Geology of the Netherlands

Geology of the Netherlands

second edition

The publication of this book is made possible through the generous support of our sponsors Energie Beheer Nederland (EBN), Ministry of Climate Policy and Green Growth, ENI Energy NL, ONE-Dyas, Element NL, COVRA N.V., Petrogas E&P NL, Wintershall Noordzee, Vermilion Energy Netherlands, Nobian, Kistos Nederland, KNGMG Royal Geological and Mining Society of the Netherlands, Aardyn, Ennatuurlijk Aardwarmte, Argo Geological Consultants, and TNO – Geological Survey of the Netherlands.

Geology of the Netherlands

second edition

Edited by

Johan H. ten Veen

Geert-Jan Vis

Jan de Jager

Theo E. Wong†

Routledge
Taylor & Francis Group
LONDON AND NEW YORK

First published in 2025 by Amsterdam University Press Ltd.

Published 2025 by Routledge
4 Park Square, Milton Park, Abingdon, Oxon OX14 4RN
605 Third Avenue, New York, NY 10158

Routledge is an imprint of the Taylor & Francis Group, an informa business

ISBN: 9789463728362 (hbk)
ISBN: 9781041180203 (pbk)
ISBN: 9781003696568 (ebk)

NUR 933

Cover design:
Nik Trabucho Alexandre, TNO – Geological Survey of the Netherlands

Design, typography and figures:
Margot Stoete & Ton Markus, Communications and Marketing, Faculty of Geosciences, Utrecht University

DOI: 10.5117/9789463728362

Contents

Applications

Effects

Dissemination

Foreword

"Denkend aan Holland
zie ik breede rivieren
traag door oneindig
laagland gaan ...
en in alle gewesten
wordt de stem van het water
met zijn eeuwige rampen
gevreesd en gehoord."

'Remembrance of Holland' by Hendrik Marsman is one of the most famous Dutch poems, first published in 1936. The poem describes the landscape in the context of the Dutch struggle against water, clearly referring to the past 1000 years since humans started controlling water and gaining land. This anthropogenic change of landscape was preceded by a constantly changing geography since the end of the last ice age, some 11,700 years ago.

The shaping of the southern Netherlands, notably characterized by its flat landscapes and intricate waterways, stems from the development of the Holocene Rhine-Meuse delta, the formation of which was driven by factors such as rising sea level, tectonic activity and sedimentary processes.

Underneath the apparently straightforward exterior lie complex geological subterranean structures. This book explores updated insights into the various rock formations and mineral deposits that make up the Dutch subsurface. From the Permian sand dunes of the north to the Cretaceous limestone hills of the south, readers will gain a deeper understanding of the forces that have shaped this small but fascinating country. This new edition of the major reference book *Geology of the Netherlands* (2007; eds Wong, Batjes and De Jager) fits in a rich tradition and echoes the significant contributions of F.J. Faber's *Geologie van Nederland* (1926).

Dutch geologists and mining engineers consistently advance scientific and technological understanding in Dutch Earth Sciences. Their contributions are particularly significant in fields such as coal mining, fossil fuel extraction (oil and gas), salt production, and more recently, in geothermal energy and storage, drawing insights from extensive field data.

This new edition describes the geology from the late Silurian to the Quaternary. The chapters on applied geology focus on various aspects of petroleum geology, coal, salt, groundwater, industrial minerals and geothermal energy. Related effects of these activities such as subsidence, induced earthquakes and storage possibilities are separately covered. A huge stream of data processed by advanced modelling tools has shed new insights on the various geological phenomena of our country that had not been reported upon earlier.

In the last decades, earth scientists have adjusted their ways of working to the changing societal and professional landscape of the 21st century. As we are starting to gain a better understanding of the impact of human activity on the planet, resources are becoming scarcer and people more numerous. Human-induced global warming is presently increasing at a rate of 0.2°C per decade with global average temperature reaching 1.45°C above pre-industrial levels in 2023. Two-thirds of the warming has occurred since 1975. Climate change has a major impact on our society: warmer temperatures are disrupting the usual balance of nature. New university courses are being introduced to allow students to gain knowledge on the earth's fundamental processes that regulate the past, present, and future dynamics of environmental systems, biodiversity, and climate. There has been an increasing push for the energy transition, in order to reduce our dependence on fossil fuels, and the Dutch government has set ambitious targets for reducing carbon emissions and increasing the use of renewable energy sources. As the energy transition progresses, the role of earth scientists is likely to change, with a greater emphasis on sustainability, geological storage and geothermal energy.

Energy transitions have been numerous in the Netherlands: the switch from wood to peat in the 17th century, the change from peat to coal in the 19th century and the dominance of oil in the 20th century are still fresh in our minds. But the transformation in the 1960s from coal to natural gas was a major game changer for Dutch society, enabled by access to cheap gas from the huge Groningen gas field in the north of the Netherlands. As a result, in 1965, the Government announced its decision to close all coal collieries in South Limburg. Energy transitions come at a cost but also with major benefits: After 60 years in which the Groningen gas production brought great prosperity, severe safety risks emerged that have resulted in the loss of the social license to operate the Groningen field. Increasing earthquakes caused by gas extraction, culminating in the 3.6 magnitude Huizinge earthquake in 2012 caused widespread damage to property. This watershed event played a central role in the decision to stop gas production in Groningen and has had an impact on the use of other fossil fuel sources: shale gas was never accepted in the Netherlands, Groningen gas production lost that acceptance between 2012 and 2018 and it is not unlikely that oil and gas production from small fields will lose it in the future. The increasingly critical attitude toward subsurface activities, at local, regional, and national levels, make procedures for new economic activities increasingly tedious and often difficult to organize. The current national government envisages a major role for carbon capture and sequestration (CCS) as a future technology, and both electricity generation and certain industrial processes with natural gas as a feedstock are likely candidates for the application of CCS technologies in the foreseeable future. But technical and commercial arguments are unlikely to convince a skeptical public, and there is often no consensus on whether extraction or injection is either justified or correct. Society decreasingly accepts that substances, like natural gas, CO_2 or wastewater should be disposed of, permanently or temporarily, in the subsurface, or that numerous wells may be drilled to facilitate deep geothermal heat extraction. Multi-stakeholder dialogues offer possible solutions for mutual trust, and an internationally proven method to ensure support among residents is the use of an independently managed area development fund.

Understanding the wide range of subsurface applications, from resource extraction to renewable energy and use of the subsurface to store CO_2, hydrogen or radioactive waste requires highly skilled and trained geologists across academia and industry. The Netherlands has a strong tradition of innovation in these areas, and it has a solid geological research base, which is fundamental to understanding earth processes and challenges.

The editors of this strongly revised second edition of *Geology of the Netherlands*, Johan ten Veen, Geert-Jan Vis, Jan de Jager and Theo Wong, have done a tremendous job in bringing together such a rich collection of papers presenting the latest knowledge of the Dutch subsurface and surface geology. Unfortunately, Theo Wong passed away before he could witness the final result. This new edition will serve yet again as a standard reference for students, professionals, and policy makers, and is an important contribution to the continued investment in Dutch geoscience skills and research that are needed for future tasks.

The current Rhine-Meuse delta will not stop defining the Dutch landscape: it will continue to evolve, increasingly influenced by climatic fluctuations, growing human population, social and engineering changes. Also, the use of the subsurface in this densely populated country will continue, supported by solid geological knowledge as presented in this updated publication about the geology of the Netherlands.

Lucia van Geuns
Strategic Advisor, Energy and Raw Materials at
The Hague Centre for Strategic Studies

Theo Edward Wong (1946-2024)

The Surinamese geologist Theo Wong, editor and co-author of both the 2007 and current version of this book, passed away on September 10, 2024, in Schoorl, the Netherlands. We could have written 'Surinamese-Dutch geologist' because he was as prominent a geologist in the Netherlands as in his home country, but as he called the book he recently wrote about his life and career *Memoires van een Surinaamse geoloog*, we will stick to his preferences. He was born on April 6, 1946, in Paramaribo, and his dream as a youngster was to become a fighter pilot. However, his hearing was not good enough, and after all, he chose for a career as a geologist. He graduated from Utrecht University and obtained a PhD with Professor Drooger on the micropaleontology of the first offshore drill cores that had been obtained in Suriname. He was deeply motivated to contribute to the development of his country and worked after his return on gold exploration at the Geological and Mining Service, later in the bauxite Industry at the Suriname Aluminium Company, and finally with Gulf Oil in offshore hydrocarbon exploration. He was also instrumental in setting up a geology and mining course at the University of Suriname in 1976.

In 1984 he got a fellowship in the Netherlands to write a revision of the stratigraphy of the Suriname coastal plain, and shortly after that, he was contracted by Petrolog to supervise exploration drilling by Amoco in the North Sea Basin. Being already an experienced hydrocarbon geologist, he was appointed in 1986 as a senior geologist by the Rijks Geologische Dienst (State Geological Survey of the Netherlands, later NITG-TNO). He became head of the mapping division of the Deep Subsurface Department, then in 1992 head of the Scientific Laboratory and finally project leader of many national and international geo-energy projects. He survived many reorganizations and stayed with TNO until his retirement in 2007.

In those years he closely collaborated with specialists from many different disciplines, including palynology, (micro)paleontology, heavy minerals, granulometry and geochemistry, which led to the publication of numerous peer-reviewed papers, especially about the Late Jurassic and Early Cretaceous hydrocarbon reservoirs in the Central North Sea Graben and the Vlieland Basin. Their interpretation completely overruled the earlier vision of NAM/Shell at that time.

His colleagues praised him as being a quiet, friendly, contemplative person, rarely searching the limelight, listening attentively, and taking the larger issues into account while leaving the details to others. He managed to considerably increase the staff of the laboratory and successfully stimulated collaboration and integration within his team.

With the support of NITG-TNO, in 1997 he was appointed as extraordinary professor in *Sedimentary Geology of the Deep Subsurface of the Netherlands* at Utrecht University. In his inaugural address *Ondergrond Nederland doorgrond?* (*Netherlands Underground Understood?*) he showed how much we still do not know about the subsurface, citing as an example how the erstwhile promising Scruff Greensand only resulted in dry wells. He successfully guided five PhD students, sponsored by TNO and industry, on different aspects of the North Sea Basin: Gesa Kuhlmann on the Plio-Pleistocene, Iwan de Lugt on the Paleogene, Allard van der Molen on Chalk, Mark Geluk on the Triassic and Henk Kombrink on the Paleozoic. He was a co-editor of the 2003 book *De Ondergrond van Nederland*, editor-in-chief of its 2007 English edition, in which he included the results of his students, and again co-editor and co-author of the present volume. He also was editor-in-chief of the *Netherlands Journal of Geosciences/Geologie en Mijnbouw*.

In spite of his busy activities in the Netherlands, he always kept a keen eye for opportunities to support his home country. In 1993 he established a committee to support geological research in Suriname and suggested strategies to upgrade his former employer the Geological and Mining Service. He also was editor-in-chief and prolific co-author of the book *The History of Earth Sciences in Suriname*, issued by the Royal Netherlands Academy of Arts and Sciences and NITG-TNO in 1998. In 2002 he was appointed as an endowed professor of geology at the Anton de Kom University of Suriname. Thanks to his excellent working relationship with Staatsolie, the Suriname State Oil Company, he managed to set up a two-year MSc course in Petroleum Geoscience at the university in 2011, which successfully continues up to the present day. Moreover, he developed intensive contacts with IAMGOLD Rosebel Gold Mines Company, which resulted in the establishment of a two-year MSc Mineral Geoscience course. With two co-authors he wrote a Dutch-language book on geology and landscape in Suriname for educational purposes, published in 2017. In this way, he managed to inspire a whole new generation of young Surinamese geologists.

We miss him every day.

Salomon Kroonenberg, Dirk Munsterman &
Waldemar Herngreen

QUATERNARY
NEOGENE
PALEOGENE
CRETACEOUS
JURASSIC
TRIASSIC
PERMIAN
CARBONIFEROUS
DEVONIAN
SILURIAN
ORDOVICIAN
CAMBRIAN
PROTEROZOIC

Introduction

<< The geological time scale with cartoons characteristic for the flora and fauna of the periods shown. Artwork by Nik Trabucho Alexandre.

DOI: 10.5117/9789463728362_intro

Since the preparation and publication of the first edition of *Geology of the Netherlands* in 2007, the world has changed considerably. In the early 21st century natural gas was the main hydrocarbon and energy source for the Netherlands – and it still is. Since 2018 however, the reduced gas production from the enormous Groningen gas field and the decline in production from small gas fields have made the Netherlands a net importer of natural gas. For the first time in many decades, the security of energy supply has emerged as a significant concern, imperilled by ongoing wars and global political conflicts. This concern has significantly boosted the new energy transition that focuses on sustainable energy resources. A transition that also affects the global search for (critical) raw materials and comes with geopolitical implications. To ensure access to a secure and sustainable supply of critical raw materials and reduce the dependence of the European Union on other states, the Critical Raw Materials Act has been implemented. The energy transition leads to a growing component of renewables in the energy mix of the Netherlands. On sunny and windy days there is even a surplus production of electricity, which currently cannot be stored adequately. Hence the growing demand for (subsurface) energy storage solutions.

During the last decades the effects of climate change have become increasingly apparent in the Netherlands, impacting flood-protection strategies, water availability and groundwater quality. At the same time the pressure on the available land surface is growing due to a growing population and changing household structures, requiring more houses to be built. That in turn increases the demand for yet another precious commodity: construction sand. In that respect, the largely Quaternary origin of the land surface of the Netherlands with many sand-rich deposits is a fortunate geological circumstance.

The abovementioned topics are just a few of a plethora of human activities requiring a thorough understanding of the geological structure and properties in our part of the world. At a time when the study of earth sciences is under pressure, earth scientists and earth science education are more important than ever.

In the light of the above, a revised edition of *Geology of the Netherlands* is timely. Although the geology hardly changed, comprehension of it has grown tremendously during the last two decades. This is partly due to new data, technological developments and legislature that help to make data availability easier and faster. More importantly, the growing need for knowledge of the subsurface on which we rely for various applications and increasingly interfere with, has initiated many new lines of research in areas, and at depths, previously underexplored. The energy transition asks for a greater understanding of the complex superposition of rock units, to be able to assess their suitability for future use. For the exploration and exploitation of aggregates (mainly for building materials) and energy (fossil, heat, solar or wind), the search for water, the storage of energy and substances, and the quest for critical raw minerals, it is crucial to understand the structure of the subsurface, its properties and the way the subsurface responds to all these activities, now and in the future.

This book covers the European part of the Kingdom of the Netherlands. A small country which is known as flat, densely populated and wedged between the North Sea and elevated hinterlands in Belgium and Germany. For much of its geological history, the country occupied such an intermediate position and has been continuously shaped by rivers and their interaction with the sea. Most of the country's present-day flat topography is cultivated and sedimentary processes are almost everywhere influenced by man: rivers are constrained by dikes; many streams are canalized; swamps, lakes and large parts of an inland sea have been converted to polders. In many places dikes and artificially maintained coastal dunes strengthen the coastline. To keep polders dry and fit for farming, pumping stations – formerly powered by the now iconic Dutch windmills – constantly transfer water into a system of water bodies, ultimately debouching into the North Sea. Water extraction since the Middle Ages has led to oxidation of peat and compaction of the upper soil layers, resulting in several metres of subsidence in vast coastal areas. Consequently, the lowest point of the Netherlands lies at 6.78 m below mean sea level. Over the last decades, ecological, cultural and scenic values of landscapes in the Netherlands have become increasingly acknowledged and appreciated, and their conservation is guaranteed through initiatives such as the European Natura 2000 protected areas network.

Most of the surface sediments of the Netherlands are of Quaternary age and about 4% of the volume of onshore Holocene deposits consists of anthropogenic deposits. Older Cenozoic, Mesozoic and Paleozoic rocks occur only locally at or near the surface along the country's eastern and southern borders. The largely siliciclastic sedimentary succession in places reaches more than 10 kilometres in thickness and is thought to rest upon metamorphic basement, which has not yet been encountered. From a vast amount of subsurface data and only a few outcrops, geologists observe that this succession is far from simple and conclude that it must have been intricately structured during several tectonic deformation phases. The Paleozoic amalgamation of tectonic plates that formed the Pangea supercontinent, was accompanied by several phases of folding, thrusting and basin formation. This continued until Carboniferous times. Later, Late Jurassic to Early Cretaceous rifting and Late Cretaceous to Paleogene inversion, together with strong deformation of the Permian Zechstein salt, had an even greater impact on the subsurface structure. Essentially, this complex subsurface structuration is the result of two important processes in geological history. One being the formation of a rock layer, the other being its alteration, either by geological processes or, in more recent times, assisted by human activity.

Previous work

This new edition of *Geology of the Netherlands* builds on a rich history of geologic publications and maps. In 1769-1771, J. le Francq van Berkhey published a detailed geological description of 'Holland'. The illustrations notably include a profile, based on shallow drilling and field observations, showing the distribution of sand, clay and peat layers down to 5 metres depth over a distance of 130 kilometres (Zagwijn, 2004). The very first geological overview map showing the entire Netherlands was made in 1821 by the German Christian Keferstein (1784-1866), an autodidact in geology. Admittedly, Jean d'Omalius d'Halloy (1783-1875), a native of Liège, had already made a geological survey map of France and the Netherlands in 1813. It was not published however, until 1822 as a result of the war against Napoleon. Both maps paid little respect to the complexity of Dutch surface geology. In 1844, Winand Staring (1808-1877) published the first geological overview map (1:800,000) of the Netherlands (*Proef eener geologische kaart van de Nederlanden*), which he produced with the help of correspondents throughout the country (Fig. 1).

Until the 20th century geological descriptions of the Netherlands only concerned the surface geology. A first insight into the deeper subsurface was provided with the start of the state-funded exploration campaign for mineral resources, directed by W.A.J.M. van Waterschoot van der Gracht. This campaign aimed at locating exploitable deposits of coal and rock salt and its results were published in detailed reports (Van Waterschoot van der Gracht, 1909, 1918). In 1956 an overview publication in English, entitled *Geological history of the Netherlands*, was edited by A.J. Pannekoek and covered both the geology and the geological resources. It was published by the Royal Netherlands Geological and Mining Society (KNGMG) and the Geological Foundation – the geological survey at that time – and was intended as an explanation to the 1:200,000 geological map, which had been prepared by the Geological Foundation on the basis of its own 1:50,000 map. In 1963, *Transactions of the Jubilee Convention* was published, following the 1962 convention on the theme 'Geology and mining in the Netherlands'. The book edited by Pannekoek in 1956 and the 1963 Transactions, give an excellent overview of the state of geological knowledge six to seven decades ago. Much of what was presented in that overview has since evolved considerably, mainly as a result of subsurface exploration for hydrocarbons, water, aggregates and heat.

Through all the exploration efforts, many insights were gained and are documented in numerous publications. A few of these publications merit to be mentioned here. A set of 1:1,000,000 depth and subcrop maps of the entire Netherlands and its offshore area was published by P. Heybroek in 1974, together with an explanatory text. It was a much-consulted paper at the time. In 1979, the Geological Survey (RGD) published the summary overview *The geology of the Netherlands* (Van Staalduinen et al., 1979). In 2003, the Netherlands Institute of Applied Geoscience (NITG), currently TNO – Geological Survey of the Netherlands, published a book in Dutch entitled *De ondergrond van Nederland* (The subsurface of the Netherlands) on the country's geology and geological resources (De Mulder et al., 2003). In 2004, the same institute published an atlas in English and Dutch compiling the results of its mapping project on the subsurface geology of the onshore Netherlands on a 1:1,000,000 scale. Duin et al. (2006) published an article including a set of depth and thickness maps of the main rock units in both the onshore and offshore Netherlands and clearly delineated the structural elements through geological time. Several years later, at a time when much more offshore data was publicly available, Kombrink et al. (2012) overhauled this publication. Many illustrations in these books, atlases and scientific publications have been used in subsurface studies and assessments and were also included in the first edition of this book.

The geological mapping of a country relies on a set of defined stratigraphic units, such as formations and members. Since the 1970s these units were defined and published in overview works such as *Toelichting bij geologische overzichtskaarten van Nederland* (Zagwijn & Van Staalduinen, 1975), *De lithostratigrafische indeling van Nederland – Formaties uit het Tertiair en Kwartair* (Weerts et al., 2000) for the Cenozoic successions, in *Stratigraphic Nomenclature of the Netherlands* (Nederlandse Aardolie Maatschappij & Rijks Geologische Dienst, 1980) and in *Stratigraphic nomenclature of the Netherlands, revision and update by RGD and NOGEPA* (Van Adrichem Boogaert & Kouwe, 1993-1997) for the deeper subsurface. These works were later converted into websites published by TNO – Geological Survey of the Netherlands. In June 2020 the pre-existing and partly

Figure 1. *Detail of the 1844 geological overview map of the Netherlands at scale 1:800,000, compiled by W.C.H. Staring. Using only a few colours, the map shows the distribution of dunes and other eolian deposits (orange), peat (grey), marine deposits (pale orange), continental deposits (yellow), and flood basin deposits (pink). The map shown here represents a digitally enhanced version, published by TNO-GDN in 2017. Figure 2 shows the 2023 geological map of the same area.*

overlapping stratigraphic nomenclature of the shallow and deep subsurface of the Netherlands were integrated and published on a new website (see www.dinoloket.nl).

Geological mapping used to focus on surface geology, and in many countries it still does. In the Netherlands, the sheer amount of subsurface data that has become available over the past decades enabled mapping down to many kilometres. The large amount of borehole data from the Dutch subsurface includes over 6000 deep boreholes and approximately 600,000 shallow boreholes. Most of these come with well-log data and some with core, or cutting material, allowing for the characterization of subsurface sediments and sedimentary rocks. Combined with 2D and 3D seismic data, these boreholes facilitate an interpretation of the underground arrangement of geologic units and faults. This arrangement, nowadays, is captured in several 3D digital geological models that target specific depth intervals. The shallow subsurface down to a depth of 50 metres is schematized by the GeoTOP model. The DGM model is a 3D stacked lithostratigraphic layer model of the entire onshore part of the Netherlands down to a depth of 1500 metres. The DGM-deep layer model is the 3D geological on- and offshore model suited for applications requiring geological information on the depth and distribution of sedimentary rocks in the deeper subsurface (1.5-7 km). Derivatives of these models are tailored at specific applications such as hydrogeology (REGIS) or geothermal potential (ThermoGIS).

In 2023, the new geological map of the Kingdom of the Netherlands was published by the Geological Survey, summarizing the state of knowledge of the Dutch subsurface (Fig. 2). Such an overview map is only released once every few decades. The last published geological map dates from 1975 and was modified over the years by several authors (e.g. Weerts et al., 2004; Schokker et al., 2010). The new paper map has a scale of 1:600,000 and offers a greatly im-

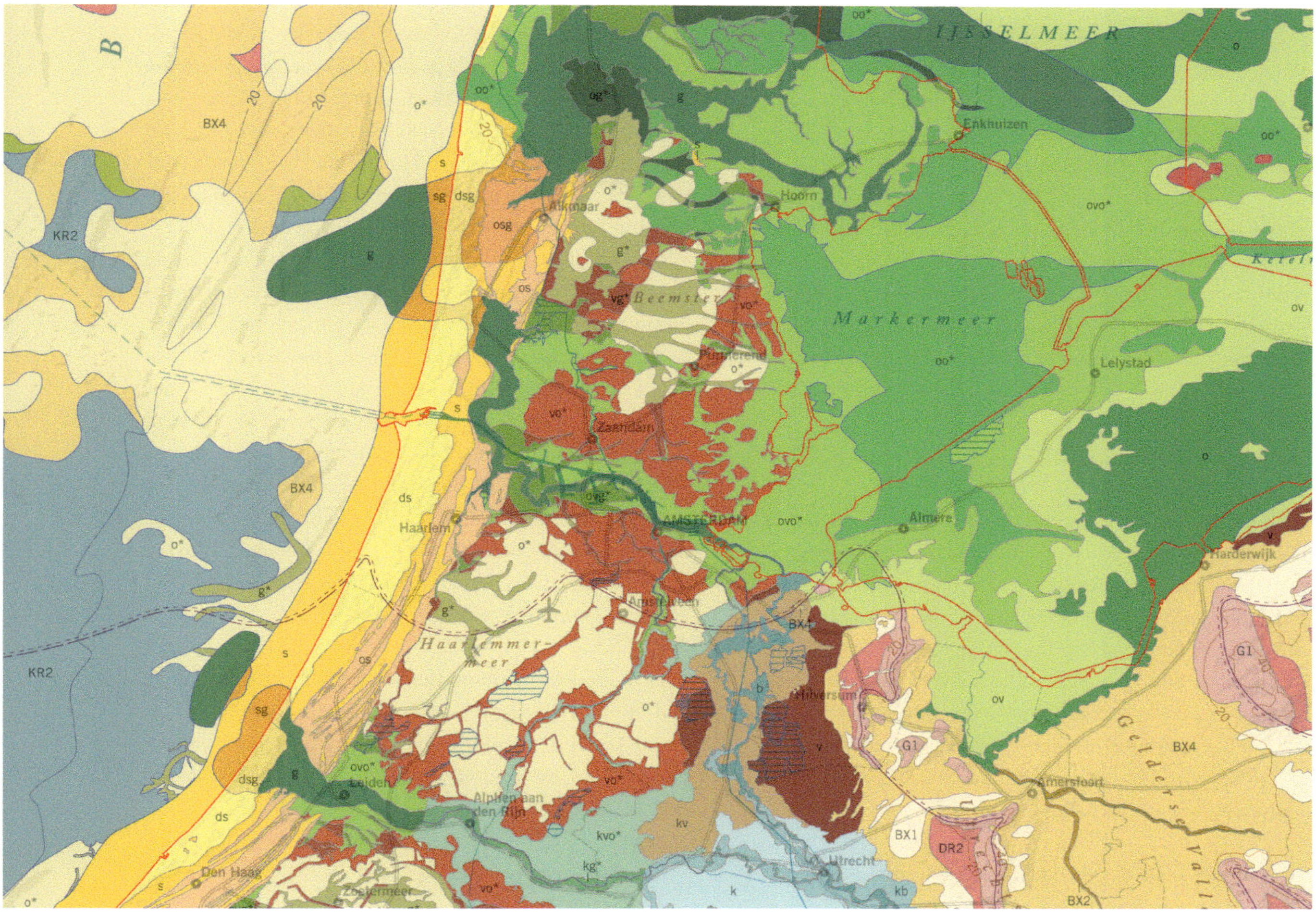

Figure 2. Detail of the 2023 geological map of the Kingdom of the Netherlands. The map shows the deposits that compose the upper few metres of the subsurface. The map units represent (combinations of) formations, members, and beds. The folded paper geological map of the Netherlands can be ordered from GDN's web shop. A PDF version is available for free (www.geologische-dienst.nl) and the digital version can be viewed at www.dinoloket.nl.

proved image of the mainland, an entirely new representation of the geology of the North Sea, and for the first time includes the Caribbean part of the Kingdom. The map shows the deposits that make up the upper metres of the subsurface and includes several countrywide cross-sections that also highlight the deeper geology. The map is based on a multitude of published data, supplemented with latest information and insights. For the Dutch mainland, this mainly involves information collected as part of the 1:50,000 geological mapping (~1950-1997) and the creation of the 3D subsurface models DGM, DGM-deep and GeoTOP (1997-present).

The book

The second edition of this book provides access to a wealth of data presented in four sections. The section **Geology and stratigraphy** deals with the state of the subsurface and describes stratigraphy, structure and rock properties. It begins with an overview of the geological development of the Netherlands (Chapter 1). The following nine chapters give a thoroughly revised and updated description of the Paleozoic to Quaternary geology of the Netherlands, including the offshore. The last three chapters of this section describe magmatic rocks (Chapter 11), salt tectonics (Chapter 12), and sediment geochemistry and mineralogy (Chapter 13). The latter two chapters are new additions compared to the first edition. The section **Applications** focuses on subsurface exploitation and describes petroleum geology (Chapter 14), peat, coal and coalbed methane (Chapter 15), salt production (Chapter 16), hydrogeology (Chapter 17), surface mineral resources (Chapter 18), geothermal energy (Chapter 19), and underground storage and sequestration (Chapter 20). The section **Effects** deals with effects of natural processes and those induced by subsurface activities and includes chapters on geological disposal of radioactive waste (Chapter

21), natural and induced seismicity (Chapter 22) and subsidence and uplift (Chapter 23). As a grand finale of this book the section **Dissemination** illustrates how data and knowledge of the subsurface are disseminated by TNO – Geological Survey of the Netherlands (Chapter 24).

This book has been written for readers with basic knowledge of earth sciences. It strives to be a comprehensive reference work for geologists, engineers, geoscience students, and all others who wish to know more about the relevance and applications of geology in the Netherlands.

Acknowledgements

In 2019, the first discussion on making this new edition of the book was held. As the original editorial team was not complete anymore (Dick Batjes passed away in 2012), editors Johan ten Veen and Geert-Jan Vis joined to strengthen the ranks. Theo Wong was the initiator of the first edition of this book and was eager to be actively involved in the creation of this second edition. We deeply regret that Theo did not live to see the result; he passed away on September 10, 2024. With his end in sight, he remarked: "They will probably place a cross behind my name".

The coming to be of this book was a lengthy process that started with a thorough review of all chapters of the first edition of the book. Many people contributed to this task and are highly acknowledged, as their work gave us insights into the necessary updates and additions for this second edition. We owe many thanks to the almost 100 authors for their elaborate work on their respective chapters. Specifically, we acknowledge the patience of the early birds who finalized their contributions a long time ago and had to wait considerable time for the final book to be published. All chapters were peer-reviewed and we are grateful for the almost fifty reviewers who gave valuable comments and suggestions for improvement.

We are much indebted to Margot Stoete and Ton Markus of the Faculty of Geosciences of Utrecht University who took care of many of the figures and the graphic design of the book. We thank TNO for permission to reproduce illustrations; Amsterdam University Press for undertaking the publishing of the book; Harry Doust for checking and adjusting the English of all the chapters; Selçuk Aksay for checking on consistency between the chapters; Anika Hurkens, Rixt Altenburg and Sytze van Heteren for their meticulous work on the references and Jenny Hettelaar for GIS support. The cover art of the book was designed by Nik Trabucho Alexandre who patiently dealt with our changing demands. We thank Jacob J. Gerrits for providing the scenic photograph for the frontispiece of the Introduction; he unfortunately passed away months before the publication of the book. Much appreciation goes to Lucia van Geuns for writing the preface, that could not have been more contemporary and in line with the objectives of this book. We are most grateful to Tirza van Daalen (director) and Michiel van der Meulen (chief geologist) of TNO – Geological Survey of the Netherlands for sponsoring this book by providing financial, administrative and management support to many of the authors and editors. The sponsors Energie Beheer Nederland, Ministry of Climate Policy and Green Growth, ENI Energy NL, ONE-Dyas, Element NL, COVRA N.V., Petrogas E&P NL, Wintershall Noordzee, Vermilion Energy Netherlands, Nobian, Kistos Nederland, KNGMG Royal Geological and Mining Society of the Netherlands, Aardyn, Ennatuurlijk Aardwarmte, Argo Geological Consultants and TNO – Geological Survey of the Netherlands are thanked for their generous financial contributions to the realization of this book.

Finally, the editors thank their families for their patience and support during the time-consuming work on this book. A special thanks to Marion ten Veen and Margriet Pluymaekers for their dedicated help in reviewing the chapter proofs.

Johan ten Veen
Geert-Jan Vis
Jan de Jager
Theo Wong†

References

d'Omalius d'Halloy, J.B.J., 1822. Essai d'une carte géologique de la France, des Pays-Bas et de quelques contrées voisines. Annales Des Mines 7: 353-376.

De Mulder, E.F.J., Geluk, M.C., Ritsema, I., Westerhoff, W.E. & Wong, T.E., 2003. De ondergrond van Nederland. TNO-NITG (Utrecht): 379 pp.

Duin, E., Doornenbal, J.C., Rijkers, R.H.B., Verbeek, J.W. & Wong, Th.E., 2006. Subsurface structure of the Netherlands; results of recent onshore and offshore mapping. Netherlands Journal of Geosciences 85 (4): 245-76. DOI: 10.1017/S0016774600023064

Heybroek, P., 1974. Explanation to tectonic maps of The Netherlands. Geologie En Mijnbouw 53 (2): 43-50.

Keferstein, C., 1821. Teutschland geognostischen-geologisch dargestellt mit Karten, Durschschnittzeichnungen, welche einer geognostischen Atlas bilden. (Weimar).

Kombrink, H., Doornenbal, J.C., Duin, E.J.T., Den Dulk, M., Ten Veen, J.H. & Witmans, N., 2012. New insights into the geological structure of the Netherlands; results of a detailed mapping project. Netherlands Journal of Geosciences 91 (4): 419-446. DOI: 10.1017/S0016774600000329

Le Francq van Berkhey, J., 1769. Natuurlyke historie van Holland.

NAM & RGD, 1980. Stratigraphic Nomenclature of the Netherlands. Verhandelingen van Het Koninklijk Nederlands Geologisch Mijnbouwkundig Genootschap 32: 77.

Pannekoek, A.J., 1956. Geologische geschiedenis van Nederland, toelichting bij de geologische overzichtskaart van Nederland op de schaal 1:20 000. Staatsdrukkerij ('s-Gravenhage): 147 pp.

Schokker, J., 2010. Geologische overzichtskaart van Nederland. TNO Bouw en Ondergrond(Utrecht).

Staring, W.C.H., 1844. Proef eener geologische kaart van de Nederlanden, 1:800.000. J. Oomkens (Groningen).

TNO-GDN, 2023. Geologische kaart van het Koninkrijk der Nederlanden 1:600 000, TNO – Geological Survey of the Netherlands (Utrecht), 1 map sheet. https://www.dinoloket.nl/en/subsurface-models/map. ISBN 9789059865142

TNO-NITG, 2004. Geological Atlas of the Subsurface of the Netherlands – onshore. Netherlands Institute of Applied Geoscience TNO (Utrecht): 104 pp.

Transactions of the Jubilee Convention, 1963a. Geology and mining in the Netherlands (1). Verhandelingen van het Koninklijk Nederlands Geologisch Mijnbouwkundig Genootschap 21 (1): 179 pp.

Transactions of the Jubilee Convention, 1963b. Geology and mining in the Netherlands (2). Verhandelingen van het Koninklijk Nederlands Geologisch Mijnbouwkundig Genootschap 21 (2): 280 pp.

Van Adrichem Boogaert, H.A. & Kouwe, W.F.P. (eds), 1993-1997. Stratigraphic nomenclature of the Netherlands; revision and update by RGD and NOGEPA (sections A-J). Mededelingen Rijks Geologische Dienst 50.

Van Staalduinen, C.J., Van Adrichem Boogaert, H.A., Bless, M.J.M., Doppert, J.W.C., Harsveldt, H.M., Van Montfrans, H.M., Oele, E., Wermuth, R.A. & Zagwijn, W.H., 1979. The geology of the Netherlands. Mededelingen Rijks Geologische Dienst 31 (2): 9-49.

Van Waterschoot van der Gracht, W.A.J.M., 1918. Eindverslag over de onderzoekingen en uitkomsten van den Dienst der Rijksopsporing van Delfstoffen in Nederland 1903-1916. Martinus Nijhoff ('s-Gravenhage): 664 pp.

Van Waterschoot van der Gracht, W.A.J.M., 1909. The deeper geology of the Netherlands and adjacent regions, with special reference to the latest borings in the Netherlands, Belgium and Westphalia. Meededeelingen van de Rijksopsporing van Delfstoffen (The Hague) 2: 437 pp.

Weerts, H.J.T., Cleveringa, P., Ebbing, J.H.J., Lang, F.D. de & Westerhoff, W.E., 2000. De lithostratigrafische indeling van Nederland. Formaties uit het Tertiair en Kwartair. Nederlands Instituut voor Toegepaste Geowetenschappen TNO (Utrecht), Report No. NITG-00-95-A: 1-86 pp.

Weerts, H.J.T., Schokker, J., Rijsdijk, K.F., Laban, C., 2004. Geologische overzichtskaart van Nederland. Nederlands Instituut voor Toegepaste Geowetenschappen TNO (Utrecht).

Weerts, H.J.T., Westerhoff, W.E., Cleveringa, P., Bierkens, M.F.P., Veldkamp, J.G. & Rijsdijk, K.F., 2005. Quaternary geological mapping of the lowlands of The Netherlands, a 21st century perspective. Quaternary International 133-134: 159-178. DOI: 10.1016/j.quaint.2004.10.011

Zagwijn, W.H., 2004. Berkhey's Treatise on the Grounds of Holland (1771): geology before the term existed. In: J.L.R. Touret & R.P.W. Visser (eds): Dutch pioneers of the Earth sciences. Royal Netherlands Academy of Arts and Sciences (KNAW, Amsterdam), Vol. 1: 1-32.

Zagwijn, W.H. & Van Staalduinen, C.J., 1975. Toelichting bij geologische overzichtskaarten van Nederland, met bijbehorende 5 kaarten (1: 600 000) en 3 overzichtsprofielen. Rijks Geologische Dienst (Haarlem): 134 pp.

1 Geological development

Jan de Jager, Kees van Ojik & Jeroen Smit*

*Corresponding author: jan.dejager@hotmail.com

ABSTRACT

The geological evolution of the Netherlands has resulted in the development of a highly structured and surprisingly varied subsurface geology beneath a deceptively flat topography. In large parts of the Dutch subsurface, more than 10 km of predominantly siliciclastic sedimentary rocks overlie the metamorphic basement. Although several major unconformities are recorded in these strata, the geological record is represented by sedimentary rocks from the Late Paleozoic onwards. The basins, platforms and highs present in the subsurface formed in response to global reorganizations of lithospheric plates. The main tectonic events that affected the area were: 1) the Caledonian and Variscan orogenies, resulting from the assembly of the Pangea supercontinent during the Paleozoic, 2) repeated rifting during the Mesozoic, related to the break-up of Pangea, 3) Alpine inversion, resulting initially from the rotation of the Iberian Peninsula and the later collision of Africa and Europe during the Late Cretaceous and Paleogene, and 4) Oligocene to Quaternary development of the European Cenozoic rift system coupled with strong, long wavelength vertical motions resulting from the opening of the Atlantic and the convergence of Africa-Europe. Notwithstanding the high degree of diversity of these events, faults are mainly parallel to each other and form predominantly NW-SE oriented structures. The general structural model is, therefore, one of repeated (oblique) reactivation of basement faults that maintain a control of the structural grain, independent of tectonic regime and stress direction. Thick Permian Zechstein salt in large parts of the subsurface was deformed during phases of extensive salt tectonics that led to structural reorganization, and (partial) decoupling of the basin fill from sub-salt faulting. The initiation of salt movement resulted in most cases from reactivation of (deep) basement faults.

<< A polder landscape with the Oukoopse Molen windmill in the Green Heart of South-Holland, east of the town of Reeuwijk. Photo: Jacob Gerrits†.

DOI: 10.5117/9789463728362_ch01

Introduction

In large parts of the Dutch subsurface, more than 10 km of largely siliciclastic sedimentary rocks overlie a poorly known metamorphic basement. These rocks are interrupted by several major unconformities but present nevertheless an overall continuous geological record from the Late Paleozoic onwards.

The most prominent structural elements in the Dutch subsurface are Late Jurassic to Early Cretaceous extensional and transtensional rift basins, viz. the Dutch Central Graben, the Broad Fourteens Basin, the West and Central Netherlands basins, the Roer Valley Graben and the Lower Saxony Basin (Figs 1.1, 1.2). However, these elements follow Early Paleozoic structural trends that over geological time have continuously been reorganized in response to changing tectonic conditions (e.g. TNO-NITG, 2004b; Duin et al., 2006; Doornenbal & Stevenson, 2010; Kombrink et al., 2012; Smit et al., 2018). The names of the structural elements as used here are defined by Kombrink et al. (2012), with each element having experienced a unique structural development and associated burial history. In most cases the structural elements are defined by their Late Jurassic to Early Cretaceous origin and development, but some structural elements developed during other time intervals, for example the Mid Netherlands Swell during the Triassic, a broad 'high' overlapping partly with the Texel-IJsselmeer High.

Many tectonic events or pulses are recorded and recognized in northwest Europe. However, formation of the

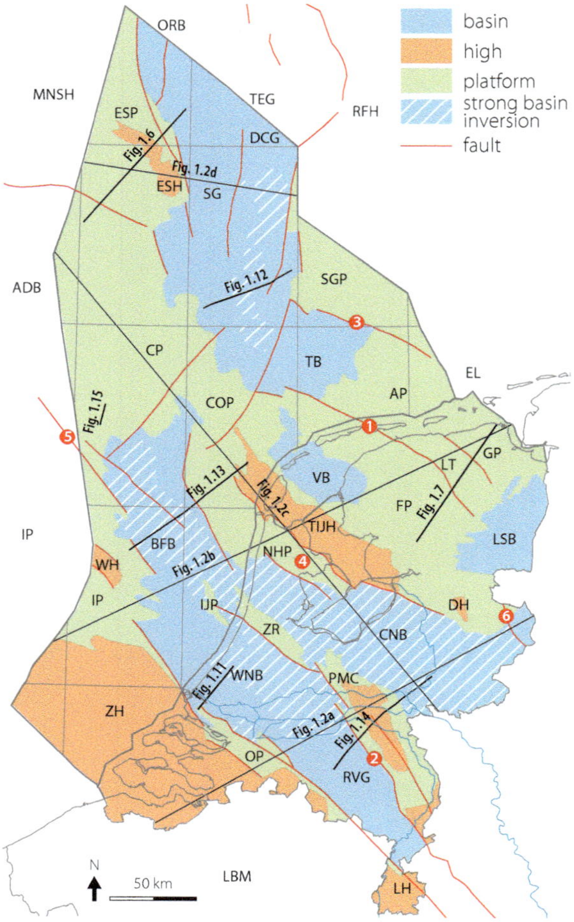

Acronym	Full name
ADB	Anglo-Dutch Basin*
AP	Ameland Platform
BFB	Broad Fourteens Basin
CNB	Central Netherlands Basin
COP	Central Offshore Platform
CP	Cleaverbank Platform
DCG	Dutch Central Graben
DH	Dalfsen High
EL	Ems Low*
ESH	Elbow Spit High
ESP	Elbow Spit Platform
FP	Friesland Platform
GP	Groningen Platform
IJP	IJmuiden Platform
IP	Indefatigable Platform
LBM	London-Brabant Massif
LH	Limburg High
LSB	Lower Saxony Basin
LT	Lauwerszee Trough
MNSH	Mid-North Sea High*
NHP	Noord Holland Platform
OP	Oosterhout Platform
ORB	Outer Rough Basin*
PMC	Peel-Maasbommel Complex
RFH	Ringkøbing-Fyn High*
RVG	Roer Valley Graben
SG	Step Graben
SGP	Schill Grund Platform
TB	Terschelling Basin
TEG	Tail End Graben*
TIJH	Texel-IJsselmeer High
VB	Vlieland Basin
WH	Winterton High
WNB	West Netherlands Basin
ZR	Zandvoort Ridge
ZH	Zeeland High

* outside Dutch territory

Figure 1.1. *Structural elements map of the Netherlands showing Mid and Late Cimmerian (Jurassic and Early Cretaceous) basins, highs and platforms. The main basement highs are formed by crystalline basement and Paleozoic rocks, generally without Rotliegend. The locations of geologic and seismic cross-sections shown in next figures are indicated (modified from Kombrink et al., 2012). 1 = Hantum Fault Zone; 2 = Peel Boundary Fault Zone; 3 = Rifgronden Fault Zone; 4 = Mid Netherlands Fault Zone; 5 = Indefatigable Fault Zone; 6 = Gronau Fault Zone. Also see Fig. 1.16 for a more detailed overview of faults in the subsurface of the Netherlands.*

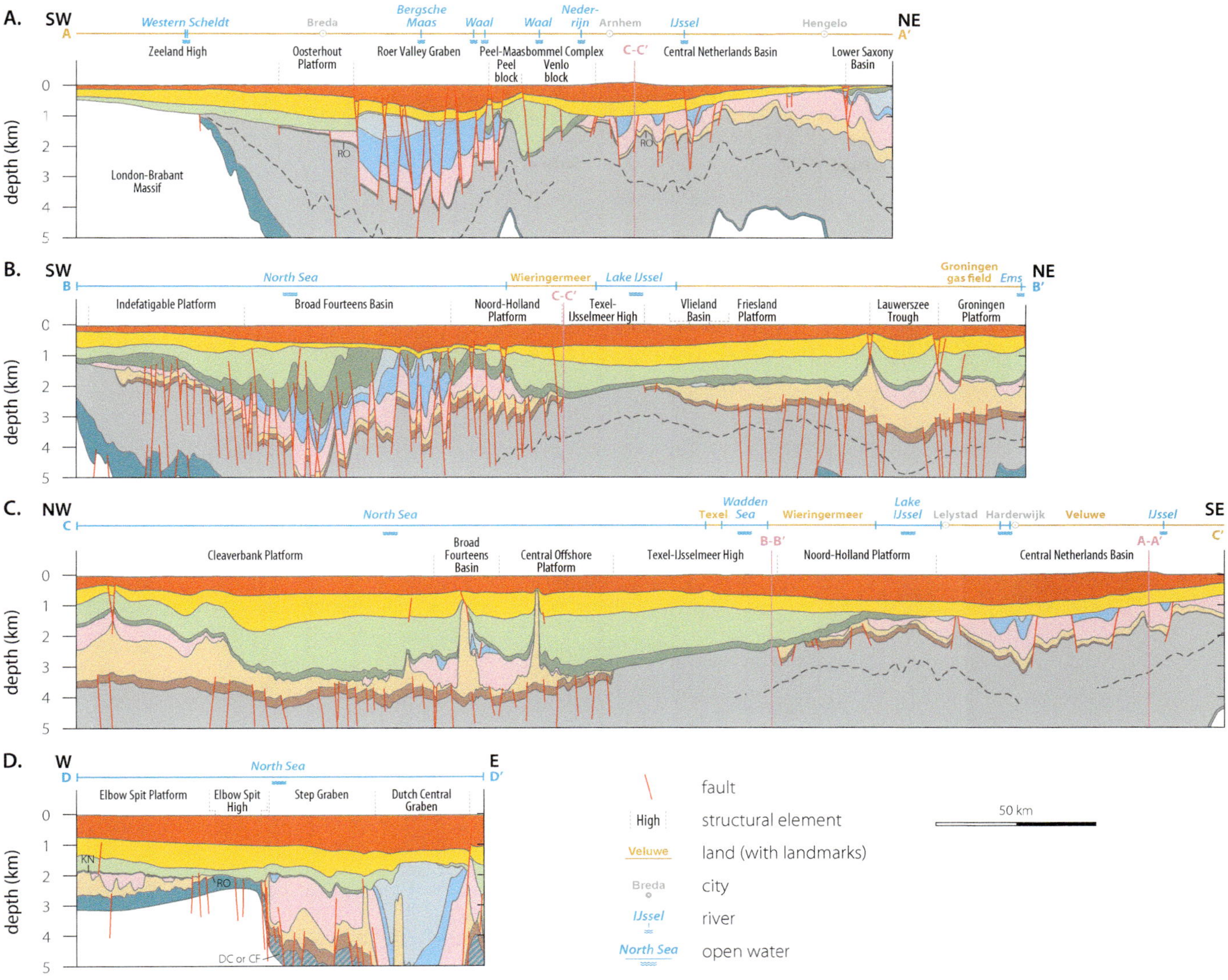

NU Upper North Sea Group – Shallow marine and continental siliciclastic deposits (Miocene – Recent)

NLM Middle North Sea Group (NM) and Lower North Sea Group (NL) – Mainly marine siliciclastic deposits (Paleocene – Miocene)

CK Chalk Group – Mainly marine limestone (Late Cretaceous – Paleocene)

KN Rijnland Group – Mainly marine siliciclastic sedimentary rock (Early Cretaceous)

S Upper Jurassic Supergroup (SG, SK, SL) – Marine, marginal marine and continental siliciclastic sedimentary rock, evaporites and limestone (Late Jurassic)

AT Altena Group – Marine siliciclastic sedimentary rock (Late Triassic – Early Jurassic)

GT Upper Germanic Trias Group (RN) and Lower Germanic Trias Group (RB) – Marine and continental siliciclastic sedimentary rock, evaporites and limestone (Triassic)

ZE Zechstein Group – Marine evaporites and limestone (Permian, Lopingian)

RO Upper Rotliegend Group – Continental siliciclastic sedimentary rock and evaporites (Permian, Guadalupian – Lopingian)

DC Limburg Group – Mainly marine and continental siliciclastic sedimentary rock, alternating with coal (upper Carboniferous, Namurian, Westphalian, Stephanian)

CL Carboniferous Limestone Group – Marine limestone (lower Carboniferous, Dinantian)

CF Farne Group – Mainly marine and continental siliciclastic sedimentary rock alternating with coal, limestone and dolomite (lower Carboniferous, Dinantian)

- - - - - base Westphalian – Base of the interval of the Limburg Group containing coal (Caumer Subgroup, Dinkel Subgroup and Hunze Subgroup; not mapped country wide)

Figure 1.2. Regional cross-sections through the Digital Geological Model-deep v5 of the Netherlands (after TNO-GDN, 2023), showing the distribution of the lithostratigraphic groups and the diverse structural styles that ensued from the various tectonic phases. See Fig. 1.1 for locations. Note that where possible the colours and abbreviations of the lithostratigraphic groups are consistently used throughout this book.

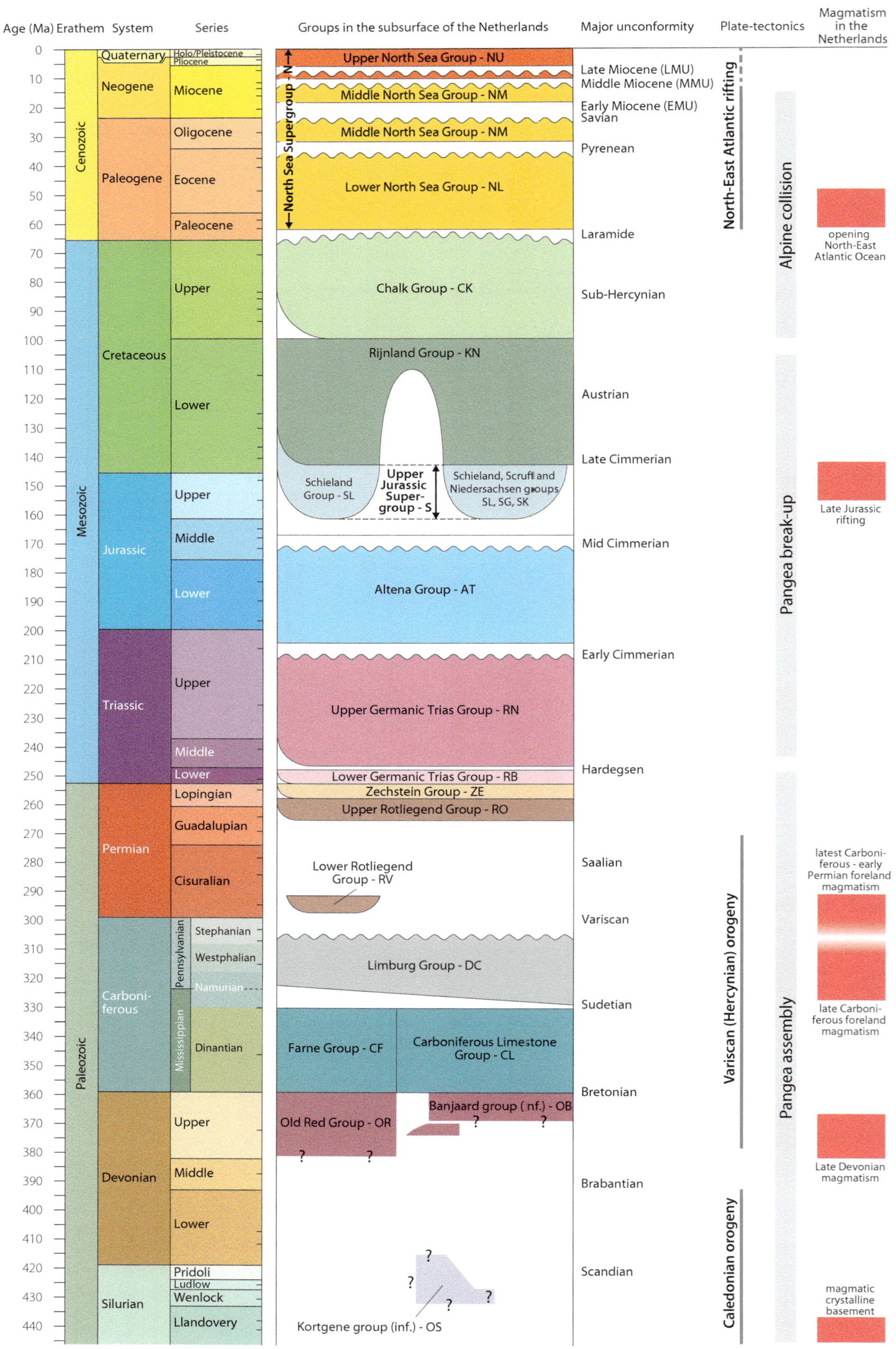

Age (Ma) Erathem System Series
Groups in the subsurface of the Netherlands
Major unconformity
Plate-tectonics
Magmatism in the Netherlands
Cenozoic
Mesozoic
Paleozoic
Quaternary
Neogene
Paleogene
Cretaceous
Jurassic
Triassic
Permian
Carboniferous
Devonian
Silurian
Holo/Pleistocene
Pliocene
Miocene
Oligocene
Eocene
Paleocene
Upper
Lower
Upper
Middle
Lower
Upper
Middle
Lower
Lopingian
Guadalupian
Cisuralian
Pennsylvanian
Mississippian
Stephanian
Westphalian
Namurian
Dinantian
Upper
Middle
Lower
Pridoli
Ludlow
Wenlock
Llandovery
North Sea Supergroup - N
Upper North Sea Group - NU
Middle North Sea Group - NM
Middle North Sea Group - NM
Lower North Sea Group - NL
Chalk Group - CK
Rijnland Group - KN
Schieland Group - SL
Upper Jurassic Super-group - S
Schieland, Scruff and Niedersachsen groups SL, SG, SK
Altena Group - AT
Upper Germanic Trias Group - RN
Lower Germanic Trias Group - RB
Zechstein Group - ZE
Upper Rotliegend Group - RO
Lower Rotliegend Group - RV
Limburg Group - DC
Farne Group - CF
Carboniferous Limestone Group - CL
Old Red Group - OR
Banjaard group (inf.) - OB
Kortgene group (inf.) - OS
Late Miocene (LMU)
Middle Miocene (MMU)
Early Miocene (EMU)
Savian
Pyrenean
Laramide
Sub-Hercynian
Austrian
Late Cimmerian
Mid Cimmerian
Early Cimmerian
Hardegsen
Saalian
Variscan
Sudetian
Bretonian
Brabantian
Scandian
North-East Atlantic rifting
Alpine collision
Pangea break-up
Variscan (Hercynian) orogeny
Pangea assembly
Caledonian orogeny
opening North-East Atlantic Ocean
Late Jurassic rifting
latest Carboniferous - early Permian foreland magmatism
late Carboniferous foreland magmatism
Late Devonian magmatism
magmatic crystalline basement

← ***Figure 1.3.*** *Paleozoic-Cenozoic chronostratigraphy of the rock succession of the Netherlands, which is divided into lithostratigraphic groups that are mostly differentiated by major unconformities that are associated with the tectonic phases. Tectonic- and magmatic phases are shown at the right together with their relation to plate-tectonic events (grey bars).*

structural grain did not occur as discrete events but responded to more or less continuously and dynamically to ongoing convergence or divergence of lithospheric plates. Nevertheless, periods of accelerated tectonic activity can be recognized (Fig. 1.3). The magmatic activity that accompanied the structural development is reviewed by Van Bergen et al. (2025, this volume).

The present chapter consists of three main sections. The first describes the structural history of the Netherlands in the light of plate-tectonic events and summarizes the overall sedimentary development. The second section describes the evolution and salient character of the main structural elements, and in the third section some remarks are made concerning the fault trends in the Dutch subsurface.

Structural history

Paleozoic assembly of Pangea

The Paleozoic structural history of Western Europe is dominated by the opening of the Iapetus and Rheic Oceans and their subsequent closure during the Caledonian and Variscan orogenic cycles, involving the three large continental plates Laurentia, Baltica and Gondwana and a number of smaller Gondwana-derived micro-plates such as Avalonia. The amalgamation of these plates occurred over a period of more than 150 Ma and finally resulted in the formation of the Pangea supercontinent at the end of the Carboniferous (Fig. 1.4, see also Torsvik & Cocks, 2017). The Netherlands, including most of its offshore territory, is located on the northeastern margin of the Avalonia microplate or terrane (Figs 1.4, 1.5) and its Paleozoic history was closely linked to it from pre-Cambrian times, when Avalonia formed part of Gondwana (e.g. Woodcock & Strachan, 2012; Herbosch et al., 2020), up to its position at the centre of Pangea during the Permian.

Due to their deep burial and limited economic interest, pre-Westphalian (ca. 320 Ma, mid-early Pennsylvanian) deposits remain under-sampled (Geluk et al., 2007; Bouroullec et al., 2018). Moreover, the structure of the older sequences is generally not well imaged on seismic reflection data and consequently is difficult to interpret and understand, although remapping (Ten Veen et al., 2019) and new seismic profiles aim to fill this gap. The early structural history of the Netherlands is therefore inferred mainly from regional data (e.g. Ziegler, 1978, 1981, 1988, 1990a,b; Glennie, 1986; Glennie & Underhill, 1998; Coward et al., 2003; Guterch et al., 2010; Pharaoh et al., 2010; Woodcock & Strachan, 2012; Smit et al., 2016, 2018; Ter Borgh et al., 2018a), upon which the following account of the plate-tectonic setting of the Netherlands is largely based.

Early Paleozoic, including Caledonian Orogeny

Avalonia formed part of the northern active margin of Gondwana during the late Precambrian and most of the Cambrian. The oldest recorded sedimentary rocks in the London-Brabant Massif are of Ediacaran age, (late Proterozoic, e.g. Verniers et al., 2002), indicating that its precursor – the London-Brabant Basins – already existed when Avalonia rifted from Gondwana during the Cambro-Ordovician. The general NW-SE trend, a prominent structural feature in the southern half of the Dutch subsurface, is likely as old as or older than the London-Brabant Basin, in line with field evidence from Wales and Ireland (e.g. Woodcock & Strachan, 2012) and the paleotectonic restoration by Smit et al. (2018). Around the Cambro-Ordovician boundary, Avalonia rifted from Gondwana toward Baltica and Laurentia, with the Rheic Ocean opening to its rear and the Iapetus Ocean being subducted below its northern margin (Pharaoh et al., 2010) along the Texel-IJsselmeer High (TIJH), the Mid North Sea High (MNSH), the Southern Uplands of England and Ireland, as well as on the opposite side of the present-day Atlantic (e.g. Verniers et al., 2002; Woodcock & Strachan, 2012; Smit et al., 2016, 2018). Collision with Baltica along the Thor or Tornquist Suture Zone occurred during the Late Ordovician-early Silurian (e.g. Berthelsen, 1998; Torsvik & Rehnström, 2003) and from the late Silurian to Early Devonian, the joint Baltica-Avalonia block collided with Laurentia to form a single Laurussian continent (Ziegler, 1990a), with the Caledonian mountain belts following the Iapetus and Thor sutures. With the closure of the Iapetus Ocean, the London-Brabant Basin was gradually inverted to form the London-Brabant Massif, starting from the early Silurian (late Llandovery) and lasting to the Early Devonian (Pragian) (Debacker et al., 2005).

The oldest recorded sedimentary rocks in the Netherlands are folded and densely compacted, fine-grained turbidites found in wells on the northern flank of the London-Brabant Massif. Houben and Vis (2021) assigned a late Silurian (Ludlow-Pridoli) to earliest Devonian (Lochkovian) age to these turbidites, penetrated in well Kortgene-1. The Silurian clastic rocks in wells O18-01 and Kortgene-1 resemble strata encountered in East Anglia and in the offshore UK wells 47/29A-1 and 53/16-I and may have been deposited as distal turbidites in a Caledo-

A. Silurian 430 Ma

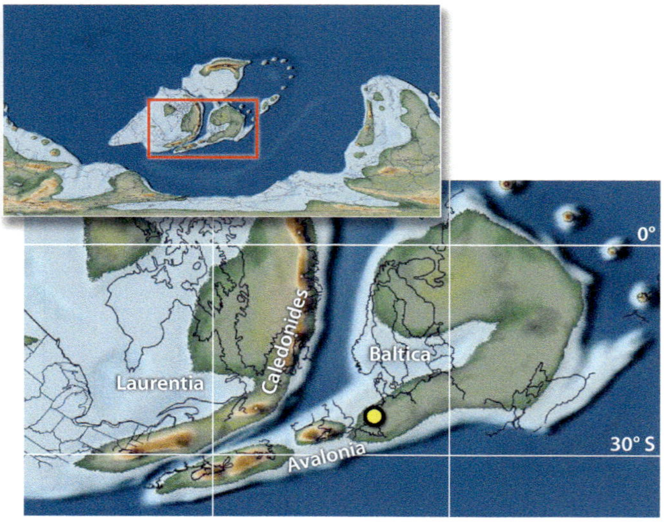

B. Devonian - early Carboniferous 375 Ma

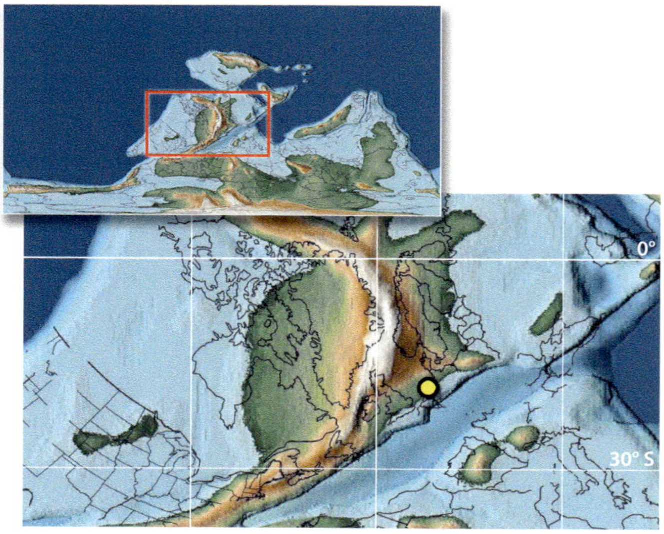

C. late Carboniferous 324 Ma

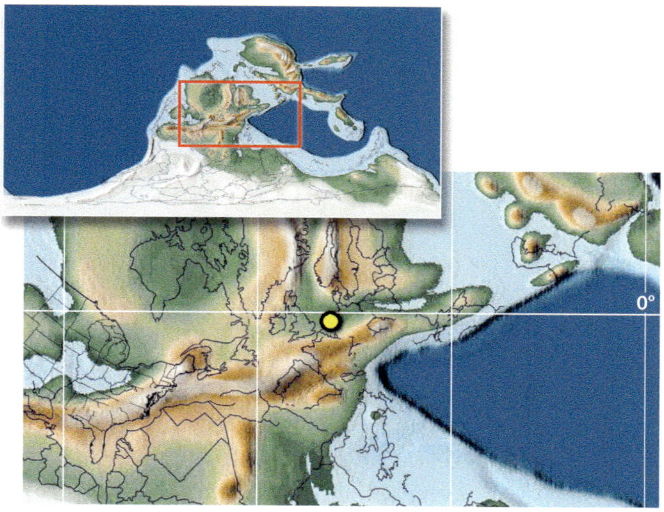

D. Late Permian 250 Ma

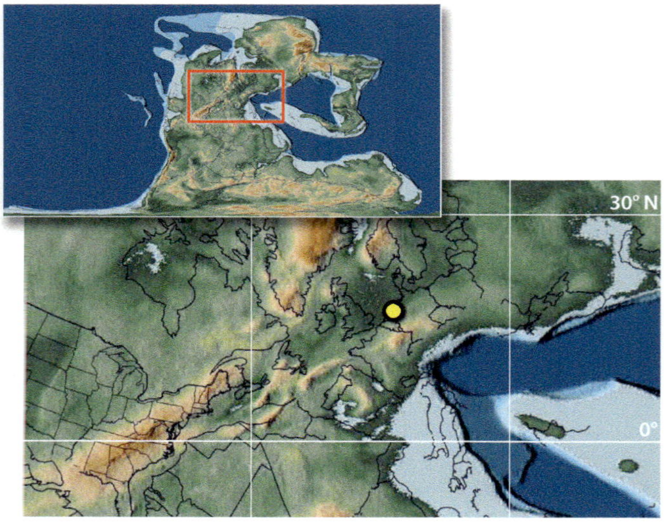

E. Late Jurassic - Early Cretaceous 137 Ma

F. Early Cenozoic 49 Ma

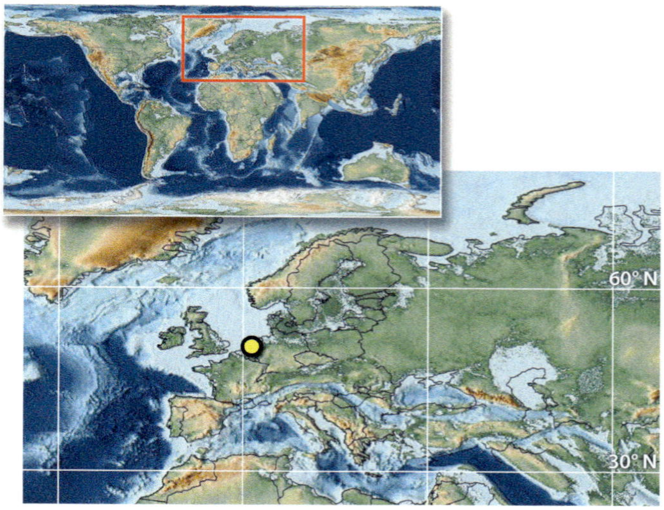

← *Figure 1.4. Global plate reconstructions from the Silurian to the early Cenozoic. During the Paleozoic (a-d) the supercontinent of Pangea was created through the collisions of Laurentia, Baltica and Gondwana, and the Caledonian and Variscan orogens developed along the plate sutures. The breaking up of Pangea during the Mesozoic (e) resulted in the formation of rift basins, while the Alpine collision of Africa (f) and Europe, caused inversion of these basins and compressional reactivation of pre-existing extensional faults. Since the early Paleozoic, the area of what is now the Netherlands (indicated in yellow) drifted from the southern hemisphere to its present location, crossing several climate zones. Source: Joint Industry Project Paleomap© (2003-2007), reproduced with permission from Prof. Christopher R. Scotese.*

nian foreland basin located between the Caledonides and the London-Brabant Massif (Pharaoh et al., 1995). During that time, the area of the current Netherlands was located at some 40° to 30° south of the equator (Torsvik & Cocks, 2017). In northern Belgium, ca. 400 Myr old (Late Silurian or Early Devonian) rock units belonging to the London-Brabant Massif are known from outcrop and well data (Ziegler, 1990a; De Vos et al., 1993).

A highly altered biotite monzogranite overlain by non-metamorphosed Devonian Old Red Sandstone is locally encountered on the Elbow Spit High in the northern Dutch offshore (well A17-01, Fig. 1.6) and represents the only direct evidence of Caledonian basement in the subsurface of the Netherlands (Frost et al., 1981; Pharaoh et al., 1995). The Old Red Sandstone was derived from the eroding Caledonides to the north (e.g. Lundmark et al., 2013).

Late Paleozoic, early Variscan extension and Variscan Orogeny

During the Middle to Late Devonian, when the area of the Netherlands had drifted further North into equatorial regions, Gondwana began to converge with Laurussia. Before the Rheic Ocean between Gondwana and Laurussia had completely closed, a short-lived oceanic basin opened to the south of Avalonia: the Rhenohercynian Ocean or Sea (see Vis et al., 2025, this volume). In SW England ophiolites of the Lizzard Complex are related to this ocean, but in the Netherlands and Belgium no traces related to this phase have been identified. The Late Devonian to Early Carboniferous post-Caledonian period is marked by widespread extension in the area north of the London-Brabant Massif (e.g. Leeder, 1982; Fraser & Gawthorpe, 1990; Kombrink et al., 2010b). Extension reached as far north as the Midland Valley of Scotland (e.g. Leeder, 1982; Fraser & Gawthorpe, 1990) and the Norwegian-Danish Basin (Kalani et al., 2020). This early Variscan extension created typical horst-and graben structures that are well known from England, Ireland and the UK North Sea, where they lie at or near the surface due to Cenozoic uplift and erosion. Deep seismic lines from the Dutch onshore show evidence of large fault blocks bounded by NNW-SSE trending faults (Fig. 1.7; Kombrink et al., 2010b; Bouroullec et al., 2019; Ten Veen et al., 2019). Late Devonian-early Carboniferous extension-related magmatism is known from the UK and Ireland (Timmerman, 2004), from the Embla Field on the northern flank of the Mid North Sea High (Lundmark et al., 2012) and as a thin layer of altered rhyolitic volcanics or quartz porphyry (Van Bergen & Sissingh, 2007) in the Old Red Sandstones of the Buchan Formation on the Elbow Spit High (well A17-01).

Thick Upper Devonian and Lower Carboniferous carbonate sequences were deposited around the London-Brabant Massif and further north on the uplifted fault blocks. In fact, the well-known pattern of basins and highs in the North Sea and the Netherlands seems to have originated from this early Variscan extension, although it remains to be seen if at that time the Central Graben was a basin or simply a large shear zone (e.g. Smit et al., 2018).

Whereas the horst-and-graben structural style is best known, the most dramatic extension may have taken place along the Caledonian Thor Suture where deep seismic refraction profiles image a 50-100 km wide zone of low P-wave velocity in the lower crust (Smit et al., 2016). This zone may represent the remnants of a Caledonian accretionary complex associated with the Thor Suture that collapsed during early Variscan extension (Fig. 1.6; Smit et al., 2016). The northern boundary of Avalonia follows the Terschelling Basin and the western margin of the Dutch Central and Step grabens (Smit et al., 2016). The northern part of the Dutch Central Graben is located on the fossil margin of Baltica, while its southern part lies on the Avalonian margin. Thus three separate provinces form the basement of the Netherlands, namely Gondwana-derived Avalonia, including the London-Brabant Massif, and the Caledonian basement in the form of the remnant Tornquist accretionary complex in the northern offshore (Smit et al., 2016, 2018; Figs 1.5, 1.6).

The London-Brabant Massif has remained a structural high since its formation in late Silurian-Early Devonian times (e.g. Pharaoh, 2018; Smit et al., 2018). Other structural highs, including the Elbow Spit High (Ter Borgh et al., 2018b), the Groningen Platform and most likely many if not all other long-lived highs in the Netherlands (e.g. Kombrink, 2008; Smit et al., 2018), have been highs since the Late Devonian-early Carboniferous early Variscan extension. Seismic data clearly show units older than the Devonian red beds of well A17-01 onlapping onto the western

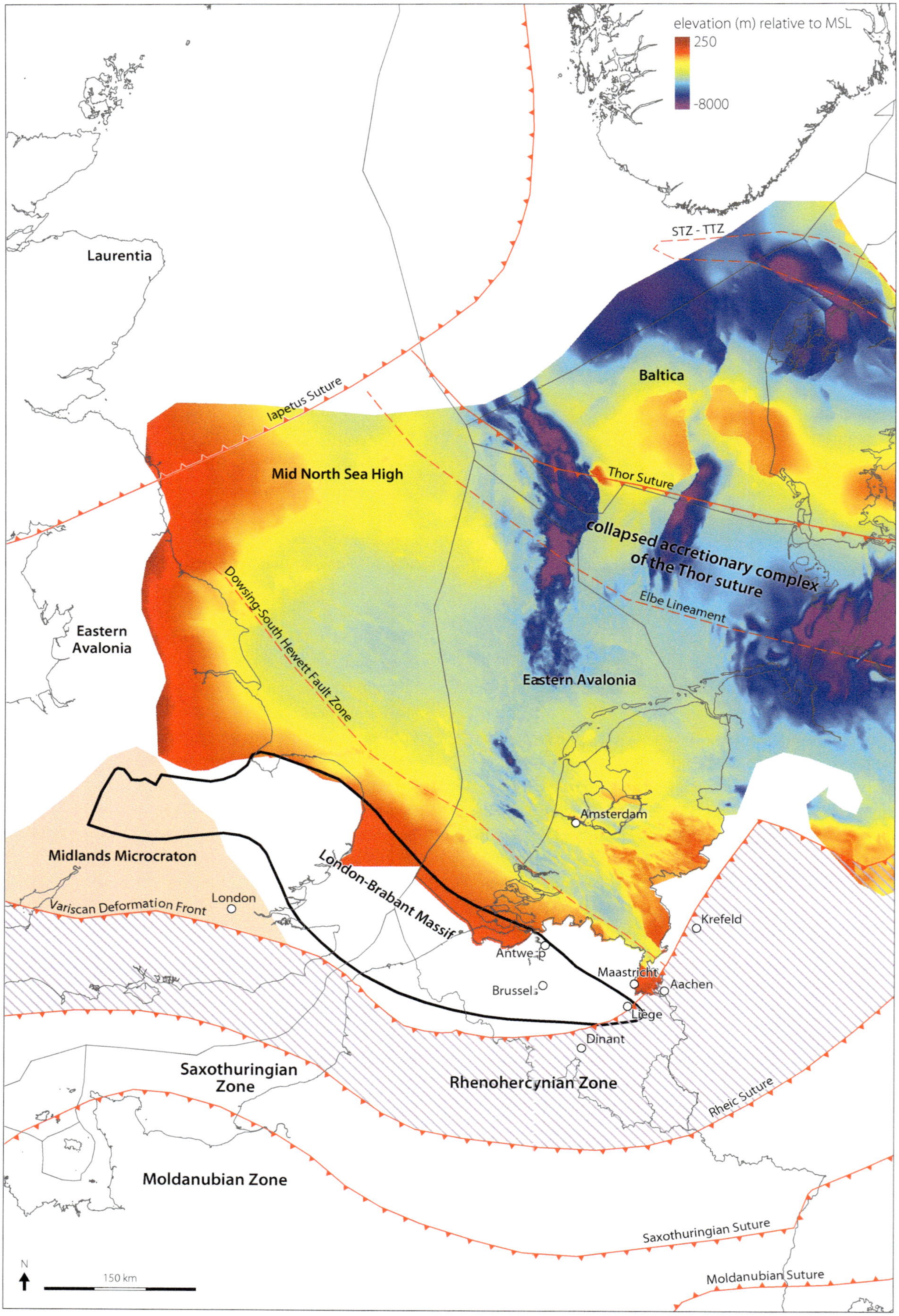

elevation (m) relative to MSL
250
-8000
STZ - TTZ
Laurentia
Iapetus Suture
Baltica
Mid North Sea High
Thor Suture
collapsed accretionary complex of the Thor suture
Elbe Lineament
Eastern Avalonia
Dowsing-South Hewett Fault Zone
Eastern Avalonia
Amsterdam
Midlands Microcraton
London-Brabant Massif
Variscan Deformation Front
London
Krefeld
Antwerp
Maastricht
Aachen
Brussels
Liège
Dinant
Saxothuringian Zone
Rhenohercynian Zone
Rheic Suture
Moldanubian Zone
Saxothuringian Suture
Moldanubian Suture
N
150 km

← ***Figure 1.5.*** *Present-day tectonic setting of Eastern Avalonia and surrounding tectonic plates and -zones. The Caledonian Thor Suture represents a collapsed accretionary complex that separates Avalonia from Baltica and which is bounded to the south by the Elbe lineament. The contact between Laurentia and the Avalonian lithosphere more or less coincides with the accretionary wedge of the Southern Uplands of England and Ireland. See text for further explanation. Modified from Smit et al. (2018) and references therein. The top Carboniferous depth map is based on DGM-deep v5, www.nlog.nl (2019) for the Netherlands area, and on Kombrink et al. (2010a) for the area outside the Netherlands.*

flank of the Elbow Spit High (Fig. 1.6). Furthermore, the Carboniferous units appear to thin onto this high as well as onto the Mid North Sea High. Source-rock maturities measured in Dinantian (lower Carboniferous, Lower to Middle Mississippian) coals in well E02-01 on the flank of the high suggest that the maximum thickness of the overlying Carboniferous succession was about 1600 m, significantly less than elsewhere (De Jager, 2007; Ter Borgh et al., 2018a).

Around the Dinantian-Namurian transition (~330 Ma), the onset of the Variscan Orogeny at the southern margin of Avalonia marked the final closure of the Rheic Ocean and the end of extension in the North Sea and surrounding areas (e.g. Zwart & Dornsiepen, 1978; Ziegler, 1990b). The amalgamation of Gondwana and the remaining micro-continents to Laurussia completed the assembly of Pangea, and the area of the Netherlands became landlocked. At that time, around 330 Ma, the Netherlands had moved into the northern hemisphere, but remained in the humid equatorial climate belt. The formation of a flexural foredeep in combination with post-rift thermal subsidence caused deepening and drowning of the Dinantian basin and the onset of clastic sedimentation (e.g. Kombrink, 2008). During the Variscan Orogeny in Namurian-Westphalian times (Late Mississippian-Middle Pennsylvanian) (Fig. 1.4), much of the Southern Permian Basin, including the Netherlands, became part of the foredeep of the fold-and-thrust belt, although it seems to have been little affected by synsedimentary tectonism. The Limburg Group consists of a largely marine and lacustrine Namurian succession, overlain by Westphalian coastal-plain and fluvial-plain deposits with coal seams and thin marine intercalations. The deposits of this group accumulated in the quickly subsiding area which is now known as the Netherlands and locally reached a thickness of 5500 m.

The front of the Variscan fold-and-thrust belt moved northwards during the late Carboniferous and its final position runs approximately E-W through Belgium, just south of the Netherlands, swinging to the northeast into Germany (Ziegler, 1990a). The northwest-directed convergence caused inversion of the Late Devonian-early Carboniferous, NE-SW oriented basins in the UK and Ireland (e.g. Bowland, Dublin, Cleveland Basins) as far north as the Midland Valley in southern Scotland during late stages of the Variscan Orogeny (late Westphalian, late Middle Pennsylvanian, ca. 310 Ma; Corfield et al., 1996). During this inversion, at least part of the mostly NW-SE oriented faults and basins in the Southern North Sea, such as the Sole Pit and Silverpit Basins, were reactivated in transpression with estimated uplifts of ca. 1 km (Corfield et al., 1996). Strike-slip reactivation of NW-SE basement faults must have occurred in the Netherlands as well but has not been reported as such, although a local NW-SE trending anticline west of the Cleaverbank Platform may very well be inversion-related (e.g. Schroot & De Haan, 2003). Long wavelength folding, probably at a (upper) crustal scale is visible in the erosion pattern on Base Permian Unconformity subcrop maps (e.g. Schroot & De Haan, 2003). In Germany and in the east of the Netherlands, Stephanian strata (Upper Pennsylvanian, Late Carboniferous) locally unconformably overlie the Westphalian (Ziegler, 1990a). These youngest Carboniferous units are red beds that were deposited after the area of the Netherlands had moved into the arid climate zone of the northern hemisphere at some 15° to 20° latitude (Torsvik & Cocks, 2016).

Permian, late-Variscan tectonics and magmatism

At the close of the Carboniferous and start of the Permian, western Europe became affected by large scale late-Variscan wrench faulting. More or less simultaneous intrusive and extrusive magmatism and thermal uplift resulted in widespread and deep erosion (Ziegler, 1990a). The NW-SE trend was reactivated again in response to this late-Variscan Stephanian-Autunian (Late Pennsylvanian- 305-290 Ma) wrench deformation while regionally a conjugate set of NE-SW to NNE-SSW faults developed (Ziegler, 1990a). The subcrop pattern of the Westphalian below the regional Base Permian Unconformity anticipates the shapes of the much younger West Netherlands Basin, Lower Saxony Basin and Lauwerszee Trough that became areas of less uplift and erosion. What is now known as the Texel-IJsselmeer High was part of a larger E-W trending uplifted zone. The Hantum Fault Zone (Fig. 1.1) at the western margin of the Lauwerszee Trough was active at this time, juxtaposing Westphalian B and C against Westphalian A. The youngest pre-Permian deposits (Westphalian D to Stephanian) are preserved in the Lower Saxony and West Netherlands basins and in a small erosional remnant on the Cleaverbank Platform. These areas experienced the least uplift during the Saalian event. Strong uplift occurred to the north of the Dutch sector on the Mid North Sea High and Ringkøbing-Fyn High.

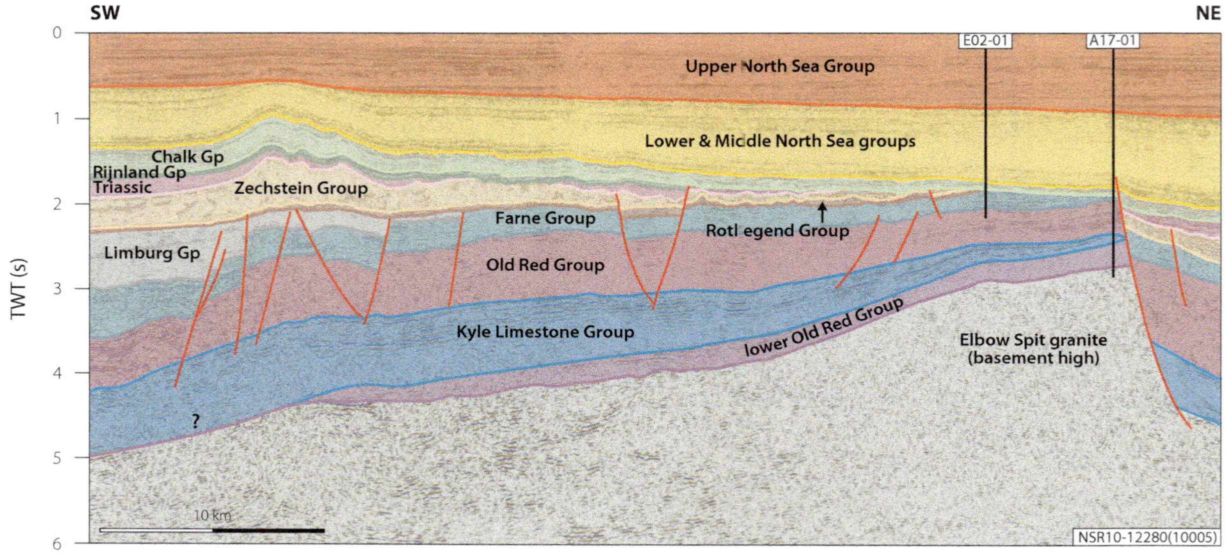

Figure 1.6. Seismic time section across the Elbow Spit High showing the onlap of Devonian deposits onto crystalline basement. For location see Fig. 1.1.

Evidence for the presence of late Carboniferous or early Permian extensional tectonics in the Dutch subsurface comes from the area along the Dutch-German offshore border north of the Wadden Islands. Seismic interpretation there has revealed a somewhat thicker lowermost section of the Upper Rotliegend located in subtle fault-bounded north-south trending grabens (Burgess et al., 2018; Doornenbal et al., 2019). Further to the east in northern Germany, narrow fault-bounded basins in which thick Lower Rotliegend volcanics and clastic sediments accumulated as syn-rift deposits had been well documented already for a long time (Ziegler, 1990a). Other occurrences suggesting active tectonism in the Netherlands during this period are thin volcanic sequences along and within the Ems Low in the easternmost onshore area, and in the Dutch Central Graben and Outer Rough Basin in the northern offshore (Geluk, 2007). In the far northern offshore, on the Step Graben, NE-SW faults may bound local late Carboniferous-early Permian extensional depocentres of the so-called Urania Graben (Ter Borgh et al., 2018b).

High values of vitrinite reflectance measured in the Westphalian in some wells on the Groningen Platform (Kettel, 1983) and in the central Netherlands (Bonté et al., 2020), in combination with data from apatite fission-track analysis (NAM data), indicate that a regional heat pulse affected this area during late Carboniferous and/or early Permian times. This heat pulse was likely related to widespread magmatism in the Northern and Southern Permian basins (Ziegler, 1990a; Benek et al., 1996; Heeremans et al., 2004; Timmerman, 2004; McCann et al., 2006), since younger strata are not affected (Bonté et al., 2020). Heat flow modelling indicates that widespread Permo-Carboniferous magmatic intrusions in the upper crust can explain the observed high maturity-depth trends. These trends cannot be explained by heating of the lithosphere from below by mantle upwelling and crustal underplating alone (Bonté et al., 2020).

After the Variscan orogeny: thermal subsidence

Regional subsidence, induced by the decay of the thermal anomaly introduced during the Permo-Carboniferous tectonic-magmatic event (Wilson et al., 2004 and papers herein) commenced while the Permian and Rotliegend clastics began to be deposited in the large E-W trending Southern Permian Basin (Van Wees et al., 2000). By then, the Netherlands was located within the arid climate zone at ca. 20° north (Fig. 1.4). The Southern Permian Basin was located at the centre of Pangea, bordered to the south by the Variscan Mountains and the stable London-Brabant Massif, and to the north by the Mid North Sea and Ringkøbing-Fyn highs. Clastic sediments entered the basin probably from all sides, but the main influx was from the south, where the Variscan mountain belt was progressively eroded. The Rotliegend clastics onlap the southern and northern basin margins and comprise conglomerates and eolian and fluviatile sandstones, passing towards the basin centre and stratigraphically upward into lacustrine siltstones and claystones, with further north also intercalations of evaporites. Rotliegend clastic sediments are proven further west in the Cygnus gas field offshore UK (Taggart & Catto, 2015), but remain unproven in Dutch waters. The forerunners of the Dutch Central Graben, Broad Fourteens Basin and Lauwerszee Trough are already subtly expressed in Rotliegend isopach patterns.

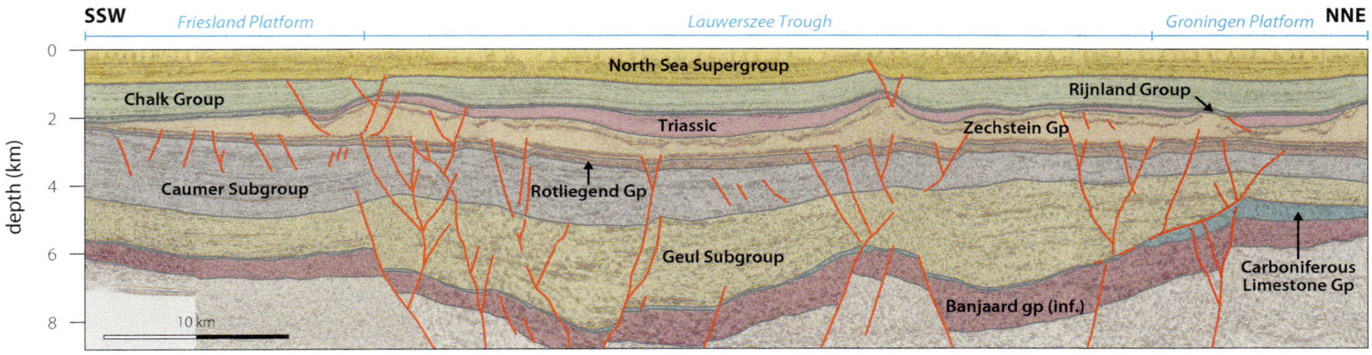

Figure 1.7. Deep seismic depth section across the eastern Netherlands. At pre-Carboniferous levels several major tilted fault blocks indicate significant early extensional faulting. For location see Fig. 1.1.

With subsidence rates exceeding sedimentation rates, a landlocked depression, probably several hundreds of metres below global sea levels, developed under an arid climate (Bouroullec & Geel, 2025, this volume). Early during the Late Permian, after an initial catastrophic infill by seawater, this depression was filled with several autocyclic cycles of the halite-dominated Zechstein sequence (Van den Belt, 2012). The autocyclic nature of the Zechstein is explained here as a response to rapid isostatic compensation and a self-sustaining process of subsidence and infill facilitated by thin, fractured crust and arid climate conditions. The original thickness of this evaporitic sequence was up to almost 1500 m in the axial parts of the Southern Permian Basin (Ten Veen et al., 2012). Where the salt sequence is well developed, in roughly the northern half of the Dutch subsurface, halokinesis played a very strong role in the subsequent structural deformation history (Ten Veen et al., 2012; Van Winden et al., 2018).

Mesozoic, Cimmerian breaking-up of Pangea

Rifting related to the Mesozoic break-up of the Pangea supercontinent commenced during the Triassic in the Arctic-North Atlantic and between Greenland and Scandinavia, and slowly propagated southwards into the Central Atlantic domain along the line of future continental break-up. An eastern branch of crustal extension propagated during the Early Triassic into the North Sea area (Ziegler, 1988, 1990a). During the Middle Triassic, this extension had reached the southern North Sea, although the degree of extension rapidly decreases southwards. Continued extension in the western rift branch resulted during the Middle Jurassic in continental break-up and the opening of the Central Atlantic Ocean. Towards the end of the Early Cretaceous, crustal separation was achieved in the North Atlantic, and rifting concentrated on the area of the Norwegian and Greenland seas (Ziegler, 1988, 1990a). By this time, rifting in the Netherlands had effectively ceased, so the North Sea Central Graben system can be considered to represent a failed rift, ongoing extension being accommodated further west. While Europe and North America broke apart, rifting in the Mediterranean domain culminated in crustal separation between Europe and Africa, forming a part of the Tethys Ocean.

During the Triassic to Early Cretaceous continental break-up, the Netherlands moved north from the arid climate zone into sub-tropical latitudes of the northern hemisphere (Fig. 1.4e).

Triassic and Early Jurassic sedimentation in the Netherlands took place under continuing subsidence. This led to broad, regular facies patterns, interrupted locally by salt movement, which first became widespread after the Early Triassic crustal extension. Fine-grained clastics of the Lower Triassic Lower Buntsandstein Subgroup were deposited in brackish to saline terrestrial and lacustrine environments and were followed by coarser-grained fluviatile and eolian clastics of the Main Buntsandstein Subgroup, which were derived from the south and split up northward (distally) into several thinner sandstone units. During the Middle Triassic, marine environments took over, leading to deposition of the claystones, carbonates and evaporites of the Röt and Muschelkalk formations. Triassic extension is well documented in the Terschelling Basin and Central Graben (Peeters et al., 2018; Van Winden et al., 2018). The Horn and Glückstadt Grabens, which cut the Ringkøbing-Fyn High and extend into the Southern Permian Basin in the German offshore, are the most significant Triassic grabens south of the Ringkøbing-Fyn High (e.g. Ziegler, 1990a; Pharaoh et al., 2010; Kilhams et al., 2018). In the Dutch subsurface, Triassic sequences thicken into the Terschelling Basin (Peeters et al., 2018) and the Dutch Central Graben and Broad Fourteens Basin, but it is difficult to relate any thickness variations across faults to syn-depositional faulting. Significant thickness variations and deposition of local sandstones (e.g. Middle Solling sandstone member) do, however, locally reflect the presence of syn-sedimentary faults in relation

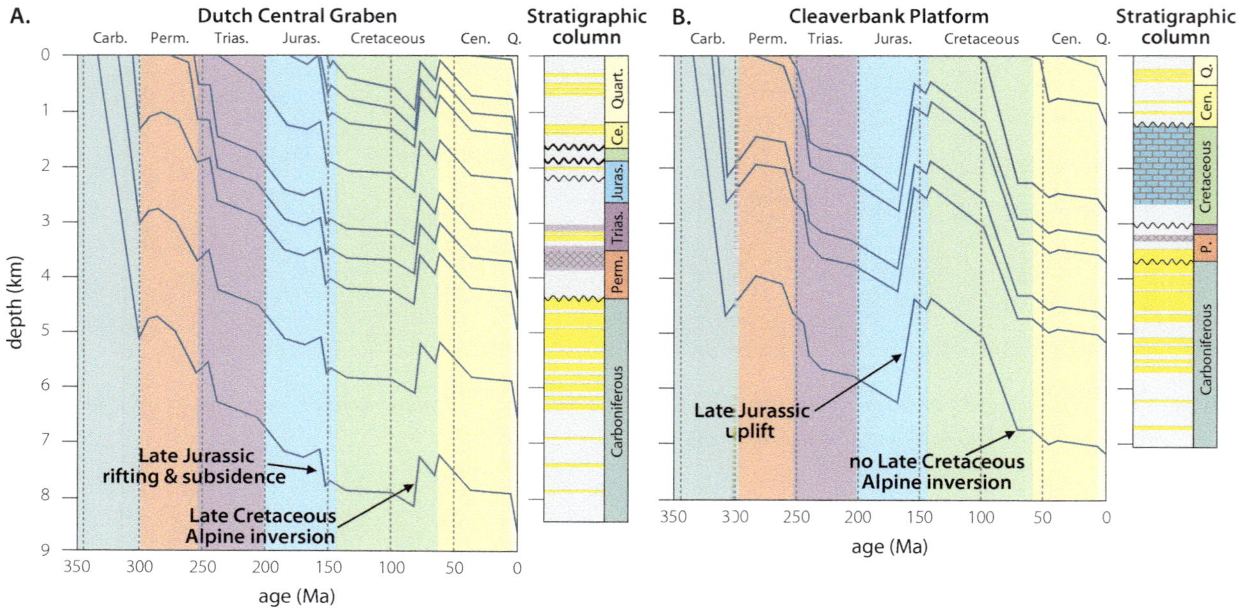

Figure 1.8. Burial graphs. a) Dutch Central Graben, showing accelerated syn-rift subsidence during the Late Jurassic and two pulses of basin inversion (uplift) during the Late Cretaceous and Cenozoic. Note the strong subsidence during the Quaternary. b) Cleaverbank Platform, showing uplift of this platform area during the Late Jurassic rifting, and no indications of Late Cretaceous uplift (inversion).

to Triassic collapse features (De Jager, 2012). In the West Netherlands Basin, to the south of the Zechstein evaporite basin, the Triassic shows only minor thickness variations across faults. Thickness variations show that most of the early salt swells formed during deposition of the Triassic Solling, Röt and Muschelkalk formations, and that piercing salt domes and associated rim-synclines developed later, during the deposition of the mainly fine-grained coastal-plain to marine clastics of the Keuper Formation (Van Winden et al., 2018). Although the onset of halokinesis is likely to have been triggered by tectonic activity, in particular where salt walls follow major sub-salt faults (Remmelts, 1995, 1996) in the Dutch Central Graben and Terschelling Basin, where the Zechstein salt reaches great thicknesses, it is difficult to directly link faults in the Triassic cover to sub-salt faults. However, it is likely that Triassic extensional faulting indeed did take place there, probably decreasing in intensity towards the south (e.g. Remmelts, 1995, 1996; Bouroullec et al., 2018; Peeters et al., 2018). No heat-flow anomaly nor any igneous activity seems to have accompanied this Triassic extension in the southern

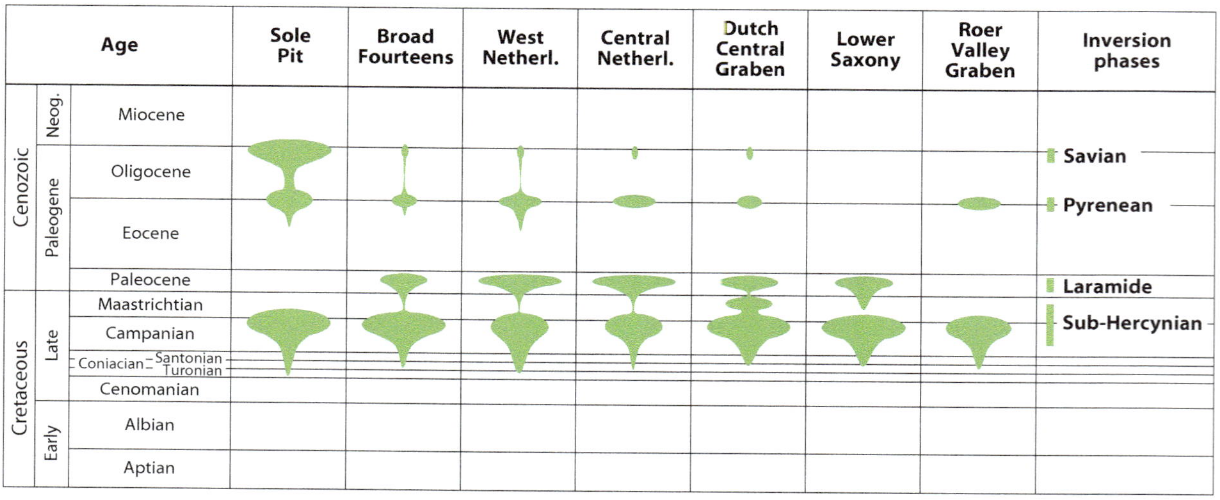

Figure 1.9. Timing and relative intensity of Alpine inversion pulses in the Dutch basins and Sole Pit Basin. Note that inversion pulses occurred simultaneously in different basins, but with varying magnitudes. Only the Sub-Hercynian pulse appears to have affected all basins. Modified from De Jager (2007).

North Sea, and uplift of rift shoulders cannot be demonstrated.

Truncation and erosion of Lower Triassic deposits below the regional Hardegsen Unconformity (also referred to as Base Solling Unconformity) at the base of the Solling Formation indicates that differential uplift took place during the transition from Early to Middle Triassic in the south of the Southern Permian Basin (McKie & Kilhams, 2025, this volume). The main uplift occurred in the Netherlands Swell, in the area of the future Texel-IJsselmeer High. No significant faulting is associated with this phase of uplift.

The Late Triassic and Early Jurassic were periods of relative tectonic quiescence. Some faulting continued in the Dutch Central Graben and locally in the Broad Fourteens Basin, but otherwise slow regional subsidence seems to have prevailed. During the Early Jurassic a wide epicontinental sea had developed in which locally more than 1800 m of fine-grained clastics of the Altena Group accumulated with, at the end of the Early Jurassic, deposition of the organic-rich shales of the Posidonia Formation (Wong, 2007).

During the Middle Jurassic, much of the North Sea, including large parts of the Dutch offshore area, experienced thermal uplift leading to erosion at the regional early Middle Jurassic Aalenian (or Mid-Cimmerian) Unconformity (Ziegler, 1990a; Underhill & Partington, 1993), and sedimentation became restricted to the rift basins. A recent study has re-evaluated this Mid-Cimmerian event based on a regional compilation of vertical motions from available well data (LaFosse & Smit, 2020). Combined with results from the ever-expanding literature, the latest global eustatic curve and modern geodynamic principles, the Mid-Cimmerian event can alternatively be explained by a combination of Late Toarcian global eustatic sea-level fall (e.g. Haq, 2018) and processes related to Late Cimmerian rifting, including rift shoulder uplift and volcanism.

During the Callovian and Oxfordian, regional sedimentation resumed. Following the Middle Jurassic crustal separation in the Central Atlantic and Tethys domains, rifting accelerated in the North Sea rift system, forming the Viking and Central grabens of Norway and the UK, the Danish Tail End Graben and the Dutch Central Graben (Fig. 1.4). Southwards the N-S trend terminated against the northern margin of Avalonia. As a result of the pervasive NW-SE structural trend that was inherited from Neoproterozoic times when Avalonia was located at the active margin of Gondwana, Jurassic crustal-extension strain was distributed over a broad area, causing transtensional development of the NW-SE trending Broad Fourteens, West Netherlands, Central Netherlands and Vlieland basins and the Roer Valley Graben. The Sole Pit Basin and the Lower Saxony Basin also resulted from this extension. Thus, the main tectonic elements in the subsurface of the Netherlands were formed mainly during Late Jurassic and Early Cretaceous Late Cimmerian rifting (Figs 1.1, 1.3). These extensional basins are filled with locally more than 2500 m of sandstones and claystones, which, depending on the basin involved, belong to the largely continental Schieland Group, the mainly marine Scruff Group or the continental to restricted marine Niedersachsen Group. Simultaneously, the adjacent platforms were uplifted and deeply truncated. Comparison of burial diagrams of the Dutch Central Graben, as an example of a Cimmerian rift basin, and of the Cleaverbank Platform as an example of a platform area, highlights the difference in burial history, which is further discussed in the section on structural elements (Fig. 1.8). The uplift culminated in the widely recognized Base Cretaceous (or Late Cimmerian) Unconformity, indicating the end of rifting. Following this, a large shallow marine basin formed with deposition of the sediments of the Rijnland Group: up to 1400 m of mainly fine-grained marine clastics with coastal sandstones (Jeremiah et al., 2010).

The regional extensional stress field was oriented E-W at the onset of Jurassic rifting, but from the middle of the Kimmeridgian onward the stress field was oriented NE-SW (Nalpas et al., 1995; Zanella & Coward, 2003; Verreussel et al., 2018). Only the Dutch Central Graben and subbasins within the Step Graben area conform to the initial E-W orientation (Ter Borgh et al., 2018b). The NW-SE trending basins that follow older structural trends were probably dextrally reactivated in transtension (see Ter Borgh et al., 2018b), although as the main fault systems were repeatedly reactivated under different stress regimes and directions, fault patterns generally do not show unambiguous evidence of oblique movements. The fault pattern seems to clearly indicate dextral displacements only locally, such as along the Rifgronden Fault Zone (Fig. 1.1) between the Terschelling Basin and Schill Grund Platform (De Jager, 2007).

Late Cretaceous to early Cenozoic, Alpine inversion

Following the mid-Cretaceous opening of the North-East Atlantic, extensional stresses became progressively concentrated on the area between Greenland, the British Isles and Norway, and tectonic activity abated rapidly in the Netherlands (Ziegler, 1988, 1990a). The Late Cretaceous was characterized by rising relative sea levels, with younger sediments overstepping the depositional limits of older sequences, until the entire area of the Netherlands was submerged in a shallow sea. The up to 1500 m of chalk deposited in this sea forms the only significant carbonate unit in the area since the Triassic.

During the Late Cretaceous, Africa-Arabia began to converge with Eurasia and the Tethys system of oceanic basins started to close. This resulted in the gradual devel-

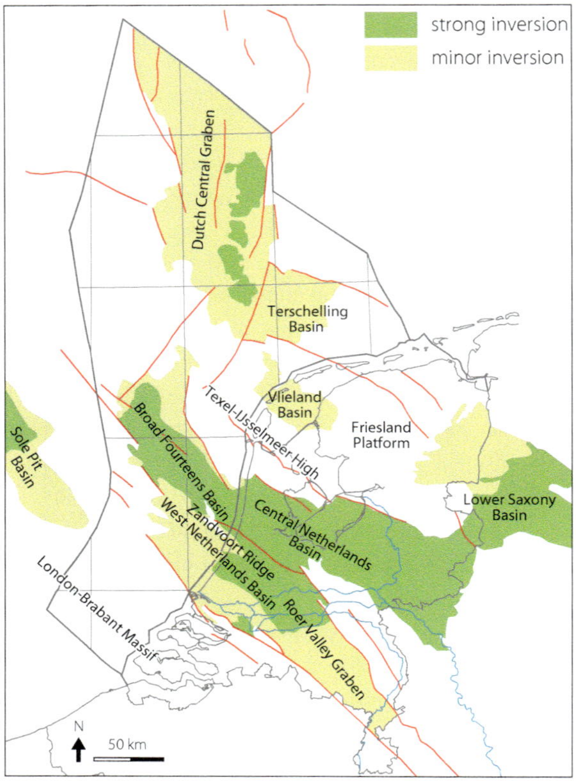

Figure 1.10. Inverted basins in the Netherlands. As a result of erosion and/or limited deposition the Upper Cretaceous chalk is thin in mildly inverted areas or absent in strongly inverted areas. Modified from De Jager (2007).

opment of the Alpine orogenic system. During the Late Cretaceous and Paleocene, increasing stresses were exerted onto its northern foreland, inducing the inversion of Mesozoic extensional basins and apparently impeding the crustal separation between Greenland and Norway, which was achieved only at the end of the Paleocene when these compressional stresses abated (Ziegler, 1988, 1990a). The Cenozoic North Sea Supergroup overlies a peneplane surface of uplifted, deformed and eroded older rocks. Above the inverted rift basins there is a clear angular unconformity, indicating that major inversion had ended by then. During the Cenozoic two minor inversion pulses occurred but, with the exception of the West Netherlands Basin and the Sole Pit Basin in the UK, they resulted in barely any erosion. In all other basins, subsidence outpaced uplift during these pulses.

Inversion-related uplift of the basins resulted in depositional thinning and in erosion of the Upper Cretaceous chalk and lower Cenozoic clastics, as well as in the local truncation of older sedimentary rocks. On the flanks of the basins, thinning of Upper Cretaceous or Lower Cenozoic deposits towards the inversion axes shows that uplift was a continuous process, albeit with several acceleration pulses. These pulses seem to have been simultaneous in most inverted basins (Fig. 1.9; De Jager, 2003) and can be correlated with increased convergence rates in the Alpine domain (Deckers & Van der Voet, 2018). Previously, the Laramide inversion pulse at the end of the Paleocene was considered to be the strongest one, but the latest views assign more importance to the Campanian Sub-Hercynian pulse. In addition, it is now concluded that in the northern offshore the Sub-Hercynian pulse may consist of two separate pulses (Deckers & Van der Voet, 2018; Van der Voet et al., 2018).

Inversion started at the onset of the Late Cretaceous with the Sub-Hercynian pulse. In most basins this was the strongest inversion pulse (Deckers & Van der Voet, 2018) and it was probably mainly related to the rotation of the Iberian Peninsula (Kley & Voigt, 2008). It consists of a stronger peak during the Campanian followed in basins in the northern offshore by a minor pulse during the Maastrichtian. Inversion resulted in uplift and reverse (transpressional) reactivation of pre-existing normal faults. Chalk sequences thin towards the inversion axes, with post-Campanian chalk onlapping unconformities in the Broad Fourteens and Central and West Netherlands basins, the Dutch Central Graben and the Roer Valley Graben (Gras, 1995; Gras & Geluk, 1999).

A second pulse of inversion, the Laramide pulse, took place during the Paleocene. It coincides with the end of chalk deposition and the beginning of deposition of the siliciclastic North Sea Supergroup. Noteworthy is that in most basins this phase resulted in the broad uplift of the old rift basins (De Jager, 2003; Deckers & Van der Voet, 2018), but only locally in reactivation of faults. In the Dutch Central Graben and the Broad Fourteens Basin, a thin layer of post-Campanian chalk has locally been preserved on top of the Sub-Hercynian unconformity. Some authors question whether any significant erosion occurred as a result of this pulse (e.g. Van Lochem, 2018), while others have observed a separate angular unconformity above the Sub-Hercynian unconformity at the base of the Cenozoic North Sea Supergroup (e.g. Van der Voet et al., 2018).

At the end of the Eocene, the Pyrenean pulse caused broad uplift of the West and Central Netherlands basins and of the Roer Valley Graben. The amplitude of this uplift decreases sharply towards the northwest in the Broad Fourteens Basin (De Jager, 2003).

An inversion pulse at the end of the Oligocene caused considerable uplift in the UK Sole Pit Basin (Van Hoorn, 1987; Pharaoh et al., 2010), but no significant inversion can be demonstrated for that time in the Netherlands. Coeval with basin inversion, the European Cenozoic Rift System formed, with the Rhône and Rhine Grabens as main elements (Ziegler, 1994; Ziegler et al., 1995). This rifting was an integral part of the late Alpine deformation in the northwestern Alpine foreland.

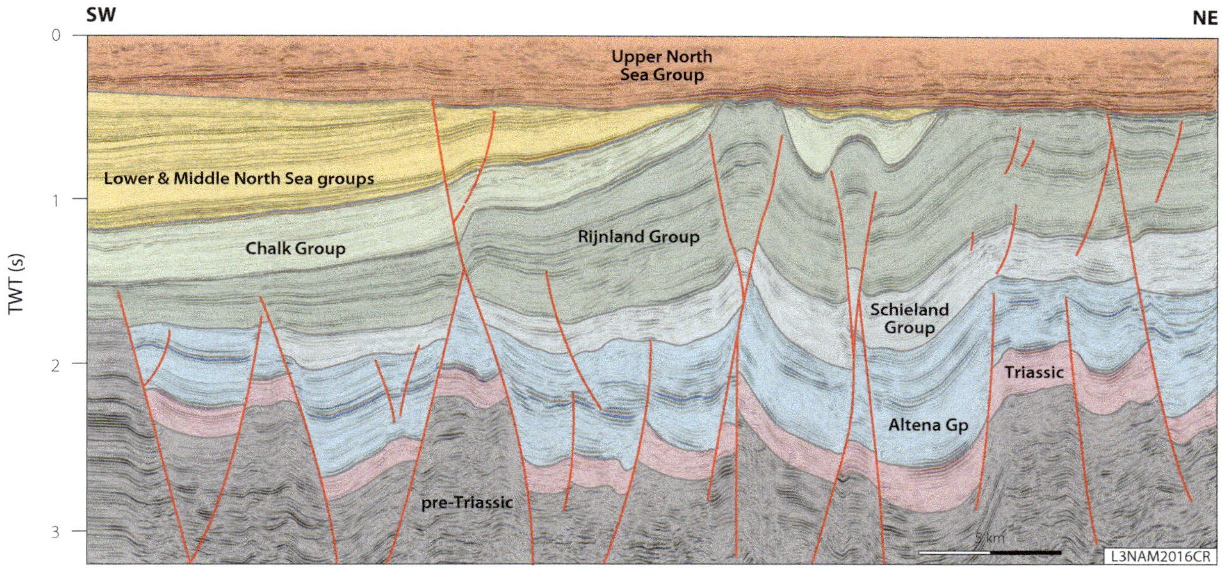

Figure 1.11. Seismic time section running from the southwestern flank of the West Netherlands Basin toward the inverted part of the basin. For location see Fig. 1.1. Note the increased thickness of the Jurassic in the basin and how the Lower Cretaceous onlaps the southwestern basin margin. Chalk deposits show depositional thinning towards the inversion axis and are truncated by the Base Lower North Sea Group Unconformity. During the Late Cretaceous inversion, positive flower structures (pop-ups) formed through transpressional reactivation of pre-existing extensional rift faults. Oligocene inversion caused broad basin uplift, without significant faulting.

The deformation related to these inversion pulses varied in intensity from basin to basin, and not all pulses can be recognized in all inverted basins (Figs 1.9, 1.10). In some cases, an older inversion unconformity may have been removed by erosion related to a later phase of uplift. In other cases, there is little doubt that an inversion pulse has left a basin unaffected. The Laramide inversion phase, for example, had little or no impact on the Lower Saxony Basin (Baldschuhn et al., 1991; Deckers, 2015).

The presence of Zechstein salt had a marked influence on how basin inversion manifests itself (De Jager, 2003). In the West Netherlands Basin, where no Zechstein salt is present, pre-existing faults were reversely reactivated forming prominent ridges or flower structures (Fig. 1.11). In the Dutch Central Graben, with more than 1 km of Zechstein salt, faults above and below the salt are entirely detached from each other, resulting in a broad uplift of post-salt deposits (Fig. 1.12). Along the northeast flank of the Broad Fourteens Basin, where the Zechstein salt is at most 200 m thick, the salt acted as a detachment surface along which the basin-fill was thrust northeastwards onto the adjacent platform (Fig. 1.13; Hooper et al., 1995; Nalpas et al., 1995).

Detailed analyses of vitrinite-reflectance, apatite fission-track and fluid-inclusion data indicate that the total uplift during the inversion nowhere exceeded 2 km, and in most places amounted to only 1 to 1.5 km (Fig. 1.8a). This is in disagreement with a reported inversion of up to 3.5 km for some Dutch basins (Dronkers & Mrozek, 1991). It seems that these higher estimates did not fully account for syn-inversion depositional thinning of Upper Cretaceous to lower Cenozoic sequences over the inverted basins (De Jager, 2003).

Late Cenozoic to Recent structural events

Oligocene rifting in the Lower Rhine Graben propagated northwards into the Netherlands, affecting areas as far as the Dutch coast (Ziegler, 1994). The Roer Valley Graben was downthrown, and up to 2000 m of Cenozoic clastics accumulated there (Van Adrichem Boogaert & Kouwe, 1993; Van den Berg, 1994; Van den Berg et al., 1994). Recent earthquakes along the main bounding faults indicate that tectonic activity continues to the present day (Houtgast, 1991; Dost et al., 2025, this volume). In the Southern North Sea Basin, westward prograding deltaic sediments supplied by the Eridanos River, which drained the Baltic Fennoscandian Shield area, were deposited during the late Neogene and Pleistocene (Sørensen & Michelsen, 1995; De Bruin et al., 2017).

Quaternary deposits to the north of the Dutch offshore are locally more than 1000 m thick, thinning towards the south. For the Dutch offshore and northwestern onshore, burial curves show a sharp increase in the rate of subsidence during the last few million years (Figs 1.8a), possibly related to flexural downwarping of the North Sea continental crust in response to the build-up of a NNW-SSE oriented compressional stress field (Kooi et al., 1989; Van Wees & Cloetingh, 1996; Cloetingh & Van Wees, 2005; Scheck-Wenderoth & Lamarche, 2005). While much of the North Sea area is still subsiding, the southeastern part of the Netherlands is rising in conjunction with uplift of the Rhenish Massif, due to the same compressional stresses. An overall review of the Miocene and younger neotectonic activity in the Netherlands was published by Van Balen et al. (2005).

Structural elements

Basins in the northern offshore and Wadden Sea

The Dutch Central Graben, the southernmost expression of the Mesozoic North Sea rift system, is situated in the northern offshore and is flanked by the shallower Step Graben and Terschelling and Vlieland basins (Figs 1.1, 1.2a). A proto-Dutch Central Graben may have been present already during Late Devonian times, when it may have formed a seaway extending from the south into Laurussia (Ziegler, 1990a). Ter Borgh et al (2018b) saw no evidence for this but did find evidence in the Westphalian units subcropping at the Base Permian Unconformity on the Step Graben, that faulting at the end of the Carboniferous was along NE-SW faults, and not along the N-S trend of the later Dutch Central Graben. Lower Permian volcanics are known from the flanks of the Southern Permian Basin (Geluk, 2007). Major salt walls, which started developing during the Triassic, closely follow the boundary faults of the Mesozoic rift basins and their development was most likely triggered by movements along the faults themselves, suggesting that downfaulting of the Dutch Central Graben was well underway already during Middle Triassic times (Remmelts, 1995, 1996; Ten Veen et al., 2012; Van Winden et al., 2018). The increased thicknesses of formations within the graben indicate that differential fault-controlled subsidence continued during Early to Middle Jurassic times (Fig. 1.12). In particular, the Middle Jurassic is thickly developed in the northern sector of the Dutch Central Graben (Heybroek, 1975; Wong, 2007; Bouroullec et al., 2018; Verreussel et al., 2025, this volume). The main rifting event, however, occurred during the Late Jurassic (Oxfordian to early Kimmeridgian), affecting northern areas somewhat earlier than areas further south. Thick, fluvial to shallow-marine sequences accumulated within the graben, while adjacent highs and platforms were uplifted and eroded (Fig. 1.8).

Rifting in the Step Graben and Terschelling Basin started during the Late Jurassic (late Kimmeridgian to Tithonian), probably coinciding with a change from E-W to NE-SW extension (Møller & Rasmussen, 2003; Verreussel et al., 2018). These basins were initially uplifted during the Mid Cimmerian event, leading to erosion of older Jurassic and Triassic deposits, then subsided later to a lesser extent than the Dutch Central Graben, resulting in the deposition of thinner Upper Jurassic sequences. The Terschelling Basin is delineated by major faults. Of these, the Hantum Fault Zone (Figs 1.1, 1.18) along its southern margin was already established during the early Permian, as can be seen from the different Westphalian units that subcrop below the Permian on either side of the fault. While there is less convincing evidence that the Rifgronden Fault Zone (Fig. 1.1) at the northern margin of the basin is also old, its

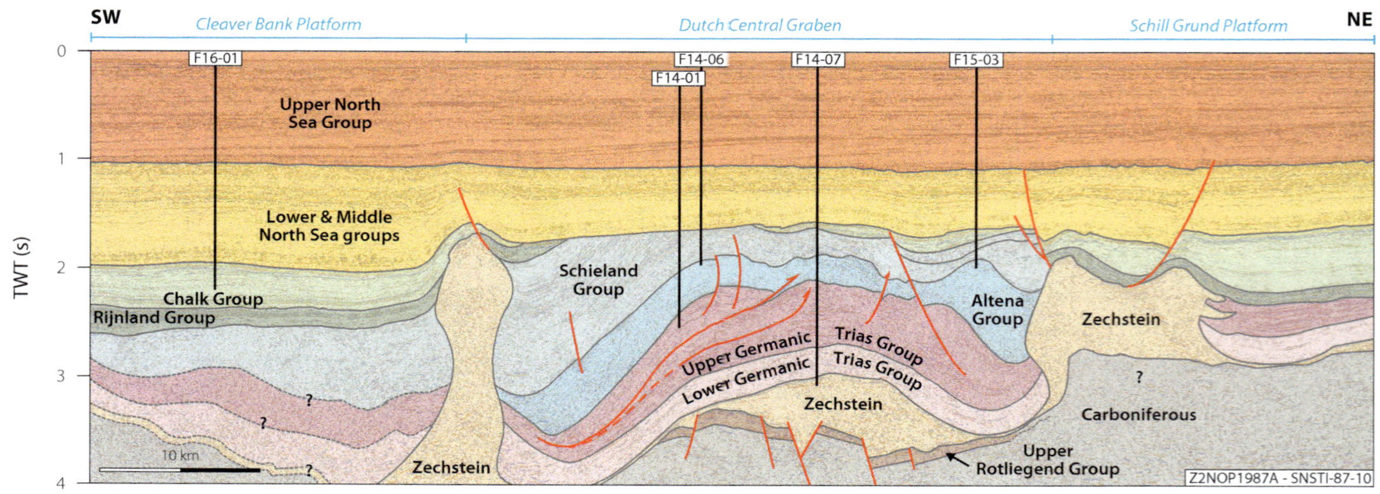

Figure 1.12. Seismic time section across the inverted Dutch Central Graben. For location see Fig. 1.1. The boundaries of the inverted Dutch Central Graben, with its thick development of Jurassic deposits, are marked by salt walls. Inversion above the Zechstein salt is characterized here by a broad dome. Only a thin veneer of mainly post-Campanian Chalk is preserved above the inverted basin.

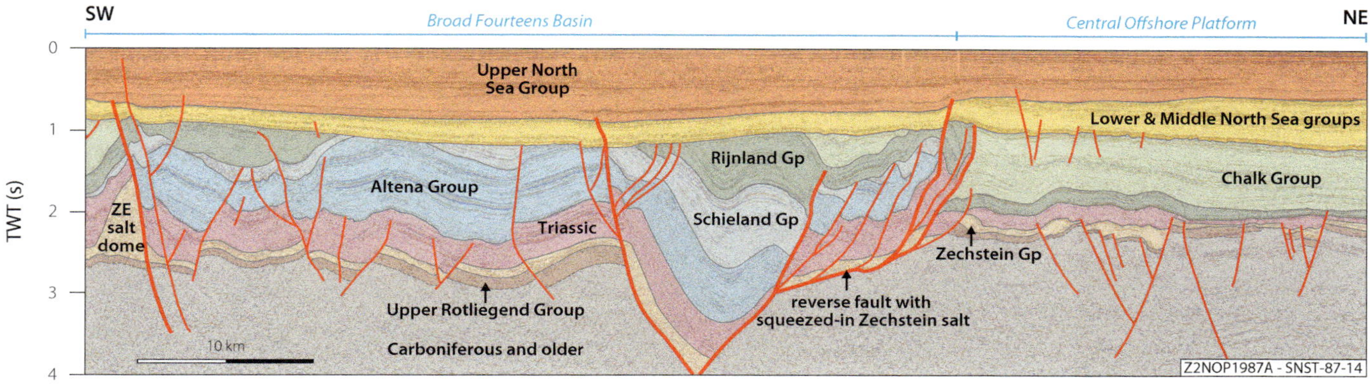

Figure 1.13. *Seismic time section across the northern sector of the Broad Fourteens Basin. For location see Fig. 1.1. Note that the Jurassic is thick within the inverted basin but absent on the adjacent platform. The inversion resulted in an impressive thrust fault along the northeast flank of the basin. Contrary to the West Netherlands Basin (Fig. 1.11), the Oligocene inversion pulse did not affect this part of the Broad Fourteens Basin.*

parallelism to the Hantum Fault Zone and its unusual direction with respect to the Late Cimmerian E-W extension speaks in favour of its origin as a reactivated pre-existing fault.

The development of the Vlieland Basin has been related to dextral shear along a NW-SE trend, linking the southern termination of the Dutch Central Graben with the Lower Saxony Basin (Herngreen et al., 1991). A significant igneous event associated with Late Cimmerian rifting led to the eruption of the Zuidwal Volcano, a major volcanic complex expressed by a marked magnetic anomaly in the Vlieland Basin (Van Bergen et al., 2025, this volume). Almost 1 km of volcanic rocks were penetrated in the Zuidwal-1 well (TD in the volcanics), which appears to have been drilled into the volcanic conduit (Perrot & Van der Poel, 1987). The local temperature anomaly that must have been associated with this igneous activity may be the reason for the limited Late Jurassic subsidence that resulted in just a thin sequence of Upper Jurassic strata below Lower Cretaceous post-rift sedimentary rocks. Investigations by TNO in 2020 (Vis, 2020; Van Bergen et al., 2025, this volume) revealed that a similar volcano erupted contemporaneously in the Dutch Central Graben: the so-called Mulciber volcano in block F16. Based on its magnetic anomaly it appears to have been somewhat smaller than the Zuidwal volcano, (see Van Bergen et al., 2025, this volume).

Several inversion pulses affected the Dutch Central Graben, decreasing in intensity towards the north (see Figs 1.9, 1.10). The most important is the Sub-Hercynian pulse, which resulted in deep truncation and erosion of pre-Campanian deposits. Within the most strongly inverted areas this erosion must have cut down to at least the Lower Cretaceous Vlieland formations. A locally developed intra-Chalk Group sandstone at the level of the Campanian unconformity (in well F17-04) is interpreted as representing the erosional products of Lower Cretaceous and possibly even Jurassic clastics. In the northwestern sector of the Dutch Central Graben Campanian and Maastrichtian pulses (Van der Voet et al., 2018) can be recognized. On the flanks of the main Sub-Hercynian inversion highs a relatively thin veneer (sometimes at the limits of seismic resolution) of post-Campanian chalk is preserved in places on top of the erosional surface. This indicates that the Laramide inversion must have been of significantly less magnitude than the Sub-Hercynian one. Van Lochem (2018) even suggested that there is no indication for Laramide inversion at all in the southern Dutch Central Graben – an angular unconformity truncating the post-Campanian Chalk Group could not be identified and mapped. In the northern Dutch Central Graben an angular Laramide unconformity has been mapped by Van der Voet et al (2018), so all in all, considering the peneplaned surface onto which the North Sea Group Supergroup was deposited, it seems that at least some Laramide uplift and erosion must have occurred.

During the early Cenozoic, the Dutch Central Graben subsided less than its surroundings, resulting in a much thinner Lower Cenozoic sequence; no marked unconformities can be observed in this sequence on seismic data (see Figs 1.12, 1.13).

Basins in the south and west of the Netherlands

The Broad Fourteens, West Netherlands and Central Netherlands basins and the Roer Valley Graben form a NW-SE trending contiguous set of Mesozoic extensional basins (Fig. 1.1). The boundary between the West and Central Netherlands basins is formed by the NW-SE striking Mid Netherlands Fault Zone (Fig. 1.1), which trends northwest from the Peel-Maasbommel Complex into the Zandvoort Ridge, the offshore IJmuiden Platform (Fig. 1.1), and ulti-

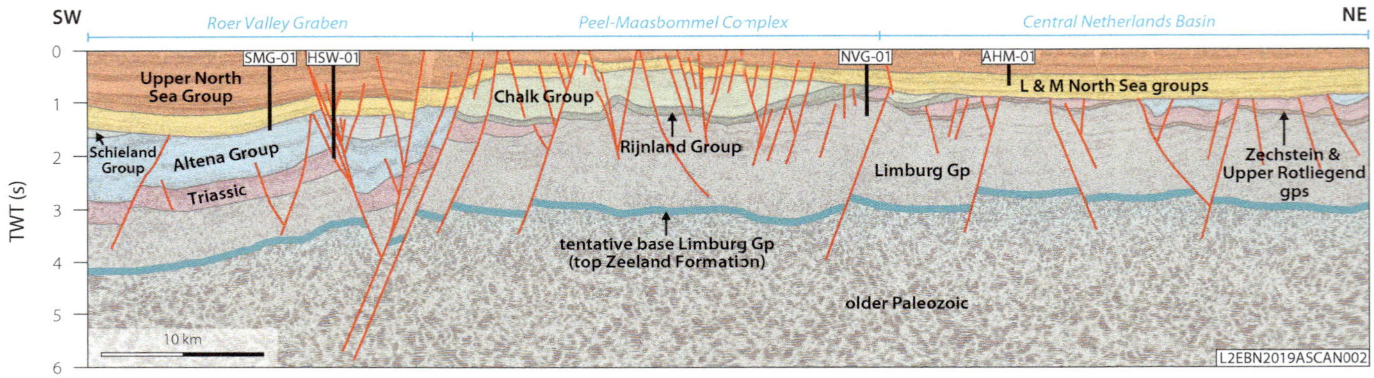

Figure 1.14. Seismic time section from the Roer Valley Graben in the southwest, across the Peel-Maasbommel Complex to the Central Netherlands Basin in the northeast. This seismic section shows the very different uplift and subsidence histories of these three structural elements. The Peel-Maasbommel Complex experienced strong uplift and erosion during the main Late Jurassic-Early Cretaceous rifting event. A thin Lower Cretaceous Rijnland Group sequence unconformably overlies Triassic deposits, but the presence of a thick Chalk Group sequence indicates that the Complex was not severely affected by Alpine inversion. The Roer Valley Graben on the other hand shows a well-developed Jurassic sequence which, as a result of Alpine uplift and erosion is unconformably overlain by lower Cenozoic deposits. The Central Netherlands Basin experienced major Alpine uplift and erosion, to the extent that it has become difficult to unravel Jurassic and Cretaceous history in any detail. For location, see Fig. 1.1.

mately into the Indefatigable Fault Zone (Fig. 1.1) in British waters. As a result of significant Cenozoic inversion, the structural evolution of this fault zone is not easily reconstructed. It is already expressed in the subcrop pattern of Westphalian units below the Base Permian Unconformity, indicating this is a major structural trend that has been repeatedly reactivated. The well Jutphaas-1 (located just southwest of the city of Utrecht) drilled through a major reverse fault with an offset of no less than 1500 m, causing a duplication of Triassic to Carboniferous stratigraphy. This faulting must have been related to Late Cretaceous (probably Sub-Hercynian) inversion, uplifting the Central Netherlands Basin more than the West Netherlands Basin. The Zandvoort Ridge must have been a tectonically shallower ridge in between the basins that experienced less Cenozoic inversion (Fig. 1.10; see Van Lochem et al., 2025, this volume).

There is no clear boundary between the Central Netherlands and Broad Fourteens basins, nor between the West Netherlands Basin and Roer Valley Graben, which are oriented in slightly different directions.

The NW-SE orientation of the Mesozoic extensional basins in the south and west of the Netherlands does not conform to the assumed E-W direction of extension during Late Cimmerian rifting. With the exception of the Broad Fourteens Basin, these basins are sub-parallel to the subcrop pattern of the Westphalian at the Base Permian Unconformity, so their localization was probably controlled by pre-existing structural elements. As in the Dutch Central Graben, increased thicknesses of Triassic sedimentary rocks in the Broad Fourteens Basin and Roer Valley Graben indicate early establishment of these basins. Seismic sections show evidence for local Early to Middle Jurassic normal faulting in the Broad Fourteens and West Netherlands basins.

In the West Netherlands Basin, which is devoid of Zechstein salt, Late Cimmerian extension resulted in the formation of a series of half-grabens filled with continental Upper Jurassic and Lower Cretaceous syn-rift clastics belonging to the Schieland Group (Kimmeridgian to Barremian; Donselaar et al., 2015). Under post-rift thermal subsidence marine waters entered the basin from the north and younger Cretaceous post-rift deposits of the Rijnland Group onlap the basin margins (Figs 1.2c, 1.11; Bodenhausen & Ott, 1981; De Jager et al., 1996). Igneous activity accompanied the rifting, as is evidenced by volcanic rocks and intrusive sills near the base of the Delfland Subgroup. During the Late Cretaceous and early Cenozoic inversion, pre-existing faults were reactivated reversely, with transpressional movements forming a series of prominent NW-SE trending flower structures (see Figs 1.2c, 1.11). Along many individual faults, a reversed sense of movement can clearly be deduced from normal offsets at deeper, and reverse offsets at shallower stratigraphic levels. Thinning of the Upper Cretaceous Chalk Group from the southeast towards the inverted basin, with several internal unconformities visible on seismic data, clearly shows that the flower structures were formed during the Late Cretaceous Sub-Hercynian inversion. During the Paleocene Laramide inversion, the West and Central Netherlands basins were both further uplifted and deeply eroded, locally down into the Triassic. The Sub-Hercynian inversion in the Central

Netherlands Basin is less well documented, as no Upper Cretaceous strata are preserved due to deep erosion during later phases. However, on the Zandvoort Ridge, forming the divide between the West Netherlands and Central basins (Fig. 1.10), chalk has been preserved and locally reaches several hundreds of metres in thickness. In the Central Netherlands Basin, several wells prove that Danian chalk overlies pre-Cretaceous successions (TNO-NITG, 2004a). During the Early Cenozoic, inversion of this basin continued. At the southwest margin of the West Netherlands Basin, the Lower Cenozoic thins towards the inverted basin (Fig. 1.11). During the Oligocene, erosion took place in the most strongly inverted sector of the basin, as well as in the Central Netherlands Basin. It is noteworthy that, although the Sub-Hercynian and Laramide inversions resulted in reverse reactivation of pre-existing faults, the Pyrenean inversion caused broad basin uplift without fault reactivation. At the end of the inversion movements, relaxation of the stress regime resulted locally in normal reactivation of faults.

The Roer Valley Graben was also affected by Sub-Hercynian inversion, which was especially strong in the northwest, while the adjacent Maas-Bommel Complex subsided considerably as evidenced by the presence of a thick package of the Upper Cretaceous Chalk Group (Geluk et al., 1994; Gras, 1995; Gras & Geluk, 1999; Deckers, 2015; Fig. 1.14). There is no evidence that Laramide inversion occurred in the Roer Valley Graben, as deposition continued without a break. The Pyrenean pulse did strongly affect the Roer Valley Graben, however (Deckers, 2015).

The Broad Fourteens Basin was affected by the same structural events as the Central and West Netherlands basins but displays a different structural style (Fig. 1.2b; Van Wijhe, 1987a,b). The main difference is caused by the presence of Zechstein salt in the northern half of the basin. Below this salt, at the level of the Upper Rotliegend Group, faulting resulted in a series of tilted fault blocks (Fig. 1.13). As structuration above the salt was decoupled from subsalt faulting, inversion of half-grabens filled with Upper Jurassic sediments, characteristic of the West Netherlands Basin, is less clearly developed. The main inversion pulse in the Broad Fourteens Basin was the Sub-Hercynian, which resulted in the development of broad NW-SE trending anticlinal arches (Dronkers & Mrozek, 1991; Nalpas et al., 1995). No angular unconformity truncates the post-Campanian chalk deposited above the Sub-Hercynian unconformity, but it is absent above the most strongly inverted areas. In addition, the Sub-Hercynian unconformity (or its correlative intra-Campanian horizon) on the platform to the northeast has been cut by a significant thrust along the northwestern boundary of the inverted Broad Fourteens Basin (Fig. 1.13) As this thrust is truncated by the unconformity at the base of the Cenozoic North Sea Supergroup but also affects Danian chalk (Kronemeijer, 2015), the thrusting must be related to Laramide inversion.

Strong Eocene and Oligocene inversion (Pyrenean and Savian pulses) resulted in uplift and erosion of older sequences of the West and Central Netherlands basins and only a thin Cenozoic cover was deposited, locally not more than 500 m. This contrasts sharply with the Cenozoic of the northern sector of the Broad Fourteens Basin, which is twice as thick. In the Roer Valley Graben the Cenozoic reaches 2000 m thickness (Geluk et al., 1994; Van den Berg, 1994; Van den Berg et al., 1994). Subsidence was particularly pronounced after the Oligocene, when crustal extension associated with the development of the Lower

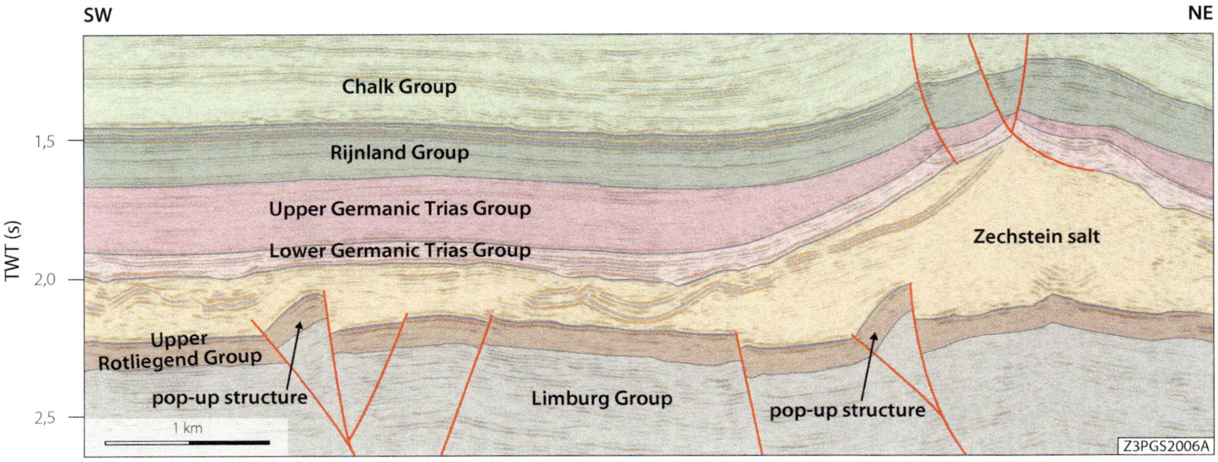

Figure 1.15. Seismic time section across the Cleaverbank Platform, showing a NW-SE trending pop-up structure bounded by a reverse fault at the Rotliegend level. The compressional nature of this structure indicates that it must have formed during Late Cretaceous to early Cenozoic inversion. For location see Fig. 1.1.

Rhine rift system propagated into the Netherlands. The Peel Boundary Fault Zone (Fig. 1.1), separating the graben from the Peel-Maasbommel Complex to the northeast, shows a post-inversion throw of up to 1 km.

Basins in the northeast of the Netherlands

The Lower Saxony Basin is an E-W trending, Late Jurassic and Early Cretaceous transtensional basin, largely located within Germany, whose westernmost part extends just into the Netherlands (Figs 1.1, 1.10), with Jurassic and Cretaceous deposits thinning westwards onto the Friesland Platform. To the north the basin is connected with the Lauwerszee Trough, which is located immediately west of the Groningen Plarform. Both the Dutch sector of the Lower Saxony Basin and the Lauwerszee Trough lack half-grabens and tilted fault blocks. The outlines of both structural elements clearly conform to subcrop patterns below the Base Permian Unconformity, indicating that their development was controlled by pre-existing structural trends. This is also clear from the Hantum and Gronau fault zones (Fig. 1.1), located west of the Lauwerszee Trough and southwest of the Lower Saxony Basin respectively, which offset Westphalian units at the Base Permian Unconformity. The Gronau Fault Zone (Fig. 1.1) continued to be active into the Cenozoic (Van Adrichem Boogaert & Kouwe, 1993).

At Base Rotliegend level the Lauwerszee Trough is only several hundreds of metres deeper than the adjacent Friesland Platform and Groningen Platform and, although it is not a very significant basin, it forms a distinct and long-lived feature, not only expressed in subcrop patterns but also in increased thicknesses of Rotliegend, Lower Cretaceous and Paleogene deposits. The latter may in part also be due to withdrawal of Zechstein salt into flanking salt walls, which grew significantly during Early Cretaceous and Cenozoic times.

The Ems Low along the German-Dutch border flanks the Groningen Platform to the east. It contains lower Permian igneous rocks that represent the western extension of much more widespread igneous activity in Germany, although they are somewhat younger (Geluk, 2007). This activity was associated with crustal extension that led to a fan-shaped system of Early Permian grabens filled with Lower Rotliegend volcanogenic clastics near Hannover (Ziegler, 1990a). In the northward offshore extension of the Ems Low, along the Dutch-German border, this extensional phase resulted in several subtle basal Rotliegend grabens containing a basal sandstone that extends further north than the younger Slochteren sandstones of the Upper Rotliegend Group (e.g. Corcoran & Lunn, 2014).

The Triassic is also thickly developed in the Ems Low, which became overprinted by the Late Cimmerian rifting associated with the Lower Saxony Basin.

Although major Late Cretaceous inversion significantly affected the German sector of the Lower Saxony Basin (Betz et al., 1987; Baldschuhn et al., 1991; Kockel, 2003), its Dutch sector was only mildly affected.

Highs and platforms

The London-Brabant Massif in the south and the Texel-IJsselmeer High in the centre of the country remained elevated structural elements during much of the geological history. The Ringkøbing-Fyn High and Mid North Sea High are located just north of the Dutch offshore and will not be discussed here. Platforms, such as the Cleaverbank, Ameland and Schill Grund platforms, have a more complete sedimentary record than the aforementioned highs. The difference with the basins which they flank is that instead of accumulating thick syn-rift sediments, they were uplifted and eroded during the Middle Jurassic and Early Cretaceous rifting phases (Fig. 1.8).

After sedimentation encroached upon the London-Brabant Massif during the Westphalian, the massif received sediments only for a relatively short period during Late Cretaceous to Paleogene times (Ziegler, 1990a). Any deposits that might have accumulated along its flanks during other periods were eroded.

The Texel-IJsselmeer High formed part of the Mid Netherlands Swell and became strongly uplifted after the Permian. Triassic units can be seen thinning towards it. Subsequent uplift and erosion during Late Cimmerian rifting (Late Jurassic to Early Cretaceous), left no rocks of latest Permian (Zechstein) to Middle Jurassic age on this high, while only remnants of the Rotliegend have been preserved in downfaulted blocks (Van Lochem et al., 2019). Thin Triassic and Lower Jurassic deposits may have accumulated there but if so, they were later eroded. Regional post-rift subsidence from Late Cretaceous times onwards accommodated younger sediments.

During long periods, sedimentation was similar on the platform areas and Mesozoic extensional basins. Only during the Mid and Late Cimmerian rifting pulses was their development different, the platforms being then uplifted and subjected to erosion. Consequently, Upper Jurassic or Lower Cretaceous syn-rift deposits are absent and Triassic and older series are unconformably overlain by a thin veneer of post-rift Lower Cretaceous sedimentary rocks and a relatively thick sequence of Upper Cretaceous chalk. Preservation of Upper Triassic to Middle Jurassic rocks was limited, and locally erosion cut down into the Zechstein. In many places, no angular unconformity associated with this hiatus can be observed on seismic sections, but well correlations clearly indicate that significant truncation and erosion has occurred. In the Mesozoic rift basins, Upper Triassic to Middle Jurassic strata reach thicknesses from hundreds of metres to more than 1 km, but on the platforms the equivalent deposits were probably less thick, as differential subsidence of the rift basins appears to have started during the Triassic. Nevertheless, hundreds of metres of Triassic and Lower to Middle Jurassic sediments were probably deposited on most platforms (Fig. 1.8b). This is suggested by the complete absence of nearshore deposits within the Lower Jurassic of the Netherlands, and also by the uniform thickness of, for example, the Toarcian Posidonia Shale. Wherever this unit is preserved, even in Germany, it always is about 30 to 35 m thick, and has very similar log expressions and oil source-rock qualities. Had the platforms been areas of non-deposition, greater variations in thickness and lithology would surely have resulted (Ziegler, 1990a).

Although the platforms and highs were not uplifted during the Alpine inversion phases, Alpine compression did affect them. On the Cleaverbank Platform, just north and northwest of the Broad Fourteens Basin, a series of pop-up structures are seen at Rotliegend level (Fig. 1.15). Such compressional pop-up structures have also been recognized on the Groningen Platform (Grötsch et al., 2011; De Jager & Visser, 2017). Even though these features occur in areas of thick Zechstein salt where it is difficult to assess the timing of faulting, their compressional nature unambiguously associates them with the Late Cretaceous and early Cenozoic inversion pulses, as these are the only post-Rotliegend compressional pulses in the Netherlands. In an area of very thin Zechstein salt, in block L13 just northeast of the Broad Fourteens Basin, a subtle but clearly expressed Riedel fault pattern developed at Base Cenozoic level suggests that dextral shear during the Pyrenean inversion phase reactivated deeper faults (Oudmayer & De Jager, 1993).

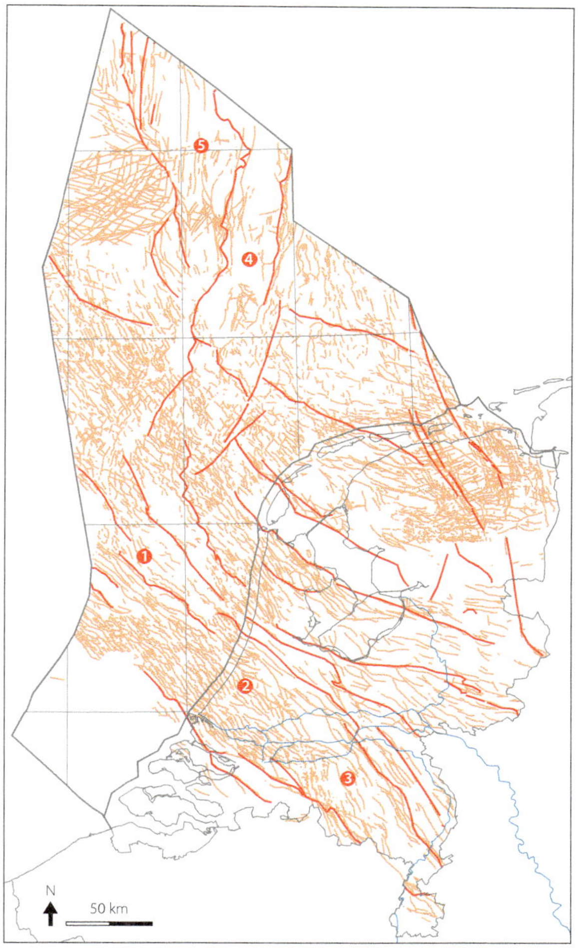

Figure 1.16. Fault map at base Rotliegend level in the Netherlands, with the major fault zones indicated in red. In many places, in particular on the structurally high platforms the pervasive NW-SE fault pattern is clearly visible. In the Late Jurassic-Early Cretaceous rift basins, reactivation of pre-existing faults affected mainly the NW-SE directed faults, as in the Broad Fourteens (1) West Netherlands (2) basins and in the Roer Valley Graben (3). In the Dutch Central Graben (4) and Step Graben (5) areas, the dominant strike of the main faults is more N-S. Source: TNO-GDN.

Fault patterns

Below the Zechstein salt, at the level of the Upper Rotliegend Group, many structural elements show a characteristic rhomboid pattern of intersecting fault trends, especially seen on the platform areas (Figs 1.16-1.18). The dominant fault system trends NW-SE and probably is pre-Caledonian in age, in analogy with the NE-trending faults in Ireland and Wales (Prigmore et al., 1997; Worthington & Walsh, 2011; Holdsworth et al., 2012; Smit et al., 2018). The Hantum and Mid Netherlands fault zones belong to this family, and show clear evidence of repeated reactivation, as do faults on the Elbow Spit Platform (Ter Borgh et al., 2018b). The second most common fault system is oriented NE-SW to ENE-WSW (Dekeyser, 1990; Oudmayer & De Jager, 1993; Schroot & De Haan, 2003). These so-called Dekeyser Faults (Van Ojik et al., 2020) are best known from the Cleaverbank Platform in the western Netherlands offshore and the UK offshore bordering area, but they are also present along the Dutch coast and possibly onshore. Typically, these faults are up to hundreds of kilometres long and, with the exception of the faults along the eastern borders of Cleaverbank Platform and Central Offshore Platform (Fig. 1.1), have low vertical and lateral offsets. Their orientation and position with respect to the Southern Central Graben suggests that these faults initiated during early Variscan extension (Smit et al., 2018), but a genetic relationship has not been established yet. Since both trends were thus already in place prior to Rotliegend and Zechstein deposition, their presence in younger intervals results from fault reactivation. The fault intersections

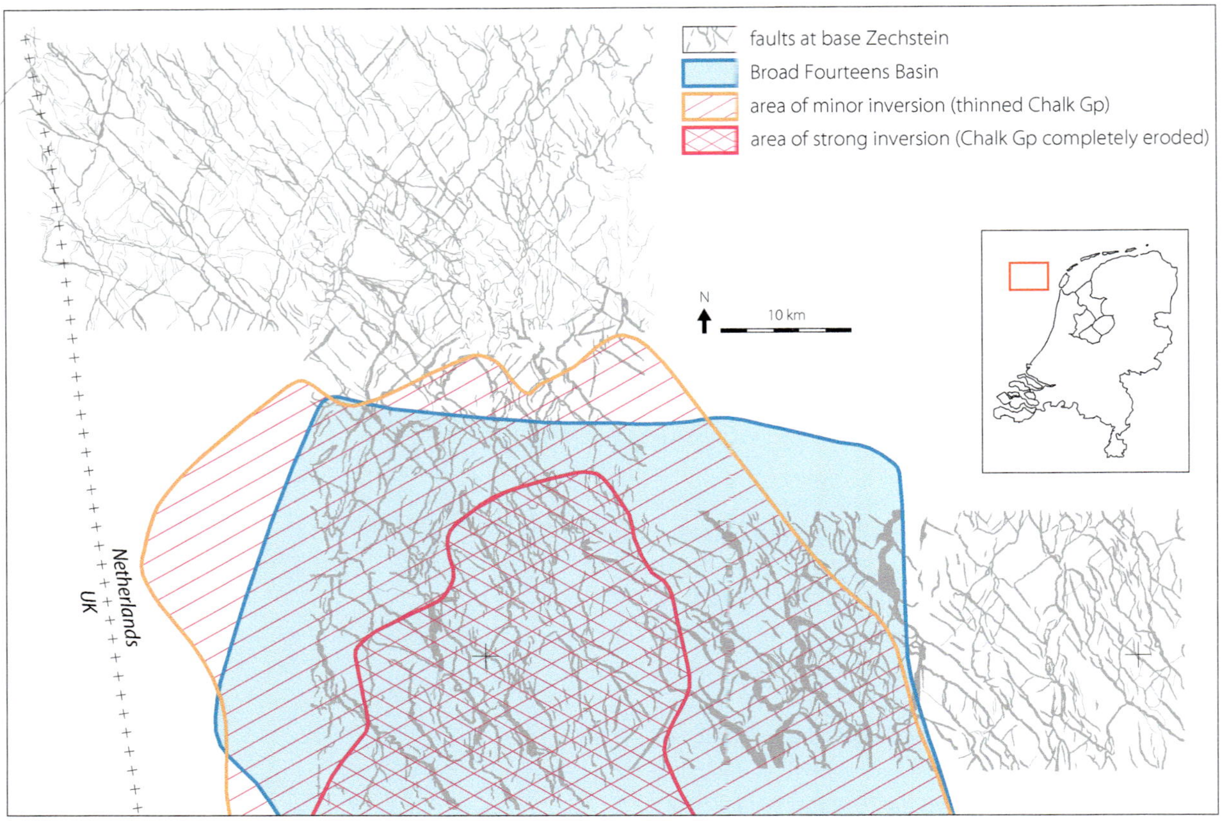

Figure 1.17. Fault pattern at Rotliegend level in the northern sector of the Broad Fourteens Basin and adjacent Cleaverbank Platform, based on good-quality 3D seismic data. Despite significant differences in structural development, there are no significant differences in fault patterns and fault density between the Broad Fourteens Basin and adjacent platform areas. Reproduced from De Jager (2007).

generally do not indicate consistent offset geometries, and it is likely that both trends were reactivated several times more or less simultaneously. Along the axis of the Dutch Central Graben N-S faults, paralleling the Late Jurassic to Early Cretaceous rift basin, dominate.

In the northern sector of the Broad Fourteens Basin and on adjacent platform areas, 3D seismic data clearly show NW-SE and NE-SW fault trends at Rotliegend level, with secondary faults trending WNW-ESE and N-S (Fig. 1.17). Apart from the better imaging of these faults on the platforms, there is no significant difference between the basins and platforms in the directions and spacing of faults. This is remarkable in view of their different structural histories. Apparently, during rifting and later inversion, pre-existing faults were reactivated, but only a few new faults formed at Rotliegend level. Trains of pop-up structures on the Cleaverbank Platform, north and northwest of the Broad Fourteens Basin, follow the NW-SE, NE-SW and WNW-ESE trends, and show that these trends were reactivated simultaneously during Alpine compression (Fig. 1.16).

In the northeast of the Netherlands, the main fault trends in the Zechstein and Rotliegend appear to have rotated slightly to NNW-SSE and ENE-WSW as compared to the western offshore. In the Lower Saxony Basin, E-W trending faults are common (Fig. 1.16). As indicated previously, the large NW-SE trending Hantum Fault Zone (Fig. 1.18) at the western boundary of the Lauwerszee Trough and the NW-SE trending Gronau Fault Zone (Fig. 1.1) along the southwestern boundary of the Lower Saxony Basin represent old fault systems. In some cases, considerable lateral displacement along individual faults seems likely, but the amount of strike-slip is mostly not more than that of the vertical offset, mainly in the order of tens to hundreds of metres for the Lauwerszee Trough area. Fault maps at Cretaceous and Cenozoic levels of this area show that both the NNW-SSE and the ENE-WSW trending faults were reactivated during the Cenozoic.

Within the West Netherlands Basin, faults affecting the Rotliegend show a characteristic anastomosing pattern of dominant WNW-ESE and NNW-SSE, and secondary N-S and E-W directions (Fig. 1.19). Dextral displacement along the main faults would be consistent with E-W Late Cimmerian extension as well as with N-S Late Cretaceous and early Cenozoic compression. However, in many of the inverted structures the Riedel patterns at the base of the Cretaceous rather appear to indicate sinistral shear.

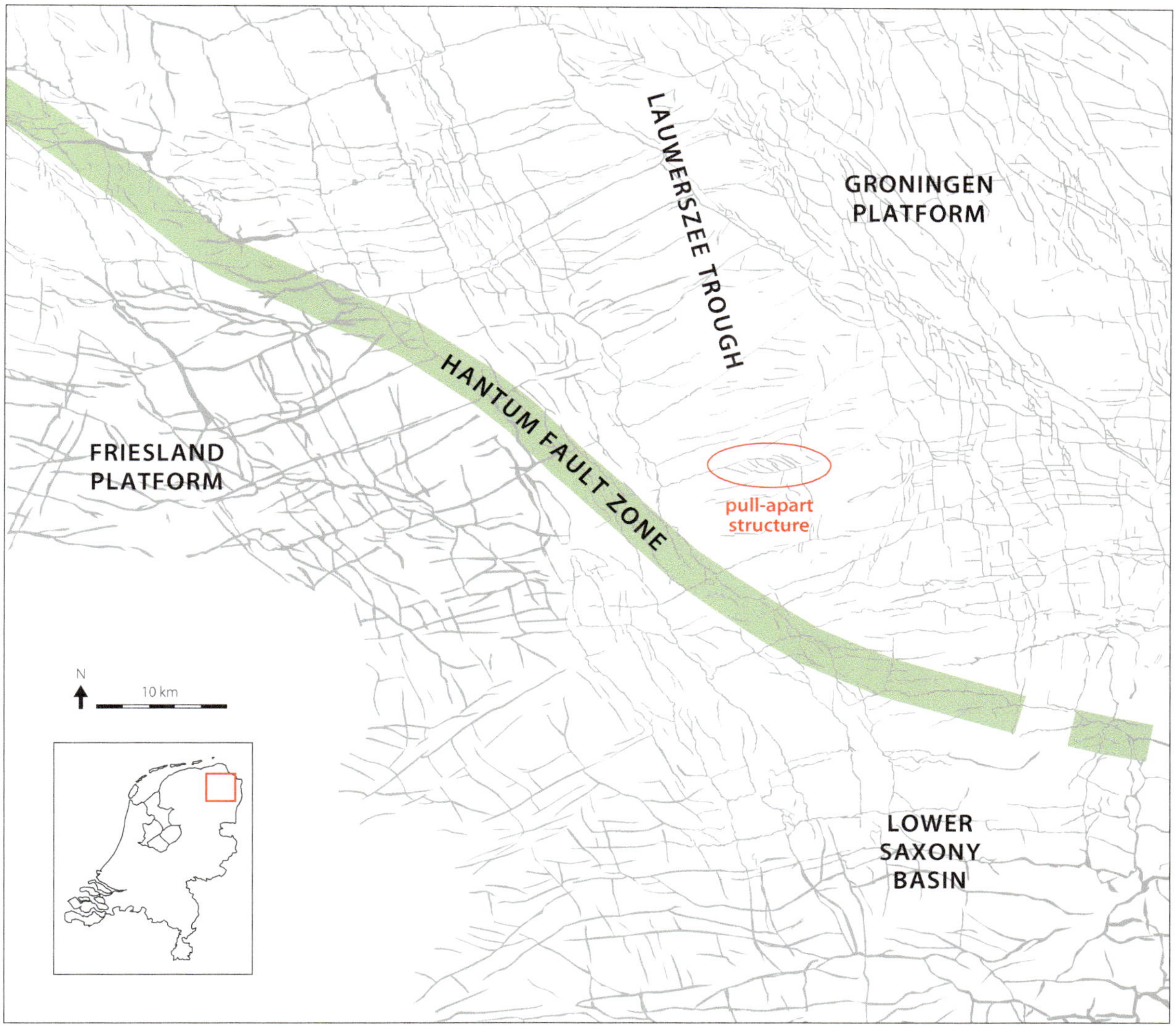

Figure 1.18. Fault pattern at Rotliegend level in part of the northeast Netherlands. The Hantum Fault Zone cannot be mapped as a through-going fault on good-quality 3D seismic data. Note the presence of a small but crisply displayed pull-apart feature within the southern Lauwerszee Trough, indicating local dextral transtensional displacement. Reproduced from De Jager (2007).

In addition, in the Pijnacker oil field in the West Netherlands Basin, an Early Cretaceous sinistral movement along WNW-ESE faults has been postulated (Racero-Baena & Drake, 1996). Dextral displacement during Alpine inversion, interpreted for this field, agrees with the regional model. However, it appears that it may not be appropriate to interpret regional stress directions from these repeatedly reactivated fault geometries. In fact, the Late Cretaceous compressional flower-structures were formed through reactivation of both NNW-SSE and WNW-ESE fault systems. What looks like an alignment of continuous flower structures at Cretaceous levels, can be seen to follow different fault trends at Rotliegend level, jumping and zigzagging from one older fault trend to the other (Fig. 1.19). Apparently, none of the pre-existing fault trends was ideally oriented to be preferentially reactivated, so present-day fault trends cannot be readily interpreted in terms of paleostress directions. As reliable indicators of the accumulated amount, or even the sense, of lateral displacement are rare, it is challenging to unravel the various evolutionary phases of individual faults.

Summary

The oldest dated sedimentary rocks in the Netherlands are of late Silurian to Early Devonian age, and the locally 10 or more kilometres thick stratigraphic record above them is largely continuous to the present day. The setting of basins, platforms and highs from the early Paleozoic onwards changed repeatedly in response to the tectonic processes that accompanied the global interaction of lithospheric

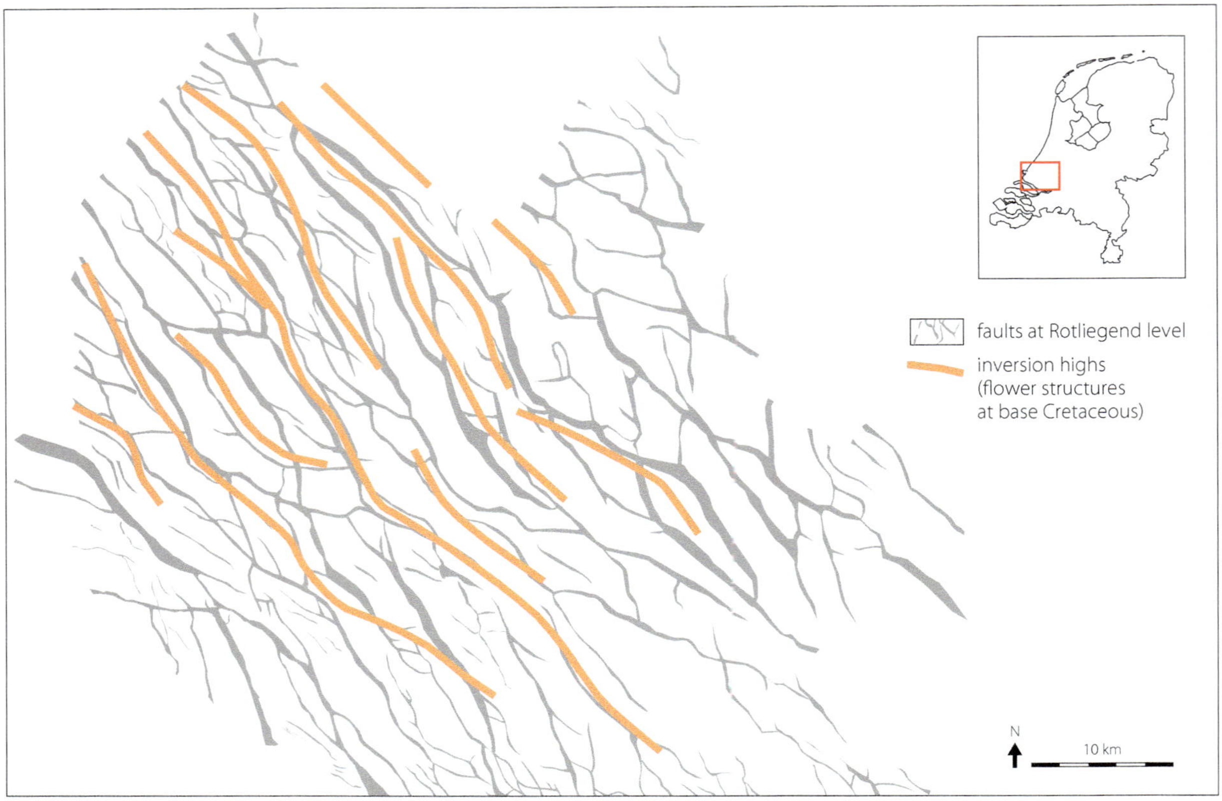

Figure 1.19. *Fault pattern at Rotliegend level in the West Netherlands Basin, based on good-quality 3D seismic data. The rhomboid pattern shows dominant WNW-ESE and NNW-SSE faults. Continuous flower structures at base Cretaceous (the inversion highs, are caused by Late Cretaceous to early Cenozoic inversion. They zig-zag from one fault trend to the other indicating that faults from both trends were reactivated during tectonic inversion. Reproduced from De Jager (2007).*

plates. Episodes of compression and mountain building alternated with phases of extension and the creation of rifted basins, whilst the present-day area of the Netherlands drifted from the southern hemisphere across equatorial regions to its current position in the temperate climate belt around 52° north.

The main plate-tectonic events that influenced the Netherlands were: i) the assembly of the Pangea supercontinent during the Paleozoic through the collisions of Avalonia, Laurentia and Baltica, resulting in the Caledonian and Variscan orogens along plate boundaries, ii) the Cimmerian breaking up of Pangea during the Mesozoic, and the accompanying formation of rift basins, and iii) the Alpine collision of Africa and Europe, causing inversion of the earlier formed Mesozoic rift basins and compressional reactivation of pre-existing extensional faults.

Periods of igneous activity, both intrusive and extrusive, accompanied some of these events. Crustal extension and thinning, accompanied by increasing heat-flow, caused uplift and erosion of platform areas adjacent to rift basins. During post-rift periods of declining heat-flow, regional subsidence occurred, while depositional areas expanded onto previously emergent platform areas.

Despite the many tectonic events, a high degree of fault parallelism can be observed. The general structural model is, therefore, one of repeated reactivation of basement faults that continued to control the structural grain, despite changes in tectonic regime and stress direction. Only rarely and by coincidence did pre-existing structural features have orientations that conformed to the regional stress of later episodes and, consequently, most faults have a component of horizontal slip.

In much of the Dutch subsurface, diapiric movements of the thick Permian Zechstein salt led to impressive salt domes of sometimes more than 3 km height, reaching very close to the present-day surface. This thick salt package has caused faulting above the salt to be decoupled from basement faulting, hampering unambiguous assessment of the timing of movements along individual faults at sub-salt levels. However, as many of the salt walls and domes follow the orientation of the main sub-salt fault systems, the salt movements were apparently triggered by basement faulting.

The tectonic forces that affected the subsurface not only shaped its present-day structural configuration, but also exerted a profound influence on the depositional history.

Acknowledgments

This chapter was originally published in the first edition of Geology of the Netherlands (2007) by permission of the Nederlandse Aardolie Maatschappij BV (NAM), Shell Internationale Petroleum Maatschappij BV (SIPM) and ExxonMobil. Reviewers Tim Pharaoh, Marten ter Borgh, Christian Brandes, and Harry Doust are thanked for their constructive edits which certainly improved the manuscript. Special thanks are due to professor Christopher R. Scotese for permission to use his plate reconstructions, to Cees van Oosterhout who prepared the reconstruction maps shown here as Figure 1.4 and to Marten ter Borgh for sharing seismic interpretations.

Digital map data

Spatial data of figures in this chapter for use in geographical information systems can be downloaded here: https://doi.org/10.5117/aup.28162868.

References

Baldschuhn, R., Best, G. & Kockel, F., 1991. Inversion tectonics in the north-west German basin. In: Spencer, A.M. (ed.): Generation, accumulation and production of Europe's hydrocarbons. Special Publication of the European Association of Petroleum Geoscientists (Oxford): 149-159.

Benek, R., Kramer, W., McCann, T., Scheck, M., Negendank, J.F.W., Korich, D., Huebscher, H.D. & Bayer, U., 1996. Permo-Carboniferous magmatism of the Northeast German Basin. Tectonophysics 266(1-4): 379-404. DOI: 10.1016/S0040-1951(96)00199-0

Berthelsen, A., 1998. The Tornquist Zone northwest of the Carpathians: An intraplate pseudosuture. GFF 120(2): 223-230. DOI: 10.1080/11035899801202223

Betz, D., Führer, F., Greiner, G. & Plein, E., 1987. Evolution of the Lower Saxony Basin. Tectonophysics 137: 127-170. DOI: 10.1016/0040-1951(87)90319-2

Bodenhausen, J.W.A. & Ott, W.F., 1981. Habitat of the Rijswijk oil province, onshore, the Netherlands. In: Illing, L.V. & Hobson, G.D. (eds): Petroleum Geology of the continental shelf of North-West Europe: Proceedings of the Second Conference. Institute of Petroleum (London): 301-309. ISBN 0-85501-656-6

Bonté, D., Smit, J., Fattah, R.A., Nelskamp, S., Cloetingh, S. & Van Wees, J.-D., 2020. Quantifying the late-to post-Variscan pervasive heat flow, central Netherlands, Southern Permian Basin. Marine and Petroleum Geology 113, 104118. DOI: 10.1016/j.marpetgeo.2019.104118

Bouroullec, R. & Geel, C.R., 2025. Permian. In: Ten Veen, J.H., Vis, G.-J., De Jager, J. & Wong, Th.E. (eds): Geology of the Netherlands, second edition. Amsterdam University Press (Amsterdam): 127-153. DOI: 10.5117/9789463728362_ch04

Bouroullec, R., Nelskamp, S., Kloppenburg, A., Fattah, R.A., Foeken, J., Ten Veen, J.H., Geel, K., Debacker, T.N. & Smit, J., 2019. Burial and Structural Analysis of the Dinantian Carbonates in the Dutch Subsurface. EBN, TNO (Utrecht), SCAN-Dinantien Project Report: 170 pp. https://www.nlog.nl/sites/default/files/2019-09/scan_dinantian_burial_and_structuration_report.pdf

Bouroullec, R., Verreussel, R.M.C.H., Geel, C.R., De Bruin, G., Zijp, M.H.A.A., Kőrösi, D., Munsterman, D.K., Janssen, N.M.M. & Kerstholt-Boegehold, S.J., 2018. Tectonostratigraphy of a rift basin affected by salt tectonics: synrift Middle Jurassic-Lower Cretaceous Dutch Central Graben, Terschelling Basin and neighbouring platforms, Dutch offshore. Geological Society of London, Special Publications 469(1): 269-303. DOI: 10.1144/SP469.22

Burgess, C.L., Clever, J.E., Corcoran, O.J., Cram, F.M., Hall, N.T. & Lunn, S.F., 2018. Seismic discrimination of an overlooked basal Rotliegend reservoir opens a new play in the Dutch offshore. First Break 36: 38-40.

Cloetingh, S. & Van Wees, J.-D., 2005. Strength reversal in Europe's intraplate lithosphere: Transition from basin inversion to lithospheric folding. Geology 33(4): 285-288. DOI: 10.1130/G21051.1

Corcoran, J. & Lunn, S.F., 2014. Old play, New Prospectivity – the Revival of the Basal Rotliegend Play, Offshore Netherlands and Germany. 76th EAGE Conference and Exhibition 2014: 1-5. DOI: 10.3997/2214-4609.20140742

Corfield, S.M., Gawthorpe, R.L., Gage, M., Fraser, A.J. & Besly, B.M., 1996. Inversion tectonics of the Variscan foreland of the British Isles. Journal of the Geological Society 153: 17-32. DOI:10.1144/gsjgs.153.1.0017

Coward, M.P., Dewey, J., Hempton, M. & Holroyd, J., 2003. Tectonic evolution. In: Evans, D., Graham, C., Armour, A. & Bathurst, P. (eds): The millennium atlas: Petroleum geology of the central and northern North Sea. Geological Society of London: 17-33.

De Bruin, G., Geel, K., Houben, S., Munsterman, D., Verweij, H., Smit, J., Janssen, N., Kerstholt-Boegehold, S. & Vandeweijer, V., 2017. MMU Unravelling the stratigraphic and structural development of the strata found underneath and above the Mid-Miocene unconformity. TNO (Utrecht), Report No. R10425: 104 pp.

De Jager, J., 2003. Inverted basins in the Netherlands, similarities and differences. Netherlands Journal of Geosciences 82(4): 339-349. DOI:10.1017/S0016774600020175

De Jager, J., 2007. Geological development. In: Wong, Th.E., Batjes, D.A.J. & De Jager, J. (eds): Geology of the Netherlands. Royal Netherlands Academy of Arts and Sciences (KNAW, Amsterdam): 5-26.

De Jager, J., 2012. The discovery of the Fat Sand Play (Solling Formation, Triassic), Northern Dutch offshore – a case of serendipity. Netherlands Journal of Geosciences 91(4): 609-619. DOI: 10.1017/S0016774600000408

De Jager, J., Doyle, M.A., Grantham, P.J. & Mabillard, J.E., 1996. Hydrocarbon habitat of the West Netherlands Basin. In: Rondeel, H.E., Batjes, D.A.J. & Nieuwenhuijs, W.H. (eds): Geology of gas and oil under the Netherlands. Springer (Dordrecht): 191-209. DOI: 10.1007/978-94-009-0121-6_17

De Jager, J. & Visser, C., 2017. Geology of the Groningen field–an overview. Netherlands Journal of Geosciences 96(5): 3-15. DOI: 10.1017/njg.2017.22

De Vos, W., Verniers, J., Herbosch, A. & Vanguestaine, M., 1993. A new geological map of the Brabant Massif. Belgium. Geological Magazine 130: 605-611. DOI: 10.1017/S0016756800020902

Debacker, T.N., Dewaele, S., Sintubin, M., Verniers, J., Muchez, P. & Boven, A., 2005. Timing and duration of the progressive deformation of the Brabant Massif, Belgium. Geologica Belgica 8(3-4): 20-34.

Deckers, J., 2015. The Paleocene stratigraphic records in the Central Netherlands and close surrounding basins: Highlighting the different responses to a late Danian change in stress regime within the Central European Basin System. Tectonophysics 659: 102-108. DOI: 10.1016/j.tecto.2015.07.031

Deckers, J. & Van der Voet, E., 2018. A review on the structural styles of deformation during Late Cretaceous and Paleocene tectonic phases in the southern North Sea area. Journal of Geodynamics 115: 1-9. DOI: 10.1016/j.jog.2018.01.005

Dekeyser, D., 1990. The Impact of 3D on the Structural Interpretation of the Barque and Clipper Fields. Shell (UK), Proprietary report.

Donselaar, M.E., Groenenberg, R.M. & Gilding, D.T., 2015. Reservoir Geology and Geothermal Potential of the Delft Sandstone Member in the West Netherlands Basin. Proceedings World Geothermal Congress, Melbourne: 9 pp.

Doornenbal, J.C. & Stevenson, A.G. (eds), 2010. Petroleum Geological Atlas of the Southern Permian Basin Area. EAGE Publications B.V. (Houten): 342 pp. ISBN 978 90 73781 61 0

Doornenbal, J.C., Kombrink, H., Bouroullec, R., Dalman, R.A.F., De Bruin, G., Geel, C.R., Houben, A.J.P., Jaarsma, B., Juez-Larré, J., Kortekaas, M., Mijnlieff, H.F., Nelskamp, S., Pharaoh, T.C., Ten Veen, J.H., Ter Borgh, M., Van Ojik, K., Verreussel, R.M.C.H., Verweij, J.M. & Vis, G.-J., 2019. New insights on subsurface energy resources in the Southern North Sea Basin area. In: Patruno, S., Archer, S.G., Chiarella, D., Howell, J.A., Jackson, C.A-L. & Kombrink, H. (eds): Cross-border themes in petroleum geology I: The North Sea. Geological Society of London, Special Publications 494: 233-268. DOI: 10.1144/SP494-2018-178

Dost, B., Ruigrok, E., Spetzler, J. & Kruiver, P., 2025. Natural and induced seismicity. In: Ten Veen, J.H., Vis, G.-J., De Jager, J. & Wong, Th.E. (eds): Geology of the Netherlands, second edition. Amsterdam University Press (Amsterdam): 793-823. DOI: 10.5117/9789463728362_ch22

Dronkers, A.J. & Mrozek, F.J., 1991. Inverted basins of The Netherlands. First Break 9: 409-418. DOI: 10.3997/1365-2397.1991019

Duin, E., Doornenbal, J.C., Rijkers, R.H.B., Verbeek, J.W. & Wong, Th.E., 2006. Subsurface structure of the Netherlands; results of recent onshore and offshore mapping. Netherlands Journal of Geosciences 85(4): 245-276. DOI: 10.1017/S0016774600023064

Fraser, A.J. & Gawthorpe, R.L., 1990. Tectono-stratigraphic development and hydrocarbon habitat of the Carboniferous in northern England. In: Hardman, R.F.P. & Brooks, J. (eds): Tectonic events responsible for Britain's oil and gas reserves. Geological Society of London, Special Publications 55: 49-86. DOI: 10.1144/GSL.SP.1990.055.01.03

Frost, R.T.C., Fitch, F.J. & Miller, J.A., 1981. The age and nature of the crystalline basement of the North Sea Basin. In: Illing, L.V. & Hobson, G.D. (eds): Petroleum Geology of the continental Shelf of North-West Europe: Proceedings of the 2nd Conference. Institute of Petroleum (London): 43-57.

Geluk, M.C., 2007. Permian. In: Wong, Th.E., Batjes, D.A.J.& De Jager, J. (eds): Geology of the Netherlands. Royal Netherlands Academy of Arts and Sciences (KNAW, Amsterdam): 63-84.

Geluk, M.C., Duin, E.J.T., Dusar, M., Rijkers, R.H.B., Van den Berg, M.W. & Van Rooijen, P., 1994. Stratigraphy and tectonics of the Roer Valley Graben. Geologie En Mijnbouw 73(2-4): 129-141.

Geluk, M C., Dusar, M. & De Vos, W., 2007. Pre-Silesian. In: Wong, Th.E., Batjes, D.A.J.& De Jager, J. (eds): Geology of the Netherlands. Royal Netherlands Academy of Arts and Sciences (KNAW, Amsterdam): 27-42.

Glennie, K.W., 1986. Development of N.W. Europe's southern Permian gas basin. In: Brooks, J., Goff, J.C. & Van Hoorn, B. (eds): Habitat of Palaeozoic Gas in N.W. Europe. Geological Society of London, Special Publications 50: 3-22.

Glennie, K.W. & Underhill, J.R., 1998. Origin, development and evolution of structural styles. In: Glennie, K.W. (ed.): Petroleum Geology of the North Sea: Basic Concepts and Recent Advances. Blackwell Science Ltd. (Oxford): 42-84.

Gras, R., 1995. Late Cretaceous sedimentation and tectonic inversion, southern Netherlands. Geologie En Mijnbouw 74 (2): 117-127.

Gras, R. & Geluk, M., 1999. Late Cretaceous – Early Tertiary sedimentation and tectonic inversion in the southern Netherlands. Geologie en Mijnbouw 78: 1-19. DOI: DOI: 10.1023/A:1003709821901

Grötsch, J., Sluijk, A., Van Ojik, K., De Keizer, M., Graaf, J. & Steenbrink J., 2011. The Groningen gas field: 50 years of exploration and gas production from a Permian dryland reservoir. Society for Sedimentary Geology (SEPM) 98: 11-33. DOI: 10.2110/pec.11.98.0011

Guterch, A., Grad, M., Krawczyk, C.M., Wybraniec, S., Ziegler, P.A. & De Vos, W., 2010. Crustal structure and structural framework. In: Doornenbal, H. & Stevenson, A.G. (eds): Petroleum Geological Atlas of the Southern Permian Basin Area. EAGE Publications B.V. (Houten): 10-23.

Haq, B.U., 2018. Jurassic sea-level variations: A reappraisal. GSA

Today 28 (1): 4-10. Scopus. DOI: 10.1130/GSATG359A.1

Heeremans, M., Timmerman, M.J., Kirstein, L.A. & Faleide, J.I., 2004. New constraints on the timing of late Carboniferous - early Permian volcanism in the central North Sea. In: Wilson, M., Neumann, E.-R., Davies, G.R., Timmerman, M.J., Heeremans, M. & Larsen, B.T. (eds): Permo-Carboniferous magmatism and rifting in Europe 223: 177-194. DOI: 10.1144/GSL.SP.2004.223.01.08

Herbosch, A., Liégeois, J.P., Gärtner, A., Hofmann, M. & Linnemann, U., 2020. The Stavelot-Venn Massif (Ardenne, Belgium), a rift shoulder basin ripped off the West African craton: Cartography, stratigraphy, sedimentology, new U-Pb on zircon ages, geochemistry and Nd isotopes evidence. Earth-Science Reviews 203, 103142. DOI: 10.1016/j.earscirev.2020.103142

Herngreen, G.F.W., Smit, R. & Wong, Th.E., 1991. Stratigraphy and tectonics of the Vlieland basin, The Netherlands. In: Spencer, A.M (ed.): Generation, accumulation and production of Europe's hydrocarbons. Special Publication of the European Association of Petroleum Geoscientists (Berlin): 175-192.

Heybroek, P., 1975. On the structure of the Dutch part of the Central North Sea Graben. In: Woodland, A.W. (ed.): Petroleum and the continental shelf of northwest Europe. Applied Science Publications (London): 339-351.

Holdsworth, R.E., Woodcock, N.H. & Strachan, R.A., 2012. Geological Framework of Britain and Ireland. In: Woodcock, N. & Strachan, R. (eds): Geological History of Britain and Ireland. Blackwell Publishing Ltd (Oxford): 19-39 pp.

Hooper, R.J., Goh, L.S. & Dewey, F., 1995. The inversion history of the northeastern margin of the Broad Fourteens Basin. In: Buchanan. J.G. & Buchanan, P.G. (eds): Basin Inversion. Geological Society of London, Special Publications 88: 307-317. DOI: 10.1144/GSL.SP.1995.088.01.17

Houben, A.J.P. & Vis, G.-J., 2021. Palynological indications for Silurian – earliest Devonian age strata in the Netherlands. Netherlands Jounal of Geosciences 100(2). DOI: 10.1017/njg.2020.20

Houtgast, G., 1991. Aardbevingen in Nederland; catalogus van aardbevingen t/m 1990: epicentra in Nederland waargenomen verschijnselen in Nederland van bevingen met het epicentrum buiten Nederland. KNMI (De Bilt), Report No. 179: 166 pp.

Jeremiah, J.M., Duxbury, S. & Rawson, P.F., 2010. Lower Cretaceous of the southern North Sea Basins: reservoir distribution within a sequence stratigraphic framework. Netherlands Journal of Geosciences 89(3-4): 203-237. DOI: 10.1017/S0016774600000706

Kalani, M., Faleide, J.I. & Gabrielsen, R.H., 2020. Paleozoic-Mesozoic tectono-sedimentary evolution and magmatism of the Egersund Basin area, Norwegian central North Sea. Marine and Petroleum Geology 122, 104642. DOI: 10.1016/j.marpetgeo.2020.104642

Kettel, D., 1983. The East Groningen Massif - Detection of an intrusive body by means of coalification. Geologie En Mijnbouw 62(1): 203-210.

Kilhams, B., Stevanovic, S. & Nicolai, C., 2018. The 'Buntsandstein' gas play of the Horn Graben (German and Danish offshore): dry well analysis and remaining hydrocarbon potential. Geological Society of London, Special Publications 469: 169-192. DOI: 10.1144/SP469.5

Kley, J. & Voigt, T., 2008. Late Cretaceous intraplate thrusting in central Europe: Effect of Africa-Europe-Iberia convergence, not Alpine collision. Geology 36(11): 839-842. DOI: 10.1130/G24930A.1

Kockel, F., 2003. Inversion structures in Central Europe – Expressions and reasons, an open discussion. Netherlands Journal of Geosciences 82(4), 351-366. DOI: 10.1017/S0016774600020187

Kombrink, H., 2008. The Carboniferous of the Netherlands and surrounding areas; a basin analysis. PhD thesis Utrecht University (the Netherlands): 184 pp.

Kombrink, H., Besly, B., Collinson, J.D., Den Hartog Jager, D.G., Drozdzewski, G., Dusar, M., Hoth, P., Pagnier, H.J.M., Stemmerik, L., Waksmundzka, M.I. & Wrede, V., 2010a. Carboniferous. In: Doornenbal, J.C. & Stevenson, A.G. (eds): Petroleum Geological Atlas of the Southern Permian Basin Area. EAGE Publications B.V. (Houten): 80-99. ISBN 978-90-73781-61-0

Kombrink, H., Doornenbal, J.C., Duin, E.J.T., Den Dulk, M., Ten Veen, J.H. & Witmans, N., 2012. New insights into the geological structure of the Netherlands; results of a detailed mapping project. Netherlands Journal of Geosciences 91(4): 419-446. DOI: 10.1017/S0016774600000329

Kombrink, H., Van Lochem, H. & Van der Zwan, K.J., 2010b. Seismic interpretation of Dinantian carbonate platforms in the Netherlands; implications for the palaeogeographical and structural development of the Northwest European Carboniferous Basin. Journal of the Geological Society 167: 99-108. DOI: 10.1144/0016-76492008-149

Kooi, H., Cloetingh, S. & Remmelts, G., 1989. Intraplate stresses and the stratigraphic evolution of the North Sea Central Graben. Geologie En Mijnbouw 68: 49-72.

Kronemeijer, G.D., 2015. The feasibility of the Chalk Play in the Dutch Broad Fourteens Basin. MSc thesis Utrecht University (the Netherlands), Wintershall.

LaFosse, M. & Smit, J., 2020. End project report: RiFa- 'Rise and Fall', the role of thermal uplift in the formation of Jurassic basins in the Dutch subsurface. Utrecht University, EBN, Wintershall Noordzee, Neptune Energy, TKI2018-05-GE: 16 pp.

Leeder, M.R., 1982. Upper Paleozoic basins of the British Isles – Caledonide inheritance versus Hercynian plate-margin processes. Journal of the Geological Society 139, 479-491.

Lundmark, A.M., Gabrielsen, R.H., Austrheim, H., Flaat, K., Strand, T. & Ohm, S.E., 2012. Late Devonian rifting in the central North Sea: Evidence from altered felsic volcanic rocks in the Embla oil field. Marine and Petroleum Geology 29: 204-218. DOI: 10.1016/j.marpetgeo.2011.06.002

Lundmark, A.M., Bue, E.P., Gabrielsen, R.H., Flaat, K., Strand, T. & Ohm, S.E., 2013. Provenance of late Palaeozoic terrestrial sediments on the northern flank of the Mid North Sea High: Detrital zircon geochronology and rutile geochemical constraints. Geological Society of London, Special Publications 386(1): 243-259. DOI: 10.1144/SP386.4

McCann, T., Pascal, C., Timmerman, M.J., Krzywiec, P., Lopez-Gomez, J., Wetzel, A., Krawczyk, C.M., Rieke, H. & Lamarche, J., 2006. Post-Variscan (end Carboniferous – Early Permian) basin evolution in Western and Central Europe. In: Gee, D.G. & Stephenson, R.A. (eds): European Lithosphere Dynamics. Geological Society of London, Memoirs 32, 355-388.

McKie, T. & Kilhams, B., 2025. Triassic. In: Ten Veen, J.H., Vis, G.-J., De Jager, J. & Wong, Th.E. (eds): Geology of the Netherlands, second edition. Amsterdam University Press (Amsterdam): 155-183. DOI: 10.5117/9789463728362_ch05

Møller, J.J. & Rasmussen, E.S., 2003. Middle Jurassic-Early Cretaceous rifting of the Danish Central Graben. Geological Survey of Denmark and Greenland Bulletin 1: 247-264. DOI: 10.34194/geusb.v1.4654

Nalpas, T., Le Douaran, S., Brun, J.P., Unternehr, P. & Richert, J.P., 1995. Inversion of the Broad Fourteens Basin (offshore Netherlands), a small scale model investigation. Sedimentary Geology 95 (3-4): 237-250. DOI: 10.1016/0037-0738(94)00113-9

Oudmayer, B.C. & De Jager, J., 1993. Fault reactivation and oblique-slip in the southern North Sea. In: Parker, J.R. (ed.): Petroleum Geology of North-West Europe: Proceedings of the 4th Conference. Geological Society of London 4: 1281-1290. DOI: 10.1144/0041281

Peeters, S., Asschert, A. & Verweij, J.M., 2018. Towards a better understanding of the highly overpressured Lower Triassic Bunter reservoir rocks in the Terschelling Basin. Geological Society of London, Special Publications 469 (1): 223-236. DOI: 10.1144/SP469.13

Perrot, J. & Van der Poel, A.B., 1987. Zuidwal - a Neocomian gas field. In: Brooks, J. & Glennie, K. (eds): Petroleum Geology of North-West Europe. Graham & Trotman (London): 325-335.

Pharaoh, T.C., 2018. The Anglo-Brabant Massif: Persistent but enigmatic palaeo-relief at the heart of western Europe. Prodeedings of the Geologists' Association 129(3): 278-328. DOI: 10.1016/j.pgeola.2018.02.009

Pharaoh, T.C., Dusar, M., Geluk, M.C., Kockel, F., Krawczyk, C.M., Krzywiec, P., Scheck-Wenderoth, M., Thybo, H., Vejbæk, O.V. & Van Wees, J.-D., 2010. Tectonic Evolution. In: Doornenbal, H. & Stevenson, A.G. (eds): Petroleum Geological Atlas of the Southern Permian Basin Area. EAGE Publications B.V. (Houten): 24-57. ISBN 978-90-73781-61-0

Pharaoh, T.C., England, R. & Lee, M.K., 1995. The concealed Caledonide basement of Eastern England and the southern North Sea - A review. Studia Geophysica et Geodaetica 39: 330-346. DOI: 10.1007/BF02295826

Prigmore, J., Butler, A. & Woodcock, N., 1997. Rifting during separation of Eastern Avalonia from Gondwana: Evidence from subsidence analysis. Geology 25(3): 203-206. DOI: 10.1130/0091-7613(1997)025<0203:RDSOEA>2.3.CO;2

Racero-Baena, A. & Drake, S.J., 1996. Structural style and reservoir development in the West Netherlands oil province. In: Rondeel, H.E., Batjes, D.A.J. & Nieuwenhuijs, W.H. (eds): Geology of gas and oil under the Netherlands. Kluwer Academic Publishers (Dordrecht): 211-228. DOI: 10.1007/978-94-009-0121-6_18

Remmelts, G., 1995. Fault-related salt tectonics in the southern North Sea, the Netherlands. In: Jackson, M.P.A., Roberts, D.G. & Snelson, S. (eds): Salt Tectonics: a Global Perspective. American Association of Petroleum Geologists Memoir 65: 261-272.

Remmelts G., 1996. Salt tectonics in the southern North Sea, the Netherlands. In: Rondeel, H.E., Batjes, D.A.J. & Nieuwenhuijs, W.H. (eds): Geology of gas and oil under the Netherlands. Royal Geological and Mining Society of the Netherlands, Kluwer Academic Publishers (Dordrecht): 143-158. DOI: 10.1007/978-94-009-0121-6_13

Scheck-Wenderoth, M. & Lamarche, J., 2005. Crustal memory and basin evolution in the Central European Basin System—new insights from a 3D structural model. Tectonophysics 397: 143-165. DOI: 10.1016/j.tecto.2004.10.007

Schroot, B.M. & De Haan, H.B., 2003. An improved regional structural model of the Upper Carboniferous of the Cleaver Bank High based on 3D seismic interpretation. In: Nieuwland, D.A. (ed.): New Insights into Structural Interpretation and Modelling. Geological Society of London, Special Publications 212: 23-37.

Smit, J., Van Wees, J.-D. & Cloetingh, S., 2016. The Thor suture zone: From subduction to intraplate basin setting. Geology 44(9): 707-710. DOI: 10.1130/G37958.1

Smit, J., Van Wees, J.-D. & Cloetingh, S., 2018. Early Carboniferous extension in East Avalonia: 350 My record of lithospheric memory. Marine and Petroleum Geology 92(1): 1010-1027. DOI: 10.1016/j.marpetgeo.2018.01.004

Sørensen, J.C. & Michelsen, O., 1995. Upper Cenozoic sequences in the southeastern North Sea Basin. Bulletin of the Geological Society of Denmark 42: 74-95.

Taggart, S. & Catto, R., 2015. New opportunities on the northern feather-edge play of the Silverpit Basin: Sedimentology and facies distribution of the Permian Lower Leman and Carboniferous Ketch Formations, Cygnus field, Southern North Sea. Conference Abstract Presented at the 8th Petroleum Geology of Northwest Europe Conference (London), 28-30 September 2015.

Ten Veen, J.H., Van Gessel, S.F. & Den Dulk, M., 2012. Thin- and thick-skinned salt tectonics in the Netherlands; a quantitative approach. Netherlands Journal of Geosciences 91(4): 447-464. DOI: 10.1017/S0016774600000330

Ten Veen, J.H., De Haan, H.B., De Bruin, G., Holleman, N. & Schöler, W., 2019. Seismic Interpretation and depth conversion of the Dinantian carbonates in the Dutch subsurface. EBN, TNO (Utrecht), SCAN-Dinantien Project Report: 170 pp.

Ter Borgh, M.M., Eikelenboom, W. & Jaarsma, B., 2018a. Hydrocarbon potential of the Visean and Namurian in the northern

Dutch offshore. In: Monaghan, A.A., Underhill, J.R., Hewett, A.J. & Marshall, J.E.A. (eds) Paleozoic plays of NW Europe. Geological Society of London, Special Publications 471: 133-153. DOI: 10.1144/SP471.5

Ter Borgh, M.M., Jaarsma, B. & Rosendaal, E.A., 2018b. Structural development of the northern Dutch offshore: Paleozoic to present. In: Monaghan, A.A., Underhill, J.R., Hewett, A.J. & Marshall, J.E.A. (eds) Paleozoic plays of NW Europe. Geological Society of London, Special Publications 471: 115-131. DOI: 10.1144/SP471.4

Timmerman, M.J., 2004. Timing, geodynamic setting and character of Permo-Carboniferous magmatism in the foreland of the Variscan Orogen, NW Europe. In: M. Wilson, E.-R. Neumann, S.J. Davies, M.J. Timmerman, M. Heeremans & B.T. Larsen (eds): Permo-Carboniferous Magmatism and Rifting in Europe. Geological Society of London, Special Publications 223: 41-74. DOI: 10.1144/GSL.SP.2004.223.01.03

TNO-NITG, 2004a. Geological Atlas of the subsurface of the Netherlands (1 : 250,000) - Explanation to map sheet IX: Harderwijk. Netherlands Institute of Applied Geoscience TNO – National Geological Survey (Utrecht): 123 pp.

TNO-NITG, 2004b. Geological Atlas of the Subsurface of the Netherlands - onshore. Netherlands Institute of Applied Geoscience TNO (Utrecht): 104 pp.

Torsvik, T.H. & Cocks, L.R.M., 2016. Earth History and Palaeogeography. Cambridge University Press: 317 pp. DOI:10.1017/9781316225523

Torsvik, T.H. & Rehnström, E.F., 2003. The Tornquist Sea and Baltica-Avalonia docking. Tectonophysics 362, 67-82. DOI: 10.1016/S0040-1951(02)00631-5

Underhill, J.R. & Partington, M.A., 1993. Jurassic thermal doming and deflation in the North Sea: implications of the sequence stratigraphic evidence. In: Parker, J.R. (ed.): Petroleum Geology of Northwest Europe: Proceedings of the 4th Conference. Geological Society of London, Petroleum Geology Conference Series 4: 337-345. DOI: 10.1144/0040337

Van Adrichem Boogaert, H.A. & Kouwe, W.F.P., 1993. Stratigraphic nomenclature of the Netherlands; revision and update by RGD and NOGEPA. Mededelingen Rijks Geologische Dienst 50: 1-24.

Van Balen, R.T., Houtgast, R.F. & Cloetingh, S.A.P.L., 2005. Neotectonics of The Netherlands: a review. In: Cloetingh, S.A.P.L. & Cornu, T.G.M. (eds): Neotectonics and Quaternary fault-reactivation in Europe's intraplate lithosphere. Quaternary Science Reviews 24: 439-454. DOI: 10.1016/j.quascirev.2004.01.011

Van Bergen, M.J. & Sissingh, W., 2007. Magmatism in the Netherlands: expression of the north-west European rifting history. In: Wong, Th.E., Batjes, D.A.J.& De Jager, J. (eds): Geology of the Netherlands. Royal Netherlands Academy of Arts and Sciences (KNAW, Amsterdam): 197-222.

Van Bergen, M.J., Vis, G.-J., Sissingh, W., Koornneef, J. & Brouwers, I., 2025. Magmatism in the Netherlands: expression of the northwest European rifting history. In: Ten Veen, J.H., Vis, G.-J., De Jager, J. & Wong, Th.E. (eds): Geology of the Netherlands, second edition. Amsterdam University Press (Amsterdam): 393-455. DOI: 10.5117/9789463728362_ch11

Van den Belt, F.J.G., 2012. Sedimentary cycles in coal and evaporite basins and the reconstruction of Palaeozoic climate. PhD thesis Utrecht University (the Netherlands): 98 pp.

Van den Berg, M.W., 1994. Neotectonics of the Roer Valley rift system. Style and rate of crustal deformation inferred from syn-tectonic sedimentation. Geologie en Mijnbouw 73: 143-156.

Van den Berg, M.W., Groenewoud, W., Lorenz, G.K., Lubbers, P.J., Brus, D.J. & Kroonenberg, S.B., 1994. Patterns and velocities of recent crustal movements in the Dutch part of the Roer Valley rift system. Geologie en Mijnbouw 73: 157-168.

Van der Voet, E., Heijnen, L. & Reijmer, J.J.G., 2018. Geological evolution of the Chalk Group in the northern Dutch North Sea: inversion, sedimentation and redeposition. Geological Magazine 156 (7): 1265-1284. DOI: 10.1017/S0016756818000572

Van Hoorn, B., 1987. Structural evolution, timing and tectonic style of the Sole Pit inversion. Tectonophysics 137(1-4): 239-284. DOI: 10.1016/0040-1951(87)90322-2

Van Lochem, H., 2018. F17-Chalk: new insights in the tectonic history of the Dutch Central Graben. In: Kilhams, B., Kukla, P.A., Mazur, S., Mckie, T., Mijnlieff, H.F. & Van Ojik, K. (eds): Mesozoic resource potential in the southern Permian Basin. Geological Society of London, Special Publications: 537-558. DOI: 10.1144/SP469.17

Van Lochem, H., Ter Borgh, M. & Mijnlief, H.F., 2019. Geological evaluation for the seismic acquisition programme for SCAN areas H (West Friesland) and I (Noordoostpolder). Report by SCAN: 32 pp.

Van Lochem, H., Vis, G.-J. & Jagt, J.W.M., 2025. Late Cretaceous. In: Ten Veen, J.H., Vis, G.-J., De Jager, J. & Wong, Th.E. (eds): Geology of the Netherlands, second edition. Amsterdam University Press (Amsterdam): 253-291. DOI: 10.5117/9789463728362_ch08

Van Ojik, K., Silvius, A., Kremer, Y. & Shipton, Z.K., 2020. Fault seal behaviour in Permian Rotliegend reservoir sequences: case studies from the Dutch Southern North Sea. Geological Society of London, Special Publications 496: 9-38. DOI: 10.1144/SP496-2018-189

Van Wees, J.-D. & Cloetingh, S.A.P.L., 1996. 3D flexure and intraplate compression in the North Sea basin. Tectonophysics 266 (1-4): 343-359.

Van Wees, J.-D., Stephenson, R.A., Ziegler, P.A., Bayer, U., McCann, T., Dadlez, R., Gaupp, R., Narkiewicz, M., Bitzer, F. & Scheck, M., 2000. On the origin of the Southern Permian Basin, Central Europe. Marine and Petroleum Geology 17 (1): 43-59. DOI: 10.1016/S0264-8172(99)00052-5

Van Wijhe, D.H., 1987. Structural evolution of inverted basins in the Dutch offshore. Tectonophysics 137: 171-219. DOI: 10.1016/0040-1951(87)90320-9

Van Wijhe, D.H., 1987b. The structural evolution of the Broad

Fourteens Basin. In: Brooks, J. & Glennie, K. (eds): Petroleum Geology of North-West Europe. Proceedings of the Third Conference. Graham & Trotman (London): 315-323. ISBN 0-08601-070-0

Van Winden, M., De Jager, J., Jaarsma, B. & Bouroullec, R., 2018. New insights into salt tectonics in the northern Dutch offshore: a framework for hydrocarbon exploration. In: Kilhams, B., Kukla, P.A., Mazur, S., Mckie, T., Mijnlieff, H.F. & Van Ojik, K. (eds): Mesozoic resource potential in the southern Permian Basin. Geological Society of London, Special Publications: 537-558. DOI: 10.1144/SP469.9

Verniers, J., Pharaoh, T.C., André, L., Debacker, T.N., De Vos, W., Everaerts, M., Herbosch, A., Samuelsson, J., Sintubin, M. & Vecoli, M., 2002. The Cambrian to mid Devonian basin development and deformation history of Eastern Avalonia, east of the Midlands Microcraton; new data and a review. In: JWinchester, J.A., Pharaoh, T.C. & Verniers, J. (eds): Palaeozoic Amalgamation of Central Europe. Geological Society of London, Special Publications 201: 47-93. DOI: 10.1144/GSL.SP.2002.201.01.04

Verreussel, R.M.C.H., Bouroullec, R., Munsterman, D.K., Dybkjær, K., Geel, C.R., Houben, A.J.P., Johannessen, P.N. & Kerstholt-Boegehold, S.J., 2018. Stepwise basin evolution of the Middle Jurassic–Early Cretaceous rift phase in the Central Graben area of Denmark, Germany and The Netherlands. In: Kilhams, B., Kukla, P.A., Mazur, S., McKie, T., Mijnlieff, H.F. & Van Oijk, K. (eds): Mesozoic resource potential in the Southern Permian Basin. Geological Society of London, Special Publications 469: 305-340. DOI: 10.1144/SP469.23

Verreussel, R.M.C.H., Munsterman, D.K., Bouroullec, R. & Houben, A.J.P., 2025. Late Jurassic - Early Cretaceous. In: Ten Veen, J.H., Vis, G.-J., De Jager, J. & Wong, Th.E. (eds): Geology of the Netherlands, second edition. Amsterdam University Press (Amsterdam): 211-251. DOI: 10.5117/9789463728362_ch07

Vis, G.-J., 2020. Nieuwe vulkaan ontdekt in Nederlandse ondergrond. https://wetenschap.nu/nieuwe-vulkaan-ontdekt-in-nederlandse-ondergrond

Vis, G.-J., Houben, A.J.P., Debacker, T.N. & Geel, C.R., 2025. The geological foundation of the Netherlands: the early Carboniferous, Devonian and older. In: Ten Veen, J.H., Vis, G.-J., De Jager, J. & Wong, Th.E. (eds): Geology of the Netherlands, second edition. Amsterdam University Press (Amsterdam): 53-93. DOI: 10.5117/9789463728362_ch02

Wilson, M., Neumann, E.-R., Davies, G.R., Timmerman, M.J., Heeremans, M. & Larsen, B.T., 2004. Permo-Carboniferous Magmatism and Rifting in Europe. In: Wilson, M., Neumann, E.R., Davies, G.R., Timmerman, M.J., Heeremans, M. & Larsen, B.T. (eds): Permo-Carboniferous Magmatism and Rifting in Europe. Geological Society of London, Special Publications 223: 1-10.

Wong, Th.E., 2007. Jurassic. In: Wong, Th.E., Batjes, D.A.J.& De Jager, J. (eds): Geology of the Netherlands. Royal Netherlands Academy of Arts and Sciences (KNAW, Amsterdam): 107-125.

Woodcock, N. & Strachan, R. 2012. Geological History of Britain and Ireland. Second edition. Blackwell Publishing Ltd (Oxford): 442 pp.

Worthington, R.P. & Walsh, J.J., 2011. Structure of Lower Carboniferous basins of NW Ireland, and its implications for structural inheritance and Cenozoic faulting. Journal of Structural Geology 33: 1285-1299. DOI: 10.1016/j.jsg.2011.05.001

Zanella, E. & Coward, M.P., 2003. Structural framework. In: Evans, D., Graham, C., Armour, A. & Bathurst, P. (eds): The Millennium Atlas: petroleum geology of the central and northern North Sea. Geological Society of London: 45-59.

Ziegler, P.A., 1978. North-Western Europe: Tectonics and basin development. Geologie en Mijnbouw 57(4): 589-626.

Ziegler, P.A., 1981. Evolution of sedimentary basins in North-West Europe. In: Illing, L.V. & Hobson, G.D. (eds): Petroleum Geology of the continental Shelf of North-West Europe: Proceedings of the Second Conference. Institute of Petroleum (London): 3-39. ISBN 0-85501-656-6

Ziegler, P.A., 1988. Evolution of the Arctic, North Atlantic and western Tethys. American Association of Petroleum Geologists, Memoir 43, 198 pp.

Ziegler, P.A., 1990a. Geological Atlas of Western and Central Europe, second edition. Shell Internationale Petroleum Mij. B.V. and Geological Society (London): 239 pp. ISBN 978-90-6644-125-5

Ziegler, P.A., 1990b. Tectonic and palaeogeographic development of the North Sea Rift system. In: Blundell, D.J. & Gibbs, A.D. (eds): Tectonic Evolution of the North Sea Rifts. Oxford Science Publications: 1-36.

Ziegler, P.A., 1994. Cenozoic rift system of western and central Europe: an overview. Geologie En Mijnbouw 73(2-4): 99-127.

Ziegler, P.A., Cloetingh, S. & Van Wees, J.-D., 1995. Dynamics of intra-plate compressional deformation: the Alpine foreland and other examples. Tectonophysics 252: 7-59.

Zwart, H.J. & Dornsiepen, U.F., 1978. The tectonic framework of Central and Western Europe. Geologie en Mijnbouw 57(4): 627-654.

2

The geological foundation of the Netherlands: the early Carboniferous, Devonian and older

Geert-Jan Vis, Alexander J.P. Houben, Timothy N. Debacker & Cornelis R. Geel*

*Corresponding author: geert-jan.vis@tno.nl

ABSTRACT

This chapter addresses the geological development of the Netherlands and its immediate surroundings during the older portion of the Paleozoic (Cambrian to early Carboniferous, 538-323 Ma). During this period northwestern Europe experienced two major phases of mountain building: the Caledonian and the Variscan phases. These laid the foundations for the geology of the Netherlands. The Caledonian phase is associated with the progressive closure of the Iapetus Ocean between Laurentia to the northwest and Baltica and Avalonia to the southeast and south, respectively. Although classically considered as a Caledonian phase, the Brabantian Deformation event is related to the progressive closure of the Rheic Ocean which existed south of Avalonia. This is represented by an angular unconformity surrounding the ancient Brabant Massif (Belgium). This unconformity separates deformed Silurian and older, (very) low-grade metamorphic units from overlying relatively undeformed non-metamorphic Middle Devonian deposits. The continued northward migration of microplates originating from Gondwana ultimately closed the Rheic Ocean, and formed the Variscan Mountains and foreland basin during the Carboniferous. It is against this backdrop that the mainly Middle Devonian to lower Carboniferous sedimentary succession of the Netherlands accumulated along the northern edge of the Rhenohercynian sea. Borehole and seismic data of this succession are still very scarce. However, over the last decade several research initiatives provided valuable new insights in this fundamental episode in the geological history of the region.

<< A tilted Famennian sandstone-shale alternation in the Trooz quarry (southeast of Liège, along the Vesdre River in Belgium). Photo: Geert-Jan Vis.

DOI: 10.5117/9789463728362_ch02

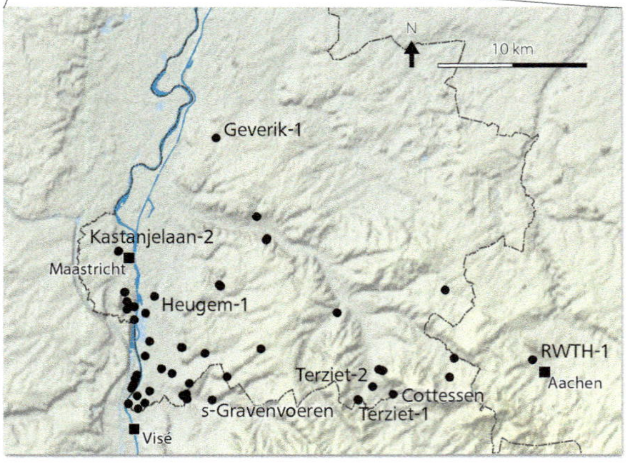

Figure 2.1. Location map showing all of the 74 wells in the Netherlands that encountered Silurian to lower Carboniferous rocks. Wells and regional names mentioned in the text are labelled. The data originate from Mining Law and non-Mining Law datasets at TNO – Geological Survey of the Netherlands. For reference some Belgian, German and UK wells are also shown. Inset shows an enlarged map of South Limburg where several coal exploration wells reached into the lower Carboniferous.

Introduction

The oldest strata encountered in the Netherlands date back to the Silurian period. In nearby Belgium, older lower Paleozoic (Cambrian to Ordovician) strata are widely encountered in outcrop and boreholes. In this chapter, the geological development during the Cambrian to early Carboniferous (539-323 Ma) of northwestern Europe is outlined and linked to the stratigraphic succession encountered in the Netherlands.

During the older part of the Paleozoic, many fundamental changes occurred, both biologically and geologically. After the first terrestrial vegetation emerged during the Ordovician (~480 Ma, Steemans et al., 2009), vascular plants radiated during the Silurian (~430 Ma) setting the stage for the spread of vegetation onto land during the Devonian. Two of the big five mass extinctions occurred during this period: the Late Ordovician (~444 Ma) and Late Devonian extinctions (375-360 Ma). Added to this there was a major tectonic reorganization along the northern margin of the Rheic Ocean.

In spite of the fundamental changes occurring during the Paleozoic, the expression of this period in the subsurface of the Netherlands is difficult to disentangle. None of the Paleozoic successions is exposed in the Netherlands, although ample outcrops are known from Belgium and Germany. In the Netherlands, pre-Serpukhovian (pre-Namurian) sedimentary rocks are only known from 74 boreholes (Fig. 2.1) or are inferred from seismic data. In the ultimate southwest and southeast of the country, this Paleozoic succession is found at relatively shallow depth (<500 m). Going northwards, the depth rapidly increases, reducing the number of boreholes and seismic surveys that reach it. Based on the interpretation of seismic data, the top and base of the lower Carboniferous limestone succession (Dinantian) have been interpreted in a patchy pattern across the onshore Netherlands (Ten Veen et al., 2019).

Interest in the occurrence of gas-prone source rocks in the northern offshore has grown over the last decades (Gerling et al., 1999; Doornenbal et al., 2019). Especially the coals in the upper Visean Elleboog, Scremerston and Yoredale formations have been considered as potential source-rocks, but others have been identified as well (Kearsey et al., 2018). About a decade ago, geothermal exploration activities increased, focussing on Upper Devonian siliciclastics and lower Carboniferous (Dinantian) limestones. Between 2017 and 2020 the Green Deal Ultra Deep Geothermal Energy programme explored the possibilities for ultra-deep, i.e. deeper than 4000 m, geothermal projects in the Netherlands, focussing on lower Carboniferous limestones. Over the last five years, activities related to the possible construction of the Einstein Telescope have generated, and continue to generate, new data for southern South Limburg and neighbouring Belgium (see Textbox 1). All these developments have generated new data and insights that are incorporated in this chapter.

Historic studies

Numerous studies have addressed the Paleozoic succession during the last 150 years. Because the Paleozoic occurs at shallow depths in and around the southeast of the Netherlands (South Limburg), most studies have focussed on that region. An overview of the most important contributions treating the pre-Serpukhovian (pre-Namurian), is given below (also see Vis & Houben, 2023).

The detailed geologic mapping by Staring (1869, map sheet 27) was the first to indicate Paleozoic rocks in or near the Netherlands. In the southeast of the Netherlands, between Maastricht and Aachen, his map shows various units which we now know as dating to the pre-Serpukhovian Paleozoic, such as 'Culm' (lower Carboniferous basinal shale), 'Bergkalk' (lower Carboniferous limestone) and 'Kolenlooze zandsteen' (Serpukhovian, Namurian sandstone). Nearly half a century later, a drilling campaign in the search for coal provided a two-dimensional component to the understanding of the distribution of the Paleozoic represented by a cross section (Van Waterschoot van der Gracht, 1918).

The first detailed Paleozoic subcrop map of South Limburg was published by Sax (1946). The mapping campaign used the enormous amount of data collected through the exploitation of upper Carboniferous coal in South Limburg. Whereas most previously published maps and profiles approach the geology from the surface, Sax (1946) compiled data from shafts and upward-directed borings from within the coal mines. The map does not include the pre-Serpukhovian Paleozoic succession, but the accompanying cross sections do indicate the presence of lower Carboniferous (Dinantian) limestones ('Kolenkalk') near Voerendaal (well SM-59). Bless et al. (1976) presented a more regionally extensive Paleozoic subcrop map for South Limburg and its surroundings. Some of the challenges posed by Bless et al. (1976) related to the distribution of Devonian, Dinantian, Namurian as well as late Westphalian strata of South Limburg, were subsequently addressed by Kimpe et al. (1978), who provide updated maps of the Paleozoic abrasion surface and its immediate cover.

The most recent Paleozoic subcrop map of South Limburg and neighbouring areas was published by the Geological Survey of the Netherlands (Rijks Geologische Dienst, 1995). This map, and the Geological Atlas of the Subsurface of the Netherlands (TNO-NITG, 1999), provide geological cross sections through South Limburg. Remarkably, these profiles show Paleozoic strata down to 6 km below sea level, while the deepest borehole (Geverik-1) only reaches 1580 m below sea level and seismic data have

Einstein Telescope: watching the universe from underground

1

From 2035 onwards the third generation gravitational wave observatory may be operational 200-300 m below the rolling land surface of South Limburg and neighbouring Belgium. Gravitational waves are ripples in space-time (analogue: ripples at the water surface after a stone was thrown into a pond) caused by some of the most violent and energetic processes in the universe. Their existence was predicted by Albert Einstein in his general theory of relativity. The strongest gravitational waves originate from colliding black holes and neutron stars and from supernovae (Einstein, 1916, 1918).

On 14 September 2015, the gravitational-wave observatory LIGO in the United States made the first physical observation of a passing gravitational wave, using laser inferometry. These undulations in space-time were caused by two colliding black holes 1.3 billion light-years away. For better triangulation (determining the origin of a gravitational wave) and higher accuracy, the third generation gravitational-wave observatory will be built in Europe and underground. Detecting gravitational-waves underground has the advantage of co-registering less noise from activities at the surface and thereby increasing the accuracy.

A situation where more or less unconsolidated rock overlies competent rock, provides ideal conditions for this purpose. In the South Limburg-Belgium region, these conditions are met. Soft Cretaceous and Cenozoic rocks overly generally competent, consolidated Upper Devonian and lower Carboniferous sedimentary rocks. Unfortunately, the region in South Limburg straddling the border with Belgium is poor in both borehole and seismic data, due to limited exploration activities for resources such as coal and hydrocarbons. These data are necessary to position the enormous underground facility, which consists of a triangular tunnel system with tunnels measuring up to 10 km in length. At the vertices, large underground caverns will hold sensitive supercooled mirror systems. To support the case for placing the Einstein Telescope in this region, various studies and a data acquisition campaign are underway. Besides the South Limburg-Belgium region, also Sardinia and possibly southeastern Germany hope to attract this high-tech project.

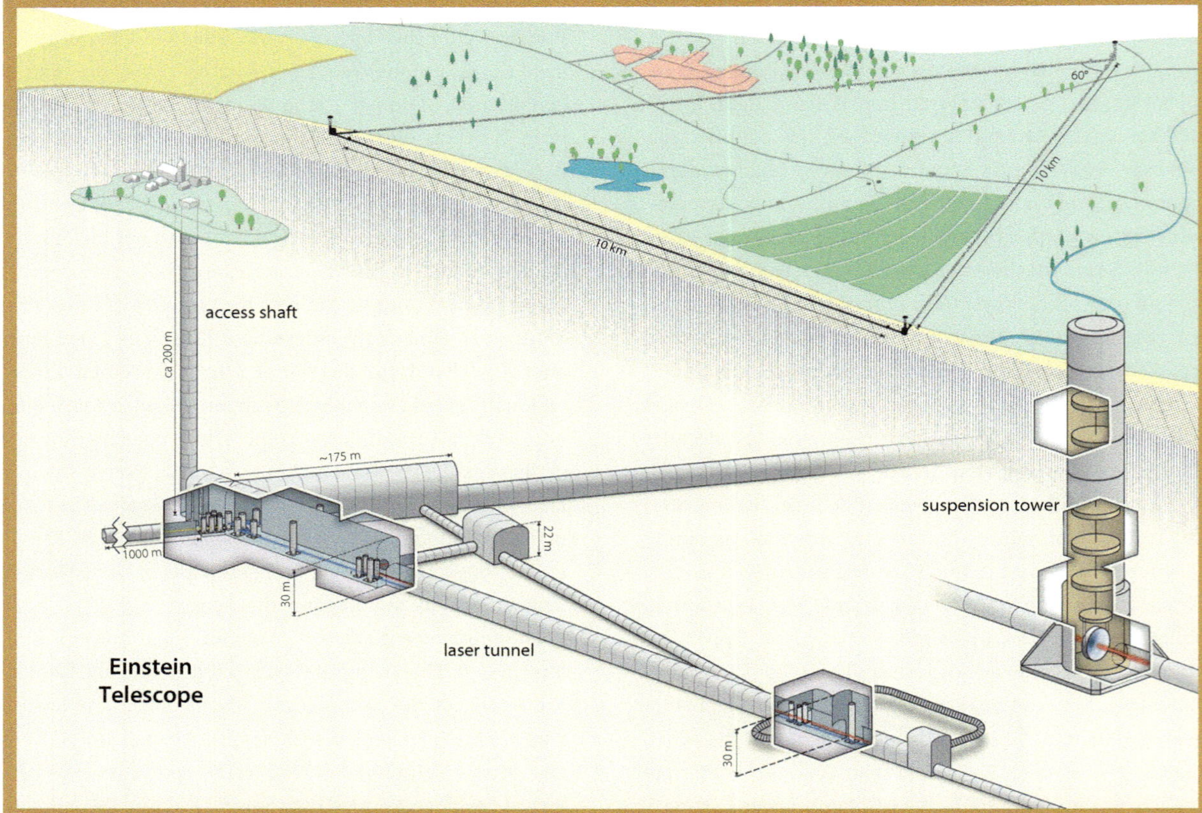

Infographic showing one of the vertices of the triangular setup of Einstein Telescope. Credits: Nikhef/Thijs Balder.

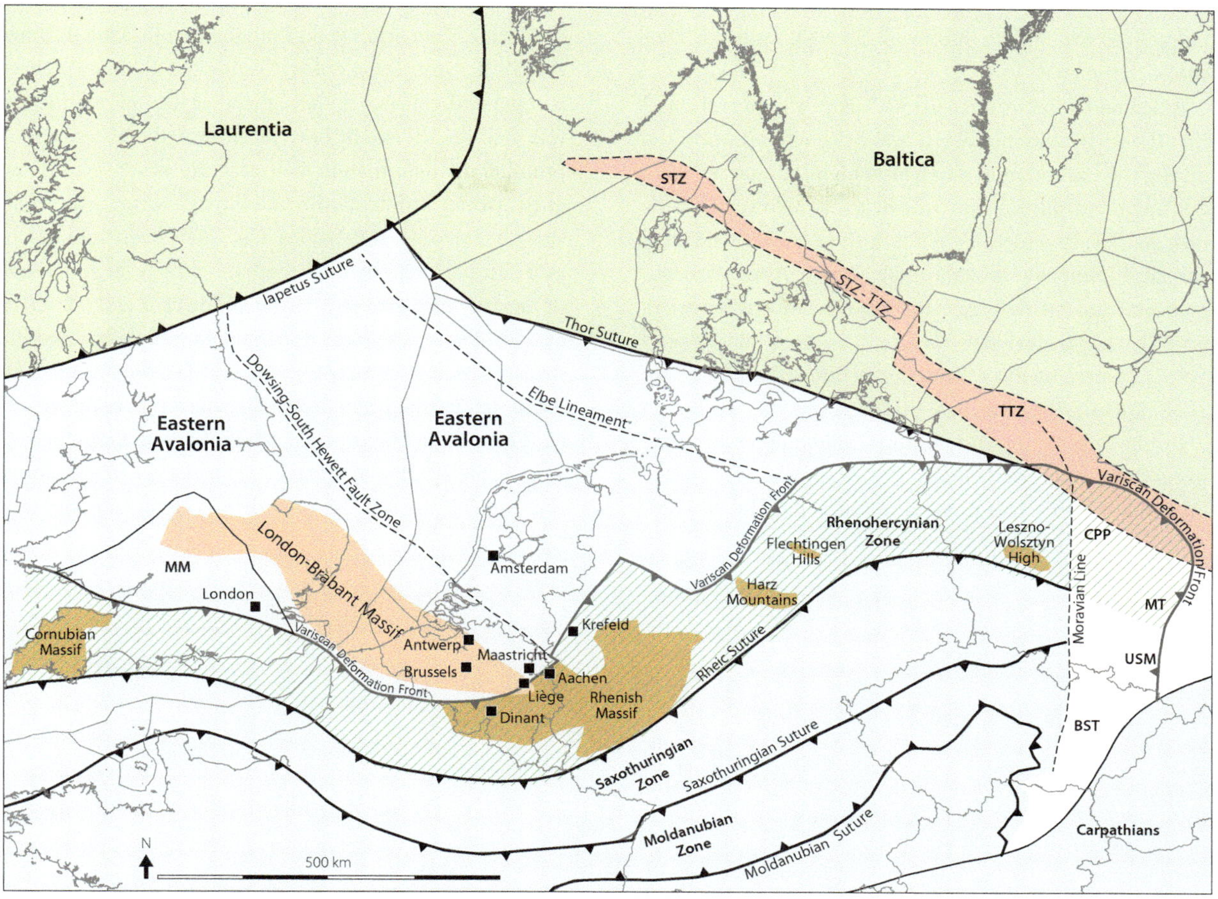

Figure 2.2. The tectonic setting of the area around Eastern Avalonia, showing the amalgamated terranes and key terminology. STZ = Sorgenfrei-Tornquist Zone; TTZ = Teisseyre-Tornquist Zone; CPP = Central Polish Paleorift; BST = Bruno-Silesian Terrane; USM = Upper Silesian Massif; MT = Malopolska Terrane; MM = Midlands Microcraton. The map is compiled using various sources (Debacker, 2001; Verniers et al., 2002; Pharaoh et al., 2006; Woodcock et al., 2007; Sintubin et al., 2009; Bełka et al., 2010).

poor resolution below the top-Paleozoic interface. Hence, some caution is advised when using these cross sections.

Updated lithostratigraphy

The limited number of wells penetrating the older Paleozoic succession in the Netherlands has hampered the definition of lithostratigraphic units. The current definitions are partially incomplete, are incorrect or are informal. During the last decades new wells became available and new studies have provided new insights. Therefore, an update of lithostratigraphic definitions is timely and is proposed in this chapter (see section Stratigraphy).

Tectonic setting

During the time-interval covered in this chapter (Cambrian to early Carboniferous), the plate-tectonic amalgamation of northwest Europe was completed (Fig. 2.2). The Netherlands was situated on continental crust of the eastern part of the microcontinent Avalonia. In the late Ordovician this microcontinent collided with Baltica, resulting in closure of the Tornquist Ocean and formation of the **Thor Suture**, and in the mid-Silurian with Laurentia, resulting in closure of the Iapetus Ocean and formation of the **Iapetus Suture** (Verniers et al., 2002; Pharaoh et al., 2006; Woodcock et al., 2007; Smit et al., 2016). Based on the timing of collision(s), the widespread occurrence of an angular unconformity overlain by red-coloured Devonian continental deposits and the paleogeographical position, this places these events within the **Caledonian cycle** (e.g. McKerrow et al., 2000). However, as explained in Textbox 2, strictly speaking true Caledonian deformation (i.e. the Scandian phase) is reserved for Silurian deformation in Scotland, Greenland and Norway, directly related to closure of the Iapetus Ocean. However, as in Belgium, where the effects of (a) late early Paleozoic deformation event(s) and the overlying unconformity can be observed in outcrop, the Netherlands occupies a south-central position within the eastern parts of Avalonia, at a considerable distance from the Iapetus suture to the NNW and the Thor suture to the ENE.

Based on faunal records, Cocks & Fortey (1982) showed that the Avalonian microplate was attached to Gondwana during the Cambrian in a southern high-latitude position, and migrated during the Ordovician to a lower latitude near Baltica, to which its fauna in the Late Ordovician is similar. Paleomagnetic data indicate that Avalonia occupied positions at 60°S in Early Ordovician Tremadocian, 40°S in Late Ordovician (Caradoc), 20°S in Silurian and 10°S in Late Devonian Famennian times (Torsvik et al., 1996; Cocks, 2000; Torsvik & Cocks, 2016).

In the late Early Ordovician, Avalonian crust rifted from Gondwana and rapidly drifted northwards. Based on faunal records, it is thought that it collided with Baltica by Late Ordovician (mid-Caradoc to mid-Ashgill) times (Vecoli & Samuelsson, 2001; Verniers et al., 2002). There is a general consensus that Tornquist Ocean subduction was towards the south but, because of the absence of a protuberant deformation belt and the absence of Tornquist Ocean subduction-related magmatism, the collision of Avalonia and Baltica has often been referred to as a soft-docking event, accompanied by a large amount of dextral displacement. Late Ordovician-early Silurian calc-alkaline magmatism documented in Belgium and the British Isles is slightly too young and far too distant to be attributed to the Avalonia-Baltica collision (e.g. Verniers et al., 2002; Linnemann et al., 2012). Nevertheless, there is clear evidence that significant deformation in the southern North Sea area (Mid North Sea High and southern

Caledonian, Scandian, Acadian, Brabantian, Ardennian – what's in a name? 2

In northwest Europe red-coloured Devonian clastic deposits unconformably overlie deformed, metamorphic lower Paleozoic deposits. This reflects a deformation event of regional extent. In the British Isles this long-recognized deformation event was defined as the **Caledonian Orogeny**, named after the Roman name for Scotland, where the effects of this orogeny were first and best documented (e.g. Suess, 1885; Haug, 1900; Stille, 1924). With the advent of plate tectonics and the recognition of the existence of different 'Caledonian' events, other names were introduced, such as **Scandian**, **Acadian**, **Brabantian** and **Ardennian**. These are explained below.

The **Caledonian Orogeny** encompasses a series of early Paleozoic to Devonian events, including intra-oceanic events, associated with the progressive closure of the Iapetus Ocean between Laurentia to the northwest and Baltica and Avalonia to the southeast (present-day orientation). The **Scandian event**, that took place from ~440 to 410 Ma, was the regionally most important Caledonian event and reflects the closure of the Iapetus Ocean and associated continent-continent collision between Laurentia to the NW and Baltica and Avalonia to the SE in mid-Silurian (Wenlock) times. Its effects are documented in Scotland, Norway and East Greenland, and include thick-skinned thrusting with high-pressure metamorphism (e.g. Gee et al., 2012). Surprisingly, there is no significant deformation along the northwestern leading edge of Avalonia that can be linked to the Scandian event (Woodcock et al., 2007, 2019).

The **Acadian event**, originally defined in the Canadian Appalachians, was introduced in the British Isles by Soper et al. (1987) in order to explain Early to Middle Devonian deformation affecting the Lower Devonian lower Old Red Sandstone, often preserved in sinistral transtensional basins in England and Wales. This deformation event took place after closure of the Iapetus Ocean and resulted in sinistral strike-slip, transtension and transpression in Scotland, Northern Ireland, England and Wales. In Wales and Northern England, sinistral transpression along the NW side of the Midlands Microcraton (Avalonia) was associated with ~400-390 Ma cleavage development.

The **Brabantian event** and **Ardennian event** were defined in Belgium at the start of the 21st century as a response to potentially erroneous long-distance correlations of deformation events based merely on timing (e.g. Debacker, 2001; Verniers et al., 2002). In Belgium, two separate angular unconformities occur between lower Paleozoic slates and red-coloured Devonian continental deposits, one occurring in northern and central Belgium, and one occurring in southern Belgium and northern France (e.g. Sintubin et al., 2009). The angular unconformity surrounding the Brabant Massif (N-Belgium) separates deformed Silurian and older, low-grade metamorphic units from overlying relatively undeformed diagenetically-altered Middle Devonian deposits. As the timing of the deformation event in the Brabant Massif coincides with the Acadian event in the British Isles (see Debacker et al., 2005), the term Acadian event was introduced also in Belgium (Van Grootel et al., 1997). However, at the start of the 21st century the Acadian event in the British Isles was considered to be related to closure of the Iapetus

parts of Ringkøbin Fyn High) and the North German-Polish Caledonides is associated with Avalonia-Baltica collision. This includes Late Ordovician-Silurian thrusting and amphibolite metamorphism around the Thor Suture and foreland basin development on the southern margin of Baltica (e.g. see De Vos et al., 2010 and references therein). Smit et al. (2016, 2018) attributed the absence of a protuberant deformation belt to Devonian-Carboniferous extension with collapse of the deformed Tornquist accretionary prism.

During the Ordovician and Silurian, the Iapetus Ocean progressively closed. From around the mid Ordovician onwards (i.e. after the Grampian Orogeny in Scotland) the Iapetus Ocean became subducted to the NW below Laurentia. The Southern Uplands accretionary prism in Scotland testifies of this subduction (e.g. Verniers et al., 2002; Woodcock et al., 2007 and references therein). The joint Avalonia-Baltica continent collided with Laurentia by Wenlock (mid-Silurian) times and gave rise to the continent Laurussia (e.g. Cocks, 2000). This collision, reflected by the Scandian deformation event, resulted in intense deformation with high pressure metamorphism in Scotland, Greenland and Norway (e.g. Gee et al., 2008, 2012), but only minor deformation along the suture between Avalonia and Laurentia (Woodcock et al., 2007, 2019). The main deformation along the northern (WSW-ENE trending) margin of Avalonia occurred later, during the Early Devonian, and was characterized by sinistral transpression, accompanied by low grade metamorphism and cleavage development at ~400-390 Ma (Soper et al., 1987;

Ocean, whereas observations in Belgium showed that the deformation within the Brabant Massif was difficult to link to either the closure of the Tornquist Ocean (between Baltica and Avalonia) or the closure of the Iapetus Ocean (between Laurentia and Avalonia-Baltica) (e.g. Debacker, 2001; Verniers et al., 2002; Sintubin & Everaerts, 2002). As a working solution, a different name, the **Brabantian event** or **Brabantian deformation phase**, was introduced (see Debacker, 2001 and Verniers et al., 2002). The angular unconformity surrounding the Ardennes Inliers (S-Belgium) separates deformed Ordovician and older, (very) low-grade metamorphic units from overlying deformed latest Silurian to Early Devonian (very) low-grade deposits, which were again deformed during the Variscan Orogeny (e.g. Sintubin et al., 2009). This event, only observed in the Ardennes Inliers now situated in the Variscan deformation belt, took place well before the Acadian and Brabantian events, and cannot be tied to a specific early Paleozoic shortening event. Hence, the term **Ardennian event** or **Ardennian deformation phase** was introduced (e.g. Verniers et al., 2002). However, this nomenclature is still under discussion (e.g. Herbosch et al., 2020; Herbosch, 2021) and there is no convincing evidence yet that the Ardennian event was compressional (Sintubin et al., 2009). The unconformity surrounding the Ardennes Inliers could also reflect a Silurian rifting event, later overprinted by the Variscan Orogeny (e.g. Sintubin et al., 2009).

Currently the Acadian deformation event is no longer considered to be associated with the closure of the Iapetus Ocean or the Tornquist Ocean, but is interpreted to be a proto-Variscan event along the northern margin of the Rheic Ocean. This interpretation reconciles the Acadian deformation in the British Isles and the Brabantian deformation in northern and central Belgium (Debacker, 2001; Woodcock et al., 2007; Sintubin et al., 2009).

Table summarizing the various terminologies on the 'Caledonian' events listed in the text.

Event name	Description
Scandian	Most important Caledonian event of regional extent reflecting the closure of the Iapetus Ocean and associated continent-continent collision in Wenlock times between Laurentia to the NW and Baltica and Avalonia to the SE. Deformation documented along Iapetus suture (Scotland, Greenland and Norway).
Acadian (in British Isles)	Deformation of lower Devonian Old Red Sandstone and underlying units in the British Isles. In Wales and Northern England sinistral transpression was associated with ~400-390 Ma cleavage development along NW side of Midlands Microcraton (Avalonia). Currently attributed to an event along the northern margin of the Rheic Ocean.
Brabantian	Deformation of Silurian and older deposits in Anglo-Brabant deformation belt, along NE side of Midlands Microcraton (Avalonia), documented in N Belgium (e.g. Brabant Massif). Deformation from Wenlock onwards, with ~400 Ma cleavage development. Currently attributed to an event along the northern margin of the Rheic Ocean.
Ardennian	Angular unconformity in N France and S Belgium between Ordovician and latest Silurian-Devonian. No convincing evidence for compressional event and potentially related to rifting event along southern side of Avalonia (Rheic margin).

Woodcock et al., 2007). This deformation event is called the Acadian Deformation event (Soper et al., 1987; see Textbox 2).

During the early Paleozoic a ~NW-SE trending basin developed along the NE side of the Midlands Microcraton (i.e. the core of the eastern part of Avalonia), called the Brabant Basin. This Brabant Basin, situated in northern Belgium and eastern England, consists of a thick, fault-bounded Cambrian succession, overlain by a thinner Ordovician succession, in turn overlain by an often very thick Silurian succession. During the Late Ordovician, around the time of Avalonia-Baltica docking, widespread slumping took place (Debacker et al., 2001; Debacker, 2012). Possibly, also the Asquempont Detachment System, an originally NE-dipping low-angle normal fault system that can be traced throughout the Brabant Massif, and that was formed between the middle Katian and the time of cleavage development (see Herbosch et al., 2008; Herbosch & Debacker, 2018, and original references therein), can be related to this time period. The Silurian succession shows evidence of foreland basin development, with a thick package of fine-grained turbidites, except where it overlies the northeastern parts of the Midlands Microcraton (Verniers et al., 2002 and references therein). The Brabant Basin was deformed from Wenlock times onwards, accompanied by Silurian foreland basin development, and resulted in the Anglo-Brabant Deformation Belt, of which the Brabant Massif in northern Belgium forms the principal outcrop area (e.g. Van Grootel et al., 1997; Sintubin, 1999; Debacker, 2001; Sintubin & Everaerts, 2002; Verniers et al., 2002; Debacker et al., 2005). An angular unconformity separates the anchizonal to epizonal, deformed units of the Anglo-Brabant Deformation Belt from overlying Givetian (Middle Devonian) and younger sedimentary deposits (De Vos et al., 1993). Deformation of the NW-trending Brabant Basin along the northeast margin of the Midlands Microcraton, with development of the Anglo-Brabant Deformation Belt, took place progressively over ~30 Myr, and culminated in cleavage development at ~400 Ma (Debacker et al., 2005 and references therein). This timing corresponds to the Acadian deformation in the British Isles, but because of the deformation belt's position and orientation at right angles to the Iapetus suture, the term Brabantian deformation event has been introduced (Debacker, 2001; Verniers et al., 2002; Sintubin et al., 2009).

The Brabantian deformation event is essentially an inversion of the pre-existing Brabant Basin, with the strongest inversion occurring in the central basin parts, surrounded by outward propagating (Silurian) foreland basins (e.g. Debacker, 2001; Sintubin & Everaerts, 2002; Debacker et al., 2005). Interestingly, along/above the northeast side of the Avalonia Microcraton, Silurian to lowermost Devonian deposits are diagenetically altered and hardly deformed (e.g. Verniers et al., 2002; cf. Welsh Borderland along NW-side of Midlands Microcraton). This implies that also in other parts of eastern Avalonia outside of areas of significant basin inversion, the presence of relatively undeformed Silurian deposits can be expected, as seen in wells Kortgene-1 and O18-01 in the southwestern Netherlands (Houben & Vis, 2021).

Deep seismic profiles in the southern North Sea (Blundell et al., 1991) show a southwest dipping deep reflector in the mantle below the Dowsing-South Hewett Fault Zone, interpreted as a possible remnant of an Ordovician intra-Avalonia subduction zone (Lee et al., 1993; Pharaoh et al., 1995; Pharaoh, 2018). The Dowsing-South Hewett Fault Zone (Fig. 2.2) may represent a suture between two different terranes within the amalgamated eastern Avalonia microplate. This terrane boundary continues to the southeast towards the Roer Valley Graben (e.g. Pharaoh et al., 2006; Pharaoh, 2018) and may have controlled later graben development.

A series of negative gravity anomalies within the Anglo-Brabant Deformation Belt are often interpreted as granitic bodies (Everaerts et al., 1996; Mansy et al., 1999), or are simply referred to as low-density bodies (Debacker, 2001; Sintubin & Everaerts, 2002). These bodies had a significant influence on Brabantian deformation and must have been present prior to this event (De Vos, 1997; Sintubin, 1999; Debacker, 2001; Verniers et al., 2002; Debacker et al., 2005). It is difficult to link these to Tornquist Ocean subduction but, if subduction-related, they may have been caused by subduction of a short-lived ocean within the eastern Avalonia terrane assembly along the Dowsing-South Hewett Fault Zone (e.g. Debacker, 2001; Verniers et al., 2002). As an alternative, Linneman et al. (2012) suggested an intracontinental origin.

During the Ordovician northward migration of Avalonia, the Rheic Ocean opened between Avalonia and Gondwana (Ziegler, 1990). After the collision of Avalonia with Baltica, the Rheic Ocean started to subduct, both southwards below the Armorican terranes and northwards below Avalonia (Franke, 2000). The ~Late Ordovician-Silurian 'Ardennian event' in Belgium is still disputed, and might even be related to Rheic extension, rather than reflecting a compressional event (see Sintubin et al., 2009; cf. Herbosch et al., 2020; Herbosch, 2021). The younger Brabantian event is regarded as the equivalent of the Acadian event, and is not related to the Iapetus Ocean closure to the north but instead represents a 'proto-Variscan' event related to progressive closure of the Rheic Ocean to the south (Debacker, 2001; Woodcock et al., 2007; Sintubin et al., 2009). Possible causes for the Brabantian and Acadian events are flat-slab subduction or impingement of another Gondwana-derived terrane (Woodcock et al., 2007).

During the Devonian and Carboniferous, the three-plate convergence gave way to a two-plate convergence (e.g. Ziegler, 1990). The latter involved the northward drift of Armorican terranes from Gondwana against Laurussia, with a major sinistral translation between Laurentia-Greenland and Fennoscandia-Baltica. During the Early Devonian a narrow (400-600 km wide) and short-lived oceanic basin opened before the Rheic Ocean had completely closed (Franke, 2000; Franke et al., 2017). This oceanic basin is referred to as Rhenohercynian Ocean (e.g. Von Raumer et al., 2009; Pharaoh et al., 2010; Franke et al., 2017). It was located south of Avalonia and north of the closing Rheic Ocean in what now is central Europe and southwest England. Multiple reconstructions of the complex plate configurations at that time have been postulated in literature (e.g. Shail & Leveridge, 2009; Von Raumer et al., 2009; Franke et al., 2017). To the north, this basin had a passive margin, while to the south its oceanic crust subducted beneath the Gondwana-derived terranes (Shail & Leveridge, 2009; von Raumer et al., 2009; Franke et al., 2017). The famous ophiolites of the Lizard Complex in southwest England are associated with early-formed oceanic crust in this basin (Kirby, 1979; Shail & Leveridge, 2009), as is the mid-ocean ridge basalt (MORB) in central Germany (Franke, 2000). In the Netherlands and Belgium no evidence for this small oceanic basin has been identified. Due to its small size and short-lived character, we here prefer the use of the term Rhenohercynian sea. The Rhenohercynian sea is not to be confused with the Rhenohercynian Zone, which is the Variscan fold-and-thrust belt that eventually formed after collision of the Variscan orogenic wedge with the passive northern margin of the Rhenohercynian sea (Fig. 2.2, Pharaoh et al., 2010).

The northward subduction of the Rheic Ocean below Avalonia was accompanied by periods of slab roll-back during the Late Devonian and Carboniferous. This resulted in a collapse of the former Tornquist Ocean accretionary prism around the Thor Suture (Smit et al., 2016, 2018). It also led to wide-spread formation of horsts and grabens around the British Isles and the southern North Sea, via reactivation of the pre-existing basement fabric (Smit et al., 2016, 2018). In addition to this back-arc extension, some expulsion of Baltica to the east may have taken place as well, causing further E-W extension in the southern North Sea area (Coward, 1993). The net result was a series of WNW-ESE trending, fault-bounded half-grabens in the southern North Sea, similar to the basins described in the English onshore (Fraser & Gawthorpe, 1990; Chadwick, 1993; Hollywood & Whorlow, 1993). The basins are separated by granite-cored horsts such as the one encountered in well A17-01 (Ziegler, 1990). Sedimentation resumed in the area of well A17-01 during the late Middle Devonian. Strong Middle Devonian to early Carboniferous block faulting, controlled by the basement fabric, has been documented from both the northern flank of the Anglo-Brabant Deformation Belt (Muchez & Langenaeker, 1993) and the Mid North Sea High (Quirk, 1993; Maynard & Dunay, 1999). Similar block-faulting is thought to exist below the Netherlands (Cameron & Ziegler, 1997) and evidence to support this can be found on good quality 3D seismic surveys in the Groningen area (Ten Veen et al., 2019) or the northern offshore (Ter Borgh et al., 2018b).

Continued subduction ultimately resulted in closure of the Rheic Ocean, with formation of the nappe complex of the Variscan Mountains during the late Carboniferous (Ziegler, 1990; Franke, 2000). The thrust front was situated south of the Brabant Massif and is represented by the Midi-Aachen Thrust (Fig. 2.2; Huis in 't Veld & Den Hartog Jager, 2025, this volume). The Rhenohercynian Zone, which includes the Ardennes and the Rhenish Massif (Fig. 2.2), has an Avalonian crust and Avalonian faunas, and was thrust upon more northerly areas of Avalonia during the late Carboniferous (Pharaoh, 1999; Franke, 2000; Oncken et al., 2000).

Climate and biotic evolution

Cambrian

The Cambrian Period (541-485 Ma) is widely known for its rapid diversification of life forms, known as the Cambrian explosion. This produced the first representatives of all modern animal phyla. Although diverse life forms prospered in the oceans, the terrestrial realm remained predominantly barren, not boasting more complex life than microbial soil crusts (Retallack, 2009). Initially stromatolites were the most abundant colonial marine organisms. The advent of grazing and bioturbation, and the proliferation of silicate-bearing sponge-like creatures like the archaeocyathids, led to the demise of stromatolites, microbial mats and associated organisms by the end of the early Cambrian (Hagadorn & Bottjer, 1997). Cambrian climatic conditions are not understood in detail. Recent oxygen-isotope analyses of phosphatic microfossils suggest temperatures that were comparable to late Mesozoic and early Cenozoic greenhouse climate conditions (20-25°C at southern high-latitudes, Hearing et al., 2018).

Ordovician

The Ordovician (485-444 Ma) marks another dramatic phase of evolutionary radiation: the Great Ordovician Biodiversification Event (GOBE, Munnecke et al., 2010). A true regime shift was the introduction of filter-feeding organisms. Groups that remained dominant during the rest of the Paleozoic arose; notably articulate brachiopods, molluscs (bivalves, gastropods and cephalopods), echino-

Figure 2.3. a) Model of a Cooksonia *plant that lived from the middle of the Silurian to the Early Devonian. Microscopic traces and spores of this plant were found in cores of Kortgene-1.* Cooksonia *model: MUSE-Science Museum of Trento (via Wikimedia Commons, CC BY-SA 3.0); b) Spore of* Cooksonia *(Kortgene-1, 1910.4 MD). These spores are also found in fossilized* Cooksonia *specimens; c) Microscopic tubular plant fragments of* Cooksonia*, that stabilized and anchored the plant (Kortgene-1, 1898.5 MD).*

derms (notably crinoids) and reef-building corals (Droser & Finnegan, 2003). The Ordovician also manifests the oldest unequivocal evidence for colonization of the terrestrial realm by arthropods (MacNaughton et al., 2002) and possibly by arachnids (spiders, Lozano-Fernandez et al., 2020). The first microscopic evidence (spores) of land-plants date back to the Early Ordovician (Steemans et al., 2009; Rubinstein et al., 2010; Strother & Foster, 2021). These plants were small, non-vascular, ephemeral and probably directly evolved from green algae (Strother & Foster, 2021).

The Early Ordovician was very warm and lacked continental-scale ice-sheets. Sea-surface temperatures as high as 45°C are suggested by some authors (Song et al., 2019). More temperate climatic conditions characterize the Middle Ordovician, with further cooling into the intense and relatively abrupt Hirnantian Glaciation (Late Ordovician). Sedimentological evidence for this glaciation is found across much of Gondwana, which was located in the high-latitude southern hemisphere (Delabroye & Vecoli, 2010; Cocks & Torsvik, 2021). Despite the polar position of Gondwana, the Hirnantian Glaciation remains remarkable given the very high atmospheric Ordovician CO_2-concentrations, which are estimated to be 14 to 16 times pre-industrial levels (Crowley & Baum, 1995). Associated with the Hirnantian Glaciation, the oldest of the 'big-five-mass-extinctions' occurred. This global extinction eliminated about 50% of marine genera and nearly 85% of marine species (Harper et al., 2014).

Silurian

The Silurian is known for the appearance of vascular land-plants, of which the first fossil records date to the middle Silurian (Wenlock, Edwards & Feehan, 1980). The earliest-known representative of this group is *Cooksonia* (Fig. 2.3). This relatively small (<10 cm high) and simple plant lacked leaves and an extensive root-system, essentially comprising a sporangium on a stem. Preferred habitats were along rivers and streams, forming micro-forests. Spores originating from *Cooksonia* have been recorded in Silurian to lowermost Devonian strata in the Netherlands (Houben & Vis, 2021, wells Kortgene-1 and O18-01; Fig. 2.1). Somewhat more complex plants that are considered the first club-mosses (lycopsids) known as *Baragwanathia* appear in the early Ludlow (Rickards, 2000). Perhaps in conjunction with the advent of land-plants, the earliest-known animals to have fully adapted to terrestrial conditions appeared during the middle Silurian, including the millipede *Pneumodesmus* (Selden & Read, 2008). In the marine realm, reef-building corals radiated during the Silurian, leading to the first extensive Paleozoic coral reefs (Copper, 2002). The early Paleozoic jawless fishes continued to thrive, and early jawed fishes, including the shark-like acanthodians and the armour-headed placoderms, evolved during the Silurian. Eurypterids were the apex predators of the Silurian oceans. These arthropods, also named sea scorpions, somewhat resemble modern horseshoe crabs.

The Silurian contains at least six large (up to 6‰) carbon isotope excursions (Melchin et al., 2020), which are interpreted to represent major carbon-cycle perturbations that are associated with minor mass extinction events and anoxia (Munnecke et al., 2010). Intense glaciations are not inferred for the Silurian, although there is sedimentological evidence for early Silurian glaciers (Díaz-Martínez & Grahn, 2007). Based on oxygen isotope data, it appears that the climate warmed immediately after the Hirnantian Glaciation, reaching a maximum towards the end of the Llandovery. The Wenlock was relatively cold and the late Silurian (Ludlow-Pridoli) is characterized by progressive warming.

Devonian

The Silurian-Devonian boundary coincides with a major carbon isotope perturbation, termed Klonk (Małkowski & Racki, 2009) which in terms of extinction-intensity was only a minor, relatively gradual event (Bambach, 2006). The drastic carbon-cycle perturbations that characterize the Silurian are no longer recorded in the Devonian (419-359 Ma).

The Devonian marks the 'mid-Paleozoic predator revolution' (Signor & Brett, 1984) and the related occupation of the free water column, the 'nekton revolution' (Klug et al., 2010). During the Devonian the first ammonoids and numerous fish groups (e.g. toothed sharks and giant placoderms) originated. More advanced vertebrates, including the first tetrapods developed (e.g. Blieck et al., 2010; Niedźwiedzki et al., 2010) and the most extensive reef complexes of the Phanerozoic (e.g. Copper & Scotese, 2003) flourished. The 'greening of land' by the diversification and spread of land plants, including the oldest true forests (e.g. Stein et al., 2012) had a profound effect on terrestrial ecosystems.

The Late Devonian is characterized by a series of biotic crises, which together comprise one of the 'big-five-mass-extinctions' (Becker et al., 2016). The most prominent of which are the Kellwasser Crisis near the Frasnian-Famennian boundary and the latest Famennian Hangenberg Crisis. These crises led to the extinction of the stromatoporoid reefs and placoderm and ostracoderm fishes (Kaiser et al., 2016). Also, the chitinozoans, stratigraphically-important marine palynomorphs, do not occur after the Devonian (Paris et al., 2000). The *Archeopteris* flora and associated *Retispora lepidophyta* spore assemblages (Streel, 1986), as observed in well Kastanjelaan-2 in Maastricht (Bless et al., 1981b), disappeared during the Hangenberg Crisis. The onset of the latter is marked by widespread black shale deposition (Hangenberg Black Shale in Europe; Kaiser et al., 2016). This black shale, which has not (yet) been identified in the Netherlands, coincided with the main extinction level for marine invertebrates (Kaiser et al., 2016).

During the Devonian, vascular land plants spread (Algeo et al., 2001), enhancing silicate weathering (Le Hir et al., 2011) and consequently leading to a long-term atmospheric CO_2-decline. Average temperatures dropped almost continuously from the Early Devonian all the way to the late Carboniferous (Buggisch et al., 2008; Joachimski et al., 2009). In spite of lacking sedimentological evidence of Devonian glaciation on Gondwana, a pronounced eustatic fall across the Frasnian-Famennian boundary is attributed to a glaciation (Isaacson et al., 2008).

Early Carboniferous

During the Mississippian (~early Carboniferous, 359-323 Ma), the Netherlands were situated near the equator (Torsvik & Cocks, 2016). The Tournaisian was a relatively warm, and largely ice-free period, starting with a major eustatic sea-level rise. Some authors link this rise to a deglaciation during the ultimate stage of the Hangenberg Crisis (Kaiser et al., 2016). The Netherlands gradually moved northward through the intertropical convergence zone, leading to the establishment of warm, monsoonal climatic conditions, with strongly seasonal run-off patterns (Falcon-Lang, 1999; Bennett et al., 2017). Following gradual cooling through the late Tournaisian and early-mid Visean (Buggisch et al., 2008), continental ice sheets started to grow in Gondwana (the UK-Brigantian Stage, Fielding et al., 2008; Limarino et al., 2014). During the following Serpukhovian (Late Mississippian, early Namurian), the ice-sheets progressively increased in volume. This increase in ice-volume was accompanied by conspicuous climatic cyclicity, ascribed to 100-kyr orbital eccentricity forcing (Waters & Condon, 2012). Climatic cycles have been described in detail for the Belgian succession by Poty (2016) who determined the onset of glacio-eustatic cyclicity already in the early Visean. Likewise, these third-order sequences have been determined in the Rhenish Kulm Basin in Germany (Herbig, 2016). Around the Early Pennsylvanian (late Namurian), maximum ice-volume stabilized, setting the stage for the glacio-eustasticy dominated depositional patterns of the Pennsylvanian and early Permian (see Bouroullec & Geel, 2025, this volume; Huis in 't Veld & Den Hartog Jager, 2025, this volume).

The Carboniferous Period is best known for its terrestrial ecosystems. Coastal peat swamps inhabited by the fern-like *Rhacophyton* had appeared by Late Devonian times, but a shift to dominance of arborescent lycophytes occurred during the early Carboniferous (Scheckler, 1986). Cyclic changes in coastal vegetation occurred in response to glacio-eustatic cycles. This is exemplified in the Visean Yoredale facies, which was deposited in the northern Dutch offshore (Ingrams et al., 2020).

Stratigraphy

Boreholes that were drilled over the last years have led to new insights regarding the pre-Serpukhovian Paleozoic stratigraphy of the Netherlands. When comparing with the previous edition of this chapter (Geluk et al., 2007), substantial additions and changes to the stratigraphy were needed. As work on Paleozoic stratigraphy is still ongoing, new insights have not completely matured and have not been ratified by the Stratigraphic Committee of the Geological Survey of the Netherlands. The revized stratigraphy below should be considered as a proposal for an updated lithostratigraphic subdivision of the Silurian, Devonian and lower Carboniferous successions in the Netherlands.

Chronostratigraphic terminology

The duration of the time interval discussed in this chapter spans approximately 200 million years. The oldest rocks within the Brabant Massif in Belgium are Cambrian (541-485 Ma) in age. The Ordovician spans the interval between 485-444 Ma, the Silurian between 444-419 Ma, the Devonian between 419-359 Ma and the older half of the Carboniferous between 359-330 Ma.

In northwestern Europe, numerous chronostratigraphic terminologies for the late Devonian and early Carboniferous are employed, particularly on the stage and substage levels (see the overviews in Hance et al., 2006a,b; Thorez et al., 2006; Poty et al., 2014; Schindler et al., 2018). Because this book follows the internationally recognized terminology as outlined by the International Commission on Stratigraphy (ICS, Gradstein & Ogg, 2020), we provide a brief overview of regionally used chronostratigraphic terms (Table 2.1). We discuss only the Devonian and early Carboniferous (Mississippian), because regional variations in older Paleozoic strata are of less importance.

According to the ICS, the Devonian System comprises three series. The Early Devonian comprises in ascending order the Lochkovian, Pragian and Emsian stages. The

Table 2.1. Overview of regionally used chronostratigraphic terminology for the Devonian and lower Carboniferous of northwest Europe. German regional chronostratigraphic units after Herbig (2016). UK units are after Waters et al. (2011). References to Belgian regional stratigraphic nomenclature are provided in the text.

	Global Series	Subseries	Regional Series	Global Stages	Global Substages	Regional Stages	UK Substages	Belgian Substages	German Substages
Carboniferous (pars.)	Pennsylvanian (pars.)	Early	Silesian (pars.)	Bashkirian (pars.)		*Namurian*	*Yeadonian*		
							Marsdenian		
							Kinder-scoutian		
							Alportian		
	Mississippian	Late		Serpukhovian			*Chokierian*		
							Arnsbergian		
							Pendleian		
		Middle	Dinantian	Visean		*Visean*	*Brigantian*	*Warnantian*	
							Asbian		
							Holkerian	*Livian*	
							Arundian	*Moliniacian*	
							Chadian		
		Early		Tournaisian		*Tournaisian*	*Courceyan*	*Ivorian*	
								Hastarian	*Hangen-bergian*
Devonian	Late			Famennian	Latest			*Strunian*	
					Late				*Dasbergian-Wocklumian*
					Middle				*Hembergian*
					Early				*Nehdenian*
				Frasnian					*Adorfian*
	Middle			Givetian					
				Eifelian			*Couvinian*		
	Early			Emsian					*Emsian*
				Pragian					*Siegenian*
				Lochkovian			*Gedinnian*		

Gedinnian is a Belgian equivalent of the Lochkovian, whereas the Pragian ICS-Stage overlaps with the German Siegenian stage. The Middle Devonian comprises the Eifelian and Givetian stages. The Couvinian is the equivalent of the Eifelian Stage in the Ardennes. The Late Devonian comprises two stages: the Frasnian and the Famennian. The German Upper Devonian Stufen I-VI, also referred to as the Adorfian, Nehdenian, Hembergian, Dasbergian, and Wocklumian stages are now also considered regional terms (Table 2.1; Schindler et al., 2018). The Famennian Stage is subdivided into four formal substages (Early, Middle, Late and Latest). The Latest Famennian corresponds to the Strunian regional substage (after Etroeungt in Belgium, Streel et al., 2006). The Hangenbergian in Germany is an equivalent to the Strunian and early Hastarian in Belgium (e.g. Paproth et al., 1983; Herbig, 2016).

The 'early' Carboniferous, according to the ICS is to be referred to as the Mississippian Series. The Mississippian consists of three subseries (Early, Middle and Late) which correspond to the Tournaisian, Visean and Serpukhovian Stages respectively. Historically, the Carboniferous succession of western Europe has been subdivided into two series. The lower part, dominated by marine depositional environments, was called the Dinantian and the upper part, dominated by terrestrial depositional environments, was called the Silesian. The Dinantian was divided into two stages; the Tournaisian and Visean. The Silesian was divided into the Namurian, Westphalian and Stephanian. This traditional scheme became obsolete with the definition of the mid-Carboniferous boundary, which corresponds to a boundary within the Silesian, corresponding roughly to the Namurian A-B boundary in traditional nomenclature (Dusar, 2006; Wagner & Winkler Prins, 2016). This also means that the Mississippian and Dinantian stages are not fully equivalent. In Belgium, Dinantian substages were defined by Conil et al. (1977); in ascending order Hastarian (early Tournaisian), Ivorian (late Tournaisian), Moliniacian (early Visean), Livian (middle Visean), and Warnantian (late Visean). Following revisions, these substages are now tied into a framework of modern biostratigraphic, lithostratigraphic, and sequence stratigraphic data (Poty et al., 2014). In Great Britain a similar subdivision on substage-level was formally proposed by George et al. (1976) and Ramsbottom et al. (1978) with a later update by Waters et al. (2011). For the Dinantian, these are in ascending order; the Courceyan (Tournaisian), the Chadian and Arundian (early Visean), the Holkerian (middle Visean) and the Asbian and Brigantian (late Visean). The lower two substages of the Namurian (Pendleian and Arnsbergian) correspond to the Serphukhovian ICS-Stage.

In this book, we use ICS-terminology and therefore mention the terms Dinantian, Namurian and Westphalian in parenthesis. For details of the Namurian development of the Netherlands, the reader is referred to Huis in 't Veld and Den Hartog Jager (2025, this volume). Because the lithostratigraphic nomenclature of the lower Carboniferous in the northern Dutch offshore is largely derived from the UK offshore, we here also refer to the regional UK Substages to describe detailed stratigraphic aspects of the Tournaisian and Visean succession.

Lithostratigraphy

In this section, some new lithostratigraphic units are introduced and some existing units are amended. As these units have not been ratified by the Stratigraphic Committee of Geological Survey of the Netherlands, and are thus still informal, they are used without capitals here (e.g. formation, not Formation).

Kortgene group – Boomvliet formation (proposed)

We propose the introduction of the Kortgene group comprising the Boomvliet formation for strata of Silurian or earliest Devonian age (Fig. 2.4). These are the oldest known strata encountered in the Netherlands (Houben & Vis, 2021), penetrated by wells Kortgene-1 and O18-01 in the southwest of the Netherlands (province of Zeeland) (Fig. 2.1) and have never been formally included in the lithostratigraphic nomenclature of the Netherlands. The names of the group and formation refer to the village and hamlet near the drilling site of well Kortgene-1.

The succession as encountered in the above mentioned wells consists of sub-horizontal dark grey shale to siltstone layers containing pyrite, alternating with intercalations of lighter grey, (very) fine grained, sandstone with poorly rounded grains (Gökdag et al., 1983; Krans, 1983). Mineralogically, the fine-grained sandstones consist of quartz (~60%) and feldspar (~9%), with a matrix consisting of clay, mica, rock fragments and clay fragments. Locally, dolomitization has occurred (Krans, 1983). The decimetre-scale sandstone layers typically have sharp basal contacts and the succession shows frequent evidence of flow (flute casts) and loading (Fig. 2.5a-d). The sediments in this succession are interpreted to have been deposited as turbidites sourced from a fine-grained source area, alternating with pelagic mud. The sediments show no signs of metamorphism but are affected by diagenesis, faulting and deformation. At least three fracture episodes have been identified based on observations in O18-01 (Swennen & Muchez, 1991). The fractures are filled up successively by ferroan dolomite, ankerite and quartz and dolomite and anhydrite satinspar (Fig. 2.5a,b; Gökdag et al., 1983). Miospores in the sedimentary rock are dark coloured and substantially affected by thermal alteration, oxic degradation and/or the imprint of diagenetic pyrite-growth structures (Houben & Vis, 2021).

The minimum age of this succession – late Silurian (Wenlock-Pridoli) to earliest Devonian (Lochkovian) – is determined based on the miospore association as recorded in core samples of Kortgene-1 (Houben & Vis, 2021). The deposits of the Kortgene group accumulated below wave-base in a marine foreland basin that started to develop over the northern part of present-day Belgium (Brabant Basin). In well Kortgene-1 the succession measures 210 m up to the bottom of the well. In Belgium, the turbiditic upper Silurian succession reaches a thickness of at least 1000 m (Verniers et al., 2001).

In UK offshore well 53/16-1, a similar succession as the one attributed here to the Boomvliet formation was encountered. The succession consists of a 223 m thick, steeply dipping grey turbiditic sandstone. The age for this succession was estimated to be pre-Carboniferous and probably latest Silurian based on the similarity with the (at that time) undated succession in O18-01 (Woodcock & Pharaoh, 1993; De Vos et al., 2010; Houben & Vis, 2021). The deposits of this newly proposed formation correspond to megasequence 3 of the Brabant Massif in Belgium (Verniers et al., 2002; Herbosch, 2021).

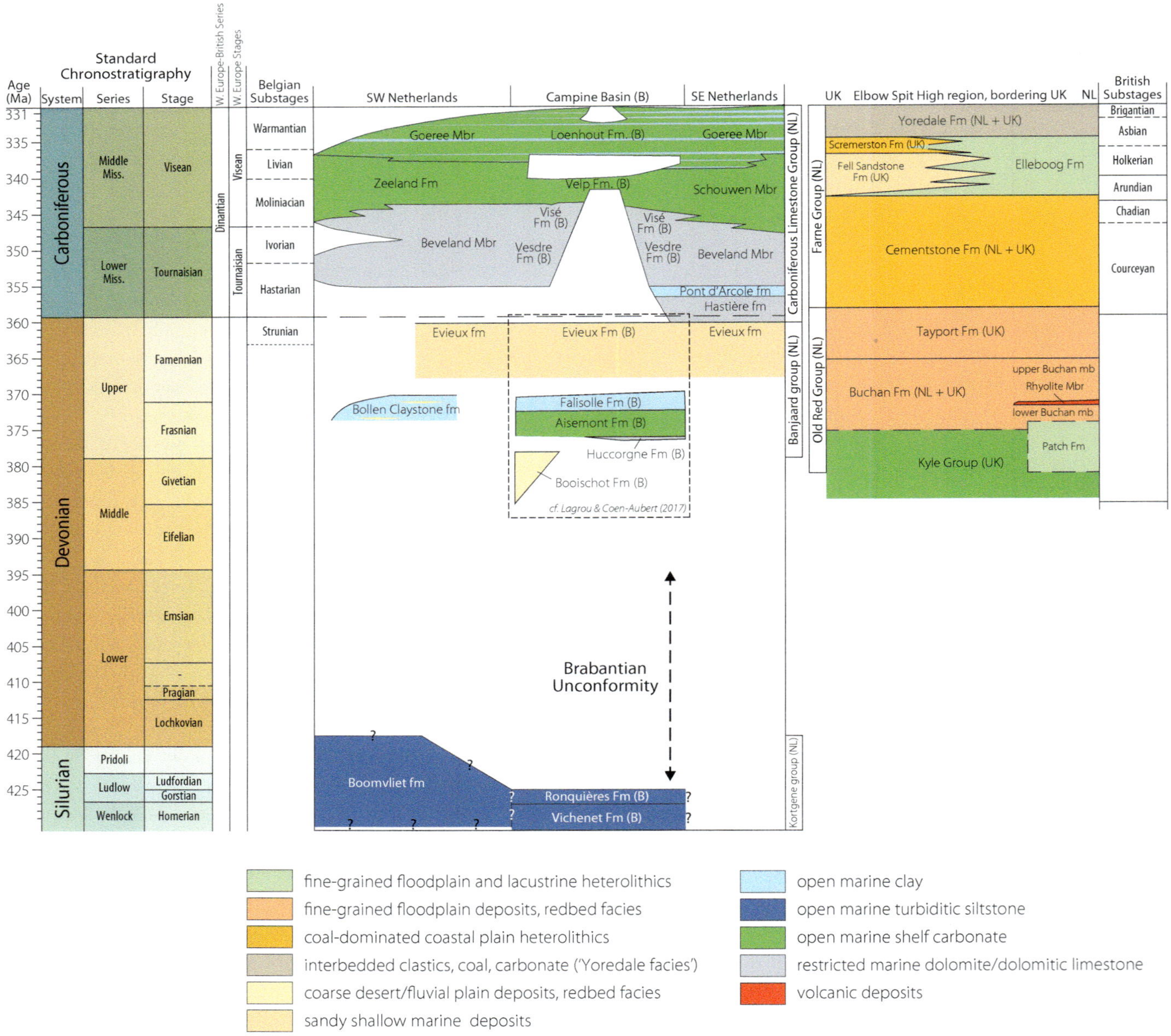

Figure 2.4. Chronostratigraphic (Wheeler) diagram for the upper Silurian to Middle Mississippian of the Netherlands. Belgian stratigraphy follows Verniers et al. (2001), Laenen (2003) and Lagrou & Coen-Aubert (2017); UK stratigraphy according to Cameron (1993b).

Banjaard group (emended)

The Banjaard group was introduced as an informal unit by Van Adrichem Boogaert & Kouwe (1994) for the upper Middle Devonian to lowermost Carboniferous, largely marine succession in the southern Netherlands. An overview of the Devonian lithological units in the Campine basin is provided by Dusar et al. (2015). Two informal formations were proposed: the Bollen Claystone formation and the overlying Bosscheveld formation. At the time of introduction only very little was known regarding the former formation, which was informally defined in Zeeland. The latter in essence represents the transition between Devonian siliciclastics and lower Carboniferous limestones. A recent re-evaluation of core material and new palynostratigraphic constraints now allow for a modification.

The Bosscheveld formation was completely cored in the Kastanjelaan-2 borehole in Maastricht and was extensively studied (Bless et al., 1981b). In addition, new boreholes drilled during the last decade near Venlo (CAL-GT-01 to CAL-GT-05) and in South Limburg close to the Belgian border (Terziet-1 (B62D1112) and -2 (B62D1113)), now provide a better appreciation of these Devonian-lowermost Carboniferous strata in the Netherlands. This leads us to abandon the Bosscheveld formation and to introduce (from old to young) the Evieux formation in the Banjaard group (Fig. 2.4) and the Hastière and the Pont d'Arcole formations in the Carboniferous Limestone Group. These modifications allow for a better synchronization with the stratigraphy of Belgium and Germany.

Bollen Claystone formation (emended)

The Bollen Claystone formation is encountered in the southwest of the Netherlands (Zeeland area, Fig. 2.4), where the formation was cored in two boreholes (Brouwershavense Gat-1 and S05-01). In these cored sections, it consists of dark shaly mudstones with some intercalated, thin, white to greenish-grey, fine-grained sandstones, which are interpreted as tempestites. The unit is in places slightly calcareous or dolomitic. Well log signatures of the unit appear fairly homogeneous and consequently, substantial limestone development is not foreseen. The Bollen Claystone was deposited on a low-energy shelf, shoreface to transitional-marine setting. Recently, miospores assignable to the genus *Diducites*, which roughly indicates a Frasnian or early Famennian age were identified in a core sample from well Brouwershavense Gat-1, at 2902.6 MD (S. Houben, pers. obs.). Based on this tentative age information, and the lithofacies character, the Bollen Claystone formation is likely to correlate to the Falisolle Formation of Lagrou & Coen-Aubert (2017), a unit that was previously referred to as the Lambermont Formation by Laenen (2003). This succession of green-grey shales becomes more sandy towards its top and spans the Frasnian-Famennian boundary (Lagrou & Coen-Aubert, 2017). In the Campine Basin, conspicuous reefal/biostromal limestones of the Aisemont Formation are encountered below the Falisolles Formation (Fig. 2.4). So far there is no indication for the presence of Frasnian limestone in the Netherlands (see also section Middle to Late Devonian), although Frasnian limestones are reported in the Visé area, just south of Maastricht (e.g. Poty & Delculée, 2011).

Evieux formation (proposed)

The Evieux formation is introduced for strata constituting alternations between red and grey-green shaly siltstones and strongly micaceous sandstones (Fig. 2.5e-h, 2.6a). The type-locality is well CAL-GT-01-S1 (2248-2708 MD = end of GR-log). The Evieux formation is derived from its namesake that is recognized in Belgium and Germany (Bultynck & Dejonghe, 2001; Laenen, 2003; Menning, 2018). Towards the top of the formation in well Kastanjelaan-2, carbonate-cemented sandstones occur, as well as mottled sediment, possibly indicating the development of soils (e.g. plant root horizons). The age of the unit is late to latest Famennian. However, the age of the base of the unit is not clearly defined (Fig. 2.4). In the Zeeland area (e.g. well Brouwershavense Gat-1) it is thought to pass downhole into the finer-grained marine siliciclastic sedimentary rocks of the Bollen Claystone formation. In South Limburg, in analogy to the Condroz Group of Belgium, a continuation to strata resembling the Montfort Formation (upper Famennian, dark-grey micaceous sandstones with abundant clayey intercalations, Bultynck & Dejonghe, 2001; Laenen, 2003) and the Esneux Formation (middle Famennian, green-coloured micaceous sand and siltstones, Bultynck & Dejonghe, 2001) is assumed to be present. The Evieux formation is also encountered in the Zeeland area (Fig. 2.4). Although a good understanding of the encountered lithofacies is lacking, similar strata appear present in wells that penetrate Devonian strata in the eastern onshore Netherlands, as far north as the Uithuizermeeden-2 well (Fig. 2.1)

The Evieux formation was deposited as part of a laterally extensive shallow marine system. This developed in the relatively narrow and shallow Rhenohercynian sea at the southern edge of the Old Red Continent. Sediments were continuously redistributed through strong along-shore currents, which are proposed to be related to a strong westward directed current within this confined basin (Paproth et al., 1986). The cyclic lithological nature was mitigated by small-scale eustatic base-level variations and, as a result, a whole spectrum of alluvial, estuarine, restricted marine (lagoonal) and shallow marine siliciclastic depositional settings may be encountered (Thorez et al., 2006).

By latest Famennian times, climatic aridification associated with the Hangenberg Crisis and a drop in sea-level led to subaerial exposure, as observed in the Kastanjelaan-2 borehole (483-468 MD). This sea-level drop may be associated with development of the Hangenberg Sandstone in Germany and Belgium (Denayer et al., 2021). This upper part of the Evieux formation in Kastanjelaan-2, correlates to the transitional strata referred to as the Dolhain, Comblain-au-Pont and Etroeungt formations in Belgium (Denayer et al., 2021).

Old Red Group

The Old Red Group comprises mainly fluvial and terrestrial sediments of Middle Devonian to early Carboniferous

(Tournaisian) age (TNO-GDN, 2022c). Sediments of this group were encountered in wells A17-01, E02-01 and E06-01 in the northern Dutch offshore (Fig. 2.1), but due to the low number of wells, the distribution of sediments of this group is not well known (Fig. 2.4). The sediments of this group are broadly equivalent to those of the Upper Old Red Group of the UK southern North Sea (Cameron, 1993b; Kearsey et al., 2015). The group has not been penetrated completely in one single well, but based upon a compilation of well data a thickness of up to 1500 m may be representative of the Mid North Sea High (Geluk et al., 2007). To the south, the group extends most likely under large parts of the Netherlands offshore and onshore, grading laterally into the marine Banjaard group.

Patch Formation

The Patch Formation was introduced by Van Adrichem Boogaert & Kouwe (1994) for a succession of grey and green-grey silty claystones, containing a few thin sandstone beds and some thin, brown, intercalated dolomite beds encountered in well A17-01 (Fig. 2.1). The depositional environment is interpreted to have been lacustrine or fluvial floodplain. Its thickness reaches nearly 300 m in well A17-01, but it is uncertain whether this formation can be recognized as a separate lithostratigraphic unit further away from well A17-01 (Fig. 2.4).

Biostratigraphic age determination is difficult for this formation, as is the case for the entire Old Red Group. A Givetian to Frasnian age is inferred (Van Adrichem Boogaert & Kouwe, 1994). The Patch Formation corresponds to the lower part of the Buchan Formation in the UK (e.g. Kearsey et al., 2015) and to the underlying limestones and mudstones of the Kyle Group in the UK (Fig. 2.4). Deposits of the Kyle Group have not been proven in wells in the Netherlands, but they have been interpreted to be present on seismic data near the Elbow Spit High (Ter Borgh et al., 2018a,b). The Kyle Group is characterized by white, buff, brownish grey, or pink limestones and thin grey shales, with an overlying, more variable unit of marine, lagoonal, and basin-marginal sabkha facies (Cameron, 1993a). The limestone contains tabulate and rugose corals, ostracods, brachiopods, bivalves, gastropods and crinoids, which demonstrate a shallow, warmwater marine environment (Marshall & Hewett, 2003). The Kyle Group is of Givetian to early Frasnian age (Cameron, 1993b).

Buchan Formation

The Buchan Formation (Fig. 2.4) comprises a succession of white to reddish sandstones, and red to red-brown and grey claystones and siltstones. The sediments are interpreted to have been deposited in a fluvial setting, where the sandstones represent stacked channel deposits. In well A17-01, nearly 140 m of felsic extrusive igneous rocks (rhyolite) occur within the formation (Rhyolite Member, Fig. 2.4). Based on the extrusive character, the radiometric age (341±30 Ma) is regarded to be unreliable (see Van Bergen et al., 2025, this volume). An older (Middle Devonian) rhyolite occurs in the Embla gas field (Norway, block 2/7; Marshall & Hewett, 2003).

In absence of biostratigraphic data from this formation in the Netherlands, a Frasnian to Famennian age is assumed. This matches with UK well 38/03-1 (Fig. 2.1), where palynological assemblages show discrete marine transgressive events of Frasnian age (Marshall & Hewett, 2003). The formation has been encountered with a thickness of over 600 m in the northwestern Dutch offshore and in the UK.

Tayport Formation

The Tayport Formation (Fig. 2.4) is composed of red to red-brown and grey claystones and siltstones, with well-developed sandstone intercalations (Fig. 2.6b.). The

← ***Figure 2.5.*** *Core photographs from selected boreholes. a) Boomvliet formation. Laminated fine-grained sandstone (grey) to siltstone (dark grey). Filled fractures are also visible, see Fig. 2.5b (O18-01, 10003 feet = ~3049 m). b) Boomvliet formation. Soft sediment deformation structures, flute casts and an intensely fractured interval resulting from three periods of fracturing according to Swennen and Muchez (1991) (O18-01, 9995.2 feet = ~3046.5 m, ruler in inches). c) Boomvliet formation. A peculiar ball-shaped sandstone in a siltstone matrix, interpreted to be the result of loading, i.e. soft sediment deformation (O18-01, 9999.4 feet = ~3048 m, finger for scale). d) Boomvliet formation. 3D view of flute casts or erosional scours on the sole of a bed, (O18-01, 9955 feet = ~3034 m). e) Cuttings from the Evieux formation illustrating the light grey, micaceous sandstone (Terziet-2 (B62D1113), 184-188 m). f) Evieux formation. Thin section of finely laminated fine-grained sandstone with laminae enriched in muscovite (m). Pores occur mostly as mouldic or oversized pores, which formed due to grain dissolution (arrows) (Terziet-2 (B62D1113), 187-188 m, 04x PPL). g) Evieux formation. Thin section of faintly laminated fine-grained litharenitic sandstone showing alignment of muscovite (m). The Authigenic Fe-dolomite crystals (d) in between the detrital quartz grains (white to black in XPL). Most of the blue colours are due to Fe-dolomite staining rather than (blue epoxy) pores (Terziet-2 (B62D1113), 151-152 m, 10x, XPL). h) Evieux formation. Thin section showing detailed image of the sandstone with pyrite (p), µm-scale microcrystals (h; possibly TiO_2). The sandstone is strongly cemented by quartz and authigenic dolomite (d), resulting in a very poor porosity (Terziet-2 (B62D1113), 149-150 m, 20x, PPL, XPL). Microphotographs (f-h) from Veeningen (2020).*

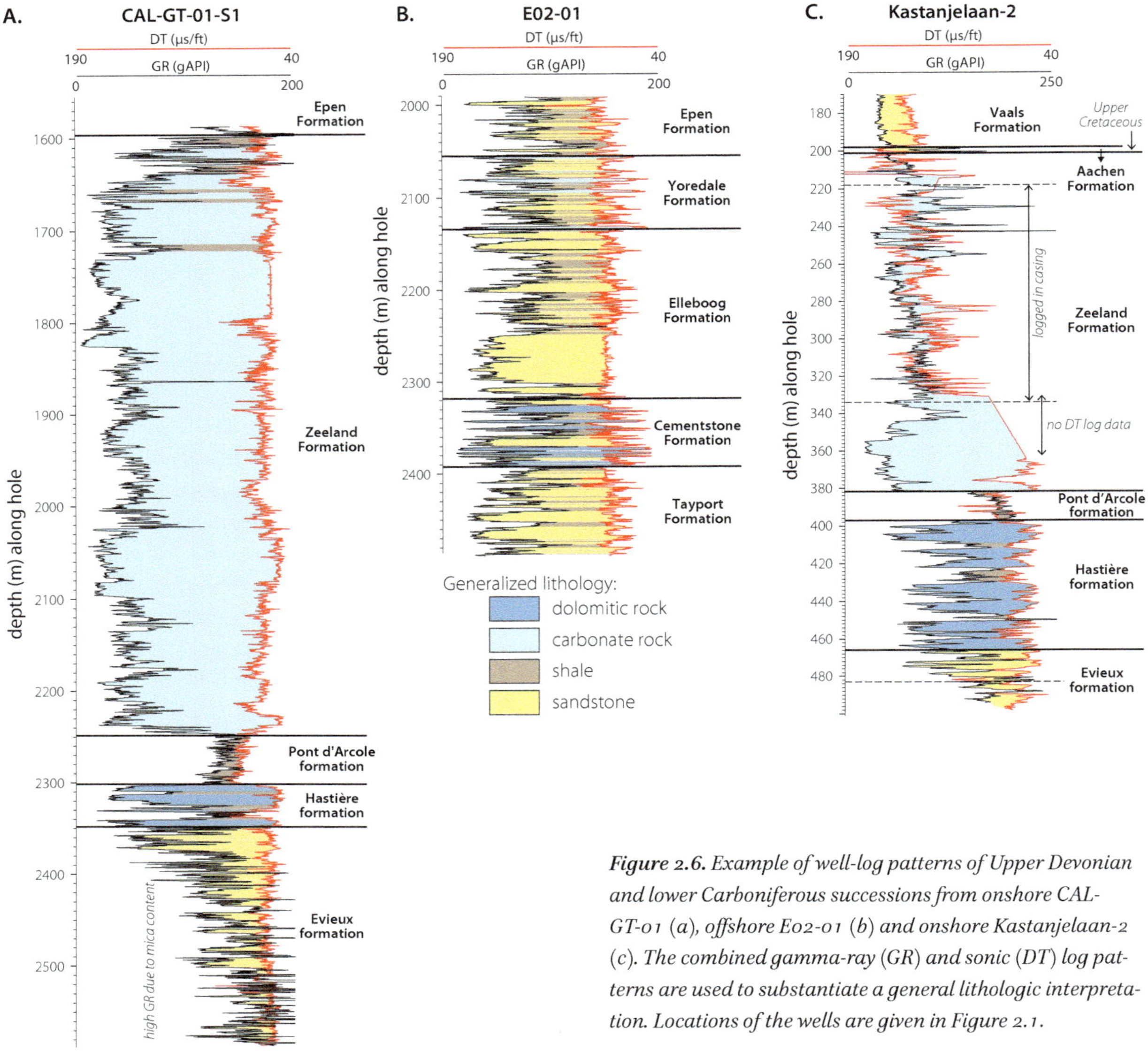

Figure 2.6. Example of well-log patterns of Upper Devonian and lower Carboniferous successions from onshore CAL-GT-01 (a), offshore E02-01 (b) and onshore Kastanjelaan-2 (c). The combined gamma-ray (GR) and sonic (DT) log patterns are used to substantiate a general lithologic interpretation. Locations of the wells are given in Figure 2.1.

presence of thin limestone beds has been inferred from wireline logs (Kearsey et al., 2015). The sandstones are thought to represent fluvial channel fills. The clay- and siltstones were deposited on a floodplain, at times well-drained, which caused a red colouration. This formation has been encountered in wells E02-01 (Fig. 2.6b) and E06-01. Based on the limited number of wells its total thickness is estimated to be at least 450 m in the Dutch sector, and 650 m in the UK sector (37/10-1). The top of the Tayport Formation is of uppermost Famennian age in well E06-01 (J.E.A. Marshall pers comm. in Kearsey et al., 2015). In UK-well 37/10-1 an earliest Tournaisian age has been assigned based on palynomorphs (McLean, 2012). The Tayport Formation correlates with the Kinnesswood Formation which is excellently exposed onshore Northumberland (UK). There, a remarkable hardpan calcrete horizon corresponds to the Devonian-Carboniferous boundary (Marshall et al., 2018), and possibly also the regressive phase of the Hangenberg Crisis (Kaiser et al., 2016). A similar phenomenon has not yet been observed in wireline log sections from offshore wells.

Farne Group

This group contains sedimentary rocks deposited in a paralic to shallow-marine environment. The variable lithology comprises claystone and sandstone, with minor coal seams and a variable amount of (dolomitic) limestone beds. The group contains three formations (Fig. 2.4), from bottom to top: the Cementstone, Elleboog and Yoredale (Cameron, 1993b; Van Adrichem Boogaert & Kouwe, 1994). The Cementstone and Yoredale formations are the equivalents of the formations defined in the UK, while the Elleboog Formation correlates with the Fell Sandstone and Scremerston formations in the UK. The group has not been penetrated entirely in one single well, but is assumed to reach a thickness between 1 and 2.5 km in the

UK sector of the North Sea (Cameron et al., 1992). Strong variations in thickness, of the order of hundreds of metres, occur over syndepositional fault zones south of the Mid North Sea High (Maynard & Dunay, 1999).

Cementstone Formation

The Cementstone Formation (Tournaisian to early Visean, Fig. 2.4) is composed of cyclically alternating carbonates, claystones and sandstones with minor coal seams (Fig. 2.6). The carbonates occur as limestones, dolomitic limestones and dolomites. The fine-grained sediments of this formation accumulated in desiccating, hypersaline lakes on an arid coastal plain (Leeder, 1992). The sandstone units represent stacked fluvial channels (Cameron, 1993b). The thickness of the formation in wells is up to 400 m (e.g. UK well 44/2-1; Cameron, 1993b).

Elleboog Formation

The Elleboog Formation (early to middle Visean, Fig. 2.4) comprises stacked fluvial sandstone bodies and claystones, with minor intercalations of coal and limestones, representing an overbank environment (Fig. 2.6b). To the west, the percentage of sandstones in the lower part increases, grading into the fluvial Fell Sandstone Formation of the UK parts of the southern North Sea (Cameron, 1993b). The sandstone is massive, pale-grey and white, fine to medium or coarse-grained and poorly sorted. It contains sporadic beds of grey, blocky to subfissile, non-calcareous mudstone up to 10 m thick and occasional beds of dark-grey, pyritic, carbonaceous mudstone.

Three sandstone units, each from 80 to over a 100 m thick, have been identified in the UK offshore blocks 43 and 44 (Cameron, 1993b). In a southward direction, the sandstones rapidly grade into mudstones with interbedded thin sandstones (e.g. E06-01). The sandstones are overlain by fine-grained deposits (UK Scremerston Formation, Kearsey et al., 2018) which comprise alternations of sandstone, siltstone, mudstone and coal, with occasional thin dolomite or limestone beds. These sediments accumulated in a range of delta-plain and overbank environments (Leeder et al., 1989, Cameron, 1993b). Their combined thickness is around 100 m.

Yoredale Formation

The Yoredale Formation (late Visean, locally earliest Serpukhovian, Fig. 2.4) comprises a cyclic alternation of marine limestones and mudstones, within a dominantly deltaic succession of claystones, sandstones and coal seams (Fig. 2.6b). The cyclicity within the limestone beds has been interpreted to reflect millennial-scale cycles, as a result of high-frequency arid-humid climatic fluctuations (Tucker et al., 2009). In the Netherlands, a thickness of 175 m was encountered in wells (e.g. A14-01, E02-01, E06-01), whereas in the UK sector, over 700 m were penetrated (41/24a-2; Cameron, 1993b).

Carboniferous Limestone Group

Sedimentary rocks interpreted as belonging to the Carboniferous Limestone Group occur in about 60 wells in the Netherlands, the majority of which are in South Limburg. At the time that the lithostratigraphic definitions for these lower Carboniferous sedimentary rocks were introduced (Van Adrichem Boogaert & Kouwe, 1994), several key wells had not been drilled or were still confidential. New data and new insights ask for an urgent revision of definitions of this group (see Mozafari et al., 2019). For the following description we rely heavily on new insights and interpretations presented by Mozafari et al. (2019). We also propose the introduction of the well-known Belgian Hastière and Pont d'Arcole formations as new units in this group in the Netherlands (Fig. 2.4).

The group generally comprises light-grey, brown and black carbonates, with variable intercalations of bedded chert and claystone, all deposited in a marine carbonate platform and slope environment. Karstic phenomena have been identified in several wells (Mozafari et al., 2019).

Hastière formation (proposed)

Well Kastanjelaan-2 was fundamental to the description of the Devonian-Carboniferous transition in South Limburg (Fig. 2.6c). The latter is thought to be located in a confined basin, the Visé-Puth Basin (Bless, 1989; Poty & Delculée, 2011) and served as the informal type-section for the Bosscheveld formation (Fig. 2.4). The fully cored 120 m thick succession below the limestones of the Zeeland Formation (500-395 MD) was completely assigned to the Bosscheveld formation by Van Adrichem Boogaert & Kouwe (1994). This cored succession comprises from bottom to top, (I) green-red coloured micaceous sandstones (500-483 MD, now included in the Evieux formation); (II) a fairly heterolithic succession of micaceous sandstones and siltstones with possible traces of soil development (mottling due to roots, concretions) and plant-remains (483-468 MD, now considered the uppermost part/top of the Evieux formation); (III) a dolomitized crinoidal bioclastic limestone (468-395 MD); and (IV) a dark grey fossiliferous shale with a distinct elevated gamma-ray signature (395-380 m). The extinction of the miospore *Retispora lepidophyta* is recorded at 447.5 MD (see Bless et al., 1981b and S. Houben pers. obs.), and serves as a good approximation for the position of the Devonian-Carboniferous boundary (Prestianni et al., 2016; Marshall et al., 2018).

This clear lithological fragmentation was also noted by Laenen (2003), who appointed the Kastanjelaan-2 well as the type-section for the Belgian Bosscheveld Formation in the Campine Basin. He placed the base of this formation

Figure 2.7. a) Hastière formation. Dolomitized crinoidal bioclastic limestone (Kastanjelaan-2; photo taken from interval 427-435 m). b) Pont d'Arcole formation. Dark grey shale with crinoid fragments (Winterswijk-1, 4434-4440 m).

at the top of the red-green coloured micaceous sandstones (i.e. at 483 MD) and subsequently included both the heterolithic as well as the crinoidal bioclastic limestone interval in the Bosscheveld Formation, which is then described as a transitional unit between the siliciclastics of the Famennian Condroz Group and the limestone-dominated strata of the Lower Carboniferous Limestone Group.

With the drilling of the California geothermal wells (CAL-GT-01 to CAL-GT-05) near Venlo, it became clear that the up-section partitioning of micaceous sandstones, bioclastic limestones and the elevated gamma-ray shales was clearly traceable on wireline-log profiles (Fig. 2.6a). This leads us to refer to the dolomitized bioclastic limestone (468-397 MD) that contains the Devonian-Carboniferous boundary as the Hastière formation (in analogy to the Ardennes where it is also applied, Denayer et al., 2021 and references therein). The Hastière formation in the Netherlands is now defined as a succession of predominantly claystone and (dolomitized) limestone, rich in crinoids, brachiopods and corals (Fig. 2.7a). The top forms at the transition to the fossiliferous dark shales, with their characteristic high-intensity gamma-ray signature. This interval is very well discernible on gamma-ray profiles. The type-section is well Kastanjelaan-2 (468-397 MD, Fig.

2.6c). The dark-grey fossiliferous shales were assigned to the Pont d'Arcole Formation by Laenen (2003). We concur with this interpretation and now propose to include this formation within the Carboniferous Limestone Group (Fig. 2.4).

Pont d'Arcole formation (proposed)

The Pont d'Arcole Formation is historically known from Belgium, where it is exposed in the type section of the Hastière Limestone, northwest of Hastière-Lavaux (Groessens, 1975; Paproth et al., 1983). In Belgium, parastratotypes are represented by wells 125E298 (Vieux-Leuze, SW Belgium) and 097E817 (Bossuit, SW Belgium) (Laenen, 2003). A third parastratotype was selected in well Kastanjelaan-2 in Maastricht (Laenen, 2003). This well has been studied in detail but the presence of an equivalent of the Pont d'Arcole Formation had not been interpreted previously, as it was estimated to be premature to establish a formal lithostratigraphic subdivision (Bless et al., 1981b).

In well Kastanjelaan-2 the proposed Pont d'Arcole formation occurs between 397 and 382 MD (Fig. 2.6c). It consists of dark grey fossiliferous (calcareous) mudstone, fissile pyritic crinoidal shale with finely dispersed plant debris and bioclastic wackestone intercalations and a generally increasing carbonate content towards the top. A crinoid-bryozoan-coral assemblage has been observed as well as the (Tournaisian) index brachiopod *Spiriferina peracuta* (Bless et al., 1981b; Poty et al., 2001; Laenen, 2003). Despite the dark colour, the organic carbon content is only 0.3-0.5 % (Bless et al., 1980). The Pont d'Arcole formation can be clearly recognized in the CAL-GT wells (Fig. 2.6a) and has been cored in well Winterswijk-1 (Fig. 2.7b). The depositional environment was open marine, below storm wave base. The base of the Pont d'Arcole formation correlates with the base of the Lower Alum Shale (Kahlenberg Formation in the Rhenish Massif area) which lies in a deeper water setting (Kulm facies) and marks a sudden flooding (Poty et al., 2001; Herbig et al., 2018). These lithofacies are fairly consistent over broad distances from southwest England to the Aachen region (Laenen, 2003; Amler & Herbig, 2006; Pharaoh et al., 2021).

The formation rests conformably on dolomitized bioclastic limestone of the Hastière formation in Kastanjelaan-2 (Fig. 2.6c). The upper boundary is placed at the base of the first massive carbonates of the Zeeland Formation, although Mozafari et al. (2019) stated that the upper boundary is gradual. The thickness is 20-30 m. This unit is widely recognized in the Dinant Synclinorium in Belgium and in Germany (North Rhine-Westphalia). It is not encountered in the southwestern Netherlands and in the Campine Basin in Belgium, where marine flooding of the London-Brabant Massif was delayed. The age is early Tournaisian.

Zeeland Formation

The Zeeland Formation (Fig. 2.4), was defined by Cameron (Cameron, 1993b) and Van Adrichem Boogaert and Kouwe (1994) for shallow water carbonate facies. However, this unit has also been used in literature for sediments in wells located in a proximal basinal environment, generating some confusion. Here we follow Mozafari et al. (2019) and apply the Zeeland Formation to generic carbonate units of lower Carboniferous age, comprising both platform and proximal basin environments. This unit is present in the Dutch onshore and offshore subsurface along the northeastern flank of the London-Brabant Massif (Fig. 2.8). The term London-Brabant Massif, sometimes also called London-Brabant Platform, refers to the late Paleozoic and younger elevated morphological feature that includes the former Anglo-Brabant Deformation Belt and the Midlands Microcraton (e.g. Pharaoh, 2018).

The base of this unit varies geographically from sharp to transitional. The contact with the underlying Pont d'Arcole formation may be transitional showing a gradual upward increase of the carbonate content, for example, in the wells Kastanjelaan-2 and Winterswijk-1. Often the Tournaisian part of this unit is missing and the Visean carbonates are in unconformable sharp contact with Devonian sedimentary rocks (e.g. Uithuizermeeden-2, Luttelgeest-1, Campine Basin and southwest Netherlands). This formation is found along the northeastern flank of the London-Brabant Massif in the Netherlands, Belgium and offshore UK, as well as in Germany close to Aachen. The same unit has been found in the isolated platforms in the central and northern Netherlands, for example, in wells Uithuizermeeden-2 and Luttelgeest-1 (Gutteridge et al., 2025). In Belgium and Germany, this unit is usually subdivided into several formations. In the Netherlands, the Zeeland Formation is subdivided into the Beveland, Schouwen and Goeree members (Fig. 2.4). The correlative basinal unit of the Zeeland shallow water carbonates is represented by the condensed Kulm facies encountered in the Rhenish Massif in Germany (Bless et al., 1976; Korn, 2010; Herbig et al., 2018; Arndt, 2021) and by proximal basin deposits found in Kastanjelaan-2, Heugem-1 and Winterswijk-1. In the subsurface of the North Sea, a transition towards the clastic units in the British offshore as found in well E06-01 is expected (Mozafari et al., 2019).

The top of the Zeeland Formation is placed at the transition from carbonate sediments to the clastics of the Epen Formation of the Limburg Group. The boundary either is represented by an unconformity with associated karst (e.g. O18-01), or, by a gradual transition to carbonaceous shales of the Geverik Member of the Epen Formation (Geverik-1 and Winterswijk-1 wells). The contact is sometimes difficult to define using gamma-ray logs, due to the similar values recorded in the upper portion of the Zeeland and

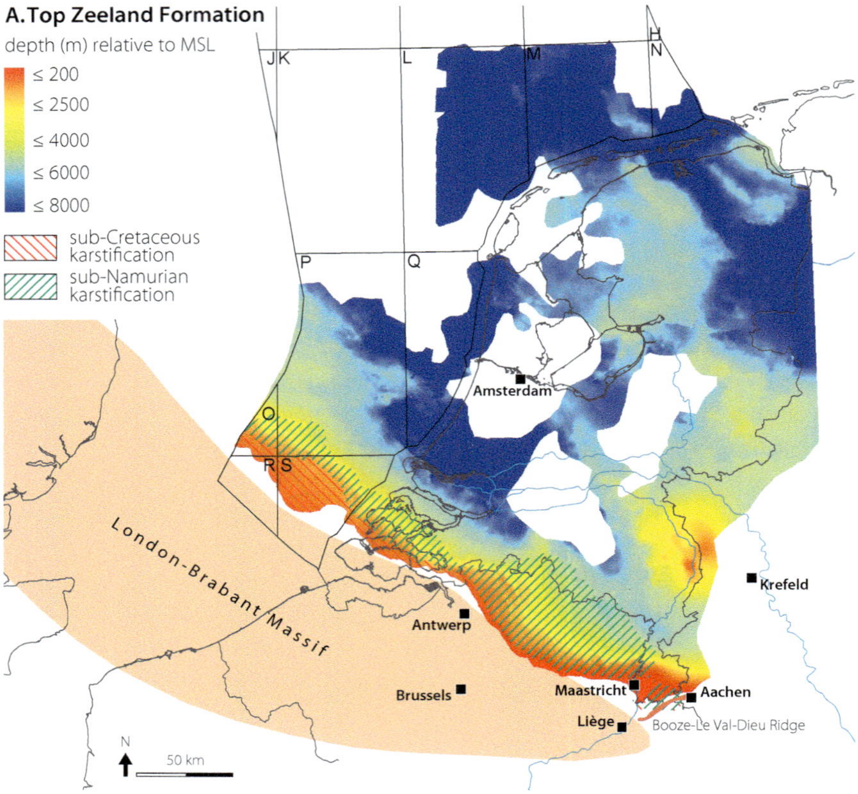

Figure 2.8. *Depth maps of the top (a) and base (b) of the lower Carboniferous (Dinantian) Zeeland Formation. White spots represent data-poor areas that have not been studied. The karstification at the base of the overlying clastic succession (sub-Namurian karstification) and the karstification related to the large hiatus between the Carboniferous succession and the overlying Upper Cretaceous succession (sub-Cretaceous karstification) are indicated in map (a). Depth data from Ten Veen et al. (2019); karstification from Mozafari et al. (2019).*

the Epen formations (Mozafari et al., 2019). In some cases, where younger unconformities erode deeply into the Carboniferous, the Zeeland Formation is also truncated (e.g. Saalian unconformity in well CAL-GT-02 or the Cretaceous Unconformity on the flank of the London-Brabant Massif). The definition of the upper boundary is also complicated by the presence of karst collapse and doline fills at the Visean to Serpukhovian contact (Dinantian to early Namurian) along the northern margin of the London-Brabant Massif (Dreesen et al., 1985, 1987; Mozafari et al., 2019; Gutteridge et al., 2025).

The age of the Zeeland Formation is Tournaisian to Visean. Depending on the location, the base of the formation is diachronous. On the isolated platforms of the central and northern Netherlands, the Tournaisian is absent and the Zeeland Formation is only Visean in age.

Beveland Member

This member consists of medium grey or brown grey to dark brown or brown-black, coarse-crystalline dolomites, often containing black organic intergrain residues. The dolomites are generally of secondary origin. In places minor grey to dark grey limestone intercalations occur, as well as minor quantities of dark brown to blackish siltstone and shaly claystone. In addition, dark beds of silicified dolomite are occasionally present (TNO-GDN, 2022a).

The onset and end of carbonate deposition appear to have taken place at different times in different parts of the basin. Both base and top of the Beveland Member are likely to be diachronous (Fig. 2.4), but their age is a matter of debate and needs to be re-examined (Mozafari et al., 2019). The top of the Beveland Member may cover a range from Tournaisian to middle Visean (late Hastarian to early Livian) in the north of the Netherlands (e.g. Uithuizermeeden-2).

The type section is defined in well S02-02 (2836-2604 MD). Van Adrichem Boogaert and Kouwe (1994) considered well Kastanjelaan-2 as an additional section (382-338 MD) but Mozafari et al. (2019) consider this well not to be representative of the Beveland Member, as it is very shaly and represents a proximal basin environment, deposited within the confined Visé-Puth Basin (Poty & Delculée, 2011).

Schouwen Member

This member comprises a succession of light to dark grey, dark- to yellowish-brown and brownish black, and light yellowish brown to dusty yellow brown limestones. The dense limestones are micritic, biosparitic to biomicritic, locally oolitic and fossiliferous. Reefal structures are known along structural highs in the Belgian Campine Basin and are also assumed to be present offshore the southwest Netherlands (e.g. Dusar et al., 2015). In places coarsely crystalline calcite veins occur. Intergranular bituminous, organic material is often present. Locally, the limestones are dolomitized, especially near fault zones, or are silicified in the leached zone in areas where later erosion truncated the Zeeland Formation (TNO-GDN, 2022d).

The age of this member is most likely early to middle late Visean (Fig. 2.4). Its boundaries are not defined, leaving uncertainty about the usage of this lithostratigraphic unit (Mozafari et al., 2019). The type section for the Schouwen Member is defined in well S02-02 (2604-2123 MD). Well Heugem-1 was considered as a parastratotype (502.7-114 MD) by Van Adrichem Boogaert & Kouwe (1994), but Mozafari et al. (2019) interpreted a rather different sedimentary environment in this well, possibly related to its position in the Visé-Puth Basin.

Goeree Member

This member comprises a succession of grey to dark grey and black limestones, thin- to thick-bedded and often partly silicified. The limestone beds often grade into calcareous and/or silicified black shales or black bedded chert towards the top. Very thin beds of tuffaceous rock occur, predominantly in the upper part of the member (TNO-GDN, 2022b).

The lower boundary is usually pinpointed by a gamma-ray increase compared to the underlying unit, due to an increase of shale and/or volcanic tuff. These gamma-ray spikes may be related to karst fills in wells along the northern margin of the London-Brabant Massif (Mozafari et al., 2019). The age of this member is mostly late Visean (Livian to Warnantian, Fig. 2.4) This member corresponds to the Goeree Formation of northern Belgium (Laenen, 2003).

Depositional history

Cambrian to Early Devonian

In the Netherlands, wells do not reach strata older than Silurian. Therefore, we outline here the older stratigraphy and geologic development based on information from surrounding countries (also see Fig. 2.9).

Deposition during the early Carboniferous, Devonian and older systems took place in a wide variety of sedimentary basins. While reconstruction of pre-Devonian sedimentary basins remains speculative for most of the area, they are better documented for the Late Devonian and early Carboniferous. Tectonostratigraphic comparisons and compilations, based on an integration of outcrop, well and geophysical data in the areas between the British Isles and the Danish-Polish Caledonides (e.g. Verniers et al., 2002; Winchester et al., 2002; Pharaoh et al., 2006; De Vos et al., 2010) serve as the main data source for interpre-

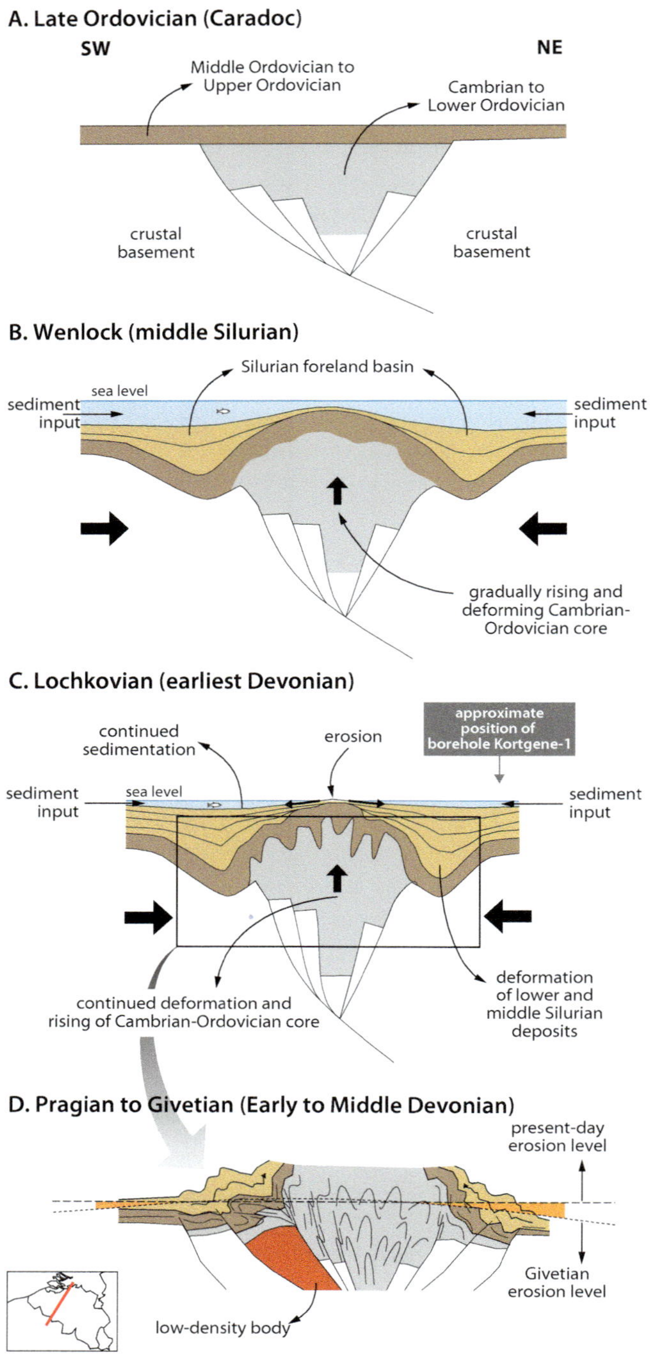

Figure 2.9. Schematic model of Debacker et al. (2005) for a four time-steps Late Ordovician to Middle Devonian evolution of the Brabant Massif, integrating the stratigraphic, sedimentological, metamorphic, geophysical, geochronological and structural data (not to scale). a) Late Ordovician (Caradoc): a basin, possibly a failed rift, filled with Cambrian to Lower Ordovician sediments overlain by more shallowly deposited Middle Ordovician sediments (cf. Verniers et al., 2002). b) Wenlock (middle Silurian): ongoing inversion of the basin, first affecting the core of the Brabant Massif. The shortening causes a rise of the core associated with a steepening of the deposits. Silurian foreland basin development is probably the result of the weight of the rising, deforming core. The sediment input is from the south in the southern foreland basin and from the north in the northern foreland basin. c) Lochkovian (earliest Devonian): continued shortening results in a continued rising and steepening of the core of the massif, which emerges, resulting in erosion of the condensed lowermost Devonian and upper Silurian deposits. The northern and southern foreland basins are now completely separated. Continued shortening leads to a spreading of deformation from the core towards the rims. Note the approximate position of borehole Kortgene-1. d) Pragian to Givetian (Early to Middle Devonian): during the Early Devonian continued shortening causes a further spreading of the deformation from the core towards the rims and results in the complete inversion of the Silurian to Early Devonian foreland basins during the late Pragian/Emsian, possibly Eifelian. At the same time, also the Cambrian core and the Ordovician sequences experienced further, final shortening. Note the lateral changes in thickness of the Silurian deposits (modified from Sintubin & Everaerts, 2002; Debacker et al., 2005).

tation of the pre-Devonian paleogeographic evolution of eastern Avalonia. Important updates and new insights on the tectonic driving mechanisms, largely based on these compilations, were published recently (e.g. Woodcock et al., 2007; Sintubin et al., 2009; Linnemann et al., 2012; Debacker et al., 2015; Smit et al., 2016, 2018; Herbosch et al., 2020).

Verniers et al. (2002) prepared a compilation of all available data on the "European Caledonides", stretching from the British Isles to the Danish-Polish Caledonides. Mostly based on observations in the Brabant Massif (Belgian part of Anglo-Brabant Deformation Belt) and surrounding areas, these authors distinguished three early Paleozoic 'megasequences' (not to be confused with sequence stratigraphic megasequences).

The Cambrian and lowermost Ordovician deposits of **megasequence 1** in the Brabant Massif are both marine and terrigenous, derived from Gondwana, as witnessed by acritarchs and other fossils (Cocks & Fortey, 1982; Cocks et al., 1997; Servais & Fatka, 1997). Their large sedimentary thickness may be related to deposition in a failed rift (e.g. Verniers et al., 2002). In Germany north of the Rhenish Massif, Lower Ordovician coarse-grained clastic schist debris with a pre-Variscan tectonic imprint was found in Middle Devonian nearshore conglomerates of the Viersen-1001 well (Ahrendt et al., 2001). This suggests that an eastern outlier of Brabant Massif-like pre-Devonian rocks may be present in the Krefeld High (Geluk et al., 2007; De Vos et al., 2010).

The core of the Ebbe Anticline in the northern Rhenish Massif contains a sedimentary record which started during the Middle Ordovician. It consists of shales and thin layers of siltstone and fine-grained sandstone followed by shales and greywackes. There are Ordovician rocks of similar composition in four isolated outcrops in the Remscheid Anticline about 50 km to the west (e.g. Verniers et al., 2002 and references therein). Within the Lippstadt High, about 60 km to the northeast of the Ebbe Anticline, graptolite-bearing pre-Devonian rocks have been identified in the Soest-Erwitte well (Clausen & Leuteritz, 1982). However, it remains unclear whether the stratigraphic gap between Ordovician 'deeper' water deposits and late Silurian shallow water shelf deposits such as the one found at the Stavelot-Venn Anticline, is sedimentary or tectonically induced, (De Vos et al., 2010). An hiatus separating uppermost Ordovician from Pridolian (upper Silurian) shales, marlstones and carbonates is known from the Ebbe, Remscheid and Müsen anticlines in Germany.

After separation from Gondwana, sediments of **megasequence 2** were deposited mainly on a shallow shelf. At the end of this sequence, in the Upper Ordovician (upper Caradoc and Ashgill), the faunal affinity of the eastern parts of Avalonia resembled that of Baltica (Giese et al., 1994, 1997; Vecoli & Samuelsson, 2001). Despite the inferred basin setting during megasequence 2, two episodes of wide-spread slumping have been documented (see Debacker, 2012 and references therein). The first took place during the Middle Ordovician (latest Dapingian-earliest Darriwilian, i.e. in the basal sequences of megasequence 2), and points to a south-dipping paleoslope (Debacker, 2012 and references therein). The second episode occurred during the Late Ordovician (late Sandbian-earliest Katian), shortly after onset of turbiditic sedimentation, and was associated locally with large-scale slumping (Debacker et al., 2001) (Fig. 2.9a). The inferred paleoslope is very poorly constrained, but is estimated to be ~north-dipping (Debacker et al., 2001; Debacker, 2012). This second episode points to basin instability around the time of Avalonia-Baltica docking (e.g. Verniers et al., 2002). Possibly, also the Asquempont Detachment System, an originally NE-dipping low-angle normal fault system that can be traced throughout the Brabant Massif (see Herbosch et al., 2008; Herbosch & Debacker, 2018, and original references therein), formed during this period. Roughly at the same time, or later during the Silurian, also the Ardennes were uplifted and eroded (Verniers et al., 2002). However, this so-called Ardennian phase, later overprinted by the Variscan deformation event (Carboniferous), cannot be tied to Avalonia-Baltica collision or another compressional plate-tectonic event, and may equally reflect Silurian extension along the northern margin of the Rheic Ocean (e.g. Sintubin et al., 2009).

North of the uplifted and eroding Ardennes Inliers, a rapidly subsiding foreland basin developed during the Silurian in the area of the future Brabant Massif (Fig. 2.9b) and its western prolongation in East Anglia (Woodcock & Pharaoh, 1993; Pharaoh et al., 1995). This occurred during **megasequence 3** of Verniers et al. (2002). Based on observations in the southern part of the Brabant Massif, it appears that during most of the Silurian turbidity currents supplied sediments from southern source areas (Verniers & Grootel, 1991; Debacker, 2001; Debacker et al., 2014 and references therein). Despite their fine-grained nature these Silurian turbidite deposits should not be regarded as distal, but rather as being sourced from a fine-grained source area, e.g. the eroding Ardennes Inliers (Debacker et al., 2014). During the Brabantian deformation phase (i.e. Acadian; e.g. Woodcock et al., 2007; Sintubin et al., 2009) inversion of this Brabant Basin formed the Anglo-Brabant Deformation Belt in the late Silurian to Early Devonian. The Cambrian core of this belt – the former possible failed rift – was progressively compressed and deformed (Fig. 2.9c), while sedimentation continued along the edges in the Silurian-earliest Devonian foreland basin (Debacker, 2001; Verniers et al., 2002; Debacker et al., 2005). Based on increased subsidence and foreland basin development, the Brabantian deformation phase already commenced around early Wenlock times (Debacker et al., 2005).

It is thought that deformation of the Cambrian-Ordovician core of the Brabant Basin was responsible for Silurian-earliest Devonian foreland basin development along its edges (Fig. 2.9c), eventually resulting in basin-wide deformation and cleavage development around 400 Ma (Sintubin & Everaerts, 2002; Verniers et al., 2002; Debacker et al., 2005). Foreland basin subsidence, degree of metamorphism and degree of Brabantian deformation decrease outwards from the Cambrian core. This is supported by observations in southwest Flanders and the Condroz Inlier, situated on the northern edge of the Midlands Microcraton, where diagenetic, undeformed Silurian to lowermost Devonian deposits occur (Verniers & Grootel, 1991; Verniers et al., 2002). Given the inferred more or less symmetrical geometry of the Brabant Basin inversion (e.g. Sintubin & Everaerts, 2002; Verniers et al., 2002; Debacker et al., 2005), a similar situation is expected in the Netherlands: non-metamorphic, very poorly deformed Silurian-lowermost Devonian deposits, except in the direct vicinity of inverted (Acadian-Brabantian) pre-existing basins. In

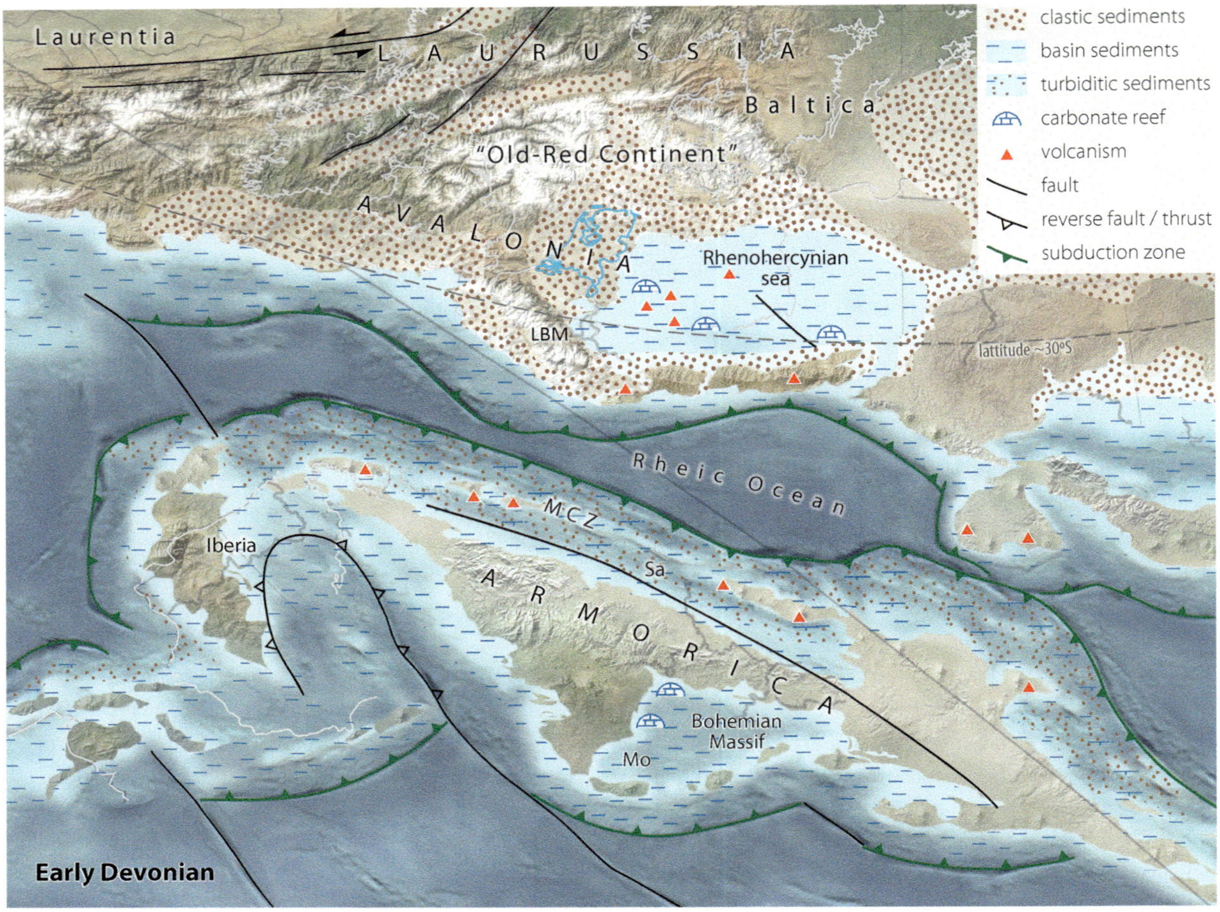

Figure 2.10. *Early Devonian paleogeographic reconstruction of the area surrounding the Netherlands (indicated with blue outline). LBM = London-Brabant Massif; MCZ = Mid-German Crystalline Zone; Mo = Moldanubian; Sa = Saxothuringian. Modified from Meschede and Warr (2019).*

fact, the diagenetically altered and (very) poorly deformed Silurian-lowermost Devonian strata of the Kortgene group encountered in wells Kortgene-1 and O18-01 (Houben & Vis, 2021) correspond to megasequence 3. In the palynological associations from well Kortgene-1 reworked Cambro-Ordovician elements were recognized, supporting the nearby orogen and development of a foreland basin over the southern Netherlands (Houben & Vis, 2021). The top of the Silurian-lowermost Devonian sequence is truncated by the Brabantian Unconformity (see Textbox 2).

Two wells with radiometric ages in the North Sea area are worth mentioning here. Well A17-01, situated on the Mid North Sea High, has a thick Upper Devonian succession overlying a granitic batholith. Seismic data show onlapping stratigraphic geometries on both sides of the granite, implying that it was in place before deposition of the sediments (see Fig. 1.6 in De Jager et al. (2025), this volume). The granite was dated using $^{40}Ar/^{39}Ar$ at 346±7 Ma (Frost et al., 1981), but is believed to have been overprinted during the early Carboniferous (Pharaoh et al., 1995), possibly as a result of burial during Late Devonian-early Carboniferous extension (Smit et al., 2018). Indeed, an unpublished U-Pb isotope age determination on zircons suggests that granite emplacement took place at 410±7 Ma, thus approximating the Silurian-Devonian boundary (Axel Gerdes, pers. com., 2022). This is further explained in Van Bergen et al. (2025, this volume). In northeast England, the Cheviot Granite was emplaced in Early Devonian time, as were many other granites in northern England (De Vos et al., 2010). The German offshore well Q1 terminated at a depth of 3804 m in muscovite-biotite schist that is probably a high-grade augengneiss showing retrograde metamorphism in greenschist facies (Frost et al., 1981). The absolute age of 415±8 Ma measured on muscovite with K-Ar is probably the age of greenschist-facies metamorphism (De Vos et al., 2010), which matches the Wenlock to Emsian age range of the progressive Brabantian deformation phase (Debacker et al., 2005).

Middle to Late Devonian

By the start of the Middle Devonian, the Brabantian deformation phase (Fig. 2.9d), which resulted in an uplifted

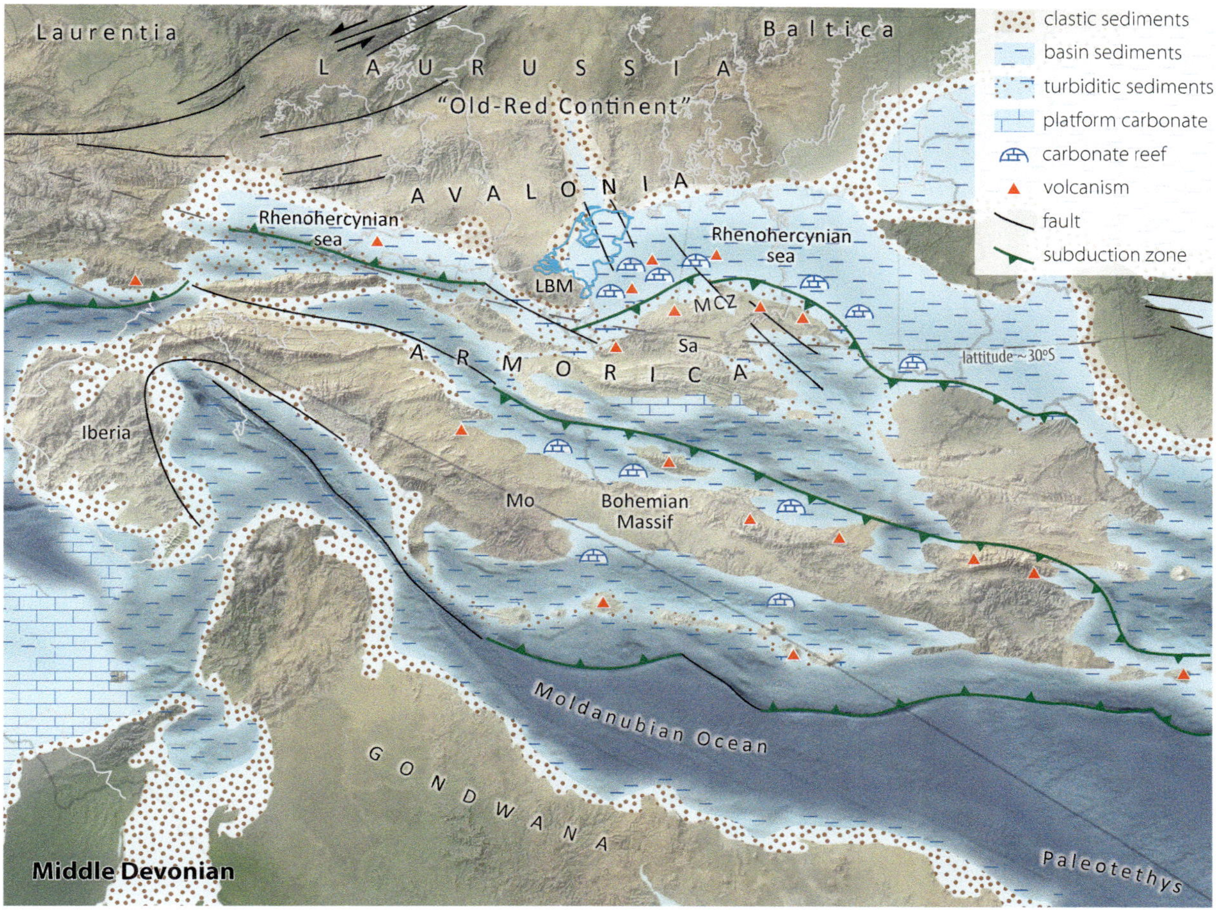

Figure 2.11. Middle Devonian paleogeographic reconstruction of the area surrounding the Netherlands (indicated with blue outline). LBM = London-Brabant Massif; MCZ = Mid-German Crystalline Zone; Mo = Moldanubian; Sa = Saxothuringian. Modified from Meschede and Warr (2019).

Brabant Massif had ended (Debacker et al., 2005). Over the course of the remaining Devonian, a complex of Gondwana-derived microcontinents such as Armorica moved northwards and docked onto Laurussia, on the southern side of Avalonia. This essentially contributed to the Variscan Orogeny in NW Europe (Figs 2.10, 2.11).

The Rhenohercynian sea opened up by Early Devonian times along the southern margin of eastern Avalonia (Figs 2.10, 2.11). The then eroding Anglo-Brabant Deformation Belt constituted a paleohigh (Fig. 2.9d), surrounded by different sub-basins of the Rhenohercynian sea (Fig. 2.11). The shape and orientation of these sub-basins were likely controlled by the pre-existing (Acadian/Brabantian and older) basement fabric. In the Campine Basin in Belgium, the wells Heibaart and Booischot (Fig. 2.1), provide insight in Middle and Upper Devonian stratigraphy (Lagrou & Coen-Aubert, 2017). In the Booischot borehole, a 400 m thick continental, mainly conglomeratic succession of the Booischot Formation overlies Caledonian (Silurian) basement rocks. The top of the Booischot Formation is dated as upper Givetian to upper Frasnian (Streel, 1965; Streel & Loboziak, 1987). This conglomeratic unit developed within a locally subsiding half-graben, close to an active fault, that coincides with the erosional boundary of the Brabant Massif. In the Heibaart borehole, located on a domal structure with stratigraphically delayed sediment cover and lower subsidence, the Caledonian basement is overlain by a few metres of the middle Frasnian Huccorgne Formation (Lagrou & Coen-Aubert, 2017). The Booischot and Huccorgne formations are overlain by the Upper Devonian marine Aisemont and Falisolle formations which are capped by Famennian sandstones of the Evieux Formation (Fig. 2.4). The Frasnian Aisemont Formation consists of nodular limestones containing corals and stromatoporoids, calcareous shales and oolitic ironstones. These reflect the transgression of the Frasnian sea over the edges of the Brabant Massif. The Falisolle Formation comprises a dominantly shaly lithology. It spans the Frasnian-Famennian boundary (Kimpe et al., 1978; Vanguestaine et al., 1983; Lagrou & Coen-Aubert, 2017). These formations have not been unequivocally recorded in the onshore Netherlands but based on seismic information, may be present.

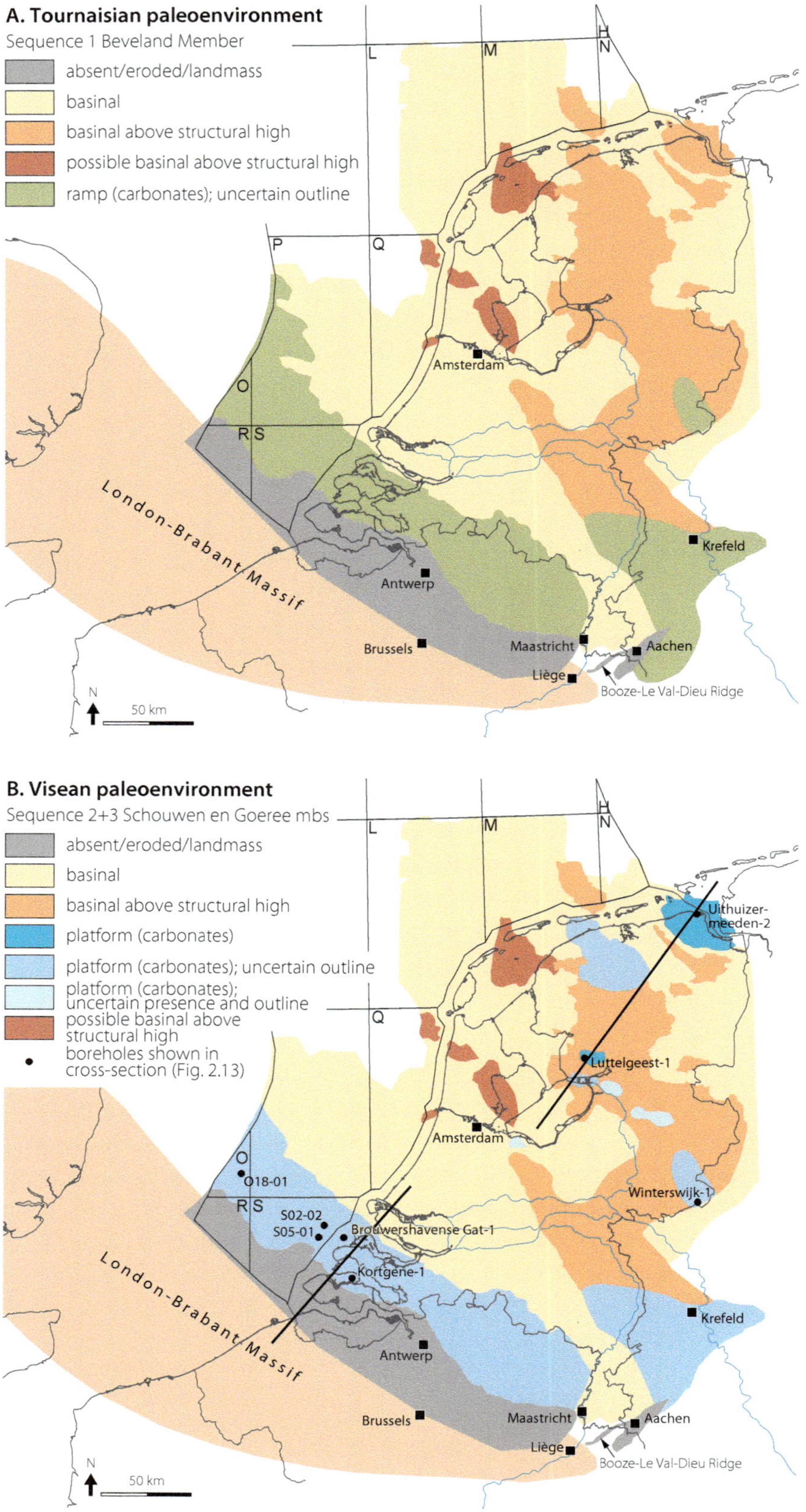

Figure 2.12. Maps depicting lower Carboniferous paleoenvironments of the Tournaisian (a) and Visean (b) during deposition of sediments representing the Zeeland Formation. Data from Mozafari et al. (2019). Lines and boreholes in (b) are used for the schematic cross section in Fig. 2.13.

In the southwest Netherlands (Zeeland area), the Middle to Upper Devonian Banjaard group rests unconformably on Silurian to lowermost Devonian turbiditic deposits of the Kortgene group (Fig. 2.4). The Bollen Claystone formation consists of shaly sediments with cyclic sandstone developments (tempestites). On wireline logs of wells in the Zeeland area, no prominent limestone can be recognized within the Banjaard group. It seems plausible that these marine strata were deposited as a shoreface complex on the northern margin of the Brabant Massif and that the reef ramp of the Brabant Massif was not located in the Zeeland area. Frasnian limestone accumulation may be expected elsewhere in the Netherlands, as illustrated by the Visé Block in Belgium (south and west of Maastricht). There, Frasnian limestones occur as strongly brecciated and karstified mogote (tower) karst. The towers are surrounded by solution collapse breccias and much younger Dinantian carbonates (Kimpe et al., 1978). Subvertical contacts between the karstified Frasnian limestones and onlapping Visean limestones are well-exposed near Visé (Poty, 1982). These Frasnian limestones are exposed at the surface because of the development of the Booze-Le Val Dieu Ridge (Ancion et al., 1943). These authors describe it as a ~5 km wide southwest to northeast trending anticlinal ridge between Visé and Aubel. Arndt et al. (2020) extend the structure towards Cottessen and Terziet in South Limburg, and further to Aachen. On this ridge, the Carboniferous Limestone Group is missing and upper Carboniferous (Namurian) or younger strata immediately overly upper Famennian strata. By Tournaisian times, the ridge formed a high separating the Visé sedimentation area (or Visé-Puth Basin in the Maastricht area) and the Namur sedimentation area to the south (Poty & Delculée, 2011).

The conglomerate in the Viersen-1001 well (see section on Cambrian to Early Devonian) on the Krefeld High stratigraphically more or less corresponds to the Booischot Formation in Belgium (Lagrou & Coen-Aubert, 2017). It is similarly covered by a Frasnian succession of reef limestones, dolomites and claystones (Birenheide, 1998). These are also encountered in well Münsterland-1 (Fig. 2.1; De Vos et al., 2010) and are regionally referred to as Massenkalk. In the east of the Netherlands (Achterhoek area), in well Winterswijk-1, 360 m of white-grey coloured sandstone is present in the lowermost part (TNO-NITG, 1998). The Frasnian reef limestones from well Münsterland-1, 25 km to the east, are absent here, leaving the Devonian stratigraphy of well Winterswijk-1 poorly understood.

A substantial Givetian to Frasnian limestone succession is suspected to be present at deeper levels below Mississippian carbonate build-ups in the Netherlands, as suggested by seismic data. More specifically, Devonian reefal build-ups are interpreted to be developed in the Groningen area and on the Luttelgeest platform (Van Hulten & Poty, 2009; Herber & De Jager, 2010; Van Hulten, 2012).

Devonian carbonates around the Mid North Sea High are referred to in UK lithostratigraphy as the Kyle Group (Cameron, 1993b; Marshall & Hewett, 2003; Kearsey et al., 2015), an interval characterized by limestone units separated by shale and sabkha facies. Two limestone units are seen in UK well 30/16- 5. The limestone is dominantly bioclastic, containing tabulate and rugose corals, ostracods, brachiopods, bivalves, gastropods and crinoids (Marshall & Hewett, 2003). The Kyle Group has been penetrated on the Auk Ridge and on the Dogger Granite High where it generates a distinct seismic reflector that can be correlated over large areas south and east of the Mid North Sea High (Arsenikos et al., 2015; Monaghan et al., 2017). This character is especially visible on relatively new long-offset seismic surveys (Ter Borgh et al., 2018b) and suggests that the Kyle limestones may be present in the Dutch northern offshore as well, despite the absence of well penetrations. In well A17-01, the carbonates were not encountered, and younger, clastic, Devonian rocks overlie granitic basement. These are assigned to the Patch Formation of Adrichem Boogaert & Kouwe (1994). In the German Q1 well, thin beds of marine Middle Devonian limestones are intercalated in the continental Old Red Sandstone (Best et al., 1983). The Kyle seismic facies can be observed only where the Devonian is not deeply buried below Mesozoic basins ((Dutch) Central Graben, Step Graben), but its distribution is thought to be widespread.

Overlying either the limestone of the UK Kyle Group or granitic basement, a thick succession of Middle Devonian to lower Carboniferous mainly fluvial, terrestrial sedimentary rocks is encountered in the Netherlands offshore, referred to as the Old Red Group (Fig. 2.4). The sediments were deposited in a massive southward prograding fluviodeltaic system, sourced from the Laurussian or Old Red Continent. Arid climate conditions due to a paleoposition in the southern hemisphere subtropical belt, caused extensive oxygenation and accumulation of red-beds. The Old Red strata have not been penetrated completely in one single well, but based on a compilation of well data thicknesses may locally be up to 1.5 km on the Mid North Sea High area (Geluk et al., 2007). To the south, the group extends most likely under large parts of the Netherlands offshore and onshore, grading laterally into the marine Banjaard group. Late Devonian transgressions were quite substantial given the intercalation of marine limestones of the Kyle Group with the Old Red fluviodeltaic system.

The nature of the lateral transition from the Old Red Group to the Banjaard group remains poorly understood. During the Famennian, the present-day onshore Netherlands was located within the Rhenohercynian sea. In this marine basin, shallow siliciclastic sediments accumulated,

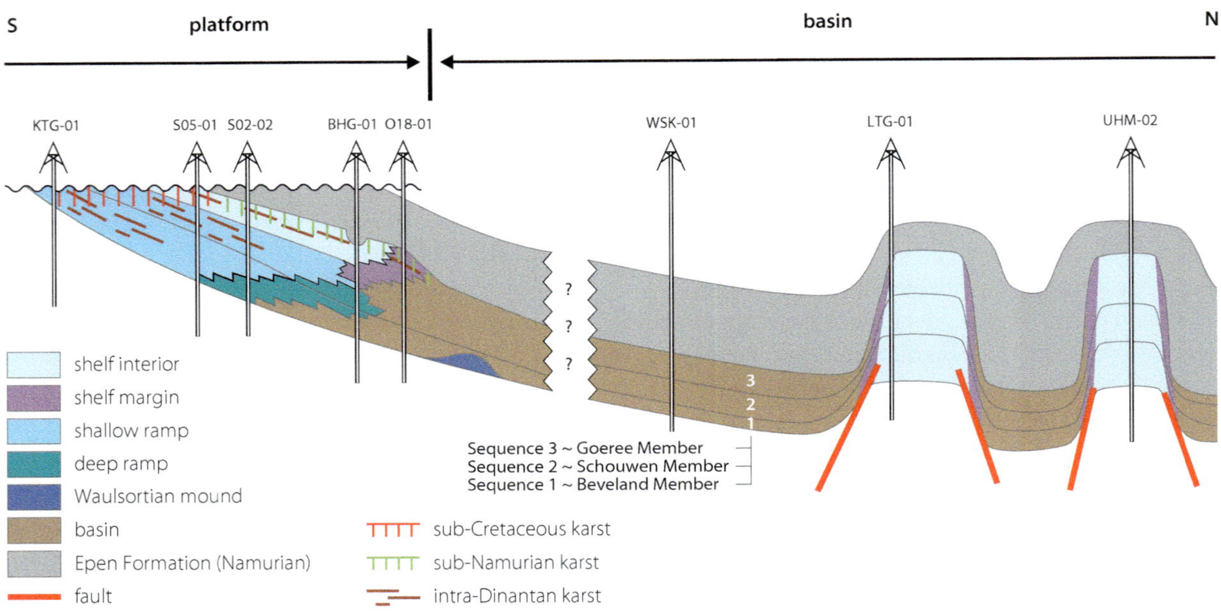

Figure 2.13. *Composite schematic north-south cross-section along projected wells (for approximate location see Fig. 2.12b), showing the sequence geometry and evolution of the attached and isolated early Carboniferous (Dinantian) carbonate systems of the Netherlands. Also note the three phases of karstification. Modified from Pickard & Gutteridge (1997). KTG-01 = Kortgene-1; BHG-01 = Brouwershavense Gat-1; WSK-01 = Winterswijk-1; LTG-01 = Luttelgeest-1; UHM-02 = Uihuizermeeden-2.*

which are thought to have been dispersed by longshore wind-driven currents (Paproth et al., 1986). The sediment is considered to be sourced from eastern source areas (Fig. 1 in Paproth et al., 1986), but Kołtonik et al. (2018) identified a Baltic provenance and also, a source from the nearby Brabant Massif cannot be excluded. These micaceous shallow marine sandstones were found in the Californië geothermal wells near Venlo and in recently drilled boreholes for the Einstein Telescope in South Limburg (Terziet-1 and -2 (B62D1112 & B62D1113) and Cottessen (B62D1181)). In wells Winterswijk-1, Uithuizermeeden-2 and Luttelgeest-1 similar strata were encountered, although lithofacies information from these wells is scarce. In well Kastanjelaan-2, near Maastricht, only the uppermost part of these Famennian sandstones was recovered, representing a regressive facies characterized by soil-development and plant-structures, possibly related to the Hangenberg regression (Kaiser et al., 2016). The nature of the transition to underlying older Devonian strata such as the carbonate-bearing Frasnian is unknown.

In wells Brouwershavense Gat-1 and S05-01, in the southwest of the Netherlands, a remarkable ~30 m thick transitional interval below the lower Carboniferous Zeeland Formation is present. This interval is characterized by a variable GR-pattern. The sedimentary rocks were cored in well Brouwershavense Gat-1 and yield an early late Famennian age (Bless, 1978). Consequently, we interpret this thin succession as corresponding to the Evieux formation. In South Limburg and wells in the eastern Netherlands, the Evieux formation transitions stepwise into the bioclastic limestone of the Hastière formation and an overlying dark grey marine shale (Pont d'Arcole formation). The latter two formations are absent in the southwest of the Netherlands, and in the Campine Basin (Belgium). In other wells in the southwest (Kortgene-1, O18-01 and S02-02), the Bollen Claystone formation is immediately overlain by the Zeeland Formation. This suggests a locally variable unconformable contact between the Middle to Upper Devonian and the lower Carboniferous successions along the London-Brabant Massif.

Early Carboniferous

During the Late Devonian to early Carboniferous extensional tectonics reactivated the pre-existing ('Caledonian') basement fabric and these reactivated basement structures controlled Late Devonian and younger sedimentation. Rifting and basement fabric reactivation took place in response to NE-SW/NNE-SSW extension that episodically occurred throughout the Late Devonian and early Carboniferous (Smit et al., 2016, 2018; Mozafari et al., 2019). In the present-day northern offshore, a large-scale coastal to fluviodeltaic system with a highly cyclic character developed at this time, first with the paralic marine Cementstone facies. This was followed during the early to middle Visean by stacked fluvial sandstone bodies and claystones, with minor intercalations of coal and limestones, that represent an overbank environment (Elleboog Formation). Towards the west, the amount of sandstone in the lower part of the formation increases and grades into the extensive fluvial sandstones of the Fell Sandstone Formation in the UK (Cameron, 1993b). Grain size decreases towards the top of the formation, the cause of which remains to be identified. The finer grained unit corresponds to the Scremerston Formation in the UK (Leeder et al., 1989, Cameron, 1993b). The shale-rich succession of the Cleveland Group in the UK (Visean to Serpukhovian) may represent the basinal environment between the fluviodeltaic system of the Farne Group and the carbonate system of the Zeeland Formation (see discussion in Kearsey et al., 2018). The well-known cyclic deposits of the Late Mississippian (late Visean) Yoredale delta occur as a lateral equivalent to the Cleveland Group; both are overlain by the coarse-grained sandstones of the Millstone Grit Formation (Kearsey et al., 2018).

The shallow marine areas around the Brabant Massif and the isolated highs in the onshore Netherlands remained outside of the influence of the fluviodeltaic system of the Farne Group and were the locus of a shallow-marine carbonate shelf environment during the Tournaisian and Visean (Fig. 2.12). The dark grey fossiliferous (calcareous) mudstone of the Pont d'Arcole formation (Fig. 2.7b) marks a major flooding (TST base, (Poty et al., 2001; Poty, 2016; Becker et al., 2021). The dark grey to black mudstones reflect an anoxic shelf setting. In a more distal setting in the Rhenohercynian sea to the east (Fig. 2.11), the base of the Pont d'Arcole formation correlates with the base of the Lower Alum Shale (Kahlenberg Formation in the Rhenish Massif) in a deeper water setting (Kulm facies). In the Netherlands, the Pont d'Arcole formation has been identified as far north as well Luttelgeest-1 and as far east as well Winterswijk-1.

The Pont d'Arcole formation is overlain by the Zeeland Formation, starting at the first massive carbonate beds. In the southeast of the Netherlands, the preservation of the lower Carboniferous succession is fragmentary and incomplete. In well Kastanjelaan-2, the transition from the Pont d'Arcole formation to the overlying carbonates of the Zeeland Formation appears to be a continuous record. Further east in South Limburg, in the area straddling the border with Belgium, the Devonian is overlain by middle Carboniferous (Namurian) sedimentary rocks. As recently demonstrated by the boreholes in Terziet (B62D1112 and B62D1113) Cottessen (B62D1181), the Tournaisian-Visean succession (Dinantian) is absent altogether, which is linked to their respective position on the Booze-Le Val Dieu Ridge (Ancion et al., 1943; Figs 2.8 and 2.9). In the Campine Basin, the Pont d'Arcole formation is absent and the base of the Tournaisian carbonates is assumed to become progressively younger (Laenen, 2003). In the southwest Netherlands and offshore towards the UK, upper Tournaisian carbonates are the oldest Carboniferous rocks present (Mozafari et al., 2019).

Recent studies of Tournaisian to Visean carbonates (Reijmer et al., 2017; Mozafari et al., 2019) have resulted in a schematic sequence stratigraphic geometry of the carbonate system (Fig. 2.13), built on previous work by Pickard & Gutteridge (1997). The correlative basinal unit of the Zeeland Formation shallow water carbonates is represented by the condensed, mostly starved siliciclastic Kulm facies encountered in Germany (Bless et al., 1976; Korn, 2010; Herbig et al., 2018; Arndt, 2021) and by proximal basin deposits found in wells Kastanjelaan-2, Heugem-1 and Winterswijk-1. In the North Sea, a transition towards marine predominantly clastic units (Farne Group) as found in well E06-01 is expected (Mozafari et al., 2019).

The carbonate system consists of three regional sequences (Fig. 2.13; Reijmer et al., 2017; Mozafari et al., 2019) that correspond to the Beveland, Schouwen and Goeree members. Within these sequences, ten depositional cycles have been recognized, which were also identified and correlated with basinal successions in the southeast and central Netherlands and with isolated carbonate platforms in the northern Netherlands (Mozafari et al., 2019). Based on biostratigraphic data, these depositional cycles are grouped into three lower order sequences (third order) of late Tournaisian to earliest Visean, early Visean and late Visean age.

Along the northern margin of the London-Brabant Massif carbonate ramps developed during late Tournaisian to mid Visean times (Fig. 2.13). Mozafari et al. (2019) proposed that deposition took place on a shallow ramp that includes cyclic peritidal carbonates sheltered behind carbonate shoals and algal reef mounds as observed on the Heibaart and Poederlee domes (e.g. Bless et al., 1981a; Muchez et al., 1987, 1990). Bioturbated packstones and wackestones were deposited in mid-ramp settings passing into a deep-water distal ramp depositional setting with inferred Waulsortian carbonate mud mounds (Fig. 2.13). The carbonate ramps sloped relatively gently at 8-9°. During the late Visean, these ramps developed into flat-topped carbonate platforms with more steeply-dipping margins at the uppermost slope (up to 30°) which passed downdip into basinal argillaceous carbonates (Kombrink et al., 2010; Ten Veen et al., 2019).

In the northern Netherlands, a number of isolated carbonate platforms (Uithuizermeeden and Luttelgeest, Fig. 2.12b) were initiated at various times during the late Tournaisian to early Visean over structural highs in an area of overall basinal sedimentation (Fig. 2.12b). These isolated platforms had relatively high depositional slopes of up to 12°, sometimes even locally 30°. Most of the platforms were built upon pre-existing tilted Devonian fault blocks, on which Bahama-type shallow platform carbonates were deposited (Mozafari et al., 2019).

An important aspect of the carbonate succession, and especially interesting for the exploration for geothermal and hydrocarbon energy sources (Mijnlieff et al., 2025, this volume), are karstification episodes. Three phases (Fig. 2.13) were identified based on core and seismic data (Mozafari et al., 2019): (1) karstification which is more or less contemporaneous with Tournaisian or Visean carbonate accumulation (intra-Dinantian karst); (2) karstification at the base of the overlying clastic succession (sub-Namurian karst) (e.g. Dusar & Lagrou, 2008); and (3) karstification related to the large hiatus between the Carboniferous succession and the overlying Upper Cretaceous succession (sub-Cretaceous karst). The vertical extent of the latter karst system reaches some 170 m beneath the Base Cretaceous Unconformity (Mozafari et al., 2019). Some wells in South Limburg (e.g. Kastanjelaan-2 and Heugem-1 and `s-Gravenvoeren in Belgium; Fig. 2.1) demonstrate complete dissolution of the upper part of the Dinantian carbonate matrix, resulting in porosities of up to 50% (Bless et al., 1981b; Batten et al., 1987; Dusar & Hogenhuis, 1997; Dusar, 2001). A well-developed paleosol including a kaolinite layer is found in association with the hiatus. The hiatus was caused by the progressing Variscan uplift in the south and the consequent development of a large foreland basin north of the emerging mountains. Early Namurian sea-level rise caused the foreland basin to be initially filled with basinal shales and turbidites of the Geverik Member. Increasing siliciclastic influx from the Variscan mountains in the south and the Fennoscandian Shield in the north resulted in a large-scale progradation, depositing a thick succession of deltaic sediments during the late Carboniferous (Westphalian). This is discussed in detail in the next chapter (Huis in 't Veld & Den Hartog Jager, 2025, this volume).

Acknowledgements

We thank Jenny Hettelaar for GIS assistance. We thank Peter Gutteridge for providing files of re-used figures and Martin Meschede from the Institut für Geographie und Geologie in Greifswald for providing the files of the paleogeographic reconstructions (Figures 2.10 and 2.11). We thank the reviewers Martin Arndt and Martin Salamon from the Geologischer Dienst NRW, Christian Brandes from the Leibniz Universität Hannover, and Michiel Dusar from the Belgian Geological Survey for their constructive and thorough reviews. Jasper Maars is thanked for discussion on the core of borehole Kastanjelaan-2.

Digital map data

Spatial data of figures in this chapter for use in geographical information systems can be downloaded here: https://doi.org/10.5117/aup.28162985.

References

Ahrendt, H., Ribbert, K.H., Vanguestaine, M. & Wemmer, K., 2001. K-Ar and acritarch dating of phyllite clasts from a resedimented Middle Devonian conglomerate in the northwestern part of the Rhenish Slate Mountains. Zeitschrift der Deutschen Geologischen Gesellschaft 152: 365-377. DOI: 10.1127/zdgg/152/2001/365

Algeo, T.J., Scheckler, S.E. & Maynard, J.B., 2001. Effects of the Middle to Late Devonian spread of vascular land plants on weathering regimes, marine biotas, and global climate. In: Gensel, P.G. & Edwards, D. (eds): Plants invade the land: evolutionary and environmental perspectives. Columbia University Press: 213-236. DOI: 10.7312/gens11160-013

Amler, M.R.W. & Herbig, H.-G., 2006. Ostrand der Kohlenkalk-Plattform und Übergang in das Kulm-Becken im westlichsten Deutschland zwischen Aachen und Wuppertal. In: Amler, M.R.W. & Stoppel, D. (eds): Stratigraphie von Deutschland VI - Unterkarbon (Mississippium). Schriftenreihe der Deutschen Gesellschaft für Geowissenschaften 41: 441-477.

Ancion, C., Van Leckwijck, W. & Ubaghs, G., 1943. A propos de la bordure méridionale du synclinal de Liège, à l'aval de Liège: la ride famennienne de Booze – Le Val-Dieu, à la limite septentrionale du plateau de Herve. Annales de la Société Géologique de Belgique 66: 299-335.

Arndt, M., 2021. 3D modelling of the Lower Carboniferous (Dinantian) as an indicator for the deep geothermal potential in North Rhine-Westphalia (NRW, Germany). Zeitschrift der Deutschen Gesellschaft Für Geowissenschaften 172: 307-324. DOI: 10.1127/zdgg/2021/0279

Arndt, M., Fritschle, T., Salamon, M., Thiel, A., 2020. Das Rhenoherzynische Becken – ein hydrothermales Reservoir für NRW und Nordwesteuropa? Scriptumonline - Geowissenschaftliche Arbeitsergebnisse aus Nordrhein-Westfalen. Geologischer Dienst NRW (Krefeld) 16: 11 pp.

Arsenikos, S., Quinn, M.F., Pharaoh, T., Sankey, M. & Monaghan, A.A., 2015. Seismic interpretation and generation of key depth structure surfaces within the Devonian and Carboniferous of the Central North Sea, Quadrants 25-44. British Geological Survey commissioned report CR/15/118: 67 pp.

Bambach, R.K., 2006. Phanerozoic biodiversity mass extinctions. Annual Review of Earth and Planetary Sciences 34: 127-155. DOI: 10.1146/annurev.earth.33.092203.122654

Batten, D.J., Streel, M., Dusar, M. & Bless, M.J.M., 1987. Late Cretaceous Palynomorphs from the boreholes Thermae 2002 (Valkenburg a/d Geul, the Netherlands) and 's-Gravenvoeren

(Belgium). Annales de la Société Géologique de Belgique 110: 47-51.

Becker, R.T., Hartenfels, S. & Kaiser, S.I., 2021. Review of Devonian-Carboniferous Boundary sections in the Rhenish Slate Mountains (Germany). Palaeobiodiversity and Palaeoenvironments 101(2): 357-420. DOI: 10.1007/s12549-020-00469-6

Becker, R.T., Königshof, P. & Brett, C.E., 2016. Devonian climate, sea level and evolutionary events: an introduction. Geological Society of London, Special Publications 423(1): 1-10. DOI: 10.1144/SP423.15

Bełka, Z., Devleeschouwer, X., Narkiewicz, M., Piecha, M., Reijers, T.J.A., Ribbert, K.-H. & Smith, N.J.P., 2010. Devonian. In: Doornenbal, J.C. & Stevenson, A.G. (eds): Petroleum geological atlas of the Southern Permian Basin area. EAGE Publications b.v. (Houten): 71-79. ISBN 978-90-73781-61-0

Bennett, C.E., Howard, A.S., Davies, S.J., Kearsey, T.I., Millward, D., Brand, P.J., Browne, M.A.E., Reeves, E.J. & Marshall, J.E.A., 2017. Ichnofauna record cryptic marine incursions onto a coastal floodplain at a key Lower Mississippian tetrapod site. Palaeogeography, Palaeoclimatology, Palaeoecology 468: 287-300. DOI: 10.1016/j.palaeo.2016.12.018

Best, G., Kockel, F. & Schöneich, H., 1983. Geological history of the southern Horn Graben. In: Kaasschieter, J.P.H. & Reijers, T.J.A. (eds): Petroleum geology of the southeastern North Sea and the adjacent onshore areas. Geologie en Mijnbouw 62: 25-33. DOI: 10.1007/978-94-009-5532-5_2

Birenheide, R., 1998. Rugose und tabulate Korallen aus der Bohrung Viersen 1001. Fortschritte in der Geologie von Rheinland und Westfalen 37: 161-213.

Bless, M.J.M., Bouckaert, J., Bouzet, P., Conil, R., Cornet, P., Fairon-Demaret, M., Groessens, E., Longerstaey, P.J., Meessen, J.P.M.T., Paproth, E., Pirlet, H., Streel, M., Van Amerom, H.W.J. & Wolf, M., 1976. Dinantian rocks in the subsurface north of the Brabant and Ardenno-Rhenish massifs in Belgium, the Netherlands and the Federal Republic of Germany. Mededelingen Rijks Geologische Dienst, Nieuwe Serie 27 (3): 81-195.

Bless, M.J.M., 1978. Het Paleozoïcum van de diepboring Brouwershavensegat-1 (NAM) tussen 1398 en 2906,8 m (No. GB 1647; 4 pp.). Rijks Geologische Dienst, Geologisch Bureau (Heerlen): 4 pp.

Bless, M.J.M., Conil, R., Defourny, P., Groessens, E., Hance, L. & Hennebert, M., 1980. Stratigraphy and thickness variations of some Strunio-Dinantian deposits around the Brabant Massif. Mededelingen Rijks Geologische Dienst, Nieuwe Serie 32: 56-65.

Bless, M.J.M., Boonen, P., Bouckaert, J., Brauckmann, C., Conil, R., Dusar, M., Felder, P.J., Felder, W.M., Gökdag, H., Kockel, F., Laloux, M., Langguth, H.R., Van der Meer Mohr, C.G., Meessen, J.P.M.T., Op het Veld, F., Paproth, E., Pietzner, H., Plum, J., Poty, E., Scherp, A., Schulz, R., Streel, M., Thorez, J., Van Rooyen, P., Vanguestaine, M., Vieslet, J.L., Wiersma, D.J., Winkler Prins, C.F. & Wolf, M., 1981a. Preliminary report on Lower Tertiary-Upper Cretaceous and Dinantian-Famennian rocks in the boreholes Heugem-1-1a and Kastanjelaan-2 (Maastricht, the Netherlands). Mededelingen Rijks Geologische Dienst 35(15): 333-415.

Bless, M.J.M., Boonen, P., Dusar, M. & Soille, P., 1981b. Microfossils and depositional environment of late Dinantian carbonates at Heibaart (northern Belgium). Annales de la Société Géologique de Belgique 104: 135-165.

Bless, M.J.M., 1989. Geological investigations around the eastern end of the Brabant Massif: historical overview. Bulletin de La Société belge de Géologie 98(2): 201-208.

Blieck, A.G.M.M., Clément, G. & Streel, M., 2010. The biostratigraphical distribution of earliest tetrapods (Late Devonian): a revised version with comments on biodiversification. Geological Society of London, Special Publications 339(1): 129-138. DOI: 10.1144/SP339.11

Blundell, D.J., Hobbs, R.W., Klemperer, S.L., Scott-Robinson, R., Long, R.E., West, T.E. & Duin, E., 1991. Crustal structure of the central and southern North Sea from BIRPS deep seismic reflection profiling. Journal of the Geological Society 148: 445-458. DOI: 10.1144/gsjgs.148.3.0445

Buggisch, W., Joachimski, M.M., Sevastopulo, G. & Morrow, J.R., 2008. Mississippian δ13Ccarb and conodont apatite δ18O records — Their relation to the Late Palaeozoic glaciation. Palaeogeography, Palaeoclimatology, Palaeoecology 268 (3): 273-292. DOI: 10.1016/j.palaeo.2008.03.043

Bultynck, P. & Dejonghe, L., 2001. Devonian lithostratigraphic units (Belgium). In: Bultynck, P. & Dejonghe, L. (eds): Guide to a revised lithostratigraphic scale of Belgium. Geologica Belgica 4 (1): 39-69. DOI: 10.20341/gb.2014.043

Cameron, N. & Ziegler, T., 1997. Probing the lower limits of a fairway: further pre-Permian potential in the southern North Sea. In: Ziegler, K., Turner, P. & Daines, A.R. (eds): Petroleum geology of the southern North Sea: future potential. Geological Society of London, Special Publications 123, 123-141. DOI: 10.1144/GSL.SP.1997.123.01.08

Cameron, T.D.J., 1993a. Carboniferous and Devonian of the Southern North Sea. In: Knox, R.W.O.B. & Cordey, W.G. (eds): Lithostratigraphic nomenclature of the UK North Sea. British Geological Survey (Nottingham): 93 pp.

Cameron, T.D.J., 1993b. 4. Triassic, Permian and pre-Permian of the Central and Northern North Sea. In: Knox, R.W.O'B. & Cordey, W.G. (eds): Lithostratigraphical nomenclature of the UK North Sea. British Geological Survey (Nottingham): 163 pp.

Cameron, T.D.J., Crosby, A., Balson, P.S., Jeffery, D.H., Lott, G.K., Bulat, J. & Harrison, D.H., 1992. United Kingdom offshore regional reports: the geology of the southern North Sea. HMSO (London) for the British Geological Survey: 152 pp.

Chadwick, R.A., 1993. Aspects of basin inversion in southern England. Journal of the Geological Society 150: 311-322. DOI: 0.1144/gsjgs.150.2.0311

Clausen, C.-D. & Leuteritz, K., 1982. Stratigraphie, Fazies und Altersstellung der paläozoischen Sedimente der Bohrung Soest-Erwitte 1/1a. Fortschritte in der Geologie von Rheinland

und Westfalen 30: 99-143.
Cocks, L.R.M., 2000. The Early Palaeozoic geography of Europe. Journal of the Geological Society 157: 1-10. DOI: 10.1144/jgs.157.1.1
Cocks, L.R.M. & Fortey, R.A., 1982. Faunal evidence for oceanic separations in the Palaeozoic of Britain. Journal of the Geological Society 139: 465-478. DOI: 10.1144/gsjgs.139.4.0465
Cocks, L.R.M. & Torsvik, T.H., 2021. Ordovician palaeogeography and climate change. Gondwana Research 100: 53-72. DOI: 10.1016/j.gr.2020.09.008
Cocks, L.R.M., Mckerrow, W.S. & Staal, C.R., 1997. The margins of Avalonia. Geological Magazine 134(5): 627-636. DOI: 0.1017/S0016756897007425
Conil, R., Groessens, E. & Pirlet, H., 1976 . Nouvelle charte stratigraphique du Dinantien type de la Belgique. Annales de la Société Géologique du Nord 96: 363-371.
Copper, P., 2002. Silurian and Devonian reefs: 80 million years of global greenhouse between two ice ages. In: Kiessling, W., Flügel, E. & Golonka, J. (eds): Phanerozoic reef patterns. SEPM Special Publication 72: 181-238. DOI: 10.2110/pec.02.72.0181
Copper, P. & Scotese, C.R., 2003. Megareefs in Middle Devonian supergreenhouse climates. In: Chan, M.A. & Archer, A.W.: Extreme depositional environments: mega end members in geologic time. GSA Special Paper 370, 209-230. DOI: 10.1130/0-8137-2370-1.209
Coward, M.P., 1993. The effect of Late Caledonian and Variscan continental escape tectonics on basement structure, Paleozoic basin kinematics and subsequent Mesozoic basin development in NW Europe. In: Parker, J.R. (ed.): Petroleum Geology of Northwest Europe. Geological Society, London, Petroleum Conference Series 4: 1095-1108. DOI: 10.1144/0041095
Crowley, T.J. & Baum, S.K., 1995. Reconciling Late Ordovician (440 Ma) glaciation with very high (14X) CO2 levels. Journal of Geophysical Research: Atmospheres 100(D1): 1093-1101. DOI: 10.1029/94JD02521
De Vos, W., 1997. Influence of the granitic batholith of Flanders on Acadian and later deformation (Brabant Massif, Belgium. In: Camelbeeck, T., Sintubin, M. & Vandycke, S. (eds): Contributions to the Belgian Symposium on Structural Geology and Tectonics. Aardkundige Mededelingen 8: 49-52.
De Vos, W., Feldrappe, H., Pharaoh, T.C., Smith, N.J.P., Vejbæk, O.V., Verniers, J., Nawrocki, J., Poprawa, P. & Bełka, Z., 2010. Pre-Devonian. In: Doornenbal, J.C. & Stevenson, A.G. (eds): Petroleum geological atlas of the Southern Permian Basin area. EAGE Publications B.V. (Houten): 59-69.
De Vos, W., Verniers, J., Herbosch, A. & Vanguestaine, M., 1993. A new geological map of the Brabant Massif. Belgium. Geological Magazine 130: 605-611. DOI: 10.1017/S0016756800020902
Debacker, T., Verniers, J., Piessens, K., Dewaele, S. & De Vos, W., 2015. Het Onder-Paleozoïcum en het Massief van Brabant. In: Borremans, M. (ed.): Geologie van Vlaanderen. Academia Press: 18-56.
Debacker, T.N., 2001. Palaeozoic deformation of the Brabant Massif within eastern Avalonia: how, when and why? PhD thesis Ghent University (Belgium): 225 pp. https://lib.ugent.be/catalog/rug01:000476399
Debacker, T.N., 2012. Folds and cleavage/fold relationships in the Brabant Massif, southeastern Anglo-Brabant Deformation Belt. Geologica Belgica 15 (1-2): 81-95.
Debacker, T.N., Dewaele, S., Sintubin, M., Verniers, J., Muchez, P. & Boven, A., 2005. Timing and duration of the progressive deformation of the Brabant Massif, Belgium. Geologica Belgica 8(3-4): 20-34.
Debacker, T.N., Sintubin, M. & Verniers, J., 2001. Large-scale slumping deduced from structural and sedimentary features in the Lower Palaeozoic Anglo-Brabant fold belt, Belgium. Journal of the Geological Society 158: 341-352. DOI: 10.1144/jgs.158.2.341
Debacker, T.N., Verniers, J., Strachan, L.J., Dumon, M. & Belmans, M., 2014. Sedimentological thickness variations within Silurian mudstone-dominated turbidite deposits and the effects on cleavage fanning (Anglo-Brabant Deformation Belt, Belgium). Journal of the Geological Society 171: 193-209. DOI: 10.1144/jgs2013-031
Delabroye, A. & Vecoli, M., 2010. The end-Ordovician glaciation and the Hirnantian Stage: a global review and questions about Late Ordovician event stratigraphy. Earth-Science Reviews 98(3-4): 269-282. DOI: 10.1016/j.earscirev.2009.10.010
Denayer, J., Prestianni, C., Mottequin, B., Hance, L. & Poty, E., 2021. The Devonian–Carboniferous boundary in Belgium and surrounding areas. Palaeobiodiversity and Palaeoenvironments 101(2): 313-356. DOI: 10.1007/s12549-020-00440-5
Díaz-Martínez, E. & Grahn, Y., 2007. Early Silurian glaciation along the western margin of Gondwana (Peru, Bolivia and northern Argentina): Palaeogeographic and geodynamic setting. Palaeogeography, Palaeoclimatology, Palaeoecology 245: 62-81. DOI: 10.1016/j.palaeo.2006.02.018
Doornenbal, J.C., Kombrink, H., Bouroullec, R., Dalman, R.A.F., De Bruin, G., Geel, C.R., Houben, A.J.P., Jaarsma, B., Juez-Larré, J., Kortekaas, M., Mijnlieff, H.F., Nelskamp, S., Pharaoh, T.C., Ten Veen, J.H., Ter Borgh, M., Van Ojik, K., Verreussel, R.M.C.H., Verweij, J.M. & Vis, G.-J., 2019. New insights on subsurface energy resources in the Southern North Sea Basin area. In: Patruno, S., Archer, S.G., Chiarella, D., Howell, J.A., Jackson, C.A-L. & Kombrink, H. (eds): Cross-border themes in petroleum geology I: The North Sea. Geological Society of London, Special Publications 494: 233-268. DOI: 10.1144/SP494-2018-178
Dreesen, R., Bouckaert, J., Dusar, M., Soille, P. & Vandenberghe, N., 1985. Collapse structures at the Dinantian - Silesian contact in the subsurface of the Campine basin, N of the Brabant Massif (N-Belgium). Congrès International de Stratigraphie et de Géologie du Carbonifère, Madrid. Compte Rendu 2, 7-14.
Dreesen, R., Bouckaert, J., Dusar, M., Soille, P. & Vandenberghe, N., 1987. Subsurface structural analysis of the late-Dinantian carbonate shelf at the northern flank of the Brabant Massif (Campine Basin, N-Belgium). Mémoires Explicatives des

Cartes Géologiques et Minières de Belgique 21: 37 pp.
Droser, M.L. & Finnegan, S., 2003. The Ordovician radiation: a follow-up to the Cambrian explosion? Integrative and Comparative Biology 43(1): 178-184. DOI: 10.1093/icb/43.1.178
Dusar, M., 2001. Karst profond dans le Viséen du sondage de recherche géologique de Fouron-le-Comte (Brabant). In: Verheyden, S. & Bernard, Ch. (eds): Actes des Journées de Spéléologie Scientifique, Han-sur-Lesse 1997 à 2000. Geological Survey of Belgium Professional Paper 2001/3(295): 110-113.
Dusar, M., 2006. Namurian. Geologica Belgica 9: 163-175.
Dusar, M. & Hogenhuis, J.E.J., 1998 . De Elvenschans te Moelingen, een thermale bron in Limburg bedreigd? Verduidelijking vanuit nieuwe geologische prospecties. In: LIKONA Jaarboek 1997. Limburgse Koepel voor Natuurstudie: 23-35.
Dusar, M. & Lagrou, D., 2008. Paleokarst-enhanced reservoir for geothermics and gas storage in Carboniferous limestone (Campine Basin, NW European Carboniferous Basin). In: Sasowsky, I.D., Feazel, C.T., Mylroie, J.E., Palmer, A.N. & Palmer, M.V. (eds): Karst from recent to reservoirs. Karst Waters Institute Special Publication 14, 43-51.
Dusar, M., Lagrou, D. & Debacker, T.N., 2015. Boven Paleozoïcum tot Mesozoïcum. In: Borremans, M. (ed.): Geologie van Vlaanderen. Academia Press: 58-102.
Edwards, D. & Feehan, J., 1980. Records of Cooksonia-type sporangia from late Wenlock strata in Ireland. Nature 287 (5777): 41-42. DOI: 10.1038/287041a0
Einstein, A., 1916. Näherungsweise Integration der Feldgleichungen der Gravitation. Sitzungsberichte der Königlich Preussischen Akademie der Wissenschaften 1: 688-696. DOI: 10.1002/3527608958.ch7
Einstein, A., 1918. Über Gravitationswellen. Sitzungsberichte der Königlich Preussischen Akademie der Wissenschaften 1: 154-167. DOI: 10.1002/3527608958.ch12
Everaerts, M., Poitevin, C., De Vos, W. & Sterpin, M., 1996. Integrated geophysical/geological modelling of the western Brabant Massif and structural implications. Bulletin de la Société belge de Géologie 105: 41-59.
Falcon-Lang, H.J., 1999. The Early Carboniferous (Courceyan–Arundian) monsoonal climate of the British Isles: evidence from growth rings in fossil woods. Geological Magazine 136(2): 177-187. DOI: 10.1017/S0016756899002307
Fielding, C.R., Frank, T.D. & Isbell, J.L., 2008. The late Paleozoic ice age – A review of current understanding and synthesis of global climate patterns. In: Fielding, C.R., Frank, T.D. & Isbell, J.L. (eds): Resolving the Late Paleozoic ice age in time and space. GSA Special Paper 441: 343-354. DOI: 10.1130/2008.2441(24)
Franke, W., 2000. The mid-European segment of the Variscides: tectonostratigraphic units, terrane boundaries and plate tectonic evolution. In: Franke, W., Haak, V., Oncken, O. & Tanner, D. (eds): Orogenic processes: quantification and modelling in the Variscan Belt. Geological Society of London, Special Publications 179, 35-61. DOI: 10.1144/GSL.SP.2000.179.01.05
Franke, W., Cocks, L.R.M. & Torsvik, T.H., 2017. The Palaeozoic Variscan oceans revisited. Gondwana Research 48: 257-284. DOI: 10.1016/j.gr.2017.03.005
Fraser, A.J. & Gawthorpe, R.L., 1990. Tectono-stratigraphic development and hydrocarbon habitat of the Carboniferous in northern England. In: Hardman, R.F.P. & Brooks, J. (eds): Tectonic events responsible for Britain's oil and gas reserves. Geological Society of London, Special Publications 55, 49-86. DOI: 10.1144/GSL.SP.1990.055.01.03
Frost, R.T.C., Fitch, F.J. & Miller, J.A., 1981. The age and nature of the crystalline basement of the North Sea Basin. In: Illing, L.V. & Hobson, G.D. (eds): Petroleum Geology of the continental shelf of North-West Europe: Proceedings of the Second Conference. Institute of Petroleum (London): 43-57.
Gee, D.G., Fossen, H., Henriksen, N. & Higgins, A.K., 2008. From the early Paleozoic platforms of Baltica and Laurentia to the Caledonide orogen of Scandinavia and Greenland. Episodes 31: 44-51. DOI: 10.18814/epiiugs/2008/v31i1/007
Gee, D.G., Janák, M., Majka, J., Robinson, P. & Van Roermund, H., 2012. Subduction along and within the Baltoscandian margin during closing of the Iapetus Ocean and Baltica-Laurentia collision. Lithosphere 5(2): 169-178. DOI: 10.1130/L220.1
Geluk, M.C., Dusar, M. & De Vos, W., 2007. Pre-Silesian. In: Wong, Th. E., Batjes, D.A.J. & De Jager, J. (eds): Geology of the Netherlands. Royal Netherlands Academy of Arts and Sciences (KNAW, Amsterdam): 27-42.
George, T.N., Johnson, G.A.L., Mitchell, M., Prentice, J.E., Ramsbottom, W.H.C., Sevastopulo, G.D. & Wilson, R.B., 1976. A correlation of Dinantian rocks in the British Isles. Geological Society of London, Special Report 7: 87 pp.
Gerling, P., Geluk, M.C., Kockel, F., Lokhorst, A., Lott, G.K. & Nicholson, R.A., 1999. 'NW European Gas Atlas' – new implications for the Carboniferous gas plays in the western part of the Southern Permian Basin. In: Fleet, A.J. & Boldy, S.A.R. (eds): Petroleum Geology of Northwest Europe: Proceedings of the 5th Conference. The Geological Society of London, Petroleum Geology Conference Series 5: 799-808. DOI: 10.1144/0050799
Giese, U., Walter, R. & Katzung, G., 1994. Detrital composition of Ordovician sandstones from the Rügen boreholes: implications for the evolution of the Tornquist Ocean. Geologische Rundschau 83: 293-308. DOI: 10.1007/BF00210546
Giese, U., Katzung, G., Walter, R. & Weber, J., 1997. The Caledonian deformation of the Brabant Massif and the Early Palaeozoic in northeast Germany: compared. Geological Magazine 134(5): 637-652. DOI: 10.1017/S0016756897007565
Gökdag, H., Van den Heuvel, P. & Zijp, F.R., 1983. The upper Palaeozoic cores from the well Kortgene-1 (Province of Zeeland) (XEE/4 Sedimentological Note 15: 39 pp. nlog.nl/brh-web/rest/brh/document/3826552441
Gradstein, F.M. & Ogg, J.G., 2020. The chronostratigraphic scale. In: Gradstein, F.M., Ogg, J.G., Schmitz, M.D. & Ogg, G.M. (eds): Geologic time scale 2020. Elsevier (Amsterdam): 21-32. DOI: 10.1016/B978-0-12-824360-2.00002-4

Groessens, E., 1974 . Distribution de conodontes dans le Dinantien de la Belgique. International Symposium on Belgian Micropaleontological Limits (Namur). Geological Survey of Belgium Special Publication 17: 193 pp.

Gutteridge, P., Ten Veen, J., Dewit, J., Garland, J., Riva, A., Mozafari, M., Jaarsma, B. & Van Melle, T., 2025 . Regional assessment of the ultra-deep geothermal (UDG) reservoir quality and potential of the Dinantian Zeeland Formation, the Netherlands. Geological Society of London, Special Publications 548 (1): SP548-2023-2106. DOI: 10.1144/SP548-2023-106

Hagadorn, J.W. & Bottjer, D.J., 1997. Wrinkle structures: Microbially mediated sedimentary structures common in subtidal siliciclastic settings at the Proterozoic-Phanerozoic transition. Geology 25(11): 1047-1050. DOI: 10.1130/0091-7613(1997)025<1047:WSMMSS>2.3.CO;2

Hance, L., Poty, E. & Devuyst, F.-X., 2006a. Tournaisian. Geologica Belgica 9(1-2): 47-53.

Hance, L., Poty, E. & Devuyst, F.-X., 2006b. Viséan. Geologica Belgica 9(1-2): 55-62.

Harper, D.A.T., Hammarlund, E.U. & Rasmussen, C.M.Ø, 2014. End Ordovician extinctions: a coincidence of causes. Gondwana Research 25(4): 1294-1307. DOI: 10.1016/j.gr.2012.12.021

Haug, E., 1900. Les géosynclinaux et les aires continentales: contribution à l'étude des transgressions et des régressions marines. Bulletin de la Société Géologique de France 3: 617-711.

Hearing, T.W., Harvey, T.H.P., Williams, M., Leng, M.J., Lamb, A.L., Wilby, P.R., Gabbott, S.E., Pohl, A. & Donnadieu, Y., 2018. An early Cambrian greenhouse climate. Science Advances 4 (5): 5690. DOI: 10.1126/sciadv.aar569

Herber, R. & De Jager, J., 2010. Oil and gas in the Netherlands - Is there a future? Netherlands Journal of Geosciences 89 (2): 119-135. DOI: 10.1017/S001677460000072X

Herbig, H.-G., 2016. Mississippian (Early Carboniferous) sequence stratigraphy of the Rhenish Kulm Basin, Germany. Geologica Belgica 19(1-2): 81-110. DOI: 10.20341/gb.2016.010

Herbig, H.-G., Salamon, M., Amler, M.R.W., Buchholz, P., Korn, D., Luppold, F.-W., Menning, M., Nesbor, H.-D., Schneider, J.W., Schultka, S., Weller, H., Weyer, D. & Wrede, V., 2018. Das Karbon in der Stratigraphischen Tabelle von Deutschland 2016. Zeitschrift der Deutschen Gesellschaft für Geowissenschaften 168(4): 483-502. DOI: 10.1127/zdgg/2017/0133

Herbosch, A., 2021. Stratigraphic correlations between the Brabant Massif and the Stavelot, Rocroi and Givonne inliers (Belgium) and geological implications. Geologica Belgica 24(3-4): 137-157. DOI: 10.20341/gb.2021.004

Herbosch, A. & Debacker, T.N., 2018. A new geological map of the outcrop areas of the Brabant Massif (Belgium). Geologica Belgica 21(1-2): 41-58. DOI: 10.20341/gb.2018.003

Herbosch, A., Debacker, T.N. & Piessens, K., 2008. The stratigraphic position of the Cambrian Jodoigne Formation redefined (Brabant Massif, Belgium). Geologica Belgica 11(3-4): 133-150.

Herbosch, A., Liégeois, J.-P., Gärtner, A., Hofmann, M. & Linnemann, U., 2020. The Stavelot-Venn Massif (Ardenne, Belgium), a rift shoulder basin ripped off the West African craton: Cartography, stratigraphy, sedimentology, new U-Pb on zircon ages, geochemistry and Nd isotopes evidence. Earth-Science Reviews 203: 103142. DOI: 10.1016/j.earscirev.2020.103142

Hollywood, J.M. & Whorlow, C.V., 1993. Structural development and hydrocarbon occurrence of the Carboniferous in the UK Southern North Sea Basin. In: Parker, J.R. (ed.): Petroleum Geology of Northwest Europe: Proceedings of the 4th Conference. The Geological Society, London, Petroleum Geology Conference Series 4: 689-696. DOI: 10.1144/0040689

Houben, A.J.P. & Vis, G.-J., 2021. Palynological indications for Silurian – earliest Devonian age strata in the Netherlands. Netherlands Jounal of Geosciences 100(2). DOI: 10.1017/njg.2020.20

Huis in 't Veld, R. & Den Hartog Jager, D., 2025. Late Carboniferous. In: Ten Veen, J.H., Vis, G.-J., De Jager, J. & Wong, Th.E. (eds): Geology of the Netherlands, second edition. Amsterdam University Press (Amsterdam): 95-125 DOI: 10.5117/9789463728362_ch03

Ingrams, S., McLean, D., Booth, M. & Bodman, D.J., 2020. Biostratigraphy and paleoecology of Asbian–Brigantian (Mississippian) miospores from Berwick-upon-Tweed, Northumberland, UK: Preliminary results. Review of Palaeobotany and Palynology 276: 104206 . DOI: 10.1016/j.revpalbo.2020.104206

Isaacson, P.E., Díaz-Martínez, E., Grader, G.W., Kalvoda, J., Babek, O. & Devuyst, F.X., 2008. Late Devonian–earliest Mississippian glaciation in Gondwanaland and its biogeographic consequences. Palaeogeography, Palaeoclimatology, Palaeoecology 268(3-4): 126-142. DOI: 10.1016/j.palaeo.2008.03.047

Joachimski, M.M., Breisig, S., Buggisch, W., Talent, J.A., Mawson, R., Gereke, M., Morrow, J.R., Day, J. & Weddige, K., 2009. Devonian climate and reef evolution: insights from oxygen isotopes in apatite. Earth and Planetary Science Letters 284(3-4): 599-609. DOI: 10.1016/j.epsl.2009.05.028

Kaiser, S.I., Aretz, M. & Becker, R.T., 2016. The global Hangenberg Crisis (Devonian–Carboniferous transition): review of a first-order mass extinction. Geological Society of London, Special Publications 423(1): 387-437. DOI: 10.1144/SP423.9

Kearsey, T., Ellen, R., Millward, D. & Monaghan, A.A., 2015. Devonian and Carboniferous stratigraphical correlation and interpretation in the Central North Sea, Quadrants 25-44. British Geological Survey commissioned report CR/15/117: 80 pp.

Kearsey, T.I., Millward, D., Ellen, R., Whitbread, K. & Monaghan, A.A., 2018. Revised stratigraphic framework of pre-Westphalian Carboniferous petroleum system elements from the Outer Moray Firth to the Silverpit Basin, North Sea, UK. Geological Society of London, Special Publications 471: 91-113. DOI: 10.1144/SP471.11

Kimpe, W.F.M., Bless, M.J.M., Bouckaert, J., Conil, R., Groessens, E., Meessen, J.P.M.T., Poty, E., Streel, M., Thorez, J. & Vanguestaine, M., 1978. Paleozoic deposits east of the Brabant Massif in Belgium and the Netherlands. Mededelingen Rijks Geologische Dienst 30(2): 37-103.

Kirby, G.A., 1979. The Lizard complex as an ophiolite. Nature 282(5734): 58-61. DOI: 10.1038/282058a0

Klug, C., Kröger, B., Kiessling, W., Mullins, G.L., Servais, T., Frýda, J., Korn, D. & Turner, S., 2010. The Devonian nekton revolution. Lethaia 43(4): 465-477. DOI: 10.1111/j.1502-3931.2009.00206.x

Kołtonik, K., Pisarzowska, A., Paszkowski, M., Sláma, J., Becker, R.T., Szczerba, M., Krawczyński, W., Hartenfels, S. & Marynowski, L., 2018. Baltic provenance of top-Famennian siliciclastic material of the northern Rhenish Massif, Rhenohercynian zone of the Variscan orogen. International Journal of Earth Sciences 107(8): 2645-2669. DOI: 10.1007/s00531-018-1628-4

Kombrink, H., Besly, B.M., Collinson, J.D., Den Hartog Jager, D.G., Drozdzewski, G., Dusar, M., Hoth, P., Pagnier, H.J.M., Stemmerik, L., Waksmundzka, M.I. & Wrede, V., 2010. Carboniferous. In: Doornenbal, J.C. & Stevenson, A.G. (eds): Petroleum Geological Atlas of the Southern Permian Basin area. EAGE Publications b.v. (Houten): 80-99.

Korn, D., 2010. Lithostratigraphy and biostratigraphy of the Kulm succession in the Rhenish Mountains. Zeitschrift der Deutschen Gesellschaft für Geowissenschaften 161(4): 431-453. DOI: 10.1127/1860-1804/2010/0161-0431

Krans, 1983. Onderzoek monsters NAM-boring Kortgene-1; Kerntrajecten 946.90-995.60, 1351-1365.10, 1720.33-1735 en 1891.50-1900 m. Rijks Geologische Dienst report GB 1888 / OP 9235. Geologisch Bureau (Heerlen): 4 pp. nlog.nl/brh-web/rest/brh/document/737173352

Laenen, B., 2003. Lithostratigrafie van het pre-Tertiair in Vlaanderen Deel II: Dinantiaan & Devoon. VITO report 2003/ETE/095: 86 pp.

Lagrou, D. & Coen-Aubert, M., 2017. Update of the Devonian lithostratigraphic subdivision in the subsurface of the Campine Basin (northern Belgium). Geologica Belgica 20(1-2): 1-13. DOI: 10.20341/gb.2016.017

Le Hir, G., Donnadieu, Y., Goddéris, Y., Meyer-Berthaud, B., Ramstein, G. & Blakey, R.C., 2011. The climate change caused by the land plant invasion in the Devonian. Earth and Planetary Science Letters 310(3): 203-212. DOI: 10.1016/j.epsl.2011.08.042

Lee, M.K., Pharaoh, T.C., Williamson, J.P., Green, C.A. & De Vos, W., 1993. Evidence on the deep structure of the Anglo-Brabant Massif from gravity and magnetic data. Geological Magazine 130: 575-582. DOI: 10.1017/S0016756800020872

Leeder, M.R., 1992. Dinantian. In: Duff, P.McL.D. & Smith, A.J. (eds): Geology of England and Wales. Geological Society (London): 207-238.

Leeder, M.R., Fairhead, D., Lee, A., Stuart, G., Clemmey, H., El-Haddaheh, B. & Green, C., 1989. Sedimentary and tectonic evolution of the Northumberland Basin. In: Arthurton, R S, Gutteridge, P, and Nolan, S C (eds): The role of tectonics in Devonian and Carboniferous sedimentation in the British Isles. Yorkshire Geological Society Occasional Publication 6: 207-223.

Limarino, C.O., Césari, S.N., Spalletti, L.A., Taboada, A.C., Isbell, J.L., Geuna, S. & Gulbranson, E.L., 2014. A paleoclimatic review of southern South America during the late Paleozoic: A record from icehouse to extreme greenhouse conditions. Gondwana Research 25(4): 1396-1421. DOI: 10.1016/j.gr.2012.12.022

Linnemann, U., Herbosch, A., Liégeois, J.-P., Pin, C., Gärtner, A. & Hofmann, M., 2012. The Cambrian to Devonian odyssey of the Brabant Massif within Avalonia: A review with new zircon ages, geochemistry, Sm-Nd isotopes, stratigraphy and palaeogeography. Earth-Science Reviews 112: 126-154. DOI: 10.1016/j.earscirev.2012.02.007

Lozano-Fernandez, J., Tanner, A.R., Puttick, M.N., Vinther, J., Edgecombe, G.D. & Pisani, D., 2020. A Cambrian–Ordovician terrestrialization of arachnids. Frontiers in Genetics 11: 182. DOI: 10.3389/fgene.2020.00182

MacNaughton, R.B., Cole, J.M., Dalrymple, R.W., Braddy, S.J., Briggs, D.E.G. & Lukie, T.D., 2002. First steps on land: Arthropod trackways in Cambrian-Ordovician eolian sandstone, southeastern Ontario, Canada. Geology 30(5): 391-394. DOI: 10.1130/0091-7613(2002)030<0391:FSOLAT>2.0.CO;2

Małkowski, K. & Racki, G., 2009. A global biogeochemical perturbation across the Silurian–Devonian boundary: Ocean–continent–biosphere feedbacks. Palaeogeography, Palaeoclimatology, Palaeoecology 276(1-4): 244-254. DOI: 10.1016/j.palaeo.2009.03.010

Mansy, J.L., Everaerts, M. & De Vos, W., 1999. Structural analysis of the adjacent Acadian and Variscan fold belts in Belgium and northern France from geophysical and geological evidence. Tectonophysics 309(1-4): 99-116. DOI: 10.1016/S0040-1951(99)00134-1

Marshall, J.E.A. & Hewett, A.J., 2003. Devonian. In: Evans, D., Graham, C., Armour, A. & Bathurst, P. (eds): The Millennium Atlas: Petroleum geology of the central and northern North Sea. Geological Society (London): 65-81.

Marshall, J.E.A., Reeves, E.J., Bennett, C.E., Davies, S.J., Kearsey, T.I., Millward, D., Smithson, T.R. & Browne, M.A.E., 2018. Reinterpreting the age of the uppermost 'Old Red Sandstone' and Early Carboniferous in Scotland. Earth and Environmental Science Transactions of The Royal Society of Edinburgh 109(1-2): 265-278. DOI: 10.1017/S1755691018000968

Maynard, J.R. & Dunay, R.E., 1999. Reservoirs of the Dinantian (Lower Carboniferous) play of the Southern North Sea. In: Fleet, A.J. & Boldy, S.A.R. (eds): Petroleum Geology of Northwest Europe: Proceedings of the 5th Conference. The Geological Society, London, Petroleum Geology Conference Series 5: 729-745. DOI: 10.1144/0050729

McKerrow, W.S., Mac Niocaill, C. & Dewey, J.F., 2000. The Caledonian orogeny redefined. Journal of the Geological Society 157: 1149-1154. DOI: 10.1144/jgs.157.6.1149

McLean, D., 2012. Palynostratigraphy of the Carboniferous and Devonian intervals in wells in southern North Sea Quadrants 36, 37, 38, 41 and 43. MB Stratigraphy Ltd. Report 89: 181 pp.

Melchin, M.J., Sadler, P.M. & Cramer, B.D., 2020. The Silurian period. In: Gradstein, F.M., Ogg, J.G., Schmitz, M.D. & Ogg, G.M. (eds): Geologic time scale 2020. Elsevier (Amsterdam): 695-732. DOI: 10.1016/B978-0-12-824360-2.00021-8

Menning, M., 2018. Die Stratigraphic Tabelle von Deutschland 2016 (STD 2016). Zeitschrift der Deutschen Gesellschaft für Geowissenschaften 169(2): 105-128. DOI: 10.1127/zdgg/2018/0161

Meschede, M. & Warr, L.N., 2019. The geology of Germany: A process-oriented approach. Springer International Publishing (Cham). DOI: 10.1007/978-3-319-76102-2

Monaghan, A.A., Arsenikos, S., Quinn, M.F., Johnson, K.R., Vincent, C.J., Vane, C.H., Kim, A.W., Uguna, C.N., Hannis, S.D., Gent, C.M.A., Millward, D., Kearsey, T.I. & Williamson, J.P., 2017. Carboniferous petroleum systems around the Mid North Sea High, UK. Marine and Petroleum Geology 88: 282-302. DOI: 10.1016/j.marpetgeo.2017.08.019

Mozafari, M., Gutteridge, P., Riva, A., Geel, K., Garland, J. & Dewit, J., 2019. Facies analysis and diagenetic evolution of the Dinantian carbonates in the Dutch subsurface. SCAN report, commissioned by the Ministry of Economic Affairs & Climate: 138 pp. https://www.nlog.nl/sites/default/files/2019-12/scan_dinantian_facies_analysis_and_diagenetic_evolution.pdf

Muchez, P. & Langenaeker, V., 1993. Middle Devonian to Dinantian sedimentation in the Campine Basin (northern Belgium): its relation to Variscan tectonism. In: Frostick, L.E. & Steel, R.J. (eds): Tectonic controls and signatures in sedimentary successions. International Association of Sedimentologists Special Publication 20, 171-181. DOI: 10.1002/9781444304053.ch10

Muchez, P., Conil, R., Viaene, W., Bouckaert, J. & Poty, E., 1987. Sedimentology and biostratigraphy of the Visean carbonates of the Heibaart (DzH1) borehole (Northern Belgium). Annales de la Société Géologique de Belgique 110(2): 199-208.

Muchez, P., Viaene, W., Bouckaert, J., Conil, R., Dusar, M., Poty, E., Soille, P. & Vandenberghe, N., 1990. The occurrence of a microbial buildup at Poederlee (Campine Basin, Belgium): biostratigraphy, sedimentology, early diagenesis and significance for early Warnantian paleogeography. Annales de la Société Géologique de Belgique 113(2): 329-339.

Munnecke, A., Calner, M., Harper, D.A.T. & Servais, T., 2010. Ordovician and Silurian sea-water chemistry, sea level, and climate: a synopsis. Palaeogeography, Palaeoclimatology, Palaeoecology 296(3-4): 389-413. DOI: 10.1016/j.palaeo.2010.08.001

Niedźwiedzki, G., Szrek, P., Narkiewicz, K., Narkiewicz, M. & Ahlberg, P.E., 2010. Tetrapod trackways from the early Middle Devonian period of Poland. Nature 463(7277): 43-48. DOI: 10.1038/nature08623

NITG, 1998. Geological Atlas of the subsurface of the Netherlands (1:250,000), Explanation to Map Sheet X Almelo–Winterswijk. Netherlands Institute of Applied Geoscience (NITG-TNO, Haarlem): 134 pp

NITG-TNO, 1999. Geological atlas of the subsurface of the Netherlands (1 : 250,000) - Explanation to Map sheet XV: Sittard-Maastricht. Netherlands Institute of Applied Geoscience TNO - National Geological Survey (Utrecht): 127 pp.

Oncken, O., Plesch, A., Weber, J., Ricken, W. & Schrader, S., 2000. Passive margin detachment during arc-continent collision (Central European Variscides). In: Franke, W., Haak, V., Oncken, O. & Tanner, D. (eds): Orogenic processes: quantification and modelling in the Variscan Belt. Geological Society of London, Special Publications 179, 199-216. DOI: 10.1144/GSL.SP.2000.179.01.13

Paproth, E., Conil, R., Bless, M.J.M., Boonen, P., Bouckaert, J., Carpentier, N., Coen, M., Delcambre, B., Deprijck, C., Deuzon, S., Dreesen, R., Groessens, E., Hance, L., Hennebert, M., Hibo, D., Hahn, G., Hahn, R., Hislaire, O., Kasig, W., Laloux, M., Lauwers, A., Lees, A., Lys, M., Op de Beeck, K., Overlau, P., Pirlet, H., Poty, E., Ramsbottom, W., Streel, M., Swennen, R., Thorez, J., Vanguestaine, M., Van Steenwinkel, M. & Vieslet, J.L., 1983. Bio- and lithostratigraphic subdivisions of the Dinantian in Belgium, a review. Annales de la Société Géologique de Belgique 106: 185-239.

Paproth, E., Dreesen, R. & Thorez, J., 1986. Famennian paleogeography and event stratigraphy of northwestern Europe. Annales de la Société Géologique de Belgique 109: 175-186.

Paris, F., Winchester-Seeto, T., Boumendjel, K. & Grahn, Y., 2000. Towards a global biozonation of Devonian chitinozoans. Courier Forschungsinstitut Senckenberg 220: 39-55.

Pharaoh, T., 2018. The Anglo-Brabant Massif: persistent but enigmatic palaeo-relief at the heart of western Europe. Proceedings of the Geologists' Association 129(3): 278-328. DOI: 10.1016/j.pgeola.2018.02.009

Pharaoh, T., England, R. & Lee, M., 1995. The concealed Caledonide basement of Eastern England and the southern North Sea - A review. Studia Geophysica et Geodaetica 39: 330-346. DOI: 10.1007/BF02295826

Pharaoh, T.C., 1999. Palaeozoic terranes and their lithospheric boundaries within the Trans-European Suture Zone (TESZ): a review. Tectonophysics 314: 17-41. DOI: 10.1016/S0040-1951(99)00235-8

Pharaoh, T., Jones, D., Kearsey, T., Newell, A., Abesser, C., Randles, T., Patton, A. & Kendall, R., 2021. Early Carboniferous limestones of southern and central Britain: Characterisation and preliminary assessment of deep geothermal prospectivity. Zeitschrift der Deutschen Gesellschaft für Geowissenschaften 172(3): 227-249. DOI: 10.1127/zdgg/2021/0282

Pharaoh, T.C., Dusar, M., Geluk, M.C., Kockel, F., Krawczyk, C.M., Krzywiec, P., Scheck-Wenderoth, M., Thybo, H., Vejbæk, O.V. & Van Wees, J.D., 2010. Tectonic evolution. In: Doornenbal, J.C. & Stevenson, A.G. (eds): Petroleum geological atlas of the Southern Permian Basin area. EAGE Publications b.v. (Houten): 24-57.

Pharaoh, T.C., Winchester, J.A., Verniers, J., Lassen, A. & Seghedi, A., 2006. The Western Accretionary Margin of the East European Craton: an overview. In: Gee, D.G. & Stephenson, R.A. (eds): European Lithosphere Dynamics. Geological

Society of London Memoirs 32: 291-312. DOI: 10.1144/GSL.MEM.2006.032.01.17
Pickard, N.A.H. & Gutteridge, P., 1997. Dinantian depositional systems and exploration potential: offshore and onshore, The Netherlands. NAM-commissioned report by Cambridge Carbonates: 596 pp. nlog.nl/brh-web/rest/brh/document/2807457197
Poty, E., 1982. Paléokarst et brèches d'effondrement dans le Frasnien moyen des environs de Visé. Leur influence dans la paléogéographie dinantienne. Annales de La Société Géologique de Belgique 105: 315-337.
Poty, E., 2016. The Dinantian (Mississippian) succession of southern Belgium and surrounding areas: stratigraphy improvement and inferred climate reconstruction. Geologica Belgica 19(1-2): 177-200. DOI: 10.20341/gb.2016.014
Poty, E. & Delculée, S., 2011. Interaction between eustacy and block-faulting in the Carboniferous of the Visé – Maastricht area (Belgium, The Netherlands). Zeitschrift der Deutschen Gesellschaft für Geowissenschaften 162(2): 117-126. DOI: 10.1127/1860-1804/2011/0162-0117
Poty, E., Aretz, M. & Hance, L., 2014. Belgian substages as a basis for an international chronostratigraphic division of the Tournaisian and Viséan. Geological Magazine 151(2): 229-243. DOI: 10.1017/S0016756813000587
Poty, E., Hance, L., Lees, A. & Hennebert, M., 2001. Dinantian lithostratigraphic units (Belgium). In: Bultynck, P. & Dejonghe, L. (eds): Guide to a revised lithostratigraphic scale of Belgium. Geologica Belgica 4(1-2): 69-93. DOI: 10.20341/gb.2014.044
Prestianni, C., Sautois, M. & Denayer, J., 2016. Disrupted continental environments around the Devonian-Carboniferous Boundary: introduction of the tener event. Geologica Belgica 19(1-2): 135-145. DOI: 10.20341/gb.2016.013
Quirk, D.G., 1993. Interpreting the Upper Carboniferous of the Dutch Cleaver Bank High. In: Parker J.R. (ed.): Proceedings of the 4th Conference on Petroleum Geology of North-West Europe. Geological Society of London, Petroleum Geology Conference series 4: 697-706. DOI: 10.1144/0040697
Ramsbottom, W.H.C., Calver, M.A., Eagar, R.M.C., Hodson, F., Holliday, D.W., Stubblefield, C.J. & Wilson, R.B., 1978. A correlation of Silesian rocks in the British Isles. Geological Society of London, Special Report 10: 81 pp.
Reijmer, J.J.G., Ten Veen, J.H., Jaarsma, B. & Boots, R., 2017. Seismic stratigraphy of Dinantian carbonates in the southern Netherlands and northern Belgium. Netherlands Journal of Geosciences 96(4): 353-379. DOI: 10.1017/njg.2017.33
Retallack, G.J., 2009. Cambrian–Ordovician non-marine fossils from South Australia. Alcheringa 33(4): 355-391. DOI: 10.1080/03115510903271066
Rickards, R.B., 2000. The age of the earliest club mosses: the Silurian Baragwanathia flora in Victoria, Australia. Geological Magazine 137(2): 207-209. DOI: 10.1017/S0016756800003800
Rijks Geologische Dienst, 1995. Geologische kaart van Zuid-Limburg en omgeving, schaal 1 : 50.000, Paleozoïcum. RGD (Haarlem): One map sheet with profile and explanation.
Rubinstein, C.V., Gerrienne, P., De la Puente, G.S., Astini, R.A. & Steemans, P., 2010. Early Middle Ordovician evidence for land plants in Argentina (eastern Gondwana). New Phytologist 188(2): 365-369. DOI: 10.1111/j.1469-8137.2010.03433.x
Sax, H.G.J., 1946. De tectoniek van het Carboon in het Zuid-Limburgsche mijngebied. Mededelingen van de Geologische Stichting C-I-1(3): 77 pp.
Scheckler, S.E., 1986. Geology, floristics and paleoecology of Late Devonian coal swamps from Appalachian Laurentia (U.S.A.). Annales de la Société Géologique de Belgique 109 (1): 209-222.
Schindler, E., Brocke, R., Becker, R.T., Buchholz, P., Jansen, U., Luppold, F.W., Nesbor, H.-D., Salamon, M., Weller, H. & Weyer, D., 2018. Das Devon in der Stratigraphischen Tabelle von Deutschland 2016. Zeitschrift der Deutschen Gesellschaft für Geowissenschaften 168(4): 447-463. DOI: 10.1127/zdgg/2017/0135
Selden, P.A. & Read, H., 2008. The oldest land animals: Silurian millipedes from Scotland. Bulletin of the British Myriapod & Isopod Group 23: 36-37.
Servais, T. & Fatka, O., 1997. Recognition of the Trans-European Suture Zone (TESZ) by the palaeobiogeographical distribution pattern of early to middle Ordovician acritarchs. Geological Magazine 134: 617-625. DOI: 10.1017/S001675689700753X
Shail, R.K. & Leveridge, B.E., 2009. The Rhenohercynian passive margin of SW England: Development, inversion and extensional reactivation. Comptes Rendus Geoscience 341(2): 140-155. DOI: 10.1016/j.crte.2008.11.002
Signor, P.W. & Brett, C.E., 1984. The mid-Paleozoic precursor to the Mesozoic marine revolution. Paleobiology 10(2): 229-245. DOI: 10.1017/S0094837300008174
Sintubin, M., 1999. Arcuate fold and cleavage patterns in the southeastern part of the Anglo-Brabant fold belt (Belgium): tectonic implications. Tectonophysics 309(1-4): 81-97. DOI: 10.1016/S0040-1951(99)00133-X
Sintubin, M. & Everaerts, M., 2002. A compressional wedge model for the Lower Palaeozoic Anglo-Brabant Belt (Belgium) based on potential field data. In: Winchester, J.A., Pharaoh, T.C. & Verniers, J. (eds): Palaeozoic Amalgamation of Central Europe. Geological Society of London, Special Publications 201: 327-343. DOI: 10.1144/GSL.SP.2002.201.01.16
Sintubin, M., Debacker, T.N. & Van Baelen, H., 2009. Early Palaeozoic orogenic events north of the Rheic suture (Brabant, Ardenne): A review. Comptes Rendus Geoscience 341(2-3): 156-173. DOI: 10.1016/j.crte.2008.11.012
Smit, J., Van Wees, J.-D. & Cloetingh, S., 2016. The Thor suture zone: From subduction to intraplate basin setting. Geology 44(9): 707-710. DOI: 10.1130/G37958.1
Smit, J., Van Wees, J.-D. & Cloetingh, S., 2018. Early Carboniferous extension in East Avalonia: 350 My record of lithospheric memory. Marine and Petroleum Geology 92(1): 1010-1027. DOI: 10.1016/j.marpetgeo.2018.01.004

Song, H., Wignall, P.B., Song, H., Dai, X. & Chu, D., 2019. Seawater temperature and dissolved oxygen over the past 500 million years. Journal of Earth Science 30(2): 236-243. DOI: 10.1007/s12583-018-1002-2

Soper, N.J., Webb, B.C. & Woodcock, N.H., 1987. Late Caledonian (Acadian) transpression in north-west England: timing, geometry and geotectonic significance. Proceedings of the Yorkshire Geological Society 46: 175-192. DOI: 10.1144/pygs.46.3.175

Staring, W.C.H., 1869 . Geologische kaart van Nederland, schaal 1 : 200.000. A.C. Kruseman (Haarlem): Map series with 28 sheets.

Steemans, P., Le Hérissé, A., Melvin, J., Miller, M.A., Paris, F., Verniers, J. & Wellman, C.H., 2009. Origin and radiation of the earliest vascular land plants. Science 324(5925): 353-353. DOI: 10.1126/science.1169659

Stein, W.E., Berry, C.M., Hernick, L.V. & Mannolini, F., 2012. Surprisingly complex community discovered in the mid-Devonian fossil forest at Gilboa. Nature 483(7387): 78-81. DOI: 10.1038/nature10819

Stille, H., 1924. Grundfragen der vergleichenden Tektonik. Gebrüder Borntraeger (Berlin): 443 pp.

Streel, M., 1965. Étude palynologique du Dévonien du sondage de Booischot (Belgique). Note préliminaire. Bulletin de la Société Belge de Géologie, de Paléontologie et d'Hydrologie LXXIII (for 1964): 172-190.

Streel, M., 1986. Miospore contribution to the upper Famennian-Strunian event stratigraphy. Annales de la Société géologique de Belgique 109(1): 75-92.

Streel, M. & Loboziak, S., 1987. Nouvelle datation par miospores du Givétien-Frasnien des sédiments non marins du sondage de Booischot (Bassin de Campine, Belgique). Bulletin de la Société Belge de Géologie 96(2): 99-106.

Streel, M., Brice, D. & Mistiaen, B., 2006. Strunian. Geologica Belgica 9(1-2): 105-109.

Strother, P.K. & Foster, C., 2021. A fossil record of land plant origins from charophyte algae. Science 373(6556): 792-796. DOI: 10.1126/science.abj2927

Suess, E., 1883. Das Antlitz der Erde. G. Freytag (Leipzig, Germany): 778 pp.

Swennen, R. & Muchez, P., 1991. Sedimentological and diagenetic study of the Dinantian carbonates and underlying siliciclastics of Placid borehole O18-1. Katholieke Universiteit Leuven research report: 128 pp. nlog.nl/brh-web/rest/brh/document/781399679

Ten Veen, J., De Haan, H., De Bruin, G., Holleman, N. & Schöler, W., 2019. Seismic interpretation and depth conversion of the Dinantian carbonates in the Dutch subsurface. SCAN Report, commissioned by the Ministry of Economic Affairs & Climate: 95 pp. https://www.nlog.nl/sites/default/files/2019-09/scan_dinantian_seismic_interpretation_and_depth_conversion_report.pdf

Ter Borgh, M.M., Eikelenboom, W. & Jaarsma, B., 2018a. Hydrocarbon potential of the Visean and Namurian in the northern Dutch offshore. In: Monaghan, A.A., Underhill, J.R., Hewett, A.J. & Marshall, J.E.A. (eds) Paleozoic plays of NW Europe. Geological Society of London, Special Publications 471: 133-153. DOI: 10.1144/SP471.5

Ter Borgh, M.M., Jaarsma, B. & Rosendaal, E.A., 2018b. Structural development of the northern Dutch offshore: Paleozoic to present. In: Monaghan, A.A., Underhill, J.R., Hewett, A.J. & Marshall, J.E.A. (eds) Paleozoic plays of NW Europe. Geological Society of London, Special Publications 471: 115-131. DOI: 10.1144/SP471.4

Thorez, J., Dreesen, R. & Streel, M., 2006 . Famennian. Geologica Belgica 9: 27-35.

TNO-GDN, 2022a. Old Red Group. In: Stratigraphic Nomenclature of the Netherlands, TNO – Geological Survey of the Netherlands. https://www.dinoloket.nl/en/stratigraphic-nomenclature/old-red-group

TNO-GDN, 2022b. Beveland Member. In: Stratigraphic Nomenclature of the Netherlands, TNO – Geological Survey of the Netherlands. https://www.dinoloket.nl/en/stratigraphic-nomenclature/beveland-member

TNO-GDN, 2022c. Schouwen Member. In: Stratigraphic Nomenclature of the Netherlands, TNO – Geological Survey of the Netherlands. https://www.dinoloket.nl/en/stratigraphic-nomenclature/schouwen-member

TNO-GDN, 2022d. Goeree Member. In: Stratigraphic Nomenclature of the Netherlands, TNO – Geological Survey of the Netherlands. https://www.dinoloket.nl/en/stratigraphic-nomenclature/goeree-member

Torsvik, T.H. & Cocks, L.R.M., 2016. Earth History and palaeogeography. Cambridge University Press: 317 pp. DOI: 10.1017/9781316225523

Torsvik, T.H., Smethurst, M.A., Meert, J.G., Van der Voo, R., McKerrow, W.S., Brasier, M.D., Sturt, B.A. & Walderhaug, H.J., 1996. Continental break-up and collision in the Neoproterozoic and Palaeozoic - a tale of Baltica and Laurentia. Earth-Science Reviews 40: 229-258. DOI: 10.1016/0012-8252(96)00008-6

Tucker, M.E., Gallagher, J. & Leng, M.J., 2009. Are beds in shelf carbonates millennial-scale cycles? An example from the mid-Carboniferous of northern England. Sedimentary Geology 214(1-4): 19-34. DOI: 10.1016/j.sedgeo.2008.03.011

Van Adrichem Boogaert, H.A. & Kouwe, W.F.P., 1994. Stratigraphic nomenclature of the Netherlands; revision and update by RGD and NOGEPA; Section B: Devonian and Dinantian (Banjaard, Old Red, Carboniferous Limestone and Farne Groups). Mededelingen Rijks Geologische Dienst 50: 1-20.

Vanguestaine, M., Declairfayt, T., Rouhart, A. & Smeesters, A., 1983. Zonation par acritarches du Frasnien Supérieur - Famennien Inférieur dans les bassins de Dinant, Namur, Herve et Campine (Dévonien Supérieur de Belgique). Annales de la Société Géologique de Belgique 106(1): 121-171.

Van Grootel, G., Verniers, J., Geerkens, B., Laduron, D., Verhaeren, M., Hertogen, J. & De Vos, W., 1997. Timing of magmatism, foreland basin development, metamorphism and inversion in the Anglo-Brabant fold belt. Geological Magazine 134: 607-616. DOI: 10.1017/S0016756897007413

Van Hulten, F.F.N., 2012. Devono-carboniferous carbonate platform systems of the Netherlands. Geologica Belgica 15(4): 284-296.

Van Hulten, F.F.N. & Poty, E., 2009. Dinantian reefs underneath the Netherlands. 71st EAGE Conference and Exhibition incorporating SPE EUROPEC 2009. Amsterdam. DOI:10.3997/2214-4609.201400477

Van Waterschoot van der Gracht, W.A.J.M., 1918. Eindverslag over de onderzoekingen en uitkomsten van den Dienst der Rijksopsporing van Delfstoffen in Nederland 1903-1916. Martinus Nijhoff ('s Gravenhage): 664 pp.

Vecoli, M. & Samuelsson, J., 2001. Quantitative evaluation of microplankton palaeobiogeography in the Ordovician-Early Silurian of the northern Trans-European Suture Zone: implications for the timing of the Avalonia-Baltica collision. Review of Palaeobotany and Palynology 115: 43-68. DOI: 10.1016/S0034-6667(01)00047-1

Veeningen, R., 2020. TNO – Terziet-01 and -02 Cuttings Petrography. PanTerra Geoconsultants B.V. (Leiderdorp) report 2000134: 49 pp (incl. Appendix).

Verniers, J. & Van Grootel, G., 1991. Review of the Silurian in the Brabant Massif, Belgium. Annales de la Société Géologique de Belgique 114(1): 163-193.

Verniers, J., Herbosch, A., Vanguestaine, M., Geukens, F., Delcambre, B., Pingot, J.-L., Belanger, I., Hennebert, M., Debacker, T., Sintubin, M. & De Vos, W., 2001. Cambrian-Ordovician-Silurian lithostratigraphic units (Belgium). In: Bultynck, P. & Dejonghe, L. (eds): Guide to a revised lithostratigraphic scale of Belgium. Geologica Belgica 4(1): 5-38. DOI: 10.20341/gb.2014.042

Verniers, J., Pharaoh, T., André, L., Debacker, T.N., De Vos, W., Everaerts, M., Herbosch, A., Samuelsson, J., Sintubin, M. & Vecoli, M., 2002. The Cambrian to mid Devonian basin development and deformation history of Eastern Avalonia, east of the Midlands Microcraton: new data and a review. In: Winchester, J.A., Pharaoh, T.C. & Verniers, J. (eds): Palaeozoic amalgamation of Central Europe. Geological Society of London, Special Publications 201, 47-93. DOI: 10.1144/GSL.SP.2002.201.01.04

Vis, G.-J. & Houben, A.J.P., 2023. Data evaluation for a new geological model of the Devonian, Carboniferous and Triassic of the southeast Netherlands (South Limburg). TNO - Geological Survey of the Netherlands Report R12252: 64 pp.

Von Raumer, J.F., Bussy, F. & Stampfli, G.M., 2009. The Variscan evolution in the External massifs of the Alps and place in their Variscan framework. Comptes Rendus Geoscience 341(2-3): 239-252. DOI: 10.1016/j.crte.2008.11.007

Wagner, R.H. & Winkler Prins, C.F., 2016. History and current status of the Pennsylvanian chronostratigraphic units: problems of definition and interregional correlation. Newsletters on Stratigraphy 49(2): 281-320. DOI: 10.1127/nos/2016/0073

Waters, C.N., Somerville, I.D., Jones, N.S., Cleal, C.J., Collinson, J.D., Waters, R.A., Besly, B.M., Dean, M.T., Stephenson, M.H., Davies, J.R., Freshney, E.C., Jackson, D.I., Mitchell, W.I., Powell, J.H., Barclay, W.J., Browne, M.A.E., Leveridge, B.E., Long, S.L. & McLean, D., 2011. A revised correlation of Carboniferous rocks in the British Isles. Geological Society of London, Special Report 26: 186 pp. DOI: 10.1144/SR26

Waters, C.N. & Condon, D.J., 2012. Nature and timing of Late Mississippian to Mid-Pennsylvanian glacio-eustatic sea-level changes of the Pennine Basin, UK. Journal of the Geological Society 169(1): 37-51. DOI: 10.1144/0016-76492011-047

Winchester, J.A., Pharaoh, T.C. & Verniers, J., 2002. Palaeozoic amalgamation of Central Europe: an introduction and synthesis of new results from recent geological and geophysical investigations. In: Winchester, J.A., Pharaoh, T.C. & Verniers, J. (eds): Palaeozoic amalgamation of Central Europe. Society of London, Special Publications 201: 1-18. DOI: 10.1144/GSL.SP.2002.201.01.01

Woodcock, N.H. & Pharaoh, T.C., 1993. Silurian facies beneath East Anglia. Geological Magazine 130: 681-690. DOI: 10.1017/S0016756800020999

Woodcock, N.H., Soper, N.J. & Miles, A.J., 2019. Age of the Acadian deformation and Devonian granites in northern England: a review. Proceedings of the Yorkshire Geological Society 62(4): 238-253. DOI: 10.1144/pygs2018-009

Woodcock, N.H., Soper, N.J. & Strachan, R.A., 2007. A Rheic cause for the Acadian deformation in Europe. Journal of the Geological Society 164(5): 1023-1036. DOI: 10.1144/0016-76492006-129

Ziegler, P.A., 1990. Geological Atlas of Western and Central Europe, second edition. Shell Internationale Petroleum Mij. B.V. and Geological Society (London): 239 pp.

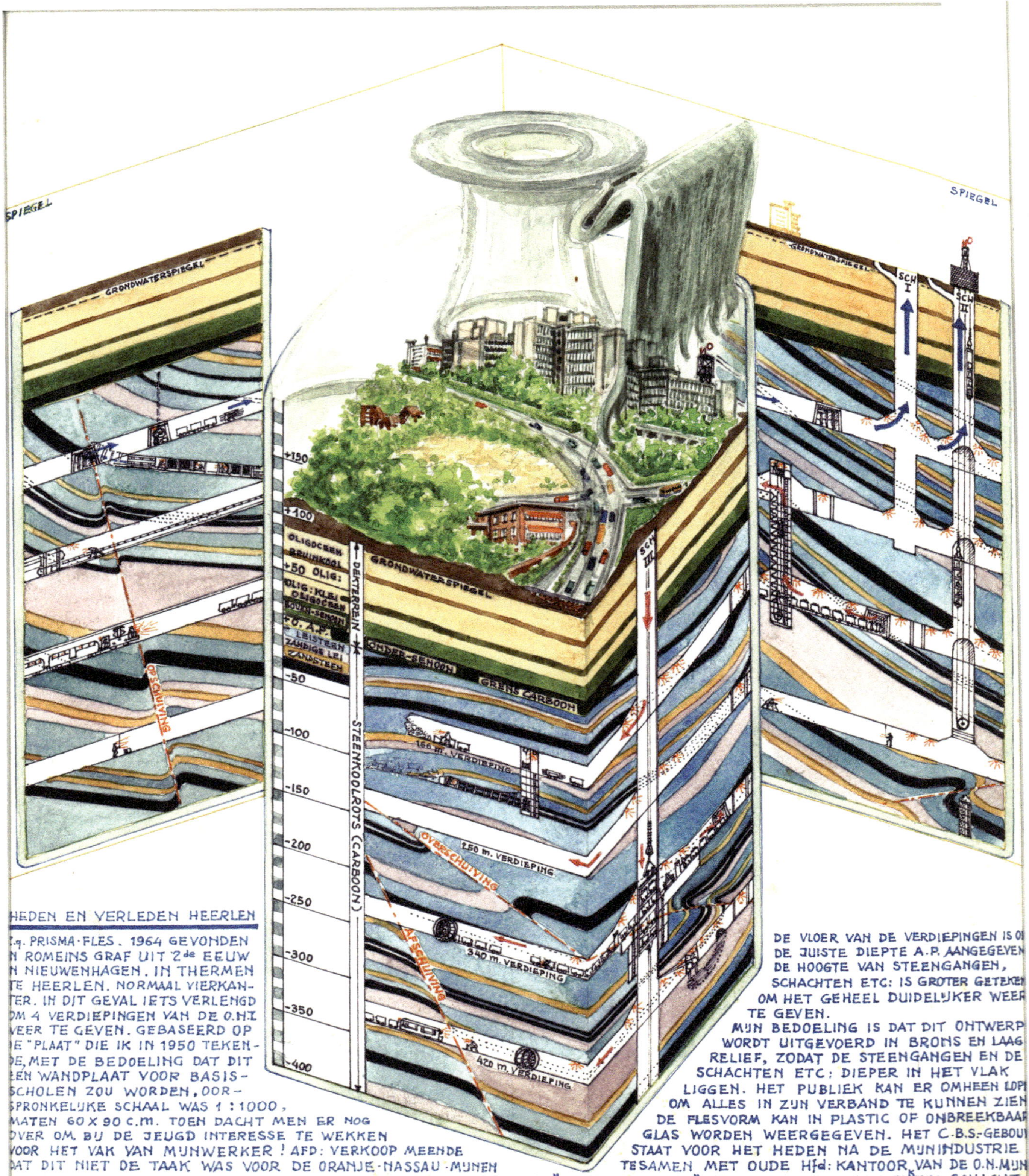

HEDEN EN VERLEDEN HEERLEN

Z.q. PRISMA·FLES. 1964 GEVONDEN IN ROMEINS GRAF UIT 2de EEUW IN NIEUWENHAGEN. IN THERMEN TE HEERLEN. NORMAAL VIERKANTER. IN DIT GEVAL IETS VERLENGD OM 4 VERDIEPINGEN VAN DE O.N.I. WEER TE GEVEN. GEBASEERD OP DE "PLAAT" DIE IK IN 1950 TEKENDE, MET DE BEDOELING DAT DIT EEN WANDPLAAT VOOR BASISSCHOLEN ZOU WORDEN, OORSPRONKELIJKE SCHAAL WAS 1 : 1000, MATEN 60 X 90 C.M. TOEN DACHT MEN ER NOG OVER OM BIJ DE JEUGD INTERESSE TE WEKKEN VOOR HET VAK VAN MIJNWERKER! AFD: VERKOOP MEENDE DAT DIT NIET DE TAAK WAS VOOR DE ORANJE-NASSAU-MIJNEN ZOIETS ALS RECLAME UIT TE GEVEN. Dhr. STEENAERT † O.A. Hfd. REDACTEUR O.N. POST, LIET DE PLAAT VERKLEINEN TOT DEKPLAAT VAN DE O.N. KALENDER. DE PLAAT VERDWEEN, MAAR ENIGE TIJD LATER BLEEK HET MIJNMUSEUM ROLDUC OVER DEZELFDE PLAAT TE BESCHIKKEN NAGETEKEND DOOR EEN LEEK DIE DE SCHACHTKOOI AAN EEN SLAGERSHAAK OPHING, HETGEEN VOOR ONS MIJNBOUWERS DESASTREUS ZOU ZIJN!

DE VLOER VAN DE VERDIEPINGEN IS OP DE JUISTE DIEPTE A.P. AANGEGEVEN DE HOOGTE VAN STEENGANGEN, SCHACHTEN ETC: IS GROTER GETEKEND OM HET GEHEEL DUIDELIJKER WEER TE GEVEN.

MIJN BEDOELING IS DAT DIT ONTWERP WORDT UITGEVOERD IN BRONS EN LAAG RELIEF, ZODAT DE STEENGANGEN EN DE SCHACHTEN ETC: DIEPER IN HET VLAK LIGGEN. HET PUBLIEK KAN ER OMHEEN LOPEN OM ALLES IN ZIJN VERBAND TE KUNNEN ZIEN. DE FLESVORM KAN IN PLASTIC OF ONBREEKBAAR GLAS WORDEN WEERGEGEVEN. HET C.B.S.-GEBOUW STAAT VOOR HET HEDEN NA DE MIJNINDUSTRIE TESAMEN MET OUDE Hfd: KANTOOR VAN DE O.N. MIJN EN "MIJNMONUMENT" SCHACHT II, DE UITTREKKENDE SCHACHT VAN DE O.N.I, WAAR OMHEEN DE OORSPRONKELIJKE OMBOUW VAN SCH. I IS AANGEBRACHT. IK BEN VAN MENING, DAT HEERLEN ALS NEDERZETTING CORIOVALLUM EN DORP ACHTERGELATEN DOOR DE ROMEINEN, DOOR DE MIJNINDUSTRIE IN HAAR GEHEEL EN DOOR DE ORANJE-NASSAU-MIJNEN IN HET BIJZONDER, GEWORDEN IS WAT HET NU IS! EEN ERESALUUT AAN DE MIJNBOUWERS DU

3 Late Carboniferous

*Richard Huis in 't Veld† and Daan den Hartog Jager**

*Corresponding author: daan.hartogjager-den@ebn.nl

ABSTRACT

A thick upper Carboniferous sequence was deposited in the Netherlands and adjacent areas. Despite intense erosion during the late Carboniferous and early Permian, it has been preserved beneath the unconformably overlying successions. The preserved thickness locally exceeds 5500 m. The sequence comprises almost exclusively siliciclastic sediments, deposited in a cratonic-molasse basin that formed as a result of the collision of the northward drifting Gondwana continent against Laurussia. The Variscan orogenic belt, the axis of this collision, provided the sediment source for the southern part of the basin. Sediments in the north were sourced from the older Caledonian mountains, which had already formed as the result of the collision of Laurentia, Baltica and Avalonia in Silurian-Devonian times. The lithology ranges from claystone, siltstone and sandstone up to conglomerate size. The units were initially deposited in a predominantly marine basin flanked by deltas, grading into a paralic basin with alluvial plains and -fans, and finally in a largely fluvial red-bed environment. Thick and extensive peat layers were formed from the Yeadonian (Early Pennsylvanian) until the early Asturian (Middle Pennsylvanian) in deltaic and paralic floodplain environments. The coal beds, which formed when this peat became deeply buried, were mined in the southern Netherlands from medieval times up until 1974 and have been the main driver for the industrial revolution in Western Europe. Combustible gas, generated from Carboniferous coals, has been the most important exploration target of the petroleum industry in the Netherlands. Since the discovery of gas in the upper Carboniferous reservoirs of the Coevorden field in 1951, numerous other discoveries have been made in the east and northeast of the Netherlands from the 1950s onward, as well as offshore from 1974 to the present.

<< Illustration by a former miner and mining engineer of the Oranje Nassau mines in South Limburg. A locally excavated Roman Prisma bottle contains the subsurface model and two mirrors are used to view the rear side. Hand drawn by Jacques Driessen (1964).

DOI: 10.5117/9789463728362_ch03

Richard Huis in 't Veld 1965-2024

Richard Huis in 't Veld, the senior author of this chapter 'Late Carboniferous', sadly passed away on 25th April 2024 after a short illness. He was only 59 years old. Many of the new figures and insights in this chapter originate with him.

For most of his life, Richard lived in or near Rotterdam, a city close to his heart. He finished his MSc Geology in Utrecht in 1988. In the midst of the oil crisis, employment for geologists was at a minimum, but undaunted, Richard and some friends started ARGO Geological Consultants, which has successfully provided them with rewarding and valuable work ever since.

His first job brought him for 9 years to NAM in Assen, where he worked mainly on the Rotliegend in the Northern Onshore, developing a taste for the Paleozoic. His next move was to Wintershall in The Hague, which at the time was actively exploring the Carboniferous on the Cleaverbank Platform. He was instrumental in several important gas discoveries both in the Netherlands and across the border in the UK; his deep knowledge of the complex geology of the local Carboniferous and its seismic expression earned him the nickname 'King of the Carboniferous'. He was the first to demonstrate the importance of the Maurits Formation as a seat seal in the Netherlands. Many of his ideas were published in Huis in 't Veld et al (2020).

In 2021 he moved on to ONE-Dyas, where he had a key role in the successful appraisal and development of gas discoveries in the N05 block. His natural curiosity and multi-discipline skills were important here: he obtained a full understanding of the regional geology also on the German side, where data can be more difficult to locate. Meanwhile, partly as a hobby, he became more interested in the geology and mining history of South Limburg, resulting in several field trips to this area. This brought new insights about the likely former presence and complete erosion of a Variscan thrust in this region, further described in the current chapter. Richard had a happy family life with wife Tanja and sons Nick and Kevin. They enjoyed traveling the world and supporting Feyenoord football club. We all miss his enthusiasm, his engagement, and his ability to understand the 'big picture'. This chapter also serves as a memorial to his legacy.

Daan den Hartog Jager

Introduction

In the Netherlands, deposits of late Carboniferous age are thicker and more widespread than those of any subsequent period. They are present below more than 95% of the country and its adjacent North Sea area, and even after major compaction, their thickness can be up to 5.5 km. This means that the upper Carboniferous is often thicker than its complete overburden (Figs 3.1-3.2).

The Carboniferous of northwest Europe has been the subject of geological studies since the 18th century. Early interest may be explained by the excellent outcrops in the region, some even within the Netherlands. More importantly, the abundance of coal as the main driver for the industrial revolution in the 19th century generated early research activities focused on the Carboniferous. For example, the landmark overview article on the Paleozoic of the Netherlands by Thiadens (1963) contains more than 20 pages on the upper Carboniferous and only one page on the Rotliegend, which at the time was not yet recognized as an important reservoir. Further details of the coal production are described by Van Bergen and Koster (2025, this volume).

Besides coal production, the Carboniferous is of major importance as a source rock for most of the gas discovered in the Netherlands, including the gas in the Groningen field (Remmelts et al., 2025, this volume). In addition, Carboniferous sandstone reservoirs were among the earliest targets for hydrocarbon exploration in the Netherlands (Van Wijhe & Bless, 1974). Coevorden was the first commercial gas field in northwest Europe, named after the nearby town which already had developed a natural gas-distribution network in 1951. The field produces commingled gas from both upper Carboniferous (upper Bolsovian-lower Asturian) fluvial sandstones and from the unconformably overlying basal Zechstein carbonates.

Offshore, the well that first tested gas from the Carboniferous was K04-01D (1974), also from upper Bolsovian-lower Asturian reservoirs. A major breakthrough followed in the period 1984-1986 with the discovery of the D15-FA field (in the lower Duckmantian), and of a number of gas fields in nearby UK quadrants 44 and 49. Overall, some 100 bcm of gas has been produced from Westphalian reservoirs in the Netherlands.

The complex fluvial architecture of Westphalian reservoirs frequently resulted in production underperformance due to poor connectivity, early water breakthrough or halite scaling. Nevertheless, in some areas these sandstones can still be considered suitable targets for CO_2-storage or geothermal exploration.

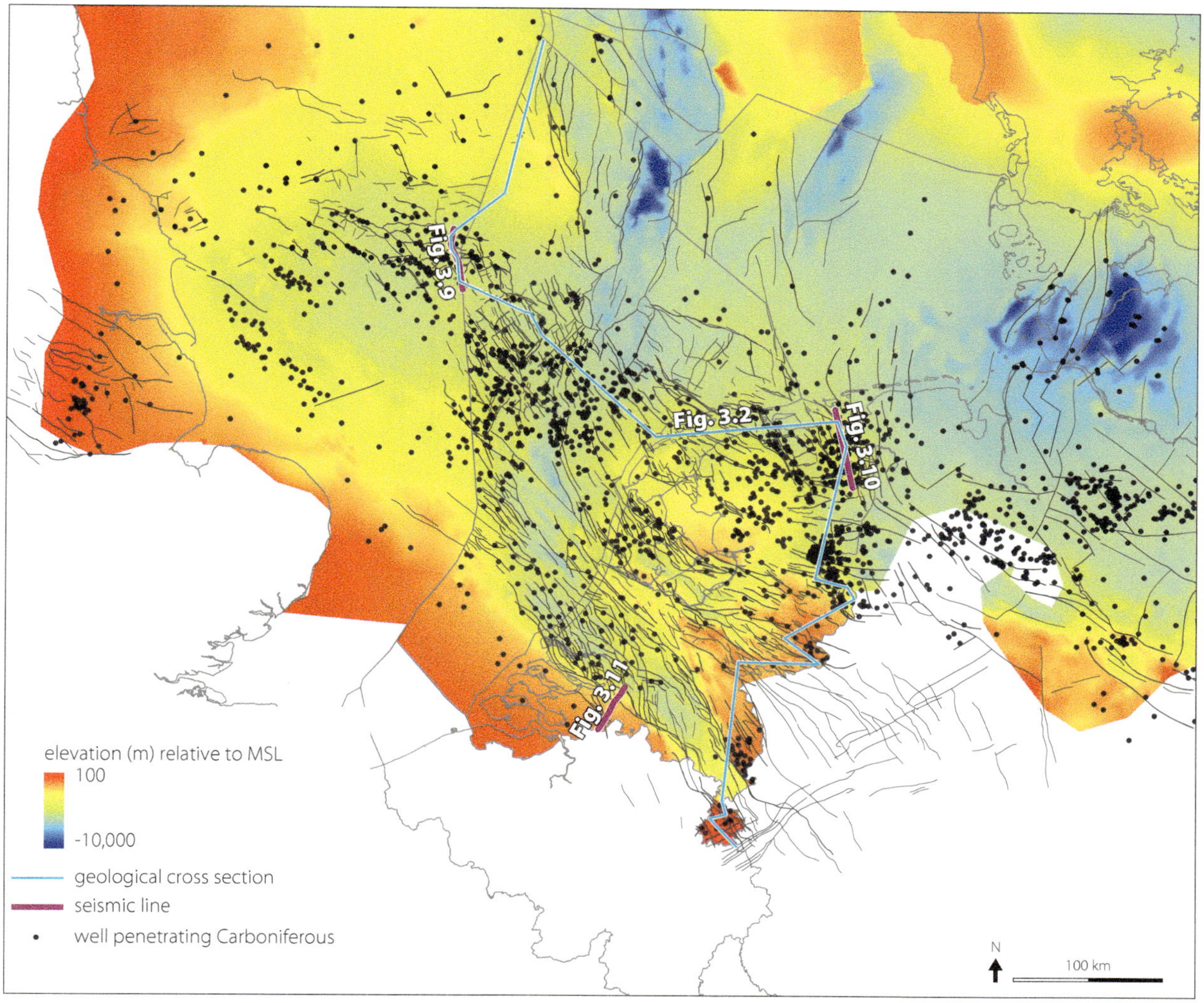

Figure 3.1. *Top Carboniferous depth map showing a depth range from >100 m + MSL in South Limburg to <10,000 m - MSL in the Dutch Central Graben from DGM-deep v5, www.nlog.nl (2019), area outside the Netherlands from Kombrink et al. (2010).*

Within the underlying Namurian, only a few small and uneconomic accumulations have been drilled in the Netherlands, for example in block E12. However, across the border in UK quadrant 43 (directly west of 44), several economic discoveries in the upper Namurian (Alportian-Yeadonian) Millstone Grit Formation, such as the Cavendish and Trent fields, indicating there might be remaining potential also in the Namurian of the Netherlands.

Despite many decades of research, numerous aspects of the upper Carboniferous remain elusive. The scarcity of clear marker beds in outcrops and wells, as well as dating uncertainties have long frustrated the construction of an unambiguous regional stratigraphic framework, such as presented in Figure 3.4. The thick clastic upper Carboniferous succession may appear monotonous, but detailed investigations reveal that rapid lateral facies changes are common.

A solid early knowledge base for the upper Carboniferous was achieved through the publication of work related to coal mine geology (e.g. Van Waterschoot van der Gracht, 1918; Trueman, 1954; Wandless, 1960; Peelcommissie, 1963; Drozdzewski et al., 1985a,b). By now, a wealth of additional publications is available, based on data from outcrop studies (e.g. David, 1990), seismic and wells, acquired during the exploration and development for natural gas both on- and offshore (e.g. Kombrink et al., 2010; Besly, 2018). This chapter is a strongly modified update of the first edition (Van Buggenum & Den Hartog Jager, 2007).

Structural setting and tectonic development

Basin development

During the late Carboniferous, the northward drifting Gondwana continent continued its collision with the Laurussia continent, which itself had been formed when

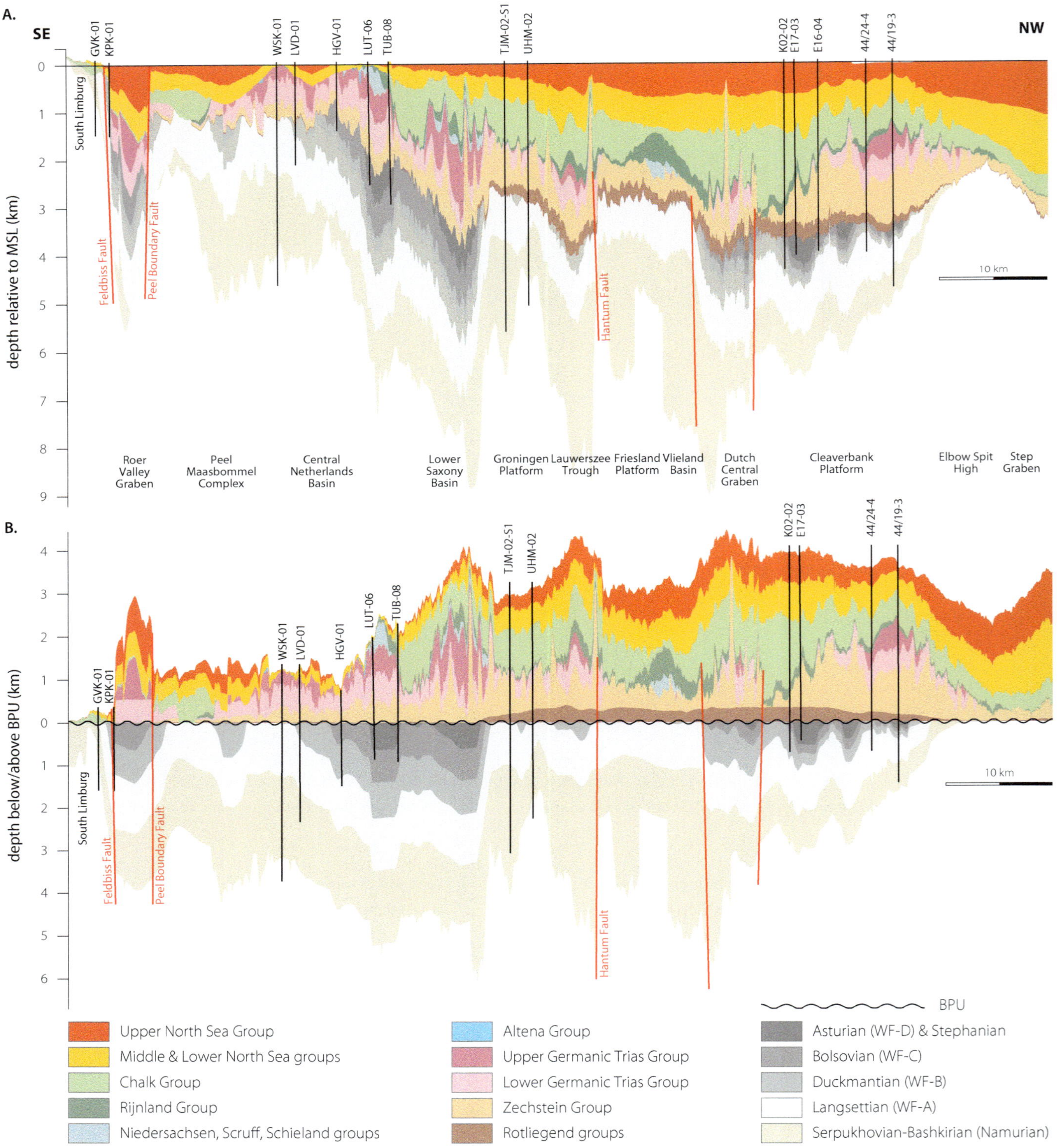

Figure 3.2. *a) Geological cross-section across the Netherlands on- and offshore, with key wells and structural elements. Note the relatively thick upper Carboniferous succession reaching locally more than 5 km. See Fig. 3.1 for section location. b) Same section as in (a), here flattened on the Base Permian Unconformity (BPU) to emphasize the upper Carboniferous thickness variation below the BPU. WF = Westphalian.*

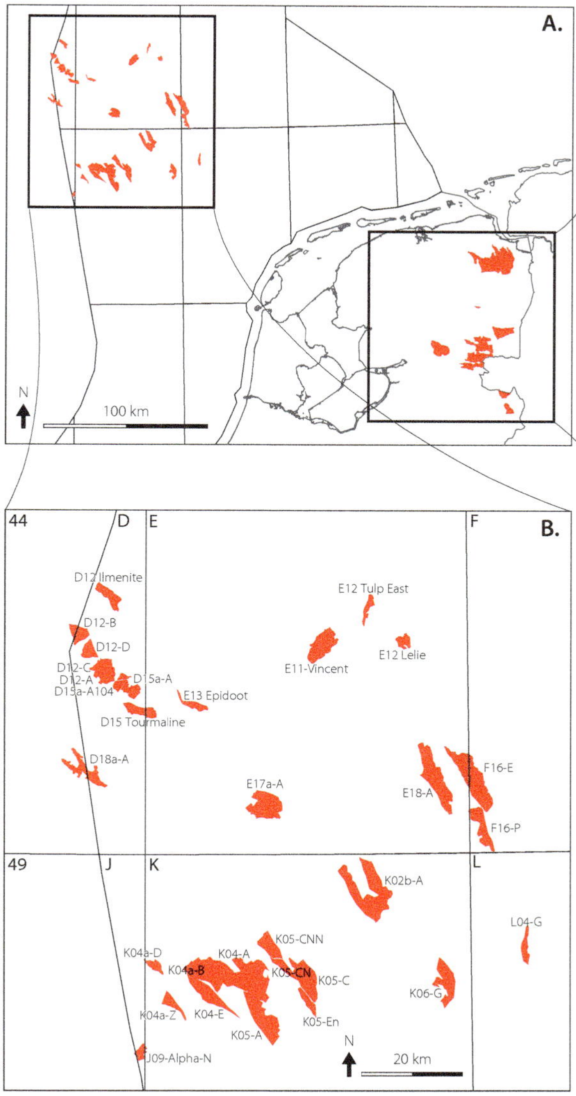

Figure 3.3. *Gas fields in the Netherlands with proven gas in Carboniferous reservoirs. a) overview map, b) offshore (Cleaverbank Platform), and c) onshore (Lower Saxony Basin area).*

the Laurentian, Baltica and Avalonia (micro-)plates collided during the Silurian-Devonian Caledonian orogeny. The collision of the African part of Gondwana with the European part of Laurussia took place near the equator and resulted in the formation of the Variscan orogenic belt (Fig. 3.5). The Variscan Mountains formed the main sediment supply to the foredeep basin north of this belt. Another supply originated from the 'passive' Caledonian hinterland north of the basin (Torsvik & Cocks, 2017).

In an early phase of the collision during the Namurian, rapid flexural subsidence took place in the southeastern part of the Carboniferous basin, with a thick succession of siliciclastic sediments ('molasse') being deposited in a foreland basin setting. This is also known as the Sudetic Phase (Stille, 1920; Schulmann et al., 2014). Sediments progressively onlapped Dinantian platform carbonates of the London-Brabant Massif and several intra-basinal highs such as the Groningen Platform (Kombrink, 2008). These highs formed during a major extension phase in the Dinantian, as result of crustal stretch along the Caledonian Thor Suture, in a back-arc setting with respect to the subducting Rheic Ocean (Von Raumer et al., 2013; Smit et al., 2016).

A thick interval of Namurian sediments within the Rheno-Hercynian Zone (Fig. 3.5; Drozdzewski, 1993) and contemporary igneous intrusions also suggest that subduction systems were present south of this zone (Ziegler, 1990; Smit et al., 2016) and support this back-arc setting. Three subduction zones, two south- and one north-dipping, associated with Variscan deformation-front sutures, have been recognized between the Rheno-Hercynian and the more southerly Moldanubian Zone (Franke, 2000; Ziegler et al., 2004; Franke et al., 2017). The northwest European upper Carboniferous basin is situated on the Avalonian northern passive margin of the northernmost subduction zone, the suture of which can be found south of the Hunsrück area, against the Mid-German Crystalline High.

During the Westphalian, the Variscan deformation front continued to migrate northward and the Dinantian-Namurian succession of the Rheno-Hercynian Zone was deformed into major thrust and nappe complexes (Ricken et al., 2000; Ziegler et al., 2004; Kombrink, 2008; Cambier & DeJonghe, 2010). The northernmost of these, named the Verviers Synclinorium in Belgium or Inde Syncline in Germany, lies close to the present-day Netherlands border in South Limburg (Figs 3.6 and 3.7), where it crops out in the city centre of Aachen as the Aachener Überschiebung (Fig. 3.7) and continues westward as the Faille du Midi in

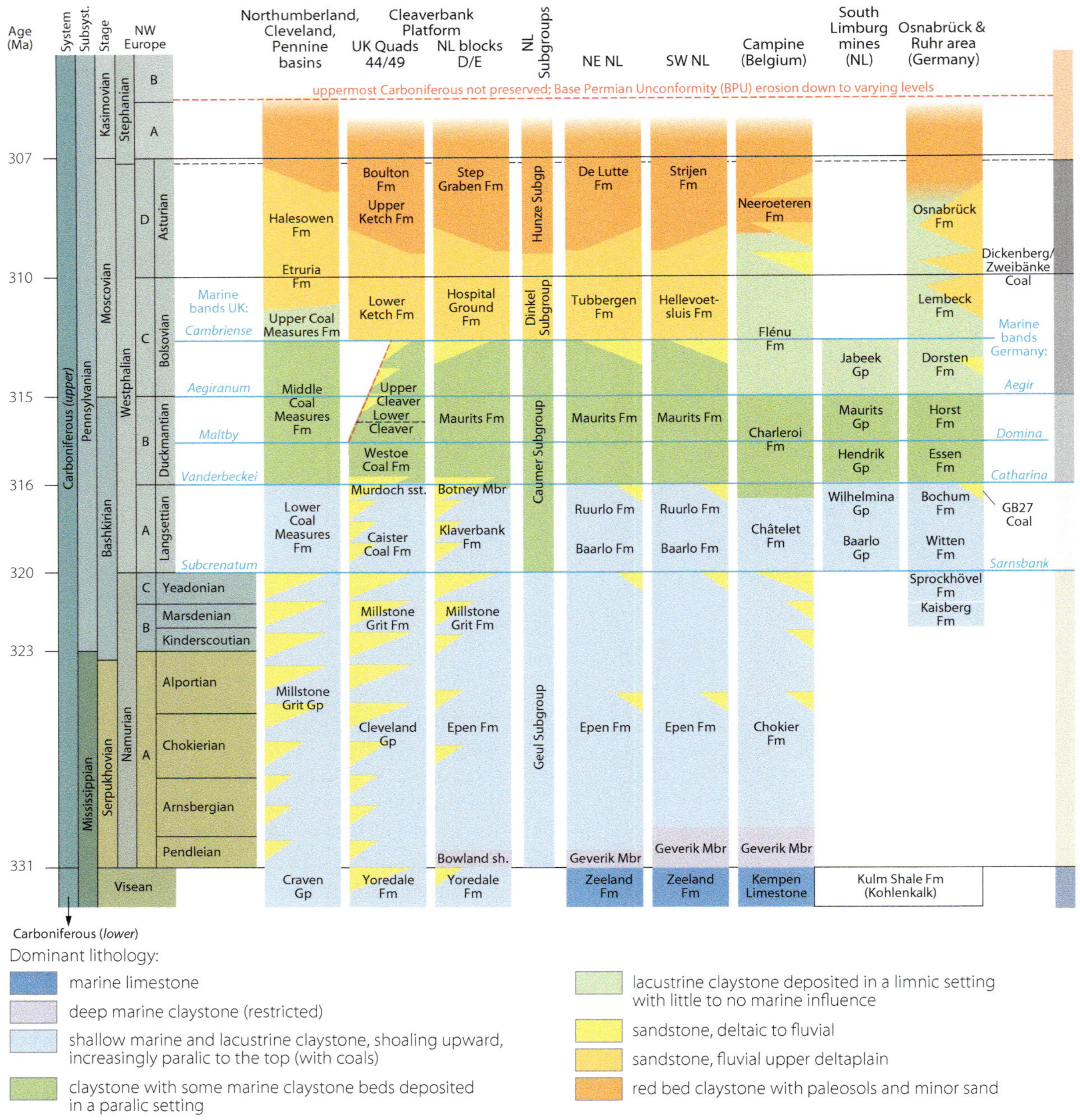

Figure 3.4. *Stratigraphic scheme with comparison of formation names used in and around the Netherlands. Absolute ages from ICS (Cohen et al., 2013). Formation names from TNO-GDN (2022), Kombrink et al. (2010), Cleal et al. (2009), Menning et al. (2006), Delmer et al. (2001) and Cameron (1993). The colours shown in the rightmost column are used throughout this volume for the corresponding stratigraphic units.*

Belgium (Fig. 3.5). North of these major Variscan thrusts, deformation decreases rapidly towards the northwest in South Limburg but extends significantly further north in the German Ruhr area, where the westernmost proven reverse fault has been drilled by well Münsterland-1. During the early formation of the thrust complexes in the Langsettian and Duckmantian, the molasse basin to the north continued to subside.

The London-Brabant Massif, in the southwest of the Netherlands Offshore, was already present during the Late Carboniferous, as indicated by thinning of both Namurian and Westphalian deposits towards it. It seems to have acted as a passive high, rather than an active source of uplift and erosion (Pharaoh, 2018; Vis et al., 2025, this volume).

In Germany, inversion-related uplift occurred from the late Bolsovian onward (Kombrink, 2008). This continued throughout the Stephanian with an increase in magnitude and culminated during the Asturian tectonic phase when a change in stress regime caused transpressional movements along NW-SE striking faults, as well as block tilting and the formation of NW-SE trending anticlines. The impact of this phase in the North Sea Basin is difficult to assess since its effects became obscured by subsequent early Permian thermal uplift and substantial erosion of Carboniferous sediments.

Permian subcrop map

The subcrop map at the Base Permian Unconformity (BPU) displays a complex erosional pattern, which resulted from the late Carboniferous Asturian inversion phase and early Permian thermal uplift and post orogenic transtension (Fig. 3.6). Several post-depositional deformation phases can be recognized.

The NW-vergent compression decreases in intensity within a short distance of the Variscan thrusts in the south east, but can be recognized as far northwest as the quadrant 44 area in the UK in the form of SW-NE striking long-wavelength open folds. This distal deformation appears to postdate the Bolsovian, as there is no evidence from the blanket 3D seismic data set for angular unconformities or syn-sedimentary tectonics around the Bolsovian-Asturian boundary. A mid-Bolsovian hiatus between the Hospital Ground (Ketch) Formation and the late Duckmantian and early Bolsovian Upper Maurits (Cleaver)

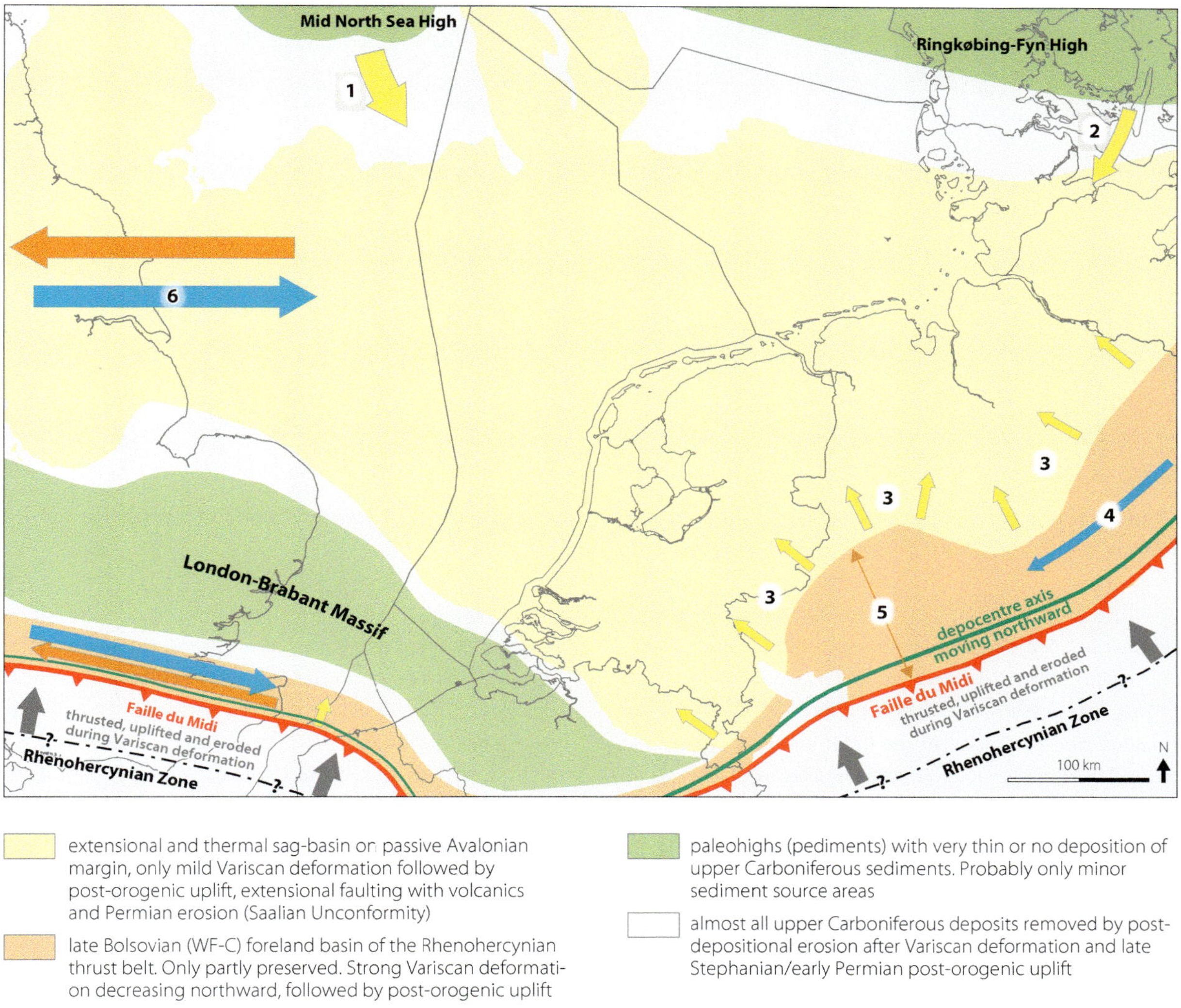

Figure 3.5. Late Carboniferous structural setting of NW Europe showing the outline of the late Carboniferous foreland basin. After Hampson et al. (1999) and Kombrink et al. (2010). 1 = Major sediment input from Caledonides in pulses from late Arnsbergian-early Asturian (Namurian delta). Reactivation and redeposition during late Bolsovian (WF-C) during lowstands (Asturian orogenic phase); 2 = Input from Baltic Shield; 3 = Distal sediment input from rising and northward migrating Variscides, major input during late Bolsovian-early Asturian from Variscan thrusts; 4 = Possible marine incursions during early Namurian highstands; 5 = Strong compressional deformation decreasing northward; 6 = Marine incursions during highstands up to late Bolsovian.

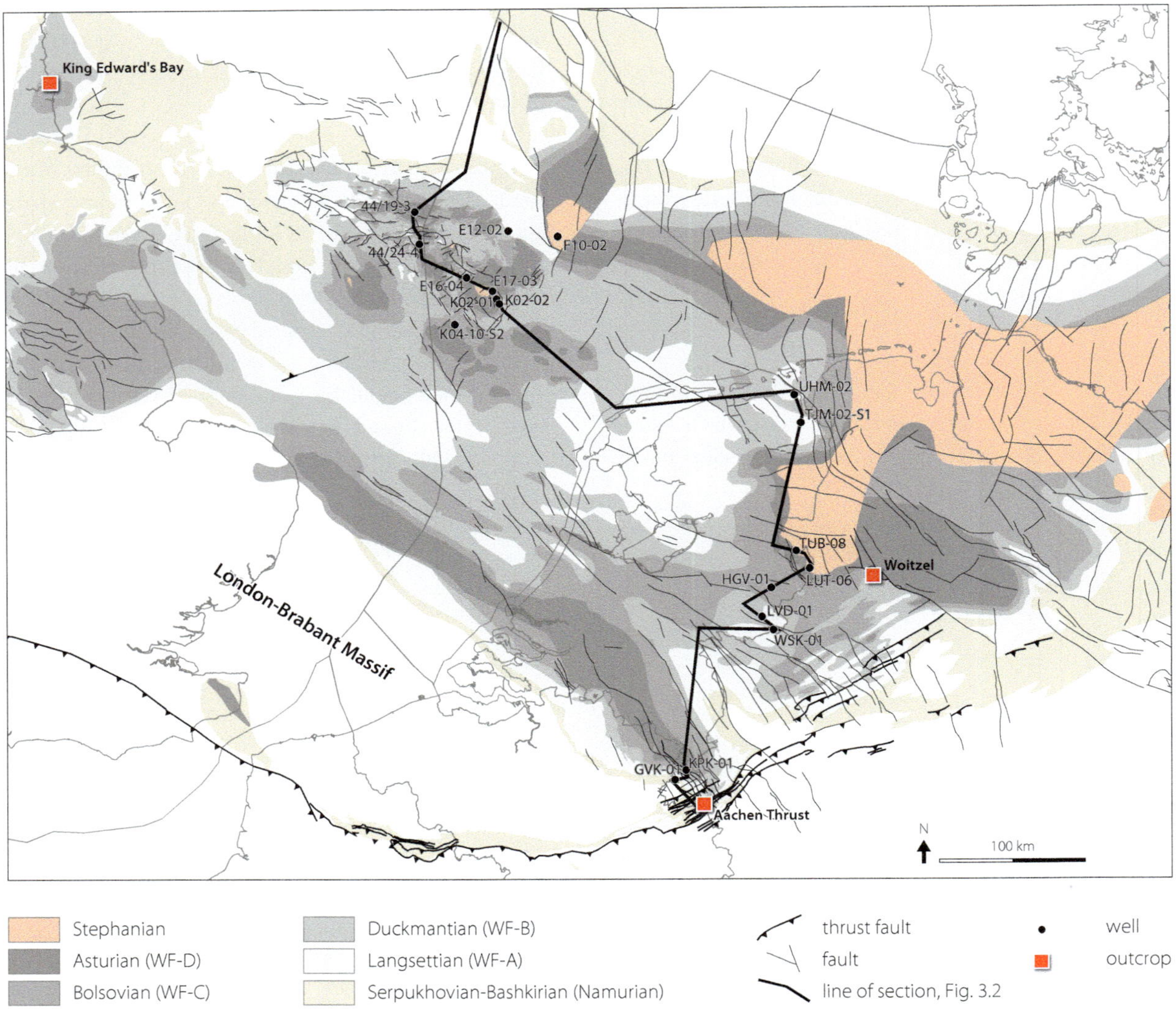

Figure 3.6. Permian subcrop map with key wells and main structural elements. Sources: Mijnlieff (2002), Van Buggenum and den Hartog Jager (2007), Kombrink et al. (2010), Roustiau et al. (2022).

Formation, can be found at the northern edge of the basin against the Mid North Sea High (e.g. McLean et al., 2005) and possibly also at the southern edge against the London Brabant Massif in the Campine Basin, where the early Asturian Neeroeteren Sandstone lies with a hiatus on underlying Bolsovian strata (Van Tongeren & Hildenbrand, 2005).

Perpendicular to the SW-NE compression, another open long wavelength folding trend with a NE-vergence, can be observed. This appears to be a distal manifestation of a late Asturian deformation trend observed north of the London Brabant Massif, but with no reverse faults and less intense folding. In the northwest of the basin, the mild Asturian folding phase is followed by extensional block faulting and tilting along W-E trending strike-slip faults, mainly in the form of north-dipping tilted fault blocks, which were uplifted and eroded during the Saalian inversion phase. Many of the gas filled traps with Bolsovian-Asturian reservoirs that have been discovered in the quadrant 44 area, like the Tyne, Wingate and Sillimanite fields, rely on such tilted fault blocks combined with intra Westphalian top- side- and seat seals (Huis in 't Veld et al., 2020).

Towards the end of the Carboniferous, the total thickness of the upper Carboniferous sedimentary succession exceeded 5500 m close to the Variscan deformation front. In areas with Langsettian-Duckmantian subcropping the BPU, it is likely that at least 1000 m of sediment was removed during early Permian uplift (Roustiau et al., 2022). A few wells (e.g. E17-03, E18-05, F10-02) have encountered Stephanian below the BPU, suggesting that Asturian and Stephanian sediments may have been deposited in the basin centre, but were subsequently eroded.

At the northern fringe of the Carboniferous basin on the Mid North Sea High, Dinantian and older sediments subcrop the BPU. The low maturity of the Dinantian coals in the Yoredale and Scremerston formations in this area indi-

Figure 3.7. Structural style at the Variscan thrust front near South Limburg with thrusts documented from the Rhenish Massif (Drozdzewski, 1993; Ricken et al., 2000). Major loading and burial by now-eroded overthrusted units explain the rapid southward increase in coal rank observed in South Limburg (e.g. Veld et al., 1996; see Textbox). Section length 26 km, near true scale (1:1). Location is given in the textbox figure. The photo shows Upper Devonian rock cropping out beneath the St. Adalbert church in Aachen.

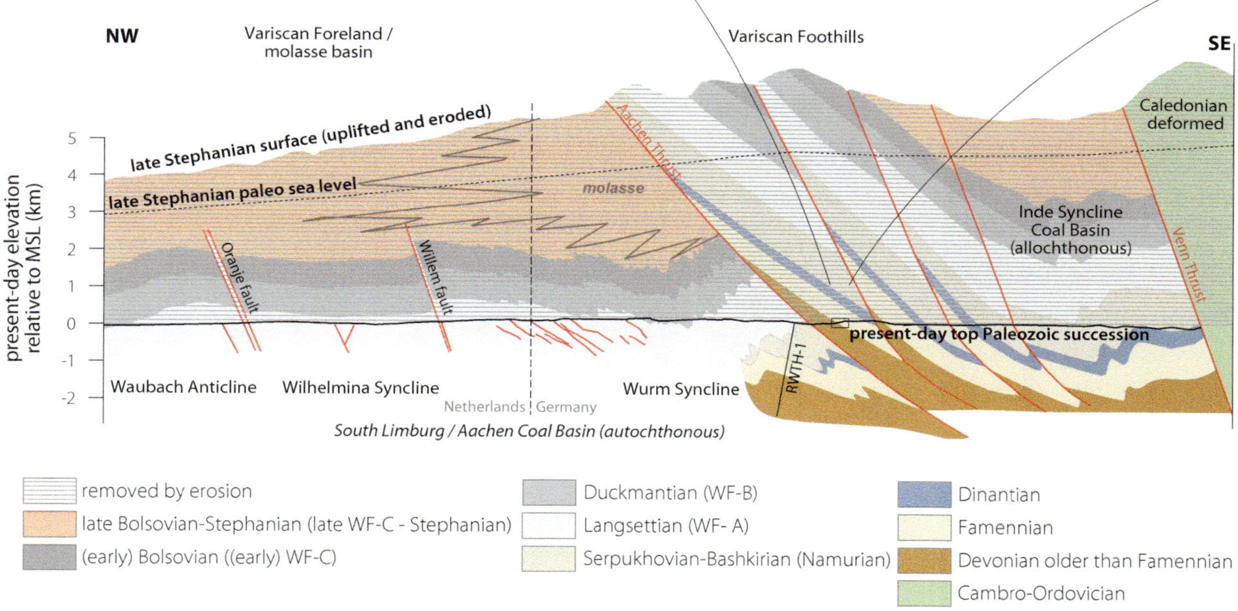

cates that the thickness of the upper Carboniferous, now fully removed, was less than 2 km (Vincent, 2015). This points to a trend of overall thinning of the upper Carboniferous succession towards the north, away from the Variscan front, so that the molasse basin in the south gradually turns into an intracratonic basin further north.

Structurally, high or low areas at the Base Permian correspond closely to later tectonic features (Fig. 3.2). High areas, such as the Friesland Platform, affected by deep erosion of Carboniferous strata during the late Carboniferous and early Permian, often reappear as horsts in later times. Low areas, less affected by such erosion, tend to reappear as grabens. Examples of such reactivated grabens are the West Netherlands Basin, the Lauwerszee Trough, the western part of the Lower Saxony Basin and the Step Graben.

During the middle and late Permian, after the early to middle Permian uplift and erosion, Rotliegend sediments covered most of the Carboniferous in the Netherlands. Excellent exposures of the Base Permian Unconformity (BPU) can be found on the English coast (Fig. 3.8) and in the German Harz area. The BPU represents in most places a time gap of some 50 to 40 Myr (Schneider & Scholze, 2018).

Upper Carboniferous unconformities, deformation and burial history

Regional seismic data in the Netherlands, both on- and offshore, show little evidence for intra-upper Carboniferous unconformities, despite the excellent coverage. Most seismic lines display a concordant pattern within the upper Carboniferous, often with a marked truncation at the Base Permian Unconformity (Figs 3.9-3.11). Although some detailed isopach maps and well data occasionally suggest a certain degree of syn-sedimentary tectonics (e.g. Dreesen et al., 1995; Kosters & Donselaar, 2003; see also Fig. 3.11), evidence is generally scarce, and the relationship may be spurious. The average rate of subsidence was high, limiting the potential to form well-expressed unconformities.

Only at the margins of the basin may erosional disconformities have formed as a result of base-level fall, possibly driven by tectonics (Jones & Glover, 2005). Incised fluvial valleys associated with such disconformities are interpreted at certain prospective reservoir levels. Examples can be found on the Cleaverbank Platform at base Caister Formation/Klaverbank Formation (late Langsettian-earliest Duckmantian) and base Ketch Formation/Hospital Ground Formation (late Bolsovian-early Asturian; Huis in 't Veld et al., 2020).

Coal grade in South Limburg

During the 20th century, significant volumes of coal were extracted from the upper Langsettian Ruurlo and Duckmantian Maurits formations in the coal mining district of South Limburg (see Van Bergen & Koster, 2025, this volume). A clear trend in coal grade is observed in a specific coalbed (GB27), situated below the Catharina Marine Band (Langsettian/Duckmantian boundary) and subcropping the Base Cretaceous Unconformity. Samples taken from this coal bed (>1 m) were analysed for vitrinite reflection (VR). Values range from (semi-)anthracite near Kerkrade in the southeast, to medium volatile coals near Geleen in the northwest (Nunna, 1960). Samples from the Namurian in well Geverik-1 yielded anthracite VR levels at a depth of 400 m (Fermont & Jegers, 1986).

The VR values correspond to a maximum burial depth of 5500 m in the southeast and 2900 m in the northeast. These values were apparently already reached during the Stephanian (Veld et al., 1996). A similar VR trend is observed in the adjacent Belgian Campine Basin (Deckers & Rombaut, 2021). This trend in Belgium is accompanied by

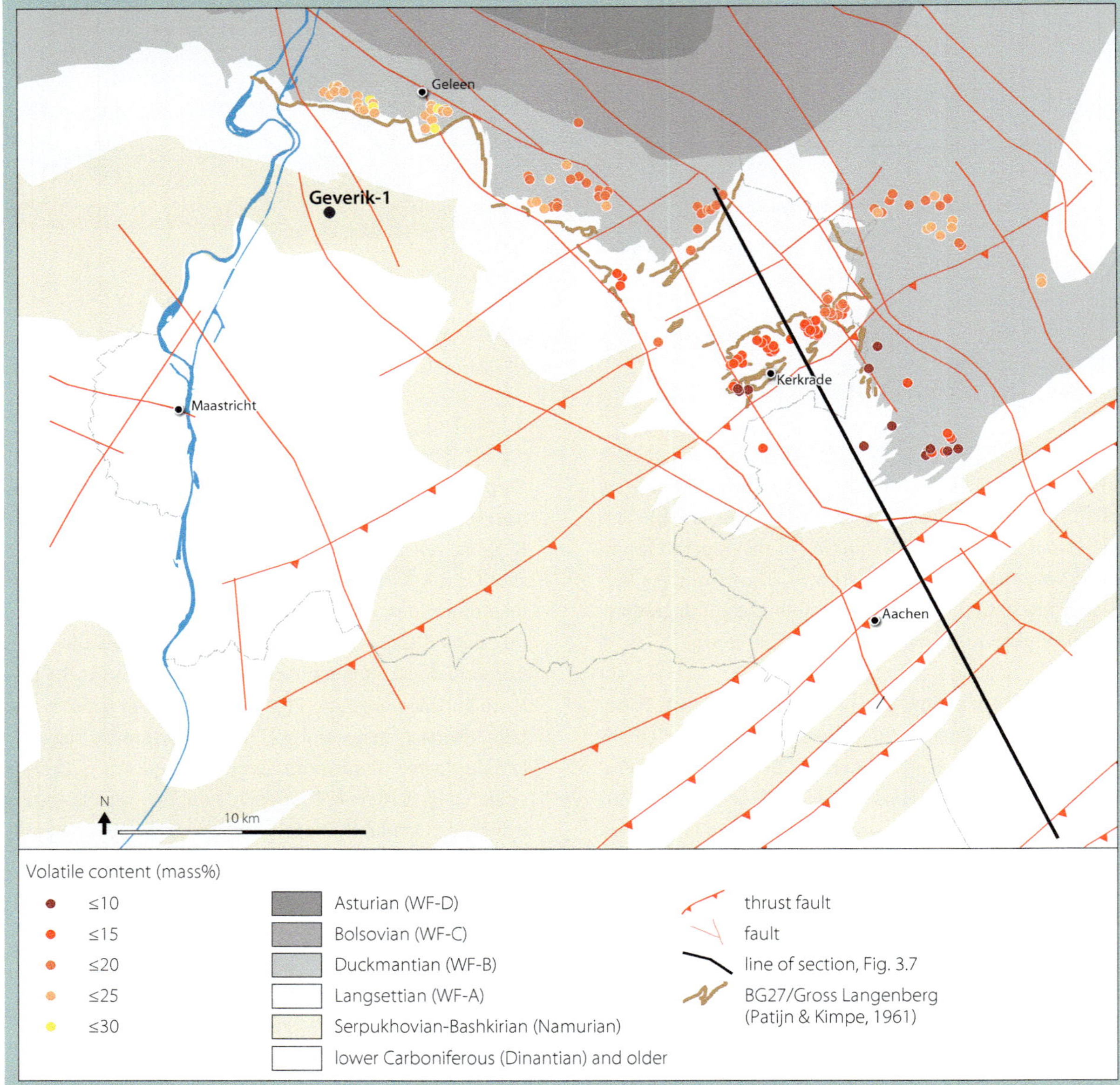

Top Carboniferous subcrop map of the South Limburg mining district (modified after Drozdzewski et al., 1985a and; RGD, 1995). Note the increasing coalification and deformation (folding) towards the southeast. The GB27 coal bed is taken from Patijn & Kimpe (1961). Volatile content from Nunna (1960).

observations on 2D seismic and well data of southward thickening successions of the upper Flénu and Neeroeteren formations (upper Bolsovian to Asturian).

These observations can be explained by rapid subsidence in a foreland basin due to loading by Variscan thrust sheets approaching from the southeast during the late Bolsovian and subsequent infill by molasse sediments (Fig. 3.6). Already during the late Bolsovian and early Asturian, older molasse sediments were eroded and redeposited further into the basin (Neeroeteren, Hellevoetsluis and Tubbergen formations). Simultaneously, on the northern fringe of the basin, the large base-level drop caused by the Variscan orogeny led to the redeposition of coarse sediments of the former Namurian-Duckmantian river system from the Mid North Sea High towards the south (Hospital Ground Formation). Later, during the Stephanian and early Permian, the remaining molasse deposits and the thrust sheets northwest of the current Faille du Midi/Aachen Thrust were eroded and redeposited to the north (Permian Slochteren Formation).

Along the northern basin margin against the Mid North Sea High, thinning at all stratigraphic levels can be observed (Fig. 3.2). A similar gradual thinning is seen in the Campine Basin (Van Tongeren & Hildenbrand, 2005), where it has been related to the presence of a forebulge (Deckers & Rombaut, 2021; Fig. 3.11). In the southern part of the basin in UK quadrant 53 (directly south of 49), the Langsettian-Duckmantian sequence is truncated below the Asturian sequence against the northern flank of the London Brabant Massif, as has been observed on seismic data (Tubb et al., 1986) and in wells (e.g. well 53/16-1). The erosional event is explained by heat flow-related uplift and associated basin-edge tectonics and is correlated with the 'Symon Unconformity' reported from the English Midlands. In the northern part of the basin at the Cleaverbank Platform, a time-equivalent disconformity has been interpreted from palynological evidence (McLean et al., 2005), but until now no angular unconformity has been observed on 3D seismic data (see Fig. 3.9). Chemostratigraphic studies on the northern margin of the basin in UK quadrant 44 (Pearce et al., 2005, 2010), where coarse-grained late Bolsovian-early Asturian fluvial sediments onlap on early Bolsovian-early Duckmantian sediments with a gradually increasing time gap to the north, support the presence of this disconformity, which is generally referred to as the Base Ketch Unconformity (Besly, 1998).

In the German on- and offshore, a Base Stephanian Unconformity has been frequently reported (e.g. David,

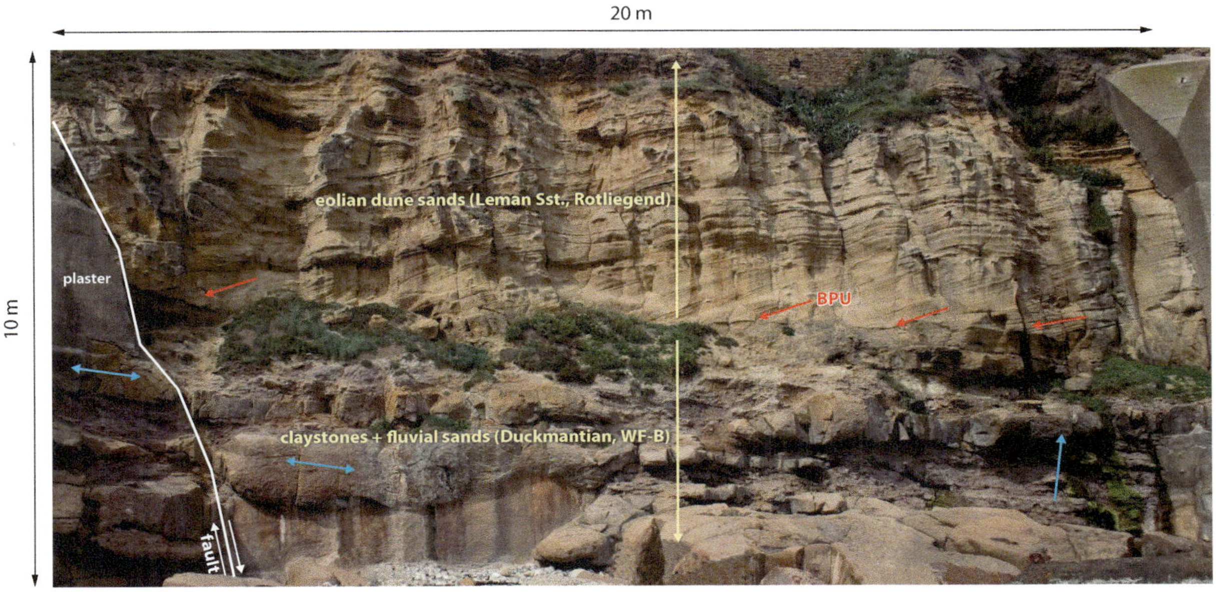

Figure 3.8. Outcrop example of the Base Permian Unconformity (BPU) at Tynemouth Priory, NE England. Here, the upper Permian eolian Leman Sandstone unconformably overlies reddened Duckmantian (Westphalian B) fluvial channel sandstones (blue arrows) and claystones, analogous to the Westoe Coal and Maurits formations. Note the approximately 5 m fault offset. Photo by Fokko van Hulten.

1990; Ziegler, 1990; Fig. 3.6). This unconformity may also be present in the Dutch northern offshore and onshore Twente. Stephanian sediments, dominantly red claystones with minor fluvial channels and carbonate paleosols (caliches) deposited in a well-drained floodplain setting, were encountered in offshore wells F10-02, E18-05, E17-03 and onshore in well De Lutte (Van Waveren et al., 2008). However, due to the seismic transparency of the Asturian-Stephanian interval, the rare preservation of Stephanian sediments below the BPU and in the absence of borehole penetrations with either unambiguous biostratigraphic dating or truncating seismic reflections, no hard evidence can be presented to support the existence of this unconformity. Also in the UK, where Stephanian sediments are encountered in boreholes and outcrops in the Shropshire, Staffordshire and Warwickshire Coalfields (Central England), the existence of an intra-Stephanian unconformity has been reviewed by Besly and Cleal (2021), who concluded that there is no evidence for a hiatus in the Asturian-Stephanian intervals and that other occurrences of such hiatuses should be revised.

The Base Permian Unconformity (Fig. 3.8) is generally the deepest coherent horizon that can be recognized on seismic data in the Dutch on- and offshore. This unconformity resulted from regional uplift of mainly early Permian age and isochore correlation and basin modelling studies with maturity evaluation, suggest that in much of the Netherlands up to 1800 m of upper Carboniferous was eroded during the early to middle Permian (Fig. 3.2b; Roustiau et al., 2022). In adjacent areas in Germany, it is estimated that up to 3000 m of upper Westphalian and Stephanian has been removed (Littke et al., 1994, 2000). From middle or late Permian times onward, upper Carboniferous rocks were steadily buried deeper and deeper. Uplift, with sometimes renewed erosion of upper Carboniferous strata, took place only in limited areas during the Late Jurassic to Early Cretaceous Cimmerian rifting and during the Late Cretaceous to early Cenozoic, Sub-Hercynian and Laramide inversion phases. Especially on the Texel-IJsselmeer High and in the southern Netherlands, the upper Carboniferous and its overburden were extensively eroded (Veld et al., 1996; Fig. 3.11).

Maturity studies on upper Langsettian coals suggest that in southern Limburg, where the Westphalian lies close to the surface, 2900 (around Sittard) to 5400 m (around Kerkrade) of upper Carboniferous may have been removed (Veld et al., 1996). The volatile content of coal decreases from around 24% (medium volatile bituminous coal) around Geleen/Sittard in the northwest to less than 9% (anthracite) around Kerkrade in the southeast (Nunna, 1960; see Textbox).

This amount of erosion would imply that a thickness increase of the upper Carboniferous of several kilometres occurs towards the south in southern Limburg, over a horizontal distance of some 40 km. While this may seem unlikely, a similar rapid increase in apparent thickness has also been observed in the UK (Besly, pers. comm.). Probably, the removed section consisted both of thrust sheets and of their eroded sediments. In addition, hot fluids expelled from Namurian and Upper Devonian claystones

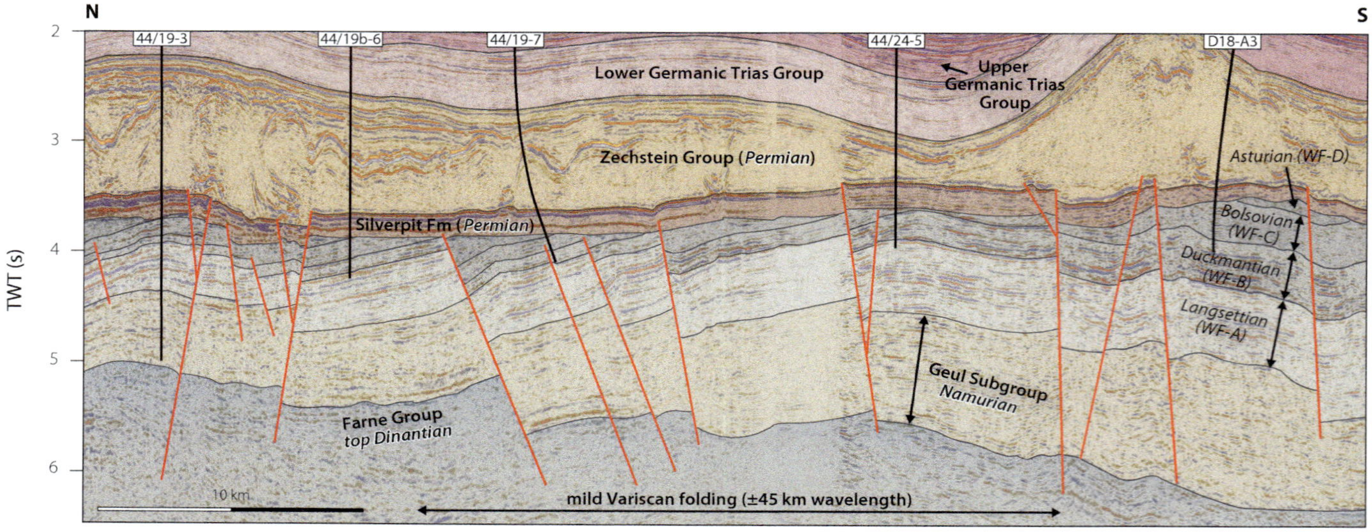

Figure 3.9. Composite seismic time section across the Cleaverbank Platform with features characteristic for the upper Carboniferous. All units show a gradual thickening towards the south. Seismic facies is representative of the depositional setting with, for example, more reflectivity in coal-bearing successions. The structural style is characterized by long-wavelength folds below a clear Base Permian Unconformity and is cut by faults from several generations (reverse faults are relatively young). For fault trends also see Fig. 3.6. Location is shown in Fig. 3.1.

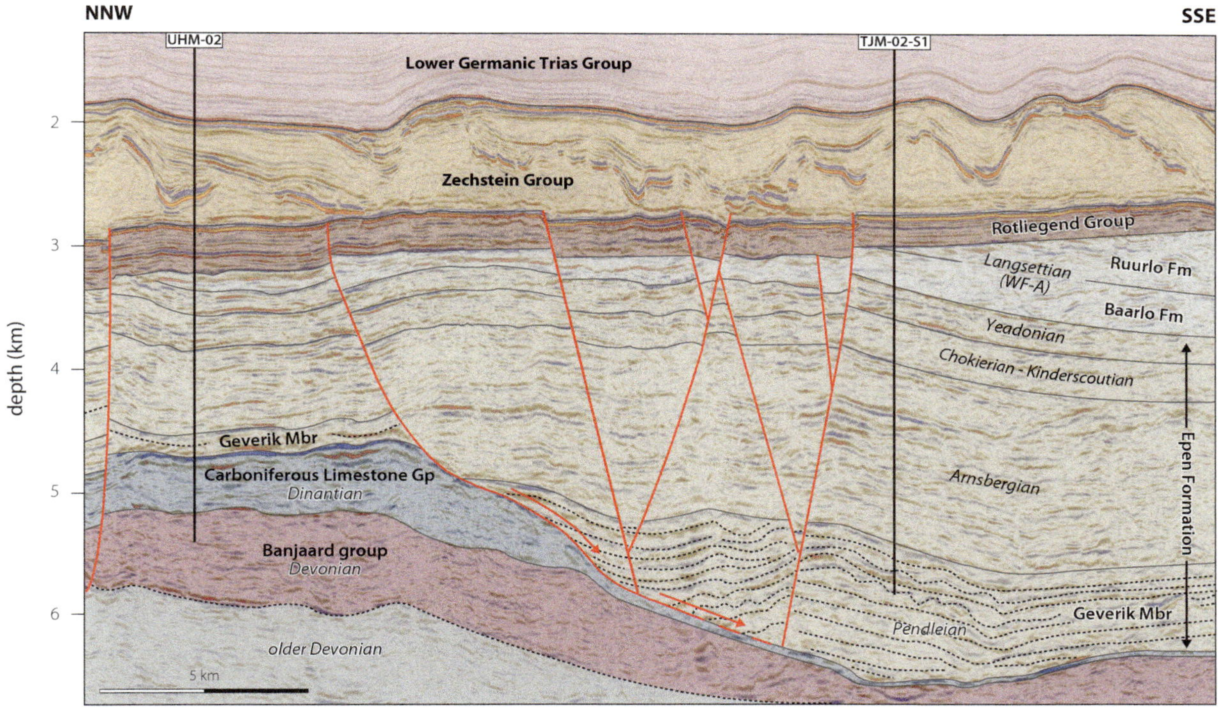

Figure 3.10. *Seismic depth section across the Groningen Platform corresponding to the northeastern part of the cross section of Fig. 3.2. The section shows progressive onlap of a lower Namurian succession against a Dinantian paleo-high, as confirmed by wells Tjuchem-2 (TJM-02) and Uithuizermeeden-2 (UHM-02). Seismic line provided by Marloes Kortekaas (EBN). Location is given in Fig. 3.1.*

due to loading of overthrusted units may have contributed to the observed increase in coalification grade.

The other area in Limburg that was prospective for coal mining, the Central Limburg Peel Horst area (Peel-commissie, 1963), also displayed an anomalously high coal rank of 4.4-11% volatile matter, despite the absence of any Variscan deformation (for example, in the Beatrix Mine east of Roermond). This is explained by the occurrence of the Erkelenz Laccolith (Erren & Bredewout, 1991; Sissingh, 2004), also probably of late Asturian-early Permian age (Teichmüller & Teichmüller, 1971).

Igneous activity

Igneous rocks are fairly common in the upper Carboniferous of the Netherlands. The first well in the country to recover oil, Corle-1, penetrated a dolerite intrusive within the Langsettian. Thiadens (1963) described seven further occurrences of igneous rock within the upper Carboniferous, and many more have been found since (Sissingh, 2004; Van Bergen et al., 2025, this volume).

Understanding the occurrence and nature of these rocks can help to improve correlation based on, for example, tuff layers (Burger, 1982) and to explain both seismic anomalies (e.g. well Winterswijk-1), and high levels of CO_2 en N_2 in gas discoveries (e.g. Kombrink et al., 2009).

Intrusives

The compilations of Sissingh (2004) and Van Bergen et al. (2025, this volume) show that intrusive rocks are rather common in the upper Carboniferous in the east of the country and in the northern offshore area. Intrusions for which radiometric age dating is available indicates a possible late Carboniferous age in wells Dwingeloo-2 and De Wijk-7, both in the province of Drenthe. These intrusions have dimensions covering several square kilometres as observed in 3D seismic data and are interpreted as relatively shallow events.

The most prominent example of an intrusive within upper Carboniferous sediments is the 72 m thick dolerite in well Winterswijk-1 between 4078-4149 m (Fig. 3.12). This dolerite intruded the lower Namurian and has a K-Ar age dating of 218 +/- 6 Ma (i.e. Late Triassic). This age is regarded unreliable, however and has possibly been reset from an older age. It causes a strong seismic amplitude anomaly, which was mistakenly interpreted as the top of the Dinantian limestone when the well was proposed (Van Bergen et al., 2025, this volume; Fig. 11.5a). Similar low angle dykes and sills have intruded into Duckmantian to Asturian rocks and were subsequently eroded by the Base Permian/Saalian unconformities. These intrusions of Stephanian to early Permian age (between 307 and 275

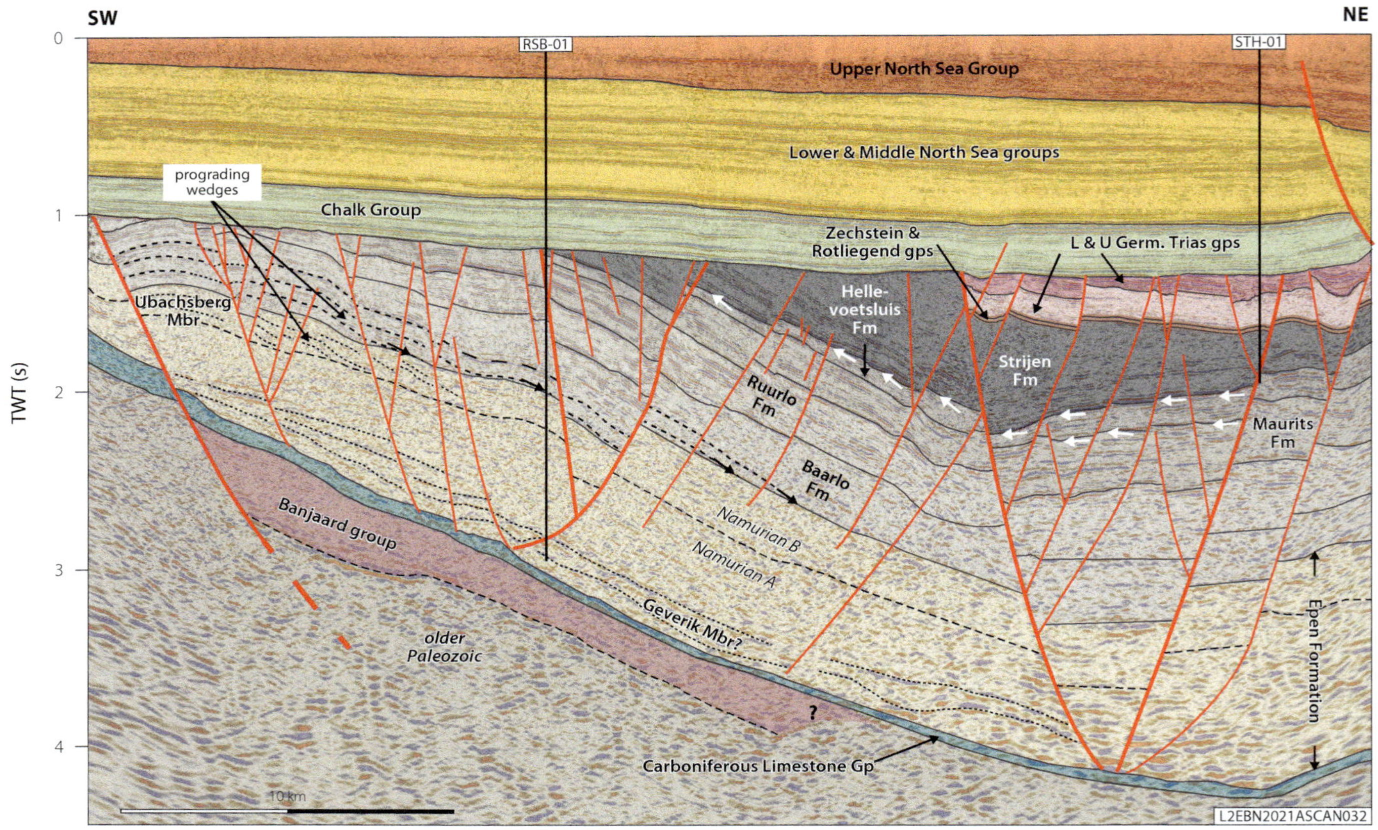

Figure 3.11. Seismic time section across the southwestern part of the Netherlands with features characteristic for the upper Carboniferous in the Campine Basin north of the London-Brabant Massif. A gradual thinning of both Namurian and Westphalian deposits towards the London-Brabant Massif in the south can be observed. An unconformity marks the change from the southward onlapping Namurian A succession to the northward prograding Namurian B succession (Ubachsberg Member of the Epen Formation). Reflector patterns indicative of prograding wedges are also present in the Langsettian successions (Baarlo Formation). The coal-bearing successions (predominantly the Ruurlo and Maurits formations) are characterized by high seismic reflectivity. Subtle unconformities (indicated by white arrows) occur at the base of the Bolsovian (Hellevoetsluis Formation) and at the base of the Asturian (Strijen Formation). The northern part of the line shows a clear Base Permian Unconformity. Both the Paleozoic and Mesozoic successions are unconformably overlain by the upper Cretaceous Chalk Group. Location is given in Fig. 3.1.

Ma, also see Bouroullec and Geel, this volume) have been identified as high impedance intervals on 2D and 3D seismic data and were encountered in several offshore wells, e.g. in blocks E18 and K02, and also in the Schill Grund Platform area (M01, G18) north of Groningen.

Extrusives

Only a single occurrence of unequivocally *extrusive* rock is known from the Carboniferous of the Netherlands onshore. In the cored interval of well Steenwijkerwold-1 (Overijssel), an 8 m thick basaltic layer is present, with soft-sediment deformation in the underlying sediments and evidence for weathering and subsequent submergence at its top (Fig. 3.20). The overlying sediments contain a well-preserved palynological assemblage, indicating a late Langsettian age of the basalt, putting the K-Ar age of 291±8 Ma in doubt. The shallow intrusives in Dwingelo-2 (and possibly that in Winterswijk-1) and the likely extrusives of Nagele-1 (Flevoland) and Haarle-1 (Overijssel) are probably related to the same volcanic phase as the extrusives in Steenwijkerwold-1, which may be linked to Variscan transtension along the proto-Texel-IJsselmeer High. This would suggest that the boundary faults of this high pre-date by far the Cimmerian tectonic phase in which they were most active. The 'Kaolin-Kohlentonsteine' described from the Langsettian-Bolsovian of the Ruhr area in Germany (e.g. Burger, 1982) and of the Netherlands (RGD, 1988) may be correlated with the same volcanic phase. Active volcanism during the Langsettian-Bolsovian has also been described from the UK, for example by Kirton (1984).

Intra-late Carboniferous extrusives also occur in the northern offshore, for example in wells A17-01 and M03-

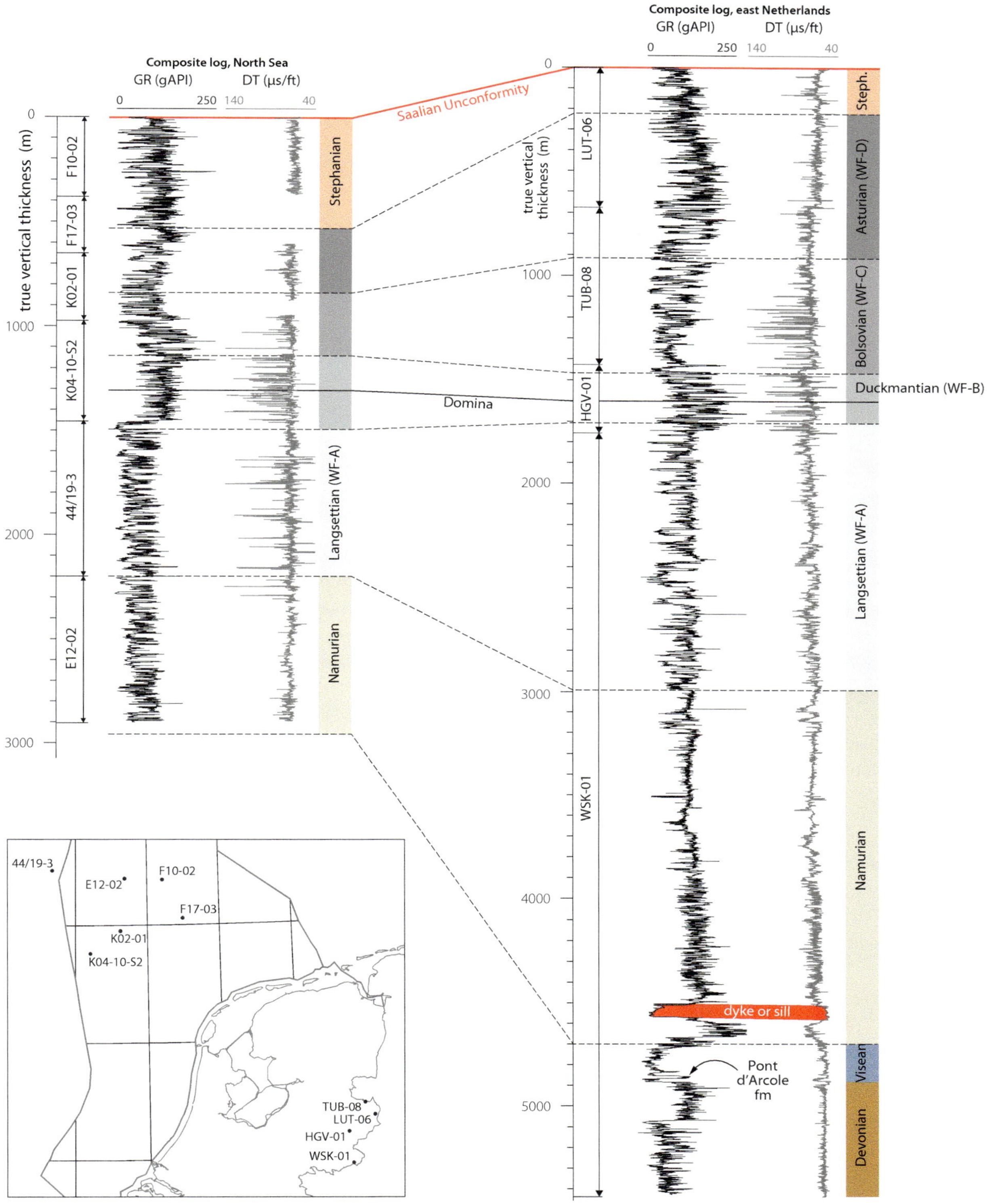

01 (Van Bergen et al., 2025, this volume). Several additional wells contain extrusives near the Carboniferous-Permian boundary, e.g. B10-02-S1, B17-04, E18-05, F04-02-A. In addition, some wells (e.g. A14-01, E12-03, E18-02, K02-02) contain igneous rocks within the Carboniferous for which an extrusive origin cannot yet be excluded. These data match observations of active Permo-Carboniferous

Figure 3.12. *Composite lithostratigraphic columns through the upper Carboniferous of the Cleaverbank Platform (left) and northeast Netherlands onshore (right), including composites of gamma-ray (GR) and sonic (DT) logs. The correlation clearly shows the thickening from the northern offshore towards the southeast where total upper Carboniferous thickness may exceed 5000 m.*

volcanism in the nearby UK, Germany, Denmark and Norway (Heeremans et al., 2004; Bouroullec & Geel, 2025, this volume).

Stratigraphy

Chrono- and lithostratigraphy

The northwest European Carboniferous has traditionally been subdivided into a carbonate-dominated lower Carboniferous and a siliciclastic upper Carboniferous succession. The actual lower-upper Carboniferous, i.e. Mississippian-Pennsylvanian chronostratigraphic boundary occurs at a stratigraphically higher level, close to the boundary between Alportian (Namurian A) and Kinderscoutian (Namurian B; Fig. 3.10; Riley et al., 1994). Both chronostratigraphic subdivision of and correlation within the upper Carboniferous are based on macroflora, palynology (pollen and spores), 'marine bands' (calcareous claystone layers with marine, brackish or fresh-water fauna), weathered volcanic ash layers ('Tonsteine') and radiometric ages of the ash layers (Menning et al., 2006). Most biostratigraphy is based on palynology, which has provided an excellent framework in the Namurian to early Bolsovian, but which becomes more difficult in the red bed intervals from the late Bolsovian upwards (McLean et al., 2005). Several other techniques have been applied to achieve a more detailed correlation in intervals barren of pollen, such as chemostratigraphy (Pearce et al., 2005) or the biostratigraphy of conchostracans, small lacustrine bivalved crustaceans that evolved quickly in restrictive environments and whose eggs spread easily through eolian and fluvial transportation (Schneider & Scholze, 2018).

The lithostratigraphic subdivision of the upper Carboniferous Limburg Group into subgroups is based on the recognition of four regressive cycles (Van Adrichem Boogaert & Kouwe, 1995). Based on a dominant or typical lithofacies characteristic, these subgroups are regionally subdivided into formations (Fig. 3.4), the boundaries of which may be diachronous. The lithology of these units ranges from claystone, siltstone and sandstone up to conglomerate. The units were deposited in a lacustrine, occasionally marine basin, bordered by deltas which prograded into a basin with alluvial plains and fans. Peat formed from the Yeadonian until the early Asturian in deltaic and floodplain environments. A basin-wide peak in peat development occurred during the late Duckmantian-early Bolsovian, with the deposition of the Maurits and Westoe Coal formations.

Regional correlation

The stratigraphy, correlation and nomenclature of the upper Carboniferous have long been matters for debate. As a result of successive conferences on Carboniferous stratigraphy in the Dutch coal-mining town of Heerlen between 1927 and 1955, some common ground was reached with the definition of three northwest European series, subdivided into stages that can be recognized basin-wide. These

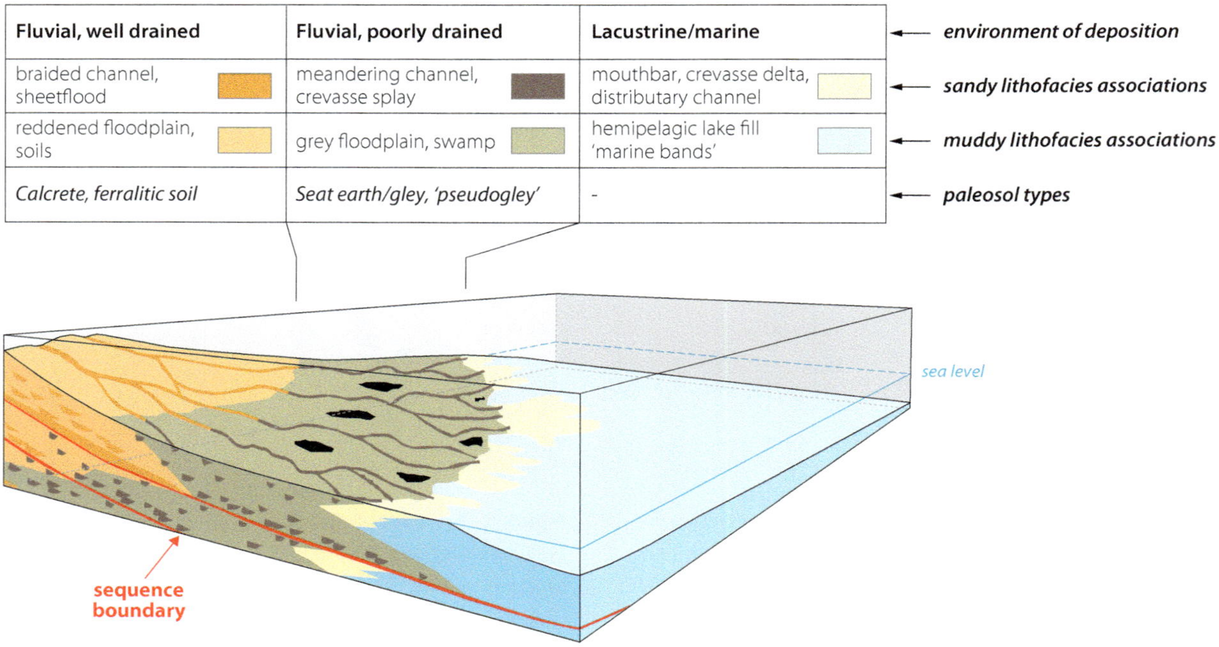

Figure 3.13. Schematic distribution of depositional facies in the upper Carboniferous of the Netherlands. Black spots indicate peat. Modified from Van Buggenum and Den Hartog Jager (2007).

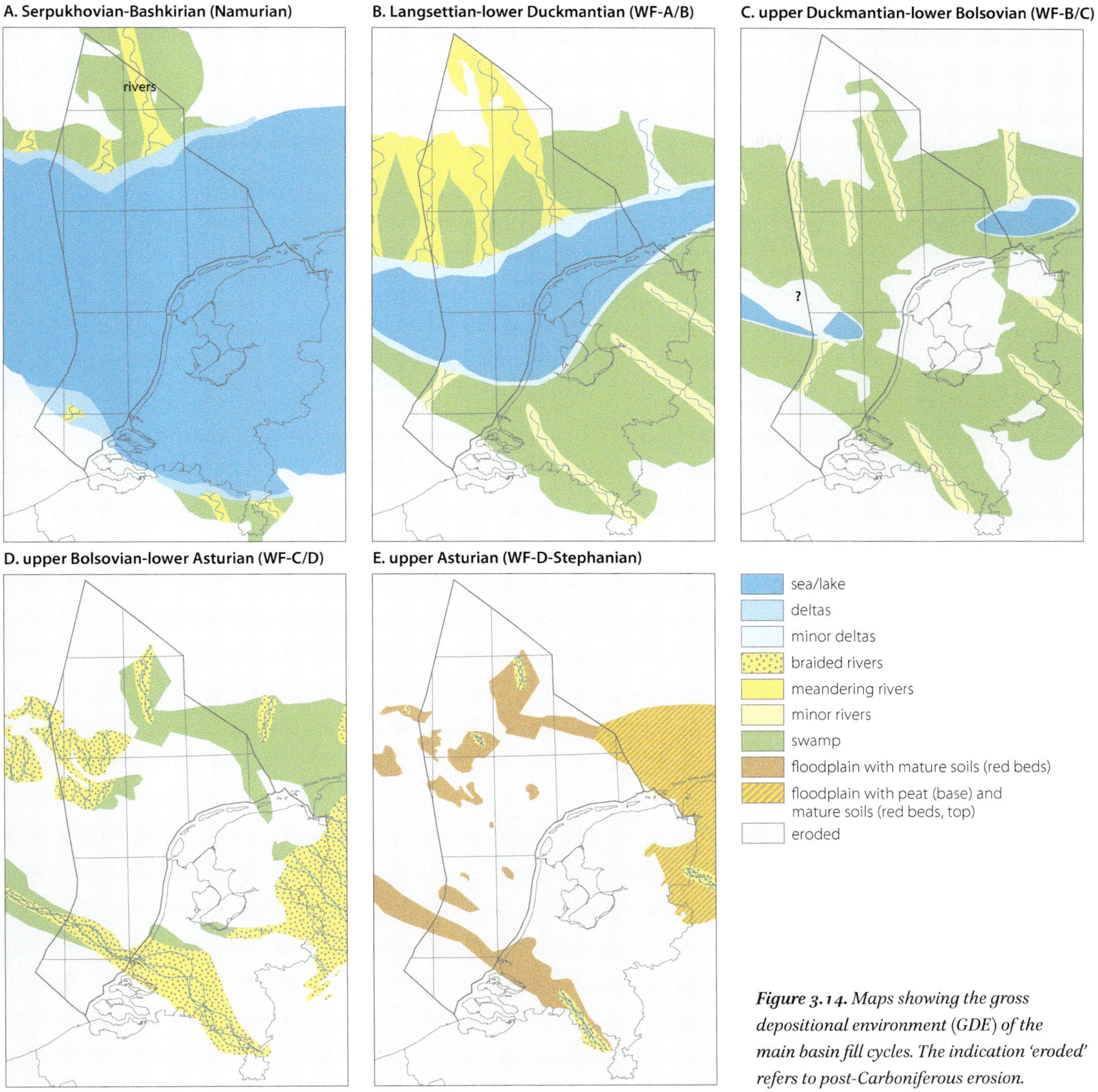

Figure 3.14. Maps showing the gross depositional environment (GDE) of the main basin fill cycles. The indication 'eroded' refers to post-Carboniferous erosion.

are from base to top: the Namurian A, B and C, now divided into seven stages: Pendleian, Arnsbergian, Chokierian, Alportian, Kinderscoutian, Marsdenian and Yeadonian; the Westphalian A, B, C and D, now renamed into Langsettian, Duckmantian, Bolsovian and Asturian; and the Stephanian A, B and C (Cameron, 1993; Menning et al., 2006).

Meanwhile, the International Commission on Stratigraphy (ICS) has adopted Russian series names. Figure 3.4 shows a comparison between local and global series names, using the official ICS absolute timescale (Cohen et al., 2013) which dates the upper Carboniferous from 331-299 Ma. The figure also shows representative stratigraphic columns and formation names for the Netherlands and surrounding areas.

Seismic character

The seismic character of the upper Carboniferous is very much associated with the presence or absence of coal beds within the sequence. The large impedance contrasts of the coal-to-shale and coal-to-sandstone interfaces generate relatively high amplitude, continuous seismic reflections, which are often grouped into easily recognisable 'reflective packages' (Evans et al., 1992; Quirk, 1993; Boscolo Gallo et al., 2020). The relatively low impedance contrasts between claystones and sandstones create low amplitude seismic reflections in overall rather 'transparent' packages. Fig. 3.9 displays an example along the UK/Dutch offshore median line showing high reflectivity coal-rich intervals in the Langsettian Klaverbank Formation (Caister Coal Formation) and the Duckmantian Maurits Formation (Westoe Coal Formation), separated by a sandstone-dominated transparent package of the Botney Member (late Langsettian-earliest Duckmantian). Due to the near absence of coals, the underlying Millstone Grit Formation (Namurian) and overlying Hospital Ground (Lower Ketch) and Step Graben formations (Upper Ketch and Boulton Formation) of upper Bolsovian to Asturian age are also relatively transparent. Similar reflective patterns in contemporaneous stratigraphic sequences are observed in the Campine Basin (Fig. 3.11).

Sedimentary cycles and sequence stratigraphy

The entire Limburg Group can be seen as a single regressive cycle. In sequence-stratigraphic terminology, this is a first-order cycle, associated with the subsidence and filling of the NW European late Carboniferous basin. At a smaller scale five second-order megacycles can be recognized which, to a certain extent, correspond to the Dutch lithostratigraphic subgroups (Figs 3.4 and 3.12). The thickness of these megacycles ranges from 100 m along the basin edges up to 2500 m close to the foredeep in the southeast (Fig. 3.17).

Following deposition of the Dinantian carbonates (Vis et al., 2025, this volume), the first-order cycle started with a locally thick succession of alternating marine and lacustrine shales with some turbiditic sands that developed into a succession of deltaic sediments toward the top of the Namurian. Thick successions occurred especially in local syn-sedimentary rift basins, for example on the Groningen Platform (Fig. 3.10). During the Westphalian, poorly drained fluvial and floodplain sediments, deposited around a lake of decreasing dimensions that mainly existed during periods of high baselevel, were succeeded by well-drained fluvial, alluvial and floodplain sediments (Figs 3.13-3.16) during low baselevel intervals. The cycle ends with a dominance of well-drained alluvial and ephemeral lake sediments of the late Asturian and Stephanian.

The Namurian megacycle of the Geul Subgroup can be recognized on both the Dutch and British sides of the basin. The observed onlap of thick packages of largely marine and lacustrine sediments onto the London-Brabant Massif in the south, the Elbow Spit High (part of the Mid North Sea High) in the north and intra basinal highs such as the Groningen Platform (Fig. 3.10) suggests that this cycle started as a result of rapid subsidence of the basin. In the northwest, the basin was subsequently filled by fluvial clastics sourced from the Caledonian mountains

Figure 3.15. Outcrop showing typical succession of fluvial environments during Bolsovian-Asturian time. Woitzel Quarry near Ibbenbüren, Germany.

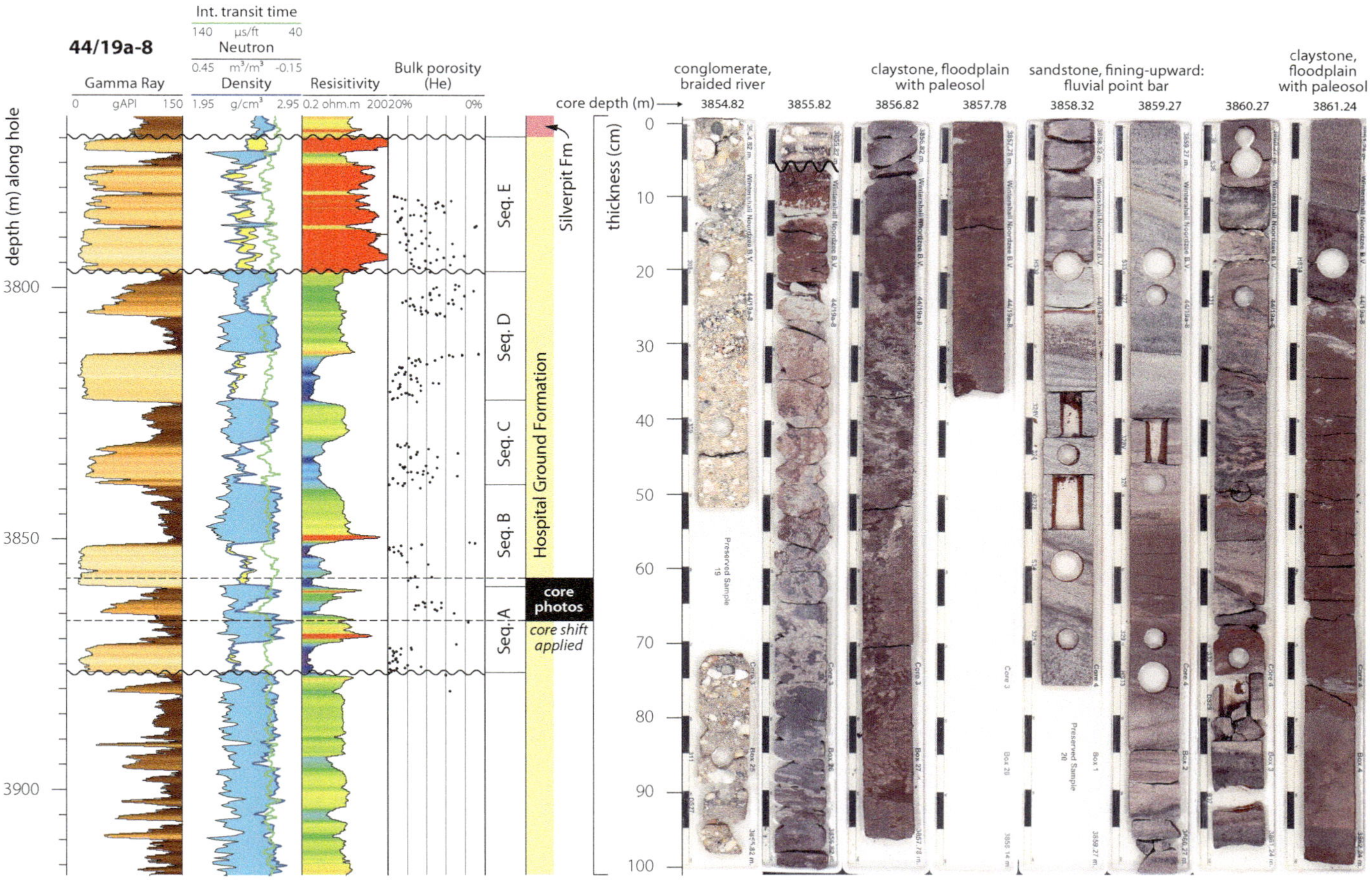

Figure 3.16. Core photographs of well 44/19a-08 on the Cleaverbank Platform showing a typical fluvial Bolsovian-Asturian succession. Cored interval is 23 m. Despite a distance of >400 km, this succession is very similar to that encountered in the Woitzel Quarry (see Fig. 3.15). Core photos from UK National Data Repository.

to the north (the Pennine River; Cliff et al., 1991). At the same time in the southeast, smaller, more distal fluvial systems developed, sourced from the rising and northward migrating Variscan Mountains (Morton et al., 2005). Proximal deposits of the southern rivers during the Namurian-Duckmantian were situated in the area now covered by the thrusts and nappes of the Rheno-Hercynian Zone (Fig. 3.7) and were partly cannibalized and redeposited in Bolsovian times. These southern sources provided the sediment for the Tubbergen Formation in NE Netherlands, the Hellevoetsluis Formation in SW Netherlands, the Neeroeteren Formation in the Campine Basin and the Osnabrück Formation in Germany.

In the northern part of the basin, three megacycles (WF-A, WF-B and WF-C in Fig. 3.12) can be recognized in the Langsettian to Bolsovian interval. They coincide more or less, with the main part of the Dutch Klaverbank Formation (including the Botney Member) and overlying Maurits Formation. The driving force for these three megacycles is thought to be the continuing development of the major southward prograding fluvial/deltaic system situated close to the UK/NL Median Line (Collinson, 2005) responding to large scale base level cycles.

The Hunze megacycle of Asturian and lower Stephanian age can be recognized on both sides of the basin. Contrary to the other cycles, it contains more sandstone at its base and more fine-grained sediments toward its top. The base coincides with the bases of the Tubbergen Formation in the east of the onshore Netherlands and of the Hospital Ground Formation (equivalent to the Lower Ketch Formation in the UK sector) in the western offshore area. The sudden increase in sand deposition, observed on both sides of the basin, is likely related to a tectonic pulse in the Variscan hinterland ('Asturian phase', Fig. 3.7; Jones & Glover, 2005). The upper part of the megacycle displays upwardly increasing evidence for dry climate conditions including for example, calcrete paleosols (well F10-02) that indicate a change in the climate as the basin drifted towards the north.

Higher-order, i.e. smaller-scale, sedimentary cycles within the megacycles may be associated with Milankovitch cyclicity. During the late Carboniferous, these cycles record waxing and waning of the Gondwana icesheet that led to global sea level fluctuations of sometimes more than 60 m. The accompanying climate variations changed the intensity of water run-off and erosion in the hinterland (Pointon et al., 2012). Such fourth- or fifth order glacio-eustatic cycles can be recognized in the Langsettian-Asturian of the offshore area and the Bolsovian-Stephanian onshore (Fig. 3.16; Huis in 't Veld et al., 2020; Baars et al., 2023). In terms of thickness, the eccentricity periodicity in the underlying Langsettian and Namurian sections has been estimated to reflect 40 and 100 m intervals. This estimate agrees with sequence-stratigraphic studies in the upper Carboniferous of the Ruhr Basin in Germany, in which approximately 75 m thick packages of alternating sand and shale, are tentatively correlated with fifth-order, short-eccentricity, (para) sequences (Süss et al., 2000).

In most of the Dutch on- and offshore, borehole penetrations of the Namurian A and B low-stand show dominantly shale deposits with some intercalated turbidites. During the Yeadonian, the turbiditic sequence passes into one dominated by prograding lowstand deltaic wedges (Fig. 3.11). Around the onset of the Westphalian, the depositional style in the Dutch on- and offshore gradually changed from a 'shelf model' in a predominantly low-stand setting into a 'ramp model' in a transgressive and high-stand setting.

Depositional response on base level cyclicity

The cyclicity observed in Westphalian sediments can be related to base level fluctuations (Quirk, 1997; Huis in 't Veld et al., 2020; Baars et al., 2023). From the Namurian until the middle Bolsovian, maximum flooding surfaces are commonly recognized as 'marine bands' (MB). These bands represent thin argillaceous layers rich in marine and sometimes lacustrine faunas that can form excellent marker beds. Examples include the Aegir MB at the base of the Bolsovian (Fig. 3.4; Jongmans, 1925; Thiadens, 1963; Van de Laar & Fermont, 1990) or the Catharina MB (also known as the Vanderbeckei MB in the UK and Quaregnon MB in Belgium) that forms the boundary between the Langsettian and Duckmantian. Some of these thin marker bands can be correlated over distances of several hundreds of kilometres, suggesting an absence of topography during at least late Namurian to early Bolsovian times.

During deposition of a 'high-stand systems tract', the relative rise of the base-level slows down. The first products of upstream erosion on the alluvial plain will be deposited downstream and prograde over the same plain, producing coarser sand layers that tend to thicken toward the top of the systems tract. Near the lake margin, small deltas may develop, which gradually prograde when the erosion intensifies. Where, in a fluvial setting, the net sedimentation of fine-grained material exceeds subsequent erosion, a 'type II sequence boundary' may develop. A condensed section with paleosols developed will mark this boundary (Van Buggenum & Den Hartog Jager, 2007; Huis in 't Veld et al., 2020).

Sedimentary development

The facies distribution maps of the main upper Carboniferous chronostratigraphic subdivisions in the areas not affected by post-Carboniferous erosion show a gradual transition from an initial largely marine and lacustrine setting to a predominantly fluvial red-bed environment (Fig. 3.14). Isochore maps of the same subdivisions indicate the considerable post-compaction thicknesses that are involved (Fig. 3.17).

Namurian

During the Namurian, a thick succession of siliciclastic sediments was deposited especially in the southeast of the basin, controlled by rapid subsidence in a foreland basin setting. Turbidites have been described from the Ruhr area and the Rhenish Massif (Ricken et al., 2000). Initially, a marine environment prevailed, but gradually it gave way to fresh-water conditions.

Parts of the basin in the south and southwest were sheltered from sand deposition by the London-Brabant Massif. In these areas, an organic-rich black shale occurs in the Pendleian (lowermost Namurian). The shale forms the lower part of the bituminous Geverik Member and has been classified as a type II source rock with a total organic carbon content of 8% (Van Balen et al., 2000). This Pendleian source rock, which developed in an anoxic marine environment, is thought to be present throughout the central and deeper parts of the basin. In northern England, the laterally equivalent Bowland Shale Formation has generated oil, for example in the East Midlands oil fields (Guion, 2008; Andrews, 2013). It may also be an under-explored source rock for plays in the North Sea Basin (Gerling et al., 1999; Besly, 2018).

In the northern part of the basin, the fluvio-deltaic system which in Dinantian times shed its sediments southward from the Caledonian Highlands, continued to be active during the Namurian (Fig. 3.14a; Collinson, 2005). In the Dutch offshore, the distal equivalent of this system is represented by lacustrine to marginal marine claystones of the Epen Formation. This sequence onlaps the Elbow Spit High (in the north) and the London-Brabant Massif in the south as observed in wells E02-01/E06-01 and S02-02/BHG-01 respectively. Its estimated paleo-water depth is hundreds of metres in the deeper parts of the basin, but much less along the basin margins (Kombrink, 2008).

During the middle part of the Namurian (Kinderscou-

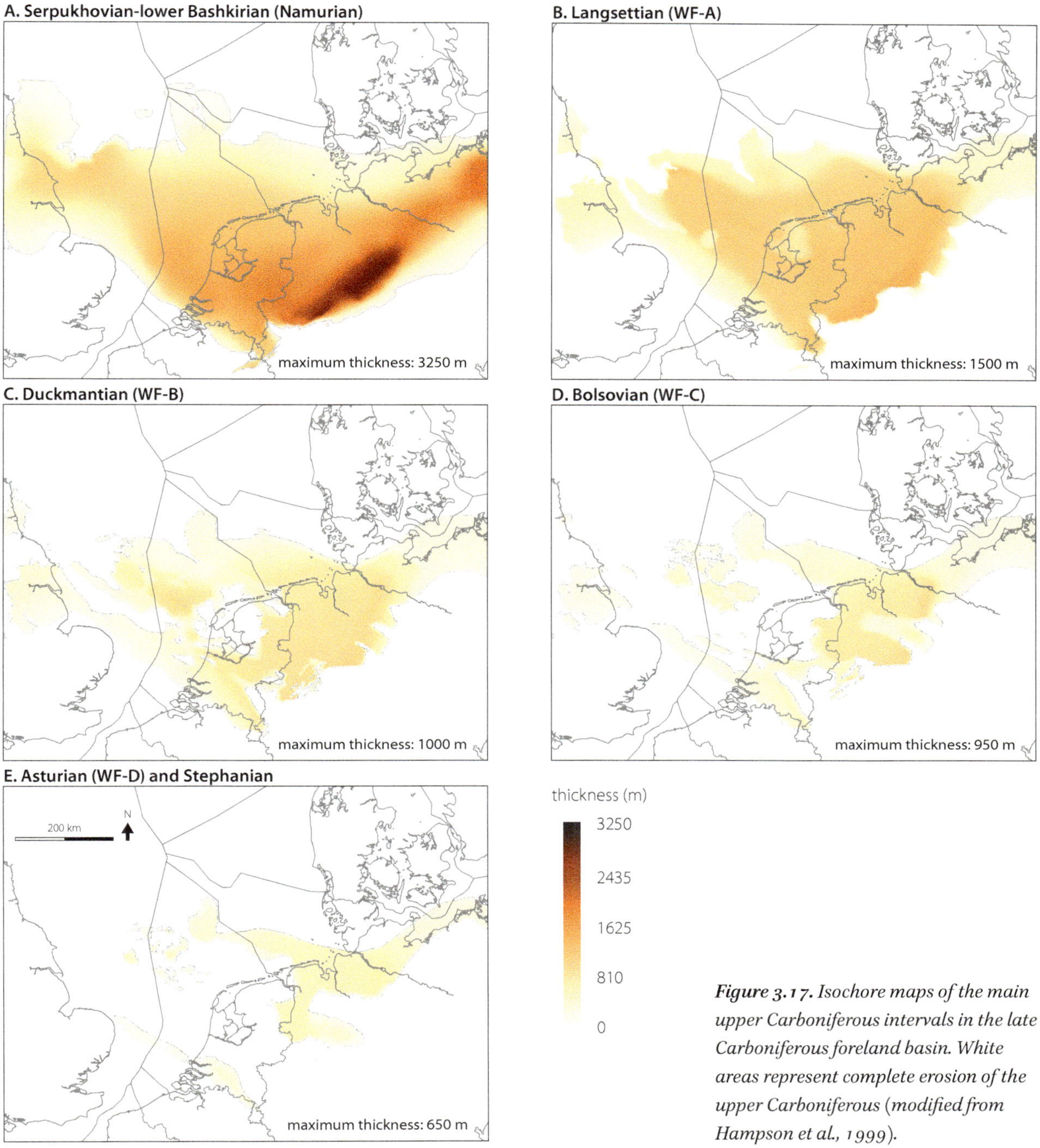

Figure 3.17. *Isochore maps of the main upper Carboniferous intervals in the late Carboniferous foreland basin. White areas represent complete erosion of the upper Carboniferous (modified from Hampson et al., 1999).*

tian-Marsdenian), the northern fluvio-deltaic systems continued to prograde towards the south where they deposited delta-front sediments of the Millstone Grit Formation (O'Mara et al., 2003; Collinson, 2005). During the Yeadonian, a similar system started to develop in the gradually shallowing southern part of the basin where delta-front sediments of the Ubachsberg Member (Epen Formation.) were deposited (Fig. 3.11).

In the onshore Netherlands, thick sections of Namurian shale with only minor amounts of sandstone were encountered in several wells, for example in Winterswijk-1 (Fig. 3.12), Tjuchem-2 (Fig. 3.10) and Nagele-1. Further south, more sandstone appears, including the Ubachsberg Member, which is present on the Peel Horst (Peelcommissie, 1963) and in the Campine Basin (Langenaeker & Dusar, 1992; Deckers & Rombaut, 2021).

Towards the end of the Namurian, sedimentation caught up with basin subsidence and much of the lacustrine shallow-water basin was transformed into a basin in which delta top- and fluvial conditions prevailed. Intercalated marine bands indicate that marine incursions continued to occur until mid-Bolsovian times. The most widespread of them form the boundaries of the upper Carboniferous stages (Fig. 3.4) and have been important correlation markers in coal exploration and mapping (see Textbox).

In the UK offshore quadrant 43, the Millstone Grit Formation forms the reservoir in the Trent and Cavendish gas fields, where it is partly developed in an incised valley-fill setting overlain by extensive sheet-like delta top sands, reworked during the early transgressive phase (Collinson, 2005). In the Dutch sector, similar sands of the delta front form the reservoir of two gas discoveries in the Yeadonian to lower Langsettian section of the Millstone Grit and Klaverbank Formation in block E12.

Westphalian

During the Westphalian, a high sedimentation rate balanced the continuing basin subsidence rate. The lake area in the basin centre gradually decreased and along the margins, early Langsettian deltaic to shallow-marine sedimentation gave way to deposition of extensive fluvial systems and adjacent marshlands during late Langsettian-Bolsovian time (Fig. 3.14). During the Asturian, the basin became more continental due to inversion tectonics related to the encroaching Variscan front. This was possibly combined with a climate shift from tropical to semi-arid as the wet, swamp-dominated environment was replaced by a dryer, floodplain-dominated setting with well drained fluvial, ephemeral-flood and fanglomerate systems and a generally lower ground water level (Van Buggenum & Den Hartog Jager, 2007; Huis in 't Veld et al., 2020).

During the *Langsettian*, the southern half of the basin was filled by distal delta systems prograding from the southeast (Langenaeker, 2000) that can be recognized as downlapping and prograding wedges in seismic data (Fig. 3.11). At the same time, the northern half of the basin was filled from the north by a similar but much larger deltaic system (Fig. 3.14b), the successor of the Pennine River System. The Langsettian consists of a series of coarsening upward cycles, the deposition of which started with a brief lacustrine transgression or a marine incursion and was followed by a period of delta progradation. Regional paleogeographic maps suggest that the marine incursions came from the east, where the eastern extension of the Variscan foredeep probably connected with the Ural Ocean (Fig. 3.5; Ziegler, 1990; DeCelles & Giles, 1996; Rippon, 1996). In the upper part of a cycle, distributary channel sandstones are intercalated with fines from the delta plain and with coal seams which originated in the coastal and delta-plain swamps. Langsettian channel sandstones form the main reservoir for the Chiswick gas field in UK block 49/4 (Smit, 2020). In the Dutch sector it has only been recorded as a secondary reservoir below the lower Duckmantian Murdoch Sandstone.

During the *Duckmantian*, fluvial sedimentation became more dominant (Fig. 3.14c). Around the shrunken central lake, a succession of coastal and floodplain fines developed with frequent intercalations of peat and occasional intercalations of fine to coarse-grained and argillaceous sands. Meandering river systems deposited fining and coarsening upward sequences up to 90 m thick. In the northern offshore, a local but distinct fluvial system developed during the late Langsettian to early Duckmantian. It gave rise to the up to 50 m thick coarse-grained stacked channel sandstones of the upper part of the Botney Member. These sandstones form the reservoirs in gas fields and gas discoveries in blocks D12, D15, E10, J03, K02 and K05 (Fig. 3.3; Schroot et al., 1997). In the nearby UK quadrant 44, similar fluvial sandstones belong to a more westerly fairway that constitute the reservoirs in the Tyne, McAdam, Caister and Murdoch gas fields. This unit is informally named Murdoch Sandstone and forms the uppermost part of the Caister Coal Formation (Huis in 't Veld et al., 2020).

During the late Duckmantian and early Bolsovian, the Carboniferous basin was dominated by floodplains and swamps with extensive peat development (Maurits and Westoe Coal formations) and relatively small meandering river systems. This thick coal- and clay rich interval has a very low sand content (low N/G) and therefore provides an excellent top seal for underlying early Duckmantian and late Langsettian reservoirs and seat seal for the overlying Bolsovian reservoirs in the D-E blocks and UK quadrant 44 (e.g. O'Mara et al., 2003; Huis in 't Veld et al., 2020). Also, in areas where these reservoirs are absent due to erosion, the Maurits/Westoe Coal formations may serve as a reliable seat- and or side-seal for overlying or juxtaposed Rotliegend reservoirs.

Provenance studies in offshore quadrant 44 show that during sea-level low stands mixing of sediment sources occurred and fluvial systems from the Variscan front interfingered with fluvial systems from the Caledonian mountains and Baltic Shield in the north (Morton et al., 2005). It appears that sediment interfingering was much more limited prior to the Bolsovian. This is shown by provenance studies of Namurian sediments in the Pennines, which appear to have been sourced mainly through the Pennine River system running from the Mid North Sea High that transported sediments possibly from East Greenland (Cliff et al., 1991). During high stands, the floodplain became periodically flooded by marine incursions, of which the Domina (Maltby) and Aegir (Duckmantian/Bolsovian boundary) marine bands are the most widespread (e.g. Kombrink et al., 2008). The last marine band recognized in the basin is the Cambriense MB in the Mid-Bolsovian (Fig. 3.4).

During the late *Bolsovian*, a pronounced fluvial system developed in the eastern Netherlands, forming the Tubbergen Formation (Figs 3.12 and 3.13). Stacked channel

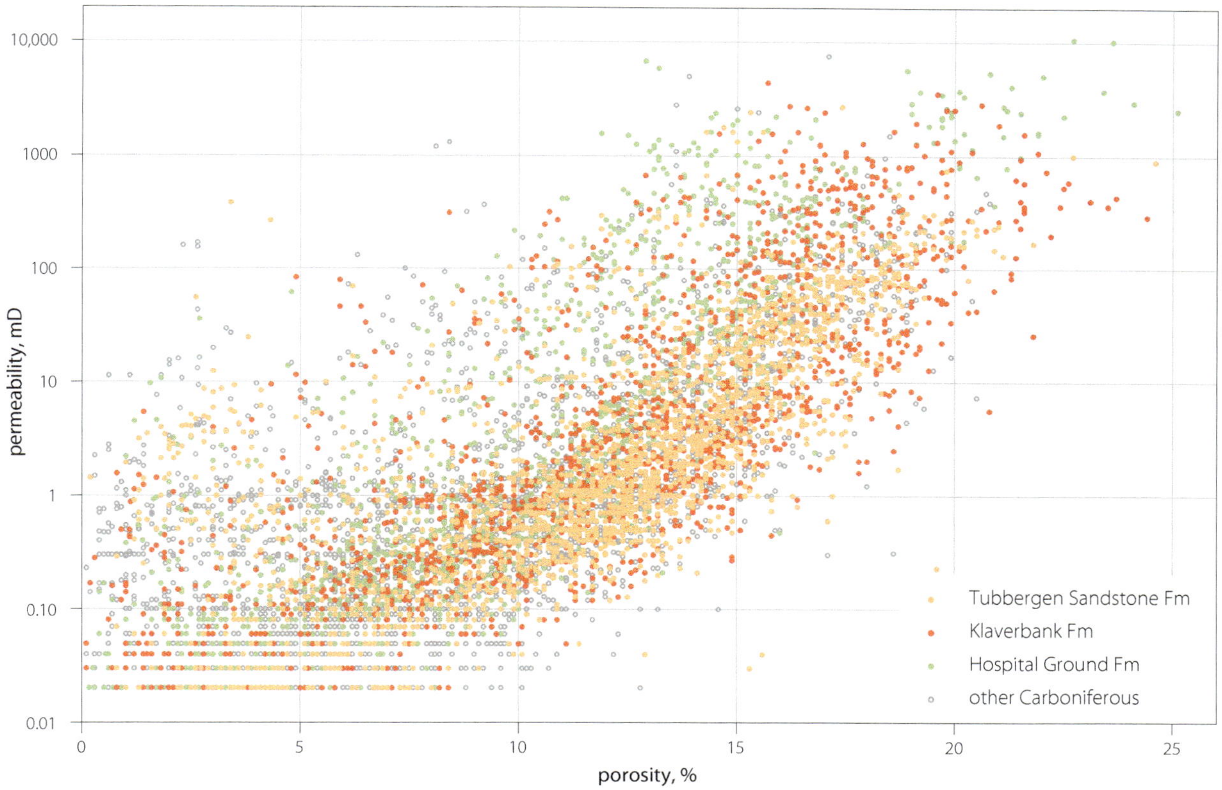

Figure 3.18. Porosity-permeability crossplot based on data from 8036 core plugs from 196 wells, both on- and offshore the Netherlands. Average porosity for all data points is 10.4%. Porosity exceeds 15% in 1493 plugs, i.e. good reservoir sandstones are not uncommon. Data provided by EBN.

sandstones of this formation are up to 100 m thick and form good reservoirs in gas fields of the Coevorden area (Kombrink et al., 2007). In the giant Groningen gas field, sandstones of the Tubbergen Formation contribute to the reservoirs where the gas column extends downward from the Rotliegend. In the central part of the basin, north of Groningen, floodplain and lacustrine claystones occasionally intercalate with relatively fine sandstone beds of meandering systems, possibly derived from the north. Red-bed facies started to develop further to the north during the late Bolsovian. In the offshore Cleaverbank Platform Area (UK quadrant 44 and the Dutch D-, E- and K-blocks) northerly sourced proximal braided channel sandstones of this red-bed facies (Hospital Ground Formation in the Netherlands, Lower Ketch Formation in the UK) form good gas reservoirs, such as in the D12-A, Sillimanite, K2-A and E17-A, Boulton, Wingate, Schooner and Ketch gas fields (Fig. 3.3; Schroot et al., 1997; Huis in 't Veld et al., 2020). Figure 3.15 shows a good example of this facies in outcrop and the core expression is shown in Figure 3.16.

During the *Asturian*, red-bed facies became dominant throughout the basin (Fig. 3.14e; Besly, 2018). The corresponding sediments consist mainly of reddish brown and greyish green, sandy and silty claystones. Mature paleosols are abundant. In the basal part, intercalated greyish green and red fluvial sandstones form channels and sheets up to 30 m thick. Pebbly sandstones and conglomerate beds are frequently incorporated within the fairly- to well-rounded and well-sorted sandstone intervals. These contemporaneous fluvial sandstones and conglomerates occur in three large areas of the basin. In the Campine area, coarse fluvial deposits of the Neeroeteren and Hellevoetsluis formations prograded along the Roer Valley Graben trend to the NW into the West Netherlands Basin in South Holland (Deckers & Rombaut, 2021). These formations and are overlain by finer-grained and more distal fluvial deposits of the Strijen formation. In the Cleaverbank Platform area, the Asturian coarse fluvial deposits belong to the Hospital Ground Formation and form the reservoirs in gas fields and gas discoveries in the blocks D18, E13, E17, K2 and K4 and adjacent UK quadrant 44 (upper part of the Lower Ketch Fm.). Overlying red bed claystones of the Step Graben Formation (Boulton Formation in the UK) of Middle-Late Asturian age form additional top- and side-seals in the D and E areas and UK quadrant 44 as demonstrated in the Wingate (Huis in 't Veld et al., 2020), and Tyne (O'Mara et al., 2003) gas fields. In the Coevorden, Tubbergen and Grolloo fields in the northeast of the Netherlands early Asturian reservoir sands belong to the Tubbergen Formation and are overlain by red claystones from the late Asturian De Lutte Formation that partially form the top seal (Van Waveren et al., 2008).

Stephanian

Only a few wells in the Netherlands, for example onshore well De Lutte-6 (Van Waveren et al., 2008) and offshore wells E17-03, E18-05 and F10-02, penetrate a Stephanian section. In all these wells, fluvial red-bed facies similar to that of the Asturian, but with better developed calcrete paleosols containing characteristic calcitic nodules, indicates a continuation of the drying-upward trend. Seismic interpretation indicates that other Stephanian sections, still undrilled (e.g. in blocks E13 and E16) may exist. They have been included on the subcrop map (Fig. 3.6).

Reservoir quality

Carboniferous reservoir quality is a key risk for gas or water production or CO_2 injection, however pockets of good reservoir (porosity >20%, permeability >1000 mD) do exist. Figure 3.18 shows a porosity-permeability cross-plot of >8000 core plug measurements from >196 wells, both on- and offshore. This figure shows that reservoir-quality sandstones can occur throughout the upper Carboniferous but are mostly found in the Dinkel Subgroup (U. Bolsovian) and Caumer Subgroup (Langsettian-L. Duckmantian). In both units, average porosity is 10-11%, and average permeability 1-2 mD. Predicting the presence of reservoir rock is difficult because of the variable Permian subcrop (Fig. 3.6) in combination with lateral variability in depositional facies (Fig. 3.13) and an often-complex diagenetic history.

Facies

The main reservoir trend in the upper Carboniferous is determined by the overall progradational nature of the fluvio-lacustrine systems. As a consequence, reservoir properties generally deteriorate from younger to older rocks and from basin fringe to basin centre (Figs 3.13 and 3.14). Multivariate analysis of petrographic data on thousands of core plugs from the Dutch upper Carboniferous has identified grain size as the main factor for porosity prediction. Larger grain size correlates with higher porosity and this means that facies distribution maps (Fig. 3.14) are key to the identification of upper Carboniferous reservoir quality.

So far, most of the commercial gas in upper Carboniferous reservoirs has been found in fluvial channels of the Klaverbank and Hospital Ground formations in the northwest offshore, and the Tubbergen Formation in the Coevorden-Groningen area (Fig. 3.3). Reservoir sandstones vary from greyish white to pinkish red (Figs 3.15 and 3.16; Kombrink et al., 2007). The sand grains are fine- to coarse, occasionally grading up to pebble and conglomerate size, moderate to poorly sorted and subangular to rounded (TNO-GDN, 2022). Quartz-arenitic sandstones dominate in the Klaverbank Formation, whereas the Tubbergen Formation is composed of a mix of lithic- and quartz arenites. Feldspar is a minor component of the reservoir sandstones, which appear to be very mature. Both formations show a distinct variability in reservoir quality, both laterally and in a downstream direction over some 50 to 100 km. In both areas, the average reservoir porosity is around 10-11%, but especially on the northern part of the Cleaverbank Platform, permeabilities are relatively high. The best reservoir intervals can have a porosity exceeding 20% and a permeability exceeding 1 Darcy (Fig. 3.18). The thickness of individual reservoir bodies ranges from 5 to 20 m (Frikken, 1999). The well-developed Asturian fluvial

Figure 3.19. *Duckmantian (WF-B) fluvial sandstone with diagenetic features shown at two different magnifications: a) Coarse to medium grained sandstone with a thin coal layer in the middle (black), siderite cement (orange) and kaolinite cement (opaque white). b) Close-up of a single pore throat, revealing authigenic quartz cement (crystals) with subsequent kaolinite cement (sugar-like texture). In this example significant reservoir quality is retained with 17% porosity and a permeability of 25 mD. Photos taken with high-resolution incident light microscopy (HOMER technique), courtesy www.geowulf.nl.*

Figure 3.20. Core photographs of an 8 metres-thick intra-Carboniferous lava flow in well Steenwijkerwold-1 (1936 m along hole). The base of the flow shows signs of soft sediment deformation (see enlargement). Undisturbed lacustrine claystone with well-preserved palynoflora that indicate a late Langsettian age overlie the weathered (fissure) top of the basaltic layer.

sandstones of the Campine Basin (Neeroeteren Formation; Kosters & Donselaar, 2003), which extend underneath the Roer Valley Graben and the West Netherlands Basin (Hellevoetsluis Formation), have hardly been explored for gas, but are currently considered as potential geothermal reservoirs.

Upper Carboniferous sandstones deposited in shallow-marine or lacustrine environments tend to be fine-grained with very low porosity. So far, in the Dutch offshore, no commercial finds have been made in such sandstones; discoveries in blocks E12 and E13 have remained undeveloped. However, the Namurian in the Netherlands is still poorly known and could locally contain better reservoirs, as suggested by wells in the UK and outcrops in the UK and Germany.

Diagenesis

Burial depth, especially paleo-burial depth, is the second most important factor in the prediction of porosity in upper Carboniferous reservoirs. The maximum temperature reached is related to the maximum depth of burial and the (paleo-)geothermal gradient and can be used as proxy for diagenetic impairment. In most of the Netherlands, maximum burial corresponds to the present-day depth (Fig. 3.1). However, in some inverted basins, the upper Carboniferous has experienced net uplift since the Late Cretaceous, and its reservoir quality can be poorer than its current depth would suggest. This is especially so when combined with the higher geothermal gradients during the Jurassic Cimmerian Phase, when burial may have been at its maximum. Examples are the Broad Fourteens Basin, the Central and West Netherlands basins and the Rheno-Hercynian Zone.

Most petrographic studies of upper Carboniferous sandstones show a complex diagenetic sequence involving cementation by quartz and sometimes carbonate, dissolution of feldspar and, most importantly, by authigenic growth of kaolinite and illite (Fig. 3.19; Van Buggenum & Den Hartog Jager, 2007). The authigenic clay is predominant where detrital clay was present in the first place. Such reservoirs may have a reasonable porosity, but a comparatively low permeability. Surprisingly, a positive correlation has been observed between quartz cement and porosity, as quartz cementation tends to take place in the cleaner sandstones where less authigenic clay is present (Fig. 3.19b). In such cases the quartz cement may form a coating around the grains, serving as fabric support and preventing compaction and grain dissolution of the reservoir rock during burial. It appears that in general, diagenesis has had a polarising effect, because layers with elevated *primary* porosity often also show increased *secondary* porosity. Finer-grained, less porous beds, especially when containing detrital clay, were at a disadvantage from the start, and often have ended up with completely occluded pores.

The depth below the Base Permian Unconformity does not stand out as an important factor in the prediction of porosity. Porosity enhancement may have occurred by leaching of upper Carboniferous rocks during the early Permian erosion, but thanks to diagenetic overprinting by other events in the following 260 Myr of burial history, the chance of finding evidence of leaching below the unconformity is small.

In parts of the northern offshore (Cleaverbank Platform, Central Graben, Schill Grund Platform), Carboniferous reservoirs may be overlain by evaporitic claystones of the Silverpit Formation, and pore space may be occluded by late halite cements. Such cementation occurs within a relatively close distance from the BPU (usually less than 10 m) and is thought to be controlled by thermodynamic processes. Cooling during phases of uplift has led to halite precipitation from the salt-saturated aquifer and rising temperature during subsequent subsidence has caused re-dissolution of previously deposited halite cements (Huis in 't Veld et al., 2020).

Acknowledgements

Many thanks to reviewers Henk Kombrink, Bernard Besly, Jan de Jager, Christian Brandes, Johan ten Veen and Geert-Jan Vis for valuable comments, and to Hans Erren for the GB27 polygon.

Digital map data

Spatial data of figures in this chapter for use in geographical information systems can be downloaded here: https://doi.org/10.5117/aup.28163108.

References

Andrews, I.J., 2013. The Carboniferous Bowland Shale gas study: geology and resource estimation. British Geological Survey (BGS) for Department of Energy and Climate Change (London): 56 pp. https://www.gov.uk/government/publications/bowland-shale-gas-study

Baars, T.F., Huis in 't Veld, R.H., Zhang, L., Koopmans, M., McLean, D., Martinius, A.W. & Abels, H.A., 2023. A cyclostratigraphic framework of the Upper Carboniferous Westoe and Cleaver formations in the southern North Sea Basin as a methodology for stratigraphic reservoir characterisation. Netherlands Journal of Geosciences 102(9). DOI: 10.1017/njg.2023.8

Besly, B.M., 1998. Carboniferous. In: Glennie, K.W. (ed.): Petroleum Geology of the North Sea: Basic Concepts and Recent Advances. Blackwell (Oxford): 104-136. DOI: 10.1002/9781444313413.ch4

Besly, B.M., 2018. Exploration and development in the Carboniferous of the Southern North Sea: a 30-year retrospective. Paleozoic Plays of NW Europe. Geological Society of London, Special Publications 471: 17-64. DOI: 10.1144/SP471.10

Besly, B.M. & Cleal, C.J., 2021. Absence of regional stratigraphic hiatus in the late Carboniferous (Asturian - Stephanian) in the northern Variscan foreland: a review of the bio- and lithostratigraphical evidence in Central England. Proceedings of the Geologists' Association 132: 50-65. DOI: 10.1016/j.pgeola.2020.09.004

Boscolo Gallo, A., Kilhams, B., Keijzer, M. & Nicolai, C., 2020. Can thin coals be resolved in seismic data? An example of the challenges for Carboniferous (Visean to Westphalian) source-rock definition across the Elbow Spit Platform (Netherlands) and the Mid North Sea High (UK) through well synthetic modelling. Geological Society of London, Special Publications 494: 215-231. DOI: 10.1144/SP494-2019-137

Bouroullec, R. & Geel, C.R., 2025. Permian. In: Ten Veen, J.H., Vis, G.-J., De Jager, J. & Wong, Th.E. (eds): Geology of the Netherlands, second edition. Amsterdam University Press (Amsterdam): 127-153. DOI: 10.5117/9789463728362_ch04

Burger, K., 1982. Kohlentonsteine als Zeitmarken, ihre Verbreitung und ihre Bedeutung für die Exploration und Exploitation von Kohlenlagerstätten. Zeitschrift Der Deutschen Geologischen Gesellschaft 133: 201-255. DOI: 10.1127/zdgg/133/1982/201

Cambier, G. & DeJonghe, L., 2010. Systematic inventory and ordering of faults in Belgium - Part 1. Professional Paper - Geological survey of Belgium 307: 1-95.

Cameron, T.D.J., 1993. Carboniferous and Devonian of the Southern North Sea. In: Knox, R.W.O.B. & Cordey, W.G. (eds): Lithostratigraphic nomenclature of the UK North Sea. British Geological Survey (Nottingham): 93 pp.

Cleal, C.J., Opluštil, S. & Thomas, B.A., 2009. Late Moscovian terrestrial biotas and palaeoenvironments of Variscan Euramerica. Netherlands Journal of Geosciences / Geologie en Mijnbouw 88(4): 181-278. DOI: 10.1017/S0016774600000846

Cliff, R.A., Drewery, S.E. & Leeder, M.R., 1991. Sourcelands for the Carboniferous Pennine river system: constraints from sedimentary evidence and U-Pb geochronology using zircon and monazite. Developments in Sedimentary Provenance Studies. Geological Society of London, Special Publications 57: 137-159. DOI: 10.1144/GSL.SP.1991.057.01.12

Cohen, K.M., Finney, S.C., Gibbard, P.L. & Fan, J.-X., 2013. The ICS International Chronostratigraphic Chart. Episodes 36(3): 199-204. DOI: 10.18814/epiiugs/2013/v36i3/002

Collinson, J.D., 2005. Dinantian and Namurian depositional systems in the southern North Sea. In: Collinson, J.D., Evans, D. Holliday, D. & Jones, N. (eds): Carboniferous hydrocarbon geology - The southern North Sea and surrounding onshore areas. Yorkshire Geological Society Occasional Publication 7: 35-56.

David, F., 1990. Sedimentologie und Beckenanalyse im Westfal C und D des nordwestdeutschen Oberkarbons. Deutsche Wissenschaftliche Gesellschaft für Erdöl, Erdgas und Kohle e.V. (DGMK): 270 pp.

DeCelles, P.G. & Giles, K.A., 1996. Foreland basin systems. Basin Research 8(2): 105-123.

Deckers, J. & Rombaut, B., 2021. Late Carboniferous differential subsidence in the Campine Basin as part of the Variscan foreland system. International Journal of Earth Sciences 110: 875-888. DOI: 10.1007/s00531-021-01996-z

Delmer, A., Dusar, M. & Delcambre, B., 2001. Upper Carboniferous lithostratigraphic units (Belgium). Geologica Belgica 4(1-2): 95-103. DOI: 10.20341/gb.2014.045

Dreesen, R., Bossiroy, D., Dusar, M., Flores Romeo, M. & Verkaeren, P., 1995. Overview of the influence of syn-sedimentary tectonics and palaeo-fluvial systems on coal seam and sand body characteristics in the Westphalian C strata, Campine Basin, Belgium. In: Whateley, M.K.G. & Spears, D.A. (eds): European Coal Geology. Geological Society of London, Special Publications 82: 215-232.

Drozdzewski, G., Bornemann, O., Kunz, E. & Wrede, V., 1985a. Beiträge zur Tiefentektonik des Ruhrkarbons. Geologisches Landesamt Nordrhein-Westfalen, Krefeld (Germany).

Drozdzewski, G., Engel, H., Wolf, R. & Wrede, V., 1985b. Beiträge zur Tiefentektonik westdeutscher Steinkohlenlagerstätten. Geologisches Landesamt von Nordrhein-Westfalen, Krefeld (Germany): 236 pp.

Drozdzewski, G., 1993. The Ruhr coal basin (Germany): structural evolution of an autochthonous foreland basin. International Journal of Coal Geology 23(1): 231-250. DOI: 10.1016/0166-5162(93)90050-K

Erren, H. & Bredewout, J.W., 1991. Model calculations on intrusive cooling and related coalification of the Peel-Erkelenz coalfield (The Netherlands and Germany). Geologie en Mijnbouw 70: 243-252.

Evans, D.J., Meneilly, A. & Brown, G., 1992. Seismic facies analysis of Westphalian sequences of the southern North Sea. Marine and Petroleum Geology 9(6): 578-589. DOI: 10.1016/0264-8172(92)90031-9

Fermont, W.J.J. & Jegers, L.F., 1986. Koolpetrografisch onderzoek aan monsters uit het Namurien van de boring Geverik ten behoeve van het OPAC-project . Rijks Geologische Dienst (Heerlen): 16 pp.

Franke, W., 2000. The mid-European segment of the Variscides; tectonostratigraphic units, terrane boundaries and plate tectonic evolution. In: Franke, W., Haak, V., Oncken, O. & Tanner, D. (eds): Orogenic Processes: Quantification and Modelling in the Variscan Belt. Geological Society of London, Special Publications 179: 35-61. DOI: 10.1144/GSL.SP.2000.179.01.05

Franke, W., Cocks, L.R.M. & Torsvik, T.H., 2017. The Palaeozoic Variscan oceans revisited. Gondwana Research 48: 257-284. DOI: 10.1016/j.gr.2017.03.005

Frikken, H.W., 1999. Reservoir-geological aspects of productivity and connectivity of gas fields in the Netherlands. PhD thesis Technische Universiteit Delft (the Netherlands) (the Netherlands): 92 pp.

Gerling, P., Kockel, F. & Krull, P., 1999. Das Kohlenwasserstoff-Potential des Präwestfals im norddeutschen Becken - Eine Synthese. DGMK Forschungsbericht 433: 106 pp.

Guion, P., 2008. Oil fields of East Midlands. Mercian Geologist 17(1): 56-60.

Hampson, G.J., Stollhofen, H. & Flint, S., 1999. A sequence stratigraphic model for the Lower Coal Measures (Upper Carboniferous) of the Ruhr district, north-west Germany. Sedimentology 46(6): 1199-1231. DOI: 10.1046/j.1365-3091.1999.00273.x

Heeremans, M., Timmerman, M.J., Kirstein, L.A. & Faleide, J.I., 2004. New constraints on the timing of late Carboniferous - early Permian volcanism in the central North Sea. In: Wilson, M., Neumann, E.-R., Davies, G.R., Timmerman, M.J., Heeremans, M. & Larsen, B.T. (eds): Permo-Carboniferous magmatism and rifting in Europe 223: 177-194. DOI: 10.1144/GSL.SP.2004.223.01.08

Huis in 't Veld, R., Schrijver, B. & Salzwedel, A., 2020. The Wingate Field, Blocks 44/23b, 44/24b and 44/19f, UK North Sea. 50th Anniversary Commemorative Volume. Geological Society of London, Memoirs 52: 288-303. DOI: 10.1144/M52-2018-75

Jones, N.S. & Glover, B.W., 2005. Fluvial sandbody architecture, cyclicity and sequence stratigraphical setting – implications for hydrocarbon reservoirs: the Westphalian C and D of the Osnabrück and Ibbenbüren area, northwest Germany. In: Collinson, J.D., Evans, D.J., Holliday, D.W. & Jones, N.S. (eds): Carboniferous hydrocarbon geology - The southern North Sea and surrounding onshore areas. Yorkshire Geological Society, Occasional Publication 7: 57-74.

Jongmans, W.J., 1925. Geologische en Paleontologische Beschrijving van het Karboon der omgeving van Epen (Limburg). Geologisch Bureau voor het Nederlandsch Mijngebied, Mededeeling 1: 31 pp.

Kirton, S.R., 1984. Carboniferous volcanicity in England with special reference to the Westphalian of the E and W Midlands. Journal of the Geological Society 141(1): 161-170. DOI: 10.1144/gsjgs.141.1.0161

Kombrink, H., 2008. The Carboniferous of the Netherlands and surrounding areas; a basin analysis. PhD thesis Utrecht Univeristy: 184 pp.

Kombrink, H., Besly, B., Collinson, J.D., Den Hartog Jager, D.G., Drozdzewski, G., Dusar, M., Hoth, P., Pagnier, H.J.M., Stemmerik, L., Waksmundzka, M.I. & Wrede, V., 2010. Carbonifer-

ous. In: Doornenbal, J.C. & Stevenson, A.G. (eds): Petroleum Geological Atlas of the Southern Permian Basin Area. EAGE Publications b.v. (Houten): 80-99.
Kombrink, H., Bridge, J.S. & Stouthamer, E., 2007. The alluvial architecture of the Coevorden Field (Upper Carboniferous), the Netherlands. Netherlands Journal of Geosciences 86 (1): 3-14. DOI: 10.1017/S0016774600021284
Kombrink, H., Della Lunga, D., Mijnlief, H.F. & Kroon, I., 2009. Nitrogen contents in the northwest Netherlands offshore area. TNO (Utrecht), unofficial report: 15 pp. https://www.nlog.nl/sites/default/files/nitrogen%20in%20the%20northern%20netherlands%20offshore.pdf
Kombrink, H., Van Os, B.J.H., Van der Zwan, C.J. & Wong, Th.E., 2008. Geochemistry of marine and lacustrine bands in the Upper Carboniferous of the Netherlands. Netherlands Journal of Geosciences 87(4): 309-322. DOI: 10.1017/S0016774600023374
Kosters, M. & Donselaar, M.E., 2003. Syn-sedimentary faulting and the formation of localized reservoir sands; Carboniferous examples from the Campine Basin, Belgium and the Dutch offshore. Petroleum Geoscience 9(4): 309-319. DOI: 10.1144/1354-079302-543
Langenaeker, V., 2000. The Campine Basin. Stratigraphy, structural geology, coalification and hydrocarbon potential for the Devonian to Jurassic. PhD thesis, Leuven University, 142 pp.
Langenaeker, V. & Dusar, M., 1992. Subsurface facies analysis of the Namurian and earliest Westphalian in the western part of the Campine Basin (N. Belgium). Geologie En Mijnbouw 71: 161-172.
Littke, R., Büker, C., Hertle, M., Karg, H., Stroetmann-Heinen, V. & Oncken, O., 2000. Heat flow evolution, subsidence and erosion in the Rheno-Hercynian orogenic wedge of Central Europe. In: Franke, W., Haak, V., Oncken, O. & Tanner, D. (eds): Orogenic Processes: Quantification and Modelling in the Variscan Belt. Geological Society of London, Special Publications 179: 231-255.
Littke, R., Büker, C., Lückge, A., Sachsenhofer, R.F. & Welte, D.H., 1994. A new evaluation of palaeo-heat flows and eroded thicknesses for the Carboniferous Ruhr basin, western Germany. International Journal of Coal Geology 26(3-4): 155-183. DOI: 10.1016/0166-5162(94)90009-4
McLean, D., Owens, B. & Neves, R., 2005. Carboniferous miospore biostratigraphy of the North Sea. Carboniferous Hydrocarbon Geology: The Southern North Sea and Surrounding Onshore Areas 7: 13-24.
Menning, M., Alekseev, A.S., Chuvashov, B.I., Davydov, V.I., Devuyst, F.X., Forke, H.C., Grunt, T.A., Hance, L., Heckel, P.H., Izokh, N.G., Jin, Y.G., Jones, P.J., Kotlyar, G.V., Kozur, H.W., Nemyrovska, T.I., Schneider, J.W., Wang, X.D., Weddige, K., Weyer, D. & Work, D.M., 2006. Global time scale and regional stratigraphic reference scales of central and west Europe, east Europe, Tethys, south China, and North America as used in the Devonian-Carboniferous-Permian Correlation Chart 2003 (DCP 2003). Palaeogeography, Palaeoclimatology, Palaeoecology 240(1-2): 318-372. DOI: 10.1016/j.palaeo.2006.03.058
Mijnlieff, H.F., 2002. Top Pre-Permian distribution map & some thematic regional geologic maps of the Netherlands. Poster presented at the Netherlands Institute of Applied Geoscience TNO - National Geological Survey, Utrecht, the Netherlands.
Morton, A.C., Hallsworth, C.R. & Moscariello, A., 2005. Interplay between northern and southern sediment sources during Westphalian deposition in the Silverpit Basin, southern North Sea. In: Collinson, J.D., Evans, D.J., Holliday, D.W. & Jones, N.S. (eds): Carboniferous hydrocarbon geology - The southern North Sea and surrounding onshore areas. Yorkshire Geological Society, Occasional Publication 7: 135-146.
Nunna, R.R., 1960. Untersuchungen über die Inkohlungsverhältnisse im Aachener Steinkohlengebiet. MSc thesis, Rheinisch-Westfaelische Technische Hochschule (RWTH).
O'Mara, P.T., Merryweather, M., Stockwell, M. & Bowler, M.M., 2003. The Trent gas field, block 43-24a, UK North Sea. In: Gluyas, J. & Hichens, H.M. (eds): United Kingdom Oil and Gas Fields, Commemorative Millennium Volume. Geological Society of London, Memoirs 20: 835-849. DOI: 10.1144/GSL.MEM.2003.020.01.70
Patijn, R.J.H. & Kimpe, W.F.M., 1961. Explanation to the map of the Carboniferous surface, the sections and the map of the overburden of the South Limburg mining District and State Mine Beatrix with the surrounding area. Mededelingen van de Geologische Stichting C I-1-4: 5-12.
Pearce, T.J., Martin, J.H., Cooper, D. & Wray, D.S., 2010. Chemostratigraphy of Upper Carboniferous (Pennsylvanian) sequences from the Southern North Sea (United Kingdom). SEPM Special Publication 94: 109-127.
Pearce, T.J., McLean, D., Wrang, J.D., Wright, D.K., Jeans, C.J. & Mearns, E.W., 2005. Stratigraphy of the Upper Carboniferous Schooner Formation, southern North Sea: chemostratigraphy, mineralogy, palynology and Sm-Nd isotope analysis. Carboniferous Hydrocarbon Geology of the Southern North Sea and Surrounding Onshore Areas 7: 165-182.
Peelcommissie, 1963. Rapport van de Peelcommissie. Koninklijk Nederlands Geologisch Mijnbouwkundig Genootschap: 133 pp.
Pharaoh, T.C., 2018. The Anglo-Brabant Massif: Persistent but enigmatic palaeo-relief at the heart of western Europe. Prodeedings of the Geologists' Association 3(129): 278-328. DOI: 10.1016/j.pgeola.2018.02.009
Pointon, M.A., Chew, D.M., Ovtcharova, M., Sevastopulo, G.D. & Crowley, Q.G., 2012. New high-precision U-Pb dates from western European Carboniferous tuffs; implications for time scale calibration, the periodicity of late Carboniferous cycles and stratigraphical correlation. Journal of the Geological Society 169(6): 713-721. DOI: 10.1144/jgs2011-092
Quirk, D.G., 1993. Interpreting the Upper Carboniferous of the Dutch Cleaver Bank High. In: Parker J.R. (ed.): Proceedings of the 4th Conference on Petroleum Geology of North-West

Europe. Geological Society of London, Petroleum Geology Conference series 4: 697-706. DOI: 10.1144/0040697

Quirk, D.G., 1997. Sequence stratigraphy of the Westphalian in the northern part of the Southern North Sea. In: K. Ziegler, P. Turner & S.R. Daines (eds): Petroleum Geology of the Southern North Sea: Future Potential. Geological Society of London, Special Publications 123: 153-168. DOI: 10.1144/GSL.SP.1997.123.01.10

Remmelts, G., Nelskamp, S., De Jager, J. & Geluk, M.C., 2025. Petroleum Geology. In: Ten Veen, J.H., Vis, G.-J., De Jager, J. & Wong, Th.E. (eds): Geology of the Netherlands, second edition. Amsterdam University Press (Amsterdam): 535-575. DOI: 10.5117/9789463728362_ch14

RGD, 1988. Rapport betreffende de stratigrafische ouderdom van een in de boring Hoogenweg-1 (NAM) gevonden vulkanische as. Rijks Geologische Dienst (Heerlen), report GB 2214a/GD 20225: 7 pp.

RGD, 1995. Geologische kaart van Zuid-Limburg en omgeving – Paleozoïcum (1:50,000). Rijks Geologische Dienst (Haarlem, the Netherlands).

Ricken, W., Schrader, S., Oncken, O. & Plesch, A., 2000. Turbidite basin and mass dynamics related to orogenic wedge growth; the Rheno-Hercynian case. In: Franke, W., Haak, V., Oncken, O. & Tanner, D. (eds): Orogenic Processes: Quantification and Modelling in the Variscan Belt. The Geological Society of London, Special Publications 179: 257-280. DOI: 10.1144/GSL.SP.2000.179.01.16

Riley, N.J., Claoué-Long, J., Higgins, A.C., Owens, B., Spears, A., Taylor, L. & Varker, W.J., 1994. Geochronometry and geochemistry of the European Mid-Carboniferous boundary global stratotype proposal, Stonehead Beck, North Yorkshire, UK. Annales de La Societe Geologique de Belgique 116(2): 275-289.

Rippon, J.H., 1996. Sand body orientation, palaeoslope analysis and basin-fill implications in the Westphalian A-C of Great Britain. Journal of the Geological Society 153(6): 881-900. DOI: 10.1144/gsjgs.153.6.0881

Roustiau, A., Bouroullec, R., Dijk, J., Den Hartog Jager, D., Nelskamp, S. & Swart, M., 2022. Play 8 Carboniferous. https://www.geodeatlas.nl/pages/play-8-carboniferous

Schneider, J.W. & Scholze, F., 2018. Late Pennsylvanian–Early Triassic conchostracan biostratigraphy: a preliminary approach. The Permian Timescale. Geological Society of London, Special Publications 450: 365-386. DOI: 10.1144/SP450.6

Schroot, B.M., De Haan, H.B. & Nio, S.D., 1997. Upper Carboniferous of the Cleaverbank High.Netherlands Institute of Applied Geoscience TNO (Haarlem), report NITG 97-138C: 42 pp.

Schulmann, K., Martínez Catalán, J. R., Lardeaux, J. M., Janoušek, V. & Oggiano, G., 2014. The Variscan Orogeny: Extent, Timescale and the Formation of the European Crust. Geological Society of London, Special Publications 405. DOI: 10.1144/SP405

Sissingh, W., 2004. Palaeozoic and Mesozoic igneous activity in the Netherlands; a tectonomagmatic overview. Netherlands Journal of Geosciences/ Geologie en Mijnbouw 83(2): 113-134. DOI: 10.1017/S0016774600020084

Smit, J., Van Wees, J.-D. & Cloetingh, S., 2016. The Thor suture zone: From subduction to intraplate basin setting. Geology 44(9): 707-710. DOI: 10.1130/G37958.1

Smit, W., 2020. Chiswick and Kew fields, Blocks 49/4a, 49/4b, 49/4c, 49/5a and 49/5b, UK North Sea. Geological Society of London, Memoirs 52: 142-150. DOI: 10.1144/M52-2018-48

Stille, H., 1920. Alter und Art der Phasen variscischer Gebirgsbildung. In: Nachrichten von der Gesellschaft der Wissenschaften zu Göttingen, Mathematisch-Physikalische Klasse. 218-224.

Süss, M.P., Drozdzewski, G. & Schäfer, A., 2000. Sequenzstratigraphie des kohlenfürenden Oberkarbons im Ruhr-Becken. Geologisches Jahrbuch Hannover A156: 45-106.

Teichmüller, M. & Teichmüller, R., 1971. Inkohlung im Rhein-Ruhr-Revier. Fortschritte Geologie Rheinland und Westfalen 19: 69-72.

Thiadens, A.A., 1963. The Palaeozoic of The Netherlands. Verhandelingen van Het Koninklijk Nederlands Geologisch Mijnbouwkundig Genootschap, Geologische Serie 21(1): 9-28.

TNO-GDN, 2022. Limburg Group. Stratigraphic Nomenclature of the Netherlands, TNO – Geological Survey of the Netherlands. https://www.dinoloket.nl/en/stratigraphic-nomenclature/limburg-group

Torsvik, T.H. & Cocks, L.R.M., 2016. Earth History and Palaeogeography. Cambridge University Press: 317 pp. DOI: 10.1017/9781316225523

Trueman, A., 1954. The Coalfields of Great Britain. Edward Arnold Publishers Ltd. London (UK): 396 pp.

Tubb, S.R., Soulsby, A. & Lawrence, S.R., 1986. Paleozoic prospects on the northern flanks of the London-Brabant Massif. Habitat of Palaeozoic Gas in N.W. Europe 23: 55-72. DOI: 10.1144/GSL.SP.1986.023.01.04

Van Adrichem Boogaert, H.A. & Kouwe, W.F.P., 1995. Stratigraphic nomenclature of The Netherlands, revision and update by RGD and NOGEPA, Section C. (Silesian). Mededelingen Rijks Geologische Dienst 50: 1-40.

Van Balen, R.T., Van Bergen, F., De Leeuw, C., Pagnier, H., Simmelink, H., Van Wees, J.-D. & Verweij, J.M., 2000. Modelling the hydrocarbon generation and migration in the West Netherlands Basin, the Netherlands. Netherlands Journal of Geosciences 79 (1): 29-44. DOI: 10.1017/S0016774600021557

Van Bergen, F. & Koster, K., 2025. Peat, coal and coalbed methane. In: J.H. Ten Veen, G.-J. Vis, J. De Jager & Th.E. Wong (eds): Geology of the Netherlands, second edition. Amsterdam University Press (Amsterdam): 577-601. DOI: 10.5117/9789463728362_ch15

Van Bergen, M.J., Vis, G.-J., Sissingh, W., Koornneef, J. & Brouwers, I., 2025. Magmatism in the Netherlands: expression of the northwest European rifting history. In: Ten Veen, J.H., Vis, G.-J., De Jager, J. & Wong, Th.E. (eds): Geology of the Netherlands, second edition. Amsterdam University Press (Amsterdam): 393-455. DOI: 10.5117/9789463728362_ch11

Van Buggenum, J.M. & Den Hartog Jager, D.G., 2007. Silesian. In:

Wong, Th.E., Batjes, D.A.J.& De Jager, J. (eds): Geology of the Netherlands. Royal Netherlands Academy of Arts and Sciences (KNAW, Amsterdam): 43-62.

Van de Laar, J.G.M. & Fermont, W.J.J., 1990. The impact of marine transgressions on palynofacies; the Carboniferous Aegir marine band in borehole Kemperkoul-1. In: Fermont, W.J.J. & Weegink, J.W. (eds): Proceedings International Symposium on Organic Petrology, Zeist. January 7-9. Mededelingen Rijks Geologische Dienst 45: 75-89.

Van Tongeren, P.C.H. & Hildenbrand, A., 2005. Subcrop kaart van de Palaeozoïsche formaties in het Kempens Bekken. Tussentijdse Rapportage. VITO: 9 pp.

Van Waterschoot van der Gracht, W.A.J.M., 1918. Eindverslag over de onderzoekingen en uitkomsten van den Dienst der Rijksopsporing van Delfstoffen in Nederland 1903-1916. Martinus Nijhoff ('s Gravenhage): 664 pp.

Van Waveren, I.M., Abbink, O.A., Van Hoof, T.B. & Van Konijnenburg - van Cittert, J.H.A., 2008. Revision of the late Carboniferous megaflora from the De Lutte-06 well (Twente, the Netherlands), and its stratigraphical implications. Netherlands Journal of Geosciences/ Geologie en Mijnbouw 87(4): 339-352. DOI: 10.1017/S0016774600023398

Van Wijhe, D.H. & Bless, M.J.M., 1974. The Westphalian of the Netherlands with special reference to miospore assemblages. Geologie en Mijnbouw 53(6): 295-328. DOI: 10.1016/S0016-7878(74)80001-280001-2

Veld, H., Fermont, W.J.J., Kerp, H. & Visscher, H., 1996. Geothermal history of the Carboniferous in South Limburg, the Netherlands. In: Rondeel, H.E., Batjes, D.A.J. & Nieuwenhuijs, W.H. (eds): Geology of gas and oil under the Netherlands. Kluwer Academic Publishers (Dordrecht): 31-43. DOI: 10.1007/978-94-009-0121-6_5

Vincent, C.J., 2015. Maturity and migration modelling of selected wells in the Central North Sea. British Geological Survey, Keyworth, Nottingham (UK). Energy And Marine Geoscience Programme Commissioned Report No. CR/15/122: 189 pp.

Vis, G.-J., Houben, A.J.P., Debacker, T.N. & Geel, C.R., 2025. The geological foundation of the Netherlands: the early Carboniferous, Devonian and older. In: Ten Veen, J.H., Vis, G.-J., De Jager, J. & Wong, Th.E. (eds): Geology of the Netherlands, second edition. Amsterdam University Press (Amsterdam): 53-93. DOI: 10.5117/9789463728362_ch02

Von Raumer, J.F., Bussy, F., Schaltegger, U., Schulz, B. & Stampfli, G.M., 2013. Pre-Mesozoic Alpine basements - Their place in the European Paleozoic framework. Geological Society of America Bulletin 125(1-2): 89-108. DOI: 10.1130/B30654.1

Wandless, A.M., 1960. The Coalfields of Great Britain: Variation in Rank of Coal. Scientific Department Coal Surveys, National Coal Board, London (UK).

Ziegler, P.A., 1990. Geological Atlas of Western and Central Europe, second edition. Shell Internationale Petroleum Mij. B.V. and Geological Society (London): 239 pp.

Ziegler, P.A., Schumacher, M.E., Dèzes, P., Van Wees, J.-D. & Cloetingh, S.A.P.L., 2004. Post-Variscan evolution of the lithosphere in the Rhine Graben area; constraints from subsidence modelling. In: Wilson, M., Neumann, E.R., Davies, G.R., Timmerman, M.J., Heeremans, M. & Larsen, B.T. (eds): Permo-Carboniferous Magmatism and Rifting in Europe. Geological Society of London, Special Publications 223: 289-317. DOI: 10.1144/GSL.SP.2004.223.01.13

4 Permian

Renaud Bouroullec & Cornelis R. Geel with a special recognition to Mark C. Geluk†*

*Corresponding author: renaud.bouroullec@tno.nl

ABSTRACT

The Permian System in the Netherlands is subdivided into the Lower Rotliegend, Upper Rotliegend and Zechstein groups. These groups were deposited in a large epicontinental basin extending from the UK to Poland. The Lower Rotliegend Group, deposited during the early Permian, unconformably overlays older strata (at the Base Permian Unconformity), is composed of volcanics and volcaniclastics and is only preserved locally in the Netherlands. After a significant erosional phase (Saalian Unconformity), the middle to upper Permian Upper Rotliegend sediments were deposited under warm and arid climatic conditions and are preserved in a large part of the Dutch sector. A broad southern and a much smaller northern lake edge depositional system composed of fluvial, eolian and playa-lake deposits are represented in the Netherlands. These systems are separated by the E-W oriented Silver Pit Lake in which clays and evaporites of the Silverpit Formation accumulated. With increasingly dry conditions, rapid marine floodings of the basin were accompanied by decreasing clastic input and thick evaporites and marginal carbonates, known as the Zechstein Group accumulated in the Netherlands and neighbouring regions. The thick layer of Zechstein rock salt is deformed into many salt bodies, sometimes localized along pre-existing faults, which often played a key role in Dutch diapirism and strongly influenced the post-Permian structural development. The depositional thickness of the Permian reaches almost 2000 m in the northern offshore. The Upper Rotliegend Group contains more than 95% of the natural gas resources in the Netherlands and it is a target for geothermal projects and CO_2 storage. The Zechstein Group contains exploitable rock salt and potassium-magnesium salts.

<< Slope carbonates of the Concretionary Limestone Member at Marsden Bay, South Shields (northeast England), equivalent to the slope facies of the Dutch Z2 Carbonate Member. Photo: Aart-Peter van den Berg van Saparoea.

DOI: 10.5117/9789463728362_ch04

Introduction

Permian strata were deposited unconformably on top of gently deformed Carboniferous strata. The base of the Permian is often referred to as the Base Permian Unconformity (or BPU) (Fig. 4.1). In large parts of the Netherlands, the Permian overlies the Westphalian sequence and, locally, the Stephanian, Namurian and Dinantian (Huis in 't Veld & Den Hartog Jager, 2025, this volume). Middle to Late Devonian rocks subcrop in the south along the Brabant Massif and in the northern part of the offshore around the Elbow Spit High. Permian rocks contain important hydrocarbon reservoirs and seals (Remmelts et al., 2025, this volume), exploitable salt deposits (Groenenberg et al., 2025, this volume), good carbon sequestration storage potential (Juez-Larré et al., 2025, this volume) and form an important deep geothermal target (Mijnlieff et al., 2025, this volume).

In the Netherlands, a major hiatus represents most of early Permian (most of the Cisuralian to the lower part of the Guadalupian) to middle Permian time (Fig. 4.2). This un-

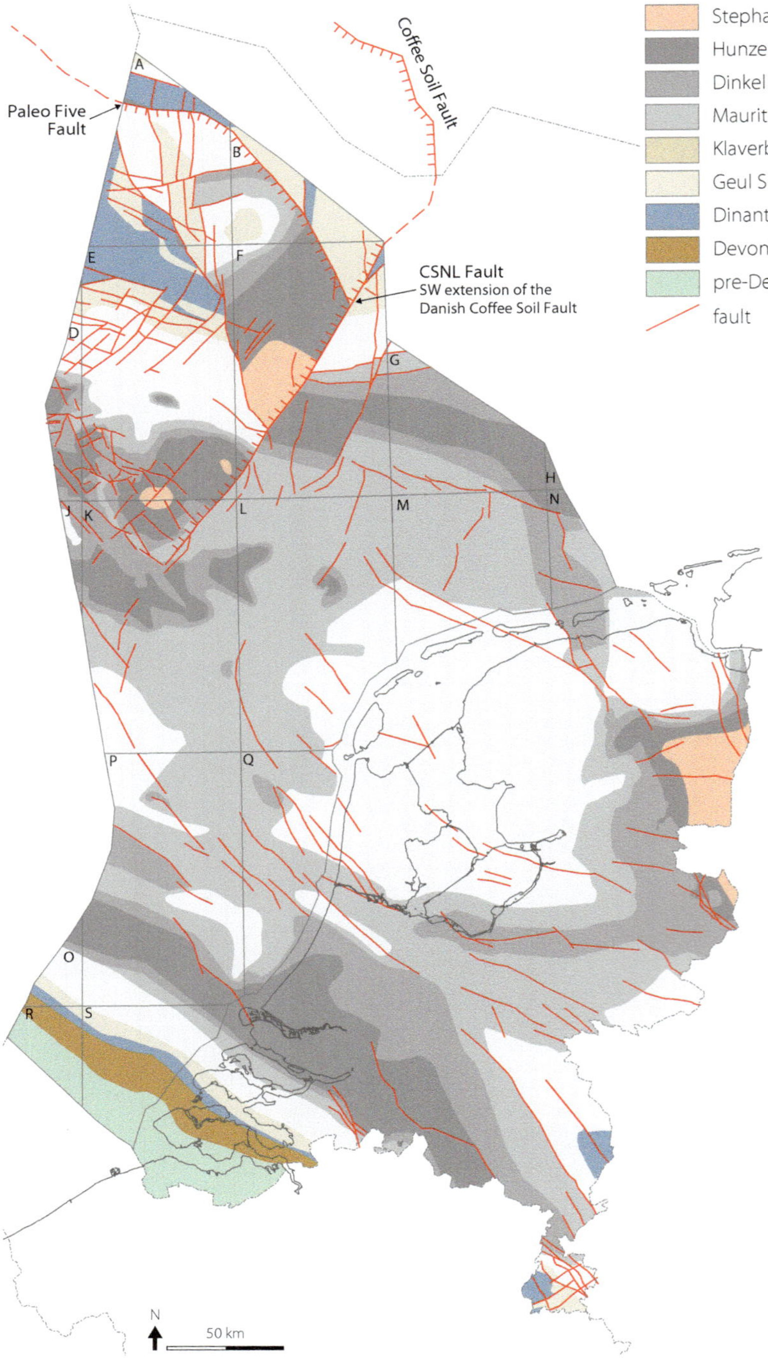

Figure 4.1. *Base Permian subcrop map based on 1) recent regional mapping in the northern part of the Dutch offshore by TNO-GDN, 2) recent well-based stratigraphic updates of the Upper Carboniferous by Roustiau et al. (2022) and 3) previously published Base Permian Subcrop maps from Mijnlieff (2002) and Kombrink et al. (2010). Note the newly mapped Paleo Five Fault, a large south-verging normal fault that extends to the UK North Sea sector and has a vertical offset of 4.5 km at top Devonian level. The Coffee Soil NL Fault (CSNL Fault) is a newly interpreted crustal scale fault that, via the German offshore (Entenschnabel), corresponds to the SW extension of the Danish Coffee Soil Fault. In the Dutch sector, Stephanian strata are preserved on the downthrown block to the northwest of the fault. This fault likely acted as a lateral ramp of the Coffee Soil and Paleo Five faults system during the Devono-Carboniferous. A stratigraphic scheme of the upper Carboniferous is provided in Huis in 't Veld, R., & Den Hartog Jager, D. (2025, this volume, Fig. 3.4).*

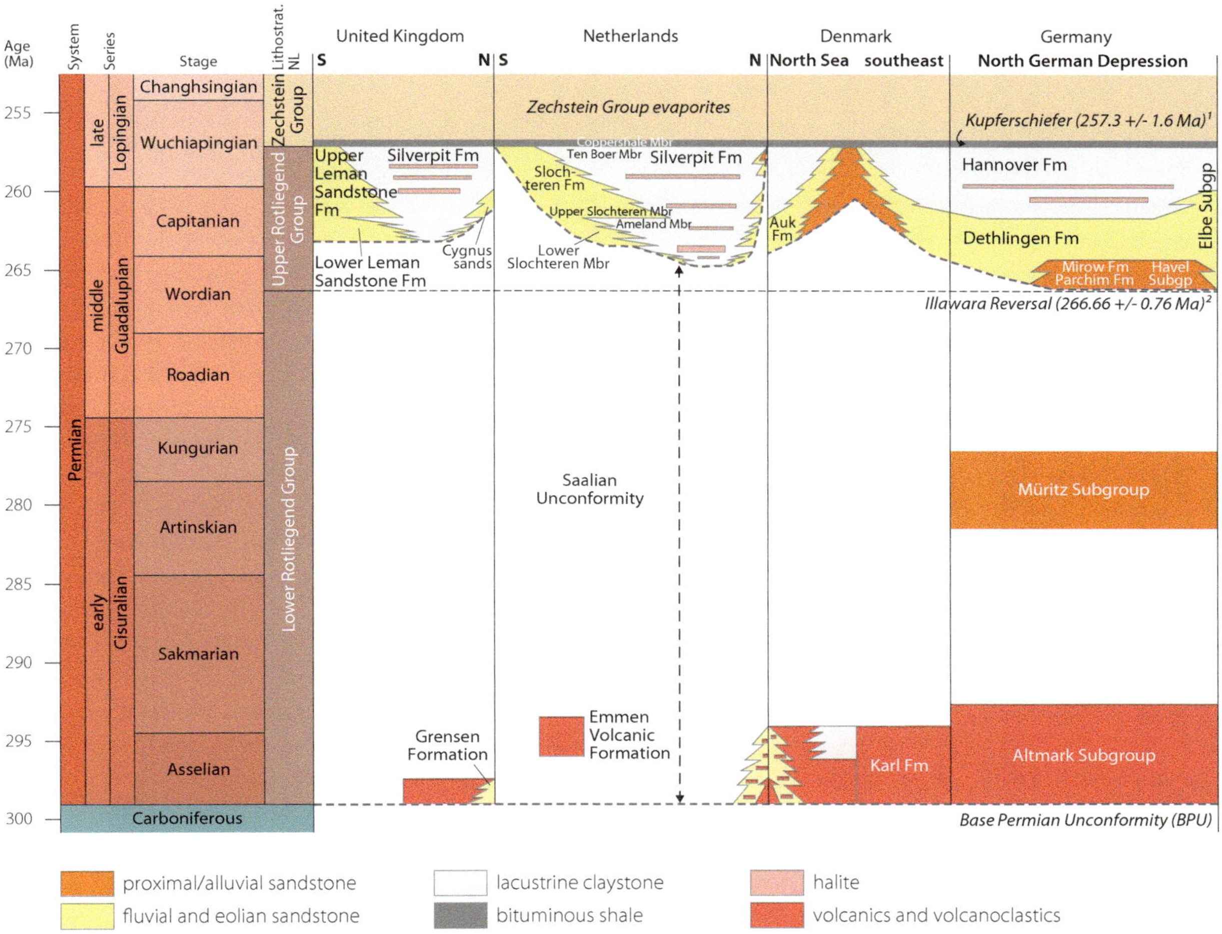

Figure 4.2. Tectonostratigraphic chart of the Permian of the UK, the Netherlands, Denmark and Germany. Modified from Gast et al. (2010). 1 = Re-Os isochron age for the Kupferschiefer (Brauns et al., 2003); 2 = age according to the geomagnetic polarity timescale for the Permian (Hounslow & Balabanov, 2016).

conformity is referred to as the Saalian Unconformity and it is often mistaken for the BPU because the Lower Rotliegend sediments deposited between those two erosional events are only sparsely preserved (Fig. 4.2). During middle and late Permian times, the Netherlands became part of the large E-W trending complex of sedimentary basins usually referred to as the Southern Permian Basin (SPB), which stretched from the UK to Poland and experienced significant thermal subsidence (Figs 4.3, 4.4; Ziegler, 1990; Geluk, 2005; Pharaoh et al., 2010). The depositional area gradually expanded during the middle and late Permian towards the London-Brabant Massif and the Rhenish Massif, which formed its southern margin, and the Mid North Sea High and Ringkøbing-Fyn High which formed its northern margin. The Lower Saxony Basin and the Dutch Central Graben formed areas of volcanic activity during the early to middle Permian (Fig. 4.7). During the middle Permian, differential subsidence of the Southern Permian Basin started. Sediment supply was mainly from the Variscan Mountains in the south and to a minor extent also from the highs north of the basin, including the Elbow Spit High.

The depositional thickness of the Permian increases from less than 50 m in the southern Netherlands to almost 2000 m in the northern offshore area. Some smaller intra-basinal paleo-highs such as the Peel-Maasbommel Complex, Dalfsen High, Winterton High and Elbow Spit High received little to no Permian sediments (Fig. 4.5). That the Texel-IJsselmeer High was a high during the Permian is still debated as it could be a younger structure. The present-day thickness of the Permian is highly variable due to 1) significant post-depositional erosion at the Mid-Cimmerian, Late-Cimmerian and Sub-Hercynian unconformities, and 2) salt dissolution and movements strongly affecting the original thickness distribution. A depth map of the base of the Permian is shown in Huis in 't Veld & Den Hartog Jager (2025, this volume). The depth of the base of the Zechstein Group ranges from less than 1 km in the eastern Netherlands to over 9 km in the Dutch Central Graben.

Stratigraphy

The Permian in the Netherlands is divided into three groups (Fig. 4.2; Van Adrichem Boogaert & Kouwe, 1994):

1. The Lower Rotliegend Group, a succession of volcanic, volcaniclastic, and clastic rocks with a geographically limited distribution.
2. The Upper Rotliegend Group, predominantly comprising sandstones in the onshore, southwestern offshore and in the northern offshore around the Elbow Spit High, and claystones with some halite beds in the E-W trending basin axis.
3. The Zechstein Group, mainly comprising marine evaporites, carbonates and claystones with minor sandstones (Zechstein Fringe Sandstone members; Fig. 4.11) in the south of the Netherlands.

The contact between the Lower and Upper Rotliegend or Zechstein groups is unconformable (Saalian Unconformity; Figs 4.2, 4.6), whereas the contact between the Upper Rotliegend and the Zechstein Group is conformable.

Traditionally, the Permian deposits have been subdivided into a Lower Permian Rotliegend and an Upper Permian Zechstein. Until recently, the former was considered to represent the early Permian, and the latter the late Permian (Menning, 1995; Plein, 1995; Glennie, 1998). Radiometric dating of the Lower Rotliegend Group in Germany documents an early Permian age (Plein, 1995; Stemmerik et al., 2000), but middle Permian ages in the Netherlands (Sissingh, 2004; Geluk, 2005). The Upper Rotliegend Group is almost completely devoid of fossils, having been deposited under hypersaline or oxidizing conditions. Here, other tools than biostratigraphy may be applied, such as chemo-stratigraphy, magneto-stratigraphy and sequence stratigraphy (George & Berry, 1994; Yang & Baumfalk, 1994; Schuurman, 1998). Rocks of the Zechstein Group, especially the carbonates and claystones, yield palynomorphs and marine fossils (bryozoa, bivalves and algae).

In 2003 the Permian Stratigraphic Committee agreed upon a three-fold subdivision of the Permian (Remane et al., 2003). Following that, the Lower Rotliegend Group, can be ascribed to the early to middle Permian, the Upper Rotliegend Group to the middle to early late Permian and the Zechstein Group to the late Permian (Fig. 4.2; Van Ojik et al., 2011). Where it is absent, the hiatus between the Upper Rotliegend and the Carboniferous in the Netherlands spans the entire early and part of the middle Permian.

A different approach to the subdivision of Permian deposits makes use of sequence stratigraphy, which aims for a better understanding of the depositional interrelations of the sediments. Sequence-stratigraphic models of the Upper Rotliegend Group have been published by George & Berry (1994), Yang & Baumfalk (1994) and Plein (1995); all models propose that wet-dry Milankovitch-driven climatic cycles with a periodicity of 100 and 200 ka controlled sedimentation. These high-resolution sequences are grouped in larger-scale cycles, with a duration of between 0.7 to 1.4 Myr. Cyclicity in the Zechstein has been described traditionally as governed by initial transgres-

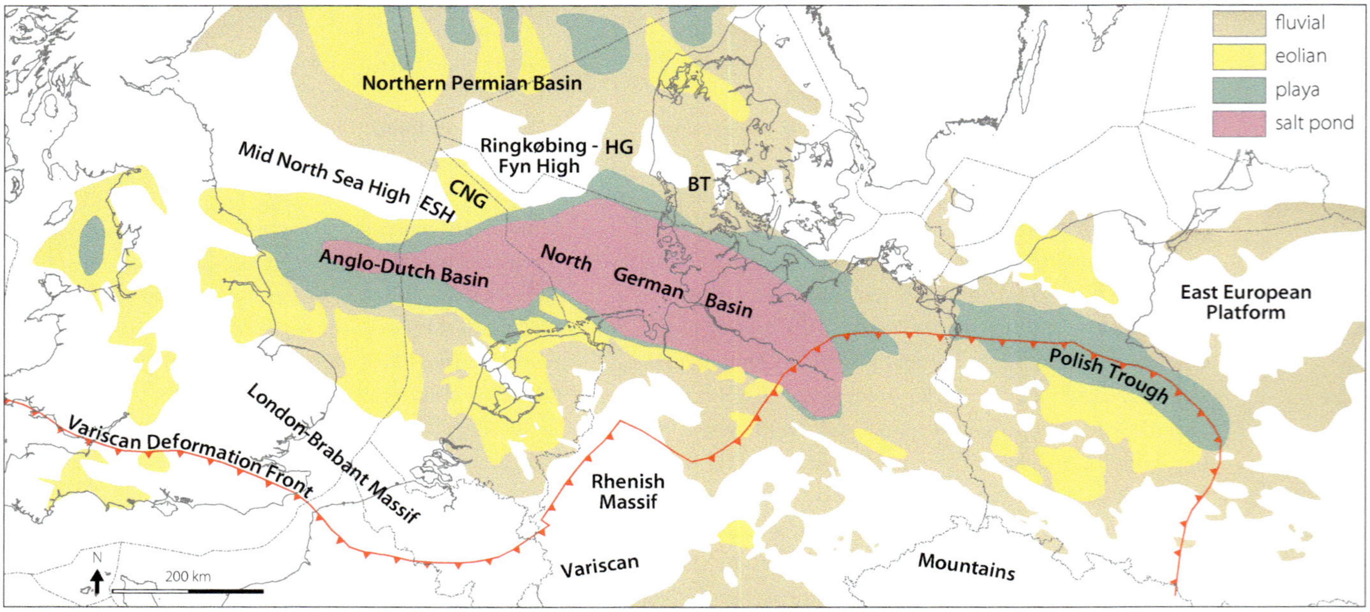

Figure 4.3. Present-day distribution and depositional environment of the Upper Rotliegend Group (late middle to early late Permian) in the Southern Permian Basin. BT = Bramble Trough; CNG = Central North Sea Graben; ESH = Elbow Spit High; HG = Horn Graben. (Modified from Geluk, 2005, 2007; De Bruin et al., 2015).

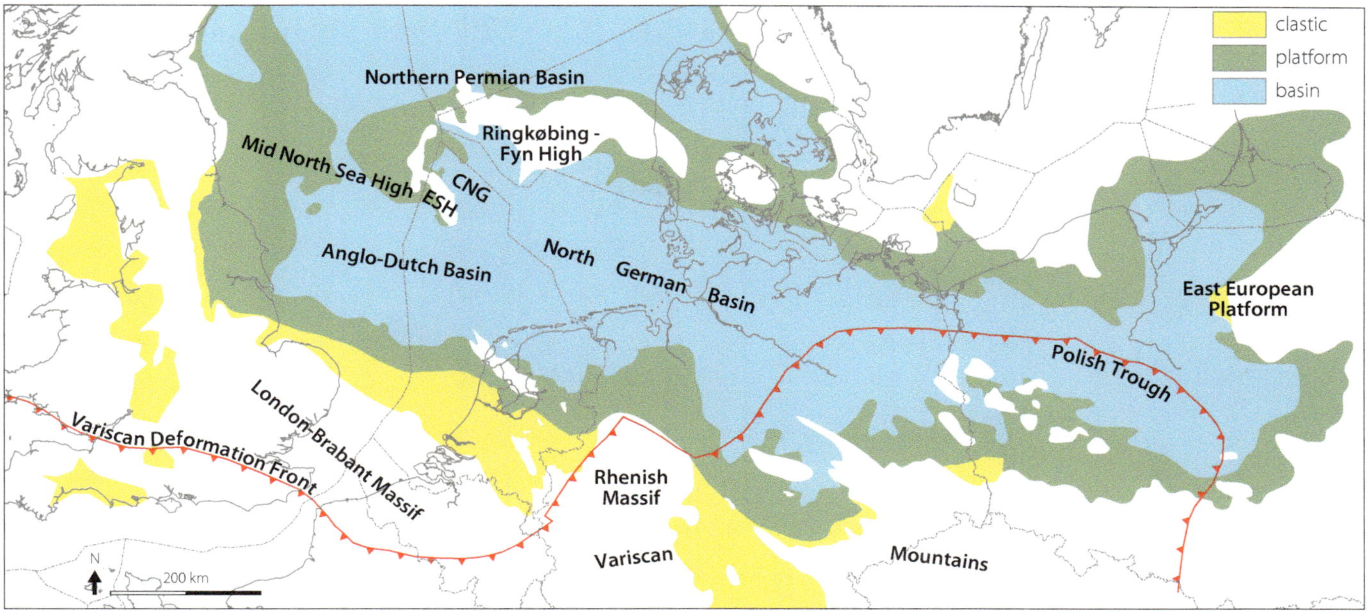

Figure 4.4. *Present-day distribution and depositional environment of the Z2 Carbonate of the Zechstein Group (late Permian) in the Southern Permian Basin. CNG = Central North Sea Graben; ESH: Elbow Spit High. (Modified from Lokhorst, 1998; Taylor, 1998; Geluk, 2005, 2007; Bouroullec et al., 2022; Peeters et al., 2023).*

sions, followed by regressive phases with evaporite deposition (Richter-Bernburg, 1955). These cycles, bounded by maximum flooding surfaces, are called genetic stratigraphic sequences (Galloway, 1989). A more modern view was published by Tucker (1991), who proposed an evaporite-carbonate sequence stratigraphy with sequence boundaries at the tops of the carbonates. The first indications for the Zechstein type cyclicity already appear in the upper part of the Rotliegend.

Regional correlations

The Permian in the Netherlands can generally be correlated with the adjacent countries without major problems (Fig. 4.2). The equivalents of the clastics, volcanics, and volcaniclastics of the Lower Rotliegend Group are known as the Karl Formation in Denmark and in the adjacent UK central North Sea (Stemmerik et al., 2000) and as the Altmark Group in Germany (Plein, 1995). The Grensen Formation (Fig. 4.2), located on the northern side of the Mid North Sea High on the eastern side of the UK offshore sector around the Flora Field and Grensen Nose, is time equivalent to the Karl Formation. This correlation is based on the K-Ar dating of 299 +/- 1.6 Ma of a volcanic extrusive unit which is unconformably overlain by the Grensen Formation (Martin et al., 2002).

The Upper Rotliegend Group correlates with the Auk Formation in the UK central North Sea and with the Silverpit and Leman formations of the UK southern North Sea (Johnson et al., 1994). The equivalent in Germany is the Elbe Subgroup (Gast et al., 2010); equivalents of the underlying formations occur in Germany and Poland (Geluk, 2005), and possibly also in Netherlands eastern offshore (Corcoran & Lunn, 2014). Equivalents of the Upper Rotliegend Group have been described as the so-called Zechstein conglomerate in northeast Belgium (De Craen & Swennen, 1992) and in the Lower Rhine coal-mining area in Germany (Schlimm & Thiermann, 1988). These are considered time equivalents of the youngest Upper Rotliegend or, alternatively, of the oldest Zechstein sediments. The Zechstein Group in the Netherlands correlates with the Zechstein Supergroup of the UK Southern North Sea and with the German Zechstein. Most of the unit names are identical. A major discrepancy, however, affects the definition of the top of the group in the different countries. In the UK, this top is picked considerably lower than in the Netherlands and Germany; as a result, rocks considered part of the Permian Zechstein in the latter two countries are assigned to the Triassic Bacton Group in the UK southern North Sea (Geluk, 2005).

Tectonic evolution

The deposition of the Permian postdates the Variscan Orogeny in western Europe. The compressive movements of this orogeny ended during the Asturian (Westphalian D) (Ziegler, 1990) and/or during the Stephanian, at which time the orogen collapsed (Arthaud & Matte,

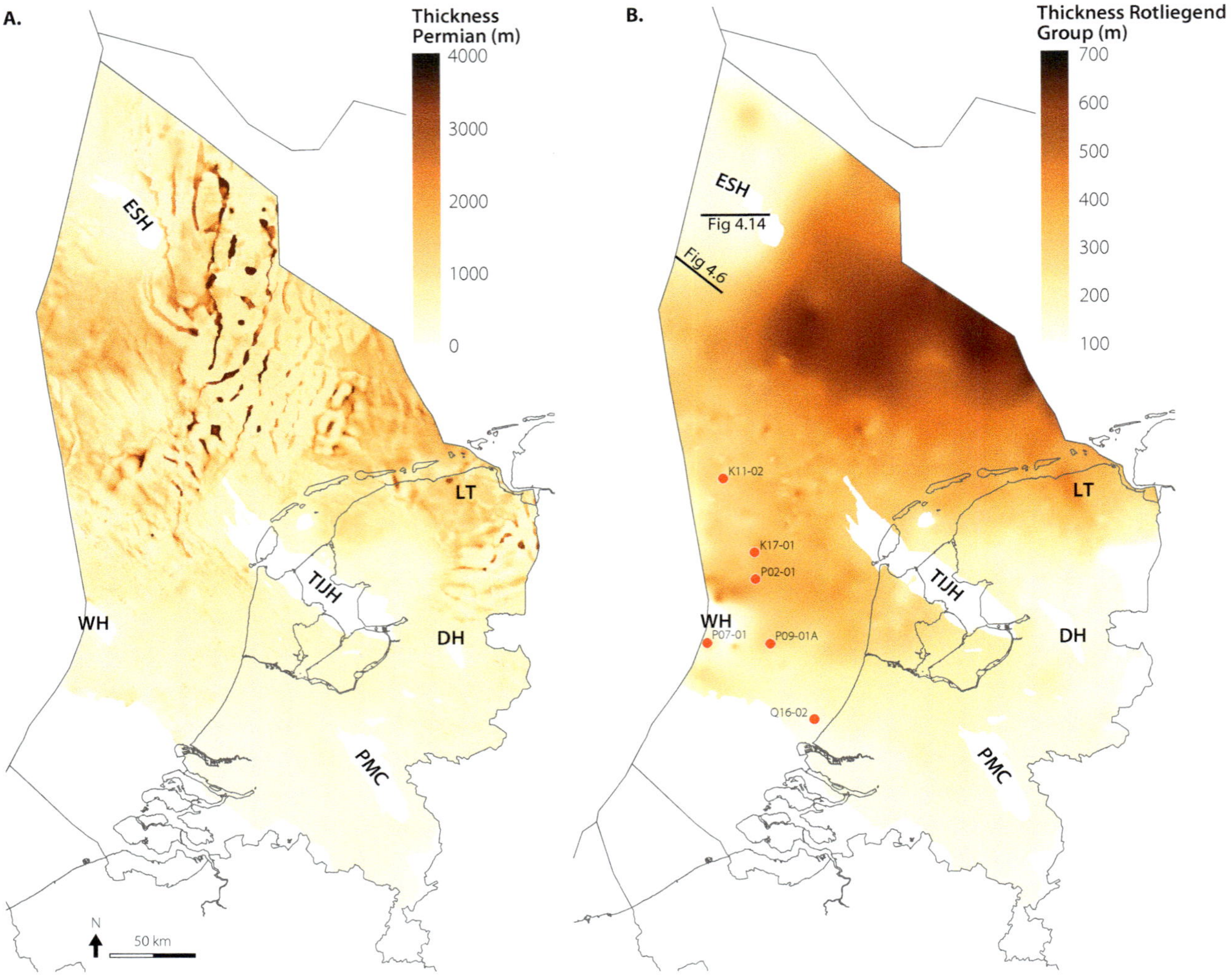

Figure 4.5. *Thickness maps of the complete Permian System (a) and Rotliegend Group (b). A Permian succession thicker than 2500 m corresponds to Zechstein salt bodies (salt diapirs and walls). DH = Dalfsen High; ESH = Elbow Spit High; LT = Lauwerszee Trough; PMC = Peel-Maasbommel Complex; TIJH = Texel-IJsselmeer High; WH = Winterton High. Note that a map for the thickness of the Zechstein Group is shown in Bouroullec & Ten Veen (2025, this volume).*

1977; Ziegler, 1989). Some of the deformation of Carboniferous strata to the north of the Variscan Front in the Netherlands post-dates the Variscan folding and has been attributed to late Carboniferous or early Permian wrench tectonics (Pharaoh et al., 2010). This late Variscan tectonic phase was accompanied by widespread plutonic and volcanic activity (Heeremans et al., 2004), which affected the Netherlands only marginally compared to Germany or Poland. The Stephanian history of the SPB is poorly understood as this was a period of uplift and erosion during the later phases of Variscan shortening. It resulted in a major unconformity between the Permian and older deposits known as the Base Permian Unconformity.

Several minor tectonic pulses took place during the late Permian (Gast, 1988; Plein, 1995; Glennie, 1998; Geluk, 1999). These are thought to herald the Permian breakup of Pangea (Vai, 2003). The pulses occurred prior to and during the deposition of the Upper Rotliegend Group (the Saalian and Altmark I to III pulses; Plein, 1995) and during deposition of the Zechstein (Tubantian I; Geluk, 1999), when a connection between the Southern Permian Basin and the Barents Sea was created. The Saalian and Altmark pulses influenced the distribution of basal Upper Rotliegend sands (Van der Baan, 1990; Van de Sande et al., 1996; NITG, 1998). The Tubantian II pulse is thought to be of compressional origin and related to the Uralian orogeny (Geluk, 2005). It marks the second, final, consolidation of Pangea at the end of the Permian (Vai, 2003).

There is evidence for considerable differential fault movement during deposition of the Z1 (Werra) Formation (Wolf, 1985; Ziegler, 1989; Geluk, 1999): the Tubantian I phase. The fault activity concentrated on the anhydrite platforms at the margins of the Zechstein basin. Loading of rapidly deposited anhydrite on a differentiated substrate is thought to have amplified these movements (Geluk, 1999). A series of half grabens and pull-apart-type basins formed, with fault offsets up to 250 m in the Central Netherlands Basin. During this faulting, the first fault – dip-closed traps in Upper Rotliegend sandstones were created, only 3 to 5 Myr after their deposition.

The syn-depositional Tubantian II movements of the upper Zechstein are more difficult to identify, but are seen in an unconformable contact between the Zechstein Upper Claystone and older Zechstein rocks. They resulted mainly in tilting and uplift and have been interpreted by Geluk (2005) as being related to mild compression. Source areas south of the basin were uplifted, and sands were shed into the basin in the western offshore area and the UK sector (Z4 or Hewett Sandstone).

Permian unconformities

The Base Permian Unconformity (BPU) is one of the most important unconformities in the regional geology of the southern North Sea (De Jager & Geluk, 2007). Where the Upper Rotliegend overlies Visean to Stephanian deposits, this hiatus represents 35 to 80 Myr (Figs 4.1, 4.2, 4.6). The BPU often represents an amalgamation of several unconformities into a single mega-unconformity (Glennie, 1998) that encompasses the entire early Permian. Where the Upper Rotliegend overlies Namurian to Stephanian deposits, as is the case in most of the Netherlands, this hiatus duration is 40 to 60 Myr. The term 'Saalian Unconformity' should be used with care since it refers to one such event, located in places where lower Permian deposits are still present above the BPU.

The Saalian Unconformity (Stille, 1924) can be distinguished from the BPU where lower Permian deposits are still present (Fig. 4.2), in the Netherlands only in small areas in the east and the north (Fig. 4.8). The time gap represented by the Saalian hiatus is poorly constrained in the Netherlands due to the lack of well penetrations and dating, but recent publications from neighbouring basins that include dating in shed some new light on the hiatus. In parts of mid-European basins, following the early Permian volcanism, sedimentation restarted in the late Permian at the time of the major geomagnetic Illawarra Re-

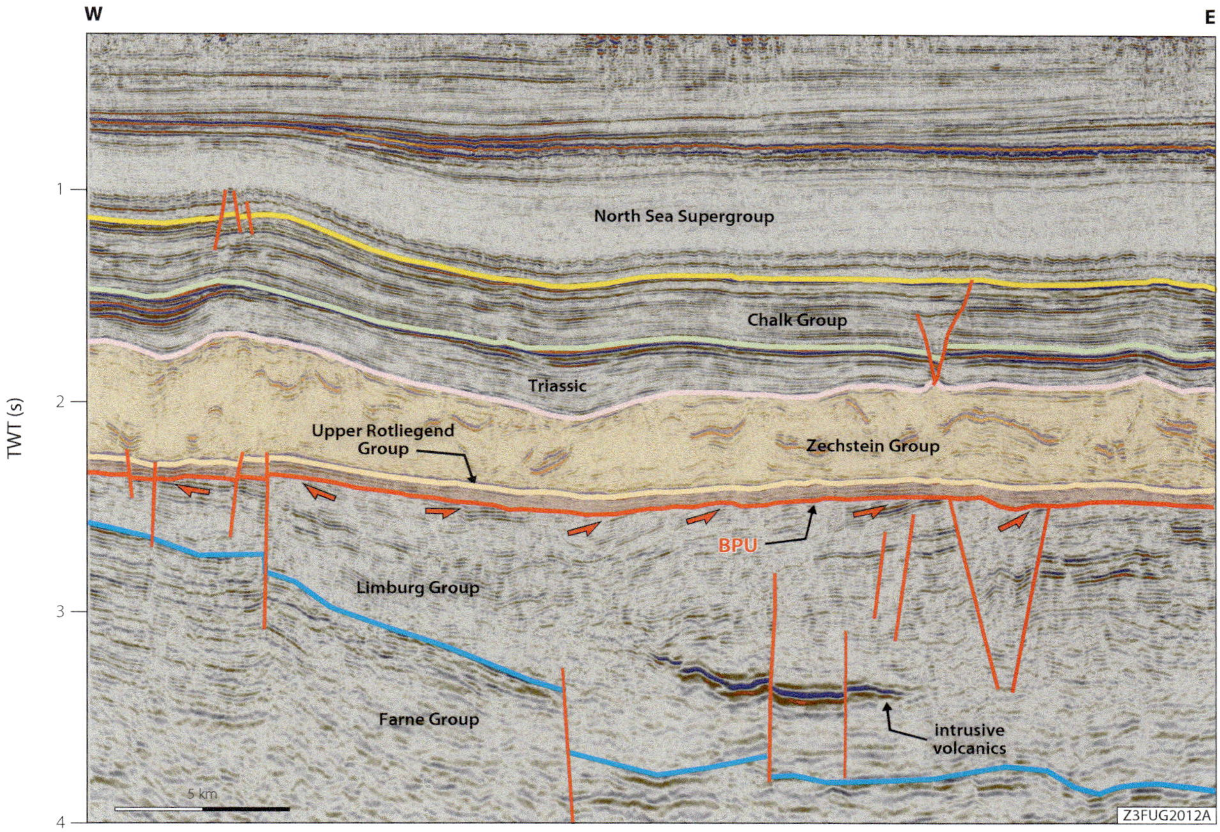

Figure 4.6. Interpreted seismic time section in the northwestern part of the Dutch offshore. The Base Permian Unconformity (BPU) here also includes the Saalian Unconformity and is clearly expressed as an angular unconformity, while the base and top of the Zechstein Group are concordant. See Fig. 4.5b for location.

versal (266.66 ± 0.76 Ma; Hounslow & Balabanov, 2018). Lützner et al. (2021) indicate that the duration of early Permian volcanism in this pull-apart Thuringian Forest Basin of central Germany was relatively short, around 4 Myr, from 300 Ma to 296 Ma and was followed by a hiatus of 25 Myr before deposition resumed in the middle to late Permian. The extent of the Saalian hiatus in the Netherlands may therefore be shorter than previously estimated (e.g. 40 Myr by Pharaoh et al., 2010).

Despite the extended period of non-deposition, the unconformities may be very difficult to identify in some areas, because of the general absence of fossils and the similarity of the red beds of the Asturian, Stephanian and Permian. Biostratigraphic zoning based on insects, especially cockroach wings, as well as amphibians, vertebrate footprints and conchostracans has been used successfully in German, Czech and Polish basins (Opluštil et al., 2017; Schneider et al., 2020), but no new data are yet available in the Netherlands.

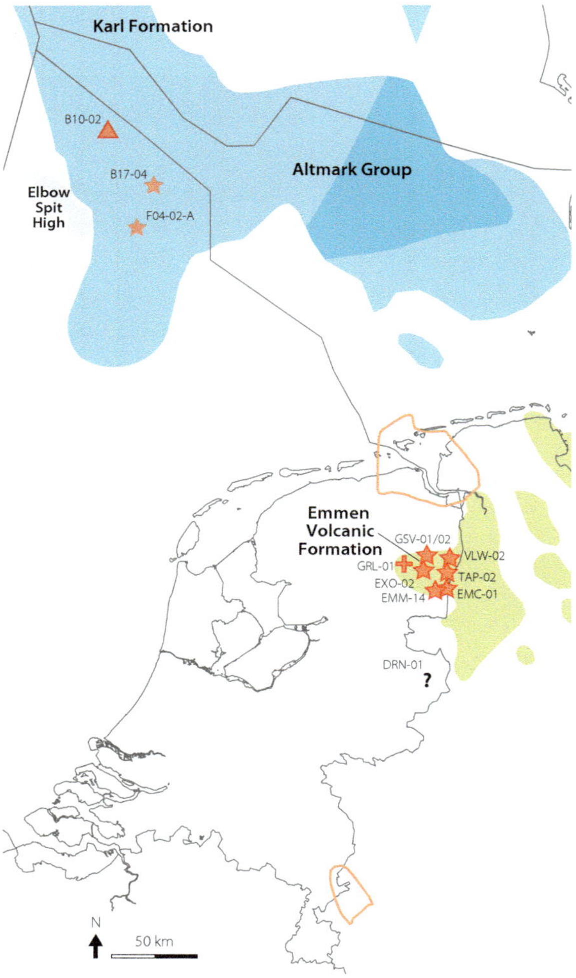

Stratigraphy

Lower Rotliegend Group

Primarily based on well data the Lower Rotliegend Group occurs in the Lower Saxony Basin, the northern part of the Dutch Central Graben, the Step Graben, and extends north-eastward into Germany and Denmark (Fig. 4.7). In the northern offshore and eastern onshore the group is represented by interbedded volcanics, volcaniclastics, and siliciclastics (Fig. 4.8). In the Lower Saxony Basin, it is represented by the Emmen Volcanic Formation. The climatic change that started in the late Westphalian from a hot tropical setting to arid conditions (red beds of Westphalian D and Stephanian) continued during the Rotliegend.

The volcaniclastics in the Lower Saxony Basin (Fig. 4.7) represent a western extension of a large area of volcanics developed in Germany (Plein, 1995) They consist of red-brown to green, spilitic, basaltic volcanics, mudstones and tephra layers. Several stacked lava flows occur in the succession. The thickness of the volcanics reaches a maximum of 80 m (NITG, 2000). The group is placed in the lower Permian in Germany based on radiometric age dating (291-289 Ma) and the lithological composition of the basalts (Plein, 1995). However, a K-Ar age for the Lower Rotliegend Group in the eastern Netherlands (well Drouwenermond-1), records a much younger volcanic pulse, which is probably incorrect (middle or late Permian, 258 ± 6 Ma; Sissingh, 2004; Van Bergen & Sissingh, 2007; Van Bergen et al., 2025, this volume).

Figure 4.7. *Locations of Lower Rotliegend Group volcanic rocks as found in wells in the Lower Saxony Basin and the northern Dutch offshore east of the Elbow Spit High. Wells and batholiths taken from Van Bergen et al. (2025, this volume). The distribution and geochemical composition of late Paleozoic volcanic rocks is taken from compilation figure 7.16 in Gast et al. (2010).*

In the Dutch Central Graben the wells B10-02, B17-04 and F04-02-A penetrate lower Permian volcanics; that is possibly also the case for wells A15-01 and F07-02. A maximum thickness of 190 m is recorded at well B10-02 (Fig. 4.8). The succession is lithologically similar to that of the Danish Karl Formation (Geluk, 2005). The lateral extent of the group is delineated towards the west, north and south by well data but remains uncertain to the east. It consists of tephra and of basaltic lava flows up to several tens of metres thick, interbedded with claystones and subordinate sandstones. De Bruin et al. (2015) identified an overall NNW-SSE-oriented depositional trend in the northern Dutch offshore with a lower sand content towards the WSW, away from the German and Danish main volcanic provinces. Stemmerik et al. (2000) identified three distinct volcanic pulses in the Permian of the Danish North Sea. The first (300-288 Ma) and second phases (281-276 Ma) both show an andesitic and rhyolitic composition and are present in the western part of the Central North Sea Graben and the Horn Graben. They are followed by a younger, basaltic, pulse (269-261 Ma, middle Permian). Although dates from the Dutch Central Graben are lacking, it is likely that the Danish volcaniclastics correlate with this younger pulse in the Netherlands.

Upper Rotliegend Group

The Upper Rotliegend Group is subdivided into two formations, the Slochteren Formation, which is mainly composed of sandstones and conglomerates and the Silverpit Formation, comprised of claystones, siltstones and evaporites (Figs 4.8, 4.9). The two formations are lateral equivalents with the Silverpit Formation deposited in the deeper, axial part of the Silver Pit Lake and the Slochteren Formation along its margins. The Upper Rotliegend Group reaches a maximum thickness of more than 700 m in the Schill Grund Platform in the eastern offshore but thins rapidly both towards the south and the north (Figs 4.8, 4.9). The oldest sediments of the group probably occur in the eastern offshore in quadrants G, H, M and N. The sequence is locally absent due to erosion on the Texel-IJsselmeer High and nearby smaller highs (Rijkers & Geluk, 1996), and because of nondeposition in most other areas such as the Elbow Spit High (Geluk, 1999; De Bruin et al., 2015).

There is strong evidence that syn-depositional fault movements affected the Upper Rotliegend Group. In the western parts of the Central Netherlands Basin, the thickness varies across major faults (RGD, 1993). It is likely that also the northern part of the West Netherlands Basin formed a fault-bounded depression, where up to 100-m-thick eolian sands occur. Other areas where faulting occurred are the eastern Netherlands (NITG, 1998), the area north of the Texel-IJsselmeer High (RGD, 1991a,b) and the Lauwerszee Trough (RGD, 1995; Fig. 4.5b). Thickness variations across faults are up to 45 m. Some of the syn-depositional faults form positive flower structures indicative of strike-slip motion, but the timing of the deformation is unclear, with some minor reactivation of pre-existing Carboniferous faults being a possibility. Based on the map-view of some of these syn-depositional faults, most are oriented NW-SE and show dextral strike slip (De Bruin et al., 2015).

The *Silverpit Formation* comprises siltstones, claystones and evaporites. In the Dutch Central Graben, tephra layers occasionally occur in its lowermost part. The formation reaches a thickness of over 700 m in the eastern offshore area and was formed in a playa lake depositional environment. In a more proximal setting, the formation is subdivided into two members, the Ameland Member and the Ten Boer Member. In the distal part of the basin, a subdivision into three members is made, i.e. the Lower and Upper Silverpit Claystone and the Silverpit Evaporite members (Fig. 4.9). The evaporitic intervals mainly comprise rock salt with subordinate anhydrite and interbedded clay-siltstones. The individual rock-salt layers form good correlation horizons throughout the Southern Permian Basin (Figs 4.8, 4.9). Their thickness reaches up to 50 m. In northern Germany and part of the adjacent Netherlands offshore (quadrants F, G), these layers have been mobilized into salt pillows. During the Triassic and Jurassic evaporite intervals locally acted as detachment horizons for syn-depositional faults and in Germany even formed diapirs. The salt layers are not composed of pure halite but contain abundant thin claystone beds since they were deposited in playa lake setting, as opposed to the marine conditions that came later during the Zechstein. The formation interfingers southward with the Slochteren Formation forming a lateral facies change and several tongues of the Silverpit Formation (Ameland and Ten Boer members) reach far to the south (Fig. 4.9).

The *Slochteren Formation* consists mainly of conglomerates and sandstones of fluvial and eolian origin. These facies can be distinguished in cores and on dipmeter logs. In the zone where the Slochteren and Silverpit formations interfinger, the former splits up into two main sandstone members, the Lower and Upper Slochteren members, separated by clay- and siltstones of the Ameland Member (Figs 4.9, 4.10a, 4.11). Eventually, both Slochteren members grade northward (in the south) and southward (in the north) into siltstones and claystones of the Silverpit Formation in the basin axis. The Lower Slochteren Member is present in the central and northern parts of the offshore and in the northern part of the onshore (Fig. 4.10a). It varies in thickness due to the presence of pre-existing

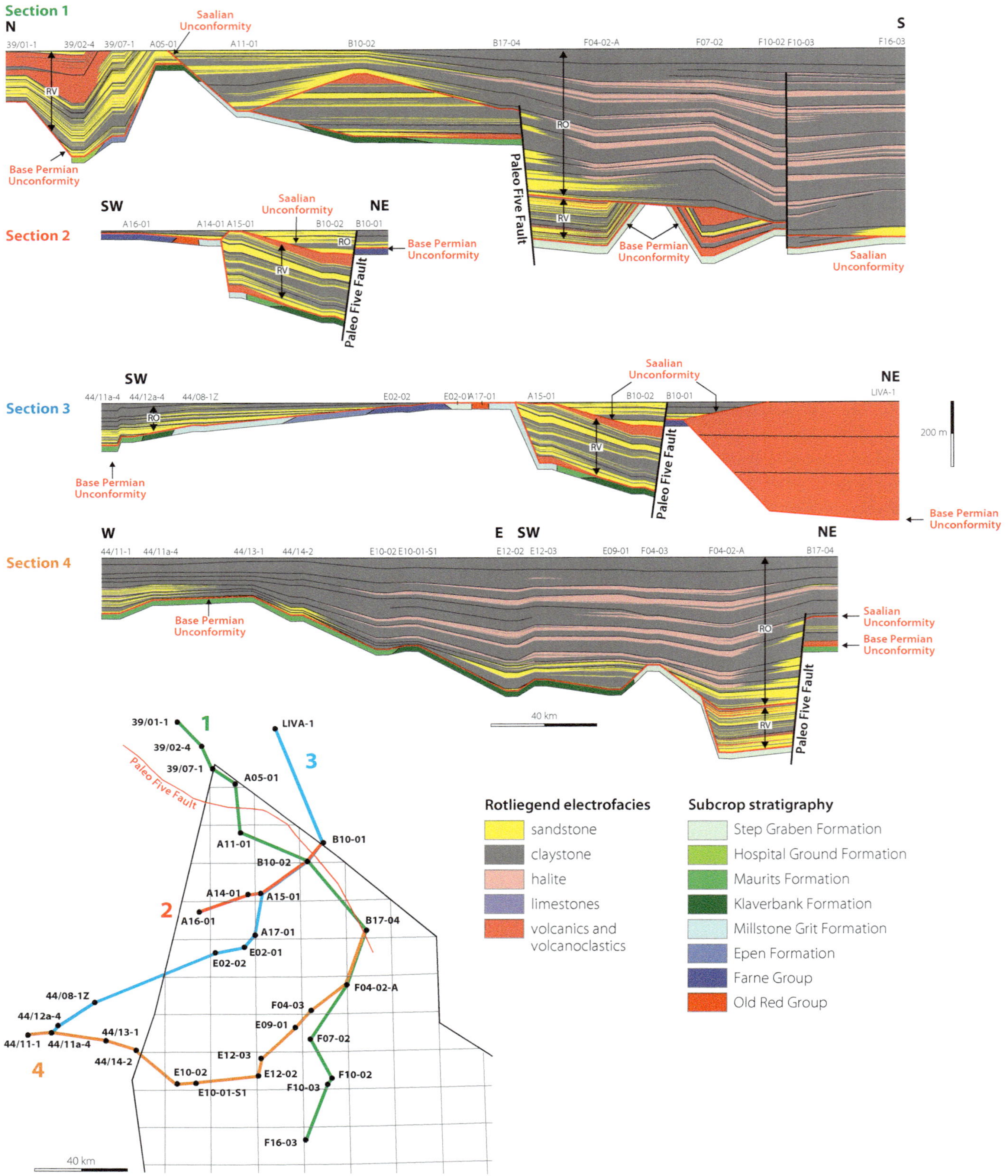

Figure 4.8. Stratigraphic correlation panels of the Lower and Upper Rotliegend groups in the Dutch Northern Offshore. Panels are flattened on the top Rotliegend, which is locally eroded in the northern part of the Dutch offshore and in the UK sector. RO = Upper Rotliegend Group; RV = Lower Rotliegend Group.

paleo-highs and -lows. The Upper Slochteren Member is well developed along the southern basin margin and is less conglomeratic than the Lower Slochteren Member. The fluvial systems of the Slochteren Formation were oriented at right angles to the basin axis with a south-to-north run-off from the southern margin of the SPB and a mainly north-to-south run-off from the northern margin.

In the southern part of the basin, the Upper Slochteren Member is partially composed of two fluvial systems located in the Dutch southern offshore and northeast onshore areas. (Fig. 4.10b). These two wadi belts were separated by a large area where eolian sediments accumulated. Detailed studies indicate that fluvial and eolian facies are interbedded in a complex way and shifted laterally through time (George & Berry, 1994; Verdier, 1996; Fryberger et al., 2011); rapid vertical and lateral facies alternations occur even on a metre-scale. Eolian sandstones vertically dominate the middle part of the formation (NITG, 2000). Within the eolian deposits a further distinction can be made between dunes, which occur predominantly in the southern central areas, and damp to dry sand flats closer to the playa lake (Glennie, 1998; Fig. 4.11). The Slochteren Formation along the northern margin of the SPB is not developed as extensively as in the southern margin (Figs 4.8, 4.10a, 4.11). There, the drainage area was limited to a much smaller area, composed of Devono-Carboniferous strata exposed on the Mid North Sea High and Elbow Spit High in the Dutch offshore. The recent Cygnus discovery (Catto et al., 2017), located 50 km to the west of the UK/Dutch border, has up to 50 m thick stacked Lower Leman Sandstone units suggesting that comparable sand-rich Slochteren sandstones may be present along the southern and eastern flanks of Elbow Spit High. Paleo-reliefs of up to 46 m formed by differential erosion of the Carboniferous subcrop are observed in the vicinity of the Cygnus Field in the UK sector.

The *Ten Boer Member* forms a transitional unit between the sandy Upper Slochteren Sandstone and the marine sediments of the Coppershale Member (or Kupferschiefer) of the basal Zechstein Group (Figs 4.9, 4.11). It marks the end of the clastic sedimentation and the onset of the Zechstein transgression, consists of shales and fine-grained sandstones and forms part of the Silverpit Formation. Locally, close to the basin edge, it may contain substantial amounts of sand. The sandstone beds are thought to have been deposited by fluvial channels and associated unconfined sheet floods in the distal part of fluvial systems (Donselaar et al., 2011). The informal Akkrum Sandstone member (equivalent to the Hyde Sandstone in the UK) which is locally present in Friesland and which is also known as the Grauliegend, can be regarded as one of the thicker sandstone beds of the Ten Boer Member.

A common, but not yet fully understood phenomenon is the Weissliegend facies. Originally described as a white to grey coloured formational unit that comprises the upper 50 to 60 m of the uppermost part of the Rotliegend sandstone (Glennie & Buller, 1983; Strömbäck & Howell, 2002). It contains mass-flow and in-situ soft-sediment deformed deposits and is thought to have formed when the Zechstein Sea flooded the Rotliegend sand sea. Whether the deformation was the result of a sudden flooding, or a gradual rise of the groundwater table remains uncertain. In any case, the Weissliegend has poor reservoir properties and is variable in thickness.

The Zechstein transgression reached the Southern Permian Basin from the Barents Sea region in the north, through a proto-Atlantic rift graben that first flooded the Northern Permian Basin (Glennie & Buller, 1983; Ziegler, 1990). The Horn Graben and the Central Graben formed N-S passages to the Southern Permian Basin which subsequently became flooded. A number of predecessors of the Zechstein transgression have been recorded in the Rotliegend and have been employed to make large-scale correlations across the basin (Legler & Schneider, 2008; Minervini et al., 2011). Based on a variety of biological, lithological, and geochemical indicators, Legler & Schneider (2008) concluded that at least some of the correlation markers represent marine ingressions.

Due to their depth of burial, the Rotliegend sediments are prone to and have been subjected to a variety of diagenetic processes. An extensive discussion of all these topics is beyond the scope of this chapter. The interested reader is referred to overview studies by Gaupp et al. (1993) and Gaupp & Okkerman (2011).

Zechstein Group

The Zechstein Group comprises five evaporite cycles (Z1-Z5), each of formation rank (Van Adrichem Boogaert & Kouwe, 1994). The Zechstein Upper Claystone Formation covers these cycles unconformably (Fig. 4.11). The lithostratigraphy of the Zechstein closely conforms to the concepts of genetic sequence stratigraphy (Galloway, 1989); three formation boundaries are picked on maximum flooding surfaces (base Z1: Coppershale Member; base Z3: Grey Salt Clay Member; base Z4: Red Salt Clay Member). These units form important correlation markers between the basin and basin-fringe deposits within the Southern Permian Basin (Geluk et al., 1996, 1997; Taylor, 1998). With respect to the distribution of the Upper Rotliegend, minor overstepping of the basin margins took place during the Zechstein (Geluk et al., 1996). The depositional thickness of the Zechstein Group increases from less than 50 m in the southern Netherlands, to over 1200 m in the northern offshore.

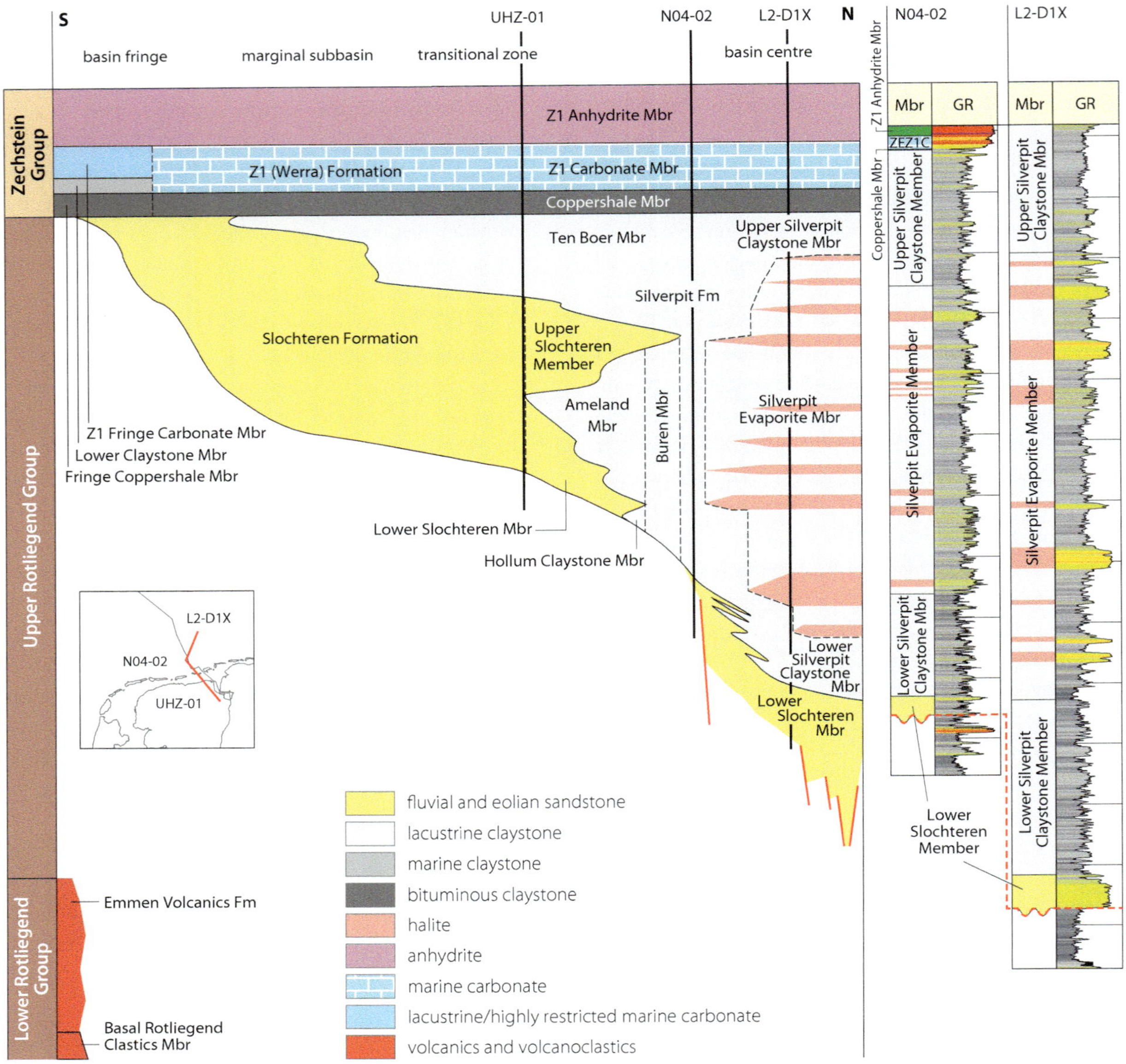

Figure 4.9. Stratigraphic subdivision of the Slochteren and Silverpit formations of the Upper Rotliegend Group in the eastern Dutch offshore and western German offshore. Modified from Doornenbal et al. (2019).

Deposition of the Z1 (*Werra*) Formation started with the Coppershale Member, a 0.5-m-thick, finely laminated claystone with a total organic-carbon content of up to 5% (Lokhorst, 1998), which represents a major marine transgression (Fig. 4.11). It was deposited in most of the Southern Permian Basin and provides an excellent marker horizon. It is found in most of the Netherlands, except for the southern onshore area. In some places (for example, in the Coevorden field) it forms a drape over ancient Carboniferous relief, with the Rotliegend missing. The Werra Formation further comprises claystone, sandstone, carbonate, anhydrite, and salt deposits, organized in several members (Figs 4.12a,b, 4.13b,d). The thickest occurrence, up to 500 m, is located in the eastern part of the Netherlands, in the province of Drenthe, where an anhydrite platform developed. The formation onlaps onto the London-Brabant Massif. In the main basin, the formation has a constant thickness of some 50 m. The N-S orientation of the anhydrite platform in the east, parallel to the occurrence of the Lower Rotliegend volcanics, points to a structural control on the outline of this platform.

The platform and slope deposits of the Z1 Carbonate consist of up to 200 m of marls and carbonates, in contrast to the 8 to 10 m of carbonate in the sediment starved basinal setting (Figs 4.12b, 4.13b). A carbonate platform was also present along the Mid North Sea and Ringkøbing-Fyn High (Taylor, 1998). Small carbonate platforms have been encountered in the southern part of the country and in adjacent parts of Germany (Visser, 1955; Fuchtbauer, 1980). Potassium-magnesium salts in the Z1 occur only in the eastern Netherlands (NITG, 1998). South of the Central Netherlands Basin, the formation is composed mainly of claystones, while in the offshore area, medium to fine-grained fluvial sandstones occur. Local sourcing of fine-grained siliciclastic material may have taken place from the Texel-IJsselmeer High (Van Adrichem Boogaert & Burgers, 1983). The carbonates and anhydrites from the Z1 cycle form important top seals for Rotliegend gas accumulations in the Netherlands (Van Ojik et al., 2021).

The Z2 (*Stassfurt*) Formation comprises a basal carbonate unit, the Z2 Carbonate, followed by the Z2 Basal Anhydrite and Z2 Salt members (Figs 4.11, 4.12c,d, 4.13a,c). In the southern onshore area, the formation is represented by anhydrite bearing claystones, while local sandstones (Z2 Fringe Sandstone Member) occur in the western offshore (Figs 4.11, 4.12c). The formation is less than 50 m thick in the southern Netherlands but locally reaches more than 700 m thick in the northern offshore due to salt movement (see Bouroullec & Ten Veen, 2025, this volume). The southern limit of the Z2 Carbonate (equivalent to the Hauptdolomit Formation in UK) lies considerably further to the north compared to the Z1 Carbonates. Within the Z2 Carbonate, three facies realms can be identified: platform, slope and basin (Figs 4.12d, 4.13a,b, 4.14; Van de Sande et al., 1996; Geluk, 2000; Bouroullec et al., 2022). The basinal deposits comprise 8 to 12 m thick, dark-coloured, bituminous, finely laminated carbonates,

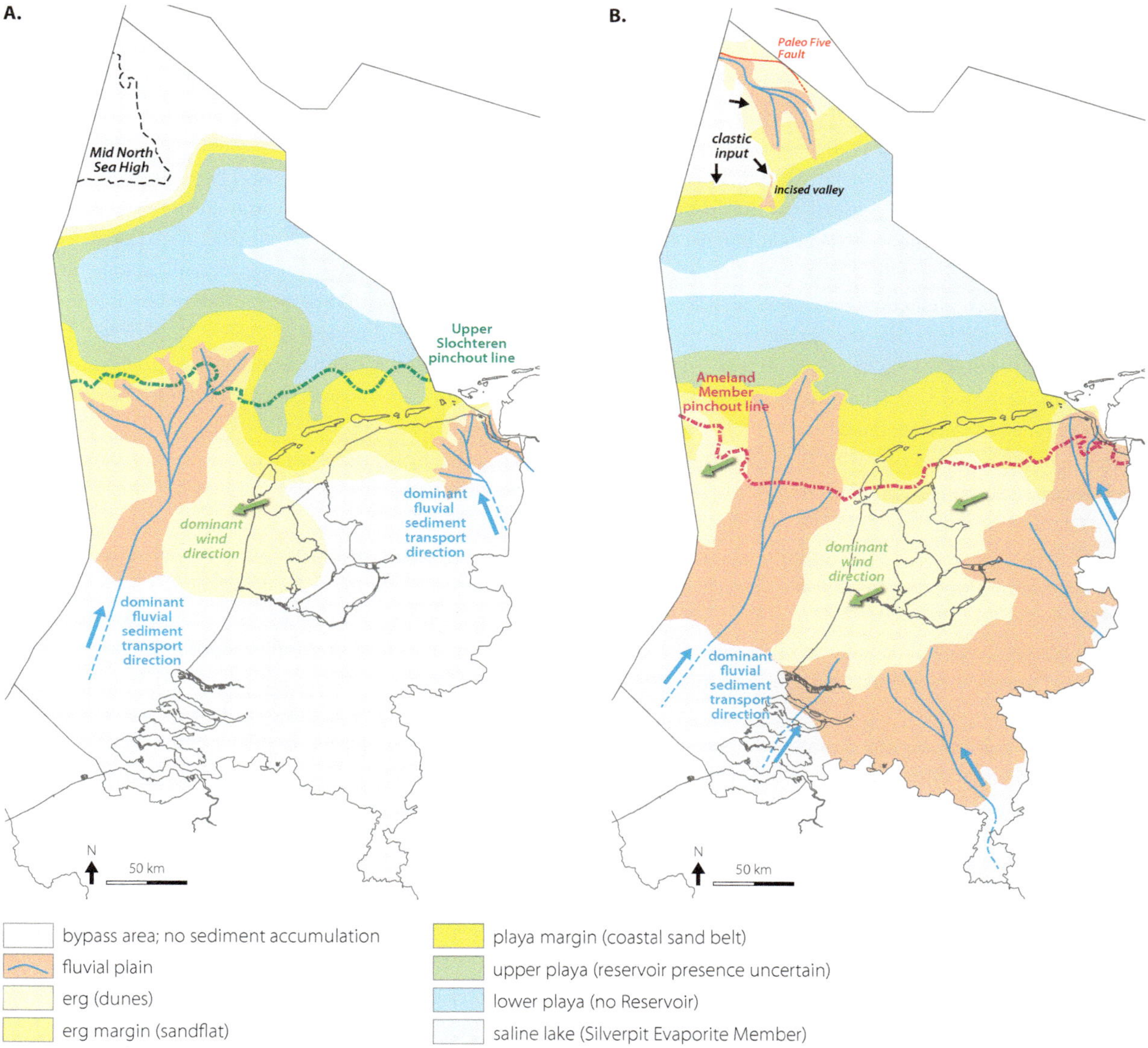

Figure 4.10. Depositional setting of a) the Lower Slochteren Member and b) the Upper Slochteren Member. Modified after Gast et al. (2010), De Bruin et al. (2015), Houben et al. (2020) and Kortekaas et al. (2023).

known as 'Stinkschiefer' (Fig. 4.13c). These rocks have source rock potential, with a total organic-carbon content of up to 1.2% (Lokhorst, 1998). Apart from the basinal deposits, high-TOC carbonates have also been found in slope and lagoonal deposits in Poland (Słowakiewicz & Gąsiewicz, 2013). Bituminous, thicker bedded slope deposits are sometimes referred to as 'Stinkkalk'. In the vicinity of the carbonate platforms slumps and turbidites of displaced shelf deposits have been identified (Amiri-Garroussi & Taylor, 1992; Van de Sande et al., 1996). The slope facies includes light-coloured limestones and dolomites, as well as of platform sediments redeposited by mass flows. In the eastern Netherlands, these facies reach a thickness of over 200 m (Van de Sande et al., 1996). The platform facies comprises a complex of oolitic, pelletoidal, bioclastic and pisolitic pack- and grainstones and finely laminated wackestones (Clark, 1986; Van der Baan, 1990; Strohmenger et al., 1996; Patruno et al., 2018). These reflect a wide range of depositional settings: sabkha, tidal flats, algal shoals, lagoons and ooid bars and shoals (Strohmenger et al., 1996; Patruno et al., 2018). 3D seismic interpretation reveals a complex outline of the platform, including isolated off-platform highs (Van de Sande et al., 1996). The shape of the Z1 Anhydrite platform controlled to a great extent the facies distribution of the Z2 Carbonate. The carbonates grade southward into anhydritic clay-siltstones, and in the eastern Netherlands, rock salt (NITG, 2000). In the eastern Netherlands, the Z2 carbonate platform complex measures some 50 to 70 km in a N-S direction, whereas in the western offshore area this is much less, between 10 and 30 km. Recent mapping of the Z2 carbonate platform in the northern Dutch offshore (Bouroullec et al., 2022) shows several isolated platforms around the Elbow Spit High and in quadrant A, at the northern tip of the Dutch offshore (Fig. 4.14). Some of these platforms show a clear platform-slope-basin geometry. The area between the platforms and the paleo-topographic high formed around the then-exposed core of the Elbow Spit High and consists of lagoonal deposits as described for the core of well A16-01 (Bouroullec et al., 2022).

The topography of the Z2 carbonates controlled the thickness and distribution of the overlying units; the thickest rock salts were deposited in the basinal zones, with only thin salt in platform areas. The depositional thickness of the Z2 Salt is more than 600 m in the basin. It is mainly composed of halite (>95%), within which three stages of deposition can be identified (Geluk, 2000; Geluk et al., 2000). During the first two stages the initial topography persisted, and platform halites graded northward into deeper-water salt complexes (intercalated halite, polyhalite and carnallite), whereas the third stage of salt deposition filled most of the remaining relief of the basin. At the top, a regionally well-developed potassium-magnesium salt unit occurs. Because of the extensive salt movement, the units overlying the Z2 Salt have become strongly deformed too, which complicates the reconstruction of their primary thickness and composition. In addition, salt may have been dissolved. Dissolution of salts is well known from the UK (Glennie, 1998; Patruno et al., 2018; Grant et al., 2019) and has been recognized in the Netherlands as well (Houben et al, 2020). It results in displacement of the overlying section. The dissolution slopes, which indicate the limits of salt removal, may be recognized on seismic.

The Z3 (*Leine*) Formation comprises the Z3 Fringe Sandstone, Z3 Fringe Claystone, Grey Salt Clay, Z3 Carbonate, Z3 Main Anhydrite and Z3 Salt members (Figs 4.11, 4.12e,f). The basal member of the formation, the Grey Salt Clay Member, forms an important regional marker within the Zechstein. Its thickness varies from 5 to 10 m. The facies pattern of the Z3 Carbonate is more complex compared to the Z2 Carbonate (Geluk, 2000; Grant et al., 2019), reflecting the levelling out of the paleo-topography by deposition of the Z2 Salt (Fig. 4.12e). In the basin centre the carbonate comprises a dark-coloured limestone of a few metres thickness only. The slope facies comprise laminated and bioturbated carbonate mudstones and silty dolomites, with a thickness of up to 40 m. The platform facies mainly consist of grey microcrystalline dolomites and algal boundstones. These are considered by Taylor (1998) to represent predominantly shallow, quiet-water, possibly lagoonal deposits. Oolitic and bioclastic grainstones locally occur in the area adjacent to the slope (Van Adrichem Boogaert & Burgers, 1983; Baird, 1993) and on the landward margin of the platform. In the southwestern offshore area, the carbonate grades into fluvial sandstones (Fig. 4.12e; Geluk et al., 1997).

The Z3 Main Anhydrite was deposited only on the northern part of the Z3 carbonate platform and slope and in the basin. The thickness of this anhydrite reaches up to 100 m, showing complex and rapid changes (NITG, 2000). Movement of the underlying Z2 Salt caused the breaking-up of this member into large slabs, known as rafts, floaters, stringers, or allochthons, which became embedded in the salt and may locally present potential drilling hazards by causing drilling fluid losses or influx of high-pressure brines or gas. The Z3 Salt is composed of a basal halite and an upper part comprising two thick potassium-magnesium salt layers. The salts encountered include beds up to 10 m thick of the rare mineral bischofite ($MgCl_2 \cdot 6H_2O$), one of the most soluble salts on earth, and kieserite, carnallite and sylvite which may be commercially exploited (Coelewij et al., 1978; Raith et al., 2017). These salts are present only in the northeastern onshore and northwestern offshore. The thickness of the Z3 Salt reaches 300 to 400 m.

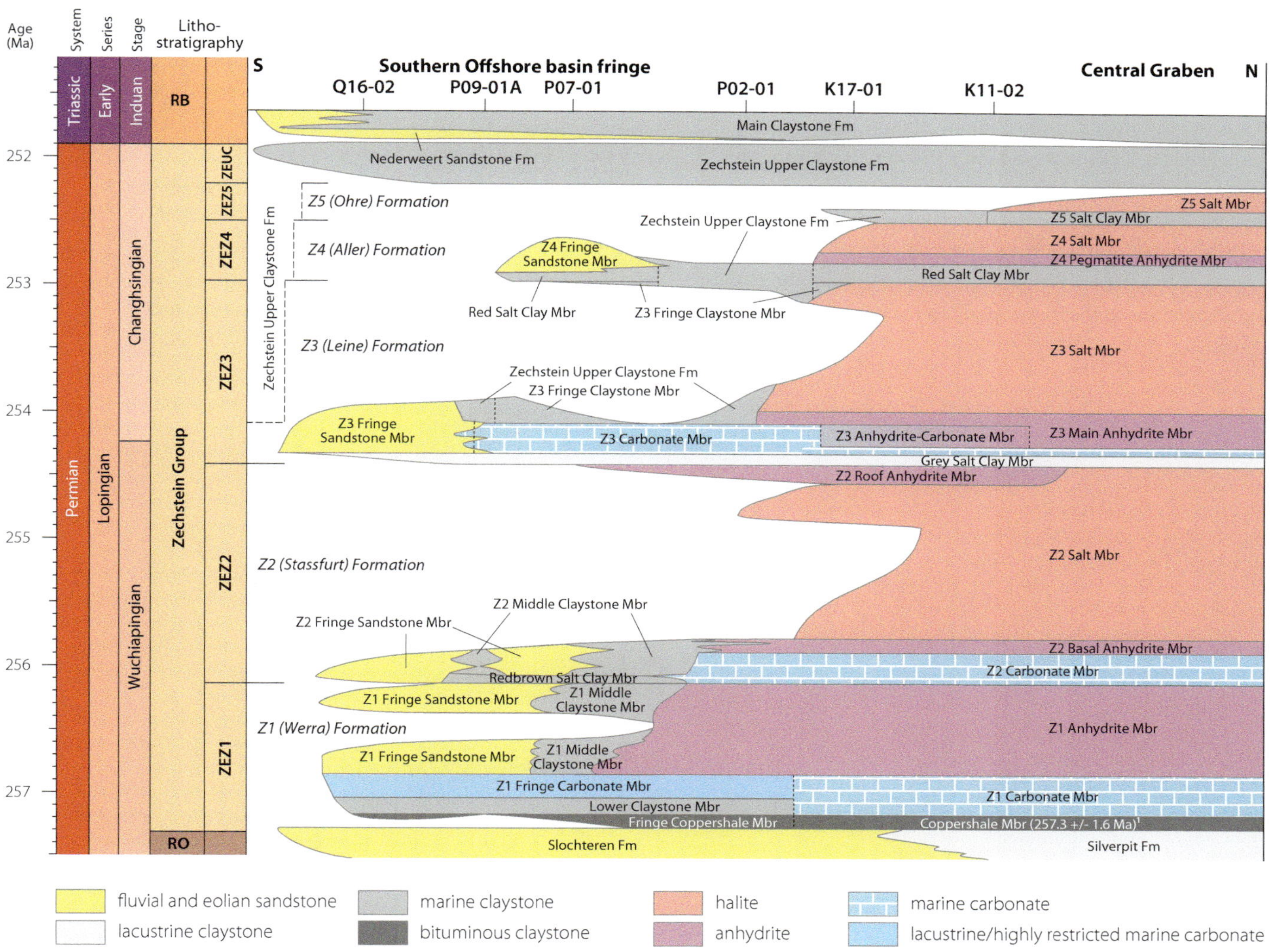

Figure 4.11. Tectonostratigraphic chart of the Zechstein Group. In the Netherlands, this group consists of five evaporite cycles (Z1-Z5). For location of the wells see Fig. 4.5b. Modified from Van Adrichem Boogaert & Kouwe (1994). 1 = Re-Os isochron age for the Coppershale (Kupferschiefer) (Brauns et al., 2003).

The Z4 (*Aller*) Formation is composed of the basal claystone of the Red Salt Clay Member and the Z4 Fringe Sandstone Member, followed by the thin Z4 Pegmatite Anhydrite and the Z4 Salt members (Fig. Figs 4.11, 4.12g). The lower two members are laterally extensive, whereas the Z4 Salt is found only in the depocentres or as isolated occurrences in half grabens (Geluk, 1999). The thickness of the salt reaches 150 m. Potassium-magnesium salts occur in the middle part of the salt; its upper part is composed of an alternation of halite and claystone. Along the southern basin margin, the formation comprises anhydritic claystones, deposited in sabkhas, and medium to coarse-grained fluvial sandstones in the southwestern offshore area (Fig. 4.12g). These sandstones are known in the UK as the Hewett Sandstone (Geluk et al., 1996).

The occurrence of the Z5 (*Ohre*) Formation is limited to the northeast of the country and to the northwestern offshore and outlines the depocentres towards the end of Zechstein sedimentation (Fig. 4.11). It comprises a basal claystone with a thickness of several metres followed by up to 15 m of halite. Younger Zechstein salts (Z6 and Z7) have been reported from northwest Germany (Best, 1989), but these are absent in the Netherlands, possibly due to nondeposition.

The Zechstein Upper Claystone Formation occurs throughout the country and accumulated disconformably on older rocks of the Zechstein and Upper Rotliegend Group (Fig. 4.11). Based upon its palynological content, the formation has been included in the Zechstein Group (Van Adrichem Boogaert & Kouwe, 1994). It is composed of red and grey anhydritic claystones and sandstones, deposited in a lacustrine to mudflat setting. Its thickness varies from 10 to 50 m.

Paleogeography

Deposition during the Permian took place north of the Variscan Mountains in the large Southern Permian Basin at paleolatitudes of around 10 to 20°N, comparable to the present-day Sahara. The mountains provided the main source of sediments that were transported northward and also prevented southern humid air masses from the Tethys Ocean from reaching the SPB (Glennie, 1998), resulting in an arid climate in the Netherlands and in neighbouring regions.

Lower Rotliegend Group

In early Permian times there was widespread intrusive igneous activity in the Netherlands (Van Bergen & Sissingh, 2007; De Bruin et al., 2015). During the middle Permian, local extrusive volcanism (Lower Rotliegend Group) occurred at different times in relation to wrenching along deep-seated faults in the Lower Saxony Basin and the Dutch Central Graben (Figs 4.6, 4.7). In the latter, minor amounts of clastic sediments were deposited. This extrusive volcanism is younger than in Germany and Poland, where it already started in late Carboniferous to early Permian times (Plein, 1995; Jakobsen et al., 2020). The

Wind and solar energy in the Permian

The Permian period was a time of great climatic change in the earth's history. During this period, the earth experienced a transition from a warm, wet climate to a much drier, desert-like climate. The main cause was the completion of the assembly of the supercontinent Pangea (see Paleomap reconstruction from Scotese (2014) below) which stretched from the North Pole to the South Pole and was surrounded by the Panthalassa Ocean. The obstruction of equatorial ocean currents and the presence of orogenic belts across nearly the whole western edge of Pangea resulted in rain shadows and persistent dry conditions.

This change in climate had a significant impact on the sedimentary deposits that were laid down during this time. Wind and sun were the most important agents affecting deposition in these vast desert-type environments.

Northeast trade winds prevailed in much of the interior of the continent. They were responsible for the erosion, transportation, and the eventual deposition of much of the sediment of the Permian age.

The sun also played a major factor in sedimentation during the Permian. Rotliegend sediments already witnessed periodic drying up and evaporation of shallow lakes as shown by the numerous mud cracks, anhydrites, and halite deposits. The evaporitic nature of the supercontinent became even more pronounced during deposition of the Zechstein evaporites, carbonates and siliciclastics.

Both wind and sun played major roles during the Permian and were instrumental in the creation of a very good reservoir, the Slochteren Sandstone and a world-class seal, the Zechstein evaporites. Today, the Netherlands are moving away from fossil fuels and toward a largely wind- and sun-powered energy supply system, bringing back in the forefront those agents that made the Dutch Permian geology what it is.

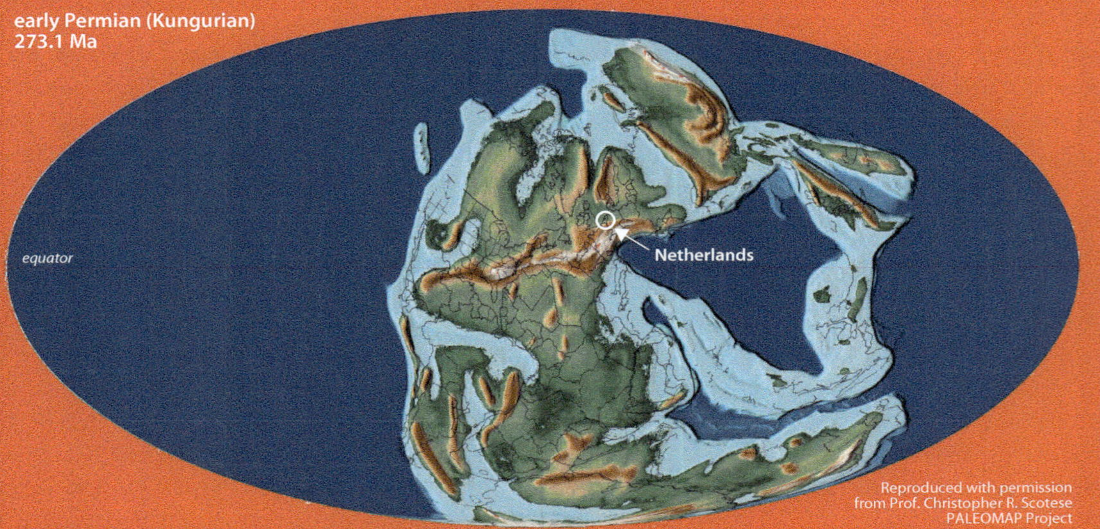

pulses are thought to have accompanied the opening of the Horn and the Central North Sea grabens (Glennie, 1997; Stemmerik et al., 2000). Prior to the deposition of the sediments of the Upper Rotliegend Group, erosion resulted in highly differentiated and locally deep truncation of the Carboniferous. The paleo-relief was not completely levelled out (Van der Baan, 1990; Van de Sande et al., 1996; De Bruin et al., 2015) and differential weathering and faulting were responsible for local relief of up to 45 m (Geluk & Mijnlieff, 2001; De Bruin et al., 2015). In the northeast of the country, the area of middle Permian volcanic activity still retained a positive relief and became covered only by the Zechstein transgression or, locally, by sediments belonging to the youngest part of the Upper Rotliegend Group (NITG, 2000).

Upper Rotliegend Group

Accumulation of sediments of the Upper Rotliegend Group started in Poland and northeast Germany during the early Permian (Upper Rotliegend I; Plein, 1995), expanded with time and reached the northern offshore of the Netherlands in the middle Permian (Fig. 4.10). Deposition did not keep pace with basin subsidence, resulting in a large inland depression lying below sea level (Glennie & Buller, 1983). In the central part of this area, a large playa lake developed where claystones and rock salt were deposited (Fig. 4.3). The bordering systems were formed by wadis with episodic run-off. With prevailing northeast trade winds, sand was deflated from fluvial deposits during dry periods and accumulated down-wind as dunes or, close to the playa lake, as damp sand flats (Glennie, 1998). In the Netherlands, sediments were supplied mainly from the southeast, where the Variscan Mountain belt was being eroded. Large amounts of sand and clay were transported northward by dryland river systems, with sands deposited in proximal settings in the south of the Netherlands and clay deposited in the distal settings in the central part of the SPB. A narrower and thinner sandy system also developed in the northern part of the Dutch offshore where Devonian and Carboniferous strata subcrop along Elbow Spit High and neighbouring paleo-highs in the UK. The Elbow Spit High likely was a sediment source and was drained southward, and possibly eastward, by ephemeral fluvial systems. The thickest Rotliegend sandstone in the Dutch offshore, >350 m, has been found in the P05 to Q04 areas. Further north, the eolian dune belt grades into playa deposits consisting of sandflats with increasing amounts of intercalated clay (Van Ojik et al., 2021).

The climate during deposition of the Upper Rotliegend Group was characterized by an alternation of wet and dry periods possibly influenced by Milankovitch cycles (Yang & Baumfalk, 1994; Howell & Mountney, 1997; Van den Belt & Van Hulten, 2011; De Jong et al., 2020). This cyclicity resulted in a periodic expansion and retreat of the lacustrine facies belts, and a complex alternation of fluvial and eolian deposits at the margins of the lake.

Zechstein Group

The claystone of the Ten Boer Member of the Upper Rotliegend Group represents an important transgressive phase, as well as a transitional phase to the overlying marine Zechstein transgression. The sudden marine transgression resulted in flooding and widespread reworking and redeposition of dune sands during the late Permian (Weissliegend facies of Glennie & Buller, 1983).

Under the prevailing dry climate, widespread cyclic deposition of evaporites took place. The controlling mechanism behind the cyclicity is thought to have been glacio-eustatic (Ziegler, 1990). Five Zechstein cycles have been identified in the Netherlands (Figs 4.11, 4.12), the bases of which are marked by transgressions. In the early cycles (Z1 to Z3), normal-marine conditions were established throughout the Southern Permian Basin, connected to the north to the Barents Sea region (Ziegler, 1990). Clastic influx was pushed back to the basin margins (Geluk et al., 1996) and carbonates were laid down as a series of prograding and aggrading sigmoidal shaped bodies, in which the facies belts and the area of maximum thickness shifted with time towards the north (Grant et al., 2019). During later cycles (Z4 and Z5) no carbonates were deposited as the conditions were permanently hypersaline.

Following the initial transgression, a starved basin up to 200 m below the paleo sea level formed. Marine deposition started with the anoxic sediments of the Coppershale Member (Sweeney et al., 1987), but at a later stage, normal-marine oxic conditions were established in the basin. The combination of an inclined basin topography, minor faulting and sea-level rise led to the establishment of locally small carbonate platforms over the southern onshore and western offshore areas (Van de Sande et al., 1996; Figs 4.12, 4.14). Around the London-Brabant Massif, extensive mudflats and sabkhas developed, whereas in the western offshore sands were deposited under fluvial to estuarine conditions. The geometry of the carbonate deposits in the western offshore suggests a relief of up to 200 m. After this transgression, the sea level fell, the connection with the Barents Sea became restricted and evaporites were deposited. A large anhydrite platform developed at the basin-ward side of this carbonate platform. In the shallow, warm water, anhydrite accumulated rapidly, in contrast to slow precipitation in the basin area north of the platform (Van der Baan, 1990).

The second Zechstein transgression did not reach as far south as the Z1 transgression (Fig. 4.12a,c). However, an extensive carbonate platform developed, with oolite shoals and lagoons (Van der Baan, 1990; Strohmenger et

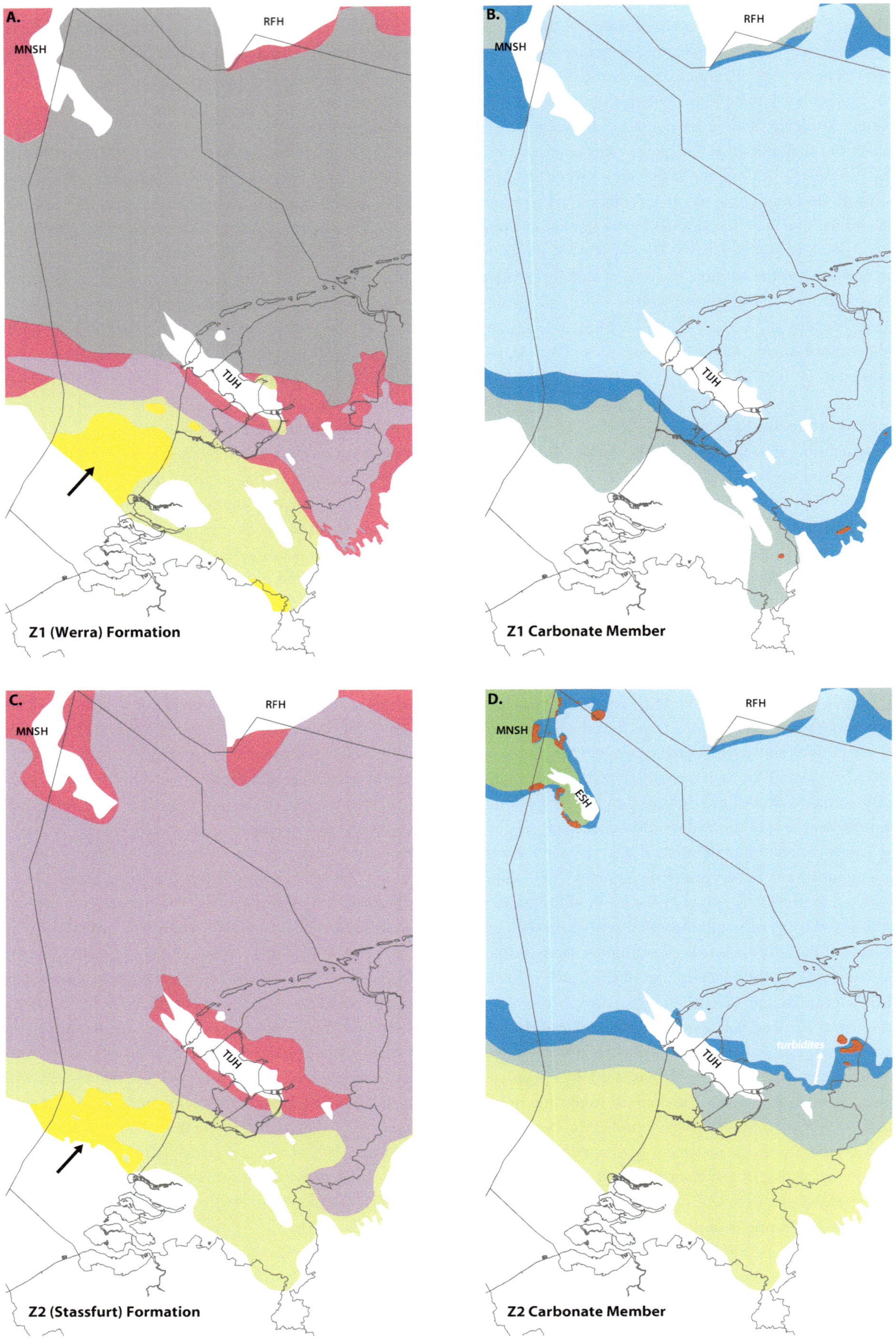
A.
RFH
MNSH
TUH
Z1 (Werra) Formation
B.
RFH
MNSH
TUH
Z1 Carbonate Member
C.
RFH
MNSH
TUH
Z2 (Stassfurt) Formation
D.
RFH
MNSH
ESH
TUH
turbidites
Z2 Carbonate Member

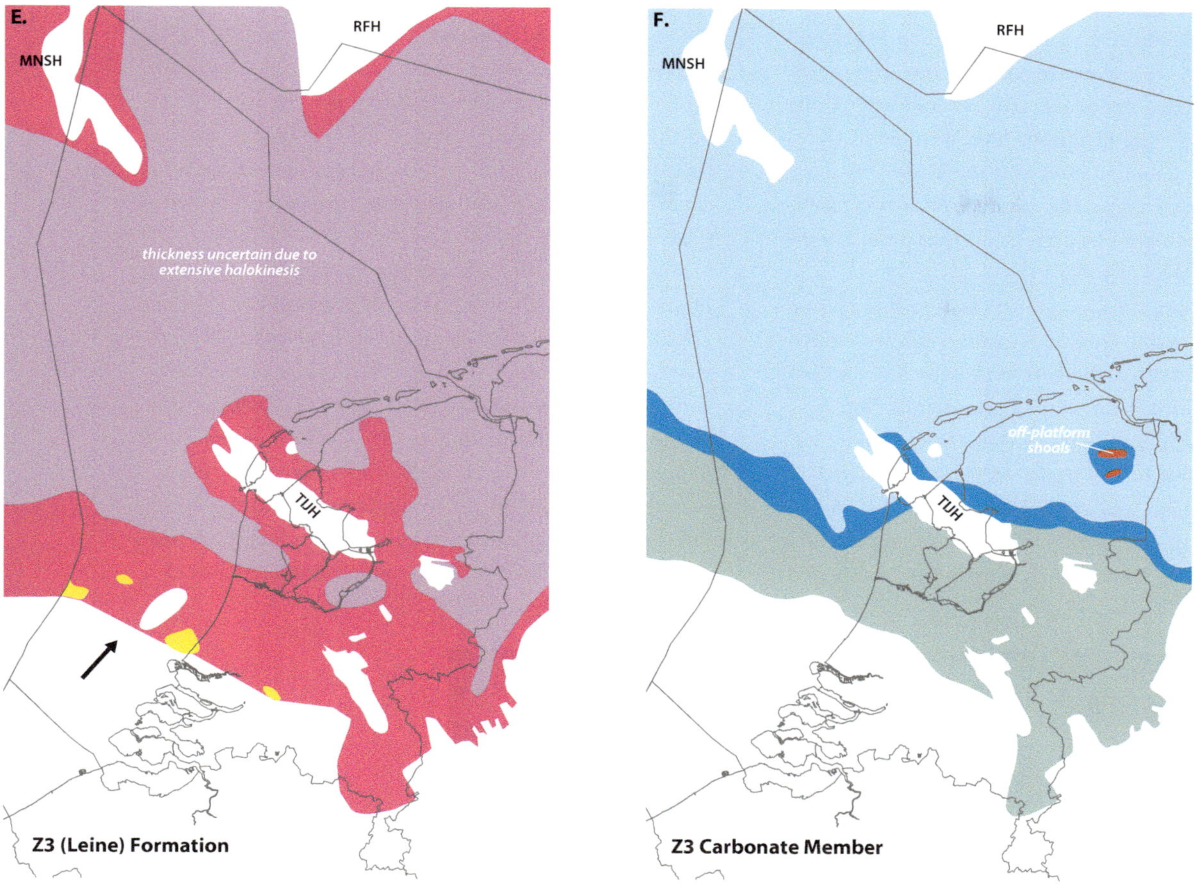

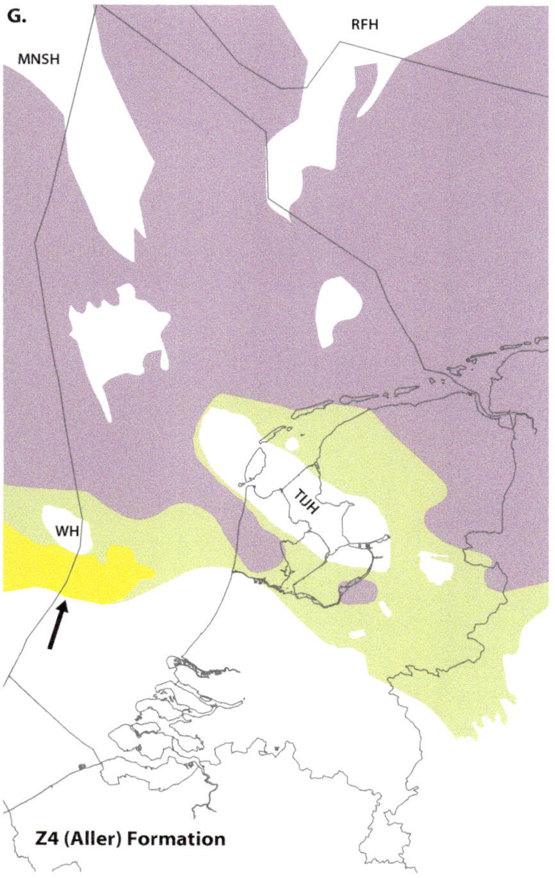

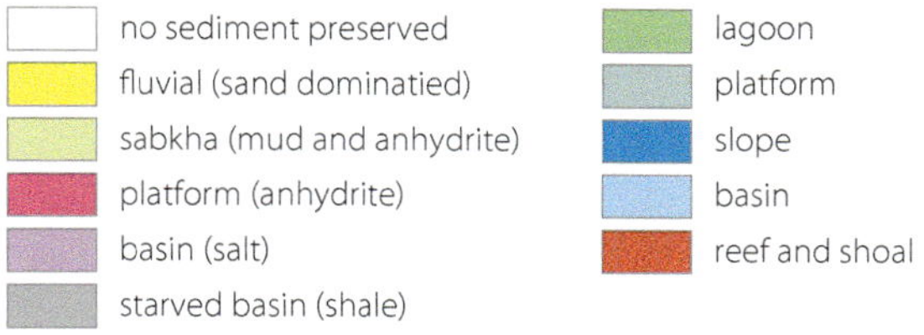

Figure 4.12. *Facies (a, c, e and g) and paleogeographic (b, d and f) maps of the Zechstein Group. a) The Z1 (Werra) Formation. Sands were shed into the basin in the western offshore (arrow). In the main basin a condensed succession of carbonates and evaporites was deposited. b) The Z1 Carbonate Member. Carbonate platforms fringed the basin both to the north and to the south. c) The Z2 (Stassfurt) Formation. A thick succession of rock salt was deposited in the centre of the SPB. d) The Z2 Carbonate Member. Well-developed platform facies are deposited in the eastern and northern part of the Netherlands. The carbonate platforms, reefs, shoals, slopes and lagoons located around and above the Elbow Spit High are from Bouroullec et al. (2022) and Peeters et al. (2023). e) The Z3 (Leine) Formation records widespread transgression, with carbonate development extending towards the margins of the basin. f) The Z3 Carbonate Member. This member has the greatest southward extension of all Zechstein carbonates and is characterized by a simpler, ramp-type development than the Z2 Carbonate. g) The Z4 (Aller) Formation is represented by thin sabkha-mudflat deposits. ESH = Elbow Spit High; MNSH = Mid North Sea High; RFH: Ringkøbing-Fyn High; TIJH: Texel-IJsselmeer High; WH: Winterton High. Modified from Geluk (2007), TNO-GDN (2022), Van Ojik et al. (2021) and Bouroullec et al. (2022).*

al., 1996; Van de Sande et al., 1996; Geluk, 2000; Fig. 4.12c, d). To the south, this platform graded into a complex of fluvial sandstones, mudflats, sabkhas and salt ponds (NITG, 1998). Lowering of the sea level resulted in widespread evaporite sedimentation and the basin was filled with up to 600 m of halite. The salt deposition was multicyclic, triggered by minor transgressions that caused salinity fluctuations (Geluk, 1995).

The third transgression records the last normal-marine salinities in the basin. This transgression reached much further southward than the previous ones and pushed fluvial deposits back as far as the basin margin (Fig. 14e, f). Less facies variation is seen within the carbonate deposits, as the relief in the basin was largely deflated to a maximum several tens of metres. A carbonate-anhydrite platform occupied most of the area on and around the Texel-IJsselmeer High, while carbonates dominated the southern Netherlands (Fig. 4.12f). Up to 10 m thick beds of bischofite and potassium-magnesium salts in the Z3 Salt indicate that partial desiccation occurred in the northeast of the country. Such thick beds likely required repeated influx of water to form in conjunction with basin-axis increase in accommodation space due to differential compaction of the Silverpit claystones.

From the fourth transgression onwards, conditions in the basin became permanently hypersaline (Ziegler, 1990). Claystones were deposited in hypersaline shallow water environments (Fig. 4.12g), but thick, massive halites and potassium-magnesium salts located in the lower part of the Z4 (Aller) Formation were probably still deposited in a marine environment. In the upper part of the formation, the alternation of saliferous claystone and thin halite indicates a gradual change to playa-type conditions. The brines, however, were still of marine origin. Deposition

Figure 4.13. *Examples of Zechstein facies as seen in cores and cuttings. a) High-energy oolite facies in ZEZ2C. Thin section of a cutting from well E02-02, 2105 m. 1 = dolomitized oolites; 2 = mouldic porosity caused by partial dissolution of oolites. Microphotograph: Bouroullec et al. (2021). b) Karstified contact (yellow dashed line) between shallow marine carbonates of ZEZ1A. 3 = ZEZ2 carbonate; 4 = large dissolution hole with anhydrite (5). Well E16-03, 3304 m. c) Bituminous, laminated, fine-grained carbonate, known as 'Stinkschiefer', with occasional anhydrite nodules. ZEZ2C, slabbed core from well D15-02, 3212 m. d) Chicken-wire anhydrite of ZEZ1A in a slabbed core from well D15-02, 3213 m.*

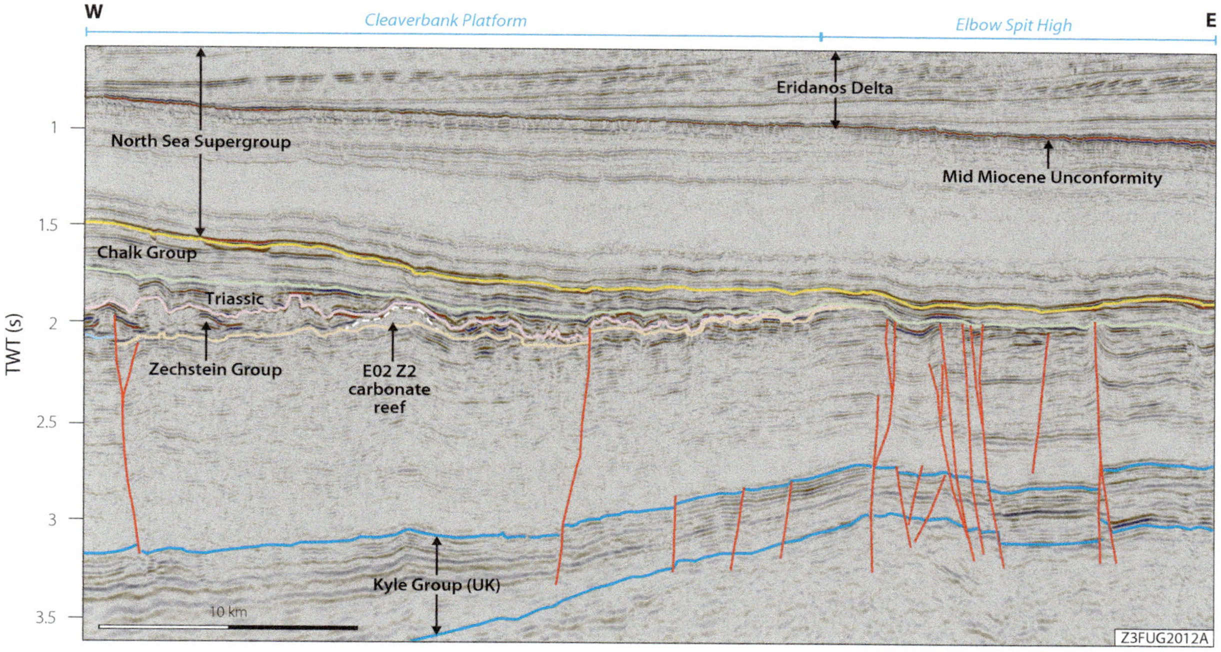

Figure 4.14. Interpreted seismic time section in the northwestern offshore. The Zechstein Group thins to the east and is absent on the Elbow Spit High. A Z2 carbonate platform, referred to as the E02 Z2 carbonate reef, can be readily identified. The platform is part of a series of carbonate platforms located around the Elbow Spit High (see Fig. 4.12d). Modified from Bouroullec et al. (2022). See Fig. 4.5b for location of the seismic line.

was governed by higher-order Milankovitch cycles, with halites representing the dry, and claystones the wet periods (Wagner, 1994; Geluk et al., 1997). The Z5 (Ohre) Formation was likewise deposited under playa conditions.
The higher Zechstein cycles Z6 and Z7, described by Best (1989) from northwest Germany, were not deposited in the Netherlands. Simultaneous with the deposition of these higher cycles, uplift and erosion affected the Netherlands, in some areas even removing the entire Zechstein (Geluk, 1999). The Zechstein Upper Claystone Formation represents a widespread development of sand- and mudflats throughout the Southern Permian Basin (Geluk et al., 1997), and represents the gradual change to continental depositional conditions in the Triassic (McKie & Kilhams, 2025, this volume).

The Permian System constitutes a major reset of the geological evolution in the Netherlands with major climatic, structural and sea level changes occurring in a relatively short amount of time (45 Myr). The occurrence of both the clastic sediments of the Rotliegend groups and the thick and mobile evaporites of the Zechstein Group has made this stratigraphic interval a major focus of research and exploration in the Netherlands and neighbouring countries since the 1950s. The combination of extensive high-quality reservoirs of the Slochteren Formation, the presence of the underlying Westphalian source rocks and the sealing capacity of the Zechstein salt units made the Rotliegend a prime and successful target for hydrocarbon exploration. Today the Permian is still the focus of research and exploration for CCS, geothermal, and hydrogen storage purposes. Many depleted Rotliegend gas fields could have a second life as a permanent storage for CO_2, but saline formation storage could also open some promising avenues for CCS in the Netherlands. The eolian facies of the Upper Rotliegend are also very attractive for geothermal exploration thanks to their high transmissivity. Underexplored areas in the Netherlands, especially where oil and gas plays were unsuccessful due to a lack of source rocks or seal, may soon become attractive for geothermal exploration and hence a new wave of seismic acquisition and exploration wells is being undertaken to fill the remaining knowledge gaps in the subsurface. Finally, Zechstein salt is also seen as a primary target for future hydrogen storage sites, both onshore and offshore within existing or to be developed salt caverns.

Acknowledgements

We would like to thank our reviewers, Clemens Visser, Daan den Hartog Jager and Fokko van Hulten for their comments and valuable input to this chapter. Thanks to Stefan Peeters and Jo Garland for their contribution on the Zechstein carbonate platform story. Finally, we would like to acknowledge the contribution of the late Mark Geluk and his pivotal research and publications on the Dutch Permian from which this chapter greatly benefitted.

Digital map data

Spatial data of figures in this chapter for use in geographical information systems can be downloaded here: https://doi.org/10.5117/aup.28163231.

References

Amiri-Garroussi, K. & Taylor, J.C.M., 1992. Displaced carbonates in the Zechstein of the UK North Sea. Marine and Petroleum Geology 9: 186-196. DOI: 10.1016/0264-8172(92)90090-2

Arthaud, F. & Matte, P., 1977. Late Paleozoic strike-slip faulting in southern Europe and northern Africa: Result of a right-lateral shear zone between the Appalachians and the Urals. Geological Society of America Bulletin 88: 1305-1320. DOI: 10.1130/0016-7606(1977)88<1305:LPSFIS>2.0.CO;2

Baird, A., 1993. An assessment of the reservoir potential of the Zechstein of the P and Q Quadrants and adjacent onshore areas, the Netherlands. Stratigraphic Services International Limited (Surrey) open-file report: 44 pp.

Best, G., 1989. Die Grenze Zechstein/Buntsandstein in Nordwest-Deutschland nach Bohrlochmessungen. Zeitschrift der Deutschen Geologischen Gesellschaft 140: 73-85. DOI: 10.1127/zdgg/140/1989/73

Bouroullec, R. & Ten Veen, J.H., 2025. Salt Tectonics. In: Ten Veen, J.H., Vis, G.-J., De Jager, J. & Wong, Th.E. (eds): Geology of the Netherlands, second edition. Amsterdam University Press (Amsterdam): 457-491. DOI: 10.5117/9789463728362_ch12

Bouroullec, R., Peeters. S., Geel, K., De Haan, H. & Garland, J., 2021. Project ZEPHYR: Hauptdolomite distribution around the Elbow Spit High, seismic mapping. Joint TNO-Cambridge Carbonates Ltd.-EBN report, EBN Kennisbank. https://www.ebn.nl/feiten-en-cijfers/kennisbank/

Bouroullec, R., Kortekaas, M., Brussée, M., Geel, C.R., Den Hartog Jager, D.G. & Peeters, S., 2022. Play 6 Zechstein. https://www.geodeatlas.nl/pages/play-6-zechstein

Brauns, C.M., Pätzold, T. & Haack, U., 2003. A Re-Os study bearing on the age of the Kupferschiefer mineralization at Sangerhausen. XVth International Congress of Carboniferous and Permian Stratigraphy, Utrecht, Abstracts: 66.

Catto, R., Taggart, S. & Poole, G., 2017. Petroleum geology of the Cygnus gas field, UK North Sea: from discovery to development. In: Bowman, M. & Levell, B. (eds): Petroleum Geology of NW Europe: 50 Years of Learning. Geological Society, London, Petroleum Conference Series 8: 307-318. DOI: 10.1144/PGC8.39

Clark, D.N., 1986. The distribution of porosity in Zechstein carbonates. In: Brooks, J., Goff, J.C. & Van Hoorn, B. (eds): Habitat of Palaeozoic gas in N.W. Europe. Geological Society of London, Special Publications 23: 121-149. DOI: 10.1144/GSL.SP.1986.023.01.09

Coelewij, P.A.J., Haug, G.M.W. & Van Kuyk, H., 1978. Magnesium-salt exploration in the northeastern Netherlands. Geologie en Mijnbouw 57: 487-502.

Corcoran, J. & Lunn, S.F., 2014. Old play, new prospectivity - the revival of the Basal Rotliegend Play, offshore Netherlands and Germany. 76th EAGE Conference and Exhibition 2014, 1-5. DOI: 10.3997/2214-4609.20140742

De Bruin, G., Bouroullec, R., Geel, K., Abdul Fattah, R., Van Hoof, T., Pluymaekers, M., Van der Belt, F., Vandeweijer, V.P. & Zijp, M., 2015. New petroleum plays in the Dutch northern offshore. TNO (Utrecht), Report No. 2015 R10920. https://www.nlog.nl/sites/default/files/2018-11/NORTHERN_OFFSHORE_TNO_report_2015_R10920_0.pdf

De Craen, M. & Swennen, R., 1992. Sedimentology and diagenesis of the ankeritized basal Zechstein conglomerate in the Campine Basin (Bree borehole, NE Belgium). Geologie en Mijnbouw 71: 145-160.

De Jager, J. & Geluk, M.C., 2007. Petroleum geology. In: Wong, Th.E., Batjes, D.A.J. & De Jager, J. (eds): Geology of the Netherlands. Royal Netherlands Academy of Arts and Sciences (KNAW, Amsterdam): 241-264.

De Jong, M.G.G., Donselaar, M.E., Boerboom, H.T.W., Van Toorenenburg, K.A., Weltje, G.J. & Van Borren, L., 2020. Long-range, high-resolution stratigraphic correlation of Rotliegend fluvial-fan deposits in the central Dutch offshore. Marine and Petroleum Geology 119: 104482. DOI: 10.1016/j.marpetgeo.2020.104482

Donselaar, M.E., Overeem, I., Reichwein, J.H.C. & Visser, C.A., 2011. Mapping of fluvial fairways in the Ten Boer Member, Southern Permian Basin. In: Grötsch, J. & Gaupp, R. (eds): The Permian Rotliegend of the Netherlands. SEPM Special Publication 98, 105-118. DOI: 0.2110/pec.11.98.0105

Fryberger, S.G., Knight, R., Hern, C., Moscariello, A. & Kabel, S., 2011. Rotliegend facies, sedimentary provinces, and stratigraphy, Southern Permian Basin UK and the Netherlands: A review with new observations. In: Grötsch, J. & Gaupp, R. (eds): The Permian Rotliegend of the Netherlands. SEPM Special Publication 98, 51-88. DOI: 10.2110/pec.11.98.0051

Fuchtbauer, H., 1980. Composition and diagenesis of a stromatolitic bryozoan bioherm in the Zechstein 1 (northwestern Germany). In: Fuchtbauer, H. & Peryt, T. (eds): The Zechstein basin with emphasis on carbonate sequences. Contributions to Sedimentology 9. 233-251.

Galloway, W.E., 1989. Genetic stratigraphic sequences in basin analysis I: architecture and genesis of flooding-surface bounded depositional units. AAPG Bulletin 73: 125-142. DOI: 10.1306/703C9AF5-1707-11D7-8645000102C1865D

Gast, R.E., 1988. Rifting im Rotliegenden Niedersachsens. Die

Geowissenschaften 6: 115-122.
Gast, R.E., Dusar, M., Breitkreuz, C., Gaupp, R., Schneider, J.W., Stemmerik, L., Geluk, M.C., Geißler, M., Kiersnowski, H., Glennie, K.W., Kabel, S. & Jones, N.S., 2010. Rotliegend. In: Doornenbal, J.C. & Stevenson, A.G. (eds): Petroleum geological atlas of the Southern Permian Basin area. EAGE Publications b.v. (Houten): 101-121.
Gaupp, R. & Okkerman, J.A., 2011. Diagenesis and reservoir quality of Rotliegend sandstones in the northern Netherlands - A review. In: Grötsch, J. & Gaupp, R. (eds): The Permian Rotliegend of the Netherlands. SEPM Special Publication 98: 193-226. DOI: 10.2110/pec.11.98.0193
Gaupp, R., Matter, A., Platt, J., Ramseyer, K. & Walzebuck, J., 1993. Diagenesis and fluid evolution of deeply buried Permian (Rotliegendes) gas reservoirs, Northwest Germany. American Association of Petroleum Geologists Bulletin 77: 1111-1128. DOI: 10.1306/BDFF8E0C-1718-11D7-8645000102C1865D
Geluk, M., 1995. Stratigraphische Gliederung der Z2-(Staßfurt-) Salzfolge in den Niederlanden: Beschreibung und Anwendung bei der Interpretation von halokinetisch gestörten Sequenzen. Zeitschrift der Deutschen Geologischen Gesellschaft 146: 458-465. DOI: 10.1127/zdgg/146/1995/458
Geluk, M., 1999. Late Permian (Zechstein) rifting in the Netherlands: models and implications for petroleum geology. Petroleum Geoscience 5: 189-199. DOI: 10.1144/petgeo.5.2.189
Geluk, M.C., 2000. Late Permian (Zechstein) carbonate-facies maps, the Netherlands. Netherlands Journal of Geosciences 79 (1): 17-27. DOI: 10.1017/S0016774600021545
Geluk, M.C., 2005. Stratigraphy and tectonics of Permo-Triassic basins in the Netherland and surrounding areas. PhD thesis Utrecht University (the Netherlands): 171 pp.
Geluk, M.C., 2007. Permian. In: Wong, Th.E., Batjes, D.A.J. & De Jager, J. (eds): Geology of the Netherlands. Royal Netherlands Academy of Arts and Sciences (KNAW, Amsterdam): 63-84.
Geluk, M.C. & Mijnlieff, H., 2001. Controls on the distribution and thickness of Permian basal Upper Rotliegend sandstones, the Netherlands - probing the limits of the Rotliegend play area. 63rd EAGE Conference and Exhibition (Amsterdam). EAGE (Houten): 522. DOI: 10.3997/2214-4609-pdb.15.P522
Geluk, M.C., Bruckner-Röhling, S. & Röhling, H.G., 2000. Salt occurrences in the Netherlands and Germany: New insights in the formation of salt basins. In: Geertman, R. (ed.): Salt 2000, Proceedings of the 8th World Salt Symposium (The Hague). Elsevier (Amsterdam): 125-130.
Geluk, M.C., Plomp, A. & Van Doorn, T.H.M., 1996. Development of the Permo-Triassic succession in the basin fringe area, southern Netherlands. In: Rondeel, H.E., Batjes, D.A.J. & Nieuwenhuijs, W.H. (eds): Geology of gas and oil under the Netherlands. Kluwer Academic Publishers (Dordrecht): 57-78. DOI: 10.1007/978-94-009-0121-6_7
Geluk, M.C., Van Wees, J.-D., Grönloh, H. & Van Adrichem Boogaert, H.A., 1997. Palaeogeography and palaeotectonics of the Zechstein Group (Upper Permian) in the Netherlands. In: Proceedings of the XIII International Congress on the Carboniferous and Permian (Krakow, Poland). Polish Geological Instiute (Warsaw) 157(2) 63-7363-75.
George, G.T. & Berry, J.K., 1994. A new palaeogeographic and depositional model for the Upper Rotliegend, offshore The Netherlands. First Break 12: 147-158. DOI: 10.3997/1365-2397.1994012
Glennie, K.W., 1997. Recent advances in understanding the southern North Sea Basin: a summary. In: K. Ziegler, P. Turner & S.R. Daines (eds): Petroleum Geology of the Southern North Sea: Future Potential. Geological Society (London): 17-29. DOI: 10.1144/GSL.SP.1997.123.01.03
Glennie, K.W., 1998. Lower Permian - Rotliegend. In: Glennie, K.W. (ed.): Petroleum geology of the North Sea - Basic concepts and recent advances. Blackwell Scientific Publishing (Oxford): 137-173. DOI: 10.1002/9781444313413.ch5
Glennie, K.W. & Buller, A.T., 1983. The Permian Weissliegend of NW Europe: the partial deformation of aeolian dune sands caused by the Zechstein transgression. Sedimentary Geology 35: 43-81. DOI: 10.1016/0037-0738(83)90069-6
Grant, R.J., Underhill, J.R., Hernández-Casado, J., Barker, S.M. & Jamieson, R.J., 2019. Upper Permian Zechstein Supergroup carbonate-evaporite platform palaeomorphology in the UK Southern North Sea. Marine and Petroleum Geology 100: 484-518. DOI: 10.1016/j.marpetgeo.2017.11.029
Groenenberg, R., Den Hartogh, M. & Fokker, P., 2025. Salt production. In: Ten Veen, J.H., Vis, G.-J., De Jager, J. & Wong, Th.E. (eds): Geology of the Netherlands, second edition. Amsterdam University Press (Amsterdam): 603-625. DOI: 10.5117/9789463728362_ch16
Heeremans, M., Timmerman, M.J., Kirstein, L.A. & Faleide, J.I., 2004. New constraints on the timing of late Carboniferous - early Permian volcanism in the central North Sea. In: Wilson, M., Neumann, E.-R., Davies, G.R., Timmerman, M.J., Heeremans, M. & Larsen, B.T. (eds): Permo-Carboniferous magmatism and rifting in Europe 223: 177-194. DOI: 10.1144/GSL.SP.2004.223.01.08
Houben, S, Bouroullec, R., De Haan, H., Geel, K., Ventra, D., Janssen, N., Verreussel, R., Peeters, S., Veenstra, F., Monaghan, A., Kearsey, T., Wakefield, O., Vane, C., Hennissen, J., Barron, H. & Stewart, B. , 2020. Paleo-Five Project: Paleozoic petroleum potential of the five countries area, Southern-Central North Sea. TNO (Utrecht), Report No. 2020 R10231: 358 pp.
Hounslow, M.W. & Balabanov, Y.P., 2018. A geomagnetic polarity timescale for the Permian, calibrated to stage boundaries. In: Lucas, S.G. & Shen, S.Z. (eds): The Permian timescale. Geological Society of London, Special Publications 450: 61-103. DOI: 10.1144/SP450.8
Howell, J. & Mountney, N., 1997. Climatic cyclicity and accommodation space in arid to semi-arid depositional systems: an example from the Rotliegend Group of the UK southern North Sea. In: Ziegler, K., Turner, P. & Daines, S.R. (eds): Petroleum geology of the southern North Sea: future potential. Geolog-

ical Society of London, Special Publication 123: 63-86. DOI: 10.1144/GSL.SP.1997.123.01.05

Huis in 't Veld, R. & Den Hartog Jager, D., 2025. Late Carboniferous. In: Ten Veen, J.H., Vis, G.-J., De Jager, J. & Wong, Th.E. (eds): Geology of the Netherlands, second edition. Amsterdam University Press (Amsterdam): 95-125. DOI: 10.5117/9789463728362_ch03

Intergeos, 1989, Buntsandstein of the Netherlands offshore. International Geoservices B.V. report EP890020 (Leiderdorp): 80 pp.

Jakobsen, F., Britze, P., Thöle, H., Jähne-Klingberg, F., Doornenbal, H. & Vis, G.-J., 2020. Harmonized stratigraphic chart for the North Sea area NL-DE-DK. GeoERA report D3: 330 pp.

Johnson, H., Warrington, G. & Stoker, S.J., 1994. Permian and Triassic of the Southern North Sea. In: Knox, R.W.O'B. & Cordey, W.G. (eds): Lithostratigraphic nomenclature of the UK North Sea. British Geological Survey (Nottingham): 141 pp.

Juez-Larré, J., Groenenberg, R., Dinkelman, D., Huijskes, T., Koornneef, J.M., Van Gessel, S.F. & Bos, C.F.M., 2025. Underground storage. In: Ten Veen, J.H., Vis, G.-J., De Jager, J. & Wong, Th.E. (eds): Geology of the Netherlands, second edition. Amsterdam University Press (Amsterdam): 729-767. DOI: 10.5117/9789463728362_ch20

Kombrink, H., Besly, B.M., Collinson, J.D., Den Hartog Jager, D.G., Drozdzewski, G., Dusar, M., Hoth, P., Pagnier, H.J.M., Stemmerik, L., Waksmundzka, M.I. & Wrede, V., 2010. Carboniferous. In: Doornenbal, J.C. & Stevenson, A.G. (eds): Petroleum geological atlas of the Southern Permian Basin area. EAGE Publications b.v. (Houten): 80-99.

Kortekaas, M., Bouroullec, R., Peeters, S., Van Unen, M., Swart, M., Den Hartog Jager, D., Wiarda, E., Nolten, M. & Beintema, K., 2023. Play 7 Rotliegend. geodeatlas.nl/pages/play-7-rotliegend

Legler, B. & Schneider, J.W., 2008. Marine ingressions into the Middle/Late Permian saline lake of the Southern Permian Basin (Rotliegend, Northern Germany) possibly linked to sea-level highstands in the Arctic rift system. Palaeogeography, Palaeoclimatology, Palaeoecology 267: 102-114. DOI: 10.1016/j.palaeo.2008.06.009

Lokhorst, A. (ed.), 1997. NW European Gas Atlas. Netherlands Institute of Applied Geosciences TNO (NITG-TNO, Haarlem): 154 pp.

Lützner, H., Tichomirowa, M., Käßner, A. & Gaupp, R., 2021. Latest Carboniferous to early Permian volcano stratigraphic evolution in Central Europe: U-Pb CA–ID–TIMS ages of volcanic rocks in the Thuringian Forest Basin (Germany). International Journal of Earth Sciences 110: 377-398. DOI: 10.1007/s00531-020-01957-y

Martin, C.A.L., Stewart, S.A. & Doubleday, P.A., 2002. Upper Carboniferous and Lower Permian tectonostratigraphy on the southern margin of the central North Sea. Journal of the Geological Society 159: 731-749. DOI: doi.org/10.1144/0016-764900-174

McKie, T. & Kilhams, B., 2025. Triassic. In: Ten Veen, J.H., Vis, G.-J., De Jager, J. & Wong, Th.E. (eds): Geology of the Netherlands, second edition. Amsterdam University Press (Amsterdam): 155-183. DOI: 10.5117/9789463728362_ch05

Menning, M., 1995. A numerical time scale for the Permian and Triassic periods: an integrated time analysis. In: Scholle, P.A., Peryt, T.M. & Ulmer-Scholle, D.S. (eds): The Permian of northern Pangea. Springer-Verlag (Berlin): 77-97. DOI: 10.1007/978-3-642-78593-1_6

Mijnlieff, H., 2002. Top Pre-Permian distribution map & some thematic regional geologic maps of the Netherlands. Poster 55th ICCP Meeting, Utrecht (Netherlands, ICCP). https://www.nlog.nl/sites/default/files/postermijnlieficcp.pdf

Mijnlieff, H., Buijze, L., Rosendaal, E., Schoof, F., Vorage, R. & Van Wees, J.-D., 2025. Geothermal energy. In: Ten Veen, J.H., Vis, G.-J., De Jager, J. & Wong, Th.E. (eds): Geology of the Netherlands, second edition. Amsterdam University Press (Amsterdam): 701-727. DOI:10.5117/9789463728362_ch19

Minervini, M., Rossi, M. & Mellere, D., 2011. Cyclicity and facies relationships at the interaction between aeolian, fluvial, and playa depositional environments in the Upper Rotliegend: Regional correlation acrosss UK (Sole Pit Basin), the Netherlands, and Germany. In: Grötsch, J. & Gaupp, R. (eds): The Permian Rotliegend of the Netherlands. SEPM Special Publication 98: 119-146. DOI: 10.2110/pec.11.98.0119

NITG, 1998. Geological Atlas of the subsurface of the Netherlands (1:250,000), Explanation to Map Sheet X Almelo–Winterswijk. Netherlands Institute of Applied Geoscience (NITG-TNO, Haarlem): 134 pp

NITG-TNO, 2000. Geological Atlas of the subsurface of the Netherlands (1:250,000) - Explanation to Map Sheet VI Veendam–Hoogeveen. Netherlands Institute for Applied Geoscience TNO - National Geological Survey (Utrecht): 152, 152 pp.

Opluštil, S., Jirásek, J., Schmitz, M. & Matýsek, D., 2017. Biotic changes around the radioisotopically constrained Carboniferous–Permian boundary in the Boskovice Basin (Czech Republic). Bulletin of Geosciences 92: 95-122. DOI: 10.3140/bull.geosci.1638

Patruno, S., Reid, W., Jackson, C.A.-L. & Davies, C., 2018. New insights into the unexploited reservoir potential of the Mid North Sea High (UKCS quadrants 35-38 and 41-43): a newly described intra-Zechstein sulphate–carbonate platform complex. In: Bowman, M. & Levell, B. (eds): Petroleum Geology of NW Europe: 50 Years of Learning. Geological Society, London, Petroleum Geology Conference Series 8: 87-124. DOI: 10.1144/PGC8.9

Peeters, S.H.J., Geel, C.R., Garland, J. & Bouroullec, R., 2023. Seismic and petrographic characterisation of the Zechstein Hauptdolomit Platforms around the Elbow Spit High, Dutch offshore. Journal of Petroleum Geology 46: 361-382. DOI: 10.1111/jpg.12842

Pharaoh, T.C., Dusar, M., Geluk, M.C., Kockel, F., Krawczyk, C.M., Krzywiec, P., Scheck-Wenderoth, M., Thybo, H., Vejbæk, O.V. & Van Wees, J.-D., 2010. Tectonic evolution. In: Doornenbal,

J.C. & Stevenson, A.G. (eds): Petroleum geological atlas of the Southern Permian Basin area. EAGE Publications b.v. (Houten): 25-57.

Plein, E., 1995. Norddeutsches Rotliegendbecken; Rotliegend-Monographie Teil II. In: Deutsche Stratigraphische Kommission (ed.): Stratigraphie von Deutschland I. Courier Forschungsinstitut Senckenberg 183: 193 pp.

Raith, A.F., Urai, J.L. & Visser, J., 2017. Structural and microstructural analysis of K–Mg salt layers in the Zechstein 3 of the Veendam Pillow, NE Netherlands: development of a tectonic mélange during salt flow. Netherlands Journal of Geosciences 96: 331-351. DOI: 10.1017/njg.2017.31

Remane, J., Cita, M.B., Dercourt, J., Bouysse, P., Repetto, F. & Faure-Muret, A. (eds), 2003: International Stratigraphic Chart. International Union of Geological Sciences: International Commission on Stratigraphy (Paris): 1 pp .

Remmelts, G., Nelskamp, S., De Jager, J. & Geluk, M.C., 2025. Petroleum Geology. In: Ten Veen, J.H., Vis, G.-J., De Jager, J. & Wong, Th.E. (eds): Geology of the Netherlands, second edition. Amsterdam University Press (Amsterdam): 535-575. DOI: 10.5117/9789463728362_ch14

Richter-Bernburg, G., 1955. Statigraphische Gliederung des Deutschen Zechsteins. Zeitschrift der Deutschen Geologischen Gesellschaft 105: 843-854. DOI: 10.1127/zdgg/105/1955/843

Rijkers, R.H.B. & Geluk, M.C., 1996. Sedimentary and structural history of the Texel-IJsselmeer High, the Netherlands. In: Rondeel, H.E., Batjes, D.A.J. & Nieuwenhuijs, W.H. (eds): Geology of gas and oil under the Netherlands. Kluwer Academic Publishers (Dordrecht): 265-284. DOI: 10.1007/978-94-009-0121-6_22

Rijks Geologische Dienst, 1991a. Geological Atlas of the subsurface of the Netherlands (1 : 250,000) - Explanation to Map Sheet I Vlieland-Terschelling. Geological Survey of the Netherlands (Haarlem): 77 pp.

Rijks Geologische Dienst, 1991b. Geological Atlas of the subsurface of the Netherlands (1 : 250,000) - Explanation to Map Sheet II Ameland-Leeuwarden. Geological Survey of the Netherlands (Haarlem): 87 pp.

Rijks Geologische Dienst, 1993. Geological Atlas of the subsurface of the Netherlands (1 : 250,000) - Explanation to Map Sheet IV Texel-Purmerend. Geological Survey of the Netherlands (Haarlem): 128 pp.

Rijks Geologische Dienst, 1995. Geological Atlas of the subsurface of the Netherlands (1 : 250,000) - Explanation to Map Sheet III Rottumeroog-Groningen. Geological Survey of the Netherlands (Haarlem): 113 pp.

Roustiau, A., Bouroullec, R., Dijk, J., Den Hartog Jager, D., Nelskamp, S. & Swart, M., 2022. Play 8 Carboniferous. https://www geodeatlas.nl/pages/play-8-carboniferous

Schlimm, W. & Thiermann, A., 1988. Perm. In: Hilden, H.D.(ed.): Geologie am Niederrhein. Geologisches Landesamt Nordrhein-Westfalen (Krefeld): 142 pp.

Schneider, J.W., Spencer, G.L., Scholze, F., Voigt, S., Marchetti, L., Klein, H., Opluštil, S., Werneburg, R., Golubev, V.K., Barrick, J.E., Nemyrovska, T., Ronchi, A., Day, M.O., Silantiev, V.V., Rößler, R., Saber, H., Linnemann, U., Zharinova, V. & Shen, S.-Z., 2020. Late Paleozoic-early Mesozoic continental biostratigraphy—links to the Standard Global Chronostratigraphic Scale. Palaeoworld 29: 186-238. DOI: 10.1016/j.palwor.2019.09.001

Schuurman, W.M.L., 1998. Carboniferous high-resolution stratigraphy the integrated approach - an example from the Cleaver Bank High Area, offshore the Netherlands. TNO report NITG-98-095-A: 3 pp.

Scotese, C.R., 2014. Atlas of Permo-Carboniferous Paleogeographic Maps (Mollweide Projection), Maps 53-64, Volumes 4, The Late Paleozoic, PALEOMAP Atlas for ArcGIS, PALEOMAP Project (Evanston, IL, USA). DOI: 10.13140/2.1.1837.8244

Sissingh, W., 2004. Palaeozoic and Mesozoic igneous activity in the Netherlands: a tectonomagmatic review. Netherlands Journal of Geosciences 83 (2): 113-134. DOI: 10.1017/S0016774600020084

Słowakiewicz, M. & Gąsiewicz, A., 2013. Palaeoclimatic imprint, distribution and genesis of Zechstein Main Dolomite (Upper Permian) petroleum source rocks in Poland: Sedimentological and geochemical rationales. In: Gąsiewicz, A. & Słowakiewicz, M. (eds): Palaeozoic climate cycles: Their evolutionary and sedimentological impact. Geological Society of London, Special Publications 376: 523-538. DOI: 10.1144/SP376.21

Stemmerik, L., Ineson, J.R. & Mitchell, J.G., 2000. Stratigraphy of the Rotliegend Group in the Danish part of the Northern Permian Basin, North Sea. Journal of the Geological Society 157 (6): 1127-1136. DOI: 10.1144/jgs.157.6.1127

Stille, H., 1924. Grundfragen der vergleichenden Tektonik. Gebrüder Borntraeger (Berlin): 443 pp.

Strohmenger, C., Antonini, M., Jäger, G., Rockenbauch, K. & Strauss, C., 1996. Zechstein 2 Carbonate reservoir facies distribution in relation to Zechstein sequence stratigraphy (Upper Permian, northwest Germany): an integrated approach. Bulletin des Centres de Recherche Exploration-Production Elf-Aquitaine 20: 1-35.

Strömbäck, A.C. & Howell, J.A., 2002. Predicting distribution of remobilized aeolian facies using sub-surface data: the Weissliegend of the UK Southern North Sea. Petroleum Geoscience 8: 237-249. DOI: 10.1144/petgeo.8.3.237

Sweeney, M., Turner, P. & Vaughan, D.J., 1987. The Marl Slate: a model for the precipitation of calcite, dolomite and sulphides in a newly formed anoxic sea. Sedimentology 34: 31-48. DOI: 10.1111/j.1365-3091.1987.tb00558.x

Taylor, J.C.M., 1998. Upper Permian - Zechstein. In: Glennie, K.W. (ed.): Petroleum geology of the North Sea - Basic concepts and recent advances. Blackwell Scientific Publishing (Oxford): 174-212. DOI: 10.1002/9781444313413.ch6

TNO-GDN, 2022. Fringe Sandstone members (ZEZ1S to ZEZ4S). In: Stratigraphic Nomenclature of the Netherlands. https://www.dinoloket.nl/en/stratigraphic-nomenclature/z1-fringe-sandstone-member (idem for z2, z3, z4)

Tucker, M.E., 1991. Sequence stratigraphy of carbonate-evaporite basins: models and application to the Upper Permian (Zechstein) of northeast England and adjoining North Sea. Journal of the Geological Society 148: 1019-1036. DOI: 10.1144/gsjgs.148.6.1019

Vai, G.B., 2003. Development of the palaeogeography of Pangaea from Late Carboniferous to Early Permian. Palaeogeography, Palaeoclimatology, Palaeoecology 196: 125-155. DOI: 10.1016/S0031-0182(03)00316-X

Van Adrichem Boogaert, H.A. & Burgers, W.F.J., 1983. The development of the Zechstein in the Netherlands. In: Kaasschieter, J.P.H. & Reijers T.J.A. (eds): Petroleum geology of the southeastern North Sea and the adjacent onshore areas. Geologie en Mijnbouw 62: 83-92. DOI: 10.1007/978-94-009-5532-5_8

Van Adrichem Boogaert, H.A. & Kouwe, W.F.P., 1994. Stratigraphic nomenclature of the Netherlands; revision and update by RGD and NOGEPA, Section D: Permian (Lower Rotliegend, Upper Rotliegend and Zechstein Groups). Mededelingen Rijks Geologische Dienst 50: 1-42.

Van Bergen, M.J. & Sissingh, W., 2007. Magmatism in the Netherlands: expression of the north-west European rifting history. In: Wong, Th.E., Batjes, D.A.J. & De Jager, J. (eds): Geology of the Netherlands. Royal Netherlands Academy of Arts and Sciences (KNAW, Amsterdam): 197-222.

Van Bergen, M.J., Vis, G.-J., Sissingh, W., Koornneef, J. & Brouwers, I., 2025. Magmatism in the Netherlands: expression of the north-west European rifting history. In: Ten Veen, J.H., Vis, G.-J., De Jager, J. & Wong, Th.E. (eds): Geology of the Netherlands, second edition. Amsterdam University Press (Amsterdam): 393-455. DOI: 10.5117/9789463728362_ch11

Van den Belt, F.J.G. & Van Hulten, F.F.N., 2011. Sedimentary architecture and palaeogeography of Lower Slochteren aeolian cycles from the Rotliegend desert-lake margin (Permian), the Markham area, Southern North Sea. In: Grötsch, J. & Gaupp, R. (eds): The Permian Rotliegend of the Netherlands. SEPM Special Publication 98: 161-176. DOI: 10.2110/pec.11.98.0161

Van der Baan, D., 1990. Zechstein reservoirs in the Netherlands. In: Brooks, J. (ed.): Classic petroleum provinces. Geological Society of London, Special Publications 50, 379-398. DOI: 10.1144/GSL.SP.1990.050.01.24

Van de Sande, J.M.M., Reijers, T.J.A. & Casson, N., 1996. Multidisciplinary exploration strategy in the northeast Netherlands Zechstein 2 Carbonate play, guided by 3D seismic. In: Rondeel, H.E., Batjes, D.A.J. & Nieuwenhuijs, W.H. (eds): Geology of gas and oil under the Netherlands. Kluwer Academic Publishers (Dordrecht): 125-142. DOI: 10.1007/978-94-009-0121-6_12

Van Ojik, K., Böhm, A.R., Cremer, H., Geluk, M.C., De Jong, M.G.G., Mijnlieff, H.F. & Nio, S.D., 2011. The rationale for an integrated stratigraphic framework of the Upper Rotliegend II depositional system in the Netherlands. In: Grötsch, J. & Gaupp, R. (eds): The Permian Rotliegend of the Netherlands. SEPM Special Publication 98: 37-48. DOI: 10.2110/pec.11.98.0037

Van Ojik, K., Den Hartog Jager, D. & Roustiau, A., 2021. Play 7 Rotliegend. https://www geodeatlas.nl/pages/play-7-rotliegend

Verdier, J.P., 1996. The Rotliegend sedimentation history of the southern North Sea and adjacent countries. In: Rondeel, H.E., Batjes, D.A.J. & Nieuwenhuijs, W.H. (eds): Geology of gas and oil under the Netherlands. Kluwer Academic Publishers (Dordrecht): 45-56. DOI: 10.1007/978-94-009-0121-6_6

Visser, W.A., 1955. The Upper Permian in the Netherlands. Leidse Geologische Mededelingen 20: 185-194. https://repository.naturalis.nl/pub/505894

Wagner, R., 1994. Stratygrafia osadów i rozwój basenu cechsztyńskiego na Niżu Polskim. Prace Państwowego Instytutu Geologicznego 146: 71 pp.

Wolf, R., 1985. Tiefentektonik des linksniederrheinischen Steinkohlengebietes. Beiträge zur Tiefentektonik westdeutscher Steinkohlenlagerstätten. Geologisches Landesamt Nordrhein-Westfalen (Krefeld) 105-167.

Yang, C.S. & Baumfalk, Y.A., 1994. Milankovitch cyclicity in the Upper Rotliegend Group of the Netherlands offshore. In: De Boer, P.L. & Smith, D.G. (eds): Orbital forcing and cyclic sequences. International Association of Sedimentologists Special Publication 19: 47-61. DOI: 10.1002/9781444304039.ch5

Ziegler, M.A., 1989. North German Zechstein facies patterns in relation to their substrate. Geologische Rundschau 78: 105-127. DOI: 10.1007/bf01988356

Ziegler, P.A., 1990. Geological Atlas of Western and Central Europe, second edition. Shell Internationale Petroleum Mij. B.V. and Geological Society (London): 239 pp.

5 Triassic

Tom McKie & Ben Kilhams*

*Corresponding author: tom.mckie@shell.com

ABSTRACT

The Triassic of the Netherlands records a complex history of episodic tectonism, changing climate and marine flooding of a continental interior basin. The section is divided into two groups separated by the Base Solling Unconformity: the latest Permian – Olenekian Lower Germanic Trias, and Olenekian – Norian Upper Germanic Trias. The lower group comprises a largely dryland continental succession of playa claystones, fluvial and eolian sandstones and lacustrine oolites that continued to fill the desiccated late Permian Zechstein sag basin. Conglomerates in the south of the Netherlands pass northwards into sandstones and offshore into claystones towards the basin centre. The section shows minor thickness changes and can be up to 800 m thick. The Upper Germanic Trias Group comprises Röt continental clastics and marine evaporites; Muschelkalk carbonate ramp and evaporite facies, and Keuper playa claystones with locally thick halite successions. The south of the Netherlands remained a fluvial clastic fringe throughout. The Upper Germanic Trias Group was profoundly influenced by Hardegsen and Cimmerian extensional events that facilitated marine ingress from the Tethys Sea, created fault-bounded depocentres for playa and evaporite accumulation and local uplift and erosion of older stratigraphy. Thickness changes are more dramatic in this interval, which can be up to 1750 m thick in the offshore Netherlands and up to 5000 m in the Glückstadt and Horn grabens. The dryland continental character of the Triassic was a persistent feature, but brief episodes of environmental disruption and wetter climatic conditions also occurred. These were expressed by the establishment of perennial lakes and rivers, with the most prominent being in the mid-Carnian and late Ladinian. Widespread marine flooding of the basin began in the latest Triassic and was the precursor to the fully marine conditions that persisted through the Early Jurassic.

<< Middle Triassic limestone of the Muschelkalk Formation in the Winterswijk quarry (east Netherlands), overlain by grey claystone of the uppermost Triassic Sleen Formation (Altena Group). The Röt Formation is visible on the quarry floor. Photo: Remco Bos.

DOI: 10.5117/9789463728362_ch05

Introduction

The Triassic of the Netherlands is stratigraphically located within the post-Variscan, late Permian to Middle Jurassic tectonic mega-cycle, bounded by the Saalian Unconformity at the base and the Mid-Cimmerian Unconformity at the top. It marks the start of the Mesozoic Era and contains a tectonic and depositional record of major geological events such as the final assembly and initial fragmentation of the Pangean supercontinent, changing climate, mass extinctions and episodic marine flooding.

The Southern Permian Basin became isolated from the northern Boreal Ocean in the latest Permian to earliest Triassic. Subsequent deposition was dominated by continental facies that blanketed and buried the former marine evaporites of the upper Permian Zechstein Group. Whilst the Early Triassic largely saw continental infilling of the Zechstein sag basin (Van Wees et al., 2000), later Early Cimmerian extensional events dissected this simple depression into a more complex mosaic of fault-bounded depocenters (Geluk, 2007) and these also opened new routes for marine ingress from the Tethys Sea to the south. These regional events created a rift network that extended across northwest Europe (Manspeizer, 1982; Coward, 1995; Štolfová & Shannon, 2009; Maystrenko et al., 2017) and provided the initial template for the ultimate fragmentation of Pangea in the Jurassic and Cretaceous.

Environmental conditions during the Triassic were harsh. The large continental landmass of Pangea maintained high temperatures, widespread aridity and intense monsoon rainfall originating from the Tethys Sea. In the early Triassic life was on a 5 Myr road to recovery following the end Permian extinction, which had been triggered by extensive and sustained volcanism centred around the Siberian large igneous province (Hallam & Wignall, 1997; Benton & Twitchett, 2003; Chen & Benton, 2012). Eruptions in the Central Atlantic Magmatic Province triggered another extinction event in the latest Triassic, so the epoch is bounded by two of the 'big five' Phanerozoic extinction events (Raup & Sepkoski, 1982; Jablonski, 1994; Van de Schootbrugge & Wignall, 2016 ; Bos et al., 2024).

The preservation of the Triassic depositional record is variable (Geluk, 2007; Pharaoh et al., 2010). Areas of rapid subsidence during Cimmerian and Jurassic rift events were protected from uplift and erosion and have a more complete record, as in the Dutch Central Graben where ca. 2 km of Triassic section is located at burial depths of ca. 7 km. However, basin-margin successions are more poorly preserved, as are areas where Middle Jurassic thermal doming, Late Jurassic rift flank uplift, base Cretaceous erosion, Alpine inversion and unroofing above rising salt diapirs all resulted in the partial or complete removal of Triassic strata. As a result, there is more stratigraphic in-

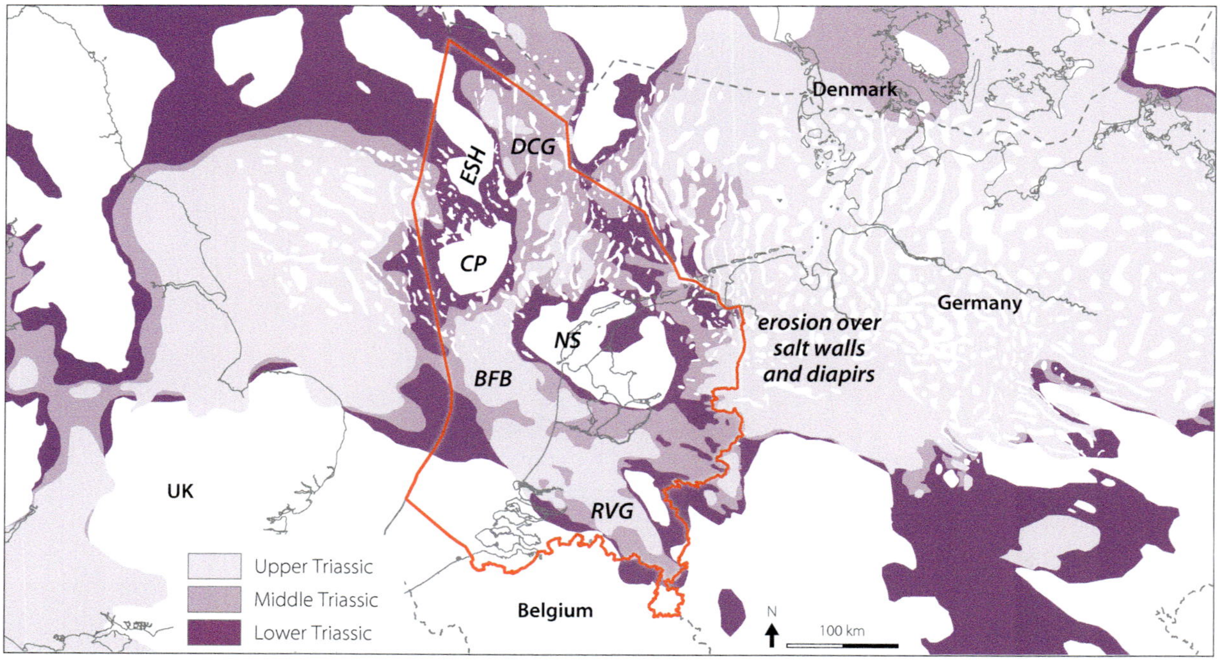

Figure 5.1. Regional distribution and preservation of Lower (Bunter), Middle (Muschelkalk and Lower Keuper) and Upper (Middle and Upper Keuper) Triassic rocks in the Netherlands onshore and offshore (outlined in red) and adjacent areas of the Southern Permian Basin. Modified after Bachmann et al. (2010). BFB = Broad Fourteens Basin, CP : Cleaverbank Platform, DCG = Dutch Central Graben, ESH = Elbow Spit High, NS = Netherlands Swell, RVG = Roer Valley Graben. Within the basin the widespread, vermicular patches represent Triassic removed over Zechstein salt walls and diapirs.

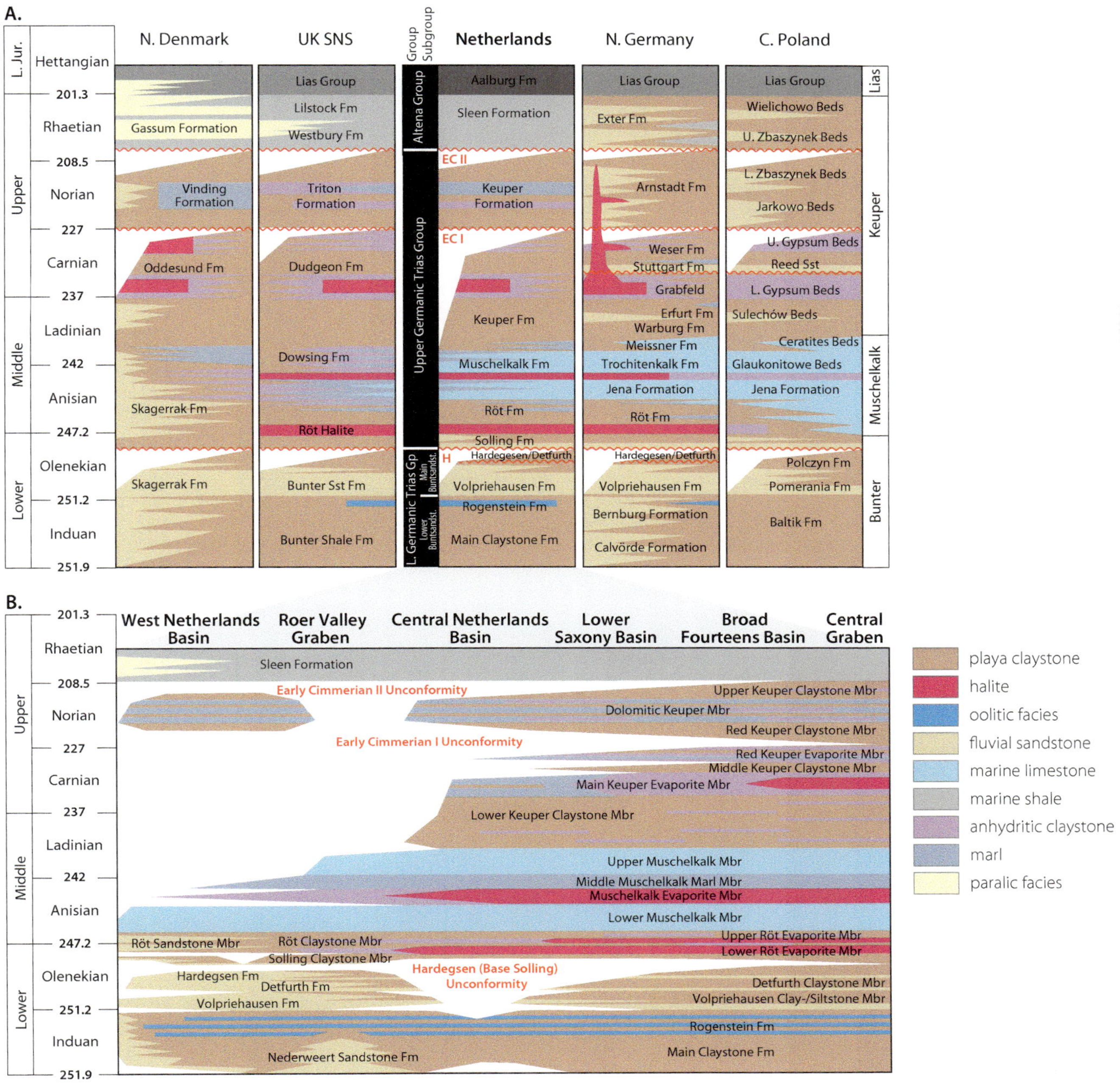

Figure 5.2. The stratigraphy and dominant facies of the Triassic succession. a) Overview of the stratigraphy in the Netherlands and comparison with adjacent areas. b) Detailed stratigraphy of the Triassic in the Netherlands (nomenclature based on TNO-GDN revision of 2020). EC I and II refer to Early Cimmerian unconformities. Timescale follows Cohen et al., (2013; updated).

formation available for the older parts of the Triassic section, with the younger parts presenting a more fragmentary record (Fig. 5.1).

Regional setting

The depositional and stratigraphic character of the Triassic in the Netherlands and across the Southern Permian Basin (Fig. 5.2) is the product of interacting tectonic, climatic and eustatic processes that operated across a wide region of northern Pangea. In addition, long-lived upland areas comprising the Variscan mountain remnants to the south, the Fennoscandian Shield to the north, and smaller UK massifs to the west functioned as sustained sediment sources throughout the Triassic, with elevated sediment yields noted in the Early Triassic (Mansouri & Hinderer, 2021) resulting from environmental perturbations. These large-scale controls ultimately dictated the nature of the Triassic stratigraphy in the Netherlands.

Tectonic events

Triassic structural trends were heavily influenced by the structural grain of the underlying basement, with Caledonian, Variscan and Permian structures exerting a strong control on fault orientation. Major Caledonian lineaments such as the Iapetus Suture and the Trans European Suture zone mark the boundaries to distinct Triassic structural styles, whilst the inherited Permian rift system formed the template for the subsequent Triassic rifts (Geluk, 2007; Crowder et al., 2020). The Early Triassic, in common with the underlying Zechstein, records deposition in a relatively quiescent, thermally subsiding sag basin, although there was initial movement on the faults that were to define the Central, Horn and Glückstadt grabens (Pharaoh et al.,

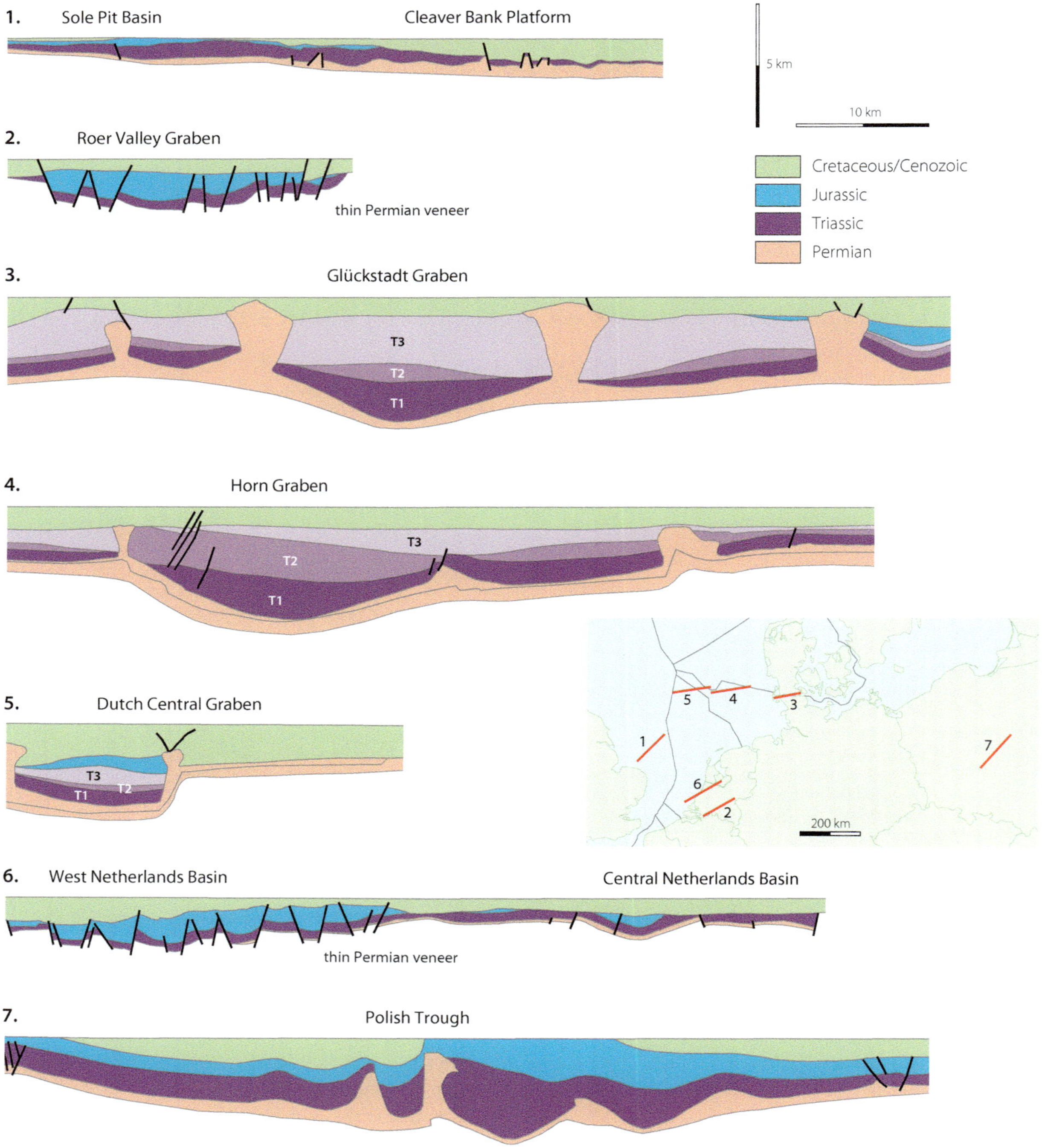

Figure 5.3. Geological cross sections of various Triassic basins in northwest Europe. The Triassic divisions T1, T2 and T3 refer to informal seismostratigraphic units. T1 = Lower and Main Buntsandstein; T2 = Solling, Röt and Muschelkalk; T3 = Keuper. Note the difference in scale between these basins, with the thickest Triassic successions associated with rim synclines in the Glückstadt and Horn grabens. Depth scale in km. Twofold vertical exaggeration. Modified after Geluk et al. (2018a).

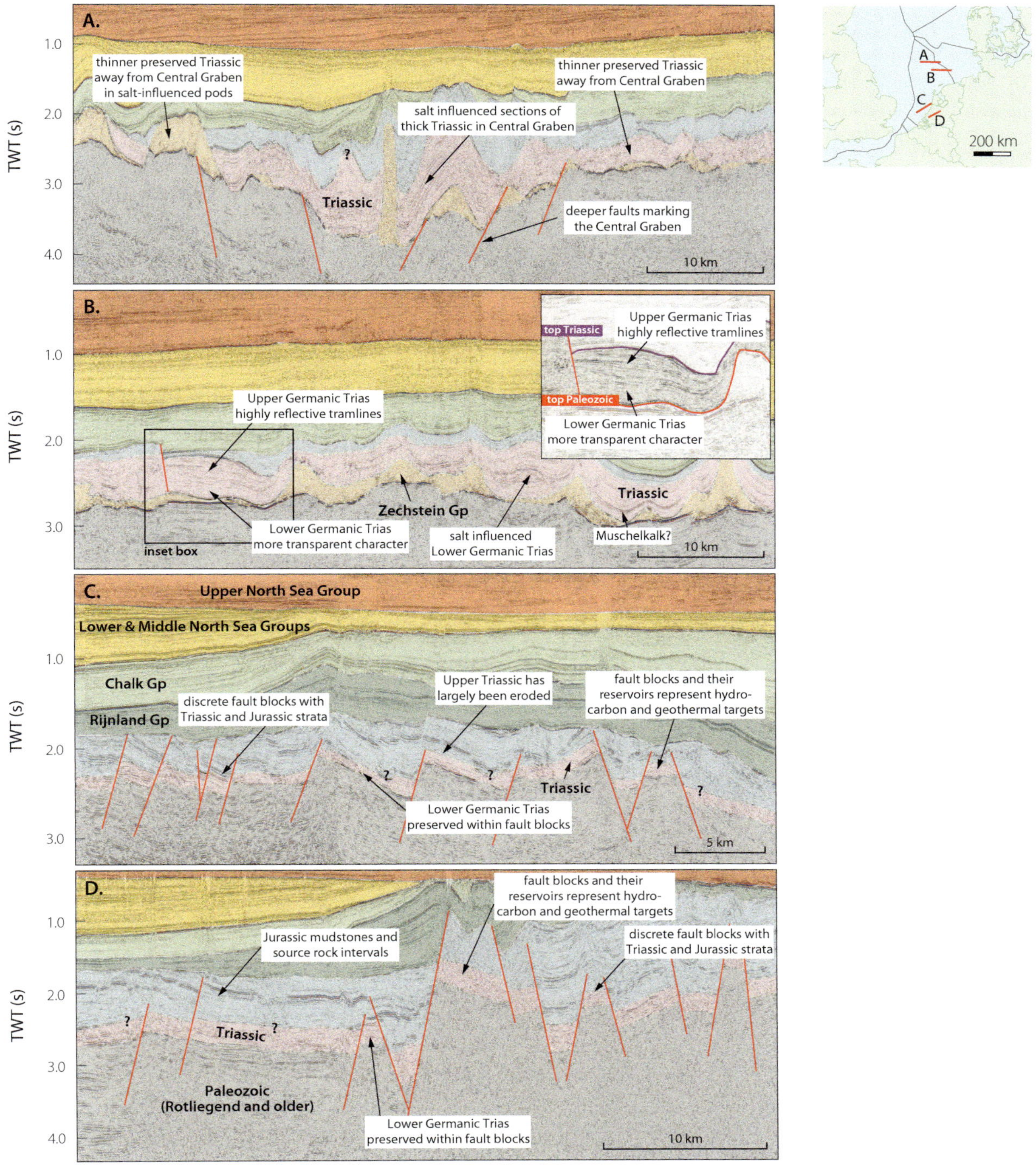

Figure 5.4. *Regional seismic lines showing the subsurface expression of the Triassic across the onshore and offshore Netherlands. a) Typical section across the Dutch Central Graben domain where a thickened Triassic section is deformed by halokinesis (for example, into rim synclines). Note that only selected deep faults are shown to illustrate the graben bounding structures that show the Triassic thickness changes across these markers. b) Typical section across the eastern Dutch offshore where halokinesis influenced localized structuration of the Triassic. The inset box highlights the seismic facies change from Lower Germanic Trias (more transparent) to Upper Germanic Trias (highly reflective). c) Section across the Broad Fourteens Basin (offshore west Netherlands) where only the Lower Triassic is preserved within a series of rotated fault blocks. d) Section through the West Netherlands Basin onshore showing a similar pattern as in c. The Triassic reservoirs in these fault blocks constitute hydrocarbon and geothermal targets.*

2010). However, widespread Zechstein salt movement was occurring in the central North Sea, while in the northern North Sea and in east Greenland half-graben structures indicate that significant extensional tectonics were active at this time but had not propagated southward into the southern North Sea region. In the late Early Triassic the Hardegsen extensional phase, comprising up to four rift pulses (Geluk & Röhling, 1997, 1999), began to reconfigure the basin. The largest of these rift pulses was responsible for the prominent unconformity at the base of the Solling Claystone (Fig. 5.2), but although its effects are expressed across a wider region than the earliest Triassic rifts in the north, the dispersed fault movements were relatively subdued. This resulted mainly in intra-Triassic unconformities rather than major footwall uplifts with erosion concentrated along a few large-displacement faults, as seen on the margins of the Jurassic rifts in the central and northern North Sea. The small-scale Hardegsen movements were, however, sufficient to set the Zechstein salt in motion and initiate the formation of salt pillows. It is possible that this widespread tectonic event, extending beyond the limits of the Southern Permian Basin, was triggered by extension that propagated from both the Boreal and Tethyan regions. Pharaoh et al., (2010) suggested that the deformation could be a response to the build-up of regional intraplate stresses, as they found evidence of warping rather than fault displacements, noting that there is little evidence of the kind of rift flank uplift that extension would predict. Minor movements persisted into the Middle Triassic, but the Late Triassic saw two major, Early Cimmerian pulses in the early Norian (base Red Keuper Claystone) and Rhaetian (base Sleen Formation), each of which generated substantial unconformities (Fig. 5.2). These are detectable across northwest Europe and mark episodes of major graben subsidence and stratigraphic thickening, particularly in the Glückstadt and Horn grabens in northwest Germany (Best et al., 1983; Geluk & Röhling, 1999; Kilhams et al., 2018b) as well as the Dutch Central Graben and Ems Low (Figs 5.3, 5.4). Cimmerian extension also triggered more substantial and widespread halokinesis of both the Zechstein and Keuper halites (Mohr et al., 2007; Warren, 2008) and the initiation of prominent salt walls and associated rim synclines (Fig. 5.3).

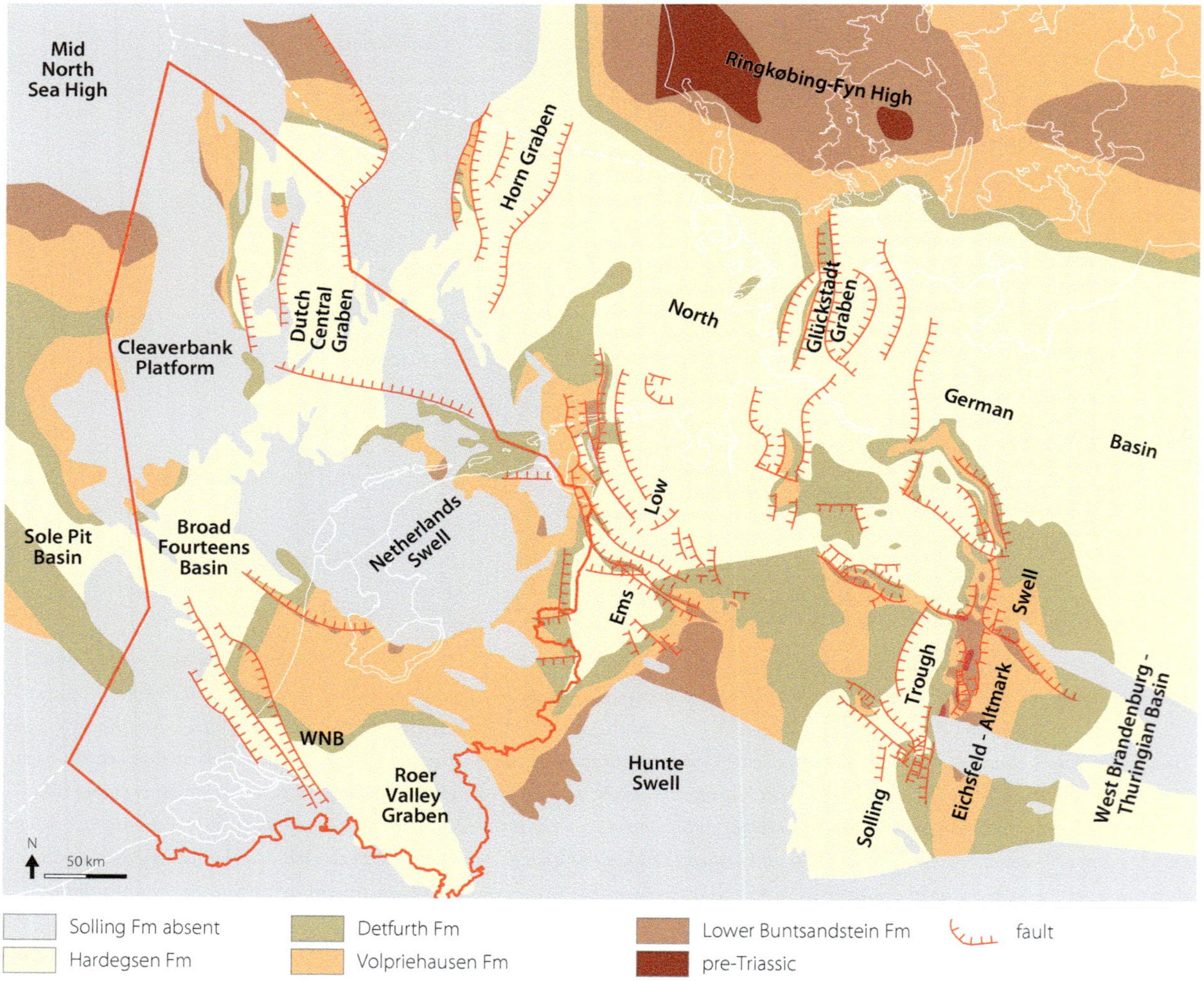

Figure 5.5. Subcrop map below the Base Solling (Hardegsen) Unconformity. Modified after Geluk (2007).

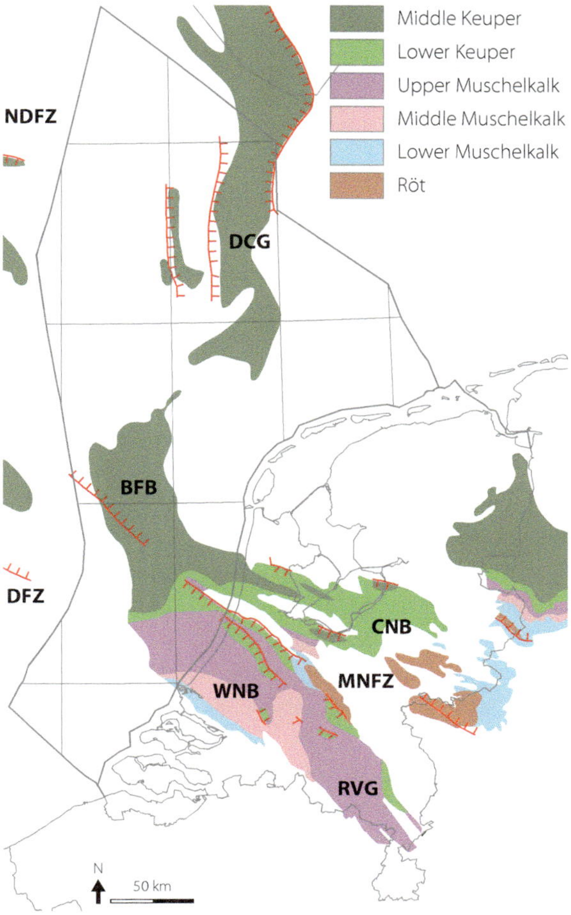

Figure 5.6. Subcrop map below the Early Cimmerian Unconformity (base Red Keuper Claystone Member or, where this unit is not present, Sleen Formation). Modified after Geluk (2007). BFB = Broad Fourteens Basin; CNB = Central Netherlands Basin; DCG = Dutch Central Graben; DFZ = Dowsing Fault Zone; MNFZ = Mid Netherlands Fault Zone; NDFZ = North Dogger Fault Zone; RVG = Roer Valley Graben; WNB = West Netherlands Basin.

These tectonic events influenced the preservation potential of the Triassic stratigraphy (Fig. 5.1). The Hardegsen Event truncated the stratigraphy on the basin margins and in areas of lower subsidence such as the Netherlands Swell and the Cleaverbank Platform (Fig. 5.5). Conversely, fault movement related to halokinesis created local accommodation for the accumulation of eolian sands, such as those seen in the L09 block offshore (De Jager, 2012). The Early Cimmerian I event also resulted in the removal of substantial stratigraphy from the basin margins, with the truncation of Lower and Middle Keuper stratigraphy in the south of the Netherlands accompanied by local removal of Muschelkalk (Fig. 5.6). Enhanced preservation in the graben areas and rim synclines led to more complete preservation of stratigraphic cycles and became sites for enhanced evaporite accumulation.

Climate

The very large size of the continental landmass of Pangea, running from pole to pole, created unique climatic conditions (Kutzbach & Gallimore, 1989; Parrish, 1993; Kidder & Worsley, 2004; Sellwood & Valdes, 2007). The absence of large open water bodies resulted in few areas where air masses could acquire moisture, creating the pronounced aridity that had persisted since the Permian assembly of the supercontinent. The physical location of the Netherlands within an equatorial arid zone and within the rain shadow of the Variscan Mountains (Kendall et al., 2003) further compounded this aridity. However, summer heating of the northern Pangean landmass created ideal conditions for the development of a very strong monsoon circulation, and this affected regions adjacent to the Tethys Sea. Monsoon precipitation over the Variscan Mountains triggered northward flowing, sediment-charged flash floods that discharged into the more persistently arid Southern Permian Basin (Van der Zwan & Spaar, 1992). This combination of monsoonal hinterland and arid basin resulted in characteristically terminal river systems (cf. Nichols, 2007) throughout the Triassic. Conditions were not always so dry, and the prevailing arid climate episodically fluctuated into 'pluvial' phases characterized by regionally warmer and wetter conditions (Trotter et al., 2015). The most pronounced of these was the Carnian Pluvial Episode (Simms & Ruffell, 1989; Hornung et al., 2007; Ruffell et al., 2016; Zhang et al., 2020) when widespread rivers and lakes appeared briefly across the region. This climate change has been attributed to atmospheric CO_2 injection by the Wrangellia large igneous province in North America and has been implicated in the explosive diversification of the dinosaurs and other key faunal groups (Bernardi et al., 2018; Dal Corso et al., 2020). Smaller-scale climate fluctuations may also have been associated with igneous activity through the Triassic, with episodes of fluvial expansion in the Southern Permian Basin apparently coinciding with major igneous activity in the Southern Siberian Traps and Angayucham Province in Alaska (Mckie, 2014; Geluk et al., 2018a; Archer et al., 2020).

Sea level

Eustatic sea level during the Permian and Triassic was generally low compared to the rest of the Phanerozoic, approximating a few tens of metres compared to modern sea levels (Haq, 2018). Levels were higher in the Induan, Olenekian and late Carnian, and lower in the late Anisian, late Norian and early Rhaetian. In many areas the European Triassic basins had subsided below sea level but were physically isolated from the global ocean (McKie, 2017). This made them prone to very rapid flooding if access to the sea was gained via tectonic openings or when sea level rise overtopped topographic sills. Conversely, this restriction also made the basins prone to being sites of evaporite accumulation. The major transgressive events in the Southern Permian Basin in the Middle Triassic (Röt and Muschelkalk formations) and Late Triassic (Rhaetian Sleen Formation) do not necessarily correspond to eustatic rises or highstands and it is therefore likely that these were at least in part controlled by tectonic movement facilitating marine access via the Silesian and Moravian Gateways in southern Poland and the more westerly Burgundy Gate (Geluk, 2005, 2007). However, once the basin was connected to the open sea, eustatic changes exerted a control on shallow marine and paralic facies architecture (Franz et al., 2015, 2018b).

Stratigraphy

The Triassic stratigraphy of the Netherlands is Germanic in style, comprising the classic three-fold subdivision of Bunter, Muschelkalk and Keuper clastic-carbonate-clastic lithologies (Fig. 5.2), although the facies composition and timing of these can differ in detail. For example, carbonate-dominated Muschelkalk lithologies become increasingly clastic-prone to the west into the UK sector of the North Sea, but conversely first appear lower in the stratigraphy to the east in Poland (Geluk, 2005; Fig. 5.2). Also, the shale-prone Rhaetian Sleen Formation in the Netherlands becomes increasingly sand-prone in the offshore UK and Germany. Areas in the south of the Netherlands, close to the former basin margin, tend to be more sand-prone in general.

Van Adrichem Boogaert & Kouwe (1994) divided the Triassic of the Netherlands into two groups separated by the Hardegsen (Base Solling) Unconformity:

- the Lower Germanic Trias Group (Fig. 5.2) comprises the Lower and Main Buntsandstein subgroups (Fig. 5.7). The former is made up of the Main Claystone Formation, with the oolitic beds of the Rogenstein Formation variably developed in the upper part. The latter comprises interbedded sandstones and shales of the Volpriehausen, Detfurth and Hardegsen formations (Fig. 5.2).
- the Upper Germanic Trias Group (Fig. 5.2) consists of an alternation of fine-grained clastics and evaporites (Solling and Röt formations; Fig. 5.7), carbonates and evaporites (Muschelkalk Formation; Fig. 5.8) overlain by a fine grained, locally anhydrite and halite-bearing section with subordinate sandstones (Keuper Formation; Fig. 5.8).

Regional correlation demonstrates the great lateral continuity of the main stratigraphic elements (Figs 5.7, 5.8), although basin-wide variations in syn-rift architecture and diachroneity of facies belts have resulted in differences in

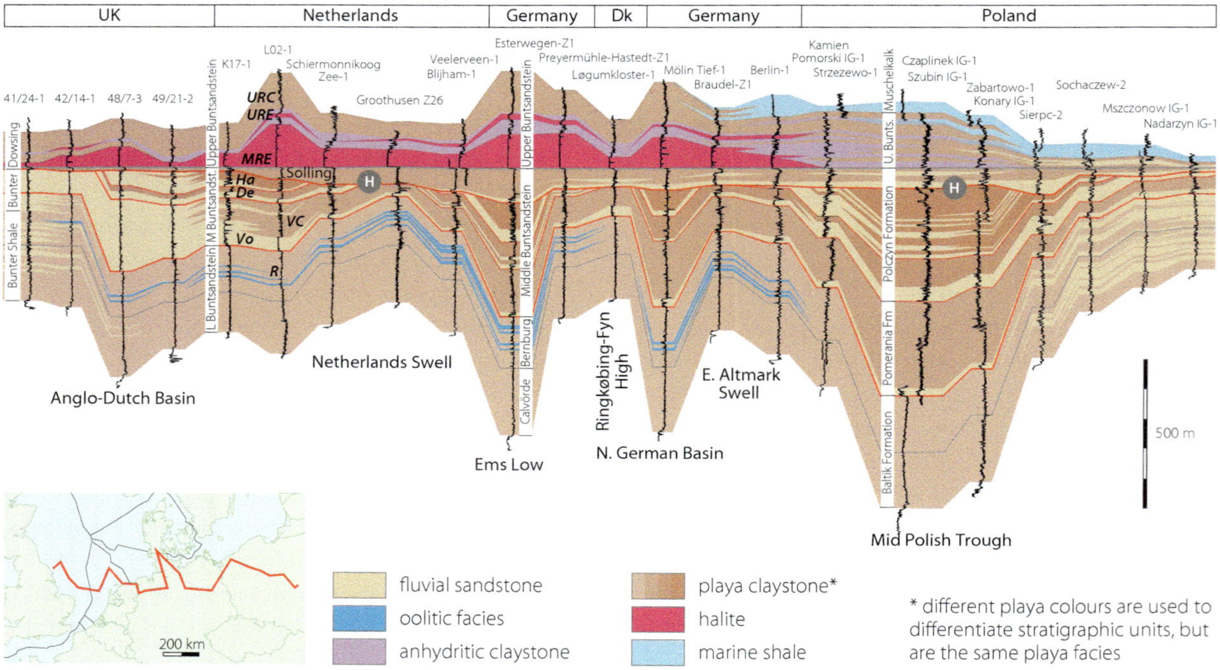

Figure 5.7. Stratigraphic correlation (based on gamma-ray logs) of the Lower, and basal Upper, Germanic Trias Group between the UK and Poland, flattened on the Base Upper Buntsandstein. Modified after Geluk (2005). R: Rogenstein Formation, Vo: Volpriehausen Formation, Vc: Volpriehausen Clay-Siltstone Member, De: Detfurth Formation, Ha: Hardegsen Formation, MRE: Main Röt Evaporite Member, URE: Upper Röt Evaporite Member, URC: Upper Röt Claystone Member. H denotes the Base Solling (Hardegsen) Unconformity.

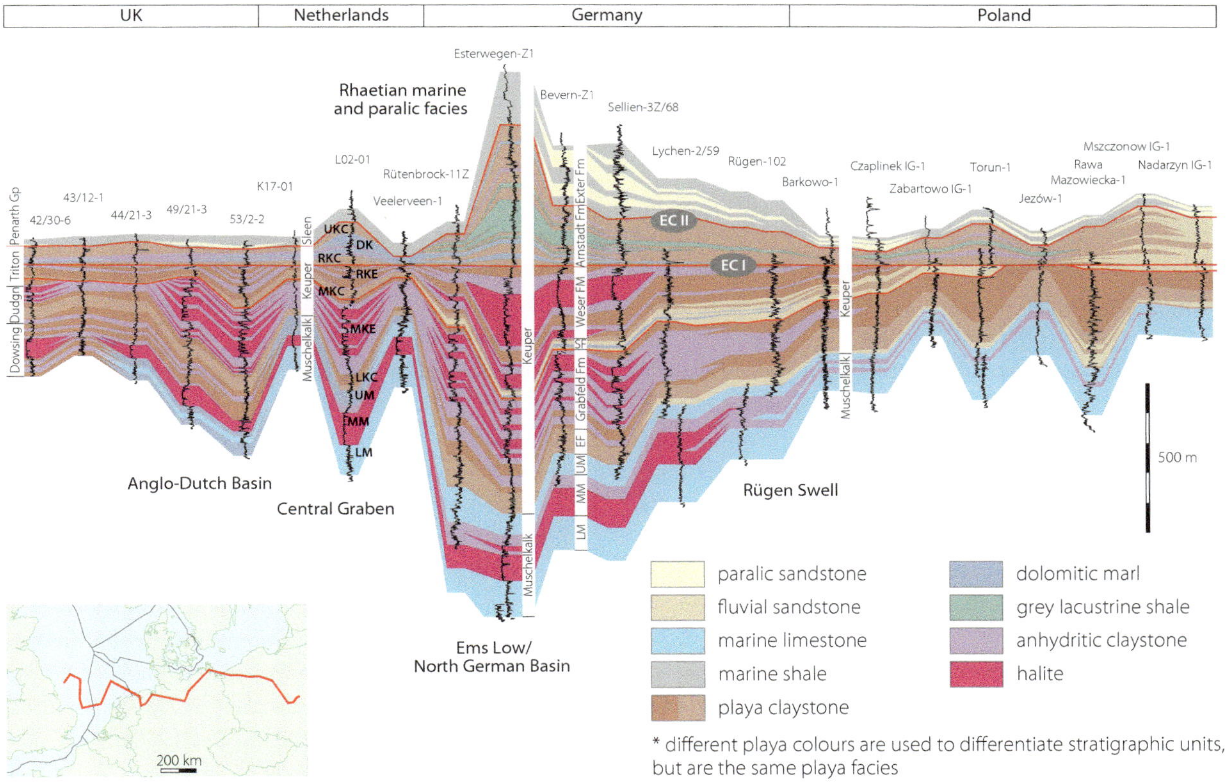

Figure 5.8. Stratigraphic correlation (based on gamma-ray logs) of the Upper Germanic Trias Group between the UK and Poland, flattened on the Early Cimmerian Unconformity I (EC I). Modified after Geluk (2005). LM: Lower Muschelkalk Member, MM: Middle Muschelkalk Evaporite Member, UM: Upper Muschelkalk Member, LKC: Lower Keuper Claystone Member, MKE: Main Keuper Evaporite Member, MKC: Middle Keuper Claystone Member, RKE: Red Keuper Evaporite Member, RKC: Red Keuper Claystone Member, DK: Dolomitic Keuper Member, UKC: Upper Keuper Claystone Member, EF: Erfurt Formation, SF: Schilfsandstein.

chosen lithostratigraphic boundaries between countries (Geluk, 2007). On seismic the lower group appears tabular and relatively transparent with few acoustic impedance contrasts. The upper group, with its more varied lithology shows greater reflectivity and greater thickness variations (Fig. 5.4).

Understanding the stratigraphic evolution of the Netherlands Triassic depends on accurate age calibration of well data with outcrops and seismic data. Age constraints are largely lacking in these continental deposits although some earlier exploration wells have palynological reports that define the broad age ranges of the lithostratigraphic units and confirm their general correlatability (albeit with demonstrable fine-scale diachroneity along many contacts, cf. Geluk, 2005; Franz et al., 2013). Some progress has been made using conchostracans, palynology, isotope stratigraphy, cyclostratigraphy and magnetostratigraphy (Szurlies, 2007; Stollhofen et al., 2008; Bachmann et al., 2010; Kürschner & Herngreen, 2010; Tanner, 2010; Szurlies et al., 2012; Heunisch & Röhling, 2016), but the correlation task is also simplified by the layer-cake nature of much of the succession across the Netherlands and the presence of recognisable litho- and seismostratigraphic motifs that can be correlated over large distances (Geluk & Röhling, 1999; Geluk, 2007; Bachmann et al., 2010).

Lower Germanic Trias Group.

The Lower Germanic Trias Group consists of the shale-prone Lower Buntsandstein Subgroup and the more sand-prone Upper Buntsandstein Subgroup. The lower subgroup consists in turn of the Main Claystone and Rogenstein formations and the upper subgroup comprises the Volpriehausen, Detfurth and Hardegsen formations (Fig. 5.2). The lithostratigraphic base of the lower subgroup is typically, and pragmatically, picked at the top of the anhydrite-prone Zechstein Upper Claystone, although the actual Permian-Triassic boundary may be higher than this (Hiete et al., 2006; McCann et al., 2008; Stollhofen et al., 2008; Szurlies et al., 2012). The top of the group is defined by the Base Solling (Hardegsen) Unconformity.

The Lower Buntsandstein Subgroup is of latest Permian to Induan age (Kozur, 1999; Szurlies et al., 2003) and is a mud-rich interval composed of stacked, subtly upward-cleaning successions of fine-grained continental sandstones to claystones (Main Claystone) overlain by claystones punctuated by common oolite beds (Rogenstein). Individual cycles are 20 to 40 m thick and can be correlated over large distances up to several 100 km (Geluk & Röhling, 1997, 1999). The cycles show few dramatic thickness variations although expansion of the stratigraphy up to 400 m in the northern and western offshore areas is indicative of locally greater subsidence rates in these areas during deposition. An area of reduced subsidence and stratigraphic thinning, the Netherlands Swell (Fig. 5.5), was present in the north-central onshore area. At the eastern side of this swell several amalgamated, up to 5 m thick, Rogenstein Formation oolite beds are present (Fig. 5.2) which elsewhere in the basin expand and split into multiple thinner beds separated by claystones. At the southern margin of the basin, in the southern Netherlands and the adjacent area in Germany (Wolburg, 1961), up to 200 m of vertically amalgamated fluvial sandstones are present (Nederweert Member), grading into conglomerates in the extreme southeast of the Netherlands (Geluk & Röhling, 1999, 2016).

The Main Buntsandstein Subgroup is a more sand-prone succession and is again organized into large-scale upward-cleaning sequences. The top of the succession is defined by the Base Solling Unconformity, but minor precursor unconformities bound the main stratigraphic elements, with the result that the Volpriehausen, Detfurth and Hardegsen formations are also tectono-stratigraphic units (Fig. 5.2). Each of these formations consists of a twofold motif comprising a basal sandstone and overlying clay-siltstones, containing a hierarchical pattern of subsidiary cleaning-upward cycles (Geluk & Röhling, 1997, 1999). In the southern Netherlands and the UK southern North Sea, the subgroup is sandstone-dominated (Cameron et al., 1992; Johnson et al., 1994; Geluk et al., 1996). These sandstones are mainly fluvial in the south and grade northward into predominantly eolian deposits with clay-siltstones representing playa-lake deposits (Fontaine et al., 1993; Ames & Farfan, 1996, Geluk et al., 2018a). The subsurface lithology and depositional architecture of these units in the Netherlands is similar to that in outcrops near Trier in the west of Germany (Mader, 1983). Subsidence was greatest in the Dutch Central Graben, where locally over 500 m of sediments were deposited. In the Sole Pit Basin and offshore western Netherlands the thickness reaches 350 m, comparable to that found in the Roer Valley Graben and Ems Low.

The Volpriehausen Unconformity, at the base of the Volpriehausen Formation, locally cuts up to several tens of metres into the Lower Buntsandstein Subgroup. The formation reaches a maximum thickness of over 200 m in the Dutch Central Graben and the Broad Fourteens Basin, is ca. 100 m thick in the Ems Low and thins towards the Netherlands Swell to less than 5 m. The thickness changes are a composite of stratigraphic thinning of individual units, onlap of the basal unconformity, diachroneity of the tops of sandstone members and erosion at the base of the Detfurth Formation (Geluk & Röhling, 1999). In southern areas of the Netherlands the Volpriehausen Formation is a sand-rich fluvial succession (Ames & Farfan, 1996), but it transitions northwards into the Volpriehausen Clay-Siltstone Member, a succession of playa siltstones and marls with subordinate fluvial sandstones and carbonate oolite beds. The sands are generally thin in this northern region (Fontaine et al., 1993), but show local thickening of up to 60 m. A possible northern provenance from the Mid North Sea High (Olivarius et al., 2017; Kortekaas et al., 2018) may contribute to this local lithostratigraphic thickening of the sandstones.

The Detfurth Formation is restricted to Early Triassic depressions (Geluk, 2007) that result from uplift and erosion prior to deposition of the Solling Formation (Fig.5.5). The depositional thickness of the formation displays considerable variation: 60-100 m in the Dutch Central Graben, 50-80 m in the Ems Low and 20-40 m in the West Netherlands Basin, Roer Valley Graben and Broad Fourteens Basin. In the West Netherlands Basin and Roer Valley Graben the formation consists entirely of sandstones. At its base the Detfurth Unconformity cuts into the Volpriehausen Formation. In the southern and western offshore areas of the Netherlands this unconformity is the most prominent in the Buntsandstein Group (Geluk et al., 1996; Geluk & Röhling, 1999).

The Hardegsen Formation consists predominantly of siltstones, with subordinate, thin sandstone beds. Significant amounts of sandstone are restricted to the basin-margin areas. Truncation by the Base Solling Unconformity (Fig. 5.5) means that only erosional remnants remain of this interval in the Ems Low and in the Dutch Central Graben (well F09-03) where up to 200 m occurs, while over 500 m have been encountered in the Horn Graben (Danish wells Pele-S-1X and Erik-1X) (Best et al., 1983; Beha et al., 2008; Kilhams et al., 2018b). In the Broad Fourteens Basin, the West Netherlands Basin and the Roer Valley Graben, thinner intervals of up to 70 m are present. During deposition of the formation, increased subsidence in the Dutch Central Graben is suggested by the thickening of individual sequences. However, in other early depocentres, such as the Broad Fourteens Basin, the West Netherlands Basin and the Roer Valley Graben, differential subsidence had apparently ceased (Geluk & Röhling, 1999).

Upper Germanic Trias Group.

The Upper Germanic Trias Group comprises the Solling, Röt, Muschelkalk and Keuper formations (Fig. 5.2). The succession is mud- and carbonate-dominated, with several prominent evaporite horizons (Fig. 5.8). The base is marked by the Base Solling (Hardegsen) Unconformity and the top by the base Rhaetian (Cimmerian II) Unconformity. The overlying, largely Jurassic Altena Group also includes the Rhaetian Sleen Formation, which will be briefly covered here to complete the Triassic narrative.

The Hardegsen event appears to have been the product of diffuse extension over a wide area of northwest Europe (Štolfová & Shannon, 2009) and forms a robust regional surface to define the base of the group (Ziegler, 1990; Geluk & Röhling, 1997, 1999; Geluk, 2005). Fault displacements appear to have been low (e.g. Radies et al., 2005) and the unconformity is draped by the Solling Formation, comprising a thin basal veneer of continental sands overlain by widespread claystones (Fig. 5.7). In the Netherlands localized accumulations of eolian sands in salt withdrawal structures associated with faults have been drilled in the offshore L09 block (De Jager, 2012; Filomena et al., 2012) and are indicative of significant eolian deflation and sand trapping in topographic lows resulting from tectonism. The thickest development of the Solling Formation occurs in the Dutch Central Graben and Ems Low, where more than 100 m accumulated. Elsewhere the section is relatively thin and rarely exceeds a few tens of metres in thickness. The facies of the Solling Formation do not differ significantly from the underlying Buntsandstein, but deposition took place in a markedly reconfigured basin which now had access to the Tethys Sea via the Silesian-Moravian and Burgundy gates (Geluk, 2005, 2007).

The Röt Formation (Fig. 5.7) comprises a lower evaporitic section and an upper part dominated by claystones and siltstones, which collectively form a succession several hundreds of metres thick that thins to a few tens of metres on the basin margin. The Main Röt Evaporite (30-300 m thick) covers a large area of the UK, Dutch, Danish and German parts of the Southern Permian Basin and records the first major access by Tethyan marine waters into the basin (Kovalevych et al., 2002), albeit represented by restricted brines rather than by open marine environments. The section is dominated by halite, with anhydritic mudstones around the fringe of the main halite bodies. An intermediate claystone separates this from the Upper Röt Evaporite; a thinner (<20 m) halite with a more localized distribution in the Central Graben and eastern Netherlands. The Upper Röt Claystone (<150 m thick) records a resumption of fine-grained clastic deposition, with sandstones (Röt fringe; Geluk et al., 1996) located on the onshore basin margin, and carbonate beds in the upper part heralding the transgression of the Muschelkalk Sea.

The Muschelkalk Formation marks a major change in environment in the Southern Permian Basin to one dominated by marine carbonates and evaporites (with fringing clastics). Although widespread and sheet-like over the Netherlands, with a layered seismic response, the Muschelkalk shows locally abrupt thickness variations on seismic in the Central and Horn grabens, suggesting that minor fault movement took place during deposition, with up to 500 m in the southern Dutch Central Graben compared to less than 250 m elsewhere (Fig. 5.8). Regionally, however, thickness changes are gradual and suggest a dominance of thermal subsidence. In the central and eastern Netherlands the section was eroded during Early Cimmerian uplift (Fig. 5.6) and the original thickness is unknown. Regionally the base of the Muschelkalk is diachronous. The boundary in the east in Poland is early Anisian while in the western part of the Southern Permian Basin it is middle Anisian. In the Netherlands the Muschelkalk stratigraphy consists of four main elements: the Lower Muschelkalk Member carbonates, Muschelkalk Evaporite and Middle Muschelkalk Marl members, and Upper Muschelkalk Member carbonates (Figs 5.2, 5.8).

The Lower Muschelkalk Member comprises widespread interbedded limestones and marls that pass westward into marls and clastics (UK Dowsing Dolomitic Formation). The Mid North Sea and Ringkøbing-Fyn highs formed a barrier to further marine penetration, and to the north of these highs the Middle Triassic dominantly comprises continental clastics of the Skagerrak Formation (Goldsmith et al., 2003; Mckie, 2014).

The Middle Muschelkalk records a restriction in access to Tethyan marine waters and the creation of evaporitic conditions: The middle Muschelkalk halite is widely distributed across the Southern Permian Basin. Local thickening, for instance along the Mid Netherlands Fault Zone (Fig. 5.6) and Dutch Central Graben (Geluk, 2007), suggest that minor synsedimentary fault movement was active. However, the total halite thickness in the Netherlands (ca. 50 m) is less than that seen in Germany (up to 500 m), suggesting a marginal location within the evaporitic basin. The locally more complete successions seen in Germany contain up to nine cycles (Röhling, 2002), but the full complement of cycles is preserved only in fault-bounded depressions. The full nine cycles are also not present in the more marginal setting of the Netherlands (Geluk, 2007).

The Upper Muschelkalk records the re-flooding of the Southern Permian Basin and the return to carbonate deposition. Again, the succession has a layer-cake character, with local thickening in the Dutch Central Graben and Ems Low. In contrast to the Lower Muschelkalk, carbonate facies do not appear to extend as far westwards (although there is limited preservation of this stratigraphic level to the west and into the UK sector; Johnson et al., 1994).

The Keuper Formation is a largely shale-prone succession with a diachronous base (Franz et al., 2013) reflecting the gradual recession of the Muschelkalk Sea. As a result, the onset of Keuper claystone deposition occurs at a lower chronostratigraphic level in the western Netherlands compared to the east (Geluk, 2007). Cimmerian fault movements and halokinesis also resulted in dramatic thickness variations across the major graben structures and into rim synclines (Frisch & Kockel, 1997). These are also sites of enhanced preservation, contrasting with large areas of former Keuper deposition which were subsequently removed by erosion in the Jurassic, Cretaceous and Cenozoic (Fig. 5.3). In the Netherlands the Keuper can be divided into lower, middle and upper divisions, with the base of the upper defined by the Early Cimmerian I Unconformity and the top by the Early Cimmerian II Unconformity (Fig. 5.8). This subdivision is informal, and differs from the definition of Lower, Middle and Upper Keuper as used in Germany (Geluk, 2007). In the south of the Netherlands much of the lower and middle divisions were removed by erosion below the Early Cimmerian I Unconformity (Fig. 5.6) whereas in the Ems Low and Central Graben up to 1 km of stratigraphy is preserved and even thicker (5 km or more) successions are present in the Glückstadt and Horn grabens in Germany (Best et al., 1983; Frisch & Kockel, 1997; Barnasch, 2009; Kilhams et al., 2018b).

The lower and middle Keuper comprises the Lower Keuper Claystone, Main Keuper Evaporite, Middle Keuper Claystone and Red Keuper Evaporite members (Fig. 5.8). All members thicken into the graben structures, indicating accelerated tectonic activity. The claystone members are composed of red coloured, locally dolomitic and anhydritic playa claystones. The evaporite members comprise halite, anhydrite and minor claystones. The Main Evaporite is halite-bearing in the Dutch Central Graben, the Ems Low, and locally in the Broad Fourteens Basin, and is anhydritic elsewhere. The Red Evaporite interval is dominated by anyhdrite with halite only present in the Dutch Central Graben. The Middle Keuper Claystone locally contains micaceous sandstones in the southern Central Graben and Ems Low areas and is interpreted to be equivalent to the more widespread Schilfsandstein seen in Germany (Geluk, 2007).

The upper Keuper comprises the Red Keuper Claystone, Dolomitic Keuper and Upper Keuper Claystone members. As a result of truncation by the Early Cimmerian I Unconformity (Fig. 5.2) the section rests on the Muschelkalk (and locally deeper into the Röt) in the south of the Netherlands and on the lower to middle Keuper towards the north and west. The most complete sections are present in the deeper graben areas in the Dutch Central Graben, Broad Fourteens Basin and along the Mid Netherlands Fault Zone (Geluk, 2007; Fig. 5.6). Lithologies are broadly similar to the lower and middle Keuper playa claystones with dolomitic and anhydritic facies developed in the West Netherlands area. The Upper Keuper Claystone Member is a darker coloured marl and claystone interval that may reflect wetter conditions during deposition, rising water tables and potentially marine ingress.

The Rhaetian Sleen Formation rests on the Early Cimmerian II Unconformity and, in the Netherlands, records shallow marine mudstone deposition. The basal section consists of a fossiliferous mudstone; the upper part contains thin sandstones. The section is relatively uniform across the Netherlands but equivalents along the basin margins to the north in Denmark (Michelsen & Clausen, 2002) comprise coal-bearing deltaic facies.

Depositional history

The prevailing aridity and monsoon climate during the Triassic resulted in the Southern Permian Basin being dominated by continental red-beds, punctuated by Middle Triassic tropical marine carbonates and latest Triassic marine shales. In the Netherlands the three-fold Germanic succession of Bunter, Muschelkalk and Keuper (Fig. 5.2) can divide the facies composition of the succession into discrete assemblages of depositional systems that were formed in response to the longer-term tectonic, climatic and eustatic conditions that operated at the time. Sedimentological observations are largely based on cores, and comparison with equivalent sections exposed onshore in outcrops and quarries beyond the Netherlands.

Early Triassic continental clastics

Continental clastics dominated the Early Triassic across a wide region of northwest Europe, framing the Netherlands within a vast continental interior and physically isolated from any significant marine influences. The pre-Hardegsen succession comprises the mud-prone Main Claystone and Rogenstein formations, overlain by the more sand-prone Volpriehausen, Detfurth and Hardegsen formations (Fig. 5.2).

The Main Claystone Formation consists of a facies type that extends beyond both the Netherlands and the Southern Permian Basin, passing northwards into the almost identical Smith Bank Formation in the central North Sea. This interval comprises heterolithic, wave and current ripple laminated deposits (crosscut by syneresis and desiccation cracks), structureless siltstones and rare coarser-grained sand stringers (Fig. 5.9a). Thick sand-prone intervals (such as the Nederweert Sandstone Formation) are restricted to a few kilometres fringing the basin margins to the south of the Netherlands, with isolated ribbons locally extending into the basin. The Roer Valley Graben was the main axis of fluvial input into the basin at this time and other regions were essentially sand-starved. A variety of origins have been interpreted for the Main Claystone, but it is generally viewed as recording deposition in ephemeral playa lakes (Paul, 1982). These were episodically filled by monsoon floods, and then became more saline through evaporation and subsequently desiccated. In the central

Figure 5.9. Early Triassic facies from cores. a) Main Claystone Formation from well G16-05, showing ripple laminated playa heterolithics (R) disrupted by desiccation cracks (D). b) Rogenstein Formation with pale coloured oolitic grainstone interbedded with darker grey lacustrine shales from well De Wijk-22, showing wave ripples (W) and root traces (R). c) Distal terminal splay facies in the Volpriehausen Formation from well G14-04, showing salt crust fabric (S), mud clast bearing splay sands (M), and eolian sands (E). Core sections are 1 m long. Source: NAM.

North Sea some structureless siltstones have been interpreted as the products of wind-blown dust (Wilkins et al., 2018) and these may also be present in the Main Claystone Formation.

Towards the top of the Main Claystone interbedded oolitic horizons (up to 5 m thick, but generally less than 1 m) collectively comprising the Rogenstein Member (Fig. 5.9b). These vary from pure oolitic limestones (Fig. 5.10) to oolite-bearing sandstones, and form widespread sheets where individual beds can be correlated over many tens of kilometres (Fig. 5.7). These deposits are generally viewed as representing shallow-water, lake-margin bar accumulations formed during episodes of sustained high lake level (Palermo et al., 2008), although a marine origin has also been proposed (Weidlich, 2007; Szulc, 2019). The oolitic facies represent a significant change in the water balance of the basin towards episodes of prolonged subaqueous conditions, albeit punctuated by repeated returns to dry playa conditions. If a marine origin is ultimately demonstrated this would imply that flooding episodes inundated the basin, although the direction of the flooding is likely to be difficult to determine. Bachmann et al. (2010) postulated a marine connection to the Boreal ocean via the central and northern North Sea. However, despite widespread well penetrations this has yet to be proven.

The Main Buntsandstein Subgroup (Volpriehausen, Detfurth and Hardegsen formations) comprises a more sand-prone interval than the Main Claystone. The formations consist of fluvial-playa cycles of interbedded sandstones

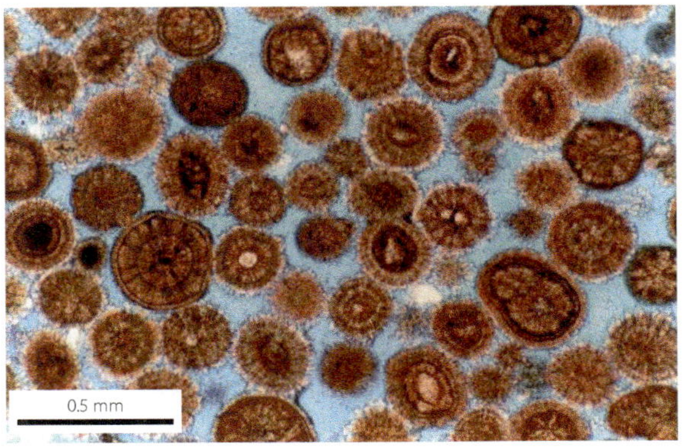

Figure 5.10. Resin-impregnated thin section with oolitic grainstone of the Rogenstein Formation. Blue colour highlights porosity. Well De Wijk-22 from the De Wijk Field. Source: NAM.

(lithic arenites) and claystones (Geluk & Röhling, 1997), with sandstones more common at the base of each cycle. Northern and eastern areas are typically more thinly bedded and mud-prone but become more sand-prone and coarser-grained towards the basin margins. In distal areas eolian-style grain flow and wind-ripple stratification is present and the associated mudstones are variably laminated, bioturbated and rooted, with sandstone laminae showing current and wave ripple lamination, soft sediment deformation and disruption by desiccation cracks and irregular salt crust fabrics (Figs 5.9c, 5.11a). Thinly bedded sandstones with intraformational mudclasts punctuate these successions. Pebbly sandstones with extraformational clasts are common in the south of the Netherlands towards the basin margin (Fig. 5.11b). The sandstones represent a spectrum of dryland depositional environments including ephemeral fluvial, eolian, terminal splay, wet and dry lake margin sandflat and ephemeral playa lake (Fig. 5. 12). Pebble-rich deposits in the south record deposition in basin margin alluvial fans or bajada (Geluk, 2005). Sparse rooting is suggestive of shrubland vegetation, and palynological data indicate the presence of drought-tolerant xerophytes (Kürschner & Herngreen, 2010). In contrast to the impoverished earliest Triassic (Main Claystone times), the Main Buntsandstein stratigraphic equivalents across the region have also yielded common evidence of vertebrate activity (Romano et al., 2020), indicating the re-establishment of continental ecosystems.

Middle Triassic terrestrial clastics, marine carbonates and evaporites

Following the Hardegsen event, fluvial and playa deposition continued from the Solling into the Röt Formation (Fig. 5.2), but with a marine influence present in the form of evaporite facies and rare marine fauna in thin bands (Kovalevych et al., 2002; Diedrich, 2009). The Anisian-Ladinian Muschelkalk records a major episode of marine flooding in the Southern Permian Basin, punctuated by the evaporitic Muschelkalk Evaporite Member during basin desiccation (Pöppelreiter, 2002; Pöppelreiter et al., 2004; Borkhataria et al., 2005, 2006).

The Solling and Röt claystone facies are similar to those described for the Main Buntsandstein (Fig. 5.11) but also contains localized, thick eolian accumulations in offshore areas (Fig. 5.13a). The Röt section is also notable for being punctuated by marine horizons and containing the Röt Halite (Fig. 5.7). This major salt body extends across a large area of the Southern Permian Basin and is composite, in the Netherlands comprising up to four halite layers separated by anhydritic mudstone intervals that have the

Figure 5.11. Early Triassic facies from cores. a) Volpriehausen Formation from well F18-11 showing interbedding of eolian beds (E), wave ripple forms (W), irregular salt crust fabrics (S) and terminal splay sandstones (T). b) More proximal, cross-stratified and mudclast-bearing fluvial sediments of the Volpriehausen Formation from well Q13-07. Core sections are 1 m long. Source: NAM.

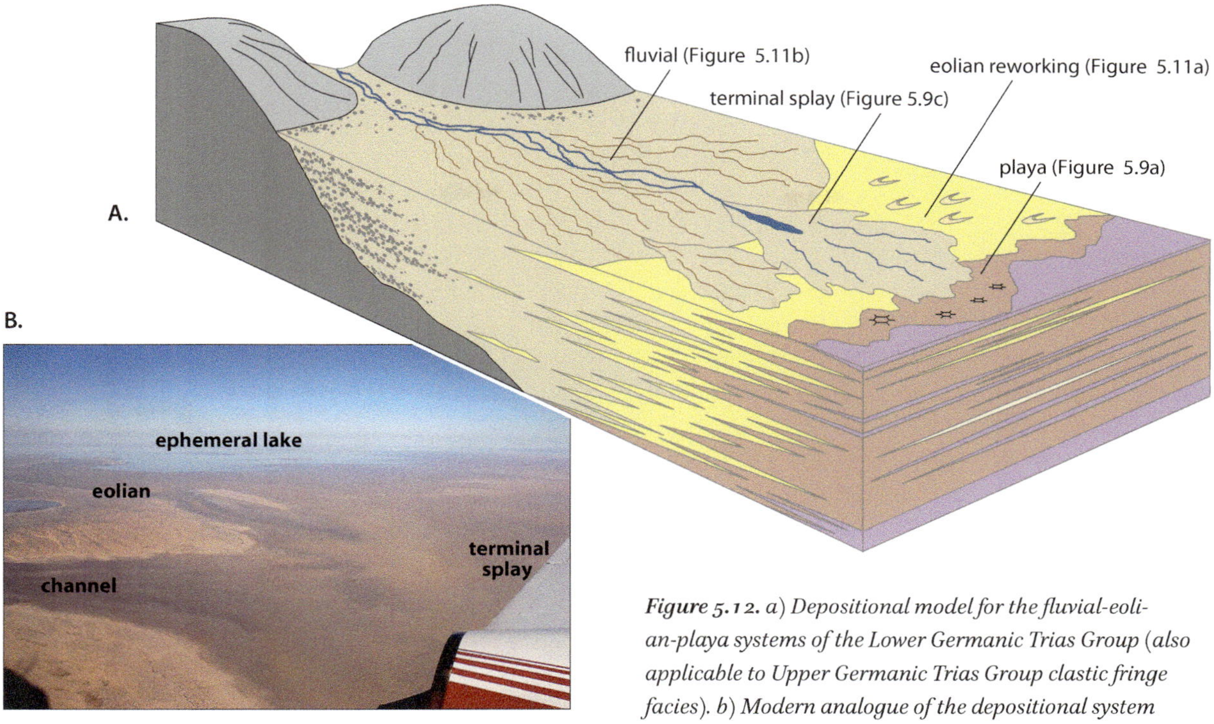

Figure 5.12. a) Depositional model for the fluvial-eolian-playa systems of the Lower Germanic Trias Group (also applicable to Upper Germanic Trias Group clastic fringe facies). b) Modern analogue of the depositional system shown in (a), Umbum Creek, Lake Eyre Basin, Australia. The channel mouth is ca. 2 km wide. Diagram modified from Geluk (2005).

chemical signature of a marine brine origin (Kovalevych et al., 2002). Quarry exposures of the Röt in the Winterswijk area (Fig. 5.14) have revealed vertebrate trackways (Diedrich, 2001) and overall conditions that were terrestrial, but which formed mud flats on the margins of an encroaching sea.

The overlying Muschelkalk carbonates comprise cyclic alternations of ooidal, peloidal and skeletal grainstones and packstones, interbedded with marls and mudstones (Fig. 5.13b). Skeletal detritus, although limited, comprises bivalve, brachiopod and ceratite shells together with crinoid debris. Tetrapod swimming traces have also been recorded from the Winterswijk area (Schulp et al., 2017). Graded beds, scours, cross-bedding, wave ripples and bioturbation are common. Both the Lower and Upper Muschelkalk have been interpreted as storm-dominated epeiric ramp deposits (Aigner, 1985; Pöppelreiter et al., 2005). The Lower Muschelkalk ramp facies in the Netherlands was mud-dominated with limited skeletal detritus (Pöppelreiter, 2002; Borkhataria et al., 2006). In contrast to this muddy ramp, the Upper Muschelkalk ramp developed ooidal and bioclastic shoals and associated back-shoal lagoons (Pöppelreiter et al., 2004; Borkhataria et al., 2006). Due to the enclosed nature of the Muschelkalk seaway it has been proposed that salinities diverged from fully marine to freshwater due to runoff from Fennoscandia and the Variscides (Korte et al., 2005; Franz et al., 2015), again probably driven by monsoon floods.

The Middle Muschelkalk is an evaporite-dominated interval dominated by halite with subsidiary anhydrite, dolomite and dolomitic marls, organized into basin wide cycles (Röhling, 2002). This interval records an episode of sustained restriction of Tethyan seawater into the western part of the Southern Permian Basin. There was sufficient inflow to generate a large halite succession due to evaporation, but this inflow was insufficient to re-establish fully marine conditions. The cyclicity would suggest that rates of recharge and evaporation varied through deposition of this interval although the reasons for this are unknown. They could record fluctuations in aridity, or tectonic activity controlling seawater inflow.

Anhydritic claystones occur around the perimeter of the Muschelkalk carbonates and these represent marginal sabkhas with clastic detritus fed by streams originating from the Variscan Mountains to the south. To the west, into the UK sector, the Muschelkalk becomes increasingly clastic dominated before ultimately transitioning into the Dowsing Dolomitic Formation (Bachmann et al., 2010).

Late Triassic evaporitic playa

The Ladinian recession of the Muschelkalk Sea was initially driven by a pluvial phase that caused the diachronous progradation of the Erfurt Formation (Lower Keuper Claystone in the Netherlands) coastal plain across the former seaway (cf. Franz et al., 2013). Following this there was a resumption of the dryland continental conditions that dominated the Carnian and Norian. The Late Triassic is

Winterswijk Quarry – a paleontological treasure trove

The quarry at Winterswijk, ca. 3 km east of the town of Winterswijk, is used to extract limestones of the Lower Muschelkalk, but in addition to its economic value it is also of great international scientific importance. The quarry represents a rare opportunity to observe Triassic rocks at outcrop in the Netherlands; the major facies transition from the largely continental Röt to marginal and fully marine Muschelkalk is exposed on the quarry floor, and the rocks display a wide variety of fossils and trace fossils that give a unique snapshot of life and death on the margins of the transgressing Muschelkalk seaway. Numerous specimens have been discovered here by both amateur and professional paleontologists, and it has been the site of discovery of a number of new species, including placodont marine reptiles and isopods.

The section is perhaps most famous for its tetrapod trackways, which are abundant at discrete levels in the Röt and the basal part of the Muschelkalk (Diedrich, 2001, 2015). They are found in association with mud-cracked claystones and microbial laminites that suggest deposition on desiccated intertidal flats and ponds on the margins of the transgressing Muschelkalk sea. In addition, Schulp et al. (2017) report a tetrapod swimming traceway, extending their environmental distribution into the marine realm.

A variety of marine vertebrate and invertebrate fossils have also been described in the lower Muschelkalk within more fully marine limestones which, in this transitional setting, are interbedded with the track-bearing subaerial facies. Three bone beds have been recorded in the lower part of the Muschelkalk section (Diedrich, 2001) containing fragments of aquatic reptiles (jawbones, vertebrae), fish (teeth, scales, skeletal material) and the skeletal remains of sharks (Diedrich, 2001; Oosterink et al., 2003; During et al., 2017). Bivalves, gastropods, brachiopods, arthropods (horseshoe crabs, insects; van Eldijk et al., 2017, 2019) and cephalopods are also recorded. Rarer fossils include lobsters (Klompmaker & Fraaije, 2011), possible jellyfish remains (Oosterink & Winkelhorst, 2013) and isopods (Schädel et al., 2020).

The marine trace fossil record in the quarry is also diverse, including rhizocorallium, cruziana, rusophycus, bifungites and radulichnus (Marchetti et al., 2019). The traces reflect the grazing, locomotion and burrowing behaviour of invertebrates that inhabited the Muschelkalk seabed. They indicate that conditions were well oxygenated and that there was a rich supply of organic detritus to feed on.

Examples of fossils and trace fossils recovered from Winterswijk quarry. a) Pachypleurosaur Anrosaurus heterodontus. *b) Palaeoniscidae fish. Photos courtesy of Nicole Klein. Scale bars are 1 cm.*

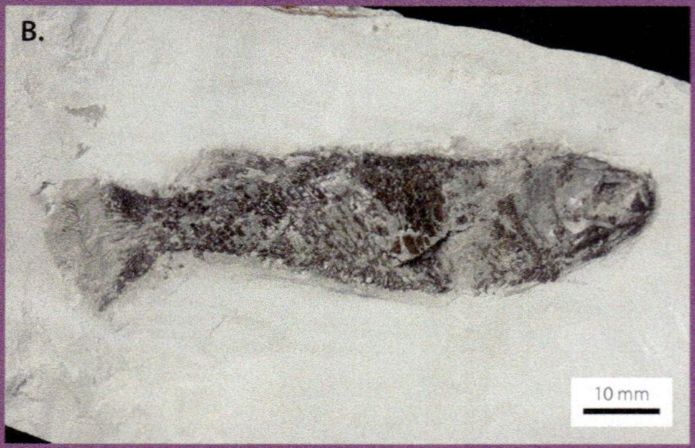

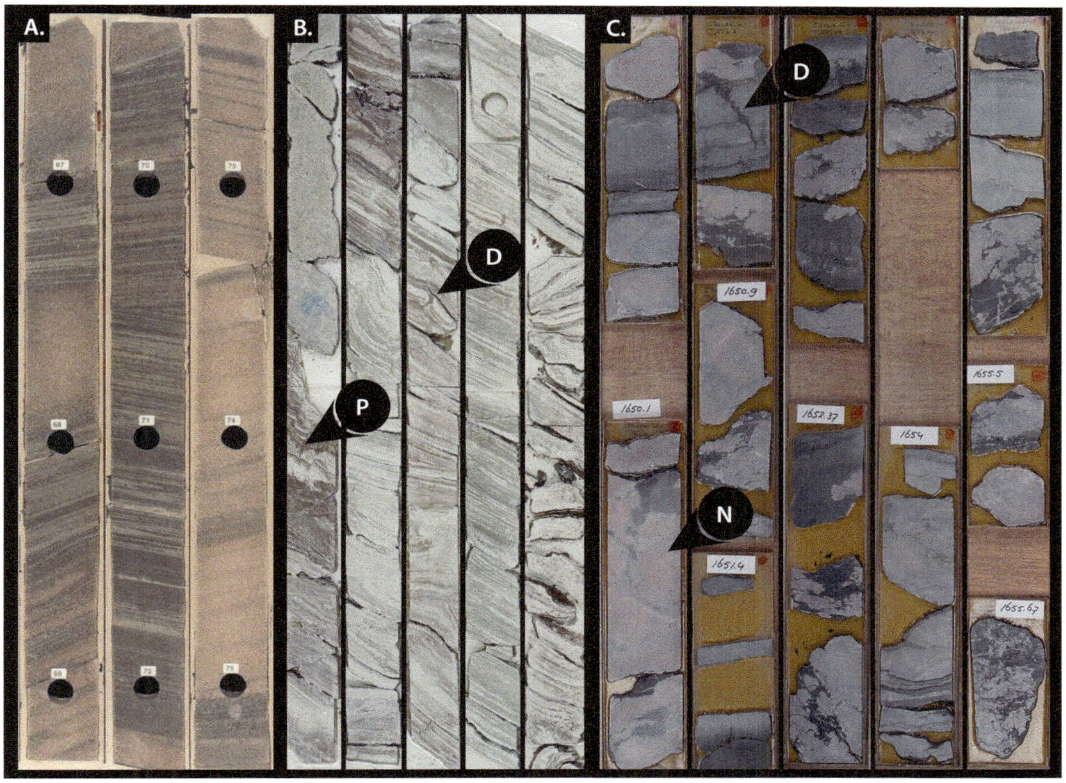

Figure 5.13. Middle and Late Triassic facies from cores. a) Eolian dune sandstones of the Solling Formation (Fat Sands; informal middle Solling Sandstone member) from well L09-10. b) Marginal limestones of the Lower Muschelkalk Member from well De Wijk-5, showing desiccation crack (D), reddened paleosol (P). c) Keuper playa claystones from well Nederweert-1 with anhydrite nodules (N) and desiccation crack (D). Core sections are 1 m long. Source: NAM.

mud-dominated over large areas of northwest Europe, but sandstones are present around the former Variscides to the south and on the fringes of Fennoscandia to the north (e.g. Hornung & Aigner, 2002). In the Netherlands there is limited preservation of the lower part of the Keuper in the south due to pervasive truncation by the Early Cimmerian unconformities. As a result, the main depositional record of the complete Keuper succession comes from offshore wells in more basinal locations, such as the Central Graben.

Anhydritic and dolomitic claystones (Fig. 5.13c) dominate the Keuper succession and are representative of similar facies seen across the offshore UK, Denmark and Germany. Well cuttings and very limited core sampling indicate that the shales are variably red and grey, with anhydrite occurring in discrete beds and as nodules (Fig. 5.12). Despite limited data available from the Netherlands these successions are interpreted to be comparable to the Keuper lithologies recognized as playa in Germany (Reinhardt & Ricken, 2000; Feist-Burkhardt et al., 2008; Vollmer et al., 2008), with similar depositional features to those seen in the Main Claystone. These claystones probably represent evaporitic playas, with the colours indicative of fluctuating groundwater conditions and the anhydrite recording near-surface evaporation of saline capillary waters of both marine and freshwater origin.

Halite bodies are present in the northern offshore Netherlands, thicken into the main graben structures (Dutch Central Graben, Broad Fourteens Basin and Ems Low) and condense laterally into clustered anhydrite beds. These occur at two distinct and correlatable horizons within the Carnian across the basin and are likely to be largely in-situ. However, it is possible that some have been extruded from underlying salt bodies as Triassic salt glaciers (Mohr et al., 2007; Bouroullec & Ten Veen, 2025, this volume). The origin of the brines is unclear, but overall the large volumes are consistent with a marine source (Warren, 2008), which would imply episodic marine brine access into the largely continental, playa dominated Keuper basin in the early and late Carnian (Aigner & Bachmann, 1989; Feist-Burkhardt et al., 2008). Marine facies in southern Germany coupled with thin marine horizons detectable in northern Germany (Stollhofen et al., 2008) suggest that this access could have been gained via the Burgundy Gate (Aigner & Bachmann, 1989).

The mid-Carnian regionally represents a departure from the typical Keuper red-bed and evaporite facies in showing a short-lived change to widespread fluvial sandstones

Figure 5.14. *Winterswijk Quarry showing the contact between the red coloured Röt Formation and grey coloured limestones of the Lower Muschelkalk Member. Photo credit: Adam McArthur. The photo at the beginning of this chapter shows a different view of the same section.*

(Schilfsandstein in Germany), grey lacustrine shales and abundant evidence of plant and animal activity (Kozur & Bachmann, 2010; Ruffell et al., 2016; Franz et al., 2018b). In the Netherlands the expression of this is limited, with sandstones only locally developed in some depocentres such as the Ems Low and the southern Dutch Central Graben. Elsewhere it is represented by the Middle Keuper Claystone. The interval records a wetter climate, more active river systems (Shukla et al., 2010) and widespread riparian vegetation. In general, however, the lithologies of the Middle Keuper Claystone seen in the Netherlands more typically form anhydritic red beds. These evaporitic (and hydrologically closed) deposits present a paradox when compared with the wetter climatic setting inferred for the Schilfsandstein rivers, which drained across multiple basins. It is possible that basinal floodplain settings may have remained semi-arid and evaporitic during this time, while the apparently wetter signature of the Schilfsandstein could be purely a reflection of intensified precipitation in a remote catchment. However, this would be in contradiction to the growing body of evidence for widespread change to wetter climate conditions across the Triassic northern hemisphere (Ruffell et al., 2016). Alternatively (and speculatively), the evaporitic Middle Keuper Claystone and Schilfsandstein fluvial sandstones may not be truly time equivalent, with the fluvial sandstones represented by a hiatal surface at either the base or top of the Middle Keuper Claystone.

Above the Early Cimmerian I tectonic event the Norian is represented by a halite-free succession of anhydritic red playa claystones. However, the medial Dolomitic Keuper, as well as containing dolomitic beds, is distinguished by a pervasive grey colouration suggesting more reducing, higher water table conditions. Equivalent sections to the east in Germany and Poland contain lacustrine intervals alternating with playa (Reinhardt & Ricken, 2000) together with isolated, thin marine horizons towards the base and top of the successions (Reinhardt & Ricken, 2000; Barth et al., 2018). To the west the UK shows more anhydritic facies (Fig. 5.8). The Dolomitic Keuper could therefore represent more sustained lacustrine conditions, or more speculatively could be analogous to the Anisian Dowsing Dolomitic Formation in representing a brackish facies associated with marine penetration from the south.

Late Triassic marine clastics

The Rhaetian Sleen Formation records the onset of the final marine flooding of the Southern Permian Basin. In the Netherlands this is marked by the marine shale of the Sleen Formation (Fig 5.2) which overlies the Early Cimmerian II Unconformity (Fig. 5.8). Equivalent sections in the UK and Germany contain more heterogenous, deltaic successions (e.g. Franz et al., 2018b). Fluvial runoff from these deltas into the enclosed basin resulted in a salinity gradient with fully marine conditions in Tethys grading into brackish and locally freshwater conditions in the Southern Permian Basin (Hallam & Shaarawy, 1982).

Paleogeographic evolution

The paleogeography of Triassic facies varied through time in response to changing climate, tectonics and sea level, all of which influenced sediment yield from fluvial catchments, the connection between basins, degree of eolian reworking of fluvial sands, vegetation cover and the distribution of marine and evaporitic facies (McKie & Williams, 2009; McKie, 2017). In addition, environmental perturbations associated with rare events such as the eruption of large igneous provinces sometimes occurred and resulted in extreme depositional responses (e.g. Mansouri & Hinderer, 2021).

The maps shown in Figures 5.15 to 5.17 were constructed using well logs, outcrop descriptions and limited seismic interpretation and, as they are interpolated between data points they give a general depiction of the depositional environment not restricted to the preserved stratigraphic fragments as shown in Figure 5.1. The maps provide the regional context of the Netherlands relative to adjacent basins so that intervals with uniform facies across the country are framed within a broader depositional setting.

The lowermost Triassic Main Claystone accumulated in a time of widespread playa deposition (Fig. 5.15a), which extended over several basins. There are no modern analogues for these vast, basin-wide playa facies (McKie, 2011), which also occur in the late Permian Silverpit Formation, and higher in the Triassic stratigraphy in the

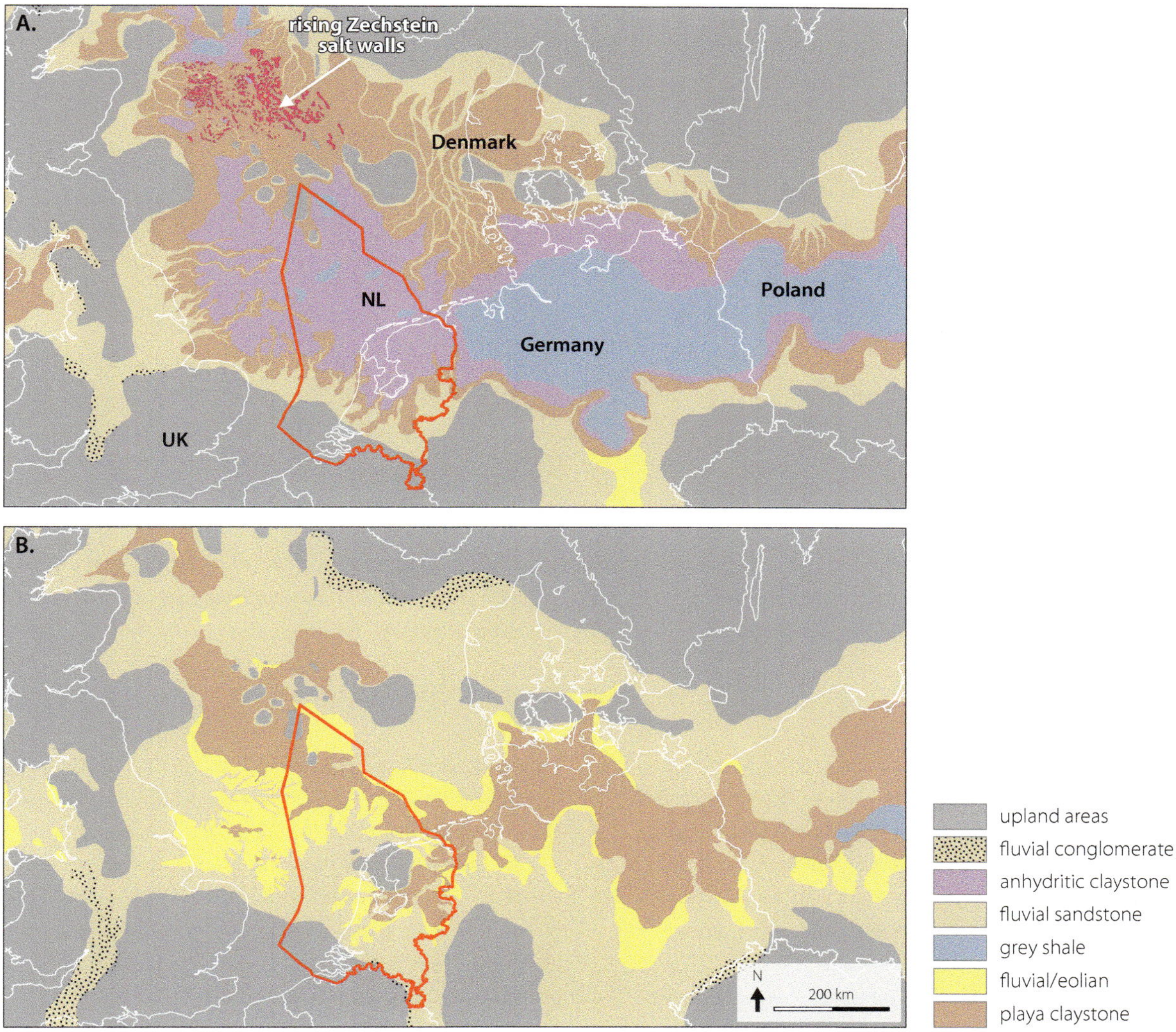

Figure 5.15. a) Main Claystone Formation paleogeography showing the position of the Netherlands (onshore and offshore outlined in red) within a vast, muddy playa system. b) Volpriehausen Formation paleogeography with basin margin fluvial systems (with eolian reworked fringes) that drained towards central playa lakes.

Keuper (Feist-Burkhardt et al., 2008) and UK Mercia Mudstone (McKie & Williams, 2009). These ancient examples are at least an order of magnitude larger in area than any present-day playa but lack comparably scaled fluvial catchments. Even with mega-monsoon floods only a small part of the ancient playa could have been inundated at any one time. Inactive areas may have been static, subject to eolian deflation or the sites of wind-blown dust accumulation. It is also possible that not all of these mud sections are playa facies and that muddy river systems transporting mud pellet aggregates were also present. These would be impossible to detect from limited core or outcrop data but have been inferred in the Mercia Mudstone (McKie & Williams, 2009) and can be observed in the present-day Australian interior. Fluvial systems can be mapped draining off the upland areas bounding the Southern Permian Basin (Fig. 5.15a) and are present in the south of the Netherlands. However, much of the north and the Dutch offshore was located within the central playa region.

The Main Buntsandstein Subgroup represents the widespread progradation of fluvial systems from the basin margin (Fig. 5.15b). These systems entered an arid basin and were subject to eolian reworking during dry seasons (e.g. Bourquin et al., 2009). The change of sediment yield compared to the muddy Main Claystone Formation was a widespread effect, but the reasons for this abrupt increase in sediment transport into the basin are unclear. Most probably it related to monsoon intensification in the hinterland, extending sand-prone river systems further into the basin (Van der Zwan & Spaar, 1992), possibly combined with episodic tectonism releasing sediment for fluvial transport. In this context the Rogenstein lakes may have been a precursor to this fluvial expansion, indicating greater volumes of water from the Variscides en-

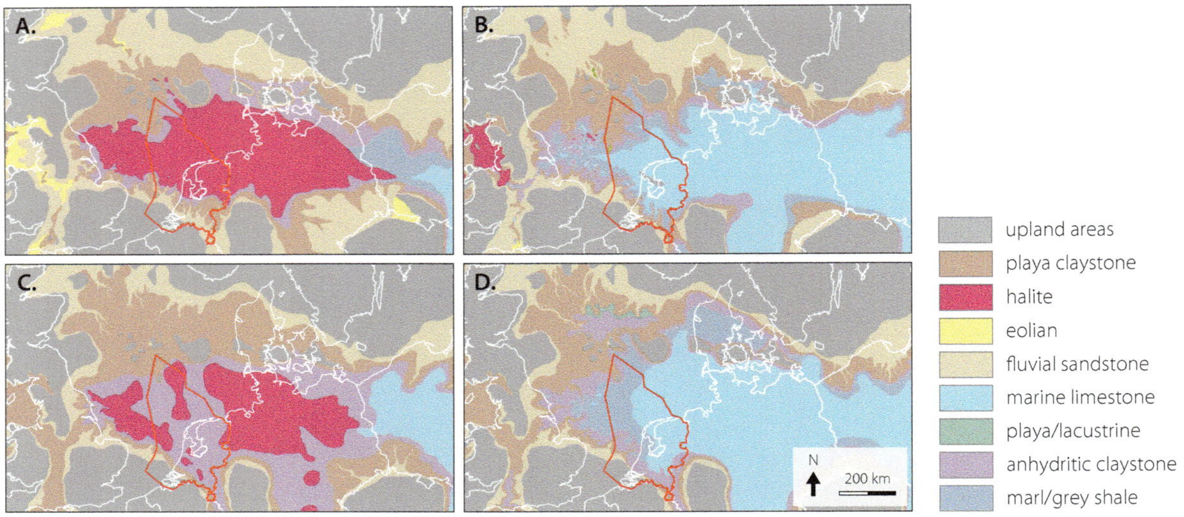

***Figure 5.16.** Middle Triassic transgressions and regressions of the Southern Permian Basin as Tethyan marine seawater gained periodic access to the basin. Restricted access led to evaporitic conditions and the concentration of brines whilst more open access led to fully marine conditions. a) Röt Halite Member evaporite basin. b) Lower Muschelkalk Member open seaway. c) Muschelkalk Evaporite Member evaporite basin. The patchy distribution of halite here is an artefact of Cretaceous erosion. d) Upper Muschelkalk Member open seaway.*

tering the basin. Fluvial transport across the Netherlands was focussed along an axis to the west of the country and extending from the southwest to the northeast. A similar axis existed during the deposition of late Permian Rotliegend (Fryberger et al., 2011) and both may be related to a long-lived antecedent drainage system derived from the Variscan Mountains.

The post-Hardegsen basin saw increased marine influence from Tethys via the Moravo-Silesian and Carpathian gates, although the initial marine penetration may have begun during deposition of the Middle Bunter (Heunisch & Röhling, 2016; Szulc, 2019). This was more fully expressed by the early Röt marine flooding event and culminated in the transgression of the Muschelkalk seaway via Poland, which subsequently expanded across the basin (Fig. 5.16). Continental sub-basins adjacent to the expanding seaway that lay below sea level were evaporitic and developed substantial halite bodies via brine seepage and overspill. The Röt Halite in the Netherlands records the early penetration of marine brines into the west of the Southern Permian Basin (Fig. 5.16a) prior to the development of open marine conditions and deposition of the Lower Muschelkalk carbonate ramps (Fig. 5.16b). The western reach of the Muschelkalk Sea extended beyond the confines of the characteristic carbonate ramp accumulations seen in the Netherlands, and grey coloured, brackish marine shales and marine evaporites can be identified to the west in the UK and Irish Sea (McKie, 2017). The Middle Muschelkalk records a reversion to conditions similar to the previous Röt evaporite (Fig. 5.16c) with ongoing seepage from Tethys into a restricted basin. During deposition of the Upper Muschelkalk (Fig. 5.16d) brackish marine penetration was directed northwards and entered the southern part of the Norwegian-Danish Basin (Michelsen & Clausen, 2002).

The Late Triassic Keuper paleogeography was influenced by the re-establishment of the vast continental playa environments that were a characteristic of the interior of Pangea. Two main episodes of halite precipitation appear to record marine seepage from Tethys in the early and late Carnian (Fig. 5.17a,c), but were not repeated in the Norian (Fig. 5.17d). The route that the marine brines took was likely to have been via the Burgundy Gate to the south. However, there is little record of marine facies beyond thin marine carbonates encased within playa and sabkha in southern Germany (locally extending into northern Germany; Stollhofen et al., 2008) and it is not fully understood how these epicontinental evaporitic environments operated (Aigner & Bachmann, 1989). Halites are restricted to the main graben areas, which may have been the location of the main brine pools or the final preservation sites of more widespread salt precipitation following repeated cycles of dissolution and reprecipitation by meteoric runoff. This regional pattern of evaporitic playa sedimentation is disrupted in the central and northern North Sea by widespread fluvial deposition, which may reflect rivers in these areas having catchments at higher and less arid latitudes (McKie & Williams, 2009).

During the middle Carnian Schilfsandstein rivers flowed across the Southern Permian Basin (Fig. 5.17b), forming lakes and deltas, before ultimately draining towards Tethys. The catchments for these southward flowing rivers were in Fennoscandia (Köppen & Carter, 2000; Paul et al.,

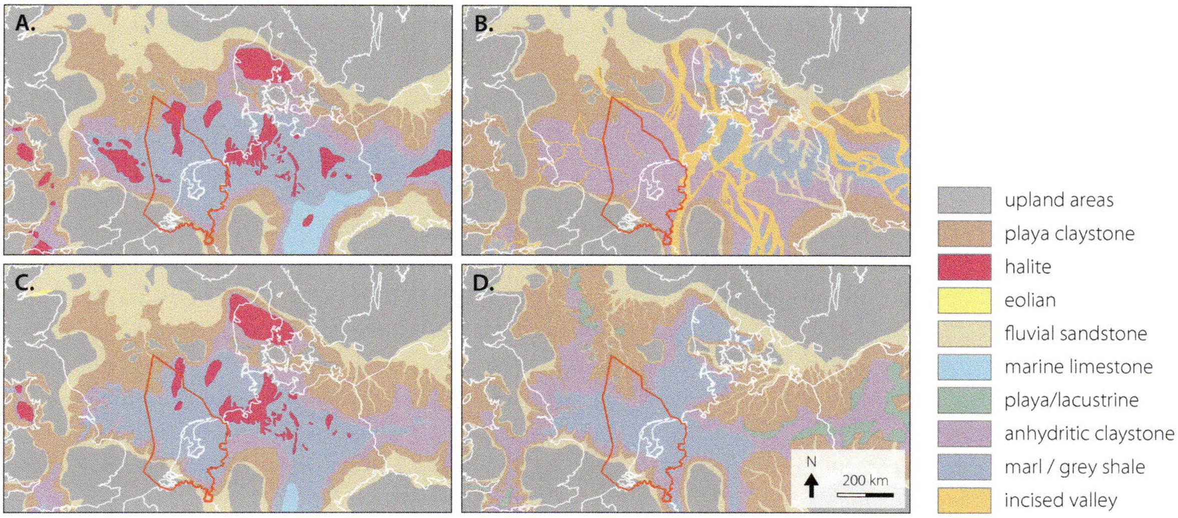

Figure 5.17. *Keuper paleogeographic evolution. a) Early Carnian (Main Keuper Evaporite Member) playa and halite bodies. Regions of grey shales and marls record the distribution of subaqueous deposits that may mark the former brine pathways that fed the precipitated halite bodies. b) Middle Carnian pluvial phase (Middle Keuper Claystone Member) with expanded fluvial drainage. The relationship between the inland delta facies mapped by Franz et al. (2018b) across Germany and the incised channels seen elsewhere is not well understood. c) Late Carnian (Red Keuper Evaporite Member) playa and halite bodies. d) Early Norian (Dolomitic Keuper Member) playa and ephemeral lacustrine facies. More persistent lacustrine water bodies appeared in the Norian across northwestern Europe but were still temporary features of the late Triassic landscape.*

2009). The architecture of this interval is potentially complex and is only beginning to be understood. Franz et al., (2018a,c) proposed that the Schilfsandstein in Germany was deposited by internally draining, distributive fluvial systems entering brackish marine water bodies that had limited connection to Tethys. These fluvial progradational cycles were episodically punctuated by transgressions that restored marine salinities. Observations from other regions suggest that more extensive, locally deeply incised, fluvial systems flowed southward and exited the basin to reach the Tethys Sea (Köppen & Carter, 2000; Geluk, 2007; Shukla et al., 2010). It is unlikely these end-member facies were synchronous and the incised river systems may be related to the basal Schilfsandstein unconformity, or to an internal sequence boundary within the Schilfsandstein itself. In the Netherlands the possible equivalents are mostly shale-prone sections (or a hiatal surface), although in the east Schilfsandstein rivers flowed through the Ems Low.

The Norian was dominated by terrestrial sedimentation (Fig. 5.17d) and lacks the halite characteristic of the Carnian playa environments. More perennial lacustrine water bodies were episodically present at this time (Fig. 5.17d) across Poland and Germany and minor marine incursions also potentially occurred (Reinhardt & Ricken, 2000; Barth et al., 2018). In the Netherlands the section is thinner and anhydritic with grey shales of the Dolomitic Keuper suggestive of higher water tables compared to the persistently anhydritic equivalents in the UK.

The Rhaetian Sleen Formation strictly belongs to the Altena Group (Van Adrichem Boogaert & Kouwe, 1994), but is briefly described here to cover the final episode of the Triassic. In the Netherlands the Sleen represents marine inundation of the basin, but there is little facies variation except towards the basin margins to the north and west where large deltaic systems prograded off Fennoscandia. These deltaic systems appear to have diluted marine waters in the basin, which had very low salinities towards Poland. In the Netherlands the Sleen Formation is most marine during the initial flooding at its base and shows a greater terrigenous input in the upper part, presumably as a result of the deltaic influence to the north.

Into the stratigraphic record

The Triassic succession that we see today did not reach its final state at the end of the Triassic. Subsequent tectonic deformation, subsidence, inversion and diagenesis resulted in continued modification of the succession into configurations that ultimately provide traps for hydrocarbons, sources for geothermal energy and shallow, exploitable sources of rock salt and limestone (Geluk et al., 2018a,b; Kilhams et al., 2018a). Diagenetic modification began shortly after deposition with the initiation of sediment compaction (Fig. 5.18a), precipitation of grain coating clays and haematite (Fig. 5.18b), and the precipitation of carbonate (mainly dolomite; Fig. 5.18b) from meteoric groundwater that flowed through the coarser grained, higher permeability beds. Further shallow burial (down to 500 m) allowed the circulation of more saline fluids which incorporated a mix of meteoric water and saline fluids expelled from deeper claystones and Zechstein evaporites. These precipitated widespread anhydrite and halite (Fig. 5.18c,d; Purvis & Okkerman, 1996). Halite cementation was pervasive adjacent to salt walls and diapirs and had a significant impact on sandstone porosity destruction (Bifani, 1986). However, later dissolution of this halite also exhumed the pore systems that had been protected from post-halite burial diagenesis, leaving an undercompacted, secondary porosity system.

Ongoing subsidence through the Early Jurassic, with minor extension localized to the Central Graben and Broad Fourteens Basin, allowed continued compaction and burial diagenesis by quartz, feldspar, ferroan dolomite and authigenic clays (mainly smectite, chlorite and illite) as burial depths and temperatures increased (Fig. 5.18e). This is a generalized list of diagenetic phases common across the basin, but different regions reached a variety of burial depths and temperatures at different times and had access to differing diagenetic fluids (including convection around Zechstein salt walls and diapirs). As a result, there is a wide range in the proportions and absolute volumes of cements precipitated in particular sub-basins or fault blocks (Van Kempen et al., 2018).

Middle Jurassic uplift related to thermal doming over a transient mantle plume in the central North Sea uplifted the northern Dutch offshore, resulting in erosion into the Triassic across the Cleaverbank Platform, Elbow Spit, Mid North Sea, Ringkøbing-Fyn and Schill Grund highs as well as uplift and erosion to Triassic level in the Terschelling Basin and Step Graben (Pharaoh et al., 2010). However, in the Sole Pit Basin and Dutch Central Graben subsidence continued, enhancing the preservation of the Triassic stratigraphy in these areas.

From the late Middle Jurassic and into the Late Jurassic accelerated rifting ensued, with extension focussed in the Central and Step grabens, West Netherlands, Broad Fourteens and Terschelling basins that took the Triassic to greater depths and higher temperatures. Rift flank uplift of the intrabasinal highs such as the Cleaverbank Platform, Elbow Spit, Friesland and Schill Grund highs further contributed to erosion of the Triassic. The widespread development of salt walls and diapirs resulted in crestal erosion above these structures (cf. Fig. 5.1). In addition, Zechstein salt walls, in conjunction with the bedded Röt Halite, began to form vertical and horizontal pressure seals that isolated Bunter fluid compartments at this time (Peeters et al., 2018).

Beginning in the Late Cretaceous, Alpine compression events caused inversion of the Sole Pit, Broad Fourteens, West and Central Netherlands basins and Central Graben (Figs 5.3, 5.4). As a result of this uplift the Triassic was locally exposed to erosion and to the action of meteoric groundwaters (Mijnlieff, 2020). Flow of meteoric waters caused the dissolution of feldspar and halite along the un-

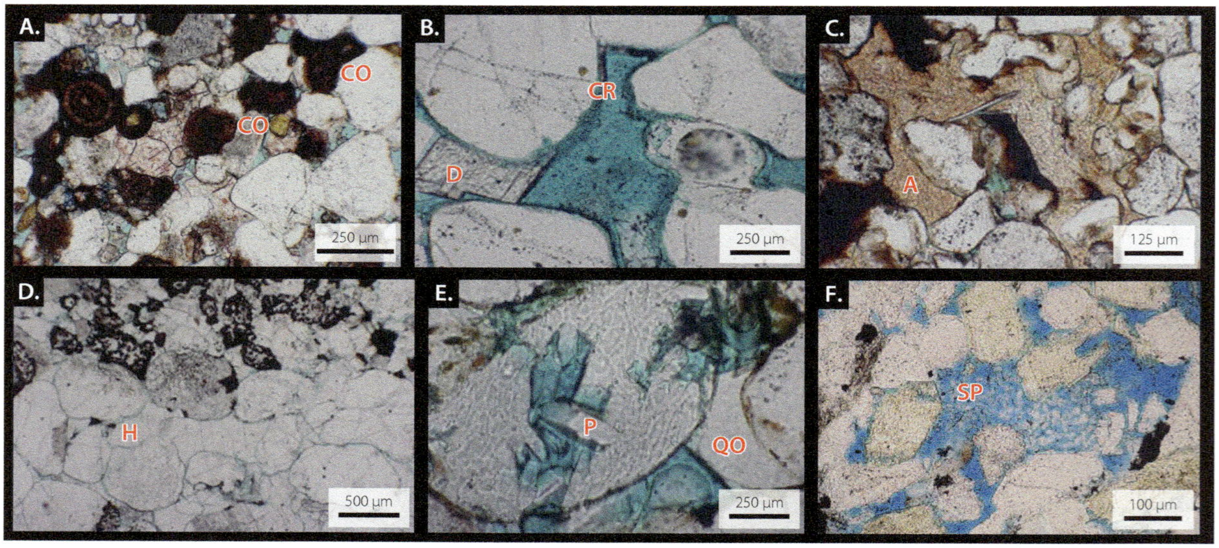

Figure 5.18. *Resin-impregnated thin section photomicrographs illustrating the diagenetic evolution of the sand-prone Lower Germanic Trias Group. a) Compaction of ductile pedogenic ooids (CO), well G14-4. b) Precipitation of grain rimming clays (CR) and non-ferroan dolomite rhombs (D), well L02-7. c) Pore-filling anhydrite (A), well L02-2. d) Pore-filling halite (H), well L02-2. e) Authigenic quartz overgrowths (QO) and prisms (P) within secondary macropore, well L04-5. f) Secondary macropore (SP) resulting from feldspar dissolution, infilled with kaolinite booklets, well Barendrecht-Ziedewij-1. Blue colour highlights porosity. Source: NAM.*

conformity surface (Fig. 5.18f), the precipitation of clays and, in the case of the De Wijk Field, the creation of a secondary porosity reservoir in the Rogenstein oolites (Fig. 5.10) accompanied by localized calcite reprecipitation (Bruijn, 1996; Goswami et al., 2018). Inversion also caused locally extensive fracturing of the now lithified Triassic. Finally, reversion to thermal sag following the latest, early Cenozoic inversion events carried the remnants of the offshore Triassic into burial again, while at the same time the onshore remained at relatively shallow depths.

Acknowledgements

This contribution owes a great debt to the pioneering and insightful work on the Triassic by Mark Geluk. He will be missed. The original manuscript also benefited greatly from the reviews by Harmen Mijnlieff and Heinz-Gerd Röhling, and editorial comments by Jan de Jager.

References

Aigner, T., 1985. Storm depositional systems. Dynamic stratigraphy in modern and ancient shallow-marine sequences. Lecture Notes in Earth Science 3: 174 pp.

Aigner, T. & Bachmann, G.H., 1989. Dynamic stratigraphy of an evaporitic-to-redbed sequence, Gipskeuper (Triassic), southwest German Basin. Sedimentary Geology 62: 5-25. DOI: 10.1016/0037-0738(89)90098-5

Ames, R. & Farfan, P.F., 1996. The environment of deposition of the Triassic Main Buntsandstein Formation in the P and Q quadrants, offshore the Netherlands. In: Rondeel, H.E., Batjes, D.A.J. & Nieuwenhuijs, W.H. (eds): Geology of gas and oil under the Netherlands. Kluwer Academic Publishers (Dordrecht): 167-178. DOI: 10.1007/978-94-009-0121-6_15

Archer, S.G., McKie, T., Andrews, S.D., Wilkins, A., Hutchison, M., Young, N., Osunde, C., Jones, J., Lang, M., Sola, B., Mouritzen, C. & Tenepalli, S., 2020. Triassic mudstones of the Central North Sea: cross-border characterisation, correlation and their palaeoclimatic significance. In: Patruno, S., Archer, S.G., Chiarella, D., Howell, L.A., Jackson, C.A.-L. & Kombrink, H. (eds): Cross-Border Themes in Petroleum Geology I: The North Sea. Geological Society of London. DOI: 10.1144/SP494-2019-61

Bachmann, G.H., Geluk, M.C., Warrington, G., Becker-Roman, A., Beutler, G., Hagdorn, H., Hounslow, M.W., Nitsch, E., Röhling, H.-G., Simon, T. & Szulc, A., 2010. Triassic. In: Doornenbal, J.C. & Stevenson, A.G. (eds): Petroleum geological atlas of the Southern Permian Basin area. EAGE Publications b.v. (Houten): 149-173.

Barnasch, J., 2009. Der Keuper im Westteil des Zentraleuropäischen Beckens (Deutschland, Niederlande, England, Dänemark): diskontinuierliche Sedimentation, Litho-, Zyklo- und Sequenzstratigraphie. PhD thesis, Martin-Luther-Universität Halle-Wittenberg (Germany): 222 pp.

Barth, G., Franz, M., Heunisch, C., Ernst, W., Zimmermann, J. & Wolfgramm, M., 2018. Marine and terrestrial sedimentation across the T-J transition in the North German Basin. Palaeogeography, Palaeoclimatology, Palaeoecology 489: 74-94. DOI: 10.1016/j.palaeo.2017.09.029

Beha, A., Thomsen, R.O. & Littke, R., 2008. Thermal history, hydrocarbon generation and migration in the Horn Graben in the Danish North Sea: a 2D basin modelling study. International Journal of Earth Sciences 97(5): 1087-1100. DOI: 10.1007/s00531-007-0247-2

Benton, M.J. & Twitchett, R.J., 2003. How to kill (almost) all life: The end-Permian extinction event. Trends in Ecology and Evolution 18: 358-365. DOI: 10.1016/S0169-5347(03)00093-4

Bernardi, M., Gianolla, P., Petti, F.M., Mietto, P. & Benton, M.J., 2018. Dinosaur diversification linked with the Carnian Pluvial Episode. Nature Communications 9: 1499. DOI: 10.1038/s41467-018-03996-1

Best, G., Kockel, F. & Schoeneich, H., 1983. Geological history of the southern Horn Graben. Geologie En Mijnbouw - Netherlands Journal of Geoscience 62: 25-33.

Bifani, R., 1986. Esmond Gas Complex. In: Brooks, J., Goff, J.C. & Van Hoorn, B. (eds): Habitat of Palaeozoic Gas in N.W. Europe. Geological Society of London, Special Publications 23: 209-221. DOI: 10.1144/GSL.SP.1986.023.01.13

Borkhataria, R., Aigner, T. & Pipping, J.C.P., 2006. An unusual, muddy, epeiric carbonate reservoir: The Lower Muschelkalk (Middle Triassic) of the Netherlands. American Association of Petroleum Geologists Bulletin 90(1): 61-89. DOI: 10.1306/08030504124

Borkhataria, R., Aigner, T., Pöppelreiter, M.C. & Pipping, J.C.P., 2005. Characterisation of epeiric 'Layer-cake' carbonate reservoirs: Upper Muschelkalk (Middle Triassic), The Netherlands. Journal of Petroleum Geology 28 (2): 119-146. DOI: 10.1111/j.1747-5457.2005.tb00076.x

Bos, R., Van Zonneveld, R.-J., Reumer, J.W.F., Vis, G.-J., Janssen, N., Everwijn, T., Sluijs, A. & Van de Schootbrugge, B., 2024. A high resolution palynological and geochemical study of the end-Triassic mass-extinction based on a new cored succession at Winterswijk (the Netherlands). Geological Magazine 161 (e17): 1-19. DOI: 10.1017/S0016756824000323

Bouroullec, R. & Ten Veen, J.H., 2025. Salt Tectonics. In: Ten Veen, J.H., Vis, G.-J., De Jager, J. & Wong, Th.E. (eds): Geology of the Netherlands, second edition. Amsterdam University Press (Amsterdam): 457-491. DOI: 10.5117/9789463728362_ch12

Bourquin, S., Guillocheau, F. & Péron, S., 2009. Braided rivers within an arid alluvial plain (example from Lower Triassic, western German Basin): recognition criteria and expression of stratigraphic cycles. Sedimentology 56: 2235-2264. DOI: 10.1111/j.1365-3091.2009.01078.x

Bruijn, A.N., 1996. De Wijk gas field (Netherlands): reservoir mapping with amplitude anomalies. In: Rondeel, H.E., Batjes, D.A.J.

& Nieuwenhuijs, W.H. (eds): Geology of gas and oil under the Netherlands. Kluwer Academic Publishers (Dordrecht): 243-253. DOI: 10.1007/978-94-009-0121-6_20

Cameron, T.D.J., Crosby, A., Balson, P.S., Jeffery, D.H., Lott, G.K., Bulat, J. & Harrison, D.H., 1992. United Kingdom offshore regional reports: the geology of the southern North Sea. HMSO (London) for the British Geological Survey: 152 pp.

Chen, Z.Q. & Benton, M.J., 2012. The timing and pattern of biotic recovery following the end-Permian mass extinction. Nature Geoscience 5: 375-383. DOI: 10.1038/ngeo1475

Cohen, K.M., Finney, S.C., Gibbard, P.L. & Fan, J.-X., 2013. The ICS International Chronostratigraphic Chart. Episodes 36(3): 199-204. DOI: 10.18814/epiiugs/2013/v36i3/002

Coward, M.P., 1995. Structural and tectonic setting of the Permo-Triassic basins of N-W Europe. In: Boldy, S.A.R. (ed.): Permian and Triassic Rifting in Northwest Europe. Geological Society of London, Special Publications 91: 7-39. DOI: 10.1144/GSL.SP.1995.091.01.02

Crowder, E., Rawlinson, N., Cornwell, D.G., Sammarco, C., Galetti, E. & Curtis, A., 2020. New insights into North Sea deep crustal structure and extension from transdimensional ambient noise tomography. Geophysical Journal International 224: 1196-1210. DOI: 10.1093/gji/ggaa475

Dal Corso, J., Bernardi, M., Sun, Y., Song, H., Seyfullah, L.J., Preto, N., Gianolla, P., Ruffell, A., Kustatscher, E., Roghi, G., Merico, A., Hohn, S., Schmidt, A.R., Marzoli, A., Newton, R.J., Wignall, P.B. & Benton, M.J., 2020. Extinction and dawn of the modern world in the Carnian (Late Triassic). Science Advances 6 (eaba0099). DOI: 10.1126/sciadv.aba0099

De Jager, J., 2012. The discovery of the Fat Sand Play (Solling Formation, Triassic), Northern Dutch offshore – a case of serendipity. Netherlands Journal of Geosciences 91(4): 609-619. DOI: 10.1017/S0016774600000408

Diedrich, C., 2009. Palaeogeographic evolution of the marine Middle Triassic marine Germanic Basin changements - with emphasis on the carbonate tidal flat and shallow marine habitats of reptiles in Central Pangaea. Global and Planetary Change 65(1-2): 27-55. DOI: 10.1016/j.gloplacha.2008.11.002

Diedrich, C., 2001. Vertebrate track-bed stratigraphy of the Röt and basal Lower Muschelkalk (Anisian) of Winterswijk (East Netherlands). Netherlands Journal of Geosciences 80: 31-40. DOI: 10.1017/S0016774600022290

Diedrich, C., 2015. Isochirotherium trackways, their possible trackmakers (?Arizonasaurus): intercontinental giant archosaur migrations in the Middle Triassic tsunami-influenced carbonate intertidal mud flats of the European Germanic Basin. Carbonates and Evaporites 30: 229-252. DOI: 10.1007/s13146-014-0228-z

During, M.A.D., Voeten, D.F.A.E., Schulp, A.S. & Ruemer, J.W.F., 2017. A possible Pararcus diepenbroeki vertebra from the Vossenveld Formation (Triassic, Anisian), Winterswijk, the Netherlands. Netherlands Journal of Geosciences 96: 63-68. DOI: 10.1017/njg.2016.7

Feist-Burkhardt, S., Götz, A.E., Szulc, J., Borkhataria, R., Geluk, M., Haas, J., Hornung, J., Jordan, P., Kempf, O., Michalík, J., Nawrocki, J., Reinhardt, L., Ricken, W., Röhling, H.G., Rüffer, T., Török, Á. & Zühlke, R., 2008. Triassic. In: McCann, T. (ed.): The Geology of Central Europe Volume 2: Mesozoic and Cenozoic. Geological Society of London 2: 749-822. DOI: 10.1144/CEV2P.1

Filomena, C.M., Stollhofen, H. & Ojik, K., 2012. High-resolution ultrasonic measurements as proxies to resolve clastic reservoir heterogeneity in a salt-cemented gas reservoir. AAPG Bulletin 96: 1197-1209. DOI: 10.1306/01021211100

Fontaine, J.M., Guastella, G., Jouault, P. & De la Vega, P., 1993. F15-A: a Triassic gas field on the easetrn limit of the Dutch Central Graben. In: Parker, J.R. (ed.): Petroleum Geology of North-West Europe: Proceedings of the 4th Conference. The Geological Society of London: 583-593. DOI: 10.1144/0040583

Franz, M., Barth, G., Zimmermann, J., Budach, I., Nowak, K. & Wolfgramm, M., 2018a. Geothermal resources of the North German Basin: exploration strategy, development examples and remaining opportunities in Mesozoic hydrothermal reservoirs. Geological Society of London, Special Publications 469. DOI: 10.1144/SP469.11

Franz, M., Henniger, M. & Barnasch, J., 2013. The strong diachronous Muschelkalk/Keuper facies shift in the Central European Basin: implications from the type section of the Erfurt Formation (Lower Keuper, Triassic) and basin-wide correlations. International Journal of Earth Sciences 102: 761-780,. DOI: 10.1007/s00531-012-0823-y

Franz, M., Kaiser, S.I., Fischer, J., Heunisch, C., Kustatscher, E., Luppold, F.W., Berner, U. & Röhling, H.G., 2015. Eustatic and climatic control on the Upper Muschelkalk Sea (late Anisian/Ladinian) in the Central European Basin. Global and Planetary Change 135: 1-27. DOI: 10.1016/j.gloplacha.2015.09.014

Franz, M., Kustatscher, E., Heunisch, C., Niegel, S. & Röhling, H.G., 2018b. The Schilfsandstein and its flora; arguments for a humid mid-Carnian episode? Journal of the Geological Society, London 176: 133-148. DOI: 10.1144/jgs2018-053

Franz, M., Nowak, K., Niegel, S., Seidel, E., Wolf, M. & Wolfgramm, M., 2018c. Deep geothermal resources of the North German Basin: The hydrothermal reservoirs of the Stuttgart Formation (Schilfsandstein, Upper Triassic. Zeitschrift der Deutschen Gesellschaft für Geowissenschaften 169: 353-387. DOI: 10.1127/zdgg/2018/0164

Frisch, U. & Kockel, F., 1997. Altkimmerische Bewegungen in Nordwestdeutschland. Brandenburgische Geowissenschaftliche Beiträge 4(1): 19-29.

Fryberger, S.G., Knight, R., Hern, C., Moscariello, A. & Kabel, S., 2011. Rotliegend facies, sedimentary provinces, and stratigraphy, Southern Permian Basin UK and the Netherlands: A review with new observations. In: Grötsch, J. & Gaupp, R. (eds): The Permian Rotliegend of the Netherlands. SEPM Special Publication 98, 51-88. DOI: 10.2110/pec.11.98.0051

Geluk, M.C., 2007. Triassic. In: Wong, Th.E., Batjes, D.A.J.& De Jager, J. (eds): Geology of the Netherlands. Royal Netherlands Academy of Arts and Sciences (KNAW, Amsterdam): 85-106.

Geluk, M.C., 2005. Stratigraphy and tectonics of Permo-Triassic basins in the Netherland and surrounding areas. PhD thesis Utrecht University (the Netherlands): 171 pp.

Geluk, M.C., Plomp, A. & Van Doorn, T.H.M., 1996. Development of the Permo-Triassic succession in the basin fringe area, southern Netherlands. In: Rondeel, H.E., Batjes, D.A.J. & Nieuwenhuijs, W.H. (eds): Geology of gas and oil under the Netherlands. Kluwer Academic Publishers (Dordrecht): 57-78. DOI: 10.1007/978-94-009-0121-6_7

Geluk, M.C. & Röhling, H.G., 1997. High-resolution sequence stratigraphy of the Lower Triassic Buntsandstein in the Netherlands and northwestern Germany. Geologie en Mijnbouw 76(3): 227-246.

Geluk, M.C. & Röhling, H.G., 1999. High-resolution sequence stratigraphy of the Lower Triassic 'Buntsandstein': a new tool for basin analysis. In: Bachmann, G.H. & Lerche, I. (eds): The Epicontinental Triassic - Zentralblatt für Geologie und Paläontology Teil I(7-8): 727-745.

Geluk, M.C. & Röhling, H.G., 2016. A tectono-stratigraphic model for the depositional history and basin development of the Permian–Early Triassic at the southern margin of the Southern Permian Basin (The Netherlands and adjacent parts of Belgium and Germany). Zeitschrift Der Deutschen Gesellschaft Für Geowissenschaften 167(2-3): 149-166. DOI: 10.1127/zdgg/2016/0073

Geluk, M.C., Kilhams, B. & Mckie, T., 2018a. Huidige inzichten in de Paleogeografie en delfstoffen van de Trias in Nederland en omringende gebieden. Grondboor en Hamer 73: 36-44.

Geluk, M.C., McKie, T. & Kilhams, B., 2018b. An introduction to the Triassic: current insights into the regional setting and energy resource potential of NW Europe. Geological Society of London, Special Publications 469: 139-147. DOI: 10.1144/SP469.1

Goldsmith, P.J., Hudson, G. & Van Veen, P., 2003. Triassic. In: Evans, D., Graham, G., Armour, A. & Bathurst, P. (eds): The Millennium Atlas: Petroleum Geology of the Central and Northern North Sea. Geological Society of London 1(9): 104-127.

Goswami, R., Seeberger, F.C. & Bosman, G., 2018. Enhanced gas recovery of an ageing field utilizing N2 displacement: De Wijk Field, The Netherlands. Geological Society of London, Special Publications 469: 237-251. DOI: 10.1144/SP469.2

Hallam, A. & Shaarawy, Z.E., 1982. Salinity reduction of the end-Triassic sea from the Alpine region into northwestern Europe. Lethaia 15: 169-178. DOI: 10.1111/j.1502-3931.1982.tb01990.x

Hallam, A. & Wignall, P.P., 1997. Mass extinctions and their aftermath. Oxford University Press.

Haq, B.U., 2018. Triassic eustatic variations reexamined. GSA Today 28: 4-9. DOI: 10.1130/GSATG381A.1

Heunisch, C. & Röhling, H. G. (eds), 2016. Early Triassic phytoplankton episodes in the Lower and Middle Buntsandstein of the Central European Basin. Zeitschrift der Deutschen Gesellschaft für Geowissenschaften 167: 227-248. DOI: 10.1127/zdgg/2016/0070

Hiete, M., Berner, U., Heunisch, C. & Röhling, H.-G., 2006. A high-resolution inorganic geochemical profile across the Zechstein-Buntsandstein boundary in the North German basin. Zeitschrift Der Deutschen Gesellschaft Fur Geowissenschaften 157(1): 77-105. DOI: 10.1127/1860-1804/2006/0157-0077

Hornung, J. & Aigner, T., 2002. Reservoir Architecture in a Terminal Alluvial Plain: An Outcrop Analogue Study (upper Triassic, Southern Germany) Part 1: Sedimentology and Petrophysics. Journal of Petroleum Geology 25(1): 3-30. DOI: 10.1111/j.1747-5457.2002.tb00097.x

Hornung, T., Brandner, R., Krystyn, L., Joachimski, M.M. & Keim, L., 2007. Multistratigraphic constraints on the NW Tethyan "Carnian Crisis". The Global Triassic.New Mexico Museum of Natural History and Science Bulletin 41: 59-67.

Jablonski, D., 1994. Extinctions in the fossil record. Philosophical Transactions - Royal Society of London, B 344: 11-17. DOI: 10.1098/rstb.1994.0045

Johnson, H., Warrington, G. & Stoker, S.J., 1994. 6. Permian and Triassic of the Southern North Sea. In: Knox, R.W.O. & Cordey, W.G. (eds): Lithostratigraphic nomenclature of the UK North Sea. British Geological Survey (Nottingham): 141 pp.

Kendall, G.C.S.C., Lake, P., Weathers III, H.D., Lakshmi, V., Althausen, J. & Alsharan, A.S., 2003. Evidence of rain shadow in the geologic record: repeated evaporite accumulation at extensional and compressional plate margins. In: Alsharan, A.S., Wood, W.W., Goudie, A.S., Fowler, A. & Abdellatif, E.M. (eds): Desertification in the Third Millennium. Swets & Zeitlinger, Lisse (the Netherlands): 45-52.

Kidder, D.L. & Worsley, T.R., 2004. Causes and consequences of extreme Permo-Triassic warming to globally equable climate and relation to the Permo-Triassic extinction and recovery. Palaeogeography, Palaeoclimatology, Palaeoecology 203: 207-237. DOI: 10.1016/S0031-0182(03)00667-9

Kilhams, B., Kukla, P.A., Mazur, S., McKie, T., Mijnlieff, H.F., Van Ojik, K. & Rosendaal, E., 2018a. Mesozoic resource potential in the Southern Permian Basin area: the geological key to exploiting remaining hydrocarbons whilst unlocking geothermal potential. Geological Society of London, Special Publications 469. DOI: 10.1144/SP469.26

Kilhams, B., Stevanovic, S. & Nicolai, C., 2018b. The 'Buntsandstein'gas play of the Horn Graben (German and Danish offshore): dry well analysis and remaining hydrocarbon potential. Geological Society of London, Special Publications 469: 169-192. DOI: 10.1144/SP469.5

Klompmaker, A.A. & Fraaije, R.H.B., 2011. The oldest (Middle Triassic, Anisian) lobsters from the Netherlands: taxonomy, taphonomy, paleoenvironment, and paleoecology. Palaeontologia Electronica 14(1): 16 pp.

Köppen, A. & Carter, A., 2000. Constraints on provenance of the central European Triassic using detrital zircon fission track data. Palaeogeography Palaeoclimatology Palaeoecology 161: 193-204. DOI: 10.1016/S0031-0182(00)00123-1

Korte, C., Kozur, H.W. & Veizer, J., 2005. δ13C and δ18O values of Triassic brachiopods and carbonate rocks as proxies for coeval seawater and palaeotemperature. Palaeogeography, Palaeoclimatology, Palaeoecology 226: 287-306. DOI: 10.1016/j.palaeo.2005.05.018

Kortekaas, M., Böker, U., Van der Kooij, C. & Jaarsma, B., 2018. Lower Triassic reservoir development in the northern Dutch offshore. Geological Society of London, Special Publications 469: 149-168. DOI: 10.1144/SP469.19

Kovalevych, V., Peryt, T.M., Beer, W., Geluk, M.C. & Halas, S., 2002. Geochemistry of Early Triassic seawater as indicated by study of the Röt halite in the Netherlands, Germany, and Poland. Chemical Geology 182(2-4): 549-563. DOI: 10.1016/S0009-2541(01)00343-6

Kozur, H., 1999. The correlation of the Germanic Buntsandstein and Muschelkalk with the Tethyan scale. In: Bachmann, G.H. & Lerche, I. (eds): The Epicontinental Triassic - Zentralblatt für Geologie und Paläontologie 7-8(1): 701-725.

Kozur, H.W. & Bachmann, G.H., 2010. The Middle Carnian Wet Intermezzo of the Stuttgart Formation (Schilfsandstein. Germanic Basin. Palaeogeography, Palaeoclimatology, Palaeoecology 290: 107-119. DOI: 10.1016/j.palaeo.2009.11.004

Kürschner, W.M. & Herngreen, G.F.W., 2010. Triassic palynology of central and northwestern Europe; a review of palynofloral diversity patterns and biostratigraphic subdivisions. In: Lucas, S.G. (ed.): The Triassic Timescale. Geological Society of London, Special Publications 334: 263-283.

Kutzbach, J.E. & Gallimore, R.G., 1989. Pangean climates: megamonsoons of the megacontinent. Journal of Geophysical Research 94: 3341-3357. DOI: 10.1029/JD094iD03p03341

Mader, D., 1983. Evolution of fluvial sedimentation in the Buntsandstein (Lower Triassic) of the Eifel (Germany). Sedimentary Geology 37: 1-84. DOI: 10.1016/0037-0738(83)90071-4

Mansouri, M. & Hinderer, M., 2021. Towards a sediment budget of Permo-Triassic successions in the Central European Basin – Implications for terrestrial perturbations around the Permian-Triassic Boundary (PTB). Zeitschrift Der Deutschen Gesellschaft Für Geowissenschaften. DOI: 10.1127/zdgg/2020/0239

Manspeizer, W., 1982. Triassic-Liassic Basins and climate of the Atlantic passive margins. Geologische Rundschau 71: 895-917. DOI: 10.1007/BF01821110

Marchetti, L., Van der Donck, H., Van Hylckama Vlieg, M. & During, M.A.D., 2019. Leaving only trace fossils -the unknown visitors of Winterswijk. Staringia 16(1): 250-257.

Maystrenko, Y.P., Bayer, U. & Scheck-Wenderoth, M., 2017. Structure and Evolution of the Glueckstadt Graben in Relation to the Other Post Permian Subbasins of the Central European Basin System. In: Soto, J.I., Flinch, J.F. & Tari, G. (eds): Permo-Triassic Salt Provinces of Europe, North Africa and the Atlantic Margins. Elsevier: 203-220. DOI: 10.1016/b978-0-12-809417-4.00010-0

McCann, T., Kiersnowski, H., Krainer, K., Vozarova, A., Peryt, T.M., Oplustil, S., Stollhofen, H., Schenider, J., Wetzel, A., Boulvain, F., Dusar, M., Torok, A., Haas, J., Tait, J. & Korner, F., 2008. Permian. In: McCann, T. (ed.): The Geology of Central Europe: Volume 1: Precambrian and Palaeozoic. Geological Society of London 1: 531-597.

Mckie, T., 2014. Climatic and tectonic controls on Triassic dryland terminal fluvial system architecture, central North Sea. In: Martinius, A.W., Ravnås, R., Howell, J.A., Steel, R.J. & Wonham, J.P. (eds): From Depositional Systems to Sedimentary Successions on the Norwegian Continental Margin. Wiley Blackwell: 19-57. DOI: 10.1002/9781118920435

McKie, T., 2017. Paleogeographic Evolution of Latest Permian and Triassic Salt Basins in Northwest Europe. In: Soto, J.I. Flinch, J.F. & Tari, G. (eds): Permo-Triassic Salt Provinces of Europe, North Africa and the Atlantic Margins. Elsevier: 159-173. DOI: 10.1016/B978-0-12-809417-4.00008-2

McKie, T., 2011. A comparisation of modern dryland depositional systems with the Rotliegend Group in the Netherlands. In: Grötsch. J. & Gaupp, R. (eds): The Permian Rotliegend of the Netherlands. SEPM Special Publication 98: 89-104. DOI:10.2110/PEC.11.98.0089

McKie, T. & Williams, B., 2009. Triassic palaeogeography and fluvial dispersal across the northwest European Basins. Geological Journal 44: 711-741.

Michelsen, O. & Clausen, O.R., 2002. Detailed stratigraphic subdivision and regional correlation of the southern Danish Triassic succession. Marine and Petroleum Geology 19(5): 563-587. DOI: 10.1016/S0264-8172(02)00028-4

Mijnlieff, H.F., 2020. Introduction to the geothermal play and reservoir geology of the Netherlands. Netherlands Journal of Geosciences 99(2): 1-19. DOI: 10.1017/njg.2020.2

Mohr, M., Warren, J.K., Kukla, P.A., Urai, J.L. & Irmen, A., 2007. Subsurface seismic record of salt glaciers in an extensional intracontinental setting (Late Triassic of northwestern Germany). Geology 35: 963-966. DOI: 10.1130/G23378A.1

Nichols, G.J., 2007. Features of fluvial systems in desiccating endorheic basins. Sedimentary Processes, Environments and Basins – a Tribute to Peter Friend. International Association of Sedimentologists, Special Publication 38: 567-587.

Olivarius, M., Weibel, R., Friis, H., Boldreel, L.O., Keulen, N. & Thomsen, T.B., 2017. Provenance of the Lower Triassic Bunter Sandstone Formation: implications for distribution and architecture of aeolian vs. fluvial reservoirs in the North German Basin. Basin Research 29(1): 113-130. DOI: 10.1111/bre.12140

Oosterink, H.W. & Winkelhorst, H., 2013. Probable remains of jellyfish (Cnidaria, Scyphozoa) from the lower Middle Triassic (Anisian) of Winterswijk, eastern Netherlands. Netherlands Journal of Geosciences 92(1): 61-67. DOI: 10.1017/S0016774600000287

Oosterink, H.W., Berkelder, W., De Jong, C., Lankamp, J. & Winkelhorst, H., 2003. Sauriens uit de Onder-Muschelkalk van Winterswijk. Staringia 11: 1-145.

Palermo, D., Aigner, T., Geluk, M.C., Pöppelreiter, M. & Pipping, K., 2008. Reservoir potential of a lacustrine mixed carbonate/siliciclastic gas reservoir: the Lower Triassic Rogenstein in the Netherlands. Journal of Petroleum Geology 31(1): 61-96. DOI: 10.1111/j.1747-5457.2008.00407.x

Parrish, J.T., 1993. Climate of the supercontinent Pangea. Journal of Geology 101: 215-233.

Paul, J., 1982. Der Untere Buntsandstein des germanischen Beckens. Geologische Rundschau 71: 795-811.

Paul, J., Wemmer, K. & Wetzel, F., 2009. Keuper (Late Triassic) sediments in Germany - indicators of rapid uplift of Caledonian rocks in southern Norway. Norwegian Journal of Geology 86: 193-202.

Peeters, S., Asschert, A. & Verweij, H., 2018. Towards a better understanding of the highly overpressured Lower Triassic Bunter reservoir rocks in the Terschelling Basin. Geological Society of London, Special Publications 469 (1): 223-236. DOI: 10.1144/SP469.13

Pharaoh, T.C., Dusar, M., Geluk, M.C., Kockel, F., Krawczyk, C.M., Krzywiec, P., Scheck-Wenderoth, M., Thybo, H., Vejbæk, O.V. & Van Wees, J.-D., 2010. Tectonic Evolution. In: Doornenbal, J.C. & Stevenson, A.G. (eds): Petroleum Geological Atlas of the Southern Permian Basin Area. EAGE Publications B.V. (Houten) : 24-57.

Pöppelreiter, M., 2002. Facies, cyclicity and reservoir properties of the Lower Muschelkalk (Middle Triassic) in the NE Netherlands. Facies 46: 119-132. DOI: 10.1007/BF02668077

Pöppelreiter, M.C., Borkhataria, R., Aigner, T. & Pipping, K., 2005. Production from Muschelkalk carbonates (Triassic, NE Netherlands): unique play or overlooked opportunity? In: Doré, A.G. & Vining, B.A. (eds): Petroleum Geology: North-West Europe and Global Perspectives - Proceedings of the 6th Petroleum Geology Conference. The Geological Society of London: 299-314.

Pöppelreiter, M., Simone, A. & Hoetz, G., 2004. Reservoir characteristics of intracontinental carbonate ramp deposits – Upper Muschelkalk, Middle Triassic, NE Netherlands. Netherlands Journal of Geosciences 83: 1-16. DOI: 10.1017/S0016774600020424

Purvis, K. & Okkerman, J.A., 1996. Inversion of reservoir quality by early diagenesis: an example from the Triassic Buntsandstein, offshore the Netherlands. In: Rondeel, H.E., Batjes, D.A.J. & Nieuwenhuijs, W.H. (eds): Geology of gas and oil under the Netherlands. Kluwer Academic Publishers (Dordrecht): 179-189.

Radies, D., Stollhofen, H., Hollmann, G. & Kukla, P., 2005. Synsedimentary faults and amalgamated unconformities: insights from 3D-seismic and core analysis of the Lower Triassic Middle Buntsandstein, Ems Trough, north-western Germany. International Journal of Earth Sciences 94: 863-875. DOI: 10.1007/s00531-005-0009-y

Raup, D.M. & Sepkoski, J.J., 1982. Mass extinctions in the marine fossil record. Science 215: 1501-1503. DOI: 10.1126/science.215.4539.1501

Reinhardt, L. & Ricken, W., 2000. The stratigraphic and geochemical record of Playa Cycles: monitoring a Pangaean monsoon-like system (Triassic, Middle Keuper, S. Germany). Palaeogeography, Palaeoclimatology, Palaeoecology 161: 205-227. DOI: 10.1016/S0031-0182(00)00124-3

Röhling, S., 2002. Der Mittlere Muschelkalk in Bohrungen Norddeutschlands: Fazies, Geochemie, Zyklo- und Sequenzstratigraphie. PhD thesis, Martin-Luther-Universität Halle-Wittenberg: 197 pp

Romano, M., Bernardi, M., Petti, F.M., Rubidge, B., Hancox, J. & Benton, M.J., 2020. Early Triassic terrestrial tetrapod fauna: a review. Earth Science Reviews 210(103331). DOI: 10.1016/j.earscirev.2020.103331

Ruffell, A., Simms, M.J. & Wignall, P.B., 2016. The Carnian Humid Episode of the late Triassic: A Review. Geological Magazine 153: 271-284. DOI: 10.1017/S0016756815000424

Schädel, M., Eldijk, T., Winkelhorst, H., Reumer, J.W.F. & Haug, J.T., 2020. Triassic Isopoda – three new species from Central Europe shed light on the early diversity of the group. Bulletin of Geosciences 95: 145-166. DOI: 10.3140/bull.geosci.1773

Schulp, A.S., Bleeker, R.W., Haarhuis, A., Van Spronsen, E., During, M.A.D., Goris, G., Kaskes, P., Matteman, Y., Winkelhorst, W. & Winkelhorst, H., 2017. A tetrapod swimming traceway from the Triassic of Winterswijk, the Netherlands. Netherlands Journal of Geosciences 96(3): 273-277. DOI: 10.1017/njg.2017.13

Sellwood, B.W. & Valdes, P.J., 2007. Mesozoic climate. In: Williams, M., Haywood, A.M., Gregory, F.J. & Schmidt, D.N. (eds): Deep-time perspectives on climate change: marrying the signal from computer models and biological proxies. The Micropalaeontological Society, Special Publications, Geological Society of London 2: 210-224. DOI: 10.1144/TMS002.10

Shukla, U.K., Bachmann, G.H. & Singh, I.B., 2010. Facies architecture of the Stuttgart Formation (Schilfsandstein, Upper Triassic), central Germany, and its comparison with modern Ganga system, India. Palaeogeography, Palaeoclimatology, Palaeoecology 297(1): 110-128. DOI: 10.1016/j.palaeo.2010.07.019

Simms, M.J. & Ruffell, A.H., 1989. Synchroneity of climatic change in the late Triassic. Geology 17: 265-268. DOI: 10.1130/0091-7613(1989)017

Štolfová, K. & Shannon, P.M., 2009. Permo-Triassic development from Ireland to Norway: basin architecture and regional controls. Geological Journal 44: 652-676. DOI: 10.1002/gj.1187

Stollhofen, H., Bachmann, G.H., Barnasch, J., Bayer, U., Beutler, G., Franz, M., Kästner, M., Legler, B., Mutterlose, J. & Radies, D., 2008. Upper Rotliegend to Early Cretaceous basin development. In: Littke, R., Bayer, U., Gajewski, D. & Nelskamp, S. (eds): Dynamics of Complex Intracontinental Basins: The Central European Basin System. Springer (Berlin): 181-210.

Szulc, J., 2019. Lower Triassic marine Buntsandstein deposits in the central European basin. Zeitschrift der Deutschen Ge-

sellschaft für Geowissenschaften 170: 311-320. DOI: 10.1127/zdgg/2019/0190
Szurlies, M., 2007. Latest Permian to Middle Triassic cyclo-magnetostratigraphy from the Central European Basin, Germany: Implications for the geomagnetic polarity timescale. Earth and Planetary Science Letters 261(3-4): 602-619. DOI: 10.1016/j.epsl.2007.07.018
Szurlies, M., Bachmann, G.H., Menning, M., Nowaczyk, N.R. & Käding, K.C., 2003. Magnetostratigraphy and high-resolution lithostratigraphy of the Permian-Triassic boundary interval in Central Germany. Earth and Planetary Science Letters 212(3-4): 263-278.
Szurlies, M., Geluk, M.C., Krijgsman, W. & Kürschner, W.M., 2012. The continental Permian–Triassic boundary in the Netherlands: Implications for the geomagnetic polarity time scale. Earth and Planetary Science Letters 317-318: 165-176. DOI: 10.1016/j.epsl.2011.11.043
Tanner, L.H., 2010. The Triassic isotope record. In: Lucas, S.G. (ed.): The Triassic Timescale. Geological Society of London, Special Publications 334: 103-118. DOI: 10.1144/SP334.5
Trotter, J.A., Williams, I.S., Nicora, A., Mazza, M. & Rigo, M., 2015. Long-term cycles of Triassic climate change: a new δ18O record from conodont apatite. Earth and Planetary Science Letters 415: 165-174. DOI: 10.1016/j.epsl.2015.01.038
Van Adrichem Boogaert, H.A. & Kouwe, W.F.P., 1994. Stratigraphic nomenclature of the Netherlands, revision and update by RGD and NOGEPA, Section E. Triassic (Lower and Upper Germanic Trias Groups). Mededelingen Rijks Geologische Dienst 50: 1-8
Van de Schootbrugge, B. & Wignall, P.B., 2016. A tale of two extinctions: Converging end-Permian and end-Triassic scenarios. Geological Magazine 153: 332-354. DOI: 10.1017/S0016756815000643
Van der Zwan, C.J. & Spaar, P., 1992. Lower to Middle Triassic sequence stratigraphy and climatology of the Netherlands, a model. Palaeogeography, Palaeoclimatology, Palaeoecology 91(3-4): 277-290. DOI: 10.1016/0031-0182(92)90072-D
Van Eldijk, T., Goris, G., Haarhuis, A., Lankamp, J., Winkelhorst, H., Reumer, J., Nel, A. & Wappler, T., 2017. New fossil insects from the Anisian (Lower to Middle Muschelkalk) from the Central European Basin (Germany and The Netherlands. PalZ 91: 185-194. DOI: 10.1007/s12542-017-0343-y
Van Eldijk, T., Heerschop, R., Haarhuis, A., Winkelhorst, H. & Klompmaker, A.A., 2019. Non-arthropod invertebrates from the Middle Triassic of Winterswijk. Staringia 16: 185-190.
Van Kempen, B.M.M., Mijnlieff, H.F. & Van der Molen, J., 2018. Data mining in the Dutch Oil and Gas Portal: a case study on the reservoir properties of the Volpriehausen Sandstone interval. Geological Society of London, Special Publications 469. DOI: 10.1144/SP469.15
Van Wees, J.-D., Stephenson, R.A., Ziegler, P.A., Bayer, U., McCann, T., Dadlez, R., Gaupp, R., Narkiewicz, M., Bitzer, F. & Scheck, M., 2000. On the origin of the Southern Permian Basin, Central Europe. Marine and Petroleum Geology 17 (1): 43-59. DOI: 10.1016/S0264-8172(99)00052-5
Vollmer, T., Werner, R., Weber, M., Tougiannidis, N., Röhling, H.G. & Hambach, U., 2008. Orbital control on Upper Triassic Playa cycles of the Steinmergel-Keuper (Norian): A new concept for ancient playa cycles. Palaeogeography, Palaeoclimatology, Palaeoecology 267: 1-16. DOI: 10.1016/j.palaeo.2007.12.017
Warren, J.K., 2008. Salt as sediment in the Central European Basin System as seen from a deep time perspective. In: Littke, R., Bayer, U., Gajewski, D. & Nelskamp, S. (eds): Dynamics of Complex Intracontinental Basins: The Central European Basin System. Springer (Berlin): 249-276.
Weidlich, O., 2007. PTB mass extinction and earliest Triassic recovery overlooked? New evidence for a marine origin of Lower Triassic mixed carbonate–siliciclastic sediments (Rogenstein Member), Germany. Palaeogeography, Palaeoclimatology, Palaeoecology 252: 259-269. DOI: 10.1016/j.palaeo.2006.11.046
Wilkins, A.D., Hurst, A., Wilson, M.J. & Archer, S., 2018. Palaeoenvironment in an ancient low-latitude, arid lacustrine basin with loessite: The Smith Bank Formation (Early Triassic) in the Central North Sea, UK Continental Shelf. Sedimentology 65: 335-339. DOI: 10.1111/sed.12382
Wolburg, J., 1961. Sedimentations-Zyklen und Stratigraphie des Buntsandsteins in NW-Deutschland. Geotektonische Forschungen 14: 7-74.
Zhang, Y., Ogg, J.G., Franz, M., Bachmann, G.H., Szurlies, M., Röhling, H.G., Minsong, L., Rolf, C. & Obst, K., 2020. Carnian (Late Triassic) magnetostratigraphy from the Germanic Basin allowing global correlation of the Mid-Carnian Episode. Earth and Planetary Science Letters 541: 116275. DOI: 10.1016/j.epsl.2020.116275
Ziegler, P.A., 1990. Geological Atlas of Western and Central Europe, second edition. Shell Internationale Petroleum Mij. B.V. and Geological Society (London): 239 pp.

6 Rhaetian to Middle Jurassic

João P. Trabucho Alexandre & Theo E. Wong†*

*Corresponding author: j.trabucho@uu.nl
Theo Wong passed away during the proof-reading stage of this chapter.

ABSTRACT

The Altena Group consists mainly of dark argillaceous mudstones deposited between the Rhaetian and the Oxfordian in the Laurasian Sea, an epeiric sea on the northwestern margin of the Tethys Ocean. The group has been subdivided into the Sleen, Aalburg, Posidonia Shale, Werkendam, and Brabant formations. Calcareous intercalations and carbonaceous mudstones characterize the Rhaetian–Toarcian part of the group, whereas the Aalenian–Oxfordian part comprises an alternation of (sandy) mudstones, sandstones, marls, and limestones. The mudstones of the lower to middle Toarcian Posidonia Shale Formation have particularly elevated organic carbon contents and constitute the main source rock for oil in the Dutch subsurface. The present-day distribution of the group is largely confined to Cimmerian basins, where it attains locally a thickness of more than 1500 m. In the northern Dutch offshore, Jurassic rocks have also been preserved outside the Dutch Central Graben in smaller basins, where accommodation was generated by salt withdrawal or along strike-slip faults. The Altena Group was also deposited on the platforms and some highs, where the Lower Jurassic succession was condensed due to limited accommodation and Early Jurassic erosion. Condensed successions on platforms and highs were subsequently removed by erosion during the Middle and Late Cimmerian uplift (Middle to Late Jurassic), resulting in the present-day patchy distribution of the group. In the Late Cretaceous, the Altena Group was also locally eroded from the Cimmerian basins due to tectonic inversion.

<< The upper Pliensbachian Penny Nab Member of the Cleveland Ironstone Formation is one of the lateral equivalents of the Aalburg Formation in Great Britain. Here, we see the Upper Striped Bed and the ironstone known as Raisdale Seam exposed at the bottom of the cliff at Jet Wyke, near Staithes, North Yorkshire, England. Photo: João Trabucho.

DOI: 10.5117/9789463728362_ch06

Introduction

The Jurassic of the Netherlands has been subdivided into the Altena, Schieland, Scruff, and Niedersachsen groups (Nederlandse Aardolie Maatschappij & Rijks Geologische Dienst, 1980; Van Adrichem Boogaert & Kouwe, 1993-1997). The last three represent approximately coeval sediments deposited between the Callovian and the Ryazanian (see Verreussel et al., 2025, this volume). In this chapter, we focus on the Altena Group, which consists mainly of dark argillaceous mudstones deposited between the Rhaetian (Late Triassic) and the Oxfordian (Late Jurassic). We will deal with the Dutch Rhaetian in this chapter because, together with the Lower and Middle Jurassic, it forms the *marine* Altena Group (see Pearson, 1970). Calcareous intercalations characterize the Rhaetian-Toarcian part of the group, whereas the Aalenian-Oxfordian part of the group consists of an alternation of (sandy) mudstones, sandstones, marls, and limestones. The group has been subdivided into five formations: the Sleen, Aalburg, Posidonia Shale, Werkendam, and Brabant formations (Nederlandse Aardolie Maatschappij & Rijks Geologische Dienst, 1980; Van Adrichem Boogaert & Kouwe, 1993-1997).

The sediments of the Altena Group and their lateral equivalents outside the Netherlands were deposited in the northern Laurasian Sea (Fig. 6.1), an epeiric sea on the northwestern margin of the Tethys Ocean surrounded by northern Pangea (Laurasia) to the west and north and by the Hispanic Corridor and southern Pangea (Gondwana) to the south (e.g. Bjerrum et al., 2001; Ruvalcaba Baroni et al., 2018). In the southern North Sea area, the water depth of the epeiric sea never exceeded a few tens of metres, except perhaps during the middle Toarcian.

Figure 6.1. Paleogeographic map of the Early Jurassic (Pliensbachian–Toarcian). The Laurasian Sea is indicated in darker blue. Map modified from Damborenea (2002) and Dera et al. (2009).

The present-day distribution of the group in the Netherlands is largely confined to Cimmerian basins (cf. Figs 6.2, 6.3), where it attains locally a thickness of more than 1500 m. These basins underwent subsidence during a protracted (Anisian–Ryazanian) interval of extensional tectonics related to the opening of the Atlantic Ocean—the Cimmerian phase (sensu Stille, 1924). The uniform character of the facies suggests that the Altena Group was originally also deposited on the platforms and maybe some of the highs, but the Lower Jurassic successions deposited there were probably condensed due to limited accommodation and Early Jurassic erosion. Sediments of the Altena Group were subsequently removed by erosion from platforms and highs during the Middle and Late Cimmerian uplift (Middle to Late Jurassic). Afterwards, due to tectonic inversion in the Late Cretaceous, the Altena Group was also locally eroded from the Cimmerian basins.

The Jurassic of the Netherlands, which lies buried in the subsurface, is the link between the outcrops of Great Britain and those of north- and southwest Germany. Lower and Middle Jurassic rocks have not been a major objective of exploration in the Southern North Sea Basin, which means that our knowledge of the Dutch Jurassic is still limited compared to surrounding countries where the rocks are exposed at the surface. For this reason, in this chapter, we make extensive reference to analogous formations in surrounding countries.

Geographic distribution

The Jurassic is one of the few pre-Quaternary systems that is represented by rock exposures at the surface of the Netherlands. Rhaetian reddish-brown and dark grey to black fine mudstones of the Sleen Formation overlie the Middle Triassic Muschelkalk Formation in the quarry near Winterswijk (Fig. 6.3; Herngreen et al., 2005; Diependaal & Reumer, 2021). Lower Jurassic (Hettangian) mudstones can be seen in brooks and ditches east of the quarry, near the German border, but the exposure is patchy and poor (Peletier & Kolstee, 1986; Witte et al., 1993; Oosterink, 2010). Lower Jurassic (Hettangian–Sinemurian) black mudstones were until recently exposed in the quarry in solution hollows in the Muschelkalk (e.g. Herngreen et al., 2005; Klompmaker & Van den Berkmortel, 2007). Near the hamlet of Kotten, again near Winterswijk, there is a rather small area in which the Middle Jurassic lies near the surface (Peletier & Kolstee, 1986; Herngreen et al., 2000; Oosterink, 2010).

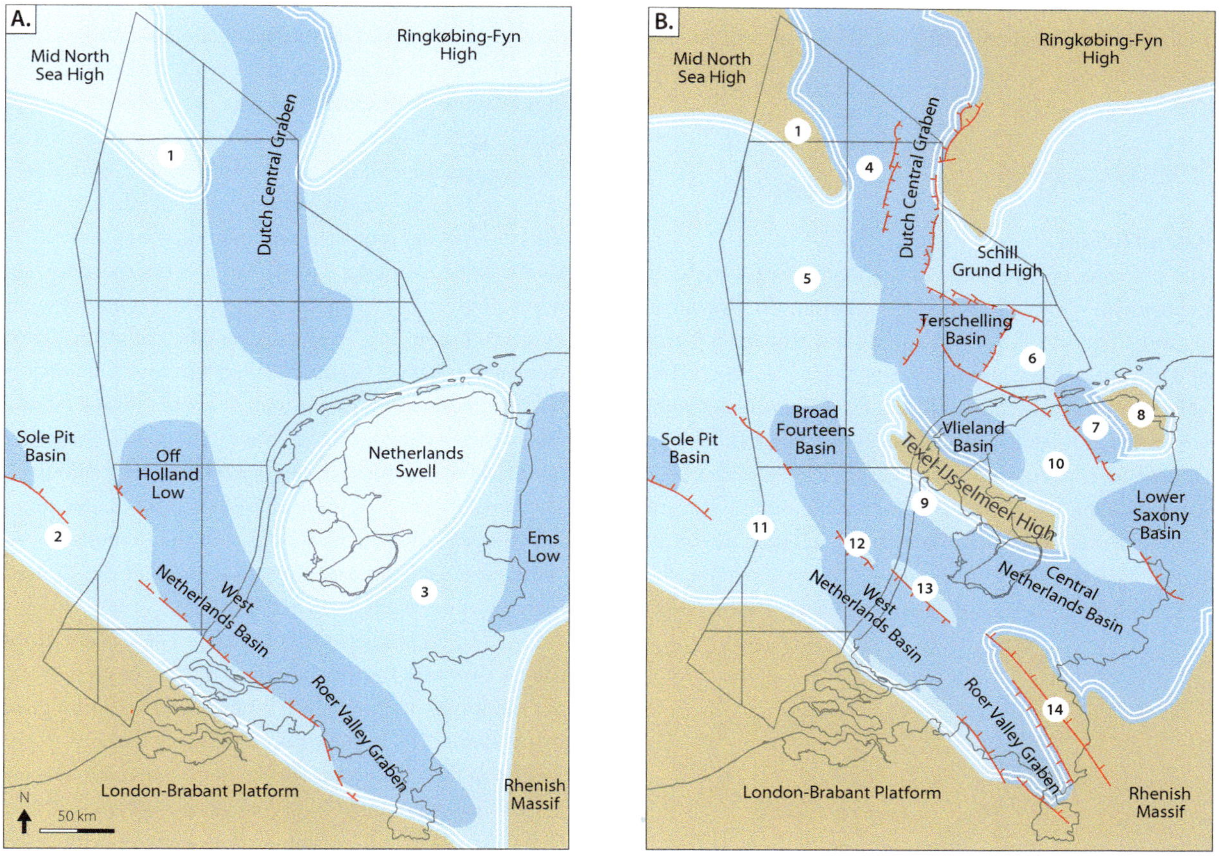

Figure 6.2. *a) Simplified paleogeographic map of Triassic to Middle Jurassic and b) Middle Jurassic to Early Cretaceous structural elements in the Netherlands (modified from Van Adrichem Boogaert & Kouwe, 1993-1997; Wong, 2007, Kombrink et al., 2012b). Brown and lightest cyan: structural highs, mostly nondepositional partly subaerial landmasses; light cyan: platforms, intermittently flooded shallow-water regions; darker blue: Cimmerian basins. Structural elements: 1 = Elbow Spit High; 2 = South Hewett Shelf; 3 = East Netherlands Platform; 4 = Step Graben; 5 = Cleaverbank Platform; 6 = Ameland Platform; 7 = Lauwerszee Trough; 8 = Groningen Platform; 9 = North Holland Platform; 10 = Friesland Platform; 11 = Winterton High; 12 = IJmuiden Platform; 13 = Zandvoort Ridge; 14 = Peel-Maasbommel Complex.*

It is clear from the above that Rhaetian and Jurassic rocks at the surface of the Netherlands are essentially an anecdotal curiosity. In the subsurface, however, representative sections of the Altena Group lie buried in Cimmerian basins, such as the West Netherlands Basin and the Dutch Central Graben (Fig. 6.2).

The present-day geographic distribution of the Altena Group is largely confined to these Cimmerian basins (cf. Figs 6.2, 6.3), where differential subsidence and erosion resulted in thicknesses varying from a few metres to more than 1500 m (Fig. 6.4). In the northern Dutch offshore, Jurassic rocks have also been preserved outside the Dutch Central Graben in smaller basins, or pods (see Hodgson et al., 1992), where accommodation was generated by salt withdrawal (Fig. 6.5 and see Fig. 5 in Van Wijhe, 1987), such as primary rim synclines (e.g. Fig. 7 in Wong et al., 1989), or in negative flower structures associated with strike-slip faults.

The stratigraphic column of platforms and highs is, by definition (see Kombrink et al., 2012b), characterized by, among other things, the absence of the Altena Group (but see Fig. 6.5). For this reason, the distinction between the two structural elements made in this chapter is paleogeographic rather than tectonostratigraphic: in the Early Jurassic, platforms were intermittently flooded shallow-water regions of the epeiric sea that were less subsident than Cimmerian basins, and highs were mostly nondepositional subaerial landmasses.

An almost completely developed lithostratigraphic section is present in the northwestern part of the Roer Valley Graben (Fig. 6.2), where the group attains a thickness of about 2000 m (Fig. 6.4). In northeast Belgium, however, the Lower Jurassic succession in the Roer Valley Graben is ca. 450 m thick and comprises the Sleen and the Aalburg formations only (Dusar et al., 2002); there, the Upper

Cretaceous lies unconformably on the Pliensbachian, and younger Jurassic rocks are absent.

Stratigraphy

Altena Group

The Jurassic of the Netherlands comprises two main depositional units. The upper, Callovian to Ryazanian (Lower Cretaceous), unit comprises the Schieland, Scruff, and Niedersachsen groups (Verreussel et al., 2025, this volume). The lower unit corresponds to the Altena Group, which has a Rhaetian to Oxfordian age and can be considered as a protracted early synrift. The Altena Group comprises the Rhaetian Sleen Formation and the Jurassic Aalburg, Posidonia Shale, Werkendam, and Brabant formations (Fig. 6.6).

The Rhaetian to middle Toarcian lower Altena Group

Figure 6.3. Distribution of the Altena Group in the Dutch subsurface. The rocks of the Altena Group are found almost exclusively in the Cimmerian basins (cf. Fig. 6.2). Places and boreholes mentioned in the text are shown in bold.

(Sleen, Aalburg, and Posidonia Shale formations) consists of up to 1000 m of epeiric-neritic dark argillaceous to calcareous mudstones with some calcareous intercalations (Fig. 6.7). In this chapter, we use the nomenclature for fine-grained sedimentary rocks proposed by Lazar et al. (2015). Similar mudstones are found in much of northwest Europe and represent sedimentation in the northern Laurasian Sea from the end of the Early Cimmerian to the onset of the Middle Cimmerian tectonic phase (~205–180 Ma). The middle Toarcian to middle Callovian–Oxfordian upper Altena Group (Werkendam and Brabant formations) consists of up to 650 m of epeiric-neritic (sandy) mudstones, sandstones, marls, and limestones. These sediments were deposited in the period up to the end of the Middle Cimmerian tectonic phase (~155 Ma). Later, in the Late Jurassic, the structural outline of the Netherlands became more complex and the stratigraphy of the Cimmerian basins more differentiated (Fig. 6.2).

Rhys (1974) proposed that the lateral equivalent of the Altena Group in the British sector of the southern North Sea comprises two units, each of which should be given group status. The lower group, chiefly argillaceous, corresponds to the Lias Group (Hettangian–Toarcian). The upper group, which Rhys (1974) named the West Sole Group (Aalenian–Callovian), consists of sandstones with lesser amounts of siltstones and finer mudstones as well as limestones.

The depth of the base of the Altena Group varies considerably, reaching maxima of more than 5000 m in the Broad Fourteens Basin and in the northern Dutch Central Graben (Fig. 6.8). On seismic reflection profiles, the base of the group is marked by a prominent high-amplitude reflector. The base of the Altena Group represents the Rhaetian transgression onto the Late Triassic erosional surface and is, therefore, unconformable: the group overlies the red-brown anhydritic beds of the Keuper Formation, the underlying Muschelkalk Formation or, locally, older units of the Upper Germanic Trias Group or even the Zechstein.

The top of the Altena Group is generally deeply eroded (Fig. 6.6), except perhaps locally in the West Netherlands Basin and Roer Valley Graben, where the succession appears to be most complete (Fig. 6.9). There, the middle Callovian to Ryazanian upper depositional unit of the Jurassic (see Verreussel et al., 2025, this volume) rests conformably on the Altena Group. Elsewhere, the group is overlain unconformably by the upper depositional unit of the Jurassic or by the Rijnland, Chalk, Lower North Sea, or Middle North Sea groups. The unconformity, which is often angular (Fig. 6.5), is sometimes referred to as the 'Mid Cimmerian Unconformity' (e.g. Burgers & Mulder, 1991; Underhill & Partington, 1993); the corresponding hiatus varies from basin to basin and exceeds 20 Ma in the Dutch part of the Lower Saxony Basin (Fig. 6.6). In strong-

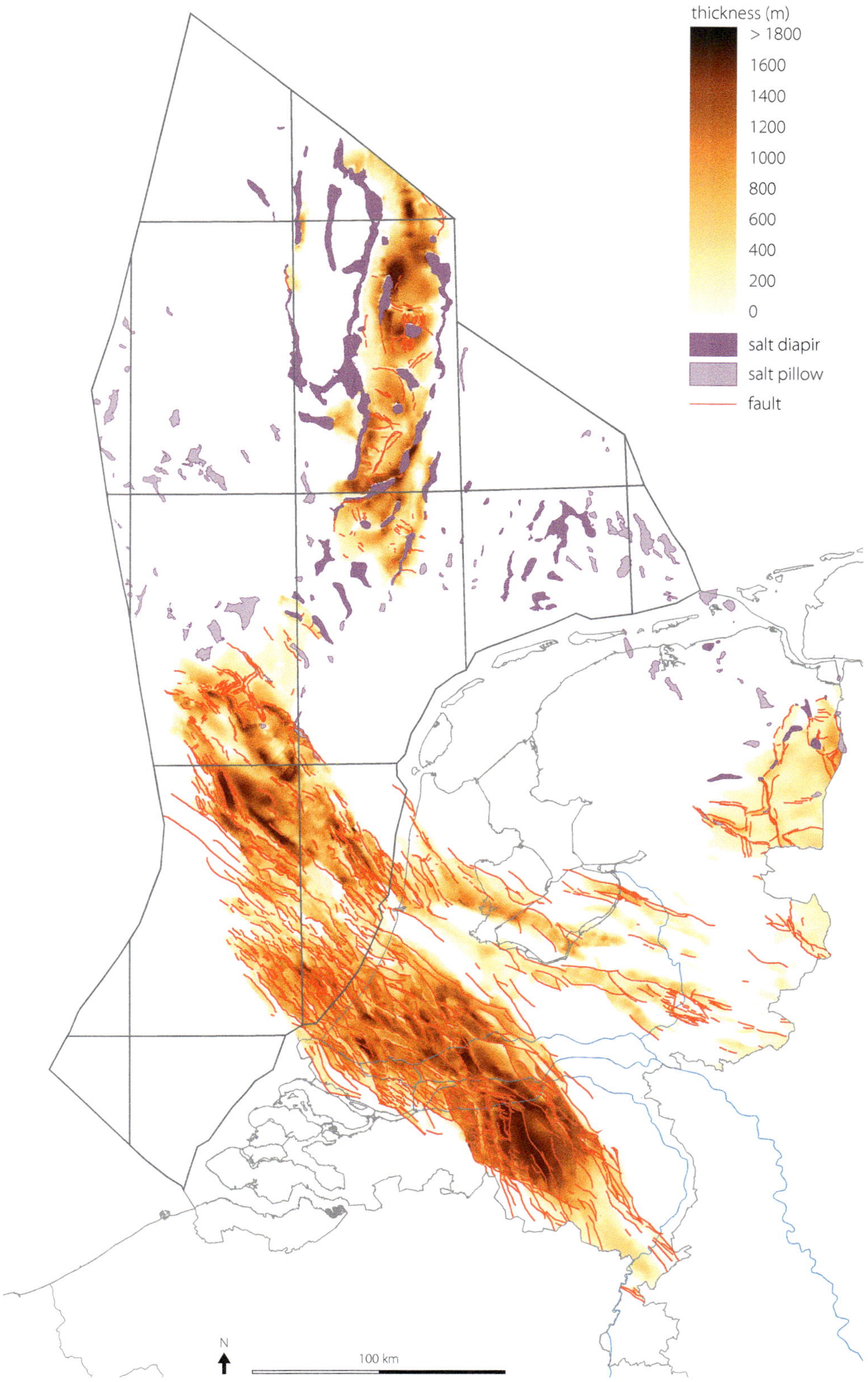

Figure 6.4. Isochore map of the Altena Group. The Altena Group reaches a thickness of more than 1500 m in the Cimmerian basins and is thickest in the Roer Valley Graben. The dominant northwest–southeast fault pattern and the distribution of salt diapirs and pillows are shown.

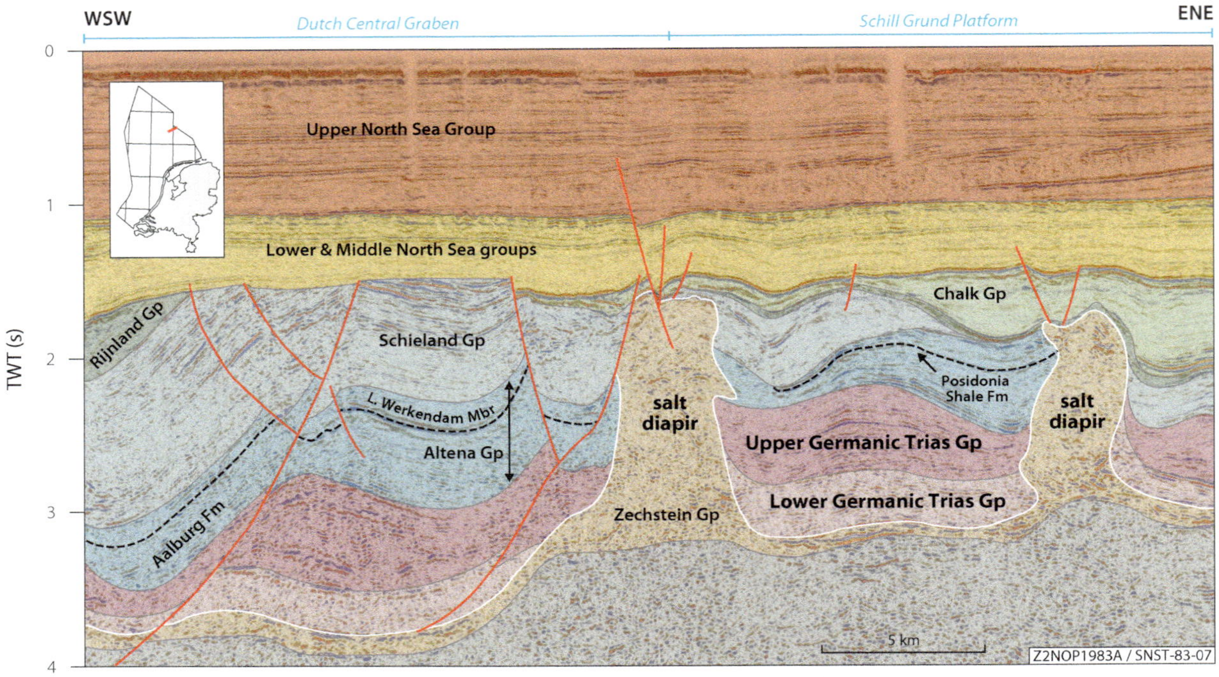

Figure 6.5. West-southwest–east-northeast seismic section across the Dutch Central Graben and Schill Grund Platform (see Fig. 6.2). Here, the lower Altena Group has been preserved outside the Cimmerian basin, on the Schill Grund Platform, in a small basin where accommodation was created by salt withdrawal.

ly inverted areas of the West Netherlands Basin, the Altena Group is overlain by the much younger Lower or Middle North Sea groups.

Sleen Formation

The shallow marine Sleen Formation consists of a succession of grey fossiliferous mudstones with reddish-brown mudstones at the top. A thin sandstone within the mudstone succession is developed locally. In logs, the formation typically displays a uniform thickness, relatively high gamma-ray values, and low sonic values (Fig. 6.9).

The geographic distribution of the formation coincides with that of the Altena Group (Fig. 6.3). The thickness of the formation normally varies between 20 and 45 m with a maximum of 157 m in the southern Dutch Central Graben. The formation lies on red-brown anhydritic and/or calcareous mudstones of the Keuper Formation, except in the eastern Netherlands, where it rests unconformably on grey-green marls and limestones of the Muschelkalk Formation or on the Röt Formation. The formation is overlain by the Aalburg Formation, and the top of the formation is defined at the base of a thin but persistent argillaceous limestone in the Aalburg Formation.

The age of the Sleen Formation is Rhaetian (Late Triassic). The formation is a lateral equivalent of the Triassic Upper Keuper Group of northwest Germany and of the Rhaetian interval of the Penarth Group of Great Britain (Lott & Warrington, 1988; Diependaal & Reumer, 2021; Hodges, 2021). In the Netherlands, as is the case in Belgium and the Danish Central Graben (see Dusar et al., 2002; Bachmann et al., 2010), this Rhaetian formation is included in the Jurassic Altena Group, rather than in the underlying Germanic Trias.

The Triassic–Jurassic boundary has been defined biostratigraphically by the lowest occurrence of the earliest forms of *Psiloceras* ammonites (see Hillebrandt et al., 2013), which means that precise recognition of the boundary in well logs, cuttings, and sometimes even in cores is normally not possible (see Lott & Warrington, 1988 for a British offshore perspective). In the British sector of the southern North Sea, the upper limit of the Triassic is placed at a facies change in the gamma-ray and sonic logs of most wells: dark grey mudstones that pass upwards into calcareous mudstones with beds of limestone.

In any case, inclusion of the Sleen Formation in the Altena Group is not without significance: the formation represents the transition from Late Triassic continental environments, through paralic environments, to the marine regime of the Early Jurassic, as well as the transition from Triassic aridity to a more humid climate (Pearson, 1970; Ahlberg et al., 2002; Bachmann et al., 2010).

Aalburg Formation

The Aalburg Formation consists of grey-green to black argillaceous to calcareous mudstones (Fig. 6.7) with thin limestone beds. The argillaceous mudstones represent

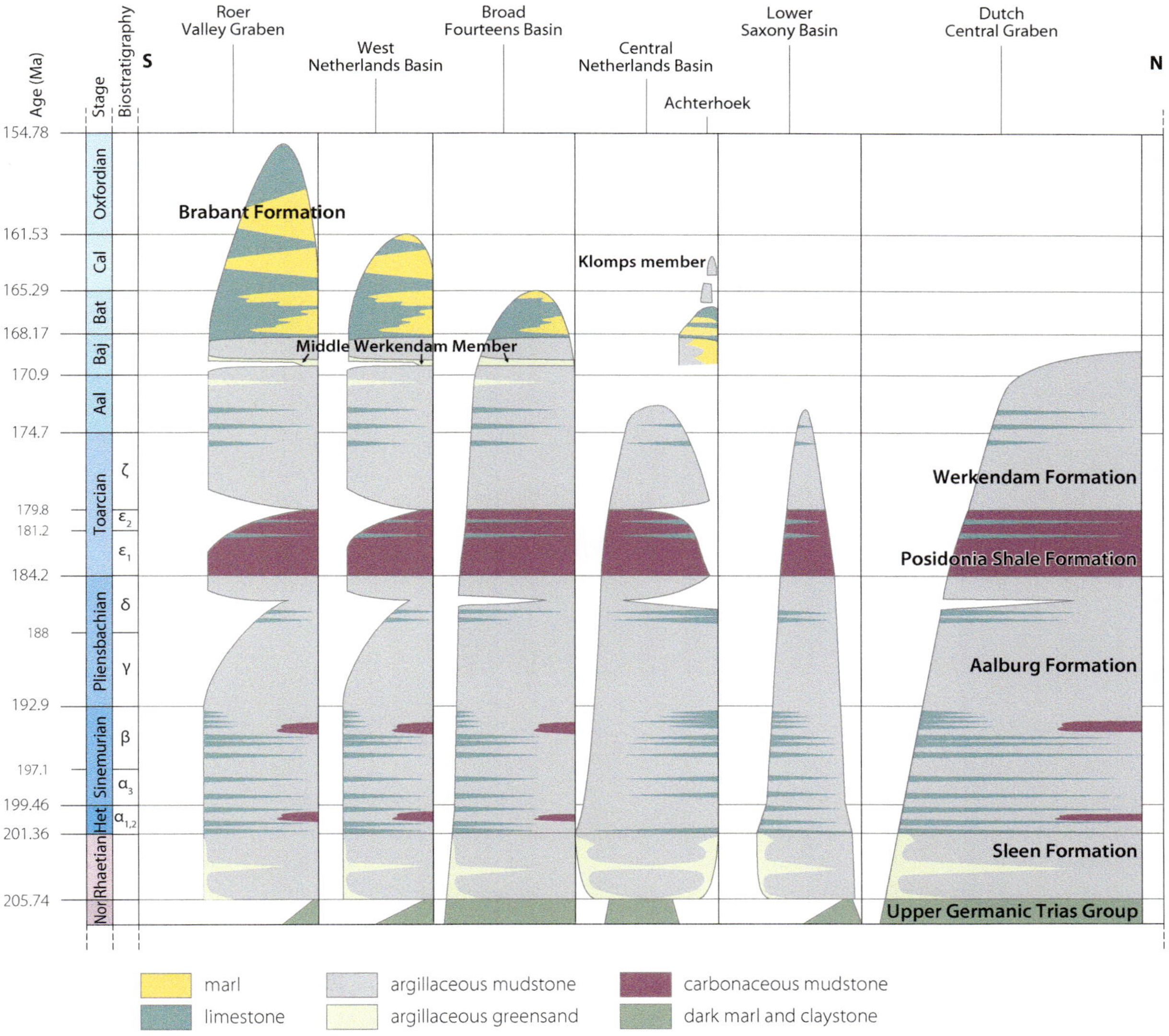

Figure 6.6. Chronostratigraphic framework of the Rhaetian to Oxfordian succession of the six larger Cimmerian, or Jurassic, basins in the Netherlands (modified from Van Adrichem Boogaert & Kouwe, 1993-1997). See Fig. 6.2 for the location of the basins. Timescale after Hesselbo et al. (2020). The top of the Altena Group is generally deeply eroded below the Middle Cimmerian Unconformity. The most completely developed lithostratigraphic section of the group is present in the northwestern part of the Roer Valley Graben. Note the consistency of the stratigraphy.

periods of greatest transgression, whereas thin limestone beds represent regression or stillstand (Donovan et al., 1979). The number of thin limestone beds increases towards the London-Brabant Platform, which formed the southern margin of the basin in the Early Jurassic (Fig. 6.5; Donovan et al., 1979; Ziegler, 1990; Bradshaw et al., 1992). A thin but persistent argillaceous limestone forms the base of the formation. In the lower part of the formation, there are a few layers of black carbonaceous mudstone (De Jager et al., 1996). Ammonites, belemnites (above lowermost Lias α3; see Fig. 6.6), and other molluscs are abundant in this formation; these are frequently pyritized. Siderite concretions are also common.

In the southern offshore, the upper part of the Aalburg Formation (upper Pliensbachian) comprises beds of calcareous silt- and sandstone with an oolitic texture. In the Lias Group of northern Germany, sideritic and chamosite/berthierine-rich ooidal ironstone beds occur sporadically in the Sinemurian and Pliensbachian succession and form useful local marker beds (Lott et al., 2010). A similar stratigraphy is found in eastern England (Powell, 2010). It seems likely that, though as yet undescribed, similar rocks may be present in the Dutch subsurface.

The geographic distribution of the formation is similar to that of the Sleen Formation (Fig. 6.3). The thickness of the Aalburg Formation can reach a maximum of 700 m.

Figure 6.7. Typical facies observed in core slabs of the Aalburg (right and centre) and Posidonia Shale (left) formations in the Dutch Central Graben. Cores are from boreholes L05-04 (left) and L05-14 (centre and right). See Fig. 6.3 for the location of the boreholes. The grey argillaceous mudstones of the Posidonia Shale Formation show very thin to thin parallel silt laminae produced by migration of current ripples on the seafloor (Trabucho Alexandre et al., 2012; Yawar & Schieber, 2017).

Particularly thick sections of the Aalburg Formation are found in the Dutch Central Graben, associated with salt withdrawal structures (rim synclines), but significant lateral thickness variations result from synsedimentary tectonics and erosion (Houben et al., 2017). The formation thins considerably towards the London-Brabant Platform, for example (see Pharaoh, 2018). These thickness variations reflect subsidence patterns that had already been established during the Early Cimmerian tectonic phase in the Middle Triassic. The formation lies concordantly on the reddish-brown mudstones at the top of the Sleen Formation and is either overlain concordantly by the Posidonia Shale Formation or discordantly by the Schieland Group, the Chalk Group, or the North Sea Supergroup.

The age of the Aalburg Formation is Hettangian to Pliensbachian or earliest Toarcian (Paltum Subzone). As is the case in Germany, the formation may be subdivided into four foraminiferal-ostracodal biostratigraphic units (Lias α-δ indicated in Fig. 6.6; Klingler, 1962).

Posidonia Shale Formation

The Posidonia Shale Formation consists of dark carbonaceous mudstones (black shales) with a few carbonate beds (Fig. 6.7). The formation was named after a bivalve genus formerly called *Posidonia* or *Posidonomya* (now *Bositra*). Total organic carbon (TOC) content is on average 5–7% (Fig. 6.10; Van Bergen et al., 2013) and type II kerogen dominates (Van Bergen et al., 2013; Song et al., 2015). Carbonate is the chief source of compositional variability both geographically and stratigraphically (Song et al., 2015; Houben et al., 2016). Erosional surfaces, grain size variations, and sedimentary structures show that the seafloor during deposition was above storm wave base (Trabucho Alexandre et al., 2012; Van Bergen et al., 2013). The formation is easily recognizable on gamma-ray and resistivity logs due to its relatively high values (Fig. 6.9; Van Bergen et al., 2013). A series of troughs in the gamma-ray profile are associated with low neutron porosities and high bulk densities. These have been interpreted as carbonate concretionary horizons (Houben et al., 2017). On seismic reflection profiles (e.g. Fig 13 in Van Bergen et al., 2013), the formation is marked by one or two continuous, very high-amplitude reflectors, which stand out from the otherwise practically transparent seismic facies of the Altena Group (Fig. 6.5). Sonic logs indicate that the strong acoustic impedance contrast is caused by the sudden change to low acoustic velocities related to the abundance of organic carbon (Fig. 6.9).

It used to be thought that the formation is absent outside the centres of Cimmerian basins (Fig 6.11; e.g. Van Adrichem Boogaert & Kouwe, 1993-1997; Van Bergen et al., 2013). In the northern Dutch offshore, however, the formation may be present outside the Dutch Central Graben in smaller basins, or pods (see Hodgson et al., 1992), where accommodation was generated due to salt withdrawal in the Early Jurassic (Fig. 6.5 and see Fig. 5 in Van Wijhe, 1987). The thickness of the formation varies between 30 and 50 m. The base of the formation

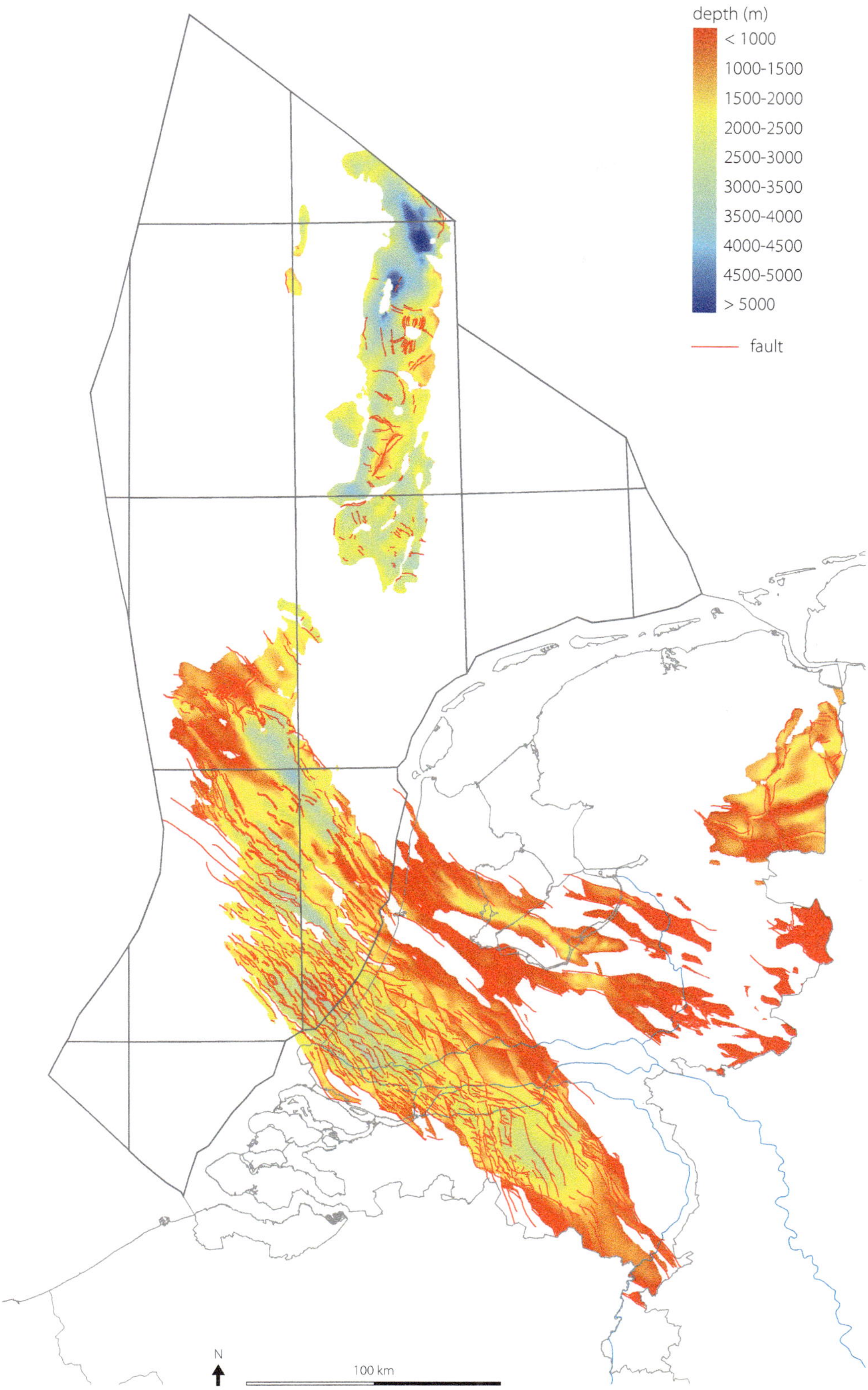

Figure 6.8. Depth map of the base of the Altena Group (Sleen Formation). The base of the group is deepest in the Broad Fourteens Basin and in the northern Dutch Central Graben.

lies presently at depths between 1 and 3 km, except in the southwestern margin of the West Netherlands Basin (3–4 km) and the northern Dutch Central Graben (>4 km; Fig. 6.12). The pyrolysis parameter and maturity indicator T_{MAX} is 420–435 °C in the southern Dutch Central Graben (L05-04) where the base of the formation is at a depth of ~2850 m (Gill, 2007). The formation lies concordantly on the Aalburg Formation and is overlain concordantly by the Werkendam Formation, except where truncated by the Middle Cimmerian Unconformity, in which case it is overlain by younger Jurassic units (e.g. Figs 6.5, 6.10).

The age of the formation is early and middle Toarcian (Lias ε1 (Tenuicostatum–Serpentinum zones) and ε2 (Bifrons Zone); Fig. 6.6), which makes it approximately laterally equivalent of the Whitby Mudstone Formation of northeast England and of the Ölschiefer of northwest Germany. The formation contains a record of the negative carbon isotope excursion associated with the Toarcian Oceanic Anoxic Event (OAE; Fig. 6.10; Trabucho Alexandre et al., 2012; Houben et al., 2021).

The Posidonia Shale Formation was raised to formation status by Van Adrichem Boogaert & Kouwe (1993-1997) to avoid problems in the subdivision into members of the Werkendam Formation where no Middle Werkendam Member can be recognized. Unlike the Sleen, Aalburg, and Werkendam formations, the newly defined Posidonia Shale Formation has kept its lithological affix 'Shale'.

Werkendam Formation

The Werkendam Formation consists of grey argillaceous to slightly calcareous mudstones with a coarser grained calcareous interval in the middle (Bajocian). The formation has, thus, been subdivided into three members (Figs 6.6, 6.9; Nederlandse Aardolie Maatschappij & Rijks Geologische Dienst, 1980). The lower member (Toarcian–lower Bajocian) contains carbonaceous beds, siderite concretions, and ironstones with an oolitic texture. Calcite concretions are more common in the upper part of the lower member. The middle member is recognizable on wireline logs by a marked increase in sonic velocity and resistivity relative to the upper member, which has uniformly low sonic velocity and resistivity profiles (Fig. 6.9).

The geographic distribution of the lower member is similar to that of the Posidonia Shale Formation (Fig. 6.11). The middle and upper members have been preserved only in the northwest–southeast-oriented Cimmerian basins of the southern Netherlands (Figs 6.2, 6.6), where the impact of Middle Jurassic uplift in the North Sea area was negligible (see Fig. 3 in Underhill & Partington, 1993). The age of the Werkendam Formation is middle Toarcian (Lias ζ; Fig. 6.6) to earliest Bathonian. The maximum thickness of the formation is 500 m. The formation lies concordantly on the Posidonia Shale Formation and is either overlain concordantly by the Brabant Formation or discordantly by the Schieland or Chalk groups or, locally, by the North Sea Supergroup.

Brabant Formation

The Brabant Formation is the uppermost unit of the Altena Group. It consists of an alternation of sandy mudstones, marls, (argillaceous) limestones, and (calcareous) sandstones that are comparable to the Cornbrash Formation of Great Britain. On seismic reflection profiles, the Brabant Formation is characterized by several parallel reflectors. The sand content increases towards the top of the formation. The relative amount of sand and the number and thickness of limestone beds have been used to further subdivide the formation into eight members (Figs 6.6, 6.9; Nederlandse Aardolie Maatschappij & Rijks Geologische Dienst, 1980; Van Adrichem Boogaert & Kouwe, 1993-1997). In the West Netherlands Basin, a calcareous sandstone is found at the top of the formation in the most complete successions. In the Roer Valley Graben, the uppermost part of the formation consists locally of oolitic limestones, representing local shoals and strandlines. In the Achterhoek (Fig. 6.3), the uppermost part of the formation consists of marls and sandy limestones with a dark grey to black calcareous mudstone at the top (the informal Klomps member).

The geographic distribution of the formation has been strongly affected by erosion. As is the case for the two upper members of the Werkendam Formation, the Brabant Formation is found only in the southern Netherlands (Fig. 6.6). The age of the formation is latest Bajocian to earliest Oxfordian. The maximum thickness of the formation is about 350 m. The formation overlies concordantly the Werkendam Formation and is either overlain discordantly by the Schieland or Chalk groups or by the North Sea Supergroup.

Structural and sedimentary development

Triassic–Jurassic tectonics and volcanism

Earth history in the latest Triassic and Early to Middle Jurassic was dominated by the progressive breakup of Pangea and the opening of the Central Atlantic Ocean. The latter was probably triggered by the massive extrusion of basaltic lava in the Central Atlantic Magmatic Province (Fig. 6.1), which took place in the Rhaetian to Sinemurian (main phase 202–195 Ma; Nomade et al., 2007). A mass extinction—one of the big five in the Phanerozoic (Raup & Sepkoski, 1982)—is coeval with this volcanism. Seafloor spreading in the Central Atlantic, as well as the opening of the Hispanic Corridor (Fig. 6.1), probably started as early

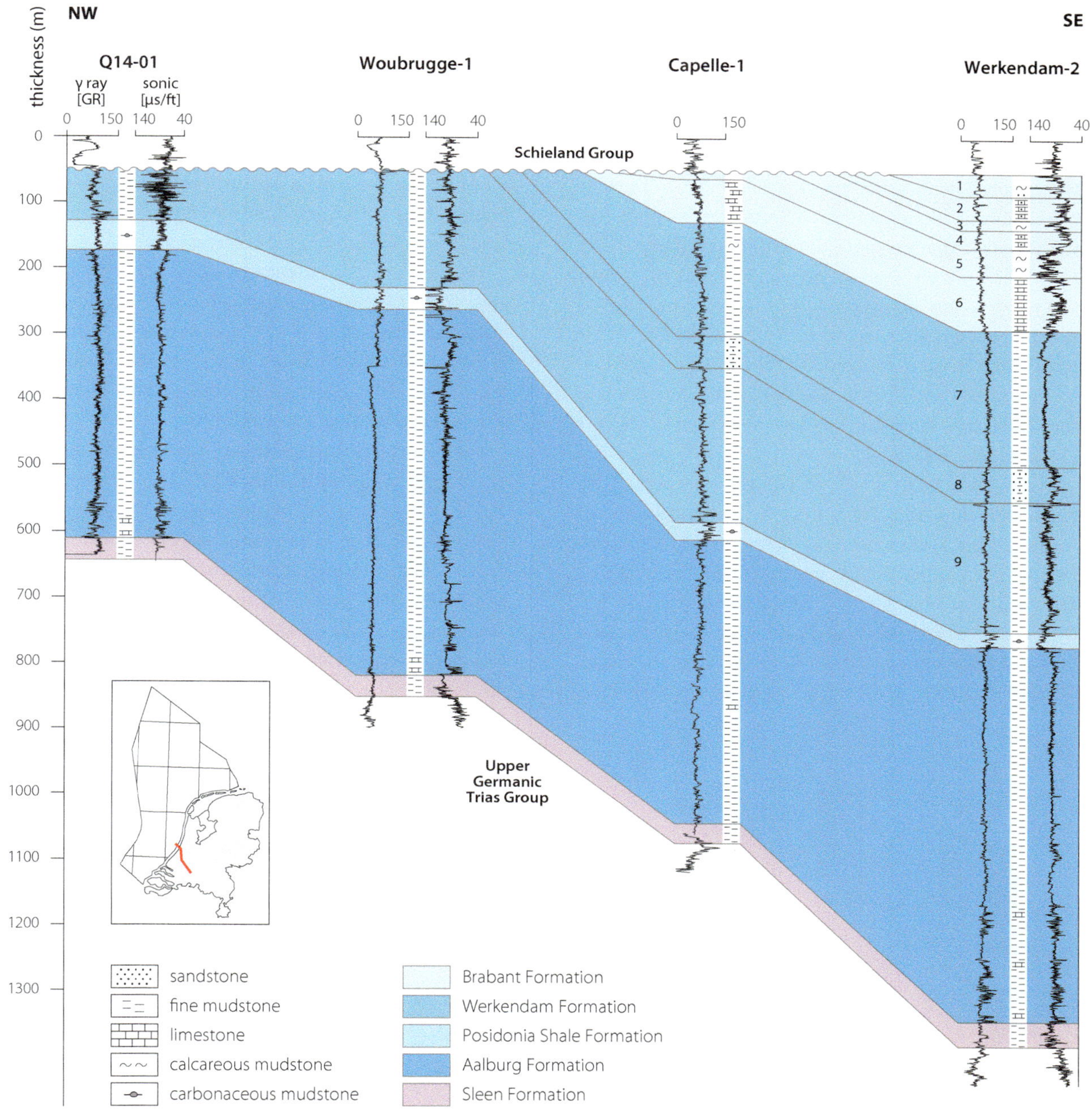

Figure 6.9. Northwest–southeast well-log correlation panel along the West Netherlands Basin (modified from Wong, 2007). Length of the panel is ~70 km. The top of the Altena Group is erosional (Middle Cimmerian Unconformity) but becomes more conformable towards the southeast, in the direction of the Roer Valley Graben, where sedimentation appears to have been more continuous. Members of the Werkendam and Brabant formations: 1 = Upper Brabant Marl Member; 2 = Upper Brabant Limestone Member; 3 = Middle Brabant Marl Member; 4 = Middle Brabant Limestone Member; 5 = Lower Brabant Marl Member; 6 = Lower Brabant Limestone Member; 7 = Upper Werkendam Member; 8 = Middle Werkendam Member; 9 = Lower Werkendam Member.

as the late Sinemurian (Sahabi et al., 2004); by the Middle Jurassic, Pangea had split into Laurasia and Gondwana, which then became separated by a Tethyan–Atlantic seaway.

The progressive breakup of Pangea changed the structural configuration of the present-day southern North Sea from a single extensive basin, the Southern Permian Basin, to a more complex pattern of basins/platforms and highs (Ziegler, 1990). In the literature, Triassic–Jurassic extensional tectonics is often treated as having taken place during four different phases: Hardegsen (Early Triassic), Early Cimmerian (Anisian–Carnian), Middle Cimmerian (Aalenian–Callovian/Oxfordian), and Late Cimmerian (Kimmeridgian–Ryazanian). All four are, however, part of a single yet protracted interval of extensional tectonics related to the opening of the Atlantic Ocean that continued into the Cretaceous and culminated in the opening of the North Atlantic Ocean (Geluk, 2005).

Latest Triassic–Early Jurassic transgression

The Rhaetian and the Hettangian were characterized by the establishment of the epeiric Laurasian Sea. Detailed correlation of the marine Sleen Formation suggests that the Rhaetian transgression, though limited (Haq, 2018), took place rapidly (Van Adrichem Boogaert & Kouwe, 1993-1997). This transgression was followed by a regression in the latest Rhaetian, when the basin shoaled. At the start of the Hettangian, however, fully marine conditions had returned to the basin, resulting in the deposition of the argillaceous limestone that has been taken as the base of the Aalburg Formation.

In the Hettangian, much of Western Europe had become an archipelago (Fig. 6.13a). As sea level continued to rise, in the Sinemurian and Pliensbachian, the Laurasian Sea became gradually connected to the Boreal Sea via the Viking Corridor (Fig. 6.1). In the southern North Sea, the gradual transgression is recorded by the progressive onlap of Lower Jurassic strata onto the London-Brabant Platform (Donovan et al., 1979; Lott et al., 2010; Pharaoh, 2018). A first connection between the Tethys and the Boreal Sea had already been established during the Rhaetian, but a more permanent connection was established by early Pliensbachian (see also Pieńkowski et al., 2021 for an example from Cardigan Bay Basin), when eustatic sea-level rise resulted in an eastward transgression that brought fully marine conditions as far east as central Poland (Pieńkowski, 2004, 2014; Lott et al., 2010).

The connection led to the development of a transitional area in the Laurasian Sea between northern colder water siliciclastic-dominated facies and southern warm-water carbonate-dominated facies. It was in this area, where Boreal and Tethyan waters met, that carbonaceous facies developed in the Toarcian (Fig. 6.13a).

The Toarcian was an interval of significant sea-level rise and transgression, but the initial rise was relatively small. The major sea-level rise occurred at the beginning of the Serpentinum Zone, when carbonaceous mudstones became widespread and burial rates decreased (Bradshaw et al., 1992; McArthur et al., 2000). In the southern North Sea, the time of maximum transgression in the Toarcian was the lower part of the Bifrons Zone (Haq, 2018). In northeast England, in the Cleveland Basin (Fig. 6.14), the

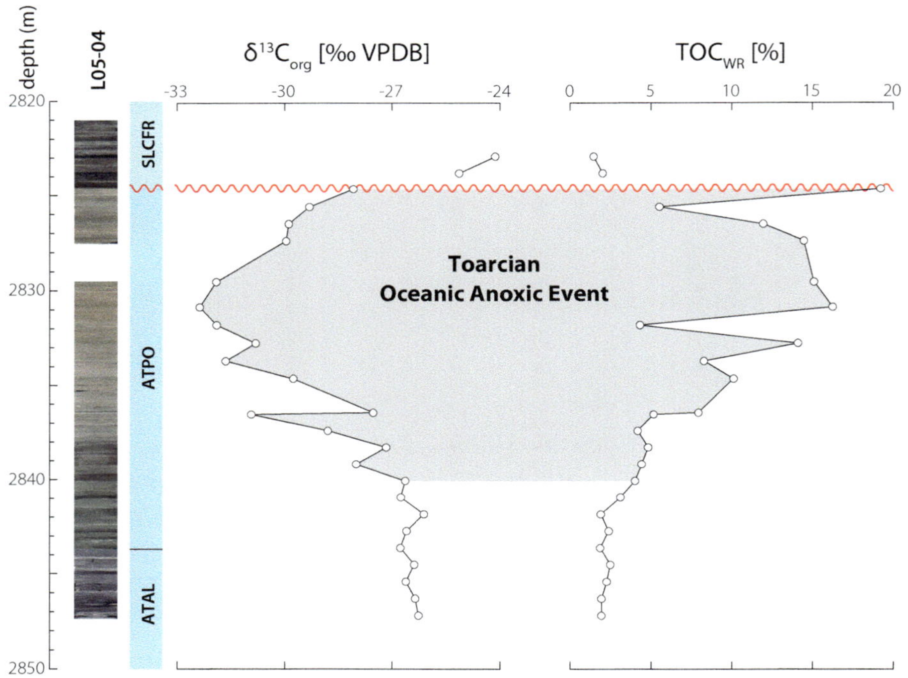

Figure 6.10. Negative organic carbon isotope excursion in borehole L05-04 in the upper Tenuicostatum and lower Serpentinum zones (southern Dutch Central Graben; Trabucho Alexandre et al., 2012). See Figs 6.3 and 6.11 for the location of the borehole. This carbon isotope excursion is associated with the Toarcian Oceanic Anoxic Event. Total organic carbon content is highest in the upper part of the excursion (recovery phase). In this area of the southern Dutch Central Graben, the Posidonia Shale Formation is overlain unconformably by the Upper Jurassic Rifgronden Member (Friese Front Formation), which is also a dark carbonaceous mudstone. The Middle Cimmerian Unconformity here, though subtle in its expression, represents a hiatus of at least 18.5 Ma. ATAL: Aalburg Formation (Pliensbachian); ATPO: Posidonia Shale Formation (Toarcian); SLCFR: Rifgronden Member (uppermost Callovian–Oxfordian).

finest Lower Jurassic sediments, which probably reflect maximum flooding and the deepest water conditions in the basin, occur in the middle of the Alum Shale Member (Pye & Krinsley, 1986; Hesselbo & Jenkyns, 1998), that is, in the lower Bifrons Zone. This interval corresponds to the upper part of the Posidonia Shale Formation in the Dutch subsurface (ε2 in Fig. 6.6). By the middle Toarcian, most of Western Europe was probably submerged (Bradshaw et al., 1992; De Graciansky et al., 1998; Hallam, 2001; Haq, 2018).

Early Jurassic subsidence and accommodation

By the end of the Triassic, the continental crust of northwest Europe was crisscrossed by a network of grabens formed by lithospheric extension related to the opening of the Atlantic Ocean (Figs 6.2, 6.13a). In the Netherlands, three major rift systems can be recognized during the Jurassic (Fig. 6.2): (i) the north–south-oriented Dutch Central Graben and Vlieland Basin system; (ii) the east–west-oriented Lower Saxony system; and (iii) the northwest–southeast-oriented block-faulted transtensional system.

The Dutch Central Graben and Vlieland Basin system included what would become the Terschelling Basin (Fig. 6.2). The Lower Saxony system extended into Germany. During the Early Jurassic, this area was situated at the western edge of the German Ems low (Fig. 6.2a). This originally north–south-oriented rift basin transformed in the Late Jurassic into the east–west-trending Lower Saxony Basin (Ziegler, 1990). The northwest–southeast-oriented block-faulted transtensional system comprised the Broad Fourteens Basin, West and Central Netherlands basins, and the Roer Valley Graben, as well as the Sole Pit Basin in British waters. The Cleveland Basin, in North Yorkshire, is an extension of this system. Permo-Triassic evaporites are generally absent in this area of the southern Netherlands, and the northwest–southeast-oriented basins form tilted half grabens.

The geographic distribution of the Lower and Middle Jurassic in the Dutch subsurface is patchy and largely restricted to these Cimmerian basins (cf. Figs 6.2b, 6.3). In many areas, basin inversion and salt tectonics resulted in the complete removal of the Jurassic before deposition of the Cretaceous (Cope et al., 1980). According to Wong (2007), the geographic distribution pattern of the Lower Jurassic is a consequence of *severe* erosion during later tectonic phases, because the lithofacies have a consistent fine-grained marine character suggestive of layer cake stratigraphy.

The interval between the Early and Middle Cimmerian phases (Norian–Toarcian) is usually described in the literature as one of tectonic quiescence characterized by regional thermal subsidence (e.g. Wong, 2007; Pieńkowski et al., 2008; Verreussel et al., 2018). Though it is true that thermal relaxation of the lithosphere continued during the Early Jurassic and controlled the regional subsidence pattern of the area of the former Southern Permian Basin (Ziegler, 1990), regional isopachs of the Lower Jurassic show that basins in northwest Europe continued to subside along fault patterns established during the Triassic (Fig. 6.2a; Ziegler, 1981, 1990).

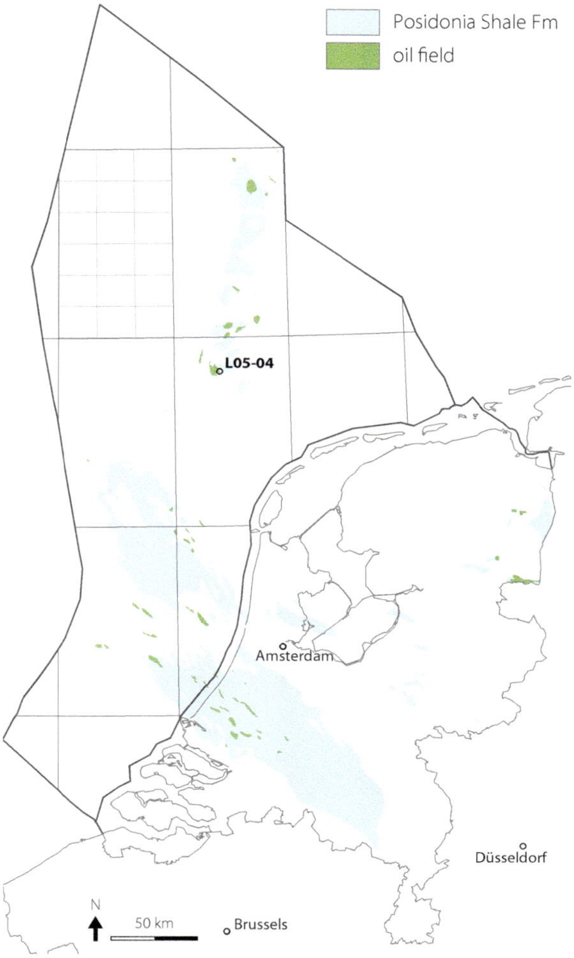

Figure 6.11. Distribution of the Posidonia Shale Formation in the Cimmerian basins. The distribution of oil fields in the Dutch subsurface is shown. The two distribution patterns largely overlap, because the formation is the main source rock for oil in the Dutch subsurface.

Many of the faults that border these Cimmerian basins were active, at least episodically, during sedimentation (Milsom & Rawson, 1989; Jenkyns & Senior, 1991; Powell, 2010; Hesselbo, 2012; Van Bergen et al., 2013). Examples from Great Britain and the British offshore are useful here (Fig. 6.14). In the Wessex Basin, for example, extensional faults were active in the Early Jurassic and resulted in several periods of active subsidence (Jenkyns & Senior, 1991). Subsidence of the Sole Pit Basin was controlled by move-

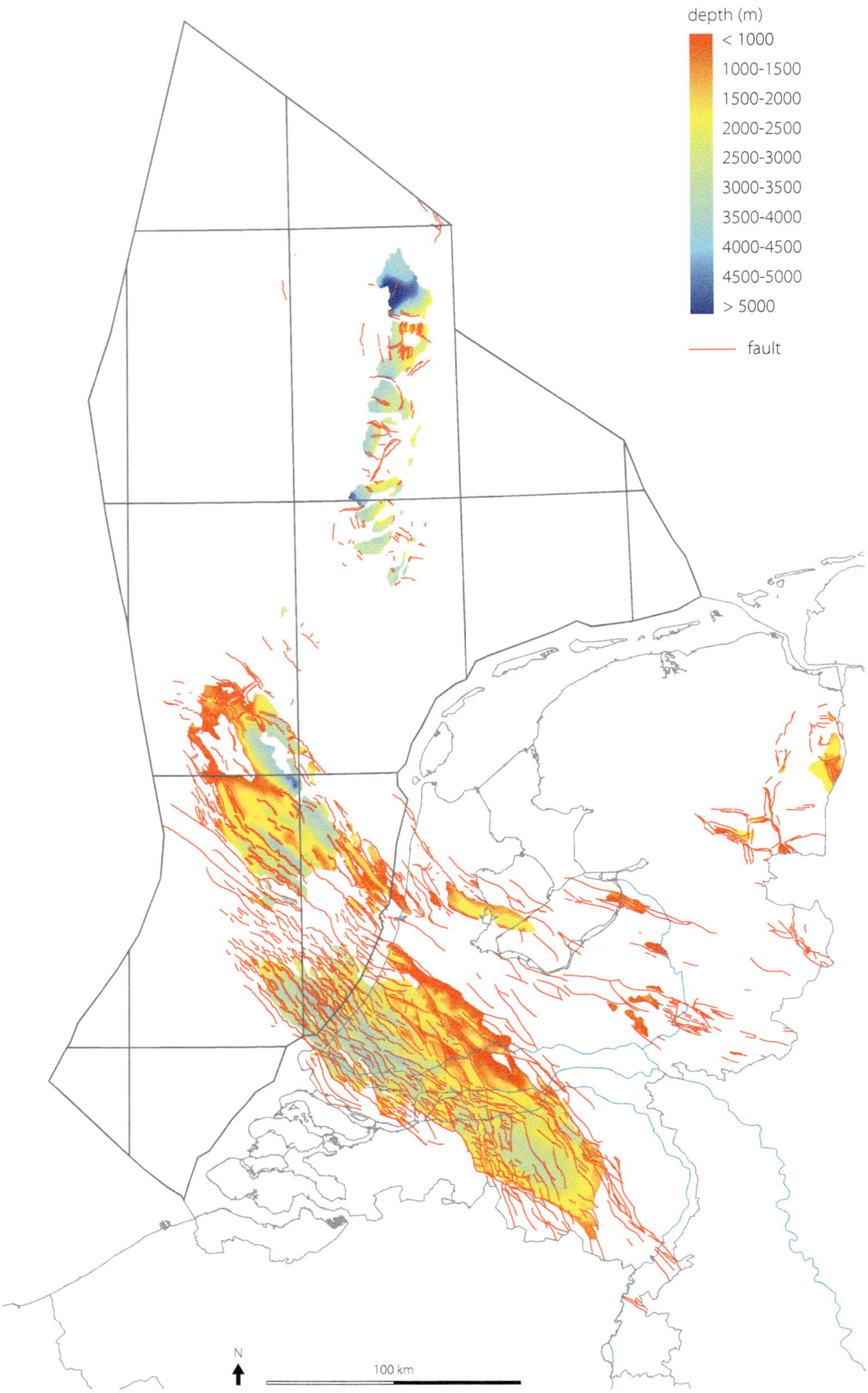

Figure 6.12. *Depth of the base of the Posidonia Shale Formation. The base of the formation is deepest in the southwestern margin of the West Netherlands Basin and in the northern Dutch Central Graben.*

ment along the Dowsing fault; the Market Weighton High, which was active during the Early Jurassic, is associated with this fault zone (Kent, 1980; Wright, 2022). Pinpointing precise times of fault movement is usually difficult, but the early Toarcian (Tenuicostatum Zone) appears to have been a period of major basin subsidence throughout England (Powell, 2010).

Evidence for synsedimentary fault movement in the Early Jurassic includes thickness changes towards and across faults (Milsom & Rawson, 1989; Jenkyns & Senior, 1991; Powell, 2010; Hesselbo, 2012). In places, considerable thinning of the Altena Group towards the highs and platforms is apparent (e.g. towards the London-Brabant Platform in the south; Donovan et al., 1979; Wong, 2007; Pharaoh, 2018; Wright, 2022). Chowns (1968) showed that two factors contributed to thickness variations in the Lias of Yorkshire: lateral differential subsidence and Early Jurassic erosion, which he correlated with uplift taking place in the basin during sedimentation.

It has been suggested that Triassic–Jurassic basins and highs were characterized by significant and negligible subsidence, respectively, and that sedimentation in each structural element was always such as to maintain a level seafloor (e.g. Sellwood & Jenkyns, 1975; Hallam & Sellwood, 1976; and see, in particular, Fig. 8 in Wignall et al., 2005). Ziegler (1981), for example, suggested that sedimentation rates in the Early Jurassic generally kept pace with subsidence rates and sea level. The thickness of the lower Altena Group, where fully preserved, reflects the general subsidence pattern.

Platforms were shallower water, less subsident, only intermittently flooded, and characterized by low sedimentation rates or even dynamic bypassing of sediment (Fig. 6.2a; Wright, 2022). On these elevated structural elements, a thinner, less carbonaceous, and more calcareous succession with occasional oolitic ironstones was deposited (Van Adrichem Boogaert & Kouwe, 1993-1997). In the absence of extensive data on the platform successions in the Netherlands, it is useful to consider examples from Great Britain (Fig. 6.14). The lower Toarcian succession of the East Midlands Shelf, for example, is less complete and the rocks have more carbonate and lower TOC content than those in the Cleveland Basin (Caswell & Coe, 2012). In the early Toarcian, the East Midlands Shelf was a shallower platform where the water column was more hospitable to benthic life than in the more subsident basin (Caswell & Coe, 2012). In southern Great Britain, the Pliensbachian–Toarcian in the coastal outcrops near Bridport is extremely condensed and carbonaceous facies are absent—probably due to erosion (Jenkyns & Macfarlane, 2021). About 40 km to the east, however, in the Winterborne Kingston borehole, the Toarcian OAE is represented by mudstones with up to 4% TOC (Jenkyns & Clayton, 1997). This suggests that low sedimentation rates and erosion were characteristic of shallower water areas of the northern Laurasian Sea; that is, on highs and platforms that subsided relatively less. In deeper water areas, however, dark carbonaceous mudstones (i.e. black shales) accumulated and formed somewhat more expanded successions (Howarth, 1992; Jenkyns & Clayton, 1997).

In conclusion, restriction of the Altena Group to the Cimmerian basins resulted from differential subsidence in the Early Jurassic, rather than solely from preservation from Middle and Upper Jurassic erosion. Erosion of Lower Jurassic mudstones from highs and platforms was probably not as severe as previously thought—the stratigraphic sequences of highs and platforms appear to have been condensed.

Pliensbachian–Toarcian oceanic anoxia and black shales

Dark carbonaceous mudstones are found in the area where Boreal and Tethyan waters met in the Laurasian Sea and characterize the lower Toarcian interval in northeast England, the Dutch subsurface, north- and southwest Germany, and the Paris Basin (Figs 6.7, 6.13a). The mudstones lose their kerogen content towards the north (e.g. Dunlin Group); eastwards, in northeast Germany and in the North Danish Basin; and towards the south, in areas dominated by Tethyan waters.

Hallam and Bradshaw (1979) were the first to suggest that these carbonaceous mudstones might represent an OAE (see Schlanger & Jenkyns, 1976). Jenkyns (1985), who studied Tethyan Lower Jurassic limestones, found a positive excursion in the carbon isotope record in the Serpentinum (or Falciferum) Zone and proposed the Toarcian OAE. The positive excursion represents a *global* increase in the rate of burial of ^{13}C-depleted organic carbon in sediments. Burial of organic carbon in the Laurasian Sea was responsible for up to ~12% of the total carbon required to explain the observed excursion (Ruvalcaba Baroni et al., 2018).

A negative excursion in organic carbon isotope ratios ($\delta^{13}C_{org}$) in the upper Tenuicostatum and lower Serpentinum zones either precedes or interrupts the positive excursion (e.g. Küspert, 1982; Jenkyns & Clayton, 1997; Hesselbo et al., 2000; Kemp et al., 2005). This negative excursion has been documented in the Dutch subsurface in the Posidonia Shale Formation (Fig. 6.10; Trabucho Alexandre et al., 2012; Houben et al., 2021). The negative excursion has been attributed to regional remineralization of organic carbon (Küspert, 1982) and sea-level fall (Röhl et al., 2001); but most authors nowadays agree that volcanism in southern Gondwana and/or release of methane from continental margin sediments were responsible for the input of ^{13}C-depleted carbon to the atmosphere and oceans (Hesselbo et al., 2000; Pálfy & Smith, 2000).

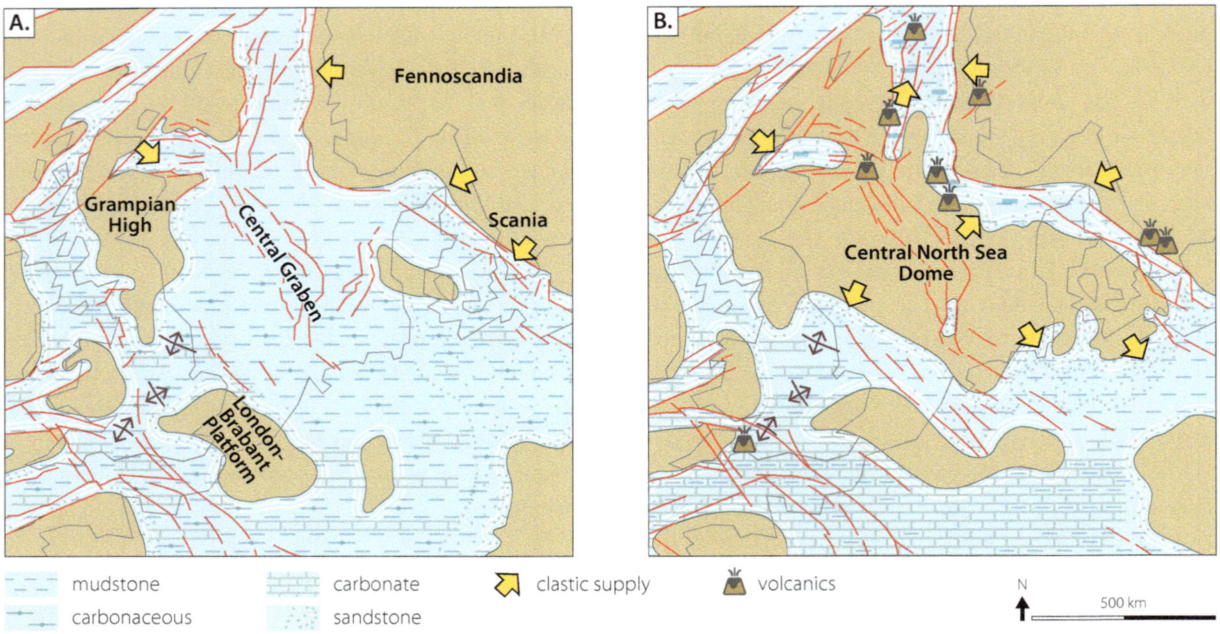

Figure 6.13. Regional paleogeographic maps. a) Early Jurassic and b) Middle Jurassic (modified from Ziegler, 1981, 1990; Bradshaw et al., 1992; Lott et al., 2010).

The increase in the rate of burial of ^{13}C-depleted organic carbon in sediments was due to oceanic anoxia (e.g. French et al., 2014; Thibault et al., 2018; Houben et al., 2021) and physiographic restriction of the northern Laurasian Sea (e.g. McArthur et al., 2008; Ruvalcaba Baroni et al., 2018; Remírez & Algeo, 2020).

Thallium isotopes indicate that oceanic anoxia started at the Pliensbachian–Toarcian boundary, some 600 ka before the negative excursion in the organic carbon isotope record (Them et al., 2018), suggesting that there is a link to volcanic activity in southern Gondwana (Pálfy & Smith, 2000), where continental flood basalts were emplaced in the Karoo-Ferrar Large Igneous Province (culmination of volcanic activity at ~183 Ma; Fig. 6.1).

Water depth in the northern Laurasian Sea has usually been overestimated (e.g. Mitchell et al., 2011; Dera & Donnadieu, 2012). Sedimentary structures in Toarcian mudstones indicate that the seafloor of the Cimmerian basins in the southern North Sea was above storm wave base (<100 m), except perhaps in the middle Toarcian (lower Bifrons Zone; Pedersen, 1985; Pye & Krinsley, 1986; Trabucho Alexandre et al., 2012, 2022; Trabucho Alexandre, 2015)—though wave base may have been deeper in the early Toarcian (Krencker et al., 2015). French et al. (2014) reported the aromatic carotenoid derivative, okenone, in a wavy laminated interval in the lower Toarcian of North Yorkshire. Okenone requires extremely shallow photic-zone euxinia and a highly restricted environment. For this reason, French et al. (2014) interpreted the sedimentary source of okenone to have been microbial mats (see O'Brien, 1990 for petrographic evidence) and the seafloor to have been *within* the euxinic photic zone.

Environments of deposition on the platforms and highs—some of which were (partly) intermittently flooded—would have been even shallower than those in the basins. This is probably reflected in the rough bathymetry (islands, shoals, and sills) invoked by Ruvalcaba Baroni et al. (2018) to explain circulation in the Laurasian Sea. According to them, the Laurasian Sea was too shallow for currents of oxygenated Tethyan water to penetrate deep into the epeiric sea, whose northern part was prone to stratification (Ruvalcaba Baroni et al., 2018). Salinity stratification, either due to runoff from the surrounding land, to rainfall, or to both (McArthur et al., 2008; Remírez & Algeo, 2020) promoted anoxia and burial of organic carbon in the basins.

As sea level rose, a southward current from the Boreal Sea into the Tethys is thought to have been established in the middle Toarcian (lower Bifrons Zone) and to have ended euxinic conditions in the Laurasian Sea (Van de Schootbrugge et al., 2020). By the middle and late Toarcian, the Werkendam Formation was deposited under more well-oxygenated conditions. In the latest Sinemurian and Pliensbachian, the Boreal Sea and Tethys were also connected but via more westerly seaways: Pliensbachian sediments in the Cardigan Bay Basin were deposited under the influence of persistent southward-flowing currents under more oxygenated conditions (Fig. 6.14; Pieńkowski et al., 2021).

Middle Jurassic uplift and erosion

Sedimentation from Aalenian to Callovian and, locally, Oxfordian was controlled by Middle Cimmerian tectonic activity, which caused significant structural differentiation (Fig. 6.2). The chief effect of this activity was the gradual uplift of the central North Sea area during the Aalenian and early Bajocian, which formed a broad dome transected by the Central Graben (Ziegler, 1981, 1990; Underhill & Partington, 1993). The uplift was approximately coeval (Bajocian–early Bathonian) with crustal separation of Gondwana from Laurasia and the opening of the Central Atlantic. After the Rhaetian–Hettangian transgression, the uplift of the central North Sea dome was arguably the most significant Jurassic paleogeographic change in northwest Europe. The uplifted terrain interrupted the seaway that had linked the Tethys to the Boreal Sea during the Early Jurassic. Domal upwarping was accompanied by volcanic activity (Bathonian–Callovian) in the central North Sea (Ziegler, 1990). Towards the end of the Bathonian, the dome began to subside. The Middle Cimmerian Unconformity cut progressively into deeper stratigraphic levels from the edges of the dome towards the margins of the Central Graben (Ziegler, 1990). Though the axial parts of the Central Graben were also uplifted and eroded, a relatively complete succession of Triassic and, locally, even Lower Jurassic sediments has been preserved in the basin. On the rift shoulders, however, erosion cut deeply into the Triassic, Permian, or even down into the basement. The products of erosion—mainly recycled Triassic sand—were shed into the adjacent subsiding basins. Exhumation (that is, rock uplift) amounted to 1500 to 2500 m, but topographic relief was much lower due to rapid erosion of the relatively poorly lithified Jurassic sediments (Ziegler, 1990).

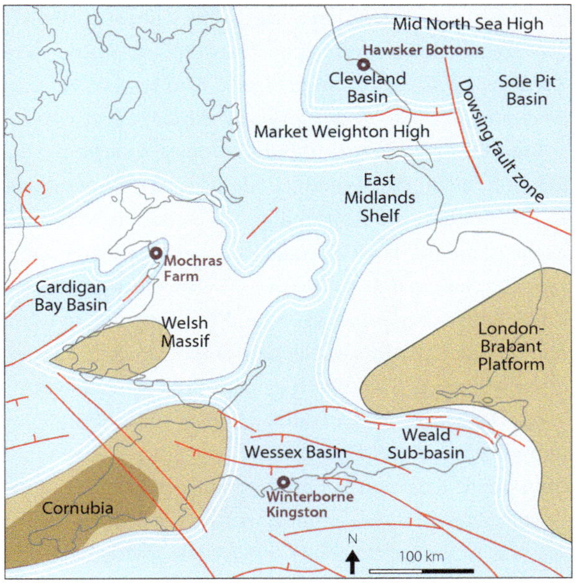

Figure 6.14. Map showing key British localities and paleogeographic features mentioned in this chapter. Map modified from Jenkyns et al. (2001) and Wright (2022).

The doming front shifted gradually southwards resulting in a considerable delay between the timing of rifting of the dome crest: middle Aalenian in the central North Sea to Kimmeridgian in North Brabant (Fig. 6.3). The extent of intra-Jurassic truncation and nondeposition decreases away from the dome (Underhill & Partington, 1993). Consequently, the corresponding hiatus at the top of the Altena Group is most prominent in the northernmost basin, the Dutch Central Graben, where epeiric sedimentation terminated in the Bajocian (Fig. 6.6). This area remained nondepositional from Bathonian to early Callovian.

While erosion also affected the Lower Saxony and Central Netherlands basins, this was not the case in the Broad Fourteens and West Netherlands basins and Roer Valley Graben in the south. Deposition of sediments of the Werkendam Formation continued into the Bajocian, with a distinct influx of marine sands along the southern margins of the basins. A facies change took place at the Bajocian–Bathonian boundary, with the deposition of sandy carbonates and marls (Cornbrash facies) of the Brabant Formation. At least three carbonate-marl cycles were deposited in the Bathonian to Oxfordian. They are preserved in the southern basins only as erosional remnants.

In the Achterhoek (Fig. 6.3), in the southeast of the Central Netherlands Basin, the sedimentary development up to the middle Bathonian shows significant similarities with the developments of the Roer Valley Graben and the German part of the Lower Saxony Basin (Herngreen et al., 1984). This suggests that the entire region was part of one sedimentary basin, which became differentiated only at the beginning of the late Bathonian. From this time until the Callovian, an open marine environment persisted in the Achterhoek and Germany, and claystones were deposited (Klomps member), whereas in the Roer Valley Graben and West Netherlands Basin deposition of shallow-water sandy carbonates continued into the Oxfordian (Fig. 6.6). In the Dutch part of the Lower Saxony Basin, most of the upper part of the Altena Group appears to have been eroded.

Paleogeography of the Laurasian Sea

Erosion of Lower Jurassic and older sediments during the early Middle Jurassic means that paleogeographic reconstructions of the northern Laurasian Sea are uncertain. The difficulty is also partly due to the paucity of records of coastal or terrestrial environments, itself evidence of the large extent of the sea (Barth et al., 2018). Even in the case of the London-Brabant Platform, which is always depicted as an island (Figs 6.2, 6.13), definitive shoreline deposits are unknown. There is no convincing evidence that the massif was an island rather than a submerged shoal, as there is little change in the character of sedimentation towards it (Bradshaw et al., 1992; and see Wright, 2022).

During the Early Jurassic, sedimentation in the North Sea area of the Laurasian Sea took place in a broad epeiric sea that has no modern equivalent (see Shaw, 1964; Hallam, 1997; Schieber, 2016). The seafloor was characterized by extremely low slopes (0.001–0.01°) over large areas and by facies belts of vast extent. This region was further characterized by a microtidal range and by dampening of tidal currents (Mitchell et al., 2011); few sediment gravity flows; and extensive storm influence (Pedersen, 1985; Krencker et al., 2015). Water depth at the time of deposition was typically a few tens of metres only (Hallam, 1975; Trabucho Alexandre et al., 2012; French et al., 2014; Trabucho Alexandre, 2015). On such broad epeiric seas, however, the correlation between distance from land and water depth that characterizes the pericontinental shelves of the Holocene is not applicable, and many fine-grained shallow-water epeiric sediments have 'deep water' characteristics (e.g. Shaw, 1964; Hallam, 1967; Schieber, 2016).

In the Early Jurassic, the main source of siliciclastic sediment for the northeastern part of the Laurasian Sea was Fennoscandia (Ziegler, 1981). The Mid North Sea High, a Paleozoic ridge northwest of the Dutch Central Graben (Figs 6.2, 6.14), was a sediment source for both the Dutch Central Graben and the Cleveland Basin (Wright, 2022). The contribution from Scotland (Grampian High) and other intrabasinal highs, such as the London-Brabant Platform, was relatively minor (Andrews et al., 1990; Hall & Bishop, 2002). In the Middle Jurassic, the central North Sea area was uplifted, and poorly lithified Jurassic sediments and Triassic sandstones were shed into the adjacent subsiding basins (Fig. 6.13b).

Cimmerian uplift of Paleozoic parent rocks controlled the provenance of Rhaetian and Lower to Middle Jurassic sediments. In the Polish Basin, Paleozoic organic matter, derived from Ordovician–Silurian marine carbonaceous mudstones and Carboniferous coals, was reworked during the Early Jurassic (Ruebsam et al., 2020). In Scotland, sand and clay mineralogy of Jurassic sedimentary rocks suggests that they were derived from the Devonian and Carboniferous (Hurst, 1985; Hall & Bishop, 2002). The high detrital kaolinite content of Rhaetian to Sinemurian mudstones in northeast Scotland has been attributed to reworking of Carboniferous regoliths, which were uplifted and eroded at the time of formation of the Viking Graben (Hurst, 1985). In Bornholm, Denmark, and Scania, southern Sweden, palynological study of the Lower and Middle Jurassic suggests that the palynomorphs were sourced from now completely eroded Carboniferous parent rocks (Nielsen & Koppelhus, 1991). Though the material is often not recognized as being reworked (Windle, 1979), Carboniferous spores are common in the Jurassic of Northern Europe (Herngreen & De Boer, 1974; Windle, 1979; Andrews et al., 1990; Nielsen & Koppelhus, 1991). According to Windle (1979), this can be explained by regional erosion of Carboniferous rocks during the Jurassic. Triassic rocks have also contributed to sedimentation in the Jurassic, particularly during the Middle Jurassic: in the Achterhoek (Fig. 6.3), for example, Keuper and Rhaetian grains occur regularly in younger Jurassic sediments (Herngreen & De Boer, 1974).

The proximal ~100 km of the journey from the sediment sources in the highs to the epeiric sea are readily explained by invoking sediment transport processes known from modern pericontinental shelves, such as hyperpycnal currents and storm-setup relaxation currents. The remaining more distal part of the journey (~500 km) towards the interior of the epeiric sea, however, requires bottom currents that do not rely on slope for propulsion (Schieber, 2016). The uniform character of Lower Jurassic strata—with limited change in facies towards and over the highs (Wright, 2022)—suggests that, despite subsidence generating more accommodation in the Cimmerian basins, the seafloor in this region of the epeiric sea was essentially flat (0.001–0.01°). Sedimentary structures in Lower Jurassic mudstones of the Cleveland Basin and the Dutch Central Graben indicate that currents in the 10–25 cm s^{-1} range (see Schieber et al., 2007) were responsible for transport and deposition of mud (O'Brien, 1990; Ghadeer & Macquaker, 2011; Trabucho Alexandre et al., 2012, 2022; Trabucho Alexandre, 2015). The Laurasian Sea was a hurricane-dominated region (Marsaglia & Klein, 1983; Krencker et al., 2015), and storm-driven bottom currents were probably responsible for most sediment transport.

Erhart's (1967) theory of bio-rhexistasy is useful in considering the timing of sediment delivery to the epeiric sea. In the Jurassic, land surrounding the Laurasian Sea as well as the islands of its archipelago were never very mountainous. The Netherlands lay about 15° south of its present latitude, between 30° and 40° N, and the climate was warm and humid. Early Jurassic climate is thought to have been globally 5 to 10 °C warmer than at present (Chandler et al., 1992; Dera & Donnadieu, 2012). During most of the time, humidity promoted a vegetation cover that protected soils from physical erosion but not from intense leaching of the dissolved products of chemical weathering into the epeiric sea (e.g. Cohen et al., 2004). Siliciclastic sediment supply to the Laurasian Sea was probably intermittent (see Brady & Bowie, 2017 for a Paleozoic example) and took place when the protective vegetation cover was reduced and argillaceous mud washed to the basin (Mander & McElwain,

2019; Slater et al., 2019). Thus, epeiric sedimentation, particularly on the platforms, was dominated by long intervals of biochemical precipitation of sediment (viz., formation of concretionary horizons under low siliciclastic mass accumulation rates) and by reworking of relict sediment by storm waves and currents.

Mechanical concentration of sulfurized organic carbon by storm-driven bottom currents has been proposed as a mechanism to explain the elevated organic carbon content of shallow-marine mudstones characterized by sedimentary structures attributed to storms (März et al., 2016; Tribovillard et al., 2019), such as those in the northern Laurasian Sea (e.g. Pedersen, 1985; Ghadeer & Macquaker, 2012; Trabucho Alexandre et al., 2012). The source of organic carbon in these mudstones was perhaps the shallow seafloor of platforms (Trabucho Alexandre et al., 2022), where erosion resulted in stratigraphic condensation and in the concentration of (carbonaceous) sediment into the subsiding basins. Storms in the northern Laurasian Sea probably promoted rapid burial of organic carbon in the basins and, thus, contributed to its preservation (Ghadeer & Macquaker, 2012).

Even though the Lower and Middle Jurassic, where present, is usually biostratigraphically complete, sedimentation in the basins was far from continuous (see Chowns, 1968). Trabucho-Alexandre (2015), for example, suggested that less than 35% of early Toarcian time is recorded in the stratigraphic succession of the Cleveland Basin (see Sloss, 1996; Kemp et al., 2018).

Economic geology

The presence and depth of burial of Lower Jurassic source rock intervals in the Netherlands, northern Germany, northern France (Paris Basin), and southern Great Britain (Weald and Wessex basins) have combined to provide charge for several small hydrocarbon discoveries that make an important contribution to the local economies (see Kombrink et al., 2012a for maps of the relevant fields). The Lower Jurassic also constitutes the main source rock interval for the Wytch Farm oilfield of Dorset, England, which is the largest onshore oilfield in northwest Europe.

In the Dutch subsurface, the Posidonia Shale Formation is the main source rock for oil (Bodenhausen & Ott, 1981). Both marine algal and land-derived organic carbon are present (type II kerogen). Several oil fields in the Broad Fourteens and West Netherlands basins and the Dutch Central Graben have been charged by this source rock (Remmelts et al., 2025, this volume). The relatively small accumulations of oil in the Netherlands may be attributed to the patchy distribution of the formation (Fig. 6.11), which has resulted from a combination of differential subsidence and later erosion. In northern Germany, the Posidonia Shale Formation (Ölschiefer Formation) lies within the oil window, making it an 'excellent target for both shale oil and shale gas exploration' (Song et al., 2015, p. 151). The lower Aalburg Formation probably also contributed to oil accumulations in the Dutch Central Graben (Wong et al., 1989) and West Netherlands Basin (De Jager et al., 1996). In the West Netherlands Basin, various members of the Brabant Formation are locally oil- and gas-bearing.

Acknowledgements

We thank Neptune Energy NL for the photos of core L05-14; Guillaume Dera for the map used to produce Fig. 6.1; Jenny Hettelaar for help with the ArcGIS maps; Nik Trabucho Alexandre for help with the figures; Grzegorz Pieńkowski and Paul Wignall for reviewing our manuscript; Harry Doust and Stephen Hesselbo for their comments and suggestions, which improved the final version of our manuscript; and the editor of our chapter, Johan ten Veen, for his comments, suggestions, and editorial work.

Digital map data

Spatial data of figures in this chapter for use in geographical information systems can be downloaded here: https://doi.org/10.5117/aup.28163258.

References

Ahlberg, A., Arndorff, L. & Guy-Ohlson, D., 2002. Onshore climate change during the Late Triassic marine inundation of the Central European Basin. Terra Nova 14: 241-248. DOI: 10.1046/j.1365-3121.2002.00416.x

Andrews, I.J., Long, D., Richards, P.C., Thomson, A.R., Brown, S., Chesher, J.A. & McCormac, M., 1990. United Kingdom offshore regional report: the geology of the Moray Firth. HMSO (London) for the British Geological Survey: 96 pp.

Bachmann, G.H., Geluk, M.C., Warrington, G., Becker-Roman, A., Beutler, G., Hagdorn, H., Hounslow, M.W., Nitsch, E., Röhling, H.-G., Simon, T. & Szulc, A., 2010. Triassic. In: Doornenbal, J.C. & Stevenson, A.G. (eds): Petroleum geological atlas of the Southern Permian Basin area. EAGE Publications b.v. (Houten): 149-173.

Barth, G., Pieńkowski, G., Zimmermann, J., Franz, M. & Kuhlmann, G., 2018. Palaeogeographical evolution of the Lower Jurassic: high-resolution biostratigraphy and sequence stratigraphy in the Central European Basin. In: Kilhams, B., Kukla,

P.A., Mazur, S., McKie, T., Mijnlieff, H.F. & Van Oijk, K. (eds): Mesozoic resource potential in the Southern Permian Basin. Geological Society of London, Special Publications 469: 341-369. DOI: 10.1144/SP469.8

Bjerrum, C.J., Surlyk, F., Callomon, J.H. & Slingerland, R.L., 2001. Numerical paleoceanographic study of the Early Jurassic Transcontinental Laurasian Seaway. Paleoceanography 16: 390-404. DOI: 10.1029/2000PA000512

Bodenhausen, J.W.A. & Ott, W.F., 1981. Habitat of the Rijswijk oil province, onshore, the Netherlands. In: Illing, L.V. & Hobson, G.D. (eds): Petroleum Geology of the continental shelf of North-West Europe: Proceedings of the Second Conference. Institute of Petroleum (London): 301-309.

Bradshaw, M.J., Cope, J.C.W., Cripps, D.W., Donovan, D.T., Howarth, M.K., Rawson, P.F., West, I.M. & Wimbledon, W.A., 1992. Jurassic. In: Cope, J.C.W., Ingham, J.K & Rawson, P.F. (eds): Atlas of palaeogeography and lithofacies. Geological Society of London, Memoir 13: 107-129. DOI: 10.1144/GSL.MEM.1992.013.01.12

Brady, M. & Bowie, C., 2017. Discontinuity surfaces and microfacies in a storm-dominated shallow epeiric sea, Devonian Cedar Valley Group, Iowa. The Depositional Record 3: 136-160. DOI: 10.1002/dep2.26

Burgers, W.F.J. & Mulder, G.G., 1991. Aspects of the Late Jurassic and Cretaceous history of The Netherlands. Geologie en Mijnbouw 70: 347-354.

Caswell, B.A. & Coe, A.L., 2012. A high-resolution shallow marine record of the Toarcian (Early Jurassic) Oceanic Anoxic Event from the East Midlands Shelf, UK. Palaeogeography, Palaeoclimatology, Palaeoecology 365-366: 124-135. DOI: 10.1016/j.palaeo.2012.09.021

Chandler, M.A., Rind, D. & Ruedy, R., 1992. Pangaean climate during the Early Jurassic: GCM simulations and the sedimentary record of paleoclimate. GSA Bulletin 104: 543-559. DOI: 10.1130/0016-7606(1992)104<0543:PCDTEJ>2.3.CO;2

Chowns, T.M., 1968. Environmental and diagenetic studies of the Cleveland Ironstone Formation of North East Yorkshire. PhD thesis University of Newcastle upon Tyne: 432 pp.

Cohen, A.S., Coe, A.L., Harding, S.M. & Schwark, L., 2004. Osmium isotope evidence for the regulation of atmospheric CO_2 by continental weathering. Geology 32: 157-160. DOI: 10.1130/G20158.1

Cope, J.C.W., Getty, T.A., Howarth, M.K., Morton, N. & Torrens, H.S., 1980. A correlation of Jurassic rocks in the British Isles. Part One: Introduction and Lower Jurassic. Geological Society of London, Special Report 14: 73 pp.

Damborenea, S.E., 2002. Jurassic evolution of Southern Hemisphere marine palaeobiogeographic units based on benthonic bivalves. Geobios 35, Supplement 1: 51-71. DOI: 10.1016/S0016-6995(02)00048-7

De Graciansky, P.-C., Jacquin, T. & Hesselbo, S.P., 1998. The Ligurian Cycle: An overview of lower Jurassic 2nd-order transgressive/regressive facies cycles in Western Europe. In: De Graciansky, P.-C., Hardenbol, J., Jacquin, T. & Vail, P.R. (eds): Mesozoic and Cenozoic sequence stratigraphy of European Basins. SEPM Special Publication 60: 467-479. DOI: 10.2110/pec.98.02.0467

De Jager, J., Doyle, M.A., Grantham, P.J. & Mabillard, J.E., 1996. Hydrocarbon habitat of the West Netherlands Basin. In: Rondeel, H.E., Batjes, D.A.J. & Nieuwenhuijs, W.H. (eds): Geology of gas and oil under the Netherlands. Kluwer Academic Publishers (Dordrecht): 191-209. DOI: 10.1007/978-94-009-0121-6_17

Dera, G. & Donnadieu, Y., 2012. Modeling evidences for global warming, Arctic seawater freshening, and sluggish oceanic circulation during the Early Toarcian anoxic event. Paleoceanography 27: PA2211. DOI: 10.1029/2012pa002283

Dera, G., Pucéat, E., Pellenard, P., Neige, P., Delsate, D., Joachimski, M.M., Reisberg, L. & Martinez, M., 2009. Water mass exchange and variations in seawater temperature in the NW Tethys during the Early Jurassic: Evidence from neodymium and oxygen isotopes of fish teeth and belemnites. Earth and Planetary Science Letters 286: 198-207. DOI: 10.1016/j.epsl.2009.06.027

Diependaal, H.J. & Reumer, J.W.F., 2021. Note on the Rhaetian fish fauna from a subrosion pipe in Winterswijk (the Netherlands), with a discussion on the validity of the genus Severnichthys Storrs, 1994. Netherlands Journal of Geosciences 100: e9. DOI: 10.1017/njg.2021.5

Donovan, D.T., Horton, A. & Ivimey-Cook, H.C., 1979. The transgression of the Lower Lias over the northern flank of the London Platform. Journal of the Geological Society 136: 165-173. DOI: 10.1144/gsjgs.136.2.0165

Duin, E.J.T., Doornenbal, J.C., Rijkers, R.H.B., Verbeek, J.W. & Wong, Th.E., 2006. Subsurface structure of the Netherlands – results of recent onshore and offshore mapping. Netherlands Journal of Geosciences 85: 245-276. DOI: 10.1017/S0016774600023064

Dusar, M., Langenaeker, V. & Wouters, L., 2002. Permian - Triassic - Jurassic lithostratigraphic units in the Campine basin and the Roer Valley Graben (NE Belgium). Geologica Belgica 4: 107-112. DOI.org/10.20341/gb.2014.047

Erhart, H., 1967. La genèse des sols en tant que phénomène géologique. Esquisse d'une théorie géologique et géochimique: biostasie en rhexistasie, second edition. Masson (Paris): 177 pp.

French, K.L., Sepúlveda, J., Trabucho-Alexandre, J., Gröcke, D.R. & Summons, R.E., 2014. Organic geochemistry of the early Toarcian oceanic anoxic event in Hawsker Bottoms, Yorkshire, England. Earth and Planetary Science Letters 390: 116-127. DOI: 10.1016/j.epsl.2013.12.033

Geluk, M.C., 2005. Stratigraphy and tectonics of Permo-Triassic basins in the Netherland and surrounding areas. PhD thesis Utrecht University (the Netherlands): 171 pp.

Ghadeer, S.G. & Macquaker, J.H.S., 2011. Sediment transport processes in an ancient mud-dominated succession: a comparison of processes operating in marine offshore settings and anoxic basinal environments. Journal of the Geological Society 168: 1121-1132. DOI: 10.1144/0016-76492010-016

Ghadeer, S.G. & Macquaker, J.H.S., 2012. The role of event beds in the preservation of organic carbon in fine-grained sediments: Analyses of the sedimentological processes operating during deposition of the Whitby Mudstone Formation (Toarcian, Lower Jurassic) preserved in northeast England. Marine and Petroleum Geology 35: 309-320. DOI: 10.1016/j.marpetgeo.2012.01.001

Gill, A.S., 2007. Organic geochemical assessment of the Posidonia Shale in the Netherlands. MSc thesis Utrecht University (the Netherlands).

Hall, A. & Bishop, P., 2002. Scotland's denudational history: an integrated view of erosion and sedimentation at an uplifted passive margin. In: Doré, A.G., Cartwright, J.A., Stoker, M.S., Turner, J.P. & White, N. (eds): Exhumation of the North Atlantic Margin: timing, mechanisms and implications for petroleum exploration. Geological Society of London, Special Publications 196: 271-290. DOI: 10.1144/GSL.SP.2002.196.01.15

Hallam, A., 1967. Editorial. In: Hallam, A. (ed.): Depth indicators in marine sedimentary environments. Marine Geology: 329-332. DOI: 10.1016/0025-3227(67)90043-6

Hallam, A., 1975. Jurassic environments. Cambridge University Press: 269 pp.

Hallam, A., 1997. Estimates of the amount and rate of sea-level change across the Rhaetian—Hettangian and Pliensbachian—Toarcian boundaries (latest Triassic to early Jurassic). Journal of the Geological Society 154: 773-779. DOI: 10.1144/gsjgs.154.5.0773

Hallam, A., 2001. A review of the broad pattern of Jurassic sea-level changes and their possible causes in the light of current knowledge. Palaeogeography, Palaeoclimatology, Palaeoecology 167: 23-37. DOI: 10.1016/S0031-0182(00)00229-7

Hallam, A. & Bradshaw, M.J., 1979. Bituminous shales and oolitic ironstones as indicators of transgressions and regressions. Journal of the Geological Society 136: 157-164. DOI: 10.1144/gsjgs.136.2.0157

Hallam, A. & Sellwood, B.W., 1976. Middle Mesozoic sedimentation in relation to tectonics in the British area. The Journal of Geology 84: 301-321. DOI: 10.1086/628197

Haq, B.U., 2018. Jurassic sea-level variations: A reappraisal. GSA Today 28 (1): 4-10. DOI: 10.1130/GSATG359A.1

Herngreen, G.F.W. & De Boer, K.F., 1974. Palynology of Rhaetian, Liassic and Dogger strata in the eastern Netherlands. Geologie en Mijnbouw 53: 343-368.

Herngreen, G.F.W., De Boer, K.F., Romein, B.J., Lissenberg, T. & Wijker, N.C., 1984. Middle Callovian beds in the Achterhoek, eastern Netherlands. In: Van Amerom, H.W.J. (ed.): S.J. Dijkstra memorial volume. Mededelingen Rijks Geologische Dienst 37: 95-123.

Herngreen, G.F.W., Van den Bosch, M. & Lissenberg, T., 2000. Nieuwe inzichten in de stratigrafische ontwikkeling van Jura, Krijt en Onder-Tertiair in de Achterhoek. Grondboor & Hamer 54: 71-90.

Herngreen, G.F.W., Van Konijnenburg-Van Cittert, J.H.A. & Oosterink, H.W., 2005. New geological data (Middle Triassic, Rhaetian-Liassic and Oligocene) of the Winterswijk quarry, the eastern Netherlands. Netherlands Journal of Geosciences 84 (4): 409-413. DOI: 10.1017/S0016774600021223

Hesselbo, S.P., 2012. Triassic-Jurassic boundary and Jurassic: Disintegrating Pangaea. In: Woodcock, N. & Strachan, R. (eds): Geological history of Britain and Ireland, second edition. Wiley-Blackwell (Oxford): 322-346. DOI: 10.1002/9781118274064.ch17

Hesselbo, S.P. & Jenkyns, H.C., 1998. British Lower Jurassic sequence stratigraphy. In: De Graciansky, P.-C., Hardenbol, J., Jacquin, T. & Vail, P.R. (eds): Mesozoic and Cenozoic sequence stratigraphy of European Basins. SEPM Special Publication 60: 561-581. DOI: 10.2110/pec.98.02.0561

Hesselbo, S.P., Gröcke, D.R., Jenkyns, H.C., Bjerrum, C.J., Farrimond, P., Morgans Bell, H.S. & Green, O.R., 2000. Massive dissociation of gas hydrate during a Jurassic oceanic anoxic event. Nature 406: 392-395. DOI: 10.1038/35019044

Hesselbo, S.P., Ogg, J.G., Ruhl, M., Hinnov, L.A. & Huang, C.J., 2020. The Jurassic Period. In: Gradstein, F.M., Ogg, J.G., Schmitz, M.D. & Ogg, G.M. (eds): Geologic time scale 2020. Elsevier (Amsterdam): 955-1021. DOI: 10.1016/B978-0-12-824360-2.00026-7

Hodges, P., 2021. A new ammonite from the Penarth Group, South Wales and the base of the Jurassic System in SW Britain. Geological Magazine 158: 1109-1114. DOI: 10.1017/S0016756820001107

Hodgson, N.A., Farnsworth, J. & Fraser, A.J., 1992. Salt-related tectonics, sedimentation and hydrocarbon plays in the Central Graben, North Sea, UKCS. In: Hardman, R.F.P. (ed.): Exploration Britain: Geological insights for the next decade. Geological Society of London, Special Publications 67: 31-63. DOI: 10.1144/GSL.SP.1992.067.01.03

Houben, A.J.P., Goldberg, T. & Slomp, C.P., 2021. Biogeochemical evolution and organic carbon deposition on the Northwestern European Shelf during the Toarcian Ocean Anoxic Event. Palaeogeography, Palaeoclimatology, Palaeoecology 565: 110191. DOI: 10.1016/j.palaeo.2020.110191

Houben, M.E., Barnhoorn, A., Wasch, L., Trabucho-Alexandre, J., Peach, C.J. & Drury, M.R., 2016. Microstructures of Early Jurassic (Toarcian) shales of Northern Europe. International Journal of Coal Geology 165: 76-89. DOI: 10.1016/j.coal.2016.08.003

Houben, S., Nelskamp, S., Goldberg, T., Verreussel, R., Janssen, N., Kerstholt, S. & Boxem, T., 2017. Hydrocarbon potential of the Lias: HYPO-Lias. TNO (Utrecht) Report No. 2017 R10380: 86 pp. https://nlog.nl/sites/default/files/2023-11/HypoLias_Report_Final.pdf

Howarth, M.K., 1992. The ammonite family Hildoceratidae in the Lower Jurassic of Britain (part 2). Monographs of the Palaeontographical Society 146: 107-192. DOI: 10.1080/25761900.20

22.12131773.
Hurst, A., 1985. The implications of clay mineralogy to palaeoclimate and provenance during the Jurassic in NE Scotland. Scottish Journal of Geology 21: 143-160. DOI: 10.1144/sjg21020143
Jenkyns, H.C., 1985. The early Toarcian and Cenomanian-Turonian anoxic events in Europe: comparisons and contrasts. Geologische Rundschau 74: 505-518. DOI: 10.1007/BF01821208
Jenkyns, H.C. & Clayton, C.J., 1997. Lower Jurassic epicontinental carbonates and mudstones from England and Wales: chemostratigraphic signals and the early Toarcian anoxic event. Sedimentology 44: 687-706. DOI: 10.1046/j.1365-3091.1997.d01-43.x
Jenkyns, H.C. & Macfarlane, S., 2022 . The chemostratigraphy and environmental significance of the Marlstone and Junction Bed (Beacon Limestone, Toarcian, Lower Jurassic, Dorset, UK. Geological Magazine 159: 357-371. DOI: 10.1017/S0016756821000972
Jenkyns, H.C. & Senior, J.R., 1991. Geological evidence for intra-Jurassic faulting in the Wessex Basin and its margins. Journal of the Geological Society 148: 245-260. DOI: 10.1144/gsjgs.148.2.0245
Jenkyns, H.C., Gröcke, D.R. & Hesselbo, S.P., 2001. Nitrogen isotope evidence for water mass denitrification during the Early Toarcian (Jurassic) oceanic anoxic event. Paleoceanography 16: 593-603. DOI: 10.1029/2000PA000558
Kemp, D.B., Coe, A.L., Cohen, A.S. & Schwark, L., 2005. Astronomical pacing of methane release in the Early Jurassic period. Nature 437: 396-399. DOI: 10.1038/nature04037
Kemp, D.B., Fraser, W.T. & Izumi, K., 2018. Stratigraphic completeness and resolution in an ancient mudrock succession. Sedimentology 65: 1875-1890. DOI: 10.1111/sed.12450
Klingler, W., 1962. Lias Deutschlands. Leitfossilien der Mikropaläontologie 3: 73-122.
Klompmaker, A.A. & Van den Berkmortel, B.J.H.M., 2007. Earliest Jurassic (Hettangian) psiloceratoid ammonites from a subrosion pipe at Winterswijk, the eastern Netherlands. Netherlands Journal of Geosciences 86: 379-388. DOI: 10.1017/S0016774600023593
Kombrink, H., Doornenbal, J.C., Duin, E.J.T., Den Dulk, M., Ten Veen, J.H. & Witmans, N., 2012. New insights into the geological structure of the Netherlands; results of a detailed mapping project. Netherlands Journal of Geosciences 91(4): 419-446. DOI: 10.1017/S0016774600000329
Kombrink, H., Ten Veen, J.H. & Geluk, M.C., 2012b. Exploration in the Netherlands, 1987-2012. Netherlands Journal of Geosciences: 403-418. DOI: 10.1017/S0016774600000317
Krencker, F.-N., Bodin, S., Suan, G., Heimhofer, U., Kabiri, L. & Immenhauser, A., 2015. Toarcian extreme warmth led to tropical cyclone intensification. Earth and Planetary Science Letters 425: 120-130. DOI: 10.1016/j.epsl.2015.06.003
Küspert, W., 1982. Environmental changes during oil shale deposition as deduced from stable isotope ratios. In: Einsele, G. & Seilacher, A. (eds): Cyclic and event stratification. Springer (Berlin): 482-501. DOI: 10.1007/978-3-642-75829-4_36
Lazar, O.R., Bohacs, K.M., Macquaker, J.H.S., Schieber, J. & Demko, T.M., 2015. Capturing key attributes of fine-grained sedimentary rocks in outcrops, cores, and thin sections: Nomenclature and description guidelines. Journal of Sedimentary Research 85: 230-246. DOI: 10.2110/jsr.2015.11
Lott, G.K. & Warrington, G., 1988. A review of the latest Triassic succession in the U.K. sector of the Southern North Sea Basin. Proceedings of the Yorkshire Geological Society 47: 139-147. DOI: 10.1144/pygs.47.2.139
Lott, G.K., Wong, Th.E., Dusar, M., Andsbjerg, J., Mönnig, E., Feldman-Olszewska, A. & Verreussel, R.M.C.H., 2010. Jurassic. In: Doornenbal, J.C. & Stevenson, A.G. (eds): Petroleum geological atlas of the Southern Permian Basin area. EAGE Publications b.v. (Houten): 175-193.
Mander, L. & McElwain, J.C., 2019. Toarcian land vegetation loss. Nature Geoscience 12: 405-406. DOI: 10.1038/s41561-019-0366-y
Marsaglia, K.M. & Klein, G.D., 1983. The paleogeography of Paleozoic and Mesozoic storm depositional systems. The Journal of Geology 91: 117-142. DOI: 10.1086/628752
März, C., Wagner, T., Aqleh, S., Al-Alaween, M., Van den Boorn, S., Podlaha, O.G., Kolonic, S., Poulton, S.W., Schnetger, B. & Brumsack, H.-J., 2016. Repeated enrichment of trace metals and organic carbon on an Eocene high-energy shelf caused by anoxia and reworking. Geology 44: 1011-1014. DOI: 10.1130/G38412.1
McArthur, J.M., Donovan, D.T., Thirlwall, M.F., Fouke, B.W. & Mattey, D., 2000. Strontium isotope profile of the early Toarcian (Jurassic) oceanic anoxic event, the duration of ammonite biozones, and belemnite palaeotemperatures. Earth and Planetary Science Letters 179: 269-285. DOI: 10.1016/S0012-821X(00)00111-4
McArthur, J.M., Algeo, T.J., Van de Schootbrugge, B., Li, Q. & Howarth, R.J., 2008. Basinal restriction, black shales, Re-Os dating, and the Early Toarcian (Jurassic) oceanic anoxic event. Paleoceanography 23: PA4212. DOI: 10.1029/2008PA001607
Milsom, J. & Rawson, P.F., 1989. The Peak Trough – a major control on the geology of the North Yorkshire coast. Geological Magazine 126: 699-705. DOI: 10.1017/S0016756800007007
Mitchell, A.J., Allison, P.A., Gorman, G.J., Piggott, M.D. & Pain, C.C., 2011. Tidal circulation in an ancient epicontinental sea: The Early Jurassic Laurasian Seaway. Geology 39: 207-210. DOI: 10.1130/G31496.1
Nederlandse Aardolie Maatschappij & Rijks Geologische Dienst, 1980. Stratigraphic nomenclature of the Netherlands. Verhandelingen van Het Koninklijk Nederlands Geologisch Mijnbouwkundig Genootschap 32: 77 pp.
Nielsen, L.H. & Koppelhus, E.B., 1991. Reworked Carboniferous palynomorphs from the Lower Jurassic of Bornholm and their palaeogeographic significance. Bulletin of the Geological Society of Denmark 38: 253-266. DOI: 10.37570/bgsd-1990-38-22

Nomade, S., Knight, K.B., Beutel, E., Renne, P.R., Verati, C., Féraud, G., Marzoli, A., Youbi, N. & Bertrand, H., 2007. Chronology of the Central Atlantic Magmatic Province: Implications for the Central Atlantic rifting processes and the Triassic–Jurassic biotic crisis. Palaeogeography, Palaeoclimatology, Palaeoecology 244: 326-344. DOI: 10.1016/j.palaeo.2006.06.034

O'Brien, N.R., 1990. Significance of lamination in Toarcian (Lower Jurassic) shales from Yorkshire, Great Britain. Sedimentary Geology 67: 25-34. DOI: 10.1016/0037-0738(90)90025-O

Oosterink, H., 2010. Jura-, Krijt- en Paleogene afzettingen in de gemeente Winterswijk. Grondboor & Hamer 64: 157-161.

Pálfy, J. & Smith, P.L., 2000. Synchrony between Early Jurassic extinction, oceanic anoxic event, and the Karoo-Ferrar flood basalt volcanism. Geology 28: 747-750. DOI: 10.1130/0091-7613(2000)28<747:SBEJEO>2.0.CO;2

Pearson, D.A.B., 1970. Problems of Rhaetian stratigraphy with special reference to the lower boundary of the stage. Quarterly Journal of the Geological Society 126: 125-150. DOI: 10.1144/gsjgs.126.1.0125

Pedersen, G.K., 1985. Thin, fine-grained storm layers in a muddy shelf sequence: an example from the Lower Jurassic in the Stenlille 1 well, Denmark. Journal of the Geological Society 142: 357-374. DOI: 10.1144/gsjgs.142.2.0357

Peletier, W. & Kolstee, H.G., 1986. Winterswijk, geologie deel I. Inleiding tot de geologie van Winterswijk. Wetenschappelijke Mededelingen K.N.N.V 175: 136 pp.

Pharaoh, T., 2018. The Anglo-Brabant Massif: Persistent but enigmatic palaeo-relief at the heart of western Europe. Prodeedings of the Geologists' Association 129: 278-328. DOI: 10.1016/j.pgeola.2018.02.009

Pieńkowski, G., 2004. The epicontinental Lower Jurassic of Poland. Polish Geological Institute Special Papers 12: 1-152.

Pieńkowski, G., 2014. The first Early Jurassic ammonite find in central Poland. Volumina Jurassica 12: 99-104.

Pieńkowski, G., Schudack, M.E., Bosák, P., Enay, R., Feldman-Olszewska, A., Golonka, J., Gutowski, J., Herngreen, G.F.W., Jordan, P., Krobicki, M., Lathuiliere, B., Leinfelder, R.R., Michalík, J., Mönnig, E., Noe-Nygaard, N., Pálfy, J., Pint, A., Rasser, M.W., Reisdorf, A.G., Schmid, D.U., Schweigert, G., Surlyk, F., Wetzel, A. & Wong, Th.E., 2008. Jurassic. In: McCann, T. (ed.): The Geology of Central Europe. Volume 2: Mesozoic and Cenozoic. The Geological Society of London: 823-922. DOI: 10.1144/CEV2P.2

Pieńkowski, G., Uchman, A., Ninard, K. & Hesselbo, S.P., 2021. Ichnology, sedimentology, and orbital cycles in the hemipelagic Early Jurassic Laurasian Seaway (Pliensbachian, Cardigan Bay Basin, UK). Global and Planetary Change 207: 103648. DOI: 10.1016/j.gloplacha.2021.103648

Powell, J.H., 2010. Jurassic sedimentation in the Cleveland Basin: a review. Proceedings of the Yorkshire Geological Society 58: 21-72. DOI: 10.1144/pygs.58.1.278

Pye, K. & Krinsley, D.H., 1986. Microfabric, mineralogy and early diagenetic history of the Whitby Mudstone Formation (Toarcian), Cleveland Basin, U.K. Geological Magazine 123: 191-203. DOI: 10.1017/S0016756800034695

Raup, D.M. & Sepkoski, J.J., 1982. Mass extinctions in the marine fossil record. Science 215: 1501-1503. DOI: 10.1126/science.215.4539.1501

Remírez, M.N. & Algeo, T.J., 2020. Paleosalinity determination in ancient epicontinental seas: A case study of the T-OAE in the Cleveland Basin (UK). Earth-Science Reviews 201: 103072. DOI: 10.1016/j.earscirev.2019.103072

Remmelts, G., Nelskamp, S., De Jager, J. & Geluk, M.C., 2025. Petroleum Geology. In: Ten Veen, J.H., Vis, G.-J., De Jager, J. & Wong, Th.E. (eds): Geology of the Netherlands, second edition. Amsterdam University Press (Amsterdam): 535-575. DOI: 10.5117/9789463728362_ch14

Rhys, G.H., 1974. A proposed standard lithostratigraphic nomenclature for the southern North Sea and an outline structural nomenclature for the whole of the (UK) North Sea. Report of the Institute of Geological Sciences 74/8. HMSO (London): 26 pp.

Röhl, H.-J., Schmid-Röhl, A., Oschmann, W., Frimmel, A. & Schwark, L., 2001. The Posidonia Shale (Lower Toarcian) of SW-Germany: an oxygen-depleted ecosystem controlled by sea level and palaeoclimate. Palaeogeography, Palaeoclimatology, Palaeoecology 165: 27-52. DOI: 10.1016/S0031-0182(00)00152-8

Ruebsam, W., Pieńkowski, G. & Schwark, L., 2020. Toarcian climate and carbon cycle perturbations – its impact on sea-level changes, enhanced mobilization and oxidation of fossil organic matter. Earth and Planetary Science Letters 546: 116417. DOI: 10.1016/j.epsl.2020.116417

Ruvalcaba Baroni, I., Pohl, A., Van Helmond, N.A.G.M., Papadomanolaki, N.M., Coe, A.L., Cohen, A.S., Van de Schootbrugge, B., Donnadieu, Y. & Slomp, C.P., 2018. Ocean circulation in the Toarcian (Early Jurassic): A key control on deoxygenation and carbon burial on the European Shelf. Paleoceanography and Paleoclimatology 33: 994-1012. DOI: 10.1029/2018PA003394

Sahabi, M., Aslanian, D. & Olivet, J.-L., 2004. Un nouveau point de départ pour l'histoire de l'Atlantique central. Comptes Rendus Geoscience 336: 1041-1052. DOI: 10.1016/j.crte.2004.03.017

Schieber, J., 2016. Mud re-distribution in epicontinental basins – Exploring likely processes. Marine and Petroleum Geology 71: 119-133. DOI: 10.1016/j.marpetgeo.2015.12.014

Schieber, J., Southard, J. & Thaisen, K., 2007. Accretion of mudstone beds from migrating floccule ripples. Science 318: 1760-1763. DOI: 10.1126/science.1147001

Schlanger, S.O. & Jenkyns, H.C., 1976. Cretaceous Oceanic Anoxic Events: causes and consequences. Geologie en Mijnbouw 55: 179-184.

Sellwood, B.W. & Jenkyns, H.G., 1975. Basins and swells and the evolution of an epeiric sea (Pliensbachian–Bajocian of Great Britain). Journal of the Geological Society 131: 373-388. DOI: 10.1144/gsjgs.131.4.0373

Shaw, A.B., 1964. Time in stratigraphy. McGraw-Hill (New York): 365 pp.

Slater, S.M., Twitchett, R.J., Danise, S. & Vajda, V., 2019. Substantial vegetation response to Early Jurassic global warming with impacts on oceanic anoxia. Nature Geoscience 12: 462-467. DOI: 10.1038/s41561-019-0349-z

Sloss, L.L., 1996. Sequence stratigraphy on the craton: caveat emptor. In: Witzke, B.J., Ludvigson, G.A. & Day, J. (eds): Paleozoic sequence stratigraphy - views from the North American Craton. Geological Society of America, Special Papers 306: 425-434. DOI: 10.1130/0-8137-2306-X.425

Song, J., Littke, R., Weniger, P., Ostertag-Henning, C. & Nelskamp, S., 2015. Shale oil potential and thermal maturity of the Lower Toarcian Posidonia Shale in NW Europe. International Journal of Coal Geology 150-151: 127-153. DOI: 10.1016/j.coal.2015.08.011

Stille, H., 1924. Grundfragen der vergleichenden Tektonik. Gebrüder Borntraeger (Berlin): 443 pp.

Them II, T.R., Gill, B.C., Caruthers, A.H., Gerhardt, A.M., Gröcke, D.R., Lyons, T.W., Marroquín, S.M., Nielsen, S.G., Trabucho Alexandre , J.P. & Owens, J.D., 2018. Thallium isotopes reveal protracted anoxia during the Toarcian (Early Jurassic) associated with volcanism, carbon burial, and mass extinction. Proceedings of the National Academy of Sciences 115: 6596-6601. DOI: 10.1073/pnas.1803478115

Thibault, N., Ruhl, M., Ullmann, C.V., Korte, C., Kemp, D.B., Gröcke, D.R. & Hesselbo, S.P., 2018. The wider context of the Lower Jurassic Toarcian oceanic anoxic event in Yorkshire coastal outcrops, UK. Proceedings of the Geologists' Association 129: 372-391. DOI: 10.1016/j.pgeola.2017.10.007

Trabucho-Alexandre, J.P., 2015. More gaps than shale: erosion of mud and its effect on preserved geochemical and palaeobiological signals. In: Smith, D.G., Bailey, R.J., Burgess, P.M. & Fraser, A.J. (eds): Strata and time: probing the gaps in our understanding. Geological Society of London, Special Publications 404: 251-270. DOI: 10.1144/SP404.10

Trabucho-Alexandre, J.P., Dirkx, R., Veld, H., Klaver, G. & De Boer, P.L., 2012. Toarcian Black Shales in the Dutch Central Graben: Record of energetic, variable depositional conditions during an oceanic anoxic event. Journal of Sedimentary Research 82: 104-120. DOI: 10.2110/jsr.2012.5

Trabucho-Alexandre, J.P., Gröcke, D.R., Atar, E., Herringshaw, L. & Jarvis, I., 2022. A new subsurface record of the Pliensbachian–Toarcian, Lower Jurassic, of Yorkshire. Proceedings of the Yorkshire Geological Society 64: pygs2022-007. DOI: 10.1144/pygs2022-007

Tribovillard, N., Koched, H., Baudin, F., Adatte, T., Delattre, M., Abraham, R. & Ferry, J.-N., 2019. Storm-induced concentration of sulfurized, marine-origin, organic matter as a possible mechanism in the formation of petroleum source-rock. Marine and Petroleum Geology 109: 808-818. DOI: 10.1016/j.marpetgeo.2019.07.003

Underhill, J.R. & Partington, M.A., 1993. Jurassic thermal doming and deflation in the North Sea: implications of the sequence stratigraphic evidence. In: Parker, J.R. (ed.): Petroleum Geology of Northwest Europe: Proceedings of the 4th Conference. Geological Society of London, Petroleum Geology Conference Series 4: 337-345. DOI: 10.1144/0040337

Van Adrichem Boogaert, H.A. & Kouwe, W.F.P., 1993-1997. Stratigraphic nomenclature of the Netherlands; revision and update by RGD and NOGEPA (sections A-J). Mededelingen Rijks Geologische Dienst 50.

Van Bergen, F., Zijp, M., Nelskamp, S. & Kombrink, H., 2013. Shale gas evaluation of the Early Jurassic Posidonia Shale Formation and the Carboniferous Epen Formation in the Netherlands. In: Chatellier, J.-Y. & Jarvie, D.M. (eds): Critical assessment of shale resource plays. American Association of Petroleum Geologists, Memoir 103: 1-24.

Van de Schootbrugge, B., Houben, A.J.P., Ercan, F.E.Z., Verreussel, R., Kerstholt, S., Janssen, N.M.M., Nikitenko, B. & Suan, G., 2020. Enhanced Arctic-Tethys connectivity ended the Toarcian Oceanic Anoxic Event in NW Europe. Geological Magazine 157: 1593-1611. DOI: 10.1017/S0016756819001262

Van Wijhe, D.H., 1987. Structural evolution of inverted basins in the Dutch offshore. Tectonophysics 137: 171-219. DOI: 10.1016/0040-1951(87)90320-9

Verreussel, R.M.C.H., Bouroullec, R., Munsterman, D.K., Dybkjær, K., Geel, C.R., Houben, A.J.P., Johannessen, P.N. & Kerstholt-Boegehold, S.J., 2018. Stepwise basin evolution of the Middle Jurassic–Early Cretaceous rift phase in the Central Graben area of Denmark, Germany and The Netherlands. In: Kilhams, B., Kukla, P.A., Mazur, S., McKie, T., Mijnlieff, H.F. & Van Oijk, K. (eds): Mesozoic resource potential in the Southern Permian Basin. Geological Society of London, Special Publications 469: 305-340. DOI: 10.1144/SP469.23

Verreussel, R.M.C.H., Munsterman, D.K., Bouroullec, R. & Houben, A.J.P., 2025. Late Jurassic - Early Cretaceous. In: Ten Veen, J.H., Vis, G.-J., De Jager, J. & Wong, Th.E. (eds): Geology of the Netherlands, second edition. Amsterdam University Press (Amsterdam): 211-251. DOI: 10.5117/9789463728362_ch07

Von Hillebrandt , A., Krystyn, L., Kürschner, W.M., Bonis, N.R., Ruhl, M., Richoz, S., Schobben, M.A.N., Urlichs, M., Bown, P.R., Kment, K., McRoberts, C.A., Simms, M. & Tomãsových, A., 2013. The Global Stratotype Sections and Point (GSSP) for the base of the Jurassic System at Kuhjoch (Karwendel Mountains, Northern Calcareous Alps, Tyrol, Austria). Episodes 36: 162-198. DOI: 10.18814/epiiugs/2013/v36i3/001

Wignall, P.B., Newton, R.J. & Little, C.T.S., 2005. The timing of paleoenvironmental change and cause-and-effect relationships during the early Jurassic mass extinction in Europe. American Journal of Science 305: 1014-1032. DOI: 10.2475/ajs.305.10.1014

Windle, T.M.F., 1979. Reworked Carboniferous spores: An example from the Lower Jurassic of northeast Scotland. Review of Palaeobotany and Palynology 27: 173-184. DOI: 10.1016/0034-6667(79)90039-3

Witte, L., Schuurman, H. & Lissenberg, Th., 1993. Microfossielen uit Jura- en Krijtontsluitingen in de oostelijke Achterhoek. Grondboor & Hamer 47: 35-40.

Wong, Th.E., 2007. Jurassic. In: Wong, Th.E., Batjes, D.A.J. & De Jager, J. (eds): Geology of the Netherlands. Royal Netherlands Academy of Arts and Sciences (KNAW, Amsterdam): 107-125.

Wong, Th.E., Van Doorn, Th.H.M. & Schroot, B.M., 1989. Late Jurassic petroleum geology of the Dutch Central North Sea Graben. Geologische Rundschau 78 (1): 319-336. DOI: 10.1007/BF01988367

Wright, J.K., 2022. The Market Weighton High in the 21st century – new understanding of a long-standing problem. Proceedings of the Yorkshire Geological Society 64. DOI: 10.1144/pygs2021-008

Yawar, Z. & Schieber, J., 2017. On the origin of silt laminae in laminated shales. Sedimentary Geology 360: 22-34. DOI: 10.1016/j.sedgeo.2017.09.001

Ziegler, P.A., 1981. Evolution of sedimentary basins in North-West Europe. In: Illing, L.V. & Hobson, G.D. (eds): Petroleum Geology of the continental shelf of North-West Europe: Proceedings of the Second Conference. Institute of Petroleum (London): 3-39.

Ziegler, P.A., 1990. Geological Atlas of Western and Central Europe, second edition. Shell Internationale Petroleum Mij. B.V. and Geological Society (London): 239 pp.

7

Late Jurassic – Early Cretaceous

Roel M.C.H. Verreussel, Renaud Bouroullec, Alexander J.P. Houben & Dirk K. Munsterman*

*Corresponding author: roel.verreussel@tno.nl

ABSTRACT

The sedimentary and structural development in the Late Jurassic-Early Cretaceous period in the Netherlands is largely governed by the Late Cimmerian rift phase and the subsequent post-rift. The rifting affected the Dutch Central Graben in the northern offshore first. East-west extension during the Callovian and Oxfordian activated the faults and salt structures that bordered the existing Triassic graben structure and created accommodation space. The basin was filled with siliciclastic non-marine and marginal marine sediments, interrupted by thick and basin-wide coal seams at the Callovian-Oxfordian boundary. In the Kimmeridgian, the extension regime changed to NE-SW and provoked the reactivation of Paleozoic NW trending faults in the subsurface of the Netherlands. As a result, accommodation space was created in several other basins and thick stacks of sediments accumulated in the hanging walls of these faults. This continued throughout the latest Jurassic until the earliest Cretaceous when movement along the faults slowed down or stalled and footwall erosion occurred in many places. During the ensuing post-rift thermal sag phase, deposition extended outside the basins onto the bordering platforms but the basins remained the most active depocentres accumulating hundreds of metres of sediment up until the Aptian. In the Aptian and Albian, the formerly prevailing siliciclastic depositional systems were gradually replaced by carbonate-dominated systems. By that time, the vast majority of the Netherlands had become fully marine.

<< Shallow marine sandstone with large concretions of the Grès de la Crèche Formation at Cap Grix-Nez (northern France), equivalent to the Terschelling Sandstone Member in the Netherlands. Photo: Bart van der Kwaak.

DOI: 10.5117/9789463728362_ch07

Introduction

Upper Jurassic and Lower Cretaceous rocks occur in large parts of the subsurface of the Netherlands both onshore and offshore, but apart from one location in the town of Losser, eastern Netherlands, outcrops are nowhere to be found. The geology of the Late Jurassic-Early Cretaceous is complex: no less than four groups, three subgroups, seventeen formations and sixty-one members have been defined within this time interval in the subsurface of the Netherlands (Van Adrichem Boogaert & Kouwe, 1993; Munsterman et al., 2012; see also Table 7.1). The main reason for this complexity is the fact that the Late Cimmerian rift phase led to the development of several sub-basins, each with its own tectonic development and depositional history (Coward et al., 2003; see also Fig. 7.1). Further complexity resulted from salt tectonics (Ten Veen et al., 2012; Bouroullec et al., 2018; Bouroullec & Ten Veen, 2025, this volume). The oldest sediments that can be attributed to the Late Cimmerian rift phase in the Netherlands are found in the offshore Dutch Central Graben and are of Middle Jurassic (Callovian) age. Because most of the Cimmerian rift phase took place during Late Jurassic and Early Cretaceous times, the short title of this chapter 'Late Jurassic-Early Cretaceous', although justified is not entirely correct. During the rift phase and shortly afterwards, the basins were filled with up to 2200 metres of siliciclastic sediments, sourced from the bordering platforms and structural highs (De Jager, 2007). All the while, Zechstein salt was being re-mobilized, faults were moving, volcanoes were active and the climate was changing from warm and wet to cooler and dry and back to wet again (Abbink et al., 2001a; Bouroullec & Ten Veen, 2025, this volume; De Jager et al., 2025, this volume; Van Bergen et al., 2025, this volume). This resulted in a wide array of different depositional environments: fluvial, lagoonal, lacustrine, marginal marine, shallow marine and deep marine. The Upper Jurassic-Lower Cretaceous deposits contain coal, oil and gas

Table 7.1. *Hierarchically arranged Upper Jurassic-Lower Cretaceous lithostratigraphic units with generalized lithologies and depositional environments. For detailed descriptions and formal definitions the reader is referred to the online stratigraphic nomenclature: dinoloket.nl/en/stratigraphic-nomenclature. For an explanation of the abbreviations of basin names: see Fig. 7.1.*

Lithostratigraphy	Lithology	Depositional environment	Basin	TMS
Rijnland Group				
Holland Formation	Marl, claystone and greensand	Open marine	**all**	**5**
Vlieland Claystone Formation	Claystone	Open marine	**all**	**4**
Vlieland Sandstone Formation	Sandstone	Coastal to offshore marine	**all**	**4**
Niedersachsen Group				
Coevorden Formation	Brownish-grey marly claystone	Lacustrine, restricted marine, lagoonal	**LSB**	**3**
Weiteveen Formation	Conglomerate, anhydritic marl, limestone	Estuarine, lacustrine, marginal marine	**LSB**	**2, 3**
Scruff Group				
Lutine Formation	Claystones and bituminous claystone	Open marine and restricted marine	**DCG, TB**	**3**
Scruff Greensand Formation	Bioturbated glauconitic sandstone	Shallow marine, shoreface	**DCG, TB**	**3**
Skylge Formation	Sandstone and claystone	Shallow marine and coastal	**DCG, TB**	**2**
Kimmeridge Clay Formation	Silty claystone with dolomite streak	Offshore distal marine	**DCG**	**1, 2**
Schieland Group, Delfland Subgroup				
Zurich Formation	Volcaniclastic, sandstone, claystone, coal	Lacustrine, coastal and deltaic	**all but LSB**	**2, 3**
Breeveertien Formation	Sandstone, siltstone, claystone and coal	Fluvial plain and coastal plain	**BFB**	**2, 3, 4**
Nieuwerkerk Formation	Sandstone, siltstone, claystone and coal	Fluvial plain and coastal plain	**WNB**	**2, 3, 4**
Schieland Group, Central Graben Subgroup				
Upper Graben Formation	Sandstone and coal	Deltaic	**DCG**	**1**
Middle Graben Formation	Sandstone, siltstone, claystone and coal	Deltaic	**DCG**	**1**
Lower Graben Formation	Sandstone, siltstone, claystone and coal	Fluvial plain and coastal plain	**DCG**	**1**
Puzzle Hole Formation	Carbonaceous claystone thin sandstone	Deltaic	**DCG**	**1**
Friese Front Formation	Claystone, siltstone and sandstone	Fluvial plain	**DCG, TB**	**1, 2**
Igneous rock group (inf.)				
Zuidwal Volcanic Formation	Massive volcanic rock and brecciated volcanic agglomerate	N/A	**VB**	N/A

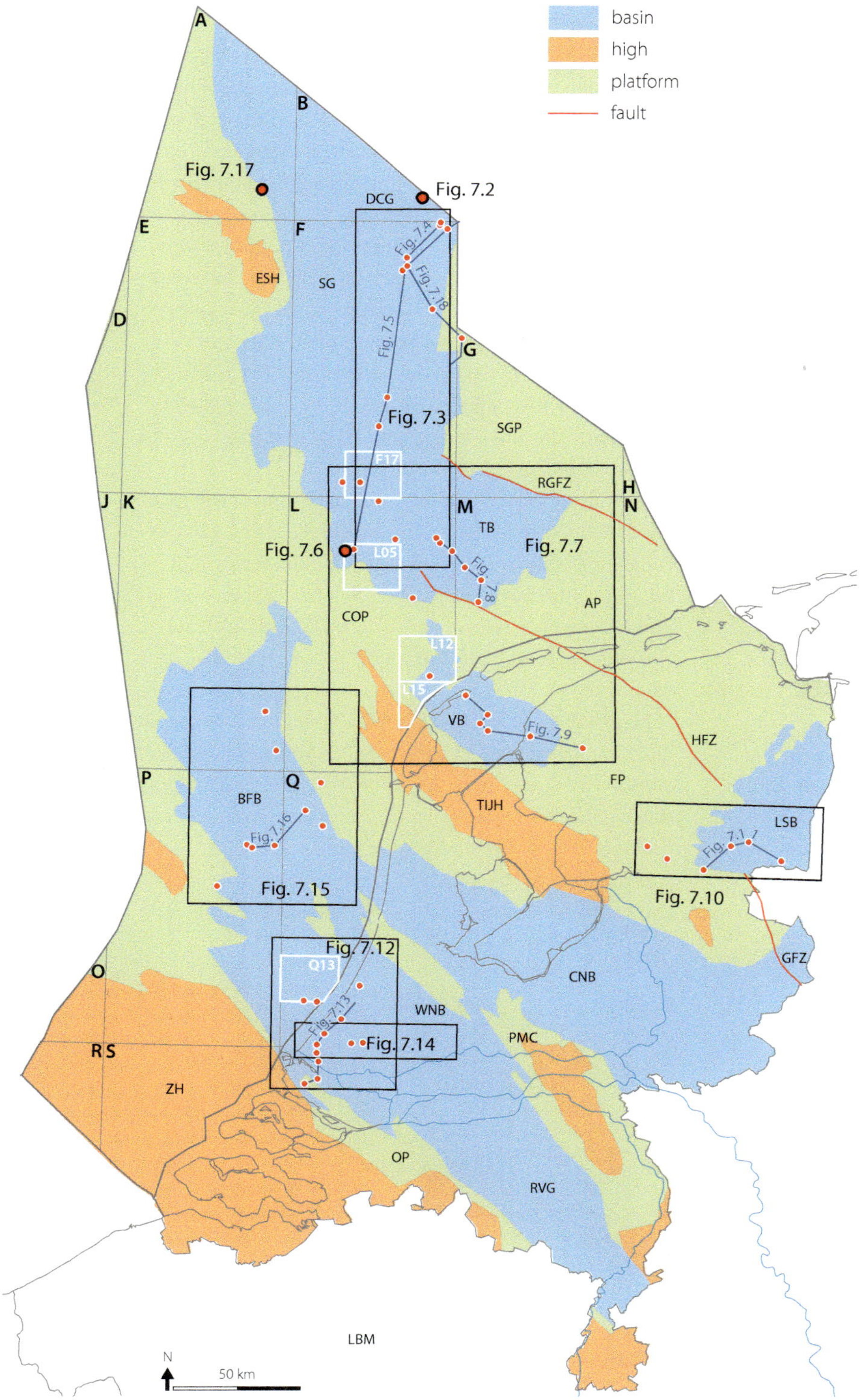

Figure 7.1. *Map of the Dutch on- and offshore displaying the structural elements. The structural elements are after Kombrink et al. (2012). Basins: BFB = Broad Fourteens Basin, CNB = Central Netherlands Basin, DCG = Dutch Central Graben, LSB = Lower Saxony Basin, SG = Step Graben, TB = Terschelling Basin, VB = Vlieland Basin, WNB = West Netherlands Basin. Highs: ESP = Elbow Spit High, LBM = London-Brabant Massif, TIJH = Texel-IJsselmeer High, PMC = Peel-Maasbommel Complex, ZH = Zeeland High. Platforms: AP = Ameland Platform, COP = Central Offshore Platform FP = Friesland Platform, OP = Oosterhout Platform. Fault Zones: RGFZ = Rifgronden Fault Zone, HFZ = Hantum Fault Zone, GFZ = Gronau Fault Zone. Thin lines with small red bullets indicate the position of well correlation panels as shown in the indicated figures. Boxes with figure numbers show the area with Wheeler diagrams for the Dutch Central Graben, Terschelling, Vlieland, Broad Fourteens, West Netherlands and the Lower Saxony basins. Large red bullets indicate the position of wells for which core material is shown in the corresponding figures.*

and have therefore been explored for these resources since the early 1960s. Most of the present-day knowledge is obtained from hydrocarbon exploration. From 2007 onward, geological characterization for geothermal energy in the Netherlands has taken off and this has resulted in renewed research activity on the Late Jurassic-Early Cretaceous interval of the onshore.

Many aspects of the Late Jurassic-Early Cretaceous geology of the Netherlands have been documented in the chapters on the Jurassic (Wong, 2007) and Cretaceous (Herngreen & Wong, 2007) in the previous edition of Geology of the Netherlands. The most comprehensive account, still very useful today, is in Van Adrichem Boogaert and Kouwe (1993). The lithostratigraphy of the interval established in Van Staalduinen et al. (1979) and Van Adrichem Boogaert and Kouwe (1993) was updated in Munsterman et al. (2012). Results of geological mapping programs were published in Duin et al. (2006) and Kombrink et al. (2012) and recent seismic intrepretation, biostratigraphic analyses, stratigraphic correlation, structural analyses and mapping results were published by Bouroullec et al. (2018), and Verreussel et al. (2018), who provide a robust tectonostratigraphic framework of the rift basins in the northern Dutch offshore.

The regional geology of the Late Jurassic-Early Cretaceous is documented in atlasses such as the Millennium Atlas (Evans et al., 2003) and the Southern Permian Basin Atlas (Doornenbal & Stevenson, 2010). Fraser et al. (2003) and Copestake et al. (2003) provided a comprehensive account of the petroleum geology of the Late Jurassic and Early Cretaceous from the central and northern North Sea and Coward et al. (2003) and Zanella and Coward (2003) described the tectonic evolution and structural framework in detail. Lott et al. (2010) and Vejbæk et al. (2010) discussed the regional development of the Jurassic to Cretaceous of the Southern Permian Basin, stretching from southeast England to Poland, while the geology of the Danish Jurassic was dealt with extensively in a series of papers in a special publication of the Geological Survey of Denmark and Greenland: Andsbjerg (2003), Andsbjerg & Dybkjaer (2003), Johannessen (2003). Verreussel et al. (2018) described the stepwise basin evolution during the Late Jurassic-Early Cretaceous and linked the geology of the Danish Upper Jurassic-Lower Cretaceous to that of the Dutch Central Graben. Jeremiah et al. (2010) provided a sequence stratigraphic framework and paleogeographic maps of the Lower Cretaceous from the southern North Sea and Patruno et al. (2021) integrated the stratigraphy of the entire northern, central and southern North Sea in a tectonostratigraphic framework. Another tectonostratigraphic framework was provided by Van Buchem et al. (2018), who described the depositional history of the Cretaceous-Danian of the Danish Central Graben in detail and compared it to that in the German and Dutch offshore sectors. Recently, publications on the geology of the German sector appeared, for example Arfai et al. (2014), Arfai & Lutz (2018) and Müller et al. (2023).

The role of salt movement in the depositional history of the Dutch Central Graben and surrounding areas has become much clearer in recent years with the recognition of e.g. turtle structures in the Dutch Central Graben and adjacent areas (Ten Veen et al., 2012; Bouroullec et al., 2018; Van Winden et al., 2018; Bouroullec & Ten Veen, 2025, this volume).

Publications focusing on biostratigraphy are primarily based on palynology: Abbink (1998); Abbink et al. (2001a, 2006); Herngreen and Wong (1989), Herngreen et al. (2000) and Verreussel et al. (2018), a Late Jurassic-Early Cretaceous ostracod stratigraphy was established in Witte and Lissenberg (1994), while nannofossil stratigraphy combined with palynology was established for the Late

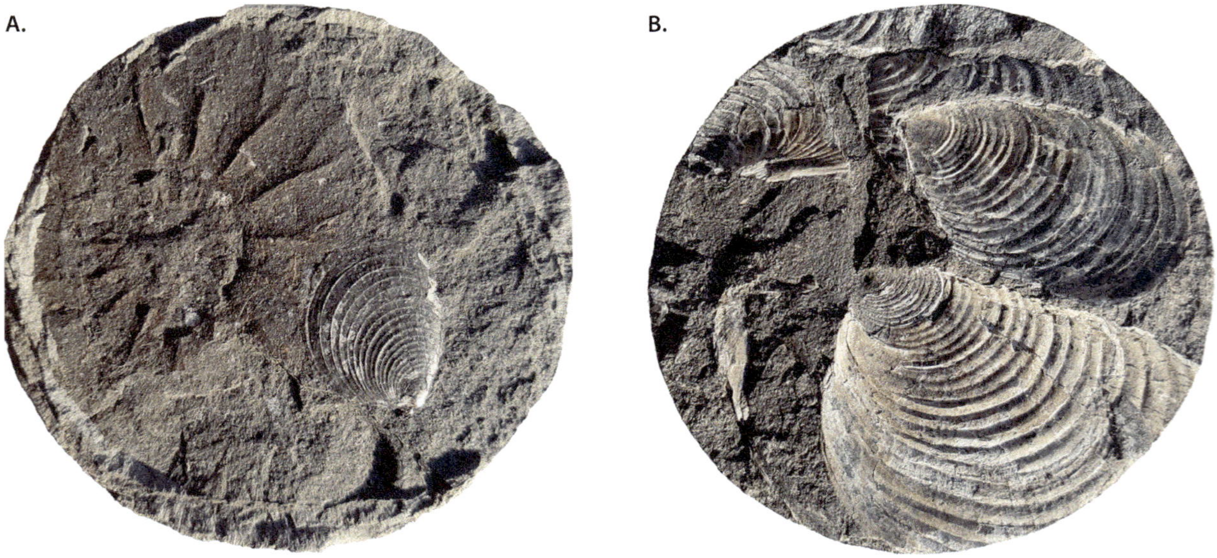

Figure 7.2. Photographs of macrofossils from the cored interval 2242-2260 m of well B18-02, northern part of the Dutch Central Graben. The cored section is part of the stratotype section (2225-2315 m) of the late Ryazanian Lutine Formation (Munsterman et al., 2012). a) Core piece from a depth of 2247.95 m showing the ammonite Lynnia icenii *and the bivalve* Buchia volgensis *from the late Ryazanian icenii ammonite zone (from Janssen et al., 2022). b) Core piece from a depth of 2256.73 m showing two well-preserved specimens of the bivalve* Buchia volgensis *(from Janssen et al., 2022). Core diameter approximately 75 mm.*

Jurassic-Early Cretaceous of the West Netherlands Basin by Jeremiah et al. (2010). Rare ammonite occurrences in the Upper Jurassic-Lower Cretaceous strata of the Netherlands are described in Abbink et al. (2001a) and Janssen et al. (2022; see also Fig. 7.2)

Key publications dealing with geological aspects of the Late Jurassic-Early Cretaceous are often only limited to one of the Dutch sub-basins and include: Geluk et al. (1995) on the Roer Valley Graben, Den Hartog Jager (1996), DeVault and Jeremiah (2002), Donselaar et al. (2015), Racero-Baena and Drake (1996), Vondrak et al., (2018), Willems et al. (2017a,b,c), Willems et al. (2020); Mijnlieff (2020) on the West Netherlands Basin, Abbink et al. (2006); Bouroullec et al. (2018) on the Dutch Central Graben, Step Graben and Terschelling Basin, Abbink et al. (2001a) on the Terschelling Basin; Beha et al. (2008); Best et al. (1983) on the Horn Graben, Wolf et al. (2018) and Heim et al. (2013) on the Schill Grund Platform, Herngreen et al. (1991), on the Vlieland Basin and Vis et al. (2018) and Rutten et al. (2020) on the Lower Saxony Basin.

Publications focusing on petroleum geological aspects of the Late Jurassic-Early Cretaceous are De Jager et al. (1996), Verweij and Simmelink (2002), De Jager and Geluk (2007), and Doornenbal et al. (2019).

Regional setting

Tectonics

The assembly of Pangea in the Paleozoic was followed by Mesozoic break-up and the development of a large rift system in Northwest Europe (Ziegler, 1990a,b; Coward et al., 2003; Zanella & Coward, 2003). Rifting started in the Middle Triassic and resulted in large and relatively wide graben structures in the North Sea and in the Norwegian Sea that were filled with predominantly non-marine sediments (Geluk, 2007; McKie & Kilhams, 2025, this volume). During Triassic rifting, salt from the Zechstein became mobile and started to influence deposition of synrift sequences in the Netherlands (De Jager, 2007; Bouroullec & Ten Veen, 2025, this volume). After a period of relative quiescence, most rift structures were blanketed with fine-grained sediments of the Lower Jurassic Altena Group (Trabucho Alexandre & Wong, 2025, this volume). During much of the Middle Jurassic, thermal doming in the North Sea affected large areas of the Dutch onshore and offshore causing regional uplift and erosion (Underhill & Partington, 1993; Husmo et al., 2003), but deposition continued in the West Netherlands Basin as it was too far away from the thermal dome area to be affected (Wong, 2007).

By Late Jurassic-Early Cretaceous times, rifting resumed in the entire North Sea area. This phase, known as the late Cimmerian rift phase (Kombrink et al., 2012) ceased in the Early Cretaceous and never developed into ocean spreading in the North Sea area (Coward et al., 2003). A failed rift complex was established in the region, comprising three branches that diverge from a triple junction in the central North Sea (Zanella et al., 2003). The southernmost branch, referred to as the Central Graben, runs across the Danish and German offshore into the Dutch offshore where it terminates against the Central Offshore Platform in the L05 block (Bouroullec et al., 2018; Verreussel et al., 2018). The Late Cimmerian rift phase developed in discrete steps, which are related to changes in the tectonic regime. In the late Kimmeridgian, the extensional regime changed from E-W towards NE-SW, related to the opening of the North-East Atlantic (Zanella et al., 2003). This long-field change in the tectonic regime had a profound impact on the evolution of the Mesozoic basins in the North Sea area. In the Netherlands and Denmark for example, basins that inherited a NW-SE early Paleozoic structural grain became active and started to accumulate thick sequences of Upper Jurassic-Lower Cretaceous deposits (Verreussel et al., 2018). This phase was characterized by active faulting and, where Zechstein salt was present by halokinesis. Deep water sediments of e.g. the Kimmeridge Clay and Farsund formations occur in large parts of the Central Graben, including the Dutch Central Graben. The phase of maximum rift activity was succeeded by a phase of transition and readjustment at the Jurassic-Cretaceous boundary (late Volgian and Ryazanian). During this transitional phase, fault movement decreased, shoulders of rotated fault blocks were eroded and the remaining depocentres were filled up to sea level and rifting came to an end. This phase is characterized by widespread erosion, which led to hiatuses and unconformities. Subsidence however, continued in most basins due to thermal sag and thick accumulations of marine shales and shoreface complexes developed during the Valanginian to Barremian post-rift sag phase.

The Late Cimmerian rift phase was accompanied by active volcanism. Tuffaceous deposits occur in the Dutch Central Graben and the Vlieland, Broad Fourteens and West Netherlands basins, but not in the Lower Saxony Basin (Van Bergen et al., 2025, this volume). On seismic images, two structures have been unambiguously identified as volcanoes, one named Mulciber in block F16 (Bouroullec et al., 2016b , Vis, 2020; Van Bergen et al., 2025, this volume) in the Dutch Central Graben, the other named Zuidwal in the Vlieland Basin (Van Bergen et al., 2025, this volume). The latter volcano was discovered during exploration for oil and gas, well Zuidwal-1 (ZDW-01) having drilled right into the volcanic conduit (Van Bergen & Sissingh, 2007). Volcanism peaked in the Volgian-Ryazanian in the north and in the Valanginian-Hauterivian in the south (Van Bergen et al., 2025, this volume).

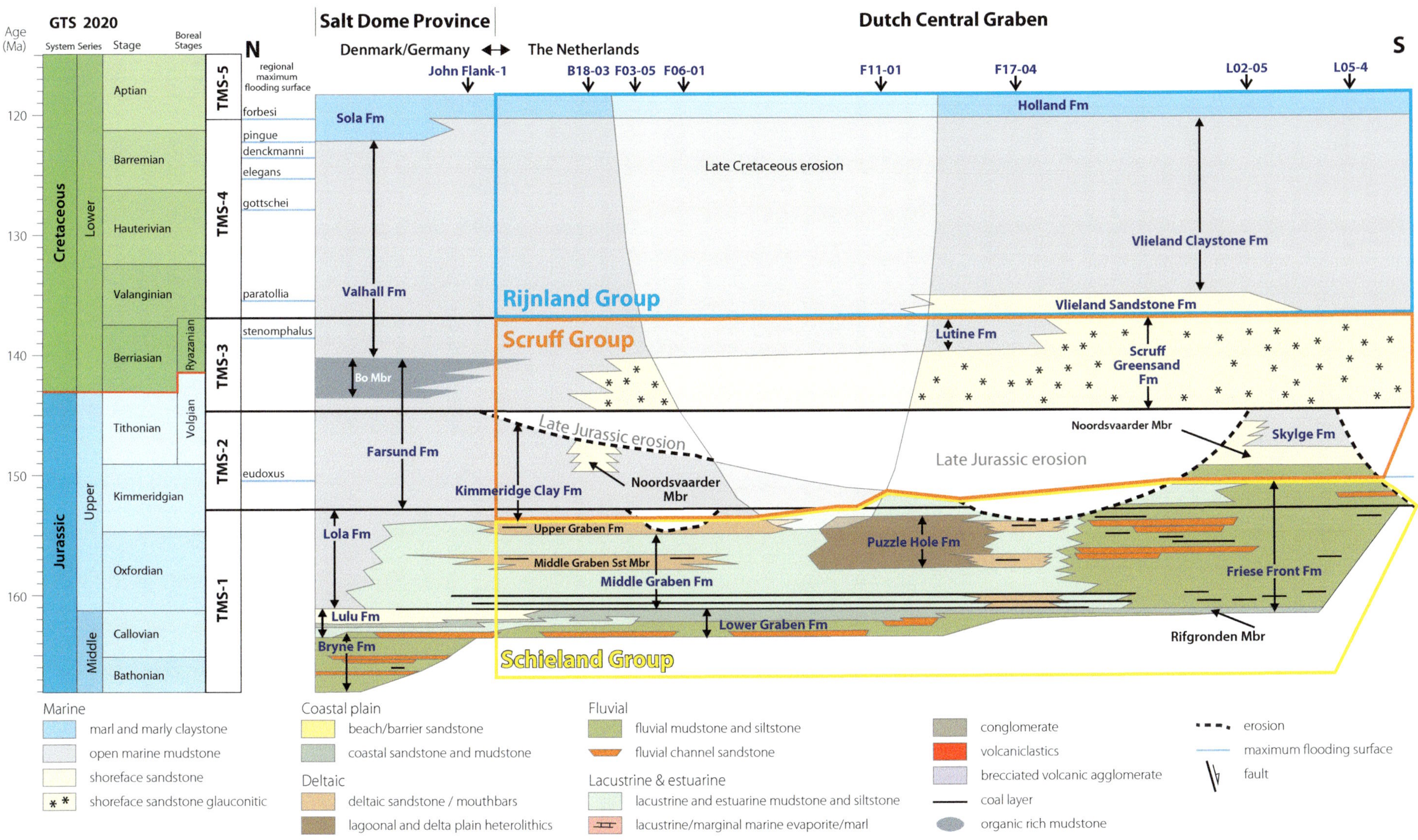

Figure 7.3. *Wheeler diagram along a N-S transect through the axis of the Dutch Central Graben and the southern tip of the Danish Salt Dome Province (see Fig. 7.1 for location). The diagram displays the generalized time and facies relationships of the Bathonian (Middle Jurassic) to Aptian (Lower Cretaceous) lithostratigraphic units. Formation names are indicated in blue, member names in black and lithostratigraphic groups are outlined in yellow, orange and blue. Relevant maximum flooding surfaces are displayed with blue lines. The black horizontal lines indicate tectonostratigraphic megasequences (TMS, see text for explanation). Key wells are shown at the top of the diagram.*

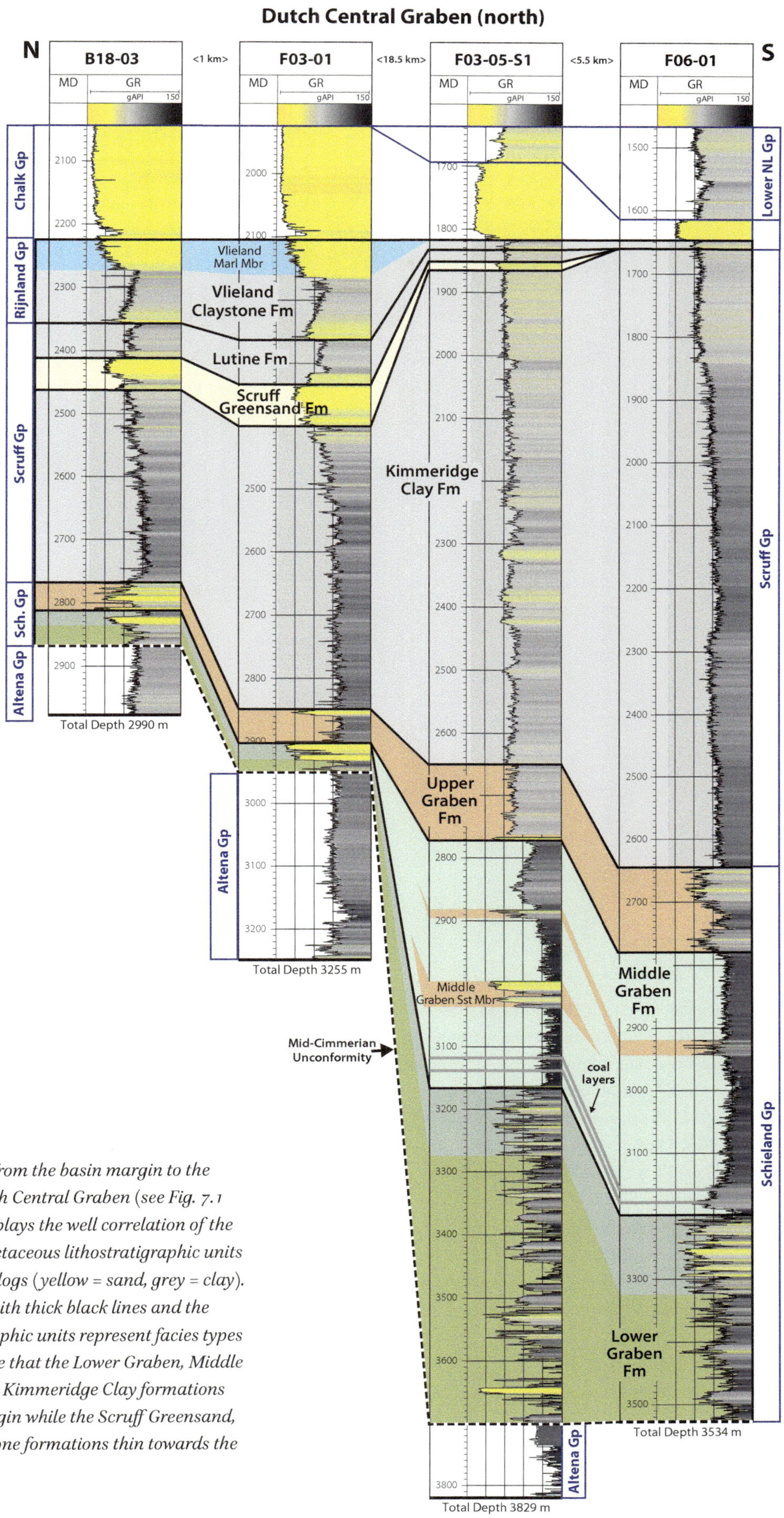

Figure 7.4. *N-S well panel from the basin margin to the centre of the northern Dutch Central Graben (see Fig. 7.1 for location). The panel displays the well correlation of the Upper Jurassic to Lower Cretaceous lithostratigraphic units based on gamma-ray (GR) logs (yellow = sand, grey = clay). Formations are indicated with thick black lines and the colours of the lithostratigraphic units represent facies types (see legend of Fig. 7.3). Note that the Lower Graben, Middle Graben, Upper Graben and Kimmeridge Clay formations thin towards the basin margin while the Scruff Greensand, Lutine and Vlieland Claystone formations thin towards the basin centre.*

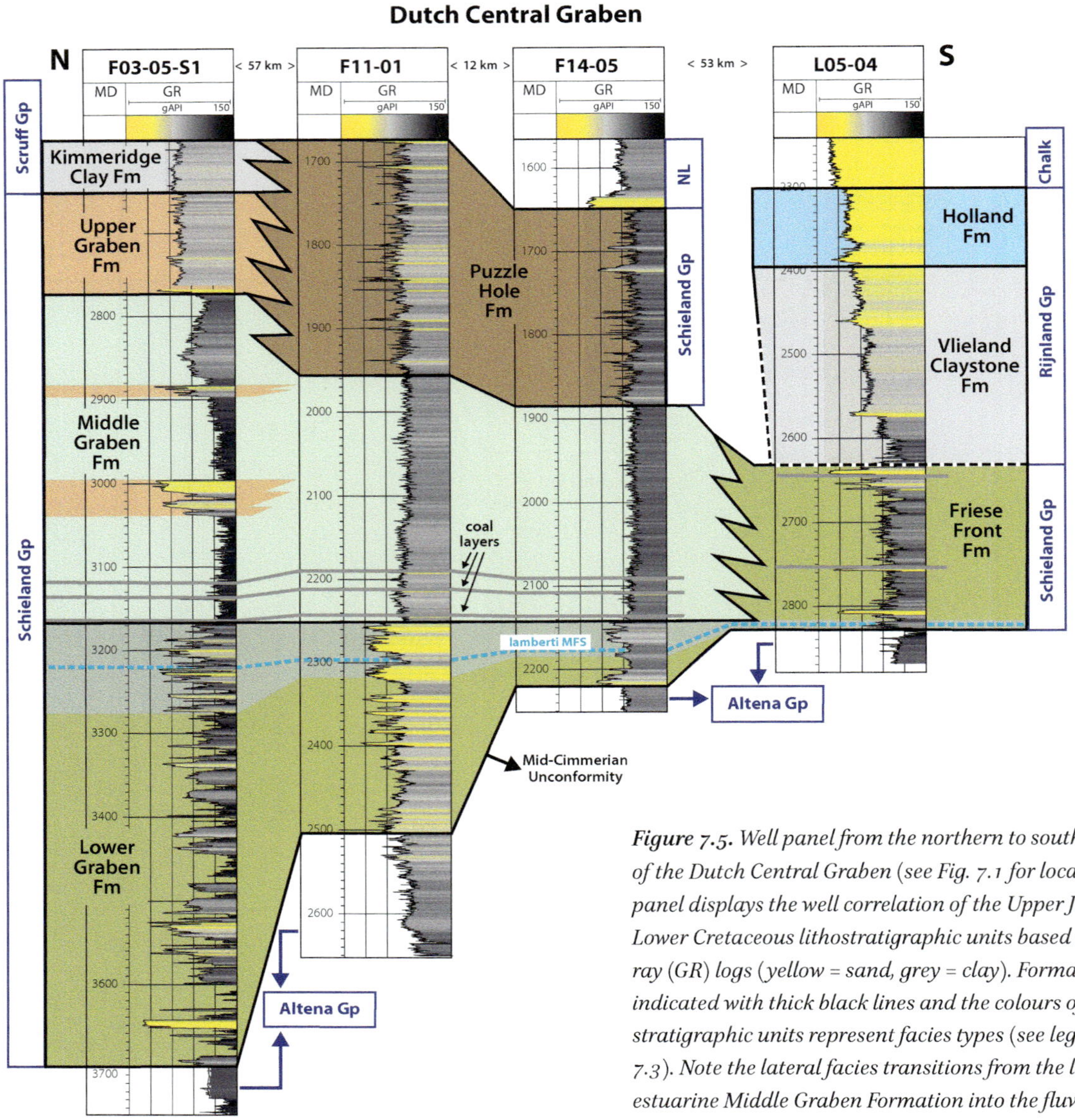

Figure 7.5. *Well panel from the northern to southern part of the Dutch Central Graben (see Fig. 7.1 for location). The panel displays the well correlation of the Upper Jurassic to Lower Cretaceous lithostratigraphic units based on gamma-ray (GR) logs (yellow = sand, grey = clay). Formations are indicated with thick black lines and the colours of the litho-stratigraphic units represent facies types (see legend of Fig. 7.3). Note the lateral facies transitions from the lacustrine/estuarine Middle Graben Formation into the fluvial Friese Front Formation and from the open marine Kimmeridge Clay Formation into the deltaic Puzzle Hole Formation. Also note the continuity of the three coal layers at the base of the Middle Graben Formation. NL = Lower North Sea Group.*

Late Jurassic-Early Cretaceous basins

In the Netherlands, nine basins contain Upper Jurassic-Lower Cretaceous strata (Fig. 7.1). Four basins are located in the Dutch offshore: the Step Graben, the Dutch Central Graben, the Terschelling Basin and the Broad Fourteens Basin. Three basins are located onshore: the Lower Saxony Basin, the Central Netherlands Basin and the Roer Valley Graben, while the Vlieland and West Netherlands basins extend from the onshore into the offshore (Fig. 7.1). Each of the nine basins displays a distinct stratigraphy and basin fill history, but overall the basins share large scale eustatic, climatic and tectonic evolutionary trends.

The Dutch Central Graben

The Dutch Central Graben is located in the northern Dutch offshore (Fig. 7.1) and runs from the B quadrant through the F quadrant to the L quadrant, where it terminates against the Central Offshore Platform. It forms the southernmost extension of the Late Jurassic rift that runs from the Viking Graben in the northern North Sea via the Tail End Graben in the Danish offshore and the German Central Graben, to the Dutch Central Graben. The structural grain of the basin is inherited from Triassic graben structures and is narrow and elongated (40 by 160 km), bounded by N-S trending faults. These faults originated in the Triassic and are associated with salt diapirs and salt walls (Bouroullec & Ten Veen, 2025, this volume) that had a major impact on the structural style and sedimentary fill of the basin. The importance of salt tectonics in the evolution of the Dutch Central Graben is expressed by the widespread occurrence of turtle structures and rim synclines (Bouroullec et al., 2018; Bouroullec & Ten Veen, 2025, this volume). Sediments from the entire Late Jurassic-Early Cretaceous interval are represented in the stratigraphic record of the basin, in the form of non-marine, estuarine and deltaic deposits of the Schieland Group (Central Graben Subgroup), marine deposits of the Scruff Group and open marine deposits of the Rijnland Group (Figs 7.3-7.5). Claystones of the Altena Group, including the Posidonia Shale Formation, subcrop underneath the Upper Jurassic (Kombrink et al., 2012; see also Fig. 7.6), while the overlying Upper Cretaceous Chalk Group is very thin or absent due to mild inversion during the Late Cretaceous.

The Step Graben

The Step Graben (SG) is situated in the northern Dutch offshore along the western side of the Dutch Central Graben (Fig. 7.1). Despite its name and in contrast to the Dutch Central Graben, it never developed into an actively subsiding basin during the Late Jurassic-Early Cretaceous and should be considered a platform area instead. The basin configuration was inherited from the Triassic early rifting expressed by N-S trending salt structures and faults (Kombrink et al., 2012; Bouroullec & Ten Veen, 2025, this volume). The Step Graben terminates against the El-

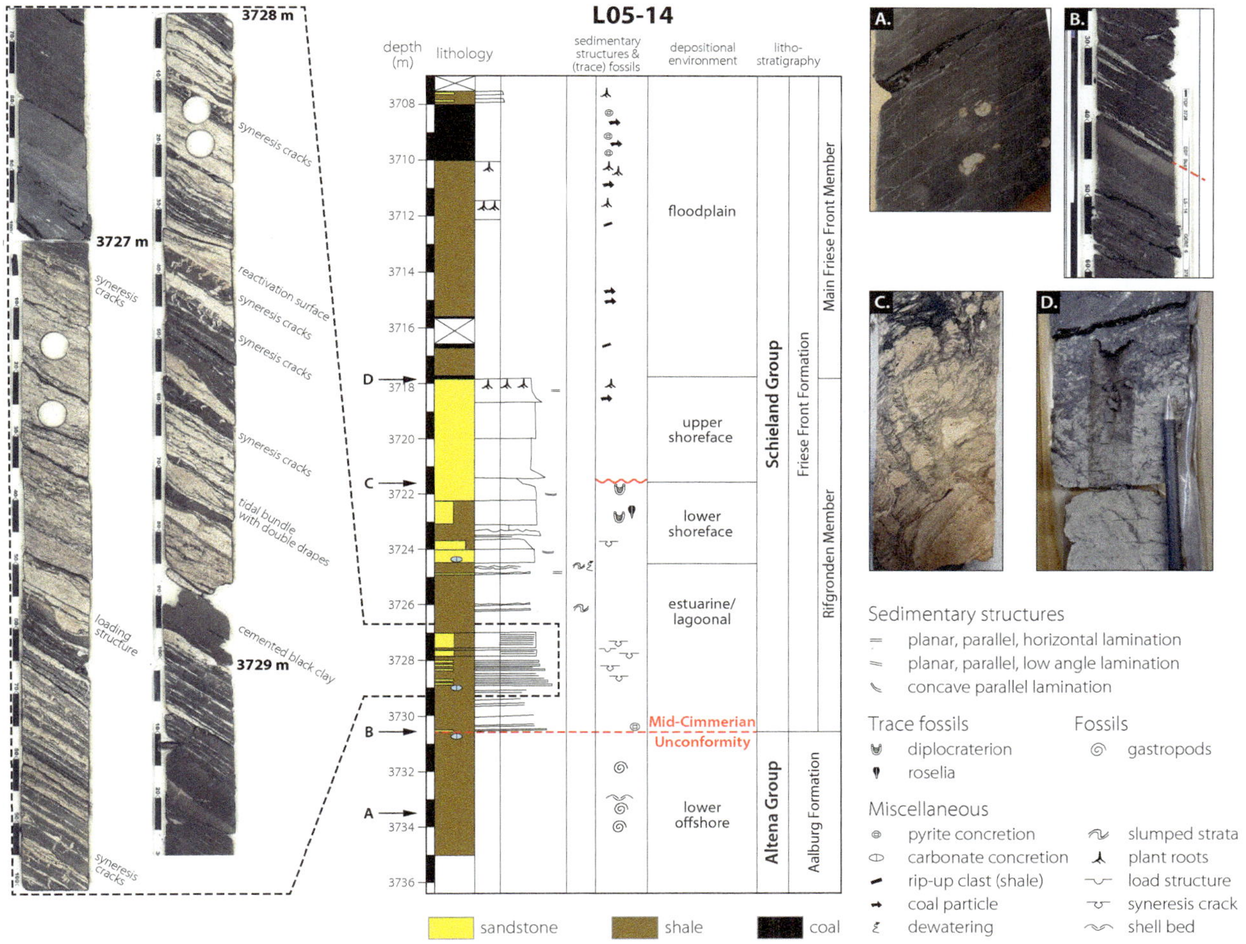

Figure 7.6. Lithological log of the cored interval 3608-3735 m of well L05-14 from the southern part of the Dutch Central Graben. The log displays the lithology, the observed sedimentary structures, the interpreted paleoenvironment and the lithostratigraphy. The far left shows a core photograph of interval 3726.65-3729.30 m with sedimentary structures annotated. To the right, more detailed core sections are shown (a-d) and their positions are indicated on the left-hand side of the lithological log. a) Cored section from a depth of 3733.6 m showing gastropod fossils within a matrix of laminated claystones from the Altena Group, b) Cored section showing the Mid-Cimmerian unconformity, indicated with a red dashed line at 3730.36 m. The unconformity separates the latest Callovian Rifgronden Member of the Friese Front Formation from the Lower Jurassic (Toarcian or older) Aalburg Formation of the Altena Group (see Trabucho-Alexandre & Wong, 2025, this volume), c) Core piece from a depth of 3721.5 m showing Diplocraterion ichnofossils. The vertical U-shaped burrows are characteristic for upper shoreface paleoenvironments, d) Core piece from a depth of 3717.9 m showing traces of plant rootlets. The rootlets indicate a transition from a marine upper shoreface paleoenvironment to a non-marine vegetated floodplain paleoenvironment.

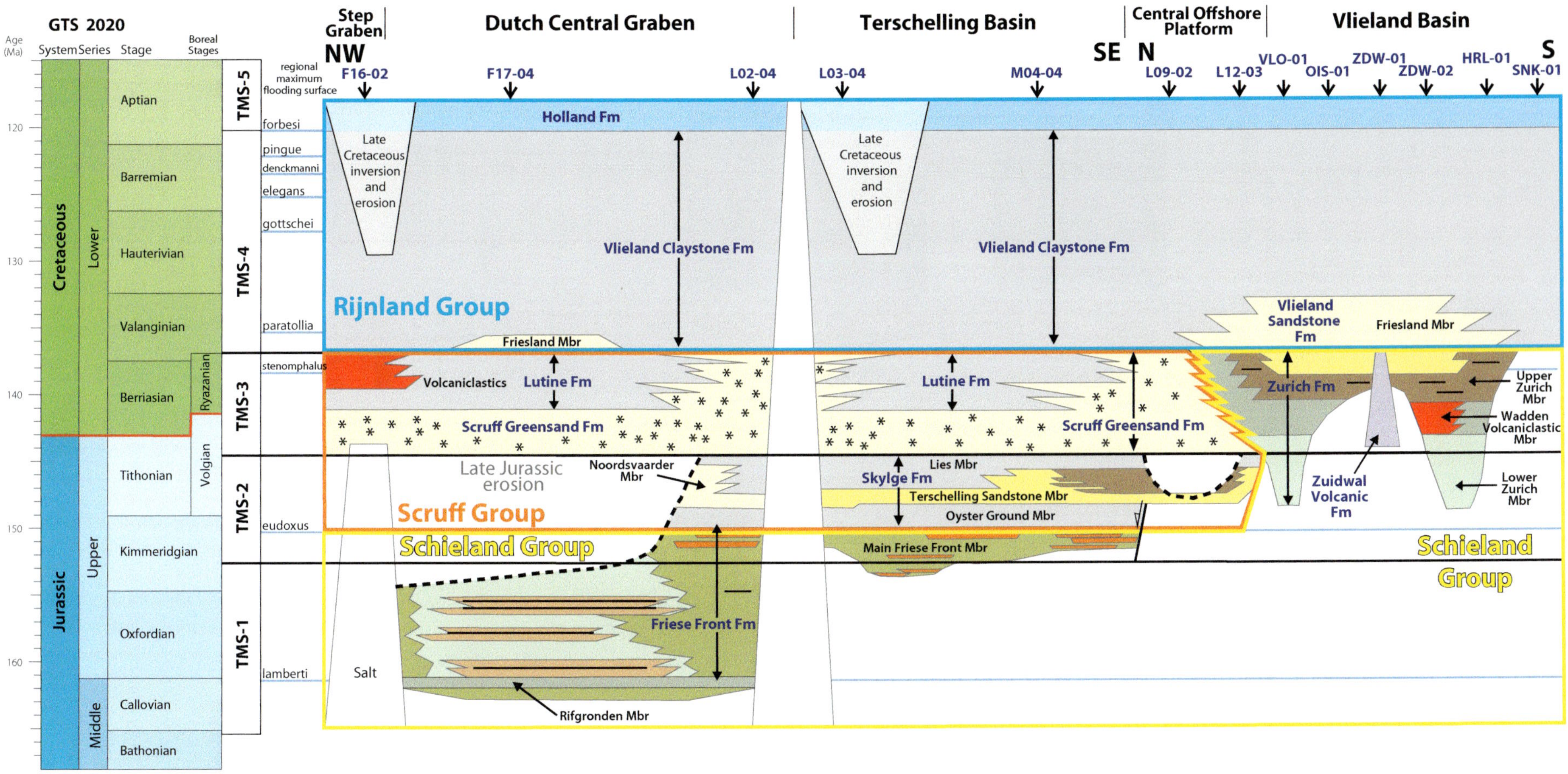

Figure 7.7. Wheeler diagram along a NW-SE transect through the southern Dutch Central Graben, the Terschelling Basin and the Vlieland Basin (see Fig. 7.1 for location). The diagram displays the generalized time and facies relationships of the Callovian (Middle Jurassic) to Aptian (Lower Cretaceous) lithostratigraphic units. Formation names are indicated in blue, member names in black and lithostratigraphic groups are outlined in yellow, orange and blue. Relevant maximum flooding surfaces are displayed with blue lines. The black horizontal lines indicate tectonostratigraphic megasequences (TMS, see text for explanation). Key wells are shown at the top of the diagram. For legend see Fig. 7.3.

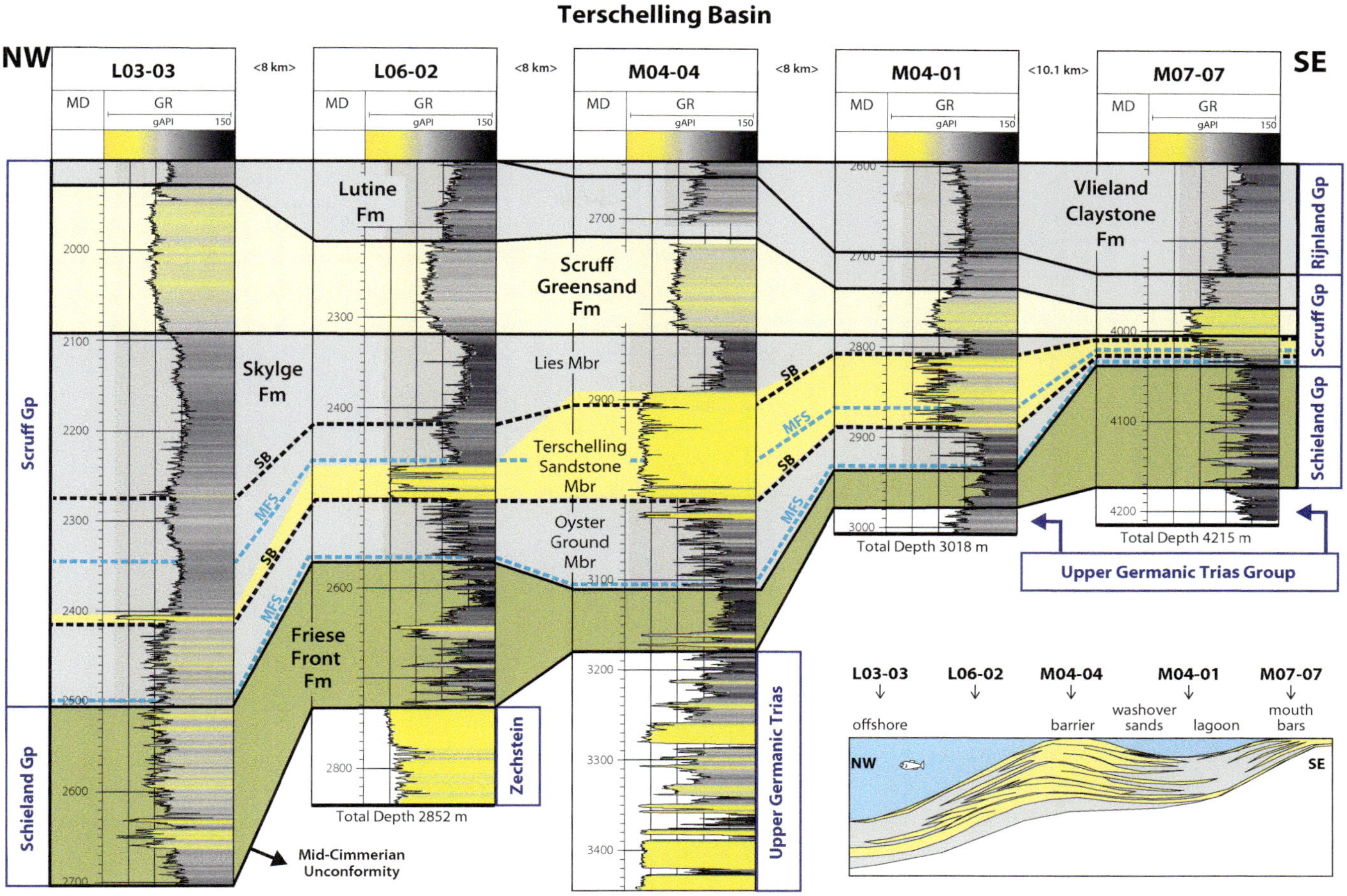

Figure 7.8. NW-SE well panel through the Terschelling Basin (see Fig. 7.1 for location). The panel displays the well correlation of the Upper Jurassic to Lower Cretaceous lithostratigraphic units based on gamma-ray (GR) logs (yellow = sand, grey = clay). Formations are indicated with thick black lines and the colours of the lithostratigraphic units represent facies types (see legend of Fig. 7.3). The blue dashed lines represent maximum flooding surfaces (MFS) and the black dashed lines represent sequence boundaries (SB). The thick amalgamated sandstones of the Terschelling Sandstone Member in M04-04 are interpreted as a stacked shoreface succession, possibly representing a coastal barrier system. Towards the northwest, the upper part of the Terschelling Member shales out rapidly. Towards the southwest, the spikey gamma-ray trend of well M04-04 is interpreted as a heterolithic back-barrier lagoonal succession (see inset diagram).

bow Spit High to the west and is connected to the Outer Rough Basin in the Danish offshore (Japsen et al., 2003) via the German Entenschnabel area (Müller et al., 2023). The Upper Jurassic-Lower Cretaceous sedimentary cover is thin due to a combination of limited accommodation space and erosion on the shoulders of the main rift system. The overlying Chalk Group is very thick (Van der Molen, 2004) because the Step Graben was not inverted during the Late Cretaceous. Below the Upper Jurassic-Lower Cretaceous, a relatively thin Upper Germanic Trias Group and a relatively thick Lower Germanic Trias Group are present (Kombrink et al., 2012). The Late Jurassic-Early Cretaceous Kimmeridgian to Ryazanian time interval is mainly represented by the Scruff Greensand and Lutine formations of the Scruff Group (Verreussel et al., 2015) with occasional presences of the Skylge Formation (Fig. 7.7).

The Terschelling Basin

The Terschelling Basin is situated in the northern L and M quadrants of the Dutch offshore (Fig. 7.1). The WNW-ESE basin trend is inherited from the middle Paleozoic (Wong, 2007). The Terschelling basin is bounded by two major fault systems; to the north by the Rifgronden Fault Zone and to the south by the Hantum Fault Zone (Duin et al., 2006; Fig. 7.1). The depocentre lies in the west, where the basin connects to the southern part of the Dutch Central Graben via a complex salt structure (Bouroullec et al., 2016a; see also Fig. 7.7). The basin is dissected by a number of NNE-SSW trending salt structures (McKie & Kilhams, 2025, this volume) and is mostly filled with sediments of late Kimmeridgian to Ryazanian age, consisting of non-marine deposits of the Schieland Group and marine deposits of the Scruff Group and the Rijnland Group (Figs 7.7, 7.8). Upper and Lower Germanic Triassic sed-

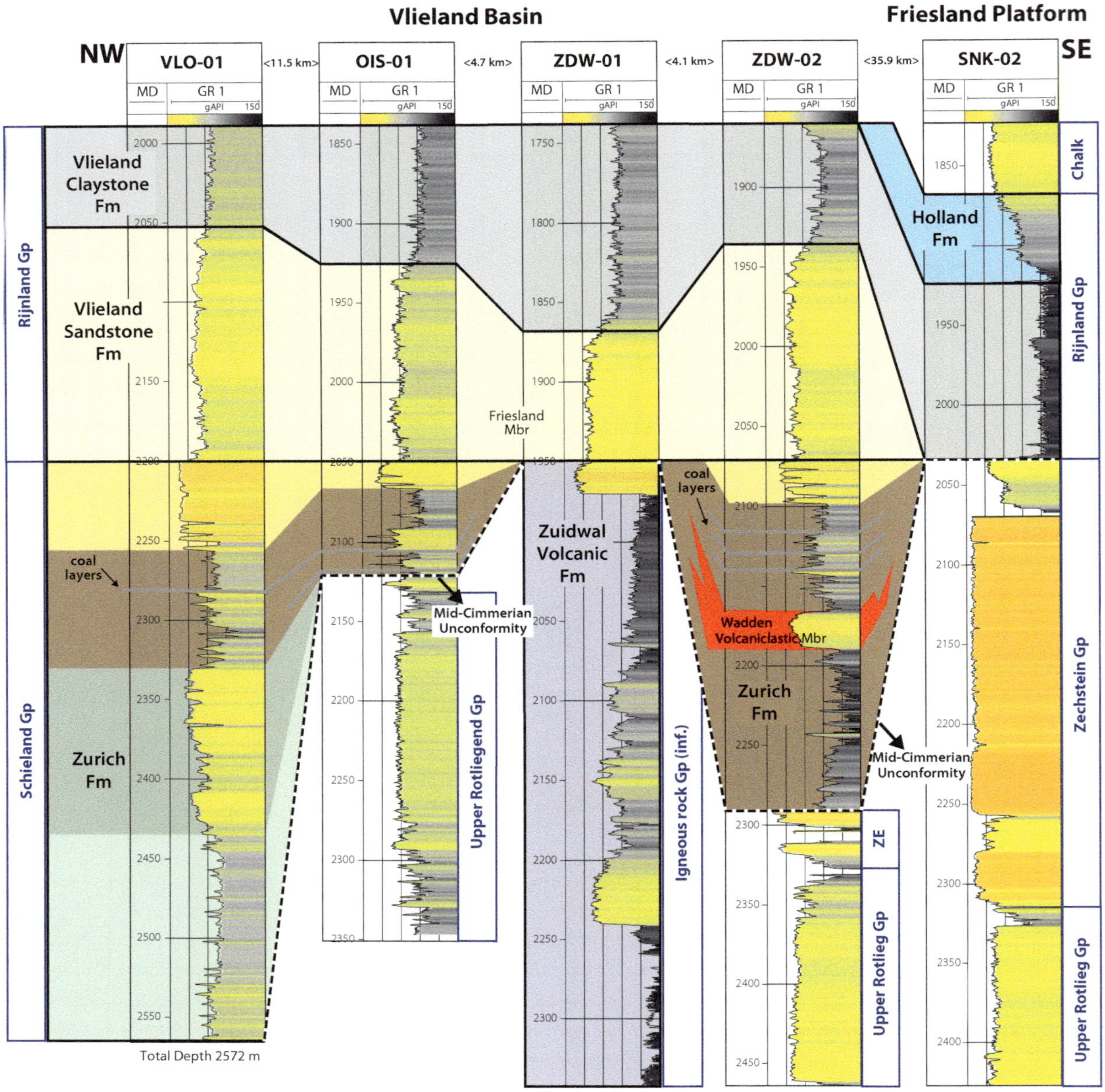

Figure 7.9. NW-SE well panel through the Vlieland Basin and on to the Friesland Platform (see Fig. 7.1 for location). The panel displays the well correlation of the Upper Jurassic to Lower Cretaceous lithostratigraphic units based on gamma-ray (GR) logs (yellow = sand, grey = clay). Formations are indicated with thick black lines and the colours of the lithostratigraphic units represent facies types (see legend of Fig. 7.3). Note that the Zurich Formation is characterized by many different facies types, including volcaniclastic sediments and coal layers.

iments lie below the interval and a thin cover of Upper Cretaceous chalk – a result of mild inversion – is situated on top (Bouroullec et al., 2016a).

The Vlieland Basin

The small Vlieland Basin trends WNW-ESE and runs from the offshore L12 and L15 blocks across the island of Vlieland via the Wadden Sea into the onshore province of Friesland (Fig. 7.1). The basin is separated from the Terschelling Basin by the Hantum Fault Zone and stands out from the other Late Jurassic-Early Cretaceous basins thanks to the prominent presence of a volcano right in its middle (Van Bergen et al., 2025, this volume; see also Fig. 7.7): The Zuidwal volcano was active in Late Jurassic-Early Cretaceous times and its tuffaceous deposits are widespread throughout the basin (Herngreen et al., 1991). The Upper Jurassic-Lower Cretaceous sequence is situated on top of the Zechstein or Upper Germanic Trias Group and is overlain by a thin cover of the Upper Cretaceous Chalk Group (Van der Molen, 2004; Duin et al., 2006; Kombrink et al., 2012). The basin is filled with predominantly marginal marine deposits of the Schieland Group associated with a thin section of the open marine Rijnland Group (Fig. 7.9).

The Lower Saxony Basin

The Lower Saxony Basin (LSB) is located onshore in the northeastern part of the Netherlands (Fig. 7.1). It trends WNW-ESE and extends into Germany where the bulk of the basin is situated, occupying an area of 300 by 80 km (Lott et al., 2010). In Germany, the basin is limited by the highs of the Pompeckj Block to the north and by the Rhenish Massif to the south (Seyfang et al., 2017) and in the Netherlands it is bordered by the Friesland Platform to the west. The Upper Jurassic-Lower Cretaceous deposits thin and wedge out onto this platform area (Fig. 7.11). Towards the south, the Gronau Fault Zone defines the boundary of the basin (Duin et al., 2006). The basin is filled with predominantly non-marine deposits of the Schieland and Niedersachsen groups and with marine deposits of the Rijnland Group (Figs 7.10, 7.11).

The Central Netherlands Basin

The Central Netherlands Basin (CNB) is a large WNW-ESE trending basin, located onshore in the central part of the Netherlands (Fig. 7.1). It was uplifted by an estimated 1500 metres during Late Cretaceous inversion (De Jager, 2003) and, as a result, the Upper Jurassic-Lower Cretaceous is only present as relatively thin erosional remnants (Duin et al., 2006) that are not discussed in detail here.

The Roer Valley Graben

The Roer Valley Graben (RVG) is a NW-SE trending basin, located onshore in the southern part of the Netherlands (Fig. 7.1). To the northwest it is connected with the West Netherlands Basin and towards Germany in the southeast it terminates against the Rhenish Massif (Kombrink et al., 2012). During Cenozoic times, the Roer Valley Graben was re-activated to connect with the Rhine Graben and it remains actively subsiding to the present day (Geluk et al., 1995; Peters & Van Balen, 2007). The basin was strongly inverted during the Late Cretaceous and as a result, most Upper Jurassic-Lower Cretaceous sediments were eroded (De Jager, 2003; Duin et al., 2006). Only in the northwestern part where the basin connects to the West Netherlands Basin, deposits from the Schieland Group are preserved.

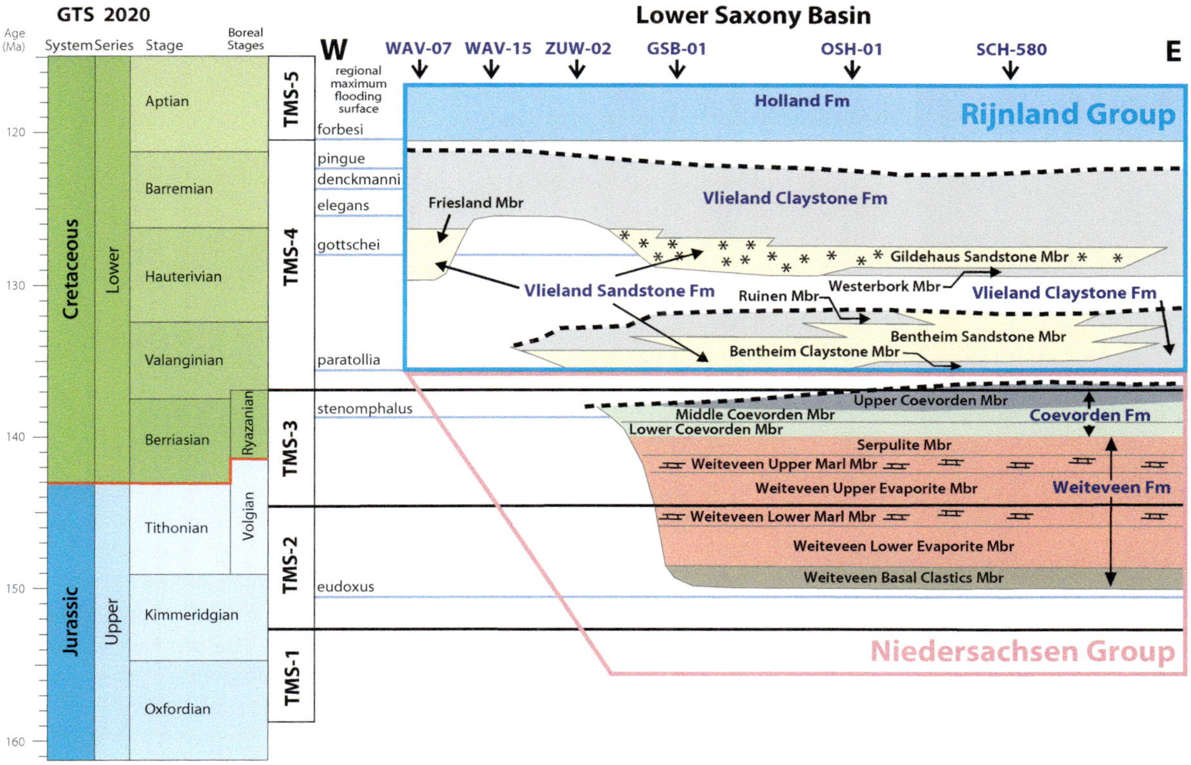

Figure 7.10. Wheeler diagram along a W-E transect through the Lower Saxony Basin (see Fig. 7.1 for location). The diagram displays the generalized time and facies relationships of the Kimmeridgian (Upper Jurassic) to Aptian (Lower Cretaceous) lithostratigraphic units. Formation names are indicated in blue, member names in black and lithostratigraphic groups are outlined in pink and blue. Relevant maximum flooding surfaces are displayed with blue lines. The black horizontal lines indicate tectonostratigraphic megasequences (TMS, see text for explanation). Key wells are shown at the top of the diagram. For legend see Fig. 7.3.

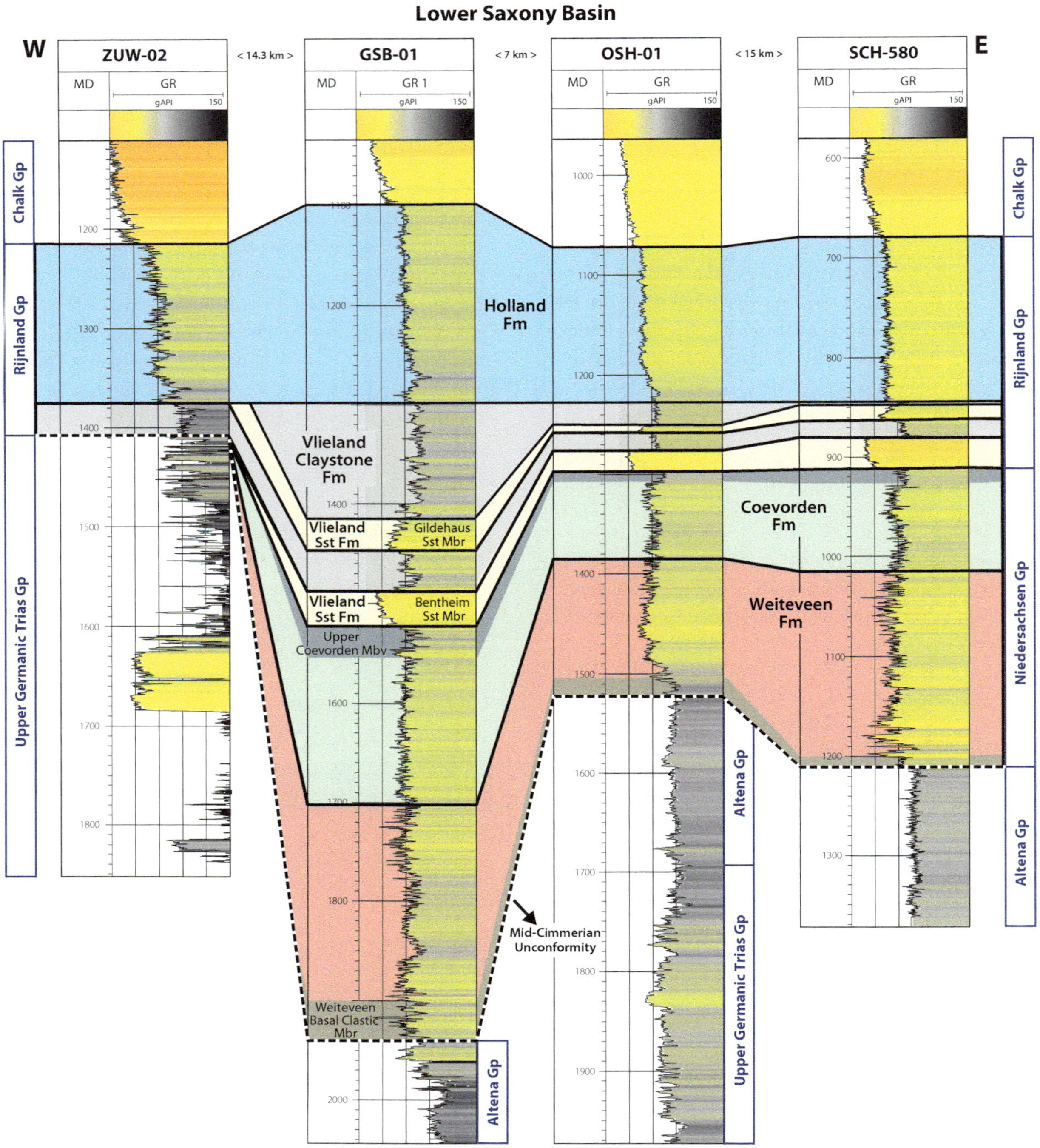

Figure 7.11. W-E well panel through the Lower Saxony Basin (see Fig. 7.1 for location). The panel displays the well correlation of the Upper Jurassic to Lower Cretaceous lithostratigraphic units based on gamma-ray (GR) logs (yellow = sand, grey = clay). Formations are indicated with thick black lines and the colours of the lithostratigraphic units represent facies types (see legend of Fig. 7.3). Two prominent sandstone units occur in the basin, the Bentheim and Gildehaus Sandstone members. The Bentheim Sandstone Member pinches out towards the west whereas the Gildehaus Sandstone Member is cut off by erosion.

The West Netherlands Basin

The West Netherlands Basin is a NW-SE trending basin that is located partially onshore and partially offshore, stretching from the provinces of Brabant and South Holland to the offshore P and Q quadrants (Fig. 7.1). The NW-SE structural grain was inherited from the Paleozoic (De Jager, 2007). The basin is virtually devoid of Zechstein Group salt (Duin et al., 2006). The basin ramps up via the intermediate Oosterhout Platform to the London-Brabant Massif in the SW and towards the NE, it is transitional into the Central Netherlands Basin. The West Netherlands Basin is dissected by an intricate pattern of faults trending

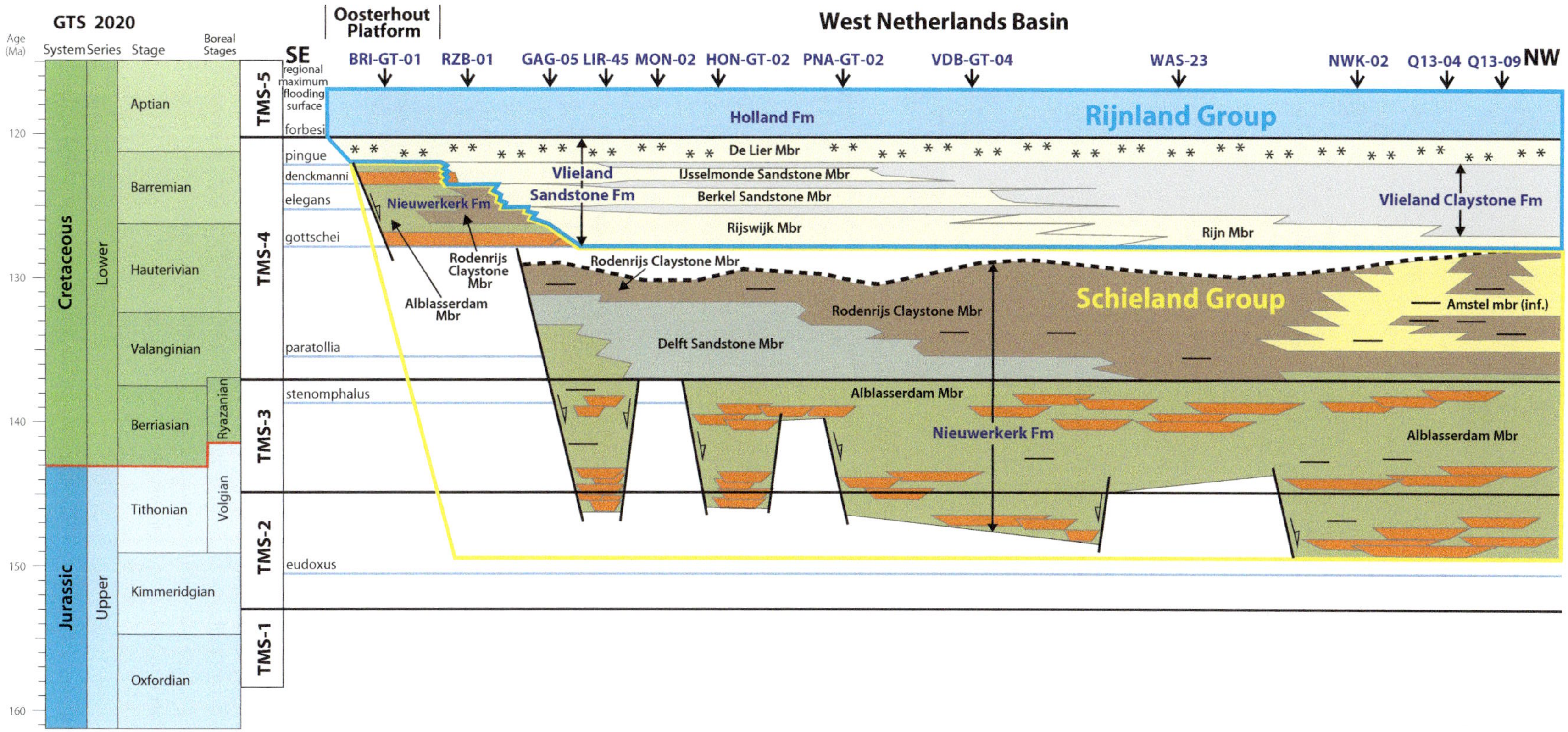

Figure 7.12. Wheeler diagram along a SE-NW transect through the West Netherlands Basin (see Fig. 7.1 for location). The diagram displays the generalized time and facies relationships of the Kimmeridigian (Upper Jurassic) to Aptian (Lower Cretaceous) lithostratigraphic units. Formation names are indicated in blue, member names in black and lithostratigraphic groups are outlined in yellow and blue. Relevant maximum flooding surfaces are displayed with blue lines. The black horizontal lines indicate tectonostratigraphic megasequences (TMS, see text for explanation). Key wells are shown at the top of the diagram. For legend see Fig. 7.3.

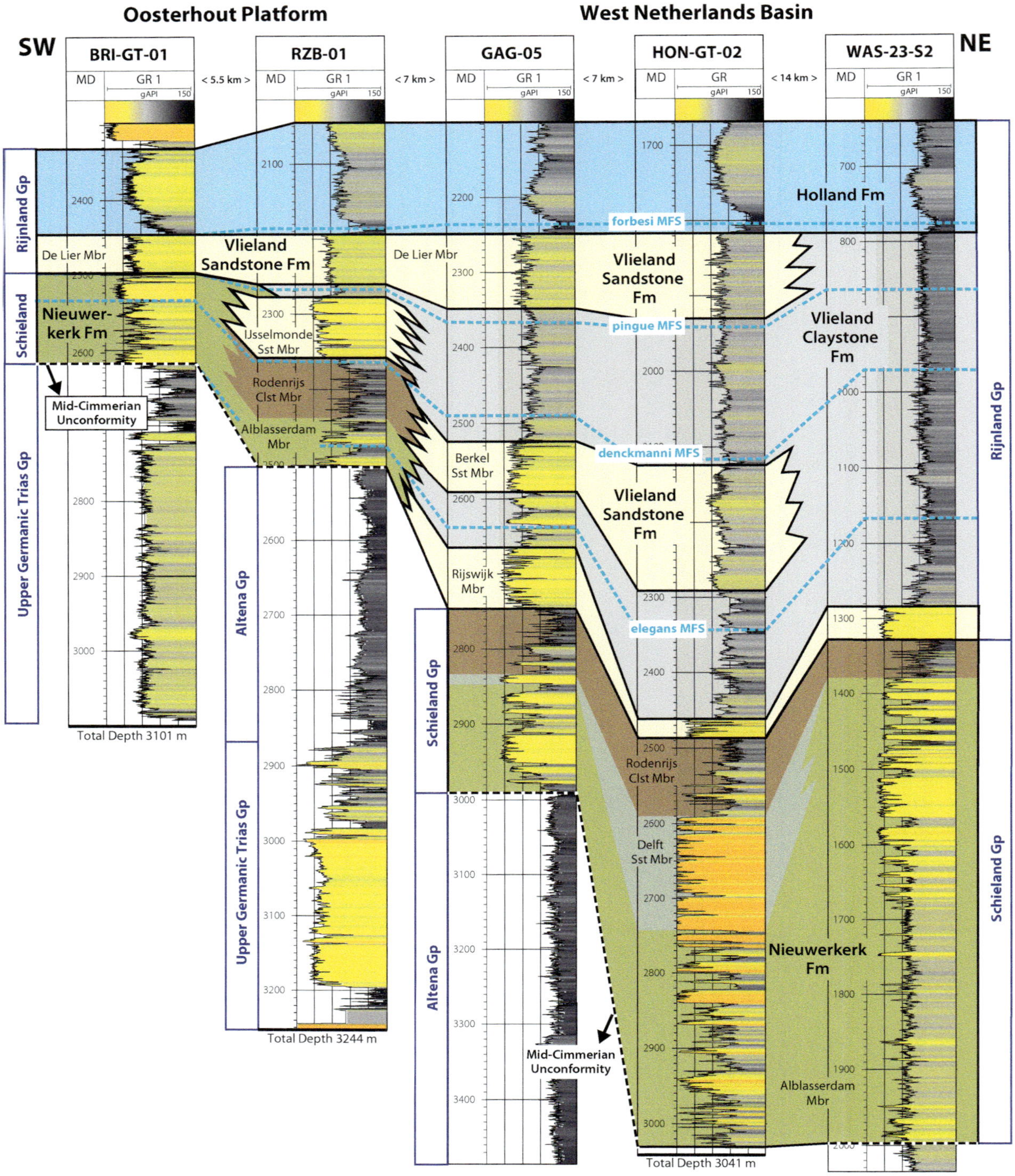

Figure 7.13. *SW-NE well panel through the West Netherlands Basin (see Fig. 7.1 for location). The panel displays the well correlation of the Upper Jurassic to Lower Cretaceous lithostratigraphic units based on gamma-ray (GR) logs (orange = coarse sand, yellow = sand, grey = clay). Formations are indicated with thick black lines and the colours of the lithostratigraphic units represent facies types (see legend of Fig. 7.3). Relevant maximum flooding surfaces are displayed with blue dashed lines. Note the lateral transition of the non-marine Nieuwerkerk Formation to the marine Vlieland Sandstone and Vlieland Claystone formations. Also note the shaling out of the IJsselmonde and Berkel Sandstone members towards the basin axis.*

NNW-SSE, NW-SE and ENE-WSW (Racero-Baena & Drake, 1996; Duin et al., 2006; also see Fig. 1.19 in De Jager et al., 2025, this volume). Due to inversion, the Upper Jurassic-Lower Cretaceous is in most parts of the basin capped by sediments of the Cenozoic North Sea Supergroup. The sequence is represented by the non-marine Schieland

Group and by the marine Rijnland Group (Figs 7.12, 7.13, 7.14).

The Broad Fourteens Basin

The Broad Fourteens Basin is located in the Dutch offshore K, L, P and Q quadrants (Fig. 7.1). Towards the south it connects to the West Netherlands Basin and towards the east, to the Central Netherlands Basin. The NNW-SSE trending structural grain of the basin is inherited from the Triassic (De Jager et al., 2025, this volume). The basin is bounded by faults and was strongly inverted in the Late Cretaceous (De Jager, 2003). As a result, much of the Upper Jurassic-Lower Cretaceous sequence has been removed by erosion from the former basin centre, where the inversion had the most profound impact (Duin et al., 2006). The thickest successions are now in the former marginal areas (De Jager, 2003). The Upper Jurassic-Lower Cretaceous succession comprises the non-marine Schieland Group and the marine Rijnland Group (Figs 7.15, 7.16).

Paleoclimate

While on a large scale, the depositional patterns of the Late Jurassic-Early Cretaceous were controlled by the tectonics of the Late Cimmerian rift phase, paleoclimate variations played an important role as well. The Jurassic Period is generally considered to have been a greenhouse world without any major glacial episodes (Jenkyns et al., 2012), but in the Middle Jurassic significantly cooler episodes, possibly including polar ice development occurred (Dera et al., 2011). Overall, the climate switched from relatively cool and humid (Callovian to Early Oxfordian) to relatively warm (middle Oxfordian to Kimmeridgian) to very warm and arid (latest Jurassic, Volgian to Early Ryazanian) and back to cool and humid again (late Ryazanian to Hauterivian) (Abbink et al., 2004a,b; Schneider et al., 2018a,b). The precise controls on these regional variations in temperature and humidity remain elusive, but the effects of both Tethyan and Boreal water-mass movements have repeatedly been considered (Dera et al., 2015; Vickers et al., 2020). In the Late Jurassic-Early Cretaceous sedimentary record of the Netherlands, continuous coal deposits developed during cooler and more humid phases. This is for instance the case for the Oxfordian Middle Graben and Friese Front formations in the Dutch Central Graben (Figs 7.3, 7.4). From the Volgian (Tithonian) onward, paleoclimate changes and warmer and more arid conditions were established. In the Lower Saxony Basin, these conditions led to strong contrasts in salinity, which are reflected in the presence of evaporites of the Weiteveen Formation (Van Adrichem Boogaert & Kouwe, 1993). In southern England, evaporites, limestones and freshwater deposits alternate in the Purbeck Formation (West, 1975). These indicate that warm and arid conditions were widespread around the Jurassic-Cretaceous boundary. In the Early Cretaceous Ryazanian (Berriasian), the warm and arid phase ended and humid conditions were established in the Netherlands and beyond (Schneider et al., 2018a,b). In the lithological record of the Broad Fourteens Basin, this is reflected by the occurrence of brackish, marginal marine and lagoonal deposits of the Bloemendaal and Neomiodon members of the Breeveertien Formation (Van Adrichem Boogaert & Kouwe, 1993).

Stratigraphy

Although the stratigraphy of the Upper Jurassic to Lower Cretaceous is complex thanks to the Late Cimmerian rift phase, many details of the stratigraphy have been revealed by biostratigraphic analyses of exploration wells drilled in the Netherlands. Biostratigraphic analyses on marine sediments generally return more precise and more reliable results than those on terrestrial sediments. As a consequence, the stratigraphy of the mostly marine middle and upper part of the Upper Jurassic-Lower Cretaceous succession is better constrained than the largely non-marine lower part. New stratigraphic methods that are independent from the fossil record, such as stable isotope- or chemo-stratigraphy, are currently being applied to counteract this shortcoming (Eldrett & Vieira, 2022; Houben et al., in prep).

Chronostratigraphy

The time interval presented in this chapter spans from the Bathonian (Middle Jurassic) to the Albian (Early Cretaceous) and consists of 11 stages, representing 68 million years of geological history according to the international Geological Time Scale GTS 2020 (100.5-168.3 Ma, Gradstein et al., 2020). For definition of the Jurassic-Cretaceous boundary, represented by the latest Jurassic Tithonian and earliest Cretaceous Berriasian stages, the international time scale is based on Tethyan reference sections and ammonite zones (Gale et al., 2020; Hesselbo et al., 2020). During that time interval, the Tethyan realm included French, Italian and southern German basins, but was isolated from the Boreal realm, which includes British, Danish and Dutch basins. As a consequence, provincialism among fossil groups was strong and it is often impossible to precisely correlate stratigraphic units across these realms. For that reason, the regional Volgian and Ryazanian stages are used for correlation purposes in the Dutch basins and in this chapter. Note that the base of the Cretaceous in the Boreal and Tethyan realms is temporally offset (Figs 7.3, 7.7, 7.10, 7.12, 7.15, 7.19). The base of the Berriasian Stage (by definition also the base of the Creta-

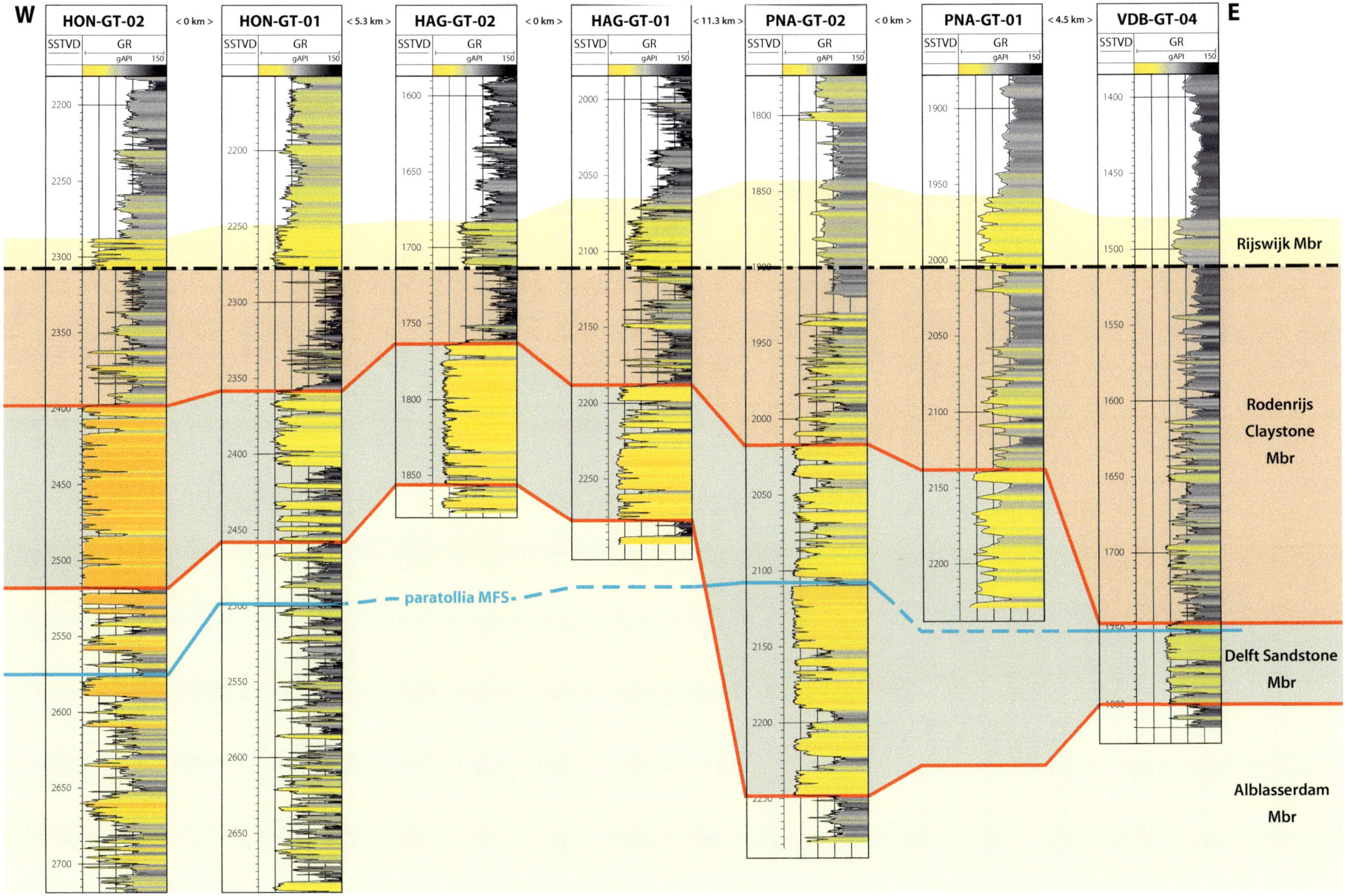

Figure 7.14. *W-E well panel displaying the correlation of the Delft Sandstone Member throughout the West Netherlands Basin based on gamma-ray (GR) logs (orange = coarse sand, yellow = sand, grey = clay).. The Delft Sandstone Member is an excellent aquifer that is targeted for geothermal energy. The Delft Sandstone Member has a diachronous relationship with both the Alblasserdam and the Rodenrijs members as is shown by its position relative to the paratollia maximum flooding surface. The correlation is based on palynological analyses (Willems et al., 2017a, 2020).*

ceous Period), is based on calpionellids, a group of calcareous microfossils. The First Appearance Datum of *Calpionella alpina,* marking the base of the *C. alpina* Subzone of the Calpionella Zone, has been selected by the Berriasian Working Group of the International Commission on Cretaceous Stratigraphy to mark the base of the Berriasian (Wimbledon et al., 2011; 2020a,b; Wimbledon, 2017; Gale et al., 2020; Hesselbo et al., 2020). This event correlates to the magnetic polarity Chron M19n.2n and corresponds to an age of 143.1 Ma. The base of the regional boreal Ryazanian Stage from the Russian Platform is defined by the base of the *ryasanensis* ammonite zone which correlates to the base of Polarity Chron M17 and corresponds to an age of 141.8 Ma (Gale et al., 2020; Hesselbo et al., 2020; Wimbledon et al., 2020a,b).

Biostratigraphy

Numerous consultancy and Geological Survey of the Netherlands reports provide detailed accounts on the age and paleoenvironments of Upper Jurassic-Lower Cretaceous deposits. Some of these reports can be found on the Netherlands Oil & Gas portal (www.nlog.nl), e.g. Bouroullec et al. (2016a,b), but the majority are not available in the public domain. Fortunately, an important part of the information derived from the studies has been incorporated in literature. The majority of the biostratigraphic information pertains to palynology (Abbink, 1998; Herngreen et al., 2000; Abbink et al., 2001a; Hoedemaeker & Herngreen, 2003; Abbink et al., 2006,Willems et al., 2017a; Verreussel et al., 2018). This discipline is ideally suited for the non-marine, marginal marine and shallow marine siliciclastic deposits that dominate the Upper Jurassic to lowermost Cretaceous of the Dutch subsurface. Palynomorphs such as dinoflagellate cysts, pollen and spores occur in a variety of environments and bridge the gap between the continental and marine realm (Jansonius & McGregor, 1996; see also Fig. 7.17). Consultancy studies on the more open marine and more calcareous deposits, such as the early Cretaceous Vlieland Sandstone Formation or the Holland Formation, have relied mostly on calcareous nannofossils. A synthesis based on this group was published in Jeremiah et al. (2010).

Sequence stratigraphy

The role of eustacy on the Upper Jurassic and Lower Cretaceous stratigraphy of the North Sea was recognized and promoted in the 1990s by authors such as Partington et al. (1993a,b) who developed a regional genetic sequence-stratigraphic framework for this time interval. Unit boundaries of their stratigraphic scheme were based on maximum flooding surfaces rather than on sequence boundaries and the stratigraphic sequences were calibrated against biostratigraphy (Partington et al., 1993b). The stratigraphic schemes were improved and further refined to fit application in specific areas such as the Outer Moray Firth (Duxbury et al., 1999; Kadolsky et al., 1999), the central and northern North Sea (Fraser et al., 2003) and the Danish Central Graben (Johannessen et al., 1996; Andsbjerg & Dybkjær, 2003). In the Netherlands, the role of eustacy is generally subordinate to the role of tectonics, but some eustatic trends are reflected in the lithological record. Within the latest Jurassic Tithonian Stage ongoing basin restriction led to an increase in provincialism that affected the evolution and distribution of environmentally sensitive fossil groups such as ammonites (Casey et al., 1977). A number of distinct maximum flooding surfaces can be recognized on wireline logs and seismic images and can be correlated across large distances. All maximum flooding surfaces are named after the ammonite zone in which they occur (Partington et al., 1993b).

The latest Callovian *Lamberti* maximum flooding surface (MFS) is reflected in the lithological record of the Dutch Central Graben by a marine flooding event near the top of the predominantly non-marine Lower Graben Formation. It heralds an overall change in depositional environment from the fluvial Lower Graben Formation to the lacustrine and estuarine Middle Graben Formation (Bouroullec et al., 2018; see also Figs 7.3, 7.5). In the southern part of the Dutch Central Graben, marine sediments of the Rifgronden Member onlap onto the partially eroded Middle Jurassic Altena Group (Fig. 7.5).

The late Kimmeridgian *Eudoxus* MFS can be traced regionally. In the Terschelling and Vlieland Basin, it marks the base of the marine Oyster Ground Member of the Skylge Formation and the estuarine/lacustrine Zurich Formation (Herngreen et al., 2000; Abbink et al., 2006; Munsterman et al., 2012; see also Figs 7.7, 7.8). In the Viking Graben and Danish Central Graben, the *eudoxus* MFS marks the transition towards the dysaerobic facies of the Kimmeridge Clay Formation (Fraser et al., 2003).

The late Ryazanian *stenomphalus* MFS is reflected in the lithological record of the Dutch Central Graben and the Terschelling Basin by the appearance of marine mudstones of the Lutine Formation (Munsterman et al., 2012; see also Figs 7.3, 7.7). In the Lower Saxony Basin, the *stenomphalus* MFS is associated with the organic-rich Upper Coevorden Member of the Coevorden Formation (Fig. 7.10) and in the Broad Fourteens Basin by the lagoonal Neomiodon Claystone Member of the Breeveertien Formation (Fig. 7.15).

The early Valanginian *paratollia* MFS more or less marks the base of the marine Rijnland Group in the Dutch Central Graben, the Terschelling Basin and the Broad Fourteens Basin (Figs 7.3, 7.7, 7.15). In the Lower Saxony Basin, the Bentheim Sandstone Member is associated with the *paratollia* MFS (Fig. 7.10). In the West Netherlands Basin, the diachronous Delft Sandstone Member of the Nieuwerkerk Formation is associated with this MFS (Willems et al., 2017a, 2020; see also Figs 7.12, 7.14).

The late Hauterivian *gottschei* MFS marks the base of the Rijswijk Member of the Vlieland Sandstone Formation in the West Netherlands Basin (Fig. 7.12).

The closely spaced early to late Barremian *Fissicostatum* and *elegans* MFSs are linked to the Berkel Sandstone Member of the Vlieland Sandstone Formation in the West Netherlands Basin (Jeremiah et al., 2010; see also Figs 7.12, 7.13). In the Danish Central Graben, the eustatic sea-level high is expressed in the organic-rich Munk Marl (Copestake et al., 2003) and in the German Lower Saxony Basin it is expressed in the bituminous Hauptblatterton (Mutterlose & Bornemann, 2000).

The *denckmanni* MFS is associated with the IJsselmonde Sandstone Member of the Vlieland Sandstone Formation in the West Netherlands Basin (Jeremiah et al., 2010; see also Figs 7.12, 7.13).

The late Barremian *pingue* MFS marks the base of the widespread De Lier Member of the Vlieland Sandstone Formation (Jeremiah et al., 2010; see also Figs 7.12, 7.13).

The early Aptian *forbesi* MFS marks the base of the Holland Formation in all basins (Figs 7.3, 7.7, 7.10, 7.12, 7.13, 7.15) and is easy to recognize on wireline logs and seismic images; the organic rich shales produce high gamma-ray readings and cause a strong contrast in acoustic impedance with the underlying sandstones. In the North Sea and Lower Saxony basins it is expressed in the organic-rich Fischshiefer (Copestake et al., 2003; Jeremiah et al., 2010). On a global scale, this maximum flooding surface is linked to the oceanic anoxic event OAE1a (Keller et al., 2011).

Lithostratigraphy

Because of the dynamic structural and climatic setting, a wide array of depositional environments developed in the Netherlands during the Late Jurassic and Early Cretaceous, including fluvial, lagoonal, lacustrine, deltaic, tidal, shoreface, offshore and deep marine environments. As a result, the stratigraphy is characterized by juxtapositions and lateral transitions. In general, the formations and members are grouped according to either the depositional setting (e.g. marine versus non-marine), the lithology (e.g.

Late Jurassic crocodile traces

When looking for Upper Jurassic or Lower Cretaceous rocks in the Netherlands you might be in for a surprise when you visit the city of Utrecht. Right at the bottom of the medieval Dom tower you will find a white limestone from the top of the Jurassic. The Dom tower is the highest church tower in the Netherlands and is mainly constructed of bricks, red sandstone and volcanic tuff from the Eifel region, but the base of the tower is clad with white oolitic limestone. The limestone is known as Portland stone and is derived from the isle of Portland in South England, where it has been quarried extensively for use as building stone. The isle also gave its name to the regional Portlandian Stage that represents the latest part of the Jurassic and earliest part of the Cretaceous. The transition from the Jurassic to the Cretaceous was characterized by a warm and arid climate and this is reflected in the rock record by the occurrence of limestones and evaporites that were deposited in shallow marine environments, often hypersaline lagoons. The characteristically abundant bivalves and oolites of the Portland stone match perfectly with such a depositional setting. Not only bivalves but also crocodiles and dinosaurs were roaming these lagoons looking for snacks and left their traces as footprints in the soft sediment. These footprints are perfectly preserved in the rocks that are now exposed in the many outcrops along the English and French coasts.

Inverted imprint of crocodile front legs from the Upper Jurassic Grés de la Crèche Formation near Boulogne-sur-Mer, northern France. The imprints are made by a crocodile belonging to a species that is characterized by five toes. The crocodile left the imprints in the soft silty sediments of a lagoon. Later, the imprints were filled with sand that was deposited on top of the partially cemented lagoon floor.

sandstone versus claystone-dominated sequences), or the geographic distribution (proximal versus distal). All major lithostratigraphic units that are defined for the Late Jurassic-Early Cretaceous are listed in Table 7.1 and the formal descriptions and definitions can be consulted online: dinoloket.nl/en/stratigraphic-nomenclature. For online listings of lithostratigraphic units from the countries bordering the Netherlands, reference is made to: litholex.bgr.de (Germany), webapps.bgs.ac.uk/lexicon/(United Kingdom) and ncs.naturalsciences.be (Belgium).

The following section provides a synopsis of the lithostratigraphic nomenclature for the Upper Jurassic and Lower Cretaceous. The lithostratigraphic units are described from old to young.

Schieland Group

The Schieland Group constitutes lithostratigraphic units that were deposited in non-marine to marginal marine depositional settings. Based on the geographic distribution, the Schieland Group is subdivided into two subgroups: the Delfland Subgroup is confined to the Vlieland, Broad Fourteens and West Netherlands basins and the Central Graben Subgroup is confined to the Dutch Central Graben and Terschelling Basin. The Schieland Group is the oldest unit of the Upper Jurassic-Lower Cretaceous succession in the Netherlands.

Delfland Subgroup

The Delfland Subgroup consists of three formations, the Zurich, the Breeveertien and the Nieuwerkerk formations (Table 7.1). The Zurich Formation is represented by non-marine deposits in the Vlieland Basin, while the Breeveertien and the Nieuwerkerk formations represent the non-marine to marginal marine base of the Upper Jurassic-Lower Cretaceous succession in the Broad Fourteens and West Netherlands basins respectively.

Zurich Formation

The Zurich Formation consists predominantly of sandy and silty claystones with thin intercalated sandstone and coal layers and includes a sequence of volcaniclastic sediments. Three members are distinguished, from old to young: the Lower Zurich, the Wadden Volcaniclastic and the Upper Zurich members. The Lower Zurich Member is dominated by claystones with few sandstone intercalations and displays weak marine influence (Herngreen et al., 2000), indicating an estuarine paleoenvironment. Based on pollen and spores, it has been assigned an early to middle Volgian age (Herngreen et al., 2000). The Wadden Volcaniclastic Member consists of variegated fine- to coarse-grained volcaniclastics with a grey tuffaceous matrix, embedded in red-brown and green clays. The Upper Zurich Member contains common coal and sandstone layers, indicative of a back-barrier lagoonal depositional setting (Van Adrichem Boogaert & Kouwe, 1993). A sandstone unit at its base in well Vlieland Oost-1 (VLO-01) may represent a lateral equivalent to (a part of) the Scruff Greensand Formation and to the Wadden Volcaniclastic Member (Figs 7.7, 7.9). The top of the Zurich Formation is often represented by more massive sandstones separated by thin claystone layers, indicating a foreshore to back-barrier depositional environment and is late Ryazanian (Herngreen et al., 2000) in age (Fig. 7.7, 7.9). The Zurich Formation reaches a thickness of more than 350 m and is limited to the Vlieland Basin. Isolated occurrences of volcaniclastic rocks are found in the Step Graben, Broad Fourteens and West Netherlands basins (Figs 7.7, 7.15, 7.16).

Breeveertien Formation

The Breeveertien Formation consists predominantly of claystones; sandstones occur mainly in the basal and the upper part of the formation. It is subdivided into seven members which are not always distinguishable. The Aerdenhout Member is the oldest and consists of claystones and channel sandstones, representing a fluvial to coastal depositional environment with an approximate late Kimmeridgian age (Van Adrichem Boogaert & Kouwe, 1993; see also Figs 7.15, 7.16). Succeeding the Aerdenhout Member are three members that share comparable depositional environments: the Fourteens Claystone Member, the Santpoort Member and the Driehuis Mottled Claystone Member. The claystone units consist predominantly of non-marine claystones with little to no marine influence which accumulated in lacustrine to estuarine depositional environments. The Santpoort Member consists of sandstones and claystones which exhibit weak marine influence, indicative of a coastal depositional environment. The age of these combined members is early to middle Volgian (Verreussel et al., 2015). At the top of the Breeveertien Formation, two partial lateral equivalents, the Bloemendaal Member and the Neomiodon Member (Figs 7.15, 7.16) occur. Palynological evidence suggests a marginal marine setting for the sandstone dominated Bloemendaal Member and a restricted marine setting for the claystone-dominated Neomiodon Member. The latter is characterized by abundant debris of the brackish water bivalve Neomiodon, indicating a lagoonal to back-barrier depositional environment (Van Adrichem Boogaert & Kouwe, 1993). The age of the Bloemendaal and Neomiodon members is late Ryazanian to earliest Valanginian (Van Adrichem Boogaert & Kouwe, 1993; Verreussel et al., 2015). Based on palynological studies, a hiatus is suggested between the Bloemendaal Member and the underlying Driehuis Mottled Claystone Member (Verreussel et al., 2015; see also Fig. 7.15). The Breeveertien Formation attains a maximum thickness of

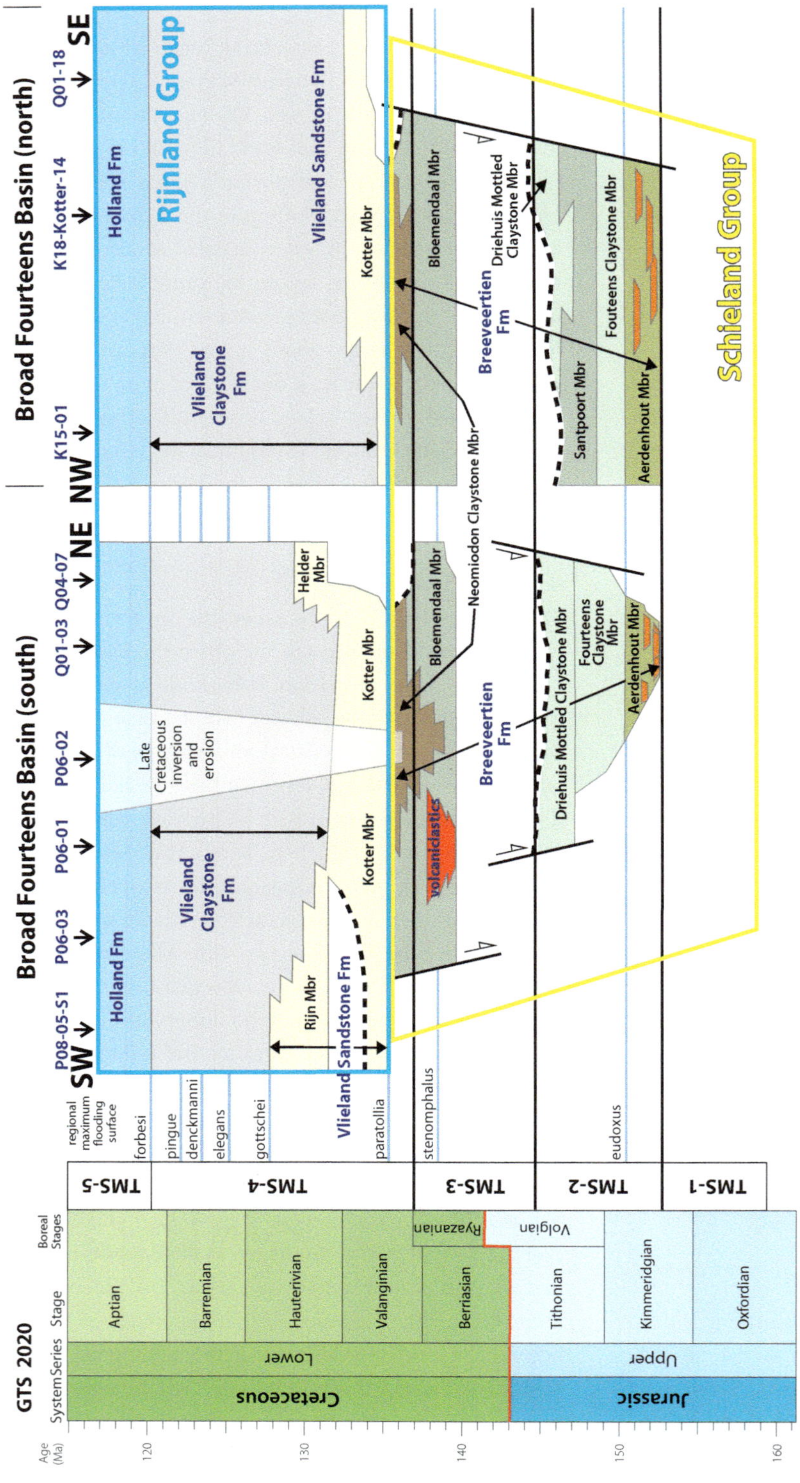

Figure 7.15. *Wheeler diagram along a SW-NE to NW-SE transect through the Broad Fourteens Basin (see Fig. 7.1 for location). The diagram displays the generalized time and facies relationships of the Kimmeridgian (Late Jurassic) to Aptian (Lower Cretaceous) lithostratigraphic units. Formation names are indicated in blue, member names in black and lithostratigraphic groups are outlined in yellow and blue. Relevant maximum flooding surfaces are displayed with blue lines. The black horizontal lines indicate tectonostratigraphic megasequences (TMS, see text for explanation). Key wells are shown at the top of the diagram. For legend see Fig. 7.3.*

640 m and is limited in distribution to the Broad Fourteens Basin.

Nieuwerkerk Formation

The Nieuwerkerk Formation consists predominantly of non-marine sandstones and variegated claystones and may attain a thickness of 1000 m (Van Adrichem Boogaert & Kouwe, 1993). The formation is limited in distribution to the West Netherlands Basin and can be considered as the slightly more proximal counterpart of the Breeveertien Formation from the Broad Fourteens Basin. Due to active faulting at the time of deposition, the age and stratigraphic thickness of the Nieuwerkerk Formation is very variable (Den Hartog Jager, 1996; Racero-Baena & Drake, 1996; DeVault & Jeremiah, 2002). For example on the Oosterhout Platform, the Hauterivian to Barremian Nieuwerkerk

Formation is juxtaposed against laterally equivalent shoreface sandstones of the Vlieland Sandstone Formation (Figs 7.12, 7.13). Three members are distinguished within the Nieuwerkerk Formation, from old to young these are the Alblasserdam Member, the Delft Sandstone Member and the Rodenrijs Member (Figs 7.13, 7.14). The Alblasserdam Member represents fluvial floodplain deposits with single and stacked channel sandstones and overbank fines. Its age is poorly known because age-diagnostic (micro-)fossils and palynomorphs are lacking, but based on its position between the underlying Brabant Formation (Trabucho Alexandre & Wong, 2025, this volume) and the overlying Delft Sandstone or Rodenrijs members, it must be younger than Oxfordian and older than Valanginian. The overlying Delft Sandstone Member is an important aquifer that is currently being targeted for geothermal energy (Mijnlieff, 2020; Willems et al., 2020; Mijnlieff et al., 2025, this volume). It consists of massive and closely spaced sandstones that probably represent meandering distributary channels in a coastal plain setting (Donselaar et al., 2015; Vondrak et al., 2018; Willems et al., 2020). The Rodenrijs Claystone Member consists of grey mudstones alternating with thin sandstones and associated coal layers. The depositional setting was coastal plain to lagoonal. In the West Netherlands Basin the Delft Sandstone Member is diachronous and may be a correlative equivalent to both the overlying Rodenrijs Member and the underlying Alblasserdam Member (Fig. 7.14). Its palynology-based age ranges from earliest Valanginian to earliest Hauterivian (Willems et al., 2017a, 2020) and it may attain a maximum thickness of 250 m. In the offshore Amstel Field in block Q13, barrier and back-barrier sandstones occur. These are likely a lateral equivalent of the Rodenrijs Member and are provisionally referred to as the informal Amstel member (Fig. 7.12).

Central Graben Subgroup

The Central Graben Subgroup includes five predominantly non-marine formations in the Dutch Central Graben and Terschelling Basin: the Lower Graben Formation, the Middle Graben Formation, the Upper Graben Formation, the Friese Front Formation and the Puzzle Hole Formation (Table 7.1). The Lower, Middle and Upper Graben formations are distributed in the middle and northern part of the Dutch Central Graben, the Puzzle Hole Formation is limited in its distribution to the middle part of the Dutch Central Graben and the Friese Front Formation occurs in the southern part of the Dutch Central Graben and in the bordering Terschelling Basin.

Friese Front Formation

The Friese Front Formation consists predominantly of carbonaceous claystones with intercalated coal beds and thin sandstones (Munsterman et al., 2012). A wide variety of depositional environments are represented, ranging from fluvial, coastal plain, fluvial channel, fluvial floodplain, estuarine and deltaic to shoreface, but on the whole, non-marine paleoenvironments prevail (Munsterman et al., 2012; Bouroullec et al., 2016b, 2018; Verreussel et al., 2018). Two members are developed, the Rifgronden Member and the main Friese Front member (Fig. 7.7). The Rifgronden Member occurs at or near the base of the formation in the southern Dutch Central Graben and displays a marine depositional environment that has been correlated to the regional J46 Lamberti maximum flooding surface in the latest Callovian (Abbink et al., 2001b; see also Fig. 7.6). The Main Friese Front Member is mostly non-marine. In block L05 at the southernmost tip of the Dutch Central Graben, it is characterized by coals and fluvial channel sands (Abbink, 1998; Abbink et al., 2006). In block F17 at the transition of the Dutch Central Graben to the Terschelling Basin, the member is characterized by three stacked sandstone units associated with coal (Fig. 7.7). These units resemble the deltaic mouth-bar sequences of the Middle and Upper Graben Formation and are possibly time equivalent, viz. Oxfordian in age (Bouroullec et al., 2016b). Within the Terschelling Basin, the main Friese Front member is characterized by fluvial channels and overbank fines and is younger than in the Dutch Central Graben: late Kimmeridgian (Verreussel et al., 2018). The Friese Front Formation is generally around 150 metres thick but may reach a maximum thickness of 480 metres.

Puzzle Hole Formation

The Puzzle Hole Formation consists of carbonaceous mudstones and siltstones, alternating with thin, fining upward sandstones and coal layers (Van Adrichem Boogaert & Kouwe, 1993). Combined with a general lack of marine indicators (Herngreen et al., 2000), a delta plain depositional environment is inferred. The formation occurs in the middle part of the Dutch Central Graben only. Towards the north, the Puzzle Hole grades into the deltaic mouthbars and estuarine claystones of the Middle and Upper Graben formations (Figs 7.3, 7.5). The Puzzle Hole Formation is limited in extent to the middle part of the Dutch Central Graben. The age of the formation is late Oxfordian to early Kimmeridgian (Verreussel et al., 2018).

Lower Graben Formation

The Lower Graben Formation is characterized by well-sorted sandstones and silty to sandy claystones, often carbonaceous with common coal layers (Van Adrichem Boogaert & Kouwe, 1993). Towards the top, the sandstones become more massive and marine influence increases (Bouroullec et al., 2018; see also Figs 7.4, 7.5). The depositional environment changed from a fluvial, quiet bay in the lower parts, to brackish lagoon and tidal shoals at

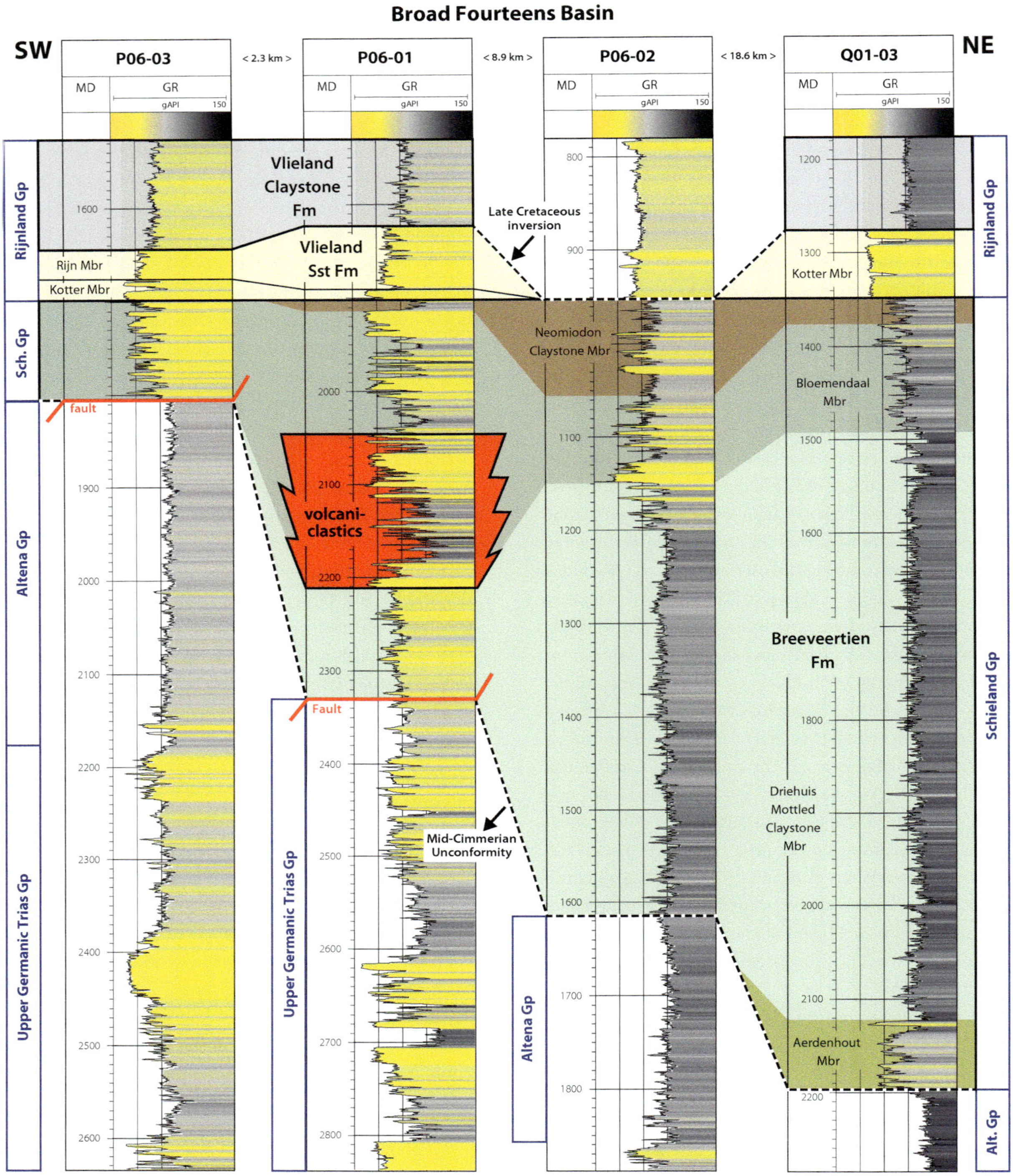

Figure 7.16. SW-NE well panel through the Broad Fourteens Basin (see Fig. 7.1 for location). The panel displays the well correlation of the Upper Jurassic to Lower Cretaceous lithostratigraphic units based on gamma-ray (GR) logs (yellow = sand, grey = clay). Formations are indicated with thick black lines and the colours of the lithostratigraphic units represent facies types (see legend of Fig. 7.3). Volcaniclastic sediments from the Zurich Formation in well P06-01 reveal the nearby presence of an active volcano. Note the large overall thickness of the non-marine Schieland Group in the basin.

the top of the formation. It reaches a maximum thickness of 500 m and is limited in its distribution to the northern and middle parts of the Dutch Central Graben. The age is middle to late Callovian and its top correlates to the Callovian-Oxfordian boundary (Verreussel et al., 2018).

Middle Graben Formation

The Middle Graben Formation is an up to 500 m thick unit that consists predominantly of carbonaceous claystones interrupted by two subordinate sandstone units (Fig. 7.14), the lower of which forms the Middle Graben Sandstone Member. These sandstones are very similar in lithology and architecture to those of the Upper Graben Formation, differing only in their more limited development. They were deposited as deltaic mouthbars in a marginal marine deltaic depositional environment, comparable to those of the Upper Graben Formation. Palynological associations from the claystones show a dominance of pollen and spores indicative of lacustrine to estuarine paleoenvironments (Herngreen et al., 2000). Three up to 3 metres thick coal seams, separated by up to 30 metres of claystone occur at the base of the formation (Figs 7.3-7. 5) and can be traced across the entire Dutch Central Graben. The thickest coal seam lies at the base and is visible as an excellent reflector on seismic images. A meandering channel feature identified within it has been attributed to a large fluvial system entering the basin from the east (Bouroullec et al., 2018). The age of the Middle Graben Formation is Oxfordian, the basal coal seam corresponds to the Callovian-Oxfordian boundary (Verreussel et al., 2018).

Upper Graben Formation

The Upper Graben Formation consists of two coarsening upward, relatively fine-grained carbonaceous sandstone units associated with coal layers, separated by fine-grained siliciclastics (Van Adrichem Boogaert & Kouwe, 1993). It is confined to the northernmost part of the Dutch Central Graben; towards the south the formation grades into the Puzzle Hole Formation (Figs 7.3, 7.4). The coarsening upward sandstone units with associated coals point to prograding mouthbars in a marginal marine deltaic depositional setting, and this is corroborated by the identification of small bird-foot deltas within the formation near the basin margin (Bouroullec et al., 2018). The age of the Upper Graben Formation is latest Oxfordian to earliest Kimmeridgian (Herngreen et al., 2000; Verreussel et al., 2018). It is succeeded by shallow marine to offshore claystones of the Kimmeridge Clay Formation (Fig. 7.4) and reaches a maximum thickness of 180 m.

Niedersachsen Group

The Niedersachsen Group differs from all other Upper Jurassic-Lower Cretaceous groups in that limestones and evaporites commonly occur. The Niedersachen Group includes two formations, the Coevorden Formation and the Weiteveen Formation and is limited in its distribution to the Lower Saxony Basin. It attains a maximum thickness of 600 m.

Weiteveen Formation

The Weiteveen Formation is characterized by anhydritic and marly claystones with halite and limestone intercalations. Coarse sands and conglomerates may occur at the base. The depositional environment is lacustrine to marginal marine. The predominance of evaporitic facies points to warm and arid climate conditions. The formation is subdivided into six members (Fig. 7.10), from old to young being the Weiteveen Basal Clastic Member (variegated conglomerates, sandstones and claystones), the Weiteveen Lower Evaporite Member (alternation of dolomitic claystones, anhydrites, halites and limestones), the Weiteveen Lower Marl Member (blue-grey marly, locally anhydritic or pebbly claystone succession), the Weiteveen Upper Evaporite Member (anhydrite, carbonate and halite), the Weiteveen Upper Marl Member (marly claystone unit with some sand, anhydrite and limestone intercalations) and the Serpulite Member (limestone-rich claystone interval, characterized by the common occurrence of fossil worm tubes known as serpulids). Towards the western and northern margins of the Lower Saxony Basin the Basal Clastic Member becomes less coarse-grained, while the Lower and Upper Evaporite members become thinner and eventually pinch out. The age of the top of the Weiteveen Formation is latest Volgian (earliest Berriasian, Hoedemaeker & Herngreen, 2003; Seyfang et al., 2017), but the base is not well established. Based on ostracods it is probably latest Kimmeridgian, (Van Adrichem Boogaert & Kouwe, 1993). The Weiteveen Formation is limited to the onshore Lower Saxony Basin and is equivalent to the Münder Formation in the German part of the basin (Erbacher et al., 2014; Schneider et al., 2018b).

Coevorden Formation

The Coevorden Formation mainly consists of marly claystones with occasional limestone beds. The depositional setting is fresh- to brackish-water lacustrine to marginal-marine lagoonal. The fresh water dominated depositional environment reflects the humid climate conditions characteristic for the earliest Cretaceous of NW Europe, often referred to as the 'Wealden' (after the comparable succession in the Weald Basin of southwest England, see e.g. Abbink et al., 2001b; Stollhofen et al., 2008). The formation is subdivided into three members, the Lower, Middle and Upper Coevorden members (Van Adrichem Boogaert & Kouwe, 1993; see also Fig. 7.10). The Lower Coevorden Member consists of grey, calcareous, occasionally sandy claystones. Limestone beds occur frequently in the basal part. The Middle Coevorden Member consists of grey, fossiliferous, silty claystones and is typically more calcareous than the other two members. The Upper Coevorden Member consists of brownish-grey, occasionally sandy claystones in which locally laminated organic-rich claystones occur, reflecting anaerobic bottom conditions (Figs 7.10, 7.11). These paper shales are the source rocks for the oil of the Schoonebeek Field (De Jager & Geluk, 2007). The age of the Coevorden Formation is late Ryazanian to earliest Valangian (Hoedemaeker & Herngreen, 2003) and it reaches a maximum thickness of 150 m. It is limited to the onshore Lower Saxony Basin and continues into Germany where the Isterberg Formation is its correlative equivalent (Erbacher et al., 2014).

Scruff Group

The Scruff Group contains lithostratigraphic units that were deposited in a marine depositional setting. It is divided into four formations: the Lutine, the Scruff Greensand, the Skylge and the Kimmeridge Clay formations (Van Adrichem Boogaert & Kouwe, 1993; Munsterman et al., 2012; see also Table 7.1). The Scruff Group is generally sandwiched between the underlying and predominantly non-marine Schieland Group and the overlying marine Rijnland Group. The Group ranges in age from late Oxfordian to late Ryazanian and units belonging to it are developed in the offshore Step Graben, Dutch Central Graben, Terschelling Basin and Vlieland Basin. It attains a maximum thickness of 850 m in the Dutch Central Graben. The deep-marine Kimmeridge Clay Formation reflects peak basin subsidence related to Late Cimmerian rifting.

Kimmeridge Clay Formation

The Kimmeridge Clay Formation consists mainly of claystones, which are siltier in its basal part. Stratigraphically up-section, dolomite streaks appear but they are less common near the basin margins. Based on the palynological content and on the predominance of claystones, an open marine offshore depositional setting has been determined (Munsterman et al., 2012). The age of the Kimmeridge Clay Formation is Kimmeridgian to middle Volgian (Verreussel et al., 2018) and it attains a maximum thickness of 1000 m. It is limited to the middle and northern part of the Dutch Central Graben and to isolated patches in the Step Graben, but does not extend into the Terschelling or Vlieland basins. The Kimmeridge Clay Formation conformably overlies the deltaic sandstones of the Upper Graben Formation (Figs 7.3, 7.4) and has correlative inter-

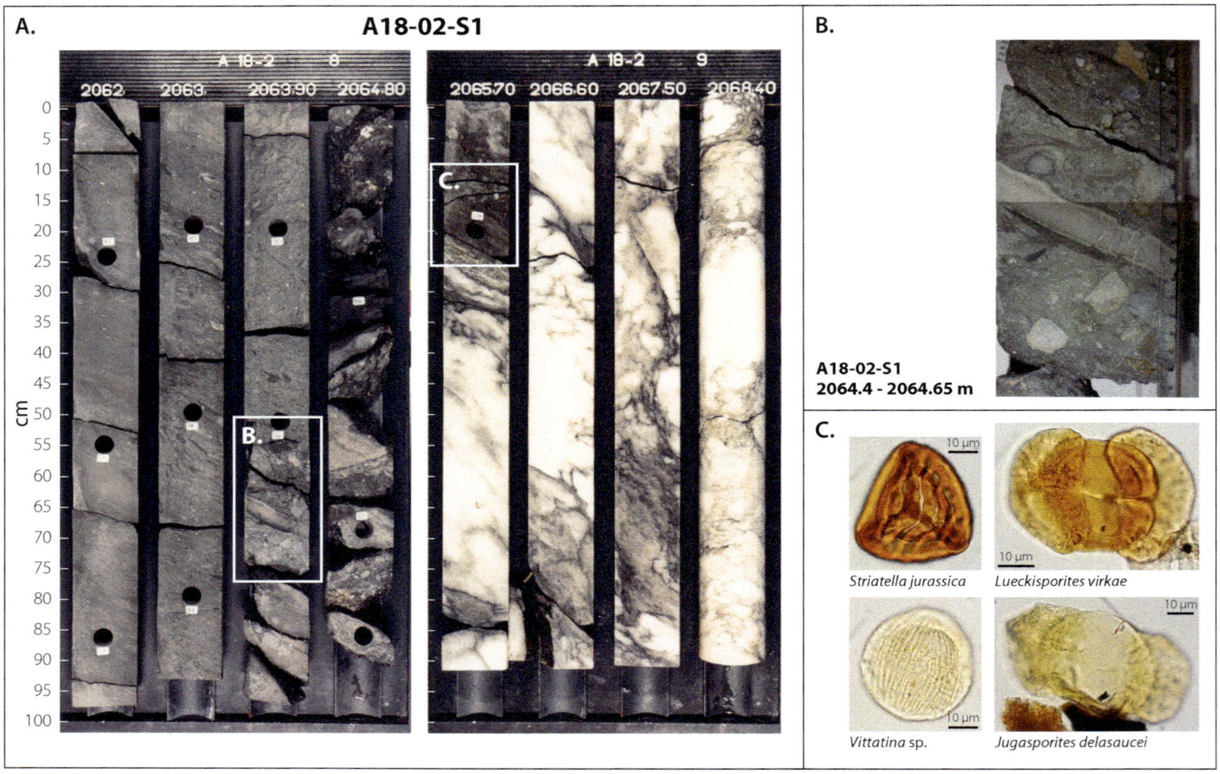

Figure 7.17. *a) Core photographs of interval 2062.0-2069.3m of well A18-02-S1 from the Step Graben in the northern Dutch offshore (see Fig. 7.1 for location; source: nlog.nl). The core displays the unconformable contact between the Permian Zechstein Group (white) and the Upper Jurassic Scruff Group (grey and black). Core photos from www.nlog.nl. b) The base of the Upper Jurassic Schieland Group is represented by sandstones and breccias of the Noordsvaarder Member. Among the large and angular clasts are limestones and dolomites from the Zechstein Group. The sandy matrix contains dinoflagellate cysts that indicate a Volgian age. c) The base of the Upper Jurassic sequence is a dark, laminated claystone that contains excellently preserved pollen and spores from the Permian, e.g.* Lueckisporites virkae, Vittatina sp. *and* Jugasporites delasaucei, *in combination with Jurassic spores such as e.g.* Striatella jurassica.

national equivalent units with widespread occurrence in the North Sea and adjacent basins, where they often display elevated total organic carbon content (Cooper et al., 1995; Van der Hoeven et al., 2022). Although the organic carbon content of the Kimmeridge Clay Formation in the Netherlands is somewhat elevated, it is not characterized as an oil-prone source rock (De Jager & Geluk, 2007; Remmelts et al., 2025, this volume).

Skylge Formation

The Skylge Formation includes the Upper Jurassic marine deposits below the Scruff Greensand Formation in the Terschelling Basin (Munsterman et al., 2012). The formation comprises discrete claystone and sandstone units which are ranked as members. These are the Oyster Ground Member, the Terschelling Sandstone Member, the Noordsvaarder Member and the Lies Member (Fig. 7.7). The early Volgian Oyster Ground Member lies at the base of the Skylge Formation in the basin centre of the Terschelling Basin and consists of parallel laminated claystones with numerous thin silty or sandy storm layers with abundant shell fragments, indicating a storm dominated lower shoreface depositional environment (Bouroullec et al., 2018; Verreussel et al., 2018). The succeeding Terschelling Sandstone Member consists of middle- to coarse-grained sandstones, often cross-bedded, with varying degrees of bioturbation. These sands mainly represent upper shoreface environments but also include back-barrier and tidal lagoon environments (Bouroullec et al., 2018) forming a stacked shoreface succession, likely representing a coastal barrier system with thick amalgamated barrier sandstone units organized in a belt. The upper sandstones thin towards the basin centre, while a more heterolithic unit represents the back-barrier lagoon (Fig. 7.8). The Heno Formation in the Danish offshore (Michelsen et al., 2003, Johannessen et al., 2010a,b) is considered to represent a correlative equivalent (Verreussel et al., 2018). The Terschelling Sandstone Member is of early Volgian to middle Volgian age (Verreussel et al., 2018), occurs in the southern part of the Terschelling Basin and reaches a maximum thickness of 150 m. The Noordsvaarder Member is time equivalent to the Terschelling Sandstone but is distributed mainly in the northern part of the Terschelling Basin and in the Dutch Central Graben. It consists of bioturbated glauconitic shoreface sandstones, comparable in composition and organization with the Scruff Greensand Formation. In the Step Graben, the Noordsvaarder Member is sometimes represented by a breccia or by an immature sandstone at the base of the Upper Jurassic sequence (Bouroullec et al., 2016b; see also Fig. 7.17). It reaches a maximum thickness of 310 m and is of early to middle Volgian age (Verreussel et al., 2018). In the northern part of the Terschelling Basin and in the southern part of the Dutch Central Graben, the Noordsvaarder Member is overlain by the Scruff Greensand Formation. Towards the centre of the Terschelling Basin, the top shales out into the Lies Member (Munsterman et al., 2012), which consists of claystones and siltstones that accumulated in a shallow marine depositional environment (Munsterman et al., 2012; see also Fig. 7.7). The age of the Lies Member is middle to late Volgian (Verreussel et al., 2018).

Scruff Greensand Formation

The Scruff Greensand Formation consists predominantly of intensely bioturbated greensands with a high glauconite content (Munsterman et al., 2012). Bedding, cross-bedding and other sedimentary features are rare, most having been obliterated by bioturbation. At the base of the formation sponge spicules are abundant and these intervals are sometimes referred to as the Scruff Spiculite Member. The sandstones were deposited in a lower shoreface to upper offshore environment (Bouroullec et al., 2018). The base of the formation is conformable in the Terschelling Basin, where it is characterized by a distinct coarsening trend following the fine-grained deposits of the underlying Lies Member of the Skylge Formation (Fig. 7.7). In other parts of the Dutch offshore the base is erosional and a hiatus separates the Scruff Greensand Formation from older units (Fig. 7.3). The Scruff Greensand Formation is often succeeded by claystones of the Lutine Formation and is limited to the northern part of the Dutch offshore, to the Step Graben, Dutch Central Graben, Terschelling Basin and the bordering platforms. The Scruff Greensand has no correlative equivalent in the German and Danish offshore as the formation shales out towards the north (Verreussel et al., 2018; see also Fig. 7.3). It reaches its maximum thickness of 180 m in the Terschelling Basin and is late Volgian to early Ryazanian in age (Verreussel et al., 2018).

Lutine Formation

The Lutine Formation consists of claystones. It is typically organic rich in the northernmost part of the Dutch offshore, where the organic-rich intervals are ranked as a separate member, the Clay Deep Member (Munsterman et al., 2012). The depositional environment is open marine, but with anoxic bottom conditions for the Clay Deep Member. The Lutine Formation is thin, with a maximum thickness of 150 m and its distribution is limited to the northern part of the Dutch offshore, the Step Graben, Dutch Central Graben and Terschelling Basin. The age of the Lutine Formation is uppermost Volgian to upper Ryazanian (Verreussel et al., 2018). The claystones of the Lutine Formation typically overlie sandstones of the Scruff Greensand Formation reflecting a regionally recorded basal Cretaceous flooding (Figs 7.3, 7.4, 7.7, 7.8). The Lutine Formation can be distinguished from the claystones of the succeeding

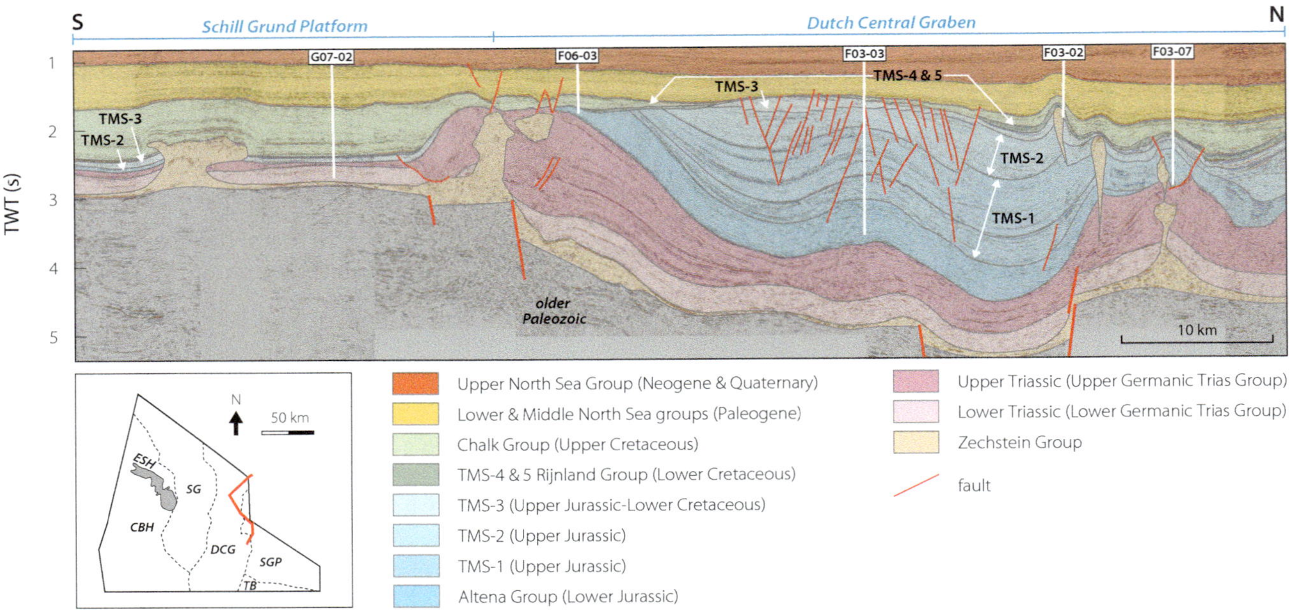

Figure 7.18. *Seismic time section through part of the Schill Grund Platform and the Dutch Central Graben showing the prominent role of salt movement in the basin development. Due to the Late Cretaceous inversion, the Chalk Group is relatively thick on the Schill Grund Platform but is virtually absent in the Dutch Central Graben. Note that TMS-1, representing the initial phase of the Late Cimmerian rifting, is very expanded in the northern part of the Dutch Central Graben.*

Vlieland Claystone Formation by a subtle kick on the wireline logs, coinciding the Base Cretaceous Unconformity (Ineson et al., 2022, see also Fig. 7.4).

Rijnland Group

All lithostratigraphic units belonging to the Rijnland Group were deposited in marine depositional environments and are of Early Cretaceous age. The Group is subdivided into one stand-alone formation, the Holland Formation, and one subgroup, the Vlieland Subgroup (Table 7.1). The Rijnland Group is widely distributed across all Mesozoic basins in the Netherlands and generally overlies other Upper Jurassic-Lower Cretaceous strata. It is stratigraphically overlain by the Upper Cretaceous Chalk Group (Van Lochem et al., 2025, this volume) and is Early Cretaceous, Valanginian to Albian in age. The Rijnland Group attains a maximum thickness of 1700 metres in the West Netherlands Basin. The widespread distribution and the predominance of layer cake type of internal organization reflects a relatively quiet tectonic phase with relative uniform subsidence following the Late Cimmerian rifting.

Holland Formation

The Holland Formation consists of marls, claystones and glauconitic sands (greensands), with locally thin layers of organic-rich shale. Greensands, represented by the Holland Greensand Member and the Spijkenisse Greensand Member, occur mainly in the West Netherlands Basin, while other members, including the Lower Holland Marl Member, Middle Holland Claystone Member and the Upper Holland Marl Member are widely distributed across all Mesozoic basins (Figs 7.3, 7.7, 7.10, 7.12, 7.15). The depositional setting was offshore marine. The age of the Holland Formation is Aptian to Albian (Jeremiah et al., 2010) and it is 100 to 500 metres thick.

Vlieland Subgroup

The Vlieland Subgroup consists of two partially laterally equivalent formations, the Vlieland Sandstone Formation and the Vlieland Claystone Formation (Table 7.1). These formations may be laterally equivalent, in which case they represent a trend in the depositional environment from proximal to distal, or they alternate stratigraphically, in which case they represent regressive-transgressive cycles. The formations may also alternate stratigraphically in which case the sandstone-claystone alternations represent regressive-transgressive cycles. These lateral and vertical facies trends are best illustrated in the West Netherlands Basin (Figs 7.12, 7.13). The age of the Vlieland Subgroup is Early Cretaceous, Valanginian to Barremian (Van Adrichem Boogaert & Kouwe, 1993).

Vlieland Claystone Formation

The Vlieland Claystone Formation consists predominantly of claystones, with subordinate marls, thin sandstones and siltstones. Eight members have been defined, mostly in order to distinguish claystone units that are intercalated within sandstone units. This is best illustrated in the Lower Saxony Basin where the Westerbork Member, Ruinen Member, Schoonebeek Member and the Bentheim Claystone Member define fine-grained lithological units associated with the Bentheim Sandstone and Gildehaus Sandstone members (Fig. 7.10). The age of the Vlieland Claystone Formation ranges from Valanginian to Barremian (Van Adrichem Boogaert & Kouwe, 1993) and the thickness varies between 200 and 600 metres. The Vlieland Claystone Formation is widely distributed across both the onshore and offshore basins.

Vlieland Sandstone Formation

The Vlieland Sandstone Formation is subdivided into thirteen members representing a wide variety of depositional environments and ages (Van Adrichem Boogaert & Kouwe, 1993). Most of the members are limited in their distribution to a single basin. For example, the Bentheim and Gildehaus Sandstone members are found only in the Lower Saxony Basin (Fig. 7.10). The Bentheim Sandstone Member is extensively exposed near the town of Bad Bentheim in Germany where it consists of massive and locally cross-bedded clean sandstones. In the subsurface of the Netherlands it consists of massive calcareous sandstones with abundant shell fragments and lignite particles. The sandstone was deposited in a coastal to offshore setting. The younger Gildehaus Sandstone Member consists of coarse-grained very glauconitic sandstones, often with abundant sponge spicules. The Friesland Member is found in the Terschelling Basin, the Broad Fourteens and Vlieland basins and on the margins of the Lower Saxony Basin (Figs 7.3, 7.7, 7.9, 7.10). The Friesland Member consists of bioturbated sandstones and represents shallow to open marine environments in a transgressive setting. In the Broad Fourteens Basin, the Vlieland Sandstone is represented by massive sandstone units like the Kotter and the Helder members, deposited in a shallow marine shoreface or barrier island depositional environment and by fining-upward transgressive sheet sandstones such as the Rijn Member. Along the southern and eastern margin of the West Netherlands Basin (Fig. 7.1) four members of the Vlieland Sandstone Formation are stacked as a series of successive shoreface complexes. From old to young these are: the Rijswijk, the Berkel Sandstone, the IJsselmonde Sandstone and the De Lier members (Figs 7.12, 7.13). The erosive base of the late Hauterivian to Barremian Rijswijk Member marks the base of the Rijnland Group in the West Netherlands Basin (Jeremiah et al., 2010). The glauconitic sandstones of the late Barremian De Lier Member cover a large part of the West Netherlands Basin and represent the top of the Vlieland Subgroup there (Jeremiah et al., 2010). The thickness of most members of the Vlieland Sandstone Formation is typically in the order of 50 to 100 metres and reaches a maximum of 200 metres for the IJsselmonde Sandstone and the Helder members in the West Netherlands and Broad Fourteens basins.

Igneous Rock

Zuidwal Volcanic Formation

The Zuidwal Volcanic Formation consists of massive volcanic rocks and brecciated volcanic agglomerates (Herngreen et al., 1991; Van Adrichem Boogaert & Kouwe, 1993). The formation is only recognized in well Zuidwal-01 (ZDW-01), which was drilled in the volcanic conduit of the Zuidwal volcano (Van Bergen et al., 2025, this volume; see also Figs 7.7, 7.9). No less than 1058 metres of the Zuidwal Volcanic Formation were penetrated until the well reached total depth at 3002 m.

Sedimentary and structural development

The geology of the Late Jurassic-Early Cretaceous interval in the Netherlands is governed to a large extent by the Late Cimmerian rift phase and salt tectonics (Bouroullec & Ten Veen, 2025, this volume). A major and widespread hiatus separates deposits of the Late Cimmerian rift phase from the underlying Middle Jurassic or older sediments (Kombrink et al., 2012). The rifting occurred in discrete steps which are reflected in the sedimentary record as genetically related accumulations of sediments or tectonostratigraphic megasequences (Bouroullec et al., 2018; Verreussel et al., 2018; see also Fig. 7.18). For each basin, the timing, style and intensity of rift activity differs, which is reflected in the stratigraphic thickness of the tectonostratigraphic megasequences and in the facies distribution, in particular in the occurrences of sandstone units (Figs 7.19, 7.20). The inherited structural grain of the basins plays an important role in the rift development and lingers on longest in the West Netherlands Basin. Five tectonostratigraphic megasequences are distinguished and described below.

TMS-1 Bathonian-early Kimmeridgian

TMS-1 represents the onset of the Late Cimmerian rift. The initial phase took place under an east-west extensional regime and had a pronounced effect on the Dutch Central Graben area (Figs 7.3, 7.4, 7., 7.7). The oldest deposits that can be ascribed to Late Cimmerian rifting in the Netherlands are middle Callovian (Verreussel et al., 2018) and are represented in the Dutch Central Graben by the Lower Graben Formation (Fig. 7.5). Hundreds of metres of sediments were deposited during TMS-1 (Verreussel et al., 2018; see also Figs 7.18, 7.20). In most other basins, such as the Terschelling Basin, the Vlieland Basin, the Lower

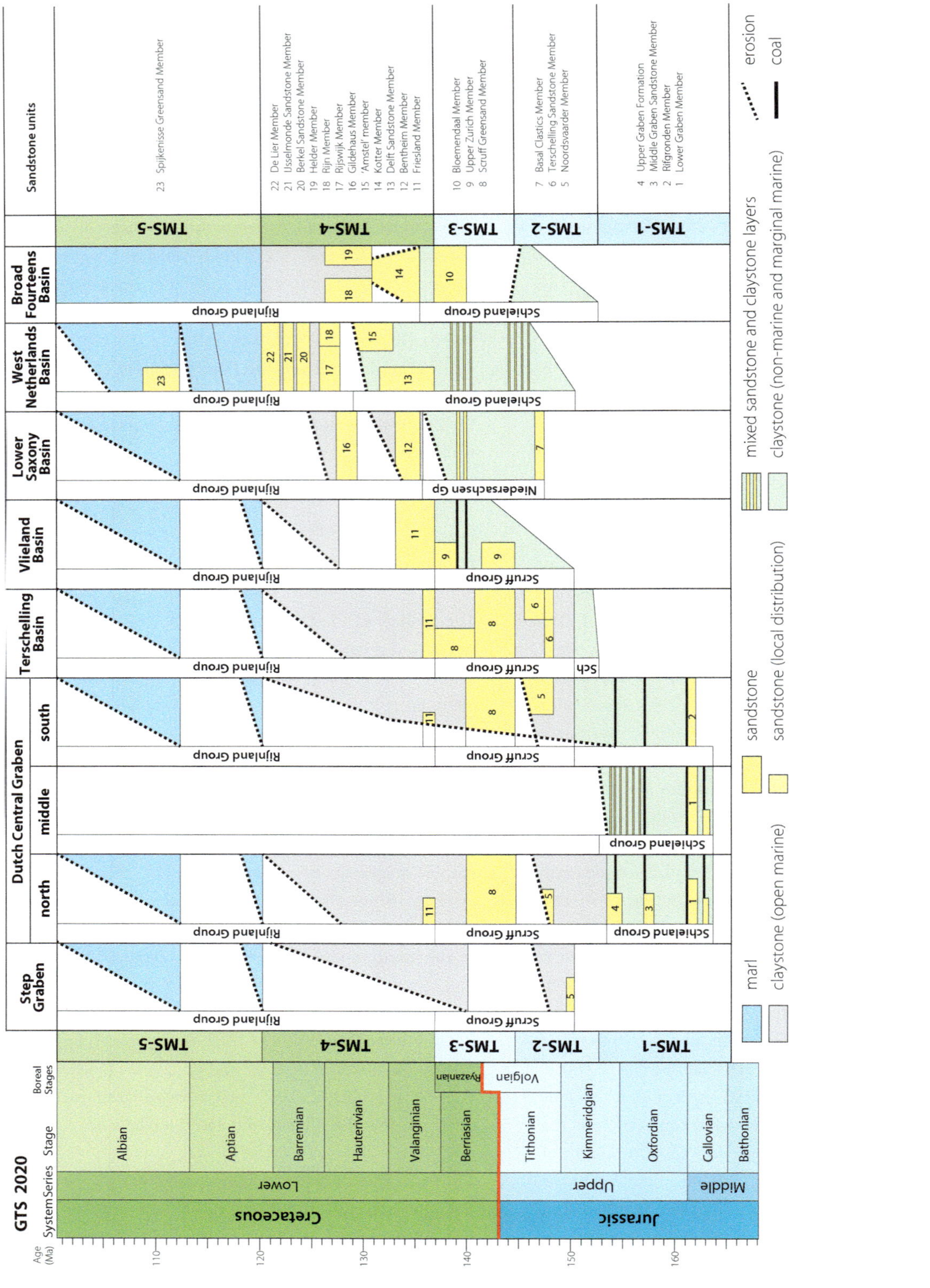

Figure 7.19. *Generalized Wheeler diagram displaying simplified time and facies relationships of the Callovian to Albian for each of the basins discussed. The different facies types from the Wheeler diagrams and the well panels are summarized into five main categories: marl, open marine claystone, sandstone, mixed sandstone and claystone layers and non-marine to marginal marine claystone.*

Saxony Basin and the Broad Fourteens Basin, Bathonian to early Kimmeridgian deposits are either missing or very thin (Wong, 2007; Van Adrichem Bogaert & Kouwe, 1993; see also Figs 7.7, 7.10, 7.12, 7.15, 7.19, 7.20).

In the Dutch Central Graben, the east-west extension triggered migration and withdrawal of Zechstein salt from the central parts of the basin towards its eastern and western margins and this led to the development of oval-shaped sub-basins that were initially disconnected from the German and Danish sectors of the Central Graben (Bouroullec et al., 2016b, 2018). Within these basins, fluvial and marginal marine sediments of the Lower Graben Formation accumulated (Fig. 7.3), the thickest accumulations of which are located in the middle and northern part of the Dutch Central Graben (Figs 7.4, 7.5). The upward transition to the deposits of the Middle Graben Formation occurred while eustatic sea level was rising and the paleoclimate was changing. This change corresponds to the Callovian-Oxfordian boundary and is marked by the occurrence of three thick and continuous coal layers, indicating that extensive swamps existed at the time (Verreussel et al., 2018; Bouroullec et al., 2018; see also Figs 7.3-7.5). During the Oxfordian, mainly claystones were deposited in lacustrine to restricted marine settings but the succession is interrupted multiple times by fluvial-deltaic sandstone units with associated coal layers, such as the Middle Graben Sandstone Member and the Upper Graben Formation (Figs 7.3-7.5). These coarsening upward units represent small-scale deltas prograding from the basin margins into the basin centre (Bouroullec et al., 2018). In the early Kimmeridigian, the Dutch Central Graben gradually developed into a deep marine basin, reflected by the organic-rich mudstones with dolomite streaks of the Kimmeridge Clay Formation. The deep marine setting continued across the boundary from TMS-1 to TMS-2.

In the Lower Saxony Basin uplift and subaerial exposure occurred during TMS-1 and led to development of a regional erosional unconformity, especially along the western and northern margins of the Lower Saxony Basin (Klassen, 1984; Betz et al., 1987).

In the West Netherlands Basin, sedimentation continued from the Bathonian to the Oxfordian without prolonged interruptions, as seen in the cyclic alterations of sand, sandy limestone and marls of the Brabant Formation (Van Adrichem Boogaert & Kouwe, 1993; Trabucho Alexandre & Wong, 2025, this volume). The Brabant Formation belongs to the Altena Group that genetically predates the Late Cimmerian rift phase (Trabucho Alexandre & Wong, 2025, this volume). The introduction of sand into the West Netherlands Basin during TMS-1 times may indicate increased erosion from the bordering highs in response to incipient tectonic activity related to the onset of Late Cimmerian rifting.

TMS-2 late Kimmeridgian-late Volgian

TMS-2 was the most active phase of the Late Cimmerian rift. All Mesozoic basins were affected and sediments accumulated in them (Figs 7.3, 7.7, 7.10, 7.12, 7.15, 7.19, 7.20). At the start of TMS-2 an important change in the extensional regime occurred; the E-W extension that characterized TMS-1 changed into an NE-SW extensional direction (Zanella & Coward, 2003; Verreussel et al., 2018). This change provoked the reactivation of fault systems that had inherited a NW-SE trending structural grain from the Paleozoic and had a profound effect on the Dutch basins. The Terschelling, Vlieland, West Netherlands and Broad Fourteens basins commenced to subside and accumulated large volumes of sediment (Figs 7.7, 7.10, 7.12, 7.15).

In the Step Graben, marine sediments of the Skylge Formation appeared for the first time in the late Kimmeridgian-early Volgian (Verreussel et al., 2015), but they are patchy and limited in extent. The same applies to the Schill Grund Platform (Fig. 7.1) on the eastern side of the Dutch Central Graben, where only thin stacks of sediment belonging to TMS-2 are found (Fig. 7.18). The base of the Upper Jurassic-Lower Cretaceous succession in the Step Graben is often represented by the Scruff Greensand or Lutine formations belonging to TMS-3 but sometimes by sandstones and breccias of the Noordsvaarder Member from TMS-2, as exemplified in a cored section from well A18-02-S1 (Fig. 7.17).

In the northern part of the N-S trending Dutch Central Graben, deep marine mudstones of the Kimmeridge Clay Formation continued to be deposited (Figs 7.3, 7.4), but the depocentres gradually shifted northward and a NW-SE trending branch of the graben developed into a more actively subsiding basin (Bouroullec et al., 2018). The basin development of the Dutch Central Graben is largely controlled by activity in the underlying salt, which is reflected in the development of turtle structures (Bouroullec & Ten Veen, 2025, this volume). At the early to middle Volgian transition, salt was evacuated from below the thick Upper Jurassic-Lower Cretaceous piles of sediment in the basin centre, with the resultant development of large-scale turtle structures (Bouroullec et al., 2018). With no more space for accommodation, these structures emerged and their tops became eroded. Alongside these turtles, for example in well L02-05, rapidly subsiding rim synclines developed and were filled in with sandy deposits of the Noordsvaarder Member (Fig. 7.3). The southernmost part of the Dutch Central Graben never developed into a deep marine basin and remained dominated by fluvial to marginal marine depositional environments (Abbink et al., 2006; see also Fig. 7.3). The NE-SW extensional regime invoked strike-slip movement on the N-S trending faults bordering the Dutch Central Graben (Peeters, 2016) which added to the complexity of the basin evolution.

In the Terschelling Basin, non-marine sediments of the Friese Front Formation lie at the base of the Upper Jurassic-Lower Cretaceous succession. These are succeeded by offshore marine mudstones of the Oyster Ground Member (Figs 7.7, 7.8). The arrival of marine sediments in the Terschelling Basin coincides with the *eudoxus* maximum flooding surface (Fraser et al., 2003; see also Fig. 7.7). On the northwestern side of the basin, erosion from the turtle structures in the Dutch Central Graben resulted in the deposition of a wedge of poorly sorted sandstones of the Noordsvaarder Member (Bouroullec et al., 2016a). On the

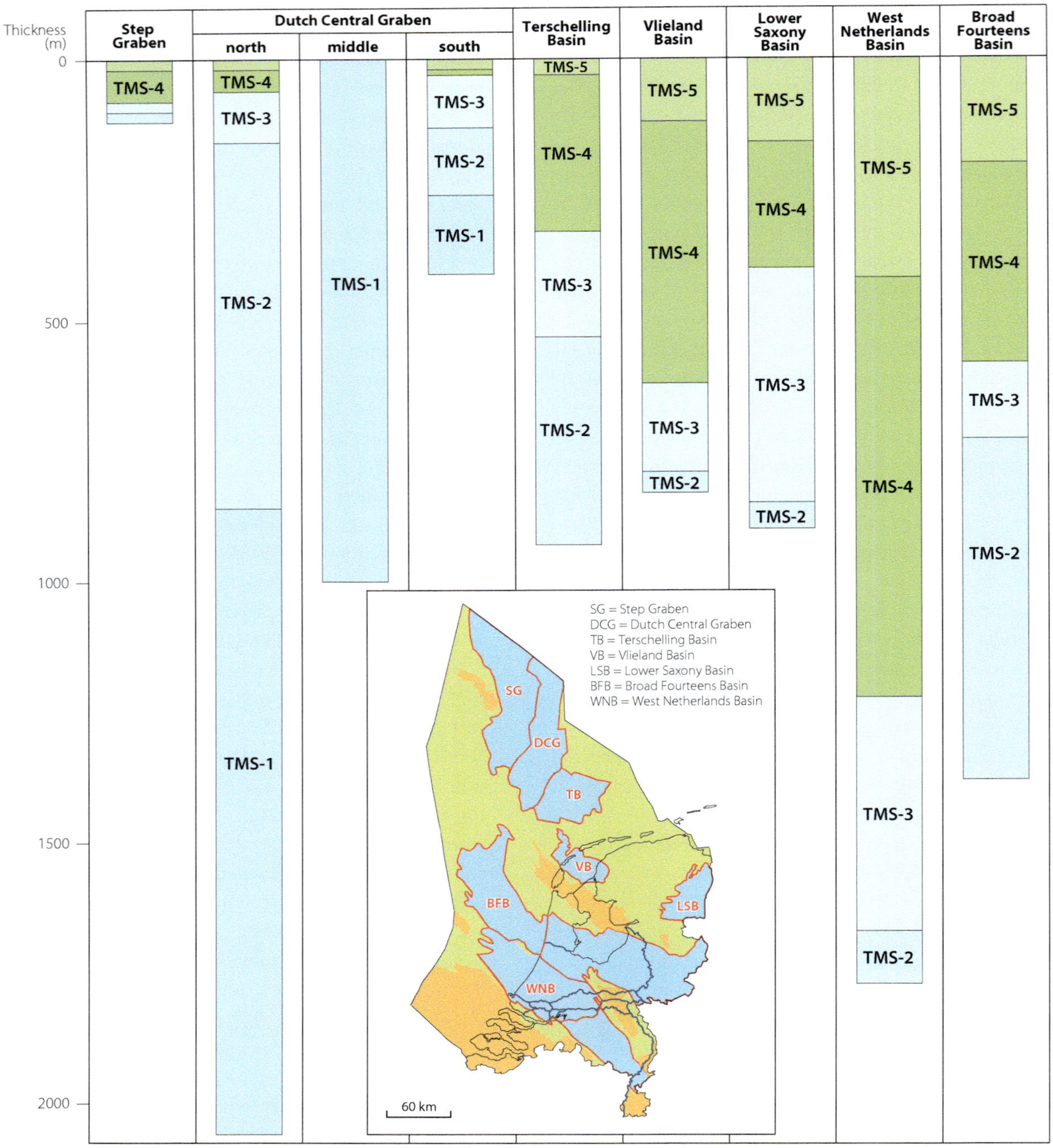

Figure 7.20. *Generalized thicknesses of tectonostratigraphic megasequences (TMS) per basin. The thicknesses provide an indication of the shift of depocentres through time.*

southeastern side of the basin close to the basin margin, log-character and palynological analyses indicate that a shoreface complex including barriers and back-barrier lagoons developed at around the same time. These sands are classified as the Terschelling Sandstone Member and onlap onto the Central Offshore Platform that borders the Terschelling Basin to the south (Figs 7.7, 7.8).

In the Vlieland Basin, the oldest sediments associated with the Late Cimmerian rift are Volgian marginal marine or estuarine claystones of the Zurich Formation (Figs 7.7, 7.9). The Hantum Fault Zone (Fig. 7.1) became an intermittent platform area with subsiding basins on both the northern (Terschelling Basin) and southern sides (Vlieland Basin). However, the accommodation space created during TMS-2 in the Vlieland Basin was considerably less (up to 100 m thick sequence) than that of the Terschelling Basin (Fig. 7.20).

In the Lower Saxony Basin, the oldest sediments associated with the Late Cimmerian rift are latest Kimmeridgian coarse-grained sandstones and breccias of the Basal Clastics Member of the Weiteveen Formation (Figs 7.10, 7.11, 7.19). At that time, the Lower Saxony Basin was isolated, which led to brackish marine conditions and eventually in the middle Volgian to the development of hypersaline conditions. Evaporites formed during this period show cyclic successions of carbonates, anhydrite, halite, and claystone (Betz et al., 1987). Warm and arid conditions and associated facies continued into the early Ryazanian, approximately halfway TMS-3 (Fig. 7.10). The general lack of siliciclastic sediments indicates that the tectonic setting was less dynamic than for instance in the Dutch Central Graben. The average thickness of TMS-2 deposits is estimated to be less than 100 m (Fig. 7.20).

In the Broad Fourteens and West Netherlands basins, the major bounding faults were re-activated in response to the E-W extension initiated at the start of TMS-2. In both basins thick Zechstein salt accumulations are absent and the basin evolution was primarily controlled by fault movements. The numerous intra-basinal faults in the West Netherlands Basin were subjected to oblique-slip faulting during TMS-2, which resulted in a series of fault blocks consisting of half-grabens, intermediate highs and relay ramps (Racero-Baena & Drake, 1996; see also Fig. 7.12). Differential subsidence acted on these fault blocks and caused large variations in the thickness of the non-marine and marginal marine units (Mijnlieff, 2020) of the Alblasserdam Member in the West Netherlands Basin (Figs 7.12, 7.13) and of the Aerdenhout, Fourteens Claystone and Driehuis Mottled Claystone members in the Broad Fourteens Basin (Figs 7.15, 7.16). Deposits of substantial thickness accumulated during TMS-2, especially in the Broad Fourteens Basin (Fig. 7.20). Note that the half-grabens were inverted into pop-up and flower structures during the Late Cretaceous inversion (Racero-Baena & Drake, 1996; De Jager, 2003; Duin et al., 2006).

TMS-3 late Volgian-Ryazanian

TMS-3 reflects the transition from the active Late Cimmerian rift phase to the post-rift sag phase and is characterized by hiatuses, unconformities and sandstone deposition (Bouroullec et al., 2018; Verreussel et al., 2018; see also Fig. 7.19).

In the Step Graben, an hiatus usually separates sediments corresponding to TMS-3 from the underlying Upper Jurassic interval (Fig. 7.19). TMS-3 is often represented by a thin succession of marine claystones of the Lutine Formation (Verreussel et al., 2015, 2018; see also Fig. 7.20). The flooding of the Step Graben and deposition of the Lutine Formation reflects a relative sea-level rise at that time.

In the Dutch Central Graben the fault activity and salt movement that characterized TMS-2 gradually came to an end and the existing relief was flattened. The boundary between TMS-2 and TMS-3 is erosional and the base of the latter consists of glauconitic sandstones of the Scruff Greensand Formation (Figs 7.3, 7.4, 7. 7, 7.8). By the end of TMS-2, the Dutch Central Graben was characterized by a rugged landscape of emergent areas in a shallow sea, situated above turtle structures in the centre of the basin and shoulders and diapir crests near the basin margin. The sands derived from the erosion of these structures filled the remaining depressions in the existing relief. In the northern part of the Dutch Central Graben, for example in well B18-03 (Fig. 7.4), the Scruff Greensand is succeeded by open marine claystones of the Lutine Formation and TMS-3 reaches a thickness of 100 m (Fig. 7.20). The presence of open marine claystones reflects high eustatic sea levels at that time (Verreussel et al., 2018).

In the Terschelling Basin, the transition from TMS-2 to TMS-3 is conformable. This is one of the few areas in the Netherlands where the Jurassic-Cretaceous boundary interval is continuously marine and is not affected by erosion (Bouroullec et al., 2016a; see also Fig. 7.7). The Scruff Greensand Formation overlies the underlying Skylge Formation and reaches thicknesses in excess of 300 metres in the basin centre (Verreussel et al., 2018). The shale content of the Scruff Greensand Formation gradually increases upward, which is reflected in increasing gamma-ray values on well logs (see e.g. well L06-02 and M04-04 in Fig 7.8). Eventually, the sandstones were replaced by claystones of the Lutine Formation, indicating a more distal depositional setting.

In the Vlieland basin marine influence is recorded for the first time in TMS-3, represented by marginal marine sandstones and volcaniclastic sediments of the Wadden Volcaniclastic and Upper Zurich members (Fig. 7.7). The age of the volcaniclastic sedimentary rocks indicates that the Zuidwal volcano was active during TMS-3. The marine sediments are followed by non-marine claystones, thin sandstones and thick coal layers, indicating a back-barrier lagoonal setting and a regressive sea-level trend. The depositional cycle of TMS-3 ends with massive coastal sandstones of the Upper Zurich Member that reach a thickness of over 50 metres in well Vlieland Oost-1 (VLO-01) and signal the end of the most active rift phase (Fig. 7.9).

In the Lower Saxony Basin, the calcareous and evaporitic deposits of the Weiteveen Formation are replaced by marly claystones and limestone beds of the Coevorden Formation in TMS-3 (Figs 7.10, 7.11). The change from evaporitic to fresh water dominated depositional environments reflects the change to the humid climate conditions of the 'Wealden', which characterized the earliest Cretaceous of NW Europe (Abbink et al., 2001b; Stollhofen et al., 2008). TMS-3 ends with the organic-rich claystones of the Upper Coevorden Member that form the source rock for the oil of the Schoonebeek Field (De Jager & Geluk, 2007; Remmelts et al., 2025, this volume).

Figure 7.21. Time-thickness map of the Holland Formation in the West Netherlands Basin. The subtle variation in thickness is caused by differential subsidence of the fault blocks. Note that values representing a large time thickness along fault blocks, are modelling artefacts.

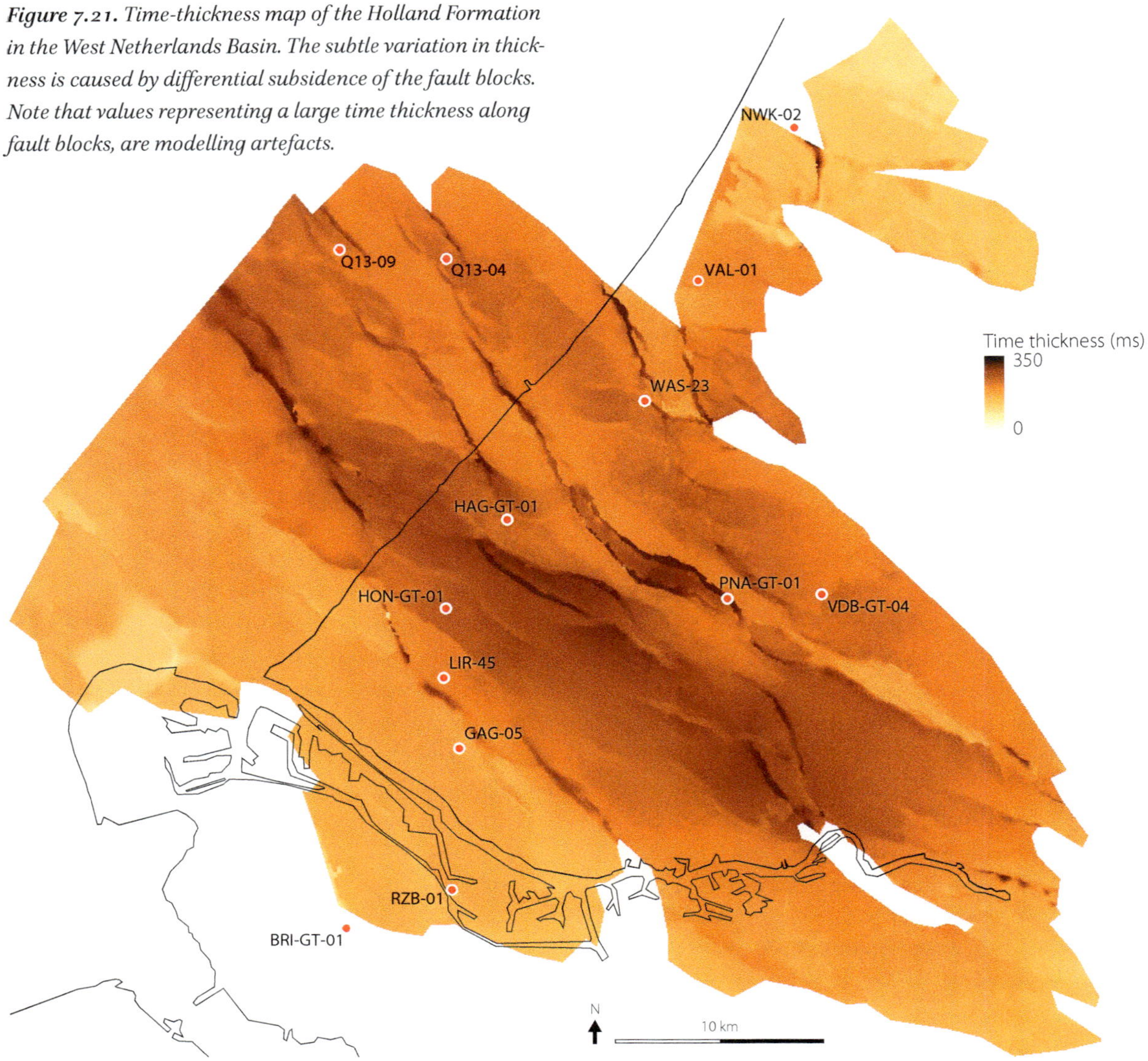

In the West Netherlands Basin, sedimentation of the non-marine Alblasserdam Member probably continued uninterruptedly throughout TMS-3 and filled the half-graben fault blocks (Fig. 7.12). However, based on the assumption that the fault activity decreased in the West Netherlands Basin during TMS-3 (Mijnlieff, 2020), it is conceivable that erosion occurred at this time and that parts of the Delft Sandstone and Rodenrijs Claystone members are separated by a time gap from the Alblasserdam Member, but so far unequivocal data to corroborate that scenario are lacking.

In the Broad Fourteens Basin erosion occurred during TMS-3 (Fig. 7.15). An hiatus separates the late Ryazanian Bloemendaal Member from the underlying Driehuis Mottled Claystone Member. Also in the Broad Fourteens Basin, a volcano must have been active at this time: volcanoclastic sediments occur in well P06-01 (Figs 7.15, 7.16; also see Van Bergen et al., 2025, this volume). The marine sandstones of the Bloemendaal Member interfinger with claystones and coals of the lagoonal Neomiodon Member. This coastal-lagoonal complex represents the transition from the active rift phase to the post-rift sag phase.

TMS-4 Valanginian-Barremian

TMS-4 represents the post-rift thermal sag following the Late Cimmerian rift and is represented by marine claystones and sandstones of the Rijnland Group that extend outside the basins onto the bordering platforms (Duin et al., 2006). In the Danish offshore, the base of TMS-4 is referred to as the Base Cretaceous Unconformity and is represented by bioturbated sandstones associated with slumped beds and with a water mass overturn from stratified and poorly oxygenated to oxygenated (Ineson et al., 2022). The slump features are linked by the authors to transgressive erosion and de-stabilization of the slopes along the basin margins.

On the Step Graben, TMS-4 is represented by a thin Vlieland Claystone Formation (Fig. 7.20). The absence of sandstones and the presence of a thin claystone unit indicates that subtle subsidence continued during TMS-4 in a relatively distal marine setting.

In the Dutch Central Graben, TMS-4 is either thin or missing due to erosion that took place during Late Cretaceous inversion (Figs 7.3, 7.4, 7.20). In wells B18-03 and F03-01, a kick in the gamma-ray log is observed at the base of the Rijnland Group, followed by a gradual increase (Fig. 7.4). This probably reflects erosion and a transgression and is equivalent to the Base Cretaceous Unconformity in the Danish offshore.

In the Terschelling Basin, claystones of the Vlieland Claystone Formation dominate TMS-4 and attain a thickness of 400 m or more (Figs 7.7, 7.20). Apparently, subsidence continued throughout the Valanginian, Hauterivian and Barremian without interruption and created substantial accommodation space. Often the distinction between the fine-grained Lutine and Vlieland Claystone formations is difficult to make, unless the thin Friesland Member of the Vlieland Sandstone Formation separates the two.

In the Vlieland Basin, the base of TMS-4 is characterized by the presence of the thick sandstone package of the Friesland Member (Figs 7.7, 7.9) which in well Vlieland Oost-1 (VLO-01) reaches a thickness of 150 metres, but which thins out rapidly towards the Terschelling Basin. The depositional setting of the Vlieland Basin was more proximal than that of the Terschelling Basin and sands coming from the Texel-IJsselmeer High accumulated there.

In the Lower Saxony Basin, TMS-4 is also characterized by a marine transgression in the Valanginian that established a shoreface complex in the area and resulted in the deposition of the Bentheim Sandstone Member (Figs 7.10, 7.11, 7.19), the main reservoir unit of the Dutch Schoonebeek oil field (De Jager, 2007) and which has also been extensively quarried in Germany for building stone (Dubelaar & Nijland, 2015). In the Hauterivian stage, a second shoreface complex developed, represented by the Gildehaus Sandstone Member. A small erosional gap is inferred between the Gildehaus Sandstone and the Bentheim Sandstone members (Vis et al., 2018; see also Figs 7.10, 7.19).

In the West Netherlands Basin, TMS-4 equivalent strata reach a thickness in excess of 700 meters and (Fig. 7.20), demonstrate that substantial TMS-3 subsidence continued for another 15 million years. There are, however, indications that the basin was affected by readjustment after TMS-3. An unconformity in the Hauterivan separates the Rijnland Group from the underlying Schieland Group (Figs 7.12, 7.14). The non-marine Alblasserdam Member that was deposited during TMS-2 and TMS-3 was succeeded by marginal marine deposits in the Valanginian. Deceleration of fault activity in TMS-3, a rise in base level and increasing marine influence led to the deposition of coastal sands of the Delft Sandstone Member, lagoonal claystones of the Rodenrijs Claystone Member and beach/barrier sandstones of the informal Amstel member. The stratigraphic relationships of these marginal marine units are complex, the Delft Sandstone Member for example is diachronous (Fig. 7.14) and the distinction between sandstones from the Delft Sandstone Member and sandstones from the Alblasserdam Member is not easy to establish based on wireline logs alone. The Schieland Group is succeeded by shallow marine sandstones of the Rijswijk and Rijn members, which are widely distributed across the basin and display erosive bases. These two members herald a change to marine conditions that persisted for the remainder of TMS-4. Note that the Oosterhout Platform, a narrow fault block near IJsselmonde and Ridderkerk (Fig. 7.1) remained in a non-marine environment for much longer (Figs 7.12, 7.13). Marine conditions reached the Oosterhout Platform in the latest Barremian when the De Lier Member onlapped onto the basin margin (Jeremiah et al., 2010; see also Figs 7.12, 7.13). The coastal sandstone units of TMS-4 display an onlapping pattern in the West Netherlands Basin. The occurrence and distribution of these units are mainly determined by eustatic sea-level variations (Jeremiah et al., 2010), but correlations between wells are not straightforward, as the marginal marine sandstones grade into finer grained sediments across short distances.

In the Broad Fourteens Basin, marine conditions were already reached at the base of TMS-4, where thick and massive sandstones of the Kotter and Helder members occur (Figs 7.15, 7.16). The provenance of these transgressive sands were probably the footwall shoulders of the large basin-bounding faults that were subjected to erosion after the active rift phase. The sandstones are succeeded by the open marine mudstones of the Vlieland Claystone Formation.

TMS-5 Aptian - Albian

In the North Sea area, TMS-5 marks the transition into the Late Cretaceous carbonate world. From a wider and more global perspective, the Aptian and Albian concluded the break-up of Pangea and paved the way for the next plate tectonic event, the alpine collision (De Jager, 2003). Large parts of the Netherlands had become fully marine by this time and the depositional areas were mostly located far away from major land masses.

In the Step Graben, the Dutch Central Graben, the Terschelling, Vlieland and Lower Saxony basins, TMS-5 is represented by thin successions of marls and claystones of the Holland Formation. A hiatus generally separates the Lower Holland Marl Member from the Albian Middle

Holland Claystone and Upper Holland Marl members (Van Adrichem Boogaert & Kouwe, 1993; see also Fig. 7.19). The hiatus has been ascribed to the Austrian tectonic phase (Herngreen & Wong, 2007) and is most evident in the Lower Saxony Basin where the entire Aptian is missing (Vis et al., 2018; see also Fig. 7.19).

In the Broad Fourteens and West Netherlands basins, the base of the Aptian is represented by the Lower Holland Marl Member, which displays very high gamma-ray readings at its base. The fossiliferous and organic-rich shale is time-equivalent to the Fischschiefer, a globally recognized black shale that reflects a regional maximum flooding (Jeremiah et al., 2010) and a global anoxic event (OAE-1a, Jenkyns, 2010). In the West Netherlands Basin, minor thickness variations in the Holland Formation can be seen on seismic images (Fig. 7.21), which indicates that differential subsidence still acted on the fault blocks within the basin.

Concluding remarks

The sedimentary rocks that were deposited during the Late Jurassic to Early Cretaceous are of economical and historical interest to Dutch society. For example, in 1648 the impressive Royal Palace of Amsterdam was constructed entirely out of the Lower Cretaceous Bentheim Sandstone, which was quarried just across the border in present day Germany. In the year 1943 oil was discovered in the small Dutch community of Schoonebeek from the same sandstone unit. For many years the Schoonebeek discovery was the largest onshore oil field in Western Europe. Not much later, Lower Cretaceous sandstones in and around Rijswijk in the western Netherlands proved oil bearing and, at the present-day, the Lower Cretaceous Delft Sandstone Member from the same area is being targeted for geothermal energy. Offshore, the Lower Cretaceous sandstones are considered for CO_2 storage.

Acknowledgements

Many current and former colleagues from the Geological Survey of the Netherlands and peers from academia contributed to knowledge of the Upper Jurassic-Lower Cretaceous geology, which is gratefully acknowledged. Industries stimulated and sponsored numerous biostratigraphic and integrated studies through which much of the present-day understanding of the Upper Jurassic-Lower Cretaceous geology came into being, which is also gratefully acknowledged. We thank Jochen Erbacher and Theo Wong† for providing helpful comments on earlier drafts of the manuscript

Digital map data

Spatial data of figures in this chapter for use in geographical information systems can be downloaded here: https://doi.org/10.5117/aup.28163300.

References

Abbink, O.A., 1998. Palynological investigations in the Jurassic of the North Sea region. PhD thesis Utrecht University (the Netherlands), Laboratory of Palaeobotany and Palynology: 192 pp.

Abbink, O.A., Callomon, J.H., Riding, J.H., Williams, P.D.B. & Wolfard, A., 2001a. Biostratigraphy of Jurassic-Cretaceous boundary strata in the Terschelling Basin, The Netherlands. Proceedings of the Yorkshire Geological Society 53(4): 275-302. DOI:10.1144/pygs.53.4.275

Abbink, O.A., Targarona, J., Brinkhuis, H. & Visscher, H., 2001b. Late Jurassic to earliest Cretaceous palaeoclimatic evolution of the Southern North Sea. Global and Planetary Change 30(3-4): 231-256. DOI:10.1016/S0921-8181(01)00101-1

Abbink, O.A., Van Konijnenburg-Van Cittert, J.H.A. & Visscher, H., 2004a. A sporomorph ecogroup model for the Northwest European Jurassic - Lower Cretaceous I: concepts and framework. Netherlands Journal of Geosciences, 83(1): 17-38. DOI:10.1017/S0016774600020436

Abbink, O.A., Van Konijnenburg-Van Cittert, J.H.A., Van der Zwan, C.J. & Visscher, H., 2004b. A sporomorph ecogroup model for the Northwest European Jurassic - Lower Cretaceous II: Application to an exploration well from the Dutch North Sea. Netherlands Journal of Geosciences 83(2): 81-92. DOI:10.1017/S0016774600020059

Abbink, O.A., Mijnlieff, H., Munsterman, D. & Verreussel, R., 2006. New stratigraphic insights in the Late Jurassic of the Southern Central North Sea Graben and Terschelling Basin (Dutch Central Graben) and related exploration potential. Netherlands Journal of Geosciences 85 (3): 221-238. DOI: 10.1017/S001677460002148X

Andsbjerg, J., 2003. Sedimentology and sequence stratigraphy of the Bryne and Lulu Formations, Middle Jurassic, northern Danish Central Graben. In: Ineson, J.R. & Surlyk, F. (eds): The Jurassic of Denmark and Greenland. Geological Survey of Denmark and Greenland Bulletin 1: 301-347.

Andsbjerg, J. & Dybkjær, K., 2003. Sequence stratigraphy of the Jurassic of the Danish Central Graben. In: Ineson, J.R. & Surlyk, F. (eds): The Jurassic of Denmark and Greenland. Geological Survey of Denmark and Greenland Bulletin 1: 265-300.

Arfai, J., Jähne, F., Lutz, R., Franke, D., Gaedicke, C. & Kley, J., 2014. Late Palaeozoic to Early Cenozoic geological evolution of the northwestern German North Sea (Entenschnabel): New results and insights. Netherlands Journal of Geosciences 93(4): 147-174. DOI:10.1017/njg.2014.22

Arfai, J. & Lutz, R., 2018. 3D basin and petroleum system modelling of the NW German North Sea (Entenschnabel). Geological Society, London, Petroleum Geology Conference Series 8: 67-86. DOI: 10.1144/PGC8.35

Beha, A., Thomsen, R.O. & Littke, R., 2008. Thermal history, hydrocarbon generation and migration in the Horn Graben in the Danish North Sea: a 2D basin modelling study. International Journal of Earth Sciences 97(5): 1087-1100. DOI: 10.1007/s00531-007-0247-2

Best, G., Kockel, F. & Schoeneich, H., 1983. Geological history of the southern Horn Graben. Geologie en Mijnbouw - Netherlands Journal of Geoscience 62: 25-33.

Betz, D., Führer, F., Greiner, G. & Plein, E., 1987. Evolution of the Lower Saxony Basin. Tectonophysics 137: 127-170. DOI: 10.1016/0040-1951(87)90319-2

Bouroullec, R. & Ten Veen, J.H., 2025. Salt Tectonics. In: Ten Veen, J.H., Vis, G.-J., De Jager, J. & Wong, Th.E. (eds): Geology of the Netherlands, second edition. Amsterdam University Press (Amsterdam): 457-491. DOI: 10.5117/9789463728362_ch12

Bouroullec, R., Verreussel, R., Boxem, T., De Bruin, G., Zijp, M., Janssen, N., Kersthold-Boegehold, S., Munsterman, D. & Korosi, D., 2016a. The COMMA Project: Understanding Jurassic sands of the complex margins of the eastern part of the Terschelling Basin during the Upper Jurassic and Lowermost Cretaceous. TNO (Utrecht), Report No. R11341: 189 pp.

Bouroullec, R., Verreussel, R., Geel, K., Munsterman, D., De Bruin, G., Zijp, M., Janssen, M., Millan, I. & Boxem, T., 2016. The FOCUS Project: Upper Jurassic sandstones: Detailed sedimentary facies analysis, correlation and stratigraphic architecture of hydrocarbon bearing shoreface complexes in the Dutch offshore. TNO (Utrecht), Report; 228 pp. https://www.nlog.nl/sites/default/files/2019-12/comma_report_full-v2016.11.30.pdf

Bouroullec, R., Verreussel, R.M.C.H., Geel, C.R., De Bruin, G., Zijp, M.H.A.A., Kőrösi, D., Munsterman, D.K., Janssen, N.M.M. & Kerstholt-Boegehold, S.J., 2018. Tectonostratigraphy of a rift basin affected by salt tectonics: synrift Middle Jurassic-Lower Cretaceous Dutch Central Graben, Terschelling Basin and neighbouring platforms, Dutch offshore. Geological Society of London, Special Publications 469(1): 269-303. DOI: 10.1144/SP469.22

Casey, R., Mesezhnikov, M.S. & Shulgina, N.I., 1977. Correlation of the Jurassic–Cretaceous boundary beds of England, Russian Platform, Subpolar Urals and Siberia. Proceedings of the USSR Academy of Sciences, Geological Series 7: 14-33.Cooper, B.S., Barnard, P.C. & Telnaes, N., 1995. The Kimmeridge clay formation of the north sea. In: Katz, B.J. (ed.): Petroleum source rocks. Springer (Berlin, Heidelberg): 89-110.

Copestake, P., Sims, A.P., Crittenden, S., Hamar, G.P., Ineson, J.R., Rose, P.T. & Tringham, M.E., 2003. Lower Creatceous. In: Evans, D., Graham, C., Armour, A. & Bathurst, P. (eds): The millennium atlas: Petroleum geology of the central and northern North Sea. Geological Society of London: 191-211.

Coward, M.P., Dewey, J., Hempton, M. & Holroyd, J., 2003. Tectonic evolution. In: Evans, D., Graham, C., Armour, A. & Bathurst, P. (eds): The millennium atlas: Petroleum geology of the central and northern North Sea. Geological Society of London : 17-33.

De Jager, J., 2003. Inverted basins in the Netherlands, similarities and differences. Netherlands Journal of Geosciences 82(4): 339-349. DOI:10.1017/S0016774600020175

De Jager, J., 2007. Geological development. In: Wong, Th.E., Batjes, D.A.J. & De Jager, J. (eds): Geology of the Netherlands. Royal Netherlands Academy of Arts and Sciences (KNAW, Amsterdam): 5-26.

De Jager, J., Doyle, M.A., Grantham, P.J. & Mabillard, J.E., 1996. Hydrocarbon habitat of the West Netherlands Basin. In: Rondeel, H.E., Batjes, D.A.J. & Nieuwenhuijs, W.H. (eds): Geology of gas and oil under the Netherlands. Springer (Dordrecht): 191-209. DOI: 10.1007/978-94-009-0121-6_17

De Jager, J. & Geluk, M.C., 2007. Petroleum geology. In: Wong, Th.E., Batjes, D.A.J.& De Jager, J. (eds): Geology of the Netherlands. Royal Netherlands Academy of Arts and Sciences (KNAW, Amsterdam): 241-264.

De Jager, J., Van Ojik, K. & Smit, J., 2025. Geological development. In: Ten Veen, J.H., Vis, G.-J., De Jager, J. & Wong, Th.E. (eds): Geology of the Netherlands, second edition. Amsterdam University Press (Amsterdam): 21-51. DOI: 10.5117/9789463728362_ch01

Den Hartog Jager, D.G., 1996. Fluviomarine sequences in the Lower Cretaceous of the West Netherlands Basin: correlation and seismic expression. In: Rondeel, H.E., Batjes, D.A.J. & Nieuwenhuijs, W.H. (eds): Geology of gas and oil under the Netherlands. Kluwer Academic Publishers (Dordrecht): 229-242. DOI: 10.1007/978-94-009-0121-6_19

Dera, G., Brigaud, B., Monna, F., Laffont, R., Pucéat, E., Deconinck, J.F., Pellenard, P., Joachimski, M.M. & Durlet, C., 2011. Climatic ups and downs in a disturbed Jurassic world. Geology 39(3): 215-218. DOI: 10.1130/G31579.1

Dera, G., Prunier, J., Smith, P.L., Haggart, J.W., Popov, E., Guzhov, A., Rogov, M., Delsate, D., Thies, D., Cuny, G. & Pucéat, E., 2015. Nd isotope constraints on ocean circulation, paleoclimate, and continental drainage during the Jurassic breakup of Pangea. Gondwana Research 27(4): 1599-1615. DOI: 10.1016/j.gr.2014.02.006

DeVault, B. & Jeremiah, J., 2002. Tectonostratigraphy of the Nieuwerkerk Formation (Delfland subgroup), West Netherlands Basin. American Association of Petroleum Geologists Bulletin 86(10): 1679-1707.

Donselaar, M.E., Groenenberg, R.M. & Gilding, D.T., 2015. Reservoir Geology and Geothermal Potential of the Delft Sandstone Member in the West Netherlands Basin. Proceedings World Geothermal Congress, Melbourne: 9 pp.

Doornenbal, J.C., Kombrink, H., Bouroullec, R., Dalman, R.A.F., De Bruin, G., Geel, C.R., Houben, A.J.P., Jaarsma, B., Juez-Larré, J., Kortekaas, M., Mijnlieff, H.F., Nelskamp, S., Pharaoh, T.C., Ten Veen, J.H., Ter Borgh, M., Van Ojik, K., Verreussel, R.M.C.H., Verweij, J.M. & Vis, G.-J., 2019. New insights on subsurface energy resources in the Southern North Sea Basin area. In: Patruno, S., Archer, S.G., Chiarella, D., Howell, J.A., Jackson, C.A-L. & Kombrink, H. (eds): Cross-border themes in petroleum geology I: The North Sea. Geological Society of London, Special Publications 494: 233-268. DOI: 10.1144/SP494-2018-178

Doornenbal, J.C. & Stevenson, A.G. (eds), 2010. Petroleum Geological Atlas of the Southern Permian Basin Area. EAGE Publications B.V. (Houten): 342 pp.

Dubelaar, C.W. & Nijland, T.G., 2015. The Bentheim Sandstone: geology, petrophysics, varieties and its use as dimension stone. In: Lollino, G., Giordan, D., Marunteanu, C., Christaras, B., Yoshinori, I. & Margottini, C. (eds): Engineering Geology for Society and Territory. Springer (Berlin) 8: 557-563 . Springer. DOI:10.1007/978-3-319-09408-3

Duin, E., Doornenbal, J.C., Rijkers, R.H.B., Verbeek, J.W. & Wong, Th.E., 2006. Subsurface structure of the Netherlands; results of recent onshore and offshore mapping. Netherlands Journal of Geosciences 85(4): 245-276. DOI: 10.1017/S0016774600023064

Duxbury, S.D.S.R., Kadolsky, D. & Johansen, S., 1999. Sequence stratigraphic subdivision of the Humber group in the outer moray firth area (UKCS, North Sea). Geological Society of London, Special Publications 152(1): 23-54.

Eldrett, J.S. & Vieira, M., 2022. An integrated carbon isotope and bio-sequence stratigraphic study of the Early Cretaceous to Paleogene, Central North Sea. Marine and Petroleum Geology 141: 105696. DOI: 10.1016/j.marpetgeo.2022.105696

Erbacher, J., Hiss, M., Luppold, F.W. & Mutterlose, J., 2014. LithoLex online database. Bundesanstalt Fur Geowissenschaften Und Rohstoffe.

Evans, D., Graham, C., Armour, A. & Bathurst, P., 2003. The Millennium Atlas: Petroleum Geology of the Central and Northern North Sea. The Geological Society of London: 389 pp.

Fraser, A., Robinson, A., Johnson, H., Underhill, J. & Kadolsky, P., 2003. Middle Jurassic to Early Cretaceous. In: Graham, C., Armour, A. & Bathurst, P. (eds): The Millennium Atlas: Petroleum geology of the Central and Northern North Sea. The Geological Society of London: 157-189.

Gale, A.S., Mutterlose, J., Batenburg, S., Gradstein, F.M., Agterberg, F.P., Ogg, J.G. & Petrizzo, M.R., 2020. The Cretaceous period. In: Gradstein, F.M., Ogg, J.G., Schmitz, M.D. & Ogg, G.M. (eds): Geologic time scale 2020. Elsevier (Amsterdam): 1023-1086. DOI: 10.1016/B978-0-12-824360-2.00026-7

Geluk, M.C., 2007. Triassic. In: Wong, Th.E., Batjes, D.A.J.& De Jager, J. (eds): Geology of the Netherlands. Royal Netherlands Academy of Arts and Sciences (KNAW, Amsterdam): 85-106.

Geluk, M.C., Duin, E.T., Dusar, M., Rijkers, R.H.B., Van den Berg, M.W. & Van Rooijen, P., 1995. Stratigraphy and tectonics of the Roer Valley Graben. Geologie en Mijnbouw 73: 129-141.

Gradstein, F. M., Ogg, J. G., Schmitz, M. D. & Ogg, G. M. (eds), 2020. Geologic time scale 2020. Elsevier (Amsterdam): 1357 pp. DOI: 10.1016/C2020-1-02369-3

Heim, S., Lutz, R., Nelskamp, S., Verweij, J.M., Kaufmann, D. & Reinhardt, L., 2013. Geological Evolution of the North Sea: Cross-border Basin Modeling Study on the Schillground High. Energy Procedia 40: 222-231. DOI: 10.1016/j.egypro.2013.08.026

Herngreen, G.F.W., Kerstholt, S.J. & Munsterman, D.K., 2000. Callovian-Ryazanian (Upper Jurassic) palynostratigraphy of the Central North Sea Graben and Vlieland Basin, the Netherlands. Mededelingen Nederlands Instituut Voor Toegepaste Geowetenschappen TNO 63: 1-99.

Herngreen, G.F.W., Smit, R. & Wong, Th.E., 1991. Upper Jurassic-Cretaceous stratigraphy of the Vlieland Basin, The Netherlands. In: Michelsen, O. & Frandsen, F. (eds): The Jurassic in the Southern Central Trough. Danmarks Geologiske Undersøgelse Series B 16: 17-19.

Herngreen, G.F.W. & Wong, Th.E., 1989. Revision of the Late Jurassic stratigraphy of the Dutch Central North Sea Graben. Geologie En Mijnbouw 68(1): 73-105.

Herngreen, G.F.W. & Wong, Th.E., 2007. Cretaceous. In: Wong, Th.E., Batjes, D.A.J.& De Jager, J. (eds): Geology of the Netherlands. Royal Netherlands Academy of Arts and Sciences (KNAW, Amsterdam): 127-150.

Hesselbo, S.P., Ogg, J.G., Ruhl, M., Hinnov, L.A. & Huang, C.J., 2020. The Jurassic Period. In: Gradstein, F.M., Ogg, J.G., Schmitz, M.D. & Ogg, G.M. (eds): Geologic time scale 2020. Elsevier (Amsterdam): 955-1021. DOI: 10.1016/B978-0-12-824360-2.00026-7

Hoedemaeker, P.J. & Herngreen, G.F.W., 2003. Correlation of Tethyan and Boreal Berriasian-Barremian strata with emphasis on strata in the subsurface of the Netherlands. Cretaceous Research 24(3): 253-275. DOI: 10.1016/S0195-6671(03)00044-2

Houben, A.J.P., Verreussel, R.M.C.H., Janssen, N.M.M., Kerstholt-Boegehold, S.J. & Munsterman, D.K., in prep. A palynostratigraphically calibrated stable carbon isotope curve for the Upper Jurassic and Lower Cretaceous from the North Sea Basin.

Husmo, T., Hamar, G.P., Høiland, O., Johannessen, E.P., Rømuld, A., Spencer, A. & Titterton, R., 2003. Lower and Middle Jurassic. In: Graham, C., Armour, A. & Bathurst, P. (eds): The Millennium Atlas: Petroleum geology of the Central and Northern North Sea. The Geological Society of London: 128-155.

Ineson, J.R., Sheldon, E., Dybkjær, K., Andersen, C., Alsen, P. & Jakobsen, F., 2022. The 'Base Cretaceous Unconformity' in a basin-centre setting, Danish Central Graben, North Sea: A cored record of resedimentation and condensation accompanying transgression and basinal overturn. Marine and Petroleum Geology 137: 105489. DOI: 10.1016/j.marpetgeo.2021.105489

Jansonius, J. & McGregor, D. C. (eds), 1996. Palynology: principles and applications. American Association of Stratigraphic Palynologists Foundation.

Janssen, N.M.M., Rogov, M.A. & Zakharov, V.A., 2022. Ryazanian (Berriasian) molluscs and biostratigraphy of the Dutch and Norwegian North Sea area (south of Viking Graben). Nether-

lands Journal of Geosciences 101(8). DOI: 10.1017/njg.2022.5
Japsen, P., Britze, P. & Andersen, C., 2003. Upper Jurassic – Lower Cretaceous of the Danish Central Graben: structural framework and nomenclature. Bulletin of the Geological Society of Denmark: 233-246.
Jenkyns, H.C., 2010. Geochemistry of oceanic anoxic events. Geochemistry, Geophysics, Geosystems 11(3).
Jenkyns, H.C., Schouten-Huibers, L., Schouten, S. & Sinninghe Damsté, J.S., 2012. Warm Middle Jurassic–Early Cretaceous high-latitude sea-surface temperatures from the Southern Ocean. Climate of the Past 8(1): 215-226. DOI: 10.5194/cp-8-215-2012
Jeremiah, J.M., Duxbury, S. & Rawson, P.F., 2010. Lower Cretaceous of the southern North Sea Basins: reservoir distribution within a sequence stratigraphic framework. Netherlands Journal of Geosciences 89(3-4): 203-237. DOI: 10.1017/S0016774600000706
Johannessen, P.N., 2003. Sedimentology and sequence stratigraphy of paralic and shallow marine Upper Jurassic sandstones in the northern Danish Central Graben. Geological Survey of Denmark and Greenland Bulletin: 367-402.
Johannessen, P.N., Dybkjær, K., Andersen, C., Kristensen, L., Hovikoski, J. & Vosgerau, H., 2010a. Upper Jurassic reservoir sandstones in the Danish Central Graben: new insights on distribution and depositional environments. Geological Society of London, Petroleum Geology Conference Series 7: 127-143. DOI: 10.1144/0070127.
Johannessen, P.N., Nielsen, L.H., Nielsen, L., Møller, I., Pejrup, M. & Andersen, T.J., 2010b . Architecture of an Upper Jurassic barrier island sandstone reservoir, Danish Central Graben: implications of a Holocene–Recent analogue from the Wadden Sea. Geological Society of London, Petroleum Geology Conference Series 7: 145-155. DOI: 10.1144/0070145.
Johannessen, P.N., Dybkjær, K. & Rasmussen, E.S., 1996. Sequence stratigraphy of Upper Jurassic reservoir sandstones in the northern part of the Danish Central Trough, North Sea. Marine and Petroleum Geology 13: 755-770.
Kadolsky, D., Johansen, S.J. & Duxbury, S., 1999. Sequence stratigraphy and sedimentary history of the Humber Group (Late Jurassic-Ryazanian) in the Outer Moray Firth (UKCS, North Sea). In: Fleet, A.J. & Boldy, S.A.R. (eds): Petroleum Geology of North-West Europe: Proceedings of the 5th Conference. The Geological Society of London: 839-860.
Keller, C.E., Hochuli, P.A., Weissert, H., Bernasconi, S.M., Giorgioni, M. & Garcia, T.I., 2011. A volcanically induced climate warming and floral change preceded the onset of OAE1a (Early Cretaceous). Palaeogeography, Palaeoclimatology, Palaeoecology 305(1-4): 43-49.
Klassen, H., 1984. Geologie des Osnabrücker Berglandes. Naturwissenschaftliches Museum (Osnabrück): 672 pp.
Kombrink, H., Doornenbal, J.C., Duin, E.J.T., Den Dulk, M., Ten Veen, J.H. & Witmans, N., 2012. New insights into the geological structure of the Netherlands; results of a detailed mapping project. Netherlands Journal of Geosciences 91(4): 419-446. DOI: 10.1017/S0016774600000329
Lott, G.K., Wong, Th.E., Dusar, M., Andsbjerg, J., Mönnig, E., Feldman-Olszewska, A. & Verreussel, R.M.C.H., 2010. Jurassic. In: JDoornenbal, J.C. & Stevenson, A.G. (eds): Petroleum Geological Atlas of the Southern Permian Basin Area. EAGE Publications B.V. (Houten): 175-193.
Michelsen, O., Nielsen, L.H., Andsbjerg, J. & Surlyk, F., 2003. Jurassic lithostratigraphy and stratigraphic development onshore and offshore Denmark. Geological Survey of Denmark and Greenland Bulletin: 147-216.
Mijnlieff, H.F., 2020. Introduction to the geothermal play and reservoir geology of the Netherlands. Netherlands Journal of Geosciences 99(2): 1-19. DOI: 10.1017/njg.2020.2
Mijnlieff, H., Buijze, L., Rosendaal, E., Schoof, F., Vorage, R. & Van Wees, J.-D., 2025. Geothermal energy. In: Ten Veen, J.H., Vis, G.-J., De Jager, J. & Wong, Th.E. (eds): Geology of the Netherlands, second edition. Amsterdam University Press (Amsterdam): 701-727. DOI:10.5117/9789463728362_ch19
Müller, S.M., Jähne-Klingberg, F., Thöle, H., Jakobsen, F.C., Bense, F., Winsemann, J. & Gaedicke, C., 2023. Jurassic to Lower Cretaceous tectonostratigraphy of the German Central Graben, southern North Sea. Netherlands Journal of Geosciences 102(4). DOI: 10.1017/njg.2023.4
Munsterman, D.K., Verreussel, R.M.C.H., Mijnlieff, H.F., Witmans, N., Kerstholt-Boegehold, S. & Abbink, O.A., 2012. Revision and update of the Callovian-Ryazanian Stratigraphic Nomenclature in the northern Dutch offshore, i.e. Central Graben Subgroup and Scruff Group. Netherlands Journal of Geosciences 91(4): 555-590. DOI: 10.1017/S001677460000038X
Mutterlose, J. & Bornemann, A., 2000. Distribution and facies patterns of Lower Cretaceous sediments in northern Germany: a review. Cretaceous Research 21(6): 733-759.
Partington, M.A., Mitchener, B.C., Milton, N.J. & Fraser, A.J., 1993a . Genetic sequence stratigraphy for the North Sea Late Jurassic and Early Cretaceous : distribution and prediction of Kimmeridgian-Late Ryazanian reservoirs in the North Sea and adjacent areas. In: Parker, J.R. (ed.): Petroleum Geology of North-West Europe, Proceedings of the 4th Conference. Geological Society of London: 347-370.
Partington, M.A., Copestake, P., Mitchener, B.C. & Underhill, J.R., 1993b . Biostratigraphic calibration of genetic stratigraphic sequences in the Jurassic-lowermost Cretaceous (Hettangian to Ryazanian) of the North Sea and adjacent areas. In: Parker, J.R. (ed.): Petroleum Geology of North-West Europe: Proceedings of the 4th Conference. Geological Society of London: 371-386.
Patruno, S., Kombrink, H. & Archer, S.G., 2021. Cross-border stratigraphy of the Northern, Central and Southern North Sea: a comparative tectono-stratigraphic megasequence synthesis. Geological Society of London, Special Publications 494 (1): 13-83. DOI: 10.1144/SP494-2020-228
Peeters, S.H.J., 2016. Mesozoic strike-slip faults in the northern Dutch offshore; new insights from seismic-and analogue mod-

elling data. MSc thesis Utrecht University (the Netherlands): 106 pp.
Peters, G. & Van Balen, R.T., 2007. Tectonic geomorphology of the northern Upper Rhine graben, Germany. Global and Planetary Change 58(1-4): 310-334. DOI: 10.1016/j.gloplacha.2006.11.041
Racero-Baena, A. & Drake, S.J., 1996. Structural style and reservoir development in the West Netherlands oil province. In: Rondeel, H.E., Batjes, D.A.J. & Nieuwenhuijs, W.H. (eds): Geology of gas and oil under the Netherlands. Kluwer Academic Publishers (Dordrecht): 211-228. DOI: 10.1007/978-94-009-0121-6_18
Remmelts, G., Nelskamp, S., De Jager, J. & Geluk, M.C., 2025. Petroleum Geology. In: Ten Veen, J.H., Vis, G.-J., De Jager, J. & Wong, Th.E. (eds): Geology of the Netherlands, second edition. Amsterdam University Press (Amsterdam): 535-575. DOI:10.5117/9789463728362_ch14
Rutten, K.W., Den Hartog Jager, D. & Vis, G.-J., 2020. Unconformity mapping in the Schoonebeek oil field, the Netherlands. First Break 38(11): 33-42. DOI: 10.3997/1365-2397.fb2020078
Schneider, A.C., Heimhofer, U., Heunisch, C. & Mutterlose, J., 2018a. From arid to humid – The Jurassic–Cretaceous boundary interval in northern Germany. Review of Palaeobotany and Palynology 255: 57-69. DOI: 10.1016/j.revpalbo.2018.04.008
Schneider, A.C., Heimhofer, U., Heunisch, C. & Mutterlose, J., 2018b. The Jurassic–Cretaceous boundary interval in non-marine strata of northwest Europe – New light on an old problem. Cretaceous Research 87: 42-54. DOI: 10.1016/j.cretres.2017.06.002
Seyfang, B., Aigner, T., Munsterman, D. & Irmen, A., 2017. An integrated workflow to assess the remaining potential of mature hydrocarbon basins: a case study from Northwest Germany (Upper Jurassic/Lower Cretaceous, Lower Saxony Basin). International Journal of Earth Sciences 106(3): 1075-1105. DOI: 10.1007/s00531-016-1354-8
Stollhofen, H., Bachmann, G.H., Barnasch, J., Bayer, U., Beutler, G., Franz, M., Kästner, M., Legler, B., Mutterlose, J. & Radies, D., 2008. Upper Rotliegend to Early Cretaceous basin development. In: Littke, R., Bayer, U., Gajewski, D. & Nelskamp, S. (eds): Dynamics of Complex Intracontinental Basins: The Central European Basin System. Springer (Berlin): 181-210.
Ten Veen, J.H., Van Gessel, S.F. & Den Dulk, M., 2012. Thin- and thick-skinned salt tectonics in the Netherlands; a quantitative approach. Netherlands Journal of Geosciences 91(4): 447-464. DOI: 10.1017/S0016774600000330
Trabucho Alexandre, J.P. & Wong, Th. E., 2025. Rhaetian to Middle Jurassic. In: Ten Veen, J.H., Vis, G.-J., De Jager, J. & Wong, Th.E. (eds): Geology of the Netherlands, second edition. Amsterdam University Press (Amsterdam): 185-209. DOI: 10.5117/9789463728362_ch06
Underhill, J.R. & Partington, M.A., 1993. Jurassic thermal doming and deflation in the North Sea: implications of the sequence stratigraphic evidence. In: Parker, J.R. (ed.): Petroleum Geology of Northwest Europe: Proceedings of the 4th Conference. Geological Society of London, Petroleum Geology Conference Series 4: 337-345. DOI: 10.1144/0040337
Van Adrichem Boogaert, H.A. & Kouwe, W.F.P., 1993. Stratigraphic nomenclature of The Netherlands; revision and update by RGD and NOGEPA, Section G. Mededelingen Rijks Geologische Dienst 50: 1-80.
Van Bergen, M.J. & Sissingh, W., 2007. Magmatism in the Netherlands:expression of the north-west European rifting history. In: Wong, Th.E., Batjes, D.A.J.& De Jager, J. (eds): Geology of the Netherlands. Royal Netherlands Academy of Arts and Sciences (KNAW, Amsterdam): 197-222.
Van Bergen, M.J., Vis, G.-J., Sissingh, W., Koornneef, J. & Brouwers, I., 2025. Magmatism in the Netherlands: expression of the northwest European rifting history. In: Ten Veen, J.H., Vis, G.-J., De Jager, J. & Wong, Th.E. (eds): Geology of the Netherlands, second edition. Amsterdam University Press (Amsterdam): 393-455. DOI: 10.5117/9789463728362_ch11
Van Buchem, F.S.P., Smit, F.W.H., Buijs, G.J.A., Trudgill, B. & Larsen, P.-H., 2018. Tectonostratigraphic framework and depositional history of the Cretaceous–Danian succession of the Danish Central Graben (North Sea)–new light on a mature area. In: Bowman, M. & Levell, B. (eds): Petroleum Geology of NW Europe: 50 Years of Learning – Proceedings of the 8th Petroleum Geology Conference. Geological Society of London: 9-46. DOI: 10.1144/PGC8.24
Van der Hoeven, I.C., Verreussel, R.M.C.H., Riboulleau, A., Tribovillard, N. & Van de Schootbrugge, B., 2022. Climate-controlled organic matter accumulation as recorded in the Upper Jurassic Argiles de Châtillon Formation, a shallow-marine counterpart of the Kimmeridge Clay Formation. Geological Magazine 1-22. DOI: 10.1017/S0016756822001121
Van der Molen, A.S., 2004. Sedimentary development, seismic stratigraphy and burial compaction of the Chalk Group in the Netherlands North Sea area. PhD thesis Utrecht University (the Netherlands): 175 pp.
Van Lochem, H., Vis, G.-J. & Jagt, J.W.M., 2025. Late Cretaceous. In: Ten Veen, J.H., Vis, G.-J., De Jager, J. & Wong, Th.E. (eds): Geology of the Netherlands, second edition. Amsterdam University Press (Amsterdam): 253-291. DOI: 10.5117/9789463728362_ch08
Van Staalduinen, C.J., Van Adrichem Boogaert, H.A., Bless, M.J.M., Doppert, J.W.C., Harsveldt, H.M., Van Montfrans, H.M., Oele, E., Wermuth, R.A. & Zagwijn, W.H., 1979. The geology of the Netherlands. Mededelingen Rijks Geologische Dienst 31(2): 9-49.
Van Winden, M., De Jager, J., Jaarsma, B. & Bouroullec, R., 2018. New insights into salt tectonics in the northern Dutch offshore: a framework for hydrocarbon exploration. In: Kilhams, B., Kukla, P.A., Mazur, S., McKie, T., Mijnlief, H.F. & Van Ojik, K. (eds): Mesozoic Resource Potential in the Southern Permian Basin. Geological Society of London, Special Publications 469: 99-117. DOI: 10.1144/SP469.9
Vejbæk, O.V., Andersen, C., Dusar, M., Herngreen, G.F.W., Krabbe,

H., Leszczyński, K., Lott, G.K., Mutterlose, J. & Van der Molen, A.S., 2010. Cretaceous. In: Doornenbal, J.C. & Stevenson, A.G. (eds): Petroleum Geological Atlas of the Southern Permian Basin Area. EAGE Publications B.V. (Houten): 194-209.

Verreussel, R., Houben, A., Munsterman, D., Janssen, N. & Kerstholt-Boegehold, S., 2015. The JUSTRAT Project: New stratigraphic framework for the Upper Jurassic-Lower Cretaceous in the southern North Sea using integrated novel techniques. TNO (Utrecht), Report No. R10343: 112 pp.

Verreussel, R.M.C.H., Bouroullec, R., Munsterman, D.K., Dybkjær, K., Geel, C.R., Houben, A.J.P., Johannessen, P.N. & Kerstholt-Boegehold, S.J., 2018. Stepwise basin evolution of the Middle Jurassic–Early Cretaceous rift phase in the Central Graben area of Denmark, Germany and The Netherlands. In: Kilhams, B., Kukla, P.A., Mazur, S., McKie, T., Mijnlieff, H.F. & Van Oijk, K. (eds): Mesozoic resource potential in the Southern Permian Basin. Geological Society of London, Special Publications 469: 305-340. DOI: 10.1144/SP469.23

Verweij, J.M. & Simmelink, H.J., 2002. Geodynamic and hydrodynamic evolution of the Broad Fourteens Basin (The Netherlands) in relation to its petroleum systems. Marine and Petroleum Geology 19(3): 339-359.

Vickers, M.L., Fernandez, A., Hesselbo, S.P., Price, G.D., Bernasconi, S.M., Lode, S., Ullmann, C.V., Thibault, N., Hougaard, I.W. & Korte, C., 2020. Unravelling Middle to Late Jurassic palaeoceanographic and palaeoclimatic signals in the Hebrides Basin using belemnite clumped isotope thermometry. Earth and Planetary Science Letters 546: 116401. DOI: 10.1016/j.epsl.2020.116401

Vis, G.-J., 2020. Nieuwe vulkaan ontdekt in Nederlandse ondergrond. https://wetenschap.nu/nieuwe-vulkaan-ontdekt-in-nederlandse-ondergrond

Vis, G.-J., Smoor, W.D., Rutten, K.W., De Jager, J. & Mijnlieff, H.F., 2018. Tectonic control on the Early Cretaceous Bentheim Sandstone sediments in the Schoonebeek oil field, The Netherlands. Geological Society of London, Special Publications 469: 435-455. DOI: 10.1144/SP469.25

Vondrak, A.G., Donselaar, M.E. & Munsterman, D.K., 2018. Reservoir architecture model of the Nieuwerkerk Formation (Early Cretaceous, West Netherlands Basin): diachronous development of sand-prone fluvial deposits. Geological Society of London, Special Publications 469: 423-434. DOI: 10.1144/SP469.18

West, I.M., 1975. Evaporites and associated sediments of the basal Purbeck Formation (Upper Jurassic) of Dorset. Proceedings of the Geologists' Association 86(2): 205-225.

Willems, C.J.L., Vondrak, A., Mijnlieff, H.F., Donselaar, M.E. & Van Kempen, B.M.M., 2020. Geology of the Upper Jurassic to Lower Cretaceous geothermal aquifers in the West Netherlands Basin – an overview. Netherlands Journal of Geosciences 99(1). DOI: 10.1017/njg.2020.1

Willems, C.J.L., Vondrak, A., Munsterman, D.K., Donselaar, M.E. & Mijnlieff, H.F., 2017a . Regional geothermal aquifer architecture of the fluvial Lower Cretaceous Nieuwerkerk Formation – a palynological analysis. Netherlands Journal of Geosciences 96(4): 319-330. DOI: 10.1017/njg.2017.23

Willems, C.J., Nick, H.M., Donselaar, M.E., Weltje, G.J. & Bruhn, D.F., 2017b. On the connectivity anisotropy in fluvial Hot Sedimentary Aquifers and its influence on geothermal doublet performance. Geothermics 65: 222-233. DOI: 10.1016/j.geothermics.2016.10.002

Willems, C.J., Nick, H.M., Weltje, G.J. & Bruhn, D.F., 2017c . An evaluation of interferences in heat production from low enthalpy geothermal doublets systems. Energy 135: 500-512. DOI: 10.1016/j.energy.2017.06.129

Wimbledon, W.A., 2017. Developments with fixing a Tithonian/Berriasian (J/K) boundary. Volumina Jurassica 15(1): 181-186.

Wimbledon, W.A., Rehakova, D., Svobodová, A., Elbra, T., Schnabel, P., Pruner, P., Sifnerova, K., Kdyr, S., Frau, C., Schnyder, J. & Galbrun, B., 2020a . The proposal of a GSSP for the Berriasian Stage (Cretaceous System): part 2. Volumina Jurassica 18(2): 119-158.

Wimbledon, W.A., Reháková, D., Svobodová, A., Elbra, T., Schnabel, P., Pruner, P., Sifnerova, K., Kdyr, S., Dzyuba, O., Schyder, J. & Galbrun, B., 2020b . The proposal of a GSSP for the Berriasian Stage (Cretaceous System): part 1. Volumina Jurassica 18(1): 53-106.

Wimbledon, W.A.P., Casellato, C.E., Reháková, D., Bulot, L.G., Erba, E., Gardin, S., Verreussel, R.M.C.H., Munsterman, D.K. & Hunt, C.O., 2011. Fixing a basal Berriasian and Jurassic/Cretaceous (J/K) boundary – Is there perhaps some light at the end of the tunnel? Revista Italiana Di Paleontologia e Stratigrafia 17(2): 295-307.

Witte, L.J. & Lissenberg, T., 1994. Ostracods from Callovian to Ryazanian strata ("Upper Jurassic") in the Central North Sea Graben (Netherlands offshore). Mededelingen Rijks Geologische Dienst 51: 1-69.

Wolf, M., Vis, A. & Asschert, A., 2018. Erosional valleys at a major Late Jurassic–Early Cretaceous unconformity offshore Germany and The Netherlands: potential reservoirs or deteriorated seals? Geological Society of London, Special Publications 469. DOI: 10.1144/SP469.4

Wong, Th.E., 2007. Jurassic. In: Wong, Th.E., Batjes, D.A.J.& De Jager, J. (eds): Geology of the Netherlands. Royal Netherlands Academy of Arts and Sciences (KNAW, Amsterdam): 107-125.

Zanella, E. & Coward, M.P., 2003. Structural framework. In: Graham, C., Armour, A. & Bathurst, P. (eds): The Millennium Atlas: Petroleum geology of the Central and Northern North Sea. The Geological Society of London: 45-59.

Ziegler, P.A., 1990a. Geological Atlas of Western and Central Europe, second edition. Shell Internationale Petroleum Mij. B.V. and Geological Society (London): 239 pp.

Ziegler, P.A., 1990b. Tectonic and palaeogeogaphic development of the North Sea Rift system. In: Blundell, D.J. & Gibbs, A.D. (eds): Tectonic Evolution of the North Sea Rifts. Oxford Science Publications: 1-36.

8

Late Cretaceous

Henk van Lochem, Geert-Jan Vis & John W.M. Jagt*

*Corresponding author: j.h.vanlochem@argo-geoscience.com

ABSTRACT

The Upper Cretaceous is one of the rare stratigraphic intervals in the Netherlands that can be studied in outcrop. The quarrying of chalky limestone and interbedded chert nodules in South Limburg has contributed to the study of these rocks and their notable fossil content. During the Late Cretaceous, a transgression which had started in the Jurassic resulted in a complete flooding of the current extent of the Netherlands. A fairly uniform succession of marine marls and limestones of the Texel and Ommelanden formations of the Chalk Group developed. This period of calm sedimentation was locally interrupted by the Sub-Hercynian (Santonian-Campanian) and Laramide (Paleocene) tectonic inversion phases. The resulting erosion removed much of the Cretaceous succession, especially in areas of the former Jurassic basins. However, deposition continued unabated in the previous Jurassic high areas, resulting in significant sediment thicknesses of more than 1800 m. The biostratigraphy and paleoenvironments of the Upper Cretaceous can be studied in detail in outcrop in South Limburg, including the Cretaceous-Paleogene boundary.

<< Fine-grained limestone with rhythmic chert layers of the Gulpen Formation overlain by the massive calcarenitic Maastricht Formation (abandoned ENCI quarry, southeast Netherlands). Photo: Geert-Jan Vis.

DOI: 10.5117/9789463728362_ch08

Introduction

Previous research

As Upper Cretaceous strata are virtually the only rocks that crop out in the Netherlands, they have attracted human interest quite early. Already during the Stone Age, Upper Cretaceous limestones in what is currently known as South Limburg (southeast Netherlands), were put to practical use. Neolithic people used intricate mining techniques to extract flint (chert) from the subsurface, which they used for knapping tools. They clearly had an understanding of the geology of the Upper Cretaceous since they preferentially mined intervals with a high chert content. In the mine field of Rijckholt-St. Geertruid this was flint level 10 of the Lanaye Member (upper Gulpen Formation; Rademakers, 1998).

From Roman times onwards these limestones have been used as building material (Engelen, 1989). For this application the miners preferred intervals of relatively soft rock, which contained only few chert nodules and which could be cut by chisels and saws. In the well-known 'caves' of Mount Saint Peter (Sint Pietersberg) for over centuries rectangular blocks of limestone were cut from the Maastricht Formation (Felder & Bosch, 2000).

The mining activities in the Mount Saint Peter also resulted in the find of the famous mosasaur jaws (Camper, 1786), which set off more scientific studies into Late Cretaceous paleontology and geology. In 1849, the 'système maestrichtien' was introduced when Dumont (1849) defined the highest stage of the Cretaceous in the Maastricht area, the only formal, international stage named after a Dutch location. In 1859, Binkhorst van den Binkhorst, published his standard work on the geology of the Cretaceous of South Limburg. A year later, part two of Staring's (1860) book 'De Bodem van Nederland' appeared, which contained a lithostratigraphic subdivision of the Cretaceous.

W.M. Felder (1975) formalized the lithostratigraphy of the Upper Cretaceous of South Limburg, while NAM & RGD (1980) defined a less detailed lithostratigraphy for the Chalk Group for that part of the Netherlands where the Upper Cretaceous is only known from boreholes. The latter subdivision was later revised in the lithostratigraphic nomenclature compiled by Van Adrichem Boogaert & Kouwe (1994), while the former was illustrated in more detail by W.M. Felder & Bosch (2000).

In South Limburg the late Felder brothers, Werner (1930-2008) and Peter Jozef ('Sjeuf'; 1928-2009), contributed much to our understanding of the stratigraphy of the Upper Cretaceous, while others have since continued to build on their legacy (Jagt et al., 2011).

Ongoing exploration for oil and gas has yielded more subsurface information on the Upper Cretaceous interval. Tectonic inversion (Fig. 8.1) in particular, during or just after the Late Cretaceous, has been and is an important topic of studies (see e.g. Van Wijhe, 1987a,b; Burgers & Mulder, 1991; Dronkers & Mrozek, 1991; Van Lochem, 2018). Duin et al. (2006) and Kombrink et al. (2012) presented a compilation of subsurface maps based on results obtained during more than 50 years of oil and gas exploration (Figs 8.2, 8.3). Vejbæk et al. (2010) placed all mapping efforts of the Cretaceous strata of the Southern Permian Basin, including the Netherlands, into a single atlas chapter.

Oil and gas exploration in Upper Cretaceous reservoirs in the Netherlands did not match the great successes of the Norwegian Ekofisk field or the Dan field in the Danish sector (Surlyk et al., 2003). Consequently, the Dutch oil industry and research institutions did not spend much time and effort on the Upper Cretaceous, aside from ongoing studies in South Limburg. Far more work was done in the Danish, Norwegian and UK sectors of the North Sea Basin. A notable exception is the PhD thesis by Van der Molen (2004), who introduced the use of seismic stratigraphy for the Chalk Group in the Dutch offshore, which yielded a better understanding of the internal structure of this group and of the timing of tectonic movements during the Late Cretaceous.

Regional geological setting

All Upper Cretaceous sedimentary rocks in the Netherlands are attributed to the Chalk Group. However, this unit also includes the Ekofisk and Houthem formations of Danian age (Van Adrichem Boogaert & Kouwe, 1994). For this reason, the present chapter on the Upper Cretaceous also includes Danian strata, the first stage of the Paleocene Epoch (Fig. 8.4). The whole interval discussed in the present chapter covers a time period of 38.8 Myr (Gale et al., 2020).

The Late Cretaceous world was utterly different from the Quaternary world (Skelton et al., 2003; MacLeod et al., 2011; Hu et al., 2012; Huber et al., 2018). While for the last three million years the earth's climate has seen an alternation of ice ages and interglacials, the Late Cretaceous can be characterized as a 'greenhouse' world. Global temperatures were much higher than today's, especially extending to higher latitudes (Frakes, 1999; Hay & Floegel, 2012; O'Brien et al., 2017). The long-term eustatic sea level was significantly higher than the current level (Haq, 2014) with the highest level (~250 m) reached during the Turonian Stage and slowly decreasing to around 200 m above the present-day level during the Danian. In general, the CO_2 content of the atmosphere was high, in the range of 500-1000 ppmv (Wang et al., 2014). Ice sheets were virtually absent, although cooling has been demonstrated at the end of the Late Cretaceous (Linnert et al., 2014, 2018).

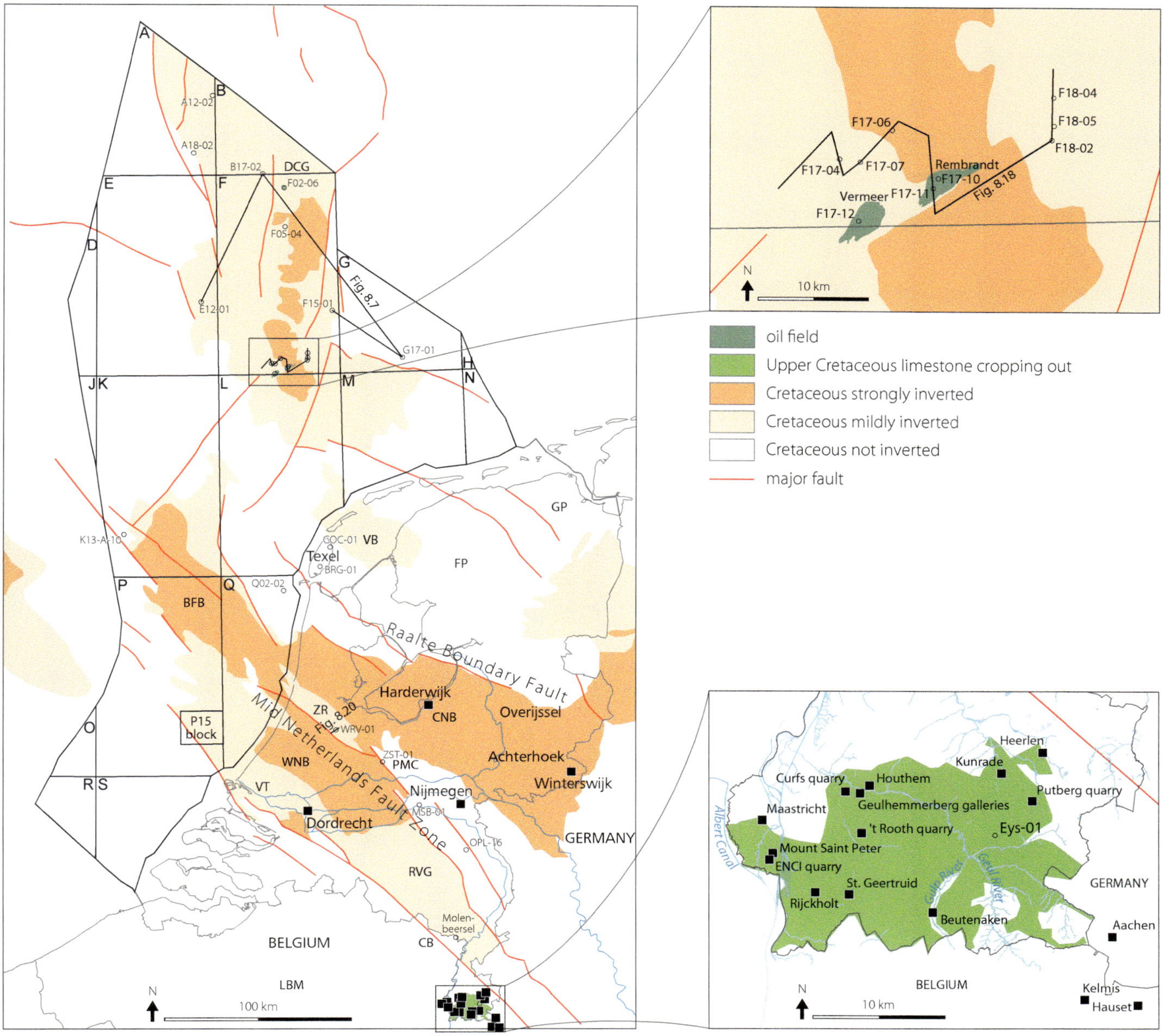

Acronym	Full name	Acronym	Full name
BFB	Broad Fourteens Basin	PMC	Peel-Maasbommel Complex
CB	Campine Basin	RVG	Roer Valley Graben
CNB	Central Netherlands Basin	VB	Vlieland Basin
DCG	Dutch Central Graben	VT	Voorne Trough
FP	Friesland Platform	WNB	West Netherlands Basin
GP	Groningen Platform	ZR	Zandvoort Ridge
LBM	London-Brabant Massif		

Figure 8.1. Inverted basins in the Netherlands. In the mildly or non inverted basins, the Upper Cretaceous chalk is thin as a result of erosion and/or limited deposition. In the strongly inverted basins, no chalk has been preserved. Locations and geological features mentioned in the text are indicated. Inverted basins modified from Kombrink et al. (2012). Eys-01 = B62B4547.

The present-day Netherlands was located at a latitude of around 45 degrees north in the Northern Mid-Latitude Warm Humid Belt according to paleoclimatic maps by Chumakov et al. (1995) and Hay & Floegel (2012).

During the Late Cretaceous the North Atlantic Ocean continued to open between West Africa and North America, while, albeit at a slower pace, the Norwegian Sea rifted between Greenland and Norway (Scotese, 2001; Pharaoh et al., 2010; Hopper et al., 2014). Meanwhile, Italy and the Balkans moved north in the Tethys Ocean and at the end of the Cretaceous started to collide with the southern margin of the European continent.

A large sea covered most of Europe (Vejbæk et al., 2010) during the Late Cretaceous (Fig. 8.5). While in western and central Europe the sea was relatively shallow and epicontinental, to the south the Tethys Ocean was consider-

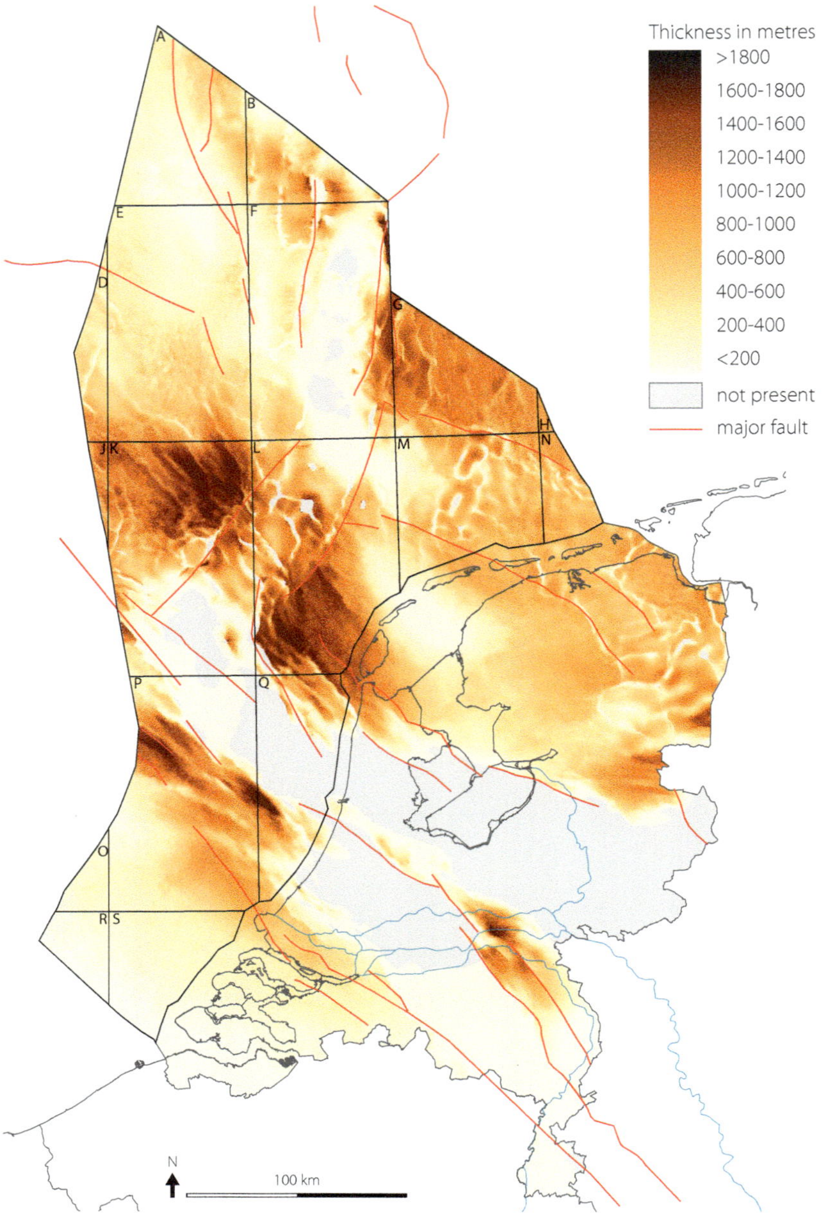

Figure 8.2. *Isochore map of the Chalk Group. Data: DGM-deep v5, www.nlog.nl* (2020).

ably deeper and in part was underlain by oceanic crust. The nearest continental landmass was Eurasia, including the Fennoscandian High and Baltic Shield, situated to the north and east. In the Late Cretaceous sea several exposed islands were present (Fig. 8.5): to the southeast lay the Mid-European Island (Wilmsen et al., 2014) which stretched from the Rhenish Massif to the Bohemian Massif, to the southwest (Bretagne) the Armorican Massif and to the west the Cornish Massif, Welsh High, Grampian High and Shetland Platform. The lack of continental sediment source areas limited the influx of siliciclastic sediments into the sea. Only locally, around the landmasses and islands, were siliciclastic, shallow-marine sediments deposited, while the largest part of the basin was filled with very fine-grained calcareous nannofossil ooze, which turned into calcareous mudstone due to lithification. In and around the Netherlands these deposits are called 'chalk' (Scholle, 1977), which consists for the largest part of the tests of coccolithophorid algae settled out of suspension, which makes these carbonate rocks very different in facies and lithology from platform and reef limestones.

During the Late Jurassic to Late Cretaceous, a large southwardly directed transgression inundated the present-day territory of the Netherlands. At the start of the Late Cretaceous, the coastline extended approximately from Dordrecht to Nijmegen, as evidenced by the presence of the shallow-marine Texel Greensand Member to the north of this line. During the Late Cretaceous (Fig. 8.5), transgression in the south continued and by Maastrichtian time the

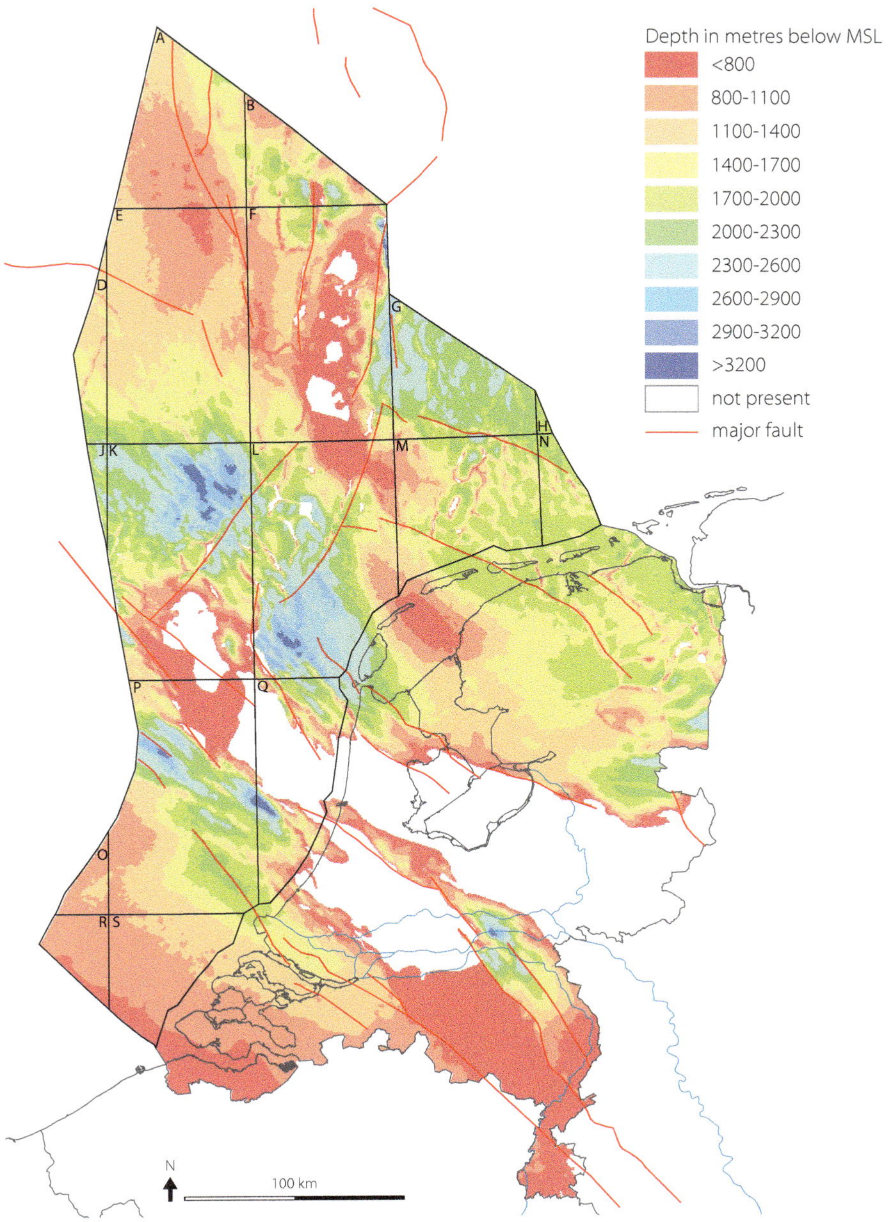

Figure 8.3. Depth map of the base of the Chalk Group. Data: DGM-deep v5, www.nlog.nl (2020).

sea had flooded the entire London-Brabant Massif (Dusar & Lagrou, 2007).

On a global scale at the Cenomanian-Turonian boundary, organic carbon-rich deposits can be found with a concomitant positive shift in carbon-isotope values (Tsikos et al., 2004). The Plenus Marl found at this boundary represents an Oceanic Anoxic Event (OAE) (Schlanger & Jenkyns, 1976), also identified as the OAE-2 or Bonarelli Event after the Livello Bonarelli, the characteristic black shales found in the central part of the Apennines in Italy. The OAE-2 represents a worldwide sea level high-stand, coupled with a rise of the oxygen-minimum zone and large-scale organic carbon deposition and burial. It occurred during an episode of global greenhouse conditions when large amounts of organic matter entered the marine sedimentary record, probably triggered by increased availability of nutrients for planktic biota. Globally widespread carbon burial and silica weathering have both been identified as important mechanisms for drawing down levels of atmospheric carbon dioxide that caused the Plenus Cold Event (Jenkyns et al., 2017).

The Cretaceous-Paleogene (K-Pg) boundary mass extinction was one of the most devastating events in the Phanerozoic history of life, both on land and in the oceans (Alvarez et al., 1980; Smit & Hertogen, 1980; Speijer et al., 2020). This is now widely acknowledged to be related to the impact of an asteroid with an estimated diameter of ~10 km at Chicxulub, Yucatán Peninsula, Mexico (Kring, 2007; Schulte et al., 2010). The impact was followed by a so-called 'impact winter', the result of the injection of

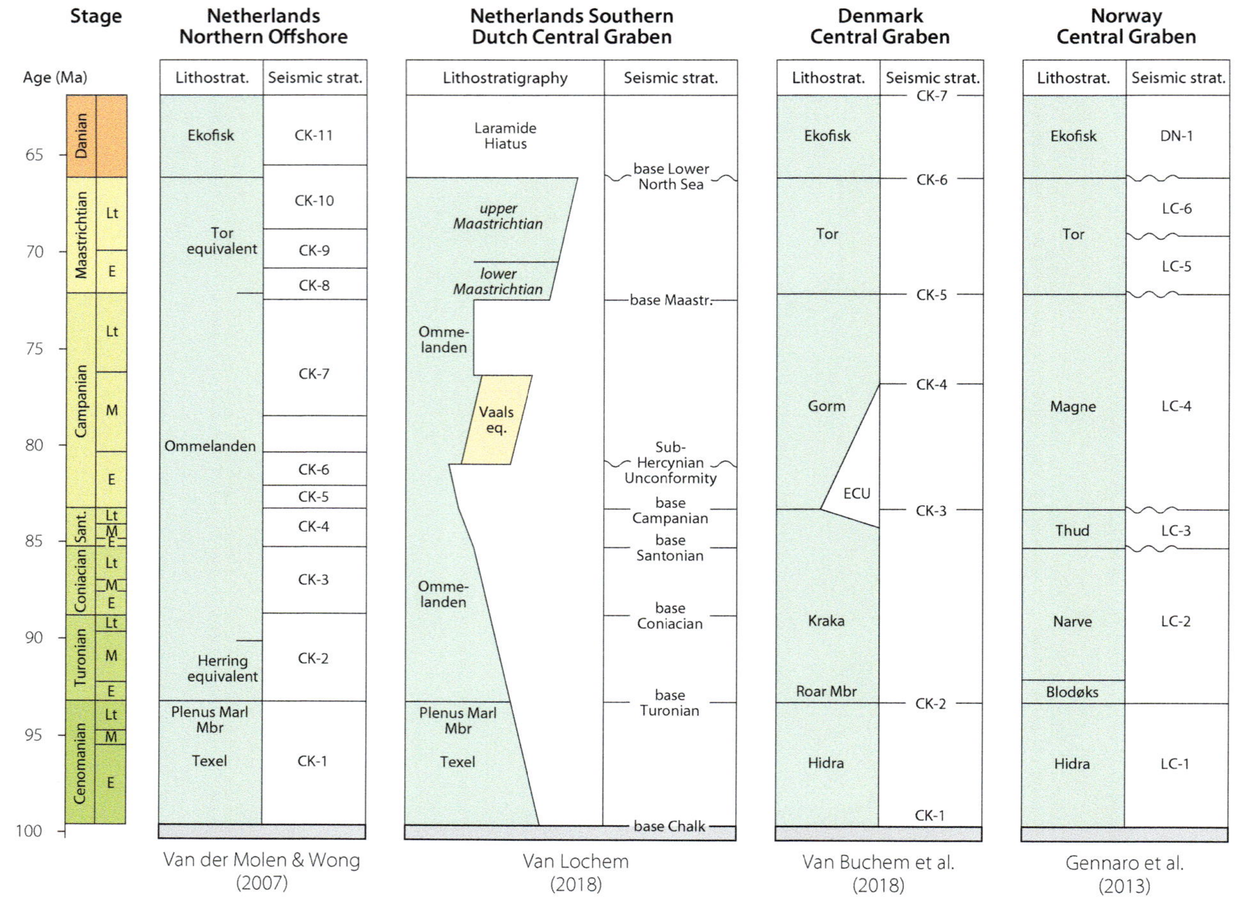

Figure 8.4. Stratigraphic scheme of the Chalk Group in the Dutch (after Van der Molen & Wong, 2007; Van Lochem, 2018), Danish (Van Buchem et al., 2018), and Norwegian (after Gennaro et al., 2013) sectors presenting (informal) lithostratigraphic units and interpreted seismic sequences. Numerical ages are from Cohen et al. (2013; updated). Modified from Van Lochem (2018). ECU = Early Campanian Unconformity.

large amounts of dust and aerosols into the stratosphere, which significantly reduced incoming solar radiation for decades. Therefore, this phase will have been a key contributory element in the extinctions of many biological clades, including most species of planktic foraminifera and many coccolithophorids but also of larger marine taxa such as ammonites and reptiles, in addition to non-avian dinosaurs and flying reptiles (Vellekoop et al., 2014; Fischer et al., 2015; Landman et al., 2015). Strata deposited just after the impact are often enriched in the element iridium (Ir), which points to an extra-terrestrial origin of the K-Pg event (Alvarez et al., 1980). The mass extinction of marine life and the deposition of impact-related sediments can also be studied in the chalk of South Limburg (Smit & Brinkhuis, 1996).

In large parts of the North Sea Basin the Danian Stage of the lower Paleocene sedimentary record consists of calcareous nannofossil mudstone that is quite similar to strata formed just prior to the K-Pg boundary. However, the fossil content of these superficially identical rocks differs significantly from that of the Upper Cretaceous levels (Hansen, 2019; Schrøder & Surlyk, 2020). Following the Danian, during the Selandian, the sedimentary development of the North Sea Basin completely changed from a carbonate system to a siliciclastic-dominated system, resulting from the Paleocene Laramide tectonic phase.

During the early Late Cretaceous (Cenomanian-Turonian) the tectonic development of the present-day Netherlands was characterized mainly by post-rift thermal subsidence, since the North Sea rift had abated in Albian times (Pharaoh et al., 2010). The lithosphere of northwest Europe was subjected to compressional intraplate stresses starting in the late Turonian-Coniacian, when extensional and transtensional Mesozoic basins began to experience tectonic inversion (Fig. 8.1; De Jager et al., 2025, this volume). Older normal faults were reactivated as reverse faults and much of the inverted basin fills were eroded.

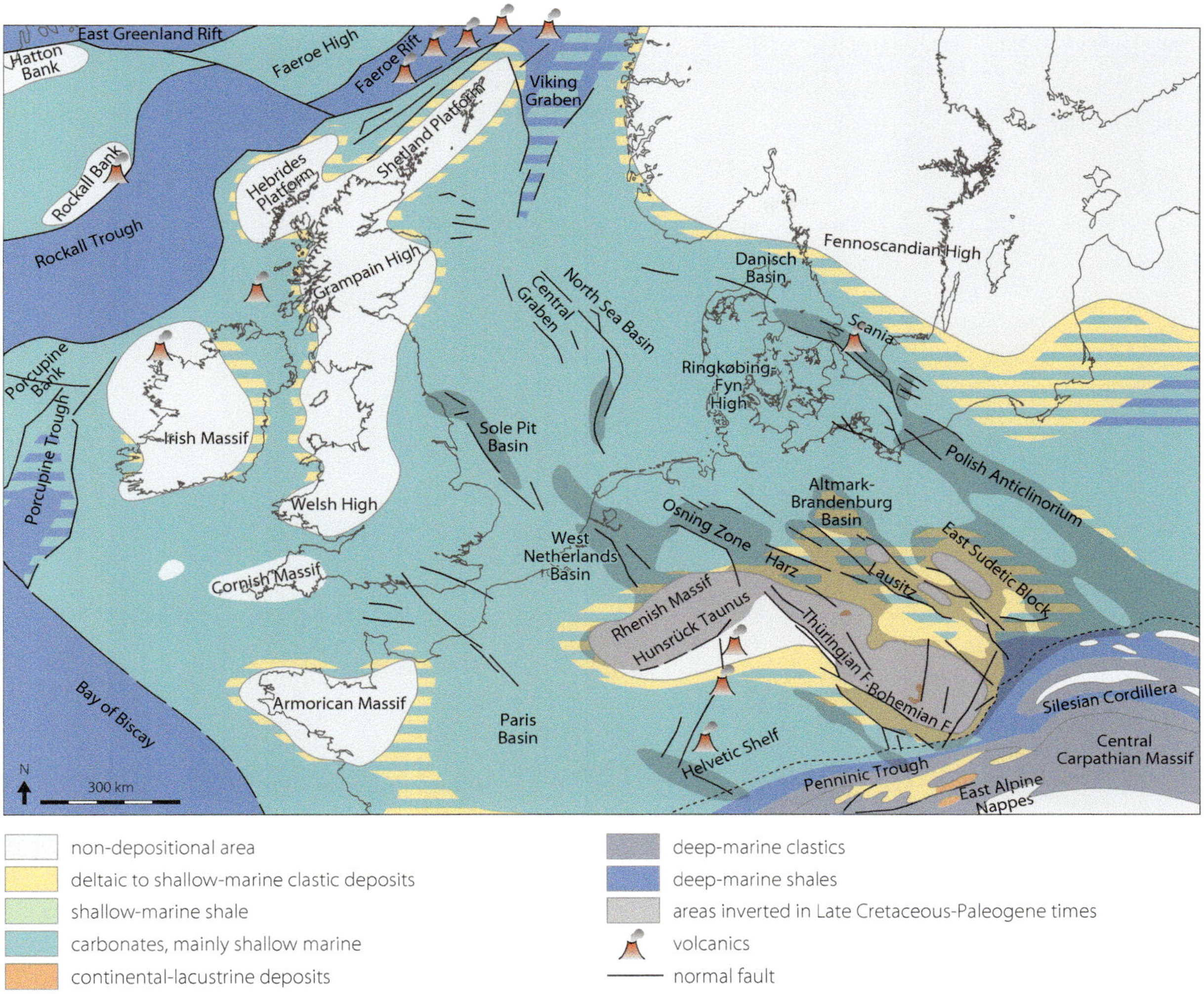

Figure 8.5. Central European paleogeography in Cenomanian to Danian times (modified from Ziegler, 1982; from Vejbæk et al., 2010).

Stratigraphy

Chalk Group

The Chalk Group consists of a succession of marine, predominantly bioclastic and in part marly limestones or calcareous mudstones (Fig. 8.6). In the southwest and southeast of the Netherlands, deposits of this group unconformably overlie strata of Paleozoic age. The stratigraphy in the southeast, in the region of South Limburg where these rocks partly crop out, follows a separate subdivision with six lithostratigraphic units which are of Santonian to Danian age (Fig. 8.6; Felder, 1975; Felder & Bosch, 2000).

Biostratigraphic interpretations of offshore sections rely on analyses of calcareous nannoplankton, planktic and benthic foraminifera and dinoflagellates in borehole cuttings and cores; data on macrofossil groups are very limited (Van Buchem et al., 2018; Eldrett et al., 2021). Ranges of key index microfossil species and, more specifically, their first and last appearance datums (FAD, LAD) can be used for correlations to other offshore and onshore sections. On the basis of foraminifera, dinoflagellates, calcareous nannofossils and diatoms Gradstein et al. (2016) distinguished 19 microfossil zones (NCF1-19) and more than one hundred events. They also used literature data to link Dinoflagellate Cyst Zones (DCZ) to these NCF zones. In all, 11 dinoflagellate zones and 39 subzones were recognized for the marine Cretaceous succession in the North Sea.

Thus, microfaunal analyses of borehole cores have yielded a robust correlative framework. Overall, depositional conditions for strata penetrated in wells may be assumed not to have differed in any significant respect from onshore sections around the North Sea (e.g. Surlyk et al., 2006, 2013; Anderskouv et al., 2007; Voigt et al., 2008; Niebuhr et al., 2011; Jelby et al., 2014; Mortimore & James, 2015; Mitchell, 2019; Wilmsen et al., 2019). In cores of the Chalk Group, ichnofossils (reflecting benthic macrofaunal activity) and skeletal elements of certain macrofossil

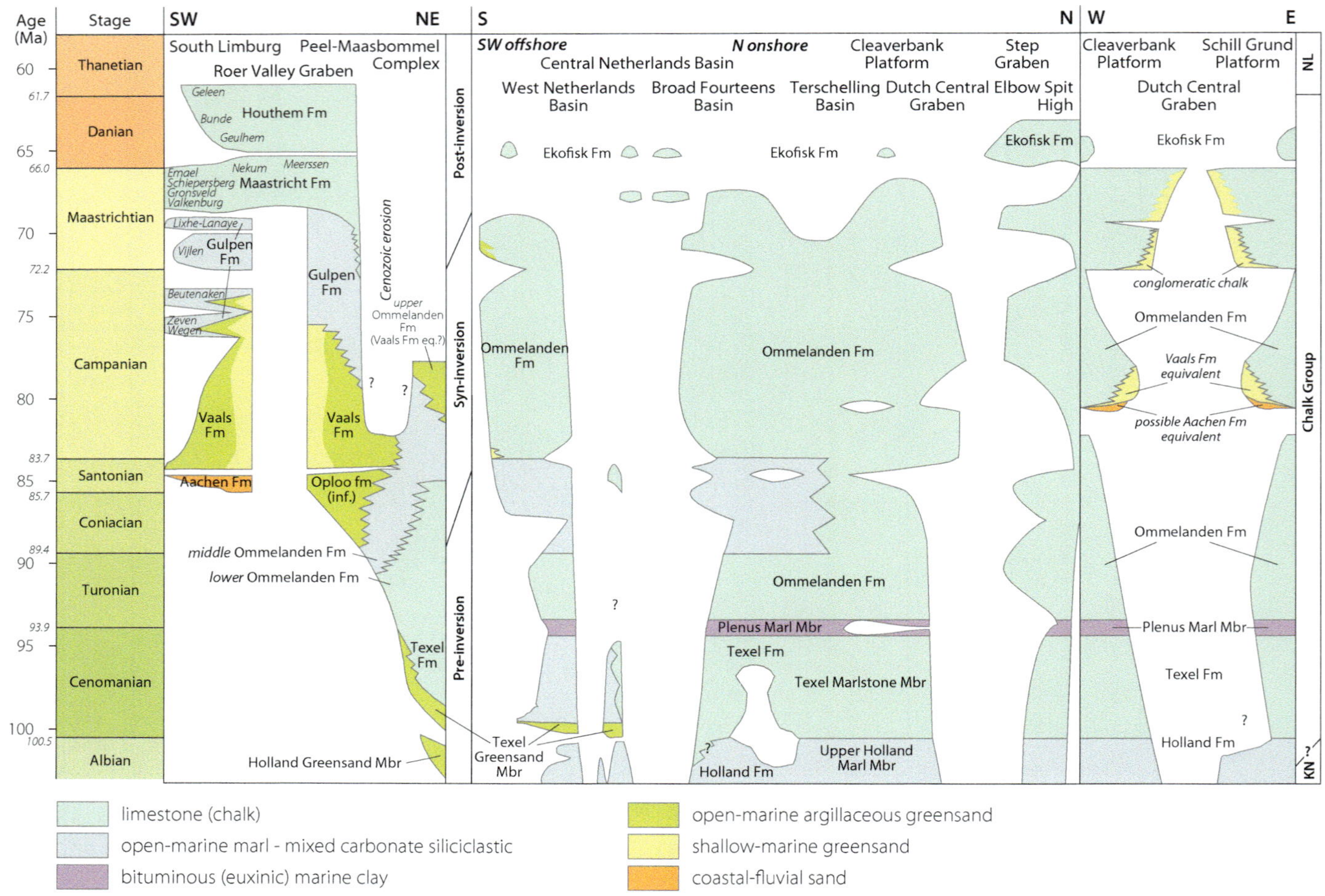

Figure 8.6. *Lithostratigraphic scheme of the Upper Cretaceous succession of the Netherlands, based on Van Adrichem Boogaert & Kouwe (1994), Gras & Geluk (1999), Jagt (1999a,b), Map sheet XIII+XIV (NITG-TNO, 2001), Vellekoop et al. (2022) and Gale et al. (2023). Lihthostratigraphy of the Dutch Central Graben (right panel) after Van Lochem (2018). KN = Vlieland Group; NL = Lower North Sea Group*

groups are occasionally noted. Key index macrofossil taxa such as ammonites, belemnites, inoceramid bivalves, brachiopods, echinoids and crinoids are also known from some observations in core and cuttings.

Texel Formation

The Texel Formation usually comprises light grey to beige and white limestones and marly chalks with some marl intercalations. Near the southern edge of its distribution, i.e. the southern part of the West Netherlands Basin and along the Peel-Maasbommel Complex, greensands form the base of the formation (Fig. 8.6). Elsewhere, in the central part of the West Netherlands Basin, and in the Central Netherlands and Broad Fourteens basins, the formation becomes increasingly marly and, in a more offshore setting north of these basins, it develops into pure limestones. The formation's thickness varies between 50 and 70 m in the northern and eastern Netherlands (Vlieland Basin, Friesland Platform, Groningen Platform and the east of the province of Overijssel) and increases southwards to 300 m (borehole Q02-02). The formation is locally exposed in the Achterhoek and is well developed in adjacent parts of Germany (Hiss et al., 2018). The sediments of this formation were deposited in a transgressive setting of relatively deep water and have been dated as Cenomanian on the basis of benthic and planktic foraminifera (Table 8.1). The formation comprises sequence CK1 of Van der Molen & Wong (2007) (Fig. 8.7).

The Texel Greensand Member (Fig. 8.6) is a highly glauconitic, calcareous sandstone with intercalated marls. It occurs only along the southern fringe of the West Netherlands Basin. The Texel Marlstone Member consists of white

Table 8.1. Index taxa for Upper Cretaceous units, arranged by groups available online at: DOI: 10.5117/aup.27108772.

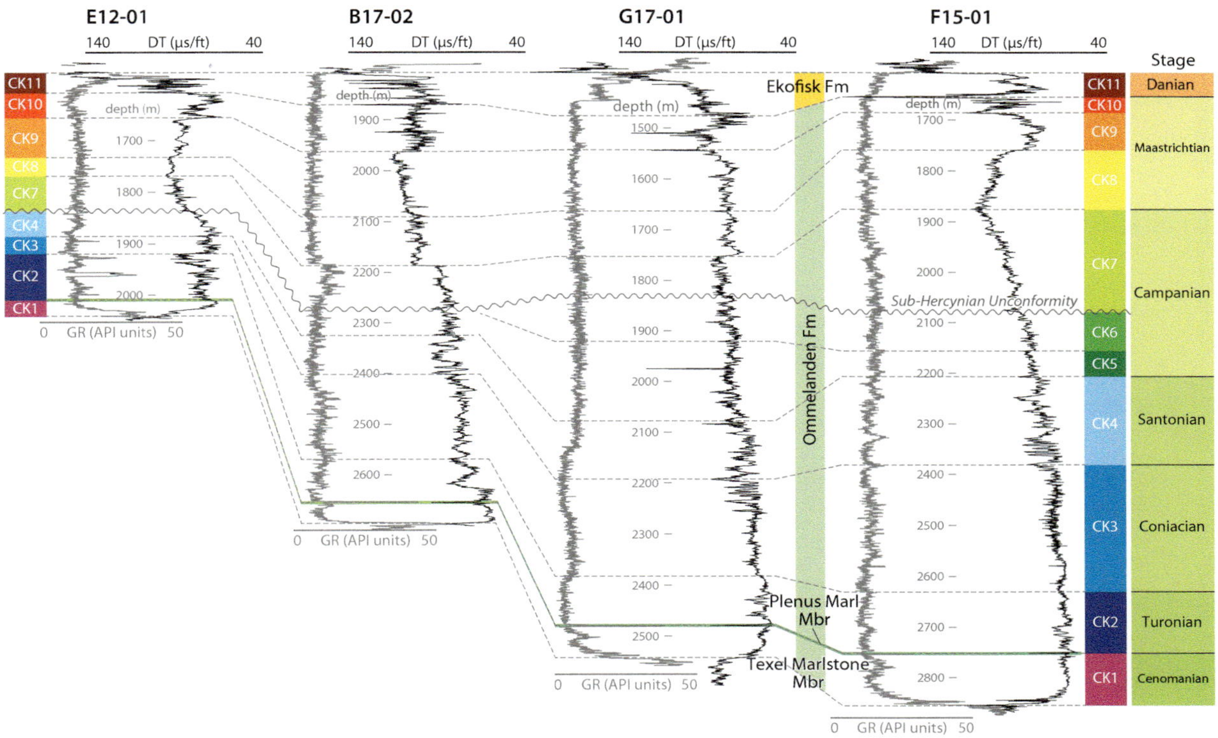

Figure 8.7. Well-log correlation panel with gamma-ray- and sonic logs of wells E12-01, B17-02, G17-01 and F15-01 (for location see Fig. 8.1). Intra-chalk seismic sequences CK1 to CK11 were tied to the wells using synthetic seismograms. The seismic sequences were subsequently dated with biostratigraphic data available in a selection of wells (modified from Van der Molen & Wong, 2007, Fig. 7). Note that the Plenus Marl Member clearly stands out in the gamma-ray and sonic logs.

to light-grey limestones and marly chalks in the northern Netherlands and of marls with marly chalks south of the Broad Fourteens Basin and in the western part of the West Netherlands Basin. The Plenus Marl Member is a dark grey, locally black, calcareous, laminated, bituminous claystone of widespread distribution. The marl represents an Oceanic Anoxic Event (OAE-2). Its thickness is usually some decimetres to a few metres and it is easily recognized on wireline logs by the characteristic high(er) gamma-ray and resistivity, and lower sonic readings (Fig. 8.7). The age of the member straddles the Cenomanian-Turonian boundary.

Ommelanden Formation

A succession of white to yellowish-white or light olive-grey, fine-grained, and in places argillaceous limestones characterizes this formation (Fig. 8.8a,b; Fig. 8.9a-d). Layers of chert nodules are locally common in its Campanian to Maastrichtian part. Along the basin edge coarse-grained, bioclastic limestones and wedges of sandstone and glauconitic sandstone occur (Gras, 1995; Gras & Geluk, 1999; Van Lochem, 2018). The formation comprises mainly hard, dense limestones (as result of compaction and cementation), but in its upper part, it tends to be softer and more chalky in texture (Fig. 8.10a). The sediments were deposited under relatively stable, low-energy conditions in carbonate-shelf and upper bathyal settings and range in age from early Turonian to Late Maastrichtian. They comprise sequences CK2-CK10 of Van der Molen & Wong (2007). The age assignment relies primarily on benthic foraminifera (genera *Stensioeina*, *Gavelinella* and *Bolivinoides*; see Table 8.1). The LADs of the planktic foraminifera *Racemiguembelina fructicosa* and *Pseudotextularia elegans* date sequence CK10 as Late Maastrichtian, but in some places, it extends into the early Danian, where the index *Globoconusa daubjergensis* has been noted.

North of the Peel-Maasbommel Complex the formation has informally been subdivided into three units (Fig. 8.6), although recent ongoing work casts doubt on this subdivision. The lowermost Turonian interval is a whitish, hard, dense limestone with high sonic readings. It is overlain by a more marly succession of Coniacian to Campanian age (NITG-TNO, 2001). The uppermost unit is less marly and, in its lower part, consists of consolidated calcarenites, grading upwards into massive chalk with abundant chert nodules.

Thicknesses up to 1500 m are reached in the offshore K and L quadrants. The thickness is greatly influenced by inversion movements whereby former Mesozoic depocentres were uplifted, resulting in a thin cover of Chalk

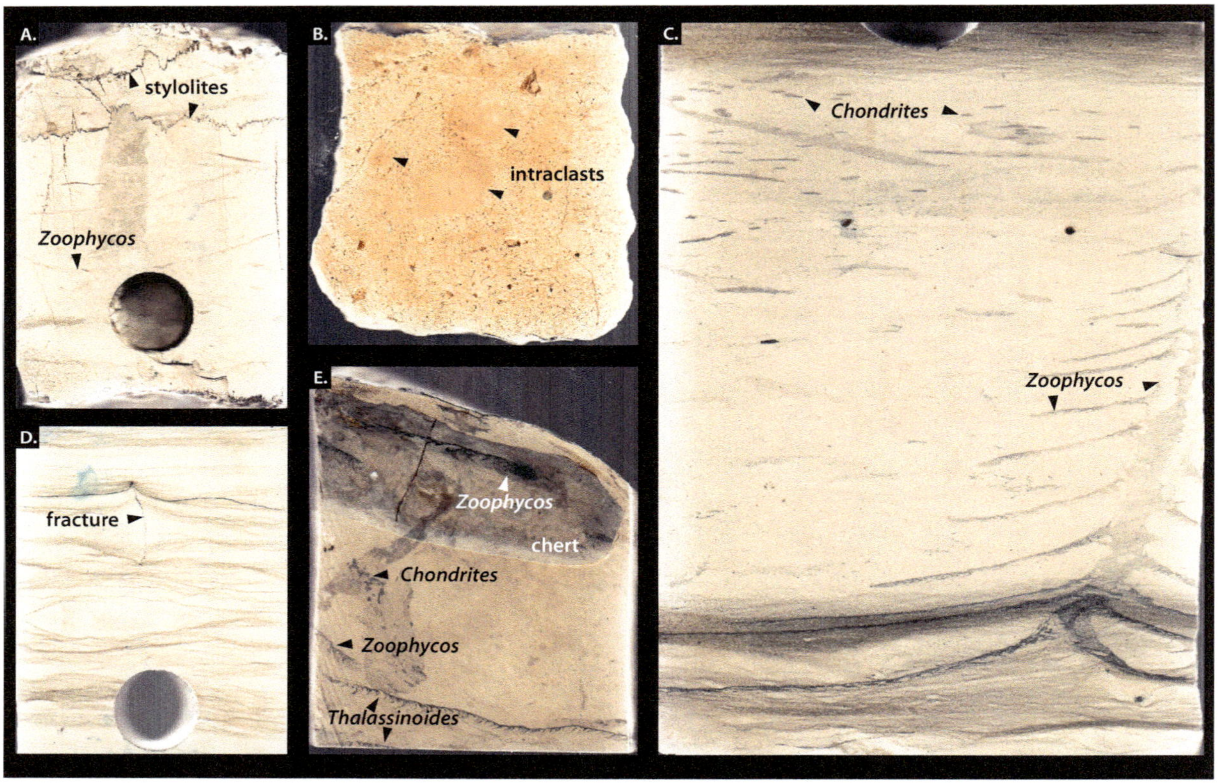

Figure 8.8. *Core photographs (from Ineson et al., 2014). a) Chalk Mudstone Association (FA-1): chalk mudstone, bio-mottled, stylolitic; note the curved concave-up* Zoophycos *traces and the discontinuous open fractures emanating from the stylolites. Ommelanden Formation, Maastrichtian (A12-02, Core 4, 2040.6 m, core width ca. 9 cm; note that Core 4 is probably inverted and is shown here in the inferred correct way-up). b) Skeletal Wackestone-Packstone Association (FA-3): skeletal packstone containing chalk mudstone intraclasts, Ommelanden Formation, lower Maastrichtian (F05-04, Core 3, 1358.75 m, core width ca. 7 cm). c) Marly Chalk Association (FA-2): slightly marly chalk showing a cross-section of the axial tube of a* Zoophycos *system (right) and* Chondrites *(top left). Ekofisk Formation, Danian (A12-02, Core 3, 2022.6 m, core width ca. 10 cm). d) Marly Chalk Association (FA-2): marly chalk showing nodular fabric produced by abundant anastomosing solution seams. Note compaction effects over an early calcite-filled discontinuous fracture. Ekofisk Formation, Danian (A12-02, Core 3, 2018.8 m, core width ca. 10 cm). e) Marly Chalk Association (FA-2): slightly marly chalk mudstone with well-defined burrows of* Thalassinoides, Zoophycos *and* Chondrites. *Note virtually uncompacted* Zoophycos *in the chert nodule. Ekofisk Formation, Danian (A18-02, Core 2, 1947.0 m, core width ca. 10 cm).*

→ ***Figure 8.9.*** *Microphotographs of thin sections of Chalk Group sedimentary rock taken in plain polarized light (PPL). a) Ommelanden Formation (Lower Maastrichtian) sandy chalk (F17-11, 1467.6 m; Van Lochem, 2018). b) Ommelanden Formation (Lower Maastrichtian) conglomeratic chalk (F17-11, 1469.9 m; Van Lochem, 2018). c) Ommelanden Formation (Upper Maastrichtian) clean chalk (F17-10, plug number: S14A core depth: 1387.85 m; Petrea & Goldberg (2013). d) Ommelanden Formation (Upper Maastrichtian) clean chalk (F17-11, 1432.6 m; Van Lochem, 2018). e) Vaals Formation equivalent (late Campanian) sandstone (F17-11, 1474.1 m; Van Lochem, 2018). f) Vaals Formation equivalent (late Campanian) sandstone (F17-12, 1437.0 m; Van Lochem, 2018). g) Vaals Formation: very fine to fine-grained glauconitic sandstone with sponge spicules and planktonic foraminifera (Eys-01 (B62B4547), 109.50 m). h) Gulpen Formation (Lixhe Member): bioclastic wackestone with planktonic foraminifera (ENCI quarry). i) Lower Maastricht Formation: highly porous bioclastic grainstone, with syntaxial overgrowth of echinoderms. The main bioclasts are echinoderms, bivalves, foraminifera, gastropods, and bryozoans (ENCI quarry). j) Houthem Formation: red algae grainstone to boundstone with bryozoans, foraminifera, and echinoderms (borehole Molenbeersel (Belgium), 1235.10 m). Images g-j kindly provided by Mateus Kroth of Utrecht University.*

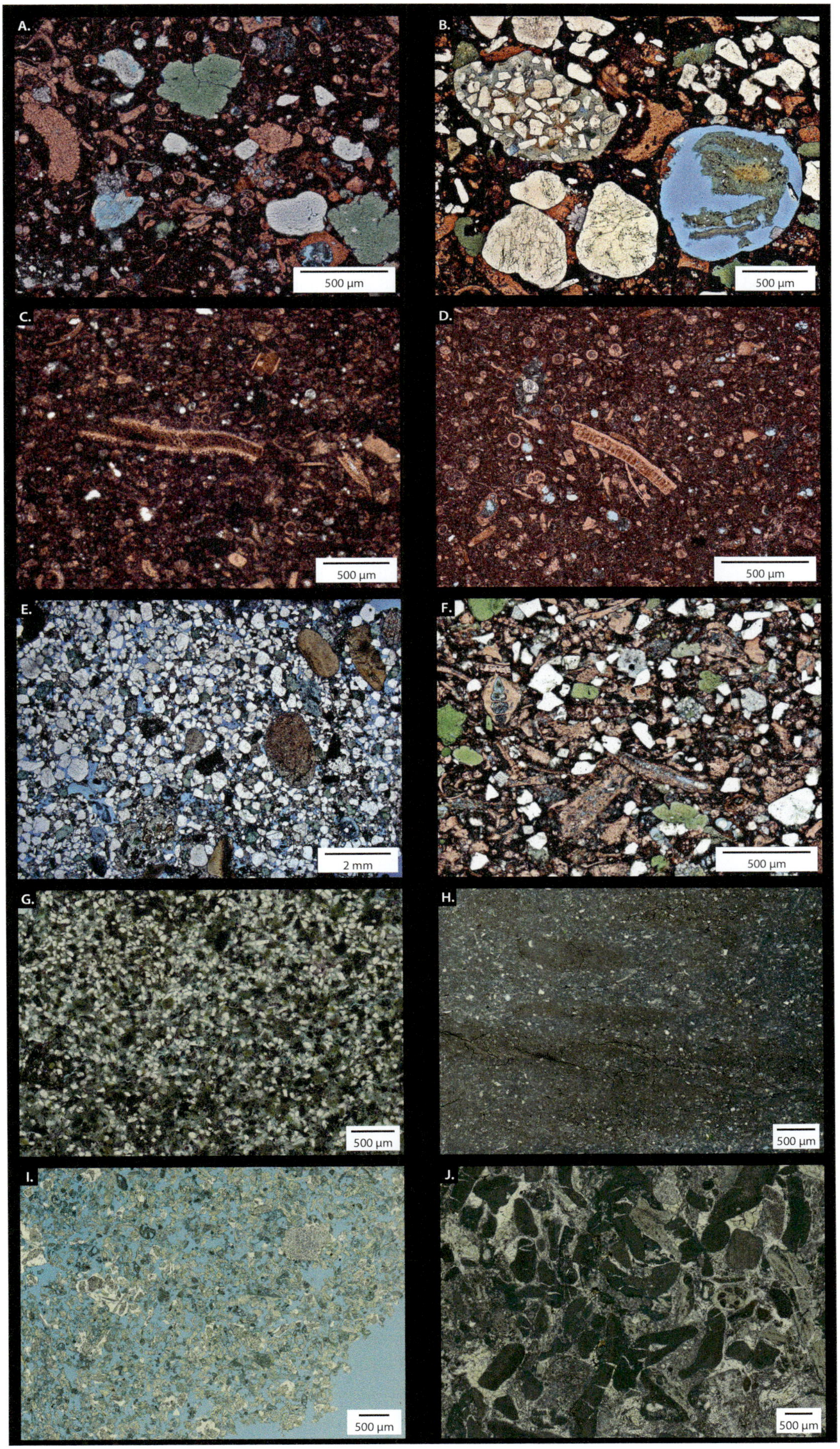
A.
500 µm
B.
500 µm
C.
500 µm
D.
500 µm
E.
2 mm
F.
500 µm
G.
500 µm
H.
500 µm
I.
500 µm
J.
500 µm

Group strata. Strong subsidence of the adjacent highs resulted in so-called marginal troughs with very thick successions. Biostratigraphic and log-correlation studies in places have identified several unconformities within the succession, which can be related to the different inversion pulses (Van Wijhe, 1987b; Herngreen et al., 1996; De Jager, 2003, 2007; Van der Molen, 2004).

Ekofisk Formation

This formation comprises white, chalky limestones with rare nodular and bedded chert layers (Fig. 8.8c-e) and thin, grey to green clay laminae; glauconite may occur in its basal part. The formation displays a characteristic gradual upward increase in transit time log readings, resulting from an increase in porosity (Fig. 8.7). The Cretaceous/Paleogene (K-Pg) boundary is placed either at a rarely observed gamma-ray peak representing the Cretaceous/Paleogene (K-Pg) boundary clay (Herngreen & Wong, 2007), or at a negative excursion in $\delta^{13}C$ values or at an interval with 'disaster taxa' among calcareous nannofossils (Eldrett et al., 2021). Near this boundary a hardground, overlain by a glauconite-rich bed, occurs locally. This boundary equates with the Fish Clay in eastern Denmark (e.g. Christensen et al., 1973; Schmitz et al., 1992; Surlyk et al., 2006) and the base of subunit IVf-7 (Meerssen Member) in South Limburg (Smit & Brinkhuis, 1996; Herngreen et al., 1998).

Figure 8.10. SEM images, both with the same level of magnification. a) Exceptionally well-preserved coccoliths (CC) from the Ommelanden Formation in the North Sea. Note that most of the grains derive from disintegrated coccoliths (well F17-10, plug number: S11A core depth: 1386.81 m, source Petrea & Goldberg, 2013). b) The only coccolith found in a sample from the Meerssen Member (former ENCI quarry, Maastricht). Note the much larger calcite crystals compared to the chalk mudstone from well F17-10.

The sediments of this formation were deposited under relatively stable, low-energy marine conditions in carbonate-shelf and upper bathyal environments. Local coral growth has been observed on submarine highs such as in the core of well F02-06 (Ineson et al., 2014). The Ekofisk Formation corresponds to sequence CK11 of Van der Molen & Wong (2007) and is dated as Danian based on foraminiferal evidence. It is not clear, however, when the onset of deposition of its sediments occurred.

The Ekofisk Formation is found in the northern and southern parts of the Dutch North Sea sector and is the equivalent of the Houthem Formation. By definition, the name Ekofisk Formation is reserved to offshore occurrences. Its thickness attains maximum values of 140 m.

Chalk Group in the (south)eastern Netherlands

In the onshore sections, both micro- and macrofaunal assemblages permit more detailed paleoenvironmental interpretations than in the offshore. From the Winterswijk area (eastern Netherlands), typical early middle Cenomanian ostracods (Witte et al., 1992) and early Cenomanian ammonites (Jagt & Oosterink, 2012) have been recorded from the so-called 'Kottense Kalk', which belongs to the Texel Formation. There are no outcrops of strata of middle Cenomanian to late Coniacian age in the Netherlands.

In the southeast of the Netherlands strata of Santonian to Danian age are present and six formations are defined (Fig. 8.6). Their lateral distribution is not clearly defined, but currently is limited to the area of South Limburg and adjacent regions, where most of these formations are found in outcrop. In general, the sedimentary rock of the Chalk Group of South Limburg has a coarser texture than its equivalents in the present-day North Sea area (Figs 8.9 h-j, 8.10, 8.11). Most formation names used in the Netherlands for this region are adopted by neighbouring Belgium and Germany. The formations have been subdivided into members (Fig. 8.6) separated by (marker) horizons (Felder, 1975; Felder & Bosch, 2000). The field characteristics on which these formations and members are subdivided is questionable (Bless et al., 1986), however, as the

Figure 8.11. *Core photographs from well Eys-01 (B62B4547, Gutteridge & Thompson, 2019). a) Aachen Formation: soil formation in smectitic Hergenrath Clay (178.6 m). b) Vaals Formation: bioturbated argillaceous sandstone with ?Teichichnus cemented by calcite (108.45 m). c) Gulpen Formation: basal clastic-rich limestone; peloidal glauconite, comminuted bioclasts and up to mm-sized detrital quartz grains (102.20 m). d) Gulpen Formation: bioturbated grainstone with bored hardground on chert nodule (100.58 m). e) Gulpen Formation: bioturbated calcarenitic limestone (78.85 m). f) Gulpen Formation: reworked lithified limestone in glauconitic limestone (71.35 m). g) Gulpen Formation: fine-grained glauconitic sandstone interval embedded in limestone (70.25 m). h) Maastricht Formation: sandy glauconitic limestone with reworked lithified burrow fills (55.90 m). i) Maastricht Formation: bioturbated calcarenitic limestone with firm ground at the top (51.10 m). j) Maastricht Formation: bioturbated calcarenitic limestone (39.90 m).*

3D-stratigraphic architecture is more complex than previously recognized. As an ongoing PhD study is likely to shed new light on the Upper Cretaceous succession in South Limburg, we limit ourselves here to presenting only the formations and noteworthy details at smaller stratigraphic scales. For reference, the members are indicated in Fig. 8.6 in italics.

For the Maastrichtian type area (South Limburg), no formal micro- or macrofaunal biozones have been proposed to date. However, certain taxa serve as proxies in correlations with biozones in the Boreal white chalk succession of northern Germany. In addition, the sequence-stratigraphic calibrated ecozonation can be used as proposed by Keutgen (2018); see also Felder (1994, 1995).

Aachen Formation

This formation has a smectitic clay unit at its base, known as the Hergenrath Clay (Fig. 8.11a). This clay was deposited as fluvial overbank deposit as evidenced by the presence of brown coal layers, root horizons and soils (Fig. 8.11a), while marine incursions with bioturbation have been recognized towards the top. The clay is overlain by a succession of well-sorted, fine-grained quartz sands, with multi-directional cross bedding, known as the Aachen and Hauset sands (Fig. 8.12). These sands are difficult to distinguish especially in small outcrops and borehole data. They are interpreted to have been deposited in a shallow-marine setting. The unit reaches a thickness of up to 60 m. The formation is rich in terrestrial plant fossils, examples being on display in Museum Valkenburg. Deposits of the Aachen Formation crop out in now disused quarries in northeast Belgium. Well-known former quarries are those of Flög (Bingeberg, near Hauset), Schampelheide and Käskorb near Kelmis (La Calamine).

The lowest part of the Aachen Formation is dated primarily on spores, pollen and dinoflagellate cysts, indicating a middle to late Santonian age (Batten et al., 1987, 1988; Knobloch et al., 1993). Very little is known of the marine fauna in this unit, as all calcareous material has been dissolved. However, in places bioturbation by decapod crustaceans is common.

Oploo formation (informal)

This informal formation is probably time-equivalent to the upper part of the Aachen Formation (Fig. 8.6) and can be found on the Peel-Maasbommel Complex (Gras & Geluk, 1999; NITG-TNO, 2001). This formation differs from the Aachen Formation by its purely marine nature. The lower part consists of a cyclic alternation of glauconitic and argillaceous sandstones, grading into bioclastic limestones. The upper part consists of fine-grained, locally argillaceous sandstones. Towards the north, the sandstones of the Oploo formation grade into marls and chalks of the Ommelanden Formation. The formation represents a marine basin-edge facies, reaches a thickness of more than 300 m and is of Coniacian to Santonian age. The geographic distribution of this unit has not been well established yet. The formation has only been penetrated in well Oploo-16 (OPL-16).

Vaals Formation

The Vaals Formation consists predominantly of partly decalcified, yellow to greyish green, fine-grained glauconitic and moderately to well-sorted argillaceous sandstone occasionally showing trough cross bedding (Fig. 8.11b). Intervals with strong bioturbation occur, with occasional mud-lined burrows. Bioclasts include echinoid spines, bivalves, bryozoans, monaxon sponge spicules and planktic foraminifera, while just across the border in Germany mosasaurid remains were found (Sachs et al., 2018). In South Limburg, the formation passes into calcareous, glauconitic silty clay to the west and northwest (Felder & Bosch, 2000). Masses of marine shelly fossils and glauconite grains are concentrated in channel fills. When complete, the succession starts with a 0.25 to 0.5 m thick basal conglomerate on top of the bioturbated Aachen Formation (Hauset Sand). The sediments were deposited mainly in a shallow-marine environment which is thought to have been interrupted by tectonic movements causing submarine erosion of major parts of the formation. Due to the syn-sedimentary movements and post-depositional erosion the thickness varies strongly but may reach up to 150 m.

Figure 8.12. Cross-bedded sandstone from the Aachen Formation with multiple clay drapes and possible tidal bundles in the lowermost set, indicating marine influence. Käskorb Quarry, Kelmis (Belgium). Source: RGD (photo PAR0014054, 1994). Estimated vertical scale: ca. 150 cm.

Age assignment of the Vaals Formation is based mainly on belemnites and ammonites (Table 8.1) that yield an (early) Campanian age (Albers, 1976; Christensen & Schmid, 1987; Kennedy & Jagt, 1995). Faunal assemblages are comparatively diverse, with infaunal and epifaunal groups (echinoids, lobsters, molluscs and echinoderms) represented in addition to ichnotaxa. The upper levels of this succession have produced early late Campanian ammonites (Fig. 8.13a), hinting at correlation with the Zeven Wegen Member (Gulpen Formation) elsewhere, as do benthic foraminifera, belemnites and crinoids (see Gale, 2018).

The Vaals Formation has received little attention since the study by Albers (1976). This formation crops out in various roadcuts in the southeast of South Limburg. Close to the now abandoned and protected Putberg quarry, near the historic farm De Dael, sand of this formation crops out below the Maastricht Formation. Deposits equivalent in age and composition to the Vaals Formation have been found in the offshore F quadrant as erosional products near Sub-Hercynian inversion highs (Figs 8.6, 8.14b-d; Van Lochem, 2018).

Gulpen Formation

The Gulpen Formation sees the onset of carbonate deposition in South Limburg. This formation is dominated by mixed siliciclastic and carbonate sediments. The basal unit is a clean white chalk (Zeven Wegen Member; Fig. 8.6). The layers above show increasing but varying amounts of glauconite, siliciclastic sediment and chert (Figs 8.9h, 8.11c-g). Towards the top of the formation, the glauconite content diminishes, but the chert content increases further into organized layers of black chert, occurring every 0.5-1.5 m (see photo at the beginning of this chapter). In the top of the lower part of the formation, the Vijlen Member (Fig. 8.6) represents the infill of erosional channels (Felder & Bless, 1994; Felder, 1997) which scoured into the top of the Carboniferous bedrock in South Limburg in two zones with a NW-SE-trending orientation. Therefore, the base of the Vijlen Member represents an unconformity. The top of the Gulpen Formation (Lanaye Member) is a clean limestone consisting of a more than 95% pure coccolithic, bioclastic silty packstone. Bioturbation is omnipresent in this formation, as are several hardgrounds. A transitional facies of the Gulpen Formation, characterized by glauconitic marls, was identified by Felder et al. (1985) in the eastern Campine area and the westerly Roer Valley Graben border zone in Belgium. These facies are provisionally indicated as 'Pre-Valkenburg strata' by these authors.

The lowermost unit of the Gulpen Formation yields typically late Campanian ammonites, belemnites, pectinid bivalves, echinoderms and brachiopods (Jagt, 1999a,b, 2000a-d; Simon & Owen, 2001; Jagt & Jagt-Yazykova, 2019), illustrating links with northern Germany, southern Belgium and eastern England. The uppermost Campanian-lowermost Maastrichtian (i.e. Beutenaken and Vijlen members) is still poorly understood, despite the fact that several key belemnite species (e.g. *Belemnella obtusa*) have been recorded. However, condensed sections and ubiquitous reworking blur the overall picture.

For the Maastrichtian succession of the Gulpen Formation (Vijlen Member and younger), ammonites and belemnites are prime index taxa (Felder & Bless, 1994; Jagt et al., 1999; Christensen et al., 2005; Keutgen et al., 2010), in addition to inoceramid bivalves and echinoids. Inoceramids include key index taxa that allow correlation with Europe, North America and South Africa (Walaszczyk et al., 2010), just prior to their extinction. The Lanaye Member represents the uppermost member of the Gulpen Formation and sees the transition from a north-temperate (Boreal) to a Tethyan setting, as documented by diverse 'newcomers' among echinoids, crinoids (Fig. 8.13c), brachiopods, cheloniid turtles and mosasaurs. These include taxa (Table 8.1) that have close relatives in North America (Atlantic Plain and Gulf Coast) and North Africa (Jagt, 1999a,b, 2000a-d, 2005; Mulder, 1999; Phillips et al., 2014).

The sediments of this formation were deposited under fully marine conditions. The formation reaches a thickness of up to 175 m. Hiatuses occur within it (Fig. 8.6) and the top of the formation is separated by a hiatus of some 0.7 Myr from the base of the overlying Maastricht Formation (Vandenberghe et al., 2004; Keutgen, 2018; Vellekoop et al., 2022). In the east of South Limburg, the Gulpen Formation can be found in some road cuttings and disused quarries, such as the Habets quarry east of the village of Beutenaken in the Gulp River valley. In the west, the formation is exposed in the now abandoned ENCI quarry south of Maastricht (see photo at the beginning of this chapter) and in the incision of the Belgian Albert Canal.

Maastricht Formation

The Maastricht Formation is less variable in lithology than the underlying Gulpen Formation and consists mainly of carbonate deposits (Fig. 8.9i). It has been informally subdivided into the Maastricht limestone facies and the Kunrade limestone facies (see for an early reference e.g. Umbgrove, 1925). In the Kunrade-Heerlen area (eastern part of South Limburg), the Kunrade limestone facies ('Kunrader Kalk') is an alternation of hard and soft light-grey, fine-grained limestone beds without chert nodules. It is considered to be the lateral, shallower and more onshore equivalent of the Maastricht limestone facies, with macrofaunas reflecting a range of sea floor conditions from firm grounds (regular echinoids, crinoids) to less consolidated substrates (echinoids). The most detailed correlation attempt to date is that of Felder & Bless (1989), who equated the 'Kunrader Kalk' with the middle Lanaye Member (Gulpen Formation) and the lower portion of the Maastricht Formation (basal Nekum Member), dating it as late Maastrichtian (Fig. 8.13l). In Belgium the 'Kunrader Kalk' has recently been raised to the rank of formation for the Campine area (see ncs.naturalsciences.be).

The Maastricht limestone facies is found in the south-western part of South Limburg, and consists of soft, coarse-grained limestone of a yellow-whitish colour with 5 to 12% chert (Fig. 8.11h-j). Felder & Bosch (2000) identified an intermediate development between the Maastricht

Figure 8.13. *Characteristic macrofossils from the Upper Cretaceous of South Limburg. a)* Patagiosites stobaei, *an ammonite from the Benzenrade Member which documents the lateral equivalence of that unit with the Zeven Wegen Member (late Campanian). b) Type specimen of the late Maastrichtian mammal (a small marsupial),* Maastrichtidelphys meurismeti, *from the abandoned ENCI quarry (Maastricht): a direct North American link (Martin et al., 2005). c) The stalked crinoid,* Dunnicrinus aequalis *(collection Oertijdmuseum, Boxtel), from the abandoned ENCI quarry (Maastricht), allows correlation with the upper Maastrichtian of Mississippi (US Gulf Coast). d) Regular echinoid,* Gauthieria gr. radiata, *with full complement of spines, apical disc and lantern,*

testifying to storm-generated obrution events during the latest Maastrichtian. e) vagile bottom fauna, such as this brittle star, is typical of all members of the Maastricht Formation. f) The heteromorph ammonite, Hoploscaphites constrictus johnjagti, *defines the uppermost Maastrichtian in Boreal settings of northwest and eastern Europe (Machalski, 2005). g) The ammonite* Sphenodiscus binckhorsti *from the uppermost Meerssen Member, directly below the K-Pg boundary at the abandoned Curfs quarry, Geulhem. h-j) The 'tegulated' inoceramid key index,* Spyridoceramus tegulatus, *typical of the Vijlen to Nekum members; this is replaced by* Tenuipteria argentea *in the uppermost Meerssen Member (Jagt & Jagt-Yazykova, 2018). k) The carnivorous snail,* Volutospina *aff. deperdita, from the Meerssen Member illustrates precursors of post-Cretaceous (Paleogene) gastropod faunas. l) The ammonite* Pachydiscus gollevillensis *allows correlation of the basal Nekum Member (Maastricht Formation) with the upper part of the Kunrade facies (middle Lanaye Member). Photo (c): Oertijdmuseum Boxtel, with permission; remaining photos by Natural History Museum Maastricht.*

and Kunrade limestone facies which they named the limestone of Schaelsberg (informal).

Towards the top of the formation, chert becomes rare, bioclastic grain size increases and large-scale trough cross bedding can be found (Brinkhuis & Smit, 1996; Zijlstra et al., 1996; Lagrou & Dreesen, 2005). The uppermost part of the formation, the Meerssen Member (Figs 8.6, 8.10b), is characterized by numerous detrital fossil layers full of algae, bryozoans, corals, etc. above hardgrounds, and does not contain chert. Lagrou & Dreesen (2005) reported mudstone to wackestone textures for the Gulpen and Maastricht formations sampled in boreholes in Belgium, indicating that deeper-marine and more offshore settings were developed towards the northwest.

The general shallowing trend in the upper Maastrichtian succession is concomitant with an increase in biodiversity at all levels reflected in seagrass meadow microhabitats and decapod crustacean burrows (Villain, 1977; Liebau, 1978; Sprechmann, 1981; Bless, 1989; Zijlstra, 1994; Schiøler et al., 1997). The lower Maastricht Formation is characterized by some obrution levels ('Lagerstätten', predominantly with echinoderms; Fig. 8.13d), testifying to severe storm activity. In addition, there was riverine/eolian input of terrestrial plants, mammals (Fig. 8.13b) and birds (Martin et al., 2005; Van der Ham et al., 2010; Field et al., 2020).

In the upper Maastricht Formation omission surfaces and hardgrounds (bored and encrusted) are present and the impact of periodic storms is visible in coquinas or lumachelles at several levels, composed mostly of serpulid worms, bivalves and gastropods (Fig. 8.13d,e). Bioherm-like structures of scleractinian corals, rudists and bryozoans became established in some places. Ammonites comprise an admixture of Boreal and Tethyan forms (Table 8.1; Fig. 8.13f, g), while the cold-water belemnite *Neobelemnella* gr. *kazimiroviensis* first appears in the Meerssen

Feathered dinosaurs from the Maastrichtian type area

Until recently, there were no records of avian dinosaurs (or birds) from the Upper Cretaceous biocalcarenites in the Maastricht region. This changed some years ago with the description of two incomplete specimens from a quarry along the southern margin of the Mount Saint Peter, south of Maastricht. One of these, *Janavis finalidens*, named by Benito et al. (2022), had been recovered from the basal Valkenburg Member (Maastricht Formation) already in 1999, but following high-resolution CT scanning at the University of Cambridge (England), it became clear how special this specimen actually was. Although the skull is not preserved, a single bone from the palate (pterygoid) is. Partly on the basis of this, it proved possible to define a new genus and species for it, closely related to the piscivorous genus *Ichthyornis* from the lower Upper Cretaceous of North America, but attaining a considerably larger size, with an estimated wingspan of 1.5 metres. *Janavis finalidens* is of great importance in constituting the youngest known toothed bird worldwide. It demonstrates that, during the late Maastrichtian, around one million years prior to the Cretaceous-Paleogene boundary, toothed birds ('stem group' birds) co-existed with very early 'modern' birds ('crown' birds) such as the partridge-sized wading bird, *Asteriornis maastrichtenis*, or 'wonderchicken' (Field et al., 2020). The type specimen of that form, recovered in the year 2000, originates from the same stratigraphic level as *Janavis finalidens*. It includes a well-preserved skull, lacking teeth, and is interpreted to have been close to the most recent common ancestor of present-day land and water fowl (chicken- and duck-like birds, or Galloanserae).

Reconstruction of the smaller-sized wading bird Asteriornis maastrichtensis *and the gull-like, tooth-bearing* Janavis finalidens (*artwork by Phillip Krzeminski*).

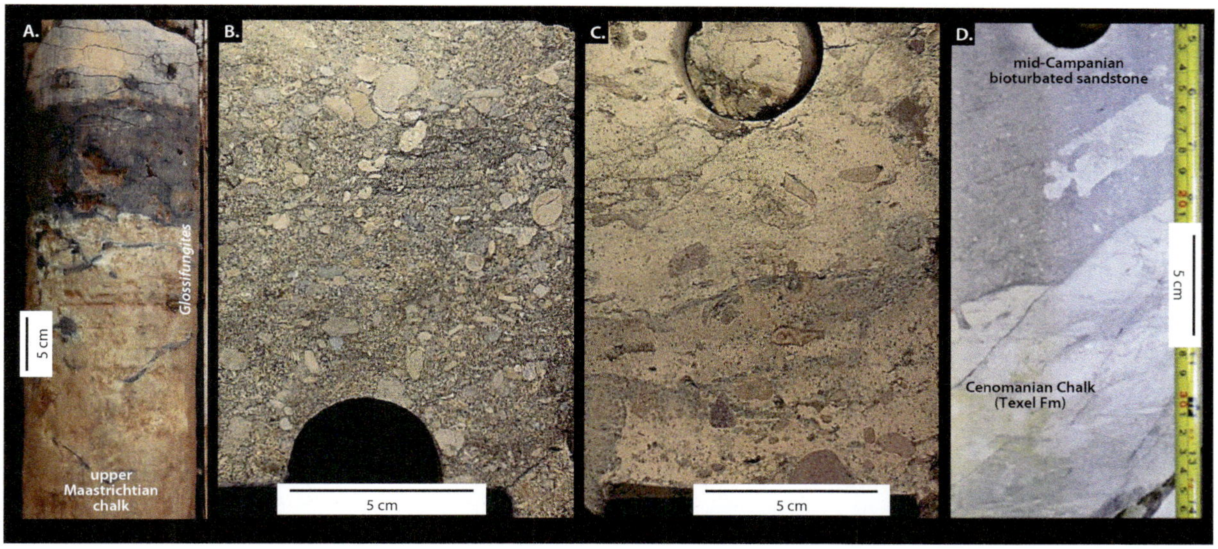

***Figure 8.14.** Core photographs (Van Lochem, 2018). a) Well F17-11 (measured depth (MD) 1420.0 m): contact between the Landen Formation and upper Maastrichtian chalk of the Ommelanden Formation. Note the* Glossifungites *ichnofossils in the chalk. b) Well F17-11 (MD 1469.8 m): clast-supported conglomerate with a chalk matrix. c) Well F17-11 (MD 1468.7 m): matrix-supported conglomerate. d) Well F17-12 (MD 1450.9 m): the Sub-Hercynian Unconformity is between the upper Campanian bioturbated sandstone (Vaals Formation equivalent) and the fractured Cenomanian chalk of the Texel Formation.*

Member (Fig. 8.6) and ranges to the K-Pg boundary (Keutgen et al., 2017).

The Maastricht Formation has yielded sharks, rays, cheloniid turtles, mosasaurs (Fig. 8.15), plesiosaurs and crocodiles (Mulder et al., 2016; Schulp et al., 2017), as well as rich associations of endo- and epibenthic invertebrates (8.13h-j,k), all indicative of clear, well-oxygenated waters. The highest part of the formation (i.e. Meerssen Member, subunit IVf-7) is now considered to be of earliest Danian (early Paleocene) age, the K-Pg boundary being equated with the Berg en Terblijt Horizon, rather than at the Vroenhoven Horizon, which is situated some three metres higher (Fig. 8.17).

In the Maastricht Formation bioturbation is omnipresent. Cryptocrystalline quartz concretions (chert) of varying sizes, morphology and abundance are concentrated mainly in layers but also occur in isolation (Zijlstra, 1987, 1994). The thickness of the formation ranges between 45 and 90 m. Sediments accumulated in a shallow-marine environment (Zijlstra, 1994). Well-known sites where this formation crops out include the disused quarries of ENCI, south of Maastricht (see photo at the beginning of this chapter), and 't Rooth, east of Maastricht near Bemelen. The transition to the overlying Houthem Formation is exposed in the abandoned Curfs quarry, northeast of Maastricht.

Houthem Formation

This formation consists of soft, light grey to beige, fine- to coarse-grained limestones with intercalated hardgrounds and indurated limestone concretions (Fig. 8.9j). The lowermost part of the formation is characterized by the occurrence of glauconite. It occurs mainly in the south of the Netherlands and is by definition only found onshore and is considered to be the lateral equivalent of the Ekofisk Formation (Fig. 8.6). The lower boundary of the Houthem Formation is the Vroenhoven Horizon, a conspicuous hardground with bioturbation at the top of the Maastricht Formation. In ascending order, this unit is subdivided into the Geulhem, exposed at the abandoned Curfs quarry (Herngreen et al., 1998; Felder & Bosch, 2000), the Bunde and the Geleen members. The latter two are known only from boreholes and mine shafts. They comprise coarse-grained, yellowish white limestone with hard bands and very coarse-grained, probably cyclic, limestones respectively.

Sediments of this formation were deposited in carbonate-shelf and upper bathyal environments. The rocks consist essentially of pelagic biogenic remains, which settled from suspension. The sedimentary rocks of the formation may locally reach a thickness of more than 100 m and are of Danian and possibly Selandian/Thanetian age. The formation crops out near the village of Houthem, northeast of Maastricht (e.g. abandoned Curfs quarry and Geul River valley).

Onshore Cretaceous-Paleogene (K-Pg) boundary

In the South Limburg area, carbonate deposition continued across the K-Pg boundary. The early Danian date of subunit IVf-7 is based on palynomorphs, calcareous

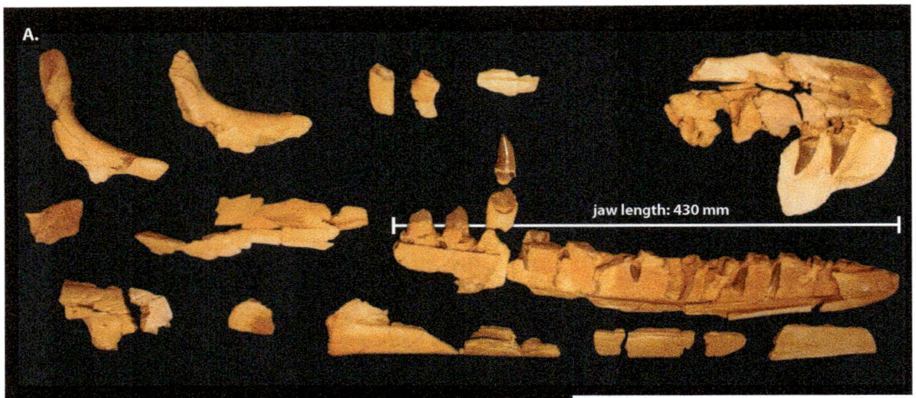

Figure 8.15. Large marine reptiles from the Maastricht area. a) Partial skull of Mosasaurus hoffmanni, *nicknamed 'Curfske', from the abandoned Curfs quarry (Geulhem; Meerssen Member, upper part of subunit IVf-6). b) The type specimen of* Prognathodon saturator *from the upper Lanaye Member (Gulpen Formation; Dortangs et al., 2002) at the former ENCI quarry, now residing on the inner square of the Natuurhistorisch Museum Maastricht. Photos by Natural History Museum Maastricht.*

nannoplankton and foraminifera (Brinkhuis & Schiøler, 1996; Kuhnt, 1996; Romein et al., 1996; Smit & Zachariasse, 1996; Witte & Schuurman, 1996), which indicate that it can be correlated with the basal Danian P0 planktic foraminifer zone, while the overlying Geulhem Member yields planktic foraminifera that tie in with the type-Danian (Moorkens, 1982). Macrofaunal elements with a link to the type-Danian (Stevns Klint and Faxe, Denmark) include echinoderms (Gravesen, 1993; Jagt, 1999a,b, 2000a-d), cirripedes, sharks and rays (Adolfssen & Ward, 2014, 2015). Vagile benthic molluscs from subunit IVf-7 and the overlying Geulhem Member include numerous taxa in common with the middle Danian ('Montien') of southern Belgium (Bignot, 1993; Jagt et al., 2013; Vellekoop et al., 2020).

The combined sections of the former Curfs quarry (Geulhem) and Geulhemmerberg galleries (Fig. 8.16) nearby constitute the most complete K-Pg boundary interval in the South Limburg area (Fig. 8.17). A 0.2-0.3 m thick, heavily indurated level capping subunit IVf-7 yielded diverse molluscan assemblages, including heteromorph ammonites (Landman et al., 2014, 2015). Numerous bivalves are still articulated or in butterfly position, and baculitid ammonites often preserve intact apertures, providing evidence that these assemblages are *in situ*. Based on gastropod faunas, Vellekoop et al. (2020) suggested that a rapid recovery followed the K-Pg extinction event.

For the K-Pg section in the Geulhemmerberg galleries, Smit & Brinkhuis (1996) presented a total picture, based on sedimentology, biostratigraphy, geochemistry and paleomagnetism. The undulating hardground (Berg en Terblijt Horizon) at the base of subunit IVf-7 (Fig. 8.17), with depressions up to 1.5 m in depth, cut into heavily

Figure 8.16. The K-Pg boundary section in the underground galleries of the Geulhemmerberg. The irregular erosion surface in the top of the Maastricht Formation is overlain by bioclastic sand and grey clay layers (IVf-7 in Fig. 8.17). Inset photograph represents 0.4 m vertically. Photos: courtesy of Rudi Dortangs.

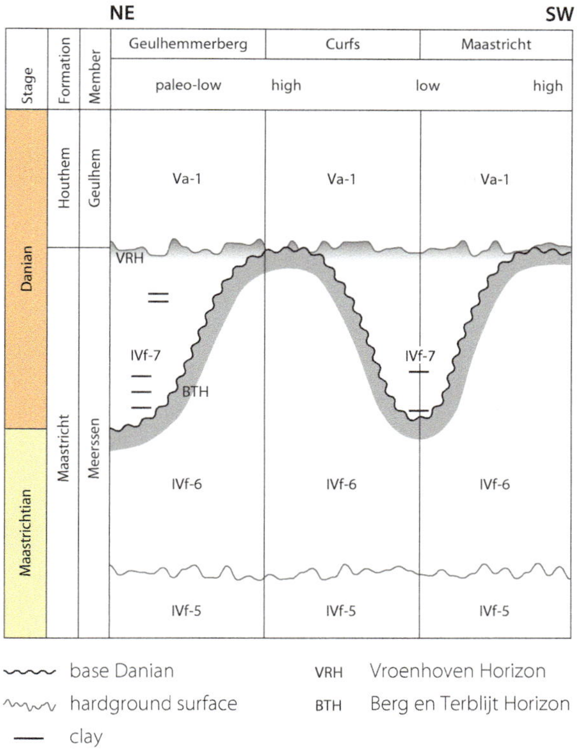

Figure 8.17. Stratigraphic relationships in the K-Pg boundary interval in South Limburg (modified from Herngreen et al., 1998 and Vellekoop et al., 2020). Codes (IVf-5, 6, 7 and Va-1) were introduced by Felder (1975).

bioturbated IVf-6 calcarenites and mark a brief period of non-deposition in the Meerssen Member. Above this, the initially coarse-grained sediments filling the paleorelief depressions represent a thinning-up and fining-up sequence, culminating in a 0.1-0.15 m thick clay level, the so-called E clay. The infilling process is inferred to have been episodic, involving decreasing storm-wave activity in waters deepening to *c.* 20 to 40 m, or alternatively a changeover to a more 'shielded' situation. The clay layers could have formed only during low-energy conditions. Just below the Berg en Terblijt Horizon, very low, yet slightly elevated iridium (Ir) contents have been observed (Smit & Brinkhuis, 1996). The time span between IVf-6 and IVf-7 sediments involved the deposition of latest Maastrichtian calcarenites, followed by K-Pg storm- or hypercane-related 'washing' of the paleoshelf and erosion of uppermost Maastrichtian and lowermost Danian sediments, inclusive of an Ir-bearing layer.

Geological history

Structural development

Chalk Group strata were deformed at various times and different scales. The earliest deformation took place during deposition. Van der Molen et al. (2005) and Van der Voet et al. (2018) recorded local slump structures in the Dutch Central Graben, while Back et al. (2011) documented a large number of gravity-driven structures like slumps and slides, which developed due to syn-depositional tectonics in the southern Danish North Sea. Smit (2018) also noted frequent fluid escape structures in this area. In comparison to the Danish sector fewer syn-sedimentary deformation structures have been found in the Dutch sector. This may in part be due to the lesser number of studies performed on the Dutch chalk interval, but it may also be linked to the shallower nature of the chalk sea in the Dutch sector (Ineson et al., 2014) which led to lower sea floor relief and, concomitantly, gravity-driven structures with a smaller lateral extent.

Syn-depositional tectonics

An important syn-depositional tectonic event was the Sub-Hercynian inversion phase, which peaked during the Campanian. Pre-inversion intervals were folded, tilted and locally faulted by this tectonic event, preferentially above pre-existing Mesozoic grabens (Van Wijhe, 1987a; Dronkers & Mrozek, 1991; De Jager, 2003). Outside these inversion areas sedimentation continued almost without interruption, leading to a continuous, horizontally stratified Cretaceous succession. In the inversion areas, islands emerged in the chalk sea due to the tectonic uplift and were subsequently eroded (Van Lochem, 2018). The pre-inversion chalk intervals, which may not have been completely lithified, may have eroded quite rapidly without leading to recognisable erosional material, as chalk easily disintegrates into its original constituent material (Buls et al., 2015). Eroded clay-rich material may also have been diluted into the surrounding chalk sea, leading to more marly chalk intervals. Siliciclastic-rich erosional material has been found around some inversion structures in the Netherlands, most being assigned to the Vaals Formation (Figs 8.4, 8.6).

In the Dutch Central Graben, seismic data show a clear unconformity between pre- and post-inversion chalk sections (Clark-Lowes et al., 1987; Van der Molen, 2004; Van der Voet et al., 2018; Van Lochem, 2018). Van Lochem (2018) presented a cored section of the Sub-Hercynian Unconformity in well F17-12, where upper Campanian sandstones overlie a fractured and eroded Texel Formation chalk section (Fig. 8.14d). 3D-seismic interpretation (Van Lochem, 2018) demonstrates that the Dutch Central Graben inversion structure has the shape of a north-south-striking anticline with, towards its core, progressively older intervals subcropping below the Sub-Hercynian Unconformity (Fig. 8.18). On the Sub-Hercynian Unconformity

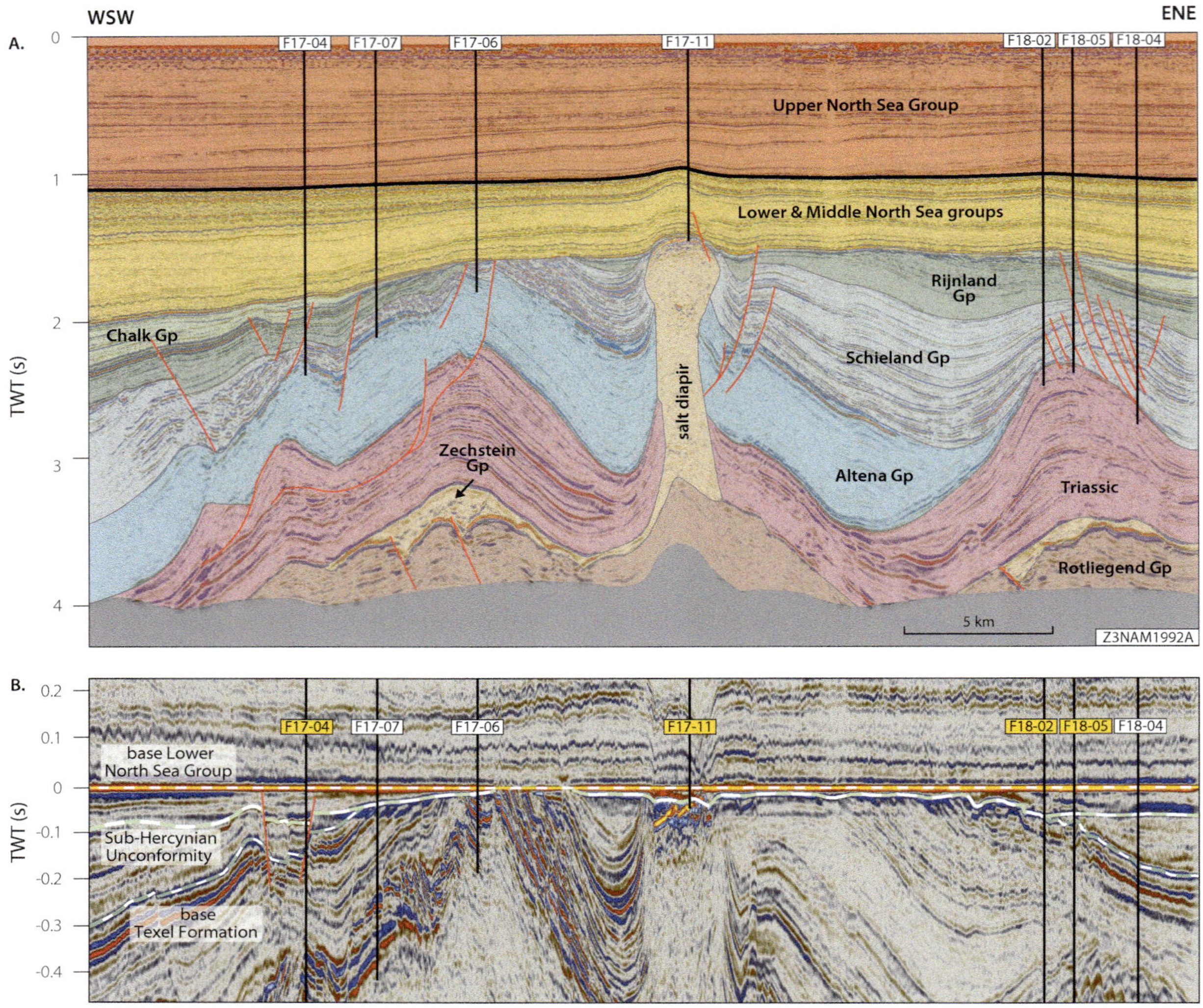

Figure 8.18. a) Interpreted seismic time section across the inverted Dutch Central Graben in blocks F17 and F18. The Upper Cretaceous is present over much of the section but is not visible at this scale. b) Same section flattened on the base Lower North Sea Group and stretched vertically to show the chalk interval. Wells indicated in yellow have a Campanian sandstone above the Sub-Hercynian Unconformity. Note that the morphology of the Sub-Hercynian Unconformity represents a fault-bounded graben at F17-04, a depression above the salt dome near F17-11, and a channel shape with a rapidly thickening succession on top near F18-02. Maastrichtian chalk is absent east of well F17-06 where highly reflective Upper Jurassic sandstone, claystone and coal beds subcrop below the base Cenozoic (from Van Lochem, 2018). For location see Fig. 8.1

surface onlapping levels of late Campanian and Maastrichtian age are found, demonstrating that most of the inversion movements had come to a halt here during the late Campanian. In the southern part of the Dutch Central Graben, the upper Maastrichtian is conformably overlain by Selandian and Thanetian deeper marine claystones of the Landen Formation. This contact was cored in well F17-11 (Fig. 8.14a); it shows a burrowed interval of Maastrichtian chalk, which can be interpreted as a marine hardground, overlain by a 0.1 m thick, glauconitic-rich claystone (Van Lochem, 2018).

The NW-SE-striking Broad Fourteens Basin (Fig. 8.1) has been the subject of several inversion-related studies during the last decades (Van Wijhe, 1987b; Hooper et al., 1995; Naplas, 1994; Bouw & Oude Essink, 2003a,b; Deckers & Van der Voet, 2018). An important difference with the Dutch Central Graben is the fact that existing Mesozoic graben faults were reactivated as reversed faults or locally as thrust faults. Zechstein salt acted as decollement surface and facilitated the formation of several low angle thrust faults (Nalpas et al., 1996). Hooper et al. (1995) estimated the shortening by inversion of the Broad Fourteens Basin at 10% and the total uplift along the inversion axis at 3500 m. The erosion that followed this uplift completely removed the Chalk Group from the centre of the Broad Fourteens Basin. At the basin's edges partially eroded sections of chalk are unconformably overlain by a chalk section of Maastrichtian age (e.g. well K13-A-10). A thin unconformable Maastrichtian and/or potentially Danian cover can be traced from the western flank towards the centre of the basin, overlying older Mesozoic sedimentary rocks (Deckers & Van der Voet, 2018). Bouw & Oude Essink (2003a,b) modelled the hydrogeological evolution of the Broad Fourteens Basin during the inversion. Fresh meteoric water infiltrated into the eroded outcropping Mesozoic sedimentary rocks and may have influenced the biodegradation of Lower Cretaceous oil fields there.

Inversion in the West Netherlands Basin is limited to the northern part of this Jurassic basin. In the southern half, which partly coincides with the Paleogene Voorne Trough, a complete and uneroded chalk succession is present (NITG-TNO, 2002). A burial study by Worum & Van Wees (2017) has shown that the maximum thickness of the Chalk Group in the centre of the basin at any moment during the Late Cretaceous was around 300 m. This figure is much smaller than the total preserved thickness of the chalk in the Voorne Trough, indicating that continuous syn-inversion sedimentation occurred there, while non-deposition or slight erosion characterized the centre of the basin, which was inverting. Thinning of the Heers Formation from the Voorne Trough in the direction of the West Netherlands Basin and truncation of the top of the Chalk Group in the same direction suggests that the basin was also inverted during the Paleocene Laramide phase (Deckers & Van der Voet, 2018). Doornenbal et al. (2019) recognized NE-SW oriented intra-chalk runoff channels, which may be related to uplift in the northeast part of offshore block P15.

The fault zone between the West Netherlands Basin and the Central Netherlands Basin to the north is characterized by reverse faulting and overthrusting of the Mesozoic succession (NITG-TNO, 2002). Directly north of this fault zone there is a ridge which has been referred to as the Zandvoort Ridge, which contains an eroded chalk section of Cenomanian to Coniacian age. Compared to the adjacent inverted basins this ridge was buried less deeply and aquifers in this structural feature may have preserved more favourable reservoir parameters due to less intense diagenesis (Pluymaekers et al., 2012). North of the Zandvoort Ridge, no chalk has been preserved in the Central Netherlands Basin, except for a thin development of the Danian-aged Houthem Formation in the area around Harderwijk (NITG-TNO, 2004) near the axis of the basin. The preservation of Danian carbonates shows that Late Cretaceous basin-wide inversion had significantly slowed down or even ended prior to the Danian and that the Central Netherlands Basin did not experience strong erosion during the Laramide phase at the end of or just after the Danian (Deckers & Van der Voet, 2018). The northeastern boundary of the Central Netherlands Basin is determined by the Raalte Boundary fault, which is a large, pre-existing Paleozoic normal fault that was reversed during Late Cretaceous inversion (NITG-TNO, 2004).

During the Cenomanian to Coniacian the Roer Valley Graben and the adjacent Campine Basin (Belgium) as well as the southern part of the Peel-Maasbommel Complex were exposed above sea level (Luijendijk et al., 2011). During this period the coastline was located on the northern part of the Peel-Maasbommel Complex in the area of Nijmegen (Gras & Geluk, 1999). During the Santonian to Campanian, a southward transgression took place. On the borders of the Peel-Maasbommel Complex, the Campine Basin and the London-Brabant Massif shallow-marine siliciclastics of the Oploo and Vaals formations accumulated. These sedimentary rocks are erosional products of the inverting Roer Valley Graben, which formed a positive feature during the Sub-Hercynian Phase (Kuyl, 1983; Bless et al., 1987; Gras, 1995; Gras & Geluk, 1999; NITG-TNO, 1999; NITG-TNO, 2001). During the Maastrichtian, the strata became more carbonate-rich as the Roer Valley Graben gradually lost its importance as a source area. Finally, during the Danian, also the Roer Valley Graben was inundated and the marine Houthem Formation formed a blanket over most of the south of the Netherlands (NITG-TNO, 2001). During the Laramide Phase a broad domal uplift of the London-Brabant Massif removed the Danian and part of the Maastrichtian successions (Deckers & Van der Voet, 2018).

In the Vlieland Basin two inversion phases can be recognized as well. Herngreen et al. (1996) concluded that thickness changes in the Coniacian to Campanian sections in wells Den Burg-1 (BRG-01) and De Cocksdorp-1 (COC-01) on the Isle of Texel could be related to the Sub-Hercynian inversion of the adjacent Vlieland Basin. The missing parts of the Maastrichtian and truncation of the Cretaceous strata reflect erosion during the Laramide Phase.

Not only did inversion tectonics have an impact on the thickness distribution of the Chalk Group, also halokinetic movements played an important role. The Chalk Group is often thinned over rising salt domes, while next to the salt structure in subsiding rim synclines the group is clearly thickened (RGD, 1993; Van der Voet et al., 2018).

Post-depositional tectonics

Following deposition, deformation of the Chalk Group continued. Due to the brittle nature of chalk extensive fracture networks can often be found. Welch et al. (2015) studied fracture geometries in chalk outcrops in the UK, while Koestler & Ehrmann (1986, 1991) mapped faults and fractures in quarries in northern Germany (Hamburg area), where chalk was pushed upwards by underlying salt diapirs (Niebuhr, 2006; Voigt et al., 2008). The development of these fractures is very important for any fluid (oil) flow to take place in the chalk, as chalk matrix permeability is generally too low (typically <10 mD) for any commercially interesting flow quantities. Van Rooijen (1989) noted that in South Limburg the ground water abstraction rates from chalk reservoirs were highly influenced by the presence of fractures.

In the North Sea Basin salt doming is the most important deformation mechanism that contributes to the creation of fracture networks in the chalk. The main chalk oil fields in the Norwegian, UK and Danish sectors rely on these fracture networks for production (Surlyk et al., 2003). In the Dutch Central Graben, where the Chalk Group lies above major salt domes and diapirs (Remmelts, 1996; Van Winden et al., 2018), fractured chalk is the reservoir of several oil fields. For the Hanze oil field average permeabilities of 200-780 mD were modelled in the course of a 3D-permeability modelling exercise, taking into account the intense macroscopic fracturing (0.8-1.2 fractures/m), as well as the microscopic fracture network and well injectivity results (Hofmann et al., 2002). For the Rembrandt and Vermeer chalk oil fields, Wang et al. (2018) built a discrete fracture network model using 3D-seismic attributes by mapping small-scale faults and lineaments, while Beller et al. (2019) used diffraction imaging to improve the fault and fracture models of these fields. Van Gent et al. (2010) used small-scale, intra-chalk faults, confined to the brittle Chalk Group, to provide a way to estimate paleostress from seismic data for specific time intervals. Based on the orientation and position and on a thickness analysis, it was concluded that these faults were formed due to the tectonic reactivation of salt structures during the latest Cretaceous.

Other large faults cutting through the Chalk Group are related to the Roer Valley Graben and the Mid-Netherlands Fault Zone. A large number of these faults may have been active up to recent times (Van Balen et al., 2005). These faults mainly have a normal-slip mode, which is in accordance with the present-day orientation of the maximum horizontal stress and the vertical orientation of the maximum stress (Van Balen et al., 2005). In and near these fault zones open fracture systems may be expected in the Chalk Group.

Karstification

Fresh water from the surface with dissolved organic acids percolating through limestone causes dissolution of the limestone. This process is called (meteoric) karstification and results in a plethora of dissolution features (Gilli, 2015). In South Limburg these features are known from quarries, boreholes and occasional sinkholes. The most

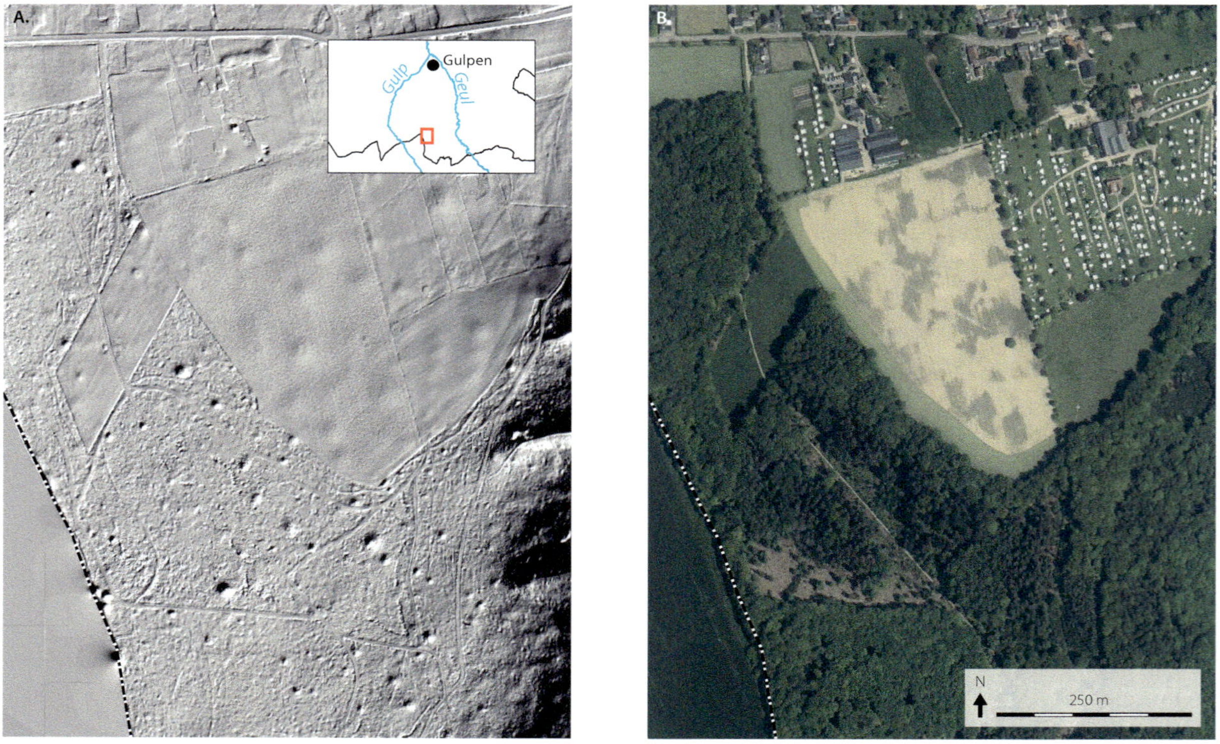

Figure 8.19. Suffosion dolines on the plateau between the rivers Gulp and Geul in South Limburg. a) LIDAR image with shaded relief effect shows pronounced and smooth dolines (AHN v3). b) Aerial photograph of the same area showing that smooth dolines are located in farmland.

commonly observed features are solution pipes, which are often filled with gravel and loess. However, in most cases they lie below a thick sedimentary cover and their presence cannot be demonstrated. In the southeast of South Limburg, the sedimentary cover is thin enough (8-20 m) to permit a topographic expression of the underlying chalk. Depressions in the sediment overlying limestone, called *suffusion dolines* can be observed using publicly available Laser Imaging Detection and Ranging data (LIDAR). A distinction can be made between dolines with a pronounced expression and those with a smooth expression (Fig. 8.19). The difference is probably the result of land-use as smooth dolines are found in farm land where ploughing and filling tend to level the dolines. Dolines with a more pronounced expression are found in forested areas.

The distribution of dolines and the underlying solution pipes in the landscape is often linked to the occurrence of fractures and cracks, as is often the case with hard, compact limestone (Jennings, 1985). In the soft and coarse-grained limestone of South Limburg, however, the dolines seem to occur more randomly, without control by fractures or cracks. This observation is also supported by several publications that state that in the Upper Cretaceous limestones of northwest Europe (chalk), solution pipes unconnected to fractures occur (Kirkaldy, 1950; Rodet, 1992; Mouriau, 1998). In the Romontbos quarry south of Maastricht, solution pipes are frequently found under the gravelly deposits of the Meuse River, while under clayey sands in the west of the quarry, no solution pipes are observed in the limestone (Juvigné & Renard, 1992; Willems et al., 2007). The presence of solution pipes seems to correlate with the presence of permeable deposits on top of the limestone.

The age of solution pipes and dolines in South Limburg is unknown and no research has been published on the subject. The frequency with which dolines form is also largely unknown, but observations of sinkholes are regularly reported from the area. The dissolution of limestone probably started tens of millions of years ago when the limestones were lifted above sea level. Data from borehole Molenbeersel in the Belgian Campine region suggest that karstification occurred shortly after the end of the Cretaceous (Swennen & Dusar, 1997). Solution pipes observed in the former ENCI quarry show that the backfill consists of thick layers of loess, which indicates that there was already a depression during the Pleistocene (2.58-0.01 Ma) when the loess was deposited.

Outside South Limburg karst features have been observed in the subsurface on the Zandvoort Ridge (Fig. 8.20), where the partially eroded chalk section shows dissolution and collapse structures. Near well Waverveen-1 (WRV-01), two 200 to 500 m wide dolines can be observed on seismic data. This karstification predates the Oligocene to lower Miocene Rupel Formation, which overlies the chalk of the Coniacian Ommelanden Formation. Wells Waverveen-1 and Zeist-1 (ZST-01) (Fig. 8.1) encountered significant losses while drilling the chalk, indicating the presence of open fractures and karst cavities. During the Cretaceous, karstification also affected the lower Carboniferous (Dinantian) limestones exposed along the London-Brabant Massif (see Vis et al., 2025, this volume).

Offshore chalk sedimentology and facies

Most geologists who do not work on the Chalk Group would consider this interval to rank amongst the most monotonous with respect to facies and sedimentology. While the bulk of the Chalk Group consists of white carbonate mudstones with few apparent sedimentological features other than interbedded chert layers or nodules, more can be said about its subtle variations. Although the Chalk Group crops out in southern South Limburg and is therefore relatively easy to study, only a few detailed sedimentological and paleoecological studies have been carried out (Zijlstra, 1994; Roep & Smit, 1996; Zijlstra et al., 1996; Lagrou & Dreesen, 2005). As work is underway to fill this knowledge gap, we here limit the description of sedimentological details and facies to the offshore Chalk Group.

The study of offshore chalk relies on rock cores for detailed sedimentological facies analysis. However, due to the lack of hydrocarbon discoveries in the Chalk Group, the core coverage is low. Preferentially, the uppermost intervals of the chalk have been cored, which has led to a bias towards study of Danian- and Maastrichtian-aged successions. Following the discovery of the Rembrandt oil field, Wintershall Noordzee B.V. and its partners commissioned a study of all released chalk cores in the Dutch offshore, as well as some onshore cores (Ineson et al., 2014). Identification of the main facies associations and the environmental model are reproduced here, with permission, from this confidential report.

Eight facies associations (Figs 8.8, 8.21) have been recognized in the cores of the Ekofisk and upper part of the Ommelanden formations, with 95% of the cored interval consisting of the Chalk Mudstone Association (FA-1; 30%), the Marly Chalk Association (FA-2; 37%) and the Skeletal Chalk Wackestone-Packstone Association (FA-3; 28%). The remaining five facies associations record more local conditions.

The Chalk Mudstone Association (Fig. 8.8a) consists of fine-grained carbonate, primarily reflecting a pelagic settling. The ubiquitous bioturbation indicates an oxygenated sea floor environment and relatively low accumulation rates, allowing for wholesale biogenic reworking of the carbonate ooze. Although the well-known, deeper-water, low-energy trace fossil assemblage *Chondrites-Planolites-Zoophycos* typifies this association, certain sections

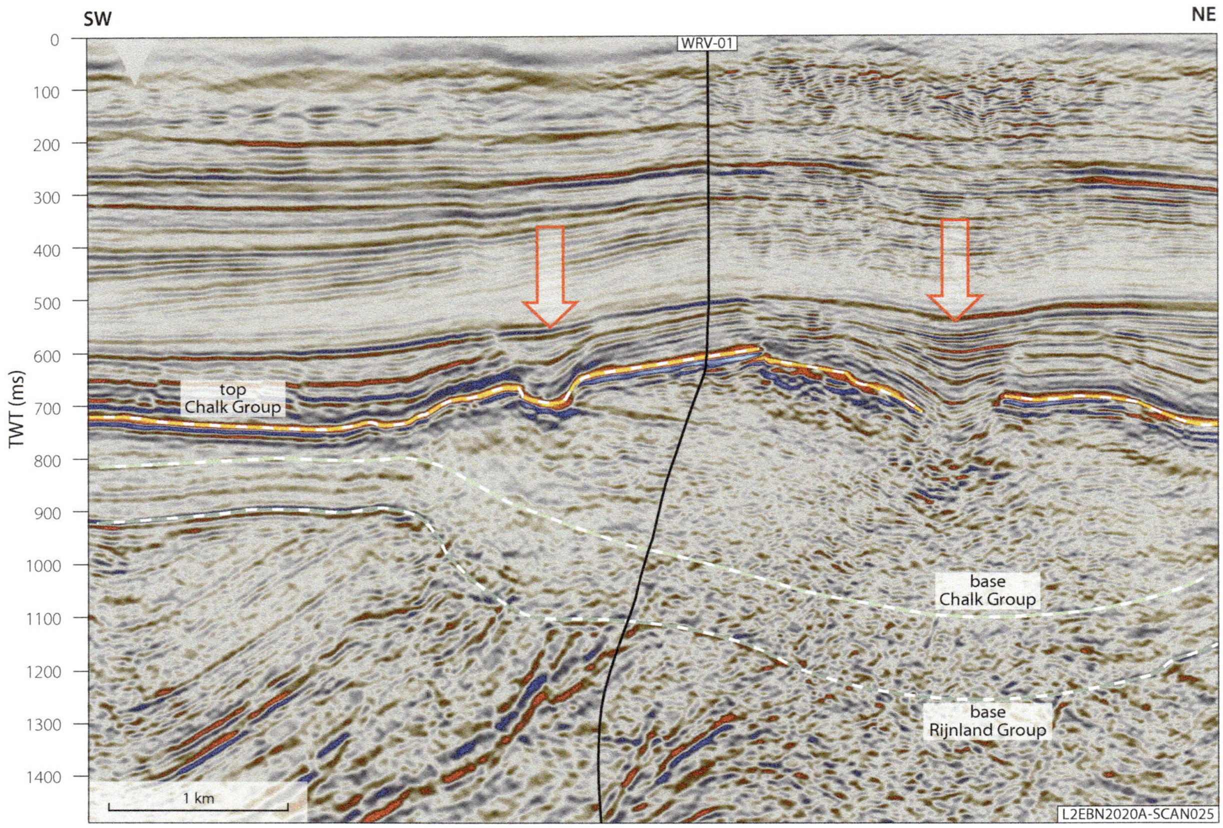

Figure 8.20. *Seismic time section showing collapse structures (arrows) at the top of the Chalk Group near well Waverveen-1 (WRV-01; for position see Fig. 8.1) on the Zandvoort Ridge.*

show well-developed *Thalassinoides* fabrics, probably indicating shallower, more hospitable conditions.

The Marly Chalk Association (Fig. 8c-e) is common in the logged cores; it is characterized by an alternation of burrowed, variably marly chalk mudstones and marlstones. It is commonly rich in chert. Two assemblages have been distinguished in this association, both named after the dominant trace fossil: the deeper-water *Zoophycos* assemblage and the more proximal *Thalassinoides* assemblage.

The Skeletal Chalk Wackestone-Packstone Association (Fig. 8.8b), with common to abundant occurrences of skeletal remains together with a predominance of *Thalassinoides*-dominated trace fossil assemblages, suggests a relatively shallow, nutrient-rich setting with moderate energy levels. Interbeds of dense bioclastic packstone suggest periodic higher-energy levels induced by increased velocity of bottom currents or possibly by deep-reaching storm waves. Accordingly, the depositional position of this facies association is interpreted to have been along submarine highs with reduced marly input and/or in intermediate shelf/ramp positions.

The remaining facies associations are related to local shallow-water or emergent conditions. It should be noted that mass-flow deposits are very rarely recognized in the cored sections.

Ineson et al. (2014) proposed two stages to explain the spatial relationships between the observed depositional environments (Fig. 8.21). Phase A depicts a distally steepened ramp environment in which a deep shelf-basinal setting is characterized by coccolith-dominated chalks with a deeper-water, *Zoophycos*-dominated trace fossil assemblage. The clay content in this setting varies temporally, controlled by regional climatic/tectonic factors. Shorewards, the chalk sea shallows to mid-shelf and becomes more nutrient-rich; the benthic fauna increases and, as the environment shifts from oligotrophic to mesotrophic calcispheres become important. This skeletal chalk belt grades into a *Thalassinoides*-burrowed more marly chalk facies, with siliciclastic mud being derived from the regional shoreline or from locally emergent highs.

Stage B (Ineson et al., 2014) schematically illustrates the effect of inversion/salt-induced uplift on the facies belts; with increasing shallowing, the developing structural high will have formed a favourable habitat for benthic invertebrates, potentially culminating in local bryozoan/coral mound development (e.g. the Hanze diapir). Further shallowing will have exposed the high area to current

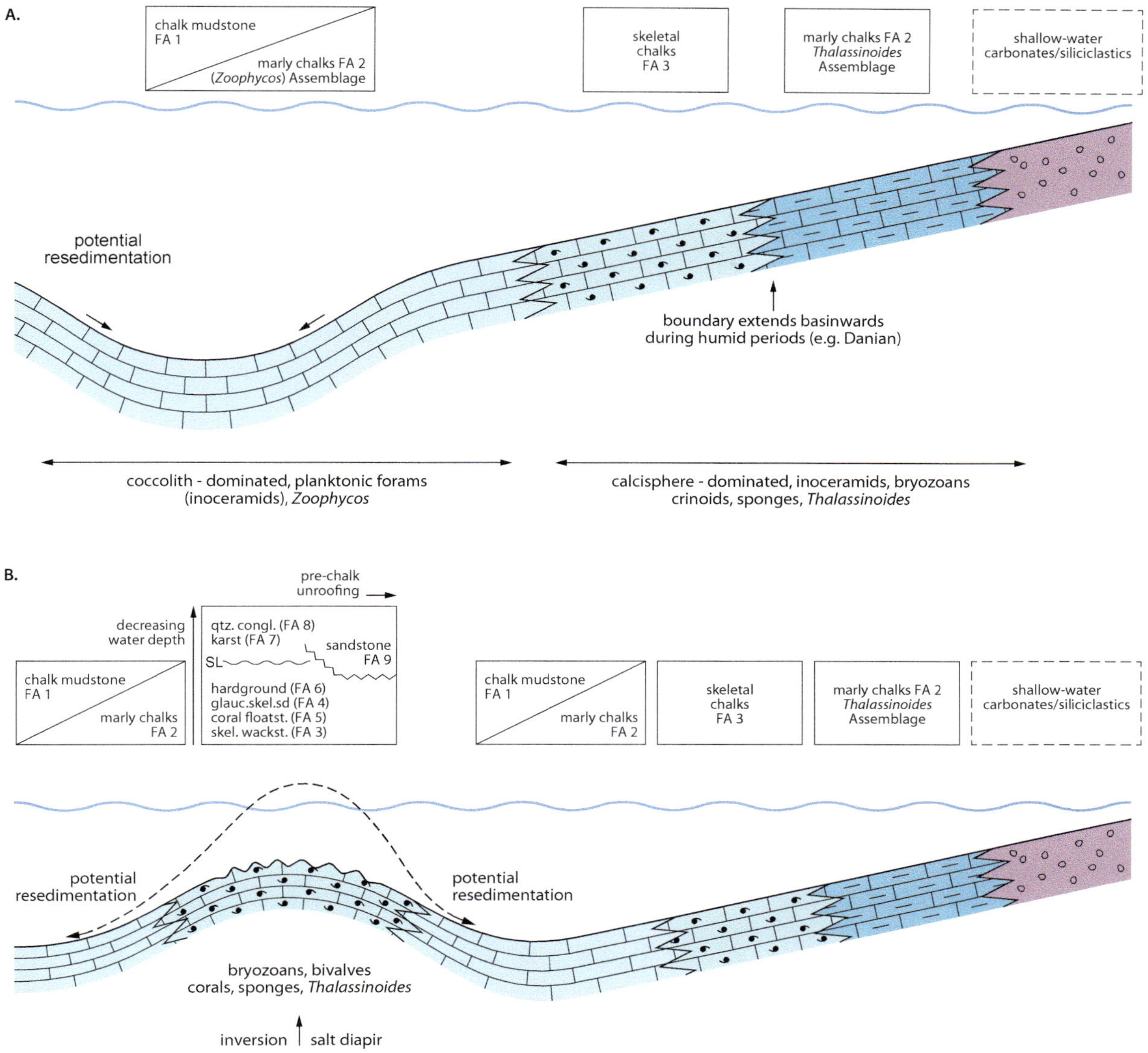

Figure 8.21. Sketches showing stages A and B for the inferred spatial relationships between depositional environments (Facies Associations (FA) 1-8) during (a) Periods of regional subsidence (Cenomanian-Santonian) and (b)Following late Campanian-Danian localized structural inversion uplift (from Ineson et al., 2014).

winnowing, concentration of skeletal sands and potentially also to hardground development. Local exposure of the high also may have resulted in karstification.

In the cores of the Chalk Group several typical, smaller-scale sedimentological features can be recognized (Fig. 8.8). Often intense bioturbation and discrete trace fossils are observed, the latter of which may be indicative for specific depositional environments. Anastomosing solution seams, also called horsetail structures, and horizontal stylolites indicate dissolution of limestone due to overburden pressure, whereby darker non-dissolving material is concentrated on the solution surfaces. In addition, chert nodules are often present, which occasionally follow burrow structures and are sometimes deposited in rhythmic vertical successions, which may be related to Milankovitch cycles (Zijlstra, 1994). The formation of these chert nodules or levels is thought to have been governed by early diagenetic bacterial redox reactions concentrating the silica particles present in the chalky sediment (Zijlstra, 1987). Chert is more common in the upper part of the Ommelanden Formation (Campanian-Maastrichtian) and in the Ekofisk Formation than in the lower part of the Ommelanden Formation and the Texel Formation (Herngreen et al., 1996).

Microfacies analyses (Fig. 8.9c) of samples from the cored upper Maastrichtian section of well F17-10 show mud-supported, bioturbated, heterogeneous wackestone textures, the sediment consisting of 97-99% calcite (Petrea & Goldberg, 2013). The main detrital components are coccoliths, foraminiferal tests (mostly planktic), calcispheres, bivalves (in particular inoceramids) and brachiopod shell fragments, ostracods, echinoderm ossicles and fragments, bryozoan colonies, calcareous sponges and sponge spicules. The main authigenic components are overgrowths on original calcite mud components and rare to common calcite sparite cements. Herngreen et al. (1996) described similar microfacies from the cores of the Den Burg-1 (BRG-01) well on Texel. These samples are from chalk of a Cenomanian to early Campanian age that formed at water depths of approximately 50-150 metres.

In well F17-10 SEM and BSEM images of the chalk (Petrea & Goldberg, 2013) document that complete or nearly complete coccolith tests are rare (Fig. 8.10a). Mostly, coccoliths are disarticulated into individual platelets, which form the bulk of this micrite mud, which is neither compacted nor cemented very well, resulting in high porosities. However, the pore throats of the micrite mud typically range between 0.5 and 2.0 µm, resulting in a low matrix permeability of 1-5 mD. Pure chalk mudstone can have a porosity of up to 45%, while admixture of bioclastic and siliciclastic material tends to decrease the total porosity in chalk wackestones in this area to 25-35% (Van Lochem & Beller, 2018).

In areas close to emergent landmasses, siliciclastic facies are present in the Chalk Group, often in the form of glauconite-rich 'greensands'. The oldest of these greensands is the Cenomanian Texel Greensand, present mainly on the southern fringe of the West Netherlands Basin and further east in well Maasbommel-1 (MSB-01) (Gras, 1995). In the Dutch Central Graben area, Van Lochem (2018) documented the presence of upper Campanian marine conglomerates and glauconitic sandstones that overlie the Sub-Hercynian Unconformity and grade upwards into lower Maastrichtian sandy chalk (Fig. 8.14b-d). The sandstone facies in this area formed when fluvial incised valleys were flooded during a transgressive phase. These valleys were eroded into emerging highs during uplift subsequent to the Sub-Hercynian tectonic phase (Fig. 8.18).

Van Lochem (2018) and Van Lochem & Beller (2018) have shown that in offshore block F17 in the interval just above the Sub-Hercynian Unconformity there is a higher content of siliciclastic material and, to a lesser degree, an increased number of skeletal grains. The siliciclastic grains comprise very fine- to medium-grained, quartz and glauconite/green clay grains (Fig. 8.9a,b,e,f) that are either dispersed through the carbonate mud matrix or concentrated within burrows, along laminae of more coarsely grained material or as component of tests of agglutinated foraminifera.

A large north-south-trending channel system was described by Van der Molen (2004) on the Schill Grund High and on the Ameland Block in the Maastrichtian and Danian interval. Surlyk et al. (2008) mapped the continuation of this feature to the north into the German offshore area. They documented channels that were filled laterally, obliquely, or longitudinally by chalk drifts and proposed that the main channel system was eroded by powerful contour-parallel bottom currents (contourites) flowing towards the southeast. Large-scale evidence for channelling due to ocean bottom currents in the chalk was found by Surlyk & Lykke-Andersen (2007) in the Danish Basin and by Hübscher et al. (2019) in the southern Baltic Sea.

Geophysical studies

Well-log correlation within the Chalk Group is quite difficult without the help of micropaleontology and/or seismic interpretation (Fig. 8.7). The gamma-ray log typically shows a continuous low value, while only the high values in the thin Plenus Marl Member stand out. The gamma-ray log readings often increase towards the base of the Texel Formation, at the transition with the underlying Rijnland Group. A clearer trend can be seen in the sonic-log with transit-times decreasing and velocities increasing from the top to the bottom of the Ommelanden Formation. It is often the interval of Turonian age which shows the highest velocity, higher than the underlying Texel Formation. Where present, also the base of the Ekofisk Formation often shows a slightly higher velocity than the underlying Ommelanden Formation.

Van der Molen (2004) used spectral frequency analyses of well logs to conclude that specific wavelength ratios between 5-7 m and 2-3 m represent Milankovitch 100-kyr eccentricity and 40-kyr obliquity cycles, respectively. From this he derived average net accumulation rates in the chalk of about 5 to 7 cm/kyr. However, it should be noted that sufficiently detailed biostratigraphic control was absent and these estimates could therefore not be corroborated. Arts (2017) recognized a 15-22 m cycle in the well-log records, which he interpreted to correspond to the 405-kyr eccentricity cycle.

Seismic stratigraphy is a powerful tool to unravel the sedimentary and large-scale tectonic development of the Chalk Group. For it to be successful a good biostratigraphic framework is needed next to high-quality seismic data. Van der Molen (2004) and Van der Molen & Wong (2007) subdivided the Chalk Group of the Dutch offshore into eleven seismic sequences (CK1-CK11; Figs 8.4, 8.7) and mapped these sequences on 2D-seismic data over most of the Dutch offshore area. In general, the seismic facies of the chalk are characterized by parallel and continuous reflections of varying amplitudes. Two megasequences were recognized, the first includes the Cenomanian to lower Campanian interval, which is separated from the second megasequence (upper Campanian to Danian) by a large uplift and erosion phase that predates the deposition of the upper Campanian. In a more detailed study by Van der Molen et al. (2005) on two 3D-seismic surveys seven sequences were mapped. They found that most of the Chalk Group is characterized by parallel and continuous reflections, indicative of pelagic sedimentation under relatively quiescent tectonic conditions. Where deposition of the Chalk Group coincided with intense tectonic activity, discontinuous and irregular seismic reflections are found. Such reflection configurations indicate reworking of sediment.

Van der Voet et al. (2018) performed a 3D-seismic sequence-stratigraphic study of the northern E and F quadrants in the Dutch sector, covering the Elbow Spit Platform, the Step Graben and the northern and central part of the Dutch Central Graben. In their study the Chalk Group was subdivided into the seven chronostratigraphic stages of the Late Cretaceous plus the Danian.

Van Lochem (2018) published the results of a seismostratigraphic study of the Chalk Group in the central and southern part of the Dutch Central Graben. The interpreted markers coincide with the boundaries of the stages of the Upper Cretaceous and, in addition, to the Sub-Hercynian Unconformity surface. This interpretation resulted in a subcrop map of the Sub-Hercynian Unconformity. The results of these studies in the Dutch North Sea compare well with studies in the Norwegian (Gennaro et al., 2013) and Danish (Van Buchem et al., 2018) sectors (Fig. 8.4). In the Danish sector the only inversion event is referred to as the Early Campanian Unconformity (Van Buchem et al., 2018) and is equivalent in timing and tectonic character to the Sub-Hercynian Unconformity in the Dutch sector.

For the Dutch onshore area the only study to make use of seismic stratigraphy was conducted by Gras & Geluk (1999). These authors subdivided the Chalk Group of the Peel-Maasbommel Complex into three seismic facies. The lowermost seismic facies I, which has a transparent character with few and discontinuous reflections, represents the sandy, shallow-marine Oploo and Vaals formations of Santonian to Campanian age. Towards the north, it grades into seismic facies II and becomes less transparent, marking a lithofacies change from siliciclastic sediments into chalk and marls of the Ommelanden Formation. Unconformably overlying is seismic facies III, which shows a fairly constant thickness of 60-80 m. This seismic facies comprises several continuous high-amplitude reflectors, parallel to the overlying Paleogene succession, representing the Gulpen, Maastricht and Houthem formations, which were deposited in a shallow-marine chalk environment.

Other types of seismic evaluations have been successfully applied to the Chalk Group, especially where 3D-seismic data are available. Since chalk is a monomineralic rock it is ideal for rock physical and seismic inversion studies to understand the acoustic properties as a function of porosity. Seismic inversion and spectral decomposition studies in the Danish sector show slump and fluid escape structures and also diagenetically altered areas (Smit, 2018). In the Dutch offshore sector seismic inversion, spectral decomposition and PaleoScan™ interpretation studies in the chalk yielded promising results (Van Lochem, personal observation), but this work has not been published yet.

The chalk has been the subject of geophysical velocity studies (Van der Molen, 2004; Van Dalfsen et al., 2006), which is of particular interest as its velocity is often clearly higher than in underlying rocks, while the velocity also increases significantly with burial depth. Van der Molen (2004) used the observed velocity changes in the chalk to estimate burial anomalies. Especially in the Dutch Central Graben the velocities proved to be lower than expected at the present burial depth, indicating that the sequence is under-compacted. This under-compaction can be related to overpressure and reduced dewatering (Verweij, 2006).

Acknowledgements

Wintershall Noordzee B.V. and its partners are thanked for releasing the information from their offshore core report (Ineson et al., 2014). Allard van der Molen provided useful suggestions for the improvement of an earlier typescript. Dominik Mock and Marie Klose of the Institute of Mineralogy of Leibniz Universität Hannover are acknowledged for producing SEM images of limestone from South Limburg. Nik Trabucho Alexandre is thanked for drawing a number of figures and we are grateful to Mateus Kroth for valuable discussion on the Upper Cretaceous stratigraphy of South Limburg and for providing images of thin sections. We appreciate our reviewers Nora Witmans and Eva de Boever of TNO – Geological Survey of the Netherlands for their valuable comments on a first draft of the manuscript.

Digital map data

Spatial data of figures in this chapter for use in geographical information systems can be downloaded here: https://doi.org/10.5117/aup.28163312.

References

Adolfssen, J.S. & Ward, D.J., 2014. Crossing the boundary: an elasmobranch fauna from Stevns Klint, Denmark. Palaeontology 57: 591-629. DOI: 10.1111/pala.12079

Adolfssen, J.S. & Ward, D.J., 2015. Neoselachians from the Danian (early Paleocene) of Denmark. Acta Palaeontologica Polonica 60: 313-338. DOI: 10.4202/app.2012.0123

Albers, H.J., 1976. Feinstratigraphie, Faziesanalyse und Zyklen des Untercampans (Vaalser Grünsand = Hervien) von Aachen und dem niederländisch-belgischen Limburg. Geologisches Jahrbuch A34: 3-68.

Alvarez, L.W., Alvarez, W., Asaro, F. & Michel, H.V., 1980. Extraterrestrial cause for the Cretaceous-Tertiary extinction. Science 208: 1095-1108. DOI: 10.1126/science.208.4448.1095

Anderskouv, K., Damholt, T. & Surlyk, F., 2007. Late Maastrichtian chalk mounds, Stevns Klint, Denmark – Combined physical and biogenic structures. Sedimentary Geology 200: 57-72. DOI: 10.1016/j.sedgeo.2007.03.005

Arts, M.C.M., 2017. Towards a cyclostratigraphic framework of the Upper Cretaceous Chalk Group in the Dutch offshore. MSc thesis Utrecht University (the Netherlands) (the Netherlands): 144 pp.

Back, S., Van Gent, H., Reuning, L., Grötsch, J., Niederau, J. & Kukla, P., 2011. 3D seismic geomorphology and sedimentology of the Chalk Group, southern Danish North Sea. Journal of the Geological Society 168: 393-406. DOI: 10.1144/0016-76492010-047

Batten, D.J., Streel, M., Dusar, M. & Bless, M.J.M., 1987. Late Cretaceous palynomorphs from the boreholes Thermae 2002 (Valkenburg a/d Geul, the Netherlands) and 's-Gravenvoeren (Belgium). Annales de la Société Géologique de Belgique 110: 47-51.

Batten, D.J., Dupagne-Kievits, J. & Lister, J.K., 1988. Palynology of the Upper Cretaceous Aachen Formation of northeast Belgium. In: Streel, M. & Bless, M.J.M. (eds): The Chalk District of the Euregio Meuse-Rhine. Selected papers on Upper Cretaceous deposits. Natuurhistorisch Museum Maastricht/Université d'État Liège: 95-103.

Beller, M., Chaouch, N., Petri, V., Veeken, P., Landa, E., Kokoshin, E., Smirnov, K. & Paleari, C., 2019. Diffraction imaging: resolving subtle seismic geo-features in the Rembrandt and Vermeer Chalk oil fields, Dutch offshore. 81st EAGE Conference and Exhibition 2019, London: 1-5. DOI: 10.3997/2214-4609.201900935

Benito, J., Kuo, P.-C., Widrig, K.E., Jagt, J.W.M. & Field, D.J., 2022. Cretaceous ornithurine supports a neognathous crown bird ancestor. Nature 612: 100-105. DOI: 10.1038/s41586-022-05445-y

Bignot, G., 1993. The position of the Montian Stage and related facies within the stratigraphic-palaeogeographic framework of NW Europe during the Danian. Contributions to Tertiary and Quaternary Geology 29: 47-59.

Binkhorst van den Binkhorst, J.T., 1859. Esquisse géologique et paléontologique des couches Crétacées du Limbourg, et plus spécialement de la craie tuffeau. Première partie. Van Osch-America et Cie. (Maastricht): 268 pp.

Bless, M.J.M., 1989. Event-induced changes in Late Cretaceous to Early Paleocene ostracode assemblages of the SE Netherlands and NE Belgium. Annales de la Société géologique de Belgique 112: 19-30.

Bless, M.J.M., Felder, P.J. & Meessen, J.P.M.Th., 1986. Late Cretaceous sea level rise and inversion: their influence on the depositional environment between Aachen and Antwerp. Annales de La Societe geologique de Belgique 109: 333-355.

Bless, M.J.M., Bouckaert, J. & Paproth, E., 1987. Fossil assemblages and depositional environments: Limits to Stratigraphical Correlations. In: Miller, J., Adams, A.E. & Wright, V.P. (eds): European Dinantian environments. Geological Journal, Special Issue 12: 61-73.

Bouw, L. & Oude Essink, G.H.P., 2003a. Development of a freshwater lens in the inverted Broad Fourteens Basin, Netherlands offshore. Journal of Geochemical Exploration 78-79: 321-325. DOI: 10.1016/S0375-6742(03)00113-4

Bouw, L. & Oude Essink, G.H.P., 2003b. Fluid flow in the northern Broad Fourteens Basin during Late Cretaceous inversion. Netherlands Journal of Geosciences: Geologie En Mijnbouw 82: 55-69. DOI: 10.1017/S0016774600022794

Brinkhuis, H. & Schiøler, P., 1996. Palynology of the Geulhemmerberg Cretaceous/Tertiary boundary section (Limburg, SE Netherlands). Geologie en Mijnbouw 75: 193-213.

Brinkhuis, H. & Smit, J., 1996. The Geulhemmerberg Cretaceous/Tertiary boundary section (Maastrichtian type area, SE Netherlands); an introduction. Geologie en Mijnbouw 75: 101-293.

Buls, T., Anderskouv, K., Fabricius, I.L., Friend, P.L., Thompson, C.E.L. & Stemmerik, L., 2015. Production of calcareous nannofossil ooze for sedimentological experiments. Journal of Sedimentary Research 85: 1228-1237. DOI: 10.2110/jsr.2015.80

Burgers, W.F.J. & Mulder, G.G., 1991. Aspects of the Late Jurassic and Cretaceous history of The Netherlands. Geologie En Mijnbouw 70: 347-354.

Camper, P., 1786. Conjectures relative to the petrifactions found in St. Peter's Mountain, near Maestricht. Philosophical Transactions of the Royal Society of London 76: 443-456. DOI: 10.1098/rstl.1786.0026

Christensen, L., Fregerslev, S., Simonsen, A. & Thiede, J., 1973. Sedimentology and depositional environment of Lower Danian Fish Clay from Stevns Klint, Denmark. Bulletin of the Geological Society of Denmark 22: 193-212.

Christensen, W.K. & Schmid, F., 1987. The belemnites of the Vaals Formation from the C.P.L. quarry at Hallembaye in Belgium – taxonomy, biometry and biostratigraphy. Geologisches Jahrbuch A94: 37 pp.

Christensen, W.K., Schmid, F. & Schulz, M.-G., 2005. Belemnitella from the Upper Maastrichtian of Hemmoor, northwest Germany. In: Schmid, F. (ed): Das Maastricht in Nordwestdeutschland – Fossilien aus der Schreibkreide von Hemmoor und Kronsmeer. Geologisches Jahrbuch A157: 23-67.

Chumakov, N.M., Zharkov, M.A., Herman, A.B., Doludenko, M.P., Kalandadze, N.N., Lebedev, E.L., Ponomarenko, A.G. & Rautian, A.S., 1995. Climatic belts of the Mid-Cretaceous time. Strati-

graphy and Geological Correlation 3: 42-63.
Clark-Lowes, D.D., Kuzemko, N.C.J. & Scott, D.A., 1987. Structure and petroleum prospectivity of the Dutch Central Graben and neighbouring platform areas. In: Brooks, J. & Glennie, K. (eds): Petroleum Geology of North-West Europe: Proceedings of the Third Conference. Graham & Trotman (London): 337-356.
Cohen, K.M., Finney, S.C., Gibbard, P.L. & Fan, J.-X., 2013. The ICS International Chronostratigraphic Chart. Episodes 36(3): 199-204. DOI: 10.18814/epiiugs/2013/v36i3/002
De Jager, J., 2003. Inverted basins in the Netherlands, similarities and differences. Netherlands Journal of Geosciences 82(4): 339-349. DOI:10.1017/S0016774600020175
De Jager, J., 2007. Geological development. In: Wong, Th.E., Batjes, D.A.J. & De Jager, J. (eds): Geology of the Netherlands. Royal Netherlands Academy of Arts and Sciences (KNAW, Amsterdam): 5-26.
De Jager, J., Van Ojik, K. & Smit, J., 2025. Geological development. In: Ten Veen, J.H., Vis, G.-J., De Jager, J. & Wong, Th.E. (eds): Geology of the Netherlands, second edition. Amsterdam University Press (Amsterdam): 21-51. DOI: 10.5117/9789463728362_ch01
Deckers, J. & Van der Voet, E., 2018. A review on the structural styles of deformation during Late Cretaceous and Paleocene tectonic phases in the southern North Sea area. Journal of Geodynamics 115: 1-9. DOI: 10.1016/j.jog.2018.01.005
Doornenbal, J.C., Kombrink, H., Bouroullec, R., Dalman, R.A.F., De Bruin, G., Geel, C.R., Houben, A.J.P., Jaarsma, B., Juez-Larré, J., Kortekaas, M., Mijnlieff, H.F., Nelskamp, S., Pharaoh, T.C., Ten Veen, J.H., Ter Borgh, M., Van Ojik, K., Verreussel, R.M.C.H., Verweij, J.M. & Vis, G.-J., 2019. New insights on subsurface energy resources in the Southern North Sea Basin area. In: Patruno, S., Archer, S.G., Chiarella, D., Howell, J.A., Jackson, C.A-L. & Kombrink, H. (eds): Cross-border themes in petroleum geology I: The North Sea. Geological Society of London, Special Publications 494: 233-268. DOI: 10.1144/SP494-2018-178
Dortangs, R.W., Schulp, A.S., Mulder, E.W.A., Jagt, J.W.M., Peeters, H.H.G. & De Graaf, D.Th., 2002. A large new mosasaur from the Upper Cretaceous of The Netherlands. Netherlands Journal of Geosciences 81: 1-8. DOI: 10.1017/S0016774600020515
Dronkers, A.J. & Mrozek, F.J., 1991. Inverted basins of The Netherlands. First Break 9: 409-418. DOI: 10.3997/1365-2397.1991019
Duin, E.J.T., Doornenbal, J.C., Rijkers, R.H.B., Verbeek, J.W. & Wong, Th.E., 2006. Subsurface structure of the Netherlands – results of recent onshore and offshore mapping. Netherlands Journal of Geosciences 85: 245-276. DOI: 10.1017/S0016774600023064
Dumont, M.A. , 1849. Rapport sur la carte géologique du Royaume. Bulletins de l'Académie Royale des Sciences, des Lettres et Des Beaux-Arts de Belgique 16: 351-373.
Dusar, M. & Lagrou, D., 2007. Cretaceous flooding of the Brabant Massif and the lithostratigraphic characteristics of its chalk cover in northern Belgium. Geologica Belgica 10: 27-38.
Eldrett, J.S., Vieira, M., Gallagher, L., Hampton, M., Blaauw, M. & Swart, P.K., 2021. Late Cretaceous to Palaeogene carbon isotope, calcareous nannofossil and foraminifera stratigraphy of the Chalk Group, Central North Sea. Marine and Petroleum Geology 124: 104789. DOI: 10.1016/j.marpetgeo.2020.104789
Engelen, F.G.H., 1989. Vuursteenwinning en toepassing. Grondboor & Hamer 43: 207-210.
Felder, P.J., 1994. Bioklasten in het Krijt uit boringen van West- en Oost-Vlaanderen. Belgische Geologische Dienst, Professional Paper 1994/3 (270). Drukkerij van het Ministerie van Economische Zaken (Brussel): 86 pp.
Felder, P.J., 1995. Bioklasten-onderzoek van Boven-Krijt en Dano-Montiaan afzettingen uit boringen in de Belgische Kempen. Belgische Geologische Dienst, Professional Paper 1994/8 (275). Drukkerij van het Ministerie van Economische Zaken (Brussel): 240 pp.
Felder, P.J., 1997. The Vijlen Chalk Member (Maastrichtian, Late Cretaceous) in the Meuse-Rhine Euregion. Annales de la Société Géologique de Belgique 119: 119-133.
Felder, P.J. & Bless, M.J.M., 1989. Biostratigraphy and ecostratigraphy of Late Cretaceous deposits in the Kunrade area (South-Limburg, SE Netherlands). Annales de la Société Géologique de Belgique 112: 31-45.
Felder, P.J. & Bless, M.J.M., 1994. The Vijlen Chalk (early Early to early Late Maastrichtian) in its type area around Vijlen and Mamelis (southern Limburg, the Netherlands). Annales de La Société Géologique de Belgique 116: 61-85.
Felder, P.J., Bless, M.J.M., Demyttenaere, R., Dusar, M., Messen, J.P.M.Th. & Robaszynski, F., 1985. Upper Cretaceous to Early Tertiary deposits (Santonian - Paleocene) in northeastern Belgium and South Limburg (The Netherlands) with reference to the Campanian - Maastrichtian. Belgische Geologische Dienst, Professional Paper 1985/1 (214). Drukkerij van het Ministerie van Economische Zaken (Brussel): 1-151 pp.
Felder, W.M., 1975. Lithostratigrafie van het Boven-Krijt en het Dano-Montien in Zuid-Limburg en het aangrenzende gebied. In: Zagwijn, W.H. & Van Staalduinen, C.J. (eds): Toelichting bij de geologische overzichtskaarten van Nederland. Rijks Geologische Dienst (Haarlem): 63-72.
Felder, W.M. & Bosch, P.W., 2000. Krijt van Zuid-Limburg. PlantijnCasparie (Rotterdam): 192 pp.
Field, D.J., Benito, J., Chen, A., Jagt, J.W.M. & Ksepka, D.T., 2020. Late Cretaceous neornithine from Europe illuminates the origins of crown birds. Nature 579: 397-401. DOI: 10.1038/s41586-020-2096-0
Fischer, V., Bardet, N., Benson, R.B.J., Arkhangelsky, M.S. & Friedman, M., 2015. Extinction of fish-shaped marine reptiles associated with reduced evolutionary rates and global environmental volatility. Nature Communications 7: 10825. DOI: 10.1038/ncomms10825
Frakes, L.A., 1999. Estimating the global thermal state from Cretaceous sea surface and continental temperature data. In: Barrera, E. & Johnson, C.C. (eds): Evolution of the Cretaceous

ocean-climate system. Geological Society of America, Special Paper 332: 49-57. DOI: 10.1130/0-8137-2332-9.49

Gale, A., Batenburg, S., Coccioni, R., Dubicka, Z., Erba, E., Falzoni, F., Haggart, J., Hasegawa, T., Ifrim, C., Jarvis, I., Jenkyns, H., Jurowska, A., Kennedy, J., Maron, M., Muttoni, G., Pearce, M., Petrizzo, M.R., Premoli-Silva, I., Thibault, N., Voigt, S., Wagreich, M. & Walaszczyk, I., 2023. The Global Boundary Stratotype Section and Point (GSSP) of the Campanian Stage at Bottaccione (Gubbio, Italy) and its Auxiliary Sections: Seaford Head (UK), Bocieniec (Poland), Postalm (Austria), Smoky Hill, Kansas (U.S.A), Tepayac (Mexico). Episodes 46: 451-490. DOI: 10.18814/epiiugs/2022/022048

Gale, A.S., 2018. An integrated microcrinoid zonation for the lower Campanian chalks of southern England, and its implications for correlation. Cretaceous Research 87: 312-357. DOI: 10.1016/j.cretres.2017.02.002

Gale, A.S., Mutterlose, J., Batenburg, S., Gradstein, F.M., Agterberg, F.P., Ogg, J.G. & Petrizzo, M.R., 2020. The Cretaceous period. In: Gradstein, F.M., Ogg, J.G., Schmitz, M.D. & Ogg, G.M. (eds): Geologic time scale 2020. Elsevier (Amsterdam): 1023-1086. DOI: 10.1016/B978-0-12-824360-2.00026-7

Gennaro, M., Wonham, J.P., Gawthorpe, R. & Sælen, G., 2013. Seismic stratigraphy of the Chalk Group in the Norwegian Central Graben, North Sea. Marine and Petroleum Geology 45: 236-266. DOI: 10.1016/j.marpetgeo.2013.04.010

Gilli, É., 2015. Karstology. Karsts, caves and springs: Elements of fundamental and applied karstology. CRC Press (Boca Raton, FL, USA): 256 pp.

Gradstein, F.M., Waters, C.N., Charnock, M., Munsterman, D.K., Hollerbach, M., Brunstad, H., Hammer, O. & Vergara, L., 2016. Stratigraphic guide to the Cromer Knoll, Shetland and Chalk Groups, North Sea and Norwegian Sea. Newsletters on Stratigraphy 49: 71-280. DOI: 10.1127/nos/2016/0071

Gras, R., 1995. Late Cretaceous sedimentation and tectonic inversion, southern Netherlands. Geologie en Mijnbouw 74 (2): 117-127.

Gras, R. & Geluk, M., 1999. Late Cretaceous – Early Tertiary sedimentation and tectonic inversion in the southern Netherlands. Geologie en Mijnbouw 78: 1-19. DOI: DOI: 10.1023/A:1003709821901

Gravesen, P., 1993. Early Danian species of the echinoid genus Tylocidaris (Cidaridae, Psychocidarinae) from eastern Denmark. Contributions to Tertiary and Quaternary Geology 30: 41-73.

Gutteridge, P. & Thompson, S., 2019. Sedimentology and microfacies of the cored interval in well Eys-1, Limburg Province, the Netherlands. Cambridge Carbonates Ltd report: 39 pp.

Hansen, T., 2019. Gastropods from the Cretaceous-Palaeogene boundary in Denmark. Zootaxa 4654: 1-196. DOI: 10.11646/zootaxa.4654.1.1

Haq, B.U., 2014. Cretaceous eustasy revisited. Global and Planetary Change 113: 44-58. DOI: 10.1016/j.gloplacha.2013.12.007

Hay, W.W. & Floegel, S., 2012. New thoughts about the Cretaceous climate and oceans. Earth-Science Reviews 115: 262-272. DOI: 10.1016/j.earscirev.2012.09.008

Herngreen, G.F.W. (ed), Eillebrecht, A.T.J.M., Gortemaker, R.E., Remmelts, G., Schuurman, H.A.H.M. & Verbeek, J.W., 1996. Upper Cetaceous Chalk Group stratigraphy near the isle of Texel, the Netherlands (a multidisciplinary approach). Mededelingen Rijks Geologische Dienst 56: 63 pp.

Herngreen, G.F.W. (ed), Schuurman, H.A.H.M., Verbeek, J.W., Brinkhuis, H., Burnett, J.A., Felder, W.M. & Kedves, M., 1998. Biostratigraphy of Cretaceous-Tertiary boundary strata in the Curfs quarry, the Netherlands. Mededelingen Nederlands Instituut voor Toegepaste Geowetenschappen TNO 61: 57 pp.

Herngreen, G.F.W. & Wong, Th.E., 2007. Cretaceous. In: Wong, Th.E., Batjes, D.A.J. & De Jager, J. (eds): Geology of the Netherlands. Royal Netherlands Academy of Arts and Sciences (KNAW, Amsterdam): 127-150.

Hiss, M., Niebuhr, B. & Teipel, U., 2018. Die Kreide in der Stratigraphischen Tabelle von Deutschland 2016. Zeitschrift der Deutschen Gesellschaft für Geowissenschaften 169: 247-266. DOI: 10.1127/zdgg/2018/0149

Hofmann, A.P., Price, A., Kaffenberger, G., Godderij, R.R.G.G. & Simpson, M., 2002. Hanze chalk oil field—the Chalk pearl in the Dutch North Sea. cp-5-00405. 64th EAGE Conference and Exhibition 2002, Florence: 211. DOI: 10.3997/2214-4609-pdb.5.P211

Hooper, R.J., Goh, L.S. & Dewey, F., 1995. The inversion history of the northeastern margin of the Broad Fourteens Basin. In: Buchanan, J.G. & Buchanan, P.G. (eds): Basin inversion. Geological Society of London, Special Publications 88, 307-317. DOI: 10.1144/GSL.SP.1995.088.01.17

Hopper, J. R., Funck, T., Stoker, M., Árting, U., Péron-Pinvidic, G., Doornenbal, H. & Gaina, C. (eds), 2014. Tectonostratigraphic Atlas of the North-East Atlantic Region. Geological Survey of Denmark and Greenland (GEUS) (Copenhagen, Denmark): 339 pp.

Hu, X., Wagreich, M. & Yilmaz, I.O., 2012. Marine rapid environmental/climatic change in the Cretaceous greenhouse world. Cretaceous Research 38: 1-6. DOI: 10.1016/j.cretres.2012.04.012

Huber, B.T., MacLeod, K.G., Watkins, D.K. & Coffin, M.F., 2018. The rise and fall of the Cretaceous Hot Greenhouse climate. Global and Planetary Change 167: 1-23. DOI: 10.1016/j.gloplacha.2018.04.004

Hübscher, C., Al Hseinat, M., Schneider, M. & Betzler, C., 2019. Evolution of contourite systems in the late Cretaceous Chalk Sea along the Tornquist Zone. Sedimentology: 1341-1360. DOI: 10.1111/sed.12564

Ineson, J., Bjerager, M. & Kristensen, L., 2014. Upper Cretaceous-Danian chalk of the Netherlands subsurface - Core descriptions, facies analysis, reservoir properties. Nationale Geologiske Undersøgelser for Danmark og Grønland (GEUS) confidential report 2014/47: 90 pp. For Wintershall Noordzee, cited with permission.

Jagt, J.W.M., 1999a. Late Cretaceous-Early Palaeogene echinoderms and the K/T boundary in the southeast Netherlands and northeast Belgium – Part 1: Introduction and stratigraphy. Scripta Geologica 116: 1-57.

Jagt, J.W.M., 1999b. Late Cretaceous-Early Palaeogene echinoderms and the K/T boundary in the southeast Netherlands and northeast Belgium – Part 2: Crinoids. Scripta Geologica 116: 59-255.

Jagt, J.W.M., 2000a. Late Cretaceous-Early Palaeogene echinoderms and the K/T boundary in the southeast Netherlands and northeast Belgium – Part 3: Ophiuroids (with a chapter on: Early Maastrichtian ophiuroids from Rügen (northeast Germany) and Møn (Denmark) by Manfred Kutschner & John W.M. Jagt). Scripta Geologica 121: 1-179.

Jagt, J.W.M., 2000b. Late Cretaceous-Early Palaeogene echinoderms and the K/T boundary in the southeast Netherlands and northeast Belgium – Part 5: Asteroids. Scripta Geologica 121: 377-503.

Jagt, J.W.M., 2000c. Late Cretaceous-Early Palaeogene echinoderms and the K/T boundary in the southeast Netherlands and northeast Belgium – Part 6: Conclusions. Scripta Geologica 121: 505-577.

Jagt, J.W.M., 2000d. Late Cretaceous-Early Palaeogene echinoderms and the K/T boundary in the southeast Netherlands and northeast Belgium – Part 4: Echnoids. Scripta Geologica 122: 181-375.

Jagt, J.W.M., 2005. Stratigraphic ranges of mosasaurs of Belgium and the Netherlands (Late Cretaceous) and cephalopod-based correlations with North America. Proceedings of the First Mosasaur Meeting. Netherlands Journal of Geosciences 84: 283-301. DOI: 10.1017/S0016774600021065

Jagt, J.W.M. & Jagt-Yazykova, E.A., 2018. Stratigraphical ranges of tegulated inoceramid bivalves in the type area of the Maastrichtian Stage (Belgium, the Netherlands). Cretaceous Research 87: 385-394. DOI: 10.1016/j.cretres.2017.05.022

Jagt, J.W.M. & Jagt-Yazykova, E.A., 2019. Late Cretaceous and Cenozoic cephalopods from the southern North Sea Basin: stocktaking and future directions. Vita Malacologica 18: 1-33.

Jagt, J.W.M. & Oosterink, H.W., 2012. Hypoturrilites uit het vroeg-Cenomanien van Winterswijk. Staringia 13: 199.

Jagt, J.W.M., Kennedy, W.J. & Machalski, M., 1999. Giant scaphitid ammonites from the Maastrichtian of Europe. Bulletin de l'Institut royal des Sciences naturelles de Belgique, Sciences de la Terre 69: 133-154.

Jagt, J.W.M., Jagt-Yazykova, E.A. & Schins, W.J.H., 2011. A tribute to the late Felder brothers – pioneers of Limburg geology and prehistoric archaeology; introduction. Netherlands Journal of Geosciences 90: 65-71. DOI: 10.1017/S0016774600001025

Jagt, J.W.M., Van Bakel, B.W.M., Cremers, G., Deckers, M.J.M., Dortangs, R.W., Van Es, M., Fraaije, R.H.B., Kisters, P.J.M., Van Knippenberg, P.H.M, Lemmens, H., Nieuwenhuis, E., Severijns, J. & Stroucken, J.W., 2013. Het Vroeg Paleoceen (Danien) van zuidelijk Limburg en aangrenzend gebied – nieuwe fauna's en nieuwe inzichten. Afzettingen 34: 198-230.

Jelby, M.E., Thibault, N., Surlyk, F., Ullmann, C.V., Harlou, R. & Korte, C., 2014. The lower Maastrichtian Hvidskud succession, Møns Klint, Denmark: calcareous nannofossil biostratigraphy, carbon isotope stratigraphy, and bulk and brachiopod oxygen isotopes. Bulletin of the Geological Society of Denmark 62: 89-104. DOI: 10.37570/bgsd-2014-62-05

Jenkyns, H.C., Dickson, A.J., Ruhl, M. & Van den Boorn, S.H.J.M., 2017. Basalt-seawater interaction, the Plenus Cold Event, enhanced weathering and geochemical change: deconstructing Oceanic Anoxic Event 2 (Cenomanian–Turonian, Late Cretaceous). In: Robinson, S.A., Heimhofer, U., Hesselbo, S.P. & Petrizzo, M.R. (eds): Mesozoic climate and oceans – a tribute to Hugh Jenkyns and Helmut Weissert. Sedimentology 64: 16-43. DOI: 10.1111/sed.12305

Jennings, J.N., 1985. Karst geomorphology. Blackwell (Oxford, UK): 293 pp.

Juvigné, E. & Renard, F., 1992. Les terrasses de la Meuse de Liège à Maastricht. Annales de la Société Géologique de Belgique 115: 167-186.

Kennedy, W.J. & Jagt, J.W.M., 1995. Lower Campanian heteromorph ammonites from the Vaals Formation around Aachen, Germany, and adjacent parts of Belgium and The Netherlands. Neues Jahrbuch für Geologie und Paläontologie - Abhandlungen 197: 275-294. DOI: 10.1127/njgpa/197/1995/275

Keutgen, N., 2018. A bioclast-based astronomical timescale for the Maastrichtian in the type area (southeast Netherlands, northeast Belgium) and stratigraphic implications: the legacy of P.J. Felder. Netherlands Journal of Geosciences 97: 229-260. DOI: 10.1017/njg.2018.15

Keutgen, N., Jagt, J.W.M., Felder, P.J. & Jagt-Yazykova, E.A., 2010. Stratigraphy of the upper Vijlen Member (Gulpen Formation; Maastrichtian) in northeast Belgium, the southeast Netherlands and the Aachen area (Germany), with special reference to belemnitellid cephalopods. Netherlands Journal of Geosciences 89: 109-136. DOI: 10.1017/S0016774600000731

Keutgen, N., Remin, Z. & Jagt, J.W.M., 2017. The late Maastrichtian Belemnella kazimiroviensis group (Cephalopoda, Coleoidea) in the Middle Vistula valley (Poland) and the Maastricht area (the Netherlands, Belgium) – taxonomy and palaeobiological implications. Palaeontologia Electronica 20.2: 1-29. DOI: 10.26879/671

Kirkaldy, J.F., 1950. Solution of the Chalk in the Mimms Valley, Herts. Proceedings of the Geologists' Association 61: 219-224. DOI: 10.1016/S0016-7878(50)80028-7

Knobloch, E., Kvaček, Z., Bůžek, Č., Mai, D.H. & Batten, D.J., 1993. Evolutionary significance of floristic changes in the Northern Hemisphere during the Late Cretaceous and Palaeogene, with particular reference to Central Europe. Review of Palaeobotany and Palynology 78: 41-54. DOI: 10.1016/0034-6667(93)90016-N

Koestler, A.G. & Ehrmann, W.U., 1986. Fault patterns in the calcareous overburden of a salt diapir: Laegerdorf, NW Germany.

Neues Jahrbuch für Geologie Und Paläontologie, Monatshefte 1986: 555-569. DOI: 10.1127/njgpm/1986/1986/555
Koestler, A.G. & Ehrmann, W.U., 1991. Description of brittle extensional features in chalk on the crest of a salt ridge (NW Germany). In: Roberts, A.M., Yielding, G. & Freeman, B. (eds): The geometry of normal faults. Geological Society of London, Special Publications 56: 113-123. DOI: 10.1144/GSL.SP.1991.056.01.0
Kombrink, H., Ten Veen, J.H. & Geluk, M.C., 2012. Exploration in the Netherlands, 1987-2012. Netherlands Journal of Geosciences 91: 403-418. DOI: 10.1017/S0016774600000317
Kring, D.A., 2007. The Chicxulub impact event and its environmental consequences at the Cretaceous–Tertiary boundary. Palaeogeography, Palaeoclimatology, Palaeoecology 255: 4-21. DOI: 10.1016/j.palaeo.2007.02.037
Kuhnt, W., 1996. Early Danian benthic foraminiferal community structures, Geulhemmerberg, SE Netherlands. Geologie en Mijnbouw 75: 163-172.
Kuyl, O.S., 1983. The inversion of part of the southern border of the Central Graben in South Limburg during the Late Cretaceous. Geologie en Mijnbouw 62: 401-408.
Lagrou, D. & Dreesen, R., 2005. Kartering en karakterisering (sedimentpetrografisch en petrofysisch) van de Krijtgesteenten in Vlaanderen. VITO (Mol, Belgium), Report 2005/MAT/R/0216: 104 pp.
Landman, N.H., Goolaerts, S., Jagt, J.W.M., Jagt-Yazykova, E.A., Machalski, M. & Yacobucci, M.M., 2014. Ammonite extinction and nautilid survival at the end of the Cretaceous. Geology 42: 707-710. DOI: 10.1130/G35776.1
Landman, N.H., Goolaerts, S., Jagt, J.W.M., Jagt-Yazykova, E.A. & Machalski, M., 2015. Ammonites on the brink of extinction: diversity, abundance, and ecology of the Order Ammonoidea at the Cretaceous/Paleogene (K/Pg) boundary. In: Klug, C., Korn, D., De Baets, K., Kruta, I. & Mapes, R.H. (eds): Ammonoid paleobiology: from macroevolution to paleogeography. Topics in Geology 44. Springer (Dordrecht): 497-553. DOI: 10.1007/978-94-017-9633-0_19
Liebau, A., 1978. Paläobathymetrische und paläoklimatische Veränderungen im Mikrofaunenbild der Maastrichter Tuffkreide. Neues Jahrbuch für Geologie und Paläontologie - Abhandlungen 157: 233-237. DOI: 10.1127/njgpa/157/1978/233
Linnert, C., Robinson, S.A., Lees, J.A., Bown, P.R., Pérez-Rodríguez, I., Petrizzo, M.R., Falzoni, F., Littler, K., Arz, J.A. & Russell, E.E., 2014. Evidence for global cooling in the Late Cretaceous. Nature Communications 5: 4194. DOI: 10.1038/ncomms5194
Linnert, C., Robinson, S.A., Lees, J.A., Pérez-Rodríguez, I., Jenkyns, H.C., Petrizzo, M.R., Arz, J.A., Bown, P.R. & Falzoni, F., 2018. Did Late Cretaceous cooling trigger the Campanian–Maastrichtian Boundary Event? Newsletters on Stratigraphy 51: 145-166. DOI: 10.1127/nos/2017/0310
Luijendijk, E., Van Balen, R.T., Ter Voorde, M. & Andriessen, P.A.M., 2011. Reconstructing the Late Cretaceous inversion of the Roer Valley Graben (southern Netherlands) using a new model that integrates burial and provenance history with fission track thermochronology. Journal of Geophysical Research 116: B06402: 1-19. DOI: 10.1029/2010JB008071
Machalski, M., 2005. Late Maastrichtian and earliest Danian scaphitid ammonites from central Europe: Taxonomy, evolution, and extinction. Acta Palaeontologica Polonica 50: 653-696.
MacLeod, K.G., Isaza Londoño, C., Martin, E.E., Jiménez Berrocoso, A. & Basak, C., 2011. Changes in North Atlantic circulation at the end of the Cretaceous greenhouse interval. Nature Geoscience 4: 779-782. DOI: 10.1038/NGEO01284
Martin, J.E., Case, J.A., Jagt, J.W.M., Schulp, A.S. & Mulder, E.W.A., 2005. A new European marsupial indicates a Late Cretaceous high-latitude transatlantic dispersal route. Journal of Mammalian Evolution 12: 495-511. DOI: 10.1007/s10914-005-7330-x
Mitchell, S.F., 2019. The Chalk Group (Upper Cretaceous) of the Northern Province, eastern England – a review. Proceedings of the Yorkshire Geological Society 62: 153-177. DOI: 10.1144/pygs2017-010
Moorkens, T.L., 1982. Foraminifera of the Montian stratotype and of subjacent strata in the "Mons Well 1969" with a review of Belgian Paleocene stratigraphy. Mémoires pour servir à l'explication des Cartes géologiques et minières de la Belgique 17: 186 pp.
Mortimore, R. & James, L., 2015. The search for onshore analogues for the offshore Upper Cretaceous Chalk of the North Sea. Proceedings of the Geologists' Association 126: 188-210. DOI: 10.1016/j.pgeola.2015.01.008
Mouriaux, P., 1998. Phénomènes de karstification dans le Calcaire grossier lutétien du nord de l'Ile-de-France et de la Picardie. Mémoires des actes du 1er Congrès franco-belge de spéléologie (Fromelennes, Ardennes): 47-69.
Mulder, E.W.A., 1999. Transatlantic latest Cretaceous mosasaurs (Reptilia, Lacertilia) from the Maastrichtian type area and New Jersey. Geologie en Mijnbouw 78: 281-300. DOI: 10.1023/A:1003838929257
Mulder, E.W.A., Jagt, J.W.M. & Stroucken, J.W., 2016. New records of latest Cretaceous neosuchian crocodyliforms from the Maastrichtian type area (southern Limburg, the Netherlands). In: Sullivan, R.M.& Lucas, S.G. (eds): Fossil Record 5. New Mexico Museum of Natural History and Science Bulletin 74: 169-172.
Nalpas, T., 1994. Inversion des grabens du sud de la mer du Nord. Données de sub-surface et modélisation analogique. PhD thesis Université Rennes (France), Memoires de Geosciences 71; 246 pp.
Nalpas, T., Richert, J.-P., Brun, J.-P., Mulder, T. & Unternehr, P., 1996. Inversion du « Broad Fourteens Basin » ou Graben de La Haye (sud de la Mer du Nord) apport de la sismique 3D. Bulletin des Centres de Recherches Exploration-Production Elf Aquitaine 20: 309-321.
Nederlandse Aardolie Maatschappij & Rijks Geologische Dienst, 1980. Stratigraphic Nomenclature of the Netherlands. Verhandelingen van Het Koninklijk Nederlands Geologisch Mijnbouwkundig Genootschap 32: 77 pp.

Niebuhr, B., 2006. Multistratigraphische Gliederung der norddeutschen Schreibkreide (Coniac bis Maastricht), Korrelation von Aufschlüssen und Bohrungen. Zeitschrift der deutschen Gesellschaft für Geowissenschaften 157: 245-262. DOI: 10.1127/1860-1804/2006/0157-0245

Niebuhr, B., Hampton, M.J., Gallagher, L.T. & Remin, Z., 2011. Integrated stratigraphy of the Kronsmoor section (northern Germany), a reference point for the base of the Maastrichtian in the Boreal Realm. Acta Geologica Polonica 61: 193-214.

NITG-TNO, 1999. Geological atlas of the subsurface of the Netherlands (1 : 250,000) - Explanation to Map sheet XV: Sittard-Maastricht. Netherlands Institute of Applied Geoscience TNO - National Geological Survey (Utrecht): 127 pp.

NITG-TNO, 2001. Geological Atlas of the subsurface of the Netherlands (1 : 250,000) - Explanation to Map Sheets XIII and XIV: Breda-Valkenswaard and Oss-Roermond + Map Sheet XIII and XIV. Netherlands Institute of Applied Geoscience TNO (NITG-TNO, Utrecht): 149 pp.

NITG-TNO, 2002. Geological Atlas of the subsurface of the Netherlands (1 : 250,000) - Explanation to map sheets VII and VIII: Noordwijk-Rotterdam and Amsterdam-Gorinchem. Netherlands Institute of Applied Geoscience TNO - National Geological Survey (Utrecht): 135 pp.

NITG-TNO, 2004. Geological Atlas of the subsurface of the Netherlands (1 : 250,000) - Explanation to map sheet IX: Harderwijk. Netherlands Institute of Applied Geoscience TNO - National Geological Survey (Utrecht): 123 pp.

O'Brien, C.L., Robinson, S.A., Pancost, R.D., Sinninghe Damsté, J.S., Schouten, S., Lunt, D.J., Alsenz, H., Bornemann, A., Bottini, C., Brassell, S.C., Farnsworth, A., Forster, A., Huber, B.T., Inglis, G.N., Jenkyns, H.C., Linnert, C., Littler, K., Markwick, P., McAnena, A., Mutterlose, J., Naafs, B.D.A., Püttmann, W., Sluijs, A., Van Helmond, N.A.G.M., Vellekoop, J., Wagner, T. & Wrobel, N.E., 2017. Cretaceous sea-surface temperature evolution: constraints from TEX86 and planktonic foraminiferal oxygen isotopes. Earth-Science Reviews 172: 224-247. DOI: 10.1016/j.earscirev.2017.07.012

Petrea, C. & Goldberg, T., 2013. Sedimentology and petrography of chalk samples from well F17-10. PanTerra Geoconsultants B.V. (Leiderdorp) report: 23 pp.

Pharaoh, T.C., Dusar, M., Geluk, M.C., Kockel, F., Krawczyk, C.M., Krzywiec, P., Scheck-Wenderoth, M., Thybo, H., Vejbæk, O.V. & Van Wees, J.D., 2010. Tectonic evolution. In: Doornenbal, J.C. & Stevenson, A.G. (eds): Petroleum geological atlas of the Southern Permian Basin area. EAGE Publications b.v. (Houten): 25-57.

Phillips, G.E., Nyborg, T. & Vega, F.J., 2014 . Icriocarcinidae: a family of portunoid crabs from the Upper Cretaceous of North America. Paläontologische Zeitschrift 88: 139-158. DOI: 10.1007/s12542-013-0190-4

Pluymaekers, M.P.D., Kramers, L., Van Wees, J.-D., Kronimus, A., Nelskamp, S., Boxem, T. & Bonté, D., 2012. Reservoir characterisation of aquifers for direct heat production: Methodology and screening of the potential reservoirs for the Netherlands. Netherlands Journal of Geosciences 91: 621-636. DOI: 10.1017/S001677460000041X

Rademakers , P.C.M. (ed.), 1998. De prehistorische vuursteenmijnen van Ryckholt - St. Geertruid. Nederlandse Geologische vereniging afdeling Limburg (Beek): 333 pp.

Remmelts, G., 1996. Salt tectonics in the southern North Sea, the Netherlands. In: Rondeel, H.E., Batjes, D.A.J. & Nieuwenhuijs, W.H. (eds): Geology of gas and oil under the Netherlands. Royal Geological and Mining Society of the Netherlands, Kluwer Academic Publishers (Dordrecht): 143-158. DOI: 10.1007/978-94-009-0121-6_13

RGD, 1993. Geological atlas of the subsurface of the Netherlands (1 : 250,000). Explanation to map sheet V Sneek- Zwolle. Rijks Geologische Dienst (Haarlem): 127 pp.

Rodet, J., 1992. La craie et ses karsts. Thèse d'Etat Centre Normand d'étude du karst et des cavités du sous-sol. In: Elbeuf & Groupe Seine-CNRS (Caen): 560 pp.

Roep, T.B. & Smit, J., 1996. Sedimentological aspects of the K/T boundary at Geulhemmerberg, Zuid Limburg, the Netherlands. Geologie en Mijnbouw 75: 119-131.

Romein, A.J.T., Willems, H. & Mai, H., 1996. Calcareous nannoplankton of the Geulhemmerberg K/T boundary section, Maastrichtian type area, the Netherlands. Geologie en Mijnbouw 75: 231-238.

Sachs, S., Hornung, J.J. & Scheer, U., 2018. Mosasaurid and plesiosaurian remains from marginal facies of the lower Campanian (Upper Cretaceous) Bottrop and Vaals formations of western Germany. Cretaceous Research 87: 358-367. DOI: 10.1016/j.cretres.2017.05.026

Schiøler, P., Brinkhuis, H., Roncaglia, L. & Wilson, G.J., 1997. Dinoflagellate biostratigraphy and sequence stratigraphy of the Type Maastrichtian (Upper Cretaceous), ENCI Quarry, The Netherlands. Marine Micropaleontology 31: 65-95. DOI: 10.1016/S0377-8398(96)00058-8

Schlanger, S.O. & Jenkyns, H.C., 1976. Cretaceous Oceanic Anoxic Events: causes and consequences. Geologie en Mijnbouw 55: 179-184.

Schmitz, B., Keller, G. & Stenvall, O., 1992. Stable isotope and foraminiferal changes across the Cretaceous–Tertiary boundary at Stevns Klint, Denmark: Arguments for long-term oceanic instability before and after bolide-impact event. Palaeogeography, Palaeoclimatology, Palaeoecology 96: 233-260. DOI: 10.1016/0031-0182(92)90104-D

Scholle, P.A., 1977. Chalk diagenesis and its relation to petroleum exploration: oil from chalks, a modern miracle? American Association of Petroleum Geologists Bulletin 61 (7): 982-1009. DOI: 10.1306/C1EA43B5-16C9-11D7-8645000102C1865D

Schrøder, A.E. & Surlyk, F., 2020. Adaptive brachiopod morphologies in four key environments of the Late Cretaceous–Danian Chalk Sea of northern Europe: A comparative study. Cretaceous Research 107: 104288. DOI: 10.1016/j.cretres.2019.104288

Schulp, A.S., Janssen, R., Van Baal, R.R., Jagt, J.W.M., Mulder, E.W.A. & Vonhof, H.B., 2017. Stable isotopes, niche partitioning and the paucity of elasmosaur remains in the Maastrichtian type area. Netherlands Journal of Geosciences 96: 29-33. DOI: 10.1017/njg.2016.20

Schulte, P., Alegret, L., Arenillas, I., Arz, J.A., Barton, P.J., Bown, P.R., Bralower, T.J., Christeson, G.L., Claeys, P., Cockell, C.S., Collins, G.S., Deutsch, A., Goldin, T.J., Goto, K., Grajales-Nishimura, J.M., Grieve, R.A.F., Gulick, S.P.S., Johnson, K.R., Kiessling, W., Koeberl, C., Kring, D.A., MacLeod, K.G., Matsui, T., Melosh, J., Montanari, A., Morgan, J.V., Neal, C.R., Nichols, D.J., Norris, R.D., Pierazzo, E., Ravizza, G., Rebolledo-Vieyra, M., Reimold, W.U., Robin, E., Salge, T., Speijer, R.P., Sweet, A.R., Urrutia-Fucugauchi, J., Vadja, V., Whalen, M.T. & Willumsen, P.S., 2010. The Chicxulub asteroid impact and mass extinction at the Cretaceous-Paleogene boundary. Science 327: 1214-1218. DOI: 10.1126/science.1177265

Scotese, C.R., 2001. Atlas of earth history, volume 1: Paleogeography. PALEOMAP Project (Arlington, Texas, USA): 52 pp.

Simon, E. & Owen, E.F., 2001. A first step in the revision of the genus Cretirhynchia Pettitt, 1950. Bulletin de l'Institut royal des Sciences naturelles de Belgique, Sciences de la Terre 71: 53-118.

Skelton, P.W., Spicer, R.A., Kelley, S.P. & Gilmour, I., 2003. The Cretaceous world. The Open University, Cambridge University Press: 360 pp.

Smit, F.W.H., 2018. Integrated seismic geomorphological analysis of syn- and post-depositional fluid migration features in the Chalk Group of the Danish North Sea. PhD thesis Technical University of Denmark: 161 pp.

Smit, J. & Brinkhuis, H., 1996. The Geulhemmerberg Cretaceous/Tertiary boundary section (Maastrichtian type area, SE Netherlands); summary of results and a scenario of events. Geologie en Mijnbouw 75: 283-293.

Smit, J. & Hertogen, J., 1980. An extraterrestrial event at the Cretaceous–Tertiary boundary. Nature 285: 198-200. DOI: 10.1038/285198a0

Smit, J. & Zachariasse, W.J., 1996. Planktic foraminifera in the Cretaceous/Tertiary boundary clays of the Geulhemmerberg (Netherlands). Geologie en Mijnbouw 75: 187-191.

Speijer, R.P., Pälike, H., Hollis, C.J., Hooker, J.J. & Ogg, J.G., 2020. The Paleogene Period. In: Gradstein, F.M., Ogg, J.G., Schmitz, M.D. & Ogg, G.M. (eds): Geologic time scale 2020. Elsevier (Amsterdam): 1088-1140. DOI: 10.1016/B978-0-12-824360-2.00028-0

Sprechmann, P., 1981. Paleocommunities and paleobathymetry of Maastrichtian sublittoral benthonic foraminifera from western Europe. Neues Jahrbuch für Geologie und Paläontologie – Abhandlungen 162: 188-230.

Staring, W.C.H., 1860. De bodem van Nederland. De zamenstelling en het ontstaan der gronden in Nederland ten behoeve van het algemeen beschreven. Deel II. A.C. Kruseman (Haarlem): 480 pp.

Surlyk, F. & Lykke-Andersen, H., 2007. Contourite drifts, moats and channels in the Upper Cretaceous chalk of the Danish Basin. Sedimentology 54: 405-422. DOI: 10.1111/j.1365-3091.2006.00842.x

Surlyk, F., Dons, T., Clausen, C.K. & Highham, J., 2003. Upper Cretaceous. In: Evans, D., Graham, C., Armour, A. & Bathurst, P. (eds): The Millennium Atlas: Petroleum geology of the central and northern North Sea. The Geological Society (London): 212-233.

Surlyk, F., Damholt, T. & Bjerager, M., 2006. Stevns Klint, Denmark: Uppermost Maastrichtian chalk, Cretaceous–Tertiary boundary, and lower Danian bryozoan mound complex. Bulletin of the Geological Society of Denmark 54: 1-48. DOI: 10.37570/bgsd-2006-54-01

Surlyk, F., Kolbye Jensen, S. & Engkilde, M., 2008. Deep channels in the Cenomanian–Danian Chalk Group of the German North Sea sector: Evidence of strong constructional and erosional bottom currents and effect on reservoir quality distribution. AAPG Bulletin 92: 1565-1586. DOI: 10.1306/07100808035

Surlyk, F., Rasmussen, S.L., Boussaha, M., Schiøler, P., Schovsbo, N.H., Sheldon, E., Stemmerik, L. & Thibault, N., 2013. Upper Campanian–Maastrichtian holostratigraphy of the eastern Danish Basin. Cretaceous Research 46: 232-256. DOI: 10.1016/j.cretres.2013.08.006

Swennen, R. & Dusar, M., 1997. Diagenesis of Late Cretaceous to Paleocene carbonates in the Rur Valley Graben (Molenbeersel borehole, NE-Belgium). Annales de la Société Géologique du Nord 5: 215-226.

Tsikos, H., Jenkyns, H.C., Walsworth-Bell, B., Petrizzo, M.R., Forster, A., Kolonic, S., Erba, E., Premoli Silva, I., Baas, M., Wagner, T. & Sinninghe Damsté, J.S., 2004. Carbon-isotope stratigraphy recorded by the Cenomanian–Turonian Oceanic Anoxic Event: correlation and implications based on three key localities. Journal of the Geological Society, London 161: 711-719. DOI: 10.1144/0016-764903-077

Umbgrove, J.H.F., 1925. Bijdrage tot de kennis der stratighaphie, tektoniek en petrographie van het Senoon in Zuid-Limburg. Leidse Geologische Mededelingen 1-1: 254-332.

Van Adrichem Boogaert, H.A. & Kouwe, W.F.P., 1994. Stratigraphic nomenclature of the Netherlands, revision and update by RGD and NOGEPA, Section H. Upper Cretaceous and Danian (Chalk Group). Mededelingen Rijks Geologische Dienst 50: 1-20.

Van Balen, R.T., Houtgast, R.F. & Cloetingh, S.A.P.L., 2005. Neotectonics of The Netherlands: a review. In: Cloetingh, S.A.P.L. & Cornu, T.G.M. (eds): Neotectonics and Quaternary fault-reactivation in Europe's intraplate lithosphere. Quaternary Science Reviews 24: 439-454. DOI: 10.1016/j.quascirev.2004.01.011

Van Buchem, F.S.P., Smit, F.W.H., Buijs, G.J.A., Trudgill, B. & Larsen, P.-H., 2018. Tectonostratigraphic framework and depositional history of the Cretaceous–Danian succession of the Danish Central Graben (North Sea)–new light on a mature area. In: Bowman, M. & Levell, B. (eds): Petroleum Geology of NW Europe: 50 Years of Learning – Proceedings of the 8th Petroleum Geology Conference. Geological Society of London: 9-46. DOI: 10.1144/PGC8.24

Van Dalfsen, W., Doornenbal, J.C., Dortland, S. & Gunnink, J.L., 2006. A comprehensive seismic velocity model for the Netherlands based on lithostratigraphic layers. Netherlands Journal of Geosciences 85: 277-292. DOI: 10.1017/S0016774600023076

Vandenberghe, N., Van Simaeys, S., Steurbaut, E., Jagt, J.W.M. & Felder, P.J., 2004. Stratigraphic architecture of the Upper Cretaceous and Cenozoic along the southern border of the North Sea Basin in Belgium. Netherlands Journal of Geosciences 83: 155-171. DOI: 10.1017/S0016774600020229

Van der Ham, R.W.J.M., Jagt, J.W.M., Renkens, S. & Van Konijnenburg-van Cittert, J.H.A., 2010. Seed-cone scales from the upper Maastrichtian document the last occurrence in Europe of the Southern Hemisphere conifer family Araucariaceae. Palaeogeography, Palaeoclimatology, Palaeoecology 291: 469-473. DOI: 10.1016/j.palaeo.2010.03.017

Van der Molen, A.S., 2004. Sedimentary development, seismic stratigraphy and burial compaction of the Chalk Group in the Netherlands North Sea area. PhD thesis Utrecht University (the Netherlands): 175 pp.

Van der Molen, A.S. & Wong, Th.E., 2007. Towards an improved lithostratigraphic subdivision of the Chalk Group in the Netherlands North Sea area – a seismic stratigraphic approach. Netherlands Journal of Geosciences 86: 131-143. DOI: 10.1017/S0016774600023131

Van der Molen, A.S., Dudok van Heel, H.W. & Wong, T.E., 2005. The influence of tectonic regime on chalk deposition: examples of the sedimentary development and 3D-seismic stratigraphy of the Chalk Group in the Netherlands offshore area. Basin Research 17: 63-81. DOI: 10.1111/j.1365-2117.2005.00261.x

Van der Voet, E., Heijnen, L. & Reijmer, J.J.G., 2018. Geological evolution of the Chalk Group in the northern Dutch North Sea: inversion, sedimentation and redeposition. Geological Magazine 156 (7): 1265-1284. DOI: 10.1017/S0016756818000572

Van Gent, H., Back, S., Urai, J.L. & Kukla, P., 2010. Small-scale faulting in the Upper Cretaceous of the Groningen block (The Netherlands): 3D seismic interpretation, fault plane analysis and regional paleostress. Journal of Structural Geology 32: 537-553. DOI: 10.1016/j.jsg.2010.03.003

Van Lochem, H., 2018. F17-Chalk: new insights in the tectonic history of the Dutch Central Graben. In: Kilhams, B., Kukla, P.A., Mazur, S., McKie, T., Mijnlieff, H.F. & Van Ojik, K. (eds): Mesozoic resource potential in the southern Permian Basin. Geological Society of London, Special Publications 469: 537-558. DOI: 10.1144/SP469.17

Van Lochem, H. & Beller, M., 2018. Maastrichtian chalk reservoir quality in the Rembrandt and Vermeer oilfields, Dutch offshore block F17. First Break 36: 60-63.

Van Rooijen, P., 1989. Grondwater in Limburg. Grondboor & Hamer 43: 377-386.

Van Wijhe, D.H., 1987a. Structural evolution of inverted basins in the Dutch offshore. Tectonophysics 137: 171-219. DOI: 10.1016/0040-1951(87)90320-9

Van Wijhe, D.H., 1987b. The structural evolution of the Broad Fourteens Basin. In: Brooks, J. & Glennie, K. (eds): Petroleum Geology of North-West Europe. Proceedings of the Third Conference. Graham & Trotman (London): 315-323.

Van Winden, M., De Jager, J., Jaarsma, B. & Bouroullec, R., 2018. New insights into salt tectonics in the northern Dutch offshore: a framework for hydrocarbon exploration. In: Kilhams, B., Kukla, P.A., Mazur, S., McKie, T., Mijnlieff, H.F. & Van Ojik, K. (eds): Mesozoic resource potential in the Southern Permian Basin. Geological Society of London, Special Publications 469: 99-117. DOI: 10.1144/SP469.9

Vejbæk, O.V., Andersen, C., Dusar, M., Herngreen, G.F.W., Krabbe, H., Leszczyński, K., Lott, G.K., Mutterlose, J. & Van der Molen, A.S., 2010. Cretaceous. In: Doornenbal, J.C. & Stevenson, A.G. (eds): Petroleum geological atlas of the Southern Permian Basin area. EAGE Publications B.V. (Houten): 195-209.

Vellekoop, J., Sluijs, A., Smit, J., Schouten, S., Weijers, J.W.H., Sinninghe Damsté, J.S. & Brinkhuis, H., 2014. Rapid short-term cooling following the Chicxulub impact at the Cretaceous–Paleogene boundary. Proceedings of the National Academy of Sciences 111: 7537-7541. DOI: 10.1073/pnas.1319253111

Vellekoop, J., Van Tilborgh, K.H., Van Knippenberg, P., Jagt, J.W.M., Stassen, P., Goolaerts, S. & Speijer, R.P., 2020. Type-Maastrichtian gastropod faunas show rapid ecosystem recovery following the Cretaceous–Palaeogene boundary catastrophe. Palaeontology 63: 349-367. DOI: 10.1111/pala.12462

Vellekoop, J., Kaskes, P., Sinnesael, M., Huygh, J., Déhais, T., Jagt, J.W.M., Speijer, R.P. & Claeys, P., 2022. A new age model and chemostratigraphic framework for the Maastrichtian type area (southeastern Netherlands, northeastern Belgium). Newsletters on Stratigraphy 55: 479-501. DOI: 10.1127/nos/2022/0703

Verweij, H., 2006. Dual hydraulic behaviour of the Chalk in the Netherlands North Sea. Journal of Geochemical Exploration 89: 405-408. DOI: 10.1016/j.gexplo.2005.11.013

Villain, J.-M., 1977. Le Maastrichtien dans sa région type (Limbourg, Pays-Bas). Étude stratigraphique et micropaléontologique. Palaeontographica A157: 1-87.

Vis, G.-J., Houben, A.J.P., Debacker, T.N. & Geel, C.R., 2025. The geological foundation of the Netherlands: the early Carboniferous, Devonian and older. In: Ten Veen, J.H., Vis, G.-J., De Jager, J. & Wong, Th.E. (eds): Geology of the Netherlands, second edition. Amsterdam University Press (Amsterdam): 53-93. DOI: 10.5117/9789463728362_ch02

Voigt, S., Wagreich, M., Surlyk, F., Walaszczyk, I., Uličný, D., Čech, S., Voigt, T., Wiese, F., Wilmsen, M., Niebuhr, B., Reich, M., Funk, H., Michalík, J., Jagt, J.W.M., Felder, P.J. & Schulp, A., 2008. Cretaceous. In: McCann, T. (ed): The Geology of Central Europe. Volume 2: Mesozoic and Cenozoic. The Geological Society of London: 923-997. DOI: 10.1144/CEV2P.3

Walaszczyk, I., Jagt, J.W.M. & Keutgen, N., 2010. The youngest Maastrichtian 'true' inoceramids from the Vijlen Member (Gulpen Formation) in northeast Belgium and the Aachen area (Germany). Netherlands Journal of Geosciences 89: 147-167. DOI: 10.1017/S0016774600000755

Wang, Y., Huang, C., Sun, B., Quan, C., Wu, J. & Lin, Z., 2014. Paleo-CO2 variation trends and the Cretaceous greenhouse climate. Earth-Science Reviews 129: 136-147. DOI: 10.1016/j.earscirev.2013.11.001

Wang, Z., Bachmann, R. & Pipping, J.C.P., 2018. F17 chalk fracture model. 80th EAGE Conference and Exhibition 2018, Copenhagen, Denmark: European Association of Geoscientists & Engineers (EAGE). DOI: 10.3997/2214-4609.201801486

Welch, M.J., Souque, C., Davies, R.K. & Knipe, R.J., 2015. Using mechanical models to investigate the controls on fracture geometry and distribution in chalk. In: Agar, S.M. & Geiger, S. (eds): Fundamental controls on fluid flow in carbonates: Current workflows to emerging technologies. Geological Society of London, Special Publications 406: 281-309. DOI: 10.1144/SP406.5

Willems, L., Rodet, J., Fournier, M., Laignel, B., Dusar, M., Lagrou, D., Pouclet, A., Massei, N., Dussart-Baptista, L., Compère, P. & Ek, C., 2007. Polyphase karst system in Cretaceous chalk and calcarenite of the Belgian-Dutch border. Zeitschrift für Geomorphologie 51: 361-376. DOI: 10.1127/0372-8854/2007/0051-0361

Wilmsen, M., Uličný, D. & Košťák, M., 2014. Cretaceous basins of central Europe: deciphering effects of global and regional processes – a short introduction. Zeitschrift der Deutschen Gesellschaft für Geowissenschaften 165: 495-499. DOI: 10.1127/1860-1804/2014/0088

Wilmsen, M., Engelke, J., Linnert, C., Mutterlose, J. & Niebuhr, B., 2019. A Boreal reference section revisited (Kronsmoor, northern Germany): high-resolution stratigraphic calibration of the Campanian-Maastrichtian boundary interval (Upper Cretaceous). Newsletters on Stratigraphy 52: 155-172. DOI: 10.1127/nos/2018/0473

Witte, L. & Schuurman, H., 1996. Calcareous benthic foraminifera across the Cretaceous/Tertiary boundary in the Geulhemmerberg (SE Netherlands). Geologie en Mijnbouw 75: 173-186.

Witte, L., Lissenberg, Th. & Schuurman, H., 1992. Ostracods from the Albian/Cenomanian boundary in the Achterhoek area (eastern part of the Netherlands). Scripta Geologica 102: 33-84.

Worum, G. & Van Wees, J.-D., 2017. High-resolution quantitative reconstruction of Late Cretaceous–Tertiary erosion in the West Netherlands Basin using multi-formation compaction trends and seismic data: implications for geothermal exploration. Acta Geodaetica et Geophysica: 243-268. DOI: 10.1007/s40328-017-0196-6

Ziegler, P.A., 1982. Geological atlas of western and central Europe. Shell Internationale Petroleum Maatschappij B.V. (The Hague), distributed by Elsevier (Amsterdam): 130 pp.

Zijlstra, H.J.P., 1987. Early diagenetic silica precipitation, in relation to redox boundaries and bacterial metabolism, in late cretaceous chalk of the Maastrichtian type locality. Geologie en Mijnbouw 66 (4): 343-355.

Zijlstra, J.J.P., 1994. Sedimentology of the Late Cretaceous and Early Tertiary (Tuffaceous) chalk of northwest Europe. PhD thesis Utrecht University (the Netherlands), Geologica Ultraiectina 119: 192 pp.

Zijlstra, J.J.P., Brouwers, M.H.M.P., Brinkhuis, H. & De Boer, P.L., 1996. Microfacies analysis of Cretaceous/Tertiary boundary sections in the quarries Geulhemmerberg and Curfs, SE Netherlands. Geologie en Mijnbouw 75: 133-151.

9

Paleogene – Neogene

Dirk K. Munsterman, Timme H. Donders, Alexander J.P. Houben, Johan H. ten Veen & Frank P. Wesselingh

*Corresponding author: dirk.munsterman@tno.nl

ABSTRACT

During the late Danian-Selandian Laramide phase, open-marine carbonate deposition of the Late Cretaceous and earliest Paleocene was replaced by clastic sediment infill of the Southern North Sea Basin. The Laramide phase, associated with domal uplift and subsidence of Mesozoic grabens, led to a break in sedimentation and reworking of Upper Cretaceous carbonates into marls. Consequently, Paleogene marine deposits are condensed in most areas. Late Paleocene to earliest Eocene uplift of basin margins caused major sand influxes into marginal marine environments with restricted circulation. In the North Sea area, global Paleogene warming culminated in near-tropical conditions and associated biota. Under maximum temperature conditions and differential subsidence, deltaic and submarine-fan sand deposition continued into the early Eocene. Cenozoic sediment input changed from the northwest during the Eocene, through northeastern sources in the Oligocene and Miocene, to dominantly southeastern and southern sources during the Pliocene and Pleistocene. The Paleocene-Eocene transition was interrupted by major volcanism, resulting in widespread ash layer deposition from volcanoes on the Greenland-Scotland ridge. From the middle Eocene onwards, regional subsidence interrupted by uplift phases led to transgression/regression patterns at the basins margins.

In the North Sea Basin, a major discontinuity formed due to the Pyrenean inversion phase that occurred just before Antarctic ice cap growth and global cooling at the onset of the Oligocene. From late Eocene to Mid Miocene, the basin experienced warmer and cooler phases, developing a rich, mostly endemic North Sea marine biota. In the early Oligocene, much of the Southern North Sea Basin drowned, and outer-neritic marine clays of the Rupel Formation (Boom Member) were deposited. During the late Oligocene through Pliocene, shallow marine sedimentation was balanced by subsidence resulting in monotonous sequences of marine clays and silts. During the Miocene Climate Optimum, peat formation was widespread at the southern margin of the North Sea Basin, followed by large-scale fluvial-deltaic deposition with local peatbogs as the climate cooled in the Late Miocene. The Upper Pliocene and Lower Pleistocene deposits are dominated by marine silty and sandy clays with ice-rafted debris, marking the first strong Northern Hemisphere glaciations, grading into shallow marine and fluvial sands towards the margins. These are overlain by predominantly sandy Pleistocene fluvial deposits. This chapter is structured around the varying tectonic and climatic factors that determined the structures of the North Sea Basin and its heterogeneous Paleogene-Neogene basin fill.

<< Laminated clay of the Belgian Boom Formation (quarry of Wienerberger in Rumst, west Belgium), equivalent to the Rupel Formation in the Netherlands. The laminations reflect Milankovitch driven sea-level fluctuations related to early Antarctic glaciations. Photo: Patrick Kiden.

DOI: 10.5117/9789463728362_ch09

Paleogene and Neogene time scale

Giovanni Arduino was a mining specialist who was the first to use the term 'Tertiary' in 1760. Based on observations in the Alps of northern Italy, Arduino proposed a subdivision of geological time into Primary, Secondary, and Tertiary periods. Later a fourth period, the Quaternary, was added. Although widely used, the 'Tertiary' was no longer recognized as a formal unit by the International Commission on Stratigraphy in the 1990s. Until 2010 the Paleogene and Neogene referred to the span of time between 65 and 1.8 million years ago (Fig. 9.1). Later, the 'dawn' of the Quaternary was put at the start of the Gelasian at 2.58 Ma (Gibbard et al., 2010).

The name Cenozoic (originally Cainozoic) Era is derived from the new (*kainos*) biota, compared to those of the Mesozoic Era (Phillips, 1841). The Cenozoic Era comprises the Paleogene (*palaios* = old, *genos* = birth), Neogene, and Quaternary periods. The subdivision of the Paleogene (66-23 Ma) and the Neogene (23-2.58 Ma) periods was originally based on the different relationship with modern fauna in marine deposits. In the Eocene, for example, only 3% of 'modern' (still extant) species are present, whereas in the Pliocene this percentage had risen to 90% (Lyell, 1833; Simmons, 2018). The Paleogene Period is subdivided into three epochs: the Paleocene, Eocene, and Oligocene, again referring to the evolution of biota (*eos* = dawn, *oligos* = little). The three Paleogene epochs are formally subdivided into nine ages. The Paleocene Epoch consists of the Danian, Selandian, and Thanetian ages, the Eocene Epoch of the Ypresian, Lutetian, Bartonian, and Priabonian ages and the Oligocene Epoch comprises the Rupelian and Chattian ages (Speijer et al., 2020). The Neogene Period presently includes the Miocene and Pliocene epochs. The Miocene includes the Aquitanian and Burdigalian (*Lower Miocene*), Langhian and Serravallian (*Middle Miocene*) and Tortonian and Messinian (*Upper Miocene*). With the recent transfer of the Gelasian to the Quaternary, the Pliocene now consists of the Zanclean (*Lower Pliocene*) and Piacenzian (*Upper Pliocene*) ages (Raffi et al., 2020). Orbital tuning of sedimentary cycles, calibrated to geomagnetic polarity- and biostratigraphic time scales has improved the resolution of the Paleogene and Neogene time scale over the last two decades (Raffi et al., 2020; Speijer et al., 2020). Astronomical age control is provided for almost all geomagnetic polarity reversals.

Introduction to the Paleogene and Neogene Southern North Sea Basin

The contours of the primordial North Sea became established in the Paleogene-Neogene (Fig. 9.2). The present-day Netherlands is located in the southern sector, but originally was positioned on the western edge of the Eurasian continent, at the approximate latitude of the Iberian Peninsula today. Over the past 60 My, the Southern North Sea Basin (SNSB) has migrated from 42° to 52° northern latitude, passing through a range of climatic zones from tropical to temperate (Van Hinsbergen et al., 2015). Throughout the Cenozoic, the Dutch part of the SNSB was located at the boundary between land and sea environments and the coastal zone was mainly located around the present province of South Limburg.

The exact position of the coastline and the types of depositional environments represented at the earth's surface are controlled by the long- and short-term interplay of change in accommodation space and sediment supply. Both accommodation changes and sediment supply responded, directly or indirectly, to the interaction of tectonics and climate. These responses are recorded in the sedimentary record by vertical variations in sedimentary facies, reflecting lateral shifts in depositional systems and the dynamic development of the paleo-landscapes.

Ongoing post-rift thermal subsidence since mid-Cretaceous times and the progressive opening of the northern Atlantic Ocean and African-European collision drove the evolution and origin of the North Sea Basin. These factors resulted in regional uplift of the British Isles, Fennoscandian Shield, Ardennes, Rhenish and Bohemian massifs around the Netherlands during the Paleogene-Neogene. These developments were responsible for the opening and closure of seaways which, together with climatic cooling, caused shifts in biotic provinces and the increasing importance of boreal elements in both fauna and flora. The developing geographic isolation of northwest Europe from southern Europe, caused by Alpine uplift and closure of epicontinental seaways, created a climatological and biological divide that precipitated extinctions and development of glacial landforms (Knox et al., 2010). As a result, the SNSB contains an almost complete geological archive of Paleogene and Neogene climate, landscape and biotic evolution, although the record is mostly contained within the subsurface. Isolated Paleogene-Neogene deposits between 200 and 1600 m thick (Fig. 9.3) crop out in restricted areas in the eastern and southern Netherlands and dip towards the Central North Sea Basin depocentre, where siliciclastic Cenozoic deposits reach over 2500 m thick (Kuhlmann et al., 2006a,b). The Cenozoic record of terrestrial environmental change is considerably more fragmented than that of the marine realm, and ages are often poorly constrained (Utescher et al., 2011).

Stable oxygen and carbon isotope records provide insight into trends in paleoclimate and carbon cycling (Zachos et al., 2008; Westerhold et al., 2020). These reveal a long-term global warming trend that started in the mid-

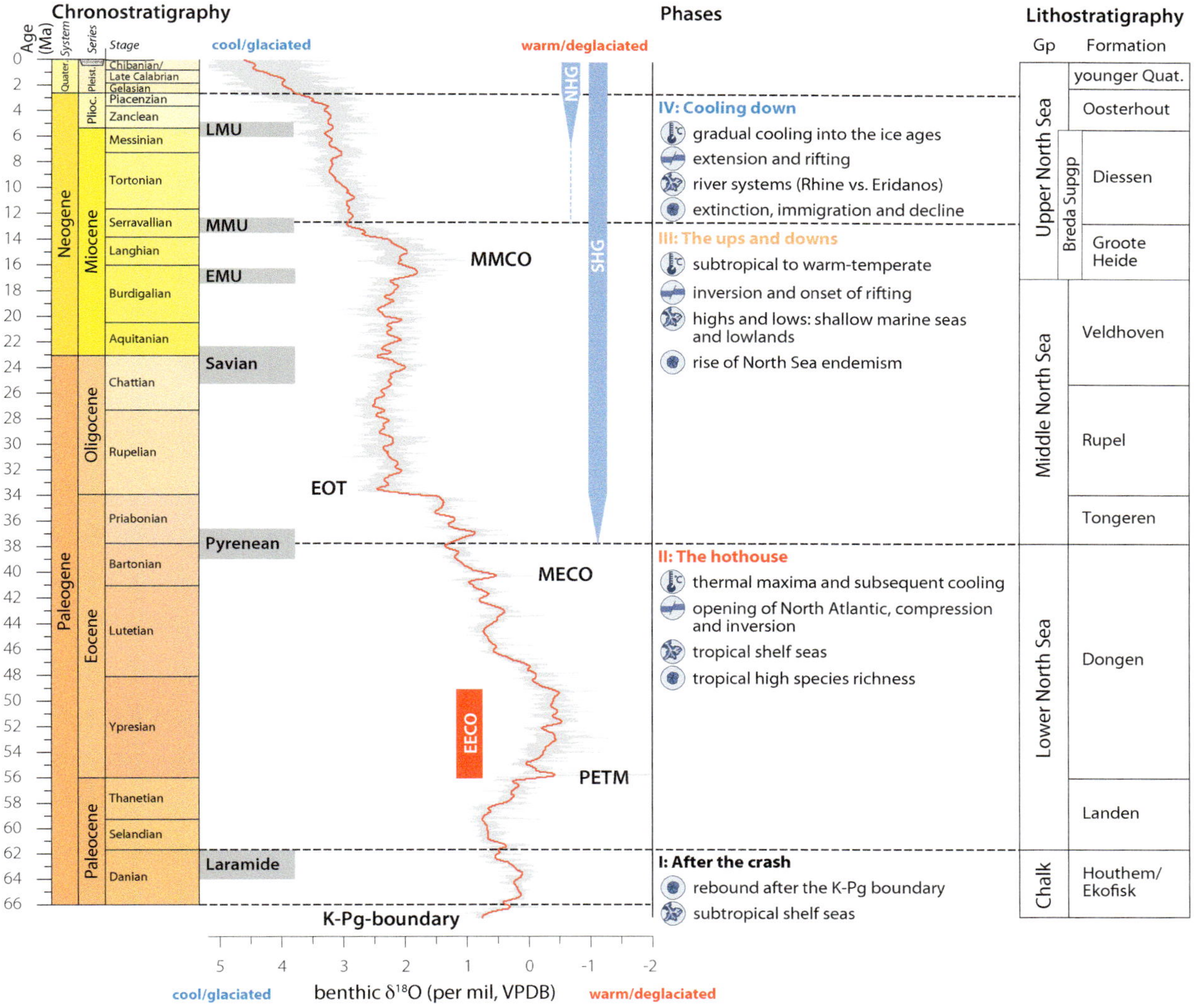

Figure 9.1. Paleogene-Neogene tectonic, climate and lithostratigraphic framework. The grey bars represent regionally recognized tectonic phases and/or unconformities. EMU – Early Miocene Unconformity, MMU – Mid-Miocene Unconformity, LMU – Late Miocene Unconformity. Phases I-IV correspond to distinct tectonic-climatic phases described in the text. The climatic framework is provided by the deep-sea benthic foraminiferal oxygen isotope compilation after Westerhold et al. (2020). Heavy (more positive) values correspond to cooler deep-sea temperatures and/or the development of continental glaciations. K-Pg = Cretaceous-Paleogene boundary; PETM = Paleocene-Eocene Thermal Maximum; EECO = Early Eocene Climatic Optimum; MECO = Middle Eocene Climatic Optimum; EOT = Eocene-Oligocene Transition; MMCO = Mid-Miocene Climatic Optimum.

dle Paleocene and culminated in the early Eocene. After peak warmth in the early Eocene this trend reversed and developed as a stepwise cooling from greenhouse to icehouse conditions, starting with the appearance of ephemeral ice-sheets on Antarctica across the Eocene-Oligocene transition (Zachos et al., 1992; Hutchinson et al., 2021). Northwest Europe and the SNSB started to cool as part of the global development of steeper and spatially more heterogeneous latitudinal climate gradients during the Cenozoic (Bijl et al., 2009; Liu et al., 2009). During the Oligocene and Miocene, numerous short-term isotope excursions mark a high degree climatic variability, largely driven by an interplay of solar insolation and oceanographic reorganizations (Palike et al., 2006; Jakobsson et al., 2007, Steinthorsdottir et al., 2021a). Progressive long-term cooling eventually culminated in the Pliocene establishment of permanent ice-sheets in the northern hemisphere, which expanded significantly at the start of the Pleistocene (Lisiecki & Raymo, 2005; Head & Gibbard, 2015).

Tectonic and depositional setting

Since the mid-Cretaceous (Van Lochem et al., 2025, this volume), the North Sea Basin has been almost continuously subsiding between the Fennoscandian Shield in the

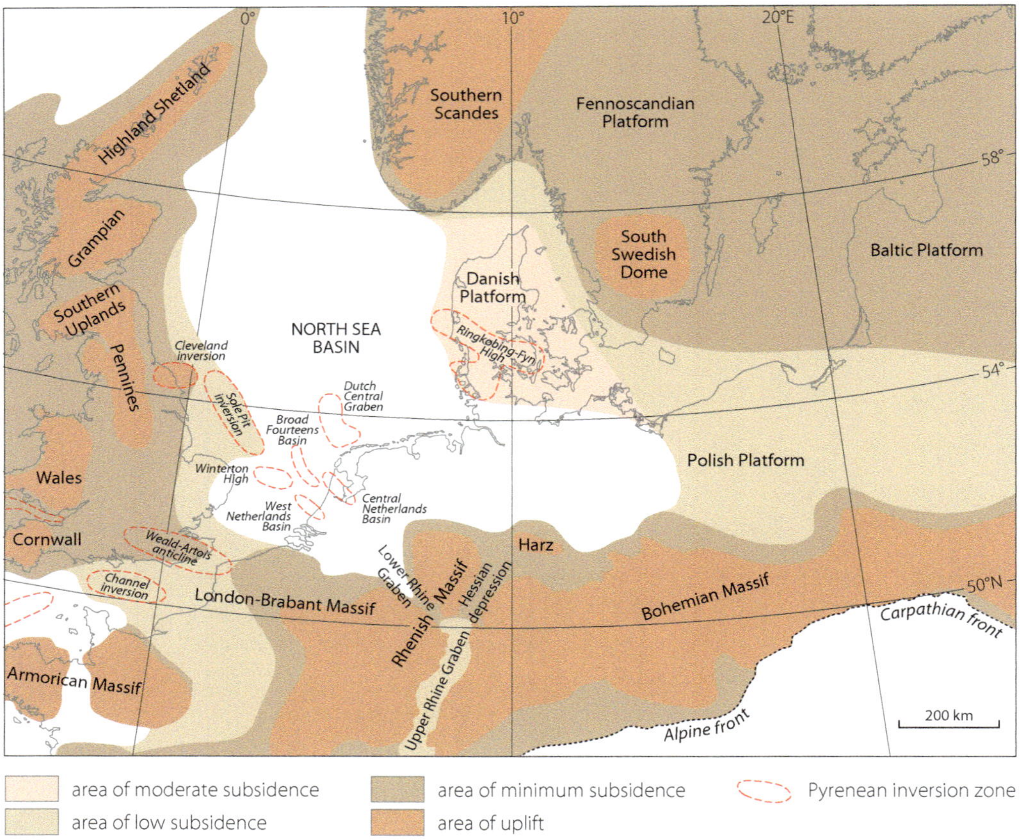

Figure 9.2. *Paleogene and Neogene structural elements in the North Sea region. Modified after Knox et al. (2010) and Gibbard & Lewin (2016).*

north, the European mainland in the south and the British Isles in the west. New seismic techniques and data have led to the realization that the predominantly subsiding post-rift thermal sag basin developed in response to gradual lithospheric cooling of underlying deeper Jurassic rift dome structures (Ziegler, 1990). Its central axis lies to the west of the Dutch Central Graben. The connection between the southern North Sea via the Channel to the Atlantic shelves of France and Ireland was opened intermittently by flooding or breaching of the Melantois-Artois High throughout the Cenozoic (Van Vliet-Lanoë et al., 1998; Gibbard & Lewin, 2003; Ziegler & Dèzes, 2007). The North Sea Basin region evolved during fragmentation of the Eurasian-North American plate and Alpine and Pyrenean orogenies. These forces caused repeated fluctuations in the magnitude and orientation of the intraplate stress field and related subtle reactivations of pre-Cenozoic structural elements (Knox et al., 2010).

The most important fracture system in the Netherlands in this regard is the approximately 1100 km long complex European Cenozoic Rift System (ECRIS; Fig. 9.4) which extends from the Southern North Sea Basin, through the Lower Rhine Graben (also called Basin or Embayment), via the Jura, Saône Graben as far as the Valencia Trough (Zagwijn, 1989; Ziegler, 1992). The Roer Valley Graben is the main tectonic feature in the Lower Rhine Graben (Geluk et al., 1994; Ziegler, 1994; Michon et al., 2003; Van Balen et al., 2005), which is bordered by the Rhenish Massif in the east and south and by the Brabant Massif in the southwest. It started to develop in the Alpine foreland and propagated northwards (and southwards) to accommodate the stress from the Alpine-Mediterranean orogenic system. During the Rupelian, the rifting progressively moved northward and the Rhine and Leine grabens developed a narrow seaway between the Alpine foredeep and the North Sea Basin (Vinken, 1988). The development of this rift system was also influenced by mantle-plume related volcanic activity, which intensified during the Oligocene and Miocene and contributed, by thermal thinning of the lithosphere, to the uplift of the Ardenno-Rhenish Massif and – in part – the Bohemian Massif (Ziegler & Dèzes, 2006, 2007).

Extensional tectonics in the Eurasian Plate attributed to the separation of North America away from Eurasia during the Paleogene resulted in the Danian Laramide phase. By latest Paleocene to earliest Eocene times, ridge-push forces exerted by the opening of the North Atlantic led to crustal separation between Greenland and Europe. This was

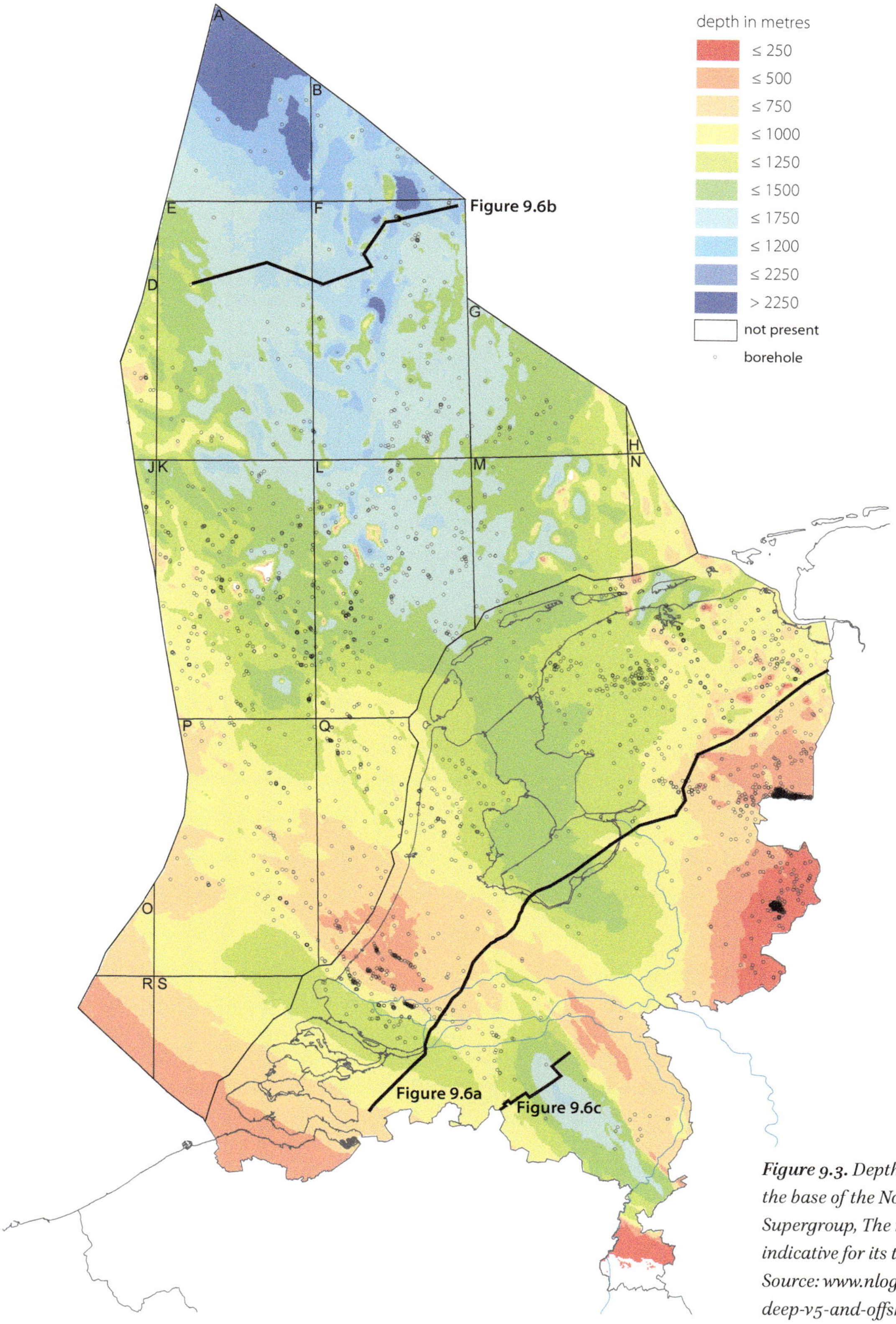

***Figure 9.3.** Depth map of the base of the North Sea Supergroup, The map is also indicative for its total thickness. Source: www.nlog.nl/en/dgm-deep-v5-and-offshore*

accompanied by major extrusive volcanism in the Norwegian Sea (Abdelmalak et al., 2016; Horni et al., 2017), which is represented by tuffaceous sediments in the Netherlands (De Wijk Member of the Dongen Formation).

In the southwestern part of present-day Europe, the Iberian Plate pushed against the southern part of France, which led to the exhumation of the Pyrenees (Pyrenean tectonic phase) in the middle to late Eocene. During this

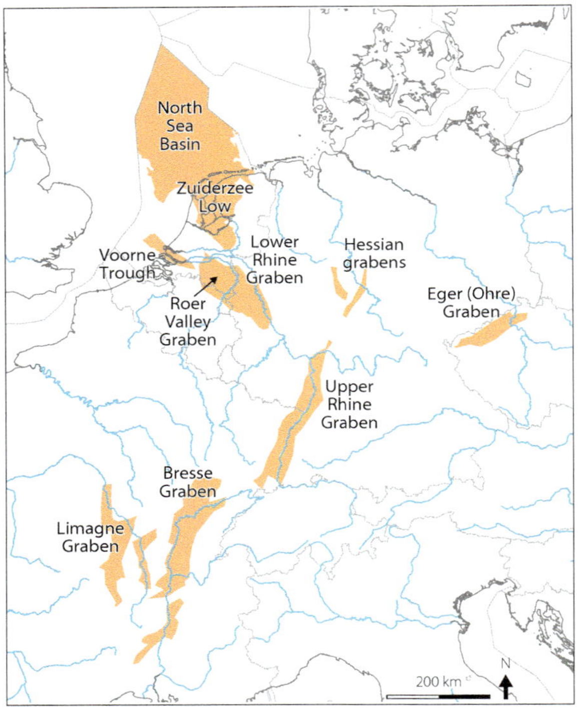

Figure 9.4. Depocentres (grabens, basins, troughs, and lows) of the European Cenozoic Rift System (ECRIS) in northwestern Europe.

Pyrenean inversion, roughly north-south directed compression is recognisable on seismic data in the southern North Sea (Nalpas et al., 1995) where a marked angular unconformity is seen in the Sole Pit area, close to the Eocene-Oligocene boundary (De Jager, 2007). This tectonic pulse caused widespread inversion (approximately 300 m) in the Mesozoic West and Central Netherlands basins, becoming less towards the Broad Fourteens Basin (Knox et al., 2010). Uplift of the Roer Valley Graben increased in southeastern direction (Deckers et al., 2016). It was estimated that the Weald Basin was uplifted by up to 1500 m during its late Eocene to Miocene inversion (Simpson et al., 1989). Massive erosional processes occurred in most of the inverted parts of the West and Central Netherlands basins. The Pyrenean phase ended in the earliest Oligocene and was followed by a widespread shallow marine transgression that even crossed the current Ardennes (Vandenberghe, 2017). An intermittent marine connection between the North Sea and the Upper Rhine Graben possibly existed via the Hessian Seaway (Berger et al., 2005; Sissingh, 2006) in the early Oligocene. In addition, a connection in the early Oligocene probably existed between the Tethys Ocean and the Atlantic via the Loire Seaway (Knox et al., 2010). Uplift and sea-level fall resulted in emergence of the Fennoscandian Platform (Knox et al., 2010). Large areas became subaerially exposed and forced the rivers seaward.

Absence of major sediment influx from the western North Sea Basin margin during the late Oligocene indicates that the land area maintained a low relief (Gibbard & Lewin, 2003, 2016). A marine connection via a Channel Seaway between the North Sea and the Atlantic is evidenced by mollusc faunas (Janssen & Gürs, 2002); coeval Lower Miocene deposits in the Paris Basin are non-marine. This implies that any threshold generated by inversion of the Central Channel and Weald basins must have been substantially reduced before the Burdigalian transgression, possibly due to the Savian tectonic phase.

During the Neogene, multiple unconformities developed due to uplift and inversion and interacted with eustatic sea-level variations. Rotational movements of the African and Iberian Plates against the Eurasian Plate led to closure of the Tethys Ocean in the Oligocene-Miocene. The related Early Miocene compression drove the orogeny of the Alps, Apennines and Jura. It was accompanied by rotational movement of the Adriatic (or Apulian) Plate on which Italy is situated and, at the same time, continued opening of the North Atlantic Ocean that enhanced regional uplift of the Fennoscandian Shield.

This interplay of plate-tectonic events resulted in an uplift and inversion phase that is referred to as the Savian phase (end Chattian-earliest Aquitanian). At the base of the Miocene, the Savian Unconformity represents a gap separating Chattian from early Aquitanian or younger deposits (Köthe, 2007). Next to its plate-tectonic origin, the Savian Unconformity corresponds to a glacio-eustatic event at the end of the Oligocene (Miller et al., 2008). The Savian Unconformity, though widespread, is less distinct in the eastern part of the North Sea Basin than in the west. Increased sediment supply in the eastern part during this time was associated with the development of a major Baltic River System, which drained a large area of Fennoscandia, the German lowlands, northern Carpathians and western Russian Platform (Gibbard & Lewin, 2016). The Baltic River flowed westwards along the present-day southern Baltic Sea (Cameron et al., 1992). Coastal progradation also took place in active basin settings such as the Lower Rhine Graben. Increased uplift in the West Netherlands Basin (De Jager, 2007), in the northern Campine Block (Munsterman & Deckers, 2020), on the Ringkøbing-Fyn High and Central Graben in the Danish sector (Rasmussen, 2009), and in the Mesozoic Sole Pit, Weald and Cleveland basins of the UK sector (Whittaker, 1985) are all related to the Savian phase, but the unconformity seems to be less well-developed in other parts of the Netherlands and Germany (Knox et al., 2010). Following the Savian phase, sedimentation recommenced and a Lower Miocene sequence of glauconite-bearing sands was deposited in the Southern North Sea Basin (Rasmussen & Dybkjær, 2014).

This sequence has a late Burdigalian, or possibly early Langhian age (Munsterman et al., 2019).

A eustatic sea-level rise associated with the Mid-Miocene Climatic Optimum (MMCO; Steinthorsdottir et al., 2021a) resulted in widespread transgression, characterized by sediment onlap on tectonic highs (marked by the Early Miocene Unconformity - EMU) and thick prograding sequences along the basin margins. In the late-middle Miocene a distinct change in the tectonic regime affected the SNSB. Accelerated uplift of Fennoscandia and Britain marked the culmination of the plate-tectonic regime that started in the Oligocene, possibly driven by mantle processes (Cloetingh et al., 2005). The uplift, coupled with a cooling in climate (Sangiorgi et al., 2021) and associated eustatic sea-level fall, led to widespread erosional processes and formation of the large-scale Mid-Miocene Unconformity (MMU). The uplift of Fennoscandia forced progradation of the fluvial-deltaic Baltic system into the southeastern embayment of the North Sea and its river system reached the Danish part of the southern North Sea in the Late Miocene (Knox et al., 2010; Rasmussen & Dybkjær, 2014; Thöle et al., 2014). In the Dutch northern offshore, time-equivalent, post-MMU, deposits are represented by a condensed sequence of Tortonian fine grained marine siliciclastics. This sequence represents toe-of-slope deposits that correspond to the distal parts of the sedimentary system that started to be deposited onto the MMU (Kuhlmann & Wong, 2008). The MMU is one of the main regional (seismic) horizons and is always characterized by stratal onlap. In the Roer Valley Graben, for instance, the MMU is expressed as a downlap surface of west-prograding clinoforms (Munsterman et al., 2019). A large-scale incision into the Paleogene successions indicates that the MMU is an erosional unconformity in the western North Sea (Cameron et al., 1992). The MMU is also associated with channel erosion at the base of the Diest Formation in Belgium (e.g. Vandenberghe et al., 2005).

Due to widespread and increasingly strong subsidence during the Neogene, the onshore Lower Rhine-Roer Valley Graben and offshore North Sea Basin became major depocentres (Munsterman et al., 2019). The timing of accelerated subsidence as well as the geometry of the depocentres could suggest that this was due to lithospheric folding under northwest-directed Alpine compression (Deckers & Louwye, 2020). During Pliocene and Pleistocene times, the central North Sea Basin also experienced accelerated subsidence, probably due to the same process of lithospheric downfolding (Van Wees & Cloetingh, 1996). This reflects the continuity of the Southern North Sea Basin as part of the Alpine foreland, together with the Ardennes and the northeastern part of the Paris Basin, which were also folded at this time (Bourgeois et al., 2007; Stoutham-er et al., 2023). Two of the three connections between the Atlantic Ocean and the Mediterranean closed completely at the end of the Miocene (~5.96-5.33 Ma) due to tectonic uplift (Ryan et al., 1973; Flecker et al., 2015; Krijgsman et al., 2018), resulting in the Messinian Salinity Crisis (Ryan, 1973; Roveri et al., 2014) and associated erosion expressed as the Late Miocene Unconformity (LMU). Seismic studies in the Roer Valley Graben (e.g. of the transect between wells Goirle and Heeswijk) show that topsets are missing and that an angular unconformity exists (Munsterman et al., 2019).

By the end of the Pliocene, plate-tectonic reorganization resulted in the Eurasian and North American landmasses to start surrounding the shallow marine polar basin and the gradual closure of the Isthmus of Panama (e.g. Keigwin, 1982; Bacon et al., 2015), linking North and South America. These gateway closures intensified Atlantic Gulf Stream circulation and set off a cascade of climatic feedbacks that ultimately led to the extensive Northern hemisphere glaciation in the Quaternary.

Stratigraphy and geological history

The stratigraphy and geological history of the Paleogene-Neogene in the Netherlands can be divided in four major phases bounded by tectonic events, described below in detail (Fig. 9.1). These phases, the late early to mid-Paleocene Laramide phase, the late Eocene Pyrenean phase and the late Middle Miocene phase (MMU) are separated by three major unconformities relevant for the Netherlands. The Savian phase strongly effected the proximal settings of the Southern North Sea Basin, including the southwestern part of the Netherlands, resulting in a Chattian-Early Miocene hiatus. More distally however, the MMU is the most prominent feature on seismic sections. In addition, an obvious trend break in climate development towards more consistent cooling is associated with the MMU. The importance is such that it led to a recent subdivision of the former Breda Formation into the Miocene Groote Heide and Diessen formations (Munsterman et al., 2019), grouped in the Breda Subgroup.

The Paleogene-Neogene successions are subdivided into a lowermost calcareous part (the Danian part of the Chalk Group) and predominately siliciclastic rocks of the Lower, Middle and Upper North Sea groups. Lithostratigraphically, the Cenozoic phases are represented by the Lower North Sea Group (interval between Laramide and Pyrenean tectonic events), the Middle North Sea Group (in between Pyrenean phase and EMU) and Upper North Sea Group (post-EMU). The alternation of mainly clay and sand and to a lesser extent marls and gravels near the southern margin of the North Sea Basin forms the basis

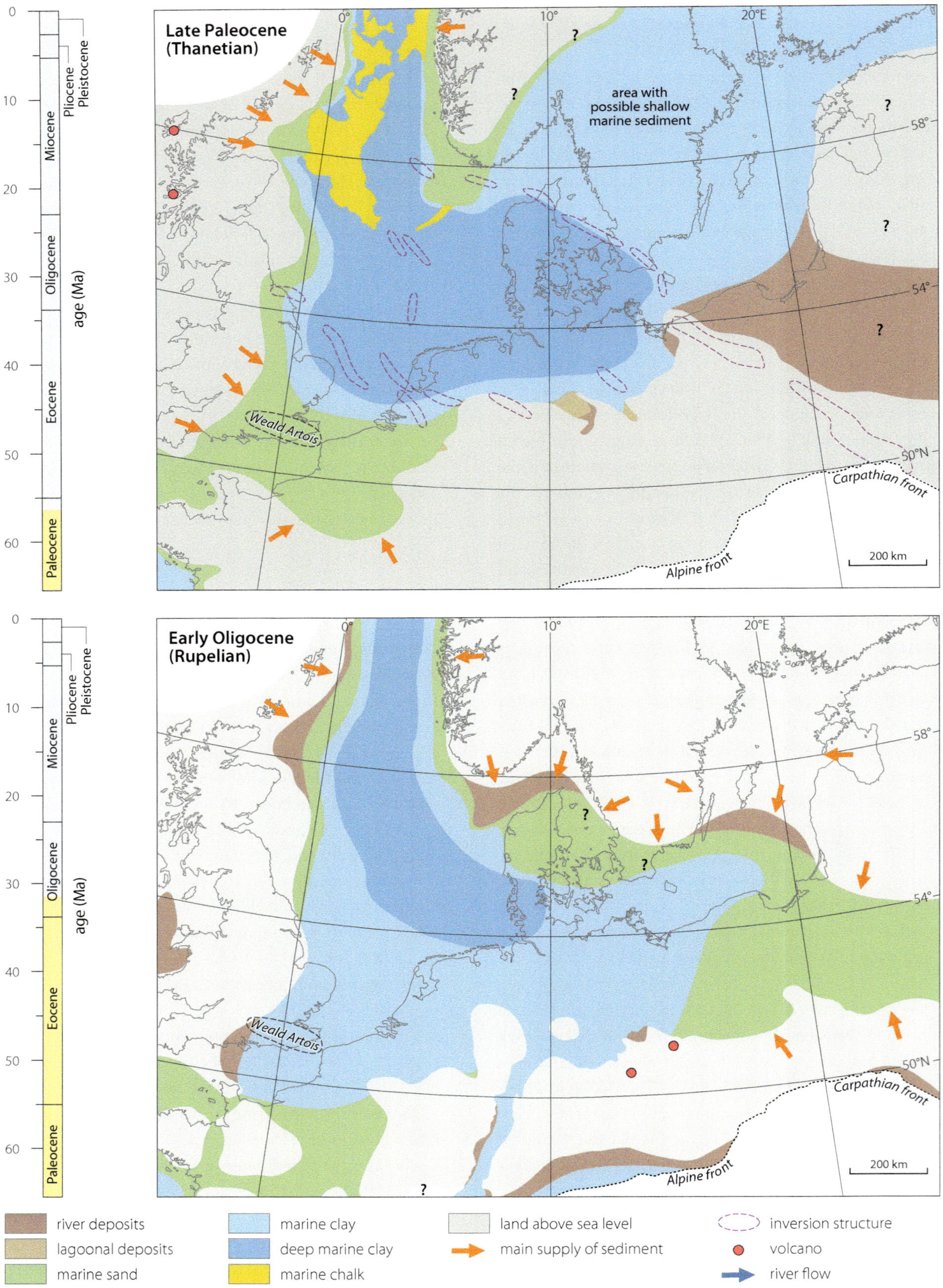

Figure 9.5. *Paleogeographic maps for a) the late Paleocene, b) the early Oligocene, c) Middle Miocene, and d) Early Pliocene. Modified after Knox et al. (2010) and Gibbard & Lewin (2016).*

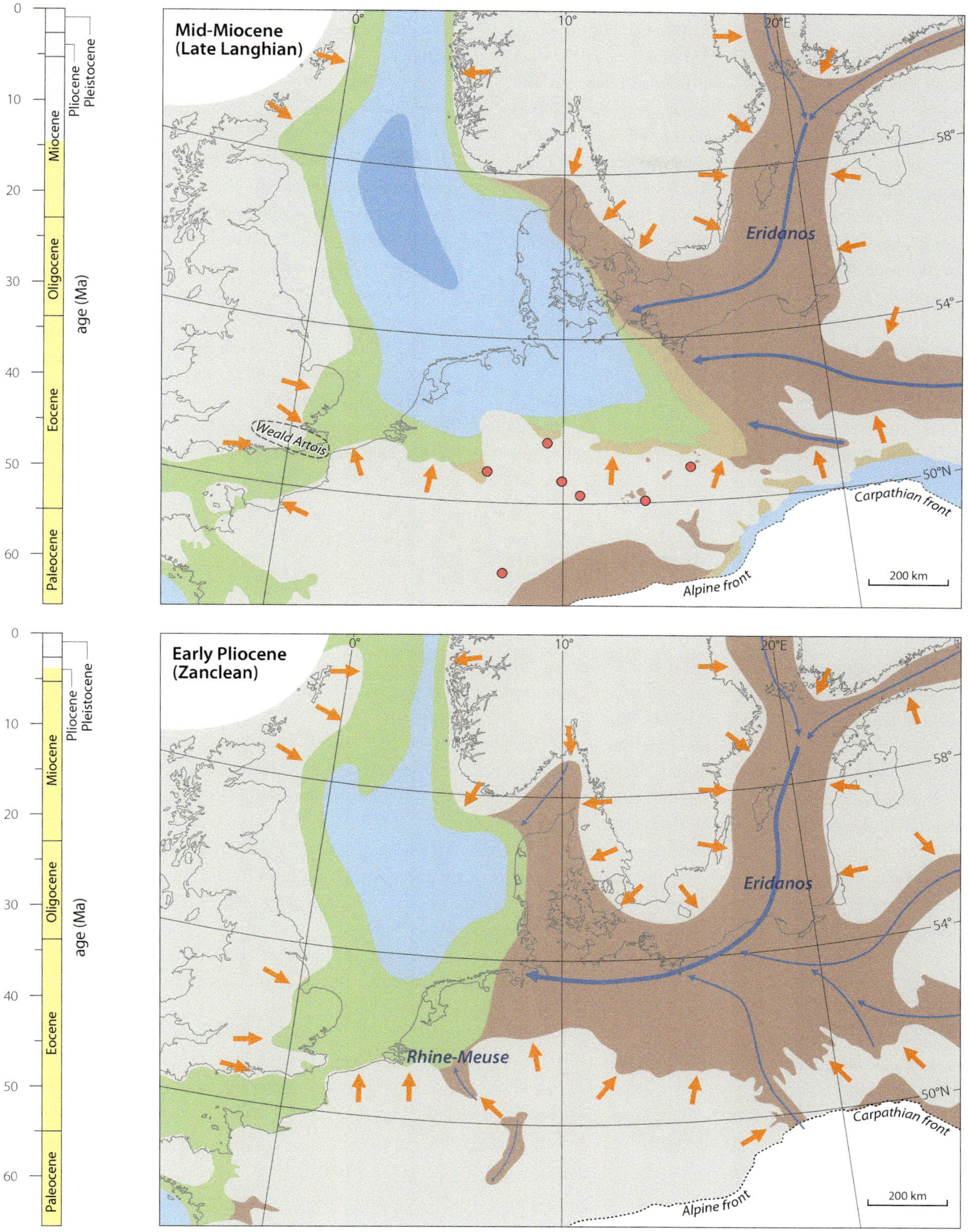

of the subdivision into formations and members. The description of the phases includes separate sections focusing on regional tectonics, climate, depositional systems, landscape, ecosystem evolution and lithostratigraphic development (Fig. 9.5).

Phase I – After the impact and Sub-Hercynian inversion (Danian, 66-62 Ma)

The Yucatan asteroid impact at 66 Ma (formerly known as the K-T event) marked the catastrophic end of the dinosaur world at the end of the Cretaceous and saw the almost instant demise of typical Mesozoic biota such as ammonites, rudists and inoceramid molluscs. The impact shell-shocked the shallow-marine tropical biota and also affected plankton groups, such as calcareous nannofossils and dinoflagellates. This was, however, followed by a rapid (partial) recovery. The early Paleocene North Sea Basin almost immediately returned to the limestone deposition that characterized the Late Cretaceous. The rapid recovery of carbonate ecosystems has been ascribed to calcium oversaturation resulting from the Yucatan impact and its aftermath (Bralower et al., 2020).

In terms of lithostratigraphy, the Paleocene chalk-type deposits are included in the Ekofisk Formation (offshore) and the Houthem Formation (onshore) of the Chalk Group (see Van Lochem et al., 2025, this volume). Deposition occurred under relatively stable, low-energy conditions in carbonate-shelf to upper bathyal environments. The sediments predominantly consist of pelagic biogenic remains, which settled from suspension with only very limited detrital input.

Life on land had a full reset with the loss of the dominant sauriid groups and the emergence of mammal dominated communities. In the Netherlands, the main paleontological record for the onset of the Paleogene is found in the marine Houthem Formation, of which a very good section used to be exposed in the Curfs quarry near Berg en Terblijt in South Limburg (Jagt et al., 2013). Rich marine faunas from the NE North Sea Basin, including the Faxe area (Denmark) and terrestrial faunas from the Mons Basin in Belgium (De Bast & Smith, 2016) provide a regional view on the ecosystem and biotic development. Top marine Mesozoic mosasauriid predators became replaced by sharks and crocodiles during the onset of the Paleocene. However, several other Mesozoic groups, including

Fossil marine mammals from the Neogene

1

The southern part of the North Sea Basin, including particular sites of the Western Scheldt and sandpits of Mill and Liessel, conceal large fossil treasures. Until approximately a decade ago, these treasures were hidden in buried seafloor sediments, but were revealed by recent sand extraction and trawling. The faunal association of marine vertebrates from the Late Miocene Diessen Formation revealed dozens of fossil taxa that were little or not known. Unique for the Netherlands and Europe!

Recent studies of these fossils show that the composition of Late Miocene whale faunas is radically different from the current ones (e.g. Post & Peters, 2023). The extant marine mammal fauna is dominated by a few species of fin whales, several dolphin species, the harbour porpoise and the grey- and harbour seal. The large sperm whale and two members of the beaked whale family are less common.

The whale faunas of the Late Miocene were even richer in genera and species than recent pendants. The former were dominated by many species of relatively small and basal fin whales (Cetotheriidae), small true fin whales, which are the ancestors of the enormous recent fin whales (Balaenopteridae), sperm whales of all shapes and sizes, with rows of extremely sharp teeth in the lower and upper jaws (Physeteroidea), a range of species of longirostral beaked whales, some of which hunted their epipelagic prey in large groups in shallow waters (Ziphiidae), and small 'river' dolphins (Pontoporiidae). These faunas also included up to half a dozen of pinniped taxa (large and small) and a seacow.

During the Pliocene, and especially by the end of the Pliocene, the highly varied Miocene faunas were replaced by faunas with a reduced number of taxa, which included the direct ancestors of the present-day North Sea fauna. Fossils of these Pliocene faunas are extremely rare worldwide, although some taxa have been described from the Oosterhout Formation (Post & Kompanje, 2010; Post & Bosselaers, 2017).

Skull with mandibles and cervical vertebrae of a yet undescribed species of beaked whale of approximately 4-5 metres body length from the Western Scheldt (collection Natuurhistorisch Museum-Natural History Museum, Rotterdam; NMR 12016).

some ammonoid and inoceramid- and trigonoid bivalve taxa briefly lingered on into the earliest Paleocene before their final extinction (Jagt et al., 2013). Echinoderms and molluscs experienced a major extinction of (sub-) tropical communities followed by rapid recovery through immigration from the northern Atlantic (Jagt, 2000; Vellekoop et al., 2019). The Houthem Formation fossils indicate shallow (sub-)tropical conditions with common algal firmgrounds. The Danian marine biota have much in common with those from the northern Paris Basin and from Denmark/southern Sweden, and indicate that the North Sea Basin at the time was an embayment of the North Atlantic (Jagt et al., 2013). During the Danian, lowland coastal habitats existed near the southern border of the Netherlands. A very rich fauna retrieved from Hainin (Belgium) represents an early Paleocene assemblage dominated by small mammals of primitive groups and insectivores that lived in forested environments with dense undergrowth at the southern margin of the North Sea Basin (De Bast & Smith, 2016). Rare fossils of *Thalassotaenia* seagrass foliage from the earliest Danian represent the first Cenozoic plants remains found in the Netherlands (Van der Ham et al., 2007).

The change from Late Cretaceous and early Paleocene chalk deposition to Paleocene-Eocene clastic deposition occurred around the transition from early-middle Paleocene (late Danian/Selandian boundary) (Steurbaut & Sztrákos, 2008). This transition has been linked to the Laramide phase and is regionally correlated with the Atlantean and Near Base Paleogene unconformities (Patruno et al., 2021). The change in sediment supply is attributed to widespread regional uplift (Knox et al., 2010) that lead to enhanced erosion and sediment supply. Hypotheses on the reason for the uplift include relaxation inversion (Nielsen et al., 2005, 2007), lithospheric folding (Deckers, 2015; Deckers & Matthijs, 2017), the Iceland mantle plume (Gale & Lovell, 2018) and a combination of these factors (e.g. Hillis et al., 2008). The uplift led to subaerial exposure of land-masses, such as the British Isles and the Brabant Massif, to the west and south of the Netherlands (Knox et al., 2010; Deckers & Matthijs, 2017), as well as in Mesozoic grabens in the Netherlands, such as the Vlieland and West Netherlands basins (Deckers & Van der Voet, 2018). Simultaneously, other areas such as the Central Netherlands Basin, Roer Valley Graben, Voorne Trough and local areas of the Broad Fourteens Basin experienced subsidence of depocentres that were filled with latest Danian continental deposits and/or Selandian greensands and marls (Deckers & Van der Voet, 2018). These marls consist mainly of reworked Cretaceous nannofossils from the Chalk Group (Steurbaut, 1998; Vandenberghe et al., 1998) that were probably eroded from the nearby highs. During the Laramide phase, the southern part of the Dutch Central Graben was an area of non-deposition (Van Lochem, 2018) or shallow marine environments, whereas the northern part was covered by marly clays of the deep marine (outer shelf to bathyal) Danish Våle Formation (Deckers & Munsterman, 2019).

Paleoclimate- and sea-level reconstructions suggest global cooling and a progressive sea-level fall through the late Paleocene (Miller et al., 2020; Speijer et al., 2020; Hollis et al., 2021). Regional biotic and/or environmental data are very limited, but the abrupt regional disappearance of the hyposaline *Braarudosphaeracea* nannofossils in conjunction with the shift from carbonate to clastic deposition suggests a reaction to reduced precipitation (Steurbaut & Sztrákos, 2008). This could be a basin-specific response to the long-term climatic developments. The contact between the Chalk Group and North Sea Supergroup is unconformable across much of the onshore Netherlands, but is quasi conformable in the deepest parts of the offshore North Sea Basin.

Phase II – Hothouse times (Paleocene-Eocene, 62-38 Ma)

Regional tectonics and paleogeography

During the Paleocene-Eocene, tectonic regimes in NW Europe were mostly driven by progressive opening and widening of the North Atlantic Ocean that led to episodic compression and inversion pulses. By early late Paleocene times, the SNSB was marginally connected to the Paris Basin to the south, but connections became severed and only a northern Atlantic connection remained (Knox et al., 2010). At the Paleocene-Eocene transition, the inception and intensification of a mantle plume between Scotland and Iceland caused extensive regional uplift across the British Isles and the SNSB (Maclennan & Jones, 2006). Volcanoclastic remnants are well documented in the North Sea Basin (Morton & Knox, 1990; Watson et al., 2017; Van Bergen et al., 2025). In the Netherlands these are known as the De Wijk Member of the Dongen Formation (previously known as the Dongen Tuffite Member).

Over the course of the Eocene, broad thermal subsidence led to voluminous clastic-dominated deposition. In addition to the offshore SNSB, other major Paleocene-Eocene depocentres include the Voorne Trough, the Zuiderzee Low (Fig. 9.6a) and the Lauwerszee Trough. The Voorne Trough is a basin located on the southern flank of the inverted part of the West Netherlands Basin referred to as the Kijkduin High (Duin et al., 2006). The origin of the Voorne Trough is linked to the Laramide phase. A seismic profile (Duin et al., 2006) shows that the Kijkduin High is a Pyrenean inversion structure that determined the cut-off and distribution of Paleogene-Eocene deposits (and is expresed as an unconformity between the Dongen and

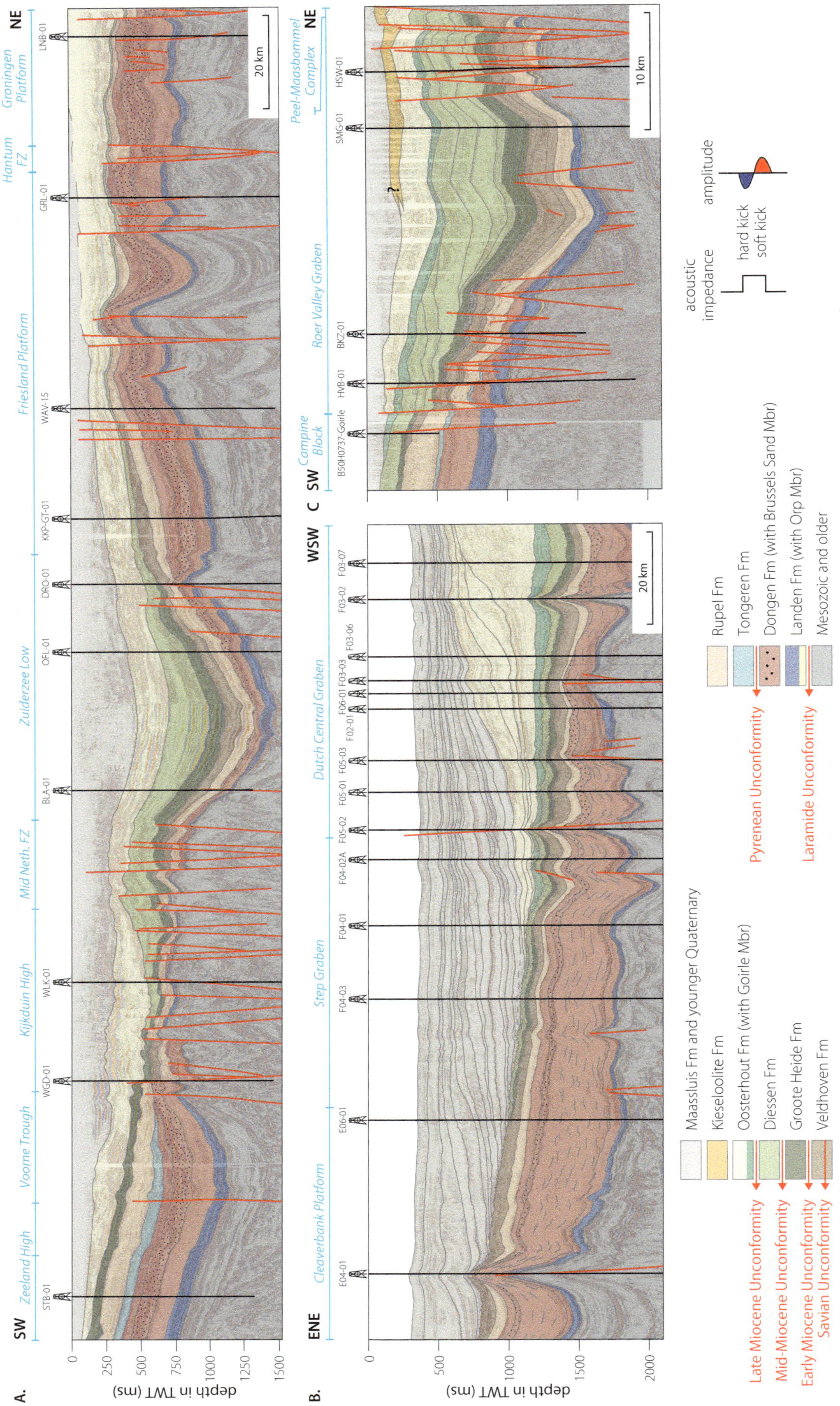

A.
SW
NE
depth in TWT (ms)
0
250
500
750
1000
1250
1500
Zeeland High
Voorne Trough
Kijkduin High
Mid Neth. FZ
Zuiderzee Low
Friesland Platform
Hantum FZ
Groningen Platform
STB-01
WGD-01
WLK-01
BLA-01
OFL-01
DRO-01
KKP-GT-01
WAV-15
GRL-01
LNB-01
20 km
B.
ENE
WSW
depth in TWT (ms)
0
500
1000
1500
2000
Cleaverbank Platform
Step Graben
Dutch Central Graben
E04-01
E06-01
F04-03
F04-01
F04-02A
F05-02
F05-01
F05-03
F02-01
F06-01
F03-03
F03-06
F03-02
F03-07
20 km
C
SW
NE
Campine Block
Roer Valley Graben
Peel-Maasbommel Complex
B50H0737-Goirle
HVB-01
BKZ-01
SMG-01
HSW-01
10 km
?
Late Miocene Unconformity
Mid-Miocene Unconformity
Early Miocene Unconformity
Savian Unconformity
Maassluis Fm and younger Quaternary
Kieseloolite Fm
Oosterhout Fm (with Goirle Mbr)
Diessen Fm
Groote Heide Fm
Veldhoven Fm
Pyrenean Unconformity
Laramide Unconformity
Rupel Fm
Tongeren Fm
Dongen Fm (with Brussels Sand Mbr)
Landen Fm (with Orp Mbr)
Mesozoic and older
acoustic impedance
hard kick
soft kick
amplitude

Rupel formations). Their original distribution was linked to the Laramide phase and thus more widespread. This cut-off can also be seen in the Voorne Trough (see Fig. 6 of Deckers & Van der Voet, 2018).

The Lauwerszee Trough is delineated by north to northwest and east to northeast trending faults that were reactivated during the Cenozoic (De Jager, 2007). It is a distinct and long-lived feature, which is clearly expressed in the pre-Permian subcrop pattern and the increased thicknesses of the Rotliegend. Thickness trends in Lower Cretaceous and Paleogene deposits are primarily due to withdrawal of Zechstein salt into flanking salt walls associated with the eastern and western (Hantum Fault Zone) boundary faults, which grew significantly during Early Cretaceous and Cenozoic times (De Jager, 2007). The Paleogene succession in this basin is up to 1000 m thick and the whole Cenozoic succession totals more than 1750 m (see Fig. 9.3).

Regional climate

Following the abrupt environmental change and ecosystem turnover at the Cretaceous-Paleogene boundary, the Paleocene was a period of relative climatic stability, not marking a major contrast with the Late Cretaceous. Polar climates were temperate and largely ice-free (Westerhold et al., 2020). The late Paleocene is characterized by a transition towards a cooler climate. Carbon-isotope data suggest that this was a phase of excess organic-burial, perhaps related to extensive uplift and consequent development of coastal swamps during the Laramide phase (Kurtz et al., 2003). The cooling trend reversed and across the Paleocene-Eocene boundary a major change in the climate occurred. The boundary coincides with the Paleocene-Eocene Thermal Maximum (PETM), the first of a series of transient warming-events that were associated with an injection of CO_2 into the ocean-atmosphere system (Dunkley Jones et al., 2013; Westerhold et al., 2020). Current hypotheses for the sources for CO_2 include the destabilization of marine methane hydrates (Dickens et al., 1995) and/or the thermogenic release associated with contemporaneous North Atlantic volcanism (Svensen et al., 2004). The carbon cycle perturbation that characterizes the PETM is also recorded in shelf environments from the North Sea (e.g. in borehole Woensdrecht-1; Sluijs et al., 2008). These so-called hyperthermal events were superimposed on a longer time-scale increase in temperature, labelled the Early Eocene Climatic Optimum (EECO). Reconstructions of atmospheric CO_2-concentrations (Rae et al., 2021) suggest that this warmest phase of the Cenozoic is associated with a coeval maximum in CO_2 concentration that lasted for most of the early Eocene (Ypresian).

The earliest well-preserved Cenozoic floras are known only from the Paleocene in Belgium (Gelinden Member) and the climatic interpretation of these floras indicates mean annual temperatures of around 16°C and high precipitation (Tanrattana et al., 2020). Comparison with floras from France show a long-term Paleocene warming trend analogous to marine-based reconstructions of temperature. Lipid paleothermometry data from early Eocene lignites from Germany and the UK (Inglis et al., 2017; Naafs et al., 2018) show warm mean annual temperatures (MAT) of up to 23-28°C, reflecting the temperature optimum at that time. The Messel Pit, near Frankfurt is a world-famous locality providing insights into the flora and fauna of the early middle Eocene (48 Ma, Lenz et al., 2015) of NW-Europe. Quantitative analyses of the vegetation composition at Messel indicate a MAT of approximately 22°C and annual precipitation rates about three times as high as at the present-day (2540 mm/yr; Grein et al., 2011). The remainder of the Eocene experienced a gradual cooling trend (e.g. Cramwinckel et al., 2018) that was only interrupted by the Middle Eocene Climatic Optimum (MECO) (Bohaty & Zachos, 2003; Bijl et al., 2010). The long-term early Eocene warmth and subsequent cooling are generally ascribed to a change from a net-excess of CO_2 outgassing from Tethyan subductive volcanism vs. CO_2-capture due to silicate weathering (Kent & Muttoni,

← ***Figure 9.6.*** *Interpreted seismic sections across the Netherlands, illustrating the distribution of Cenozoic lithostratigraphic units in various structural elements (blue italic text) as discussed in text. Key wells (mostly energy exploration wells) used for lithostratigraphic correlation are indicated. Note that (only) vertical scales are equal. a) NE-SW section crossing the main structural elements onshore the Netherlands, illustrating how the presence, thickness and internal arrangement of the Paleogene-Neogene units was strongly controlled by several phases of subsidence and inversion of the main sedimentary basins. b) Seismic section in the northern offshore E and F blocks, showing eastward thinning and truncation of the pre-Eocene section (with internal stratal arrangement indicated with dashed lines), a relatively thin sequence of Eocene-Miocene rocks, and the strong progradational stratigraphic arrangement of seismic sequences S1-S13 (after Kuhlmann & Wong, 2008) belonging to the Pliocene-Pleistocene Eridanos shelf-edge delta system. The lower boundary of the Eridanos interval is the Mid-Miocene Unconformity (MMU sensu Kuhlmann, 2004). Note that the upper/younger part of this system continues into the Pleistocene and is discussed in Busschers et al. (2025, this volume). c) Seismic section across the Roer Valley Graben, showing the relatively thick development of Paleogene and Neogene sequences. The Diessen Formation shows several stacked clinoform sets that were deposited in a time of strong differential vertical movements that resulted in considerable deepening of the graben in Miocene times.*

2008). Over the course of the Eocene, CO_2-concentrations declined to levels that eventually allowed for full-scale Antarctic glaciation across the Eocene-Oligocene Transition (DeConto & Pollard, 2003; Pearson et al., 2009).

Depositional systems, landscape and ecosystem evolution

After the Danian the shallow marine carbonates were replaced by siliciclastic deposits. Most of the time the Southern North Sea Basin was a subtropical-tropical shelf sea bordered by coastal areas in the southern Netherlands and Belgium. Low diversity North Atlantic marine faunas from the middle Paleocene eventually gave way to the rich tropical marine biota of the middle Eocene. At the same time, life on land diversified and very rich tropical to subtropical lowland biota emerged around the southern shores of the North Sea Basin in the early to middle Eocene. During the PETM, the warming and accompanying hydrological changes caused the shallow seas of the SNSB to experience widespread anoxia (Kender et al., 2012).

For this time interval sparse and exclusively marine fossils have been recorded from outcrops and boreholes in Twente and Zeeuws-Vlaanderen as well as from some coal mine shafts in Limburg. Furthermore, Eocene fossils deposited from Quaternary rivers draining the Flemish hinterland have been found on beaches in Zeeland. In adjacent regions, such as northern Belgium, Denmark and the eastern UK, an extensive sedimentary record provides detailed insights into the regional biotic- and ecosystem evolution through this hothouse phase.

In the Paleocene Gelinden flora, mostly thermophilic *Fagaceae* and *Lauraceae* made up the evergreen forest with palms and a few conifers. Terrestrial mammal faunas diversified as did bird faunas. Along the shores of the southern North Sea large flightless birds filled the vacant niches of the large terrestrial predators, while arboreal birds diversified together with the diversification of forests (Mayr & Smith, 2019). The late Paleocene exchange with North American terrestrial biota shaped the subsequent faunas.

The early Eocene tropical climate supported coastal mangrove forests with (*Nypa*) palms and tropical ferns (*Acrostichum*). There is no direct evidence from the Netherlands, but mangrove fossils known from Belgium and Southern England reveal a very rich lowland flora (Kvaček, 2010). From the London Clay in the southern UK over 350 species of plants have been identified. Fossil forest remains from Hoegaarden (Belgium) show that swamp forests with cypress trees, so typical of Neogene lignites, were already present in the lowlands of the SNSB during the EETM (Fairon-Demaret et al., 2003)).

Benthic marine communities during the Selandian-Thanetian (mid-late Paleocene) were relatively species-poor in the Southern North Sea Basin, which at that time extended into the northern Paris Basin. The Ypresian (early Eocene) saw the rise of rich tropical marine and terrestrial biota. The presence of nummulitids (larger foraminifera that host endosymbionts) in the Ieper Group of Belgium indicate very warm shallow marine conditions and most likely some connection towards more southerly Atlantic basins. A very widespread transgression resulted in the almost entire drowning of the North Sea Basin and the Netherlands must have been fully overlain by sea.

The early Lutetian is represented by the fossiliferous lower part of the Brussels Sand Member in the shallow subsurface of Zeeuws-Vlaanderen. In the region between Gent and Brugge this unit is known as the Oedelem Member and crops out extensively. Fossil shells and shark teeth eroded from this formation are well known from beaches such as at Cadzand. Truly tropical marine faunas occurred in the middle Eocene (Lutetian-Bartonian) as shown for example by the occurrence of very large *Campanile* gastropods in western Belgium and the Dutch offshore sector (Wesselingh et al., 2013). During the Bartonian, marine ecosystems in the Netherlands were dominated by fine-grained sandy and clayey sea floors and little of the fauna was preserved here. The middle to late Eocene floras regionally show a decline in diversity due to climatic cooling (Kvaček, 2010), but they are not preserved in the Netherlands.

Lithostratigraphy

The interacting climate, tectonic and depositional regimes resulted in the formation of a series of units that are most distinct in the coastal regions, but which shale-out towards the central North Sea where they become amalgamated within the Lower North Sea Group. The lithological succession of the Lower North Sea Group consist of an alternation of predominantly marine grey sands, sandstones and clays. The total thickness is up to ~650 m. In large parts of the LNSG the lower boundary is characterized by an unconformity (Laramide tectonic phase), expressed as a sharp lithologic break marking the top of the calcareous Chalk Group. Elsewhere, the group rests with a distinct unconformity on even older deposits. The upper boundary is defined by unconformably overlying deposits of the Middle North Sea Group or younger units and corresponds to the Pyrenean tectonic phase. The Lower North Sea Group comprises the Landen and Dongen formations.

Landen Formation

In the southern part of North Brabant and in Limburg (basin fringe), the Landen Formation (Fig. 9.7) starts with dark green-grey, fine-grained glauconitic sand (Orp Member) followed upward by light-grey marl (Gelinden Member) with dark green-grey clay at its base (Liessel Member). These become less limey and more sandy up-

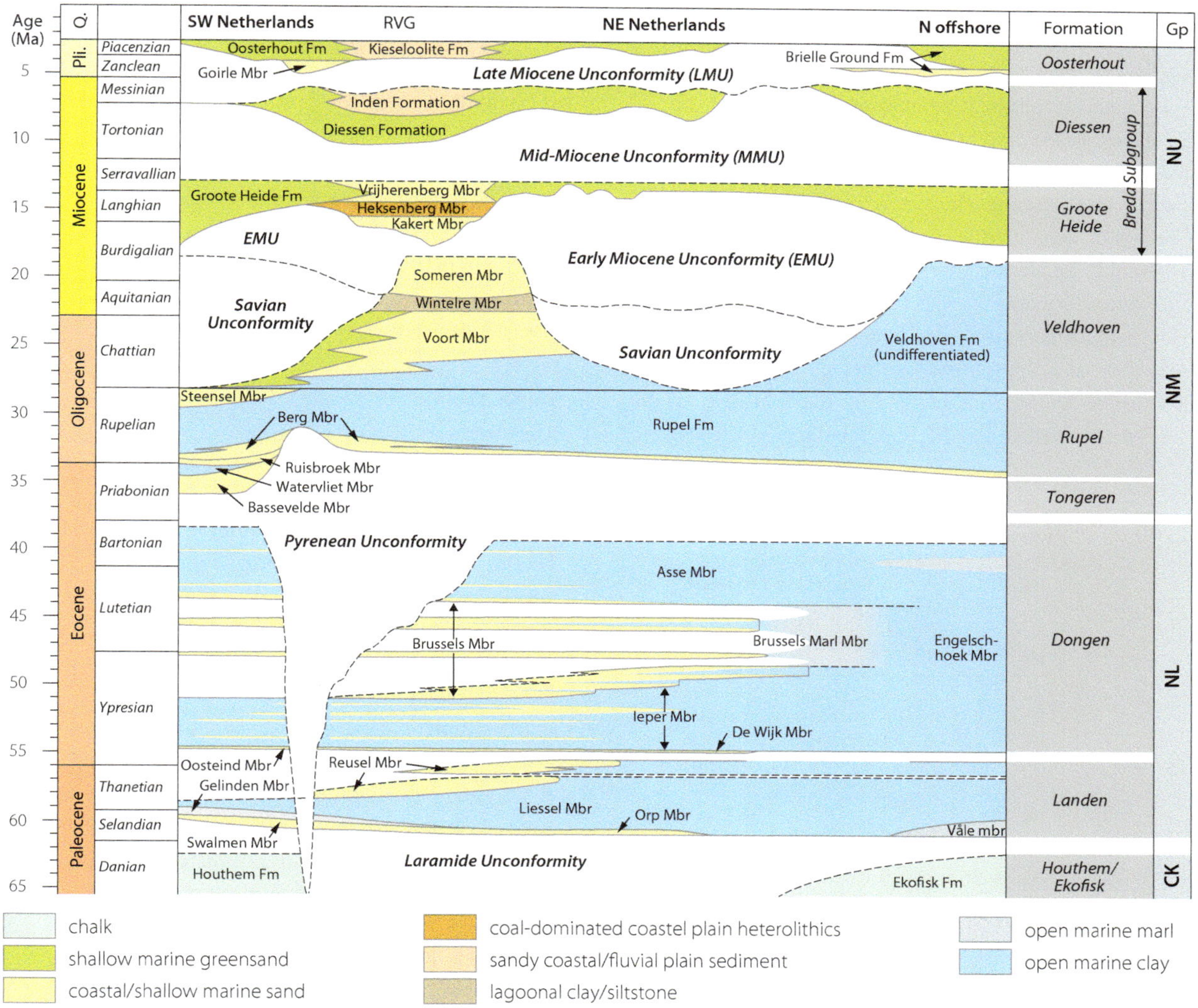

Figure 9.7. Stratigraphic scheme (Wheeler diagram) of the Lower (NL), Middle (NM) and Upper (NU) North Sea groups in the Netherlands (with exception of the Quaternary interval). The scheme corresponds to seismic lines shown in Fig. 9.6. Hiatuses are linked to major unconformities as discusses in text. The figure is partly based on Van Adrichem Boogaert & Kouwe (1997) and De Mulder et al. (2003). Stratigraphic codes are conform the Dutch Stratigraphic Nomenclator (www.dinoloket.nl). CK = Chalk Group; RVG = Roer Valley Graben.

wards and in turn are overlain by sandy clay with fine-grained sandy intercalations (Reussel Member). Locally, in the north of Limburg, variegated clay and minor sand occur. Further north (more distally), the formation consists exclusively of grey to greenish-grey clays with local marl intercalations (Liessel Member), particularly in the basal part. Initially, the overall depositional setting was shallow-marine and later changed to open-marine conditions. The maximum water depth may have reached several hundreds of metres in the central parts of the basin. An exception is the Swalmen Member at the base of the Landen Formation that is locally present in the south of the Netherlands and possibly in the Central Netherlands Basin (Deckers, 2015) and which was deposited in a lagoonal, brackish to non-marine environment. The upper part of the formation consists of mid-late Paleocene regressive sands up to 300 m thick, probably related to NW-SE oriented prograding deltaic lobes, which locally led to brackish conditions.

Recently the Våle member, as lateral equivalent of the Våle Formation in Denmark and Norway and the Maureen Formation in the UK, was proposed to represent the basal late Danian-Selandian interval of the Landen Formation (Deckers & Munsterman, 2019; Fig. 9.8). In an area that spans over 100 km in the northernmost offshore Netherlands, marly claystones to marlstones form a transitional interval up to over 30 metres thick, sandwiched between the calcareous coccolithic mudstones of the Ekofisk Formation and the claystones of the Liessel Member (Landen Formation). The latest Danian and Seelandian successions in the North Sea Basin contain significant amounts of reworked Cretaceous fossils, due to the erosional effects of the Laramide phase (Clemmensen & Thomsen, 2005). At the southern margin of the SNSB, abundant reworked fossils are recorded in the Gelinden Member (Vandenberghe et al., 1998). The newly named Våle member is the lateral, deep marine equivalent of the shallow marine Orp and Gelinden members (Fig. 9.8). The Landen Formation unconformably overlies the chalks of the Ekofisk and Houthem formations or older units. The lower boundary is expressed as a sharp break in lithology and log signature. The upper boundary is marked by the transition to the sand-prone Oosteind or De Wijk members of the Dongen Formation, which overlie the clays of the Landen Formation with a sharp contact. In the southern Netherlands, the boundary is less evident because sands of the Reussel Member are intercalated with sands of the Oosteind Member (Dongen Formation) and a distinction can only be made on biostratigraphical grounds. Particularly southeast of the Tilburg-Den Bosch area, the Dongen Formation is absent below the Pyrenean Unconformity. Here, the Landen Formation is unconformably covered by sandy deposits of the Tongeren Formation or the Rupel Formation.

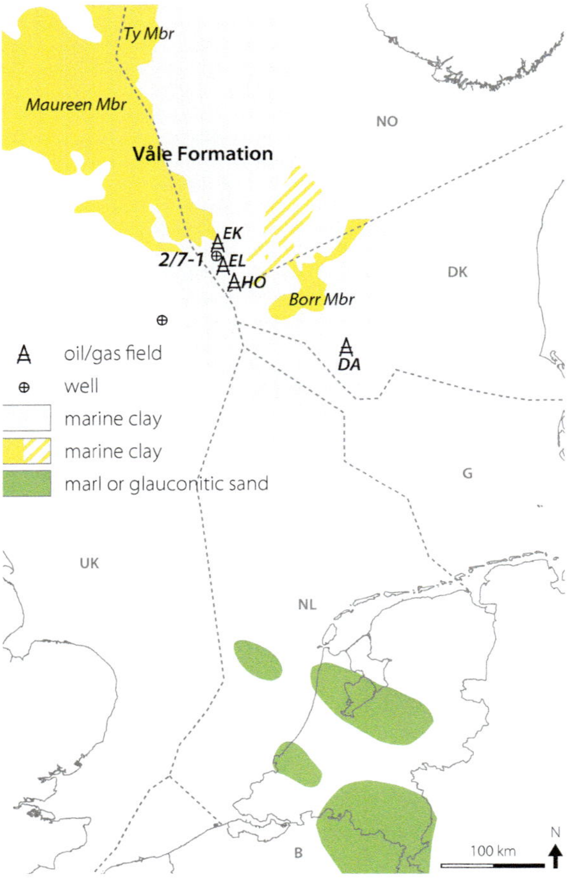

Figure 9.8. The North Sea Basin with the middle Paleocene distal Våle Formation and the more proximal Orp and Gelinden members. In the Dutch northern offshore, the marine clay (grey color) is proposed to represent the Våle member within the Landen Formation. The location of the sandy members of the Våle Formation is indicated in yellow and based on Brunstad et al. (2013). The geographic extents of the middle Paleocene sandy units in the Netherlands units were taken from Deckers & Van der Voet (2018).. The location of several of the major hydrocarbon fields in the Chalk Group (EK= Ekofisk, EL=Eldfisk, Ho=Hod, DA= Dan) and reference well 2/7-1 are indicated.

Dongen Formation

The lowermost part of this formation is represented by the De Wijk Member, which consists of tuffaceous clays and silts that grade into fine-grained sands in more proximal positions of the middle and southern parts of the Netherlands. The depositional setting is inner to outer neritic with a transgressive basal part of glauconitic fine sands (Oosteind Member). The Ieper Member comprises dark-grey, green and brown, slightly calcareous clays, with intercalated glauconitic sands. Distally these sands grade into marly sediments. Apart from near-shore and inner-neritic conditions also estuarine conditions are interpreted for the Lutetian Brussels Sand Member Above the Brussels Sand, dark greenish-grey and blue-grey clays of the Asse Member indicate a return to full marine conditions. In a more proximal setting, the Maldegem Formation, the Belgian lateral equivalent of the Asse Member, includes sandy intercalations. In distal settings the Brussels Sand Member is absent and a continuous sequence of marine clays exists (Engelschhoek Member). The age of the Dongen Formation is early-middle Eocene and its thickness is up to 850 m.

Generally, the lowermost boundary is characterized by a sharp transition from the clay of the Liessel Member to the sandy or tuffaceous base of the Dongen Formation. The latter appears to be synchronous with the PETM, during which significant flooding occurred through thermal expansion of sea-water, and which can be recognized by the acme of the dinoflagellate cyst *Apectodinium* (Sluijs et al., 2008). Outside the major depocentres, the PETM is manifested as a thin (1-2 m thick) but distinct clay layer that rests on top of the Landen Formation, but in the major depocentres, the lithological signature of the PETM flooding is less prominent due to a combined increase in accommodation space and sediment supply. Also, in the southern Netherlands, where the underlying Landen Formation has a regressive sandy top, the boundary is less clear. In the easternmost parts of the provinces of Gelderland and Overijssel, the formation unconformably overlies Mesozoic deposits. In areas where the succession is virtually complete, the upper boundary of the formation is marked by an usually clear transition from its upper argillaceous part (Asse Member) to the unconformably overlying sands of the Berg Member of the Rupel Formation. Around the eroded Kijkduin High and its north-western extension into the North Sea, an unconformable contact with the overlying formations exists. Locally the Asse Member may be absent and calcareous sands of the Brussels Sand Member are directly overlain by slightly calcareous to non-calcareous sands of the Berg Member (Rupel Formation; Fig. 9.7). Due to erosion during the Pyrenean phase, multiple lithostratigraphic units may be missing. Northward increase of erosion towards the Roer Valley Graben (Deckers et al., 2015) is shown by progressive removal of the Asse and Brussels Sand members and eventually erosion down into Cretaceous successions (Fig. 9.6). Similar erosion can be seen on the southern edge of West Netherlands Basin. In those areas where the Berg Member is absent, the Boom Member rests directly on the clays of the Asse Member (Dongen Formation) and in these cases, determination of the boundary on lithological grounds may be difficult. However, on wire-line logs the Boom Member shows a somewhat higher gamma-ray response compared to clays of the Lower North Sea Group.

Phase III – Ups and downs of the middle Cenozoic (late Eocene-Middle Miocene, 38-12 Ma)

Regional tectonics and paleogeography

Phase III is bounded by the Pyrenean and MMU tectonic phases. The Pyrenean inversion affected the West and Central Netherlands basins, the southern part of the Broad Fourteens Basin (Van Wijhe, 1987) and the Roer Valley Graben (Deckers, 2015) and was restricted to the southern North Sea (Michon et al., 2003). The Pyrenean inversion caused broad basin uplift. Most of the inverted parts of the West and Central Netherlands basins were subjected to erosion during the late Oligocene Savian inversion as well. Uplift and erosion associated with the Pyrenean inversion is in the order of 400 m, whereas the effects of the Savian inversion are approximately 100 m. Broad uplift of the West Netherlands Basin occurred at the Mid-Netherlands Fault Zone, forming the Kijkduin High in the late Eocene. The Savian inversion, which is very conspicuous in the West Netherlands Basin (De Jager, 2007; Fig. 9.6), barely influenced the Broad Fourteens Basin and had no effect in the Roer Valley Graben (Munsterman & Deckers, 2022). The strong inversion of the Mesozoic West and Central Netherlands basins during the Pyrenean and Savian phases leaves a preserved Cenozoic succession of 500 m maximally, i.e. far less compared to the 1000 m in the northern part of the non-inverted Broad Fourteens Basin (Knox et al., 2010).

The Roer Valley Rift System (RVRS) developed from the late Oligocene onward and consists of several structural elements (Geluk et al., 1994). To the northwest, the rift system steps over to the West Netherlands Basin. At a regional scale, the Roer Valley Rift System is delineated by NW-SE and NNW-SSE faults that represent reactivated Variscan structures (e.g. Ziegler, 1990). Blocks of intermediate subsidence or uplift flank the central Roer Valley Graben on both sides. In the southwest, these areas are the Eastern and Western Campine Blocks, while the Venlo, Peel and Köln Blocks are recognized in the northeast (Geluk et al., 1994). The Campine Block is separated from the subsiding Roer Valley Graben by the Feldbiss Fault System (Dusar et al., 2001). The deepest part of the Venlo-Peel Block is often referred to as the Venlo Graben (Van den Berg, 1994). The Peel Block has been uplifted ~1000 m along the NW-SE oriented Peel Boundary Fault. The Tegelen Fault separates the Peel Block from the Venlo Block (Van Adrichem Boogaert & Kouwe, 1997). The Viersen Fault is the principal displacement zone that separates the Venlo Block from the Krefeld Block to the northeast. Before rifting started, the Roer Valley Rift System area subsided uniformly until it was interrupted by middle-late Eocene inversion movements and erosion related to the Pyrenean tectonic phase (Letsch & Sissingh, 1983). In the Chattian, rifting became evident in the southern Netherlands (Demyttenaere, 1989; Geluk et al., 1994) and resulted in differential subsidence of the various blocks in the Roer Valley Rift System (Michon et al., 2003; Munsterman & Brinkhuis, 2004). The rifting was accompanied by sinistral strike-slip faulting that led to a southeastward translation of the Roer Valley Graben relative to the Peel Block (Geluk et al., 1994). A strong eustatic sea-level drop apparently occurred at the Rupelian-Chattian transition (e.g. Hardenbol et al., 1998; Van Simaeys et al., 2004) and led to substantial erosion of the top sets of the latest Rupelian highstand system tracts.

Śliwinska et al. (2014) showed that the base of the Chattian is synchronous with the Oxygen index-2a cooling event around 28.3 Ma (Vis et al., 2016). Hence the recorded unconformity at the base of the Chattian deposits in the North Sea Basin is not related to a glacio-eustatic sea-level fall linked to the Oi-2b cooling event, because calcareous nannofossils indicate a younger (early Chattian) age for this cooling (Van Adrichem Boogaert & Kouwe, 1993; Clausen et al., 2012; Śliwinska et al., 2014). Following the base-Chattian tectonic event, a progressive, probably eustatic, sea-level rise led to widespread marine transgression. Warm-water faunas with affinities to those of both the Atlantic and Tethys oceans (Janssen, 1979) indicate renewed connections with the marine domain to the west and south. Connection with the Tethys Ocean via the Upper Rhine Graben can be ruled out, as paralic and non-marine environments prevailed in the graben throughout Chattian times (Sissingh, 2006). However, a connection with the eastern Paratethys via the eastward

Moravian Seaway seems possible, as sediments of Chattian age in southwest Poland consist of marine glauconitic sands (Knox et al., 2010). The Roer Valley Graben and the Lower Rhine Graben developed as nearshore depocentres during the Chattian (Figs 9.7, 9.9). In the northern, western and eastern parts of the Netherlands, the marine Chattian (Voort Member) is believed to be largely absent, either due to non-deposition, synsedimentary uplift or later erosion (Zagwijn, 1989; Fig. 9.7). In large parts of the western Campine area, most of the (upper) Chattian and Aquitanian strata have been removed during both the Savian and EMU phases and sometimes only a single gravel bed remains at the base of the Miocene (Louwye, 2005; Munsterman & Deckers, 2020). This hiatus is known from large parts of the North Sea Basin and is related to the Savian tectonic phase (Utescher et al., 2000; Wong et al., 2001; Verbeek et al., 2002; Munsterman & Brinkhuis, 2004; Knox et al., 2010; Rasmussen et al., 2010; Eidvin et al., 2014; Dybkjær et al., 2020). Around the Oligocene/Miocene boundary a glacio-eustatic sea-level fall coupled with the Savian phase (e.g. Beddow et al., 2018; Miller et al., 2020; Steinthorsdottir et al., 2021b) was most likely responsible for development of restricted marine conditions in the strongest subsiding areas. Non-deposition or erosion occurred in the surrounding highs such that isolated (sub)basins were created. The younger EMU is mainly seen as a regional onlap/baselap surface and represents a small hiatus that covers the mid-Burdigalian (Munsterman et al., 2019). Outside the main depocentres, the EMU is often combined with the Savian Unconformity and the hiatus spans the Chattian-Early Miocene period. In the Achterhoek, for instance, the Miocene sequence directly overlies the Rupel Formation. During the subsequent Burdigalian-Langhian sea-level rise, the area of marine sedimentation increased compared to the Chattian. Marginal to shallow marine successions of the Groote Heide Formation were deposited over most of the Netherlands except the southwesternmost parts. During the Neogene, progressive uplift of the Ardennes and the Rhenish and Bohemian massifs resulted in increased clastic influx into the North Sea Basin.

Regional climate

Global climate in this period changed from a warm, 'greenhouse' world with very little ice towards an 'icehouse' world characterized by significant ephemeral Antarctic ice-sheets and sea-level fluctuations (Zachos et al., 2008; Westerhold et al., 2020). In the early Oligocene, the first substantial Cenozoic ice build-up on Antarctica caused shallow marine and lowland areas to experience notable orbitally-controlled glacioeustatic sea-level oscillations. These led to distinct and laterally persistent metre-scale alternations of shallow marine silt- and claystones in the Rupelian deposits at the southern margin of the SNSB (Abels et al., 2006). The early Oligocene temperature decline is reflected in the paleobotanical record of western Europe by a ~2.5°C cooling and increased seasonal contrast (Utescher et al., 2015). Chattian pollen-based data and $\delta^{18}O$ values of fossil bivalve shells and shark teeth show further evidence for cooling in both the marine and terrestrial environments (Walliser et al., 2016), as well as some evidence for associated precipitation decline (Utescher et al., 2015). Towards the Miocene, climate experienced a renewed warming phase.

At the end of the late Eocene-Middle Miocene phase III, a prominent period of global warmth named the Miocene Climatic Optimum (MCO) occurred. This period lasted from 17 to 14.5 Ma. The MCO world became 7-8°C warmer than present day with atmospheric CO_2 concentrations of ~500-600 parts per million (Steinthorsdottir et al., 2021a,b), possibly even more. Most regional and local climate information is based on fossil assemblages but few recent studies use microbial cell membrane lipids to reconstruct surface water temperatures. Such 'lipid paleothermometer' data, for instance, have been used in reconstructions from the west coast of Ireland and suggest sea surface temperatures of over 26°C during the MCO (Sangiorgi et al., 2021). Records from the Danish North Sea (Herbert et al., 2020) also suggest that Miocene sea surface temperatures were 11-20°C higher than modern mean annual temperatures at ~55°N. Terrestrial climate reconstructions from marginal marine deposits in the SE Netherlands show equally high temperatures (Donders et al., 2009). A MCO record from Denmark based on plant fossil data suggests that maximum mean annual temperatures varied mostly between 17-20°C and that there was little or no winter frost (Larsson et al., 2011). These temperatures are much higher than at present but remain significantly lower than the lipid-based estimates, which only show some indication for a bias to the summer season (Sangiorgi et al., 2021). Precipitation was generally high, exceeding 1200 mm/year (Larsson et al., 2011). All available records show that long-term temperature decline started at about 14 Ma and that, within the age uncertainties, several cooling phases were associated with increases in Antarctic ice build-up and global-scale reorganization of carbon cycling and ocean circulation. Despite such cooling events, the stratigraphic records in the southern Netherlands (Deckers & Munsterman, 2020) and Denmark (Rasmussen, 2004) indicate renewed transgression of the margins of the North Sea Basin after the MCO in the Serravallian.

Depositional systems, landscape and ecosystem evolution

During the Priabonian-Rupelian the preceding rich tropical marine faunas were entirely replaced by lower diverse communities. From the Chattian onwards, an increasingly rich and endemic marine North Sea biota evolved that culminated in the very rich 'paratropical' middle Miocene faunas. Subtropical lowland swamp ecosystems started to develop in the Lower Rhine Graben to the east of Limburg (Huhn et al., 1997; Utescher et al., 2021). Important sites covering this phase in the Netherlands are isolated occurrences of the Goudberg Member around Valkenburg and Rupel Formation outcrops near Meerssen (Limburg) and Winterswijk (Herngreen et al., 2005). The middle Miocene is particularly well known from outcrops near Winterswijk. Important regional information on the Priabonian-Rupelian comes from extensive outcrops in northern Belgium. Mine shafts around Krefeld (Germany) have disclosed important marine Chattian successions. Miocene terrestrial and coastal deposits are found in the lignite mines of Nordrhein-Westfalen, while extensive fossiliferous marine localities are known from the Antwerp region and from isolated outcrops and boreholes in northern Germany and Denmark (Schwarzhans, 2010; Larsson et al., 2011; Utescher et al., 2015).

The Priabonian-Rupelian transition in the Southern North Sea Basin is marked by a series of Milankovitch related transgressions and regressions culminating in widespread outer-shelf marine settings of the Boom Clay covering the entire Netherlands. Initially, an intermediate diverse para-tropical biota (Marquet, 2016) reminiscent of the earliest Oligocene faunas developed. It is found in the Belgian Grimmertingen Sands (a lateral equivalent of the Dutch Tongeren Formation), which are preserved in some of the mine shafts in Limburg. In overlying deposits, diversity levels drop (Marquet & Herman, 2012) and successions contain low diverse near-shore communities with intermittent tropical cerithiid snail communities dominating the brackish coastal zone (Fairon-Demaret et al., 2003; Marquet, 2016). During the Rupelian, much of the Southern North Sea Basin drowned and extensive deeper marine clays of the Rupel Formation (Boom Member) were deposited that contain mollusc, fish and microfossil assemblages representing neritic environments. Faunas were not particularly diverse (Marquet & Herman, 2012). During this interval, most of the Netherlands became covered by the Rupel Sea. A major terrestrial faunal transition known as the 'grand coupure' extinction phase, took place around 33.5 Ma when endemic European mammal species became extinct and new groups immigrated from Asia. It is preserved in the famous Hoogbutsel locality in the Belgian province of Brabant (Smith & Smith, 2003). Forested coastal wetlands existed at that time around the North Sea with a wealth of fish, reptilians and amphibians, birds and mammals.

The Rupelian-Chattian boundary marks a major shift in ecosystems and biota. In general, it is represented by a regional unconformity that is attributed to renewed tectonic activity coupled with a glacially induced sea-level fall (Oi-2a event of Miller et al., 2008). The cooler conditions of the Chattian are reflected in more drought-tolerant taxa and an overall decline in evergreen broadleaved elements. From the Chattian to the Middle Miocene, extensive coastal lowland swamp ecosystems developed at times in the Lower Rhine Graben in Nordrhein-Westfalen, and some isolated *Taxodium* macrofossils are known from isolated outcrops in the Sittard area. Following the base Chattian tectonic event, a progressive – probably eustatic – sea-level rise led to a marine transgression. Warm-water faunas with affinities to those of both the Atlantic and Paratethys (Janssen, 1979) indicate renewed connections with marine waters to the west and south. A connection of the SNSB to the eastern Paratethys via the Moravian Seaway seems possible. Widespread shelfs developed with very rich shallow marine algal, seagrass and soft bottom communities that included a rich endemic North Sea marine mollusc fauna (Welle, 1997; 1998) as well as Sirenians and archaic whales (Diedrich, 2012). The rich subtropical marine assemblages with their very diverse endemic invertebrate and fish lineages continued through the Early Miocene into the Middle Miocene when they diversified even further (Janssen, 1984; Schwarzhans, 2010). During the Burdigalian stage additional tropical marine groups (molluscs, fish) that originated from more southerly Atlantic regions such as the Aquitaine Basin in France appeared in the North Sea Basin, suggesting a possible direct marine connection to the south (Janssen, 2001; Schwarzhans, 2010). Rare Paratethyan fish species in these Middle Miocene faunas imply limited connection with the Middle Miocene (Badenian) Paratethyan seas, yet no evidence for large scale faunal exchange exists. The Middle Miocene marine biota were very species rich: in the famous locality of Miste near Winterswijk, over 500 species of molluscs are found (Janssen, 1984).

Later in the Burdigalian, a coastal lowland system with freshwater marshes and swamps (Ville Formation) prograded from Germany into the Netherlands and preserves the remains of rich assemblages of subtropical and warm temperate plants and animals, including straight-tusked elephants and tapirs. The subsidence of extensive wetlands resulted in thick lignite deposits in the Lower Rhine Graben that extend into the southeast of the Netherlands. Near Kerkrade in South Limburg, Middle Miocene browncoals of the Ville Formation contain a rich palynoflora (Manten, 1958), but most information in the Netherlands comes from boreholes. The peat-forming taxa included wetland forest trees that are presently limited to SE Asia and the SE USA. The lowland swamp environments were dominated by cypress (*Glyptostrobus* and *Taxodium* species) and umbrella pine (*Sciadopitys*, currently confined to

Japan) together with other extant conifer taxa and diverse angiosperm shrubs and trees such as ash and oaks (Figueiral et al., 1999; Utescher et al., 2017, 2021; Mosbrugger et al., 2022). From the German lignite mines extensive flora and fauna has been collected that show a lowland subtropical swamp biota with intermittent shallow marine species such as dolphins (Mörs, 2002). The lignite swamp forests were inhabited by diverse fish, amphibians and reptiles including aquatic and terrestrial crocodiles and turtles, as well as very diverse mammal communities that included a range of small mammals, commonly beavers, but also larger-sized groups such as primates, carnivores, deer, horses and elephants.

Lithostratigraphy

Phase III is delimited by the Pyrenean tectonic phase and the Middle Miocene Unconformity (MMU) and is represented by the entire Middle North Sea Group and the lower part of the Upper North Sea Group. This interval includes the Tongeren, Rupel, Veldhoven, and Groote Heide formations (Figs 9.1, 9.7). The thickness of this interval is minimally 700 m and decreases towards the northwest, i.e. the centre of the Southern North Sea Basin.

The Tongeren Formation

The new Dutch Stratigraphic Nomenclature Online (Dinoloket, 2021) integrated two earlier versions for the deep (Van Adrichem Boogaert & Kouwe, 1997) and shallow subsurface (Rijsdijk et al., 2005) of the Geological Survey of the Netherlands. The distribution of the Tongeren Formation was originally restricted to the southern part of the province of Limburg (SE Netherlands) but has been extended to the southern part of the province of Zeeland (SW Netherlands). The depositional setting of the Tongeren Formation is interpreted successively as a shallow-marine and lagoonal to coastal-plain. In the SE part of the Netherlands the formation is split into the Klimmen Member (clayey sands) followed by the Goudsberg Member (clays). In the southwestern part of the Netherlands a tripartition into the Bassevelde (grey sands), Watervliet (clays) and Ruisbroek (green-grey glauconitic sands) members is made. The Tongeren Formation unconformably overlies the Dongen Formation or older deposits. The upper boundary is sharp when overlain by fluvial sands and gravels or eolian sediments of the Upper North Sea Group, but gradual when overlain by transgressive sands and clays of the Rupel Formation. The thickness is up to 60 m and the age is Priabonian to earliest Rupelian. In the southwestern part of the Netherlands, the unit contains reworked middle Eocene sediments suggesting that the Tongeren Formation developed during the Pyrenean tectonic phase, south of the main inversion axis of the Kijkduin High. Also in Belgium just to the southwest of the Netherlands, Priabonian sands were deposited during the Pyrenean inversion as supported by changes in clay mineral associations (Saeys et al., 2004).

Rupel Formation

The Rupel Formation has recently been updated and revised (see www.dinoloket.nl). The lithology is described as heavy, dark brown-grey silty clays, rich in pyrite and relatively glauconite-poor. Towards both base and top, the clays grade into silts and rather abruptly into sands along the southern basin margin (Vis et al., 2016). The basal sands are transgressive (Berg Member). In Flanders, the Berg Member passes northwestwards into the Belsele-Waas Member, which is silty. This may explain the local occurrence of the Berg Member at the base of the Rupel Formation, where flint pebbles and phosphorite nodules commonly occur. In the SE part of the Netherlands these sands are followed by laminated clays (Kleine Spouwen Member) and white sands with clay layers (Waterval Member). The clayey Boom Member contains septarian carbonate concretions often documented from Belgian quarries (De Craen et al., 1999). Middle to outer neritic marine conditions are interpreted for the clays (water depths over 200 m). At the top of the formation an alternation of sands and silty clays grade upward into sands (Steensel Member). The latter unit is of shallow to marginal marine origin. The thickness of the formation is up to 125 m and the age is early Oligocene (Rupelian). The Rupel Formation unconformably overlies the Tongeren Formation, the Dongen Formation or older sediments. In those areas where the sandy development at the base is missing, the clays of the Rupel Formation rest directly on the Asse or Engelsche Hoek members, which complicates definition of the boundary. However, on wire-line logs, the Rupel clays show a somewhat higher gamma-ray response compared to the clays of the Lower North Sea Group. Biostratigraphic data may assist in establishing this lower boundary. In the central and eastern Netherlands, and in parts of the northeastern offshore the Rupel Formation is conformably (or mildly unconformably) overlain by the sands and silty clays of the Veldhoven Formation. In the southern Netherlands the latter has a typical, thin clay interval at its base. The upper limit is usually sharp in those parts of the Netherlands where the Rupel Formation is partly eroded (e.g. on the Middle Netherlands High between the cities of Nijmegen and Leiden) and is overlain by the Groote Heide Formation.

Veldhoven Formation

The Veldhoven Formation overlies either the Steensel Member or Boom Member of the Rupel Formation with a minor unconformity. The base of the formation is marked by a distinct clay layer. It is overlain by the Groote Heide

Formation from which it is separated by the Early Miocene Unconformity. Outside the Roer Valley Graben and the Venlo Block, the EMU represents a distinct and sometimes angular unconformity that encompasses the late Burdigalian (Munsterman et al., 2019). The age of the Veldhoven Formation is Chattian-Early Miocene. The thickness may reach over 400 m. In the Roer Valley Graben the formation predominantly comprises light grey to green grey, very fine-grained, micaceous sand (Voort Member) followed by (dark) green-grey sandy clay (Wintelre Member) that coarsens upward into sands (Someren Member). In general, the depositional setting was shallow-marine. In contrast to earlier interpretations, the increase in clay of the Wintelre Member (formerly Veldhoven Clay) is now attributed to a shallowing towards restricted marine conditions instead of an increase in sea-level. Munsterman & Deckers (2022) claim that these clays can be distinguished from clays intercalated in the Voort Member by a higher concentration of shells, as is recorded in the lithological description of the Groote Heide borehole. Dinoflagellate cyst assemblages are here dominated by a single genus indicating extreme depositional environments, with salinities that deviate from normal marine conditions, most probably due to periodic minor connection to the Atlantic Ocean. The glacioeustatic sea-level fall of 50-70 m around the Oligocene/Miocene boundary most likely limited sea coverage to the strongest subsiding areas, where deposition of the Wintelre Member took place. Non-deposition or erosion occurred in the surrounding highs, hence creating an isolated (sub)basin (e.g. Miller et al., 2020; Steinthorsdottir et al., 2021b). The most recent depositional interpretation places the Wintelre Member around the Oligocene/Miocene boundary as a restricted marine facies related to this sea level fall in concert with the Savian inversion tectonics (Dybkjær & Rasmussen, 2007; Dybkjær et al., 2012; Munsterman & Deckers, 2022). This is in accordance with the eustacy based model of Hager et al. (1998) and disproves the sedimentological 'dynamic' infill model put forward by Schäfer et al. (2005) and Wong et al. (2007). The North Sea Basin probably became a (semi-)enclosed basin during the latest Oligocene-earliest Miocene, with a narrow connection to the North Atlantic Ocean between the Shetland Isles and Norway (Rasmussen et al., 2008; Knox et al., 2010).

Groote Heide Formation

The recognition of an intraformational hiatus within the former Breda Formation (known as the Mid-Miocene Unconformity) led to a decision to split the formation. As a result, the Breda Formation was promoted to Subgroup, of which the part between the EMU and MMU was defined as the Groote Heide Formation and the part between the MMU and LMU as the Diessen Formation (Munsterman et al., 2019; Figs 9.1, 9.7). The age of the Groote Heide Formation is late Burdigalian-Serravallian and the thickness is up to 200 m. The formation comprises greyish to blackish green silts and very fine to medium sands that can be (very) glauconitic, calcareous, and locally micaceous or organic. Sandy to moderately silty clays are also present. The presence of glauconite is very characteristic of the formation, except for the deposits that developed in coastal settings. In the province of Zeeland and in the western part of North Brabant, moderately fine- to moderately coarse-grained glauconite-rich sands occur. On the Peel Horst, i.e. the northern rift shoulder of the Roer Valley Graben, the deposits are generally fine-grained and include mica flakes. Within the Roer Valley Graben and Venlo Slenk the formation also contains organic material. In the eastern part of the Netherlands the formation has sand-clay alternations with varying glauconite content and goethite and phosphorite concretions. In the offshore centre of the North Sea Basin, the Middle Miocene sequences are often very condensed into glauconitic clays (Rasmussen & Dybkjær, 2014).

The Groote Heide Formation was deposited in a shallow-marine environment, except at its easternmost edges where (near)coastal sedimentary facies occur. The glauconitic content of this coastal facies is markedly lower than in the remainder of the formation. In the southeastern part of the Netherlands shallow marine fine sands of the Kakert Member form the base of the formation. In the northern part of the Roer Valley Graben and in the Venlo Graben, the gravelly Elsloo Bed (Zagwijn & Van Staalduinen, 1975; Kuyl, 1980) marks the transition of the Someren Member to the Kakert Member and represents a winnowed interval corresponding to the EMU. In the Achterhoek, the base of the Miocene is represented by the Miste Bed that unconformably overlies the Rupel Formation and contains reworked phosphoritic pebbles and nodules and contains Chattian and Early-to-middle Middle Miocene reworked taxa (Janssen, 1984). In the Limburg area, the Kakert Member is succeeded by pure quartz sands of the Heksenberg Member (locally known as 'silver sands') that were deposited in a coastal setting. At the lower boundary of the member, sandy coastal prisms with thin lignite layers are locally present that correlate with the Morken Seam of the Ville Formation (Fig. 9.9). The Belgian Genk Sands are considered equivalent to the Dutch Heksenberg sands (Vandenberghe et al., 1998) and both are laterally equivalent to regionally widespread lignite seams of the Frimmersdorf Seam (Van Adrichem Boogaert & Kouwe, 1997). Following the Serravallian transgression, the Heksenberg Member passes up into brownish-yellow to green-grey fine sands and silts that belong to the Vrijherenberg Member, laterally equivalent to the Neurath Sand in Germany (Prinz et al., 2017). In the

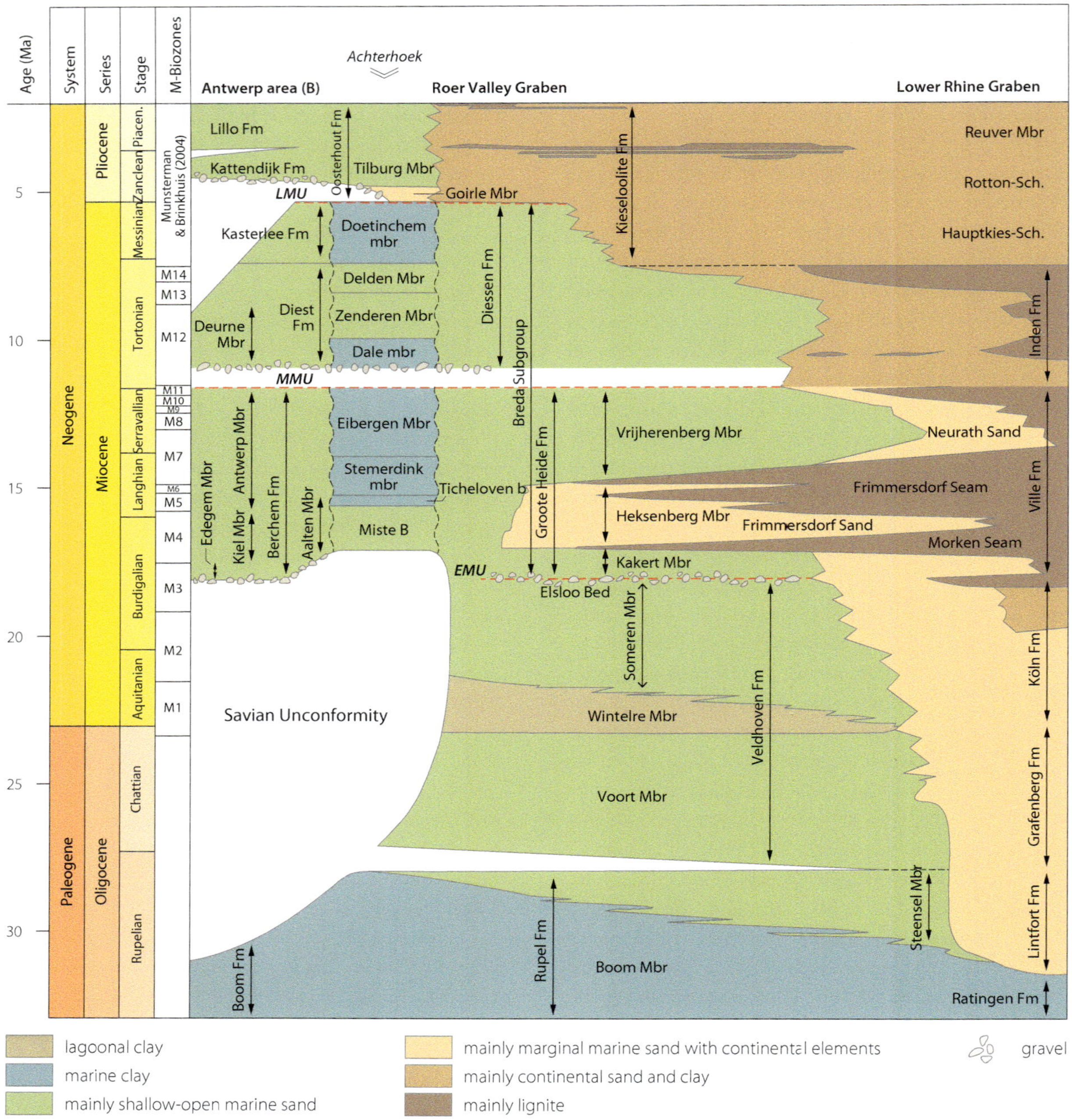

Figure 9.9. *Stratigraphic scheme (Wheeler diagram) of the Oligocene-Early Pliocene of the Antwerp area (Belgium), the Achterhoek, the Roer Valley Graben (the Netherlands) and its continuation into the Lower Rhine Graben (Germany). The lithostratigraphy is modified from Van Adrichem Boogaert & Kouwe (1997), Deckers & Louwye (2019), Munsterman et al. (2019), Vandenberghe & Louwye (2020), and Everaert et al. (2020). Everaert et al. (2020), and Munsterman et al. (2024). EMU = Early Miocene Unconformity; MMU = Mid-Miocene Unconformity; LMU = Late Miocene Unconformity.*

central part of these sands a maximum flooding surface is interpreted, which is rich in shells and glauconite, and is expressed by a gamma-ray peak (Deckers & Munsterman, 2020).

The distribution of sedimentary facies in combination with foreset-bottomset geometries observed on seismic data suggest a delta plain (east) to delta front (west) setting. An unconformity (EMU) is usually present at the base of the unit. On the relatively high margins of the Roer Valley Rift System, such as the Peel-Maasbommel High and the Campine area, the formation truncates the Veldhoven and Rupel formations (mainly Boom Member) or older deposits. The gamma-ray log shows lower values at the base of the formation, due to the low concentration of glauconite. In the Peel region and in part of the Roer Valley Graben the lower boundary of the formation is difficult to establish due to the gradual transition of the slightly glauconitic beds of the Veldhoven Formation (Someren Member) into the greensands of the Groote Heide Formation (Deckers & Munsterman, 2020). Above the basal interval, an upward decrease in glauconite content is compensated by an upward increase in clay content within the Groote Heide Formation, which masks the upward reduction in gamma-ray values that is observed elsewhere. The MMU marks the top of the Groote Heide Formation and is characterized in all wells by a very distinctive positive gamma-ray log shift immediately followed by a sharp decrease, presumably reflecting reworking of glauconite. This gamma-ray spike suggests that the MMU represents a late-Middle Miocene hiatus after which sedimentation resumed in the early Tortonian. In Belgium, basal gravels are deposited at this time. In the centre of the Roer Valley Graben the upper boundary of the Groote Heide Formation is associated with a shift in colour of the glauconitic clays from dark green to greenish-grey. In more proximal settings, the MMU is associated with a lithological change from strong sandy loam to silty fine-grained sands. The formation is present in most of the Dutch subsurface. In the eastern part of the Netherlands (Achterhoek), the Groote Heide Formation is divided into the Aalten (sands passing into clays), Stemerdink (silty layered brownish-grey clay) and Eibergen (stiff blackish clays) members (Munsterman et al., 2024). However, it is absent in small areas in the extreme eastern, southeastern and southwestern parts of the country, on the Kijkduin High (Fig. 9.6a) and in the northwestern offshore.

Phase IV – Late Neogene cooling (Late Miocene-Early Pleistocene, 12-2.5 Ma)

Regional tectonics and paleogeography

The onset of phase IV is constrained by a distinct change in the tectonic regime of the North Sea area. Regional tectonic movements led to an increased tilting of the Fennoscandian Platform and accelerated uplift of Britain (Japsen et al., 2002; Anell et al., 2010). This resulted in divergence of the Baltic river system, which started to feed a large-scale delta stretching from the present-day Baltic region into the Dutch part of the North Sea Basin. The entire fluvio-deltaic system is referred to as the Eridanos (Overeem et al., 2001) or Southern North Sea delta (SNS). It supplied large amounts of freshwater and sediment into the shelf sea during the Neogene and Early Pleistocene (Overeem et al., 2001). In the Late Miocene the delta was located along the southern Norwegian and Danish shore. Western progradation continued with a lateral shift of depocentre towards the Dutch northern offshore during the Pliocene-Pleistocene (De Bruin et al., 2017). Whereas during the Miocene sediments laid down in the Danish North Sea sector were predominantly sourced from Fennoscandia, in the Pliocene this was overtaken by Central European sources (Rasmussen & Dybkjær, 2014). The southwestward Eridanos extention during the Gelasian is clearly discernible from seismic data that show progradational sigmoidal and oblique shelf-prism clinoforms that downlap onto the MMU (Ten Veen et al., 2018; Busschers et al., 2025, this volume). Hence, in this phase the eastern boundary of the North Sea Basin shifted westwards and the depositional basin considerably narrowed. The SNSB became separated from a shallow basin to the southeast known as the East German-Polish Basin, which closely represents the present-day configuration. Locally, fast-growing salt diapirs surrounded by rim synclines likely contributed to failure of delta slopes and mass transport deposition and an irregular seafloor topography (Benvenuti et al., 2012; Thöle et al., 2016).

The Mid-Miocene Unconformity forms the lower (seismo)stratigraphic boundary of the Eridanos delta. With the exception of the Danish offshore section, the central axis of the Roer Valley Graben and the Venlo Graben in the Netherlands, it represents a hiatus covering a substantial part of the Serravallian. In Denmark, a 3 m thick condensed section composed of deeper marine hemipelagic clays spans the several million years that coincide with the MMU elsewhere (Rasmussen & Dybkjær, 2014). In the Venlo Graben the MMU time span is represented by a thick coarsening upward sequence dominated by a coastal to restricted-marine facies of the latest Serravallian-earliest Tortonian age (Munsterman et al., 2019).

In the Serravallian, uplift of the London-Brabant Massif resulted in non-deposition and formation of erosional channels with basal gravels and rounded flints in the southernmost North Sea Basin (Demyttenaere, 1989; Wouters & Vandenberghe, 1994; Van Vliet-Lanoë et al., 2002; Louwye & Laga, 2008). Massive reworking of Middle Miocene Antwerp Sands in the Late Miocene Diest Formation is recorded in the Campine area (Vandenberghe et al., 2014). Uplift along the Weald-Artois Axis (Fig. 9.2) led to closure of the Channel Seaway (Knox et al., 2010) that linked the SNSB to the eastern Atlantic, further isolating the basin during the Late Miocene and ending the few indications for direct southern connections in the late Burdigalian and Langhian. Compared to the Groote Heide Formation, the Diessen Formation is much thicker on the southern margin of the North Sea Basin. Deckers & Louwye (2020) suggested that this might be explained by the start of Late Miocene lithospheric folding in that area that continued into Pliocene-Pleistocene times (Van Wees & Cloetingh, 1996). In addition to folding, upward mantle flows resulted in uplift of the Rhenish Massif during the Late Miocene (Garcia-Castellanos et al., 2000; Van Balen et al., 2000). After the Middle to Late Miocene transition, this uplift instigated the progradation of the proto-Rhine fluvial-deltaic Inden Formation (Schäfer et al., 2005) and the overlying latest Tortonian Kieseloolite Formation (Munsterman et al., 2019) from Germany into the southeastern Netherlands. The Inden Formation passes westwards into the shallow marine Diessen Formation. On seismic data, the latter is characterized by multiple stacked clinoforms that indicate a westward shelf-delta progradation into the Roer Valley Graben/Lower Rhine Graben during the Late Miocene and Early Pliocene. Similar progradational geometries are observed in the time-equivalent Diest Formation in the Campine Basin (Vandenberghe et al., 2014). These observations suggest that the proto-Rhine delta front became the most important contributor of sediment here. Additionally, the Eridanos system was a source for background sedimentation here, albeit with a rather small contribution (Verhaegen, 2019). The input of southerly-sourced sediments (Meuse-Ardennes and proto-Rhine) strongly increased from the Early Miocene to the Late Miocene and Pliocene (Adriaens, 2015; Verhaegen et al., 2019). During the Late Miocene, besides the Roer Valley Graben, the Zuiderzee Low (Fig. 9.4; Duin et al., 2006) developed as a second depocentre. A direct connection to the Alpine forelands was established in the Late Pliocene and is traceable in unstable mineral associations with a proto-Rhine origin (Kemna & Westerhoff, 2007).

In many seismic profiles through the Netherlands it is evident that topset beds of clinoforms of the Diessen Formation are missing and are directly overlain by either parallel or downlapping (inclined and concave-downward) strata of the Oosterhout Formation (Fig. 9.6c). This angular unconformity is defined as the LMU (Munsterman et al., 2019) and is inferred to represent a hiatus due to considerable erosion.

Regional climate

Following the MCO, strengthening of the North Atlantic Current caused divergent climate developments on either side of the Atlantic Ocean since the Late Miocene. Long-term Miocene cooling and humidity decrease were clearly stronger in NW Europe (Donders et al., 2009; Sangiorgi et al., 2021) than at similar latitudes along the Atlantic coast of North America (Kotthoff et al., 2014) and the east of Eurasia (Utescher et al., 2015). Climate reconstructions based on plant macrofossils from German basins place the most pronounced cooling between 12 and 10 Ma. The diversity decrease of subtropical plants most likely reflects decline of winter temperature minima, due to their sensitivity of frost. The plant macrofossil records further indicate a decline in Neogene precipitation in response to long-term cooling that seems to lag behind the major temperature transitions (Mosbrugger et al., 2005; Utescher et al., 2011). In the Netherlands, the only quantitative reconstruction of Miocene terrestrial climate is based on analysis of the Groote Heide borehole in the Venlo Graben. Here, integrated marine and terrestrial microfossil and lipid-based paleothermometry analyses record a clear cooling phase between 12 and 11.2 Ma followed by a relatively stable period and a gradual cooling trend between 9.5 and 7.4 Ma. The quantitative reconstructions suggest a 3-5°C mean annual temperature decline during each of these cooling phases, associated with impoverishment of the terrestrial flora (Donders et al., 2009).

The Pliocene Climatic Optimum (3.26-3.0 Ma) is characterized by global temperatures that were 2-4°C higher compared to the present (Haywood et al., 2013). Locally, lipid paleothermometry indicates that the SNSB was 1-2°C warmer on land, with marine anomalies of up to 10°C (Dearing Crampton-Flood et al., 2018, 2020). This optimum was followed by 'precursor' Northern Hemisphere glaciations that resulted in variable sedimentary facies in the SNSB. The tectonically-driven isolation of the Arctic Ocean and intensification of the North Atlantic Current, increased moisture distribution northward (Haug & Tiedemann, 1998). This led to an enhanced fresh water budget in the Arctic Ocean. The additional fresh water is thought to have facilitated the formation of sea ice, causing a positive ice-albedo feedback. In turn, Arctic waters flowing back into the North Atlantic could have become less cold and salty, hence interrupting and slowing the Ocean Conveyor belt as a global heat pump to North Atlantic regions (Haug et al., 2004). The resulting globally cooler conditions and initiation of Northern Hemisphere ice-sheet growth during the Late Pliocene (Bintanja & Wal, 2008) developed into periodic glaciations covering large parts of the Northern Hemisphere during the Middle and Late Pleistocene (De Schepper et al., 2014). The M2 glacial at 3.3 Ma is globally the most pronounced early glaciation and is associated with a regional unconformity in the shallow marine deposits in the onshore western Netherlands (Dearing Crampton-Flood et al., 2020). Coupled records of such marine and terrestrial environments show contemporaneous floral extinction and climate cooling with marine ingressions of arctic water masses during this and subsequent glacial periods (Kuhlmann et al., 2006a; Donders et al., 2018, 2009).

Depositional systems, landscape and ecosystem evolution

During the second half of the Neogene, the progressive disappearance of (sub)tropical and endemic taxa drove declining marine and terrestrial biodiversity levels, while the occurrence of Pacific immigrants in the Pliocene shaped the modern marine faunas. Macrofossils from terrestrial biota have been dredged from the Liessel and Langenboom localities and small outcrops in the Venlo area in Limburg, and are extensively exposed in lignite mines of adjacent Nordrhein-Westfalen, such as the well-known Hambach locality. Most microfossil data rely on mixed samples from airlift boreholes and there are very few continuously cored intervals.

The presence of palm fossils in the Lower Rhine Graben shows that frost conditions were still rare during the Miocene (Van der Burgh, 1984). Pollen records from the Miocene seams include many so-called 'form' taxa for which the exact botanical affinity is not known. Most of these taxa disappeared during the Miocene cooling events, suggesting that they have tropical and subtropical affinities (Donders et al., 2009). In comparison, Pliocene lignite seams contain an essentially modern flora but, compared to today, have higher quantities of angiosperm wetland taxa such as poplar (*Populus* spp.), tupelo (*Nyssa*) and sweetgum (*Liquidambar*), characteristic of present-day hardwood swamps in the southeastern United States (Zagwijn, 1960; Westerhoff et al., 2020). Upland vegetation from outside the wetland areas is much less known but analyses from shallow marine deposits show that conifers were surprisingly common regionally (Larsson et al., 2011; Dearing Crampton-Flood et al., 2020). Many conifer taxa like pine (*Pinus*), spruce (*Picea*) and fir (*Abies*) but also hemlock (*Tsuga*) – which presently is confined to North America – are today characteristic of cooler, mountainous regions. Leached, acidic soils that formed during extended warm periods might explain the abundance of conifers that are more tolerant to such conditions. Other 'cool' vegetation elements like birch (*Betula*) and heath (*Ericaceae*) are indeed rare in the Neogene and increased in abundance only in the Early Pleistocene (Donders et al., 2018).

Marine fossils from the Late Miocene-Pliocene time interval are known from a number of outcrops and extraction sites in the Netherlands, including Langenboom and Liessel in North Brabant (Peters, 2009; Peters & Peters, 2013) and from boreholes throughout the country (e.g. Sliggers & Leeuwen, 1987). Rich fossil faunas are known from beaches and dredged samples in the Westerschelde area (Moerdijk et al., 2010; Post & Reumer, 2016). As with the flora, faunas during the Late Miocene-Early Pliocene were warm-temperate in character. Most of the truly tropical taxa that flourished in the Middle Miocene – such as conids and cowries – had disappeared. Endemic mollusc and fish lineages continued to dominate the marine biota and there was an increase of transatlantic fish species during the Tortonian, indicating oceanic dispersal (Schwarzhans, 2010), but overall diversity was considerably lower than during the Middle Miocene. Rich marine mammal and fish (including shark) faunas have been recovered from Upper Miocene sediments at Liessel and the Westerschelde (Marx et al., 2016, 2019; Post et al., 2018; Bisconti et al., 2019, 2020; Peters et al., 2021; Bosselaers & Munsterman, 2022) The occurrence of very large whales indicates that shallow to open marine conditions extended over the southern and southeastern Netherlands during the Tortonian. Dinoflagellate cyst assemblages indicate nutrient-rich water, associated with prograding fluvial deltaic river systems (Inden and Kieseloolite formations) from the southeast. During Early Pliocene times shallow to open marine conditions persisted well towards the east as is shown by cetacean and marine bird assemblages (including albatross) recovered from the Langenboom locality (Wijnker et al., 2008). The arrival of Pacific immigrant species such as clams (*Macoma*), whelks (*Neptunea*) and mussels (*Mytilus*) started to shape the modern faunas. Proposed connections through the Channel region during Late Miocene and Pliocene (Van Vliet-Lanoë et al., 2002) are not in agreement with the largely endemic and northern faunas represented in the North Sea Basin, which at that time probably was an embayment of the North Atlantic. During the Late Pliocene a progressive loss of species around cooling events, most notably thermophilous and endemic taxa, drove a strong decline in marine diversity, analogous to changes on land. Invertebrate faunas became dominated by species of Pacific origin.

Lithostratigraphy

Sediments deposited during tectonic phase IV include the Diessen, Inden, Kieseloolite and Oosterhout formations. Figure 9.9 shows a summary of these post-MMU sediments in the Netherlands.

Diessen Formation

The Diessen Formation is placed in between the MMU and LMU and is Tortonian to Messinian in age. With the LMU as upper boundary, locally the upper part of the former Breda Formation was assigned to the newly defined Goirle Member of the Oosterhout Formation. The lithology of the Diessen Formation changes from sandy loam at its base into silty fine-grained sand upwards and has a moderate glauconite content. This coarsening upward trend is associated with colours shifting from green-grey to grey as the concentration of mica increases upward. Traces of molluscs and sea urchin spines are present. Coincident with a shift to lower grain size there is a sharp increase in epidote and amphibole relative to tourmaline and metamorphic minerals. Upwards, the heavy mineral composition again becomes increasingly rich in tourmaline and metamorphic minerals. The succession in the deepest part of the Roer Valley Graben (well Sint Michielsgestel-1) consists of grey-greenish clays with glauconite. The sediments were deposited in the prodeltaic environment of a fluvio-deltaic system and reach a thickness up to 500 m. On seismic data the base is detectible as widespread downlap surface

Dinoflagellate cysts

2

Dinoflagellates are free-living, single-celled organisms that inhabit aquatic environments. The name is derived from the Greek word *dinos* (rotation) and the Latin word *flagellum* (small whip). Their food strategy may be autotrophic or heterotrophic, i.e. indicating that their energy is obtained by photosynthesis or from organic material, respectively. Mixture of endosymbiosis and parasitism also occurs. Dinoflagellates are particularly common in marine nearshore and neritic settings and populations are most varied in shallow to open marine conditions. The organism is characterized by the presence of a swimming (motile) cell with two flagella: a transverse flagellum that encircles the cell and is ribbonlike, and a longitudinal flagellum that is cylindrical and trails the cell posteriorly (left figure). Combined actions of the flagella move motile cells through water in a corkscrew manner. Most living dinoflagellate cells range in diameter from 5 to 100 μm. The life cycle of many dinoflagellates includes at least one cyst stage, during which the organism is dormant/resting and life processes are suspended. Some produce long-term resting cysts (hypnozygotes) of which the wall is composed of a fossilizable organic material (dinosporin), calcium carbonate or, very rarely, silica (Fensome & Munsterman, 2020). These cysts can be well-conserved in sediments and occur frequent in the rock record. Cyst fossils go back to the Middle Triassic and can be used to study the evolution of dinoflagellates. The intricate Jurassic to Cenozoic organic-walled cyst taxonomy allows a refined marine biostratigraphy (right figure). Fossil cyst analysis is an excellent application tool in marine geology for age assessment, reconstruction of paleoenvironment and paleoclimatic evolution.

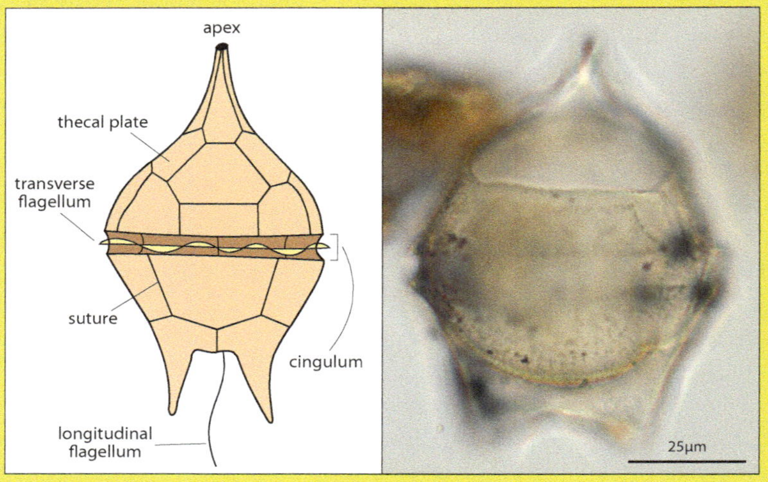

Left: General morphology of a motile dinoflagellate cell showing the position of the two flagella. Right: Light microscope photograph of a* Deflandrea phosphoritica *cyst, which in the Dutch North Sea Basin is an age-diagnostic taxon for the top of the late Oligocene.

of the overlying clinoforms, which corresponds to a gamma-ray peak in log data. This suggests that the MMU can be interpreted as unconformity (or disconformity). The arrangement of clinoforms on seismic profiles in the Roer Valley Graben is indicative for multiple phases and directions of progradation and/or avulsion of a large fluvio-deltaic system. In the Goirle borehole, marine dinoflagellate cysts become very rare to absent at the top of the Diessen Formation, indicating an upward shallowing to marginal marine and possibly even subaerial conditions. Based on foraminiferal analysis in well Sint Michielsgestel-1, infra- to epineritic conditions (water depths between less than 100 m at the base and ca. 10-30 m at the top of this unit) were interpreted (Doppert et al., 1975). The LMU spans a period equal to at least all of the missing topset beds of the Diessen Formation clinoforms. Despite the fact that a hiatus cannot be identified on the basis of palynological criteria, across the unconformity all Miocene dinoflagellate cyst marker taxa disappeared from the assemblages. In the Oud-Turnhout borehole (Belgium) a hiatus is recorded as the Lower Pliocene Kattendijk Formation is missing and the Upper Pliocene (Piacenzian) directly overlies the Upper Miocene Diest Formation (Louwye & De Schepper, 2010). In the southeastern part of the Netherlands the formation interfingers with the mainly continental deposits of the Inden and Kieseloolite formations. These transitions are marked by erosional scours and shifts to more coarse-grained deposits. In the northeastern part of the country the Diessen Formation is in lateral contact with the nearshore to continental beds of the Peize Formation. In the eastern part of the Netherlands (Achterhoek) it is divided into four units (Munsterman et al., 2024): the informal Dale member (dark grey-green clayey silts and silty clays), followed by the Zenderen Member (green glauconitic fine sands), the Delden Member (sharp increase in coarse-grained sands) and, finally the informal Doetinchem member (loam, sandy, silty clays).

Inden Formation

The Inden Formation is described as dark grey medium to very coarse sand, locally gravelly, with minimal calcareous content and dark grey to brown clay, which is slightly sandy and silty. In the Rur Block (Germany) interbedded, locally clayey lignite seams compose the Tortonian Upper Seam that reaches up to 30 m in thickness (Schafer et al., 2005). The depositional environment is interpreted as coastal lowland (proto-Rhine), including swamp (lignite), meandering channel (sand), residual channel (clay), crevasse (sand) and floodplain (clay) sub-environments (Boersma et al., 1981; Abraham, 1994; Schäfer et al., 2004, 2005). The lower boundary is defined at a sharp contact with finer glauconitic shallow-marine sand and clay of the Diessen Formation. The upper boundary commonly shows a thick clay layer and a sharp contact with overlying coarser fluvial sand and gravel (Waubach Member, Kieseloolite Formation). The age is early late Miocene (Tortonian) and the thickness reaches up to 140 m. The distribution is limited to the eastern part of the Roer Valley Graben.

Kieseloolite Formation

The depositional conditions were fluvial (proto-Rhine and -Meuse) (Boenigk, 1978; Gliese & Hager, 1978; Westerhoff, 2009), including floodplain and swamp to tidally influenced coastal plain with back-barrier areas and beach ridges. Sediments were sourced from the Ardennes and Lower Rhine Graben. The lithology comprises whitish to brownish grey, very fine to very coarse sand (quarzitic) and gravel (granules to boulders) together with thick intercalated clay beds and associated peat/lignite layers. The formation is split into the Brunssum Member (stiff greyish brown to dark brown clay, slightly sandy to very silty, with intercalated layers of fine to medium sand and lignite) and the Waubach Member (white to light grey medium to very coarse sand, very gravelly, partially lithified in coarsening-upward sequences with poorly developed clay layers). The lower boundary is a sharp contact with either fine, glauconitic shallow-marine sand of the Diessen Formation or dark grey medium to very coarse and gravelly sand of the Inden Formation. Laterally, it is transitonal to glauconitic shallow-marine sand with marine molluscs of the Oosterhout Formation. The upper boundary is generally a transition into micaceous fluvial sands (Waalre Formation), but may be sharp where this fluvial unit overlies the Reuver Member clay beds. There is a conformable, gradual transition into very fine to fine eolian and local river sand of the Stramproy Formation, which has thinner clay beds and a locally sharp contact with coarse gravelly fluvial sand of the Kreftenheye and Beegden formations. The age is Late Miocene to Pliocene and the thickness is up to 200 m. Distribution is limited to the southeastern part of the Netherlands (the province of Limburg and eastern part of Brabant).

Oosterhout Formation

The recently redefined base of the Oosterhout Formation is now located at an angular unconformity defined as the LMU (Munsterman et al., 2019). In some parts of the Netherlands the LMU is directly overlain by parallel-layered sands of the Goirle Member of the Oosterhout Formation. In these instances, the unconformity is marked by a sharp decrease in gamma-ray readings. The Goirle sands were previously assigned to the former Breda Formation. Sands in the upper levels of the Diessen Formation are relatively enriched in glauconite and the sudden lower gamma-ray values are due to waning concentrations of glauconite in the Goirle Member. This change can be seen in the colour of the sands, which changes from green into light grey and/or white. These lightly coloured sands are followed by dark green-grey, glauconitic, slightly silty sands including clays, defined as Tilburg Member, and interpreted as a marine flooding surface. In the Antwerp area and by extension probably also in the neighbouring parts of Dutch Zeeland, Deckers & Louwye (2020) described a >10 m deep gully incision at the base of the Belgian Kattendijk Formation which is the lateral equivalent to the Tilburg Member.

In general, the lithology of the Oosterhout Formation consists of light grey-green very fine to medium sand rich in shells and can locally be clayey and glauconitic. At the top, dark grey to greyish-brown silty or sandy clays occur. At this level several shell banks (crags) are present. Locally, yellow to reddish-brown medium grained fossiliferous sands are encountered that are partly cemented by iron(hydr)oxides. The depositional setting is shallow-marine and includes estuarine, deltaic, beach and nearshore facies. Landward, to the southeast the formation is transitional to and interfingers with coastal-plain and fluvial deposits of the Kieseloolite Formation. The exact transition of the Oosterhout Formation into the overlying Maassluis Formation is a subject of ongoing study that focuses on depositional systems and biotic evolution at the Pliocene-Pleistocene transition.

The Phase IV sedimentary succession of the Dutch northern offshore onlaps onto the MMU (Fig. 9.6b). Above the latter a toe-of-slope facies of the early Eridanos (mid-Late Miocene; Kuhlmann et al., 2006a) and a relatively thick Pliocene-Pleistocene late Eridanos shelf-edge delta sequence (De Bruin et al., 2017; Busschers et al., 2025, this volume) corresponds to the Diessen and Oosterhout formations, respectively. In this area, the Brielle Ground and Westkapelle Ground formations (Rijsdijk et al., 2005) which have a slightly more distal depositional setting (shelf to slope) compared to the Oosterhout Formation, denote the Pliocene part of the westward prograding Eridanos interval. Offshore the western part of the Netherlands, the Brielle Ground Formation also shows influences from the Rhine/Meuse system. Further to the east, the formation has a sharp contact with coarser fluvial and coastal sand and gravel (Waalre and Peize formations) or passes gradually into eolian or local river sand (Stramproy Formation). The thickness is up to 150 m and the age is Pliocene.

Acknowledgements

Many collaborators and colleagues at various Geological Surveys (Utrecht, Brussels, Krefeld, Copenhagen, Nottingham, Hannover and Oslo), research institutes, companies, consultants, and universities (including MSc and PhD students) at many places in northwest Europe are recognized for inspiring discussions on the Paleogene and Neogene interval of the Dutch part of the North Sea Basin in the past decades. Jef Deckers (VITO, Belgium) and Torsten Utescher (University of Bonn, Germany) are highly acknowledged for their constructive reviews, improving the manuscript. Klaas Post is thanked for his contribution and review of the textbox "Neogene fossil marine mammals from the Netherlands". Native speaker Harry Doust is thanked for the adjustments made to the English text and for tinkering with structure and syntax. Phil Gibbard and John Lewin are thanked for sharing their paleogeographical maps.

Digital map data

Spatial data of figures in this chapter for use in geographical information systems can be downloaded here: https://doi.org/10.5117/aup.28163597.

References

Á Horni , J., Hopper, J.R., Blischke, A., Geisler, W.H., Stewart, M., McDermott, K. & Árting, U., 2017. Regional distribution of volcanism within the North Atlantic Igneous Province. Geological Society, London, Special Publications 447(1): 105-125. DOI: 10.1144/SP447.18

Abdelmalak, M.M., Meyer, R., Planke, S., Faleide, J.I., Gernigon, L., Frieling, J. & Myklebust, R., 2016. Pre-breakup magmatism on the Vøring Margin: Insight from new sub-basalt imaging and results from Ocean Drilling Program Hole 642E. Tectonophysics 675: 258-274. DOI: 10.1016/j.tecto.2016.02.037

Abels, H.A., Van Simaeys, S., Hilgen, F.J., De Man, E. & Vandenberghe, N., 2006. Obliquity-dominated glacio-eustatic sea level change in the early Oligocene; evidence from the shallow marine siliciclastic Rupelian stratotype (Boom Formation, Belgium). Terra Nova 19(1): 65-73. DOI: 10.1111/j.1365-3121.2006.00716.x

Abraham, M., 1994. Untersuchungen zur sedimentologischen Entwicklung der fluviatilen Deckschichten (Miozän, Pliozän) der Rheinischen Braunkohle. Bonner Geowissenschaftliche Schriften 15: 1-27.

Adriaens, R., 2015. Neogene and Quaternary clay minerals in the southern North Sea. PhD thesis KU Leuven (Belgium), 47: 273 pp.

Anell, I., Thybo, H. & Stratford, W., 2010. Relating Cenozoic North Sea sediments to topography in southern Norway: The interplay between tectonics and climate. Earth and Planetary Science Letters 300: 19-32. DOI: 10.1016/j.epsl.2010.09.009

Bacon, C.D., Silvestro, D., Jaramillo, C., Smith, B.T., Chakrabarty, P. & Antonelli, A., 2015. Biological evidence supports an early and complex emergence of the Isthmus of Panama. Proceedings of the National Academy of Sciences 112(19): 6110-6115. DOI: 10.1073/pnas.1423853112

Beddow, H.M., Liebrand, D., Wilson, D.S., Hilgen, F.J., Sluijs, A., Wade, B.S. & Lourens, L.J., 2018. Astronomical tunings of the Oligocene-Miocene transition from Pacific Ocean Site U1334 and implications for the carbon cycle. Climate of the Past 14(3): 255-270. DOI: 10.5194/cp-14-255-2018

Benvenuti, A., Kombrink, H., Ten Veen, J.H., Munsterman, D.K., Bardi, F. & Benvenuti, M., 2012. Late Cenozoic shelf delta development and Mass Transport Deposits in the Dutch offshore area – results of 3D seismic interpretation. Netherlands Journal of Geosciences 91(4): 591-608. DOI: 10.1017/S0016774600000391

Berger, J.-P., Reichenbacher, B., Becker, D., Grimm, M., Grimm, K., Picot, L., Storni, A., Pirkenseer, C., Derer, C. & Schaefer, A., 2005. Paleogeography of the Upper Rhine Graben (URG) and the Swiss Molasse Basin (SMB) from Eocene to Pliocene. International Journal of Earth Sciences 94(4): 697-710. DOI: 10.1007/s00531-005-0475-2

Bijl, P.K., Houben, A.J., Schouten, S., Bohaty, S.M., Sluijs, A., Reichart, G.J. & Brinkhuis, H., 2010. Transient Middle Eocene atmospheric CO2 and temperature variations. Science 330(6005): 819-821. DOI: 10.1126/science.1193654

Bijl, P.K., Schouten, S., Sluijs, A., Reichart, G.J., Zachos, J.C. & Brinkhuis, H., 2009. Early Palaeogene temperature evolution of the southwest Pacific Ocean. Nature 461(7265): 776-779. DOI: 10.1038/nature08399

Bintanja, R. & Wal, R.S.W., 2008. North American ice-sheet dynamics and the onset of 100,000-year glacial cycles. Nature 454(7206): 869-872. DOI: 10.1038/nature07158

Bisconti, M., Munsterman, D.K., Fraaije, R.H.B., Bosselaers, M.E.J. & Post, K., 2020. A new species of rorqual whale (Cetacea, Mysticeti, Balaenopteridae) from the Late Miocene of the Southern North Sea Basin and the role of the North Atlantic in the paleobiogeography of Archaebalaenoptera. PeerJ 8(8315). DOI: 10.7717/peerj.8315

Bisconti, M., Munsterman, D.K. & Post, K., 2019. A new balaenopterid whale from the late Miocene of the Southern North Sea Basin and the evolution of balaenopterid diversity (Cetacea, Mysticeti). PeerJ 7(6915). DOI: 10.7717/peerj.6915

Boenigk, W., 1978. Gliederung der altquartären Ablagerungen in der Niederrheinischen Bucht. Fortschritte in der Geologie von Rheinland und Westfalen 28: 135-212.

Boersma, J.R., Van Gelder, A., De Groot, T.A.M. & Puigdefabregas, C., 1981. Formen fluviatiler Sedimentation in neogenen und jungeren Ablagerungen im Braunkohlentagebau Frechen (Niederrheinischen Bucht). Fortschritte in der Geologie von Rheinland und Westfalen 29: 275-307.

Bohaty, S.M. & Zachos, J.C., 2003. Significant Southern Ocean warming event in the late middle Eocene. Geology 31(11): 1017-1020. DOI: 10.1130/G19800.1

Bosselaers, M. & Munsterman, D.K., 2022. The discovery of a Balaenomorpha (Persufflatius renefraaijeni n. gen., n. sp.) from the upper Miocene of the Netherlands sheds new light on the cranial anatomy of archaic rorqual relatives. Geodiversitas 44(30): 933-973. DOI: 10.5252/geodiversitas2022v44a30

Bourgeois, O., Ford, M., Diraison, M., Veslud, C.L.C.D., Gerbault, M., Pik, R. & Bonnet, S., 2007. Separation of rifting and lithospheric folding signatures in the NW-Alpine foreland. International Journal of Earth Sciences 96(6): 1003-1031. DOI: 10.1007/s00531-007-0202-2

Bralower, T.J., Cosmidis, J., Heaney, P.J., Kump, L.R., Morgan, J.V., Harper, D.T. & Vajda, V., 2020. Origin of a global carbonate layer deposited in the aftermath of the Cretaceous-Paleogene boundary impact. Earth and Planetary Science Letters 548: 116476. DOI: 10.1016/j.epsl.2020.116476

Brunstad, H., Gradstein, F., Lie, J.E., Hammer, Ø., Munsterman, D.K., Ogg, G. & Hollerbach, M., 2013. Stratigraphic guide to the Rogaland group, Norwegian North Sea. Norwegian North Sea 46(2): 137-286.

Busschers, F.S., Cohen, K., Wesselingh, F.P., Schokker, J., Bakker, M., Van Balen, R. & Van Heteren, S., 2025. Quaternary. In: Ten Veen, J.H., Vis, G.-J., De Jager, J. & Wong, Th.E. (eds): Geology of the Netherlands, second edition. Amsterdam University Press (Amsterdam): 333-391. DOI: 10.5117/9789463728362_ch10

Cameron, T.D.J., Crosby, A., Balson, P.S., Jeffery, D.H., Lott, G.K., Bulat, J. & Harrison, D.H., 1992. United Kingdom offshore regional reports: the geology of the southern North Sea. HMSO (London) for the British Geological Survey: 152 pp.

Clausen, O.R., Śliwińska, K.K. & Gołędowski, B., 2012. Oligocene climate changes controlling forced regression in the eastern North Sea. Marine and Petroleum Geology 29: 1-14. DOI: 10.1016/j.marpetgeo.2011.10.002

Clemmensen, A. & Thomsen, E., 2005. Palaeoenvironmental changes across the Danian-Selandian boundary in the North Sea Basin. Palaeogeography, Palaeoclimatology, Palaeoecology 219(3-4): 351-394. DOI: 10.1016/j.palaeo.2005.01.005

Cloetingh, S.A.P.L., Ziegler, P.A., Beekman, F., Andriessen, P.A.M., Matenco, L., Bada, G. & Sokoutis, D., 2005. Lithospheric memory, state of stress and rheology: neotectonic controls on Europe's intraplate continental topography. Quaternary Science Reviews 24(3-4): 241-304. DOI: 10.1016/j.quascirev.2004.06.015

Cramwinckel, M.J., Huber, M., Kocken, I.J., Agnini, C., Bijl, P.K., Bohaty, S.M. & Sluijs, A., 2018. Synchronous tropical and polar temperature evolution in the Eocene. Nature 559(7714): 382-386. DOI: 10.1038/s41586-018-0272-2

De Bast, E. & Smith, T., 2016. The oldest Cenozoic mammal fauna of Europe: implication of the Hainin reference fauna for mammalian evolution and dispersals during the Paleocene. Journal of Systematic Palaeontology 15(9): 741-785. DOI: 10.1080/14772019.2016.1237582

De Bruin, G., Geel, K., Houben, S., Munsterman, D., Verweij, H., Smit, J., Janssen, N., Kerstholt-Boegehold, S. & Vandeweijer, V., 2017. MMU Unravelling the stratigraphic and structural development of the strata found underneath and above the Mid-Miocene unconformity. TNO (Utrecht), Report No. R10425: 104 pp.

De Craen, M., Swennen, R., Keppens, E., Macaulay, C.I. & Kiriakoulakis, K., 1999. Bacterially mediated formation of Carbonate concretions in the Oligocene Boom Clay of Northern Belgium. Journal of Sedimentary Research 69(5): 1098-1106. DOI: 10.2110/jsr.69.1098

De Jager, J., 2007. Geological development. In: Wong, Th.E., Batjes, D.A.J. & De Jager, J. (eds): Geology of the Netherlands. Royal Netherlands Academy of Arts and Sciences (KNAW, Amsterdam): 5-26.

De Mulder, E.F.J., Geluk, M.C., Ritsema, I., Westerhoff, W.E. & Wong, Th.E., 2003. De ondergrond van Nederland. Wolters-Noordhof (Groningen): 379 pp.

De Schepper, S., Gibbard, Ph.L., Salzmann, U. & Ehlers, J., 2014. A global synthesis of the marine and terrestrial evidence for glaciation during the Pliocene Epoch. Earth-Science Reviews 135: 83-102. DOI: 10.1016/j.earscirev.2014.04.003

Dearing Crampton-Flood, E., Noorbergen, L.J., Smits, D., Boschman, R.C., Donders, T.H., Munsterman, D.K., Ten Veen, J.H., Peterse, F., Lourens, L. & Sinninghe Damsté, J.S., 2020. A new age model for the Pliocene of the Southern North Sea Basin: a multi-proxy climate reconstruction. Climate of the Past 16: 523-541. DOI: 10.5194/cp-16-523-2020, 2020

Dearing Crampton-Flood, E., Peterse, P., Munsterman, D. & Sinninghe Damsté, J.S., 2018. Using tetraether lipids archived in North Sea Basin sediments to extract North Western European Pliocene continental air temperatures. Earth and Planetary Science Letters 490: 193-205. DOI: 10.1016/j.epsl.2018.03.030

Deckers, J., 2015. The Paleocene stratigraphic records in the Central Netherlands and close surrounding basins: Highlighting the different responses to a late Danian change in stress regime within the Central European Basin System. Tectonophysics 659: 102-108. DOI: 10.1016/j.tecto.2015.07.031

Deckers, J. & Louwye, S., 2020. Late Miocene increase in sediment accommodation rates in the Southern North Sea Basin. Geological Journal 55(1): 728-736. DOI: 10.1002/gj.3438

Deckers, J. & Louwye, S., 2019. A reinterpretation of the ages and depositional environments of the lower and middle Miocene stratigraphic records in a key area along the southern margin of the North Sea Basin. Geological Magazine 156(3): 525-532. DOI: 10.1017/S0016756817000991

Deckers, J. & Matthijs, J., 2017. Middle Paleocene uplift of the Brabant Massif from central Belgium up to the southeast coast of England. Geological Magazine 154(5): 1117-1126. DOI: 10.1017/S001675681600073X

Deckers, J. & Munsterman, D., 2019. Introducing the Våle Formation in the southern central North Sea Basin, offshore Netherlands. Marine and Petroleum Geology 101: 512-518. DOI: 10.1016/j.marpetgeo.2018.12.032

Deckers, J. & Munsterman, D., 2020. Middle Miocene depositional evolution of the central Roer Valley Rift System. Geological Journal 55(9). DOI: 10.1002/gj.3799

Deckers, J. & Van der Voet, E., 2018. A review on the structural styles of deformation during Late Cretaceous and Paleocene tectonic phases in the southern North Sea area. Journal of Geodynamics 115: 1-9. DOI: 10.1016/j.jog.2018.01.005

Deckers, J., Vandenberghe, N., Lanckacker, T. & de Koninck, R., 2016. The Pyrenean inversion phase in northern Belgium: an example of a relaxation inversion? International Journal of Earth Sciences 105(2): 583-593.

DeConto, R.M. & Pollard, D., 2003. Rapid Cenozoic glaciation of Antarctica induced by declining atmospheric CO2. Nature 421(6920): 245-249. DOI: 10.1038/nature01290

Demyttenaere, R., 1989. The Post-Paleozoic geological history of north-eastern Belgium. Mededelingen van de Koninklijke Academie Voor Wetenschappen, Letteren En Schone Kunsten van België 51(4): 51-81.

Dickens, G.R., O'Neil, J.R., Rea, D.K. & Owen, R.M., 1995. Dissociation of oceanic methane hydrate as a cause of the carbon isotope excursion at the end of the Paleocene. Paleoceanography 10(6): 965-971. DOI: 10.1029/95PA02087

Diedrich, C.G., 2012. Eocene (Lutetian) Shark-Rich Coastal Paleoenvironments of the Southern North Sea Basin in Europe: Biodiversity of the Marine Fürstenau Formation Including Early White and Megatooth Sharks. International Journal of Oceanography: 1-22. DOI: 10.1155/2012/565326

Donders, T.H., Van Helmond, N.A.G.M., Verreussel, R., Munsterman, D., Ten Veen, J., Speijer, R.P., Weijers, J.W.H., Sangiorgi, F., Peterse, F., Reichart, G., Sinninghe Damste, J.S., Lourens, L., Kuhlmann, G. & Brinkhuis, H., 2018. Land–sea coupling of early Pleistocene glacial cycles in the southern North Sea exhibit dominant Northern Hemisphere forcing. Climate of the Past 14(3): 287. DOI: 10.5194/cp-14-397-2018

Donders, T.H., Weijers, J.W.H., Munsterman, D.K., Kloosterboer-van Hoeve, M.L., Buckles, L.K., Pancost, R.D., Schouten, S., Sinninghe Damste, J.S. & Brinkhuis, H., 2009. Strong climate coupling of terrestrial and marine environments in the Miocene of northwest Europe. Earth and Planetary Science Letters 281(3-4): 215-225. DOI: 10.1016/j.epsl.2009.02.034

Doppert, J.W.C., Ruegg, G.H.J., Van Staalduinen, C.J., Zagwijn, W.H. & Zandstra, J.G., 1975. Formaties van het Kwartair en Boven Tertiair in Nederland. In: Zagwijn, W.H. & JVan Staalduinen, J.C. (eds): Toelichting bij de geologische overzichtskaarten van Nederland. Rijks Geologische Dienst (Haarlem) 1: 11-56.

Duin, E., Doornenbal, J.C., Rijkers, R.H.B., Verbeek, J.W. & Wong, Th.E., 2006. Subsurface structure of the Netherlands; results of recent onshore and offshore mapping. Netherlands Journal of Geosciences 85(4): 245-276. DOI: 10.1017/S0016774600023064

Dunkley Jones, T., Lunt, D.J., Schmidt, D.N., Ridgwell, A., Sluijs, A., Valdes, P.J. & Maslin, M., 2013. Climate model and proxy data constraints on ocean warming across the Paleocene–Eocene Thermal Maximum. Earth-Science Reviews 125: 123-145. DOI: 10.1016/j.earscirev.2013.07.004

Dusar, M., Rijpens, J., Sintubin, M. & Wouters, L., 2001. Plio-Pleistocene fault pattern of the Feldbiss fault system (southern border of the Roer Valley Graben, Belgium) based on high resolution reflection seismic data. Netherlands Journal of Geosciences 80(3/4): 79-94.

Dybkjær, K., King, C. & Sheldon, E., 2012. Identification and characterisation of the Oligocene–Miocene boundary (base Neogene) in the eastern North Sea Basin — based on dinocyst stratigraphy, micropalaeontology and δ13C-isotope data. Palaeogeography, Palaeoclimatology, Palaeoecology 363-364: 11-22. DOI: 10.1016/j.palaeo.2012.08.007

Dybkjær, K. & Rasmussen, E.S., 2007. Organic-walled dinoflagellate cyst stratigraphy in an expanded Oligocene-Miocene boundary section in the eastern North Sea Basin (the Frida-1 well, Denmark) and correlation from basinal to marginal areas. Journal of Micropalaeontology 26(1): 1-17. DOI:10.1144/jm.26.1.1

Dybkjær, K., Rasmussen, E.S., Eidvin, T., Grøsfjeld, K., Riis, F., Piasecki, S. & Śliwińska, K.K., 2020. A new stratigraphic framework for the Miocene – Lower Pliocene deposits offshore Scandinavia: a multiscale approach. Geological Journal 56(3): 1699-1725. DOI: 10.1002/gj.3982

Eidvin, T., Riis, F. & Rasmussen, E.S., 2014. Oligocene to Lower Pliocene deposits of the Norwegian continental shelf, Norwegian Sea, Svalbard, Denmark and their relation to the uplift of Fennoscandia: A synthesis. Marine and Petroleum Geology 56: 184-221. DOI: 10.1016/j.marpetgeo.2014.04.006

Everaert, S., Munsterman, D.K., De Schuiter, P.J., Bosselaers, M., Van Boeckel, J., Cleemput, G. & Bor, T.J., 2020. Stratigraphy and palaeontology of the lower Miocene Kiel Sand Member (Berchem Formation) in temporary exposures in Antwerp

(northern Belgium). Geologica Belgica 23(3-4): 167-198. DOI: 10.20341/gb.2020.025
Fairon-Demaret, M., Steurbaut, E., Damblon, F., Dupuis, C., Smith, T. & Gerrienne, P., 2003. The in situ Glyptostroboxylon forest of Hoegaarden (Belgium) at the Initial Eocene Thermal Maximum (55 Ma). Review of Palaeobotany and Palynology 126(1): 103-129. DOI: 10.1016/S0034-6667(03)00062-9
Fensome, R.A. & Munsterman, D.K., 2020. Dinoflagellates. In: Gradstein, F.M., Ogg, J.G., Schmitz, M.D. & Ogg. G.M. (eds): The Geological Time Scale 2020. Elsevier Science Publishing (Amsterdam) 1: 99-108. DOI: 10.1016/C2020-1-02369-3
Figueiral, I., Mosbrugger, V., Rowe, N.P., Ashra, A.R., Utescher, T. & Jones, T.P., 1999. The Miocene peat-forming vegetation of northwestern Germany: an analysis of wood remains and comparison with previous palynological interpretations. Review of Palaeobotany and Palynology 104(3-4): 239-266. DOI: 10.1016/S0034-6667(98)00059-1
Flecker, R., Krijgsman, W., Capella, W., de Castro Martíns, C., Dmitrieva, E., Mayser, J.P. & Yousfi, M.Z., 2015. Evolution of the Late Miocene Mediterranean–Atlantic gateways and their impact on regional and global environmental change. Earth-Science Reviews 150: 365-392. DOI: 10.1016/j.earscirev.2015.08.007
Gale, A.S. & Lovell, B., 2018. The Cretaceous–Paleogene unconformity in England: Uplift and erosion related to the Iceland mantle plume. Proceedings of the Geologists' Association 129(3): 421-435. DOI: 10.1016/j.pgeola.2017.04.002
Garcia-Castellanos, D., Cloetingh, S. & Van Balen, R., 2000. Modelling the Middle Pleistocene uplift in the Ardennes–Rhenish Massif: thermo-mechanical weakening under the Eifel? Global and Planetary Change 27(1-4): 39-52. DOI: 10.1016/S0921-8181(01)00058-3
Geluk, M.C., Duin, E.J.T., Dusar, M., Rijkers, R.H.B., Van den Berg, M.W. & Van Rooijen, P., 1994. Stratigraphy and tectonics of the Roer Valley Graben. Geologie En Mijnbouw 73(2-4): 129-141.
Gibbard, P.L. & Lewin, J., 2003. The history of major rivers of southern Britain during the Tertiary. Journal of the Geological Society 160(6): 829-845. DOI: 10.1144/0016-764902-137
Gibbard, P.L. & Lewin, J., 2016. Filling the North Sea Basin: Cenozoic sediment sources and river styles. Geologica Belgica 19(3/4): 1-17. DOI: 10.20341/gb.2015.017
Gibbard, P.L., Head, M.J. & Walker, M.J.C., 2010. Formal ratification of the Quaternary System/Period and the Pleistocene Series/Epoch with a base at 2.58 Ma. Journal of Quaternary Science 25(2): 96-102. DOI: 10.1002/jqs.1338
Gliese, J. & Hager, H., 1978. On brown coal resources in the Lower Rhine Embayment (West Germany). Geologie En Mijnbouw 57(4): 517-525.
Grein, M., Utescher, T., Wilde, V. & Roth-Nebelsick, A., 2011. Reconstruction of the middle Eocene climate of Messel using palaeobotanical data. Neues Jahrbuch Fur Geologie Und Palaontologie-Abhandlungen 260(3): 305-318. DOI: 10.1127/0077-7749/2011/0139
Hager, H., Vandenberghe, N., Van den Bosch, M., Abraham, M., Von der Hocht, F., Rescher, K., Laga, P., Nickel, E., Verstraelen, A., Leroi, S. & Van Leeuwen, R.J.W., 1998. The geometry of the Rupelian and Chattian depositional bodies in the Lower Rhine district and its border area: implications for Oligocene lithostratigraphy. Bulletin of the Geological Society of Denmark 45: 53-62.
Hardenbol, J., Thierry, J., Farley, M.B., Jacquin, T., De Graciansky, P.-C. & Vail, P.R., 1998. Mesozoic and Cenozoic sequence chronostratigraphic framework of European basins. In: Graciansky, P.-C., Hardenbol, J., Jacquin, T. & Vail, P.R. (eds): Mesozoic and Cenozoic Sequence Stratigraphy of European Basins. (Tulsa) 1: 3-13.
Haug, G.H. & Tiedemann, R., 1998. Effect of the formation of the Isthmus of Panama on Atlantic Ocean thermohaline circulation. Nature 393(6686): 673-676.
Haug, G.H., Tiedemann, R. & Keigwin, L.D., 2004. How the Isthmus of Panama put ice in the Arctic. Oceanus 42: 94-97.
Haywood, A.M., Hill, D.J., Dolan, A.M., Otto-Bliesner, B.L., Bragg, F., Chan, W.L. & Zhang, Z., 2013. Large-scale features of Pliocene climate: results from the Pliocene Model Intercomparison Project. Climate of the Past 9(1): 191-209. DOI: 10.5194/cp-9-191-2013
Head, M.J. & Gibbard, P.L., 2015. Early–Middle Pleistocene transitions: linking terrestrial and marine realms. Quaternary International 389: 7-46. DOI: 10.1016/j.quaint.2015.09.042
Herbert, T.D., Rose, R., Dybkjær, K., Rasmussen, E.S. & Śliwińska, K.K., 2020. Bihemispheric Warming in the Miocene Climatic Optimum as Seen From the Danish North Sea. Paleoceanography and Paleoclimatology 35(10): e2020PA003935. DOI: 10.1029/2020PA003935
Herngreen, G.F.W., Van Konijnenburg-Van Cittert, J.H.A. & Oosterink, H.W., 2005. New geological data (Middle Triassic, Rhaetian-Liassic and Oligocene) of the Winterswijk quarry, the eastern Netherlands. Netherlands Journal of Geosciences 84 (4): 409-413. DOI: 10.1017/S0016774600021223
Hillis, R.R., Holford, S.P., Green, P.F., Doré, A.G., Gatliff, R.W., Stoker, M.S., Thompson, K., Turner, J.P., Underhill, J.R. & Williams, G.A., 2008. Cenozoic exhumation of the southern British Isles. Geology 36(5): 371-374. DOI: 10.1130/G24699A.1
Hollis, C.J., Naeher, S., Clowes, C.D., Naafs, B.D.A., Pancost, R.D., Taylor, K.W.R., Dahl, J., Li, X., Ventura, G.T. & Sykes, R., 2021. Late Paleocene CO2 drawdown, climatic cooling, and terrestrial denudation in the southwest Pacific. Climate of the Past Discussion 18:1295-1320. DOI: 10.5194/cp-18-1295-2022.
Huhn, B., Utescher, T., Ashraf, A.R. & Mosbrugger, V., 1997. The peat-forming vegetation in the Middle Miocene Lower Rhine Embayment, an analysis based on palynological data. Mededelingen Nederlands Instituut voor Toegepaste Geowetenschappen TNO.
Hutchinson, D.K., Coxall, H.K., Lunt, D.J., Steinthorsdottir, M., Boer, A.M., Baatsen, M. & Zhang, Z., 2021. The Eocene–Oligocene transition: a review of marine and terrestrial proxy data,

models and model–data comparisons. Climate of the Past 17(1): 269-315. DOI: 10.5194/cp-17-269-2021

Inglis, G.N., Collinson, M.E., Riegel, W., Wilde, V., Farnsworth, A., Lunt, D.J., Valdes, P., Robson, B.E., Scott, A.C., Lenz, O.K., David, B., Naafs, A. & Pancost, R.D., 2017. Mid-latitude continental temperatures through the early Eocene in western Europe. Earth and Planetary Science Letters 460: 86-96. DOI: 10.1016/j.epsl.2016.12.009

Jagt, J.W.M., 2000. Late Cretaceous-Early Palaeogene echinoderms and the K/T boundary in the southeast Netherlands and northeast Belgium-Part 5: Asteroids. Scripta Geologica 121: 377-503.

Jagt, J.W.M., Van Bakel, B.W.M., Cremers, G., Deckers, M.J.M., Dortangs, R.W., Van Es, M., Fraaije, R.H.B., Kisters, P.J.M., Van Knippenberg, P.H.M, Lemmens, H., Nieuwenhuis, E., Severijns, J. & Stroucken, J.W., 2013. Het Vroeg Paleoceen (Danien) van zuidelijk Limburg en aangrenzend gebied – nieuwe fauna's en nieuwe inzichten. Afzettingen 34: 198-230.

Jakobsson, M., Backman, J., Rudels, B., Nycander, J., Frank, M., Mayer, L. & Moran, K., 2007. The early Miocene onset of a ventilated circulation regime in the Arctic Ocean. Nature 447(7147): 986-990. DOI: 10.1038/nature05924

Janssen, A.W., 1979. Over de molluskenfauna van de Nuculaklei en de relatieve ouderdom van enkele midden-oligocene afzettingen in Belgisch Limburg. Afzettingen 1:10-18.

Janssen, A.W., 1984. Mollusken uit het Mioceen van Winterswijk - Miste. Stichting Uitgeverij Koninklijke Nederlandse Natuurhistorische Vereniging 36.

Janssen, A.W., 2001. The age of the North Sea Basin Hemmoorian (Miocene); holoplanktonic molluscan evidence. Aardkundige Mededelingen 11: 45-50.

Janssen, A.W. & Gürs, K., 2002. Notes on the systematics, morphology and biostratigraphy of fossil holoplanktonic Mollusca, 12. On the identity of Hyalea perovalis Von Koenen, 1882 (Mollusca, Gastropoda, Euthecosomata) from the Early Miocene of the North Sea Basin. Basteria 66(4/6): 143-148.

Japsen, P., Bidstrup, T. & Lidmar-Bergström, K., 2002. Neogene uplift and erosion of southern Scandinavia induced by the rise of the South Swedish Dome. In: Doré, A.G., Cartwright, J.A., Stoker, M.S., Turner, J.P. & White, N. (eds): Exhumation of the North Atlantic margin: timing, mechanism and implications for petroleum exploration. Geological Society of London, Special Publications 196: 183-207.

Keigwin, L., 1982. Isotopic paleoceanography of the Caribbean and East Pacific: role of Panama uplift in late Neogene time. Science 217(4557): 350-353.

Kemna, H.A. & Westerhoff, W.E., 2007. Remarks on the palynology-based chronostratigraphical subdivision of Pliocene terrestrial deposits in NW-Europe. Quaternary International 164: 184-196. DOI: 10.1016/j.quaint.2006.10.017

Kender, S., Stephenson, M.H., Riding, J.B., Leng, M.J., Knox, R.W.B., Peck, V.L. & Jamieson, R., 2012. Marine and terrestrial environmental changes in NW Europe preceding carbon release at the Paleocene–Eocene transition. Earth and Planetary Science Letters 353: 108-120. DOI: 10.1016/j.epsl.2012.08.011

Kent, D.V. & Muttoni, G., 2008. Equatorial convergence of India and early Cenozoic climate trends. Proceedings of the National Academy of Sciences 105(42): 16065-16070. DOI: 10.1073/pnas.0805382105

Knox, R.W.O.B., Bosch, J.H.A., Rasmussen, E.S., Heilmann-Clausen, C., Hiss, M., De Lugt, I.R., Kasińksi, J., King, C., Köthe, A., Słodkowska, B., Standke, G. & Vandenberghe, N., 2010. Cenozoic. In: Doornenbal, J.C. & Stevenson, A.G. (eds): Petroleum geological atlas of the Southern Permian Basin area. EAGE Publications b.v. (Houten): 211-223.

Köthe, A., 2007. Cenozoic biostratigraphy from the German North Sea sector (G-11-1 borehole, dinoflagellate cysts, calcareous nannoplankton). Zeitschrift Der Deutschen Gesellschaft Fur Geowissenschaften 158(2): 287-327. DOI: 10.1127/1860-1804/2007/0158-0287

Kotthoff, U., Greenwood, D.R., McCarthy, F.M.G., Müller-Navarra, K., Prader, S. & Hesselbo, S.P., 2014. Late Eocene to middle Miocene (33 to 13 million years ago) vegetation and climate development on the North American Atlantic Coastal Plain (IODP Expedition 313, Site M0027). Climate of the Past 10(4): 1523-1539. DOI: 10.5194/cp-10-1523-2014

Krijgsman, W., Capella, W., Simon, D., Hilgen, F.J., Kouwenhoven, T.J., Meijer, P.T. & Flecker, R., 2018. The Gibraltar corridor: Watergate of the Messinian salinity crisis. Marine Geology 403: 238-246. DOI: 10.1016/j.margeo.2018.06.008

Kuhlmann, G., Langereis, C.G., Munsterman, D., Van Leeuwen, R.-J., Verreussel, R., Meulenkamp, J.E. & Wong, Th.E., 2006a. Chronostratigraphy of Late Neogene sediments in the Southern North Sea Basin and paleoenvironmental interpretations. Palaeogeography, Palaeoclimatology, Palaeoecology 239: 426-455. DOI: 10.1016/j.palaeo.2006.02.004

Kuhlmann, G., Langereis, C.G., Munsterman, D., Van Leeuwen, R.-J., Verreussel, R., Meulenkamp, J.E. & Wong, Th.E., 2006b. Intergrated chronostratigraphy of the Piocene-Pleistocene interval and its relation to the regional stratigraphical stages in the southern North Sea region. Netherlands Journal of Geosciences 85 (1): 19-35. DOI: 10.1017/S0016774600021405

Kuhlmann, G. & Wong, Th.E., 2008. Pliocene paleoenvironment evolution as interpreted from 3D-seismic data in the southern North Sea, Dutch offshore sector. Marine and Petroleum Geology 25: 173-189. DOI: 10.1016/j.marpetgeo.2007.05.009

Kurtz, A.C., Kump, L.R., Arthur, M.A., Zachos, J.C. & Paytan, A., 2003. Early Cenozoic decoupling of the global carbon and sulfur cycles. Paleoceanography 18: 1090. DOI: 10.1029/2003PA000908

Kuyl, O.S., 1980. Toelichting bij de geologische kaart van Nederland 1 : 50.000, Blad Heerlen (62 W oostelijke helft, 62 O westelijke helft). Rijks Geologische Dienst (Haarlem): 206 pp.

Kvaček, Z., 2010. Forest flora and vegetation of the European early Palaeogene – a review. Bulletin of Geosciences 85(1): 63-76. DOI: 10.3140/bull.geosci.1146

Larsson, L.M., Dybkjær, K., Rasmussen, E.S., Piasecki, S., Utescher, T. & Vajda, V., 2011. Miocene climate evolution of northern Europe: A palynological investigation from Denmark. Palaeogeography, Palaeoclimatology, Palaeoecology 309(3-4): 161-175. DOI: 10.1016/j.palaeo.2011.05.003

Lenz, O.K., Wilde, V., Mertz, D.F. & Riegel, W., 2015. New palynology-based astronomical and revised 40Ar/39Ar ages for the Eocene maar lake of Messel (Germany). International Journal of Earth Sciences 104(3): 873-889. DOI:10.1007/s00531-014-1126-2

Letsch, W.J. & Sissingh, W., 1983. Tertiary stratigraphy of The Netherlands. Geologie En Mijnbouw 62(2): 305-318.

Lisiecki, L.E. & Raymo, M.E., 2005. A Pliocene-Pleistocene stack of 57 globally distributed benthic δ18O records. Paleoceanography 20(PA1003). DOI: 10.1029/2004PA001071

Liu, L., Yu, L., Pearl, D.K. & Edwards, S.V., 2009. Estimating species phylogenies using coalescence times among sequences. Systematic biology 58(5): 468-477. DOI:10.1093/sysbio/syp031

Louwye, S., 2005. The Early and Middle Miocene transgression at the southern border of the North Sea Basin (northern Belgium. Geological Journal 40(4): 441-456. DOI: 10.1002/gj.1021

Louwye, S. & De Schepper, S., 2010. The Miocene–Pliocene hiatus in the Southern North Sea Basin (northern Belgium) revealed by dinoflagellate cysts. Geological Magazine 147(5): 760-776. DOI: 10.1017/S0016756810000191

Louwye, S. & Laga, P., 2008. Dinoflagellate cyst stratigraphy and palaeoenvironment of the marginal marine Middle and Upper Miocene of the eastern Campine area, northern Belgium (Southern North Sea Basin). Geological Journal 43(1): 75-94. DOI: 10.1002/gj.1103

Lyell, C., 1833. Principles of Geology, being an attempt to explain the former changes of the earth's surface, by reference to causes now in operation. London (John Murray): 398 pp.

Maclennan, J. & Jones, S.M., 2006. Regional uplift, gas hydrate dissociation and the origins of the Paleocene–Eocene Thermal Maximum. Earth and Planetary Science Letters 245(1-2): 65-80. DOI: 10.1016/j.epsl.2006.01.069

Manten, 1958. Palynology of the Miocene browncoal mined at Haanrade (Limburg, Netherlands. Acta Botanica Neerlandica 7(3): 445-488.

Marquet, R., 2016. A revision of the gastropod fauna of the Boom Clay Formation (Rupelian, Early Oligocene) in Belgium, 1. Vetigastropoda and Caenogastropoda to Eulimoidea. Cainozoic Research 16(1): 13-33.

Marquet, R. & Herman, J., 2012. Reinvestigation of the invertebrate fauna of the Boom Clay Formation and the Ruisbroek Sand Member (Oligocene, Rupelian) of Belgium, with the description of a new lithostratigraphic unit: the Sint Niklaas Phosphorite Bed. Cainozoic Research 9(1): 101-120.

Marx, F.G., Bosselaers, M.E.J. & Louwye, S., 2016. A new species of Metopocetus (Cetacea, Mysticeti, Cetotheriidae) from the Late Miocene of the Netherlands. PeerJ 4(1572). DOI: 10.7717/peerj.1572

Marx, F.G., Post, K., Bosselaers, M. & Munsterman, D.K., 2019. A large Late Miocene cetotheriid (Cetacea, Mysticeti) from the Netherlands clarifies the status of Tranatocetidae. PeerJ 7(6426). DOI: 10.7717/peerj.6426

Mayr, G. & Smith, T., 2019. New Paleocene bird fossils from the North Sea Basin in Belgium and France. Geologica Belgica 22(1-2): 35-46. DOI: 10.20341/gb.2019.003

Michon, L., Van Balen, R.T., Merle, O. & Pagnier, H., 2003. The Cenozoic evolution of the Roer Valley rift system integrated at a European scale. Tectonophysics 367 (1-2): 101-126. DOI: 10.1016/S0040-1951(03)00132-X00132-X

Miller, K.G., Browning, J.V., Aubry, M.P., Wade, B.S., Katz, M.E., Kulpecz, A.A. & Wright, J.D., 2008. Eocene–Oligocene global climate and sea-level changes: St. Stephens Quarry, Alabama. Geological Society of America Bulletin 120(1-2): 34-53.

Miller, K.G., Browning, J.V., Schmelz, W.J., Kopp, R.E., Mountain, G.S. & Wright, J.D., 2020. Cenozoic sea-level and cryospheric evolution from deep-sea geochemical and continental margin records. Science Advances 6(20): 1346.

Moerdijk, P.W., Janssen, A.W., Wesselingh, F.P., Peeters, G.A., Pouwer, R., Nieulande, F.A.D. & Ter Poorten, J.J., (eds) 2010. De fossiele schelpen van de Nederlandse kust. Nederlands Centrum voor Biodiversiteit Naturalis (Leiden): 332 pp.

Mörs, T., 2002. Biostratigraphy and paleoecology of continental Tertiary vertebrate faunas in the Lower Rhine Embayment (NW-Germany). Netherlands Journal of Geosciences 81(2): 177-183. DOI: 10.1017/S0016774600022411

Morton, A.C. & Knox, R.O.B., 1990. Geochemistry of late Palaeocene and early Eocene tephras from the North Sea Basin. Journal of the Geological Society 147(3): 425-437. DOI: 10.1144/gsjgs.147.3.0425

Mosbrugger, V., Gee, C.T., Belz, G. & Ashraf, A.R., 2022. Three-dimensional reconstruction of an in-situ Miocene peat forest from the Lower Rhine Embayment, northwestern Germany—new methods in palaeovegetation analysis. Palaeogeography, Palaeoclimatology, Palaeoecology 110(3-4): 295-317. DOI: 10.1016/0031-0182(94)90089-2

Mosbrugger, V., Utescher, T. & Dilcher, D.L., 2005. Cenozoic continental climatic evolution of Central Europe. Proceedings of the National Academy of Sciences 102(42): 14964-14969.

Munsterman, D.K. & Deckers, J., 2020. The Oligocene/Miocene boundary in the ON-Mol-1 and Weelde boreholes along the southern margin of the North Sea Basin, Belgium. Geologica Belgica 23(3-4). DOI: 10.20341/gb.2020.007

Munsterman, D.K., Ten Veen, J.H., Menkovic, A., Deckers, J., Witmans, N., Verhaegen, J., Kerstholt-Boegehold, S.J., Van de Ven, T. & Busschers, F.S., 2019. An updated and revised stratigraphic framework for the Miocene and earliest Pliocene strata of the Roer Valley Graben and adjacent blocks. Netherlands Journal of Geosciences 98(8). DOI: 10.1017/njg.2019.10

Munsterman, D.K., Van den Bosch, M., Wesselingh, F.P., Helwerda, M. & Busschers, F.S., 2024. A proposal for an updated and revised stratigraphical framework of the Miocene in the

Achterhoek (eastern Netherlands). Netherlands Journal of Geosciences 103(7). DOI: 10.1017/njg.2024.3

Munsterman, D.K. & Brinkhuis, H., 2004. A southern North Sea Miocene dinoflagellate cyst zonation. Netherlands Journal of Geosciences 83(4): 267-285. DOI:10.1017/S0016774600020369

Munsterman, D.K. & Deckers, J., 2022. Biostratigraphic ages and depositional environments of the upper Oligocene to lower Miocene Veldhoven Formation in the central Roer Valley Rift System (SE Netherlands-NE Belgium). Netherlands Journal of Geosciences 101(6). DOI:10.1017/njg.2022.3

Naafs, B.D.A., Rohrssen, M. & Inglis, G.N., 2018. High temperatures in the terrestrial mid-latitudes during the early Palaeogene. Nature Geosciences 11: 766-771.

Nalpas, T., Le Douaran, S., Brun, J.P., Unternehr, P. & Richert, J.P., 1995. Inversion of the Broad Fourteens Basin (offshore Netherlands), a small-scale model investigation. Sedimentary Geology 95 (3-4): 237-250. DOI: 10.1016/0037-0738(94)00113-9

Nielsen, S.B., Stephenson, R. & Thomsen, E., 2007. Dynamics of Mid-Palaeocene North Atlantic rifting linked with European intra-plate deformations. Nature 450: 1071-1074. DOI:10.1038/nature06379

Nielsen, S.B., Thomsen, E., Hansen, D.L. & Clausen, O.R., 2005. Plate-wide stress relaxation explains European Paleocene basin inversions. Nature 435(7039): 195-198. DOI:10.1038/nature03599

Overeem, I., Weltje, G.J., Bishop-Kay, C. & Kroonenberg, S.B., 2001. The Late Cenozoic Eridanos delta system in the Southern North Sea Basin: a climate signal in sediment supply? Basin Research 13: 293-312. DOI:10.1046/j.1365-2117.2001.00151.x

Palike, H., Norris, R.D., Herrle, J.O., Wilson, P.A., Coxall, H.K., Lear, C.H. & Wade, B.S., 2006. The heartbeat of the Oligocene climate system. Science 314(5807): 1894-1898. DOI:10.1126/science.1133822

Patruno, S., Kombrink, H. & Archer, S.G., 2021. Cross-border stratigraphy of the Northern, Central and Southern North Sea: a comparative tectono-stratigraphic megasequence synthesis. Geological Society of London, Special Publications 494 (1): 13-83. DOI:10.1144/sp494-2020-228

Pearson, P.N., Foster, G.L. & Wade, B.S., 2009. Atmospheric carbon dioxide through the Eocene–Oligocene climate transition. Nature 461(7267): 1110-1113 .

Peters, N., (ed.) 2013. Van reuzenhaai tot Chalicotherium. Fossielen uit Mill-Langenboom. Oertijdmuseum De Groene Poort (Boxtel): 158 pp.

Peters, N., Munsterman, D.K. & Post, K., 2021. A late Miocene balaenopterid petrotympanic from Liessel (The Netherlands). Cainozoic Research 21(2): 165-172

Peters, N. & Peters, H., (eds) 2009. Brabant tussen walvissen en mastodonten: fossielen uit Liessel. Nationaal Beiaard- en Natuurmuseum (Asten) & Oertijdmuseum De Groene Poort (Boxte)l: 111 pp.

Phillips, J., (ed.) 1841. Figures and Descriptions of the Palaeozoic Fossils of Cornwall, Devon and East Somerset. Longman, Brown, Green and Longmans (London): 231 pp.

Post, K. & Bosselaers, M., 2017. Cetacean fossils from a 1961 expedition at the Schelde estuary, province of Zeeland, the Netherlands. Cainozoic Research 17(1): 11-21.

Post, K. & Kompanje, E.J.O., 2010. A new dolphin (Cetacea, Delphinidae) from the Plio-Pleistocene of the North Sea. Deinsea 14: 1-12

Post, K., Munsterman, D.K., Peters, M.E., Hille, P. & Reumer, J., 2018. Lederschildpadden, reuzenhaaien en La Plata dolfijnen zwommen ooit in het zuidelijk deel van het Noordzeebekken (Nu: Westerschelde. Cranium 35(2): 37-42

Post, K. & Peters, N., (eds) 2023. De fossiele zeezoogdieren van Nederland. GBU (Urk): 112 pp.

Post, K. & Reumer, J.W., 2016. History and future of paleontological surveys in the Westerschelde Estuary (Province of Zeeland, the Netherlands. Deinsea 16: 1-9.

Prinz, L., Schäfer, A., McCann, T., Utescher, T., Lokay, P. & Asmus, S., 2017. Facies analysis and depositional model of the Serravallian-age Neurath Sand, Lower Rhine Basin (W Germany). Netherlands Journal of Geosciences 96(3): 211-231. DOI: 10.1017/njg.2016.51

Rae, J.W., Zhang, Y.G., Liu, X., Foster, G.L., Stoll, H.M. & Whiteford, R.D., 2021. Atmospheric CO_2 over the past 66 million years from marine archives. Annual Review of Earth and Planetary Sciences 49: 609-641.

Raffi, I., Wade, B.S. & Palike, H., 2020. The Neogene Period. In: Gradstein, F.M., Ogg, J.G., Schmitz, M.D. & Ogg, G.M. (eds): The Geologic Time Scale 2020. Elsevier (Boston) 2: 1141-1215. DOI: 10.1016/B978-0-12-824360-2.00029-2

Rasmussen, E.S., 2009. Neogene inversion of the Central Graben and Ringkobing-Fyn High, Denmark. Tectonophysics 465(1-4): 84-97. DOI:10.1016/j.tecto.2008.10.025

Rasmussen, E.S., 2004. The interplay between true eustatic sea-level changes, tectonics and climatic changes: what is the dominating factor in sequence formation of the Upper Oligocene - Miocene succession in the eastern North Sea Basin, Denmark. Global and Planetary Change 41(1): 15-30. DOI:10.1016/j.gloplacha.2003.08.004

Rasmussen, E.S. & Dybkjær, K., 2014. Patterns of Cenozoic sediment flux from western Scandinavia: discussion. Basin Research 26(2): 338-346. DOI:10.1111/bre.12024

Rasmussen, E.S., Heilmann-Clausen, C., Waagstein, R. & Eidvin, T., 2008. The Tertiary of Norden. Episodes 31(1): 66-72. DOI:10.18814/epiiugs/2008/v31i1/010

Rasmussen, E.S., Dybkjær, K. & Piasecki, S., 2010. Lithostratigraphy of the Upper Oligocene – Miocene Succession of Denmark. Geological Survey of Denmark and Greenland Bulletin 22: 92.

Rijsdijk, K.F., Passchier, S., Weerts, H.J.T., Laban, C., Van Leeuwen, R.J.W. & Ebbing, J.H.J., 2005. Revised Upper Cenozoic stratigraphy of the Dutch sector of the North Sea Basin: towards an integrated lithostratigraphic, seismostratigraphic and al-

lostratigraphic approach. Netherlands Journal of Geosciences 84(2): 129-146. DOI:10.1017/S0016774600023015
Roveri, M., Flecker, R., Krijgsman, W., Lofi, J., Lugli, S., Manzi, V. & Stoica, M., 2014. The Messinian Salinity Crisis: Past and future of a great challenge for marine sciences. Marine Geology 352: 25-58. DOI:10.1016/j.margeo.2014.02.002
Ryan, W.B., Dewey, J.F., Pitman, W.C. & Bonnin, J., 1973. Plate tectonics and the evolution of the Alpine system. Geological Society of America Bulletin 84(10): 3137-3180. DOI:10.1130/0016-7606(1973)84<3137:PTATEO>2.0.CO;2
Ryan, W.B.F., 1973. Geodynamic implications of the Messinian crisis of salinity. In: Drooger, C.W. (ed.): Messinian events in the Mediterranean. North-Bolland Publishing Company (Amsterdam): 26-38.
Saeys, R., Verheyen, A. & Vandenberghe, N., 2004. A rapid clay-mineral change in the earliest Priabonian of the North Sea Basin? Netherlands Journal of Geosciences 83(3): 179-185. DOI:10.1017/S0016774600020242
Sangiorgi, F., Quaijtaal, W., Donders, T.H., Schouten, S. & Louwye, S., 2021. Middle Miocene temperature and productivity evolution at a Northeast Atlantic shelf site (IODP U1318, Porcupine Basin): Global and regional changes. Paleoceanography and Paleoclimatology 36(7): e2020PA004059. DOI:10.1029/2020PA004059
Schäfer, A., Utescher, T., Klett, M. & Valdivia-Manchego, M., 2005. The Cenozoic Lower Rhine Basin – rifting, sedimentation, and cyclic stratigraphy. International Journal of Earth Sciences 94: 621-639. DOI:10.1007/s00531-005-0499-7
Schäfer, A., Utescher, T. & Mörs, T., 2004. Stratigraphy of the Cenozoic Lower Rhine Basin, northwestern Germany. Newsletters in Stratigraphy 40(1-2): 73-110. DOI:10.1127/0078-0421/2004/0040-0073
Schwarzhans, W., (ed.) 2010. The otoliths from the Miocene of the North Sea Basin. Backhuys Publishers (Leiden) & Margraf Publishers (Weikersheim): 352 pp.
Simmons, M.D., 2018. Sir Charles Lyell. In: Simmons, M.D. (ed.) Great geologists. (Abington): 39-41.
Simpson, I.R., Gravestock, M., Ham, D., Leach, H. & Thompson, S.D., 1989. Notes and cross-sections illustrating inversion tectonics in the Wessex Basin. Geological Society, London, Special Publications 44: 123-129.
Sissingh, W., 2006. Syn-kinematic palaeogeographic evolution of the West European Platform: correlation with Alpine plate collision and foreland deformation. Netherlands Journal of Geosciences 85(2): 131-180. DOI:10.1017/S0016774600077933
Sliggers, B.C. & Leeuwen, R.J.W., 1987. Mollusc biozonation of the Miocene in the south-eastern Netherlands and correlation with the foraminiferal biostratigraphy. Mededelingen van de Werkgroep voor Tertiaire en Kwartaire Geologie 24(1-2): 41-57.
Śliwinska, K.K., Heilmann-Clausen, C. & Thomsen, E., 2014. Correlation between the type Chattian in NW Europe and the Rupelian–Chattian candidate GSSP in Italy. STRATI 2013, Springer: 283-286. DOI: 10.1007/978-3-319-04364-7_57
Sluijs, A., Brinkhuis, H., Crouch, E.M., John, C.M., Handley, L., Munsterman, D.K. & Dickens, G.R., 2008. Eustatic variations during the Paleocene-Eocene greenhouse world. Paleoceanography 23(4). DOI: 10.1029/2008PA001615
Smith, T. & Smith, R., 2003. Terrestrial mammals as biostratigraphic indicators in upper Paleocene-lower Eocene marine deposits of the Southern North Sea Basin. Geological Society of America Special Papers 369: 513-520.
Speijer, R.P., Pälike, H., Hollis, C.J., Hooker, J.J. & Ogg, J.G., 2020. The Paleogene Period. In: Gradstein, F.M., Ogg, J.G., Schmitz, M.D. & Ogg, G.M. (eds): Geologic time scale 2020. Elsevier (Amsterdam): 1088-1140. DOI: 10.1016/B978-0-12-824360-2.00028-0
Steinthorsdottir, M., Coxall, H.K., Boer, A.M., Huber, M., Barbolini, N., Bradshaw, C.D., N.J., B., Feakins, S.J., Gasson, E., Henderiks, J., Holbourn, A.E., Kiel, S., Kohn, M.J., Knorr, G., Kürschner, W.M., Lear, C.H., Liebrand, D., Lunt, D.J., Mörs, T., Pearson, P.N., Pound, M.J., Stoll, H. & Strömberg, C.A.E., 2021a. The Miocene: The Future of the Past. Paleoceanography and Paleoclimatology 36(4): e2020PA004037. DOI:10.1029/2020PA004037
Steinthorsdottir, M., Jardine, P.E. & Rember, W.B., 2021b. Near-future pCO2 during the hot Mid Miocene Climatic Optimum. Paleoceanography and Paleoclimatology 36(1): e2020PA003900. DOI:10.1029/2020PA003900
Steurbaut, E., 1998. High-resolution holostratigraphy of Middle Paleocene to Early Eocene strata in Belgium and adjacent areas. Palaeontographica 247(5-6): 91-156.
Steurbaut, E. & Sztrákos, K., 2008. Danian/Selandian boundary criteria and North Sea Basin–Tethys correlations based on calcareous nannofossil and foraminiferal trends in SW France. Marine Micropaleontology 67(1-2): 1-29. DOI:10.1016/j.marmicro.2007.08.004
Stouthamer, E., Cohen, K.M. & Hoek, W.Z., (eds) 2023. De vorming van het land. Geologie en geomorfologie. Perspectief Uitgevers (Bilthoven) 9: 448 pp.
Svensen, H., Planke, S., Malthe-Sørenssen, A., Jamtveit, B., Myklebust, R., Rasmussen Eidem, T. & Rey, S.S., 2004. Release of methane from a volcanic basin as a mechanism for initial Eocene global warming. Nature 429(6991): 542-545. DOI:10.1038/nature02566
Tanrattana, M., Boura, A., Jacques F.M., B., Villier, L., Fournier, F., Enguehard, A., Cardonnet, S., Voland, G., Garcia, A., Chaouch, S. & Franceschi, D., 2020. Climatic evolution in Western Europe during the Cenozoic: insights from historical collections using leaf physiognomy. Geodiversitas 42(11): 151-174. DOI: 10.5252/geodiversitas2020v42a11
Ten Veen, J., De Bruin, G., Donders, T., Verweij, H. & Geel, K., 2018. Shallow gas traps in the Cenozoic Southern North Sea delta, offshore Netherlands. First Break 36: 30-36.
Thöle, H., Gaedicke, C., Kuhlmann, G. & Reinhardt, L., 2014. Late Cenozoic sedimentary evolution of the German North Sea – A seismic stratigraphic approach. Newsletters on Stratigraphy

47(3): 299-329. DOI: 10.1127/0078-0421/2014/0049

Thöle, H., Kuhlmann, G., Lutz, R. & Gaedicke, C., 2016. Late Cenozoic submarine slope failures in the southern North Sea–Evolution and controlling factors. Marine and Petroleum Geology 75: 272-290. DOI:10.1016/j.marpetgeo.2016.04.028

Utescher, T., Ashraf, A.R., Kern, A.K. & Mosbrugger, V., 2021. Diversity patterns in microfloras recovered from Miocene brown coals of the lower Rhine Basin reveal distinct coupling of the structure of the peat-forming vegetation and continental climate variability. Geological Journal 56: 768-785. DOI:10.1002/gj.3801

Utescher, T., Bondarenko, O.V. & Mosbrugger, V., 2015. The Cenozoic Cooling—Continental signals from the Atlantic and Pacific side of Eurasia. Earth and Planetary Science Letters 415: 121-133. DOI:10.1016/j.epsl.2015.01.019

Utescher, T., Bruch, A.A., Micheels, V., Mosbrugger, V. & Popova, S., 2011. Cenozoic climate gradients in Eurasia — a palaeo-perspective on future climate change? Palaeogeography, Palaeoclimatology, Palaeoecology 304(3-4): 351-358. DOI:10.1016/j.palaeo.2010.09.031

Utescher, T., Erdei, B., Hably, L. & Mosbrugger, V., 2017. Late Miocene vegetation of the Pannonian Basin. Palaeogeography, Palaeoclimatology, Palaeoecology 467: 131-148. DOI:10.1016/j.palaeo.2016.02.042

Utescher, T., Mosbrugger, V. & Ashraf, A.R., 2000. Terrestrial climate evolution in Northwest Germany over the last 25 million years. Palaios 15(5): 430-449. DOI:10.1669/0883-1351(2000)015<0430:TCEING>2.0.CO;2

Van Adrichem Boogaert, H.A. & Kouwe, W.F.P., 1993. Stratigraphic nomenclature of The Netherlands, revision and update by RGD and NOGEPA. Mededelingen Rijks Geologische Dienst 50: 1-24.

Van Adrichem Boogaert, H.A. & Kouwe, W.F.P., 1997. Stratigraphic nomenclature of the Netherlands, revision and update by RGD and NOGEPA, Section I, Tertiary. Mededelingen Rijks Geologische Dienst (Haarlem) 50: 1-39.

Van Balen, R.T., Houtgast, R.F. & Cloetingh, S.A.P.L., 2005. Neotectonics of The Netherlands: a review. In: Cloetingh, S.A.P.L. & Cornu, T.G.M. (eds): Neotectonics and Quaternary fault-reactivation in Europe's intraplate lithosphere. Quaternary Science Reviews 24: 439-454. DOI: 10.1016/j.quascirev.2004.01.011

Van Balen, R.T., Houtgast, R.F., Van der Wateren, F.M., Vandenberghe, J. & Bogaart, P.W., 2000. Sediment budget and tectonic evolution of the Meuse catchment in the Ardennes and the Roer Valley Rift System. Global and Planetary Change 27 (1-4): 113-129. DOI:10.1016/S0921-8181(01)00062-5

Van Bergen, M.J., Vis, G.-J., Sissingh, W., Koornneef, J. & Brouwers, I., 2024. Magmatism in the Netherlands: expression of the north-west European rifting history. In: Ten Veen, J.H., Vis, G.-J., De Jager, J. & Wong, Th.E. (eds): Geology of the Netherlands, second edition. Amsterdam University Press (Amsterdam): 393-455. DOI: 10.5117/9789463728362_ch11

Van den Berg, M.W., 1994. Neotectonics of the Roer Valley rift system; style and rate of crustal deformation inferred from syn-tectonic sedimentation. Geologie En Mijnbouw 73: 143-156.

Van der Burgh, J., 1984. Some palms in the Miocene of the Lower Rhenish Plain. Review of Palaeobotany and Palynology 40: 359-374.

Van der Ham, R.W.J.M., Konijnenburg-van Cittert, J.A.H. & Indeherberge, L., 2007. Seagrass foliage from the Maastrichtian type area (Maastrichtian, Danian, NE Belgium, SE Netherlands. Review of Palaeobotany and Palynology 144(3-4): 301-321. DOI:10.1016/j.revpalbo.2006.07.008

Van Hinsbergen, D.J.J., De Groot, L.V., Van Schaik, S.J., Spakman, W., Bijl, P.K., Sluijs, A., Langereis, C.G. & Brinkhuis, H., 2015. A paleolatitude calculator for paleoclimate studies. PLoS ONE 10(6): e0126946. DOI:10.1371/journal.pone.0126946

Van Lochem, H., 2018. F17-Chalk: new insights in the tectonic history of the Dutch Central Graben. Geological Society, London, Special Publications 468: 537-558. DOI: 10.1144/SP469.17

Van Lochem, H., Vis, G.-J. & Jagt, J.W.M., 2025. Late Cretaceous. In: Ten Veen, J.H., Vis, G.-J., De Jager, J. & Wong, Th.E. (eds): Geology of the Netherlands, second edition. Amsterdam University Press (Amsterdam): 253-291. DOI: 10.5117/9789463728362_ch08

Van Simaeys, S., De Man, E., Vandenberghe, N., Brinkhuis, H. & Steurbaut, E., 2004. Stratigraphic and Paleoenvironmental Analysis of the Rupelian and Chattian in the Type Region: evidence from dinoflagellate cysts, foraminifera and calcareous nannofossils. Palaeogeography, Palaeoclimatology, Palaeoecology 208: 31-58. DOI: 10.1016/j.palaeo.2004.02.029

Van Vliet-Lanoë, B., Laurent, M., Hallégouët, B., Margerel, J.P., Chauvel, J.J., Michel, Y., Moguedet, G., Trautman, F. & Vauthier, S., 1998. Le Mio-Pliocène du Massif armoricain. Données nouvelles. Comptes Rendus de l'Académie des Sciences - Series IIA - Earth and Planetary Science 326(5): 333-340. DOI: 10.1016/S1251-8050(98)80303-0

Van Vliet-Lanoë, B., Vandenberghe, N., Laurent, M., Laignel, B., Lauriat-Rage, A., Louwye, S., Mansy, J.-L., Mercier, D., Hallégouët, B., Laga, P., Laquement, F., Meilliez, F., Michel, Y., Moguedet, G. & Vidier, J.-P., 2002. Palaeogeographic evolution of northwestern Europe during the Upper Cenozoic. Geodiversitas 24(3): 511-540.

Van Wees, J.-D. & Cloetingh, S.A.P.L., 1996. 3D flexure and intraplate compression in the North Sea basin. Tectonophysics 266 (1-4): 343-359.

Van Wijhe, D.H., 1987. Structural evolution of inverted basins in the Dutch offshore. Tectonophysics 137: 171-219. DOI: 10.1016/0040-1951(87)90320-9

Vandenberghe, N., 2017. Tectonic and climatic signals in the Oligocene sediments of the Southern North-Sea Basin. Geologica Belgica 20(3-4): 105-123. DOI:10.20341/gb.2017.007

Vandenberghe, N., Harris, W.B., Wampler, J.M., Houthuys, R., Louwye, S., Adriaens, R., Vos, K., Lanckacker, T., Matthijs, J.,

Deckers, J., Verhaegen, J., Laga, P., Westerhoff, W. & Munsterman, D., 2014. The implications of K-Ar glauconite dating of the Diest Formation on the paleogeography of the Upper Miocene in Belgium. Geologica Belgica 17(2): 161-174. DOI: 10.20341/gb.2020.010

Vandenberghe, N., Laga, P., Louwye, S., Vanhoorne, R., Marquet, R., De Meuter, F., Wouters, K. & Hagemann, H.W., 2005. Stratigraphic interpretation of the Neogene marine – continental record in the Maaseik well (49W0220) in the Roer Valley Graben, NE Belgium. Memoirs of the Geological Survey of Belgium 52: 1-39.

Vandenberghe, N., Laga, P., Steurbaut, E., Hardenbol, J. & Vail, P.R., 1998. Tertiary sequence stratigraphy at the southern border of the North Sea Basin in Belgium. In: De Graciansky, P.-C., Hardenbol, J., Jacquin, T. & Vail, P.R. (eds): Mesozoic and Cenozoic Sequence Stratigraphy of European Basins. SEPM, Special Publications (Claremore) 1(60): 119-154. DOI: 10.2110/pec.98.02.0119.

Vandenberghe, N. & Louwye, S., 2020. An introduction to the Neogene stratigraphy of northern Belgium: present status. Geologica Belgica 23(3-4): 97-112. DOI: 10.20341/gb.2020.008

Vellekoop, J., Woelders, L., Sluijs, A., Miller, K.G. & Speijer, R.P., 2019. Phytoplankton community disruption caused by latest Cretaceous global warming. Biogeosciences 16(21): 4201-4210. DOI:10.5194/bg-16-4201-2019

Verbeek, J.W., De Leeuw, C.S., Parker, N. & Wong, Th.E., 2002. Characterisation and correlation of Tertiary seismostratigraphic units in the Roer Valley Graben. Netherlands Journal of Geosciences 81(2): 159-166. DOI: 10.1017/S0016774600022393

Verhaegen, J., 2019. Quantitative sediment provenance and volumetric reconstruction: application to the Miocene of the Southern North Sea Basin (PhD study KU Leuven.).

Verhaegen, J., Weltje, G.J. & Munsterman, D., 2019. Workflow for analysis of compositional data in sedimentary petrology: provenance changes in sedimentary basins from spatio-temporal variation in heavy-mineral assemblages. Geological Magazine 156(7): 1111-1130. DOI: 10.1017/S0016756818000584

Vinken, R., 1988. The Northwest European Tertiary Basin. Results of the IGCP Project No. 124. Geologisches Jahrbuch A100: 508.

Vis, G.-J., Verweij, J.M. & Koenen, M., 2016. The Rupel Clay Member in the Netherlands: towards a comprehensive understanding of its geometry and depositional environment. Netherlands Journal of Geosciences 95 (3): 221-251. DOI: 10.1017/njg.2016.25

Walliser, E.O., Lohmann, G., Niezgodzki, I., Tütken, T. & Schöne, B., 2016. Response of Central European SST to atmospheric pCO2 forcing during the Oligocene—A combined proxy data and numerical climate model approach. Palaeogeography, Palaeoclimatology, Palaeoecology 459: 552-569. DOI:10.1016/j.palaeo.2016.07.033

Watson, D., Schofield, N., Jolley, D., Archer, S., Finlay, A.J., Mark, N. & Watton, T., 2017. Stratigraphic overview of Palaeogene tuffs in the Faroe–Shetland Basin, NE Atlantic margin. Journal of the Geological Society 174(4): 627-645. DOI:10.1144/jgs2016-132

Welle, J., 1997. Oligozäne Mollusken aus dem Schacht 8 der Bergwerksgesellschaft Sophia Jacoba bei Erkelenz (Niederrheinische Bucht). Teil 1: Lamellibranchia und ScaphopodaLamellibranchia und Scaphopoda 4: 1-137.

Welle, J., 1998. Die Molluskenfauna des Rupeltons im Tagebau Amsdorf westlich von Halle, Systematik und Paläoökologie. Münstersche Forschungen zur Geologie Paläontologie 85: 137-187.

Wesselingh, F., Janse, A.C., Vervoenen, M. & Van Nieulande, F.A.D., 2013. New records of giant campanilid gastropods (Mollusca) from the Southern North Sea Basin: implications for Eocene and Quaternary palaeogeography. Netherlands Journal of Geosciences 92(2-3): 159-164. DOI:10.1017/S0016774600000093

Westerhoff, W.E., 2009. Stratigraphy and sedimentary evolution - The Lower Rhine-Meuse system during the Late Pliocene and aerly Pleistocene (Southern North Sea Basin). PhD thesis Vrije Universiteit Amsterdam (the Netherlands): 168 pp.

Westerhoff, W.E., Donders, T.H., Trabucho Alexandre, N. & Busschers, F.S., 2020. Early Pleistocene Tiglian sites in the Netherlands: A revised view on the significance for quaternary stratigraphy. Quaternary Science Reviews 242: 106417. DOI: 10.1016/j.quascirev.2020.106417

Westerhold, T., Marwan, N., Drury, A.J., Liebrand, D., Agnini, C., Anagnostou, E. & Zachos, J.C., 2020. An astronomically dated record of Earth's climate and its predictability over the last 66 million years. Science 369(6509): 1383-1387. DOI: 10.1126/science.aba6853

Whittaker, A., (ed.) 1985. Atlas of onshore sedimentary basins in England and Wales: Post-Carboniferous tectonics and stratigraphy. Blackie, Glasgow & London: 71 pp.

Wijnker, E., Bor, T.J., Wesselingh, F.P., Munsterman, D.K., Brinkhuis, H., Burger, A.W., Vonhof, H.B., Post, K., Hoedemakers, K., Janse, A.C. & Taverne, N., 2008. Neogene stratigraphy of the Langenboom locality (Noord-Brabant, the Netherlands). Netherlands Journal of Geosciences 87(2): 165-180. DOI: 10.1017/S0016774600023209

Wong, T.E., De Lugt, I.R., Kuhlmann, G. & Overeem, I., 2007. Tertiary. In: Wong, Th.E., Batjes, D.A.J.& De Jager, J. (eds): Geology of the Netherlands. Royal Netherlands Academy of Arts and Sciences (KNAW, Amsterdam): 151-172.

Wong, Th.E., Parker, N. & Horst, P., 2001. Tertiary sedimentary development of the Broad Fourteens area, the Netherlands. Netherlands Journal of Geosciences 80(1): 85-94. DOI: 10.1017/S0016774600022186

Wouters, L. & Vandenberghe, N., 1994. Geologie van de Kempen. Division of Geology, KU Leuven, Belgium.

Zachos, J.C., Breza, J.R. & Wise, S.W., 1992. Early Oligocene ice-sheet expansion on Antarctica: Stable isotope and sedimentological evidence from Kerguelen Plateau, southern Indian Ocean. Geology 20(6): 569-573.

Zachos, J.C., Dickens, G.R. & Zeebe, R.E., 2008. An early Cenozoic perspective on greenhouse warming and carbon-cycle dynamics. Nature 451: 279-283. DOI:10.1038/nature06588

Zagwijn, W.H., 1989. The Netherlands during the Tertiary and the Quaternary: A case history of Coastal Lowland evolution. In: Van der Linden, W.J.M., Cloetingh, S.A.P.L., Kaasschieter, J.P.K., van de Graaff, W.J.E., Vandenberghe, J. & van der Gun, J.A.M. (eds): Coastal Lowlands: Geology and Geotechnology. Springer Netherlands (Dordrecht): 107-120. DOI: 10.1007/978-94-017-1064-0_6

Zagwijn, W.H., 1960. Aspects of the Pliocene and Early Pleistocene vegetation in the Netherlands. PhD thesis Leiden University. Mededelingen van de Geologische Stichting, Serie C-III-1 5: 78 pp.

Zagwijn, W.H. & Van Staalduinen, C.J., 1975. Toelichting bij geologische overzichtskaarten van Nederland, met bijbehorende 5 kaarten (1 : 600 000) en 3 overzichtsprofielen. Rijks Geologische Dienst (Haarlem): 134 pp.

Ziegler, P.A., (ed.) 1990. Geological Atlas of Western and Central Europe. Shell Internationale Petroleum Maatschappij (Den Haag) & Geological Society Publishing House (Bath) 2: 239 pp.

Ziegler, P.A., 1992. European Cenozoic rift system. Tectonophysics 208 (1-3): 91-111. DOI:10.1016/0040-1951(92)90338-7

Ziegler, P.A. & Dèzes, P., 2006. Crustal evolution of Western and Central Europe. Geological Society of London, Memoirs 32: 43-56. DOI:10.1144/GSL.MEM.2006.032.01.03

Ziegler, P.A., 1994. Cenozoic rift system of western and central Europe: an overview. Geologie En Mijnbouw 73(2-4): 99-127.

Ziegler, P.A. & Dèzes, P., 2007. Cenozoic uplift of Variscan Massifs in the Alpine foreland: Timing and controlling mechanisms. Global and Planetary Change 58(1-4): 237-269. DOI:10.1016/j.gloplacha.2006.12.004

Amkreutz, L. & Van der Vaart-Verschoof, S. (eds), 2022. Doggerland - Lost world in the North Sea. Sidestone Press: 210 pp.

10

Quaternary

Freek Busschers*, Kim Cohen, Frank Wesselingh, Marcel Bakker, Jeroen Schokker, Ronald van Balen & Sytze van Heteren

*Corresponding author: freek.busschers@tno.nl

ABSTRACT

During the Quaternary (last 2.58 Myr) polycyclic and progressive climate shifts controlled depositional systems in the southern North Sea region and shaped today's Dutch landscapes. Continuous sediment supply from surrounding regions and strong basin subsidence resulted in thick depositional sequences that recorded information on successive regional climate conditions, glaciations, tectonic movement, landscapes and ecosystems.

During the Early Pleistocene, the Baltic River System acted as the principal sediment source for the Southern North Sea Basin, which was rapidly filled and consequently progressively shallowed. During the last 1 Myr the glacial climatic cycles intensified significantly, ushering in a mode of stronger and longer Fennoscandian ice-sheet expansion. This caused the demise of the Baltic River System, which in turn allowed the Rhine-Meuse River System to expand. As the Middle Pleistocene progressed, the Fennoscandian ice sheets repeatedly reached far into the North Sea region. During glaciation maxima in the Elsterian (ca. 450 ka) and Saalian (ca. 160 ka) land ice even covered northern parts of the onshore Netherlands. Along the ice-sheet fronts, ice and meltwater widely eroded, reworked and deformed pre-existing substrates and created a diverse topography that controlled river-valley courses and transgressive pathways, both in glacial and postglacial times. Across the southwestern shoulder of the North Sea Basin, an erosional connection with the English Channel was established that progressively deepened over time. Late Pleistocene (Weichselian) glacial advances were confined to the northern Dutch North Sea. Postglacial transgression, culminating in the present highstand, resulted in laterally extensive Holocene coastal-deltaic onlap.

Marine and terrestrial life in the region changed composition throughout the Quaternary: modern marine faunal and terrestrial floral assemblages emerged during the Mid-Pleistocene Revolution (from 1.2 to 0.8 Ma). Terrestrial faunas underwent spectacular turnovers, culminating in the Late Pleistocene megafaunal extinction. Humans became a dominant force in the last few millennia, in the heavily water-managed Holocene coastal plain and in inland areas of Pleistocene sediments.

<< Holocene transgression as exposed in a 20-m-deep construction pit (Kruisplein, Rotterdam). Lowstand fluvial succession (Kreftenheye Formation) with soil formation, overlain (near knife) by gyttja/basal peat and topped by estuarine clay. Photo: Geert-Jan Vis.

DOI: 10.5117/9789463728362_ch10

The Quaternary research tradition of the Netherlands

1

For many centuries, the challenges that come with living in a low-lying delta with unconsolidated sediments have provided a motivation to explore and understand the subsurface. As early as 1605, in search of high-quality water to brew beer, Pieter Pieterszn Ente managed to drill a borehole to a depth of 73 m in the centre of Amsterdam in only twenty-one days. Unfortunately, just brackish water was encountered, but the recovered sediment samples were described in considerable detail. The associated report contains one of the oldest borehole descriptions to such depth known worldwide. In those same early days, foundation demands for increasingly large stone buildings on Amsterdam's soft soil triggered the need to collect further information on the shallow subsurface. In 1655, for example, the Amsterdam City Hall (presently the Royal Palace) was founded on 13,659 wooden piles, all resting on Pleistocene sand at 12 m below NAP (the Dutch ordnance datum, about MSL).

In the nineteenth century, the search for potable groundwater in Dutch cities intensified. Pieter Harting, professor of natural history, noticed that shell-rich sediments were present at considerable depths below the surface in multiple cities in the central and western Netherlands. The uppermost shell-rich bed contained a shallow-marine molluscan association with species that now occur in the Bay of Biscay. He concluded that the North Sea must once have reached further inland, at a time when climate was warmer than today. To denote the shell-rich sediments at this stratigraphic level he introduced the term Eemian (Harting, 1874, 1875), named after the river Eem in the central part of the country, which marks the landward margin of the shell bed. Internationally, the term is still widely used to denote the Last Interglacial (equating to Substage 5e in the global marine isotope stratigraphy; Lisiecki & Raymo, 2005).

In the mid-twentieth century the well-preserved Dutch Quaternary record became subject to intensifying research efforts. At its location just south of the maximum extent of the most extensive Pleistocene glaciations and sufficiently north to document strong climatic, paleogeographic and biotic changes throughout the Quaternary, the Netherlands region appeared ideal for emerging integrated climato-, chrono- and biostratigraphic approaches. The work of various prominent researchers of that time, including I.M. van der Vlerk, F. Florschütz and W.H. Zagwijn (see photo below) coincided with the establishment of a whole range of new research methods, including palynology and radiocarbon dating, and led to a wealth of new information. Van der Vlerk & Florschütz (1950, 1953) presented a subdivision of the Pleistocene, based on floral and faunal associations (Hooghiemstra & Richards, 2022). Zagwijn (1960, 1961, 1963) further expanded and detailed this scheme. It combines the climatostratigraphic research, mainly based on palynology, and mineralogical analyses that have formed the basis for a chronostratigraphic subdivision (Hooghiemstra & Hoek, 2019). The glacial and interglacial periods that Zagwijn defined were subsequently adopted in large parts of Europe and were tied to the paleomagnetic and marine isotope record (e.g. Zagwijn, 1985; Van Kolfschoten & Gibbard, 1998; Zagwijn, 1998). Their names still appear on modern regional and global correlation charts (Cohen & Gibbard, 2019).

Waldo H. Zagwijn in quarry Maalbeek in 1992 (Hooghiemstra & Hoek, 2019).

Following pioneering work (e.g. Oele, 1969), the number of studies addressing the Dutch offshore stratigraphic record increased in the 1980s and 1990s, profiting from improving and expanding seismic-surveying capabilities (e.g. Cameron et al., 1986; Laban, 1995). Up to today, however, correlation of the offshore and onshore Quaternary stratigraphy and records remains challenging (Fig. 10.1).

Introduction

Following a lengthy period of obliquity-forced, 41-kyr climatic cycles alternating between relatively warm and temperate phases, northwestern Europe first started to experience very deep glacial minima between 2.6 and 2.4 Ma (e.g. Lisiecki & Raymo, 2005; Fig. 10.1). The first occurrence of such a cold glacial minimum in the global marine isotope record roughly coincides with the Gauss-Matuyama magnetic reversal at 2.58 Ma. It marks the onset of the Quaternary Period or the Pleistocene Epoch. Ample evidence of ice-sheet development over Fennoscandia at this time comes from iceberg scour marks in marine sedimentary records of northwestern Europe, easily recognisable on time slices of 3D seismic surveys (Ruddiman et al., 1986; Sosdian & Rosenthal, 2009; Rea et al., 2018). Although the onset of large-scale glaciation in the Northern Hemisphere was driven by a complex interconnected set of (conditional) factors and feedback mechanisms, global oceanographic change resulting from closure of the Panama isthmus in the latest Pliocene is regarded a critical factor (Mudelsee & Raymo, 2005).

The Dutch terrestrial record of the onset of the Quaternary shows evidence of phases with steppe-like vegetation (Zagwijn, 1957, 1960; Donders et al., 2018), large-scale faunal turnover (Meloro et al., 2008) and massive increases in sediment supply by the dominant river systems of the time. It also marks the beginning of the disappearance of many thermophile species that dominated the area before the Quaternary. Following continuing alternations of temperate and cold climatic phases after 2.4 Ma, the intensity of cold-climate conditions strongly increased from about 1.2 million years ago onwards, the start of the Mid-Pleistocene Revolution. At the same time, cycles of ice build-up in the Northern Hemisphere lengthened (Fig. 10.1) shown by the spacing of the Marine Isotope Stages (MIS). This resulted in periods of ice-sheet advance into previously temperate latitudes of the Northern Hemisphere, including the present-day Dutch territory, and large changes in sea level.

Under intensively varying climatic conditions of the Quaternary, a thick, complex succession of marine, deltaic, fluvial and glacigenic sediments accumulated in the Dutch part of the North Sea Basin (Fig. 10.2). This succession reflects the evolution of several terrestrial sedimentary systems and their associated coastal and shallow-marine equivalents, with clear evidence for cyclic landward and seaward shifts of the coastline. It also contains a wealth of information on the development of ecosystems and biota. Vast parts of the succession were deposited by the Baltic River System, which had been functioning since the Miocene. By the time of the Pliocene and Early Pleistocene, its depocentre was just offshore the northern Netherlands (Ottesen et al., 2018). Once it terminated as a river system at the end of the Early Pleistocene, the Rhine-Meuse River and episodically functioning depositional systems of the Fennoscandian ice-sheet margin became the main sedimentary drivers in the Netherlands and its offshore region. This new situation characterized the remainder of the Pleistocene and Holocene.

The exceptional near-surface preservation of the Quaternary sediments has generated a great deal of interest from both applied and fundamental scientific fields. Geoscientific study started as early as the seventeenth century, evolved into various disciplinary branches and has helped answer many applied questions, for example on building material, groundwater resources and the surface stability of soft soils. The Dutch Quaternary record has hence been subject to intense mapping and research efforts, particularly during the second half of the twentieth century (see Textbox 1).

With its technological and digital progress, the 21st century heralded a new phase for Quaternary research. Greatly increased amounts of accumulated and newly collected data, a focus on multidisciplinary core analysis, improved seismic acquisition and a broadening suite of dating techniques have upgraded the insight into the depositional evolution and its tectonic, geographic and paleoenvironmental contexts. Diverse scientific studies of the Quaternary were, among others, spawned from increased activities regarding exploration for aggregate and groundwater resources, for thermal energy storage in aquifers, for transportation and energy-transition infrastructure (e.g. tunnels, windfarms), for coastal safety (nourishments), and for legally required archeological prospection and investigation related to plans that will disturb the geologic record. The decision within the Geological Survey of the Netherlands to move to nationwide digital mapping and modelling, and its increasing involvement in cross-border mapping collaborations, also brought considerable new research impetus (Van der Meulen et al., 2013). Academic research output increased manyfold and involved inter-institutional collaboration. Herein, global marine isotope stratigraphy has provided an essential framework for referring to and discussing the particular age of regionally recognized stratigraphic units, events and stages. As such, traditionally established regional chronostratigraphic naming schemes remain. They are applied in parallel with international nomenclature, with alignments and correlations from time to time re-evaluated. With the raised awareness of ongoing global climate change and issues of biodiversity loss, the Quaternary archive is more than ever used to assess past conditions and responses to climate and environmental change that can be relevant to address projected sea-level rise and the development of sustainable energy sources. These issues are expected to be a major

Table 10.1. Codes for lithostratigraphic units used by TNO-GDN (status 2024).

Code	Stratigraphic unit
NUAA	Anthropogenic deposits
NUAAES	Plaggen soil
NUAAOM	Reworked ground
NUAAOP	Made ground
NUAP	Appelscha Formation
NUAPWE	Weerdinge Member
NUAU	Aurora Formation
NUBB	Brown Bank Formation
NUBE*	Beegden Formation
NUBERO	Rosmalen Bed
NUBEWY	Wijchen Bed
NUBG	Brielle Ground Formation
NUB*	Breda Subgroup
NUBV	Batavier Formation
NUBX	Boxtel Formation
NUBXBS	Best Member
NUBXDE	Delwijnen Member
NUBXKO	Kootwijk Member
NUBXLM	Liempde Member
NUBXSC	Schimmert Member
NUBXSI	Singraven Member
NUBXTI	Tilligte Member
NUBXWI	Wierden Member
NUDB	Dogger Bight Formation
NUDBBC	Botney Cut Member
NUDBBO	Bolders Bank Member
NUDBDB	Dogger Bank Member
NUDBVO	Volans Member
NUDBWG	Well Ground Member
NUDN	Drachten Formation
NUDR	Drente Formation
NUDRGI	Gieten Member
NUDRSC	Schaarsbergen Member
NUDRUI	Uitdam Member
NUDRUIOO	Oosterdok Bed
NUEC	Echteld Formation
NUEE	Eem Formation
NUEG	Egmond Ground Formation
NUHS	Holset Formation
NUHT	Heijenrath Formation
NUIE	Inden Formation
NUKI*	Kieseloolite Formation
NUKK	Kreekrak Formation
NUKR	Kreftenheye Formation
NUKROC	Ockenburg Member
NUKRTW	Twello Member
NUKRWE	Well Member
NUKRWY	Wijchen Bed
NUKRZU	Zutphen Member
NUKW	Koewacht Formation
NUMH	Markhams Hole Formation
NUMS	Maassluis Formation
NUNA	Naaldwijk Formation
NUNASC	Schoorl Member
NUNAWA	Walcheren Member
NUNAWAAL	Almere Bed
NUNAWAYE	IJe Bed
NUNAWAYS	IJsselmeer Bed
NUNAWAZU	Zuiderzee Bed
NUNAWO	Wormer Member
NUNAWOBE	Bergen Bed
NUNAWOVE	Velsen Bed
NUNAZA	Zandvoort Member
NUNI	Nieuwkoop Formation
NUNIBA	Basisveen Bed
NUNIFL	Flevomeer Bed
NUNIGR	Griendtsveen Member
NUNIHO	Hollandveen Member
NUOO*	Oosterhout Formation
NUOS	Outer Silver Pit Formation
NUPE	Peelo Formation
NUPENI	Nieuwolda Member
NUPZ	Peize Formation
NUPZBA	Balk Member
NUSB	Southern Bight Formation
NUSBBL	Bligh Bank Member
NUSBBU	Buitenbanken Member
NUSBIG	Indefatigable Grounds Member
NUSBTB	Terschellinger Bank Member
NUSK	Smith's Knoll Formation
NUST	Sterksel Formation
NUSY	Stramproy Formation
NUSYHO	Hoogcruts Member
NUUA	Urania Formation
NUUAWH	Well Hole Member
NUUAWM	Western Mud Hole Member
NUUR	Urk Formation
NUURLI	Lingsfort Member
NUURTY	Tijnje Member
NUURVE	Veenhuizen Member
NUVI	Ville Formation
NUWA	Waalre Formation
NUWAHO	Hoogerheide Member
NUWATE	Tegelen Member
NUWAWO	Woensdrecht Member
NUWB	Woudenberg Formation
NUWG	Westkapelle Ground Formation
NUWS	Winterton Shoal Formation
NUYG	IJmuiden Ground Formation
NUYR	Yarmouth Roads Formation
NUYRAL	Alkaid Member

* Subdivision not indicated

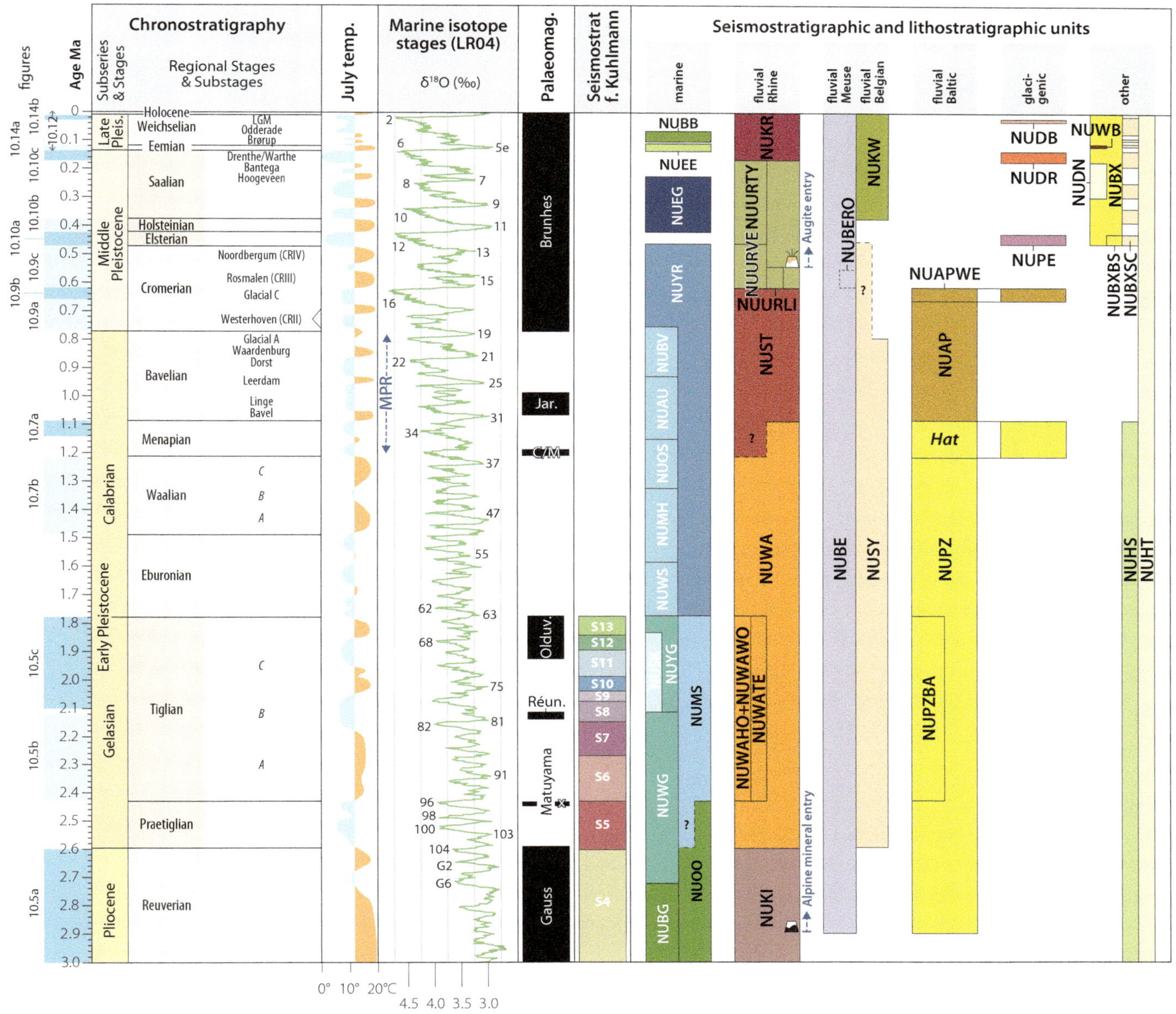

Figure 10.1. *Subdivision of the latest Pliocene and Pleistocene with, from left to right, chronostratigraphic subseries and stages (Cohen & Gibbard, 2019; complemented with stages and substages from regional schemes for northwestern Europe), pollen-based July temperature curve (Zagwijn, 1992, 1998; age model partly revised), deep-ocean oxygen isotope sequence (Marine Isotope Stages – MIS; Lisiecki & Raymo, 2005), paleomagnetic record (Channell et al., 2016), offshore seismostratigraphic scheme (Kuhlmann & Wong, 2008), and formally defined stratigraphic units in the on- and offshore domains of the Netherlands (TNO-GDN, 2023d). Age ranges of seismostratigraphic units were taken from Sha et al. (1991), Cameron et al. (1992), Gatliff et al. (1994), Stoker et al. (2011) and Westerhoff et al. (2003). Age ranges of lithostratigraphic units were primarily taken from Zagwijn (1996b), Westerhoff et al. (2003), Westerhoff et al. (2020), Busschers et al. (2007, 2008), Schokker et al. (2005), and Kars et al. (2012), but some changes to these ages are presented in this chapter. Although the figure suggests uniform ages and continuity between minimum and maximum ages, diachroneity and hiatuses occur throughout the sedimentary sequences. Late Pleistocene and Holocene (not indicated here) units are shown in Fig. 10.12. See Table 10.1 for lithostratigraphic codes. Paleomagnetic abbreviations: Jar. = Jamarillo; C/M = Cobb Mountain Event; Olduv. = Olduvai; Réun. = Réunion Event; X = X Event. Other abbreviations: MPR = Mid-Pleistocene Revolution; Hat = Hattem beds (informal).*

driver for further geological research in the forthcoming decades.

The next sections describe the state-of-the-art knowledge on the Quaternary geology of the southern North Sea region. The initial focus is on the general depositional setting, as controlled by tectonics, sediment sources and processes. Stage-wise overviews of the temporal evolution then follow for the Early Pleistocene, Middle Pleistocene, Late Pleistocene and Holocene. In these overviews, we explore the interaction between terrestrial, marine and glacigenic depositional systems in the study area, as well as their external drivers.

Setting of the southern North Sea region

Regional tectonics

The Netherlands is situated in the south-southeastern sector of the North Sea Basin (Ziegler, 1994). The modern neotectonic situation became established in the Early Miocene. Fairly strong subsidence in the North Sea Basin has characterized much of the Neogene-Quaternary phase and is consistent with widespread uplift around this region (e.g. Ardennes/Rhenish Shield, Britain and southwestern Scandinavia). While Alpine orogeny-related plate-tectonic stress fields are endogenic drivers of the subsidence and uplift (Michon et al., 2003), considerable further vertical movement of the lithosphere has been caused by isostastic adjustments (Kooi et al., 1998). Increased rates of sediment supply from surrounding uplands, attributed to the intensified cold-climatic swings from the onset of the Early Pleistocene (Fig. 10.1), outpaced tectonic and isostasy-driven subsidence rates and thus facilitated progradational basin fill (Arfai et al., 2018; Ottesen et al., 2018). The main offshore Quaternary depocentre, labelled here as Dogger Basin, is confined to a relatively narrow zone that is orientated north-northwest to south-southeast (Lamb et al., 2018; Ottesen et al., 2018). In this basin, which corresponds to the Mesozoic Step Graben, Dutch Central Graben and Terschelling Basin (De Jager et al., 2025, this volume, their Fig. 1.1), Quaternary sediments reach a maximum thickness of 1150 to 1250 m (Figs 10.2, 10.3a). A secondary depocentre, labelled here as the Southern Bight Basin, extends offshore westward from the coast of Holland to the boundary with the British continental shelf (Fig. 10.2). In this basin, which occupies roughly the same area as the Mesozoic Broad Fourteens Basin, Quaternary sediments reach a maximum thickness of 800 to 900 m (Fig. 10.2). Further southwest and also in landward directions, the Quaternary gradually thins.

Onshore, subordinate depocentres are the Zuiderzee Basin in the northwestern Netherlands and the Roer Valley Rift System in the southern Netherlands and northeastern Belgium, the latter connecting to the Lower Rhine Graben in Germany. A maximum Quaternary sediment thickness of 200 to 300 m is reached in the central part of the Roer Valley Graben (RVG in Fig. 10.3d). Vertical motions of structural blocks of the Roer Valley Rift System are controlled by normal faulting, mainly along the Feldbiss and Peel Boundary Fault Zones that bound the Roer Valley Graben to the south and north, respectively. These active fault zones have a geomorphological scarp expression and record historic earthquake activity. The most recent major earthquake (Mw 5.4) in the Peel Boundary Fault Zone occurred in 1992 (see Dost et al., 2025, this volume). Paleoseismology revealed evidence of larger earthquakes (Mw ca. 6.8) at the end of the Last Glacial, attributed to contemporaneous glacio-isostatic adjustment of the crustal-stress regime (Van Balen et al., 2019, 2021, 2024). Fluvial deposition interacted with the regional subsidence and uplift dynamics of the graben and its shoulder blocks, leading to repeated shifts of valley positions and preservation of terrace flights (see Textbox 2). Throughout the Quaternary, this has controlled the routing, preservation and erosion of sediment produced in the Rhine and Meuse hinterland through the hinge zone of the North Sea Basin towards northwestern depocentres (Van Balen et al., 2005; Boenigk & Frechen, 2006). Since the end of the Last Glacial, fault-zone activity and glacio-isostatically enhanced subsidence have also modified paleovalley gradients, thus controlling sedimentation as well as preservation of Rhine and Meuse fluvial deposits and morphology (Cohen et al., 2002; Van Balen et al., 2005; Busschers et al., 2007; Woolderink et al., 2019).

Sediment sources, transport agents and stratigraphy

Two major fluvial systems supplied sediment to the Dutch part of the North Sea Basin during the Quaternary. These are the Baltic (or Eridanos) River System sourced from the east-northeast, and the Rhine-Meuse System sourced from the south. The Baltic River System was by far the largest sediment supplier for most of the Quaternary. Small-scale fluvial (Scheldt and other systems), fluvioperiglacial and eolian sediments only make up a small part of the sediment record but are highly significant in the shallowest, youngest part of the succession and in basin areas temporarily vacated by the larger rivers. Glacigenic sediments occur at distinct levels in the record of the North Sea Basin. Although subglacial erosion (valleys) and glaciotectonic structures had a major impact on landscape evolution, glacigenic sediments are of lesser importance in terms of sediment volume. Tidal and open-marine processes have reworked the fluvial and glacigenic record to varying degrees.

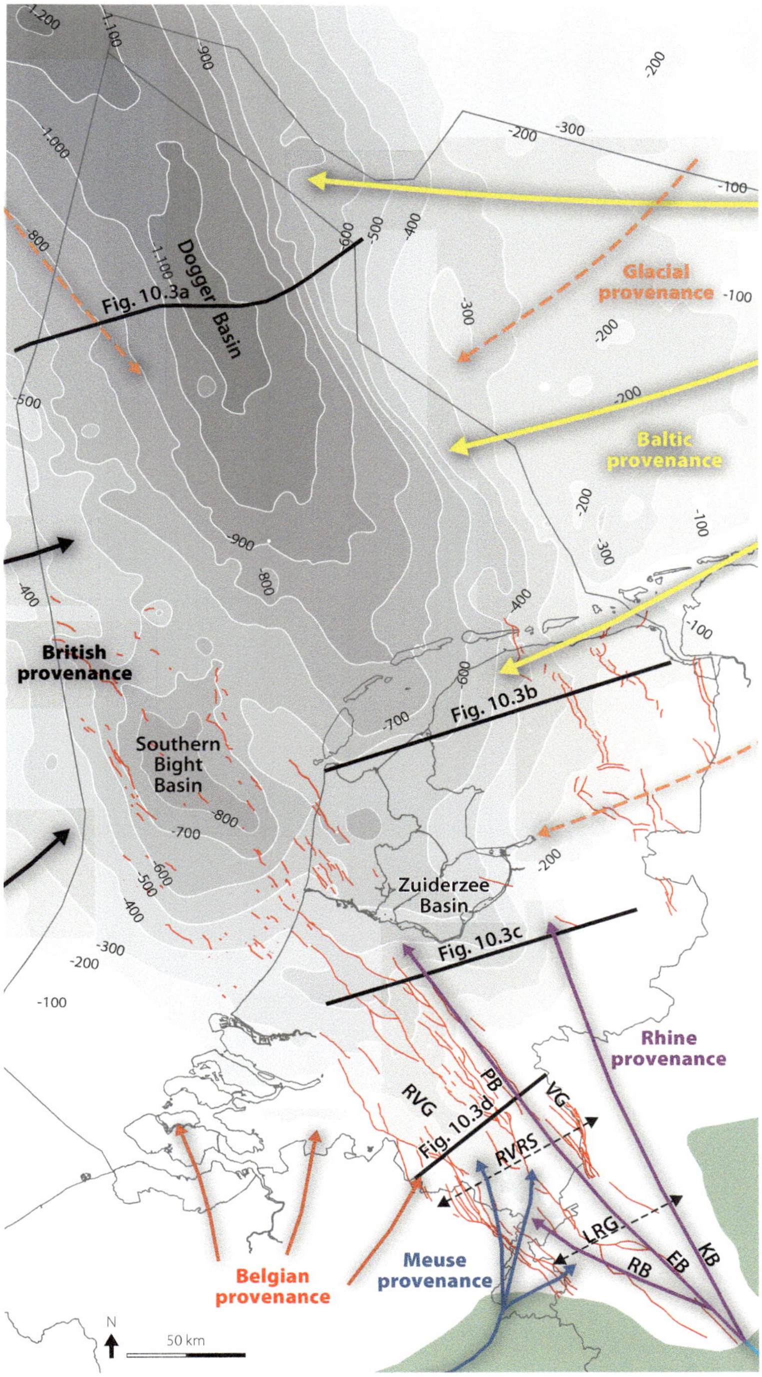

Figure 10.2. Depth map of the base of the Quaternary (isolines: depth below modern MSL) with main tectonic elements, faults (in red) sediment-transport pathways (Kooi et al., 1998; Van Balen et al., 2005; Westerhoff, 2009; Cohen et al., 2022; TNO-GDN, 2023b). Depths above -100 m MSL not indicated. LRG = Lower Rhine Graben, RB = Roer Block, EB = Erft Block, KB = Köln Block, RVRS = Roer Valley Rift System, RVG = Roer Valley Graben, PB = Peel Block, VG = Venlo Graben.

Baltic River System

The Baltic River System had a large source area east-northeast of the Netherlands (Fig. 10.2). From the Late Miocene onwards until the early Middle Pleistocene, this river system supplied vast amounts of sediment to the eastern part of the North Sea Basin (Peize and Appelscha formations and their offshore equivalents; Figs 10.1, 10.3). Its source area comprised two major domains: (I) the Baltic region and (II) the Central Uplands of Germany including the Harz, Erzgebirge, Teutoburger and Thüringer Wald, Bohemia and the northern Carpathians. Quaternary Baltic River formations in the Netherlands reach thicknesses of 50 to 200 m onshore (TNO-GDN, 2023c) and up to 900 m in the offshore region (Overeem et al. 2001; Kuhlmann et al., 2006a,b; Arfai et al., 2018). Offshore, seismic data have revealed 100 to 300 m thick sets of clinothems that represent prograding shelf-edge delta fronts of the Baltic River System (Fig. 10.3a; Kuhlmann et al., 2006a,b; Patruno et al., 2020).

The onshore Baltic River sediments are composed most-

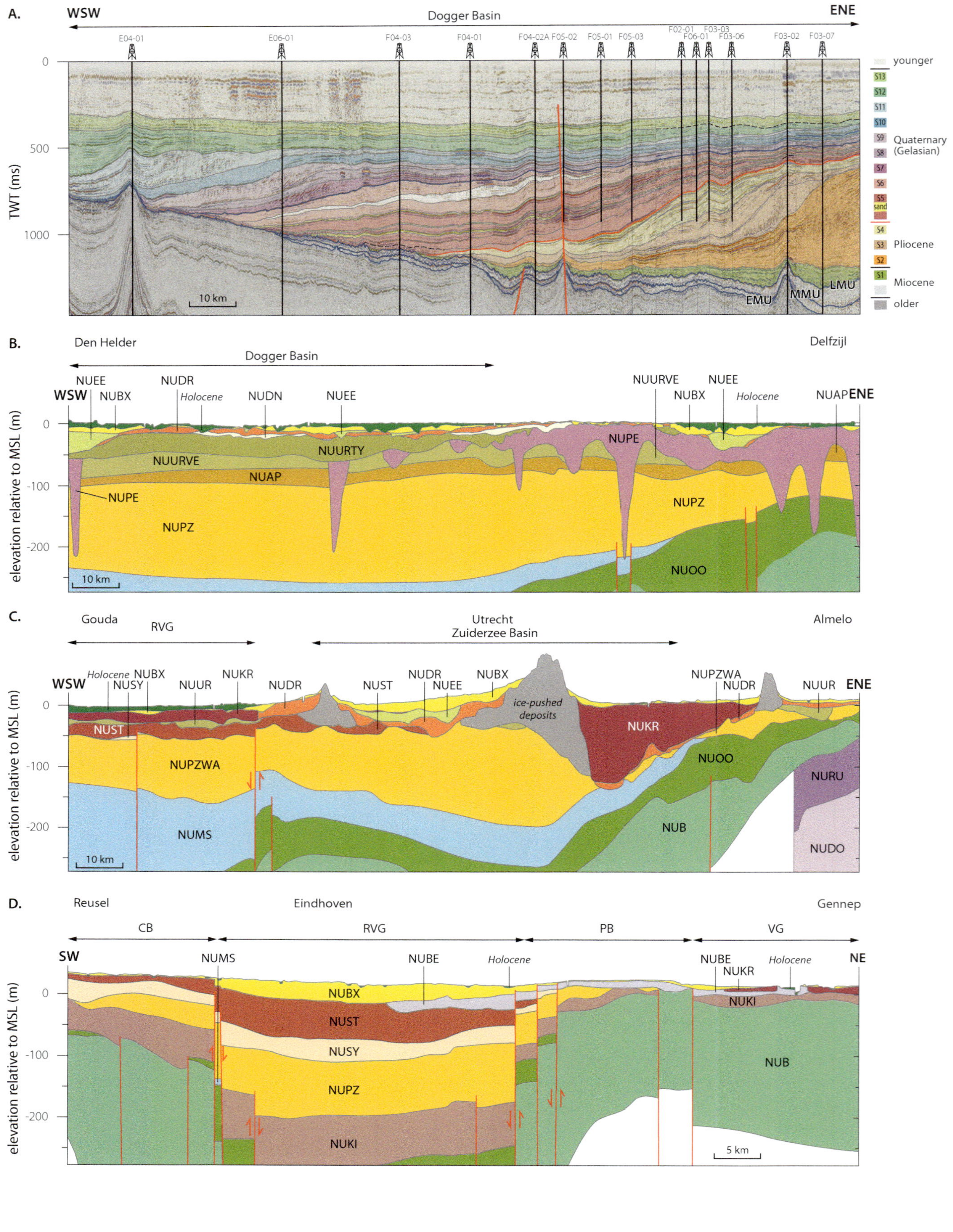
A.
WSW
Dogger Basin
ENE
E04-01
E06-01
F04-03
F04-01
F04-02A
F05-02
F05-01
F05-03
F02-01
F06-01
F03-03
F03-06
F03-02
F03-07
TWT (ms)
0
500
1000
10 km
EMU
MMU
LMU
younger
S13
S12
S11
S10
S9
S8
S7
S6
S5
sand
S4
S3
S2
S1
Quaternary (Gelasian)
Pliocene
Miocene
older
B.
Den Helder
Dogger Basin
Delfzijl
WSW
ENE
NUEE
NUBX
NUDR
Holocene
NUDN
NUEE
NUURVE
NUBX
NUEE
Holocene
NUAP
NUURVE
NUURTY
NUAP
NUPE
NUPE
NUPZ
NUPZ
NUOO
elevation relative to MSL (m)
0
-100
-200
10 km
C.
Gouda
RVG
Utrecht
Zuiderzee Basin
Almelo
WSW
ENE
Holocene
NUBX
NUSY
NUUR
NUKR
NUDR
NUST
NUDR
NUEE
NUBX
NUPZWA
NUDR
NUUR
ice-pushed deposits
NUST
NUKR
NUPZWA
NUOO
NURU
NUMS
NUB
NUDO
elevation relative to MSL (m)
0
-100
-200
10 km
D.
Reusel
Eindhoven
Gennep
CB
RVG
PB
VG
SW
NE
NUMS
NUBE
Holocene
NUBE
NUKR
Holocene
NUBX
NUKI
NUST
NUSY
NUPZ
NUKI
NUB
elevation relative to MSL (m)
0
-100
-200
5 km

← ***Figure 10.3.*** *Four sections (a-d) representative of the Quaternary stratigraphic succession of the North Sea region in the on- and offshore areas of the Netherlands (see Fig. 10.2 for section locations and Table 10.1 for explanation of lithostratigraphic codes). a) Interpreted composite seismic time section (modified from Ten Veen et al., 2020) through the Dogger Basin in the northern part of the Dutch North Sea illustrating westward-prograding seismostratigraphic units S1 to S13 (Kuhlmann & Wong, 2008) belonging to the Pliocene-Pleistocene shelf-edge delta of the Baltic River System. Units S1-S4 represent the Upper Miocene-Pliocene part of this system, directly overlying the Late Miocene Unconformity (LMU; see also Munsterman et al., 2025, this volume). Units S5-13 represent the Gelasian sequence. Seismostratigraphic units S5-13 are tentatively correlated to the Westkapelle and IJmuiden Ground formations (Fig. 10.1). The top of the interpreted part of the sequence is of Eburonian age (unit S13; Kuhlmann & Wong, 2008) and is marked by a major unconformity. b) Section showing lithostratigraphy of the southernmost part of the Dogger Basin and adjacent basin-marginal zone in the northern Netherlands (DGM model; TNO-GDN, 2023c). Volumetrically, the stratigraphy is dominated by Baltic sediments (NUPZWA) and by an alternation of Rhine sediments (NUURVE, NUURTY) with glacigenic sediments from the Elsterian (NUPE) and Saalian (NUDR) glaciations. The presence of Elsterian tunnel valleys, here reaching depths more than 200 m, is striking. Marine sediments and sediments of local origin are present on top of the Saalian and locally Elsterian glacigenic series. c) Section showing lithostratigraphy of the central Netherlands (DGM model; TNO-GDN, 2023c). The section crosses the northwestern part of the Roer Valley Rift System (west) and the southeastern part of the Zuiderzee Basin (centre), terminating on the basin margin (east). Aside from Rhine, marine and local stratigraphic units, this section contains the southernmost occurrence of Baltic sediments (here included in units NUPZ and NUWA). Also visible is a strong imprint of erosion and deposition associated with the Saalian glaciation (e.g. subglacial formation and postglacial infill of the IJssel Valley glacial basin with sediments of unit NUKR), as well as glaciotectonic ice-pushed deposits. d) Section showing lithostratigraphy of the Roer Valley Rift System in the southeastern Netherlands (DGM model; TNO-GDN, 2023c). A thick stack of Quaternary Rhine, Meuse, Belgian and local fluvial units, overlying Pliocene and Miocene shallow marine units, is present in the Roer Valley Graben. The Quaternary sediment thickness on the adjacent blocks is limited.*

ly of whitish, non-calcareous sand, commonly gravelly and rich in transparent quartz. In the basin, sediments primarily consist of fine-grained sand and silt. Large-scale glaciation of the Baltic area at the end of the Early Pleistocene beheaded the Baltic drainage basin (Bijlsma, 1981; Hall & Van Boeckel, 2020), significantly reducing its size. Its northern domain became restricted to the Baltic depression (today's Baltic Sea, including the Finnish and Bothnian Gulfs). Ice-rafted erratics derived from Fennoscandia (Hattem beds), embedded in the uppermost part of Baltic River sediments, mark the onset of this change (Fig. 10.1). A shrunken fluvial system remained, in which precursors of the Oder, Elbe and Weser Rivers drained westward towards the North Sea Basin. Their sediments differ from those of their earlier Baltic counterparts thanks to a much larger share of gravel components sourced from the central German uplands (e.g. lydite) and by topaz from Bohemia in the heavy-mineral fraction.

Rhine-Meuse River System

Late Miocene uplift of the Rhenish Massif instigated progradation of the proto-Rhine fluvial-deltaic Inden Formation and the overlying latest Tortonian Kieseloolite Formation into Germany and the southeastern Netherlands (Schäfer et al., 2005; Munsterman et al., 2019, 2025, this volume). The Rhine-Meuse River System expanded its catchment considerably in the period immediately preceding the Early Pleistocene, capturing Upper Rhine Graben and Alpine headwater reaches that have proven to be critical to the hydrological regimes, sediment composition and magnitude of supply throughout the Quaternary. Its deposits are contained in terrace flights at the basin margin, stacked fluvial successions at the basin rim and estuarine/deltaic successions in the basin proper (Waalre, Sterksel, Urk, Kreftenheye and Beegden formations and offshore equivalents; Figs 10.1, 10.3). The combination of lowland topography with little dividing relief, differential tectonic motions (grabens and shoulders), and increasing base-level fluctuations (due to intensification of glacial-interglacial climate and sea-level cycles) allowed the Rhine-Meuse System to shift in the basin, feeding several depocentres. Pre-Quaternary Rhine and Meuse sediments had been primarily derived from the strongly weathered sediment cover of the Rhenish Shield and Ardennes (Boenigk, 2002). This is evident from quartz-rich gravels and coarse sands with stable heavy-mineral components and low carbonate contents (Kieseloolite Formation; Figs 10.1, 10.3). The expansion of the system altered the Rhine sediment mineralogy, adding an unstable heavy-mineral assemblage characterized by dominant epidote, green hornblende and garnet (Westerhoff, 2009; Hülscher et al., 2018; Fig. 10.1), and also changed zircon fission-track data towards a spectrum dominated by Cenozoic cooling ages (Tatzel et al., 2015). Furthermore, carbonate and mica-flake content increased. The gradual increase of iron-stained sands and quartzitic gravel through the course of the Pleistocene reflects accelerating bedrock incision into the Rhenish Shield and Ardennes where they were

The Meuse terrace sequence

2

The Meuse River originates in the northeastern part of the Paris Basin in France, crosses the Ardennes in Belgium, and converges with the Rhine in the Netherlands. Younger courses merge with the Rhine in the central Netherlands, whereas Early and Middle Pleistocene courses were confluent further south. The paleogeographic and incisional history of the Meuse is heavily affected by tectonic uplift of the Ardennes-Rhenish Massif and by rifting in the Roer Valley Rift System.

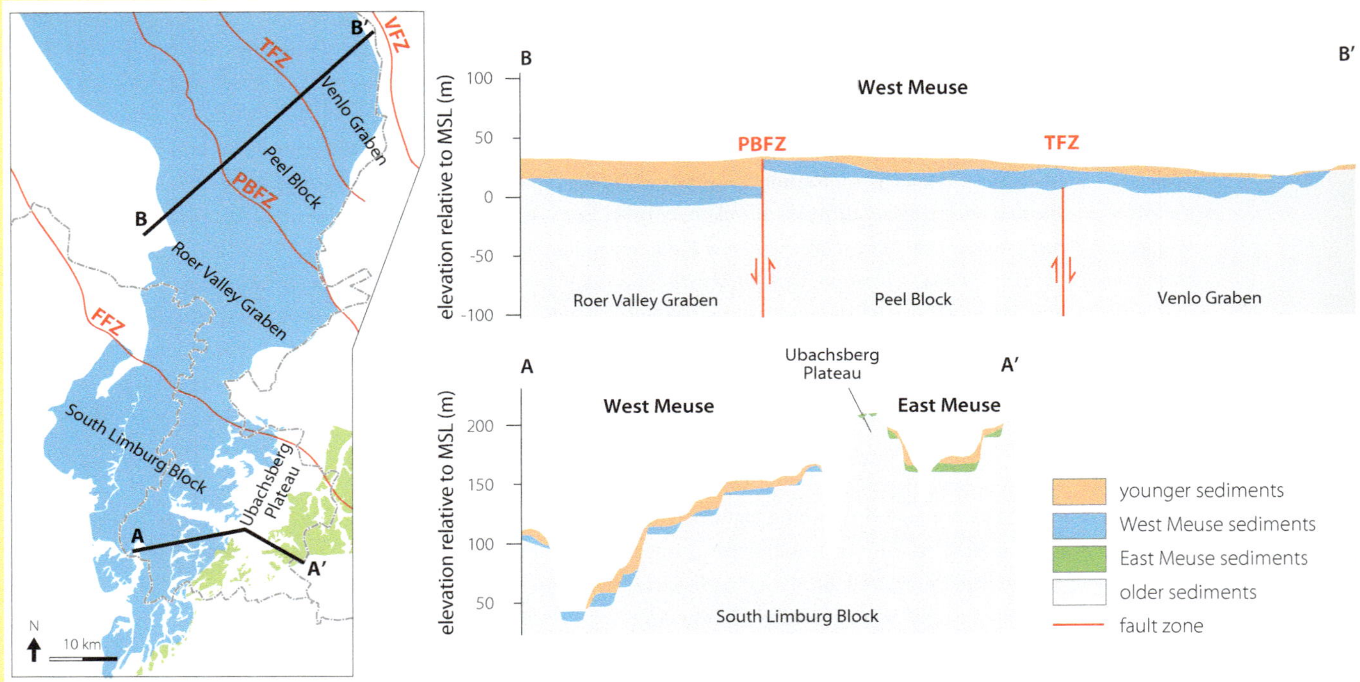

Distribution of deposits from the East Meuse (green) and West Meuse (blue) and composite Meuse terrace sequence. Schematized cross-sections illustrate differences between a tectonically uplifting (terraced) area in the south (A-A') and tectonically stable and subsiding areas in the north (B-B'). Abbreviations: FFZ = Feldbiss Fault Zone; PBFZ = Peel Boundary Fault Zone; TFZ - Tegelen Fault Zone; VFZ = Viersen Fault Zone.

crossed by Rhine and Meuse rivers. These proximal Ardennes and Rhenish Shield clasts (e.g. 'Taunus' quartzite) dominate the gravel fraction. While Alpine fine sand fractions (including heavy minerals) reached the North Sea Basin, Alpine gravels did not since they were not abraded during fluvial transport but became trapped upstream in the Upper Rhine Graben.

Rhine sediments younger than about 500 kyr (halfway through the Middle Pleistocene) are enriched in volcanic minerals such as clino-pyroxenes ('augite') and brown hornblende, introduced into the system by increased Eifel eruptions during that time (Zagwijn, 1985; Van den Bogaard & Schmincke, 1990; Boenigk & Frechen, 2006; Gallant et al., 2014; Fig. 10.1). The occurrence of these minerals is of marked stratigraphic benefit in separating older from younger depositional units (Van Kolfschoten & Turner, 1996; Zagwijn, 1996b; Fig. 10.1). Deposits that are exclusive to the Meuse River (i.e. upstream from its confluence with the Rhine) are found in the extreme south of the country, concentrated along an Early Pleistocene East Meuse Valley (which extended into Germany) and the younger West Meuse Valley (see Textbox 2). These sediments lack the Alpine and Eifel-volcanic Rhine components but share stable heavy-mineral components of the Ardennes-Rhenish Shield and Vosges bedrock hinterlands (Zonneveld, 1974; Westerhoff, 2008). The Meuse sediments are relatively gravelly, rich in flints and Ardennes-specific indicator gravels (e.g. 'Revinian' quartzite). Downstream of the confluence of the Rhine and Meuse, the Meuse signature, which is dominated by stable minerals, is entirely diluted by the Rhine sediment contribution with its high concentrations of unstable heavy minerals

In between the Ardennes s.s. and the Feldbiss Fault Zone of the Roer Valley Rift System, on the South Limburg Block, the Meuse Valley has a well-developed terrace staircase consisting of ca. 31 levels (Zonneveld, 1974; Felder & Bosch, 1989; Van den Berg, 1996; Van den Berg & Van Hoof, 2001). Individual levels are vertically separated by 4 to 9 m. The oldest levels probably date back to the Pliocene. The terraces formed in response to climatic oscillations, but they have been preserved because of tectonic uplift causing long-term incision. During cold climatic conditions, the Meuse was an aggrading braided river with high peak discharges and high sediment loads. During temperate conditions it took on a meandering style and during warm periods the river was likely stable, like at present. The incisional steps were formed during cold-warm and warm-cold transitions, reflecting a temporary imbalance of discharge and sediment load (Vandenberghe, 2015). Terraces were formed first along the East Meuse Valley course, and later, following a capture, along the West Meuse Valley course. For most of the Quaternary, the confluence with the Rhine was located in the subsiding Roer Valley Graben in between the Feldbiss Fault Zone and the Peel Boundary Fault Zone, where sediments accumulated as a buried stack. In the middle Saalian (250 ka; MIS 8), the Meuse started to shift its course away from the graben. From the late Saalian onward (150 ka; MIS 6), it formed terraces along its current course on the uplifting Peel Block and Venlo Graben, north of the Peel Boundary Fault Zone. Here, relatively small terraces were formed during post-LGM climatic oscillations and associated river style changes (Woolderink et al., 2019).

One of the first-dated terrace remnants was at Maastricht-Belvédère, an important Paleolithic hominin site. Using thermoluminescence dating of burnt flint artefacts, an age of ca. 250 ka (MIS 7) was established for the Caberg-3 terrace level (Huxtable & Aitken, 1985; Vandenberghe, 1993). This terrace is the first in which the heavy-mineral assemblage lacks Vosges hornblende. Its absence is attributed to the loss of the Upper Moselle tributary to the Rhine system (Losson & Quinif, 2001). Paleomagnetic analyses, U-Th dating of cements, pollen biostratigraphy, luminescence dating, cosmogenic isotope dating, and up- and downstream correlations have provided increasingly precise local age control for terraces on the South Limburg Block (Van Balen et al., 2000; Schaller et al., 2004; Van Balen et al., 2021; Da Silva Guimarães et al., 2024). The Meuse terraces have been used to determine uplift rates of the South Limburg Block (Van Balen et al., 2000; Van den Berg & Van Hoof, 2001) and displacement rates on the major fault zones (Houtgast et al., 2002; Van Balen et al., 2021).

from the Alps and Upper Rhine Graben. (Busschers et al., 2007).

Glacigenic systems

Sediments belonging to Quaternary glacigenic systems occur in the northern half of the Netherlands and below the Dutch part of the North Sea (Van den Berg & Beets, 1987; Laban & Van der Meer, 2011). These sediments comprise tills, fluvioglacial outwash and glaciolacustrine fines, which are thickest in infills of subglacial valleys and glacial basins (Peelo, Drente and Dogger Bight formations and in those upper parts of the Peize and Appelscha formations that have glacigenic input; Figs 10.1, 10.3). The glacial sediments are primarily derived from erosion and reworking of older sediments and are mixed with Fennoscandian gravel (crystalline fragments, flints and chalk) and other reworked elements. As a whole, glacigenic systems may not have left large sediment volumes in the North Sea Basin, but they did have a severe impact on the evolution of landscapes and depocentres in the form of subglacial erosion (i.e. tunnel valleys and glacial basins), glaciotectonic deformation and sedimentation (i.e. push moraines and sandurs; Van den Berg & Beets, 1987; Bakker & Van der Meer, 2003). This strongly affected the evolution of the pathways and positions of successive fluvial and marine systems.

Evidence for grounded ice sheets in the northern and central North Sea dates back as far as MIS 100 (2.53 Ma) while the Fennoscandian and British ice sheets first became confluent ca. 1.9 Ma ago (Bendixen et al., 2018; Rea et al., 2018). In the on- and offshore part of the northern Netherlands, at least five levels of glacial evidence have been identified to date (Fig. 10.1), each marked by a set of erosional, sedimentary and/or glaciotectonic features.

Small-scale fluvial, fluvioperiglacial and eolian systems

Small-scale fluvial, fluvioperiglacial and eolian sedimentary systems produced deposits (primarily belonging to the Stramproy, Drachten, Koewacht and Boxtel formations) that are intercalated with, and overlie sediment left by the larger-scale 'main' depositional systems (Figs 10.1, 10.3). These deposits are dominated by medium- and fine-grained sands, loam and local peat strata. Mainly they lack carbonate and contain a heavy-mineral spectrum that is more stable than the mineralogy of the Rhine sediments. The sediments were primarily deposited under periglacial climatic conditions in areas beyond the limits of the Rhine-Meuse activity at the time (Westerhoff, 2009). The thickest record of these sediments occurs in the Roer Valley Graben (Schokker et al., 2005; Westerhoff, 2009), as infills of glacially eroded topography (Kasse et al., 2022), directly below the late Saalian till (Zagwijn, 1973) and as infills of some Late Pleistocene river valleys (for example the Scheldt Valley; Kiden, 2006; De Clercq et al., 2018; Kasse et al., 2020). In major parts of the on- and offshore Netherlands, the uppermost section of the Quaternary sediment sequence is formed by a 1- to 2-m-thick eolian sand sequence, referred to as 'coversands' (Kasse et al., 2007). In Limburg, this upper Quaternary sequence is primarily represented by loess that is part of the northwestern fringe of the European loess belt (Meijs et al., 2013; Lehmkuhl et al., 2021).

Reworking by tidal and open-marine systems

The North Sea Basin is mostly portrayed as a sink that has trapped riverine and glacially transported sediment produced under terrestrial conditions north and east of the basin. Extensive shelf-edge deltaic units, formed especially in the Early Pleistocene, predate the major glaciations. The Dutch marine realm was farthest away from source areas. It was a place where, once deposited, such terrestrial sediments were reworked during transgressive and highstand conditions. Where rapid subsidence quickly moved sediment to depths beyond the reach of marine erosive processes, reworking was limited to single highstands. Farthest south, close to the basin edge, repeated reworking during successive highstands has resulted in a fragmentary and highly discontinuous stratigraphy. Open-marine transport and deposition, causing some of the reworking, were dominated by tidal and wave processes. Sand-wave fields and other large-scale marine bedforms displace sands over large areas but small distances, and transgressive surfaces typically leave coarse wave-formed lags. Marine muds indicate a strongly mixed provenance far away from modern-day river mouths. They are derived in part from eroding British cliffs and from seabed scouring of Belgian Paleogene clay (especially in the Late Pleistocene and Holocene). All of these fines are tidally circulated in the North Sea and are relevant to both the nearshore and offshore realms (Beets & Van der Spek, 2000; Rijsdijk et al., 2005).

Closer to continuously migrating shorelines, tidal currents have caused local but volumetrically more significant erosion and (re)deposition. Because of rapidly reorganising and partly cyclical coastal landscapes, stacked generations of channel sediments are common (Van der Spek, 1994). Although tidal and other coastal units dominate the Holocene record, their preservation potential has typically been limited. Pre-Eemian coastal deposits are rare.

Evolution of southern North Sea sedimentary systems and landscapes

Early Pleistocene Subseries, Gelasian Stage (2.58-1.8 Ma)

Climate, sea level, ice sheets and biota

Stable-isotope records of North Atlantic planktonic foraminifers, further referred to as the marine oxygen isotope record, show that the chronostratigraphic Gelasian Stage was characterized by an obliquity-forced cyclicity (41 kyr) of alternating relatively warm (to temperate) and cool climatic phases (Fig. 10.1). Very low $^{18}O/^{16}O$ isotope values indicate that several cycles were characterized by very cold climatic conditions (or glacial minima). Although average temperatures were generally lower than in the preceding Pliocene, the markedly higher frequency of these very cold climatic conditions makes the Gelasian critically different. The earliest phase of this stage (2.6-2.4 Ma) is characterized by a distinct triplet of such glacial minima (MIS 100, 98 and 96; Fig. 10.1), also known as the Praetiglian Regional Stage in northwestern Europe (Zagwijn, 1998; Cohen & Gibbard, 2019; Fig. 10.1). This interval is marked by the first extensive development of steppe-like vegetation (Zagwijn, 1957, 1960) and faunal turnover (Meloro et al., 2008) and was accompanied by extensive ice-sheet development over Fennoscandia and associated North Sea iceberg-rafting events (Rea et al., 2018; Løseth et al., 2022). It effectively marks the onset of the ice-age period.

Fossil pollen records indicate that for most of the following Tiglian Regional Stage, vegetation patterns changed from boreal forest during cool climatic phases, periodically with permafrost (e.g. the Beerse Glacial of Kasse, 1993; Kasse & Bohncke, 2001), to deciduous forest during the more temperate phases. Even during these warmer times, many thermophile species that existed in the area during the Pliocene did not return or became incrementally regionally extinct (Zagwijn, 1998; Donders et al., 2018; Westerhoff et al., 2020). The deciduous forests were characterized by a diverse arboreal vegetation that included exotic taxa such as *Eucommia, Tsuga* and *Pterocarya* (Westerhoff et al., 2020; Fig. 10.4).

The climate variations of the Tiglian Regional Stage are also reflected in successions of warm temperate to arctic palynological and molluscan assemblages in marine sediment (Meijer et al., 2006; Donders et al., 2018). Turnover events (including extinction/extirpation) were frequent, leading to overall impoverishment of the fauna. At the

Figure 10.4. Artist impression of the Tegelen landscape and biota along the Rhine River, with rich floodplain forests including thermophile taxa such as the Tegelen macaque (Macaca sylvanus florentina). *Drawing E.J. Bosch, Naturalis.*

same time, the proportion of species with a Pacific origin increased (Slupik et al., 2007; Preece et al., 2020).

On the basis of extensive work near the Tiglian type locality in the Venlo Graben and in the Roer Valley Graben in the southeastern Netherlands (see also Textbox 1), Zagwijn (1963, 1992) suggested that during the Tiglian only two cold intervals occurred, which he referred to as 'Tiglian B' and 'Tiglian C4' (Fig. 10.1). This vision was formally incorporated into the chronostratigraphic subdivision and is still widely used as a standard throughout northwestern Europe. Westerhoff et al. (2020), however, showed that the Tiglian record in the Roer Valley Rift System includes more cold-climate intervals, reflected in peaks of non-arboreal pollen or 'Tiglian B' associations, than had hitherto been identified. The mismatch between the number of observed Tiglian climate oscillations in the terrestrial record (given the meaning of incorporated information) and the many more isotopic oscillations that are reported from the ocean-floor, as well as the link to periods of grounded icebergs that are observed in marine records from the North Sea Basin (Rea et al., 2018), make this interval a focus of ongoing study.

Sedimentary systems and paleogeography

As in the Late Pliocene situation (Fig. 10.5a), the Gelasian North Sea was an embayment open to the Atlantic Ocean to the northwest, but without a marine passage towards the southwest (Fig. 10.5b,c). The Baltic River System entered the marine area from the east, while the Rhine-Meuse System entered from the south. As a result of the marked increase in climate-driven sediment supply, the early Gelasian, and in particular the Tiglian Regional Stage (Fig. 10.5b), was dominated by extensive deposition of fluvial sediments in the present onshore region (Peize and Waalre formations) and large-scale progradation of their equivalent shelf-delta complexes in the offshore (upper part of the Oosterhout Formation; Maassluis, Westkapelle Ground and IJmuiden Ground formations; Fig. 10.1; Overeem et al., 2001; Arfai et al., 2018). The deeper parts of the contemporaneous North Sea had initial water depths up to 300 m but became rapidly filled by shelf deposits arranged in amalgamated clinothems (Fig. 10.3a; Arfai et al., 2018; Lamb et al., 2018; Rea et al., 2018). As a consequence of the associated progradation (Fig. 10.5), the marine depositional area decreased in size and became shallower, while depositional rates increased by an order of magnitude compared to the preceding Neogene period (Kooi et al., 1998; Knox et al., 2010). The Gelasian Stage can be regarded as a period of under-accommodation as clinoform progradation resulted in rapid basin infill. As a result, the depocentre and the coastlines shifted towards the north to northwest (Funnell, 1996; Ottesen et al., 2018).

During the Gelasian, the Rhine followed alternate courses through the Roer Valley Graben, over the Venlo Graben and through the Lower Rhine Graben (Kieseloolite and Waalre formations; Fig. 10.1). Towards the late Gelasian, the southern Roer Valley Graben was abandoned and deposition was concentrated in the northern part of the Roer Valley Graben and in an adjacent area towards the west (Fig. 10.5c). Abandonment of the southern Roer Valley Graben initiated a period of deposition of the Campine River System (which gradually extended northward during latter periods of the Early Pleistocene). Extensive floodplains and meandering channels characterized the fluvial systems during the warm(er) climatic intervals of the Gelasian (Fig. 10.6). These environments graded into estuarine and shallow-marine settings to the west and northwest (Kasse, 1988; Westerhoff, 2009; Fig. 10.5b,c). Towards the end of the Gelasian, continued progradation of the Baltic fluviodeltaic system to the west-southwest resulted in interfingering with the Rhine-Meuse System in the central Netherlands, as reflected in spatio-temporal alternations of the Waalre and Peize formations. Throughout the Gelasian, the Meuse (Beegden Formation) flowed through the East Meuse valley (Fig. 10.5b,c; Textbox 2).

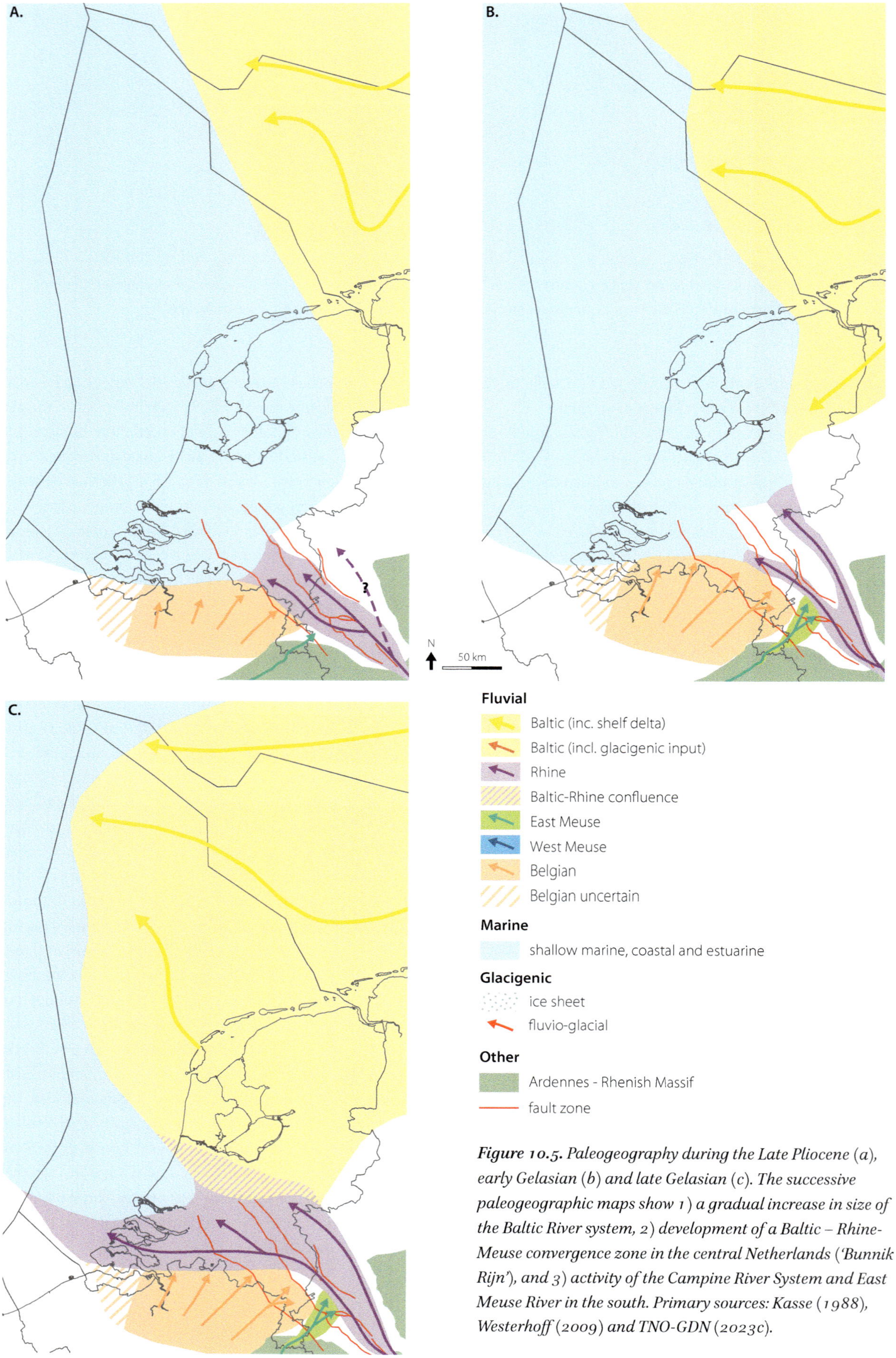

Figure 10.5. *Paleogeography during the Late Pliocene (a), early Gelasian (b) and late Gelasian (c). The successive paleogeographic maps show 1) a gradual increase in size of the Baltic River system, 2) development of a Baltic – Rhine-Meuse convergence zone in the central Netherlands ('Bunnik Rijn'), and 3) activity of the Campine River System and East Meuse River in the south. Primary sources: Kasse (1988), Westerhoff (2009) and TNO-GDN (2023c).*

Figure 10.6. a) Tiglian oxbow lake fill with bedded clay being excavated in the Maalbeek pit reprinted from Westerhoff et al. (2020) with permission from Elsevier. b) Detail of Tiglian flood-basin clay (Waalre Formation) in the Hoher Stall pit (photo: W. de Weger).

Early Pleistocene Subseries, Calabrian Stage (1.8-0.8 Ma)

Climate, sea level, ice sheets and biota

Like the earlier Gelasian, the Calabrian Stage was globally characterized by glacial-interglacial cycles (Fig. 10.1). However, from the onset of the Calabrian by ca. 1.8 Ma, particularly cold glacial minima occurred more frequently than before. Phases with steppe-like vegetation and cryoturbation, indicating temporal permafrost conditions, left their mark in the terrestrial record. In the regional chronostratigraphy (Fig. 10.1), the resulting glacial complex is known as the Eburonian. Provisional correlation with the marine isotope record aligns this complex with the minima MIS 54 to 46, for which evidence of iceberg rafting and grounding has been observed in marine sequences of the North Sea Basin (Rea et al., 2018). In the terrestrial record, the Eburonian is succeeded by a generally warmer period, referred to as the Waalian Regional Stage, with pollen indicative of forest cover. It is provisionally aligned to MIS 47 to 37 and lasted until about 1.2 Ma (Fig. 10.1).

During the last 400 kyr of the Calabrian Stage (ca. 1.2-0.8 Ma), the periodicity of the Earth's glacial-interglacial climate cycle began to change significantly, a phenomenon known as the Mid-Pleistocene Revolution or the Early-Middle Pleistocene Transition (Head & Gibbard, 2005; Maslin & Ridgwell, 2005; Berends et al., 2021; Head, 2021). This time interval saw intensification of cold periods and cycles of ice-sheet growth and decay lengthening to 100 kyr. The cold intervals were characterized by gradual cooling and contrasting rapid terminations (Fig. 10.1), mirrored in global sea-level minima some 100-140 m lower than at present, and maxima up to ca. 10 m above present sea level. The Mid-Pleistocene Revolution is the first post-Gelasian phase during which major ice-sheet expansion occurred in the North Sea region, although the extents of these ice sheets are not well-constrained (Ottesen et al., 2018; Rea et al., 2018; Hughes et al., 2020). In the terrestrial record onshore and in the regional stratigraphic scheme, the onset of the Mid-Pleistocene Revolution is provisionally correlated to MIS 34 (Margari et al., 2023; Fig. 10.1) and associated with the Menapian Regional Stage, which has intervals characterized by a non-arboreous palynology (Zagwijn, 1960; Zagwijn & De Jong, 1984; De Jong, 1988).

The marine oxygen isotope record shows that this phase was followed by pronounced alternations of very warm interglacials and colder climatic phases (Bavelian Regional Stage; Fig. 10.1). Owing to their intensity, some of these interglacials are referred to as 'super-interglacials' (e.g. MIS 31; Oliveira et al., 2017; Tzedakis et al., 2017). In the onshore record, three warm phases are reported within this period: the Bavel, Leerdam and Waardenburg interglacials. Originally, Zagwijn (1996b) considered the latter to be part of the Mid-Pleistocene Cromerian Regional Stage. In light of the international decision to align the onset of the Middle Pleistocene with the Brunhes-Matuyama reversal

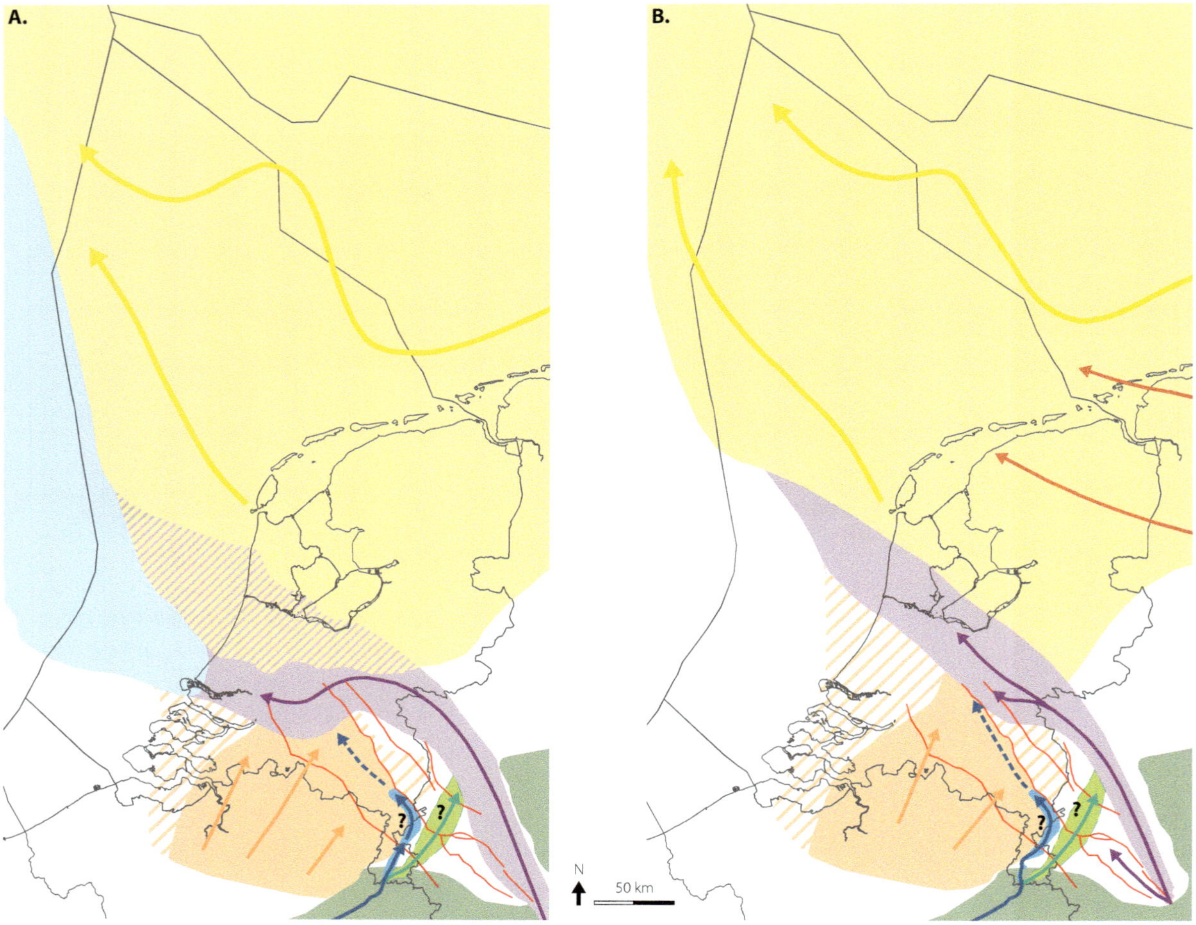

Figure 10.7. Paleogeography during a) the middle Calabrian (Waalian Regional Stage) and b) late Calabrian Stage (Menapian to earliest Bavelian Regional Stages). The successive paleogeographic maps show 1) a gradual demise of the Baltic River System, 2) first evidence of Fennoscandian influence, and 3) expansion of the Campine River System into the Roer Valley Rift System (see Fig. 10.5 for legend). Primary sources: Kasse (1988), Westerhoff (2009) and TNO-GDN (2023c).

(e.g. Chibanian Global Boundary Stratotype Section and Point; Suganuma et al., 2021), our preference is to reassign the Waardenburg interglacial and its overlying 'Glacial A' to the Bavelian (Fig. 10.1). Current insight favours correlation of the Bavel interglacial to MIS 31, Leerdam to MIS 25 and Waardenburg to MIS 21 (Fig. 10.1). Palynologically, these interglacials are the last to have assemblages that include the 'typical' Early Pleistocene taxa *Eucommia* and *Tsuga* (Zagwijn & De Jong, 1984; De Jong, 1988; Preece & Parfitt, 2012). Contemporaneously, critical turnovers (extinctions and arrivals of new species) in molluscan assemblages took place (Meijer, 1988, 1991), along with immigration of Eurasian mammal species (Van Kolfschoten, 2001).

Sedimentary systems and paleogeography

At the onset of the Calabrian, the southern North Sea had shallowed considerably compared to the beginning of the Pleistocene, following major sediment infill (Kuhlmann et al., 2006a,b; Ottesen et al., 2018). At the same time, increasingly accentuated glacial-interglacial sea-level cycles deepened lowstand levels. This combination set the stage for major change in terms of landscape development and fluvial-marine configurations (Fig. 10.7a,b).

During the cold-climate Eburonian (Fig. 10.1), marine influence was absent from the present-day Dutch onshore, but during the Waalian, marine conditions returned to the western Netherlands (Zagwijn, 1975; Kasse, 1988; Fig. 10.7a). The Rhine repositioned itself eastwards to the Lower Rhine Graben, largely abandoning the Roer Valley Rift System (Westerhoff, 2009), and continuing its convergence with the Baltic River System in the central Netherlands. This led to the subsequent expansion of the Campine River System (Stramproy Formation). Contradicting views exist on the position of the Meuse during this period. Westerhoff et al. (2008) suggested the East Meuse Valley was still an active branch during this phase and correlations to the Holzweiler Formation in Germany point to active deposition up to the early Bavelian Regional Stage (Boenigk, 2002; Boenigk & Frechen, 2006). Other age

Figure 10.8. *Artist impression of a typical Mid-Pleistocene interglacial landscape with rich floodplain forests where early humans* (Homo heidelbergensis) *lived alongside thermophile taxa such as forest elephant, rhinoceros and hippopotamus. Drawing E.J. Bosch, Naturalis.*

models suggest that the East Meuse Valley was abandoned earlier, around the Gelasian-Calabrian transition (Zagwijn, 1975; Kuyl, 1980; Kasse, 1988; Van den Berg & Van Hoof, 2001). As new dating of the East Meuse sediments is needed to solve this puzzling issue, both scenarios are presented in the paleogeographic reconstruction for this period (Fig. 10.7a).

The intensified periodic glaciation of Fennoscandia during and following the Mid-Pleistocene Revolution triggered major reorganization of sediment delivery and depositional system interactions in the North Sea area at large (Bijlsma, 1981; Gibbard, 1988). Basin shallowing in the northern and central North Sea facilitated the expansion of Scandinavian and British-Irish ice sheets onto the North Sea shelf (Sejrup et al., 1995; Batchelor et al., 2019). The northern Danish subsurface bears sedimentary-stratigraphic indications that the Fennoscandian ice sheet first extended that far south by ca. 1.2-1.0 Ma (Houmark-Nielsen, 2004), correlating to the Menapian periglacial indicators in the Netherlands. Seismic records from the northern and central North Sea feature the Crenulate Marker, attributed to shelf glaciation at this time (Buckley, 2017; Ottesen et al., 2018). Sejrup et al. (1995) and Ottesen et al. (2014) suggested that initiation of the Norwegian Channel Ice Stream at the northern fringe of the basin could be of the same age, but more recent results point to a likely younger initiation around 0.8 Ma (Løseth et al., 2022).

The Fennoscandian ice-sheet expansions transformed the hinterland of the Baltic River drainage basin. This upstream change is reflected in reworked erratics in the top of Baltic River units in the Netherlands (Hattem beds, Peize Formation; Zandstra, 1971; Fig. 10.1), suggesting proximity of an ice sheet during the Menapian glaciation (Fig. 10.7b). Remnant tributaries of the Baltic River System (the Weser and Elbe Rivers) draining Central Germany and Bohemia remained active as sediment suppliers from the east (Appelscha Formation), but with strongly reduced sediment supply. During the Menapian, the Rhine remained in its position outside of the Roer Valley Rift System. Towards the northwest the Rhine-Meuse System started to dominate the shrinking Baltic River System (Fig. 10.7b). Shallow-marine deposition was restricted to the northwestern offshore (just northwest of the map domain of Fig. 10.7b; Outer Silver Pit and/or Aurora formations; Fig. 10.1).

At the end of the Menapian or in the early Bavelian, the Rhine-Meuse System reestablished a course through the Roer Valley Graben (Zagwijn, 1985; Kasse, 1988; Fig. 10.7b; Zagwijn, 1996b). From this moment onwards, the Rhine-Meuse deposits show considerable coarsening (Sterksel Formation; Kasse, 1988). This trend is attributed to a combination of frost weathering and intensified periglacial processes in hinterland and valleys, which favoured the production of coarse sediment during the cold climatic phases that are reported for this period (Linge, Dorst and Glacial A glacials; Fig. 10.1). Increased uplift rates in the Rhenish Massif and Ardennes were fluvially countered by bedrock incision (Van Balen et al., 2000; Meyer & Stets, 2002; Rixhon & Demoulin, 2018) and further added to bedload reaching the Netherlands. Vast volumes of coarse-grained sand and gravel ended up in the fluvial domains.

Mid-Pleistocene Subseries (773-130 ka)

Climate, sea level, ice sheets and biota

The North Atlantic isotope record shows that during the Middle Pleistocene (0.78-0.13 Ma) climatic cyclicity was dominated by 100-kyr periodicity, with an overprint of shorter obliquity- and/or precession-paced warm to cool climatic phases (Fig. 10.1). In general, each 100-kyr cycle began with a very warm interglacial phase followed by stepwise cooling (alternating interstadials and stadials) that culminated in a very cold glacial minimum, of which MIS 16, 12 and 6 were the most pronounced. The marine oxygen isotope record shows that MIS 19 and 11c were the most strongly developed interglacials (Gibbard & Lewin,

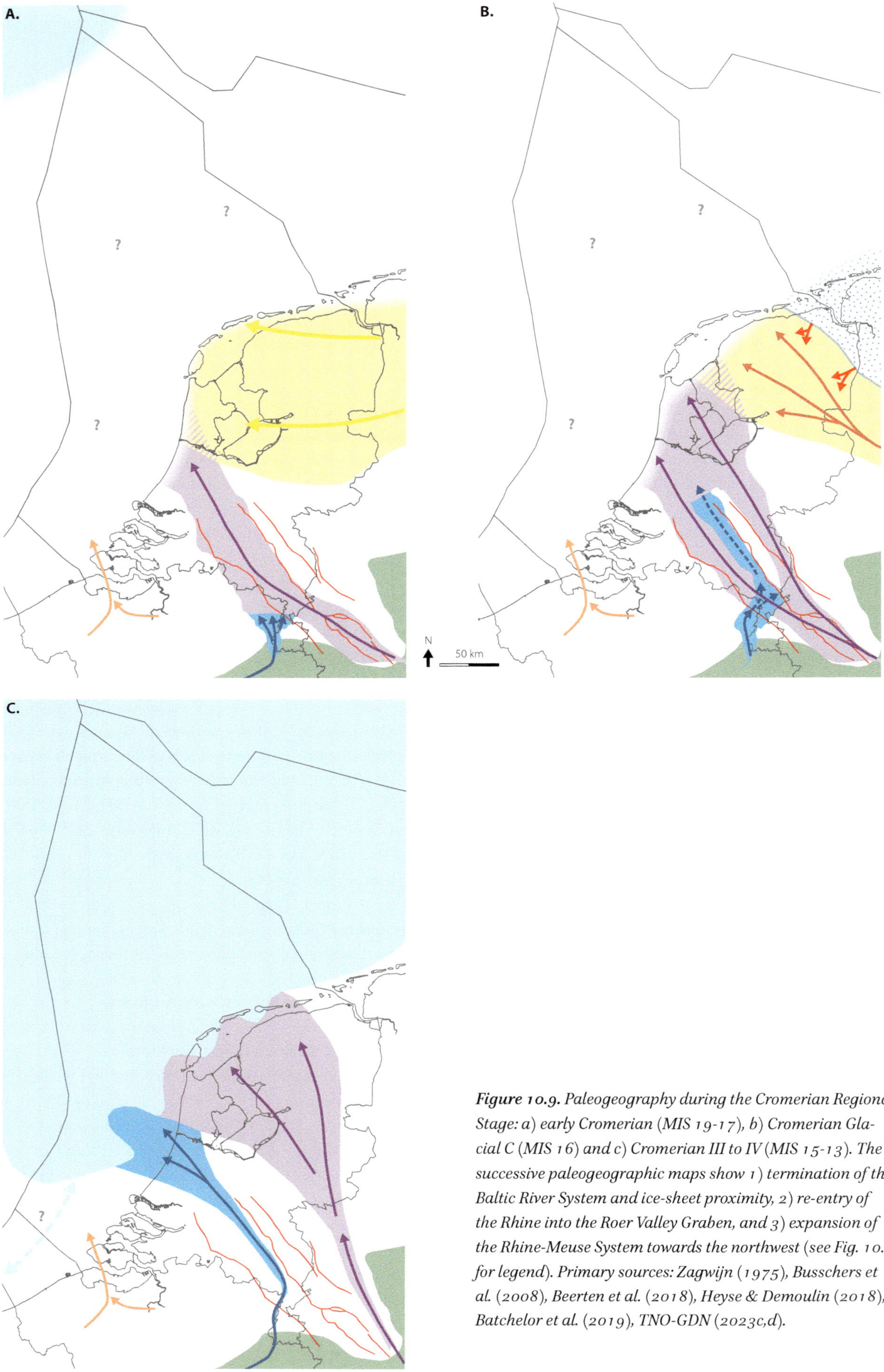

Figure 10.9. Paleogeography during the Cromerian Regional Stage: a) early Cromerian (MIS 19-17), b) Cromerian Glacial C (MIS 16) and c) Cromerian III to IV (MIS 15-13). The successive paleogeographic maps show 1) termination of the Baltic River System and ice-sheet proximity, 2) re-entry of the Rhine into the Roer Valley Graben, and 3) expansion of the Rhine-Meuse System towards the northwest (see Fig. 10.5 for legend). Primary sources: Zagwijn (1975), Busschers et al. (2008), Beerten et al. (2018), Heyse & Demoulin (2018), Batchelor et al. (2019), TNO-GDN (2023c,d).

2009; Tzedakis et al., 2017). Sea level oscillated semi-synchronously with the glacial-interglacial cycles, producing interglacial highs within metres of modern-day levels and sea-level drops of more than 100 m during glacial minima.

On the basis of pollen analysis of fine-grained sediments in fluvial and estuarine sequences of the Sterksel, Urk and Beegden formations, Zagwijn & Zonneveld (1956), Zagwijn et al. (1971) and Zagwijn (1996b) defined several Middle Pleistocene interglacial and glacial intervals (Fig. 10.1). The pollen records indicate that the warmest phases were characterized by deciduous forest vegetation. During cooler conditions, boreal forest and/or steppe-like vegetation dominated the region (Zagwijn, 1996b; Fig. 10.8).

Within the Cromerian three interglacials are recognized (Fig. 10.1). A critical difference with older interglacial periods is that their pollen spectra lack the thermophile tree taxa *Eucommia* and *Tsuga*. Details of vegetation successions into the interglacials differ considerably between the interglacials (Zagwijn, 1996b; Westerhoff et al., 2003). The Cromerian was followed by the Elsterian glacial (MIS 12) and Holsteinian interglacial (MIS 11). Holsteinian pollen spectra are the last to include the thermophile taxon *Pterocarya*. The longer Saalian complex (MIS 10-6) completes the Middle Pleistocene, with three glacials and two interglacials dominated by boreal forest (Fig. 10.1).

The coldest climatic phases were characterized by open vegetation types and/or polar desert conditions, development of continuous permafrost, and extensive ice-sheet advance. Known advances of the Fennoscandian and British ice sheets close to or into Dutch territory took place during the Cromerian Glacial C (MIS 16), the Elsterian (MIS 12) and the late Saalian (MIS 6) (Van den Berg & Beets, 1987; Laban & Van der Meer, 2011; Batchelor et al., 2019; Hughes et al., 2020). In the central and northern North Sea multiple advances of the Middle Pleistocene ice sheets led to the formation of a marked glacigenic unconformity (Upper Regional Unconformity; Ottesen et al., 2014, 2018).

Sedimentary systems and paleogeography

Middle Pleistocene up to the Elsterian (773-478 ka)

During the earliest Cromerian, the Rhine was positioned in the Roer Valley Graben (Sterksel Formation) (Fig. 10.9a). The Meuse and Rhine converged in the southern part of the graben. Towards the northwest, the Rhine appears to have merged with remaining easterly rivers (Appelscha Formation) and their depositional systems jointly extended into the marine realm (Yarmouth Roads and Batavier formations). The southwest of the Netherlands and adjacent parts of Belgium began to be incised by local river systems, while the Scheldt and its tributaries started developing the Flemish Valley in the uplifting Campine Block (Heyse & Demoulin, 2018; Fig. 10.9a).

For Glacial C (MIS 16) in the middle of the Cromerian (Fig. 10.9b), specific information comes from fluvioglacial deposits (Weerdinge Member, Appelscha Formation) exposed in the Emmerschans quarry in the northeastern Netherlands (Ruegg & Zandstra, 1977; Zagwijn, 1985). The Weerdinge Member is regarded as a glacial outwash and ice-marginal river product, indicating a contemporaneously active ice front to the northeast in northern Germany and its offshore area that predates more prominent glaciation of the same area during the Elsterian (MIS 12). The Rhine deposited significant volumes of coarse-grained sediments in the Roer Valley Graben and Venlo Graben (Sterksel Formation, Fig. 10.9b). Their compositional signature matches key levels in Rhine terrace flights in the German part of the Roer Valley Rift System record (Boenigk, 1978; Zagwijn, 1985), for which age constraints point to MIS 16 (Upper Terrace 3, cf. Boenigk & Frechen, 2006). A similar age correspondence follows from recent cosmogenic dating (Beerten et al., 2018; Vandermaelen et al., 2022) of morphostratigraphically equivalent terrace levels along the Meuse in Belgian borderland (Gullentops et al., 2001). During MIS 16 Glacial C, the Rhine-Meuse System of the southern and west-central Netherlands and the ice-marginal outwash river of the northeast were likely confluent in the northwestern Netherlands (Fig. 10.9b).

During or shortly after Glacial C, the Rhine part of the depositional system abandoned the Roer Valley Graben, although the Meuse System kept using this route for the remainder of the Cromerian (Fig. 10.9b,c). Late Cromerian interglacial depositional evidence is available both from Meuse deposits in the Roer Valley Graben (Rosmalen Member, Beegden Formation) and from Rhine deposits traced through the central and northern Netherlands (Veenhuizen Member, Urk Formation). These areas host the respective palynological type sites for Cromerian III and Cromerian IV (Zagwijn, 1985, 1996b), correlated to MIS 15 and MIS 13 respectively (Fig. 10.1). A minor cold period in between explains the compositional differences between the associated units. A review of geological mapping evidence indicates that no major glacial interval separates the two interglacials, in contrast to the suggestion of Zagwijn (1985, 1996b). Therefore, both the Weerdinge Member ('Glacial C' of Zagwijn, 1986, 1996b) and the upper part of Sterksel Formation ('Glacial B' of Zagwijn, 1986, 1996a), are attributed to a single major northwest European glaciation episode, for which the name 'Glacial C' is retained and the correlation to MIS 16 upheld (Fig. 10.1).

In the late Cromerian (MIS 15-13), the Rhine followed a relatively easterly course through the central Netherlands (Veenhuizen Member; Zagwijn, 1975; Busschers et al.,

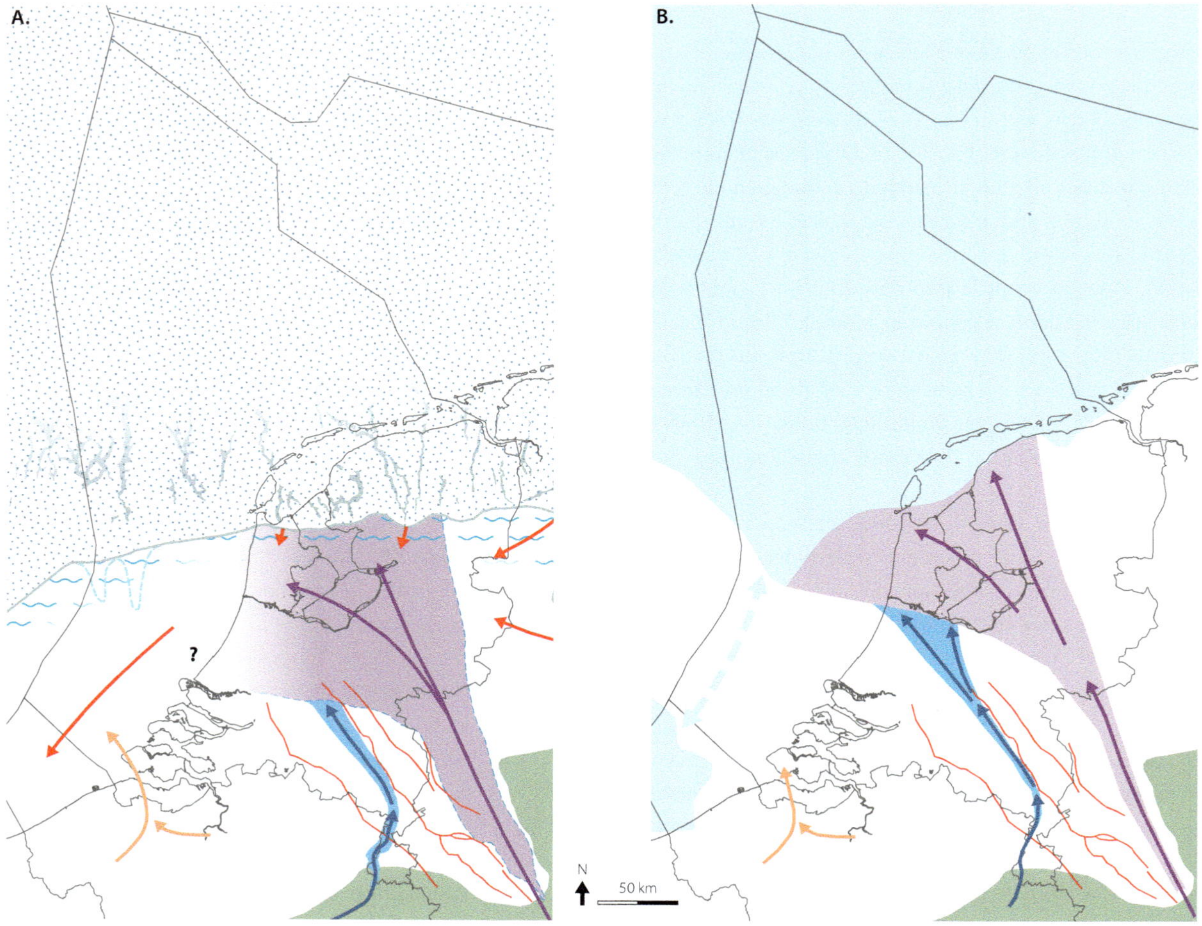

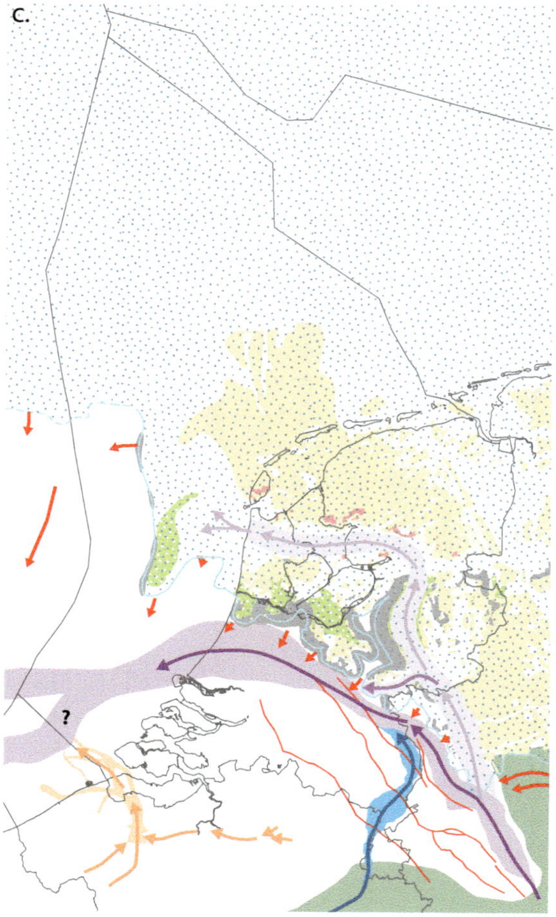

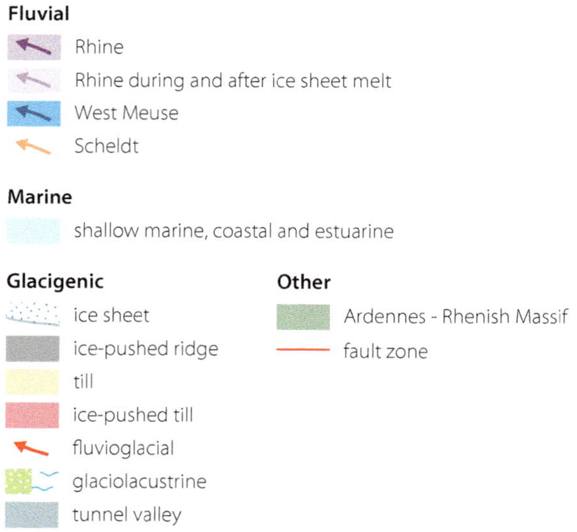

***Figure 10.10.** Paleogeography during a) the Elsterian glaciation (MIS 12), b) Holsteinian interglacial (MIS 11) and c) Saalian glaciation (MIS 6). The successive paleogeographic maps show 1) periods of large-scale glaciation of both the on- and offshore sector, 2) major (glaciation-controlled) changes in landscape topography and position of depocentres, and 3) fluvial and marine exchange between the North Sea Basin and the Channel area. Late Saalian (c) position of the Rhine near question mark and the dotted ice-sheet position in the west are uncertain. Primary sources: Zagwijn (1975), Laban (1995), Busschers et al. (2008), Moreau et al. (2012), Batchelor et al. (2019), Cartelle et al. (2021), Breuer et al. (2023), TNO-GDN (2023c) and TNO-GDN (2023d).*

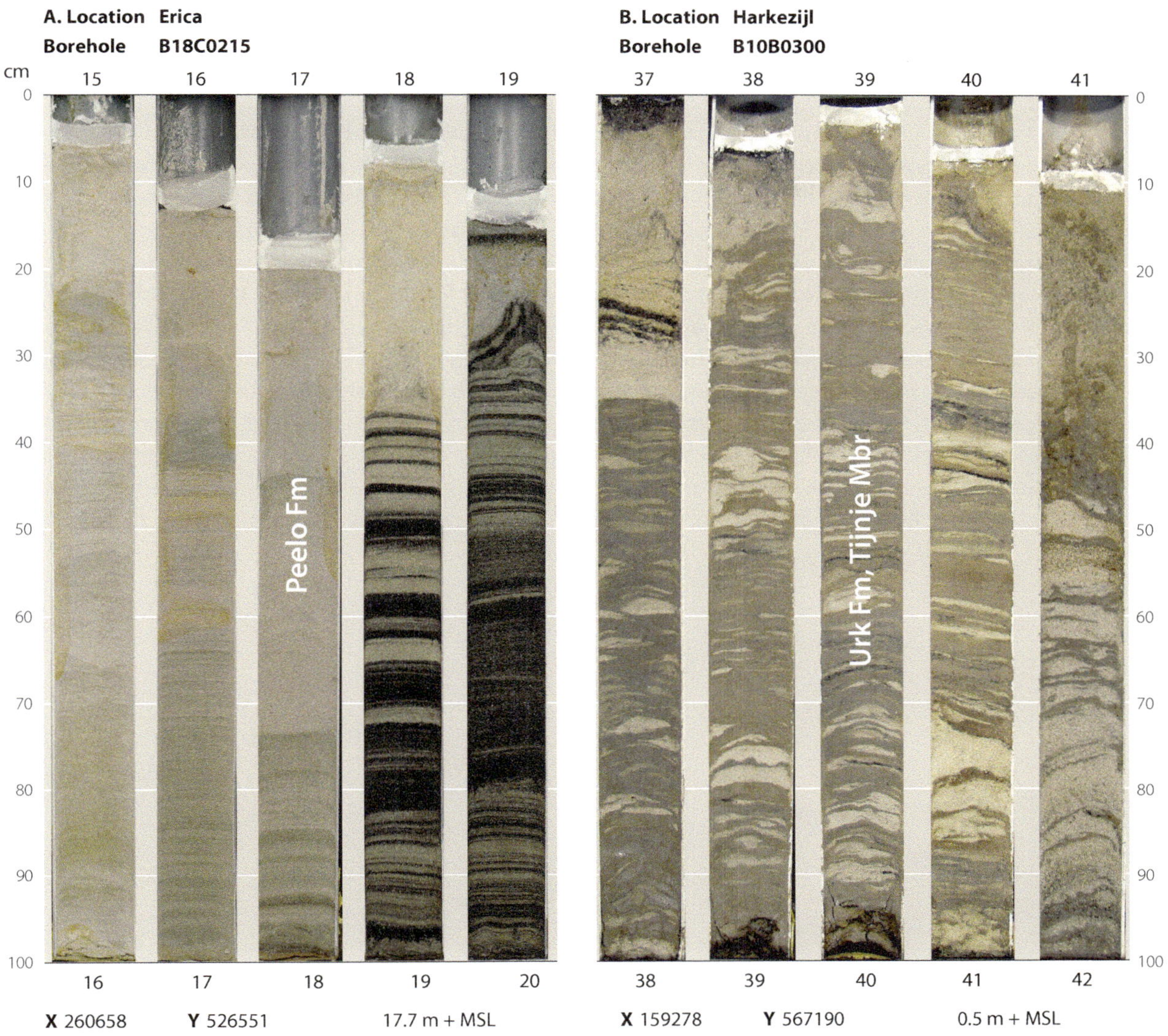

Figure 10.11. a) Elsterian glaciolacustrine sands with parallel lamination and climbing ripples in a core from the northeastern Netherlands (Peelo Formation). The base of the core shows high concentrates of organic fragments reworked from underlying peat layers of the Urk Formation (Veenhuizen Member). b) Holsteinian tidal sediments (heterolithic bedding) in a core from the northern Netherlands (Tijnje Member, Urk Formation). Coordinates in Rijksdriehoeksstelsel. Core photos: TNO-GDN, 2023d.

2008), widening into a coastal-deltaic plain with estuarine mouths aligned northwest, grading into shallow-marine deposits (Fig. 10.9c) and continuing as part of the Yarmouth Roads Formation in the offshore domain (Zagwijn, 1996b). Interglacial mollusc assemblages from the coastal and estuarine levels have a Lusitanian signature suggesting early marine exchange between the North Sea and English Channel (Wesselingh et al., 2023). The MIS-13 attribution of Cromerian-IV interglacial sites within the Veenhuizen Member is corroborated by the presence of clino-pyroxene (augite) in the sediments: a heavy mineral released by Eifel volcanic systems active since around 0.5 Ma (Fig. 10.1). Dating of the entry of augite was established using tephra layers in sequences bearing assemblages of small mammal species along the Middle Rhine in Germany (Van Kolfschoten, 1990; Van Kolfschoten & Turner, 1996; Gallant et al., 2014).

Late Saalian glacial landforms and sediments

3

Glacial deposits of the late Saalian glaciation (Drente Formation), show variable compositions and landforms. Till sheets formed in the northern part of the Netherlands (Gieten Member) and the adjacent part of the offshore sector as the ice advanced over fine-grained sediments (Fig. 10.10c). Thicker accumulations of till are limited to standstill and terminal locations where ice-marginal deposition and the formation of small-scale glaciotectonic ridges took place. Elongated, parallel ridges of up to 70 km length developed in the northeastern Netherlands, consisting of till and deformed preglacial material. This major geomorphological feature is known as the Hondsrug megaflute complex, formed below an ice stream that drained the ice sheet from the northwest (Van den Berg & Beets, 1987).

In the central Netherlands, the ice sheet advanced over coarse-grained fluvial sediments. Meltwater drainage from the permeable glacier bed resulted in coupling of the ice sheet with the substrate, and in the formation of large ice-pushed ridges (or push moraines) and associated glacial basins (Fig. 10.10c). The ice-pushed ridges consist of stacked (thrust) sheets of preglacial and fluvioglacial sediments (see figures a and b below). In the central Netherlands and the North Sea, these are mainly Pleistocene gravelly sands with few clay and loam beds. In the eastern Netherlands, Neogene units have been glaciotectonised as well. Fluvioglacial outwash was deposited prior to and during

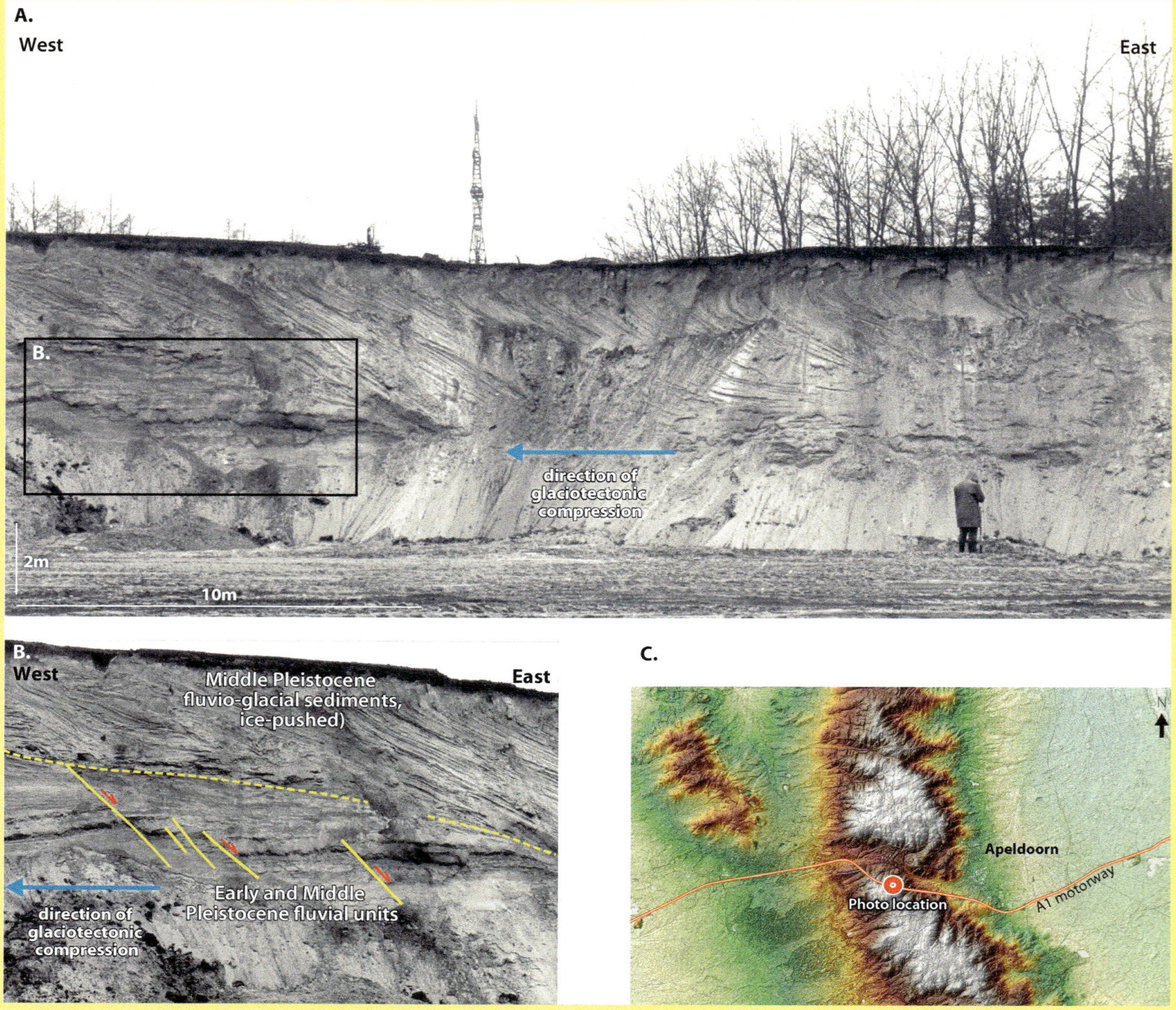

a) Exposure of glaciotectonied sediments visible during construction of the A1 motorway west of Apeldoorn in 1970 (photo taken by Van der Veen; archive TNO-GDN). b) Close-up of glaciotectonic sequence with low-angle thrusts and relaxation structures (photo taken by Van der Veen; archive TNO-GDN). c) Lidar DTM (AHN4) of the Veluwe ice-pushed ridge with the location of the exposure shown in a and b.

the formation of ice-pushed ridges (Schaarsbergen Member). Till patches are present in the glacial basins, on the proximal flanks of the ice-pushed ridges and partly in ice-pushed position. During deglaciation, the initial infill of the glacial basins comprised fine-grained glaciolacustrine sediments and coarser-grained mass-flow deposits (Uitdam Member), the latter originating from the steep slopes, flanking ice-pushed ridges (Van Leeuwen et al., 2000; Beets & Beets, 2003).

After deglaciation, the remaining lows formed depositional sinks for marine, coastal and estuarine sediments of Eemian age and fluvial and eolian sediments of Weichselian age, respectively (e.g. Fig. 10.15). Currently the basins are completely filled, their remnants buried in the subsurface. This is also true for the toes of the ice-pushed ridges in the east and the lower half of the ice-pushed ridges in the central Netherlands. From Amsterdam westward and further below the North Sea (Fig. 10.10c), the ice-pushed ridges are entirely covered by younger sediments. Geophysical data show that postglacial denudation has lowered the crests of the ice-pushed ridges by about 15 m. The eroded sediments have been redistributed in the basins and on the ridge flanks as alluvial fans (Boxtel Formation; Busschers et al., 2007). Other preserved features related to the Saalian glaciation include drumlins, eskers, buried tunnel valleys and deposits of ice-dammed lakes, including kettle-hole fills.

Artefacts found in ice-pushed sediments testify to the late Middle Pleistocene presence of early humans (Stapert, 1987). Aligned series of hand-dug pits on the ice-pushed ridges themselves are relicts of early medieval iron industry. These pits follow outcrops of iron-bearing concretions and reflect the strike of the ice-pushed strata. Currently the ice-pushed ridges provide major groundwater reserves. Hydrogeological aspects include a marked anisotropic character of the aquifers arising from their specific structural architecture.

Middle Pleistocene from Elsterian to Mid-Late Pleistocene Transition (478-130 ka)

The Elsterian (MIS 12) was a glacial period that began with severe cooling and sea-level lowering, followed by far-reaching Fennoscandian ice-sheet advance (Fig. 10.10a). Eventually, this ice sheet covered the northern Netherlands and all but the southernmost parts of the Dutch and English sectors of the North Sea, while coalescing with British ice cover over East Anglia (Anglian glaciation in the British scheme). This arguably most extensive glaciation to ever cover the North Sea region left a vast network of subglacial tunnel valleys that cut deeply into the substrate (Fig. 10.3b). Towards the east, they are marked by considerable outwash deposits at their southern terminations. The irregular network geometry and ups and downs in thalweg depth indicate that overpressured meltwater was the main agent responsible for tunnel-valley formation, with steady-state conditions and catastrophic outbursts fed by supra- and englacial meltwater as key mechanisms (Praeg, 2003; Benvenuti et al., 2018; Kirkham et al., 2024).

Sediments of the Elsterian glaciation are grouped into the Peelo Formation (Fig. 10.1). Usually, the tunnel-valley infill has an erosional lag at the base. It consists of a heterogeneous unit of gravel, sand and diamicton that may contain rock fragments of Neogene to Cretaceous age. The lag is overlain by sandy deposits. Seismic imagery commonly shows characteristic northward-dipping clinoforms. Their precise mode of origin is debated, but they likely were formed synchronously with headward erosion of the valleys during or immediately following ice-front recession (Huuse & Lykke-Andersen, 2000; Praeg, 2003; Moreau & Huuse, 2014). Subsequently, glaciolacustrine conditions prevailed, during which glacial clay ('pottery clay') and fine-grained sands were deposited preferentially in the tunnel valleys and other topographic lows. The clay can be very stiff, reflecting postdepositional dewatering and overcompaction from loading by the younger late Saalian ice sheet.

In the Dutch onshore and offshore, the southern limit of Elsterian glaciation is effectively defined by the southernmost extent of tunnel valleys. In and around the Netherlands, it maps out as a remarkably straight ice-sheet margin (Fig. 10.10a), linking up to more irregular British and German ice-front maxima constrained from tunnel-valley tips and patchy tills (Batchelor et al., 2019; Breuer et al., 2023). Along the Dutch Elsterian margin, glaciotectonic deformation appears to be rare, which is in sharp contrast to the Dutch Saalian margin. This may indicate that it was bounded and held up by a large freshwater body, with a calving ice front and tunnel valleys releasing abundant sediment-laden subglacial water. In the east of the country, remnants of thick glaciolacustrine outwash complexes of Elsterian age have been identified (Peelo Formation, Fig. 10.11a). Erosion during subsequent times makes it difficult, however, to reconstruct the situation westward and to demonstrate lateral continuity of lakes bounded by the ice front. Exceptionally large proglacial lake systems have been suggested, with tentatively projected shoreline positions in the southern Netherlands, but the absence of positively identifiable glaciolacustrine facies in the central Netherlands (e.g. in Saalian ice-pushed complexes), makes it more likely that Elsterian glacial lakes were restricted to zones closer to the ice front. Whether or not spillage of ice-lake waters initiated strong erosion in the Dover Strait, and to what degree it explains cliff and erosional seafloor features over there, is also still a matter of debate (e.g. Gibbard, 1995; Gupta et al., 2007; Hijma et al., 2012; Gupta et al., 2017).

The Elsterian Rhine River System must have merged with easterly proglacial drainage systems some distance south of the ice front in the central Netherlands. It built out coarse-grained braidplain deposits that comprise parts of the Urk Formation (Fig. 10.10a). The exact extent of this braidplain is unknown. From the south, the Meuse braidplain (Beegden Formation) merged with the Rhine ice-marginal system. South of the Rhine braidplain, eolian and fluvio-eolian deposition prevailed (Best Member, Boxtel Formation; Schokker et al., 2005, 2007; Kars et al., 2012; Meijs et al., 2013).

During the Holsteinian interglacial (MIS 11), the Rhine ran a northward course (Fig. 10.10b) towards estuarine and shallow-marine areas in the northern Netherlands (Tijnje Member, Urk Formation). Here, the Rhine record has a complex architecture comprising alternating channel sands and flood-basin clays and peats, (Fig. 10.11b; Bosch, 1990; Lee et al., 2012), analogous to the Holocene coastal prism (see section on the Holocene). This complex holds a palynological record of several warm phases associated with interglacial deciduous forests. The earliest and most widespread of these has been linked to interglacial MIS 11 on the basis of relative intensity and duration as well as the match with traditional Holsteinian pollen spectra (Ehlers, 2011; Stephan, 2014; Tzedakis et al., 2022). A second warm level, although originally less widespread or less well preserved than the basal level, occurs higher in the complex. It is separated from the basal level by coarse-grained cold-stage Rhine sands suggesting that it might correlate to MIS 9. Offshore, this ensemble passes into estuarine and shallow-marine facies of the Egmond Ground Formation (Zagwijn, 1983; Laban, 1995). During the Holsteinian, the Meuse maintained a course through the Roer Valley Graben (Schokker et al., 2005), with estuarine reaches projected in the northwestern Netherlands (Busschers et al., 2008).

During cooling phases following the Holsteinian, the Rhine-Meuse System adopted a new course through the central Netherlands (Fig. 10.10c). Early to middle Saalian (MIS 10-7) Rhine deposits (Urk Formation) are encountered in areas that later were glaciotectonised during the Drenthe glaciation (ca. 160-150 ka, within MIS 6; Busschers et al., 2008). These deposits contain intervals bearing Paleolithic artefacts assigned to MIS 7 and early MIS 6 ('Rhenen industry'; Stapert, 1987; Van Balen & Busschers, 2010; Niekus et al., 2023). Paleolithic human presence during MIS 7 (243-190 ka) is also evident from the Meuse Valley upstream (Maastricht Belvédère, on the Caberg-3 Terrace; see Textbox 2). Estuarine and shallow-marine sediments were deposited in the northwestern Netherlands and offshore area during MIS 9 and possibly MIS 7 (Tijnje Member and Egmond Ground Formation; Meijer et al., 2021).

Saalian sediments belonging to local depositional systems are widespread both in the north and the south of the country. In the north, the Drachten Formation (Fig. 10.1) comprises fluvial, fluvioperiglacial and eolian sands with organic intercalations, overlying the aforementioned Rhine complex of the Urk Formation. Two organic levels record the periodic return of boreal forest during the Hoogeveen and Bantega interstadials (Fig. 10.1), which likely correlate to MIS 7e (ca. 240 ka) and MIS 7c (ca. 215 ka), respectively. Luminescence dating of quartz grains from the upper parts of the Drachten Formation gave an age of ca. 170 ka (Kars et al., 2012) corresponding to MIS 6, just prior to glaciation of the area. In the southern Netherlands, a similar package of fluvial, fluvioperiglacial and eolian sands that accumulated in the Roer Valley Graben along with lacustro-eolian silts, stretches further back in time (Boxtel Formation; Schokker et al., 2005, 2007; Kars et al., 2012). Multiple involuted intercalations encountered in the Boxtel Formation indicate multiple freezing and thawing cycles. The depositional sequence in the Roer Valley Graben holds several peaty beds with pollen assemblages attributed to Holsteinian and early Saalian interglacials (MIS 11 and 9), and to the Hoogeveen and Bantega interstadials (MIS 7) also encountered farther north. Even farther south, loess blanketing of the landscape occurred repeatedly during Saalian stadials (in MIS 10, 8 and 6) alternating with soil formation during interglacials and interstadials (in MIS 9 and 7; Meijs et al., 2013).

Towards the end of the Saalian, during the Drenthe glaciation (MIS 6, ca. 170-140 ka), a major sector of the Fennoscandian ice sheet extended into the Netherlands. The glaciation had a major influence on the paleogeographic developments, sedimentary records and landforms (also see Textbox 3). The lobate ice front advanced from the northeast and eventually covered the northern and central Netherlands (Van den Berg & Beets, 1987). As it overcame temporal stillstands, the ice sheet overrode and deformed its own proglacial outwash and associated ice-marginal river deposits. Once the ice reached basin-rim areas with their coarser-grained substrate of earlier Rhine-Meuse and easterly river formations, it glaciotectonised, froze up and bulldozed that substrate into a series of pronounced ice-

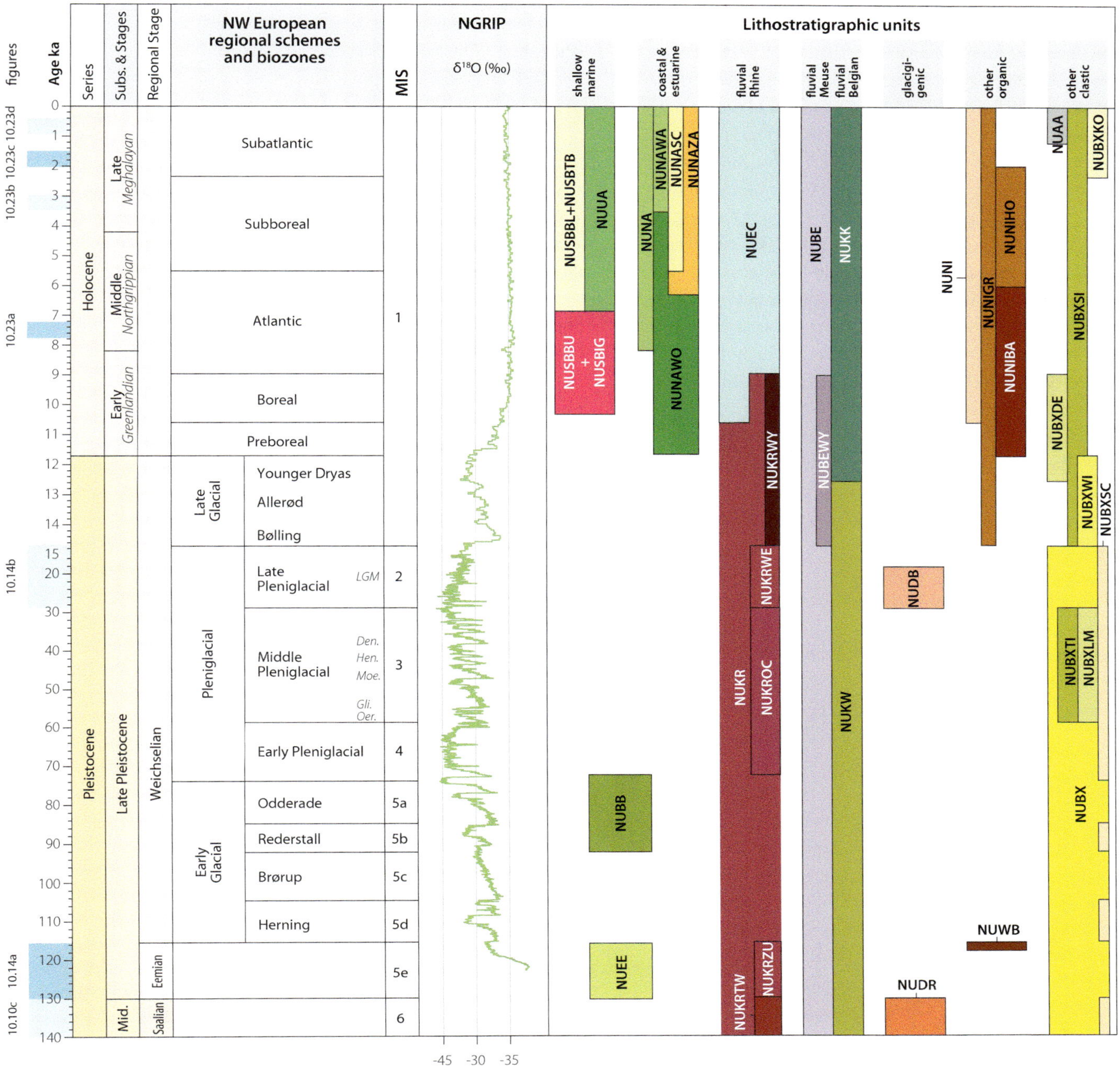

Figure 10.12. Subdivision of the Late Pleistocene and Holocene, with from left to right chronostratigraphic subseries and stages (Cohen & Gibbard, 2019) complemented with regional (northwest European) stages and substages, northwest European regional schemes and biozones, deep-ocean oxygen isotope stages (Lisiecki & Raymo, 2005), NGRIP oxygen isotope record (North Greenland Ice Core Project members, 2004), and formally defined stratigraphic units in the on- and offshore domains of the Netherlands (TNO-GDN, 2023d). See Table 10.1 for explanation of lithostratigraphic codes. Age ranges indicate minimum and maximum ages of each unit; diachroneity and hiatuses, present throughout the stratigraphic column, are not shown.

pushed ridges (Van der Wateren, 1995; Bakker & Van der Meer, 2003; Fig. 10.10c) and associated glacial basins. The ridges are especially well developed in the central Netherlands where they form the Utrechtse Heuvelrug and Veluwe push-ridge complexes (see also Textbox 3).

By ca. 160 ka (Busschers et al., 2008), the ice sheet had reached it southernmost extent in the central Netherlands, continuing into Germany as well as the Netherlands offshore area (Laban, 1995; Moreau et al., 2012; Cartelle et al., 2021; Fig. 10.10c). The topographic prominence of the ridge complex forced the Rhine-Meuse System to adopt a westward proglacial course. During the final stage of the glaciation, major incision and dissection occurred by the ice-marginal Rhine-Meuse River System (Busschers et al., 2008; Rijsdijk et al., 2013). This is the first incised lowstand valley of the Rhine-Meuse System that can be positively traced westward to the Dover Strait (Fig. 10.10c).

During ice-sheet disintegration (ca. 150-140 ka), series of (interconnected) lakes developed in the newly exposed glacial basins. The Rhine shifted its course to a position north of the glaciotectonic ridge complexes and began depositing sequences of fine-grained fluviolacustrine sediments (Twello Member, Kreftenheye Formation) in the largest glacial basin in the east (IJssel glacial basin; Fig. 10.10c). These fine sediments were gradually buried by coarse-grained sands. Through time, the lake transformed into a river valley. Via the IJssel glacial basin, the Rhine continued westward through the central Netherlands, shaping a deeply incised valley partly filled with fluvial sand of the Kreftenheye Formation (Peeters et al., 2016). The position of the offshore continuation of this valley is still unknown (Fig. 10.10c).

Late Pleistocene Subseries (130-11.7 ka)

Climate, sea level, ice sheets and biota

The Late Pleistocene marks the final, full interglacial-glacial cycle of the Quaternary (Fig. 10.1). It started with the Eemian interglacial (ca. MIS 5e), when temperatures and global MSL reached slightly higher values than those of today (Medina-Elizalde, 2013; Long et al., 2015; Spratt & Lisiecki, 2016). The Eemian vegetation succession is different from that of the Holocene, especially in the later part of the Eemian (Zagwijn, 1961, 1996a; Helmens, 2013). The pollen assemblages, particularly the presence of *Taxus*, indicate that temperatures during the peak Eemian climate were slightly higher than during the Holocene climatic optimum. The Eemian interglacial was followed by the Weichselian Early Glacial with a succession of cool to temperate interstadials (notably Brørup and Odderade; Fig. 10.12) with *Betula-Pinus-Picea* forests, and cold stadials characterized by tundra vegetation and discontinuous permafrost (Zagwijn, 1961; Van der Meer et al., 1984; Behre, 1989; Helmens, 2013; Kasse et al., 2022). During the warmer phases of the Weichselian Early Glacial, global MSL was positioned ca. 10-20 m below the present-day level during interstadials and even lower during stadials (Medina-Elizalde, 2013; Spratt & Lisiecki, 2016).

Rapid and severe cooling, accompanied by extreme sea-level lowering, occurred at the onset of the Weichselian Early Pleniglacial (MIS 4), with polar desert conditions becoming established. This interval also saw extensive Fennoscandian ice sheet expansion, but it was more limited than during the latest part of the Weichselian. The Early Pleniglacial also marked the transition from the diverse warm-temperate faunas to the mammoth fauna that flourished during the Middle Pleniglacial (MIS 3). During the Middle Pleniglacial the climate was generally milder than in the Early Pleniglacial, with tundra- and steppe-dominated landscapes and common but non-pervasive permafrost (Van Huissteden et al., 2003; Busschers et al., 2007). Coeval ice-sheet build-up over Fennoscandia was limited at this time (Mangerud et al., 2011; Batchelor et al., 2019) and sea level was positioned several tens of metres lower than that of the Weichselian Early Glacial (MIS 5d-a).

The Greenland ice sheet recorded repeated alternations of very cold stadials and cool-temperate interstadials (Dansgaard-Oeschger events), which particularly affected the North Atlantic and northwest European regional climate strongly (Fig. 10.12). In the Dutch Weichselian sequence, these interstadials (e.g. Moershoofd, Hengelo and Denekamp interstadials) are evident from peat beds that are dominated by tundra vegetation (Zagwijn, 1961; Van der Hammen & Wijmstra, 1971; Van der Meer et al., 1984; Van Huissteden, 1990; Vandenberghe & Van der Plicht, 2016). Botanical macrofossil research indicates that *Betula nana* and *Salix polaris* were common in the tundra landscapes. For the duration of the Pleniglacial, closed boreal forests were unable to return to the Netherlands. For extensive periods, stadial-interstadial vegetation cover was dominated by abundant grasses, sedges and some herbaceous vegetation. The southern North Sea region formed part of a mammoth-steppe ecosystem that supported a wealth of smaller and larger mammals. Large grazers (mammoth, horse, bison, woolly rhinoceros) were particularly common while hyena was abundant and other predators such as cave bear and cave lion were also present (Fig. 10.13).

Modern humans may have first arrived in the region during this time (Higham et al 2011). Through the course of MIS 3, megafaunal species began to disappear. The ranges of their typical habitats contracted mostly towards the northern Asian realm before they ultimately became extinct (see Textbox 4).

The cooling that followed MIS 3 culminated in the Weichselian Late Pleniglacial (MIS 2), when the Netherlands and North Sea floor experienced polar desert conditions. This phase includes the Last Glacial Maximum (LGM), ca. 21 ka (Fig. 10.12). Between 29 and 23 ka, Fennoscandian and British ice-sheet sectors advancing across the North Sea reached local maximum extents, forming the now submerged Dogger Bank glaciotectonic ridge in the central North Sea (Laban, 1995; Roberts et al., 2018; Emery et al., 2019). In regions further south, polar desert conditions prevailed. After about 19 ka and accelerating from 14.7 ka onwards, warming caused the land ice to recede. Newly exposed terrain was rapidly occupied by tundra to taiga ecosystems. The warming continued into the early Holocene, preceded by the Younger Dryas Stadial

Figure 10.13. Artist impression of the MIS 3 mammoth steppe, a dry periglacial landscape home to a typical ice-age megafauna including mammoths, whooly rhinoceros and hyenas. Drawing E.J. Bosch, Naturalis.

(12.9-11.7 ka), when a North American meltwater pulse disrupted the North Atlantic meridional overturning and subarctic conditions returned briefly.

Sedimentary systems and paleogeography

Eemian and Weichselian Early Glacial (MIS 5)

At the onset of the Eemian interglacial, the rising sea level of the North Sea encountered a diverse topography with ice-pushed ridges and deep glacial basins that strongly affected the inundation history (Fig. 10.14a). The coastline configuration of the Eemian sea was different from the present one, with peninsular promontories as well as embayments extending far inland. The Rhine flowed northward through today's IJssel Valley, triggering extensive floodplain deposition (Zutphen Member, Kreftenheye Formation; Fig. 10.12; Busschers et al., 2007; Peeters et al., 2016). It debouched into a shallow-marine embayment (Fig. 10.14a). The Meuse River broadly followed its modern course and discharged into the North Sea tens of kilometres further inland than its present-day mouth.

During the Eemian highstand, which was coeval with pollen zone E5 (*Carpinus* zone; Zagwijn, 1983; Long et al., 2015; Cohen et al., 2022; Kasse et al., 2022; late MIS 5e), coastal marine conditions prevailed in large parts of the northern and western Netherlands, including the Amsterdam and Amersfoort basins (Fig. 10.14a). These basins became filled with generally shell-rich shallow-marine and lagoonal deposits of the Eem Formation (Fig. 10.15). The topography resulting from the previous Saalian glaciation provided varied habitats (Cleveringa et al., 2000; Van Leeuwen et al., 2000). A species-rich nearshore fauna with common warm seagrass-associated mollusc species lived in embayments that experienced different salinity regimes depending on basin configuration and the spatiotemporally varying amount of river inflow. As a consequence of the warm-temperate conditions, vegetation on land was lush, the landscape stabilized and extensive soil formation took place. Organic-rich units attributed to the Boxtel Formation were formed in peat bogs that developed in areas with a high groundwater table (Schokker et al., 2004).

At the end of the Eemian (pollen zone E6), regression enabled peat of the Woudenberg Formation to form along the coast in more protected areas (Fig. 10.15). As a result of ongoing and severe climate cooling as well as sea-level fall at the end of the Eemian, the North Sea probably became disconnected from the English Channel region, with a single ocean connection in the northern North Sea remaining. This oceanic connection drove strong cooling of marine habitats in the North Sea even though summer temperatures on land may have remained moderately high. This created a setting somewhat comparable to today's Hudson Bay (Waajen et al. 2024). Cold, boreal marine faunas, with walrus and beluga, dominated the southern North Sea at the time.

During this period, the Rhine continued to follow a course through the IJssel Valley and the province of Noord-Holland in the northwest, forming an extensive deltaic and prodeltaic depositional complex centred below the present day coast (Kreftenheye Formation; Busschers et al., 2007; Hijma et al., 2012; Peeters et al., 2016). The Meuse remained approximately in the same position as during the Eemian, although later erosion has fragmented its depositional record (Busschers et al., 2007). Seismic data and shallow vibrocores from the North Sea demonstrate that deposition farther offshore was dominated by fine-grained sand and clay under alternating shallow-marine and lagoonal conditions (Brown Bank Formation; Fig. 10.12; Laban, 1995; Busschers et al., 2007; Hijma et al., 2012; Eaton et al., 2020; Missiaen et al., 2021; Waajen et al., 2024).

Beyond the major river valleys, alternations between small-scale fluvial, fluvioperiglacial, lacustrine and peat strata dominate Early Glacial successions, reflecting transitions between warmer and cool to cold climatic phases (Zagwijn, 1961; Van den Berg & Den Otter, 1993; Kasse et al., 2022). Closer to the Saalian ice-pushed ridges, the imprints of these MIS 5 climatic variations are also visible

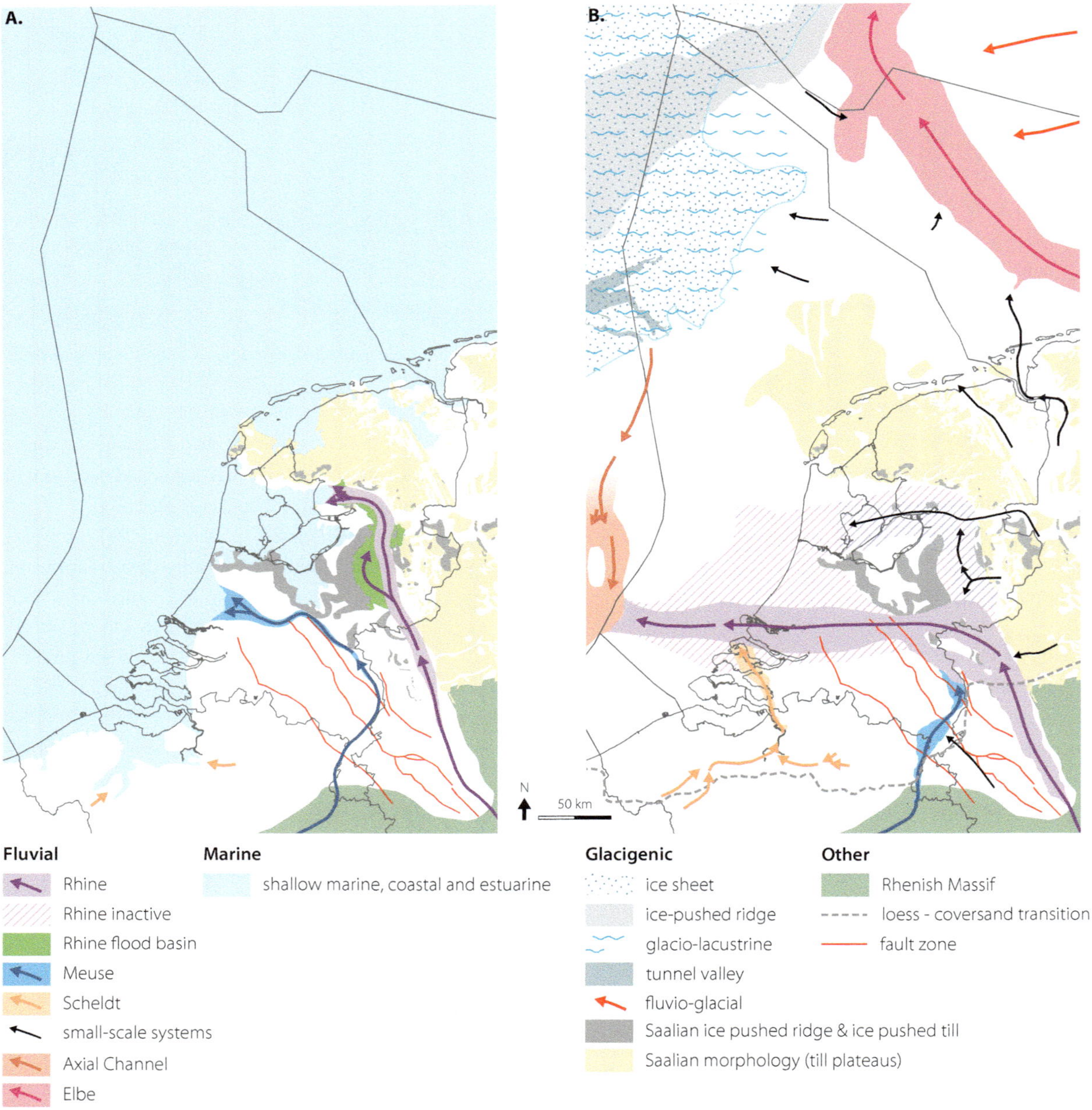

Figure 10.14. *Paleogeography during a) the Eemian (MIS 5e) and b) the Weichselian Late Pleniglacial (MIS 2). The successive maps show 1) a complex of Eemian marine embayments and protruding capes, in the central and northern Netherlands strongly affected by inherited glacial relief (a), 2) initial separation and eventual joining of Rhine and Meuse (b), and 3) glaciation of the northern offshore sector towards the end of the Late Pleistocene (b). Primary sources: Busschers et al. (2007), Hijma et al. (2012); Peeters et al. (2016, De Clercq et al. (2018), Batchelor et al. (2019), Hepp et al. (2019), Cartelle et al. (2021), Lehmkuhl et al. (2021), Cohen et al. (2022), TNO-GDN (2023c) and TNO-GDN (2023d).*

in alluvial fan deposits as alternations between coarse-grained sediments, eolian sands, loess and peat horizons (Van der Meer et al., 1984; Palstra et al., 2021).

Weichselian Early and Middle Pleniglacial (MIS 4-3)

During the Weichselian Early Pleniglacial, climate cooling and sea-level lowering resulted in incision and large-scale reworking of Eemian and Weichselian Early Glacial marine sediments in the valleys of the Rhine-Meuse (Kreftenheye Formation) and other smaller rivers (Koewacht and Boxtel formations; Van Huissteden et al., 2003; De Clercq et al., 2018). These valleys largely followed the same courses as during the preceding interglacial but extended their downstream ranges onto the dry North Sea floor, becoming confluent towards Dover Strait (Busschers et al., 2007; Hijma et al., 2012; Peeters et al., 2016). Extremely cold and dry conditions towards the end of the Weichselian Early Pleniglacial resulted in a phase of large-scale eolian transport and deposition of sand (Schokker & Koster, 2004) and, in the southeasternmost province of Limburg, of loess (Meijs, 2011). Large ice-wedge casts in sediments from this period testify to the presence of continuous permafrost at a time of very low temperatures (Huijzer & Vandenberghe, 1998; Fig. 10.16).

Halfway through the Middle Pleniglacial, major changes in the course of the Rhine occurred. As a result of avulsion in the upstream part of the IJssel Valley, related to ongoing climate-driven aggradation and a strong gradient advantage, the Rhine adopted a new course to the south of the ice-pushed ridges of the central Netherlands (Fig. 10.14b). Here the river became confluent with the Meuse. By ca. 38 ka, the northern route was fully abandoned, possibly also in response to initial glacio-isostatic uplift resulting from the advance of the growing Fennoscandian and British ice sheets (Busschers et al., 2007; Peeters et al., 2016).

The precursor of the Scheldt ran northward and merged with the Rhine-Meuse close to today's harbour city of Rotterdam (Slupik et al., 2013; Koewacht Formation). The Rhine-Meuse System resumed its course into the present-day southern North Sea (Kreftenheye Formation), where it connected with the Thames and was deflected southward through Dover Strait into the Channel River System (Toucanne et al., 2010; Hijma et al., 2012). The latter debouched into the Atlantic Ocean near the present-day shelf break between Brittany and southern Ireland. Severe seasonality, with massive meltwater pulses during the spring, resulted in strong sediment supply and redeposition.

Outside the realm of the large rivers, sediments from MIS 3 filled glacial basins in the central Netherlands as well as subsiding areas in the south where they have been well-studied. In the Nordhorn and Hengelo Glacial Basins, up to 20-m-thick sequences of small-scale fluvial sediments and organic deposits occur (Tilligte Member, Boxtel Formation). These sediments provide information on paleoclimate, paleoecological conditions and changes in fluvial styles (Van der Hammen & Wijmstra, 1971; Van Huissteden, 1990; Berrittella, 2017). The sedimentary sequence in the Roer Valley Graben forms a complex alternation of small-scale fluvial, wet-eolian and lacustrine deposits. Grain-size distribution, sedimentary structures and malacological content of thick calcareous loam beds (Liempde Member, Boxtel Formation; Schokker et al., 2007) indicate the presence of shallow pools filled with loess-like sediments in a sub-arctic climate. The sedimentary sequence in this area testifies to generally wet, tundra-like conditions with repeated cryoturbation phases (Deeben et al., 2010).

Weichselian Late Pleniglacial (MIS 2)

Extreme climate cooling at the onset of the Weichselian Late Pleniglacial resulted in major expansion of the Fennoscandian and British ice sheets. The ice sheets advanced into the central North Sea, creating the Cleaver Bank and Dogger Bank highs through glaciotectonics (Laban, 1995; Roberts et al., 2018; Emery et al., 2019), as well as into the southwestern Baltic (Batchelor et al., 2019). The southernmost ice-margin positions were obtained between 27 and 21 ka (Figs 10.1, 10.14b).

Ice-pushed ridges and outwash complexes formed along maximum and recessional ice margins. Some of these are evident from seafloor topography and from subsurface features visible in shallow seismics. Additional evidence comes from Dogger Bank cores. The subsurface features are mapped as the Dogger Bight Formation (Fig. 10.12), which includes the Bolders Bank Member (till), the Well Ground Member (sandy outwash deposits) and the Dogger Bank Member (clayey glaciolacustrine deposits). Northeast of the Dutch sector the Elbe established a southeast-northwest oriented incised valley (Hepp et al., 2019; Fig. 10.14b). In the northwest of the Dutch Sector, a British ice lobe advanced farthest south, leaving tunnel valleys such as Outer Silver Pit (Fig. 10.14b) and Botney Cut. These valleys were partially infilled during deglaciation (Botney Cut and Volans members) and still receive fine sediment from suspension today. In general, modern high-resolution bathymetric charts reveal many well-preserved details of the last glacial action in the region (lineations, fields of boulder-sized erratics), overprinted to varying degrees by shallow-marine bedform fields formed by subsequent Holocene tidal and wave action.

During the Weichselian Late Pleniglacial, the Rhine-Meuse System followed an east-west orientated course (Fig. 10.14b) and deposited coarse-grained sands and gravel (Kreftenheye and Beegden formations; Fig. 10.12). The Scheldt joined with the Rhine-Meuse System close to the present-day coastline. Further west, the joint Rhine-Meuse-Scheldt System coalesced with the north-south orientated Axial Channel (Fig. 10.14b). The Axial Channel also received copious amounts of meltwater from the ice front to the north (Cohen et al., 2014; De Clercq et al., 2018). It connected with the Thames and was deflected southward through Dover Strait into the Channel River System (Toucanne et al., 2010; Hijma et al., 2012; García-Moreno, 2017; De Clercq et al., 2018).

In its peripheral position relative to the Fennoscandian and British ice-mass centres, the southern parts of the North Sea Basin experienced glacio-isostatic uplift (forebulge updoming). The northern Netherlands was part of the main centre of uplift, which temporarily inverted the general subsidence regime of the Netherlands depocentre (Kiden et al., 2002; Busschers et al., 2007). Because the uplift occurred during the cold-period sea-level lowstand, its effect was mainly reflected in the changing behaviour of fluvial drainage systems like the Rhine-Meuse, notably as southward deflections away from the uplift centre and as a dominance of incision (Busschers et al., 2007). During deglaciation (ca. 25-11 ka), glacio-isostatic adjustment led to enhanced subsidence (forebulge collapse). This is recorded in the Rhine-Meuse System by marked coarse-grained aggradation in a widening active braidplain (ca. 18-15 ka) and eventually (ca. 13-10 ka) by a northward lateral deflection (Busschers et al., 2007; Hijma et al., 2009). Coevally, forebulge collapse also led to an increase in paleoseismic activity in the Roer Valley Rift System (Van Balen et al., 2019, 2021). At diminishing rates, glacio-isostatic subsidence continued into the Holocene, as is evident from south-to-north differences in postglacial relative sea-level rise (Kiden et al., 2002).

Beyond the active rivers, polar desert conditions prevailed between 29 and 16 ka. As a consequence of the extremely cold and dry climatic conditions, the vegetation cover was very limited and wind became an impor-

Constraining the age of Europe's megafaunal extinction

4

The southern North Sea is globally renowned for its very rich Quaternary vertebrate faunas. Many of the numerous fossils have been recovered during bottom trawling, a common fishing practice. In recent decades, they have also been dredged from offshore sand-extraction pits. Sand from these pits has been used for coastal reinforcements (nourishments) and for construction. Sand nourishments are an increasingly popular faunal resource to a large and active community of citizen scientists. The richest beach trajectories include those of Maasvlakte and Maasvlakte 2, two Rotterdam harbour extensions, and the Sand Motor mega-nourishment (21.5 million m^3) slightly further north. Most fossils (and Paleolithic artefacts) found there come from the Late Pleistocene Kreftenheye Formation (Amkreutz & Van der Vaart-Verschoof, 2022).

The majority of vertebrate faunal elements shown in the photo below are typical of the mammoth steppe and include species such as mammoth, woolly rhinoceros, hyena, horse and deer (Mol et al., 2006). Cold marine vertebrate taxa such as walrus, bearded seal and beluga are rare, but ecologically compatible. Fossil mollusc taxa, such as the boreal *Astarte*, are frequently found. Warm-temperate terrestrial taxa originating from interglacial Eemian intervals, such as forest elephant, rhinoceros and hippo, are poorly represented, but thermophilous marine mollusc species, including *Acanthocardia tuberculata* are common. Finally, Early and Middle Pleistocene faunas, rare on the beaches, have been trawled from the Eastern Scheldt and from some offshore areas, making the North Sea one of the most productive fossil-rich regions worldwide.

The continued collection of fossils and artefacts is contributing majorly to our understanding of late Quaternary landscapes, biotic evolution and hominin occupation of the Southern North Sea Basin (Amkreutz & Van der Vaart-Verschoof, 2022). However, difficulties of age constraining hamper detailed understanding of biotic successions. The ongoing debate on the timing and causes of the mammal extinction around the end of the last ice age invokes rapid climate deterioration and/or human overkill as primary drivers (Stuart, 2021). According to the present consensus, the timing of extinction of many taxa within western Eurasia was not sudden but spanned the entire Late Pleistocene and Early Holocene, making a single climatic driver untenable. However, direct and indirect evidence for extensive demise related to hominid hunting is extremely scarce and fails to match observed extinction patterns. Erecting a detailed and correct chronostratigraphic framework for the occurrence and disappearance of the ice-age fauna is a prerequisite to improve the understanding of the drivers of the demise of many great mammals. For this, the southern North Sea fossil record provides many opportunities.

Photo: P. Wildschut, D. Mol.

tant sediment-transport agent. In a broad area known as the European Sand Belt, sand sheets and dune fields were formed (Wierden Member, Boxtel Formation; Figs 10.12, 10.16, 10.17). These so-called coversands have been studied in detail, focussing on stratigraphic, sedimentological and palynological aspects, and have also been dated extensively by luminescence and radiocarbon methods (Kasse et al., 2007; Kolstrup, 2007; Vandenberghe et al., 2013; Bazelmans et al., 2021; Van Hateren et al., 2022). In areas farther south, including Belgium, Germany and neighbouring regions of France, loess deposits (Schimmert Member, Boxtel Formation) rather than coversands accumulated as wind-blown sediment blankets that successively colluviated (Fig. 10.4b; Meijs, 2011; Lehmkuhl et al., 2021). Pockets of loess are also known from the flanks of the central Netherlands ice-pushed ridges.

Weichselian Late Glacial (onset MIS 1)

The Late Glacial stage comprises the last interstadial and stadial of the Weichselian (14.7-11.7 ka; Fig. 10.12). The transition from full glacial conditions to warmer ones is recorded in the coversand sequence, at various well-described sites, notably those of Lutterzand (Van der Hammen & Wijmstra, 1971; Van Huissteden, 1990; Kasse, 2002; Fig. 10.17) and Usselo (e.g. Van Geel et al., 1989; Kolstrup, 2007).

These coversand records reflect both eolian deposition during cold and/or dry phases and organic beds and associated soils (with charcoal fragments) representing warmer and/or wetter times. The Usselo Soil is a characteristic Late Glacial Allerød marker horizon. It is typified by the presence of dung-beetle burrow casts (Fig. 10.18a), indicating that herds of large herbivores roamed the landscape. In wet low-lying areas, including pingo remains, organic deposits from the same age accumulated. Where preserved, they show a dominance of open vegetation with mainly herbs and grasses, and *Betula* as well as *Pinus* in favourable places. Recently, remains from a Late Glacial stand of pine forest have been uncovered in Leusden in the central Netherlands (Fig. 10.18b). Dated wood samples have Allerød and early Younger Dryas ages, suggesting that the trees did not survive the abrupt climatic and environmental shift from the mild Allerød into the cold and dry Younger Dryas Stadial (Bazelmans et al., 2021). During the second part of this last cold interval, before the onset of the Holocene, large fields of low, eolian dunes developed (upper part of the Wierden Member). These dunes are characteristic elements in the modern landscape of the eastern and southern Netherlands.

The climatic amelioration that marked the end of the Late Glacial and onset of the Holocene (ca. 15-9 ka) had a strong influence on channel styles and floodplain terraces. It is also reflected in paleosols belonging to the river valleys upstream of the Rhine-Meuse delta and those buried beneath the delta itself. In Late Pleniglacial times (up to 15 ka), the river style had been braided, with active multichannel beds up to a few kilometres wide. These saw high spring regimes of meltwater discharge generated by a hinterland with frozen subsoils, little vegetation and plenty of sediment input. Through the Late Glacial (ca. 15-12 ka), climate conditions improved overall. The permafrost disintegrated, the discharge regime changed from peaked to more regular, vegetation could become established in hinterland and floodplains, and sediment loads decreased. Most notably, the active width of riverbed reduced and the channel style was transformed from braided to meandering (Kasse, 1995; Vandenberghe, 1995; Hoek et al., 2017; Fig. 10.17). Abandoned stretches of braidplain became floodplain terrace. They silted up in a fluvial system that was increasingly meandering in nature (Wijchen Bed in the Kreftenheye and Beegden formations). Newly formed terraces stabilized after regular flooding events ceased.

The Younger Dryas cold period (12.9-11.7 ka) initiated a temporal return to seasonally frozen subsoils, which interrupted the transformation from braided to meandering conditions, temporarily widening riverbeds that failed to reach the widths of Pleniglacial times (Fig. 10.17). These wider beds enabled wind to pick up sediment, creating inland dune systems along the rivers. Following this short break, during the onset of Early Holocene warming, the transformation to a single-thread narrow channel with a meandering style resumed (Fig. 10.17) and floodplain forests returned. The latter influenced the formation of soils on terraces and accelerated the maturation of floodplain dunes. They also contributed to further narrowing of the river course into a single channel. This situation was reached by 10.5 ka in the southeastern Netherlands Meuse Valley (Woolderink et al., 2019) and by 9.5 ka in the Rhine-Meuse Valley of the western Netherlands (Hijma et al., 2009; 2011). By that time, the Holocene was in full swing.

Holocene Series (11.7 ka – present)

Climate, sea level and human influence

The current, Holocene Series is a relatively stable, warm and humid interglacial that followed the cold Younger Dryas Stadial that marked the end of the Pleistocene. In and around the Netherlands, rising temperatures had a profound effect on sea level (Fig. 10.19), groundwater tables, vegetation, indigenous fauna and, importantly, the human population. Abundant evidence of this effect and of the associated changes in paleogeography has been preserved in the shallow Dutch subsurface record, both onshore and offshore.

Early on in the Holocene (Preboreal biozone, 11.7-10.6 ka, Fig. 10.12), warming was rapid. Through glacio-

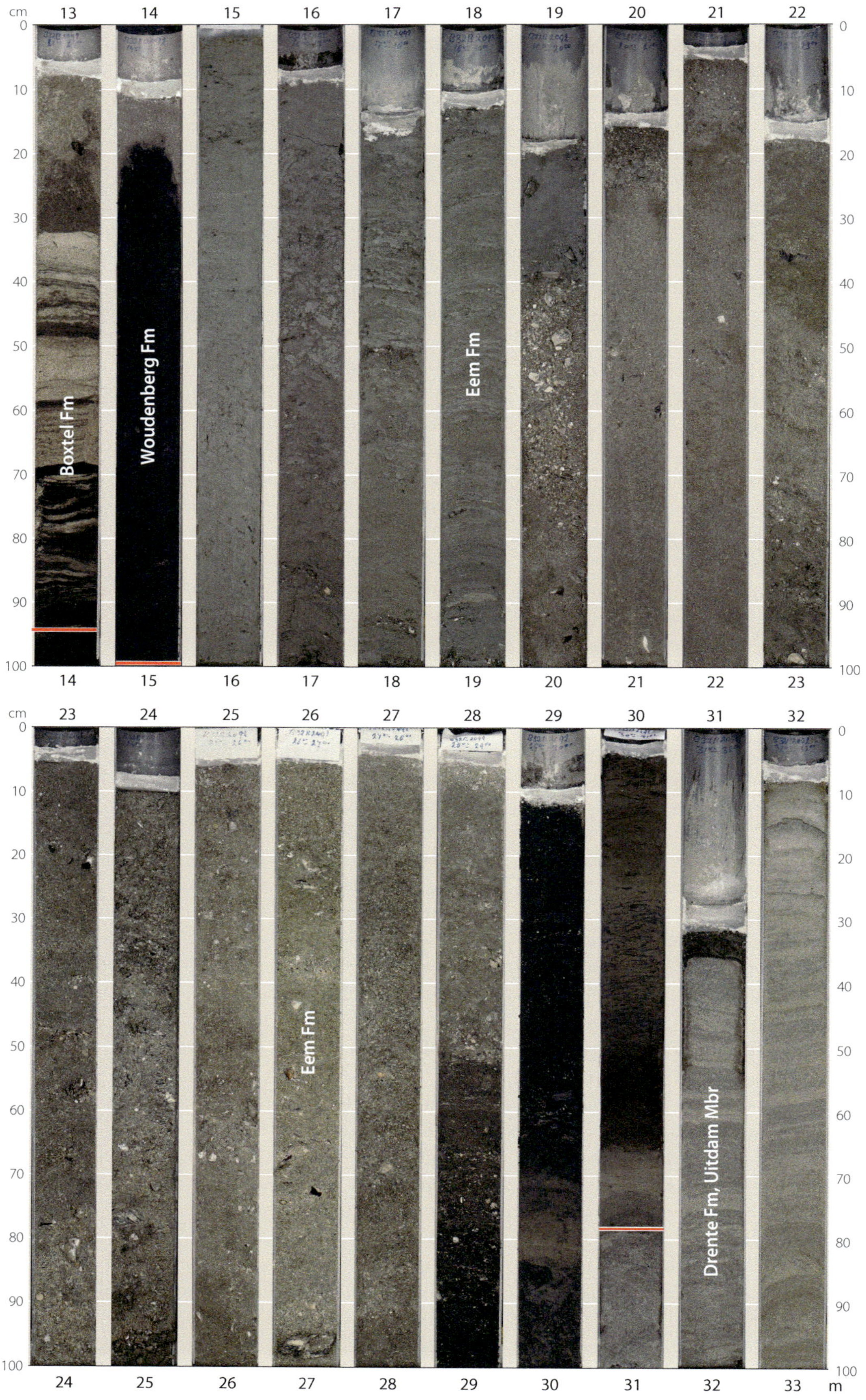

Location **Amersfoort**
Borehole **B32B2092**

X 157127 **Y** 463605

2.9 m + MSL

← ***Figure 10.15.*** *Latest Saalian, Eemian and Weichselian succession in the Amersfoort glacial basin (core B32B2092; Kasse et al., 2022). Meltwater sand of late Saalian age (Drente Formation) at the base of the core is overlain by an organic unit (gyttja/peat; Eem Formation) reflecting early Eemian lacustrine situation with rising water table owing to nearby sea-level rise. The organic deposits are partly reworked and are covered by a coarsening-upward sequence of loam and fine sand, indicating a brackish to salt-water subtidal environment (the embayment shown in Fig. 10.14a). This is followed by 8 m of shell-bearing coarse-grained sand deposited under highly energetic shallow-marine conditions during the Eemian highstand. The shelly sand is separated from overlying loam and clay by a sharp boundary at 19.4 m that formed in response to shallowing or partial closure of the marine basin. The lithological transition from clay at the top of the Eem Formation to peat of the Woudenberg Formation witnesses a gradual climate cooling and the onset of sea-level fall. The subsequent change to fully terrestrial conditions at ca. 14 m depth (sandy and loamy, partly organic-rich deposits; Boxtel Formation) marks major climate cooling at the onset of the Weichselian Early Glacial. Coordinates in Rijksdriehoeksstelsel. Core photos: TNO-GDN, 2023d.*

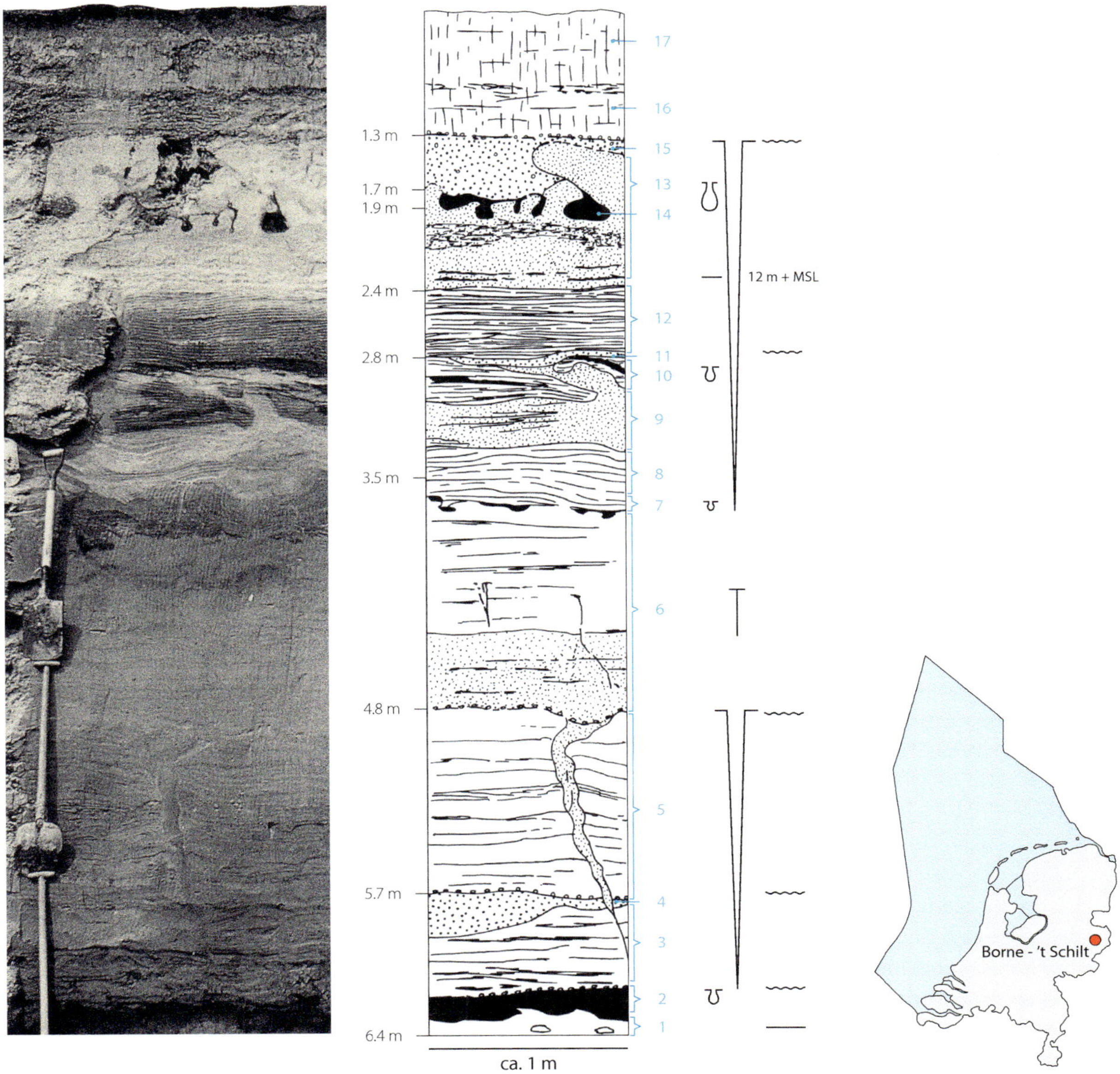

Figure 10.16. *Weichselian sedimentary sequence and associated ice-wedge casts, frost cracks and involutions in outcrop Borne-'t Schilt in the eastern Netherlands (Van den Berg & Den Otter, 1993). Left: outcrop photo; right: interpreted sedimentary log. Layer 1: Saalian glacial diamicton; 2: Eemian peat; 3-5: Early Glacial fluvial and lacustrine deposits; 6-14: Middle-Pleniglacial fluvio-eolian and lacustrine deposits; 15: Late Pleniglacial fluvial deposits, capped by a desert pavement; 16: Late Glacial eolian deposits; 17: Holocene brook deposits. Two levels with large ice-wedge casts and involutions (unit 5 and units 13-15) indicate periods of permafrost degradation at the end of MIS 4 and MIS 2, respectively.*

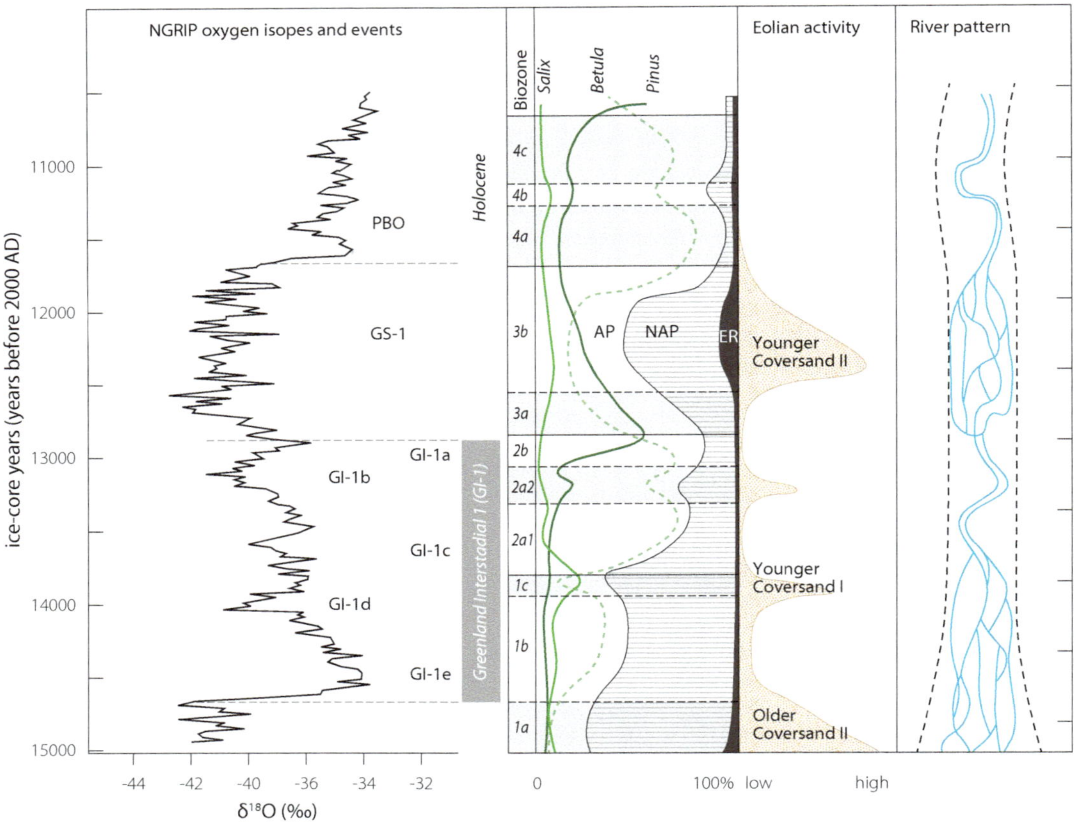

Figure 10.17. Schematic overview of Weichselian Late Glacial coversand deposition and river style changes in relation to climate and vegetation development (Hoek & Bohncke, 2002; Lowe et al., 2008; Hoek et al., 2017). NGRIP oxygen isotope record from North Greenland Ice Core Project members (2004). Biozonation according to Hoek & Bohncke (2002). GS-1 = Younger Dryas, GI-1 = Bølling / Allerød, AP = Arboreal Pollen, NAP = Non-Arboreal Pollen, ER = Ericales. Younger Coversand I + II equal the Wierden Member of the Boxtel Formation. © Cambridge University Press, reproduced with permission.

eustasy and glacio-isostatic subsidence, climate-driven deglaciation caused regional relative sea level to rise tens of metres at rates well exceeding 1 m per century. This sea-level change is recorded in basal peats that cap the Pleistocene paleosurface in many locations offshore and below the near- and onshore coastal plain. The lowest such areas of the Pleistocene landscape, more than 60 m below present-day MSL, were submerged soon after the Younger Dryas ended. By 9.2 ka (Boreal biozone, Fig. 10.12), the sea had flooded vast parts of the present marine basin, creating ample accommodation space in areas that had been erosional or non-depositional before. By 8.5 ka (early Atlantic biozone, Fig. 10.12), when relative sea level had risen to 15-20 m below current MSL, the North Sea shoreline had moved to positions close to its modern one, even landward from it at some places (Vos, 2015). The Holocene succession offshore and onshore shows effects of short-lived events superimposed on the main long-term trend: relative sea level (Hijma & Cohen, 2010) that was still rising about 1 m per century. Around 7 ka, only a few Pleistocene promontories remained along a gradually straightening coastline. By then, Doggerland, with its fertile lowlands occupying the North Sea during times of low sea level, was drowned.

The regional climate of northwestern Europe and the North Atlantic Ocean reached its natural temperature optimum by 7-6 ka (middle of Atlantic biozone, Fig. 10.12). Lowland organic deposits (e.g. lake muds), fine-grained Rhine-Meuse channel fills and buried paleosols as well as upland peat bogs provide spatially consistent records of climate change impacting shallow groundwater systems as well as surficial drainage (e.g. Hoek & Bohncke, 2002; Makaske & Maas, 2023).

By 6 ka, climate-driven deglaciation had ceased and climate variability as such stopped being a major factor in landscape and depositional change. With eustatic sea-level rise no longer a principal influence, only tectonic subsidence and compaction continued to affect regional sea level. Decelerating but persistent relative sea-level rise, governed mainly by remnant glacio-isostasy remained to drive the establishment of the Dutch coastal plain. Its indirect influence on sediment supply, especially from the eroding shoreface, explains the south-to-north diachronic turnovers between transgressive and regressive phases in coastal behaviour. As inlets opened and closed, and tidal basins expanded and filled, adjacent coastlines adjusted accordingly. An anastomosing network of meandering and straight river branches supplied sediment on the landward side of these tidal basins, repeatedly triggering avulsions (Törnqvist, 1993; Hijma et al., 2011; Stouthamer et al., 2011). Extensive peatlands covered ever larger tidal and fluvial areas with poor drainage conditions. Seaward of the steepening shoreface, fully marine processes became dominant, with tidal ridges and sand waves occupying extensive areas of residual sand transport.

Around 5 ka (early Subboreal biozone, Fig. 10.12), the rate of relative sea-level rise had dropped to 0.15 m per century. A further reduction to just a few centimetres per century followed in the Late Holocene. This overall deceleration is reflected in the depositional record of the onshore coastal prism (Fig. 10.19): a wedge-shaped sediment body that extends from the lower shoreface to the landward edge of marine or lowland fluvial (Rhine-Meuse and Scheldt) influence and is typically thickest near the present-day coastline. Along with spatiotemporally varying sediment supply and human influence, even this modest sea-level change has been an influential driver of sedimentation and erosion.

Agriculture and the associated deforestation commenced in the Netherlands by 7.5 ka (Neolithization, Louwe Kooijmans, 2005), spreading westward from the eastern and southeastern entry points of Rhine and Meuse. It influenced flow regimes and riverbank vegetation along distributaries that formed and decayed in response to avulsions. By 5 ka, agriculture was also practiced close to the contemporaneous coastline, on innermost beach ridges and in the lowland delta plain. Along small-scale rivers, it even spread northeast into extensive upland with less fertile substrate. After 3.5 ka (Bronze Age), when forest clearing intensified and ploughing was introduced in the Rhine-Meuse hinterlands, sediment loads of these main rivers began to increase. Around 2 ka (Roman times), they carried twice as much suspended sediment compared to their natural states before (Hoffmann et al., 2007), dramatically increasing deposition in estuaries and along rivers, and triggering major channel-network alteration downstream and upstream in the delta (Stouthamer et al., 2011; Pierik et al., 2017b; 2018b). Simultaneously, digging of ditches to drain coastal-plain bogs and reclaim peatland for agriculture increased (Van Bergen & Kosters, 2025, this volume). This resulted in peat loss, in the coastal plain as well as inland. Part of the deforestation-related mud ended up covering depressions that formed by excavation and oxidation of peat. Around 1 ka (medieval times), people started to construct interconnected dykes along the rivers, tidal inlets and lagoon shores. This strategy fixed the channel network of the delta and its tidal river mouths and stopped deposition in large former flood-basin areas. Relatively narrow embanked floodplains remained along the delta rivers. Greater volumes of mud were carried fur-

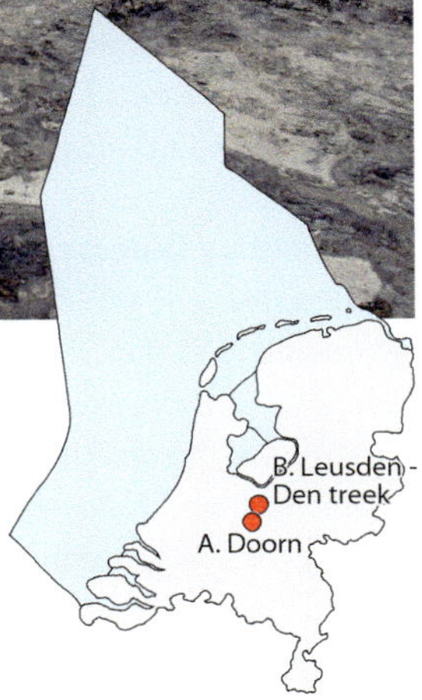

Figure 10.18. a) Usselo soil at a key site in Doorn (Photo: J. Schokker). b) Excavation trench in Leusden-Den Treek with exceptionally well-preserved Late Glacial pine-tree remnants (Bazelmans et al., 2021). Photo: RCE/J. Bazelmans. The pine-tree remnants are covered by eolian sediments (Younger Coversand II; see Fig. 10.17). See inset map for locations.

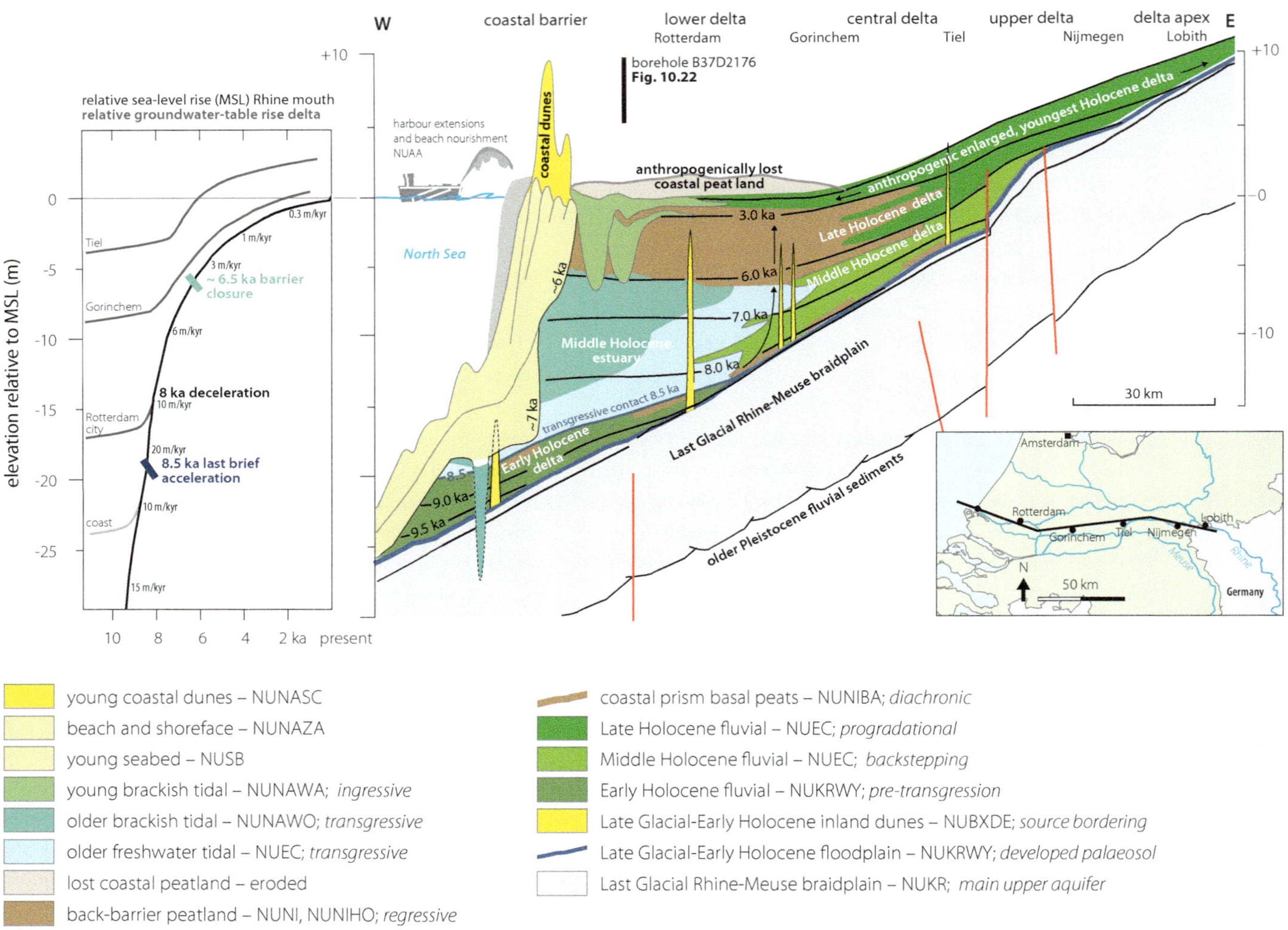

Figure 10.19. Longitudinal architecture of the Holocene coastal prism, showing the infill of the main fluvial basin in response to changes in relative sea-level rise. Left panel plots Holocene relative sea-level rise at the Rhine mouth and coeval groundwater-table rise in the delta inland (modified from Hijma & Cohen, 2011). Because the seaward reaches of shoreface sediments (NUNAZA) are difficult to distinguish from young seabed sediments, they are denoted in the same colour.

ther downstream than before and were delivered to estuarine reaches where the fine sediment settled in narrow zones between channels and dykes. Since the twentieth century, man-managed volumes of relocated sediment (dredging, nourishment, building) in the urbanized lowlands and its waters have grown substantially, outweighing natural sediment transport.

Sedimentary systems and paleogeography

To describe and explain the preserved Holocene record, we subdivided the Dutch Holocene sequence into three domains with differing lithostratigraphic sequence types, Pleistocene topography and substrate lithology: the offshore, onshore lowland/coastal, and onshore upland (Fig. 10.20). This tripartition is clearly visible on the recently published Geological map of the Netherlands (TNO-GDN, 2023d).

The offshore domain is dominated by open-marine units that are part of the Southern Bight and Urania formations. They occur either directly on top of Pleistocene or older sediments, or are separated from these sediments by a mostly reworked tidal sequence (Naaldwijk Formation) with a local peat bed (Basisveen Bed, Nieuwkoop Formation). See Figures 10.12 and 10.19 to 10.22.

The lowland/coastal domain is dominated by a very large Middle- to Late Holocene coastal prism that extends across roughly half of the country (Naaldwijk, Nieuwkoop, Echteld and Kreekrak formations; Fig. 10.12). The coastal prism can be further divided into a southwestern (Zeeland province), central (Holland coast and back-barrier areas including the Rhine-Meuse delta and central Netherlands lagoon) and northern part (Wadden Sea and coastal plain). Its predominantly wave-formed and tidal sediments (Naaldwijk Formation) cover and alternate with

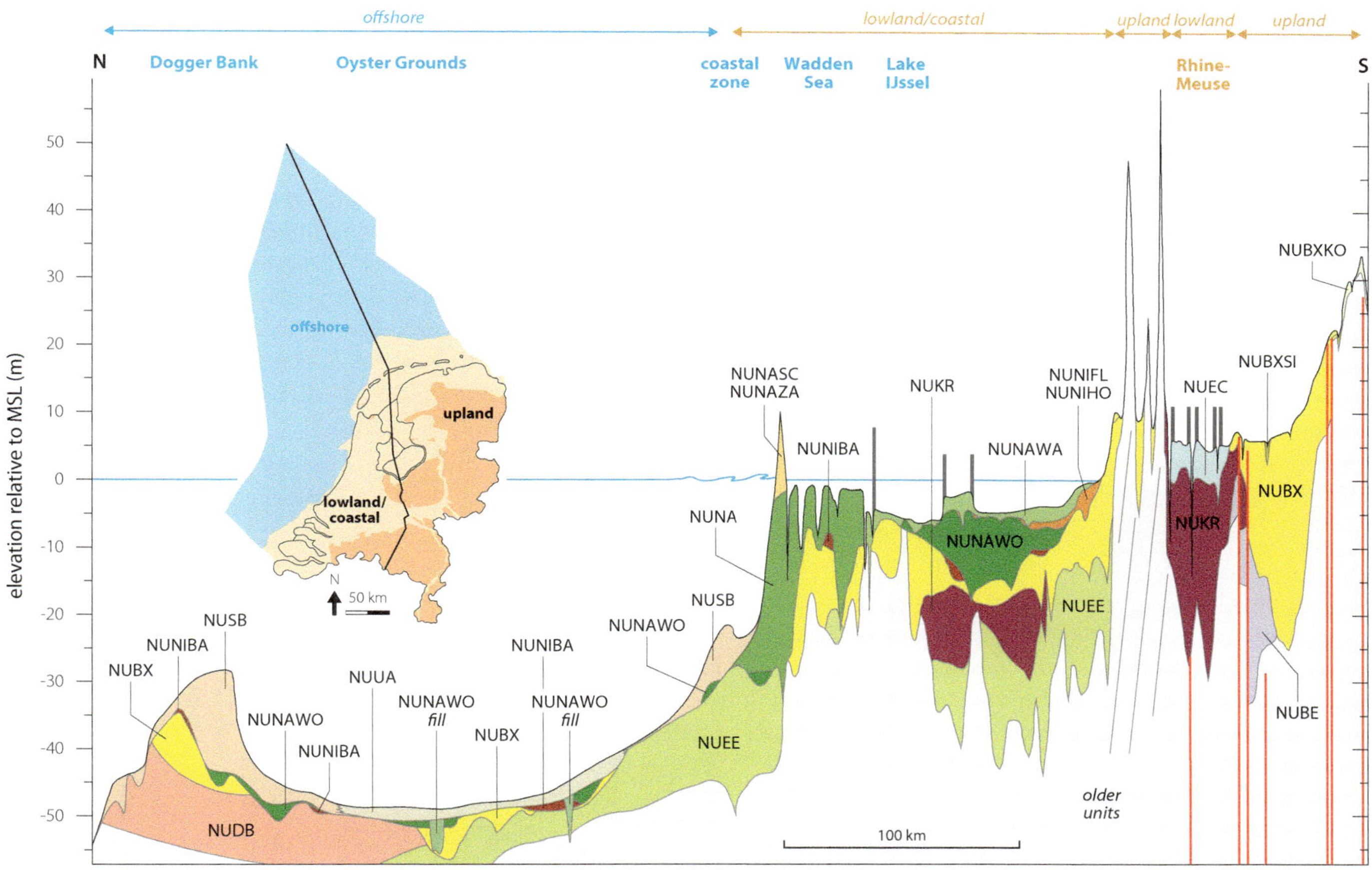

Figure 10.20. Section from the northernmost point in the Dutch part of the North Sea to the Pleistocene upland at the south-central border with Belgium (see inset map for location). It shows a bipartition of depth settings for the Nieuwkoop and Naaldwijk formations between the offshore and lowland/coastal zones. In places such as below the Wadden Sea, where the Wormer and Walcheren members are directly stacked and the Hollandveen Member is missing, these clastic tidal units are grouped in the Naaldwijk Formation (undifferentiated). See Table 10.1 for explanation of lithostratigraphic codes.

extensive horizons of peat (Nieuwkoop Formation). Near the coast, fluvial influence is mostly limited to indirect, estuarine imprints. Moving inland, the prevalence of riverine deposits (Echteld Formation of Rhine-Meuse origin; Kreekrak Formation of Scheldt origin) increases, but their total volume is an order of magnitude smaller than that of the Naaldwijk Formation (Figs 10.19, 10.20).

The Holocene depositional volumes in the upland domain are minor compared to the offshore and lowland/coastal domains. They comprise sediments of brook valley systems (Singraven Member, Boxtel Formation) and sand drifts (Kootwijk Member, Boxtel Formation), the Rhine-Meuse System (youngest parts of Beegden and Kreftenheye formations), and inland peats formed by blanket bogs (Griendtsveen Member, Nieuwkoop Formation; Fig. 10.12). Because they contain abundant organic indicators of climate and paleoenvironment, easily datable soils and paleosols are an especially informative part of the Holocene geological record (TNO-GDN, 2023a).

Onshore, the subsurface record is complete enough to enable high-resolution paleogeographic mapping (Berendsen & Stouthamer, 2001; Vos et al., 2020 ; Fig. 10.23) and detailed reconstruction of the changing depositional settings (Beets et al., 1996a). Offshore, the Holocene succession is more fragmentary and is dominated by open-marine units that overlie heavily eroded coastal deposits or Pleistocene substrate.

Early Holocene (Greenlandian: 11.7-8.2 ka)

Aside from inland depositional activity independent of rising base level, the first unit that started developing in the Early Holocene was the Basisveen Bed, reflecting a rising water table. It was soon followed by the Wormer Member (Naaldwijk Formation) with its early lagoonal clay (Velsen Bed), heterogeneous tidal-channel fills and muddy tidal-flat deposits. Offshore, these units are located as deep as 65 m below MSL. They cap terrestrial surfaces of land where hunter-gatherer populations lived, known as Doggerland, where a hunter-gatherer population lived (Coles, 1998; Amkreutz & Van der Vaart-Verschoof, 2021). This was the first area in present-day Dutch territory to be affected by postglacial transgression. In just a few thousand years, 11.7-8.5 ka, the expanding North Sea left a transgressive record of extremely rapid water-table and sea-level rise (Figs 10.19, 10.20, 10.21).

In seaward parts of the coastal lowland, the Early Holocene sequence indicates that incipient aggradation was being initiated as base level started to rise, even before the sea reached this area (Fig. 10.19). As they flooded, four Weichselian drainage basins of different size (from south to north: Lower Scheldt, Rhine-Meuse, Vecht and Boorne-Hunze areas) began to fill up. Intact sections of the sloping Pleistocene paleosurface are capped by peat (Basisveen Bed). Profiting from its diachronous nature, this peat has been dated to reconstruct water-level rise and marine submergence (e.g. Van de Plassche, 1982; Hijma & Cohen, 2019). Its compositional variability is used to distinguish brackish, freshwater and rainwater-fed environments in paleogeographic mapping. Commonly thick tidal facies of the Wormer Member (including Velsen Bed; Fig. 10.20), with alternating sand and mud, overlie and incise much of this peat as well as some of the Pleistocene underlying it.

The fluvial lowland record further inland is mostly limited to the Rhine-Meuse Valley (Berendsen & Stouthamer, 2001; Hijma & Cohen, 2011). The Late Glacial transition in river style, with its width reduction and regime change, continued. It was completed in the Early Holocene and the eventual meander belts occupy only a small area of the ice-age braidplain width (Hijma et al., 2009; Woolderink et al., 2019). Multiple immature paleosols are part of a fine-grained floodplain succession associated with the newly established meandering channel system (Wijchen Bed; Figs 10.12, 10.19, 10.22; see also photo at beginning of this chapter). Deeper than ca. 5 m below MSL, this unit is typically capped by the Basisveen Bed (Figs 10.19-10.22). The basal peats of the Rhine-Meuse delta are woody and clayey and formed in the freshwater-dominated flood basins of rivers. Eutrophic reed peats and mesotrophic sedge peats are dominant further away from contemporaneous channel systems (Bos et al., 2012).

The uplands underwent little activity limited to small rivers and brook systems (Singraven Member , Boxtel Formation), aside from soil formation at the undulating Late Pleistocene surface.

Middle Holocene (Northgrippian: 8.2-4.2 ka)

Offshore, the Middle Holocene was dominated by ravinement through wave and tidal action impacting the drowned Pleistocene land surface. Many of the Early Holocene sediments were reworked along with the top of the underlying Pleistocene (Fig. 10.20). This process left a spatially variable lag deposit enriched with shell and local gravels, depending on the nature of the eroded sediment. Gravel-rich lags (Indefatigable Grounds Member, Southern Bight Formation) overlie glacigenic units at the Cleaver Bank and offshore Texel, the southernmost Wadden island. Gravel-bearing lags rich in shells and shell fragments (Buitenbanken Member, Southern Bight Formation) occupy a southwest-northeast belt extending from the Belgian sector up to the Rhine-Meuse mouth, indicating reworking of Late Pleistocene Rhine-Meuse deposits. Shelly lags with local peat and clay pebbles (base of the Bligh Bank and Terschellinger Bank members of the Southern Bight Formation), widespread elsewhere, represent reworking of both Pleistocene sands and gravel and Early Holocene transgressive clay, sand and peat. As the North Sea bathymetry and tidal circulation approached their highstand equilibrium, by 6 ka, large parts of these lag deposits were being covered in turn by open-marine sand with a much lower shell content (Bligh Bank and Terschellinger Bank members). Ever since their formation, the variable thickness of these units has been a function of sand-wave and megaripple migration in response to (residual) tidal transport and combined wave-current flow.

In the lowland/coastal domain, an extensive coastal prism grew by 9.0-8.0 ka during Middle Holocene time (Beets & Van der Spek, 2000; De Haas et al., 2019), in the southwestern and central sectors, and by ca. 7.0 ka in the north (Kiden et al., 2002; Hijma & Cohen, 2011; Vos, 2015; Meijles et al., 2018). Incipient barrier shorelines and back-barrier plains developed from 8.5 ka onward (Rieu et al., 2005; Fig. 10.19). The general succession for this phase – seen in all transgressed drainage systems and the southwest, centre and north alike – was that of a landward-migrating transition from incipient, freshwater-dominated terrestrial wetlands with low-energy conditions (fens, fringes of inland lagoons: with formation of Basisveen Bed) to increasingly dynamic inter- and subtidal environments in brackish and saline estuaries and tidal basins. These basins were connected to tidal inlets with direct marine influence (Wormer Member). In the lower Rhine-Meuse, a phase of tidally influenced freshwater conditions left transgressive subaqueous muds (Echteld Formation) wedged between wetland peats at its base and brackish estuarine clastics on top (Figs 10.19, 10.20). Near the coastal barrier, Middle Holocene tidal basins were deepest, and tidal channels developed where tidal sediment import was largest, allowing development of fully saline tidal-flat sand and mud complexes (Wormer Member; Figs 10.19-10.22, 10.23a). Because of growing tidal prisms, partly associated with increasing tidal range, the share of sand in these deposits increased. Simultaneously, the barrier complex matured into widening and lengthening barrier spits, no longer backstepping and with modest eolian dune capping. The narrow oldest preserved beach barriers of this stage mark the inland fringe of the Holland coastal barrier complex (central sector). They may have had counterparts in the southwest but the latter have been lost to Late Holocene erosion. It is likely that counterparts in the north never existed. Under continued transgression, this part of the barrier coast matured into an island chain rather than an elongated amalgamated barrier complex.

In the millennia that followed the initial establishment of the barrier complex, adjacent beach-dune sections and associated back-barrier areas underwent complex diachronous development, with different architectural outcomes. In the southwestern and central sectors, back-barrier accommodation space had filled with tidal and fluvio-tidal sediments by 6.0-4.0 ka: sediment supply was rapidly catching up with decelerating sea-level rise. This occurred contemporaneously with a switch from retrogradational to progradational shoreline behaviour. Accretion of beach ridges widened the barrier complex (Zandvoort Member, Naaldwijk Formation). The transition from transgressive to stabilized and regressive overall conditions affected the barrier and back-barrier areas and was not a synchronous event. It dates back to 7.0-6.5 ka in the southwest and the Rhine-Meuse mouth (Figs 10.19, 10.23b), and to 4.0 ka in northwest-central (Bergen inlet) section. A switch between transgression and regression never materialized in the northern sector (Wadden Sea), which ended up being stabilized by man.

In the central sector, the accreting barrier complex widened to several kilometres and thickened to some 20 m between 7.0 and 4.0 ka (Fig. 10.19). The cross-sectional areas of tidal inlets decreased in response to an interplay with tidal-prism reduction and the associated filling of tidal basins (youngest parts of the Wormer Member). Upon tidal-inlet closure through spit accretion, tidal basins transformed into increasingly sheltered salt marshes and freshwater coastal-plain peatlands, favouring development of the Hollandveen Member (Nieuwkoop Formation; Fig. 10.22; Beets & Van der Spek, 2000; see also Van Bergen & Kosters, 2025, this volume). Aided by poor drainage, peat gradually grew to about 1 m above contemporaneous mean high-tide level. Eutrophic and mesotrophic reed and sedge peat had the upper hand early on, but over time, woody peat started to develop in areas of the lower Rhine-Meuse delta and the central delta interior (Fig. 10.19), as

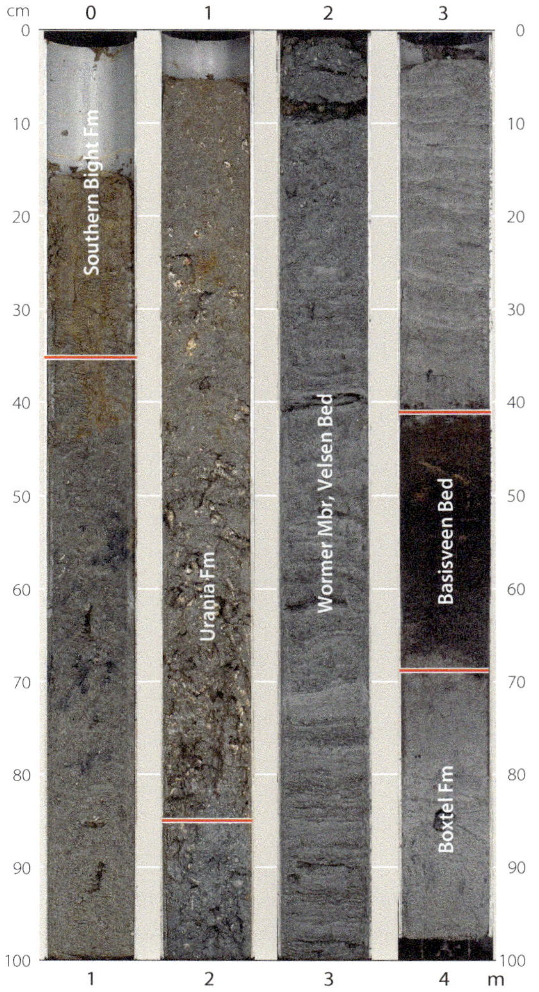

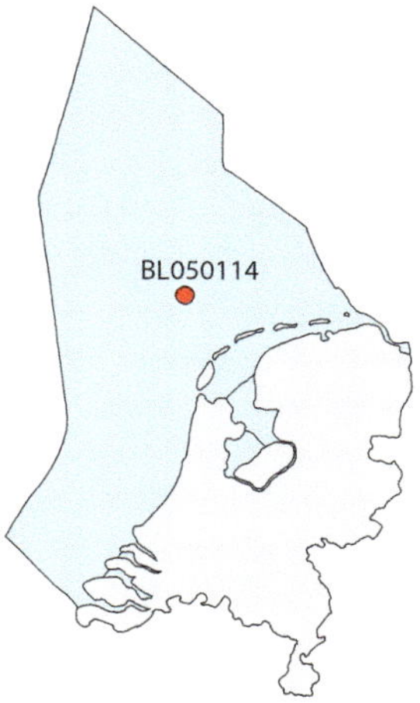

Location **Oyster Grounds**
Borehole **BL050114**
X 598814 **Y** 5955857 39.6 m - MSL

Figure 10.21. Core from the Oyster Grounds in the offshore domain, showing the Basisveen Bed followed by lagoonal clay of the Velsen Bed, indicating Early Holocene submergence of the Pleistocene surface (top Boxtel Formation). Coordinates in UTM31-WGS84. Core photos: TNO-GDN, 2023d.

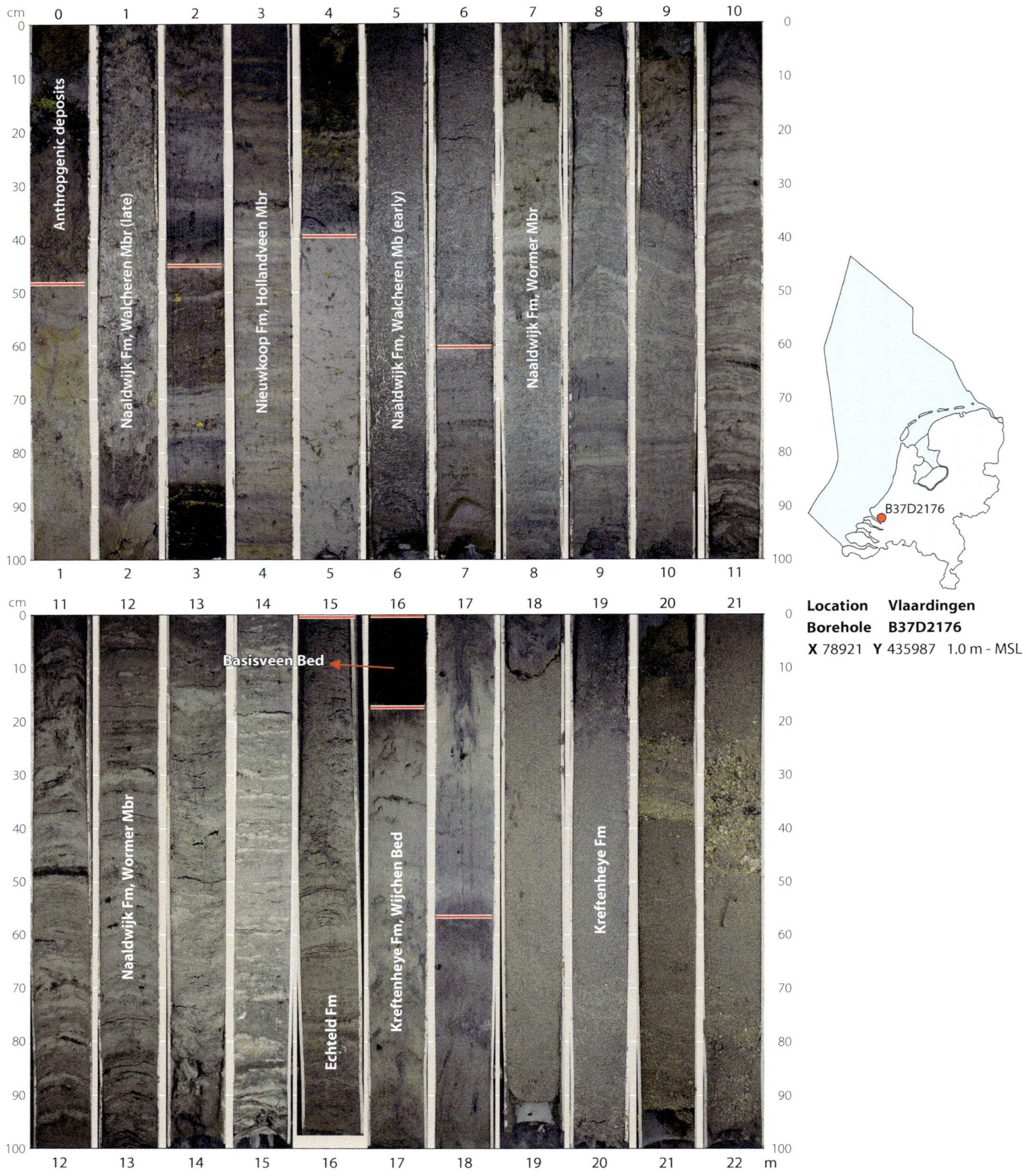

Figure 10.22. Core from the central coastal lowland, showing the Basisveen Bed that marks submergence of the Pleistocene upland (Kreftenheye Formation). The overlying Naaldwijk Formation shows distinct phases in Holocene tidal-basin development. Coordinates in Rijksdriehoeksstelsel. Core photos: TNO-GDN, 2023d.

arboreal wetlands replaced reed marsh and sedge fens (Van Asselen et al., 2017). Sizable fen-rimmed, peat-calving lakes also developed, particularly in the north-central Netherlands (Pons & Wiggers, 1960). Part of the eroded material accumulated nearby as organic detritus (Flevomeer Bed, Nieuwkoop Formation).

The barrier-island chain and large tidal basin of the Wadden Sea shows that the northern shoreline did not become regressive. When by 5.0 ka sea-level rise reduced to background rates (higher in the Wadden Sea than farther south because of differential subsidence regimes), the island chain occupied a position just a few kilometres north of its current position (Oost et al., 2012). Cyclic sediment exchanges between open sea and tidal basin have since kept most inlets open. Barrier configuration has been governed by longshore as well as ebb- and flood-tidal currents and storm-related overwash, and was constrained by inlet positions, island heads and tails (a function of substrate lithology and paleotopography). Continuously rejuvenating islands shifted slowly to their current positions (Fig. 10.23b). Rather than changing into ubiquitous peatland, the Wadden Sea back-barrier tidal basin remained open. In the Late Holocene, it would even increase in size in the northwest.

Early in the Middle Holocene (8.5-6.0 ka), development of the Rhine-Meuse delta was characterized by repeated backstepping, with an important role for avulsions triggered by downstream sea-level rise (channel belts in the Echteld Formation). Thereafter, backstepping lessened and was replaced by nodal avulsion (6.0-3.0 ka), in response to decelerating sea-level rise and coastal barrier development (Stouthamer et al., 2011). Longitudinally, the inland apex of the delta established itself in the east-central Netherlands (Fig. 10.19). In both planform and cross-section, the avulsions left an anastomosing mosaic of cross-cutting channel-belt units embedded in flood-basin clay and woody peat (Törnqvist, 1993). Local lacustrine gyttja and sandy crevasse splays formed as well (Echteld Formation). A major avulsion, commencing at 6.3 ka, shifted the main Rhine branch from its original southerly mouth that was shared with the Meuse to an independent one with a more northerly position (Fig. 10.23a,b). This new branch steadily developed into a progradational tidal river mouth, with sandy promontory beach ridges adhering to the originally linear coastal barrier, and a small pro-deltaic muddy body forming just offshore (Naaldwijk Formation). Throughout the delta, the assemblage of sandy channel systems, clayey proximal overbanks and peaty intercalated beds has been radiocarbon-dated extensively. Extensive ^{14}C-dating has documented the activation and deactivation of avulsing channels through time (Cohen et al., 2012). It has enabled a separate characterization of developments in main river-trunk channels and secondary distributaries. The excellent dating record has also been crucial in the development of similarly constrained chronologies for flood-basin environmental evolution and prehistoric occupancy (Berendsen & Stouthamer, 2001; Louwe Kooijmans, 2005; Toonen et al., 2012; 2015; Van Asselen et al., 2017).

Beyond the coastal plain and the Rhine-Meuse floodplain, above the contemporaneous mean high tide and independent of sea level, extensive fenlands and bogs blanketed the Pleistocene topography in areas of poor drainage, groundwater seepage and early deforestation. Mesotrophic sedge peat formed in the lowest parts of brook valleys and in seepage zones, oligotrophic moss peat accumulated on topographic sills in the inherited Pleistocene relief (Griendtsveen Member, Nieuwkoop Formation). The few remnants of peat developed in these wet upland environments are much less widespread than fluvial deposits of local streams (Singraven Member) from the same Middle Holocene times.

Late Holocene – the dawn of human impact (Meghalayan: 4.2-1.5 ka)

Onshore in the lowland, tidal-inlet evolution continued to have a strong influence on erosion and deposition in the northwest and north, while from about 3 ka onwards avulsions and estuarine development altered the Rhine-Meuse delta network majorly once again. Fluvial and tidal channel systems tend to be wider than their Middle Holocene counterparts, owing to relative longevity at a time of stable sea level. Peat continued to accumulate, both in the coastal plain and in the upland. At its peak, almost half of the Netherlands was peatland. In the back-barrier area of the southwestern, western and northwestern plain, freshwater wetland thrived, especially in areas away from established main channels and behind a seaward-growing coastal barrier that impeded the landward flow of saline groundwater (Fig. 10.23b,c). Reed and sedge peat formed mainly in tidal basins and estuaries, woody peat in riverine environments, and moss peat where groundwater had no influence. Domes of this oligotrophic peat, with local stands of oak, reached up to 3 m above regional water level. In the north, peatlands were formed inland from the Wadden Sea and were separated from intertidal back-barrier areas by a brackish supratidal salt-marsh zone (Vos, 2015; Vos & Knol, 2015).

From about 2.5 ka onward, intensifying human presence and especially peatland cultivation started to leave its mark in the coastal record (Van Bergen & Kosters, 2025, this volume). Swamp forests situated on peaty substrates and separating freshwater-dominated tidal rivers from boggy coastal-plain interiors, were cleared for timber and fuel, while ditches were dug to drain these lands for agriculture (Van Dinter, 2013; Pierik et al., 2018b). With the natural levees cultivated and cleared of forests, and dykes not yet in place, floodwaters could invade the cleared and drained lands deeper and more frequently than before. This resulted in the blanketing of fringes of Middle- to Late Holocene coastal and deltaic peatlands with clays: brackish tidal in the west (Walcheren Member, Naaldwijk Formation) and fluvial in the central delta (Echteld Formation) (Figs 10.19, 10.22). Feedback processes involving intruding tidal channels and retrograding lakeshores, as well as human peat extraction for salt and fuel, eroded and removed considerable parts of Late Holocene peat stock and this was further exacerbated by later land use in polders.

The progradational shoreface and beach-ridge system of the Holland coastline in the western Netherlands reached a width of more than 10 km during the Late Holocene (Fig. 10.23c). Radiocarbon dating of shells in beach and shoreface facies has made it possible to reconstruct this growth in detail (Beets et al., 1992; Cleveringa, 2000). Early during the progradation phase, marine scouring was able to remobilize sand from a wide zone of drowned, gently sloping Pleistocene substrate that was still well within reach of waves. As the coastline migrated seaward, this zone of erodible sediment became narrower and shoreface slopes steepened. Because of later erosion, older parts of the regressive wedge have been preserved more completely than younger and thus more seaward ones. The sequence is dominated by shoreface facies but also includes associated beach facies, both belonging to the Zandvoort Member (Figs 10.19, 10.20) and an eolian cap (Schoorl Member, Naaldwijk Formation). It is particularly complete and informative at the mouth of the Middle to Late Holocene Rhine that was gradually abandoned and simultaneously sanded up between 2.0 and 1.0 ka (Van de Plassche, 1982; Beets & Van der Spek, 2000; De Haas et al., 2019). In between subparallel supratidal beach ridges that mimic former coastlines, linear swales accommodated marshes that filled the lows with clayey tidal deposits and peat.

Barrier widening affected the width of remaining tidal inlets downdrift to the north (Fig. 10.23c). There, previously large tidal basins shrunk and silted up, and adjacent coastal peatland expanded across increasingly vegetated tidal flats (Pons & Wiggers, 1960; Beets et al., 1996b; Vos, 2015). Closure of the originally major Bergen inlet occurred around 3.5 ka, as is evident from barrier-spit remnants and a clayey tidal-inlet fill (Bergen Bed, part of the Wormer Member). After this, the only remaining interruptions of the Holland coastal barrier were the outlets of Meuse and Rhine and the Oer-IJ inlet west of Amsterdam. This latter inlet formed a western connection between lowland lakes and small rivers in the central Netherlands and the North Sea. Other local rivers with catchments in the eastern Netherlands had rerouted northward to the Wadden Sea when the Bergen inlet just north of the Oer-IJ began closing. They drained into the North Sea through the Vlie tidal inlet just north of Texel. From 3.0 ka onward, this inlet gradually increased its influence, servicing a tidal basin that enlarged into the central Netherlands (Fig. 10.23c), thus tapping into the lagoon-like peat-fringed freshwater lakes that had been depositing organic detritus (Flevomeer Bed; Fig. 10.20). Around 2.0 ka, the Oer-IJ inlet closed off naturally, its demise being expedited when the central peat lakes connected to the Vlie inlet captured discharge. From 3.0 to 2.0 ka, these lakes had grown and become interconnected, receiving water and sediment from a minor Rhine branch (Angstel-Vecht, ca. 2.6 ka; Bos, 2010). From 2.0 ka onwards (Vos, 2015), freshwater lacustrine conditions made way for increasingly brackish circumstances (Fig. 10.23d) with deposition of fine silts and humic clays (Almere Bed, Naaldwijk Formation).

Also around 2.0 ka, barrier erosion and shoreline retreat commenced along the Holland coastline of the central sector, following an earlier shift towards transgression in the southwest (Beets & Van der Spek, 2000). This new behaviour stood in marked contrast to the barrier widening observed over millennia before. Shoreface deepening and steepening narrowed the source area where wave-base orbitals could harvest Pleistocene sand, while decreases in the efficiency of longshore sediment supply were an additional cause for the change. In the southwest, the shoreline has been set back the strongest (Fig. 10.23cd), with no surviving direct evidence for prior prograding barriers (Vos, 2015). Pleistocene sand sources below the shoreface in the southwest were reworked and exhausted earlier than in the central sector, and the Middle to Late Holocene barrier system may have been a comparatively narrow complex anyway. More importantly, southwesterly shoreface sands were gradually cannibalized by northward-directed longshore currents to supply the more prominent Holland barrier, adding to sediment brought along by local waves and by the Rhine and Meuse. Despite this barrier cannibalization and retreat, the southwestern coastal plain was continuously protected by a narrow barrier from 4.2 to 1.5 ka, with just a single small river outlet of the Scheldt (Fig. 10.23c). A vast back-barrier peatland produced enormous volumes of organic deposits (Hollandveen Member) up to ca. 1.5 ka (Vos, 2015; Pierik et al., 2017a).

In the north, the Wadden Islands system matured during the Late Holocene, migrating landward (Zandvoort Member) over a Middle Holocene back-barrier depositional substrate (Naaldwijk Formation; Fig. 10.20). Crescentic complexes of coastal dune ridges also form a substantial part of the barrier islands (Schoorl Member). The back-barrier tidal basin continued to grow as the northwesterly Vlie inlet reached southward and connected to more and more peat lakes in the central Netherlands (Fig. 10.23c). Other Wadden Sea inlet systems also developed ingressions into coastal-plain fenlands, in part eroding and in part burying their peat with clastics (Walcheren Member). Dwelling mounds show that this development was coeval with human settlement and cultivation of supratidal terrain (Vos & Knol, 2015).

With respect to avulsion history and sedimentation, Late Holocene development of the Rhine-Meuse delta and estuaries contrasted with the Middle Holocene (Gouw & Erkens, 2007; Stouthamer et al., 2011). Up to 3.0 ka, avulsions mainly affected short-lived secondary channels in the central delta. These secondary channels rejoined established tidal outlets of the northerly Rhine and southerly Meuse, and did not affect the main discharge routing. Between 3.0 and 1.5 ka, the frequency of avulsions increased. First, central-delta avulsions created Rhine distributaries (Linge and Merwede Rivers) that connected to the Meuse estuary (Rotterdam), and upper-delta avulsions (Nederrijn River) stepped back the delta apex (Fig. 10.19) to the modern position (Fig. 10.23c,d). This development is associated with volumetrically doubled deposition of fine sediment in response to increased anthropogenic hinterland supply. Consequently, the deltaic Rhine-Meuse deposition (Echteld Formation) expanded downstream *and* upstream, as did estuarine deposition in the Meuse estuary (Walcheren Member). After 2.0 ka a second wave of avulsions created three further Rhine connections to the Rotterdam estuary (central delta: Waal; lower delta: Lek and Hollandse IJssel) and a new northward one (delta apex: Gelderse IJssel). Of these avulsions, the ones in the lower delta were due to pre-embankment human impacts in the lower and central delta itself (Pierik et al., 2018b). Avulsions in the central delta and apex are attributed to increased sedimentation, with additional influence of flooding regime and inherited topography (Cohen et al., 2009; Pierik et al., 2017b). They strongly diminished discharge to the former Rhine outlet, and its promontory of beach ridges eroded quickly.

The sandy uplands of the Late Holocene were gradually deforested. The terrain was turned into arable land (Celtic Fields). Upon exhaustion, it was either abandoned to let secondary forests appear, or was left overgrazed to develop short-lived dune fields of drift sand (Kootwijk Member; Pierik et al., 2018a), heaths and peat bogs. Once established, these mossy peat bogs (Griendtsveen Member) tended to expand gradually across larger plateaus and drainage divides (Quik et al., 2023), but little remains today. Sedge peats were topographically more constrained, accumulating locally over seepage areas and in overgrown brook valleys, as well as more regionally along the rim of the coastal plain. Paleoenvironmental key records contained within remnant bog peats indicate that climate became cooler and wetter around 2.8 ka (Van Geel et al., 1980).

Late Holocene, youngest millennia with abundant human impacts (1.5 ka – present)

From ca. 1.5 ka onward, human-induced deposition and erosion strongly affected lowland and upland landscapes, starting to exceed natural processes in extent and magnitude (Pierik et al., 2017b). The first such human action drove increased tidal incursion into peaty back-barrier areas in the southwestern Netherlands and the Rhine-Meuse estuary (Vos, 2015), the central Netherlands lagoon (Pons & Wiggers, 1960; Van Popta et al., 2020) and the northeastern Netherlands (Dollard; Vos & Knol, 2015). Between 1.5 and 0.5 ka, erosion created new estuarine and lagoonal water bodies (Fig. 10.23d) in which considerable sub-, inter- and supratidal deposition took place (Walcheren Member).

Storm surges started to breach coastal barriers, creating new inlets that were further enlarged by subsequent extreme events. Depletion of sediment supply from lower-shoreface erosion inhibited the natural ability of barrier systems to fix such event-based damage. More importantly, human activities in the peaty coastal plain created circumstances and feedback mechanisms (like those in the Meuse estuary around 2.0 ka) that accelerated marine incursion. Over time, tidal erosion reached progressively further inland, commonly into areas with highly vulnerable organic substrate. Most of the Late Holocene southwestern peat was eroded, surviving only where it was buried under tidal mud (Fig. 10.22, upper metres). Following land loss during storm surges, areas marked by rapid accretion turned into salt marshes, many of which were successfully reclaimed from 0.8 ka onwards.

In the southwest, successive storm surges between 1.5 and 0.6 ka created four new main estuarine sea arms, servicing the Scheldt and Rhine-Meuse systems. They were interconnected through numerous tie channels. In the northwest, a series of storm surges between 1.5 and 0.2 ka enlarged the connection between the Wadden Sea and the central Netherlands lagoon (allowing a second inlet to connect to it) and made the lagoon itself bigger as well while turning it increasingly brackish (Almere, Zuiderzee and IJe beds of the Walcheren Member). In the northeast, several incursions were created during storm surges and then silted up again between 0.5 and 0.2 ka. As in the southwest, the Ems estuary at the German border was especially prone to such changes.

The presence and actions of humans affected lowland and waterbodies of the central and northern coastal plains as well. To minimize the loss of lowland to erosion, many dykes were built along vulnerable parts of the coastal plain and along rivers. These early structures, dating mostly from 1.3 to 0.8 ka, protected peatland. The strategy of raising dykes was particularly successful in the lower Rhine-Meuse delta, where the channel network

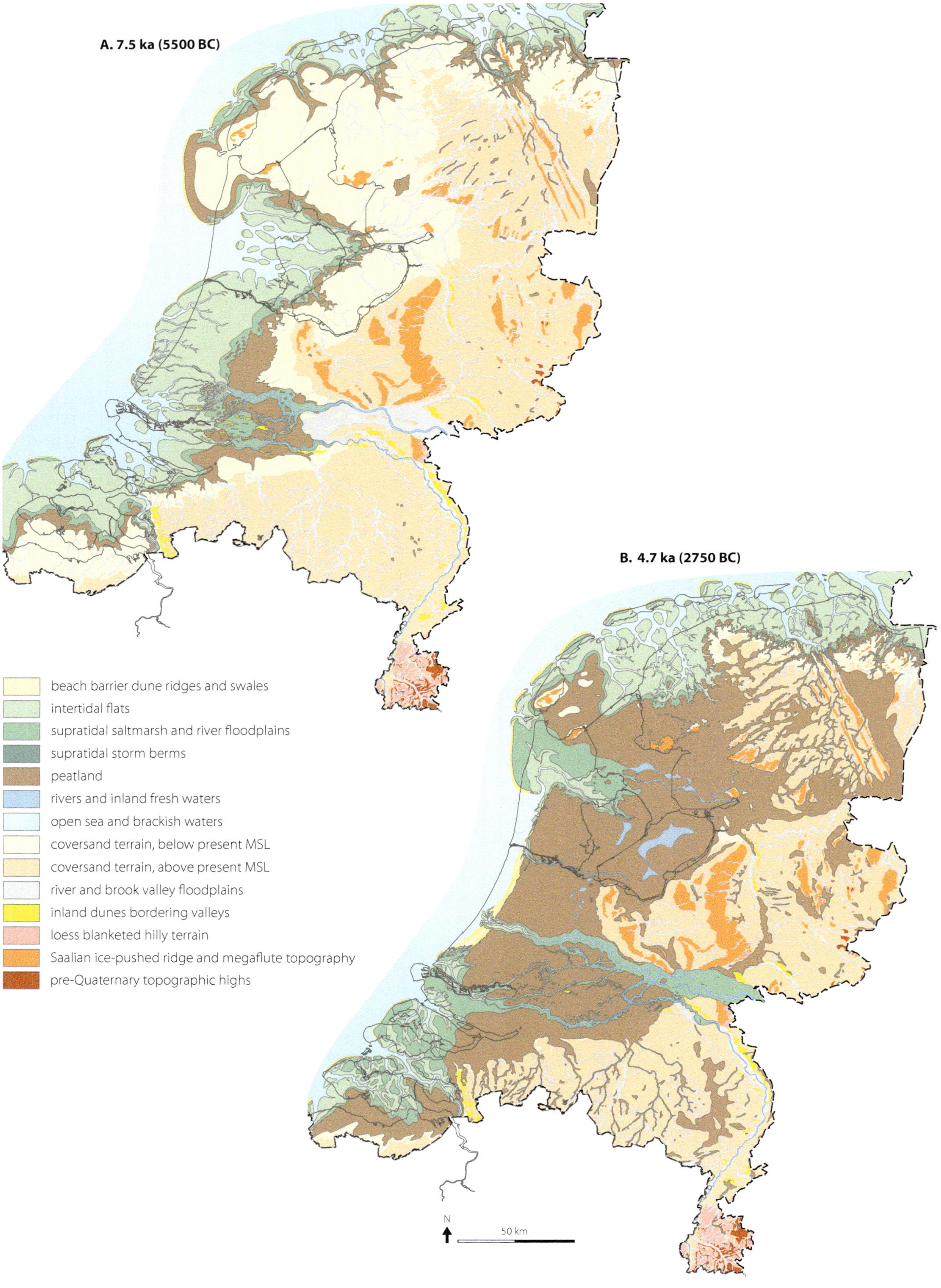
A. 7.5 ka (5500 BC)
B. 4.7 ka (2750 BC)
beach barrier dune ridges and swales
intertidal flats
supratidal saltmarsh and river floodplains
supratidal storm berms
peatland
rivers and inland fresh waters
open sea and brackish waters
coversand terrain, below present MSL
coversand terrain, above present MSL
river and brook valley floodplains
inland dunes bordering valleys
loess blanketed hilly terrain
Saalian ice-pushed ridge and megaflute topography
pre-Quaternary topographic highs
N
50 km

C. 2.5 ka (500 BC)

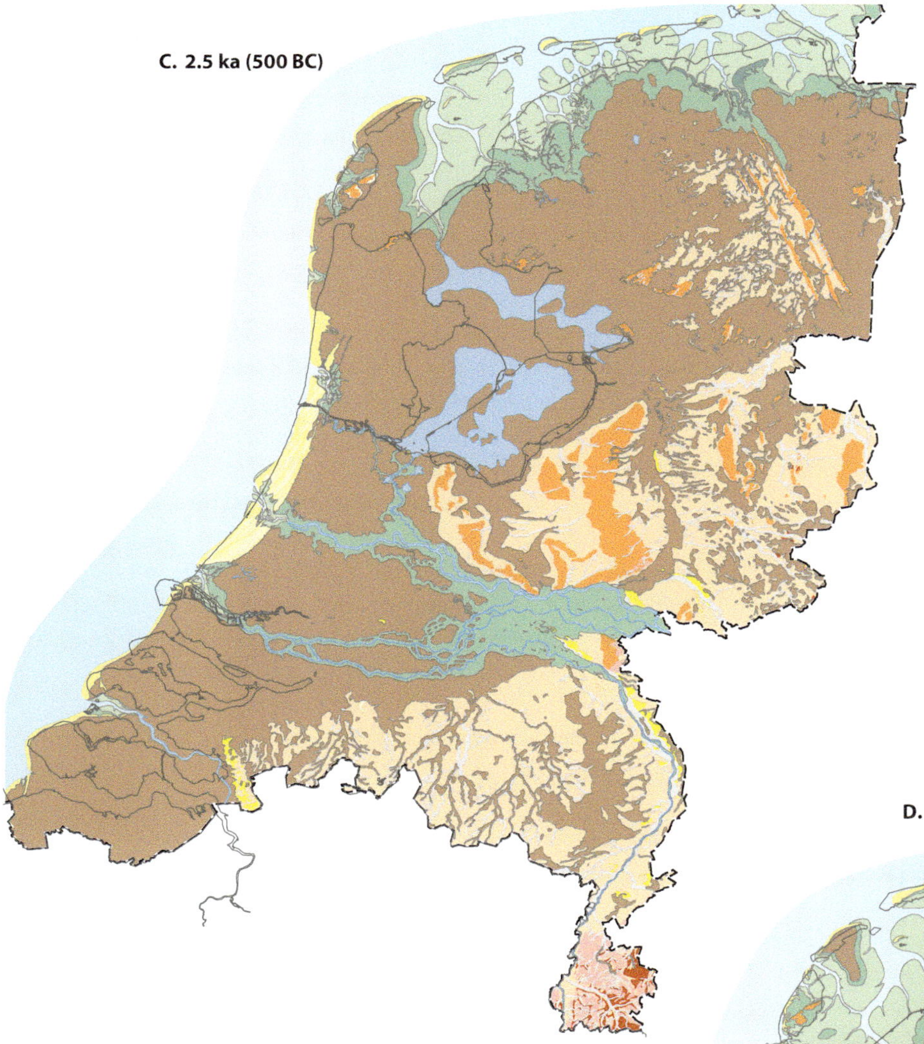

Figure 10.23. *Paleogeography during the a) Holocene middle Atlantic (7.5 ka), b) late Subboreal (4.7 ka), c) early Subatlantic (2.5 ka) and d) late Subatlantic (1.2 ka). The main developments shown are submergence of the Pleistocene landscape, closure of tidal basins, peatland expansion, inland-lake formation and growth, and barrier breaching and tidal-basin expansion. Modified from Vos et al. (2020).*

D. 1.2 ka (800 AD)

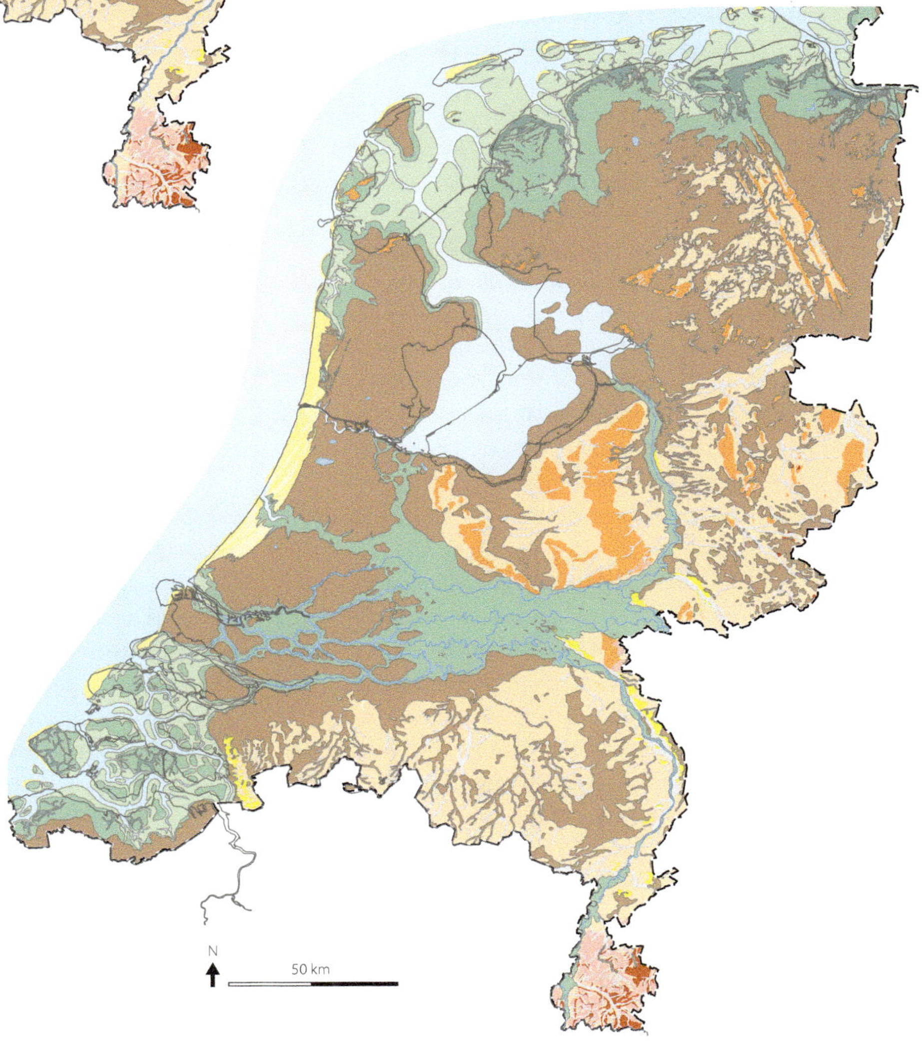

became essentially fixed. Further engineering measures between 0.3 and 0.1 ka also stopped meandering, removed mid-channel bars and shoals, fixed and deepened navigable channels, and regulated discharge division.

While human society protected vast parts of the lowlands against flooding and erosion with dykes and other defences (groynes, revetments, stone armouring) from medieval times onwards, they also dug into this protected land as part of cultivation and resource use. This has lowered the surface elevation to well below original levels of regular inundation, such as those associated with daily high tide or seasonal flooding. In the fifteenth to eighteenth century, Late Holocene peat (Hollandveen Member) was extracted to its base (top of Wormer Member) and water filled the resulting depressions in the west-central coastal plain. Wave-driven downwind shore erosion made the resulting lakes grow through time. Where technically possible, they were pumped dry after 0.4 ka, introducing a new form of land reclamation.

In addition to surface lowering, a lack of sediment accretion increased lowland vulnerability to flooding. Geologically, polder segmentation and widespread embankment turned extensive parts of the lowland into non-depositional areas at a time when the Rhine-Meuse system was at its muddiest (Gouw & Erkens, 2007; Cohen et al., 2009; Toonen et al., 2015; Pierik et al., 2017b). Instead of dispersing across floodplains, much of the suspended fluvial mud was now carried further downstream and delivered to tidal river and estuarine reaches. Upstream, sedimentation became restricted to zones between the river channel and flanking dykes, where rapidly accumulating clay was being extracted. Illustrative of the vulnerability of embanked areas, the St. Elizabeth's Flood in the fifteenth century helped earlier formed incursions to break poorly maintained dykes of the lower delta and turn a large and populated medieval polder into a tidal water body. With it, direct inner-estuarine connections established themselves between Rhine-Meuse System and the estuarine sea arms farther southwest. A century later, another another major storm surge (St. Elizabeth's Flood) established the Western Scheldt. This new, dominant sea arm owes its name to rerouting of the Scheldt, triggered as a large embanked inland area was lost.

In the upland, medieval agriculture altered the relief by interfering with small river systems, by creating raised fields that capped the nutrient-poor natural surface with manure-fertilized sediment, and by overgrazing that induced sand drifts (Kootwijk Member). Other human impact involving the geological record includes gravel, sand and clay mining (in quarries) as well as peat extraction (superficially). Their intensity increased gradually up to the twentieth century.

In the twentieth century, reclamation of lakes would take on a new dimension, extending to the enormous Zuiderzee lagoon. In 1916, a deadly storm surge impacted the central coastal lowland, triggering extensive modifications that shaped the land as we know it today. Construction of a closure dam between the Wadden Sea and Zuiderzee, finalized in the 1930s, turned the latter into a large freshwater lake and its deposits represent the youngest lithostratigraphic unit (IJsselmeer Bed, Walcheren Member). In the absence of tides, a substantial area of lagoon floor could be reclaimed, turning land that had been lost over previous millennia into fertile agricultural soil.

In today's lowland polders with peaty substrate and ditches, man-managed land subsidence occurs at rates that are an order of magnitude greater than Late Holocene natural subsidence rates (Fokker et al., 2025, this volume). Oxidation, consolidation and compaction have left their mark. Because natural sedimentation is excluded from these low-lying polder areas, substantial areas of raised ground have been created for infrastructure and urban development (anthropogenic deposits; Fig. 10.22). They are needed to allow the population to live and move around in areas as low as 7 m - MSL.

From present to future

The management of land and water continues today, and intensifies with lengthening human presence. Owing to ongoing climate change and accelerating sea-level rise, storm surges and floods are expected to become increasingly severe and frequent. To limit their negative impact on society, two strategies have been deployed. First, triggered by the 1916 event and even more by the storm-surge disaster that hit the southwest in 1953, hard protective engineering structures have been built in the southwest and northwest at unprecedented scales. These constructions drastically modified patterns of erosion and deposition in tidal basins that were allowed to stay open, such as the Western Scheldt, as well as in partly or fully enclosed counterparts. The second strategy sees the seabed, shoreface and beach-dune complex as important linked assets that need to be managed in an adaptive way. During millennia of coastal erosion, southwestern and central coastal sections became straightened, developing a prominent beach-parallel foredune in many places. This foredune, fronting either a belt of parabolic dunes (Schoorl Member) or low-lying polders, has been increasingly managed by man to keep it high and continuous enough. Here and elsewhere, the protection afforded by this dune system has been increasingly compromised over the twentieth century as sand sources for semi-natural maintenance dried up. Since the 1990s, large-scale sand nourishments have become common practice. In nourishing the upper shoreface and the beaches, this soft engineering measure mimics the natural process of sand supply from inner shelf to shoreface to beach and dune that had been in play until well into the Late Holocene, but which was no longer functioning. Every year, this requires some 12 million m³ of sand dredged from extraction pits 2 to 6 m into the seabed, beyond the 20 m depth contour. The nourished beaches now form a rich source of stray Quaternary fossils, including hominin remains, and artefacts (Textbox 4; Amkreutz & Van der Vaart-Verschoof, 2021; Niekus et al., 2023). Sand has also been extracted for raising, consolidating and levelling of construction sites onshore. The recent Maasvlakte 2 harbour extension required extraction (Fig. 10.19) from seabed pits as deep as 20 m, reaching well into Pleistocene units. Material dredged for harbour maintenance follows the opposite route, ending up in offshore dumping areas. In total, about one billion m³ of North Sea sand has been displaced so far.

Onshore, quarry-based gravel, sand and clay extraction as well as peat mining have been reduced, either due to exhaustion of resources or for environmental reasons. Sand and gravel extraction was relocated downriver. Along lowland sections of the main rivers and in the central Netherlands lagoon, subaqueous sand and gravel mining has multiplied since the middle of the twentieth century, reshaping landscapes and modifying sediment-transport regimes. Because of upstream weirs and dams in hinterland tributaries and due to reforestration, the twentieth century has seen a dramatic drop in fluvial suspended-sediment load, which was returned to values equal to those before prehistoric impacts. Depleted of sediment, the trunk channel Waal in the upper delta is geologically deepening itself. Socio-economically, this is felt during summer droughts as it lessens discharge in other Rhine-Meuse branches. Intensifying high-discharge events in the Rhine and Meuse during the late twentieth century have necessitated dyke upgrades and modifications, including relocations to increase room for the river between the dykes. This is designed to keep peak flooding levels below nationally defined maximum threshold risk levels.

Humans have left their marks in various, sometimes surprising kinds of ways. They have changed, for example, geochemical signatures of riverine sand and clay, estuarine mud and the active seabed, and have also introduced new particle types. In the Southern Bight and Urania formations offshore, the active layer from the twentieth and twenty-first century can now be identified by looking at pollutants and microplastics.

During the past century alone, humans have completely reshaped sedimentary landscapes and environments in the Netherlands and had as much influence as during the thousand years before that. By doing so, they have strongly modified how depositional and erosional processes can respond to sea-level and climate changes, in the parts of the landscape where man allows so. Human intervention has been inducing severe and rapid changes in the coastal and fluvial systems well beyond the natural thresholds seen during the Quaternary.

Acknowledgements

Reviews by Cees Kasse and Phil Gibbard significantly improved this chapter. Research by the late Wim Westerhoff highly inspired the Pleistocene paleogeographical maps presented. Editors Geert-Jan Vis and Johan ten Veen are thanked for their oversight on text and figure quality. Nik Trabucho Alexandre is thanked for his graphic support.

Digital map data

Spatial data of figures in this chapter for use in geographical information systems can be downloaded here: https://doi.org/10.5117/aup.28163615.

References

Arfai, J., Franke, D., Lutz, R., Reinhardt, L., Kley, J. & Gaedicke, C., 2018. Rapid Quaternary subsidence in the northwestern German North Sea. Scientific Reports 8: 11524. DOI: 10.1038/s41598-018-29638-6

Bakker, M.A.J. & Van der Meer, J.J.M., 2003. Structure of a Pleistocene push moraine revealed by GPR: the eastern Veluwe Ridge, The Netherlands. In: Bristow, C.S. & Jol, H.M. (eds): Ground penetrating radar in sediments. Geological Society of London, Special Publications 211: 143-151. DOI: 0.1144/GSL.SP.2001.211.01.12

Batchelor, C.L., Margold, M., Krapp, M., Murton, D.K., Dalton, A.S., Gibbard, P.L., Stokes, C.R., Murton, J.B. & Manica, A., 2019. The configuration of Northern Hemisphere ice sheets through the Quaternary. Nature Communications 10: 3713. DOI: 10.1038/s41467-019-11601-2

Bazelmans, J., Van Balen, R., Bos, J., Brinkkemper, O., Colenberg, J., Doeve, P., Van Geel, B., Hakbijl, T., Van Hateren, H., Hoek, W.Z., Huisman, H., Jansma, E., Kasse, C., Van Os, B., Van der Plicht, H., Schokker, J., Van der Putten, N. & Van der Woude, J., 2021. Environmental changes in the late Allerød and early Younger Dryas in the Netherlands: a multiproxy high-resolution record from a site with two Pinus sylvestris populations. Quaternary Science Reviews 272: 107199. DOI: 10.1016/j.quascirev.2021.107199

Beerten, K., Dreesen, R., Janssen, J. & Van Uytven, D., 2018. The Campine Plateau. In: Demoulin, A. (ed.): Landscapes and landforms of Belgium and Luxembourg. Springer International Publishing (Cham): 193-214. DOI: 10.1007/978-3-319-58239-9_12

Beets, C.J. & Beets, D.J., 2003. A high resolution stable isotope record of the penultimate deglaciation in lake sediments below the city of Amsterdam, The Netherlands. Quaternary Science Reviews 22: 195-207. DOI: 10.1016/S0277-3791(02)00089-6

Beets, D.J. & Van der Spek, A.J.F., 2000. The Holocene evolution of the barrier and back-barrier basins of Belgium and the Netherlands as a function of late Weichselian morphology, relative sea-level rise and sediment supply. Netherlands Journal of Geosciences 79: 3-16. DOI: 10.1017/S0016774600021533

Beets, D.J., Van der Valk, L. & Stive, M.J.F., 1992. Holocene evolution of the coast of Holland. Marine Geology 103: 423-443. DOI: 10.1016/0025-3227(92)90030-L

Beets, D.J., Fischer, M.M. & De Gans, W. (eds), 1996a . Coastal studies on the Holocene of the Netherlands. Mededelingen Rijks Geologische Dienst 57: 268 pp.

Beets, D.J., Roep, Th.B. & Westerhoff, W.E., 1996b . The Holocene Bergen lnlet: closing history and related barrier progradation. In: Beets, D.J., Fischer, M.M. & De Gans, W. (eds): Coastal studies on the Holocene of the Netherlands. Mededelingen Rijks Geologische Dienst 57: 97-131.

Behre, K.-E., 1989. Biostratigraphy of the last glacial period in Europe. Quaternary Science Reviews 8: 25-44. DOI: 10.1016/0277-3791(89)90019-X

Bendixen, C., Lamb, R.M., Huuse, M., Boldreel, L.O., Jensen, J.B. & Clausen, O.R., 2018. Evidence for a grounded ice sheet in the central North Sea during the early Middle Pleistocene Donian Glaciation. Journal of Geological Society 175: 291-307. DOI: 10.1144/jgs2017-073

Benvenuti, A., Šegvić, B. & Moscariello, A., 2018. Tunnel valley deposits from the southern North Sea – material provenance and depositional processes. Boreas 47: 625-642. DOI: 10.1111/bor.12292

Berends, C.J., Köhler, P., Lourens, L.J. & Van de Wal, R.S.W., 2021. On the cause of the mid-Pleistocene transition. Reviews of Geophysics 59: e2020RG000727. DOI: 10.1029/2020RG000727

Berendsen, H.J.A. & Stouthamer, E., 2001. Palaeogeographic development of the Rhine-Meuse delta, The Netherlands. Koninklijke van Gorcum B.V. (Assen): 268 pp.

Berrittella, C., 2017. Wetland methane emissions during Last Glacial climate warming. PhD thesis Vrije Universiteit Amsterdam (the Netherlands): 192 pp.

Bijlsma, S., 1981. Fluvial sedimentation from the Fennoscandian area into the North-West European Basin during the Late Cenozoic. In: Van Loon, A.J. (ed.): Quaternary geology: a farewell to A.J. Wiggers. Geologie en Mijnbouw 60: 337-345.

Boenigk, W., 1978. Gliederung der altquartären Ablagerungen in der Niederrheinischen Bucht. Fortschritte in der Geologie von Rheinland und Westfalen 28: 135-212.

Boenigk, W., 2002. The Pleistocene drainage pattern in the Lower Rhine Basin. Netherlands Journal of Geosciences 81: 201-209. DOI: 10.1017/S0016774600022447

Boenigk, W. & Frechen, M., 2006. The Pliocene and Quaternary fluvial archives of the Rhine System. Quaternary Science Reviews 25(5-6): 550-574. DOI: 10.1016/j.quascirev.2005.01.018

Bos, I.J., 2010. Architecture and facies distribution of organic-clastic lake fills in the fluvio-deltaic Rhine-Meuse system, The Netherlands. Journal of Sedimentary Research 80: 339-356. DOI: 10.2110/jsr.2010.035

Bos, I.J., Busschers, F.S. & Hoek, W.Z., 2012. Organic-facies determination: a key for understanding facies distribution in the basal peat layer of the Holocene Rhine-Meuse delta, The Netherlands. Sedimentology 59: 676-703. DOI: 10.1111/j.1365-3091.2011.01271.x

Bosch, J.H.A., 1990. Landijs, zee en rivieren als geologische "opbouwwerkers" van het Noorden. Grondboor & Hamer 44: 90-94.

Breuer, S., Bebiolka, A., Noack, V. & Lang, J., 2023. The past is the key to the future – considering Pleistocene subglacial erosion for the minimum depth of a radioactive waste repository. E&G Quaternary Science Journal 72: 113-125. DOI: 10.5194/egqsj-72-113-2023

Buckley, F.A., 2017. A glaciogenic sequence from the Early Pleistocene of the Central North Sea. Journal of Quaternary Science 32: 145-168. DOI: 10.1002/jqs.2867

Busschers, F.S., Kasse, C., Van Balen, R.T., Vandenberghe, J., Cohen, K.M., Weerts, H.J.T., Wallinga, J., Johns, C., Cleveringa, P. & Bunnik, F.P.M., 2007. Late Pleistocene evolution of the Rhine-Meuse system in the Southern North Sea Basin: imprints of climate change, sea-level oscillation and glacio-isostacy. Quaternary Science Reviews 26: 3216-3248. DOI: 10.1016/j.quascirev.2007.07.013

Busschers, F.S., Van Balen, R.T., Cohen, K.M., Kasse, C., Weerts, H.J.T., Wallinga, J. & Bunnik, F.P.M., 2008. Response of the Rhine-Meuse fluvial system to Saalian ice-sheet dynamics. Boreas 37: 377-398. DOI: 10.1111/j.1502-3885.2008.00025.x

Cameron, T.D.J., Laban, C., Mesdag, C.S. & Schüttenhelm, R.T.E., 1986. Indefatigable Sheet 53°N-02°E, 1:250 000 series. British Geological Survey and Rijks Geologische Dienst. Ordnance Survey (Southampton) for the British Geological Survey.

Cameron, T.D.J., Crosby, A., Balson, P.S., Jeffery, D.H., Lott, G.K., Bulat, J. & Harrison, D.H., 1992. United Kingdom offshore regional reports: the geology of the southern North Sea. HMSO (London) for the British Geological Survey: 152 pp.

Cartelle, V., Barlow, N.L.M., Hodgson, D.M., Busschers, F.S., Cohen, K.M., Meijninger, B.M.L. & Van Kesteren, W.P., 2021. Sedimentary architecture and landforms of the late Saalian (MIS 6) ice sheet margin offshore of the Netherlands. Earth Surface Dynamics 9: 1399-1421. DOI: 10.5194/esurf-9-1399-2021

Channell, J.E.T., Hodell, D.A. & Curtis, J.H., 2016. Relative paleointensity (RPI) and oxygen isotope stratigraphy at IODP Site U1308: North Atlantic RPI stack for 1.2-2.2 Ma (NARPI-2200) and age of the Olduvai Subchron. Quaternary Science Reviews 131: 1-19. DOI: 10.1016/j.quascirev.2015.10.011

Cleveringa, J., 2000. Reconstruction and modelling of Holocene coastal evolution of the western Netherlands. PhD thesis

Utrecht University, Geologica Ultraectina 200: 198 pp.

Cleveringa, P., Meijer, T., Van Leeuwen, R.J.W., De Wolf, H., Pouwer, R., Lissenberg, T. & Burger, A.W., 2000. The Eemian stratotype locality at Amersfoort in the central Netherlands: a re-evaluation of old and new data. Geologie en Mijnbouw / Netherlands Journal of Geosciences 79: 197-216. DOI: 10.1017/S0016774600023659

Cohen, K.M. & Gibbard, P.L., 2019. Global chronostratigraphical correlation table for the last 2.7 million years, version 2019 QI-500. Quaternary International 500: 20-31. DOI: 10.1016/j.quaint.2019.03.009

Cohen, K.M., Stouthamer, E. & Berendsen, H.J.A., 2002. Fluvial deposits as a record for neotectonic activity in the Rhine–Meuse delta, The Netherlands. Geologie en Mijnbouw / Netherlands Journal of Geosciences 81: 389-405. DOI: 10.1017/S0016774600022678

Cohen, K.M., Stouthamer, E., Hoek, W.Z., Berendsen, H.J.A. & Kempen, H.F.J., 2009. Zand in banen: zanddieptekaarten van het Rivierengebied en het IJsseldal in de provincies Gelderland en Overijssel. Provincie Gelderland (Arnhem): 130 pp.

Cohen, K.M., Stouthamer, E., Pierik, H.J. & Geurts, A.H., 2012. Digitaal basisbestand paleogeografie van de Rijn-Maas delta. DANS Data Station Archaeology: Dataset/database v.5. DOI: 10.17026/dans-x7g-sjtw

Cohen, K.M., Gibbard, P.L. & Weerts, H.J.T., 2014. North Sea palaeogeographical reconstructions for the last 1 Ma. Netherlands Journal of Geosciences 93: 7-29. DOI: 10.1017/njg.2014.12

Cohen, K.M., Cartelle, V., Barnett, R., Busschers, F.S. & Barlow, N.L.M., 2022. Last Interglacial sea-level data points from Northwest Europe. Earth System Science Data 14: 2895-2937. DOI: 10.5194/essd-14-2895-2022

Coles, B.J., 1998. Doggerland: a speculative survey. Proceedings of the Prehistoric Society 64: 45-81. DOI: 10.1017/S0079497X00002176

Da Silva Guimarães, E., Busschers, F.S., Kasse, C., Van Haren, T., Menkovic, A. & Van Balen, R.T., 2024. Quaternary tectonic and climatic forcing on the spatio-temporal evolution of the Meuse fluvial terrace staircase. Geomorphology 459: 109270. DOI: 10.1016/j.geomorph.2024.109270

De Clercq, M., Missiaen, T., Wallinga, J., Zurita Hurtado, O., Versendaal, A., Mathys, M. & De Batist, M., 2018. A well-preserved Eemian incised-valley fill in the Southern North Sea Basin, Belgian Continental Shelf – Coastal Plain: Implications for northwest European landscape evolution. Earth Surface Processes and Landforms 43: 1913-1942. DOI: 10.1002/esp.4365

De Haas, T., Van der Valk, L., Cohen, K.M., Pierik, H.J., Weisscher, S.A.H., Hijma, M.P., Van der Spek, A.J.F. & Kleinhans, M.G., 2019. Long-term evolution of the Old Rhine estuary: Unravelling effects of changing boundary conditions and inherited landscape. The Depositional Record 5: 84-108. DOI: 10.1002/dep2.56

De Jager, J., Van Ojik, K. & Smit, J., 2025. Geological development. In: Ten Veen, J.H., Vis, G.-J., De Jager, J. & Wong, Th.E. (eds): Geology of the Netherlands, second edition. Amsterdam University Press (Amsterdam): 21-51. DOI: 10.5117/9789463728362_ch01

De Jong, J., 1988. Climatic variability during the past three million years, as indicated by vegetational evolution in northwest Europe and with emphasis on data from The Netherlands. Philosophical Transactions of the Royal Society of London B 318: 603-617. DOI: 10.1098/rstb.1988.0025

Deeben, J., Hiddink, H., Huisman, D.J., Müller, A., Schokker, J. & Wallinga, J., 2010. Middle Palaeolithic artefact migration due to periglacial processes; a geological investigation into near-surface occurrence of Palaeolithic artefacts (Limburg-Eastern Brabant coversand region, the Netherlands). Netherlands Journal of Geosciences 89: 35-50. DOI: 10.1017/S0016774600000809

Donders, T.H., Van Helmond, N.A.G.M., Verreussel, R., Munsterman, D., Ten Veen, J., Speijer, R.P., Weijers, J.W.H., Sangiorgi, F., Peterse, F., Reichart, G.-J., Sinninghe Damsté, J.S., Lourens, L., Kuhlmann, G. & Brinkhuis, H., 2018. Land–sea coupling of early Pleistocene glacial cycles in the southern North Sea exhibit dominant Northern Hemisphere forcing. Climate of the Past 14: 397-411. DOI: 10.5194/cp-14-397-2018

Dost, B., Ruigrok, E., Spetzler, J. & Kruiver, P., 2025. Natural and induced seismicity. In: Ten Veen, J.H., Vis, G.-J., De Jager, J. & Wong, Th.E. (eds): Geology of the Netherlands, second edition. Amsterdam University Press (Amsterdam): 793-823. DOI: 10.5117/9789463728362_ch22

Eaton, S.J., Hodgson, D.M., Barlow, N.L., Mortimer, E.E.J. & Mellett, C.L., 2020. Palaeogeographical changes in response to glacial–interglacial cycles, as recorded in Middle and Late Pleistocene seismic stratigraphy, southern North Sea. Journal of Quaternary Science 35: 760-775. DOI: 10.1002/jqs.3230

Ehlers, J., 2011. Das Eiszeitalter. Spektrum (Heidelberg). 367 pp.

Emery, A.R., Hodgson, D.M., Barlow, N.L.M., Carrivick, J.L., Cotterill, C.J. & Phillips, E., 2019. Left high and dry: Deglaciation of Dogger Bank, North Sea, recorded in proglacial lake evolution. Frontiers in Earth Science 7: 234. DOI: 10.3389/feart.2019.00234

Felder, W.M. & Bosch, P.B., 1989. Geologische kaart van Zuid-Limburg en omgeving, 1:50.000. Afzettingen van de Maas. Rijks Geologische Dienst (Haarlem).

Fokker, P.A., Candela, T.G.G., Erkens, G., Hanssen, R.F., Kooi, H., Koster, K. & Van Leijen, F.J., 2025. Subsidence. In: Ten Veen, J.H., Vis, G.-J., De Jager, J. & Wong, Th.E. (eds): Geology of the Netherlands, second edition. Amsterdam University Press (Amsterdam): 825-847. DOI: 10.5117/9789463728362_ch23

Funnell, B.M., 1996. Plio-Pleistocene palaeogeography of the Southern North Sea Basin (3.75-0.60 Ma). Quaternary Science Reviews 15: 391-405. DOI: 10.1016/0277-3791(96)00022-4

Gallant, C.E., Candy, I., Van den Bogaard, P., Silva, B.N. & Turner, E., 2014. Stable isotopic evidence for Middle Pleistocene

environmental change from a loess-paleosol sequence: Kärlich, Germany. Boreas 43: 818-833. DOI: 10.1111/bor.12065
García-Moreno, D., 2017. Origin and geomorphology of Dover Strait and southern North Sea palaeovalleys and palaeo-depressions. PhD thesis Ghent University (Belgium): 313 pp.
Gatliff, R.W., Richards, P.C., Smith, K., Graham, C.C., McCormac, M., Smith, N.J.P., Long, D., Cameron, T.D.J., Evans, D., Stevenson, A.G., Bulat, J. & Ritchie, J.D., 1994. The geology of the central North Sea (United Kingdom Offshore Regional Report). HMSO (London) for the British Geological Survey: 118 pp.
Gibbard, P.L., 1988. The history of the great northwest European rivers during the past three million years. Philosophical Transactions of the Royal Society of London B: 318: 559-602. DOI: 10.1098/rstb.1988.0024
Gibbard, P.L., 1995. The formation of the Strait of Dover. In: Preece, R.C. (ed.): Island Britain: a Quaternary perspective. Geological Society Special of London, Special Publications 96, 15-26. DOI: 10.1144/GSL.SP.1995.096.01.03
Gibbard, P.L. & Lewin, J., 2009. River incision and terrace formation in the Late Cenozoic of Europe. Tectonophysics 474: 41-55. DOI: 10.1016/j.tecto.2008.11.017
Gouw, M.J.P. & Erkens, G., 2007. Architecture of the Holocene Rhine-Meuse delta (the Netherlands) - A result of changing external controls. Netherlands Journal of Geosciences 86: 23-54. DOI: 10.1017/S0016774600021302
Gullentops, F., Bogemans, F., De Moor, G., Paulissen, E. & Pissart, A., 2001. Quaternary lithostratigraphic units (Belgium). Geologica Belgica 4: 153-164. DOI: 10.20341/gb.2014.051
Gupta, S., Collier, J.S., Palmer-Felgate, A. & Potter, G., 2007. Catastrophic flooding origin of shelf valley systems in the English Channel. Nature 448: 342-346. DOI: 10.1038/nature06018
Gupta, S., Collier, J.S., Garcia-Moreno, D., Oggioni, F., Trentesaux, A., Vanneste, K., De Batist, M., Camelbeeck, T., Potter, G., Van Vliet-Lanoë, B. & Arthur, J.C.R., 2017. Two-stage opening of the Dover Strait and the origin of island Britain. Nature Communications 8: 15101. DOI: 10.1038/ncomms15101
Harting, P., 1874. De bodem van het Eemdal. Verslagen en Mededeelingen der Koninklijke Akademie van Wetenschappen, Afdeeling Natuurkunde. Tweede Reeks 8: 282-290.
Harting, P., 1875. Le système Éemien. Archives Néerlandaises des Sciences Exactes et Naturelles 10: 443-454.
Head, M.J., 2021. Review of the Early–Middle Pleistocene boundary and Marine Isotope Stage 19. Progress in Earth and Planetary Science 8: 50. DOI: 10.1186/s40645-021-00439-2
Head, M.J. & Gibbard, P.L., 2005. Early-Middle Pleistocene transitions: an overview and recommendation for the defining boundary. In: Head, M.J. & Gibbard, P.L. (eds): Early-Middle Pleistocene transitions: The land-ocean evidence. Geological Society of London, Special Publications 247: 1-18. DOI: 10.1144/GSL.SP.2005.247.01.0
Helmens, K.F., 2014 . The Last Interglacial-Glacial cycle (MIS 5-2) re-examined based on long proxy records from central and northern Europe. Quaternary Science Reviews 86: 115-143. DOI: 10.1016/j.quascirev.2013.12.012
Hepp, D.A., Romero, O.E., Mörz, T., De Pol-Holz, R. & Hebbeln, D., 2019. How a river submerges into the sea: a geological record of changing a fluvial to a marine paleoenvironment during early Holocene sea level rise. Journal of Quaternary Science 34: 581-592. DOI: 10.1002/jqs.3147
Heyse, I. & Demoulin, A., 2018. The Flemish Valley: response of the Scheldt drainage system to climatic and glacio-eustatic oscillations. In: Demoulin, A. (ed.): Landscapes and landforms of Belgium and Luxembourg. Springer International Publishing (Cham): 297-311. DOI: 10.1007/978-3-319-58239-9_18
Hijma, M.P. & Cohen, K.M., 2010. Timing and magnitude of the sea-level jump preluding the 8200 yr event. Geology 38: 275-278. DOI: 10.1130/G30439.1
Hijma, M.P. & Cohen, K.M., 2011. Holocene transgression of the Rhine river mouth area, The Netherlands/Southern North Sea: palaeogeography and sequence stratigraphy. Sedimentology 58: 1453-1485. DOI: 10.1111/j.1365-3091.2010.01222.x
Hijma, M.P. & Cohen, K.M., 2019. Holocene sea-level database for the Rhine-Meuse Delta, The Netherlands: Implications for the pre-8.2 ka sea-level jump. Quaternary Science Reviews 214: 68-86. DOI: 10.1016/j.quascirev.2019.05.001
Hijma, M.P., Cohen, K.M., Hoffmann, G., Van der Spek, A.J.F. & Stouthamer, E., 2009. From river valley to estuary: the evolution of the Rhine mouth in the early to middle Holocene (western Netherlands, Rhine-Meuse delta). Netherlands Journal of Geosciences 88: 13-53. DOI: 10.1017/S0016774600000986
Hijma, M.P, Cohen, K.M., Roebroeks, W., Westerhoff, W.E. & Busschers, F.S., 2012a. Pleistocene Rhine–Thames landscapes: geological background for hominin occupation of the southern North Sea region. Journal of Quaternary Science 27: 17-39. DOI: 10.1002/jqs.1549
Hijma, M.P., Cohen, K.M., Roebroeks, W., Westerhoff, W.E. & Busschers, F.S., 2012b . Corrigendum: Pleistocene Rhine–Thames landscapes: geological background for hominin occupation of the southern North Sea region. Journal of Quaternary Science 27: 336. DOI: 10.1002/jqs.2545
Hoek, W.Z. & Bohncke, S.J.P., 2002. Climatic and environmental events over the Last Termination, as recorded in The Netherlands: a review. Netherlands Journal of Geosciences 81: 123-137. DOI: 10.1017/S001677460002062X
Hoek, W.Z., Lammertsma, E.I., Bohncke, S.J.P., Bos, J.A.A., Bunnik, F., Kasse, C., Schokker, J. & Westerhoff, W., 2017. Lateglacial and early Holocene vegetation development and fluvial system changes in the northern Meuse valley, the Netherlands: a review of palynological data. Netherlands Journal of Geosciences 96: 93-114. DOI: 10.1017/njg.2017.4
Hoffmann, T., Erkens, G., Cohen, K.M., Houben, P., Seidel, J. & Dikau, R., 2007. Holocene floodplain sediment storage and hillslope erosion within the Rhine catchment. The Holocene 17: 105-118. DOI: 10.1177/0959683607073287
Hooghiemstra, H. & Hoek, W.Z., 2019. Waldo Heliodoor Zagwijn (1928-2018): the instigator and architect of European chron-

ostratigraphy. Netherlands Journal of Geosciences 98: e7. DOI: 10.1017/njg.2019.9

Hooghiemstra, H. & Richards, K., 2022. Frans Florschütz as founding father of pollen analysis in the Netherlands, and expansion of palynology into the tropics. Grana 61: 241-255. DOI: 10.1080/00173134.2022.2089226

Houmark-Nielsen, M., 2004. The Pleistocene of Denmark: a review of stratigraphy and glaciation history. Developments in Quaternary Sciences 2: 35-46. DOI: 10.1016/S1571-0866(04)80055-1

Houtgast, R.F., Van Balen, R.T., Bouwer, L.M., Brand, G.B.M. & Brijker, J.M., 2002. Late Quaternary activity of the Feldbiss Fault Zone, Roer Valley Rift System, the Netherlands, based on displaced fluvial terrace fragments. Tectonophysics 352: 295-315. DOI: 10.1016/S0040-1951(02)00219-6

Hughes, P.D., Gibbard, P.L. & Ehlers, J., 2020. The 'missing glaciations' of the Middle Pleistocene. Quaternary Research 96: 161-183. DOI: 10.1017/qua.2019.76

Huijzer, B. & Vandenberghe, J., 1998. Climatic reconstruction of the Weichselian Pleniglacial in northwestern and Central Europe. Journal of Quaternary Science 13: 391-417. DOI: 10.1002/(SICI)1099-1417(1998090)13:5<391::AID-JQS397>3.0.CO;2-6

Huuse, M. & Lykke-Andersen, H., 2000. Overdeepened Quaternary valleys in the eastern Danish North Sea: morphology and origin. Quaternary Science Reviews 19: 1233-1253. DOI: 10.1016/S0277-3791(99)00103-1

Huxtable, J. & Aitken, J., 1985. Thermoluminescence dating results for the Paleolithic site Maastricht–Belvédère. Mededelingen Rijks Geologische Dienst 39: 41-44.

Kars, R.H., Busschers, F.S. & Wallinga, J., 2012. Validating post IR-IRSL dating on K-feldspars through comparison with quartz OSL ages. Quaternary Geochronology 12: 74-86. DOI: 10.1016/j.quageo.2012.05.001

Kasse, C., 1988. Early-Pleistocene tidal and fluviatile environments in the southern Netherlands and northern Belgium. PhD thesis Vrije Universiteit Amsterdam (the Netherlands): 190 pp.

Kasse, C., 1993. Periglacial environments and climatic development during the Early Pleistocene Tiglian stage (Beerse Glacial) in northern Belgium. Geologie en Mijnbouw 72: 107-123.

Kasse, C., 1995. Younger Dryas cooling and fluvial response (Maas River, the Netherlands). Geologie en Mijnbouw 74: 251-256.

Kasse, C., 2002. Sandy aeolian deposits and environments and their relation to climate during the Last Glacial Maximum and Lateglacial in northwest and central Europe. Progress in Physical Geography: Earth and Environment 26: 507-532. DOI: 10.1191/0309133302pp350ra

Kasse, C. & Bohncke, S.J.P., 2001. Early Pleistocene fluvial and estuarine records of climate change in the southern Netherlands and northern Belgium. In: Maddy, D., Macklin, M.G. & Woodward, J.C.: River basin sediment systems: archives of environmental change. Balkema Publishers (Lisse): 171-193. DOI: 10.1201/9781439824672.pt3

Kasse, C., Vandenberghe, D., De Corte, F. & Van den Haute, P., 2007. Late Weichselian fluvio-aeolian sands and coversands of the type locality Grubbenvorst (southern Netherlands): sedimentary environments, climate record and age. Journal of Quaternary Science 22: 695-708. DOI: 10.1002/jqs.1087

Kasse, C., Woolderink, H.A.G., Kloos, M.E. & Hoek, W.Z., 2020. Source-bordering aeolian dune formation along the Scheldt River (southern Netherlands – northern Belgium) was caused by Younger Dryas cooling, high river gradient and southwesterly summer winds. Netherlands Journal of Geosciences 99: e13. DOI: 10.1017/njg.2020.15

Kasse, C., Van der Woude, J.D., Woolderink, H.A.G. & Schokker, J., 2022. Eemian to Early Weichselian regional and local vegetation development and sedimentary and geomorphological controls, Amersfoort Basin, The Netherlands. Netherlands Journal of Geosciences 101: e7. DOI: 10.1017/njg.2022.4

Kiden, P., 2006. De evolutie van de Beneden-Schelde in België en Zuidwest-Nederland na de laatste ijstijd. Belgeo 3: 279-294. DOI: 10.4000/belgeo.12025

Kiden, P., Denys, L. & Johnston, P., 2002. Late Quaternary sea-level change and isostatic and tectonic land movements along the Belgian–Dutch North Sea coast: geological data and model results. Journal of Quaternary Science 17: 535-546. DOI: 10.1002/jqs.709

Kirkham, J.D., Hogan, K.A., Larter, R.D., Arnold, N.S., Ely, J.C., Clark, C.D., Self, E., Games, K., Huuse, M., Stewart, M.A., Ottesen, D. & Dowdeswell, J.A., 2024. Tunnel valley formation beneath deglaciating mid-latitude ice sheets: Observations and modelling. Quaternary Science Reviews 323: 107680. DOI: 10.1016/j.quascirev.2022.107680

Knox, R.W.O.B., Bosch, J.H.A., Rasmussen, E.S., Heilmann-Clausen, C., Hiss, M., De Lugt, I.R., Kasińksi, J., King, C., Köthe, A., Słodkowska, B., Standke, G. & Vandenberghe, N., 2010. Cenozoic. In: Doornenbal, J.C. & Stevenson, A.G. (eds): Petroleum geological atlas of the Southern Permian Basin area. EAGE Publications b.v. (Houten): 211-223.

Kolstrup, E., 2007. Lateglacial older and younger coversand in northwest Europe: chronology and relation to climate and vegetation. Boreas 36: 65-75. DOI: 10.1111/j.1502-3885.2007.tb01181.x

Kooi, H., Johnston, P., Lambeck, K., Smither, C. & Molendijk, R., 1998. Geological causes of recent (~100 yr) vertical land movement in the Netherlands. Tectonophysics 299: 297-316. DOI: 10.1016/S0040-1951(98)00209-1

Kuhlmann, G. & Wong, Th.E., 2008. Pliocene paleoenvironment evolution as interpreted from 3D-seismic data in the southern North Sea, Dutch offshore sector. Marine and Petroleum Geology 25: 173-189. DOI: 10.1016/j.marpetgeo.2007.05.009

Kuhlmann, G., Langereis, C.G., Munsterman, D., Van Leeuwen, R.-J., Verreussel, R., Meulenkamp, J.E. & Wong, Th.E., 2006a. Intergrated chronostratigraphy of the Pliocene-Pleistocene interval and its relation to the regional stratigraphical stages

in the southern North Sea region. Netherlands Journal of Geosciences 85 (1): 19-35. DOI: 10.1017/S0016774600021405
Kuhlmann, G., Langereis, C.G., Munsterman, D., Van Leeuwen, R.-J., Verreussel, R., Meulenkamp, J.E. & Wong, Th.E., 2006a. Chronostratigraphy of Late Neogene sediments in the Southern North Sea Basin and paleoenvironmental interpretations. Palaeogeography, Palaeoclimatology, Palaeoecology 239: 426-455. DOI: 10.1016/j.palaeo.2006.02.004
Kuyl, O.S., 1980. Toelichting bij de geologische kaart van Nederland 1:50,000, Blad Heerlen (62W oostelijke helft, 62O westelijke helft). Rijks Geologische Dienst (Haarlem): 206 pp.
Laban, C., 1995. The Pleistocene glaciations in the Dutch sector of the North Sea. A synthesis of sedimentary and seismic data. PhD thesis University of Amsterdam (the Netherlands): 194 pp.
Laban, C. & Van der Meer, J.J.M., 2011. Pleistocene glaciation in The Netherlands. In: Ehlers, J., Gibbard, P.L. & Hughes, D. (eds): Quaternary glaciations - extent and chronology: a closer look. Developments in Quaternary Sciences 15: 247-260, 745. DOI: 10.1016/B978-0-444-53447-7.00020-9
Lamb, R.M., Harding, R., Huuse, M., Stewart, M. & Brocklehurst, S.H., 2018. The early Quaternary North Sea Basin. Journal of the Geological Society 175: 275-290. DOI: 10.1144/jgs2017-057
Lee, J.R., Busschers, F.S. & Sejrup, H.P., 2012. Pre-Weichselian Quaternary glaciations of the British Isles, The Netherlands, Norway and adjacent marine areas south of 68°N: implications for long-term ice sheet development in northern Europe. Quaternary Science Reviews 44: 213-228. DOI: 10.1016/j.quascirev.2010.02.027
Lehmkuhl, F., Nett, J.J., Pötter, S., Schulte, P., Sprafke, T., Jary, Z., Antoine, P., Wacha, L., Wolf, D., Zerboni, A., Hošek, J., Marković, S.B., Obreht, I., Sümegi, P., Veres, D., Zeeden, C., Boemke, B., Schaubert, V., Viehweger, J. & Hambach, U., 2021. Loess landscapes of Europe – Mapping, geomorphology, and zonal differentiation. Earth-Science Reviews 215: 103496. DOI: 10.1016/j.earscirev.2020.103496
Lisiecki, L.E. & Raymo, M.E., 2005. A Pliocene-Pleistocene stack of 57 globally distributed benthic δ18O records. Paleoceanography 20: PA1003. DOI: 10.1029/2004PA001071
Long, A.J., Barlow, N.L.M., Busschers, F.S., Cohen, K.M., Gehrels, W.R. & Wake, L.M., 2015. Near-field sea-level variability in northwest Europe and ice sheet stability during the last interglacial. Quaternary Science Reviews 126: 26-40. DOI: 10.1016/j.quascirev.2015.08.021
Løseth, H., Nygård, A., Batchelor, C.L. & Fayzullaev, T., 2022. A regionally consistent 3D seismic-stratigraphic framework and age model for the Quaternary sediments of the northern North Sea. Marine and Petroleum Geology 142: 105766. DOI: 10.1016/j.marpetgeo.2022.105766
Losson, B. & Quinif, Y., 2001. La capture de la Moselle – Nouvelles données chronologiques par datations U/Th sur spéléothèmes . Karstologia 37: 29-40. DOI: 10.3406/karst.2001.2469
Louwe Kooijmans, L.P., Van den Broeke, P.W., Fokkens, H. & Van Gijn, A.L. (ed.), 2005. The prehistory of the Netherlands. Amsterdam University Press (Amsterdam): 844 pp.
Lowe, J.J., Rasmussen, S.O., Björck, S., Hoek, W.Z., Steffensen, J.P., Walker, M.J.C. & Yu, Z.C., 2008. Synchronisation of palaeoenvironmental events in the North Atlantic region during the Last Termination: a revised protocol recommended by the INTIMATE group. Quaternary Science Reviews 27: 6-17. DOI: 10.1016/j.quascirev.2007.09.016
Makaske, B. & Maas, G.J., 2023. Different hydrological controls causing variable rates of Holocene peat growth in a lowland valley system, north-eastern Netherlands; implications for valley peatland restoration. The Holocene 33: 960-974. DOI: . 10.1177/09596836231169985
Mangerud, J., Gyllencreutz, R., Lohne, Ø. & Svendsen, J.I., 2011. Glacial history of Norway. In: Ehlers, J., Gibbard, P.L. & Hughes, P.D. (eds): Quaternary glaciations – extent and chronology: a closer look. Developments in Quaternary Sciences 15: 1108 pp.
Margari, V., Hodell, D.A., Parfitt, S.A., Ashton, N.M., Grimalt, J.O., Kim, H., Yun, K.-S., Gibbard, P.L., Stringer, C.B., Timmermann, A. & Tzedakis, P.C., 2023. Extreme glacial cooling likely led to hominin depopulation of Europe in the Early Pleistocene. Science 381: 693-699. DOI: 10.1126/science.adf4445
Maslin, M.A. & Ridgwell, A.J., 2005. Mid-Pleistocene revolution and the 'eccentricity myth'. In: Head, M.J. & Gibbard, P.L. (eds): Early-Middle Pleistocene transitions: The land-ocean evidence. Geological Society of London, Special Publications 247: 19-34. DOI: 10.1144/GSL.SP.2005.247.01.02
Medina-Elizalde, M., 2013. A global compilation of coral sea-level benchmarks: Implications and new challenges. Earth and Planetary Science Letters 362: 310-318. DOI: 10.1016/j.epsl.2012.12.001
Meijer, T., 1988. Mollusca from the borehole Zuurland-2 at Brielle, The Netherlands (an interim report. Mededelingen van de Werkgroep voor Tertiaire en Kwartaire Geologie 25: 49-60.
Meijer, T., 1991. Molluscan investigation of ice-pushed Pleistocene deposits near Wageningen,The Netherlands. In: Ruegg, G.H.J. (ed.): Geology and archeology of ice-pushed Pleistocene deposits near Wageningen. Mededelingen Rijks Geologische Dienst 46: 55-64.
Meijer, T., Cleveringa, P., Munsterman, D.K. & Verreussel, R.M.C.H., 2006. The Early Pleistocene Praetiglian and Ludhamian pollen stages in the North Sea Basin and their relationship to the marine isotope record. Journal of Quaternary Science 21: 307-310. DOI: 10.1002/jqs.956
Meijer, T., Pouwer, R., Cleveringa, P., De Wolf, H., Busschers, F.S. & Wesselingh, F.P., 2021. Fossil molluscs from borehole Hollum (Ameland, the Netherlands) constrain three successive Quaternary interglacial marine intervals in the Southern North Sea Basin. Netherlands Journal of Geosciences 100: e13. DOI: 10.1017/njg.2021.2
Meijles, E.W., Kiden, P., Streurman, H.-J., Van der Plicht, J., Vos, P.C., Gehrels, W.R. & Kopp, R.E., 2018. Holocene relative

mean sea-level changes in the Wadden Sea area, northern Netherlands. Journal of Quaternary Science 33: 905-923. DOI: 10.1002/jqs.3068

Meijs, E.P.M., 2011. The Veldwezelt site (province of Limburg, Belgium): environmental and stratigraphical interpretations. Netherlands Journal of Geosciences 90 (2-3): 73-94. DOI: 10.1017/S0016774600001037

Meijs, E.P.M., Van Peer, Ph. & De Warrimont, J.P.L.M.N., 2013. Geomorphologic context and proposed chronostratigraphic position of Lower Palaeolithic artefacts from the Op de Schans pit near Kesselt (Belgium) to the west of Maastricht. Netherlands Journal of Geosciences 91: 137-157. Cambridge Core. DOI: 10.1017/S0016774600001554

Meloro, C., Raia, P., Carotenuto, F. & Barbera, C., 2008. Diversity and turnover of Plio-Pleistocene large mammal fauna from the Italian Peninsula. Palaeogeography, Palaeoclimatology, Palaeoecology 268: 58-64. DOI: 10.1016/j.palaeo.2008.08.002

Meyer, W. & Stets, J., 2002. Pleistocene to Recent tectonics in the Rhenish Massif (Germany). In: Schäfer, A & Siehl, A. (eds): Rift tectonics and syngenetic sedimentation - the Cenozoic Lower Rhine Graben and related structures. Netherlands Journal of Geosciences 81: 217-221. DOI: 10.1017/S0016774600022460

Michon, L., Van Balen, R.T., Merle, O. & Pagnier, H., 2003. The Cenozoic evolution of the Roer Valley rift system integrated at a European scale. Tectonophysics 367 (1-2): 101-126. DOI: 10.1016/S0040-1951(03)00132-X00132-X

Missiaen, T., Fitch, S., Harding, R., Muru, M., Fraser, A., De Clercq, M., Garcia Moreno, D., Versteeg, W., Busschers, F.S., Van Heteren, S., Hijma, M.P., Reichart, G.-J. & Gaffney, V., 2021. Targeting the Mesolithic: Interdisciplinary approaches to archaeological prospection in the Brown Bank area, southern North Sea. Quaternary International 584: 141-151. DOI: 10.1016/j.quaint.2020.05.004

Mol, D., Post, K., Reumer, J.W.F., Van der Plicht, J., De Vos, J., Van Geel, B., Van Reenen, G., Pals, J.P. & Glimmerveen, J., 2006. The Eurogeul—first report of the palaeontological, palynological and archaeological investigations of this part of the North Sea. In: Storer, J.E. (ed.): Third International Mammoth Conference, Dawson, Yukon. Quaternary International 142-143: 178-185. DOI: 10.1016/j.quaint.2005.03.015

Moreau, J. & Huuse, M., 2014. Infill of tunnel valleys associated with landward-flowing ice sheets: The missing Middle Pleistocene record of the NW European rivers? Geochemistry, Geophysics, Geosystems 15: 1-9. DOI: 10.1002/2013GC005007

Moreau, J., Huuse, M., Janszen, A., Van der Vegt, P., Gibbard, P.L. & Moscariello, A., 2012. The glaciogenic unconformity of the southern North Sea. In: Huuse, M., Redfern, J., Le Heron, D.P., Dixon, R.J., Moscariello, A. & Craig, J. (eds): Glaciogenic reservoirs and hydrocarbon systems. Geological Society of London, Special Publications 368: 99-110. DOI: 10.1144/SP368

Mudelsee, M. & Raymo, M.E., 2005. Slow dynamics of the Northern Hemisphere glaciation. Paleoceanography 20: PA4022. DOI: 10.1029/2005PA001153

Munsterman, D.K., Ten Veen, J.H., Menkovic, A., Deckers, J., Witmans, N., Verhaegen, J., Kerstholt-Boegehold, S.J., Van de Ven, T. & Busschers, F.S., 2019. An updated and revised stratigraphic framework for the Miocene and earliest Pliocene strata of the Roer Valley Graben and adjacent blocks. Netherlands Journal of Geosciences 98: e8. DOI: 10.1017/njg.2019.10

Munsterman, D.K., Donders, T.H., Houben, A.J.P., Ten Veen, J.H. & Wesselingh, F.P., 2025. Paleogene-Neogene. In: Ten Veen, J.H., Vis, G.-J., De Jager, J. & Wong, Th.E. (eds): Geology of the Netherlands, second edition. Amsterdam University Press (Amsterdam): 293-331. DOI: 10.5117/9789463728362_ch9

Niekus, M.J.L.Th., Cohen, K.M., Johansen, L. & Stapert, D., 2023. Opgezogen, opgespoten en opgeraapt - Vuurstenen en hardstenen artefacten van de Zandmotor en hun sedimentaire context. Rapportage Archeologische Monumentenzorg 276: 168 pp.

North Greenland Ice Core Project members, 2004. High-resolution record of Northern Hemisphere climate extending into the last interglacial period. Nature 431: 147-151. DOI: doi.org/10.1038/nature02805

Oele, E., 1969. The Quaternary geology of the Dutch part of the North Sea, north of the Frisian isles. Geologie En Mijnbouw 48: 467-480.

Oliveira, D., Sanchez Goñi, M.F., Naughton, F., Polanco-Martínez, J.M., Jimenez-Espejo, F.J., Grimalt, J.O., Martrat, B., Voelker, A.H.L., Trigo, R., Hodell, D., Abrantes, F. & Desprat, S., 2017. Unexpected weak seasonal climate in the western Mediterranean region during MIS 31, a high-insolation forced interglacial. Quaternary Science Reviews 161:1-17. DOI: 10.1016/j.quascirev.2017.02.013

Oost, A.P., Hoekstra, P., Wiersma, A., Flemming, B., Lammerts, E.J., Pejrup, M., Hofstede, J., Van der Valk, B., Kiden, P., Bartholdy, J., Van der Berg, M.W., Vos, P.C., De Vries, S. & Wang, Z.B., 2012. Barrier island management: Lessons from the past and directions for the future. Ocean & Coastal Management 68: 18-38. DOI: 10.1016/j.ocecoaman.2012.07.010

Ottesen, D., Dowdeswell, J.A. & Bugge, T., 2014. Morphology, sedimentary infill and depositional environments of the Early Quaternary North Sea Basin (56°-62°N). Marine and Petroleum Geology 56: 123-146. DOI: 10.1016/j.marpetgeo.2014.04.007

Ottesen, D., Batchelor, C.L., Dowdeswell, J.A. & Løseth, H., 2018. Morphology and pattern of Quaternary sedimentation in the North Sea Basin (52-62°N). Marine and Petroleum Geology 98: 836-859. DOI: 10.1016/j.marpetgeo.2018.08.022

Ottesen, D., Stewart, M., Brönner, M. & Batchelor, C.L., 2020. Tunnel valleys of the central and northern North Sea (56°N to 62°N): Distribution and characteristics. Marine Geology 425: 106199. DOI: 10.1016/j.margeo.2020.106199

Overeem, I., Weltje, G.J., Bishop-Kay, C. & Kroonenberg, S.B., 2001. The Late Cenozoic Eridanos delta system in the Southern North Sea Basin: a climate signal in sediment supply? Basin Research 13: 293-312. DOI: 10.1046/j.1365-2117.2001.00151.x

Palstra, S.W.L., Wallinga, J., Viveen, W., Schoorl, J.M., Van den Berg, M. & Van der Plicht, J., 2021. Cross-comparison of last glacial radiocarbon and OSL ages using periglacial fan deposits. Quaternary Geochronology 61: 101128. DOI: 10.1016/j.quageo.2020.101128

Patruno, S., Scisciani, V., Helland-Hansen, W., D'Intino, N., Reid, W. & Pellegrini, C., 2020. Upslope-climbing shelf-edge clinoforms and the stepwise evolution of the northern European glaciation (lower Pleistocene Eridanos Delta system, U.K. North Sea): When sediment supply overwhelms accommodation. In: Pellegrini, C., Patruno, S., Helland-Hansen, W. & Steel, R.J. (eds): Clinoforms and clinothems: Fundamental elements of basin infill. Basin Research 32: 224-239. DOI: 10.1111/bre.12379

Peeters, J., Busschers, F.S., Stouthamer, E., Bosch, J.H.A., Van den Berg, M.W., Wallinga, J., Versendaal, A.J., Bunnik, F.P.M. & Middelkoop, H., 2016. Sedimentary architecture and chronostratigraphy of a late Quaternary incised-valley fill: A case study of the late Middle and Late Pleistocene Rhine system in the Netherlands. Quaternary Science Reviews 131: 211-236. DOI: 10.1016/j.quascirev.2015.10.015

Pierik, H.J., Cohen, K.M., Vos, P.C., Van der Spek, A.J.F. & Stouthamer, E., 2017a. Late Holocene coastal-plain evolution of the Netherlands: the role of natural preconditions in human-induced sea ingressions. Proceedings of the Geologists' Association 128: 180-197. DOI: 10.1016/j.pgeola.2016.12.002

Pierik, H.J., Stouthamer, E. & Cohen, K.M., 2017b. Natural levee evolution in the Rhine-Meuse delta, the Netherlands, during the first millennium CE. Geomorphology 295: 215-234. DOI: 10.1016/j.geomorph.2017.07.003

Pierik, H.J., Stouthamer, E., Schuring, T. & Cohen, K.M., 2018a. Human-caused avulsion in the Rhine-Meuse delta before historic embankment (The Netherlands). Geology 46: 935-938. DOI: 10.1130/G45188.1

Pierik, H.J., Van Lanen, R.J., Gouw-Bouman, M.T.I.J., Groenewoudt, B.J., Wallinga, J. & Hoek, W.Z., 2018b. Controls on late-Holocene drift-sand dynamics: The dominant role of human pressure in the Netherlands. The Holocene 28: 1361-1381. DOI: . Controls on late-Holocene drift-sand dynamics: The dominant role of human pressure in the Netherlands

Pons, L. & Wiggers, A.J., 1960. De Holocene wordingsgeschiedenis van Noordholland en het Zuiderzeegebied - The holocene genesis of the province of North-Holland and the Zuyder Sea region – Deel II. Tijdschrift van Het Koninklijk Nederlandsch Aardrijkskundig Genootschap 67: 3-57.

Praeg, D., 2003. Seismic imaging of mid-Pleistocene tunnel-valleys in the North Sea Basin—high resolution from low frequencies. Journal of Applied Geophysics 53: 273-298. DOI: 10.1016/j.jappgeo.2003.08.001

Preece, R.C. & Parfitt, S.A., 2012. The Early and early Middle Pleistocene context of human occupation and lowland glaciation in Britain and northern Europe. Quaternary International 271: 6-28. DOI: 10.1016/j.quaint.2012.04.018

Preece, R.C., Meijer, T., Penkman, K.E.H., Demarchi, B., Mayhew, D.F. & Parfitt, S.A., 2020. The palaeontology and dating of the 'Weybourne Crag', an important marker horizon in the Early Pleistocene of the Southern North Sea Basin. Quaternary Science Reviews 236: 106177. DOI: 10.1016/j.quascirev.2020.106177

Quik, C., Van der Velde, Y., Candel, J.H.J., Steinbuch, L., Van Beek, R. & Wallinga, J., 2023. Faded landscape: unravelling peat initiation and lateral expansion at one of northwest Europe's largest bog remnants. Biogeosciences 20: 695-718. DOI: 10.5194/bg-20-695-2023

Rea, B.R., Newton, A.M.W., Lamb, R.M., Harding, R., Bigg, G.R., Rose, P., Spagnolo, M., Huuse, M., Cater, J.M.L., Archer, S., Buckley, F., Halliyeva, M., Huuse, J., Cornwell, D.G., Brocklehurst, S.H. & Howell, J.A., 2018. Extensive marine-terminating ice sheets in Europe from 2.5 million years ago. Science Advances 4: eaar8327. DOI: 10.1126/sciadv.aar83

Rieu, R., Van Heteren, S., Van der Spek, A.J.F. & De Boer, P.L., 2005. Development and preservation of a mid-Holocene tidal-channel network offshore the western Netherlands. Journal of Sedimentary Research 75: 409-419. DOI: 10.2110/jsr.2005.032

Rijsdijk, K.F., Passchier, S., Weerts, H.J.T., Laban, C., Van Leeuwen, R.J.W. & Ebbing, J.H.J., 2005. Revised Upper Cenozoic stratigraphy of the Dutch sector of the North Sea Basin: towards an integrated lithostratigraphic, seismostratigraphic and allostratigraphic approach. Netherlands Journal of Geosciences 84: 129-146. DOI: 10.1017/S0016774600023015

Rijsdijk, K.F., Kroon, I.C., Meijer, T., Passchier, S., Van Dijk, T.A.G.P., Bunnik, F.P.M. & Janse, A.C., 2013. Reconstructing Quaternary Rhine–Meuse dynamics in the southern North Sea: architecture, seismo-lithofacies associations and malacological biozonation. Journal of Quaternary Science 28: 453-466. DOI: doi.org/10.1002/jqs.2627

Rixhon, G. & Demoulin, A., 2018. The picturesque Ardennian valleys: Plio-Quaternary incision of the drainage system in the uplifting Ardenne. In: Demoulin, A. (ed.): Landscapes and landforms of Belgium and Luxembourg. Springer International Publishing (Cham): 159-175. DOI: 10.1007/978-3-319-58239-9_10

Roberts, D.H., Evans, D.J.A., Callard, S.L., Clark, C.D., Bateman, M.D., Medialdea, A., Dove, D., Cotterill, C.J., Saher, M., Cofaigh, C.Ó., Chiverrell, R.C., Moreton, S.G., Fabel, D. & Bradwell, T., 2018. Ice marginal dynamics of the last British-Irish Ice Sheet in the southern North Sea: Ice limits, timing and the influence of the Dogger Bank. Quaternary Science Reviews 198: 181-207,. DOI: 10.1016/j.quascirev.2018.08.010

Ruddiman, W.F., Raymo, M. & McIntyre, A., 1986. Matuyama 41,000-year cycles: North Atlantic Ocean and northern hemisphere ice sheets. Earth and Planetary Science Letters 80: 117-129. DOI: 10.1016/0012-821X(86)90024-5

Ruegg, G.H.J. & Zandstra, J.G., 1977. Pliozäne und pleistozäne gestauchte Ablagerungen bei Emmerschans (Drenthe, Nieder-

lande). Mededelingen Rijks Geologische Dienst, Nieuwe Serie 28: 66-99.
Schäfer, A., Utescher, T., Klett, M. & Valdivia-Manchego, M., 2005. The Cenozoic Lower Rhine Basin – rifting, sedimentation, and cyclic stratigraphy. International Journal of Earth Sciences 94: 621-639. DOI: 10.1007/s00531-005-0499-7
Schaller, M., Von Blanckenburg, F., Hovius, N., Veldkamp, A., Van den Berg, M.W. & Kubik, P.W., 2004. Paleoerosion rates from cosmogenic 10Be in a 1.3 Ma terrace sequence: Response of the River Meuse to changes in climate and rock uplift. The Journal of Geology 112: 127-144. DOI: 10.1086/381654
Schokker, J. & Koster, E.A., 2004. Sedimentology and facies distribution of Pleistocene cold-climate aeolian and fluvial deposits in the Roer Valley Graben (southeastern Netherlands). Permafrost and Periglacial Processes 15: 1-20. DOI: 10.1002/ppp.477
Schokker, J., Cleveringa, P. & Murray, A.S., 2004. Palaeoenvironmental reconstruction and OSL dating of terrestrial Eemian deposits in the southeastern Netherlands. Journal of Quaternary Science 19: 193-202. DOI: 10.1002/jqs.808
Schokker, J., Cleveringa, P., Murray, A.S., Wallinga, J. & Westerhoff, W.E., 2005. An OSL dated Middle and Late Quaternary sedimentary record in the Roer Valley Graben (southeastern Netherlands). Quaternary Science Reviews 24: 2243-2264. DOI: 10.1016/j.quascirev.2005.01.010
Schokker, J., Weerts, H.J.T., Westerhoff, W.E., Berendsen, H.J.A. & Den Otter, C., 2007. Introduction of the Boxtel Formation and implications for the Quaternary lithostratigraphy of the Netherlands. Netherlands Journal of Geosciences 86 (3): 197-210. DOI: 10.1017/S0016774600077805
Sejrup, H.P., Aarseth, I., Haflidason, H., Løvlie, R., Bratten, Å., Tjøstheim, G., Forsberg, C.F. & Ellingsen, K.L., 1995. Quaternary of the Norwegian Channel: glaciation history and palaeoceanography. Norsk Geologisk Tidsskrift 75: 65-87.
Sha, L.P., Schwarz, C., Maenhaut Van Lemberge, V., Cameron, T.D.J., Zöllmer, V., Konradi, P., Laban, C., Streif, H. & Schüttenhelm, R.T.E., 1991. Quaternary sedimentary sequences in the Southern North Sea Basin. In: L.P. Sha (ed.): Final discipline report of the project: The modelling and dynamics of the Quaternary geology of the southern North Sea and their applications to environmental protection and industrial developments. Rijks Geologische Dienst (Haarlem): 135 pp. DOI: 10.13140/RG.2.1.4118.2803
Sirocko, F., Knapp, H., Dreher, F., Förster, M.W., Albert, J., Brunck, H., Veres, D., Dietrich, S., Zech, M., Hambach, U., Röhner, M., Rudert, S., Schwibus, K., Adams, C. & Sigl, P., 2016. The ELSA-Vegetation-Stack: Reconstruction of Landscape Evolution Zones (LEZ) from laminated Eifel maar sediments of the last 60,000 years. Global and Planetary Change 142: 108-135. DOI: 10.1016/j.gloplacha.2016.03.005
Slupik, A.A., Wesselingh, F.P., Janse, A.C. & Reumer, J.W.F., 2007. The stratigraphy of the Neogene-Quaternary succession in the southwest Netherlands from the Schelphoek borehole (42G4-11/42G0022) – a sequence-stratigraphic approach. Netherlands Journal of Geosciences 86: 317-332. DOI: 10.1017/S0016774600023556
Slupik, A.A., Wesselingh, F.P., Mayhew, D.F., Janse, A.C., Dieleman, F.E., Van Strydonck, M., Kiden, P., Burger, A.W. & Reumer, J.W.F., 2013. The role of a proto-Schelde River in the genesis of the southwestern Netherlands, inferred from the Quaternary successions and fossils in Moriaanshoofd Borehole (Zeeland, the Netherlands). Netherlands Journal of Geosciences 92: 69-86. DOI: 10.1017/S0016774600000299
Sosdian, S. & Rosenthal, Y., 2009. Deep-sea temperature and ice volume changes across the Pliocene-Pleistocene climate transitions. Science 325: 306-310. DOI: 10.1126/science.1169938
Spratt, R.M. & Lisiecki, L.E., 2016. A Late Pleistocene sea level stack. Climate of the Past 12: 1079-1092. DOI: 10.5194/cp-12-1079-2016
Stapert, D., 1987. A progress report on the Rhenen industry (Central Netherlands) and its stratigraphical context. Palaeohistoria 29: 219-243. DOI: 10.1201/9781003079446
Stephan, H.-J., 2014. Climato-stratigraphic subdivision of the Pleistocene in Schleswig-Holstein, Germany and adjoining areas. E&G Quaternary Science Journal 63: 3-18. DOI: 10.3285/eg.63.1.01
Stoker, M.S., Balson, P.S., Long, D. & Tappin, D.R., 2011. An overview of the lithostratigraphical framework for the Quaternary deposits on the United Kingdom continental shelf. Research Report RR/11/03. British Geological Survey (Keyworth): 48 pp.
Stouthamer, E., Cohen, K.M. & Gouw, M.J.P., 2011. Avulsion and its implications for fluvial-deltaic architecture: insights from the Holocene Rhine-Meuse delta. In: Davidson, S.K., Leleu, S. & North, C.P. (eds): From river to rock record: The preservation of fluvial sediments and their subsequent interpretation. SEPM Special Publication 97: 215-232. DOI: 10.2110/sepm-sp.097.215
Stuart, A.J., 2021. Vanished giants – the lost world of the ice age. University of Chicago Press: 288 pp.
Suganuma, Y., Okada, M., Head, M.J., Kameo, K., Haneda, Y., Hayashi, H., Irizuki, T., Itaki, T., Izumi, K., Kubota, Y., Nakazato, H., Nishida, N., Okuda, M., Satoguchi, Y., Simon, Q. & Takeshita, Y., 2021. Formal ratification of the Global Boundary Stratotype Section and Point (GSSP) for the Chibanian Stage and Middle Pleistocene Subseries of the Quaternary System: the Chiba Section, Japan. Episodes 44: 317-347. DOI: 10.18814/epiiugs/2020/020080
Ten Veen, J., Peeters, S., Den Dulk, M. & Geel, K., 2020. Regional understanding of the Cenozoic "Shallow Gas" play of blocks F4 & F5. TNO - Geological Survey of the Netherlands Joint Industry Project final report 28-04-2020: 157 pp. https://www.nlog.nl/media/2903
TNO-GDN, 2023a. The Digital Geological Model: DGM (BRO DGM v2.2). TNO - Geological Survey of the Netherlands. https://www.dinoloket.nl/en/the-digital-geological-model-dgm

TNO-GDN, 2023b. Geologische kaart van het Koninkrijk der Nederlanden 1 : 600 000, TNO - Geological Survey of the Netherlands (Utrecht), 1 map sheet. https://www.dinoloket.nl/en/subsurface-models/map. ISBN 9789059865142

TNO-GDN, 2023c. Stratigraphic Nomenclature of the Netherlands, TNO – Geological Survey of the Netherlands. https://www.dinoloket.nl/en/stratigraphic-nomenclature

TNO-GDN, 2023d. Borehole lithology, stratigraphic data and core photographs. TNO - Geological Survey of the Netherlands. https://www.dinoloket.nl/en/subsurface-data

Toonen, W.H.J., Kleinhans, M.G. & Cohen, K.M., 2012. Sedimentary architecture of abandoned channel fills. Earth Surface Processes and Landforms 37: 459-472. DOI: 10.1002/esp.3189

Toonen, W.H.J., Winkels, T.G., Cohen, K.M., Prins, M.A. & Middelkoop, H., 2015. Lower Rhine historical flood magnitudes of the last 450 years reproduced from grain-size measurements of flood deposits using End Member Modelling. Catena 130: 69-81. 10.1016/j.catena.2014.12.004

Törnqvist, T.E., 1993. Holocene alternation of meandering and anatomosing fluvial systems in the Rhine–Meuse delta (central Netherlands) controlled by sea-level rise and subsoil erodibility. Journal of Sedimentary Petrology 63: 683-693. DOI: 10.1306/D4267BB8-2B26-11D7-8648000102C1865D

Toucanne, S., Zaragosi, S., Bourillet, J.-F., Marieu, V., Cremer, M., Kageyama, M., Van Vliet-Lanoë, B., Eynaud, F., Turon, J.-L. & Gibbard, P.L., 2010. The first estimation of Fleuve Manche palaeoriver discharge during the last deglaciation: Evidence for Fennoscandian ice sheet meltwater flow in the English Channel ca 20-18 ka ago. Earth and Planetary Science Letters 290: 459-473. DOI: 10.1016/j.epsl.2009.12.050

Tzedakis, P.C., Crucifix, M., Mitsui, T. & Wolff, E.W., 2017. A simple rule to determine which insolation cycles lead to interglacials. Nature 542: 427-432. DOI: 10.1038/nature21364

Tzedakis, P.C., Hodell, D.A., Nehrbass-Ahles, C., Mitsui, T. & Wolff, E.W., 2022. Marine Isotope Stage 11c: An unusual interglacial. Quaternary Science Reviews 284: 107493. DOI: 10.1016/j.quascirev.2022.107493

Van Asselen, S., Cohen, K.M. & Stouthamer, E., 2017. The impact of avulsion on groundwater level and peat formation in delta floodbasins during the middle-Holocene transgression in the Rhine-Meuse delta, The Netherlands. The Holocene 27: 1694-1706. DOI: 10.1177/0959683617702

Van Baelen, A., Raczynski-Henk, Y., De Kort, J.W., Huisman, H., Van Os, B., Versendaal, A.J., Wallinga, J., Meijs, E.P.M. & Deeben, J., 2017. Onderzoek naar de stratigrafie van de löss-sequentie en de daarin aangetroffen artefacten op De Kaap bij St. Geertruid. Rapportage Archeologische Monumentenzorg (RAM) 236: 88 pp.

Van Balen, R.T. & Busschers, F.S., 2010. Human presence in the central Netherlands during early MIS 6 (~170-190 Ka): evidence from early Middle Palaeolithic artefacts in ice-pushed Rhine-Meuse sediments. Netherlands Journal of Geosciences 89: 77-83. DOI: 10.1017/S0016774600000834

Van Balen, R.T., Houtgast, R.F., Van der Wateren, F.M., Vandenberghe, J. & Bogaart, P.W., 2000. Sediment budget and tectonic evolution of the Meuse catchment in the Ardennes and the Roer Valley Rift System. Global and Planetary Change 27 (1-4): 113-129. DOI: 10.1016/S0921-8181(01)00062-5

Van Balen, R.T., Houtgast, R.F. & Cloetingh, S.A.P.L., 2005. Neotectonics of The Netherlands: a review. In: Cloetingh, S.A.P.L. & Cornu, T.G.M. (eds): Neotectonics and Quaternary fault-reactivation in Europe's intraplate lithosphere. Quaternary Science Reviews 24: 439-454. DOI: 10.1016/j.quascirev.2004.01.011

Van Balen, R.T., Bakker, M.A.J., Kasse, C., Wallinga, J. & Woolderink, H.A.G., 2019. A Late Glacial surface rupturing earthquake at the Peel Boundary fault zone, Roer Valley Rift System, the Netherlands. Quaternary Science Reviews 218: 254-266. DOI: 10.1016/j.quascirev.2019.06.033

Van Balen, R.T., Kasse, C., Wallinga, J. & Woolderink, H.A.G., 2021. Middle to Late Pleistocene faulting history of the Heerlerheide fault, Roer Valley Rift System, influenced by glacio-isostasy and mining-induced displacement. Quaternary Science Reviews 268: 107111. DOI: 10.1016/j.quascirev.2021.107111

Van Balen, R.T., Lapperre, R.E., Woolderink, H.A.G., Wallinga, J. & Kasse, C., 2024. Magnitudes and surface rupture lengths of paleo-earthquakes at the NW-part of the Peel Boundary fault zone, Roer Valley Rift System. Tectonophysics 879: 230322. DOI: 10.1016/j.tecto.2024.230322

Van Bergen, F. & Koster, K., 2025. Peat, coal and coalbed methane. In: Ten Veen, J.H., Vis, G.-J., De Jager, J. & Wong, Th.E. (eds): Geology of the Netherlands, second edition. Amsterdam University Press (Amsterdam): 577-601. DOI: 10.5117/9789463728362_ch15

Van den Berg, M.W., 1996. Fluvial sequences of the Maas: a 10 Ma record of neotectonics and climatic change at various timescales. PhD thesis University Wageningen (the Netherlands): 181 pp.

Van den Berg, M.W. & Beets, D.J., 1987. Saalian glacial deposits and morphology in The Netherlands. In: Van der Meer, J.J.M. (ed.): Tills and glaciotectonics: proceedings of an INQUA symposium on genesis and lithology of glacial deposits. Balkema (Rotterdam): 235-251.

Van den Berg, M.W. & Den Otter, C., 1993. Toelichtingen bij de geologische kaart van Nederland 1:50.000. Blad Almelo Oost / Denekamp (28O/29). Rijks Geologische Dienst (Haarlem): 240 pp.

Van den Berg, M.W. & Van Hoof, T., 2001. The Maas terrace sequence at Maastricht, SE Netherlands: evidence for 200 m of late Neogene and Quaternary surface uplift. In: Maddy, D., Macklin, M.G. & Woodward, J.C. (eds): River basin sediment systems: archives of environmental change. Balkema (Lisse): 45-86. DOI: 10.1201/9781439824672.PT2

Vandenberghe, D.A.G., Derese, C., Kasse, C. & Van den haute, P., 2013. Late Weichselian (fluvio-)aeolian sediments and Holocene drift-sands of the classic type locality in Twente (E Netherlands): a high-resolution dating study using optically stimulated luminescence. Quaternary Science Reviews 68: 96-113. DOI: 10.1016/j.quascirev.2013.02.009

Vandenberghe, J., 1993. Changing fluvial processes under changing periglacial conditions. Zeitschrift für Geomorphologie Neue Folge, Supplementband 88: 17-28.

Vandenberghe, J., 1995. Timescales, climate and river development. Quaternary Science Reviews 14: 631-638. DOI: 10.1016/0277-3791(95)00043-O

Vandenberghe, J., 2015. River terraces as a response to climatic forcing: Formation processes, sedimentary characteristics and sites for human occupation. Quaternary International 370: 3-11. DOI: 10.1016/j.quaint.2014.05.046

Vandenberghe, J. & Van der Plicht, J., 2016. The age of the Hengelo interstadial revisited. Quaternary Geochronology 32: 21-28. DOI: 10.1016/j.quageo.2015.12.004

Van den Bogaard, P. & Schmincke, H.-U., 1990. Die Entwicklungsgeschichte des Mittelrheinraumes und die Eruptionsgeschichte des Osteifel-Vulkanfeldes. In: Schirmer, W. (ed.): Rheingeschichte zwischen Mosel und Maas. Deuqua-Führer 1 (Hannover): 166-190.

Van de Plassche, O., 1982. Sea-level change and water-level movements in the Netherlands during the Holocene. Mededelingen Rijks Geologische Dienst, Nieuwe Serie 36: 93 pp.

Van der Hammen, T. & Wijmstra, T.A., 1971. The Upper Quaternary of the Dinkel valley. (Twente, Eastern Overijssel, The Netherlands). Mededelingen Rijks Geologische Dienst, Nieuwe Serie 22: 55-214.

Vandermaelen, N., Beerten, K., Clapuyt, F., Christl, M. & Vanacker, V., 2022. Constraining the aggradation mode of Pleistocene river deposits based on cosmogenic radionuclide depth profiling and numerical modelling. Geochronology 4 : 713-730. DOI: 10.5194/gchron-4-713-2022

Van der Meer, J.J.M., Slotboom, R.T. & De Vires-Bruynsteen, I.M.E., 1984. Lithology and palynology of Weichselian alluvial fan deposits near Eerbeek. The Netherlands. Boreas 13: 393-402. DOI: 10.1111/j.1502-3885.1984.tb00955.x

Van der Meulen, M.J., Doornenbal, J.C., Gunnink, J.L., Stafleu, J., Schokker, J., Vernes, R.W., Van Geer, F.C., Van Gessel, S.F., Van Heteren, S., Van Leeuwen, R.J.W., Bakker, M.A.J., Boogaard, P.J.F., Busschers, F.S., Griffioen, J., Gruijters, S.H.L.L., Kiden, P., Schroot, B.M., Simmelink, H.J., Van Berkel, W.O., Van der Krogt, R.A.A., Westerhoff, W.E. & Van Daalen, T.M., 2013. 3D geology in a 2D country: perspectives for geological surveying in the Netherlands. Netherlands Journal of Geosciences 92 (4): 217-241. DOI: 10.1017/S0016774600000184

Van der Spek, A.J.F., 1994. Large-scale evolution of Holocene tidal basins in the Netherlands. PhD thesis Utrecht University (the Netherlands): 191 pp.

Van der Vlerk, I.M. & Florschütz, F., 1950. Nederland in het IJstijdvak. W. de Haan (Utrecht): 287 pp.

Van der Vlerk, I.M. & Florschütz, F., 1953. The palaeontological base of the subdivision of the Pleistocene in the Netherlands. Verhandelingen van de Koninklijke Nederlandse Akademie van Wetenschappen, Afdeling Natuurkunde, Eerste Reeks 20: 3-58.

Van der Wateren, F.M., 1995. Structural geology and sedimentology of push moraines - Processes of soft sediment deformation in a glacial environment and the distribution of glaciotectonic styles. Mededelingen Rijks Geologische Dienst 54: 168 pp.

Van Dinter, M., 2013. The Roman Limes in the Netherlands: how a delta landscape determined the location of the military structures. Netherlands Journal of Geosciences 92: 11-32. DOI: 10.1017/S0016774600000251

Van Geel, B., Bohncke, S.J.P. & Dee, H., 1980. A palaeoecological study of an upper Late Glacial and Holocene sequence from "De Borchert", The Netherlands. Review of Palaeobotany and Palynology 31: 367-448. DOI: 10.1016/0034-6667(80)90035-4

Van Geel, B., Coope, G.R. & Van der Hammen, T., 1989. Palaeoecology and stratigraphy of the Lateglacial typesection at Usselo (The Netherlands). Review of Palaeobotany and Palynology 60: 25-129. DOI: 10.1016/0034-6667(89)90072-9

Van Hateren, J.A., Kasse, C., Van der Woude, J., Schokker, J., Prins, M.A. & Van Balen, R.T., 2022. Late Weichselian and Holocene climatic and local controls on aeolian deposition inferred from decomposing grain size-shape distributions. Quaternary Science Reviews 287: 107554. DOI: 10.1016/j.quascirev.2022.107554

Van Huissteden, J., 1990. Tundra rivers of the Last Glacial: sedimentation and geomorphological processes during the Middle Pleniglacial in Twente, Eastern Netherlands. Mededelingen Rijks Geologische Dienst 44: 3-138.

Van Huissteden, J., Gibbard, P.L. & Briant, R.M., 2001 . Periglacial fluvial systems in northwest Europe during marine isotope stages 4 and 3. Quaternary International 79: 75-88. DOI: 10.1016/S1040-6182(00)00124-5

Van Kolfschoten, T., 1990. The evolution of the mammal fauna in the Netherlands and the middle Rhine Area (Western Germany) during the late Middle Pleistocene. Mededelingen Rijks Geologische Dienst 43: 1-69.

Van Kolfschoten, T., 2001. Pleistocene mammals from the Netherlands. Bollettino Della Società Paleontologica Italiana 40: 209-215.

Van Kolfschoten, T. & Gibbard, P.L., 1998. The dawn of the Quaternary: an introduction. In: Van Kolfschoten, Th. & Gibbard, P.L. (eds): The Dawn of the Quaternary - Proceedings of the SEQS-EuroMam Symposium 1996. Mededelingen Rijks Geologische Dienst 60: 13-17.

Van Kolfschoten, T. & Turner, E., 1996. Early Middle Pleistocene mammalian faunas from Kärlich and Miesenheim I and their biostratigraphical implications. In: Turner, C. (ed.): The early Middle Pleistocene in Europe. Balkema (Rotterdam): 227-253. DOI: 10.1201/9781003077879-16

Van Leeuwen, R.J.W., Beets, D.J., Bosch, J.H.A., Burger, A.W., Cleveringa, P., Van Harten, D., Herngreen, G.F.W., Kruk, R.W., Langereis, C.G., Meijer, T., Pouwer, R. & De Wolf, H., 2000. Stratigraphy and integrated facies analysis of the Saalian and

Eemian sediments in the Amsterdam-Terminal borehole, the Netherlands. Netherlands Journal of Geosciences 79: 161-196. DOI: 10.1017/S0016774600023647

Van Popta, Y.T., Cohen, K.M., Vos, P.C. & Spek, Th., 2020. Reconstructing medieval eroded landscapes of the north-eastern Zuyder Zee (the Netherlands): a refined palaeogeographical time series of the Noordoostpolder between A.D. 1100 and 1400. Landscape History 41: 27-56. DOI: 10.1080/01433768.2020.1835180

Vos, P.C., 2015. Origin of the Dutch coastal landscape: long-term landscape evolution of the Netherlands during the Holocene, described and visualized in national regional and local palaeogeographical map series. PhD thesis Utrecht University (the Netherlands), Barkhuis, Groningen: 369 pp.

Vos, P., Van der Meulen, M., Weerts, H. & Bazelmans, J., 2020. Atlas of the Holocene Netherlands - landscape and habitation since the last ice age. Amsterdam University Press (Amsterdam): 96 pp.

Vos, P.C. & Knol, E., 2015. Holocene landscape reconstruction of the Wadden Sea area between Marsdiep and Weser - Explanation of the coastal evolution and visualisation of the landscape development of the northern Netherlands and Niedersachsen in five palaeogeographical maps from 500 BC to present. Netherlands Journal of Geosciences 94: 157-183. DOI: 10.1017/njg.2015.4

Waajen, I.M., Busschers, F.S., Donders, T.H., Van Heteren, S., Plets, R., Wallinga, J., Hennekam, R., Reichart, G.-J., Kinnaird, T. & Wagner-Cremer, F., 2024. Late MIS5a in the southern North Sea: new chronostratigraphic insights from the Brown Bank Formation. Journal of Quaternary Science 39: 408-420. DOI: 10.1002/jqs.3592

Wageningen Environmental Research, 2023. BRO Soil and groundwater map, 2023-01. Wageningen University and Research. https://www.dinoloket.nl/en/bro-soil-map

Wesselingh, F.P., Meijer, T., Harting, R., Bakker, M. & Busschers, F.S., 2023. New marine warm-temperate molluscan assemblage demonstrates warm conditions during the Middle Pleistocene of the North Sea Basin. Netherlands Journal of Geosciences 102: e3. DOI: 10.1017/njg.2023.1

Westerhoff, W.E., 2008. The Rhine - a major fluvial record. Netherlands Journal of Geosciences 87: 3-5. DOI: 10.1017/S0016774600023994

Westerhoff, W.E., 2009. Stratigraphy and sedimentary evolution - The Lower Rhine-Meuse system during the Late Pliocene and Early Pleistocene (Southern North Sea Basin). PhD thesis Vrije Universiteit Amsterdam (the Netherlands): 168 pp.

Westerhoff, W.E., Wong, Th.E. & De Mulder, E.F.J., 2003. Opbouw van de ondergrond. In: De Mulder, E.F.J., Geluk, M.C., Ritsema, I.L., Westerhoff, W.E. & Wong, Th.E. (eds): De ondergrond van Nederland. Wolters-Noordhoff (Groningen/Houten): 247-352.

Westerhoff, W.E., Kemna, H.A. & Boenigk, W., 2008. The confluence area of Rhine, Meuse, and Belgian rivers: Late Pliocene and Early Pleistocene fluvial history of the northern Lower Rhine Embayment. Netherlands Journal of Geosciences 87: 107-125. DOI: 10.1017/S0016774600024070

Westerhoff, W.E., Donders, T.H., Trabucho Alexandre, N. & Busschers, F.S., 2020. Early Pleistocene Tiglian sites in the Netherlands: A revised view on the significance for quaternary stratigraphy. Quaternary Science Reviews 242: 106417. DOI: 10.1016/j.quascirev.2020.106417

Woolderink, H.A.G., Kasse, C., Cohen, K.M., Hoek, W.Z. & Van Balen, R.T., 2019. Spatial and temporal variations in river terrace formation, preservation and morphology in the Lower Meuse Valley, the Netherlands. Quaternary Research 91 (2): 548-569. DOI: 10.1017/qua.2018.49

Zagwijn, W.H., 1957. Vegetation, climate and time-correlations in the Early Pleistocene of Europe. Geologie en Mijnbouw, Nieuwe Serie 19: 233-244.

Zagwijn, W.H., 1960. Aspects of the Pliocene and Early Pleistocene vegetation in the Netherlands. PhD thesis Leiden University (the Netherlands). Mededelingen van de Geologische Stichting, Serie C-III-1 5: 78 pp.

Zagwijn, W.H., 1961. Vegetation, climate and radiocarbon datings in the Late Pleistocene of the Netherlands. Part I: Eemian and Early Weichselian. Mededelingen van de Geologische Stichting, Nieuwe Serie 14: 15-45.

Zagwijn, W.H., 1963. Pleistocene stratigraphy in the Netherlands based on changes in vegetation and climate. Verhandelingen van Het Koninklijk Nederlands Geologisch Mijnbouwkundig Genootschap. Geologische Serie 21: 173-196.

Zagwijn, W.H., 1973. Pollenanalytic studies of Holsteinian and Saalian beds in the northern Netherlands. Mededelingen Rijks Geologische Dienst, Nieuwe Serie 24: 139-156.

Zagwijn, W.H., 1975. De paleogeografische ontwikkeling van Nederland in de laatste drie miljoen jaar. Geografisch Tijdschrift 9: 181-201.

Zagwijn, W.H., 1983. Sea-level changes in the Netherlands during the Eemian. In: Van den Berg, M.W. & Felix, R. (eds): Special issue in the honour of J.D. de Jong. Geologie en Mijnbouw 62: 437-450.

Zagwijn, W.H., 1985. An outline of the Quaternary stratigraphy of the Netherlands. Geologie en Mijnbouw 64 (1): 17-24.

Zagwijn, W.H., 1986. Nederland in het Holoceen. Staatsuitgeverij ('s Gravenhage): 46 pp.

Zagwijn, W.H., 1992. The beginning of the Ice Age in Europe and its major subdivisions. Quaternary Science Reviews 11: 583-591. DOI: 10.1016/0277-3791(92)90015-Z

Zagwijn, W.H., 1996a. The Cromerian Complex Stage of the Netherlands and correlation with other areas of Europe. In: Turner, C. (ed.): The early Middle Pleistocene in Europe. Balkema (Rotterdam): 145-172. DOI: 10.1016/0277-3791(92)90015-Z

Zagwijn, W.H., 1996b. An analysis of Eemian climate in western and central Europe. Quaternary Science Reviews 15: 451-469. DOI: 10.1016/0277-3791(96)00011-X

Zagwijn, W.H., 1998. Borders and boundaries: a century of stratigraphical research in the Tegelen–Reuver area of Limburg (The

Netherlands). In: Van Kolfschoten, Th. & Gibbard, P.L. (eds): The Dawn of the Quaternary - Proceedings of the SEQS-EuroMam Symposium 1996. Mededelingen Rijks Geologische Dienst 60: 19-34.

Zagwijn, W.H. & De Jong, J., 1984. Die Interglaziale von Bavel und Leerdam und ihre stratigraphische Stellung im niederländischen Früh-Pleistozän. In: Van Amerom, H.W.J. (ed.): S.J. Dijkstra memorial volume. Mededelingen Rijks Geologische Dienst 37: 155-169.

Zagwijn, W.H. & Zonneveld, J.I.S., 1956. The interglacial of Westerhoven. Geologie en Mijnbouw, Nieuwe Serie 18: 37-46.

Zagwijn, W.H., Van Montfrans, H.M. & Zandstra, J.G., 1971. Subdivision of the "Cromerian" in the Netherlands; pollen analysis, palaeomagnetism and sedimentary petrology. Geologie en Mijnbouw 50: 41-58.

Zandstra, J.G., 1971. Geologisch onderzoek in de stuwwal van de oostelijke Veluwe bij Hattem en Wapenveld. Mededelingen Rijks Geologische Dienst, Nieuwe Serie 22: 215-260.

Ziegler, P.A., 1994. Cenozoic rift system of western and central Europe: an overview. Geologie En Mijnbouw 73(2-4): 99-127.

Zonneveld, J.I.S., 1974. The terraces of the Maas (and the Rhine) downstream of Maastricht. In: L'évolution quaternaire des bassins fluviaux de la Mer du Nord méridionale - Centenaire de la Société Géologique de Belgique. Annales de la Société Géologique de Belgique : 133-157.

5 cm

11

Magmatism in the Netherlands: expression of the northwest European rifting history

Manfred J. van Bergen, Geert-Jan Vis, Wim Sissingh, Janne M. Koornneef & Isaac Brouwers*

*Corresponding author: M.J.vanBergen@uu.nl

ABSTRACT

The presence of igneous rocks in about 100 wells drilled for hydrocarbons, mainly concentrated in the eastern provinces of the Netherlands, the western onshore and offshore areas and in the northern offshore testify to the significance of magmatic activity in the geological history of the Netherlands. Intrusive rocks, mainly emplaced into Carboniferous and younger sedimentary rocks dominate over lavas and other extrusive products. Prominent exceptions are two buried Late Jurassic volcanoes. The Zuidwal volcano under the Waddenzee consists of thick extrusive deposits and associated subsurface features that collectively define a well-developed volcanic centre. The recently discovered Mulciber volcano is a complex on the rim of the Central Graben below the North Sea.

Radiometric ages have been determined on samples from 25 wells, but most of the currently available results are inaccurate due to ubiquitous effects of alteration. Despite this limitation, additional stratigraphic, geochemical and petrological evidence indicates that magmatism in the Netherlands was largely periodic and followed a pattern of intermittent activity in the North Sea region and adjacent areas in late Paleozoic, Mesozoic and Cenozoic times.

Samples from ca. 15 core intervals studied in detail suggest that a large majority of igneous rocks in the Dutch subsurface have a mafic composition and a moderate to high alkali content. Geochemical signatures are consistent with a within-plate tectonic setting throughout the successive episodes of magmatism, of which the Carboniferous – early Permian and Late Jurassic periods were the most productive. The lithospheric rifting processes that have dominated the northwest European geological history since Paleozoic times provided favourable conditions for melt generation, the emplacement of intrusive magma bodies and associated volcanism.

<< Porphyritic trachyte lava in the upper Carboniferous Step Graben Formation in well F04-02-A (offshore, Dutch Central Graben). Photo: Geert-Jan Vis.

DOI: 10.5117/9789463728362_ch11

Introduction

Igneous rocks are not readily associated with the low-lying, mostly flat landscapes of the Netherlands where surface outcrops of any hard rock are rare and restricted to the rolling countryside near the eastern border and to the south-easternmost part of the country. The only noticeable igneous rocks are building materials that were imported from surrounding countries over many centuries. Columnar basalts, traditionally used for the reinforcement of dikes, are well-known examples.

Nonetheless, the long geological history of the lowlands has known episodes of strong magmatic activity, as is evidenced by a surprisingly rich hidden record in the subsurface across the country, both on land and below the sea. Striking illustrations are the buried Zuidwal and Mulciber volcanoes, which were active in Late Jurassic times. In strong contrast, volcanic activity is ongoing in the Caribbean Netherlands at the present day. The latest eruption of the Quill volcano on the island of Statia was around 250 AD, and that of Mount Scenery on Saba as recent as in 1640 AD or shortly before.

The first discovery of igneous rock in the Dutch subsurface dates back to 1923 when a dolerite intrusion was encountered in an exploration well near Corle in the east of the country (Tesch, 1925, 1928; Tomkeieff & Tesch, 1931). Several years later, similar intrusive rocks were found in the Meddeho and Hupsel wells nearby (Tesch & Van Voorthuysen, 1944). In all of these cases, the intrusion was encountered in Carboniferous shales at current depths of between 957 and 1320 m. Since then, a diversity of intrusive and extrusive igneous rocks has been found in numerous wells drilled mainly by the hydrocarbon industry, both onshore and offshore.

Based on published, unpublished and newly available data, the number of wells in which pre-Quaternary igneous material was identified has grown from 71 (Van Bergen & Sissingh, 2007) to 101 (this work), while the number of radiometrically dated samples increased from 23 to 33. This update presents a thoroughly revized overview of igneous intrusive material (sills, dykes, larger bodies) and extrusive deposits (lavas, ash and other pyroclastic rocks) reported from onshore and offshore exploration wells in publicly available sources, ranging from hand-written notes in internal company documents with core descriptions to peer-reviewed publications. Indirect geophysical evidence for the subsurface presence of igneous rocks is included as well.

In comparison to the preceding overviews (Sissingh, 2004; Van Bergen & Sissingh, 2007) several adjustments have been made. The most conspicuous change concerns the re-evaluation of ages assigned to individual occurrences in the sources consulted. This issue is particularly relevant, given the difficulty of reliably dating magmatic rock without information on the mode of emplacement and the local geological context, a typical drawback in studies of drill-core samples. Since the large majority of radiometric ages is based on conventional K-Ar dating of whole-rock samples, and virtually all of this material underwent significant post-emplacement alteration, the reported ages are of limited value and mostly significantly too young. The restricted number of ages obtained by Ar-Ar methods, generally considered more accurate, should be considered with caution as well, as only a few meet an acceptable quality standard. Furthermore, (bio)stratigraphic age constraints for host sedimentary rocks have little relevance in cases where petrography is inadequate to determine if an interval of solid igneous rock represents effusive material (lava) or a subvolcanic equivalent (sill or dyke). This uncertainty applies to a fair number of cases.

Taking these limitations into account, the available information has been scrutinized and re-interpreted, resulting in an improved subdivision into distinct episodes for Paleozoic, Mesozoic and Cenozoic magmatic activity in the Netherlands. Supporting evidence comes from geochemical signatures of Dutch igneous rocks including an extensive set of new data compiled and discussed in this work. Additional petrographic evidence is illustrated by optical microscope images, representative for the distinct rock types. Coherence from inferred chronology, compositional properties and geographic distribution yields new time frames of recurrent igneous activity, which match the general pattern of magmatism in neighbouring parts of northwest Europe and can be linked to a common history of episodic large-scale rift tectonics.

Finally, more attention is devoted to Cenozoic magmatic activity, including Paleocene manifestations related to the opening of the North-East Atlantic Ocean and the distribution of pyroclastic material in the Netherlands emitted by Quaternary explosive volcanism abroad.

Rock names mentioned throughout this work are not always based on a rigorous application of petrographic or geochemical classifications because appropriate information is sometimes lacking. Instead, the terminology of the original sources is maintained, except for a few cases of obviously incorrect labelling. Qualifications of intrusive or extrusive modes of emplacement should also be taken with care, as interpretations sometimes hinge on poor or conflicting evidence.

Crystalline basement

The oldest igneous rocks in the Netherlands belong to the crystalline basement and were encountered only in offshore well A17-01 on the Elbow Spit High (Fig. 11.1).

According to Frost et al. (1981), the rocks can be classified as biotite monzogranite containing heavily altered biotite and oligoclase (Fig. 11.2). The authors reported a $^{40}Ar/^{39}Ar$ age of 346±7 Ma, obtained from micas, but this should be regarded as a minimum in view of the degree of alteration. The considerably older U-Pb zircon age of 410±7 Ma (A. Gerdes, pers. comm. 2022) is probably a better approximation of the intrusion age. The granite is covered by Devonian sediments and was presumably emplaced during or shortly after the Caledonian Orogeny.

The A17-01 granite is one of 22 occurrences of igneous rocks in the North Sea Basin that can be assigned to the Caledonian basement complex, which further contains a variety of metamorphic rocks (Frost et al., 1981). It is one of the few locations where basement has been encountered in the far eastern Avalonia terrain, close to the Thor Suture. The latter represents the structural border with Precambrian crust of the Baltica paleo-continent (see Vis et al., 2025, this volume) and geophysical modelling suggests that the suture originated from the south-westward underthrusting of Baltica lower crust below the Avalonian margin prior to collision of the continental blocks (Pharaoh et al., 2006; Lyngsie & Thybo, 2007). The A17-01 well is situated in a 50-100-km-wide band running parallel to this boundary, marking a crustal domain characterized by abnormally low seismic velocity and interpreted as remnants of a collapsed Caledonian accretionary complex (Smit et al., 2016).

Available $^{40}Ar/^{39}Ar$ ages of these metamorphic basement rocks cluster around 450 to 440 Ma (latest Ordovician or earliest Silurian) and fall within the Caledonian radiometric dates of Britain and Scandinavia (Frost et al., 1981, and references therein). Many of the obtained ages likely represent overprints of earlier phases of metamorphism or deformation related to the Caledonian Orogeny or older, Precambrian orogenies. Post Caledonian overprints show less consistency and can be attributed to local thermal or tectonic events. One such event around 350 Ma may explain the younger $^{40}Ar/^{39}Ar$ age of the A17-01 monzogranite relative to its intrusion age.

Devonian

The A17-01 well (Fig. 11.3; Table 11.1) also contained Devonian igneous rocks consisting of altered rhyolitic volcanics or quartz porphyry. These rocks are intercalated in non-metamorphic Old Red Sandstone, which covers the basement and was dated at 341±30 Ma with unknown method and reliability (see Sissingh, 2004). However, if the extrusive mode of emplacement and stratigraphic position are correct, this early Carboniferous radiometric age is too young. In the southern end of the Norwegian sector of the North Sea (Fig. 11.3), highly altered quartz porphyries in the Embla oil field have been dated at 374±3 Ma (zircon U-Pb age). They are thought to represent multiple early Famennian alkali rhyolite eruptions with an intraplate compositional signature that form a bimodal suite together with transitional basalts (Lundmark et al., 2012, 2018). These isolated examples of volcanic activity in the North Sea area accompanied the tensional tectonics characteristic for the Devonian development of northwest and central Europe. Devonian bimodal alkaline volcanics also occur in the Cornwall and Rhenish basins, as well as in the more southerly Central Armorican-Saxothuringian Basin. In contrast, calcalkaline volcanism and emplacement of post-orogenic granites accompanied the development of Old Red basins in the northern British Isles (Ziegler, 1990).

Carboniferous and Permian

There is ample evidence for Carboniferous and Permian magmatic activity in the form of igneous intrusive and effusive rocks in ca. 30 wells in the Netherlands subsurface, both onshore and offshore (Fig 11.1). The majority of these occurrences were found in the eastern part of the country. Two igneous episodes have been distinguished but it is worth emphasizing that radiometric age constraints are poor, making a chronological subdivision tenuous. Rocks here denoted as (late) Carboniferous are dominated by mafic intrusions (dolerite, gabbro), while the extrusive nature of subordinate rocks referred to as 'basalts' is not always certain. On the other hand, 'Rotliegend volcanics' of latest Carboniferous-early Permian age are mostly effusive. Both rhyolitic and basaltic varieties occur but compositional information is often lacking. The allocation of igneous rock intervals from individual wells to one of these igneous episodes in the description below and Table 11.1 should be taken with care, particularly for the intrusions that lack independent stratigraphic age control.

Carboniferous

Igneous rocks with a radiometrically determined Carboniferous age are restricted to onshore wells around the eastern Texel-IJsselmeer High (Fig. 11.1). A basaltic lava from Nagele-1 (NAG-01) has been K-Ar dated at 327±8 Ma, which is consistent with its occurrence in the Epen Formation of the Limburg Group (Sissingh, 2004), i.e. near the lower and upper Carboniferous transition. An extrusive basaltic rock in Westphalian sediments in Steenwijkerwold-1 (SWD-01), however, was K-Ar dated as early Permian, in obvious conflict with its alleged extrusive nature (Sissingh, 2004; core photo in Huis in 't Veld & Den Hartog Jager, 2025, this volume). Gabbroic intrusive rocks

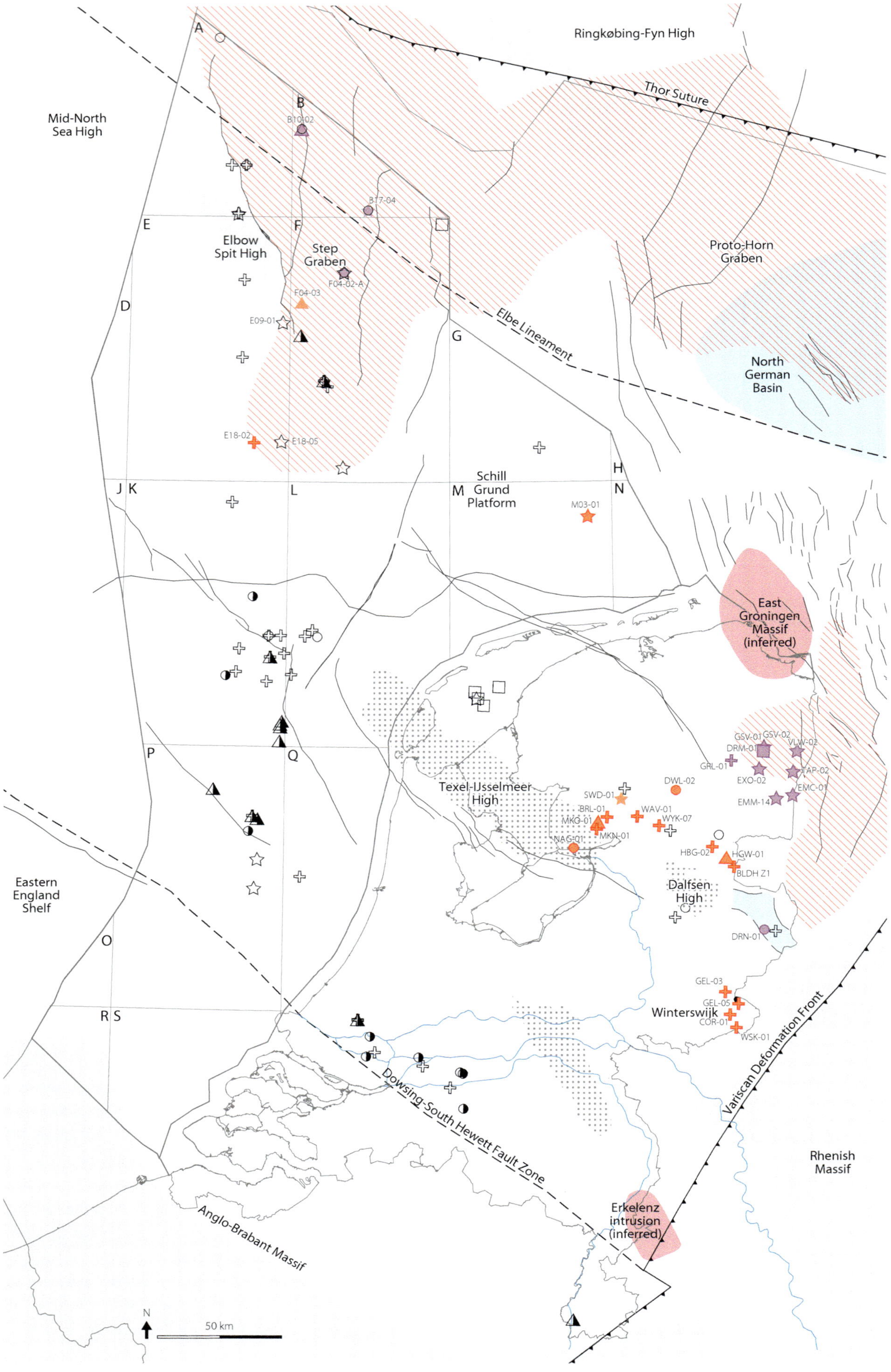
Ringkøbing-Fyn High
Thor Suture
Mid-North Sea High
Elbow Spit High
Step Graben
Proto-Horn Graben
Elbe Lineament
North German Basin
Schill Grund Platform
East Groningen Massif (inferred)
Texel-IJsselmeer High
Dalfsen High
Winterswijk
Eastern England Shelf
Dowsing-South Hewett Fault Zone
Variscan Deformation Front
Rhenish Massif
Erkelenz intrusion (inferred)
Anglo-Brabant Massif
A
B
D
E
F
G
H
J
K
L
M
N
O
P
Q
R
S
B10-02
B17-04
F04-02-A
F04-03
E09-01
E18-02
E18-05
M03-01
GSV-01
GSV-02
DRM-01
VLW-02
GRL-01
EXO-02
TAP-02
EMC-01
EMM-14
DWL-02
SWD-01
BRL-01
WAV-01
WYK-07
MKO-01
MKN-01
NAG-01
HBG-02
HGW-01
BLDH Z1
DRN-01
GEL-03
GEL-05
COR-01
WSK-01
N
50 km

← *Figure 11.1. Well locations in the Netherlands where igneous rocks, attributed to the Carboniferous and latest Carboniferous-early Permian ('lower Rotliegend') episodes of magmatic activity, have been found. Wells with igneous material from other or unknown periods are shown for comparison. Structural elements and faults from early Permian times are based on Gast et al. (2010) and Pharaoh et al. (2010). Details on the rocks encountered in the wells are listed in Table 11.1.*

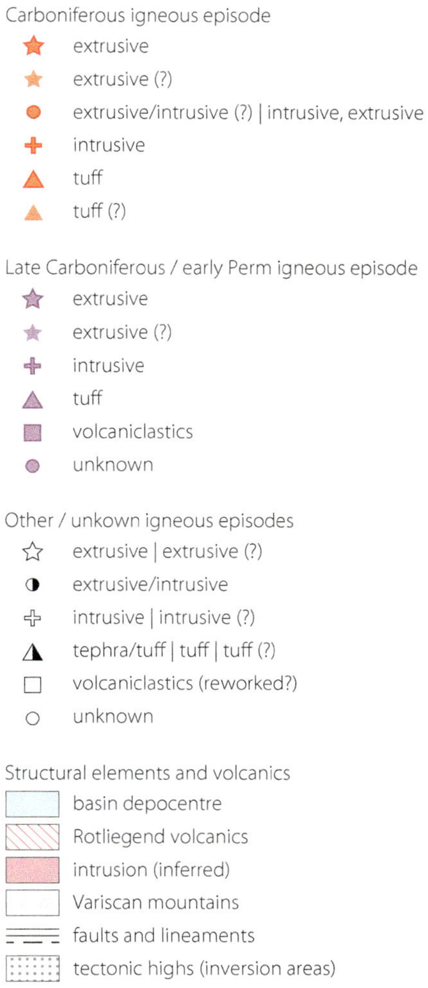

in Dwingelo-2 (DWL-02) (Thiadens, 1963; Eigenfeld & Eigenfeld-Mende, 1986) form a sill-like body of several kilometres in length according to seismic data. K-Ar dating yielded an age of 322±15 Ma. Based on the consistent mafic (gabbroic) character and geographic clustering of these intrusions, it is reasonable to suppose that all of the rocks in the east of the Texel-IJsselmeer High and Dalfsen High are manifestations of this Carboniferous igneous episode despite their generally much younger K-Ar ages.

A similar affinity is probably also valid for some 15 igneous bodies in well Wanneperveen-1 (WAV-01, Fig. 11.1), which were originally interpreted as basalts, dolerites and olivine-dolerites intruded into lower Permian (British Autunian stage) shales and siltstones (Kimpe, 1953). However, in a later revision of the biostratigraphy, Van Amerom (1972) assigned an upper Carboniferous age (Westphalian-A) to these sedimentary rocks. On the basis of petrographic similarity, Dixon et al. (1981) proposed that most, if not all of the igneous rocks in Wanneperveen-1 represent Permian flows equivalent to those present in North Sea wells in the German, Danish and UK sectors around the Ringkøbing-Fyn High. K-Ar dating of a sample from 2035-2030.5 m depth yielded an age of 217±20 Ma (Sissingh, 2004), but this is inconsistent with Rotliegend (or older) volcanics and thus reflects the inaccuracy of the method for these rocks. Hence, if (at least some of) the igneous intervals in Wanneperveen-1 are extrusive, the biostratigraphic age of the intercalated sediments favours a late Carboniferous rather than a 'Rotliegend' magmatic episode. Available compositional evidence is inconclusive as will be discussed below.

Doleritic intrusions in wells around Winterswijk (Fig. 11.1) have not been dated except for an olivine dolerite in Winterswijk-1 (WSK-01; Figs 11.4d, 11.5a), which yielded an (erroneous) K-Ar age of 218±6 Ma. They include the nearby Corle 'melaphyre' and Gelria-3 (GEL-03) 'leucophyre' intrusives, originally referred to as 'dolerites' (Tomkeieff & Tesch, 1931; Tesch & Van Voorthuysen, 1944), and the Gelria-5 (GEL-05) 'dolerite', and are presumably representatives of the same (late) Carboniferous magmatic episode as that of the igneous intervals on the eastern Texel-IJsselmeer High.

Eigenfeld & Eigenfeld-Mende (1986) used petrographic criteria to infer that 'Permo-Carboniferous' mafic intrusive rocks in the eastern Netherlands, including those in Wanneperveen-1, Dwingelo-2, De Wijk-7 (WYK-07) (Fig. 11.6a), the undated intrusions in the wells around Winterswijk (Gelria-3, Corle-1 (COR-01)), and (equally altered) equivalents encountered in wells across the border in Germany all belong to a single magmatic province. An olivine-gabbroic dyke in Balderhaar-Z1 (see Table 11.1b) may well be an example that would fit into this group. These authors proposed that these events could be genetically related to a laccolithic intrusion ('Bramsche Massif') in the southwestern part of the Lower Saxony Basin, inferred from a sizeable gravity and magnetic anomaly coinciding with high thermal maturity of sedimentary organic matter. However, since the maturation process may have been related to the Mesozoic basin evolution instead, a magmatic intrusion as explanation for the anomaly is questionable (Senglaub et al., 2006). In view of the poor absolute age constraints, a Carboniferous rather than Permian affinity is favoured.

Table 11.1. Pre-Cenozoic igneous rocks encountered in wells in the subsurface of the Netherlands: a) Radiometrically dated igneous rocks (33 ages, 25 wells); b) Igneous rocks without radiometric age dates, but with interpretation of igneous episode (42 wells); c) Igneous rocks without reliable age (41 wells). A more elaborate version of this table, including remarks is available as online supplementary material. Igneous episode is interpreted from geochronological and stratigraphic evidence (letter code with question mark indicates an expected but unconfirmed episode): C = Carboniferous; D = Devonian; J = Jurassic; LC-EP = late Carboniferous-early Permian. Emplacement is based on petrographic and stratigraphic evidence; note that original information is not always conclusive (letter code with question mark indicates uncertainty): ? = undefined; E/E? = extrusive (possibly); E/I? = extrusive or intrusive; I,E = intrusive and extrusive; I/I? = intrusive (possibly); T/T? = tuff (possibly); V/V (rew.?) = volcaniclastic/ possibly reworked volcaniclastic. Rock type/composition is based on the original report where possible; note that the compositions listed here do not follow a rigid classification scheme and must be used with care. NCP = Netherlands Continental Platform; MD = metres along hole; T.D. = Total Depth; CWL = Composite Well Log. A more elaborate online version is available at: DOI: 10.5117/ aup.27114895.

Table 11.1. (a)

Location	Short name	Well name	Igneous interval(s) (MD)	Dated interval (MD)/ Sample depth	Age (Ma)
NCP	A17-01	A17-01	2157.0-2295.0	?	341±30
NCP	A17-01	A17-01	3013.0-3044.0 (=T.D.)	~3043.0	Min. 346±7
NCP	A17-01	A17-01	3013.0-3044.0 (=T.D.)	unknown	410±7
Andel-4	AND-04	Andel-4	1651-1657	1651.4-1652.1 or 1657.1-1658.0	133±2
Baarlo-1-S1	BRL-01-S1	Baarlo-1-S1	1772.5-1775.0; 1780.0-1782.0; 1792.0-1794.0	1775.6	266±18
Berkel-1-S1	BRK-01-S1	Berkel-1-S1	2952-2970	~2964.66	214±25?
De Wijk-7	WYK-07	De Wijk-7	2443-2486	2456.1	155±4
De Wijk-7	WYK-07	De Wijk-7	2684.0-2691.0	2690.5	289±7
Drouwenermond-1-S1	DRM-01-S1	Drouwenermond-1-S1	3920.5-3953.5	3921.0	258±6
Dwingelo-2	DWL-02	Dwingelo-2	3719.0-3799.7 (=T.D.)	3788.8	322±15
NCP	E06-01	E06-01	2405-2428	2415	161±4
NCP	E06-01	E06-01	2405-2428	2420	183±4
NCP	E12-03	E12-03	3653-3682; 3788-3875 (=T.D.)	3678	221±23
NCP	E18-02	E18-02	4432-4460	4439	274±7
NCP	F03-07	F03-07	?	~2928.05	236±6
NCP	F04-02-A	F04-02-A	4630-4656 (=T.D.)	4650-4656	165±2
NCP	F10-01	F10-01	3426-3460 (=T.D.)	3450.25	99±5
NCP	F10-01	F10-01	3426-3460 (=T.D.)	~3449-3460	>127
NCP	F10-01	F10-01	3426-3460 (=T.D.)	~3449-3460	151±2
Giessendam-1	GSD-01	Giessendam-1	1190-1218; 1350-1380	~1220	125±25
Hardenberg-2-S1	HBG-02-S1	Hardenberg-2-S1	3376.5-3441.0	3410.6	290±16
NCP	K12-05-S1	K12-05-S1	3057-3292	~3121-3133	152±7
NCP	L13-03	L13-03	877-909, 1604-1645; 1808-1821; 1931-1943	~1820	101±1
Loon-op-Zand-1	LOZ-01	Loon-op-Zand-1	2454-2457.5; 2572-2610; 2682-2694	?	132±3
NCP	M03-01	M03-01	4038-4055	4036-4055	188.5±8.4
Nagele-1	NAG-01	Nagele-1	2772.0-2776.0	2774.0	327±8
NCP	Q07-02	Q07-02	3405-3440	~3450 (?)	95±2 to 106±2
Steenwijkerwold-1	SWD-01	Steenwijkerwold-1	1937.0-1944.5	1940.5	291±8
Wanneperveen-1	WAV-01	Wanneperveen-1	2017-2019; 2028-2028.5; 2064-2069.5	2034.0	217±20?
Winterswijk-1	WSK-01	Winterswijk-1	4078-4149	4150.0	218±6
Zuidwal-1	ZDW-01	Zuidwal-1	1944-3002	?	144±1
Zuidwal-1	ZDW-01	Zuidwal-1	1944-3002	?	145
Zuidwal-1	ZDW-01	Zuidwal-1	1944-3002	?	152±3

Method	Igneous episode (interpreted)	Stratigraphy of host/ intruded interval	Emplacement	Rock type/composition as reported	Reference(s)
?	D	Buchan Fm	E	Rhyolite-Rhyodacite	Sissingh (2004)
$^{40}Ar/^{39}Ar$	D	'Basement'	I	Biotite monzo-granite	Frost et al. (1981)
U-Pb	D	'Basement'	I	Biotite monzo-granite	Axel Gerdes, pers. com. (2022)
$^{40}Ar/^{39}Ar$	J	Werkendam Fm	E/I?	Nephelinite, basanite	Dixon et al. (1981)
K-Ar	C	Ruurlo Fm	I	Dolerite ('diabase')	Sissingh (2004)
K-Ar	J	Werkendam Fm	I	Essexite or theralite	Sissingh (2004); Dixon et al. (1981); Van der Sijp (1953)
K-Ar	C	Maurits Fm	I	Olivine gabbro	Sissingh (2004); Eigenfeld & Eigenfeld-Mende (1986)
K-Ar	C	Maurits Fm	I	Gabbro	Sissingh (2004)
K-Ar	LC-EP?	Emmen Volcanic Fm	V	Mixed clastic and igneous	Sissingh (2004)
K-Ar	C	Baarlo Fm	I,E	Olivine gabbro	Sissingh (2004); Eigenfeld & Eigenfeld-Mende (1986)
K-Ar	?	Elleboog Fm	I	No data	Sissingh (2004)
K-Ar	?	Elleboog Fm	I	No data	Sissingh (2004)
K-Ar	?	Millstone Grit Fm	I	Dolerite or microgabbro	CWL on NLOG.nl; Peniguel et al. (1992)
K-Ar	C	Klaverbank Fm	I	Gabbro	Sissingh (2004)
K-Ar	?	Lower Graben Fm	V (rew.?)	No data	Sissingh (2004); Herngreen (1985)
K-Ar	LC-EP	Step Graben Fm	E?	Porphyritic rhyolite lava/ micro-granodiorite	Slater (1980); Kuijper (1991)
K-Ar	J	Zechstein Gp	I	Lamprophyre	Kuijper (1991); Sissingh (2004)
$^{40}Ar/^{39}Ar$ whole-rock	J	Zechstein Gp	I	Potassic alkaline lamprophyre	Latin (1990)
$^{40}Ar/^{39}Ar$ kaersutite	J	Zechstein Gp	I	Potassic alkaline lamprophyre	Latin (1990)
K-Ar	J	Nieuwerkerk Fm	E/I?	'Basalt'	Rieffe & Van Lil (1992); Sissingh (2004)
K-Ar	C	Ruurlo Fm	I	Olivine gabbro	Sissingh (2004)
K-Ar	J	Zechstein Gp	I	Alkali-rich lamprohyre	Well K12-05-S1 on NLOG.nl
$^{40}Ar/^{39}Ar$	J	Zechstein Gp	I	Resembling South Holland olivine nephelinites	CWL on NLOG.nl; Dixon et al. (1981)
$^{40}Ar/^{39}Ar$	J	Aalburg Fm	E/I?	Nephelinite, basanite	Dixon et al. (1981)
K-Ar	C	Ruurlo Fm	E	Pyroxene basalt with trachytic texture	Meadows & Hrycyszyn (1992)
K-Ar	C	Epen Fm	E/I?	Thin dolerite-like intrusive bodies (?)	RGD (1993); Sissingh (2004)
$^{40}Ar/^{39}Ar$	J	Solling Fm	I	Undersaturated basalt	Dixon et al. (1981)
K-Ar	C	Ruurlo Fm	E?	'Basalt'	Sissingh (2004)
K-Ar	C	Limburg Gp	I	Dolerite, (olivine-) gabbro	Sissingh (2004); Kimpe (1953)
K-Ar	C	Epen Fm	I	Olivine dolerite	Sissingh (2004); NITG (1998)
$^{40}Ar/^{39}Ar$	J	Zuidwal Volcanic Fm	E	(Mafic) phonolite, phonolitic basanite	Dixon et al. (1981)
K-Ar	J	Zuidwal Volcanic Fm	E	Trachyte	Harrison et al. (1979)
$^{40}Ar/^{36}Ar$-$^{40}K/^{36}Ar$	J	Zuidwal Volcanic Fm	E	Trachyte, phonolite, leucitite	Perrot & Van der Poel (1987)

Table 11.1. (b)

Location	Short name	Well name	Igneous interval(s) (MD)	Igneous episode (interpreted)
Andel-2	AND-02	Andel-2	1355-1421	J?
NCP	B10-02-S1	B10-02-S1	3780-3810	LC-EP
NCP	B17-04	B17-04	4376-4391	LC-EP
Balderhaar Z1 (Germany)	BLDH Z1	Balderhaar Z1 (Germany)	3608.4-3748.3; 3845.0-3847.0; 3863.0-3866.0	C
Berkel-1-S1	BRK-01-S1	Berkel-1-S1	2863-2875; 2933-2984	J
Corle-1	COR-01	Corle-1	970-983	C?
Deurningen-1	DRN-01	Deurningen-1	1468-1471	LC-EP
Emmen-14	EMM-14	Emmen-14	4087-4153	LC-EP
Emmer-Compascuum-1(-S1)	EMC-01(-S1)	Emmer-Compascuum-1(-S1)	3935-4014	LC-EP
Exloo-2-S2	EXO-02-S2	Exloo-2-S2	4190-4200	LC-EP
NCP	F04-03	F04-03	4206-4250; 4262-4288; 4375-4403	C
NCP	F16-02	F16-02	1662-1701	J
Gasselternijveen-1	GSV-01	Gasselternijveen-1	3736-3766	LC-EP
Gasselternijveen-2	GSV-02	Gasselternijveen-2	3909.5-3941.0 (=T.D.)	LC-EP
Gelria-3 (Hupsel)	GEL-03	Gelria-3 (Hupsel)	1320.7-1321.7	C?
Gelria-5 (Meddeho)	GEL-05	Gelria-5 (Meddeho)	957-958; 1070-1070.45	C?
Grollo-1	GRL-01	Grollo-1	3513-3525.5	LC-EP
Heinenoord-1	HEI-01	Heinenoord-1	2213-2261	J
Hoogenweg-1	HGW-01	Hoogenweg-1	3134	C
NCP	K12-G-05-S1	K12-G-05-S1	3917-3971; 4011-4095; 4098-4104	J
NCP	K14-FA-103	K14-FA-103	3215-3240 or 3439-3459 (cored sections)?	J
NCP	K15-13	K15-13	3687-3691	J
NCP	K18-01	K18-01	2184-2436 (interval with tuffs)	J
NCP	K18-02-A	K18-02-A	2170-2190 (interval with tuffs)	J
NCP	K18-03	K18-03	1905-1927 (interval with tuffs)	J
NCP	K18-05	K18-05	2550-2645 (interval with tuffs)	J
Marknesse Oost-1	MKO-01	Marknesse Oost-1	1864-1882	C
Marknesse-1	MKN-01	Marknesse-1	1908-1918	C
NCP	P02-06	P02-06	1366-1560 (interval with tuffs)	J
NCP	P06-01	P06-01	2060-2180 (interval with tuffs)	J
NCP	P06-B-01	P06-B-01	1595-1640 (possible tuff layer)	J
NCP	P09-08-S1	P09-08-S1	3060 (breccia: 3049-3071)	J
NCP	P12-08	P12-08	2665 (breccia: 2659-2702)	J
Slenk-1	SLK-01	Slenk-1	2274.5-2352.5	J
Ter Apel-2-S1	TAP-02-S1	Ter Apel-2-S1	4534.2-4601.5	LC-EP
Vlagtwedde-2	VLW-02	Vlagtwedde-2	4140-4165	LC-EP
Zuidwal-2	ZDW-02	Zuidwal-2	2167-2189	J
Zuidwal-3	ZDW-03	Zuidwal-3	2055-2080	J
Zuidwal-A-01(-S1)	ZDW-A-01(-S1)	Zuidwal-A-01(-S1)	1950-2060; 1950-2130 (S1)	J
Zuidwal-A-02	ZDW-A-02	Zuidwal-A-02	2048-2245	J
Zuidwal-A-04	ZDW-A-04	Zuidwal-A-04	2662-2835	J
Zuidwal-A-05	ZDW-A-05	Zuidwal-A-05	2060-2260	J

Stratigraphy of host/ intruded interval	Emplacement	Rock type/composition as reported	Reference(s)
Werkendam Fm	E/I?	Olivine nephelinite, basanite	Kuijper (1991)
Lower Rotliegend Gp	T	Volcanic tuff	Sissingh (2004)
Lower Rotliegend Gp	E?	?	CWL on NLOG.nl
Westphalian A	I	Dolerite (gabbroic rock containing abundant olivine and feldspar laths, as well as magnetite, glass and biotite)	Sissingh (2004); Anonymous (1992)
Werkendam Fm	T	No data	Sissingh (2004)
Ruurlo Fm	I	Dolerite	Tomkeieff & Tesch (1931)
Emmen Volcanic Fm	?	?	CWL on NLOG.nl
Emmen Volcanic Fm	E	No data	Sissingh (2004)
Emmen Volcanic Fm	E	Basalt and tephra	Sissingh (2004); NITG (2000)
Emmen Volcanic Fm	E	No data	Sissingh (2004)
Maurits Fm	T?	No data	CWL on NLOG.nl; De Bruin et al. (2015)
New Fm; Mulciber	E	Tuff grading into "basalt" and tuffaceous siltstone; porphyritic (basaltic-) andesitic extrusive rock	Lester (1985)
Emmen Volcanic Fm	E	No data	Sissingh (2004)
Emmen Volcanic Fm	E	No data	Sissingh (2004)
Ruurlo Fm	I	Dolerite ("diabase")	Tesch & Van Voorthuysen (1944)
Ruurlo Fm	I	Dolerite ("diabase")	Tesch & Van Voorthuysen (1944)
Emmen Volcanic Fm	I	Dolerite ("diabase")	Sissingh (2004)
Nieuwerkerk Fm	E/I?	Alkali basalt	Helmers (1991)
Maurits Fm	T	Sanidine-containing ash	Van Buggenum & Den Hartog Jager (2007)
Zechstein Gp	I	No data	Neptune Energy
Slochteren Fm	E/I?	Trachybasalt	Dixon et al. (1981)
Slochteren Fm	I	No data	CWL on NLOG.nl
Breeveertien Fm	T	No data	Sissingh (2004)
Breeveertien Fm	T	No data	CWL on NLOG.nl
Breeveertien Fm	T	No data	Sissingh (2004)
Breeveertien Fm	T	No data	Sissingh (2004)
Limburg Gp	T	Tuff with augite	Rieffe & Van Lil (1992)
Ruurlo Fm	I	Extrusive rock according to CWL	Sissingh (2004)
Nieuwerkerk Fm	T	No data	Sissingh (2004)
Nieuwerkerk Fm.	T	No data	Sissingh (2004)
Vlieland Sandstone Fm	T?	No data	King et al. (1985)
Breeveertien Fm	E?	Olivine basalt	Sissingh (2004); Galavazi (1999); NLOG.nl
Vlieland Sandstone Fm	E?	Lamprophyric	Sissingh (2004); Galavazi (1999); NLOG.nl
Wadden Volcaniclastic Mbr	V	Volcanic breccia (CWL)	Herngreen et al. (1991)
Emmen Volcanic Fm	E	No data	Sissingh (2004)
Emmen Volcanic Fm	E	No data	Sissingh (2004)
Wadden Volcaniclastic Mbr	V	Volcanic tuff (CWL)	Herngreen et al. (1991)
Wadden Volcaniclastic Mbr	V	Volcanic breccia (CWL)	Herngreen et al. (1991)
Wadden Volcaniclastic Mbr	V	Volcanic breccia (CWL)	CWL on NLOG.nl
Wadden Volcaniclastic Mbr	V	Volcanic breccia (CWL)	CWL on NLOG.nl
Wadden Volcaniclastic Mbr	V	Volcanic breccia (CWL)	CWL on NLOG.nl
Wadden Volcaniclastic Mbr	V	Volcanic breccia (CWL)	CWL on NLOG.nl

Table 11.1. (c)

Location	Short name	Well name	Igneous interval(s) (MD)
NCP	A05-01	A05-01	3930-3971.8 (=T.D.)
NCP	A14-01	A14-01	2718-2775
NCP	A15-01	A15-01	3591-3753
NCP	A15-01	A15-01	3753-3773; 3790-3883
NCP	B10-02-S1	B10-02-S1	3810-3972 (=T.D.)
NCP	B17-04	B17-04	4300-4376
Barendrecht-Ziedewij-1	BRTZ-01	Barendrecht-Ziedewij-1	3092-3096
Berkel-2	BRK-02	Berkel-2	1758-1784
Coevorden-17-S1	COV-17-S1	Coevorden-17-S1	?
De Blesse-1-S1	BLS-01-S1	De Blesse-1-S1	2014-2070
De Wijk-15-S1	WYK-15-S1	De Wijk-15-S1	?
NCP	E09-01	E09-01	3290-3345
NCP	E18-05	E18-05	~4318-4328
NCP	F04-02-A	F04-02-A	4400-4425; 4490-4515
NCP	F04-02-A	F04-02-A	4425-4533.5
NCP	F07-02	F07-02	3783-3927
NCP	F10-02	F10-02	2010-3441 (interval with tuffs)
NCP	F10-02	F10-02	4055-4061
NCP	F10-02	F10-02	4061-4090
NCP	F16-02	F16-02	1738-1872 (interval with tuffs)
NCP	F16-02	F16-02	1872-1985
NCP	G17-02	G17-02	4295-4310
Haarle-1	HLE-01	Haarle-1	1480-1497
Hellendoorn-1	HLD-01	Hellendoorn-1	1455-1493 (=T.D.)
Heugem-1	HEU-01	Heugem-1	397
IJsselmonde-64(-S1)	IJS-64(-S1)	IJsselmonde-64(-S1)	2242-2263; 2260-2268 (S1)
NCP	K02-02	K02-02	4556-4588
NCP	K09-01	K09-01	3567-3571
NCP	K12-05	K12-05	2897-2902
NCP	K12-05-S1	K12-05-S1	2919-2923
NCP	K12-B-04	K12-B-04	4103.5-4110.5 (tuff band in this interval)
NCP	K12-B-06	K12-B-06	3875-3888 (within cored interval?)
NCP	K12-D-01	K12-D-01	3760-3765, 3854-3862, 3989-3997
NCP	K12-D-03	K12-D-03	3880-3890
NCP	K12-G-08	K12-G-08	3804-3810; 3881-3906, 4082-4248
NCP	K12-K-01	K12-K-01	3795-3917 (three intercalated salt layers)
NCP	K15-04-S2	K15-04-S2	3230, 3237, 3344, 3338, 3350, 3242, 3248, 3250, 3217, 3226, 3414, 3436, 3568
NCP	L10-06	L10-06	3835-3870
NCP	L10-31	L10-31	3610-3638
NCP	L10-L-05-S1(?)	L10-L-05-S1(?)	5164, 5176, 5194
Oldenzaal-2	OLZ-02	Oldenzaal-2	269-274
Overkamp	B34G1431	Overkamp	142.0-142.5
NCP	P06-10	P06-10	3200-3227
NCP	P06-B-01	P06-B-01	?
Sprang-Capelle-1	SPC-01	Sprang-Capelle-1	2219-2223
Sprang-Capelle-1	SPC-01	Sprang-Capelle-1	2357-2377
Sprang-Capelle-1	SPC-01	Sprang-Capelle-1	2598-2606.5
Werkendam-1	WED-01	Werkendam-1	1553.5-1554.5

Stratigraphy of host/intruded interval	Emplacement	Rock type/composition as reported	Reference(s)
Lower Rotliegend Gp	?	Volcanics in claystone	CWL on NLOG.nl
Epen Fm	I	Porphyry?	Sissingh (2004)
Lower Rotliegend Gp	?	?	CWL on NLOG.nl
Lower Rotliegend Gp & Millstone Grit Fm	I	Gabbro, micro-granodiorite	Kuijper (1991)
Lower Rotliegend Gp	?	?	CWL on NLOG.nl
Lower Rotliegend Gp	?	?	CWL on NLOG.nl
Strijen Fm	I	Dolerite	Van Buggenum & Den Hartog Jager (2007)
Nieuwerkerk Fm	I	No data	Sissingh (2004)
?	?	Uncertain; no public data	Sissingh (2004)
Ruurlo Fm	I	Gabbroic (plagioclase, pyroxene, olivine in groundmass of chloritic serpentinitic clays with local magnetite and calcite)	Sissingh (2004); NLOG.nl
?	I	No data	Sissingh (2004)
Millstone Grit Fm	E?	Porphyry?	Sissingh (2004)
Limburg Gp	E?	Basaltic	Kuijper (2003)
Lower Rotliegend Gp	E?	Claystone with volcanic fragments	CWL on NLOG.nl
Lower Rotliegend Gp	?	Volcanics (?) in siltstone/sandstone	CWL on NLOG.nl
Lower Rotliegend Gp	T?	Tuff (?) in claystone/siltstone/sandstone	CWL on NLOG.nl
Lower Zechstein salt	T?	No data	CWL on NLOG.nl
Lower Rotliegend Gp	I?	Leucocratic, equigranular, quartz, amphibole, feldspar	CWL on NLOG.nl; De Bruin et al. (2015)
Lower Rotliegend Gp	?	?	CWL on NLOG.nl
Zechstein Gp	E	Anhydrite with tuff layers	Lester (1985)
Zechstein Gp	E	Acidic porphyritic volcanic rock containing augite, biotite, apatite; volcaniclastic sediment	Lester (1985)
Limburg Gp	I	Dolerite	Kuijper (1991)
Maurits Fm	I	Dolerite (diabase')	Sissingh (2004)
Ruurlo Fm	?	No data	Sissingh (2004)
Zeeland Fm	T	No data	Bless et al. (1981)
Werkendam Fm	E/I?	Basanitic	Sissingh (2004)
Klaverbank Fm	I?	No data	CWL on NLOG.nl
Z3 Salt Mbr	E/I?	Olivine basalt	CWL on NLOG.nl
Zechstein Gp	I	Lamprophyre	Kuijper (1991)
Zechstein Gp	I	Alkali-rich lamprohyre	Neptune Energy
Zechstein Gp	T	No data	Neptune Energy
Silverpit Fm	I	No data	Neptune Energy
Slochteren Fm	I	Andesite, olivine dolerite	Neptune Energy
Slochteren Fm	?	No data	Neptune Energy
Zechstein Gp	I	Lamprophyre	Neptune Energy
Zechstein Gp	I	Lamprophyre	Neptune Energy
Lower and Upper Germanic Trias gps	I?	No data	Sissingh (2004)
Slochteren Fm	I	Lamprophyre-type (?)	Neptune Energy
Slochteren Fm	I	Lamprophyre-type (cutting thin section 3630 mMD)	Neptune Energy
Limburg Groep	?	High plagioclase content (14-60 wt.%)	Neptune Energy
Coevorden Fm	I	Dolerite (hornblende 'diabase')	Van Voorthuysen (1944)
Volpriehausen Fm (?)	T	Tephra?	water well (pers. com. M. van den Bosch, 2022)
Detfurth Fm	E/I?	Mafic rock	Sissingh (2004)
?	I	Lamprophyre	Kuijper (1991)
Werkendam Fm	I	Contains nepheline and biotite	CWL on NLOG.nl
Sleen Fm	I	Contains nepheline and biotite	CWL on NLOG.nl
Hardegsen Fm	I	Contains nepheline and biotite	CWL on NLOG.nl
Brabant Fm	I	Dolerite	Sissingh (2004)

Widespread kaolinite-coal-tonstein beds in the Limburg Group represent altered tuffaceous horizons in the former coal-mining district of the south-eastern Netherlands, providing evidence for explosive volcanism during the late Carboniferous (Lippolt et al., 1984). A 5-cm-thick ash layer at 3134 m depth in well Hoogenweg-1 (HGW-01) (Huis in 't Veld & Den Hartog Jager, 2025, this volume) may be a northern equivalent.

Figure 11.2. Microphotographs and the single available core sample of the oldest igneous rock encountered in the Dutch subsurface (well A17-01): a) Altered biotite granite (PPL) (depth 2985 m); b) Same in XPL; c) Core sample of the same rock of ~5 cm wide. (estimated depth 2985 m). Images (a-b) are taken from De Vos et al. (2010) and were originally supplied by Frost (Conoco) and Miller (University of Cambridge).

In the offshore, dated and undated igneous intervals in Carboniferous sediments have been encountered mostly in the Step Graben area and the Schill Grund Platform (quadrants A, E, F and M; see Fig. 11.1). Putative extrusive lavas and ash/tuff layers would be consistent with emplacement during the Carboniferous episode (wells E09-01, E18-05, F04-02A), whereas a K-Ar age of 188.5±8.4 Ma for a basaltic rock in the Ruurlo Formation (well M03-01; Meadows & Hrycyszyn, 1992) is too young.

Considering the stratigraphic relationships of the extrusive products, most of the magmatic bodies encountered in wells in the Netherlands would appear to have a late Carboniferous age. Given the similar compositional character and foreland position (just north of the Variscan front), they likely correspond to manifestations of Namurian-Westphalian magmatism cropping out in the British Isles, in particular with alkaline and locally subalkaline basaltic flows, sills and dykes in the English Midlands and Derbyshire (Timmerman, 2004; Wilson et al., 2004). Probably, there also is a temporal and compositional overlap with the long-lived magmatism in the Midland Valley of Scotland where mostly mildly alkaline magmas were produced during intermitted pulses throughout Carboniferous times (Monaghan & Pringle, 2004; Upton et al., 2004; Monaghan & Parrish, 2006).

Latest Carboniferous – early Permian

Igneous rocks representing 'Rotliegend volcanics' with reliable, radiometrically determined ages around the Carboniferous-Permian boundary (299 Ma; Gradstein et al., 2004) are lacking in the database documented here (Table 11.1). This is in spite of the widespread occurrence of 'Rotliegend volcanics' and intrusives in the Southern Permian Basin (Fig. 11.7) (cf. Glennie, 1997, and references therein; cf. Scheck & Bayer, 1999), which extend over some 1700 km from England across the North Sea through northern Germany into Poland (Ziegler, 1990; Gast et al., 2010).

According to zircon U-Pb and whole-rock Ar-Ar ages of igneous intervals in wells in the neighbouring northeast German Basin system, North Sea Central Graben and Danish North Sea, 'Rotliegend volcanics' were emplaced in a narrow time window between 303±10 and 293±2 Ma (Heeremans et al., 2004; Breitkreuz et al., 2007). The thickest occurrences of these volcanics may have been associated with caldera subsidence (Benek et al., 1996). In North Sea wells, volcanic intervals from the lower Permian have been referred to as the Inge Volcanics Formation in the UK Central Graben and as the Karl Formation in the Danish Central Graben and Norwegian-Danish Basin.

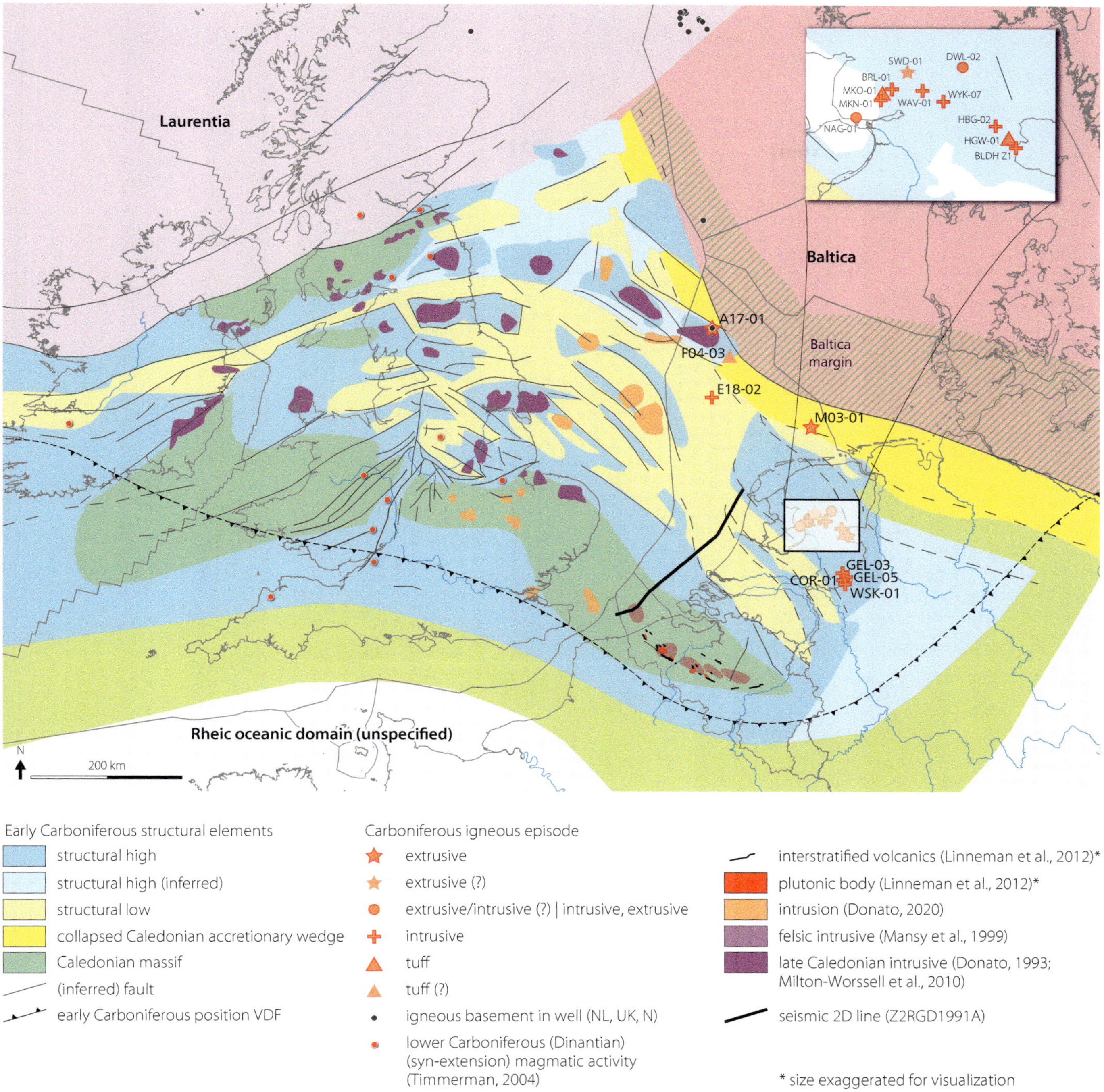

Figure 11.3. *Onshore and offshore locations where the oldest igneous rocks in the Netherlands were recovered from drill cores. In well A17-01, a pre-Devonian basement granite and a presumably late Devonian extrusive felsic rock were encountered. In other labelled wells, intrusive or extrusive igneous rocks with an assumed (late) Carboniferous age were found. Red dots indicate locations of Carboniferous magmatism in Britain and Ireland (Timmerman, 2004) and black dots indicate wells where igneous basement rock was found in the northern North Sea (Frost et al., 1981; Slagstad et al., 2011; Lundmark et al., 2012; Riber et al., 2015). The horst-graben type structural framework at the end of the early Carboniferous and the approximate margin of the Variscan Deformation Front (VDF) are after Smit et al. (2018). Late Caledonian, mostly granitic intrusions, inferred from gravity anomalies after Donato (1993, 2020), Mansy et al. (1999), Linneman et al. (2012) and Milton-Worssell et al. (2010).*

A.
2505
2506
2507
2508
2509
2510
10 cm
B.
10 cm
G.
10 cm
C.
5 cm
D.
5 cm
E.
5 cm
F.
5 cm

Corresponding deposits in the Dutch subsurface onshore are represented by undated rocks of the Emmen Volcanic Formation (Fig. 11.7), which consist of basaltic lava flows and pyroclastic rocks, usually intercalated between clastic sedimentary rocks of the Limburg Group (below) and basal layers of the Zechstein Group (above). Reaching a maximum thickness of 66 m (TNO-GDN, 2022), they can be considered as a westerly extension of 'Rotliegend volcanics' in the adjacent German Ems Graben (Figs 11.1 & 11.7), which have been described as spilitized, mildly alkaline andesites with subordinate basaltic and rhyolitic rocks (Eckhardt, 1979; cf. Plein, 1995, and references therein). Volcaniclastics in the sidetrack of Drouwenermond-1 (DRM-01-S1) with a K-Ar age of 258±6 Ma may also belong to this suite. Near the Dutch Central Graben, a porphyritic rhyolite in well F04-02A (Figs 11.1 & 11.4e) and rocks in wells within and near the A and B quadrants (Figs 11.1 & 11.7) represent time-equivalent magmatic rocks. These volcanics are almost 150 m thick and consist of pyroclastic rocks and lavas up to tens of metres thick (Geluk, 1997).

Dixon et al. (1981) reported the presence of a bimodal association of lower Permian volcanics along the northern and western flanks of the Ringkøbing-Fyn High and from within the Horn Graben, west of Denmark. Strongly altered basalts contain primary biotite and, on the basis of immobile trace elements, have been classified as transitional between olivine tholeiites and mildly alkaline basalts. The silicic rocks may represent silicified trachytes, although their original chemical composition is unclear. Stemmerik et al. (2000) suggested that Permian volcanism in the Danish part of the North Sea took place during two separate events dated 276-281 Ma and 261-269 Ma, thus postdating the early Permian volcanism in the Southern Permian Basin (Fig. 11.8). However, Heeremans et al. (2004) argued that these K-Ar ages might be too young, based on an Ar-Ar age of 299±3 Ma, which the authors obtained on a basalt sample from well 39/2-4 (Fig. 11.7) on the adjacent western flank of the UK Central Graben. Thin intervals of (strongly altered) alkaline intra-plate basalts have also been reported from the Norwegian Embla oil field (well 2/7-23S, Fig. 11.3), and potentially correlative basalts from the nearby Argyll and Auk oil fields in the UK sector of the Central North Sea (Lundmark et al., 2012, 2018, and references therein). A sample from the Danish offshore well C-1 in the eastern North Sea near the Jylland Peninsula (Fig. 11.7), yielded a high-resolution ion microprobe U-Pb zircon age of 303±10 Ma (Breitkreuz et al., 2007).

Different periods of extensive intrusive and extrusive magmatism accompanied the development of the Oslo Rift (Fig. 11.7; Neumann et al., 1992, 2004). According to U-Pb dating (zircon, perovskite, titanite), initial mainly mafic magmatism at ca. 300 Ma was followed by a ca. 40 Myr long interval of mafic and felsic intrusive and extrusive activity (Pedersen et al., 1995; Dahlgren et al., 1996; Corfu & Dahlgren, 2008; Corfu & Larsen, 2020). The rocks vary in composition between basaltic and granitic, all having a strongly alkaline character.

A comprehensive study of the widespread late Carboniferous-Permian magmatism in the northeast German Basin showed that thick rhyolitic rocks and ignimbrites volumetrically dominate over intermediate and basaltic varieties in this part of the Southern Permian Basin (Benek et al., 1996). Zircon U-Pb dating yielded ages between 302±3 and 297±3 Ma for the volcanic activity in this area, thus straddling the Carboniferous-Permian boundary (Breitkreuz & Kennedy, 1999). The authors also reported that these rocks included populations of inherited zircons with ages reflecting reworked Cadomian (650-550 Ma) and older Gondwanan elements. Manifestations of corresponding latest Carboniferous-earliest Permian volcanism and dyke and sill emplacements on the British Isles include the voluminous Whin Sill Complex (mostly high-Fe subalkaline basalts) of northeast England and the extensive dyke swarm and sill in the Midland Valley of Scotland (tholeiitic quartz dolerites). Coeval intrusive rocks in southwest England have a granitic composition and their emplacement was probably accompanied by extrusion of rhyolitic lavas (Timmerman, 2004).

In summary, although many of the Carboniferous and Permian igneous rocks discovered in the Netherlands subsurface lack unequivocal constraints on their composition, timing and mode of emplacement, they tend to fit well into the distribution and setting of equivalent intervals in onshore and offshore wells across northwestern Europe.

← ***Figure 11.4.*** *Core photographs showing examples of igneous rocks from the Dutch subsurface: a) Volcanic agglomerate from the Upper Jurassic Zuidwal volcano (well Zuidwal-1, upper 0.4 m of interval 2505-2511 m); b) Olivine gabbro in the upper Carboniferous Ruurlo Formation (well Hardenberg-2-S1, 3408.2-3408.6 m), photo: NAM; c) Volcanic agglomerate from the Upper Jurassic-Lower Cretaceous Zuidwal volcano (well Zuidwal-1, 3001.0-3001.2 m); d) Olivine dolerite in the upper Carboniferous Epen Formation (well Winterswijk-1, 4149.0-4149.3 m), photo: NAM; e) Porphyritic trachyte lava in the upper Carboniferous Step Graben Formation (well F04-02-A, 4652.27-4652.77 m); f) Dolerite sill or dyke in the Middle Jurassic Brabant Formation (well Werkendam-1, 1554.70-1554.97 m); g) Alkali basaltic rock in the Upper Jurassic Nieuwerkerk Formation (well Heinenoord-1, 2239.31-2240.20 m), photo: NAM.*

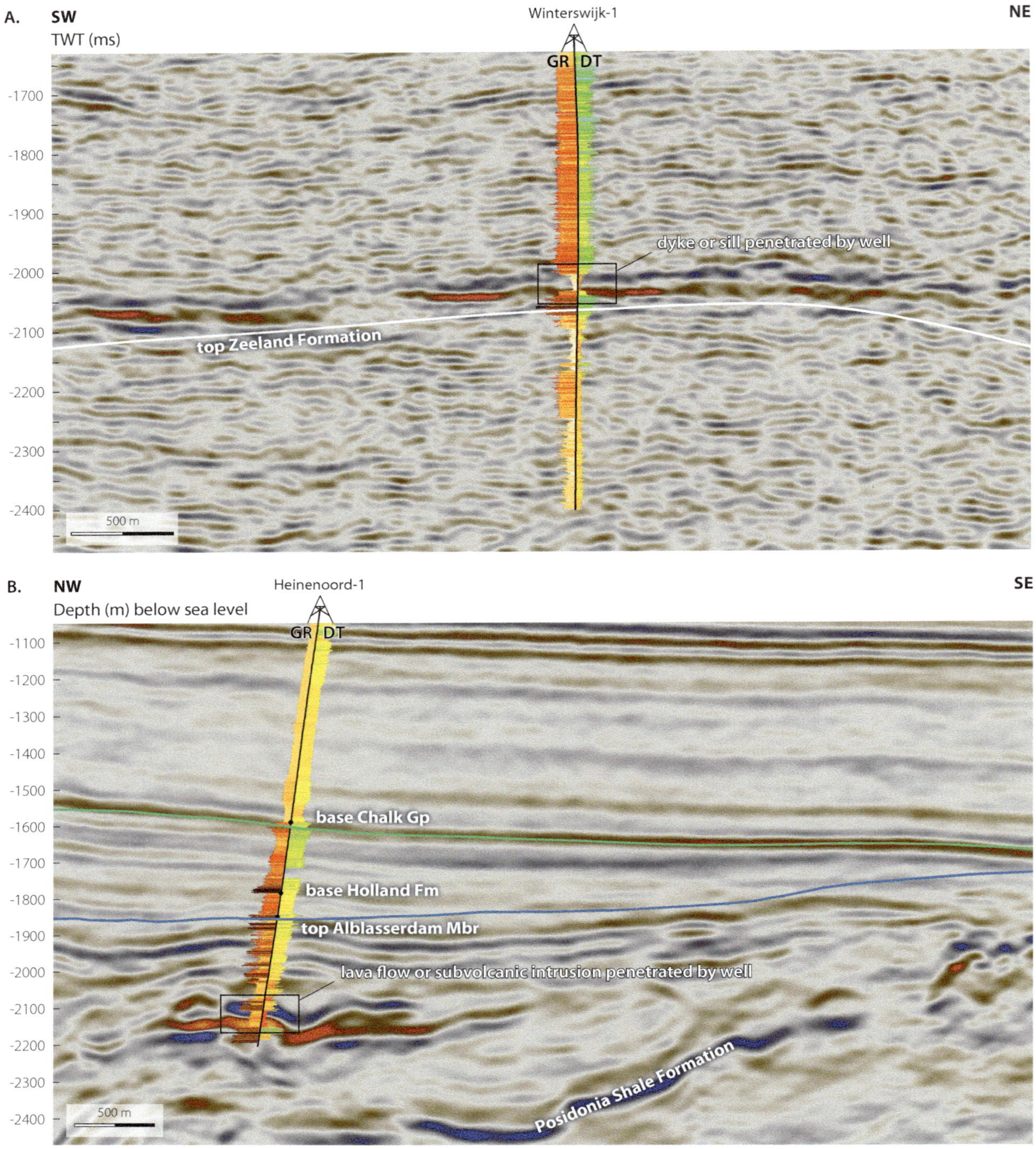

Figure 11.5. Seismic profiles in which direct observations confirm the presence of intrusive magmatic rock (sill or dyke) as high-amplitude reflectors in seismic data: a) Well Winterswijk-1 where the upper Carboniferous Epen Formation hosts an interval of olivine dolerite (see Fig. 11.4d); seismic survey L2NAM1975L, line 1011; b) Well Heinenoord-1 where the Upper Jurassic Nieuwerkerk Formation contains an interval of alkali basaltic rock (see Fig. 11.4g); seismic image from a depth-migrated merged 3D seismic dataset for the West Netherlands Basin (TNO). GR = gamma-ray log, DT = sonic log.

Jurassic

In the Dutch subsurface, igneous rocks with a Late Jurassic age have been found offshore in wells along the rim of the Dutch Central graben (Fig. 11.9), in the Broad Fourteens Basin (K, L, P, Q-quadrants) and in the Vlieland Basin (Zuidwal volcano), but also in onshore wells in the West Netherlands Basin (near Andel, Berkel, Giessendam, Heinenoord and Loon op Zand). They represent the southernmost expression of Jurassic magmatism in and around the

North Sea Central Graben System (Fig. 11.9). Details on the rifting history are given in De Jager et al. (2025, this volume).

The igneous intervals encountered in wells in the West Netherlands Basin are mostly relatively mafic rocks with a silica-undersaturated affinity, referred to as nephelinites, basanites, theralites/essexites and alkali basalts in original descriptions (Table 11.1).

Fine-grained rocks in the Andel-2 and Andel-4 and Loon op Zand-1 wells (AND-02 and LOZ-01 shown in Fig. 11.6e,f) have been described as intrusions in the middle and lower Jurassic, respectively. A sample from Andel-4 (AND-04) has been $^{40}Ar/^{39}Ar$ dated at 133±2 Ma and one from Loon op Zand-1 at 132±3 (Dixon et al., 1981). Because of a partial overprint at <120 Ma the emplacement ages are likely older. The Loon op Zand-1 and Andel-4 rocks are highly altered glassy olivine nephelinites, containing pseudomorphed olivine and clinopyroxene phenocrysts in a groundmass with the same mineral phases, as well as kaersutitic hornblende, biotite and apatite. The Andel-2 samples are probably basanites with pseudomorphed olivine, plagioclase and possibly nepheline. Several observations cast doubt on the alleged intrusive nature as dykes or sills in the Andel and Loon op Zand wells. The strongly porphyritic texture and abundance of vesicles, as well as the presence of a thin layer of baked sediment at bottom sides only, as core descriptions suggest, are consistent with an origin as lava flows. As the true emplacement ages may well fall in the late Jurassic band (Fig 11.8), an extrusive emplacement cannot be excluded at least for the rocks in the Andel wells.

The Ar-Ar ages correspond reasonably well with a K-Ar age of 125±25 Ma obtained on a body of (biotite-bearing?) olivine basalt in the Tithonian-Berriasian sandstone-shale succession of the Alblasserdam Member, which was sampled in Giessendam-1 (GSD-01) (Sissingh, 2004). The thin basaltic layers found in this well may have had an extrusive origin but the K-Ar age is not sufficiently reliable to confirm this. Dixon et al. (1981) suggested that the Berkel-1 (BRK-01) hornblende basalt intruded in Jurassic sediments (Van der Sijp, 1953) may be a fine-grained essexite or theralite, and is possibly also similar in age to the Andel and Loon op Zand rocks.

A nearly 40 m thick interval of relatively fresh olivine-rich alkali basalt in well Heinenoord-1 (HEI-01, Fig. 11.5b) has been studied in some detail (Helmers, 1991; Frezzotti, 1992; this work). The porphyritic rock (Figs 11.4g, 11.6d), carrying abundant clinopyroxene and olivine phenocrysts, shows petrographic similarities with the olivine nephelinites in Loon op Zand-1 (Fig. 11.6f). Although the basaltic rock is considered to represent a near-surface sill, emplacement as a lava flow cannot be excluded, given its stratigraphic position in the upper Jurassic-lower Cretaceous Nieuwerkerk Formation, its textural appearance (Fig. 11.4g) and the absence of a radiometric age date. Based on the whole-rock composition, the basalt classifies as tephrite or basanite. Helmers (1991) reported the presence of accessory rhönite, a rare high-Ti alumino-silicate, which is usually only present in silica-undersaturated mafic to intermediate igneous rocks.

In summary, the available information suggests that the intervals of igneous rocks in the onshore wells of the West Netherlands Basin were emplaced in a Jurassic shallow-marine succession of predominantly fine-grained mudstones and minor silt- and sandstones, either as near-surface subvolcanic intrusions or as lava flows.

Igneous rocks from wells in the offshore F, K, L, P and Q quadrants have been described as strongly silica-undersaturated basaltic intrusions (Dixon et al., 1981; Latin, 1990; Latin et al., 1990b). According to phenocryst assemblages and bulk-rock compositions, the rocks from K12-05-S1, K14-FA-103 (in the border areas of the Broad Fourteens Basin) and F10-01 (near the rim of the Central Graben) share a potassium and volatile-rich character. The first constitutes a ca. 130 m thick lamprophyric intrusion in salt of the Zechstein Group, the second a thin fine-grained trachybasaltic layer in Upper Permian sandstone and the third a >30 m thick potassic alkaline lamprophyre, also intercalated in Zechstein salt. Typical phenocrysts include Ti-rich augite and amphibole (kaersutite), while minor olivine and biotite have been found as well.

The other offshore occurrences (L13-03 and those in the P and Q quadrants further to the south) are mafic intrusions with a more nephelinitic character, compositionally and petrographically comparable to the rocks of the West Netherlands Basin mentioned above. Tuffaceous layers in late Jurassic sediments in the region of the Broad Fourteens Basin may indicate that alkaline magmatism of this episode was also accompanied by explosive eruptive activity.

The large variation in radiometric ages available for this period (Table 11.1, Fig. 11.8) underscores the uncertainties inherent to the dating methods used. The ca. 50 Ma age range obtained by different methods on virtually the same rock sampled in F10-01 is an instructive example. The plateau $^{40}Ar/^{39}Ar$ age of 151±2 Ma, obtained on separated kaersutite, is considered to represent the most robust result (Latin, 1990). A within-errors identical whole-rock K-Ar age of 152±7 Ma, obtained on fresh material from the thick lamprophyre in K12-05-S1 (Krueger, 1982), together with consistent results for the Zuidwal volcanic centre below the Waddenzee and biostratigraphic age constraints for the Mulciber volcanic centre on the rim of the Dutch Central Graben (see below) further limit the timing of Mesozoic magmatism in the Netherlands to a relatively short late Jurassic episode. The other reported ages (Table

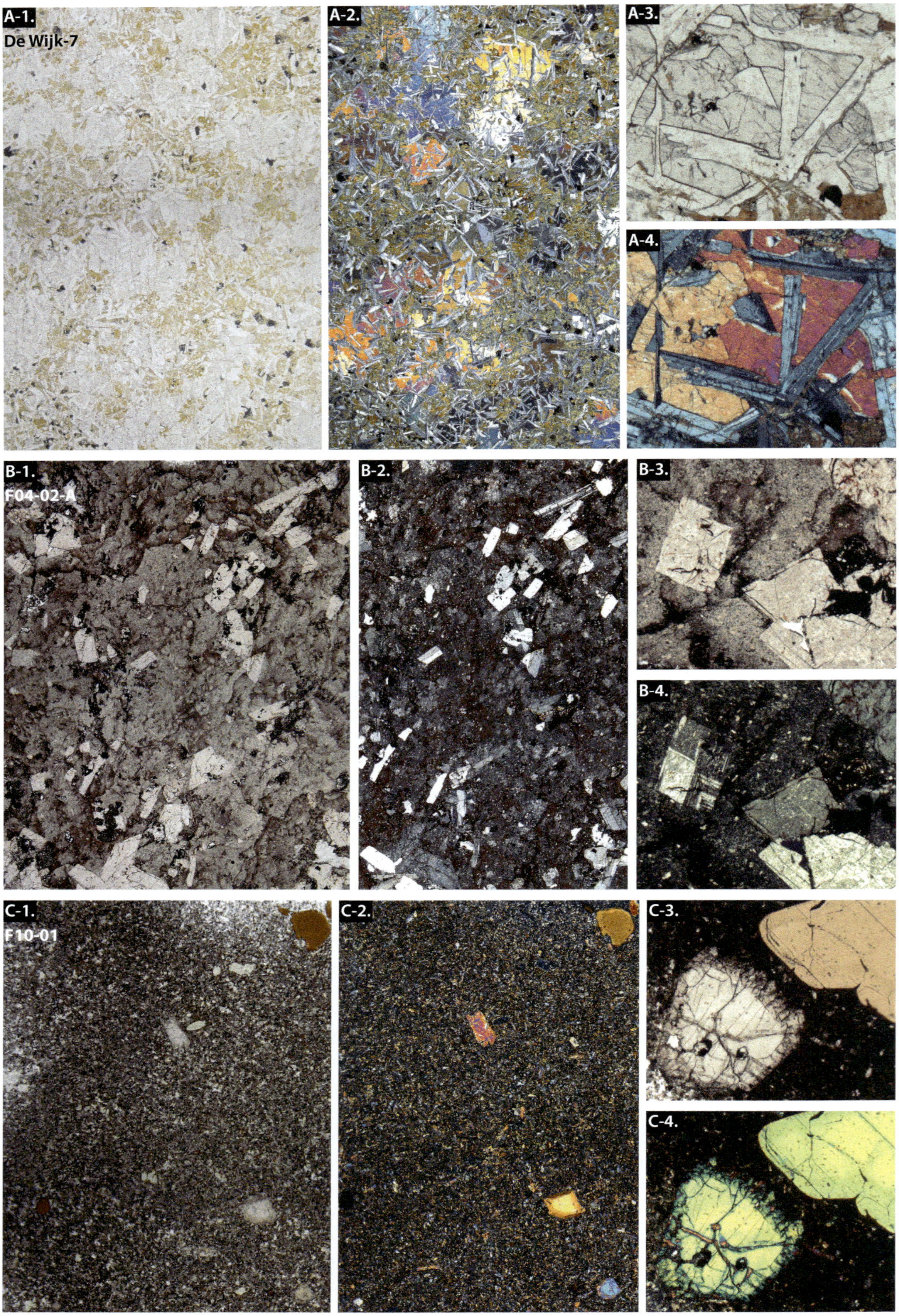
A-1.
De Wijk-7
A-2.
A-3.
A-4.
B-1.
F04-02-A
B-2.
B-3.
B-4.
C-1.
F10-01
C-2.
C-3.
C-4.

11.1), a few older but mostly considerably younger, are ambiguous and may not represent magma emplacement ages.

Buried volcanoes

Two prominent geomorphic structures have been interpreted as buried volcanoes in Mesozoic sedimentary basins offshore: the 'Zuidwal volcano' below the Wadden Sea, southeast of the island of Vlieland, and 'Mulciber volcano' below the North Sea, some 100 km northwest of the island of Terschelling. Their outlines correspond to gravity highs and positive geomagnetic anomalies (Fig. 11.10). Considering the size of the geomagnetic anomalies and their volumes based on 3D seismic reflection data, both systems fall in the category of large (>5km³), composite shield and caldera volcanoes (cf. Bischoff et al., 2021).

The Zuidwal volcanic centre

The 'Zuidwal volcano', identified during petroleum exploration below the Wadden Sea (Cottençon et al., 1975), is located north of the Texel-IJsselmeer High in the Vlieland Basin (Fig. 11.9). Cores from well Zuidwal-1 (ZDW-01) revealed volcanic agglomerates between 1950 m and the total depth of 3000 m (Fig. 11.4a, c). The volcanics are unconformably overlain by gas-bearing Valanginian sandstone. Details on the geological setting, exploration geophysics, structural evolution and reservoir characteristics are provided in Perrot & Van der Poel (1987) and Herngreen et al. (1991). Magma may have reached the surface along Permo-Carboniferous and Cimmerian faults that were opened during the Late Jurassic. Based on geophysical data the agglomerates represent a neck, which forms part of a dome-like structure. This is illustrated by the isopach map of 'Upper Jurassic' units (Delfland Subgroup and Kimmeridge Clay Formation), which shows a division of the Vlieland Basin into two sub-basins with thinning around the dome (Fig. 11.11a). Perrot & Van der Poel (1987) interpreted local aeromagnetic anomalies in terms of a circular caldera-type body with a central volcanic conduit (Fig. 11.11c). They hypothesized that after a major

*← **Figure 11.6.** Microphotographs of thin sections of representative igneous rocks from well cores in the Dutch onshore and offshore, taken in plain polarized light (PPL) and in cross polarized light (XPL). Indicated ages are best estimates and in most cases are not constrained by accurate radiometric dating (see Table 11.1 and Fig. 11.8). **De Wijk-7, depth 2445 m:** a-1) Gabbroic intrusive rock (dolerite) largely consisting of ophitic intergrowths of plagioclase laths and large pyroxene crystals and containing small opaque grains of Fe-Ti oxide (PPL); a-2) Same in XPL; a-3) Detail showing intergrown clinopyroxene and elongated plagioclase (PPL); a-4) Same in XPL. **F04-02-A, depth 4650 m:** b-1) Porphyritic trachyte showing euhedral plagioclase phenocrysts set in an altered fine-grained matrix (PPL); b-2) Same in XPL; b-3) Detail showing plagioclase cluster with Fe-Ti oxide (PPL); b-4) Same in XPL; note the complex twinning in the plagioclase on the left. **F10-01, depth 3451 m:** c-1) Porphyritic basanite (referred to as 'lamprophyre' by Latin, 1990) with phenocrysts of clinopyroxene with brown Ti-rich amphibole (e.g. in the upper right corner) and minor biotite in a crystalline matrix (PPL); c-2) Same in XPL; c-3) Magnified detail showing clinopyroxene phenocryst (left) with somewhat darker outgrowths on its rim, as well as a brownish amphibole megacryst with concentric colour zoning and smoothly curved outlines suggesting resorption (PPL); c-4) Same in XPL. **Heinenoord-1, depth 2249.3 m:** d-1) Porphyritic rock with an alkali-basaltic composition, containing abundant phenocrysts of olivine and clinopyroxene, chromium spinel and rhönite in a fine crystalline matrix (PPL); d-2) Same in XPL; d-3) Magnified detail (PPL) showing micro-phenocrysts of partly serpentinized olivine (green rims and crack fringes); d-4) Same in XPL. **Andel-2, depth 1367.8 m:** e-1) Strongly altered porphyritic alkalibasalt with a fine-grained groundmass (PPL); white spots (black in e-2) represent former phenocrysts (probably clinopyroxene and or amphibole), which are now voids sometimes partly filled with carbonate; the F10-01 basanite (see c) may well be a much fresher equivalent; e-2) Same in XPL; e-3) Detail showing the abundance of small feldspar laths in the groundmass and a completely altered microphenocryst (PPL); e-4) Same in XPL. **Loon op Zand-1, depth 2589.5 m:** f-1) Strongly altered porphyritic alkalibasaltic rock with ghosts of clinopyroxene and olivine phenocrysts set in a fine-grained groundmass (PPL); f-2) Same in XPL; f-3) Detail showing carbonated phenocryst, presumably olivine (PPL); f-4) Same in XPL. **Zuidwal-1, depth 1948.5 m:** g-1) Altered porphyritic trachyandesite with phenocrysts of elongated plagioclase and euhedral nepheline set in a crystal-rich groundmass made up of the same minerals, (altered) biotite flakes and Fe-Ti oxides (PPL); g-2) Same in XPL; note the low birefringent to nearly isotropic appearance and euhedral outlines of nepheline crystals (XPL); **Zuidwal-1, depth 1957.9 m:** g-3) Magnified detail with plagioclase phenocrysts (PPL); g-4) Same in XPL; note the strong zoning in the plagioclase (XPL). **Zuidwal-1, depths unknown:** h-1) Leucitophyre with abundant pseudomorphs after leucite, now consisting of a fine aggregate of brownish carbonate (sample ZU6); h-2) Cluster of brown titaniferous clinopyroxene (salite) and small greenish clinopyroxene (aegirine augite) in the groundmass in leucite basanite (PPL) (sample ZU5); h-3) Phenocryst of brown titaniferous clinopyroxene (salite) rimmed by green clinopyroxene (aegirine augite) (PPL) (sample ZU5). Images h are from Latin (1990). Approximate image widths: 16.5 mm for panels 1 and 2 of a, b, c, d and e; 9.5 mm for panels 1 and 2 of f and g; 3.3 mm for panels 3 and 4 of a, b, c, d, e, f, g; 1.5 mm for panels h-1 and h-3; 3.6 mm for panel h-2.*

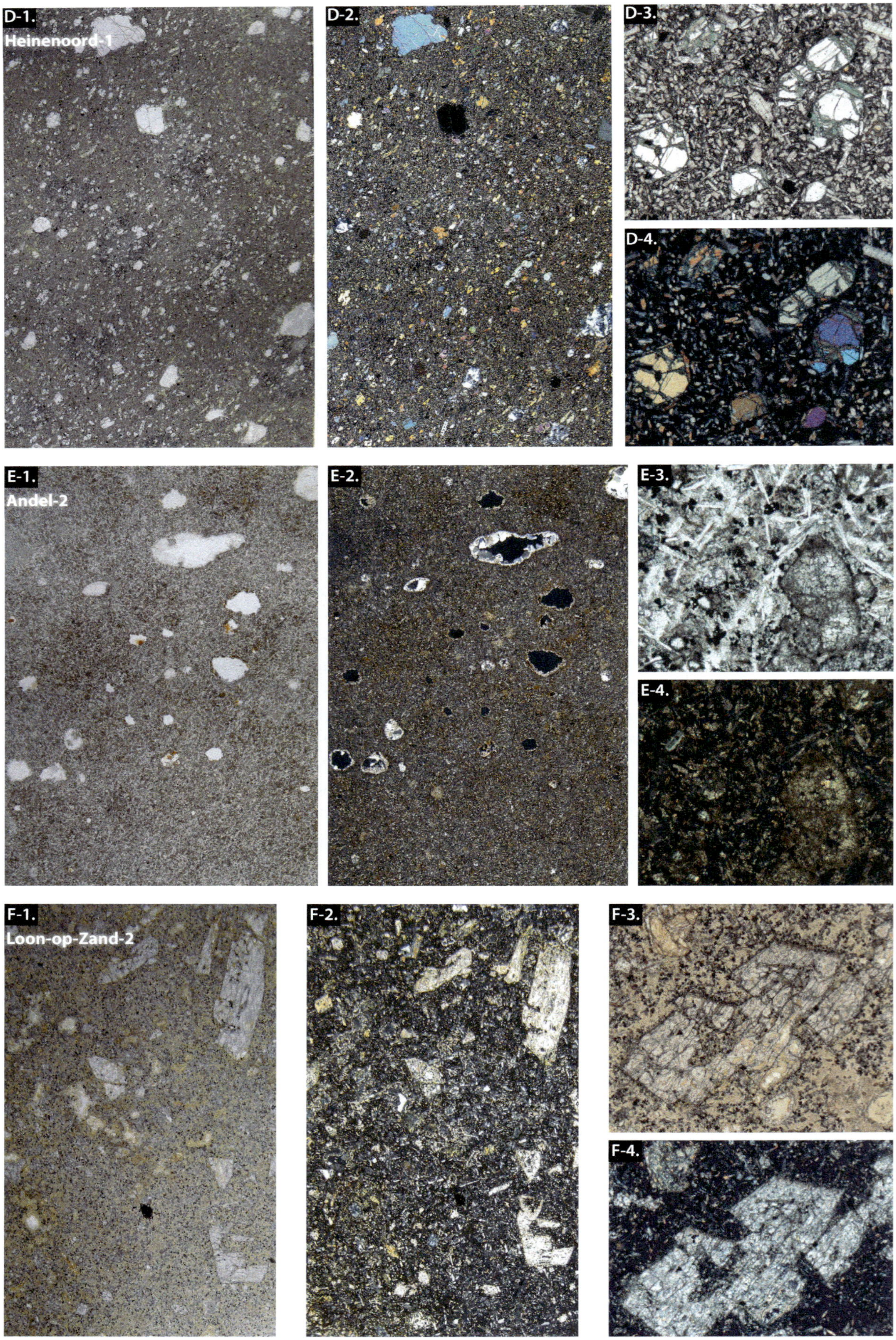
D-1.
Heinenoord-1
D-2.
D-3.
D-4.
E-1.
Andel-2
E-2.
E-3.
E-4.
F-1.
Loon-op-Zand-2
F-2.
F-3.
F-4.

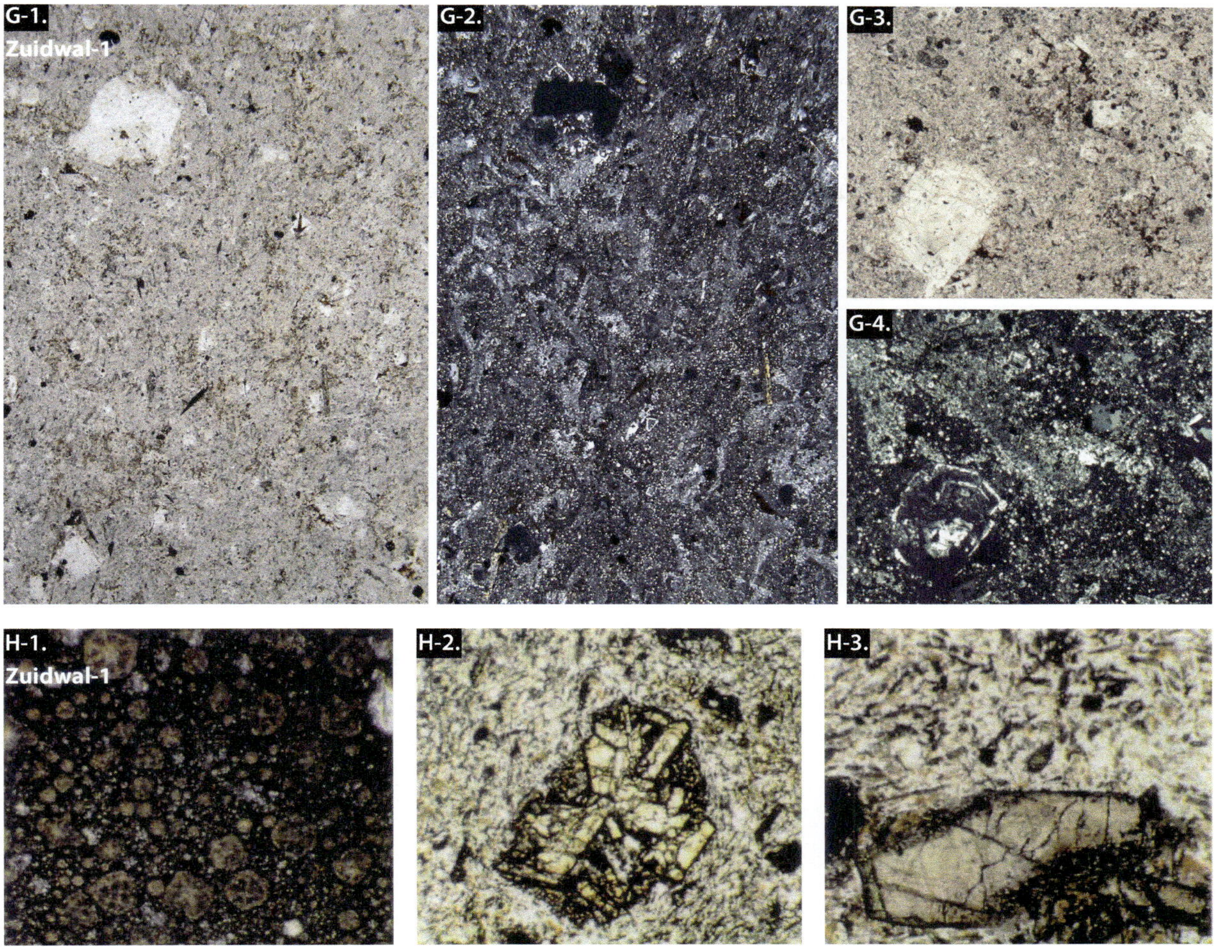

eruption, the magma chamber must have collapsed, and that any volcanic islands were largely removed by Late Jurassic and Cretaceous erosion. The Wadden Volcaniclastic Member of the Delfland Subgroup, 78 m thick in Slenk-1 (SLK-01) and also found in Zuidwal-2 and Zuidwal-3 (ZDW-02, -03), consists of fine- to coarse-grained volcaniclasts in a tuffaceous matrix (Fig. 11.4a, c). Because of the weathered appearance and proximity to the dome, these sediments probably represent erosion products rather than primary volcanic deposits (Herngreen et al., 1991).

Radiometric dating of Zuidwal-1 yielded inconsistent results. Initial K-Ar ages obtained on four samples (3000-1950 m) ranged between 119±2 Ma and 92±2. These are probably only minimum ages, considering the extensive alteration (Jeans et al., 1977; Harrison et al., 1979). An expected true age of >120 Ma was supported by a K-Ar date of 145 Ma obtained on a sample of unknown origin (pers. comm. G. Flacelière, in Jeans et al., 1977). Subsequent ^{40}Ar/^{39}Ar dating yielded a virtually identical age of 144±1 Ma with a minor overprint between about 120 and 90 Ma, suggesting that the younger K-Ar ages reflect argon loss (Dixon et al., 1981). Finally, Perrot & Van der Poel (1987) reported an age of 152±3 Ma, based on the ^{40}Ar/^{36}Ar-^{40}K/^{36}Ar method. These oldest ages (Late Jurassic) are presumably closest to the true emplacement age.

Documented petrographic descriptions of Zuidwal samples are not uniform. Dixon et al. (1981) inferred that the eight altered rocks they examined originally were phonolite samples, biotite pyroxenites, a phonolitic basanite and a rock rich in pseudomorphed leucite respectively (cf. Fig. 11.6g, h). The agglomerate matrix attached to one of the phonolites contained leucite basanite and leucite tephrite clasts. Rock types referred to as trachytes, phonolites and leucitites from observations on 12 samples (Perrot & Van der Poel, 1987) are consistent with this description. On the other hand, Harrison et al. (1979) described finely crystalline to partly glassy trachyte as the most common rock type, together with less abundant, heavily altered lava of probably basic composition and pieces of originally glassy vesicular pumice and minette. They noted that sphene is relatively abundant and do not record the presence of primary feldspathoids or their pseudomorphs. According to Dixon et al. (1981) it is conceivable that the Zuidwal volcanics represent one or more cycles of trachyte-phonolite eruptions and that they included subordinate amounts of more primitive lavas such as basanites and tephrites.

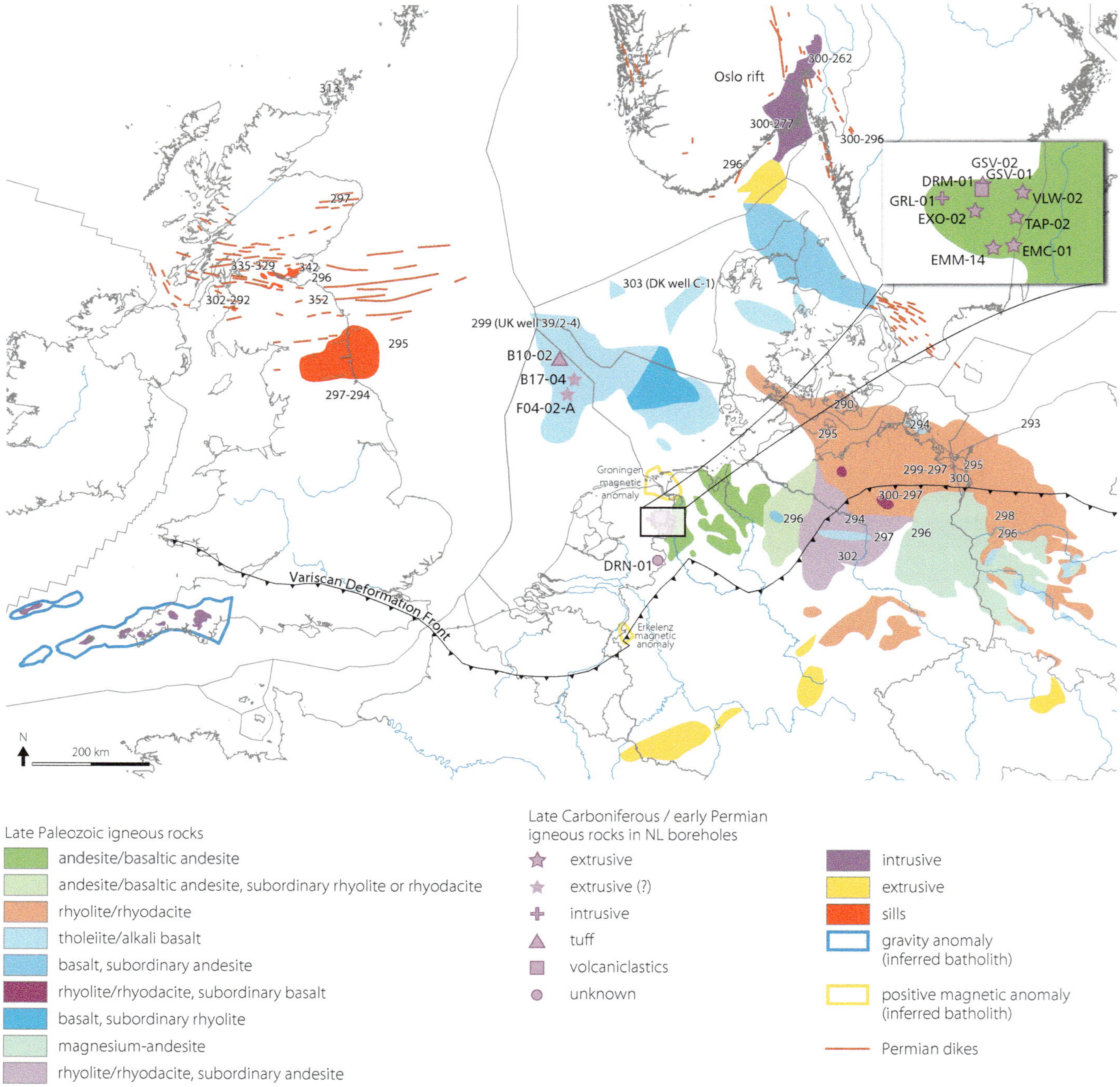

Figure 11.7. *Distribution of latest Carboniferous-early Permian ('lower Rotliegend') igneous rocks in northwest Europe and locations of wells in the Netherlands with intrusive and extrusive igneous rocks, volcaniclastics and tuffs from this period. Numbers refer to radiometric ages in Ma obtained by U-Pb zircon, $^{39}Ar/^{40}Ar$ and by other high-quality dating methods. Information on the Whin Sill Complex: Hamilton and Pearson (2011), Timmerman (2004); Midland Valley of Scotland: Monaghan and Pringle (2004); Orkney: Lundmark et al. (2011); Mid North Sea High: Lundmark et al. (2012), Heeremans et al. (2004); Horn Graben: Breitkreuz et al. (2007); northeast German Basin: Breitkreuz and Kennedy (1999); Oslo Rift: Pedersen et al. (1995), Corfu and Dahlgren (2008), Corfu and Larsen (2020); late Paleozoic igneous rocks: Gast et al. (2010; Fig. 7.16); additional Permain igneous rocks: Van Bergen & Sissingh (2007); East Groningen Massif: Bredewout (1989); Cornubian batolith (Cornwall): Searl et al., (2024); Watts et al., (2024).*

Taking effects of secondary alteration into account, the available geochemical data (see below) indicate that the Zuidwal volcanics originally represent a typical ultra-potassic, silica-undersaturated rock series, including leucite-bearing phonolitic members (Fig. 11.6g, h), which probably originated during the residence of a more mafic, mantle-derived, clinopyroxene-rich tephritic-basanitic parent magma in a relatively shallow subvolcanic reservoir.

The Mulciber volcanic centre

The existence of a buried igneous centre in the subsurface of the Dutch sector of the North Sea (block F16, Fig. 11.12) had been suspected for some time from the existence of a positive magnetic intensity anomaly around Amoco exploration well F16-02 (Sissingh, 2004) and from the presence of volcanic material in core samples (Lester, 1985). An integrated regional sedimentary and stratigraphic study in the area around the well (Bouroullec et al., 2018) ultimately led to the identification of the Mulciber volcanic centre, made public by the Geological Survey of the Netherlands in 2020. The volcano was named after the Roman god of fire and volcanoes.

Sidewall core samples and borehole logs from well F16-02 (Fig. 11.12a) include tuff and tuffaceous claystone grading into solid volcanic rock in the succession between 1702 and 1662.5 m depth, as well as tuffaceous claystone interbedded with anhydrite, and volcaniclastics between 1983 and 1737 m (Lester, 1985). Petrographic information provides insight into the mode of emplacement and composition. Five pieces of igneous rock sampled between 1672 and 1662.5 m were interpreted as basaltic-andesitic and andesitic lavas. This was based on their porphyritic, vesicular nature and the mineral content, which includes clinopyroxene, plagioclase, Fe-oxide, apatite and a matrix of devitrified glass that hosts the same phases, often together with biotite. Due to strong alteration, most of the original assemblage has been replaced by secondary minerals. The presence of biotite (and locally amphibole) suggests an alkaline affinity of the rock suite. Although it has been proposed that the igneous material of this interval represents a continuation of the Zuidwal volcanics into the Central North Sea Graben (Lester, 1985), it is now clear that the Mulciber volcanic centre was the source.

The magnetic anomaly in block F16 is similar to that of the Zuidwal volcano (Fig. 11.10b). A strongly reflective interval with a high acoustic impedance contrast, mapped on 3D seismic data a few kilometres to the west of well F16-02 (Fig. 11.12b), indicates the presence of layers of solid rock, as can be readily inferred from the sonic log (Fig. 11.12a). Its spatial distribution coincides with that of the magnetic anomaly (Fig. 11.12b), and its reflections can be traced to the well (Fig. 11.12c, d), so it probably represents the same lavas as found in the sidewall cores in F16-02. The morphology of the uppermost seismic reflector resembles that of a smoothed volcano, likely due to subaerial weathering and erosion prior to burial by younger sediments.

Biostratigraphic constraints indicate that the upper interval of igneous rocks (1702-1662.5 m) is latest Jurassic to Early Cretaceous in age, while the overlying claystone (1662.5-1659 m) is Late Hauterivian to Early Barremian (Lester, 1985; Bouroullec et al., 2018). Samples from the sidewall cores were completely used for petrographic analyses and igneous material is no longer available for radiometric age determination.

The position of strong seismic reflectors at a considerably deeper level (3300 m) west of well F16-02 relative to that near the well and the volcanic interval (ca. 1660 m), marks the effect of salt tectonics (Fig. 11.12d). The seismic data suggest that faulting along the Dutch Central Graben was active during magmatic activity and that, rock salt (Zechstein) moved upward along the fault-related weakness zones in the overburden, pushing the succession of igneous rocks to higher levels at the well location (Fig. 11.12c).

Below the igneous rocks of Jurassic age in well F16-02, evaporites of the Zechstein Group host an interval with felsic volcanics between 1985 and 1738 m. Considering the stratigraphic position, these are probably of late Permian age. Strongly altered material at 1975-1965 m represents a porphyritic extrusive rock with clinopyroxene, biotite and glassy matter, presumably originally a dacitic lava.

Jurassic igneous rocks in adjacent areas

The triple junction between the Moray Firth, the Viking Graben and the Central Graben (northern North Sea) has been known as a centre of voluminous volcanism in middle to late Jurassic times since its discovery in 1970 (Fig. 11.13). An over 3000 m thick succession of basaltic lavas ('Forties Volcanics') constitutes a volcanic field that covers an area of about 12000 km^2, extending eastward across the southernmost part of the Viking Graben (Woodhall & Knox, 1979). The volcanics are also known as the Forties Igneous Province and the Rattray Volcanics amongst others. Based on well data and seismic reflection profiles, Smith and Ritchie (1993) proposed that at least three major volcanic centres existed, whereas from a recent re-interpretation using 3D-seismic data Quirie et al. (2019) inferred that the volcanics were sourced in fissure eruptions from linear vents. The rocks were emplaced as subaerial lavas and volcaniclastics, subaqueous basaltic hyaloclastics and reworked material and as subordinate intrusive units (Quirie et al., 2019). The basalts are silica-undersaturated, belong to the alkaline series and contain abundant phenocrysts of olivine and clinopyrox-

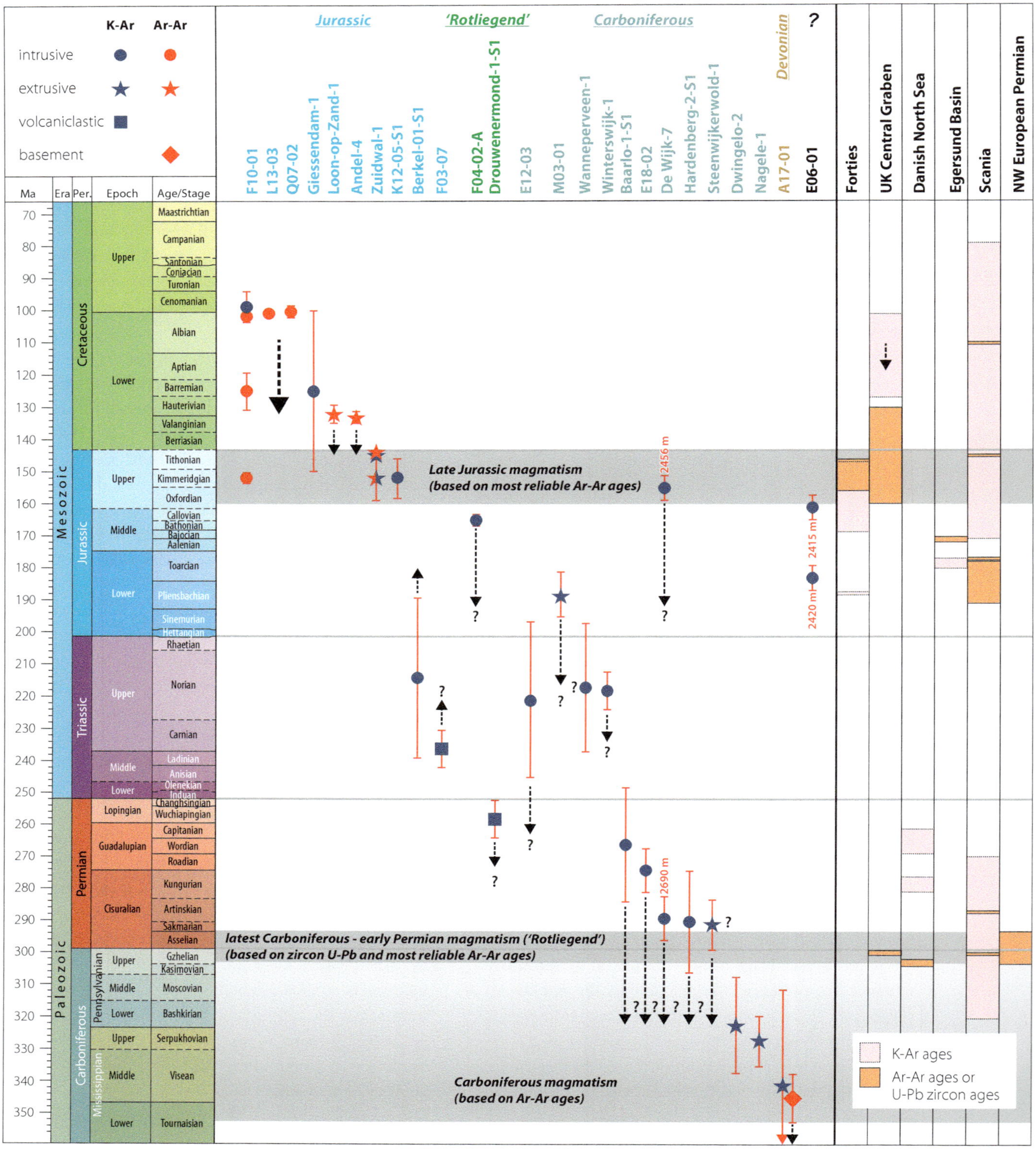

Figure 11.8. *Stratigraphic overview of radiometrically dated igneous and volcaniclastic rocks from wells in the Netherlands, plotted against periods of magmatic activity in surrounding northwest European areas (see Table 11.1 for references). Red bars indicate error margins, black arrows point towards expected more correct ages.*

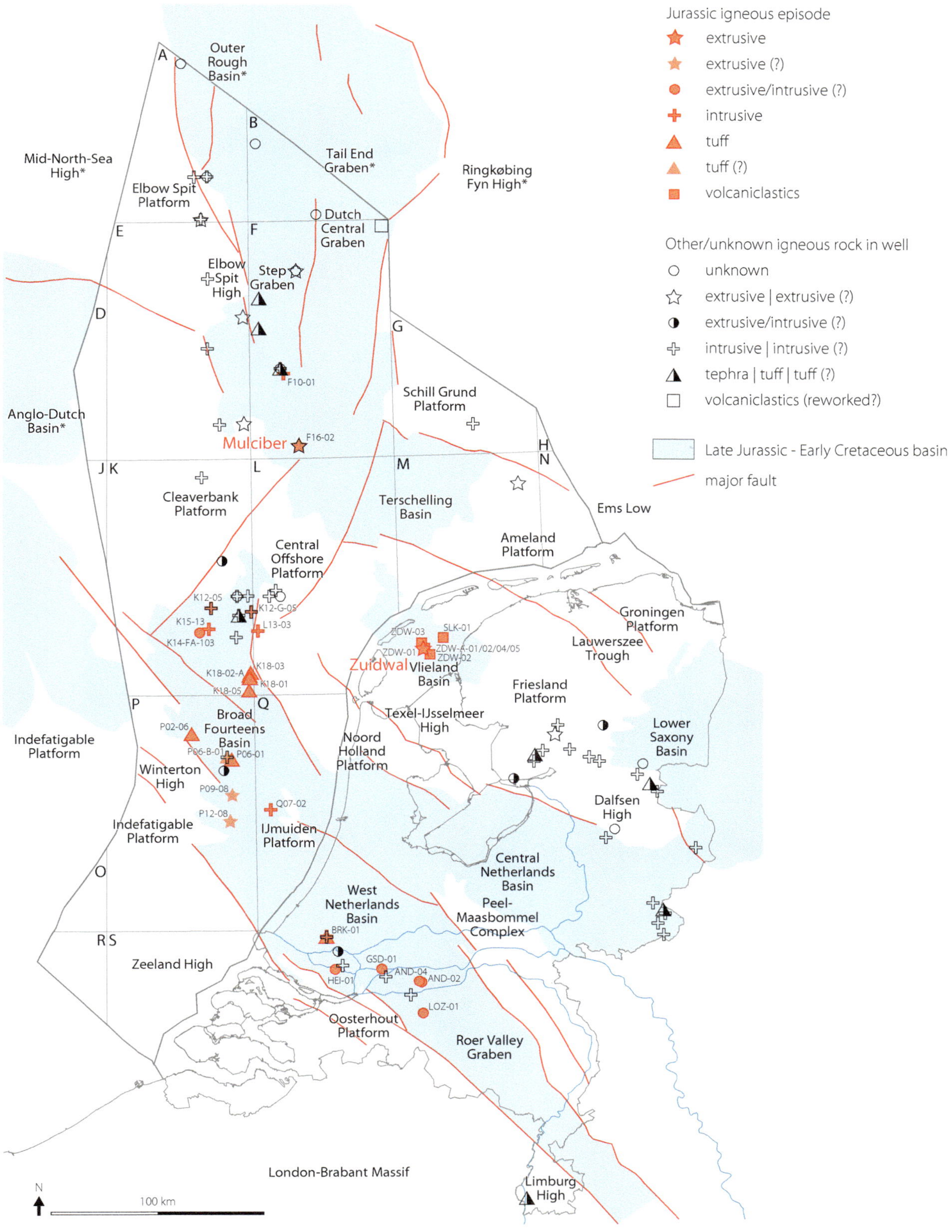

Figure 11.9. Wells located in the Netherlands in which igneous rocks attributed to the Late Jurassic episode of magmatic activity have been found. Wells with igneous material from other or unknown periods are shown for comparison. Structural elements are those from Late Jurassic-Early Cretaceous times (Pharaoh et al., 2010; Kombrink et al., 2012; Hopper et al., 2014). Details on the rocks in each well are listed in Table 11.1.

Figure 11.10. *a) Bouguer gravity map of the Netherlands; b) Local variations in the Earth's magnetic field. Both maps were compiled by TNO-GDN from various surveys (data can be found on NLOG.nl). Wells with igneous rocks (Table 11.1) and locations of the Zuidwal and Mulciber volcanic centres are shown for comparison.*

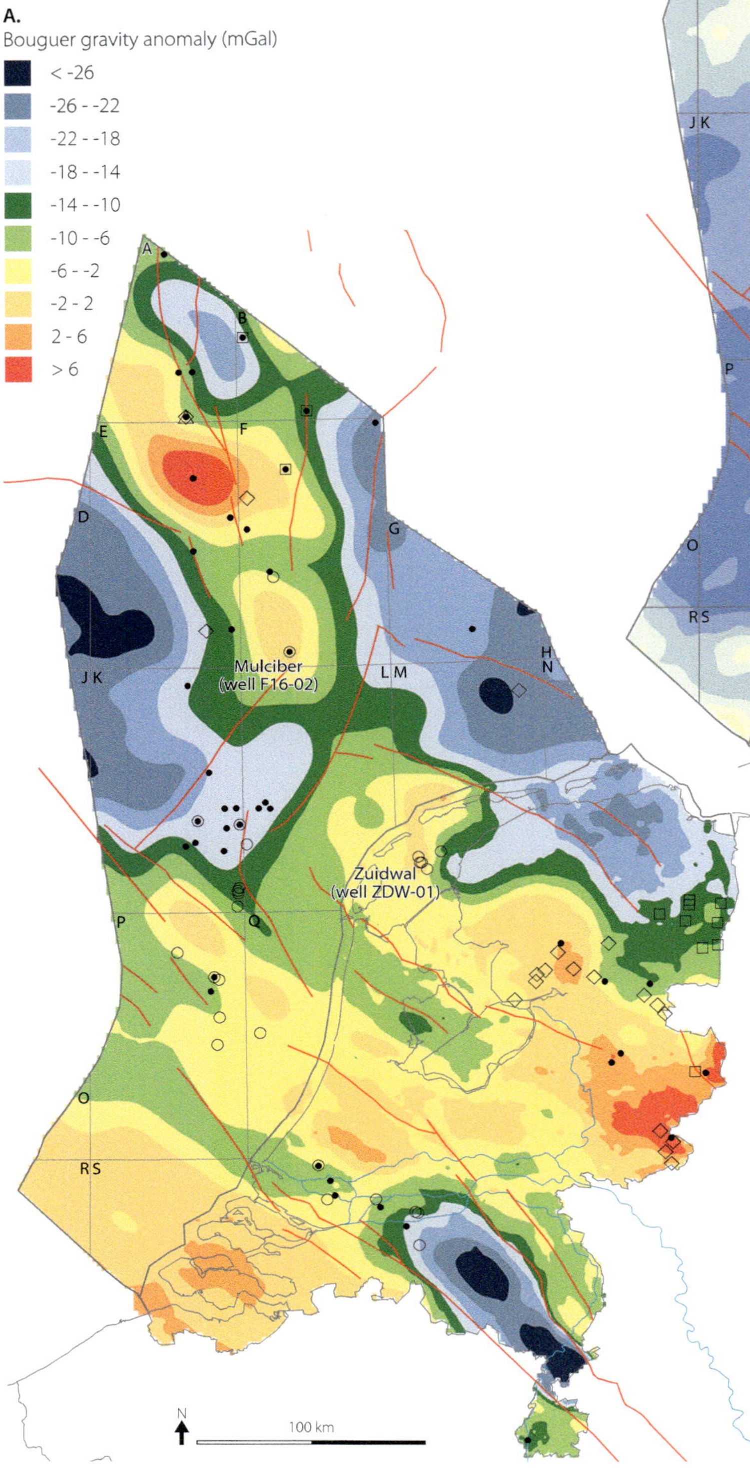

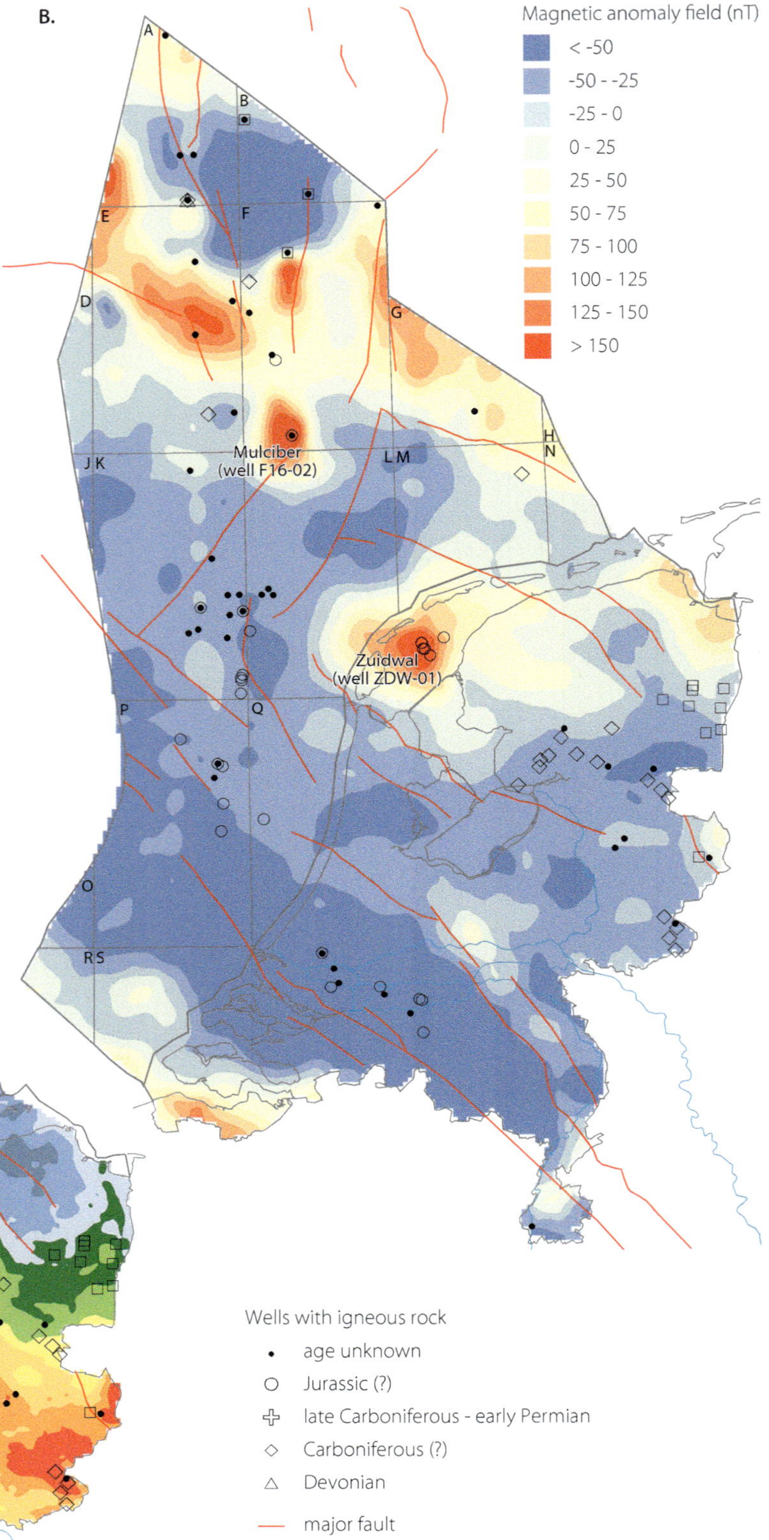

ene (Gibb & Kanaris-Sotiriou, 1976; Dixon et al., 1981; Latin & Waters, 1992). Most are fairly primitive, whereas more evolved hawaiites and mugearites have been found as well. The precise age and duration of the effusive event is still elusive as there is disagreement between radiometric and biostratigraphic age dates (see discussions in Latin, 1990; Quirie et al., 2020). Stratigraphic evidence based on palynology, suggests a middle Jurassic age (Bajocian-Callovian), while scrutiny of available radiometric dating results (Latin, 1990) favours an upper Jurassic age (Kimmeridgian-Tithonian), based on ages of 153±4 Ma and 148±2 Ma obtained on a $^{40}Ar/^{39}Ar$ stepwise degassing study of samples from well 21/03b-3 (Ritchie et al., 1988).

The Central North Sea igneous province on the western margin of the UK Central Graben (Fig. 11.13) is made up of mostly subaerial lava flows, tuffs and epiclastic rocks, together with several intrusions in the Zechstein interval. The rocks share a silica-undersaturated and potassium-rich character and carry mineral assemblages that include primary biotite and amphibole (Latin et al., 1990b). Fine-grained extrusive rocks in well 30/16-A13Y (sample SH1), labelled as kaersutite leucitite, contain Ti-rich clinopyroxene and amphibole as phenocrysts, whereas severely altered, fine-grained and largely aphyric rocks in well 9/14b-1 (sample AH2) were tentatively classified as basanite (Latin, 1990). Two intervals with intrusive varieties in well 29/25-1, referred to as mafic biotite phonolite (Dixon et al., 1981) or potassic lamprophyre (Latin, 1990), contain phlogopite and pseudomorphs of clinopyroxene and olivine phenocrysts. From an irregular $^{40}Ar/^{39}Ar$ age plateau of a phlogopite sample, the oldest age of 157 Ma is probably closest to that of the emplacement (Latin, 1990). All of these ultrapotassic rocks belong to the same series and share similar settings as those of the Dutch Central Graben and the Zuidwal volcanic centre.

On the western rim of the Egersund Basin in the Norwegian sector of the North Sea, nephelinite lavas have been found interbedded with Jurassic sediments, forming a sequence several hundred metre thick (well 17/9-1; Fig. 11.13). They are porphyritic vesicular rocks with large clinopyroxenes and pseudomorphs after olivine, which are set in a fine-grained or glassy groundmass. High contents of incompatible trace elements confirm the alkaline nature of these rocks (Dixon et al., 1981; Latin & Waters, 1992). A lower succession, cored in the same well, intruded into Lower Jurassic and/or Triassic sediments (Furnes et al., 1982; Latin et al., 1990b). These are strongly undersaturated mafic potassic-alkaline rocks, similar to the lamprophyres of the Central Graben area, with phlogopite-rich mineral assemblages and textures that resemble alnöites (Latin, 1990). Conventional K-Ar ages obtained on phlogopite in the intrusive rocks yielded 180-177 Ma (Furnes et al., 1982) and a $^{40}Ar/^{39}Ar$ plateau age on phlogopite a younger age of 170±2 Ma (Latin, 1990). Together with the Bajocian-Bathonian age of the sedimentary rocks intercalated with the lavas, this would make the Egersund magmatic event somewhat older than the most reliable radiometric ages of Jurassic magmatism in the Netherlands and UK sectors of the North Sea.

Numerous volcanic necks and plugs, associated with minor remnants of lava flows and pyroclastic deposits represent Jurassic mafic alkaline magmatic rocks with nephelinite-basanite composition in central Scania (south Sweden, Fig. 11.13). Whole-rock determinations produced a K-Ar age range between 171 and 79 Ma, and a $^{40}Ar/^{39}Ar$ range in three age groups between 191 and 110 Ma, but the younger results are probably compromized by secondary Ar loss (Tappe et al., 2016, and references therein). However, two $^{40}Ar/^{39}Ar$ plateau ages obtained on anorthoclase feldspar megacrysts yielding a weighted average of 176.7±0.5 Ma (2-sigma), probably provide the current best estimate for Jurassic magmatism along the southwestern margin of the Baltic Shield. A Middle Jurassic olivine-biotite gabbro with a K-Ar age of 166±4 Ma was encountered in a borehole off the west coast of Cornwall (Harrison et al., 1979), while Jurassic smectitic clays in southern and eastern England have been interpreted as alteration products of volcanic air-fall, part of which may originate from volcanic centres in the North Sea (Bradshaw, 1975; Jeans et al., 1977, 2000).

In summary, the extensive database on Mesozoic magmatism in the Netherlands encountered in wells (Table 11.1) discloses clear systematic patterns in the onshore and offshore distribution, timing, composition and mode of emplacement of the igneous rocks. Despite a usually high degree of alteration, the available petrographic and geochemical evidence indicates that all of the rocks sampled have a silica-undersaturated alkaline signature. Most of the occurrences solidified as sills or dykes from mafic mantle-derived magmas, whereas a minority represents lava flows or tuffaceous intervals indicative of explosive eruptive activity. The buried volcanoes of Zuidwal and Mulciber form complex igneous edifices. Petrographically the rocks are often porphyritic, carrying clinopyroxene and olivine (or pseudomorphs) as dominant phenocrysts, while assemblages of primary silicate phases may further include amphibole, biotite, nepheline, leucite and plagioclase. Parental magmas are tephritic/basanitic with a more sodic affinity around the West Netherlands Basin and a more potassic character in the areas of the Vlieland Basin, Broad Fourteens Basin and Dutch Central Graben. Tuff-bearing intervals are mostly reported in logs from wells near the Broad Fourteens Basin. Available results from conventional absolute age dating span a range from Late Triassic to Late Cretaceous but in many cases relia-

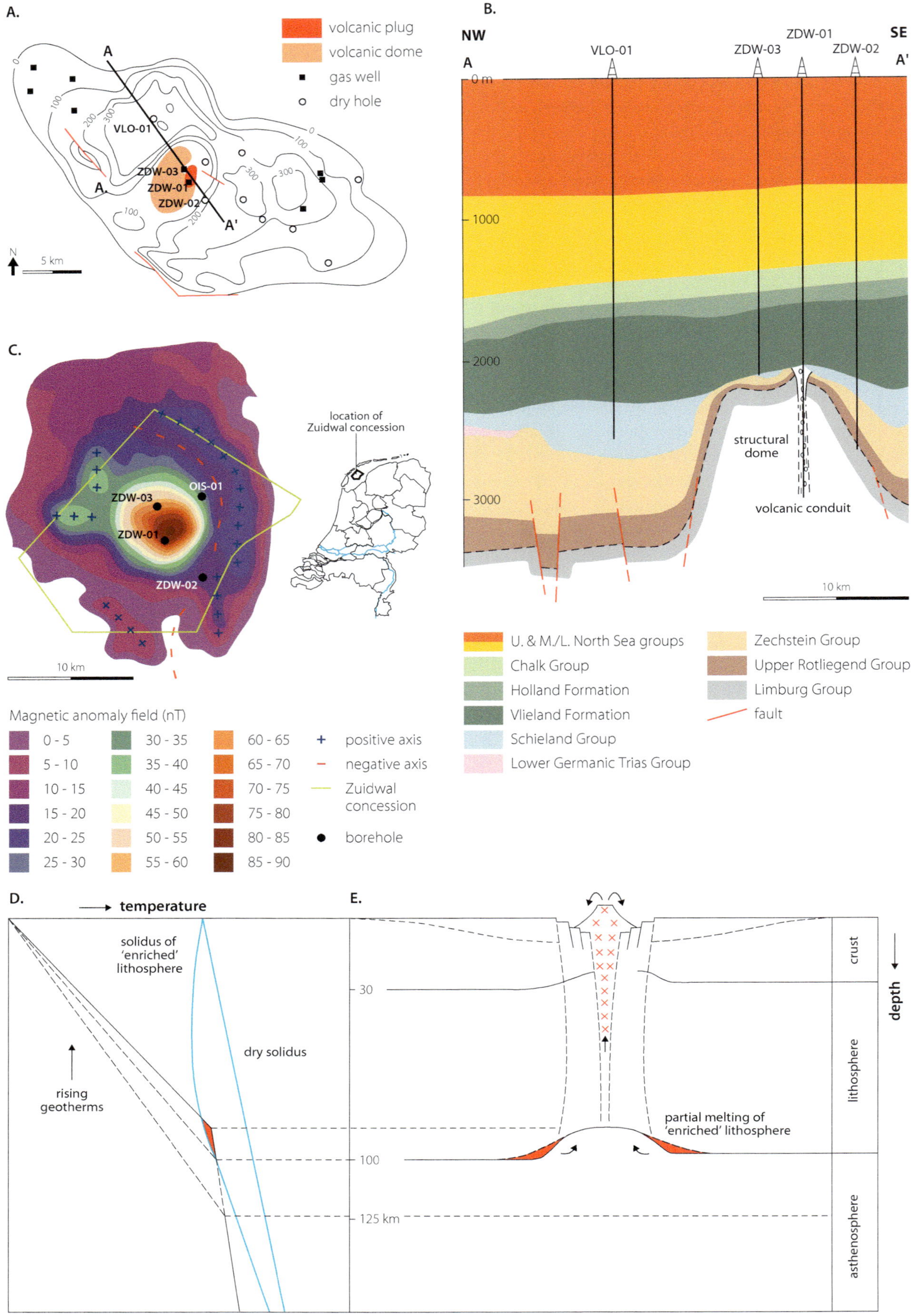

A.
volcanic plug
volcanic dome
gas well
dry hole
A
A'
VLO-01
ZDW-03
ZDW-01
ZDW-02
0
100
200
300
N
5 km
B.
NW
SE
A
A'
VLO-01
ZDW-03
ZDW-01
ZDW-02
0 m
1000
2000
3000
structural dome
volcanic conduit
10 km
U. & M./L. North Sea groups
Chalk Group
Holland Formation
Vlieland Formation
Schieland Group
Lower Germanic Trias Group
Zechstein Group
Upper Rotliegend Group
Limburg Group
fault
C.
location of Zuidwal concession
OIS-01
ZDW-03
ZDW-01
ZDW-02
10 km
Magnetic anomaly field (nT)
0 - 5
5 - 10
10 - 15
15 - 20
20 - 25
25 - 30
30 - 35
35 - 40
40 - 45
45 - 50
50 - 55
55 - 60
60 - 65
65 - 70
70 - 75
75 - 80
80 - 85
85 - 90
positive axis
negative axis
Zuidwal concession
borehole
D.
temperature
solidus of 'enriched' lithosphere
dry solidus
rising geotherms
E.
30
100
125 km
partial melting of 'enriched' lithosphere
crust
lithosphere
asthenosphere
depth

← ***Figure 11.11.*** *The Zuidwal volcanic centre (for location, see Fig. 11.1): a) Isopach map (metres) of the 'Upper Jurassic' in the southern part of the Vlieland Basin (modified from Herngreen et al., 1991); b) Cross section showing the dome structure of the Zuidwal volcano (modified from Herngreen et al., 1991) c) Aeromagnetic map (modified from Perrot & Van der Poel, 1987; cf. Fig. 11.10b); d) and e) Tentative model for mantle melting below the Zuidwal volcano in response to stretching and thinning of the continental lithosphere, illustrated by connecting a hypothetical temperature-depth (pressure) phase diagram for mantle rock (d) to a simplified cross section through the local crust (modified from Herngreen et al., 1991; see text for details).*

bility is compromized by post-crystallization disturbance. Considering the most robust age dates only, Mesozoic magmatism is probably limited to a Late Jurassic interval.

Paleogene-Neogene

The lower Eocene marine tuffaceous clay of the De Wijk Member (formerly Basal Dongen Tuffite Member) is widely distributed in the subsurface of the Netherlands (Table 11.2). Equivalent tephra deposits are widespread in offshore and onshore stratigraphic records of the UK, Germany and Denmark (Fig. 11.14), where locally tens to hundreds of individual air fall layers have been identified. This early Eocene ash interval has been recognized in large parts of NW Europe and is the expression of an episode of major explosive volcanism (Knox & Morton, 1988; Morton & Knox, 1990; Larsen et al., 2003; Stokke et al., 2020). Given the regional distribution, the interval represents a conspicuous marker in the North Sea Basin, which is well detectible in wire-line logs (Jacqué & Thouvenin, 1975; Knox & Morton, 1988).

The ashfalls were initially bimodal (mixed basaltic to rhyolitic) but subsequent main stage series were predominantly basaltic, as inferred from key sections in northwest Denmark (e.g. Larsen et al., 2003; Stokke et al., 2020). Tephra deposits or their alteration products with largely similar geochemical signatures, have also been described from wells and clay pits in northern Germany (e.g. Obst et al., 2015 and references therein) and from offshore wells in the North Sea area (Morton & Knox, 1990). The ashes probably originated from volcanic centres in the North Atlantic Igneous Province (NAIP), formerly referred to as the 'Thulean igneous province', which comprises intrusive and extrusive rocks encountered in Greenland, Iceland, the Faroe Islands, the UK and Ireland as well as vast submarine areas of the adjacent continental margin (Á Horni et al., 2017). Massive flood basalts and sill intrusions were emplaced between 63 and 52 Ma (maximum activity between 56 and 54 Ma), accompanying the separation of Greenland from Eurasia and the opening of the North-East Atlantic Ocean (Storey et al., 2007). The ash eruptions were particularly explosive and voluminous. The most powerful single event in the succession (dated at 54.04±0.14 Ma; Chambers et al., 2003) ejected ca. 1200 km^3 of ash material with a fall-out distance up to some 1900 km from the source, i.e. as far south as the Austrian Alps (Egger & Brückl, 2006). In the absence of evidence from tephrochronology, it is assumed that the ash particles in the De Wijk Member originate from this eruption (Table 11.2). The magnitude and explosivity of the NAIP eruptions are exceptional for basaltic systems and may have been enhanced by hydromagmatic interaction in a shallow marine environment (Larsen et al., 2003; Stokke et al., 2020). The intrusion of dykes in the North Sea area (see below) may have further contributed to the widespread deposition of tephra in the North Sea Basin through explosive interaction between rising magma and sea water (Wall et al., 2010).

The large-scale igneous activity that marks the NAIP is related to a plume of upwelling hot mantle material arriving at the base of the lithosphere below Greenland, the subsequent continental break-up and the initiation of ocean floor spreading (Morgan, 1971; Saunders et al., 1997; Torsvik et al., 2001). The postulated plume is currently centred below Iceland. Although the mantle plume concept successfully explains the origin of the NAIP, there is ongoing debate as to the shape, size and depth of the mantle source, the location of first impingement, the areal extent of the geological effects, and the stability of its position (see review of Meyer et al., 2007). Alternatives for a classical hot-plume hypothesis have been proposed as well (e.g. Foulger & Anderson, 2005; Hole & Natland, 2020). This episode of intense volcanism had a global environmental impact as it is thought to have triggered the onset of the Paleocene-Eocene Thermal Maximum (PETM) at ca. 56 Ma (e.g. Storey et al., 2007).

The NAIP encompasses the British-Irish Paleogene Igneous Province (BIPIP), which consists of igneous centres and dyke swarms in Scotland, England, (Northern-)Ireland and the Irish Sea (Fig. 11.14; Wright et al., 1971). Dyke swarms also extend into the Southern North Sea (Kirton & Donato, 1985; Brown et al., 1994), as inferred from magnetic anomalies and seismic stratigraphy (see compilation in Fig. 11.14). Since their discovery, several authors have reported evidence for WNW-ESE striking dykes in the Southern North Sea, reaching as far east as the Netherlands offshore (Gauer et al., 2004; Underhill, 2009; Wall et al., 2010; Hernandez Casado & Underhill, 2013; Kortekaas et al., 2018; Carver et al., 2023; Engie/Neptune Energy unpublished internal studies). The dyke swarms appear as

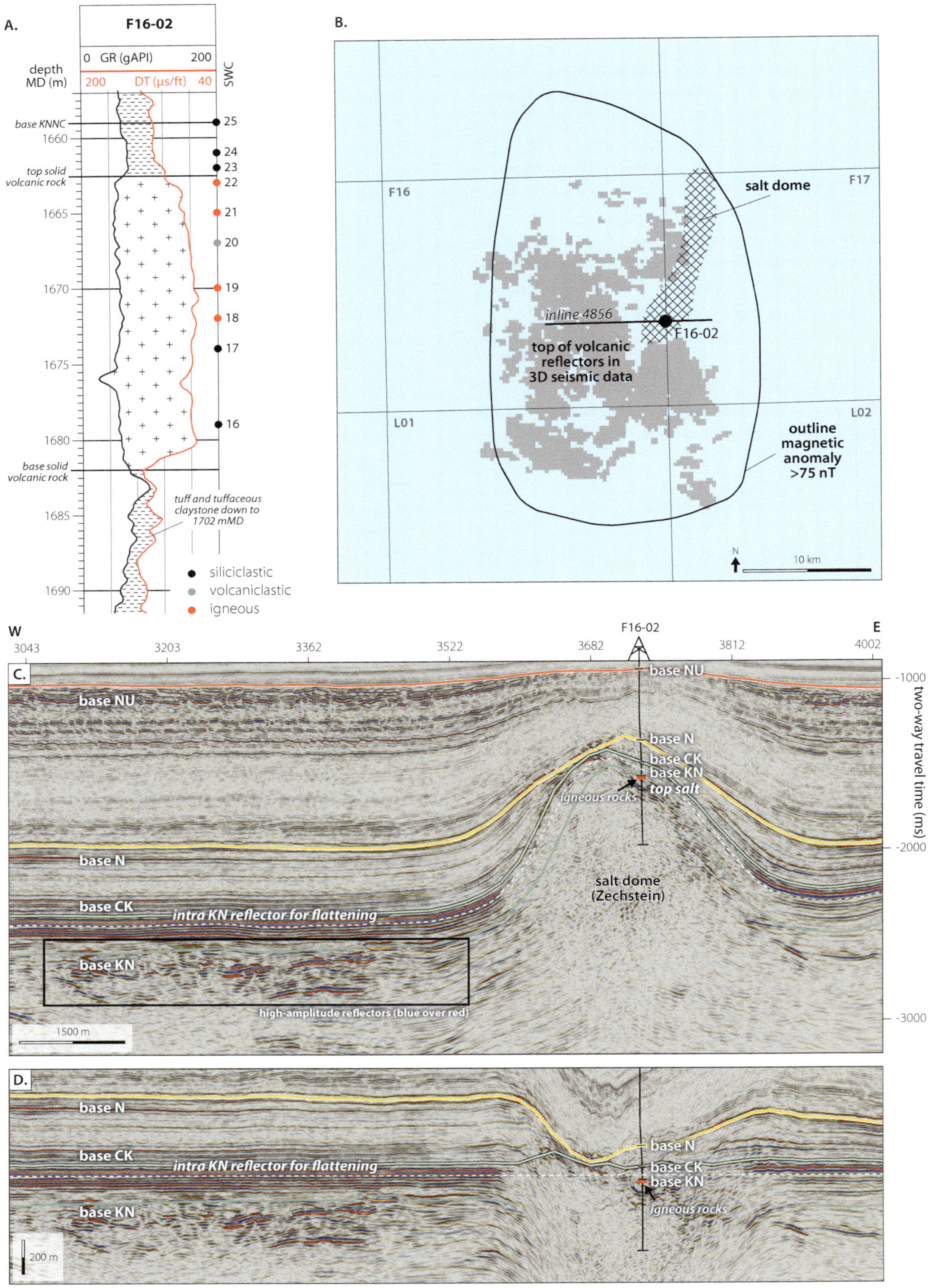
A.
F16-02
0 GR (gAPI) 200
200 DT (µs/ft) 40
depth MD (m)
SWC
base KNNC
1660
top solid volcanic rock
1665
1670
1675
1680
base solid volcanic rock
1685
1690
tuff and tuffaceous claystone down to 1702 mMD
25
24
23
22
21
20
19
18
17
16
siliciclastic
volcaniclastic
igneous
B.
F16
F17
L01
L02
salt dome
inline 4856
F16-02
top of volcanic reflectors in 3D seismic data
outline magnetic anomaly >75 nT
N
10 km
W
E
F16-02
3043
3203
3362
3522
3682
3812
4002
C.
base NU
base N
base CK
base KN
top salt
igneous rocks
intra KN reflector for flattening
salt dome (Zechstein)
high-amplitude reflectors (blue over red)
1500 m
-1000
-2000
-3000
two-way travel time (ms)
D.
200 m

← *Figure 11.12. The Mulciber volcanic centre (for location, see Fig. 11.9): a) Volcanic rock, tuff and tuffaceous claystone in well F16-02, as inferred from geophysical borehole logging and identified in sidewall cores (SWC). GR = gamma-ray log, DT = sonic log, MD = Measured Depth, KNNC = Vlieland Claystone Formation; b) The area around well F16-02 showing the coincidence of the magnetic anomaly (also see Fig. 11.10b) and a seismic reflector representing layers of solid (volcanic) rock with a high acoustic impedance; c) Seismic line across well F16-02, penetrating an underlying salt dome (3D survey Z3WIN2003B, inline 4856); d) Flattened version of the seismic line to illustrate the correlation between the volcanic deposits in well F16-02 and the seismic reflector further to the west. NU = Upper North Sea Group, N = North Sea Supergroup, CK = Chalk Group, KN = Rijnland Group.*

noisy sub-vertical zones on seismic data but the deep roots of the igneous bodies remain elusive.

The age of the dykes in the Southern North Sea is not well established but can be inferred from their onshore equivalents in Scotland and England. Interpretation of conventional K-Ar dates of a large set of samples from the Cleveland Dyke Echelon, which stretches from the Mull intrusive complex in western Scotland to the north-east coast of England, yielded an emplacement age of 59.3±2.0 Ma (Mitchell et al., 1989). Stratigraphic relationships point to a consistent age of 58-54 Ma (Underhill, 2009; Wall et al., 2010) or 59-58 Ma (Carver et al., 2023), i.e. near to the Paleocene-Eocene boundary for the Southern North Sea dykes that form the south-easterly continuation of the Cleveland and associated dykes. Thoroughly screened and recalibrated Ar-Ar and U-Pb dates of various igneous rocks return an age cluster between 63 and 58 Ma for the entire BIPIP (Wilkinson et al., 2017).

Dyke swarms of similar age to those of the BIPIP and Southern North Sea also occur on the north-American side of the spreading zone, i.e. along the east coast of Greenland (Kirton & Donato, 1985; Wilkinson et al., 2017). In the Southern North Sea, the dykes may reflect the regional stress pattern in the Paleogene as they do not coincide with prominent faults at sub-salt level (Underhill, 2009). On the other hand, Wall et al. (2010) noted that dykes have the same trend as regional basement lineaments. The dykes occur preferentially in synclines where Zechstein salt has flowed away (Brown et al., 1994; Underhill, 2009; Wall et al., 2010). Based on seismic interpretation, Hernandez Casado & Underhill (2013) suggested that Paleogene igneous activity triggered mobility of the Zechstein salt in the Southern North Sea, which played a major role in shaping the structural configuration. Carver et al. (2023) tentatively inferred a minimum of at least three intrusive episodes spread over a period of about one million years. Linear channel-like features, visible at the top of the Upper Cretaceous Chalk (Brown et al., 1994; Carver et al., 2023), may reflect catastrophic dewatering, compaction and consequent extensional collapse as a thermal effect of dyke intrusion. Based on 3D seismic imaging, Wall et al. (2010) identified various types of craters at the top of the Chalk.

The dyke intrusions may have influenced petroleum prospectivity in various ways. Some authors (e.g. Kirton & Donato, 1985; Dewey & Windley, 1988) refer to the coincidence of the NW-SE trending dyke systems in Scotland and northern England with a broad zone of 'devolatilized Westphalian coals' that crosses the North Sea and extends to the Netherlands towards the Ruhr Graben. Fracturing around and above the dykes could have permitted gas migration from the Carboniferous source rocks into Triassic reservoirs, thus explaining the linear orientation of Triassic gas fields in the Southern North Sea (Underhill, 2009; Kortekaas et al., 2018). It has also been proposed that intrusion of the Paleogene dykes had an influence on the composition of subsurface gas, such as a rise in CO_2 or N_2 content (Verweij et al., 2016).

Volcanism also affected sediment composition in the North Sea Basin during the early Oligocene (Rupelian), in particular that of the Boom Member ('Boom clay'), a 50 to 75-m-thick marine shelf sediment present over large parts of the Netherlands subsurface (Table 11.2). According to petrographic observations on samples from outcrops in Belgium the sediment contains volcanic heavy minerals and up to 10% of argillized volcaniclasts in a clayey matrix that is thought to be the altered volcanic ash of trachytic and basaltic composition (Zimmerle, 1993). Plausible sources are volcanic centres of the Siebengebirge and Hocheifel that produced basaltic and evolved lavas and pyroclastics with alkaline affinities (Wedepohl et al., 1994; Jung et al., 2006, 2012). They were active from ca. 30-22 and 44-35 Ma, respectively, according to $^{40}Ar/^{39}Ar$ dating (Fekiacova et al., 2007; Przybyla et al., 2018). Direct ash falls and/or reworking of weathered volcanic soils were possible modes of transport to the North Sea area. Admixture of ash from the BIPIP cannot be excluded, as the main sea current along the British uplands at the time was counterclockwise (Vandenberghe et al., 2014).

Several layers containing ash particles have been found intercalated in shallow marine clay of the upper Oligocene to lower Miocene Veldhoven Formation in boreholes in the Winterswijk region (Table 11.2; Burger, 1979, 1992; Van den Bosch, 2015). The presence of 'Eifel titanite' and 'basaltic hornblende' suggests a link with explosive volcanism in the Eifel. This volcanism around the Rhine-Roer-Hessian graben triple junction in the Rhenish Massif started in the Eocene and lasted well into the Quaternary, with episodes of increased activity in the Late Oligocene

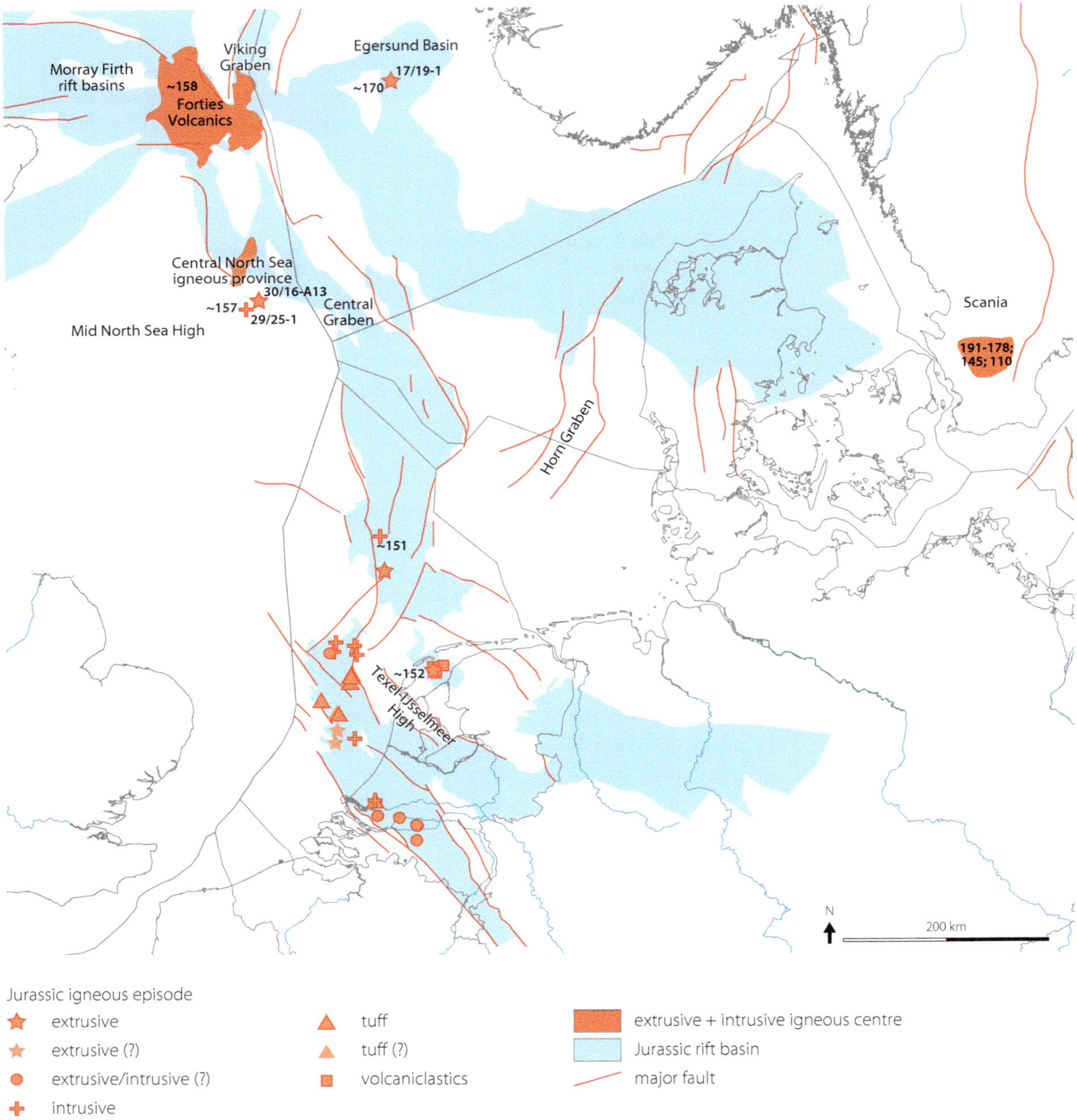

Figure 11.13. Distribution of Jurassic igneous rocks in northwest Europe and wells in the Netherlands that penetrated intrusive and extrusive rocks, volcaniclastics and tuffs from this period. Well names can be found in Fig. 11.9.

and Miocene (Sissingh, 2003; Ziegler & Dèzes, 2005, and references therein). It accompanied the development of the Rhine rift system and has been associated with the rise of a mantle plume and related thermal thinning of the mantle-lithosphere (Ritter et al., 2001).

Quaternary

Tephra layers originating from volcanic centres in the Eifel or Iceland are locally intercalated in Quaternary sedimentary deposits, providing a rare opportunity for tephrostratigraphic correlations in the Netherlands.

Middle Pleistocene fluvial terrace sequences in the Middle and Lower Rhine Basin in Germany record the onset of volcanism in the eastern Eifel about 600 ky ago through the appearance of volcanic heavy minerals that were initially dominated by hornblende and later by clinopyroxene (Boenigk & Frechen, 2006). In the Netherlands, a similar change in the heavy mineral spectrum is used to separate fluviatile deposits of the Sterksel Formation from the overlying Urk Formation (Preusser, 2008), in view of

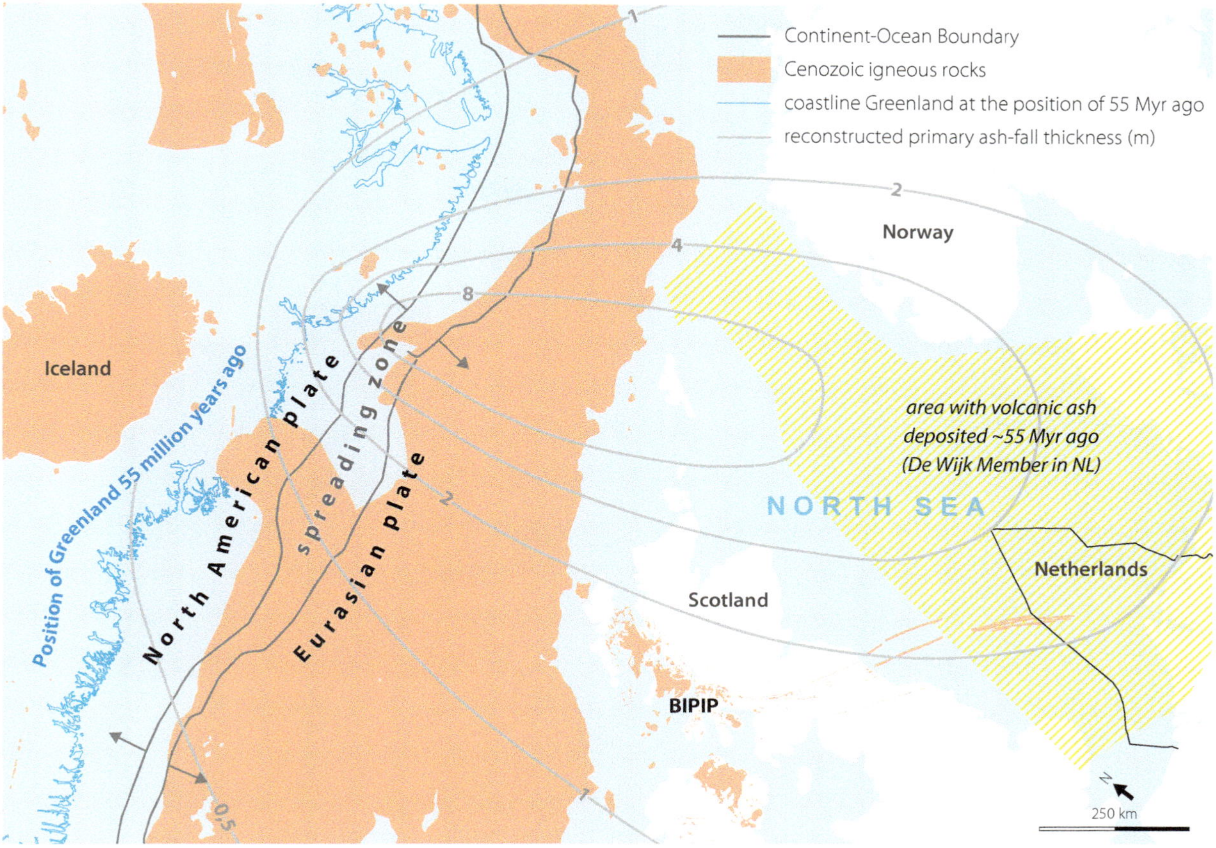

Figure 11.14. The North-East Atlantic Ocean at the onset of breakup (ca. 55 Ma, after Gaina et al., 2014) and the distribution of associated basaltic rocks (after Á Horni et al., 2017). Also shown are reconstructed isopach contours for primary ash fall produced by explosive volcanism that accompanied the rifting process (based on Knox & Morton, 1988; Knox, 1997). In the Netherlands, these ashes form a constituent of the De Wijk Member of the Dongen Formation. Geophysical evidence suggests that rifting-related dykes extend as far as the Dutch offshore (see text). Compilation of dikes in the North Sea is based on: Underhill (2009), Kirton & Donato (1985), Kortekaas et al. (2018), Youri Poslawski/Neptune Energy. BIPIP = British and Irish Paleogene Igneous Province.

the fact that the Lingsfort Member, the lowermost member of the Urk Formation, is thought to correlate with the oldest Middle Terrace sediments (MT I) in the Lower Rhine area (Zagwijn, 1985), where the increased influx of volcanic material is also recorded (Boenigk & Frechen, 2006). This is a rare example of the use of magmatic products for stratigraphic purposes in the Netherlands.

The loess succession of the Middle Pleistocene-Lower Holocene Boxtel Formation in South Limburg contains several tephras with an origin in the Eifel region, taking observations at locations in the immediate vicinity into account: the Rocourt Tephra, the Eltville Tephra and the Laacher Sea Tephra (Fig. 11.15, Table 11.2).

The Rocourt Tephra (Fig. 11.15b) was produced by a basaltic alkaline volcano with an unknown location presumably in the West Eifel. The tephra is marked by a blocky shape of altered glass shards, suggesting a phreatomagmatic character of the eruption, with a mineral assemblage comprising clinopyroxene, brown amphibole, orthopyroxene (enstatite) and Cr-spinel in decreasing order of abundance (Pouclet et al., 2008). The combination of high- to middle-pressure enstatite and aluminous clinopyroxene with low-pressure fassaitic clinopyroxene is rather unique and makes the tephra easy to recognize. Stratigraphic dating points to an age between 90.3 and 74 ka (Pouclet et al., 2008). Rocourt tephra has been found across large parts of eastern Belgium, including in loess sections near Kesselt, Vroenhoven en Veldwezelt, just west of Maastricht (Mees & Meijs, 1984; Pouclet et al., 2008; Meijs, 2011).

The Weichselian Eltville Tephra (Fig. 11.15c) erupted from an unidentified volcanic centre in the West or East Eifel and has been recognized in loess regions of western and central Germany, southern Netherlands and eastern Belgium, where two to five individual ash layers are part of an interval with a total thickness ranging between a few mm to 20 cm (Meijs et al., 1983; Pouclet & Juvigne, 2009; Zens et al., 2017). The assemblage of mafic vol-

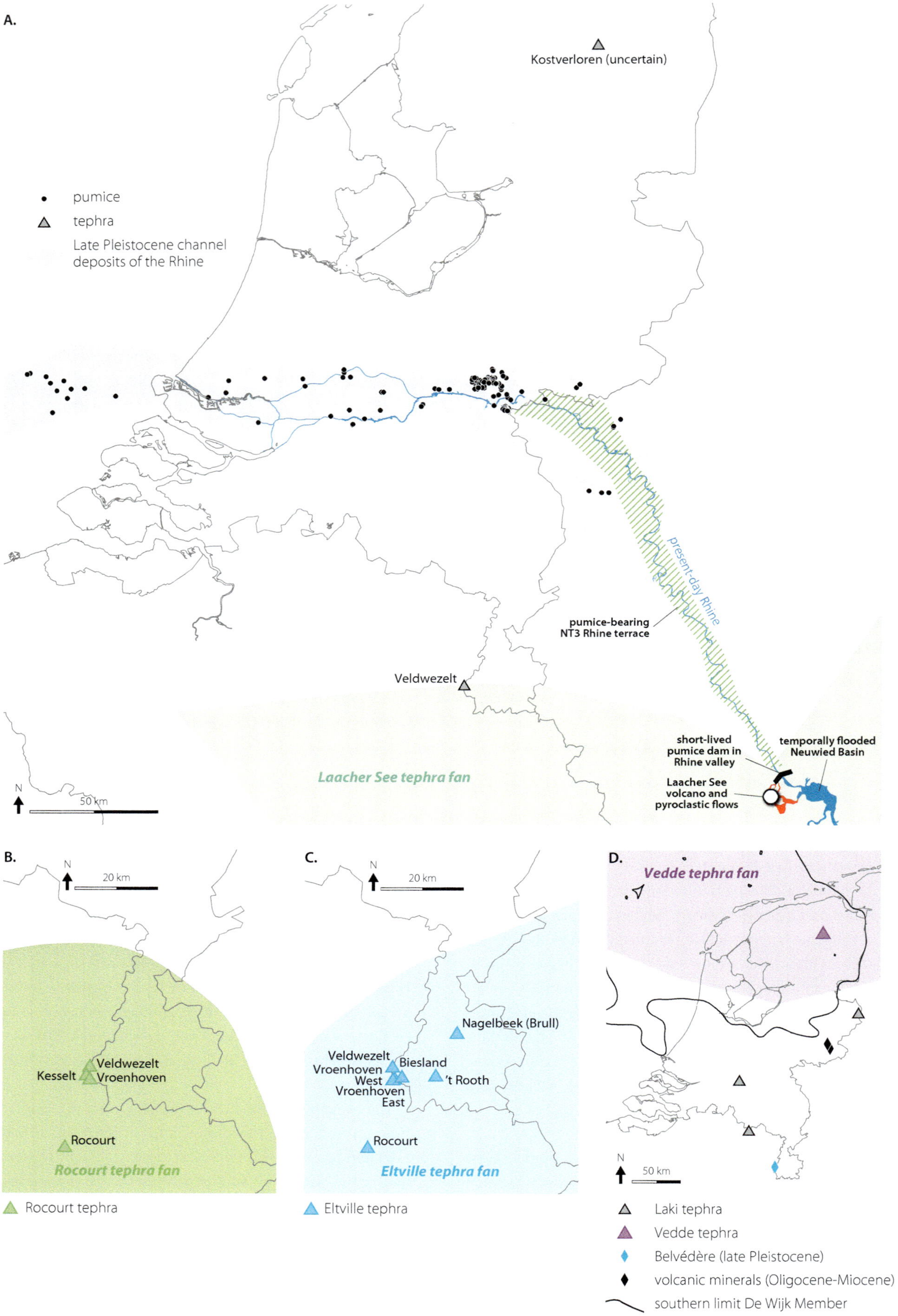
A.
Kostverloren (uncertain)
pumice
tephra
Late Pleistocene channel deposits of the Rhine
present-day Rhine
pumice-bearing NT3 Rhine terrace
Veldwezelt
short-lived pumice dam in Rhine valley
temporally flooded Neuwied Basin
Laacher See volcano and pyroclastic flows
Laacher See tephra fan
N
50 km
B.
N
20 km
Veldwezelt
Kesselt
Vroenhoven
Rocourt
Rocourt tephra fan
Rocourt tephra
C.
N
20 km
Nagelbeek (Brull)
Veldwezelt
Biesland
Vroenhoven West
't Rooth
Vroenhoven East
Rocourt
Eltville tephra fan
Eltville tephra
D.
Vedde tephra fan
N
50 km
Laki tephra
Vedde tephra
Belvédère (late Pleistocene)
volcanic minerals (Oligocene-Miocene)
southern limit De Wijk Member

← *Figure 11.15. Cenozoic volcanic material in the Netherlands: a) Reworked Laacher See pumice dispersed along the (Late Pleistocene) Rhine River and schematic distribution of air-fall tephra ejected from the Laacher See eruption centre (pumice compilation: K.M. Cohen (2015, 2022: various databases TNO-GDN, UU, literature); tephra and details: Schmincke et al. (1999), Park & Schmincke (2009); Late Pleistocene channel modified from Cohen et al. (2012); b) Approximate distribution of air-fall tephra from the Rocourt event (between 90.3 and 74 ka) and locations where it has been found in sediments (tephra: Pouclet & Juvigne (2009); locations are from: Mees & Meijs (1984), Meijs (2011), Juvigné (1999)); c) Approximate distribution of air-fall tephra from the 25.6-23.2 ka Eltville event and locations where it has been found in sediments (see also Fig. 11.16a) (tephra: Zens et al. (2017); locations are from: Meijs (1980), Juvigné & Semmel (1981), Meijs et al. (1983), Meijs (2011), Juvigné & Semmel (1981)); d) Approximate distribution of crypto-tephra from the 12.1 ka BP Vedde eruption and locations where crypto-tephra from the Vedde and 1783-84 AD Laki eruptions has been found in sediments. The distribution of the De Wijk Member containing Paleocene-Eocene tephra or tuff is indicated based on TNO-GDN (2023). (tephra: Davies et al. (2005); volcanic minerals: Burger (1979, 1992); Laki: Cremer et al. (2010), Andronikov et al. (2016); Belvédère: Meijs (1985)).*

canic minerals is dominated by clinopyroxene, olivine and brown amphibole, whereas titanite, apatite and traces of phlogopite, prehnite, haüyne, tridimite, nepheline and K-feldspars have been identified as well (Juvigné & Semmel, 1981; Meijs et al., 1983; Pouclet & Juvigne, 2009). The tephra is a prominent stratigraphic marker with an age of 25.6-23.2 ka obtained from luminescence dating (Zens et al., 2017). Meijs (1980, 2011), Meijs et al. (1983) and Pouclet & Juvigné (2009) reported its presence in quarries and sections around Maastricht (Nagelbeek, 't Rooth (Fig. 11.16a), Biesland, Lixhe, Veldwezelt, wall of the Albert Canal near Vroenhove).

The Late Pleistocene Laacher See Tephra (Fig. 11.15a) is a prominent stratigraphic marker in large parts of Central Europe. Three main events of violent explosive activity from a centre in the East Eifel Volcanic Field produced ash clouds fanning in different directions and fall-out up to more than 1000 km from the source (Wörner & Schmincke, 1984; Van den Bogaard & Schmincke, 1990). Owing to the phonolitic composition of glass shards, the dominance of clinopyroxene, amphibole and sphene (titanite) as heavy minerals and sanidine and plagioclase as light phases, the Laacher See Tephra is easily distinguishable from ashes with a similar age. The chemical and mineralogical composition changed in the course of time from highly differentiated to mafic phonolite. Recent high-precision dendrochronological and ^{14}C dating yielded an eruption age of 13,006 ± 9 calibrated years before present (Reinig et al., 2021).

Blockage of the Rhine River by abundant pyroclastic flows and rapid deposition of tephra fall near the crater, followed by breaching of the pumice dam and catastrophic flooding, discharged large amounts of tephra into the lower Middle Rhine Valley (Schmincke et al., 1999). Further reworking explains why Rhine sediments in the Netherlands contain Laacher See pumice granules from the Late Allerød onwards (Busschers et al., 2007; Autin, 2008; Fig. 11.16b). Layers with glassy pumice fragments containing phenocrysts of augite and 'basaltic hornblende' were encountered in river sediments from several locations near Arnhem (Crommelin, 1963; Erkens, 2009).

Airborne Laacher See tephra has been identified in the form of volcanic heavy minerals (clinopyroxene and brown amphibole) in the loess succession of the Veldwezelt section that also contains the Rocourt and Eltville tephras (Meijs, 2011). Trace-element signatures of Late Pleistocene eolian cover sands (Lutterzand in the easternmost part of the Netherlands and sites near Lommel in NE Belgium) point to the presence of a volcanic component that may also be derived from the Laacher See Tephra (Andronikov et al., 2016).

Layers with distal (crypto-)tephra have also been identified in a Late Pleistocene peat and gyttja infill of a pingo-remnant at Kostverloren, in the province of Drenthe (Fig 11.15d; Davies et al., 2005). Geochemical fingerprinting identified glass shards of one horizon as the rhyolitic component of the bimodal (basaltic-rhyolitic) Vedde Ash (mid-Younger Dryas), an important regional stratigraphic marker in the North Atlantic, the Norwegian Sea, and the adjacent land area, dated at 12 ka (e.g. Lohne et al., 2013). These Vedde Ash particles are fall-out from an ash cloud thought to have originated from a major explosive eruption in southern Iceland (probably from Katla Volcano), which drifted in south-easterly directions over northern Britain, southern Scandinavia and western Russia. Two other tephras in the Kostverloren succession could not be geochemically characterized but one has tentatively been attributed to the Laacher See Tephra (Davies et al., 2005).

A Holocene micro-tephra has been preserved in the form of glass shards in a sedimentary succession dominated by gyttja, filling in a historic scour-hole lake near Haarsteeg (Fig. 11.15d; Noord Brabant). From chronostratigraphic evidence the tephra corresponds to the large 1783-1784 AD volcanic eruption of the Laki volcano in Iceland (Cremer et al., 2010). This event produced a dry sulphuric fog across large parts of Europe (Thordarson & Self, 1993) with a severe environmental impact. Brugmans (1784) provided a remarkably detailed historic account on

Table 11.2. Cenozoic tephra in the Dutch subsurface with sources and dated eruption events. Note that ages are mostly based on dating of equivalent deposits outside the Netherlands (see references). Emplacement is based on available evidence.

Eruptive event/marker	Location	Well/sample depth (m)	Age	Dating method
Laki eruption	Haarsteegse Wiel (scour lake)	Northern basin (HA-07-05)/ 2.60-2.70	1783-1784 CE	Age-depth relationships and interpolation between various historic events recorded in lake sediments
Vedde Tephra – probably from Katla Volcano	Kostverloren Veen (peat bog)	N/A	12,066±42 cal yr BP	Tephrochronology (Kostverloren Veen) and bayesian age-depth modelling of AMS ^{14}C dates from Icelandic lake sediments hosting Vedde-ash
Laacher See eruption	Rhine-Meuse delta	N/A	13,006±9 cal yr BP	^{14}C and dendrochronology
Eltville Tephra	South Limburg loess	N/A	25.6-23.2 ka	Bayesian age-depth modelling of luminescence ages of minerals above and below the tephra in loess-paleosol sequences between eastern Belgium and central Germany
Rocourt Tephra	South Limburg loess	N/A	90.3-74 ka	Stratigraphic bracketing with thermoluminescence dating of underlying limestone clasts and ^{14}C dating of an overlying peat layer
Tephra in Veldhoven Formation	Subsuface Winterswijk region (East Netherlands)	Beltrum ('100') B34D0232/ 131.5-136.5	28-16 Ma (?)	Age estimate based on formation age
		Lievelde ('101') B41B0073/ 92.5-93.5	28-16 Ma (?)	Age estimate based on formation age
Tephra in Boom Member of Rupel Formation	Onshore and offshore subsurface Netherlands	Holostratotype (B51D0127, Veldhoven-01) and other wells	44-35, 30-22 Ma	Inferred from $^{40}Ar/^{39}Ar$ ages of possible centres in Siebengebirge and Hocheifel (Germany)
North East Atlantic Province tephra	Onshore and offshore subsurface north Netherlands	Holostratotype (BL020070, L02-04) and other wells	54.04 ± 0.14 Ma	$^{40}Ar/^{39}Ar$ age of '+19 ash' from the Fur Formation (Denmark), presumably equivalent to De Wijk ash

***Figure 11.16.** a) Thin layer of Eltville tephra in a loess succession in quarry 't Rooth (Province of Limburg). Luminescence dating at locations in Germany and Belgium yielded an age of 25.6-23.2 ka for the tephra (Zens et al., 2017). Diameter of coin: 1.9 cm. Photograph courtesy of S.B. Kroonenberg; b) Pumice in fluvial deposits of the Rhine (Kreftenheye Formation, 'terrace X-deposits', location unknown). The pumice particles represent reworked material from pyroclastic-flow deposits of the Laacher See Volcano, which erupted 13,006±9 cal yr BP according to dendrochronological and radiocarbon age dates obtained from subfossil trees buried by pyroclastic deposits near the eruption centre (Reinig et al., 2021). Photograph from Berendsen & Stouthamer (2001), lacquer-peel made by J. van der Staay (Geological Survey of the Netherlands). Frame size: ~100 x 60 cm.*

Source region	Host formation	Chronostratigraphy of host interval	Emplacement	Deposit type/composition	Reference(s)
Iceland	Echteld Formation	Holocene	Airfall	Micro-tephra (glass shards)	Cremer et al. (2010)
Iceland	Nieuwkoop Formation	Pleistocene	Airfall	Cryptotephra (rhyolitic glass shards)	Davies et al. (2005); Lohne et al. (2013)
East Eifel	Kreftenheye Formation	Pleistocene	Reworked airfall	Reworked felsic pumice particles	Verbraeck (1984); Autin (2008); Meijs (2011); Reinig et al. (2021)
West or East Eifel volcanic field	Boxtel Formation	Pleistocene	Airfall	Tephra layer containing mafic volcanic minerals	Meijs et al. (1983); Zens et al. (2017)
Eifel (location unknown)	Boxtel Formation	Pleistocene	Airfall	Reworked tephra containing altered glass shards and a mineral assemblage including orthopyroxene	Mees & Meijs (1984); Pouclet et al. (2008); Meijs (2011)
Eifel (?)	Veldhoven Formation	Oligocene-Miocene	Airfall?	Igneous minerals	Burger (1992); Van den Bosch (2015)
Eifel (?)	Veldhoven Formation	Oligocene-Miocene	Airfall?	Igneous minerals	Burger (1979); Van den Bosch (2015)
Siebengebirge and/ or Hocheifel (?)	Rupel Formation	Rupelian	Airfall or reworked airfall	Volcanic heavy minerals and altered trachytic-basaltic volcanic ash	Zimmerle (1993); Fekiacova et al. (2007); Przybyla et al. (2018)
NE Atlantic	Dongen Formation (De Wijk Member)	Eocene	Airfall	Distal ash	Chambers et al. (2003)

an array of adverse effects in the Netherlands. The 2010 eruption of Eyjafjallajökull volcano illustrates that explosive eruptions on Iceland occasionally have produced fallout from ash clouds in the Netherlands in recent times.

Indirect evidence for other magmatic intrusions

Apart from igneous rocks encountered in hydrocarbon wells and seismic data and the evidence for Paleogene dykes extending in the southern North Sea, the presence of intrusive magmatic rocks in the Dutch subsurface has also been inferred from geophysical anomalies and localized heating effects on organic matter. The presence of two sizeable intrusive bodies has been postulated along the border with Germany (Fig. 11.1). A coalification anomaly in upper Carboniferous sediments from wells in the Ems estuary (northeastern Groningen) and the adjacent part of Germany defines the East Groningen Massif (Kettel, 1983). A positive magnetic anomaly and structural contours of the top of the Rotliegend, inferred for Jurassic times, outline the shape of this anomaly (Fig. 11.7, also visible in Fig. 11.10b). This, in combination with the timing of the Zuidwal, Andel and Loon op Zand igneous activity, led Kettel (1983) to assume that the East Groningen Massif represents an intrusive body emplaced around the Jurassic-Cretaceous transition. It cannot be excluded, however, that the inferred intrusion is older and could have an early Permian age (Van Wijhe et al., 1980).

The Erkelenz intrusion near the Peel Boundary Fault Zone of the Roer Valley Graben in the province of Limburg and neighbouring Germany has been identified from geophysical surveys (Fig. 11.7). A pronounced positive magnetic anomaly, in combination with a minor residual gravity anomaly, favours a granitic body as interpretation (Bredewout, 1989). The initial intrusion temperature was 800 ± 100 °C, based on cooling modelling and vitrinite reflection data collected from local coal seams (Erren & Bredewout, 1991). The timing of the intrusion is unknown, but fission-track data and the degree of coalification in overlying sediments point to a heat pulse between late Carboniferous and mid-Cretaceous times (Teichmüller & Teichmüller, 1971; Bredewout, 1989). Assuming a granitic composition, a Permian age of the intrusion is plausible, given the widespread manifestations of 'Rotliegend' acidic magmatism in northwest Europe, particularly near the Variscan front. In the Campine Basin, across the southwestern boundary fault of the Roer Valley Graben, a magnetic anomaly near Bilzen (Dusar & Langenaeker, 1992), and a gravimetric anomaly near the city of Maastricht (see Textbox) also hint at the possible presence of intrusive bodies. A continuous positive pre-Permian residual gravity anomaly largely follows the Dutch onshore and offshore rift pattern and could signal the presence of high-density magmatic intrusions in the crust, but Moho shallowing cannot be ruled out as an alternative explanation (Dirkzwager et al., 2000).

Indirect geophysical evidence points to the presence of buried, mainly granitic intrusions in the domain of the

Anglo-Brabant Massif below the Southern North Sea Basin (e.g. Rijkers & Duin, 1994). They must be pre-Devonian and are probably equivalent to igneous rocks encountered in onshore drill holes and outcrops in east England and Belgium (e.g. Pharaoh et al., 1993; De Vos et al., 2010). Similar geophysical interpretations in areas around the Mid North Sea high (Donato et al., 1983; Kimbell & Williamson, 2015) also suggest that buried granites within the Paleozoic basement may be more widespread than the single occurrence in the A17-01 well. Structural highs such as the Dogger High may well have a granite core. It is of interest to note that the decay of uranium, thorium and potassium isotopes in granitic rocks produces radioactive heat that may drive hydrothermal circulation long after

A granite intrusion below Maastricht?

Nearly a hundred years ago, the front page of a newspaper from The Hague reported on a surprising discovery in a 330 m deep borehole in the city of Maastricht (De Avondpost, 1 February 1930). In a search for new drinking water resources, subsurface water with a temperature of 18°C was sampled, about eight degrees higher than expected. According to a laboratory in Berlin it had medicinal qualities and could be bottled similarly to what was then already practiced in the nearby town of Spa. The well (Kastanjelaan-1) produced the local Trega mineral water until 1960.

In the 1980s two new boreholes were drilled in an attempt to identify a successor to this well: Kastanjelaan-2 and Heugem-1. They were cored over great lengths and investigated in detail under supervision of Martin Bless of the Natural History Museum of Maastricht (Bless et al., 1981). Borehole Heugem-1 drew particular attention, as a temperature of 20-21°C was recorded at 500 m depth. Even more striking was the high degree of coalification in samples of the Dinantian limestone host rock. Measured vitrinite reflectance of 5-7% (and a >10% outlier) fall in the (meta-)anthracite range, suggesting temperatures up to 300°C and burial down to 10 km. A sequence of strong tectonic burial, followed by uplift, could have generated such a high coalification but does not fit with the regional geodynamic history.

It is also conceivable that a granitic intrusion is responsible for the heat. A gravity low beneath Maastricht suggests its presence (Mansy et al., 1999). Interestingly, a magnetic anomaly around Bilzen just west of the city across the Belgian border (Dusar & Langenaeker, 1992) has characteristics similar to the Erkelenz anomaly, which has been linked to a hidden granitic body (Bredewout, 1989), possibly related to a Permo-Carboniferous intrusion *(see main text)*. Unfortunately, drilling could not conclusively confirm this for Bilzen, as borehole Martenslinde ended in Cambrian quartzite at only 300 m depth (Dusar & Langenaeker, 1992).

If a magma body intruded underneath Maastricht in, say, Permo-Carboniferous times, it will have cooled down and solidified long ago. This could explain the anomalous coalification, but seemingly not the elevated temperature of the subsurface water. However, ancient granitic rock bodies often continue to produce heat today, as they are rich in the radioactive isotopes ^{235}U, ^{238}U, ^{232}Th and ^{40}K with half-lives between 700 and 14,000 million years. Upward flowing fluids could transfer this heat to the surface through steep faults. Since fault locations are poorly known in the Maastricht region, could the warm water hint at a hidden fault? On the other hand, how certain is the presence of a granite body? Anomalous coalification ranks have also been recorded in Dinantian shales and carbonates from the Thermae-2000 and Thermae-2002 wells (Wolf & Bless, 1987).

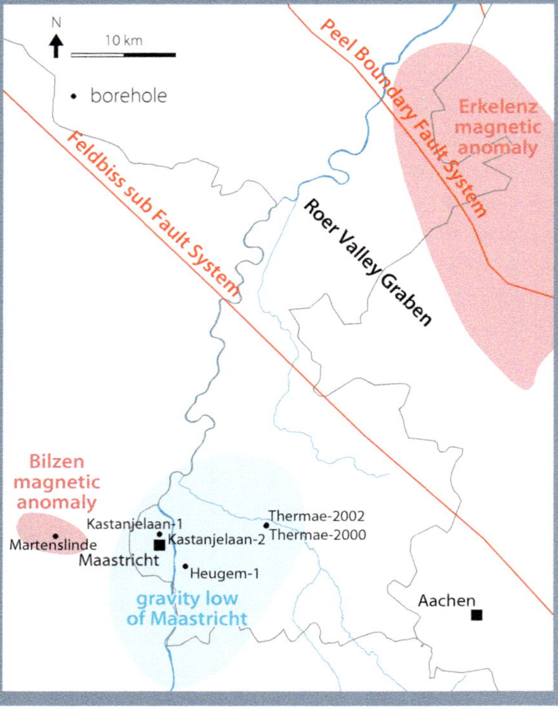

Obvously, further research is needed to test the validity of a Trega water-granite connection.

the intrusion event. This may be sufficient to generate local areas of anomalously high organic maturity, or remobilization of Zechstein salts and carbonates (Parnell, 1988).

Geochemical signatures

The petrographic observations show that secondary alteration is a ubiquitous feature in all the Paleozoic and Mesozoic igneous rocks listed in Table 11.1. Hence, many of the reported geochemical compositions of bulk-rock samples are expected to reflect modifications of original signatures to a certain extent (Fig. 11.17). Alteration usually affects mobile elements such as Si, Mg, Ca, K, Na, Sr, Rb, Ba, which may either be added or removed. Standard classifications and petrogenetic interpretations based on these elements will therefore not always be unambiguous. Instead, inferences from relatively immobile minor and trace elements such as Ti, P, Zr, Nb, Y and the rare-earth elements (REE) may be more trustworthy, as these are least sensitive to alteration processes. Compositions of the igneous rocks in the Dutch subsurface are compiled in Table 11.3. The available data show a clear geochemical distinction between Paleozoic and Mesozoic occurrences, as is illustrated in Figs 11.18 and 11.21.

Carboniferous-early Permian

Carboniferous-early Permian magmatic rocks (in Wanneperveen-1, De Wijk-7, Dwingelo-2, Corle-1, Gelria-3, F04-02-A) are subalkaline to mildly alkaline, showing compositional variability, both between different locations and among samples from individual wells. In a TAS classification diagram (wt.% SiO_2 vs Na_2O+K_2O; Fig. 11.18a) they tend to concentrate in the basalt and trachybasalt fields. It should be noted that at least part of the scatter in this diagram is attributable to effects of secondary alteration. The few analysed samples with more evolved compositions (trachyte-trachydacite) are from offshore well F04-02-A (rim of the Dutch Central Graben) and onshore well De Wijk-7 (east of Texel-IJsselmeer High). The additional presence of predominantly basaltic rocks in the latter well signals a bimodal distribution, which is also seen in the Horn Graben (Danish offshore) and in Carboniferous-early Permian sills and dykes in Scotland and southern Sweden (Kirstein et al., 2006). Immobile elements further constrain the geochemical affinity of the Dutch Permo-carboniferous rocks. In a Nb/Y vs. Zr/Ti discrimination diagram (Fig. 11.18c) the mafic rocks straddle the basalt-alkali basalt boundary, whereas the more felsic rocks plot close to the trachyte-trachyandesite-rhyolite+dacite junction. Samples from offshore wells in other sectors of the North Sea show a similar distribution. In contrast, the available data also indicate that the Dutch onshore

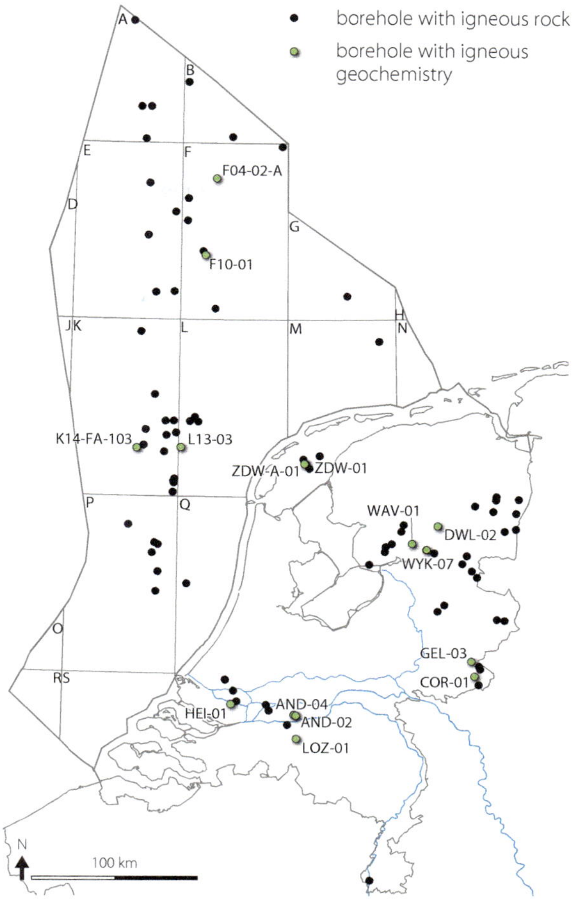

Figure 11.17. Locations of wells with geochemical analyses on core samples of igneous rocks(see Table 11.3).

occurrences are distinct, as they tend to have lower concentrations in TiO_2 and P_2O_5 (Fig. 11.19c), as well as in Zr, Nb, Y, LREE (light rare-earth elements) than Rotliegend rocks from neighbouring North Sea areas if only the relatively mafic samples (>4 wt.% MgO) are considered. Rare-earth-element patterns display a modest LREE enrichment and a relatively flat HREE (heavy rare-earth elements) (Fig. 11.20a). These trends are comparable to those of the subalkaline basalts in Central Graben UK well 39/2-4, but LREE/HREE ratios and total REE concentrations tend to be lower than in the majority of equivalent igneous rocks of northwestern Europe. A final point to note is the significant variation in major and trace element contents of the intrusive rocks from Wanneperveen-1 (Table 11.3a). Several Wanneperveen-1 samples, as well as the Dwingelo-2 rock, are marked by very high MgO (>15 wt.%), presumably due to accumulation of olivine phenocrysts under magmatic conditions.

Abundant geochemical data on Carboniferous-early Permian igneous occurrences encountered in drill cores and outcrops in northwest Europe and available for com-

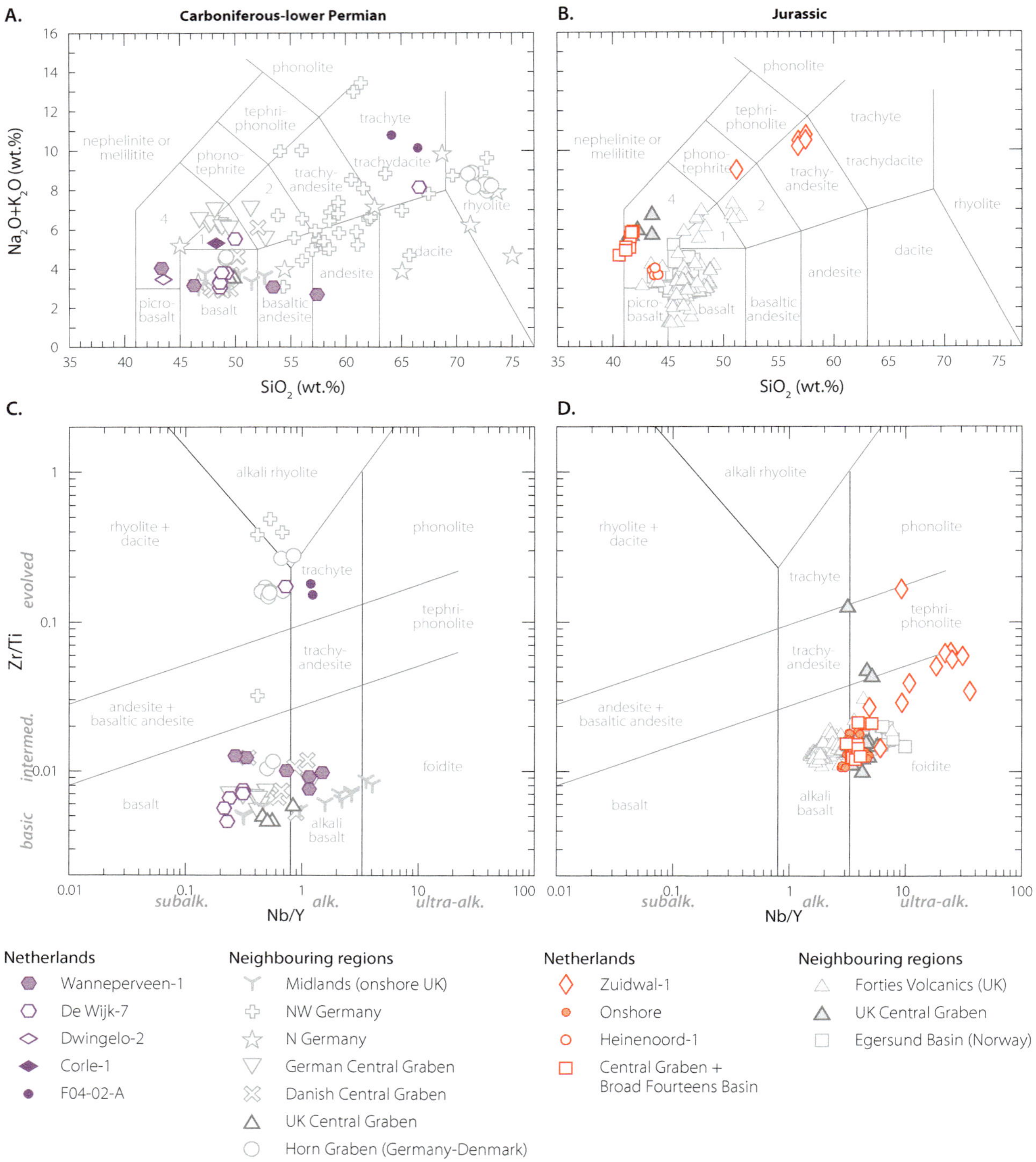

Figure 11.18. *Major and trace-element classification diagrams showing Carboniferous-lower Permian and Jurassic igneous rocks from Dutch exploration wells onshore and offshore, in comparison to rocks from neighbouring regions in northwestern Europe. See Table 11.3 for data sources of the Dutch rocks. Data for neighbouring regions are from Latin (1990), Eckhardt (1979), Aghabawa (1993), Kirstein et al. (2006). (a) and (b) TAS classification diagrams (Le Maitre et al., 2002) for relatively fresh (LOI<6 wt.%) Carboniferous-lower Permian and Jurassic rocks, respectively; (c) and (d) Nb/Y-Zr/Ti classification diagrams (Pearce, 1996); note the subalkaline to mildly alkaline character and the bimodal distribution of the Carboniferous-lower Permian rocks and the (ultra-)alkaline signatures of the Jurassic rocks. 1 = trachybasalt, 2 = basaltic trachyandesite, 3 = tephrite or basanite.*

parison have been obtained from intrusive and extrusive rocks from mainland England and Scotland, the region around the Mid North Sea High, Denmark and offshore surroundings, northern Germany, the Kattegat area, the Oslo Graben and Scania in southern Sweden (Dixon et al., 1981; Latin, 1990; Aghabawa, 1993; Benek et al., 1996;

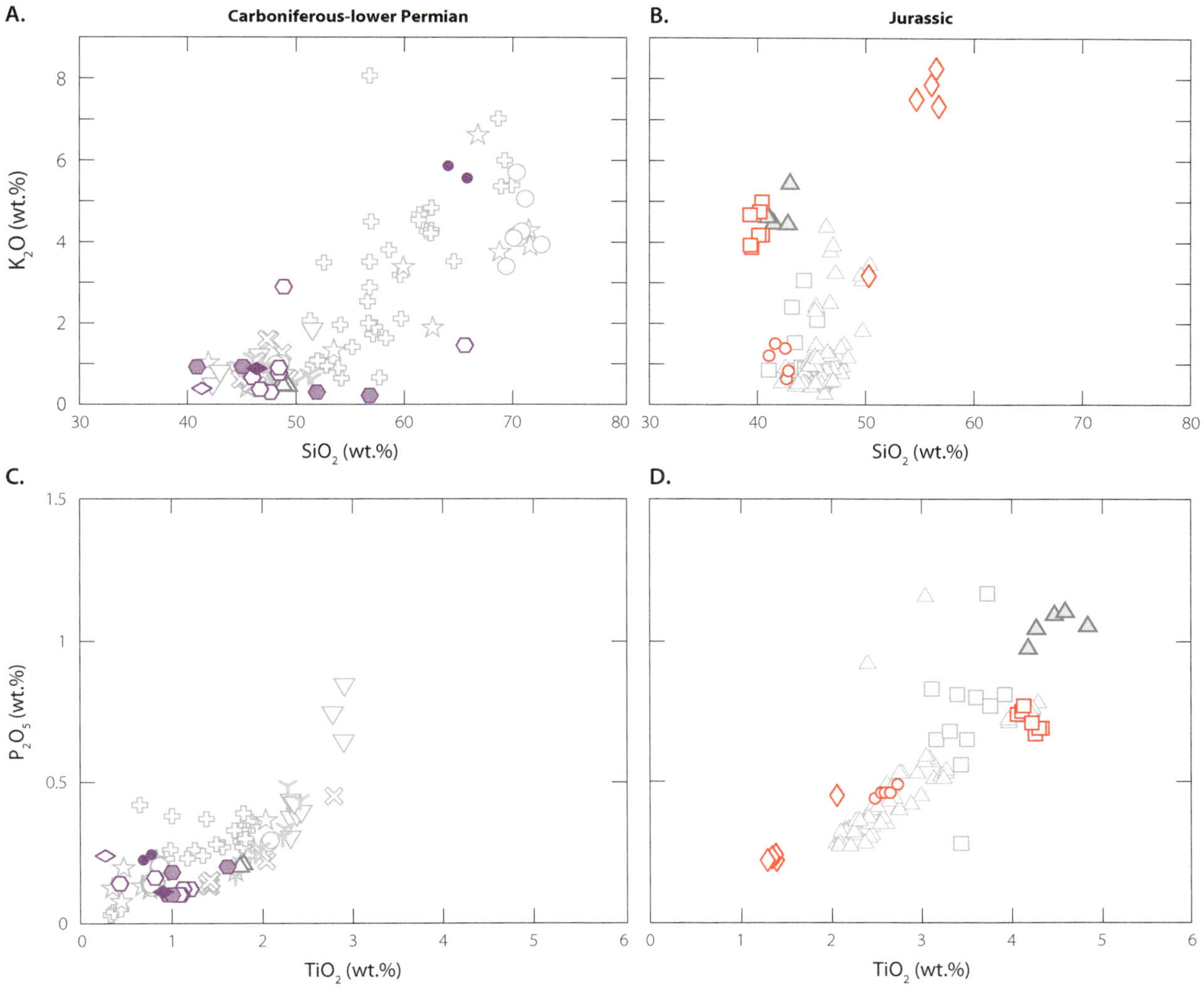

Figure 11.19. Major and minor element variation diagrams showing compositional differences between Carboniferous-lower Permian and Jurassic igneous rocks from Dutch exploration wells onshore and offshore, in comparison with rocks from neighbouring regions in northwestern Europe: a) and b) SiO_2-K_2O diagram for Carboniferous-lower Permian and Jurassic rocks, respectively; c) and d) TiO_2-P_2O_5 diagrams illustrating variations of minor oxides relatively insensitive to the effects of secondary alteration. For legend and data sources, see Fig. 11.18.

Heeremans et al., 2004; Obst et al., 2004; Kirstein et al., 2006). They are thus representative for the foreland of the Variscan orogeny.

Mafic rocks from different geographic locations in the North Sea area show significant compositional variation, while diversity in geochemical affinity may also be seen in lava flows, sills or dykes sampled in individual wells. The Central Graben area is illustrative, as most of the lavas are mildly alkaline trachybasalts-tephrites/basanites that are associated with subordinate amounts of subalkaline and highly alkaline basalts or transitional varieties. The ca. 299 Ma old samples from well 39/2-4 on the western flank of the UK Central Graben are an example of subalkaline (tholeiitic) basalts with low abundances of incompatible trace elements, including the light rare-earth elements (REE). They deviate from normal mid-ocean-ridge basalts (N-MORB) in being more enriched in light REE and more depleted in heavy REE (Heeremans et al., 2004).

Immobile-trace-element signatures confirm a 'within-plate' tectonic affinity for virtually all the Carboniferous-early Permian magmatic rocks in the Variscan foreland region (Fig. 11.21). Basalts in the northeast German Basin are an exception, as they show a wide diversity in tectonic discrimination diagrams (Benek et al., 1996).

The trachytic lava in F04-02-A (Fig. 11.1) is most likely the only 'Rotliegend' igneous rock analysed from the Dutch subsurface, since compositional data on rocks from the Emmen Volcanic Formation in onshore wells are lacking and the studied intrusions in the eastern part of the Netherlands presumably have a Carboniferous age. It is reasonable to assume that the igneous rocks of the Emmen Volcanic Formation are similar to (sub-)volcanics in

Table 11.3. Geochemical compositions of igneous rocks in onshore and offshore wells in the Netherlands: a) Inventory of major and trace element compositions; b) Rare-earth-element concentrations. Source of data and analytical techniques applied: 1 = Latin (1990), major elements by XRF on glass disks, trace elements by XRF on pressed powder pellets, LOI represents weight loss after ignition at 1100°C, all XRF results reported on anhydrous basis, REEs determined by ICP-AES (results reported on hydrous basis); 2 = Helmers (1991, internal report), major elements by XRF on glass disks, trace elements by XRF (no further information); 3 = Eigenfeld & Eigenfeld-Mende (1986), analytical method not reported; 4 = this work (VU Amsterdam), major elements by XRF on glass disks, trace elements by XRF on pressed powder pellets, REEs determined by solution ICP-MS (Klaver et al., 2018), major element totals based on subtracted LOI. [1] All Fe is expressed as FeO or Fe_2O_3, except when data for both oxides are given; n.a. = not available; <d.l. = below detection limit. Well names for Carboniferous samples from Latin (1990) are inferred from Aghabawa (1993); Latin (1990) listed these samples as 'Permian'. Sample codes and rock types taken from the references.

Igneous episode	LATE JURASSIC												
Well name	F10-01							K14-FA-103			L13-03		
Depth (m)	n.a.	n.a.	n.a.	n.a.	n.a.	3451	3458	n.a.	n.a.	n.a.	n.a.	n.a.	n.a.
Sample code	**PL1/5**	**PL1/6**	**PL1/10**	**PL1/11**	**PL1/12**	**F10-6**	**F10-7**	**DL1**	**DL2**	**DL2A**	**001-1A**	**001-1B**	**001-1C**
Rock type	Lamprophyre	Lamprophyre	Lamprophyre	Lamprophyre	Lamprophyre	Foïdite	Foïdite	Hawaiite-Basanite(?)	Basanite(?)	Basanite(?)	?	?	?
Reference	**1**	**1**	**1**	**1**	**1**	**4**	**4**	**1**	**1**	**1**	**1**	**1**	**1**
Major elements (wt.%)													
SiO_2	39.47	40.47	40.44	40.22	40.13	39.37	39.34	32.56	42.78	51.11	39.95	39.88	39.05
Al_2O_3	14.41	14.13	14.59	14.76	14.45	13.90	14.11	18.73	18.28	19.78	15.15	14.82	15.33
Fe_2O_3 [1]	14.04	13.92	14.65	14.50	14.64	13.89	13.85	36.90	27.23	16.46	13.22	12.61	12.50
FeO [1]	-	-	-	-	-	-	-	-	-	-	-	-	-
MgO	11.80	10.80	10.51	10.67	10.74	11.28	10.58	1.87	1.78	1.26	17.78	18.15	20.30
CaO	9.11	9.44	7.20	7.55	8.78	8.67	7.23	1.61	1.65	0.90	3.83	4.33	1.28
Na_2O	0.67	0.76	0.71	0.67	0.78	0.77	0.83	0.34	0.49	0.43	1.09	1.16	0.65
K_2O	3.87	4.18	4.99	4.75	4.18	3.93	4.68	0.38	1.27	2.28	2.60	2.73	3.26
TiO_2	4.33	4.26	4.06	4.11	4.30	4.22	4.13	3.49	3.48	3.64	3.80	3.69	3.93
MnO	0.28	0.27	0.24	0.24	0.24	0.28	0.22	1.35	0.96	0.64	0.12	0.12	0.12
P_2O_5	0.69	0.67	0.74	0.75	0.69	0.71	0.77	0.87	0.87	1.36	0.79	1.00	1.61
TOTAL	98.66	98.89	98.12	98.22	98.93	97.02	95.74	98.10	98.79	97.85	98.34	98.50	98.01
LOI	3.7	3.4	4.2	3.9	3.2	3.75	4.05	20.4	16.8	7.8	8.5	8.8	9.1
H_2O^+	-	-	-	-	-	-	-	-	-	-	-	-	-
CO_2	-	-	-	-	-	-	-	-	-	-	-	-	-
Trace elements (ppm)													
Ni	43	44	34	39	46	38	31	39	56	36	139	138	123
Cr	27	23	19	22	32	24	17	270	264	252	378	361	379
V	484	476	459	461	501	411	387	398	346	267	397	381	392
Sc	35	34	26	24	36	37	32	22	17	9	35	32	34
Cu	52	66	90	81	69	84	55	9	7	8	20	18	24
Zn	119	112	80	114	94	117	80	47	23	10	84	79	39
Sr	651	796	1174	895	883	613	880	1432	1180	6114	589	577	156
Rb	62	57	61	68	63	60	63	19	54	84	25	26	56
Zr	325	320	321	323	318	303	312	434	443	456	345	338	335
Nb	98	96	103	102	94	95	103	154	156	164	115	111	117
Ba	530	614	814	808	760	443	669	142	265	387	724	685	845
Pb	5	3	4	4	2	1.7	1.3	3	7	1	12	21	426
Th	7	6	5	7	6	13	15	6	11	-	8	8	15
La	87	81	81	83	67	102	101	117	108	333	113	99	96
Ce	177	176	182	173	160	154	162	217	198	674	201	180	160
Nd	67	71	69	70	67	67	69	76	65	260	76	69	55
Y	28	28	27	28	27	26	25	35	40	32	30	36	30

Table 11.3 a) Continued

LATE JURASSIC																		
Zuidwal-1 (ZDW-01)										ZDW-A1	Loon-op-Zand-1 (LOZ-01)							
n.a.	n.a.	n.a.	n.a.	n.a.	n.a.	n.a.	n.a.	1958	2180	1950	n.a.	n.a.	n.a.	n.a.	n.a.	n.a.	n.a.	2590	2590
ZU1	**ZU2**	**ZU3**	**ZU4**	**ZU5**	**ZU6**	**ZU7**	**ZU8**	**ZDW-2**	**ZDW-4**	**ZDW-5**	**NAM 2/2**	**NAM 2/3**	**NAM 2/7**	**NAM 2/8**	**DL3**	**DL4**	**DL5**	**LOZ-11**	**LOZ-12**
Phonolite	Phonolite	Phonolite	Phonolite	Phonolite-Basanite	Phonolite	Lamprophyre(?)	Lamprophyre(?)	Trachyandesite	Altered andesite	Trachyte	Basanite	Basanite	Basanite	Basanite	Basanite	Basanite	Basanite	Altered basaltic andesite	Altered basaltic andesite
1	**1**	**1**	**1**	**1**	**1**	**1**	**1**	**4**	**4**	**4**	**1**	**1**	**1**	**1**	**1**	**1**	**1**	**4**	**4**
52.78	56.08	56.50	56.74	50.30	55.75	28.51	28.75	54.69	55.95	44.10	39.04	40.71	40.60	35.43	38.05	37.42	43.44	37.44	36.31
26.71	25.45	25.56	24.33	17.71	18.94	15.73	16.76	23.85	13.41	22.02	17.13	18.14	18.32	16.34	16.35	17.16	18.68	16.03	16.27
7.77	4.39	4.77	4.49	8.38	4.14	18.02	17.54	4.20	7.72	10.53	12.40	12.58	13.26	14.18	12.14	14.18	14.06	12.56	12.23
-	-	-	-	-	-	-	-	-	-	-	-	-	-	-	-	-	-	-	-
0.67	0.66	0.80	0.78	4.78	5.72	9.61	9.74	0.90	5.57	1.08	6.05	6.16	6.60	7.18	6.41	7.03	7.74	6.26	5.81
0.65	0.63	0.62	0.62	6.43	1.69	16.37	16.49	0.46	3.65	1.51	15.29	12.36	11.75	17.24	17.36	14.34	9.27	10.24	10.75
1.14	2.49	1.89	3.31	5.67	4.99	0.62	0.41	2.51	1.32	0.75	1.70	2.19	1.70	1.40	1.73	1.57	1.83	1.81	1.59
7.14	7.86	8.25	7.33	3.17	7.28	3.69	3.84	7.50	4.30	5.81	0.93	0.75	0.77	0.74	0.71	0.68	0.81	0.75	0.71
1.52	1.39	1.38	1.34	2.06	0.33	3.15	3.15	1.29	1.28	1.89	3.61	3.94	4.04	3.48	3.48	3.62	3.91	3.31	3.56
0.37	0.02	0.03	0.02	0.17	0.12	0.49	0.47	0.03	0.12	0.55	0.20	0.17	0.16	0.22	0.21	0.18	0.12	0.14	0.15
0.27	0.22	0.24	0.23	0.45	0.03	0.37	0.41	0.22	0.24	0.42	0.96	1.01	1.02	0.88	0.89	0.92	1.00	0.87	0.92
99.03	99.18	100.04	99.18	99.14	99.00	96.56	97.56	95.65	93.56	88.66	97.30	98.02	98.21	97.10	97.32	97.11	100.88	89.41	88.30
8.5	5.0	4.7	4.7	4.0	6.0	21.5	21.4	4.92	6.35	10.80	14.6	13.0	12.5	15.4	15.4	13.0	9.8	10.86	12.22
-	-	-	-	-	-	-	-	-	-	-	-	-	-	-	-	-	-	-	-
-	-	-	-	-	-	-	-	-	-	-	-	-	-	-	-	-	-	-	-
21	50	44	59	40	39	267	254	50	112	34	142	147	148	188	181	181	151	161	119
4	3	5	4	9	120	1881	2098	5.8	313	50	274	303	303	271	294	302	312	211	238
157	218	203	200	135	238	480	484	166	159	236	377	406	421	363	375	396	439	290	341
-	-	-	-	-	13	30	28	0.4	18	5.4	27	36	37	35	29	31	38	34	38
802	11	294	12	143	5	182	446	11	14	26	53	56	55	46	49	54	58	47	56
137	145	175	164	137	92	213	200	165	131	841	112	79	85	103	64	119	79	94	74
2634	3086	2995	3043	2501	1992	650	377	2934	621	533	895	1027	930	826	949	852	823	704	727
201	176	191	165	150	64	170	176	167	161	186	27	20	21	19	20	19	24	23	21
459	522	511	450	423	329	282	287	457	206	440	293	316	323	265	293	291	311	249	256
331	319	307	302	464	157	63	63	302	83	248	113	116	117	99	103	109	109	98	96
1630	1982	1862	2162	3616	1469	3215	1488	2253	933	851	976	1013	1033	952	854	905	1064	871	727
11	12	12	12	4	9	3	4	12	14	64	9	6	6	8	3	5	4	4.0	2.8
26	21	20	18	13	15	-	-	41	15	36	6	6	7	3	5	5	8	13	14
90	72	64	93	137	109	59	34	131	77	236	94	88	89	88	89	94	92	102	93
180	152	135	175	190	178	62	53	160	99	304	177	166	181	190	168	180	157	141	145
53	43	32	42	30	45	20	19	44	38	95	74	60	57	72	64	70	57	53	57
18	13	14	12	13	17	18	19	10	17	23	33	27	27	31	30	31	24	20	20

Table 11.3 a) Continued

Igneous episode	LATE JURASSIC									
Well name	Andel-2 (AND-02)			Andel-4 (AND-04)		Heinenoord-1 (HEI-01)				
Depth (m)	n.a.	n.a.	1368	n.a.	n.a.	2236,5	2238,5	2237	2242	2249
Sample code	DL6	DL7	AND-19	DL8	DL9	HEI-1	HEI-1	HEI-16	HEI-17	HEI-18
Rock type	Alkali basalt-Basanite	Alkali basalt-Basanite	Trachy-basalt	Basanite	Basanite	Alkali basalt	Alkali basalt	Basanite	Basanite	Basanite
Reference	1	1	4	1	1	2	2	4	4	4
Major elements (wt.%)										
SiO_2	42.23	39.63	39.27	41.07	40.17	41.18	41.74	42.67	42.78	42.97
Al_2O_3	19.85	20.70	15.50	19.73	17.82	12.34	12.79	12.69	12.86	13.29
Fe_2O_3 [1])	13.64	14.34	11.17	12.01	14.15	10.84	11.00	11.17	11.36	11.32
FeO [1])	-	-	-	-	-	-	-	-	-	-
MgO	5.66	5.40	5.40	4.43	7.14	12.57	12.22	12.94	12.15	12.32
CaO	7.99	9.01	7.12	11.43	9.02	11.59	12.12	12.24	12.06	11.85
Na_2O	2.28	1.43	2.35	1.62	1.80	2.34	2.23	2.16	2.91	3.08
K_2O	2.38	2.45	2.11	1.70	1.71	1.18	1.48	1.36	0.61	0.81
TiO_2	3.61	3.74	2.84	4.77	4.28	2.48	2.55	2.59	2.65	2.73
MnO	0.21	0.23	0.17	0.18	0.17	0.17	0.17	0.17	0.18	0.17
P_2O_5	0.77	0.81	0.63	0.99	0.89	0.44	0.46	0.46	0.46	0.49
TOTAL	98.63	97.73	86.56	97.93	97.14	95.12	96.76	98.45	98.02	99.03
LOI	16.5	18.2	13.72	15.9	15.0	3.07	2.43	2.75	3.60	3.28
H_2O^+	-	-	-	-	-	-	-	-	-	-
CO_2	-	-	-	-	-	-	-	-	-	-
Trace elements (ppm)										
Ni	126	172	125	190	218	-	-	243	247	220
Cr	236	283	178	645	517	-	-	587	593	333
V	388	401	274	522	461	-	-	248	253	250
Sc	36	34	31	53	43	-	-	38	36	34
Cu	53	55	48	70	66	-	-	61	91	77
Zn	157	132	96	71	135	-	-	77	76	75
Sr	732	631	1165	1302	1430	1360	970	1595	692	722
Rb	56	52	51	51	44	30	40	39	11	13
Zr	383	395	273	524	447	150	160	161	170	170
Nb	110	115	83	135	120	-	-	56	57	60
Ba	587	702	621	1058	943	-	-	592	573	698
Pb	8	11	0.5	5	4	<10	<10	<d.l.	0.7	0.9
Th	9	9	16	7	5	<10	15	15	10	10
La	96	109	86	110	95	-	-	62	60	62
Ce	177	193	142	214	184	-	-	93	89	93
Nd	68	75	61	82	76	-	-	39	39	40
Y	33	34	25	34	29	20	20	20	19	19

Table 11.3 a) Continued

CARBONIFEROUS – EARLY PERMIAN																				
F04-02-A		Wanneperveen-1 (WAV-01)								De Wijk-7 (WYK-07)								Dwingelo-2 (DWL-02)	Corle-1 (COR-01)	Hupsel (GEL-03)
4650	4654	2070	2015	2030-2070	2030-2070	2030-2070	2030-2070	2030-2070	2030-2070	2690	2440-2460	2440-2460	2440-2460	2440-2460	2440-2460	2445	2464	3797	970	1320
F04-8	**F04-9**	**WAV**	**WAV**	**NAM 1/1**	**NAM 1/4**	**NAM 1/13**	**NAM 1/17**	**NAM 1/19**	**NAM 1/21**	**WYK-7**	**NAM 3/1**	**NAM 3/4**	**NAM 3/8**	**NAM 3/12**	**NAM 3/11**	**WYK-14**	**WYK-15**	**DWL**	**COR**	**GEL-3**
Trachyte	Trachyte	Olivine gabbro	Gabbro	Dolerite	Dolerite	Dolerite	Picritic dolerite	Picritic dolerite	Picritic dolerite	Olivine gabbro	Dolerite	Dolerite	Dolerite	Dolerite	Dolerite (Syenite)	Olivine gabbro	Olivine gabbro	Olivine gabbro	Gabbro/ melaphyre	Leucophyre
4	**4**	**3**	**3**	**1**	**1**	**1**	**1**	**1**	**1**	**3**	**1**	**1**	**1**	**1**	**1**	**4**	**4**	**3**	**3**	**3**
64.17	65.92	40.87	45.15	51.94	56.82	45.82	45.08	44.15	43.86	45.99	47.95	47.66	48.88	48.44	65.60	46.67	48.44	41.32	46.38	37.61
16.67	15.58	10.45	16.23	17.42	17.00	16.92	12.66	11.67	11.06	16.14	16.51	16.21	15.97	16.25	14.35	15.62	16.33	11.07	15.44	18.95
7.09	6.24	7.21	3.79	8.35	7.94	12.97	13.61	14.58	15.22	4.81	12.07	12.25	11.44	11.90	4.39	11.83	11.95	3.70	4.69	3.82
-	-	9.11	7.84	-	-	-	-	-	-	6.52	-	-	-	-	-	-	-	8.79	8.03	9.69
0.29	0.27	15.31	8.25	6.74	6.15	9.76	16.22	18.51	19.00	8.11	7.48	7.02	8.64	8.62	2.02	7.75	8.47	19.88	7.38	4.22
0.67	0.69	7.12	3.99	9.44	8.06	6.45	6.15	5.31	5.07	9.70	11.51	11.71	6.92	9.36	3.44	10.81	10.08	6.78	8.49	5.29
4.90	4.46	2.88	5.29	2.68	2.44	3.95	2.14	2.17	1.80	2.51	2.63	2.68	2.52	2.96	6.57	2.78	2.88	2.90	4.23	2.69
5.84	5.54	0.92	1.88	0.30	0.22	0.90	0.93	0.65	0.65	0.66	0.28	0.30	2.89	0.77	1.45	0.37	0.90	0.39	0.88	0.38
0.78	0.69	1.00	0.49	1.00	0.95	2.24	1.61	1.65	1.53	0.42	0.94	1.12	1.20	1.08	0.81	1.06	1.10	0.26	0.90	0.24
0.02	0.04	trace		0.12	0.10	0.18	0.14	0.15	0.15		0.14	0.16	0.33	0.18	0.10	0.16	0.20	-	-	-
0.24	0.22	0.18	0.19	0.10	0.10	0.34	0.20	0.22	0.23	0.14	0.08	0.12	0.12	0.10	0.16	0.10	0.10	0.24	0.11	0.08
100.67	99.65	95.05	93.1	98.10	99.78	99.52	98.76	99.06	98.57	95.00	99.59	99.23	98.92	99.67	98.89	97.15	100.45	95.33	96.53	82.97
1.35	1.49			3.9	3.7	8.3	4.6	5.6	6.0		5.5	3.4	3.1	2.4	1.9	2.97	2.22	-	-	-
-	-	2.77	2.94	-	-	-	-	-	-	2.31	-	-	-	-	-	-	-	3.00	2.02	13.03
-	-	1.80	2.77	-	-	-	-	-	-	0.94	-	-	-	-	-	-	-	1.56	0.70	3.90
6.4	6.2	-	-	102	90	101	430	548	682	-	136	121	103	136	9	123	128	-	-	-
2.7	3	-	-	344	334	228	611	675	804	-	309	267	209	264	0	220	216	-	-	-
17	9.4	-	-	287	292	345	263	266	232	-	264	293	298	261	36	230	221	-	-	-
13	9.8	-	-	38	43	33	22	20	13	-	34	35	30	27	10	29	28	-	-	-
4.4	6.8	-	-	-	-	-	-	-	-	-	-	-	-	-	-	111	124	-	-	-
20	28	-	-	-	-	-	-	-	-	-	-	-	-	-	-	75	71	-	-	-
78	61	-	-	210	211	333	305	294	239	-	152	150	164	173	134	137	179	-	-	-
185	175	-	-	8	6	27	32	21	22	-	7	7	30	16	25	9.3	17	-	-	-
434	432	-	-	97	94	189	133	114	121	-	59	76	82	71	501	60	54	-	-	-
46	41	-	-	7	6	17	20	15	15	-	5	7	7	7	36	4.4	4.3	-	-	-
496	517	-	-	121	74	240	287	209	279	-	296	156	805	159	192	328	308	-	-	-
5.0	3.8	-	-	-	-	-	-	-	-	-	-	-	-	-	-	1.6	1.1	-	-	-
15	14	-	-	-	-	-	-	-	-	-	-	-	-	-	-	2.5	2.5	-	-	-
60	56	-	-	0	0	10	8	10	9	-	4	6	14	4	39	10	11	-	-	-
102	84	-	-	26	28	44	36	32	26	-	10	21	19	17	82	12	8.8	-	-	-
52	46	-	-	12	11	28	15	15	11	-	10	8	10	11	37	9.0	8.1	-	-	-
51	47	-	-	20	20	28	20	18	18	-	18	21	21	21	60	17	16	-	-	-

Table 11.3 b) Rare-earth-element concentrations in igneous rocks from offshore and onshore wells in the Netherlands

Igneous episode	**LATE JURASSIC**											**CARBONIFEROUS – EARLY PERMIAN**				
Well name	**F10-01**				**Zuidwal-1 (ZDW-01)**		**Loon-op-Zand-1 (LOZ-01)**		**Andel-2 (AND-02)**	**Heinenoord-1 (HEI-01)**		**F04-02-A**	**Wanneperveen-1 (WAV-01)**	**De Wijk-7 (WYK-07)**		
Depth (m)	n.a.	n.a.	3451	3458	n.a.	2180	n.a.	2590	1368	2237	2249	4654	2030-2070	2440-2460	2445	2464
Sample code	**PL1/6**	**PL1/10**	**F10-6**	**F10-7**	**ZU5**	**ZDW-4**	**NAM2/8**	**LOZ-12**	**AND-19**	**HEI-16**	**HEI-18**	**F04-9**	**NAM1/1**	**NAM3/8**	**WYK-14**	**WYK-15**
Rock type	Lamprophyre	Lamprophyre	Foïdite	Foïdite	Phonolite-Basanite	Altered andesite	Basanite	Altered basaltic andesite	Trachy-basalt	Basanite	Basanite	Trachyte	Dolerite	Dolerite	Olivine gabbro	Olivine gabbro
Reference	**1**	**1**	**4**	**4**	**1**	**4**	**1**	**4**	**4**	**4**	**4**	**4**	**1**	**1**	**4**	**4**
La	78	82	78	83	93	69	59	77	84	46	48	30	10	5.3	8.6	7.9
Ce	159	157	159	168	159	113	121	143	161	88	93	71	23	13	19	18
Pr	18	16	18	19	16	11	14	15	18	10	11	9.0	2.8	1.9	2.6	2.4
Nd	70	63	69	71	50	39	51	54	66	40	41	37	13	10	12	11
Sm	11	10	12	12	7.7	6.1	8.5	8.7	11	7.3	7.6	8.3	2.5	2.1	3.1	2.9
Eu	3.3	3.0	3.5	3.4	2.2	1.7	2.4	2.6	3.4	2.3	2.4	2.1	0.82	0.84	1.1	1.1
Tb	-	-	1.3	1.3	-	0.70	-	0.97	1.4	0.91	0.94	1.3	-	-	0.68	0.61
Gd	9.1	8.4	9.3	9.0	5.6	4.9	6.7	6.8	9.3	6.4	6.5	8.2	3.0	3.1	3.9	3.7
Dy	6.4	6.2	6.4	6.3	4.4	3.8	5.4	5.0	7.0	4.8	4.9	7.6	3.6	3.5	4.3	3.9
Ho	1.1	1.1	1.1	1.1	0.69	0.72	0.86	0.88	1.3	0.86	0.88	1.6	0.66	0.64	0.91	0.83
Er	2.9	2.7	2.7	2.8	1.5	2.0	2.1	2.2	3.3	2.1	2.2	4.6	1.9	1.8	2.6	2.3
Tm	-	-	0.37	0.38	-	0.30	-	0.29	0.44	0.29	0.29	0.72	-	-	0.38	0.34
Yb	2.1	2.1	2.2	2.3	1.5	1.9	1.78	1.8	2.7	1.7	1.8	4.7	1.9	1.8	2.4	2.1
Lu	0.31	0.30	0.31	0.32	0.21	0.29	0.25	0.25	0.39	0.24	0.25	0.71	0.28	0.26	0.35	0.31

the nearby northwest German Ems-Weser area, where a large number of 'spilitized diabases' show a dominance of intermediate compositions and a wide range between basaltic and rhyolitic rock types (Eckhardt, 1968, 1979). The least altered rocks define a continuous subalkaline to mildly alkaline series between basaltic and rhyolitic compositions, comparable to those of lower Rotliegend volcanics from other locations in northern Germany (Benek et al., 1996).

Overall, the Dutch occurrences fit into the general characteristics of Carboniferous-lower Permian igneous rocks of northwestern Europe, which are marked by a dominance of mafic compositions locally showing a bimodal distribution with felsic rocks, variable alkali contents ranging from subalkaline (tholeiitic) to highly alkaline, subtle variations in incompatible trace-element contents and a prominent within-plate signature. The geochemical signatures are consistent with derivation of relatively small magma volumes from variable low degrees of partial melting of heterogeneous mantle sources in local independent systems, whereas crystal fractionation and contamination/melting of crustal components may have further added to the compositional diversity (e.g. Aghabawa, 1993; Benek et al., 1996; Heeremans et al., 2004; Neumann et al., 2004; Obst et al., 2004; Kirstein et al., 2006).

Late Jurassic

In comparison to those of other periods, Late Jurassic igneous rocks are geochemically well documented. Major and trace-element compositions have been reported on multiple samples from offshore wells F10-01 (sample PL-1), K14-FA-103, L13-03, Zuidwal-1, and from onshore wells Loon-op-Zand-1, Andel-2, Andel-4 and Heinenoord-1 (Dixon et al., 1981; Eigenfeld & Eigenfeld-Mende, 1986; Latin, 1990; Helmers, 1991; this work). Taking the effects of alteration into account, all of these rocks are silica-undersaturated and are marked by high contents of alkalis. With the exception of the samples from the Zuidwal volcano, which represent more evolved members of an ultrapotassic suite, they are relatively mafic. The least altered rocks (LOI <5 wt.%) of Heinenoord-1 (West Netherlands Basin) and F10-01 (Rim Dutch Central Graben) illustrate the compositional diversity of the late Jurassic mantle-derived magmas. The analysed samples from both locations plot in the tephrite-basanite field of the TAS classification diagram (Fig. 11.18b), but the Heinenoord-1 rock is more sodic (nephelinitic) and has a higher Na_2O/K_2O ratio, higher CaO and lower TiO_2 and P_2O_5 contents than F10-01, which has an ultrapotassic affinity. Mafic samples from the other locations may well represent magmas with similar compositions but strong alteration likely

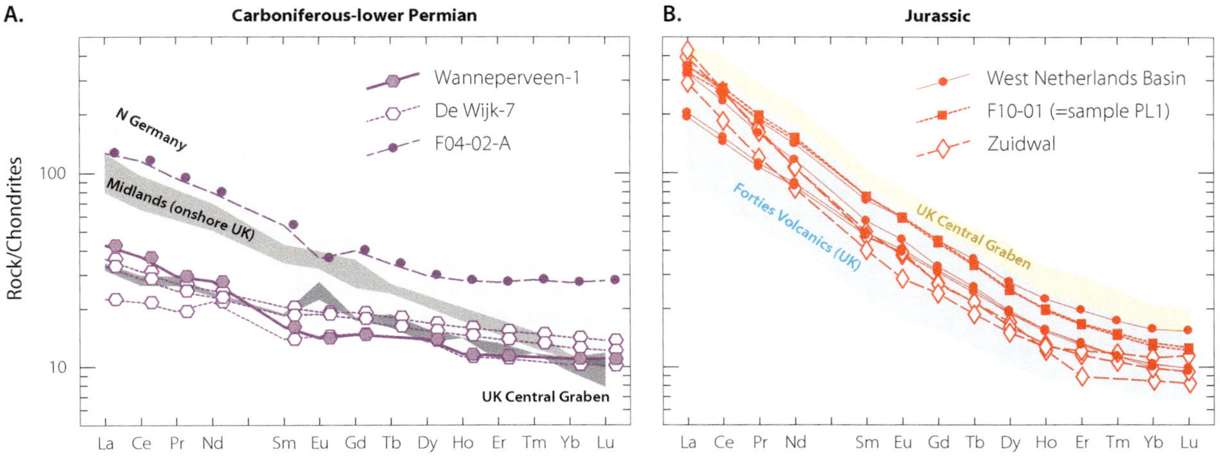

Figure 11.20. *Chondrite-normalized REE patterns for Carboniferous-lower Permian (a) and Jurassic (b) igneous rocks from Dutch exploration wells (data in Table 11.3b) in comparison with those for rocks from neighbouring regions in northwestern Europe. All Dutch rocks are basaltic except for the trachytic F04-02-A sample. Note the relative flat trends of the subalkaline-alkaline Carboniferous-lower Permian rocks and steep parallel trends of the ultra-alkaline Jurassic rocks. Data sources as in Fig. 11.18.*

affects the accuracy of their TAS classification. In the Nb/Y vs Zr/Ti diagram (Fig. 11.18d), all of the mafic rocks plot near the foidite-alkali basalt boundary whereas the Zuidwal rock series trends towards the tephri-phonolite field.

The geochemical compositions of the Dutch onshore and offshore occurrences mirror the differences between the subalkaline-alkaline, predominantly mafic rocks of the Forties Volcanics (or Rattray Volcanics) at the triple junction of the North Sea continental rift system (Gibb & Kanaris-Sotiriou, 1976; Woodhall & Knox, 1979; Dixon et al., 1981; Fall et al., 1982; Ritchie et al., 1988; Latin, 1990; Latin et al., 1990a; Latin & Waters, 1992; Quirie et al., 2019, 2020), and the more alkaline mafic centres in the west Central Graben (Latin, 1990; Latin et al., 1990a; Latin & Waters, 1992) (Fig. 11.13). Typically, the Central Graben tephrites-basanites are most enriched in alkalis and incompatible minor and trace elements (e.g. TiO_2, P_2O_5, Nb, Zr, REE) and contain lower amounts of Cr and Ni, which compares well with the ultrapotassic signature of F10-01 and the Zuidwal volcano. The nephelinitic alkali-basaltic rock of Heinenoord-1 has a closer affinity with the relatively alkaline varieties of the Forties Volcanics.

This regional difference in geochemical affinity also appears in REE patterns. In all cases they show almost straight trends with no Eu anomalies and enrichment of light REEs over heavy REEs (Fig. 11.20b), but the LREE/HREE ratios tend to be higher in the Central Graben area (including F10-01) than in the Forties region and Heinenoord-1. Similar systematics in Zr/Nb and Ce/Y ratios (Fig. 11.21b) suggest that if a common mantle source is assumed (Latin et al., 1990b), the latter group was generated by a higher degree of partial melting. Immobile-trace-element signatures further confirm a 'within-plate' tectonic affinity for all of the mafic rocks in these regions, as is illustrated in the Nb/Y-Ti/Y tectonic discrimination diagram of Fig. 11.21).

Magmagenesis and rifting

The inventory of igneous rocks in the Dutch subsurface provides insight into the intimate link between magmagenesis and rifting that has existed throughout the post-Devonian geological evolution of the (proto) North Sea region. A returning point of discussion concerns the geodynamic forces that controlled the rifting processes and to what extent signals from associated magmatism discriminate between different options. Possible involvement of mantle plumes is a subject of ongoing debate for both the Carboniferous-early Permian and the Jurassic episodes. In this section, some key issues on the two main magmatic episodes are summarized and briefly explained in a northwest European context, with emphasis on a petrological/geochemical perspective. The Cenozoic rifting period will not be treated since Dutch territories only host airborne products from volcanic activity outside the country.

Rifts can be distinguished into two main groups based on models for their origin and evolution: 'passive' rifts, developed as a result of lithospheric extension driven by far-field stresses (e.g. McKenzie & Bickle, 1988), and 'active' rifts, originated from thermal upwelling of the asthenosphere (White & McKenzie, 1995). In either case, volcanism accompanying rifting in continental settings is marked by a sequence ideally comprising the initial production of minor amounts of alkaline magma with a lithospheric imprint, followed by more voluminous subalkaline magmas predominantly derived from upwelling mantle rock as stretching and/or thinning of continental lithosphere provides room for asthenosphere to rise. This adiabatic melting of asthenospheric mantle occurs when the accompanying perturbation of the local geotherm advances in such a way that it intersects the mantle solidus (i.e. the melting point is reached at a given pressure) and partial melting starts. The composition of the melt that ultimately escapes from a solid residue is a function of the original chemical and mineral composition of the mantle source, depth and degree of partial melting.

In a scenario where continental rifting is actively driven by deep mantle processes, a plume of solid mantle rock rises diapirically to the surface since it has a lower density than surrounding mantle, being hotter and/or different in composition. Impingement of a plume head from below may heat up, uplift and erode the lithosphere, which eventually can lead to tensional failure and continental breakup (e.g. Courtillot et al., 1999). Melting of mantle rock may occur in the centre of the rising plume by adiabatic decompression at higher potential temperature and/or in adjacent domains through heat transfer when a plume head cannot pass a physical barrier and spreads out below it. If the amount of melt produced becomes large enough it may escape to the surface and produce voluminous basaltic volcanism.

Carboniferous-early Permian

Magmatic activity occurred throughout Carboniferous and early Permian times over a wide region of the Variscan foreland in northwest Europe (Figs 11.3 and 11.7). The basement is a heterogeneous blend of relatively small crustal domains with Neoarchean to Neoproterozoic ages and low to high-grade metamorphic Caledonian belts, presumably bounded by major zones of deep-reaching faults. Hence, intrinsic variations in lithospheric thicknesses, thermal histories and other local controls explain differences in timing and nature of magma-producing events observed in individual areas. Decompression-induced low-degree partial melting of predominantly asthenospheric mantle rock is thought to have accompanied lithospheric extension, thinning and basin formation (see below for details on this mechanism). Magmatism was mostly marked by within-plate geochemical signatures, but locally with more arc-type imprints.

Early Carboniferous

Early Carboniferous (Dinantian) extension-related magmatism and basin formation from Ireland to Poland was driven by geodynamic events such as closure of the Rhenohercynian Ocean, accretion of a magmatic arc and amalgamation of microcontinents. In Ireland and the UK, where early Carboniferous magmatism peaked in the Visean, intrusive and extrusive rocks are mildly alkaline-transitional in composition (e.g. Smedley, 1986; Timmerman, 2004; Upton et al., 2004). Widespread manifestations of Namurian-Westphalian magmatic activity in the UK, centred in the Midland Valley of Scotland, the English Midlands and Derbyshire, includes alkaline and subalkaline varieties as well (Timmerman, 2004).

Minor volumes of tholeiitic to alkaline basaltic Visean volcanic rocks and Namurian-Westphalian dolerite sills with a typical within-plate signature in Derbyshire are thought to represent products of low-degree melting in isolated pockets of a heterogeneous mantle magma. They managed to reach the surface along fractures in attenuated crust (Macdonald et al., 1984; Timmerman, 2004).

Late Carboniferous

A similar scenario is conceivable for the presumably (late) Carboniferous dolerites found in Dutch onshore wells (e.g. Wanneperveen-1, De Wijk-7). Their geochemical signature tends to be mostly subalkaline but does not provide a uniform assessment of the tectonic setting. Discrimination diagrams suggest a within-plate signature for some of the rocks and magma derivation from a mantle source with possibly a modest subduction-related ('volcanic-arc') chemical imprint for others (Fig. 11.21). This diversity hints at small-scale heterogeneity of mantle sources that may include an earlier (Devonian?) back-arc spreading domain, perhaps related to the northward subduction of Rheic ocean crust.

Latest Carboniferous – early Permian

A conspicuous feature of the latest Carboniferous-early Permian period of voluminous magmatism is that it started almost simultaneously in different provinces over an extensive region north of the Variscan front. Magmatism coincided with a period of wrench-related lithospheric deformation in response to a fundamental change in the regional stress field that affected western and central Europe at the end of the Variscan orogenic activity (Ziegler, 1990) and is generally associated with lithospheric stretching and rifting.

Dated volcanic and plutonic rocks from the Oslo Graben, Scania, the North Sea area, Scotland, northern England and the northeast German Basin show that magmatic activity peaked in a rather narrow time interval around ca. 300 Ma, while centres in the internal Variscides as far south as Iberia and Italy have comparable ages (Heeremans et al., 2004; Neumann et al., 2004; Timmerman, 2004; Wilson et al., 2004; and references in caption to Fig. 11.7).

Bulk-rock compositions, often showing a bimodal distribution, are much more diverse than in later periods. In particular, the large abundance of intermediate and acid

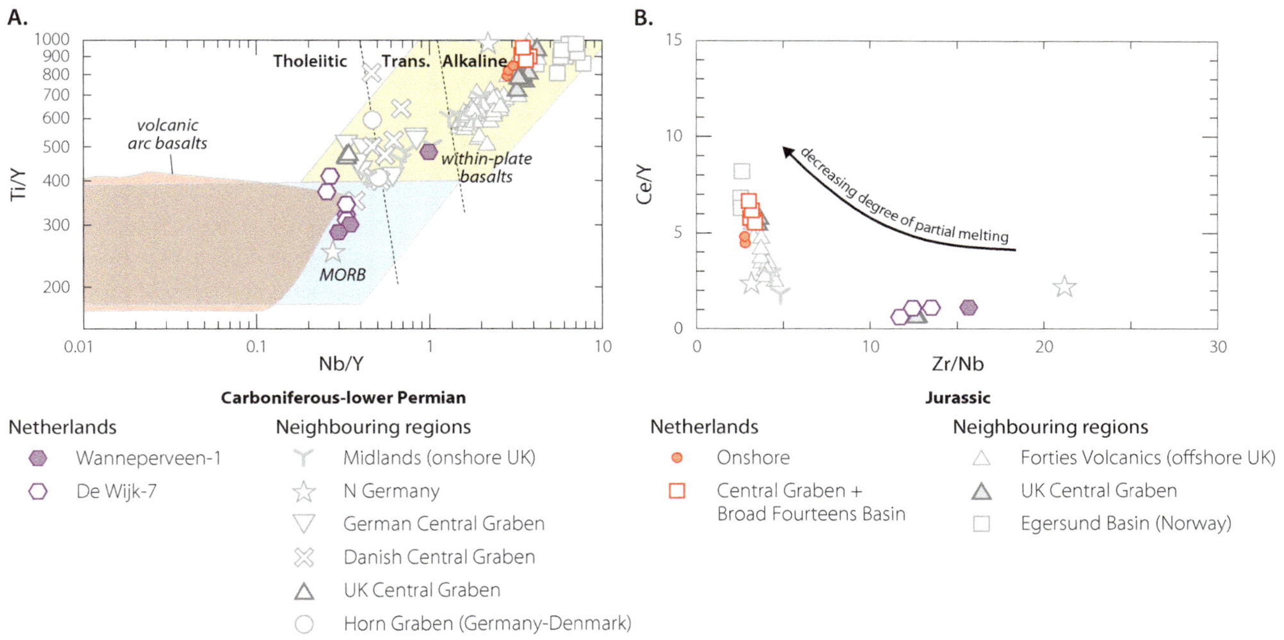

Figure 11.21. Comparison of magmagenetic controls: a) Tectonic Nb/Y-Ti/Y discrimination diagram after Pearce (1982), illustrating an alkaline, within-plate signature of the Dutch Jurassic rocks and their difference with the transitional Carboniferous-lower Permian rocks, some of which trend towards the MORB or volcanic arc field; b) Zr/Nb vs. Ce/Y diagram for samples from Dutch onshore and offshore exploration wells, suggesting that the Jurassic mantle-derived magmas originated by lower degrees of melting than the Carboniferous-lower Permian magmas, assuming a reasonably similar source. Only relatively fresh samples (LOI<5 wt.%) of rocks with >5 wt.% MgO are shown. Data sources as in Fig. 11.18.

varieties is noticeable. Although there tends to be a regional north-south trend from strongly alkaline in the Oslo Rift to mildly alkaline in the Variscan foredeep and calcalkaline in the Rhenohercynian orogenic belt (e.g. references in Ziegler, 1990), rock types vary widely both on regional and local scales. For example, basalts encompass the entire spectrum from highly alkaline to tholeiitic in the Oslo Graben (Neumann et al., 2004), while tholeiitic as well as alkali-rich basalts have also been found in North Sea wells (Heeremans et al., 2004).

'Rotliegend' volcanic rocks in Danish offshore and onshore wells and in surrounding areas are dominated by basaltic flows and pyroclastic ash flow deposits, while associated intermediate and acidic volcanic rocks are present in subordinate amounts (Aghabawa, 1993). There is a general tendency towards increasing alkalinity and degree of silica undersaturation from the Kattegat region in the east to the Central Graben in the west. The Horn Graben region is marked by a bimodal alkali/transitional basalt-rhyolite/rhyodacite suite. The trachytes from the Dutch Central Graben (well F04-02-A) fit in this scheme. Petrological modelling suggests that 'Rotliegend' basalts in the Danish North Sea and surroundings originate from variable-degree melting at different depths of a heterogeneous mantle or from separate mantle sources, including a depleted asthenospheric domain and an enriched lithospheric domain (Aghabawa, 1993).

Based on trace-element and isotopic signatures, Neumann et al. (2004) suggested that the main mantle source of magmatism in the Oslo Rift, Scania and possibly also the North Sea was similar to a Prevalent Mantle-type (PREMA) component residing in the lithospheric mantle. Melting was probably induced by local decompression and thinning of lithosphere in response to regional stretching north of the Variscan front, although the PREMA affinity implies that involvement of a mantle plume cannot be discarded.

Similar scenarios have been proposed for the northeast German Basin (e.g. Breitkreuz & Kennedy, 1999), where different structural domains and a heterogeneous basement added to geochemical diversity on a relatively small scale (Benek et al., 1996). Crustal thinning and block faulting facilitated the production of large volumes of intrusive and extrusive rocks. Magmas with a calcalkaline character could have been derived from a pre-existing subduction-influenced basaltic magma source (cf. Benek et al., 1996).

Locally, the thermal perturbation associated with Carboniferous-Permian magmatism may have been more pronounced than in Jurassic times, as supported by the higher degrees of melting (~10%) inferred from trace-element signatures of the North Sea samples studied by Latin et al. (1990b). According to the data in Eckhardt (1979), particularly the Zr/Nb and Ce/Y ratios, a similar melting regime probably affected the Ems-Weser basalts in northwest Germany (and perhaps also the Dutch Permian volcanics) (Fig. 11.21b). A stronger thermal anomaly is also consistent with the widespread generation of the acidic magmas, which can be explained as the products of anatexis of the lower crust, possibly provoked by heat input from underplated basalt (Breitkreuz & Kennedy, 1999).

Ernst & Buchan (1997) noticed that the dyke swarms in the Oslo Rift, Scania, northeast England and Scotland radiate from a common triple junction in Denmark, which would localize the axis of a deep-mantle plume. From the large areal extent, volume, brevity of the activity interval and the convergent dyke swarms, Torsvik et al. (2008) connected the ~300 Ma igneous episode to that of a typical Large Igneous Province (LIP), sourced by a deep-sourced mantle plume below the Skagerrak. They proposed that mantle-derived basaltic magma originated at this centre and propagated over some 1000 km to Scotland and Northern England, where it formed the voluminous dike swarms and sills of the Midland Valley Rift, the Great Whin Sill and surrounding areas.

However, geochemical and tectonic evidence for a plume model is weak. In a geochemical study on dyke and sill intrusions across northern Europe, Kirstein et al. (2006) discussed a relationship between the depth of melting and lithosphere thickness, which strongly varies between the different areas and is difficult to reconcile with a region-wide plume control. Furthermore, radiogenic-isotope and trace-element signatures are not inconsistent with lithospheric mantle as a major source of the basaltic magma prior to decompression melting of the underlying asthenosphere, which would also be more in line with a local lithospheric extension rather than a single major plume as principal control (e.g. Neumann et al., 2004). Kirstein et al. (2006) suggested that the apparent dyke orientation is a relic of larger tectonic features (emplaced sub-parallel to the Caledonian trend, terrane boundaries in the Pre-Cambrian basement, or coeval extensional structures), and concluded that there is no evidence for a thermally anomalous mantle plume during Carboniferous-Permian magmatism in northern Europe.

Late Jurassic

The relation between late Jurassic magmatism and rifting in the Netherlands is best illustrated in conjunction with the igneous manifestations in other parts of the North Sea Basin (Fig. 11.13). Variations in timing, location, volume and composition allow the construction of a coherent magmagenetic framework for the entire region (Latin et al., 1990a, b; Latin & Waters, 1992). The Forties basaltic province (or Rattray Volcanic Province), situated at the triple junction between the Viking Graben, Central Graben and Moray Firth rift basins, can be seen as the focal point of the magmatic activity within the main rift system. Two phases of activity, possibly in a series of fissure eruptions from linear vents, produced an up to 1.5 km thick succession of subaerial basaltic lavas and hyaloclastites in a terrestrial-lacustrine environment over an area of ca. 7400 km^2, and ended with a minor intrusive suite (Quirie et al., 2019, 2020).

The occurrences in the Dutch Central Graben and Egersund Basin are much less voluminous, are located on the flanks of the main rift system or in minor sub-basins and are more silica-undersaturated and alkaline (ultrapotassic of nephelinitic) than the Forties basalts. All of these magmas were produced during the syn-rift phase of basin development, but none were derived from an asthenospheric mantle source similar to that producing mid-ocean-ridge basalts. Instead, the mantle sources were variably enriched in incompatible elements and have often been referred to as OIB-type (Oceanic Island Basalt) or E-MORB-type (Enriched Mid Ocean Rich Basalt) mantle without, however, necessarily implying a similar setting. The Forties basalts (mildly alkaline and mildly silica-undersaturated) represent the largest-degree melts of the entire sedimentary rock succession in the North Sea Basin. They did originate from asthenospheric mantle, but their trace-element and isotope signatures indicate that an enriched component is involved as well, either as heterogeneities in the asthenosphere or in the form of mixed-in lithospheric melt. The other magmas (highly alkaline and strongly silica-undersaturated), including those of the Netherlands, must have formed by lower-degree melting of a volatile- and incompatible-trace-element-enriched region of the mantle, which had remained isolated from the asthenosphere for hundreds of millions of years. Therefore, these magmas originated in a separate source, of which continental lithosphere is the most plausible domain (Latin & Waters, 1992).

The Zuidwal volcano likely represents a case of extension-driven melt generation in the continental lithosphere, as is illustrated in the generalized model of Fig. 11.11d, e (cf. Herngreen et al., 1991), wherein a schematic cross section of the magma system and a hypothetical P-T diagram visualize the potential controls. Given the small size of the Vlieland pull-apart basin, it is improbable that decompression was sufficient to cause melting of dry asthenosphere. Therefore, a more realistic scenario is that magma formed mainly at the base of the lithosphere, which must have been enriched in volatiles (H_2O and

CO_2) and incompatible elements in order to shift its solidus to sufficiently lower temperature for melting to occur. The nature and timing of this enrichment are unknown but may be related to an older melt-infiltration event on a regional scale. Localized stretching of the lithosphere, controlling the development of this basin, may have led to uplift of the asthenosphere-lithosphere boundary to a point where the geotherm crossed the solidus. The inferred limited degree of melting would be in line with the small size of the basin. The evolved compositions of the Zuidwal volcanics (Fig. 11.18b) and indications for a caldera structure (Perrot & Van der Poel, 1987) suggest that mantle-derived melt accumulated and fractionated in a crustal magma reservoir at relatively shallow depth before and during the period of eruptive activity.

Magmagenetic interpretations are consistent with the rift setting. The Forties basalts (Fig. 11.13) occur in a region that experienced the maximum lithospheric stretching at the rift triple junction. Major zones of weakness in the lithosphere may have further facilitated their rise to the surface. This would explain the only scattered volcanic occurrences in the adjacent Viking and North Sea Central grabens, where the degree of extension was slightly smaller. It is conceivable that asthenospheric melts were produced here as well, but that these solidified at depth and were not able to reach the surface. Melt generation in the Netherlands and on the flanks of the Central Graben, where the amount of stretching was much smaller, may have been facilitated by special conditions connected to the intersection of the graben with pre-existing and long-lived stable highs such as the Texel-IJsselmeer High and the Mid North Sea High (Fig. 11.13). At these locations, enriched continental lithosphere could have been partially molten in contact with small amounts of asthenospheric melts to create magma batches sufficiently voluminous to reach the surface.

Stratigraphic information from well data (e.g., detection of unconformities, absence of syn-rift sediments, erosion of pre-rift sediments), has been used to argue for a regional thermally driven doming event in Jurassic times, with the Forties basaltic province as focal point (Ziegler, 1990; Underhill & Partington, 1993). If valid for Zuidwal and the other Jurassic alkaline occurrences of the Netherlands, such a scenario would imply that localized low-degree partial melts originated in response to this regional heat pulse and/or uplift, and that ascent to shallow crustal levels was facilitated by crustal-scale faulting.

A model of active rifting driven by impingement of a mantle plume predicts that doming will be followed by volcanism and subsequent or simultaneous rift formation (e.g. Courtillot et al., 1999). Based on chronostratigraphic data and assuming a middle Jurassic age for North Sea volcanism, it has been contended that the sequence of events follows this scheme and supports a scenario of doming and deflation of a transient, warm plume head or 'blob' (Underhill & Partington, 1993). However, from a detailed paleogeographic study in the area of the Forties basaltic province, Quirie et al. (2020) suggested that subsidence already started before the onset of volcanism. Also, the amount and areal extent of regional uplift associated with a plume control have been contested and, if radiometric ages are taken into account, there are other uncertainties as to the chronology of magmatism relative to the main phase of faulting (Latin, 1990).

The entire set of available radiometric ages might suggest a southward migration of magmatism from the Forties basaltic province towards the Netherlands in the course of time, possibly connected to propagation of the rift system in the same direction. It may also point to northwards motion of the European plate over a fixed hot-spot (Latin et al., 1990a,b), which would be consistent with a plume model. However, if only the most reliable $^{40}Ar/^{39}Ar$ age dates on fresh samples are considered (see Fig. 11.8), the apparent geographic systematics of a hot-spot trail disappears, as ages tend to cluster around 160-150 Ma in the igneous provinces from the Forties triple junction to the Netherlands (Latin, 1990). This crude absolute age bracket places the timing of North Sea magmatism in the late Jurassic instead of the middle Jurassic inferred from stratigraphic interpretations. Consequently, magmatic activity would not precede but largely coincide with the main phases of extension and subsidence, which seems to discard a key argument against the passive rifting model (cf. Latin, 1990).

Also, the observed volume of basalts produced at the triple junction is much lower than modelling of decompression melting of a hot mantle plume, rising beneath stretched and thinned lithosphere in a rift setting, predicts (White & Latin, 1993; White & McKenzie, 1995). Significant uncertainty, inherent to some of the assumptions and input parameters on which this modelling approach relies, makes outcomes not always conclusive. In particular, the amount of stretching and associated vertical movements in the North Sea area have been subject of considerable debate (White & Latin, 1993 and references therein). Finally, available Sr-Nd-Pb isotope data are not sufficiently diagnostic (see discussion in Latin, 1990). The isotope ratios point to a source that experienced time-integrated enrichment relative to MORB, but candidate mantle components could have been tapped from a deep mantle domain by a rising plume or could have resided in the subcontinental lithospheric mantle.

In summary, there is no compelling evidence from geochemical signatures, petrological constraints and radiometric age dating that Jurassic magmatism in the North

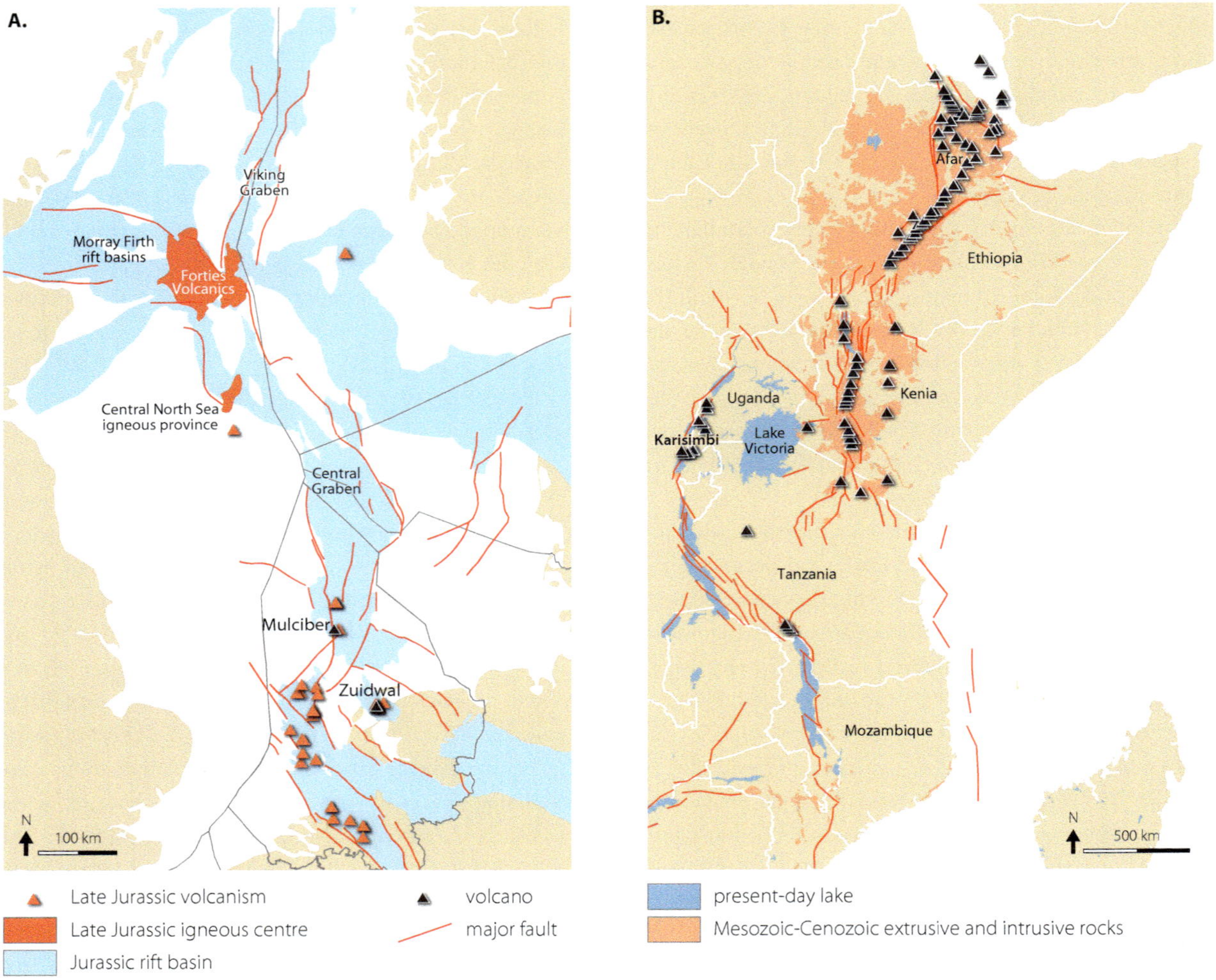

Figure 11.22. Linear distribution of known magmatic manifestations in the Jurassic of the North Sea Basin (a) compared with the chain of recently active volcanoes in the rift zone of East Africa (b). Note the difference in scale. Rift basins and faults in the North Sea region from Guterch et al. (2010), Kombrink et al. (2012) and Hopper et al. (2014). Faults, igneous rocks and volcanoes in the East African Rift Zone: Natural Environment Research Council, BGS Dataset ID: 13607585 and Global Volcanism Program (2023).

Sea region was controlled by the arrival of a major, deeply sourced mantle plume. A passive-rifting model adequately explains most of the observations, although minor effects from a modest, short-lived hot spot cannot be excluded, since the largest magma volume, degrees of partial melting and contributions of asthenospheric mantle material have all been inferred for the basalts in the triple-junction region. For the onshore and offshore subsurface occurrences in Dutch territory (Dutch Central Graben, Broad Fourteens Basin, West Netherlands Basin, Vlieland Basin), the subcontinental mantle lithosphere is the most likely source of the magmas. Melting can be explained in the context of passive stretching, thinning and rifting in the continental setting. If a distant mantle plume played a role in melt generation at all, it may only have acted as source of heat or as propagator of additional uplift.

Continental rifting and magmatism in East Africa: present-day analogue of the Jurassic North Sea region?

The magmatic manifestations from the Jurassic of the North Sea region show conspicuous similarities to those in the currently active rift zone of East Africa, both in distribution and in composition (Fig. 11.22). Parallels in fault patterns, volcanism and geological dimensions between the North Sea Graben system and the Kenyan-North Tanzanian segment of the East African Rift were already noticed by Dixon et al. (1981).

A row of (recently) active volcanoes in East Africa, all situated on continental basement, generally follow the branched rift structure from north to south. The northern termination in the Afar region and surrounding areas represents a triple junction (Fig. 11.22b) and can be seen as an evolution between continental rifting and sea floor

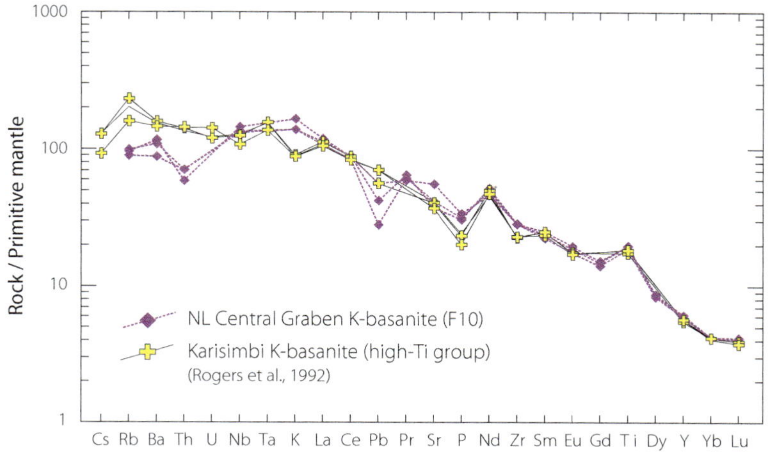

Figure 11.23. *Primordial-mantle normalized incompatible trace-element patterns illustrating similarity between the Jurassic alkaline rocks in Dutch onshore and offshore wells with Quaternary to Recent volcanics in the Virunga Volcanic Province and the Toro Ankole area, situated in the western arm of the East African Rift. Overlapping trends for K-basanites from the F10 well in the Dutch Central Graben (Table 11.3) and Karisimbi Volcano (Rogers et al., 1992) may indicate common petrogenetic controls.*

spreading in the Red Sea and Gulf of Aden. Magmatism in the East African Rift System (EARS) is generally marked by a systematic compositional evolution in space and time. While volcanics in a given sector tend to become less alkaline and more silica-saturated with time, there is also a gradual increase in alkalinity and silica under-saturation towards the south (Rooney, 2020 and references therein). This can be explained by differences in magma-contributing mantle sources: initial highly alkaline and silica-undersaturated magmas are derived from lithospheric subcontinental mantle, whereas continued lithospheric thinning ultimately allows melting of convecting upper mantle and the production of more silica-saturated, transitional to subalkaline magmas.

In this framework, the Forties volcanic field of the North Sea area, situated at the Moray Firth-Central North Sea Graben-Viking Graben triple junction (Fig. 11.22a) would have its counterpart in the Afar region. The tectonic setting, relatively high magma volumes and compositional affinity suggest a contribution from asthenospheric mantle. It is of interest to note that there is wide geochemical and geophysical support to regard magmatism in the Afar region (and possibly along the entire EARS) as a manifestation of mantle plume activity (e.g. Marty et al., 1996; Chang et al., 2020).

Quaternary to Recent volcanism of the Virunga Volcanic Province (Uganda-Rwanda-Congo border) in the western branch of the East African Rift system and the Toro Ankole Volcanic Province (west Uganda) produced alkali-rich rocks with compositions (Rogers et al., 1998; Pitcavage et al., 2021) very similar to the basanites, nephelinites and more evolved products of the Central Graben, Zuidwal and Dutch onshore occurrences, as is illustrated by incompatible trace-element patterns (Fig. 11.23). The parental magmas are thought to have originated from the subcontinental lithospheric mantle that had been variably modified by metasomatic agents and thus acquired a mineralogical composition that deviates substantially from common lherzolitic assemblages. Phlogopite is usually a prominent source component responsible for the (ultra) potassic nature of the volcanics. Results of melting experiments on phlogopite-clinopyroxenite rock (Lloyd et al., 1985) have shown that such a source can produce magma with compositions very close to those of the Central Graben (Latin, 1990). Also, metasomatically induced variability in the source mineralogy, inferred for the Ugandan rocks (Pitcavage et al., 2021), compares well with the compositional differences seen in the rocks from the Dutch wells on a similar spatial scale (e.g. Heinenoord-1, Zuidwal-1, F10-01). The volcanic setting of the western branch of the East African Rift system thus provides an attractive analogue for the manifestations of Jurassic volcanism in the Dutch subsurface. It is conceivable that Karisimbi or any other volcano in the Virunga area is a present-day equivalent of Zuidwal and Mulciber.

Concluding remarks

The igneous rocks encountered in ca. 100 wells in the Dutch subsurface represent a sequence of intermittent magmatic activity that fits well in the region-wide pattern of post-Caledonian melting events recorded in northwest Europe. There is no doubt about the relationship between magmatism and periods of extension and rifting since late Paleozoic times, but evidence for an accompanying role of mantle-plume activity as a driving force remains controversial, partly due to insufficient constraints from absolute age dating. This episodic magmatism in northwest Europe tends to be related to specific periods in the opening history of the North-East Atlantic Ocean (e.g. Woodhall & Knox, 1979; Ziegler, 1988). Since stretching, thinning and rifting did not lead to a full breakup and transition to seafloor spreading and MORB-type magmatism, the North

Sea region is a perfect example of a failed continental rift system (Ziegler, 1992).

The geographically widespread Carboniferous-Permian magmatism was voluminous and compositionally diverse, presumably because melting and assimilation of crustal components was promoted by the mobilization of large amounts of hot mantle-derived magma. In contrast, Jurassic magmatism was much less abundant, is predominantly basaltic in composition, and is typically confined to the Mesozoic rift system with the Forties (Rattray) triple junction as focal point. The preceding Triassic rifting was probably not accompanied by magmatic activity. The Cenozoic dyke swarms inferred to be present in the Southern North Sea are likely related to the final stages in the opening of the Northeast Atlantic.

Acknowledgements

We thank Neptune Energy (now ENI), Total Energies and NAM for supplying new data and photos. Pieter Vroon, Roel van Elsas, Bouke Lacet and Graham Hagen-Peter are thanked for their laboratory assistance at the VU Amsterdam. Kim Cohen of Utrecht University is thanked for data and feedback on Quaternary tephra distributions. David Latin helped us with interpreting sample codes in his PhD thesis, and Axel Gerdes shared details on the U-Pb age determination of the granite in well A17-01. Rixt Altenburg is thanked for mapping the distribution of igneous rocks of Mulciber in seismic data. We thank Johan ten Veen for a comprehensive review of an earlier draft of this chapter.

Digital map data

Spatial data of figures in this chapter for use in geographical information systems can be downloaded here: https://doi.org/10.5117/aup.28163870.

References

Á Horni, J., Hopper, J.R., Blischke, A., Geisler, W.H., Stewart, M., McDermott, K. & Árting, U., 2017. Regional distribution of volcanism within the North Atlantic Igneous Province. Geological Society, London, Special Publications 447(1): 105-125. DOI: 10.1144/SP447.18

Aghabawa, M.A., 1993. Petrology and geochemistry of the Rotliegendes Volcanic Rocks in Denmark and their tectonic implications. In: Stouge, S. & Vejbaek, O.V. (eds): Dynamisk/stratigrafisk analyse af Palaeozoikum i Denmark. EFP-89; Omrade 1: Olie og Naturgas 3, DGU Kunderapport. 351 pp.Andronikov, A.V., Van Hoesel, A., Andronikova, I.E. & Hoek, W.Z., 2016. Trace element distribution and implications in sediments across the allerød–younger dryas boundary in the Netherlands and Belgium. Geografiska Annaler: Series A, Physical Geography 98(4): 325-345. DOI: 10.1111/geoa.12140

Anonymous, 1992. Schichtenverzeichnis Balderhaar Z1. Preussag Erdöl und Erdgas GmbH, Provided by Landesamt für Bergbau, Energie und Geologie, LBEG (Hannover): 63 pp.

Autin, W.J., 2008. Stratigraphic analysis and paleoenvironmental implications of the Wijchen Member in the lower Rhine-Meuse Valley of the Netherlands. Netherlands Journal of Geosciences 87(4): 291-308. DOI: 10.1017/S0016774600023362

Benek, R., Kramer, W., McCann, T., Scheck, M., Negendank, J.F.W., Korich, D., Huebscher, H.D. & Bayer, U., 1996. Permo-Carboniferous magmatism of the Northeast German Basin. Tectonophysics 266(1-4): 379-404. DOI: 10.1016/S0040-1951(96)00199-0

Berendsen, H.J.A. & Stouthamer, E., 2001. Palaeogeographic development of the Rhine-Meuse delta, The Netherlands. Koninklijke van Gorcum B.V. (Assen): 268 pp.

Bischoff, A., Planke, S., Holford, S. & Nicol, A., 2021. Seismic Geomorphology, Architecture and Stratigraphy of Volcanoes Buried in Sedimentary Basins. In: Németh, K. (ed.): Updates in Volcanology - Transdisciplinary Nature of Volcano Science. IntechOpen: 1-34. DOI: 10.5772/intechopen.95282

Bless, M.J.M., Boonen, P., Bouckaert, J., Brauckmann, C., Conil, R., Dusar, M., Felder, P.J., Felder, W.M., Gökdag, H., Kockel, F., Laloux, M., Langguth, H.R., Van der Meer Mohr, C.G., Meessen, J.P.M.T., Op het Veld, F., Paproth, E., Pietzner, H., Plum, J., Poty, E., Scherp, A., Schulz, R., Streel, M., Thorez, J., Van Rooyen, P., Vanguestaine, M., Vieslet, J.L., Wiersma, D.J., Winkler Prins, C.F. & Wolf, M., 1981. Preliminary report on Lower Tertiary-Upper Cretaceous and Dinantian-Famennian rocks in the boreholes Heugem-1-1a and Kastanjelaan-2 (Maastricht, the Netherlands). Mededelingen Rijks Geologische Dienst 35(15): 333-415

Boenigk, W. & Frechen, M., 2006. The Pliocene and Quaternary fluvial archives of the Rhine System. Quaternary Science Reviews 25(5-6): 550-574. DOI: 10.1016/j.quascirev.2005.01.018

Bouroullec, R., Verreussel, R.M.C.H., Geel, C.R., De Bruin, G., Zijp, M.H.A.A., Kőrösi, D., Munsterman, D.K., Janssen, N.M.M. & Kerstholt-Boegehold, S.J., 2018. Tectonostratigraphy of a rift basin affected by salt tectonics: synrift Middle Jurassic-Lower Cretaceous Dutch Central Graben, Terschelling Basin and neighbouring platforms, Dutch offshore. Geological Society of London, Special Publications 469(1): 269-303. DOI: 10.1144/SP469.22

Bradshaw, M.J., 1975. Origin of montmorillonite bands in the Middle Jurassic of Eastern England. Earth and Planetary Science Letters 26(2): 245-252.

Bredewout, J.W., 1989. The character of the Erkelenz intrusive as derived from geophysical data. Geologie en Mijnbouw 68(4): 445-454.

Breitkreuz, C. & Kennedy, A., 1999. Magmatic flare-up at the

Carboniferous/Permian boundary in the NE German Basin revealed by SHRIMP zircon ages. Tectonophysics 302(3-4): 307-326.

Breitkreuz, C., Kennedy, A., Geissler, M., Ehling, B.-C., Kopp, J., Muszynski, A., Protas, A. & Stouge, S., 2007. Far Eastern Avalonia: its chronostratigraphic structure revealed by SHRIMP zircon ages from Upper Carboniferous to Lower Permian volcanic rocks (drill cores from Germany, Poland and Denmark). GSA Special Paper 423: 173-190.

Brown, G., Platt, N.H. & McGrandle, A., 1994. The geophysical expression of Tertiary dykes in the southern North Sea. First Break 12(3): 137-145.

Brugmans, S.J., 1784. Natuurkundige verhandeling over een zwavelagtigen nevel den 24 Juni 1783 in de provintie van Stad en Lande en naburige landen waargenomen. Petrus Doekema (Groningen): 74 pp.

Burger, A.W., 1979. Sedimentpetrologisch onderzoek van boring Lievelde 41 B/73 (No. 645). Rijks Geologische Dienst (Haarlem).

Burger, A.W., 1992. Sedimentpetrologisch onderzoek aan boring Beltrum 34D/232 (No. 960). Rijks Geologisch Dienst (Haarlem).

Busschers, F.S., Kasse, C., Van Balen, R.T., Vandenberghe, J., Cohen, K.M., Weerts, H.J.T., Wallinga, J., Johns, C., Cleveringa, P. & Bunnik, F.P.M., 2007. Late Pleistocene evolution of the Rhine-Meuse system in the southern North Sea basin: imprints of climate change, sea-level oscillation and glacio-isostacy. Quaternary Science Reviews 26: 3216-3248. DOI: 10.1016/j.quascirev.2007.07.013

Carver, F., Cartwright, J., McGrandle, A., Kirkham, C. & Pryce, E., 2023. The continuation of the Mull Dyke Swarm into the Southern North Sea. Journal of the Geological Society 180(6): jgs2023-039. DOI: 10.1144/jgs2023-039

Chambers, L.M., Pringle, M., Fitton, G., Larsen, L.M., Pedersen, A.K. & Parrish, R., 2003. Recalibration of the Palaeocene–Eocene boundary (P-E) using high precision U-Pb and Ar-Ar isotope dating. Presented at the EGS-AGU-EGU Joint Assembly, Nice.

Chang, S.-J., Kendall, E., Davaille, A. & Ferreira, A.M.G., 2020. The Evolution of Mantle Plumes in East Africa. Journal of Geophysical Research: Solid Earth 125(12): e2020JB019929. DOI: 10.1029/2020JB019929

Cohen, K.M. & Stouthamer, E., 2012. Digitaal Basisbestand Paleogeografie van de Rijn-Maas Delta. DANS Data Station Archaeology 5. DOI: 10.17026/dans-x7g-sjtw

Corfu, F. & Dahlgren, S., 2008. Perovskite U-Pb ages and the Pb isotopic composition of alkaline volcanism initiating the Permo-Carboniferous Oslo Rift. Earth and Planetary Science Letters 265(1): 256-269. DOI: 10.1016/j.epsl.2007.10.019

Corfu, F. & Larsen, B.T., 2020. U-Pb systematics in volcanic and plutonic rocks of the Krokskogen area: Resolving a 40 million years long evolution in the Oslo Rift. Lithos 376-377: 105755. DOI: 10.1016/j.lithos.2020.105755

Cottençon, A., Parant, B. & Flacelière, B., 1975. Lower Cretaceous Gas-fields in Holland. In: Woodland, A.W. (ed.): Petroleum and the continental shelf of northwest Europe. Applied Science Publications (London) 1-30: 403-412

Courtillot, V., Jaupart, C., Manighetti, I., Tapponnier, P. & Besse, J., 1999. On causal links between flood basalts and continental breakup. Earth and Planetary Science Letters 166(3): 177-195. DOI: 10.1016/S0012-821X(98)00282-9

Cremer, H., Bunnik, F.P.M., Donders, T.H., Hoek, W.Z., Koolen-Eekhout, M., Koolmees, H.H. & Lavooi, E., 2010. River flooding and landscape changes impact ecological conditions of a scour hole lake in the Rhine-Meuse delta, The Netherlands. Journal of Paleolimnology 44: 789-801. DOI: 10.1007/s10933-010-9452-2

Crommelin, R.D., 1963. Een vulkanisch sediment in de ondergrond van de Betuwe. Netherlands Journal of Geosciences-Geologie en Mijnbouw 42(4): 112-113.

Dahlgren, S., Corfu, F. & Heaman, L.M., 1996. U-Pb time constraints, and Hf and Pb source characteristics of the Larvik plutonic complex, Oslo Paleorift. Geodynamic and geochemical implications for the rift evolution. VM Goldschmidt Conference. Journal of Conference Abstracts 1 (Cambridge Publications).

Davies, S.M., Hoek, W.Z., Bohncke, S.J.P., Lowe, J.J., Pyne-O'Donnell, S. & Turney, C.S.M., 2005. Detection of Lateglacial distal tephra layers in the Netherlands. Boreas 34: 123-135. DOI: 10.1080/03009480510012944

De Bruin, G., Bouroullec, R., Geel, K., Abdul Fattah, R., Van Hoof, T., Pluymaekers, M., Van der Belt, F., Vandeweijer, V.P. & Zijp, M., 2015. New petroleum plays in the Dutch northern offshore. TNO (Utrecht), Report No. 2015 R10920. https://www.nlog.nl/sites/default/files/2018-11/NORTHERN_OFFSHORE_TNO_report_2015_R10920_0.pdf

De Jager, J., Van Ojik, K. & Smit, J., 2025. Geological development. In: Ten Veen, J.H., Vis, G.-J., De Jager, J. & Wong, Th.E. (eds): Geology of the Netherlands, second edition. Amsterdam University Press (Amsterdam): 21-51. DOI: 10.5117/9789463728362_ch01

De Vos, W., Feldrappe, H., Pharaoh, T.C., Smith, N.J.P., Vejbæk, O.V., Verniers, J., Nawrocki, J., Poprawa, P. & Bełka, Z., 2010. Pre-Devonian. In: Doornenbal, J.C. & Stevenson, A.G. (eds): Petroleum Geological Atlas of the Southern Permian Basin Area. EAGE Publications B.V. (Houten): 59-69.

Dewey, J.F. & Windley, B.F., 1988. Palaeocene-Oligocene tectonics of NW Europe. In: Morton, A.C. & Parsons, L.M. (eds): Early Tertiary volcanism and the opening of the NE Atlantic. Geological Society of London, Special Publications 39: 25-31.

Dirkzwager, J.B., Stephenson, R.A. & Legostaeva, O.V., 2000. The pre-Permian residual gravity field for the Dutch onshore and adjacent offshore. Global and Planetary Change 27(1-4): 53-66.

Dixon, J.E., Fitton, J.G. & Frost, R.T.C., 1981. The tectonic significance of post-Carboniferous igneous activity in the North Sea

Basin. In: Illing, L.V. & Hobson, G.D. (eds): Petroleum Geology of the continental Shelf of North-West Europe: Proceedings of the 2nd Conference. Institute of Petroleum (London): 121-137.

Donato, J.A., 1993. A buried batholith and the origin of the Sole Pit Basin, UK Southern North Sea. Journal of the Geological Society 150: 255-258.

Donato, J.A., Martindale, W. & Tully, M.C., 1983. Buried granites within the Mid North Sea High. Journal of the Geological Society 140: 825-837.

Donato, J.A., 2020. Gravity modelling across two postulated granite batholiths within the UK onshore East Midlands Shelf. UK Onshore Geophysical Library: 14 pp.

Dusar, M. & Langenaeker, V., 1992. De oostrand van het Massief van Brabant, met beschrijving van de geologische verkenningsboring te Martenslinde. Belgische Geologische Dienst: 22 pp.

Eckhardt, F.J., 1979. Der permische Vulkanismus Mitteleuropas. Geologisches Jahrbuch D35: 3-84.

Eckhardt, F.J., 1968. Vorkommen und Petrogenese spilitisierter Diabase des Rotliegenden im Weser-Ems-Gebiet. Geologisches Jahrbuch 85: 227-264.

Egger, H. & Brückl, W., 2006. Gigantic volcanic eruptions and climatic change in the early Eocene. International Journal of Earth Sciences 95(6): 1065-1070. DOI: 10.1007/s00531-006-0085-7

Eigenfeld, R.W.F. & Eigenfeld-Mende, I., 1986. Niederländische permokarbone basische Magmatite als Fortsetzung der spilisierten Effusiva in NW-Deutschland. Mededelingen Rijks Geologische Dienst 40(1): 11-21.

Erkens, G., 2009. Sediment dynamics in the Rhine catchment - Quantification of fluvial response to climate change and human impact. PhD thesis Utrecht University, Netherlands Geographical Studies 388: 278 pp.

Ernst, R.E. & Buchan, K.L., 1997. Giant radiating dyke swarms: their use in identifying pre-Mesozoic large igneous provinces and mantle plumes. In: Large Igneous Provinces: Continental, Oceanic, and Planetary Flood Volcanism. American Geophysical Union: 297-334.

Erren, H. & Bredewout, J.W., 1991. Model calculations on intrusive cooling and related coalification of the Peel-Erkelenz coalfield (The Netherlands and Germany). Geologie En Mijnbouw 70(3): 243-252.

Fall, H.G., Gibb, F.G.F. & Kanaris-Sotiriou, R., 1982. Jurassic volcanic rocks of the northern North Sea. Journal of the Geological Society 139(3): 277-292. DOI: 10.1144/gsjgs.139.3.0277

Fekiacova, Z., Mertz, D.F. & Renne, P.R., 2007. Geodynamic Setting of the Tertiary Hocheifel Volcanism (Germany), Part I: 40Ar/39Ar geochronology. In: Ritter, J.R.R. & Christensen, U.R. (eds): Mantle Plumes: A Multidisciplinary Approach. Springer (Berlin, Heidelberg): 185-206. DOI: 10.1007/978-3-540-68046-8_6

Foulger, G.R. & Anderson, D.L., 2005. A cool model for the Iceland hotspot. Journal of Volcanology and Geothermal Research 141(1): 1-22. DOI: 10.1016/j.jvolgeores.2004.10.007

Frezzotti, M.L., 1992. Fluid and melt inclusion study (No. XUM/090; 15 pp.). Nederlandse Aardolie Maatschappij B.V. (Assen): 15 pp.

Frost, R.T.C., Fitch, F.J. & Miller, J.A., 1981. The age and nature of the crystalline basement of the North Sea Basin. In: Illing, L.V. & Hobson, G.D. (eds): Petroleum Geology of the continental Shelf of North-West Europe: Proceedings of the 2nd Conference. Institute of Petroleum (London): 43-57.

Furnes, H., Elvsborg, A. & Malm, O.A., 1982. Lower and Middle Jurassic alkaline magmatism in the Egersund sub-basin, North Sea. Marine Geology 46(1-2): 53-69.

Gaina, C., Medvedev, S., Torsvik, T.H., Koulakov, I. & Werner, S.C., 2014. 4D Arctic: A Glimpse into the Structure and Evolution of the Arctic in the Light of New Geophysical Maps, Plate Tectonics and Tomographic Models. Surveys in Geophysics 35(5): 1095-1122. DOI: 10.1007/s10712-013-9254-y

Galavazi, M.A., 1999. Study of volcanic fragments in cuttings samples of the lower Delfland, wells P9-8 and P12-8 (No. G307b-1). PanTerra Geoconsultants B.V. for Clyde Petroleum Exploration B.V.: 7 pp. https://www.nlog.nl/brh-web/rest/brh/document/605736932

Gast, R.E., Dusar, M., Breitkreuz, C., Gaupp, R., Schneider, J.W., Stemmerik, L., Geluk, M.C., Geißler, M., Kiersnowski, H., Glennie, K.W., Kabel, S. & Jones, N.S., 2010. Rotliegend. In: Doornenbal, J.C. & Stevenson, A.G. (eds): Petroleum Geological Atlas of the Southern Permian Basin Area. EAGE Publications B.V. (Houten): 100-121.

Gauer, M.B., Upton, A.J. & McGrandle, A., 2004. Identification and interpretation of igneous sills in the Zechstein of the Southern North Sea Basin. Extended Abstract presented at PETEX: 23-25 pp.

Geluk, M.C., 1997. Palaeogeographic maps of Moscovian and Artinskian: contributions from the Netherlands. In: Crasquin-Soleau, S. & De Wever, P. (eds): Peri-Tetheys stratigraphic correlations. Geodiversitas 19: 229-234.

Gibb, F.G.F. & Kanaris-Sotiriou, R., 1976. Jurassic igneous rocks of the Forties Field. Nature 260: 23-25.

Glennie, K.W., 1997. Recent advances in understanding the southern North Sea Basin: a summary. In: Ziegler, K., Turner, P. & Daines, S.R. (eds): Petroleum Geology of the Southern North Sea: Future Potential. Geological Society of London: 17-29. DOI: 10.1144/GSL.SP.1997.123.01.03

Global Volcanism Program, 2023. [Database] Volcanoes of the World. Smithsonian Institution: v. 5.1.2. DOI: 10.5479/si.GVP.VOTW5-2023.5.1

Gradstein, F.M., Ogg, J.G. & Smith, A.G., 2004. A Geologic Time Scale 2004. Cambridge University Press (Cambridge): 589 pp.

Guterch, A., Grad, M., Krawczyk, C.M., Wybraniec, S., Ziegler, P.A. & De Vos, W., 2010. Crustal structure and structural framework. In: Doornenbal, J.C. & Stevenson, A.G. (eds): Petroleum Geological Atlas of the Southern Permian Basin Area. EAGE Publications B.V. (Houten): 10-23.

Hamilton, M. & Pearson, G., 2011. Precise U-Pb Age for the Great Whin Dolerite Complex, N.E. England and Its Significance: 495-507. DOI: 10.1007/978-3-642-12496-9_26

Harrison, R.K., Jeans, C.V. & Merriman, R.J., 1979. Mesozoic igneous rocks, hydrothermal mineralisation and volcanogenic sediments in Britain and adjacent regions. Bulletin of the Geological Survey of Great Britain 70: 57-69.

Heeremans, M., Timmerman, M.J., Kirstein, L.A. & Faleide, J.I., 2004. New constraints on the timing of late Carboniferous - early Permian volcanism in the central North Sea. In: Wilson, M., Neumann, E.-R., Davies, G.R., Timmerman, M.J., Heeremans, M. & Larsen, B.T. (eds): Permo-Carboniferous magmatism and rifting in Europe 223: 177-194. DOI: 10.1144/GSL.SP.2004.223.01.08

Helmers, H., 1991. Ultramafic alkalibasalt of the Heinenoord-1 drilling. Vrije Universiteit for NAM (Assen): 29 pp.

Hernandez Casado, J. & Underhill, J.R., 2013. Consequences of intrusions for sub-salt, sub-basalt exploration and development. 75th EAGE Conference and Exhibition (London). EAGE (Houten): cp-348-00615. DOI: 10.3997/2214-4609.20130141

Herngreen, C.F.W., 1985. Resultaten van het palynologisch onderzoek van de kernen 1 en 2 van boring F3-7 (No. 2167-A; 2 pp.). Rijks Geologische Dienst: 2 pp.

Herngreen, G.F.W., Smit, R. & Wong, Th.E., 1991. Stratigraphy and tectonics of the Vlieland basin, The Netherlands. In: Spencer, A.M. (ed.): Generation, accumulation and production of Europe's hydrocarbons. Special Publication of the European Association of Petroleum Geoscientists (Berlin) 1: 175-192.

Hole, M.J. & Natland, J.H., 2020. Magmatism in the North Atlantic Igneous Province; mantle temperatures, rifting and geodynamics. Earth-Science Reviews 206: 102794. DOI: 10.1016/j.earscirev.2019.02.011

Hopper, J. R., Funck, T., Stoker, M., Árting, U., Péron-Pinvidic, G., Doornenbal, H. & Gaina, C. (eds), 2014. Tectonostratigraphic Atlas of the North-East Atlantic Region. Geological Survey of Denmark and Greenland (GEUS) (Copenhagen, Denmark): 339 pp.

Huis in 't Veld, R. & Den Hartog Jager, D., 2025. Late Carboniferous. In: Ten Veen, J.H., Vis, G.-J., De Jager, J. & Wong, Th.E. (eds): Geology of the Netherlands, second edition. Amsterdam University Press (Amsterdam): 95-125. DOI: 10.5117/9789463728362_ch03

Jacqué, M. & Thouvenin, J., 1975. Lower Tertiary tuffs and volcanic activity in the North Sea. In: Woodland, A.W. (ed.): Petroleum and the continental shelf of northwest Europe. Applied Science Publications (London) 1-30: 455-465.

Jeans, C.V., Merriman, R.J. & Mitchell, J.G., 1977. Origin of Middle Jurassic and Lower Cretaceous fuller's earths in England. Clay Minerals 12: 11-44.

Jeans, C.V., Wray, D.S., Merriman, R.J. & Fisher, M.J., 2000. Volcanogenic clays in Jurassic and Cretaceous strata of England and the North Sea Basin. Clay Minerals 35(1): 25-55.

Jung, C., Jung, S., Hoffer, E. & Berndt, J., 2006. Petrogenesis of Tertiary Mafic Alkaline Magmas in the Hocheifel, Germany. Journal of Petrology 47(8): 1637-1671. DOI: 10.1093/petrology/egl023

Jung, S., Vieten, K., Romer, R.L., Mezger, K., Hoernes, S. & Satir, M., 2012. Petrogenesis of Tertiary Alkaline Magmas in the Siebengebirge, Germany. Journal of Petrology 53(11): 2381-2409. DOI: 10.1093/petrology/egs047

Juvigné, E., 1999. Téphrostratigraphie du Quaternaire en Belgique. Geologica Belgica 2/3: 73-87. DOI: 10.20341/gb.2014.011

Juvigné, E. & Semmel, A., 1981. Un tuf volcanique semblable à l'Eltviller Tuff dans les loess de Hesbaye (Belgique) et du Limbourg néerlandais. E&G Quaternary Science Journal 31(1): 83-90. DOI: 10.3285/eg.31.1.08

Kettel, D., 1983. The East Groningen Massif - Detection of an intrusive body by means of coalification. Geologie En Mijnbouw 62(1): 203-210.

Kimbell, G.S. & Williamson, J.P., 2015. A gravity interpretation of the Central North Sea. British Geological Survey (Nottingham, UK), Commissioned Report No. CR/15/119: 75 pp.

Kimpe, W.F.M., 1953. Doleritic and gabbroic instrusives in the Autunian (Lower Permian) of the boring Wanneperveen 1, eastern Netherlands. Geologie en Mijnbouw 15: 57-65.

King, A.D., Marshall, K.L. & Burton, C.A., 1985. Well P/6-B1 Stratigraphical Report (Interval 1545-1850 m). Paleoservices Ltd. for Mobil Producing Netherlands Inc., Report No. 1523: 18 pp. https://www.nlog.nl/brh-web/rest/brh/document/910480758

Kirstein, L.A., Davies, G.R. & Heeremans, M., 2006. The petrogenesis of Carboniferous–Permian dyke and sill intrusions across northern Europe. Contributions to Mineralogy and Petrology 152(6): 721-742. DOI: 10.1007/s00410-006-0129-9

Kirton, S.R. & Donato, J.A., 1985. Some buried Tertiary dykes of Britain and surrounding waters deduced by magnetic modelling and seismic reflection methods. Journal of the Geological Society 142(6): 1047-1057. DOI: 10.1144/gsjgs.142.6.1047

Klaver, M., Blundy, J.D. & Vroon, P.Z., 2018. Generation of arc rhyodacites through cumulate-melt reactions in a deep crustal hot zone: Evidence from Nisyros volcano. Earth and Planetary Science Letters 497: 169-180. DOI: 10.1016/j.epsl.2018.06.019

Knox, R.W.O., 1997. The late Paleocene to early Eocene ash layers of the Danish Mo-Clay (Fur Formation): stratigraphic and tectonic significance. In: Thomsen, E. & Pedersen, S.A.S. (eds): Geology and palaeontology of the Mo-clay. Aarhus Geoscience 6: 7-11.

Knox, R.W.O. & Morton, A.C., 1988. The record of early Tertiary N Atlantic volcanism in sediments of the North Sea Basin. Geological Society Special Publication 39: 407-419.

Kombrink, H., Doornenbal, J.C., Duin, E.J.T., Den Dulk, M., Ten Veen, J.H. & Witmans, N., 2012. New insights into the geological structure of the Netherlands; results of a detailed mapping project. Netherlands Journal of Geosciences 91(4): 419-446. DOI: 10.1017/S0016774600000329

Kortekaas, M., Böker, U., Van der Kooij, C. & Jaarsma, B., 2018.

Lower Triassic reservoir development in the northern Dutch offshore. Geological Society of London, Special Publications 469: 149-168. DOI: 10.1144/SP469.19
Krueger, 1982. Potassium-Argon age determination of a sample from borehole K12-05-S1. Krueger Enterprises Inc., Geochron Laboratories Division (Cambridge), Report No. R-6006: 1 pp. https://www.nlog.nl/brh-web/rest/brh/document/3881042201
Kuijper, R.P., 1991. Petrology of a dolerite in Netherlands offshore well G/17-2. Scripta Geologica 97 (Journal Article): 33-97.
Kuijper, R.P., 2003. Petrography of igneous rocks in E18-5. Goal Olie- en Gasexploratie B.V. for Wintershall Noordzee B.V., Report No. 2003/025: 14 pp. https://www.nlog.nl/brh-web/rest/brh/document/3831683293
Larsen, L.M., Fitton, J.G. & Pedersen, A.K., 2003. Paleogene volcanic ash layers in the Danish Basin: compositions and source areas in the North Atlantic Igneous Province. Lithos 71(1): 47-80. DOI: 10.1016/j.lithos.2003.07.001
Latin, D.M., Dixon, J.E. & Fitton, J.G., 1990a. Rift-related magmatism in the North Sea basin. In: Blundell, D.J. & Gibbs, A.D. (eds): Tectonic Evolution of the North Sea Rifts. Oxford Science Publications (Oxford): 101-144.
Latin, D.M., Dixon, J.E., Fitton, J.G. & White, N., 1990b. Mesozoic magmatic activity in the North Sea Basin: implications for stretching history. Tectonic Events Responsible for Britain's Oil and Gas Reserves 55: 207-227.
Latin, D.M., 1990. The relationship between extension and magmatism in the North Sea Basin. PhD thesis, University of Edinburgh: 404 pp.
Latin, D.M. & Waters, F.G., 1992. Basaltic magmatism in the North Sea and its relationship to lithospheric extension. Tectonophysics 208(1-3): 77-90. DOI: 10.1016/0040-1951(92)90337-6
Le Maitre, R. W., Streckeisen, A., Zanettin, B., Le Bas, M. J., Bonin, B. & Bateman, P. (eds), 2002. Igneous Rocks: A Classification and Glossary of Terms: Recommendations of the International Union of Geological Sciences Subcommission on the Systematics of Igneous Rocks. Cambridge University Press (Cambridge): 2nd ed. DOI: 10.1017/CBO9780511535581
Lester, M.J., 1985. Biostratigraphical analysis of the Amoco F/16-2 Well, offshore Netherlands. Amoco Netherlands Petroleum Company (Surrey): 25 pp.
Linnemann, U., Herbosch, A., Liégeois, J.P., Pin, C., Gärtner, A. & Hofmann, M., 2012. The Cambrian to Devonian odyssey of the Brabant Massif within Avalonia: A review with new zircon ages, geochemistry, Sm-Nd isotopes, stratigraphy and palaeogeography. Earth-Science Reviews 112: 126-154. DOI: 10.1016/j.earscirev.2012.02.007
Lippolt, H.J., Hess, J.C. & Burger, K., 1984. Isotopische Alter von pyroklastischen Sanidinen aus Kaolin-Kohlentonsteinen als Korrelationsmarken für das mitteleuropäische Oberkarbon. Fortschritte in Der Geologie von Rheinland Und Westfalen 32: 119-150.
Lloyd, F.E., Arima, M. & Edgar, A.D., 1985. Partial melting of a phlogopite-clinopyroxenite nodule from south-west Uganda: an experimental study bearing on the origin of highly potassic continental rift volcanics. Contributions to Mineralogy and Petrology 91(4): 321-329. DOI: 10.1007/BF00374688
Lohne, Ø.S., Mangerud, J. & Birks, H.H., 2013. Precise 14C ages of the Vedde and Saksunarvatn ashes and the Younger Dryas boundaries from western Norway and their comparison with the Greenland Ice Core (GICC05) chronology. Journal of Quaternary Science 28(5): 490-500. DOI: 10.1002/jqs.2640
Lundmark, A.M., Gabrielsen, R.H., Austrheim, H., Flaat, K., Strand, T. & Ohm, S.E., 2012. Late Devonian rifting in the central North Sea: Evidence from altered felsic volcanic rocks in the Embla oil field. Marine and Petroleum Geology 29: 204-218. DOI: 10.1016/j.marpetgeo.2011.06.002
Lundmark, A.M., Gabrielsen, R.H. & Brown, J.F., 2011. Zircon U-Pb age for the Orkney lamprophyre dyke swarm, Scotland, and relations to Permo- Carboniferous magmatism in northwestern Europe. Journal of the Geological Society 168: 1233-1236.
Lundmark, A.M., Gabrielsen, R.H., Strand, T. & Ohm, S.E., 2018. Repeated post-Caledonian intra-cratonic rifting in the central North Sea: Evidence from the volcanic record in the Embla oil field. Marine and Petroleum Geology 92: 505-518. DOI: 10.1016/j.marpetgeo.2017.11.018
Lyngsie, S.B. & Thybo, H., 2007. A new tectonic model for the Laurentia-Avalonia-Baltica sutures in the North Sea: A case study along MONA LISA profile 3. Tectonophysics 429(3-4): 201-227. DOI: 10.1016/j.tecto.2006.09.017
Macdonald, R., Gass, K.N., Thorpe, R.S. & Gass, I.G., 1984. Geochemistry and petrogenesis of the Derbyshire Carboniferous basalts. Journal of the Geological Society 141: 147-159.
Mansy, J.L., Everaerts, M. & De Vos, W., 1999. Structural analysis of the adjacent Acadian and Variscan fold belts in Belgium and northern France from geophysical and geological evidence. Tectonophysics 309(1-4): 99-116. DOI: 10.1016/S0040-1951(99)00134-1
Marty, B., Pik, R. & Gezahegn, Y., 1996. Helium isotopic variations in Ethiopian plume lavas: nature of magmatic sources and limit on lower mantle contribution. Earth and Planetary Science Letters 144(1): 223-237. DOI: 10.1016/0012-821X(96)00158-6
McKenzie, D. & Bickle, M.J., 1988. The Volume and Composition of Melt Generated by Extension of the Lithosphere. Journal of Petrology 29(3): 625-679. DOI: 10.1093/petrology/29.3.625
Meadows, N. & Hrycyszyn, A., 1992. A sedimentological and petrographical evaluation of 17.90 m of core from Bow Valley Industries LTD. Well M/3-1, southern North Sea, Netherlands sector. Geochem Group NAM (Assen), Report No. 5435/0004: 91 pp. https://www.nlog.nl/brh-web/rest/brh/document/1784665012
Mees, R.P.R. & Meijs, E.P.M., 1984. Note on the presence of pre-weichselian loess deposits along the Albert Canal near Kesselt

and Vroenhoven (Belgian Limbourg). Geologie en Mijnbouw 63(1): 7-11.

Meijs, E., Mücher, H.J., Ouwerkerk, G., Romein, A. & Stoltenberg, H., 1983. Evidence of Presence of the Eltville Tuff Layer in Dutsch and Belgian Limbourg and the Consequences for the Loess Stratigraphy. Eiszeitalter Und Gegenwart: Jahrbuch Der Deutschen Quartärvereinigung 33: 59 pp.

Meijs, E.P.M., 2011. The Veldwezelt site (province of Limburg, Belgium): environmental and stratigraphical interpretations. Netherlands Journal of Geosciences 90 (2-3): 73-94. DOI: 10.1017/S0016774600001037

Meijs, E.P.M., 1980. A short note on the presence of the Eltville-tuff layer in the surroundings of Maastricht. Geologie en Mijnbouw 59(4): 409-410.

Meijs, E.P.M., 1985. Loess stratigraphical research at the Palaeolithic site Maastricht-Belvédère. Mededelingen Rijks Geologische Dienst 39(1): 31-34.

Meyer, R., Wijk, J. & Gernigon, L., 2007. The North Atlantic Igneous Province: A review of models for its formation. The Geological Society of America, Special Paper 430: 525-552.

Milton-Worssell, R.J., Smith, K., McGrandle, A., Watson, J. & Cameron, D., 2010. The search for a Carboniferous petroleum system beneath the Central North Sea. Petroleum Geology Conference Series 7: 57-75. DOI: 10.1144/0070057

Mitchell, J.G., Rands, P.N. & Ineson, P.R., 1989. Perturbation of the K-Ar age system in the Cleveland dyke, U.K.: Evidence of an Early Eocene age for barite mineralisation in the Magnesian Limestone of County Durham. Chemical Geology: Isotope Geoscience Section 79(1): 49-64. DOI: 10.1016/0168-9622(89)90006-7

Monaghan, A.A. & Pringle, M.S., 2004. 40Ar/39Ar geochronology of Carboniferous-Permian volcanism in the Midland Valley, Scotland. Geological Society, London, Special Publications 223(1): 219-241. DOI: 10.1144/GSL.SP.2004.223.01.10

Monaghan, A.A. & Parrish, R.R., 2006. Geochronology of Carboniferous–Permian magmatism in the Midland Valley of Scotland: implications for regional tectonomagmatic evolution and the numerical time scale. Journal of the Geological Society 163(1): 15-28. DOI: 10.1144/0016-764904-142

Morgan, W.J., 1971. Convection Plumes in the Lower Mantle. Nature 230(5288): 42-43. DOI: 10.1038/230042a0

Morton, A.C. & Knox, R.O.B., 1990. Geochemistry of late Palaeocene and early Eocene tephras from the North Sea Basin. Journal of the Geological Society 147(3): 425-437. DOI: 10.1144/gsjgs.147.3.0425

Neumann, E.-R., Olsen, K.H., Baldridge, W.S. & Sundvoll, B., 1992. The Oslo Rift: a review. Tectonophysics 208(1-3): 1-18. DOI: 10.1016/0040-1951(92)90333-2

Neumann, E.-R., Wilson, M., Heeremans, M., Spencer, E.A., Obst, K., Timmerman, M.J. & Kirstein, L., 2004. Carboniferous-Permian rifting and magmatism in southern Scandinavia, the North Sea and northern Germany: a review. Permo-Carboniferous Magmatism and Rifting in Europe 223: 11-40.

NITG, 1998. Geological Atlas of the subsurface of the Netherlands (1:250,000), Explanation to Map Sheet X Almelo–Winterswijk. Netherlands Institute of Applied Geoscience (NITG-TNO, Haarlem): 134 pp.

NITG, 2000. Geological Atlas of the subsurface of the Netherlands (1:250,000), Explanation to Map Sheet VI Veendam Hoogeveen. Netherlands Institute of Applied Geoscience (NITG-TNO, Utrecht): 152 pp.

Obst, K., Ansorge, J., Matting, S. & Hüneke, H., 2015. Early Eocene volcanic ashes on Greifswalder Oie and their depositional environment, with an overview of coeval ash-bearing deposits in northern Germany and Denmark. International Journal of Earth Sciences 104(8): 2179-2212. DOI: 10.1007/s00531-015-1203-1

Obst, K., Hammer, J., Katzung, G. & Korich, D., 2004. The Mesoproterozoic basement in the southern Baltic Sea: insights from the G 14-1 off-shore borehole. International Journal of Earth Sciences 93(1): 1-12. DOI: 10.1007/s00531-003-0371-6

Park, C. & Schmincke, H., 2009. Apokalypse im Rheintal. Spektrum Der Wissenschaft 2: 78-87.

Parnell, J., 1988. Migration of biogenic hydrocarbons into granites: a review of hydrocarbons in British plutons. Marine and Petroleum Geology 5(4): 385-396. DOI: 10.1016/0264-8172(88)90032-3

Pearce, J.A., 1996. A users guide to basalt discrimination Diagrams. In: Wyman, D.A. (ed.): Trace element geochemisty of volcanic rocks: Application for massive sulphide exploration. Geological Association of Canada, Short Course Notes 12: 79-113.

Pearce, J.A., 1982. Trace element characteristics of lavas from destructive plate boundaries. In: Thorpe, R.S. (ed.): Orogenic andesites and related rocks. John Wiley and Sons (Chichester, England): 528-548.

Pedersen, L.E., Heaman, L.M. & Holm, P.M., 1995. Further constraints on the temporal evolution of the Oslo Rift from precise U-Pb zircon dating in the Siljan-Skrim area. Lithos 34(4): 301-315. DOI: 10.1016/0024-4937(94)00014-S

Peniguel, G., Acheriteguy, J., Bignoumba, J., Dugue, O., Euvrard, B., Gout, C., Janisek, J.M., Mathis, B., Meyer, A., Neumann, C., Nicolas, G. & Philippe, B., 1992. E/12-3 (The Netherlands offshore – Block E12) – Laboratory studies. Société Nationale Elf Aquitaine Production (France), Report No. 92/013: 59 pp.

Perrot, J. & Van der Poel, A.B., 1987. Zuidwal - a Neocomian gas field. In: Brooks, J. & Glennie, K. (eds): Petroleum Geology of North-West Europe. Graham & Trotman (London): 325-335.

Pharaoh, T.C., Dusar, M., Geluk, M.C., Kockel, F., Krawczyk, C.M., Krzywiec, P., Scheck-Wenderoth, M., Thybo, H., Vejbæk, O.V. & Van Wees, J.-D., 2010. Tectonic Evolution. In: Doornenbal, J.C. & Stevenson, A.G. (eds): Petroleum Geological Atlas of the Southern Permian Basin Area. EAGE Publications B.V. (Houten): 24-57.

Pharaoh, T.C., England, R. & Lee, M.K., 1995. The concealed Caledonide basement of Eastern England and the southern North

Sea - A review. Sudia Geophysica et Geodaetica 39: 330-346. DOI: 10.1007/BF02295826

Pharaoh, T.C., Winchester, J.A., Verniers, J., Lassen, A. & Seghedi, A., 2006. The Western Accretionary Margin of the East European Craton: an overview. In: Gee, D.G. & Stephenson, R.A. (eds): European Lithosphere Dynamics. Geological Society of London Memoirs 32: 291-312. DOI: 10.1144/GSL.MEM.2006.032.01.17

Pitcavage, E., Furman, T., Nelson, W.R., Kalegga, P.K. & Barifaijo, E., 2021. Petrogenesis of primitive lavas from the Toro Ankole and Virunga Volcanic Provinces: Metasomatic mineralogy beneath East Africa's Western Rift. Lithos 396-397: 106192. DOI: 10.1016/j.lithos.2021.106192

Plein, E., 1995. Norddeutsches Rotliegendbecken; Rotliegend-Monographie Teil II. Stratigraphie von Deutschland I. Courier Forschungsinstitut Senckenberg 183.

Pouclet, A. & Juvigne, E., 2009. The Eltville Tephra, a late Pleistocene widespread tephra layer in Germany, Belgium and the Netherlands; symptomatic compositions of the minerals. Geologica Belgica 12(1-2): 93-103.

Pouclet, A., Juvigné, E. & Pirson, S., 2008. The Rocourt Tephra, a Widespread 90-74 ka stratigraphic marker in Belgium. Quaternary Research 70(1): 105-120. DOI: 10.1016/j.yqres.2008.03.010

Preusser, F., 2008. Characterisation and evolution of the River Rhine system. Netherlands Journal of Geosciences 87(1): 7-19.

Przybyla, T., Pfänder, J.A., Münker, C., Kolb, M., Becker, M. & Hamacher, U., 2018. High-resolution 40Ar/39Ar geochronology of volcanic rocks from the Siebengebirge (Central Germany)—Implications for eruption timescales and petrogenetic evolution of intraplate volcanic fields. International Journal of Earth Sciences 107(4): 1465-1484. DOI: 10.1007/s00531-017-1553-y

Quirie, A.K., Schofield, N., Hartley, A., Hole, M.J., Archer, S.G., Underhill, J.R., Watson, D. & Holford, S.P., 2019. The Rattray Volcanics: Mid-Jurassic fissure volcanism in the UK Central North Sea. Journal of the Geological Society 176(3): 462-481. DOI: 10.1144/jgs2018-151

Quirie, A.K., Schofield, N., Jolley, D.W., Archer, S.G., Hole, M.J., Hartley, A., Watson, D., Burgess, R., Pugsley, J.H., Underhill, J.R. & Holford, S.P., 2020. Palaeogeographical evolution of the Rattray Volcanic Province, Central North Sea. Journal of the Geological Society 177(4): 718-737. DOI: 10.1144/jgs2019-182

Reinig, F., Wacker, L., Jöris, O., Oppenheimer, C., Guidobaldi, G., Nievergelt, D., Adolphi, F., Cherubini, P., Engels, S., Esper, J., Land, A., Lane, C., Pfanz, H., Remmele, S., Sigl, M., Sookdeo, A. & Büntgen, U., 2021. Precise date for the Laacher See eruption synchronizes the Younger Dryas. Nature 595(7865): 66-69. DOI: 10.1038/s41586-021-03608-x

Riber, L., Dypvik, H. & Sørlie, R., 2015. Altered basement rocks on the Utsira High and its surroundings, Norwegian North Sea. Norwegian Journal of Geology 95(1). DOI: 10.17850/njg95-1-04

Rieffe, E.C. & Van Lil, R., 1992. Optical analyses of igneous rock samples from 45 wells in the Netherlands and the Dutch part of the continental shelf. MSc thesis. Vrije Universiteit Amsterdam/NAM (Assen): 41 pp.

Rijkers, R. & Duin, E., 1994. Crustal observations beneath the southern North Sea and their tectonic and geological implications. Tectonophysics 240(1-4): 215-224.

Rijks Geologische Dienst, 1993. Geological atlas of the subsurface of the Netherlands (1 : 250 000). Explanation to Map sheet V Sneek- Zwolle. Rijks Geologische Dienst (Haarlem) 127.

Ritchie, J.D., Swallow, J.L., Mitchell, J.G. & Morton, A.C., 1988. Jurassic ages from intrusives and extrusives within the Forties igneous province. Scottish Journal of Geology 24(1): 81-88. DOI: 10.1144/sjg24010081

Ritter, J.R.R., Jordan, M., Christensen, U.R. & Achauer, U., 2001. A mantle plume below the Eifel volcanic fields, Germany. Earth and Planetary Science Letters 186(1): 7-14.

Rogers, N.W., De Mulder, M. & Hawkesworth, C.J., 1992. An enriched mantle source for potassic basanites: evidence from Karisimbi volcano, Virunga volcanic province, Rwanda. Contributions to Mineralogy and Petrology 111(4): 543-556. DOI: 10.1007/BF00320908

Rogers, N.W., James, D., Kelley, S.P. & De Mulder, M., 1998. The Generation of Potassic Lavas from the Eastern Virunga Province, Rwanda. Journal of Petrology 39(6): 1223-1247. DOI: 10.1093/petroj/39.6.1223

Rooney, T.O., 2020. The Cenozoic magmatism of East Africa: Part V – Magma sources and processes in the East African Rift. Lithos 360-361: 105296. DOI: 10.1016/j.lithos.2019.105296

Saunders, A.D., Fitton, J.G., Kerr, A.C., Norry, M.J. & Kent, R.W., 1997. The north Atlantic igneous province. In: Mahony, J.J. & Coffin, M.F. (eds): Large Igneous Provinces: Continental, Oceanic and Planetary. Geophysical Monograph-American Geophysical Union (Washington, D.C.) 100: 45-94.

Scheck, M. & Bayer, U., 1999. Evolution of the Northeast German Basin – inferences from a 3D structural model and subsidence analysis. Tectonophysics 313(1-2): 145-169. DOI: 10.1016/S0040-1951(99)00194-8

Schmincke, H.-U., Park, C. & Harms, E., 1999. Evolution and environmental impacts of the eruption of Laacher See Volcano (Germany) 12,900 a BP. Quaternary International 61(1): 61-72. DOI: 10.1016/S1040-6182(99)00017-8

Searle, M.P., Shail, R.K., Pownall, J.M., Jurkowski, C., Watts, A.B. & Robb, L.J., 2024. The Permian Cornubian granite batholith, SW England; Part 1: Field, structural, and petrological constraints. Geological Society of America Bulletin 136 (9-10), 4301-4320. DOI: 10.1130/B37457.1

Senglaub, Y., Littke, R. & Brix, M.R., 2006. Numerical modelling of burial and temperature history as an approach for an alternative interpretation of the Bramsche anomaly, Lower Saxony Basin. International Journal of Earth Sciences 95: 204-224. DOI: 10.1007/s00531-005-0033-y

Sissingh, W., 2004. Palaeozoic and Mesozoic igneous activity

in the Netherlands: a tectonomagmatic review. Netherlands Journal of Geosciences 83 (2): 113-134. DOI: 10.1017/S0016774600020084
Sissingh, W., 2003. Tertiary paleogeographic and tectonostratigraphic evolution of the Rhenish Triple Junction. Palaeogeography, Palaeoclimatology, Palaeoecology 196(1-2): 229-263. DOI: 10.1016/S0031-0182(03)00320-1
Slagstad, T., Davidsen, B. & Daly, J.S., 2011. Age and composition of crystalline basement rocks on the Norwegian continental margin: offshore extension and continuity of the Caledonian–Appalachian orogenic belt. Journal of the Geological Society 168(5): 1167-1185. DOI: 10.1144/0016-76492010-136
Slater, T.H.A., 1980. Potassium-Argon age dating and petrographic analysis of a core piece sample from the Union Oil F/4-2A well, Dutch North Sea (No. 4503P/C; 3 pp.). Robertson Research International Limited (Gwynedd, Wales): 3 pp.
Smedley, P.L., 1986. The relationship between calc-alkaline volcanism and within-plate continental rift volcanism: evidence from Scottish Palaeozoic lavas. Earth and Planetary Science Letters 77(1): 113-128. DOI: 10.1016/0012-821X(86)90137-8
Smit, J., Van Wees, J.-D. & Cloetingh, S., 2016. The Thor suture zone: From subduction to intraplate basin setting. Geology 44(9): 707-710. DOI: 10.1130/G37958.1
Smit, J., Van Wees, J.-D. & Cloetingh, S., 2018. Early Carboniferous extension in East Avalonia: 350 My record of lithospheric memory. Marine and Petroleum Geology 92(1): 1010-1027. DOI: 10.1016/j.marpetgeo.2018.01.004
Smith, K. & Ritchie, J.D., 1993. Jurassic volcanic centres in the Central North Sea. In: Parker, J.R. (ed.): Petroleum Geology of North-West Europe: Proceedings of the 4th Conference. Geological Society of London: 519-531.
Stemmerik, L., Ineson, J.R. & Mitchell, J.G., 2000. Stratigraphy of the Rotliegend Group in the Danish part of the Northern Permian Basin, North Sea. Journal of the Geological Society 157 (6): 1127-1136. DOI: 10.1144/jgs.157.6.1127
Stokke, E.W., Liu, E.J. & Jones, M.T., 2020. Evidence of explosive hydromagmatic eruptions during the emplacement of the North Atlantic Igneous Province. Volcanica 3(2): 227-250. DOI: 10.30909/vol.03.02.227250
Storey, M., Duncan, R.A. & Swisher, C.C., 2007. Paleocene-Eocene Thermal Maximum and the Opening of the Northeast Atlantic. Science 316(5824): 587-589. DOI: 10.1126/science.1135274
Tappe, S., Smart, K.A., Stracke, A., Romer, R.L., Prelević, D. & Van den Bogaard, P., 2016. Melt evolution beneath a rifted craton edge: 40Ar/39Ar geochronology and Sr-Nd-Hf-Pb isotope systematics of primitive alkaline basalts and lamprophyres from the SW Baltic Shield. Geochimica et Cosmochimica Acta 173: 1-36. DOI: 10.1016/j.gca.2015.10.006
Teichmüller, M. & Teichmüller, R., 1971. Inkohlung im Rhein-Ruhr-Revier. Fortschritte Geologie Rheinland und Westfalen 19: 69-72.
Tesch, P., 1925. Over een intrusie in het Carboon van oostelijk Gelderland. Verslag Vergadering van 22 Maart 1924. Geologie Sectie Geologisch Mijnbouwkundig Genootschap van Nederland en Koloniën 3(4).
Tesch, P., 1928. On the occurrence of igneous rocks in the Dutch Carboniferous. Le Congres pour l'Avancement des Études de Stratigrafie Carbonifère, Compte Rendu: 731-732.
Tesch, P. & Van Voorthuysen, J.H., 1944. Nog drie intrusies in het Carboon van Oost-Gelderland. Geologie en Mijnbouw 5: 56-57.
Thiadens, A.A., 1963. The Palaeozoic of The Netherlands. Verhandelingen van Het Koninklijk Nederlands Geologisch Mijnbouwkundig Genootschap, Geologische Serie 21(1): 9-28.
Thordarson, Th. & Self, S., 1993. The Laki (Skaftár Fires) and Grímsvötn eruptions in 1783-1785. Bulletin of Volcanology 55(4): 233-263. DOI: 10.1007/BF00624353
Timmerman, M.J., 2004. Timing, geodynamic setting and character of Permo-Carboniferous magmatism in the foreland of the Variscan Orogen, NW Europe. In: Wilson, M., Neumann, E.R., Davies, S.J., Timmerman, M.J., Heeremans, M. & Larsen, B.T. (eds): Permo-Carboniferous Magmatism and Rifting in Europe. Geological Society of London, Special Publications 223: 41-74. DOI: 10.1144/GSL.SP.2004.223.01.03
TNO-GDN, 2022. Emmen Volcanic Formation. Stratigraphic Nomenclature of the Netherlands, TNO – Geological Survey of the Netherlands. https://www.dinoloket.nl/en/stratigraphic-nomenclature/emmen-volcanic-formation
TNO-GDN, 2024. De Wijk Member. Stratigraphic Nomenclature of the Netherlands, TNO – Geological Survey of the Netherlands. https://www.dinoloket.nl/en/stratigraphic-nomenclature/de-wijk-member
Tomkeieff, S. & Tesch, P., 1931. On a dolerite in the Dutch Carboniferous. Geological Magazine 68(5): 231-236.
Torsvik, T.H., Mosar, J. & Eide, E.A., 2001. Cretaceous–Tertiary geodynamics: a North Atlantic exercise. Geophysical Journal International 146(3): 850-866. DOI: 10.1046/j.0956-540x.2001.01511.x
Torsvik, T.H., Smethurst, M.A., Burke, K. & Steinberger, B., 2008. Long term stability in deep mantle structure: Evidence from the ~300 Ma Skagerrak-Centered Large Igneous Province (the SCLIP). Earth and Planetary Science Letters 267(3): 444-452. DOI: 10.1016/j.epsl.2007.12.004
Underhill, J.R., 2009. Role of intrusion-induced salt mobility in controlling the formation of the enigmatic Silverpit Crater, UK Southern North Sea. Petroleum Geoscience 15(3): 197-216. DOI: 10.1144/1354-079309-843
Underhill, J.R. & Partington, M.A., 1993. Jurassic thermal doming and deflation in the North Sea: implications of the sequence stratigraphic evidence. In: Parker, J.R. (ed.): Petroleum Geology of Northwest Europe: Proceedings of the 4th Conference. Geological Society of London, Petroleum Geology Conference Series 4: 337-345. DOI: 10.1144/0040337
Upton, B.G.J., Stephenson, D., Smedley, P.M., Wallis, S.M. & Fitton, J.G., 2004. Carboniferous and Permian magmatism in Scotland. Geological Society of London, Special Publications 223(1):

195-218. DOI: 10.1144/GSL.SP.2004.223.01.09
Van Amerom, H.W.J., 1972. Das karbonische Alter der Tiefbohrung Wanneperveen 1, eine Revision. Geologie En Mijnbouw 51(4): 491-495.
Van Bergen, M.J. & Sissingh, W., 2007. Magmatism in the Netherlands:expression of the north-west European rifting history. In: Wong, Th.E., Batjes, D.A.J.& De Jager, J. (eds): Geology of the Netherlands. Royal Netherlands Academy of Arts and Sciences (KNAW, Amsterdam): 197-222.
Van Buggenum, J.M. & Den Hartog Jager, D.G., 2007. Silesian. In: Wong, Th.E., Batjes, D.A.J.& De Jager, J. (eds): Geology of the Netherlands. Royal Netherlands Academy of Arts and Sciences (KNAW, Amsterdam): 43-62.
Van den Bogaard, P. & Schmincke, H.-U., 1990. Die Entwicklungsgeschichte des Mittelrheinraumes und die Eruptionsgeschichte des Osteifel-Vulkanfeldes. In: Schirmer, W. (ed.): Rheingeschichte zwischen Mosel und Maas. Deuqua-Führer 1 (Hannover): 166-190.
Van den Bosch, M., 2015. Lithostratigraphy of the Oligocene in the Almelo-Winterswijk area, eastern Netherlands. Eburon Publishers (Delft): 164 pp.
Van der Sijp, J.W.C.M., 1953. Intrusive rocks in the Berkel well. Geologie en Mijnbouw 15: 65-66.
Van Voorthuysen, J.H., 1944. Hoornblendediabaas-intrusie in het Wealden van Oost-Nederland. Geologie en Mijnbouw 6: 24-26.
Van Wijhe, D.H., Lutz, M. & Kaasschieter, J.P.H., 1980. The Rotliegend in The Netherlands and its gas accumulations. Geologie En Mijnbouw 59(1): 3-24.
Vandenberghe, N., De Craen, M. & Wouters, L., 2014. The Boom clay geology - from sedimentation to present-day occurrence - a review. Memoirs of the Geological Survey of Belgium 60: 1-76.
Verbraeck, A., 1984. Toelichtingen bij de Geologische kaart van Nederland, schaal 1 : 50.000, bladen Tiel West (39W) en Tiel Oost (39O). Rijks Geologische Dienst (Haarlem).
Verweij, J.M., Nelskamp, S. & Hegen, D., 2016. Geochemical composition and origin of natural gas in onshore and offshore Netherlands TNO (Utrecht), Report No. R10599: 182 pp.
Vis, G.-J., Houben, A.J.P., Debacker, T.N. & Geel, C.R., 2025. The geological foundation of the Netherlands: the early Carboniferous, Devonian and older. In: Ten Veen, J.H., Vis, G.-J., De Jager, J. & Wong, Th.E. (eds): Geology of the Netherlands, second edition. Amsterdam University Press (Amsterdam): 53-93. DOI: 10.5117/9789463728362_ch02
Wall, M., Cartwright, J., Davies, R. & McGrandle, A., 2010. 3D seismic imaging of a Tertiary Dyke Swarm in the Southern North Sea, UK. Basin Research 22: 181-194.
Watts, A.B., Xu, C., Searle, M.P., Jurkowski, C. & Shail, R.K., 2024. The Permian Cornubian granite batholith, SW England; Part 2: Gravity anomalies, structure, and state of isostasy. Geological Society of America Bulletin 136 (9-10), 4381-4397. DOI: 10.1130/B37459.1
Wedepohl, K.H., Gohn, E. & Hartmann, G., 1994. Cenozoic alkali basaltic magmas of western Germany and their products of differentiation. Contributions to Mineralogy and Petrology 115(3): 253-278. DOI: 10.1007/BF00310766
White, N. & Latin, D., 1993. Subsidence analyses in the North Sea 'triple-junction'. Journal of the Geological Society 150(3): 473-488.
White, R.S. & McKenzie, D., 1995. Mantle plumes and flood basalts. Journal of Geophysical Research 100(B9): 17543-17585. DOI: 10.1029/95jb01585
Wilkinson, C.M., Ganerød, M., Hendriks, B.W.H. & Eide, E.A., 2017. Compilation and appraisal of geochronological data from the North Atlantic Igneous Province (NAIP). Geological Society of London, Special Publications 447(1): 69-103. DOI: 10.1144/SP447.10
Wilson, M., Neumann, E.-R., Davies, G.R., Timmerman, M.J., Heeremans, M. & Larsen, B.T., 2004. Permo-Carboniferous Magmatism and Rifting in Europe. In: Wilson, M., Neumann, E.-R., Davies, G.R., Timmerman, M.J., Heeremans, M. & Larsen, B.T. (eds): Permo-Carboniferous Magmatism and Rifting in Europe. Geological Society of London, Special Publications 223: 1-10.
Wolf, M. & Bless, M.J.M., 1987. Coal-petrographic investigations on samples from the boreholes Thermae 2000 and Thermae 2002 Valkenburg a/d Geul, The Netherlands. Annales de La Société Géologique de Belgique 110: 77-84.
Woodhall, D. & Knox, R.W.O., 1979. Mesozoic volcanism in the northern North Sea and adjacent areas. Bulletin of the Geological Survey of Great Britain 70: 34-56.
Wörner, G. & Schmincke, H.-U., 1984. Mineralogical and Chemical Zonation of the Laacher See Tephra Sequence (East Eifel, W. Germany). Journal of Petrology 25(4): 805-835. DOI: 10.1093/petrology/25.4.805
Wright, J.E., Hull, J.H., McQulllan, R. & Arnold, S.E., 1971. Irish Sea investigations 1969-70. Natural Environment Research Council.
Zagwijn, W.H., 1985. An outline of the Quaternary stratigraphy of The Netherlands. Geologie en Mijnbouw 64 (1): 17-24.
Zens, J., Zeeden, C., Römer, W., Fuchs, M., Klasen, N. & Lehmkuhl, F., 2017. The Eltville Tephra (Western Europe) age revised: Integrating stratigraphic and dating information from different Last Glacial loess localities. Palaeogeography, Palaeoclimatology, Palaeoecology 466: 240-251. DOI: 10.1016/j.palaeo.2016.11.033
Ziegler, P.A., 1990. Tectonic and palaeogeogaphic development of the North Sea Rift system. In: Blundell, D.J. & Gibbs, A.D. (eds): Tectonic Evolution of the North Sea Rifts. Oxford Science Publications (Oxford): 1-36.
Ziegler, P.A., 1988. Evolution of the Arctic, North Atlantic and western Tethys. American Association of Petroleum Geologists Memoir. 43: 198 pp.
Ziegler, P.A., 1992. North Sea rift system. Tectonophysics 208(1-3): 55-75. DOI: 10.1016/0040-1951(92)90336-5
Ziegler, P.A. & Dèzes, P., 2005. Evolution of the lithosphere in the

area of the Rhine Rift System. International Journal of Earth Sciences 94(4): 594-614.
Zimmerle, W., 1993. On the lithology and provenance of the Rupelian Boom Clay in northern Belgium, a volcaniclastic deposit. Bulletin de La Société Belge de Géologie 102: 91-103.

12 Salt tectonics

Renaud Bouroullec & Johan H. ten Veen*

*Corresponding author: renaud.bouroullec@tno.nl

ABSTRACT

Salt tectonics refers to structural deformation involving bodies composed of evaporite minerals. The Dutch subsurface is characterized by a thick sequence of Zechstein salt deposited during the Late Permian. Its deformation in response to several tectonic phases had a pronounced influence on Meso-Cenozoic tectonostratigraphic and depositional evolution. Salt layers of the Zechstein Group are the most prominent and regionally significant intervals that were affected by salt tectonics in the Netherlands. Other salt layers in the middle to early late Permian Silverpit Formation and of Triassic age (e.g. Main Röt Evaporite Member) were affected by more local and less extensive salt tectonics than the Zechstein and are of secondary importance. The main salt tectonic structures observed in the Dutch sector are due to large volumes of Zechstein salt that were mobilized laterally or vertically, feeding large salt bodies and allowing basins to form in response to a combination of salt withdrawal, extensional- and contractional tectonic forces. These growth basins include mini basins, sub-basins, rim synclines, rollover anticlines and collapse grabens. Successive crustal-scale tectonic events induced the salt tectonics in the Netherlands, localizing salt movements, enhancing salt migration, and deforming pre-existing salt bodies (e.g. through diapir squeeze). In this chapter we discuss the main salt tectonic processes, the original Zechstein salt thickness and the five types of salt structures observed in the Dutch subsurface (salt pillows, diapirs, walls, rollers, and sheets), including their geometry, their geographic distribution, their kinematics and their effects on associated structures such as growth fault/raft systems, crestal faults, turtle structures, caprock and collapse grabens. The Meso-Cenozoic sediment pathways, sediment distribution/preservation and the overall basin architecture were also affected by accommodation changes resulting from salt withdrawal and migration.

In the Netherlands, limited syn-depositional faulting during the Tubantian I phase (Geluk, 1999) caused local thickness variations within the Zechstein Group. The actual movement of Zechstein salt started as early as the Early Triassic but most of the significant salt tectonics occurred between the Middle Triassic and the Paleogene in response to a combination of gravitational forces (halokinetics) and tectonic events. Since the Paleogene, salt tectonic intensity has diminished but salt bodies and their crestal faults affect several processes in the shallow overburden up until the Quaternary.

<< Rhythmic millimetre-scale calcite (brown) and gypsum-bearing (white) laminites interpreted as seasonal varves formed from hemipelagic settling and correlatable over a hundred kilometres. Photo: Zoltan Sylvester.

DOI: 10.5117/9789463728362_ch12

Introduction

Knowledge about salt tectonics in the Dutch subsurface has grown rapidly since the seminal work of Heybroek (1975). The work carried out by the oil and gas industry as well as academic research on salt basins worldwide, including those of the North Sea Basin has provided the framework for salt tectonics analysis in the Netherlands (Clark-Lowes et al., 1987; Wong et al., 1989; Nalpas et al., 1995; Remmelts, 1995, 1996; Nalpas, 1996; Geluk, 2005; De Jager, 2012; Nelskamp et al., 2012; Ten Veen et al., 2012; Bouroullec et al., 2016, 2017, 2018; Peeters, 2016; Van Winden et al., 2018). The Zechstein salt in the Dutch subsurface displays a wide variety of salt body types and associated structures. 'Salt body' is a general term referring to any individual salt feature. In the Dutch sector the main salt body types are salt pillows, salt diapirs and salt walls but locally salt rollers and -sheets are observed as well (Fig. 12.1). Salt bodies and structures formed during multiple periods and their distribution depends on a combination of factors, including 1) the depositional thickness and lithology of the original (mother) salt layer, 2) the burial depth of the salt, 3) the topography of the subsalt strata (e.g. faulted pre-salt), 4) the sediment loading rate and direction, 5) the structural evolution of basins, platforms and highs (e.g. rifting, inversion), 6) the presence of pre-existing salt bodies formed during previous tectonic- or basin fill phases and 7) the occurrence and amount of erosion. All these factors will be described in this chapter.

Rheology and geomechanics of evaporites

Rheology is the study of deformation of matter and in geology deals with the flow of solid rock visible only over geological time. The rheological behaviour of evaporites and particularly rock salt (halite, NaCl) plays a major role in how salt basins develop. Geomechanics, i.e. the mechanics of rocks, is a broader term that describes the brittle behaviour of rocks and tackles the formation of faults and fractures. Both rheology and geomechanics play a role in salt tectonics, especially in the case of Zechstein salt deformation. The Zechstein Group is a layered geological interval composed of various lithologies (halite, anhydrite, carbonate, siliciclastics) with variable mechanical behaviour (Bouroullec & Geel, 2025, this volume). Halite layers are the weakest, while anhydrites, carbonates and siliciclastics are stronger. Such mechanical layering plays an important role in the way the internal deformation of salt layers and bodies occurs (Rowan et al., 2019). Salt weakness is extremely variable and many concepts and quantifications based on studies in the Netherlands are now used worldwide (Urai, 1983; Urai et al., 1986, 2008; Schléder & Urai, 2005). A thorough summary in Hudec and Jackson (2007) testifies to the current understanding of salt behaviour at multiple scales.

Halite behaves in a ductile fashion when deformed or when mixed with higher-pressured fluids. The short-term fracture strength of halite is as high as that of concrete.

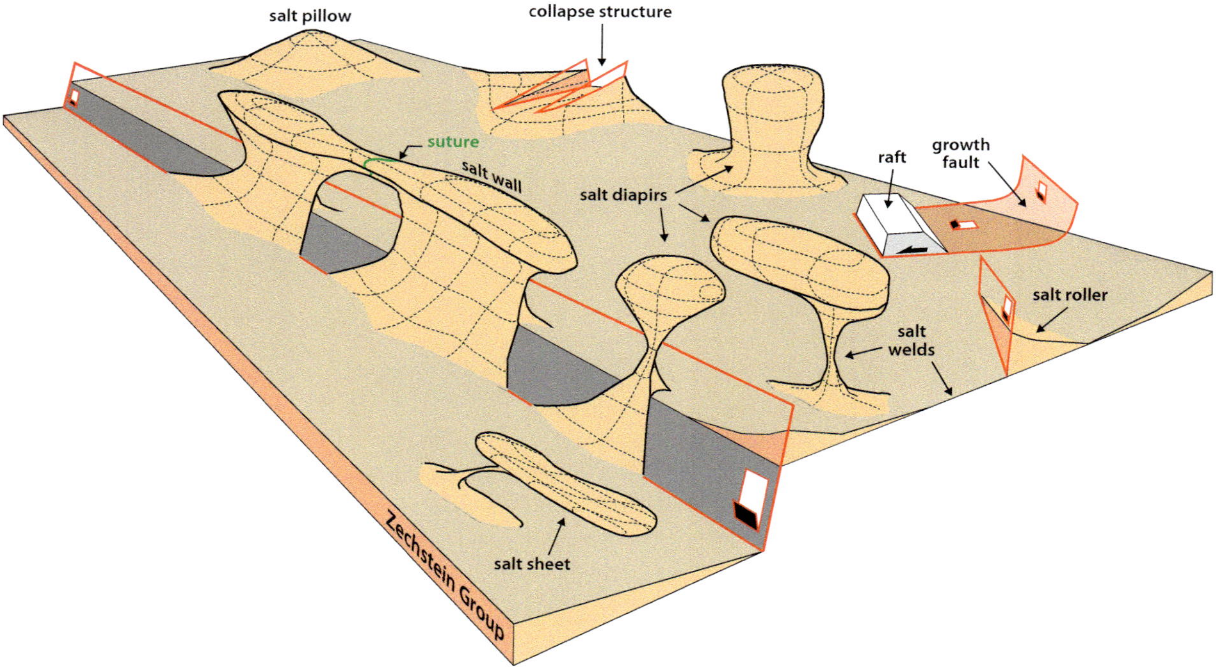

Figure 12.1. *The main salt body types in the Dutch subsurface. See text for explanation.*

However, the yield strength for long-term creep of halite is near zero, which means that salt will creep plastically even if shear stress is very low. Various processes such as dislocation glide, dislocation creep, solution-precipitation creep and water-assisted dynamic recrystallization play a role in the microstructural deformation of halite (Urai et al., 1986, 2008). Grain size (typically around 10 mm in diapiric halite), impurities, stress paths, temperature and fluid chemistry also play a role (Rowan et al., 2019). Differential stress in naturally deforming halite generally ranges between 0.5 and 2 MPa but can be as high as 5 MPa at shallow levels (Spiers & Carter, 1998; Schléder & Urai, 2005). Anhydrite ($CaSO_4$) is stronger than halite but weakens rapidly when burial temperatures reach 100° to 200°C. Deformation is aided by low effective stress caused by highly pressurized pore fluids sealed within encasing rock salt (Van Gent et al., 2011). Therefore, anhydrite is weaker than most rock types and forms effective decollements. For example, in the Dutch subsurface many faults are detached onto the Triassic Keuper evaporites. Thick anhydrite layers can also form stringers (see below for definition) when embedded in thick halite layers. If shortened, anhydrite stringers are folded, while they may be subject to boudinage while extended. Gypsum ($CaSO_4 \cdot 2H_2O$) is not as weak as halite but is still a ductile rock type and can also act as an effective detachment. However, the rates of pressure solution in compacting gypsum are typically three to four orders of magnitude lower than in halite under similar conditions (Spiers et al., 1990). Carnallite (hydrated potassium-magnesium chloride) and Bischofite (hydrated magnesium chloride) are much weaker than halite and are often referred to as squeezing salts by drilling engineers. These salts have much lower viscosities than halite and flow 100-1000 time faster under a given differential stress (Urai, 1983; Urai et al., 1986). For example, at 60°C, carnallite has a viscosity as much as 10^6 times lower than that of halite (Van Keken et al., 1993). For a differential stress of 10 MPa at a temperature of 60°C, carnallite creeps about 100 times faster than halite; and bischofite creeps about 10,000 times faster than halite (Urai et al., 1986).

Interbeds of various lithologies within halite layers will deform in specific manners when salt flows, depending on their own lithology, degree of cementation, overpressure, thickness, and burial. For example, in the Dutch sector, the "Z3 stringers" composed of anhydrite, carbonates and clay, are locally well imaged on seismic data since they are relatively thick (40-50 m in average and up to 150 m). These stringers are often folded or deformed as boudin structures and locally have perturbed salt flow (Van Gent et al., 2011). The mechanical behaviour of both carbonate and anhydrite is different from rock salt. Under conditions relevant for deformation of stringers, rock salt tends to become ductile while carbonates and anhydrites tend to be relatively more competent with folding being common; hence the stringer does not deform in a fully brittle fashion (Van Gent et al., 2011). This basically means the depth or confining pressure for the transition between brittle and ductile deformation is larger for carbonates and anhydrite than for rock salt. It also means that at greater depths, the rate of ductile deformation is lower for carbonate and anhydrite than for rock salt at similar flow stress.

Salt tectonics processes

Current understanding of salt tectonic processes has significantly benefited from technical progress made during the decades of hydrocarbon exploration. Access to 3D seismic data and improvement of seismic imaging and analysis techniques allows for better seismic and structural interpretation and has changed the view on how salt diapirs behave within the main extensional grabens in NW Europe (e.g. Mohr et al., 2005).

There is consensus that the main reason salt mobilizes and migrates in the subsurface over geological time scales is because of its mechanical weakness. However, since the 1930s the ideas on salt flow triggering mechanism(s), rheology, timing and modes of salt structure formation have changed. The movement of salt under the influence of gravity is referred to as halokinesis (Trusheim, 1957). One of the oldest ideas was that low-density salt penetrates higher-density overburden sediments solely due to differential loading (salt piercement model of Trusheim, 1957), and that salt buoyancy (lava lamp-like deformation) was the main process involved in salt tectonics. After the 1980s, the buoyancy process fell out of favour as an explanation for salt tectonics and the now-established – and still evolving concepts – consider the combination of a brittle and competent overburden and differential loading as the main process driving salt flow.

After reaching as little as 200-300 m of burial depth, the confining pressure and diagenetic processes affecting siliciclastic sediments increases to the point that they become mechanically stronger than salt layers. With further compaction the sedimentary overburden lithifies, increasing further the strength of the supra-salt sediments (sediments located above the salt layer or body). Contrastingly, the salt weakens only slightly with increasing burial and temperature. With continued burial, the salt layers become the weakest rock and, at a geological time scale, will tend to flow and deform under gravitational displacement and thermal loading forces (Hudec & Jackson, 2007). Active tectonics increases the likelihood of salt structure formation as it triggers differential loading (e.g. Vendeville &

Figure 12.2. a) Zechstein Group thickness map (in metres) in which six areas (see numbers) with distinct thickness trend are indicated (see text for explanation). b) Top Zechstein time structure map (depth in s) overlain by polygons indicating the location of salt pillows, diapirs and walls in the northern part of the Dutch offshore. Modified from Bouroullec et al. (2017).

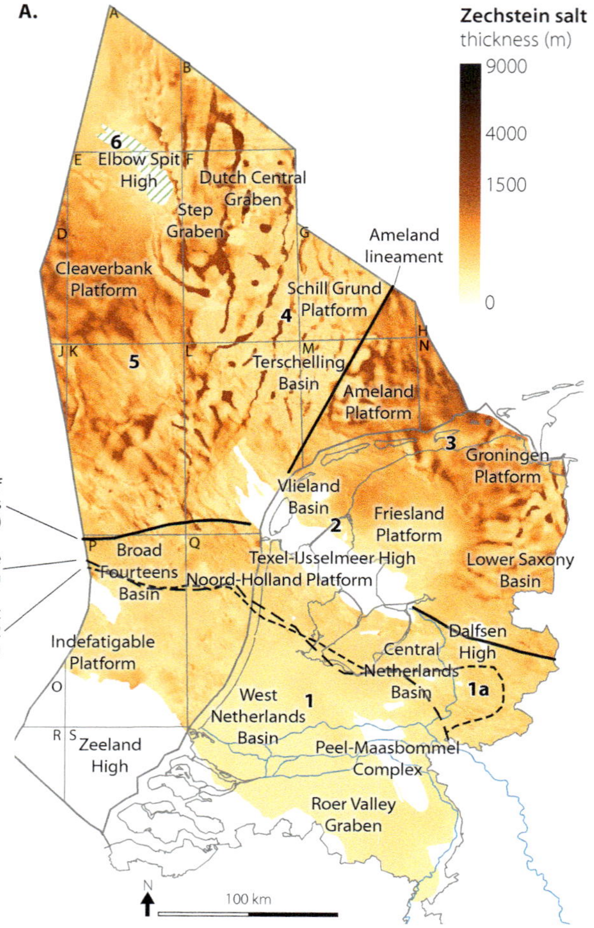

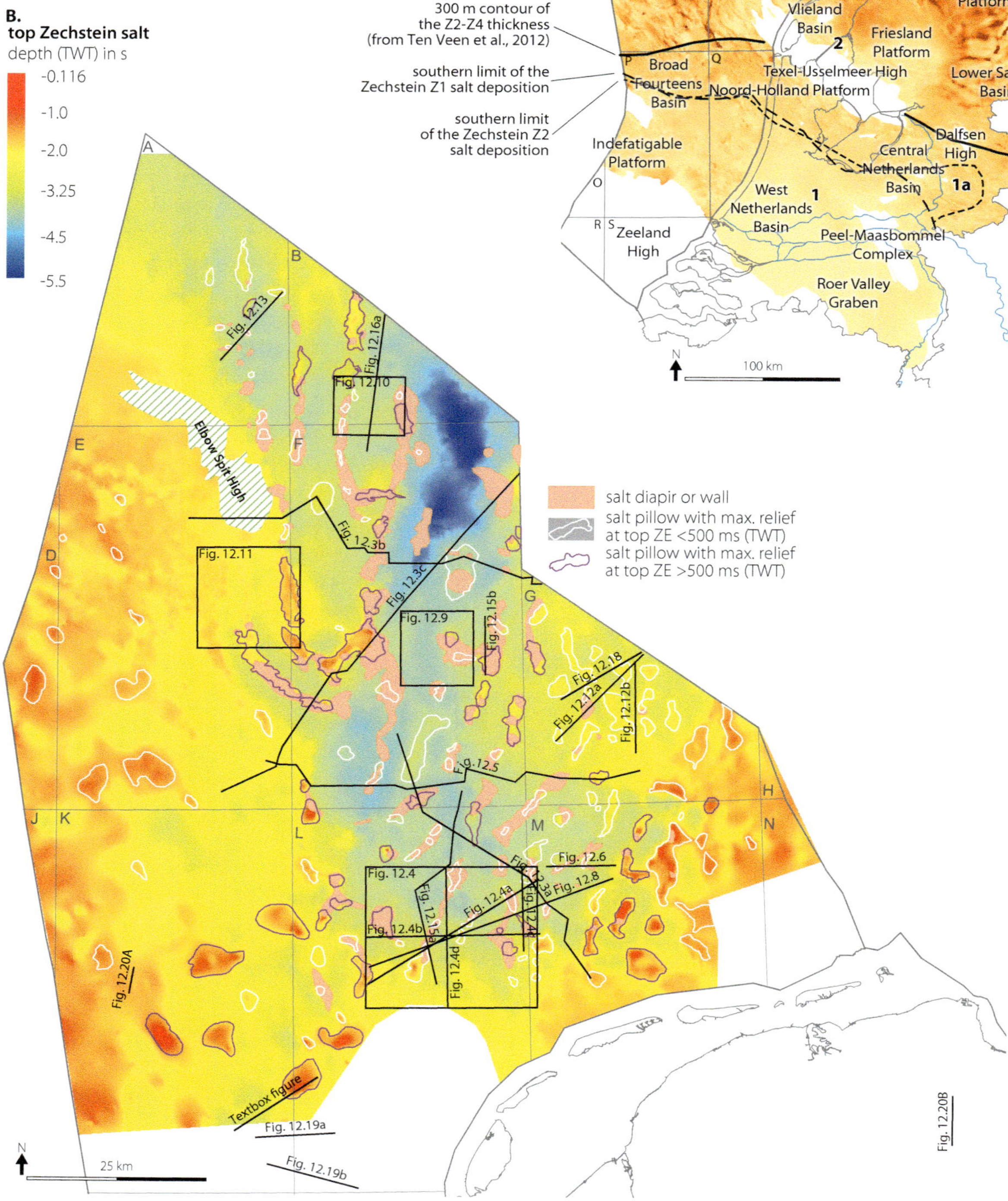

Jackson, 1992; Vendeville, 2002) and associated lateral salt flow (migration). The term halotectonics or salt tectonics covers all salt mobilization processes where tectonic forces are also involved and is more appropriate for the Netherlands given its long and complex tectonic history. All these processes allow the salt to migrate, forming salt bodies (salt pillows and/or diapirs) and ridges (salt walls) separated by depressions where thicker sedimentary successions could accumulate (Fig. 12.1). Continued sedimentation in these depressions increases differential loading, further favouring salt flow until salt supply from the source layer is cut off, due either to the salt volume running out or the creation of a physical barrier (e.g. fault, or collapse of the salt body feeder).

In the Dutch sector only a small number of salt bodies are allochthonous, meaning that they were emplaced as sub-horizontal or moderately dipping sheet-like salt diapirs, at a stratigraphic level above the autochthonous (mother-salt) layer (Zechstein).

The salt bodies can often reach shallow depth and even extrude at the seafloor or free surface, forming salt sheets or salt glaciers such as those that can be seen at present in the Zagros Mountains of Iran. Zechstein salt glaciers that extruded during the Late Triassic have been recognized in Germany (Mohr et al., 2007). In the Netherlands, most salt bodies are buried under a thick Cenozoic sequence and show little signs of deformation post-dating the Alpine orogeny. Differential subsidence around salt bodies may have led to formation of crestal faults that play an important role in hydrocarbon migration (e.g. shallow gas).

Zechstein salt distribution in the Netherlands

The combined effects of burial, temperature and (for the Dutch subsurface) external forces such as differential loading, extension and compression, ensure that thick salt flows in various fashions to form salt bodies that have various sizes and geometries. As pointed out by Geluk (2007a) for the Netherlands it is extremely important to differentiate between areas with thick salt and those without salt (Fig. 12.2a), since the structural styles of these areas are completely different during phases of extensional or compressional tectonics (De Jager, 2007). Salt flow was more intense during periods that coincide with known tectonic phases, but the intensity differs considerably between structural element types (e.g. Ten Veen et al., 2012).

The partitioning of the Southern Permian Basin into several basins and highs during the Triassic and Jurassic was accentuated by intense salt tectonics, located primarily along fault-bounded basin margins (Heybroek, 1975; Clark-Lowes et al., 1987; Wong, 2007). No evidence of Permian-age salt tectonics has been described in the Netherlands. Salt tectonics is dominant in rift basins such as the Dutch Central Graben, the Step Graben, the Terschelling and Broad Fourteens basins and their associated platforms (Fig. 12.2a). These basins have a horst and graben configuration and are bounded by large syn-sedimentary faults and salt diapirs or salt walls that were repeatedly reactivated during Mesozoic and Cenozoic rift and inversion events and in response to differential loading and erosional events. Basin compartmentalization and sub-basin formation were focused predominantly in the Dutch Central Graben, Step Graben and Terschelling Platform, where salt withdrawal involved thick salt layers. In the Broad Fourteens Basin compartmentalization occurred to a lesser extent since the original salt layer was thinner, as discussed below. Salt tectonics also took place in the Lower Saxony Basin, Groningen Platform, Friesland platform and around the Lauwerszee Trough (Strozyk et al., 2014).

The present-day thickness of the Zechstein Group increases from less than 50 m in the southern part of the Netherlands to more than 1200 m in the northern offshore (Geluk, 2007a). It is composed of five evaporite cycles (Z1-Z5), which may comprise fringe sandstone, carbonates, anhydrites, halite or claystone (for detail, see Bouroullec & Geel, 2025, this volume). The thickest halite layers are the Stassfurt Halite (Z2) and Leine Halite members, up to 600 m and 300 m thick, respectively. Locally, movements can result in salt bodies that can reach heights of 8 km. The main Zechstein unit deforming into salt bodies is the Stassfurt Halite (Z2), which has the highest overall percentage of halite (>95%) and forms the most important top seal for the Upper Rotliegend gas reservoirs (Geluk, 2007a; Ten Veen et al., 2012). The depositional Zechstein salt geometry was of a horizontal nature with possibly some local thickening on the downthrown sides of pre-existing faults. Based on numerical restoration Ten Veen et al. (2012) estimated that the original depositional thickness of the main Zechstein salt ranged from a few hundred metres in the southern part of the Dutch on- and offshore to an average of 500-700 m in the northern half of the offshore, reaching a maximum thickness of 900 m in the northern part of the present-day Dutch Central Graben and on the northeastern part of the Step Graben and thickening toward the German offshore sector.

It should be considered, however, that the amount of salt observed in many Dutch basins is only a portion of the originally deposited salt since it can be subject to intense dissolution and/or withdrawal and may have been eroded during local or regional basin evolution. Since the Jurassic, erosional events have occurred frequently in the Dutch sector. For instance, during the Middle Jurassic Mid-Cimmerian erosional phase (Ziegler, 1990; Underhill & Partington, 1993; Surlyk & Ineson, 2003; Bouroullec et

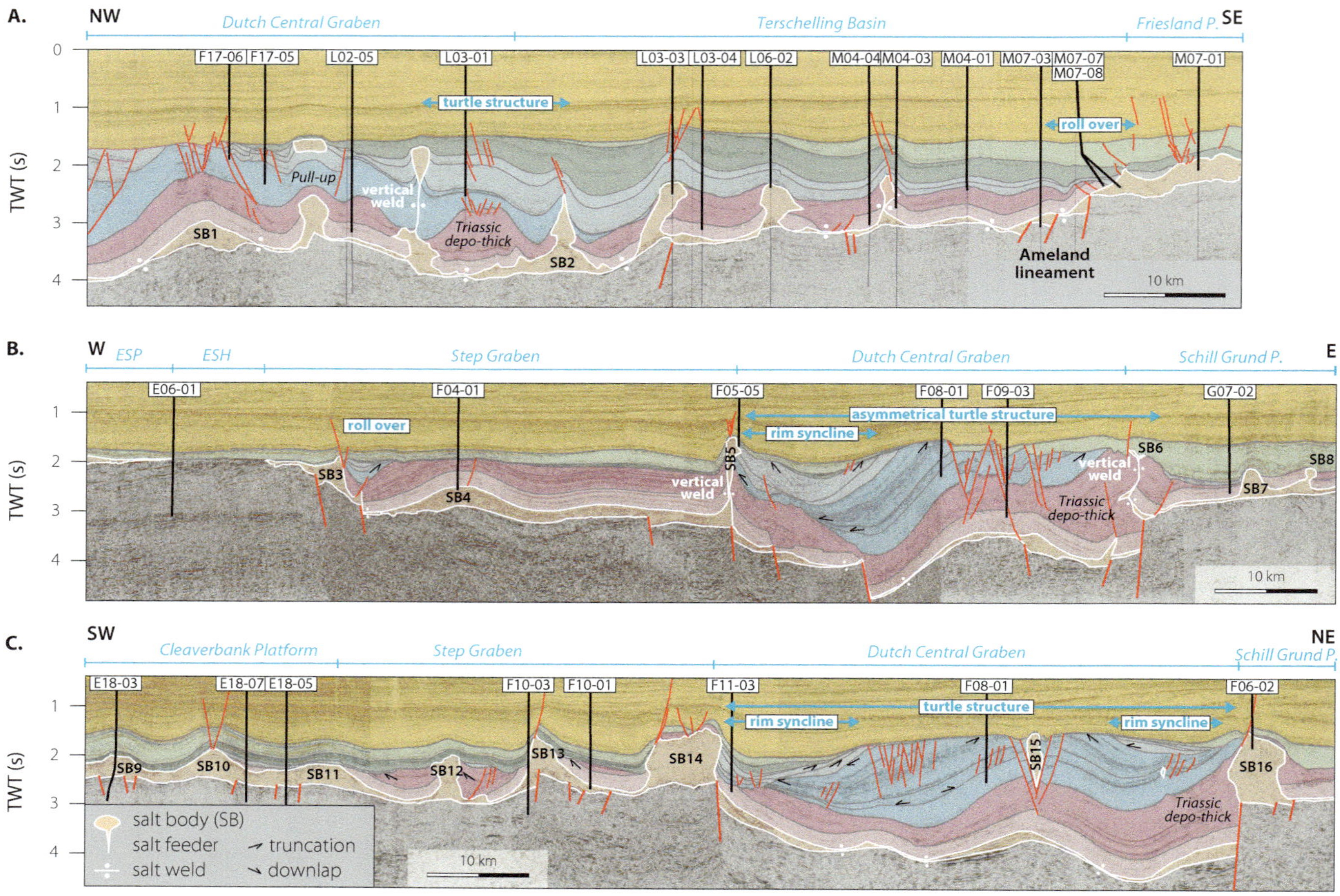

Figure 12.3. Interpreted seismic time sections across the northern part of the Dutch offshore highlighting the type of salt bodies (pink polygons) observed. Each panel is a composite from various 2D and 3D seismic datasets, selected such that key wells are intercepted. The pre-Zechstein interval is not interpreted. a) NW-SE trending panel from the southern part of the Dutch Central Graben to the Friesland Platform through the Terschelling Basin. b) W-E trending panel from the western part of the Elbow Spit Platform to the Schill Grund Platform across the Elbow Spit High, the Step Graben and the Dutch Central Graben. c) SW-NE-trending panel from the Cleaverbank Platform to the Dutch Central Graben through the Step Graben. See text for further explanation and Fig. 12.2b for location of the seismic sections. Modified from Bouroullec et al. (2016, 2019). ESP = Elbow Spit Platform; ESH = Elbow Spit High.

al., 2018; Verreussel et al., 2018) shallow salt bodies located along the flanks and within the Jurassic rift system were significantly eroded (see Verreussel et al., 2025, this volume). Throughout parts of the Dutch sector, there is evidence that during continued flow salt bodies at one time may have reached shallow depths or even the paleo-free surface, forming glaciers, as observed in Germany (Mohr et al., 2007). The presence of cap rocks at the top of salt bodies (see discussion on cap rock below) also indicates that salt dissolution was likely a common phenomenon. Cartwright et al. (2001) indicated that lateral dissolution of salt bodies could have taken place in the North Sea basins resulting in the removal of as much as 40% of the original Zechstein volumes. These dissolution phenomena and erosional events complicate attempts to reconciliate the present-day volume of Zechstein salt preserved in the subsurface with salt-related stratigraphic and basinal geometry observed throughout the Dutch subsurface. Often, a considerable volume of salt needs to have been lost in order to explain the extreme basinal geometries observed in the northeastern part of the Dutch sector. In this area, at least 1000 m of salt must have been deposited and as much as 1300 m or more if significant dissolution occurred. These values exceed the earlier 900 m estimate of Ten Veen et al. (2012).

Present-day salt thickness can be approximated by looking at the variations of the overall Zechstein Group thickness (Fig. 12.2a) which, to a large extent, is related to migration of Zechstein Z2 salt. The salt migrated and withdrew in multiple directions, forming different types of salt bodies and associated basins and sub-basins (see below). At the present-day six Zechstein Group thickness areas can be differentiated in the Netherlands:

1. The first area (Area 1 in Fig. 12.2a) extends NE to SW across the southern Dutch on- and offshore. This area is located north of the Zeeland High and includes the Indefatigable Platform, the West Netherlands Basin, the Roer Valley Graben, the Peel-Maasbommel Complex and the southern part of the Central Netherlands Platform. This area shows little to no thickness change and no salt bodies have been identified. The area is located south of the approximate southern limit of Zechstein halite deposition (Fig. 12.2a) and is devoid of salt tectonic features. A small area in the eastern part of the Central Netherlands Basin (Area 1a, Fig. 12.2a) shows some thickness variations that are likely related to Z1 salt movement since the southern depositional limit of Z2 halite is farther to the north.
2. The Zechstein Group is very thin or absent in the area to the north (Area 2 in Fig. 12.2a), This area is in the vicinity of the Texel-IJsselmeer High and includes the northern part of the Central Netherlands Platform, Friesland Platform, the Vlieland Basin and the Noord-Holland Platform. The Texel-IJsselmeer High was a high during the deposition of the Zechstein Group and the limited thickness of the salt (mainly Z2 and limited Z1) allows for minor movement into small salt pillows only. In the area located between the southern depositional limits of Z2 and Z1 salt (Fig. 12.2a) and the original 300 m Z2-Z4 thickness (from Ten Veen et al., 2012) the salt is considered to be too thin to have been subject to significant diapirism and this is shown by the lack of salt bodies. Rather, this area was subject to minor salt migration and salt-induced differential subsidence.
3. The third area (Area 3 in Fig. 12.2a) is in the central eastern part of the Netherlands, spanning from the onshore Lower Saxony Basin via the offshore Ameland Platform to the eastern part of the Schill Grund Platform. Strozyk et al. (2014) illustrated the salt tectonic style of this region and highlighted the presence of Z3 stringers (see below for description of stringers). Overall, this area consists of a relatively thick (700 to 1400 m) Zechstein Group with a predominance of large salt pillows and salt diapirs. The transition in thickness between Area 2 and 3 is smooth, in contrast to the distinct and linear transition between the third and fourth areas (referred to as the Ameland lineament; Fig. 12.2a). This feature was likely an inherited pre-Zechstein fault that affected the depositional thickness of the salt as well as the later deposition and preservation of the Triassic sequence (see Fig. 12.3a, in particular around wells M07-07 and M07-08).
4. The fourth area (Area 4 in Fig. 12.2a) is in the central part of the Dutch offshore and comprises the Terschelling Basin, the western part of the Schill Grund Platform, the Dutch Central Graben and the Step Graben. This area shows by far the most extreme post-depositional Zechstein Group thickness variation with large zones where the salt withdrew and migrated toward preferential locations forming large and numerous salt diapirs and salt walls above or in the vicinity of rift bounding faults (Fig. 12.3; Strozyk et al., 2012). This is one of the few areas that contains disconnected salt stocks, i.e. salt bulbs with salt welds beneath them. Note that this is not the norm in the Netherlands since most salt pillows and diapirs have relatively thick bases or stems. The resulting growth basins and sub-basins were formed during Middle Triassic to Early Cretaceous rifting phases and comprise turtle back structures with rim synclines, roll-over anticlines and intra-rift sub-basins. Some of the salt diapirs located in this area were squeezed and became disconnected during the Middle to Late Triassic in response to thin-skin extension around the paleo basin margin (see below for additional information).
5. The fifth area (Area 5 in Fig. 12.2a) is located west of areas two and four and extends from the Broad Fourteens Basin in the south to the Cleaverbank Platform (see Fig. 12.3c). This southern part of the area is dominated by Alpine compressional features such as large folds, inverted pre-existing normal faults and thrusts that are occasionally salt-cored (Nalpas & Brun, 1993). Prior to compression, the original Mesozoic geometries of these salt bodies were likely to have been similar to the ones found in Area four and they show a predominance of rift-related salt withdrawal and migration into salt diapirs and walls. The central part of Area five is dominated by a relatively thick salt layer composed of numerous salt pillows and some rare salt bodies. The eastward transition between Area five and four is rather smooth and is demarcated by an increasing number of salt diapirs and collapsed salt bodies (see below). The northern part of Area 5 transitions smoothly into Area 6 with a decreasing Zechstein thickness towards the Elbow Spit Platform and High.
6. Area 6 is in the northwestern part of the Dutch offshore around the Elbow Spit High (Figs 12.2a, 12.3b). In this area the Zechstein Group is very thin or absent. A few small isolated Early to Middle Triassic mini basins (pods) exist in the eastern part of this area, similar to those in the UK offshore sector (Penge et al., 1999; Stewart & Clark, 1999). Mini basins are small intra-salt basins largely surrounded by and subsiding into relatively thick salt areas (e.g. deep in the Gulf of Mexico; Pilcher et al., 2011; Bouroullec & Weimer, 2017). Such mini basins are not clearly observed elsewhere in the Dutch sector except for some areas located between salt pillows on the Groningen Platform (Raith et al., 2016).

Salt bodies, associated structures and phenomena in the Netherlands

More than 230 salt bodies primarily consisting of Zechstein salt pillows, salt diapirs and salt walls (Figs 12.1 and 12.2b; Groenenberg et al., 2025, this volume, their Figure 16.4) have been recognized in the Netherlands. For the Dutch onshore, the term "dome" has been historically widely used as a synonym for salt diapir or -wall in the hydrocarbon industry as well as in the salt mining and -production industry. Other salt body types observed in the Dutch subsurface include salt rollers, i.e. overthickened

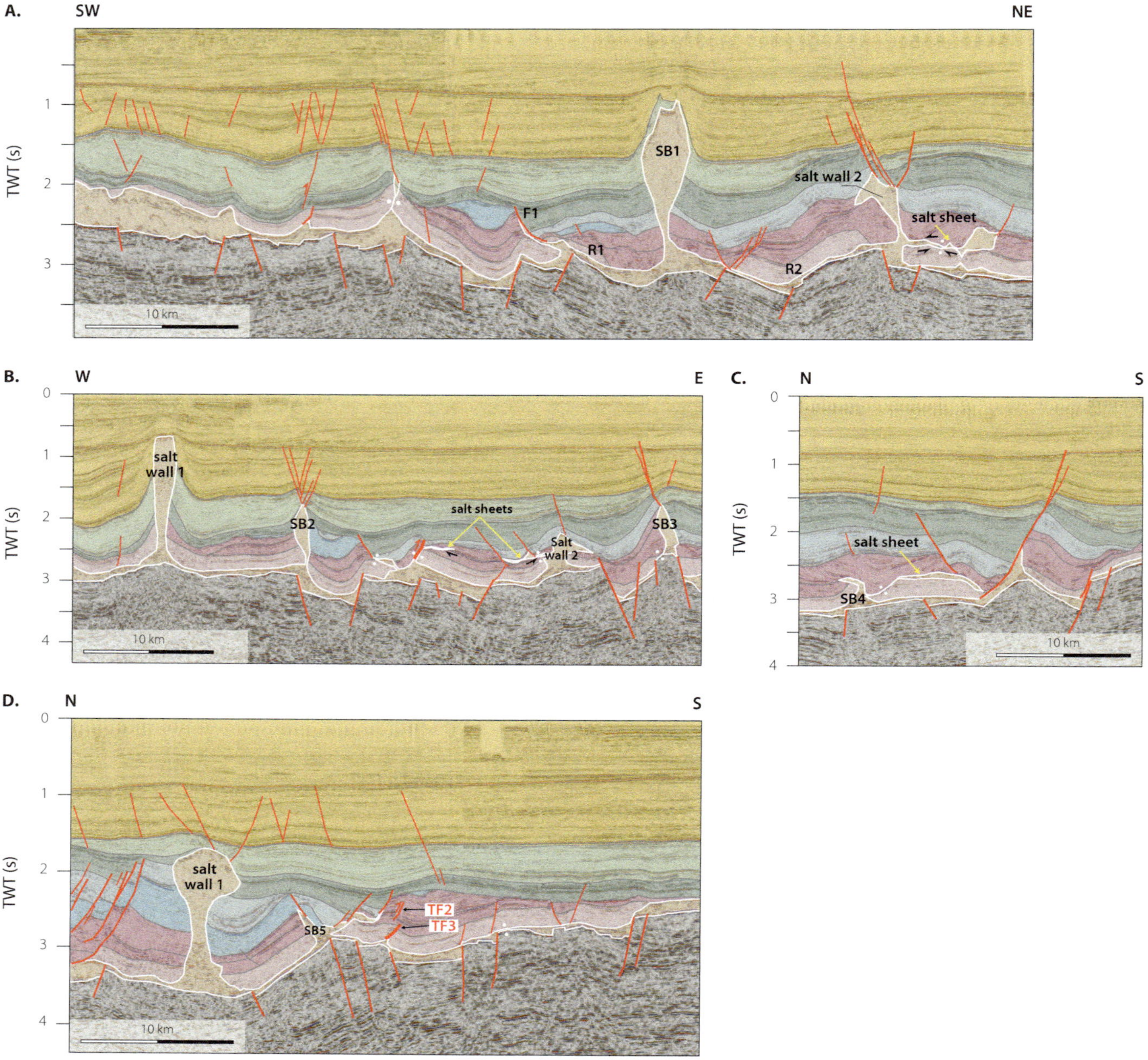

Figure 12.4. Interpreted seismic time sections located in the southern part of the Terschelling Basin and the northern part of the Friesland Platform. Zechstein salt is shown as pink polygons, faults as red lines. Note the series of loaded salt sheets in sections (a), (b), and (c) as well as collapsed diapirs in section (a), (b), and (d). Section A shows a zone around salt body 1 (SB1) where the Lower Triassic is absent, likely due to rafting. Note the rugose nature of the base Zechstein Group with its large number of normal faults and a large flexure that demarcates the Hantum Fault Zone that strikes SE-NW from the southern part of the Lauwerszee Trough to the southern part of the Terschelling Basin. See Fig. 12.2b for location of the seismic sections and Fig. 12.3c for explanation of salt symbology. Modified from Bouroullec et al. (2017).

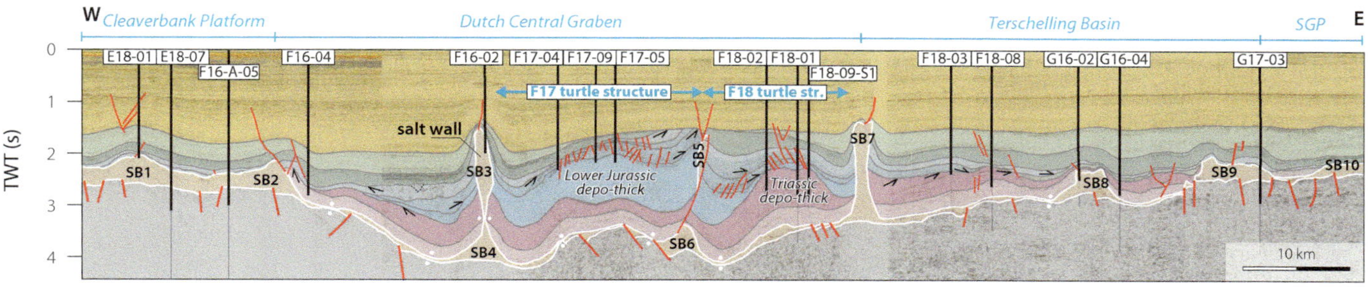

Figure 12.5. Interpreted composite seismic time section across the Dutch Central Graben showing turtle structures and associated salt bodies. The F17 and F18 turtle structures are located in the central and eastern part of the rift basin and are separated by a squeezed salt wall. See Fig. 12.2b for location and Fig. 12.3c for explanation of salt symbology. Modified from Bouroullec et al. (2017).

asymmetric pyramid-shaped salt bodies on the footwall side of syn-depositional faults (SB3 in Figs 12.1, 12.3b) and small loaded salt sheets that are most likely loaded paleo-salt glaciers (Figs 12.1, 12.4a-c). Figures 12.2b and 12.3 highlights the omnipresence of salt pillows and diapirs in the northern part of the Dutch offshore. The salt pillows are present across most of this part of the Dutch sector but the highest relief pillows (more than ~500 m of relief) are predominantly related to Mesozoic rift structures around the Step Graben, the Dutch Central Graben and the Terschelling Basin (Fig. 12.3b,c). Salt diapirs and salt walls are also predominantly organized along Mesozoic rift trends and their formation is closely related to the tectonostratigraphic evolution of the same basins. The exact geometry of the salt bodies is difficult to interpret using time seismic data since salt velocities are much higher than those of the surrounding sediments. Since high-velocity salt absorbs a lot of the acoustic energy, good quality seismic imaging below salt bodies is often difficult to obtain. In the past few decades radical improvements in subsalt seismic data imaging technologies have helped exploration in salt basins such as the Gulf of Mexico and the South Atlantic margins. Most seismic surveys acquired in the Netherlands, however, only provide limited illumination below salt diapirs such that the exact geometry and width of salt stems/feeders is often open to debate. Because of the high salt velocities, velocity pull-up structures below thick salt bodies are commonly seen on Dutch time seismic data or sub-optimally migrated depth data. Subsalt geometries are therefore unrealistically represented as can be seen for the salt bodies between wells F17-05 and L02-05 in Figure 12.3a.

Zechstein salt migration also triggered the formation of several associated structures such as syn-sedimentary faults, crestal faults, salt welds, salt stems, salt feeders, growth fault/raft systems, salt sheets, mini basins, rim synclines, turtle structures, collapse graben and collapse diapirs that are described in more detail below.

Salt pillows

A salt pillow (Fig. 12.1) is a subcircular upwelling of salt that has a concordant overburden (Trusheim, 1960) and a low-amplitude reflection at the upper surface of the salt layer. Dutch salt pillows (e.g. SB1, 2 and 7-11 in Fig. 12.3; salt bodies SB1, 2, 5, 6, 8-10 in Fig. 12.5) are primarily observed on the platforms such as the Schill Grund- (Fig. 12.3b), Cleaverbank- (Fig. 12.3b), Friesland-, Groningen- and Ameland platforms (Fig. 12.3a). Here the amount of differential sediment loading was limited compared to the Mesozoic rift basins. The salt pillows vary in shape from circular to elongated in map view, are up to 15 km in diameter or length and vary from gentle dome-like structures with a few tens of metres of relief to concave upward or A-shape salt bodies with up to 1200 m of relief. Within the pillows, the overall thickness of the Zechstein salt can reach 1500 m (Ten Veen et al., 2012). In cross-sectional view the salt pillows are often asymmetric due differential loading and later deformation related to rifting and compression. The base Zechstein can be faulted below the salt pillows, which often played a role in localizing the flow of salt into those salt bodies. Such relationships between salt bodies and faulting in pre-salt strata (commonly referred to as 'basement' in salt tectonic terms) is common in many salt basins such as the Gulf of Mexico (Bouroullec & Weimer, 2017) or Angola (Jackson & Hudec, 2005). See also the basement section below.

Salt diapirs

A salt diapir (Fig. 12.1) is a mass of salt that has flowed in a ductile manner and appears to have discordantly pierced or intruded into the overburden. The term was first coined in 1907 by the Romanian geologist Ludovic Mrazek to describe a salt formation that intrudes due to the rise of lighter rocks into denser rocks. Diapirs commonly start to rise vertically, driven by buoyancy contrasts between the relatively low-density salt and the denser overburden rocks (Bouroullec & Weimer, 2017; Jackson

& Hudec, 2017). In extensional settings the overburden commonly thins and fractures, allowing the mobile salt to rise further. As the diapir becomes taller the driving force can be taken-up by differential loading. Halokinetic sequences around the deforming salt diapirs develop with a wide range of stratigraphic geometries, architectures and deposits that depend on the syn-tectonic depositional environments, sedimentary flux in relation to the salt deformation rate and preservation potential. The map view geometry of the diapir bodies in the Dutch subsurface can greatly vary, from circular to elongated (Fig. 12.2b). The diameter (or main axis) of salt diapirs ranges from 1 to 15 km and their height from 2 to 8 km. With a planform aspect ratio of 2 or more the salt diapirs are commonly referred to as salt walls (see below). The cross-sectional view of many salt diapirs in the Dutch sector is often difficult to evaluate due to seismic data imaging challenges. However, well data and depth-domain seismic data from the North Sea, the Gulf of Mexico and the African West Coast show that narrowing of the salt diapirs' core (feeder or stem) is common. A salt stem refers to the narrow part of the salt body connecting the source salt to the shallower salt body. Most salt bodies in the Dutch sector have a salt stem, though with variable narrowing ratios (Figs 12.4b,e; 12.3a, c; 12.5). The width of the stems is difficult to estimate but vertical separation (welding; see definition below) of both sides of salt bodies can locally be observed (Figs 12.3a,b; 12.4b; 12.13). In general, most salt diapirs in the Netherlands did not intrude into sediments above the Cretaceous interval and are buried by Cenozoic strata. Occasionally, however, the North Sea Group was intruded or affected (Figs 12.3b; 12.4a,b; 12.5; 12.19) especially in the eastern part of the Dutch onshore, where many salt diapirs reach shallow depths and are being solution mined (Groenenberg et al., 2025, this volume). The shapes of salt diapirs in the Netherlands are often related to the interplay of tectonic deformation (rifting and compression) and halokinetic deformation. Hence many diapirs have complex geometries due to multiple phases of deformation, with evidence of diapiric rise, fall (collapse), squeezing/shortening, strike slip deformation and erosion (see Evolution of salt tectonics section below).

Salt walls

A salt wall (Fig. 12.1) is an elongate upwelling of diapiric salt that forms parallel sinuous rows (Trusheim, 1960). In the Dutch subsurface, less than twenty structures can be referred to as salt walls. They are predominantly located within and along rift basin faults (Figs 12.2, 12.5). In their short axis they have a similar cross-sectional view as in salt diapirs but can extend up to 65 km in length. Salt wall systems are often organized along regional trends and are often seen above the main rift bounding fault systems (see Fig. 12.3b). Most salt walls were formed along the newly formed rift system from the amalgamation of salt diapirs that formed during the Triassic as individual salt bodies, but which later collided and merged into elongated salt walls (Fig. 12.5). The identification of sutures (zone of collision between different salt bodies) within Dutch salt walls has not been successful due the subsequent defor-

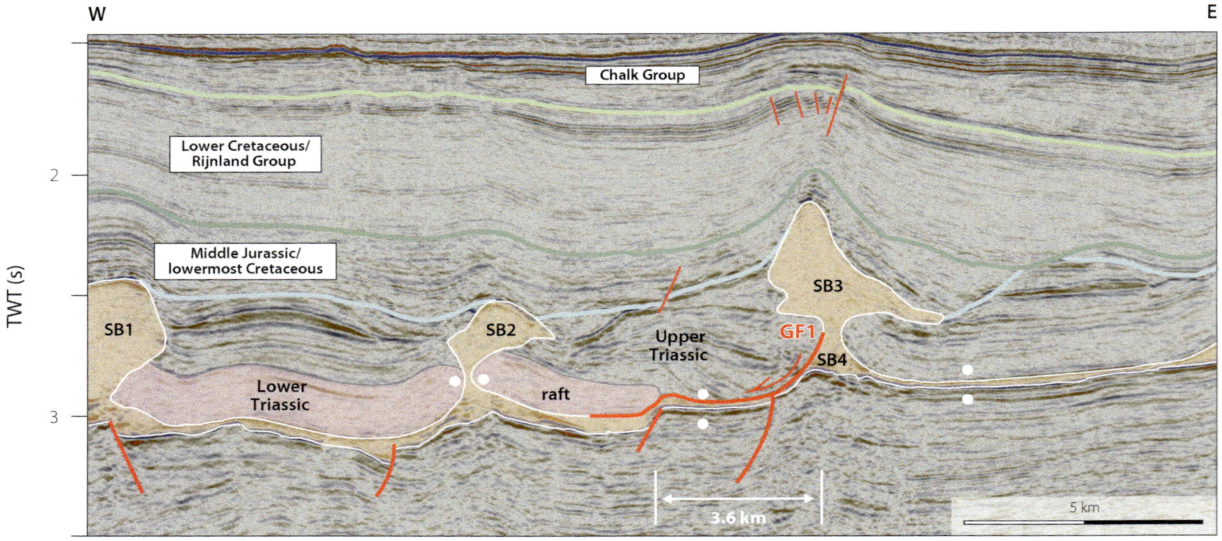

Figure 12.6. Interpreted seismic time section from a 3D seismic survey in the eastern part of the Terschelling Basin showing a growth fault/raft system that was active during the Middle-Late Triassic as shown by the rollover on the downthrown side of the growth fault GF1 (modified from Bouroullec et al., 2017). The Lower Triassic raft translated westward on top of the mother salt over a distance of 3.6 km. Three salt bodies (SB1-3) were squeezed during the deposition of the Upper Cretaceous and SB2 was also shortened during the Late Triassic in response to the upward extension during gravitational gliding. See Fig. 12.2b for location.

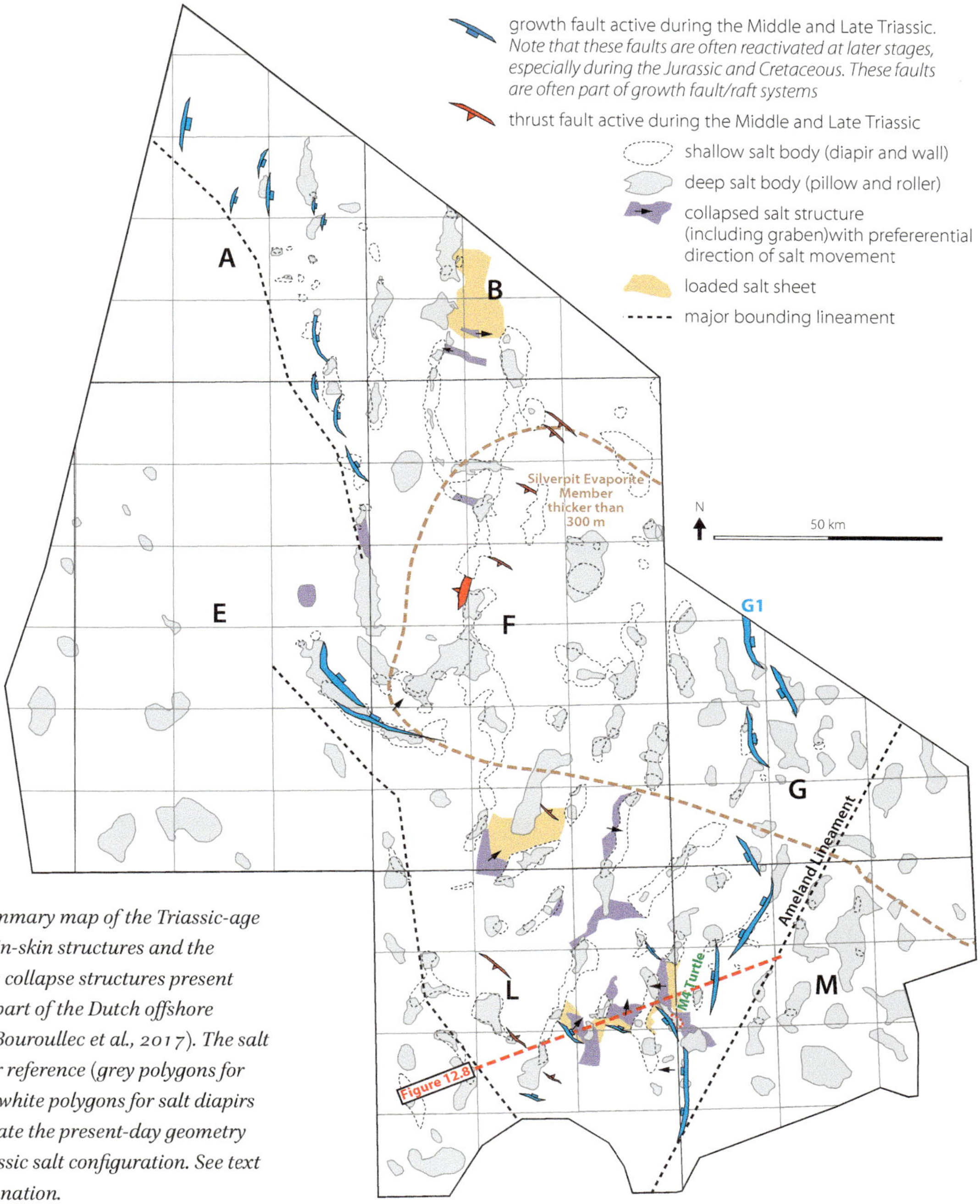

Figure 12.7. Summary map of the Triassic-age gravitational thin-skin structures and the Triassic-Jurassic collapse structures present in the northern part of the Dutch offshore (modified from Bouroullec et al., 2017). The salt bodies shown for reference (grey polygons for salt pillows and white polygons for salt diapirs and walls) indicate the present-day geometry and not the Triassic salt configuration. See text for further explanation.

mation that has blurred the picture, and seismic imaging challenges. Especially during the Alpine contractional phases many salt bodies in the Netherlands were shortened and squeezed.

Salt rollers

A salt roller (Fig. 12.1) is a low-amplitude, asymmetric salt structure composed of two flanks, a gently dipping flank with a conformable stratigraphic contact with the overburden, and a steeply dipping flank with a normal-faulted contact with the overburden (Bally, 1981). Many large faults detaching on the Zechstein salt have a salt roller on their footwall side (see SB3 in Fig. 12.3; SB2 in Fig. 12.5; SB4 in Fig. 12.12; SB1 and SB2 in Fig. 12.18) and are often present on the footwall side of growth faults (Rouby et al., 2003; Jackson & Lewis, 2012) as is the case for salt body SB4 in Figure 12.6.

Salt sheets

A salt sheet (Fig. 12.1) is a sub-horizontal salt body that originated by salt expansion from a salt-diapir configuration at or near the seafloor. Salt sheets can amalgamate and form salt canopies. These salt body types have become better understood since the 1980s, when improvements

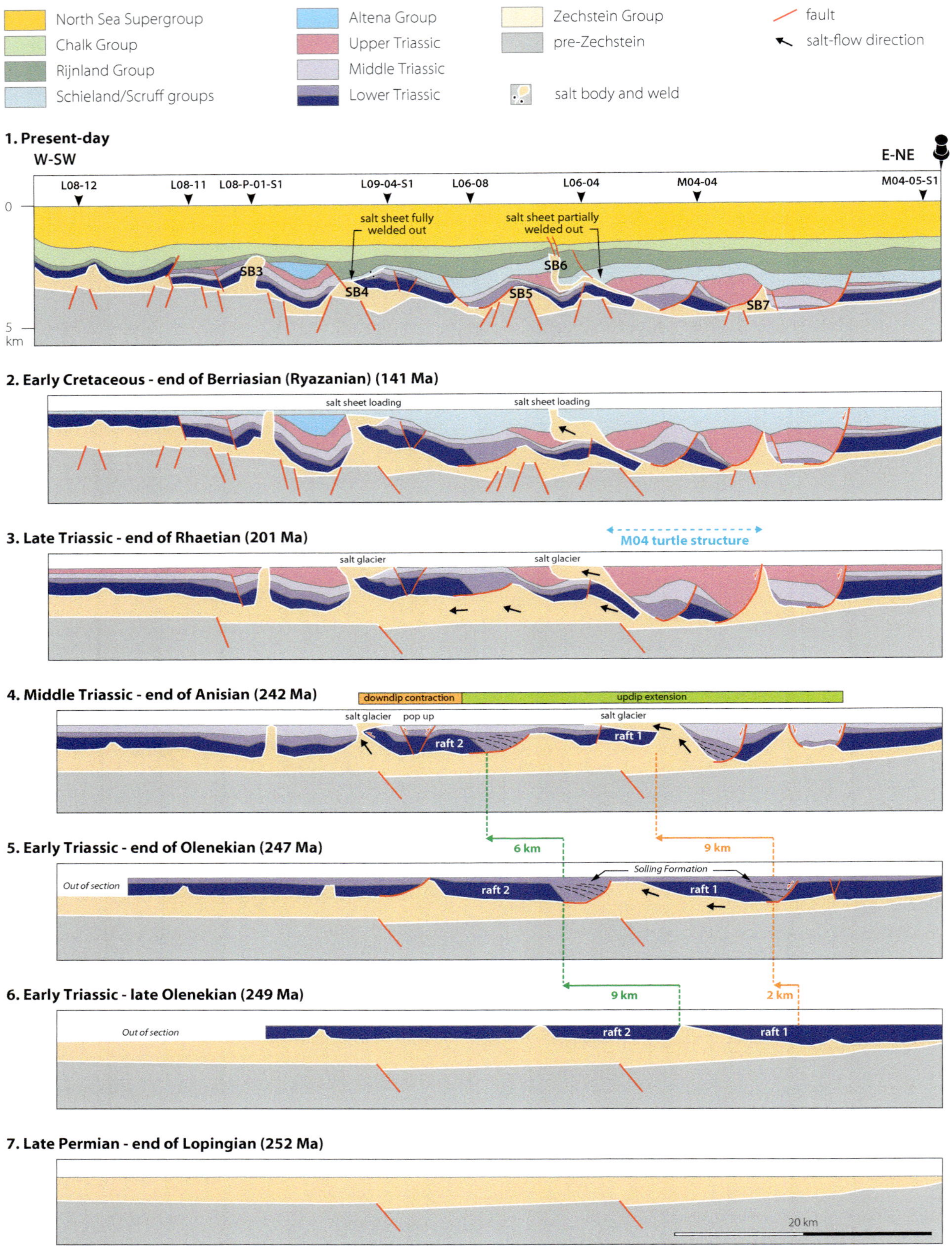

Figure 12.8. 2D palinspastic restoration of a SW-NE trending section from the Friesland Platform to the Ameland Platform through the southern part of the Terschelling Basin. The section intercepts blocks L08, L09, L06 and M04 and was constructed from 2D and 3D time seismic data that was depth converted and restored using 2D Move software (Petroleum Experts Ltd). Most of the wells displayed are located on the section, but wells L08-12, L08-11, L09-04-S1 and L06-08 were projected from 2-4 km distance. See text for further explanation and Fig. 12.7 for location. Modified from Bouroullec et al. (2017).

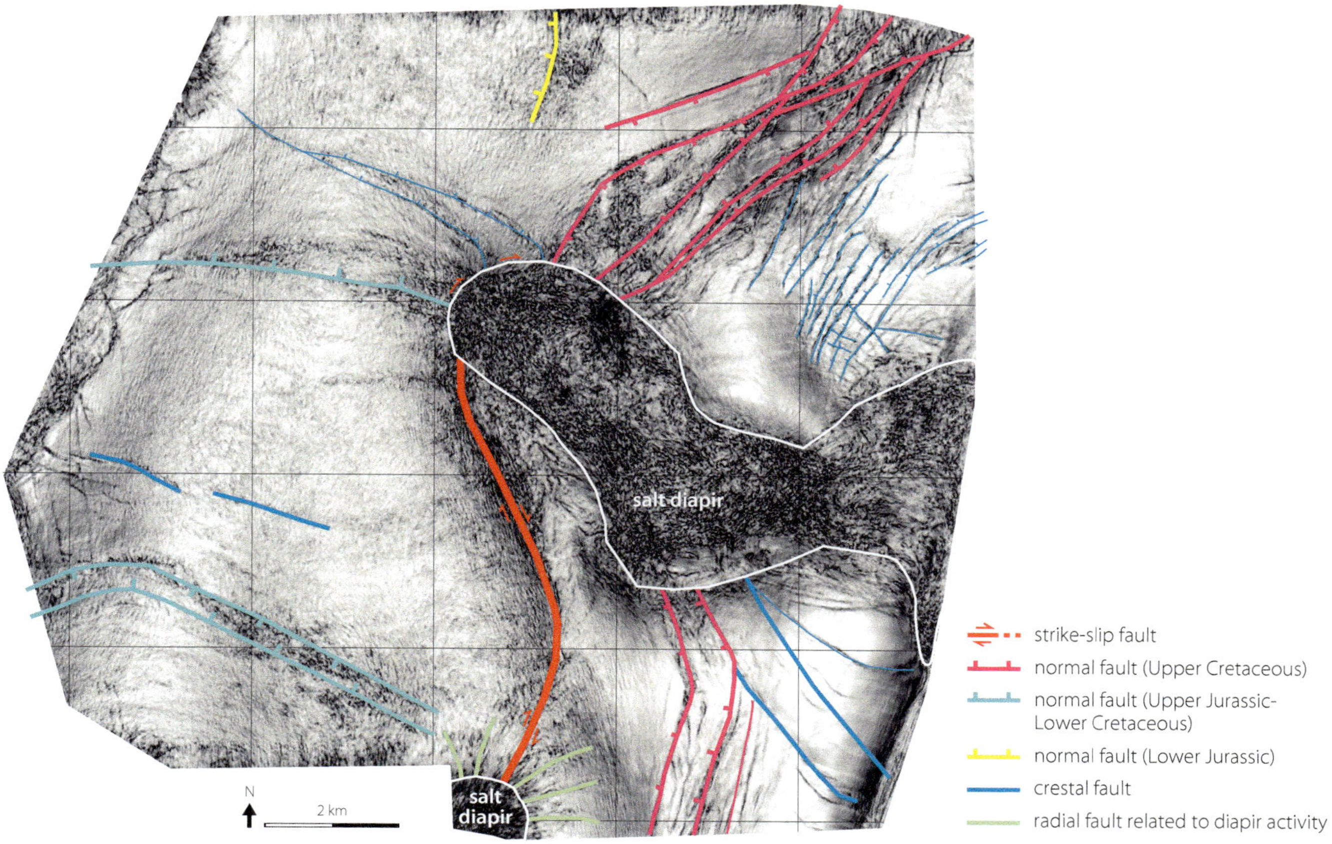

Figure 12.9. *Similarity attribute map of the base Altena Group in block F11. Note the deflection of the Cretaceous-age strike slip systems around the central salt body. See Fig. 12.2b for location. Modified from Peeters (2016).*

in drilling and seismic technologies led to a surge of interest in allochthonous salt on deep-water passive margins. In the past thirty years, thousands of salt sheets in more than thirty-five basins have been recognized worldwide. Recently, several small sheets have also been interpreted in the Dutch sector (Bouroullec et al., 2017), mainly in the southern part of the offshore L quadrant along the southern margin of the Terschelling Basin (Fig. 12.4a-c). These salt sheets are located downdip of growth fault/raft systems and were mainly loaded by Middle to Late Triassic strata. They are interpreted as resulting from excessive shortening of pre-existing salt diapirs. During gravitational gliding (see section on growth fault/raft systems below) the updip thin-skinned extension was mainly balanced by shortening of pre-existing (Early to Middle Triassic) salt diapirs instead of reverse faulting. This explains why only a few Triassic-age reverse faults are observed in the Dutch sector (see below). Instead, diapirs were horizontally squeezed and salt migrated upward, occasionally extruding at the free surface to form exposed salt masses (glaciers, see Mohr et al., 2007) that were later either completely or partially eroded and/or dissolved. Occasionally the remaining parts of these salt sheets were loaded by Triassic or Jurassic sediments and were deformed again to accommodate the resulting differential subsidence (Fig. 12.16). The remaining salt was loaded by Middle to Upper Triassic deposits and migrated into peripheral salt bodies or was partially to entirely welded out. The presence of salt welds related to such loaded salt sheets is often difficult to distinguish from stratigraphic geometry. In Germany Zechstein salt can form wings when salt is remobilized laterally from a diapir into adjacent stratigraphic positions, protruding itself into weak Triassic strata such as the Main Röt Evaporite Member (Hudec & Jackson, 2006; Pharaoh et al., 2010). Such salt wings have not been recognized yet in the Dutch sector. As lateral intrusions require salt to compete with gravity, they are mechanically more likely to occur with the help of external (compressional or extensional) forces. One way to distinguish between intruded (salt wing or salt-wing intrusion) or extruded (loaded salt sheet) salt systems is to analyse the sub-and supra-salt stratigraphy, with intrusion favouring concordant geometry while extrusion favours angular disconformities. For more information regarding loaded salt sheets see the section below on the evolution of salt tectonics.

Salt welds

A salt weld (Fig. 12.1) is a thin or narrow salt interval that forms when a salt layer becomes very thin due to salt movement, dissolution, or removal by faulting such that the overburden and the underlying sub-salt strata become effectively welded together (Jackson & Cramez, 1989). Salt welds may also develop vertically when the squeezing is horizontal, bringing the sides of salt diapirs in contact with each other. Horizontal salt welds have been interpreted in the Dutch sector of the North Sea where, especially in the central parts of the rift basins, where most or all the available mobile mother salt has withdrawn laterally (see Figs 12.3-12.6). Locally, such as in the southern part of the Terschelling Basin, a combination of salt welds and growth fault/raft systems places the Middle or Upper Triassic strata directly above Rotliegend strata, without erosional processes being involved (see Fig. 12.3a, between wells L06-02 and M04-04; around SB1 in Fig. 12.4a; Figs 12.6, 12.8). Salt welding in the Dutch subsurface may often be incomplete, meaning that some tens of metres of (probably) deformed and fractured Zechstein strata (likely the mechanically stronger lithologies such as the carbonates and caprocks) may remain on the weld surface after most of the ductile salt has migrated away. This situation differs from the relatively clean salt welds found in other basins (e.g. Gulf of Mexico; Bouroullec & Weimer, 2017). If such welds become faulted later, they may provide important zones for hydrocarbon migration. It is difficult to determine ahead of drilling whether an apparent salt weld contains any evaporites or non-evaporitic residue (Jackson & Hudec, 2017).

Growth fault/raft systems

A growth fault/raft system (Fig. 12.1) is an association of 1) an extensional listric fault that detaches on a sub-horizontal salt layer (here the Zechstein Group) and 2) one or

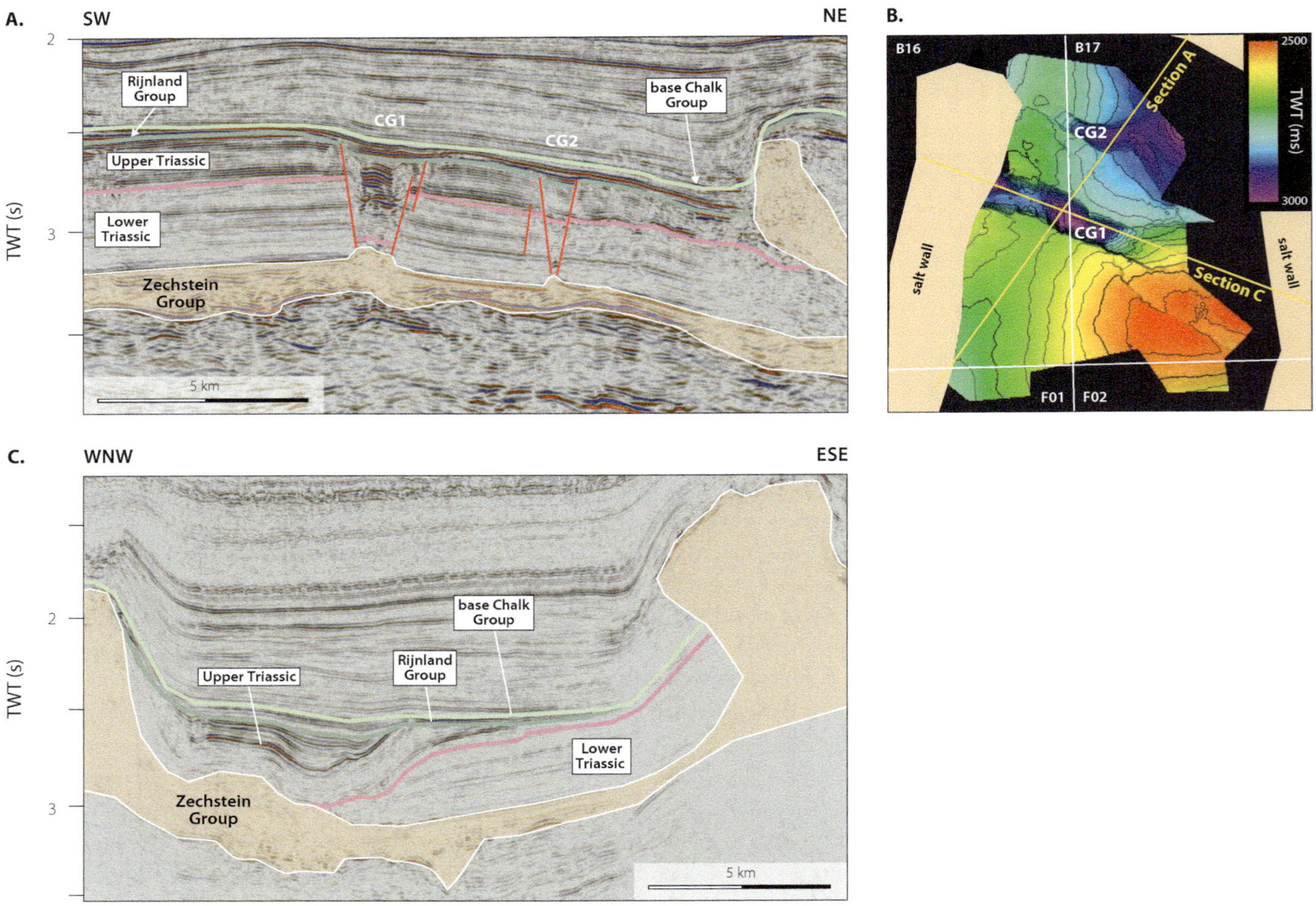

Figure 12.10. Two WNW-SSE trending collapse grabens (CG 1 and CG2) are present spanning the B16 and B17 blocks. a) Interpreted time seismic section showing a strike view across the collapse grabens (see (b) for location). b) Time structure map of the top Lower Triassic horizon (see Fig. 12.2b for location). c) Interpreted time seismic section showing a dip view along the axis of the collapse graben CG1 (See (b) for location). Note that the western part of the graben is filled with Upper Triassic strata which is indicative of salt withdrawal that fed a shallow salt body located toward the WNW in block B16. Modified from Bouroullec et al. (2017).

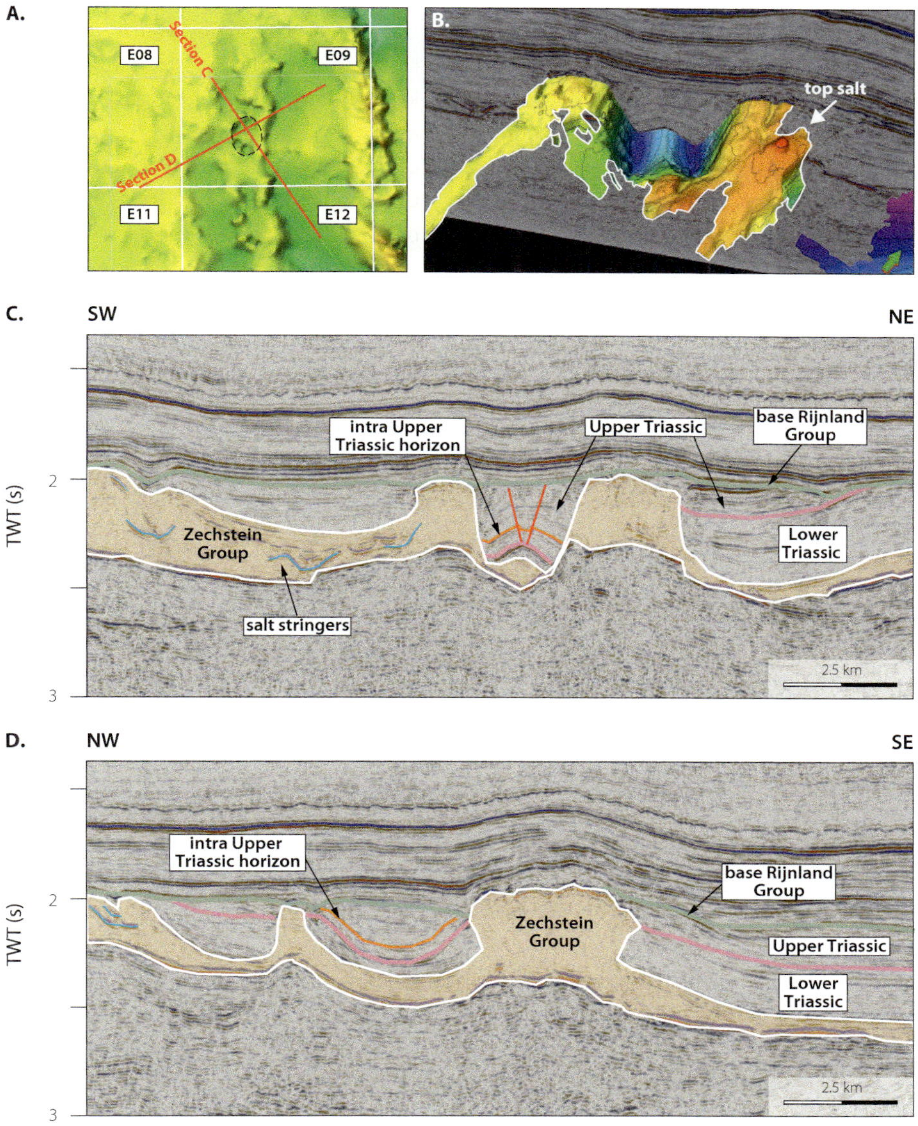

Figure 12.11. *Collapsed salt body on the Cleaverbank Platform in block E09. a) Location map (see Fig. 12.2b for location). The position of the mini basin that formed on a collapsed salt body is encircled with a black dashed line. b) Perspective view, looking towards the NW of the E09 collapse mini basin and showing the top Zechstein salt surface with a reference strike seismic line. c) SW-NE seismic section showing the narrow mini basin configuration with Upper Triassic growth wedges along the margins. Blue lines highlight salt stringers. d) NW-SE seismic section showing the Upper Triassic depocentre in the central part of the mini basin. Modified from Bouroullec et al. (2017).*

a series of stratal blocks referred to as rafts (here blocks containing Lower Triassic rocks). A growth fault is a listric syn-sedimentary fault that detaches on a ductile layer (in this case mainly autochthonous but occasionally allochthonous salt layers) and allows for a large growth stratigraphic wedge (roll-over) to form on its hanging wall side. A raft is fault-block located below or downdip of the main stratigraphic wedge and is formed by extreme extension over a horizontal to gently dipping detachment. These rafts can be separated from their footwall stratal equivalent by distances up to 16 km, as can be seen on the Congo-Cabinda continental margin (Rouby et al., 2002, 2003). Note that the usage of the term raft differs in the above case from that commonly used to describe salt stringers, which are blocks of non-halite rock within a salt structure. The growth fault and the rafts within this system are separated by trough-like depocentres of younger syn-kinematic strata (Burollet, 1975; Rouby et al., 2002, 2003).

Several growth fault/raft systems that were active during the Middle to Late Triassic as shown by the age of growth strata have been recently identified in the Dutch sector (Bouroullec et al., 2017). They have been identified in the Schill Grund Platform, the eastern and southern part of the Terschelling Basin and along the western flank of the Step Graben (Figs 12.6-12.8). The amount of extension was limited in the western part of the Dutch offshore (Step Graben) and was at its maximum in the Terschelling Basin where up to 12 km of raft gliding/translation have been deduced from palinspastic restoration (Fig. 12.8). This amount is comparable to the horizontal translation observed in classical raft tectonics in west Africa (14 km in Rouby et al., 2002) or the Gulf of Mexico (11 km in Bouroullec & Weimer, 2017; 8 km in Jackson & Hudec, 2017). Lower Triassic rafts were transported westward by gravitational gliding toward the paleo-basin centre located at that time in the eastern part of the offshore F quadrant. Several eastward-directed gravitational slides on the eastern part of the Schill Grund Platform indicate that gravitational gliding occurred at the western margin of the paleobasin, i.e. heading towards the German sector of the North Sea as well. De Jager (2012) showed a similar salt-tectonic geometry associated with the Triassic Fat Sand Play (Solling Formation) which is now interpreted as being part of a growth fault/raft system.

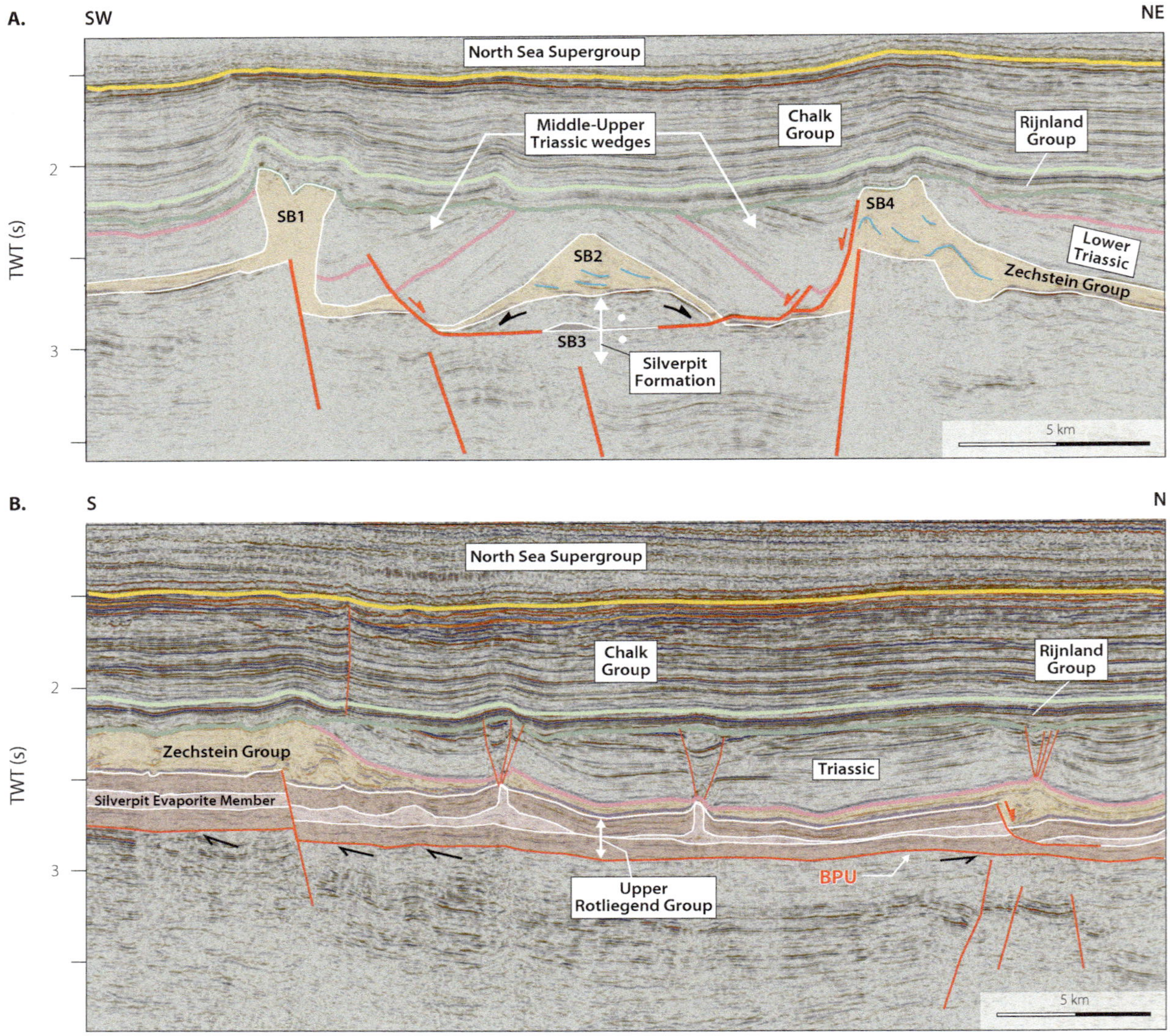

Figure 12.12. Interpreted time seismic sections on the Schill Grund Platform illustrating salt tectonic features associated with Rotliegend-age salt layers of the Silverpit Evaporite Member (see Fig. 12.2b for location). a) Two Upper Triassic growth faults (thick red lines) are detached on the thick halite layers of the Silverpit Formation (Rotliegend Group), i.e. on a deeper stratigraphic level than the Zechstein Group. Blue lines highlight salt stringers. Modified from Bouroullec et al. (2017). b) Salt pillows and diapirs formed in the Silverpit Evaporite Member. Based on the formation of collapse grabens above these diapirs, the salt movement and formation of the deep salt diapirs is likely Triassic in age. Note the growth fault on the northern side of the section showing Zechstein-age syn-sedimentary fault activity detaching onto the Silverpit Evaporite Member of the Upper Rotliegend Group. Modified from Peeters (2016). BPU = Base Permian Unconformity.

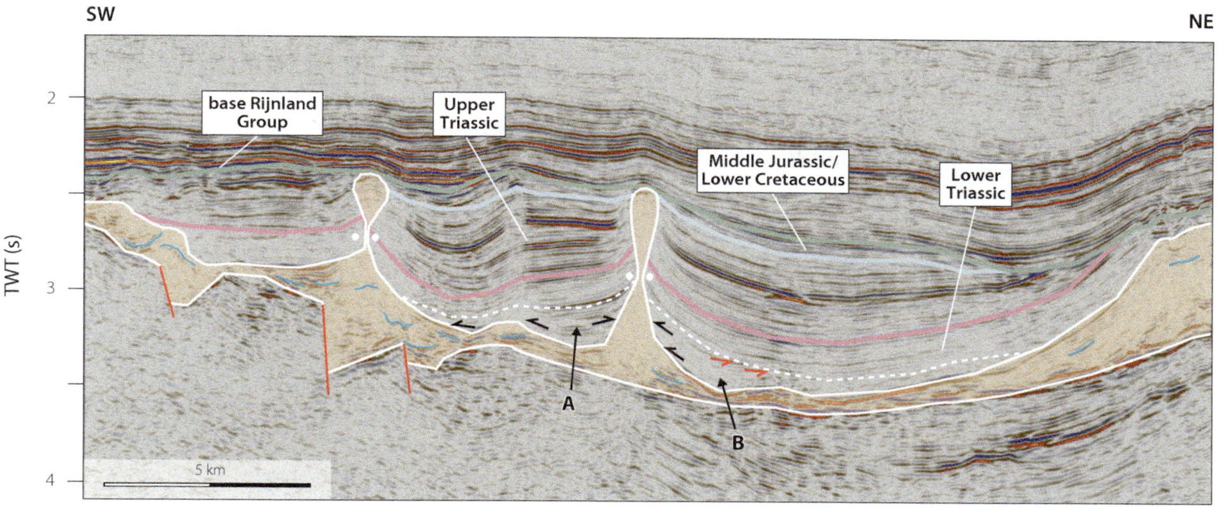

Figure 12.13. Interpreted 2D time seismic sections showing salt-related Early Triassic mini basins (A and B). Basal and internal onlaps (black arrows) and truncations (red arrows) are only observed in the Lower Triassic within these mini basins. This line is crossing the northern parts of the A15 and A13 blocks. Blue lines highlight salt stringers. See Fig. 12.2b for location. Modified from Bouroullec et al. (2017).

Crestal faults

Crestal faulting often occurs above salt pillows, diapirs and walls in the areas defined as crests, which are the anticlines formed in the syn- or post-kinematic strata above salt bodies. These faults are mainly linear in cross section but can be organized radially or longitudinally in the case of circular or elongated salt bodies, respectively. The faults connect downward to the salt body's crest and the degree of upward penetration into shallow stratigraphic levels depends on the deformation of the salt, the rock properties of the suprasalt (e.g. chalk has more crestal faults above salt bodies), the burial history and the local and regional tectonics. Many salt bodies in the Dutch subsurface have crestal faults (Figs 12.3, 12.4, 12.5, 12.19) that range from low-displacement to large-displacement and are associated with supra-salt grabens that formed during salt body collapse (e.g. Salt wall 2 in Fig. 12.4a, Salt wall 2 and SB2, SB3 in Fig. 12.4b, Salt wall 1 in Fig. 12.4d; see section below on collapse salt bodies). Some salt bodies, for instance on the Cleaverbank Platform and the Schill Grund High (Figs 12.3b,c), where faults are mainly observed below major unconformities (e.g. Mid-Cimmerian Unconformity), as well as those on rift shoulders and platforms, are devoid of crestal faults. This suggests that most crestal faults were initiated during the early stages of extensional salt tectonics (in the Triassic and Jurassic). In later stages, salt bodies were often shortened during the Cenozoic Alpine compression and show reactivated, often rotated outward-directed faults or new associated crestal faults. Some shallow faults located above salt bodies can be wrongly interpreted as crestal faults in cross sectional view but appear to be related to large-scale strike slip systems when seen on 3D seismic data (Fig. 12.10). Such flower structures, formed during Mesozoic or Cenozoic transpressional regimes, were accommodated by brittle deformation in the sub- and supra-salt section and by ductile movement within the salt bodies (Peeters, 2016). The presence of such multilayered detached deformation is the reason why the identification of even large strike slip systems in the Netherlands is challenging.

Turtle structures

Turtle structures (or turtle-structure anticlines) are mounded strata lying between salt diapirs, typically consisting of a flat base and a rounded crest. The structure may have a low salt pillow core above which a thick sedimentary sequence thins laterally. Turtle structures form between diapirs whose flanks have subsided because of regional extension or between salt structures whose withdrawal basins migrate and widen through time (Trusheim, 1960).

Several turtle structures are present in the Dutch offshore (Figs 12.3, 12.5, 12.8). They are often large and formed within the Mesozoic rifts in response to extension and outward salt migration towards the basin margins. All turtle structures have a unique kinematic setting formed by local parameters such as: 1) the amount and characteristics of original salt available, 2) the timing of sediment loading, and 3) the timing of Zechstein salt welding at the basin-centres. These parameters control the turtle structure inversion and lateral depocentre shift toward the basin flanks in later stages (see below 'Evolution of salt tectonics' for additional discussion). Depocentre shifts highlight the change of the main loci of subsidence and loading from the axial position of the turtle towards

Glacial tunnel valleys and salt domes

In a strip across the Broad Fourteens Basin, the northern part of the Netherlands onshore and northern Germany, Quaternary (Elsterian), glacial tunnel valleys appear to partly coincide with the location of salt diapirs and -walls, suggesting that at these sites increased and large-scale glacial erosion of pre-glacial sediments occurred. For the offshore Broad Fourteens Basin, Wenau & Alves (2020) showed that glacial valleys follow the strike of a Zechstein salt wall and its crestal faults. Changes in downstream tunnel valley orientation coincide with the strike direction of these salt-induced faults.

One possible explanation is that sediments affected by crestal faults above a salt diapir (see figure) offered less resistance to erosion, which facilitated the formation of Elsterian glacial tunnel valleys and the development of subglacial dewatering pathways during periods of ice sheet advance (Wenau & Alves, 2020). Renewed glacial erosion during subsequent ice ages is evidenced by the observation that Elsterian sediments are thinner above the salt domes of Anloo and Schoonlo (Bregman & Smit, 2012).

In an additional explanation, it is assumed that open fault systems are spots with higher heat flow, which could increase basal melting rates of the ice (Bregman & Smit, 2012). Ice streams may then be 'pulled towards' these favourable pathways since sliding will be facilitated by higher geothermal fluxes. Above the salt diapir, a higher salinity content of the subglacial groundwater may also have had an impact on the above-mentioned processes by lowering the freezing point of the ice. Although most of the salt diapirs are not shallow enough to have played a role in diverting tunnel valley courses, the presence of diapirs may have triggered forebulging in front of the advancing glaciers due to their relative rigidity compared to the surrounding unconsolidated sediments. This may have focused glacier surge to the less elevated areas above and around salt diapirs. Differential compaction and subsidence probably created some undulating surface relief of the pre-glacial landscape making the salt diapirs areas of preferential incision and erosion.

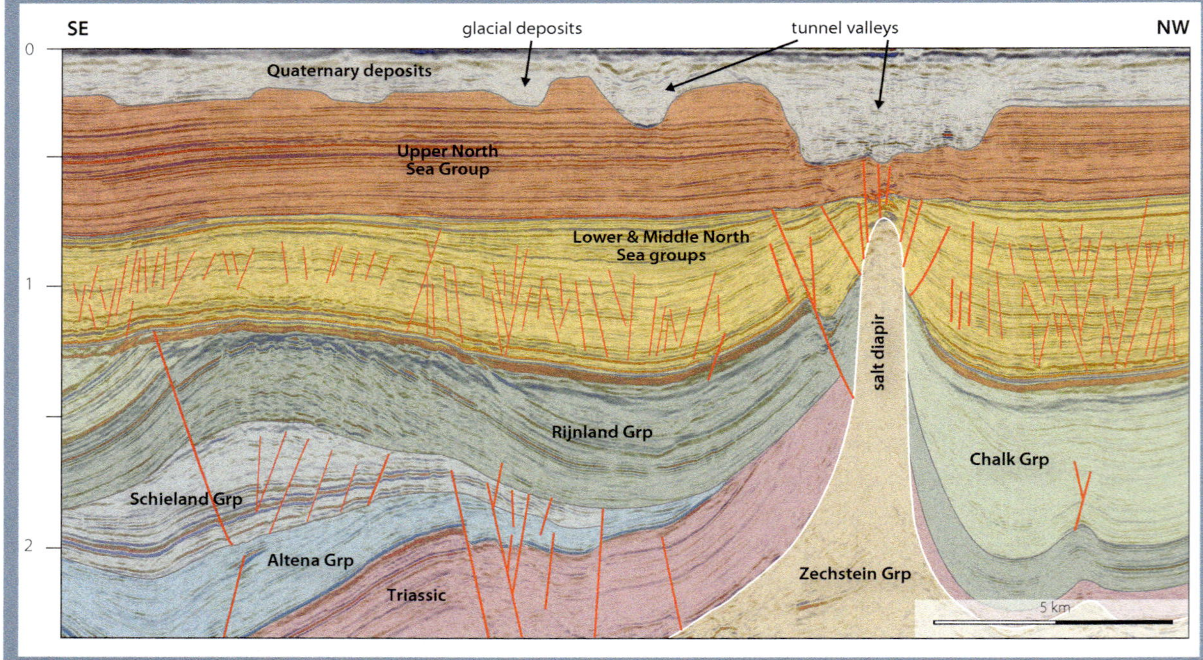

Relationship between salt body-related crestal faulting and glacial tunnel valleys in the offshore Broad Fourteens Basin (K quadrant, see Fig. 12.2b for location). The seismic interpretation of the tunnel valleys is modified from Wenau and Alves (2020).

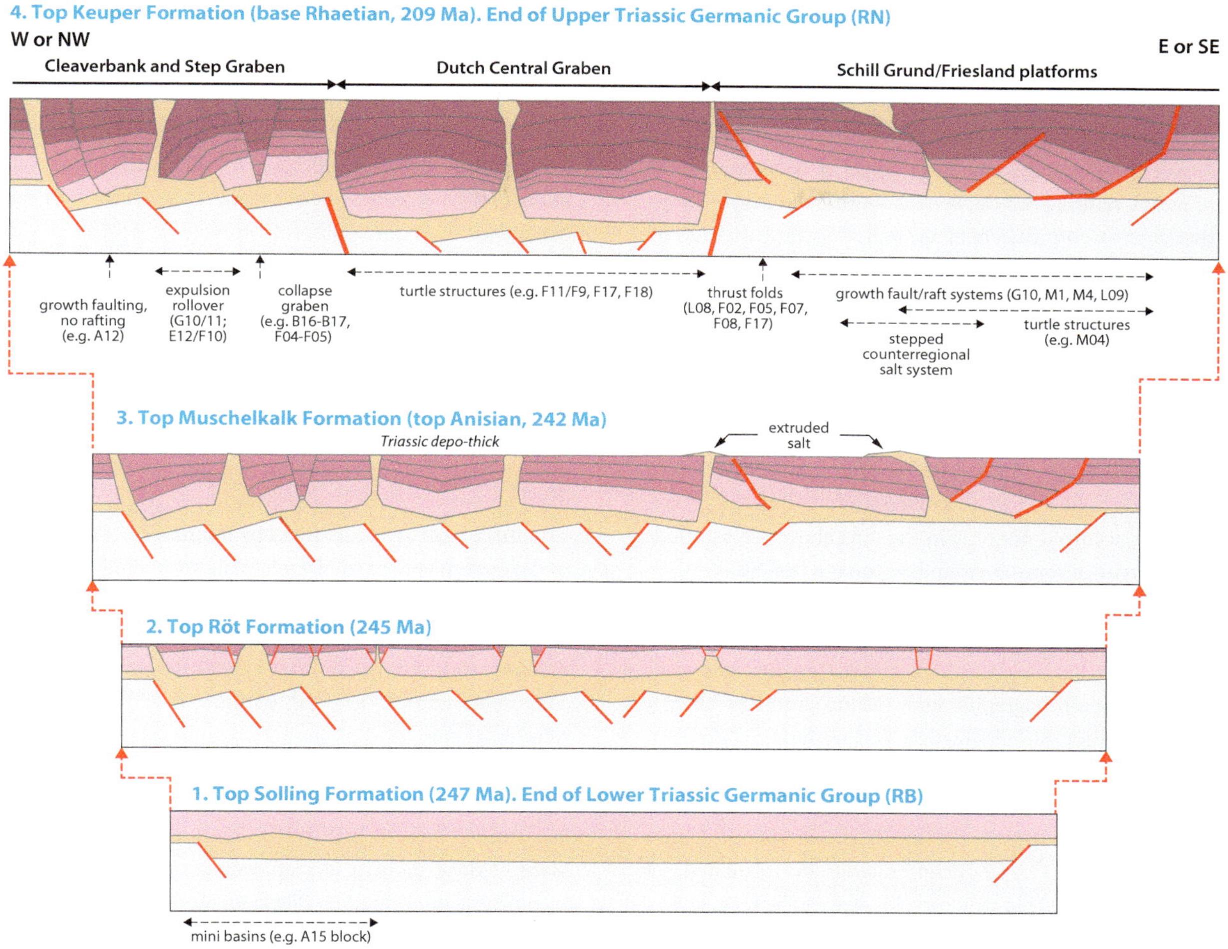

Figure 12.14. Conceptual kinematic model showing the structural evolution during stages of the Triassic, following the deposition of the Solling, Röt, Muschelkalk and Keuper formations in the vicinity of the Dutch Central Graben. This cross section summarizes the different types of salt tectonic structures observed in the area and at specific locations (see text for explanation). The approximate orientation of the sections is from W-E to NW-SE; the sections are not to scale. Modified from Bouroullec et al. (2017).

its margins, following the direction of salt migration. For illustration, Figure 12.8 shows a structural restoration of one of the turtle structures (M04 Turtle) located in the southern part of the Terschelling Basin.

Distinguishing between individual turtle structures in the Dutch offshore is not straightforward since many neighbouring turtle structures are genetically related to each other. Additionally, the salt migration that controls the formation of turtle structures is rarely unidirectional and may have changed direction during the formation of specific structures, for instance due to salt volume constraints during different phases of rifting. Such dynamics make Dutch turtle structures very complex and evaluation of observed structures and depositional trends requires full 3D structural restoration. To date, such a task has not been carried out and would likely be a challenge for years to come. Figure 12.3c shows a typical turtle structure geometry that comprise:s 1) a Lower to Upper Jurassic TMS-1 basin-centered depo-thick that formed when the Zechstein salt evacuated from the axis of the Dutch Central Graben, and 2) marginal rim synclines with stratigraphic wedges of Upper Jurassic to Lower Cretaceous (TMS-2 and 3; Verreussel et al., 2025, this volume) deposits that formed when the salt migrated and fed two large salt diapirs at the rift edges (SB 4 in Fig. 12.3c). The period during which the depocentres shift from axis to margin is referred to as turtle structure inversion and indicates when the mother salt, located below the initially subsiding basin, runs out due to complete salt withdrawal. From this time onward, the salt can only migrate laterally and the turtle structure 'inverts'. This creates an initial point of contact between sub- and supra-salt strata and forces the salt to further withdraw laterally, resulting in the downward and outward rotation of the flanks of the basin and in the formation of rim synclines.

Rim synclines (also referred to as peripheral sinks) are depressions that formed along or around a salt body when salt migrated into that salt body. In this chapter, rim-synclines refer to sub-basins that form in the late stage of turtle structure development along the bounding salt diapirs or walls. They are the result of salt withdrawal along the flanks of a growing salt pillow, diapir of wall and consist of strata thinning or thickening toward the salt body. In the Dutch Central Graben, the initial inward rotation of pre-kinematic strata such as the Lower to Middle Triassic along the basin margins during the initial stages of turtle structure formation was so drastic that the later-developing rim synclines were eroded down to older Jurassic units and locally down to Lower Triassic (e.g. around the location of well F11-03 in Fig. 12.3). Such local salt-tectonically-controlled erosion permitted vast amounts of Mesozoic sediments to be eroded and redeposited within the rift basin. This occurred not only in the area above the salt bodies but in large zones surrounding the salt bodies.

Recent work by Bouroullec et al. (2018) showed that in extreme cases the central part of the Dutch Central Graben (e.g. Fig. 12.3b), which was the original depocentre, became a topographic high (island) and was likely a source area for cannibalization and rerouting of Jurassic sediments toward neighbouring Late Jurassic and Early Cretaceous rim synclines. The F17 and F18 turtles are noticeable structures in the southern part of the Dutch Central Graben (Fig. 12.5) and underlie several oil and gas fields. The M04 turtle structure, located in the Terschelling Basin, formed as a final stage of a growth fault/raft system which explains why Lower Triassic rocks are absent from its core due to earlier rafting toward the west (Fig. 12.8).

Caprocks

A caprock is a dissolution interval located at the crest of a salt body. If the top of a salt body reaches depths at or near the free surface halite will be dissolved leaving a residue of insoluble anhydrite that is later altered to gypsum, calcite, and sulfur. The cap rock may be intermixed with deformed and cemented rocks from the overlying strata transported upward with the migrating salt. As solution takes place in the circulating (shallow) water zone, it is likely that deeply buried domes with cap rock may have been located at shallow depth at some former time and subsequently buried, or else deeper fluid migration may have occurred to trigger the formation of caprock at the crest or along the shoulders of the salt bodies. In that context the Mid-Cimmerian Unconformity played a significant role in bringing the Zechstein salt into contact with Upper Jurassic and Lower Cretaceous reservoir rocks. Locally, the dissolution of vast amounts of halite led to the formation of depressions that controlled Late Jurassic sediment pathways. The formation of these depressions may also have been enhanced by salt body collapse (see below) which occurred during the same period. Verreussel et al. (2018) showed that palynofloras occurring in the basal part of cap rock successions indicate a latest Callovian to earliest Oxfordian age and a marine depositional setting. This suggests that these diapirs were partly dissolved, collapsed, and subsequently flooded and filled in with TMS-1 age sediments (see Verreussel et al., 2025, this volume). In the Netherlands several Zechstein caprocks were drilled and a Scruff Group (Upper Jurassic-Lower Cretaceous) play on the Schill Grund Platform was proved to be economic (G16-A discovery).

Collapse structures

Collapse structures are circular, elongated or graben-like depocentres (Fig. 12.1) above a salt layer or diapir resulting from salt withdrawal or dissolution. Seventeen collapse structures have been identified by Bouroullec et al. (2017) in the central part of the Dutch offshore (notably in quadrants B, F, L). They range from 2 to 20 km in length and are filled with Triassic and Jurassic pre- and syn-kinematic strata relative to the timing of salt movement. For these collapse structures, deep salt withdrawal rather than dissolution is suggested as the most applicable mechanism. The geometry of the collapse structures is normal to the direction of salt migration, with unidirectional salt flow favouring elongated graben-like depressions (Fig. 12.10). Some of these grabens formed over salt pillows of the Silverpit Evaporite Member (Fig. 12.12b) rather than those of the Zechstein Group. These collapsed salt bodies provided new sediment pathways by turning the areas above salt bodies from topographic highs to topographic lows in a similar fashion to the fluvial systems of the Paradox Basin in Utah (Banham & Mountney, 2013). For example, in well A18-02-S1 sediments of the Noordsvaarder Member include Zechstein pebbles that are located directly above the salt. The area is interpreted as an incised valley that was flooded and filled with marine sandstones during the earliest Volgian (Bouroullec et al., 2016).

Mini basins

Mini basins are small intrasalt basins largely surrounded by and subsiding into relatively thick salt. Those in the Dutch subsurface can be grouped into three categories.

The Lower Triassic mini basins: supra-salt stratigraphy is primarily concordant with the upper surface of the Zechstein pillows, except in few locations such as the eastern part of the northern offshore (Fig. 12.13) and the Cleaverbank Platform (Fig. 12.20a), where several small thin ponded mini basins with Lower Triassic growth strata are observed. These mini basins formed directly above the autochthonous Zechstein.

The Middle to Upper Triassic mini basins: these basins also formed over autochthonous Zechstein salt on the rift platforms but distinguish themselves by their greater

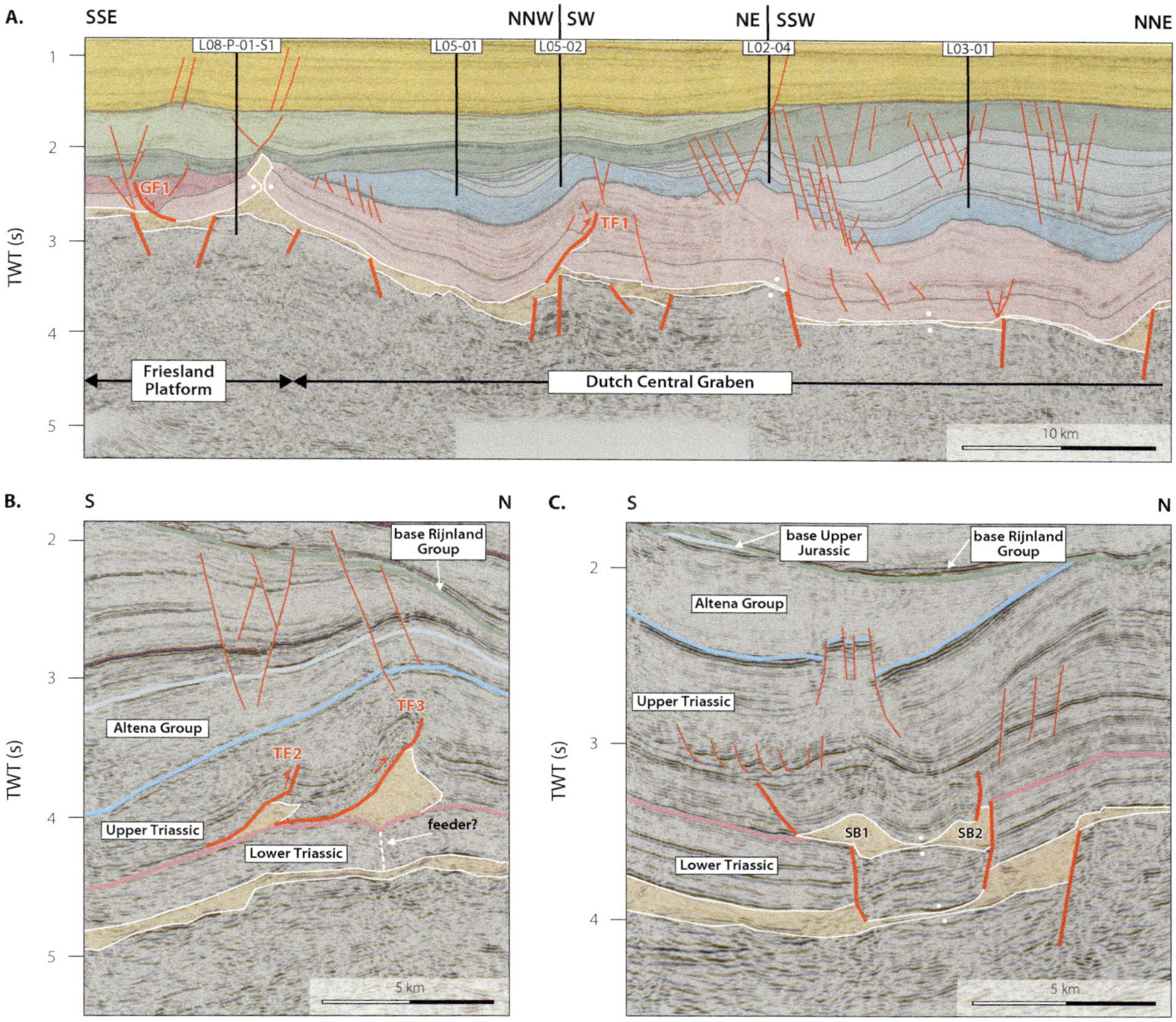

Figure 12.15. Features of Triassic shortening in the Dutch offshore. a) Interpreted seismic time section from the Friesland Platform to the southern part of the Dutch Central Graben. b) Upper Triassic contraction in the F02 Block. Two thrust faults (TF2 and TF3) are observed that contain a salt core (beige polygons). Either, the salt core contains remobilized Zechstein salt injected upward from its original position or, more likely, the salt represents structurally over-thickened Röt Evaporite that forms a shallow detachment, possibly connected to the Zechstein salt layer further towards the south (ramp and flat geometry). c) Intra Upper Triassic salt features in the F09 Block. This salt sheet has not been drilled and is composed of two salt bodies (SB1-2 indicated with beige polygons) that are located stratigraphically above the top of the Lower Triassic. In the northern part of the Dutch offshore, folding and thrusting during the Middle to Late Triassic is interpreted as downdip shortening, balancing some of the thin-skin extensional tectonics (growth fault/raft systems) that took place along the outer rims of the Triassic early rift basins. See Fig. 12.2b for location. Modified from Bouroullec et al. (2017).

thickness and circular shape (Fig. 12.11). They form over large salt pillows that collapsed during the early phase of rifting (Middle to Late Triassic). Similar mini basins (or pods) are observed in the UK North Sea sector (Penge et al., 1999; Stewart & Clark, 1999).

The Mid Triassic to Early Cretaceous mini basins: these basins formed over large allochthonous salt bodies (Fig. 12.16) that formed close to the free surface due to shortening of their stems resulting in increased diapiric rise during the Middle to Late Triassic. These allochthonous salt bodies, which may have locally been extruded during the Mid Triassic, were then loaded to form intrasalt mini basins.

Salt stringers

Salt stringers are deformation features within a salt body and refer to layers or fragments of rock consisting mainly of minerals other than halite (e.g. anhydrite or carbonate) (Figs 12.20, 12.11, 12.12, 12.13, 12.16a). Most of

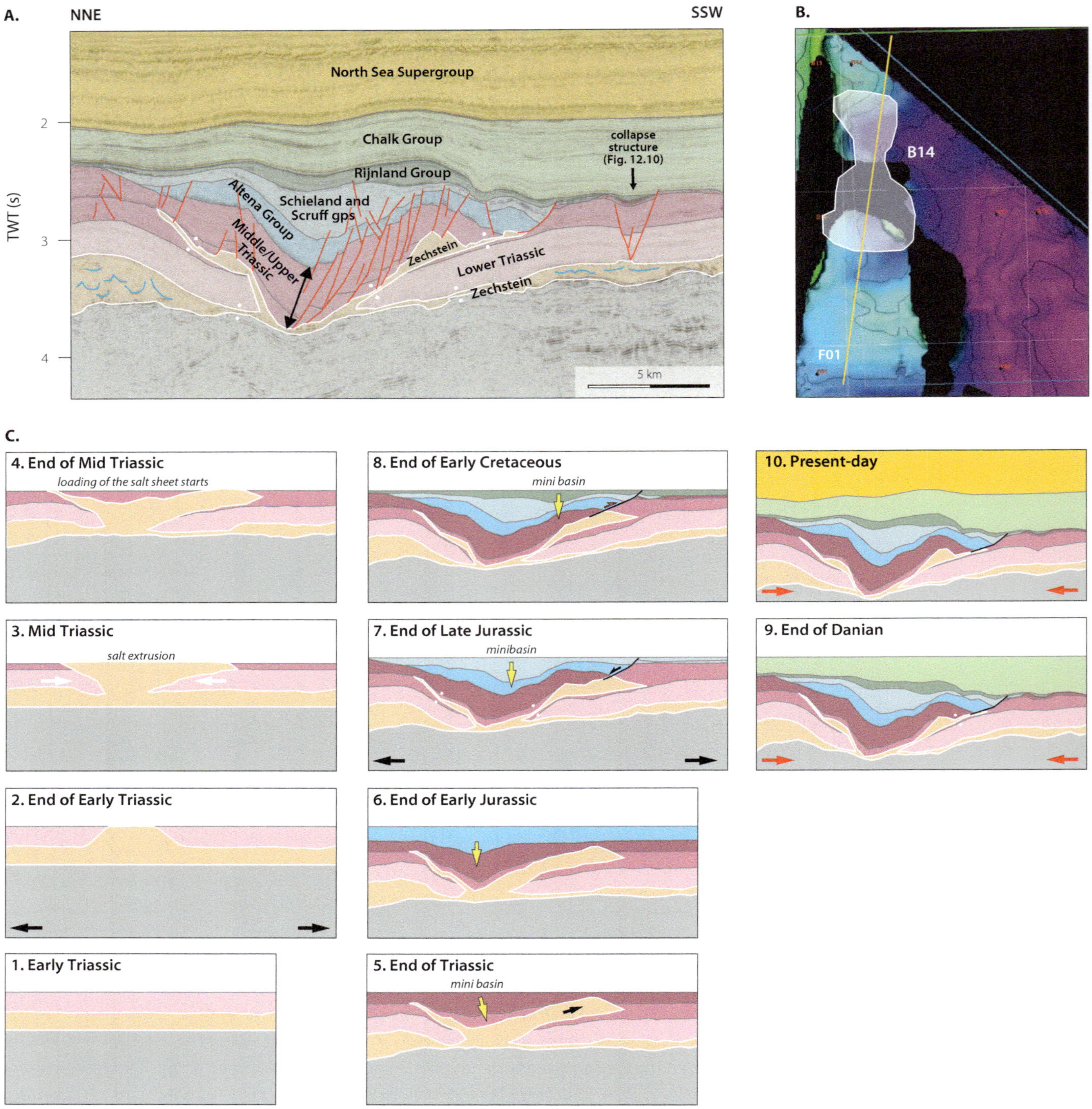

Figure 12.16. Squeezed and loaded salt diapir in the B14-B17 area in the northern part of the Dutch offshore. a) NNE-SSW interpreted seismic time section showing the morphology of a squeezed salt body. Part of the deformation is related to Cenozoic shortening but most of the intra-Triassic deformation is related to the shortening of a pre-existing salt body during the early part of the Middle Triassic. This salt body was later loaded by Upper Triassic and Jurassic sediments. Blue lines highlight salt stringers. See Fig. 12.3c for explanation of salt symbology and inset map (b) and Fig. 12.2b for location of the seismic section. b) Location map. The white polygon shows the extent of the squeezed and loaded salt body that stretches 20 km in N-S direction. c) Conceptual structural restoration of the section shown in (a). Not to scale and without major erosion or dynamic decompaction taken into account. Modified from Bouroullec et al. (2017).

the stringers identified within the Zechstein of the Dutch subsurface are Z3 anhydrite-carbonate (Z3AC) stringers (Geluk, 2007a; Van Gent et al., 2011; Strozyk, 2017). The Z3AC lithologies are assumed to have formed as continuous layers across large parts of the Zechstein Basin, especially in the southern part of the Southern Permian Basin currently located in the subsurface of the onshore (Geluk, 2007a). Further north (basinward), the thickness of Z3AC decreases. Stringers have been identified from the Cleaverbank Platform in the west as far as the Ameland Platform in the east and are well developed in the Waddenzee area (Strozyk et al., 2012; Strozyk, 2017). The apparent lack of stringers in other parts of the Dutch subsurface is likely due to the difficulty of imaging internal heterogeneities within the Zechstein Group. Stringers clearly stand out on seismic data because of their different acoustic properties and, where visible, the geometry of the stringers can give an indication of the severity of salt flow. They are found in many wells as well: Van Gent et al. (2011) showed that the Z3AC stringers are on average 40-50 m thick and have a complex structure dominated by boudinage and folding (Fig. 12.3; Hoetz et al., 2011). They also indicated that these stringers form blocky thickness anomalies and may be up to 150 m thick. They are commonly broken into mappable fragments of varying sizes but are most likely still arranged along the deformed boundary between the Z2 and Z3 salt, generally following the shape of top Zechstein (Strozyk, 2017). In some areas the salt and the imbedded Z3AC stringers are almost undeformed (for example at the Friesland Platform), while in other areas they are very fragmented (for example at the Vlieland Platform; Fig. 12.2a). Recent research (SODM, 2020) concluded that the internal heterogeneities and their effects on stress distribution within the salt are not understood well enough to allow salt solution mining (Groenenberg et al., 2025, this volume) and storage of liquids and gasses (such as hydrogen). For the energy transition in the coming decades, research will be needed to de-risk the presence and effects of salt stringers to ensure the safe development of storage of fuel-based energy carriers (natural gas, oil and likely hydrogen in the near future), feedstocks (nitrogen) and disposal of nuclear waste in salt domes.

Non-Zechstein salt tectonics

Besides the dominant Zechstein salt units other stratigraphic intervals were subject to and initiated salt-related deformation in the Netherlands. These include the Silverpit Formation of the Upper Rotliegend Group and evaporitic intervals in the Triassic, especially the thicker Main Röt Evaporite (see McKie & Kilhams, 2025, this volume).

The Silverpit lake depocentre contains the thickest halite units of the Silverpit Evaporite Member and was located in the eastern part of the Dutch sector around the quadrants F and G (see Fig. 12.7). In this area, several anomalously thick but small salt pillows and rollers have been identified (Peeters, 2016; Bouroullec et al., 2017). Locally these salt layers were used as detachment surfaces for Mesozoic growth faults (see SB3 in Fig. 12.12) and they reach deeper than the ductile Zechstein layer, probably because at that time the Zechstein salt was welded out and was not able to act as a detachment.

Occasionally, Triassic salt layers can also form salt bodies and form detachment horizons for extensional and contractional faults. Figure 12.15b shows two salt-cored thrusts that were active during the Late Triassic. These thrusts are interpreted as the downdip compressional part of gravitational gliding systems that were active during the Triassic. Growth fault/raft systems were active around the basin margins while the basin centre was affected by thin-skin compressional structures that occasionally detached into shallower ductile layers (multilayer salt tectonics). Distinguishing between intruded Zechstein salt into Triassic weak layers (e.g. salt wing model), buried salt glaciers and deformed in-situ Triassic salt is not always easy and only detailed seismic and geochemical analysis may help.

Role of base salt geometry on salt tectonics

The geometry of pre-salt strata (referred to as 'basement' in salt tectonics) played an important role in the location and the way salt flows into preferential zones, influencing the location of diapirism. Many salt bodies are located close or above basement (sub-Zechstein) faults (Figs 12.3, 12.4, 12.5, 12.8 and 12.12). Either the basement faults were present prior to the deposition of the mother salt or formed during and/or after the deposition of the salt layer. Both scenarios have been described from the Dutch subsurface. Barabasch et al. (2019) interpreted disrupted Z3 stringers above base salt faults in the Friesland Platform and the overthickening of the supra-stringer salt layers (Z3 and Z4) as evidence for syn-sedimentary tectonics during Zechstein deposition. These subtle structures are difficult to identify in many parts of the Netherlands because they have been overprinted by later salt tectonics. Geluk (2007a) described syn-rift salt fill of Permian half graben- and pull-apart basins that formed during the Tubantian extensional phase, which is seen as a relaxation phase of the Variscan orogeny. Where the depositional thickness of salt changes across fault zones it may have later controlled salt flow and differential loading and resulted in the localization of salt pillows in the vicinity of basement faults. Tectonic events post-dating the salt deposition also af-

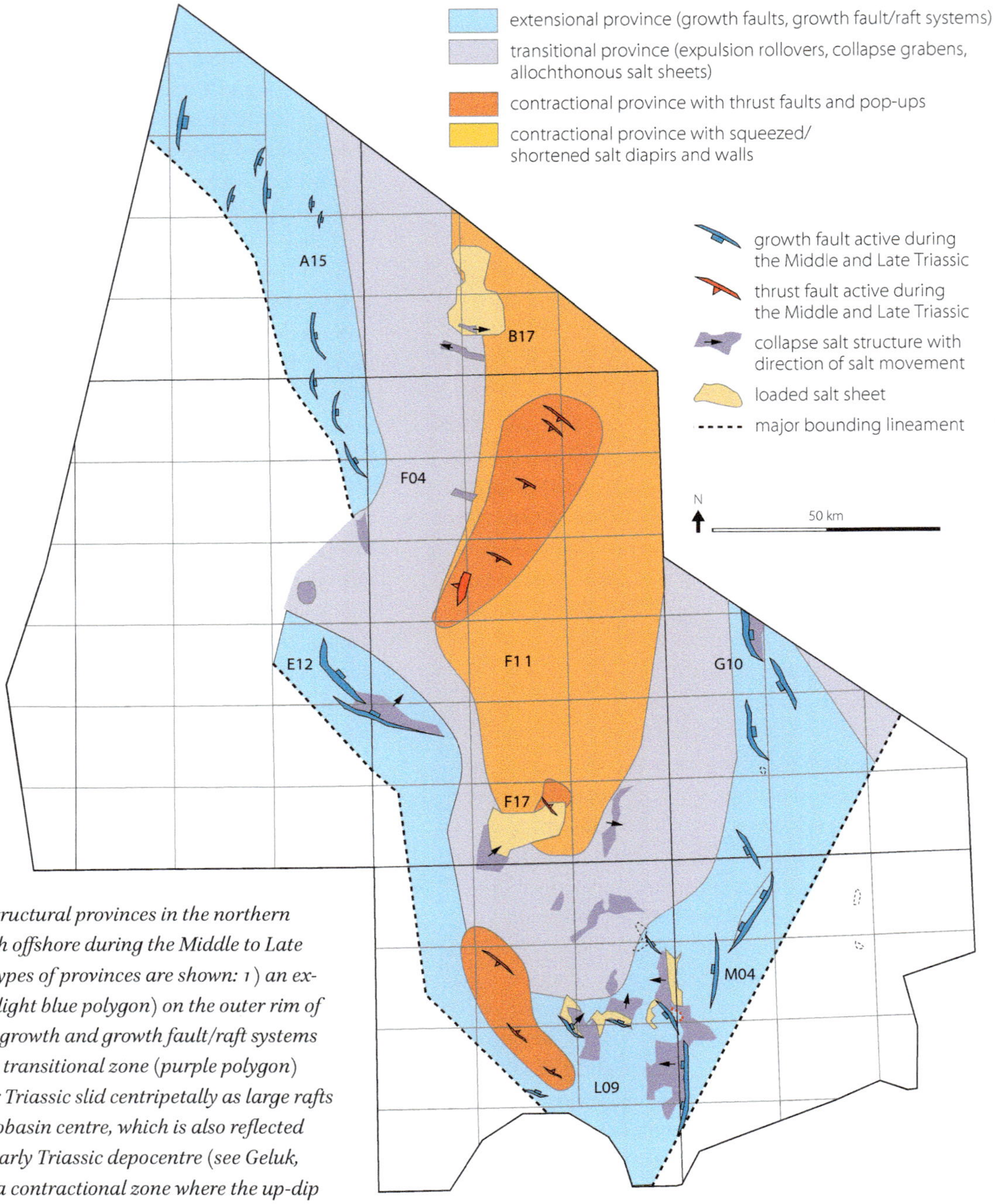

Figure 12.17. Structural provinces in the northern part of the Dutch offshore during the Middle to Late Triassic. Three types of provinces are shown: 1) an extensional zone (light blue polygon) on the outer rim of the basin where growth and growth fault/raft systems were active, 2) a transitional zone (purple polygon) where the Lower Triassic slid centripetally as large rafts toward the paleobasin centre, which is also reflected by the thickest Early Triassic depocentre (see Geluk, 2007b), and 3) a contractional zone where the up-dip extension was accommodated by pre-existing salt body shortening (light orange polygon) and minor thrusting (dark orange polygons). Some of the salt bodies were squeezed to the extent that salt extruded at the paleo-free surface and formed salt glaciers that were later loaded as salt sheets (pale orange polygons). Note the eastward dip of some of the faults in quadrant G, which form the base to eastward-slid stratal blocks. Also note the small area with thrusting in the middle part of quadrant L, which could be related to pre-existing base Zechstein topography ('basement' faults) that may have limited gravitational gliding over long distances and resulted in a small fold belt in this area. Modified from Bouroullec et al. (2017).

fected the preferential directions of salt flow around basement faults, the loci of differential accommodation and hence sediment loading, and deformation of salt bodies in the vicinity of basement faults.

Evolution of salt tectonics in the Dutch subsurface

In this section we discuss the main geological events and phenomena related to salt tectonics that took place in the Netherlands since the Permian. Much of the following is based on offshore studies and publications. Figure 12.21

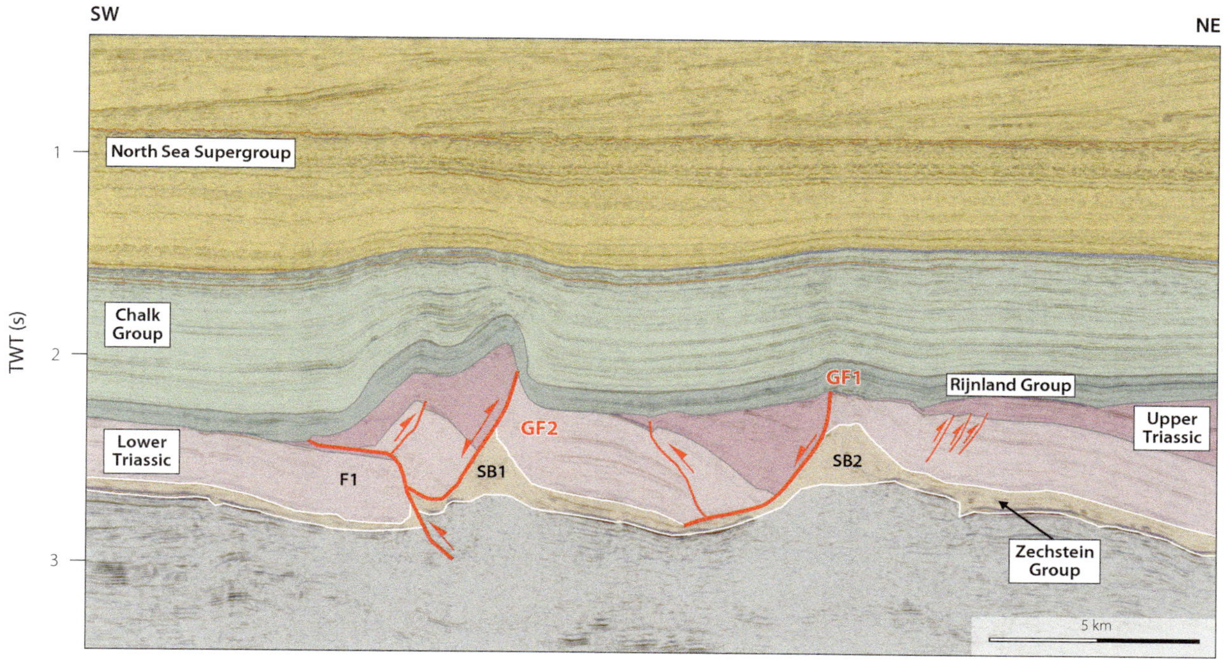

Figure 12.18. Interpreted seismic time section from a 3D survey showing an example of inversion of Triassic structures during Alpine compression. A pre-existing Triassic growth fault (GF2) was reversed during the depositional time of the Chalk and lower Cenozoic and forms one of the sides of a pop-up structure. Note the local blind thrusts in the core of the pop-up and on both sides of growth fault GF1. See Fig. 12.2b for location. Modified from Bouroullec et al. (2017).

summarizes most of the salt evolution described in the text below.

Permian

I. The first significant event was the deposition of thick halites of the Silverpit Formation (Upper Rotliegend Group; see Fig. 4.2 of Bouroullec & Geel, 2025, this volume) during the Capitanian (middle Permian). These individual halite layers are up to 40 m thick in the central part of the basin (e.g. well G13-01).

II. The Zechstein Group was deposited in a large part of the Netherlands during the Wuchiapingian and Changhsingian (late Permian). Its thick salt members, in particular the up to 1300 m thick Z2 halite (Fig. 12.2a), were deposited in the northern part of the onshore and in most of the central and northern parts of the offshore. The original salt thickness was variable across the Dutch sector (see above; Ten Veen et al., 2012) and along pre-existing fault blocks (section 1 in Fig. 12.14).

III. In contrast to the UK (Stewart, 2007), no evidence of Permian-age salt tectonics has been described for the Netherlands. It is however conceivable that some of the thick halite layers of the Silver Pit Formation may have been buried deeply enough to start flowing into salt pillows, but such observations have not been made so far.

Triassic

IV. Overall, the Early Triassic is a period without salt tectonics in the Netherlands (Geluk, 2007b). Recent work by Bouroullec et al. (2017), however, shows that some small growth mini basins (pods) formed in the northwestern part of the Dutch sector in the A15 and A13 Blocks (Figs 12.13, 12.20a). In the rest of the Dutch offshore the Lower Triassic thickness is more uniform and reflects a homogeneous subsidence (mainly thermal in origin), with Buntsandstein reflectors unaffected by syn-depositional faulting (Ziegler, 1990; Hoffmann & Stiewe, 1994, Geluk, 2007b; Ten Veen et al., 2012). This makes the Lower Triassic a very useful pre-kinematic reference unit ('rail-track') across most of the Dutch salt province as is illustrated in Figures 12.3, 12.4 and 12.6.

V. The Middle Triassic is a period of great importance for the Dutch salt tectonic history since it sets the stage for many other events that followed. Three notable events can be highlighted:

   i. Thickening of the Solling Formation into listric faults (Fig. 12.8) suggests that locally gravitational sliding must have started to play a role already during the latest Olekenian. In the early Anisian (around 247 to 245 Ma), at the onset of rifting in the Dutch sector, small amounts of extension in the sedimentary basins allowed local faulting and

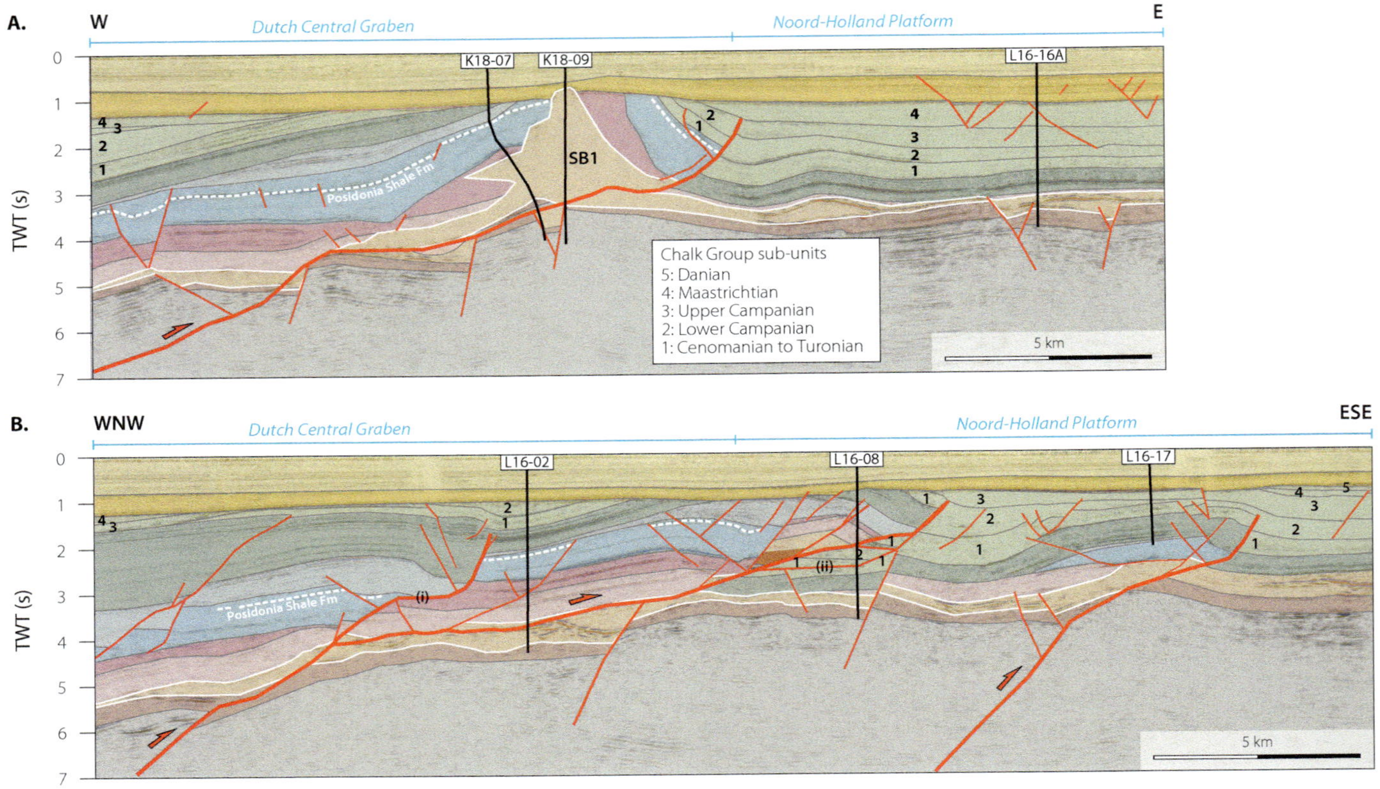

Figure 12.19. Interpreted seismic time sections in the southern part of the Dutch offshore (blocks K18-L16). a) W-E oriented section across a large thrust fault (thick red line) The core of the hanging wall is occupied by a large salt body (SB1). The presence of a Lower Jurassic stratigraphic wedge on the eastern side of the salt body indicates that the thrust occupied a normal fault that was originally associated with Jurassic rifting. This salt was located in the vicinity of the rift bounding fault on the eastern side of the Broad Fourteen Basin and was remobilized during reactivation in the Late Cretaceous. Unconformities between sub units of the Chalk Group attest to multiple periods of Late Cretaceous to early Cenozoic inversion. Note the repeated section of Carboniferous (Limburg Group) in well K18-07. b) Section just south of section (a) showing two thrust faults (thick red lines), one of which is interpreted as a duplex with multiple branches (thinner red lines). Note that the flat and ramp geometries of the larger thrust faults mainly detach in the Zechstein salt. The branches also use Middle-Upper Triassic mechanically weak layers (likely in-situ salt layers of the Main Röt Evaporite Member) or weak levels in the Texel Formation of the Chalk Group (likely the Plenus Marl Member). These detachments are indicated with (i) and (ii), respectively. See Fig. 12.2b for location of the lines. Modified from Peeters et al. (2021).

extension within the suprasalt overburden to take place, locally breaking strata into blocks and allowing salt to flow into the newly formed grabens (sections 1 and 2 in Fig. 12.14). This is also well documented in the Gulf of Mexico and on the west African continental margins (Schultz-Ela, 1992; Jackson et al., 1994; Mauduit et al., 1998), allowing several salt pillows to form. The Middle Triassic shows more thickness variation than the lower Triassic but less than the Upper Triassic, which indicates that the intensity of salt tectonics was not yet at its peak during this period (Fig. 12.14).

ii. During the Anisian two main Triassic salt-prone units were deposited, namely the Main Röt Evaporite and the Muschelkalk Evaporite members.

iii. Gravitational sliding intensified at the end of the Anisian period with a predominance of extension on the outer-rim of the Triassic basin (eastern part of quadrant A, southwestern part of F; northern part of L, northwestern part of M and G) (Bouroullec et al., 2017; see Fig. 12.7). Growth fault/raft systems formed in the same regions during this period with a vergence toward the Permian and Lower Triassic paleo-basin axis that was located in the quadrant F (Figs 12.6, 12.7, 12.8, 12.12, 12.14) and continuing farther northeast towards Germany. Fault G1 in Figure 12.7 dips to the east and is indicative of eastward gravitational gliding. The best developed growth fault/raft systems are observed in the L and M quadrants and are associated with salt sheets that represent squeezed and extruded salt from pre-existing salt diapirs that formed at the onset of the Middle Triassic (Fig. 12.8). With the predominance of extensional gravitational systems on the outer rim of the paleo-basin, one would expect a large concentration of reverse faults in the down-dip/basin axis area (quadrant F). However, only a few of that age have been observed (TF1-3 in Fig. 12.15). One example is seen in Figure 12.15a where up-dip extension during the Late Triassic (growth fault GF1) is balanced downdip by a thrust fault (TF1) located below well L05-02. TF1 detached onto the Zechstein salt and has a narrow salt body on its footwall side that is interpreted as a shortened and overthrusted pre-existing salt diapir that formed around the depositional time of the Röt Formation. Figure 12.15b shows examples of thrusts faults that detach on shallower (Triassic) salt units. Bouroullec et al. (2017) suggested that most of the balancing was accommodated by squeezing of pre-existing salt diapirs and -walls (SB6 in Fig. 12.3b; Salt wall 2, SB 4 and SB5 in Fig. 12.4; SB2 in Fig. 12.6; SB4 and SB6 in Fig. 12.8; Fig. 12.14; step 3 of Fig. 12.16). This increased the upward migration of the salt in those areas and locally allowed salt to extrude at the free surface (in a similar fashion as the salt glaciers in the Zagros Mountains in Iran). Such extruded salt bodies are represented by a few remaining (non-eroded) loaded salt sheets in the south of the paleo-basin (quadrant L) and in the north of the Dutch onshore in quadrant B (see Figs 12.4a-c, 12.7, 12.8 and step 3 of Fig. 12.14).

VI. The Late Triassic was a period of intense salt tectonics that led to the large variation in stratigraphic thicknesses observed throughout the Dutch sector. Large stratigraphic wedges formed on top of rollover anticlines in growth faults/raft systems. Salt migrated toward rift bounding faults where it formed large salt diapirs and -walls. Concurrently, Triassic axial depocentres developed and evolved into turtle structures as salt withdrawal continued (e.g. Figs 12.3, 12.8, 12.14). Several salt sheets that were emplaced during the Middle Triassic became subsequently loaded by thick Upper Triassic (and younger) strata that locally accumulated in mini basins above the originally squeezed and extruded salt masses (Figs 12.4d, 12.16).

VII. During the Early Jurassic salt tectonics declined and only minor and long wavelength salt-related thickening features are observed. The Altena Group is often thicker in the axis of the Dutch Central Graben, which is indicative for continued salt migration from the basin axis toward its margins but this took place in a more homogeneous manner that during the Middle/Late Triassic or during the upcoming Middle Jurassic-Early Cretaceous phases (Figs 12.3c, 12.5, 12.16). This can be related to a relatively quiet phase in the Dutch rifting history and the decrease of accumulation rate that triggered less differential loading on pre-existing salt bodies.

VIII. The Middle Jurassic is missing from most of the Dutch salt province due to the Mid-Cimmerian erosional phase which eroded large portions of the older strata (as well as shallow paleo-salt features). This holds especially for rift shoulders represented by the Step Graben and Terschelling Basin (which were not yet loci of Jurassic deposition at that time), Cleaverbank Platform and Schill Grund Platform. The Middle Jurassic Werkendam Member is only locally present in the Roer Valley Graben, the West Netherlands Basin, the Broad Fourteens Basin and the Dutch Central Graben. The role salt tectonics played during this period is therefore unknown.

IX. The latest part of the Middle Jurassic (Callovian) to the late Middle Volgian (TMS-1 and TMS-2 in Figs 12.3 and 12.5) consists of a growth stratigraphic interval highly affected by the continued salt withdrawal from the Dutch Central Graben axis toward its margins (Bouroullec et al., 2016, 2017, 2018; Verreussel et al., 2018, 2025, this volume). The depositional area widened during the Kimmeridgian with platform areas such as the Step Graben and the Terschelling Basin becoming part of the larger rift system thanks to a change from E-W to NE-SW rifting (Bouroullec et al., 2017; Verreussel et al., 2018). During the Late Jurassic-Early Cretaceous, crustal extension accelerated across the North Sea rift system driven by the reorganization of the stress field due to the opening of the Arctic-North Atlantic rift system and the related clockwise, westwards rotation of Laurentia-Greenland relative to Eurasia (Ziegler, 1988, 1990; Torsvik et al., 2002). This reorganization resulted in NW-SE-trending transtensional basins that formed along the southern margin of the Southern Permian Basin. Outside of the rift basins large areas were uplifted, exposed and eroded (Figs 12.3b,c, 12.5, 12.8). Evidence that salt bodies acted as sediment pathways during the latter part of this period has been documented on the Schill Grund and Cleaverbank platforms (Bouroullec et al., 2017, 2018, 2019). Most of the contemporaneous sediment pathways located on the rift shoulders toward the basinal sinks were, however, originally located away from the still growing salt bodies. When some of those salt bodies collapsed, due to the lack of salt flowing into their locations (salt welding out

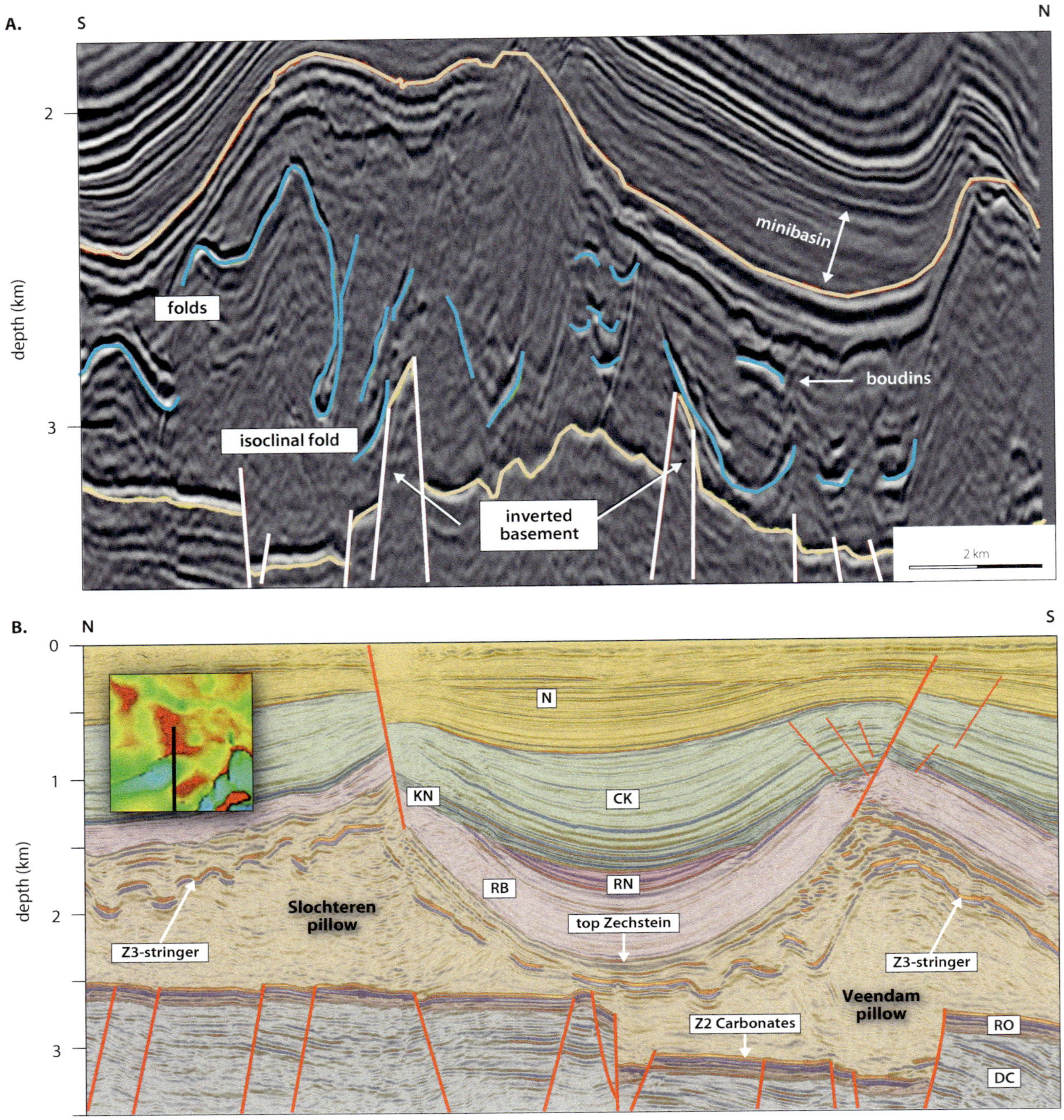

Figure 12.20. Zechstein salt stringers. a) N-S seismic depth section through the Cleaverbank Platform. Stringers are shown as blue lines. Note that the isoclinally folded stringers are located approximately above the basement-block bounding faults, whereas the boudins occur above the fault block. Location shown in Fig. 12.2b. Reprinted from Van Gent et al. (2011) with permission from Elsevier. b) N-S profile through the Slochteren and the Veendam salt pillows (see Fig. 12.2b and inset map showing top Zechstein for location). The Z3 stringer became separated (boudinaged) under the mini basin and was folded and thrusted in the salt pillows. Seismic interpretation based on Raith et al. (2016). N = North Sea Supergroup; CK = Chalk Group; KN = Rijnland Group; RN = Upper Germanic Trias Group; RB = Lower Germanic Trias Group; RO = Upper Rotliegend Group; DC = Limburg Group.

in neighbouring regions), they occasionally became sediment sinks (collapse structures) or acted as sediment conduits to more distal zones, especially where collapse salt walls are located in the vicinity of the rift basins. Most of the turtle structures observed in the Dutch offshore entered their inversion stage during this period. The timing of inversion spans from the Late Triassic (Turtle M4 in Fig. 12.8), through the Kimmeridgian (F17 and F18 turtles in Fig. 12.5) to the Middle Volgian (F09 turtle in, Fig. 3b, c). Hence, most of the younger Late Jurassic to Early Cretaceous accumulations in the Dutch Central Graben are in rim synclines around these turtle structures.

X. The late Volgian (Late Jurassic) and Ryazanian (earliest Early Cretaceous) (TMS-3; Bouroullec et al., 2018; Verreussel et al., 2025, this volume) was a period of continued salt tectonics but is distinguished from the previous period by the presence of an erosional event at its onset. This event eroded large amounts of older strata as well as salt bodies in the rift basins. Especially on the rift shoulders this erosion reached down to Permo-Triassic strata (Fig. 12.3a). Following this erosional event, the Scruff Greensand Formation and Lutine Formation were deposited in the basins and on their neighbouring platforms. During the Late Cretaceous, these deposits were often eroded, especially on the platforms and uplifted parts of the rift and are only preserved at the paleo-basin margins where accommodation was greater.

XI. After the rifting ceased, deposition of the Early Cretaceous Rijnland Group testifies to a period of relative quiescence during which only regional thermal subsidence occurred. The Rijnland Group often drapes salt bodies and is thicker between pre-existing salt bodies (Figs 12.3a, 12.4a). Most of the thickness changes around salt bodies are more likely to have been caused by differential subsidence rather than reactivation of salt features. In the northern Dutch offshore, where the original salt was thickest, a few collapsed salt bodies show thicker Rijnland Group developments, suggesting that Zechstein salt was still actively withdrawing from the northern tip of the Dutch Central Graben during this period (Fig. 12.16). No new diapirs formed during this period.

XII. During the Late Cretaceous to Danian deposition of the Chalk Group multiple salt features were reactivated. The presence or absence of thick Zechstein salt plays an important role in the architecture of inverted basins (De Jager, 2007). During multiple phases of the Alpine Orogeny, shortening was expressed as a series of large, uplifted zones in the older rift basins (Figs 12.3, 12.4a,c, 12.8, 12.15a, 12.16a), shortened and squeezed salt bodies (SB5 and SB6 in Fig. 12.3b; SB7 in Fig. 12.5) and thrust faults (Figs 12.18, 12.19). Many of the largest inverted structures were focussed along pre-existing rift-bounding faults with associated salt bodies (e.g. SB1 in Fig. 12.19 that coincides with the northern margin of the Broad Fourteens Basin). In the Dutch subsurface, large thrust faults in the shallow (post-Permian) sequence are often expressed as thin-skinned ramp- and flat thrust systems detaching onto the Zechstein (mother) salt (Fig. 12.19). Reactivation and shortening of pre-existing salt bodies resulted in substantial folding of chalk layers and older supra salt strata (Figs 12.3, 12.4a,b, 12.5, 12.18, 12.19). Such folding allowed for intense faulting and fracturing of the chalk above salt-cored anticlines. The most significant phases of shortening were the Sub-Hercynian (Campanian) and Laramide (Mid Paleocene) phases that triggered major local uplift (especially of the paleo-basin centres) and erosion. Some of the rift bounding faults were inverted during this period and can be salt-cored (SB1 in Fig. 12.19). No extrusion of salt associated with these thrusts has been documented. During the Alpine orogeny some salt bodies were squeezed and their crests uplifted and eroded. Some local intra-Chalk Group carbonate and siliciclastic syn-tectonically reworked sediments, equivalent to the onshore Aachen and Vaals formations, were deposited on the flanks of inverted basins (Van Lochem et al., 2025, this volume). Also, contourites formed in the depocentres located between elevated suprasalt areas and mass transport complexes (including slides and slumps) formed on the steeper slopes (Van der Voet et al., 2018). Multiple transpressional faults were active during this period in the Netherlands (Nalpas et al., 1995; De Jager, 2007; Peeters, 2016). During the transpressive deformation salt bodies along pre-existing rifting-bounding faults were further squeezed, with narrowing of salt stems and formation of vertical welds.

Mid-Paleocene to Present: many faults active during this period are detached on the Zechstein salt bodies, although no significant evidence for salt flow is recognized during this period. During the Cenozoic, the main structural element in the southern North Sea is the Central Graben, which is interpreted as an intracratonic sag basin (Huuse & Clausen, 2001). This is the area with greatest accommodation during the deposition of the late Cenozoic southern North Sea delta. During the Quaternary, a combination of rapid subsidence in the centre of the basin and uplift at the UK and Fennoscandian margins is attributed to intra-plate stresses (Overeem et al., 2001). Although several mechanisms have been proposed to explain Quaternary subsidence, compaction and load-induced subsidence alone explain about 75% of the observed subsidence (Arfai et al., 2018). Consequently, in the basin centre salt bodies became rapidly buried below a thick (more than 1 km) pile of shallow marine sediments. Despite their deep burial and the fact that salt bodies are not found shallower than strata as young as the Middle Miocene they have a significant influence on the thickness distribution of the Cenozoic interval, which often shows anticlinal domes atop- and flexural depressions around salt bodies. Differential compac-

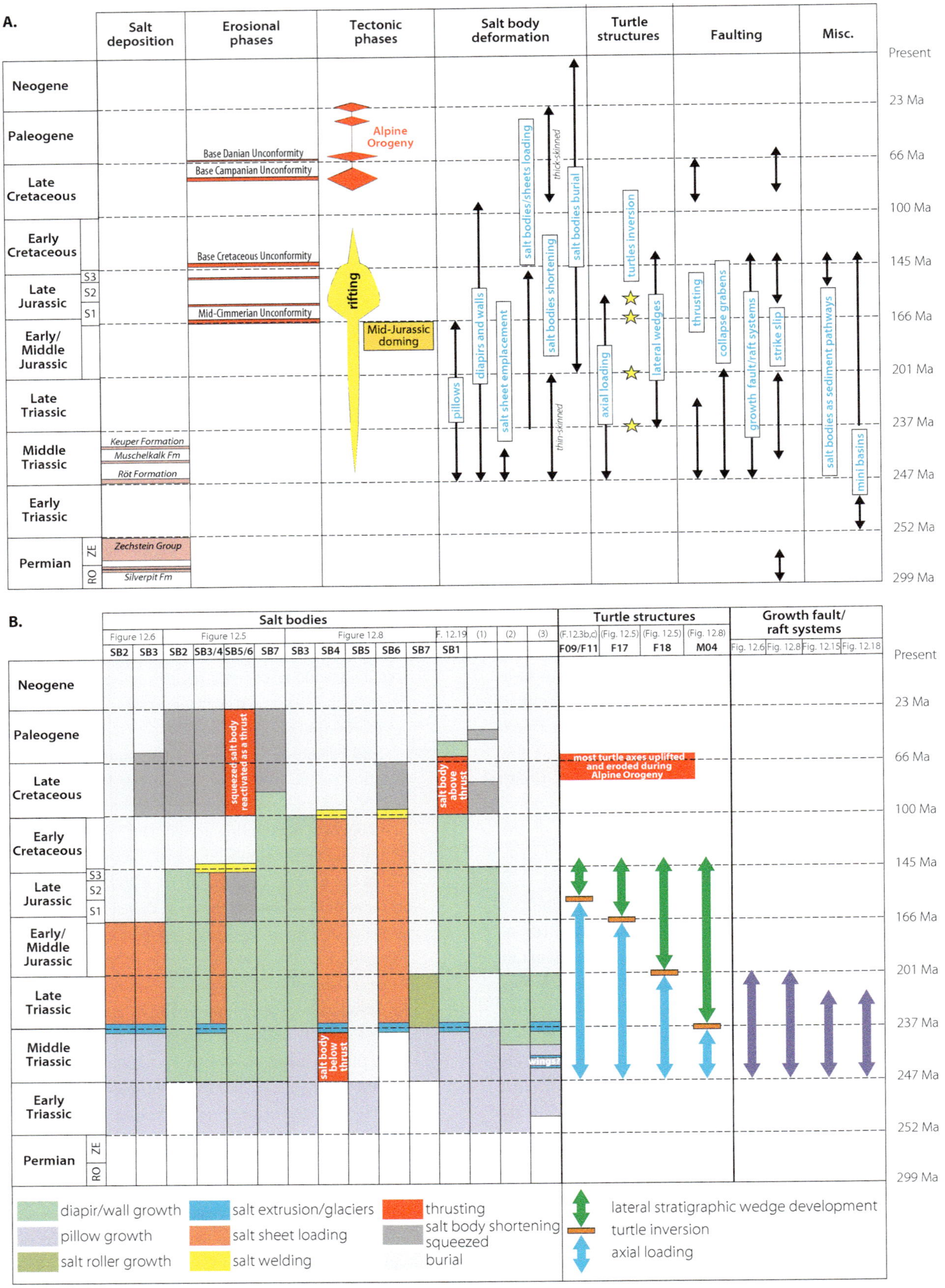
A.
Salt deposition
Erosional phases
Tectonic phases
Salt body deformation
Turtle structures
Faulting
Misc.
Present
23 Ma
66 Ma
100 Ma
145 Ma
166 Ma
201 Ma
237 Ma
247 Ma
252 Ma
299 Ma
Neogene
Paleogene
Late Cretaceous
Early Cretaceous
Late Jurassic
Early/ Middle Jurassic
Late Triassic
Middle Triassic
Early Triassic
Permian
S3
S2
S1
ZE
RO
Base Danian Unconformity
Base Campanian Unconformity
Base Cretaceous Unconformity
Mid-Cimmerian Unconformity
Keuper Formation
Muschelkalk Fm
Röt Formation
Zechstein Group
Silverpit Fm
Alpine Orogeny
rifting
Mid-Jurassic doming
pillows
diapirs and walls
salt sheet emplacement
salt bodies/sheets loading
thick-skinned
salt bodies shortening
thin-skinned
salt bodies burial
axial loading
turtles inversion
lateral wedges
thrusting
collapse grabens
growth fault/raft systems
strike slip
salt bodies as sediment pathways
mini basins
B.
Salt bodies
Figure 12.6
Figure 12.5
Figure 12.8
F. 12.19
(1)
(2)
(3)
SB2
SB3
SB2
SB3/4
SB5/6
SB7
SB3
SB4
SB5
SB6
SB7
SB1
squeezed salt body reactivated as a thrust
salt body above thrust
salt body below thrust
wings?
Turtle structures
(F.12.3b,c)
(Fig. 12.5)
(Fig. 12.5)
(Fig. 12.8)
F09/F11
F17
F18
M04
most turtle axes uplifted and eroded during Alpine Orogeny
Growth fault/ raft systems
Fig. 12.6
Fig. 12.8
Fig. 12.15
Fig. 12.18
diapir/wall growth
pillow growth
salt roller growth
salt extrusion/glaciers
salt sheet loading
salt welding
thrusting
salt body shortening
squeezed
burial
lateral stratigraphic wedge development
turtle inversion
axial loading

tion related to thickness variations is one of the explanations for the (relative) up-doming of Cenozoic sediments. The resulting anticlinal structures favour migration and trapping of gas in shallow intervals. Hence most of the shallow gas discoveries in the Netherlands are located above salt bodies (see Remmelts et al., 2025, this volume). Eastward of the intracratonic sink area, i.e. in the northeastern part of the Netherlands and NW Germany, the thickness of the Cenozoic interval is less and several salt diapirs reach up to 100 m below the surface. These shallow depths make such salt diapirs favourable sites for salt production (see Groenenberg et al., 2025, this volume).

Synthesis and outlook

Salt tectonics has shaped the Dutch geology in significant ways since the Triassic. The Netherlands contains one of the most complex salt tectonic provinces in the world since it involves multiple salt layers, tectonic phases and erosion events. The complex structural styles resulting from this multi-stage salt tectonics have often favoured the migration, trapping and sealing of economically significant oil and gas volumes by providing enhanced accommodation for the deposition of thick reservoirs above and around salt bodies. For the coming decades, in which the role of the subsurface in the energy transition will be crucial, an improved understanding of salt tectonics in the Dutch on- and offshore will remain an important topic. Especially for CCUS a good understanding of the behavior of faults that detach onto salt layers and bodies will be crucial. Ongoing and future exploitation of salt onshore will also benefit from a good understanding of shallow salt bodies and layers. Current activities include rock salt and magnesium salt mining, storage of natural gas, nitrogen and diesel, whereas future activities will include storage of hydrogen, compressed air (CAES) and, although still only theoretically feasible, the disposal of nuclear waste. Geothermal activity in the Dutch onshore also targets many stratigraphic intervals that are affected by salt tectonics and new research will help to further reduce the risk of future developments in these structurally challenging areas.

← ***Figure 12.21.*** *Tectonic summary charts. a) Timing of main salt tectonic features and related phenomena in the Netherlands that formed since the Permian. Yellow stars indicate specific periods of turtle structure inversion. b) Timing of specific salt bodies, turtle structures and growth fault/raft systems shown in this chapter's figures and in three representative publications: (1) Van Winden et al. (2018); (2) De Jager (2012); (3) Ten Veen et al. (2012).*

Acknowledgements

We would like to express our deepest gratitude to the numerous colleagues and peers who, through their work and observations, made important contributions to the content of this chapter. A special thanks goes to Susanne Nelskamp, Stefan Peeters, Geert de Bruin and Sander Osinga. An earlier version of the manuscript was reviewed by Heijn van Gent, who made valuable suggestions for improvement, for which we are highly grateful. Another review was provided by Janos Urai, the godfather of salt mechanics, who sadly passed away in 2023. His passion and expertise on the salt topic greatly enriched this chapter. His legacy will continue to inspire us.

Digital map data

Spatial data of figures in this chapter for use in geographical information systems can be downloaded here: https://doi.org/10.5117/aup.28163951.

References

Arfai, J., Franke, D., Lutz, R., Reinhardt, L., Kley, J. & Gaedicke, C., 2018. Rapid Quaternary subsidence in the northwestern German North Sea. Scientific Reports 8: 11524. DOI: 10.1038/s41598-018-29638-6

Bally, A.W., 1981. Thoughts on the tectonics of folded belts. Thrust and Nappe Tectonics: London, Geological Society, Special Publication 9: 13-32. DOI: 10.1144/GSL.SP.1981.009.01.03

Banham, S. & Mountney, N., 2013. Evolution of fluvial systems in salt-walled mini-basins: A review and new insights. Sedimentary Geology 296: 142-166. DOI: 10.1016/j.sedgeo.2013.08.010

Barabasch, J., Urai, J., Raith, A. & De Jager, J., 2019. The early life of a salt giant: 3D seismic study on syntectonic Zechstein salt and stringer deposition on the Friesland Platform, Netherlands. Zeitschrift Der Deutschen Gesellschaft Für Geowissenschaften 170. DOI: 10.1127/zdgg/2019/0186

Bouroullec, R. & Geel, C.R., 2025. Permian. In: Ten Veen, J.H., Vis, G.-J., De Jager, J. & Wong, Th.E. (eds): Geology of the Netherlands, second edition. Amsterdam University Press (Amsterdam): 127-153. DOI: 10.5117/9789463728362_ch04

Bouroullec, R., Nelskamp, S., Kloppenburg, A., Fattah, R.A., Foeken, J., Ten Veen, J.H., Geel, K., Debacker, T.N. & Smit, J., 2019. Burial and Structural Analysis of the Dinantian Carbonates in the Dutch Subsurface. EBN, TNO (Utrecht), SCAN-Dinantien Project Report: 170 pp. https://www.nlog.nl/sites/default/files/2019-09/scan_dinantian_burial_and_structuration_report.pdf

Bouroullec, R., Osinga, S., Neldskamp, S., Goldberg, T., Zijp, M., De Bruin, G., Verreussel, R., Janssen, N., Houben, A., Kerst-

holt-Boegehold, S., Van Hoof, T., Peeters, S. & Halter, M., 2017. The STEM Project: Salt tectonics early movement, Dutch offshore. TNO (Utrecht), Report No. 2017 R11242: 203 pp.
Bouroullec, R., Verreussel, R., Geel, K., Munsterman, D., De Bruin, G., Zijp, M., Janssen, M., Millan, I. & Boxem, T., 2016. The FOCUS Project: Upper Jurassic sandstones: Detailed sedimentary facies analysis, correlation and stratigraphic architecture of hydrocarbon bearing shoreface complexes in the Dutch offshore. TNO (Utrecht), Report; 228 pp. https://www.nlog.nl/sites/default/files/2019-12/comma_report_full-v2016.11.30.pdf
Bouroullec, R., Verreussel, R.M.C.H., Geel, C.R., De Bruin, G., Zijp, M.H.A.A., Kőrösi, D., Munsterman, D.K., Janssen, N.M.M. & Kerstholt-Boegehold, S.J., 2018. Tectonostratigraphy of a rift basin affected by salt tectonics: synrift Middle Jurassic-Lower Cretaceous Dutch Central Graben, Terschelling Basin and neighbouring platforms, Dutch offshore. Geological Society, London, Special Publications 469 (1): 269-303. DOI: 10.1144/SP469.22
Bouroullec, R. & Weimer, P., 2017. Geometry and kinematics of Neogene allochthonous salt systems in the Mississippi Canyon, Atwater Valley, western Lloyd Ridge, and western DeSoto Canyon protraction areas, northern deep-water Gulf of Mexico. AAPG Bulletin 101: 1003-1034. DOI: 10.1306/09011609186
Bregman, E. & Smit, F., 2012. Genesis of the Hondsrug - A Saalian Megaflute. Drenthe, The Netherlands - Aspiring European Geopark. Report Utrecht University (commissioned by the Province of Drenthe): 115 pp.
Burollet, P., 1975. Tectonique en radeaux en Angola. Bulletin de la Société Géologique de France 17: 503-504.
Cartwright, J., Stewart, S. & Clark, J., 2001. Salt dissolution and salt-related deformation of the Forth Approaches Basin, UK North Sea. Marine and Petroleum Geology 18 (6): 757-778. DOI: 10.1016/S0264-8172(01)00019-8
Clark-Lowes, D.D., Kuzemko, N.C.J. & Scott, D.A., 1987. Structure and petroleum prospectivity of the Dutch Central Graben and neighbouring platform areas. In: Brooks, J. & Glennie, K. (eds): Petroleum Geology of North-West Europe: Proceedings of the Third Conference. Graham & Trotman (London): 337-356.
De Jager, J., 2007. Geological development. In: Wong, Th.E., Batjes, D.A.J. & De Jager, J. (eds): Geology of the Netherlands. Royal Netherlands Academy of Arts and Sciences (KNAW, Amsterdam): 5-26.
De Jager, J., 2012. The discovery of the Fat Sand Play (Solling Formation, Triassic), Northern Dutch offshore – a case of serendipity. Netherlands Journal of Geosciences 91 (4): 609-619. DOI: 10.1017/S0016774600000408
Geluk, M.C., 1999. Late Permian (Zechstein) rifting in the Netherlands: models and implications for petroleum geology. Petroleum Geoscience 5: 189-199. DOI: 10.1144/petgeo.5.2.189
Geluk, M.C., 2005. Stratigraphy and tectonics of Permo-Triassic basins in the Netherland and surrounding areas. PhD thesis Utrecht University (the Netherlands): 171 pp.
Geluk, M.C., 2007a. Permian. In: Wong, Th.E., Batjes, D.A.J. & De Jager, J. (eds): Geology of the Netherlands. Royal Netherlands Academy of Arts and Sciences (KNAW, Amsterdam): 63-84.
Geluk, M.C., 2007b. Triassic. In: Wong, Th.E., Batjes, D.A.J. & De Jager, J. (eds): Geology of the Netherlands. Royal Netherlands Academy of Arts and Sciences (KNAW, Amsterdam): 85-106.
Groenenberg, R., Den Hartogh, M. & Fokker, P., 2025. Salt Production. In: Ten Veen, J.H., Vis, G.-J., De Jager, J. & Wong, Th.E. (eds): Geology of the Netherlands, second edition. Amsterdam University Press (Amsterdam): 603-625. DOI: 10.5117/9789463728362_ch16
Heybroek, P., 1975. On the structure of the Dutch part of the Central North Sea Graben. In: A.W. Woodland (ed.): Petroleum and the continental shelf of northwest Europe. Applied Science Publications (London): 339-351.
Hoetz, H.L.J.G., Steenbrink, J., Bekkers, N., Vogelaar, A. & Luthi, S.M., 2011. Salt-induced stress anomalies: an explanation for variations in seismic velocity and reservoir quality. Petroleum Geoscience 17 (4): 385-396. DOI: DOI: 10.1144/1354-079311-002
Hoffmann, N. & Stiewe, H., 1994. Neuerkenntnisse zur geologisch-geophysikalischen Modellierung der Pritzwalker Anomalie im Bereich des Ostelbischen Massivs. Zeitschrift Für Geologische Wissenschaften 22: 161-171.
Hudec, M.R. & Jackson, M.P.A., 2006. Advance of allochthonous salt sheets in passive margins and orogens. American Association of Petroleum Geologists Bulletin 90: 1535-1564,. DOI: 10.1306/05080605143
Hudec, M.R. & Jackson, M.P.A., 2007. Terra infirma: Understanding salt tectonics. Earth-Science Reviews 82: 1-28. DOI: 10.1016/j.earscirev.2007.01.001
Huuse, M. & Clausen, O.R., 2001. Morphology and origin of major Cenozoic boundaries: eastern Danish North Sea. Basin Research 13 (1): 17-41. DOI: 10.1046/j.1365-2117.2001.00123.x
Jackson, C.A.-L. & Lewis, M.M., 2012. Origin of an anhydrite sheath encircling a salt diapir and implications for the seismic imaging of steep-sided salt structures, Egersund Basin, Northern North Sea. Journal of the Geological Society 169 (5): 593-599. DOI: 10.1144/0016-76492011-126
Jackson, M.P.A. & Cramez, C., 1989. Seismic recognition of salt welds in salt tectonics regimes. Gulf of Mexico Salt Tectonics, Associated Processes and Exploration Potential: Society of Economic Paleontologists and Mineralogists Gulf Coast Section, 10th Annual Research Conference Program and Abstracts: 66-71.
Jackson, M.P.A. & Hudec, M.R., 2005. Stratigraphic record of translation down ramps in a passive-margin detachment. Journal of Structural Geology 27: 889-911. DOI: 10.1016/j.jsg.2005.01.010
Jackson, M.P.A. & Hudec, M.R., 2017. Salt tectonics : principles and practice. Cambridge University Press: 494 pp. DOI.org/10.1017/9781139003988
Jackson, M.P.A., Vendeville, B.C. & Schultz-Ela, D.D., 1994. Structural dynamics of salt systems. Annual Review of Earth

and Planetary Sciences 22: 93-117. DOI: 10.1146/annurev.ea.22.050194.000521
Mauduit, T., Gaullier, V. & Brun, J.-P., 1998. On the asymmetry of turtle-back growth anticlines. Marine and Petroleum Geology 19: 1219-1230. DOI: 10.1016/s0264-8172(97)00053-6
McKie, T. & Kilhams, B., 2025. Triassic. In: Ten Veen, J.H., Vis, G.-J., De Jager, J. & Wong, Th.E. (eds): Geology of the Netherlands, second edition. Amsterdam University Press (Amsterdam): 155-183. DOI: 10.5117/9789463728362_ch05
Mohr, M., Kukla, P.A., Urai, J.L. & Bresser, G., 2005. Multiphase salt tectonic evolution in NW Germany: seismic interpretation and retro-deformation. International Journal of Earth Sciences 94 (5-6): 917-940. DOI: 10.1007/s00531-005-0039-5
Mohr, M., Warren, J.K., Kukla, P.A., Urai, J.L. & Irmen, A., 2007. Subsurface seismic record of salt glaciers in an extensional intracontinental setting (Late Triassic of northwestern Germany). Geology 35: 963-966. DOI: 10.1130/G23378A.1
Nalpas, T., 1996. Inversion des Grabens du sud de la Mer du Nord. Données de sub-surface et modélisation analogique. Géosciences Rennes, University of Rennes (France): 267 pp.
Nalpas, T. & Brun, J.P., 1993. Salt flow and diapirism related to extension at crustal scale. Tectonophysics 228: 349-362. DOI: 10.1016/0040-1951(93)90348-N
Nalpas, T., Le Douaran, S., Brun, J.P., Unternehr, P. & Richert, J.P., 1995. Inversion of the Broad Fourteens Basin (offshore Netherlands), a small scale model investigation. Sedimentary Geology 95 (3-4): 237-250. DOI: 10.1016/0037-0738(94)00113-9
Nelskamp, S., Verweij, J.M. & Witmans, N., 2012. The role of salt tectonics and overburden in the generation of overpressure in the Dutch North Sea area. Netherlands Journal of Geosciences 91 (4): 517-534. DOI: 10.1017/S0016774600000366
Overeem, I., Weltje, G.J., Bishop-Kay, C. & Kroonenberg, S.B., 2001. The Late Cenozoic Eridanos delta system in the Southern North Sea Basin: a climate signal in sediment supply? Basin Research 13: 293-312. DOI: 10.1046/j.1365-2117.2001.00151.x
Peeters, S.H.J., 2016. Mesozoic strike-slip faults in the northern Dutch offshore; new insights from seismic-and analogue modelling data. MSc thesis Utrecht University (the Netherlands): 106 pp.
Penge, J., Munns, J.W., Taylor, B. & Windle, T.M.F., 1999. Rift-raft tectonics: examples of gravitational tectonics from the Zechstein basins of northwest Europe. In: Fleet, A.J. & Boldy, S.A.R. (eds): Petroleum Geology of North-West Europe: Proceedings of the 5th Conference. The Geological Society (London): 201-213. DOI:10.1144/0050201
Pharaoh, T.C., Dusar, M., Geluk, M.C., Kockel, F., Krawczyk, C.M., Krzywiec, P., Scheck-Wenderoth, M., Thybo, H., Vejbæk, O.V. & Van Wees, J.-D., 2010. Tectonic Evolution. In: Doornenbal, H. & Stevenson, A.G. (eds): Petroleum Geological Atlas of the Southern Permian Basin Area. EAGE Publications B.V. (Houten): 24-57.
Pilcher, R.S., Kilsdonk, B. & Trude, J., 2011. Primary basins and their boundaries in the deep-water northern Gulf of Mexico: Origin, trap types, and petroleum system implications. American Association of Petroleum Geologists Bulletin 95: 219-240. DOI: 10.1306/06301010004
Raith, A.F., Strozyk, F., Visser, J. & Urai, J.L., 2016. Evolution of rheologically heterogeneous salt structures: a case study from the NE Netherlands. Solid Earth 7: 67-82. DOI: 10.5194/se-7-67-2016
Remmelts, G., 1995. Fault-related salt tectonics in the southern North Sea, the Netherlands. In: M.P.A. Jackson, D.G. Roberts & S. Snelson (eds): Salt Tectonics: a Global Perspective. American Association of Petroleum Geologists Memoir 65: 261-272.
Remmelts, G., 1996. Salt tectonics in the southern North Sea, the Netherlands. In: Rondeel, H.E., Batjes, D.A.J. & Nieuwenhuijs, W.H. (eds): Geology of gas and oil under the Netherlands. Royal Geological and Mining Society of the Netherlands, Kluwer Academic Publishers (Dordrecht): 143-158. DOI: 10.1007/978-94-009-0121-6_13
Rouby, D., Guillocheau, F., Robin, C., Bouroullec, R., Raillard, S., Castelltort, C. & Nalpas, T., 2003. Rates of deformation of an extensional growth fault/raft system (offshore Congo, West African Margin) from combined accommodation measurements and 3-D restoration. Basin Research 15: 183-200. DOI: 10.1046/j.1365-2117.2003.00200.x
Rouby, D., Raillard, S., Guillocheau, F., Bouroullec, R. & Nalpas, T., 2002. Kinematics of a growth fault/raft system on the West African margin using 3-D restoration. Journal of Structural Geology 24: 783. DOI:10.1016/S0191-8141(01)00108-0
Rowan, M.G., Urai, J.L., Fiduk, J.C. & Kukla, P.A., 2019. Deformation of intrasalt competent layers in different modes of salt tectonics. Solid Earth 10 (3): 987-1013. DOI: 10.5194/se-10-987-2019
Schléder, Z. & Urai, J.L., 2005. Microstructural evolution of deformation-modified primary halite from the Middle Triassic Röt Formation at Hengelo, the Netherlands. International Journal of Earth Sciences 94: 941-956,. DOI: 10.1007/s00531-005-0503-2
Schultz-Ela, D.D., 1992. Restoration of cross-sections to constrain deformation processes of extensional terranes. Marine and Petroleum Geology 9: 372-388. DOI: 10.1016/0264-8172(92)90049-K
Schuster, D.C., 1995. Deformation of allochthonous salt and evolution of related salt-structural systems, eastern Louisiana Gulf Coast. In: Jackson M.P.A., Roberts, D.G., & Snelson, S. (eds): Salt tectonics: A global perspective. American Association of Petroleum Geologists Memoir (Tulsa, OK) 65: 177-198.
SODM, 2020. Conclusions and recommendations of the over-pressured caverns and leakage mechanisms project (KEM-17). SODM Report: 16 pp.
Spiers, C., Schutjens, P., Brzesowsky, R.H., Peach, C., Liezenberg, J. & Zwart, H., 1990. Experimental determination of constitutive parameters governing creep of rocksalt by pressure solution. Geological Society, London, Special Publications 54: 215-227.

DOI: 10.1144/GSL.SP.1990.054.01.21
Spiers, C.J. & Carter, N.L., 1998. Microphysics of rocksalt flow in nature. In: Aubertin, M. & Hardy, H.R. (eds): The Mechanical Behavior of Salt: Proceedings of the 4th Conference.
Stewart, S.A., 2007. Salt tectonics in the North Sea Basin: a structural style template for seismic interpreters. In: A.C. Ries, R.W.H. Butler & R.H. Graham (eds): Deformation of the Continental Crust: The Legacy of Mike Coward. Geological Society of London, Special Publication 272: 361-396. DOI: 10.1144/GSL.SP.2007.272.01.19
Stewart, S.A. & Clark, J.A., 1999. Impact of salt on the structure of the Central North Sea hydrocarbon fairways. In: Fleet, A.J. & Boldy, S.A.R. (eds): Petroleum Geology of North-West Europe: Proceedings of the 5th Conference. The Geological Society (London): 189-200. DOI: 10.1144/0050179
Strozyk, F., 2017. The Internal Structure of the Zechstein Salt and Related Drilling Risks in the Northern Netherlands. In: Permo-Triassic Salt Provinces of Europe, North Africa and the Atlantic Margins. 115-128. DOI: 10.1016/B978-0-12-809417-4.00006-9
Strozyk, F., Urai, J.L., Van Gent, H., De Keijzer, M. & Kukla, P.A., 2014. Regional variations in the structure of the Permian Zechstein 3 intra-salt stringer in the Northern Netherlands: 3D seismic interpretation and implications for salt tectonic evolution. Interpretation 2 (4): 1-17. DOI: 10.1190/INT-2014-0037.1
Strozyk, F., Van Gent, H., Urai, J.L. & Kukla, P.A., 2012. 3D seismic study of complex intra-salt deformation: An example from the Upper Permian Zechstein 3 stringer, western Dutch offshore. Geological Society of London, Special Publications 363: 489-501. DOI: 10.1144/sp363.23
Surlyk, F. & Ineson, J.R., 2003. The Jurassic of Denmark and Greenland: key elements in the reconstruction of the North Atlantic Jurassic rift system. Geological Survey of Denmark and Greenland Bulletin 1: 9-20. DOI: 10.34194/geusb.v1.4644
Ten Veen, J.H., Van Gessel, S.F. & Den Dulk, M., 2012. Thin- and thick-skinned salt tectonics in the Netherlands; a quantitative approach. Netherlands Journal of Geosciences 91 (4): 447-464. DOI: 10.1017/S0016774600000330
Torsvik, T.H., Carlos, D., Mosar, J., Cocks, L.R.M. & Malme, T., 2002. Global reconstructions and North Atlantic paleogeography 440 Ma to recent. BATLAS–Mid Norway Plate Reconstruction Atlas with Global and Atlantic Perspectives. Geological Survey of Norway (Trondheim): 18-39.
Trusheim, F., 1957. Über Halokinese und ihre Bedeutung für die strukturelle Entwicklung Norddeutschlands. Zeitschrift Der Deutschen Geologischen Gesellschaft 112: 150-163.
Trusheim, F., 1960. Mechanism of Salt Migration in Northern Germany. American Association of Petroleum Geologists Bulletin 44 (9): 1519-1540.
Underhill, J.R. & Partington, M.A., 1993. Jurassic thermal doming and deflation in the North Sea: implications of the sequence stratigraphic evidence. In: Parker, J.R. (ed.): Petroleum Geology of Northwest Europe: Proceedings of the 4th Conference. Geological Society of London, Petroleum Geology Conference Series 4: 337-345. DOI: 10.1144/0040337
Urai, J.L., 1983. Water Assisted Dynamic Recrystallization and Weakening in Polycrystalline Bischofite. Tectonophysics 96: 125-157. DOI: 10.1016/0040-1951(83)90247-0
Urai, J.L., Schléder, Z., Spiers, C.J. & Kukla, P.A., 2008. Flow and transport properties of salt rocks. In: Littke, R., Bayer, U., Gajewski, D. & Nelskamp, S. (eds): Dynamics of Complex Intracontinental Basins: The Central European Basin System. Springer-Verlag (Berlin Heidelberg): 277-290. DOI:10.1007/978-3-540-85085-4_5
Urai, J.L., Spiers, C.J., Zwart, H.J. & Lister, G.S., 1986. Weakening of rock salt by water during long-term creep. Nature 324: 554-557,. DOI: 10.1038/32455
Van der Voet, E., Heijnen, L. & Reijmer, J.J.G., 2018. Geological evolution of the Chalk Group in the northern Dutch North Sea: inversion, sedimentation and redeposition. Geological Magazine 156 (7): 1265-1284. DOI: 10.1017/S0016756818000572
Van Gent, H., Urai, J.L. & De Keijzer, M. de, 2011. The internal geometry of salt structures - A first look using 3D seismic data from the Zechstein of the Netherlands. Journal of Structural Geology 33: 292-311. DOI: 10.1016/j.jsg.2010.07.005
Van Keken, P.E., Spiers, C.J., Van Den Berg, A.P. & Muyzert, E.J., 1993. The effective viscosity of rocksalt: implementation of steadystate creep laws in numerical models of salt diapirism. Tectonophysics 225: 457-476.
Van Lochem, H., Vis, G.-J. & Jagt, J.W.M., 2025. Late Cretaceous. In: Ten Veen, J.H., Vis, G.-J., De Jager, J. & Wong, Th.E. (eds): Geology of the Netherlands, second edition. Amsterdam University Press (Amsterdam): 253-291. DOI: 10.5117/9789463728362_ch08
Van Winden, M., De Jager, J., Jaarsma, B. & Bouroullec, R., 2018. New insights into salt tectonics in the northern Dutch offshore: a framework for hydrocarbon exploration. In: Kilhams, B., Kukla, P.A., Mazur, S., McKie, T., Mijnlief, H.F. & Van Ojik, K. (eds): Mesozoic Resource Potential in the Southern Permian Basin. Geological Society of London, Special Publications 469: 99-117. DOI: 10.1144/SP469.9
Vendeville, B.C., 2002. A new interpretation of Thrusheim's classic model on saltdiapir growth. Transactions Gulf Coast Association of Geological Societies 52: 943-952.
Vendeville, B.C. & Jackson, M.P.A., 1992. The fall of diapirs during thin-skinned extension. Marine and Petroleum Geology 9: 354-371. DOI: 10.1016/0264-8172(92)90048-J
Verreussel, R.M.C.H., Bouroullec, R., Munsterman, D.K., Dybkjær, K., Geel, C.R., Houben, A.J.P., Johannessen, P.N. & Kerstholt-Boegehold, S.J., 2018. Stepwise basin evolution of the Middle Jurassic–Early Cretaceous rift phase in the Central Graben area of Denmark, Germany and The Netherlands. In: Kilhams, B., Kukla, P.A., Mazur, S., McKie, T., Mijnlieff, H.F. & Van Oijk, K. (eds): Mesozoic resource potential in the Southern Permian Basin. Geological Society of London, Special Publications 469: 305-340. DOI: 10.1144/SP469.23

Verreussel, R.M.C.H., Munsterman, D.K., Bouroullec, R. & Houben, A.J.P., 2025. Late Jurassic - Early Cretaceous. In: Ten Veen, J.H., Vis, G.-J., De Jager, J. & Wong, Th.E. (eds): Geology of the Netherlands, second edition. Amsterdam University Press (Amsterdam): 211-251. DOI: 10.5117/9789463728362_ch07

Wenau, S. & Alves, T.M., 2020. Salt-induced crestal faults control the formation of Quaternary tunnel valleys in the southern North Sea. Boreas 49 (4): 799-812. DOI: 10.1111/bor.12461

Wong, Th.E., 2007. Jurassic. In: Wong, Th.E., Batjes, D.A.J. & De Jager, J. (eds): Geology of the Netherlands. Royal Netherlands Academy of Arts and Sciences (KNAW, Amsterdam): 107-125.

Wong, Th.E., Van Doorn, Th.H.M. & Schroot, B.M., 1989. Late Jurassic petroleum geology of the Dutch Central North Sea Graben. Geologische Rundschau 78 (1): 319-336. DOI: 10.1007/BF01988367

Ziegler, P.A., 1988. Evolution of the Arctic, North Atlantic and western Tethys. American Association of Petroleum Geologists Memoir 43: 198 pp.

Ziegler, P.A., 1990. Geological Atlas of Western and Central Europe, second edition. Shell Internationale Petroleum Mij. B.V. and Geological Society (London): 239 pp.

13

Sediment mineralogy and geochemistry

Jasper Griffioen, Marita Felder, Nicolaas Molenaar, Job Spijker & Alwina Hoving*

*Corresponding author: jasper.griffioen@tno.nl

ABSTRACT

Insight in the mineralogical and the geochemical composition of Dutch geological sediments is needed for environmental management of the subsurface. A comprehensive overview of these compositions is given in this chapter. The first part of this chapter deals with the mineralogy of the Paleozoic and Mesozoic sedimentary rocks where properties of the rocks are discussed per stratigraphic period. The information presented is mainly based on studies that had a commercial interest in exploring and exploiting the associated natural resources, which gives some indication of what has been studied. The detrital versus diagenetic origin of minerals is considered as fully as possible. The second part covers the Cenozoic, unconsolidated sediments where both the mineralogical and the geochemical compositions are considered. The studies that are summarized were diverse in scope and dealt with the soil and deeper Cenozoic sediments. The sediment geochemical composition is first considered based on results of factor analysis, which has frequently been applied in sediment-geochemical studies. Next, individual elemental relationships are presented for several minor elements. Then the mineralogy is briefly considered for quartz, feldspars, heavy minerals, clay minerals and carbonates, whereby notable differences as well as similarities are discussed. Finally, four specific features are discussed that are worth to be summarized from an environmental perspective: bog iron ores, trace elements in pyrite, sulphur in peat, and lead and its isotopes in soils.

<< A freshly excavated clay deposit (1 m ruler for scale) with redox layering (Tegelen Member, Maalbeek quarry, southeast Netherlands). Photo: Jasper Griffioen.

DOI: 10.5117/9789463728362_ch13

Introduction

The objective of this chapter is to present a comprehensive overview of the mineralogical and geochemical properties of Dutch geological sediments, where geochemical refers to the elemental composition. Staring (1856) and contemporaries already paid attention to the mineralogical and geochemical composition of Dutch sediments and soils. The occurrences of shells, bog iron ore and reactive Fe minerals were frequently addressed in his monograph on Dutch soils. Since then, mineralogical and geochemical analysis of Dutch geological sediments, soils and rocks has been performed to characterize these solid geo-materials for a broad range of purposes. Examples refer to: provenance and geological stratigraphy (Baak, 1936; Schuttenhelm & Laban, 2005), properties of oil and gas (and geothermal) reservoirs (Gaupp & Okkerman, 2011), soil genesis and biogeochemical cycling (Zuur, 1936; Kooistra, 1978; Mol et al., 2003), background composition and the natural range of variation (Tebbens et al., 2000; Van der Veer, 2006), anthropogenic enrichment and impact of land use (Spijker et al., 2005; Van Gaans et al., 2007), weathering rates in relation to critical acid loads (Van der Salm, 2001), the geo-availability and reactive transport of trace elements (Van der Grift & Griffioen, 2008; Zhang & Oliver, 2009), production of drinking water (Saaltink et al., 2003; Antoniou et al., 2013), subsurface disposal of radioactive waste (Koenen & Griffioen, 2016; Hoving et al., 2017, 2020), geo-scientific mapping in general (Huisman et al., 2000; Goldberg et al., 2021).

Overviews on the geochemistry and mineralogy of Dutch sediments are limitedly present. The geochemical soil atlas of the Netherlands was presented as a national reference by Mol et al. (2012) and primarily addresses the elemental composition of shallow soils. Griffioen et al. (2016) recently wrote an extensive review of the mineralogical composition of Cenozoic sediments in which fresh sediments and suspended matter in marine, riverine and limnological environments were also addressed. To date, there are no overviews for the geochemical composition of Cenozoic, Mesozoic and Paleozoic sediments.

The full range from Paleozoic to Cenozoic sediments is addressed in this chapter. The set-up of the overview for the Paleozoic and Mesozoic sediments differs from that for the Cenozoic sediments. For the first, qualitative, summarising descriptions of the mineralogy are presented per geological unit for the entire Dutch subsurface; both on- and offshore. For the Cenozoic, quantitative descriptions of the geochemical and mineralogical properties of the different types of sediments are provided supported with statistical or other quantitative data. Here, focus is on the deposits on land, which have been studied much more intensively than in the offshore. Some attention is also paid to anthropogenic contamination of soils and sediments with trace metals. Four specific, geochemical topics are highlighted that are relevant from an environmental perspective and have not been summarized before: bog iron ores, trace elements in pyrite, sulphur speciation in peat and isotope geochemistry of lead.

Approach

For this overview, existing literature was consulted and original data were used. The mineralogy and petrography of the Paleozoic and Mesozoic sedimentary rocks were mainly studied with thin section petrography for quantitative and descriptive analyses of the components, including their mineralogy, texture and paragenesis. In addition, XRD was applied for the minerology of the bulk rock and, in particular, the clay fraction. Thin section petrography and SEM analysis have the advantage that the detrital and authigenic minerals and components can be distinguished and their spatial and temporal relationship can be determined. Moreover, the effects of diagenesis, including changes in the mineralogy and texture, on the sediment and the reservoir-aquifer quality can be reconstructed. This is done by the interpretation of the spatial relationships of grains, cementing minerals and pores. Open, geochemical data from Paleozoic and Mesozoic sedimentary rocks are only limitedly available for the Dutch subsurface. Most geochemical data are company confidential data, which could therefore not be included in this review.

The most common methods used in studying the mineralogical composition of the Quaternary sediments have been optical microscopy and X-ray diffraction (XRD). The first was traditional in the past and has become uncommon in the last decades. It especially refers to analysis of heavy minerals for stratigraphic and provenance studies. XRD-analysis has transformed into a quantitative approach in more recent years, including XRD-analysis of clay assemblages (e.g. Zeelmaekers, 2011). About 300 XRD-analyses of geological sediment and soil samples were available from recent research projects and literature as old as Favejee (1951). As applied to Dutch sediments, the most common methods used to study the bulk geochemical composition have been X-ray fluorescence analysis (XRF) and ICP analysis following aqua regia destruction. Other methods that have been used intensively are thermographimetric analysis and CS-elemental analysis, especially to characterize the organic matter and carbonate contents. Various other methods have been applied, including various selective and sequential extraction techniques. More than 15,000 geochemical analyses were available from research projects over the past 30 years. Most of these projects were performed by TNO –

Geological Survey of the Netherlands and its predecessors, while several were performed by research institute KWR (or its predecessor KIWA) for drinking water companies, and others by PhD students.

The mineralogical and geochemical compositions of sediments are strongly interlinked but cannot be set equal to each other as the chemical elements may be distributed among various minerals major or not. The mineralogical composition of Dutch, Quaternary sediments has frequently been deduced from geochemical analysis using chemometric methods (e.g. Van Gaans et al., 2011; Griffioen et al., 2012). When translating the geochemical composition to the mineralogical composition of sediments, one should bear in mind that such approaches are based on assumptions. One such assumption is that carbonates are found as calcite and aragonite, which neglects the occurrence of dolomite, siderite and also ankerite. However, other studies have shown that these carbonates are also present in Dutch sediments (e.g. Salomons, 1975; Huisman & Kiden, 1998). Another major assumption is that sulphur in Cenozoic sediments is predominantly found as pyrite, which may not always be true. Some studies showed that gypsum and jarosite are present in the soil of reclaimed marine polders (Van Dam & Pons, 1972; Miedema et al., 1973).

Sediment mineralogy of Paleozoic and Mesozoic sedimentary rocks

Introduction

The Dutch Paleozoic and Mesozoic subsurface demonstrates a complex geological history and contains siliciclastic, carbonate and evaporite deposits (De Jager et al., 2025, this volume). Lithologies often occur in separate stratigraphic intervals and in a framework of permeable (mostly sandstones but also carbonates) and impermeable (shales and evaporites) lithologies. The present-day mineralogical composition and petrophysical properties of sediments in general are the result of provenance, depositional processes and early to late diagenetic processes. The burial history, and resulting factors such as maximum burial depth, temperature and effective pressure reached, residence time at depth and uplift and depressurising, are external factors influencing burial diagenesis. Internal factors are the initial mineralogical composition of the clastic as well as carbonate sediments in combination with their texture. As a consequence, the mineralogical composition can always be expected to be highly variable and the rocks in the Dutch subsurface are indeed no exception.

The main petrophysical properties relevant for both hydrocarbon and geothermal reservoirs are porosity and permeability, which affect their storage and flow capacity. The roughness of pore surfaces and wettability play an important role and are governed by the mineralogical composition of detrital grains and diagenetic cements. The mineralogical composition is of great importance for geothermal exploitation since induced fluid flow or induced temperature changes will cause fluid-rock interactions that will change the chemistry of the pore fluids, which may have adverse consequences. The sedimentary and also diagenetic components may dissolve as a consequence of induced physical-chemical changes, such as changes due to CO_2 sequestration (e.g. Laenen et al., 2004), which may cause precipitation of minerals along the fluid flow either within the sediments or in the technical installations. In addition, dissolved minerals may be nutrients to bacterial growth. The mineralogical composition of shales has an effect on their ductility, which may be relevant for their sealing capacity. Some other examples of the importance of understanding mineralogical and textural properties of the subsurface on hydrocarbon exploitation and geothermal production are the presence of quartz and carbonate cements preventing sand production, formation damage caused by authigenic clay minerals and the presence of minerals reactive to introduced fluids such as feldspars and clay minerals.

Mineralogical studies of Paleozoic and Mesozoic sediments in the Netherlands have virtually only been carried out where the mineralogy (rock salt) and rock properties (hydrocarbon and geothermal reservoirs) are of commercial interest. In the Netherlands and neighbouring countries, Permian and to a lesser degree Triassic sandstones have been most intensively studied petrographically and petrophysically (e.g. Glennie et al., 1978; Hancock, 1978; Gaupp & Okkerman, 2011; Olivarius et al., 2015), because they contain most of the economic gas and oil reservoirs. Relatively recently, interest in Triassic sandstones rose because they form potential geothermal reservoirs in parts of the onshore. Less is known about upper Carboniferous (Namurian-Stephanian) sandstones, simply because they contain less hydrocarbons. Carboniferous sandstones are often rather tight and low permeable, with a few exceptions of the Coevorden field (Kombrink et al., 2007) and in the Dutch-British border area (Huis in 't Veld et al., 2020). Carbonates have locally been the subject of studies where they form reservoirs in the lower Carboniferous (Gökdag, 1982), Zechstein (Clark, 1986; Reijers, 2012), Muschelkalk (Pöppelreiter et al., 2004, 2005; Borkhataria et al., 2005, 2006) and Late Cretaceous chalk (Scholle, 1977; Taylor & Lapré, 1987; Hjuler, 2007). Evaporites and shales were of interest when forming regional seals and/or source rocks (Wei & Swennen, 2022; Remmelts et al., 2025, this volume). Rocksalt deposits have only rarely been cored and described petrographically (e.g. wells Barradeel Salt-1; Pieterburen-1; see www.nlog.nl for information on deep oil, gas and geothermal wells), partly because of their mechanical and chemical instability. In parts of the Netherlands, sufficient petrographic research has been published for some of these mentioned economic stratigraphic units to provide a robust overview. This includes the West Netherlands Basin, The Roer Valley Graben, Groningen and Friesland, and offshore areas south from the E and F quadrants. Summary maps of areas with hydrocarbon interest of the individual formations can be found in Remmelts et al. (2025, this volume). Very little is known for other areas, including the upper Carboniferous coal-bearing sediments in the Limburg area, as there was apparently a lack of interest in the mineralogy of their inorganic components, and as mines were closed already before 1975. For areas without sufficient data only basic descriptions based on the general literature and models can be given and observations from adjacent areas such as the Carboniferous in the Campine Basin in Belgium (e.g. Bertier et al., 2008). Most studies that form the basis of this chapter are single-well industry studies published on the NLOG website (www.nlog.nl/datacenter/brh-overview).

Detrital composition and clastic source areas

The supply of siliciclastic detritus or its absence determines whether clastic sediments or carbonates and evaporites were deposited. Hybrid deposits are rarely present in the Dutch subsurface; one of them is the Triassic Rogenstein (Palermo et al., 2008; McKie & Kilhams, 2025, this volume). Carbonates (Tournaisian-Visean, Triassic Muschelkalk, Zechstein, Cretaceous chalk) and evaporites (Zechstein, Middle Triassic, locally Jurassic and Cretaceous) formed or precipitated only when there was little or no siliciclastic sand and/or clay influx. Carbonates and evaporites formed in situ, biogenically or chemically respectively, but were also locally eroded and reworked.

The provenance, also influenced by climatic and tectonic activity, determines the initial mineralogical composition of the sediment. The main clastic source areas were the London Brabant Massif in the southwest, the Rhenish Massif in the southeast, the Vindelician-Bohemian Massif in the far southeast, the Mid North Sea High (MNH) in the northwest, and the Ringkøbing-Fyn High (RFH) area in the northeast (De Jager et al., 2025, this volume). Detrital input from the Fennoscandian Massif is uncertain for the Dutch subsurface. These areas remained the main source areas during the Devonian until the Late Triassic at least. The Carboniferous Variscan orogeny formed a main tectonic activity phase and in front of the deforming Variscan Mountains, a late Carboniferous foreland basin was filled with a thick succession of molasse-type clastic deposits. After the Variscan orogeny, the source areas became increasingly denuded and as a consequence were less prolific clastic source areas. Related to these key changes in the source areas, also the detrital composition changed accordingly (Fig. 13.1 and discussions in next sections).

Not only the overall paleogeographic setting, but also local factors were of influence on the detrital composition and depositional processes. Permo-Carboniferous volcanism provided a local source of detritus (Van Bergen et al., 2025, this volume), while diapiric movements of Zechstein evaporites caused both local uplift and deformation of overlying younger deposits and also created accommodation space (De Jager et al., 2025, this volume). Rifting periods and associated block faulting determined the location of depocentres, and local erosion of intrabasinal older deposits provided an additional source of sediments (Verreussel et al., 2025, this volume). Rapid lateral variations in the amount of subsidence and uplift played an important role in the inverted West Netherlands and Broad Fourteens basins and in the Dutch Central Graben, making prediction of reservoir properties difficult. Maximum (paleo-)burial depths, and thus the effects of diagenesis, can change quickly laterally, and consequently the mineralogical composition and petrophysical properties are difficult to extrapolate. Not only sediment input from the hinterland but also reworking of older deposits within the basin are important for understanding the primary mineralogy and lateral changes of clastic sediments in the Netherlands. In places, changes in accommodation space and local uplift caused synsedimentary diagenesis resulting in hardgrounds, for instance in the Upper Cretaceous Chalk Group (Molenaar & Zijlstra, 1997).

Diagenesis

Grey paleosols in the Namurian-lower Westphalian are often related to vegetated paleosurfaces and coal beds, and red aridisols in the upper Westphalian-Stephanian reflect a gradual aridization of the paleoclimate. In fluvial and eolian deposits of the Upper Rotliegend Group and the Lower Germanic Trias Group, early diagenesis started with clay infiltration and clay grain coating development, nodular calcrete and dolocrete development in calcareous paleosols, and anhydrite and possibly gypsum crystal growth in fine-grained floodplain-playa deposits. Early marine or terrestrial diagenesis was important in several stratigraphic units. Early marine diagenesis comprises marine hardground development by carbonate cementation, phosphate accumulation and glauconite development in Jurassic deposits and the formation of marine hardgrounds by calcite cementation in chalk. Such hardgrounds are often linked to local diapiric uplift.

The present-day burial depths of the stratigraphic units relevant for hydrocarbon and geothermal exploration and exploitation vary quite significantly. For the Cretaceous, the maximum burial depth ranges from 750 m in

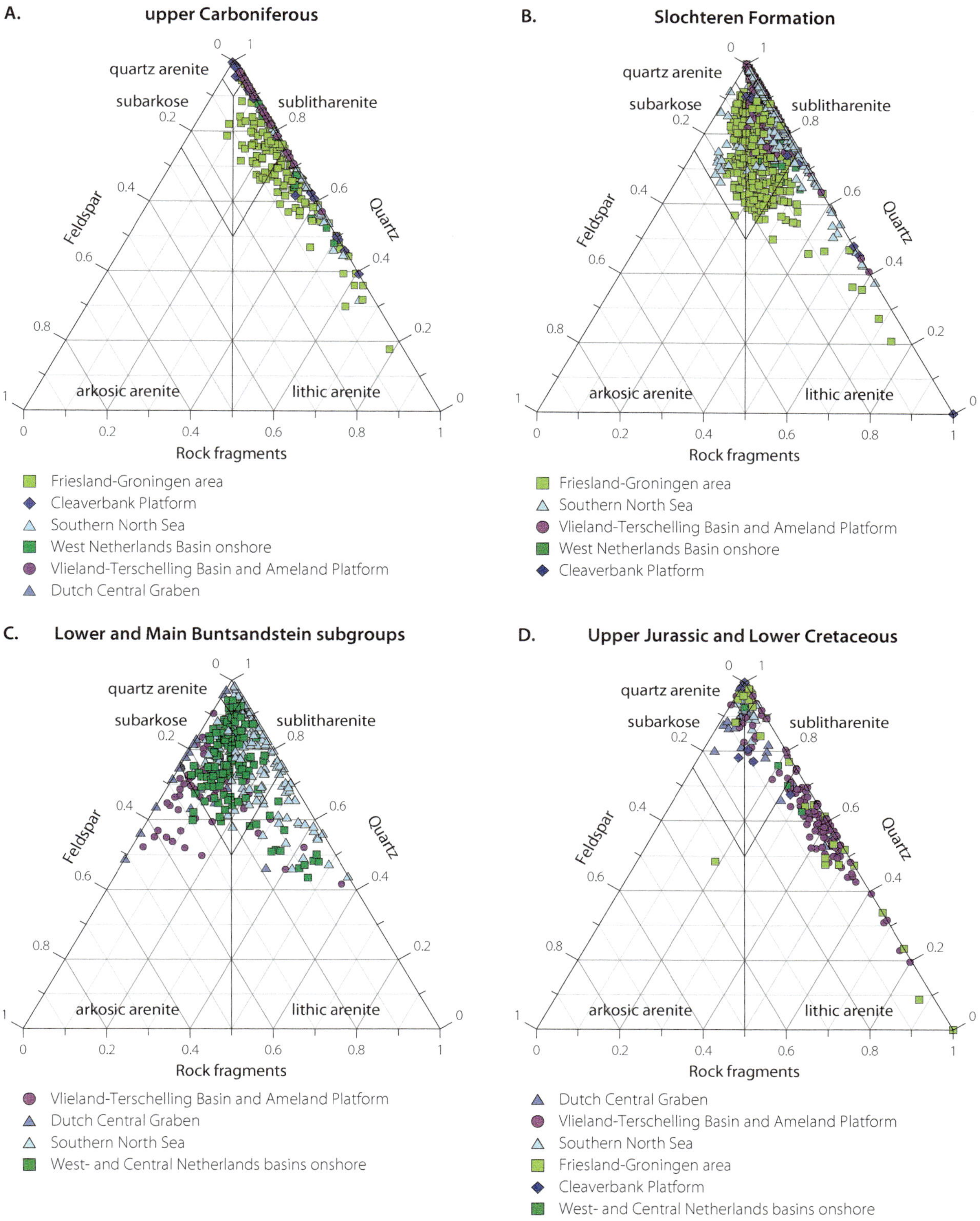

Figure 13.1. Detrital composition of arenites and wackestones for four different sandstone units in the Dutch subsurface. a) Upper Carboniferous. b) Slochteren Formation. c) Lower and Main Buntsandstein subgroups. d) Upper Jurassic and Lower Cretaceous. Detrital quartz is monocrystalline and polycrystalline quartz; feldspars comprise orthoclase, albite and minor microcline; rock fragments comprise extrabasinal fragments from magmatic and metamorphic rocks, Permo-Carboniferous volcanic rock fragments and intrabasinal sedimentary grains, including clay-shale and glauconite grains. See main text for discussion.

the southern part of the Netherlands to more than 2750 m in parts of the offshore K- and L-quadrants (Duin et al., 2006). Upper Permian and Triassic sandstones have been buried from about 2000 m down to more than 5000 m in the Dutch Central Graben (Duin et al., 2006). The burial depth of upper Carboniferous sandstones is in most places between 3000 to far more than 6000 m depth.

Burial diagenesis modified the original composition and texture of sandstones, carbonates, evaporites and clays. The more reactive minerals they contained originally, the more the mineralogy and also the texture changed due to diagenesis. A main and usually first process in sandstones is mechanical compaction which reduces the initial intergranular porosity and permeability (Paxton et al., 2002).

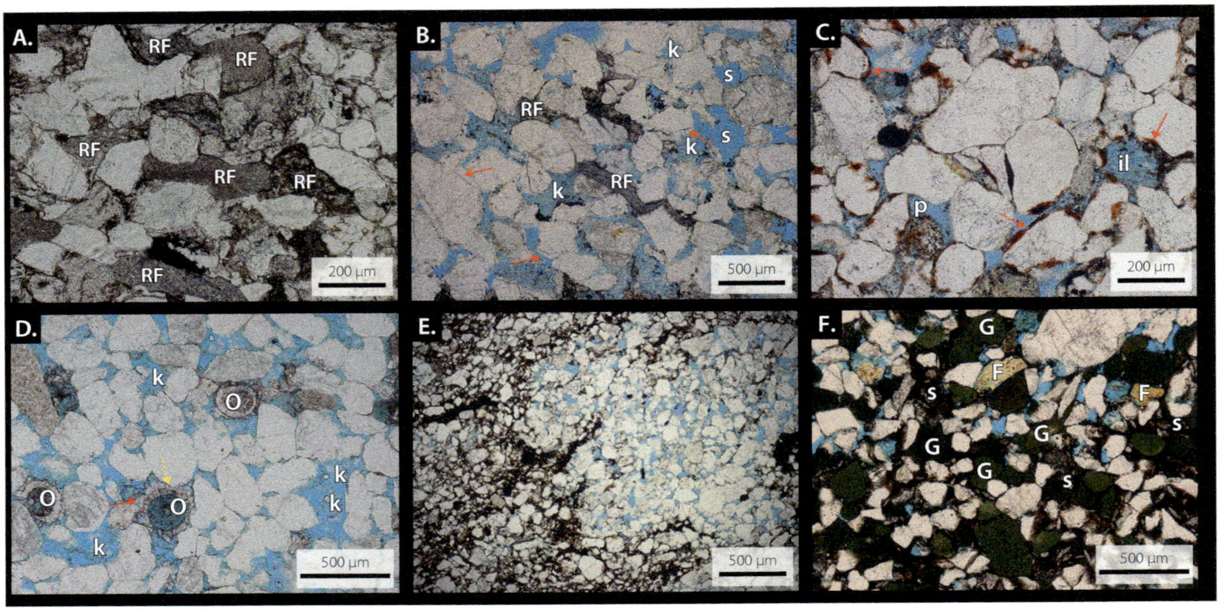

Figure 13.2 Thin sections of sandstone units in the Dutch subsurface (blue resin shows porosity). Note that a) and b) illustrate two endmembers (see Fig. 13.1) of the range in detrital compositions (respectively lithic and quartz arenites) and the effect on diagenesis and final petrophysical properties of upper Carboniferous sandstones. a) In this lithic arenite, ductile rock fragments (RF) are abundant, most of which were deformed during burial compaction that significantly reduced the intergranular pores. Sample from the Bolsovian (WF-C), Lower Saxony Basin. b) Quartz arenite in which the amount of rock fragments was limited and the degree of compaction was therefore also limited. The dispersed rock fragments (RF) were nevertheless deformed by compaction. Dissolution of detrital feldspars caused precipitation of authigenic kaolinite (k) in the secondary pores. Some other detrital grains dissolved completely leaving oversized secondary pores (s). In parts of the sandstones with clean quartz grains, quartz cement overgrowths (red arrows) developed. Sample from the Asturian (WF-D), Cleaverbank Platform. c) Typical example of Permian Slochteren Formation sandstone (North Sea). Most of the detrital quartz grains are partly covered with hematite-stained red clay cutans (red arrows). Some feldspar grains dissolved leaving authigenic kaolinite and illite (il) clay minerals and authigenic feldspars in the secondary pores. The sandstone was compacted but ample intergranular pores (p) remained open. d) Example of a sandstone from the Triassic Volpriehausen Formation with dominant quartz grains and dispersed carbonate ooids (o). The quartz grains have thin and often incomplete quartz overgrowths (white arrows) whereas the ooids have rims of dolomite cement which precipitated in multiple phases (yellow and red arrows). Authigenic kaolinite (k) occurs in oversized pores after dissolution of detrital feldspar. The kaolinite formed loosely stacked booklets randomly oriented. e) Example of marine sandstone (Alblasserdam Member, Lower Cretaceous) with intense bioturbation. Some of the burrows have a sand infill that is slightly larger grained, better sorted and without a clayey matrix but with ample intergranular pores, whereas the adjacent bioturbated sandstone is matrix rich and poorly sorted. Bioturbation increased the heterogeneity of sandstones. Note that the quartz grains have virtually no quartz overgrowths, probably because the burial depth was insufficient here. f) Jurassic Greensand (Scruff Greensand Formation, Dutch Central Graben), characteristic of many Upper Jurassic and Lower Cretaceous greensands. The green grains (G), consisting mainly of the mineral glauconite, were deformed by compaction, depending on the mechanical stability of the surrounding grains. Some intrabasinal glauconite-quartz silt fragments and clay grains occur with euhedral siderite crystals (s) in the clay matrix. Authigenic siderite crystals also occur in the dispersed clay matrix. The greensand was bioturbated, leaving parts with clayey matrix and other parts with little or no matrix and with intergranular pores. Some phosphate grains with intragranular pores occur, derived from marine hardgrounds. Some other grain types largely dissolved leaving secondary pores.

Thereafter, carbonate, quartz and clay mineral cementation further reduce the porosity and lithify the sandstones. In particular clay mineral cements reduce the permeability. Locally, anhydrite and halite partly fill remaining pores. In the process, pore fluids were either expelled and/or became overpressured. Precipitation of diagenetic minerals and dissolution of components are impacted by the composition of the pore fluids which continuously changed through rock-fluid interactions, and on the availability of nucleation sites for precipitation (e.g. Ajdukiewicz & Larese, 2012; Molenaar & Felder, 2018).

In the following sections, the detrital mineralogical composition and the post-depositional mineralogical and textural changes of the main commercially interesting sediments in the Netherlands are briefly summarized.

Carboniferous

In the south of the Netherlands, the lower Carboniferous Dinantian (Tournaisian-Visean) is composed of shallow-marine fossiliferous limestones, dolostones and shales (e.g. wells S02-02, Q18-01, Californie-GT-1). Their basic lithology is discussed in Huis in 't Veld & Den Hartog Jager (2025, this volume) and Vis et al. (2025, this volume). Generally, these carbonates are tightly calcite cemented or dolomitized (e.g. Reijmer et al., 2017) and have low porosity and permeability. Recently, these carbonates became of increased interest as deep geothermal reservoirs, because of exposure-related karstic porosity (Van Hulten, 2012; Jaarsma et al., 2013, 2018; Mozafari et al., 2019), fracture permeability (Van Oversteeg et al., 2014) or their interaction (e.g. Californië geothermal wells, published on NLOG). In the transition to the terrestrial Namurian deposits, organic-rich shales of lower Namurian Geverik Member of the Epen Formation, have shale gas potential (Wei et al., 2021).

Namurian and lower Westphalian clastic deposits were deposited under humid climate conditions, while during the late Westphalian and Stephanian the climate became more arid (Peyser & Poulsen, 2008; Tabor & Poulsen, 2008). The sandstones are fluvial to deltaic channel-fill deposits (e.g. Kombrink et al., 2007, 2008), whereas the shales and coals are paralic to floodplain related and locally marine (Dreesen et al., 1995; Huis in 't Veld & Den Hartog Jager, 2025, this volume). The detrital composition is highly variable ranging from lithic arenites and arkosic arenites to quartz arenites, with low feldspar contents. These are partly regional variations due to minor provenance differences, but also within each region the composition is highly variable, in particular with respect to the amount of rock fragments (Fig. 13.1a). The sandstones in the Lower Saxony Basin and Friesland-Groningen area contain more feldspars and more metamorphic rock fragments, including schist and phyllite rock fragments than other areas (Becker et al., 2017). Volcanic rock fragments are rare. The detrital feldspar content is predominantly below 5 vol%. In some wells a significant fraction of detrital feldspars is replaced by kaolinite (well Deurningen-1), indicating that the feldspar content at the time of deposition was higher. Mica is another common minor detrital component in the upper Carboniferous sandstones. The detritus was derived from the, partly synsedimentary evolving, Variscan Mountains and from reworking of the foreland molasse in the south (Remmelts et al., 2025, this volume). However, also quartz arenites occur, mainly in the Cleaverbank Platform, with clastic supply from the north.

Depositional environment-related early diagenetic cement in the upper Carboniferous sandstones is siderite in lacustrine and paralic environments and pyrite in marine beds and in intervals with plant remains, such as coals. Locally in upper Westphalian and Stephanian sandstones, hematite-stained dolomite rhombohedra occur, which are thought to represent partly dissolved and oxidized former iron-rich dolomites related to surface exposure and weathering below the Base-Permian Unconformity (e.g. Besly et al., 1993).

Schist and phyllite rock fragments as well as mica grains became ductile during burial, resulting in a high degree of mechanical compaction or even complete framework collapse (Fig. 13.2a). In sandstones with common ductile lithic grains and mica, the degree of compaction is therefore high, with very high deformation of lithic grains leading in some cases to pseudomatrix development. The high degree of compaction explains the low porosity and permeability of most of the sublitharenites and lithic arenites (e.g. Bertier et al., 2008).

The main post-compactional cements are quartz and carbonates, both in highly variable amounts in each stratigraphic unit (Fig. 13.3). In sandstones with very small amounts of ductile lithic grains, the long period of burial below 2 km depth has not caused more quartz cementation. For instance, the high reservoir quality sandstones in the Cleaverbank Platform area have quartz cement contents below 8 vol% (e.g. wells D12-05, D15-FA-101, E18-3, Fig. 13.2b). Higher contents only occur in very few individual wells, such as K18-9. Overall, the range of quartz cement and carbonate contents overlaps in all regions and stratigraphic units in the Dutch subsurface.

Kaolinite and illite are the dominant clay minerals in sandstones (Fig. 13.4) and in shales interbedded in the sandstones, with variable amounts of illite/smectite and minor chlorite. Smectite is rare to virtually absent due to burial illitization (e.g. Lanson et al., 1996; Molenaar and Felder, 2018). Both, thin section point-count data and XRD clay fraction data (Fig. 13.4) show no indication of a systematic increase or decrease of any clay mineral rel-

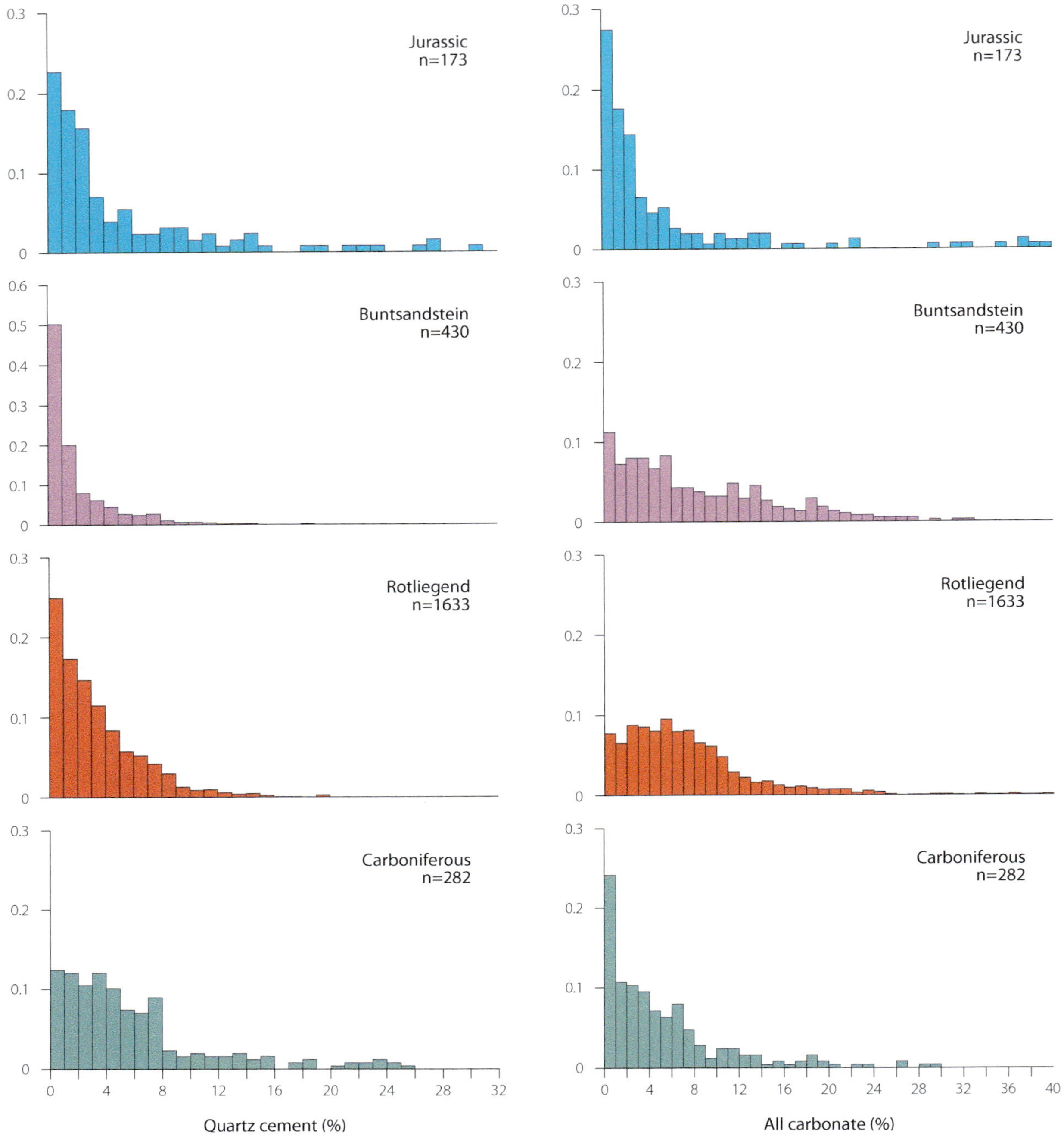

Figure 13.3. *Left: Frequency of point-counted quartz cement content in vol% for the sandstone units. Right: Frequency of the content in vol% of all carbonates. Note that the total carbonate content is displayed, because grains and cements were often not distinguished in the point counts.*

ative to younger stratigraphic units. This reflects similar provenance areas through time. Some stratigraphic levels in the Upper Rotliegend Group have higher chlorite contents, mainly as authigenic chlorite rim cement, whereas Jurassic deposits have high amounts of illite/smectite.

Permian

The Permian Upper Rotliegend Group (including the Slochteren Formation with Lower and Upper Slochteren members) is mainly composed of fluvial and eolian sandstones with interbedded playa shales, all deposited under hot, arid desert conditions (Grötsch & Gaupp, 2011). The sandstones are mainly sublitharenites and minor subarkoses, lithic arenites and quartz arenites (Fig. 13.1b). The main source of the siliciclastic detritus in the Permian deposits were the Variscan Mountains towards the south, and the eastern part of the Southern Permian Basin from which eolian deposits were mainly derived (Gaupp & Okkerman,

2011; Grötsch & Gaupp, 2011). Frequent local reworking is proven by common intrabasinal clay grains, which range from clay to granule size, dolomite grains (Molenaar & Felder, 2019) and occasional anhydrite grains (Henares et al., 2014). These intrabasinal grains can be the dominant type of rock fragments and had significant effect on diagenesis, more so than extrabasinal rock fragments, which include metamorphic and magmatic rock fragments. The latter were derived from the Variscan Mountains in the hinterland. Volcanic rock fragments occur especially in the northeast of the Netherlands in the Friesland-Groningen area and the adjacent offshore area (well Zeerijp-3A; NAM, 2017) as well as in northwest Germany (Hancock, 1978; Gaupp & Okkerman, 2011). These volcanic rock fragments are most likely derived from Carboniferous to lower Permian volcanic rocks from the Ems Graben area and in Germany (Hancock, 1978; Van Bergen et al., 2025, this volume). The feldspar content and composition (orthoclase and albite, rare microcline) show a regional change in the Slochteren Formation. The feldspar content is generally higher in the Friesland-Groningen area and the adjacent offshore area than in the North Sea towards the west (Fig. 13.1b), where K-feldspar is the dominant feldspar (e.g. wells L16-16A, K12-17). Both plagioclase and K-feldspar occur in the Groningen area (e.g. well Zeerijp-3A).

The Slochteren Formation sandstones are mostly orange to red, and are similar to sand in many modern deserts. Due to the hot, arid paleoclimate during the Permian, chemical weathering at the surface was minimal during times of low net deposition. Instead, dust and detrital clay particles with some iron oxides, now hematite, could infiltrate and settle in the sediment and cover detrital grains, whereby the eolian and fluvial sands obtained the typical red to orange colour (Fig. 13.2c; Seemann, 1982; Pye & Tsoar, 1987; Molenaar & Felder, 2018). Such hematite-stained clay-coated siliciclastic grains occur in large parts of the Slochteren sandstones and are similar to sand deposits in many modern deserts.

The main intergranular cements in the Slochteren Formation sandstones are dolomite and quartz, which are both wide-spread. Locally anhydrite and rarely halite form the cement. Clay mineral cements (illite and chlorite) are also common. All of these cements precipitated after mechanical compaction. Compaction was enhanced by the presence of clay grains, shale clasts and carbonate grains. Dolomite has a distinct negative impact on the reservoir quality by filling intergranular pores and obstructing pore connectivity. Usually, non-ferroan dolomite predates ferroan dolomite, ankerite and siderite (Pye & Krinsley, 1986; Vincent et al., 2018; Molenaar & Felder, 2019; Miocic et al., 2020). Dolomite is commonly dispersed throughout sandstone and also occurs in thick amalgamated sandstone intervals and is not related with shale intervals or shale interbeds.

The origin of dolomite has lately been discussed (Vincent et al., 2018; Molenaar & Felder, 2019; Miocic et al., 2020). That faulting and influx of Zechstein fluids as postulated by Vincent et al. (2018) causing dolomite cementation is considered unlikely, because dolomite is also the dominant carbonate cement in areas with few or no faults. Dolomite is even dominant in the Lower Slochteren Member sandstones, where hundreds of metres of shales from the Silverpit Formation separate the Rotliegend from the Zechstein, and in areas without Zechstein salt.

Dolomite has been interpreted as cement in sandstones that formed during early diagenesis due to evaporation (Amthor & Okkerman, 1998). Some authors consider that the dolomite finds its origin as cement in fluids expelled from adjacent shales (Miocic et al., 2020). Dolomite occurs as an authigenic early diagenetic-pedogenic component in fine-grained playa deposits, which were reworked during subsequent fluvial and eolian activity (Molenaar & Felder, 2019). In most sandstones, part of the dolomite present comprises detrital grains, reworked from the fine-grained playa deposits, and only part is cement. The cement occurs as syntaxial overgrowths on the detrital carbonate grains and formed after compaction (Molenaar & Felder, 2019). Dolomite grains became pressure dissolved at grain contacts yielding carbonate for local dolomite cementation as overgrowths on detrital dolomite grains, thereby obscuring the original grain shape. Other carbonate grains dissolved completely leaving secondary pores.

Quartz cement can locally be abundant in the Slochteren Formation sandstones, but usually is present with less than 5-9 vol% (Fig. 13.3). In most cases, it can be clearly linked to the well-established model of clay-induced siliciclastic grain dissolution (i.e. chemical compaction) as silica source (Sibley & Blatt, 1976; Walderhaug & Bjørkum, 2003; Čyžienė et al., 2006) in combination with the need of nucleation sites for quartz cement precipitation (Pittman et al., 1992; Ajdukiewicz & Larese, 2012). In the Slochteren Formation, grain rims of illite and chlorite cements retarded or even inhibited quartz cementation (Busch et al., 2020).

Illite is an important diagenetic mineral in the sandstones (Gaupp et al., 1993; Gaupp & Okkerman, 2011; Molenaar & Felder, 2019). A few percent of illite cement can already have a significant negative effect on the reservoir quality. How much depends on the origin and texture of the illite. Some authigenic illite occurs in secondary pores after partial or complete dissolution of detrital K-feldspar grains, which has little if any effect on porosity and permeability (Molenaar & Felder, 2019). In some stratigraphic levels, illite cement forms rims around detrital grains with crystals oriented perpendicular-oblique to

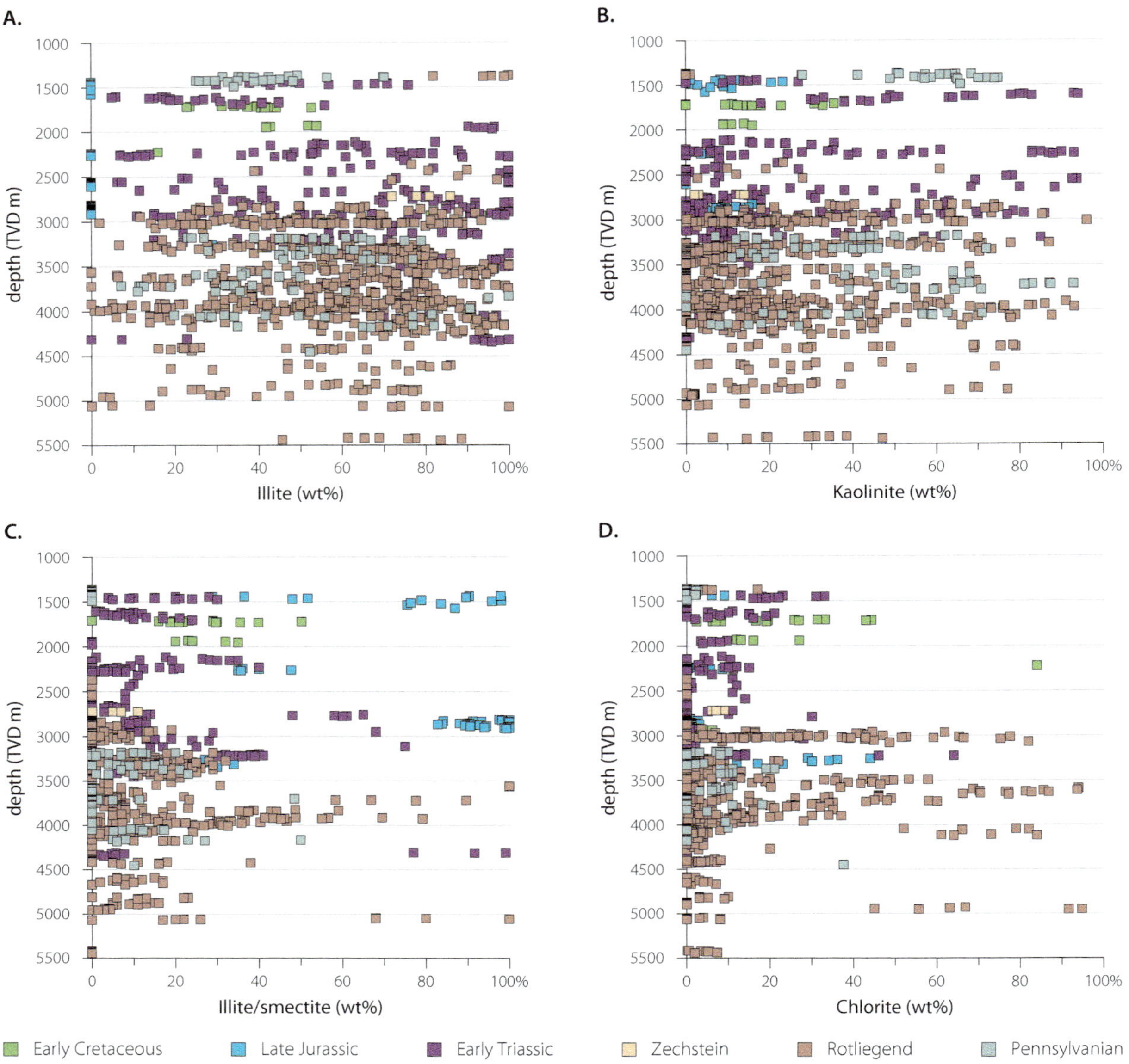

Figure 13.4. Clay fraction XRD results versus core depths, showing clay mineral percentages, mainly from sandstones but also from interbedded shales. Note that the maximum burial depth may have been larger than at present, but insufficient information is available to use it for the individual datasets in this plot. The mineral contents are given in weight%. Note the clay mineral contents show a wide range. a) Illite. b) Kaolinite. c) Illite-smectite. d) Chlorite (mostly clinochlore).

grain surfaces and may even bridge and fill pores completely. This authigenic illite cement significantly decreases the permeability (Gaupp & Okkerman, 2011; Molenaar & Felder, 2018; Busch et al., 2020). The bladed, and in particular fibrous, illite crystals are friable and could cause formation damage when dislodged by induced fluid flow. The origin of the illite rim cement was formerly considered to be linked to fluid flow along open faults derived from the Zechstein or Carboniferous and to illitization of kaolinite (Pye & Krinsley, 1986; Gaupp et al., 1993; Lanson et al., 1996). More likely, a local origin is associated with diagenesis of the detrital clay coatings around siliciclastic grains, causing precipitation of illite during burial after compaction (Seemann, 1982; Molenaar & Felder, 2019). Geochemical work done by Platt (1994) supports the view that the Zechstein derived formation waters only played a role in some horst blocks, where the Zechstein is juxtaposed to sandstones of the Upper Rotliegend Group. However, this is contradicted by the wide-spread regional occurrence of the illite cement.

Kaolinite is another common diagenetic mineral in the Slochteren Formation sandstones (Pye & Krinsley, 1986; Gaupp et al., 1993; Gaupp & Okkerman, 2011), but is absent in some areas, such as in well P02-07. It usually occurs with up to 10 vol% as randomly oriented thin booklets of crystals in grain-shaped secondary pores after dissolution of plagioclase feldspars (Fig. 13.2c), and rarely as larger clusters. The detrital clay assemblage in both sandstones and shales of the Slochteren Formation shows variable amounts of illite, kaolinite, chlorite and minor illite/smectite, similar to the detrital clay minerals found in Carboniferous and Triassic sandstones (Fig. 13.4). Smectite is usually not preserved in the Permian deposits. Increased chlorite contents can clearly be linked to the volcanic province in the east of the Netherlands (see also Gaupp & Okkerman, 2011).

The shales of the Silverpit Formation, overlying and partly laterally equivalent to the Slochteren Formation, are composed of silty-sandy shales and shaly silt- and sandstones with intervals of anhydrite and halite (Glennie et al., 1978; McKie, 2011; Bouroullec & Geel, 2025, this volume). The shale is mainly composed of illite and kaolinite. The silt- to sand-sized fractions contain quartz, dolomite, anhydrite and feldspar.

In the biggest part of the Southern Permian Basin, the Zechstein succession is dominated by evaporites, mainly anhydrite, halite and sylvite, and carbonates. It forms the impermeable seal for hydrocarbons in the underlying sandstones of the Slochteren Formation (Remmelts et al., 2025, this volume). Sandstones occur along the southern margin of the basin (Bouroullec & Geel, 2025, this volume) with similar mineralogy as sandstones in the underlying Slochteren Formation. Zechstein carbonates include oolithic- and calcimicrobial limestones and dolostones. The dolostones, usually with anhydrite, are either of synsedimentary early diagenetic origin (Peryt & Scholle, 1996) or may have formed primarily in an arid setting without calcium carbonate precursors under the influence of bacteria or as evaporitic precipitate (e.g. Bouroullec & Geel, 2025, this volume). The anhydrite could be a primary in situ precipitate formed by dehydration of primary gypsum near the paleosurface or at elevated temperature during burial. The topmost part of the Zechstein Group and Lower Germanic Trias Group are composed of shales of which the mineralogy is largely unknown.

Triassic

The sandy oolitic limestones (grainstones) and oolitic sandstones of the Rogenstein Formation (Lower Buntsandstein Subgroup) have been discussed and described in detail by Palermo et al. (2008). The ooids are mostly calcite and the sandstone is calcite cemented. Similar calcite ooids and calcite cemented ooid grainstone fragments also occur in the younger siliciclastic sandstones of the Main Buntsandstein Subgroup, e.g. in the Volpriehausen Formation and in the lower Detfurth Sandstone Member (e.g. well Q01-27, Fig. 13.2d). They were probably not eroded from the underlying Rogenstein Member but from laterally existing ephemeral lakes with ooids, analogue to the situation in the German Basin (Roman, 2004). Dolomitic ooids and partly dissolved ooids indicate replacement or dissolution of originally calcite ooids during diagenesis in some areas. In the Lower Germanic Trias sandstones, the ooids are usually slightly deformed and have pressure-dissolved contacts, yielding carbonate for carbonate cementation after or during late phases of compaction (Fig. 13.2d).

In many aspects, the mineralogical composition of the detrital and authigenic sandstone of the Main Buntsandstein Subgroup, Solling and Röt formations is similar to that of the Slochteren Formation sandstones. Differences are the low content of lithic grains derived from outside the Southern Permian Basin, the high feldspar content (Fig. 13.1c) and a higher total carbonate content in the sandstones of the Lower and Upper Germanic Trias Group. The detrital composition shows regional differences with enhanced feldspar contents in areas such as the West Netherlands Basin and Roer Valley Graben, the Dutch Central Graben and the Schill Grund High due to regional differences in provenance (Fig. 13.1c). The Broad Fourteens Basin and Friesland Platform have the lowest feldspar content and exhibit a wide range in the content of lithic components. The sandstones are mainly subarkoses, sublitharenites and minor arkosic arenites and lithic arenites with intraclasts and some quartz arenites. Intraclasts include calcite and dolomite ooids and intrabasinal dolomite grains and shale clasts, some with dolomite derived from eroded fine-grained playa deposits.

Sandstones are variably cemented by quartz, dolomite, kaolinite, anhydrite, halite and locally illite (see also Füchtbauer, 1967; Olivarius et al., 2015). The cementation and clay mineral composition are similar to the Slochteren Formation sandstones (Figs 13.3, 13.4), but anhydrite and halite are more abundant in sandstones of the Main Buntsandstein Subgroup, whereas illite is less common. The composition of the clay fraction shows the same variation as in the Slochteren Formation. Halite cement is only common in some areas. Most of the halite and anhydrite cements occur in compacted fabrics (e.g. wells M01-04 and L05-12; Füchtbauer, 1967) and thus formed after burial compaction, and not early as published by Purvis & Okkerman (1996). Halite cementation is probably related to local salt doming, lateral flow of halite supersaturated formation fluids and cooling of these brines (Dronkert & Remmelts, 1996). This preferentially cemented the coarser grained laminae. Authigenic feldspar (orthoclase and albite) is a minor but common cement that, apart from a few exceptions is largely absent in the Slochteren Formation sandstones. This probably reflects minor differences in provenance. The authigenic feldspar forms thin overgrowths on detrital feldspar grains.

The main carbonate interval of the Triassic is the Muschelkalk Formation, comprising shallow-marine limestones, dolostones and marls (Borkhataria et al., 2005, 2006; McKie & Kilhams, 2025, this volume) with intercalated evaporites. Evaporites also occur in the overlying Keuper Formation, indicating prevailing hot arid paleoclimate conditions (McKie & Kilhams, 2025, this volume). Dolomitization of limestones occurred during early diagenesis where meteoric water or more likely, evaporitic brines periodically flushed the sediment (Pöppelreiter et al., 2004; Borkhataria et al., 2006).

Jurassic and Lower Cretaceous

The Jurassic and Lower Cretaceous deposits are very similar with respect to their depositional setting (Herngreen et al., 2003; Trabucho Alexandre & Wong, 2025, this volume; Verreussel et al., 2025, this volume), detrital composition and diagenetic mineralogy. The Lower Jurassic and younger shale intervals are mainly marine deposits. Some shale intervals are organic rich and are important source rocks (notably the Toarcian Posidonia Shale Formation). In the Jurassic there are no significant carbonates (Trabucho Alexandre & Wong, 2025, this volume). The main early diagenetic mineralogical changes in shales and sandstones are represented by the formation of pyrite and siderite (well G16-A-01-S1; Drost & Korenromp, 2009). The pyrite is mostly bound to plant remains or other organic matter. Illite, illite/smectite and glauconite are the main detrital clay minerals in the clay fraction of sandstones and in the shales. Kaolinite and chlorite are often absent or form minor detrital components (less than 1-2 vol%), except in some sandstones in the Schieland Group which contain up to 15 vol% kaolinite.

During Jurassic and Cretaceous times, the relief in the hinterland was low. Clastic sediment input mainly came from the south and southeast, and from local tectonic highs or highs created by salt diapirism. This resulted in a low content of extrabasinal metamorphic and magmatic rock fragments. The depositional environments of the sandstones range from fluvial, coastal, to shallow marine. Sandstone intervals occur in the Upper Jurassic Aerdenhout Member and the Scruff Group, (including the Terschelling Sandstone, the Scruff Spiculite, the Scruff Greensand and the Schill Grund members), and also in the Lower Cretaceous Vlieland Subgroup. The main compositional difference of these sandstones is the highly variable glauconite grain content. Some of the sandstones have a high content of detrital matrix through bioturbation. The sandstone composition ranges from quartz arenites and quartz wackes (Vlieland Sandstone Formation, Aerdenhout Member), to sublitharenites (Fig. 13.1d) and wackestones (Fig. 13.2e), to glauconitic sandstones, i.e. so-called greensands (Fig. 13.2f). The content of detrital feldspar is commonly very low (e.g. in the Scruff Group, and Vlieland Subgroup). The amount of extrabasinal rock fragments derived from the hinterland is usually less than ~3%, but intrabasinal rock fragments and grains can be abundant. The reservoir quality of the coastal, paralic and marine sandstones is strongly related to the depositional facies.

The fluvial and coastal shallow-marine sandstones are quartz rich, and represent sublitharenites to quartz arenites that have little or no detrital matrix except where they were modified by pedogenesis and bioturbation (Fig. 13.2e). The marine sandstones are often bioturbated and contain detrital clay matrix, turning arenites into wackes (e.g. well Tietjerksteradeel-901, K18-Kotter-14). Locally, the sandstones contain intrabasinal shale clasts and carbonate grains, including fossil fragments (e.g. wells G16-06-S1, K18-Kotter-14). Most sandstones typically contain variable amounts of glauconite (e.g. well L06-02). These so-called greensands are often bioturbated to a variable degree and therefore contain detrital clay matrix. Feldspar- and extrabasinal rock fragment contents are very low.

The greensands are shallow-marine deposits and the intrabasinal grains are of marine origin, many of which are reworked material from sediments deposited during periods of low net sedimentation and hardground development. Greensands occur in the Scruff Group (the Scruff Spiculite, the Scruff Greensand and the Schill Grund members). They are composed of a mixture of detrital grains including quartz, feldspar, carbonate, fossil shells, and glauconite, phosphate bone fragments and phosphate intraclasts from reworked phosphatic marine hardgrounds (Gallois & Owen, 2017). Shell fragments are locally replaced by phosphate and/or by chert. Grains from local diapirically uplifted areas and from evaporites eroded at the sea floor are common and sometimes dominant. For instance, sandstones of the Scruff Group contain euhedral shaped detrital quartz grains with abundant anhydrite inclusions (well G16-06A). These quartz grains are locally the dominant detrital grain type. Occasionally carbonate breccias (dolostone and limestone clasts, e.g. well G16-A-01-S1) are intercalated in the sandstone, derived from diapiric caprocks. Spiculites are part of the Scruff Greensand Formation in the Dutch Central Graben (Herngreen et al., 2003; Abbink et al., 2006). They are also associated with diapiric highs and occur at the top of these structures.

The dominant diagenetic process in Jurassic sandstones is mechanical compaction resulting in grain deformation during burial, which affected the texture and petrophysical properties. The amount of glauconite grains is thereby crucial. During burial, glauconite grains deform plastically, which results in an enhanced mechanical compaction (Fig. 13.2f). Consequently, high glauconite contents significantly decrease porosity and permeability. Some glauconite-rich sandstones also have intergranular siderite cement, and euhedral siderite crystals and crystal clusters occur dispersed in the bioturbated detrital matrix. Besides siderite, also calcite and Fe-dolomite cements, and early diagenetic pyrite occur in the detrital matrix. Quartz overgrowth cement is restricted to matrix-free and matrix-poor sandstones. In particular, the quartz arenites are quartz cemented but also in matrix-free parts of wackestones. These quartz-rich arenites were less compacted because of the rigid quartz grains. Carbonate cements and locally also quartz cement is present in sandstones or parts of the sandstone with low glauconite content. Anhydrite cement is rare and only occurs in de Aerdenhout and Terschelling Sandstone members (e.g. well K18-Kotter-14). Halite cement is absent. Because of the dependency of these cements on the bioturbated texture, the sandstones are often highly heterogeneous with respect to cements and mineralogy.

Besides the common high content of glauconite grains, the clay mineralogy of the Jurassic and Lower Cretaceous sandstones stands out because these sandstones contain much more illite/smectite, which is often the dominant clay mineral, and far less illite than the older strata (Figs 13.4a,b). The kaolinite content is very low in Jurassic sandstones.

Upper Cretaceous

At the transition to the Late Cretaceous, the sediments became marlier and eventually fine-grained pelagic limestones were deposited (the Chalk Group; Van Lochem et al., 2025, this volume) often with nodular chert beds (Zijlstra, 1987). These Upper Cretaceous carbonates predominantly consist of chalk composed of low-Mg calcite. The more proximal coastal chalk onshore the Netherlands is composed of coarser grained skeletal debris whereas the more distal chalk is composed of pelagic carbonate mud (Molenaar & Zijlstra, 1997; Van Lochem et al., 2025, this volume). The chalk is composed of micritic limestone derived from microfossils and microfossil debris (mainly coccolith tests) and fine crystalline calcite cement which precipitated partly as overgrowths on the microfossils (Hjuler, 2007). Porosity declines rapidly even with shallow burial depths because chalk typically has high rates of compaction (Scholle, 1977). With increasing burial, low-Mg calcite coccoliths became unstable (e.g. Neugebauer, 1974; Hjuler, 2007) because of the small size of the composing skeletal parts. Aragonite and high-Mg calcite skeletal debris, which are more common in shallower paleo-coastal areas, dissolved resulting in secondary micropores and calcite cementation. Precipitation of low-Mg calcite cement mainly occurred as overgrowths on still stable low-Mg calcite fossil debris or on cement fringes of coarser grained fossil debris (e.g. Mapstone, 1975; Hjuler, 2007). In addition, pervasive stylolites developed in and along marlier interbeds, starting already at burial depths of several hundreds of metres (Egeberg & Saigal, 1991; Lind, 1993). Part of the microfossils, e.g. diatoms and radiolaria, was composed of metastable opal-A, the dissolution of which led to the development of layers of nodular chert. The silica precipitated often along burrows and/or filling burrows, probably influenced by microbial activity. Burial diagenesis thus resulted in increasing crystal grain size (Lind, 1993) and modification of the original carbonates into low-Mg calcite, mainly by local mass conservation.

Typically, marine hardgrounds developed during periods of non-deposition due to early marine cementation of the chalk layers at or closely below the paleo seafloor (Molenaar & Zijlstra, 1997). Such well-cemented hardgrounds form low-porous and low-permeable interbeds in the Chalk Group succession.

Comparison of Cretaceous and older stratigraphic units

Most remarkable in the studied Paleozoic and Mesozoic sandstone intervals is the similarity of detrital compositions, which resulted in similar diagenetic features and authigenic minerals that formed during burial. Nevertheless, subtle regional differences in composition in upper Carboniferous, Permian and Lower Triassic sandstones reflect input from different source areas in the south and in the southeast. Gradual unroofing of the southern source area explains the differences in contents of metamorphic rock fragments between Carboniferous and younger deposits. Also differences in the intensity and frequency of intrabasinal reworking affected the detrital composition and influenced diagenesis locally and regionally.

Compositional variation at local scales is related to different sedimentary processes and selective deposition of grain types, which tend to occur within narrow size fractions. The latter resulted in slight differences in detrital composition at the scale of laminae, beds and depositional cycles. Variations within stratigraphic intervals are mainly the result of changes in types and amounts of intrabasinal detritus. Intrabasinal grains are carbonate, shale, anhydrite clasts, and siliciclastic grains with clay coatings, the latter in Permian-Triassic fluvial-eolian sandstones. In Jurassic-Cretaceous deposits another type of intrabasinal grains is locally important comprising glauconite, phosphate and carbonate from eroded marine hardgrounds and euhedral quartz grains with evaporite inclusions from eroded evaporites. The content of these grains depends on the presence and erosion of paleosols, ephemeral lakes, and fine-grained playa deposits, as well as on the activity of the fluvial channel belts and lateral changes of these belts. In marine Jurassic deposits this depends on local uplift related to diapiric activity.

The clay mineral assemblages are similar throughout the stratigraphy and show no depth trends. Still the quantities of individual clay minerals vary widely even within a single stratigraphic interval. The same can be stated for the cement mineralogy. The overlapping assemblages of detrital and diagenetic minerals and properties clearly reflect the result of a multivariate controlled diagenesis. As example for the multitude of factors that influence cementation, quartz cement can be mentioned. Quartz cement mainly precipitated as overgrowths on quartz grains when clean quartz surfaces were available and when the temperature was sufficient. The amount and distribution of detrital clay matrix, clay grains, clay grain coatings, and other cements that formed earlier than quartz cement can therefore inhibit quartz cement precipitation partly or completely, and locally or pervasively. Changes in the detrital composition of a few percent can thereby have a significant effect on the amount of quartz cement. The intrabasinal clay grains and clay grain coatings in sandstones of the Slochteren Formation and Lower Germanic Trias Group vary stratigraphically and locally. This indicates that the tectonic setting did not play a significant role for quartz cementation. The presence of quartz cement inhibited further compaction, mainly chemical, thereby preserving porosity.

This complex interplay of regional and local parameters implies that the prediction of the composition (and thus also the petrophysical properties) of sandstones from extrapolation of known cases, as would be desirable for hydrocarbon and geothermal purposes, is difficult.

Geochemistry and mineralogy of Cenozoic sediments

Introduction

The Cenozoic formations consist of clastic sediments as well as peat layers. The Paleogene formations are dominated by marine clays while the Neogene formations are mainly composed of marine sands and clays (Munsterman et al., 2025, this volume). The Quaternary formations mostly contain fluvial sands but significant marine and glacial clayey to sandy deposits are present in parts of the Netherlands (Busschers et al., 2025, this volume). Eolian and limnological deposits are also present but mostly at smaller scales and so are peat layers. In terms of terrestrial provenances, two major systems must be recognized: the Eridanos system from the east and the Rhine system from the south (Table 13.1). The deposits of the Eridanos system originated from the Baltic area and have been assigned to the Lower Pleistocene Peize Formation (Overeem & Kroonenberg, 2002). The sediments of the Middle Pleistocene Appelscha Formation also have an eastern provenance, but they originate from northern Germany and the Central German Uplands (i.e. Mittelgebirge in Central Germany). The provenance area of the Rhine system grew kind of abruptly towards the Alps at the onset of the Pleistocene. This had important effects on the sediment mineralogy, in particular the heavy minerals (Westerhoff, 2009). The Meuse River system is a much smaller system as the river discharge has been considerably smaller and frequently it was part of the Rhine system. Its importance for the southern Netherlands is, however, large. The Scheldt system plays a role in the SW Netherlands, and local Belgian rivers leading to the Stramproy Formation played a role during the Middle Pleistocene (Busschers et al., 2025, this volume).

It is important to realize that considerable amounts of Quaternary sediments originate from the reworking of older sediments. Parts of the glacial deposits consist of locally eroded sediments that may have been fluvial or glacial themselves. For example, some of the periglacial sediments at the foot of the ice-pushed ridges originate from glacial ridges. Mineralogically, they are highly comparable, although sorting processes have resulted in some differences. Sediments from the Boxtel Formation frequently have a local origin as well. Another example is that sandy sediments from the Kreftenheije Formation in the provinces of North and South Holland partly consist of reworked sediments from the marine Eem Formation (Busschers et al., 2007), resulting in high carbonate contents. Syn- and post-depositional diagenetic processes and weathering altered the geochemical and mineralogical composition of Cenozoic sediments, although they have

Table 13.1. Provenance areas of the Dutch fluvial Cenozoic formations.

Provenance area	Lithostratigraphic formation
Rhine	Echteld, Kreftenheye, Urk, Sterksel, Waalre
Meuse	Beegden
Belgian rivers, including Scheldt	Kreekrak, Koewacht, Stramproy
Eastern – northern Germany and Mittelgebirge	Appelscha
Eastern – Eridanos (Baltic)	Peize
Pre-Alpine Rhine	Kieseloolite, Inden

not been as intensive as for the Paleozoic and Mesozoic sediments largely for reasons of time and burial depth. Ultimately, this has led to a situation where the mineralogical and geochemical composition of individual Cenozoic geological units is not unique.

The geochemical and mineralogical characteristics of the Cenozoic formations have not been surveyed at a national scale as systematically as their geological characteristics. Consequently, systematic descriptions at the level of geological formations cannot be made. The TopIntegraal project of TNO – Geological Survey of the Netherlands has systematically collected geochemical data for the first 30-40 m below surface for about half the Netherlands until 2024. In the overview below, these new results are combined with previous results.

Sediment geochemistry

Factor analysis

Factor analysis (FA) or principal component analysis has frequently been applied to Dutch sediment geochemical data sets (Hakstege et al., 1993; Huisman, 1998; Tebbens et al., 1998; Mol, 2002; Spijker, 2005; Van Helvoort et al., 2005; Heerdink & Griffioen, 2007, 2012; Vermooten et al., 2011; Klein et al., 2015). It provides insight into a) the minerals that are dominantly present and b) the association of trace elements with main elements. Table 13.2 summarizes the findings of studies that characterize individual geological units based on total element analysis extended with some element contents according to aqua regia destruction (Spijker, 2005) and 0.43 M HNO_3 extraction (Mol, 2002). The factor analysis performed was not carried out in the same manner for all studies and this may affect the detailed results. Despite this, the number of factors needed to describe the geochemical variability within the data sets in adequate detail lies between 2-5 with an average of 3.46 ± 1.24. Only twice, more than 5 factors are required. The total explained variance varied between 0.66 and 0.97 with an average of 0.85 ± 0.06. When a small number of factors was present, one factor (or two) typically consist of a combination of two factors also observed in the models having 3-5 factors.

The most prominent factor identified in many (25 of 37) of the FA models, and which then usually explained most of the variance, is the Al-factor. In all but one of the other FA models, the factor explaining most of the variance is akin to this first one, being defined by both Al and Ca or both Al and S. The first features 8 times in total and the last 6 times. Another major factor is a Ca-factor. This factor features 21 times in total, always in combination with the first Al or Al-S factor. In 6 additional models, Na is a co-defining variable on the Ca-factor. Two further factors that emerged frequently in the 37 FA models are an S-dominated factor (occurring 20 times) and a factor defined by Na and/or K (occurring 13 times). Finally, the heavy minerals and trace metals factor types emerged more than four times. The geochemical compositions of the various Cenozoic geological formations are thus characterized by a small number of factor types. They are discussed below based on their defining variables and geological parentage. First, the four, major factor types are discussed that have a singular explanation. Subsequently, three major factor types are discussed that are intrinsically composed of two or three of these factor types. Finally, some minor factor types are briefly mentioned.

Al-factor

The variable defining this factor is Al; in addition, K is frequently present as co-defining variable with 24 scores out of 25. Other variables that contribute in more than half of the cases are the clay and silt contents, the major elements Ti, Fe, Mg, and the trace elements Ba, Cr, Ni, Pb, Zn, Nb, Rb, V and Y. Ca-loadings on this factor are always below 0.6. The Al-factor is found for both carbonate-poor and carbonate-rich, sandy to clayey formations from various sedimentary origins. This factor most likely represents the variability in the content of Al-silicates as an overall group. Iron may be excluded from this factor or not depending on the presence of reactive Fe minerals in addition to association of Fe with Al-silicate minerals. The prominence of K points in particular to illite, muscovite and/or K-feldspar, where the first is associated with the clay fraction and the last two with the sand and silt fractions (Boggs, 2009). Also Pb, when not present in ore minerals, is known to be mainly hosted by K-aluminosilicates, specifically K-feldspar, where the content ranges between 25-100 mg kg^{-1} (Heier & Taylor, 1959; Heinrichs et al., 1980; Herron & Matteson, 1993), while the average Pb content in the upper continental crust is 17 mg kg^{-1} (e.g. McLennan, 2001; Rudnick & Gao, 2014). Barium is typically an accessory element in K-feldspar (Deer et al., 2013), further illustrating the dominance of sedimentary Ba in aluminosilicates that define this factor.

Ca-factor

The Ca-factor is frequently, but not always, co-defined by Sr and Mn, where the factor loadings are usually higher for the first than for the last. High loadings of Al, Na, K and Pb are never observed on this factor; Fe, Mg, P and As appear as co-variables in about one third of the cases, other variables just a few times. The Ca-factor factor plays a role in data sets for both carbonate-rich formations such as present in the Holocene part of the Netherlands, as well as in carbonate-poor formations in the Pleistocene part. This factor clearly represents carbonate content as controlling factor for the related element variability. Several important observations can be made on some specific element

Table 13.2. Factors as deduced from factor analysis or principal component analyses applied to 37 data sets of single geological units or soils. Note that the Kedichem Formation is now part of the Waalre Formation. The Schoorl, Walcheren and Wormer members are part of the Naaldwijk Formation, the Singraven Member is part of the Boxtel Formation. Factors that were found three times or less are not included.

Factor (n observed)	Variance explained	Elements typically associated (loading mostly >0.6)	Minerals represented	Distinguished in geological units (formations and members)
Al (25)	0.20-0.55	Al, Fe, K, Ti, Ba, Cr, Ni, Pb, Nb, Rb, V, Y, Zn, Ga, (Mg)	Clay minerals, mica	Echteld, Pleistocene fluvial, Boxtel, Drente, Oosterhout, Naaldwijk, Walcheren Member, Holocene Beegden, acid sandy soils, Holocene marine soils
Al (25)	0.07-0.40	Ca, Mn, Sr, (Fe, Mg, P)	Ca-carbonates	Waalre (before: Kedichem), Holocene Beegden, Singraven Member, buried Boxtel, Sterksel, Stramproy, Waalre, Kreftenheye, Drente, Eem, Naaldwijk, Holocene marine soils
S (20)	0.07-0.21	S, org. matter, As	(Fe-)sulphides, (atmospheric S binding to soil org. matter)	Boxtel, Sterksel, Stramproy, Waalre, Kreften-heye, Drente, Oosterhout, Naaldwijk, Echteld, Waalre (before: Kedichem), acid sandy soils
Na (13)	0.02-0.28	Na, Ba, (K)	Na(,K)-feldspar	Waalre (before: Kedichem), Sterksel, Waalre, Urk, Peelo, Eem, Naaldwijk, Holocene Beegden, Holocene marine soils
Al/Ca + Na (4) + S (4)	0.33-0.64	Al, Fe, Mn, Ca, Mg, K, P, Cr, Ni, Pb, Sr, Zn	Clay minerals, mica, Ca-carbonates	Boxtel, Drachten, Urk, Peelo, Naaldwijk
Al/S (6)	0.31-0.71	Al, Fe, Mn, Mg, K, Ti, P, S, As, Cr, Ni, Pb, Zn	Clay minerals, mica, Fe-sulphides	Echteld, Eem, Wormer Member, Naaldwijk
Ca/Na (8)	0.17-0.38	Ca, Sr, Na	Ca-carbonate, Na-feldspar	Echteld, Urk, Drente, Oosterhout, Walcheren Member
Heavy minerals (8)	0.04-0.14	Ti, Fe, Cr, Nb, Zr, Y	Heavy minerals	Boxtel, Oosterhout, Drente, Schoorl Member, Echteld, Waalre (before: Kedichem), acid sandy soils
Trace metals (5)	0.13-0.20	Cu, Ni, Zn, Th	Anthropogenic contamination?	Boxtel, Singraven Member, Drachten, Stramproy, Echteld

associations relating to carbonate content, diagenesis and weathering:

1. Sr is absent for four, sandy, carbonate-poor formations but present in the rest, including many carbonate-rich marine formations (except the calcareous Urk Formation in the northern Netherlands). This difference may be due to lack of dominance of detrital carbonates for the first cases.
2. When Mn is absent, the factor is present in the Holocene Naaldwijk Formation and its members where Ca-carbonate contents (including shell fragments) are highest in both clayey and sandy sediments. This may indicate a difference between dominance of detrital Ca carbonates versus more diagenetic carbonates incorporating Mn under reduced conditions as commonly present in the Dutch subsurface (Griffioen et al., 2013).
3. Fe and P show parallel behaviour while As is also frequently present with or without S. This may refer to diagenetic precipitation of siderite, vivianite and pyrite (Postma, 1982; Huisman, 1998; Hartog et al., 2005). High factor loadings for these elements are observed for all four types of sedimentary environments. In the Netherlands, shallow groundwater is frequently saturated for siderite, rhodochrosite and vivianite (Griffioen, 1994; Griffioen et al., 2013) suggesting that secondary precipitation of these minerals is probable across the country.
4. The relevance of this factor for Mg may be controlled by secondary enrichment of dolomite relative to calcite and aragonite following leaching of the last two under paleohydrological conditions (Griffioen et al., 2013, 2016).

The differences in elemental associations among the different data sets point to a sedimentological, detrital signal (as more often observed for the marine formations) versus a post-sedimentary diagenetic or leaching signal (as is more often observed for the periglacial, glacial and also fluvial formations) for the Ca-carbonate factor.

S-factor

This factor is often co-defined by organic matter and As and is found in fluvial and marine units as well as in the

Drente and Boxtel formations. It is notable that this factor occurs at a lower frequency in marine formations for which it appears as a Al/S factor instead (see below). The S-factor is complemented with some of the following variables in one or two data sets out of the 20 in which this factor is found: Fe, Mn, Ca, Mg, P, Ti, or clay, and the trace elements Cr, Cu, Pb, Zn, U, or V. Sulphur is presumably present as pyrite since presence of gypsum is negligible in the shallow subsurface of the Netherlands (Griffioen et al., 2013, 2016). The strong association with organic matter (as reductant) and with As (as impurity in sulphides) also points to diagenetic sulphides as controlling minerals. The general absence of Fe as co-defining variable for this factor implies that sulphides are of minor importance for explaining the variability in Fe, probably because in the formations studied Fe is for the greater part associated with other minerals like Al-silicates, heavy minerals and Fe-bearing carbonates (as indicated earlier). Most of the associated trace elements are well scavenged by sulphides but their limited presence in this factor indicate that this is not the major control for these trace elements.

The emergence of the S-factor in the FA models for data sets from the Pleistocene part of the Netherlands is peculiar as the formations concerned have not been subject to marine conditions after deposition (with the exception of the Pliocene Oosterhout Formation). Therefore, the source of S cannot be seawater-SO_4 but must have been fresh river water on a geological time scale and/or recent atmospheric deposition and agricultural activities (cf. Griffioen et al., 2008).

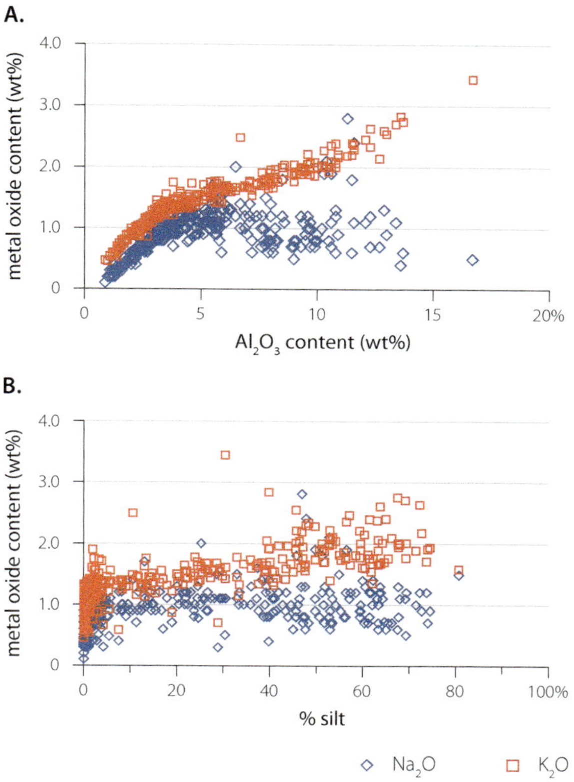

Figure 13.5. The differences in correlation between the total content of Na_2O or K_2O with that of Al_2O_3 (a) and the silt content (b) for samples from the Naaldwijk Formation in the western Netherlands illustrating the presence of an independent Na-factor.

Na- (and K-) factor

The Na factor shows up 13 times independently from an Al-bearing factor. Frequently, K and Ba are associated with this factor and a comparable factor that contains K and Ba without Na is found just once. When Ba appears as significant contributor to this factor, it is not strongly correlated to the Al-factor or vice versa. Feldspars are usually the most dominant mineralogical hosts for Na in sediments or rock. Barium is a typical accessory in K-feldspars (Deer et al., 2013). This factor thus specifically refers to Na and K-feldspars as far as their content varies independently of that of the overall group of Al silicates. Silt or clay are only once predominant, which implies that the relationship with grain size is weak. Figure 13.5 illustrates that there is a linear relationship between Na or K with Al_2O_3 up to 5% Al_2O_3 indicating that these elements are associated with Al-silicates. A second linear relationship with lower slope is found for K above 5% but not for Na. No relationship with the silt fraction is found for Na, and only a weak one for K, which indicates the independence of especially Na to grain size. Some other elements appear as co-definers once (Zn, Sr, Rb, Nb, Y) or twice (Zr) making this a rather unimportant factor for the trace elements as well. This factor is found within about half the data sets of the fluvial and marine formations and not for the data sets of the Boxtel Formation.

Al/Ca-factor

This factor is mostly found in data sets that generated a 2-factor model. The Al/Ca-factor is then co-defined by many of the other variables considered, which accounts for the high explained variances by this factor. Sodium is notably present in half of the data sets while S is present in the other half. This factor type thus combines three of the above factor types. Iron and Mn are always present as co-defining variables. Other co-defining variables that appear in nearly all models are silt, Mg, K, P, Cr, Ni, Pb and Sr. Many main and trace elements are thus present in this factor.

Mostly, this factor emerges in FA models for calcareous-poor, sandy formations with low contents of pyrite and other reactive Fe-minerals, as well as in the model for the Naaldwijk Formation in Zeeland, which is characterized by being rich in all reactive compounds. This factor presumably represents the main dichotomy between 'sterile' quartz-rich sand deposits on the one hand and deposits rich in Al-silicates and reactive minerals that all correlate together on the other hand. For the first, this factor can be viewed as describing variability due to the 'impurity' of the sand, whereas it describes the 'dilution' by quartz sand for the Naaldwijk Formation as the second. At both extremes, the need (in terms of explained variance) and opportunity (in terms of available resolution) to have a larger number of factors that recognize independent variability among the various non-quartz fractions appears to be absent.

Al/S factor

This elemental association especially co-defines factor models for the marine Eem and Naaldwijk formations and the buried Echteld Formation just once. Iron, Mg, P, Cr, Ni and Pb are always associated with this factor and K, As, Zn, clay and silt almost always. The depositional or postdepositional marine environment for this factor type indicates the presence of seawater-SO_4 for the reduction to pyrite or other reduced S minerals. For these data sets, organic matter is also often incorporated in this Al/S-factor with high explained variances, probably due to a grain-size effect where solids other than quartz increase with decreasing grain size together with a strong association between clay minerals and sedimentary organic matter.

Ca/Na-factor

This factor is defined by Ca, Na and also Sr with loadings for Al, Fe, clay and silt always below 0.6. Manganese occurs as co-defining element in this factor in half of the factor models. Magnesium, Fe and P, which frequently co-define the Ca-factor, are usually irrelevant in this factor. Potassium, As, Ba and Zn co-define once and Cu is also noted once. This makes this factor type unimportant for the trace elements. The Ca/Na factor is only found in data sets where the separate Ca- and Na/K-factors are absent while the Al-factor is present. It mostly emerges for calcareous formations of fluvial, glacial and marine origin. This factor is thought to describe the parallel variability in carbonate as well as feldspar content. Since the elements Mg, P and Fe are not strongly associated with this factor, diagenetic carbonates are not reflected in this factor, whereas detrital Ca-carbonate is.

Other factors

A few additional factors appear in the factor models, and always more than once. One such factor is defined by Zn and some other trace elements (Cu, Pb, Cr, Ba, Ni, Th, Y). It may be related to anthropogenic contamination as it occurs in formations close to the surface with acid to neutral groundwater pH or young floodplain sediments, i.e. vulnerable to anthropogenic contamination and soil leaching of trace metals (Schröder et al., 2005; Fest et al., 2007). The other factors refer to:

1. organic matter with or without clay
2. U as single-element factor
3. Nb with Cr and Zr or with Y
4. a combination of silt, Ti and Fe.

The last two factors refer to heavy minerals where associations with rutile and zircon are evident (Deer et al., 2013). These factors are limited in their appearance because the related elements were not always selected in the data set studied.

Trace element composition

The results of the factor analysis point to some common associations among the major and trace elements. These associations may be recognized as a baseline, i.e. a linear relationship between the trace element and a major element, especially Al, that refers to the natural, geochemical composition of the sediments (Huisman et al., 1997; Mol et al., 2012). We will highlight some of these to illustrate

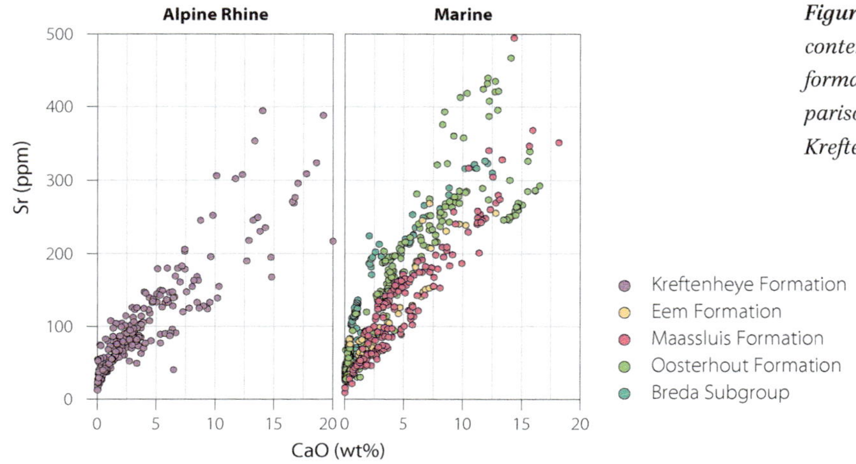

Figure 13.6. Comparison between the total contents of CaO and Sr for pre-Holocene formations with a marine origin. For comparison results are also shown for the fluvial Kreftenheye Formation.

noteworthy similarities and differences in the element composition that can be explained by differences in sedimentary provenance, diagenetic processes or anthropogenic pollution. The data presented concern total element contents as measured with XRF.

Strontium

High total contents of CaO as measured by XRF analyses are indicative for high Ca carbonate contents in clastic sediments. The minor element that is most strongly correlated with CaO is Sr as illustrated in Figure 13.6 for sediments with a marine provenance. The Sr content varies in Ca carbonates but it is overall more abundant in aragonite than in calcite as the larger Sr^{2+} fits better in the aragonite crystal lattice (Carpenter & Lohman, 1992; Deer et al., 2013). The scatter on the cross-plots of Figure 13.6 indicates varying Sr contents in Ca carbonate and varying calcite to aragonite ratios (see section of Carbonates). In Figure 13.6, a steeper slope in the series of data points is noted at low Sr and CaO contents. This slope may be in-

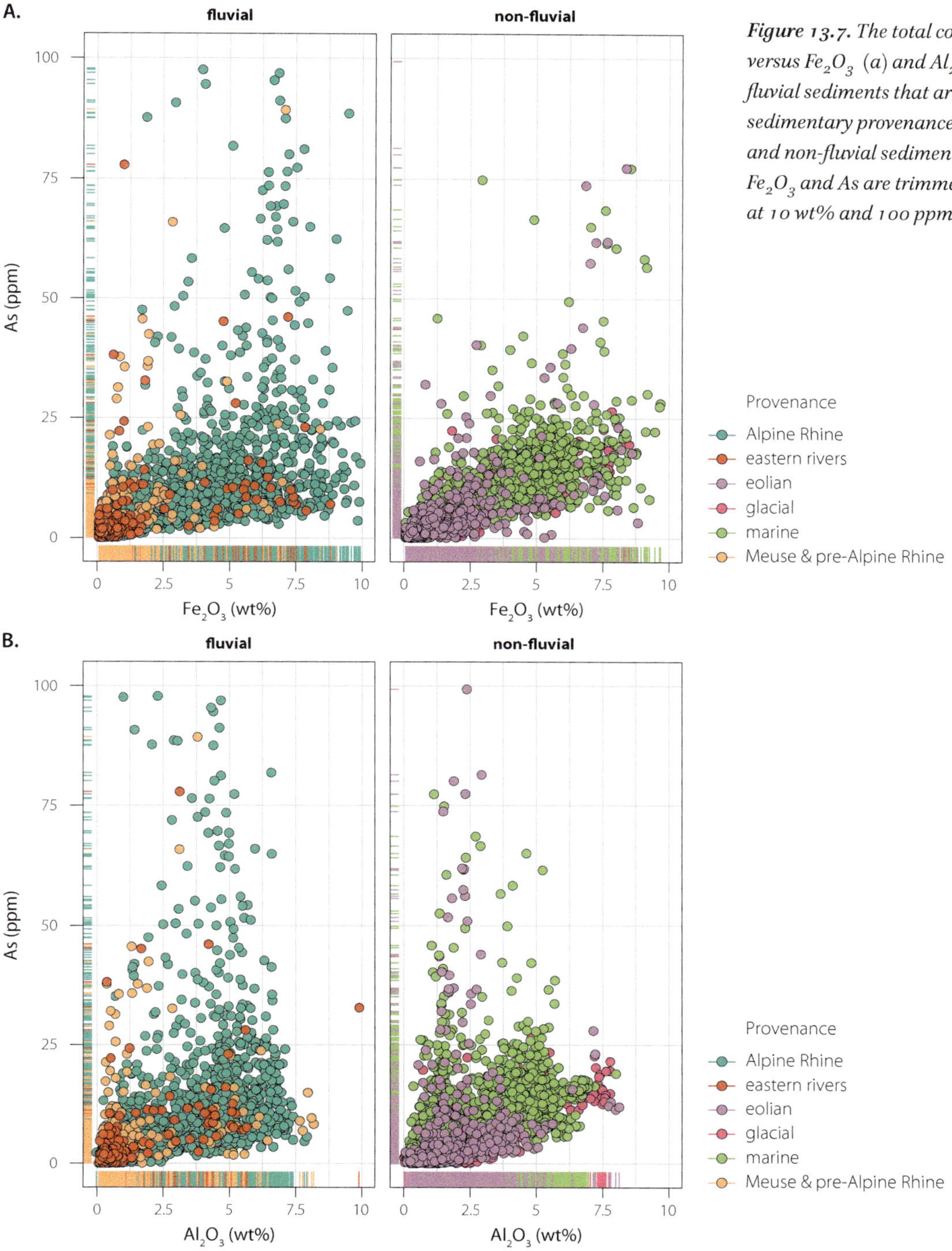

Figure 13.7. The total contents of As versus Fe_2O_3 (a) and Al_2O_3 (b) for fluvial sediments that are classified by sedimentary provenance (cf. Table 13.1), and non-fluvial sediments. The axes of Fe_2O_3 and As are trimmed for visibility at 10 wt% and 100 ppm, respectively.

dicative for Sr in feldspars as Sr can be an important substitution for Ca, Na and K in feldspars (Deer et al., 2013). A systematic difference among the marine formations seems present with higher Sr contents for the Breda Subgroup and Oosterhout Formation. The trend for the Breda Subgroup is more curved compared to the other datasets. The data scattering for the fluvial Kreftenheye Formation is comparable, although some high CaO samples possess low Sr.

Arsenic

Arsenic (As) may be present in Dutch sediments at such high contents that the sediment is considered to be naturally polluted with it (Vermooten & Gunnink, 2008). The evaluation of the factor analysis models revealed that As is often in the same factor as S with or without Al and Fe. Figure 13.7 shows that As follows a baseline with considerable variability when plotted against Al_2O_3 or Fe_2O_3, where the correlation with Fe is somewhat better. This is expected since As may be associated with Fe sulphides and oxyhydroxides. Especially marine and Alpine Rhine sediments are enriched with As, with contents up to 100 mg/kg. This is due to secondary enrichments (Vermooten & Gunnink, 2008). Especially bog iron ores are notorious for their high As contents as discussed below.

Trace elements Pb, Ni and Zn

For the visualization of anthropogenic elements, we selected Pb, Ni, and Zn. Overall, the collected samples seldomly refer to sites with anthropogenic pollution from point sources. Total contents from XRF analysis are intercompared in Figures 13.8-13.10 where each heavy metal is plotted against Al_2O_3. For all plots, a linear baseline can be recognized at the underside of the cloud of data points. Anthropogenic influences on the variability of these three heavy metals occur mostly in the soil zone, for which reason we examined the elemental composition of our samples in the following depth intervals: 0-1.2 m, 1.2-10 m and more than 10 m below surface, which reflects decreasing influence from anthropogenic contamination at the surface.

For lead (Pb), the y-axis in Figure 13.8 is trimmed to a maximum value of 100 ppm Pb for clarity. Sandy eolian sediments (Boxtel Formation) show the most enrichments relative to the baseline for the shallow soil zone. Enrichments are also found in marine soils (Naaldwijk Formation) and to a minor extent in the Al_2O_3-rich (i.e. clayey) fluvial soils (Echteld Formation). These Pb enrichments are far less frequently found in deeper layers although they do occur. Depletions relative to the baseline are also found in these deeper layers.

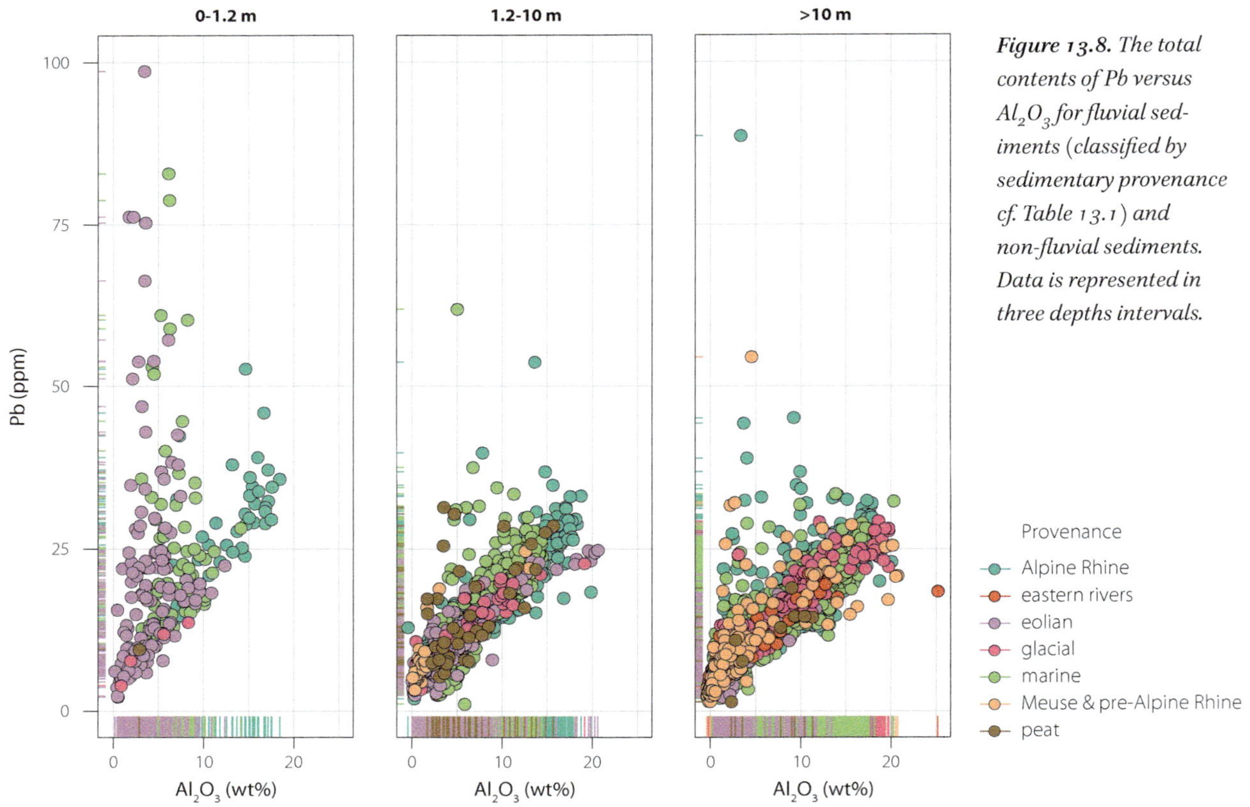

Figure 13.8. The total contents of Pb versus Al_2O_3 for fluvial sediments (classified by sedimentary provenance cf. Table 13.1) and non-fluvial sediments. Data is represented in three depths intervals.

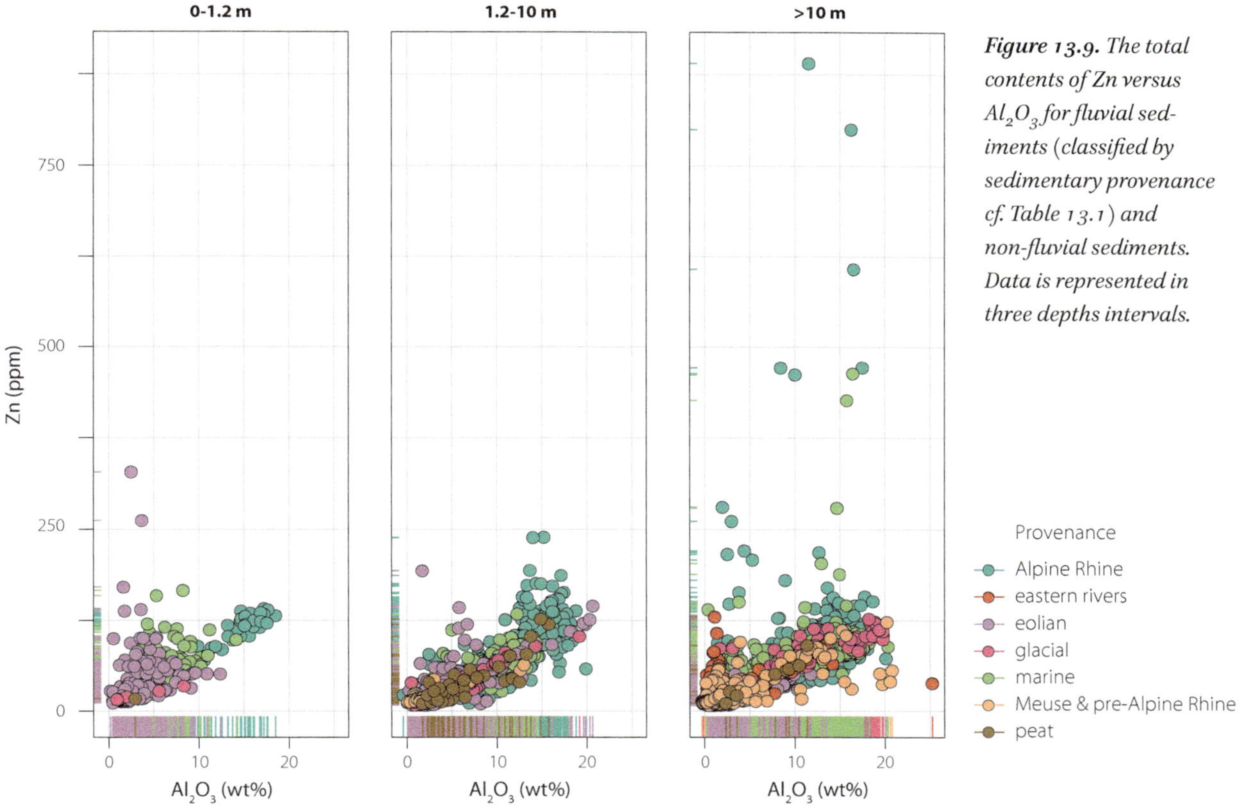

Figure 13.9. *The total contents of Zn versus Al_2O_3 for fluvial sediments (classified by sedimentary provenance cf. Table 13.1) and non-fluvial sediments. Data is represented in three depths intervals.*

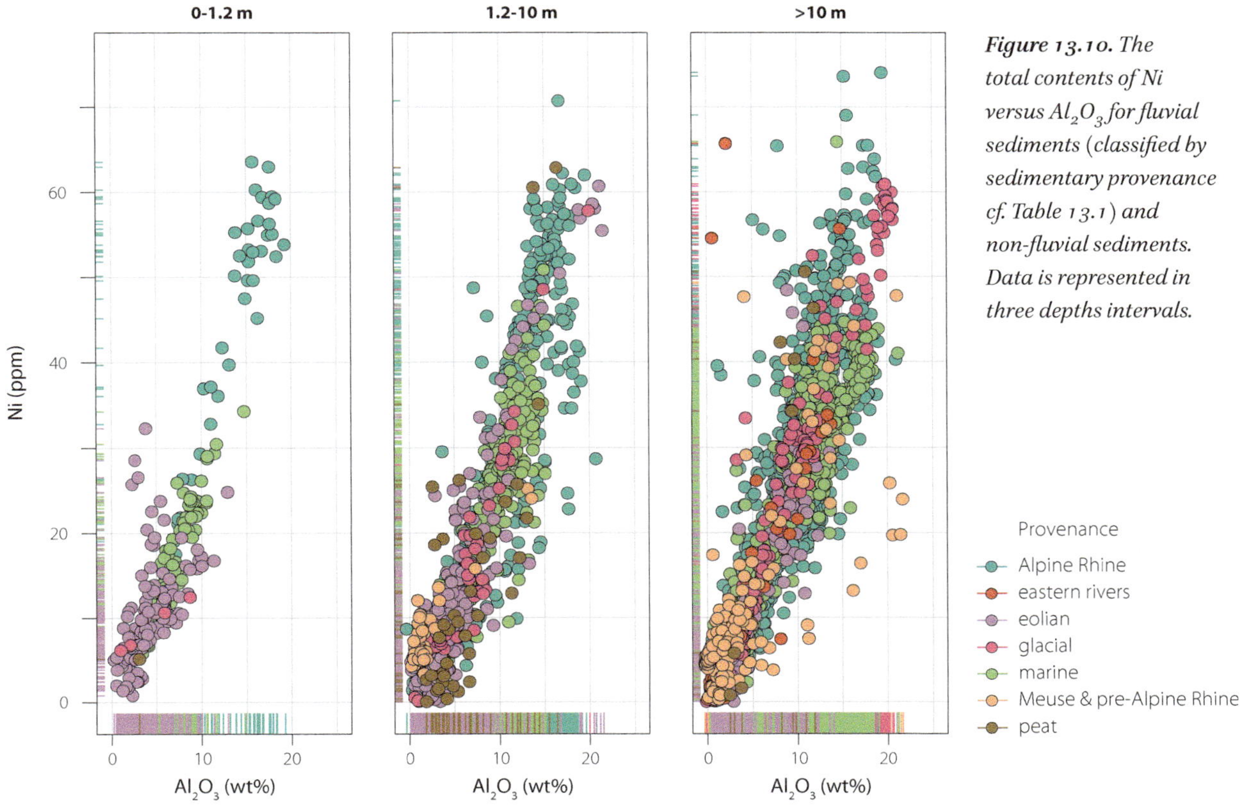

Figure 13.10. *The total contents of Ni versus Al_2O_3 for fluvial sediments (classified by sedimentary provenance cf. Table 13.1) and non-fluvial sediments. Data is represented in three depths intervals.*

The patterns for Zn are broadly similar to those for Pb (Fig. 13.9). The relative enrichments of Zn in the soil layer (0-1.2 m) seem less strong than for Pb, but are occasionally strong in the deepest layer (> 10 m). There is more scatter in the data for the deepest layer, which needs to be explored in more detail.

For Ni, the y-axis in Figure 13.10 is trimmed to a maximum of 75 ppm for clarity. For the upper soil layer, a good correlation with Al_2O_3 is observed, with few exceptions. More data scatter is observed for the deeper layers, showing both depletions and minor enrichments. The depletion in Ni for sediments with a Meuse/pre Alpine Rhine origin is striking and might be indicative for another compositional group.

Gallium

Gallium as chemical element is part of the boron family like Al. The chemical similarity of Ga and Al implies that Ga, being a trace metal, commonly substitutes for Al in Al-bearing silicate minerals. The close geochemical association is also indicated in Figure 13.11 in which a strong linear relationship is shown for Dutch sediments. However, two separate relationships can be recognized. Spijker (2005) showed that this must be attributed to an analytical artefact since the recovery of certain elements (including Ga) changed over time due to changes in instruments in the laboratory of TNO – Geological Survey of the Netherlands. This example illustrates the importance of quality control before data become geo-scientifically interpreted since it is inevitable that data come with errors.

Sediment mineralogy

Recently, research on the mineralogy of Cenozoic sediments in the Netherlands was extensively reviewed by Griffioen et al. (2016). The text below summarizes several

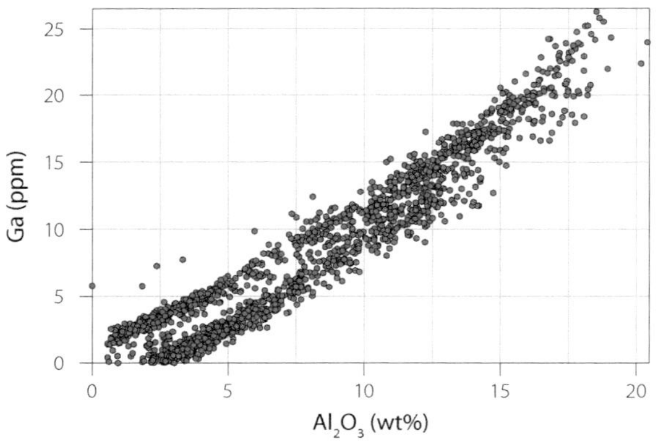

Figure 13.11. The total contents of Ga versus Al_2O_3 for a broad suite of Dutch sediment samples. See text for details.

of the results of that review and includes some additional findings. The original review was made from the perspective of the depositional environment, whereas here we primarily look at mineral type.

Quartz and feldspars

Quartz and feldspars are common minerals in Dutch sediments. Quartz is often the most abundant mineral in sandy sediment with usual percentages of 60-90%. It is also abundantly present in clayey sediments with contents of usually 25-55%. Feldspar minerals are also commonly present. The percentages in which they are present vary between different formations where selective weathering may play a role apart from amount deposited. The specific dissolution rate of Ca-plagioclase is highest and that of K-feldspar lowest, with Na-feldspar showing an intermediate behaviour. Consequently, stronger depletion of Ca-plagioclase upon prolonged weathering in sediments or soils is expected.

Usual percentages for both (Na,K)-alkali feldspars and (Ca,Na)-plagioclase feldspars vary several percent around 5%, largely irrespective whether one deals with sand or clay (cf. Fig. 13.5). The reason is that feldspar minerals are strongly associated with the silt fraction due to their hardness. The total Na_2O content may be used as a proxy for the Na-feldspar content. It commonly increases with Al_2O_3 content up to 5.0% Al_2O_3 and is invariant at higher contents of Al_2O_3 (Huisman & Kiden, 1998). Huisman (1999) concluded for the studied fluvial sediments that the Kreftenheye, Urk and Sterksel formations are richest in Na-feldspar, the Neogene Kieseloolite and Peize formations are poorest, and the Waalre and Stramproy formations have intermediate contents, as illustrated by Figure 13.12. This conclusion can be broadened as the fluvial sediments with an Alpine Rhine provenance and the marine sediments show the widest range in Na_2O content while the fluvial sediments with other provenances and the glacial sediments show a much smaller range. It must be realized that Na_2O content as proxy for the Na feldspar content works less well in clayey sediment with Na-rich pore water where the Na occupancy of the cation exchange complex contributes substantially to the Na_2O content, as is evidenced by the wider scattering for marine sediments at high Al_2O_3 content.

For K_2O, the initial linear relationship with Al_2O_3 can be attributed to K-feldspar and the further increase in content to clay minerals, in particular illite that is rich in K and commonly present in tens of percents within the clay mineral assemblage of Dutch sediments (cf. Fig. 13.5). It is noteworthy that similar relationships are observed for many Cenozoic sediments in the Dutch subsurface (Fig. 13.13). An important exception is glauconite-bearing marine sand of the Breda Subgroup as discussed later.

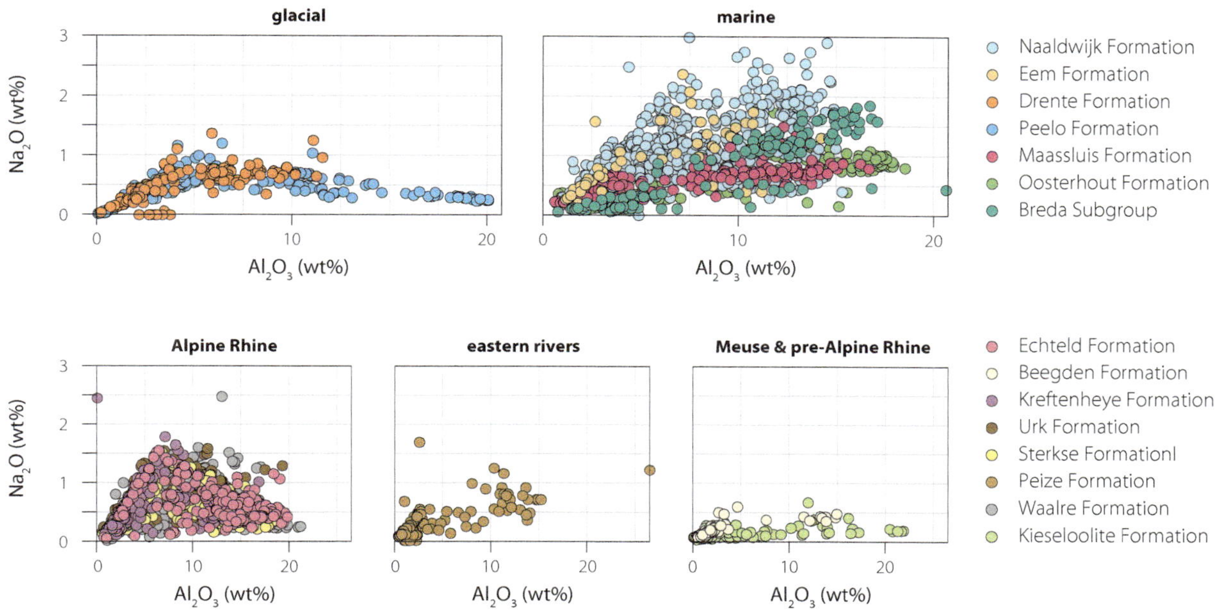

Figure 13.12. The relationships between the total contents of Al_2O_3 and Na_2O for Cenozoic sediments with different sedimentary origins and belonging to different lithostratigraphic formations.

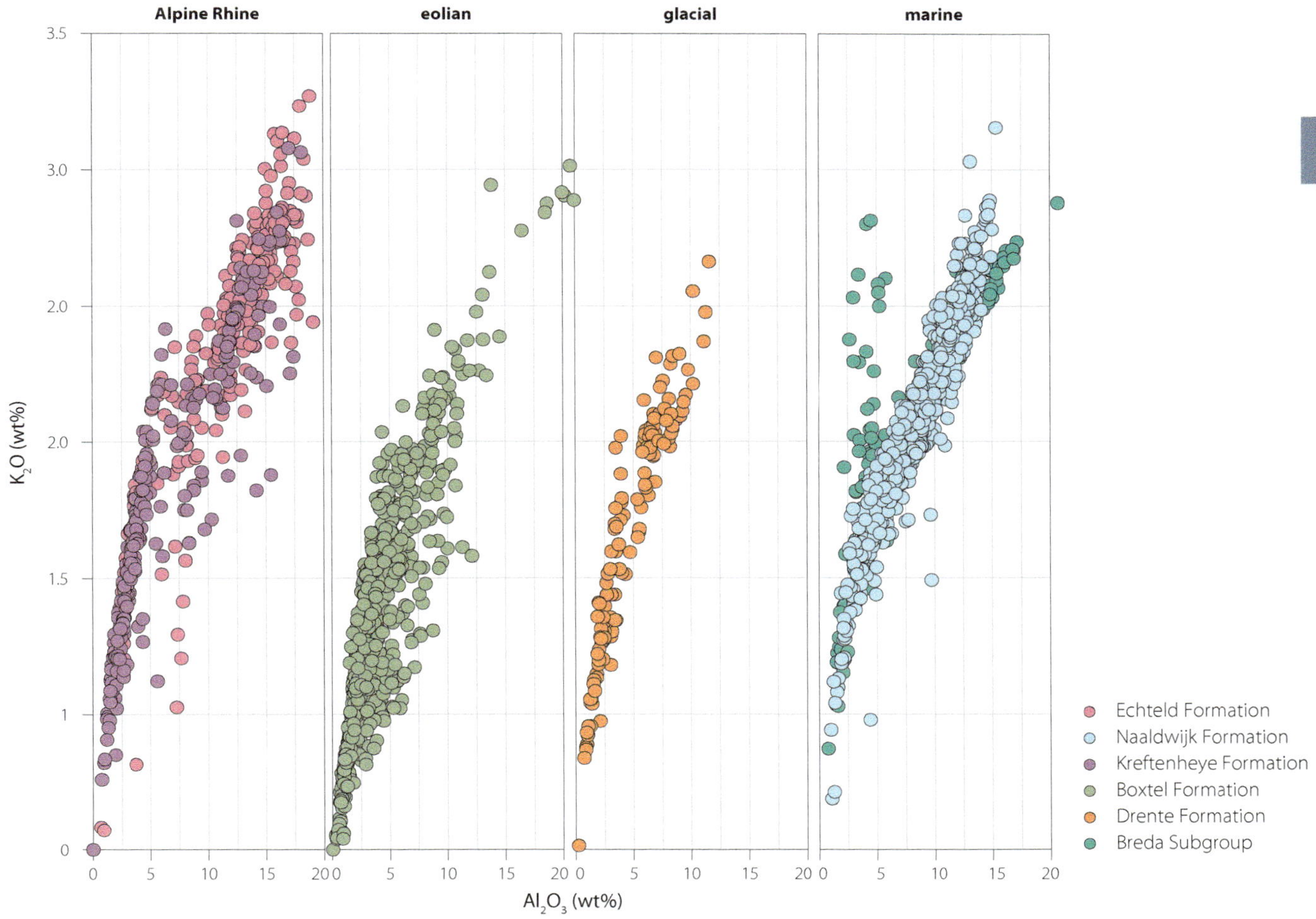

Figure 13.13. The relationship between the total contents of Al_2O_3 and K_2O for Cenozoic sediments with different sedimentary origins and belonging to different lithostratigraphic formations.

Van Baren (1934) pointed out that sediments having an eastern provenance (i.e. the fluvial Peize and Appelscha formations) are dominated by microcline as K-feldspar whereas those with a Rhine provenance by orthoclase as K-feldspar and albite as Na-feldspar.

Heavy minerals

Heavy minerals are defined as those minerals that have a density exceeding 2.85 g/cm^3 (Boggs, 2009). In absolute contents, they contribute less than 1 to 2% to the sand fraction (Druif, 1927; Van Andel, 1950; Wartel, 1977). Sand with higher percentages of heavy minerals is found in a beach environment due to dune collapse and washing out of lighter minerals (Crommelin & Slotboom, 1945). Despite their limited occurrence, heavy minerals have been intensively studied since Edelman (1933) as a tool to increase the understanding of the stratigraphy of Dutch sediments based on their provenance and paleogeography.

The concept of mineral associations has been widely used to denote the particular combination of minerals that characterize a sediment. Here, two kinds of associations are distinguished: 1) a stable association with minerals such as zircon, rutile, anatase and tourmaline with or without the metamorphic group index minerals (staurolite, kyanite, chloritoid, sillimanite and andalusite) and 2) an unstable association with epidote, garnet, alterite, saussurite and hornblende. The heavy mineral assemblages are presented in Table 13.3. The deposits of the eastern rivers contain both stable and unstable assemblages. The Beegden Formation, being built from Meuse sediments, is dominated by a stable assemblage and has Vosges hornblende, anatase and chloritoid as particular guide minerals. The Rhine deposits show two shifts in assemblage (Tatzel et al., 2017). A first shift from the so-called pre-Rhine system containing a stable assemblage, to the Rhine system where the Alps became part of the drainage area and a dominance of the unstable assemblage and also a minor admixture of the stable Meuse assemblage. This is combined with the presence of mica flakes, an increase in red-brown particles (i.e. feldspar and sandstone fragments), and decreasing quartz content. After the second shift from the Urk Formation onwards, additional volcanic minerals from the Eifel appear, in particular augite, while the Meuse stable mineral assemblage disappears. The Belgian Stramproy Formation has predominantly stable heavy mineral assemblages.

Table 13.3. *Heavy mineral assemblages as valid for various Cenozoic geological formations (based on Griffioen et al., 2016). X and m denote dominantly and minorly present, respectively. HMA means heavy mineral assemblage.*

Formation/ Member	origin	augite (Eifel volcanics)	garnet	epidote	hornblende	saussurite	saussurite/ alterite	Vosges hornblende	Meuse chloritoid	zircon	rutile	anatase	tourmaline	staurolite	sillimanite	other metam. min.
		unstable								stable				stable/ metamorphic		
Schimmert (BX)	loess		X	X	m					X	X		m	m		m
Kreftenheye	Rhine	X	X	X	X		X									
Eem	marine	m	X	X	X	X							m	m		m
Urk	Rhine	X	X	X	green		X									
Appelscha	eastern		X	X	X								X	X		X
Sterksel	Rhine+Meuse		X	X	X		X			+ stable Meuse HMA						
Beegden	Meuse		X	X				X	X	X	X	X	X			X
Stramproy	Belgian									X			X	X		X
Waalre	Rhine+Meuse		X	X	X		X			+ stable Meuse HMA						
Peize	eastern		X	X	X								X	X	X	X
Maassluis	marine		X	X	X		X									
Kieseloolite	pre-Rhine									X			X	X		X
Oosterhout	marine		m	X	X								m			m
Breda Subgp	marine		m	X	X								m			m

The loess deposits from Limburg, classified as the Schimmert Member (part of the Boxtel Formation) are different from the underlying formations and the recent Meuse deposits. They are dominated by zircon and epidote and are also rich in garnet and rutile. It makes them comparable to the Tertiary group as distinguished by Schuttenhelm & Laban (2005) for seabed sands near the eastern English coast. The heavy mineral composition of the cover sands is a reflection of the underlying pre-Weichselian formations (Crommelin, 1964). In the northern region garnet, tourmaline and metamorphic minerals are dominantly present, similar to what is found in the eastern river deposits. Epidote-saussurite and hornblende-pyroxene groups are more important in the central Dutch region, indicating a Rhine signature. High percentages of tourmaline and metamorphic minerals are found in the southern region as in the Beegden, Stramproy and Kieseloolite formations (cf. Table 13.3).

Clay minerals

Clay minerals are abundant in Cenozoic sediments of the Dutch subsurface. Different types of clay minerals have distinct properties, which are important for the sediment's chemical and physical behaviour. These clay minerals can be indicative of climatic changes. In the Dutch subsurface, the most common clay minerals are of the smectite, illite and kaolinite groups (Fig. 13.14). In addition to these 'pure' clay minerals, mixed-layered illite-smectites are also abundant in Dutch sediments. Chlorite and vermiculite are usually present in minor amounts but can be important as indicator for depositional origin. Clay minerals can form through the following features: neoformation, transformation and inheritance (Galán & Ferrell, 2013). Important factors for the type of clay minerals formed are climate, time, parent material, topography, soil-profile type, transport processes and burial diagenesis (Singer, 1984; Hillier, 1995). In the Netherlands, most clay minerals present in Cenozoic sediments are inherited from pre-existing parent rock or weathered material and have not been transformed or newly formed in the soils and sediments after deposition (Breeuwsma, 1985; Tebbens et al., 1998). Clay minerals in sediments that are buried deeper and experienced diagenesis can be altered by illitization, the transformation of smectite or kaolinite to illite (e.g. Lanson et al., 1996; Molenaar & Felder, 2018). Another transformation mechanism specific for glacial tunnel valleys in Pleistocene sediments in the North Sea is *in-situ* smectitization of illite during a cold climate (Šegvić et al., 2016). Griffioen et al. (2016) give an extensive literature review of the mineralogy of fresh and Cenozoic sediments in the Netherlands, which is summarized below.

Fluvial clays

The dominant clay minerals in fluvial Rhine and Meuse clays are illite, smectite and vermiculite, with minor amounts of kaolinite and chlorite (e.g. Breeuwsma, 1985; Zeelmaekers, 2011; Adriaens, 2014). The clay mineral content varies over time in the Rhine clays (Fig. 13.14). The total illite to total smectite ratio increases from less than one (average 0.7) for the Lower Pleistocene Rhine-Meuse-Scheldt clays, to equal to one for the Mid-Holocene Rhine-Meuse clays, to more than one (around 1.5) for Lower Holocene Rhine and Lower to Mid-Holocene Scheldt clays. Huisman and Kiden (1998) reported that for the lower units of the Waalre Formation (Early Pleistocene Rhine) the clay is dominated by well-crystallized smectite, illite and kaolinite, whereas for the upper units it is dominated by illite and kaolinite. Such large variations in clay mineralogy may be attributed to drastic climate- or provenance changes in time (Tebbens et al., 1998; Alizai et al., 2012; Limmer et al., 2012). Scheldt-derived sediments of the Kreekrak Formation have a smectite-dominated clay-mineralogical composition and low contents of illite (e.g. Huisman & Kiden, 1998). One sample from the Peize Formation, originating from the Eastern rivers, has a higher kaolinite content compared to the other fluvial clays.

Alluvial brook clays from the eastern part of the Netherlands, which are stratigraphically part of the Singraven Member of the Boxtel Formation, have smectite contents that are significantly higher than in sediments from the rivers Rhine and Meuse. The smectites must originate from deposits present in the catchments of these brooks. Possible candidates are Paleogene and/or Neogene marine deposits or Cretaceous carbonates (Breeuwsma, 1985). Clays with such high smectite contents have very specific properties, such as high swelling capacities and cation exchange capacity, and K fixation.

Paleogene and Neogene marine formations

The clay mineralogy of Paleogene and Neogene marine deposits is highly variable along depth profiles. Overall, clays of the Ieper Member (Dongen Formation; NLDO in Fig. 13.14) contain larger fractions of smectite than other marine formations. Kuhlmann et al. (2004) and Zeelmaekers (2011) explain the shifts in clay mineral assemblage over time by changes in provenance. Climate factors seem to affect the contribution of sediments from specific hinterlands, but the climate dependency on weathering of clay minerals is believed to be of minor importance. The high smectite fraction in the Ieper Member clay can be attributed to a contribution of basic pyroclastic material related to the opening of the North Atlantic and connectivity with the North Sea (cf. Zeelmaekers, 2011; Nielsen et al., 2015). The shifts in the clay mineral assemblage of the Late Pliocene to Early Pleistocene record are attributed to

European geochemical monitoring

Geological materials are usually heterogeneous at various scales. Characterization of this heterogeneity is thus also needed at various scales. In the past decades, two large geochemical programmes were performed at the European continental-scale by geochemists employed at European geological surveys and some other organizations under the umbrella of EuroGeoSurveys: the FOREGS programme running in 1997-1999 and the GEMAS programme running in 2008-2009 (De Vos et al., 2005; Salminen et al., 2005; Reimann et al., 2014a,b; Demetriades et al., 2021). The soil samples were not taken at known contaminated sites, in the immediate vicinity of industry or power plants, railway lines, etc. The data thus provide insight in the baselines: the composition is resulting from parent material, its weathering, atmospheric deposition and agricultural use.

The FOREGS programme (weppi.gtk.fi/publ/foregsatlas) includes ca. 3500 samples for solid media: topsoil, subsoil, floodplain sediment, stream sediment and humus. Further, it contains ca. 800 stream water samples. For the soil sampling sites, this averages to approximately one site per 5700 km^2. Top and subsoil, and floodplain sediment samples were analysed at various laboratories across Europe by X-ray fluorescence, mixed acid extraction with ICP-MS, aqua regia extraction with ICP-AES, Hg analyser and a granulometric method for total organic carbon (TOC).

The GEMAS project assessed the element variation in productive soil at the European scale. Topsoil and subsoil samples were collected from two land-use categories at an average density of 1 site per 2500 km^2 each. 2024 grazing land soils were sampled and defined as 'land under permanent grass cover' and 2108 agricultural arable soils that are regularly ploughed were sampled. All collected soil samples were prepared in the same laboratory, and subsequently analysed for the same suite of elements and physico-chemical parameters in the same laboratory. The following analyses have been performed until now: 1. aqua regia extraction with ICP-MS analysis, 2. total element content with X-ray fluorescence, 3. mobile metal ion cold leach (MMI), 4. lead isotope ratios, 5. soil pH in 0.01 M $CaCl_2$ solution, 6. total organic carbon, total carbon, total sulphur, 7. cation-exchange capacity at prevailing soil pH, 8. mid-infrared spectra, 9. texture (i.e. sand, silt, clay) and 10. partitioning coefficients for selected elements.

The results show that the distribution patterns are primarily related to geochemical variation of large lithological units. For many elements, a striking difference in composition is observed between northern and southern Europe that coincides with the extension of the Weichselian glaciation. Contamination reflecting urbanized, industrialized areas and regions of intensive agriculture can better be recognized by MMI contents than total element contents. Some cities (e.g, London, Paris) cause anthropogenic trace element anomalies (e.g. Au, Pb, Hg) in their vicinity. Few samples (most of which taken in vineyards) have such high contents that they pose a toxic risk for soil organisms. Oppositely, several minor nutrients (like Cu, Zn) show such low contents across large areas of Europe that trace element deficiency is of concern.

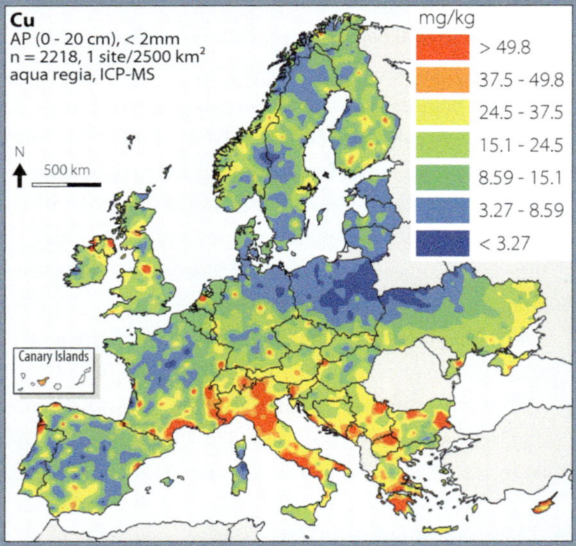

Spatial distribution of copper content in European agricultural soils (based on aqua regia destruction) as established from the GEMAS dataset using kriging (Reimann et al., 2014a; Demetriades et al., 2021). © BGR, 2014.

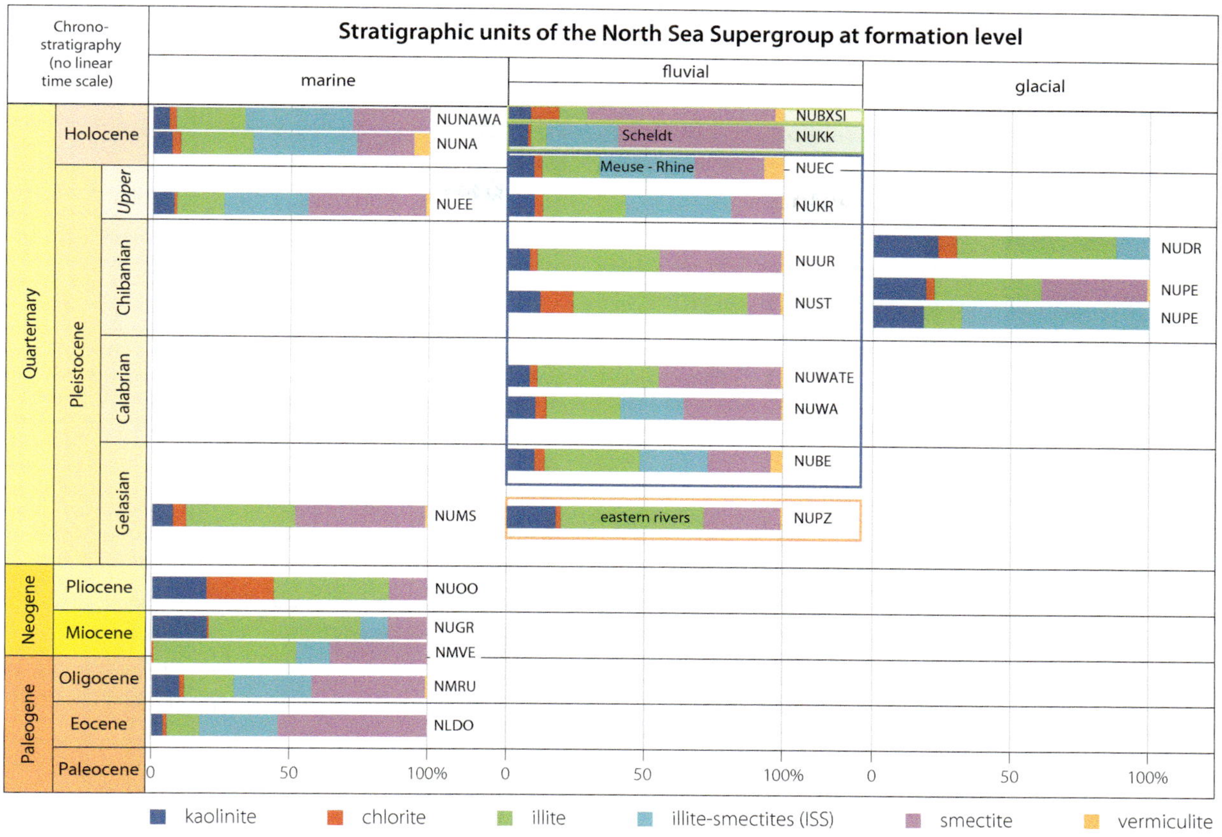

Figure 13.14. *Averaged clay mineral assemblages of the <2 μm size fraction in different Cenozoic formations. The clay mineralogy data in this figure originate from several studies: Favejee (1951), CTI-TNO (1970), Breeuwsma & Zwijnen (1984), Breeuwsma (1985), Breeuwsma et al. (1987), Heederik et al. (1989), Irion & Zollmer (1999), Kuhlmann et al. (2004), Adriaens (2014), Koenen & Griffioen (2014), Hemes et al. (2016), Dijkstra et al. (2020) and several unpublished studies performed at TNO. The data was obtained by X-ray diffraction (XRD) using different methodologies. Not all methods are suitable to determine mixed-layer illite-smectites (ISS). When ISS is not specifically analysed it is included in both the smectite and illite contents. Also vermiculite is not measured in each method and can therefore not be compared between studies using different methods. NLDO = Dongen Formation, NMRU = Rupel Formation, NMVE = Veldhoven Formation, NUGR = Groote Heide Formation, NUOO = Oosterhout Formation, NUMS = Maassluis Formation, NUPZ = Peize Formation, NUBE = Beegden Formation, NUWA = Waalre Formation, NUWATE = Tegelen Member of NUWA, NUST = Sterksel Formation, NUPE = Peelo Formation, NUDR = Drente Formation, NUUR = Urk Formation, NUKR = Kreftenheye Formation, NUEE = Eem Formation, NUNA = Naaldwijk Formation, NUNAWA = Walcheren Member of NUNA, NUEC = Echteld Formation, NUKK = Kreekrak Formation, NUBXSI = Singraven Member of the Boxtel Formation.*

shifts in the river-borne supply from Scandinavian sources. Illite and chlorite dominate from Scandinavian sources during cold periods, whereas smectite from southern sources dominates during warmer conditions (Griffioen et al., 2016). Apart from differences in mineral assemblages between formations they can also be recognized inside one formation. In the Belgian Boom Clay, Zeelmaekers (2011) observed a higher kaolinite fraction for the more clayey intervals compared to the more silty intervals, following a Milankovitch cyclic pattern. Koenen & Griffioen (2014) observed a similar trend for the clays from the Boom Member (Rupel Formation; NMRU in Fig. 13.14) with more kaolinite/smectite mixed layers and smectite in the clayrich samples.

Recent marine and estuarine sediments

Fluvial and marine clay mineralogies of Dutch Quaternary sediments exhibit some distinct differences. Smectite content in marine clays is similar to that in fluvial clays but has a higher iron content compared to the fluvial smectites. The higher iron-rich smectite content in the marine sediments is interpreted to originate from mixed input of sediment supplied by Dutch rivers and sediment transported northwards via the English Channel along the coast that has a high iron-rich smectite content (Breeuwsma, 1985).

The mineral composition along the entire Dutch coast is highly uniform. There are no significant differences between the muds from the Wadden Sea, the Eastern and

Western Scheldt estuaries, the North Sea floor and suspended material in the tidal inlets. The results of clay-mineralogy (Favejee, 1951) and heavy-mineral studies (Crommelin, 1940, 1943) in the Wadden Sea and the estuaries in Zeeland indicate that the sediments are predominantly marine in origin. Transport studies of marine sediments along the Dutch coast (De Groot, 1963; Terwindt, 1977; Eisma, 1981) all show that the amount of mud originating from the English Channel exceeds the discharge of fine material by the Rhine and Meuse. For the marine part of the Scheldt estuary the clay mineralogy was found to be identical with that of the mudflats along the Belgian coastal zone (Zeelmaekers, 2011; Adriaens et al., 2018).

Glauconite

Glauconite is mineralogically related to clay minerals but unlike other clay minerals in Dutch sediments its origin is primarily autochthonous and not detrital. The mineral glauconite is an iron potassium phyllosilicate of the mica group which has a characteristic green colour. High content of authigenic glauconite is an indicator of low sedimentation rates in a marine environment and is often associated with marine transgression. It is used for sedimentological and sequence stratigraphic interpretations as well as K-Ar dating studies (Adriaens et al., 2018).

The presence of glauconite in marine sand can be chemically recognized by higher K_2O to Al_2O_3 ratios compared to this ratio in glauconite-free sand (Huisman & Kiden, 1998). High K_2O contents in the Breda Subgroup are clearly visible in Figure 13.13 which can be attributed to a high glauconite content. Such high K_2O to Al_2O_3 ratios are also visible in a few samples from the Oosterhout Formation. Very high glauconite contents in the Breda Subgroup are likely due to in situ production whereas lower contents in younger sediments such as the Oosterhout Formation are due to reworking of these layers and mixing with sediments from the hinterland (Vandenberghe et al., 2014).

Carbonates

The carbonate state of sediments is important in two different ways. Firstly, it is used to stratigraphically classify the sediments and to characterize their provenance. Secondly, the presence of carbonate minerals exerts a control on the carbonate chemistry and pH of soil moisture and groundwater. It is an attribute that is routinely considered in borehole descriptions using the HCl effervescence test and visual observations on especially shells and their fragments. The effervescence test is only indicative for the presence of calcite or aragonite but not for dolomite or siderite. Whereas calcium carbonates are well studied in Dutch Cenozoic sediments (Griffioen et al., 2016), considerably less is known about Fe- and Mg-bearing carbonates.

Table 13.4 presents an updated overview of the carbonate state of Cenozoic Dutch geological formations. It essentially refers to the state with respect to calcite and aragonite except for the Waalre Formation for which a remark about siderite is made. Generally, the marine formations are calcareous but the carbonate state varies among

Table 13.4. General characterization of carbonates of Neogene and Quaternary formations outside the soil zone (based on De Mulder et al., 2003; Klein & Griffioen, 2010; Klein et al., 2015; Griffioen et al., 2016 and a query of the DINO-database in November 2021).

Formation	Description
Naaldwijk	Very often calcareous; may contain shells and shell fragments
Echteld	Mostly calcareous but also non-calcareous; sporadically fresh water shells
Boxtel	Mostly carbonate-poor or non-calcareous and sometimes rich
Kreftenheye	Mostly calcareous but also non-calcareous
Eem	Mostly calcareous and often shells; locally rich in shells
Drente	Often non-calcareous or carbonate-rich
Drachten	Mostly non-calcareous
Peelo	Carbonate-poor sand and carbonate-rich clay
Urk	Non-calcareous to carbonate-rich
Appelscha	Mostly non-calcareous
Sterksel	Partly calcareous
Beegden	Often non-calcareous
Stramproy	Mostly non-calcareous or carbonate-poor
Peize	Mostly non-calcareous
Waalre	Often non-calcareous or carbonate-rich; contains siderite-bearing clay layers
Kieseloolite	Mostly non-calcareous
Maassluis	Mostly carbonate-rich with shells
Oosterhout	Often carbonate-rich; rich in shells
Breda Subgp	Variable; locally with shells

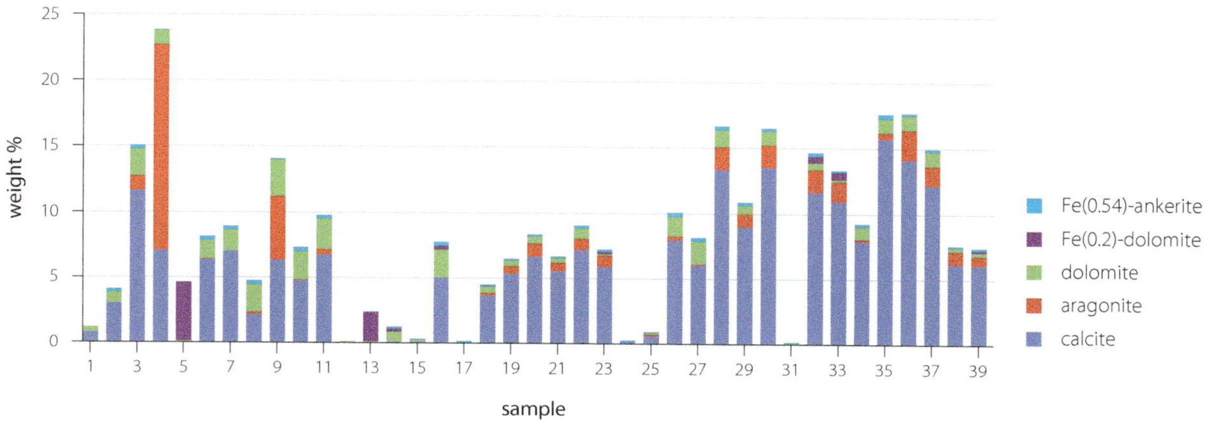

Figure 13.15. *Contents of various carbonate minerals in marine sediments from the surficial Naaldwijk Formation based on XRD-analyses of samples from 9 boreholes in Walcheren, province of Zeeland.*

non-marine formations. Sediments may be non-calcareous upon deposition or they may have been decalcified post-depositionally during pedogenesis or diagenesis. This may have happened under paleohydrological conditions, as is suggested by regional differences in pH and carbonate saturation state of groundwater in the Pleistocene part of the Netherlands (Griffioen et al., 2013) or more recently, following impoldering activities (for example of supratidal flats). A limited number of XRD analyses indicates that both calcite and aragonite are present in marine and other deposits. This is illustrated in Figure 13.15 that indicates a considerable heterogeneity in carbonate mineralogy of the marine Naaldwijk Formation sampled within a limited area in Walcheren (province of Zeeland).

Dolomite is also a common detrital mineral in Dutch sediments. Fractions of around 10% of the total Ca carbonate content have been reported for fluvial, marine and glacial sediments and also in loess (Griffioen et al., 2016). Dolomite tends to become relatively enriched during decalcification (Van der Sleen, 1912; Ritsema & Groenenberg, 1993) as the dissolution rate of dolomite is much lower than that of calcite and aragonite. It thus becomes preserved for a longer period than pure Ca carbonate minerals.

Siderite has been frequently detected in various kinds of geological sediments (Griffioen et al., 2016). It occurs in traces in suspended matter of the Rhine and Scheldt rivers (Crommelin, 1943; Wartel, 1977), which brings forward that it might be detrital when it does not become oxidized post-depositionally. However, a diagenetic origin is more likely as can be illustrated by scanning electron microscopy or just visual observations from borehole samples. The appearance of siderite varies from coatings on Ca carbonate grains via micrometer-size nodules and white substance in gyttja (which are organic-rich lacustrine sediments) to large concretions of ca. 10 cm in clay layers (Huisman, 1998; Griffioen et al., 2016). XRD-analysis as well as EDX elemental analysis indicate that the diagenetic precipitate may also be a carbonate solid-solution containing Fe, Ca, Mn and Mg, which is referred to as ferroan dolomite and ankerite instead of siderite (see Fig. 13.15). The visual observations prove that sufficient time was available for precipitation of Fe-bearing carbonates in Cenozoic sediments, despite the fact that siderite precipitation is kinetically very slow at earth surface conditions and that Dutch groundwater is frequently supersaturated with siderite, which indicates a lack of thermodynamic equilibrium (Griffioen et al., 2013).

Specific features

Bog iron ores

Bog iron ores deserve special attention as they regularly occur at local scale and have a geochemical composition that deviates from regular clastic sediments. They are typically found in sandy deposits at the bottom of peat layers, in alluvial deposits in brook valleys and within peat layers as tubular crusts or powder (Van Bemmelen & Reinders, 1901; Reinders, 1902; Booij, 1986; Landuydt, 1990). Their diagenesis has always been linked to exfiltration of Fe-containing groundwater in low-lying areas. They were mined in Roman times, the Early Middle Ages and the 17th-19th centuries in three major areas: the Vecht (province of Overijssel), Montferland and the Veluwe (Joosten, 2004).

The bog iron ores are notorious for their high contents of trace elements, that get bound during groundwater exfiltration (Van Pruissen & Zuurdeeg, 1988; Joosten, 2004). This is illustrated in Table 13.5, in which two geochemical data sets are summarized. As expected, the Fe content is high and other main metals are relatively low. The P content is also high as PO_4 binds strongly to Fe-oxyhydroxides or forms Fe-hydroxyphosphates during precipitation at high PO_4 to Fe groundwater ratios. The statistics for the trace elements may be compared to the Dutch, aqua regia-based baseline AW2000 value ('Regeling bodemkwaliteit', 2007) and to the Dutch intervention value, above which one speaks of severely contaminated soil or sediment ('Circulaire Bodemsanering', 2013). Both values were derived for sediment with ≤2% clay and ≤2% organic matter. For regular soils and sediments, aqua regia-based contents are often lower than XRF-based contents (e.g. Salminen et al., 2008). However, this difference may be small or absent for bog iron ores as much of the solids will be well soluble in aqua regia. A direct comparison, therefore, seems appropriate. The average contents for trace elements lie above the AW2000 values except for Cu. The averages also lie above the intervention value for As, Ba and Cr, which illustrates the polluted nature of bog iron ore due to secondary enrichments. Within the framework of the Dutch Soil Protection Law, their occurrences may be classified as natural pollution sites. This avoids their classification as anthropogenically contaminated sites and individual parties are not held responsible for the pollution. Bog iron ore fragments are also brought to the surface at the coastal nourishment site of the Sand Motor (Pit et al., 2017). Their surficial occurrence in this recreational area lies within acceptable limits with respect to human toxicological exposure (Van Bruggen et al., 2014).

Trace elements in pyrite

Another topic that deserves attention when it comes to the presence of trace elements is the elemental composition of pyrite. It is widely recognized that trace elements that form sulphide minerals, are present in Fe-sulphides as impurities (e.g. Morse, 1994). These trace elements may become mobilized when pyrite gets oxidized, especially in relation to denitrification. This may induce negative environmental side effects when the concentration becomes high compared to environmental standards. For the Netherland it turns out that this is especially the case for Ni and As (Van Beek et al., 1989; Zhang et al., 2009).

Selective extraction techniques have been used to characterize the elemental composition of pyrite and this has been applied a few times to Dutch sediments. Figure 13.16 summarizes the findings as cumulative frequency distribution plots of the molar ratios between the trace metals of interest and Fe as main constituent of pyrite. The plots illustrate that the ratios are highest for As, Zn, Ni, Cu and Co and vary within two orders of magnitude. For Pb and Mo, the ratios are about an order of magnitude smaller and have a smaller range. The ratio is lowest for Cd but varies three orders of magnitude. Considering the fact that Ni and As show high ratios while they have lower environmental standards compared to Zn and Co, it is not surprising that Ni and As have received the most attention. Copper also has a high ratio but is more strongly sorbed to oxides and humic acids, which makes it less mobile than the other four trace elements. Referring to the earlier discussed results of the factor analysis of geochemical data sets of Dutch sediments, it is worth to point out that an S-factor is frequently observed that contains As but often no other trace elements. This indicates that the trace metals of concern are not majorly bound by sulphides but that specific mobilization is a topic of environmental concern.

Table 13.5. Average elemental composition and associated standard deviation of bog iron ore as calculated from 15 XRF-analyses presented by Joosten (2004) and 18 XRF-analyses of bog iron ore as identified in the DINO database. Values below detection limit were set to half the detection limit when calculating the statistics.

Data set	Main elements (wt%)								
	Fe_2O_3	SiO_2	Al_2O_3	MnO	CaO	MgO	Na_2O	K_2O	P_2O_5
Joosten (2004)	71.5±13.8	22.5±11.6	1.7±0.8	1.8±2.0	1.1±1.2	< 0.1	0.1±0.1	0.4±0.2	3.0±1.7
DINO dbase	32.1±11.6	37.4±15.3	3.97±1.93	1.18±1.18	0.74±0.55	0.07±0.11	0.20±0.10	0.61±0.27	1.1±0.3
	Trace elements (mg/kg)								
	As	Ba	Co	Cr	Cu	Ni	Sr	V	Zn
Joosten (2004)	95.4±90.4	1415±1641	29.3±20.1		< 1	67.9±61.4	123.4±149.8	77.5±68.4	91.6±73.0
DINO dbase	303±322	449±266		47.1±21.9	10.4±2.1	20.7±5.6	55.0±26.1	40.1±21.9	63.6±32.8
AW2000[1]	11.4	49.0	4.3	29.7	19.3	10.9		27.4	59.0
Intervention[1]	43.5	237	54.0	42.1[2]	91.8	31.1		85.7[3]	303

1 note that the AW2000 and intervention values are based on aqua regia destruction and not total element analysis according to XRF;
2 assuming Cr(VI);
3 indicative value

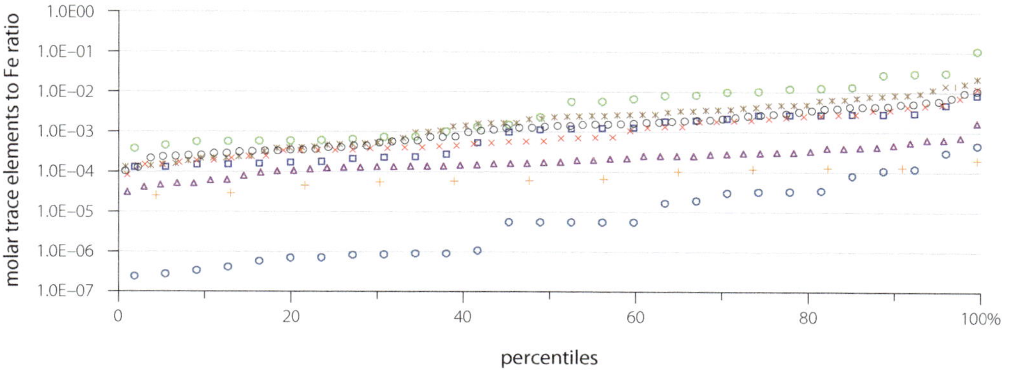

Figure 13.16. Cumulative frequency diagrams for the molar ratio of trace metals to Fe as based on selective extraction of pyrite. Data originates from Griffioen & Broers (1993), Broers & Buijs (1997) and Roskam & Griffioen (2011).

Sulphur in peat

In the Netherlands, peat has been poorly studied mineralogically and geochemically. Many peatlands in the Netherlands cope with subsidence and water quality problems. Drainage of these areas causes the peat to oxidize (e.g. Van Bergen & Koster, 2025, this volume), resulting in subsidence (Fokker et al., 2025, this volume), acidification and eutrophication. The form in which sulfur is originally present in peat, as pyrite (FeS_2), iron monosulfide (FeS), solid elemental sulfur or organic sulfur, influences the dynamics of sulfur cycling and thereby oxidation products and consequences such as acidification or the release of nutrients (Wieder & Lang, 1988; Mandernack et al., 2000). Thus, pyrite does not only play an important role in As and Ni release as pointed out in the previous section.

Studies on sulfur in Dutch peatlands show that sulfur can be present in very high amounts up to 8-9 wt% (Klein et al., 2015). S:Fe molar ratios in peat vary considerably with an average of 22, indicating that in addition to its presence as iron sulfides, sulfur is likely also present as organic sulfur (data from DINO database). The amount of organic sulfur in organic matter can vary considerably with S:C ratios between 1:30 and 1:1000 as measured in different types of peat (Kohnen et al., 1989; Price & Casagrande, 1991; Tipping et al., 2016). It is therefore not possible to estimate sulfur speciation from total organic carbon content or total element analysis of S. In a limited study on the Hollandveen Member of the Nieuwkoop Formation, at four different locations, sulfur was found to be predominantly present in the form of pyrite and organic sulfur species (Kim, 2017). FeS and sulfate were present in very small quantities and the amount of elemental sulfur was negligible.

It is suggested that in peat having had both fluvial and marine input, pyrite is the dominant form of sulfur, while in peatlands having marine influence but lacking freshwater input, organic sulfur is predominant (Kim, 2017). This is thought to be related to the availability of dissolved iron (Fe) at the time of the peat formation. In peat formed in brackish water, dissolved sulfate was abundant from the seawater, while the freshwater input brought Fe-rich suspended matter, and together they eventually formed pyrite (FeS_2) under reducing conditions. Whether such regional differences in sulfur speciation have an influence on potential acidification or eutrophication as a result of drainage or droughts has not been investigated for the Netherlands.

Lead and its isotopes

Lead (Pb) and its isotopes deserve special attention as soil contamination with lead frequently occurs in the Netherlands and there is no threshold value below which bioaccessible Pb can be assumed not toxic when ingested by children, especially (EFSA, 2010). Soil contamination originates from for example white lead ($(PbCO_3)_2{\cdot}Pb(OH)_2$) that was used in paint and from atmospheric deposition resulting from use of leaded gasoline. Historical contamination with Pb is thus widespread in both urban and rural soils. The origin of lead in soils can be traced by analyzing its isotopes. Walraven and coworkers performed a series of studies on the isotope geochemistry of lead in Dutch soils and lake sediments. Firstly, they showed that for unpolluted soils, the four main lithologies distinguished – clay, sand, peat and loess – have distinct natural isotope signatures (Walraven et al., 2013a). This variation can be explained by the soil Al and Zr contents, where Al is indicative for the Al-silicate minerals and Zr for the proportion of U and/or Th containing primary minerals that have a more radiogenic Pb isotope composition. No significant difference was found between marine and fluvial clays, which points to the dominance of a similar sedimentary provenance in the Pb isotope signal.

Anthropogenic Pb contamination of the soil is recognized from the Pb isotope composition, where the sources and their intensity vary for rural soils between atmospheric deposition from coal burning (or galena), incinerator ashes, gasoline, and the application of animal manure and fertiliser salts (Walraven et al., 2013b). For these soils, the highest contamination is found in the Randstad area and near the Dutch borders and the lowest contamination occurs in the coastal dunes and forest areas in the Pleistocene part of the Netherlands. At the national scale, the agricultural contamination of the soil with Pb is not distinctly different from that caused by atmospheric deposition. The history of atmospheric lead deposition can be well recognized in lake sediments from undisturbed sites and the extent of Pb deposition from gasoline can be identified in the vicinity of motorways (Walraven et al., 2014a,b). Lead is a trace metal that is more immobile than, for example, Ni or Zn. However, leaching of Pb to groundwater may have occurred in acid forest soils for which Pb cannot be assumed to be immobile.

Acknowledgments

Pauline van Gaans is thanked for earlier discussions on the factor analysis of geochemical datasets. Eva Hoekstra is thanked for her database activities on the geochemical sediment analyses. The reviewers Mark Hollanders, Bas van der Grift and Rudy Swennen, and the editor Jan de Jager are acknowledged for their constructive comments on an earlier version of this chapter.

References

Abbink, O.A., Mijnlieff, H., Munsterman, D. & Verreussel, R., 2006. New stratigraphic insights in the Late Jurassic of the Southern Central North Sea Graben and Terschelling Basin (Dutch Central Graben) and related exploration potential. Netherlands Journal of Geosciences 85 (3): 221-238. DOI: 10.1017/S001677460002148X

Adriaens, R., 2014. Staalname en analyse TNO/Deltares. Staalname en analyse TNO/Deltares. Doctoraatsproject KU Leuven: 14 pp.

Adriaens, R., Zeelmaekers, E., Fettweis, M., Vanlierde, E., Vanlede, J., Stassen, P., Elsen, J., Srodon, J. & Vandenberghe, N., 2018. Quantitative clay mineralogy as provenance indicator for recent muds in the southern North Sea. Marine Geology 398: 48-58. DOI: 10.1016/j.margeo.2017.12.011

Ajdukiewicz, J.M. & Larese, R.E., 2012. How clay grain coats inhibit quartz cement and preserve porosity in deeply buried sandstones: Observations and experiments. American Association of Petroleum Geologists Bulletin 96: 2091-2119. DOI: 10.1306/02211211075

Alizai, A., Hillier, S., Clift, P.D., Giosan, L. & Hurst, A., 2012. Clay mineral variations in Holocene terrestrial sediments from the Indus Basin. Quaternary Research 77: 368-381. DOI: 10.1016/j.yqres.2012.01.008

Amthor, J.E. & Okkerman, J., 1998. Influence of early Diagenesis on Reservoir Quality of Rotliegende Sandstones, Northern Netherlands. American Association of Petroleum Geologists Bulletin 82 (12): 2246-2265. DOI: 10.2110/pec.11.98.0193

Antoniou, E.A., Stuyfzand, P.J. & Breukelen, B.M., 2013. Reactive transport modeling of an aquifer storage and recovery (ASR) pilot to assess long-term water quality improvements and potential solutions. Applied Geochemistry 35: 173-186. DOI: 10.1016/j.apgeochem.2013.04.009

Baak, J.A., 1936. Regional petrology of the southern North Sea. PhD thesis, Agricultural University, Wageningen (the Netherlands): 128 pp.

Becker, I., Wüstefeld, P., Koehrer, B., Felder, M. & Hilgers, C., 2017. Porosity and permeability variations in a tight gas sandstone reservoir analogue, Westphalian D, Lower Saxony Basin, NW Germany: Influence of depositional setting and diagenesis. Journal of Petroleum Geology 40 (4): 363-389. DOI: 10.1111/jpg.12685

Bertier, P., Swennen, R., Lagrou, D., Laenen, B. & Kemps, R., 2008. Palaeo-climate controlled diagenesis of the Westphalian C & D fluvial sandstones in the Campine Basin (north-east Belgium). Sedimentology 55: 1375-1417. DOI: 10.1111/j.1365-3091.2008.00950.x

Besly, B.M., Burley, S.D. & Turner, P., 1993. The late Carboniferous 'Barren Red Play' of the Silver Pit area, Southern North Sea. Geological Society of London, Special Publications 4 (1): 727-740. DOI: 10.1144/0040727

Boggs, S., 2009. Petrology of Sedimentary Rocks. Cambridge University Press: 600 pp.

Booij, A.H., 1986. IJzeroer in Drenthe. Ontstaan, voorkomen, winning en gebruik. In: Nieuwe Drentsche Volksalmanak. Van Gorcum (Assen): 66-87.

Borkhataria, R., Aigner, T. & Pipping, J.C.P., 2006. An unusual, muddy, epeiric carbonate reservoir: The Lower Muschelkalk (Middle Triassic) of the Netherlands. American Association of Petroleum Geologists Bulletin 90 (1): 61-89. DOI: 10.1306/08030504124

Borkhataria, R., Aigner, T., Pöppelreiter, M.C. & Pipping, J.C.P., 2005. Characterisation of epeiric 'Layer-cake' carbonate reservoirs: Upper Muschelkalk (Middle Triassic), The Netherlands. Journal of Petroleum Geology 28 (2): 119-146. DOI: 10.1111/j.1747-5457.2005.tb00076.x

Breeuwsma, A., 1985. Kleimineralogische en chemische karakteristieken van zeeklei, rivierklei en beekklei. Stichting voor bodemkartering (STIBOKA), Report No. 1869: 34 pp.

Breeuwsma, A., Balkema, W. & Zwijnen, P., 1987. Kleimineralogische onderzoek van enkele Kwartaire en Tertiaire afzettingen. Stichting voor bodemkartering (STIBOKA), Report No. 1980: 30 pp.

Breeuwsma, A. & Zwijnen, P., 1984. Kleimineralogische samenstelling van Nederlandse aquifers. Stichting voor bodemkartering (STIBOKA), Report No. 1816: 13 pp.

Broers, H.P. & Buijs, E.A., 1997. De herkomst van sporenmetalen en arseen in het waterwingebied Oostrum. TNO Report No. TNO-NITG97- 198-A: 40 pp.

Busch, B., Hilgers, C. & Adelmann, D., 2020. Reservoir quality controls on Rotliegend fluvio-aeolian wells in Germany and the Netherlands, Southern Permian Basin - Impact of grain coatings and cements. Marine and Petroleum Geology 112: 104075. DOI: 10.1016/j.marpetgeo.2019.104075

Busschers, F.S., Kasse, C., Van Balen, R.T., Vandenberghe, J., Cohen, K.M., Weerts, H.J.T., Wallinga, J., Johns, C., Cleveringa, P. & Bunnik, F.P.M., 2007. Late Pleistocene evolution of the Rhine-Meuse system in the southern North Sea basin: imprints of climate change, sea-level oscillation and glacio-isostacy. Quaternary Science Reviews 26: 3216-3248. DOI: 10.1016/j.quascirev.2007.07.013

Busschers, F.S., Cohen, K., Wesselingh, F.P., Schokker, J., Bakker, M., Van Balen, R. & Van Heteren, S., 2025. Quaternary. In: Ten Veen, J.H., Vis, G.-J., De Jager, J. & Wong, Th.E. (eds): Geology of the Netherlands, second edition. Amsterdam University Press (Amsterdam): 333-391. DOI: 10.5117/9789463728362_ch10

Carpenter, S.J. & Lohman, K.C., 1992. Sr/Mg ratios of modern marine calcite: empirical indicators of ocean chemistry and precipitation rate. Geochemica et Cosmochimica Acta 56: 1837-1849. DOI: 10.1016/0016-7037(92)90314-9

Circulaire bodemsanering per 1 juli 2013, 2013. Staatscourant nr. 16675: 58 pp.

Clark, D.N., 1986. The distribution of porosity in Zechstein carbonates. In: Brooks, J., Goff, J.C. & Van Hoorn, B. (eds): Habitat of Palaeozoic gas in N.W. Europe. Geological Society of London, Special Publications 23: 121-149. DOI: 10.1144/GSL.SP.1986.023.01.09

Crommelin, R.D., 1964. A contribution to the sedimentary petrology and provenance of young Pleistocene coversand in The Netherlands. Geologie En Mijnbouw 43: 389-402.

Crommelin, R.D., 1940. De herkomst van het zand van de Waddenzee. Tijdschrift van het Koninklijk Nederlands Aardrijkskundig Genootschap 57: 347-361.

Crommelin, R.D., 1943. De herkomst van het waddenslib met korrelgrootte boven 10 micron. Verhandelingen van het Geologisch Mijnbouwkundig Genootschap. Geologische Serie 13: 299-333.

Crommelin, R.D. & Slotboom, G., 1945. Een voorkomen van granaatzandlagen op het strand van Goeree. Tijdschrift van het Koninklijk Nederlands Aardrijkskundig Genootschap LXII: 142-147.

CTI-TNO, 1970. Samenstelling en eigenschappen van 31 kleisoorten, Deel II. Centraal Technisch Instituut TNO, Report No.70-04032: 121 pp.

Čyžienė, J., Molenaar, N. & Šliaupa, S., 2006. Clay-induced pressure solution as a Si source for quartz cement in sandstones of the Cambrian Deimena Group. Geologija 53: 8-21.

De Groot, A.J., 1963. Mangaantoestand van nederlandse en duitse holocene sedimenten in verband met slibtransport en bodemgenesese. Verslagen Landbouwkundige Onderzoekingen 69 (7): 168 pp.

De Jager, J., Van Ojik, K. & Smit, J., 2025. Geological development. In: Ten Veen, J.H., Vis, G.-J., De Jager, J. & Wong, Th.E. (eds): Geology of the Netherlands, second edition. Amsterdam University Press (Amsterdam): 21-51. DOI: 10.5117/9789463728362_ch01

De Mulder, E.F.J., Geluk, M.C., Ritsema, I., Westerhoff, W.E. & Wong, Th.E., 2003. De ondergrond van Nederland. Wolters-Noordhof (Groningen): 379 pp.

De Vos W., Tarvainen T. (eds), Salminen R., Reeder S., De Vivo B., Demetriades A., Pirc S., Batista M.J., Marsina K., Ottesen R.-T., O'Connor P.J., Bidovec M., Lima A., Siewers U., Smith B., Taylor H., Shaw R., Salpeteur I., Gregorauskiene V., Halamic J., Slaninka I., Lax K., Gravesen P., Birke M., Breward N., Ander E.L., Jordan G., Duris M., Klein P., Locutura J., Bel-lan A., Pasieczna A., Lis J., Mazreku A., Gilucis A., Heitzmann P., Klaver G., Petersell V. (2006). Geochemical Atlas of Europe. Part 2 - Interpretation of Geochemical Maps, Additional Tables, Figures, Maps, and Related Publications. Geological Survey of Finland (Espoo, Finland): 690 pp.

Deer, W.A., Howie, R.A. & Zussmann, J., 2013. An introduction to the rock-forming minerals, third edition. Mineralogical Society (London): 498 pp.

Demetriades, A., Reimann, C., Birke, M., Négrel, P., Ladenberger, A., Tarvainen, T. & Sadeghi, M., 2021. GEMAS: Geochemistry of European agricultural and grazing land soil. European Geologist 52. DOI: 10.5281/zenodo.5770107

Dijkstra, H., De Boer, C., Griffioen, J. & Koenen, M., 2020. Workflow to evaluate the risk of mineral scaling in a HT-ATES system and application to a potential site in Middenmeer, The Netherlands, TNO, Report No. 2020 R10437: 56 pp.

Dreesen, R., Bossiroy, D., Dusar, M., Flores Romeo, M. & Verkaeren, P., 1995. Overview of the influence of syn-sedimentary tectonics and palaeo-fluvial systems on coal seam and sand body characteristics in the Westphalian C strata, Campine Basin, Belgium. In: Whateley, M.K.G. & Spears, D.A. (eds): European Coal Geology. Special Publications Geological Society of London 82: 215-232.

Dronkert, H. & Remmelts, G., 1996. Influence of salt structures on reservoir rocks in Block L2, Dutch continental shelf. In: Rondeel, H.E., Batjes, D.A.J. & Nieuwenhuijs, W.H. (eds): Geology of gas and oil under the Netherlands. Kluwer Academic Publishers (Dordrecht): 159-166. DOI: 10.1007/978-94-009-0121-6_14

Drost, G.I.A. & Korenromp, M.H.A., 2009. Sediment-petrographical analysis of the Delft Sandstone. BSC thesis, Technical University Delft (the Netherlands): 122 pp.

Druif, J.H., 1927. Over het ontstaan der Limburgsche löss in verband met haar mineralogische samenstelling. PhD thesis, Rijksuniversiteit Utrecht (the Netherlands): 348 pp.

Duin, E., Doornenbal, J.C., Rijkers, R.H.B., Verbeek, J.W. & Wong, Th.E., 2006. Subsurface structure of the Netherlands; results of recent onshore and offshore mapping. Netherlands Journal of Geosciences 85 (4): 245-276. DOI: 10.1017/S0016774600023064

Edelman, C.H., 1933. Petrologische provincies in het Nederlandse Kwartair. PhD thesis Universiteit van Amsterdam (the Netherlands): 104 pp.

EFSA, 2010. Scientific opinion on lead in food. Eur. Food Safety Auth 8 (4): 1570. DOI: 10.2903/j.efsa.2010.1570

Egeberg, P.K. & Saigal, G.C., 1991. North Sea chalk diagenesis: cementation of chalks and healing of fractures. Chemical Geology 92: 339-354. DOI: 10.1016/0009-2541(91)90078-6

Eisma, D., 1981. Supply and deposition of suspended matter in the North Sea. In: Special publication of the International Association of Sedimentologists 5: 415-428.

Favejee, J.Ch.L., 1951. The origin of the 'Wadden' mud. Mededelingen Landbouw Hogeschool Wageningen 51: 113-141.

Fest, E.P.M.J., Temminghoff, E.J.M., Griffioen, J., Van der Grift, B. & Van Riemsdijk, W.H., 2007. Groundwater chemistry of Al under Dutch acid sandy soils: effects of land use and depth. Applied Geochemistry 22: 1427-1438. DOI:10.1016/j.apgeochem.2007.02.002

Fokker, P.A., Candela, T.G.G., Erkens, G., Hanssen, R.F., Kooi, H., Koster, K. & Van Leijen, F.J., 2025. Subsidence. In: Ten Veen, J.H., Vis, G.-J., De Jager, J. & Wong, Th.E. (eds): Geology of the Netherlands, second edition. Amsterdam University Press (Amsterdam): 825-847. DOI: 10.5117/9789463728362_ch23

Füchtbauer, H., 1967. Der Einfluss des Ablagerungsmilieus auf die Sandsteindiagenese im mittleren Buntsandstein/Influence of depositional environment on diagenesis of the Triassic Bunter sandstones. Sedimentary Geology 1: 159-179.

Galán, E. & Ferrell, R.E., 2013. Genesis of clay minerals. Developments in clay science 5: 83-126. DOI: 10.1016/B978-0-08-098258-8.00003-1

Gallois, R. & Owen, H., 2017. The stratigraphy of the Upper Greensand Formation (Albian, Cretaceous) of the Isle of Purbeck, Dorset, UK. Acta Geologica Polonica 67: 405-424. DOI:10.1515/AGP-2017-0016

Gaupp, R., Matter, A., Platt, J., Ramseyer, K. & Walzebuck, J., 1993. Diagenesis and fluid evolution of deeply buried Permian (Rotliegendes) gas reservoirs, Northwest Germany. American Association of Petroleum Geologists Bulletin 77: 1111-1128. DOI: 10.1306/BDFF8E0C-1718-11D7-8645000102C1865D

Gaupp, R. & Okkerman, J.A., 2011. Diagenesis and reservoir quality of Rotliegend sandstones in the northern Netherlands - A review. In: Grötsch, J. & Gaupp, R. (eds): The Permian Rotliegend of the Netherlands. SEPM Special Publication 98: 193-226. DOI: 10.2110/pec.11.98.0193

Glennie, K.W., Mudd, G.W. & Nagtegaal, P.J.C., 1978. Depositional environment and diagenesis of Permian Rotliegendes sandstones in Leman Bank and Sole Pit areas of the UK southern North Sea. Journal of the Geological Society 135 (1): 25-34.

Gökdag, H., 1982. Some diagenetic aspects and origin of porosity in the Dinantian (Early Carboniferous) carbonates in the wells Heugem 1a and Kastanjelaan 2 (Maastricht, the Netherlands). Publicaties van Het Natuurhistorisch Genootschap in Limburg 31 (1-4): 50-53.

Goldberg, T., Hennekam, R., Wasch, L., Reichart, G.J., Rach, O., Stammeier, J.A. & Griffioen, J., 2021. Suitability of calibrated X-ray fluorescence core scanning for environmental geochemical characterisation of heterogeneous sediment cores. Applied Geochemistry 125: 104824. DOI: 10.1016/j.apgeochem.2020.104824

Griffioen, J., 1994. Uptake of Phosphate by Fe-hydroxides during Seepage in Relation to Saline/Fresh Groundwater Displacements. Environmental Science & Technology 28: 675-681. DOI: 10.1021/es00053a022

Griffioen, J., Klaver, G. & Westerhoff, W., 2016. The mineralogy of suspended matter, fresh and Cenozoic sediments in the fluvio-deltaic Rhine-Meuse-Scheldt-Ems area, the Netherlands: An overview and review. Netherlands Journal of Geosciences 95 (1): 23-107. DOI: 10.1017/njg.2015.32

Griffioen, J., Klein, J. & Van Gaans, P.F.M., 2012. Reaction capacity characterisation of shallow sedimentary deposits in geologically different regions of the Netherlands. Journal of Contaminant Hydrology 127: 30-46. DOI: 10.1016/j.jconhyd.2011.04.001

Griffioen, J., Passier, H. & Klein, J., 2008. Comparison of selection methods to deduce natural background levels for groundwater units. Environmental Science & Technology 42: 4863-4869. DOI: 10.1021/es7032586

Griffioen, J., Vermooten, S. & Janssen, G.J.A., 2013. Geochemical and palaeohydrological controls on the composition of shallow groundwater in the Netherlands. Applied Geochemistry 39: 129-149. DOI: 10.1016/j.apgeochem.2013.10.005

Griffioen, J., Broers, H.P. & Buijs, E.A., 1993. Characterization of sediment reactivity: the feasibility of sequential extraction techniques. TNO, Report No. OS-93-65: 15 pp.

Grötsch, J. & Gaupp, R. (eds), 2011. The Permian Rotliegend of the Netherlands. SEPM Special Publication 98: 374 pp. DOI: 10.2110/pec.11.98

Hakstege, A.L., Kroonenberg, S.B. & van Wijck, H., 1993. Geochemistry of Holocene clays of the Rhine and Meuse rivers in the central-eastern Netherlands. Geologie En Mijnbouw 71 (4): 301-315.

Hancock, N.J., 1978. Possible causes of Rotliegend sandstones diagenesis in northern West Germany. Journal of the Geological Society 135 (1): 35-40.

Hartog, N., Griffioen, J. & Bergen, P.F., 2005. Depositional and paleohydrogeological controls on the reactivity of organic matter and other reductants in aquifer sediments. Chemical Geology 216: 113-131. DOI: 10.1016/j.chemgeo.2004.11.006

Heederik, J. P. (ed.), 1989. Geothermische reserves Centrale Slenk - Exploratie en evaluatie. TNO-DGV, Report No. OS89-18: 177 pp.

Heerdink, R. & Griffioen, J., 2007. De associatie van sporenele-

menten met hoofdcomponenten in sedimentafzettingen in de geotop van Noord-Brabant en het noorden van Limburg. TNO Bouw en Ondergrond, Report No. 2007-U-R1013/A.
Heerdink, R. & Griffioen, J., 2012. Geochemische karakterisering van de geotop van Noord-Nederland (hoofdgebied 5). Gehaltes en associaties van hoofd- en sporenelementen. TNO Energie, Report No. TNO2012R10136: 277 pp.
Heier, K.S. & Taylor, S.R., 1959. Distribution of Li, Na, K, Rb, Cs, Pb and Tl in southern Norwegian pre-Cambrian alkali feldspars. Geochemica et Cosmochimica Acta 15: 284-304.
Heinrichs, H., Schulz-Dobrick, B. & Wedepohl, K.H., 1980. Terrestrial geochemistry of Cd, Bi, Tl, Pb, Zn and Rb. Geochimica et Cosmochimica Acta 44 (10): 1519-1534.
Hemes, S., Desbois, G., Klaver, J. & Urai, J.L., 2016. Microstructural characterisation of the Ypresian clays (Kallo-1) at nanometre resolution, using broad-ion beam milling and scanning electron microscopy. Netherlands Journal of Geosciences 95 (3): 293-313. DOI: 10.1017/njg.2016.16
Henares, S., Bloemsma, M.R., Donselaar, M.E., Mijnlieff, H.F., Redjosentono, A.E., Veldkamp, H.G. & Weltje, G.J., 2014. The role of detrital anhydrite in diagenesis of aeolian sandstones (Upper Rotliegend, The Netherlands): Implications for reservoir-quality prediction. Sedimentary Geology 314: 60-74. DOI: 10.1016/j.sedgeo.2014.10.001
Herngreen, G.F.W., Kouwe, W.F.P. & Wong, Th.E., 2003. The Jurassic of the Netherlands. Geological Survey of Denmark and Greenland Bulletin 1: 217-230. DOI: 10.34194/geusb.v1.4652
Herron, M.M. & Matteson, A., 1993. Elemental composition and nuclear parameters of some common sedimentary minerals. Nuclear Geophysics 7: 383-406.
Hillier, S., 1995. Erosion, sedimentation and sedimentary origin of clays. In: Origin and mineralogy of clays. Springer (Berlin, Heidelberg): 162-219. DOI: 10.1007/978-3-662-12648-6_4
Hjuler, M.L., 2007. Diagenesis of Upper Cretaceous onshore and offshore chalk from the North Sea area. PhD thesis, DTU Environment, Technical University of Denmark (Lyngby): 71 pp.
Hoving, A.L., Sander, M., Bruggeman, C. & Behrends, T., 2017. Redox properties of clay-rich sediments as assessed by mediated electrochemical analysis: Separating pyrite, siderite and structural Fe in clay minerals. Chemical Geology 457: 149-161. DOI: 10.1016/j.chemgeo.2017.03.022
Hoving, A.L., Sander, M., Frederickx, L., Dugulan, A.I., Bruggeman, C. & Behrends, T., 2020. What does mediated electrochemistry reveal about regional differences in the redox properties of Boom Clay? Applied Geochemistry 120: 104681. DOI: 10.1016/j.apgeochem.2020.104681
Huis in 't Veld, R., Schrijver, B. & Salzwedel, A., 2020. The Wingate Field, Blocks 44/23b, 44/24b and 44/19f, UK North Sea. 50Th Anniversary Commemorative Volume. Geological Society of London, Memoirs 52: 288-303. DOI: 10.1144/M52-2018-75
Huis in 't Veld, R. & Den Hartog Jager, D., 2025. Late Carboniferous. In: Ten Veen, J.H., Vis, G.-J., De Jager, J. & Wong, Th.E. (eds): Geology of the Netherlands, second edition. Amsterdam University Press (Amsterdam): 95-125. DOI: 10.5117/9789463728362_ch03
Huisman, D.J. & Kiden, P., 1998. A geochemical record of Late Cenozoic sedimentation history in the southern Netherlands. Geologie En Mijnbouw 76 (4): 277-292. DOI: 10.1023/A:1003212721020
Huisman, D.J., Vermeulen, F.J.H., Baker, J., Veldkamp, A., Kroonenberg, S.B. & Klaver, G.T., 1997. A geological interpretation of heavy metal concentrations in soils and sediments in the southern Netherlands. Journal of Geochemical Exploration 59 (3): 163-174. DOI: 10.1016/S0375-6742(97)00018-6
Huisman, D.J., Weijers, J.P., Dijkshoorn, L. & Veldkamp, A., 2000. Spatial prediction of the variability of Early Pleistocene subsurface sediments in the Netherlands - Part 2: Geochemistry. Netherlands Journal of Geosciences 79 (4): 381-390. DOI: 10.1017/S0016774600021892
Huisman, D.J., 1998. Geochemical characterization of subsurface sediments in the Netherlands. PhD thesis, Agricultural University Wageningen (the Netherlands): 177 pp.
Huisman, D.J., 1999. Bulk geochemie als karteringsgereedschap. TNO-NITG, Report No. NITG 99-1-B:
Irion, G. & Zöllmer, V., 1999. Clay mineral associations in fine-grained surface sediments of the North Sea. Journal of Sea Research 41: 119-128. DOI: 10.1016/S1385-1101(98)00041-0
Jaarsma, B., Brolsma, M.J., Hoetz, G. & Lutgert, J.E., 2013. Exploring Dinantian carbonates in the SNS – new data offering new insights. 75th EAGE Conference & Exhibition Incorporating SPE EUROPEC (London). EAGE (Houten): cp-348-00382.
Jaarsma, B., Heijnen, N., Veldkamp, H. & Schavemaker, Y., 2018. Dinantian Carbonates Are Hot! 80th EAGE Conference and Exhibition 2018(1), 1-5. European Association of Geoscientists & Engineers: 1-5.
Joosten, I., 2004. Technology of early historical iron production in the Netherlands. PhD thesis Vrije Universiteit Amsterdam (the Netherlands), Geoarchaeological and Bioarchaeological Studies 2: 133 pp.
Kim, J., 2017. Speciation of Sulfur in Peat in Four Locations in the Northwestern Netherlands. BSc thesis University Utrecht (the Netherlands): 32 pp.
Klein, J. & Griffioen, J., 2010. Organisch materiaal, kalk en zwavel in sediment: uitwerking van analyses naar gehalten, TNO, Report No. TNO-034-UT-2010-02327/A.
Klein, J., Van Gaans, P. & Griffioen., J., 2015. Geochemische karakterisering van de geotop van Holland (gebied 1b en 1c), TNO, Report No. TNO 2015 R 10785: 91 pp.
Koenen, M. & Griffioen, J., 2014. Mineralogical and geochemical characterization of the Boom Clay in the Netherlands. COVRA N.V., Report No. OPERA-PU-TNO521-1: 106 pp.
Koenen, M. & Griffioen, J., 2016. Charaterisation of the geochemical heterogeneity of the Rupel Clay Member in the Netherlands. Netherlands Journal of Geosciences 95 (Journal Article): 269-281. DOI: 10.1017/njg.2016.6
Kohnen, M.E.L., Damsté, J.S., Ten Haven, H.L. & De Leeuw,

J.W., 1989. Early incorporation of polysulphides in sedimentary organic matter. Nature 341 (6243): 640-641. DOI: 10.1038/341640a0

Kombrink, H., Bridge, J.S. & Stouthamer, E., 2007. The alluvial architecture of the Coevorden Field (Upper Carboniferous), the Netherlands. Netherlands Journal of Geosciences 86 (1): 3-14. DOI: 10.1017/S0016774600021284

Kombrink, H., Leever, K.A., Van Wees, J.-D., Van Bergen, F., David, P. & Wong, Th.E., 2008. Late Carboniferous foreland basin formation and Early Carboniferous stretching in Northwestern Europe - Inferences from quantitative subsidence analyses in the Netherlands. Basin Research 20: 377-395. DOI: 10.1111/j.1365-2117.2008.00353.x

Kooistra, M.J., 1978. Soil development in recent marine sediments of the intertidal zone in the Oosterschelde – the Netherlands. A soil micromorphological approach. PhD thesis University of Amsterdam (the Netherlands): 187 pp.

Kuhlmann, G., De Boer, P.L., Pedersen, R.B. & Wong, Th.E., 2004. Provenance of Pliocene sediments and paleoenvironmental changes in the southern North Sea region using Samarium–Neodymium (Sm/Nd) provenance ages and clay mineralogy. Sedimentary Geology 171: 205-226. DOI: 10.1016/j.sedgeo.2004.05.016

Laenen, B., Van Tongeren, P.C.H., Dreesen, R. & Dusar, M., 2004. Carbon dioxide sequestration in the Campine Basin and the adjacent Roer Valley Graben (North Belgium): an inventory. In: Baines, S.J. & Worden, R.H. (eds): Geological storage of carbon dioxide. Special Publication Geological Society of London 233: 193-210.

Landuydt, C.J., 1990. Micromorphology of iron minerals from bog ores of the Belgian Campine area. Developments in Soil Science 19: 289-294. DOI: 10.1016/S0166-2481(08)70340-4

Lanson, B., Beaufort, D., Berger, G., Baradat, J. & Lacharpagne, J.C., 1996. Illitization of diagenetic kaolinite-to-dickite conversion series: late-stage diagenesis of the Lower Permian Rotliegend Sandstone reservoir, offshore of the Netherlands. Journal of Sedimentary Research 66 (3): 501-518. DOI: 10.1306/D4268392-2B26-11D7-8648000102C1865D

Limmer, D.R., Köhler, C.M., Hiller, S., Moreton, S.G., Tabrez, A.R. & Clift, P.D., 2012. Chemical weathering and provenance evolution of Holocene- Recent sediments from the western Indus Shelf, Northern Arabian Sea inferred from physical and mineralogical properties. Marine Geology 326-328: 101-115. DOI: 10.1016/j.margeo.2012.07.009

Lind, I.L., 1993. Stylolites in chalk from Leg 130, Ontong Java Plateau. Proceedings of the Ocean Drilling Program Scientific Results 130: 445-451. DOI: 10.2973/odp.proc.sr.130.006.1993

Mandernack, K.W., Lynch, L., Krouse, H.R. & Morgan, M.D., 2000. Sulfur cycling in wetland peat of the New Jersey Pinelands and its effect on stream water chemistry. Geochimica et Cosmochimica Acta 64: 3949-3964. DOI: 10.1016/S0016-7037(00)00491-9

Mapstone, N.B., 1975. Diagenetic history of a North Sea chalk. Sedimentology 22: 601-614.

McKie, T., 2011. A comparisation of modern dryland depositional systems with the Rotliegend Group in the Netherlands. In: Grötsch. J. & Gaupp, R. (eds): The Permian Rotliegend of the Netherlands. SEPM Special Publication 98: 89-104. DOI:10.2110/PEC.11.98.0089

McKie, T. & Kilhams, B., 2025. Triassic. In: Ten Veen, J.H., Vis, G.-J., De Jager, J. & Wong, Th.E. (eds): Geology of the Netherlands, second edition. Amsterdam University Press (Amsterdam): 155-183. DOI: 10.5117/9789463728362_ch05

McLennan, S.M., 2001. Relationship between the trace element composition of sedimentary rocks and upper continental crust. Geochemistry Geophysics Geosystems 2: 2000GC000109. DOI: 10.1029/2000GC000109

Miedema, R., Jongmans, A.G. & Slager, S., 1973. Micromorphological observations on pyrite and its oxidation products in four Holocene alluvial soils in the Netherlands. Soil Microscopy, Proceedings of the 4th International Working-Meeting on Soil Micromorphology, The Limestone Press (Kingston): 772-794.

Miocic, J.M., Girard, J.-P., Schöner, R. & Gaupp, R., 2020. Mudstone/sandstone ratio control on carbonate cementation and reservoir quality in Upper Permian Rotliegend sandstones, offshore the Netherlands. Marine and Petroleum Geology 115: 104293. DOI: 10.1016/j.marpetgeo.2020.104293

Mol, G., 2002. Soil acidification monitoring in the Netherlands. PhD thesis Utrecht University (the Netherlands): 151 pp.

Mol, G., Spijker, J., Van Gaans, P. & Romkens, P., 2012. Geochemische bodematlas van Nederland. Wageningen Academic Publishers (Wageningen): 276.

Mol, G., Vriend, S.P. & Van Gaans, P.F.M., 2003. Feldspar weathering as the key to understanding soil acidification monitoring data; a study of acid sandy soils in the Netherlands. Chemical Geology 202: 417-441. DOI: 10.1016/j.chemgeo.2002.07.001

Molenaar, N. & Felder, M., 2018. Clay Cutans and the Origin of Illite Rim Cement: An Example from the Siliciclastic Rotliegend Sandstone in the Dutch Southern Permian Basin. Journal of Sedimentary Research 88 (5): 641-658. DOI: 10.2110/jsr.2018.33

Molenaar, N. & Felder, M., 2019. Origin and distribution of dolomite in Permian Rotliegend siliciclastic sandstones (Dutch Southern Permian Basin). Journal of Sedimentary Research 89: 1055-1073. DOI:10.2110/jsr.2019.58

Molenaar, N. & Zijlstra, J.J.P., 1997. Differential early diagenetic low-Mg calcite cementation and rhythmic hardground development in Campanian-Maastrichtian chalk. Sedimentary Geology 109 (3-4): 261-281. DOI: 10.1016/S0037-0738(96)00064-4

Morse, J.W., 1994. Interactions of trace metals with authigenic sulfide minerals: implications for their bioavailability. Marine Chemistry 46: 1-6. DOI: 10.1016/0304-4203(94)90040-X

Mozafari, M., Gutteridge, P., Riva, A., Geel, K., Garland, J. & Dewit, J., 2019. Facies analysis and diagenetic evolution of the Dinantian carbonates in the Dutch subsurface. SCAN report, commissioned by the Ministry of Economic Affairs & Climate: 138 pp. https://www.nlog.nl/sites/default/files/2019-12/scan_dinantian_facies_analysis_and_diagenetic_evolution.pdf

Munsterman, D.K., Donders, T.H., Houben, A.J.P., Ten Veen, J.H. & Wesselingh, F.P., 2025. Paleogene-Neogene. In: Ten Veen, J.H., Vis, G.-J., De Jager, J. & Wong, Th.E. (eds): Geology of the Netherlands, second edition. Amsterdam University Press (Amsterdam): 293-331. DOI: 10.5117/9789463728362_ch9

Neugebauer, J., 1974. Some aspects of cementation in chalk. In: Hsu, K. & Jenkyns, H.C. (eds): Pelagic Sediments: on Land and under the Sea. Int. Ass. Sedimentologists Spec. Pub: Vol. 1, 149-176.

Nielsen, O.B., Rasmussen, W.E.S. & Thyberg, B.I., 2015. Distribution of clay minerals in the northern North Sea basin during the Paleogene and Neogene: A result of source-area geology and sorting processes. Journal of Sedimentary Research 85: 562-581. DOI: 10.2110/jsr.2015.40

Olivarius, M., Weibel, R., Hjuler, M.L., Kristensen, L., Mathiesen, A., Nielsen, L.H. & Kjøller, C., 2015. Diagenetic effects on porosity–permeability relationships in red beds of the Lower Triassic Bunter Sandstone Formation in the North German Basin. Sedimentary Geology 321: 139-153. DOI: 10.1016/j.sedgeo.2015.03.003

Overeem, I. & Kroonenberg, S.B., 2002. De Amazone in de Noordzee. Natuur en Techniek 70 (1): 66-71.

Palermo, D., Aigner, T., Geluk, M.C., Pöppelreiter, M. & Pipping, K., 2008. Reservoir potential of a lacustrine mixed carbonate/siliciclastic gas reservoir: the Lower Triassic Rogenstein in the Netherlands. Journal of Petroleum Geology 31: 61-96. DOI: 10.1111/j.1747-5457.2008.00407.x

Paxton, S.T., Szabo, J.O., Ajdukiewicz, J.M. & Klimentidis, R.E., 2002. Construction of an intergranular volume compaction curve for evaluating and predicting compaction and porosity loss in rigid-grain sandstone reservoirs. AAPG Bulletin 86: 2047-2067. DOI: 10.1306/61EEDDFA-173E-11D7-8645000102C1865D

Peryt, T.M. & Scholle, P.A., 1996. Regional setting and role of meteoric water in dolomite formation and diagenesis in an evaporite basin: studies in the Zechstein (Permian) deposits of Poland. Sedimentology 43: 1005-1023.

Peyser, C.E. & Poulsen, C.J., 2008. Controls on Permo-Carboniferous precipitation over tropical Pangaea A GCM sensitivity study. Palaeogeography, Palaeoclimatology, Palaeoecology 268: 181-192. 10.1016/j.palaeo.2008.03.048

Pit, I.R., Griffioen, J. & Wassen, M.J., 2017. Environmental geochemistry of a mega beach nourishment in the Netherlands: monitoring freshening and oxidation processes. Applied Geochemistry 80: 72-89. DOI: 10.1016/j.apgeochem.2017.02.003

Pittman, E. D., Larese, R. E., Heald, M. T., Houseknecht, D. W. & Pittman, E. D. (eds), 1992. Clay coats: occurrence and relevance to preservation of porosity in sand-stones. Origin, Diagenesis, and Petrophysics of Clay Minerals in Sandstones. SEPM Special Publication 47: 241-255. DOI: 10.2110/pec.92.47.0241

Platt, J.D., 1994. Geochemical evolution of pore waters in the Rotliegend (Early Permian) of northern Germany. Marine and Petroleum Geology 11 (1): 66-78. DOI: 10.1016/0264-8172(94)90010-8

Pöppelreiter, M.C., Borkhataria, R., Aigner, T. & Pipping, K., 2005. Production from Muschelkalk carbonates (Triassic, NE Netherlands): unique play or overlooked opportunity? In: Doré, A.G. & Vining, B.A. (eds): Petroleum Geology: North-West Europe and Global Perspectives - Proceedings of the 6th Petroleum Geology Conference. The Geological Society (London): 299-314.

Pöppelreiter, M.C., Simone, A. & Hoetz, G., 2004. Reservoir characteristics of intracontinental carbonate ramp deposits - Upper Muschelkalk, Middle Triassic, NE Netherlands. Netherlands Journal of Geosciences 83 (1): 1-16. DOI: 10.1017/S0016774600020424

Postma, D., 1982. Pyrite and siderite formation in brackish and fresh water swamp sediments. American Journal of Science 282: 1151-1183.

Price, F.T. & Casagrande, D.J., 1991. Sulfur distribution and isotopic composition in peats from the Okefenokee Swamp, Georgia and the Everglades, Florida. International Journal of Coal Geology 17: 1-20. DOI: 10.1016/0166-5162(91)90002-Z

Purvis, K. & Okkerman, J.A., 1996. Inversion of reservoir quality by early diagenesis: an example from the Triassic Buntsandstein, offshore the Netherlands. In: Rondeel, H.E., Batjes, D.A.J. & Nieuwenhuijs, W.H. (eds): Geology of gas and oil under the Netherlands. Kluwer Academic Publishers (Dordrecht): 179-189. DOI: 10.1007/978-94-009-0121-6_177

Pye, K. & Krinsley, D.H., 1986. Diagenetic carbonate and evaporite minerals in Rotliegend aeolian sandstones of the southern North Sea: their nature and relationship to secondary porosity. Clay Minerals 21: 443-457. DOI: 10.1180/claymin.1986.021.4.03

Pye, K. & Tsoar, H., 1987. The mechanics and geological implications of dust transport and deposition in deserts with particular reference to loess formation and dune sand diagenesis in the northern Negev, Israel. Geological Society, London, Special Publications 35: 139-156. DOI: 10.1144/GSL.SP.1987.035.01.1

Regeling bodemkwaliteit, 2007. Staatscourant nr. 247: 91 pp.

Reijers, T.J.A., 2012. Sedimentology and diagenesis as 'hydrocarbon exploration tools'in the Late Permian Zechstein-2 Carbonate Member (NE Netherlands). Geologos 18 (3): 163-195. DOI: 10.2478/v10118-012-0009-x

Reijmer, J.J.G., Ten Veen, J.H., Jaarsma, B. & Boots, R., 2017. Seismic stratigraphy of Dinantian carbonates in the southern Netherlands and northern Belgium. Netherlands Journal of Geosciences 96 (4): 353-379. DOI: 10.1017/njg.2017.33

Reimann, C., Demetriades, A., Filzmoser, P. & O'Connor, P. (eds), 2014a. Chemistry of Europe's Agricultural Soils, Part B. Geologisches Jahrbuch Reihe B (102).

Reimann, C., Demetriades, A., Filzmoser, P. & O'Connor, P. (eds),

2014b. Chemistry of Europe's Agricultural Soils, Part A. Geologisches Jahrbuch Reihe B (103).

Reinders, G., 1902. Mededeeling omtrent de verspreiding van het deels poedervormig deels pijpvormig ijzeroer in de provinciën Groningen en Drente. Verhandelingen der Koininklijke Akademie van Wetenschapopen (Amsterdam) 9 (2): 3-18.

Remmelts, G., Nelskamp, S., De Jager, J. & Geluk, M.C., 2025. Petroleum Geology. In: Ten Veen, J.H., Vis, G.-J., De Jager, J. & Wong, Th.E. (eds): Geology of the Netherlands, second edition. Amsterdam University Press (Amsterdam): 535-575. DOI: 10.5117/9789463728362_ch14

Ritsema, C.J. & Groenenberg, J.E., 1993. Pyrite oxidation, carbonate weathering, and gypsum formation in a drained potential acid sulfate soil. Soil Science Society of America Journal 57: 968-976. DOI: 10.2136/sssaj1993.03615995005700040015x

Roman, A., 2004. Sequenzstratigraphie und Fazies des Unteren und Mittleren Buntsandsteins im östlichen Teile des Germanischen Beckens (Deutschland, Polen), PhD thesis University of Halle-Wittemberg (Germany): 266 pp.

Roskam, G. & Griffioen, J., 2011. Een verkennende studie naar sporenelementen in pyriet in Nederlandse geologische afzettingen, TNO, Report No. TNO-060-UT-2011-02229: 26 pp.

Rudnick, R.L. and Gao, S., 2014. Composition of the Continental Crust. In: Holland, H.D. and Turekian, K.K. (eds): Treatise on Geochemistry. Elsevier (Oxford): 1-51. DOI: 10.1016/B978-0-08-095975-7.00301-6

Saaltink, M.W., Ayora, C., Stuyfzand, P.J. & Timmer, H., 2003. Analysis of a deep well recharge experiment by calibrating a reactive transport model with field data. Journal of Contaminant Hydrology 65: 1-18. DOI: 10.1016/S0169-7722(02)00236-X

Salminen, R. (ed.), Batista, M.J., Bidovec, M., Demetriades, A., De Vivo, B., De Vos, W., Duris, M., Gilucis, A., Gregorauskiene, V., Halamic, J., Heitzmann, P., Lima, A., Jordan, G., Klaver, G., Klein, P., Lis, J., Locutura, J., Marsina, K., Mazreku, A., O'Connor, P.J., Olsson, S.Å., Ottesen, R.-T., Petersell, V., Plant, J.A., Reeder, S., Salpeteur, I., Sandström, H., Siewers, U., Steenfelt, A. & Tarvainen, T., 2005. Geochemical Atlas of Europe. Part 1: Background Information, Methodology and maps. Geological Survey of Finland.

Salminen, R., Gregorauskiene, V. & Tarvainen, T., 2008. The normative mineralogy of 10 soil profiles in Fennoscandia and north-western Russia. Applied Geochemistry 23: 3651-3665. DOI: 10.1016/j.apgeochem.2008.09.007

Salomons, W., 1975. Chemical and isotopic composition of carbonates in recent sediments and soils from western Europe. Journal of Sedimentary Petrology 45: 440-449.

Scholle, P.A., 1977. Chalk diagenesis and its relation to petroleum exploration: oil from chalks, a modern miracle? American Association of Petroleum Geologists Bulletin 61 (7): 982-1009. DOI: 10.1306/C1EA43B5-16C9-11D7-8645000102C1865D

Schröder, T.J., Hiemstra, T., Vink, J.P.M. & Van der Zee, S.E.A.T.M., 2005. Modeling of the solid-solution partitioning of heavy metals and arsenic in embanked flood plain soils of the rivers Rhine and Meuse. Environmental Science & Technology 39: 7176-7184. DOI: 10.1021/es048154s

Schuttenhelm, R.T.E. & Laban, C., 2005. Heavy minerals, provenance and large scale dynamics of seabed sands in the Southern North Sea: Baak's (1936) heavy mineral study revisited. Quaternary International 133 (1): 179-193. DOI: 10.1016/j.quaint.2004.10.012

Seemann, U., 1982. Depositional facies, diagenetic clay minerals and reservoir quality of Rotliegend sediments in the southern Permian Basin (North Sea): A review. Clay Minerals 17: 55-67.

Šegvić, B., Benvenuti, A. & Moscariello, A., 2016. Illite-smectite-rich clay parageneses from Quaternary tunnel valley sediments of the Dutch southern North Sea–mineral origin and paleoenvironment implications. Clays and Clay Minerals 64: 608-627. DOI: 10.1346/000986016820500438

Sibley, D.F. & Blatt, H., 1976. Intergranular pressure solution and cementation of the Tuscarora orthoquartzite. Journal of Sedimentary Research 46: 881-896. DOI: 10.1306/212F7081-2B24-11D7-8648000102C1865D

Singer, A., 1984. The paleoclimatic interpretation of clay minerals in sediments—a review. Earth-Science Reviews 21: 251-293. DOI: 10.1016/0012-8252(84)90055-2

Spijker, J., 2005. Geochemical patterns in the soils of Zeeland. Natural variability versus anthropogenic impact. PhD thesis Utrecht University (the Netherlands): 205 pp.

Spijker, J., Vriend, S.P. & Van Gaans, P.F.M., 2005. Natural and anthropogenic patterns of covariance and spatial variability of minor and trace elements in agricultural topsoil. Geoderma 127: 24-35. DOI: 10.1016/j.geoderma.2004.11.002

Staring, W.C.H., 1856. De bodem van Nederland. De zamenstelling en het ontstaan der gronden in Nederland ten behoeve van het algemeen beschreven. Deel I. Kruseman (Haarlem): 441 pp.

Tabor, N.J. & Poulsen, C.J., 2008. Palaeoclimate across the Late Pennsylvanian–Early Permian tropical palaeolatitudes: A review of climate indicators, their distribution, and relation to palaeophysiographic climate factors. Palaeogeography, Palaeoclimatology, Palaeoecology 268: 293-310. DOI: 10.1016/j.palaeo.2008.03.052

Tatzel, M., Dunkl, I. & Eynatten, H., 2017. Provenance of Palaeo-Rhine sediments from zircon thermochronology, geochemistry, U/Pb dating and heavy mineral assemblages. Basin Research 29: 396-417. DOI: 10.1111/bre.12155

Taylor, S.R. & Lapré, J.F., 1987. North Sea chalk diagenesis: its effect on reservoir location and properties. Conference on Petroleum Geology of North West Europe 3: 483-495.

Tebbens, L.A., Veldkamp, A. & Kroonenberg, S.B., 2000. Natural composition variation of the River Meuse (Maas) suspended load: a 13 ka bulk geochemical record from the upper-Kreftenheye and Betuwe Formations in northern Limburg. Netherlands Journal of Geosciences 79 (4): 391-409. DOI: 10.1017/S0016774600021909

Tebbens, L.A., Veldkamp, A. & Kroonenberg, S.B., 1998. The impact of climate change on the bulk and clay geochemistry of fluvial residual channel infillings: the Late Weichselian and Early Holocene River Meuse sediments (The Netherlands). Journal of Quaternary Science 13: 345-356. DOI: 10.1002/(SICI)1099-1417(199807/08)13:4<345::AID-JQS384>3.0.CO;2-B

Terwindt, J.H.J., 1977. Mud in the Dutch delta area. Geologie & Mijnbouw 56: 203-210.

Tipping, E., Somerville, C.J. & Luster, J., 2016. The C:N:P:S stoichiometry of soil organic matter. Biogeochemistry 130: 117-131. DOI: 10.1007/s10533-016-0247-z

Trabucho Alexandre, J.P. & Wong, Th. E., 2025. Rhaetian to Middle Jurassic. In: Ten Veen, J.H., Vis, G.-J., De Jager, J. & Wong, Th.E. (eds): Geology of the Netherlands, second edition. Amsterdam University Press (Amsterdam): 185-209. DOI: 10.5117/9789463728362_ch06

Van Andel, T.H., 1950. Provenance, transport and deposition of Rhine sediments : a heavy mineral study on river sands from the drainage area of the Rhine. PhD thesis, Rijksuniversiteit Groningen (the Netherlands): 129 pp.

Van Baren, F.A., 1934. Het voorkomen en de betekenis van kali-houdende mineralen in Nederlandse gronden. Landbouwuniversiteit Wageningen: 120 pp.

Van Beek, C.G.E.M., Hettinga, F.A.M. & Straatman, R., 1989. Release of heavy metals in groundwater due to manure spreading. In: D.L. Miles (ed.): Proceeding of the 6th Int. Symp. on Water–Rock Interaction. Balkema Publishers (The Netherlands): 703-706.

Van Bemmelen, J.M. & Reinders, G., 1901. Twee nieuwe vindplaatsen van moerasijzersteen in en onder veen. Verhandelingen der Koininklijke Akademie van Wetenschapopen (Amsterdam) 9: 406-418.

Van Bergen, F. & Koster, K., 2025. Peat, coal and coalbed methane. In: Ten Veen, J.H., Vis, G.-J., De Jager, J. & Wong, Th.E. (eds): Geology of the Netherlands, second edition. Amsterdam University Press (Amsterdam): 577-601. DOI: 10.5117/9789463728362_ch15

Van Bergen, M.J., Vis, G.-J., Sissingh, W., Koornneef, J. & Brouwers, I., 2025. Magmatism in the Netherlands: expression of the northwest European rifting history. In: Ten Veen, J.H., Vis, G.-J., De Jager, J. & Wong, Th.E. (eds): Geology of the Netherlands, second edition. Amsterdam University Press (Amsterdam): 393-455. DOI: 10.5117/9789463728362_ch11

Van Bruggen, M., Swartjes, F.A., Janssen, P.J.C.M., Pit, I.R., Griffioen, J. & Spijker, J., 2014. Beoordeling gezondheidsrisico's van arseen op de Zandmotor. RIVM, Universiteit Utrecht, Deltares, briefrapport no. 2014-0063: 42 pp.

Van Dam, D. & Pons, L.J., 1972. Micropedological observations on pyrite and its pedological reaction products. Proc. Int. Symp. On Acid Sulphate Soils, 169-196. (Wageningen): 169-196.

Van der Grift, B. & Griffioen, J., 2008. Modelling assessment of regional groundwater contamination due to historic smelter emissions of heavy metals. Journal of Contaminant Hydrology 96: 48-68. DOI: 10.1016/j.jconhyd.2007.10.001

Van der Salm, C., 2001. Assessment of the regional variation in weathering rates of loess and clay soils in the Netherlands. Water, Air, and Soil Pollution 131: 217-243. DOI: 10.1023/A:1011912120842

Van der Sleen, W.G.N., 1912. Bijdrage tot de kennis der chemische samenstelling van het duinwater in verband met de geo-mineralogische gesteldheid van den bodem. PhD thesis University of Amsterdam (the Netherlands): 157 pp.

Van der Veer, G., 2006. Geochemical soil survey of the Netherlands. Atlas of major and trace elements in topsoil and parent material; assessment of natural and anthropogenic enrichment factors. PhD thesis Utrecht University (the Netherlands): 345 pp.

Van Gaans, P.F.M., Griffioen, J., Mol, G. & Klaver, G.T., 2011. Geochemical reactivity of subsurface sediments as potential buffer to anthropogenic inputs: a strategy for regional characterization in the Netherlands. Journal of Soils and Sediments 11: 336-351. DOI: 10.1007/s11368-010-0313-4

Van Gaans, P.F.M., Spijker, J., Vriend, S.P. & De Jong, J.N., 2007. Patterns in soil quality: Natural geochemical variability versus anthropogenic impact in soils of Zeeland, The Netherlands. International Journal Geographical Information Science 21: 569-587. DOI: 10.1080/13658810601064900

Van Helvoort, P.J., Filzmoser, P. & Van Gaans, P.F., 2005. Sequential factor analysis as a new approach to multivariate analysis of heterogeneous geochemical datasets: an application to a bulk chemical characterization of fluvial deposits (Rhine-Meuse delta, The Netherlands. Applied Geochemistry 20: 2233-2251. DOI:10.1016/j.apgeochem.2005.08.009

Van Hulten, F.F.N., 2012. Devono-carboniferous carbonate platform systems of the Netherlands. Geologica Belgica 15 (4): 284-296.

Van Lochem, H., Vis, G.-J. & Jagt, J.W.M., 2025. Late Cretaceous. In: Ten Veen, J.H., Vis, G.-J., De Jager, J. & Wong, Th.E. (eds): Geology of the Netherlands, second edition. Amsterdam University Press (Amsterdam): 253-291. DOI: 10.5117/9789463728362_ch08

Van Oversteeg, K., Lipsey, L., Pluymaekers, M., Van Wees, J.-D., Fokker, P. & Spiers, C., 2014. Fracture Permeability Assessment in Deeply Buried Carbonates and Implications for Enhanced Geothermal Systems: Inferences from a Detailed Well Study at Luttelgeest-01, the Netherlands. Proceedings Thirty-Eighth Workshop on Geothermal Reservoir Engineering, Stanford University, Stanford, California.

Van Pruissen, F.G.M. & Zuurdeeg, B.W., 1988. Hoge metaalgehalten in ijzeroerknollen in de Nederlandse bodem. Milieutechniek 3: 84-91.

Vandenberghe, N., Harris, W.B., Wampler, J.M., Houthuys, R., Louwye, S., Adriaens, R., Vos, K., Lanckacker, T., Matthijs, J., Deckers, J., Verhaegen, J., Laga, P., Westerhoff, W. & Munsterman, D., 2014. The implications of K-Ar glauconite dating of the Diest Formation on the paleogeography of the Upper Miocene in Belgium. Geologica Belgica 17 (2): 161-174. DOI: 10.20341/gb.2020.010

Vermooten, S., Griffioen, J., Heerdink, R. & Visser, A., 2011. Geochemische karakterisering van de geotop van Zeeland. TNO, Report No. TNO-034-UT-2010-02397: 138 pp.

Vermooten, S. & Gunnink, J., 2008. Arsenic in the Dutch coastal provinces. Netherlands national Committee of the International Association of Hydrologists. Proceedings Seminar (Utrecht): 93-101.

Verreussel, R.M.C.H., Munsterman, D.K., Bouroullec, R. & Houben, A.J.P., 2025. Late Jurassic - Early Cretaceous. In: Ten Veen, J.H., Vis, G.-J., De Jager, J. & Wong, Th.E. (eds): Geology of the Netherlands, second edition. Amsterdam University Press (Amsterdam): 211-251. DOI: 10.5117/9789463728362_ch07

Vincent, B., Waters, J., Witkowski, F., Daniau, G., Oxtoby, N., Crowley, S. & Ellam, R., 2018. Diagenesis of Rotliegend sandstone reservoirs (offshore Netherlands): The origin and impact of dolomite cements. Sedimentary Geology 373 (Journal Article): 272-291. DOI: 10.1016/j.sedgeo.2018.06.012

Vis, G.-J., Houben, A.J.P., Debacker, T.N. & Geel, C.R., 2025. The geological foundation of the Netherlands: the early Carboniferous, Devonian and older. In: Ten Veen, J.H., Vis, G.-J., De Jager, J. & Wong, Th.E. (eds): Geology of the Netherlands, second edition. Amsterdam University Press (Amsterdam): 53-93. DOI: 10.5117/9789463728362_ch02

Walderhaug, O. & Bjørkum, P.A., 2003. The effect of stylolite spacing on quartz cementation in the Lower Jurassic Stø Formation, southern Barents Sea. Journal of Sedimentary Research 73: 146-156. DOI: 10.1306/090502730146

Walraven, N., Van Gaans, P.F.M., Van der Veer, G., VanOs, B.J.H., Klaver, G., Vriend, S.P., Middelburg, J.J. & Davies, G.R., 2013b. Lithologically inherited variation in Pb isotope ratios in sedimentary soils in The Netherlands. Applied Geochemistry 37: 228-241.

Walraven, N., Van Gaans, P.F.M., Van der Veer, G., VanOs, B.J.H., Klaver, G., Vriend, S.P., Middelburg, J.J. & Davies, G.R., 2013b. Lithologically inherited variation in Pb isotope ratios in sedimentary soils in The Netherlands. Applied Geochemistry 37: 228-241.

Walraven, N., Van Os, B.J.H., Klaver, G., Middelburg, J.J. & Davies, G.R., 2014a. Reconstruction of historical atmospheric Pb using Dutch urban lake sediments: A Pb isotope study. Science of the Total Environment 484: 185-195. DOI: 10.1016/j.scitotenv.2014.02.062

Walraven, N., Van Os, B.J.H., Klaver, G., Middelburg, J.J. & Davies, G.R., 2014b. The lead (Pb) isotope signature, behaviour and fate of traffic-related lead pollution in roadside soils in The Netherlands. Science of the Total Environment 472: 888-900. DOI: 10.1016/j.scitotenv.2013.11.110

Wartel, S., 1977. Composition, transport and origin of sediments in the Schelde estuary. Geologie & Mijnbouw 56: 219-233.

Wei, W., Littke, R. & Swennen, R., 2021. Depositional environment and characteristics of organic matter of Namurian Shale, Namur Synclinorium and Campine Basin (Belgium and the S-Netherlands). Abstract Book 7th Geologica Belgica Meeting (Tervuren): 176-177.

Wei, W. & Swennen, R., 2022. Sedimentology and lithofacies of organic-rich Namurian Shale, Namur Synclinorium and Campine Basin (Belgium and S-Netherlands). Marine and Petroleum Geology 138: 105553. DOI: 10.1016/j.marpetgeo.2022.105553

Westerhoff, W.E., 2009. Stratigraphy and sedimentary evolution - The Lower Rhine-Meuse system during the Late Pliocene and Early Pleistocene (southern North Sea Basin). PhD thesis Vrije Universiteit Amsterdam (the Netherlands): 168 pp.

Wieder, R.K. & Lang, G.E., 1988. Cycling of inorganic and organic sulfur in peat from Big Run Bog, West Virginia. Biogeochemistry 5: 221-242. DOI: 10.1007/BF02180229

Zeelmaekers, E., 2011. Computerized qualitative and quantitative clay mineralogy. Introduction and application to known geological cases. PhD thesis Katholieke Universiteit Leuven (Belgium): 414 pp.

Zhang, Y.C., Slomp, C.P., Broers, H.P., Passier, H.F. & Van Cappellen, P., 2009. Denitrification coupled to pyrite oxidation and changes in groundwater quality in a shallow sandy aquifer. Geochimica et Cosmochimica Acta 73: 6716-6726. DOI: 10.1016/j.gca.2009.08.026

Zijlstra, H.J.P., 1987. Early diagenetic silica precipitation, in relation to redox boundaries and bacterial metabolism, in late cretaceous chalk of the Maastrichtian type locality. Geologie En Mijnbouw 66 (4): 343-355.

Zuur, A.J., 1936. Over de bodemkundige gesteldheid van de Wieringermeer. Landsdrukkerij (Den Haag): 113 pp.

PI-G14-PB

14

Petroleum geology

Gijs Remmelts, Susanne Nelskamp, Jan de Jager & Mark C. Geluk†*

Corresponding author: susanne.nelskamp@tno.nl

ABSTRACT

A great variety of proven hydrocarbon plays and trap styles is present in the sedimentary succession in the subsurface of the Netherlands. This sedimentary succession started to develop in the middle Paleozoic and has been deformed by several periods of tectonic structuration.

In much of the subsurface, the thick Permian Zechstein salt provides an effective seal between a prolific Paleozoic gas system and a mixed, oil and gas-prone Mesozoic hydrocarbon system.

The dominating Rotliegend play is formed by a near-ideal superposition of i) a thick upper Carboniferous (Westphalian) succession with abundant coal measures as the source rocks for gas, ii) well developed fluvial and eolian reservoir sandstones of the Permian Upper Rotliegend Group, and iii) the excellent seal of the Zechstein salt. The giant Groningen field is by far the largest representative of this play and represents two thirds of the total recoverable Dutch gas reserves, containing an initial recoverable gas volume of ca. 2800×10^9 m^3. Production of the Groningen gas was very profitable for decades, but eventually the impact of production induced seismicity resulted in a premature cessation of production.

Triassic plays are second in importance with respect to proven gas accumulations. Relatively minor gas reserves occur in Permian Zechstein carbonates, Upper Jurassic and Lower Cretaceous sandstones, Upper Cretaceous chalk as well as in shallow, unconsolidated sands of Cenozoic age.

Producible oil occurs within Upper Jurassic and Lower Cretaceous rift-basin sediments in a variety of sandstone reservoirs and trap styles. Minor amounts of oil have also been produced from the Upper Cretaceous chalk and the Triassic sandstones. The exploration of the Dutch hydrocarbon plays is in a mature stage and many of the current gas fields will be depleted in the next decade. The urge for energy transition is growing and moreover diminishing public acceptance is restricting exploration and production activities. Consequently, exploration and production are concentrating on the offshore area. The role of fossil fuels in the energy system will decrease significantly in the coming decades and is expected to cease (almost) completely in 2050.

<< G14-B production platform (now Eni Energy Netherlands B.V.), producing gas from the Triassic Volpriehausen Formation in the Dutch northeastern offshore. Photo: Neptune Energy.

DOI: 10.5117/9789463728362_ch14

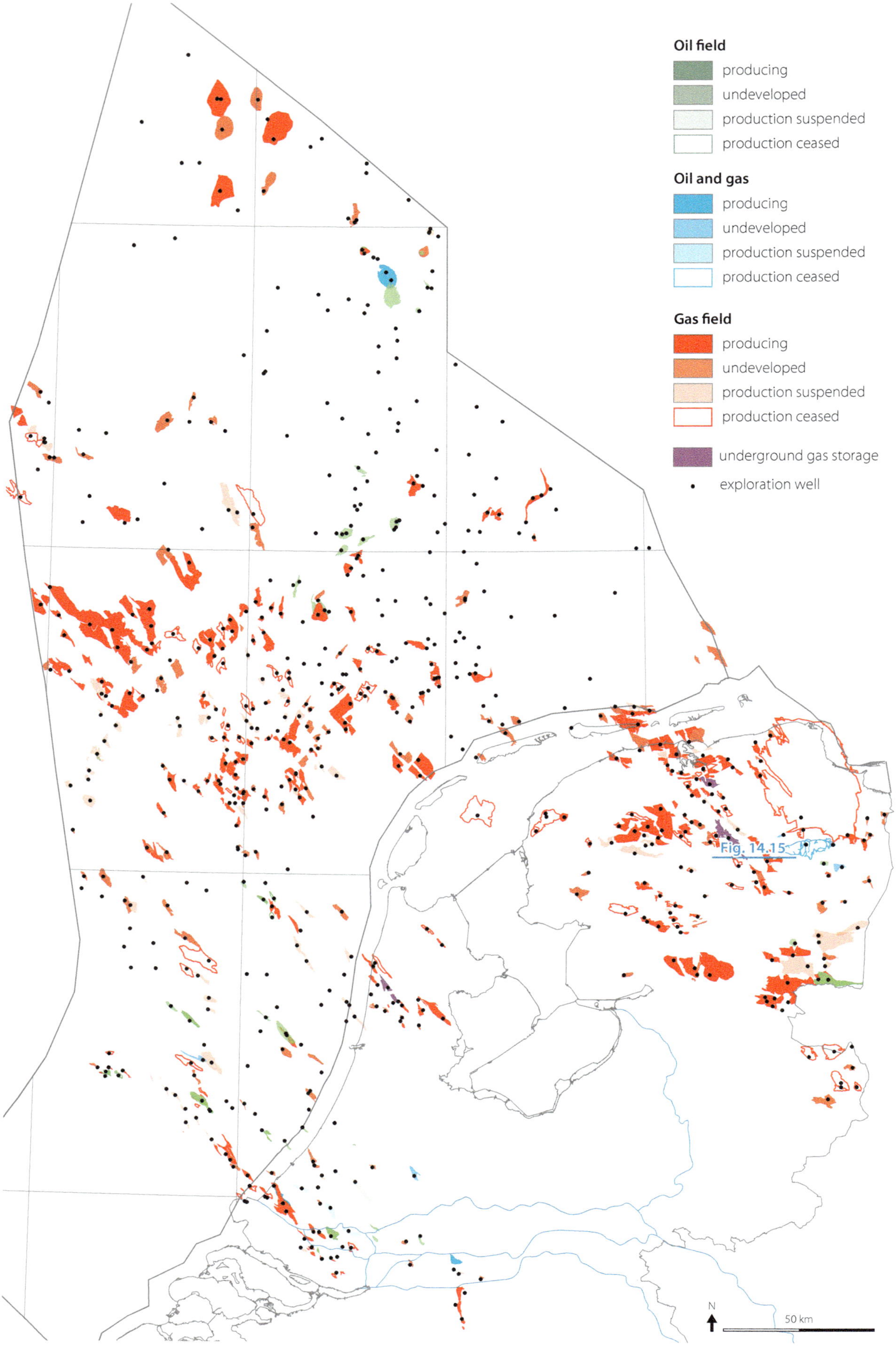
Oil field
producing
undeveloped
production suspended
production ceased
Oil and gas
producing
undeveloped
production suspended
production ceased
Gas field
producing
undeveloped
production suspended
production ceased
underground gas storage
exploration well
Fig. 14.15
N
50 km

Introduction

In terms of natural resources, The Netherlands is a natural gas country and the history of Dutch natural gas is dominated by the giant Groningen field in the northeast of the country (Fig. 14.1). The Netherlands has produced over 3500 x 10^9 m^3 of natural gas since the first gas field came on stream in 1951. Approximately 2100 x 10^9 m^3 (60%) of that volume came from the Groningen field.

Since the discovery of the Groningen field in 1959, the Netherlands has developed into the main gas 'hub' of Western Europe. At the time all exploration activities were focussed on finding oil as the domestic gas market was deemed to be saturated with so-called 'town gas'. In fact, before the 1960s, encountering gas was considered an exploration risk. Once the size of Groningen was recognized, the European energy supply changed fundamentally with natural gas dominating the Dutch energy market as well as much of the European market. The discovery of the Groningen field close to a large potential market served as a magnet for the oil & gas industry to explore the rest of the onshore and later also the North Sea (Van de Graaff, 2001).

For 60 years the Groningen field outnumbered other gas fields in terms of reserves- and production. These other fields were therefore called 'small fields' as they were at least 40 to even 1000 times smaller than the Groningen field. Groningen originally contained almost 2800 x 10^9 m^3 producible gas. The second largest gas field (Annerveen – which adjoins the Groningen field) contains just over 70 x 10^9 m^3 of gas (2.5% of the Groningen field). Most of the small fields initially contained in between 1 to 10 x 10^9 m^3 of gas. In total 499 gas discoveries have been made of which 386 were developed as fields and for another 26 development is pending (Ministry of Economic Affairs and Climate Policy (MEACP, 2022)). Using the Groningen field as a swing producer to accommodate seasonal variations in demand enabled all the other fields to optimize their production and economics. This so-called 'small fields policy' (*kleineveldenbeleid*) strongly contributed to the success of the development of the oil and gas resources in the country. Most gas is contained in upper Permian sandstones of the Upper Rotliegend Group (Bouroullec & Geel, 2025, this volume). Other hydrocarbon plays, ranging from Carboniferous to Quaternary in age, are present as well (Fig. 14.2).

Since the late 1980s, however, gas production has induced over 1500 earth tremors with magnitude >1 (larger than 1 Richter) in the Groningen field (Dost et al., 2025, this volume) and eventually these earth tremors resulted in serious damage to many houses and buildings and, more importantly, in a severe feeling of insecurity amongst the people living above the gas field. The 2012 Huizinge earthquake with a magnitude of 3.6 on the Richter scale (M_L= 3.6) was the largest one ever registered and was recognized as the tipping point. The government committed to an early closure of the gas field in 2023/2024 to end the seismic threat. Consequently, approximately 500 x 10^9 m^3 of technically recoverable gas will be left stranded. Depletion of the remaining gas resources from the small fields is expected to last at least for another two decades. Natural gas is considered the preferred fossil energy source in the transition towards a sustainable energy system thanks to its relatively low carbon footprint compared to the other fossil fuel types.

The Netherlands has a modest share of oil as well. The Schoonebeek oil field is the largest onshore oil field of western Europe. Other oil fields occur onshore in the western Netherlands and in the Dutch offshore (Fig. 14.1). Up to 2022, 52 oil discoveries were reported (MEACP, 2022). Of these, 23 oil fields have been developed, of which 10 are still producing. Another four may come on stream within five years. Development of the other discoveries is presently not economically viable. Remaining reserves and contingent resources (according to the SPE Petroleum Resource Management System (www.spe.org/en/industry/reserves) add up to a modest 34.8 million m^3 of oil.

Throughout this paper reference is made to structural units that are shown in Figure 14.3. Locations of individual fields and wells mentioned in the text can best be located on the maps of www.nlog.nl. This website shows a complete and updated overview of these data.

Data

As a result of exploration and development activities, a wealth of subsurface data is available. Currently, a total of ca. 64,000 km^2 (ca. 65%) of the total Dutch on- and offshore surface territory is covered with high-quality 3D seismic data (Fig. 14.4). Significant parts of the remaining 35% are covered by 2D surveys (www.nlog.nl). This makes the Netherlands rank amongst the countries that have the seismically best imaged subsurface geology in the world (Dessens, 1996). The seismic data set allows a good appreciation of the vast variety of geological settings and

← ***Figure 14.1.*** *Gas and oil fields identified by status as well as underground gas storage locations and exploration wells. Total reserves are dominated by the giant Groningen gas field in the northeast onshore. (Gas and oil fields from www.nlog.nl, retrieved on 13.12.2022).*

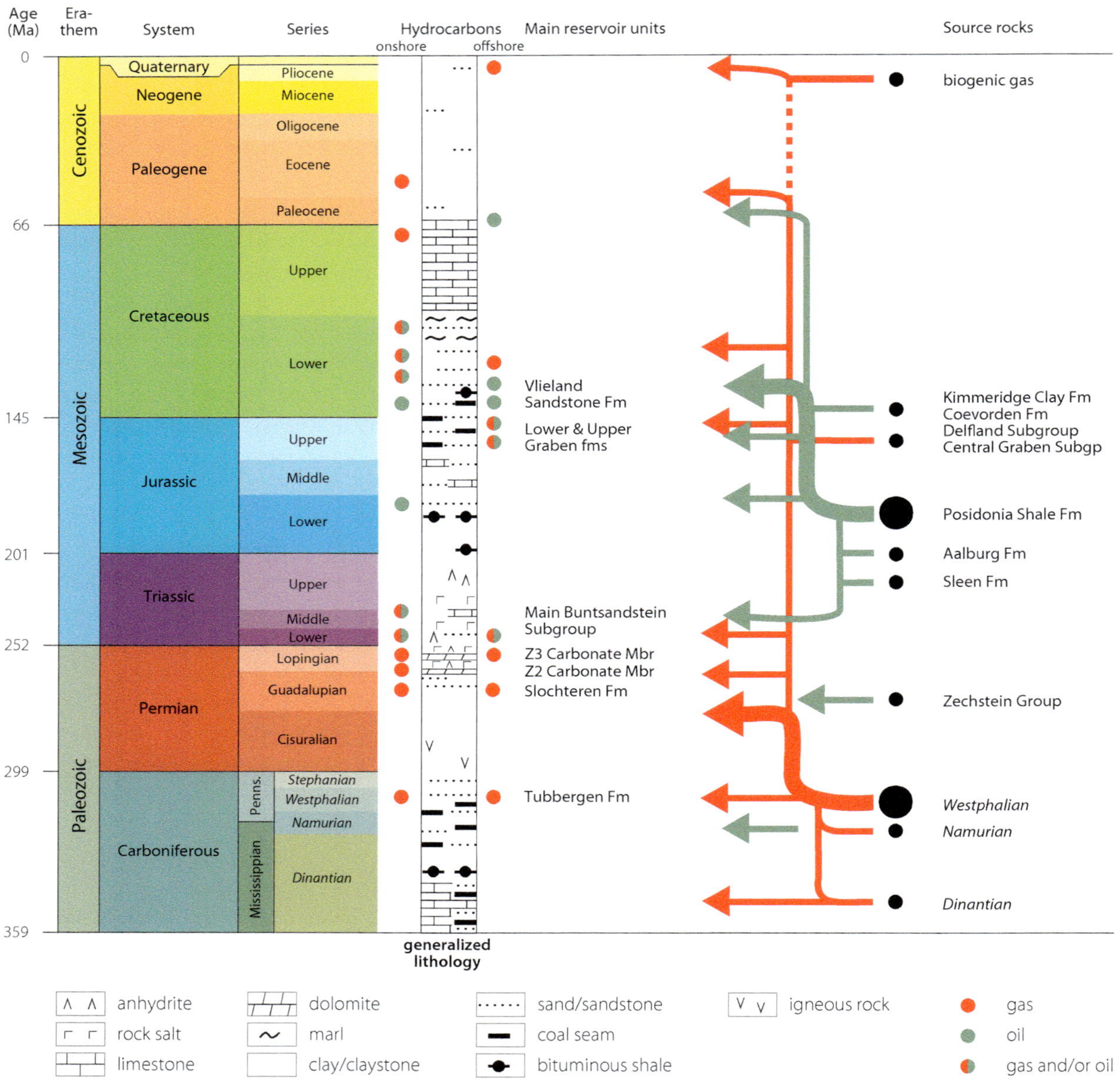

Figure 14.2. Hydrocarbon systems in the Dutch subsurface. Arrows show from which source rocks the main reservoirs have been charged with gas and/or oil. The Upper Permian Zechstein salt, present in much of the subsurface, provides a regional seal between a Paleozoic gas and a Mesozoic oil and gas system. Not acknowledged here is that with time probably >95% of the generated hydrocarbons escaped into the biosphere (see Textbox). Modified from De Jager and Geluk (2007).

structural styles in the Dutch subsurface. On January 1st, 2022, the oil and gas exploration & production (E&P) industry had drilled a total of 1417 exploration wells (Fig. 14.5), 498 appraisal wells and 2382 production wells (MEACP, 2022). All seismic data, well-log measurements and rock samples (including core analysis) are made publicly accessible after a five-year confidentiality period. Much of the data is freely downloadable at www.nlog.nl. The quest for geothermal energy, the plans for underground storage of carbon-dioxide as well as the foreseen need for subsurface storage of hydrogen and compressed air, they all benefit substantially from these data.

Exploration and production activities over the years

A brief summary of the history of oil and gas exploration in the Netherlands follows below. Reference is made to the more extensive overviews published by Knaap & Coenen (1987), Breunese & Rispens (1996), Glennie (2001), Breunese et al., (2010), Kombrink et al. (2012) and Doornenbal et al. (2019). An annual update on E&P activities is published by the Ministry of Economic Affairs and Climate Policy, under the title 'Natural resources and geothermal energy in the Netherlands' (available from www.nlog.nl/en/annual-reports).

The first Dutch oil was found in 1923, near Corle, close to the German border, where 1.5 barrels of oil (240 l) were recovered from Zechstein anhydrites and Carboniferous sandstones (Fig. 14.5). A blowout in 1938 at Bad Bentheim in Germany, ca. 50 km NE of Corle, indicated the presence of gas at Zechstein level (Knaap & Coenen, 1987). The second indication for oil was found in 1938, during an exhibition in The Hague where the Bataafsche Petroleum Maatschappij, an operating company of Shell, drilled a demonstration well. At a depth of 460 m, the drillers unexpectedly encountered oil stains. The first commercial oil discovery dates from 1943 in the Lower Cretaceous sandstones of the Schoonebeek field in the southeast of the province of Drenthe at a depth of ca. 800 m. With an initial in-place volume of ca. 1 billion barrels (160 x 10^6 m^3), this is the largest onshore oil field in Western Europe. The Schoonebeek field was closed-in in 1996 as it became uneconomic after 25% of the viscous oil (25° API) had been produced. However, after a shut-in period of 15 years, production resumed in 2011 with the application of steam solution (solution gas drive) so as to deliver another 10-15% (NAM, 2008).

After World War II, exploration resumed with further oil discoveries in the West Netherlands Basin (e.g. the Rijswijk, IJsselmonde, Wassenaar, Ridderkerk and Rotterdam fields) and gas discoveries in the east of the country (e.g. the Coevorden, De Wijk, Wanneperveen and Tubbergen fields). Until the end of the 1950s Zechstein carbonates and Lower Cretaceous sandstones were considered the most prospective hydrocarbon reservoirs.

The giant Groningen gas field, which would change the exploration outlook forever, was discovered in 1959 with the Slochteren-1 well. It has repeatedly been reported that the sheer size of this field was initially not recognized. The discovery well's target was a small structural closure at the level of the basal Zechstein carbonates in which gas had already been encountered in southeast Drenthe. Unexpectedly, the underlying sandstones of the Slochteren Formation (Upper Permian) at approximately 3 km depth turned out to contain gas as well. Only after several

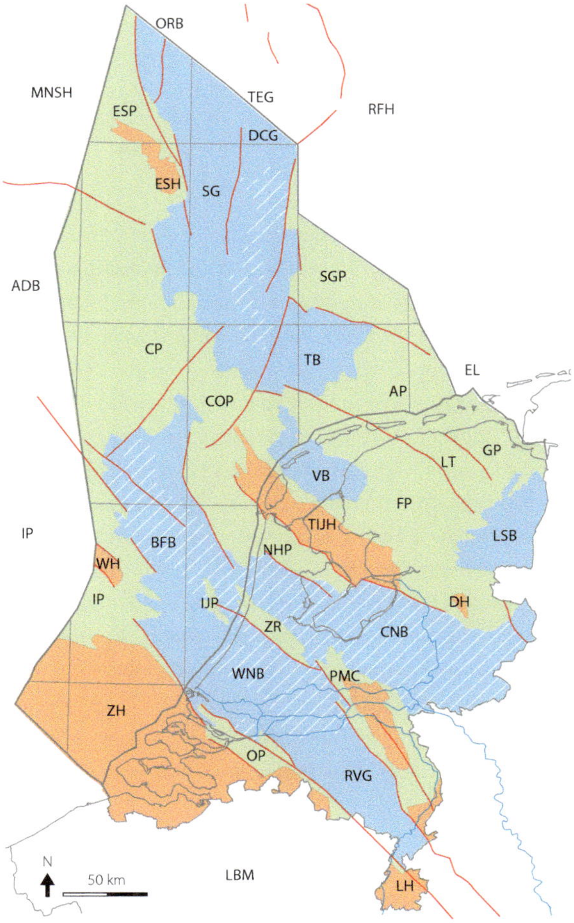

Acronym	Full name
ADB	Anglo-Dutch Basin*
AP	Ameland Platform
BFB	Broad Fourteens Basin
CNB	Central Netherlands Basin
COP	Central Offshore Platform
CP	Cleaverbank Platform
DCG	Dutch Central Graben
DH	Dalfsen High
EL	Ems Low*
ESH	Elbow Spit High
ESP	Elbow Spit Platform
FP	Friesland Platform
GP	Groningen Platform
IJP	IJmuiden Platform
IP	Indefatigable Platform
LBM	London-Brabant Massif
LH	Limburg High
LSB	Lower Saxony Basin
LT	Lauwerszee Trough
MNSH	Mid-North Sea High*
NHP	Noord Holland Platform
OP	Oosterhout Platform
ORB	Outer Rough Basin*
PMC	Peel-Maasbommel Complex
RFH	Ringkøbing Fyn High*
RVG	Roer Valley Graben
SG	Step Graben
SGP	Schill Grund Platform
TB	Terschelling Basin
TEG	Tail End Graben*
TIJH	Texel-IJsselmeer High
VB	Vlieland Basin
WH	Winterton High
WNB	West Netherlands Basin
ZH	Zeeland High
ZR	Zandvoort Ridge

* outside Dutch territory

Figure 14.3. *Main Mesozoic structural elements of the Dutch subsurface (after de Jager et al. 2025, this volume).*

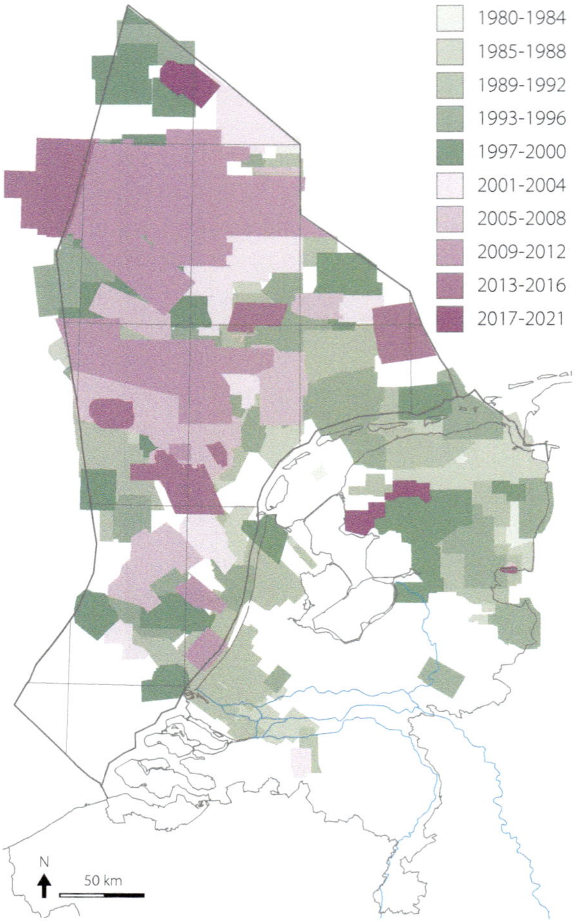

Figure 14.4. Coverage of 3D seismic surveys. In total ca. 64,000 km² of 3D seismic data has been acquired, covering ca. 65% of the Dutch on- and offshore areas.

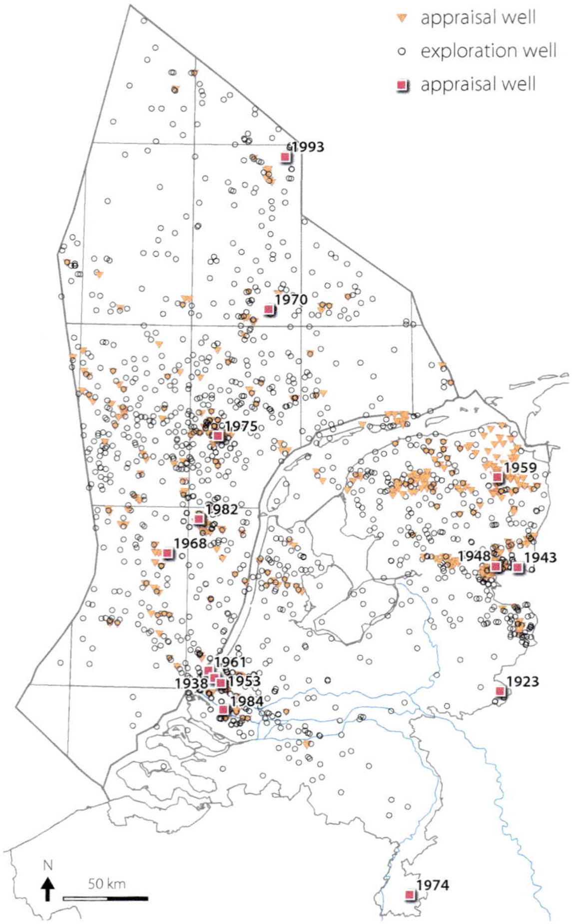

Figure 14.5. Hydrocarbon exploration and appraisal wells. A total of 1417 exploration wells and 498 appraisal wells had been drilled by the end of 2021. Also shown are the milestones in the Dutch exploration and production history (after PGK, 1993). 1923 = Corle oil show; 1938 = The Hague oil show; 1943 = Schoonebeek oil discovery (on stream in 1945); 1948 = Coevorden gas discovery (first production in 1951); 1953 = Rijswijk oil discovery (first production in 1954); 1959 = Slochteren gas discovery (Groningen field on stream in 1963); 1961 = first offshore well (Kijkduin Zee-1); 1968 = first offshore gas discovery (P6); 1970 = first offshore oil discovery (F18); 1974 = last coal mine in Zuid-Limburg closed; 1974 = first offshore gas production (L10); 1982 = first offshore oil production (Q1); 1984 = Botlek gas discovery in Rijswijk concession; 1987 = discovery of shallow gas in A18; 1993 =- F3 oil field on stream; 2017 = discovery Basal Slochteren Clastics (Upper Rotliegend Group).

wells were drilled it was realized that the small structure that was targeted, was in fact much larger with a length of more than 40 km and a width of almost 30 km. With initial recoverable reserves of ca. 2800 x 10^9 m^3 of gas, the Groningen gas field is by far the largest of Europe and it contained two thirds of the total recoverable gas resources in the Netherlands. Following the discovery of the Groningen field, the first careful steps into the offshore were made in the early 1960s with the near-coastal well Kijkduin Zee-1. Legislation to regulate offshore exploration activities became effective in the mid-sixties, the first offshore gas discovery was in block P6 in 1968 and was followed in 1970 by the first oil discovery in block F18. Offshore gas production started in 1975 from the large L10 Rotliegend gas field and the first offshore oil was delivered in 1982 from the Cretaceous in block Q1. The introduction of 3D seismic in the 1980s very much improved the exploration efficiency. In 1987 the first gas in Cenozoic reservoirs was discovered (A18). However, it took another twenty years before reservoirs from this play came on stream. Recently, recoverable gas was discovered in the Basal Slochteren Sandstones in the N blocks, which is an extension of the well-known Rotliegend play (Burgess et al., 2018). Although the first discovery (N05-01, 2017) appears very promising, the potential of this play remains uncertain.

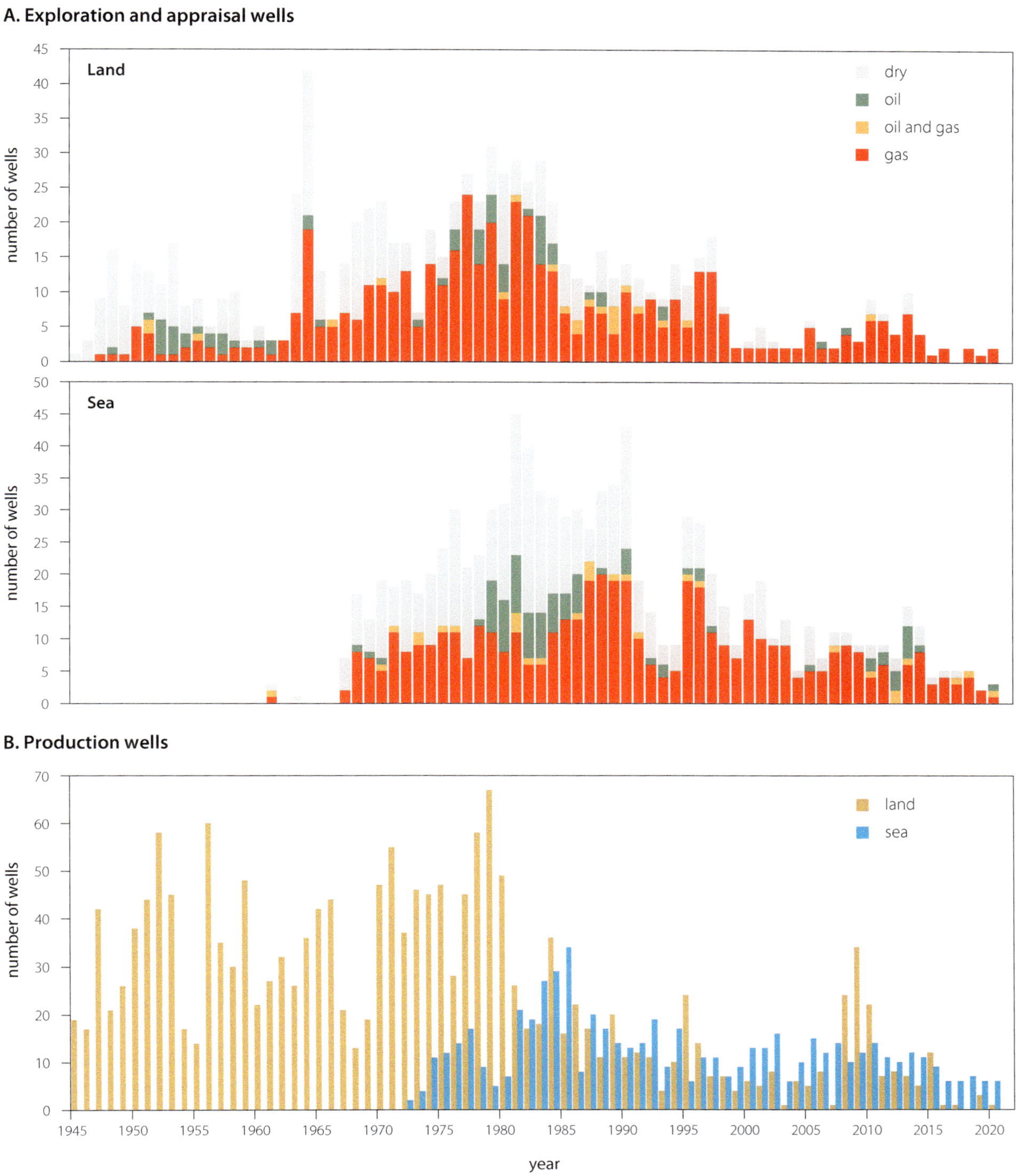

Figure 14.6. The onshore and offshore drilling sequence since 1946 reflects the exploration and on- and offshore appraisal history (a) and the overview of the consequent production wells (b). Source: MEACP (2022).

Drilling activity since 1946 reflects the E&P history as summarized above. Drilling started onshore with relatively low success ratios (Fig. 14.6a) with increased activity in de mid-sixties following the discovery of the Groningen field. By the end of the sixties offshore drilling commenced. Activities remained high during the 20th century with peak periods from 1975-1985 onshore and slightly later offshore (1985-1997). Since 2000, drilling activity has strongly declined, especially onshore. Most recent exploration can be categorized as near-field drilling, i.e. drilling either blocks adjacent to producing fields or prospects well within reach of existing infrastructure in order to lower the economic hurdle. This reflects the maturity of the Dutch petroleum province. Numerous production wells were

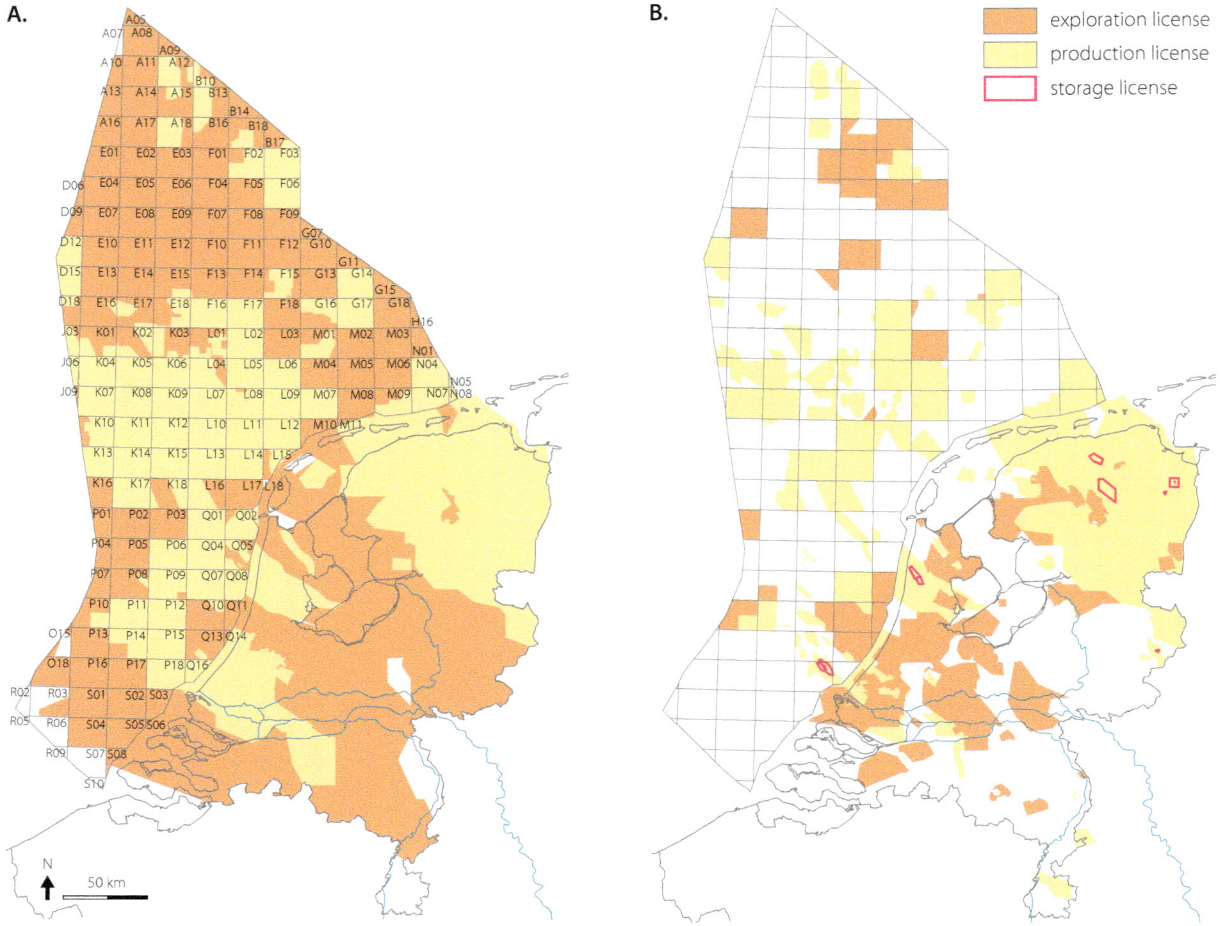

Figure 14.7. a) All license areas ever issued onshore and offshore the Netherlands. Total licensed area is 94,912 km². This equals 96.3% of the total Dutch onshore and offshore territory. b) License situation as of April 2023. Source: www.nlog.nl.

drilled up to the eighties to develop the large Groningen, Annerveen and Schoonebeek fields (Fig. 14.6b).

Since the beginning of the 21st century 111 gas discoveries have been made in the Netherlands, of which 30 are onshore and 81 offshore. Oil discoveries are restricted to 5 offshore and 1 onshore. Most of these gas discoveries have already been put on stream (77% offshore and 87% onshore). The Rotliegend play is dominant with 54% of all the gas discoveries since 2000. The Triassic play represents 34% and the remaining 12% are Carboniferous, Lower Cretaceous and Zechstein discoveries. Most discoveries are of modest size (less than 2 x 10^9 m^3 recoverable gas). Of the oil finds, only one onshore and one offshore field is currently producing. Economically viable development of the other four offshore oil discoveries has not yet been proven.

Over the years, almost the entire Dutch onshore and offshore acreage has been covered by exploration and/or production licences (Fig 14.7a). Successful exploration licences have matured into production licences, while less successful licences were (partly) relinquished, as were production licences after depletion of the relevant oil and gas fields. This resulted in the current licence situation (April 2023, Fig. 14.7b).

Production

The Dutch cumulative gas production up to January 1st, 2022, amounted to 3.589 x 10^9 m^3 of gas and 944 million barrels (150 x 10^6 m^3) of oil. Remaining proven reserves are 150 x 10^9 m^3 of gas and 69 million barrels (11 x 10^6 m^3) of oil (Fig. 14.8). Production has come from 221 gas fields, with an annual production of some 19× 10^9 m^3. This is 44% of the present total Dutch gas consumption of ca. 43 x 10^9 m^3 per year. The annual oil production of almost 5.6 million barrels (0.9 x 10^6 m^3) from 10 fields is relatively modest.

In 2018, the Netherlands became a net gas importer (Van Geuns et al., 2017; MEACP, 2022) as the Dutch government decided to rapidly decrease production from the Groningen gas field with a planned complete cessation of production in 2024. Ever since the seventies the Groningen field was used as a swing producer to match the (sea-

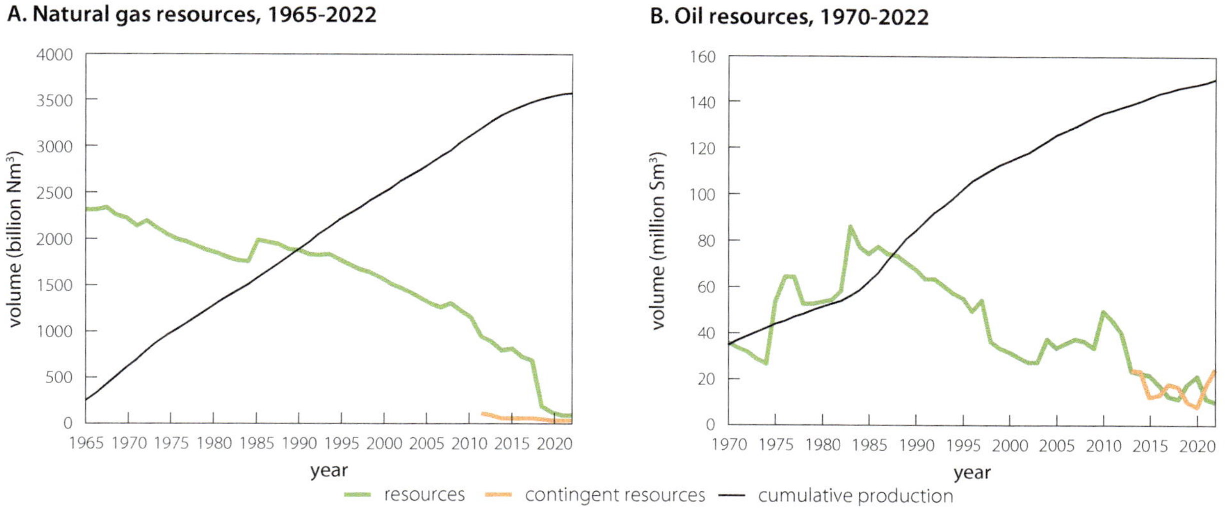

Figure 14.8. a) Natural gas reserves and cumulative production from 1965-2022 in billion Nm³ as of January 1, 2022 (MEACP, 2022). The strong reduction in natural gas reserves in 2018 with some 500 x 10⁹ m³ results from the government's decision to cease gas production from the Groningen field – i.e. this reflects a 'de-booking' phenomenon rather than a failure in the production of gas. Contingent resources registered since 2013. b) Oil reserves and cumulative production (1970-2022) in million Sm³ as of January 1, 2022 (MEACP, 2022). Contingent resources registered since 2013.

sonal) variation in gas demand by adjusting its production rate. Due to ongoing production and the corresponding pressure decline in the Groningen gas field, its capacity to act as a swing producer during extreme cold weather spells was decreasing. To provide additional swing production capacity, three underground gas storage facilities (Grijpskerk, Norg and Alkmaar) were therefore installed in 1996. More recently, this swing capacity was increased with implementation of the Bergermeer and the Zuidwending facilities. Storage in the latter is in several salt caverns rather than in originally gas-bearing sandstone or carbonate reservoir rocks (see Juez-Larré et al., 2025, this volume). Currently, with the cessation of the Groningen production, all swing production capacity is provided by these underground gas storage facilities. The remaining part of the gas demand will have to be covered by import of gas through international pipelines or by ship (LNG).

From a commercial perspective, two types of gas are distinguished in the Netherlands: gas with a low heating value ('low cal' gas or Groningen quality gas) and a nitrogen content of ca. 10-12% and gas with a high heating value ('high cal' gas) that contains little or no non-hydrocarbon components. There are separate pipeline systems for the two different gas types. The low-cal gas is predominantly distributed to Dutch households, buildings, and greenhouses, while high-cal gas is mainly for export and industrial use. An overview of the composition of the gases in the Netherlands is presented later in this chapter and in more detail in the Northwest European Gas Atlas (Lokhorst, 1997).

In 2021 the total hydrocarbon production provided the Dutch State with an income of a mere 32 million Euros (MEACP, 2022). In previous years, these revenues have been significantly higher. Between 2006 and 2014 annual revenues were on average around €11 billion, adding up to a grand total of €417 billion since 1965 (CBS, 2019). Gas revenues are expected to decrease significantly due to the decreasing production. However, extremely high gas prices in 2022 resulted in a much higher revenue.

Legislation

On January 1st, 2003, a new Mining Act (*Mijnbouwwet*) became effective, governing exploration and exploitation of oil and gas for both the onshore and offshore. Licences are mandatory for all exploration and production activities, as well as for the underground storage of hydrocarbons or other substances. The Mining Act also covers mining installations and pipelines. A production licence may be obtained once production is deemed economically viable. However, before production may start, ministerial consent to the mandatory production plan is required. Further details on implementing the Mining Act are elaborated in a separate decree (*Mijnbouwbesluit*) and regulation (*Mijnbouwregeling*).

The petroleum industry and society

In the densely populated and low-lying country of the Netherlands, there are increasing concerns that exploration and production activities may negatively influence the environment, or the well-being or comfort of the people. The main concerns are about surface subsidence, earth tremors, operations in environmentally sensitive areas and any activities to do with shale gas.

Subsidence

The depletion of oil or gas fields may cause surface subsidence due to the compaction of reservoir rocks as the pressure in the reservoir declines. For most gas fields in the Netherlands this subsidence is well below 0.1 m. Above some of the larger and thicker reservoirs, subsidence may reach between 0.1 and 0.3 m, with an expected maximum subsidence of 0.46 m at the centre of the Groningen field (NAM, 2020; VEN B.V, 2020). Subsidence due to oil production is very minor as the reservoir pressure decrease is much lower as compared to gas fields.

Land surface subsidence affects surface water run-off and reduces the distance to the groundwater table. Especially in polders and other onshore areas in the north and west of the Netherlands that lie below sea-level, this may be a serious issue. Depending on the situation, if surface subsidence exceeds 0.05-0.1 m, mitigating measures may have to be taken by the local Waterboard (*Waterschap*). For built structures such as houses, subsidence caused by gas production is not expected to cause any structural damage as the vertical subsidence is small in relation to the horizontal extent (Van Staalduinen et al., 2019; Huijgen et al., 2020). The cause of structural damage often lies in the upper part of the subsurface, particularly in areas underlain by surface peat alternating with more solid rock types such as sandstones. Although the physics of surface subsidence is well understood, this does not necessarily re-assure the

Most gas escapes to the atmosphere

Combustible gas is unquestionably the main commodity obtained from the Dutch subsurface. It is estimated that over 95% of all gas discovered in the Netherlands was generated from Carboniferous (Westphalian) coals, with the remainder originating from other source rocks such as the Jurassic Posidonia Shale.

Based on thickness maps, the total volume of upper Carboniferous (Namurian and Westphalian) deposits in the Dutch on- and offshore is estimated to be 231,000 km³. This corresponds to an average thickness of 2.35 km. Average coal percentages per stage are estimated between 0.1 and 2.1% (in-house studies by Nederlandse Aardolie Maatschappij). These percentages can be considered conservative; other studies report higher estimates of up to 3.5% (e.g. Van Wijhe & Bless, 1974; Hedemann, 2001). However, those estimates are based on onshore observations from the coal mines along the southern basin margin. Studies of the depositional environment suggest lower average coal percentages for the present-day offshore area.

Based on a back-of-the-envelope calculation of rock, coal and combustible gas volumes for the upper Carboniferous, an estimate was made of the total volume of gas in the subsurface of the Netherlands that was generated over geological time. Using an average density of coal of 1.5 g/cm³ (i.e. low-volatile bituminous to anthracitic coal) and the rule-of-thumb that 1000 kg of this type of coal generated some 136 m³ of combustible gas (Rightmire et al., 1984), a volume of 300×10^{12} m³ of generated combustible gas can be estimated. Less than 1.5% of this volume has been proven in discovered accumulations, so it is safe to assume that by far the bulk (>95%) of the gas ever generated has leaked away to the surface or may still be escaping through surface seeps.

Estimated rock, coal and combustible gas volumes for the upper Carboniferous in the Dutch on- and offshore (after Van Buggenum & Den Hartog Jager, 2007).

Stratigraphy	Rock volume (10^3 km³)	Coal % (av. est.)	Coal volume (km³)	Coal weight (10^{12} kg)	Combustible gas generated (10^{12} m³)
Asturian (WF-D)	3	0.1	3	5	1
Bolsovian (WF-C)	7	1.6	114	170	23
Duckmantian (WF-B)	23	2.1	477	715	97
Langsettian (WF-A)	82	1.0	820	1230	167
Namurian	116	0.1	116	174	24
Total	231		1530	2294	312

public. Reasons for this include the physical complexity of this process, combined with a lack of trust in the oil & gas industry and in many cases also a lack of trust in the government.

The subsidence caused by hydrocarbon production is monitored at regular intervals (2-5 years) and the environmental implications are carefully evaluated. Prior to the start of production, predictions of expected future subsidence are required for all onshore oil and gas fields as part of the mandatory production plans.

Induced earthquakes

Gas production will inevitably cause a drop of the pore pressure in the reservoir that it is being produced from. This pressure drop will in turn cause changes in the in-situ stress field in the subsurface. Such production induced changes in subsurface stresses can in some cases give rise to earthquakes and this is called induced seismicity. The induced earthquakes in the Netherlands seem to be correlated with faults within or at the boundaries of producing gas fields. A reactivation of existing zones of weakness is, as in the case of natural earthquakes, considered the most likely mechanism. Some small fields in the provinces of North-Holland, Drenthe, Friesland and Groningen have experienced earthquakes, but by far the most severe case in the Netherlands is undoubtedly the Groningen field where, since the late 1980s over 1500 events of $M_L>1$ related to gas production have occurred. To date, the strongest registered earth tremor in the Groningen field was the Huizinge earthquake ($M_L=3.6$) in 2012. It was this Huizinge event which fundamentally changed perception about the safety of gas production from the Groningen field. For the people living on top of the Groningen field who are directly affected by these induced earthquakes. It resulted in a severe feeling of discomfort, fear and insecurity. Consequently, public opposition against onshore gas production has strongly increased since the 2012 Huizinge event. Eventually, after years of protest and unsuccessful attempts by the operator and the Dutch government to mitigate this problem, it has led to the premature cessation of production of the Groningen gas field with approximately 500 billion Nm^3 of producible gas still in the reservoir.

Operations in sensitive areas

The Mining Act stipulates the permits and related procedures for exploration and production operations. Other acts also play a role in the permitting and execution of these operations: the *Wet Ruimtelijke Ordening* (Spatial Planning Act), *Wet Milieubeheer* (Environmental Act) and the *Natuurbeschermingswet* (Nature Protection Act) – recently replaced by the *Omgevingswet* (Environment and Planning Act). Permitting requirements vary according to the nature of the area involved. Onshore, the areas with the highest level of protection include ecologically designated areas, 'silence preservation areas', soil-conservation areas, birdland habitat-directive areas and nature-protection areas. Offshore, restrictions apply in ecologically designated areas, shipping lanes and military exercise areas. Moreover, for all new onshore drilling or production locations a building permit issued by the local municipality is required. For activities in sensitive areas a location permit from the Ministry of Economic Affairs, coupled to an environmental impact assessment (in Dutch: MER, *Milieu Effect Rapportage*) must be obtained. Other permits may be required for nature-protection and environmental issues. For all exploration wells an environmental permit from the Ministry of Economic Affairs is needed and production may only commence (or proceed in the case of fields already developed) after the Ministry's consent to the submitted production plan (comparable to a summary field development plan). Together with the application for a production permit, the operator must also submit a comprehensive field development plan. During all permitting procedures there is an option to file comments and objections. For activities in or near sensitive areas it may take several years before all permits are secured.

The exploration and production of gas from below the Wadden Sea (UNESCO World Heritage Site since 2009) has met with strong objections and received considerable media attention. In 1984 exploration in the Wadden Sea was put on hold and only after extensive studies concerning the environmental impact, Parliament endorsed exploration and production in 2004 under strict conditions. The main concern was that surface subsidence due to gas production might result in the drowning of the tidal flats with their unique ecological system. To prevent the possible damage to the tidal flats, gas production is constrained by a maximum rate of subsidence which may be compensated by the natural sedimentation rate of the Wadden Sea system. If the observed subsidence rate is too high, gas production will be reduced to stay within the agreed limit.

Shale gas

Spurred by the commercial success of shale gas in the United States, several licence applications for shale gas exploration were submitted in 2008 to the Ministry of Economic Affairs. These plans to explore for shale gas quickly gave rise to significant public opposition, most of which was related to the use of hydraulic fracking that is needed to produce gas at economic rates from low permeability rocks. Fracking requires large quantities of frack fluids to be injected under high pressure into the target layer. Environmental concerns centred around the composition of the chemical additives to the frack fluid, and the risks these might pose to people and the environment in the event of accidental spills. Also concerns were about the

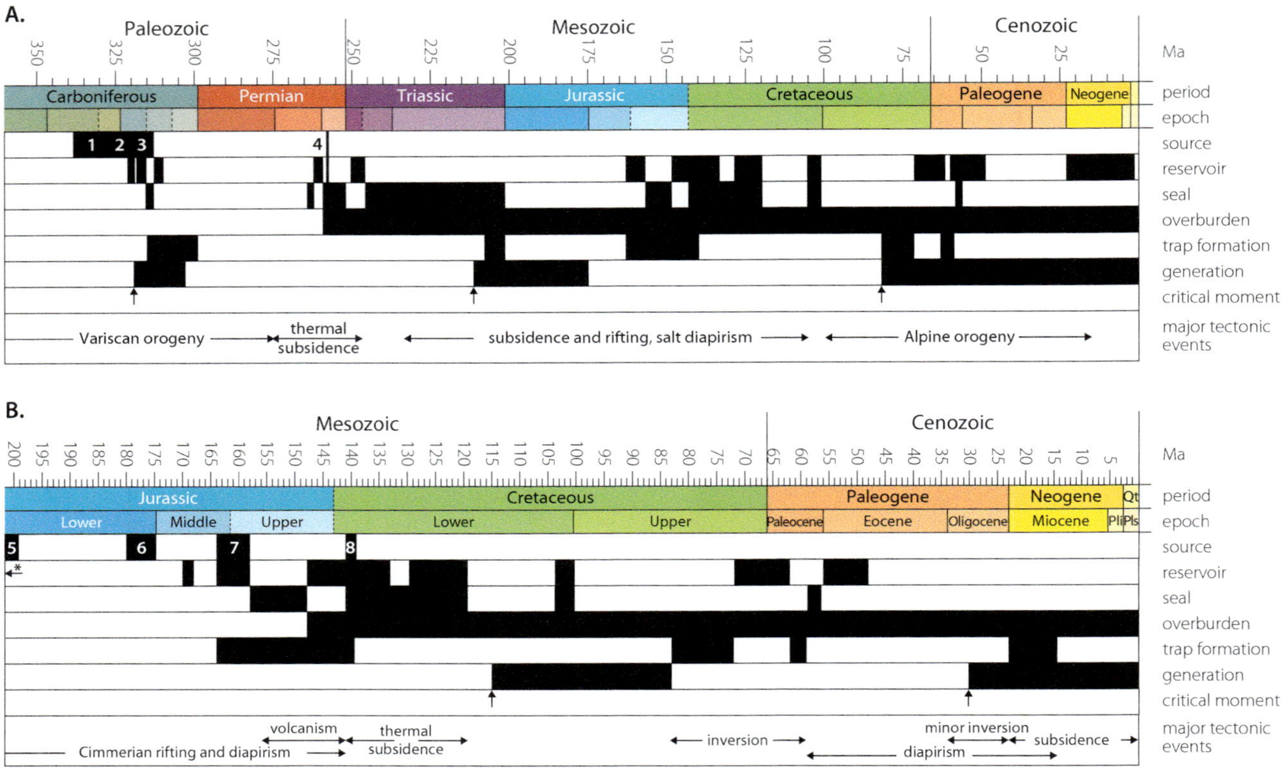

*Figure 14.9. Petroleum system element chart (PSE) for (a) Paleozoic source rocks and (b) Mesozoic source rocks. For both petroleum systems the graphs show the timing of deposition of the main elements (source, reservoir, seal) as well as trap formation, generation and related critical moment related to the main tectonic events. The Paleozoic systems consist of 1) lower Carboniferous coals (Elleboog and Yoredale formations), 2) Namurian shales (Geverik Member and Epen Formation), 3) Westphalian coals (Baarlo, Ruurlo, Klaverbank and Maurits formations), 4) Zechstein Carbonates (Z1 Coppershale and Z2 Carbonate members). The Mesozoic systems consist of 5) The Sleen and lowermost Altena formations, 6) Posidonia Shale Formation, 7) Middle to Upper Jurassic coals (Middle and Upper Graben formations and Rifgronden Member), 8) Lower Cretaceous marine shales (Lutine Formation, northern offshore) and Lower Cretaceous lacustrine shales (Coevorden Formation, eastern onshore). Cross-fault migration into Triassic reservoirs indicated with *.*

risk of induced earth tremors and potential risks to public health from the emissions associated with shale gas extraction. In addition, the spatial implications of the necessary dense spacing of drilling pads and consequent activities (traffic, drilling, and fracking operations) were a concern. After an extensive period of public consultations, debates, and an environmental impact assessment (MEA, 2015), in 2015 the ministry decided to put a moratorium on shale-gas exploration (De Boer, 2015; Kamp, 2015) which became irrevocable in 2018 (Kamerstukken 32849 en 33529 nr. 126).

Hydrocarbon source rocks

Hydrocarbon source rocks are sedimentary rocks that contain high percentages of organic matter (generally >2 wt% C_{org}, Tissot & Welte, 2013). This organic matter is termed kerogen and is derived from plants and other biomass (bacteria, archaea, animals). Source rocks can be classified depending on the origin of the kerogen and the related type of hydrocarbon that may be generated by it. Type I kerogen is mostly derived from algal biomass, deposited under oxygen-deficient freshwater conditions, such as in lake deposits. It is characterized by a high initial hydrogen/carbon ratio with a low oxygen/carbon ratio and produces mostly oil during maturation. Type II kerogens were mostly deposited under marine dys- to anoxic conditions and are often derived from algal or plankton biomass. They are characterized by intermediate H/C and low O/C ratios and generate mostly oil but also gas at higher maturities. Type III kerogen is derived from terrestrial plant matter such as coal and is characterized by high O/C ratios and lower H/C ratios. Type III kerogen has only very limited oil generation potential, but high natural gas generation potential (e.g. Killops & Killops, 2005).

In the Dutch subsurface gas plays are volumetrically and economically much more important than oil plays. With

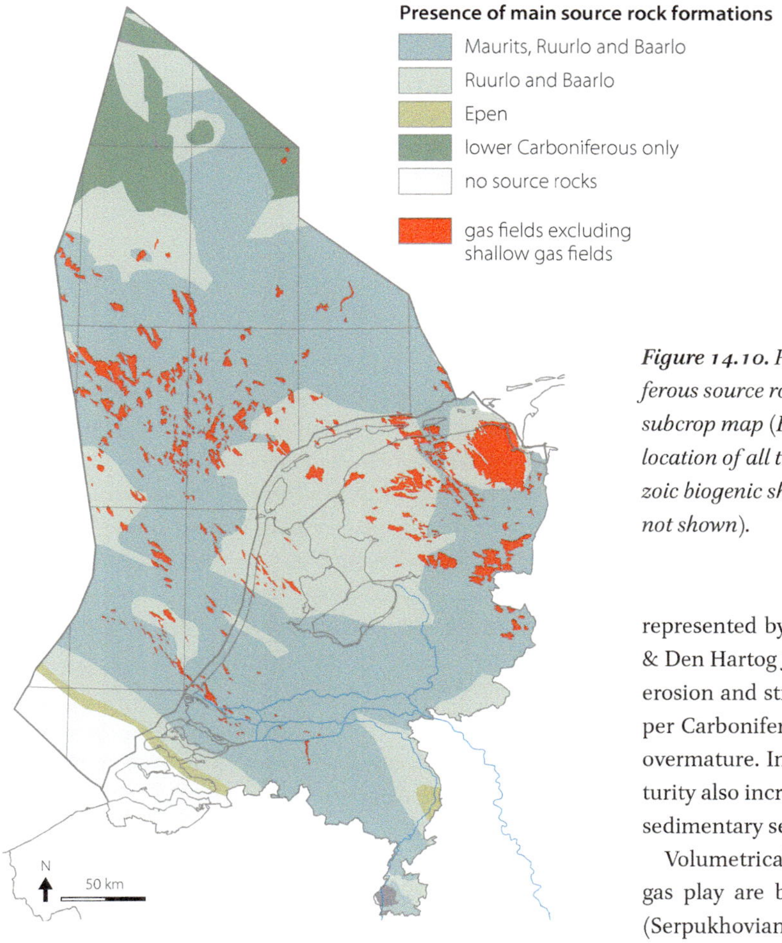

Figure 14.10. *Present-day distribution of the main Carboniferous source rock intervals based on the Base Permian subcrop map (Bouroullec & Geel, 2025, this volume). The location of all thermogenic gas fields is shown in red (Cenozoic biogenic shallow gas fields in the northern offshore are not shown).*

respect to the age of the source rocks, reservoirs and seals of these gas plays, the dominant hydrocarbon system is of Paleozoic age (Fig. 14.9a). Where the thick Permian Zechstein salt is present, it provides an effective seal between this system and the overlying oil plays which belong almost entirely to the Mesozoic hydrocarbon systems (Fig. 14.9b). The Cenozoic shallow gas plays are thought to be predominantly of biogenic origin with only minor contributions from thermogenic gas.

Gas: source rocks and generation

The principal source rocks for gas are the upper Carboniferous (Westphalian) coals and carbonaceous shales that underlie most of the Dutch territory (Fig. 14.10; Lokhorst, 1997; Gerling et al., 1999; Breunese et al., 2010; Doornenbal et al., 2019). The depositional thickness of the upper Carboniferous ranges from more than 5.5 km in the south to less than 2 km in the northern offshore. The cumulative thickness of the upper Carboniferous coals is estimated to be several tens of metres. Carboniferous peats accumulated from the late Namurian (Yeadonian) to the late Westphalian (early Asturian). Most coal-forming peats accumulated during the Westphalian B (late Duckmantian) and early Westphalian C (early Bolsovian) and are represented by the Maurits Formation (see Huis in 't Veld & Den Hartog Jager, 2025, this volume). Due to differential erosion and structural complexity, the maturity of the upper Carboniferous coals varies spatially from immature to overmature. In areas with thick preserved sequences, maturity also increases significantly from top to bottom of the sedimentary sequence.

Volumetrically minor contributions to the Paleozoic gas play are believed to originate from basal Namurian (Serpukhovian, Epen Formation) organic-rich shales (Lokhorst, 1997; NITG, 1998; Gerling et al., 1999) and coals of the Elleboog and Yoredale formations (late Viséan; e.g. Ter Borgh et al., 2018). In most places these Dinantian and Namurian source rocks became overmature during deep burial prior to the main Mesozoic uplift and erosion phases. However, in the northernmost part of the offshore territory and along the London-Brabant Massif in the south, these Namurian source rock intervals are presently in the gas generation window and could be viable source rocks in areas where Westphalian coals are absent or immature (Fig. 14.10).

Gas-prone source rocks also occur in several Mesozoic intervals, mostly as coals in the Upper Jurassic and Lower Cretaceous Delfland Subgroup in the West Netherlands and Broad Fourteens basins, and in the Central Graben Subgroup of the Dutch Central Graben and Terschelling Basin.

Gas generation in the main depocentres started during the late Carboniferous and was widespread until the Middle Jurassic. At that time source rocks of the lower Carboniferous and at the base of the Westphalian already reached over-maturity in the West Netherlands and Broad Fourteens basins and in the Dutch Central Graben. After the Middle Jurassic, a distinction must be made between rift basins and adjoining platforms and highs (Fig. 14.3).

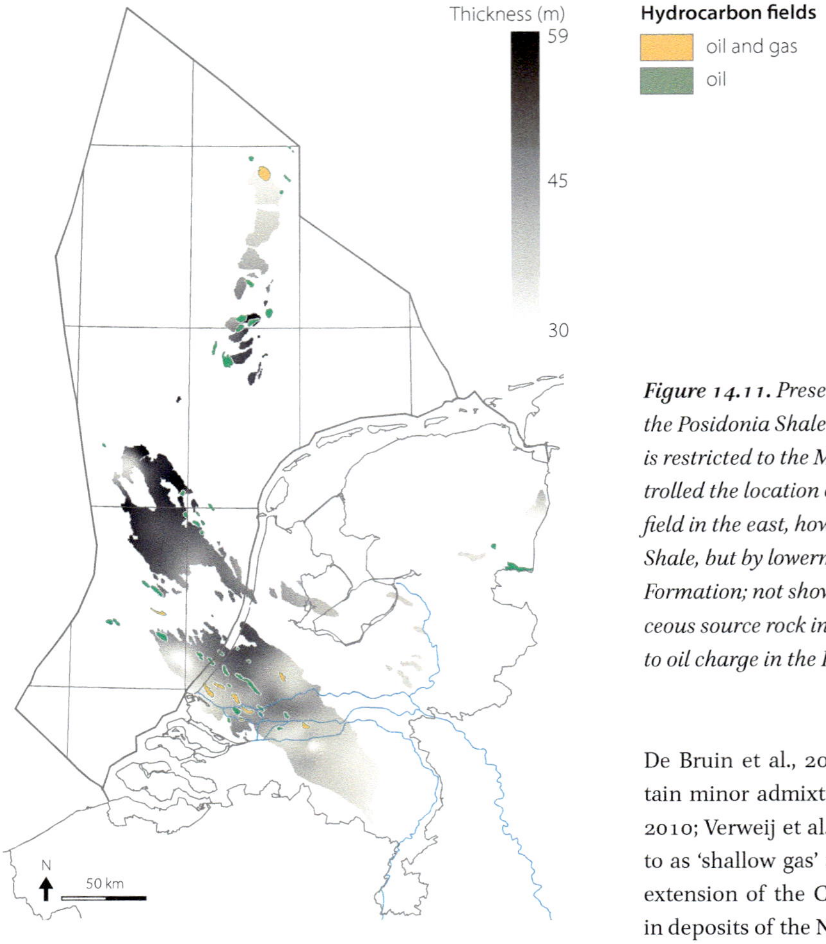

Figure 14.11. Present-day distribution and thickness map of the Posidonia Shale Formation. The present-day distribution is restricted to the Mesozoic rift basins and has clearly controlled the location of the oil fields. The large Schoonebeek field in the east, however, was not sourced by the Posidonia Shale, but by lowermost Cretaceous source rocks (Coevorden Formation; not shown). Other Upper Jurassic to Lower Cretaceous source rock intervals are believed to have contributed to oil charge in the Dutch Central Graben area.

During the Late Jurassic to Early Cretaceous, increased subsidence within the rift basins resulted in accelerated hydrocarbon generation. Gas generation ceased during the Late Cretaceous due to inversion-related uplift and declining heat flow. In parts of these inverted rift basins, the subsequent burial was not strong enough to revive hydrocarbon generation, resulting in the so-called burial anomaly where no hydrocarbon generation occurred since the Late Cretaceous (Pluymaekers et al., 2012; Nelskamp et al., 2021) At the basin margins, for example on the southwest margin of the West Netherlands Basin where uplift due to inversion was limited and was followed by strong subsidence, gas charge from Westphalian coals resumed during the Cenozoic and continues until the present day (De Jager et al., 1996). The platforms and highs, on the other hand, were uplifted during the Late Jurassic rifting, interrupting hydrocarbon generation. Where subsequent re-burial caused temperatures at the Westphalian source-rock levels to exceed the maximum temperatures reached earlier, gas generation resumed.

Apart from the thermogenic gas plays described above, there are also gas accumulations at depths shallower than 1000 m that are mainly of microbial origin (biogenic gas, De Bruin et al., 2022), although some of them may contain minor admixtures of thermogenic gas (Pletsch et al., 2010; Verweij et al., 2018). These occurrences are referred to as 'shallow gas' and are mostly located in the northern extension of the Central Graben area, typically occurring in deposits of the Neogene Eridanos Delta. Several of these shallow gas fields have recently been taken into production (Doornenbal et al., 2019). A detailed overview of the occurrences of 'shallow gas' in the Dutch offshore region was recently published by Ecclestone et al. (2021).

Oil: source rocks and generation

The Lower Jurassic Posidonia Shale Formation is the main oil-prone source rock in the Netherlands. This marine, Type-II source rock is only preserved in the Late Jurassic rift basins of the Netherlands and northern Germany. Time-equivalents also occur in Belgium, Luxemburg, and the French Paris Basin (Schistes Carton) as well as in northern England (Jet Rock of the Whitby Mudstone Formation, Lott et al., 2010; Song et al., 2015).

The Posidonia Shale Formation has a gross thickness of approximately 36 m with, locally, thicknesses of up to 46 m (Fig. 14.11). It has an average total organic carbon content (TOC) of 10%, which can locally reach up to 17%,with a Hydrogen Index (HI) of up to 700 mg HC/g TOC. It forms a very distinct interval throughout the subsurface of the Netherlands and is easily recognizable on seismic data and wire-line logs (Van Adrichem Boogaert & Kouwe, 1993). The Posidonia Shale Formation shows a strong cyclic signal and little lateral heterogeneity with respect to thickness. Correlation across basins can easily be achieved using stable carbon isotope measurements on the organic material (Houben et al., 2021). Total organic carbon (TOC) content is generally higher in the lower half of the formation, as indicated by both laboratory measurements and calculated TOC logs. Geochemical proxies, biomarkers and palynological facies all point towards a sharp chemocline and anoxic to euxinic water-column conditions as the main driving factors for TOC enrichment. Locally increased subsidence in salt-rim synclines gave rise to a very high TOC contents. This overall homogeneous depositional character and laterally traceable TOC-trends suggest that the deposition occurred in an extensive marine basin that was susceptible to stratification and enhanced preservation of organic matter.

The present-day distribution (Fig. 14.11) as well as the maturity of the Posidonia Shale Formation are strongly influenced by uplift and erosion. On the Mesozoic platform areas, i.e. the uplifted shoulders of the main Jurassic rift basins, most of the Lower Jurassic sediments were eroded (Fig. 14.3; Trabucho Alexandre & Wong, 2025, this volume). In addition, Late Cretaceous basin inversion resulted in uplift along the axes of the basins. Consequently, the onshore and the southern offshore region basin fills show a so-called burial anomaly. This means that the past burial depths and related temperatures were higher than those observed at present (e.g. Pluymaekers et al., 2012; Nelskamp et al., 2021). In regions affected by this burial anomaly, little to no hydrocarbon generation can be expected since the uplift, making them higher exploration risk areas.

The large Schoonebeek oil field was not sourced from the Posidonia Shale, but from Lower Cretaceous lacustrine source rocks of the Coevorden Formation. In the Netherlands, these algal, Type-I source rocks only occur in the onshore Lower Saxony Basin (Binot et al., 1991; NITG, 2000; Amberg et al., 2022). As the name 'Lower Saxony' indicates, the main extent of this basin is in Germany where the source rocks belong to the Bückeberg Formation (or Wealden Shale, e.g. Froidl et al., 2021).

Source rocks with similar characteristics to the Posidonia Shale Formation occur in the Lower Jurassic Aalburg and uppermost Triassic Sleen Formations. However, the limited geochemical analyses suggest significantly lower oil generation potential compared to the Posidonia Shale (Houben et al., 2017). Due to their similar geochemical footprint the contribution of these source rocks is unknown. The bituminous Clay Deep Member of the Lower Cretaceous Lutine Formation in the northern part of the Dutch Central Graben also has oil-generating potential. However, the areal and stratigraphic extent of this member is much smaller than those of the time equivalent, very productive, hot shales in the British, Danish and Norwegian sectors of the North Sea (e.g. Petersen et al., 2017; Verreussel et al., 2018, 2025, this volume). Middle Jurassic coals, with relatively high Hydrogen/Carbon ratios, may also have contributed to the oil fields in the Danish offshore (Petersen & Hertle, 2018). The equivalent formations in the Dutch Central Graben area are the Middle and Upper Graben formations which contain coaly intervals with similar organic geochemical properties (Doornenbal & Stevenson, 2010).

Other oil-prone source rocks occur in the Permian Coppershale- and Z2 Carbonate members. Despite numerous oil shows in Zechstein reservoirs, these source rocks have given rise to minor oil accumulations only locally (e.g. Stadskanaal, Gieterveen and E13-01). This is not only because of their limited thickness, but also because any oil expelled from these source rocks and originally trapped in Zechstein or Rotliegend reservoirs, has been subsequently flushed out by the abundant gas charge from the Westphalian below. The presence of H_2S in Zechstein gas reservoirs as well as high condensate content in Rotliegend gas is linked to the hydrocarbons generated from these Zechstein source rocks.

Carboniferous oil-prone source rocks occur in the basal Namurian organic-rich shales of the Geverik Member which has been studied for shale oil/gas potential (Van Bergen et al., 2013; Zijp et al., 2015). These basinal organic-rich shales have been encountered in deep wells (e.g. Winterswijk-1, Luttelgeest-1, Uithuizermeeden-2) in the Groningen area and in the southeastern Netherlands onshore (Geverik-1, Californie-GT-01). The Geverik Member can be up to 50 m thick and has TOC contents of up to 9% (Bouw & Lutgert, 2012). In much of the Netherlands the Geverik Member is post-mature for oil due to its great burial depth (>5 km; Fig. 14.26) and vitrinite reflectance values of over 2%. However, there are clear exceptions to this. In the Geverik-1 and Californie-GT-01 wells in Limburg the Geverik Member was encountered at depths of less than 1 km TVD and ~1.5 km TVD, respectively. However, vitrinite reflectance measurements show maturities above 2% in these areas. In the Limburg area the maturity of the Geverik Member is therefore controlled by its past burial, uplift and erosional history rather than the present burial depth (see also maturity map in Zijp et al., 2015). In the UK, the equivalent Bowland Shale is the source rock for the earliest oil discoveries, including the Hardstoft (1919) and Eakring (1939) discoveries (Kombrink et al., 2010) and has been investigated in the context of shale gas exploration (Andrews, 2013).

Based on regional paleogeographic reconstructions, source rocks for oil may also occur in Dinantian and Devonian deposits (Cameron & Ziegler, 1997). However, in most places the present-day (or maximum) burial depth will render them post-mature for oil and gas generation.

Hydrocarbon plays

A hydrocarbon play is used here to describe a region in which the oil or gas occurrences are controlled by the same set of geological circumstances. The term is widely used in the realm of exploration to categorize hydrocarbon resources. Proven gas and oil plays in the Netherlands range from Westphalian to Quaternary in age.

Devonian and lower Carboniferous plays are regarded as speculative because of their poor reservoir quality and the post-maturity of their source kitchens. The Namurian Epen Formation is prospective for (shale) gas, but this play has never been effectively explored.

The Rotliegend play is by far the most important gas play, with the Groningen field accounting for over 60% of the Dutch initial gas reserves. Excluding the Groningen field, approximately 65% of all the initial resources in the so-called 'small fields' are found in the Rotliegend as well. The cumulative amount of produced gas and the remaining economically producible Dutch gas resources amount to 3724 x 10^9 m^3 (MEACP, 2022). Recoverable gas reserves for the Groningen field were reduced by 470 billion m^3 due to the premature cessation of production. Safety issues related to the induced seismicity in the depleting Groningen field prompted this closing-in (MEACP, 2018).

The most important oil play consists of the Jurassic and lower Cretaceous sands that were charged from the Jurassic Posidonia shale. Cumulative produced oil and economically producible remaining oil resources are reported to amount to 190.7 million Sm^3 (MEACP, 2022).

For oil, there is a marked variation in the recovery factors (e.g. 12% to 70% in the West Netherlands Basin and 35% for Schoonebeek), but for gas, the recovery factors generally vary between 70% and 90%, and in some cases are as high as 95% (e.g. as originally predicted for the Groningen field prior to the decision to close-in the field). For gas in tight or in very shallow reservoirs, or in reservoirs with low connectivity (e.g. in the Westphalian), recovery efficiencies are notably lower.

Westphalian play

The Westphalian is present throughout most of the Dutch subsurface and is locally up to 5.5 km thick (Huis in 't Veld & Den Hartog Jager, 2025, this volume). Gas fields producing from Westphalian sandstones occur in the eastern Netherlands and adjacent parts of Germany, as well as in the Cleaverbank Platform in the northwestern offshore and in the adjacent British sector (Figs 14.3, 14.12). Gas charge is derived from Westphalian coals, with on the Cleaverbank Platform also a minor contribution from Namurian source rocks (Gerling et al., 1999; Breunese et al., 2010; Besly, 2018). Traps are formed by tilted fault block closures below the Base Permian Unconformity. Where Rotliegend sandstones directly overlie this unconformity, gas will migrate into the Rotliegend reservoir and the Westphalian sandstones only contain gas when the height of the trap (or gas column) exceeds the thickness of the Rotliegend, e.g. in the Groningen field. In the east of the Netherlands where no Rotliegend sandstones overlie the unconformity, the Zechstein salt forms a near perfect top seal for the Westphalian traps. On the Cleaverbank Platform the claystones and evaporites of the Silverpit Formation form the seal. Locally, intra-Westphalian shales and faults may also form seals (Figs 14.3, 14.12).

A key uncertainty for the Westphalian play is the presence of good reservoir sandstones within the traps at the level of the Base Permian Unconformity. Due to early Permian uplift and erosion, different rocks subcrop at this unconformity. A good understanding of facies distribution within the Westphalian and mapping of intra-Westphalian horizons is therefore a pre-requisite to pursue this play (Quirk, 1993; Quirk & Aitken, 1997). Fluvial sandstone reservoirs occur at distinct stratigraphic levels within the Westphalian, with the most important reservoir units occurring in the Tubbergen Formation. These Westphalian C-D (Bolsovian-Asturian) sandstones are gas-bearing in several fields. Approximately 65% of the total volume of gas produced from Carboniferous reservoirs is from onshore fields, mainly in the east of the Netherlands. The largest of these is the Coevorden field, discovered in 1948 at a depth of ca. 2800 m. Since its discovery, approximately 20 x 10^9 m^3 gas has been produced from the Carboniferous. In addition, 12 x 10^9 m^3 has been produced from the Zechstein Carbonates and a minor 140 million m^3 from the Triassic (Production plan Coevorden, NAM, 2018). Other significant Carboniferous fields in the area are Dalen, Tubbergen and Hardenberg. The upper Carboniferous reservoir sands were derived from erosion of the rising Variscan mountain belt to the southeast. They wedge-out northwestward showing an overall decrease in reservoir potential in that direction. The typically sheet-like stacked sandstones have net-to-gross ratios as high as 50% (Kombrink et al., 2007). The uppermost Carboniferous (Westphalian D and Stephanian) comprises red beds, with low net-to-gross ratios of 10 to 20%. Porous conglomeratic sandstones occur, but their distribution is hard to predict. The reservoir architecture of the Westphalian sandstones and the faulted nature of the traps in the east of the Netherlands result in strong compartmentalization of the gas fields.

Offshore Carboniferous fields in the Cleaverbank Platform-area have contributed ca. 35% of the gas produced from Carboniferous reservoirs in the Netherlands. In this offshore area, gas-bearing fluvial channel sandstones occur

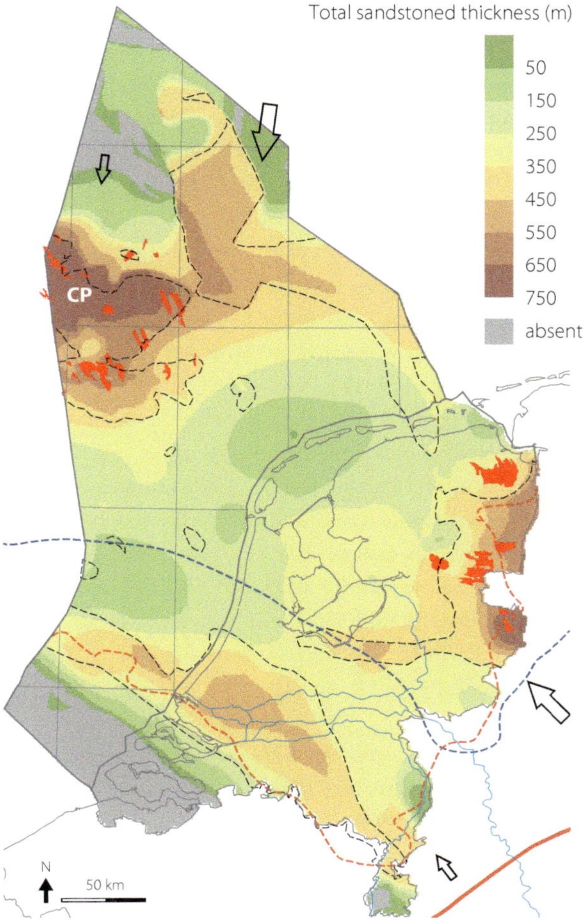

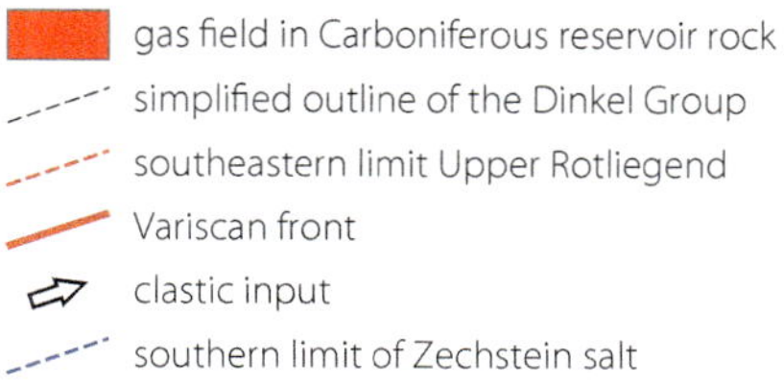

Figure 14.12. Westphalian play map. Distribution and net sandstone thickness of the Westphalian reservoir rocks. The best reservoirs occur mostly in the sandstone-prone Westphalian C and D (Dinkel Group). On the Cleaverbank Platform (CP), gas has also been discovered in Westphalian A and B reservoirs. Modified from Roustiau et al. (2022).

mainly in the Westphalian A and B (Botney Member of the Klaverbank Formation, Caister and Murdoch sandstones; Huis in 't Veld & Den Hartog Jager, 2025, this volume) and, as in the Dutch onshore, also in Westphalian C and D deposits (Hospital Ground and Step Graben formations). The fluvial systems that deposited these sands were sourced from the north (see Fig. 14.12) and their reservoir quality deteriorates towards the south. The sandy facies may extend further eastward to the Schill Grund High, at the southern margin of the Ringkøbing-Fyn High, although there are insufficient well data to substantiate this. The Westphalian A and B deposits generally have low net-to-gross ratios. The main fields are located in the northern K and southern D and E quadrants, at depths of around 3500 to 4000 m, north of the pinch-out of the Rotliegend sandstones, where the evaporite bearing Silverpit shales generally form an excellent seal. The largest Westphalian field is E17a-A, with an initial gas-in-place volume of ca. 18 x 10^9 m^3 of which 10 x 10^9 m^3 may be produced (Neptune, 2021). The trend of Westphalian gas fields continues into the UK sector of the North Sea, with fields like Orca, Caister, Murdoch, Ketch and Schooner and several other discoveries (Mijnssen, 1997). The fields in E18, F16 and the northern K quadrant produce commingled gas from the basal Rotliegend sandstone and the Hospital Ground Formation. Although most of the Carboniferous gas fields are approaching end-of-field life, new discoveries may still be possible, as shown by the cross-border Sillimanite Field (D12-B). Some developments have shown significant upside, like the E17a-A Field. In 2014 a promising gas discovery was made in well E11-01 in sandstones of reasonable quality in the Hospital Ground Formation. However, as for several other discoveries in the D9, E12, E13 blocks, development of E11-01 was judged to be uneconomic.

Where the Westphalian sandstones are gas-bearing, they mostly have fair to good porosities (average 9%, maximum 20%) and permeabilities (average 1-2 mD, locally >100 mD) and good initial production rates of about 1 x 10^6 m^3/day or more can generally be achieved (Van Buggenum & Den Hartog Jager, 2007). However, their poor reservoir connectivity often results in limited connected volumes per well accompanied by rapidly declining flow rates, thus seriously hampering the economic viability of offshore Westphalian gas accumulations.

Rotliegend play

Even excluding the giant Groningen field, the Slochteren Formation of the Upper Rotliegend Group is volumetrically by far the most important gas reservoir in the Netherlands. The Rotliegend play consists of an ideal superposition of 1) the prolific Westphalian source rocks for gas, 2) the thick Slochteren sandstone reservoirs, and 3) the near-perfect seal of the Zechstein salt (Glennie & Provan, 1990; Grötsch & Gaupp, 2011). Most traps form simple asymmetrical tilt-block configurations (Fig. 14.13). The Slochteren reservoir sandstones are present in the classical Rotliegend play fairway which runs E-W from the Groningen field to the offshore K and L blocks. To the north the Slochteren sandstones shale-out into the Silverpit Formation. The southern boundary of the play is determined by the depositional limit of either the Slochteren reservoir or the Zechstein seal (Fig. 14.14). Two exceptions are the P6 an Q10 gas fields which are not sealed by Zechstein salt but by the distal Zechstein Fringe deposits. Although most gas in the Upper Rotliegend reservoirs was sourced from Westphalian coal measures (Van Wijhe et al., 1980; Glennie, 1998), in some areas (northern Netherlands, northwest offshore) the basal Namurian hot shales and/or the lower Carboniferous coals (e.g. Gerling et al., 1999) potentially contributed as well.

The sands in the Slochteren Formation were mainly derived from the Variscan Mountains to the south (Verdier, 1996; Geluk, 2007). They show marked differences in reservoir quality. Well-sorted and clean eolian sandstones form the best reservoirs, but there are also packages of up to tens of metres of mainly fluvial sandstones with locally conglomeratic intervals that are of fair reservoir quality. The sandstones of the Upper Slochteren Member shale out towards the north into the Ten Boer Member (Bouroullec & Geel, 2025, this volume). Depending on the amount of sand present, the Ten Boer Member acts either as a waste zone or as seal for the underlying Slochteren sandstones. Even where the Ten Boer is not a top seal, it forms a lateral seal where it is juxtaposed against Slochteren reservoirs across faults (Fig. 14.16). Its sealing potential improves with increasing shale content, which suggests that a sealing shale-gouge formed along the fault planes. In some gas accumulations, the gas-water contact extends even below the mapped spill point controlled by the Ten Boer Member lateral seal, indicating that faults in clean sandstones can also be sealing. It is therefore likely that factors other than shale gouge, such as cataclasis and/or diagenesis, also play a role in determining whether a fault is sealing (Leveille et al., 1997).

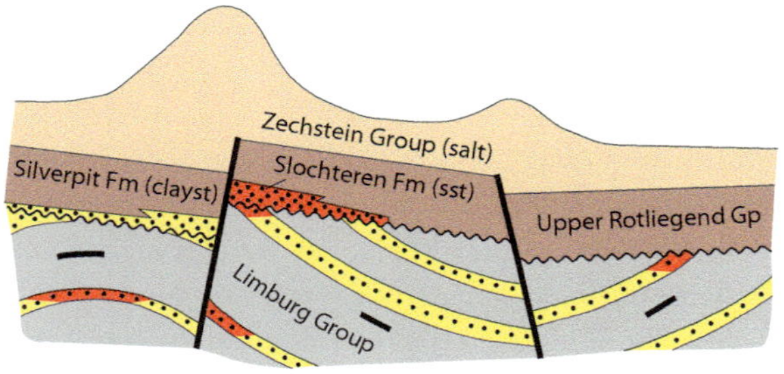

Figure 14.13. Hydrocarbon trapping styles in the Westphalian. Most gas (red) is found directly below the Base Permian Unconformity, where Slochteren sandstones are thin or absent. At intra-Westphalian levels only sub-commercial gas volumes have been encountered to date. Modified from De Jager and Geluk (2007).

Below the Silverpit shales, basal sandstones of the Lower Slochteren Member are locally present. Thickness variations of these sandstones are controlled by the paleotopography at the time of deposition. This paleotopography, was in turn controlled by faulting and the type of formation sub-cropping the Base Permian Unconformity. The shaly parts of the Westphalian A and B were less resistant to erosion than the sandy parts. The eroded shaley areas turned into lows where sand accumulated, while the more resistant areas formed ridges, on top of which little or no sand was deposited (Geluk & Mijnlieff, 2001; Mijnlieff & Geluk, 2011).

Most Rotliegend gas reserves occur in the northeast of the Netherlands and in the offshore K and L quadrants (Central Offshore Saddle, the southern sector of the Cleaverbank Platform and the northern sector of the Broad Fourteens Basin). This E-W belt of Rotliegend gas fields extends westward into the British Indefatigable and Sole Pit areas. Additional Rotliegend fields occur in the Vlieland Basin and the northwestern sector of the Central Netherlands Basin (Fig. 14.13).

The giant Groningen field in the northeastern onshore is by far the largest gas field (Roels, 2001; Verberg, 2001; Fig. 14.1). Other large accumulations are Annerveen, located directly south of the Groningen field (Veenhof, 1996) and the Ameland complex of fields located directly north of Groningen, below the Wadden Sea (Crouch et al., 1996). Both fields have a gas-initially-in-place (GIIP) volume of ca. 75 x 10^9 m^3. It is noteworthy that the Ameland complex appears to be underfilled. Structural reconstruction shows that Late Cretaceous inversion enlarged the Ameland trap. Apparently, subsequent charge was insufficient to completely fill it.

Marked variations in the nitrogen contents in gas in the northeast of the Netherlands are attributed to variations in charge history. Traps that have access to recent gas charge are characterized by very low nitrogen contents (<5%). On the Friesland Platform, where charge modelling indicates that the last phase of gas charge occurred prior to Jurassic rifting, nitrogen levels were as high as 25%. Residual oil is often reported from cores cut in Slochteren sandstones. This is interpreted to indicate an early oil charge from intra Zechstein and/or Coppershale source rocks. The oil was later flushed out by excessive amounts of gas, filling the majority of trapping structures with gas down to spill point. The only live oil found so far in Slochteren sandstones is in Midlaren in northeast Drenthe, where a 120-m oil column was encountered below a gas cap. The oil is juxtaposed across a fault against Z2 Carbonate. The carbonates may have allowed leakage of gas but formed a lateral seal to oil.

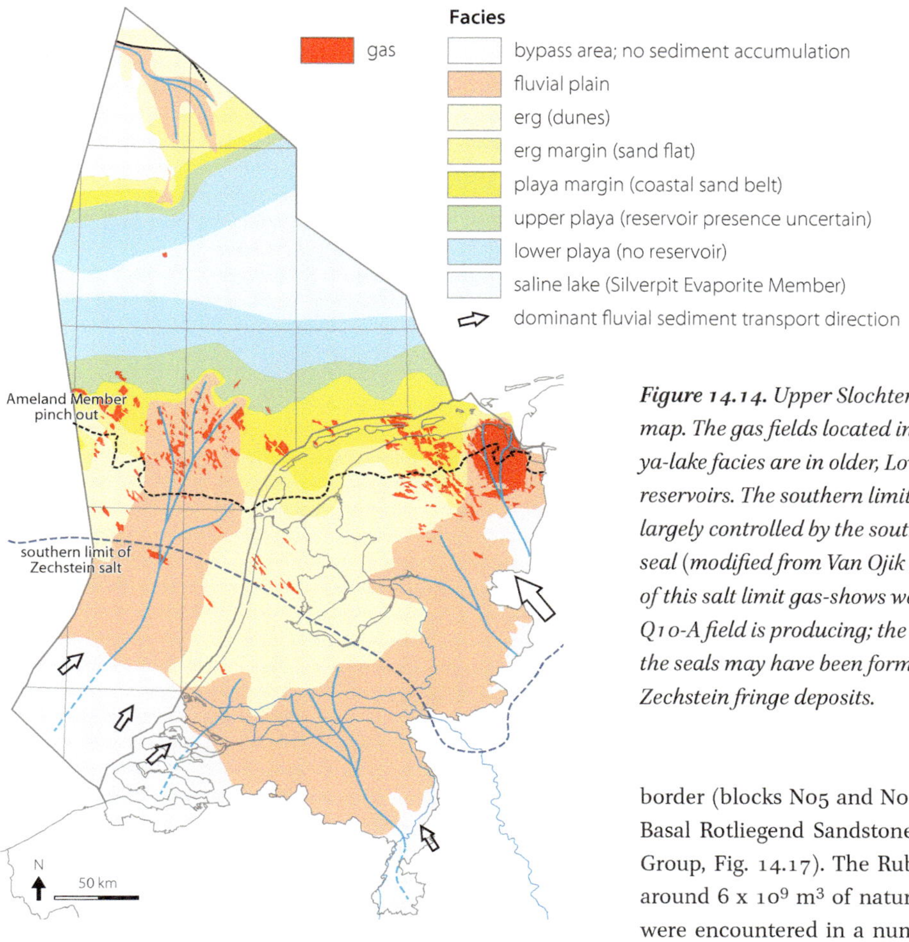

Figure 14.14. Upper Slochteren (Upper Rotliegend) play map. The gas fields located in the northern area of playa-lake facies are in older, Lower Slochteren sandstone reservoirs. The southern limit of Rotliegend gas fields is largely controlled by the southern extent of the Zechstein top seal (modified from Van Ojik et al., 2023). In four wells south of this salt limit gas-shows were encountered, but only the Q10-A field is producing; the other reservoirs are tight. Here, the seals may have been formed by clay layers within the Zechstein fringe deposits.

In the offshore K and L quadrants, K8-FA is the largest field with a GIIP of some 64 x 10^9 m^3 (NAM, 2016b). In these offshore fields the traps are also simple tilt blocks with the overlying thick Zechstein evaporites forming the top seal. As the Ten Boer shales are much thinner than in the northeast Netherlands, there is less potential for an additional gas column height due to fault juxtaposition of the Ten Boer Member against Slochteren sands. Locally, where the Zechstein salt seal is breached by erosion, salt welding and/or by faulting, basal Triassic shales form an effective alternative top seal (e.g. in K15-FC and K15-FK). Variations in gas-water contacts across different field compartments, and gas columns below conventional spill points, indicate that sealing faults are also present in these offshore fields. This has also been observed in the western continuation of the Rotliegend play in the UK sector of the southern North Sea (Leveille et al., 1997).

In the northern K quadrant and adjacent areas, basal Slochteren sandstones (see Bouroullec & Geel, 2025, this volume), sealed by the overlying Silverpit Formation also contain commercial gas accumulations. The most significant is the Markham field in the J3/J6 area, straddling the UK-Netherlands border. The recent discoveries of the Ruby and Turkoois fields, both straddling the Dutch German border (blocks N05 and N04), proved the potential of the Basal Rotliegend Sandstones Member (Upper Rotliegend Group, Fig. 14.17). The Ruby field is expected to contain around 6 x 10^9 m^3 of natural gas. These basal sandstones were encountered in a number of legacy wells. Although they were gas bearing, the uncertainty of their presence, relatively high nitrogen content and a questionable seal quality make this a high-risk play (Corcoran & Lunn, 2014). The ability to discriminate these sands on modern 3D seismic data opened the play which is claimed to potentially contain over 30 x 10^9 m^3 (Burgess et al., 2018; Fig. 14.17). According to the paleogeographical study of the Rotliegend by De Bruin et al. (2015), this basal reservoir sand may also be present along the northern margin of the Southern Permian Basin (northern A-, B- and F quadrants), but this play has not yet been tested.

Also in the western sector of the Central Netherlands Basin, Rotliegend gas fields such as Bergen, Groet, Bergermeer, Schermer and Middelie have been brought to production (Van Lith, 1983). Even at significant depths, reasonable porosities (>15% at 4000 m) occur in the better reservoir levels of the Slochteren Formation. The main deterioration of reservoir potential is by early carbonate cementation or the late growth of fibrous illite. The latter is associated with deep paleo-burial, in particular within the inverted portion of the Broad Fourteens Basin. Growth of illite fibres in the pores has little effect on porosity but significantly reduces permeability. The presence of formation waters rich in ions of clay minerals is conducive

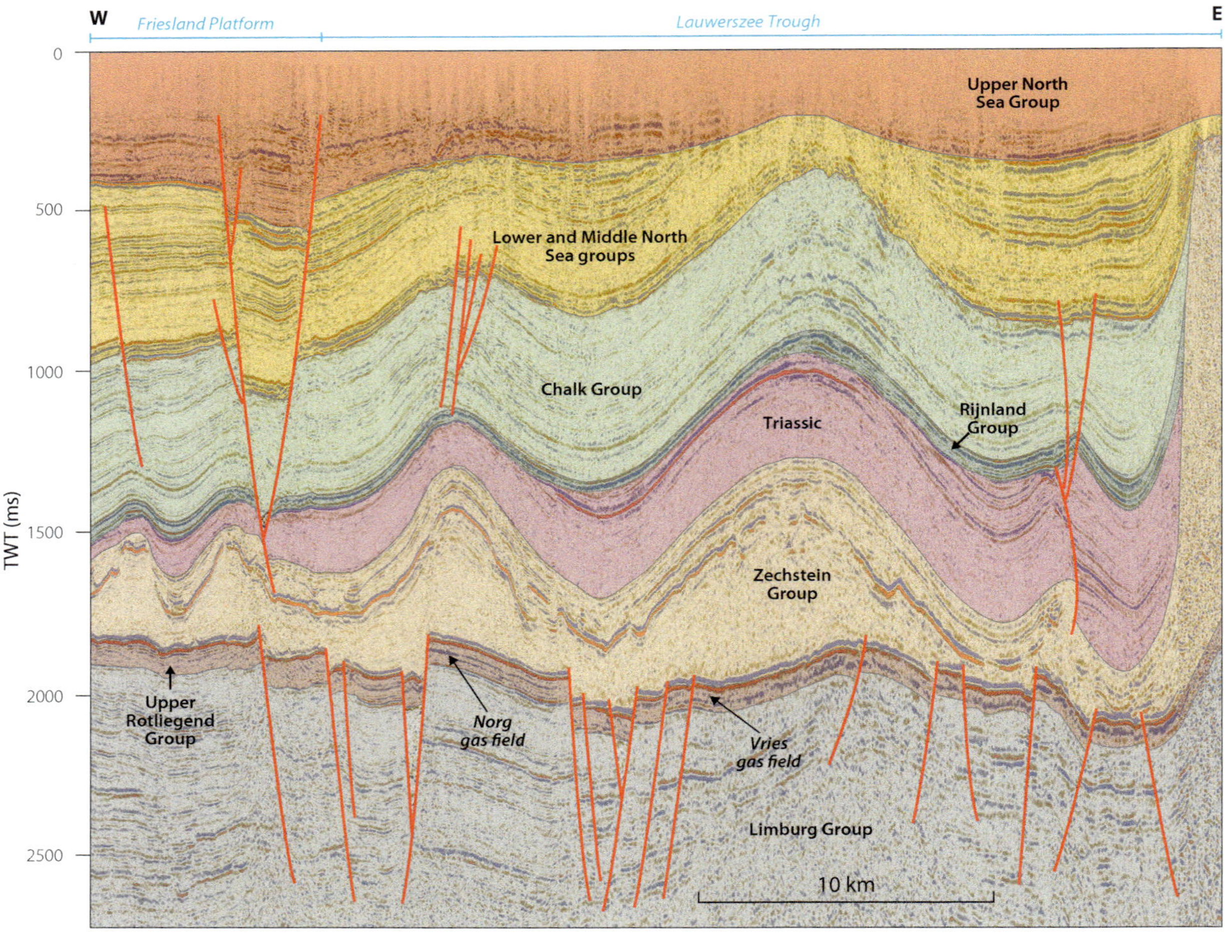

Figure 14.15. Seismic line in the northeast Netherlands (Friesland Platform to Lauwerszee Trough) showing the setting of the Norg en Vries gas fields in fault blocks in the Upper Rotliegend Group. For location see Fig. 14.1.

to illite diagenesis. This is likely to be the case where for example the Slochteren is juxtaposed against the shaly Westphalian (Gaupp et al., 1993). However, where the Slochteren was already gas-bearing prior to deep burial, growth of diagenetic illite has been inhibited. Therefore, to be able to predict reservoir quality, a good understanding of depositional trends, as well as of timing of trap formation, gas charge and temperature and burial history, is required. Low permeabilities and resultant uneconomic production rates are linked to depositional facies, such as in fine-grained, argillaceous wadi sands (Frikken, 1999).

Zechstein play

Gas fields producing from the Zechstein Group mostly occur in platform carbonates. These carbonates developed in marine settings along the Southern Permian Basin margin and are sealed by Zechstein salt and anhydrite (Van der Baan, 1990; Van de Sande et al., 1996; Geluk, 2007; Bouroullec & Geel, 2025, this volume). In addition, some minor clastics (so-called fringe sandstones) along the southern extremities of the basin, host a number of gas fields (Fig. 14.18). The traps are predominantly fault-dip closures, with some four-way dip closures. As with the Rotliegend play, gas charge to the Zechstein was from the Westphalian and Namurian. Contributions from Zechstein oil-prone source rocks have locally resulted in condensate-rich gas. Oil shows in the Zechstein have been recorded in numerous wells in the northeast of the Netherlands (e.g. fields Stadskanaal and Gieterveen). Analyses indicate that these shows are derived from intra-Zechstein source-rock intervals and from the Coppershale Member. In the northwestern offshore (E13-01) minor amounts of oil were encountered as well. The most likely source rock for this oil is the so-called Stinkkalk (or Stinkschiefer), which is the basinal equivalent of the Z2 Carbonate Member (De Jager & Geluk, 2007; Bouroullec & Geel, 2025, this volume).The best Zechstein reservoirs with porosities of ca. 14% occur in the Z2 Carbonate and, less frequently, in the Z3 Carbonate members. A single accumulation in the Z1 Carbonate Member (Q05-A) came on stream late 2004 (Geluk, 2000; MEA, 2006). The highest porosities occur in the shelf-edge facies that developed above the paleo-highs formed by the underlying Z1 Anhydrite. Subaerial exposure of the platform occurred occasionally and karstification locally enhanced reservoir quality. The adjacent slope facies consists mainly of redeposited platform sediments. Small-scale reefs occur locally, but they are restricted to the Z1 Carbonate (Geluk, 2007). The basinal facies of the carbonates comprises fine-grained limestones and dolomites. The Zechstein carbonates experienced a complex diagenetic history, including cementation, leaching and dolomitization, which has resulted sometimes in strong lateral variations in their porosity and permeability. Especially in the tight slope- and basinal facies with porosities <4-5%, such as encountered in the Dalen, Emmen and Coevorden fields, economic gas production rates largely depend on the presence of fractures (Frikken, 1999).

The first offshore gas discovery, in 1968, was in Zechstein dolomites in block P6 (Van der Poel, 1989). Various additional wells in the P and Q quadrants and in the southernmost part of the K quadrant have subsequently tested gas, but gas is produced from a Zechstein reservoir below a larger Triassic gas accumulation only in block P6. One of the Zechstein fields in the onshore part of the Central Netherlands Basin is the Alkmaar field which is now used

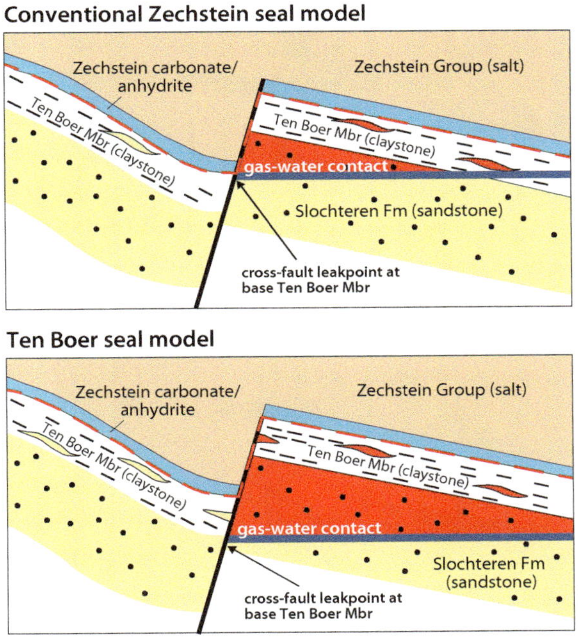

Figure 14.16. Ten Boer sealing concept. The regional top seal for the Rotliegend is the base of the Zechstein (red line in upper image). Where the Ten Boer Member is sufficiently shaly, it provides a seal in fault juxtaposition (red line in lower image), even though intra-Ten Boer Member sandstones may be gas-bearing, and on the same pressure gradient as the gas in the Slochteren sandstones. Modified from De Jager and Geluk (2007).

for underground gas storage (Juez-Larré et al., 2025, this volume).

Along the northern margin of the Southern Permian Basin, in the area of the Elbow Spit and Mid-North Sea highs, the Zechstein also comprises slope and platform carbonates as proven by well E02-02. Although quite thin, carbonate build-ups in the Zechstein may have better reservoir quality further north. In this context the recent discovery of gas in the Pensacola prospect along the northern margin of the basin on the UK side is supportive (jpt.spe.org/shell-makes-gas-discovery-at-pensacola, 2023).

A special case is the gas accumulation discovered in 1985 by well G16-01 on the Schill Grund High, where a leached Z3 Carbonate and an overlying Upper Jurassic sandstone form a single composite reservoir above a Zechstein salt dome. The Lower Cretaceous Vlieland Claystone Formation provides the seal.

In the West Netherlands Basin, the four Zechstein cycles are developed as siliciclastic basin-margin facies ('fringe sandstones'- Fig. 14.18). The best reservoir was found in offshore block P10, where a sandstone interval reaches a thickness of ca. 125 m. Since no Zechstein evaporites are present here to provide a top seal, predominantly sub-commercial quantities of gas have been encountered in the area. An exception is the Zechstein reservoir unit of the Q10-A field that recently came on stream in combination with Rotliegend and Triassic sandstone reservoirs. Recent exploration activities for Zechstein reservoirs are limited to the Friesland area where several small fields, often associated with gas-bearing reservoir rocks in Rotliegend and/or Vlieland Sandstones were discovered.

Triassic plays

Triassic sandstones are widely distributed in the Netherlands and form the second most important gas reservoir, next to the Rotliegend sandstones. Producing Triassic fields occur in the east of the Netherlands, in the West Netherlands Basin and in the northern and western offshore (Fig. 14.19). The top seal is in most cases formed by Triassic shales. Where the Triassic is truncated by the Base Cretaceous Unconformity, Lower Cretaceous shales form the top seal. The Solling Claystone Member seals accumulations in the Main Buntsandstein (Spain & Conrad, 1997), whereas the clayey Muschelkalk and Keuper formations seal accumulations in the Röt. The gas was mainly sourced from Westphalian coal measures, and partly also from the basal Namurian hot shales (Fontaine et al., 1993; De Jager et al., 1996; De Jager & Geluk, 2007) and is trapped in a range of styles. In the northern sector of the Dutch subsurface, the underlying Zechstein salt forms a barrier that prevented Westphalian gas from reaching Triassic reservoirs. Only where this regional seal is breached by salt withdrawal or by faulting, could Westphalian gas charge Triassic reservoirs. In the West Netherlands Basin, Zechstein salt is absent and the Triassic sandstones form the first well-sealed reservoir above the Westphalian. The Triassic sandstones were deposited in fluvial and eolian settings and were derived from the Variscan mountains towards the south. In the south, the Main Buntsandstein Subgroup forms a thick massive reservoir package. To the north this package splits into several thinner sandstone units, such as the sandstones of the Volpriehausen and Detfurth formations, which are separated by clay- and siltstones that form regional seals (Van Adrichem Boogaert & Kouwe, 1994; Ames & Farfan, 1996; Geluk, 2005; Bachmann et al., 2010; McKie & Kilhams, 2025, this volume).

The first Triassic discovery was the Coevorden field in 1948. This tilted fault block was found to be gas-bearing in the Main Buntsandstein Subgroup and in the Upper Triassic Muschelkalk, Keuper and Röt, as well as in the underlying upper Carboniferous. The Buntsandstein is also a proven gas reservoir in the eastern Netherlands. The Wanneperveen and De Wijk fields (Gdula, 1983; Bruijn, 1996; Pipping et al., 2001) produce from reservoirs that are sealed by shales of the Lower Cretaceous Vlieland Claystone Formation. Reservoir characteristics are enhanced as a result of leaching below the Base Cretaceous Unconformity. Even leached anhydritic shales of the Lower Triassic Rogenstein Formation are productive in these fields. Channel sands at the base of the Vlieland Sandstone Formation locally form part of the complex traps.

In the West Netherlands Basin, exploration for Triassic gas accumulations only effectively started in 1982. The typical trap in this play consists of a Late Jurassic horst block in which the reservoir is sealed vertically by Upper Triassic evaporitic shales and laterally by Upper Triassic to Lower Jurassic shales (Fig. 14.20; De Jager et al., 1996). Lateral seal risks exist where fault-throws are so large that Triassic sandstones are juxtaposed against the Jurassic Delfland sequence (Delft Sandstone Member). Gas discoveries line up along the southwestern border of the West Netherlands Basin on a WNW-ESE trending structural trend consisting of tilted fault blocks or inverted flower structures. The best reservoirs occur along its southwestern margin, with excellent gas production rates of several million cubic metres per day being reached in some fields. Towards the north, reservoir quality deteriorates rapidly due to erosion of the best reservoir section in the upper part of the Main Buntsandstein Subgroup, in combination

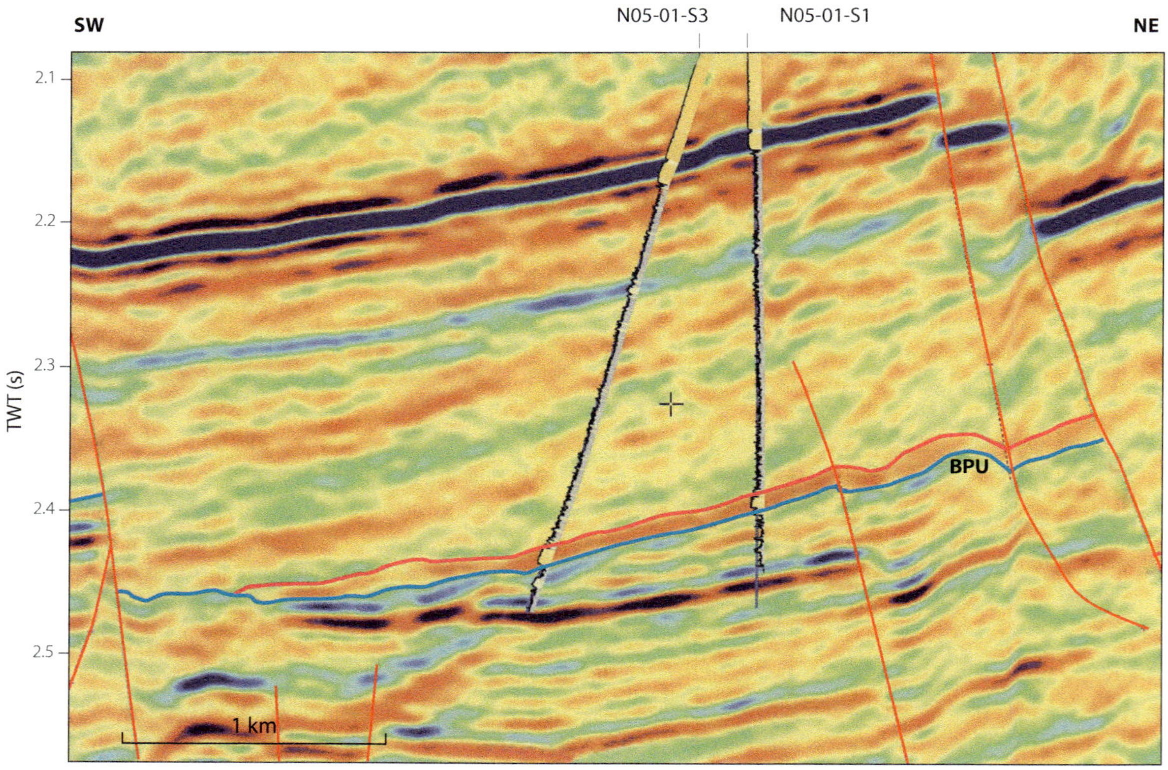

Figure 14.17. Seismic time section (freely available at www.nlog.nl) showing the overall setting and the tilt-block trap style of the Ruby field, a basal Slochteren clastics (Upper Rotliegend Group) gas discovery made in 2017. Gamma-ray logs of N05-01-S1 and N05-01-S3 show that this thin basal sandstone unit (yellow in the log) directly overlies the Base Permian Unconformity (BPU; in light blue). A high acoustic impedance contrast is associated with these reservoir sandstones (modified from Burgess et al., 2018 and Doornenbal et al., 2019).

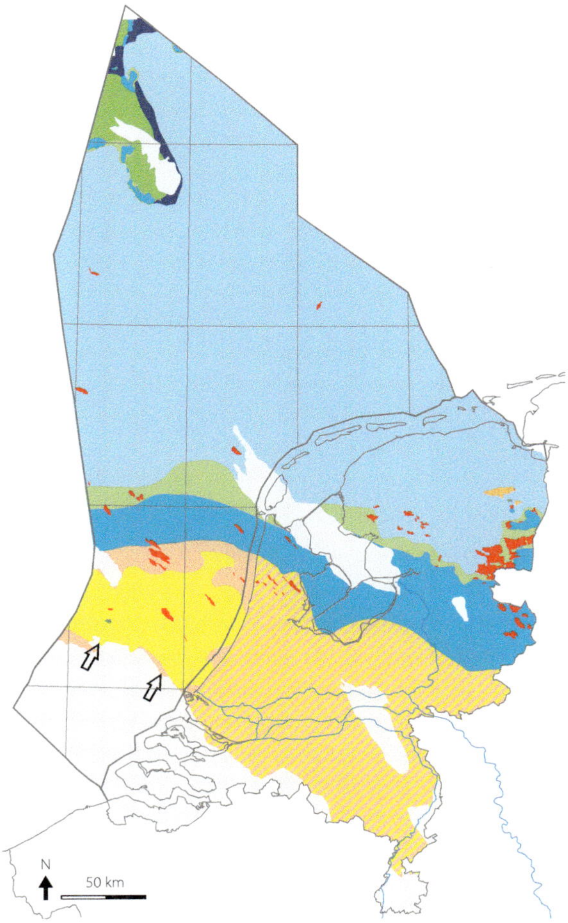

Figure 14.18. Zechstein play map showing the facies distribution of the Z2 Carbonate and the location of prospective Z1 to Z4 fringe sandstones in yellow. Gas accumulations in the central offshore are located outside the distribution area of the Z2 carbonate reservoirs and reside in the Z3 Carbonate. Modified from Bouroullec et al. (2022).

with a general increase in pre-inversion burial. This deep burial in inverted basins caused diagenetic reduction of porosity and permeability as can be seen in a number of fields in blocks P02 and P12 (Van Hulten, 2009). Furthermore, because of tectonic inversion during the Late Cretaceous, traps in the central and northern part of the West Netherlands Basin changed their structural configuration or were even destroyed, while new ones formed. At the same time, following uplift and declining heat flow most of the Carboniferous and the Jurassic source rocks stopped generating hydrocarbons in this part of the basin (De Jager et al., 1996). A timing problem thus exists for new traps that formed during tectonic inversion. In the southern part of the West Netherlands Basin, where no or very little inversion occurred, gas generation continues to the present day.

In the Jurassic basins in the South (West Netherlands Basin, Central Netherlands Basin and Broad Fourteens Basin) gas-bearing Triassic sandstone reservoirs occur along the WNW-ESE structural trend. Although the Triassic play in the West Netherlands Basin is largely a gas play, in several accumulations (e.g. Papekop, Pernis West, Spijkenisse Oost, Waalwijk, Zuid Botlek, Oud Beijerland-Noord, Q16-Maasmond and offshore in the P10/P11 blocks) an oil leg was found underneath the gas, while the Ottoland structure predominantly contains oil. The oil has been matched to the source rocks of the Lower Jurassic Posidonia Shale, with probably some contribution from the Lower Jurassic Aalburg and uppermost Triassic Sleen formations, and possibly also from Carboniferous rocks (De Jager et al., 1996). This oil migrated across faults into the Triassic reservoirs from younger source rocks in adjoining down-thrown blocks. Where the Posidonia Shale has not been downfaulted to below the Triassic reservoir, as is the case for some oil-bearing structures, it is very unlikely that it could have charged these traps. In those cases, the oil is most probably entirely sourced from the Aalburg and Sleen Formations (Van Balen et al., 2000).

In the northern offshore, sandstones of the Volpriehausen and Detfurth formations are reservoirs for gas fields located in the northern Vlieland Basin, Terschelling Basin, Schill Grund High and southern Dutch Central Graben area. In 1992, the L5-FA field was the first sizeable Triassic field to come on stream in the northern offshore. This field lies on the NE-SW structural trend that dominates the Southern Central Graben and the Terschelling Basin. In these blocks most traps are related to halokinesis of the underlying Zechstein salt, and occur in turtleback anticlines, against salt walls or as fault-dip closures (Fig. 14.21; see also Bouroullec & Ten Veen, 2025, this volume). Examples of these trap types are the economic accumulations in L2, F15 (Fontaine et al., 1993) and G14, G16, and G17.

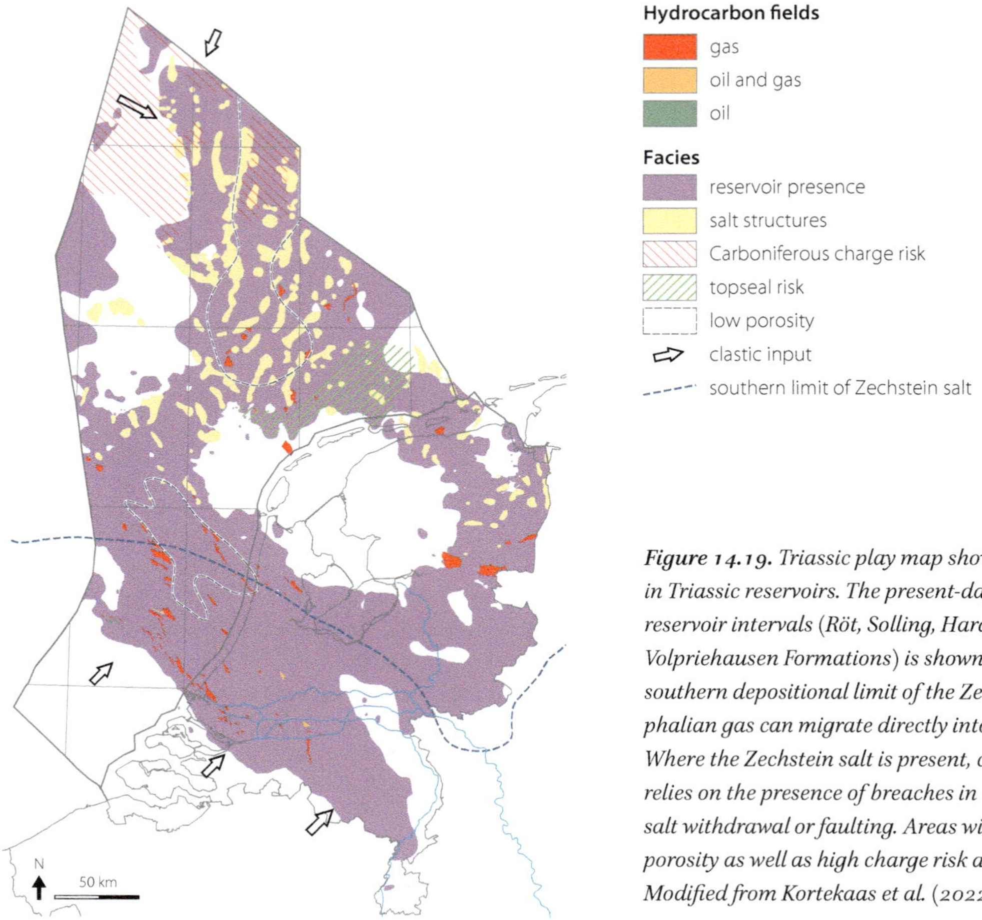

Figure 14.19. *Triassic play map showing gas and oil fields in Triassic reservoirs. The present-day distribution of the reservoir intervals (Röt, Solling, Hardegsen, Detfurth and Volpriehausen Formations) is shown in purple. South of the southern depositional limit of the Zechstein salt, Westphalian gas can migrate directly into Triassic reservoirs. Where the Zechstein salt is present, charge into the Triassic relies on the presence of breaches in the Zechstein due to salt withdrawal or faulting. Areas with weak top seal, low porosity as well as high charge risk are indicated on the map. Modified from Kortekaas et al.* (2022).

Salt walls in the Terschelling Basin and southern Dutch Central Graben seal-off separate pressure cells that contain Triassic reservoirs (Crepieux et al., 1998). In some of these cells, reservoir pressures equal the minimum horizontal stress, thus causing the seal to be breached. Deeper culminations in these pressure cells may be protected against seal failure.

Sandstones of the Triassic middle Solling Sandstone Member form an exceptional reservoir in the L9 area. Well L09-07, drilled in 1992, found this hitherto unknown reservoir to be gas-bearing (Geluk, 2005; De Jager, 2012). Subsequent development drilling of the L09-FD, FF and FI complex of fields established total recoverable gas reserves of ca. 31 x 10^9 m^3 (NAM, 2014) in the so-called 'Fat Sand' reservoir. The 'Fat Sand' is up to 125 m thick and has a net-to-gross ratio of close to 100% with average permeabilities in the Darcy range. Gas production rates from this extremely prolific reservoir are typically several million cubic metres per day per well. To the north, in block L6, the same reservoir is salt-plugged. The unusually large thickness of the Solling Sandstone in part of block L9 implies a unique depositional setting. Extensive core coverage indicates that the reservoir mainly consists of eolian sandstones, with locally some water-laid sediments. Well- and seismic data show that the reservoir package is wedge-shaped and thickens into a listric normal fault that detaches onto the top of the Zechstein salt, adjacent to a salt wall (De Jager, 2012). The eolian sandstones are clearly preserved in a syn-depositional half-graben, the development of which appears to be related to Early Triassic extension. This development has proven to be rather local, and no further discoveries of these 'Fat Sands' have been made since the discoveries in L9 (De Jager, 2012).

Salt plugging of reservoirs is a serious risk to the Triassic play in the northern offshore (Purvis & Okkerman, 1996). Salt-plugged reservoirs have been encountered near salt walls and along fault planes and are often characterized by a phase reversal of the seismic response at top reservoir. Careful study of seismic amplitudes may indicate whether good reservoirs are present or not.

Because of the economics-driven focus on near-field exploration, recent Triassic discoveries are concentrated around known fields. Most discoveries are in the onshore part of the West Netherlands Basin in the Rotterdam area and offshore in the P10-P11 and G16 clusters. All discoveries are limited in volume (between 0.5 and 2.5 10^9 m^3). To date, no Triassic discoveries have been made on the Cleaverbank Platform area. Because of deep erosion, the

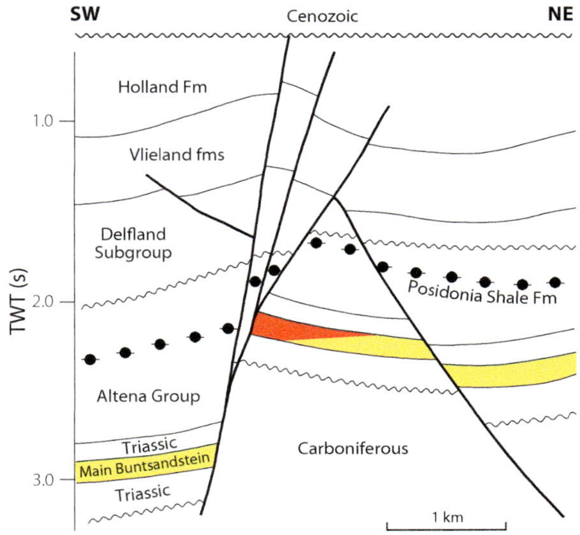

Figure 14.20. Example of a trap in the Triassic of the West Netherlands Basin (Wassenaar Deep gas field). The Main Buntsandstein reservoirs (yellow) have younger Triassic shales as top seal and are laterally sealed by Lower Jurassic shales. Modified from De Jager and Geluk (2007).

Triassic is in many places truncated by the Base Cretaceous Unconformity. Associated leaching has resulted in excellent reservoirs, with average porosities as high as 27%. The lack of exploration success in the Cleaverbank Platform area is attributed to the absence of charge through windows in the thick sealing claystones and evaporites of the Rotliegend and Zechstein. Kortekaas et al. (2022) suggested a northern provenance of reservoir sands in the Step Graben to the west of the Central Graben. This resembles the locally sourced fluvial deposits in the North German Basin as described by Olivarious et al. (2017). Charge for any trapping structures in the Cleaverbank Platform area may come from Namurian shales or Carboniferous coals, which are likely to be present at depth and in the gas window (Schroot et al., 2006; De Bruin et al., 2015; Ter Borgh et al., 2018). Migration paths may be along faults, through windows in the salt seal or, as recently demonstrated, along volcanic dykes (Underhill, 2009; Kortekaas et al., 2018). Given the attractive trapping configuration, migration is the main risk due to the presence of extensive Zechstein salt layers. Kortekaas et al. (2018) have identified 42 leads in their regional study. However, none of these has been drilled yet.

Upper Jurassic and Lower Cretaceous plays

Upper Jurassic and Lower Cretaceous oil plays are restricted to the Cimmerian rift basins, such as the Dutch Central Graben, Broad Fourteens, Lower Saxony and West- and Central Netherlands basins. The reservoirs occur in clastic syn-rift and early post-rift deposits and show rapid facies changes ranging from continental to marine. There is also a great variation in trap styles. These plays are much more oil-prone than the other Dutch plays. This is mainly because of oil charge from the Lower Jurassic Posidonia Shale, which is only present in the extensional Late Jurassic-Early Cretaceous rift basins (Fig. 14.11). Most of the gas present is derived from the Westphalian, and in the Dutch Central Graben also from Jurassic coal-bearing sequences and/or deeper Jurassic shales of the Aalburg Formation.

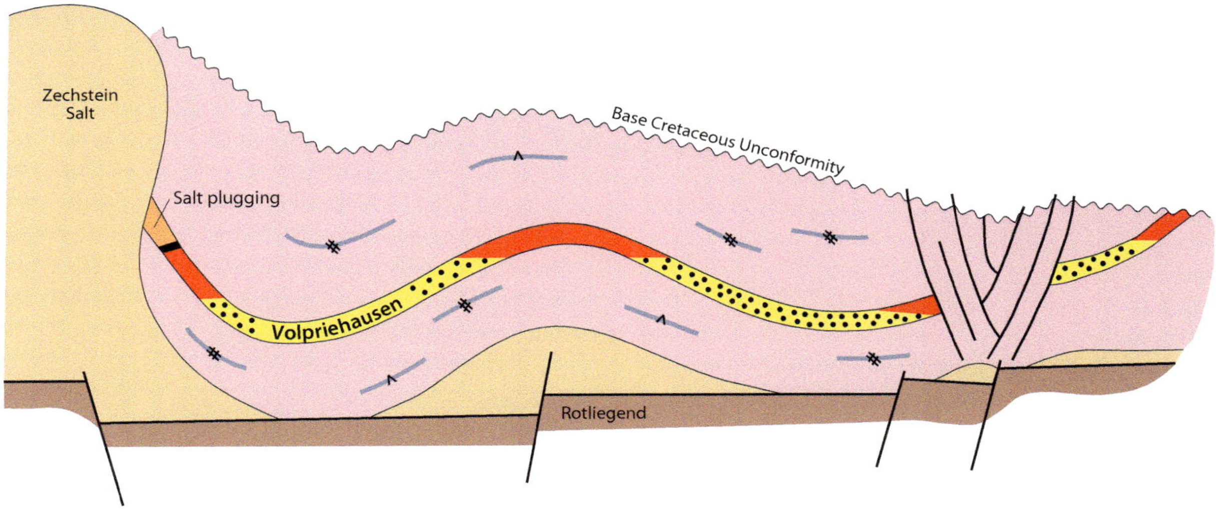

Figure 14.21. Trapping styles in the Triassic of the northern offshore. Most gas (red) is trapped in four-way dip-closed structures above salt swells or in turtle-back structures. Where the Triassic is truncated at the Base Cretaceous Unconformity, Lower Cretaceous marine shales may provide the seal. Salt-plugged reservoirs may provide lateral stratigraphic seals. Modified from De Jager and Geluk (2007).

In the east of the Netherlands, the Schoonebeek field has produced a total of ca. 250 million barrels (40 x 10^6 m^3) of oil from the Lower Cretaceous Bentheim Sandstone. The accumulation at a depth of ca. 800 m is exceptional, as it is sourced from lacustrine algal Type-I source rocks of Early Cretaceous age (Coevorden Formation). These source rocks have a limited distribution in the Netherlands but extend into the Lower Saxony Basin in Germany (Binot et al., 1991). The relatively low gravity of the oil of 25° API reflects the low maturity of the source rocks. Geochemical data indicate that the oil is an early expulsion product, without any sign of biodegradation. The field was discovered in 1943 and produced conventionally from 1947 until 1996. In 2011 the western part of the field was re-developed with a low-pressure gravity-assisted steam flood. The eastern part of the field is permanently abandoned (De Keijzer et al., 2009).

Gas fields at the level of the Lower Cretaceous Vlieland sandstones on the Friesland Platform in the northeast Netherlands are, amongst others, Tietjerksteradeel, Wanneperveen, Leeuwarden, Harlingen (Van den Bosch, 1983) and De Wijk. Additional discoveries have been made in the intermediate areas where Vermilion drilled several gas-bearing structures such as Vinkega, De Hoeve, Langezwaag and Diever. The main difficulty of this play is to predict the presence of the Vlieland Sandstone, the distribution of which is somewhat erratic and cannot readily

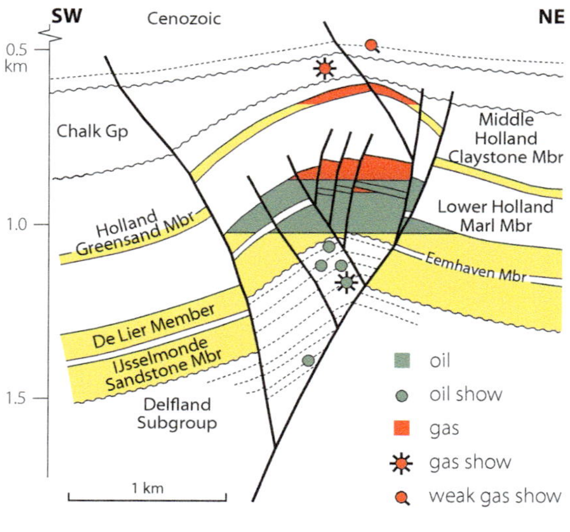

Figure 14.22. Example of a hydrocarbon trap in the Upper Jurassic and Lower Cretaceous in the West Netherlands (IJsselmonde-Ridderkerk field). Several stacked reservoirs are present, the shallower of which contain more gas and the deeper more oil. The complex anticlinal structure is a so-called 'flower structure' formed during strike-slip faulting related to inversion of the West Netherlands Basin and the Broad Fourteens Basin. Modified from De Jager and Geluk (2007).

be interpreted from the seismic data. Many of these fields have stacked reservoirs in which the Rotliegend, Zechstein and Vlieland Sandstone are all hydrocarbon bearing.

In the West Netherlands Basin, an intensive drilling programme in the 1950s led to the discovery of the Rijswijk, Pijnacker, De Lier, IJsselmonde, Wassenaar, Zoetermeer and Moerkapelle oil fields. Exploration from the late 1970s to early 1990s resulted in further discoveries, both onshore (Berkel, Barendrecht, Rotterdam, Pernis and Pernis West) and offshore (P8, P9, P11, P12, P15, and Q13). Rotterdam is the largest field, with initial reserves of ca. 90 million barrels (14 x 10^6 m^3). Both gas and oil are present in sandstones of the Upper Jurassic to Lower Cretaceous syn-rift and Lower Cretaceous post-rift deposits (Bodenhausen & Ott, 1981). Many of the older fields have been abandoned or closed-in over the last few years, and currently oil production is only from Rotterdam and the offshore fields Q13a-Amstel and the P15-Rijn. The best reservoirs are the post-rift Rijswijk, Berkel and IJsselmonde sandstones that were deposited as coastal barrier complexes overlying the Late Cimmerian Unconformity (Den Hartog Jager, 1996). Additional reservoirs are the younger De Lier and Holland Greensand sandstones as well as the older synrift Delfland Subgroup. Many of the reservoirs have oil columns with a gas cap, with the younger reservoirs generally containing more gas (De Jager et al., 1996). The presence of gas is often a downgrading factor, as it reduces the potential oil volumes in the traps, while the gas volumes are insufficient to justify development as gas fields.

Some of the fields contain very heavy oil with API gravities of 13° to 20°, resulting in low recovery factors (e.g. Moerkapelle and Wassenaar). The low API gravity is the result of biodegradation at the onset of the Cenozoic. Because of tectonic inversion, reservoirs were then at or near the surface, where bacteria had access to fresh meteoric waters (De Jager et al., 1996).

Traps are mainly four-way dip-closures along anticlinal trends that formed in response to the Late Cretaceous and early Cenozoic inversion along the southwestern basin margin (Fig. 14.22; Racero-Baena & Drake, 1996). Further north, in the more strongly inverted sectors of the West Netherlands and Central Netherlands basins, the target reservoirs have been eroded. Although the rapid facies variations within the Upper Jurassic and Lower Cretaceous suggest a potential for stratigraphic traps, these have yet to be encountered.

The Upper Jurassic and Lower Cretaceous oil and gas plays of the Broad Fourteens Basin are essentially the same as in the West Netherlands Basin, with the same source rocks and reservoir sandstones. All known trapped hydrocarbons are located along the northeastern margin of the inverted basin in Q1 (Helm, Hoorn, Helder, Haven and Halfweg fields; Roelofsen & De Boer, 1991) and in K18 and

L16 (Kotter and Logger fields; De Jong & Laker, 1992; Goh, 1993). These oil fields have ceased production. In the inverted areas of the Broad Fourteens Basin, as in the West Netherlands Basin, there is a timing problem regarding hydrocarbon generation and trap formation. In these areas, the Carboniferous source rocks became mature in the Late Jurassic or the Cretaceous, prior to the main phase of trap formation, which is Late Cretaceous or even younger.

In the Dutch Central Graben, several oil discoveries were made in Upper Jurassic and Lower Cretaceous sandstones (Wong et al., 1989; Wong, 1991). The largest field is F3-FB, where several sandstones in the Upper Jurassic Central Graben Group contain light oil/condensate (ca. 50° API) and gas. The volume of oil initially in place in this field is ca. 100 million barrels (16 x 10^6 m^3). Other oil accumulations are in F3-FA, F14-FA, F17-FA, F17-FB, F18-FA, L1-FB, L2-FA and L5-FA. Of these only F3-FA was developed as a gas-condensate producer. Gas caps are present in F3-FB and F18-FA. Most of these hydrocarbons have been generated by the Posidonia Shale, with minor contributions from the Aalburg and Sleen formations and from coals in the Central Graben Group.

The traps in the graben occur either in either four-way dip-closures in salt-related turtle-back structures (F3-FB, Fig. 14.23) or in tilted fault blocks. The southern sector of the graben has been more strongly inverted, resulting in greater structural complexity and compartmentalization of the traps. Reservoir distribution is another complicating factor in that the syn-tectonic sedimentary sequences have been deposited in fluvial, deltaic and lagoonal environments. This is especially the case in the southern part of the graben, where the sand-to-shale ratios vary greatly, and where individual sands seem to have a limited lateral extent.

In the Terschelling Basin only one small gas discovery has been made in block L6. This lack of success is attributed mainly to a lack of lateral and top seals. Gas charge may be limited as well, as migration paths from the Westphalian through the Zechstein evaporites are tortuous at best. Coals within the Central Graben Group are not very extensive but may nevertheless have contributed some gas charge. In the Wadden Sea area of the Vlieland Basin, the large Zuidwal field produces gas from the Vlieland Sandstone Formation in a drape structure over an Upper Jurassic volcanic complex (Perrot & Van der Poel, 1987; Herngreen et al., 1991).

Upper Cretaceous chalk play

The Chalk Group forms the reservoir in several major oil fields in the Norwegian and Danish sectors of the North Sea (e.g. Ekofisk, Dan, Gorm and Skjold). In the Netherlands, the first and so far only economic oil accumulation in the chalk is the Hanze field, which was discovered

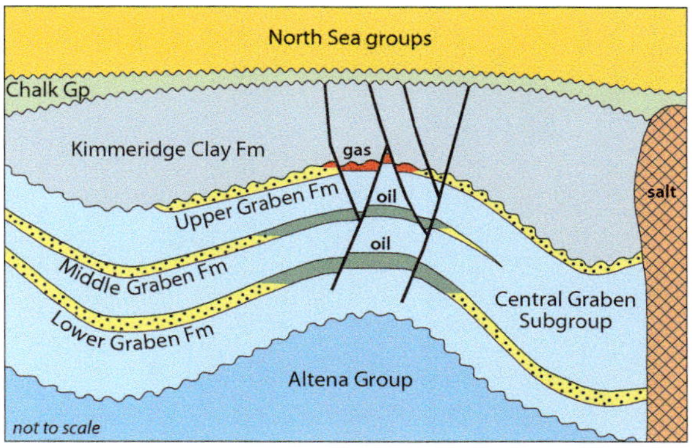

Figure 14.23. The F3-FB field in the northern part of the Dutch Central Graben. The three main Upper Jurassic sandstone reservoirs are stacked four-way dip-closed turtle-back structures. The reservoirs are compartmentalized by the faults and have different hydrocarbon-water contacts. Modified from De Jager and Geluk (2007).

in 1996 in offshore block F2 in the north of the Dutch Central Graben. The Ommelanden and Ekofisk formations constitute the reservoirs. This field is the largest producing oil field in the offshore Netherlands with a STOIIP of 20.5 x 10^6 m^3 (Dominik et al., 2000; Dana Petroleum, 2014). The success of the Hanze field initiated an increased interest in the chalk play (Van der Molen, 2004; Van der Molen et al., 2005; Guasti et al., 2009, 2010). In several structurally valid tests, oil shows have been encountered in the chalk, in particular in the Dutch Central Graben area. This includes the F17-Rembrandt and F17-Vermeer oil accumulations discovered in 2012 (Van Lochem, 2018) as well as the F06b-Snellius and F06b-Zulu oil accumulations (both discovered in 2014). So far none of these discoveries has come on stream. In most of these cases the oil is likely to have been derived from the Posidonia Shale. The Upper Maastrichtian and Danian parts of the Chalk Group in general have a fair reservoir quality for oil and gas reservoirs, due to the presence of fractures in the very fine-grained chalk. These fracture networks resulted from diapirism in the underlying Zechstein salt pushing the chalk upwards thus forming domal structures.

The only onshore gas field that has produced from the Chalk Group is the Harlingen field (Van den Bosch, 1983), but due to severe production-induced subsidence, it ceased production in 2008. Subsidence significantly exceeded predictions in both rate and amount (23 cm since the start of production in 1988). Strong subsidence is assumed to be the result of pore collapse and pressure solution within very porous layers of the reservoir (De Waal et al., 2016). In the West Netherlands Basin, gas has been tested from the Maastrichtian and Danian parts of the Chalk Group in the IJsselmonde structure. In the De Wijk field, chalk is assumed to be gas bearing as well (NAM, 2016a).

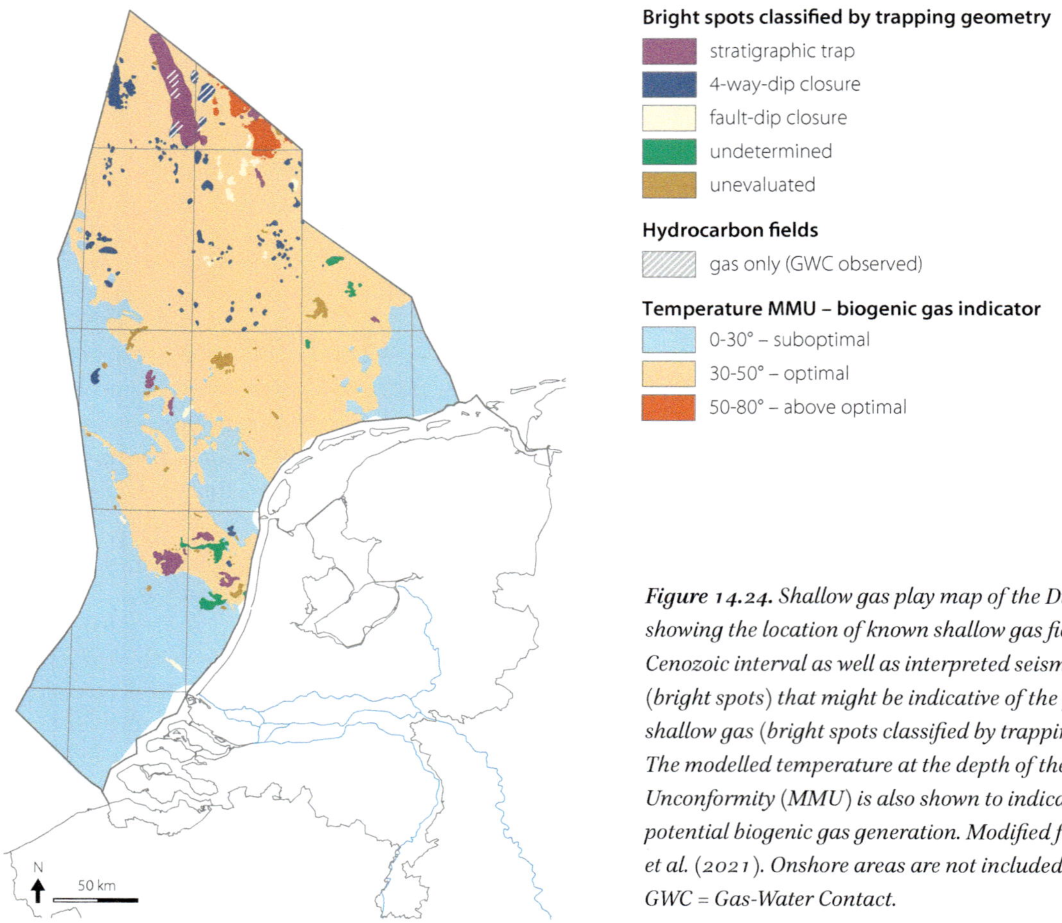

Figure 14.24. Shallow gas play map of the Dutch offshore showing the location of known shallow gas fields in the Cenozoic interval as well as interpreted seismic anomalies (bright spots) that might be indicative of the presence of shallow gas (bright spots classified by trapping geometry). The modelled temperature at the depth of the Mid Miocene Unconformity (MMU) is also shown to indicate zones of potential biogenic gas generation. Modified from Ecclestone et al. (2021). Onshore areas are not included in this study. GWC = Gas-Water Contact.

The Upper Maastrichtian and Danian parts of the Chalk Group in general have a fair reservoir quality for oil and gas due to the presence of fractures in the very fine-grained chalk. These fracture networks resulted from diapirism of the underlying Zechstein salt pushing the chalk upwards and forming domal structures. Although the reservoir characteristics of the Chalk Group are generally only poor to fair, they are comparable to those of the chalk in the Danish and Norwegian sectors of the North Sea. The lack of success in the chalk in the Netherlands is possibly related to the relatively shallow depth of burial and to a limited charge. The shallow burial has resulted in a reduced sealing capacity of the Paleogene above the somewhat overpressured chalk reservoirs, causing leaky traps.

Cenozoic plays (shallow gas)

The main Cenozoic gas accumulations occur offshore in the A and B quadrants at depths of 300 to 800 m in Plio-Pleistocene sediments of the westward prograding Eridanos shelf-edge delta system. Because of the shallow depth this offshore play is often pragmatically referred to as 'shallow gas play' (Fig. 14.24). Strong amplitude anomalies or bright spots and deeper velocity pull-downs on seismic profiles clearly indicate the presence of gas in subtle four-way dip structures with a structural relief of only a few tens of metres. Structures are often situated above underlying diapirs, either due to Cenozoic salt rise or differential compaction (see also Bouroullec & Ten Veen, 2025, this volume). The more porous silty-sandy layers form the reservoirs while intercalated clays provide the necessary sealing capacity. Many of the shallow gas accumulations seem to be associated with gas chimneys, suggesting that leakage has taken place (Schroot & Schüttenhelm, 2003; Schroot et al., 2005; Ten Veen et al., 2014) from a deep thermogenic gas source, while gas composition and isotopes indicate a mostly biogenic source. De Bruin et al. (2022) argued that in some cases these gas chimneys may be artefacts of transmission and attenuation of the seismic signal below the shallow accumulations rather than indicating actual leakage.

The Plio-Pleistocene gas-bearing silts- and sandstones had long been recognized on seismic and had been encountered in wells drilled to deeper objectives. Eight shallow gas accumulations were discovered in the late eighties-early nineties. Although potentially economic production rates were tested from the silty sands in the topsets of prograding shelf-edge delta deposits in 1987 (well A18-02-S2), development was still slow for a number of reasons. Pipeline infrastructure did not yet reach these remote fields, sand production from these unconsolidated sands posed a technical challenge, and the need for early compression formed an economic hurdle. It thus took some twenty years before the first shallow-gas discovery (A12-FA) came on stream in 2007. The A12-FA field ranks amongst the better producing gas fields in the Netherlands with production rates around 3 million Nm^3/day from six wells. Currently production also comes from A18-A, B13-A and F02-Pliocene, while A15-A, B10-A and B16-A are expected to be developed in the near future. For B17-A development seems not viable. The reduction of pore volume by production-induced reservoir compaction has had a positive impact on the reservoir pressure and therefore recovery. The potentially negative effect of compaction on reservoir quality, i.e. reduction in permeability, is negligible (Van den Boogaard & Hoetz, 2012).

Gas in onshore Cenozoic sediments is of minor importance and is mostly associated with gas accumulations in traps with stacked reservoirs. Examples of these settings are the IJsselmonde structure in the West Netherlands Basin and the De Wijk field in Drenthe. In the IJsselmonde field small amounts of gas have been tested from the Eocene Oosteind Member, which lies some 30 m above the base of the Lower North Sea Group and has average porosities ranging from 34 to 39%. The De Wijk Member produced over a short period of time but was shut in due to sand and clay influx. In the De Wijk field the gas trapped in the tuffaceous sediments of the De Wijk Member is associated with underlying accumulations in the Liessel Member, Lower Cretaceous and Triassic reservoirs. Although the deeper parts of the gas field have been developed, the reservoir compaction of the unconsolidated De Wijk Member was very high and the consequent subsidence was such that production had to cease after only a small fraction of the GIIP had been produced. Recently production from the De Wijk Member was resumed in combination with the injection of nitrogen. This should drive the natural gas towards the production wells and at the same time keep the pressure from depleting (enhanced gas recovery). This way of producing has been successfully tested in an earlier phase in the underlying reservoirs of the De Wijk field (NAM, 2016a).

Gas compositions

The composition of the hydrocarbon and non-hydrocarbon components of natural gas are influenced by both the source rock and by secondary processes (e.g. mixing, migration fractionation, secondary cracking, biodegradation, water washing). Several studies have evaluated the gas composition of the Dutch on- and offshore (Lokhorst, 1997; De Jager & Geluk, 2007; Doornenbal & Stevenson, 2010). More recent studies have updated these compilations with newly released data and interpretations with respect to potential gas sources and secondary processes (Verweij et al., 2018).

Hydrocarbon gas

The main differences in hydrocarbon-gas composition are linked to the mixing of gasses expelled from different source rocks as well as to differences in source rock maturity. The gas-dryness ratio (C1/C2+) of the Carboniferous, Rotliegend and Triassic reservoirs, as well as that of the Jurassic and Cretaceous reservoirs in the Vlieland Basin, Friesland Platform and Terschelling Basin, varies between 10 and 50. This mostly reflects variations in the maturity of the upper Carboniferous coals that sourced the gas. The most prominent exception is in the northern part of the Cleaverbank Platform, where the gas-dryness ratio suggests a much higher source-rock maturity than is measured on local, unaltered coal samples. This may be related to the local presence of igneous intrusions in the area, to longer migration distances or to the contribution of a more mature, deeper source rock such as lower Carboniferous coals. Some of the gas in the Zechstein reservoirs shows differences compared to other reservoirs, in isotopic composition as well as in the dryness ratio. This is most likely caused by secondary processes such as thermogenic sulphate reduction (TSR), as well as by source mixing. Low gas-dryness ratio of gas in Triassic reservoirs in the West Netherlands Basin indicates mixing of gas sourced from the Carboniferous with younger oil-associated gas. Similarly, the percentage of C2+ components in gas from Jurassic and Cretaceous reservoirs in the West Netherlands Basin, Broad Fourteens Basin and Dutch Central Graben is also high and related to oil-associated gas. This oil-associated gas is probably sourced from the Lower Jurassic Posidonia Shale Formation or from Upper Jurassic coals (De Jager et al., 1996).

Important non-hydrocarbon components in gas accumulations in the Netherlands, are nitrogen (N_2), carbon dioxide (CO_2) and hydrogen sulphide (H_2S). While N_2 and CO_2 mainly influence the calorific value of the gas, H_2S poses serious safety and health risks during drilling as well as during production, and significantly increases development costs.

Nitrogen

In Dutch gas accumulations nitrogen is the most abundant non-hydrocarbon component. The nitrogen content is highly variable and ranges from less than 1% to – locally – more than 85%. These ranges are also apparent on the larger scale of the Southern Permian Basin/Central European Basin System (Lokhorst, 1997; Gras & Clayton, 1998; Gerling et al., 1999; Krooss et al., 2005, 2008; Pletsch et al., 2010; Kotarba & Nagao, 2015). In general, sources of nitrogen in natural gas accumulations include atmospheric nitrogen, organic nitrogen in organic matter, such as peat, lignite, coal, anthracite, and dispersed organic matter in clastic and carbonate rocks, inorganic nitrogen (ammonium) in mineral phases of sedimentary and metamorphic rocks, inorganic nitrogen in nitrate salts (KNO_3; $NaNO_3$), and even the deep mantle (Littke et al., 1995; Krooss et al., 2005; Jurisch et al., 2012; Hoefs, 2015). Main potential sources of N_2 in the Netherlands include:

1. Organic sources: Westphalian coal measures, of which the most important coal-bearing source rock is the Maurits Formation, are the principal source rocks for Dutch hydrocarbon accumulations and are considered also to be the main source for nitrogen. The present-day distribution and thickness of these formations varies laterally due to uplift and erosion related to the Variscan Orogeny (Fig. 14.9a) and the depositional facies of the Westphalian A and B show regional variations as well. Additional sources of nitrogen are dispersed organic matter in Namurian and Westphalian shales and secondary sources include coals of the Dinantian Yoredale Formation (in the northern offshore), Zechstein source rock intervals (Copper Shale, ZEZ2C), as well as younger gas-prone source rocks, such as coals in the Mesozoic Delfland Subgroup and Central Graben Group, as well as the oil-prone Jurassic source rocks;
2. Inorganic sources include release of inorganic nitrogen from clay minerals in Namurian and Westphalian shales through ammonium-potassium exchange, from inorganic-nitrogen bearing minerals present in volcanic rocks in the Lower Rotliegend Group (extending from the Ems Low in Germany into the Lower Saxony Basin in the northeastern Netherlands) and from volcaniclastics in the Upper Rotliegend (Fig. 4 in Gaupp & Okkerman, 2011);
3. Magmatic sources: potential magmatic sources are listed by Van Bergen & Sissingh (2007) and Paleogene sources are described by Hernandez Casado & Underhill (2013), Underhill (2009), Wall et al. (2010). Magmatism is further discussed in Van Bergen et al. (2025, this volume).

Gas accumulations in Upper Rotliegend and Carboniferous reservoirs with a N_2 content exceeding 10% occur in the Province of Friesland, in the Lauwerszee Trough, in Groningen and the adjacent part of the offshore Schill Grund Platform (including Ameland wells and the H16-01 well). The maximum N_2 content in the Lauwerszee Trough is less than on the Groningen and Friesland Platforms. Both nitrogen and hydrocarbons in the Groningen gas field probably originates from Westphalian sources in basins surrounding the Groningen field, while in the Lauwerszee Trough area the charge from Westphalian sources is local.

N_2-rich gas accumulations containing more than 15% N_2 occur in the northeastern onshore part of the Friesland Platform (Upper Rotliegend, Rijnland Group, De Wijk Member), in Zechstein 2 carbonates in the onshore Lower Saxony Basin, and in the northwestern offshore in the D & E quadrants of the Cleaverbank Platform, Step Graben (and locally in K17-02, P02-03 in the Broad Fourteens Basin). The N_2-rich gases in the Upper Rotliegend and Rijnland Groups in the Friesland Platform are probably related to timing and/or distance of migration. The other N_2-rich gas accumulations seem to be related to specific local conditions, such as:

1. Rapid and intense heating of organic and inorganic nitrogen sources due to magmatic activity, especially in Step Graben and on the Cleaverbank Platform. Of special interest in this respect is the possible influence of deep-seated Paleogene dikes.
2. Intra-Zechstein processes, including Thermogenic Sulfate Reduction in the Lower Saxony Basin and the Broad Fourteens Basin, which explain why N_2 rich gases in Zechstein 2 carbonates also contain H_2S.
3. Intra-reservoir conditions in the De Wijk Member, related to the mineral composition of tuffites in combination with the possible influence of fluid flow and microbial activity at the shallow burial depth of this North Sea Supergroup unit.

Low to very low N_2 contents have been measured in gas samples from the central and western onshore and the southern offshore parts of the Netherlands (Roer Valley Graben, West Netherlands Basin, Indefatigable Platform and southernmost part of the Broad Fourteens Basin), as well as from the southern part of the Cleaverbank Platform (K, J03, J06 blocks) and Central Offshore Platform. This suggests that low N_2 contents may be related to recent gas charge from low maturity source rocks. However, low nitrogen content also occurs in combination with relatively dry and possibly more mature hydrocarbon gases. Alternatively, the nitrogen content might also be related to facies characteristics of the source rock, in combination with varying maturity.

Hydrogen sulphide

Many Zechstein gas accumulations in the eastern Netherlands as well as in the Broad Fourteens Basin and Friesland Platform contain hydrogen sulphide resulting in so-called sour gas. The H_2S content of these sour gas fields is commonly below 5 Mol% but can reach up to more than 30 Mol% (e.g. Vlagtwedde: 45 Mol%).

The most common sources of sulphur in the Dutch subsurface are:

1. Dissolved sulphate in formation water, originating from anhydrite in the Zechstein and Upper Germanic Triassic;
2. Accumulations of oil and wet gas, and their associated sources. For oil, the sources include Type II source rocks, such as Zechstein Copper Shale and ZEZ2C (Stinkkalk/Stinkschiefer), Jurassic source rocks and Namurian Geverik Shale. For wet gas, the main sources considered here are the Westphalian coals and organic matter rich Carboniferous shales;
3. Carbonate reservoirs (carbonates of the Zechstein and Upper Germanic Triassic groups, Dinantian carbonates, Cretaceous Chalk Group).

H_2S is mainly generated by Thermochemical Sulphate Reduction (TSR), which strongly accelerates at temperatures above ca. 120°C (Orr, 1977). Consequently, significant quantities of sour gas occur mainly where the reservoir is buried sufficiently deep, and where some of the above-mentioned sulphur sources are present. However, the complex interplay of all controls on TSR and other H_2S-generating processes, such as generation from carbonate source rocks and biochemical sulphate reduction, hampers accurate prediction of H_2S concentrations in undrilled prospects (Mittag-Brendel, 2000; Krooss et al., 2008).

Carbon dioxide

Carbon dioxide occurs naturally in the subsurface, often in large volumes (e.g. Barnes et al., 1978; Wycherley et al., 1999; Baines & Worden, 2004; Pearce et al., 2004). A large number of possible sources of CO_2 have been identified worldwide (Wycherley et al., 1999; Baines & Worden, 2004). The large spatial variation in CO_2 content in natural gas accumulations in the on- and offshore Netherlands results from several factors, such as different organic and inorganic sources for CO_2, high solubility of CO_2 in formation fluids and its high reactivity. CO_2 percentages in Dutch gas accumulations range up to 77%, with a large spatial variation. Sweet gas with CO_2 <1% occurs in about half of the wells. High to very high CO_2 contents of 10-20% and >20%, have been encountered at 7 and 10 well locations, respectively.

Both organic and inorganic sources of CO_2 have been identified in Dutch gas accumulations; gas accumulations may contain a mixture of both sources:

1. Organic source: especially during early maturation both gas and oil-prone source rocks generate CO_2. Given the widespread occurrence of especially gas-prone source rocks, organic CO_2 is an important contributor. Gas samples containing CO_2 of organic origin are from different structural settings and stratigraphic units. Average CO_2 contents vary between 0.77 and 3.34 Mol%.
2. Inorganic source:
   a. CO_2 in gas accumulations in Zechstein carbonates (wells Geesbrug-1, Q02-03, Den Velde-2, -03 and Middelie-302) originated locally from dissolution of the carbonate rock, The high CO_2 content in well Geesbrug-1 may be due to the presence of H_2S and related carbonate dissolution by acidic fluids (Lokhorst, 1997; Gras & Clayton, 1998; Liu et al., 2014).
   b. Gas accumulations with more than 20% CO_2 are spatially related to Early Jurassic-Early Cretaceous rift-related magmatism and Paleogene magmatism. The high heat flow might have increased CO_2 generation from inorganic carbonate sources, with a possible contribution of direct CO_2 outgassing from the magmatic source;

Mixed organic-inorganic origin of CO_2 occurs where an organic origin is predominant, such as on the Cleaverbank Platform (in Upper Rotliegend reservoirs), while a predominantly inorganic origin is spatially related to the presence of igneous intrusions (Broad Fourteens and West Netherlands basins).

Future potential

The Dutch government's policy is to gradually transform to a CO_2 free energy system by 2050. As long as natural gas is needed domestic production is preferred over import. For that reason offshore production will be stimulated, but at the same time new exploration licences onshore will no longer be granted, no new licences for production from the Wadden Sea area will be issued and shale gas exploration and production will not be permitted anymore.

Proven plays

The Netherlands clearly is a mature hydrocarbon province. With most of the current licences covered with 3D seismic data and with many wells drilled, the main prospective fairways have already been well explored. Yet, new discoveries continue to be made, although there are indications of a reduced exploration efficiency or 'creaming' (Fig. 14.25). From 1970 to 2000 the average discovery rate was 2 x 10^9 m^3 per well, but from 2000 to 2020 this rate decreased to only 0.8 x 10^9 m^3 per well. The exact reason needs to be examined. It may be due either to risk-averse exploration with a focus on near-field developments or to the exhaustion of the prospect portfolio. Note that there is also a remarkable change in exploration activities from 32 wells per year in the period 1970-1997, via 9 wells per year

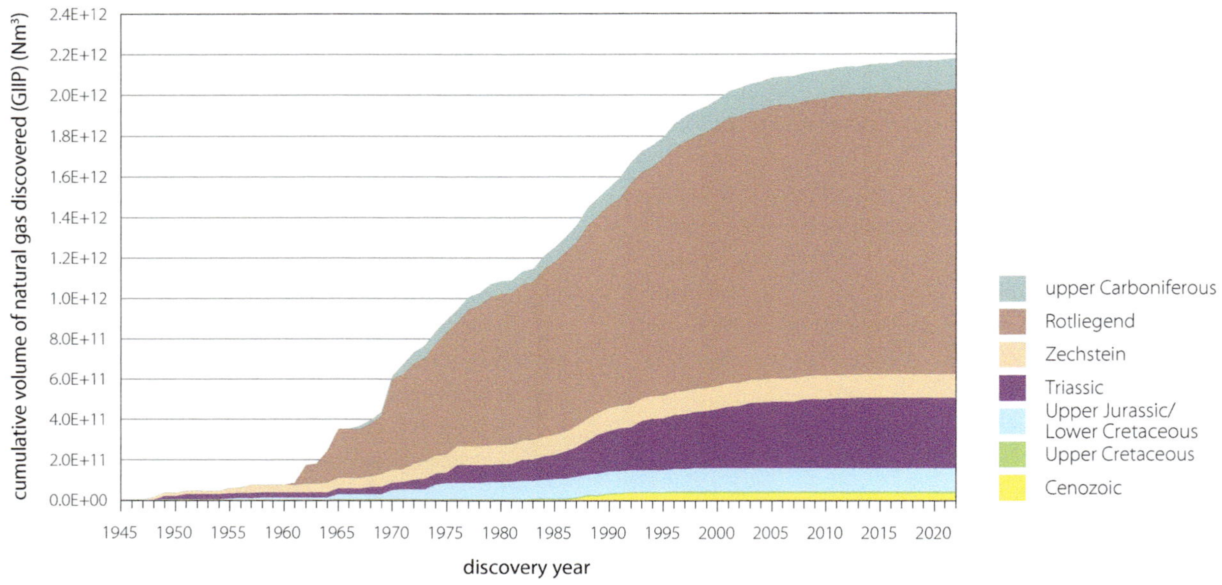

Figure 14.25. *Creaming curve of gas resources of the small fields (excluding Groningen). Note that the contribution of the Upper Cretaceous is minor. Source: TNO – Geological Survey of the Netherlands.*

from 1997-2015, decreasing to only 2-5 wells per year in the period 2015-2022 (Fig. 14.6).

The steady pace of new discoveries over the last two decades of the 20th century certainly partly resulted from the application of 3D seismic technology, which has revealed traps that hitherto were unnoticed. More recently advances in seismic processing, such as pre-stack depth migration, have further improved the accuracy of predictive subsurface models. In addition, geological concepts continue to be improved, resulting in the identification of new play opportunities and trapping styles. The future potential for new gas discoveries in the Netherlands, based on identified prospects in proven plays, as reported by the Ministry of Economic Affairs and Climate Policy (MEACP, 2022), is estimated to be approximately 80 x 10^9 m^3 of gas. Based on a scenario with four exploration wells per year and subsequent field development, a total production of 56 x 10^9 m^3 could potentially be realized in the next 25 years. These future reserves represent the addition of risked volumes of identified prospects in proven plays. These numbers do not include the unidentified potential that may result from identification of new plays and prospects, for example through the application of new technologies or extensions to proven play areas.

In a densely explored area such as the Netherlands, ongoing exploration continues to lead to new geological insights. The great variety of structural styles and potential reservoir-seal pairs only improves the chances that unidentified plays and prospects can be recognized. New technological developments are likely to provide increased accuracy to the predictions and a better understanding of play and prospect risks. A break-through to improve gas production from tight reservoirs may not only allow the exploitation of many accumulations that are currently uneconomic but will also broaden the scope for exploration. Similarly, a break-through in permeability prediction (e.g. from seismic data) would result in the de-risking of prospects that are currently considered high risk. And, last but not least, higher gas prices, lower economic thresholds or cheaper drilling- and production technologies may render presently non-commercial accumulations economic in the future.

New or speculative plays

In most of the Netherlands the Devonian occurs at depths greater than 5 or 6 km and is a speculative target at best. Dinantian carbonates on the northern flank of the London-Brabant Massif and in well Winterswijk-1 in the east of the Netherlands have been found tight and water-bearing. Nevertheless, some potential remains in karstified and fractured platform-edge carbonates sealed by Namurian shales that may contain source rocks (Van Hulten & Poty, 2008; Jaarsma et al., 2013). In the P blocks exploration activities focus on this play, but no wells have been drilled yet. The 53/14a-2 exploration well just across the median line in UK waters with a similar Dinantian target was disappointing.

Wells in northern Belgium have indicated good porosities and Darcy scale permeabilities in karstified Dinantian carbonates near Loenhout, east of Antwerp. The Loenhout gas storage project was developed in this carbonate reservoir. The same Dinantian carbonates are the target of a successful geothermal project in Mol (Belgium) and of a short-lived geothermal project near Venlo (Limburg). Though these Dinantian carbonates are directly overlain by the organic-rich Geverik Member, this has not given rise to any gas accumulations. Even though, the Geverik Member was a potential exploration target for shale-gas.

In the northern offshore, Viséan-Namurian sandstones of reservoir-quality are assumed to be present and Viséan-Namurian coals and shales are considered promising source rocks to charge these reservoirs in the northern part of the Dutch Central Graben and the adjacent Step Graben. The presence of a deep, mature Paleozoic source rock is supported by hydrocarbon shows and vitrinite reflectance data. In the southern E and F quadrants, charge may also occur laterally from upper Carboniferous Westphalian coals (Ter Borgh et al., 2018). Intra-Westphalian gas accumulations must rely on intra-formational seals and on lateral seals by favourable juxtaposition across faults. This concept is proven by a small and uneconomic gas discovery in block Q13. Also elsewhere in the Netherlands, intra-Westphalian shales occasionally seal small gas columns. Although no regional Westphalian seals have been identified, the 'marine bands,' which form laterally extensive but thin shale intervals, could prove to be effective as intra-formational seals.

The recent discovery of the Ruby and Turkoois fields straddling the Dutch German border (block N05 and N04, Fig. 14.17) proved the potential of the Basal Rotliegend Sandstone Member (Upper Rotliegend Group) in that area. This play certainly needs more appraisal drilling, but this discovery also focussed attention on possible gas accumulations in Rotliegend sandstones along the northern margin of the Southern Permian Basin in the A and B quadrants (De Bruin et al., 2015). Prospectivity in that area relies on finding structurally valid and well-sealed traps. A risk is the absence of Westphalian source rocks. However, Namurian organic-rich shales and Viséan coals (if present) display source-rock quality in nearby German and British wells.

Elsewhere, remaining Rotliegend potential is likely to be present in down-faulted traps relying on laterally sealing faults as was proven in the north of the Netherlands onshore (Corona, 2005). A technological break-through in the production of gas from tight reservoirs (permeability <1 mD) would open up additional potential, although as with the anticipated production of shale gas the economics may not be favourable.

Two potential Zechstein hydrocarbon objectives were proposed by Geluk (2000). Firstly, Z1 reefs, which could be present along the Dutch Central Graben, on the Mid-North Sea High and in the eastern Netherlands, as well as above topographic base-Zechstein highs elsewhere. Secondly, Z2 carbonates in slump and mass-flow deposits in the western offshore, in front of a carbonate platform that existed in the southern K and L quadrants. Furthermore, Z2 off-platform highs may occur in this area. The recent Pensacola gas discovery in Zechstein Carbonates in the UK part of the Southern North Sea may open up this play. For successful exploration of these plays, 3D seismic interpretation and high-resolution sequence stratigraphy form a prerequisite.

Above the Zechstein salt, the Triassic is in many places intricately structured, and several potential reservoir objectives are present. New prospects may be present in the A and B quadrants formed by fluvial reservoir rocks in local depocentres with local sourcing from the north. These hydrocarbons would have to come from the lower Carboniferous coals of the Elleboog and Yoredale formations or the Namurian shales.

Several oil discoveries in the Upper Jurassic and Lower Cretaceous of the Dutch Central Graben are sub-economic because of compartmentalization related to the presence of sealing faults in combination with a complex depositional reservoir architecture. Improved understanding of their depositional setting may lead to a better assessment of risks and to improved prediction of reservoirs, thus creating possibilities for development of these accumulations.

Finally, the shallow gas play has proven to be a valid gas play, but still many areas need further exploration to get a better grip on its presence and the gas saturations.

Unconventional gas or shale gas

The shale gas rush in the USA which started around 2005, quickly spread to other parts of the world, including the Netherlands. The first exploration licence that explicitly targeted shale gas in the Netherlands was applied for in 2009. Shale gas occurs in tight, organic-rich (low permeability) rocks which are of regional or sub-regional extent and, if gas-bearing do not have a well-defined hydrocarbon-water contact. Targets in the Netherlands were the Namurian Geverik Member in the Carboniferous, the Epen Formation and the Jurassic Posidonia shales (Van Bergen et al., 2013; Zijp et al., 2015, Fig. 14.26).

The Geverik member is a relatively thin (15-25 m) siliceous and bituminous, carbonate-rich black shale with a high TOC of 8-9% (Van Balen et al., 2000; Bouw & Lutgert, 2012; Zijp et al., 2015). In the Netherland this source rock was penetrated by 12 wells, but it is assumed to underlie large parts of the country. The main source of information on this source rock interval is well Geverik-1 which was cored over more than 1200 metres and included the entire Geverik Member. Due to its depth of burial its resource potential is assumed to be mainly restricted to the southern

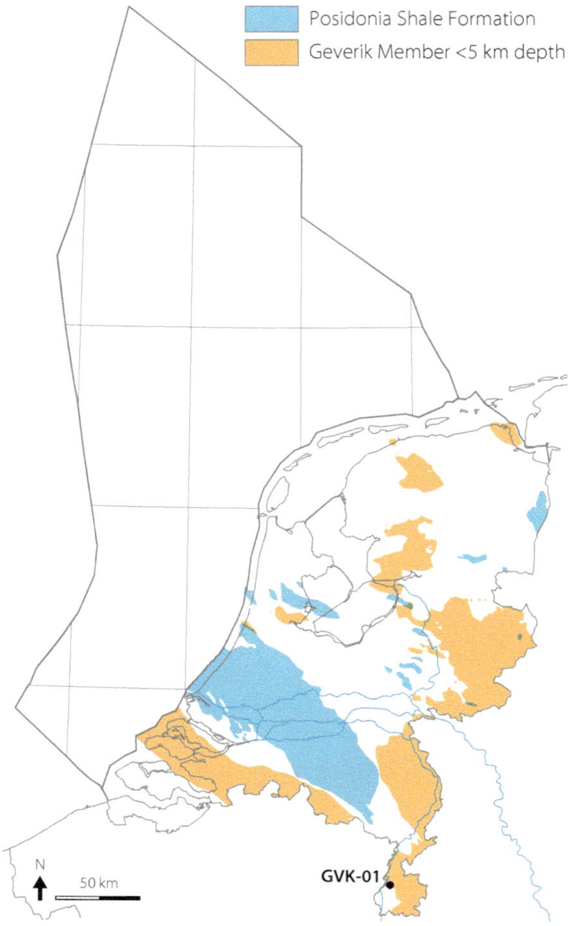

Figure 14.26. Onshore areas with shale gas/oil exploration potential for the Lower Jurassic Posidonia Shale Formation and the Namurian Geverik Member. For the Geverik Member only the distribution at a depth of less than 5000 m is shown as deeper burial is considered not to be prospective. Offshore shale gas production is assumed not to be prospective for economic reasons (from www.nlog.nl).

margin of the Carboniferous Basin where burial depths are less than 5000 m (Fig. 14.26).

The Jurassic Posidonia shale (Toarcian, 182-180 Ma) is the main oil source rock in the Netherlands. The formation has been through drilled over 100 times, both on and offshore, but never as the primary target of a well. For that reason, few samples and measurements are available. Extensive outcrop analogue and well sampling studies have shown variable potential (e.g. Song et al., 2015; Ter Heege et al., 2015; Houben et al., 2016, 2018).

Shale gas resource estimates are poorly constrained due to the lack of data. Regional studies (Zijp, 2012, 2017; Zijp et al., 2017) have estimated recoverable reserves of 200 up to 500 x 10^9 m^3. Herber & De Jager (2010) estimated a mere 10-20 x 10^9 m^3 for the Geverik Member as surface constraints were likely to prevent drilling a large number of wells.

Development of unconventional reservoirs requires the drilling of a many closely spaced, preferably horizontal wells to drain the tight rocks. Ultimate recovery per well is significantly lower than from conventional gas fields (at least 20-50 times, Herber & De Jager, 2010). Moreover, in most cases large scale hydraulic fracking operations are unavoidable. Despite these techno-economic challenges, a number of licences to explore for shale gas were applied for. In 2018, based on public opposition the Dutch government decided that exploration for shale gas was not permitted (Kamerstukken 32849 and 33529 nr. 126) and to this day the shale gas play is yet unproven as no exploration well has been spudded.

Acknowledgements

This paper is an updated version of the chapter published by de Jager and Geluk in 2007. We regret that Mark Geluk passed away in 2018. We hereby gratefully acknowledge his contribution to the development of the play concepts presented here. These concepts are based on the work by several generations of geoscientists. Comments on this manuscript by Bert Manders and Evert van de Graaff are gratefully acknowledged.

Digital map data

Spatial data of figures in this chapter for use in geographical information systems can be downloaded here: https://doi.org/10.5117/aup.28164023.

References

Amberg, S., Back, S., Sachse, V. & Littke, R., 2022. Numerical 3D modelling of burial and temperature history, source rock maturity, and hydrocarbon generation in the onshore north-eastern Netherlands. International Journal of Earth Sciences 111: 1033-1055. DOI: 10.1007/s00531-022-02168-3

Ames, R. & Farfan, P.F., 1996. The environment of deposition of the Triassic Main Buntsandstein Formation in the P and Q quadrants, offshore the Netherlands. In: Rondeel, H.E., Batjes, D.A.J. & Nieuwenhuijs, W.H. (eds): Geology of gas and oil under the Netherlands. Kluwer Academic Publishers (Dordrecht): 167-178. DOI: 10.1007/978-94-009-0121-6_15

Andrews, I.J., 2013. The Carboniferous Bowland Shale gas study: geology and resource estimation. British Geological Survey (BGS) for Department of Energy and Climate Change (London): 56 pp. https://www.gov.uk/government/publications/bowland-shale-gas-study

Bachmann, G.H., Geluk, M.C., Warrington, G., Becker-Roman, A., Beutler, G., Hagdorn, H., Hounslow, M.W., Nitsch, E., Röhling, H.-G., Simon, T. & Szulc, A., 2010. Triassic. In: Doornenbal, J.C. & Stevenson, A.G. (eds): Petroleum geological atlas of the Southern Permian Basin area. EAGE Publications b.v. (Houten): 149-173.

Baines, S.J. & Worden, R.H., 2004. The long-term fate of CO2 in the subsurface: natural analogues for CO2 storage. Geological Storage of Carbon Dioxide. Geological Society of London, Special Publications 233: 59-85. DOI: 10.1144/GSL.SP.2004.233.01.06

Barnes, I., Irwin, W.P. & White, D.E., 1978. Global distribution of carbon dioxide discharges and major zones of seismicity. U.S. Geological Survey (USGS), Water-Resources Investigations Report No. 78-39: 12 pp. DOI: 10.3133/wri7839

Besly, B.M., 2018. Exploration and development in the Carboniferous of the Southern North Sea: a 30-year retrospective. Paleozoic Plays of NW Europe. Geological Society of London, Special Publications 471: 17-64. DOI: 10.1144/SP471.10

Binot, F., Gerling, P., Hiltmann, W., Kockel, F. & Wehner, H., 1991. The Petroleum System in the Lower Saxony Basin. In: Spencer, A.M. (ed.): Generation, accumulation and production of Europe's hydrocarbons. Special Publication of the European Association of Petroleum Geoscientists (Berlin): 121-139. DOI: 10.1007/978-3-642-77859-9_11

Bodenhausen, J.W.A. & Ott, W.F., 1981. Habitat of the Rijswijk oil province, onshore, the Netherlands. In: Illing, L.V. & Hobson, G.D. (eds): Petroleum Geology of the continental shelf of North-West Europe: Proceedings of the Second Conference. Institute of Petroleum (London): 301-309.

Bouroullec, R. & Geel, C.R., 2025. Permian. In: Ten Veen, J.H., Vis, G.-J., De Jager, J. & Wong, Th.E. (eds): Geology of the Netherlands, second edition. Amsterdam University Press (Amsterdam): 127-183. DOI: 10.5117/9789463728362_ch04

Bouroullec, R., Kortekaas, M., Brussée, M., Geel, C.R., Den Hartog Jager, D.G. & Peeters, S., 2022. Play 6 Zechstein. https://www.geodeatlas.nl/pages/play-6-zechstein

Bouroullec, R. & Ten Veen, J.H., 2025. Salt Tectonics. In: Ten Veen, J.H., Vis, G.-J., De Jager, J. & Wong, Th.E. (eds): Geology of the Netherlands, second edition. Amsterdam University Press (Amsterdam): 457-491. DOI: 10.5117/9789463728362_ch12

Bouw, S. & Lutgert, J., 2012. Shale plays in the Netherlands. European Unconventional Resources Conference and Exhibition 2012. Society of Petroleum Engineers - SPE/EAGE: 623-634. DOI: 10.2118/152644-MS

Breunese, J.N., Andersen, J.H., Brinkman, S., Jagosiak, P., Karnin, W.-D., Karnkowski, P.H., Kombrink, H., Messner, J., Mijnlieff, H., Olsen, S.B., Peryt, T.M., Piske, J., Poprawa, P., Roelofsen, J.W., Stoker, S.J., Smith, N.J.P., Swann, G., Waksmundzka, M.I. & Veldkamp, J.G., 2010. Reserves and production history. In: Doornenbal, J.C. & Stevenson, A.G. (eds): Petroleum Geological Atlas of the Southern Permian Basin Area. EAGE Publications B.V. (Houten): 270-281.

Breunese, J.N. & Rispens, F.B., 1996. Natural gas in the Netherlands: Exploration and development in historic and future perspective. In: Rondeel, H.E., Batjes, D.A.J. & Nieuwenhuijs, W.H. (eds): Geology of Gas and Oil under the Netherlands. Springer Netherlands (Dordrecht): 19-30. DOI: 10.1007/978-94-009-0121-6_4

Bruijn, A.N., 1996. De Wijk gas field (Netherlands): reservoir mapping with amplitude anomalies. In: Rondeel, H.E., Batjes, D.A.J. & Nieuwenhuijs, W.H. (eds): Geology of gas and oil under the Netherlands. Kluwer Academic Publishers (Dordrecht): 243-253. DOI: 10.1007/978-94-009-0121-6_20

Burgess, C.L., Clever, J.E., Corcoran, O.J., Cram, F.M., Hall, N.T. & Lunn, S.F., 2018. Seismic discrimination of an overlooked basal Rotliegend reservoir opens a new play in the Dutch offshore. First Break 36: 38-40.

Cameron, N. & Ziegler, T., 1997. Probing the lower limits of a fairway: further pre-Permian potential in the southern North Sea. In: Ziegler, K., Turner, P. & Daines, A.R. (eds): Petroleum Geology of the Southern North Sea: Future Potential. Geological Society of London, Special Publication 123: 123-141. DOI: 10.1144/gsl.sp.1997.123.01.08

CBS, 2019. Aardgasbaten. Statistics Netherlands (CBS). https://www.cbs.nl/nl-nl/nieuws/2019/22/aardgasbaten-uit-gaswinning-bijna-417-miljard-euro

Corcoran, J. & Lunn, S.F., 2014. Old play, New Prospectivity - the Revival of the Basal Rotliegend Play, Offshore Netherlands and Germany. 76th EAGE Conference and Exhibition 2014: 1-5. DOI: 10.3997/2214-4609.20140742

Corona, F.V., 2005. Fault trap analysis of the Permian Rotliegend gas play, Lauwerszee Trough, NE Netherlands. Geological Society of London, Petroleum Geology Conference Series 6 (1): 327-335. DOI: 10.1144/0060327

Crepieux, N., Sacleux, M. & Mathis, B., 1998. Influence of the pressure on the petroleum system. Example from the Triassic in the Netherlands Central Graben. In: Mitchell, A. & Grauls, D. (eds): Overpressures in Petroleum Exploration. Elf EP-Editions Memoir 22: 123-132.

Crouch, S.V., Baumgartner, W.E.L., Houlleberghs, E.J.M.J. & Walzebuck, J.P., 1996. Development of a tight gas reservoir by a multiple fracced horizontal well: Ameland-204, the Netherlands. In: Rondeel, H.E., Batjes, D.A.J. & Nieuwenhuijs, W.H. (eds): Geology of gas and oil under the Netherlands. Kluwer Academic Publishers (Dordrecht): 93-102. DOI: 10.1007/978-94-009-0121-6_9

Dana Petroleum, 2014. Production plan F2. https://www.nlog.nl

De Boer, R., 2015. Tussen hoogmoed en hysterie, vijf jaar strijd tegen schaliegas in Nederland. Veen Media: 350 pp.

De Bruin, G., Bouroullec, R., Geel, K., Abdul Fattah, R., Van Hoof, T., Pluymaekers, M., Van der Belt, F., Vandeweijer, V.P. & Zijp, M., 2015. New petroleum plays in the Dutch northern offshore. TNO (Utrecht), Report No. 2015 R10920. https://www.nlog.nl/

sites/default/files/2018-11/NORTHERN_OFFSHORE_TNO_report_2015_R10920_0.pdf
De Bruin, G., Ten Veen, J., Wilpshaar, M., Versteijlen, N., Geel, K., Verweij, H. & Carpentier, S., 2022. Origin of shallow gas in the Dutch North Sea - Seismic versus geochemical evidence. Interpretation 10: 1F-T211. DOI: 10.1190/INT-2021-0081.1
De Jager, J., 2012. The discovery of the Fat Sand Play (Solling Formation, Triassic), Northern Dutch offshore – a case of serendipity. Netherlands Journal of Geosciences 91 (4): 609-619. DOI: 10.1017/S0016774600000408
De Jager, J., Doyle, M.A., Grantham, P.J. & Mabillard, J.E., 1996. Hydrocarbon habitat of the West Netherlands Basin. In: Rondeel, H.E., Batjes, D.A.J. & Nieuwenhuijs, W.H. (eds): Geology of gas and oil under the Netherlands. Springer (Dordrecht): 191-209. DOI: 10.1007/978-94-009-0121-6_17
De Jager, J. & Geluk, M.C., 2007. Petroleum geology. In: Wong, Th.E., Batjes, D.A.J. & De Jager, J. (eds): Geology of the Netherlands. Royal Netherlands Academy of Arts and Sciences (KNAW, Amsterdam): 241-264.
De Jager, J., Van Ojik, K. & Smit, J., 2025. Geological development. In: Ten Veen, J.H., Vis, G.-J., De Jager, J. & Wong, Th.E. (eds): Geology of the Netherlands, second edition. Amsterdam University Press (Amsterdam): 21-51. DOI: 10.5117/9789463728362_ch01
De Jong, M.G.G. & Laker, N., 1992. Reservoir modeling of the Vlieland Sandstone of the Kotter Field (Block K18b), offshore, The Netherlands. Geologie En Mijnbouw 71 (2): 173-188.
De Keijzer, M., Boscolo, A., Evers, T., Foster, A., Den Hartog Jager, D.G., Jelgersma, F., De Mik, F., Shokoya, O. & Vandeweyer, T., 2009. Redevelopment of the Schoonebeek Oil Field, Netherlands – Tackling a Giant with the Necessary Detail. 61st EAGE Conference and Exhibition (Helsinki). EAGE (Houten): cp-127-00248. DOI: 10.3997/2214-4609.201400255
De Waal, H., Muntendam-Bos, A.G. & Van Thienen-Visser, K., 2016. Lessons from larger than expected subsidence due to production of halite and natural gas in Fryslân. 50th US Rock Mechanics/geomechanics symposium, Houston (USA). American Rock Mechanics Association (ARMA): 10 pp.
Den Hartog Jager, D.G., 1996. Fluviomarine sequences in the Lower Cretaceous of the West Netherlands Basin: correlation and seismic expression. In: Rondeel, H.E., Batjes, D.A.J. & Nieuwenhuijs, W.H. (eds): Geology of gas and oil under the Netherlands. Kluwer Academic Publishers (Amsterdam): 229-242. DOI: 10.1007/978-94-009-0121-6_19
Dessens, C.W.M., 1996. The role of oil and gas in the Dutch energy policy. In: Rondeel, H.E., Batjes, D.A.J. & Nieuwenhuijs, W.H. (eds): Geology of Gas and Oil under the Netherlands. Kluwer Academic Publishers (Dordrecht, the Netherlands): 7-10. DOI: 10.1007/978-94-009-0121-6_2
Dominik, W.R., Gralla, P., Rose, J., Schlichter, C. & Smit, R., 2000. Hanze Field – the Southernmost Chalk Oil Discovery in the North Sea. 62nd EAGE Conference and Exhibition (Glasgow). EAGE (Houten): cp-28-00009. DOI: 10.3997/2214-4609-pdb.28.A11
Doornenbal, J.C., Kombrink, H., Bouroullec, R., Dalman, R.A.F., De Bruin, G., Geel, C.R., Houben, A.J.P., Jaarsma, B., Juez-Larré, J., Kortekaas, M., Mijnlieff, H.F., Nelskamp, S., Pharaoh, T.C., Ten Veen, J.H., Ter Borgh, M., Van Ojik, K., Verreussel, R.M.C.H., Verweij, J.M. & Vis, G.-J., 2019. New insights on subsurface energy resources in the Southern North Sea Basin area. In: Patruno, S., Archer, S.G., Chiarella, D., Howell, J.A., Jackson, C.A-L. & Kombrink, H. (eds): Cross-border themes in petroleum geology I: The North Sea. Geological Society of London, Special Publications 494: 233-268. DOI: 10.1144/SP494-2018-178
Doornenbal, J.C. & Stevenson, A.G. (eds), 2010. Petroleum Geological Atlas of the Southern Permian Basin Area. EAGE Publications B.V. (Houten): 342 pp.
Dost, B., Ruigrok, E., Spetzler, J. & Kruiver, P., 2025. Natural and induced seismicity. In: Ten Veen, J.H., Vis, G.-J., De Jager, J. & Wong, Th.E. (eds): Geology of the Netherlands, second edition. Amsterdam University Press (Amsterdam): 793-823. DOI: 10.5117/9789463728362_ch22
Ecclestone, M., Ten Veen, J., Brussée, M., Korevaar, S., Nelskamp, S., Peeters, S. & Versteijlen, N., 2021. Play 1 Cenozoic. https://www.geodeatlas.nl/pages/play-1-cenozioc
Fontaine, J.M., Guastella, G., Jouault, P. & De la Vega, P., 1993. F15-A: a Triassic gas field on the easetrn limit of the Dutch Central Graben. In: Parker, J.R. (ed.): Petroleum Geology of North-West Europe: Geological Society of London, Petroleum Geology Conference series 4: 583-593. DOI: 10.1144/0040583
Frikken, H.W., 1999. Reservoir-geological aspects of productivity and connectivity of gas fields in the Netherlands. PhD thesis Technische Universiteit Delft (the Netherlands): 92 pp.
Froidl, F., Littke, R., Baniasad, A., Zheng, T., Röth, J., Böcker, J., Hartkopf-Fröder, C. & Strauss, H., 2021. Peculiar Berriasian "Wealden" Shales of northwest Germany: Organic facies, depositional environment, thermal maturity and kinetics of petroleum generation. Marine and Petroleum Geology 124: 104819. DOI: 10.1016/j.marpetgeo.2020.104819
Gaupp, R., Matter, A., Platt, J., Ramseyer, K. & Walzebuck, J., 1993. Diagenesis and fluid evolution of deeply buried Permian (Rotliegendes) gas reservoirs, Northwest Germany. American Association of Petroleum Geologists Bulletin 77: 1111-1128. DOI: 10.1306/BDFF8E0C-1718-11D7-8645000102C1865D
Gaupp, R. & Okkerman, J.A., 2011. Diagenesis and reservoir quality of Rotliegend sandstones in the northern Netherlands - A review. In: Grötsch, J. & Gaupp, R. (eds): The Permian Rotliegend of the Netherlands. SEPM Special Publication 98: 193-226. DOI: 10.2110/pec.11.98.0193
Gdula, J.E., 1983. Reservoir geology, structural framework and petrophysical aspects of the De Wijk gas field. Geologie En Mijnbouw 62 (1): 191-202. DOI: 10.1007/978-94-009-5532-5_18
Geluk, M.C., 2000. Late Permian (Zechstein) carbonate-facies maps, the Netherlands. Netherlands Journal of Geosciences 79 (1): 17-27. DOI: 10.1017/S0016774600021545
Geluk, M.C., 2005. Stratigraphy and tectonics of Permo-Triassic

basins in the Netherland and surrounding areas. PhD thesis Utrecht University (the Netherlands): 171 pp.

Geluk, M.C., 2007. Permian. In: Wong, Th.E., Batjes, D.A.J. & De Jager, J. (eds): Geology of the Netherlands. Royal Netherlands Academy of Arts and Sciences (KNAW, Amsterdam): 63-84.

Geluk, M.C. & Mijnlieff, H., 2001. Controls on the distribution and thickness of Permian basal Upper Rotliegend sandstones, the Netherlands - probing the limits of the Rotliegend play area. 63rd EAGE Conference and Exhibition (Amsterdam). EAGE (Houten): 522. DOI: 10.3997/2214-4609-pdb.15.P522

Gerling, P., Geluk, M.C., Kockel, F., Lokhorst, A., Lott, G.K. & Nicholson, R.A., 1999. 'NW European Gas Atlas' - new implications for the Carboniferous gas plays in the western part of the Southern Permian Basin. In: Fleet, A.J. & Boldy, S.A.R. (eds): Petroleum Geology of Northwest Europe: Proceedings of the 5th Conference. The Geological Society of London, Petroleum Geology Conference Series 5: 799-808. DOI: 10.1144/0050799

Glennie, K.W., 1998. Lower Permian - Rotliegend. In: Glennie, K.W. (ed.): Petroleum geology of the North Sea - Basic concepts and recent advances. Blackwell Scientific Publishing (Oxford): 137-173. DOI: 10.1002/9781444313413.ch5

Glennie, K.W., 2001. Exploration activities in the Netherlands and North-West Europe since Groningen. Netherlands Journal of Geosciences 80 (1): 33-53. DOI: 10.1017/S0016774600022150

Glennie, K.W. & Provan, D.M.J., 1990. Lower Permian Rotliegend reservoir of the Southern North Sea gas province. In: Brooks, J. (ed.): Classic Petroleum Provinces. Geological Society of London, Special Publication 50: 399-416. DOI: 10.1144/GSL.SP.1990.050.01.25

Goh, L.S., 1993. The Logger Field: Geology and Reservoir characterization. In: Aasen, J.O., Buller, A.T., Hjelmeland, O, Holt, R.M., Kleppe, J. & Torsæter, O. (eds): North Sea oil and gas reservoirs - III, Proceedings 3rd North Sea Oil and Gas Reservoirs Conference (Trondheim). Kluwer (Dordrecht): 75-93. DOI: 10.1017/S0016756800013704

Gras, R. & Clayton, C.J., 1998. Non-hydrocarbon components of Carboniferous-sourced gas in the Southern Permian Basin, northwest Europe. Petroleum Geoscience 4 (2): 147-156. DOI: 10.1144/petgeo.4.2.147

Grötsch, J. & Gaupp, R. (eds), 2011. The Permian Rotliegend of the Netherlands. SEPM Special Publication 98: 374 pp. DOI: 10.2110/pec.11.98

Guasti, E., Abbink, O., Souto Carneiro Echternach, M., Verweij, J.M. & Witmans, N., 2009. Unravelling the petroleum system in the Chalk by means of microfossils. 78th EAGE Conference and Exhibition (Amsterdam). EAGE (Houten): cp-127-00476. DOI: 10.3997/2214-4609.201400483

Guasti, E., Abbink, O., Van Den Berg, Y., Nelskamp, S., Kunakbayeva, G. & Verweij, H., 2010. The Chalk in the Netherlands, a very poor play or a very poorly understood play? Petex Conference (London).

Hedemann, H.A., 2001. Die Bedeutung des Oberkarbons fur die Kohlenwasserstoffvorkommen im Nordseebecken. Erdol Und Kohle, Erdgas, Petrochemie 33 (6): 255-266.

Herber, R. & De Jager, J., 2010. Oil and gas in the Netherlands - Is there a future? Netherlands Journal of Geosciences 89 (2): 119-135. DOI: 10.1017/S001677460000072X

Hernandez Casado, J. & Underhill, J.R., 2013. Consequences of intrusions for sub-salt, sub-basalt exploration and development. 75th EAGE Conference and Exhibition (London). EAGE (Houten): cp-348-00615. DOI: 10.3997/2214-4609.20130141

Herngreen, G.F.W., Smit, R. & Wong, Th.E., 1991. Stratigraphy and tectonics of the Vlieland basin, The Netherlands. In: Spencer, A.M. (ed.): Generation, accumulation and production of Europe's hydrocarbons. Special Publication of the European Association of Petroleum Geoscientists (Berlin) 1: 175-192.

Hoefs, J., 2015. Stable isotope geochemistry, 7th edition. Springer International Publishing Switzerland: 404 pp. DOI: 10.1007/978-3-319-19716-6

Houben, A.J.P., Goldberg, T. & Slomp, C.P., 2021. Biogeochemical evolution and organic carbon deposition on the Northwestern European Shelf during the Toarcian Ocean Anoxic Event. Palaeogeography, Palaeoclimatology, Palaeoecology 565: 110191. DOI: 10.1016/j.palaeo.2020.110191

Houben, A.J.P., Nelskamp, S., Goldberg, T., Verreussel, R.M.C.H., Janssen, N.M.M., Kerstholt, S. & Boxem, T.A.P., 2017. Hydrocarbon Potential of the Lias: HYPO-Lias. TNO report R10380: 86 pp. https://www.nlog.nl/sites/default/files/2023-11/Hypo-Lias_Report_Final.pdf

Houben, M.E., Barnhoorn, A., Peach, C.J. & Drury, M.R., 2018. Potential permeability enhancement in Early Jurassic shales due to their swelling and shrinkage behavior. International Journal of Coal Geology 196: 115-125. DOI: 10.1016/j.coal.2018.07.003

Houben, M.E., Barnhoorn, A., Wasch, L., Trabucho Alexander, J., Peach, C.J. & Drury, M.R., 2016. Microstructures of Early Jurassic (Toarcian) shales of northern Europe. International Journal of Coal Geology 165: 76-89. DOI: 10.1016/j.coal.2016.08.003

Huijgen, M., Remmelts, G. & Roholl, J., 2020. Bodemdaling door gaswinning veroorzaakt geen schade. Bodem 3: 26-28.

Jaarsma, B., Brolsma, M.J., Hoetz, G. & Lutgert, J.E., 2013. Exploring Dinantian carbonates in the SNS – new data offering new insights. 75th EAGE Conference & Exhibition Incorporating SPE EUROPEC (London). EAGE (Houten): cp-348-00382.

Juez-Larré, J., Groenenberg, R., Dinkelman, D., Huijskes, T., Koornneef, J.M., Van Gessel, S.F. & Bos, C.F.M., 2025. Underground storage. In: Ten Veen, J.H., Vis, G.-J., De Jager, J. & Wong, Th.E. (eds): Geology of the Netherlands, second edition. Amsterdam University Press (Amsterdam): 729-767. DOI: 10.5117/9789463728362_ch20

Jurisch, S.A., Heim, S., Krooss, B.M. & Littke, R., 2012. Systematics of pyrolytic gas (N2, CH4) liberation from sedimentary rocks: Contribution of organic and inorganic rock constituents. International Journal of Coal Geology 89: 95-107. DOI: 10.1016/j.coal.2011.11.004

Kamp, H.G.J., 2015. Brief van de Minister van Economische Zaken inzake schaliegas (Parliamentary papers, session year 2014-2015 No. 33952, no. 32). Ministerie van Economische Zaken (The Hague, Netherlands). https://zoek.officielebekendmakingen.nl/kst-33952-32.html

Killops, S. & Killops, V., 2005. Introduction to organic geochemistry, 2nd ed. Blackwell Science Ltd. (Oxford): 393 pp. DOI:10.1002/9781118697214

Knaap, W.A. & Coenen, M.J., 1987. Exploration for oil and natural gas. In: Visser, W.A., Zonneveld, J.I.S. & Van Loon, A.J. (eds): Seventy-five years of geology and mining in the Netherlands (1912-1987). Royal Geological and Mining Society of the Netherlands (KNGMG, The Hague): 207-230.

Kombrink, H., Besly, B., Collinson, J.D., Den Hartog Jager, D.G., Drozdzewski, G., Dusar, M., Hoth, P., Pagnier, H.J.M., Stemmerik, L., Waksmundzka, M.I. & Wrede, V., 2010. Carboniferous. In: Doornenbal, J.C. & Stevenson, A.G. (eds): Petroleum Geological Atlas of the Southern Permian Basin Area. EAGE Publications B.V. (Houten): 80-99.

Kombrink, H., Bridge, J.S. & Stouthamer, E., 2007. The alluvial architecture of the Coevorden Field (Upper Carboniferous), the Netherlands. Netherlands Journal of Geosciences 86 (1): 3-14. DOI: 10.1017/S0016774600021284

Kombrink, H., Ten Veen, J.H. & Geluk, M.C., 2012. Exploration in the Netherlands, 1987-2012. Netherlands Journal of Geosciences 91 (4): 403-418. DOI: 10.1017/S0016774600000317

Kortekaas, M., Böker, U., Van der Kooij, C. & Jaarsma, B., 2018. Lower Triassic reservoir development in the northern Dutch offshore. Geological Society of London, Special Publications 469: 149-168. DOI: 10.1144/SP469.19

Kortekaas, M., Dalman, R., Foeken, J., Hanemaaijer, L., Den Hartog Jager, D.G. & Van der Kraan, V., 2022. Play 5 Triassic. https://www.geodeatlas.nl/pages/play-5-triassic

Kotarba, M.J. & Nagao, K., 2015. Molecular and isotopic compositions and origin of natural gases from Cambrian and CarboniferousLower Permian reservoirs of the onshore Polish Baltic region. International Journal of Earth Sciences 104: 241-261. Doi; 10.1007/s00531-014-1063-0

Krooss, B.M., Friberg, L., Gensterblum, Y., Hollenstein, J., Prinz, D. & Littke, R., 2005. Investigation of the pyrolytic liberation of molecular nitrogen from Palaeozoic sedimentary rocks. International Journal of Earth Sciences 94 (5-6): 1023-1038. DOI: 10.1007/s00531-005-0012-3

Krooss, B.M., Plessen, B., Machel, H.G., Lueders, V. & Littke, R., 2008. Origin and distribution of non-hydrocarbon gases. In: R. Littke, U. Bayer, D. Gajewski & S. Nelskamp (eds): Dynamics of Complex Intracontinental Basins: The Central European Basin System. Springer-Verlag (Berlin): 433-458.

Leveille, G.P., Knipe, R., More, C., Ellis, D.D., Jones, G. & Allinson, G.J., 1997. Compartmentalization of Rotliegendes gas reservoirs by sealing faults, Jupiter Fields area, southern North Sea. In: Ziegler, K., Turner, P. & Daines, S.R. (eds): Petroleum Geology of the Southern North Sea: Future Potential. Geological Society of London, Special Publication 123: 87-104. DOI: 10.1144/GSL.SP.1997.123.01.06

Littke, R., Krooss, B.M., Idiz, E. & Frielingsdorf, J., 1995. Molecular nitrogen in natural gas accumulations: Generation from sedimentary organic matter at high temperatures. American Association of Petroleum Geologists Bulletin 79 (3): 410-430. DOI: 10.1306/8D2B1548-171E-11D7-8645000102C1865D

Liu, Q., Jin, Z., Wu, X., Liu, W., Gao, B., Zhang, D., Li, J. & Hu, A., 2014. Origin and carbon isotope fractionation of CO2 in marine sour gas reservoirs in the Eastern Sichuan Basin. Organic Geochemistry 74: 22-32. DOI: 10.1016/j.orggeochem.2014.01.012

Lokhorst, A. (ed.), 1997. NW European Gas Atlas. Netherlands Institute of Applied Geosciences TNO (NITG-TNO, Haarlem): 154 pp.

Lott, G.K., Wong, Th.E., Dusar, M., Andsbjerg, J., Mönnig, E., Feldman-Olszewska, A. & Verreussel, R.M.C.H., 2010. Jurassic. In: Doornenbal, J.C. & Stevenson, A.G. (eds): Petroleum Geological Atlas of the Southern Permian Basin Area. EAGE Publications B.V. (Houten): 175-193.

McKie, T. & Kilhams, B., 2025. Triassic. In: Ten Veen, J.H., Vis, G.-J., De Jager, J. & Wong, Th.E. (eds): Geology of the Netherlands, second edition. Amsterdam University Press (Amsterdam): 155-183. DOI: 10.5117/9789463728362_ch05

MEA, 2006. Oil and gas in the Netherlands. Annual review 2006 en prognosis 2007-2016. Ministerie van Economische Zaken (The Hague): 122 pp. https://www.nlog.nl/en/archive

MEA, 2015. PlanMER Structuurvisie Schaliegas, (No. B02047.000182.0100). Ministry of Economic Affairs (The Hague). https://zoek.officielebekendmakingen.nl/blg-548921.pdf

MEACP, 2018. Kamerstuk 33529-678. Ministry of Economic Affairs and Climate Policy (The Hague). https://zoek.officielebekendmakingen.nl/kst-33529-678.html

MEACP, 2022. Natural resources and geothermal energy in the Netherlands 2021. Annual Review Ministry of Economic Affairs and Climate Policy (The Hague). Ministry of Economic Affairs and Climate Policy (The Hague): 159 pp. https://www.nlog.nl/en/archive

Mijnlieff, H.F. & Geluk, M.C., 2011. Palaeotopography-governed sediment distribution - a new predictive model for the Permian Upper Rotliegend in the Dutch sector of the Southern Permian Basin. In: Grötsch, J. & Gaupp, R. (eds): The Permian Rotliegend of the Netherlands. SEPM Special Publication 98: 147-159. DOI: 10.2110/pec.11.98.0147

Mijnssen, F.C.J., 1997. Modelling of sandbody connectivity in the Schooner Field. In: K. Ziegler, P. Turner & S.R. Daines (eds): Petroleum Geology of the Southern North Sea: Future Potential. Geological Society of London, Special Publication 123: 169-180.

Mittag-Brendel, E., 2000. Schwefelwasserstoff-Genese in den Zechstein-Erdgaslagerstätten NW-Deutschlands. PhD thesis Universität Hannover (Germany).

NAM, 2008. Production plan Schoonebeek oilfield. www.nlog.nl

NAM, 2014. Production plan L09. https://www.nlog.nl
NAM, 2016a. Production plan De Wijk. https://www.nlog.nl
NAM, 2016b. Production plan K8-FA. https://www.nlog.nl
NAM, 2018. Production plan Coevorden. https://www.nlog.nl
NAM, 2020. Bodemdaling door aardgaswinning - Statusrapport 2020 en Prognose tot het jaar 2080. Report No. EP202011201629: 58 pp.
Nelskamp, S., Foeken, J., Hanemaaijer, L. & Den Hartog Jager, D.G., 2021. Play 9a Source rocks - Westphalian coals. https://www.geodeatlas.nl/pages/play-9a-sr-westphalian
Neptune, 2021. Production plan E17a-A. https://www.nlog.nl
NITG, 1998. Geological Atlas of the subsurface of the Netherlands (1:250,000), Explanation to Map Sheet X Almelo–Winterswijk. Netherlands Institute of Applied Geoscience (NITG-TNO, Haarlem): 134 pp.
NITG, 2000. Geological Atlas of the subsurface of the Netherlands (1:250,000), Explanation to Map Sheet VI Veendam Hoogeveen. Netherlands Institute of Applied Geoscience (NITG-TNO, Utrecht): 152 pp.
Olivarius, M., Weibel, R., Friis, H., Boldreel, L.O., Keulen, N. & Thomsen, T.B., 2017. Provenance of the Lower Triassic Bunter Sandstone Formation: implications for distribution and architecture of aeolian vs. fluvial reservoirs in the North German Basin. Basin Research 29 (S1): 113-130. DOI: 10.1111/bre.12140
Orr, W.L., 1977. Geologic and geochemical controls in the distribution of hydrogen sulphide in natural gas. In: Campos, R. & Goni, J. (eds): Advances in organic geochemistry. Enadisma (Madrid): 571-597.
Pearce, J., Czernichowski-Lauriol, I., Lombardi, S., Brune, S., Nador, A., Baker, J., Pauwels, H., Hatziyannis, G., Beaubien, S. & Faber, E., 2004. A review of natural CO2 accumulations in Europe as analogues for geological sequestration. Geological Society of London, Special Publications 233 (1): 29-41. DOI: 10.1144/GSL.SP.2004.233.01.04
Perrot, J. & Van der Poel, A.B., 1987. Zuidwal - a Neocomian gas field. In: Brooks, J. & Glennie, K. (eds): Petroleum Geology of North-West Europe. Graham & Trotman (London): 325-335.
Petersen, H.I. & Hertle, M., 2018. A Review of the Coaly Source Rocks and Generated Petroleums in the Danish North Sea: An Underexplored Middle Jurassic Petroleum System? Journal of Petroleum Geology 41 (2): 135-154. DOI: 10.1111/jpg.12697
Petersen, H.I., Hertle, M. & Sulsbrück, H., 2017. Upper Jurassic–lowermost Cretaceous marine shale source rocks (Farsund Formation), North Sea: kerogen composition and quality and the adverse effect of oil-based mud contamination on organic geochemical analyses. International Journal of Coal Geology 173: 26-39. DOI: 10.1016/j.coal.2017.02.006
PGK, 1993. Synopsis: Petroleum Geology of the Netherlands. In: Rondeel, H.E., Batjes, D.A.J. & Nieuwenhuijs, W.H. (eds): Geology of Gas and Oil under the Netherlands. Kluwer (Dordrecht, the Netherlands): 1-20.
Pipping, J.C.P., Carlson, T., Frikken, H.W. & Vellinga, P.M., 2001. Sedimentary cycles are key to improve reservoir performance in carbonates, Triassic Lower Muschelkalk – De Wijk gas field, The Netherlands. extended abstract P523. Amsterdam.
Pletsch, T., Appel, J., Botor, D., Clayton, C.J., Duin, E., Faber, E., Górecki, W., Kombrink, H., Kosakowski, P., Kuper, G., Kus, J., Lutz, R., Mathiesen, A., Ostertag-Henning, C., Papiernek, B. & Van Bergen, F., 2010. Petroleum Generation and Migration. In: Doornenbal, J.C. & Stevenson, A.G. (eds): Petroleum Geological Atlas of the Southern Permian Basin Area. EAGE Publications B.V. (Houten): 224-253.
Pluymaekers, M.P.D., Kramers, L., Van Wees, J.-D., Kronimus, A., Nelskamp, S., Boxem, T. & Bonté, D., 2012. Reservoir characterisation of aquifers for direct heat production: Methodology and screening of the potential reservoirs for the Netherlands. Netherlands Journal of Geosciences 91 (4): 621-636. DOI: 10.1017/S001677460000041X
Purvis, K. & Okkerman, J.A., 1996. Inversion of reservoir quality by early diagenesis: an example from the Triassic Buntsandstein, offshore the Netherlands. In: Rondeel, H.E., Batjes, D.A.J. & Nieuwenhuijs, W.H. (eds): Geology of gas and oil under the Netherlands. Kluwer Academic Publishers (Dordrecht): 179-189. DOI: 10.1007/978-94-009-0121-6_177
Quirk, D.G., 1993. Interpreting the Upper Carboniferous of the Dutch Cleaver Bank High. In: Parker J.R. (ed.): Proceedings of the 4th Conference on Petroleum Geology of North-West Europe. Geological Society of London, Petroleum Geology Conference series 4: 697-706. DOI: 10.1144/0040697
Quirk, D.G. & Aitken, J.F., 1997. The structure of the Westphalian in the northern part of the southern North Sea. In: Ziegler, K., Turner, P. & Daines, S.R. (eds): Petroleum Geology of the Southern North Sea: Future Potential. Geological Society Special Publication (London): Vol. 123, 143-152.
Racero-Baena, A. & Drake, S.J., 1996. Structural style and reservoir development in the West Netherlands oil province. In: Rondeel, H.E., Batjes, D.A.J. & Nieuwenhuijs, W.H. (eds): Geology of gas and oil under the Netherlands. Kluwer Academic Publishers (Dordrecht): 211-228. DOI: 10.1007/978-94-009-0121-6_18
Rightmire, C. T., Eddy, G. E. & Kirr, J. N. (eds), 1984. Coalbed Methane Resources of the United States. American Association of Petroleum Geologists: 378 pp.
Roelofsen, J.W. & De Boer, W.D., 1991. Geology of the Lower Cretaceous Q-1 oil-fields, Broad Fourteens basin, The Netherlands. In: Spencer, A.M. (ed.): Generation, accumulation and production of Europe's hydrocarbons. Special Publication of the European Association of Petroleum Geoscientists (Berlin) 1: 203-216.
Roels, H.J.M., 2001. Groningen field, past, present and future. Netherlands Journal of Geosciences 80 (1): 12-14. DOI: 10.1017/S001677460002206X
Roustiau, A., Bouroullec, R., Dijk, J., Den Hartog Jager, D.G., Nelskamp, S. & Swart, M., 2022. Play 8 Carboniferous. https://www.geodeatlas.nl/pages/play-8-carboniferous
Schroot, B.M., Klaver, G.T. & Schüttenhelm, R.T.E., 2005. Surface

and subsurface expressions of gas seepage to the seabed - examples from the Southern North Sea. Marine and Petroleum Geology 22: 499-515. DOI: 10.1016/j.marpetgeo.2004.08.007

Schroot, B.M. & Schüttenhelm, R.T.E., 2003. Expressions of shallow gas in the Netherlands North Sea. Netherlands Journal of Geosciences 82 (1): 91-105. DOI: 10.1017/S0016774600022812

Schroot, B.M., Van Bergen, F., Abbink, O.A., David, P., Van Eijs, R. & Veld, H., 2006. Hydrocarbon potential of the Pre-Westphalian in the Netherlands on- and offshore - report of the PetroPlay project. Report No. NITG-05-155-C. TNO Built Environment and Geosciences (Utrecht): 1-436 pp.

Song, J., Littke, R., Weniger, P., Ostertag-Henning, C. & Nelskamp, S., 2015. Shale oil potential and thermal maturity of the Lower Toarcian Posidonia Shale in NW Europe. International Journal of Coal Geology 150-151: 127-153. DOI: 10.1016/j.coal.2015.08.011

Spain, D.R. & Conrad, C.P., 1997. Quantitative analysis of top-seal capacity offshore Netherlands, southern North Sea. Geologie En Mijnbouw 76: 217-226. DOI: 10.1023/A:1003056609771

Ten Veen, J.H., Verweij, J.M., De Bruin, G. & Donders, T., 2014. First steps toward maturing the shallow gas play - results of an integrated exploration workflow. International Petroleum Technology Conference (Doha, Qatar): IPTC-17563-MS. DOI: 10.2523/IPTC-17563-MS

Ter Borgh, M.M., Eikelenboom, W. & Jaarsma, B., 2018. Hydrocarbon potential of the Visean and Namurian in the northern Dutch offshore. In: Monaghan, A.A., Underhill, J.R., Hewett, A.J. & Marshall, J.E.A. (eds) Paleozoic plays of NW Europe. Geological Society of London, Special Publications 471: 133-153. DOI: 10.1144/SP471.5

Ter Heege, J., Zijp, M.H.A.A., Nelskamp, S., Douma, L., Verreussel, R., Ten Veen, J., De Bruin, G. & Peters, R., 2015. Sweet spot identification in underexplored shales using multidisciplinary reservoir characterization and key performance indicators: Example of the Posidonia Shale Formation in the Netherlands. Journal of Natural Gas Science and Engineering 27 (2): 558-577.

Tissot, B.P. & Welte, D.H., 1984. Petroleum Formation and Occurrence, second edition. Springer Berlin (Heidelberg): 699 pp. DOI: 10.1007/978-3-642-87813-8

Trabucho Alexandre, J.P. & Wong, Th. E., 2025. Rhaetian to Middle Jurassic. In: Ten Veen, J.H., Vis, G.-J., De Jager, J. & Wong, Th.E. (eds): Geology of the Netherlands, second edition. Amsterdam University Press (Amsterdam): 185-209. DOI: 10.5117/9789463728362_ch06

Underhill, J.R., 2009. Role of intrusion-induced salt mobility in controlling the formation of the enigmatic Silverpit Crater, UK Southern North Sea. Petroleum Geoscience 15 (3): 197-216. DOI: 10.1144/1354-079309-843

Van Adrichem Boogaert, H.A. & Kouwe, W.F.P., 1993. Stratigraphic nomenclature of the Netherlands, revision and update by RGD and NOGEPA; Section F: Lower and Middle Jurassic (Altena Group). Mededelingen Rijks Geologische Dienst 50: 1–20.

Van Adrichem Boogaert, H.A. & Kouwe, W.F.P., 1994. Stratigraphic nomenclature of the Netherlands, revision and update by RGD and NOGEPA; Section E, Triassic. Mededelingen Rijks Geologische Dienst, 50, 1-28.

Van Balen, R.T., Van Bergen, F., De Leeuw, C., Pagnier, H., Simmelink, H., Van Wees, J.-D. & Verweij, J.M., 2000. Modelling the hydrocarbon generation and migration in the West Netherlands Basin, the Netherlands. Netherlands Journal of Geosciences 79 (1): 29-44. DOI: 10.1017/S0016774600021557

Van Bergen, F., Zijp, M.H.A.A., Nelskamp, S. & Kombrink, H., 2013. Shale Gas Evaluation of the Early Jurassic Posidonia Shale Formation and the Carboniferous Epen Formation in the Netherlands. American Association of Petroleum Geologists, Memoir 103: 1-24. DOI: 10.1306/134017221H53468

Van Bergen, M.J. & Sissingh, W., 2007. Magmatism in the Netherlands: expression of the north-west European rifting history. In: Wong, Th.E., Batjes, D.A.J. & De Jager, J. (eds): Geology of the Netherlands. Royal Netherlands Academy of Arts and Sciences (KNAW, Amsterdam): 197-222.

Van Bergen, M.J., Vis, G.-J., Sissingh, W., Koornneef, J. & Brouwers, I., 2025. Magmatism in the Netherlands: expression of the northwest European rifting history. In: Ten Veen, J.H., Vis, G.-J., De Jager, J. & Wong, Th.E. (eds): Geology of the Netherlands, second edition. Amsterdam University Press (Amsterdam): 393-455. DOI: 10.5117/9789463728362_ch11

Van Buggenum, J.M. & Den Hartog Jager, D.G., 2007. Silesian. In: Wong, Th.E., Batjes, D.A.J. & De Jager, J. (eds): Geology of the Netherlands. Royal Netherlands Academy of Arts and Sciences (KNAW, Amsterdam): 43-62.

Van de Graaff, W.J.E., 2001. Preface – Special issue: Groningen Catalyst for the North West European Oil and Gas Industry. Netherlands Journal of Geosciences/Geologie En Mijnbouw 80 (1): 3-4.

Van de Sande, J.M.M., Reijers, T.J.A. & Casson, N., 1996. Multidisciplinary exploration strategy in the northeast Netherlands Zechstein 2 Carbonate play, guided by 3D seismics. In: Rondeel, H.E., Batjes, D.A.J. & Nieuwenhuijs, W.H. (eds): Oil and Gas under the Netherlands. Kluwer (Dordrecht): 125-142. DOI: 10.1007/978-94-009-0121-6_125

Van den Boogaard, M. & Hoetz, H.L.J.G., 2012. Shallow Gas Play in the Netherlands Takes Off. EAGE 74th EAGE Conference and Exhibition Incorporating EUROPEC 2012 (Kopenhagen). EAGE (Houten): cp-293-00768. DOI: 10.3997/2214-4609.20148764

Van den Bosch, W.J., 1983. The Harlingen Field, the only gas field in the Upper Cretaceous Chalk of The Netherlands. Geologie En Mijnbouw 62 (1): 145-156.

Van der Baan, D., 1990. Zechstein reservoirs in the Netherlands. In: Brooks, J. (ed.): Classic Petroleum Provinces. Geological Society of London, Special Publication 50: 379-398. DOI: 10.1144/GSL.SP.1990.050.01.2

Van der Molen, A.S., 2004. Sedimentary development, seismic

stratigraphy and burial compaction of the Chalk Group in the Netherlands North Sea area. PhD thesis Utrecht University (the Netherlands): 175 pp.

Van der Molen, A.S., Dudok van Heel, H.W. & Wong, Th.E., 2005. The influence of tectonic regime on chalk deposition: examples of the sedimentary development and 3D-seismic stratigraphy of the Chalk Group in the Netherlands offshore area. Basin Research 17: 63-81. DOI: 10.1111/j.1365-2117.2005.00261.x

Van der Poel, A.B., 1989. A case study on the hydrocarbon geology of Upper Permian (Zechstein-3) carbonates in licence P6, the Netherlands' offshore. Geologie En Mijnbouw 68 (3): 285-296.

Van Geuns, L., Juez-Larré, J. & De Jong, S., 2017. Van exporteur naar importeur. De verander(en)de rol van aardgas in Nederland. White Paper TNO & The Hague Centre for Strategic Studies: 28 pp. https://publications.tno.nl/publication/34625382/IkMSR6/TNO-2017-exporteur.pdf

Van Hulten, F.F.N., 2009. Brief history of petroleum exploration in the Netherlands. 15-16.

Van Hulten, F.F.N. & Poty, E., 2008. Geological factors controlling Early Carboniferous carbonate platform development in the Netherlands. Geological Journal 43 (2-3): 175-196. DOI: 10.1002/gj.1104

Van Lith, J.G.J., 1983. Gas fields of Bergen consession, The Netherlands. Geologie En Mijnbouw 62 (1): 63-74.

Van Lochem, H., 2018. F17-Chalk: new insights in the tectonic history of the Dutch Central Graben. In: Kilhams, B., Kukla, P.A., Mazur, S., McKie, T., Mijnlieff, H.F. & Van Ojik, K. (eds): Mesozoic resource potential in the southern Permian Basin. Geological Society of London, Special Publications 469: 537-558. DOI: 10.1144/SP469.17

Van Ojik, K., Kortekaas, M., Bouroullec, R., Peeters, S., Van Unen, M., Swart, M., Den Hartog Jager, D.G., Wiarda, E., Nolten, M. & Beintema, K., 2023. Play 7 Rotliegend. https://www.geodeatlas.nl/pages/play-7-Rotliegend

Van Staalduinen, P.C., Terwel, K.C. & Rots, J.G., 2019. Onderzoek naar de oorzaken van bouwkundige schade in Groningen. Methodologie en case studies ter duiding van de oorzaken. Delft University of Technology, Report No. CM-2018-01: 221 pp.

Van Wijhe, D.H. & Bless, M.J.M., 1974. The Westphalian of the Netherlands with special reference to miospore assemblages. Geologie En Mijnbouw 53 (6): 295-328.

Van Wijhe, D.H., Lutz, M. & Kaasschieter, J.P.H., 1980. The Rotliegend in The Netherlands and its gas accumulations. Geologie En Mijnbouw 59 (1): 3-24.

Veenhof, E.N., 1996. Geological aspects of the Annerveen gas field, the Netherlands. In: Rondeel, H.E., Batjes, D.A.J. & Nieuwenhuijs, W.H. (eds): Geology of gas and oil under the Netherlands. Kluwer Academic Publishers (Dordrecht): 79-92. DOI: 10.1007/978-94-009-0121-6_79

VEN, 2020. Bodemdaling Statusrapport 2020 - Drenthe Overijssel Friesland, versie 1.1. Vermilion Energy Netherlands B.V. (VEN): 101 pp.

Verberg, G.H.B., 2001. Groningen, Gasunie and the gas market. Netherlands Journal of Geosciences 80 (1): 1-15. DOI: 10.1017/S0016774600022071

Verdier, J.P., 1996. The Rotliegend sedimentation history of the southern North Sea and adjacent countries. In: Rondeel, H.E., Batjes, D.A.J. & Nieuwenhuijs, W.H. (eds): Geology of gas and oil under the Netherlands. Kluwer Academic Publishers (Dordrecht): 45-56. DOI: 10.1007/978-94-009-0121-6_45

Verreussel, R.M.C.H., Bouroullec, R., Munsterman, D.K., Dybkjær, K., Geel, C.R., Houben, A.J.P., Johannessen, P.N. & Kerstholt-Boegehold, S.J., 2018. Stepwise basin evolution of the Middle Jurassic–Early Cretaceous rift phase in the Central Graben area of Denmark, Germany and The Netherlands. In: Kilhams, B., Kukla, P.A., Mazur, S., McKie, T., Mijnlieff, H.F. & Van Oijk, K. (eds): Mesozoic resource potential in the Southern Permian Basin. Geological Society of London, Special Publications 469: 305-340. DOI: 10.1144/SP469.23

Verreussel, R.M.C.H., Munsterman, D.K., Bouroullec, R. & Houben, A.J.P., 2025. Late Jurassic - Early Cretaceous. In: Ten Veen, J.H., Vis, G.-J., De Jager, J. & Wong, Th.E. (eds): Geology of the Netherlands, second edition. Amsterdam University Press (Amsterdam): 211-251. DOI: 10.5117/9789463728362_ch07

Verweij, J.M., Nelskamp, S.N., Ten Veen, J.H., De Bruin, G., Geel, K. & Donders, T.H., 2018. Generation, migration, entrapment and leakage of microbial gas in the Dutch part of the Southern North Sea Delta. Marine and Petroleum Geology 97: 493-516. DOI: 10.1016/j.marpetgeo.2018.07.034

Wall, M., Cartwright, J., Davies, R. & McGrandle, A., 2010. 3D seismic imaging of a Tertiary Dyke Swarm in the Southern North Sea, UK. Basin Research 22: 181-194. DOI: 10.1111/j.1365-2117.2009.00416.x

Wong, Th.E., 1991. Petroleum geology of the Dutch Central North Sea Graben. In: O. Michelsen & F. Frandsen (eds): The Jurassic in the Southern Central Trough. Danmarks Geologiske Undersøgelse, Series B: Vol. 16: 36-40. DOI: 10.34194/serieb.v16.7091

Wong, Th.E., Van Doorn, Th.H.M. & Schroot, B.M., 1989. Late Jurassic petroleum geology of the Dutch Central North Sea Graben. Geologische Rundschau 78 (1): 319-336. DOI: 10.1007/BF01988367

Wycherley, H., Fleet, A. & Shaw, H., 1999. Some observations on the origins of large volumes of carbon dioxide accumulations in sedimentary basins. Marine and Petroleum Geology 16: 489-494. DOI: 10.1016/S0264-8172(99)00047-1

Zijp, M.H.A.A., Nelskamp, S., Schovsbo, N., Tougaard, L. & Bocin-Dumitriu, A., 2017. Resource Estimation of Eighty-Two European Shale Formations. Unconventional Resources Technology Conference (URTeC). DOI 10.15530-urtec-2017-2686270

Zijp, M.H.A.A., Nelskamp, S., Verreussel, R. & Ter Heege, J., 2015. The Geverik Member of the Carboniferous Epen Formation, Shale Gas Potential in Western Europe. Presented at the International Petroleum Technology Conference (Doha, Qatar). DOI: 10.2523/IPTC-18410-MS

15 Peat, coal and coalbed methane

Frank van Bergen & Kay Koster*

Corresponding author: frank.vanbergen@tno.nl

ABSTRACT

Multiple stratigraphic intervals in the subsurface of the Netherlands contain layers of peat and coal. These include (pre-)Holocene peat, Miocene lignite and upper Carboniferous (Westphalian) bituminous coal and anthracite, all of which have played important roles as energy sources in the country. In the Netherlands, total peat mining between the 13th and 20th centuries reached ca. 9.4 x 10^9 m^3, while peak annual lignite production reached 1.5 Mt. Total production of bituminous coal and anthracite in the 20th century amounted to 568 Mt. Production of these energy sources ceased during the 20th century. Current resources of lignite are estimated to be at least 1700 Mt; estimates for bituminous and anthracite coal resources shallower than 1500 m below surface are between 4000 and 38000 Mt. At the turn of the 21st century, interest in coalbed methane, i.e. gas that is contained in bituminous coal started to increase. Estimates for the theoretically recoverable volumes of coalbed methane shallower than 1500 m below surface are between 7 and 107 x 10^9 m^3. The possibility to combine the production of this gas while simultaneously storing CO_2 within the coal (considered as a sink for the CO_2) was investigated but is technologically challenging. While the subsurface coal and coalbed methane resources in the Netherlands are significant, there is no desire to exploit them currently, at a time when dependency on fossil fuels is diminishing in order to avoid further increase in CO_2 emissions. Substantial volumes of peat are still present in the subsurface, but its organic nature and high porosity causes land subsidence and brings challenges to the built environment. Furthermore, peat oxidation induced by drainage, especially in agricultural areas, is estimated to contribute 4.246 Mt/yr or ca. 2% of the Netherlands' annual CO_2 emissions. The remaining peat has the potential to emit 2.0 Gt CO_2 into the atmosphere and regardless of the energy content of the organic rich layers, this does not fit in a sustainable energy supply system. It is therefore important to increase knowledge and understanding of the characteristics and behaviour of these layers in the Dutch subsurface.

<< Coal miner in the Oranje Nassau mine (Heerlen, southeastern Netherlands, 1952-1953). Photo: © Nico Jesse/Nederlands Fotomuseum.

DOI: 10.5117/9789463728362_ch15

Introduction

The world has been relying on peat and coal for a large portion of its energy demands for centuries. In the Netherlands, large scale commercial peat mining for energy production commenced in the 14th century (Van Dam, 2001), but gradually became replaced by coal as a main source of energy from the 19th century onwards (Gerding et al., 2015). Despite a significant shift over the last decades towards gas and renewables for electricity production, 14% of the electricity in 2021 was still generated from coal-fired power plants, with an associated CO_2 emission of 11.2 Megaton (CBS, 2022a,b). Although the usage of coal has been phasing out in the Netherlands, it is likely that globally the world will continue to rely on coal for several decades to come. In its 2020 World Energy Outlook, the International Energy Agency predicts a relatively stable demand for coal until 2030, albeit lower than the 2019 demand (IEA, 2020). Currently it seems very unlikely that further exploitation of the natural coal and peat resources in the Netherlands as fossil fuel will ever be reconsidered, given the likely impact that exploitation would have on society and the associated emission of greenhouse gases. However, the subsurface is envisaged to play a key role in the energy transition for extraction of heat and storage of fluids and gases. Organic rich layers are an integral part of the subsurface that need to be taken into account, as they show different responses to temperature and pressure changes compared to other rocks and sediments. Understanding of the nature of these layers and prediction of their behaviour under changing conditions will therefore remain of utmost importance. Peat, coal and coalbed methane are consecutively described by briefly discussing their formation, their exploitation history and the chapter concludes with a high-level resource assessment.

Peat

Peat is one of the most peculiar lithologies encountered in the shallow subsurface of the Netherlands. Its organic composition makes it suitable for various uses, but very prone to degradation. The reclamation of vast peatlands has boosted the country's economy for centuries as an energy source for domestic and industrial use, as agricultural land, and even as a reservoir for sea salt mining. Nevertheless, its utilization, has led to severe land subsidence due to volumetric loss by mining, decomposition by microbial oxidation following drainage for land reclamation, and by peat compaction through reduced pore water pressures and anthropogenic loading. Land subsidence results in higher flood risks and damage to buildings and infrastructure. In addition, the oxidation of shallow peat beds is a prime source of greenhouse gas emissions, and its weak organic structure brings engineering challenges to the built environment.

Formation and composition

Peat forms under waterlogged anoxic conditions in wetlands, where organic matter production from decaying vegetation exceeds decomposition by biochemical processes. In the Netherlands, wetlands of Holocene age can be divided into low-lying fens and elevated bogs (Verhoeven, 1992). Fens develop under influence of groundwater and are situated in delta and coastal plains, brook valleys, and areas with groundwater seepage. Bogs are elevated above regional groundwater levels and are primarily precipitation fed. Remnants of bogs can be found in pre-Holocene elevated substrates, as well as superimposed on coastal fens (Pons, 1992). Peat of Holocene age typically has organic matter dry mass percentages of 20 to 90% (Den Haan & Kruse, 2006; Koster, 2017;

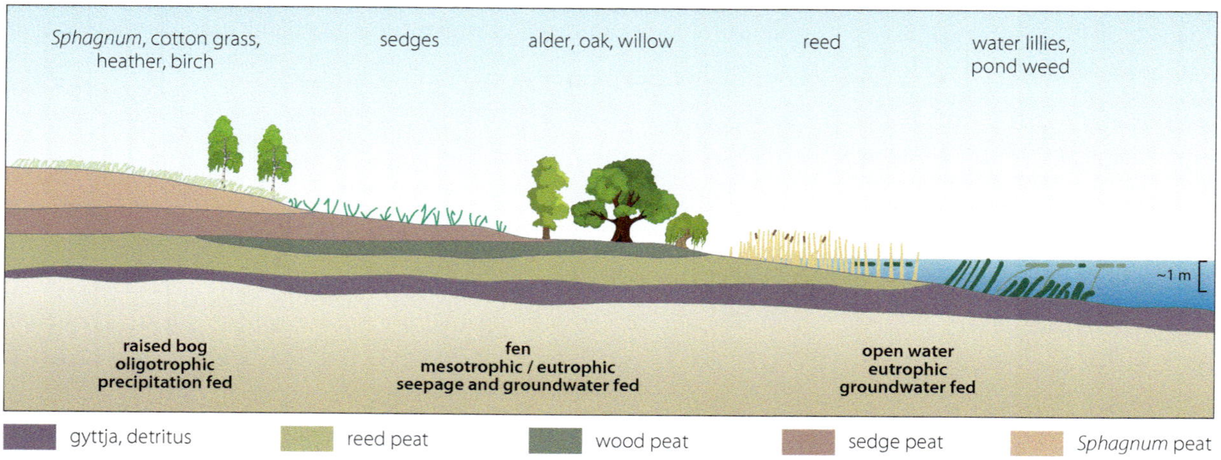

Figure 15.1. Schematic diagram of the depositional setting of various peat types (after Hobbs, 1986).

Koster et al., 2018a) and in pristine settings has porosities up to 95%, owing to the low specific weight of organic matter (Landva, 2006). The high porosity of peat substantially decreases during compression by loading of younger, higher density sediments, leading to substantial volumetric loss (Van Asselen et al., 2009). This process occurs naturally and has influenced for instance sedimentation patterns of the Rhine and Meuse rivers during the Holocene, as the loading of peat by clastic sediments has provided locally 40% additional accommodation space (Van Asselen, 2011). Peat often contains clastic admixtures that are naturally introduced by water, air, or bioturbation, thereby changing its composition and reducing its porosity. The content of clastic sediments in peat beds often increases at its vertical and horizontal boundaries, thereby reflecting transitions in depositional environments (Koster et al., 2018b).
The type of peat-forming vegetation depends mostly on the nutrient level of the accommodating water and water depth (Den Held et al., 1992). Major peat-forming vegetation in the Netherlands during the Holocene were i) eutrophic reed peat in deeper water (ca. 1 m) influenced by fluvial and tidal systems, ii) eutrophic to mesotrophic wood peat (alder, oak, willow) in shallow (ca. 0.1 m) slightly less nutrient-rich fluvial dominated zones, iii) mesotrophic sedge peat in areas of seepage and in shallow zones (ca. 0.3 m) distal to rivers, and iv) oligotrophic peat domes that are precipitation fed and primarily consist of mosses and heather.

Organic lake deposits are also encountered in the subsurface of the Netherlands. Gyttja is a brown to greenish fine-grained organic deposit that accumulates in fresh eutrophic standing water too deep to accommodate peat forming vegetation (ca. >1 m). It consists of decayed organic constituents present in the water column such as algae, water plants and humic acids, and can have a white-yellowish appearance when formed in water saturated with calcium carbonate. Although gyttja is very amorphous, it can contain macroscopically visible remains of water flora and fauna. Detritus found in areas where peat beds were eroded, mechanically broken down during transport and subsequently deposited consists of small peat, plant and/or wood fragments and is typically encountered at former lake shores, such as at the edge of the former Almere lake system (Menke et al., 1998). Detritus often contains admixed clastic sediments and in contrast to peat is not very cohesive.

Geological setting

Most peat currently found in the subsurface of the Netherlands formed during the Holocene, although shallow peat beds of Pleistocene age are locally present as well

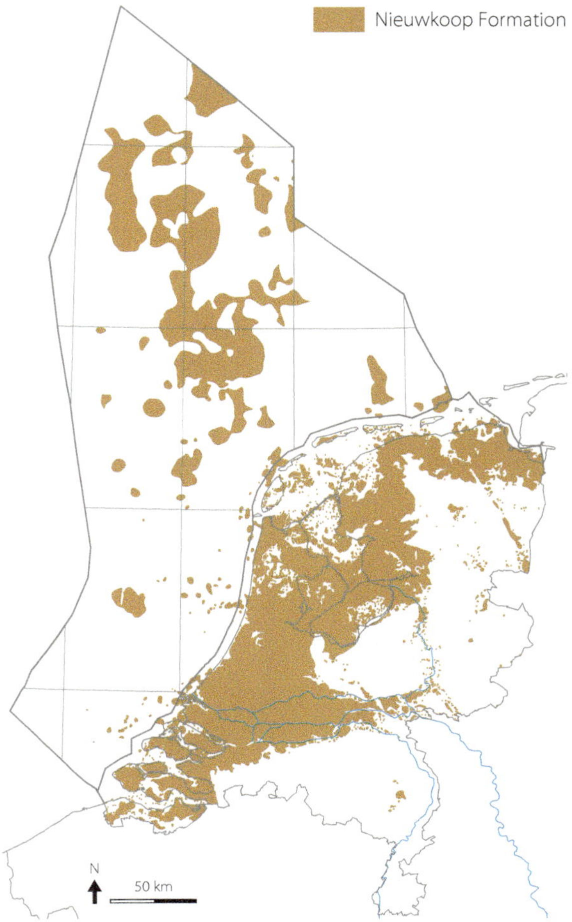

Figure 15.2. Present-day distribution of Holocene peat of the Nieuwkoop Formation in the Netherlands (TNO-GDN, 2022d). The original distribution of peat was much wider (see e.g. Vos, 2015) but through the centuries has been significantly affected by excavation activities that completely removed peat from certain areas.

(Schokker et al., 2007; Peeters et al., 2015). In general, two areas of Holocene peat formation are distinguished based on their hydrological regimes (Pons, 1992): the coastal and delta plains where fen and bog peat formed under influence of relative sea-level rise, and the higher elevated pre-Holocene substrates where fen peat formed in swampy brook-valleys and bog peat on plateaus under influence of seepage and precipitation (Fig. 15.1).

The formation of a basal peat bed (Nieuwkoop Formation; Basisveen Bed; see Figs 15.2, 15.3) lying directly on top of the Pleistocene surface commenced around 10,900 year BP, when this surface drowned under influence of relative sea-level rise (Jelgersma, 1961; Hijma & Cohen, 2010). These basal peat beds are situated around 23 to 24 m below present mean sea-level. Offshore, this basal peat bed is encountered even deeper, at depths up to ca. 60 m below mean sea-level. There, basal peat beds have an age of ca. 12,700 to 13,800 years BP (Lippmann et al., 2021), coinciding with the Late Glacial Allerød interstadial. Basal peat consists primarily of eutrophic and mesotrophic peat formed in fens, locally alternating with organic deposits and oligotrophic bogs, the latter primarily at locations of groundwater seepage (Bos et al., 2012). This peat layer is overlain by clastic tidal basin and fluvial deposits of the Naaldwijk Formation and Echteld Formation respectively (TNO-GDN, 2022a,c). These tidal basin and fluvial deposits themselves are also locally embedded with thin peat layers that formed during relatively short intervals of non-deposition in the Early Holocene. The formation of the up to metres thick vast peat beds that are found in the shallow subsurface of the present-day coastal plains com-

The vanished peat

During many millennia large parts of the Netherlands were covered with swamps, in which metres thick peat beds accumulated. At present, most of these swamps and the peat beds are gone. For peat to form, a high groundwater table is needed to prevent oxidation of vegetation remains. Natural conditions in the Netherlands favour peat formation, because of its low-lying near-coastal position and abundant precipitation. Even further away from the coast, large-scale peat growth in swamps and raised bogs occurred.

Peat swamps provide unfavourable conditions for humans to live in. Initially, swamps were used merely for hunting and performing rituals. The soggy conditions of peat swamps made them difficult to access and unsuitable for agricultural practices. Archeological finds indicate that already from the Bronze Age onwards, over 3000 years ago, people started to construct infrastructural works to allow for access into the swamps. During the Iron Age, and later in Roman Times, technological advancements allowed people to manage peatlands, by draining and cultivating them. However, by draining the peat, the vegetational remains were exposed to oxygen and started to degrade. This marked the onset, on a local scale, of the disappearance of peat.

It was not until the Middle Ages that the vanishing of peat commenced on large regional scales, when it was mined for fuel and salt, and drained and dug to create agricultural lands. Today, almost all swamps have disappeared. The few that remain are now in use as nature reserve areas, such as the national parks 'De Groote Peel' and 'Weerribben-Wieden'. Although these swamps are preserved, their water dynamics and therefore peat growth are still heavily controlled by humans.

In the low-lying coastal areas most of the peat has been exploited and the thick peat layers that once characterized the area are now strongly reduced. This can best be seen underneath centuries old earthen structures that very locally shielded-off the peat from human-induced degradation. Examples are for instance embankments, but also dwelling mounds ('terpen').

The image shows two samples of the same peat layer: on the left a sample taken from a dug exposure and on the right one from a cored borehole (Vos, 2011). On the left, the peat has vanished and been degraded to faint black lines embedded within clay. On the right the peat was shielded-off by a dwelling mound, which preserved it from degradation. It has a thickness of some decimetres. Note the thin clay bed within the peat layer, indicative of a flood event and visible in both samples. It can be used to correlate the vanished with the fresh peat bed. The diameter of the core is 3 cm. To prevent the remaining peat in the Netherlands from vanishing it is important to keep the remaining swamps and peat beds waterlogged.

menced around 6000-5500 year BP (Nieuwkoop Formation; Hollandveen Member), when the open tidal basins transformed into an enclosed fresh water fed area (Beets & Van der Spek, 2000). This was the result of the maturation of a coastal-barrier complex under influence of decelerating relative sea-level rise, closing-off the inland from saline water and sediments, thereby providing freshwater accommodation space for peat formation. Locally, the peat of the Hollandveen Member was elevated in rainwater fed peat bogs and furthermore, in the Flevoland and IJsselmeer area, within the peatlands of the Hollandveen Member, the regional Flevomeer lake system developed. Within this lake system, gyttja and detritus were deposited, stratigraphically interfingering with peat beds of the Hollandveen Member (Nieuwkoop Formation; Flevomeer Bed).

Peat formation in brook valleys and on elevated plateaus (Nieuwkoop Formation; Griendtsveen Member) commenced in two phases. Locally, mesotrophic peat started to develop as early as the relatively warm Allerød and Bølling interstadials during the Late Glacial (Quik et al., 2021). However, largescale peat formation developed relatively late, around 5600 years BP, simultaneously to the widespread peat formation in the coastal plain that followed the maturation of the barrier complex. The latter most likely influenced the local drainage conditions on the higher elevated plateaus as well (Quik et al., 2021). The second phase of formation of the Griendtsveen Member took place in less nutrient-rich conditions. It often consists of oligotrophic bogs, with an amorphous highly organic base layer (*gliede*), locally alternating with mesotrophic fen peat.

Historical development and economics

Inhabitants of the coastal plains of the Netherlands have been exploiting Holocene peat on a local scale for millennia, inevitably contributing to subsidence and thereby precipitating a series of floods that repeatedly destroyed arable land and settlements. For instance, peat areas adjacent to the former estuaries of the Scheldt and Meuse rivers were already artificially being drained during the Late Iron Age around 2500 years BP and the consequent subsidence due especially to peat oxidation, led to series of severe floods and the eventual abandonment of the areas until Roman times (Vos & Van Heeringen, 1997; Vos, 2017).
Large scale more centrally organized reclamation of the peat areas commenced between 800-1000 AD (Van Dam, 2001). Since then, this has resulted in a loss of 19.8 x 10^9 m^3 of peat from the coastal plain alone, of which the majority (14.6 x 10^9 m^3) has disappeared through oxidation (Erkens et al., 2016). The natural process of peat oxidation and the combustion of excavated peat as a fuel emitted total 2.02-5.72 Gton CO_2 (Erkens et al., 2016).

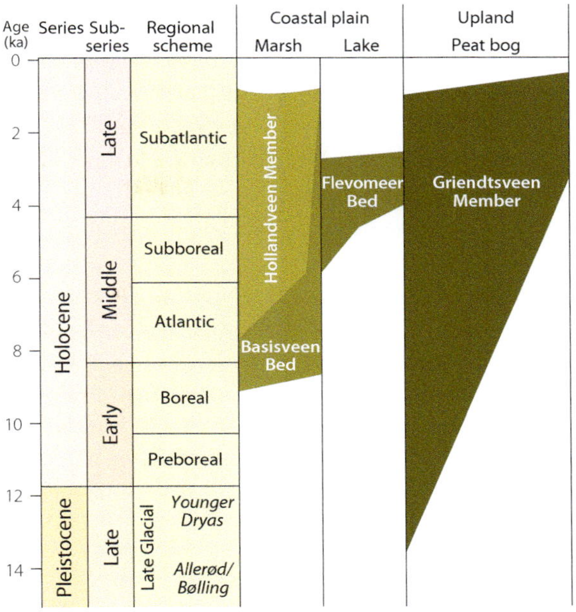

Figure 15.3. Chronostratigraphy and lithostratigraphy of Holocene peat units in the Netherlands (from Vos, 2015).

During the early Middle Ages, peat was primarily mined at the higher elevated plateaus in the north eastern Netherlands, in parallel to north western Germany and northern Flanders. This *turf* (brick-sized blocks of dried peat) was used predominantly as a fuel. Eventually, production shifted from northern Flanders and western Brabant to the northern Netherlands, Holland, and Utrecht (Fig. 15.4). In the provinces of Friesland, Groningen, Drenthe and Overijssel, about 1420 km^2 of peat bogs were excavated for the production of turf. This turf is estimated to represent a total of 3.5 x 10^9 Gigajoules of energy (Gerding, 1995).

Until the 14th century, farmers exploited the extensive peatlands in the coastal plains for the production of grain. However, subsidence resulted in a relative groundwater level rise, making circumstances unsuitable for agriculture (Van Dam, 2001). Instead, the remnants of the subsiding peat bogs were mined for fuel, even below regional ground-water levels, turning the areas into open water, which expanded in vast lakes during storm surges (Borger, 1992). Large scale land subsidence in the coastal plains as a result of mainly peat mining and peat drainage for land reclamation led to centuries of flooding, especially in areas adjacent to estuaries and tidal inlets (Vos, 2015). In parallel to the peatland reclamations, peatland flooded by saline water was also mined and combusted to extract trapped sea salt (*selnering*), which remained as a residue (De Kraker, 2006).

Local authorities were aware of the relation between peat mining, subsidence, and floods. The counts of the provinces of Holland and Zeeland for instance demanded in 1404 AD the termination of all mining and drainage activities in the peatlands surrounding Dordrecht (which were primarily related to the sea salt mining industry), imposing as punishment the cutting off one's right hand (Tuinstra, 1951). Despite such repressive measures, the area continued to subside and was struck in 1421 AD by the St. Elisabeth's flood, one of the most devastating floods recorded in the Netherlands, from both landscape and socio-economic points-of-view.

Between 1700 and 1900 AD, most of the remaining raised bogs on the elevated plateaus were mined for fuel or were burned to create arable lands. Major peat mining for turf production gradually stopped between the two World Wars. Based on annual production data, it is estimated that 9.4 x 10^9 m^3 of turf was produced from 1200 to 1950, production from the Peel area not included (Leenders, 1987; Fig. 15.4).

Present-day challenges

At present, ca. 15 x 10^9 m^3 of Holocene peat remains in the coastal and delta plains of the Netherlands, often at shallow depths, bringing new challenges to society (Koster et al., 2018b). Its high porosity and mechanically weak organic structure make it very compactable, causing it to be the most undesirable soil layer from an engineering perspective (Ngan-Tillard et al., 2010; Van Asselen et al., 2018). Yet, the Netherlands is burdened with a substantial construction task, especially in the heavily populated peat-rich coastal plains, to address the housing shortage, increase railroad capacity and deploy subsurface infrastructure to facilitate the energy transition. Activities that include the construction of buildings as well as the installation of below- and above-ground infrastructures must all take the disadvantageous peat properties into account. Koster et al. (2016) for instance, calculated that the overburden of an earthen embankment adjacent to the Markermeer reduced the thickness of an underlying peat layer by between 35 and 95% and this has demanded periodic heightening of the embankment in the recent past to compensate for the elevation loss. Ngan-Tillard et al. (2010) emphasized that during construction of line infrastructures such as roads in peat areas, the heterogeneous character of the subsurface should always be taken into account, in order to avoid differential settlement, damage, and eventually high maintenance costs.

Another challenge is related to global climate change. The Dutch government has committed to the Paris climate agreement to reduce CO_2 emissions by 49% in 2030 and 95% in 2050 (UNTC, 2016). Shallow and aerated peat layers have been identified as substantial contributors to our annual CO_2 emissions budget. It is estimated that microbial oxidation of peat organic carbon leads to a CO_2 emission of 4246 Mt/yr, amounting to ca. 2% of all the Netherlands' annual CO_2 emissions (Kuikman et al., 2005; Arets et al., 2019), while the remaining peat layers in the coastal plains have the potential to emit 2.0 Gt CO_2 when completely oxidized (Koster et al., 2018b). To minimize this process it is critical to keep peat waterlogged, especially in agricultural areas (Weideveld et al., 2021).

To meet the stated climate goals, a transition in land use in agricultural areas may be inevitable to prevent the remaining peat from further degradation that might be exacerbated by climate change. It is predicted that the severity and frequency of extreme droughts in northwestern Europe will increase in the next decades (Spinoni et al., 2018). Droughts are expected to accelerate peat shrinkage and oxidation by diminishing soil moisture content and raising soil temperatures (Kechavarzi et al., 2010), increasing CO_2 emissions even more, as well as enhancing mechanical breakdown of organic matter. Indeed, pilots have

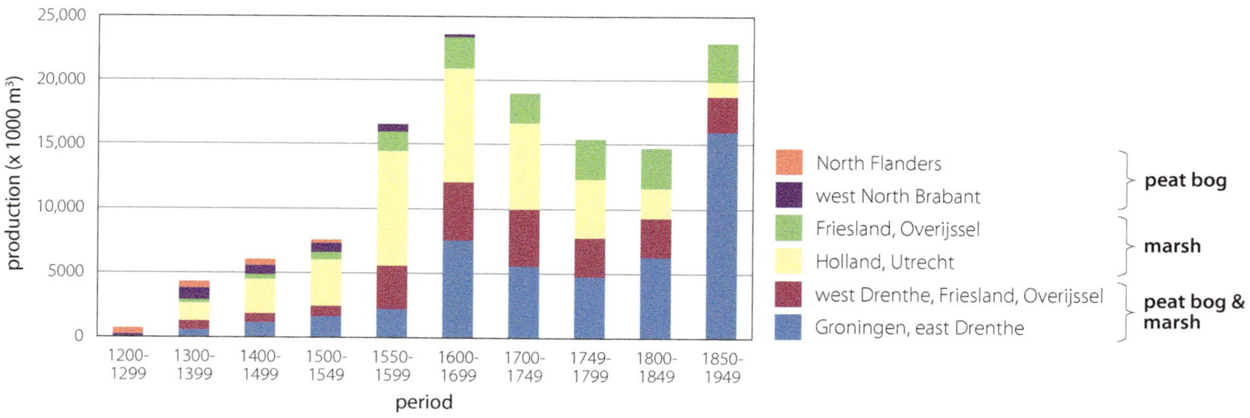

Figure 15.4. Volumes of peat mined in the Netherlands and the north of Flanders (Belgium), per period and per region. Note the differences in the duration of the time periods. Figures indicate average yearly production and do not include the Peel area (east North Brabant and northern Limburg). After data of Leenders (1987). Modified from Van Bergen et al. (2007).

started in recent years to actively keep groundwater levels high in peaty areas (Hoogland et al., 2020; Weideveld et al., 2021), indicating that a transition in agricultural use of peatlands is feasible.

High groundwater levels will in the best case only slow down peat oxidation. However, studies from the peat-rich and heavily subsiding Sacramento-San Joaquin delta (United States) have shown that shifting land use from meadows to wet rice-agriculture completely stopped subsidence and strongly reduced CO_2 emissions (Deverel et al., 2016). Although a transition to rice paddies is probably not realistic for the Netherlands, the Sacramento-San Joaquin case proves that a drastic shift in land use can prevent the remaining peat from oxidizing. However, while through a transition to wet agriculture the remaining peat is preserved, the peat does not regain its natural carbon sink function (Chmura et al., 2003). To stop oxidation and transform peat into carbon sinks, peat areas ought to be inundated again, allowing them to form naturally. Although groundwater levels will still be managed, this situation would approaches natural conditions that prevailed in the coastal plains and higher sandy areas during the major part of the Holocene.

Coal

Coal is generally brown to black and consists of an aggregate of mainly organic constituents called macerals, which are associated with minor to moderate amounts of mineral matter (Karayiğit & Köksoy, 1994; Taylor et al., 1998). Differences between types of coal result from variations in peat forming environments, as well as on different degrees of coalification. With increasing pressure and temperature during burial, peat changes with time first into lignite, then into bituminous coal and finally into anthracite (Fig. 15.5). During this coalification process, hydrocarbons are generated and expelled biogenically and thermogenically.

Carboniferous coal deposits of Westphalian (Pennsylvanian) age, which underlie most of the Netherlands, are the main source rock for natural gas accumulations, with an estimated contribution of more than 95%. The giant Groningen field with its initial recoverable reserves of ca. 2900 x 10^9 m^3 of gas (81% methane; Lokhorst et al., 1997) has also been sourced from Westphalian coals. Given the large volumes of gas-mature Westphalian coal in the subsurface it can be estimated that far more gas has been generated and expelled in the geological past than is currently accumulated in traps in the subsurface of the Netherlands. Probably only 1.5% or less of the total generated volume of gas through time has been trapped, the remainder having escaped to the surface during geological history (Van Buggenum & Den Hartog Jager, 2007; Remmelts et al., 2025, this volume). Part of the generated methane has not been expelled but is retained within the coal itself. Depending on the (technical) process that releases the gas from the coal, this resource is referred to as coalmine methane or coalbed methane.

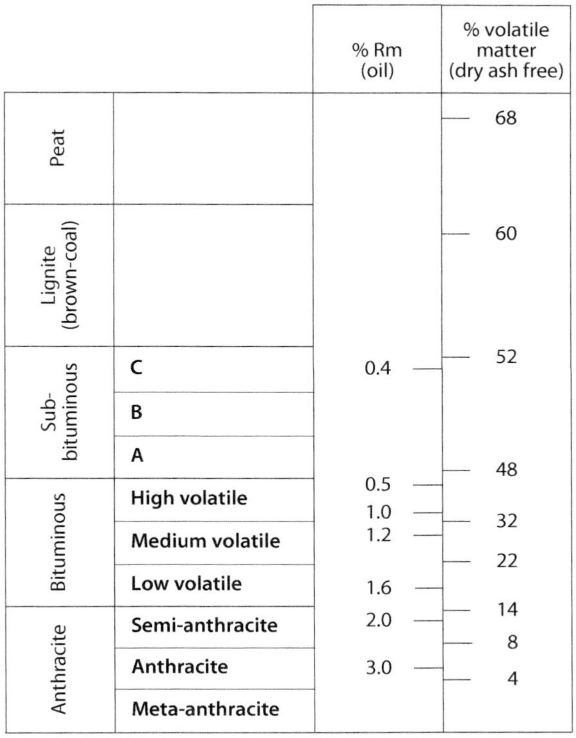

Figure 15.5. Overview of ASTM (American Society for Testing and Materials) coal rank classification (after Stach et al., 1982; Taylor et al., 1998). The stage of coalification is indicated by the rank of the coal, most commonly expressed in percentage vitrinite reflectance (% Rm) or in percentage of volatile matter (dry ash free, daf). Coalification is the progressive change in composition and structure of peat deposits and any other carbon components within sediments during burial, as a result of various interrelated physical, chemical and biological processes (Levine, 1993). The nature of these processes changes significantly with the different stages of coalification (Stach et al., 1982; Taylor et al., 1998). With advancing coalification, biochemical processes become less active (anaerobic microbial processes have some influence until the formation of lignite). During further burial, only physical factors (pressure, temperature) play a role (Van Krevelen, 1993). Lignite has a higher density, lower porosity and lower moisture content than peat. Further coalification of lignite into bituminous coal, and finally into anthracite, occurs as a result of loss of water and thermally controlled chemical alteration of the organic matter. These processes result in volumetric shrinkage due to physical compaction. Modified from Van Bergen et al. (2007).

Lignite

Geological setting

The lignite, or brown coal, deposits in the province of Limburg are part of extensive lignite deposits that were formed during the Miocene in the lower Rhine Embayment north of the Eifel and Ardennes uplands (Fig. 15.6) and it is therefore relevant to include an evaluation of the deposits outside the Netherlands. Lignite deposits reach a thickness of 100 m in Germany (Zagwijn & Hager, 1987; Van der Burgh et al., 1988; NITG-TNO, 1999; Gąsiewicz et al., 2010; Stock et al., 2015; Fig. 15.7), implying by analogy with studies of peat layers in alluvial floodplains that have already compacted up to 43% within a few centuries (Van Asselen et al., 2010) that its original thickness must have been even 2-3 times larger. Further compaction is expected with conversion to lignite. Locally up to 300 m thick peats were deposited to form the main seam (Hager, 1993; Stock et al., 2015) These lignite deposits are part of the Ville Formation in Germany that contains, from oldest to youngest, the Morken, Frimmersdorf and Garzweiler seams (Stock et al., 2015; Fig. 15.7). Towards the west, these deposits include more and more sand intercalations which in effect split a seam into several lignite beds that can be defined as members (Zagwijn & Hager, 1987). The Ville Formation grades laterally in South Limburg into unconsolidated sand with three to four lignite beds belonging to the Middle Miocene Heksenberg Member (Fig. 15.8; Engelen, 1987). This member is up to 100 m thick with its upper and lower limits defined by the Frimmersdorf 1 and 2 coal seams (Munsterman et al., 2019). Individual zones with lignite exceed 15 m (NITG-TNO, 1999; TNO-GDN, 2022b) but decrease markedly towards the west and northwest as they continue into Belgium and North Brabant. Thinner seams continue into the Late Miocene, e.g. within the Kieseloolite Formation.

Historical development and economics

In Germany, in the area between Cologne, Bonn and Aachen, small-scale lignite mining started during the 16th century. In South Limburg, small-scale, though illegal, exploitation commenced between 1865 and 1868 near Eygelshoven (Engelen, 1989a). The first mining conces-

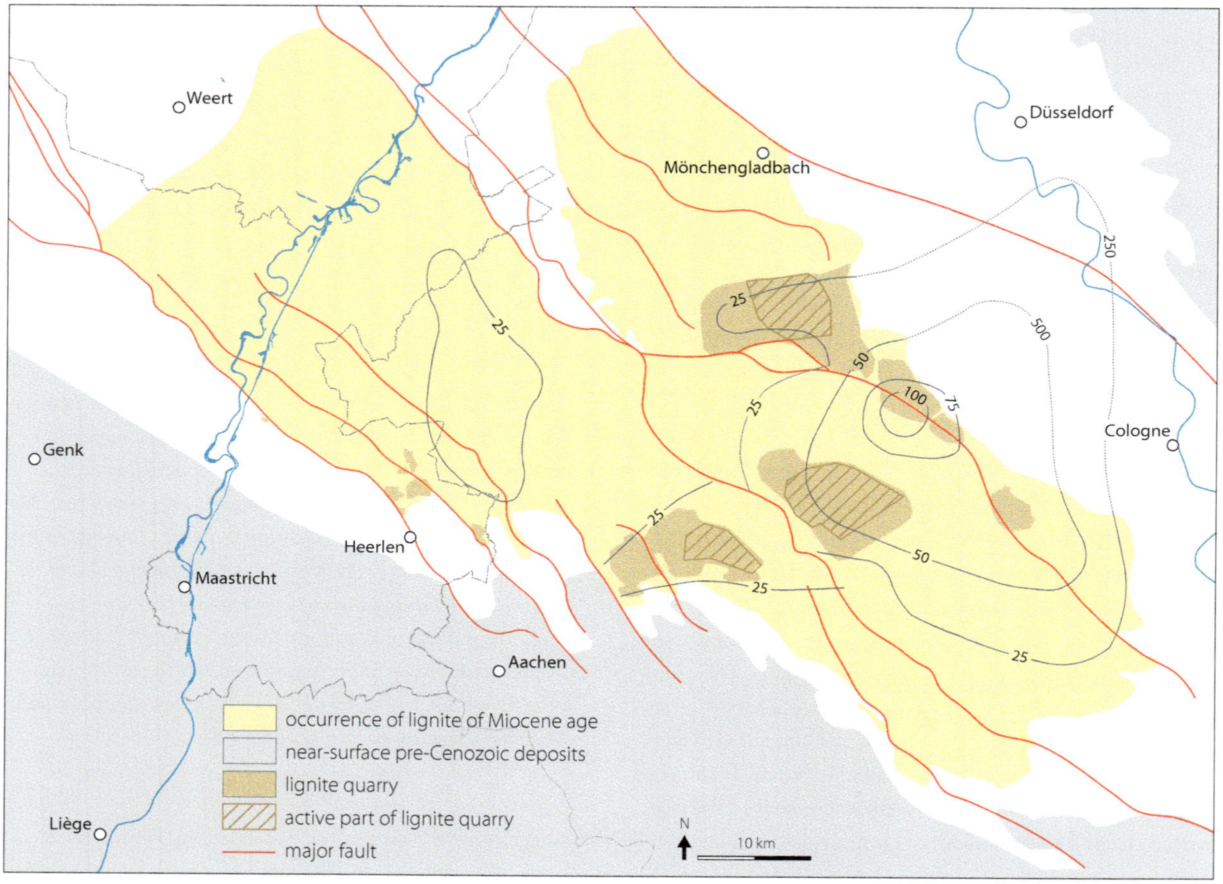

Figure 15.6. *The occurrence and thickness (in metres) of Miocene lignite deposits in the Lower Rhine Graben (modified from Zagwijn & Hager, 1987; GD-NRW, 2023; in-house data of TNO-GDN, 2023). Bounds of lignite quarries in the Netherlands after Jongmans & Van Rummelen (1930) and in Germany after T. Römer (commons.wikimedia.org/w/index.php?curid=11539190 used under CC BY-SA 2.0).*

Figure 15.7. Picture of the opencast lignite mine of Garzweiler (Germany), exposing the Neogene sedimentary infill of the Lower Rhine Embayment and with the three regional lignite seams indicated. The location of the mine is given in Figure 15.6. Photo: Jan Stafleu.

sion was granted in 1906 near Heerlerheide, but the exploitation was not successful (Engelen, 1987). It took until 1917, when fuel became scarce during the First World War, before new lignite extraction started in several villages near Heerlen. Dutch lignite production peaked at about 1.5 Mt/yr around the end of the war (1918-1919). By 1921 production had decreased to less than 0.13 Mt. In the following years most quarries ceased operations and only the Carisborg I quarry was kept open (NITG-TNO, 1999). Renewed production commenced during the Second World War and lasted until 1962. The quarries 'de Energie' and 'de Herman' were brought back to production and 'de Anna' was opened (NITG-TNO, 1999). Production reached over 0.2 Mt/yr during these last years (Engelen, 1989a). The lignite resources in the Netherlands are calculated to exceed 1700 Mt (Van der Burgh et al., 1988). However, despite their relatively shallow depth, any future mining is unlikely because of environmental concerns and population density.

The lignite of the Lower Rhine Embayment in Germany continues to be mined intensively, providing cheap fuel for the large power plants in the area between Cologne and Aachen, not far from the border with the Netherlands. It is mined in very large open pits to a depth of 399 m below the original surface (293 m below sea level) in the Hambach quarry (Eifelnatur.de, 2023). The annual lignite production from the Garzweiler, Neurath and Hambach open pit mines declined from 95.2 Mt in 2015 (BGR, 2016) to 51.4 Mt in 2021 (BGR, 2022). The original recoverable reserves were estimated to be about 50,000 Mt (Van Montfrans et al., 1988) or 55,000 Mt (Hager, 1993; DEBRIV, 2023) of which around 30,000 Mt was considered technically recoverable. The production of lignite in the area is expected to cease in 2030 (DEBRIV, 2023). During mining, millions of cubic metres of groundwater are pumped off and discharged. As a consequence, the groundwater system changed from a condition with upward seepage into a situation with downward infiltration (NITG-TNO, 2001; De Vries, 2007; Stuurman et al., 2007).

Bituminous coal and anthracite

Geological setting

Carboniferous sedimentary rocks of the Netherlands and surrounding areas accumulated in an east-west striking

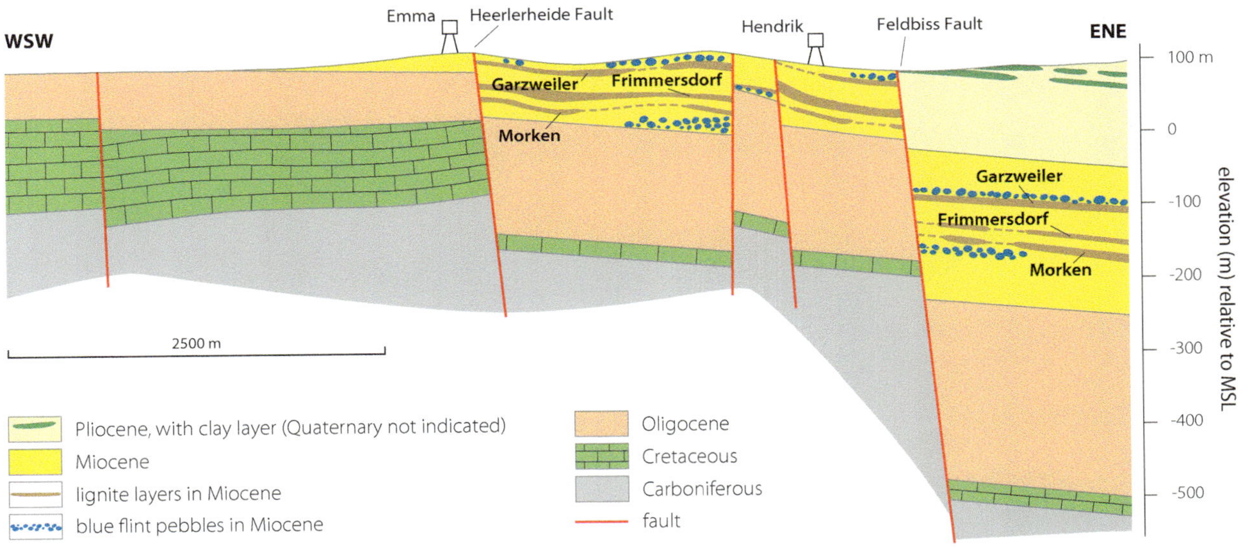

Figure 15.8. Cross section through Cenozoic strata near Heerlen in the northeastern part of South Limburg, showing the depth of the lignite-bearing Miocene silver sands (adapted from Jongmans & Van Rummelen, 1930; Pannekoek, 1956; De Jong & Van der Waals, 1971). For location of the Emma and Hendrik mines see Figure 15.12.

basin over a period of 60 million years (359-299 Myr) and have a present-day (thus compacted) maximum thickness of 5.5 km (Van Buggenum & Den Hartog Jager, 2007; Kombrink, 2008). These deposits contain numerous organic rich layers that, depending on their maturity, are classified as anthracite or bituminous coal (Figs 15.5, 15.9). At the same location both types can occur with the anthracite buried deeper than the bituminous coal. In this chapter both are referred to as coal. The main coal-bearing formations in the Netherlands were part of this succession and were deposited during the late Carboniferous (mainly during the Westphalian) in the Northwest European Coal Basin. During a phase of deformation and uplift at the end of the Carboniferous about 2000 m of Carboniferous strata were eroded across large parts of the country, leading to the current configuration of the Carboniferous subcrop below the basal Permian unconformity (Pagnier et al., 1987; Van Buggenum & Den Hartog Jager, 2007; Huis in 't Veld & Den Hartog Jager, 2025, this volume). At present, the coal-bearing deposits are found at shallow depth in the southernmost Netherlands but deepen northward to more than 3 km, where offshore the top of the Carboniferous can be as deep as 10 km in the northern Central Graben (De Bruin et al., 2015).

The Epen Formation of Namurian age consists mainly of lacustrine, marine, and deltaic shale with some sandstone intervals that locally are overlain by coal beds. Namurian aged coal beds are generally not well developed in the Netherlands, whereas they are well represented further to the east in Germany and Poland (Kombrink, 2008). The main coal-bearing deposits belong to the Baarlo, Ruurlo and Maurits formations of Langsettian to lower Bolsovian age (±321-311 Ma; Huis in 't Veld & Den Hartog Jager, 2025, this volume). The Westphalian sedimentary succession reflects continued aggradation at or near emergent conditions, comprising deltaic to fluvio-lacustrine deposits with numerous coal layers in the early Westphalian that give way to well drained fluvial sediments (eventually red beds) in the late Westphalian (Kombrink, 2008). These formations have a total thickness up to 3000 m, were deposited in a cyclic river-dominated environment and have coal contents between 1.0 and 2.1% (Van Buggenum & Den Hartog Jager, 2007). Other authors have given higher estimates of up to 3.5% based on onshore observations along the southern basin margin (e.g. Van Wijhe & Bless, 1974; Hedemann et al., 1984). Facies maps suggest lower average coal percentages for the offshore (Huis in 't Veld & Den Hartog Jager, 2025, this volume). The younger fluvial sandstone and shale sequence of the Bolsovian and

Figure 15.9. Mining equipment (kolenschaaf) in a pillar in the Oranje Nassau Mines taken around 1952. In the background a Westphalian coal seam is visible with the typical layering and break-out patterns of coal. © Nico Jesse / Nederlands Fotomuseum.

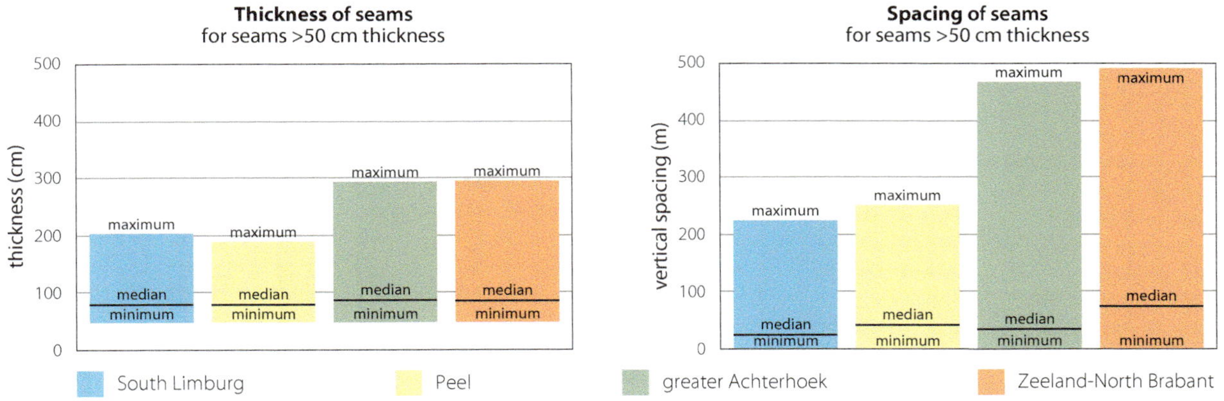

Figure 15.10. Thickness (>50 cm thickness) and spacing of Westphalian coal seams in the areas shown in Figure 15.12 (South Limburg: depth <1500 m; Peel: depth <1500 m; greater Achterhoek/south Overijssel area: depth <2000 m; Zeeland-Noord Brabant: depth 1500-2000 m). The number of evaluated coal seams varies per area from 21 encountered in 2 wells in Zeeland-North Brabant, to well over 100 seams in other areas. For a limited number of coal-exploration wells, the coal-seam thickness was known from cores, otherwise geophysical well logs (sonic, density) were used (from Van Bergen et al., 2000).

Asturian locally contains thin coal seams (Van Adrichem Boogaert & Kouwe, 1995). The rank of these coal seams is mainly bituminous, but anthracite occurs locally, especially in South Limburg where most likely a major Variscan thrust sequence has been eroded (NITG-TNO, 1999; Huis in 't Veld & Den Hartog Jager, 2025, this volume).

Coal-clastic cycles, also known as 'cyclothems', are metre- to decametre-scale alternations of (marginal) marine shales and terrestrial shales, sands and coal beds (Weller, 1930; Greb et al., 2008; Van den Belt, 2012). These cycles may (Veevers & Powell, 1987; Klein & Willard, 1989) or may not (Wilkinson et al., 2003) have resulted from glacio-eustatic sea level fluctuations in the Milankovitch frequency band. Van den Belt (2012) suggested that carbon storage in the late Paleozoic terrestrial system was controlled by climate rather than by local sedimentary conditions. Core analysis indicates that a distinct two-fold cyclicity at wavelengths of ~255 m (~395 ky) and ~60 m (93 ky) is present, as is shown in Carboniferous cores, indicating long and short-eccentricity control (Van den Belt, 2012). Based on Markov analysis, para-sequence thicknesses in the Westphalian deposits were quantified with cycle thicknesses for the Langsettian and Duckmantian at around 10 m, while not all cycles contain coal (Pagnier et al., 1987). Average duration of individual cyclothems with highly variable thicknesses was ~21 ky (Van den Belt, 2012). The estimated coal-accumulation rate in the Netherlands was 5-10 m/My for the Langsettian and Duckmantian (ca. 320-316 Ma) and increased abruptly to ~25 m/My at 315 Ma. Spacing between the individual seams as measured in wells is shown in Figure 15.10. Based on five coal-exploration wells drilled in the 1980s in the Achterhoek and in South Limburg, the mean thickness of Westphalian coal seams (of >50 cm) is 1 m (RGD, 1986). According to Van Bergen et al. (2000) the median thickness of such seams in the areas mentioned is 88 and 80 cm, respectively. Individual coal-layer thicknesses reach up to 3 m in the greater Achterhoek area, or, in terms of cumulative thickness, even up to 3.5 m (RGD, 1986; Pagnier et al., 1987; Fig. 15.11).

Historical development and economics

South Limburg

The lower part of the Westphalian coal measures lies at or near the surface in the valley of the river Worm north of the German city of Aachen, where it has been mined from Roman times onwards in simple open pits (Visser & Zonneveld, 1987; NITG-TNO, 1999). Coal has been mined in the Kerkrade area since medieval times (Bless et al., 1984). The Annales Rodenses of 1113 AD, the chronicles of Kloosterrade (Rolduc Abbey), situated in the Worm Valley of Kerkrade, contain a transfer of land named 'kalculen'. The etymology is thought to refer to a small open pit coal mining area, which could be regarded as the oldest official recordings of coal production in Europe (Moonen, 2012). Underground mining in galleries started early in the 14th century (Visser & Zonneveld, 1987; NITG-TNO, 1999). The Limburg dialect word *Koul*, (e.g. the Sittard district Kemperkoul or the small hamlet of Koulen, near Valkenburg) refers to such small open pits that were dug to produce near-surface coal or to store coal or gravel until the 18th century. In the 16th century, when exploitation began to require more technical and financial resources, an industry developed that was based on the early capitalist system (Raedts, 1971; Engelen, 1989b; NITG-TNO, 1999). At the beginning of the 18th century, the abbey of Kloosterrade (Rolduc) began developing coal mines. By the

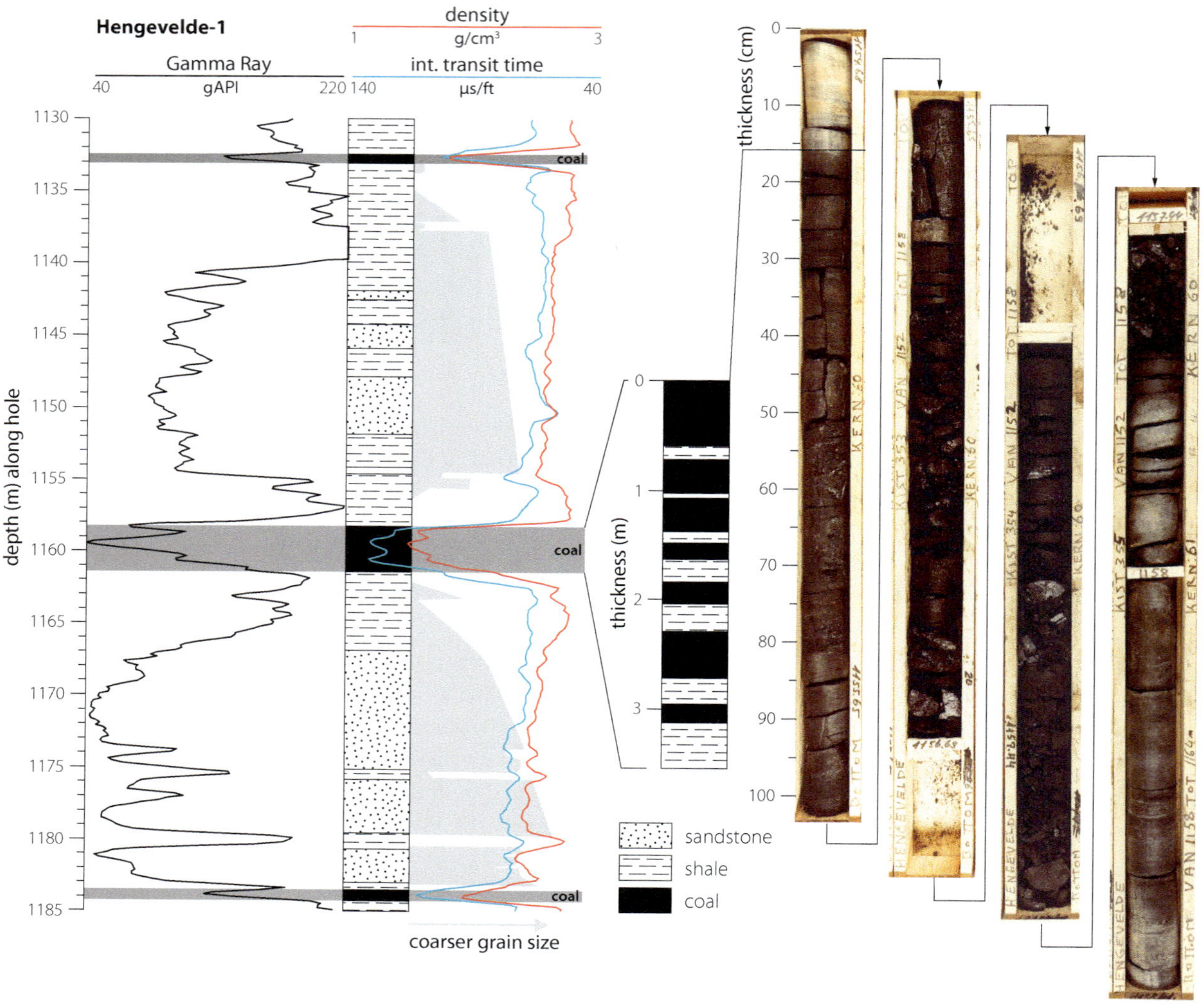

Figure 15.11. Log response with core description and core pictures of the thickest coal seam encountered in the well Hengevelde-1, where a cumulative thickness of 356 cm is encountered. Core analysis indicates that this coal seam (named seam VI) is actually composed of smaller coal (grey) and shale (brown) intervals as shown by the picture in the centre. This coal is classified as a high volatile bituminous coal based on a vitrinite reflectance (Rm) of 0.77%, which is not fully consistent with the measured volatile matter content of 51.62% (dry ash free). Ash was 15.88% and sulphur 1.27% (dry basis). The layers in this section dip at 8-25°. The alternation of sandstone, shale coal, shale, sandstone, shale and coal are expressions of (non-ideal) cyclothems. Core photos from Rijks Geologische Dienst.

middle of that century mining was conducted from small shafts and with small mining corridors from the western wall of the Worm Valley where the Rolduc Abbey is situated. The first shaft mining started during the French occupation in 1797 on the nearby Kerkrade Plateau, after the mines were nationalized. Following the invention of the steam engine, it became possible to handle the vast amounts of ground water and to deepen the mine shafts.

However, during the French occupation of 1794-1815, mining came to a near standstill as a result of seizure and incompetent management (NITG-TNO, 1999). During the French period, all natural resources became the property of the State under the Napoleonic law of 1810 and under this law, concessions for exploitation were granted by the State. Exploitation in the state-owned mines was unsuccessful after 1815. The Domaniale concession was therefore leased to a private enterprise in 1845. Administration of commercial coal production in South Limburg started in 1847, but mining became truly established at the end of the 19th century when the arrival of a railway line improved access to eastern South Limburg (NITG-TNO, 1999). During this period, the large Oranje-Nassau concession was awarded, and the Oranje-Nassau I mine became operational in 1899.

From 1902 onwards the 'State Collieries' (Staatsmijnen) Wilhelmina, Emma, Hendrik and Maurits were established in the substantial areas not yet covered by earlier concessions. The new concessions were repeatedly extended dur-

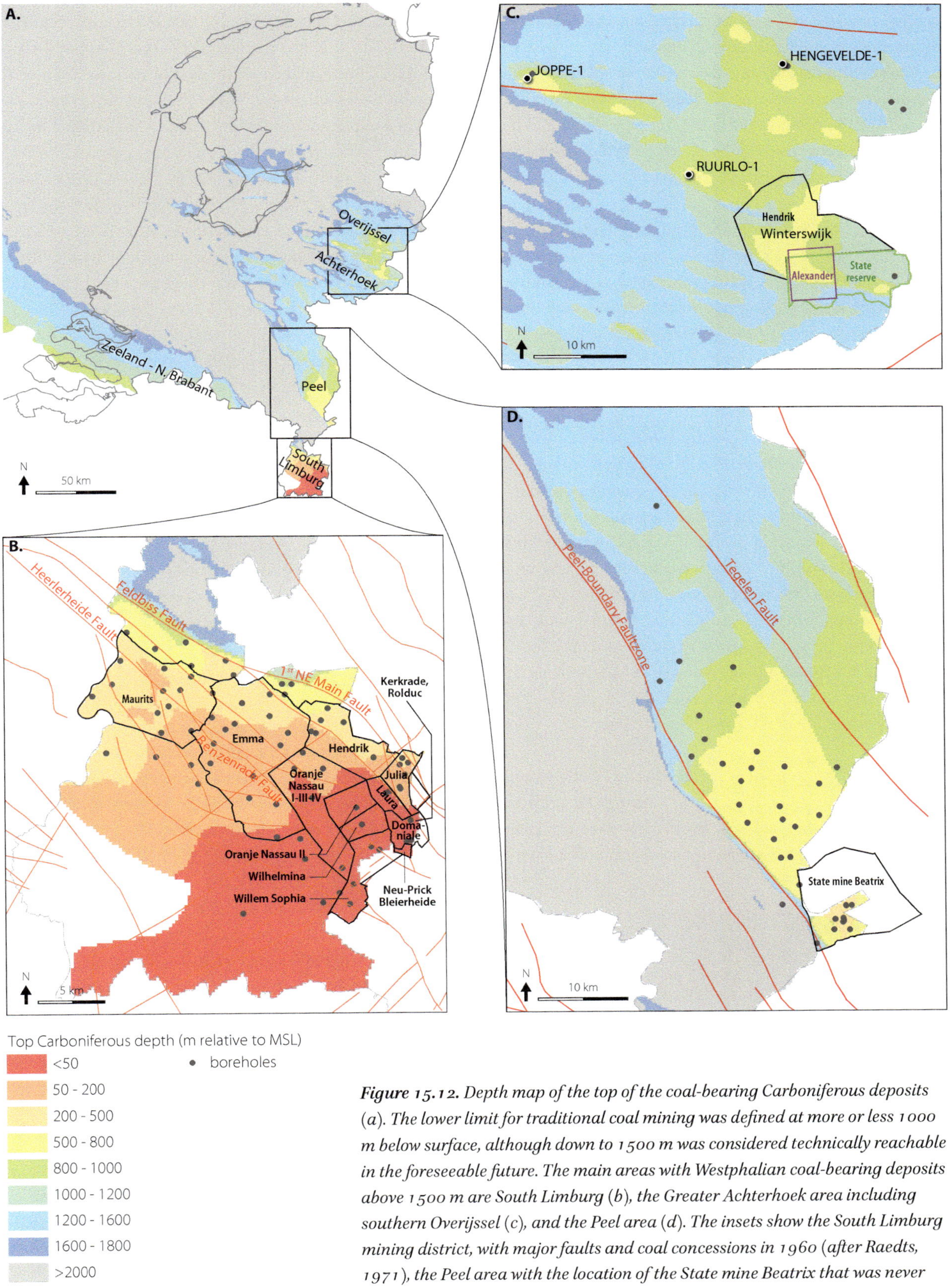

Figure 15.12. *Depth map of the top of the coal-bearing Carboniferous deposits (a). The lower limit for traditional coal mining was defined at more or less 1000 m below surface, although down to 1500 m was considered technically reachable in the foreseeable future. The main areas with Westphalian coal-bearing deposits above 1500 m are South Limburg (b), the Greater Achterhoek area including southern Overijssel (c), and the Peel area (d). The insets show the South Limburg mining district, with major faults and coal concessions in 1960 (after Raedts, 1971), the Peel area with the location of the State mine Beatrix that was never taken into production and the Achterhoek/southern Overijssel area with the former concession for coal and salt exploration. The application for the Alexander concession by private companies had to be withdrawn after it had been established a State reserve by law in 1924 (Visser, 1987).*

ing the first half of the 20th century. Smaller enterprises opened mines along the German border.

Coal mining in Limburg took place between the Benzenrade Fault and the First NE Main Fault (Fig. 15.12). As a result of the north to northeast dip of bedding and because of faulting, the depths at which exploitation took place varied considerably. The bottom of a shaft of the Domaniale mine was only 22 m below surface level, whereas a shaft of the State Colliery Hendrik extended down to 1008 m. A total of 248 coal seams with a cumulative coal thickness of over 73 m (IGCP 166, 1980) is present. Approximately 65 of these seams are exploitable, with thicknesses ranging roughly from 50 to 150 cm (Kuyl, 1980).

After the global economic crisis in the late twenties and early thirties, the Dutch coal-mining industry experienced a boom that lasted until after World War II. Twelve surface facilities were in operation, with a combined annual production of about 12.6 Mt (Engelen, 1989b). In the early 1960s, when production was peaking, the mines provided work for more than 28,000 workers underground and nearly 23,000 at the surface (Westen, 1971). Interest in coal outside South Limburg diminished rapidly after the discovery of the giant Groningen gas field in 1959 (Van Tongeren, 1987). This discovery, together with low world-energy prices and rising exploitation costs, led to the collapse of coal mining industry in the Netherlands (Raedts, 1971; Van Tongeren, 1987). The period from 1967-1974 saw the rapid closure of the mines.

Coal production stopped with the closure of the Oranje Nassau mine in 1974 (Fig. 15.13). During the previous 128 years, an estimated 582 Mt of coal were produced from thirteen mines (four state-owned and nine private),

Figure 15.13. Picture of the last piece of produced coal, in custody of OneDyas, extracted from the Oranje-Nassau I mine on 31 December 1974, showing the centimetre-scale internal fracture (cleat) system of the coal. Production from this last operational mine in the Netherlands reached 427,000 tons coal in that last year. Photo: Richard Huis in 't Veld.

all in South Limburg (Fig. 15.14; Westen, 1971; Stuffken, 1987). This is only a fraction of the coal volumes present in the subsurface and enormous coal resources remain in Limburg and elsewhere. In the Peel and Achterhoek, numerous exploration wells for coal have been drilled but no mining has taken place (Peelcommissie, 1963; RGD, 1986). During the 1980s, interest in coal was renewed, induced by the second oil crisis and also triggered by new techniques such as underground coal gasification. This led to an inventory of coal deposits down to a depth of 1500 m (RGD, 1986; Pagnier et al., 1987; Van Tongeren, 1987), the results of which indicate that the fault block between the Tegelen Fault and the Peel-Boundary Faultzone, north of the former mining area, contains a coal resource of 591 Mt (RGD, 1986). Underground coal gasification was considered as early as 1946 (Visser, 1987), but no applications have been made, although elsewhere, especially in the United States, demonstration projects have been conducted.

Peel

In view of the coal resources immediately across the border in Germany, the State Service for the Exploration of Mineral Resources (ROD) carried out a drilling campaign between 1903 and 1916 to investigate the occurrence of exploitable quantities of coal in the Peel area (Fig. 15.12). Based on the results from 13 deep wells and many shallow boreholes, the ROD calculated that reserves of about 2500 Mt of coal exist above 1500 m below mean sea level in seams of exploitable thickness (Van Waterschoot van der Gracht, 1918). During the Second World War, a gravimetric survey was carried out in the Peel and adjacent areas (De Sitter et al., 1949; Van Weelden, 1957) and halfway through the 20th century, the State Collieries drilled 12 wells (NITG-TNO, 2001). In 1952, exploration intensified after establishment of the advisory Peelcommissie. It included the surveying of 368 km of seismic data and the drilling of eight additional wells. The committee calculated that a geological reserve of approximately 2900 Mt of coal in seams over 50 cm thick is present in the evaluated areas but observed that the net cumulative coal thickness in the area is less than in South Limburg (Peelcommissie, 1963). Meanwhile, the construction of two shafts for the new Beatrix State Colliery in the eastern Peel area commenced in 1952; it was subsequently suspended in 1962 as a result of a change in energy policy (Peelcommissie, 1963; NITG-TNO, 2001). In 1984, further investigations were carried out by the Geological Survey (RGD, 1986; Pagnier et al., 1987). Seven seismic lines with a total length of 115 km were acquired, covering about 220 km^2. The results of this research confirmed the conclusions of the Peelcommissie. They also showed that the Carboniferous to the north of the Peel area is represented by less productive coal seams of the Namurian and Westphalian

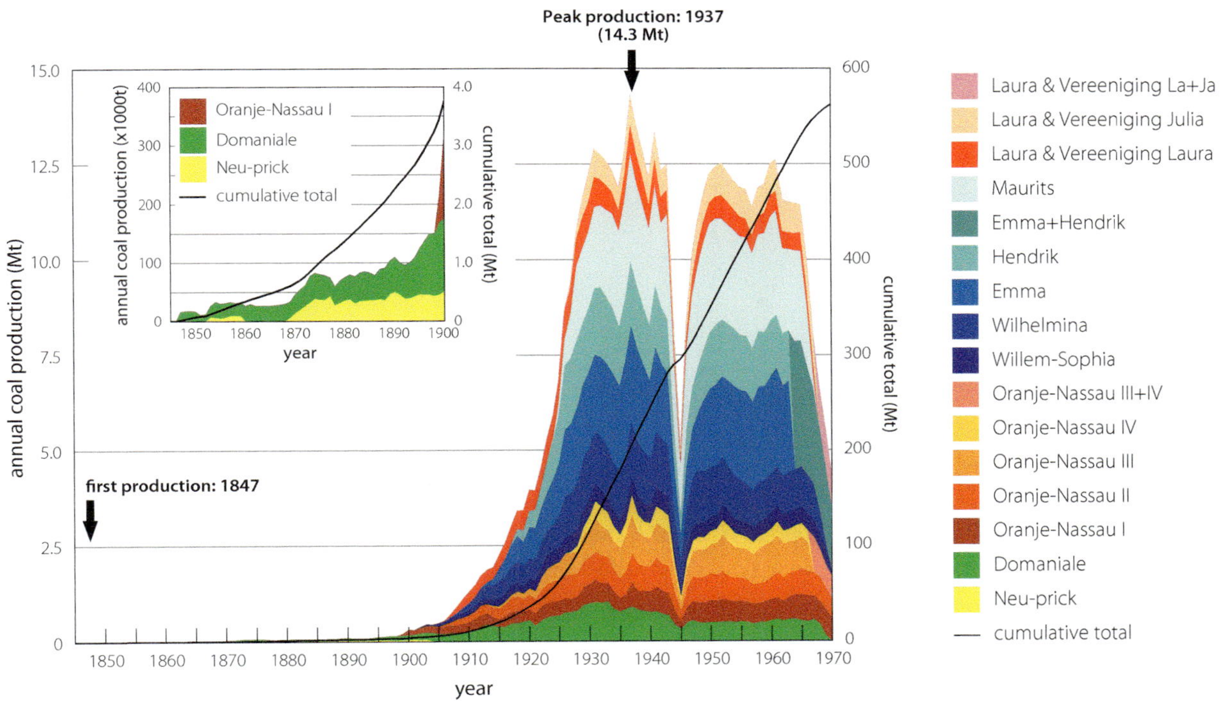

Figure 15.14. *Net annual coal production from 1847 to 1970 from mines in South Limburg (after data of Westen, 1971). Data up to 1929 without coal sludge; since 1929 including coal sludge. Extrapolation of the declining production curves indicates a cumulative production of ca. 8.5 Mt in the period from 1971 until the closure of the last mine in 1974. Modified from Van Bergen et al. (2007).*

A (NITG-TNO, 2001). The latest evaluation of the Peel area covered twelve fault blocks, some of which are located in Germany, but which fall under a Dutch concession according to an agreement reached in 1959 (Peelcommissie, 1963). The resources of the area (P50: 3300 Mt; Fig. 15.15a; Van Bergen et al., 2000) are slightly higher than those reported earlier by the Peelcommissie (1963).

Achterhoek (and southern Overijssel)

A concession for coal and rock-salt exploitation in the Achterhoek was granted in 1930, after Carboniferous coal measures had been identified near Winterswijk by the State Service for the Exploration of Mineral Resources (Fig. 15.12). The area of this 'Gelria' concession is greater than 100 km². At that time, the reserves of bituminous coal in this area at depths between 900 and 1500 m were estimated at about 360 Mt. However, mining never took place because of high costs and the risk involved, not to mention an economic slump (Visser, 1987).

In the 1980s, further investigations were carried out by the Geological Survey (RGD, 1986; Pagnier et al., 1987). Within the framework of this study, seismic lines with a total length of 79 km were aquired and 375 km of previously existing lines were re-evaluated. The resources of three coal seams from the top of the Carboniferous to 1500 m, in a triangular area of 157 km², between the coal exploration wells Joppe-1, Hengevelde-1 and Ruurlo-1 were estimated at 693 Mt (RGD, 1986). For the greater Achterhoek area, including south Overijssel, the total coal resources shallower than 1500 m were estimated to be about 9100 Mt in the latest inventory (P50; Fig. 15.15a; Van Bergen et al., 2000).

Zeeland and North Brabant

Exploration wells were drilled throughout the Netherlands in search of coal by the Rijks Opsporingsdienst Voor Delfstoffen just before World War I. During this campaign well Woensdrecht-01 drilled down to a depth of 1205 m, but found that coal measures were missing. Since then, a number of wells drilled into the Carboniferous in this area, but further to the north encountered coal seams, mostly at depths greater than 1500 m. Given the large area, the total resources are substantial but the seams lie too deep and are too widely spaced.

Coalbed methane

At the turn of the 21st century attention to the coal resources of the Netherlands resumed, this time not for exploitation of the coal itself but for the gas that is contained in the coal seams. Coalbed methane, or CBM, is the

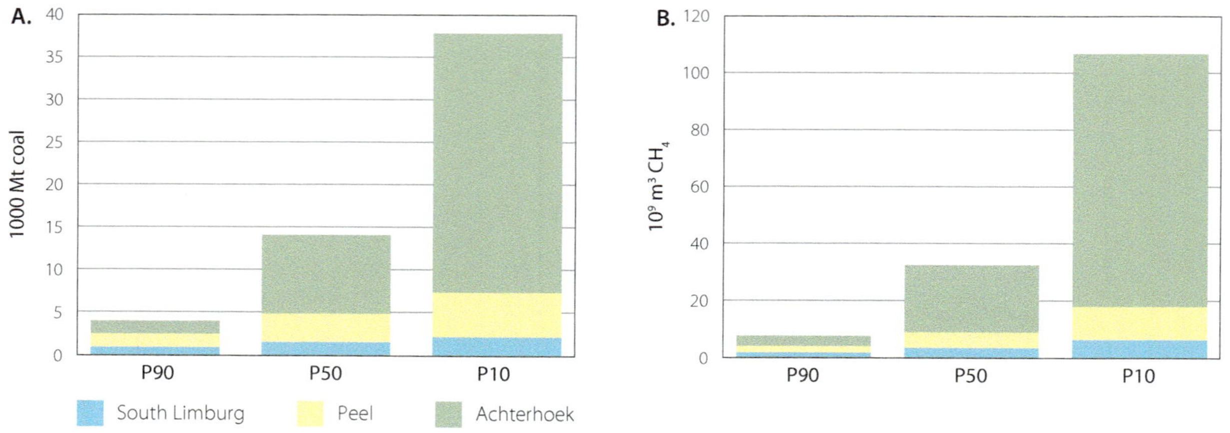

Figure 15.15. Calculated quantities expressed in probability values, of (a) coal resources and (b) potentially producible (technically recoverable) coalbed methane (CBM) in the Netherlands from Westphalian coals shallower than 1500 m depth (after Van Bergen et al., 2000). The (theoretically) technically recoverable amount of CBM depends on high net coal thickness (Pashin & Hinkle, 1997), while producibility depends on several critical factors, including hydrodynamics, depositional setting and coal distribution, tectonic and structural setting, coal rank and gas generation, permeability, and gas content (Scott, 2002). In the calculations by Van Bergen et al, (2000) these dependencies were represented by a completion and a recovery factor. The completion factor will be high in thick, closely spaced seams, and low in thin, sparsely spaced seams. The recovery factor is the percentage of gas that can be produced from the contributing coals. In CBM production as applied in the United States, this depends strongly on the pressure drop that can be achieved in the coal seams by the pumping of large volumes of water; normally 20 to 80% of the gas originally in place can be recovered. The South Limburg area has potentially the highest producible methane content per square kilometre due to its high net cumulative coal thickness, but the potential for total producible methane is highest in the large, greater Achterhoek area (including southern Overijssel) and fairly high in the medium-sized Peel area. In addition to the inventory down to 1500 m, methane resources for the depth interval between 1500 and 2000 m are largely due to potentially large amounts of coal and the high pressure. Estimated resources in the greater Achterhoek and Zeeland and North Brabant areas amount to 50 and 118 × 10^9 m³ gas respectively (P50 values). However, the producibility of these deeper resources is technically and economically questionable at present because the permeability of deeper coal seams is expected to be too low and the costs of drilling too high (Van Bergen et al., 2000; Scott, 2002). Modified from Van Bergen et al. (2007).

natural gas that is retained in coal in the subsurface. It is known as mine gas, or coal mine methane in the mining industry, where it is dangerous due to its explosive nature. From the 1970s onwards interest has grown in coalbed methane as a fuel and a related industry has developed in the United States as a mainstream natural gas producer (Saulsberry et al., 1996; Ayers, 2002). This has also triggered interest in many of the European countries with coal resources, because the Carboniferous deposits in western Europe show many similarities to those producing from Variscan foredeep basin in the USA, such as the Black Warrior Basin in Alabama (Fails, 1996; Pashin, 1998). In Europe, particular interest has been raised in the production of CBM while simultaneously storing CO_2 in the coal seams. This could become a climate-friendly way of producing energy as discussed below. To evaluate the scope for exploration for CBM, an update of the coal resources in the Netherlands was made for the relatively shallow occurrences of the Carboniferous in the year 2000 (Fig. 15.15; Van Bergen et al., 2000). Deep coal seams were considered to be of limited economic significance.

Although coal seams in the subsurface generally contain large amounts of gas (mainly methane), there is at present no interest in producing coalbed methane as the Netherlands are shifting away from fossil fuel for its energy production. This is, however, a relatively new development, as methane has been considered as a 'transition fuel' in the 1990s-2000s, that could support a society with lower CO_2 emissions by replacing coal-fired by gas-fired power plants. Efforts were therefore undertaken to investigate the potential of CBM production in the Netherlands. Following the success in coalbed-methane production in the United States, the Peer well was drilled in 1992 in the Campine Basin in northeastern Belgium, close to South Limburg. This remains the only well close to the Netherlands that was drilled and tested for coalbed methane production. Only a limited amount of CH_4 was produced because the inflow of water in the well was too high to reduce the pressure sufficiently to allow substantial gas desorption. Wenselaers et al. (1996) concluded that these unfavourable hydrological conditions were due to the proximity of the well to a major open fault (the

'Donderslagbreuk'). According to Wenselaers et al. (1996) perspectives for coalbed methane production could exist in the Campine Basin, with more than enough coal and gas present, despite the low productivity observed. Nevertheless, no further CBM wells have been drilled until present.

The investigations into the potential for CBM production in the Netherlands did trigger major research efforts to understand the presence of this fuel in the context of the geology and reservoir conditions. Fails (1996) established that the depositional setting and coal distribution of the Eastern Ruhr Basin are favourable for CBM. As the coal basins in the Netherlands resemble the Eastern Ruhr Basin in the sense that the coal-bearing deposits are of the same age and the basins are located on a similar position with respect to the Variscan front, it could be assumed that their depositional setting and coal distribution were suitable for CBM production. However, the tectonic history and structural setting in the Netherlands is in general more complex than CBM producing basins in the United States and this affects the production potential. The basins are heavily faulted and locally also folding occurs, especially near the Variscan front (e.g. in South Limburg). Still, tectonic fault blocks of substantial size with continuous coal seams are present. During the geological history of parts of the Netherlands, in South Limburg, coal seams have been buried deeper than their present depth. The Carboniferous coal in the Netherlands is bituminous or anthracitic, indicating that large volumes of thermogenic gas have been generated during coalification. Contributions from secondary biogenic gas generation have so-far not been reported and seem unlikely due to the depth of the coal. In coal at around 1000 m below surface, permeability of 0.1 to 1 (maximal about 5) m is to be expected in the Netherlands which is on the low end of the permeability range (0.5 and 100 m) established for the most highly productive wells (Scott, 2002). Permeability in coal beds is determined by its fracture (cleat) system, which is in turn largely controlled by the tectonic/structural regime (Fig. 15.13; Scott, 2002) and is strongly dependent on present-day in situ stress orientation. The permeability in the aforementioned Peer well was less than 1 m in Langsettian and Duckmantianseams at depths between 850 and 1250 m, i.e. about 450 m above their maximum paleo-burial depth (Wenselaers et al., 1996). In view of the lack of coal-mining activities in the Netherlands since 1974 and of coal exploration since 1985, the gas content can only be estimated by indirect methods. Stuffken (1957) investigated the methane content of coals in South Limburg on the basis of data from mine ventilation air and found a U-shaped relation between rank and gas content, with a maximum at about 25 wt% V.M. However, it is difficult to convert these data from ventilation air into gas content per in situ coal volume. The majority of the gas is not in the pore system of the coal, as is for example the case in sandstones, but is adsorbed onto the coal surface. Resource estimates for CBM are therefore in general based on the sorption capacity of coal. The multiple parametres that influence this are generally interpreted to depend on pressure, temperature and coal characteristics (e.g. Kim, 1977; Yang & Saunders, 1985; Levy et al., 1997; Bustin & Clarkson, 1998). Coal characteristics that affect sorption capacity are maceral composition, rank, ash content and moisture content (e.g. Ettinger et al., 1966). Opposing effects of pressure and temperature result in a maximum in gas-sorption capacity around a depth of 1500 m (Fig. 15.16).

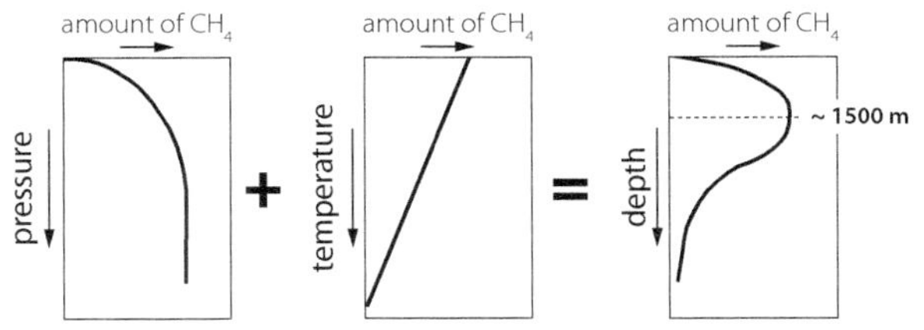

Figure 15.16. Effects of increasing pressure, temperature and depth on the amount of methane adsorbed on coal, assuming a normal geothermal gradient and hydrostatic pressure comparable to those in the Carboniferous of the Ruhr Basin in Germany (Freudenberg et al., 1996). This figure shows a linear decrease (after Kim, 1977; Levy et al., 1997; Bustin & Clarkson, 1998), but an exponential decrease is also reported (Yang & Saunders, 1985). Modified from Van Bergen et al. (2007).

Coppens (1967) showed that the sorption capacity of coal samples from the Belgian part of the Campine Basin varies, depending on rank, between 5 and 9 m^3/t. A small set of isotherms is available for the Netherlands from core samples that were taken from wells in the Achterhoek area (Krooss et al., 2002). Sorption capacities of the investigated dry coals were measured to be about 15 m^3/t, while the sorption capacities of moist coal were found to be lower (Krooss et al., 2002). No rank effect could be discerned on these samples, probably due to the small number of samples (Krooss et al., 2002).

The resources of coalbed methane have been inventoried in several studies based mainly on isotherm data and on relations between rank and gas content. However, the hydrological conditions should be considered in follow-up studies as hydrodynamics strongly affect CBM producibility (Scott, 2002; Van Bergen et al., 2007). An estimation by Wolf et al. (1997) indicated very high resources of 770 x 10^9 m^3 down to a depth of 2000 m and 1400 x 10^9 m^3 in deeper seams. However, the negative impact of increased temperatures below 2000 m was not taken into consideration in this evaluation, and the resources are therefore probably overestimated (Fig. 15.16).

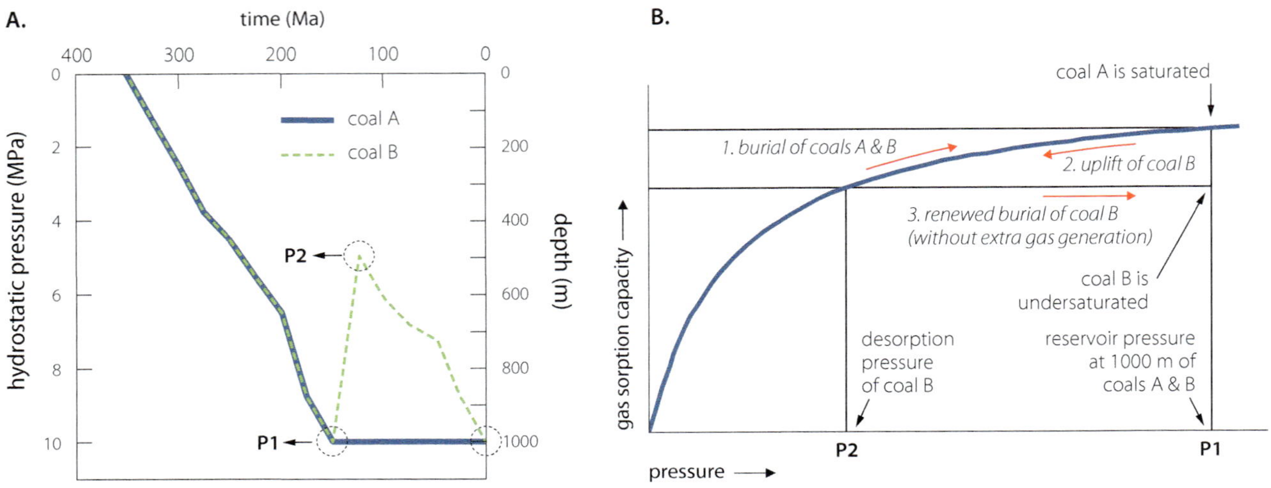

Figure 15.17. In tectonically active areas, coal seams were often buried deeper during geological history than at present. As a result of irreversible compaction during burial, present-day reservoir properties, such as porosity and permeability are affected by paleo-burial depth effects. Uplift can also possibly result in degassing of coal beds. This figure shows the effect of burial history on the amount of adsorbed methane. a) Coals A and B were buried to depths of 1000 m. While coal A remained at that depth, coal B experienced a phase of uplift, after which it was reburied to 1000 m. b) During the uplift, coal B released some of its adsorbed methane. Assuming no renewed gas charge, coal B will not adsorb more methane during reburial, and hence becomes undersaturated (adapted from McElhiney et al., 1993). Undersaturation of coal implies that it holds a lower gas content at a certain pressure and temperature than can be expected on the basis of its sorption capacity (Scott et al., 1994; Van Bergen et al., 2000; McCants et al., 2001). Modified from Van Bergen et al. (2007).

The technically recoverable quantity of such gas in the Netherlands was estimated to range from 5 to 27 x 10^9 m^3 (GAPS, 1994; NITG-TNO, 1999). Van Bergen et al. (2000) focused on the resources within Westphalian coal seams shallower than 1500 m, and calculated slightly higher total recoverable resources of 32 x 10^9 m^3 for the three evaluated areas (P50 value; Fig. 15.15b). In the latter inventory also the impact of tectonic uplift on the gas content of the coal, which is highly relevant depending on burial history of the coal seams was taken into account (Fig. 15.17).

Apart from the earlier mentioned interest in CBM as a transition fuel, efforts were undertaken in the beginning of the century to investigate the possibility of combining CBM production with CO_2 injection for storage in the underground coal seams. At that time this was considered to have potential to reduce CO_2 emissions while producing a relatively clean fossil fuel. Gas adsorption has proven stable through geological time, and the risk of future CO_2 release was considered low. Calculations of the amounts of CO_2 that could potentially be stored in coal seams shallower than 1500 m in the Netherlands ranged from 35 to almost 600 Mt (Van Bergen et al., 2000; Hamelinck et al., 2001), up to 3 or 4 times the total annual emission of CO_2 in the Netherlands at the time. Under the assumptions prevalent at that time, it was concluded that CO_2 sequestration in coal could be economically feasible in the Netherlands if the technology proved to be applicable and infrastructural, societal and environmental issues were fully addressed (Hamelinck et al., 2001). It was expected that incentives would be required to make this technology attractive for industry. Several demonstration sites were developed worldwide, e.g. in Canada, the United States, Poland, and Japan (Gunter et al., 1997, 1998; Erickson & Jensen, 2000; Schoeling & McGovern, 2000; Van Bergen et al., 2006) but it appeared both commercially and technically challenging to scale-up these demonstrations. Most of these sites have experienced permeability reduction in the coal seams, generally thought to be caused by swelling of the coal as the result of CO_2 injection as also observed in the laboratory. Understanding and managing the development of swelling with time was considered one of the most critical factors but appeared to be difficult (Van Bergen, 2009).

Acknowledgements

This chapter is based on the chapter in the 2007 edition of the book 'Geology of the Netherlands'. The co-authors of that chapter, Henk Pagnier and Peter van Tongeren, are acknowledged for their contributions. Richard Huis in 't Veld and Daan den Hartog Jager are thanked for reviewing and contributing to the text of this chapter. Armin Menkovic of the TNO-Geological Survey of the Netherlands, is thanked for his help in delineating the lignite distribution in the Roer Valley Graben. The editors of this second edition are thanked for their valuable suggestions. Sanneke van Asselen and Petra David are acknowledged for their thorough reviews.

Digital map data

Spatial data of figures in this chapter for use in geographical information systems can be downloaded here: https://doi.org/10.5117/aup.28164248.

References

Arets, E.J.M.M., Kolk, J.W.H., Hengeveld, G.M., Lesschen, J.P., Kramer, H., Kuikman, P.J. & Schelhaas, M.J., 2019. Greenhouse gas reporting of the LULUCF sector in the Netherlands. Methodological background, update 2019. Wageningen Environmental Research, WOt-technical report No. 146: 111 pp. DOI: 10.18174/472433

Ayers, W.B.Jr., 2002. Coalbed gas systems, resources, and production and a review of contrasting cases from the San Juan and Powder River basins. American Association of Petroleum Geologists Bulletin 86: 1853-1890. DOI: 10.1306/61EEDDAA-173E-11D7-8645000102C1865D

Beets, D.J. & Van der Spek, A.J.F., 2000. The Holocene evolution of the barrier and back-barrier basins of Belgium and the Netherlands as a function of late Weichselian morphology, relative sea-level rise and sediment supply. Netherlands Journal of Geosciences 79: 3-16. DOI: 10.1017/S0016774600021533

BGR, 2016. Energiestudie: Reserven, Ressourcen und Verfügbarkeit von. Bundesanstalt für Geowissenschaften und Rohstoffe (BGR, Hannover), Report No. 20: 180 pp. DOI: 10.25928/es-2021

BGR, 2022. BGR Energiestudie 2021: Daten und Entwicklungen der deutschen und globalen Energieversorgung. Bundesanstalt für Geowissenschaften und Rohstoffe (BGR, Hannover), Report No. 24: 175 pp. DOI: 10.25928/es-2021

Bless, M.J.M., Bouckaert, J., Finger, J.A.M. & Paproth, E., 1984. Oorsprong en winning van steenkool langs Henne, Samber, Maas en Worm. Geofiles (Tervuren): 68 pp.

Borger, G.J., 1992. Draining-digging-dredging; the creation of a new landscape in the peat areas of the low countries. In: Verhoeven, J.T.A. (ed.): Fens and bogs in the Netherlands: vegetation, history, nutrient dynamics and conservation. Springer (Dordrecht): 131-172.

Bos, I.J., Busschers, F.S. & Hoek, W.Z., 2012. Organic-facies determination: a key for understanding facies distribution in the basal peat layer of the Holocene Rhine-Meuse delta, The Netherlands. Sedimentology 59: 676-703. DOI: 10.1111/j.1365-3091.2011.01271.x

Bustin, R.M. & Clarkson, C.R., 1998. Geological controls on coalbed methane reservoir capacity and gas content. International Journal of Coal Geology 38: 3-26.

CBS, 2022a. Meer elektriciteit uit kolen en hernieuwbare bronnen minder uit gas. Statistics Netherlands (CBS). https://www.cbs.nl/nl-nl/nieuws/2022/36/meer-elektriciteit-uit-kolen-en-hernieuwbare-bronnen-minder-uit-gas

CBS, 2022b. Welke sectoren stoten broeikasgassen uit? Statistics Netherlands (CBS). https://www.cbs.nl/nl-nl/dossier/dossier-broeikasgassen/welke-sectoren-stoten-broeikasgassen-uit-

Chmura, G.L., Anisfeld, S.C., Cahoon, D.R. & Lynch, J.C., 2003. Global carbon sequestration in tidal, saline wetlands soils. Global Biogeochemical Cycles 17: 1-12.

Coppens, P.L., 1967. Synthèse des Propriétés Chimiques et Physiques des Houilles Belges. Inichar Publication (Liège): 216 pp.

De Bruin, G., Bouroullec, R., Geel, K., Abdul Fattah, R., Van Hoof, T., Pluymaekers, M., Van der Belt, F., Vandeweijer, V.P. & Zijp, M., 2015. New petroleum plays in the Dutch northern offshore. TNO (Utrecht), Report No. 2015 R10920. https://www.nlog.nl/sites/default/files/2018-11/NORTHERN_OFFSHORE_TNO_report_2015_R10920_0.pdf

De Jong, J.D. & Van der Waals, L., 1971. Depositional environment and weathering phenomena of the white Miocene sands of southern Limburg (the Netherlands). Geologie En Mijnbouw 50 (3): 417-424.

De Kraker, A.M.J., 2006. Flood events in the southwestern Netherlands and coastal Belgium, 1400-1953. Hydrological Sciences Journal 51: 913-929. DOI: 10.1623/hysj.51.5.913

De Sitter, L.U., Tesch, P., Van der Drift, J.B. & Jongmans, W.J., 1949. Eindverslag van het geophysische onderzoek in ZO-Nederland door de Geophysische Dienst der Staatsmijnen. Mededelingen van de Geologische Stichting C-1-3-1: 372 pp.

De Vries, J.J., 2007. Groundwater. In: Wong, Th.E., Batjes, D.A.J. & De Jager, J. (eds): Geology of the Netherlands. Royal Netherlands Academy of Arts and Sciences (KNAW, Amsterdam): 295-315.

DEBRIV, 2023. Braunhohlereviere. Bundesverband Braunkohle. https://braunkohle.de/media/daten-und-fakten

Den Haan, E.J. & Kruse, G.A.M., 2006. Characterisation and engineering properties of Dutch peats. In: Tan, E.T.S., Phoon, K.K., Hight, D.W., Leroueil, S. & Balkema, A.A. (eds): Second International Workshop on Characterisation and Engineering Properties of Natural Soils 3, 2101-2133.

Den Held, A.J., Schmitz, M. & Wirdum, G., 1992. Types of terrestrializing fen vegetation in the Netherlands. In: Verhoeven, J.T.A. (ed.): Fens and bogs in the Netherlands: vegetation, history, nutrient dynamics and conservation. Springer (Dordrecht): 237-321.

Deverel, S.J., Ingrum, T. & Leighton, D., 2016. Present-day oxidative subsidence of organic soils and mitigation in the Sacramento-San Joaquin Delta, California, USA. Hydrogeology 24: 569-586. DOI: 10.1007/s10040-016-1391-1

Eifelnatur.de, 2023. Bruinkoolgroeve Hambach. https://eifelnatur.

de/Niederl%E4ndisch/Seiten/Bruinkoolgroeve%20Hambach.html
Engelen, F.H.G., 1987. Resources at the surface. In: Visser, W.A., Zonneveld, J.I.S. & Van Loon, A.J. (eds): Seventy-five years of geology and mining in the Netherlands (1912-1987). Royal Geological and Mining Society of the Netherlands (KNGMG, The Hague): 125-136.
Engelen, F.H.G., 1989a. De exploitatie van bruinkool. Grondboor & Hamer 43 (5/6): 343-344.
Engelen, F.H.G., 1989b. De exploitatie van steenkool. Grondboor & Hamer 43 (5/6): 349-352.
Erickson, D. & Jensen, J.R., 2000. CO2 sequestration in an unminable coalbed – San Juan Basin, Colorado, USA. In: Williams, D., Durie, B., McMullan, P., Paulson, C. & Smith, A. (eds): Proceedings of the 5th International Conference on Greenhouse Control Technologies (Cairns): 589-592.
Erkens, G., Van der Meulen, M.J. & Middelkoop, H., 2016. Double trouble: subsidence and CO2 respiration due to 1,000 years of Dutch coastal peatlands cultivation. Hydrogeology Journal 24 (3): 551-568. DOI: 10.1007/s10040-016-1380-4
Ettinger, I., Eremin, I., Zimakov, B. & Yanovskaya, M., 1966. Natural factors influencing coal sorption properties. I. Petrography and sorption properties of coals. Fuel 45: 267-275.
Fails, T.G., 1996. Coalbed methane potential of some Variscan foredeep basins. Coalbed Methane and Coal Geology. Geological Society of London, Special Publication 109: 13-26.
Freudenberg, U., Lou, S., Schlüter, R., Schütz, K. & Thomas, K., 1996. Main factors controlling coalbed methane distribution in the Ruhr District, Germany. In: R. Gayer & I. Harris (eds): Coalbed Methane and Coal Geology. Geological Society Special Publication (London): Vol. 109, 165-179. DOI: 10.1144/GSL.SP.1996.109.01.06
GAPS, 1994. Coalbed methane potential of the Netherlands (Warmond) [GAPS Nederland B.V.]. 60 pp.
Gąsiewicz, A., Bos, C., Czapowski, G., Evans, D., Górecki, W., Hajto, M., Holliday, D.W., Holloway, S., Jones, N.S., Kasiński, J.R., Kępińska, B., Kramers, L., Laenen, B., Lott, G.K., Lokhorst, A., Mathiesen, A., May, F., Rowley, W.J., Schmidt, S., Sedlacek, R., Seibt, P., Simmelink, E., Tarkowski, R., Uliasz-Misiak, B., Van Wees, J.D., Wildenborg, A., Wolfgramm, M. & Wong, T.E., 2010. Applied geology. In: Doornenbal, J.C. & Stevenson, A.G. (eds): Petroleum Geological Atlas of the Southern Permian Basin Area. EAGE Publications B.V. (Houten): 282-299.
GD-NRW, 2023. Informationssystem Rohstoffkundliche Übersichtskarte von NRW 1:500,000. Geologischer Dienst Nordrhein-Westfalen. https://www.gd.nrw.de/pr_kd_rohstoffkarte-500000.php
Gerding, M.A.W., 1995. Vier eeuwen turfwinning. De verveningen in Groningen, Friesland, Drenthe en Overijssel tussen 1550 en 1950. PhD thesis Landbouwuniversiteit Wageningen (the Netherlands): 512 pp. DOI: 10.18174/136833
Gerding, M.A.W., Karel, E.H.K. & De Vries, G.E., 2015. The history of the peat manufacturing industry in The Netherlands: peat moss litter and active carbon. Mires and Peat 10: 1-9. DOI: 10.19189/MaP.2015.OMB.174
Greb, S.F., Pashin, J.C., Martino, R.L. & Eble, C.F., 2008. Appalachian sedimentary cycles during the Pennsylvanian: Changing influences of sea level, climate, and tectonics. Geological Society of America Special Paper 441: 235-248. DOI: 10.1130/2008.2441
Gunter, W.D., Gentzis, T., Rottenfusser, B.A. & Richardson, R.J.H., 1997. Deep coalbed methane in Alberta, Canada: a fuel resource with the potential of zero greenhouse gas emissions. Energy Conversion and Management 38: 217-222.
Gunter, W.D., Wong, S., Cheel, D.B. & Sjostrom, G., 1998. Large CO2 sinks: Their role in the mitigation of greenhouse gases from an international, national (Canadian) and provincial (Alberta) perspective. Applied Energy 61: 209-227.
Hager, H., 1993. The origin of the Tertiary lignite deposits in the Lower Rhine region, Germany. International Journal of Coal Geology 23 (1-4): 251-262.
Hamelinck, C.N., Faaij, A.P.C., Ruijg, G.J., Jansen, D., Pagnier, H.J.M., Bergen, F. van, Wolf, K.H., Barzandji, O., Bruining, H. & Schreurs, H., 2001. Potential for CO2 Sequestration and Enhanced Coalbed Methane production in the Netherlands. Netherlands Agency for Energy and the Environment (Novem, Utrecht), Report No. 2ECBM01.01: 105 pp.
Hedemann, H.-A., Schuster, A., Stancu-Kristoff, G. & Lösch, J., 1984. Die Verbreitung der Kohlenflöze des Oberkarbons in Nordwest-deutschland und ihre stratigraphische Einstufung. Fortschritte in Der Geologie von Rheinland Und Westfalen 32: 39-88.
Hijma, M. & Cohen, K.M., 2010. Timing and magnitude of the sea-level jump preluding the 8200 yr event. Geology 38 (3): 275-278. DOI: 10.1130/G30336.1
Hobbs, N.B., 1986. Mire morphology and the properties and behavior of some British and foreign peats. Quarterly Journal of Engineering Geology 19: 7-80. DOI: 10.1144/GSL.QJEG.1986.019.01.02
Hoogland, F., Roelandse, A.S., de la Loma González, B., Waterloo, M.J., Mooij, P.W., Verhagen, S.A. & Velstra, J., 2020. Investigating the effectiveness of drain infiltration to minimize peat oxidation in agricultural fields in Flevoland, the Netherlands. Proc. IAHS 382: 747-753. DOI: 10.5194/piahs-382-747-2020
Huis in 't Veld, R. & Den Hartog Jager, D., 2025. Late Carboniferous. In: Ten Veen, J.H., Vis, G.-J., De Jager, J. & Wong, Th.E. (eds): Geology of the Netherlands, second edition. Amsterdam University Press (Amsterdam): 95-125. DOI: 10.5117/9789463728362_ch03
IEA, 2020. International Energy Agency 2020 World Energy Outlook. https://www.iea.org/reports/world-energy-outlook-2020/outlook-for-energy-demand#abstract
IGCP, 1980. World Coal Fields. Rijks Geologische Dienst (Haarlem).
Jelgersma, S., 1961. Holocene sea level changes in the Netherlands. PhD thesis University Leiden (the Netherlands): 100 pp.

Jongmans, W.J. & Van Rummelen, F.H., 1930. Het voorkomen van bruinkool en bruinkoolformatie in Zuid-Limburg in verband met den bouw van het steenkoolgebied. Geologisch Bureau voor het Nederlandsche Mijngebied te Heerlen, Jaarverslag over 1930: 25-59.

Karayiğit, A.I. & Köksoy, M., 1994. Origin of coal. In: Kural O. (ed.): Coal: resources, properties, utilization, pollution. Mining Faculty, Istanbul Technical University: 494 pp.

Kechavarzi, C., Dawson, Q. & Leeds-Harrison, P.B., 2010. Physical properties of low-lying agricultural peat soils in England. Geoderma 154: 196-202. DOI: 10.1016/j.geoderma.2010.01.002

Kim, A.G., 1977. Estimating methane content of bituminous coalbeds from adsorption data. United States, Bureau of Mines, Report No. 8245: 22 pp.

Klein, G. & Willard, D.A., 1989. Origin of Pennsylvanian coal-bearing cyclothems of North America. Geology 17: 152-155.

Kombrink, H., 2008. The Carboniferous of the Netherlands and surrounding areas; a basin analysis. PhD thesis Utrecht University (the Netherlands): 184 pp.

Koster, K., 2017. 3D characterization of Holocene peat in the Netherlands. Implications for coastal-deltaic subsidence. PhD thesis Utrecht University (the Netherlands): 185 pp.

Koster, K., Erkens, G. & Zwanenburg, C., 2016. A new soil mechanics approach to quantify and predict land subsidence by peat compaction. Geophysical Research Letters 43 (20): 10792-10799. DOI: 10.1002/2016GL070207

Koster, K., De Lange, G., Harting, R., De Heer, E. & Middelkoop, H., 2018a. Characterizing void ratio and compressibility of Holocene peat with CPT for assessing coastal-deltaic subsidence. Quarterly Journal of Engineering Geology and Hydrogeology 51: 210-218. DOI: 10.1144/qjegh2017-014

Koster, K., Stafleu, J., Cohen, K.M., Stouthamer, E., Busschers, F.S. & Middelkoop, H., 2018b. Three-dimensional distribution of organic matter in coastal-deltaic peat: Implications for subsidence and carbon dioxide emissions by human-induced peat oxidation. Anthropocene 22: 1-9. DOI: 10.1016/j.ancene.2018.01.001

Krooss, B.M., Van Bergen, F., Gensterblum, Y., Siemons, N., Pagnier, H.J.M. & David, P., 2002. High-pressure methane and carbon dioxide adsorption on dry and moisture-equilibrated Pennsylvanian coals. International Journal of Coal Geology 51: 69-92. DOI: 10.1016/S0166-5162(02)00078-200078-2

Kuikman, P.J., Van den Akker, J.J.H. & De Vries, F., 2005. Emissie van N2O en CO2 uit organische landbouwbodems. Alterra (Wageningen), Report No. 1035.2: 66 pp.

Kuyl, O.S., 1980. Toelichting bij de geologische kaart van Nederland 1:50,000, Blad Heerlen (62W oostelijke helft, 62O westelijke helft). Rijks Geologische Dienst (Haarlem): 206 pp.

Landva, A., 2006. Characterization of Escuminac peat and construction on peatland. In: Tan, E.T.S., Phoon, K.K., Hight, D.W. & Leroueil, S. (eds): Second International Workshop on Characterisation and Engineering Properties of Natural Soils 3: 2135-2192.

Leenders, K.A.H.W., 1987. De diffusie van een techniek. De vergraving van het veen in de Nederlanden (1150-1950). Tijdschrift van de Belgische Vereniging Aardkundige Studies 56: 197-216.

Levine, J.R., Johnson, P.W. & Beamish, B.B., 1993. High pressure microgravimetry provides a viable alternative to volumetric method in gas sorption studies on coal. Proceedings 1993 International Coalbed Methane Symposium (Tuscaloosa): 187-195. DOI: 10.1016/S0016-2361(97)00078-100078-1

Levy, J., Day, S.J. & Killingley, J.S., 1997. Methane capacity of Bowen Basin coals related to coal properties. Fuel 74: 1-7.

Lippmann, T.J.R., in 't Zandt, M.H., Van der Putten, N.N.L., Busschers, F.S., Hijma, M.P., van der Velden, P., de Groot, T., van Aalderen, Z., Meisel, O.H., Slomp, C.P., Niemann, H., Jetten, M.S.M., Dolman, H.A.J. & Welte, C.U., 2021. Microbial activity, methane production, and carbon storage in Early Holocene North Sea peats. Biogeosciences 18 (19): 5491-5511. DOI: 10.5194/bg-18-5491-2021

Lokhorst, A. (ed.), 1997. NW European Gas Atlas. Netherlands Institute of Applied Geosciences TNO (NITG-TNO, Haarlem): 154 pp.

McCants, C.Y., Spafford, S. & Stevens, S.H., 2001. Five-Spot Production Pilot on Tight Spacing: Rapid Evaluation of a Coalbed Methane Block in the Upper Silesian Coal Basin, Poland. Proceedings of the 2001 International Coalbed Methane Symposium (Tuscaloosa, AL): 193-204.

McElhiney, J.E., Paul, G.W., Young, G.B.C. & McCartney, J.A., 1993. Reservoir engineering aspects of coalbed methane. In: Law, B.E. & Rice, D.D. (eds): Hydrocarbons from coal. American Association of Petroleum Geologists Bulletin Studies in Geology 38: 361-372.

Menke, U., Van de Laar, E. & Lenselink, G., 1998. De geologie en bodem van Zuidelijk Flevoland. Flevobericht 415: 94 pp.

Moonen, H., 2012. Kalculen, mijnbouw in dagbouw? https://www.demijnen.nl/actueel/artikel/kalculen-mijnbouw-dagbouw

Munsterman, D.K., Ten Veen, J.H., Menkovic, A., Deckers, J., Witmans, N., Verhaegen, J., Kerstholt-Boegehold, S.J., Van de Ven, T. & Busschers, F.S., 2019. An updated and revised stratigraphic framework for the Miocene and earliest Pliocene strata of the Roer Valley Graben and adjacent blocks. Netherlands Journal of Geosciences 98 (8). DOI: 10.1017/njg.2019.10

Ngan-Tillard, D., Venmans, A. & Slob, E., 2010. Total engineering geology approach applied to motorway construction and widening in the Netherlands: Part I: A pragmatic approach. Engineering Geology 114: 164-170. DOI: 10.1016/j.enggeo.2010.01.002

NITG-TNO, 1999. Geological atlas of the subsurface of the Netherlands (1:250,000) - Explanation to Map sheet XV: Sittard-Maastricht. Netherlands Institute of Applied Geoscience TNO - National Geological Survey (Utrecht): 127 pp.

NITG-TNO, 2001. Geological Atlas of the subsurface of the Netherlands (1 : 250,000) - Explanation to Map Sheets XIII and XIV:

Breda-Valkenswaard and Oss-Roermond + Map Sheet XIII and XIV. Netherlands Institute of Applied Geoscience TNO (NITG-TNO, Utrecht): 149 pp.

Pagnier, H.J.M., Pestman, P.J. & van Tongeren, P.C.H., 1987. Recent coal exploration in the Netherlands. In: J.W. Martin & S.P. Acarome (eds): Proceedings of the 13th Annual Underground Coal Gasification Symposium. US Deptartment of Energy (Laramie, WY): 151-162.

Pannekoek, A.J., 1956. Geologische geschiedenis van Nederland, toelichting bij de geologische overzichtskaart van Nederland op de schaal 1:20 000. Staatsdrukkerij- en Uitgeverijbedrijf (Den Haag): 154 pp.

Pashin, J.C., 1998. Stratigraphy and structure of coalbed methane reservoirs in the United States: an overview. International Journal of Coal Geology v. 35: 207-238.

Pashin, J.C. & Hinkle, F., 1997. Coalbed Methane in Alabama. Alabama Geological Survey Circular 192: 71. DOI: 10.1016/S0166-5162(97)00013-X00013-X

Peelcommissie, 1963. Rapport van de Peelcommissie. Koninklijk Nederlands Geologisch Mijnbouwkundig Genootschap: 133 pp.

Peeters, J., Busschers, F.S. & Stouthamer, E., 2015. Fluvial evolution of the Rhine during the last interglacial-glacial cycle in the southern North Sea basin: A review and look forward. Quaternary International 357: 176-188. DOI: 10.1016/j.quaint.2015.03.013

Pons, L.J., 1992. Holocene peat formation in the lower parts of the Netherlands. In: Verhoeven, J.T.A. (ed.): Fens and bogs in the Netherlands: vegetation, history, nutrient dynamics and conservation. Springer (Dordrecht): 7-80.

Quik, C., Van der Velde, Y., Harkema, T., Van der Plicht, H., Quik, J., Van Beek, R. & Wallinga, J., 2021. Using legacy data to reconstruct the past? Rescue, rigour and reuse in peatland geochronology. Earth Surface Processes and Landforms 46: 2607-2631. DOI: 10.1002/esp.5131

Raedts, C.E.P.M., 1971. De opgang en teleurgang van de Limburgse steenkoolindustrie (Een historisch overzicht). Geologie En Mijnbouw 50 (2): 105-118.

Remmelts, G., Nelskamp, S., De Jager, J. & Geluk, M.C., 2025. Petroleum Geology. In: Ten Veen, J.H., Vis, G.-J., De Jager, J. & Wong, Th.E. (eds): Geology of the Netherlands, second edition. Amsterdam University Press (Amsterdam): 535-575. DOI:10.5117/9789463728362_ch14

RGD, 1986. Eindrapport project inventarisatieonderzoek Nederlandse kolenvoorkomens, eerste fase 1981-1985. Rijks Geologische Dienst (Heerlen), Report No. GB 2107: 62 pp.

Saulsberry, J.L., Schafer, P.S. & Schraufnagel, R.A., 1996. A guide to coalbed methane reservoir engineering. Gas Research Institute (Chicago), Report No. GRI-94/0397: 330 pp.

Schoeling, L. & McGovern, M., 2000. Pilot test demonstrates how CO2 injection enhances coalbed methane recovery. Petroleum Technology Digest: 14 pp.

Schokker, J., Weerts, H.J.T., Westerhoff, W.E., Berendsen, H.J.A. & Den Otter, C., 2007. Introduction of the Boxtel Formation and implications for the Quaternary lithostratigraphy of the Netherlands. Netherlands Journal of Geosciences 86 (3): 197-210. DOI: 10.1017/S0016774600077805

Scott, A.R., 2002. Hydrogeologic factors affecting gas content distribution in coal beds International Journal of Coal Geology 50: 363-387. DOI: 10.1016/S0166-5162(02)00135-000135-0

Scott, A.R., Kaiser, W.R. & Ayers, W.B., Jr, 1994. Thermogenic and secondary biogenic gases, San Juan basin, Colorado and New Mexico – Implications for coalbed gas producibility. American Association of Petroleum Geologists Bulletin 78: 1186-1209. DOI: 10.1306/A25FEAA9-171B-11D7-8645000102C1865D

Spinoni, J., Vogt, J.V., Naumann, G., Barbosa, P. & Dosio, A., 2018. Will drought events become more frequent and severe in Europe? International Journal of Climatology 38: 1718-1736. DOI: 10.1002/joc.5291

Stach, E., Mackowsky, M.-Th., Teichmüller, M., Taylor, G.H., Chandra, D. & Teichmüller, R., 1982. Stach's Textbook of Coal Petrology, third edition. Gebrüder Borntraeger (Berlin): 535 pp.

Stock, A.T., Littke, R., Lücke, A., Zieger, L. & Thielemann, T., 2015. Miocene depositional environment and climate in western Europe: The lignite deposits of the Lower Rhine Basin, Germany. International Journal of Coal Geology 157: 2-18. DOI: 10.1016/j.coal.2015.06.009

Stuffken, J., 1987. Coal mining. In: Visser, W.A., Zonneveld, J.I.S. & Van Loon, A.J. (eds): Seventy-five years of geology and mining in the Netherlands (1912-1987). Royal Geological and Mining Society of the Netherlands (KNGMG, The Hague): 153-160.

Stuffken, J., 1957. De mijngasafgifte van kolenlagen. PhD thesis Technische Hogeschool Delft (the Netherlands).

Stuurman, R., Baggelaar, P., Van Oostrom, N., Van der Meulen, M.J. & Oude Essink, G.H.P., 2007. Schetsen van het Nederlandse grondwatersysteem in 2050. TNO (Utrecht), Report No. TNO 2007-U-R0225/B: 156 pp.

Taylor, G.H., Teichmüller, M., Davis, A., Diessel, C.F.K., Littke, R. & Robert, P., 1998. Organic Petrology. Gebrüder Borntraeger (Berlin): 704 pp.

TNO-GDN, 2022a. Echteld Formation. In: Stratigraphic Nomenclature of the Netherlands, TNO – Geological Survey of the Netherlands. https://www.dinoloket.nl/en/stratigraphic-nomenclature/echteld-formation

TNO-GDN, 2022b. Heksenberg Member. In: Stratigraphic Nomenclature of the Netherlands, TNO – Geological Survey of the Netherlands. https://www.dinoloket.nl/en/stratigraphic-nomenclature/heksenberg-member

TNO-GDN, 2022c. Naaldwijk Formation. In: Stratigraphic Nomenclature of the Netherlands, TNO – Geological Survey of the Netherlands. https://www.dinoloket.nl/en/stratigraphic-nomenclature/naaldwijk-formation

TNO-GDN, 2022d. Nieuwkoop Formation. In: Stratigraphic Nomenclature of the Netherlands, TNO – Geological Survey of the Netherlands. https://www.dinoloket.nl/en/stratigraphic-nomenclature/nieuwkoop-formation

Tuinstra, U., 1951. Bijdrage tot de kennis van holocene landschapsontwikkeling in het noordwesten van Noordbrabant. PhD thesis University of Amsterdam: 139 pp.

UNTC, 2016. Paris Agreement. United Nations Treaty Collection, Report No. 54113: 257 pp. treaties.un.org/doc/Publication/UNTS/Volume%203156/Part/volume-3156-I-54113.pdf

Van Adrichem Boogaert, H.A. & Kouwe, W.F.P., 1995. Stratigraphic nomenclature of The Netherlands, revision and update by RGD and NOGEPA, Section C. (Silesian). Mededelingen Rijks Geologische Dienst 50: 1-40.

Van Asselen, S., 2011. The contribution of peat compaction to total basin subsidence: Implications for the provision of accommodation space in organic-rich deltas. Basin Research 23: 239-255. DOI: 10.1111/j.1365-2117.2010.00482.x

Van Asselen, S., Erkens, G., Stouthamer, E., Woolderink, H.A.G., Geeraert, R.E.E. & Hefting, M.M., 2018. The relative contribution of peat compaction and oxidation to subsidence in built-up areas in the Rhine-Meuse delta, The Netherlands. Science of the Total Environment 636: 177-191. DOI: 10.1016/j.scitotenv.2018.04.141

Van Asselen, S., Stouthamer, E. & Smith, N.D., 2010. Factors controlling peat compaction in alluvial floodplains: A case study in the cold-temperate Cumberland Marshes, Canada. Journal of Sedimentary Research 80 (2): 155-166. DOI: 10.2110/jsr.2010.015

Van Asselen, S., Stouthamer, E. & Van Asch, T.W.J., 2009. Effects of peat compaction on delta evolution: A review on processes, responses, measuring and modeling. Earth Science Reviews 92: 35-51. DOI: 10.1016/j.earscirev.2008.11.001

Van Bergen, F., 2009. Effect of coal matrix swelling on enhanced coalbed methane production. A field and laboratory study. PhD thesis Utrecht University (the Netherlands): 196 pp.

Van Bergen, F., Pagnier, H.J.M., David, P. & Krooss, B.M., 2000. Inventory of potential volumes of methane extraction and carbon dioxide storage in coallayers in the Dutch subsurface. TNO contribution to the 'Feasibility study of combined Coalbed Methane production and Carbon Dioxide storage in the Netherlands'. TNO (Utrecht): 1-87 pp.

Van Bergen, F., Pagnier, H.J.M. & Van Tongeren, P.C.H., 2007. Peat, coal and coalbed methane. In: Wong, Th.E., Batjes, D.A.J. & De Jager, J. (eds): Geology of the Netherlands. Royal Netherlands Academy of Arts and Sciences (KNAW, Amsterdam): 265-282.

Van Bergen, F., Pagnier, H. & Krzystolik, P., 2006. Field experiment of enhanced coalbed methane-CO2 in the Upper Silesian Basin of Poland. Environmental Geosciences 13 (3): 201-224. DOI: 10.2113/gseegeosci.13.3.201

Van Buggenum, J.M. & Den Hartog Jager, D.G., 2007. Silesian. In: Wong, Th.E., Batjes, D.A.J. & De Jager, J. (eds): Geology of the Netherlands. Royal Netherlands Academy of Arts and Sciences (KNAW, Amsterdam): 43-62.

Van Dam, P.J.E.M., 2001. Sinking Peat Bogs: Environmental Change in Holland, 1350-1550. Environmental History 6 (1): 32-45. DOI: 10.2307/3985230

Van den Belt, F.J.G., 2012. Sedimentary cycles in coal and evaporite basins and the reconstruction of Palaeozoic climate. PhD thesis Utrecht University (the Netherlands): 98 pp.

Van der Burgh, J., Rooijen, P. & Amerom, H.W.J., 1988. Bruinkool, 20 miljoen jaar geschiedenis van een energiebron. Rijks Geologische Dienst (RGD, Heerlen): 47 pp.

Van Krevelen, D.W., 1993. Coal: typology, physics, chemistry, constitution, third edition. Elsevier (Amsterdam): 979 pp.

Van Montfrans, H.M., De Graaf, L.W.S., Mourik, J.M. & Zagwijn, W.H., 1988. Delfstoffen en Samenleving. Rijks Geologische Dienst (RGD, Haarlem): 83 pp.

Van Tongeren, P.C.H., 1987. Renewed interest in coal. In: Visser, W.A., Zonneveld, J.I.S. & Van Loon, A.J. (eds): Seventy-five years of geology and mining in the Netherlands (1912-1987). Royal Geological and Mining Society of the Netherlands (KNGMG, The Hague): 231-242.

Van Waterschoot van der Gracht, W.A.J.M., 1918. Eindverslag over de onderzoekingen en uitkomsten van den Dienst der Rijksopsporing van Delfstoffen in Nederland 1903-1916. Martinus Nijhoff ('s Gravenhage): 664 pp.

Van Weelden, A., 1957. History of gravity observations in the Netherlands. Verhandelingen van Het Koninklijk Nederlands Geologisch Mijnbouwkundig Genootschap. Geologische Serie 18: 305-308.

Van Wijhe, D.H. & Bless, M.J.M., 1974. The Westphalian of the Netherlands with special reference to miospore assemblages. Geologie En Mijnbouw 53 (6): 295-328. DOI: 10.1016/S0016-7878(74)80001-280001-2

Veevers, J.J. & Powell, C.M.A., 1987. Late Palaeozoic glacial episodes in Gondwanaland reflected by transgressive-regressive depositional sequences in Euramerica. Geological Society of America Bulletin 98 (4): 475-487.

Verhoeven, J.T.A., 1992. Introduction. In: Verhoeven, J.T.A. (ed.): Fens and bogs in the Netherlands: vegetation, history, nutrient dynamics and conservation. Springer (Dordrecht): 1-6. DOI: 10.1007/978-94-015-7997-1

Visser, W.A., 1987. The Gelria concession. In: Visser, W.A., Zonneveld, J.I.S. & Van Loon, A.J. (eds): Seventy-five years of geology and mining in the Netherlands (1912-1987). Royal Geological and Mining Society of the Netherlands (KNGMG, The Hague): 147-152.

Visser, W.A. & Zonneveld, J.I.S., 1987. Introduction. In: Visser, W.A., Zonneveld, J.I.S. & Van Loon, A.J. (eds): Seventy-five years of geology and mining in the Netherlands (1912-1987). Royal Geological and Mining Society of the Netherlands (KNGMG, The Hague): 9-26.

Vos, P.C. & Van Heeringen, R.M., 1997. Holocene geology and occupation history of the Province of Zeeland (SW Netherlands). Mededelingen Nederlands Instituut Voor Toegepaste Geowetenschappen TNO 59: 5-109.

Vos, P.C., 2017. Het ontstaan van Westland-Delftland, gebaseerd op paleolandschappelijk onderzoek en getijsysteemkennis. Delftse Archeologische Rapporten 130: 95.

Vos, P.C., 2011. Verfijning van de archeologische verwachtingskaart voor het pilotgebied Kantens - Stitswerd (Noord Groningen). Deltares (Utrecht), Report No. 1204018-000-BGS-0004: 30 pp.

Vos, P.C., 2015. Origin of the Dutch coastal landscape: long-term landscape evolution of the Netherlands during the Holocene, described and visualized in national regional and local palaeogeographical map series. PhD thesis Utrecht University (the Netherlands), Barkhuis, Groningen: 369 pp.

Weideveld, S.T.J., Liu, W., Van den Berg, M., Lamers, L.P.M. & Fritz, C., 2021. Conventional subsoil irrigation techniques do not lower carbon emissions from drained peat meadows. Biogeosciences 18: 3881-3902. DOI: 10.5194/bg-18-3881-2021

Weller, J.M., 1930. Cyclical sedimentation of the Pennsylvanian period and its significance. Journal of Geology 38: 97-135.

Wenselaers, P., Dusar, M. & Van Tongeren, P.C.H., 1996. Steenkoollaag methaangaswinning in het Kempisch kolenbekken 'Het proefproject te Peer'. Ministerie van de Vlaamse Gemeenschap, afdeling Natuurlijke Rijkdommen en Energie (Brussel): 67 pp.

Westen, J.M.J., 1971. Statistisch overzicht van productie, bezetting en prestaties van de Limburgse steenkolenmijnen. Geologie En Mijnbouw 50: 312-319.

Wilkinson, B.H., Merrill, G.K. & Kivett, S.J., 2003. Stratal order in Pennsylvanian cyclothems. Geological Society of America Bulletin 115: 1068-1087. DOI: 10.1130/B25329.1

Wolf, K.-H.A.A., Westerink, H.H.E., Van Delft, P.T.P. & Bruining, J., 1997. Coalbed methane production in the Netherlands. An inventory. Delft University of Technology, Report No. TUD-TA/PF-97.004: 73 pp.

Yang, R.T. & Saunders, J.T., 1985. Adsorption of gases on coals and heat-treated coals at elevated temperature and pressure. Fuel 64: 616-620. DOI: 10.1016/0016-2361(85)90043-2

Zagwijn, W.H. & Hager, H., 1987. Correlations of the continental and marine neogene deposits in the southeastern Netherlands and the Lower Rhine district. Mededelingen van de Werkgroep Voor Tertiaire En Kwartaire Geologie 24 (1/2): 59-78.

16 Salt production

Remco Groenenberg, Peter Fokker & Marinus den Hartogh*

*Corresponding author: remco.groenenberg@tno.nl

ABSTRACT

In the Netherlands salt deposits are present within the Permian and Triassic intervals. The salt units in the upper Permian Zechstein Group attain especially great thicknesses (up to 1000 m) in the north of the country and the adjacent offshore areas. Salt is produced by solution mining in the eastern and northern Netherlands: rock salt from the Triassic Röt Formation and the Permian Zechstein Group and magnesium salt from the Zechstein Group. Total production amounts to 5.5-6.7 Mt/yr of rock salt and 0.25-0.30 Mt/yr of magnesium salt. Exploitation takes place between appx. 500 and 3000 m, the mining of Permian rock salt in the Barradeel concessions being the deepest worldwide. The rock salt is mainly used for the manufacturing of chlorine; most of the magnesium salt is used to produce magnesium oxide. The lifecycle of a salt mining concession can be divided into five phases: exploration, development and production, suspension, abandonment, and after-care. Optionally, a sixth phase, in-between production and suspension, can be to use the salt caverns (large brine-filled cavities in the salt that develop during production) for storage of a gas (natural gas, nitrogen, hydrogen, air) or a liquid (oil products). The environmental effects of salt mining are largely comparable to the mining effects of oil and gas operations in the Netherlands. Mining can result in ground subsidence, excess noise levels from drilling and production processes and production of saline wastewater. These effects also include the requirement for pipelines and cables installation as well as the (temporary) change of mostly agricultural land usage, hence changing the landscape to a limited extent.

<< Nobian salt storage in Hengelo (eastern Netherlands).
Photo: Nobian.

DOI: 10.5117/9789463728362_ch16

Introduction

Salt belongs to a group of chemically deposited minerals which crystallize by evaporation of salt-rich lake or sea water (called 'brine') in a lagoon, lake or shallow sea and which has limited or no connection to the ocean most of the time. Their chemical compositions reflect the original composition of water, although the brine changes in composition after precipitation of certain salts. Fresh influxes of seawater from the ocean at times, re-dissolution of salt near or at surface due to rainfall (and precipitation elsewhere) and temporary aerial exposure locally change the natural order of precipitation of the minerals in time or creates iterating cycles, the evidence of which can be observed in the stratigraphic record as sequences of evaporites. These sequences of evaporites thus result from variations in the chemistry of the brine from which they were deposited. The word salt is used as a general name for evaporitic salts, like gypsum/anhydrite ($CaSO_4 \cdot 2H_2O/CaSO_4$), halite (NaCl), sylvite (KCl), carnallite ($KClMgCl_2 \cdot 6H_2O$), kainite ($KClMgSO_4 \cdot 3H_2O$), kieserite ($MgSO_4 \cdot H_2O$) and bischofite ($MgCl_2 \cdot 6H_2O$). Many more salts exist, but usually not in large volumes or concentrations from natural deposition. The largest known and also economically most important salt occurrences on earth are of marine origin.

The upper Permian Zechstein salt is the most widespread and thickest salt formation in the Netherlands and in Western Europe. It contains rock salt, potassium-magnesium salts and thick sulphates (Bouroullec & Geel, 2025, this volume). Apart from the Zechstein, salt occurs in the Permian Upper Rotliegend Group, in the Triassic Röt, Muschelkalk and Keuper formations (McKie & Kilhams, 2025, this volume) and locally in the Upper Jurassic Weiteveen Formation. Salt has been exploited in the Netherlands since 1919, by means of solution mining. Currently, in three clusters of concession areas, i.e. the Twenthe-Rijn cluster, the Barradeel-Havenmond cluster and the Adolf van Nassau cluster, rock salt is mined from a single concession (Fig. 16.1; Table 16.1).

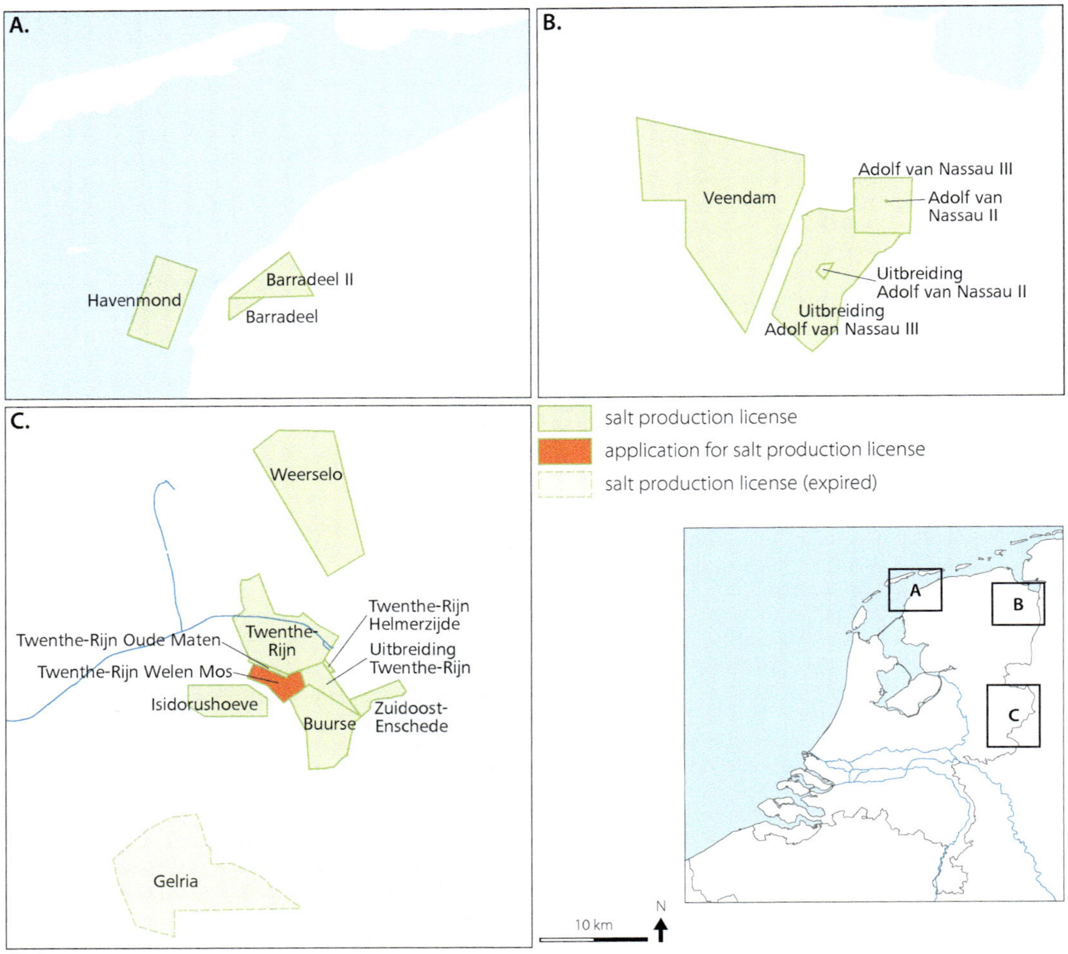

Figure 16.1. Overview of salt concessions in the Netherlands (reproduced from MEACP, 2022).

Table 16.1. Salt production and salt concessions in the Netherlands.

Concession	Owner	Salt formation	Depth	Annual production[1]
Adolf van Nassau[2]	Nobian	Z2 Salt (diapir)	500-1600 m	2600 (2735) kt NaCl
Barradeel	Frisia	Z2 Salt (layered)	2500-3000 m	158 (174) kt NaCl
Buurse	Nobian	Röt Salt (layered)	300-400 m	42 (1440) kt NaCl[3]
Havenmond	Frisia	Z2 (layered)	2600-3000 m	65 (0) kt NaCl
Isidorushoeve	Nobian	Z1 Salt (pillow)	600-1100 m	-[4]
Twenthe-Rijn[5]	Nobian	Röt Salt (layered)	300-550 m	2643 (2710) kt NaCl
Veendam	Nedmag	Z3 Salt (pillow)	1400-1800 m	246 (296) kt $MgCl_2$
Weerselo	Nobian	Z1 Salt (complex)	400-1000 m	-[4]
Zuidoost-Enschede	SGW[6]	Z1 Salt (pillow)	-[4]	-[4]

1 Production 2020 (2019)
2 Including Uitbreiding Adolf van Nassau
3 Salt production stopped in 1953
4 No salt production has taken place
5 Including Uitbreiding Twenthe-Rijn, Twenthe-Rijn Helmerzijde, and Twenthe-Rijn Oude Maten
6 Salzgewinnungsgesellschaft Westfalen

History

Rock salt was discovered in the eastern part of the Netherlands during the 1880s in a drinking-water well (Wassmann & Brouwer, 1987). Later, during the drilling campaign of the *Dienst der Rijksopsporing van Delfstoffen* (National Agency for Exploration of Minerals) between 1903 and 1923, more deposits of rock salt and of potassium-magnesium salts were discovered in the east of the Netherlands (Van Waterschoot van der Gracht, 1909, 1918). In 1919, rock-salt production in the Buurse concession (Fig. 16.1c) from solution mining of salt belonging to the Triassic Röt Formation was started by the *Koninklijke Nederlandse Zoutindustrie* (KNZ, now Nobian) following heavy taxation on German salt exports. In 1936 production started in the Twenthe-Rijn concession (Fig. 16.1c), located close to the newly constructed Twentekanaal, a canal that connects the three largest cities of the Twente region (Almelo, Hengelo and Enschede) to the national network of rivers and canals and which allowed better transport facilities. In the same period, the Gelria concession (Fig. 16.1c) was granted for the exploitation of rock salt and coal. To protect the underlying Carboniferous coal reserves, the production of rock salt was restricted to conventional room and pillar mining, which made it uneconomic. The production of rock salt in the Buurse concession was discontinued in 1953 and salt production in the east of the Netherlands was continued in the Twenthe-Rijn area only. In 2012 the Isidorushoeve concession was granted for salt extraction from Z1 Zechstein salt (Bouroullec & Geel, 2025, this volume), following the gradual depletion of the Twenthe-Rijn Röt salt concession.

In the north of the Netherlands several salt domes were discovered while drilling for oil and gas during and after the Second World War; others were inferred based upon gravity anomalies (Bentz, 1947; Mulder, 1950). More salt domes and pillows were interpreted from seismic surveys during the 1950s to 1980s. This resulted in the granting of new concessions for the production of rock salt to KNZ (Adolf van Nassau in 1954 with an extension in 1967 and Weerselo concessions in 1967) and of potassium-magnesium salts to Billiton (now Nedmag; Veendam concession, 1980). The presence of magnesium salts within some of the salt pillows was also proven thanks to drilling of gas wells from the 1960s onwards, when the pure $MgCl_2$ salt bischofite was found exclusively in this area between the towns of Veendam and Slochteren in the concession Veendam (Fig. 16.1b).

In 1996 the company Frisia started its salt mining operation near Harlingen-Franeker in the Barradeel concession (granted in 1991 with an extension in 2003 – Fig. 16.1a, Table 16.1). In 2012 the Havenmond Concession was granted. At the Havenmond concession (Fig. 16.1a), salt is extracted from below the Wadden Sea. The first Havenmond well was drilled from the salt factory site (at Harlingen harbour) with a deviation to 3 km offshore where the cavern is being developed. Mining from the Havenmond concession commenced in 2020. Note that carnallite salt layers were identified at the top of the Zechstein Z2 salt but they are not commercially exploited here.

Salt mining lifecycle

The lifecycle of a salt mining concession, and its associated caverns, can be divided into six phases: 1) Exploration and field design; 2) Development & Production; 3) Storage

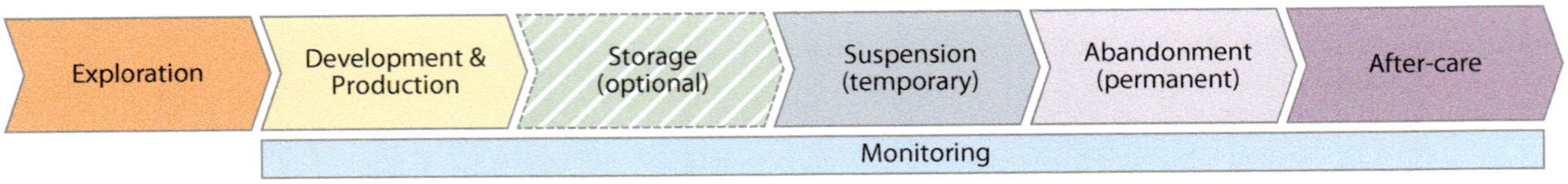

Figure 16.2. The lifecycle of a salt mining concession can be divided into six phases: 1 = Exploration; 2 = Development & Production; 3 = Storage (optional); 4 = Suspension (temporary); 5 = Abandonment; 6 = After-care. The (optional) storage phase only applies when the caverns are used for storage of gas or liquids.

(optional); 4) Suspension; 5) Abandonment; and 6) After-care (see Fig. 16.2).

In the first phase, the subsurface must be explored to confirm the presence of rock salt volume and structures in which caverns can be developed. This must be matched with suitable above-ground locations where facilities can be constructed, and where the necessary infrastructure is present or can be developed. With this aim an exploration permit is applied for. When granted, exploration work will be performed. Examples of exploration work are seismic data acquisition and interpretation and the drilling of an exploration well (including coring, logging and analyses and interpretation of these data). When the presence of the salt deposit has been confirmed, the application for the concession (mining permit) starts. At the same time the cavern field and its infrastructure will be designed. When the concession has been granted and the field design completed, the extraction plan (*winningsplan*) is sent to the authorities for approval and the environmental permits are applied for. After the permits have been granted the second phase of the field lifecycle starts.

In the second phase, drilling and construction of brine transportation and processing infrastructure is executed and production from the caverns starts. During this phase, a cavern is gradually increased in size by continuous leaching, while control measures and monitoring are carried out to ensure that it develops in a geomechanically stable fashion. In this process the amount of salt dissolved in the extracted brine gradually increases, until it reaches a fully salt-saturated level. Production continues until the cavern reaches its final shape according to the extraction plan.

In the optional third phase, gas or liquid storage can be implemented. After a period of production, when the cavern has reached a certain volume, it can be repurposed for storage by converting the former brine well to a storage well and replacing (a part of) the brine in the cavern with a gas (e.g. natural gas, nitrogen, hydrogen, air) or liquid (e.g. oil products). This process is called 'debrining'. This optional third phase can last for several decades, during which the stored product is injected and withdrawn in a cyclic manner depending on market dynamics (Juez-Larré et al., 2025, this volume) or remains untouched for a long time (in the case of storage for strategic reserves)

Afterwards, the stored product must be removed from the cavern, and brine (or water) must be injected again, either to restart a phase of production (second phase), or to safely suspend the mining operation (fourth phase) and initiate abandonment procedures (fifth phase). Note that suspension is a temporary measure applied to a cavern and can be undone. Suspension of a cavern involves removing the brine production string, shutting off the well, periodically in case of pressure build-up, and, when deemed safe, installing a cement plug downhole. Abandonment is commonly done at cavern and/or well pad level, is permanent and involves removing all surface infrastructure (well sites, brine transport and processing infrastructure, etc.) and returning the land into its original state.

After abandonment, the sixth and final phase is that of after-care, during which the surface continues to be monitored for years at least until site-specific subsidence cannot be measured or is constrained within the range of natural subsidence of 1-2 mm per year. In the next sections, the individual phases of the salt mining lifecycle are further detailed.

Exploration

The first phase in the lifecycle of a salt mining concession is the exploration phase. The most important topics for exploration are to confirm 1) the presence of salt, 2) the surface situation and 3) the distance to necessary infrastructure. A large enough salt volume should also be available for the long-term economic viability of the mining operation.

The geometry, volume and depth of the salt structures are important parametres to determine if mining is technically feasible and economically viable. Large volumes and shallow depths are favourable from a mining cost perspective. Proximity to sea-harbours can however favour deeper salt targets as well, limiting the need for long land pipelines to harbour-bound salt plants. This can save costs on salt transport to international clients who can be reached by larger ships, such as in the case of the Barradeel concession, with its production plant located in the Harlingen harbour. When locations with promising salt structures are identified, the surface situation is assessed. Agricultural land, or in combination with commercial estates represent favourable surface situations for the development of a salt mine. Protected areas (nature, drinking water) and residential areas are less suitable for mining activities. Another important topic is the distance to necessary infrastructure. For regular solution mining this means access to water, electricity, pipelines for water and brine and a salt plant to process the brine to salt, or a large body of brackish or salt water (e.g. sea, estuary) to dispose all of the remaining brine. Note that this practice is currently not allowed under Dutch law since it is seen as a waste of natural resources, therefore brine production in the Netherlands is always coupled with salt production. Salt plants are in general located near canals or seaports to enable transport by ship. In the case of cavern development for storage of methane, hydrogen or compressed air the distance to main gas and electricity infrastructure is also important.

For the first step of the exploration phase assessment of all existing subsurface data is done and a new evaluation is carried out, for example involving reprocessing and reinterpretation of existing seismic data. Once the most promising location has been identified, drilling of an exploration well is often the logical next step. Core material from such a well is preferred since it enables the direct assessment of the quality and creep properties of the salt and increases understanding of the stratigraphy and internal structure of the salt deposits.

Salt structures and occurrence

On a geological timescale, and when buried below 500 m, salt behaves in a visco-plastic way, whereas other sedimentary rocks show brittle deformation. Rock salt compacts already during early stages of burial to a tight mass with a constant density of 2168 kg/m^3, slightly depending on the quantity of impurities such as anhydrite, carbonate and clay layers. Other sediments show an increase in density with depth owing to cementation and the reduction of pore volume as a function of overburden and pore pressures. Consequently, in near-surface positions, where sand and clay typically show densities of 1200 to 1400 kg/m^3, halite is relatively heavy, whilst below 500 m it is lighter than the surrounding rocks. This results in an unstable situation. Under the right conditions, the salt rises while solid rocks sink (Trusheim, 1963; Kockel, 1990; Remmelts, 1995). This process of flowing and rising salt is called diapirism. Salt will initially flow mainly in a lateral sense, forming what is known as salt pillows (Fig. 16.3). Above a pillow, the sedimentary cover does not become pierced, but in the next stage a pillow can evolve into a salt diapir. The Zechstein salt in the Netherlands has formed many diapirs (Fig. 16.4), which indicates that the boundary conditions for diapirism have been met on a wide scale in the geological past. However, in the northwest of the Netherlands the salts remain almost undisturbed. A more in-depth explanation of salt tectonics, and its role in the geological history of the Dutch subsurface can be found in Bouroullec & Ten Veen (2025, this volume). Salt can be mined from different types of salt structures, and different internal tectonic styles can be recognized in them. In literature a separation is commonly made between bedded (layered) salt, salt pillows and salt diapirs, which represent increasing states of salt deformation (Fig. 16.3; Trusheim, 1963).

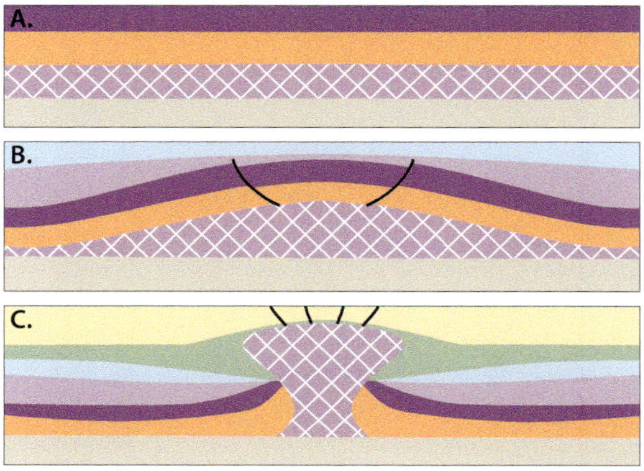

Figure 16.3. Types of salt structures. a) Original salt layer, end Middle Triassic. b) Salt pillow, Late Jurassic. c) Salt diapir, present-day situation. Modified after Trusheim (1963).

In ***bedded salt*** the top and bottom of the layer are parallel to the interbedded elastic carbonate and anhydrite beds, reflecting the original depositional geometry. Salt layers may be discontinuous due to faults with significant offset, especially in areas with relatively thin depositional salt. In most cases the faults will be salt smeared, in similar fashion to clay smeared faults, by which process the faults will be barriers to fluid migration. In contrast to paleo basement highs where the salt is often thin, basinal areas can have salt thicknesses that generally exceed the throw of most faults. Even with bedded salt the internal individual salt layers can be deformed due to salt lateral migration, expressed by flow folds with horizontal axial planes (Lotze, 1957). They occur on the decimetre to decametre-scale. Potassium-magnesium salts, with interbedded layers of halite, usually show strong folding. With the exception of possible faults, exploration in bedded salt can be relatively simple due to the often large extent of continuous (semi-)horizontal layers. Possible trends in depth or thickness can be interpreted from one well to the next. The salt occurrences in the Twenthe-Rijn, Buurse and Friesland (Barradeel/Havenmond) concessions are examples of layered or bedded deposits (Fig. 16.1; Table 16.1). In the Twenthe-Rijn and Buurse concessions, 50 to 100 m of rock salt occur in the Triassic Röt Formation at depths of 200 to 600 m (Harsveldt, 1980, 1986). In the Friesland

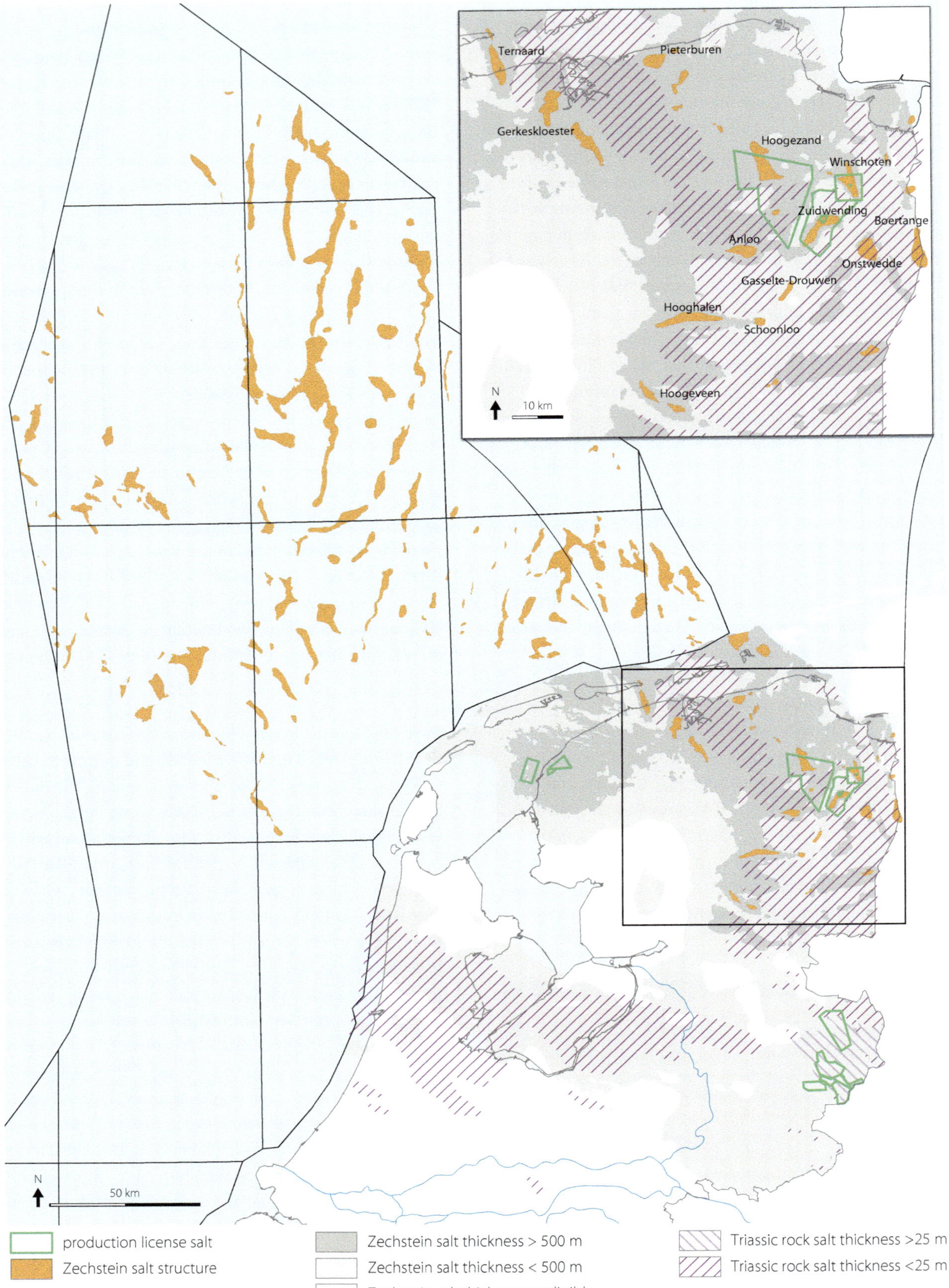

Figure 16.4. *Map displaying salt occurrences of the Permian (Zechstein) and the Triassic (Röt) in the Netherlands. Zechstein salt diapirs are displayed in orange. Elongated salt walls (e.g. in the northern offshore) are related to major sub-salt faults, whereas circular diapirs developed at intersections of fault trends.*

concessions, bedded rock salt occurs in the Z2 (Stassfurt) and Z3 (Leine) formations. Salt is only exploited from the Z2 where it has a thickness of up to 650 m and occurs at depths of 2200 to 3000 m, which makes it the deepest salt exploited in the world.

Salt pillows are formed by a local thickening of the salt due to horizontal salt flow. In central areas of the Zechstein basin (in the central North Sea, the south of Denmark, and the north of Germany) the cores of the pillows are composed of the Z2 (Stassfurt) salt; in marginal areas (in the east of the Netherlands and the middle of Germany) the Z1 (Werra) salt deformed into thinner and thicker zones, the thicker of which form salt pillows. Extensional faulting commonly affects the beds over the crest of the pillow and are referred as crestal faults. The thickening of salt into a salt pillow is usually accompanied by a series of stacked, large-scale flow folds in the salt, with horizontal axial planes. Interbedded carbonate and anhydrite layers within the pillow are folded and their axial planes are generally parallel to the pillow's roof. Superimposed upon such large-scale folds one commonly finds a series of smaller secondary folds. In the late stage of pillow formation, the direction of salt flow may change from mainly horizontal to more vertical, forming salt diapirs (Lotze, 1957). In the northern and eastern parts of the Netherlands, a series of Zechstein salt pillows is present at relatively shallow depth (500-800 m). These salt pillows were brought close to the surface thanks to a combination of deep erosion and limited burial during the Late Jurassic to Early Cretaceous accompanied by uplift during the Cenozoic. This is in contrast to rift basin areas (e.g. West Netherlands Basin, Broad Fourteens Basin or Dutch Central Graben). Exploration in pillows is a bit more complex compared to that in bedded salt, because the salt layer thickness varies. A good coverage of the pillow by seismic lines is important to interpret the thickness variation. After that, it is possible to interpret trends from well to well. For example, the Isidorushoeve concession (Fig. 16.1) contains a salt pillow with a maximum thickness of 400 metres in a depth range between 500 and 1100 metres (Fig. 16.5). In areas around the Dutch Central Graben, pillows of upper Permian (Silver Pit Formation, Upper Rotliegend Group, Bouroullec & Ten Veen, 2025, this volume) as well as of Triassic (Röt, Muschelkalk) salts have been identified. The Veendam concession (Fig. 16.1) covers two Zechstein salt pillows. Both contain thick layers of carnallite and bischofite in the Z3 salt (Coelewij et al., 1978). It is the only known occurrence of primary bischofite in western Europe.

In ***salt diapirs*** the movement of the salt changed from mainly horizontal to mainly vertical, and the salt deforms or even pierces the cover beds. The salt flow results in folds within the diapir with vertical axial planes and horizontal fold axes as well as in curtain folds with vertical fold axes. In the late stage of diapirism these latter folds usually will have deformed the previous generation of folds with horizontal fold axes, but due to the extremely high strains in the diapir, the older deformation phase may be indistinguishable. Along the margins of diapirs, friction between non viscous and viscous rocks can create additional complex fault systems and associated fracture patterns (Jackson et al., 1990). For this reason, wells aiming at targets located below the salt (subsalt) are usually drilled through the diapir itself or at a safe distance from it in order to to avoid such structurally complex zones. Sometimes the salt flow and migration may even envelop blocks of the overlying succession (as is observed in e.g. the Gulf of Mexico and West Africa). The salt in a diapir rarely moves at a uniform rate (Kupfer, 1976). Shear zones, known as spines, separate sectors moving at higher speed, from those moving more slowly. The development of salt diapirs normally results in a complicated internal structure, with folds of many different scales and orientations (Richter-Bernburg, 1980). Compressional tectonics during the Late Cretaceous and Cenozoic accelerated the rate of diapirism, or

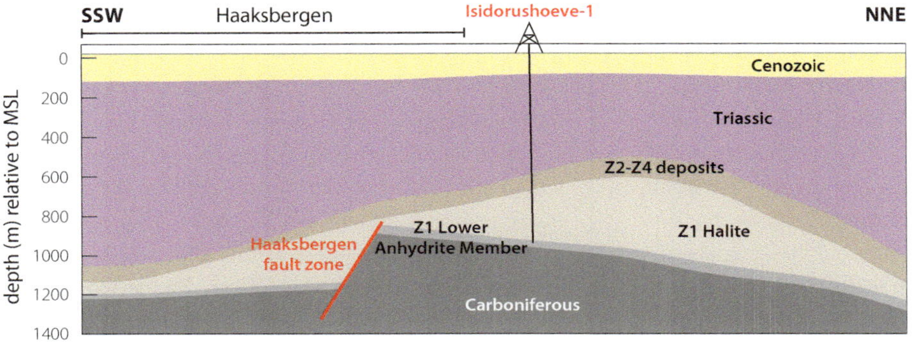

Figure 16.5. *Cross-section through the Haaksbergen salt pillow on the Isidorushoeve Concession of Nobian in the northeast of the Netherlands. Location of well Isidorushoeve-1 is indicated. See Fig. 16.1 for location.*

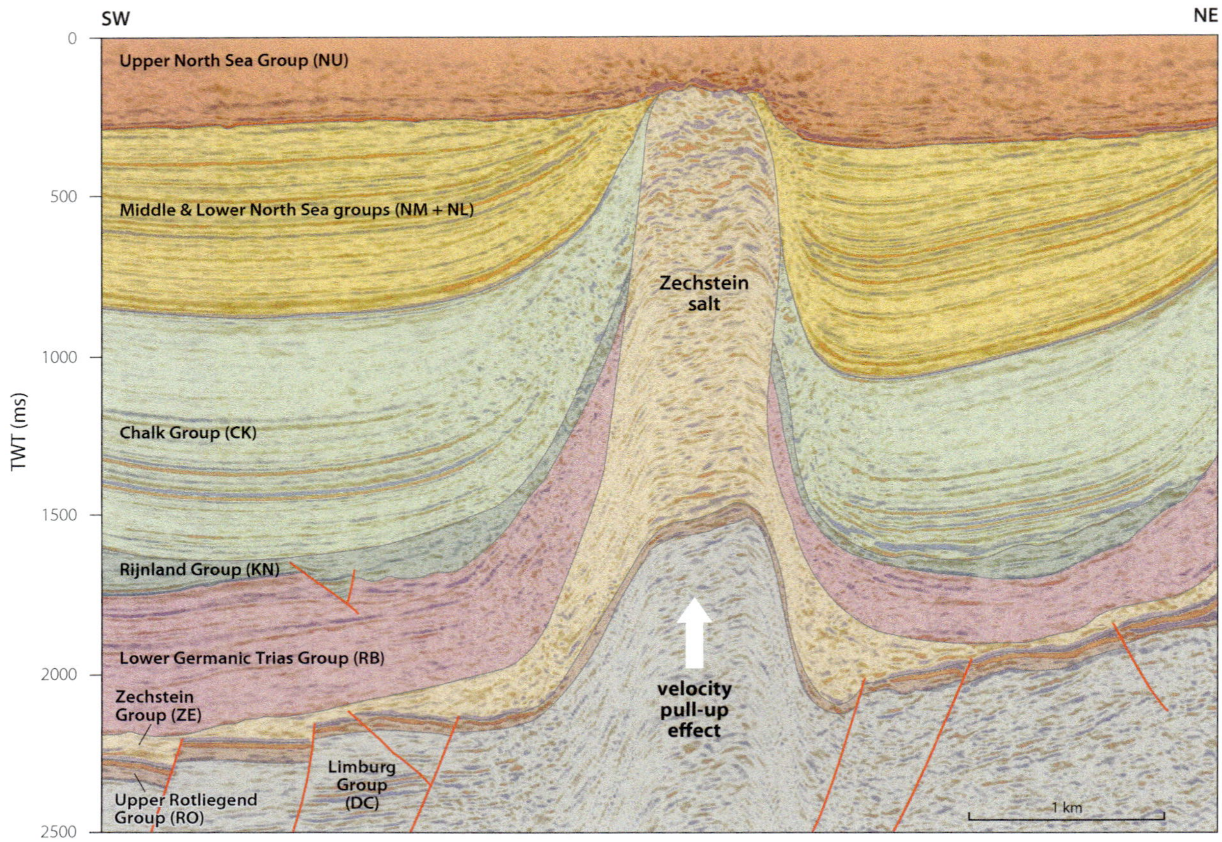

Figure 16.6. *Seismic time section of the Zuidwending salt dome in the Adolf van Nassau concession. This dome is the shallowest in the northern Netherlands, reaching to ca. 100 m below the surface. Thickness variations in adjacent sediments result from periods of main salt movement. See inset map in Fig. 16.4 for location of the Zuidwending salt dome.*

strongly deformed some of the diapirs. An example of this is shown in Figure 16.6, which displays a seismic cross-section through the Zuidwending salt dome.

The shapes and the internal structures of individual salt bodies are complex; reference is made to descriptions by De Boer (1971), Richter-Bernburg (1972, 1980), Kupfer (1976), Bornemann (1991) and Geluk (1995). Salt diapirs outcropping in the Iranian Dasht-e-Kavir (Jackson et al., 1990) and Zagros Mountains (Kent, 1979) and also in the Spanish Pyrenees (Wagner et al., 1971) present analogues for buried salt structures in the Netherlands (Geluk, 1998, 2000). A generalized simplified model of Zechstein salt diapirs is that the central part of the dome contains mostly Z2 salt and that the marginal parts are made up of younger salt units. This is shown to be true for structures such as the Winschoten and Zuidwending diapirs (Harsveldt, 1980, 1986). In the Pieterburen diapir, however, the central part is made up of younger salts and the Z2 salt occurs at the margin (Harsveldt, 1980; RGD, 1995); in Germany this occurs at Gorleben (Bornemann, 1991) and Hanigsen-Wathlingen (Schachl, 1987).

Exploration of salt diapirs is relatively complex compared to exploration of bedded salt or salt pillows. This is caused by two risks related to cavern development in a salt diapir: 1) the location of the salt dome flank and 2) the presence of non-halite layers within the dome. Caverns should be developed at a safe distance from the salt dome flank (typical values are 150 or 200 metres). However, it is not easy to precisely determine the salt dome flank position due to the (semi-)vertical layering (difficult to interpret on seismic data) and the often heavy folding and faulting in the overburden layers near the salt dome flank. Another risk is the presence of non-halite layers within the dome. Often these are not easy to recognize on seismic data. Due to the complex structure identifying trends from one well to the other may not provide a reliable prediction for the geology of the next well. Ground penetration radar (GPR) within newly drilled boreholes is regularly used within salt wells to improve the reliability of the geological interpretation within the salt diapir.

Several salt domes in the Netherlands reached shallow depth and are close to the surface. The shallowest salt diapir is the Zuidwending diapir, where the top of the caprock (insoluble residue of the salt formed by groundwater dissolution, see below) is at a present-day depth of 100 to 110 m below mean sea level (Fig. 16.6). Other noticeable shallow diapirs are the Schoonlo (120 m), Pieterburen (190 m), Hooghalen (250 m), Onstwedde (250 m) and Gasselte-Drouwen (350 m) salt diapirs (see insert map in Fig. 16.4 for locations of diapirs). In the western offshore block K9, a salt diapir almost reaches to the seafloor (Giesen & Mesdag, 1995). Some of the Dutch salt diapirs have reached the surface at some point in their geological history, mainly during the Mid/Late Triassic, Late Jurassic or Early Cretaceous, and became subject to later erosion as well as near-surface dissolution of salt by ground or surface water (Bouroullec & Ten Veen, 2025, this volume). As a result, a caprock composed of the insoluble residue of the salt formed (Batsche & Klarr, 1980). This caprock may locally include thick anhydrite floaters and remnant suprasalt sediments such as Jurassic claystones, as dated using palynologiocal analysis in the Dutch offshore G Block (Verreussel et al., 2018). The thickness of the caprock shows considerable variation between salt domes, from 20 m in the Winschoten salt dome to over 100 m in the Pieterburen salt dome. Also, within a single salt diapir, this thickness may vary strongly in relation to the percentages of insolubles in the subcropping salt succession and their structural style (Balzer, 2000a,b, 2001). In the Adolf van Nassau concession, two salt diapirs are present. The shallow Zuidwending diapir has a typical crescent shape in map view and is interpreted to have been formed by the amalgamation of two individual diapirs that eventually formed a salt wall (Figs 16.6, 16.7).

Production

The techniques applied in solution mining depend upon the geological characteristics of the salt and particularly upon its internal structure. After a general introduction on solution mining, the mining of rock salt and potassium-magnesium salts will be discussed.

In the past, salt was produced from natural brine springs by evaporation. In the coastal areas of the Netherlands, it was produced by the burning of peat wetted by salt water.

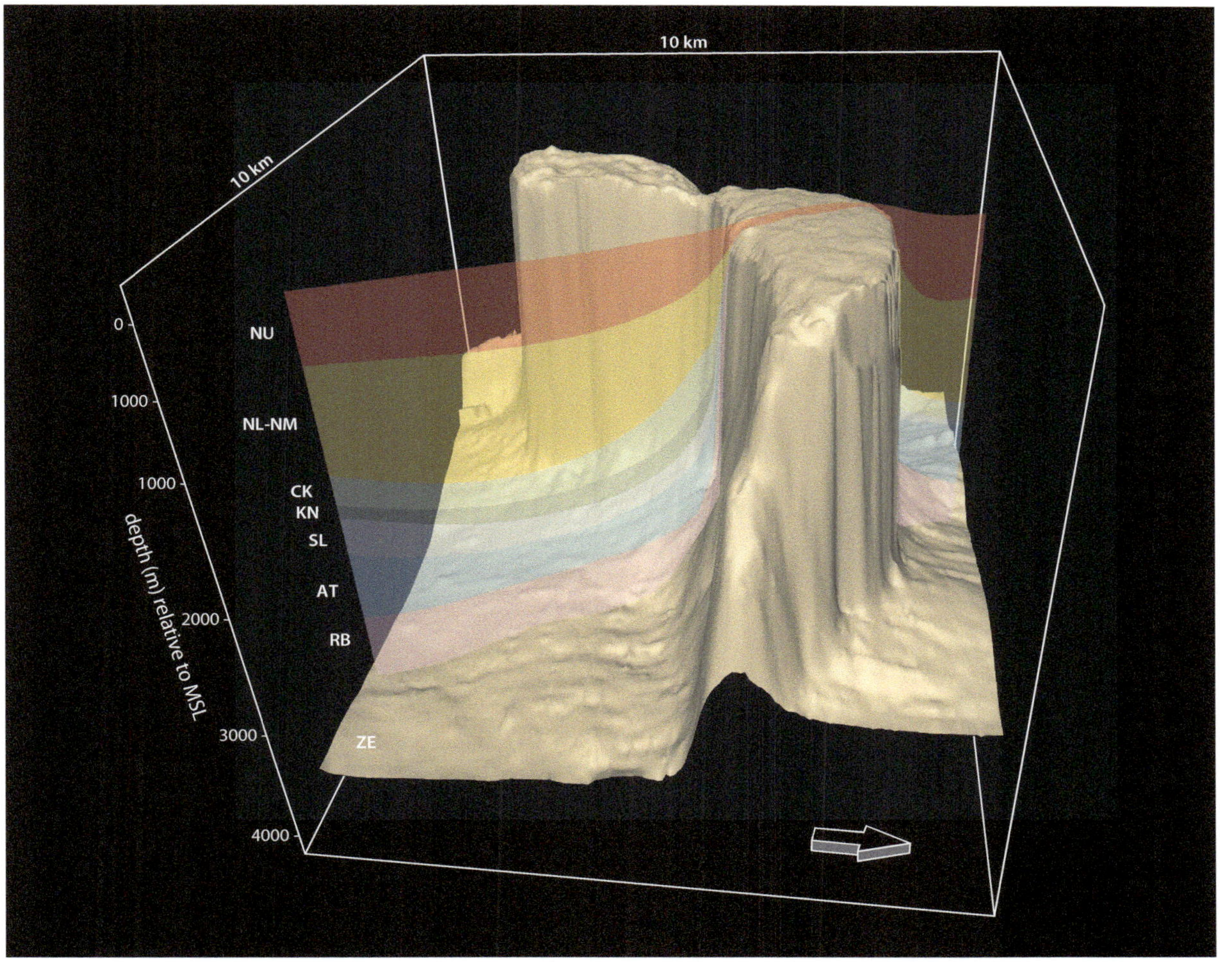

Figure 16.7. 3D geometric model of the Zuidwending salt diapir, viewed obliquely from above. The structure is an amalgamation of two individual domes with a marked difference in strike. The height of the dome is ca. 3 km above the base of the Zechstein salt layer. The salt diapir covers an area of ca. 6 x 6 km. See Fig. 16.4 for location and Fig. 16.6 for explanation of the stratigraphic codes.

The peat was then burned, water was added to the ashes and subsequently the ashes were removed from the water and the water evaporated to produce the salt. Solution mining started by the pumping of natural salt-saturated water (Jeremic, 1994). In this process, which was never applied in the Netherlands, the produced water was replenished by fresh or unsaturated groundwater in an uncontrolled manner. As dissolution of salt occurred mostly at the top of the salt layer, the overlying rocks became unsupported. These phenomena produce local ground subsidence at the surface above the dissolution zone. This process requires systematic monitoring since uncontrolled dissolution can produce large-scale collapse and even create sinkholes at the surface. This type of dissolution may also occur naturally by groundwater flow, whereby salt is dissolved and flushed away. In the Netherlands, controlled solution mining is applied, using several methods to regulate salt dissolution, raw brine production and cavern development. These methods involve considerations regarding cavern integrity and long-term stability of the cavern walls and roof, surface subsidence as well as post-abandonment monitoring.

Solution mining of rock salt may be designed as a single or multiple-well operation, depending on the nature of the salt deposits and the shape of the cavern to be developed. Two modes of operation can be distinguished, i.e. the indirect and the direct circulation mode. The indirect circulation mode is also called 'top injection method', whereby the fresh water is injected into the upper part of the cavern (Fig. 16.8). The water is flushed downward to the cavern bottom, increasing in salt concentration by dissolution of solid salt on its way. The brine is produced from near the base of the cavern. The advantage of this mode is that it produces saturated brine at an early stage of cavern development. A disadvantage is that the cavern primarily develops at the top, where the brine is dilute. Due to temperature changes in the well bore, moreover, crystallization of salt or gypsum in the supersaturated brine may cause clogging of the tubing (the free hanging pipe in the well, used to produce the brine). In the direct circulation or 'bottom injection' method, fresh water is injected at the cavern bottom, and saturated brine is produced from the upper part of the cavern. This method results in a cavern in which the brine produced is somewhat diluted by

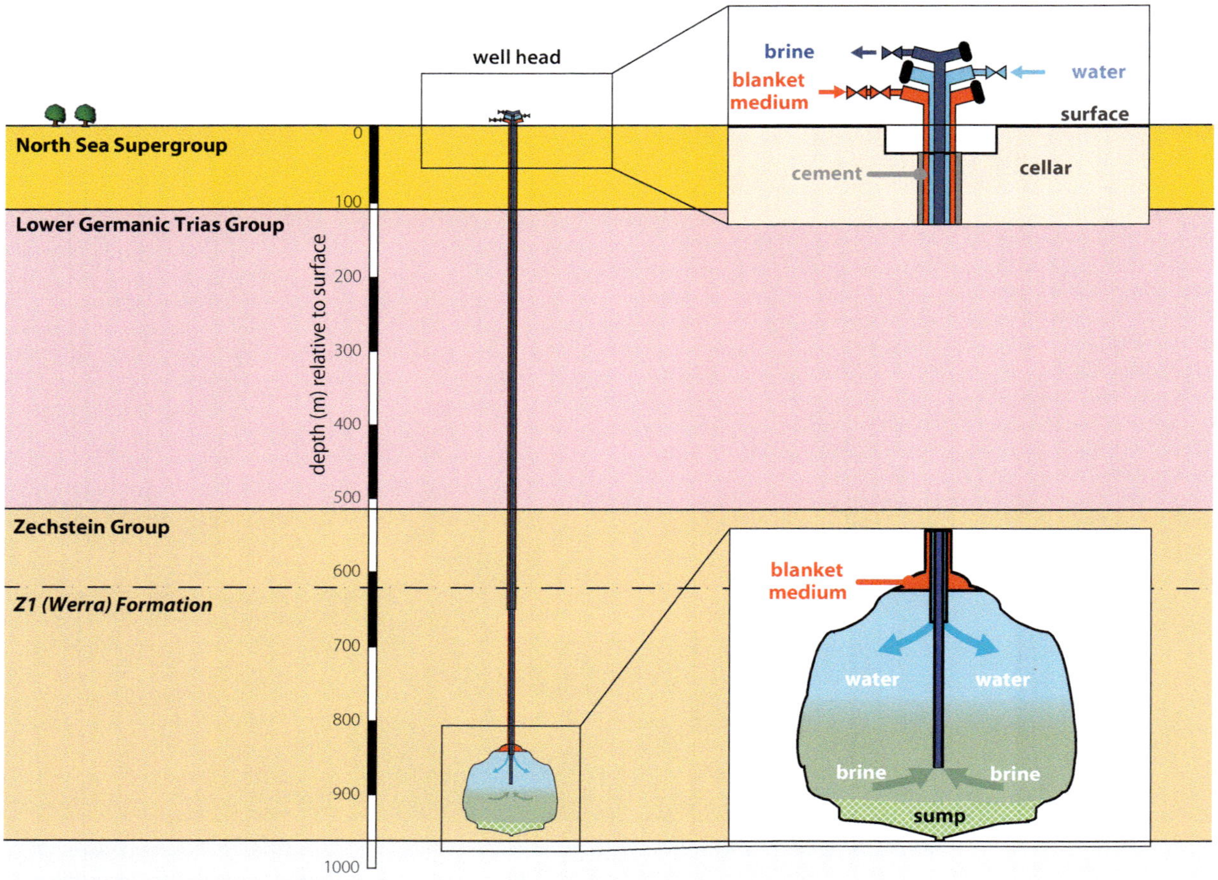

Figure 16.8. Cross-section of a planned cavern at a depth of 800-950 m in the Haaksbergen salt pillow of the Isidorushoeve Concession. The leaching will be controlled by the application of a blanket fluid in such a way that a stable conical roof will develop.

the injection buoyancy plume. Thus, clogging of tubulars is prevented. In the past, mining started using this direct circulation mode, transitioning to the indirect circulation mode later. Nowadays, the direct circulation mode is hardly used and indirect circulation (top injection) is used from the start.

The final shape of a cavern is controlled by the original thickness and geometry of the salt layer, the depths of the injection and production tubulars, the mode of operation, the water throughput, the specific dissolution rate of each salt layer and the setting depth of the blanket fluid. Cavern development preferentially takes place in an upward vertical direction as fresh water is lighter (1000 kg/m^3) than saturated brine (1200 kg/m^3). In order to allow controlled vertical and lateral development of a cavern, a so-called blanket fluid may be injected (Fig. 16.8). This liquid or gas should not dissolve salt nor mix (too much) with brine, and it should float on brine. Commonly, hydrocarbon-based fluids like oil or LPG are applied (700-900 kg/m^3), but compressed air or nitrogen (both in gas phase) can be used as well. Cavern lifetime amounts to tens of years depending on salt layer thickness and mining method. Sequential cavern development, i.e. the leaching of the void by applying subsequent leaching steps, is simulated and subsequently monitored by time-lapse acoustic measurements (sonar). These measurements are acquired using a wireline-mounted probe, which is lowered into the well bore. Comparison with data of former surveys or with results of computer simulations enables the verification of compliance with planned cavern development.

The integrity and long-term stability of a cavern are important issues with respect to potential surface subsidence. Maximum allowable cavern dimensions are determined by finite-element modelling. When the leaching is executed within these given dimensions the cavern wall and roof can be considered stable. The mechanical behaviour of the salt is described by means of a constitutive law, which gives the relation between stress and deformation and which incorporates the visco-plastic, time-dependent, non-linear behaviour of salt. On a scale of weeks to decades, salt behaves like a viscous fluid upon application of an external stress field. The resulting 'creep' is temperature (i.e. depth) dependent and has a large impact on cavern operations. Over time, creep results in a homogeneous stress field around a cavern, which is a prerequisite for the preservation of cavern integrity and long-term stability. Brine production is closely monitored by mass balancing and chemical analysis. In addition, parameters like flow, pressure, temperature, saturation, etc. are recorded continuously to provide information on the actual development of a cavern and on any deviation from default values. Periodically, the surface subsidence is measured to assure compliance with planned cavern development.

Shallow depth salt production (300-600 m), e.g. Twenthe-Rijn concession

Since production started in 1936, over 550 wells have been drilled and over 100 million tonnes (Mt) of rock salt have been mined. Initially, wells were drilled in the direct vicinity of the evaporation plant, south of the city of Hengelo. The mining method evolved over time; in the past salt was extracted from caverns which each had two or three wells at a spacing of 40 m. Currently caverns are developed from single wells. Cavern development starts at the base of the salt. Once the initial cavern has developed, production is increased and the cavern is further developed, laterally and vertically, into a cylindrical void (Fig. 16.9). The maximum height of a cavern depends on the depth of the salt deposit (more specifically on the distance between the top of the salt and the base of the Cenozoic) and the bulking factor, that is the ratio between loose and consolidated rock material in the overburden in a case of collapse. When the cavern is developed within this maximum height, which generally ranges between 20 and 40 metres, it will not cause a sinkhole even in the unlikely event of a collapse.

Caverns are spaced such that between caverns a safety pillar of proper size is maintained to guarantee cavern integrity and to minimize creep-induced subsidence. Thin dolomitic claystones subdivide the Röt salt into the layers A to D (from base to top) and may hamper cavern development. Salt production takes place from layers A, B and (partially) C. On top of each cavern a safety roof is maintained in salt layer C to ensure cavern integrity and stability and to prevent future surface subsidence. Without a safety roof, as was the case with caverns developed before 1980, mechanical failure of the overlying strata by gravity force may occur. This may result in an upward vertical growth of the cavern, and a resulting surface subsidence of several millimetres per year. Once the collapse zone in the overburden reaches the base of the unconsolidated Cenozoic (at 60 to 100 m depth in the case of the Twenthe-Rijn) and depending on the height of the remaining void, severe subsidence, up to the formation of a sinkhole, may occur. In the Netherlands, this has happened only once, in 1991. On that occasion a collapsing cavern reached the surface and a sinkhole of about 30 metres diameter and 4.5 metres depth was formed. By limiting the height of a cavern, the formation of a sinkhole is prevented, because the material falling from the cavern roof to the cavern bottom occupies a greater volume than the original rock layer. Therefore, a collapsing cavern gradually decreases in height. By using an empirically determined so-called bulking factor Nobian limits the height of the cavern such that the remaining cavern height will be zero when it reaches a pre-defined limit below the base of the unconsolidated layers. This migration process, whereby a collapsing cavern

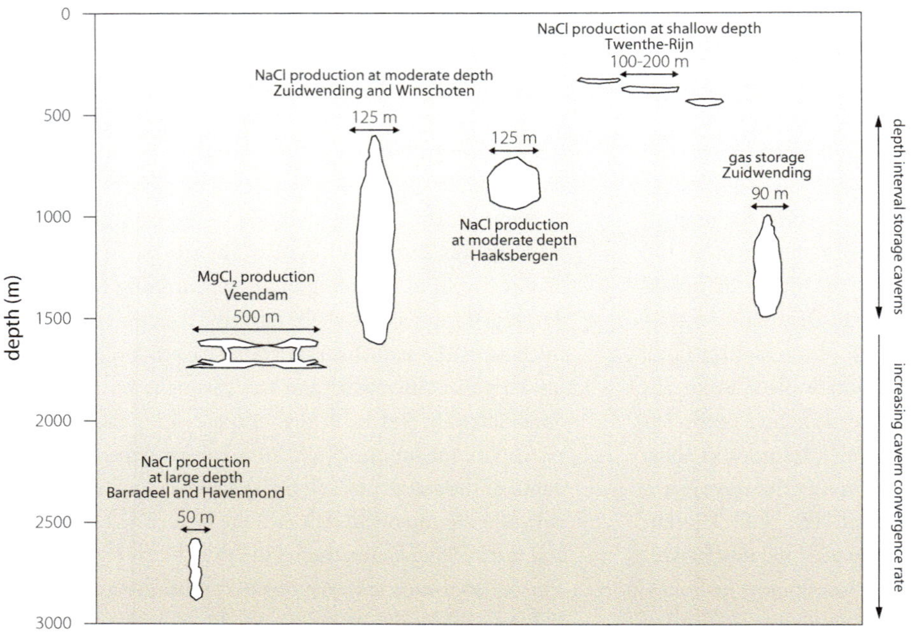

Figure 16.9. Typical cavern geometries, depths and dimensions of salt mining and gas storage concessions in the Netherlands.

gradually migrates upward to the surface, may last 10-20 years. The critical area, i.e. the zone of surface subsidence, is limited to an area that is defined by the 45-degree angle of draw relative to the base of the Cenozoic (Wassmann, 1980, 1983, 1993). When the caverns are developed according to the current safety guidelines, sinkhole formation is prevented and subsidence due to salt creep is limited to a maximum around 1.0 mm/yr, which is in line with natural subsidence rates in large parts of the Netherlands. Therefore, negative surface effects are not expected.

Moderate depth salt production (400-1600 m), e.g. Adolf van Nassau and Isidorushoeve concessions

Solution mining at moderate depth is carried out from the Zechstein salt using one well per cavern. Presently, production is from 5 wells in the Winschoten salt diapir and from 8 wells in the Zuidwending salt diapir (Figs 16.4, 16.9), from the Zechstein Z2 salt layer. The roofs of the caverns are at depths between 650 and 800 m (Winschoten salt diapir) and 450 and 500 m (Zuidwending salt diapir). The actual heights of the caverns vary between 400 and 700 m (Winschoten salt diapir) and between 400 and 900 m (Zuidwending diapir). The well spacing is approx. 250 m in a line configuration. Cumulative production is over 100 Mt of rock salt since the beginning of the production in the 1950s (Winschoten diapir) and 1960s (Zuidwending diapir) respectively. Surface subsidence due to cavern convergence amounts to 1 to 3 mm/yr. As in the Hengelo brine field, a sphere of influence is defined in this case by a 45-degree angle of draw relative to the deepest point of a cavern. In the coming years the mining company Nobian will develop a new brinefield in the Isidorushoeve concession, near Haaksbergen, to supply brine to the Hengelo salt plant. The salt to be mined is a Zechstein Z1 salt pillow. Cavern development wells will be drilled with a 300 m spacing. The cavern depth is determined by the depth of the base of the Z1 salt. The permitted cavern height is 125 m, in order to ensure a safe long-term cavern abandonment.

Large depth salt production (1600-3000 m), e.g. Friesland concessions

The world's deepest solution mine, operated in the Friesland concessions of Barradeel and Havenmond, feeds a vacuum salt factory near the town of Harlingen. Two caverns, 500 m apart, have been in production since 1995 and new caverns at larger distances are under construction. Halite is extracted at 2500-3000 m depth from a 300 m interval in the bedded Z2 salt. Minor faults affect the base and top of the salt. The top 30 m of the Z2 salt is formed by a carnallitic layer. By applying a gasoline or LPG-based blanket fluid the cavern development is kept well below this layer, to avoid carnallite coming in contact with the cavern brine, which would result in contamination of the brine and in loss of the cavern for further production. Subsurface temperatures at cavern depths are as high as 105°C. As a result, the viscosity of the salt is low, giving a high cavern convergence. This effect is greatly increased by the 25-30 MPa pressure difference between the cavern pressure and the far-field lithostatic pressure (60-65 MPa) at a cavern depth of 2500-3000 m. The high temperatures do not influence the solubility of NaCl, but significantly increase its rate of dissolution. Where it takes usually one to two years to create sufficient wall surface in a cavern in order to saturate the brine to acceptable levels for production, the deep caverns become saturated in a few months, which is a great advantage for salt production. Although the process of solution mining at large depth is similar to that at moderate depth, the behaviour of caverns strongly differs. At moderate depth, most caverns (at near final volume) have a convergence rate that is only 0.05 to 0.1% of the dissolution rate; in the deep caverns, the convergence rate becomes equal to the dissolution rate within a few years. The cavern shapes and sizes are frequently measured with sonar (3D imaging technique). Although their shapes change somewhat with time, their volumes remain almost constant. Hence the term 'squeeze mining' is often used for this type of deep solution mining to reflect the fact that the expected increase in cavern volume by dissolution of salt is completely compensated by the surrounding salt flowing inward at a fast rate, thereby effectively squeezing the cavern.

Solution mining of potassium-magnesium salt, e.g. Veendam concession

Solution mining for potassium and magnesium salts differs greatly from that of rock salt. They are exclusively mined from bedded salt or salt pillows. In diapirs these salts are strongly folded and could be extracted only by conventional dry (room-and-pillar) mining, as is being done in many potassium salt mines in Germany. In the Veendam concession, the magnesium component is the main target of mining, while the potassium component is currently not produced. The mined magnesium salts occur in the Z3 salt. Three evaporation cycles have been recognized in this salt, and the potassium and magnesium salts are separated by layers of halite. The first, lower, cycle contains most of the magnesium salts and all of the bischofite; the second and third cycles contain only carnallite. The depth ranges from 1300 to 1800 m (Fig. 16.10). The layers are dipping up to about 30 degrees at the mining location. The halite layers occur in all wells with equal thickness and composition, but the magnesium salts are interbedded with thin layers of halite at a centimetre to metre-scale. This interbedding is thought to reflect daily or seasonal variation in the concentration of the depositional brine.

During production the magnesium salts are dissolved preferentially, leaving many less dissolvable overhangs on the cavern walls. Eventually some overhangs will fall into the caverns, occasionally damaging production tubings. Sonar surveys, made in the initial stages of cavern development in the 1980s, showed a very irregular cavern shape, but were largely inconclusive in determining the precise shapes and volumes of the caverns due to many reflections from the rock-salt balconies. Presently, the cavern system has a labyrinthine architecture, comparable to a natural cave system formed by karstification. Non-saturated brines (to bischofite or carnallite) find their way to those parts of a cavern that are the easiest to dissolve and most easily accessible by density flow. The total interconnected areas in the cavern field (called the cavern cluster) have a span of more than two kilometres. Salt dissolution is mainly in the updip direction, since water and brine change-out is driven by density flow.

In 2022 the Veendam pillow had 15 wells, 2 of which are abandoned. Injection and production intermittently take place through the same well or via separate wells. The cavern cluster will be reduced in volume by allowing salt creep (mainly the bischofite) to leave as little cavern brine behind as possible before abandoning the connecting wells. Some 90 cm of maximum ground subsidence is expected in a few decades. This excludes overlapping

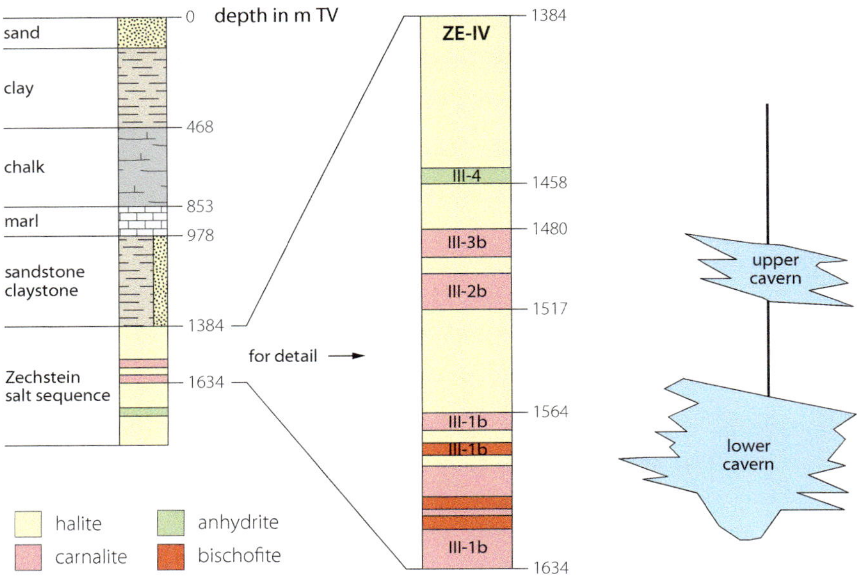

Figure 16.10. Stratigraphy and cavern development in the Veendam concession. Cavern shapes reflect dip and solubility of the salt layers. The upper cavern is no longer solution-mined from the year 2000 onwards. Depths are indicative due to the pillow shape of the salt.

subsidence by gas production from the Groningen and Annerveen gas fields. The (lateral) cavern growth for the same salt-production rate is much larger for bischofite than for halite because: i) the brine, which replaces the dissolved salt contains much more salt (500 kg $MgCl_2/m^3$ versus 300 kg $NaCl/m^3$), ii) bischofite contains more water than solids (850 kg H_2O/m^3 and 750 kg $MgCl_2/m^3$) whereas rock salt contains 2168 kg $NaCl/m^3$ (and no water) and iii) the cumulative thickness of the Mg-salt layers is ca. 40 to 50 m, compared to hundreds of metres in rock salt.

Fresh water is injected 30 to 60 m below a rock salt layer into a cavern filled with fully saturated brine (1350 kg/m^3 at bischofite saturation). The buoyancy effects make the water rise in the cavern in a turbulent plume, which causes intense mixing with the existing brine. The brine-water mixture is almost saturated with $MgCl_2$/when it reaches the rock salt and can only dissolve negligible amounts of NaCl. Hence the rock salt acts as a natural barrier to solution mining. The slightly $MgCl_2$-undersaturated brine will flow, usually following the stratification updip, to the highest spot in the cavern, where bischofite or carnallite will be dissolved. Subsequently, the saturated brine will flow down-dip to the deepest point of the cavern, where it can be produced. If the salts were immobile, or if the cavern pressure could be kept at almost lithostatic, the caverns would continue to grow. However, the magnesium salts have a low viscosity and they flow towards the caverns, driven by the weight of the overburden. Since the magnesium salts occur in layers, salt movement occurs mainly in and through these layers. The salts are squeezed towards the caverns, in a similar fashion to tooth paste in a tube. This rate of salt flow is such that the brine volume of the total cavern field remains limited during production: the cavern growth by dissolution is balanced by the cavern convergence.

Suspension and abandonment

Once a cavern has reached its maximum size, depending on practical, permitting or legislative considerations and once no further use is seen for storage of fluid or sequestration of waste material, a cavern usually becomes inactive. If the cavern was used for storage of oil or gas, these get extracted to the maximum possible, and replaced with water or brine from an active solution mining cavern in the vicinity. Sometimes the free-hanging leaching or production tubings are removed during this phase, after which a monitoring phase starts, usually lasting several years or decades.

During the monitoring phase, the cavern can be monitored for pressure build-up and temperature rise (due to heating by the surrounding rocks, especially relevant for the caverns at larger depth, i.e. 1000-3000 m, which were significantly cooled during the production phase). Repeated sonar measurements can help to prove that a cavern is stable in terms of roof fall in cases where the cavern roof consists of mixed halite-anhydrite-clay interfaces, which sometimes give roof falls due to stress build-up in the non-creeping materials and slow brine migration into clays that can degrade clay and shale strengths. In case of $MgCl_2$ mining, the sonar measurements cannot be used to measure the labyrinth type of cavern system for volume or extent, as it only sees a central core of the cavern, which adds no information.

The overall objective of cavern abandonment is to make sure that the cavern (or well) does not leak at elevated pressures, for instance caused by a leak path due to bad cementation of the casing. If this is established, the cavern can be closed by abandonment of the well, for example by placing barriers such as cement plugs over the salt roof section. The final abandonment (including more cement plugs to prevent cross flow of overburden fluids in the future) and the removal of the top section to return the well site to its original state (usually farmland) coincides with the full removal of the well site. Once the brine pressure reaches close-to-lithostatic pressures due to salt creep (and a bit of thermal expansion) the cavern is believed to be stable at human timescales. On geological time scale, but also on a scale of centuries, the cavern brine will slowly migrate out of the cavern due to the small excess pressure of brine towards rock stresses at the cavern roof, basically a buoyancy effect caused by the low density of brine with respect to the surrounding salts. When the brine pressures reach lithostatic stress level, it is likely that salt permeability increases, either at a crystal-interface level or by the creation of micro- or macro-fractures.

Which mechanism dominates is still a matter of scientific debate, but most likely it will depend on the pressurization rate, the presence of embedded fracture barriers like anhydrite (not present in domal salt, but usually in bedded salt) and the potential of crystallization in micro-annuli due to cooling of brine when moving upwards. This effect is strong for magnesium brines but almost absent for sodium chloride brine. Especially for higher caverns within halite the cavern field-closure and aftercare is now prepared following a risk-based approach in order to prevent any lagging surface effects on the longer term.

Pressurization tests in (small) sodium chloride test caverns have indicated that leak-off (a stabilization of pressure and continued injection) occurs around lithostatic pressure, but that no macro-scale fracture is created that would cause a massive brine outflow and large pressure drop. Such a macro-scale fracture was seen in 2018 in the Nedmag cavern system for $MgCl_2$ production where about 100 000 m^3 of brine leaked off before stopping due to suspension of the fracture at brine levels of about 60 bars less than original. This cavern system is still in operation and operating normally. Pressure levels will however not be increased anymore until final abandonment.

The general opinion is that all (or most) of the brine will leak from the salt layers (towards some overlying aquifer) in time. The time frame will, however, differ from cavern to cavern and from salt structure to salt structure. Shallow caverns in relatively cold salt will take thousands of years (or more) to leak-off half of the remaining brine, assuming the brine pressure remains close to lithostatic levels. For the deeper (warmer) salt and more creep-prone magnesium salts, the time frame may be centuries. Combined with political and public concerns that subsidence may occur at faster rates, or that fracture processes speed up the process potentially long after the mining operator left the mining site, the present vision on deep and magnesium salt mining is to allow salt creep to occur at a reduced pressure, in order to reduce the cavern volumes to the practical minimum before abandoning the well and cavern. This will create additional subsidence (and associated costs) at an early phase, but at least the consequences are clear and measurable during a phase when the operator is still mining or accountable.

Environmental impacts

The environmental effects of salt mining are largely comparable to the mining effects of oil and gas operations in the Netherlands, in the sense of creating subsidence, noise levels from drilling and production processes, saline wastewater volumes, the requirement for pipelines and cables (usually requiring cooperation from landowners, both private and public) and the (temporal) change of mostly agricultural land use into mining use, hence changing the landscape to a limited extent.

Subsidence

Salt mining leads to subsidence due to the (mainly) viscous flow of salt towards the caverns. The cavern fluid pressure is always lower than the lithostatic stress level to prevent leakage of fluids by permeation or fracturing, but in most it is kept lower for practical reasons. Solution mining for halite usually takes place at the lowest possible injection pressure and is mainly dictated by friction in the injection and production pipelines, including the surface pipelines for production towards the brine tanks. In case of fast flowing salt due to elevated depth and temperature or the presence of magnesium salts, a higher pressure may be preferred by choking the production line. This is at the expense of higher energy costs for injection, but it can reduce the salt creep. The shallow caverns near Hengelo have negligible salt creep rates at halmostatic fluid pressure. Exceptions are some caverns predating 1950, which were basically overmined, leaving too little salt between the caverns and hence loading the remaining salt (sometimes called pillars from analogy with conventional mining). The overburden load rests on the pillars resulting in creep at measurable rates. Subsidence is a few mm per year at these locations. A few caverns were also overmined at the roof, leaving the clay overburden exposed to the brine, resulting in roof-falls into the caverns. One cavern created a sinkhole at surface. Another cavern caused a local depression (subsidence of about one metre). Around 40 other caverns are earmarked to have the potential to have significant roof falls and creating significant additional subsidence on top of salt-creep induced subsidence. These caverns are monitored regularly. They are being backfilled with sulphates and carbonates from the brine purification plant. The sequence of the caverns selected for backfilling is determined by risk level.

For mining from the salt domes (Zuidwending and Winschoten by Nobian) and the storage of liquids, the pressure in caverns is usually kept close to brine hydrostatic (halmostatic) levels, whereas storage is usually in shallow caverns with limited salt creep. Storage of gasses at Zuidwending and Winschoten (natural gas, nitrogen and most likely also hydrogen and compressed air in the future) is usually performed at pressures that are dictated by demand. In high demand the pressures are low (down to 10-30% of lithostatic), whereas in low demand (or high production), the gas pressure can be as high as 85% of lithostatic (depending on permits, pump and pipeline ratings and safety analyses). The salt creep related cavern convergence mainly takes place during periods of low cavern pressure. Typical cavern convergence rates for the gas storage caverns in Zuidwending (Veendam) are 0.5%-1% per year for caverns with volumes of about 1 million m^3. The subsidence at both Zuidwending and Winschoten domal cavern sites amounts to a few mm per year at the well head centre, with a bowl extending to about 2 km radius. The subsidence so far (in 2022) is about 8 cm since start production for salt production and gas storage combined.

The Frisia caverns converge rapidly, so that a maximum volume is reached during solution mining. The caverns are at 2500 m depth and 100°C salt temperature. This makes viscosity low, combined with large difference between lithostatic stress and cavern pressure. The subsidence on land from 5 caverns amounts to some 30 cm, with partially overlapping bowls with a radius of about 4 km each. At present salt extraction is shifted to a cavern below the Wadden Sea. Subsidence will be measured with subsurface anchors and two GPS stations above water (on poles). Only at the sea dike and the harbour site direct measurement is possible (via levelling, InSAR and GPS), but only 5% or less of the maximum subsidence is expected on some of these locations. The subsidence bowl, although expected to fill itself by natural sand transport, is expected to reach a total of ca. 1 metre in the decades to come (Fig. 16.11).

For the caverns on the Nedmag concession, the bischofite salt causes 80-90% of the cavern convergence, with the bischofite at 70°C, so close to its melting (self-dissolution) point of about 110°C. The caverns are kept under artificial pressure to limit the salt creep and cavern convergence during the active water injection stage in order to maximize the retention time of the brine and maximize the exposure time to bischofite, maximizing the purity of $MgCl_2$ brine (minimizing levels of SO_4, K and Na in the brine). In doing so, several of the original 1970-1980s caverns have interconnected to form a single labyrinth type of cavern (Visser, 2018). The subsidence amounts to 1-3 cm per year (depending on the cavern pressure and volume), reached about 60 cm by 2024 and is projected to increase to about 90 cm by 2045 (Fig. 16.12). The radius of the total bowl is about 3 km. New wells are expected to be drilled at a distance of about 2 km from the heart of the existing bowl, so that the deepest point of the existing bowl is hardly affected by the new mining activities.

Developments on subsidence (land surface) measuring

Traditionally the subsidence is being measured by bi-annual or even less frequent levelling campaigns, where hundreds of 'solid' points anchored on shallow foundation poles in the land or bolted to well-founded houses or civil structures are manually measured. Advantage of the method is that shallow surface effects, that are influenced by fluctuating water saturation and ground water levels or ploughing of the fields can be discarded. Effect of water extraction from shallow wells will still be visible, however. The downside is that short term effects are not seen and the levelling campaign is expensive (and time consuming, usually taking a month or two).

GPS beacons allow daily measurements from a satellite, the height being compared with other GPS beacons (of other operators or from governmental organizations). The GPS is inaccurate on a daily basis due to fluctuations in the atmosphere, but on a monthly scale can be accurate to within a few mm. The downside is that a GPS station is

Land-surface monitoring from space

1

InSAR measures the altitude changes of hard surfaces, like roads or roofs. Soft or non-constant reflectors (like farmland) are usually filtered out. Normally, there are sufficient roads or farmhouses to give a reasonable impression of subsidence, such that the resolution in built-up areas (towns) is very good. InSAR is relatively cheap, including the semi-manual interpretation by a specialized company. It can give data on (approximately) a monthly basis and can also give data in retrospect (from about the year 2016 onwards). The downside is that road and roof surfaces react somewhat to groundwater levels and air temperature, so that there is slightly more seasonal fluctuation compared to levelling. If subsidence is several mm per year, this inaccuracy is small over the period of two (or more) years, assuming that the natural compaction of the top 1-10 m of the subsurface is small, like in most places in Groningen and Friesland, where most peat layers have been removed (or have been eroded) in the past centuries.

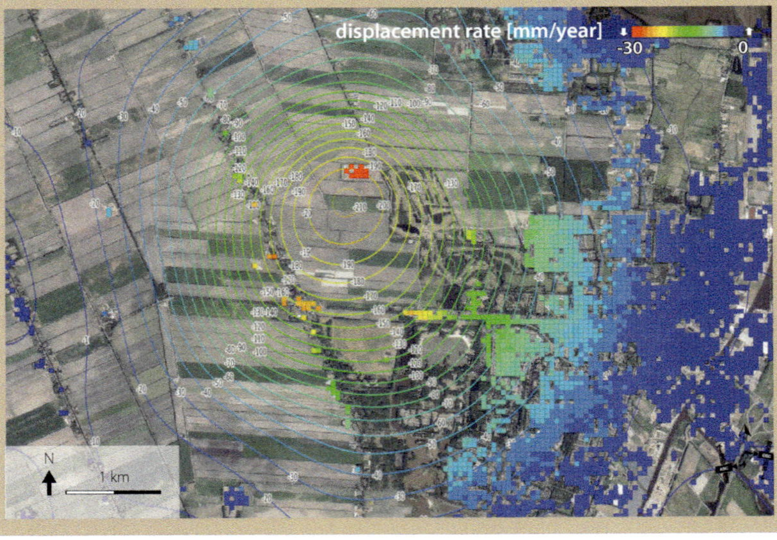

Example of a contour map of subsidence (concentric coloured lines) based on InSAR data (coloured squares).

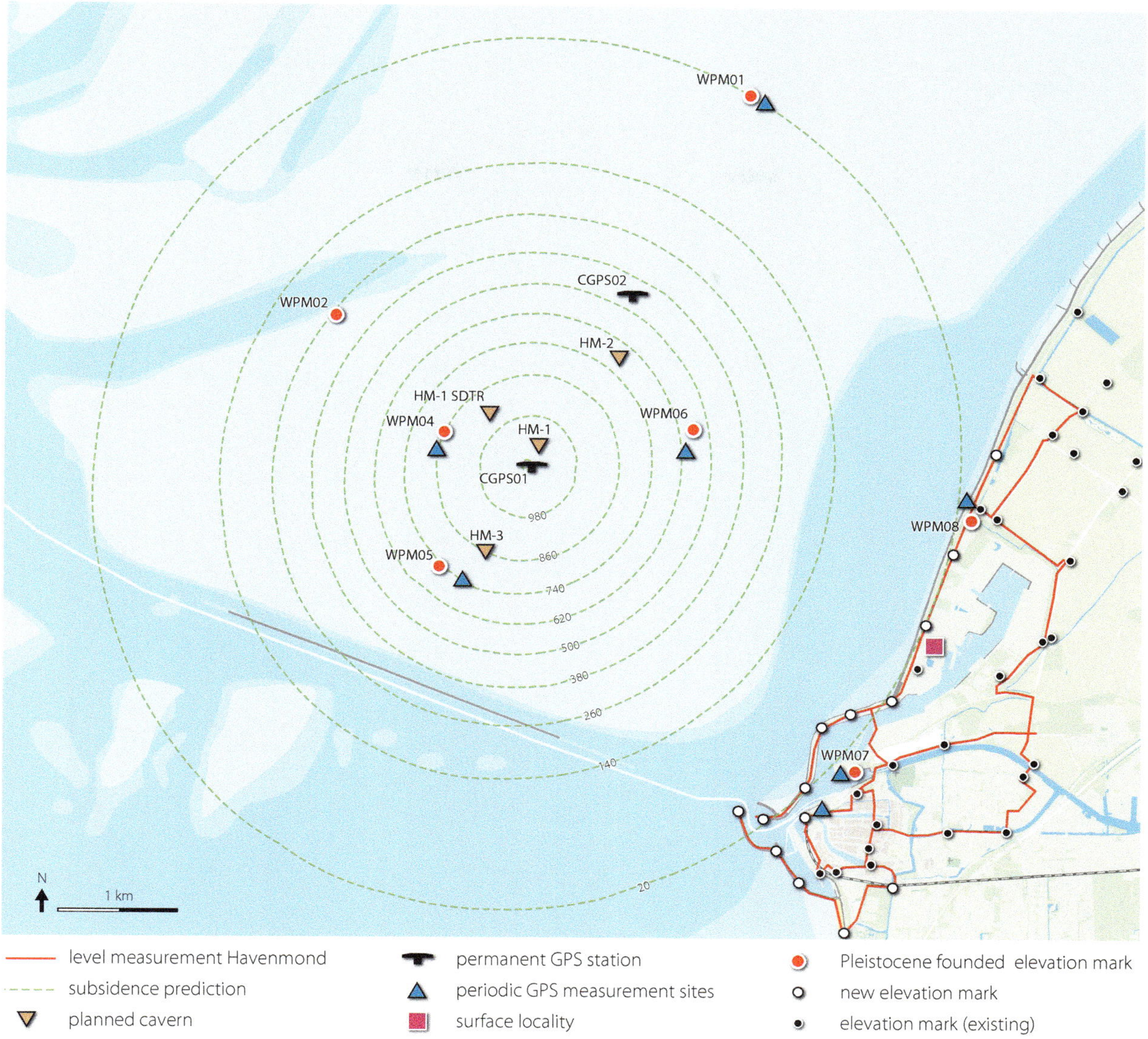

***Figure 16.11.** Frisia Wadden Sea subsidence bowl prediction (in mm contours) for 2050.*

costly and requires electric power and maintenance. Usually, the GPS is placed at a mining site or on a large structure (office building, church, etc).

A relatively new development is the use of InSAR (Interferometric Synthetic Aperture Radar) whereby satellites measure the earth surface altitude (or more accurately it's relative change in altitude in time with respect to a reference point). So far, in salt mining it has not replaced levelling but is seen as complementary.

Damage to infrastructure, farmland or environment due to salt mining subsidence

Mining related subsidence, with the exception of sinkholes, is usually very gradual and it is generally assumed that infrastructure (houses, offices, pipelines, viaducts, roads) are not damaged by the subsidence itself (including bending, tilting and horizontal strain that is associated with the subsidence; Geurts et al., 2021). Indirect damage may be possible if the groundwater table is not properly or not timely adjusted in cases where tens of centimetres of subsidence are involved. The water table adjustments are usually the prerogative of the Water Boards (and within city limits sometimes the municipality). So far, water tables could be adjusted properly at all of the salt mining locations in the Netherlands.

Earthquake damage by gas extraction from the Groningen gas field has changed the legislation for this specific field, so that damage that could have resulted from earthquakes or (uneven) subsidence is assumed to be related to gas production, unless there is proof of a different cause for the damage. The salt extraction fields of Veendam, Zuidwending and Winschoten overlap with the Groningen gas field (effect zone), but formally do not have the burden of

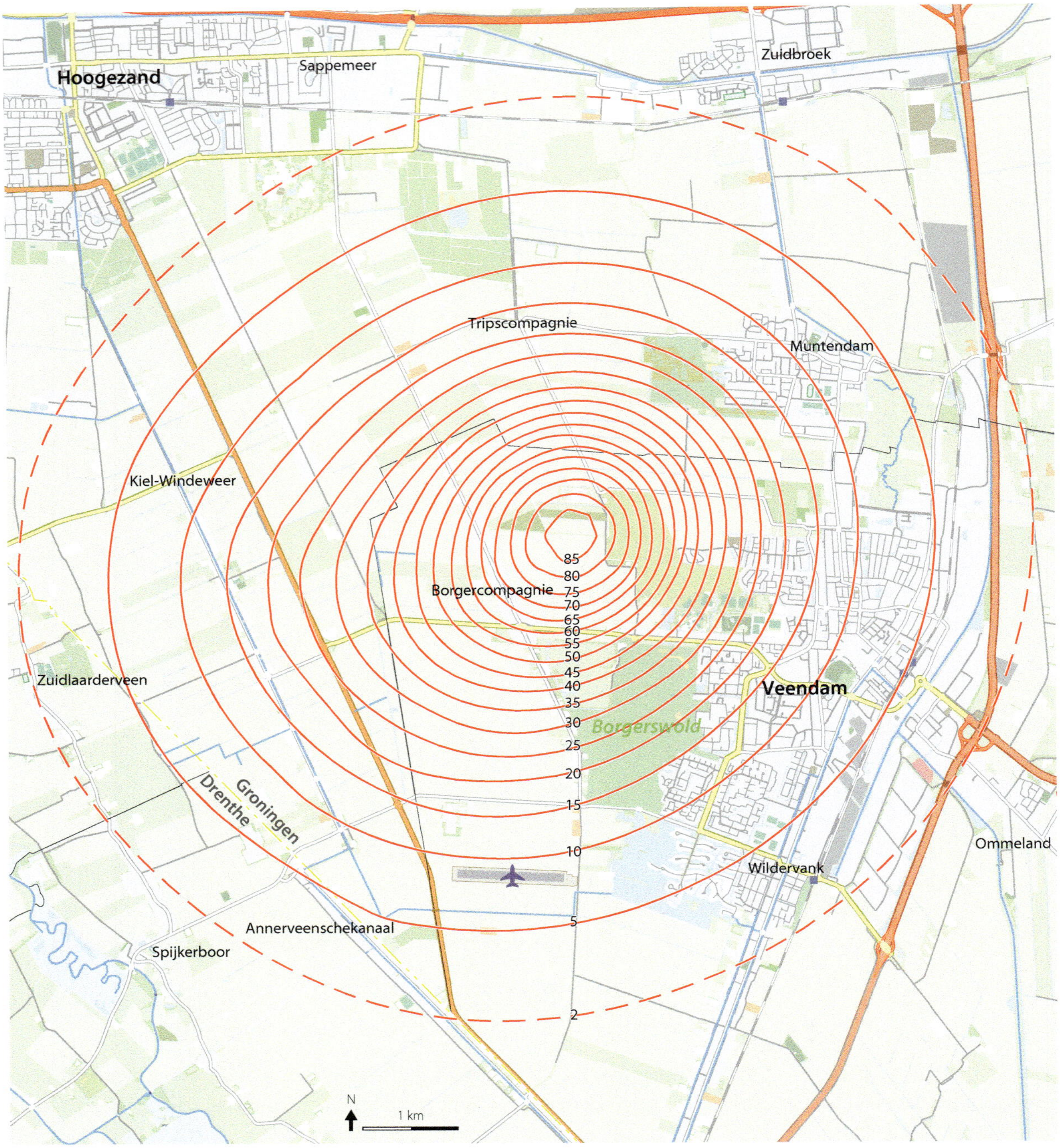

Figure 16.12. Nedmag subsidence bowl in cm, contours as expected in 2050, according to extraction plan 2018.

proof. It is not sure how this will develop in law and legislature in the future, after the closure of the Groningen gasfield (2023), and also with respect to the Annerveen gas field (which overlaps with the Veendam and Zuidwending subsidence fields).

For the mine sites close to the Wadden Sea (only Frisia for salt extraction), subsidence (as sea-level rise) may cause a slight increase in salt sea water volumes that seep in (Dutch: *kwel*) and makes agriculture more difficult or less diverse. As of 2021 effects appear not to be severe (at 35 cm maximum subsidence).

Subsidence under the Wadden Sea is much more debated, from what appears to be a political-environmental point of view. Salt extraction was moved here, because of public concerns about subsidence on land, but environmental groups that wish to protect the Wadden Sea (Natura 2000 nature reserve) have objected to any mining at all below the Wadden Sea. Environmental studies have shown that the maximum subsidence itself is not as important as the subsidence bowl volumes per year. The bowl will attract sand from the normal sand and silt flux that moves (starting at the river influx from the Belgian and Dutch rivers Scheldt, Rhine, Meuse) via the coast northwards, partially via the Wadden Sea. Too much subsidence (from salt and gas extraction) may cause a disruption of the natural sand fluxes, by which the low tide islands in the Wadden Sea may no longer (or to a lesser extent) be available for seals to rest and birds to feed. Although the expected 'sand-hunger' is much less than the sand flux, negative effects cannot be ruled out entirely. Monitoring is hence required for the wild life. Due to the highly fluctuating circumstances in the Wadden Sea (especially after storms with a lot of sand and silt displacement) wildlife populations fluctuate considerably, so that it is not easy to distinguish natural effects from mining effects. Sand suppletion via dredging in deeper North Sea waters and deposition in the Wadden Sea can in principle supplement shortages in sand, when it is believed that (temporarily) too much sand is trapped in subsidence bowls.

Seismicity

In contrast to gas extraction, earthquakes (anything above M=2.5) are unknown to the solution mining industry, most likely due to the viscous behaviour of salt and the limited associated shear stresses. Some light tremors (the largest measured M=1.3 near the Winschoten salt dome (Fig. 16.4; KNMI, 2022) have occurred around salt extraction areas. Two possible causes have been identified, the first being movement along existing fault planes above or next to the salt structure, possibly induced by the salt mining. The second cause could be related to collapse of parts of the cavern roof. Given the public concern from mainly the Groningen field earthquakes (mainly those exceeding M=2), all the salt mining operators have installed micro-seismic monitors in shallow wells in order to be able to measure and (more precisely) locate smaller tremors. An example is the Zuidwending micro-seismic monitoring network installed by Nobian and EnergyStock to measure and locate micro-seismic activity in the vicinity of the salt mining and gas storage concessions in Zuidwending (see Fig. 16.13). Certainly, for Groningen and Friesland this also allows for a (better) discrimination of the most likely trigger of the earthquakes: salt extraction or gas extraction. No major tremors or earthquakes (M>2.0) are known to have been caused by solution mining globally. For the minor tremors, there is far less evidence, as they may not have been noticed at the ground surface, perhaps have not been published in literature or press, or have not been measured beyond noise level by national seismic stations at large distances from the mining operation site.

Environmental footprint

The indirect effect of solution mining is that it requires energy to remove water by evaporation in a vacuum evaporation plant, assuming that solid salt is required for trade

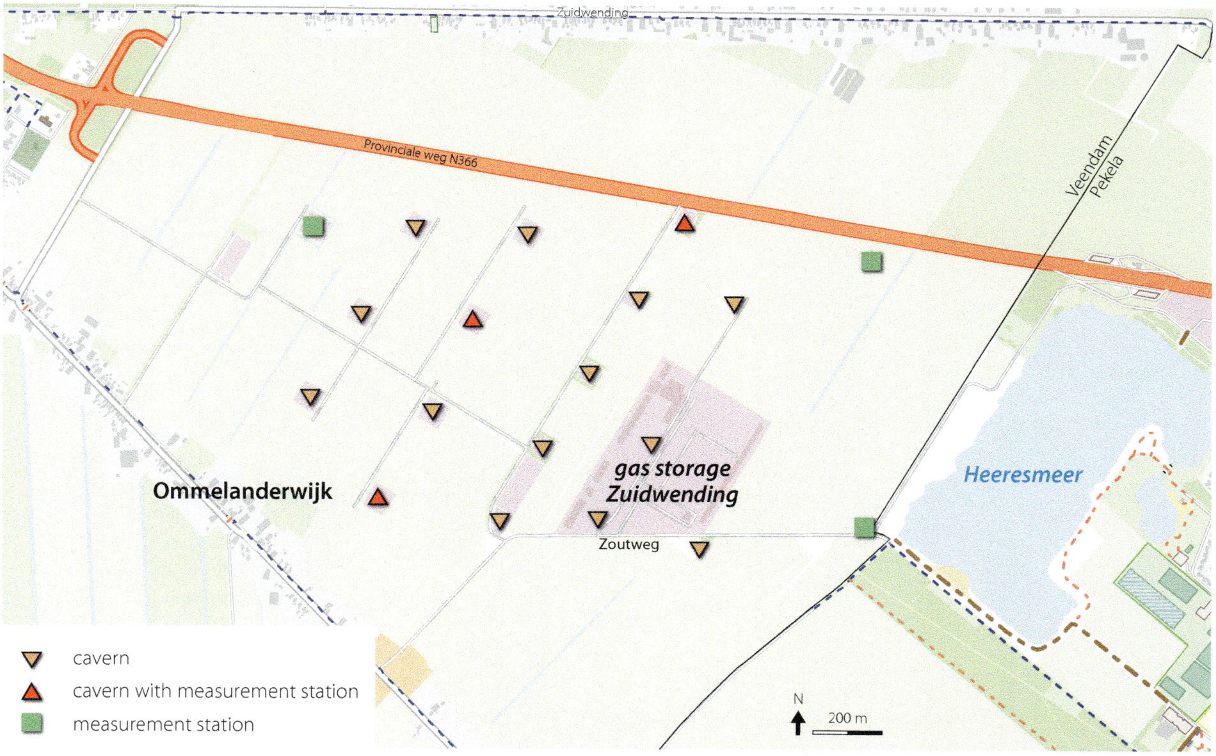

Figure 16.13. Micro-seismic monitoring network to measure and locate small tremors in the vicinity of the salt mining and gas storage concessions in Zuidwending (Nobian, personal communication).

and transport to external clients. As conventional (excavated) salt is less pure, it may require higher purification costs or energy consumption from the client, usually a chemical plant. Which of the two salt mining methods results in a lower CO_2 footprint per ton of salt or final product is unknown at present. Part of brine is sometimes used as brine without having the intermediate step of creating salt, usually in chemical plants that are associated with the mining company, but only if brine transport is possible by pipeline. In that case, brine is mined, purified and consumed by, for example, a soda ash plant. More than 95% of the $MgCl_2$ brine of Nedmag is processed as brine. Only 5% or less is sold as bischofite salt in the shape of flakes or pellets.

The evaporation cycles of salt use some 90-95% of the inflowing brine to produce commercial salt. Some brine is purged (bled off) due to rising levels of bromine and potassium in the last stage of evaporation, which may off-set quality of salt for some or most of the clients. These brines are allowed to be disposed of in the sea at the Harlingen and Delfzijl salt plants or are being fed into existing caverns again (Hengelo salt plant). Prior to evaporation, the levels of magnesium and sulphates are reduced by precipitation of gypsum and magnesium carbonates. These solids are separated and usually disposed of or reinjected into the cavern system, where they precipitate. The solids are usually too impure and of too little volume to be used commercially.

From brine to salt

2

In a vacuum evaporation plant, water is evaporated from the saturated brine in multiple stages to obtain pure solid salt. The evaporation stages are usually four to five repeated cycles in the vacuum evaporation plant, where the vapour of one stage is used to heat the next boiler (stage) that holds a lower vapour pressure. This subsequent boiler has a lower boiling point of the brine, ending at close to vacuum pressure in the last boiler (hence the name vacuum salt). In this process the evaporation energy is therefore reused three to four times. The energy source for evaporation is dry steam from steam boilers. The boilers are heated with gas, but also alternative sources of energy are being used. Frisia Salt is using energy from the neighbouring waste incineration plant (Omrin Harlingen), Nedmag is partly using hydrogen and biogas for its processes, while Nobian is using steam generated by Twence (Hengelo salt plant) and EEW and Eneco (Delfzijl salt plant). Excavated salt from conventional mining contains much more insolubles, like anhydrite, clay and potassium, that cannot be removed to a large percentage (without dissolving and re-crystallization by evaporation). The impurities create additional costs, energy or waste streams with the client chemical plants, for which reasons they usually choose vacuum salt.

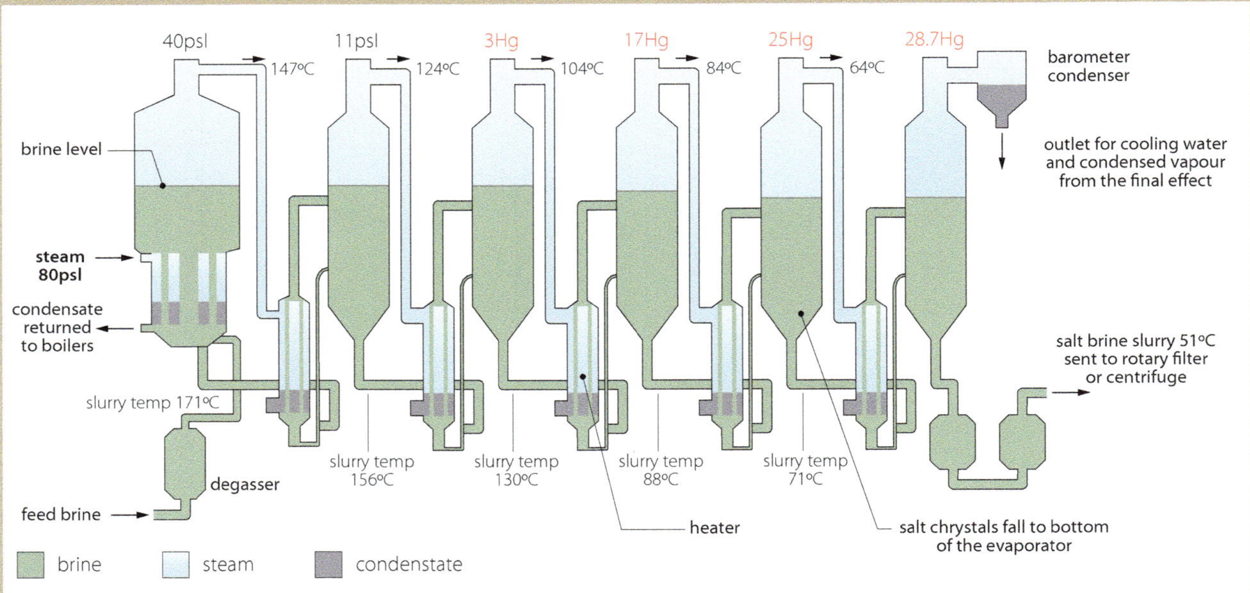

Schematic diagram of the multi-stage vacuum evaporation process to extract pure salt from saturated brine. Redrawn after a figure on the website of the Salt Association (UK).

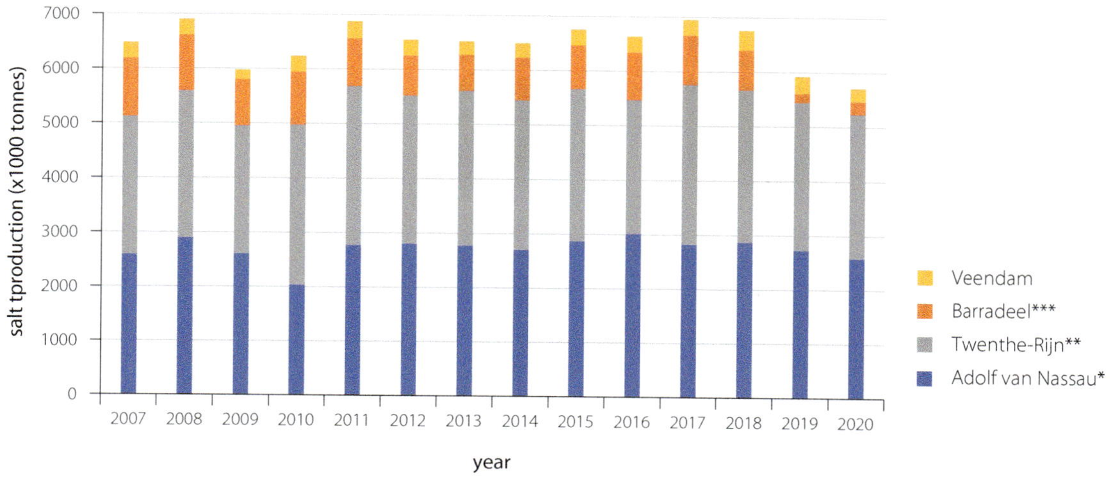

*incl. concession Uitbreiding Adolf van Nassau III
**incl. concessions Uitbreiding Twenthe-Rijn, Twenthe-Rijn Helmerzijde and Twenthe-Rijn Oude Maten
***incl. concessions Barradeel II and Havenmond

Figure 16.14. *Salt production figures 2007-2020 (reproduced from MEACP, 2022).*

Economic aspects

Approximately 70% of the Dutch NaCl production of 5.46 Mt/yr (2020, see Table 16.1) is used as feedstock for the production of chlorine in the chemical industry, which is indispensable for the manufacturing of PVC, textiles, aluminium, soap and detergents. Other uses of salt include applications in food and beverages, water treatment, licks for livestock, tanneries and road de-icing. Apart from chemical pureness and constant high quality, the cost of transportation is an important issue for most end-users. This is the reason why salt plants usually have direct access to navigable water.

About 0.25-0.30 Mt/yr of $MgCl_2$ is produced, of which the main part is used as brine for the production of high-grade magnesium oxide (MgO) and magnesium hydroxide ($Mg(OH)_2$). The remainder is sold as brine or salt and transported by ship or by road. The magnesium oxide is sold and used for the production of heat-resistant bricks, which in turn are used in iron and cement furnaces. The brine is used in many applications in the cement industry, and in manure treatment, but also as de-dusting and de-icing material. The process of creating MgO with $MgCl_2$ and Dolime (burnt dolomite) $MgCaO_2$, also results in a $CaCl_2$ brine stream that is sold as brine or processed towards solid salt.

Figure 16.14 gives an overview of the salt production (NaCl and $MgCl_2$) in the period 2007-2020, based on figures provided by operators. During this period the salt production has been fairly constant, between 6 and 7 million tons per year. In 2009, 2019 and 2020, the total salt production fell to slightly under 6 million tons. In 2009 and in 2020 this was due to the economic crisis and the Corona Pandemic respectively, while in 2019, the main reason was that the production from the Barradeel production licenses decreased compared to previous years (MEACP, 2021).

Acknowledgements

The authors like to acknowledge Alexander Nagelhout and Abel Jan Smit for their feedback on the initial manuscript, Tjeerd Koopmans and Renaud Bouroullec for their reviews of the manuscript, and Isis van Wetten and Alma Hollen for support in obtaining production data and creating visual content.

Digital map data

Spatial data of figures in this chapter for use in geographical information systems can be downloaded here: https://doi.org/10.5117/aup.28164287.

References

Balzer, D., 2000a. Analysis and interpretation of saliniferous Zechstein structures (Upper Permian) in subrosive facies – examples from the Subhercynian basin (Germany). In: Geertman, R.M. (ed.): Proceedings of the 8th World Salt Symposium. Elsevier (Amsterdam), Vol. 1: 66-71.

Balzer, D., 2000b. Lithostratographie, Fazies, Strukturbau und subrosive Entwicklung des Hutgesteins über der Allertal-Salzstruktur zwischen Alleringersleben und Beendorf (Sachsen-Anhalt, Bundesrepublik Deutschland). Geologisches

Jahrbuch A 154: 3-85.
Balzer, D., 2001. Geologische Interpretation von Salinar (Zechstein) und Hutgesteinsbohrungen im Bereich der Kaverne Schönebeck. Glückauf-Forschungshefte 137: 122-131.
Batsche, H. & Klarr, K., 1980. Beobachtungen und Gedanken zur Gipshutgenese. Fifth International Symposium on Salt (Hamburg), Vol.1: 9-19.
Bentz, A., 1947. Geotektonische Karte von Nordwestdeutschland. Amt für Bodenforschung (Hannover-Celle).
Bornemann, O., 1991. Zur Geologie des Salzstocks Gorleben nach Bohrergebnissen. Bundesamt für Strahlenschutz Schriften 4/91 (Salzgitter): 67 pp.
Bouroullec, R. & Ten Veen, J.H., 2025. Salt Tectonics. In: Ten Veen, J.H., Vis, G.-J., De Jager, J. & Wong, Th.E. (eds): Geology of the Netherlands, second edition. Amsterdam University Press (Amsterdam): 457-491. DOI: 10.5117/9789463728362_ch12
Coelewij, P.A.J., Haug, G.M.W. & Van Kuijk, H., 1978. Magnesium-salt exploration in the northeastern Netherlands. Geologie En Mijnbouw 57: 487-502.
De Boer, H.U., 1971. Gefügeregelung in Salzstöcken und ihren Hüllgesteinen. Kali Und Steinsalz 5: 403-425.
Geluk, M.C., 1995. Stratigraphische Gliederung der Z2-(Staßfurt-) Salzfolge in den Niederlanden: Beschreibung und Anwendung bei der Interpretation von halokinetisch gestörten Sequenzen. Zeitschrift Der Deutschen Geologischen Gesellschaft 146: 458-465.
Geluk, M.C., 1998. Internal tectonics of salt structures. Journal of Seismic Exploration 7: 237-251.
Geluk, M.C., 2000. Steps towards successful prediction of the internal tectonics of salt structures. In: Geertman, R.M. (ed.): Proceedings of the 8th World Salt Symposium. Elsevier (Amsterdam): 125-130.
Geurts, C.P.W., Pluymaekers, M.P.D. & Rots, J.G., 2021. Schade aan gebouwen door diepe bodemdaling en -stijging, TNO, Report No. 2021 R10325B
Giesen, R. & Mesdag, C., 1995. De zoutkoepel van K9. Grondboor & Hamer 49 (1): 11-16.
Harsveldt, H.M., 1980. Salt resources in The Netherlands as surveyed mainly by AKZO. Proceedings of the 5th Information Symposium on Salt 1. Northern Ohio Geological Society: 65-81.
Harsveldt, H.M., 1986. The Netherlands. In: Dunning, F.W. & Evans, A.M. (eds): Mineral deposits of Europe. The Institute of Mining and Metallurgy and The Mineralological Society (London) 3: 113-116.
Jackson, M.P.A., Cornelius, R.R., Craig, C.H., Gansser, A., Stöcklin, J. & Talbot, C.J., 1990. Salt diapirs of the Great Kavir, Central Iran. Geolological Society of America Memoir 177: 139 pp.
Jeremic, M.L., 1994. Rock mechanics in Salt Mining. CRC Press (Boca Raton): 532 pp. DOI: 10.1201/9781003077589
Juez-Larré, J., Groenenberg, R., Dinkelman, D., Huijskes, T., Koornneef, J.M., Van Gessel, S.F. & Bos, C.F.M., 2025. Underground storage. In: Ten Veen, J.H., Vis, G.-J., De Jager, J. & Wong, Th.E. (eds): Geology of the Netherlands, second edition. Amsterdam University Press (Amsterdam): 729-767. DOI: 10.5117/9789463728362_ch20
Kent, P.E., 1979. The emergent Hormuz salt plugs of southern Iran. Journal of Petroleum Geology 2: 117-144.
KNMI, 2022. Aarbevingscatalogus website. https://www.knmi.nl/kennis-en-datacentrum/dataset/aardbevingscatalogus
Kockel, F., 1990. Morphology and genesis of northwest German salt structures. Proceedings of the Symposium of Diapirism with Special Reference to Iran. Geological Survey of Iran (Tehran), 2: 229-249.
Kupfer, D., 1976. Shear zones inside Gulf Coast salt stocks help to delineate spines of movements. American Association of Petroleum Geologists Bulletin 60: 1434-1447.
Lotze, F., 1957. Steinsalz und Kalisalze, I. Teil. Bornträger (Berlin): 465 pp.
McKie, T. & Kilhams, B., 2025. Triassic. In: Ten Veen, J.H., Vis, G.-J., De Jager, J. & Wong, Th.E. (eds): Geology of the Netherlands, second edition. Amsterdam University Press (Amsterdam): 155-183. DOI: 10.5117/9789463728362_ch05
MEACP., 2021. Natural resources and geothermal energy in the Netherlands 2020. Annual Review Ministry of Economic Affairs and Climate Policy (The Hague): 175 pp.
MEACP, 2022. Natural resources and geothermal energy in the Netherlands 2021. Annual Review Ministry of Economic Affairs and Climate Policy (The Hague). Ministry of Economic Affairs and Climate Policy (The Hague): 159 pp. https://www.nlog.nl/en/archive
Mulder, A.J., 1950. De zoutpijler van Schoonlo. Geologie En Mijnbouw 12 (6): 169-176.
Remmelts, G., 1995. Fault-related salt tectonics in the southern North Sea, the Netherlands. In: Jackson, M.P.A., Roberts, D.G. & Snelson, S. (eds): Salt Tectonics: a Global Perspective. American Association of Petroleum Geologists Memoir 65: 261-272.
RGD, 1995. Geological Atlas of the subsurface of The Netherlands (1 : 250 000). Explanation to map sheet III Rotumeroog-Groningen. Rijks Geologische Dienst (Haarlem): 113 pp.
Richter-Bernburg, G., 1972. Saline deposits in Germany: a review and general introduction to the excursion. Hannover: Unesco: 275-287.
Richter-Bernburg, G., 1980. Interior structures of salt bodies. Bulletin des Centres de Recherches Exploration-Production elf aquitaine 4: 373-389.
Schachl, E., 1987. Kali- und Steinsalzbergwerk Niedersachsen-Riedel der Kali und Salz AG, Schachtanlage Riedel – Zechsteinstratigraphie und Innenbau des Salzstockes von Wathlingen Hänigsen. Internationales Symposium Zechstein 1987, Exkursionsführer I (Hannover/Kassel): 69-100.
Trusheim, F., 1963. Mechanism of salt migration in Northern Germany. American Association of Petroleum Geologists Bulletin 44: 1519-1540.
Van Waterschoot van der Gracht, W.A.J.M., 1909. The deeper geology of the Netherlands and adjacent regions, with special

reference to the latest borings in the Netherlands, Belgium and Westphalia. In: Memoir Government Institute for the Geological Exploration of the Netherlands 2. (The Hague): 437 pp.
Van Waterschoot van der Gracht, W.A.J.M., 1918. Eindverslag over de onderzoekingen en uitkomsten van den Dienst der Rijksopsporing van Delfstoffen in Nederland 1903-1916. Martinus Nijhoff ('s Gravenhage): 664 pp.
Verreussel, R.M.C.H., Bouroullec, R., Munsterman, D.K., Dybkjær, K., Geel, C.R., Houben, A.J.P., Johannessen, P.N. & Kerstholt-Boegehold, S.J., 2018. Stepwise basin evolution of the Middle Jurassic–Early Cretaceous rift phase in the Central Graben area of Denmark, Germany and The Netherlands. In: Kilhams, B., Kukla, P.A., Mazur, S., McKie, T., Mijnlieff, H.F. & Van Oijk, K. (eds): Mesozoic resource potential in the Southern Permian Basin. Geological Society of London, Special Publications 469: 305-340. DOI: 10.1144/SP469.23
Visser, J., 2018. Hydraulic connections between Nedmag caverns Underlying mechanism, chance of new connections and possible consequences.
Wagner, G., Mauthe, F. & Mensink, H., 1971. Der Salzstock von Cardona in Nordspanien. Geologische Rundschau 60: 970-996.
Wassmann, T.H., 1980. Mining subsidence in Twente, East Netherlands. Geologie En Mijnbouw 59 (3): 225-231.
Wassmann, T.H. & Brouwer, M.S., 1987. The mining of Rock Salt. In: Visser, W.A., Zonneveld, J.I.S. & Van Loon, A.J. (eds): Seventy-five years of geology and mining in the Netherlands (1912-1987). Royal Geological and Mining Society of the Netherlands (KNGMG, The Hague): 137-146.
Wassmann, Th.H., 1983. Cavity Utilization in the Netherlands. Proceedings Sixth International Salt Symposium. Salt Institute (Alexandria, Virginia) II: 191-201.
Wassmann, Th.H., 1993. Mining Subsidence above Cavities Created by Solution Mining of Rock Salt. Proceedings Seventh International Salt Symposium. Elsevier (Amsterdam): 425-431.

17

Hydrogeology

Hanneke Verweij, Gualbert Oude Essink, Niels Hartog & Willem Jan Zaadnoordijk*

*Corresponding author: hanneke.verweij@outlook.com

ABSTRACT

Groundwater is a crucial source of fresh water for domestic, agricultural and industrial uses, and for maintaining aquatic- and groundwater-dependent ecosystems. It is also of importance for identifying and securing sustainable use of energy resources and subsurface storage sites. Groundwater interacts with different parts of system earth, including the atmosphere, surface water, soil, geological environment and with many geological processes. This chapter provides an overview of groundwater systems in the Netherlands and the Dutch continental shelf at depths from the surface down to about 5 km. The overview shows how the groundwater systems are shaped by the interplay between natural mechanisms operating on various geological timescales and the impact of anthropogenic activities. Important mechanisms involved in the development of groundwater flow systems include differences in the elevation of the groundwater table (its topographic relief), contrasts in groundwater density and hydromechanical interaction of groundwater with geologic media. The topography-driven groundwater flow systems in the coastal dunes, Pleistocene ice-pushed ridges, and the southeastern part of the country contain important fresh groundwater resources of meteoric origin. These resources occur largely in unconsolidated sedimentary sequences of Holocene and Pleistocene to Neogene age. Natural and anthropogenic factors explain the Holocene history of salinization and seepage in the coastal zone. The large transboundary topography-driven groundwater flow system in the southeast of the Netherlands has developed since Miocene times. It induced freshening of groundwater to relatively great depths and cooling of subsurface temperatures. Case studies show the effects of shallow and deep fault zones on flow and chemical conditions of groundwater. Groundwater in older, pre-Paleogene to Carboniferous units outside the realm of topography-driven flow mostly consists of highly saline brines. Groundwater in these units also shows high overpressures in the northern offshore and northern and northeastern part of the Netherlands, while close to hydrostatic pressures prevail in the southern onshore and adjacent offshore area. This spatial difference reflects the differences in burial history and hydrogeological framework.

<< Infiltration pond in the coastal dunes (Meijendel, The Hague). Groundwater abstraction is supported by artificial recharge through infiltration ponds supplemented with pre-treated Meuse River water. Photo: Hanneke Verweij.

DOI: 10.5117/9789463728362_ch17

Introduction

Hydrogeology concerns the occurrence, movement, and properties of groundwater, its mechanical, chemical and thermal interaction with sediments and porous and/or fractured rocks (geological environment) and with other pore fluids, as well as the transport of energy, solutes and particulate matter by moving groundwater. The flow of groundwater, transport of solutes, transport of heat and rock deformation are coupled processes (Person et al., 1996; Ingebritsen et al., 2006). Groundwater is the subsurface part of the hydrosphere. It includes all subsurface aqueous fluids, such as those of meteoric, marine, syn-sedimentary, or even magmatic origin (Ingebritsen et al., 2006). Groundwater is interconnected and interacts with surface, near-surface and above surface components of the hydrosphere (such as water stored in rivers, lakes, seas, soils, and atmosphere) and with the biosphere. In addition, anthropogenic activities may impact groundwater to a greater or lesser extent. Groundwater is present in the pores and fractures of the entire sedimentary sequence of the onshore and offshore Netherlands. The depth range of groundwater considered here extends from the land surface down to about 5 km, i.e. from Quaternary to Carboniferous stratigraphic units (Table 17.1).

The development of hydrogeological research in the Netherlands since the 19th century has largely been driven by the ever-expanding societal needs to solve practical groundwater-related problems in the densely populated Dutch delta with its polders below sea level, its strong interactions between surface water, groundwater and the sea, and the various increasing impacts of anthropogenic activities on the groundwater system (De Vries, 1982, 2004, 2007). Especially the demand for fresh groundwater resources for domestic, agricultural and industrial uses, the need for land drainage and concerns about sea-water

Table 17.1. Generalized hydrostratigraphy of the Netherlands onshore and offshore (Fm = Formation, Mbr = Member).

Age (Ma)	Chronostratigraphy		Lithostratigraphy	Hydrostratigraphy	
				Aquifers/reservoirs	Aquitards/seals
23-0	Cenozoic	Quaternary Neogene	**Upper North Sea Group-NU**	Sand fms & mbrs	Clay fms & mbrs
66-23		Paleogene	**Middle & Lower North Sea groups-NM&NL** Dongen Fm Landen Fm	Brussels Sand Mbr Sand in southern area	Clay and Silt mbrs Clay in northern area
100.5-66	Mesozoic	Late Cretaceous	**Chalk Group-CK**	Post-inversion upper part Chalk (Ekofisk Fm, Houthem Fm)	Older deeply buried Chalk
145-100.5		Early Cretaceous	**Rijnland Group-KN**	Sandstone mbrs of Vlieland Sandstone Fm	Claystone mbrs of Vlieland Claystone Fm
163.5-145		Late Jurassic	**Schieland, Scruff, Niedersachsen groups** (SL, SG, SK)	Various sandstone mbrs	Various claystone and marl mbrs
201-174		Early Jurassic	**Altena Group-AT** Werkendam Fm Posidonia Shale Fm Sleen and Aalburg fms	Sandstone mbr	Claystone mbrs Organic-matter rich claystones Claystones
247-201		Middle & Late Triassic	**Upper Germanic Trias Group-RN** Keuper Fm Muschelkalk Fm Röt Fm Solling Fm	Basal Solling Sandstone Mbr	Claystone mbrs Marl, Claystone, Evaporite mbrs Evaporite and Claystone mbrs
252-247		Early Triassic	**Lower Germanic Trias Group-RB** Main Buntsandstein Subgroup Main Buntsandstein Subgroup Lower Buntsandstein Subgroup	Sandstones of Hardegsen Fm Sandstone mbrs Detfurth & Volpriehausen fms	Detfurth Claystone Mbr Main Claystone Fm
260-252	Paleozoic	Permian (Lopingian)	**Zechstein Group-ZE**	Carbonate mbrs	Salt mbrs
273-260		Permian (Guadalupian)	**Upper Rotliegend Group-RO**	Sandstone mbrs of Slochteren Fm	Clay-siltstone mbrs of Silverpit Fm Evaporite mbr of Silverpit Fm
323-299		Carboniferous-Pennsylvanian (Nam.-Westph.-Steph.)	**Limburg Group-DC**	Sandstone fms of Dinkel Subgroup	Claystone fms of Hunze and Geul subgroups Claystones and coal-rich fms of Caumer Subgroup
359-323		Carboniferous-Mississippian (Dinantian)	**Carboniferous Limestone Group-CL**	Carbonates of Zeeland Fm	

intrusion and salinization have always been important research topics at universities and knowledge institutes, as well as at water supply companies, consultancy firms, provinces and waterboards. Until recently, only relatively short time scales were taken into account in most hydrogeological research. In the last few decades, however, the scope of hydrogeological research has expanded in response to urgent questions concerning climate change and the need for the energy transition. The need for a transition to a more sustainable energy supply has induced growth in research and development to assess subsurface groundwater conditions in support of the efficient and safe supply of geothermal energy (Mijnlieff et al., 2025, this volume), storage of energy and CO_2 (Juez-Larré et al., 2025, this volume), disposal of nuclear waste (Neeft et al., 2025, this volume), use of aquifer thermal energy storage (ATES; at less than 500 m) and geothermal energy (at 1.5-3 km) (Mijnlieff et al., 2025, this volume). Note that geothermal resources, potential storage sites for CO_2 and energy, in addition to oil and gas and rock salt resources in the Dutch subsurface all occur well below the depths of fresh groundwater resources. The present-day multiple uses of the subsurface on-and offshore and their foreseen future growth, as well as environmental and societal concerns related to these activities, have increased the need for better knowledge and understanding of hydraulic properties of the subsurface and the physical and chemical groundwater conditions at greater depths than those conventionally studied. Hydraulic properties of the deeper subsurface as well as the current physical and chemical properties of the groundwater and associated flow conditions are increasingly seen to reflect their development in response to past and ongoing geological processes and water-rock interactions on geological timescales.

Important questions concerning the impact of climate change, related sea-level rise, and the energy transition not only require assessment and process-based knowledge and understanding of present-day hydrogeological conditions. It also requires forecasting of hydrogeological properties and flow conditions at different spatial and time scales.

The three major types of groundwater flow systems described in this chapter are the topography-driven system (also known as gravity-driven system), the hydromechanical-driven system (also known as compaction-driven system) and the variable density-driven system. The current characteristics of these groundwater systems are the result of the interplay between the natural evolution of the systems driven by processes operating on various geological timescales at thousands to tens and even hundred millions of years and the impact of anthropogenic activities driven by societal needs at short timescales of years to often not more than decades. Continued multiple short-time anthropogenic activities have led to important large-scale physical and chemical impacts on the natural groundwater systems and the occurrence of artificially managed groundwater systems.

The emphasis in this chapter is on highlighting the relationship between geology and hydrogeology at different time and spatial scales, incorporating the impact of past and present anthropogenic activities on the subsurface and its contained groundwater from the land surface to great depths. The main focus is on regional and larger spatial scales. The previous version of this chapter (De Vries, 2007) and overview books on groundwater of the Netherlands (Dufour, 1998, 2000) were limited to shallow groundwater systems down to depths of 100s of metres. The current overview has been expanded to include 'deep' hydrogeology down to depths of 5 km, including the hydrogeology of the Dutch offshore. The overview incorporates the growth in knowledge and understanding of hydrogeology based on the considerable increase in publicly available data and – quantitative – information on the sedimentary sequences and their contained pore water in the subsurface of both the Netherlands itself and the Dutch part of the continental shelf. The successive sections describe the development of hydrogeological knowledge and understanding, the present-day hydrogeological framework, groundwater flow systems and hydrochemistry of groundwater systems onshore, and deep hydrogeology. The term 'deep' hydrogeology is not very well defined and here we use it to describe groundwater systems located outside the onshore groundwater resources that are used and studied for fresh water supply and water management purposes.

The Appendix provides background information on general concepts and principles used in this chapter.

Development of hydrogeological knowledge and understanding

Anthropogenic interventions in landscape and water systems were undertaken many centuries before groundwater research and understanding started to develop in the 19th century. Since 1000 AD interventions that have had a major impact on current groundwater flow systems include those related to landscape and water systems for meeting societal needs, such as the need for suitable agricultural land, flood protection, reclamation of land, industry, urbanization and water supply (Van de Ven, 1993; Huisman, 2004; Vos, 2015; Koster, 2017). The continued draining of peat areas to make land habitable and suitable for agricultural purposes as well as for peat extraction for use as fuel and salt production have led to large-scale lowering of groundwater levels, compaction, and surface sub-

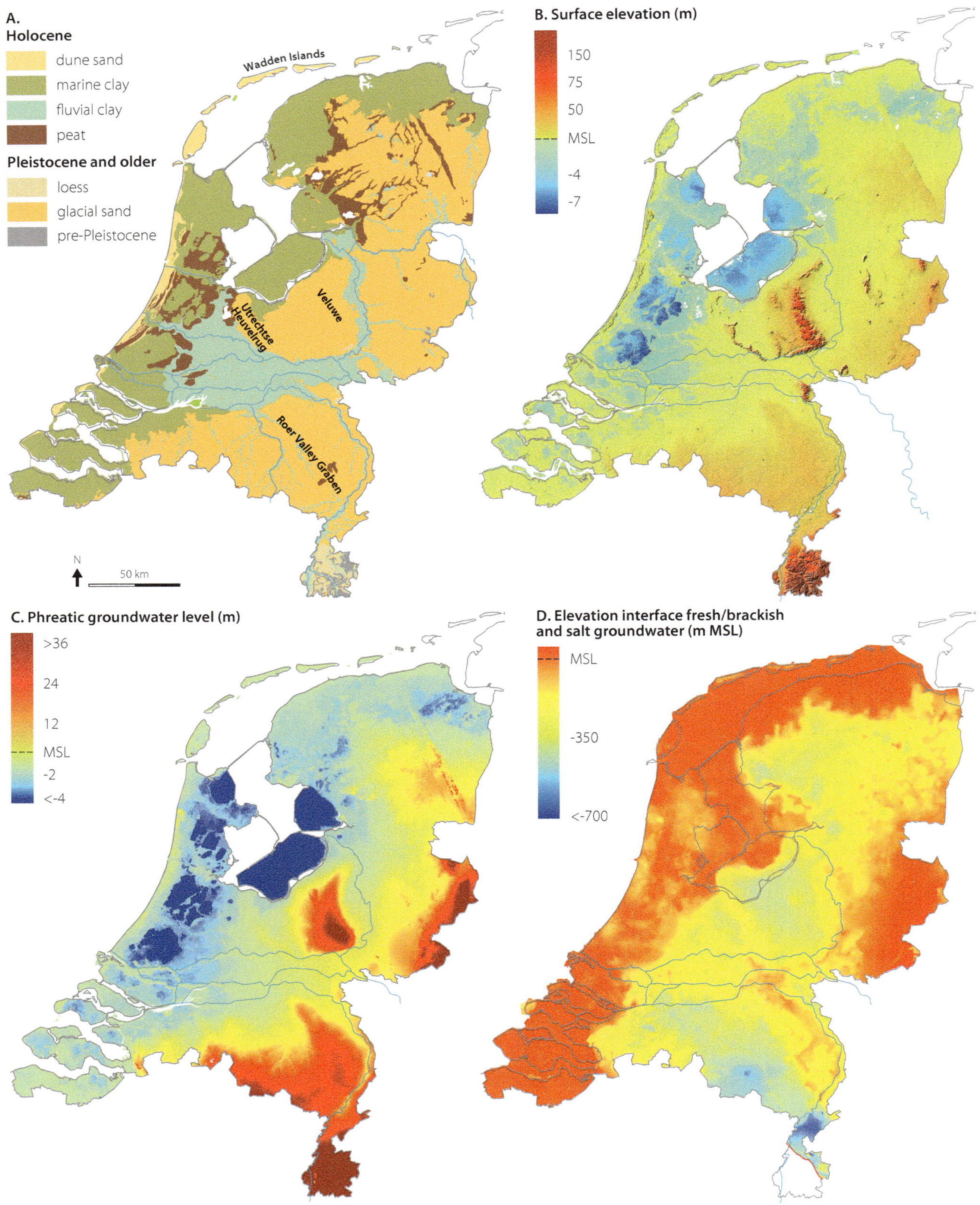

Figure 17.1. Schematic maps of: a) surface geology (reprinted from Stafleu et al., 2019) and b) elevation of the Netherlands (AHN.3), both showing a strong correlation with c) the phreatic groundwater level (Delsman et al., 2020) and d) depth of the fresh-salt groundwater interface (map based on data from Stuurman & Oude Essink (2007) and www.grondwatertools.nl). The area south of the Feldbiss fault is not mapped here. Fresh groundwater has a chloride concentration <0.15 g/l; saline groundwater has a chloride concentration >1 g/l.

sidence (Koster, 2017). Peat exploitation and excavation created lakes which subsequently were reclaimed in the 16th and 17th century, while several natural large lakes were reclaimed in the 19th and 20th century, including the Haarlemmermeer around 1850 and IJsselmeer polders between 1942-1968. Largely as a result of the cumulative effects of these anthropogenic activities, about 25% of the land surface of the entire country, and 50% of the coastal zone are nowadays below mean sea level (Fig. 17.1b). The deepest polders lie at more than 6 m below mean sea level. At present continued draining of these low-lying areas remains necessary for agricultural purposes and to prevent inundations of the land. Peat-related subsidence is ongoing (Koster, 2017; Fokker et al., 2025, this volume). In addition, the Netherlands continues to subside on geological time and spatial scales related to mechanisms such as compaction of buried pre-Holocene geological units due to mechanical loading (Kooi, 2000), and tectonic and isostatic mechanisms (Kooi et al., 1998; Hijma & Kooi, 2018).

Another important ongoing impact of anthropogenic activity on the groundwater system concerns the extraction of fresh groundwater for water supply, which first started in 1853 in the dunes west of Amsterdam, followed by The Hague and Rotterdam in 1874. Since then supply has been predominantly based on groundwater. Fresh groundwater lenses below the dunes were heavily exploited until 1940; thereafter artificial recharge and recovery was used more and more (Huisman, 2004; Geelen et al., 2017). Through time, fresh groundwater from Pleistocene elevated areas in the eastern and southeastern part of the country (Fig 17.1 a-b) was increasingly used for water supply as well.

Second half 19th century

Until the mid-19th century, there was no knowledge about the spatial extent of the fresh groundwater resources and no knowledge or understanding of the underlying processes and conditions controlling the extent of the resources. Despite the increasingly innovative technical water works, there also was no clear understanding of the groundwater flow processes involved in the drainage of the land. Groundwater only began to emerge as a distinct research topic in the late 19th century (De Vries, 1982). Darcy's Law, as developed by the French engineer Henry Darcy in 1856 (Darcy, 1856), can be considered as the basis for understanding groundwater flow, showing that flow is driven by differences in hydraulic head. Badon Ghijben delivered a Dutch contribution by proposing the principle of a fresh-water lens in the dunes, floating on water of sea-water density, assuming theoretical hydrostatic equilibrium between fresh water and salt water, i.e. no flow conditions, as well as isotropic permeability of the subsurface (Drabbe & Badon Ghijben, 1889; Herzberg, 1901). According to

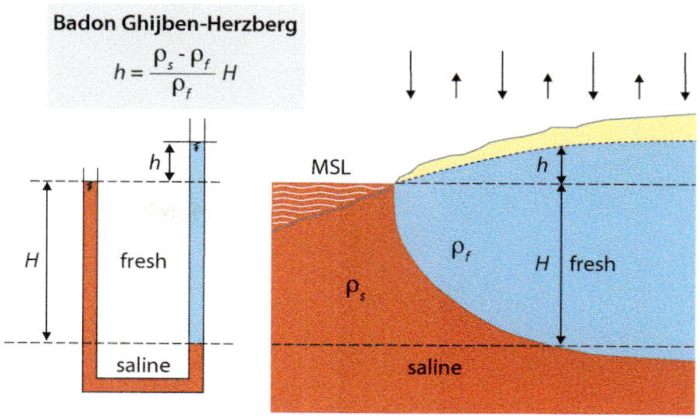

Figure 17.2. Badon Ghijben-Herzberg principle: Under the assumption of a hydrostatic pressure distribution, the pressures of fresh groundwater (indicated in blue) and saline groundwater (indicated in red) at the fresh-salt interfaces are equal, and the depth of the fresh-salt groundwater interface (H) below MSL can be related to the density differences of the fresh and salt groundwater and the elevation of the groundwater table above MSL (reprinted from Oude Essink, 2001a with permission from Elsevier).

this so-called Badon Ghijben-Herzberg principle, the ratio between water-table elevation (h) above mean sea level (MSL) and the depth of the freshwater-saltwater interface (H) below MSL in coastal aquifers is about 40 (Fig. 17.2). Pennink (1905) demonstrated the existence of a 2D flow pattern towards a drainage channel based on field measurements and experiments. He also proposed, already at that time, to apply artificial recharge of the dunes with river water. His plans were only implemented a half century later (De Vries, 1982).

First half 20th century

Research in the first half of the 20th century was dominated by engineering approaches to the solving of specific groundwater flow problems in relation to groundwater extraction, large reclamation works (Zuiderzee polders), infrastructural works and management of groundwater levels for agriculture. The engineering approach involved the development of mathematical-analytical-equations to describe the flow of groundwater through a hydrogeological schematic version of the shallow – unconsolidated – subsurface in combination with simple boundary conditions. The physicist Lorentz (1913) combined Darcy's flow equation with the continuity equation to develop a differential equation for steady-state groundwater flow. Another early example is the equation of De Glee (1930) describing steady-state flow to a pumped well in an aquifer with leakage across an overlying aquitard. The De Glee equation continued to be used for calculating aquifer and aquitard properties and for determining sustainable groundwater extractions (Kruseman et al., 2000). Additional examples

Use and protection of fresh groundwater resources

Fresh groundwater is a vital resource for the supply of fresh water for domestic, industrial and agricultural uses and is of importance for maintaining groundwater-dependent eco-systems. About 60% of the drinking water supply in the Netherlands is produced at about 200 locations from fresh groundwater resources, including 6% derived from river bank infiltration and 1% from natural dune water (CBS, 2021; VEWIN, 2022). Most fresh groundwater is extracted from aquifers in the eastern and southeastern part of the Netherlands.

The country faces increased stress from climate change, land-use transitions, extraction from and contamination and competitive use of fresh groundwater resources. Sea-level rise and land subsidence in the coastal zone are expected to increase salinization of fresh groundwater systems. The dry summers of 2018, 2019, 2020, and 2022 caused a decrease in recharge, increased groundwater extractions, and a lowering of groundwater tables.

Anthropogenic activities have significantly impacted the quantity and quality of fresh groundwater resources. The groundwater table has declined in the order of decimetres outside the coastal zone over the last 60 years due to groundwater extraction, drainage, and land use changes (Van Bakel et al., 2017; Witte et al., 2019; Vermeulen & Op den Kelder, 2020). Extraction from the groundwater resources in the dunes along the western coast is supported by artificial recharge with pre-purified river water from the Rhine and Meuse (Geelen et al., 2017). Natural groundwater resources in the dunes of the islands of Vlieland and Schiermonnikoog can keep up with increasing drinking water demand, while other islands rely largely on mainland drinking water.

Competitive anthropogenic uses of fresh groundwater resources include storage and abstraction of heat or cold in aquifers at depths of less than 500 m, which have increased by almost 40% between 2015 and 2020 (VEWIN, 2022). Mining activities in relation to energy supply at greater depth may have consequences for the sustainability of fresh groundwater resources.

The Netherlands' spatial planning strategy emphasizes the importance of drinking water supply and mining activities, with national and supplementary spatial groundwater reserves and protection measures to ensure long-term security of public drinking water supply. Water and subsurface are both implemented as a leading principle in spatial planning (Harbers & Heijnen, 2022). Since the publication of Strong (2018), there has been an increase in national programmes and research projects concerning groundwater and sustainability challenges, such as Delta programmes (e.g. 2021, 2023) and the integral groundwater study (Deltares, 2023).

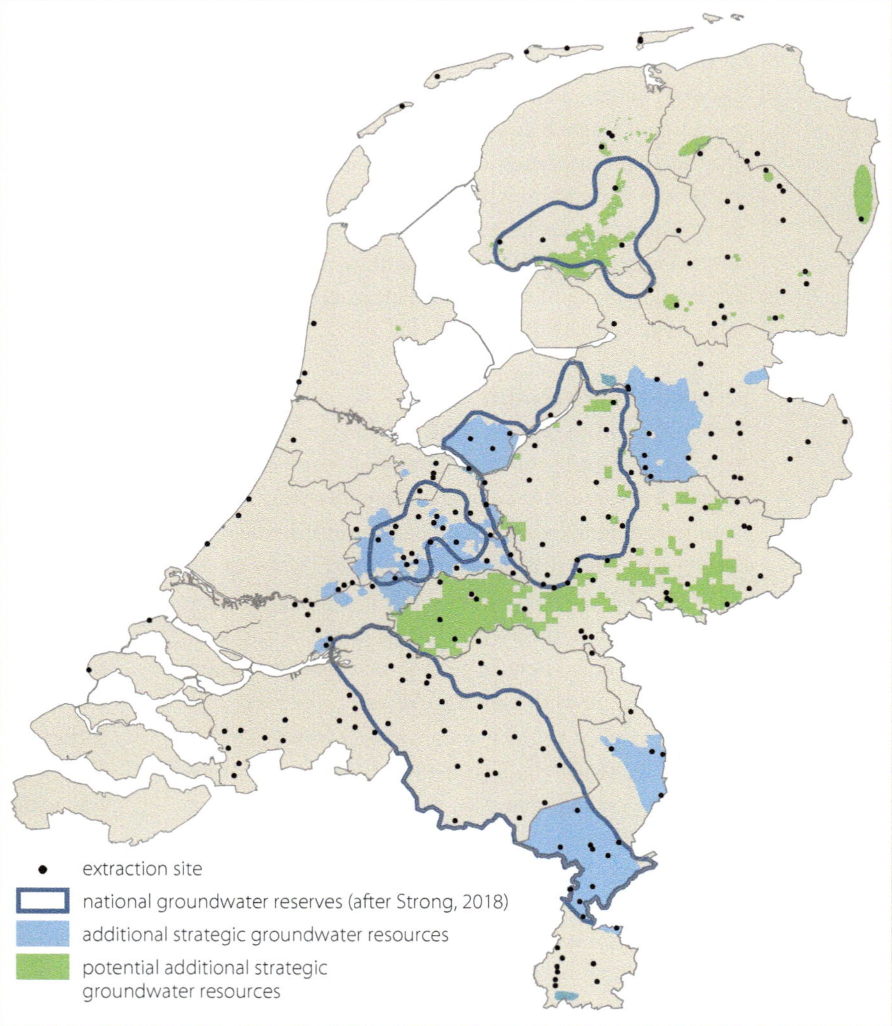

Groundwater reserves and additional groundwater resources (modified from: VEWIN, 2022).

of important analytical solutions include the equations of Hooghoudt (1937, 1940) and Edelman (1947) that were developed for solving drainage and infiltration problems, and the equation of Mazure (1936) describing the flow of groundwater below a dike driven by the difference in water level on either side of the dike. These examples of early applied research taken from De Vries (1982) illustrate the strong relation of the research in response to societal needs of public water supply and water management in the coastal zone. From these early days onward Dutch researchers have played and still play an important role in developing analytical solutions for a wide range of groundwater flow conditions (Bruggeman, 1999; Bakker & Post, 2022). Bruggeman (1999) compiled and published analytical solutions for more than 1000 groundwater flow problems, including several new problems identified by himself. Bakker & Post (2022) outline the use of practical applications of novel analytical groundwater modelling approaches.

Second half 20th century

After 1945 large-scale programmes on groundwater research were initiated to improve general water management in the Netherlands (Krul, 1962). Important early initiatives included the establishment of the National Archive for groundwater level data TNO in 1948 and the national inventory of groundwater levels executed in 1952-1956, which was used to compile maps of the depth of the water table in summer and winter that are important for agricultural water management (Visser, 1958). A national monitoring network for groundwater quality was established between 1979 and 1992 (Van Duijvenboode, 1993). The importance of geological knowledge of the structure and composition of the subsurface and its geological history for groundwater research was recognized by some researchers early on (Krul, 1940, 1962; Ernst & De Ridder, 1960). Since about the 1970s the geological knowledge delivered by the Geological Survey of the Netherlands (RGD) has been integrated increasingly in hydrogeological research at local to national scales (Jelgersma & Visser, 1972; Breeuwer & Jelgersma, 1973; De Vries, 1974; Jelgersma, 1977). Increasing environmental awareness, including the role of groundwater therein since the early 1970s has helped drive the change from mostly hydraulic engineering approaches and mathematical descriptions towards a more comprehensive earth science based approach to hydrogeological research. This is apparent in the national groundwater mapping, monitoring and data management programme that was executed by the Groundwater Survey TNO in 1970-1989. The resulting Groundwater Maps of the Netherlands and the explanatory reports provide a comprehensive overview of fresh groundwater resources, geology, hydrostratigraphy, hydraulic properties, groundwater levels, patterns of groundwater flow, and general chemical characteristics of the groundwater.

An important boost in the development of Dutch hydrogeological research was the establishment of an earth science based education and research programme 'Hydrogeology and Geographical Hydrology' at the Vrije Universiteit Amsterdam, starting in the 1960s. The concept of groundwater flow systems as integral part of earth system sciences was extended and adapted to the Dutch situation (Engelen, 1986; Engelen & Kloosterman, 1996) and was broadened into an integrated hydrological systems analysis, including natural and man-made systems of interacting surface and groundwater. The concept was applied in the 'National analysis of regional groundwater flow systems' (1991-1995; e.g. Kloosterman, 1993), which included paleo- and present-day flow patterns at various scales, trends in groundwater quality, and impacts on groundwater-dependent ecosystems as important study objectives. It demonstrated and further established the role of hydrogeology as a distinct geoscience discipline within earth system science.

End 20th century – Recent

In addition to the ongoing focus of research on – often artificially managed – fresh groundwater systems to ensure e.g. supply of good quality water for domestic, agricultural, and industrial uses, new fields of application of groundwater research have emerged since about the 1980s. They include topics such as groundwater-dependent ecosystems; geological storage and disposal sites (radioactive waste); subsurface storage, recovery and extraction of water and thermal energy; exploration and production performance of natural resources other than groundwater (geothermal energy, oil and gas) and their impact on groundwater conditions and ground surface movement; the impact of land use change and, more recently, the impact of climate change and foreseen increases in multiple use of the subsurface related to the energy transition. Extension and improvement of monitoring networks (Van Bracht, 2001; Broers, 2002) and the development of statistical and modelling methods for analysing the monitoring data (Van Geer, 1987; Gehrels, 1999; Zaadnoordijk et al., 2019) have been of key importance for mapping and understanding groundwater flow systems. The merger of the Groundwater Survey TNO and Dutch Geological Survey (RGD) into one institute in 1997 resulted in the development of a publicly available integrated data and information system with hydrogeological as well as geological data (Dutch national subsurface database DINO (www.dinoloket.nl). Since 2018, an additional Key register of the Subsurface has been developed, standardizing public information about the subsurface and making it publicly available (basisregistratieondergrond.nl). Since the establishment of the Mining Act of the Netherlands 2003, a wealth of data and information from the petroleum industry have been made publicly available on TNO's oil and gas portal (www.nlog.nl). The broadening of scope, as well as the increase in available national databases and digital capabilities and the incorporation of new technologies and concepts, as well as theories and tools from other earth science disciplines, has stimulated the development of the systems approach of groundwater research (Stuyfzand, 1993, 1999; Zijl, 1999; Griffioen et al., 2003) as well as specialization and the emergence of subdisciplines. Recognition of the impact of groundwater extraction, the associated decline of groundwater levels and, in coastal zones, salinization on groundwater-dependent ecosystems emerged in the 1990s. Ecohydrology has also rapidly developed as a separate discipline (Klijn & Witte, 1999; Batelaan et al., 2003; Stofberg et al., 2016) and has influenced and has been integrated into groundwater research and management since that time (Griffioen et al., 2003; Geelen et al., 2017).

Other examples of key subdisciplines related to specific natural Dutch circumstances are coastal hydrogeology, exchange between surface water and groundwater, as well as surface subsidence due to drainage and pumping. Coastal hydrogeology continues to be a key subdiscipline for research in the Netherlands. Dune water management/groundwater extraction and natural and human-induced causes of observed distributions of fresh, brackish, and saline groundwater have stimulated groundwater research of dunes and adjacent polder areas since the end of the 19th century (Dubois, 1903; Versluys, 1931; De Vries, 1981; Appelo & Geirnaert, 1991; Stuyfzand, 1993). More recent process-based focus is on modelling the role of past and future sea-level changes and sea-water intrusion on salinization of the groundwater (Oude Essink, 1996; Post et al., 2003; De Louw et al., 2010; Pauw et al., 2012; Delsman et al., 2014).

Some new applications involve research into the realm of deep hydrogeology. Since the 1980s such research includes investigating the feasibility of geological disposal of radioactive waste in Cenozoic mudstones, especially the Oligocene Boom Clay, or in Zechstein rock (Neeft et al., 2025, this volume). These programmes also included compilation of hydrogeological properties and groundwater flow and transport assessments. Because of the long timescales involved in radioactive waste disposal, these groundwater flow studies have not only considered current conditions but also have involved studies of paleo groundwater flow and transport processes in order to forecast future groundwater conditions on a geological timescale of up to 1 million years (Griffioen & Wildenborg, 2016). The recognition of the influence of hydrodynamics on petroleum systems was an important trigger to start research on pressure and groundwater systems at greater depths and covering large – geological – time and spatial scales in the Netherlands onshore and offshore in the late 1980s (Verweij, 1999; Verweij & Simmelink, 2002; Verweij, 2003). Research advanced rapidly since the late 1990s when data on the deep subsurface became increasingly available. Both hydrogeology- and petroleum geoscience-based approaches were used to characterize and explain current and past pressure and groundwater flow conditions. For example, present-day characteristics of the sedimentary sequences, such as sediment diagenetic distributions, could be used to identify current and past groundwater flow conditions (Verweij, 2003). Basin modelling tools from the petroleum industry were used to quantify different processes (such as sedimentary loading and unloading, tectonic compression and water table change) that affect fluid pressure and groundwater systems during the evolution of a sedimentary basin (Verweij, 2003; Verweij et al., 2011; Nelskamp et al., 2012; Verweij et al., 2012). The tools have also proved very useful in forecasting spatial distributions of present-day key properties in the subsurface, such as porosity, permeability and temperature (e.g. Nelskamp & Verweij, 2012).

Hydrogeological framework

The present-day hydrogeological framework includes the spatial distribution, thickness, dip, and hydraulic properties of hydrogeological units (aquifers, aquitards) and the location of geological structures and tectonic elements important for groundwater flow.

Hydrogeological framework onshore and offshore

The subsurface of onshore and offshore Netherlands consists of different structural elements (sedimentary basins, grabens, platforms, highs), each with a characteristic history of sedimentation, uplift, non-deposition, and erosion (Fig. 17.3). The complex geological history manifests itself not only in the stratigraphy and structure, but also in properties such as porosity, permeability, compressibility and the geochemical composition of stratigraphic units and fault zones, as well as the properties and flow condi-

→ ***Figure 17.3.*** *a) Map of onshore and offshore Netherlands showing location of structural elements (after Kombrink et al., 2012). b) SW-NE geological cross section showing main features of the present-day stratigraphic buildup and structural framework (after TNO-GDN, 2023). It highlights the regional differences in stratigraphic buildup between basins and platforms/highs and differences between the fault-dominated southwest with deep reaching faults cross-cutting Carboniferous to Recent strata, and the northeast with its salt-dominated structural framework.*

A.

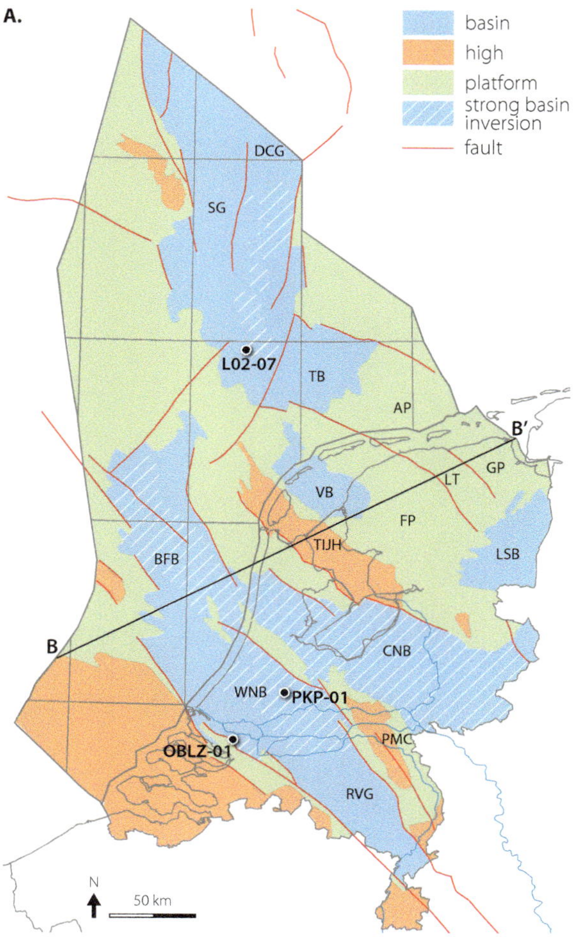

Acronym	Full name
AP	Ameland Platform
BFB	Broad Fourteens Basin
CNB	Central Netherlands Basin
DCG	Dutch Central Graben
FP	Friesland Platform
GP	Groningen Platform
LSB	Lower Saxony Basin
LT	Lauwerszee Trough
PMC	Peel-Maasbommel Complex
RVG	Roer Valley Graben
SG	Step Graben
TB	Terschelling Basin
TIJH	Texel-IJsselmeer High
VB	Vlieland Basin
WNB	West Netherlands Basin

fault
High – structural element
IJssel – river
North Sea – open water
50 km

B.

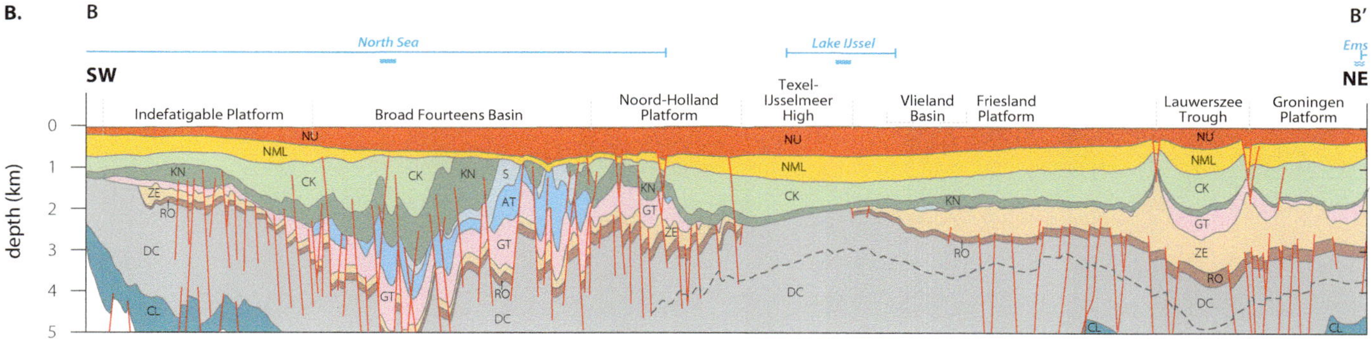

- **NU** Upper North Sea Group – Shallow marine and continental siliciclastic deposits (Miocene – Recent)
- **NLM** Middle North Sea Group (NM) and Lower North Sea Group (NL) – Mainly marine siliciclastic deposits (Paleocene – Miocene)
- **CK** Chalk Group – Mainly marine limestone (Late Cretaceous – Paleocene)
- **KN** Rijnland Group – Mainly marine siliciclastic sedimentary rock (Early Cretaceous)
- **S** Upper Jurassic Supergroup (SG, SK, SL) – Marine, marginal marine and continental siliciclastic sedimentary rock, evaporites and limestone (Late Jurassic)
- **AT** Altena Group – Marine siliciclastic sedimentary rock (Late Triassic – Early Jurassic)
- **GT** Upper Germanic Trias Group (RN) and Lower Germanic Trias Group (RB) – Marine and continental siliciclastic sedimentary rock, evaporites and limestone (Triassic)
- **ZE** Zechstein Group – Marine evaporites and limestone (Permian, Lopingian)
- **RO** Upper Rotliegend Group – Continental siliciclastic sedimentary rock and evaporites (Permian, Guadalupian – Lopingian)
- **DC** Limburg Group – Mainly marine and continental siliciclastic sedimentary rock, alternating with coal (upper Carboniferous, Namurian, Westphalian, Stephanian)
- **CL** Carboniferous Limestone Group – Marine limestone (lower Carboniferous, Dinantian)
- **CF** Farne Group – Mainly marine and continental siliciclastic sedimentary rock alternating with coal, limestone and dolomite (lower Carboniferous, Dinantian)
- - - - - base Westphalian – Base of the interval of the Limburg Group containing coal (Caumer Subgroup, Dinkel Subgroup and Hunze Subgroup; not mapped country wide)

Figure 17.4. *Spatial distribution of porosity and permeability of the Triassic Lower Volpriehausen Sandstone Member in onshore and offshore Netherlands. The continental sandstones were deposited in a fluvial setting in the southern area and in an eolian setting in the northern offshore. The maps show a large spatial variation in porosity and permeability. The lowest porosity and permeability values occur in the southern basins and in parts of the northern grabens. (maps retrieved in February 2022 from www.nlog.nl, and slightly modified). Permeability k = 1 mD corresponds to approximately 1.0E-15 m² and 1.0E-8 m/s).*

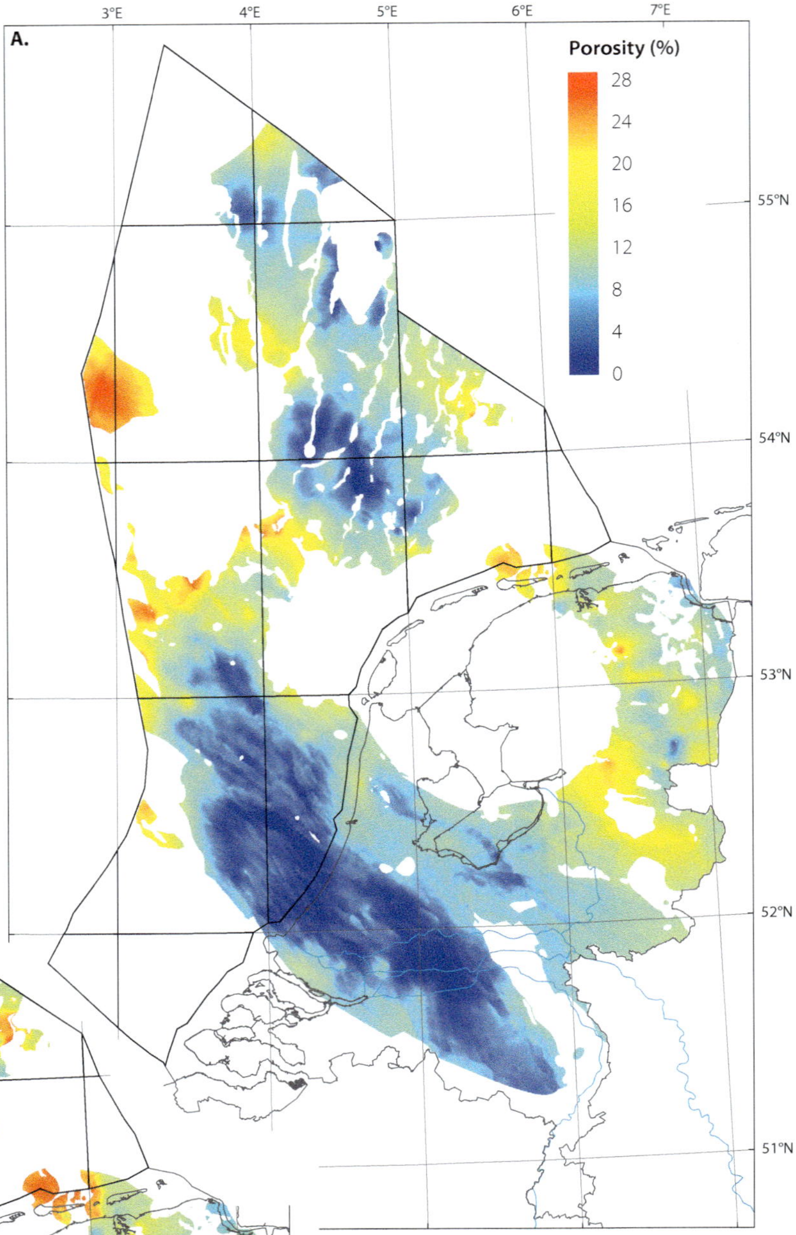

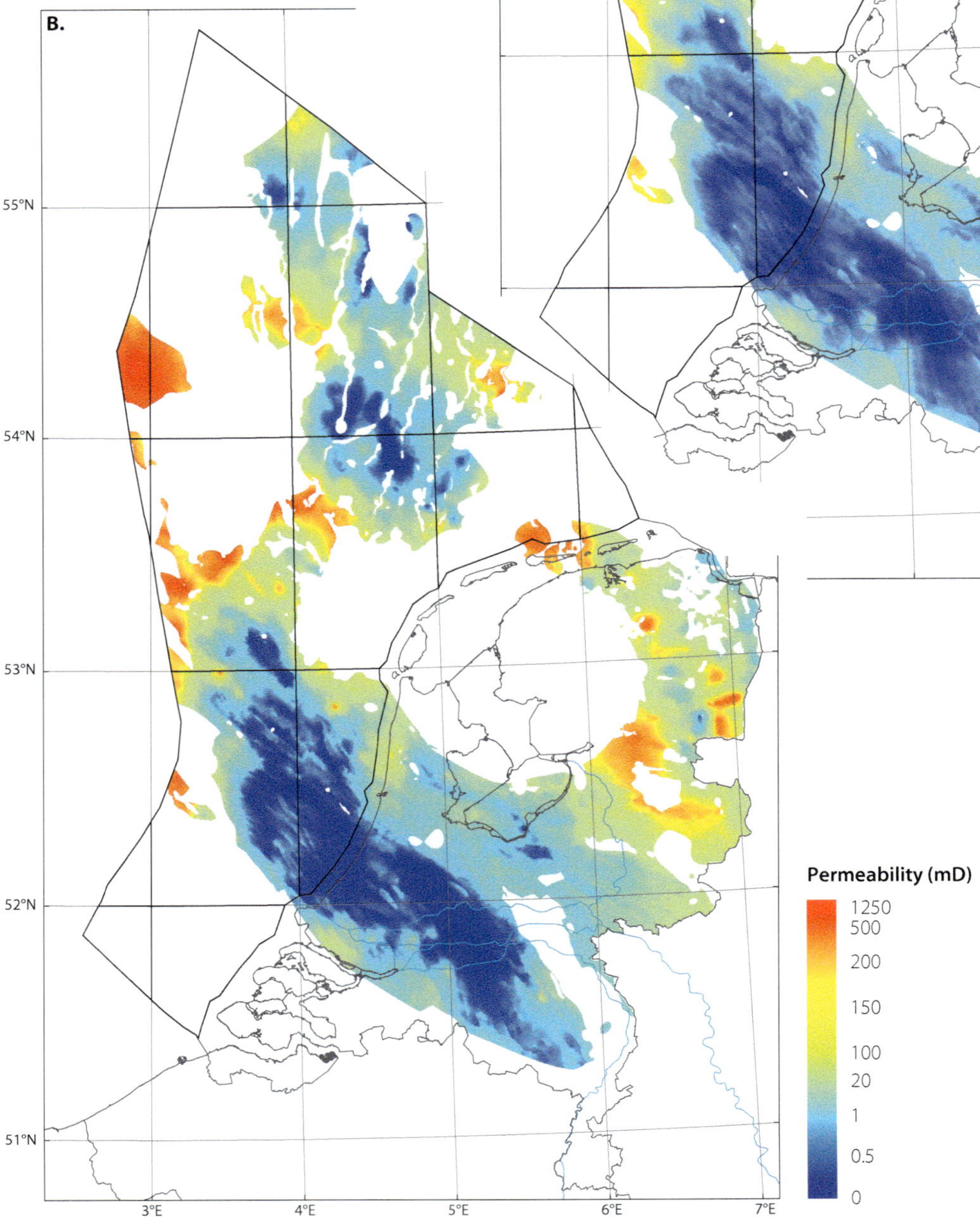

tions of the pore waters they contain. During most of the Permian to Quaternary geological history, the Netherlands was located at the southern edge of large sedimentary basins and this has directly affected the facies distribution of the deposited sediments. Generally, more coarse-grained clastic sediments were deposited in the southern part of the onshore and offshore Netherlands. This facies distribution has resulted in generally more permeable facies in the south and an increasing number of intercalated layers of poor permeability towards the northern offshore part of the area. Clear present-day differences in facies between the south and the north are apparent in, for example, the Permian Upper Rotliegend Group (sandstones of the Slochteren Formation in the south and mudstones and evaporites of the Silverpit Formation in the north), Permian Zechstein Group (clastic sediments in the south versus evaporites in the north), Upper Germanic Trias Group (salts largely absent in the south), Cretaceous Rijnland Group (sandstones concentrated in the southern half of the area) and the Paleogene Lower North Sea Group (sandy in southern onshore). Stratigraphic groups of large areal extent are the Limburg, Upper Rotliegend, Zechstein, Rijnland, and Chalk groups, and the Lower and Upper North Sea groups. The presence of the Germanic Trias and Jurassic Altena groups is more scattered. The main permeable units forming aquifers and reservoirs are the sandstone members of the Limburg Group and Slochteren Formation of the Upper Rotliegend Group, carbonate members of the Zechstein Group, sandstone members of the Lower Germanic Trias Group, Solling Formation of the Upper Germanic Trias Group, sandstone members of the Schieland, Scruff and Niedersachsen groups, Vlieland Sandstone Formation of the Rijnland Group, Cenozoic sands (in southern onshore) and Quaternary sands of the Upper North Sea Group (Table 17.1). Pre-Cenozoic hydrogeologic units consist of consolidated rocks and the aquifer units are confined. Porosity and permeability maps of the main Cretaceous, Triassic, and Permian sandstone aquifers in the onshore and offshore Netherlands are publicly available at the oil and gas portal of TNO. The spatial distribution of geological units, and therefore of aquifers and aquitards, is not restricted by the boundary between the onshore and offshore. Figure 17.4 shows the spatial distribution of the Triassic Lower Volpriehausen Sandstone Member, including the variation of the porosity and permeability of this sandstone aquifer.

Poorly permeable units of considerable lateral extent include the mudstones of the Limburg Group, the mudstones and evaporites of the Silverpit Formation and the extremely poorly permeable Salt members of the Zechstein Group as well as evaporite members of the Upper Germanic Trias Group which have a more restricted distribution. Mud-rich deposits of low matrix permeability occur throughout the entire stratigraphic sequence, separating the main permeable hydrogeological units (Table 17.1). Hitherto there have been no property maps available for poorly permeable units.

The post-Zechstein structural development in the northern Netherlands and adjacent offshore was affected by salt movement and the creation of large salt structures (Bouroullec & Ten Veen, 2025, this volume). The salt structures and numerous deep faults disrupt the hydraulic continuity of especially the pre-Cenozoic hydrogeological units add (Fig. 17.3).

Detailed shallow hydrogeological models onshore

The dominant features of the present-day structural geological and lithostratigraphy of the subsurface of the Netherlands are incorporated in hydrogeological models. The regional hydrogeological model of the Netherlands REGISII (available at DINOloket.nl; Stafleu et al., 2025, this volume) is a 3D layer model describing hydraulic properties down to a depth of about 500 m, with a maximum depth of 1200 m below sea level in the Roer Valley Graben. The model covers the depth ranges of topography-driven groundwater flow and is based on the digital geological model of the Netherlands (DGM). The lithostratigraphic units in DGM are subdivided into one or more hydrogeological units (aquifers and aquitards), while the location of faults is taken into account in determining the lateral extent of the units. The faults themselves and their hydraulic properties are not included in the model. The lithostratigraphic units are characterized by hydraulic conductivity, transmissivity, and hydraulic resistance (Fig. 17.5). REGISII distinguishes 130 units of Late Cretaceous to Holocene age. The unconsolidated Neogene and Quaternary units of the Upper North Sea Group are mapped nationwide. Older Cenozoic and Mesozoic units are only included in the southwestern and eastern parts of the Netherlands and southern part of Limburg.

The upper, Holocene, hydrogeological unit in the Dutch coastal lowland is composed of marine, lagoonal, peat, and fluvial deposits with a maximum thickness of 25 metres that accumulated behind a series of coastal barriers and dunes. The Holocene unit is generally considered to act as a semi-confining layer overlying Pleistocene fluvial sedimentary formations (Kreftenheye, Urk, Sterksel, Waalre, and Peize formations) that incorporate important sandy aquifers separated by clayey aquitards. The underlying Early Pleistocene (Gelasian) marine Maassluis and Pliocene marine Oosterhout formations are composed of a sequence of sandy aquifers and clayey aquitards. The Miocene Breda aquitard (BRk1) forms the base layer of the REGISII model in the coastal zone. In the northern coastal area, subglacial tunnel valleys (Peelo Formation) reaching

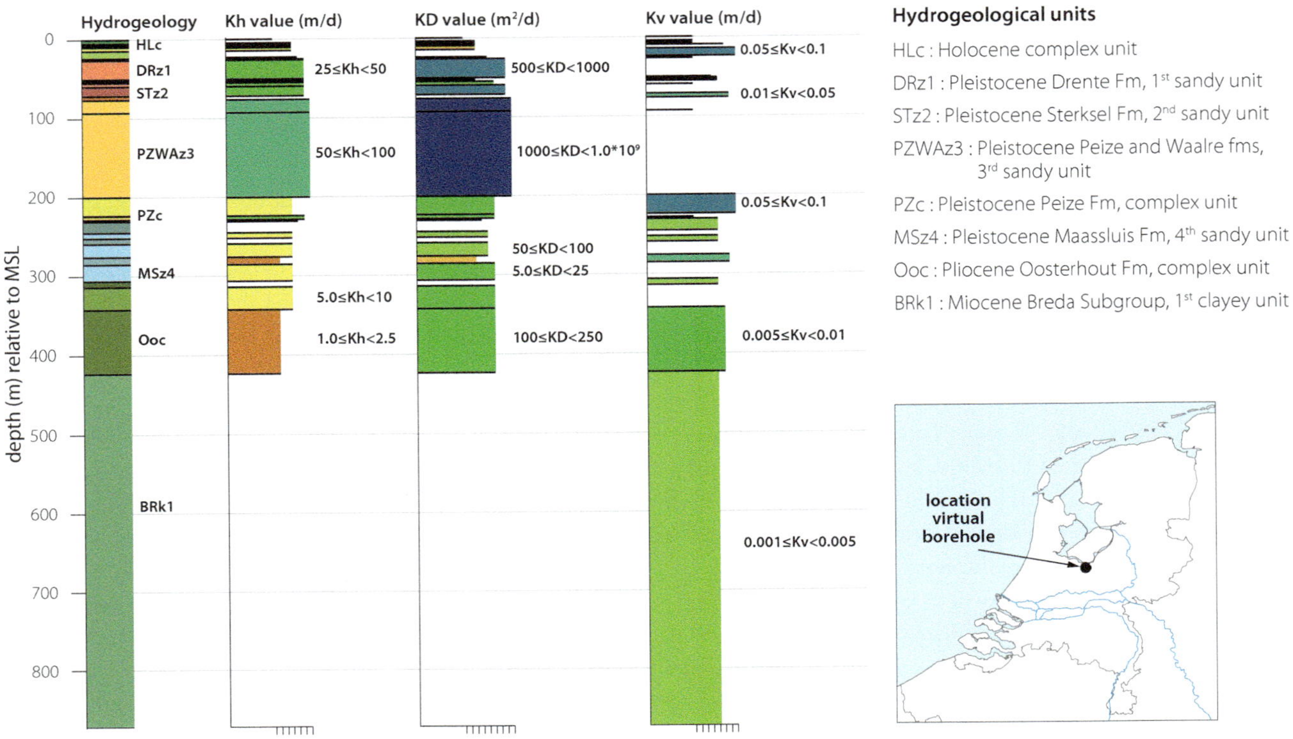

Figure 17.5. Example of Miocene-Recent stratigraphy of aquifers, aquitards and complex hydrogeological units in onshore Netherlands, based on hydrogeological model REGISII v2.2. The hydrogeological units are coded according to the corresponding stratigraphic unit (capitals) plus a hydrogeological classification: z = sandy unit (aquifer), k = clayey unit (aquitard), or c = complex hydrogeological unit (mixed). Kh = horizontal hydraulic conductivity; Kv = vertical hydraulic conductivity; KD = transmissivity.

depths of 300 m disrupt the lateral continuity of Pleistocene fluvial and marine Oosterhout units (Fig. 17.6). The glacial and fluvioglacial clays of the Peelo and also the Drente Formation form important aquitards in this area (Fig. 17.6).

The NW-SE hydrogeological cross-section (Fig. 17.7) shows the large variation in thickness and hydrogeological properties of the third sandy aquifer of the Waalre and Peize formations (PZWAz3) and – especially – in the Miocene Breda units. A large thickness of aquifer quality of the Breda Subgroup (BRz1; 5.0 m/day ≤ K_h < 10 m/day) is found in the central part of the Roer Valley Graben (RVG), while the Breda Subgroup is of an overall aquitard nature (BRk1; 0.001 m/day ≤ K_v < 0.005 m/day and hy-

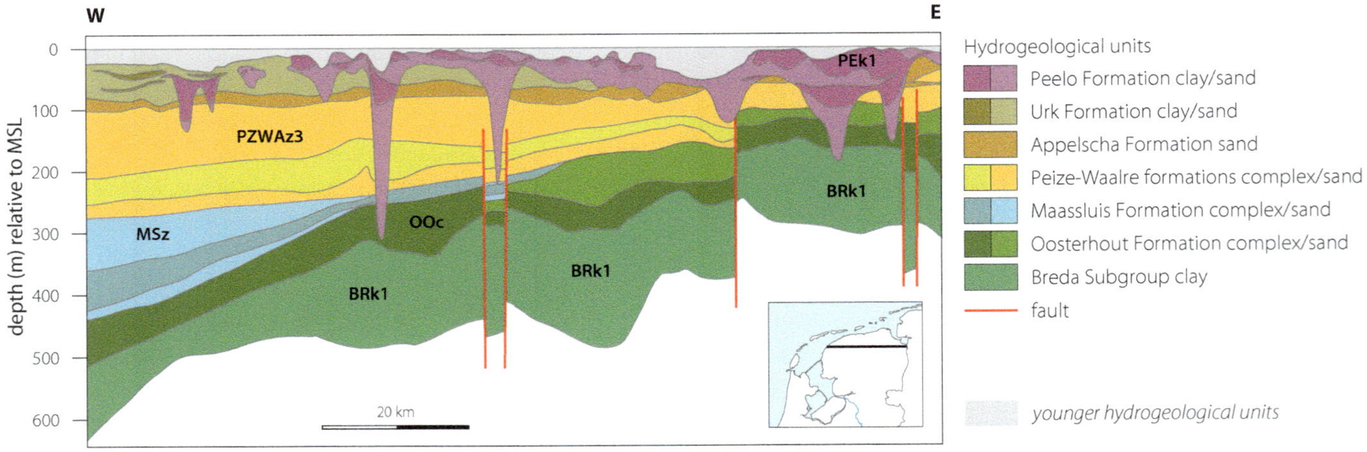

Figure 17.6. W-E hydrogeological cross section through the northern part of the Netherlands, showing the subglacial valleys and their hydrogeological infill, belonging to the Peelo Formation (purple PE hydrogeological units). Extracted from REGISII v2.2.

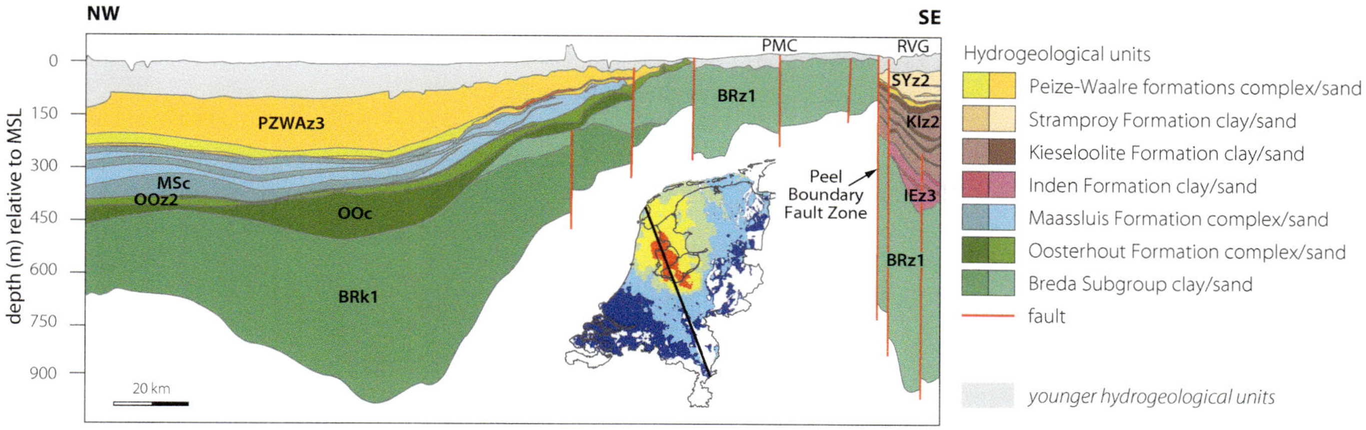

Figure 17.7. *NW-SE hydrogeological cross section across the Netherlands, showing north-westward increase in thickness of the third sandy unit of the Peize and Waalre formations (PZWAz3), and great thickness of the Breda Subgroup in the central part of the Netherlands (Central Netherlands Basin; clayey hydrogeological unit BRk1) and in the southeast (Roer Valley Graben (RVG); sandy hydrogeological unit BRz1) (extracted from REGISII v2.2). Inset map shows the distribution of the third sandy hydrogeological unit of the Peize and Waalre formations (PZWAz3) and its lateral variation of hydraulic conductivity from about 5 m/day in the SW (dark blue area) to 50-100 m/day in NW of the Netherlands (yellow-red area). PMC = Peel-Maasbommel Complex.*

draulic resistance c = 10,000 days-1 million days) in the depocentre of the Central Netherlands Basin. Figure 17.7 also shows the large displacements of hydrogeological units along the boundary fault of the Roer Valley Graben. The total thickness of the Pliocene to Holocene hydrogeological units overlying the Breda units increases from 0 m in the south and east to more than 500 m in the north-west, reflecting the position of the Netherlands at the south-eastern limit of the subsiding Southern North Sea Basin. The Pleistocene fluvial hydrogeological units reach the surface in the eastern part of the Netherlands, where also the Pleistocene glacial and periglacial units are close to the surface in this area. The Plio-Pleistocene aquifers in the east and south of the Netherlands contain the most important groundwater resources for public drinking water supply.

Additional hydrogeological units of different nature occur in the southernmost part of the Netherlands. Here, Pleistocene loess deposits (part of the Boxtel Formation) at the surface form an aquitard covering consolidated Cretaceous Chalk aquifer units (Maastricht and Gulpen formations) or older Cenozoic clastic sedimentary sequences. The Chalk aquifer is a dual porosity-permeability aquifer with a high permeability due to fissures and small-scale faults (De Wit, 1988).

The public web-based geographical information system developed to support the development of geothermal energy, ThermoGIS (Mijnlieff et al., 2025, this volume), provides additional information on aquifers in the onshore Netherlands. It includes, amongst other things, depth, thickness, porosity, and permeability maps of 29 potential aquifers of Carboniferous to Paleogene age.

Groundwater flow systems onshore introduction

The most well-known and well-studied type of natural flow in onshore settings is topography-driven cross-formational groundwater flow that is related to spatial differences in the elevation of the groundwater table, with its pattern of hierarchical sets of flow systems of different orders of magnitude in lateral extent and depth of penetration (Tóth, 1963, 2009; Zijl, 1999). In each system flow is from an infiltration or recharge area towards one or more discharge areas. A system is local if recharge and discharge areas are contiguous. In a recharge area the groundwater flows downwards, and the associated groundwater head decreases with depth. In the discharge area, the groundwater flow is upwards towards the water table and groundwater heads increase with depth. Recharge of groundwater systems refers to the addition of water to the groundwater system, either naturally (by infiltration of net precipitation or surface water) or artificially (by e.g. managed aquifer recharge). Topography-driven groundwater flow systems form the subsurface part of the conventional hydrological cycle. The term gravity-driven groundwater flow is more comprehensive and describes flow driven by differences in potential energy of groundwater rather than hydraulic head (see Appendix).

A water table (also known as phreatic groundwater level) has existed in subaerial parts of onshore and offshore Netherlands during different geologic time periods since Permian times (Verweij, 2003). Supra-regional topography-driven flow systems recharged from topographically high regions south of the Netherlands have probably been active repeatedly and for long periods of time. The evolution of the current supra-regional flow system in southern and southeastern Netherlands started in the Miocene. During the Quaternary local and regional topography-driven flow systems developed in the rather flat lowland area of onshore and offshore Netherlands and were added to this supra-regional system. Elsterian and Saalian glaciations modified the flat topography, creating glacial features such as ice-pushed ridges (Busschers et al., 2025, this volume). The coastal dunes that developed during the last 1000 years of the Holocene are characterized by their great mobility and undergo continuous shifts in their location seawards and landwards with respect to the current coastline and changes to their morphology (Vos, 2015). Topography-driven flow could start to develop in the dunes as soon as they were formed. The present geometry and topographic relief of the coastal dunes reflect their Holocene development, but have now in large part been stabilized by vegetation and sand nourishment. Repeated changes of climate and related changes of sea level in the Quaternary have exerted an important influence on the regional extent and depth of penetration of the topography-driven flow systems, as well as the subsurface distribution of fresh, brackish and salt water. Changes in topographic relief of the groundwater table, i.e. in boundary conditions of groundwater flow systems, take place at very different time and spatial scales, from very rapid changes related to meteorological conditions (dry and wet events) and anthropogenic activities to much longer-term changes in topographic relief of the land surface and climate. It takes time for the groundwater flow system to adjust to changes in boundary conditions, depending on the spatial scale of the flow system and hydraulic properties of the subsurface. Poorly permeable hydrogeologic units will increase the adjustment time, while large regional flow systems will adjust more slowly than small local systems. Present-day regional groundwater flow systems are often in transient state, i.e. not yet in balance with present-day boundary conditions.

Nation-wide differences in the elevation of the groundwater table in the Netherlands today (Fig. 17.1c) are related to the largely flat topography of its ground surface (Fig. 17.1b) in combination with the net groundwater replenishment from – especially – net precipitation in areas with relatively little drainage and surface waters in the Pleistocene inland part of the country and the drainage of the low-lying areas. This drainage is strongly influenced by anthropogenically controlled fixed surface water heads in the polders located below sea level in the coastal zones. Locally to regionally the topography of the groundwater table is affected by groundwater extractions. The current fresh groundwater of meteoric origin reaches its greatest depths in the Holocene coastal dunes (tens of metres depth), the Pleistocene ice-pushed hills (Veluwe and Utrechtse Heuvelrug) in the central part of the country (up to a few hundred metres depth), and in the supra-regional groundwater flow system in the southeastern part of the country (≥600 m) (Fig. 17.1d). These fresh parts of the groundwater flow systems occur largely in unconsolidated sedimentary sequences of dominantly Holocene and Pleistocene to Neogene age.

Local and regional groundwater flow systems onshore

Natural recharge of the groundwater systems is dominated by infiltration of net precipitation (precipitation minus evapotranspiration). The average annual precipitation in the Netherlands of about 800 mm is approximately evenly distributed over the year. The evapotranspiration is about 600 mm and is highest during spring and summer. As a consequence, the annual surplus of precipitation is concentrated in autumn and winter.

Modelled interaction between groundwater and surface water for an average year (2004) concerning the amount of net precipitation, indicates that discharge of groundwater into streams, rivers, drainage channels and as overland flow occurs in most of the country during most of the year. Surface water infiltrates into the groundwater only in limited areas (e.g. riverbank infiltration Rhine) during part of the year (Hendriks et al., 2012). The discharge of groundwater to surface water in the coastal zone with its poorly permeable Holocene clay and peat layers is largely controlled by the managed surface water levels in the polder areas. Most of the net precipitation in these areas is discharged rapidly after a short subsurface passage along an extensive network of draining watercourses, and there is very little or no recharge of groundwater in underlying aquifers from local precipitation. The main recharge of groundwater by net precipitation occurs in the central, eastern and southeastern Pleistocene part of the Netherlands and in the coastal dunes (Fig. 17.8). These areas are most vulnerable to periods of meteoric droughts as exemplified by the lowering of groundwater tables since the dry summer of 2018 resulting from multiple relatively dry periods during the spring and summer (e.g. Van den Eertwegh et al., 2021).

The topographic relief of the groundwater table (Fig.17.1c) shows that groundwater flow is directed from relatively high areas in the central, southeastern and eastern parts of the Netherlands towards the low-lying coastal zone (Fig. 17.1b), where the groundwater table is very close to the surface. It lies at greater depth below ground level in local recharge areas in the Holocene coastal dunes and in recharge areas of larger scale in the elevated Pleis-

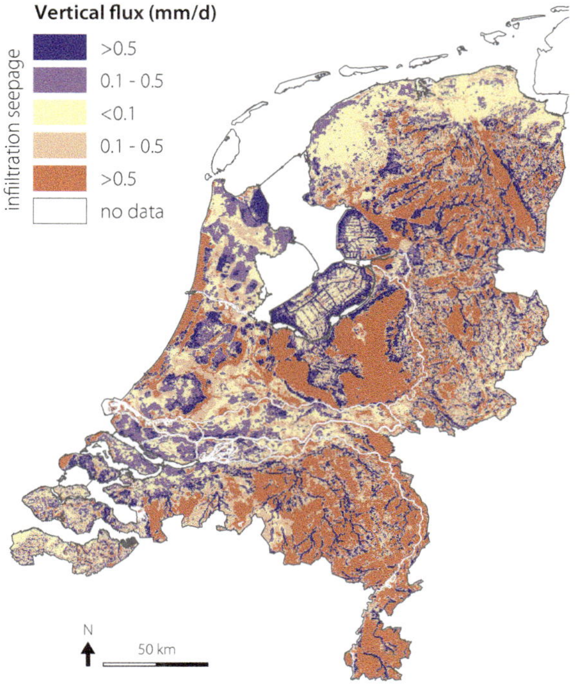

Figure 17.8. Seepage and infiltration fluxes based on calibrated groundwater model calculation (reprinted with permission from De Louw, 2013).

tocene parts (at 15 m to several tens of metres in Veluwe ice-pushed ridge). Around the borders of the Veluwe, the groundwater table is at shallow depth and its level is controlled in the north by the water level of Lake IJssel and along the other borders by water levels of rivers. In the recharge areas in the elevated ice-pushed ridges a shallow, lower-order drainage system is absent, while fluctuations of the deeper groundwater level are of low frequency and relatively large amplitudes in response to fluctuations in net precipitation because it takes longer for infiltrated precipitation to move through the unsaturated zone and because of the larger distance to surface waters along the borders (Fig. 17.9). The recharged groundwater is discharged in part along the edges of the ice-pushed ridges by rivers and anthropogenic drainage systems and also by regional flow into low-lying polders (Fig. 17.10). Already in the 17th-19th century, good quality groundwater was tapped and guided artificially (through so-called 'sprengen') into little streams along the eastern and southern borders of the Veluwe to provide a constant supply of good quality water for the production of paper and for driving water-powered paper mills (Renes et al., 2002). Nowadays these groundwater-fed streams in the Veluwe and Utrechtse Heuvelrug are highly valued for their associated ecosystems and cultural history (Renes et al., 2002). The fluctuations of the shallow groundwater table along the edges of the ice-pushed ridges are of high frequency with small amplitudes due to the shallow groundwater table and the proximity of surface water (Gehrels, 1999; Van Engelenburg et al., 2017; Fig. 17.9).

Groundwater flow in the eastern part of the Veluwe is influenced by ice-pushed tilted and discontinuous clay and silt layers (Gehrels, 1999) and is reflected in stepwise changes of the phreatic groundwater level. Groundwater recharge and groundwater levels decreased in the Veluwe area in the period between 1851-2016 (Van Huijgevoort et al., 2020). Change in land use was the main cause of this decrease in recharge, while in more recent years the extraction of groundwater contributed relatively more. Since 1950 groundwater levels in the Veluwe have also been influenced by reclamation of the Flevopolders. This induced a hydraulic head decline of 4 to 6 m in these polders (Gehrels, 1999) which, after time, propagated into the Veluwe area. Gehrels showed that it took about 25 years for the reclamation-induced head decline to reach the centre of the Veluwe area.

Groundwater flow in the slightly undulating north and westward sloping Pleistocene areas outside the ice-pushed ridges is characterized by deeper regional systems between higher-order topographic elements and by shallow, local or intermediate systems related to lower-order topographic elements such as brook valleys and small streams and ditches (Engelen & Kloosterman, 1996). The area is drained by a hierarchical stream system that expands and contracts with the seasonal variation of the groundwater table. This fluctuation thus regulates the number of channels that participate in the drainage process, so that the drainage density increases with increasing net precipitation (De Vries, 1974, 1994). Natural variation in recharge and also the variation in drainage level management both exert a large and direct influence on shallow groundwater flow systems in sandy phreatic aquifers (Vissers, 2005; Vissers & Van der Perk, 2008). Several approaches have been applied in order to separate the impact of natural meteoric variations on groundwater from variations due to anthropogenic influences (such as changes in water management, groundwater extraction, land use). Examples include statistical analysis of time series of groundwater level data, groundwater flow modelling, or a combination of both (Gehrels, 1999; Zaadnoordijk et al., 2019; Zaadnoordijk & Lourens, 2019). Figure 17.11 illustrates the separation of the fluctuation of the groundwater level into the contribution of various influences for a monitoring well near the river Meuse using time series modelling (Zaadnoordijk et al., 2019). The evaporation and to a lesser extent the river Meuse, which is partially controlled by weirs, cause a distinct seasonal fluctuation of the groundwater level.

The net precipitation in the coastal dunes has created a phreatic groundwater level located above sea level and a

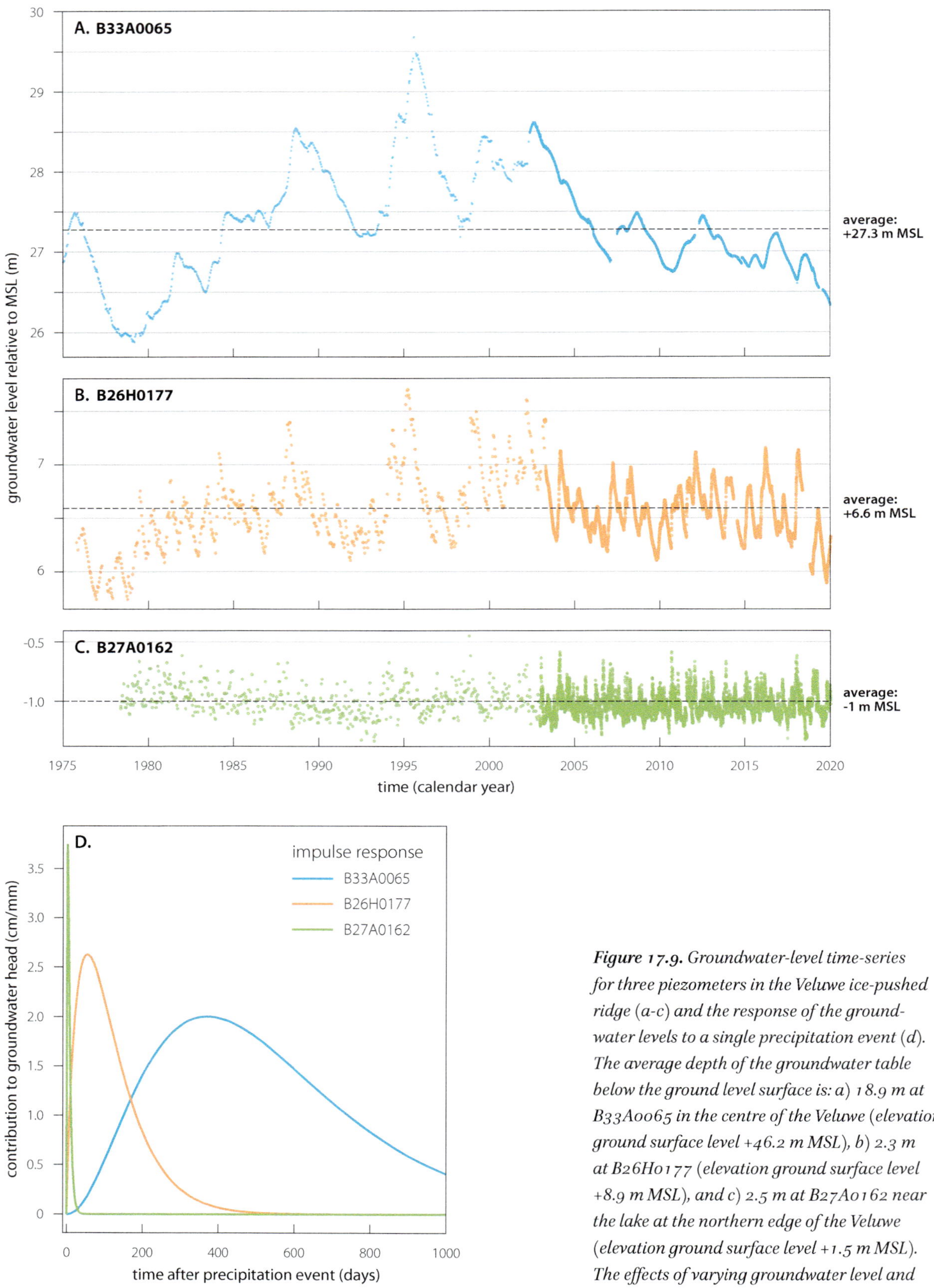

Figure 17.9. Groundwater-level time-series for three piezometers in the Veluwe ice-pushed ridge (a-c) and the response of the groundwater levels to a single precipitation event (d). The average depth of the groundwater table below the ground level surface is: a) 18.9 m at B33A0065 in the centre of the Veluwe (elevation ground surface level +46.2 m MSL), b) 2.3 m at B26H0177 (elevation ground surface level +8.9 m MSL), and c) 2.5 m at B27A0162 near the lake at the northern edge of the Veluwe (elevation ground surface level +1.5 m MSL). The effects of varying groundwater level and distance to controlling surface water result in different responses of the hydraulic head to precipitation (d) (source: www.grondwatertools.nl/gwsinbeeld).

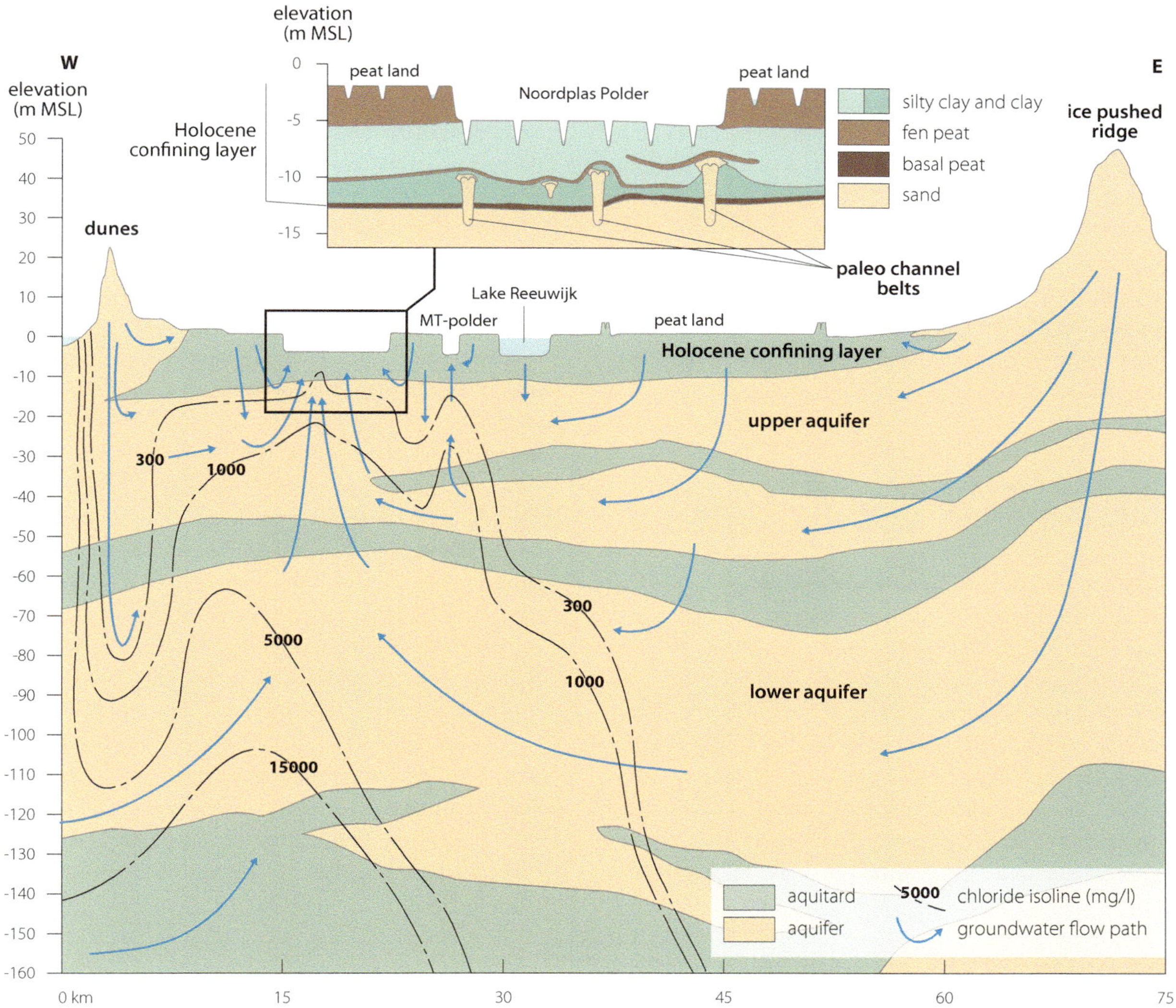

Figure 17.10. Regional W-E cross section perpendicular to the North Sea from the coastal dunes to the Utrechtse Heuvelrug (ice-pushed ridge), showing patterns of groundwater flow and chloride isolines (reprinted from De Louw et al., 2010 with permission from Elsevier).

freshwater lens floating on saline groundwater. The fresh groundwater flows and discharges towards the edges of the dunes both seaward and towards the coastal lowland. Submarine fresh groundwater discharge to the North Sea has been detected several hundred metres from the coastline, for example north of the island of Terschelling (Auken et al., 2010). In the Amsterdam dunes, water tables reach up to 7 m above sea level and the 5 km wide freshwater lens reaches a maximum depth of about 120 m (Geelen et al., 2017). Under natural conditions, the fresh groundwater lenses in the dunes at 21 locations along the Dutch North Sea coast reach depths of 16 to 140 m for water table levels of 0.35 to 8 m above sea level and correspond to a Badon Ghijben-Herzberg ratio (Fig. 17.2) ranging between 6 and 46, and 18.6 on average (Stuyfzand, 2017). However, the observed depths of the fresh groundwater lenses below the dunes are often shallower than suggested by the theoretical Badon Ghijben-Herzberg ratio (Fig. 17.2), probably as a result of poorly permeable clay layers, hampering vertical flow (Stuyfzand, 2017). The Dutch coast is protected by continuous addition of sand to counter the ongoing erosion of the coast. This sand supply has an impact on the spatial extent and pattern of groundwater flow in adjacent dunes. This is especially apparent at the location of the intensely studied mega-scale sand nourishment (sand motor/sand engine) constructed in 2011 at the coast near the Hague. Freshwater lenses formed in small dunes that developed at the nourishment itself (Huizer et al., 2016; Huizer, 2019) while groundwater tables in the adjacent water supply dunes are artificially kept at the same level by a newly installed drainage system.

The managed surface water levels, and therefore the groundwater tables, lie below sea level in the polders of the coastal lowland. This induces a continued subregional flow of groundwater towards the low-lying polder areas from the relatively higher regions, such as coastal dunes in the west, ice-pushed ridges in the east, and from the less deep-lying polders to the deeper ones, as well as from infiltrating sea water (Fig. 17.10; De Louw et al., 2010). Figure 17.12 shows the existence of very small local flow systems within the polders. Upward seepage of groundwater into the polders includes flow from the Pleistocene aquifers into the overlying Holocene units (Fig. 17.12). Vertical seepage of brackish groundwater from the Pleistocene aquifers through paleochannel deposits influences the chemical composition in the Holocene units.

Coastal flow systems; salinization and saline seepage

History of salinization and saline seepage

From early investigations in the 1850s up to the present-day, studies have been carried out to address the possible natural and human-induced causes of observed – often complex – distributions of fresh, brackish, and saline groundwater in Holocene and Pleistocene units (Dubois, 1903; Versluys, 1931; De Vries, 1981; Appelo & Geirnaert, 1991; Meinardi, 1991; Stuyfzand, 1993; Oude Essink, 1996; Post et al., 2003; Post, 2004; Delsman et al., 2014). Fresh groundwater is defined by chloride (Cl) concentrations below 150 mg/l, while saline groundwater is typically defined by concentrations above 1000 mg/l, with intermediate concentrations being considered brackish in the Netherlands (Dufour, 2000). An important focus was on the role of – Holocene – sea-level changes and sea-water intrusion on salinization of the groundwater. The chloride concentration of sea water is about 19000 mg/l.

Various theories have been developed during the last decades about the occurrence of saline and brackish groundwater in the Dutch coastal aquifers. The physical processes under consideration are advection, dispersion, molecular diffusion and possibly also chemical osmosis (Volker, 1961; Meinardi, 1973; Volker & Van der Molen, 1991; Post et al., 2003; Post, 2004). More recently, a clearer integrated picture has been constructed based on modelling (De Louw, 2013; Delsman, 2015; De Louw et al., 2019). Saline groundwater in the subsurface of the Dutch coastal area originates from syn-sedimentary groundwater in marine deposits of Pliocene to Early-Pleistocene origin as well as from transgressions of the sea during the Holocene (Post et al., 2003; Delsman et al., 2014). During the Holocene rise of sea levels, sea water migrated into the coastal plains while the coastline moved inland (Vos, 2015). During these transgressions, density currents allowed sea water to infiltrate into the underlying aquifers relatively quickly. Post and Kooi (2003) calculated, using conceptual models, that this process of density-depend-

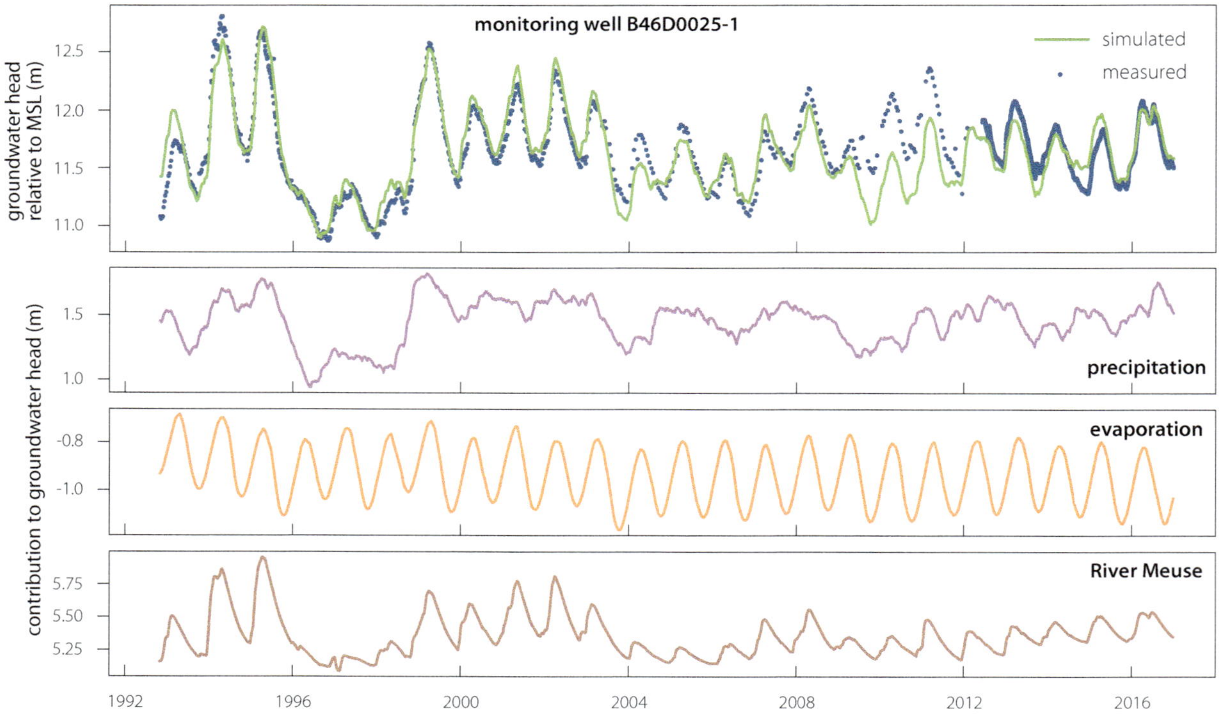

Figure 17.11. Time series model of groundwater heads (blue) measured at monitoring well B46D0025-1 (filter depth at 0.95-1.95 m MSL; ground surface at 13.8 m MSL) near the river Meuse in the eastern part of the Netherlands separating the fluctuation of the head (green) into contributions of precipitation (violet), evaporation (orange) and river level (brown).

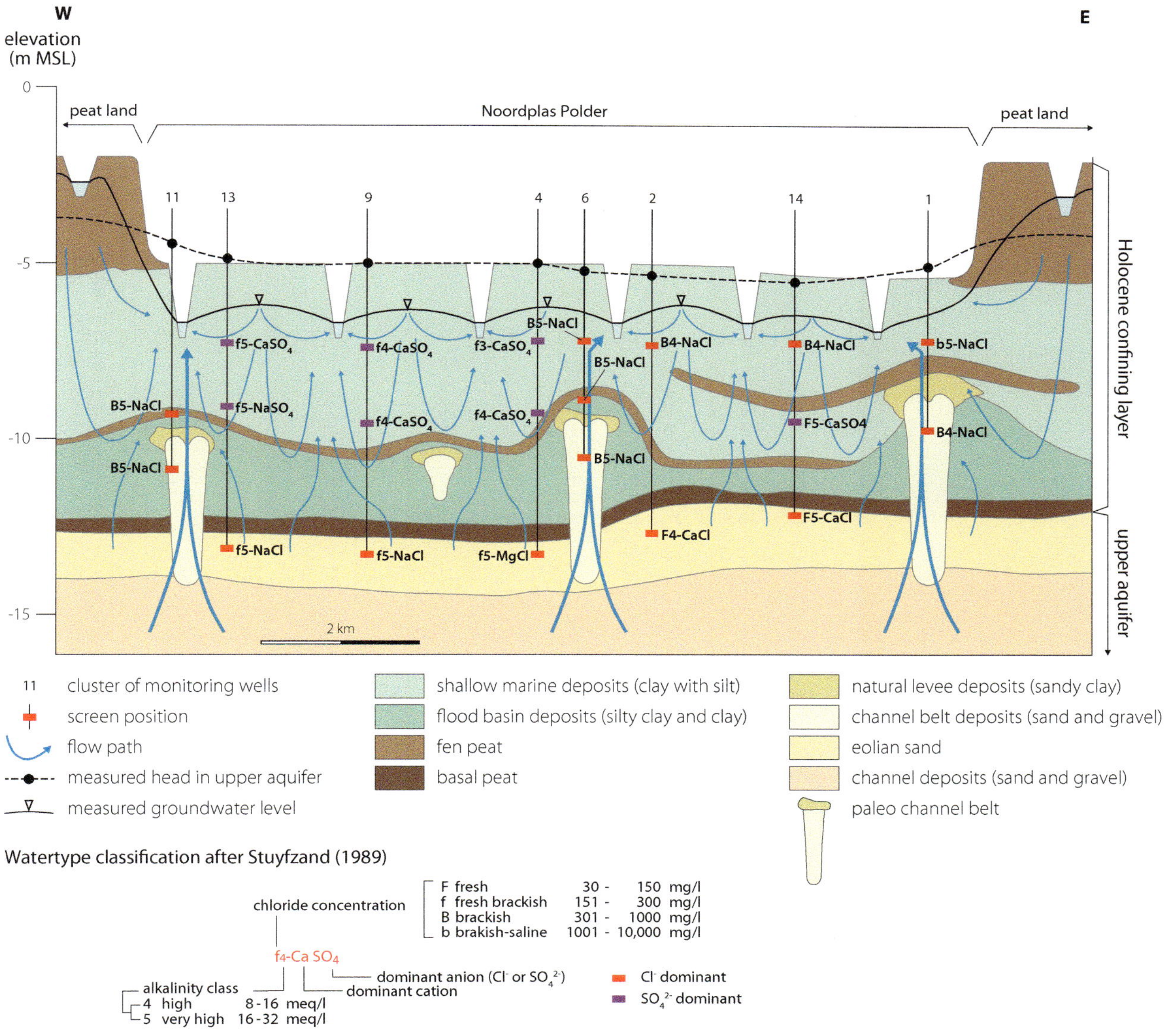

Figure 17.12. Local W-E cross section across a polder showing patterns of shallow groundwater flow systems in the Holocene succession, including focused vertical upward flow of brackish groundwater from Pleistocene aquifers. See Fig. 17.10 for location of cross section through the Noordplas polder (reprinted from De Louw et al., 2010 with permission from Elsevier).

ent free convection can make a 100 m thick aquifer system saline within one to two centuries. This is very fast on a geological time scale. Figure 17.13a-d shows the paleogeographic development of the Netherlands in four phases. The maximum transgression took place some 6 000 years ago (Fig. 17.13a; 3850 BC) and its distribution corresponds with the area where brackish-saline groundwater is found within 100 m depth (compare Figs 17.13a and 17.13e). Shortly after this event, dunes evolved and protected the coastal area of the western Netherlands from the sea. More inland, peat in a fresh water environment could grow. No further transgressions of the sea have taken place in the western part of the Netherlands and precipitation with a fresh signature and fresh river water could infiltrate into the subsurface. As a consequence, fresh groundwater reaches much deeper levels in the subsurface than in the southwestern most (Zeeland) and northern part of the Netherlands (compare Figs 17.13b and 17.13e). The latter areas have been under the influence of the sea until very recently, 800 to 1100 AD. Some 2000 years ago (Fig. 17.13c), Zeeland was temporarily covered with peat, but because of its extraction coupled with drainage by the Romans, it became a tidal area with salt marshes again. The peat eroded and most of Zeeland came under the influence of the sea for at least a few centuries during which the upper groundwater system became saline again. Very low-permeable clay layers, such as those in the Boom clay (Rupel Formation), hindered deeper infiltration of salt groundwater. Pre-Holocene fresh groundwater can still be found underneath brackish to saline groundwater (Van

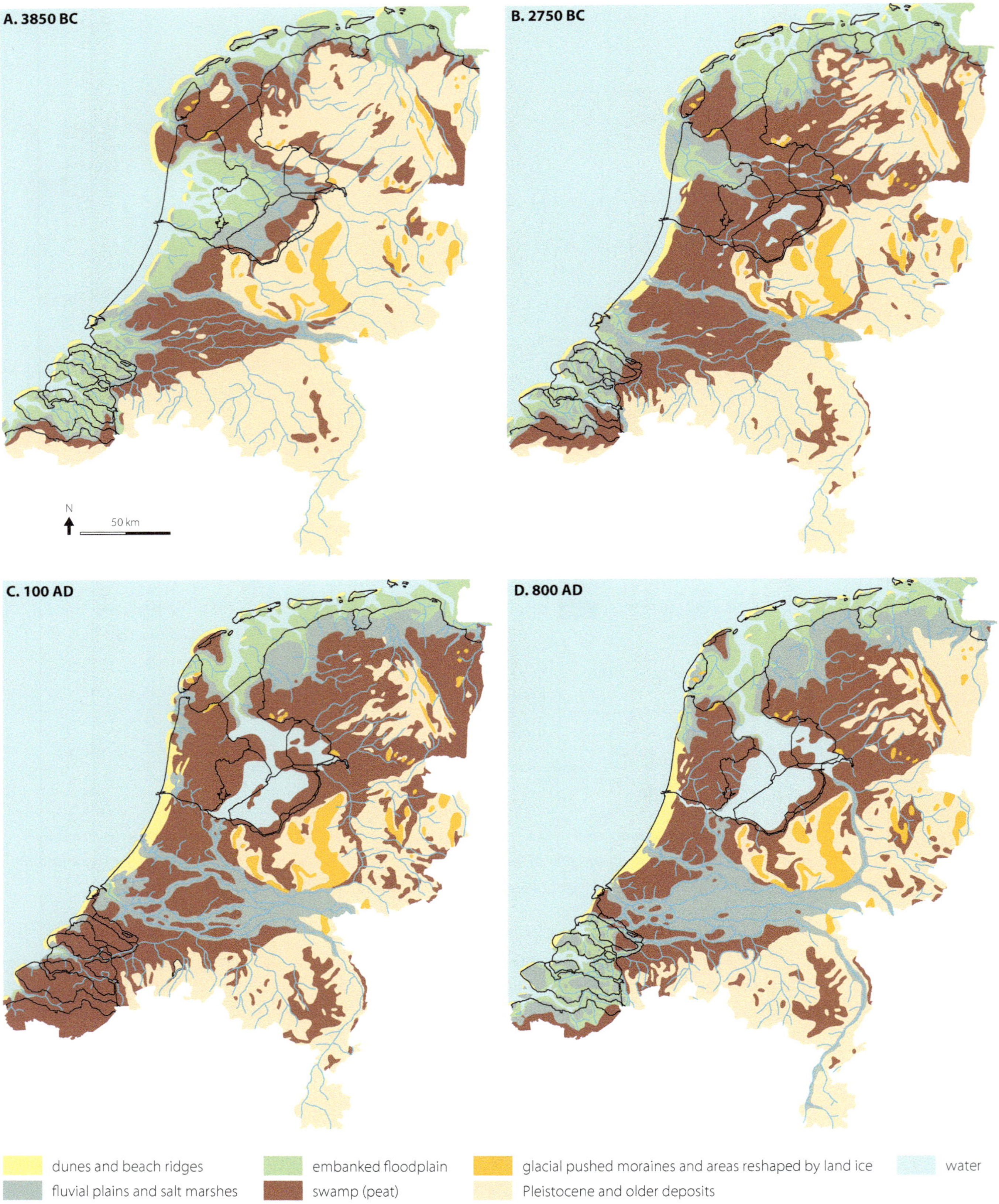

Figure 17.13. *Paleogeographic development of the Netherlands in four phases (after Vos, 2015): a) 3850 BC; b) 2750 BC; c) 100 AD; d) 800 AD; e) depth (in metres relative to mean sea level) of the brackish-saline (limit 1 g Cl/l) groundwater interface (Delsman et al., 2020). The distribution of salt water reflects the paleogeographic evolution of the coastal zone (as first visualized and described by De Louw, 2013).*

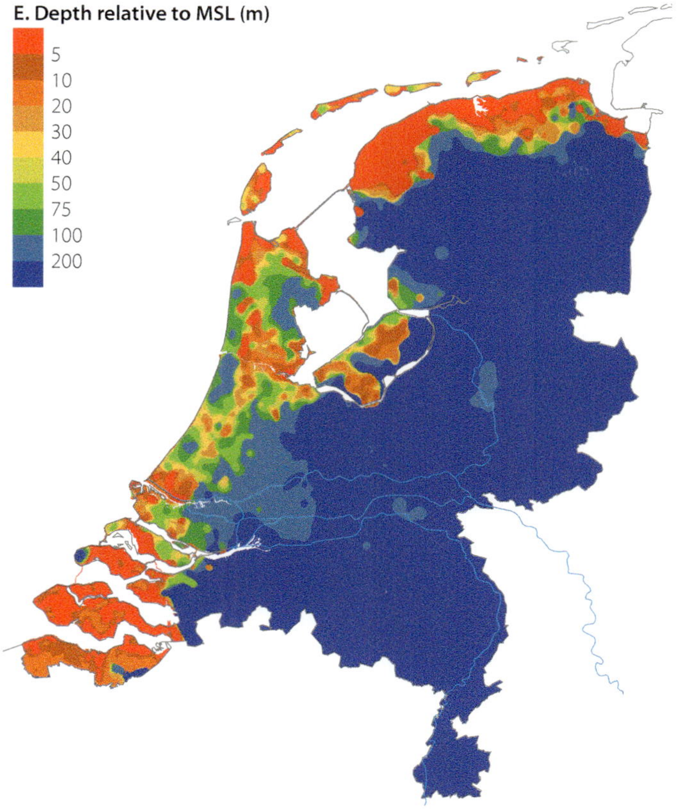

Baaren et al., 2018). The relatively recent transgression areas (southwestern and northern Netherlands) are characterized by very shallow brackish-saline groundwater (<5 m below ground surface; compare Figs 17.13d and 17.13e). Transgressions up to 3800 BC and the associated infiltration of salt water and peat formation that followed led to infiltration of freshwater and determined the characteristic fresh-salt distribution of the western part of the Netherlands. Over the past millennium, human activities have become more prominent in the area and changes to the land surface affected the groundwater system: dewatering of peaty and clayey soils, land subsidence caused by loading, peat oxidation as well as compaction and shrinkage of clay (Fokker et al., 2025, this volume) and the creation of polders. The artificially lowered water table relative to surrounding land has initiated groundwater flow in polder areas, transporting saline groundwater to shallower depths while fresh groundwater started to flow to the surface. In the coastal zone, over time, a significant head difference developed between the high mean sea level and the controlled surface water level. Many low-lying polders are currently experiencing upward groundwater flow (seepage) with a higher salinity (Fig. 17.13e; Oude Essink, 2001b). Delsman et al. (2014) simulated this paleohydrogeological evolution of groundwater salinity in a 2D section perpendicular to the coastline, through the Haarlemmermeer polder and Horstermeer, using a density-dependent groundwater model over the last 8500 years (Fig. 17.14). Since the creation of the deep reclaimed land, the brackish-saline transgression water flows back up to the surface, resulting in saline seepage in the polders. This salinization process is slow and the fresh-salt distribution is far from being in balance with the new boundary conditions in the west of the country. Older, deeper, and therefore saltier groundwater will continue to reach the surface water during the coming centuries, in a process referred to as 'autonomous salinization'. Special phenomena are salty boils, which are vents that connect the underlying aquifer and the surface water or ground level through the confining top layer (De Louw et al., 2010; Fig. 17.12). In the Noordplas Polder (Fig. 17.12), more than 60% of the salt load takes place via boils (De Louw et al., 2011) and this is even more than 80% in the Haarlemmermeer polder (Delsman, 2015).

Current and future impact of salinization

Salinization of the surface water system in the coastal zone affects agriculture, while decline in fresh groundwater reserves may affect domestic and industrial water supply. Under normal conditions, watercourses in low-lying areas are flushed with fresh water frequently and dispose of excess salt. Until recently shortages of fresh water during dry summers were infrequent, but under climate change conditions will probably happen more often (Ter Maat et al., 2014; Haasnoot et al., 2018). Modelling demonstrated that an anticipated relative sea-level rise accelerates the ongoing saltwater intrusion (e.g. Oude Essink et al., 2010). Climate change related sea-level rise, as well as continued land subsidence, are expected to jeopardize the fresh groundwater system even more in the future.

During the last decade, the subsurface is considered to contribute to an adaptation strategy towards a more sustainable and resilient freshwater supply (Strong, 2018) and has started to become a priority within the Dutch National Delta Programme (Deltaprogramma, 2021). Pilots have been set up to store surplus surface water in the subsurface during the wet season and extract it when needed during dry periods. Though the concept of Managed Aquifer Recharge or Aquifer Storage and Recovery is not new (Sprenger et al., 2017; Dillon et al., 2019), its application in a doable, cost-efficient way on the land-use scale by farmers in the Netherlands is (Delsman et al., 2018; Oude Essink et al., 2018; Acacia Water, 2019). Various measures are available to mitigate groundwater salinization (Oude Essink, 2001b; Sprenger et al., 2017; Dillon et al., 2019), several of which are shown in Figure 17.15. The large sand nourishments along the coast and development of fresh groundwater bodies in it (Huizer et al., 2016; Huizer, 2019) could also potentially delay the inflow of saline groundwater.

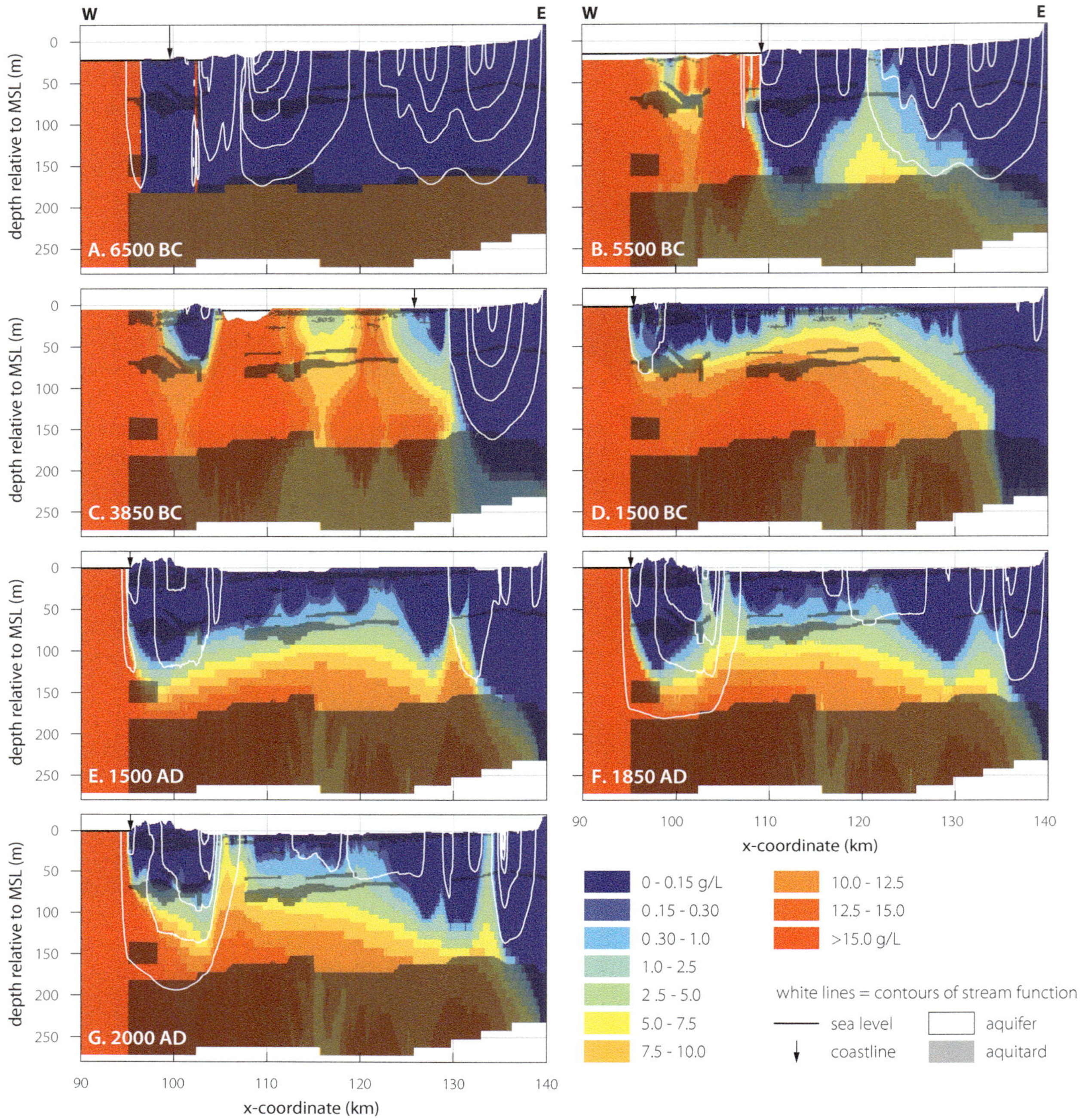

Figure 17.14. The development of the fresh-salt distribution during the Holocene for a cross-section perpendicular to the coast, through the dunes, polders (Haarlemmermeerpolder, Horstermeerpolder) and ice-pushed ridge (Utrechtse Heuvelrug), by courtesy of Delsman et al. (2014).

Supra-regional groundwater systems onshore

The tectonic and depositional history of the Roer Valley Rift System has played an important role in the development of a supra-regional transboundary flow system in the south-eastern part of the Netherlands since Miocene times. The Roer Valley Rift System, which extends into Germany and Belgium (Fig. 17.16) comprises the northern part of the Campine block in the south, the Roer Valley Graben in the centre, the Peel Block, and the Venlo Block in the northeast. The Roer Valley Graben is separated from these blocks by the deep-reaching NW-SE striking Peel Boundary Fault Zone in the north and Feldbiss Fault Zone in the south. The last rifting phase of the Roer Valley Rift System started during the Late Oligocene and continues today (Houtgast & Van Balen, 2000; Van Balen et al., 2005). At about the same time uplift of the Rhenish massif at the south-eastern end of the graben started and then accelerated during mid-late Miocene. A package of sediments of 1200 m of Late Oligocene to Quaternary age has accumulated in the subsiding graben.

The boundary fault zones of the Roer Valley Graben have clearly influenced the flow pattern of both very shallow local systems as well as the larger and deeper reaching systems. The long-time evolution of this large system introduced freshening of the groundwater to relatively great depths and cooling of the subsurface temperatures. Local extraction of groundwater, as well as extraction of groundwater in adjacent parts of Germany and Belgium have had an important impact on the hydraulic head distributions. In addition, changes in groundwater flow and hydraulic head due to mining activities have induced surface movements. Case-studies that illustrate these processes are presented below.

Impact of faults

Reduction of lateral permeability along faults in unconsolidated hydrogeological units may result from juxtaposition of hydrogeological units with different hydraulic properties, as well as processes such as clay smearing, grain-scale mixing, and iron oxide precipitation (Bense, 2004; Bense et al., 2013).

Studies of hydraulic heads in unconsolidated aquifers around the near-surface part of boundary fault zones of the Roer Valley Graben identified fault-related steps in groundwater level indicating that the faults act as lateral hydraulic barriers (Ernst & Ridder, 1960; Stuurman & Atari, 1997; Bense et al., 2003; Bense, 2004; Lapperre et al., 2019, 2022). The faults hamper, but do not completely prevent, phreatic lateral groundwater flow from adjacent topographically higher footwall areas (for example the Peel Block) towards topographically lower hanging wall areas in the graben (Lapperre et al., 2022). The reduced hydraulic conductivity of the fault zone in comparison to that of the adjacent hydrogeological units forces the groundwater to flow upwards at the footwall side of the fault zone (Lapperre et al., 2022; Fig. 17.17). Seepage along the Peel Boundary Fault zone to the surface has resulted in fault-related wet areas (in Dutch: *wijstgronden*) characterized by iron-rich groundwater and related specific groundwater-dependent ecosystems. Broers et al. (2021) found evidence from the groundwater age-dating study that the Peel Boundary Fault Zone also restricts groundwater flow from the Peel block into the graben at the somewhat greater depth of 100-150 m (Fig. 17.18).

Evolution of transboundary groundwater flow system

The recharge areas of the supra-regional transboundary flow system are located in Belgium and Germany in the Campine Block and the Ardennes-Rhenish Massif. Several hierarchical nested systems are part of this transboundary system (Engelen & Kloosterman, 1996). An important part of the flow is concentrated in the Roer Valley Graben (Vermeulen & Op den Kelder, 2020). At present, the fresh-brackish groundwater interface reaches its greatest depth (>600 m) in the southeast of the graben, while in its central part it is related to flow from the Campine recharge area. The thickness of the brackish groundwater layer can reach several hundred metres (Zuurdeeg et al., 1989). The ages of groundwater at shallow depths (<25

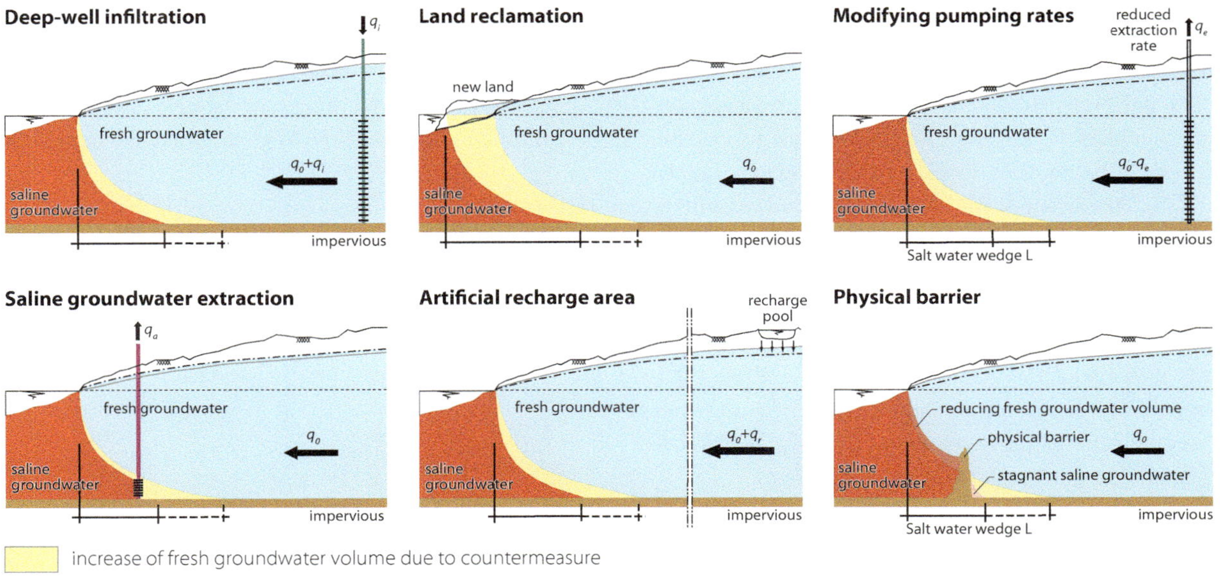

Figure 17.15. Concepts of countermeasures to control landward salt water intrusion (reprinted from Oude Essink, 2001a with permission from Elsevier).

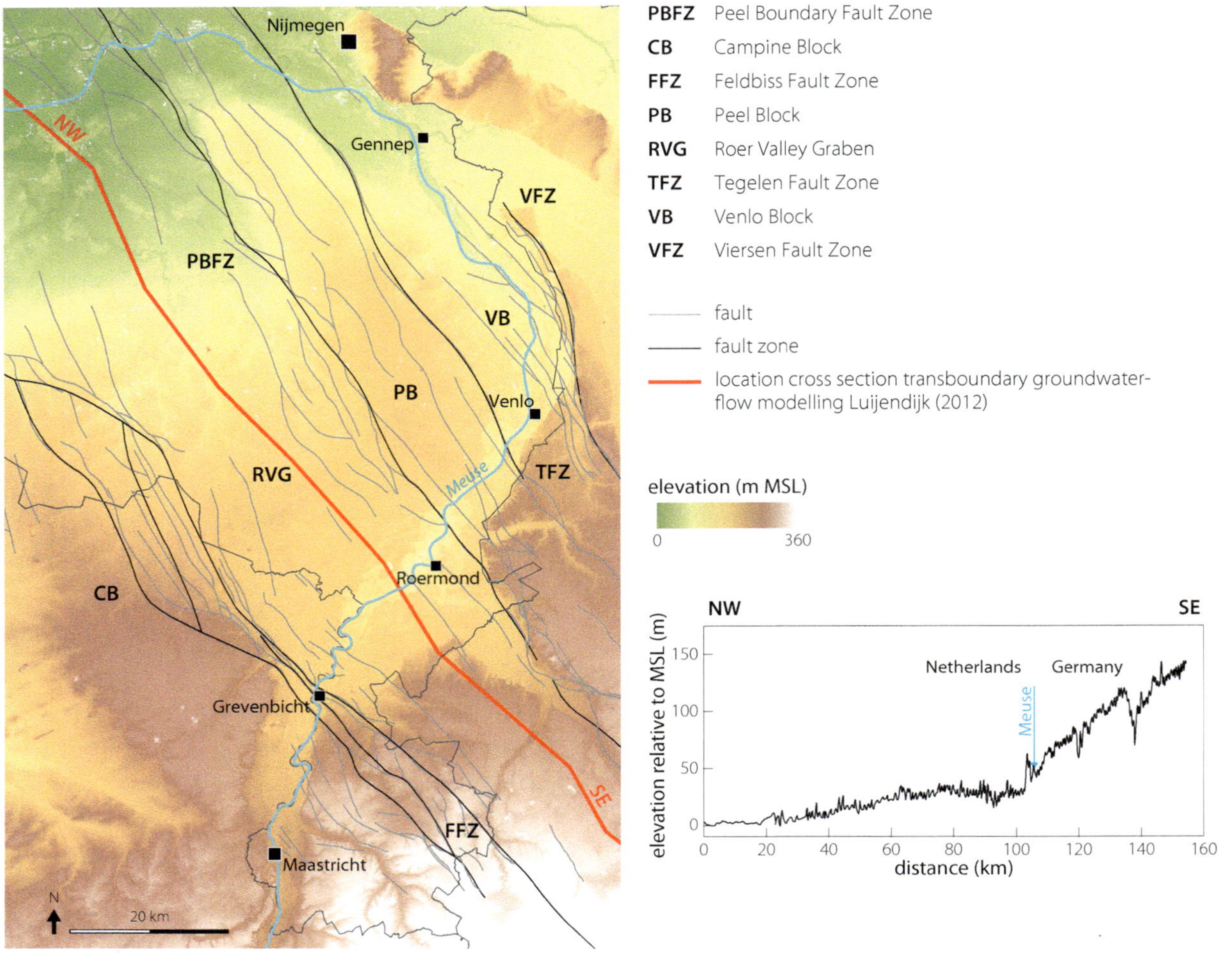

Figure 17.16. Roer Valley Rift system in the south-eastern part of the Netherlands (modified from Woolderink et al., 2019). The figure shows the location of the cross section used for transboundary groundwater-flow modelling in the graben (see Fig. 17.19) and the elevation of the ground surface along the cross section (reprinted with permission from Luijendijk, 2012).

m) are mostly less than 60 years (Broers et al., 2021). Multi-tracer age-dating of the sampled mixed water from public supply well fields clearly show an increase in age along the southeast to northwest flow path below two laterally extensive aquitards in the graben (Broers et al., 2021; Fig. 17.18). Pleistocene ages of more than 25000 years BP are observed in the northwestern part of the graben. This confirms previously revealed Pleistocene 14C ages of groundwater in discharge areas of the transboundary system in the province of Brabant (Stuurman et al., 1990). Hence the recharge of the groundwater system and the subsequent evolution of the groundwater flow occurred under very different boundary conditions (climatic, sea level, topography of water table) than the current ones.

Luijendijk (2012) used numerical modelling code RIFT2D to simulate the evolution of transboundary groundwater flow since Miocene times (17 Ma) along a 2D SW-NE cross-section through the central part of the Roer Valley Graben and extension into Germany (Fig. 17.16). The applied hydrogeological model is a layered one with a maximum depth of about 4 km depth and includes Permian to Quaternary hydrogeological units. The simulated groundwater flow is driven by groundwater table gradients

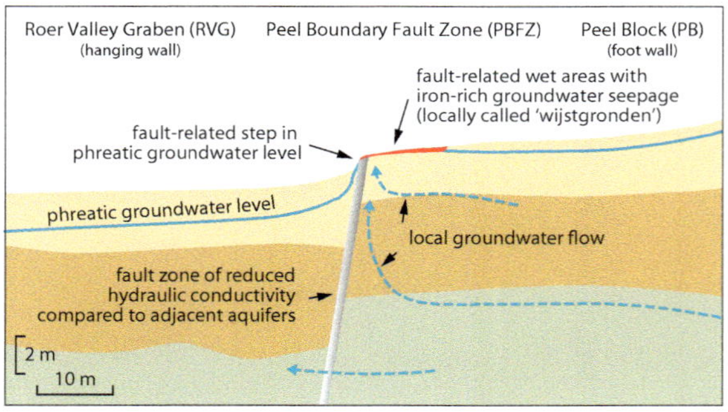

Figure 17.17. Impact of reduced lateral permeability of the Peel Boundary Fault Zone on the pattern of very shallow groundwater flow (modified from Lapperre et al., 2019).

related to a topographic relief of 130 m, sediment compaction and buoyancy forces generated by salinity and temperature differences. The simulation results (Fig. 17.19) show that most flow is concentrated in the upper 500 m. It also indicates that freshening of groundwater in Miocene and older marine sedimentary sequences in the graben can be explained by supra-regional flow that penetrated the sediments following the Pliocene-Pleistocene shift from marine to continental conditions in the area. The infiltration of relatively cool meteoric water caused a significant cooling of on average 14°C compared to conductive temperatures in the graben and is in accordance with observed relatively low temperatures in the central and southeastern parts of the graben. The simulated heating by upward flow in the regional discharge area in the NW was very minor (<1°C). Positive temperature anomalies measured at wells at depths of 1-1.5 km at the northwestern end of the Roer Valley Graben cannot be explained quantitatively by 2D flow simulations, by variations in thermal conductivity with depth (Luijendijk, 2012) nor by the presence of magmatic rocks in the subsurface (Stegers et al., 2018).

Impact of groundwater extraction

Measured groundwater heads in the provinces of Noord Brabant and Limburg over the years 2005-2020 show decreasing trends of heads of 1cm/year to 10 cm/year (Zaadnoordijk et al., 2021). A concentration of trend slopes of <-10 cm/year occur in the SE of the Roer Valley Graben. Variations in precipitation and evaporation are an important cause for the decreasing trends. It is unclear to what extent increased groundwater extractions have contribut-

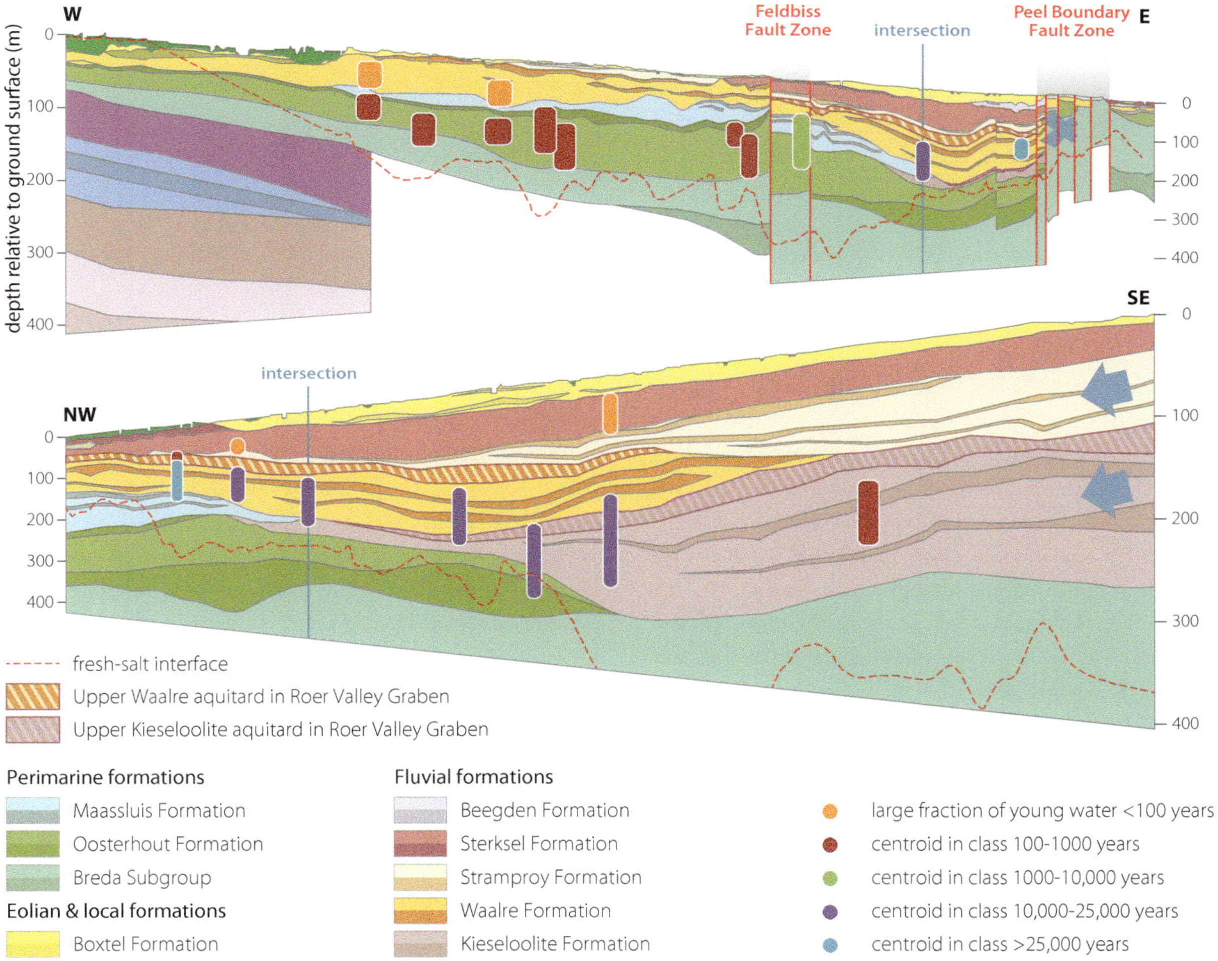

Figure 17.18. Relation between hydrogeological framework, groundwater-flow system and groundwater age in the Roer Valley Graben (NW-SE cross section) and perpendicular to the graben (W-E cross section). Groundwater age at shallow depths is mostly less than 100 years. The W-E cross section shows that the groundwater is much younger to the west of the graben (<1000 years) compared to groundwater at about the same depths in the graben (>1000 years reaching >25,000 years close to the Peel Boundary Fault Zone). Groundwater beneath two extensive aquitards in the graben increases in age in the direction of groundwater flow (blue arrow) from SE to NW. Modified from Broers et al. (2021).

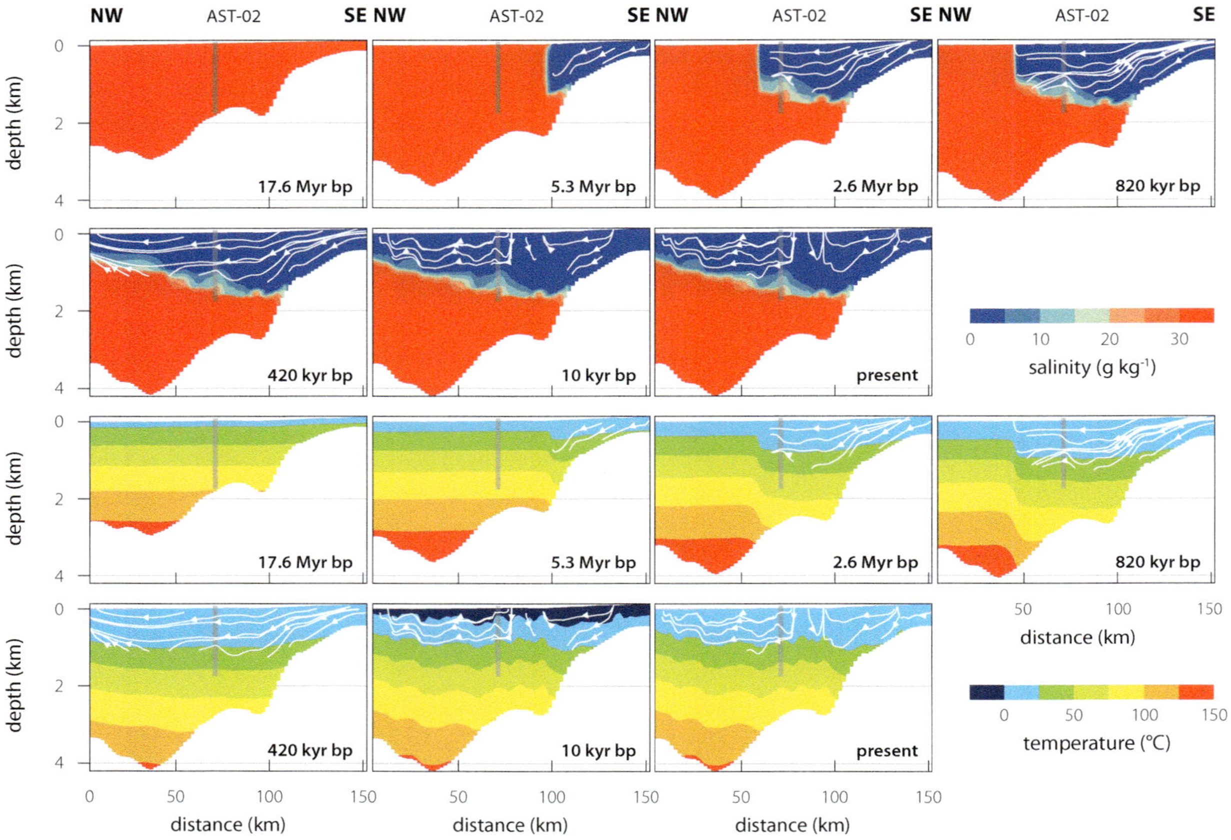

Figure 17.19. Evolution of modelled groundwater flow (white lines), salinity and temperature in the Roer Valley Graben from Late Miocene to present-day (reprinted with permission from Luijendijk, 2012). For location see Fig. 17.16.

ed. Vermeulen & Op den Kelder (2020) used the IBRAHYMv2 model to study the water balance and the changes through time of the impact of total groundwater extractions by the Dutch provinces of Limburg and Noord Brabant, German Waterboard Erftverband, Belgium, as well as extractions related to the open-pit lignite mines Inden and Hambach in Germany on the groundwater system in the Roer Valley Graben. The base of the applied hydrogeological model in the Dutch part of the model is the Oligocene marine Boom Member. The thick Miocene Breda Subgroup in the model consists of poorly permeable marine sands, in which lignite layers are present. Groundwater extractions from the Inden pit mine are mainly located below these high resistance lignite aquitards. The hydrogeological framework of the model includes lateral resistances to groundwater flow of the major faults that vary spatially and with depth. The assigned resistance for the Peelrand Fault ranges from 10,000 days (first and second aquifer) up to 100,000 days (Kieseloolite Formation) to 10,000 days in the Breda Subgroup. The resistance of the Feldbiss Fault varies from 1000 days (first and second aquifer) to 10,000 days for deeper aquifers. Vermeulen & Op den Kelder (2020) pointed out that it was difficult to determine the actual resistance values due to a lack of groundwater head data on either side of the faults. The computed results of a water balance for the Roer Valley Graben indicate that the Graben is for 41% fed by recharge from net precipitation, for 20% by infiltration of water from the river Meuse and for 26% by lateral inflow of groundwater into the Roer Valley Graben. This lateral inflow occurs from the south, the southwest crossing the Feldbiss fault, and from the northeast crossing the Peel Boundary Fault. Most of this lateral inflow enters from the south of the Graben, where a ca 400 metre thick aquifer intercalated with aquitards crops out at the surface. Model computations of the groundwater dynamics on a yearly basis from 1955 to the end of 2018 show that groundwater extractions in the Netherlands and Germany mutually influence drawdown of hydraulic heads (Vermeulen & Op den Kelder, 2020). The effect of extractions and lowering of hydraulic heads in the confined aquifer below the aquitards may extend laterally over large distances of 10 to 100 km. The order of magnitude of the response time of changes in hydraulic head in the confined aquifer over a distance of L= 10 km can be estimated from L^2S_S/K (see Appendix for explanation) and properties of the confined aquifer. The response time varies between 50 and 1000 days, assuming K= 1-2 m/day and S_S =1.0E-5m^{-1} to 1.0E-6m^{-1}. The model calculations of Vermeulen & Op den Kelder (2020) show that open pit mines in Germany affect the total drawdown of the groundwater level in Limburg for 40% and in Noord Brabant for 25%. The extractions in Limburg influence the drawdown locally in Germany up to about 30%, and in Noord Brabant for about 25%. The extractions in Noord Brabant affect most of the drawdown for half of Limburg for up to 80% and the total drawdown in Germany for about 5 to 10% (Vermeulen & Op den Kelder, 2020). The computed main drawdown occurs in the deeper layers and confirms earlier findings (Stuurman & Vermeulen, 1996; see also Dufour, 1998).

The influence of extractions in Belgium and by Erftverband on drawdown in neighbouring countries were computed to be significantly less.

Impact of coal mining

Carboniferous coals in the southern part of the province of Limburg were mined until the mid-1970s. Dewatering of the mines lowered the mine water levels to 400-800 m MSL (Projectgroup GS-ZL, 2016). This long-term mining of coal induced land subsidence (Pöttgens, 1985). A large part of which was caused by pumping of groundwater from the galleries and the rocks surrounding them. After the mining of coal ended and pumping stopped, groundwater flow entered the mines again and started to increase the water levels in the mines. Small shallow seismic events (depth <10 km) have been observed in the former coal mining area since 1985, which according to Muntendam-Bos et al. (2021) most likely were associated with post-mining groundwater flow and rising mine water levels.

Caro Cuenca (2012) and Caro Cuenca et al. (2013) used satellite radar interferometry (InSAR) data to study vertical land surface movements between 1992 and 2009. They identified land surface uplifts in Limburg that showed a strong correlation with rising mine water levels at three old shafts, suggesting that groundwater causes the uplift. In addition, the identified uplifts were found to be strictly confined to areas bounded by faults, including the major NW-SE trending faults parallel to the Roer Valley Graben as well as faults orthogonal to the NW-SE faults. Observed strong hydraulic head differences across the faults suggest that the faults act as hydraulic barriers, hampering lateral flow across the fault (Caro Cuenca, 2012).

Hydrochemistry of groundwater systems

The hydrochemistry of groundwater systems includes both the chemical composition of groundwater and the groundwater quality. The chemical composition refers to the concentration of the different chemical components and compounds in the groundwater. The quality of groundwater concerns the concentration of chemical components in relation to the standards set for a specific purpose, such as production of drinking water, irrigation and groundwater-dependent nature. Understanding the natural and anthropogenic causes of the chemical composition of groundwater systems is a prerequisite for water management aimed at protecting fresh groundwater resources and groundwater-dependent ecosystems now and in the future. The major cations and anions that describe the chemical composition include Na, K, Ca, Mg, Cl, HCO3, SO4, NO3, as well as NH4, PO4, Fe, Mn and the measure for acidity of the water (pH). In addition, the groundwater also contains a wide variety of natural and anthropogenic trace elements and metals. The chemical composition of groundwater in a certain flow system results from a complex set of influences, such as initial composition of syn-sedimentary pore water, initial natural or anthropogenically influenced composition of the water recharging the system, the location and residence time of the groundwater in the flow system, temperature, transport of chemical compounds along the flow path, mixing of groundwaters and the geochemical and mineralogical composition and reactivity of the sediments. Differences in time and spatial scales as well as in flux of the different groundwater flow systems have an important impact on the chemical composition. Hydrochemical classifications have been developed to enable process-based interpretation of measured chemical compositions (Stuyfzand, 1999, 2005; Graf Pannatier et al., 2000). In addition, natural or anthropogenic chemical elements and components including stable and radioactive isotopes are used to estimate the age of groundwater and patterns of groundwater flow (Visser, 2009). In turn a combination of groundwater age-dating with measurement of the concentration of chemical components can be used to assess the change in water composition in time (e.g. Broers et al., 2021; Fig. 17.18). A database of chemical composition and types of groundwater in the Netherlands is publicly available from the national database (www. grondwatertools.nl).

Hydrochemical classification

A hydrochemical classification is based on a combination of parametres. The classification of water types included in the Dutch national database is a slight adaptation of the Stuyfzand classification (Graf Pannatier et al., 2000). Stuyfzand (1999, 2005) developed a widely used hydrochemical classification and system analysis which includes information on salinity, most important cation/anion and total hardness. Stuyfzand defined a hydrochemical groundwater system, or hydrosome as a coherent 3D unit of groundwater with a specific origin (for example,

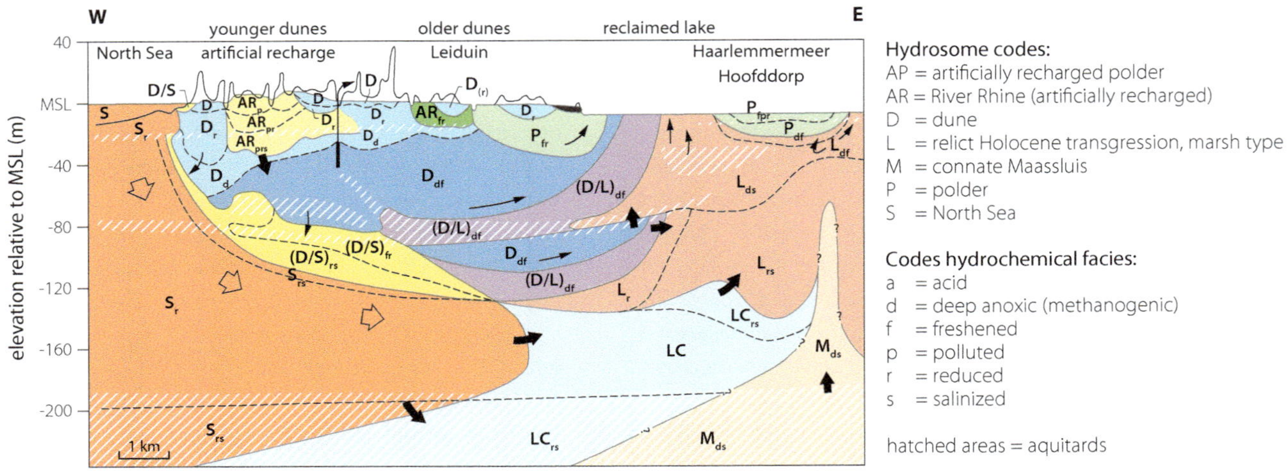

Figure 17.20. Cross section through the coastal dunes to the east of Amsterdam, showing the areal distribution of hydrosomes with their hydrochemical facies as based on 200 well-distributed samples. Arrows indicate flow direction of groundwater (reprinted from Mendizabal et al., 2011 with permission from Springer Nature; after Stuyfzand, 1993).

infiltrated precipitation, sea water, river water; Fig. 17.20). Within a certain hydrosome the chemical composition varies in time and space due to changes in recharge composition and in flow patterns and to chemical processes between water and the porous medium. Based on variations in chemical composition a hydrosome is subdivided into hydrochemical zones (facies). A facies is characterized by indices, such as a base exchange index, an age index, redox index, alkalinity index, pollution index (Stuyfzand, 1999; Mendizabal et al., 2011). The hydrochemical system therefore addresses the spatial distribution of groundwater bodies with specific origins (hydrosomes) and the characteristic hydrochemical zones (facies) within each hydrosome. A groundwater flow system may include various hydrosomes and a hydrosome may even exist, as paleo groundwater without flow (Stuyfzand, 1999). The change in chemical composition in the direction of groundwater flow can be small to large, from polluted to unpolluted, from acidic to basic, from oxic to anoxic, from no to significant base exchange and from fresh to brackish (Stuyfzand, 1999). The evolution in time and space of a fresh groundwater body and of hydrochemical facies due to changes in boundary conditions proceeds more slowly in comparison with adjustment of the groundwater flow system.

Role of mineralogy, geochemistry and reactivity

Organic matter and the mineralogical composition of sediments (Griffioen et al., 2016; Van Bergen & Kosters, 2025, this volume) influence the chemical composition of groundwater and the distribution of contaminants in the groundwater flow system. This is because mineralogical composition plays an important role in the reactivity of sediments (Van Gaans et al., 2011; Griffioen et al., 2013), affecting processes such as mineral dissolution and precipitation, exchange of cations and anions, sorption of organic substances, redox processes such as denitrification and pyrite oxidation. These reactive processes influence groundwater contamination patterns, because of their buffering capacity (STOWA, 2022). STOWA (2022) describes geochemical buffering processes of important reactive components, such as organic material, calcium carbonate, iron hydroxides, iron sulfides (pyrite), iron carbonates (siderite), clay minerals, glauconite and aluminium hydroxides. Organic material plays a role in redox reactions (denitrification, Mn(IV) reduction, Fe(III) reduction, sulfate reduction, methanogenesis), exchange of cations and anions and adsorption of organic and inorganic pollutants. Degradation of marine derived organic matter as a natural source of nutrients (NH_4, PO_4) is likely more intense than that of peat and remnants of terrestrial plants (Griffioen et al., 2013). Marine deposits are rich in calcium carbonates, while the concentration in fluvial deposits varies between poor to rich (Griffioen et al., 2016). Calcium carbonate acts as a buffer against acidification, involving dissolution of calcium carbonate in groundwater. Iron sulfides (especially pyrite FeS_2) are an important buffer against oxidation. Oxygen and nitrate reduce pyrite producing Fe, SO_4 and H. Pyrite is quite common in the Dutch subsurface, especially in marine deposits. Clay minerals are common in the subsurface and play a role in exchange of cations and anions.

There is a regional difference in reactivity of sedimentary sequences between the Holocene lowland and Pleistocene region: the Holocene marine sediments are most reactive and Upper Pleistocene eolian sands the least (Griffioen et al., 2013). Peat is also reactive with regard to redox chemistry but less on carbonates. Fluvial and glacial deposits show intermediate reactivity. Degradation of marine derived organic matter is likely to be most intense in the Holocene deposits.

The chemical composition of groundwater in the Holocene confining layer in the Noordplas Polder (Fig. 17.12) provides a nice example of the influence of infiltrating water on water-rock interaction in the reactive Holocene layer, and the effects of influx of chloride-rich groundwater into the Holocene from the underlying aquifer (De Louw, 2013). The infiltrating dilute, oxic or nitrate-contaminated, infiltrating waters from the surface of the agricultural polders oxidize pyrite, which in turn produce sulfate and protons. The resulting acid groundwater enhances dissolution of calcium carbonates in the Holocene layer. These two processes produce the calcium sulfate water type in the Holocene layer (Fig. 17.12). The local occurrence of a sodium chloride water type in the Holocene layer is explained by strong vertical upward flow of chloride-rich groundwater from the underlying aquifer through paleochannels (De Louw, 2013).

Groundwater quality

The Dutch Drinking Water Decree prescribes the limits for chemical parametres for drinking water. Important chemical quality parametres of groundwater for both drinking water production and groundwater-dependent eco-systems include salinity, hardness, iron (and manganese), methane and a variety of contaminants.

Salinity: The main parametre used to describe salinity is chloride concentration. The limit of 150 mg/l for fresh groundwater is also the maximum value allowed according to the Dutch Drinking Water Decree.

Hardness: Hardness, the quality parametre of the summed concentrations of calcium (Ca) and magnesium (Mg) is an important groundwater quality parametre for both drinking water production and groundwater dependent nature areas. These concentrations primarily depend on the equilibrium resulting from groundwater acidity interacting with carbonate, mainly calcite containing sediments. Since the equilibrium concentrations increase with increasing partial carbon dioxide pressure, hardness is typically elevated in groundwater that has been affected by organic matter degradation, for example due to the impact of oxidation processes in the root zone during recharge or by the occurrence of methanogenesis in organic rich deeply anoxic strata. Depending on the depositional conditions of the sediments, the original carbonate content varies. Typically, marine sediments contain more calcite but the actual carbonate content also depends on how much of the carbonate has already been leached. For example, in the main infiltration areas, such as the Utrechtse Heuvelrug carbonates have been leached out to a large extent and therefore the associated groundwater has lower hardness and higher acidity. A value of 1 mmol/l is used as an indicator concentration for hardness (Ca + Mg) for when softening is required for drinking water production.

Iron (and manganese): Iron (II) is considered an important groundwater quality parametre as it may cause discolorations to e.g. rust staining on clothing, but it is also known for causing well clogging problems when iron containing groundwater is mixed during extraction with more oxic groundwater. Manganese (II) is typically present at a tenth of the iron concentrations and, as with iron, is unwanted mainly because it causes black staining. Maximum concentrations considered for iron and manganese in drinking water (Dutch Drinking Water Decree) are 0.2 mg/l and 0.05 mg/l respectively. Iron(II) concentrations in fresh groundwater in Dutch aquifers range from absent up to 50 mg/l and are largely controlled by the mobilization through reduction of hydrous ferric oxides (HFO). The latter is determined by the reactivity of sedimentary organic matter (Hartog et al., 2004) and its precipitation in carbonate phases once these reach saturation (Van Beek et al., 2021). Subsurface iron removal has proven itself as a valuable technique for preventing well clogging for several of the drinking water production sites in the Netherlands, as well as in aquifer storage and recovery operations (Zuurbier et al., 2016; Van Beek et al., 2020).

Methane: Methane frequently occurs in groundwater in the Netherlands, with the highest concentrations typically found at relatively shallow depths near organic-rich (peat) strata. Removal of methane from drinking water is relatively difficult, however, at drinking water production sites with abstracted methane concentrations up to 50 mg/l, the stripping of methane may provide a useful energy source (Cirkel et al., 2015). The vast majority of methane in the shallow, fresh part of the groundwater system is of shallow biogenic origin (Cirkel et al., 2015; Schout, 2020; Schout et al., 2024). However, pathways have been shown to exist that may allow thermogenic methane to leak from greater depths through either natural pathways or from oil and gas wells and these may locally impact shallow fresh groundwater systems, even though that risk is considered to be relatively low (SodM, 2019).

Contaminants: A wide range of anthropogenic activities, such as agriculture and industry, pose threats to the quality of groundwater in the Netherlands. Many of the compounds that impact groundwater quality are considered to be contamination due to their nature and/or their concentration. A recent nationwide analysis of shallow groundwater showed that 85% of groundwater samples taken at 10 m depth contain chemical compounds from anthropogenic sources (Van Loon et al., 2020). The most abundant are pesticides and pharmaceuticals, encountered in respectively 60% and 35% of the sampled locations. Pharmaceuticals occur mostly close to infiltrating surface water or in urban areas. In addition, a range of other pollutants were found in 70% of the sampled locations. Nitrogen and phosphorus form part of fertilizers used in agriculture. An excess of these components may leach together with infiltrating excess precipitation into the groundwater. Nitrogen, occurring as nitrate in the groundwater, is a threat for drinking water quality (quality standard is 50 mg/l). The concentration of nitrate below agricultural areas is higher than in other regions, especially in sandy areas (Fraters et al., 2020). Groundwater contaminants may react with the sediments in the aquifer. Nitrate in contact with pyrite in the sediments may induce pyrite oxidation by nitrate reduction as shown by several studies in the province of Noord Brabant, an area seriously affected by agricultural pollution (Broers, 2002; Visser, 2009). This denitrification process will lower the concentration of nitrate in the groundwater but may also release secondary contaminants such as arsenic, nickel and zinc enclosed in pyrite (Visser, 2009). Observed contamination of groundwater by anthropogenic activities has been followed by studies to improve strategies for groundwater quality monitoring by Broers (2002) and to differentiate between geogenetic and anthropogenic origins of chemical components in groundwater (Griffioen et al., 2013).

Groundwater contamination is increasingly caused by contaminants that have been introduced relatively recently, although historic soil contamination from several decades ago is still causing considerable contamination in the Netherlands (e.g. Van Leeuwen et al., 2020). In the past, groundwater quality threats to drinking water production were mainly considered to originate from surface levels. The increasing use of the subsurface also poses a risk from multiple anthropogenic activities that take place below the ground surface. These include activities at depths of fresh groundwater resources, for example aquifer thermal energy storage (ATES) systems (Bonte et al., 2011; Bonte, 2013; Zuurbier et al., 2013). Risks from activities at greater depth may be related to conventional (oil and gas) exploration and production and to more recent and future activities, as described by Remmelts et al. (2025, this volume) and Neeft et al. (2025, this volume), such as geothermal energy production, subsurface hydrogen storage or CO_2 sequestration (Schout, 2020).

Deep hydrogeology onshore and offshore

Introduction

Current properties and flow conditions of groundwater systems in the realm of deep hydrogeology in onshore and offshore Netherlands are quite different from onshore shallow groundwater occurrences. The differences include chemical and physical properties, as well as past and present mechanisms involved in creating the properties and flow conditions (Verweij, 2003; Verweij et al., 2011, 2012). These include stress related mechanisms, such as sedimentary, tectonic and glacial loading and unloading, in addition to the topographic relief of groundwater tables. The climatic and geodynamic history largely control the distribution in time and space of the mechanisms. Both hydrogeology- and petroleum-based approaches have been used to reconstruct, and in part also quantify, the evolution of different mechanisms and their effect on the evolution of pore pressure, hydrochemical and groundwater flow conditions (Bouw & Oude Essink, 2003; Verweij, 2003; Verweij et al., 2011; Nelskamp et al., 2012; Verweij et al., 2012; Peeters et al., 2018). These studies revealed for example, that different times in the Permian to recent history a water table was able to establish itself in subaerial parts of the onshore and offshore and in subaerial regions close to the Netherlands southern and south eastern borders. Supra-regional topography-driven flow systems with recharge areas in the highs located south of the Netherlands have probably been active repeatedly and for long periods of time, e.g. during the Permian, the Late Jurassic-Early Cretaceous syn-rift period and from the Miocene to the present-day. Sedimentary loading played an important role during most of the post-Carboniferous evolution. During each geologic time period, including the current one, different subsidence and uplift histories have acted simultaneously on the groundwater (Fig. 17.21). As a result, different pressure and groundwater flow systems have coexisted and interacted laterally and vertically in the past and coexist today. Table 17.1 shows the main aquifers and aquitards in onshore and offshore Netherlands. It is good to realize that the porosity and permeability of these hydrogeologic units at their present-day depths reflect their burial history and the associated history of physical and chemical processes. This may lead to relatively low porosity and permeability in hydrogeologic units in which past maximum burial depth exceeded the present-day one. This can be observed in inverted parts of basins (Fig. 17.21), such as for example in Triassic and Jurassic units in the West Netherlands Basin (Nelskamp & Verweij, 2012). The evolution of flow and the chemical interaction of groundwater with the mineralogical and geochemical composition of the subsurface have been important in creating the current chemical compositions of groundwater. At greater depth important reactive geological units of influence on the hydrochemical composition are the coal-bearing Carboniferous Limburg Group, salt and carbonates of the Zechstein Group and Upper Germanic Trias Group, and the organic matter rich Posidonia shale. Higher temperatures in the deeper subsurface enhance chemical reactions. The result

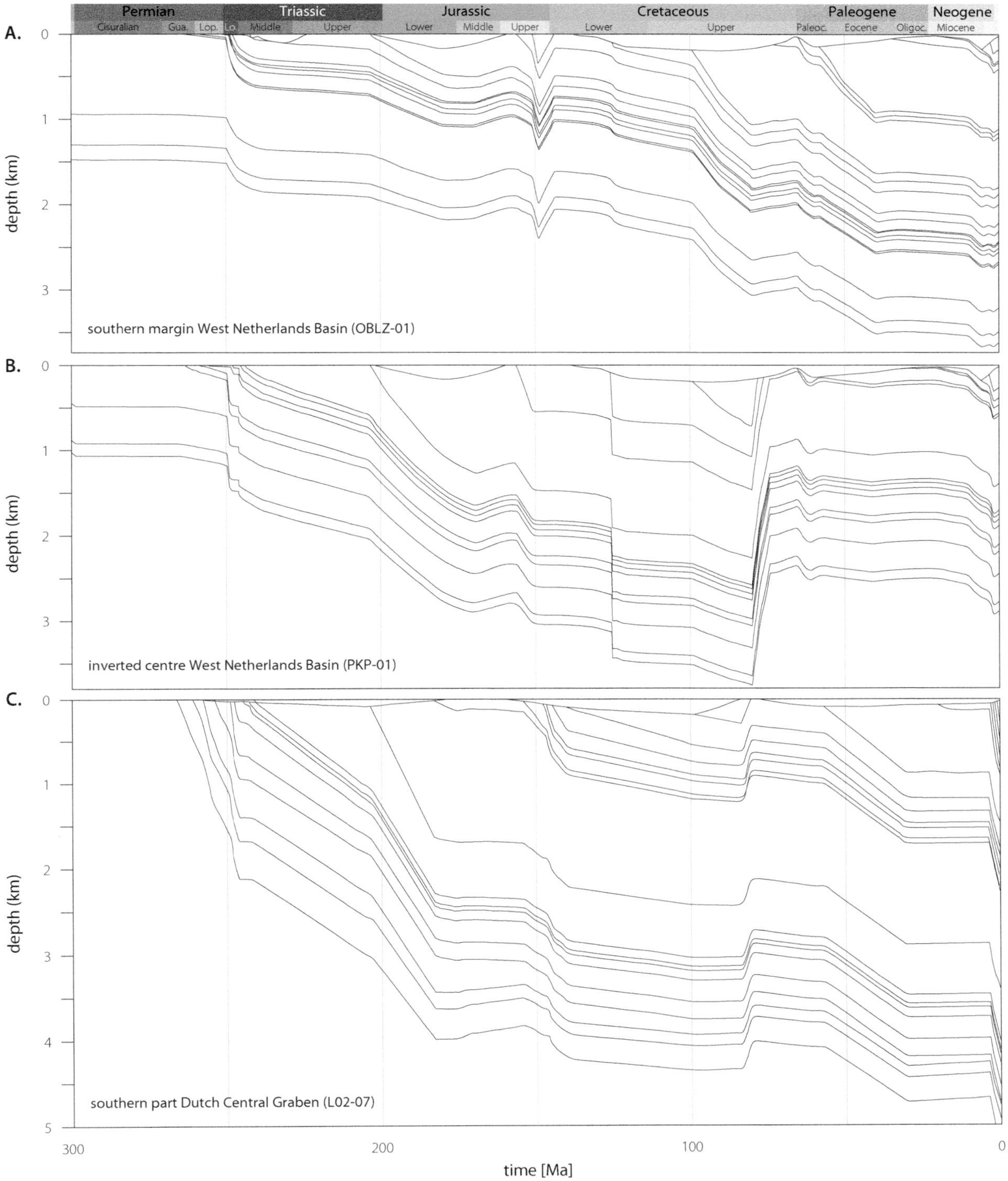

Figure 17.21. 1D extractions of 3D burial-history modelling at the southern margin (a), and inverted centre (b) of the onshore West Netherlands Basin, and (c) at the southern part of the offshore Dutch Central Graben (reprinted from Verweij et al., 2012 with permission from Cambridge University Press). For location of the wells see Fig. 17.3.

of water-rock interaction and paleo fluid flow are apparent today in chemical composition of groundwater and in diagenetic features. These diagenetic features are often used to reconstruct paleo fluid flow (e.g. Lee et al., 1989; Verweij, 2003; Gaupp & Okkerman, 2011).

Knowledge of groundwater properties (pressure, salinity, chemical composition, density) in the realm of deep hydrogeology is mainly based on measurements in reservoirs taken at oil and gas wells since the 1960s. The number of measurements of groundwater properties is very limited in comparison with those available for the fresh groundwater flow systems, and their distribution is clustered in reservoirs, depths and areas of interest for oil and gas industry (Remmelts et al., 2025, this volume). Depths of these measurements mostly exceed 1500-2000 m. In general, present-day natural flow conditions at these depths can be considered to be stagnant in comparison with time scales usually considered in studying onshore natural and managed shallow groundwater systems. Some measurements at depths of less than 1 km are available in the northernmost offshore where shallow gas fields occur. Present-day salinity and pore water pressure distributions are used hereafter as basis to describe deep hydrogeology.

Salinity

Salinity values derived from water chemistry data, petrophysical log analyses and analyses of pressure gradients exceed average sea-water salinity of 35 g/l (salinity expressed here as total dissolved solids, TDS; Verweij, 2018). There is a large vertical and lateral variation in salinity at depths exceeding 500 m, from about 50 g/l to even more than 300 g/l. Water samples show that most groundwaters are chloride dominated brines (Griffioen et al., 2016). The presence of evaporites of the Upper Germanic Trias, Zechstein groups and Silverpit Formation as well as diffusive transport seem to play important roles in controlling the salinity (Griffioen et al., 2016). The least saline groundwaters occur in the southern onshore and offshore. These areas are outside the area of distribution of evaporite sequences, and in addition have been affected by meteoric flushing multiple times during geologic history (Verweij, 2003), including during more recently since Miocene times (Luijendijk, 2012). The most saline brines are found in the northern onshore area and in the offshore, where the evaporites occur. Dissolution of halite can lead to large increases in sodium and chloride. This will initially enhance groundwater salinity near salt layers and salt structures. The subsequent movement of the resulting brine through aquifers may spread brines further. High salinity brines are not restricted to groundwater at great depths, as shown by the chloride-dominated brine (chloride content = ca 74 g/l) in a Paleogene aquifer at a depth of 582 m in the northeast of the Netherlands. There, the saline nature was explained by rock salt dissolution through groundwater contact with a salt structure (Glasbergen, 1981). Large variations in salinity are especially apparent in Jurassic and Triassic aquifers and reservoirs at the same depth of measurement in offshore basins and grabens (e.g. Terschelling Basin and Dutch Central Graben) that are bounded by large Zechstein salt structures. The lateral contacts of aquifers and reservoirs with Zechstein salt structures explain the observed salinity variations. Supporting evidence for the impact of Zechstein salt on properties of adjacent aquifers and reservoirs is the observed widespread occurrence of salt plugging/halite cementation of Triassic sandstones close to Zechstein salt structures which affects their porosity and permeability (e.g. Dronkert & Remmelts, 1996; Peeters et al., 2018).

Pore pressure and flow conditions

The main focus in this section is on the distribution of natural – pre-production – pore water pressures. Figures 17.22 and 17.23 provide an overview of pore water pressures in onshore and offshore Netherlands. The figures show cross plots of pore pressures versus depth relative to a standard hydrostatic gradient which corresponds to an increase of pressure with depth for a groundwater density of 1020 kg/m^3. In reality, the hydrostatic pressure (p) is determined by the actual density of the groundwater (ρ) at the depth of pressure measurement (z) in the aquifer $p = \rho gz$). The actual – salinity-related – density of the groundwater shows large variations in all aquifers. Relatively low densities occur in the southern onshore and offshore. The groundwater density in the Roer Valley Graben ranges between 1020-1090 kg/m^3 at the points of pressure measurements. Because brines are present in most pre-Cenozoic aquifers in the northern on- and offshore areas, the groundwater density is often much higher and can reach values of up to about 1200 kg/m^3. The magnitude of overpressure at a certain depth is the difference between the measured pressure and the hydrostatic pressure at that depth. Hence, pore water overpressures calculated from the cross plots reflect the effect of in situ pore water density as well as other causes of overpressuring. Sometimes only pressures measured in a petroleum fluid were available. These fluid pressures also incorporate the effect of the density contrast between a petroleum fluid and pore water. There is a wide spatial variation in magnitudes of overpressure on different scales both horizontally and vertically in Neogene to Carboniferous aquifers and reservoirs in the on- and offshore (Figs 17.22 and 17.23). The overpressures vary from near hydrostatic pressures to almost lithostatic pressures (lithostatic pressure = pressure of rocks and fluids of geologic units overlying the aquifer). Overpressure values corresponding to virtual hydraulic heads of 100s of metres to several kilometres above mean sea level are common. The observed overpressures reach much higher values than observed in – paleo – artesian conditions in topography-driven groundwater flow systems in the onshore Netherlands. Glasbergen (1985), for example, reported magnitudes of hydraulic head in the order of metres to about 12 metres above ground surface in Limburg,

The foregoing outline of onshore shallow groundwater systems includes examples of the physical interaction of groundwater with geologic media through hydromechanical coupling due to anthropogenic activities, such as pumping and drainage of groundwater. These activities have induced compaction of sediments and land surface subsidence. Comparable interactions are also active in response to geologic mechanisms operating at geologic timescales. Important geological mechanisms are stress-related processes, such as sedimentary, glacial, tectonic loading and unloading as well as processes that change the pore fluid volume, such as fluid generation by gas generation, dehydration of minerals, and aquathermal pressuring. The increase in load on a water-saturated rock pore may lead to reduction in pore volume (compaction), water flow from the pore, or increase in pore-water pressure. Compaction of water-saturated sediments requires that water be expelled. If the flow rate from the pore cannot keep pace with the loading rate acting on the water-saturated rock pore, the pore-water pressure will bear part of the increase in load and the pressure will increase, i.e. will create overpressuring. Lithologies with high matrix compressibility and low permeability such as clays, are especially susceptible to load-induced overpressuring. In such lithologies the increase in overpressure is accompanied by slowdown or even stop of compaction (known as compaction disequilibrium).

Once the overpressure generating mechanism is no longer active, the overpressure may dissipate over time. The rate of overpressure dissipation by water flow depends on the hydrogeologic and hydromechanical characteristics of the fluid saturated stratigraphic unit itself in relation to its position in the regional hydrogeologic framework. For example, the time required to dissipate pore water overpressure for 1D vertical flow through a stratigraphic unit of thickness L is determined by the ratio: L^2S_s/K (see Appendix). Overpressures generated by past mechanisms may persist for a very long time if lateral and vertical flow is hydraulically restricted. Stratigraphic units of low porosity and permeability such as mudstones, that overly an overpressured stratigraphic unit will restrict dissipation of the overpressure by vertical flow, increasingly so for thicker units. Stratigraphic units of extreme low porosity and permeability, such as the Zechstein and Triassic evaporites/salts, and fault zones of very low permeability will severely restrict flow and dissipation of overpressures.

Studies involving hydrodynamic analysis covering geological time and spatial scales have provided insights into observed pore fluid and pore water overpressure distributions in relation to the hydrogeological framework and burial history. The descriptions below are largely derived from these studies (Verweij et al., 2011, 2012; Verweij & Hegen, 2015).

Regional differences in the hydrogeological framework and the burial history of reservoirs/aquifers (the history of sedimentation and erosion after deposition of the aquifers; Fig. 17.21) lead to a regional subdivision of the Netherlands on- and offshore into southern and northern areas with different characteristics, important for pressure generation and dissipation. Regional differences in hydrogeological framework include the southward changing of facies towards more porous and permeable lithologies and differences in geological structures, dominated by large salt structures in the north and deep reaching faults in the south (Fig. 17.3). There are important regional differences in burial history (Fig. 17.21), including the concentration of major Late Cretaceous-Early Paleogene uplift and erosion in the south (e.g. of West Netherlands, Broad Fourteens, Central Netherlands, Lower Saxony basins, Roer Valley Graben) and to a minor extent in the centre of the Dutch Central Graben, and the overall northward increase of subsidence and associated depositional thickness of Paleogene and Neogene sediments (Munsterman et al., 2025, this volume). In the northernmost offshore subsidence and sedimentation rates accelerated and reached especially high magnitudes in the Early Pleistocene (up to 800 m/Myr), while sedimentation rates remained high (>200 m/Myr) since then.

Spatial distribution of pressure and flow conditions

The observed pore water pressures in on- and offshore demonstrate a clear difference between overpressure conditions in the southern and northern areas, reflecting differences in burial history and hydrogeological framework. In the southern onshore (Roer Valley Graben, West Netherlands Basin, Central Netherlands Basin) and adjacent offshore area, pressures close to hydrostatic prevail (Fig. 17.22), the only exceptions being small local pockets of minor overpressuring in Triassic reservoirs. In the northern offshore pore waters in most pre-Paleogene stratigraphic units are overpressured (Fig. 17.23).

In general, the competition between pressure generating and pressure dissipating mechanisms operating during a specific time interval controls the overpressure distribution. The relatively rapid late Cenozoic sedimentary loading plays an important role in explaining present-day overpressure distribution in Paleogene mudstones, Chalk Group, Rijnland Group and Jurassic sandstones in the northern offshore. This is in accordance with previous findings concerning the overpressures in the North Sea outside the Dutch sector (e.g. Gaarenstoom et al., 1993; Winefield et al., 2005). Late Cenozoic vertical loading by sedimentation is probably not the only mechanism involved in overpressure generation in the deeper and older Triassic, Upper Rotliegend and Carboniferous units in northern offshore basins and grabens and in Triassic and older units in northern and northeastern part of the onshore. Gas generation in Carboniferous source rocks and the entrapment of gas in these reservoirs (Remmelts et al., 2025, this volume) may have played a role, as well as

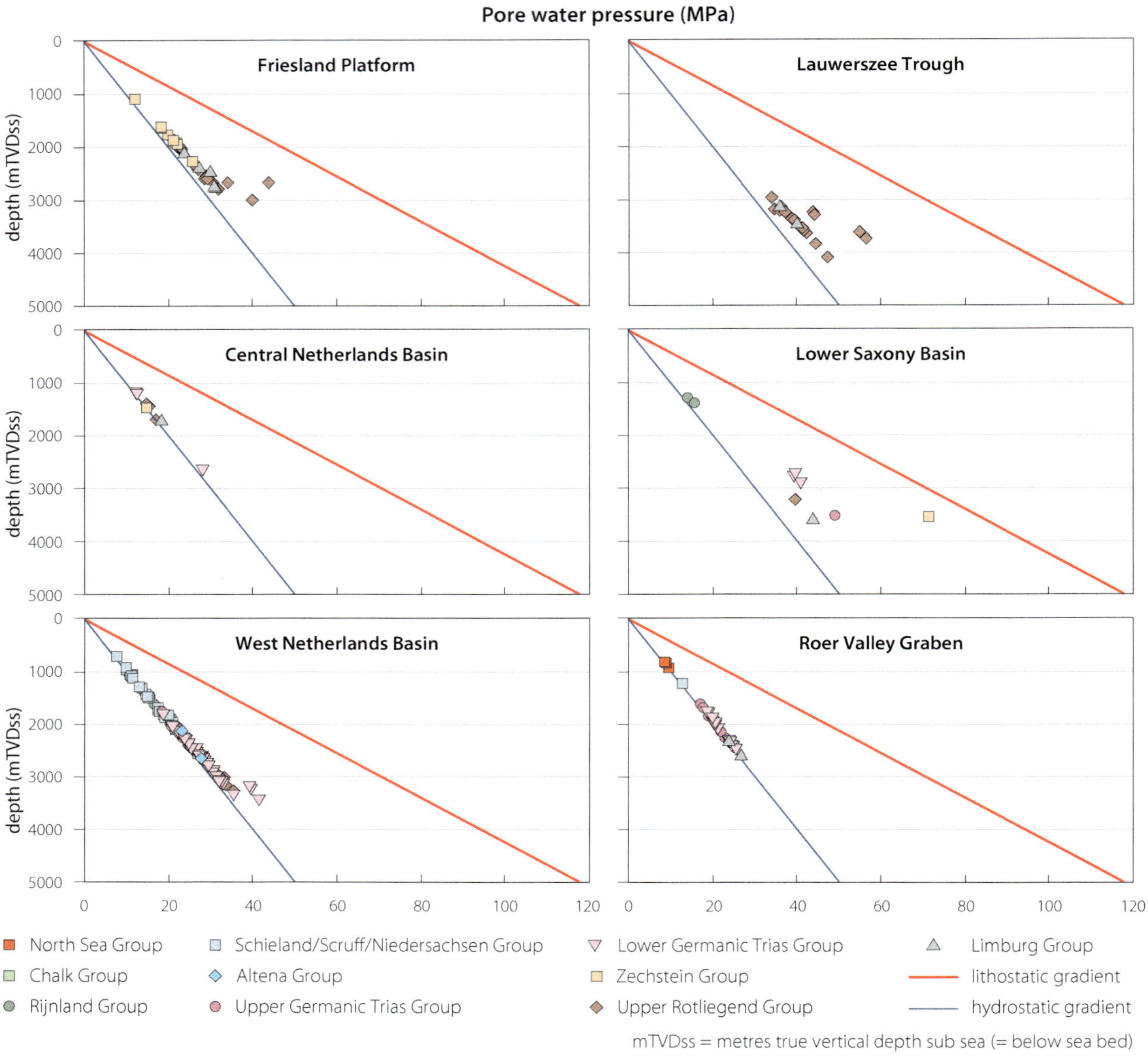

Figure 17.22. Multiwell cross plot of pore-water pressures versus depth for the main stratigraphic units and for important structural elements in the onshore Netherlands. Close-to-hydrostatic pressures prevail in the Roer Valley Graben, West Netherlands Basin and Central Netherlands Basin, while overpressures occur in the Lower Saxony Basin, Lauwerszee Trough and Friesland Platform. Standardized hydrostatic gradient is 10 MPa/km, i.e. increase in vertical pressure due to the weight of groundwater for water density of 1020 kg/m³; standardized lithostatic gradient is 23 MPa/km, i.e. increase in vertical weight of rock and groundwater. The location of the structural elements is shown in Fig. 17.3.

stress-related mechanisms other than vertical loading by sedimentation, such as tectonic and glacial loading and in-situ chemical diagenetic processes. The effect of Elsterian and Saalian glacial loading and unloading in the north eastern part of the Netherlands, for example, was studied by Amberg et al. (2022) using 3D basin modelling. They documented pressure buildup during glaciations and also the remaining overpressures in Mesozoic and older units today.

Severely hydraulically restricted conditions affect overpressure generation and are important in long-time preservation of overpressured conditions. More detailed outlines of regional overpressure and flow conditions are given below.

Cenozoic North Sea groups

There are only a very limited number of pore pressure measurements from oil and gas wells available for the Cenozoic sequences. Most measurements concern the Neogene and Quaternary sediments in the northernmost offshore. These include intercalated silty-sandy and clayey-silty sediments of the Southern North Sea Delta of

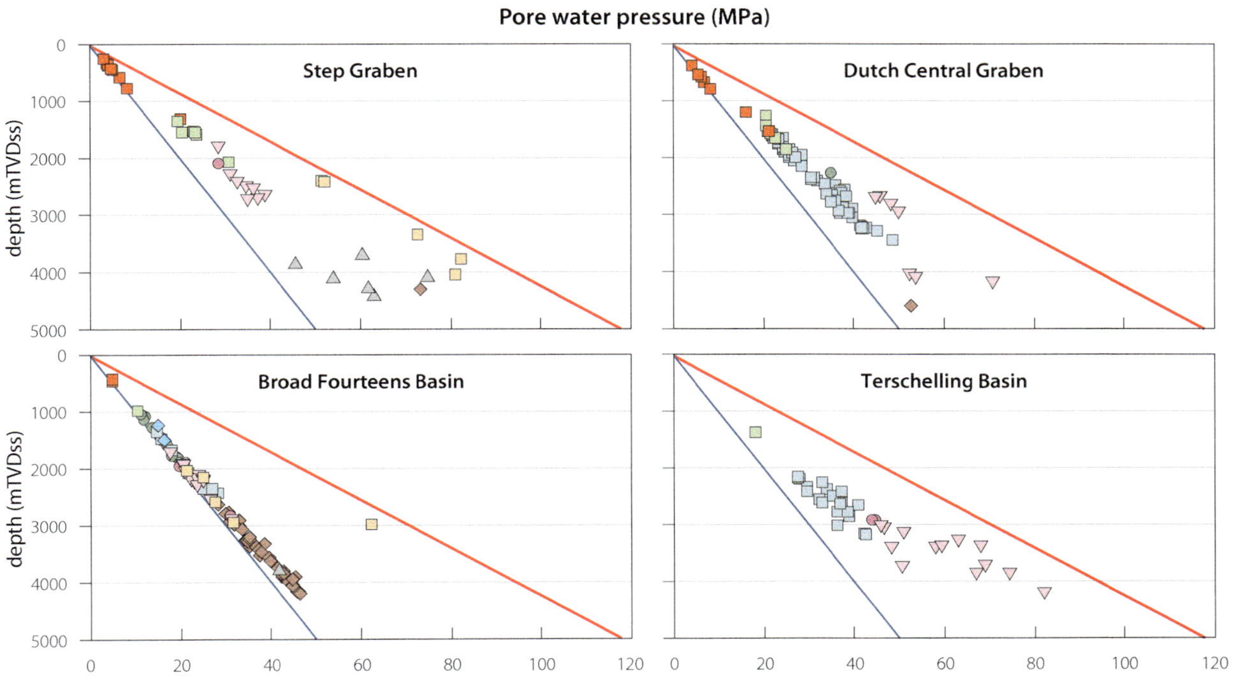

Figure 17.23. Multiwell cross plot of pore water pressures versus depth for the main stratigraphic units and for important structural elements in offshore Netherlands. Overpressured to severely overpressured conditions occur in Cretaceous and older units in the Terschelling Basin, Dutch Central Graben and Step Graben. In the Broad Fourteens Basin most observed pressures are near hydrostatic, and only mild overpressures occur at greater depth, with the exception of one observed severely overpressured Zechstein unit (see Fig. 17.22 for legend, and Fig. 17.3 for location of structural elements).

mostly Pliocene to Pleistocene age (Ten Veen et al., 2013). The porosity of the delta sediments is high (>20% to about 40%). The calculated – vertical – permeability of the intra-delta clay/silt sediments (2.8E-20 m^2 to 1.1E-18 m^2) is several orders of magnitude smaller than the measured – lateral – permeability of the silty/sandy sediments (2.5E-16 m^2 to 4.5E-12 m^2; Verweij, 2018). The delta reaches its greatest thickness and depth in the northeastern most part of the Dutch offshore (base delta at >1200 m depth). It is capped by 300-400 m of predominantly continental sediments of Late Pleistocene and Holocene age. The measured pore-water pressures in the Southern North Sea Delta are normal (i.e. hydrostatic) or close to normal. This suggests that the competition between pressure generating mechanisms (such as ongoing sedimentary loading and in-situ generation of microbial gas), compaction of the delta sedimentary sequences and pressure dissipating mechanisms (dewatering of the sediments by lateral and vertical flow) are in equilibrium.

Figure 17.23 shows a large stepwise change in measured pore water pressures between the normally pressured sediments of the Southern North Sea Delta and overpressures in pre-Paleogene sequences in the Step and Central Grabens. The figure also indicates that measured pressures are missing from the Paleogene section.

Data on mudweights used during drilling of the Cenozoic sequences are more widely available than measured pore fluid pressures and these were used as proxies for pore fluid pressure in Paleogene mudstones. The observed change of mudweights with depth suggests that slightly overpressured conditions occur in Paleogene mudstones in the northern offshore at depths greater than around 1000 m (Verweij & Hegen, 2015). Pliocene to Quaternary sedimentation rates in the northernmost offshore were calculated to be high enough to explain the occurrence of overpressure in the thick package of low permeability Paleogene mudstones (Verweij, 2019). The southward decreasing thickness and increasing sand content of the Paleogene sedimentary sequences reduces the likelihood of overpressuring of these sequences in the southern offshore and in the onshore. This is confirmed by the much lower mudweights used for drilling the Paleogene in the West Netherlands and Broad Fourteens basins in comparison with those in the northern offshore.

Basin modelling of the history of pressure and groundwater flow in response to Quaternary sedimentation rates in the offshore Broad Fourteens Basin have provided calculated vertical flow rates through Cenozoic sequences in the order of 0.02-0.1 mm/yr and lateral flow rates through Quaternary aquifers of more than 2 mm/yr (Verweij,

2003). Kooi (2000) modelled groundwater flow and natural land subsidence due to compaction of Cenozoic sedimentary sequences in the coastal area of the Netherlands in response to sediment loading during the Late Pleistocene and Holocene. The modelling showed that lateral loading-related flow through the sandy layers of the sand-clay sequence exerts a first-order control on the simulated current compaction-driven land subsidence rates of less than 0.1 mm/year.

Cretaceous Chalk and Rijnland groups

Significant overpressures exist in the Chalk Group in large parts of the Dutch offshore. Pore water overpressures increase northward from about 2.2 MPa at the northeastern border of the Broad Fourteens Basin to 4.4 MPa south of the Dutch Central Graben, to 5.8 MPa in the southern part of the Dutch Central Graben and finally to 9.9 MPa in the Step Graben in the northernmost part of the offshore (Fig. 17.23). The lateral increase in overpressure shows a strong relation with the northward increasing burial depth of the top of the Chalk, reaching >2 km, and high Pleistocene to present-day burial rates (Fig. 17.21). The observed overpressures indicate that the Chalk Group is not able to dewater rapidly enough to keep pace with the rapid ongoing sedimentary loading. The laterally extensive low permeability Paleogene mudstones delay the dissipation of overpressure in the Chalk Group through vertical groundwater flow. The overpressuring of the Chalk Group can be expected to be maintained for longer periods of time towards the north because of the northward increasing thickness and decreasing depth-related permeability of the Paleogene mudstones.

The largely argillaceous Rijnland Group includes a number of sandstone units, mostly in the southern part of the area. Figures 17.22 and 17.23 show that these sandstones are normally to only slightly overpressured in the onshore and southern part of the North Sea. One overpressure value (Fig. 17.23) and also mudweight proxies suggest that more overpressured conditions prevail in sandstones of the Friesland Member of the Rijnland Group in the southern part of Dutch Central Graben..

Upper Jurassic-Lower Cretaceous Schieland/Scruff groups

Pore water overpressures occur in sandstone aquifers of the Upper Jurassic-Lower Cretaceous Schieland/Scruff Groups in the Dutch Central Graben and Terschelling Basin (Fig. 17.23). The overpressures show a different lateral variation in comparison with the northward increasing values observed in the Chalk, exceeding 10 MPa in the Terschelling Basin, southernmost Central Graben and northernmost part of the graben. The maximum overpressures in the – inverted – centre of the graben are lower and reach about 6 MPa. Lateral flow through the Upper Jurassic aquifers is restricted by large fault zones and salt structures that form hydraulic barriers. The overpressure distribution in the sandstone aquifers of the Schieland/Scruff groups can largely be explained by a balance of rapid Late Cenozoic sedimentary loading and the preservation of the generated pressure which is controlled largely by vertical dewatering of the aquifers. This in turn is related to the distribution and variation in thickness of the overlying low permeable layers. The pore water overpressures exceeding 10 MPa in the aquifers in the Terschelling Basin and Dutch Central Graben are maintained by the overlying mudstones of the Upper Jurassic Group and Rijnland Group, the lower tight parts of the Chalk Group and the laterally extensive Paleogene mudstones. The total thickness of these poorly permeable units in the inverted centre of the graben is reduced due to erosion and the original reduced depositional thickness of the Paleogene mudstones and this has probably allowed overpressures to dissipate more rapidly by vertical groundwater flow. This is reflected in the observed lower overpressure values of 6 MPa.

Germanic Trias groups

Overpressured conditions in sandstone reservoirs/aquifers of the Germanic Trias groups are more severe than encountered in younger groups. The highest pore water overpressures in the northern offshore occur in the Terschelling Basin (reaching about 40 MPa), southern Dutch Central Graben (about 28 MPa) and directly south of the graben (about 18 MPa; Fig. 17.23). Peeters et al. (2018) studied the possible causes of the large variation of observed overpressures in the Terschelling Basin. They found that overpressures are generally ≥30 MPa, where laterally continuous evaporite/salt units of the Upper Germanic Trias Group are on top of the reservoirs and aquifers. The overpressures are <15 MPa in parts of the basin where this top hydraulic barrier is missing and reservoirs are capped by Upper Jurassic mudstones or Lower Cretaceous layers. The Upper Triassic has largely been eroded from the platform areas to the west, east and south of the basin and graben and no such severely overpressured compartments occur in the Triassic reservoirs in those locations. Nelskamp et al. (2012) used basin modelling to study the overpressure generation history in the Triassic in the Terschelling Basin and Dutch Central Graben. The simulation results indicate that timing of salt movement and closure of lateral flow barriers in Terschelling Basin and Dutch Central Graben control the overpressure generation history. The results also suggest that preservation of overpressures in such closed compartments can be maintained for very long periods of geological time.

Interestingly pore water overpressures have also been encountered in sandstone aquifers of the Germanic Trias

Group in the onshore Lower Saxony Basin (Fig. 17.22) in the northeast of the Netherlands. The pore water overpressures occur in aquifers lying on top of Zechstein salt that are restricted laterally by Zechstein salt structures or sealing faults and are capped by evaporite/salt units of the Upper Germanic Trias Group. The overpressures exceed 10 MPa with a maximum of 13.6 MPa in such hydraulically restricted compartments. Pressures are close to normal in onshore areas where the Upper Triassic evaporite hydraulic barrier is missing.

Permian Zechstein Group

Fluid overpressures and pore water overpressures in the Zechstein Group (Fig. 17.22 and Fig. 17.23) show magnitudes from less than 1 MPa to near lithostatic values. The location of the pressure measurements in the Zechstein Group (enclosed in evaporites, or situated at the top or bottom of the group) significantly influences the magnitude of the overpressure. The highest overpressures occur in units (such as carbonate and anhydrite units, including stringers) completely enclosed in evaporites. Overpressures in carbonates enclosed by evaporites in the Zechstein Group may reach very high values also in the onshore (e.g. in Lower Saxony Basin, Fig. 17.22).

Permian Upper Rotliegend Group

The Slochteren sandstone units of the Upper Rotliegend Group change gradually northward in the offshore into the poorly permeable siltstones and evaporitic mudstones of the Silverpit Formation. These poorly permeable units are a regional hydraulic barrier for lateral flow. The evaporites of the Zechstein provide the regional top hydraulic barrier of the sandstones. Both the Slochteren sandstone units and the Zechstein evaporites are missing from the Texel IJsselmeer High. The overpressure values are high to very high along the offshore northern limits of the deeply buried Lower Slochteren sandstones, where the sandstones become increasingly faulted and become intercalated with mudstones that form barriers to flow. Extremely high fluid overpressures have been observed in the southern part of the Dutch Central Graben (60 MPa) and in the Terschelling Basin (38 MPa).

Overpressures in the Slochteren sandstone aquifers and reservoirs located below the Zechstein salt hydraulic barrier in the northern provinces of the Netherlands and adjacent offshore show a general decreasing trend towards the Texel IJsselmeer High, where the Zechstein is missing. The decrease of overpressures away from the severely overpressured conditions in the Dutch Central Graben and Terschelling Basin often occurs in a stepwise manner related to the numerous faults that intersect the sandstone aquifers and restrict its lateral hydraulic continuity. This is apparent, for example, along a cross section just north of the Wadden Islands (Fig. 17.24) and in the Lauwerszee

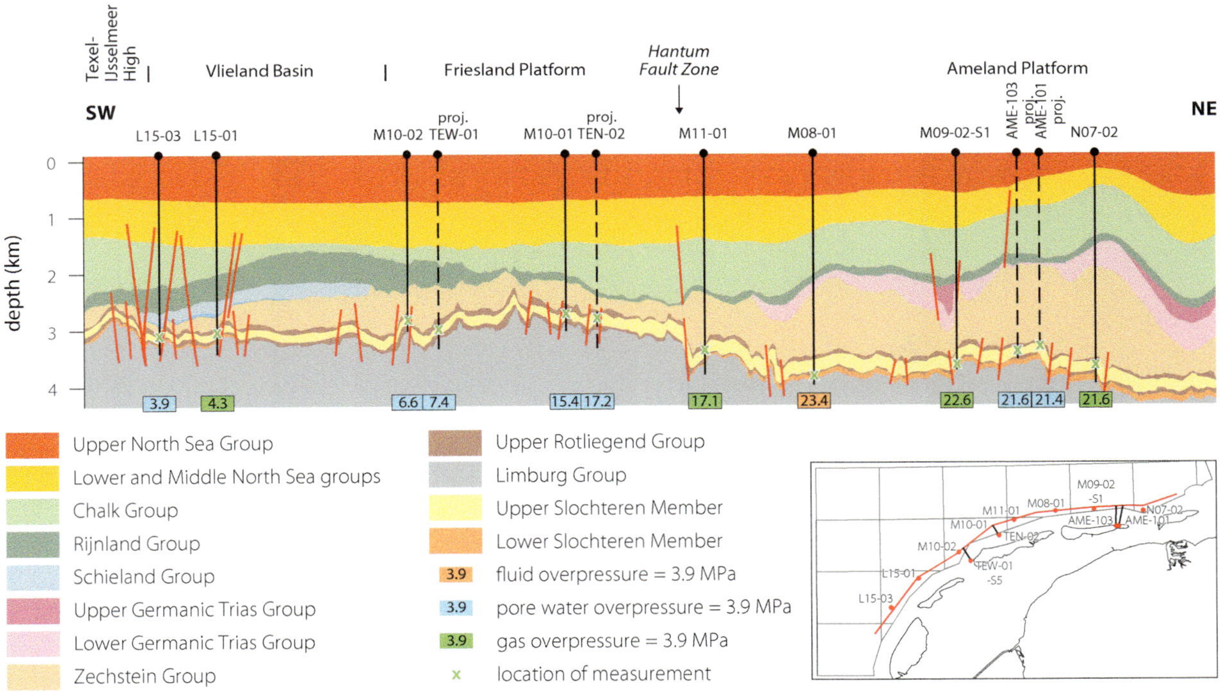

Figure 17.24. SW-NE cross section along Wadden islands showing southward stepwise decrease of pore water overpressures in the faulted Slochteren sandstone aquifers and reservoirs below the regional Zechstein top seal towards the Texel IJsselmeer High (modified from Verweij & Hegen, 2015).

Trough in the northwestern part of which, where very high overpressures are maintained in pressure compartments bounded by faults. Pressure differences of different magnitudes exist across faults in the Trough (Corona, 2005; Corona et al., 2010) and pore water overpressures generally decrease stepwise from about 20 MPa in the northwest to about 4 MPa in the southeast. In the provinces of Friesland and Groningen, to the west and east of the Trough, overpressuring dissipates. In the Lower Saxony Basin fluid overpressures vary between about 6.5 and 9.8 MPa in the Slochteren sandstones and are higher than those observed to the north and west of this basin, but are lower than those in overlying pressure compartments in Triassic sandstones at the same location.

Carboniferous Limburg and Limestone groups

Measured pore water overpressures in sandstone units of the Carboniferous Limburg Group are relatively scarce and scattered. The highest magnitudes were observed at depths >3700 m in the northernmost offshore (>23 MPa in Step Graben) and south-eastern offshore (about 24 MPa). Overpressures are much lower in platform areas west and southwest of the Step Graben (with pore water overpressures varying between about 6 and 10.5 MPa).

Pressure measurements in onshore aquifers and reservoirs of the Carboniferous Limburg Group are largely limited to the Friesland Platform and Lower Saxony Basin. Pore water overpressures on the Friesland Platform range between about 2.5 and 3.5 MPa, while for the Lower Saxony Basin, pore they range between 2 and 8 MPa.

Two wells encountered overpressured pore water conditions of about 13 and 19 MPa in the carbonates of the Carboniferous Limestone Group at depths of about 5 km at the Texel IJsselmeer High and in the northern part of Groningen, respectively.

Impact of faults

The above-described overpressure distributions show that faults/fault zones act as important hydraulic barriers or even seals for groundwater flow on geological timescales. It should be realized, however, that fault zones may show time-dependent hydraulic behaviour: the hydraulic behaviour of faults may be different during production (oil, gas, hot water) in comparison with their observed impact on pre-production overpressure distributions (e.g. Muggeridge & Smalley, 2008; Wibberley et al., 2017). The induced large changes in pressure and pressure differences across faults during production may force flow across faults that were interpreted to be sealing based on pre-production across-fault pressure differences. In turn, faults with permeabilities not low enough to maintain overpressure differences on geological timescales, may act as hydraulic barriers on production timescales if fault permeability is significantly lower than that of the reservoir or aquifer (Wibberley et al., 2017). The differences in fault behaviour can be related to the short periods of production, of the order of 10 to 100 years, in combination with the much larger changes in pressure created during production in comparison with those associated with geological processes, such as sedimentary loading.

Impact of anthropogenic activities

Anthropogenic activities in the realm of deep hydrogeology disturb the prevailing characteristics of the groundwa-

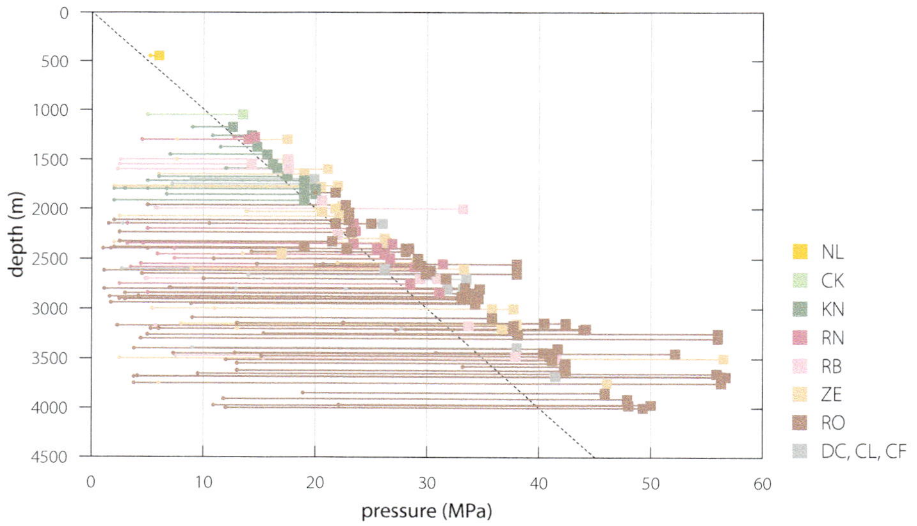

Figure 17.25. Cross plot of pressure versus depth for gas fields in different stratigraphic units. The plot shows the production-induced pressure changes in the gas fields from the initial pressures at the start of production of the gas fields to the final depleted pressure projected at the end of the fields' lifetimes (from Buijze et al., 2023). Codes in the legend refer to stratigraphic groups.

ter system and hydrogeologic framework that have developed over geologic timescales covering often 10s to 100s of million years. The pore water pressures described above refer to undisturbed pre-production pressures. Production of oil and gas during the last 60 years has resulted in major pressure depletion of the originally normally to highly overpressured accumulations: a decrease of several 10s of MPa to even up to 10% of the original gas pressure is not uncommon (Fig. 17.25; NAM, 2020; Buijze et al., 2023). It is known that pressure depletion in gas/oil fields has several effects, including 1) flow of pore water from the water bearing part of a reservoir towards the overlying field and associated reduction of the original pore water pressure in that water bearing part as well as reduction of pore pressures in aquifers/aquitards hydraulically connected to the produced reservoir, 2) compaction of reservoir and connected aquifers/aquitards, 3) ground surface subsidence (Fokker et al., 2025, this volume) and 4) seismicity (Dost et al., 2025, this volume). These production-induced depleted pressures are highly out of equilibrium with pore water pressures in surrounding rocks and this continues after production stops. The quantitative evolution of pore pressure and flow in time and space in response to the depleted reservoir pressures is only poorly known, especially in relation to the long time evolution after production stops. Evidence of flow towards a depleted gas/oil field during production is apparent from, for example: presence of co-produced water (e.g. Van der Molen et al., 2019); pressure interference between neighbouring gas/oil fields; and ground surface subsidence outside producing gas fields (Fokker et al., 2018; NAM, 2020). Subsidence studies executed in the Wadden Sea area included the investigation into the lateral and vertical pore water decline due to depletion of the producing Ameland and Wadden Sea gas fields in order to to better understand the geometry and volume of surface subsidence (e.g. NAM, 2015). These studies identified a delay in pressure drop in the water leg, creating a pressure difference between the produced gas field and the water bearing part of the reservoir. The authors relate the pressure difference to the hampering of vertical pressure equilibration due to the presence of layering and residual gas in the water bearing part of the reservoir.

A simplified regional modelling approach was used to study the evolution of flow and pressure redistribution for different scenarios during production and after abandonment of the large Groningen gas field (TNO & Deltares, 2022). The modelling results showed a continuing decline of pore pressures in aquifers connected to the field over modelled periods of 100-500 years. Decline of pressures was found to be concentrated in aquifers to the NW and W of the field. The pore pressure depletion in these aquifers and associated aquifer compaction were calculated to result in additional surface subsidence above the depleted aquifers (TNO & Deltares, 2022). Mass balance analyses further showed that the assumed volume of connected aquifers is insufficient to repressurize the depleted Groningen gas field by flow from the aquifers. Hence, low pressures are expected to persist for a very long time in the field area and will continue to influence pressures in surrounding rocks.

Ground surface subsidence above produced reservoirs and depleted aquifers (Fokker et al., 2025, this volume) may have consequences for management of shallow groundwater systems in onshore Netherlands now and in the future (Deltares, 2021). This is especially of importance in the coastal zones, because of the delicate balance between ground surface level and groundwater table. After all a change in height and geometry of the groundwater table directly affects groundwater flow and the balance between fresh and salt groundwater.

Additional potential effects of conventional (oil and gas) exploration and production on shallow groundwater systems include leakage of gas along wellbores. Schout (2020) identified that 1 out of the investigated 29 abandoned wells was leaking thermogenic methane due to well integrity failure. He also measured elevated methane concentrations in groundwater 500 m downstream of a blowout that had taken place in 1965.

Besides conventional (oil and gas) exploration and production other anthropogenic activities, such as geothermal energy production, subsurface hydrogen storage or CO_2 sequestration will disturb the virgin deep hydrogeology to a greater or lesser extent now and in the future. Geothermal energy production (Mijnlieff et al., 2025, this volume) induces flow, changes of pore pressure, temperature and thereby affects chemistry of pore water in the produced aquifer and beyond.

Synthesis

- This chapter shows the close relation between geology and hydrogeology and the impact of anthropogenic activities on groundwater systems.
- Detailed geology-based hydrogeologic models (REGISII) with assigned static properties of aquifers and aquitards cover the depth ranges of topography-driven groundwater flow in onshore Netherlands. Maps of selected aquifer units with static properties are available for the onshore and offshore at greater depths.
- Current characteristics of groundwater flow systems are the result of the interplay between the natural evolution driven by processes (such as creation of fresh groundwater table, sea-water intrusion, sedimentary and glacial loading and unloading, water-rock interac-

tion) operating on geological timescales of thousands to tens and even hundred millions of years and the impact of multiple anthropogenic activities operating at timescales of years to tens of years.

- Topography-driven groundwater flow systems developed in mostly unconsolidated sedimentary sequences of Holocene and Pleistocene to Neogene age.
- Topography-driven systems contain important fresh groundwater resources.
- Most groundwater spatially outside the topography-driven systems are saline to saline brines. Groundwater in the realm of deep hydrogeology shows high overpressures in the northern offshore and northern and north-eastern part of the Netherlands.
- Past and present anthropogenic activities (such as extraction of water and fossil fluids, drainage, surface water management, land use, mining) have impacted the subsurface and the physical and chemical properties of its contained groundwater from surface to depth of several kilometres, e.g.:
  - o Drainage of groundwater and associated lowering of groundwater tables in coastal low land areas have resulted in salinization and saline seepage;
  - o Extraction of groundwater for water supply and for dewatering for mining purposes have changed groundwater pressure, groundwater head and flow patterns; such changes in pressure travel fast and over large lateral distances in hydraulically continuous confined aquifers;
  - o Shallow fresh groundwater resources have been contaminated chemically from anthropogenic sources with e.g. fertilizers, pesticides, pharmaceuticals;
  - o Pressure depletion due to fossil fluid extraction at great depth has resulted in ground surface subsidence and associated changes in near-surface groundwater flow conditions.
- The broadening of scope of hydrogeology since about the 1980s especially in relation to climate change (sea-level changes, extreme dry and wet events) and the energy transition (increase in multiple use of the subsurface) have triggered research and application of hydrogeology at greater depths and on larger – geological – timescales than those conventionally studied for fresh water supply and water management purposes.
- Data, concepts and tools from petroleum geoscience and industry enable the assessment of the history and current physical and chemical properties of the hydrogeologic framework and groundwater systems, especially in the realm of deep hydrogeology.
- The foreseen future developments in relation to climate change, energy transition and population growth increase stress on sustainability of the quantity and quality of fresh groundwater resources.
- Current and future multiple uses of the shallow subsurface in the context of fresh groundwater resources and the use of the subsurface at greater depth, as well as environmental and social concerns related to these activities require better understanding of hydrogeological properties and processes, such as:
  - o Integrity of the subsurface:
    - hydrogeological models beyond the areas covered by REGISII;
    - hydraulic properties of aquitards/seals and faults at all depth ranges;
    - dynamic nature of hydraulic properties;
  - o Chemical composition of groundwater, including groundwater in the realm of deep hydrogeology;
  - o Hydrogeological properties and processes at timescales of importance for the provision of advice concerning integration of surface water and groundwater management with spatial planning at the surface and subsurface.
- Data, skills, concepts, tools and technologies other than traditionally used are needed in order to acquire this knowledge and understanding.

Acknowledgements

The constructive and insightful reviews by Victor Bense and Elco Luijendijk are gratefully acknowledged. We also thank Harry Doust for his careful improvement of English grammar and editor Geert-Jan Vis for his help in preparing good quality figures. Peter Vos and Jenny Hetelaar are thanked for preparing the generalized paleogeographic maps.

Digital map data

Spatial data of figures in this chapter for use in geographical information systems can be downloaded here: https://doi.org/10.5117/aup.28164308.

References

Acacia Water, 2019. Spaarwater. Rendabel en duurzaam agrarisch watergebruik en waterbeheer in de verziltende Waddenregio. Hoofdrapport 2016-2018: 81 pp.

Amberg, S., Sachse, V., Littke, R. & Back, S., 2022. Influence of Quaternary glaciations on subsurface temperatures, pore pressures, rock properties and petroleum systems in the onshore northeastern Netherlands. Netherlands Journal of Geosciences 101: e10. DOI: 10.1017/njg.2022.6

Appelo, C.A.J. & Geirnaert, W., 1991. Processes accompanying the intrusion of salt water. In: De Breuck, W.D. (ed.): International Contributions to Hydrogeology. Verlag Heinz Heise (Hannover, Germany): 291-303.

Auken, E., Bosch, A., Courtens, C., Elderhorst, W., Euwe, M., Gunnink, J., Hinsby, K., Jansen, J., Johnsen, R., Kok, A., Lebbe, L., De Louw, P., Noorlandt, R., Oude Essink, G., Pedersen, J., Rasmussen, P., Scheer, W., Siemon, B., Sonnenborg, T., Sulzbacher, H., Ullmann, A., Vandenbohede, A. & Wiederhold, H., 2010. CLIWAT: a transnational project about climate change and coastal groundwater in the North Sea Region. Proceedings SWIM21 (Azores, Portugal): 89-92.

Bakker, M. & Post, V., 2022. Analytical Groundwater Modeling: Theory and Applications using Python. CRC Press (London): 242 pp. DOI: 10.1201/9781315206134

Batelaan, O., De Smedt, F. & Triest, L., 2003. Regional groundwater discharge: phreatophyte mapping, groundwater modelling and impact analysis of land-use change. Journal of Hydrology 275: 86-108. DOI: 10.1016/S0022-1694(03)00018-0

Bense, V.F., Van Balen, R.T. & De Vries, J.J., 2003. The impact of faults on the hydrogeological conditions in the Roer Valley Rift System: an overview. Netherlands Journal of Geosciences 82 (1): 41-54. DOI: 10.1017/S0016774600022782

Bense, V.F., 2004. The hydraulic properties of faults in unconsolidated sediments and their impact on groundwater flow: a study in the Roer Valley Rift System and adjacent areas in the Lower Rhine Embayment. PhD thesis Vrije Universiteit Amsterdam (the Netherlands): 143 pp.

Bense, V.F., Gleeson, T., Loveless, S.E., Bour, O. & Scibek, J., 2013. Fault zone hydrogeology. Earth-Science Reviews 127: 171-192. DOI: 10.1016/j.earscirev.2013.09.008

Bonte, M., 2013. Impacts of shallow geothermal energy on groundwater quality; A hydrochemical and geomicrobial study of the effects of ground source heat pumps and aquifer thermal energy storage. PhD thesis Vrije Universiteit Amsterdam (the Netherlands): 175 pp.

Bonte, M., Stuyfzand, P.J., Van Den Berg, G.A. & Hijnen, W.A.M., 2011. Effects of aquifer thermal energy storage on groundwater quality and the consequences for drinking water production: A case study from the Netherlands. Water Science and Technology 63 (9): 1922-1931. DOI: 10.2166/wst.2011.189

Bouroullec, R. & Ten Veen, J.H., 2025. Salt Tectonics. In: Ten Veen, J.H., Vis, G.-J., De Jager, J. & Wong, Th.E. (eds): Geology of the Netherlands, second edition. Amsterdam University Press (Amsterdam): 457-491. DOI: 10.5117/9789463728362_ch12

Bouw, L. & Oude Essink, G.H.P., 2003. Fluid flow in the northern Broad Fourteens Basin during Late Cretaceous inversion. Netherlands Journal of Geosciences: Geologie En Mijnbouw 82 (1): 55-69. DOI: 10.1017/S0016774600022794

Breeuwer, J.B. & Jelgersma, S., 1973. An east-west geo-hydrological section across the Netherlands. Verhandelingen van Het Koninklijk Nederlands Geologisch Mijnbouwkundig Genootschap 29: 105-106.

Broers, H.P., 2002. Strategies for regional groundwater quality monitoring. PhD thesis Utrecht University (the Netherlands): 231 pp.

Broers, H.P., Sultenfus, J., Aeschbach, W., Kersting, A., Menkovich, A., de Weert, J. & Castelijns, J., 2021. Paleoclimate signals and groundwater age distributions from 39 public water works in the Netherlands; Insights from noble gases and carbon, hydrogen and oxygen isotope tracers. Water Resources Research 57 (7): e2020WR029058. DOI: 10.1029/2020WR029058

Bruggeman, G., 1999. Analytical solutions of geohydrological problems. Developments in Water Science 46. Elsevier (Amsterdam): 956 pp.

Buijze, L., Veldkamp, H. & Wassing, B., 2023. Comparison of hydrocarbon and geothermal energy production in the Netherlands: Reservoir characteristics, pressure and temperature changes, and implications for fault reactivation. Netherlands Journal of Geosciences 102: e7. DOI: 10.1017/njg.2023.6

Busschers, F.S., Cohen, K., Wesselingh, F.P., Schokker, J., Bakker, M., Van Balen, R. & Van Heteren, S., 2025. Quaternary. In: Ten Veen, J.H., Vis, G.-J., De Jager, J. & Wong, Th.E. (eds): Geology of the Netherlands, second edition. Amsterdam University Press (Amsterdam): 333-391. DOI: 10.5117/9789463728362_ch10

Caro Cuenca, M., 2012. Improving radar interferometry for monitoring fault-related surface deformation; Applications for the Roer Valley Graben and coal mine induced displacements in the southern Netherlands. PhD thesis Delft University of Technology, Department of Geoscience and Remote Sensing: 164 pp.

Caro Cuenca, M., Hooper, A.J. & Hanssen, R.F., 2013. Surface deformation induced by water influx in the abandoned coal mines in Limburg, The Netherlands observed by satellite radar interferometry. Journal of Applied Geophysics 88: 1-11. DOI: 10.1016/j.jappgeo.2012.10.003

CBS, 2021. Watergebruik bedrijven en particuliere huishoudens; nationale rekeningen. Statistics Netherlands (CBS). https://www.opendata.cbs.nl/statline.

Cirkel, G., Hartog, N., De la Loma Gonzales, B. & Stuyfzand, P.J., 2015. Methaan in ondiep Nederlands grondwater: verbinding met de diepe ondergrond? H2O, Tijdschrift Voor Watervoorziening en Afvalwaterbehandeling (online): 1-7.

Corona, F.V., 2005. Fault trap analysis of the Permian Rotliegend gas play, Lauwerszee Trough, NE Netherlands. Geological Society, London, Petroleum Geology Conference Series 6 (1): 327-335. DOI: 10.1144/0060327

Corona, F.V., Davis, J.S., Hippler, S.J. & Vrolijk, P.J., 2010. Multi-fault analysis scorecard: testing the stochastic approach in fault seal prediction. Geological Society, London, Special Publication 347: 317-332. DOI : 10.1144/SP347.18

Darcy, H., 1856. Les fontaines publiques de la ville de Dijon. Victor Dalmont (Paris): 647 pp.

De Glee, G.J., 1930. Over grondwaterstroomingen bij wateronttrekking door middel van putten. PhD thesis Technische Hoogeschool Delft (the Netherlands): 175 pp.

De Louw, P.G.B., 2013. Saline seepage in deltaic areas. Preferential groundwater discharge through boils and interactions between thin rainwater lenses and upward saline seepage. PhD thesis Vrije Universiteit Amsterdam (the Netherlands): 198 pp.

De Louw, P.G.B., Oude Essink, G.H.P., Delsman, J.R., Van Baaren, E.S., America, I. & Van Engelen, J., 2019. Het langetermijngeheugen van de zoet-zoutverdeling. Stromingen 33: 43-60.

De Louw, P.G.B., Oude Essink, G.H.P., Stuyfzand, P.J. & Van der Zee, S.E.A.T.M., 2010. Upward groundwater flow in boils as the dominant mechanism of salinization in deep polders, The Netherlands. Journal of Hydrology 394: 494-506. DOI: 10.1016/j.jhydrol.2010.10.009

De Louw, P.G.B., Van der Velde, Y. & Van der Zee, S.E.A.T.M., 2011. Quantifying water and salt fluxes in a lowland polder catchment dominated by boil seepage: a probabilistic end-member mixing approach. Hydrology and Earth System Science 15: 2101-2117. DOI: 10.5194/hess-15-2101-2011

De Vries, J.J., 1974. Groundwater flow systems and stream nets in the Netherlands: a groundwater-hydrologic approach to the functional relationship between the drainage system and the geological and climatical conditions in a Quaternary accumulation area. PhD thesis Vrije Universiteit Amsterdam (the Netherlands): 226 pp.

De Vries, J.J., 1982. Anderhalve eeuw hydrologisch onderzoek in Nederland (1830-1980). Rodopi (Amsterdam): 195 pp.

De Vries, J.J., 2004. From speculation to science: the founding of groundwater hydrology in the Netherlands. In: Touret, J.L.R. & Visser, R.P.W. (eds): Dutch pioneers of the Earth sciences. Royal Netherlands Academy of Arts and Sciences (KNAW, Amsterdam): 139-164.

De Vries, J.J., 2007. Groundwater. In: Wong, Th.E., Batjes, D.A.J. & De Jager, J. (eds): Geology of the Netherlands. Royal Netherlands Academy of Arts and Sciences (KNAW, Amsterdam): 295-315.

De Vries, J.J., 1981. Fresh and salt groundwater in the Dutch coastal area in relation to geomorphological evolution. Geologie En Mijnbouw 60: 363-368.

De Vries, J.J., 1994. Dynamics of the interface between stream nets and groundwater systems in lowland areas, with reference to stream net evolution. Journal of Hydrology 155: 39-56.

De Wit, R.G., 1988. De invloed van discontinuïteiten op de grondwaterstroming in kalksteenformaties van Zuid Limburg: Inventarisatie, modellering en correlatie. MSc thesis; Technische Universteit Delft (the Netherlands): 157 pp.

Delsman, J., Oude Essink, G., Huizer, S., Bootsma, H., Mulder, T., Zitman, P., Romero Verastegui, B. & Janssen, G., 2020. Actualisatie zout in het NHI - Toolbox NHI zoet-zout modellering en landelijk model. Deltares Report No. 11205261-003-BGS-0001 (Utrecht, the Netherlands): 98 pp.

Delsman, J.R., 2015. Saline groundwater – surface water interaction in coastal lowlands. PhD thesis Vrije Universiteit Amsterdam (the Netherlands): 194 pp.

Delsman, J.R., Hu-a-ng, K.R.M., Vos, P.C., De Louw, P.G.B., Oude Essink, G.H.P., Stuyfzand, P.J. & Bierkens, M.F.P., 2014. Paleo-modeling of coastal saltwater intrusion during the Holocene: An application to the Netherlands. Hydrology and Earth System Sciences 18 (10): 3891-3905. DOI: 10.5194/hess-18-3891-2014

Delsman, J.R., Van Boekel, E., Reinhard, S., Te Winkel, T., Van Loon, A., Bartholomeus, R.P., Mulder, M., Massop, H., Polman, N. & Schasfoort, F., 2018. Regioscan Zoetwatermaatregelen - Verkennen van het perspectief van kleinschalige zoetwatermaatregelen voor de regionale zoetwateropgave. Stichting Toegepast Onderzoek Waterbeheer (Amersfoort). Report No. STOWA 2018-13: 135 pp.

Delta Programme, 2023. National Delta Programme 2023: speed-up, connect and reconstruct. 108 pp. https://www.english.deltaprogramma.nl

Deltaprogramma, 2021. Nationaal Deltaprogramma 2021. Koersvast Werken Aan Een Klimaatbestendig Nederland: 140 pp.

Deltares, 2021. Indirecte schade-effecten van diepe bodemdaling en – stijging bij het Groningen gasveld en gasopslag Norg. Deltares (Utrecht). Report No. 11207096-002-BGS-0001: 66 pp.

Deltares, 2023. Integrale Grondwaterstudie Nederland. Module 1: landelijke analyse. Deltares (Utrecht). Report No. 11208092-001-BGS-0001: 195 pp.

Dillon, P.J., Stuyfzand, P.J., Grischek, T., Lluria, M., Pyne, R.D.G., Jain, R.C., Bear, J., Schwarz, J., Wang, W., Fernandez, E., Stefan, C., Pettenati, M., Van der Gun, J., Sprenger, C., Massmann, G., Scanlon, B.R., Xanke, J., Jokela, P., Zheng, Y., Rossetto, R., Shamrukh, M., Pavelic, P., Murray, E., Ross, A., Valverde, J.P.B., Nava, A.P., Ansems, N., Posavec, K., Ha, K., Martin, R. & Sapiano, M., 2019. Sixty years of global progress in managed aquifer recharge. Hydrogeology Journal 27: 1-30.

Dost, B., Ruigrok, E., Spetzler, J. & Kruiver, P., 2025. Natural and induced seismicity. In: Ten Veen, J.H., Vis, G.-J., De Jager, J. & Wong, Th.E. (eds): Geology of the Netherlands, second edition. Amsterdam University Press (Amsterdam): 793-823. DOI: 10.5117/9789463728362_ch22

Drabbe, J. & Badon Ghijben, W., 1889. Nota in verband met de voorgenomen putboring nabij Amsterdam. Tijdschrift Koninklijk Instituut van Ingenieurs (Verhandelingen 1888/1889): 8-22.

Dronkert, H. & Remmelts, G., 1996. Influence of salt structures on reservoir rocks in Block L2, Dutch continental shelf. In: Rondeel, H.E., Batjes, D.A.J.& Nieuwenhuijs, W.H. (eds): Geology of gas and oil under the Netherlands. Kluwer Academic Publishers (Dordrecht): 159-166. DOI: 10.1007/978-94-009-0121-6_14

Dubois, E., 1903. Feiten ter opsporing van de bewegingsrichting en den oorsprong van het grondwater onzer zeeprovincien (Facts for the detection of the direction of migration and origin of the groundwater in our sea provinces). Verslagen en Mededelingen Koninklijke Nederlandse Akademie van Wetenschappen, Afdeling Wis- en Natuurkunde 12 (1): 187-212.

Dufour, F.C., 1998. Grondwater in Nederland: Onzichtbaar water waarop wij lopen. Nederlands Instituut voor Toegepaste Geowetenschappen TNO (Delft): 265 pp.

Dufour, F.C., 2000. Groundwater in the Netherlands – Facts and figures. Nederlands Instituut voor Toegepaste Geowetenschappen TNO (Delft): 96 pp.

Edelman, J.H., 1947. De berekening van grondwaterstroomingen. PhD thesis Technische Hoogeschool Delft (the Netherlands): 77 pp.

Engelen, G.B., 1986. Water systems in the South-eastern part of The Netherlands. In: Engelen, G.B. & Jones, G.P. (eds): Developments in the analysis of groundwater flow systems. IAHS Publication 163: 205-224.

Engelen, G.B. & Kloosterman, F.H., 1996. Hydrological systems analysis. Kluwer (Dordrecht): 152 pp.

Ernst, L.F. & De Ridder, N.A., 1960. High resistance to horizontal ground-water flow in coarse sediments due to faulting. Geologie En Mijnbouw 39: 66-85.

Fokker, P.A., Van Leijen, F.J., Orlic, B., Van der Marel, H. & Hanssen, R.F., 2018. Subsidence in the Dutch Wadden Sea. Netherlands Journal of Geosciences 97 (3): 129-181. DOI: 10.1017/njg.2018.9

Fokker, P.A., Candela, T.G.G., Erkens, G., Hanssen, R.F., Kooi, H., Koster, K. & Van Leijen, F.J., 2025. Subsidence. In: Ten Veen, J.H., Vis, G.-J., De Jager, J. & Wong, Th.E. (eds): Geology of the Netherlands, second edition. Amsterdam University Press (Amsterdam): 825-847. DOI: 10.5117/9789463728362_ch23

Fraters, B., Hooijboer, A.E.J., Vrijhoef, A., Plette, A.C.C., Van Duijnhoven, N., Rozemeijer, J.C., Gosseling, M., Daatselaar, C.H.G., Roskam, J.L. & Begeman, H.A.L., 2020. Landbouwpraktijk en waterkwaliteit in Nederland; toestand (2016-2019) en trend (1992-2019). De Nitraatrapportage 2020 met de resultaten van de monitoring van de effecten van de EU Nitraatrichtlijn actieprogramma's. Rijksinstituut voor Volksgezondheid en Milieu, RIVM-rapport 2020-0121: 232 pp.

Gaarenstroom, L., Tromp, R.A.J., De Jong, M.C. & Brandenburg, A.M., 1993. Overpressures in the Central North Sea: implications for trap integrity and drilling safety. In: Parker, J.R. (ed.): Petroleum Geology of North-West Europe, Geological Society, London, Petroleum Geology Conference Series 4: 1305-1313. DOI: 10.1144/0041305

Gambolati, G. & Teatini, P., 2021. Land subsidence and its mitigation. The Groundwater Project. Guelph (Ontario, Canada): 92 pp.

Gaupp, R. & Okkerman, J.A., 2011. Diagenesis and reservoir quality of Rotliegend sandstones in the northern Netherlands - A review. In: Grötsch, J. & Gaupp, R. (eds): The Permian Rotliegend of the Netherlands. SEPM Special Publication 98: 193-226. DOI: 10.2110/pec.11.98.0193

Geelen, L.H.W.T., Kamps, P.T.W.J. & Olsthoorn, T.N., 2017. From overexploitation to sustainable use, an overview of 160 years of water extraction in the Amsterdam dunes, the Netherlands. Journal of Coastal Conservation 21: 657-668. DOI: 10.1007/s11852-016-0452-x

Gehrels, J.C., 1999. Groundwater level fluctuations: Separation of natural from anthropogenic influences and determinations of groundwater recharge in the Veluwe area, the Netherlands. PhD thesis Vrije Universiteit Amsterdam (the Netherlands): 269 pp.

Glasbergen, P., 1981. Extreme salt concentrations in deep aquifers in The Netherlands. Science of the Total Environment 21: 251-259. DOI: 10.1016/0048-9697(81)90157-1

Glasbergen, P., 1985. The origin of groundwater in Carboniferous and Devonian aquifers at Maastricht. Geologie En Mijnbouw 64 (2): 123-129.

Graf Pannatier, E., Broers, H.P., Venema, P. & Van Beusekom, G., 2000. A new process-based hydrogeochemical classification of groundwater Application to the Netherlands national monitoring network. Nederlands Instituut voor Toegepaste Geowetenschappen TNO (Delft), Report No. NITG 00-143-B: 57 pp.

Griffioen, J., Felder, M., Molenaar, N., Spijker, J. & Hoving, A., 2025. Sediment mineralogy and geochemistry. In: Ten Veen, J.H., Vis, G.-J., De Jager, J. & Wong, Th.E. (eds): Geology of the Netherlands, second edition. Amsterdam University Press (Amsterdam): 493-533. DOI: 10.5117/9789463728362_ch13

Griffioen, J., Klaver, G. & Westerhoff, W.E., 2016a. The mineralogy of suspended matter, fresh and Cenozoic sediments in the fluvio-deltaic Rhine-Meuse-Scheldt-Ems area, the Netherlands: An overview and review. Netherlands Journal of Geosciences 95 (1): 23-107. DOI: 10.1017/njg.2015.32

Griffioen, J., Notenboom, J., Schraa, G., Stuurman, R.J., Runhaar, H. & Wirdum, G., 2003. Systeemgericht grondwaterbeheer. Wolters-Noordhoff (Groningen): 189 pp.

Griffioen, J., Vermooten, S. & Janssen, G.J.A., 2013. Geochemical and palaeohydrological controls on the composition of shallow groundwater in the Netherlands. Applied Geochemistry 39: 129-149. DOI: 10.1016/j.apgeochem.2013.10.005

Griffioen, J., Verweij, J.M. & Stuurman, R.J., 2016b. The composition of groundwater in Palaeogene and older formations in the Netherlands, A synthesis. Netherlands Journal of Geosciences 95 (3): 349-372. DOI: 10.1017/njg.2016.19

Griffioen, J. & Wildenborg, T. (2016). Editorial: special issue on the Rupel Clay related to research on subsurface disposal of radioactive waste. Netherlands Journal of Geosciences, 95(3): 215 DOI: 10.1017/njg.2016.27

Haasnoot, M., Bouwer, L., Diermanse, F., Kwadijk, J.C.J., Spek, A. J.F., Oude Essink, G.H.P., Delsman, J.R., Weiler, O., Mens, M.J.P., Ter Maat, J., Van der Kaaij, T., Plieger, R., Huismans, Y., Sloff, K., Mosselman, E. & Lenselink, G., 2018. Mogelijke gevolgen van versnelde zeespiegelstijging voor het Deltaprogramma. Een verkenning. Deltares (Utrecht), Report No. 11202230-005-0002: 43 pp.

Harbers, M. & Heijnen, V.L.W.A., 2022. Kamerbrief Water en Bodem sturend. https://open.overheid.nl/documenten/ronl-c35e65eba0903d738ae26dab222462337b0d8de7/pdf

Hartog, N., Van Bergen, P.F., De Leeuw, J.W. & Griffioen, J., 2004. Reactivity of organic matter in aquifer sediments: geological and geochemical controls. Geochimica et Cosmochimica Acta 68 (6): 1281-1292. DOI: 10.1016/j.gca.2003.09.004

Hendriks, D., Broers, H.P., Van Ek, R. & Hoogewoud, J., 2012. Dynamiek van uitwisseling grond- en oppervlaktewater in Nederland. H2O 10: 27-30.

Herzberg, A., 1901. Die Wasserversorgung einiger Nordseebader. Journal für Gasbeleuchtung und Wasserversorgung 44: 815-819.

Hijma, M.P. & Kooi, H., 2018. Bodemdaling in het kustfundament en de getijdenbekkens: door geologische processen en menselijke activiteiten. Deltares (Utrecht), Report No. 11200538-008-ZKS-0001: 63 pp.

Hooghoudt, S.B., 1937. Bijdragen tot de kennis van eenige natuurkundige grootheden van den grond. No. 6. Bepaling van de doorlatendheid in gronden van de tweede soort: theorie en toepassingen van de kwantiatieve strooming van het water in ondiep gelegen grondslagen, vooral in verband met ontwaterings- en infiltratievraagstukken. Bodemkundige instituut te Groningen. Verschenen in de reeks Verslagen van landbouwkundige onderzoekingen 43 (13): 461-676.

Hooghoudt, S.B., 1940. Bijdragen tot de kennis van eenige natuurkundige grootheden van den grond. No. 7. Algemeene beschouwing van het probleem van de detailontwatering en de infiltratie door middel van parallel loopende drains, greppels, slooten en kanalen. Bodemkundige Instituut te Groningen 46 (14B): 515-707.

Houtgast, R.F. & Van Balen, R.T., 2000. Neotectonics of the Roer Valley Rift System, the Netherlands. Global and Planetary Change 27 (1-4): 131-146. DOI: 10.1016/S0921-8181(01)00063-7

Huisman, P., 2004. Water in the Netherlands; managing checks and balances. Netherlands Hydrological Society (NHV) (Utrecht, the Netherlands): 132 pp.

Huizer, S., 2019. Fresh groundwater in large beach nourishments. Growth of fresh groundwater resources in coastal areas. PhD thesis University Utrecht (the Netherlands). Utrecht Studies in Earth Sciences 183: 152 pp.

Huizer, S., Oude Essink, G.H.P. & Bierkens, M.F.P., 2016. Fresh groundwater resources in a large sand replenishment. Hydrology and Earth System Sciences 20: 3149-3166. DOI: 10.5194/hess-20-3149-2016

Ingebritsen, S.E., Sanford, W.E. & Neuzil, Ch.E., 2006. Groundwater in Geologic Processes, second edition. Cambridge University Press: 536 pp.

Jelgersma, S., 1977. De geologische grondslag van het geohydrologisch onderzoek in Nederland. H2O 9: 543-550.

Jelgersma, S. & Visser, W.A., 1972. Hydrogeological maps of the Netherlands, scale 1 : 1,500,000. Geologie En Mijnbouw 51 (1): 7-10.

Juez-Larré, J., Groenenberg, R., Dinkelman, D., Huijskes, T., Koornneef, J.M., Van Gessel, S.F. & Bos, C.F.M., 2025. Underground storage. In: Ten Veen, J.H., Vis, G.-J., De Jager, J. & Wong, Th.E. (eds): Geology of the Netherlands, second edition. Amsterdam University Press (Amsterdam): 729-767. DOI: 10.5117/9789463728362_ch20

Klijn, F. & Witte, J.-P.M., 1999. Eco-hydrology: Groundwater flow and site factors in plant ecology. Hydrogeology Journal 7: 65-77. DOI: 10.1007/s100400050180

Kloosterman, F.H., 1993. National hydrological systems analysis. Report 2. The regional groundwater flow systems around the ice-pushed ridge complexes of the Veluwe and Utrecht (in Dutch). TNO-IAG (Delt), Report No. OS 93-41.

Kombrink, H., Doornenbal, J.C., Duin, E.J.T., Den Dulk, M., Ten Veen, J.H. & Witmans, N., 2012. New insights into the geological structure of the Netherlands; results of a detailed mapping project. Netherlands Journal of Geosciences 91 (4): 419-446. DOI: 10.1017/S0016774600000329

Kooi, H., 2000. Land subsidence due to compaction in the coastal area of The Netherlands: The role of lateral fluid flow and constraints from well-log data. Global and Planetary Change 27 (1-4): 207-222. DOI: 10.1016/S0921-8181(01)00067-4

Kooi, H., Johnston, P., Lambeck, K., Smither, C. & Molendijk, R., 1998. Geological causes of recent (~100 yr) vertical land movement in the Netherlands. Tectonophysics 299: 297-316. DOI: 10.1016/S0040-1951(98)00209-1

Koster, K., 2017. 3D characterization of Holocene peat in the Netherlands. Implications for coastal-deltaic subsidence. PhD thesis Utrecht University (the Netherlands): 185 pp.

Krul, W.F.J.M., 1940. De invloed van mariene transgressies op de samenstelling van het grondwater in Nederland. Geologie en Mijnbouw 2 (4): 66-67.

Krul, W.F.J.M., 1962. Development of geo-hydrological knowledge and research in the Netherlands. Geologie en Mijnbouw 41: 175-180.

Kruseman, G.P., De Ridder, N.A. & Verweij, J.M., 2000. Analysis and evaluation of pumping test data. Second completely revised edition. ILRI (Wageningen, the Netherlands), Publication 47: 377 pp.

Lapperre, R.E., Kasse, C., Bense, V.F., Woolderink, H.A.G. & Van Balen, R.T., 2019. An overview of fault zone permeabilities and groundwater level steps in the Roer Valley Rift System. Netherlands Journal of Geosciences 98 (e5). DOI: 10.1017/njg.2019.4

Lapperre, R.E., Bense, V.F., Kasse, C. & Van Balen, R.T., 2022. Temporal and spatial variability of cross-fault groundwater level differences: the impact of fault-induced permeability reduction, precipitation and evaporanspiration. Hydrogeology Journal 30: 1233-1257. DOI: 10.1007/s10040-022-02465-w

Lee, M.J., Aronson, J.L. & Savin, S.M., 1989. Timing and conditions of Permian Rotliegend sandstone diagenesis, southern North Sea: K-Ar and oxygen isotopic data. American Association of Petroleum Geologists Bulletin 73: 195-215. DOI: 10.1306/703C9B0E-1707-11D7-8645000102C1865D

Lorentz, H.A., 1913. Grondwaterbeweging in de nabijheid van bronnen. De Ingenieur 28: 24-26.

Luijendijk, E., 2012. The role of fluid flow in the thermal history of sedimentary basins: Inferences from thermochronology and numerical modeling in the Roer Valley Graben, southern Netherlands. PhD thesis Vrije Universiteit Amsterdam (the Netherlands): 198 pp.

Mazure, J.B., 1936. Geohydrologische gesteldheid van de Wieringermeer. Rapporten en Mededelingen Zuiderzeewerken (5): 67-131.

Meinardi, C.R., 1973. Het zoutwatervoorkomen in de ondergrond van de lage gedeelten van Nederland. H2O 6: 454-460.

Meinardi, C.R., 1991. The origin of brackish groundwater in the lower parts of The Netherlands. In: Breuck, W.D. (ed.): Hydrogeology of salt water intrusion: A selection of SWIM papers. Verlag Heinz Heise (Hannover, Germany): 271-290.

Mendizabal, I., Stuyfzand, P.J. & Wiersma, A.P., 2011. Hydrochemical system analysis of public supply well fields, to reveal water-quality patterns and define groundwater bodies: The Netherlands. Hydrogeology Journal 19: 83-100. DOI: 10.1007/s10040-010-0614-0

Mijnlieff, H., Buijze, L., Rosendaal, E., Schoof, F., Vorage, R. & Van Wees, J.-D., 2025. Geothermal energy. In: Ten Veen, J.H., Vis, G.-J., De Jager, J. & Wong, Th.E. (eds): Geology of the Netherlands, second edition. Amsterdam University Press (Amsterdam): 701-727. DOI: 10.5117/9789463728362_ch19

Muggeridge, A.H. & Smalley, P.C., 2008. A diagnostic toolkit to detect compartmentalization using time-scales for reservoir mixing. SPE Paper 118323. Paper presented at the Abu Dhabi International Petroleum Exhibition and Conference, Abu Dhabi, UAE, November 2008.

Munsterman, D.K., Donders, T.H., Houben, A.J.P., Ten Veen, J.H. & Wesselingh, F.P., 2025. Paleogene-Neogene. In: Ten Veen, J.H., Vis, G.-J., De Jager, J. & Wong, Th.E. (eds): Geology of the Netherlands, second edition. Amsterdam University Press (Amsterdam): 293-331. DOI: 10.5117/9789463728362_ch9

Muntendam-Bos, A.G., Hoedeman, G., Polychronopoulou, K., Draganov, D., Weemstra, C., van der Zee, W., Bakker, R.R. & Roest, H., 2022. An overview of induced seismicity in the Netherlands. Netherlands Journal of Geosciences 101: e1. DOI:10.1017/njg.2021.14

NAM, 2015. Update of methodology of predicting gas- and aquifer pressures in the Waddenzee development area (by Seeberger, F.C.) NAM (Assen, the Netherlands), Report No. EP201505216677: 19 pp.

NAM, 2020. Bodemdaling door aardgaswinning - Statusrapport 2020 en Prognose tot het jaar 2080. NAM (Assen, the Netherlands), Report No. EP202011201629: 58 pp.

Neeft, E.A.C., Bartol, J., Vuorio, M. & Vis, G.-J., 2025. Geological disposal of radioactive waste. In: Ten Veen, J.H., Vis, G.-J., De Jager, J. & Wong, Th.E. (eds): Geology of the Netherlands, second edition. Amsterdam University Press (Amsterdam): 769-791. DOI: 10.5117/9789463728362_ch21

Nelskamp, S. & Verweij, J.M., 2012. Using basin modeling for geothermal energy exploration in the Netherlands - an example from the West Netherlands Basin and Roer Valley Graben TNO (Utrecht), Report No. TNO-060-UT-2012-00245: 113 pp.

Nelskamp, S., Verweij, J.M. & Witmans, N., 2012. The role of salt tectonics and overburden in the generation of overpressure in the Dutch North Sea area. Netherlands Journal of Geosciences 91 (4): 517-534. DOI: 10.1017/S0016774600000366

Neuzil, C.E., 1995. Abnormal pressures as hydrodynamic phenomena. American Journal of Science 295: 742-786. DOI: 10.2475/ajs.295.6.742

Neuzil, C.E., 2015. Interpreting fluid pressure anomalies in shallow intraplate argillaceous formations. Geophysical Research Letters 42: 4801-4808. DOI: 10.1002/2015GL064140

Oude Essink, G.H.P., 1996. Impact of sea-level rise on groundwater flow regimes, a sensitivity analysis for the Netherlands. PhD thesis Delft University of Technology (the Netherlands): 411 pp.

Oude Essink, G.H.P., 2001a. Improving fresh groundwater supply - problems and solutions. Ocean & Coastal Management 44 (5-6): 429-449. DOI: 10.1016/S0964-5691(01)00057-6

Oude Essink, G.H.P., 2001b. Salt water intrusion in a three-dimensional groundwater system in the Netherlands: A numerical study. Transport in Porous Media 43: 137-158. DOI: 10.1023/A:1010625913251

Oude Essink, G.H.P., Pauw, P.S., Van Baaren, E.S., Zuurbier, K.G., De Louw, P.G.B., Veraart, J.A., MacAteer, E., Van der Schoot, M., Groot, N., Cappon, H., Waterloo, M.J., Hu-A-Ng, K.R.M. & Groen, M.M.A., 2018. GO-FRESH: Valorisatie kansrijke oplossingen voor een robuuste zoetwatervoorziening, Deltares (Utrecht): 177 pp

Oude Essink, G.H.P., Van Baaren, E.S. & De Louw, P.G.B., 2010. Effects of climate change on coastal groundwater systems: A modeling study in the Netherlands. Water Resources Research 46 (10): 16. DOI: 10.1029/2009WR008719

Pauw, P., De Louw, P.G.B. & Oude Essink, G.H.P., 2012. Groundwater salinisation in the Wadden Sea area of the Netherlands: quantifying the effects of climate change, sea-level rise and anthropogenic interferences. Netherlands Journal of Geosciences 91 (3): 373-383. DOI: 10.1017/S0016774600000500

Peeters, S., Asschert, A. & Verweij, J.M., 2018. Towards a better understanding of the highly overpressured Lower Triassic Bunter reservoir rocks in the Terschelling Basin. Geological Society of London, Special Publications 469 (1): 223-236. DOI: 10.1144/SP469.13

Pennink, J.M.K., 1905. Over de beweging van grondwater. De Ingenieur 20 (30): 482-492.

Person, M., Raffensperger, J.P., Ge, S. & Garven, G., 1996. Basin-scale hydrogeologic modeling. Reviews of Geophysics 34: 61-87. DOI: 10.1029/95RG03286

Peters, E. & Verweij, J.M., 2012. An investigation of fault sealing in the Dutch offshore based on virgin overpressures. TNO (Utrecht), Report No. TNO-060-UT-2011-02158: 46 pp.

Post, V.E.A., 2004. Groundwater salinization processes in the coastal area of the Netherlands due to transgressions during the Holocene. PhD thesis Vrije Universiteit Amsterdam (the Netherlands): 138 pp.

Post, V.E.A. & Kooi, H., 2003. Rates of salinization by free convection in high-permeability sediments: Insights from numerical modeling and application to the Dutch coastal area. Hydrogeology Journal 11: 549-559. DOI: 10.1007/s10040-003-0271-7

Post, V.E.A., Van der Plicht, H. & Meijer, H.A.J., 2003. The origin of brackish and saline groundwater in the coastal area of the Netherlands. Netherlands Journal of Geosciences - Geologie en Mijnbouw 82: 133-147. DOI: 10.1017/S0016774600020692

Pöttgens, J.J.E., 1985. Bodemhebung durch ansteigendes Grubenwasser [Uplift as a result of rising minewaters]. The Development Science and Art of Minerals Surveying. Proceedings VIth International Congress (Harrogate, United Kingdom): International Society for Mine Surveying 2: 928-938.

Projectgroup GS-ZL, 2016. After effects from mining South Limburg. Summary report with Bow-Tie analysis. Projectgroup 'Naijlende gevolgen van de steenkolenwinning in Zuid Limburg' (Aachen, Deventer): 166 pp.

Remmelts, G., Nelskamp, S., De Jager, J. & Geluk, M.C., 2025. Petroleum Geology. In: Ten Veen, J.H., Vis, G.-J., De Jager, J. & Wong, Th.E. (eds): Geology of the Netherlands, second edition. Amsterdam University Press (Amsterdam): 535-575. DOI: 10.5117/9789463728362_ch14

Renes, J., Meijer, J. & De Poel, K.R., 2002. Het Veluwse Sprengenlandschap; een cultuurelement. Alterra (Wageningen) en Stichting tot behoud van de Veluwse Sprengen en Beken: 255 pp.

Schout, G., 2020. Origin, fate and detection of methane leaking from the deep subsurface into groundwater and soil. PhD thesis Utrecht University (the Netherlands): 202 pp..

Schout, G., Griffioen, J., Hartog, N., Eggenkamp, H.G.M. & Cirkel, D.G., 2024. Methane occurrence and origin in Dutch groundwater: from shallow aquifers to deep reservoirs. Netherlands Journal of Geosciences 103: e24. DOI 10.1017/ njg.2024.20

SodM, 2019. De integriteit van onshore putten in Nederland. State Supervision of Mines (Den Haag, Netherlands): 22 pp.

Sprenger, C., Hartog, N., Hernández, M., Vilanova, E., Grützmacher, G., Scheibler, F. & Hannappel, S., 2017. Inventory of managed aquifer recharge sites in Europe: historical development, current situation and perspectives. Hydrogeology Journal 25: 1909-1922. DOI: 10.1007/s10040-017-1554-8

Stafleu, J., Busschers, F.S., Van der Meulen, M.J., Den Dulk, M., Gunnink, J.L., Maljers, D., Hummelman, J.H., Schokker, J., Vernes, R.W., Stam, J., Dabekaussen, W., Ten Veen, J.H., Doornenbal, H., Kars, R. & De Bruijn, R., 2025. Geomodelling. In: Ten Veen, J.H., Vis, G.-J., De Jager, J. & Wong, Th.E. (eds): Geology of the Netherlands, second edition. Amsterdam University Press (Amsterdam): 849-893. DOI: 10.5117/9789463728362_ch24

Stafleu, J., Van der Meulen, M.J., Gunnink, J.L., Maljers, D., Hummelman, J., Busschers, F.S., Schokker, J., Vernes, R.W., Doornenbal, H., Den Dulk, M. & Ten Veen, J.H., 2019. Systematic 3D subsurface mapping in the Netherlands. In: K.E. MacCormack, R.C. Berg, H.A.J. Russell, H. Kessler & L.H. Thorleifson (eds): Synopsis of Current Three-Dimensional Geological Mapping and Modelling in Geological Survey Organizations. Alberta Energy Regulator / Alberta Geological Survey (AER/AGS), Special Report 112 (3): 179-190.

Stegers, D.P.M., Edelman, S. & Pieterse, K., 2018. Geological study into the geothermal potential near fault zones in the Roer Valley Graben and West Netherlands Basin. T&A Survey B.V., Amsterdam (the Netherlands). Report No. ODE6425: 87 pp.

Stofberg, S.F., Oude Essink, G.H.P., Pauw, P.S., De Louw, P.G.B., Leijnse, A. & Van der Zee, S.E.A.T.M., 2016. Fresh Water Lens Persistence and Root Zone Salinization Hazard Under Temperate Climate. Water Resources Management 31 (2): 689-702. DOI: 10.1007/s11269-016-1315-9

STOWA, 2022. Geochemisch buffervermogen van de ondergrond - Beschikbare data, 3D Modellen en trendanalyse. STOWA rapport No. 2022-27: 77 pp.

Strong, 2018. Structuurvisie Ondergrond. Ministerie Infrastructuur en Waterstaat & Ministerie Economische Zaken en Klimaat: 116 pp.

Stuurman, R.J. & Vermeulen, P., 1996. Grensoverschrijdende grondwaterstromen in de Centrale en Roerdalslenk. Spraakwater (Periodiek van TNO Grondwater en Geo-Energie) 4: 6-9.

Stuurman, R.J. & Atari, R.H., 1997. The groundwater flow situation around the 'Wijstgronden' near the village of Uden (in Dutch). Netherlands Institute of Applied GeosciencesTNO (Delft, the Netherlands). Report No. 97-212(a).

Stuurman, R.J. & Oude Essink, G.H.P., 2007. Naar een uniforme landelijke inrichting van het KRW-grondwatermeetnet Zoet-Zout? TNO (Delft), Report No. 2007-U-R0490/B: 27 pp.

Stuurman, R.J., Van der Meij, J.L., Engelen, G.B., Biesheuvel, A. & Van Zadelhoff, F.J., 1990. De hydrologische systeemanalyse van West Brabant en omgeving. Een integraal onderzoek naar de structuur en dynamiek van watersystemen en de processen die deze bedreigen. Dienst Grondwaterverkenning TNO (Delft), Report No. OS 90 25 A: 195 pp.

Stuyfzand, P.J., 1993. Hydrochemistry and hydrology of the coastal dune area of the western Netherlands. PhD thesis Vrije Universiteit Amsterdam (the Netherlands): 366 pp.

Stuyfzand, P.J., 1999. Patterns in groundwater chemistry reflecting groundwater flow. Groundwater as a Geologic Agent Special Issue of Hydrogeology Journal 7: 15-27.

Stuyfzand, P.J., 2005. Mapping groundwater bodies with artificial or induced recharge, by determination of their origin and chemical facies. Proceedings of the 5th International Symposium on Management of Aquifer Recharge ISMAR 5, 839-850. (Berlin, Germany): 839-850.

Stuyfzand, P.J., 2017. Observations and analytical modeling of freshwater and rainwater lenses in coastal dune systems. Journal of Coastal Conservation 21: 577-593. DOI: 10.1007/s11852-016-0456-6

Ten Veen, J., Verweij, J.M., Donders, T., Geel, K., De Bruin, G., Munsterman, D., Verreusel, R., Daza Cajigal, V., Harding, R. & Cremer, H., 2013. Anatomy of the Cenozoic Eridanos Delta Hydrocarbon System. TNO (Utrecht), Report No. TNO 2013 R10060: 217 pp.

Ter Maat, J., Van der Vat, M., Hunink, J.C., Haasnoot, M., Prinsen, G., Visser, M.H., Boderie, P., Van Ek, R., Maarse, M., Van der Sligte, R., Verheij, H. & Wesselius, C., 2014. Effecten van maatregelen voor de zoetwatervoorziening in Nederland in de 21e eeuw (No. 1209141-001-VEB-0004; 215 pp.). Deltares (Delft, Netherlands): 215 pp.

TNO & Deltares, 2022. KEM-19 - Evaluation of post-abandonment fluid migration and ground motion risks in subsurface exploitation operations in the Netherlands. Final report Phase 2 Simulations and case studies of fluid migration, leakage, and ground motion threats in selected hydrocarbon areas in the Netherlands. KEM-19 Report: 161 pp

TNO-GDN, 2023. Geologische kaart van het Koninkrijk der Nederlanden 1 : 600 000, TNO - Geological Survey of the Netherlands (Utrecht), 1 map sheet. https://www.dinoloket.nl/en/subsurface-models/map. ISBN 9789059865142

Tóth, J., 1963. A theoretical analysis of groundwater flow in small drainage basins. Journal of Geophysical Research 68 (16): 4795-4812.

Tóth, J., 2009. Gravitational systems of groundwater flow: Theory, Evaluation, Utilization. Cambridge University Press (Cambridge, UK): 297 pp.

Van Baaren, E.S., Delsman, J.R., Karaoulis, M., Pauw, P.S., Vermaas, T., Bootsma, H., De Louw, P.G.B., Oude Essink, G.H.P., Dabekaussen, W., Gunnink, J.L., Dubelaar, W., Menkovic, A., Siemon, B., Steuer, A. & Meyer, U., 2018. FRESHEM Zeeland - FREsh Salt groundwater distribution by Helicopter ElectroMagnetic survey in the Province of Zeeland. Deltares (Utrecht, the Netherlands): Technical Report No. 1209220-000-BGS-0030: 140 pp. Baaren, E.S., Delsman, J.R., Karaoulis, M., Pauw, P.S., Vermaas, T., Bootsma, H., De Louw, P.G.B., Oude Essink, G.H.P., Dabekaussen, W., Gunnink, J.L., Dubelaar, W., Menkovic, A., Siemon, B., Steuer, A. & Meyer, U., 2018. FRESHEM Zeeland - FREsh Salt groundwater distribution by Helicopter Electro-Magnetic survey in the Province of Zeeland. Deltares (Utrecht, the Netherlands): Technical Report No. 1209220-000-BGS-0030: 140 pp.

Van Bakel, J., Van den Akker, C., Querner, E., Rot, G., Schouten, G., Straathof, N., Vaarkamp, W., Witte, J.P. & Zaadnoordijk, W.J., 2017. Zicht op Achtergrondverlaging. Nederlandse Hydrologische Vereniging (NHV) (Wageningen, the Netherlands). NHV Report: 60 pp.

Van Balen, R.T., Houtgast, R.F. & Cloetingh, S.A.P.L., 2005. Neotectonics of The Netherlands: a review. In: Cloetingh, S.A.P.L. & Cornu, T.G.M. (eds): Neotectonics and Quaternary fault-reactivation in Europe's intraplate lithosphere. Quaternary Science Reviews 24: 439-454. DOI: 10.1016/j.quascirev.2004.01.011

Van Beek, C.G.E.M., Cirkel, D.G., De Jonge, M.J. & Hartog, N., 2021. Concentration of Iron(II) in Fresh Groundwater Controlled by Siderite, Field Evidence. Aquatic Geochemistry 27: 49-61. DOI: 10.1007/s10498-020-09390-y

Van Beek, C.G.E.M., Hofman-Caris, C.H.M. & Zweere, G.J., 2020. Drinking water treatment and chemical well clogging by iron(II) oxidation and hydrous ferric oxide (HFO) precipitation. Journal of Water Supply-Research and Technology-AQUA 69 (5): 427-437.

Van Bergen, F. & Koster, K., 2025. Peat, coal and coalbed methane. In: J.H. Ten Veen, G.-J. Vis, J. De Jager & Th.E. Wong (eds): Geology of the Netherlands, second edition. Amsterdam University Press (Amsterdam): 1-2. DOI: 10.5117/9789463728362_ch15

Van Bracht, M.J., 2001. Made to Measure. Information requirements and groundwater level monitoring networks.(PhD thesis Vrije Universiteit Amsterdam (the Netherlands): 211 pp.

Van de Ven, G.P., 1993. Leefbaar laagland - Geschiedenis van de waterbeheersing en landaanwinning in Nederland. Uitgeverij Matrijs (Utrecht): 455 pp.

Van den Eertwegh, G.A.P.H., De Louw, P.G.B., Witte, J.P.M., Van Huijgevoort, M., Bartholomeus, R., Van Deijl, D., Van Dam, J.C., Hunink, J., America, I., Pouwels, J., Hoefsloot, P. & De Wit, J., 2021. Droogte in zandgebieden van Zuid, Midden- en Oost-Nederland. Het verhaal – analyse van droogte 2018 en 2019 en bevindingen [Eindrapport Projectteam Droogte zandgronden Nederland (Fase 3), KnowH2O, KWR, Deltares, WUR, HSS, FWE]. 179 pp.

Van der Molen, J., Peters, E., Jedari-Eyvazi, F. & Van Gessel, S.F., 2019. Dual Hydrocarbon-geothermal energy exploitation: potential synergy between the production of natural gas and warm water from the subsurface. Netherlands Journal of Geosciences 98: e12. DOI: 10.1017/njg.2019.11

Van Duijvenboode, W., 1993. Groundwater quality monitoring in The Netherlands. In: W.M. Alley (ed.): Regional groundwater quality. Van Nostrand Reinhold (New York): 515-535.

Van Engelenburg, J., Hueting, R., Rijpkema, S., Teuling, A.J., Uijlenhoet, R. & Ludwig, F., 2017. Impact of changes in groundwater extractions and climate change on groundwater-dependent ecosystems in a complex hydrogeological setting. Water Resource Management 32: 259-272.

Van Gaans, P.F.M., Griffioen, J., Mol, G. & Klaver, G.T., 2011. Geochemical reactivity of subsurface sediments as potential buffer to anthropogenic inputs: a strategy for regional characterization in the Netherlands. Journal of Soils and Sediments 11: 336-351. DOI: 10.1007/s11368-010-0313-4

Van Geer, F.C., 1987. Applications of Kalman filtering in the analysis and design of groundwater monitoring networks. PhD thesis Delft University of Technology (the Netherlands): 161 pp.

Van Huijgevoort, M.H.J., Voortman, B.R., Rijpkema, S., Nijhuis, K.H.S. & Witte, J.-Ph.M., 2020. Influence of Climate and Land Use Change on the Groundwater System of the Veluwe, The Netherlands: A Historical and Future Perspective. Water 12 (10): 16. DOI: 10.3390/w12102866

Van Leeuwen, J.A., Hartog, N.J., Gerritse, J., Gallacher, C., Helmus, R., Brock, O., Parsons, J.R. & Hassanizadeh, S.M., 2020. The dissolution and microbial degradation of mobile aromatic hydrocarbons from a Pintsch gas tar DNAPL source zone.

Science of the Total Environment 722: 13. DOI: 10.1016/j.scitotenv.2020.137797

Van Loon, A., Pronk, T., Raterman, B. & Ros, S., 2020. Grondwater kwaliteit Nederland 2020. Anorganische parameters, bestrijdingsmiddelen, farmaceutica en overige verontreinigende stoffen in de grondwatermeetnetten van de provincies. Watercycle Research Institute (KWR, the Netherlands). Report No. KWR 2020.067: 145 pp.

Vermeulen, P. & Op den Kelder, T., 2020. The Impact of Groundwater extractions in the Roer Valley Graben. Deltares (the Netherlands). Report No. 11204053-002-BGS-0001: 111 pp.

Versluys, J., 1931. Subterranean water conditions in the coastal regions of the Netherlands. Economic Geology 26: 65-95.

Verweij, J.M., 1999. Application of fluid flow systems analysis to reconstruct the post Carboniferous hydrogeohistory of the onshore and offshore Netherlands. Marine and Petroleum Geology 16 (6): 561-579. DOI: 10.1016/S0264-8172(98)00074-9

Verweij, J.M., 2003. Fluid flow systems analysis on geological timescales in onshore and offshore Netherlands, with special reference to the Broad Fourteens Basin. PhD thesis Vrije Universiteit Amsterdam (the Netherlands): 278 pp.

Verweij, J.M., Nelskamp, S.N., Ten Veen, J.H., De Bruin, G., Geel, K. & Donders, T.H., 2018. Generation, migration, entrapment and leakage of microbial gas in the Dutch part of the Southern North Sea Delta. Marine and Petroleum Geology 97: 493-516. DOI: 10.1016/j.marpetgeo.2018.07.034

Verweij, J.M., Simmelink, E. & Underschultz, J., 2011. Pressure and fluid flow systems in the Permian Rotliegend in the Netherlands onshore and offshore. In: J. Grötsch & R. Gaupp (eds): The Permian Rotliegend of the Netherlands. SEPM Special Publication: Vol. 98, 247-263. ISBN 978-1-56576-300-5

Verweij, J.M. & Simmelink, H.J., 2002. Geodynamic and hydrodynamic evolution of the Broad Fourteens Basin (The Netherlands) in relation to its petroleum systems. Marine and Petroleum Geology 19 (3): 339-359. DOI: 10.1016/S0264-8172(02)00021-1

Verweij, J.M., Simmelink, H.J., Underschultz, J. & Witmans, N., 2012. Pressure and fluid dynamic characterisation of the Dutch subsurface. Netherlands Journal of Geosciences 91 (4): 465-490.

Verweij, H., 2018. 3D spatial variation in salinity of shallow to deep groundwater in the Dutch on- and offshore. Oral presentation at the 20th EGU General Assembly, 11 April 2018, Vienna, Austria. Geophysical Research Abstracts Vol.20 EGU2018-16732.

Verweij, J.M., 2019. Integrated approaches to characterize Cenozoic clay-rich layers and investigate their role in regional groundwater flow; Dutch case studies. In: Oral presentation 46th IAH Congress Malaga 22-27 September 2019.

Verweij, J.M. & Hegen, D., 2015. Integrated pressure information system for the onshore and offshore Netherlands). TNO (Utrecht). Report No. TNO 2015 R10056: 83 pp.

VEWIN, 2022. Dutch drinking water statistics 2022. From source to tap. Association of Dutch Water Companies (VEWIN): 161 pp.

Visser, A., 2009. Trends in groundwater quality in relation to groundwater age. Netherlands Geographical Studies 384. PhD thesis Utrecht University (the Netherlands): 188 pp.

Visser, W.C., 1958. De landbouwwaterhuishouding van Nederland. Commissie landbouwwaterhuishouding Nederland-TNO (the Netherlands): 159 pp.

Vissers, M.J.M., 2005. Patterns of Groundwater quality in sandy aquifers under environmental pressure. PhD thesis Utrecht University (the Netherlands): 144 pp.

Vissers, M.J.M. & Van der Perk, M., 2008. The stability of groundwater flow systems in unconfined sandy aquifers in the Netherlands. Journal of Hydrology 348: 292-304.

Volker, A., 1961. Source of brackish ground water in Pleistocene formations beneath the Dutch polder land. Economic Geology 56: 1045-1057.

Volker, A. & Van der Molen, W.H., 1991. The influence of groundwater currents on diffusion processes in a lake bottom: an old report reviewed. Journal of Hydrology 126 (3-4): 159-169. DOI: 10.1016/0022-1694(91)90153-9

Vos, P.C., 2015. Origin of the Dutch coastal landscape: long-term landscape evolution of the Netherlands during the Holocene, described and visualized in national regional and local palaeogeographical map series. PhD thesis Utrecht University (the Netherlands), Barkhuis, Groningen: 369 pp.

Wibberley, C.A.J., Gonzalez-Dunia, J. & Billon, O., 2017. Faults as barriers or channels to production-related flow: insights from case studies. Petroleum Geoscience 23: 134-147. DOI: 10.1144/petgeo2016-057

Winefield, P., Gilham, R. & Elsinger, R., 2005. Plumbing the depths of the Central Graben: towards an integrated pressure, fluid and charge model for the Central North Sea HPHT play. Geological Society, London, Petroleum Geology Conference Series 6 (1): 1301-1315. DOI: 10.1144/0061301

Witte, J.P.M., Zaadnoordijk, W.J. & Buyse, J.J., 2019. Forensic hydrology reveals why groundwater tables in the province of Noord Brabant (The Netherlands) dropped more than expected. Water 11 (3): 478. DOI: 10.3390/w11030478

Woolderink, H.A.G., Kasse, C., Cohen, K.M., Hoek, W.Z. & Van Balen, R.T., 2019. Spatial and temporal variations in river terrace formation, preservation, and morphology in the Lower Meuse Valley, The Netherlands. Quaternary Research 91 (2): 548-569. DOI: 10.1017/qua.2018.49

Zaadnoordijk, W.J., Buma, J., Vernes, R., Beerten, K., Rogiers, B., Slenter, C., Jaritz, R., Linder, B. & Broers, H.P., 2021. 3D visualization of cross-border patterns of groundwater depletion [Report on Deliverable 3.5 WP3 of GeoERA project H3O-Plus]. 89 pp.

Zaadnoordijk, W.J., Bus, S.A.R., Lourens, A. & Berendrecht, W.L., 2019. Automated time series modeling for piezometers in the National database of the Netherlands. Groundwater 57 (6): 834-843.

Zaadnoordijk, W.J. & Lourens, A., 2019. Groundwater dynamics in the aquifers of the Netherlands. TNO Geological Survey of the Netherlands. (Utrecht). Report No. TNO 2019 R12031: 59 pp.

Zijl, W., 1999. Scale aspects of groundwater flow and transport systems. Hydrogeology Journal 7 (1): 139-150.

Zuurbier, K.G., Hartog, N. & Stuyfzand, P.J., 2016. Reactive transport impacts on recovered freshwater quality during multiple partially penetrating wells (MPPW-)ASR in a brackish heterogeneous aquifer. Applied Geochemistry 71: 35-47. DOI: 10.1016/j.apgeochem.2016.05.013Get rights and content

Zuurbier, K.G., Hartog, N., Valstar, J., Post, V.E.A. & Van Breukelen, B.M., 2013. The impact of low temperature seasonal aquifer thermal energy storage (SATES) systems on chlorinated solvent contaminated groundwater: Modeling of spreading and degradation. Journal of Contaminant Hydrology 147: 1-13. DOI: 10.1016/j.jconhyd.2013.01.002

Zuurdeeg, B.W., Coenegracht, Y.M.A. & Van Gaans, P.F.M., 1989. Geothermische reserves in de Centrale Slenk Nederland - Geochemische aspecten van het formatiewater (No. OS 89-20). Dienst Grondwaterverkenning TNO & Rijksuniversiteit Utrecht.

Appendix: General concepts

A.1 Hydrogeological framework

Properties of the rock framework of importance for groundwater flow include porosity, permeability and compressibility of geological units and geological structures and tectonic elements such as unconformities and fault and fracture zones. The spatial distribution of these properties controls the lateral and vertical extent of hydrogeological units (aquifers, aquitards). In groundwater practice in the Netherlands aquifers are mostly characterized by hydraulic conductivity K (m/day) and transmissivity KD or T (m^2/day), where D is thickness of the aquifer, and aquitards are characterized by vertical hydraulic conductivity K_v (m/day) and hydraulic resistance c (= K_v/D, expressed in days). Hydraulic conductivity K=k ($\rho g/\mu$) includes the rock matrix permeability and properties of the contained groundwater (density ρ and viscosity μ). In geosciences dealing with greater depths and depth ranges, permeability of the rock matrix k (expressed in m^2 or Darcy) is commonly used instead of hydraulic conductivity K. This is because of the larger spatial variation of density and viscosity of the groundwater at these depths and its associated effect on hydraulic conductivity. In petroleum geoscience, stratigraphic units with aquifer properties are referred to as reservoirs in case the pore fluids are petroleum and aquitards are often indicated as seals or caprocks.

The expression "confined aquifer" is used for permeable water-saturated hydrogeological units overlain by less permeable units (aquitards). Unconfined aquifers, also known as phreatic aquifers or water table aquifers, are permeable water-saturated hydrogeologic units with the groundwater table as its upper boundary. The water phase in the water-saturated part of the subsurface is continuous. It will extend, for example, from an aquifer into an adjacent aquitard. The water-saturated subsurface can be considered to be hydraulically continuous allowing 3D cross-formational flow.

The geochemical composition of hydrogeological units (treated in Griffioen et al., 2025, this volume) exerts an important control on the chemical composition of groundwater together with groundwater flow and the composition of water recharging the groundwater.

A.2 Groundwater flow

The flow of groundwater is proportional to its potential energy gradient, which under the assumption that groundwater is incompressible (i.e. assumption of constant groundwater density), is often expressed as hydraulic head gradient in hydrogeological practice. Hydraulic head (h, h= h_p + h_z) is the energy per unit weight of groundwater. It is composed of the pressure head (h_p = p/ρg, where p = groundwater pressure, ρ=groundwater density, g = acceleration due to gravity) and the elevation head (h_z) representing the gravitational potential energy of the groundwater at a certain location. Flow as based on Darcy's Law (Darcy, 1856) can be described by: q = -K (dh/dl), where q volumetric flow rate per unit area; K = hydraulic conductivity; dh/dl = hydraulic head gradient. If the groundwater density is not constant, flow is also influenced by the differences in density. In coastal zones of the Netherlands salinity-related density differences are particularly important. Also, at greater depth significant spatial variations in density occur due to differences in salinity and/or temperature of groundwater.

Changes in hydraulic head need time to propagate through a groundwater system. The change of groundwater pressures (hydraulic heads) in time and space by the flow is incorporated in groundwater flow equations by the so-called specific storage (S_s = $\rho g(\alpha + n\beta)$, where ρ = density of the groundwater, α = compressibility of the rock matrix, β = compressibility of the groundwater, n = porosity). The response time is controlled by L^2S_s/K, i.e. the specific storage S_s, hydraulic conductivity K, and spatial scale L, assuming 1D flow and uniform magnitude of the ratio S_s/K (e.g. Neuzil, 2015). The response time is large for large hydrogeologic systems of small hydraulic conductivity and large specific storage. In contrast, changes in pressure will propagate fast in a confined hydrogeologic units of relatively large hydraulic conductivity and small specific storage (compressibility) values. The dissipation of overpressures by multiphase flow (e.g. in the presence of gas) is also affected by capillary pressures of the geologic framework (Peters & Verweij, 2012).

A.3 Topography-driven groundwater flow

The water table forms the top boundary of nested topography-driven groundwater flow systems, where in each local, regional or supra-regional flow system flow is directed from the recharge area towards one or more spatially related discharge or seepage areas (Tóth, 1963, 2009; Engelen & Kloosterman, 1996) (see also section 'Groundwater flow systems onshore')

A.4 Compaction-driven flow and hydromechanical coupling

Interaction of groundwater with geologic media through hydromechanical coupling is also an important mechanism for explaining the hydrogeology of the Netherlands. Hydromechanical coupling involves the causal effects of changes in groundwater pressure on deformation of sediments and rocks, and vice versa the effects of deformation on the pressure and flow of groundwater (Neuzil, 1995; Ingebritsen et al., 2006; Gambolati & Teatini, 2021). The following two examples illustrate this interaction. Consider a certain hydrogeologic unit in which the total load (stress) of the overlying layers (the overburden) acting on the unit is balanced by pore water pressure in the unit and effective stress acting on the unit's rock matrix. When groundwater is extracted from the unit by pumping or drainage, the hydraulic head and pore pressure decrease. As a consequence, part of the overburden load supported by the pore water decreases, while part of the overburden load acting on the rock matrix, the effective stress, increases. The increase in effective stress results in reduction of the pore volume and thickness of the unit, i.e. results in compaction. The subsurface compaction of the hydrogeologic unit propagates directly to the surface resulting in ground surface subsidence. Changes in total stress acting on a hydrogeologic unit due to anthropogenic activities (loading by heavy buildings or infrastructure) or geologic processes such as sedimentary loading or erosional unloading may lead to changes in effective stress, changes in pore pressure and water flow (also known as compaction-driven flow). Changes in effective stress, pore pressure and flow are highly interconnected. In groundwater practice, focus is on the flow part of this relation. The interactions of groundwater with geologic media through hydromechanical coupling operate from shallow to great depths. It is the underlying mechanism responsible for various observed phenomena, such as aquifer compaction and surface subsidence in response to reduction of groundwater pressure due to extraction of water by pumping or drainage, compaction and reduction of porosity and permeability of sediments and rocks due to sedimentary loading, compaction flow and occurrence of significant overpressures. Overpressures are pressures in excess of hydrostatic pressure, and represent actually artesian conditions. Significant overpressures occur at greater depth below the topography-driven groundwater flow systems in onshore Netherlands and in the offshore.

18

Surface mineral resources

Michiel J. van der Meulen, Joris J. Dijkstra, Tjeerd P.F. Koopmans, Hans S. Pietersen, Jelte Stam & Denise Maljers*

**Corresponding author: michiel.vandermeulen@tno.nl*

ABSTRACT

Dutch Quaternary and Neogene unconsolidated clastic deposits constitute major resources of aggregates and clay that are used in building and construction. Fine sand is available all over the country (including offshore), and is used mainly for filling, coastal nourishment and land reclamation. Coarse sand, which roughly occurs in the southeastern half of the country, is mainly used in concrete and masonry mortars. Gravel occurs in the southeasternmost part of the country, and is used in concrete and a variety of unbound applications. In addition to home-produced sand and gravel, the Netherlands uses significant amounts of imported and secondary aggregates. High-grade silica sand of Miocene age, occurring in the southeasternmost parts of the country, is used in several process industries. Clay is used for the production of structural ceramics (bricks and rooftiles), dyke building and maintenance, landfill covering and landscaping. Most ceramic clay is extracted from Holocene overbank deposits of the central/eastern Rhine-Meuse Delta. Clay for other applications is available all over the Holocene fluvial and coastal lowlands, except where peat is prevalent. Dutch carbonate resources include Cretaceous chalk in the southeasternmost part of the country, Triassic limestone and dolomite in the central-east, and Holocene shell deposits in the northern and southwestern tidal basins. Until recently, the chalk has been extracted to produce cement. The Triassic carbonates are grinded to powder and used in a variety of industrial and agricultural applications. The shells are used as drainage, paving and insulation material, or grinded for use in animal fodder.

<< Silica sand (*zilverzand*) of the Heksenberg Member in the De Groot quarry in Abdissenbosch (southeast Netherlands). Photo: Rudolf Bak.

DOI: 10.5117/9789463728362_ch18

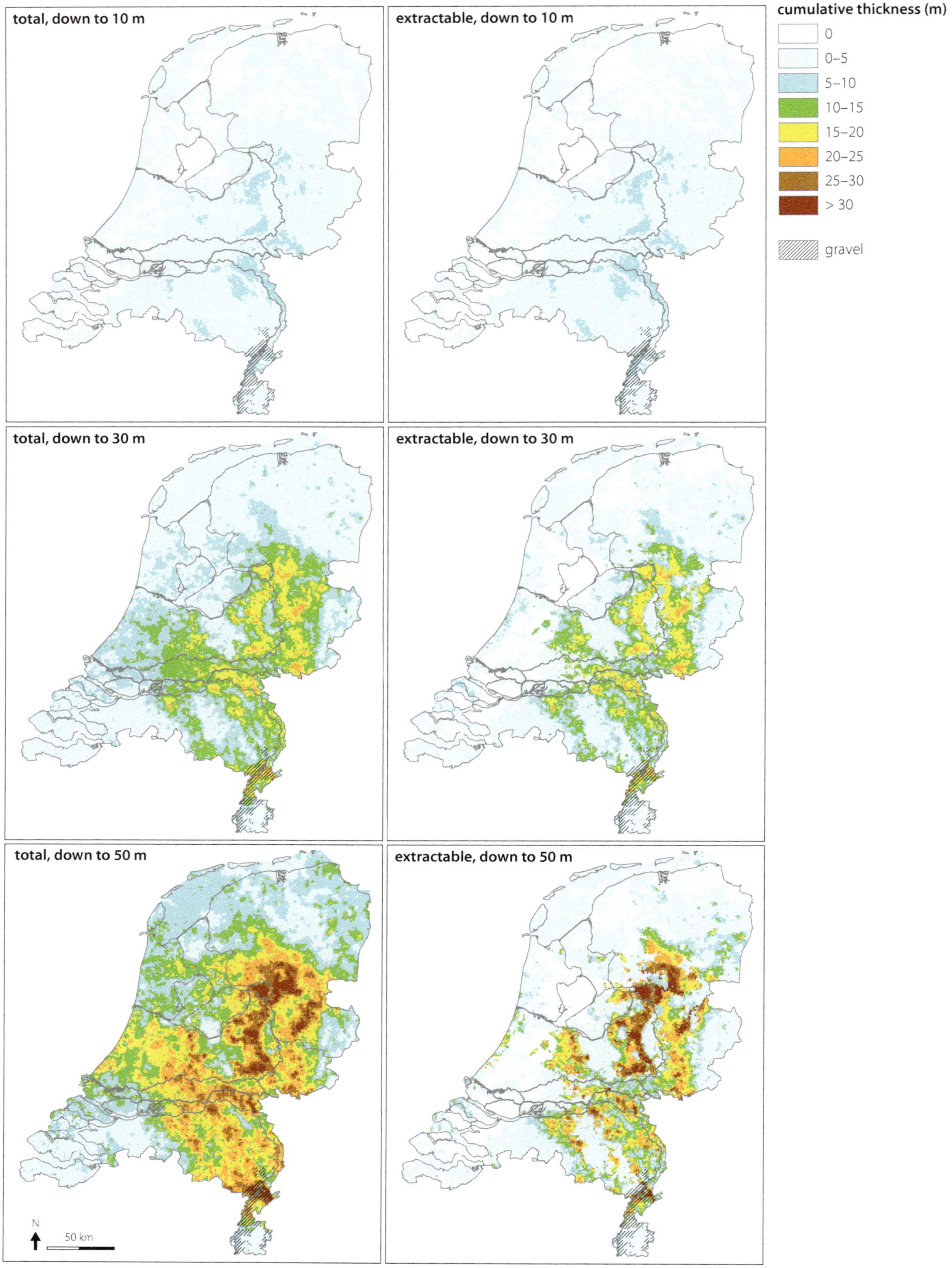
total, down to 10 m
extractable, down to 10 m
total, down to 30 m
extractable, down to 30 m
total, down to 50 m
extractable, down to 50 m
cumulative thickness (m)
0
0–5
5–10
10–15
15–20
20–25
25–30
> 30
gravel
N
50 km

Introduction

Definitions

The production of geological resources in the Netherlands is governed by two laws. The Mining Act (Mijnbouwwet; Anonymous, 2002) applies to any use of the subsurface, including mining in the broadest sense, at depths greater than 100 m (or deeper than 500 m for geothermal energy). Over the past decades, this has mainly concerned the exploration and production of hydrocarbons and salt, but it has been extended to the storage of CO_2 (amendment of 2011) and geothermal energy (amendment of 2022). Classical underground mining has not taken place in the Netherlands since domestic coal production was terminated in 1975.

Surface mining, i.e. quarrying, excavating and dredging operations, is governed by the Environment and Planning Act (*Omgevingswet*; Anonymous, 2016). The legal distinction between surface and deep mining in Dutch policy recognizes the category of 'surface mineral resources' (*oppervlaktedelfstoffen*), the subject of the present chapter. It is based on depth of occurrence only and does not correspond to any common mineral categorization. Currently exploited Dutch surface mineral resources include aggregates, clay, silica sand and carbonates, all of which are industrial minerals, used mainly in construction. The only industrial mineral produced in the Netherlands that is not included in the surface mineral resource category is rock salt, which is treated in a separate chapter (Groenenberg et al., 2025, this volume).

Constitutionally, the Netherlands proper includes three municipalities in the Caribbean (the islands of Bonaire, Sint Eustatius and Saba), while the Kingdom of the Netherlands contains three additional constituent countries (the Caribbean islands of Aruba, Curaçao and Sint Maarten). Currently known mineral resources in the Caribbean parts of the Kingdom include building materials (in all six islands); gold (mined in Aruba from the early 1800s till the early 1900s); phosphate (mined in Aruba and Curaçao from the late 1800s till the early 1900s); copper (mined in Curaçao in the 1700s); and sulphur (mined in Saba from late 1800s till the early 1900s). The present chapter focusses on the Netherlands in Europe, including the Dutch sector of the North Sea and excludes the geologically unrelated Caribbean parts of the Kingdom.

Geological setting

The shallow subsurface of the Netherlands consists largely of unconsolidated Quaternary clastic deposits, which are discussed in Busschers et al. (2025, this volume). Their general physiographic setting is determined by uplifted sediment source areas in the catchments of the rivers Rhine, Meuse, Scheldt, and Eridanos (the Eocene-Pleistocene Baltic river system; Overeem et al., 2001), together with a few smaller rivers in the south and east of the country, as well as by the subsiding North Sea Basin in the west and north. Accordingly, Quaternary sediments generally grade from coarse-grained to fine-grained in a northwestward direction. Glaciations affected deposition, erosion and geomorphology repeatedly through the effects on sea level and vegetation, as well as directly by advancing ice sheets. Occurrences of Mesozoic and older solid rock at or near the surface are limited to the province of Limburg (southeast Netherlands) and the easternmost parts of the provinces of Gelderland and Overijssel (mid-east Netherlands), where they are related to the exhumation of the Ardennes massif towards the south and the Münsterland Cretaceous Basin towards the east, respectively. In both areas, Mesozoic carbonates are quarried for a variety of purposes or have been up until recently.

Aggregates

Introduction

Aggregate is the catch-all term for granular matter used in construction, either as such (e.g. as constructional fill, railway ballast, in slit drainage systems and for coastal nourishment) or with some binder (e.g. in concrete, asphalt, masonry mortars, etc.). In terms of mass and volume, they represent the most-used mineral resource worldwide (UNEP, 2019). The Netherlands is more or less self-sufficient in sand, resources of which are abundant (Fig. 18.1). However, as resources of gravel are comparatively limited and there is hardly any crushable rock at or near the surface, the Netherlands imports about 75% of these coarser aggregates.

Sometimes confusingly, the terms sand and gravel have different meanings in geology and in the extractive industry. In geology, they refer to either nominal grain-size fractions, or clastic sedimentary deposits having sand or gravel

← ***Figure 18.1.*** *Resources of coarse sand and gravel down to 10, 30 and 50 m below the surface, based on a nationwide 3D lithological model (Van der Meulen et al., 2005, 2007). The left panel shows the total cumulative resource thickness within the depth range indicated in the upper left corner. The right panels show cumulative extractable resource thickness. Aggregates are considered extractable according to criteria explained in the text. Grey hatching in the southeasternmost part of the country indicates the occurrence of gravel. The abundance of fine sand is more or less inversely proportional to that of coarse sand and gravel at depths >10 m in most of the country.*

as the dominant fraction. Regardless of the sediment classification scheme of choice, 2 mm is the limit between the sand and gravel fractions (cf. Wentworth, 1922; see Van der Meulen et al., 2005, 2007 and 2009; Jones et al., 2021 for a discussion). In industry, sand and gravel refer to specific types of aggregate products that are each, depending on the application, defined primarily by grain-size distribution. Processing of natural sand and gravel (classification, sieving) is generally required in order to meet specifications. Depending on the application, the lower grain size limit of gravel and corresponding crushed aggregates varies between 4 mm and 8 mm while the upper limit varies between 16 mm and 32 mm. Sand has a lower limit of 63 to 125 µm and an upper limit of 4 to 8 mm (Van der Meulen et al., 2005). In addition to grain size, aggregate specifications may include criteria for shape, physical or mineralogical properties of grains, or for bulk properties such as the friction angle.

For permitting and reporting purposes, the wide range of aggregate products is categorized into gravel and crushed rock, concreting and masonry sand, sand for sand-lime bricks, filling sand, sand for coastal nourishment, sand for land reclamation, and secondary aggregates. These aggregate types are discussed below in terms of applications, supplies and resources. Silica sand, which is used for its chemical composition in the process industry rather than for its physical properties in construction, does not belong to the aggregates family and is discussed separately.

Figure 18.2 shows Dutch aggregate consumption from 1980 to 2020. A clear correlation with the overall performance of the Dutch economy (expressed as GDP growth) reveals some of the main drivers behind minerals consumption and this is also evident in the statistics for individual commodities discussed below. Five major crises occurred in the time span represented in Figure 18.2. The global debt crisis of the early 1980s (Bakker, 1999) resulted in a marked decrease of aggregates consumption. The crisis was followed by almost two decades of more or less continuous growth in aggregate consumption, affected only by the mild European recession of 1991-1993 (e.g. Roy & Kemme, 2020). Aggregates consumption peaked in 2001, after which the Dutch economy came to a virtual standstill during the global recession of the early 2000s (Roy & Kemme, 2020) and aggregate consumption levels dropped accordingly. The Dutch economy was then hit by the Great Recession of 2007-2009 and the ensuing European Sovereign Debt Crisis (2009-2014; Kickert, 2012;

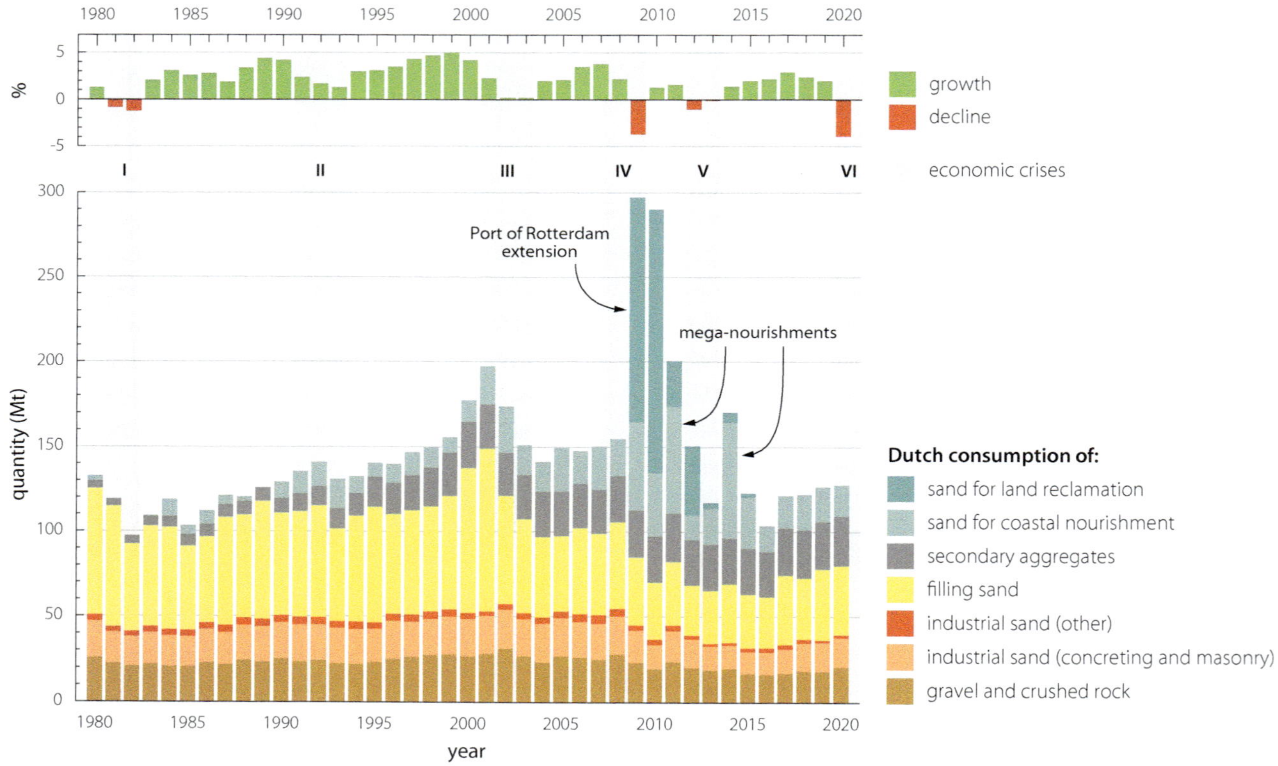

Figure 18.2. Lower panel: the consumption of aggregates in the Netherlands from 1980 to 2020. Upper panel: GDP-growth per annum, with an indication of economic crises: (I) global debt crisis, (II) European recession, (III) global recession, (IV) global financial crisis, (V) European sovereign debt crisis (VI) impacts of the Corona Pandemic. Production figures compiled from data provided by the Ministry of Infrastructure and the Environment (1980-2009), and MWH Global and H2H (2009-2020).

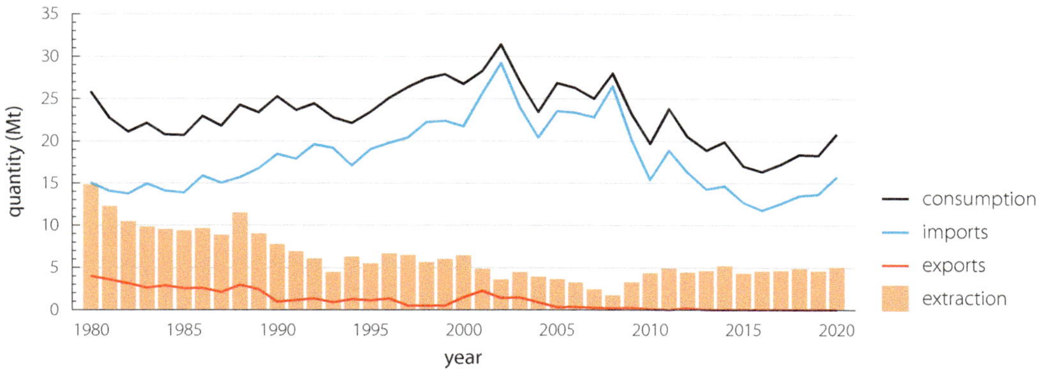

Figure 18.3. Dutch gravel and crushed rock supplies from 1980 to 2020. Sources: see Fig. 18.2.

Ureche-Rangau & Burietz, 2013). These events, however, coincided with three major national projects discussed below which demanded vast amounts of marine aggregates. As a result the consumption of sand for land reclamation and coastal nourishment peaked at his time, while that of all other aggregates declined and stabilized below the levels of 1990-2010. The effects of the Corona Pandemic crisis on the economy and minerals consumption have yet to become evident.

Another clear trend is the increasing share of secondary aggregates, the use of which was promoted by national policy during the last decades of the previous century and which has led to a stable market for their use (discussed in detail in the section on secondary aggregates). In principle, all recyclable aggregates are reused, amounting to a significant substitution of primary aggregates.

Gravel and crushed rock

Applications and demand

The Dutch building industry uses around 19 Mt/yr of gravel and crushed rock, mainly in the production of concrete and asphalt, but also for unbound applications (e.g. railway construction and maintenance, and hydraulic engineering). About one fourth of the demand is met with home-produced gravel, almost all of which is extracted in the province of Limburg in the southeast of the country (Fig. 18.1). The long-term trends in demand are similar to those discussed for aggregates in general (Fig. 18.3). However, production steadily declined from 2000 to about 2008 in response to a government policy to gradually decrease production to the approximate level of provincial self-supply, in response to environmental concerns and because of local resistance against extraction (Van der Meulen, 2005). Production subsequently stabilized at the current level. Meanwhile, gravel imports rose accordingly (Van der Meulen et al., 2003, 2005), stabilising at the current level of about 5 Mt/yr, mainly from Germany, Belgium and the British sector of the North Sea. About 6 Mt/yr of crushed rock aggregates are imported from Belgium, Germany, Scotland, France and Norway.

All current Dutch gravel extraction permits are tied to *Maaswerken*, a project that combines river widening and aggregates extraction (Van der Meulen et al., 2006). They will expire in the foreseeable future, which means that home production of gravel will come to an end unless new permits are granted.

Resources

The main Dutch gravel resources are located in the province of Limburg (Fig. 18.1), where they have been deposited by the river Meuse from the Early Pleistocene onwards. As a result of syn-sedimentary uplift, they have been preserved in terraces that cover the flanks of the river valley. In the provinces of Gelderland, Utrecht, North-Brabant and Overijssel, i.e. roughly in the southeastern half of the country, gravel occurs as admixtures and intercalations in coarse Pleistocene sands, where they constitute by-products of concreting and masonry sand production (<1 Mt/yr). There are some isolated, relatively minor surficial occurrences of gravel and gravely sand in the Dutch sector of the North Sea. The largest of these is located in the area of the Cleaver Bank (Fig. 18.4) and was formed as an end moraine during the last glaciation and subsequently marine-reworked. The estimated amount of gravel in place is a mere 30 Mt (Laban, 2002), resolved at the level of UNFC Prospective Projects class (UNECE, 2019).

Concreting and masonry sand (*beton- en metselzand*)

Applications and demand

Concreting and masonry sand refers to a range of industrially produced coarse sands. About 75% is used for concrete production and about 15% in masonry mortars. The remaining ~10% is used for a wide range of purposes such as the production of asphalt and bricks, and for unconsolidated applications such as drains and filters. Each of these

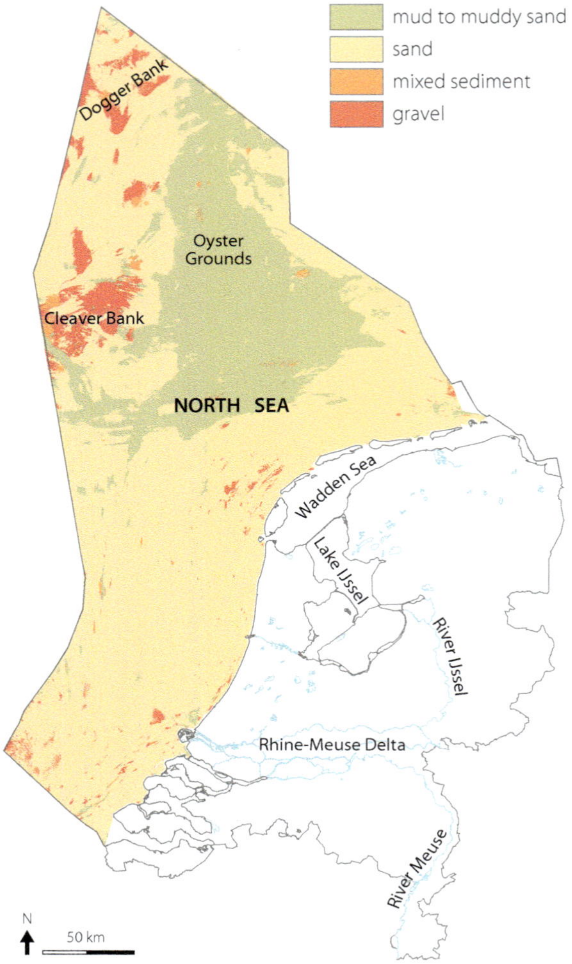

Figure 18.4. As the 3D lithological model used for Fig. 18.1 does not cover the Dutch offshore, a seabed sediment map for the Dutch sector of the North Sea Continental Shelf is provided as an indication of marine aggregate resources. Source: www.dinoloket.nl/en/subsurface-models/map (select seabed sediment).

applications has specific requirements for the grain-size distribution (Fig. 18.5) and, to some extent, for the grain shape of the aggregates. Producers have to sieve and blend natural aggregates in order to meet these requirements.

The demand for concreting and masonry sand shows more or less the same long-term trends as for aggregate in general: averaging at about 21 Mt/yr between 1980 and 2008, then dropping and stabilising at the current level of about 15 Mt/yr (Fig. 18.6). During the last 40 years sources of supply have shifted considerably. From the early 1980s until the late 1990s, a national policy was in place to approximately balance home production with demand by issuing sufficient permits to maintain the required supplies. However, due to the geological scarcity of sand suitable for concrete applications, the total production proved to be too little to meet the demand of Dutch concrete producers. Therefore, about 7 Mt/yr of coarse concreting and masonry sand were imported from Germany, while a similar amount of home-produced finer concreting and masonry sand was exported to Flanders (Belgium), which has limited coarse-sand resources.

Around 2000, the national government abandoned regulating the production of concreting and masonry sand (Van der Meulen, 2005). Home production decreased significantly and net imports rose to about 10 Mt/yr in 2003, remaining at such unprecedented high levels for 6 years. Only around 2015 has the balance between imports and exports been restored at about 5 Mt/yr, albeit under different economic circumstances and under a liberalized national permitting policy.

Resources

Concreting and masonry sand is mostly produced from Pleistocene fluvial deposits in the southern and eastern parts of the Netherlands (Kreftenheye, Urk, Beegden, Sterksel, Appelscha, Peize and Waalre formations; Busschers et al., 2025, this volume). These deposits protrude westward below the Rhine and Meuse riverplains and into the North Sea. However, since they are overlain by westward thickening, fine-grained Holocene flood and tidal basin deposits, and also display downstream fining, their exploitability and suitability for concreting applications in the coastal and North Sea areas is debatable (RWS,

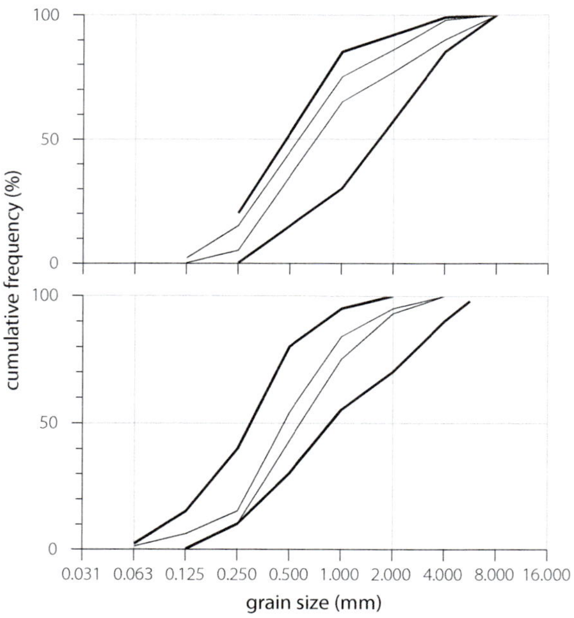

Figure 18.5. Grain-size distributions of the most important types of mortar sand (lower panel) and concrete sand (upper panel). Thick lines show the official minimum and maximum requirements (NEN, 1991, 1999); thin lines delimit the actual distributions on the market (RWS, 1998 and unpublished data).

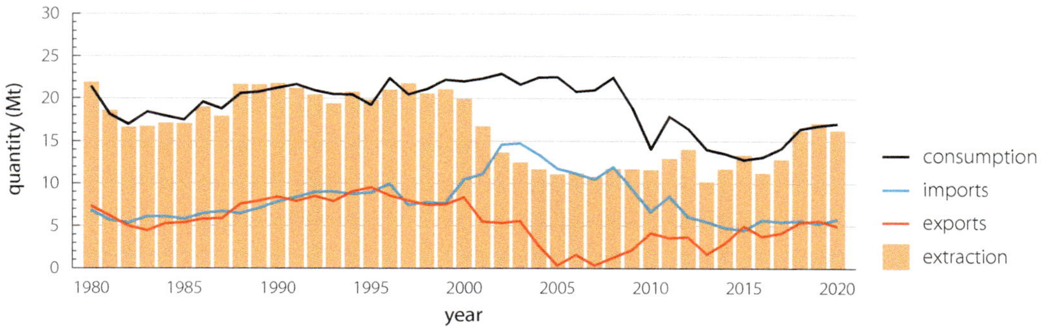

Figure 18.6. Dutch concreting and masonry sand supplies from 1980 to 2020. Sources: see Fig. 18.2.

2004). Coarse-sand extraction sites exist throughout the eastern and southern Netherlands, but since much of the sand is shipped to its destination, most of it is extracted in the vicinity of major rivers.

Sand for sand-lime bricks (*kalkzandsteenzand*)

Sand-lime bricks are produced from a mixture of sand (~90%), lime (~7%) and water, which is moulded under mechanical pressure and then hardened under steam pressure and heating. They are mainly used for inside building walls. Sand for sand-lime bricks has a median grain size between 150 and 400 µm. Ten to 30% in the grain-size distribution has to be finer than 125 µm and an equal share coarser than 500 µm. Natural sands which meet these requirements are quite common in the Netherlands. The production is about 1.5 Mt/yr (Fig. 18.7).

Filling sand (*ophoogzand*)

Applications and demand

The western and northern parts of the Netherlands have a largely muddy or peaty subsurface that is prone to compaction and subsidence. Building in these areas therefore requires large amounts of filling sand as a foundation (De Jong & De Mulder, 1998). The current yearly demand is about 25 x 10^6 m^3, all of which is extracted in the country (Fig. 18.8).

The share of sand from the Dutch sector of the North Sea rose from approximately 10% in 1980 to about 50% in the early 2000s (Stolk & Seeger, 1999; Fig. 18.6). This increase is due to a policy of phasing out the extraction from the densely populated coastal provinces of North Holland and South Holland, and more recently from the province of North Brabant (southern Netherlands). After 2009 the relative share of marine sand for offshore use increased significantly, especially in connection with the extension of the Port of Rotterdam and the mega-nourishments discussed below.

Resources

In contrast to the other aggregates discussed above, filling sand has virtually no grain-size constraints and includes all sands which are unsuitable (i.e. too fine-grained) to be used for higher-value applications. Fine sand is available all over the Netherlands. The Dutch Holocene comprises a wide range of fine-grained sand bodies, deposited in sedimentary environments ranging from fluviatile to shallow marine. Resources in the North Sea are vast (Fig. 18.4).

In Limburg, hoggin (*stol*) is used for about 7% of the provincial filling-material requirement. It is extracted from fine-grained soils which developed on local Miocene-Pliocene fluviatile pebbly sands. The pebbly sand-loam mixture is relatively cohesive and is unsuitable for concreting and masonry sand production, but apt for use as filling or road-foundation material.

Sand for coastal nourishment (*suppletiezand*)

The current Dutch coastal defence programme, initiated in 1990, involves monitoring the coastline and coastal processes, and instigating measures at places where the coast is being eroded (Brand et al., 2021). Sustained coastal erosion is countered by the application of sand on

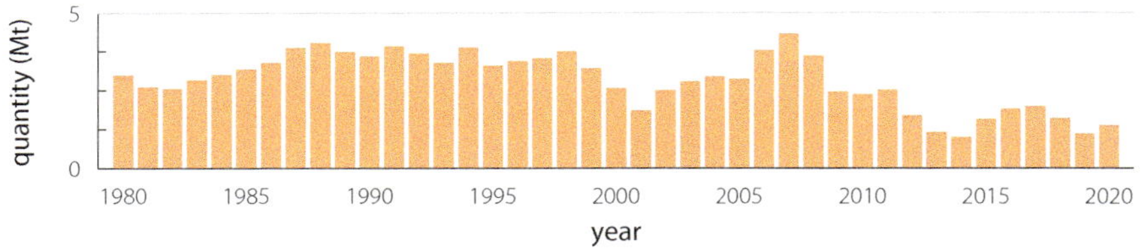

Figure 18.7. Production of sand for sand-lime bricks in the Netherlands from 1980 to 2020. Import and export figures are unavailable but are expected to be negligible. Sources: see Fig. 18.2.

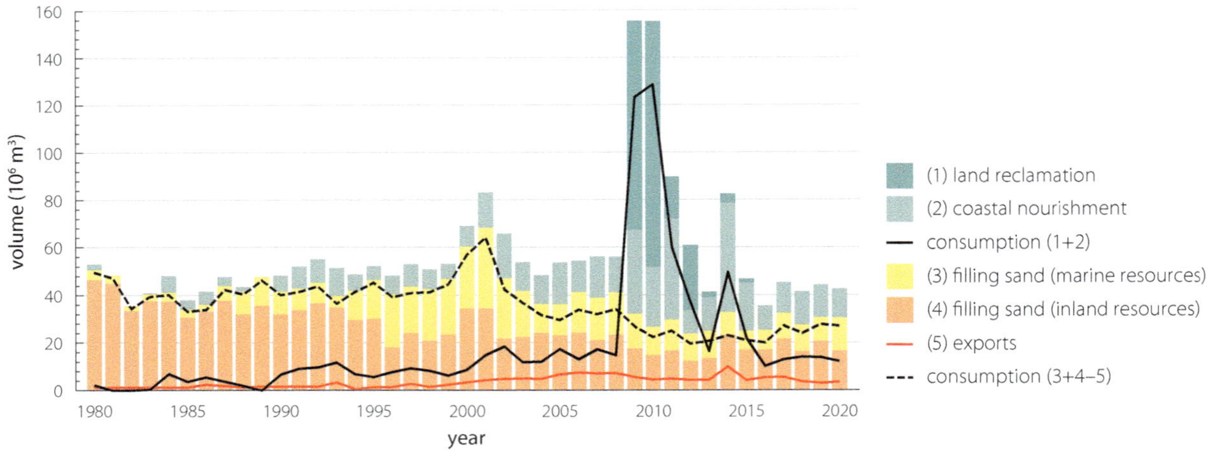

Figure 18.8. Dutch supplies of filling sand, sand for coastal nourishment and sand for land reclamation from 1980 to 2020. Amounts of these aggregates are usually given in volume rather than in weight units. Sources: see Fig. 18.2.

the beach or shoreface. The programme requires approximately 13 x 10^6 m^3 of sand per year (Fig. 18.8), which is extracted from the inner shelf as near as possible to the nourishment site. On top of that, one-off large coastal reinforcement and land reclamation projects may double or triple the use. Over the past decade, the Dutch coastal nourishment program was supplemented with two large individual coastal protection projects. In 2011, an innovative nourishment scheme was implemented that involved emplacing 21.5 x 10^6 m^3 of sand near The Hague (Stive et al., 2013). The resulting body of sand, the 'Sand Motor' (*Zandmotor*), reinforces a thin section of coastal dunes that was identified as a weaker zone in Dutch coastal defence policy. The sand is also intended to become eroded and dispersed along the coast via long-shore current sediment transport. Between 2013 and 2015, 30 million m^3 of sand were emplaced to reinforce the Hondsbossche and Pettemer sea defence, a large seawall erected in 1880 that has been used to protect a section of coastline without significant coastal dunes (Oost et al., 2015).

Sand for land reclamation (*landaanwinningszand*)

Between 1980 and 2009, the sand used to reclaim land from the sea was included in the sand for coastal nourishment statistics. However, when the Port of Rotterdam established its most recent seaward extension, *Maasvlakte II*, the 200 · 10^6 m^3 of sand it took to reclaim 20 km^2 of land in 2009 and 2010 (Fig. 18.8) was administered as a separate category. The port extension sits on the largest Dutch anthropogenic deposit created in a single project.

Secondary aggregates

Since the early 1990s, the substitution of primary by secondary aggregates has been stimulated by the Dutch national government through financial and regulatory instruments and, leading by example, by making use of secondary aggregates in national infrastructure projects. Since the mid-1990s, legislation and standards have been developed to enable the environmentally safe use of waste materials and industrial by-products in construction (e.g. VROM, 1995, 2007; Broers et al., 2002; Dijkstra et al., 2019). Reusing secondary aggregates reduces the consumption of the above primary aggregates in several applications, especially where filling is involved.

The share of secondary materials in the total provision has risen from ~7 Mt/yr (6%) in 1980 to ~32 Mt/yr (20%) in the early 2000s and stayed at approximately that level ever since (Fig. 18.2). Figure 18.9 is a breakdown of the secondary aggregate category in Figure 18.2 and considers the main types of secondary aggregates that are re-used in the Netherlands. Most of these materials, including granulated construction and demolition waste (CDW), steel slag and municipal solid waste incineration bottom ash (MWSI bottom ash), are currently used in unbound applications, replacing filling sand and coarse aggregate to an extent that should be accounted for in the total demand (see also Dijkstra et al., 2019). Typical applications include road foundations, landscaping, noise barriers, or ballast material in water works and coastal protection. Colliery spoil (not included in Fig. 18.9) is occasionally used for hydraulic engineering and as filling or paving material.

The strong rise in the use of secondary aggregates in the 1990s can mostly be attributed to CDW recycling. In 1985, the production of CDW was about 12 Mt/yr, of which 50% was landfilled and 50% was recycled (Hendriks & Pietersen, 2000; CLO, 2020). The steepest increase was coeval with the ban on landfilling of re-usable CDW in 1997 (VROM, 1997). The production of CDW stabilized in the years 2000-2014 at about 24 Mt/yr (CLO, 2020). Some other secondary aggregates in Figure 18.9 show a growth

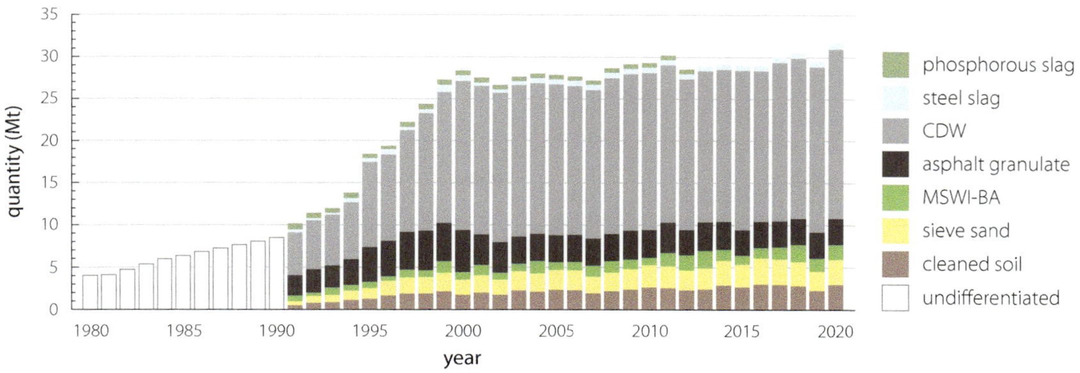

Figure 18.9. Main categories of secondary aggregates and their usage in the period 1980-2016 in the Netherlands. CDW = granulated construction and demolition waste. MSWI-BA = Municipal Solid Waste Incineration Bottom Ash, which is the main ash fraction originating from the combustion of household waste; sieve sand is the fine fraction from granulated construction and demolition waste; cleaned soil is thermally or wet-treated contaminated soil. Sources: Hendriks et al. (1999), RWS (2005, 2022), Bijleveld et al. (2015), Van Lieshout & Nusselder (2016), I&W (2017), 't Hoen (2017), CLO (2020), Tata Steel (2020).

as well. Most importantly, the use of MSWI bottom ash in infrastructural works doubled from 0.75 Mt in 1990-2000 to 1.5 Mt in 2010-2020 (RWS, 2005, 2022).

The use of secondary aggregates in bound applications such as concrete is relatively low. Replacement of concreting and masonry sand in bound applications is estimated to have been ~1 Mt/yr on a consumption of about 20 Mt/yr in the years 2001-2010 (about 5% replacement) and includes for instance sieve sand from construction and demolition waste and sand from asphalt granulate (MWH, 2012). Replacement of gravel and crushed stone by recycled concrete aggregate in new concrete amounted to a stable 0.5-0.7 Mt/yr in the years 2009-2016 (about 4% replacement, 't Hoen, 2017). In line with the Paris agreement (UN, 2016), the national government and the concrete sector expressed the ambition to achieve 100% circularity of the concrete production (*Betonakkoord*; Anonymous, 2018). Based on the ratio between the production volume of new concrete and the availability of recycled concrete aggregate, this could in theory result in a substitution of 15-20% of primary aggregate (Anonymous, 2018).

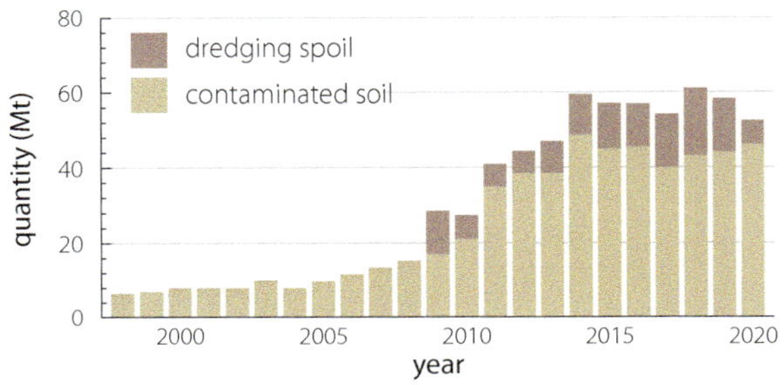

Figure 18.10. Amounts of contaminated soil and freshwater dredging spoil reused in the Netherlands. Source: RWS (2022).

Asphalt contains ~80-90 % m/m aggregate (sand, gravel, crushed rock; Bak et al., 2018). Asphalt granulate from infrastructural works is either recycled in unbound applications, replacing primary aggregate such as filling sand, or is recycled in new asphalt. MSWI fly ash is used as filler in asphalt, replacing sand. MSWI bottom ash has come to be treated and processed in such a way that its mineral fraction can substitute primary aggregate in concrete (e.g. Keulen, 2016).

Besides industrial wastes and by-products, increasing amounts of excavated contaminated soil and dredging spoil are partially replacing primary aggregates in unbound constructions (Fig. 18.10). Between 2010 and 2020, 91-95% of all excavated contaminated soil and 67-85% of freshwater dredging spoil was beneficially re-used in earthworks such as embankments and noise barriers (RWS, 2022), partially replacing filling sand. The remainder is too contaminated and is either treated, immobilized or landfilled. The (re-)use of freshwater dredging spoil peaked in 2018, when it was used in dyke maintenance (*Houtribdijk*, Lelystad) and in a large land reclamation project (*Marker Wadden*). However, it has sharply decreased following completion of these projects and also in response to the 'PFAS crisis', i.e. the widespread contamination of sediments with polyfluor-alkyl substances (PFAS) above the recognized safe limits (RWS, 2022).

Over the past decade, about 20% of the total re-use of contaminated soil and dredging spoil consists of backfilling sand and gravel extraction pits ('t Hoen, 2018). Although this practice can be viewed as the most literal replacement of primary aggregate by secondary materials, it cannot be attributed unambiguously to the substitution of primary aggregates in construction.

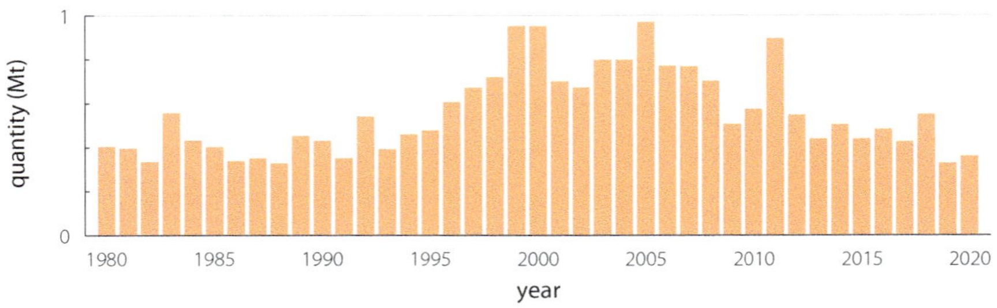

Figure 18.11. The production of silica sand in the Netherlands from 1980 to 2020. Sources: see Fig. 18.2.

Silica sand (*zilverzand*)

Applications and demand

Silica sand is a quartz resource employed in several industries (Van der Meulen et al., 2009). Various types of silica sand, with minimum SiO_2 percentages ranging from 97.0 to 99.8%, are used in the glass industry (~55%), the chemical industry (15%; the production of, e.g. water glass, zeolites and carborundum (SiC)), the ceramic industry (~20%) and foundries (~10%). Lower-grade silica sands are used when a white colour is desired for, e.g. masonry, the sand joints in pavements, or the production of sand-faced bricks. Table 18.1 shows the quality requirements for the main applications of silica sands in terms of quartz, iron, aluminium and titanium contents, and loss on ignition. Additional quality parameters, not shown in the table, include the grain-size distribution, and the chromium (Cr) and heavy-mineral contents.

Natural silica sands usually do not meet the industrial requirements and have to be upgraded. Impurities related to specific grain-size fractions such as clay are removed by washing or wet screening. Grain coatings of iron minerals such as limonite can be removed by attrition (scrubbing), or acid leaching (both cold and hot). Froth flotation and magnetic- or gravity separation methods are used to remove discrete contaminant minerals. The silica sands currently extracted in the Netherlands usually require only washing, or the combination of washing and attrition.

The production of silica sand was around 0.4 Mt/yr in the 1980s and has risen to around 0.8 Mt/yr in the early 2000s (Fig. 18.11). The rise was largely an increasing use of somewhat lesser-grade silica sands, which are more abundant. After that, the production of silica sand shows the economically driven trends shown in Figure 18.2 for aggregates. There are no accurate import and export figures. Exports to Germany are known to be significant and silica sand is imported from Belgium and Germany.

Resources

Currently exploited silica sands of the Miocene Groote Heide Formation (Munsterman et al., 2025, this volume) of South Limburg, near the town of Heerlen (Figs 18.12, 18.13), rank among the purest in Europe. The high quartz contents are the result of a combination of a quartz-rich source material, sorting in a coastal depositional environment and pedogenic processes (Van der Meulen et al., 2009). The silica sand is often intercalated with brown-coal layers, which may have acted as adsorption filters reducing the mineral content of percolating groundwater.

High-grade silica sand is used to produce silicon metal, which in its turn is listed as a critical raw material, i.e. a raw material of high economic importance with a high risk of supply disruption due to a dependency of imports from a very limited number of producing countries worldwide and lack of substitutes. The European Critical Raw Materials Act (EC, 2023) defines several types of actions

Table 18.1. Matrix of five categories of silica sand (1-5) and their applications in the structural ceramic industry (A), foundries (B), low-grade glass (C), fine ceramics (D), high-grade glass I, water glass, zeolites (F) and carborundum (SiC) (G). The quality per category is given in terms of minimum quartz (SiO_2) and maximum iron (Fe_2O_3), aluminium (Al_2O_3) and titanium (TiO_2) contents, as well as the maximum loss on ignition (LOI; %m/m). Modified from Feenstra & Mulder (2003).

	Applications							Chemical characterization				
Categories	A	B	C	D	E	F	G	SiO_2	Fe_2O_3	Al_2O_3	TiO_2	LOI
1					■	■	■	99.80	0.01	0.05	0.02	0.05
2				■	■	■	■	99.70	0.02	0.08	0.03	0.10
3				■	■	■		99.50	0.03	0.14	0.04	0.25
4		■	■					99.0	0.1	0.2	0.1	0.5
5	■	■						97.0	0.5	0.5–0.10	0.3	1.0

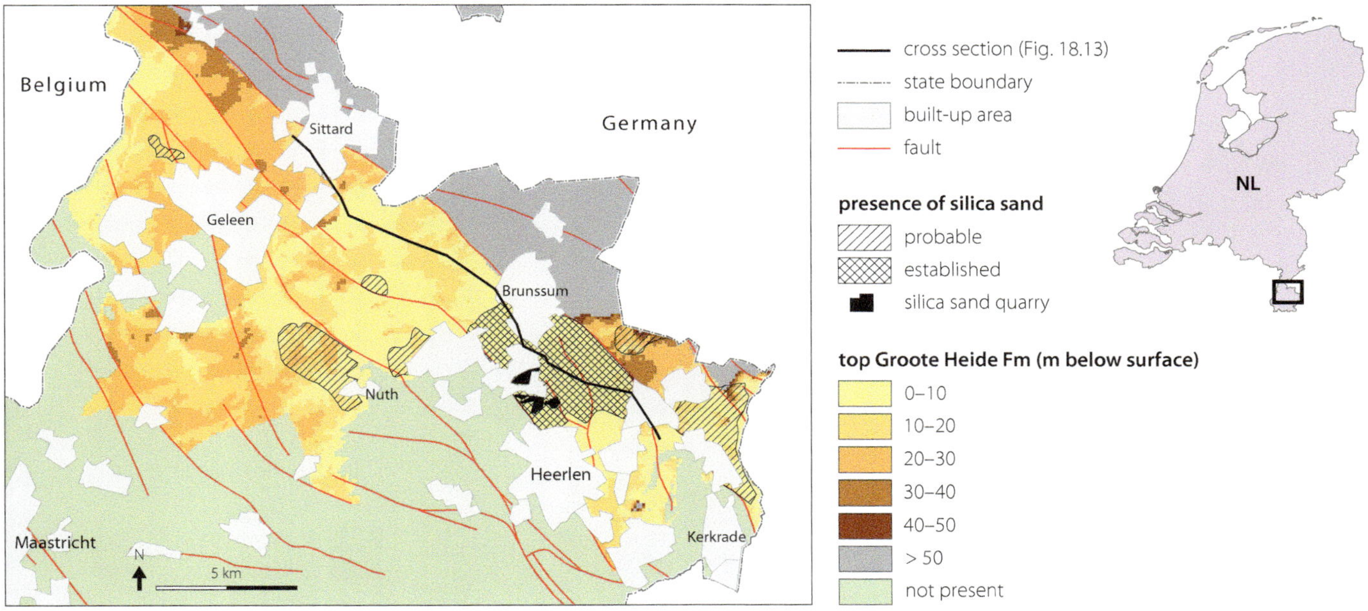

Figure 18.12. The distribution of silica sand in South Limburg. The top of the Groote Heide Formation is shown because silica sand in the area occurs in this stratigraphic unit (more detail in Fig. 18.13). Modified from Van der Meulen et al. (2009), reprinted with permission.

to secure supplies, one of which is to revive mining in Europe. This act will in its turn require the Dutch government to treat high-grade silica sand as a strategic resource and ensure some form of planning protection that is now not in place. Such policy would not only need to apply to the known, high-grade resources in the province of Limburg, but also to lower-grade sands that occur locally in the eastern and southern parts of the country (Van der Meulen et al., 2009).

Clay

Applications and demand

The structural ceramic industry uses an estimated amount of 1.5×10^6 m^3/yr of clay, mainly for the production of bricks and roof tiles (Fig. 18.14). Traditionally, bricks were the most important construction material in the Netherlands, but nowadays, they are mainly used for outer walls (Van der Meulen et al., 2007). Customer demands for certain colours or textures are important to the industry. These and other properties, such as behaviour during drying and firing, as well as the strength of the end product, are determined by the clay chemistry and mineralogy, and the sand and organic-matter contents (Van der Zwan, 1990; Van Wijck, 1997; Table 18.2). In order to obtain the desired clay mixtures, producers add sands or clays from other localities to their local stocks. Limited amounts of clay are imported for this purpose, mainly from Germany.

Clay (which in this case also includes loam) is also used for the building and maintenance of dykes, the covering of landfills, and landscaping. The material specifications for

Table 18.2. Approximate requirements for clay in the structural ceramic industry and for dyke maintenance (modified from Van der Zwan, 1990). Note that clay in the industrial definition may include fine-grained loam and silt, if they are sufficiently cohesive. Anonymous (2001) defined clay raw materials (in both applications) as having clay percentages between 17.5 and 35. Clays meeting this criterion are shown in Fig. 18.12.

Application	Share of grain-size fraction (% m/m)					C-org	$CaCO_3$
	Clay	< 10 µm	63–250 µm	> 250 µm	Sand	(% m/m)	(% m/m)
Roof tiles		50–57	< 20	< 40		< 3	< 25
Bricks		40–42	< 20	< 40		< 3	< 25
Extruded products		40–55	< 20	< 40		< 3	< 25
Dyke maintenance	18–40				< 40	< 3	

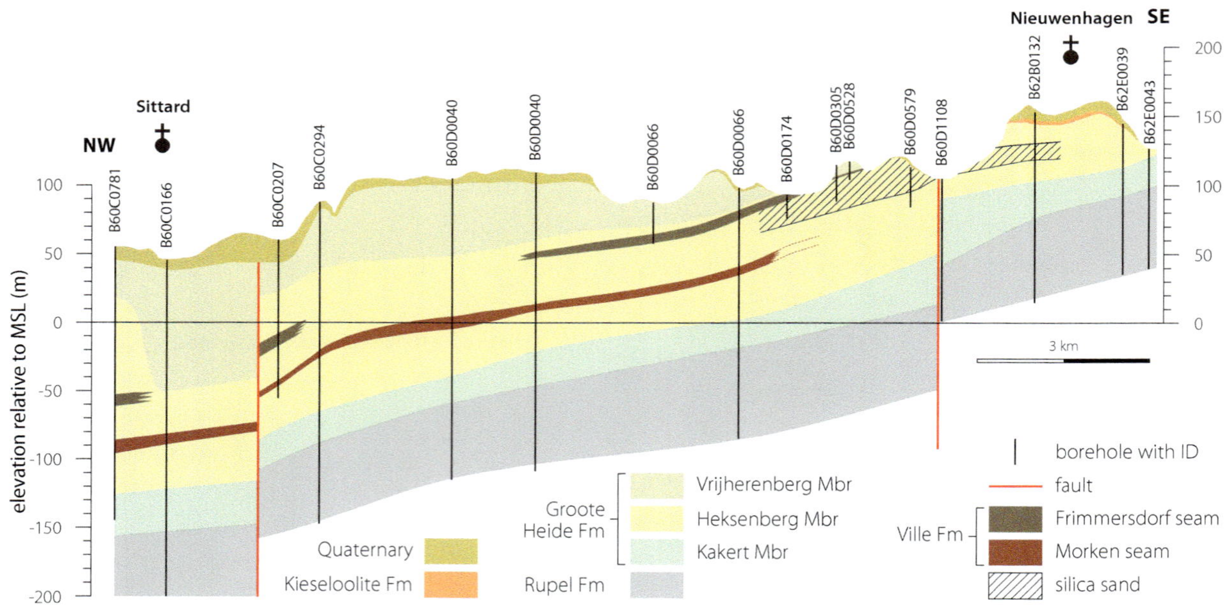

Figure 18.13. Profile through South Limburg (see Fig. 18.12 for profile line), showing the occurrence of silica sand in the Heksenberg Member of the Groote Heide Formation. Silica sand formation and the significance of the Frimmersdorf and Morken lignite seams are discussed in Van der Meulen et al. (2009), from which this figure is modified. Reprinted with permission.

such applications are related to the desired cohesiveness and impermeability. Up till the mid-1990s dyke maintenance required only relatively small amounts of clay, in the order of 0.5 to $1 \cdot 10^6$ m^3/yr, but at that time, the demand rose spectacularly as a consequence of a major river-dyke upgrading programme, which was undertaken after two winters of exceptionally high waters and flood risk (Fig. 18.14). There is no differentiation in clay use statistics after 2002, but consumption peaks in that period are most probably in the 'other' category as well.

Resources

Quaternary overbank deposits of the rivers Meuse and Rhine and Holocene peri-marine clays in the coastal provinces provide the bulk of the Dutch clay resources (Fig. 18.15). Currently, the structural ceramic industry mainly uses the fluviatile clays, while the peri-marine clays are only used in the province of Groningen. Other exploited resources include Oligocene marine clays in the province of Gelderland, Quaternary brook deposits in the provinces of North Brabant, Gelderland and Overijssel, and Pleistocene eolian silt deposits in Limburg (loess).

Most of the ceramic clay extraction sites are located in the embanked floodplains of the central/eastern Rhine-Meuse Delta (RWS, 2001; Van der Meulen et al., 2007). Dutch rivers were progressively embanked from approximately 1000 AD onwards. Within the confinement of the dykes, the main river channel was initially free to migrate and able to deposit sand throughout the embanked floodplains (Middelkoop, 1997). From approximately 1850 AD

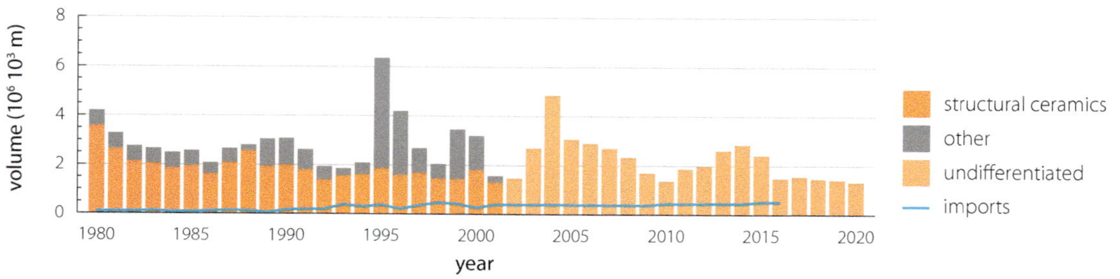

Figure 18.14. The extraction of clay for the structural ceramics industry and other applications (dyke building/maintenance, landscaping, lining of landfills) from 1980 to 2020. Reliable import figures for 2002-2020 are not available. Limited amounts of ceramic clay are imported, exports (not shown) are negligible. Amounts of clay are usually given in volume rather than in weight units. Sources: see Fig. 18.1.

Figure 18.15. The distribution of superficial fine-grained deposits in the Netherlands.

onwards, however, the river channels were fixed by a combination of groyne construction, riverbed armouring and/or strengthening of natural levees. From that time, clay became the dominant overbank deposit in the embanked floodplains, while sand transport and sedimentation became limited to the river channel. These particular clay deposits have become the main resource for the structural ceramic industry and amount to about two-thirds in the total consumption. For dyke maintenance and the covering of landfills, locally extracted clay is used wherever possible.

Carbonates

Limestone and dolomite

Applications and demand

The only Dutch Portland cement production plant (ENCI), operational from 1926 until 2020, was located near the town of Maastricht in South Limburg. The plant operated a large quarry until 2018, where between 1.2-1.6 Mt/yrof chalk was extracted over the past decades (Fig. 18.16). The factory used various secondary materials such as blast furnace slag and coal fly ash for its pozzolanic properties, i.e. the ability to react with calcium hydroxide and form calcium silicates. After the permit to quarry chalk ended in 2018, the factory continued to produce cement until 2020 from clinker imported from Belgium.

About 0.3 Mt/yr of chalk, limestone and dolomite is quarried for the production of limestone powder in Limburg and eastern Gelderland (Fig. 18.16). Limestone powder is mainly used in agriculture (fertilizing, fodder), as a filler in ceramic products, paperware and asphalt, and for the desulphurisation of flue gases.

Chalk in Limburg has been used as dimension stone since the Roman era. Nowadays, only a minor amount of chalk is quarried for this purpose, mainly for reconstruction works. The Netherlands depends almost completely on imports of dimension stone, mainly from Belgium and Germany. Between 2000 and 2021, Dutch annual imports of dimension stone varied between 0.3 and 3.1 Mt (Statistics Netherlands).

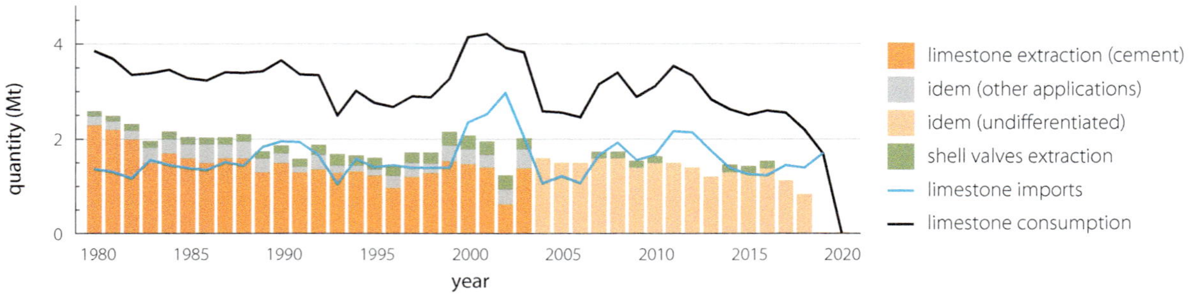

Figure 18.16. *Dutch carbonate supplies 1980 to 2020 (excluding dimension stone). Exports (not shown) are negligible. Shell extraction data are incomplete. Sources: see Fig. 18.1.*

Resources

Upper Cretaceous chalk deposits in Limburg form the most common pre-Cenozoic naturally exposed rocks in the Netherlands, and are the main carbonate resource (Fig. 18.17). The chalk was deposited as bioclastic carbonate muds, silts and sands in a shallow-marine basin that covered large portions of northwestern Europe during the Late Cretaceous and earliest Cenozoic (Zijlstra, 1994). The chalk has a very high $CaCO_3$ content (up to 99%); the only significant siliceous component being flint, which occurs in nodules or layers. The flint was used to make human tools as early as 6000 BP, and now is a by-product of limestone quarries, mainly used for road foundations, path pavements and hydraulic engineering.

Middle Triassic Muschelkalk carbonates quarried in Gelderland are an alternation of nodular limestone and dolomite, deposited in peri-marine environments. They have been quarried east of Winterswijk in the eastern Netherlands since 1933. The material is mainly used in road building and agriculture.

Shells

Mollusc shells are used in drains (~22%), path pavements (~20%), animal fodder (~16%), thermal and moisture insulation (~13%), and in several minor applications such as foundation material and the production of lime mortar (De Graaf et al., 2000). All applications except lime-mortar production make use of cockle shells (*Cerastoderma edule*), which endure intensive use and mechanical processing. When crushed, the combination of concentric growth increments and radial ribs results in approximately equidimensional fragments: a useful property for fodder production and path pavements.

Shell extraction was monitored from 1980 to 2002, but more recent statistics are sparse and less reliable. The current demand for shells is estimated at 150 x 10^3 m^3/yr. In the 1990s, extraction rose from about 180 to 290 x 10^3 m^3/yr (Fig. 18.16). About 70% of this amount was extracted from Holocene deposits of the Waddenzee and its inlets. The remaining 30% is mainly extracted from the Pleistocene of the Westerschelde estuary and its prodelta. For reasons of sustainability, the annually permitted volumes in these areas are limited to the estimated average yearly growth increments of the exploitable shell stocks (cf. Beukema & Cadée, 1999). An estimated amount of about 10 x 10^3 m^3/yr of cockle shells is obtained from seafood producers.

Some historically exploited resources

Peat that has accumulated in Quaternary paralic, fluviatile and lacustrine environments and has been exploited from Roman times up to the mid-20th century (Van Bergen & Koster, 2025, this volume). Historically, dried peat has mainly been used as fuel. In Roman to medieval times paralic peat has been extracted and burned to produce salt from the ashes. The current demand for dried peat in the production of horticultural compost etc., is almost completely met by imports, mainly from Germany. Between 2000 and 2021, Dutch annual imports of 'turf and derived products' varied between 1.8 and 2.9 Mt (Statistics Netherlands).

Bog ore. In the central and southeastern parts of the country, iron has been extracted from bog ore, iron (hydr-) oxide concretions, mainly from Pleistocene ice-pushed ridges in Gelderland (see Griffioen et al., 2025, this volume), and from siderite nodules (e.g. from the upper Carboniferous of southernmost Limburg; Laban et al., 1988; Joosten, 2004). Artisanal iron production started in prehistoric times and reached a high in medieval times. Since 1926, the Netherlands has been producing iron and steel from imported iron ore in its steelworks at the Port of IJmuiden located east of Amsterdam. Between 2000 and 2021, Dutch annual net imports of 'iron ore, iron, semi-products and scrap iron' varied between 2.7 and 8.1 Mt (Statistics Netherlands).

Till. Pleistocene subglacial tills in the northern and eastern parts of the country (Busschers et al., 2025, this volume) have been used for dyke building, in particular

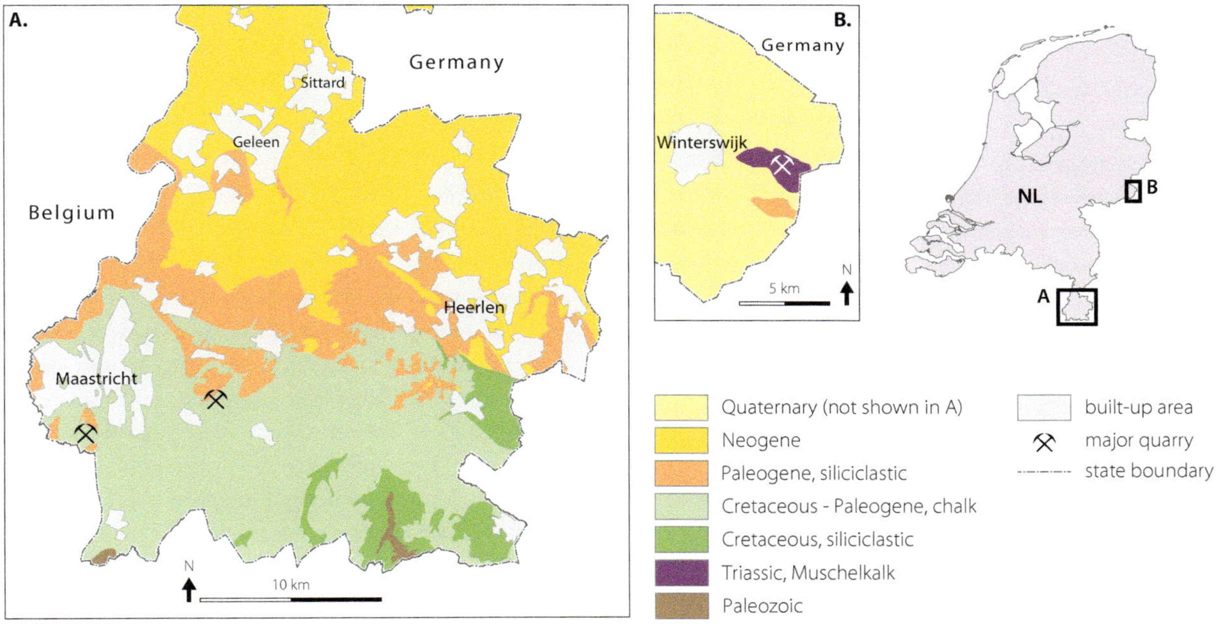

Figure 18.17. Carbonate resources: a) Pre-Quaternary sub-crop map of South Limburg showing Cretaceous to Paleogene chalk (source: RGD, 1989); b) Simplified geological map of Winterswijk and surroundings showing the occurrence of the Musschelkalk dolomite and limestone formation (source: TNO-GDN, 2023).

for the '*Afsluitdijk*', which separates Lake IJssel from the Wadden Sea (Fig. 18.4). Its construction, in 1927-1932, required 13.5 x 10^6 m^3 of till and 23 x 10^6 m^3 of sand.

Phosporite nodules. In eastern Gelderland and Overijssel, Oligocene phosphorite nodules occur in the shallow subsurface, either as a primary constituent of marine deposits, or reworked in Pleistocene moraines (Laban, 1988). These nodules were processed into fertilizer during the First and Second World Wars, when all imports were restricted. The current demand for phosphate in chemical fertilizers is about 50 kt/yr (Agricultural Economics Research Institute, Wageningen).

Diatomites are mainly used in various industrial filters, as grinding and polishing material, as a reactive silica source and for the production of paint, matches and dynamite (Vos & De Wolf, 1988). The only known Dutch exploitable resource, a Holocene lake deposit near Renkum, southern Gelderland, was virtually depleted in a 40-year period around the turn of the 19th century (Laban, 1987). Current import figures for diatomite are unavailable but are expected to be low.

Flint occurring as intercalations and nodules in the chalk of South Limburg was mined extensively during Neolithic times (7000-4000 BP). A total of about two thousand 12-m deep shafts were dug in an area of about 8 acres along the eastern slope of the Meuse valley near the current village of Ryckholt (Felder et al., 1998). These shafts were discovered by the Belgian archaeologist Marcel de Puydt in 1891 and the first one was cleared in 1914. Flint was used to make fire, and to produce axes, scrapers, spearheads, and other types of blades. From Roman times onwards, flint was a by-product of chalk mining and used as building stone.

Exploration and production

Exploration for aggregates

Exploration by Dutch sand and gravel extraction companies is aimed at estimating the tonnage of specific mixtures of sand and/or gravel that can be produced at a site. This requires a high-density drilling campaign (of up to one drilling per hectare) and grain-size analyses of samples per metre drilled. Within such a dense drilling grid, additional geophysical techniques are considered to have limited added value and are seldom used for commercial exploration. Another important exploration objective is to identify cohesive intercalations of clay or peat which may inhibit exploitation with dredging devices. The exploration depths vary between approximately 20 and 40 m, depending on the geological situation and the local policy concerning the depth of extraction pits.

The exploration results are interpreted in terms of the estimated costs of the removal of the overburden and the subsequent extraction and processing of the commodity. These costs, together with those for obtaining the concession area, the required permits, transportation etc., determine the price of the commodity and the viability of the project.

The Dutch government has occasionally commissioned national aggregates exploration studies for policy and planning purposes, or in connection with large-scale hydraulic-engineering projects. Such projects include the exploration for concreting and masonry sand in the North Sea and the combination of hydraulic measures in sand or gravel extraction along the major rivers (RWS, 2004; Van der Meulen et al., 2006). For planning and pre-prospection purposes, TNO – Geological Survey of the Netherlands provides aggregate and other surface mineral resources information online (delfstoffenonline.nl; Van der Meulen et al., 2007, 2013; Maljers et al., 2015, 2020).

While exploration for sand and gravel requires more or less standard geological and sometimes geophysical techniques, exploration for ceramic clay and silica sand may involve various types of specialized chemical and mineralogical analyses (Brijsse & Casier, 1997; Walda, 2000; Van der Meulen et al., 2007). Their interpretation involves the context of production processes and is not discussed here further.

Production of sand, gravel and clay

Most of the sand and gravel produced in the Netherlands is dredged from extraction pits or from open water. Dry extraction, with equipment like draglines, bulldozers and shovels, is used for clay and only occasionally for sand and gravel (e.g. silica sand in South Limburg). Both mechanical (digging), and hydraulic (suction) processes are used for wet extraction. Gravel is usually dredged mechanically, suction being less suited for reasons of wear. Suction dredgers are preferred for sand extraction because of their high production rates and the possibility to use pipelines for transport.

The bucket dredger is one of the oldest types of dredgers. It features a chain of buckets, supported by a ladder, which fill while scraping the bottom (Fig. 18.18a). Bucket dredgers are still in use for coarse-sand and gravel extraction in Limburg. Their main advantage is the possibility to dredge very accurately. Disadvantages are the limitations in depth (down to 20 m) and the noisy operation. Since 1980 a few bucket dredgers have been modernized to reduce the noise and enable extraction from depths down to 35 m.

The wire grab dredger is a versatile excavator suitable for depths down to 100 m (Fig. 18.18b). For sand and gravel extraction, it has largely been replaced by the backhoe dredger which has a higher production rate and allows for more accurate dredging. Backhoe dredgers are based on a digging action towards the machine and can reach depths down to 15 m (Fig. 18.18c).

A stationary suction dredger is the most common device to extract sand from pits (Fig. 18.18d). Suction caus-

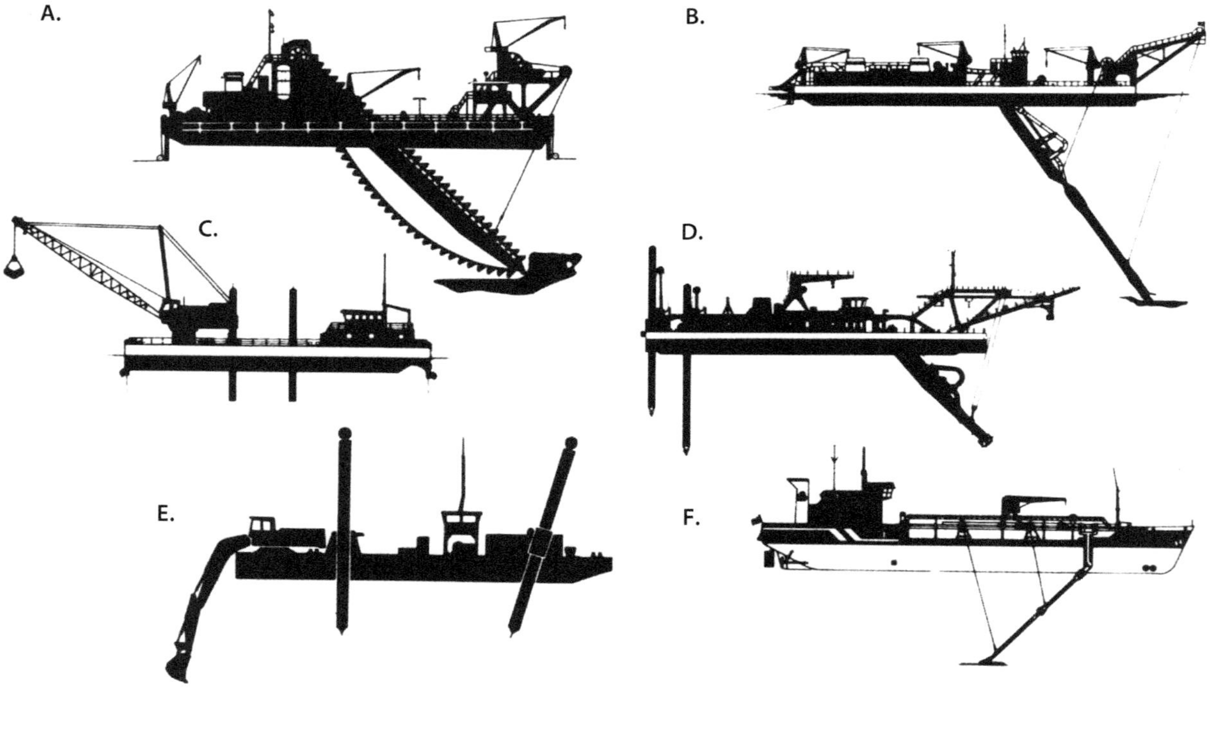

Figure 18.18. Dredging vessels used for sand and gravel extraction: a) bucket dredger; b) wire grab dredger; c) backhoe dredger; d) stationary suction dredger; e) cutter suction dredger; f) trailing suction hopper dredger. Modified from Donze (1990).

Figure 18.19. Floating aggregates classification unit (Geertjesgolf sandpit, Winssen, Netherlands). Photo by Rudolf Bak.

es breaching, which generates a flow of sand towards the suction pipe. The process may be disturbed by cohesive sediments, like peat or clay, but such materials can be loosened or penetrated by a suction pipe equipped with water jets. The standard suction dredger is limited to a depth of 30 m, but equipment capable of dredging down to 70 m is available.

A cutter suction dredger dredges cohesive materials using a combination of a cutter head and hydraulic transport (Fig. 18.18e). Cutter suction dredgers are used for the construction and maintenance of harbours and navigation channels, and occasionally for the extraction of sand and gravel.

The trailing suction hopper dredger is a self-propelled vessel which scrapes a thin layer from the bottom and stores the dredged material in its hold (Fig. 18.18f). This device can only be used in deep, open waters. Advantages are the high dredging accuracy, high production rates and the capability to operate in rough waters. Filling sand from the North Sea is extracted with medium-sized hoppers.

After extraction, gravel and sand (except filling sand) are classified by sieving and hydrocyclonage. Floating classification units can be mounted on the dredging vessel or on a separate pontoon (Fig. 18.19), while larger or more sophisticated plants are either mounted next to sand or gravel pits or at central locations where sand from several extraction sites is processed (Fig. 18.20).

Policy context and outlook

Under the Environmental and Planning Act (Anonymous, 2006), extraction permits are issued by provincial authorities or the Ministry of Infrastructure and Water Management for State Waters (*Rijkswateren*). Areas where minerals extraction is considered acceptable are outlined in mineral-extraction plans, which are issued as a part of, or in accordance with regional development plans (*streekplannen* or *omgevingsplannen*). Tree-cutting, noise, turbidity and other environmental impacts associated with extraction may require additional permits from local municipal authorities. Application procedures tend to be time-consuming and may take well over five years.

There is a scale issue in the distribution of burden and benefits that is peculiar to mining: while stable supplies of raw materials for the construction industry are important enough to be considered a matter of national importance, the impacts of their extraction affect only the site and its immediate surroundings. Minerals extraction has been challenged by ever-growing societal resistance since

Figure 18.20. Aggregates classification plant near Gendt, Netherlands. Photo by Rudolf Bak.

the 1960s. As a result, provincial administrations have become increasingly reluctant to grant extraction permits and, by the mid-1970s the supply of gravel, concreting and masonry sand and clay for the structural ceramic industry became a matter of national concern (Ike, 2000; Van der Meulen, 2005). The problem was taken up by the Ministries of Transport, Public Works and Water Management, which started preparing for mineral planning guidance at a national level (Anonymous, 1996). As a first step, the Ministry and the provincial administrations agreed on a set of assignments (*taakstellingen*, i.e. amounts of product for which extraction permits are to be granted) for the extraction of gravel, concreting and masonry sand, and clay during the 1980s.

The next step, in the late 1980s, was to prepare the first national policy plan on surface mineral extraction. It stated that the exploitation of surface mineral resources requires planning and co-ordination, as well as better embedding in other policy fields, especially those related to spatial planning (Anonymous, 1989). Policy from then on aimed at sustainable exploitation of the surface mineral resources in order to meet the demand for construction and building materials, at the economical use of such materials and at the maximum use of renewable and secondary materials (Anonymous, 1996, 2001).

Production assignments, though abandoned for clay in 1989, remained the primary policy instrument in mineral planning for about two decades. Up to 1998, the production levels of concreting and masonry sand were kept in approximate balance with the national demand.

However, in 2002 the national government decided to take on a lesser role in mineral supply planning and left supply provision to the market. Mineral planning became a matter of spatial planning and has not been the subject of a specific national policy plan since then. It is mandatory that minerals extraction is conducted in a socially responsible manner and, where possible, contributes to the realization of nature, recreation, lakeside housing and water management.

Future developments

After about two decades of *laissez-faire*, Dutch minerals policy seems to have reached a turning point. The Dutch government now aspires to better consider subsurface resources in spatial planning together with water, under the motto 'water and subsurface guiding' (*water en bodem sturend*; Anonymous, 2022). In addition to geologically determined factors such as ground conditions, implementing this policy requires government to consider geological resources when making spatial plans. The government also defined a housing assignment (*woningbouwopgave*) to counter a persistent housing crisis, which has raised the obvious question whether the considerable amounts of raw materials required for this purpose are available or not. Finally, even though the European Critical Raw Materials Act only concerns one of the minerals treated in this chapter, it has put minerals back on the policy agenda. The question now is whether this is going to add up to better management and stewardship of vital geological resources in the Netherlands.

Acknowledgements

D.A.J. Batjes†, T.P. Bide, J.W. Broers, C.W. Dubelaar, A.L. Hakstege, D.J. Harrison, and M.W.I.M. van Heijst are thanked for their contributions to/comments on the first and revised second editions of this chapter. J. 't Hoen, L.A.C. van der Voort and S. Mergenthal are thanked for sharing data and valuable insights. R.L. Bak is thanked for providing imagery.

References

Anonymous, 1989. Gegrond ontgronden. Tweede Kamer, zitting 1988-1989, 21199/1: 180 pp.

Anonymous, 1996. Structuurschema Oppervlaktedelfstoffen, Deel 4, Planologische kernbeslissing. Ministerie van Verkeer en Waterstaat (Den Haag): 49 pp.

Anonymous, 2001. Tweede Structuurschema Oppervlaktedelfstoffen, Deel 1, Ontwerp Planologische kernbeslissing. Ministerie van Verkeer en Waterstaat (Den Haag): 280 pp.

Anonymous, 2002. Mijnbouwwet. Staatsblad van het Koninkrijk der Nederlanden 542: 57 pp.

Anonymous, 2016. Omgevingswet. Staatsblad van het Koninkrijk der Nederlanden 156: 105 pp.

Anonymous, 2018. Betonakkoord: 20 pp. https://www.betonakkoord.nl

Anonymous, 2022. Water en bodem sturen; brief van de Minister en Staatssecretaris van Infrastructuur en Waterstaat. Tweede Kamer, zitting 2022-2023, 27652/592: 34 pp.

Bak, I., Overmars, L. & Van der Kruk, T., 2018. LCA-Achtergrondrapport voor Nederlandse Asfaltmengsels. Ecochain (Amsterdam): 75 pp.

Bakker, B.B., 1999. Crisis and Recovery. In: Watson, C.M., Bakker, B.B., Martijn, J.K. & Halikias, I. (eds): The Netherlands: Transforming a market economy. IMF (Washington, DC): 4-15. DOI: 10.5089/9781557758040.084

Beukema, J.J. & Cadée, G.C., 1999. An estimate of the sustainable rate of shell extraction from the Dutch Wadden Sea. Journal of Applied Ecology 36: 49-58. DOI: 10.1046/j.1365-2664.1999.00380.x

Bijleveld, M.M., Bergsma, G.C., Krutwagen, B.T.J.M. & Afman, M.A., 2015. Meten is weten in de Nederlandse bouw: milieu-impacts van Nederlandse bouw- en sloopactiviteiten in 2010. CE Delft (Delft), publication nr. 14.2746.25: 108 pp.

Brand, E., Ramaekers, G. & Lodder Q., 2021. Dutch experience with sand nourishments for dynamic coastline conservation – An operational overview. Ocean & Coastal Management 217: 106008. DOI: 10.1016/j.ocecoaman.2021.106008

Brijsse, Y.H.G. & Casier, C., 1997. De bepaling van het drooggedrag van kleigrondstoffen. Klei, Glas, Keramiek 18(4): 19-26.

Broers, J.W., Pietersen, H.S. & Smits, R.G., 2002. Secondary raw materials in the Dutch building industry – an overview of policy, current research and practices. Third European Conference on Mineral Planning (ECMP '02) – Conference Transcript and Field Trip Guide. Geological Survey of North Rhine–Westphalia (Krefeld): 161-167.

Busschers, F.S., Cohen, K., Wesselingh, F.P., Schokker, J., Bakker, M., Van Balen, R. & Van Heteren, S., 2025. Quaternary. In: Ten Veen, J.H., Vis, G.-J., De Jager, J. & Wong, Th.E. (eds): Geology of the Netherlands, second edition. Amsterdam University Press (Amsterdam): 333-391. DOI: 10.5117/9789463728362_ch10

CLO, 2020. Bouw- en sloopafval: vrijkomen en verwerking, 1985-2018. Statistics Netherlands (Den Haag), PBL (Den Haag), RIVM (Bilthoven), WUR (Wageningen). https://www.clo.nl/indicatoren/nl0147-bouw--en-sloopafval.

De Graaf, W.J., Van Veen, E., Smeets, J.H.J.M., Seijdel, R.R., Boonekamp, H.A.L. & Van der Meulen, M.J., 2000. Marktonderzoek Schelpen – Een verkenning van de schelpenmarkt in de komende 15 jaar. Dienst Weg- en Waterbouwkunde (Delft), Publicatiereeks Grondstoffen 2000/05: 87 pp.

De Jong, B. & De Mulder, E.F.J., 1998. Construction material in the Netherlands: Resources and policy. In: Bobrowsky P.T. (ed.): Aggregate Resources – A global perspective. A.A. Balkema (Rotterdam): 203-214.

Dijkstra, J.J., Comans, R.N.J., Schokker, J. & Van der Meulen, M.J., 2019. The geological significance of novel anthropogenic materials: deposits of industrial waste and by-products, Anthropocene 28: 100229. DOI: 10.1016/j.ancene.2019.100229

Donze, M, 1990. Aquatic pollution and dredging in the European Community: shaping the environment. Delwel Publishers (Den Haag): 184 pp.

EC, 2023. Proposal for a regulation of the European Parliament and of the Council establishing a framework for ensuring a secure and sustainable supply of critical raw materials and amending Regulations (EU) 168/2013, (EU) 2018/858, 2018/1724 and (EU) 2019/102. European Commission (Brussels): 94 pp.

Feenstra, L. & Mulder, E., 2003. Project Zilverzand – Deelproject toetsingsnorm. TNO (Apeldoorn), report R 2003/454: 70 pp.

Felder, P.J., Rademakers, P.C.M. & De Grooth, M.E.Th., 1998. Excavations of prehistoric flint mines at Rijckholt-St. Geertruid (Limburg, The Netherlands) by the Prehistoric Flint Mines Working Group of the Dutch Geologic Society, Limburg Sec-

tion. Archäologische Berichte 12: 86 pp.
Griffioen, J., Felder, M., Molenaar, N., Spijker, J. & Hoving, A., 2025. Sediment mineralogy and geochemistry. In: Ten Veen, J.H., Vis, G.-J., De Jager, J. & Wong, Th.E. (eds): Geology of the Netherlands, second edition. Amsterdam University Press (Amsterdam): 493-533. DOI: 10.5117/9789463728362_ch13
Groenenberg, R., Den Hartogh, M. & Fokker, P., 2025. Salt Production. In: Ten Veen, J.H., Vis, G.-J., De Jager, J. & Wong, Th.E. (eds): Geology of the Netherlands, second edition. Amsterdam University Press (Amsterdam): 603-625. DOI:10.5117/9789463728362_ch16
Hendriks, Ch.F., Nijkerk, A.A. & Van Koppen, A.E., 1999. De Bouwcyclus. Æneas (Best): 231 pp.
Hendriks, Ch.F. & Pietersen, H.S., 2000. Sustainable Raw Materials: Construction and Demolition Waste. RILEM (Cachan), report 22: 201 pp.
I&W, 2017. Rechtsoordeel status bijproduct LD-staalslak. Ministerie van Infrastructuur en Milieu (Den Haag): 7 pp.
Ike, P., 2000. De planning van ontgrondingen. PhD thesis Groningen University (the Netherlands). Geo Press (Groningen): 489 pp.
Jones, L., Terrington, R., Hulbert, A., Van der Meulen, M.J., Maljers, D., Stafleu, J., Mee, K., Marchant, B.P., Mankelow, J.M. & Bide, T.P., 2021. Building and construction. In: Turner, A.K., Kessler, H. & Van der Meulen, M.J. (eds): Applied multidimensional geological modeling: Informing sustainable human interactions with the shallow subsurface. John Wiley & Sons (Hoboken, NJ, USA): 573-593. DOI: 10.1002/9781119163091.ch24
Joosten, I., 2004. Technology of early historical iron production in the Netherlands. PhD thesis Vrije Universiteit Amsterdam (the Netherlands), Geoarchaeological and Bioarchaeological Studies 2: 133 pp.
Keulen, A., Van Zomeren, A., Harpe, P., Simans. H.A.E. & Brouwers, H.J.H., 2016. High performance of treated and washed MSWI bottom ash granulates as natural aggregate replacement within earth-moist concrete. Waste Management 49: 83-95. DOI: 10.1016/j.wasman.2016.01.010
Kickert, W., 2012. How the Dutch government responded to financial, economic and fiscal crisis. Public Money & Management 32(6): 439-443. DOI: 10.1080/09540962.2012.728784
Laban, C., 1987. Diatomeeënaarde nog niet op. Grondboor & Hamer 41: 31-35.
Laban, C., 1988. Fosforietknollen ooit gewonnen als delfstof. Grondboor & Hamer 42: 33-38.
Laban, C., 2002. Geologisch onderzoek grindgebied Klaverbank. TNO (Utrecht), report NITG 01-003-A: 28 pp.
Laban, C., Kars, H. & Heidinga, A., 1988. IJzer uit eigen bodem. Grondboor & Hamer 42: 1-11.
Maljers, D., Stafleu, J., Van der Meulen, M.J. & Dambrink, R.M., 2015. Advances in constructing regional geological voxel models, illustrated by their application in aggregate resource assessments. Netherlands Journal of Geosciences 94(3): 257-270. DOI: 10.1017/njg.2014.46
Maljers, D., Van der Meulen, M.J., Stafleu, J., Stam, J., Dabekaussen, W. & Gunnink, J.L., 2020. National 3D models as input for decision-support systems. GSA Abstracts with Programs 52(6). DOI: 10.1130/abs/2020AM-355677
Middelkoop, H., 1997. Embanked floodplains in the Netherlands. PhD thesis Utrecht University (the Netherlands). Geographical Studies 224: 352 pp.
MWH, 2012. Productie en verbruik van beton- en metselzand en (gebroken) grind in 2010; stand van het Zand XVI Lint aan het Grind XIV. MWH (Delft): 111 pp.
NEN, 1991. NEN 3835:1991 NL, Mortels voor metselwerk van stenen, blokken of elementen van baksteen, kalkzandsteen, beton en gasbeton. NEN (Delft): 16 pp.
NEN, 1999. NEN 5905:1997/A1:1999 NL, Toeslagmaterialen voor beton – Materialen met een volumieke massa van ten minste 2000 kg/m^3. NEN (Delft): 1 p.
Oost, A.P., Cado van der Lelij, J.A., De Bel, M., Oude Essink, G.H.P. & Löffler, M.A.M., 2015. The usability of the sand motor concept. Deltares (Delft), report 1221025-000-ZKS-0009: 49 pp.
Overeem, I., Weltje, G.J., Bishop-Kay, C. & Kroonenberg, S.B., 2002. The Late Cenozoic Eridanos delta system in the southern North Sea Basin: a climate signal in sediment supply? Basin Research 13: 293-312. DOI: 10.1046/j.1365-2117.2001.00151.x
RGD, 1989. Geologische kaart van Zuid-Limburg en omgeving (1 : 50.000). Rijks Geologische Dienst (Haarlem): 4 map sheets.
Roy, S. & Kemme, D.M., 2020. The run-up to the global financial crisis: A longer historical view of financial liberalization, capital inflows, and asset bubbles. International Review of Financial Analysis 69: 101377. DOI: 10.1016/j.irfa.2019.101377
RWS, 1998. Inventarisatie van kwaliteit en kwantiteit van betonzand in de markt. Dienst Weg- en Waterbouwkunde (Delft), Publicatiereeks Grondstoffen 1998/09: 27 pp.
RWS, 2001. Ruimte voor Rijntakken: inventarisatie kleivoorkomens. Dienst Weg- en Waterbouwkunde (Delft), Publicatiereeks Grondstoffen 2001/11: 24 pp.
RWS, 2002. Handboek grondonderzoek grote projecten: geologisch onderzoek voorafgaand aan grootschalig grond- of baggerwerk op de bodemgesteldheid en eventuele opbrengsten aan zand, grond en klei. Dienst Weg- en Waterbouwkunde (Delft), Publicatiereeks Grondstoffen 2002/26: 132 pp.
RWS, 2004. Beton- en metselzand uit de Noordzee? Dienst Weg- en Waterbouwkunde (Delft), Publicatiereeks Grondstoffen 2004/1: 38 pp.
RWS, 2005. Registratie productie en afzet secundaire grondstoffen: inventarisatie gegevens 1989-2003. Dienst Weg- en Waterbouwkunde (Delft), Publicatiereeks Grondstoffen 2004/10: 48 pp.
RWS, 2022. Afvalverwerking in Nederland, jaarlijkse gegevens tot en met 2022. Rijkswaterstaat (Den Haag): 76 pp.
Stive, M.J.F., De Schipper, M.A., Luijendijk, A.P., Aarninkhof, S.G.J., Van Gelder-Maas, C., Van Thiel de Vries, J.S.M., De Vries, S., Henriquez, M., Marx, S. & Ranasinghe, R., 2013. A new alternative to saving our beaches from local sea-level rise: the

sand engine. Journal of Coastal Research 29: 1001-1008. DOI: 10.2112/JCOASTRES-D-13-00070.1

Stolk, A. & Seeger, P. 1999. Managing marine sand extraction in the Netherlands; anticipating challenges. In: Fuchs, P.E.K., Smith, M.R. & Arthur, M.J. (eds): Mineral Planning in Europe. Proceedings of the 2nd European Conference on Mineral Planning, Harrogate (UK). The Institute of Quarrying (Nottingham): 213-219.

Tata Steel, 2020. Factsheet converterslak. Tata Steel (IJmuiden): 2 pp.

't Hoen, J., 2017. Rapportage Monitoring bouwgrondstoffen, 2015-2016. H2H Advies (Arnhem): 16 pp.

't Hoen, J., 2018. Rapportage Toepassing grond en bagger in diepe plassen. H2H Advies (Arnhem): 8 pp.

TNO-GDN, 2023. Geologische kaart van het Koninkrijk der Nederlanden 1 : 600 000, TNO - Geological Survey of the Netherlands (Utrecht), 1 map sheet. https://www.dinoloket.nl/en/subsurface-models/map. ISBN 9789059865142

UN, 2016. The Paris Agreement. United Nations (Geneva): 54 pp.

UNECE, 2019. United Nations Framework Classification for Resources. Update 2019. ECE ENERGY SERIES No. 61: 28 pp.

UNEP 2019. Sand and sustainability: Finding new solutions for environmental governance of global sand resources. GRID-Geneva, United Nations Environment Programme (Geneva): 56 pp.

Ureche-Rangau, L. & Burietz, A., 2013. One crisis, two crises...the subprime crisis and the European sovereign debt problems. Economic Modelling 35: 35-44. DOI: 10.1016/j.econmod.2013.06.026

Van Bergen, F. & Koster, K., 2025. Peat, coal and coalbed methane. In: Ten Veen, J.H., Vis, G.-J., De Jager, J. & Wong, Th.E. (eds): Geology of the Netherlands, second edition. Amsterdam University Press (Amsterdam): 577-601. DOI:10.5117/9789463728362_ch15

Van der Meulen, M.J., 2005. Sustainable mineral development: possibilities and pitfalls illustrated by the rise and fall of Dutch mineral planning guidance. In: Marker, B.R., Petterson, M.G., McEvoy, F. & Stephenson, M.H. (eds): Sustainable minerals operations in the developing world. Geological Society of London, Special Publications 250: 225-232: DOI: 10.1144/GSL.SP.2005.250.01.20

Van der Meulen, M.J., Koopmans, T.P.F. & Pietersen H.S., 2003. Construction raw materials policy and supply practices in Northwestern Europe. In: Degryse, P. & Elsen, J. (eds): Industrial Minerals – Resources, Characteristics and Applications. Leuven University, Geology Group, Aardkundige Mededelingen 13: 19-30.

Van der Meulen, M.J., De Kleine, M.P.E., Veldkamp, J.G., Dubelaar, C.W. & Pietersen, H.S., 2004. The sand extraction potential of embedded land surface lowering in the Netherlands. Netherlands Journal of Geosciences 83(2): 147-151. DOI: 10.1017/S0016774600020102

Van der Meulen, M.J., Van Gessel, S.F. & Veldkamp, J.G., 2005. Aggregate resources in the Netherlands. Netherlands Journal of Geosciences 84(3): 379-387. DOI: 10.1017/S0016774600021193

Van der Meulen, M.J., Rijnveld, M., Gerrits, L.M., Joziasse, J., Van Heijst, M.W.I.M. & Gruijters, S.H.L.L., 2006. Handling sediments in Dutch river management: the planning stage of the Maaswerken river widening project. Journal of Soils and Sediments 6: 163-172. DOI: 10.1065/jss2006.06.165

Van der Meulen, M.J., Maljers, D., Van Gessel, S.F. & Gruijters, S.H.L.L., 2007. Clay resources in the Netherlands. Netherlands Journal of Geosciences 86(2): 117-130. DOI: 10.1017/S001677460002312X

Van der Meulen, M.J., Westerhoff, W.E., Menkovic, A., Gruijters, S.H.L.L., Dubelaar, C.W. & Maljers, D., 2009. Silica sand resources in the Netherlands. Netherlands Journal of Geosciences 88(3): 147-160. DOI: 10.1017/S001677460000086X

Van der Meulen, M.J., Doornenbal, J.C., Gunnink, J.L., Stafleu, J., Schokker, J., Vernes, R.W., Van Geer, F.C., Van Gessel, S.F., Van Heteren, S., Van Leeuwen, R.J.W., Bakker, M.A.J., Boogaard, P.J.F., Busschers, F.S., Griffioen, J., Gruijters, S.H.L.L., Kiden, P., Schroot, B.M., Simmelink, H.J., Van Berkel, W.O., Van der Krogt, R.A.A., Westerhoff, W.E. & Van Daalen, T.M., 2013. 3D geology in a 2D country: perspectives for geological surveying in the Netherlands. Netherlands Journal of Geosciences 92 (4): 217-241. DOI: 10.1017/S0016774600000184

Van der Zwan, J.T., 1990. Inventarisatie van kleivoorkomens in Nederland geschikt voor de grof-keramische industrie en de dijkbouw. Dienst Weg- en Waterbouwkunde (Delft), report TW-R-89-4: 36 pp.

Van Lieshout, M. & Nusselder, S., 2016. Update prioritering handelingsperspectieven verduurzaming betonketen 2016. CE Delft (Delft), publication nr. 16.2G75.110: 103 pp.

Van Wijck, J.H., 1997. Grofkeramische kleien in Nederland. Klei, Glas, Keramiek 18(10): 14-17.

Vos, P.C. & De Wolf, H., 1988. Geologie en diatomeeën. Grondboor & Hamer 42: 57-68.

VROM, 1995. Bouwstoffenbesluit bodem- en oppervlaktewaterenbescherming. Staatsblad van het Koninkrijk der Nederlanden 567: 92 pp.

VROM, 1997. Besluit Stortverbod Afvalstoffen. Staatsblad van het Koninkrijk der Nederlanden 665: 8 pp.

VROM, 2007. Besluit Bodemkwaliteit. Staatsblad van het Koninkrijk der Nederlanden 469: 173 pp.

Walda, E.J., 2000. Thermomechanische Analyse (TMA) in de keramische industrie, deel 3. Klei, Glas, Keramiek 21(6): 6-7.

Wentworth, C.K., 1922. A scale of grade and class terms for clastic sediments. The Journal of Geology 30: 377-392. DOI: 10.1086/622910

Zijlstra, J.J.P., 1994. Sedimentology of the Late Cretaceous and Early Tertiary (Tuffaceous) chalk of northwest Europe. PhD thesis Utrecht University (the Netherlands), Geologica Ultraiectina 119: 192 pp.

19

Geothermal energy – from potential plays to successful growth

Harmen Mijnlieff, Loes Buijze, Eveline Rosendaal, Frank Schoof, Radboud Vorage &, Jan-Diederik van Wees*

*Corresponding author: harmen.mijnlieff@tno.nl

ABSTRACT

In the last decade and a half the Netherlands has seen a significant increase in the application of geothermal energy production. The most prolific geothermal reservoirs are Lower Cretaceous-Upper Jurassic fluvial sandstones of the Delft Sandstone Member in the West Netherlands Basin and the Permian predominantly eolian and fluvial Slochteren Formation in the northern half of the country. Runners-up are reservoirs of the Triassic terrestrial Main Buntsandstein Subgroup and the Lower Cretaceous marine Vlieland Sandstone Formation. The Carboniferous Limestone Group was the target reservoir for two geothermal systems in the Roer Valley Graben. The heat produced is primarily used in the horticultural sector, which has played a pivotal role in the initiation and first development of geothermal energy in the Netherlands. Developing geothermal energy is a complex multi-issue exercise that involves numerous surface and subsurface uncertainties. The matching of heat supply to demand is just one of the surface uncertainties. Main uncertainties from the subsurface geological point of view are reservoir quality and potential for seismicity, while main points of attention from an operational point of view are well integrity, productivity and injectivity maintenance and scaling. Geothermal development has been aided by various financial and other supportive governmental measures. The sector is maturing by integrating experience, (well) design standards, seismicity protocols, innovations and lessons learnt in multistakeholder networks. These include operators, service industry, national and local governments, academia and research/knowledge-institutes and -centres. The initial focus on the horticultural sector has widened to include district heating and light industry applications. Additionally, the single system operator model has evolved into a portfolio operator model. The aim of all this is to achieve the ambitious target of supplying a large part of the Dutch heat demand with geothermal heat by 2050.

<< Drill rig at the location of the Monster geothermal doublet (west Netherlands).
Photo courtesy of HVC.

DOI: 10.5117/9789463728362_ch19

Introduction

Since 2007 the Netherlands has seen a significant development from the identification of a small inventory of prospective areas without active geothermal systems (Lokhorst & Wong, 2007) to 27 realized geothermal systems at 1-1-2022. Nineteen of the 27 systems were in operation in 2021 (Ministry of Economic Affairs and Climate Policy - MEACP, 2022; Fig. 19.1) adding up to 6.3 PJ of heat produced, equal to 176 x 10^6 m^3 of (Groningen quality) natural gas and a contribution to CO_2 emission reduction of approximately 0.35 Megaton.

Geothermal exploration studies in the Netherlands started in the late 1970s (Visser & Heederik, 1987). In this era of geothermal exploration and development five relatively shallow (<1 km) spa or balneology development wells and one deep (ca. 2 km) research/exploratory borehole, the well Asten-GT-02, were drilled. The latter for the purpose of assessing the suitability of lower Cenozoic sediments for direct heat application and high temperature storage. Asten-GT-02 failed because the transmissivity encountered was insufficient for economically feasible heat production or high temperature storage (Dufour & Heederik, 2019). The well was used in the next decades as a groundwater monitoring well. The disappointing results of the Asten well, combined with low hydrocarbon prices, halted the development of geothermal exploitation for two decades.

Current geothermal developments are contributing to the transition from fossil fuels to sustainable energy sources and the ambition to reach the Paris Agreement goals (UN, 2015). The Dutch government aims to have a CO_2 neutral energy system in 2050. Geothermal energy production is proven to have a relatively low CO_2 footprint, taking into account the whole project chain (Gonzales, 2017) and is well suited to the future energy mix. In 2050 the ambition of the Dutch Government is to produce 80 PJ/yr from systems at production temperatures <120°C and some 60 PJ/yr from systems with production temperatures >120°C, in total comprising some 14% of the domestic heat demand (MEACP, 2018). Moreover, the ambition of the geothermal sector is to ramp up the geothermal production to 200 PJ in the year 2050 (Stichting Platform Geothermie et al., 2018; Fig. 19.2). Besides the climate policy ambition there are other drivers, as for example in the horticultural sector, where the urge to be independent of volatile and high gas prices and to replace them with a stable and sustainable heat price was felt. Additionally, the transition to renewable energy sources is becoming more and more a condition for the placement of horticultural products in the market.

All of the geothermal systems presently producing are for direct use of heat, mostly for the heating of greenhouses (Fig. 19.3), which suits the local high heat demand of the Dutch horticulture sector with its numerous greenhouses. The first companies that utilized geothermal energy systems were predominantly horticulturalists; only one district heating company was among the pioneers. The Van den Bosch tomato nursery was the first to realize a doublet. Its example was quickly followed by its colleague innovators in the vicinity and in other parts of the country. They proved that geothermal energy is a successful al-

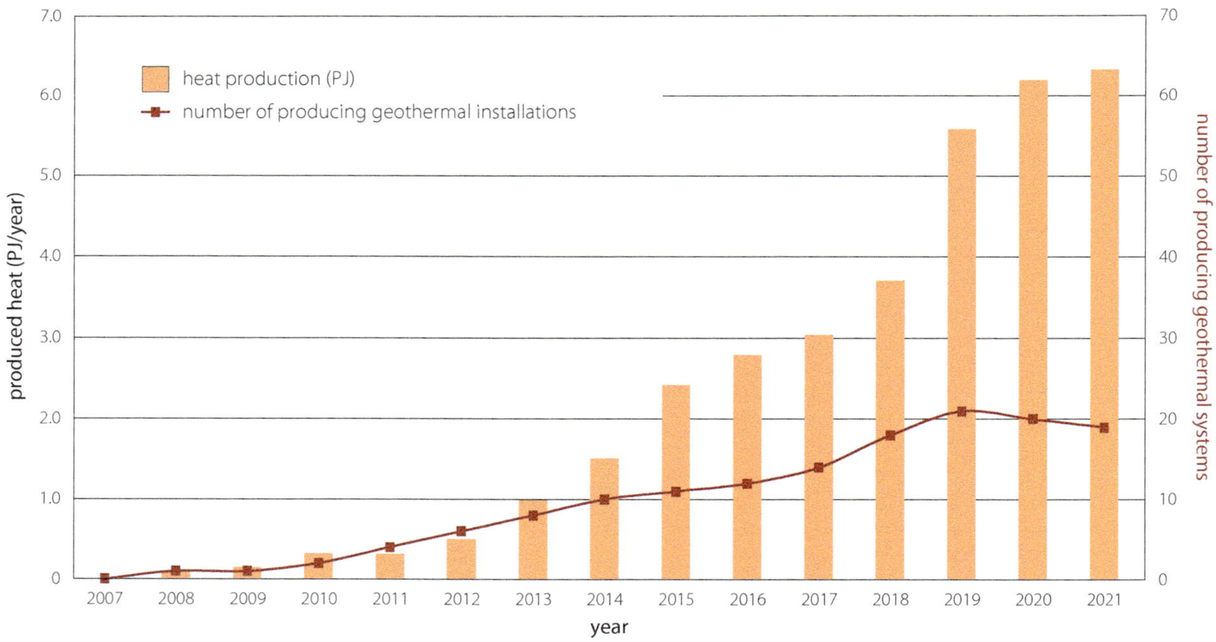

Figure 19.1. The development of geothermal heat production and number of geothermal systems (status date 1-1-2022).

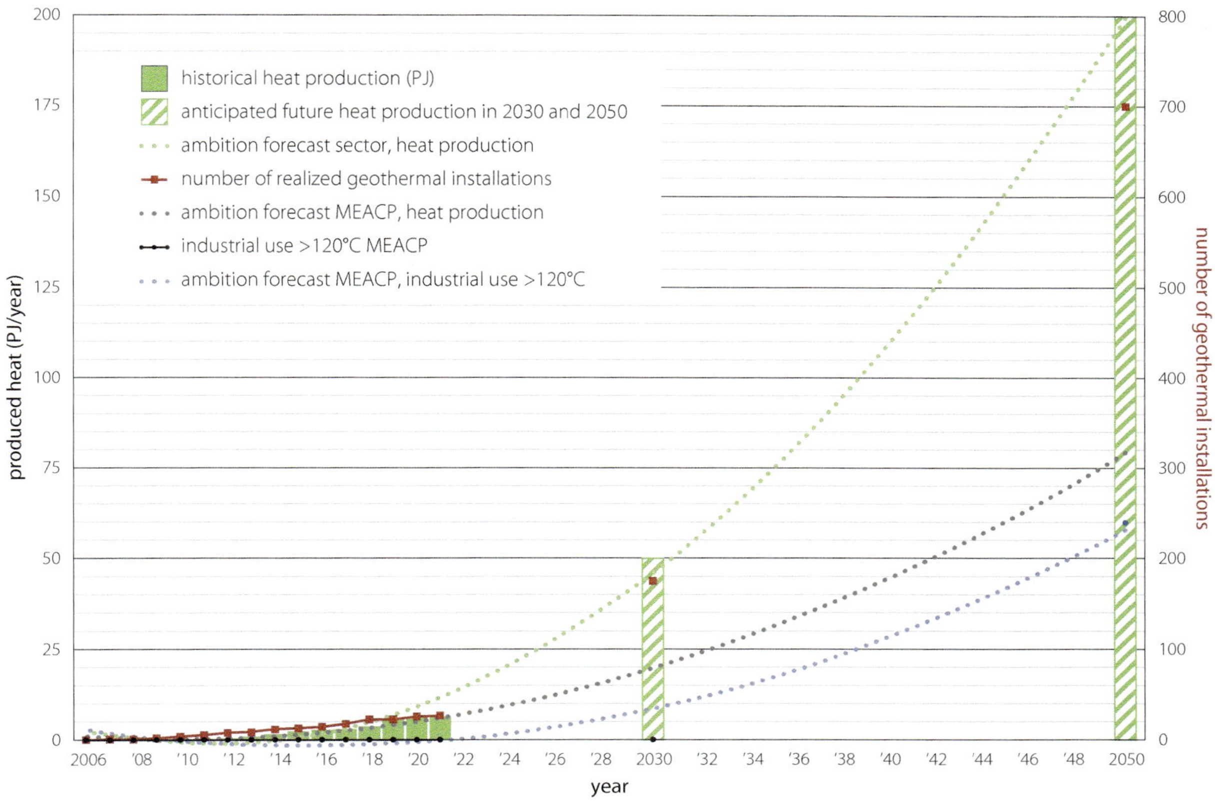

Figure 19.2. Graph of the ambitions set by the geothermal sector in the Masterplan 2018 and the Ministry of Economic Affairs and Climate Policy (MEACP, 2018) including the historical production figures up to status date 1-1-2022.

ternative for the gas fired heat production. Subsequently, other horticultural entrepreneurs entered the sector as 'early adaptors'. Presently, geothermal operators have become diverse with the arrival of portfolio operators, the entrance of traditional oil & gas companies, the participation of the state owned EBN company (www.ebn.nl) and a

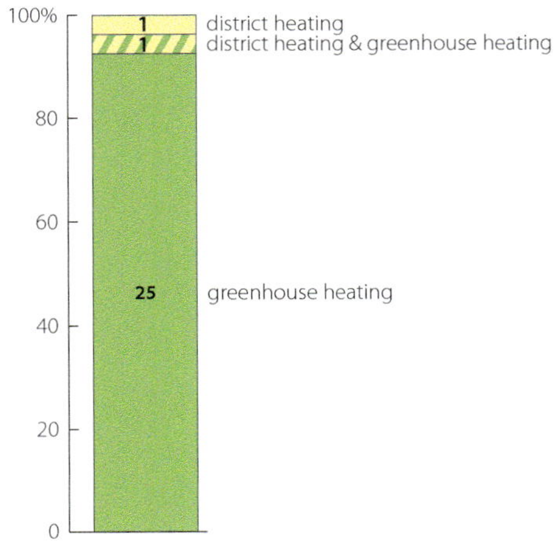

Figure 19.3. Utilization of heat from the Dutch geothermal systems (MEACP, 2022, status date 1-1-2022; numbers refer to amount of realized doublets).

shift from direct use in the greenhouse sector towards the urban environment. This change of operators, whose main aim and core business is to produce and sell heat, will change the geothermal landscape in the coming decade.

The Heerlen mine-water project realized in 2006 is regarded (legally) as a geothermal project although it is better characterized as a thermal energy storage project (TES) according Verhoeven et al. (2014). The thermal energy is stored within shafts and galleries of an abandoned mine in upper Carboniferous strata and is deemed a different type of 'geothermal reservoir' which will not be discussed in this chapter.

The technology adaptation graph (Fig. 19.4) illustrates that the sector is still in the early stage of its development, but it is eager to grow (Stichting Platform Geothermie et al., 2018). However, as emphasized in a number of publications by the Dutch State Supervision of Mines (SodM, 2017, 2021) the journey towards a mature sector will encounter hurdles. To ensure a safe, socially and environmentally responsible sector, the sharing, preservation and growth of practical, scientific, and operational knowledge and experience are key. Various initiatives have been taken by the government and the geothermal sector to ensure this growth. Studies have been executed, legislation is being improved, and standards and methodologies are being developed. For example (sectoral) guidelines for geothermal well design and for community engagement have been issued.

At present, geothermal development in the Netherlands is in a transition phase towards a more mature position in sustainable heat supply. In this chapter an overview is presented of geothermal developments in the Netherlands, emphasizing the uncertainties and hurdles encountered, as well as the lessons learned. It will also shed some light on perceived future perspectives. It is written from a geo-scientific point of view, but will address both subsurface and surface development challenges as well as the related technical, economic and environmental challenges.

Uncertainties in both supply and demand have significant negative impacts on the business case of a project. Despite these uncertainties, the development of geothermal projects in the last one and a half decades has been generally successful. The following technical, economic, legislative, and policy factors spurred this development:

- High and volatile gas price;
- Environmental awareness and peer pressure;
- Targeted and fit-for-purpose financial support and subsidy schemes like Market introduction, Energy Innovation in the horticultural sector (MEI) and Stimulating Sustainable Energy production (SDE), together with the guarantee scheme for exploration risk (RNES);
- Supportive governmental licensing policy towards geothermal exploration and production;
- Supportive governmental policy towards geothermal research and data acquisition;
- Probably the most important factor: the window of opportunity in especially the horticultural sector in combination with active and dedicated greenhouse entrepreneurs.

The policy incentives listed above were installed with the aim of directly or indirectly reducing the *uncertainty* of an element in the project maturation trajectory, aiming to increase the bankability of the project. For example:

- The Exploration Guarantee Fund (RNES), the Market Introduction of Energy Innovations in the horticultural sector (MEI) and a Feed-In Premium scheme for produced sustainable energy products (SDE) give more comfort to parties who finance the project
- Installation of a 'temporarily policy' that allows production under an exploration license prior to award of a production license. This reduces the 'legal risk' to the operator
- More and better data to reduce the geotechnical uncertainty. Government funded projects include:
  - o 'KennisAgenda' projects have increased the knowledge base of the geothermal sector and have provided the first steps towards detailed research project descriptions and initiatives
  - o Research on Ultra Deep Geothermal targets based on existing data (UDG)
  - o Reprocessing of existing vintage 2D-seismic data and acquisition of new 2D-seismic data (SCAN project)
- Resource assessment and integration of geothermal resources in urban heating:
  - o High temperature heat storage
  - o Embedding in heat networks and
  - o Portfolio development

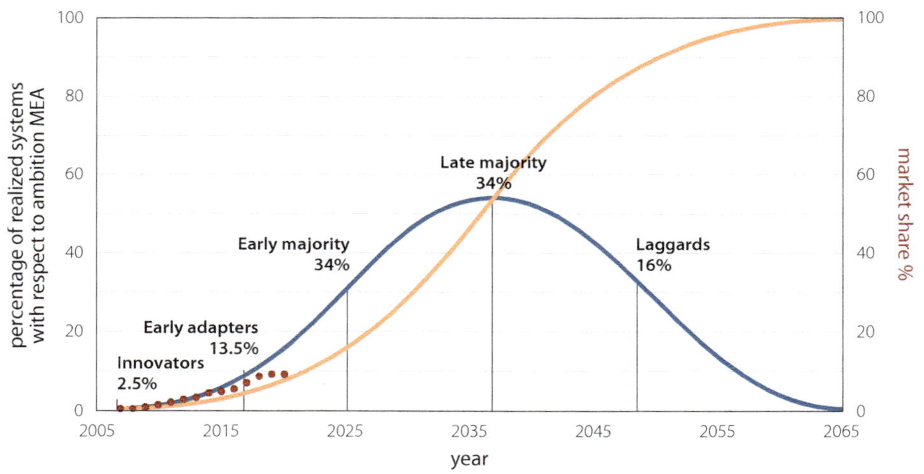

Figure 19.4. Categories of adopters in the Dutch geothermal arena (background graph after Rogers, 1962) projected 100% of market share in accordance to the MEACP ambition of ca. 260 projected conventional geothermal systems in 2050. The Dutch geothermal sector started with the innovators in 2007 and entered the early adaptors class in 2017 (red dotted line).

Reading guide

This chapter is organized as follows. In the first section, we introduce the Dutch geothermal setting and highlight the key role the availability of comprehensive subsurface data sets and information in the public domain will play in geothermal growth. This section also describes the different geothermal plays that are currently being developed, the geothermal production systems and the geothermal licensing system. In the second section the uncertainties and potential hazards related to geothermal sources are described. Geothermal prospects are marked by significant subsurface uncertainty and associated project development risks, both during the exploration and exploitation phases. In the third section we address the challenges currently faced by operators in relation to uncertainties on the supply and the demand side of geothermal heat. In the fourth section, challenges related to organizational and public engagement aspects, financing, policy and legal frameworks are described. The last section highlights some future developments and sheds some light on the Dutch geothermal resource potential.

In geothermal nomenclature the words 'aquifer' and 'reservoir' are used for the subsurface rock interval in which the hot geothermal water/brine resides, is produced from and is subsequently injected into. Although they are largely synonymous, here we use the term reservoir, as the term aquifer is commonly used in the very shallow subsurface domain in relation to fresh water resources.

Dutch geothermal setting

A wealth of subsurface data is available thanks to over 60 years of hydrocarbon exploration and production. It includes excellent seismic coverage and imaging of the subsurface to some 4 km depth (available from www.nlog.nl). About 40% of the Dutch onshore area is covered with 3D-seismic surveys. As the exploration risk is lowest in the data-rich areas, geothermal exploration started in the 3D-seismic covered West Netherlands Basin, where also many boreholes have been drilled. Outside the 3D-seismic covered area numerous 2D-seismic lines cross-cut the country, but despite this there are still areas of poor data coverage that carry a higher exploration risk. The governmental seismic data acquisition and reprocessing program (*Seismische Campagne Aardwarmte Nederland*: SCAN) aims to improve the coverage of these areas with a dedicated programme of 2D-seismic lines complemented with boreholes for reservoir data-acquisition to even further reduce the exploration risk.

All available subsurface data are used for regional maps prepared by TNO – Geological Survey of the Netherlands (Digital Geological Model-deep or DGM-deep v5.0, 2019). Next to the depth, thickness and distribution maps of the main stratigraphic intervals, special products have been made based on the regional mapping. One of the products, ThermoGIS is geared towards mapping the geothermal potential (Kramers et al., 2012; Pluymaekers et al., 2012; Van Wees et al., 2012; ThermoGIS, 2019; Vrijlandt et al., 2019). It is web-based and draws on the regional mapping results to calculate, amongst other things, Heat In-Place (HIP), geothermal power potential and, given certain technical and economic cut-offs, resource potential. As such ThermoGIS-v2.1 provides overall insight into the resource potential of geothermal plays in the Dutch subsurface (Mijnlieff et al., 2019, 2020; Van Wees et al., 2020).

Geothermal plays

In the maturation of a geothermal project the geological evaluation starts from a broad basin wide assessment (Limberger et al., 2014) within which geothermal plays characterize reservoirs that have a set of similar denominators (Moeck, 2014; Breede et al., 2015; Mijnlieff et al., 2020; Van Wees et al., 2020).

The Dutch geothermal reservoirs belong to a sequence of intracontinental sedimentary basin formation stages, marked by limited tectonics and absence of magmatic activity (Bonté et al., 2012; Békési et al., 2020; De Jager et al., 2025, this volume). Heat flow is largely conductive and significant hot spots are not found, so depth, facies and burial history, affecting reservoir temperatures and flow properties, largely determine the geothermal potential. In our approach, a geothermal play is defined by the presence of water and sedimentary formations with suitable flow properties that have comparable geological characteristics. As such the Dutch geothermal play is a hydrothermal one, specifically defined as a Hot Sedimentary Aquifer play (HSA; Breede et al., 2015), in which either the matrix or the fractures/faults form the dominant permeability system. This differentiation is essential because the seismicity hazard of geothermal systems in fracture/fault plays is higher than in matrix permeability plays (Buijze et al., 2019). Dutch geothermal plays are then divided into subplays according to the lithostratigraphic label of the reservoir sequence (Mijnlieff et al., 2020). The various geothermal reservoir-seal pairs are depicted in Figure 19.5, which shows top and base seals in relation to the stratigraphic column. The permeable sandstones of the Permian, Triassic, Upper Jurassic, Lower Cretaceous and Paleogene, plus the fractured carbonates of the lower Carboniferous form the proven sub-plays. Most units occur under large parts of the Netherlands, but their depth (and thus temperature), thickness and permeability vary greatly. The distribution of these geothermal reservoirs with their expected potential is depicted in Figure 19.6. Porous and to some extent permeable rocks of upper Carboniferous, Zechstein and Upper Cretaceous age may mature into proven sub-plays in the future. From a geological point of view the geothermal plays and reservoir distribution are described in the stratigraphic chapters of this book (Ten Veen et al., 2025) and definitions of the stratigraphic units can be found online in the Dutch Stratigraphic Nomenclature at www.dinoloket.nl/en/stratigraphic-nomenclature. The geothermal plays are described in Mijnlieff (2020). Their geothermal potential was mapped and assessed regionally in ThermoGIS (www.thermogis.nl; Vrijlandt et al., 2019; Van Wees et al., 2020 ; Stafleu et al, 2025, this volume).

The production temperature and reservoir transmissivity are of prime importance to the geothermal potential estimate of prospects and their application domain (Gudmundsson et al., 1985). The reservoir temperature

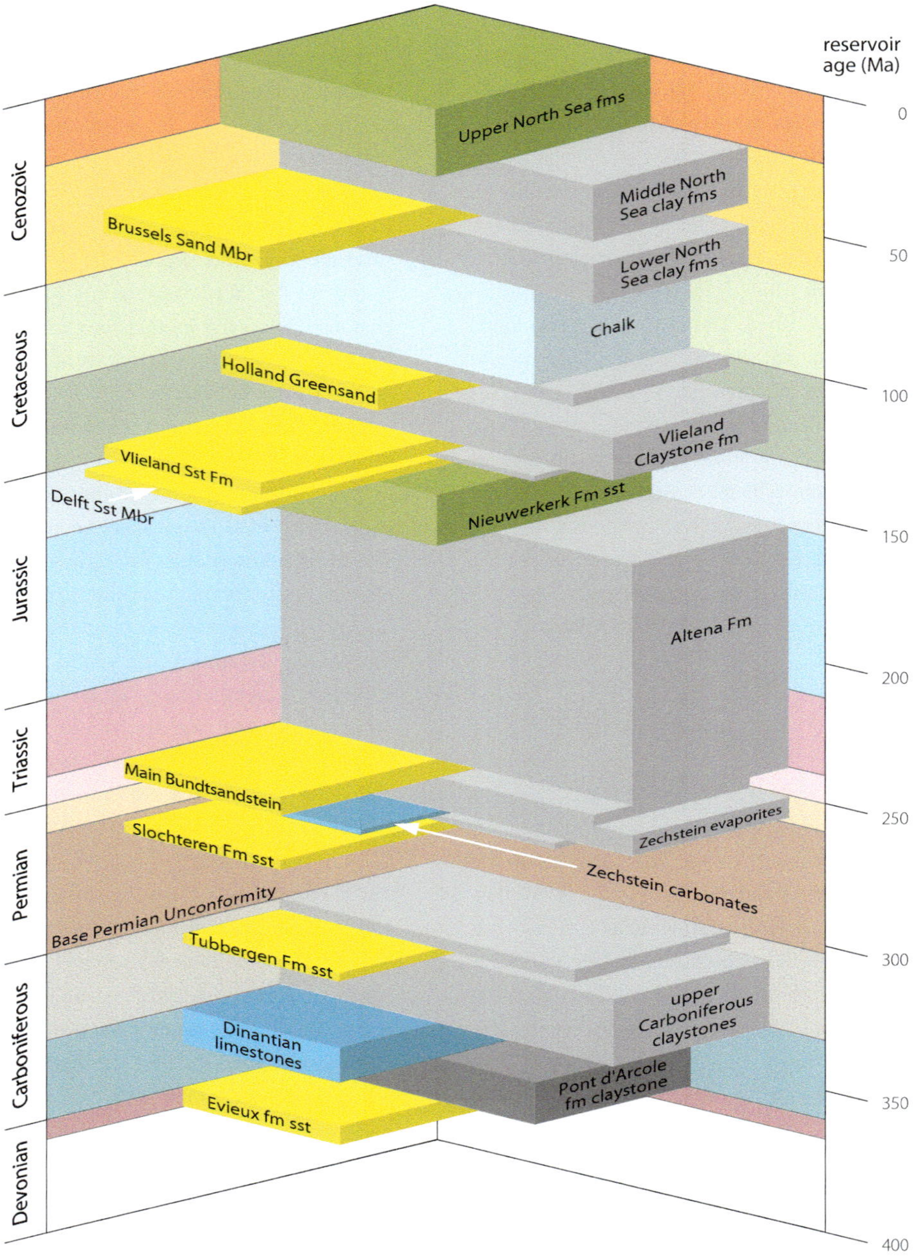

Figure 19.5. Generalized stratigraphic column of the Netherlands showing the geothermal reservoirs and sealing/aquitard packages: sandstone reservoirs in yellow, carbonate reservoirs in blue. The extent to the left of these reservoir 'boxes' gives an indication of the reservoir quality. The grey boxes denote the sealing clay-rich formations and their extent to the right gives an indication of the seal quality. The green and light blue boxes include poor quality reservoir and sealing lithologies. Note that the vertical scale is in millions of years, not stratigraphic thickness in metres.

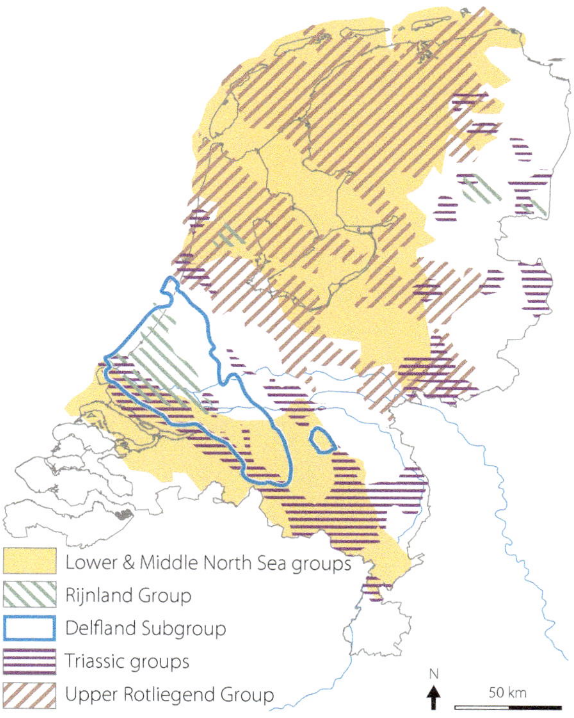

Figure 19.6. *Distribution of the main perceived productive geothermal reservoirs in the Netherlands (after P50 Power maps of the various geothermal plays from ThermoGIS, 2019, consulted in February 2022).*

is relatively easy to predict because all geothermal plays, fields or reservoirs, lie within a certain depth domain, while the average geothermal gradient is 31°C/km (Bonté et al., 2012), although more recent insights suggest a higher average geothermal gradient for the Netherlands. More detailed temperature estimates which take into account lateral and vertical variations in the thermal gradient are presented in Bonté et al. (2012) and Békési et al. (2020). Deviation, from the overall straight gradient line within certain intervals is caused, amongst others, by variations in thermal conductivity between the rock types and paleo-surface temperature changes related to the latest glacial period (Gies et al., 2021). Transmissivity, the product of thickness and permeability, is a key performance indicator for successful geothermal exploration, since it is linearly proportional to the achievable flow rate in the reservoir (e.g. Van Wees et al., 2012).

A geothermal project aims to harvest geothermal heat from the subsurface through one or more production systems. Initiation is linked to a (company) decision and is followed by a series of staged investment decisions towards maturation of the project to production. Within the Netherlands geothermal systems and projects are classified as shallow, deep, or ultradeep which correspond to specific depth domains which in turn grossly define temperature domains. The latter are largely related to specific direct use purposes. Electricity generation from geothermal energy is presently not implemented in the Netherlands as production temperatures are barely sufficient for efficient conversion. Suitable temperatures occur at depths in excess of 4 km where reservoir properties generally are poor. In the future electricity generation using Organic Rankine Cycle (ORC)-systems might be an option as they become progressively more efficient (albeit that even for a temperature of 180°C efficiency will never exceed 20%).

Figure 19.7 shows the depth and stratigraphic distribution of Dutch geothermal systems. It shows that individual geothermal reservoir at the various locations occur at different depths, as is evident from the top reservoir depth maps (DGM-deep v5). It also shows that the main targets are situated at depths ranging from 1600 to 2800 m, at which production temperatures vary from 60 to 100°C. Finally, it shows that most geothermal systems are developed in the prolific Delft Sandstone reservoir of the Nieuwerkerk Formation of Late Jurassic to Early Cretaceous. Runners-up are geothermal systems in the Slochteren Formation (Upper Rotliegend Group) of Permian age. As yet under-represented are the Vlieland Sandstone Formation, North Sea Supergroup and Triassic Main Buntsandstein reservoirs. In addition, two geothermal systems have been developed within a lower Carboniferous carbonate reservoir. They belong to the fracture/fault play, in contrast to all other systems which are assigned to the clastic Hot Sedimentary Aquifer matrix permeability play.

Table 19.1. *Depth classification of geothermal systems in the Netherlands.*

Depth denominator	Depth domain (m)	Temperature domain (°C)	Direct use domain
Shallow	500-1500	25-55	Greenhouses and well-insulated housing districts. Often combined with heat pump technology
Deep	1500-4000	55-125	Greenhouses and district heating. Optionally combined with heat pump technology to realize deeper cooling to increase the geothermal heat production and simultaneously resulting in lower injection temperatures
Ultradeep	>4000	>125	Light industry like food and paper

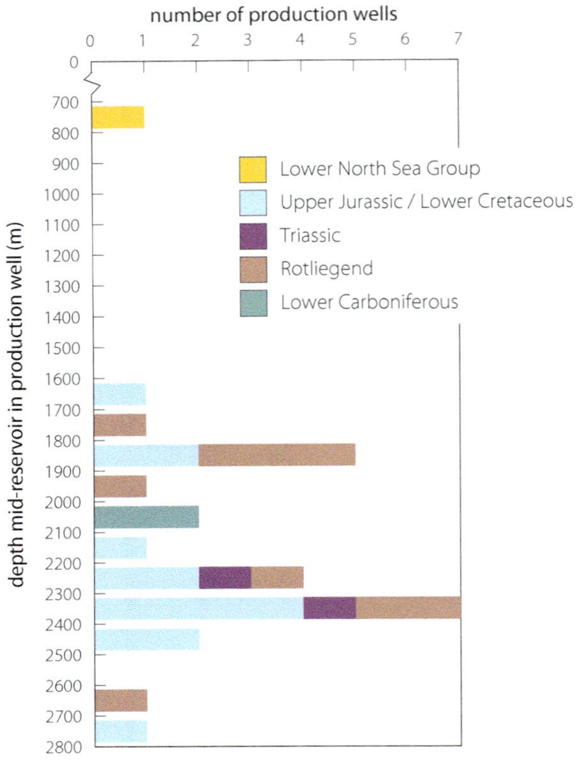

Figure 19.7. *The depth and reservoir lithostratigraphy of Dutch geothermal systems (status date 1-1-2022).*

Geothermal production systems

Most geothermal systems in the Netherlands are doublets. A doublet consists of a production and injection well, both completed in the same reservoir interval (the geothermal source) at a subsurface distance at reservoir level of up to approximately 1.5 km (Fig. 19.8). Dutch triplets presently have one production well and two injection wells. The production is executed as follows:

1. Hot geothermal brine is brought to the surface by using a pump, an Electrical Submersible Pump (ESP) placed in the production well at depths between 300 and 700 m;
2. Heat is extracted from the geothermal brine in a surface heat exchanger;
3. Finally, the cooled brine is re-injected into the reservoir via the injection well, often facilitated by a pump at surface.

Since the heat extraction from the well proceeds faster than the heat replenishment from the earth, a cold waterfront will form around the injection well in the reservoir and will eventually reach the production well (Fig. 19.9a). The moment it does so is referred to as thermal breakthrough. During production a pressure difference is imposed between the well and the reservoir. By definition, the pressure difference is negative in the production well and positive in the injection well (Fig. 19.9b). The area in which the original reservoir temperature and pressure are influenced, can be called the *subsurface area of influence* of the geothermal system.

Geothermal licensing

In the onshore Netherlands (Fig. 19.10) there are presently 96 exploration and 33 production licenses (status date 1-1-2022), including the ones in application status (MEACP, 2022). Geothermal projects targeting reservoirs deeper than 500 m are subject to Dutch Mining Act (*Mijn-*

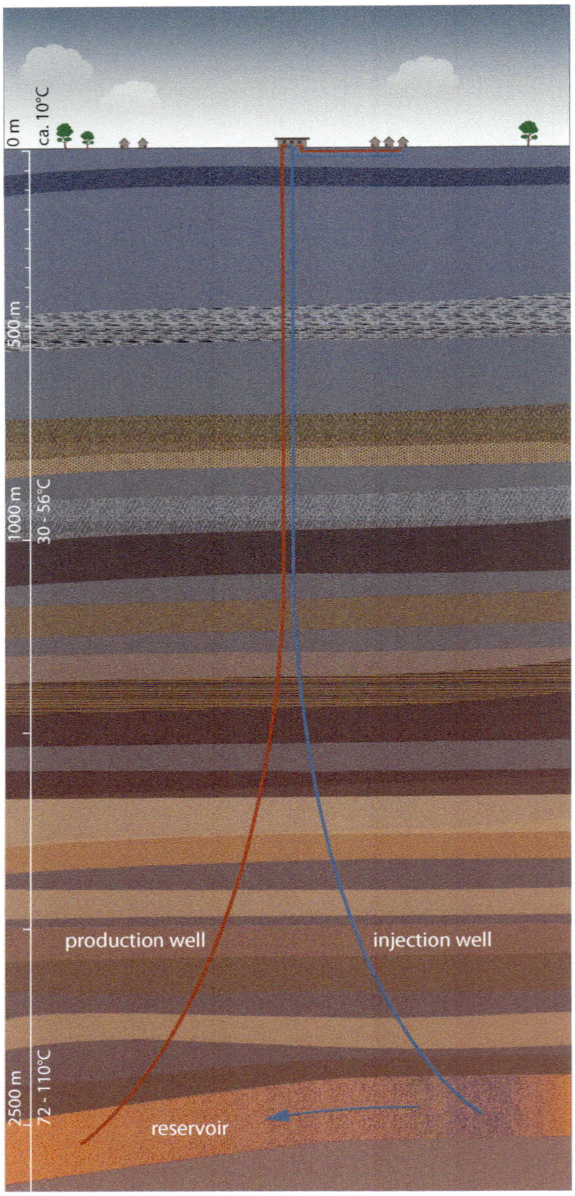

Figure 19.8. *Sketch of a typical doublet configuration with a production well (red) and an injection well (blue) targeting the same deeply buried reservoir. In this case a reservoir at about 2.5 km depth with an approximate temperature of 70 °C.*

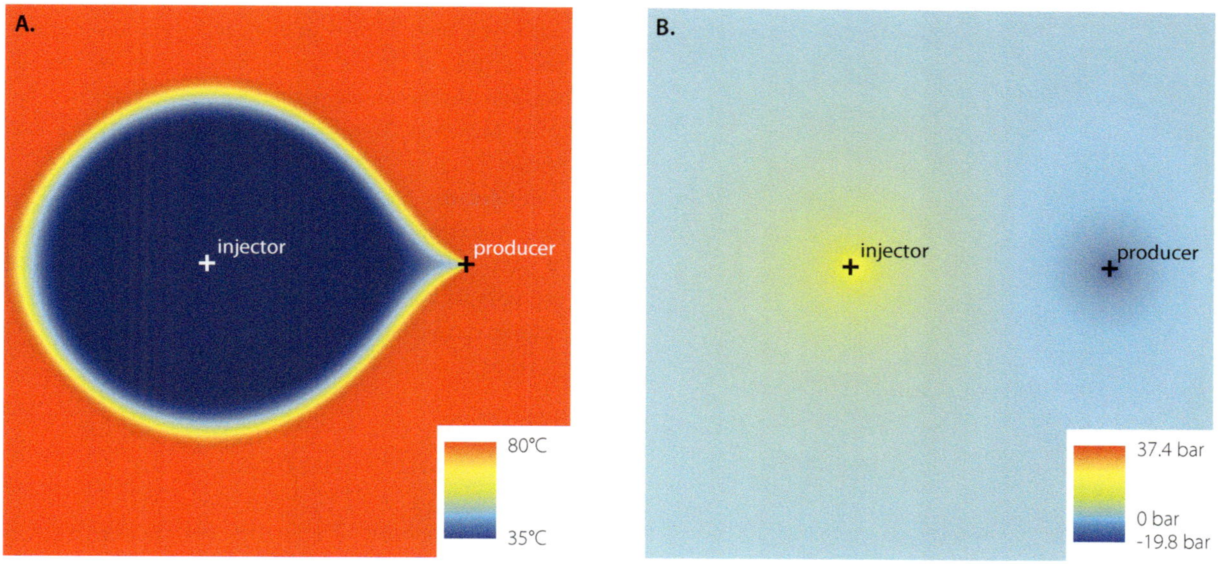

Figure 19.9. Approximation of the subsurface area of influence of a geothermal doublet in an homogeneous reservoir. a) The extent of the cold waterfront in the reservoir at thermal breakthrough. b) The extent of the pressure perturbation (ΔP in bar) in the reservoir. Modelled with DoubletCalc2D (DoubletCalc, 2014).

bouwwet), which means that separate licenses are required for geothermal exploration and production. Licenses are concentrated in areas with prominent presence of the horticultural sector like South Holland or the north of North Holland (Fig. 19.11). Actually, as shown in Figure 19.3, the majority of past geothermal development has been for greenhouse heating. However, the number of exploration licenses overlapping urban areas is growing rapidly (Fig. 19.11) and is in line with the ambition to supply geothermal heat to residential areas. In general, an exploration licence is granted for a depth domain from 500 m to infinite depth, although in some cases it is truncated at 4000 m to separate geothermal exploration at conventional depths from that for ultra-deep targets. Production licenses are usually restricted to specified intervals, in order to prevent the possibility of interference between vertically stacked geothermal systems. In such cases the upper and/or lower, license boundary is generally defined as the top or base

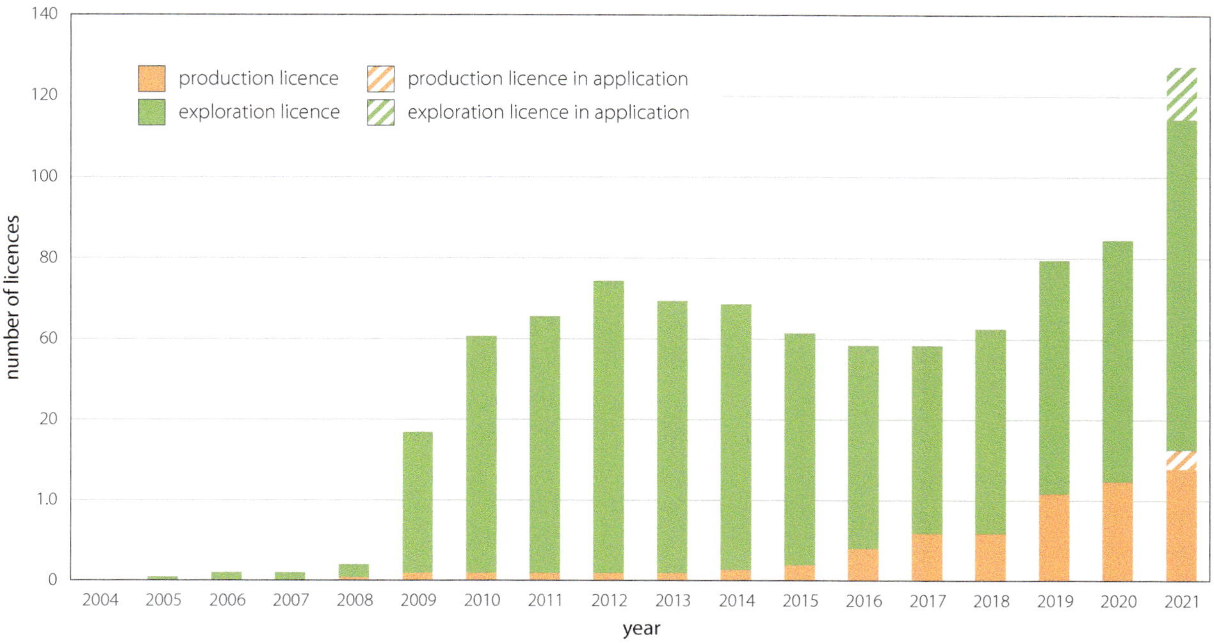

Figure 19.10. Geothermal license development between 2004 and 2021 (MEACP, 2022).

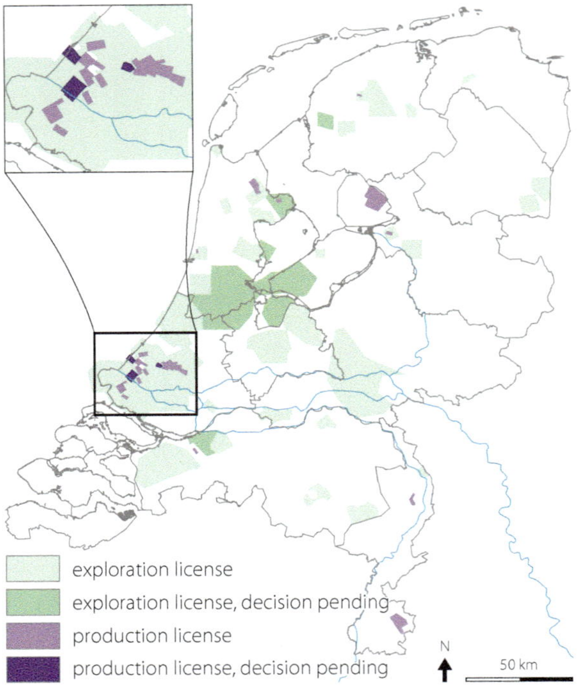

Figure 19.11. Geothermal license position at status date 01-01-2022.

of an impermeable lithostratigraphic interval that acts as a laterally continuous aquitard (Fig. 19.12; TNO, 2014a). Occasionally an absolute depth domain is licensed when it is evident that no vertical interference is possible within the licenced area.

Complementary to the licencing process, regulations demand reports detailing the process from incipient project idea to production, for example the drilling plan must be submitted to the mining authority (SodM) and the anticipated production plan (*Winningsplan*) to the Ministry of Economic Affairs and Climate Policy (MEACP) before the actual wells are drilled. The Mining Act and regulations have been revised in 2023 to better accommodate geothermal energy. The procedures in this revised Mining Act, and its subordinate regulations, are depicted in Figure 19.13 including the responsible authorities and advisory bodies.

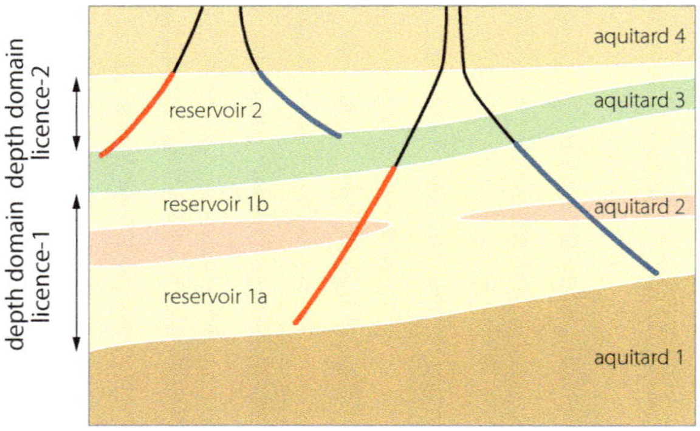

Figure 19.12. Vertical license delineation strategy on basis of laterally continuous lithostratigraphic units which act as aquitards (from TNO, 2014a).

Besides the Mining Act, the Environment and Planning Act (Omgevingsvergunning) applies to the impact of a geothermal system at the surface. Permit application can start as soon as a project location has been found and must be completed before any on-site activities start.

Managing the geothermal source; uncertainties and hazards

In the development of a project, uncertainties are supposed to reduce with maturation through time (Fig. 19.14; Van Wees et al., 2020). At project conception the uncertainties are large, but decrease gradually as geological, geophysical, and geochemical studies are executed prior to the drilling of a well. The well results record the actual reservoir parameters and productivity potential and consequently what the most likely production capacity of the system will be.

Low risk during the 'operation and maintenance' phase depends on thorough pre-operational studies, modelling, monitoring, and control during operation of the behaviour of the reservoir and reservoir fluids in relation to the imposed pressure (ΔP) and temperature (ΔT) differences and their impact on scaling, well integrity, and seismicity. In the Netherlands and neighbouring countries early production has seen unanticipated problems like casing and screen failure, ESP failure, scaling and in a few special cases induced seismicity. This emphasizes that a 'perception of low risk' in the operational stage as depicted in Figure 19.14 is only valid when the behaviour of the reservoir, the wells and the facilities under the imposed operational conditions are thoroughly known.

Uncertainty in reservoir flow properties and performance

Despite the wealth of subsurface data and knowledge of the Dutch subsurface (available from www.nlog.nl) pre-drill estimates of the geothermal power at any particular location still carry uncertainty related to the geology and the amount of relevant data. The primary geologic parameters for calculating geothermal power and energy are reservoir thickness, the net-to-gross ratio, porosity, permeability, reservoir depth, geothermal gradient or reservoir temperature and geothermal brine salinity as well as derived parameters like viscosity, density and fluid heat capacity. Also important are permeability anisotropy and rock thermal properties such as heat capacity, thermal

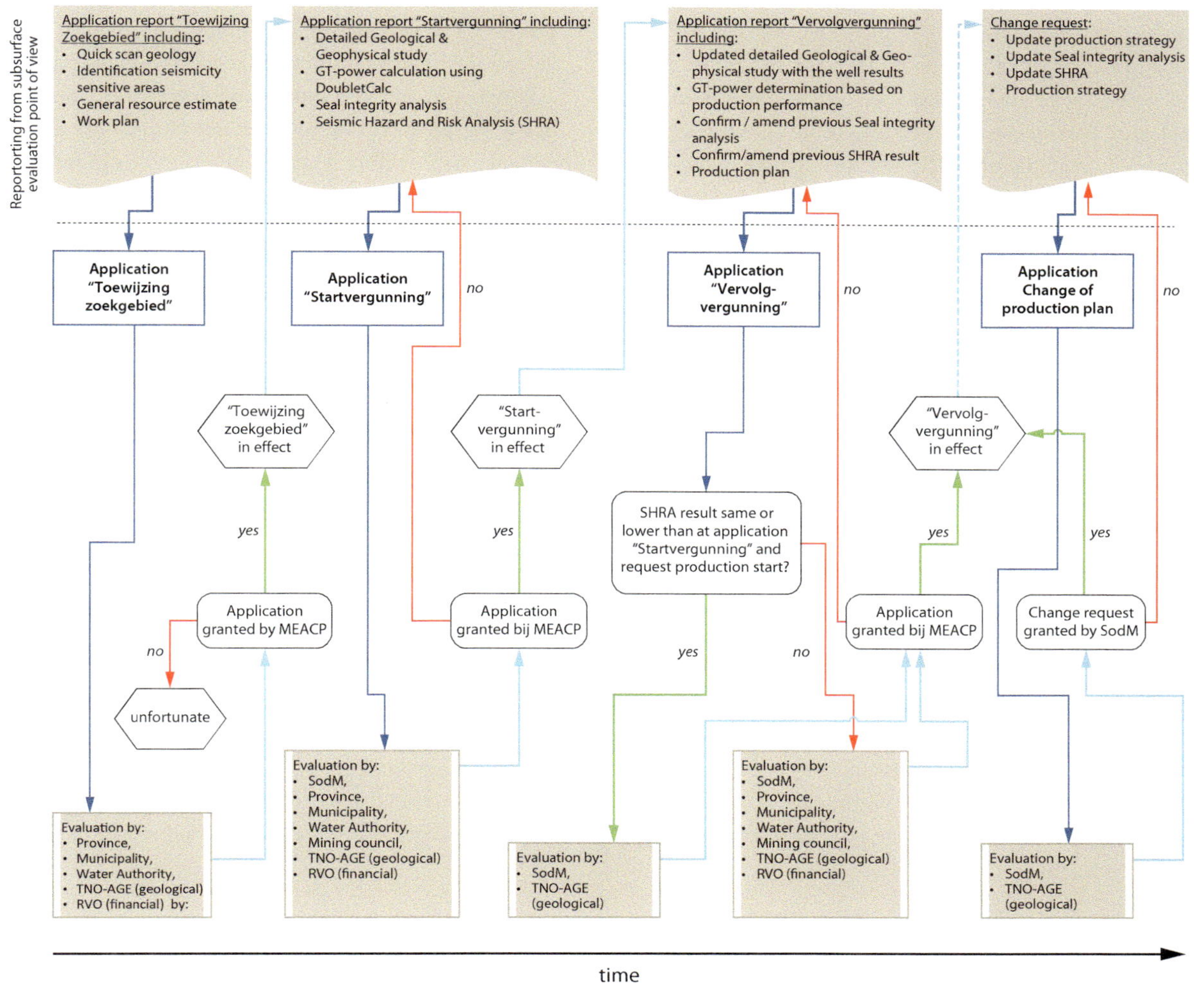

Figure 19.13. *Steps to take in mining licensing for geothermal energy under the revised Mining Act (since 2023). The top row lists reports to be filed; the row in bold shows the subsequent application steps. The lowest row shows the evaluation and advisory bodies (MEACP or SodM, shown as rectangles with rounded corners). A requested element (diamond shape) is subsequently granted or denied. Toewijzing zoekgebied can be seen as exploration license phase 1, Startvergunning as exploration license phase 2 and Vervolgvergunning is similar to production license. TNO-AGE = Advisory Group for Economic Affairs (part of TNO - Geological Survey of the Netherlands); RVO = Netherlands Enterprise Agency.*

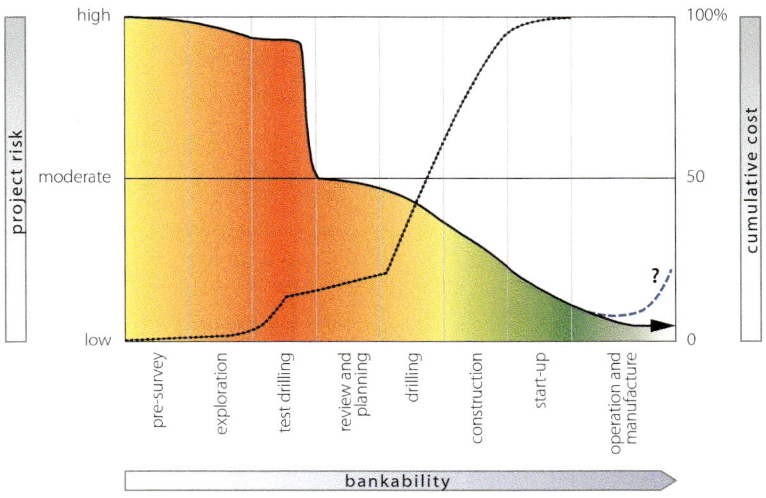

Figure 19.14. *Project risk/uncertainty during the progressive project phases from incipient thoughts to producing system (after ESMAP 2012).*

conductivity and diffusivity, both of the reservoir and of the surrounding rocks. An early estimate of the geothermal power range for a location can be calculated from the geological parameters and their uncertainty range coupled with the intended production engineering parameters and operational settings, for instance using the free DoubletCalc1D software tool (Van Wees et al., 2012; DoubletCalc, 2014; Fig. 19.15). The resulting flow rate and geothermal power are represented by an expectation curve. Threshold expectation values can be compiled on maps representing most likely (P50), the low (P90) and high (P10) estimates of the nominal thermal power as presented in ThermoGIS (Fig. 19.16). Subsequently, these most likely, low, and high values can be multiplied by the expected yearly amount of full load hours. This results in low, medium and high case yearly energy production estimates.

Over the lifetime of the project the yearly energy production estimates can be displayed as a production profile. Figure 19.17 gives an example of such a production profile for a geothermal system anticipated to produce 0.1 PJ of geothermal heat per year (in the most likely case). Uncertainty in geological parameters is reflected in a bandwidth around the most likely production profile. In addition, due to uncertainty in timing of the thermal breakthrough, the total energy produced during the systems' lifetime can be variable. The cumulative heat production in this profile defines the project resource with its uncertainty range.

Better and more comprehensive geological mapping based on progressively more data, information and experience will result in a better understanding of the reservoirs and the geothermal plays. It will reduce the geologic uncertainty but also the exploration/play uncertainty, ultimately to a level inherent to the subsurface reservoir typed (Van Wees et al., 2020).

Operational regimes and environmental hazards and risks

Nothing in the world comes free of risk and the same applies for geothermal energy production. Those related to geothermal energy production include corrosion and scaling, system integrity, groundwater pollution, subsidence and induced seismicity.

Subsidence

In systems that rely on circulation of the formation water there is in general no pressure depletion in the reservoir because the produced geothermal water is re-injected in the same reservoir. The cooling of the rock volume around the injection well due to injection of cooled water causes a shrinkage of the reservoir of less than 0.1%, but it chang-

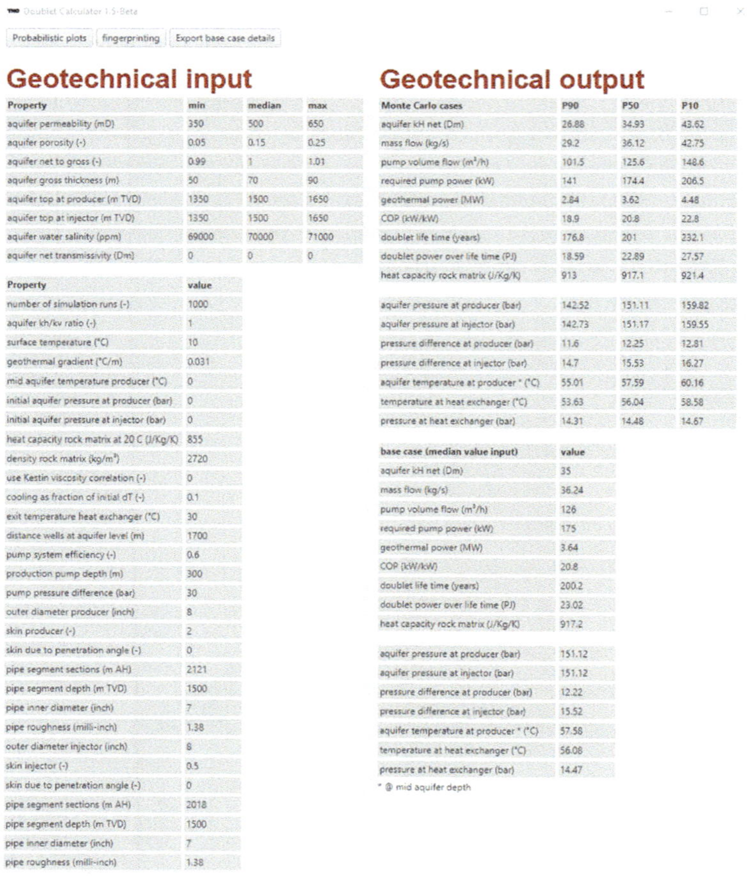

Doublet Calculator 1.5-Beta

Probabilistic plots | fingerprinting | Export base case details

Geotechnical input

Property	min	median	max
aquifer permeability (mD)	350	500	650
aquifer porosity (-)	0.05	0.15	0.25
aquifer net to gross (-)	0.99	1	1.01
aquifer gross thickness (m)	50	70	90
aquifer top at producer (m TVD)	1350	1500	1650
aquifer top at injector (m TVD)	1350	1500	1650
aquifer water salinity (ppm)	69000	70000	71000
aquifer net transmissivity (Dm)	0	0	0

Property	value
number of simulation runs (-)	1000
aquifer kh/kv ratio (-)	1
surface temperature (°C)	10
geothermal gradient (°C/m)	0.031
mid aquifer temperature producer (°C)	0
initial aquifer pressure at producer (bar)	0
initial aquifer pressure at injector (bar)	0
heat capacity rock matrix at 20 C (J/Kg/K)	855
density rock matrix (kg/m³)	2720
use Kestin viscosity correlation (-)	0
cooling as fraction of initial dT (-)	0.1
exit temperature heat exchanger (°C)	30
distance wells at aquifer level (m)	1700
pump system efficiency (-)	0.6
production pump depth (m)	300
pump pressure difference (bar)	30
outer diameter producer (inch)	8
skin producer (-)	2
skin due to penetration angle (-)	0
pipe segment sections (m AH)	2121
pipe segment depth (m TVD)	1500
pipe inner diameter (inch)	7
pipe roughness (milli-inch)	1.38
outer diameter injector (inch)	8
skin injector (-)	0.5
skin due to penetration angle (-)	0
pipe segment sections (m AH)	2018
pipe segment depth (m TVD)	1500
pipe inner diameter (inch)	7
pipe roughness (milli-inch)	1.38

Geotechnical output

Monte Carlo cases	P90	P50	P10
aquifer kH net (Dm)	26.88	34.93	43.62
mass flow (kg/s)	29.2	36.12	42.75
pump volume flow (m³/h)	101.5	125.6	148.6
required pump power (kW)	141	174.4	206.5
geothermal power (MW)	2.84	3.62	4.48
COP (kW/kW)	18.9	20.8	22.8
doublet life time (years)	176.8	201	232.1
doublet power over life time (PJ)	18.59	22.89	27.57
heat capacity rock matrix (J/Kg/K)	913	917.1	921.4
aquifer pressure at producer (bar)	142.52	151.11	159.82
aquifer pressure at injector (bar)	142.73	151.17	159.55
pressure difference at producer (bar)	11.6	12.25	12.81
pressure difference at injector (bar)	14.7	15.53	16.27
aquifer temperature at producer * (°C)	55.01	57.59	60.16
temperature at heat exchanger (°C)	53.63	56.04	58.58
pressure at heat exchanger (bar)	14.31	14.48	14.67

base case (median value input)	value
aquifer kH net (Dm)	35
mass flow (kg/s)	36.24
pump volume flow (m³/h)	126
required pump power (kW)	175
geothermal power (MW)	3.64
COP (kW/kW)	20.8
doublet life time (years)	200.2
doublet power over life time (PJ)	23.02
heat capacity rock matrix (J/Kg/K)	917.2
aquifer pressure at producer (bar)	151.12
aquifer pressure at injector (bar)	151.12
pressure difference at producer (bar)	12.22
pressure difference at injector (bar)	15.52
aquifer temperature at producer * (°C)	57.58
temperature at heat exchanger (°C)	56.08
pressure at heat exchanger (bar)	14.47

* @ mid aquifer depth

Figure 19.15. *Output screen of DoubletCalc version 1.5 (Beta) showing the geologic input value ranges and typical system architecture and operational setting in the left column. The stochastic and deterministic (base case scenario) calculation results are in the right column (see www.nlog.nl/tools).*

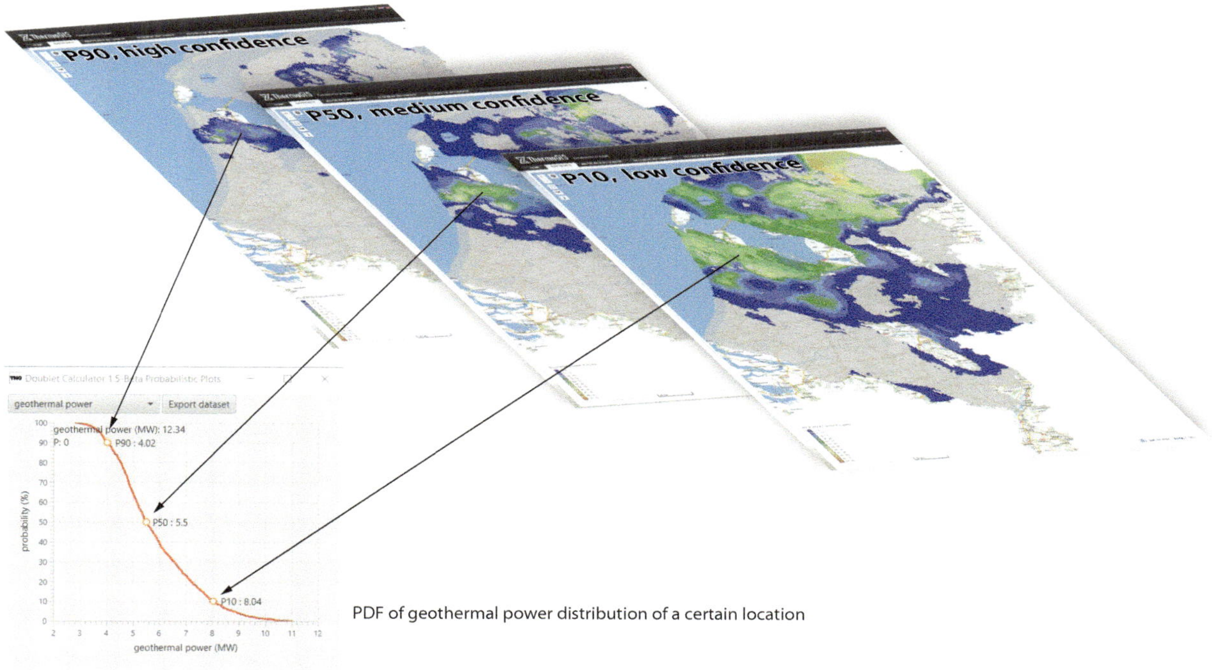

Figure 19.16. *Three maps (P90, P50 and P10) of the probabilistic geothermal power potential in the Netherlands. This is the result of the stochastic calculation using ranges of the geologic parameters of the 2D reservoir maps: permeability, thickness, net-to-gross ratio, depth and salinity. PDF = Probability Density Function.*

es the local stress field in the *subsurface area of influence*. The shrinkage may lead to subsidence from a few millimeters to a maximum of 2 cm in the centre of the subsidence bowl (Fokker et al., 2015). All subsidence modelling results of operating Dutch geothermal systems included in the 'production plan' evaluations at 1-1-2022, indicate that the effect of cooling of the reservoir at surface is in the order of a few to tens of millimetres over the licensed period of some 35 years.

Induced seismicity and seal failure

The imposed temperature and injection pressure changes on the reservoir lead to stress changes in the subsurface that potentially might reactivate existing faults in the subsurface area of influence of the geothermal system. If such fault reactivation occurs it may induce a-seismic or seismic activity. If the seismic events can be felt at the surface it may cause anxiety to the public or damage to infrastructure and housing.

A balanced mass-flow with the same production and injection volume is obligatory. Pressure communication between the injection and production well is ensured by for example interference testing. Therefore, on average there is no pressure depletion in the reservoir and no stress changes due to pressure depletion are anticipated. In the close vicinity of the injection and production wells, pressure changes do occur, but they typically decay logarithmically with distance from the well in the matrix of the porous reservoir and cause only local stress changes (e.g. Buijze et al., 2022; Kivi et al., 2022). This is in contrast to gas production where pressure in the entire reservoir is decreased by up to several hundred bar over a field's lifetime, (Roholl et al., 2021; Buijze et al., 2023).

Substantial cooling is expected to occur in the reservoir as the reinjection temperature is often several tens of degrees lower than the in-situ reservoir temperature. The lateral extent of the cooled reservoir can attain several hundreds of metres, or even more than one kilometre, depending on the subsurface injection-production well distance, the flow rate, porosity and reservoir thickness (Buijze et al., 2022, 2023). Part of the top and base seals that are in contact with the reservoir also cool due to heat diffusion. The resultant shrinkage will induce stress changes in the reservoir and surrounding rocks. These cooling-induced stress changes may cause slip along pre-existing faults when the fault strength is exceeded, primarily within the cooled zone where stress changes are largest (e.g. Van Wees et al., 2019). Cooling-induced stress changes dominate the imposed effect of the pressure changes in the matrix permeability dominated Dutch systems (Buijze et al., 2019, 2022; Kivi et al., 2022).

In fracture dominated systems however, elevated pressures at an injection well diffuse rapidly and directionally through the pre-existing fracture and fault network and can cause induced seismic events. This holds especially for hydraulic stimulations where large volumes are inject-

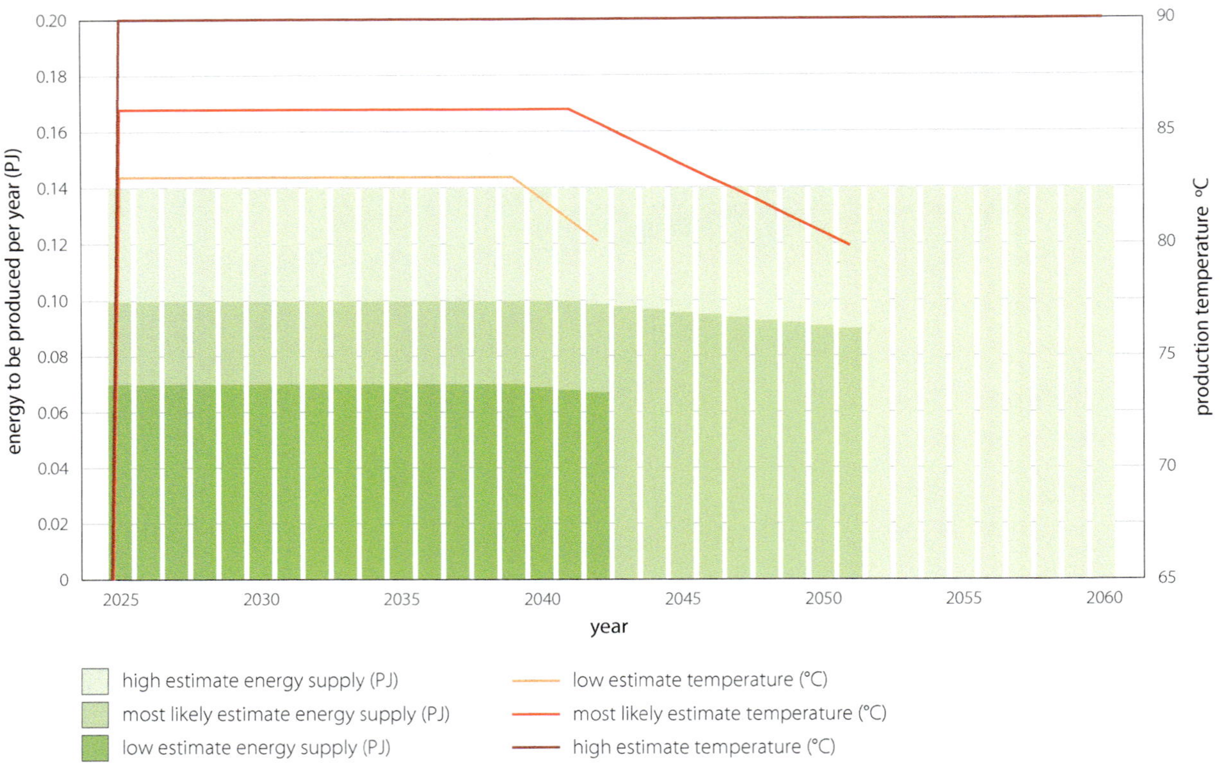

Figure 19.17. *Example of a future production profile of a geothermal system. The low value of the expected geothermal power estimate times the amount of full load hours per year results in the 'low estimate' of energy production. The same calculation is for the most likely and high estimates. Additionally, the expected production temperature profiles are given in low, most likely, and high estimates. Production in this case ceases when the production temperature drops below the minimum threshold of some 80°C.*

ed under relatively high pressures (Wassing et al., 2014; Gaucher et al., 2015). Such stimulations have not yet been put into practice in the Netherlands. The injection pressure stress perturbation normally fades away quickly when pumps are stopped, however in fracture permeable systems larger seismic events sometimes follow such a shut-down as pressure diffusion is still ongoing. The inferred pressure changes at hypocentre locations can often be small, reflecting a close to critical state of stress in these systems.

A worldwide inventory of geothermal production systems (Buijze et al., 2019) demonstrated that geothermal systems that rely on matrix permeable systems induced no seismic events with magnitudes of more than 2.0 on the Richter scale. In contrast, seismicity occasionally occurs when production and injection of water is in fault/fracture related permeability systems and even more often when it is in Enhanced Geothermal Systems (EGS). All but two Dutch geothermal systems produce from reservoirs with predominantly matrix permeability (Mijnlieff, 2020) and no seismicity has been recorded which could be unambiguously linked to such systems (Muntendam-Bos et al., 2022). In the local seismic monitoring network that was installed in the surroundings of the two systems producing from faulted and fractured (and maybe partly karstified) carbonate reservoirs sub-feelable seismic events were recorded, although the largest ($<M_L = 1.7$, (Baisch & Vörös, 2018) could potentially have been felt at the surface.

Seismic hazard and risk analysis is performed before activities commence. Screening takes into account a set of key elements that may cause seismicity (Baisch et al., 2016). Depending on the outcome of the screening, additional studies and measures which may include a detailed location-specific seismicity evaluation and the installation of a monitoring and control system to ensure safe and responsible operations.

In addition to balanced production it is mandatory that the geothermal water does not escape and leak into other reservoirs through seal failure. An injection protocol (SodM, 2013, 2019; MEACP, 2021) has been established in order to define a safe operational space with respect to imposed pressure and temperature changes on the subsurface reservoir. SodM considers that unsolicited and hazardous events like seal failure or induced seismicity can be mitigated by applying safety limits in the operational space combined with monitoring to validate operational conformance. Operational constraints define the project's 'installed capacity', which may then be lower than what would be technically achievable if limits were not imposed (Fig. 19.18). Additionally, limiting the ΔP and ΔT may not only contribute to mitigation of environmental hazards, but also to sustainable performance of the system during its lifetime. For example, limitations in ΔT may prevent certain scaling. If these uncertainties are addressed in a comprehensive manner with proper methodologies to monitor, control, learn, and adapt, geothermal projects can flourish.

Corrosion and scaling

In the production phase operational issues rooted in geological causes may arise. The geothermal brines circulating in the system are highly saline and corrosive. Elements in the brine can react with the steel and corrode the piping, impacting the well integrity (TNO, 2016). The change of pressure and temperature of the circulating brine when its heat is harvested disturbs its geochemical equilibrium and chemical reactions will take place. An example includes precipitation of solids like calcium carbonate onto the steel piping of the well (TNO, 2013).

Currently operational systems (except the shallow North Sea Supergroup system (Fig. 19.7)) have salinities between 100,000 and 300,000 ppm NaCl eq. Figure 19.19 illustrates the amount of salt that precipitates from 100 ml of approx. 180,000 ppm salinity geothermal brine. Sodium and chloride dominate the brine, however other ions are also present in lower concentrations. Additionally, various gases, including methane, carbon dioxide and nitrogen are dissolved at reservoir depth (solution gas). The water-gas ratio is in the order of 0.3 to 1.0 m^3/m^3 in the hydrocarbon producing provinces (Dijkstra et al., 2020) but the exact value varies per location, depth and reservoir. In the production well, the temperature and the pressure decrease in the brine's path to the surface and when the pressure drops below the bubble point, solution gas escapes from the brine. When the gas escapes the geochemical equilibrium in the brine is distorted and the acidity is changed, favouring chemical reactions.

The brine-piping contact, the degassing, temperature and pressure changes and subsequent brine chemistry change may cause corrosion of the piping and pumps and/or precipitation of minerals. A few years after the start of the first projects this became clear and repairs to several existing wells and alterations to the well designs were made. Lead is seen to react (redox) with the casing, corroding the steel and forming a lead residue in the injection well (TNO, 2014b). Carbonate scaling on the ESP rotors was observed. In some cases the mineral deposits include naturally occurring radioactive material (NORM). These are caught in the filters or accumulate in the system and have to be removed periodically. Nowadays, most systems use inhibitors to effectively mitigate and minimize the corrosion and scaling issues, use glass fiber reinforced epoxy (GRE) piping for their surface installations or use GRE and coated materials for their sub-surface casing/piping.

System integrity (leakage)

To ensure balanced flow it is essential to keep the geothermal brine confined and contained within the reservoir interval and in the primary loop that comprises the (two) wells and the surface installation, including the heat exchanger. This is also of prime importance to minimize the pollution risk of fresh groundwater aquifers (Hartog,

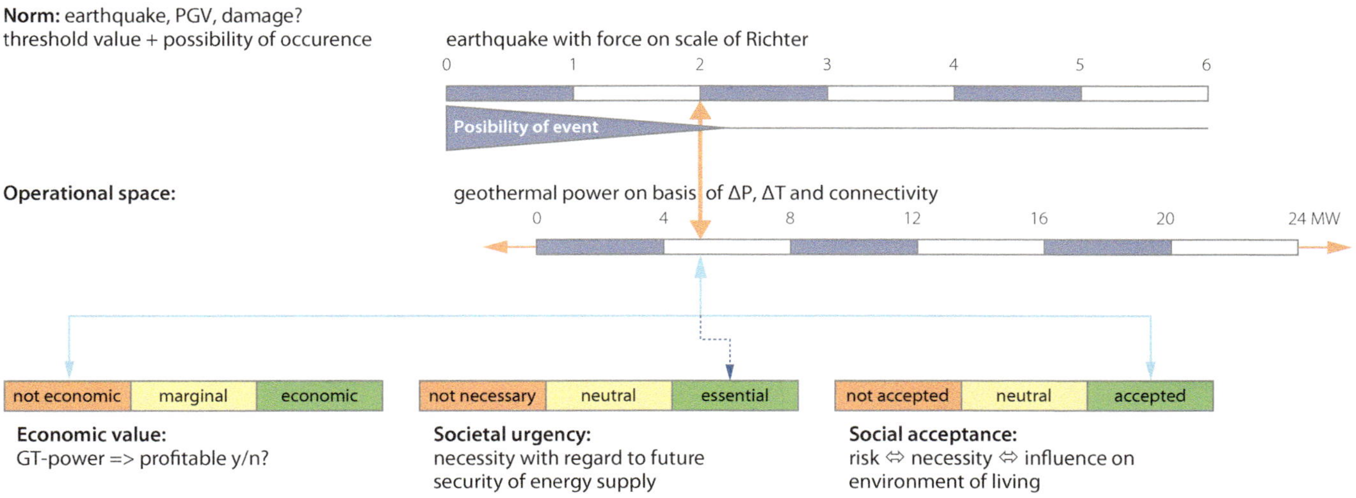

Figure 19.18. Operational space defined by (i) the threshold value of an earthquake both in magnitude and chance of occurrence, (ii) the installed power of that geothermal system given the maximum injection pressure (ΔP), (iii) the minimum injection temperature (ΔT) matching the seismicity norm, and subsequently (iv) the relation to economic value, societal urgency and social acceptance. PGV = Peak Ground Velocity.

2016). Various processes may cause the well barriers, conductor, casing or liners to fail. As the geothermal brine is highly saline and corrosive, the metal piping may corrode and ultimately lead to a leakage pathway. Similarly, flow of water with suspended particles (sand) may result in erosion of the piping, while turbulent flow with local degassing may cause pitting, which may eventually result in holes in the piping. In addition, stresses exerted on the piping by ductile lithologies like rock salt may deform it to the point that it partially breaks and creates a leakage pathway.

The choice of material and well design is essential to reduce the corrosion, pitting, and deformation processes. The improved well design, which was adopted in 2020 as an industry standard, also proposes improved monitoring options (Geothermie Nederland, 2021). The annulus between the first and the second barrier is to be monitored continuously. A compromised production or injection tubing can then be adequately identified. Secondly, adequate periodical monitoring and inspection of the installation can indicate if, for instance, the wall thickness or the shape of the piping is reduced and when they become critical. All these measures and more are detailed by the operator in a well integrity management system (WIMS).

Figure 19.19. *Example of the amount of salt precipitated from 100 ml of geothermal brine from the Koekoekspolder geothermal system after evaporating the water. The reddish colour of the salt precipitate indicates that the geothermal brine is rich in iron, which forms iron oxide when in contact with the atmosphere (photo courtesy of Viola Lucza).*

Matching supply and demand: organizational and public engagement aspects

In the horticultural sector geothermal heat replaces local gas fired heat and power plants or boilers. Although this heat source was reliable, flexible, and efficient it has major drawbacks, like the volatile pricing, geopolitical issues and CO_2 footprint. Nevertheless, in horticulture the CO_2 contributes to the growth of the plants and thus needs to be available. The use of the co-produced solution gas in the heat and power plants and boilers partly satisfies the horticultural CO_2 demand. As the required geothermal temperature for greenhouses is relatively low, in the order of 70 to 50°C, geothermal heat from the Dutch subsurface is very well suited to match the demand. Moreover, it can provide a stable year-round base load, independent of weather conditions and as such it provides the user with a future-proof alternative energy source. Additionally, the transition to renewable energy sources can be a bonus to the placement of horticultural products in the market.

A project initiative in the horticultural sector is mostly taken by an individual or a cluster of greenhouse companies that purchase the heat and become project-owners at the same time. This way the offtake risk is small, in contrast to the uncertainty of producers who sell heat to third parties and have to rely on 'heat purchase agreements'.

The first geothermal system exclusively designed to provide heat for the urban environment is the 'The Hague' geothermal doublet, drilled in 2010. The wells were successfully tested and proved able to match the pre-drill anticipated heat demand. However, due to a crisis in the housing market the demand was not developed in time and the company faced bankruptcy (Haagse aardwarmte, 2021). However, at the end of the 2020s, the local heat demand was back to pre-drill expectation and the geothermal system production started to deliver heat to residential, office and hospital buildings early 2022.

About forty percent of the Dutch energy consumption is used for direct heat (CBS & TNO, 2020). This includes heating residential, recreational and business buildings, as well as greenhouses and industrial processes. The first three building types and to a lesser extent the greenhouses have a seasonal heat demand profile. The demand is low in summer and high in winter months, resulting in a typical curved demand profile (Fig. 19.20). Where the geothermal system matches the full heat demand, it operates at full capacity during the winter months and at low capacity during the summer months. From both the project economics

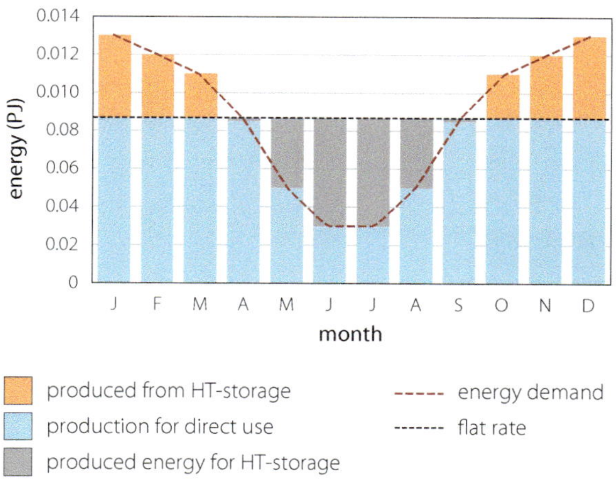

Figure 19.20. An ideal supply and demand graph for a combined deep geothermal system and a HT-ATES facility. The red dotted line depicts a typical profile of energy demand through the year. It displays the 'bath tub' profile as a consequence of the monthly varying heat demand; low in summer and high in winter. The black dashed line is the ideal flat rate supply profile from a deep geothermal system. Overproduction in the summer months with respect to the demand is buffered in the HT-ATES facility (grey bars) to be produced in winter (orange bars).

and the production and reservoir technical point of view this is sub-optimal. Geothermal systems perform better technically and are more economical when they produce at a stable rate year-round. The combination of a deep geothermal system with a seasonal high temperature heat buffer system is a solution to mismatches in monthly heat demand and supply. When the system operates year-round at full capacity, the excess heat production can be stored in shallow geologic formations in summer and later produced during the winter period to match the higher heat demand (Fig. 19.20). In 2021 in the north of North Holland a High Temperature Aquifer Thermal Energy storage (HT-ATES) pilot system that is linked to a deep geothermal system (www.heatstore.eu/) was realized.

In addition to the seasonal demand-supply match challenges, there can also be a mismatch in energy capacity (PJ), temperature of the geothermal supply and requirements on the demand side. In all cases a heat pump which can raise the production temperature of the heat produced by the system may cover the gap. Alternatively, it can produce more energy from the primary loop by lowering the injection temperature while delivering heat at the desired temperature level. Heat pumps are frequently applied in Dutch geothermal systems because they are a cost-efficient way to raise the energy output of deep geothermal systems. Moreover, a heat pump is an essential component in shallow geothermal systems (~500-1500 m depth) as they provide the required temperature levels for direct use applications. In the future use of geothermal in high temperature district heating networks or industrial applications, the use of industrial heat pumps is foreseen (Dijkstra et al., 2020).

Another challenge, as seen in the 'The Hague' project, is the alignment of geothermal heat supply and district heat network demand. Ideally the demand should be in line with the maximum capacity that the geothermal source can deliver at the start of production. For projects in the horticultural sector this appears to be fairly well matched due to clustering of demand and optimization of available buffer capacity. However, connecting customers to a heating network can lead to a timing mismatch, for instance when demand lags behind the supply offered by the geothermal resource (Fig. 19.21). In addition, the temperature provided by the geothermal system may be insufficient for the traditional heat grids, as the latter are generally designed for 90°C or more. Finally, project developers are confronted with the chicken or egg problem: "who is developing first?". Building a heat district network with sufficient offtake and developing a geothermal project both carry large pre-execution uncertainties in relation to the energy demand or supply profile, and the large stakeholder group involved.

Resource potential

The Dutch subsurface holds ample geothermal resources to match the ambition of the geothermal sector or the MEACP as depicted in Figure 19.2. The first quantitative geothermal resource assessment for the Netherlands was published by Kramers et al. (2012). Building on that approach an update was presented in 2020 (Mijnlieff et al., 2020). First, the Heat In-Place is calculated based on geological maps (Muffler & Cataldi, 1978). Subsequently the technical potential is determined by extracting the relevant input data from the ThermoGIS map set to compute the geothermal capacity in MW at every grid location. The economic potential is evaluated using the geothermal capacity together with an economic model and subsequently, the resource is be matched to the local heat demand. Geothermal heat is only of value when heat demand exists at more or less the same location, so the matching is preferably done on a local level where a heat transition strategy is to be developed.

Once the resource potential is mapped and estimated, the resources can be classified with respect to their maturity in (potential) development. The various hurdles in the project maturation have to be addressed. There might be obstructions in the maturation that lead to a

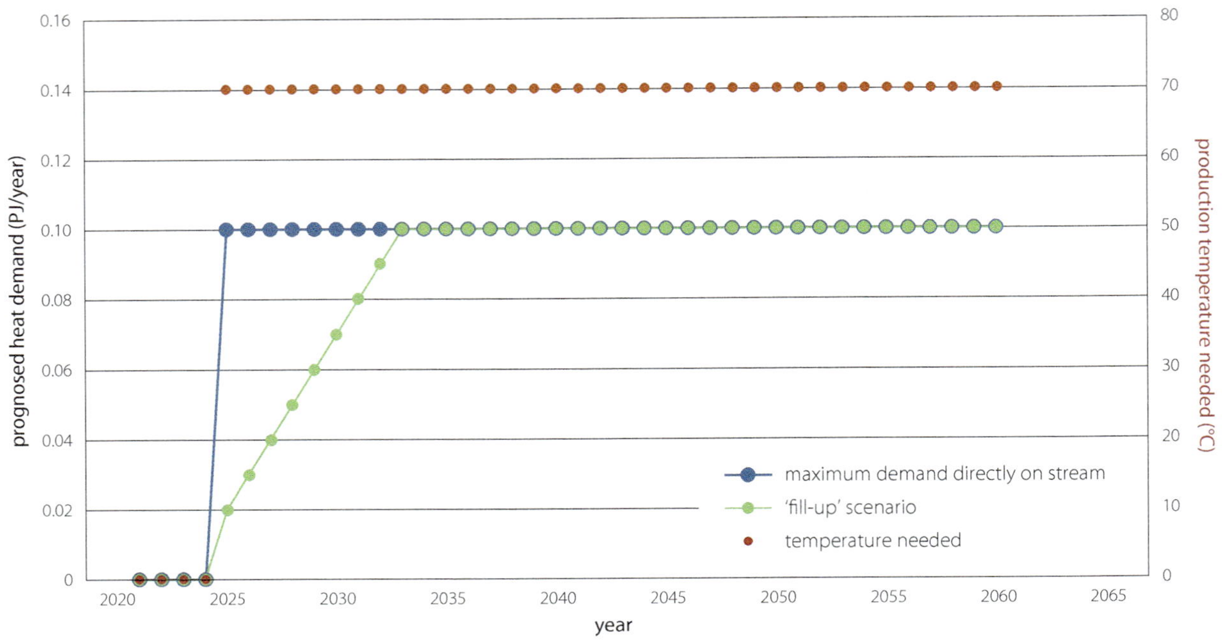

Figure 19.21. *Example of expected heat demand profiles in 'energy volume' and temperature. The blue line represents a system for which the heating demand matches the geothermal capacity profile from the start. The green line illustrates a case where the heat demand starts at zero and increases ('fills up') over a period of 10 years to its maximum. This may reflect the development of the heat network only after it is proven that the geothermal source works as expected.*

prolonged development time or even prohibit the source to be developed at that location. Examples of resource development hurdles involve overlapping drinking water protection areas, nature reserves and seismicity sensitive areas, but may also include economic hurdles (e.g. areas that are too far away from the demand location). The installed capacity together with development hurdles and perceived risks define the project categorization in terms of economic, social, environmental, and societal viability (UNFC, 2019).

Next to hurdles at the surface the subsurface uncertainty has an impact on the project maturation speed. Van Wees et al. (2020) illustrated that this hurdle can effectively be lowered by an orchestrated exploration approach where learnings, data and information are shared to lower the geological exploration risk. Using this play-based exploration approach the geothermal sector ambition to produce 50 PJ in 2030 could be efficiently matched.

Subsurface spatial planning

As the heat demand is quite localized (cities, light industries and greenhouse clusters), it is expected that geothermal systems will interfere with each other more than is presently the case. 3D or even 4D (taking time in consideration) subsurface spatial planning will be necessary. Orchestrated development of neighbouring systems, to optimize heat recovery and efficiency and simultaneously ensure safe operations, is expected to become common practice in geothermal subsurface domains. A prerequisite for a long-term joint production strategy between neighbouring geothermal projects is the presence of a linked heat distribution system at the surface, which can ensure that the security of supply is not compromised for individual operators in a jointly developed geothermal field.

Support measures and policy

The previous sections illustrated that developing the geothermal resource is a trajectory riddled with uncertainties. Governmental support measures soften the sharp edges of some of these uncertainties.

Financial support measures

The development of a geothermal system is a capital-intensive project. The drilling of deep boreholes, the construction of the surface installation and the distribution network may cost well over 30 million euros (EBN&GNL, 2021). The combination of high development costs and project uncertainties in, for example, the exploration and production phases, the cost price, the licensing and the offtake, represent major challenges in developing geothermal projects. The current legal, regulatory, financial and market frameworks are not optimised for initiating new heat value chains where offtake and infrastructure still need to be acquired and built.

On average the cost price of geothermal heat per PJ is/was higher than the equivalent gas price, which in the Netherlands (still) is the reference and in addition, the uncertainties and risks discussed in the previous sections negatively influence the business case. To support the development of geothermal energy several policy measures have been taken and some of these will be discussed.

Often, it takes a long time to reach financial closure, typically three years or more after the application for an exploration license. Convincing the different financiers that there is enough certainty for the project to be executed according to plan and the result will match pre-drill expectations despite all the uncertainties, proves to be complex. Three governmental financial support measures have been installed in the Netherlands in order to reduce the financial risk to the owner and investors. Firstly, the investment subsidy, secondly the exploration risk scheme and thirdly the feed-in premium scheme SDE (*Stimulering Duurzame Energie & Klimaattransitie*).

Dumas et al. (2019) visualized financial and risk mitigation support schemes with respect to the maturity of the geothermal development. In the Netherlands the development of support measures largely coincides with the maturity sequence. In the start-up phase (2006-2013), in addition to the project financing from banks, the investment costs of the first ten systems were supported by the Market Introduction Energy Innovation (MEI) subsidy grant, special loans from the Ministry of agriculture and the agricultural organization LTO. The exploration risk mitigation scheme, a public risk insurance, was installed by the Ministry of Economic Affairs in 2010 and updated in 2013. It offers an insurance against disappointing well productivity and/or injectivity and therefore lower than expected realized geothermal power, caused by less favourable geologic conditions than used in the pre-drill P90 power estimate (Mijnlieff et al., 2013).

In 2013 the geothermal energy production was incorporated in the feed-in premium scheme '*Stimulering Duurzame Energie*' (SDE) (Mijnlieff et al., 2013; RVO, 2021). Upon admittance of a project to the SDE scheme, a budget is reserved, based on the pre-drill expected capacity of the geothermal project times a fixed amount of full load hours during a 15 years period. The government pays-out the unprofitable top (GJ cost price of geothermal heat minus GJ cost price of natural gas) of the energy cost price per kWh when produced over the lifetime of the project with a maximum of 15 years.

A recent governmental financial initiative is the state participation in geothermal projects by EBN, which may take a 20 to 40% share in geothermal projects as has been done for decades in the gas exploration and production sector.

Regulations

The Dutch Mining Act was designed for the exploration and production of oil, gas, and salt. In 2003 the law was amended to include geothermal projects in the legal framework. The law stipulates that all data retrieved from the subsurface must be delivered to the MEACP. After a 5 years confidentiality period the data become public. This ensures the possibility that geothermal projects can use almost all available subsurface data in order to minimize the exploration risk.

Geothermal energy operates in the same subsurface environment as oil and gas production, and the (oil and gas based) legal and governance frameworks (Mining Act, Environment and Planning Act, Heat Act) used to misfit the development of geothermal energy in a number of important aspects. For example, for economic reasons and subsidy rules, production of heat needs to start shortly after wells are drilled and the surface facility is ready. However, the exploration licence did contain a permit for the production of the resource at that moment. This, in combination with long lead times related to review and approval of the production license and development plan, showed that application and granting procedures were unpractical and uneconomical for geothermal development. The revised Mining Act of 2023 takes care of this (Fig. 19.13) and, in addition, geothermal applicants will need to prove that there is a demand for heat in the application area through indicative heat offtake arrangements.

Future outlook

The so-called Masterplan Geothermal Energy (Stichting Platform Geothermie et al., 2018) outlines a number of bottlenecks in the development of the sector. It also lists innovations for the benefit of risk reduction, safety, optimization of the geothermal value chain, increasing the effectiveness of geothermal projects and cost reduction. Future research activities are likely to address topics in these areas. One focus area is the standardization of, amongst others, the design of wells, surface systems materials, and reporting. Second focus area is technical innovation, execution of which demands that funds are made available, The Netherlands Enterprise Agency (RVO) subsidizes innovation projects in the applied technical domain through subsidies. Next to these national funds there are numerous European Union calls for innovative research and demonstration projects with an international (EU) component. The list of innovation possibilities is almost endless; a number of spearhead innovation domains are described below.

The Koekoekspolder geothermal system

The Koekoekspolder geothermal system is one of the first systems developed in the Netherlands. The wells of the doublet were drilled in 2011 and production started in 2012. The doublet comprises a production and injection well and the surface installation comprising the heat exchanger. The geothermal heat is carried by water from the reservoir at 1950 m depth to the surface via the production well. After extracting the heat in the heat exchanger, the cooled-down formation water is re-injected via the injection well into the same reservoir, but at a distance of approximately 1500 m from the producer. In the first ten years the system produced more than 1,600,000 GJ of geothermal heat, the equivalent of 50 million cubic metres of natural gas. The produced heat directly feeds into a cluster of greenhouses where tomatoes, cucumbers, peppers, strawberries and more flourish on the geothermal heat.

The oblique aerial photo shows the facilities and greenhouse complex in a green rural area in the close vicinity of the towns of IJsselmuiden and Kampen. The present-day temperate coastal climate contrasts sharply with the paleo-environment in which the reservoir sediments were deposited. At the time of deposition of these sediments, some 258 million years ago during the Permian period, the area was part of an extensive desert, somewhat like the present-day Sahara. Sand was blown in from the northeast, forming extensive sand sheets which in time were stacked on top of each other ultimately forming the about 100 m-thick Upper Slochteren Member reservoir.

The reservoir productivity and injectivity is moderate, yet in the past production years the injectivity increased, resulting in better performance. The doublet yields a yearly heat production of up to 0.18 PJ. The success of this project and an increased local heat demand warranted the extension of the doublet to a triplet configuration. The photo shows the rig drilling of the third well in 2022, to boost future production and add heat to the future energy mix of this horticultural region.

Photo courtesy of R. Vorage.

Reduction of pre-drill mining risk

To ramp up the deployment of geothermal resources, it has long been considered that the Netherlands stood to benefit effectively from the wealth of existing oil and gas data (worth over 50 billion euros in terms of original acquisition costs), combined with common practice methodologies from extensive R&D and industrial experience (Van Wees et al., 2012, 2017). However, in recent years it has become apparent that such an approach is significantly less than optimal. This conclusion relates largely to the following pre-drill stage challenges:

Data-poor areas: Geothermal operations in the Netherlands have been developed primarily in two areas from reservoirs that have been well-characterized from previous conventional geo-resource (mostly oil and gas) exploitation. To meet the targets of the Dutch Climate Agreement (Klimaatakkoord, 2019), however, the existing resource base needs to be extended to areas where the data density is relatively low, and/or where existing information indicates that reservoirs are not suitable for sustainable oil and gas production, but which may be suitable for geothermal production. In order to decrease the uncertainty and exploration risk, data resolution needs to be enhanced by reprocessing existing data and the acquisition of new data (SCAN data acquisition programme). In addition, unconventional geothermal resources in lithologies and at depth levels not so far studied in detail need to be considered, including Ultra Deep Geothermal (UDG) resources at a depth range of 4-8 km.

Unknown/less-known specific properties and processes: Geothermal energy exploitation involves knowledge of rock properties and reservoir processes that may be different from those relevant for conventional oil and gas exploitation. Such properties and processes relate, for instance, to diagenetic transformation of reservoir properties during geological history, to in-situ stress conditions and thermo-mechanical properties essential to the establishment of stress changes related to cooling that may have an impact on safety. Predictions of the lateral variability of properties (i.e. porosity, permeability, geochemical, mechanical, thermal etc) can benefit considerably from high resolution understanding of structural and sedimentation processes and the link with external forcing factors (e.g. tectonics, climate) at the time of deposition.

The play-based portfolio approach for unlocking Dutch geothermal energy (Van Wees et al., 2020) has facilitated the development of various nationally funded Public-Private Partnerships for the investigation of data-poor areas. One example is the MEACP funded SCAN project (SCAN, 2024) that focuses on reprocessing of vintage seismic data, acquisition of 2D reflection seismic lines and, drilling of ~10 scientific wells in an exploration campaign. Other examples include the LEAN project (discontinued) that aimed at developing the data-poor Utrecht area, the Ultra Deep Geothermal Project (UDG, 2018) and the EU funded DGE-rollout project which targets the lower Carboniferous Dinantian reservoir (DGE-Rollout, 2024). These efforts will greatly improve regional subsurface resolution and information in data-poor areas, create extensive new datasets, know-how and provide the basis for an active and lively research and development collaboration.

Induced seismicity

One of the hazards associated with altering the subsurface pressure or temperatures during subsurface mining activities (see section on induced seismicity and seal failure) is induced seismicity. Induced seismicity related to gas production is a prominent issue in the societal and policy domain. With the type of geothermal systems and plays in the Netherlands (predominantly matrix permeable reservoirs) the hazard and risks are, for now regarded as very low outside naturally or induced pre-stressed areas. However, to reduce the risks further a Seismic Hazard and Risk Analysis (SHRA) is prescribed prior to drilling and operation, together with a fit for purpose seismic monitoring and response (M&R) system. Periodic review and update of the SHRA and M&R systems is required, based on the data and experiences in the field, coupled with scientific progress in the field of understanding and modelling seismicity.

Enhancing technical and environmental performance

Heat demand is sometimes located in places where due to poor flow performance known reservoirs cannot be developed by conventional techniques, (so-called marginal reservoirs). In addition, there is growing concern related to environmental safety and sustainability of geothermal developments, for instance where induced seismicity may result in early suspension or failure of system performance. For such situations alternative development approaches to deliver affordable energy need to be explored. At the same time there is a need to augment the environmental robustness of the development strategy in the face of pre-drill uncertainty of in-situ stress and fluid chemistry. The WarmingUP programme (19 million euros), part of the Multiple-year Mission-driven Innovation Program (MMIP) and the EU cofunded Geothermica project RESULT (NL budget 12 million euros) are focused on these subjects.

Alternative well designs

The productivity/injectivity of a well is directly related to the in- or out-flow surface area of the well-reservoir interface. In conventional wells this is the surface area of the borehole cylinder through the reservoir; the larger the better. To increase the inflow area, wells can be deviated to horizontal well trajectories through the reservoir. Another option is to drill multi-laterals or horizontal wells, a technique used extensively in the oil and gas industry and which potentially can be implemented in geothermal drilling (Ungemach et al., 2019). Similarly, the well-reservoir interface can be increased by stimulating the reservoir (see next section) or alternatively by radial jetting (Den Boogert, 2018; Reinsch et al., 2021). Implementation of these techniques varies from being already in execution to the design phase of a demonstration project, such as planned in the European Geothermica project RESULT.

Stimulation

The pressure required to drive flow in the reservoir is largely dependent on the permeability in the near vicinity of the wellbore, which can be improved by hydraulic or chemical stimulation (Gaucher et al., 2015; Maurer et al., 2020). Hydraulic and chemical stimulation has been proven successful in enhancing the technical performance of geothermal production systems in fractured reservoirs (Wassing et al., 2014, 2021; Gaucher et al., 2015; Maurer et al., 2020), but it also increases the risk of induced seismicity, which may result in a suspension or abortion of the development (e.g. Häring et al., 2008). In the Netherlands, neither hydraulic nor extensive chemical stimulation has been pursued yet.

Table 19.2. Definitions pertaining to geothermal resources used in this paper.

Term	Definition / Description
Geothermal power	The 'installed' or 'nameplate' capacity of a geothermal system. Power has the dimension of Watt (W) or Joule per second (J/s) and is generally reported in terms of MegaWatt (MW).
Geothermal play	A geothermal play can be defined as the area where a certain type of geothermal reservoir is present which can be described and evaluated in a uniform and reproducible manner and where data and knowledge is acquired that can be used to evaluate geothermal prospects in the same play resulting in reduction of the 'exploration' uncertainty of that prospect. Within the Netherlands two play types can be defined using the dominant permeability system as diagnostic criterium: - The dominant play is the matrix permeability dominated 'Hot Sedimentary Aquifer', 'Hydrothermal' or 'Intra-cratonic Conductive play' (Mijnlieff 2020). - The fault/fracture permeability dominated 'Hot Sedimentary Aquifer', 'Hydrothermal' or 'Intra-cratonic Conductive play'.
Geothermal project	A project is a defined development or geothermal operation which provides the basis for economic evaluation and decision-making. In the early stages of evaluation, including exploration, the project might be defined only in conceptual terms, whereas more mature projects will be defined in significant detail. A geothermal project comprises one or more geothermal systems. It is a term with an economic connotation; generally an incentive to harvest geothermal heat from the subsurface through one or more geothermal systems linked to a company policy decision and related investment decision.
Geothermal field	"A geothermal field in the 'Hot Sedimentary Aquifer play' is defined as a distinct (mappable) subsurface area where a porous and permeable reservoir is present bounded at top and base by an impermeable interval (seal/aquitard) and is laterally bounded by impermeable zones like reservoir pinch-outs or faults with adequate off-set. Essentially, it would be a subsurface domain in which there is sufficient pressure communication and if exploited with multiple geothermal systems these systems would interfere with each other in a positive or negative manner" (from Mijnlieff, 2020).
Geothermal system	"The geothermal system is the link between the geothermal source, the amount of heat in the ground, and the geothermal resource, the amount of heat which can be delivered to the user. In the Dutch context, most geothermal systems to date are doublets but at status date 1-1-2022 also some triplets are operational. A doublet consists of a production and injection well, both completed in the same reservoir (the geothermal source) at a subsurface distance of up to 1.5 km (Figure 19.8). At surface the two wells are linked via a heat-exchanger, which captures the heat, the geothermal product, from the primary, salt water loop and transfers it to the secondary, fresh water loop." (adapted from Mijnlieff, 2020). The triplets have one production well and two injection wells.
Geothermal Energy Resources (UNFC 2016)	Geothermal Energy Resources are the cumulative quantities of Geothermal Energy Products that will be extracted from the Geothermal Energy Source, from the Effective Date of the evaluation forward, measured or evaluated at the Reference Point. Quantity of energy in the form of heat which are typically reported in Joule (J) or kiloWatt hours (kWh).
Geothermal Energy Source (UNFC 2016)	In the geothermal context, the Renewable Energy Source is the thermal energy contained in a body of rock, sediment and/or soil, including any contained fluids, that is available for extraction and conversion into energy products. This source is termed the Geothermal Energy Source and is equivalent to the terms 'deposit' or 'accumulation' used for solid minerals and fossil fuels. The Geothermal Energy Source results from any influx to, outflux from or internal generation of energy within the system over a specified period of time.

Closed loop systems

The present geothermal systems in the Netherlands are typically open systems, in which the working fluid, the geothermal water, is extracted from one well and injected in another well in the same deep reservoir flowing in the subsurface through a permeable rock. Closed systems are gaining interest as they do not interact directly with the subsurface. In general, closed loop systems are a development option for non-permeable, tight rock at any depth and/or at shallower depth where interference with groundwater is to be avoided. In a closed loop system the working fluid (energy carrier) remains in the well, and heat transfer is by conduction only, as is the method of choice for shallow Borehole Thermal Energy Systems (BTES). Deep closed loop systems have also been developed as workovers for deep (hydrocarbon) boreholes, marked by a coaxial design (Kohl et al., 2002). Although the performance for such a coaxial system is significantly lower than for a doublet in some cases it may be an attractive alternative to the abandonment of wells (Lokhorst & Van Wees, 2005). Alternatively, the subsurface heat exchanger design in a closed loop system can be significantly enhanced by one or multiple connecting laterals (Holmes et al., 2021). In 2019 a pilot project in Canada by EAVOR technically proved the concept (TNO, 2021), and economic economic feasibility is under investigation.

High temperature heat storage

High temperature heat storage in a relatively shallow reservoir (in the interval between 200 and approximately 1000 m depth) is a new technique for the Netherlands. This technique is well suited for balancing the swing in winter-summer heat demand. As such it can increase the applicability and thus potential of geothermal in especially the built environment.

Metal harvesting from geothermal fluid

The geothermal production water contains metal ions that are often of high value. For example, lithium is present in most geothermal brines in the Netherlands in low concentrations of some tens of milligrams per litre. Worldwide tests have been performed to extract lithium from geothermal brines but none has yet evolved to a sustainable commercial development. With the high demand for lithium for batteries, renewed interest in lithium extraction from geothermal brine has stimulated various research projects, amongst which the European funded EuGeLi project (EuGeLi, 2019). In this project battery-grade, lithium carbonates were produced from geothermal brines from the Soultz-sous-Forêts geothermal plant in France (Jungmann et al., 2025).

Acknowledgements

We thank Serge van Gessel in providing the basic script for Figure 19.5. We also thank Ivan Das for discussions on the financial part of geothermal projects. The paper benefitted a lot from the suggestions and remarks of the reviewers Barbara Cox, Hans Veldkamp and Harry Doust.

Digital map data

Spatial data of figures in this chapter for use in geographical information systems can be downloaded here: https://doi.org/10.5117/aup.28164332.

References

Baisch, S., Koch, C., Stang, H., Pittens, B., Drijver, B. & Buik, N., 2016. Defining the Framework for Seismic Hazard Assessment in Geothermal Projects V0.1. Kennisagenda Technical Report 161005: 65 pp.

Baisch, S. & Vörös, R., 2018. Earthquakes near the Californie Geothermal site: August 2015-November 2018. Qcon GmbH, archive no. CLGG005: 41 pp.

Békési, E., Struijk, M., Bonté, D., Veldkamp, H., Limberger, J., Fokker, P.A., Vrijlandt, M. & van Wees, J.-D., 2020. An updated geothermal model of the Dutch subsurface based on inversion of temperature data. Geothermics 88: 101880. DOI: 10.1016/j.geothermics.2020.101880

Bonté, D., Van Wees, J.-D. & Verweij, J.M., 2012. Subsurface temperature of the onshore Netherlands: new temperature dataset and modelling. Netherlands Journal of Geosciences 91 (4): 491-515. DOI: 10.1017/S0016774600000354

Breede, K., Dzebisashvili, K. & Falcone, G., 2015. Overcoming challenges in the classification of deep geothermal potential. Geothermal Energy Science 3: 19-39. DOI: 10.5194/gtes-3-19-2015

Breunese, J.N., Andersen, J.H., Brinkman, S., Jagosiak, P., Karnin, W.-D., Karnkowski, P.H., Kombrink, H., Messner, J., Mijnlieff, H., Olsen, S.B., Peryt, T.M., Piske, J., Poprawa, P., Roelofsen, J.W., Stoker, S.J., Smith, N.J.P., Swann, G., Waksmundzka, M.I. & Veldkamp, J.G., 2010. Reserves and production history. In: Doornenbal, J.C. & Stevenson, A.G. (eds): Petroleum Geological Atlas of the Southern Permian Basin Area. EAGE Publications b.v. (Houten): 270-281.

Buijze, L., van Bijsterveldt, L., Cremer, H., Paap, B., Veldkamp, H., Wassing,, B.B.T., van Wees, J-D., van Yperen, G.C.N., ter Heege, J.H. & Jaarsma, B., 2019. Review of induced seismicity in geothermal systems worldwide and implications for geothermal systems in the Netherlands. Netherlands Journal of Geosciences 98, e13. DOI: 10.1017/njg.2019.6

Buijze, L., Fokker, P.A. & Wassing, B.B.T., 2022. Quantification of induced seismicity potential of geothermal operations: Analytical and numerical model approaches. Scientific Report Published in the framework of WarmingUp Theme 4B, TNO (Utrecht): 112 pp.

Buijze, L., Veldkamp, H. & Wassing, B., 2023. Comparison of hydrocarbon and geothermal energy production in the Netherlands: reservoir characteristics, pressure and temperature changes, and implications for fault reactivation. Netherlands Journal of Geosciences 102: e7. DOI: 10.1017/njg.2023.6

CBS & TNO, 2020. Warmtemonitor 2019. https://www.cbs.nl/nl-nl/achtergrond/2020/35/warmtemonitor-2019

De Jager, J., Van Ojik, K. & Smit, J., 2025. Geological development. In: Ten Veen, J.H., Vis, G.-J., De Jager, J. & Wong, Th.E. (eds): Geology of the Netherlands, second edition. Amsterdam University Press (Amsterdam): 21-51. DOI: 10.5117/9789463728362_ch01

Den Boogert, R., 2018. Testing the applicability of radial jetting for Dutch geothermal projects. Extending the geothermal potential in the Netherlands with the help of radial jetting. For the Dutch Knowledge Agenda/Kennisagenda Aardwarmte: 22 pp.

DGE-Rollout, 2024. DGE-ROLLOUT - Roll-out of Deep Geothermal Energy in NWE. https://vb.nweurope.eu/projects/project-search/dge-rollout-roll-out-of-deep-geothermal-energy-in-nwe/

Dijkstra, H., Dinkelman, D., Hanegraaf, M., Veldkamp, H. & Van Wees, J.-D., 2020. White Paper Duurzaamheid van Geothermie in Warmtenetten, TNO (Utrecht): 36 pp.

DoubletCalc, 2014. DoubletCalc v1.4.3 software. 17 December 2014 version, including Java 7 script. https://www.nlog.nl/en/tools.

Dufour, F.C. & Heederik, J.P., 2019. Early Geothermal Exploration in the Netherlands 1980 - 2000. Proceedings European Geothermal Congress (The Hague, the Netherlands).

Dumas, P., Garabetian, T., Le Guénan, T., Kępińska, B., Kasztelewicz, A., Karytsas, S., Siddiqi, G., Lupi, N., Syidov, F., Nador, A., Kaufhold, J., Boissavy, C., Yildirim, C., Bozkurt, C., Kujbus, A., Spyridonos, E., Oztekin, R. & Link, K., 2019. Risk Mitigation and Insurance Schemes Adapted to Geothermal Market Maturity: The Right Scheme for my Market. Proceedings European Geothermal Congress (The Hague, the Netherlands).

EBN & GNL, 2021. Integraal Kostprijsreductie Programma Aardwarmte. Whitepaper status december 2021: 27 pp. https://www.ebn.nl/wp-content/uploads/2022/10/Whitepaper-Integraal-Kostprijsreductie-Programma-Aardwarmte-2021-2.pdf

EuGeLi, 2019. EuGeLi project. https://www.eramet.com/en/activities/innovate-design/eugeli-project

Fokker, P.A., Van der Veer, E.F. & Van Wees, J.-D., 2015. Surface movement induced by a geothermal well doublet. World Geothermal Congress (Melbourne, Australia).

Gaucher, E., Schoenball, M., Heidbach, O., Zand, A., Fokker, P.A., Van Wees, J.-D. & Kohl, T., 2015. Induced seismicity in geothermal reservoirs: A review of forecasting approaches. Renewable and Sustainable Energy Reviews 52: 1473-1490. DOI: 10.1016/j.rser.2015.08.026

Geothermie Nederland, 2021. Industriestandaard Duurzaam putontwerp - factsheet. https://geothermie.nl/images/bestanden/20210315_Industriestandaard_Duurzaam_Putontwerp-final_GNL_verkort.pdf

Gies, C., Struijk, M., Békési, E., Veldkamp, H. & Van Wees, J.-D., 2021. An effective method for paleo-temperature correction of 3D thermal models: A demonstration based on high resolution datasets in the Netherlands. Global and Planetary Change 199: 103445. DOI: 10.1016/j.gloplacha.2021.103445

Gonzalez, L., 2017. Research sustainability Geothermal Wells in the Netherlands, Energy pay-back time, Life cycle analysis and social cost benefit analysis. https://www.kasalsenergiebron.nl/content/user_upload/Presentatie_171208_Advies_in_Duurzaamheid_-_Energetische_terugverdientijd.pdf

Gudmundsson, J.S., Freeston, D.H. & Lienau, P.J., 1985. The Lindal Diagram. Geothermal Resources Council Transactions 9 Part I: 151-156.

Haagse aardwarmte, 2021. FAQ page. https://haagseaardwarmte.nl

Häring, M.O., Schanz, U., Ladner, F. & Dyer, B.C., 2008. Characterisation of the Basel 1 enhanced geothermal system. Geothermics 37 (5): 469-495. DOI: 10.1016/j.geothermics.2008.06.002

Hartog, N., 2016. Risico's van Geothermie voor Grondwater BTO rapport. KWR Watercycle Research Institute, Report No. BTO 2016.077.

Holmes, M., Toews, M., Jenkins, J. & Sepulveda, N., 2021. Multilateral Closed-Loop Geothermal Systems as a Zero-Emission Load-Following Resource. Using the Earth to Save the Earth - 2021 Geothermal Rising Conference, Transactions - Geothermal Resources Council 45: 25-52.

Jungmann, M., Walter, B.F., Eiche, E., Giebel, R.J. & Kolb, J., 2025. The source of lithium in connate fluids: Evidence from the geothermal reservoir at Soultz-sous-Forêts, Upper Rhine Graben, France. Journal of Geochemical Exploration 270: 107641. DOI: 10.1016/j.gexplo.2024.107641

Kivi, I.R., Pujades, E., Rutqvist, J. & Vilarrasa, V., 2022. Cooling-induced reactivation of distant faults during long-term geothermal energy production in hot sedimentary aquifers. Scientific Reports 12 (1): 2065. DOI: 10.1038/s41598-022-06067-0

Klimaatakkoord, 2019. https://www.klimaatakkoord.nl/documenten/publicaties/2019/06/28/klimaatakkoord

Kohl, T., Brenni, R. & Eugster, W., 2002. System performance of a deep borehole heat exchanger. Geothermics 31 (6): 687-708. DOI: 10.1016/S0375-6505(02)00031-7

Kramers, L., Van Wees, J.-D., Pluymaekers, M.P.D., Kronimus, A. & Boxem, T., 2012. Direct heat resource assessment and subsurface information systems for geothermal aquifers; the Dutch perspective. Netherlands Journal of Geosciences 91 (4): 637-649. DOI: 10.1017/S0016774600000421

Limberger, J., Calcagno, P., Manzella, A., Trumpy, E., Boxem, T., Pluymaekers, M. & Van Wees, J.-D., 2014. Assessing the prospective resource base for enhanced geothermal systems in Europe. Geothermal Energy Science 2 (1): 55-71. DOI: 10.5194/gtes-2-55-2014

Lokhorst, A. & Van Wees, J.-D., 2005. Technical and economical feasibility of a deep Borehole Heat Exchanger (DBHE) in LTG-01. TNO (Utrecht), Report No. NITG 05-068-B510. 05-06: 510 pp.

Lokhorst, A. & Wong, Th.E., 2007. Geothermal Energy. In: Wong, Th.E., Batjes, D.A.J. & De Jager. J. (eds): Geology of the Netherlands. Royal Netherlands Academy of Arts and Sciences (KNAW, Amsterdam): 341-346.

Maurer, V., Gaucher, E., Grunberg, M., Koepke, R., Pestourie, R. & Cuenot, N., 2020. Seismicity induced during the development of the Rittershoffen geothermal field, France. Geotherm Energy 8: 1-5. DOI: 10.1186/s40517-020-0155-2

MEACP, 2018. Beleidsbrief Geothermie. https://open.overheid.nl/documenten/ronl-0778844c-5a07-4b2e-b77a-0026152a825d/pdf

MEACP, 2021. Reactie op toezichtsignaal Integriteit van het geothermisch reservoirs. https://www.sodm.nl/sectoren/geothermie/documenten/brieven/2021/01/19/toezichtsignaal-integriteit-afsluitende-laag-geothermie.

MEACP, 2022. Natural resources and geothermal energy in the Netherlands, 2021 Annual review. An overview of exploration, production and underground storage. https://www.nlog.nl/

sites/default/files/2022-07/annual_report_2021_natural_resources_and_geothermal_energy.pdf

Mijnlieff, H.F., 2020. Introduction to the geothermal play and reservoir geology of the Netherlands. Netherlands Journal of Geosciences 99(2): 1-19. DOI: 10.1017/njg.2020.2

Mijnlieff, H., Ramsak, P., Lako, P., Smeets, J., In 't Groen, B. & Veldkamp, H., 2013. Geothermal energy and support schemes in The Netherlands. Proceedings European Geothermal Congress: 1-8

Mijnlieff, H.F., Obdam, A.N.M., Van Wees, J.-D., Pluymaekers, M.P.D. & Veldkamp, J.G., 2014. DoubletCalc 1.4 manual, English version for DoubletCalc 1.4.3. TNO Report R11396; 54 pp.

Mijnlieff, H., Van Kempen, B., Tolsma, S., De Vries, C., Martins, J.E., Veldkamp, H., Struijk, M., Vrijlandt, M. & Van Wees, J.-D., 2019. Dutch geothermal resource reporting. A first attempt of Dutch nationwide geothermal resource using the UNFC resource classification system. Status date January 2019. Proceedings European Geothermal Congress (The Hague, the Netherlands).

Mijnlieff, H., Van Kempen, B., Tolsma, S., De Vries, C., Esteves Martins, J., Veldkamp, H., Struijk, M., Vrijlandt, M. & Van Wees, J.-D., 2020. The Dutch Geothermal Resource Base: Classified Using UNFC Resource Classification System and its Potential to meet The Dutch Geothermal Ambition. Proceedings World Geothermal Congress 2020+1 (Reykjavik, Iceland).

Moeck, I.S., 2014. Catalog of geothermal play types based on geologic controls. Renewable and Sustainable Energy Reviews 37: 867-882. DOI: 10.1016/j.rser.2014.05.032

Muffler, P. & Cataldi, R., 1978. Methods for regional assessment of geothermal resources. Geothermics 7 (2-4): 53-89. DOI: 10.1016/0375-6505(78)90002-0

Muntendam-Bos, A.G., Hoedeman, G., Polychronopoulou, K., Draganov, D., Weemstra, C., Van der Zee, W., Bakker, R.R. & Roest, H., 2021. An overview of induced seismicity in the Netherlands. Netherlands Journal of Geosciences 101: e1. DOI: 10.1017/njg.2021.14

Pluymaekers, M.P.D., Kramers, L., Van Wees, J.-D., Kronimus, A., Nelskamp, S., Boxem, T. & Bonté, D., 2012. Reservoir characterisation of aquifers for direct heat production: Methodology and screening of the potential reservoirs for the Netherlands. Netherlands Journal of Geosciences 91 (4): 621-636. DOI: 10.1017/S001677460000041X

Reinsch, T., Blöcher, G., Bruhn, D., Zotzmann, J., Wittig, V., Thorbjornsson, I.O., Kaldal, G., Hoogland, K., Peters, E., Latham, J.-P., Petrauskas, S., Šliaupa, S. & Nick, H., 2021. Novel Productivity Enhancement Concept for a Sustainable Utilization of a Geothermal Resource -The H2020 SURE Project. Proceedings World Geothermal Congress 2020+1 (Reykjavik, Iceland).

Rogers, E.M., 1962. Diffusion of Innovations. The Free Press of Gleneoe: 367 pp.

Roholl, J.A., Brunner, L.G., Versteijlen, N., Hettelaar, J. & Wilpshaar, M., 2021. Deterministische hazard analyse voor geïnduceerde seismiciteit. TNO (Utrecht), Rapport No. R10977: 41 pp.

RVO, 2021. Stimulering Duurzame Energieproductie en Klimaattransitie (SDE++). https://www.rvo.nl/subsidies-financiering/sde

SCAN, 2024. SCAN Aardwarmte. https://scanaardwarmte.nl.

SodM, 2013. Protocol bepaling maximale injectiedrukken bij aardwarmtewinning – versie 2. Staatstoezicht op de Mijnen (Den Haag): 2 pp.

SodM, 2017.Staat van de Sector Geothermie. Staatstoezicht op de Mijnen (Den Haag): 25 pp.

SodM, 2019. Protocol ter bepaling maximale injectiedrukken bij aardwarmtewinning - versie 3. Staatstoezicht op de Mijnen (Den Haag): 3 pp.

SodM, 2021. Evaluatie aanbevelingen Staat van de Sector Geothermie. Staatstoezicht op de Mijnen (Den Haag): 32 pp.

Stafleu, J., Busschers, F.S., Van der Meulen, M.J., Den Dulk, M., Gunnink, J.L., Maljers, D., Hummelman, J.H., Schokker, J., Vernes, R.W., Stam, J., Dabekaussen, W., Ten Veen, J.H., Doornenbal, H., Kars, R. & De Bruijn, R., 2025. Geomodelling. In: Ten Veen, J.H., Vis, G.-J., De Jager, J. & Wong, Th.E. (eds): Geology of the Netherlands, second edition. Amsterdam University Press (Amsterdam): 849-893. DOI: 10.5117/9789463728362_ch24

Stichting Platform Geothermie, Stichting WarmteNetwerk, DAGO & EBN, 2018. Masterplan Aardwarmte in Nederland. Een brede basis voor een duurzame warmtevoorziening. https://www.ebn.nl/feiten-en-cijfers/kennisbank/het-masterplan-aardwarmte-nederland/

Ten Veen, J. H., Vis, G.-J., De Jager, J. & Wong, Th. E. (eds), 2025. Geology of the Netherlands, second edition. Amsterdam University Press (Amsterdam): 909 pp. DOI: 10.5117/9789463728362

ThermoGIS., 2019. ThermoGIS V2.1. https://www.thermogis.nl/en

TNO, 2013. Geothermal energy – Scaling potential with cooling and CO2 degassing. TNO (Utrecht), Report No. R11661R11661: 19 pp.

TNO, 2014a. Bepaling begrenzing Winningsvergunning aardwarmte. TNO (Utrecht), Report No. AGE 14-10.050: 11 pp.

TNO, 2014b. Lead deposition in geothermal installations. TNO (Utrecht), Report No.R11416: 49 pp.

TNO, 2016. Corrosion in Dutch geothermal systems. TNO (Utrecht), Report No. R10160: 105 pp.

TNO, 2021. Eavor Loop Audit Report. TNO (Utrecht), Report No. R11549: 18 pp.

UDG, 2018. C-217 Green Deal ultradiepe geothermie: 11 pp. https://www.greendeals.nl/sites/default/files/downloads/GD217-dealtekst-Ultradiepe-Geothermie.pdf

UN, 2015. Paris Agreement: 27 pp. https://unfccc.int/files/essential_background/convention/application/pdf/english_paris_agreement.pdf

UNECE, 2016. Specifications for the application of the United Nations Framework Classification for Fossil Energy and Mineral Reserves and Resources 2009 (UNFC-2009) to Geothermal Energy Resources: 28 pp.

UNECE, 2019. United Nations Framework Classification for Resources. Update 2019. ECE ENERGY SERIES No. 61: 28 pp.
Ungemach, P., Antics, M. & Davau, M., 2019. Subhorizontal well architecture and geosteering navigation enhance well performance and reservoir evaluation. A field validation. Proceedings, 44th Workshop on Geothermal Reservoir Engineering.
Van Wees, J.-D., Kronimus, A., Van Putten, M., Pluymaekers, M.P.D., Mijnlieff, H., Van Hooff, P., Obdam, A. & Kramers, L., 2012. Geothermal aquifer performance assessment for direct heat production – Methodology and application to Rotliegend aquifers. Netherlands Journal of Geosciences 91 (4): 651-665. DOI:10.1017/S0016774600000433
Van Wees, J.-D., Pluymaekers, M., Bonté, D., Van Gessel, S. & Veldkamp, H., 2017. Unlocking Geothermal Energy from Mature Oil and Gas Basins: A Success Story from the Netherlands. In: Şimşek, Ş. (ed.): Perspectives for Geothermal Energy in Europe 1: 187-214. DOI: 10.1142/9781786342324_0007
Van Wees, J.-D., Pluymaekers, M., Osinga, S., Fokker, P.A., Van Thienen-Visser, K., Orlic, B., Wassing, B., Hegen, D. & Candela, T., 2019. 3D Mechanical Analysis of Complex Reservoirs: a novel mesh-free approach. International Geophysical Journal 219 (2): 1118-1130. DOI: 10.1093/gji/ggz352
Van Wees, J.-D., Veldkamp, H., Brunner, L., Vrijlandt, M., De Jong, S., Van Langen, C., Heijnen, N. & Peijster., J., 2020. Accelerating geothermal development with a play-based portfolio approach. Netherlands Journal of Geosciences 99: e5. DOI: 10.1017/njg.2020.4
Verhoeven, R., Willems, E., Harcouët-Menou, V., De Boever, E., Hiddes, L., Op 't Veld, P. & Demollin, E., 2014. Minewater 2.0 Project in Heerlen the Netherlands: Transformation of a Geothermal Mine Water Pilot Project into a Full Scale Hybrid Sustainable Energy Infrastructure for Heating and Cooling. Energy Procedia 46: 58-67. DOI: 10.1016/j.egypro.2014.01.158
Visser, W.A. & Heederik, J.P., 1987. Geothermal Energy. In: Visser, W.A., Zonneveld, J.I.S. & Van Loon, A.J. (eds): Seventy-five years of geology and mining in the Netherlands (1912-1987). Royal Geological and Mining Society of the Netherlands (KNGMG, The Hague): 243-258.
Vrijlandt, M.A.W., Struijk, E.L.M., Brunner, L.G., Veldkamp, J.G., Witmans, N., Maljers, D. & Van Wees, J.-D., 2019. ThermoGIS-v2.1 update: a renewed view on geothermal potential in the Netherlands. Proceedings European Geothermal Congress (The Hague, the Netherlands).
Wassing, B.B.T., Van Wees, J.-D. & Fokker, P.A., 2014. Coupled continuum modeling of fracture reactivation and induced seismicity during enhanced geothermal operations. Geothermics 52: 153-164. DOI: 10.1016/j.geothermics.2014.05.001

20

Underground storage

Joaquim Juez-Larré, Remco M. Groenenberg, Dorien Dinkelman, Thijs D. Huijskes, Joris M. Koornneef, Serge F. van Gessel, and Christian F.M. Bos.*

*Corresponding author: joaquim.juezlarre@tno.nl

ABSTRACT

The subsurface of the Netherlands has been a prolific source of raw materials and energy, key to the country's economic, technological and welfare development. However, due to the significant decline in domestic hydrocarbon production, especially natural gas, and despite the increase in geothermal projects, the subsurface is transitioning from being a provider of energy and resources to a provider of storage services, such as underground natural gas storage. This shift aligns with the current transition to renewable energy, increased energy imports, and global environmental and climate challenges.

Recent studies on the future energy system of the Netherlands point to the need for a (large) underground storage capacity, which can also facilitate the permanent storage of residues from energy production and conversion (e.g. wastewater and CO_2). This need arises from traditional variations in energy demand, such as seasonal heat demand, and the increased share of fluctuating energy production from renewable sources like wind and solar. Aboveground facilities can provide short-term storage, but the surface space required for mid-to-long-term storage would be extensive, making underground storage solutions essential.

In the Netherlands, there is a rich and broad knowledge of the subsurface derived from long-standing and large-scale exploration and production activities for natural resources. This knowledge allows for the potential use of large underground volumes in depleted oil and gas fields, aquifers, and engineered cavities. Coupled with ongoing research and development of various storage technologies and the growing need for flexibility services to balance the system and secure energy supply, this makes it feasible and attractive to store large amounts of energy (GWh to TWh) underground in the form of gas, liquid, heat, and mechanical power.

Current research focuses on identifying suitable geological formations on- and offshore and assessing the technical, economic, societal, and environmental feasibility for safe, efficient, and timely deployment of underground storage. Pilot and demonstration projects are crucial for scaling up and commercializing underground storage technologies (e.g. CO_2, hydrogen). It is important to realize that the successful integration of underground storage into the energy landscape requires informing and involving local communities and authorities at all stages of the project. Only in this way can underground storage contribute to a smooth transition towards a clean and climate-neutral future energy system.

<< The Porthos site in the Port of Rotterdam (Maasvlakte, west Netherlands) where CO_2 will be compressed and subsequently transported to an injection platform some 20 km off the coast. Photo: ©PorthosCO_2.

DOI: 10.5117/9789463728362_ch20

Introduction

Several chapters in this book describe the role of the Dutch subsurface as a source of natural resources such as fossil fuels, geothermal energy, groundwater, salt, aggregates and clay. With the traditional oil and gas exploration and production activities (E&P) in decline, one may conclude that the future role of the subsurface is dwindling, giving way to renewable energy sources that, with the exception of geothermal energy, are largely harvested above ground. This chapter addresses the important role the Dutch subsurface can play in large-scale underground storage of fuel-based energy carriers, mechanical energy, heat and CO_2, for the successful realization of energy transition ambitions and greenhouse gas emission-reduction targets.

Since the discovery of the giant Groningen gas field in 1959, and subsequent discoveries of a large number of other gas fields onshore and offshore (the so-called small fields), the Netherlands has had access to a relatively low-cost and reliable source of energy (Fig. 20.1). It has also been one of the largest energy exporters in Europe (Correljé et al., 2003; Breunese et al., 2005; Doornenbal et al., 2019). It was almost sixty years later, in 2018, that the Netherlands became a net importer of natural gas for the first time (Fig. 20.1). This was due to restrictions imposed on the natural gas production from the Groningen field related to induced seismicity and to the continued natural decline in domestic production from the small fields both on- and offshore (Van Geuns et al., 2017). Until that time, the unique swing production capacity of the Groningen field had provided most of the flexibility required by securing the seasonal balance between natural gas supply and demand, not only for the Netherlands, but also for parts of Germany and Belgium (Van den Beukel & Van Geuns, 2019). From the start of production restrictions for the Groningen field in 2014 to 2022, the percentage

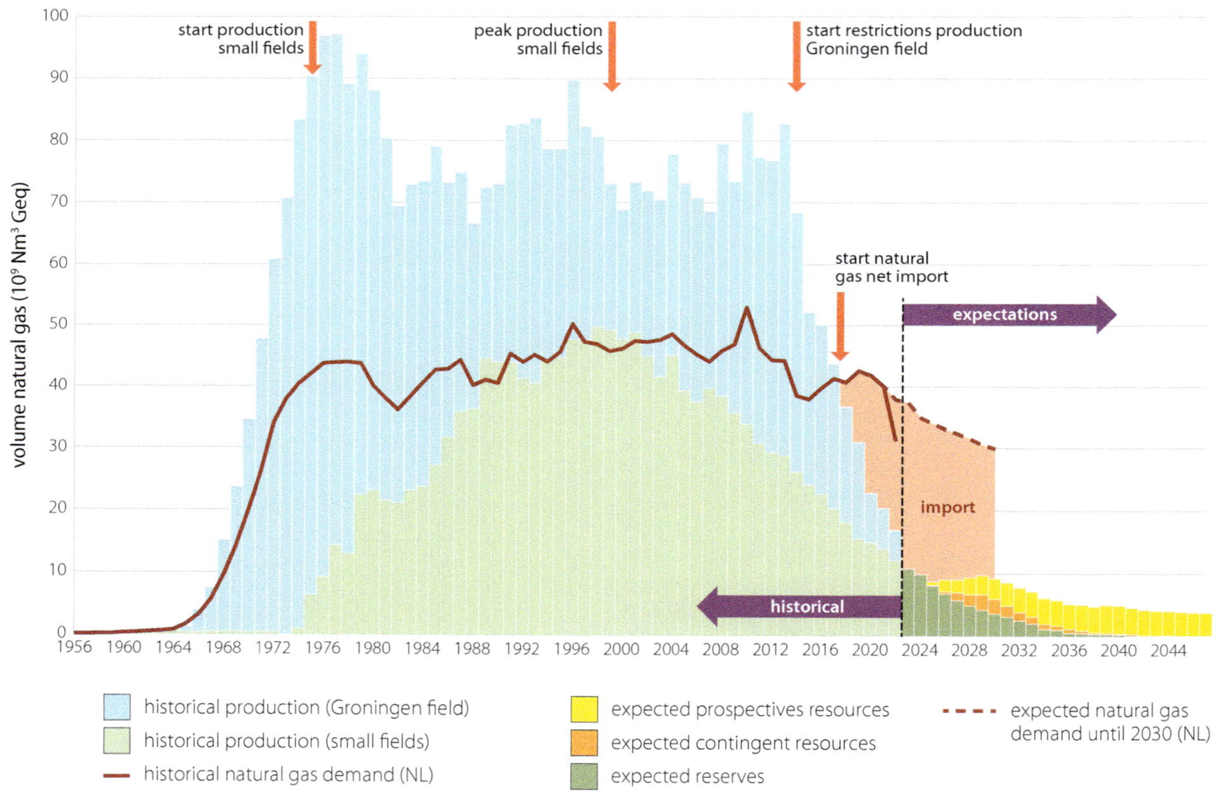

Figure 20.1. *History of domestically produced natural gas from the Groningen field (blue bars) and the Small Fields (light green bars) in the Netherlands. The corresponding expected domestic natural gas production from the Small Fields is shown in three categories (green/orange/yellow bars; MEACP, 2023) and estimated according to the international Petroleum Resources Management System (PRMS, 2018). The historical and expected demand of natural gas in the Netherlands is represented by a red line (www.pbl.nl/kev). Due to high natural gas prices in 2022, significantly less natural gas was consumed than predicted. It is difficult to make new forecasts for natural gas demand due to the current high uncertainty (www.pbl.nl/kev). All types of natural gas are converted to the Groningen equivalent (Geq) based on their gross-heating-value (GHV). The Groningen gas has a GHV of 35.17 MJ/Nm³ (Megajoules per normal cubic metre; 'Normal' refers to the reference conditions: 0°C and 101.325 kPa).*

of natural gas stored in the Netherlands relative to annual gas demand increased from 5-10% to ~32% (CBS, 2022). Currently, natural gas is stored in five large underground storage sites, which together have a maximum storage capacity of ~14 x 10^9 m^3, corresponding to 140 TWh (agsi.gie.eu, gas.kyos.com).

The 2021 annual consumption in the Netherlands clearly shows the still high dependence on hydrocarbons, with an annual primary energy consumption of 840 TWh (3024 PJ), the majority of which comes from natural gas (44%) and oil (37%), and the rest from renewables (10%), coal (6%), nuclear energy (1%) and others (2%). Although total domestic consumption of natural gas has been relatively constant at around 40 x 10^9 m^3 (390 TWh) for the past 50 years, there is a large uncertainty about how fast consumption will decline in the future (Fig. 20.1; www.pbl.nl/kev). Following the decision to completely close the Groningen field (MEACP, 2020, 2022b, 2024), the need to continue matching natural gas supply and demand in the Netherlands will depend for some time on the use of the current underground storage capacity, remaining production from small gas fields, reduction in domestic consumption, imports from other countries and the use of other energy sources such as coal-fired and nuclear power plants. And all this will be needed before natural gas can be partly replaced by renewable energy sources (MEACP, 2019b; ENCO, 2020; Van den Beukel & Van Geuns, 2020; Patrahau & Van Geuns, 2021; MEACP, 2022a; Van den Berg et al., 2022). With the increase in electrification in many sectors of our society and the plans set out in the Dutch climate agreement, in which solar and wind power are expected to grow significantly producing up to 70% of electricity by 2030 (MEACP, 2019a), this transformation represents the most significant restructuring of the Dutch energy system since the introduction of natural gas in the 1960s. Today about 71% of the natural gas consumed is still used for the production of electricity and for heating purposes (CBS, 2022).

With underground storage of natural gas gaining more importance today due to the developments concerning security of gas supply in Europe (Patrahau et al., 2022), there are three main questions about the future need for any type of underground (energy) storage in the Netherlands. First, what will be the expected volume and type of storage needed (short cycle, seasonal and/or strategic) for commodities such as heat, electricity and any alternative energy carriers, in addition to natural gas? Second, to what extent will underground storage be integrated with other aboveground energy storage solutions and flexibility options at regional, national and international level? And how will this interfere with other subsurface activities? (e.g. Carbon Capture and Storage, oil and gas production, geothermal). Third, how will the life cycle of underground storage sites be assessed and monitored to ensure their timely, safe and responsible development, operation and decommissioning to guarantee societal acceptance?

These and other questions are currently being addressed by many (research) groups and studies (e.g. Afman & Rooijers, 2017; EBN & Gasunie, 2018; Van Gessel et al., 2018; Bos, 2019; Berenschot & Kalavasta, 2020; EP, 2020; Groenenberg et al., 2020a; Scheepers et al., 2020; Sijm et al., 2020; Winters et al., 2020; Eising, 2021; Guidehouse & Berenschot, 2021; Van Gessel et al., 2021; Mendrinos et al., 2022; NSE, 2022; Sprenkeling et al., 2022; Van Gessel et al., 2022; www.gasopslagnederland.nl). The results of some of these scenario studies show a wide range in the estimated future storage capacities that may be required, depending on how energy flexibility solutions are implemented and integrated into the energy system of the future. They also take into account uncertainties about the amount of renewable generation capacity, CO_2-free production capacity, demand management and interconnection capacities between the European and global energy markets.

The subsurface can be key in providing capacities for storage of electricity and heat at different time scales and volumes. Electricity from intermittent solar and wind production can be converted into mechanical energy, hydrogen or derivatives and be stored in large volumes underground. In addition, underground heat storage from power-to-heat, waste heat, geothermal and solar thermal energy can help reduce natural gas consumption and speed up the energy transition in the heating sector. Besides energy storage, the subsurface can also be used for the cyclic storage of feedstocks (e.g. nitrogen) and the permanent storage (disposal) of by-products such as waste water, radioactive waste and CO_2. Underground disposal of radioactive solid waste is not discussed here, as it is discussed in Neeft et al. (2025, this volume).

Like other countries with a long history of hydrocarbon exploration and production, the Netherlands has the advantage of having an extensive source of data and knowledge of the subsurface, much of which is publicly available (www.nlog.nl). Over the past century, a large amount of well data (e.g. well logs and cores, as well as pressure and temperature measurements) and 2D- and 3D-seismic data have been collected, enabling the mapping and characterization of a wide range of geological formations. This knowledge has opened the possibility to investigate the potential for deploying underground storage technologies. Despite significant advances have been made in the energy sector since the publication of the previous edition of this chapter (Bos, 2007), only the large potential of underground storage of natural gas is being fully exploited today (Juez-Larré et al., 2016). However, the storage potential of the Dutch subsurface is very large and further deployment will not be limited by the subsurface itself, but by other

Qué será, será… when two kings meet and the energy transition in the Netherlands

On June 14th 2023, His Majesty King Willem-Alexander of the Netherlands paid a two-day visit to His Majesty King Felipe VI of Spain. The King travelled to Andalusia together with Rob Jetten, Minister for Climate and Energy Policy and a group of representatives of Dutch companies and knowledge institutes. In essence, the reason for this trip was the same as that of many of the 3.7 million Dutch tourists that visit Spain every year: the Spanish sun. The King, however, did not enjoy the sun on the white sandy beaches of the Costa del Sol, but instead together with King Felipe visited one of the solar/wind parks that will become part of the largest green hydrogen factory in Europe. Green hydrogen is likely to be an important energy carrier in the future energy system. With this historic visit, the Netherlands confirmed the ambitions of the Spanish energy company Cepsa to export green hydrogen to the Port of Rotterdam. The production plant and the hydrogen export route are expected to be operational by 2026-2027. The transport of hydrogen will take place in the form of ammonia through an armada of large vessels.

Cepsa has signed a collaboration agreement with Yara Clean Ammonia, a Norwegian multinational world leader in the field of green ammonia and with Gasunie, the Dutch national gas company and leader in gas transportation and infrastructure that connects the Port of Rotterdam with other European industrial clusters. This collaboration offers the Netherlands the possibility to create new import-export corridors for green energy to help secure energy supply in the current complex geopolitical situation around oil and natural gas import in Europe.

For the development of new sustainable energy markets such as that of hydrogen, the Netherlands must establish international collaboration to exchange knowledge, stimulate supply and demand, and build infrastructure for the production, import, transportation, storage and distribution of green energy. With the presence of King Willem-Alexander, the Netherlands shows a strong commitment to the green hydrogen revolution not only internationally but also nationally. This is why the King also attends important events on green hydrogen in the Netherlands. He presided over the inauguration of the first salt cavern storage test facility for green hydrogen in Zuidwending (HyStock project) in 2019, and spoke at the 'Wind meets gas' conference in 2021 on the importance of hydrogen for the Netherlands and the energy transition. He also visited students working with hydrogen at TU Delft (2021), the hydrogen cluster at the industrial park Kleefse Waard in Arnhem, the Port of Rotterdam and various green hydrogen projects in the North Sea (2022) including PosHYdon, a pilot project for offshore hydrogen production from wind power. Currently it is hard to say how fast the energy transition will take place in the Netherlands, as the current dependence on fossil fuels remains high (87%). We all hope for a swift and smooth passage, but just like in the famous song by Doris Day "Qué Será, Será, Whatever Will Be, Will Be, but is the (energy) future's not ours to see? Qué Será, Será...

Ammonia tanker Yara Nauma with, among others, their Majesties the King of Spain, Felipe VI, and the King of the Netherlands, Willem-Alexander, Mr. Maarten Wetselaar (CEO Cepsa), Ms. Monica Andres (EVP Yara Europe) and Mr. Michael Schlaug (Plant manager Yara Sluiskil). Copyright: Rijksvoorlichtingsdienst.

factors such as future energy needs, the availability of storage technology solutions with favourable economic and environmental performance and the level of societal acceptance. It is important to realize that the energy system in the Netherlands is becoming more complex and competitive than in the past. The increase in energy imports and the implementation of more renewable sources of energy and types of storage technologies may imply a major change in how the future role of the subsurface should be viewed, that is, as less of a molecule provider (with limited competition) and more of a service provider for storage (with extensive competition).

This chapter uses a top-down approach to look into the different types of underground storage technologies and their coupling to available/potential underground reservoirs in the Netherlands. The first part of the chapter presents the main state-of-the-art underground storage technologies available today, dividing them into two large groups: cyclic and long-term. For each, the main drivers and incentives for development and the services they can provide, are described in order to understand why underground storage is a crucial element in the energy transition and climate and environmental issues, and how it relates to aboveground storage and energy flexibility options. The second part of the chapter looks at the most relevant storage technologies for the Netherlands according to the type of underground reservoir and the commodities they can store. This includes a description of the general technical background and past, present and future implementations and developments in specific geological formations. Finally, the chapter concludes with a summary and outlook of the main factors to be considered for the future development of underground (energy) storage in the Netherlands.

Aboveground and underground storage technologies and rationale for application

Aboveground and underground storage technologies can be subdivided into two main categories, depending on the time frame involved: 'cyclic (temporary) storage' or 'long-term (permanent) storage'. Cyclic storage is mainly used for the temporary storage of energy and feedstocks with charge and discharge periods ranging from hours to months. Long-term storage is for the disposal of (by)products to minimize environmental impact or mitigate climate change. The storage technologies belonging to each category can support one or multiple services, each of which requires a total storage volume and a given injectivity or withdrawal capacity.

Cyclic storage

Cyclic storage of energy or feedstock can take place aboveground in tanks, pipelines, batteries, supercapacitors and underground in natural reservoirs and engineered cavities. However, underground energy storage has the main advantage that it generally has a larger volumetric capacity and lower unit technical cost than aboveground storage. Typical applications for underground storage are:

- Ensuring the supply of energy for peak and seasonal heat (and cold) demand.
- Balancing supply and demand with the increased share of variable energy sources in the energy mix.
- Serving international energy trade or large-scale energy flows between sectors in combination with arbitrage, i.e. the economic exploitation of price differences.
- Creating strategic reserves for possible prolonged periods of supply disruption that cannot be replaced by domestic and/or imported resources.
- More efficiently using the large-scale supply of constant (baseload) heat sources (e.g. geothermal, waste heat) and seasonal heat sources (e.g. solar) within local heat networks.

The future demand for cyclic energy storage will be influenced by many interrelated factors, such as:

- The configuration, predictability and intermittency of primary energy supply (base load production, variable renewable energy generation, imports).
- The characteristics of the final demand (type of energy carriers needed per sector, demand patterns, flexibility of demand, dependencies between sectors).
- The way in which supply and demand are linked (energy networks, conversions).
- The development of technology, energy markets, energy storage hubs and energy and climate policies (innovations, technology costs, energy prices, controls, subsidies).
- External influences such as weather, climate and geopolitical developments.
- Risks associated with the technologies and societal perception and acceptance.

Table 20.1 shows the most relevant types of flexibility solutions for the current gas, electricity and heat grids in the Netherlands. The table clearly shows that besides underground energy storage for natural gas and heat, there are aboveground storage solutions, which are used depending on their associated capacities, demand, prevailing legal and regulatory regimes, and cost-benefit ratio (Bos, 2019).

In general, above- and underground cyclic storage technologies can be typically categorized based on their send-out capacity (i.e. power in Watt) and discharge time

Table 20.1. Flexibility solutions and system levels per type of commodity (natural gas, electricity and heat) in the Netherlands. Transmission System (TS) and Distribution System (DS) are the infrastructures to deliver natural gas, electrical power or heat at a national and regional level, respectively.

Type commodity	General characteristics	Flexibility solutions (system level)	Description
(Natural) gas	The total domestic natural gas production from gas fields, (underground) gas storage sites, import, conversion, transportation/ distribution and biogas allow matching demand.	Line-pack (TS)	Using pressure differences in the existing transmission pipelines to store gas.
		Ramping (TS)	Adjusting the production from primary natural gas reservoirs (most reservoirs contain high-caloric* gas, but the Groningen reservoir and a few other reservoirs contain low-caloric* natural gas).
		UGS (TS)	Adjusting the withdrawal or injection rates of Underground Gas Storage sites (UGS, both high-caloric and low-caloric storage sites).
		Gas to power – G2P (TS)	Conversion of natural gas to electricity.
		Conversion of hi-cal natural gas to lo-cal gas (TS)	Conversion of hi-cal natural gas to lo-cal gas by adding nitrogen. In the Netherlands, nitrogen is captured from the air (Ommen, Wieringermeer, Pernis and Zuidbroek) and in one location stored in an underground salt cavern (Heiligerlee).
		Power to gas – P2G (TS)	Conversion of electricity to hydrogen gas or to syngas.
		International pipelines (TS)	- Import of natural gas through interconnection (pipelines) with foreign suppliers to ensure seasonal balancing (e.g. imported gas is supplied directly or used to fill the underground gas storage sites in summer, when demand is low). - Import of foreign LNG to LNG terminals using LNG carriers. LNG terminals in the Netherlands.
		Liquefied natural gas (LNG) (TS)	Gate Terminal, Rotterdam, capacity 8-12 billion m^3/yr (www.gateterminal.com). Eemshaven (floating storage regasification Unit), capacity 8 billion m^3/yr (www.eemsenergyterminal.nl).
		Biogas	Biogas is a mixture of gases (methane, carbon dioxide and hydrogen sulphide) produced from, among other things, sludge, waste from landfills, garden waste, vegetable and fruit residues, and animal residual products such as cow manure. The biogas is purified and dried to achieve the same quality as natural gas.
Electricity	Matching electricity supply and demand requires that the electricity grid is fully balanced, every second and in every sub-system of the grid (i.e. from low to high voltage systems).	Ramping (TS+DS)	Standby dispatchable power supply sources.
		Interconnectivity (TS)	Connection to other countries provided by TenneT (the national Transmission System Operator, TSO) and international trade to Balancing Responsible Parties (BRPs) for supply and demand matching.
		Storage at different scales	Batteries, capacitors, flywheel and potentially in the future e.g. Underground Hydrogen Storage (UHS), Pumped Hydro Storage (PHS), Compressed Air Energy Storage (CAES).
		Curtailment	Reducing or restricting energy delivery from an energy production capacity.
		Demand-response	Portion of the demand that can be reduced, increased or shifted in a specific period of time.
Heat	Heat grids are generally local (district heating) and often restricted to one or more heat sources (e.g. waste heat; geothermal)	Ramping up/down (DS)	Increasing/decreasing energy output in response to demand. Each heat production technology has different ramping rates and preferred operational modes (e.g. waste incinerator, Combined Heat and Power plant (CHP), geothermal installation).
		Gas to Power to Heat – G2P/P2H conversion (DS)	Supplying additional heat to the heat grid by burning gas or by using power to heat (i.e. electric boiler, heat pump) when heat demand is high.
		Aboveground thermal energy storage	Heat storage in tanks is very common in heating grids to balance heat supply and demand and to decouple heat demand from supply. Water is very often used as a storage medium.
		Underground shallow low-temperature thermal energy storage, borehole heat exchanger (heat and cold storage) (DS)	Mainly used on scales from individual houses to large buildings and complexes (e.g. offices, hospitals). And often in combination with a heat pump. This type of heat storage can be done in various geological formations, or even in buried steel vessels.
		Deeper high-temperature aquifer thermal energy storage – ATES (DS).	- Heat grids use aquifers to store heat seasonally. In summer, when heat demand is low, hot water is injected into the aquifer. In winter, when heat demand is high, the hot water is produced from the aquifer.
		Heat pump	- Heat pump can be used in combination with heat networks.

* Natural gas with a low content of heavier hydrocarbons and higher contents of gases such a nitrogen are called low-caloric (lo-cal) gas. High-caloric (hi-cal) gases (with high content of methane and heavier hydrocarbons) are considered when the gross-heating-value (GHV) is above ~39 MJ/Nm^3.

(hours), which together determine the total storage capacity (amount of energy in Watt hour, Wh) that could be delivered in a full cycle. These technical characteristics determine the specific services that a particular technology can provide (Fig. 20.2).

Underground storage technologies score high in both capacity and discharge duration in comparison to aboveground storage technologies (Fig. 20.2). Underground storage of hydrocarbons and hydrogen can potentially provide the largest flexibilities for centralized power and heat balancing, with peak send-out capacities of hundreds of MW to tens of GW (thermal energy) for hours to months. In the case of hydrogen, which must first be produced from natural gas (steam reforming) or water and electricity (electrolysis), the energy stored underground can be used as fuel or for the generation of electricity through a Combined Cycle Gas Turbine (CCGT). Another large scale (underground) storage technology is the Pumped Hydro Energy Storage (PHES), which uses the gravitational potential energy of water to generate electricity. It can potentially deliver a capacity of hundreds of MW to a few GW for periods of up to a few weeks (i.e. up to 100 GWh) and is normally used for ancillary services (electricity grid balancing) and arbitrage, the latter being the energy stored during lower-priced hours and delivered during higher-priced hours. This is a typical aboveground technology often seen in water dams in mountainous areas. However, in regions with no topography, this can also be achieved by building an elevated water basin or by combining a large aboveground water basin (e.g. pit or river) and an underground reservoir (e.g. abandoned mineshafts or excavated lined rock caverns and tunnels). Large amounts of electricity needed during peak-off hours can also be stored in underground cavities in the form of mechanical energy using Compressed Air Energy Storage (CAES). CAES systems, with potential capacities of tens to hundreds of MW for more than a week (i.e. ~10 GWh), are designed to be competitive in delivering a suite of flexibility services that are valued by utility companies, owners of generation assets and grid operators.

For the seasonal storage of heat, there are remarkable differences between the amount of energy that can be stored by above- and underground technologies. Since water is the preferred carrier for heat transport and storage, the amount of energy that an aboveground heat storage can deliver is generally on the order of a few kW to ~100

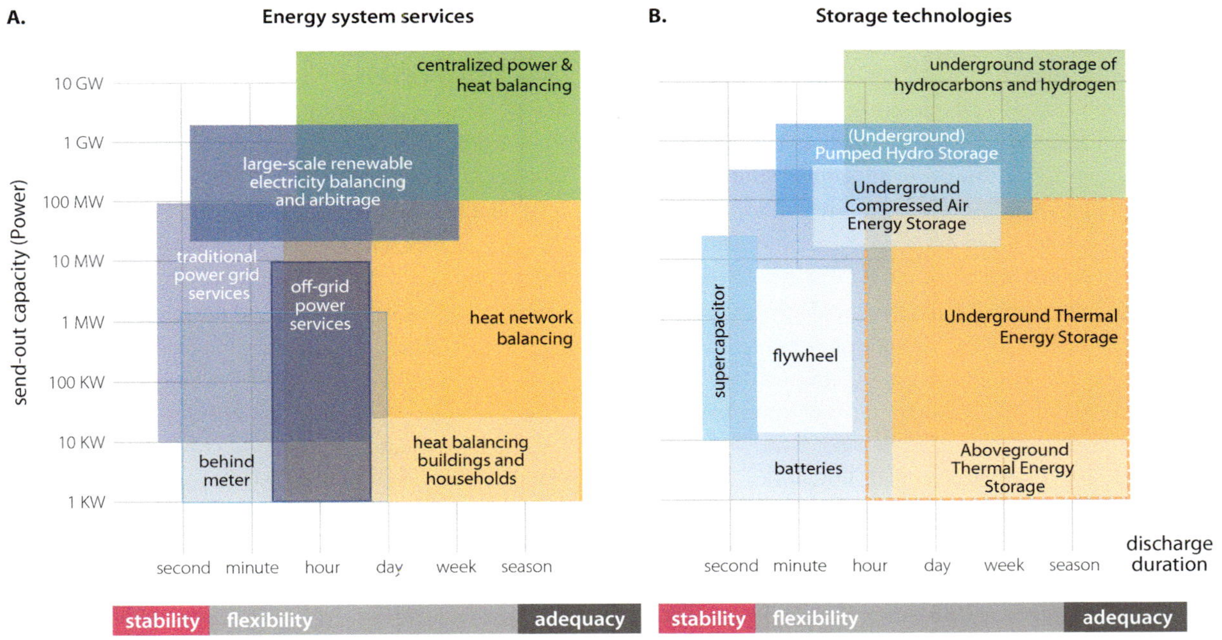

Figure 20.2. a) Energy system (balancing) services mapped according to their power (Watt) and relevant timescale for discharge. The types of services can be classified as stability, flexibility and adequacy based on their duration: stability is defined as the response to very short and fast fluctuations (especially in the power system); flexibility represents response to load and supply variations up to the seasonal timescale; adequacy (security of supply) determines the ability to adapt to long-term trends and emergencies. These three types of services are referred to in the text with the general term 'flexibility'. b) Above- and underground storage technologies that can provide the energy system services. The colour code indicates to which grid the energy system service and storage technology belongs: electricity (blue), gas (green) and heat (orange). Modified from IEA and Groenenberg et al., (2020b).

MW for several hours or days (i.e. up to several GWh). This is suitable as a daily or weekly buffer for buildings and local heat networks, but underground heat storage can supply up to 100 MW of capacity and, depending on the volume stored, can supply heat up to several months.

Besides pumped hydro and thermal storage at surface, there are other important aboveground storage technologies such as batteries, flywheels and supercapacitors. Although they can offer a considerable power range (between 1 kW and >100 MW) to provide grid stability and/or flexibility for traditional power grids, off-grids and behind meter services, their deliverability period is mostly limited to less than one day. Recent development of very large battery stacks may result in an extension of this period to a few days, albeit against very high costs and surface area requirements.

Permanent (long-term) storage

Permanent storage generally applies to substances that need to be disposed or safely stored in large quantities for long periods (thousands of years) as they can be harmful to the environment or climate (e.g. waste water, filtrated residues and CO_2). For this, the underground is the preferred storage medium. For the screening of potential underground reservoirs, a large volumetric capacity is often a critical factor, as well as the long-term integrity of the reservoir and seal, among others (EBN & Gasunie, 2018; Verhoef et al., 2020).

Waste waters produced during the extraction of petroleum and natural gas (so-called production waters) are often reinjected into the same (or similar) underground reservoir, or into nearby deep aquifers to minimize the environmental impact, and occasionally used for enhanced oil recovery (EOR; Liang et al., 2018). This is because these waters are brines that are more saline than seawater and may still contain small amounts of hydrocarbons and radioactive elements, as well as traces of chemicals used during the production process. Other types of water waste are the filtrate residues derived from water treatment facilities, which can be readily and safely disposed of in shallow aquifers a few hundred metres deep.

In the case of CO_2 storage, or so-called carbon sequestration, the goal is to isolate CO_2 from the atmosphere long enough for it to become part of the geological carbon cycle (Ringrose, 2020). Carbon Capture and Storage (CCS) is considered an essential technology for a timely achievement of Greenhouse Gas (GHG) emission reduction targets. It can be applied to centralized fossil-based energy production facilities and CO_2-intensive industries (such as steel, ammonium (fertilizer), cement plants, refineries and hydrogen production) whose emissions cannot be easily reduced in the short-term in any other way. Direct air capture of CO_2 is also an option under development for the longer term, which could result in net negative emissions when sequestered (Ringrose, 2020). CO_2 can be (re)injected for enhanced oil recovery (EOR). Rather than sequestering CO_2 for climate considerations, this is a commercial application where the costs of CO_2 capture and injection are offset by revenue from the extra oil produced.

Geological potential for underground storage in the Netherlands

The development of underground storage in the Netherlands, as elsewhere, will be determined by the need for specific storage services and by the availability of suitable underground formations. Underground storage can be accomplished in either 'natural reservoirs' or 'engineered cavities' (Muhammed et al., 2022). The term 'natural reservoirs' refers to porous reservoirs, such as depleted gas fields and aquifers. Engineered cavities represent constructed underground voids, which are either created by solution mining in salt formations, excavation and tunnelling in rock formations, or by drilled wells. For these two main underground storage options, there is a wide range of technologies for the storage of different types of commodities and their current degree of technical development or implementation in the Netherlands and abroad (Fig. 20.3). In this chapter we show that the Netherlands holds a strong position in the significant amount of research being carried out on many types of underground storage technologies and, as a result, is experiencing a notable increase in the number of storage projects in natural reservoirs and salt caverns. Only the use of engineered cavities in hard rock (i.e. crystalline and metamorphic rock) is limited, but this is due to the generally great depth of the basement in the Netherlands, and the large abundance of overlying geological formations with good reservoir properties (Rotliegend to Upper North Sea Group; Fig. 20.4).

Today, depleted gas fields in Rotliegend sandstones and Zechstein carbonates are the most commonly used reservoir types for the cyclic storage of natural gas and permanent storage of production water in the Netherlands (Figs. 20.3 and 20.4). Along with sandstones of the Lower Triassic, they are also emerging as potential candidates for large scale permanent storage of CO_2 (Fig. 20.4). Deep aquifers in the Rotliegend and Triassic are also being considered as candidates for CO_2 storage, while shallow aquifers (up to 200 m depth) in the Quaternary are already widely used for low-temperature (<25°C) aquifer thermal energy storage (LT-ATES; Fig. 20.4). Shallow aquifers have also been used for the storage of filtrate residues left after the purification of water for drinking purposes. Slightly deeper aquifers in the North Sea Supergroup are being considered as thermal storages of water at higher temperatures

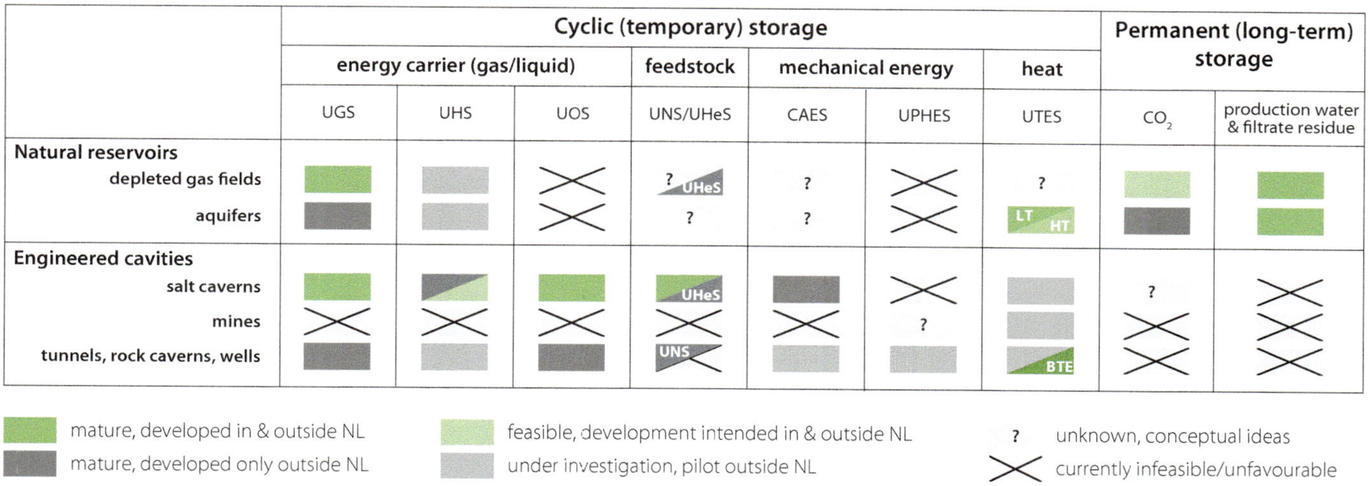

Figure 20.3. Overview of various underground reservoirs and storage technologies that are used or considered for application in the Netherlands and/or elsewhere. Abbreviations: UGS = Underground (natural) Gas Storage, UHS = Underground Hydrogen Storage; UOS = Underground Oil Storage; UNS = Underground Nitrogen Storage; UHeS = Underground Helium Storage; CAES = Compressed Air Energy Storage; UPHES = Underground Pumped Hydro Energy Storage; UTES = Underground Thermal Energy Storage; BTE = Borehole Thermal Storage; HT/LT = High/Low Temperature (LT <25°C; HT >25°C).

(HT-ATES). The stable nature and large abundance of salt formations of the Zechstein and Triassic groups have long been used for the storage of a wide variety of fuel-based energy carriers, such as natural gas and oil, and nitrogen for feedstocks (Figs 20.3, 20.4). Salt caverns may also see the development of Compressed Air Energy Storage (CAES) and Underground Hydrogen storage (UHS) in the near future. Regarding the lower Carboniferous (Dinantian) carbonates, since the 1980s many research efforts have been made to support the construction of an Underground Pumped Hydro Energy Storage (UPHES) facility in the southeast of the Netherlands (Limburg) (Fig. 20.4). In the following sections of this chapter, both cyclic and permanent underground storage technologies along with their corresponding most suitable geological groups are described in more detail for the Netherlands.

Cyclic underground storage of energy carriers, feedstocks and mechanical energy

Underground storage facilities for cyclic (temporary) storage of energy carriers (in gaseous or liquid form), mechanical power or feedstock share several common features. They all require an underground porous reservoir (depleted oil or gas field, aquifer) or cavity (one or several salt or lined rock caverns), and one or more wells connecting the underground reservoir to the aboveground processing facilities. Gas processing includes compression, expansion, drying and cleaning. The storage facilities are typically interconnected with a supplier (production facilities, import terminals) and the consumers (households, industry, power plants) via a transmission and distribution (pipeline) network. During periods of excess supply and/or low demand, a fuel-based energy carrier, compressed air or feedstock is injected (charged) into the underground storage, which then becomes a pressurized gas container. It is withdrawn (discharged) from the storage during peak demand or as needed, depending on its use and purpose. In the case of underground gas storage, the total volume of gas/liquid contained in the reservoir is divided into two volumes: the cushion volume (cv) and the working volume (wv). The cushion volume is the volume of gas/liquid that must remain in the reservoir to maintain sufficient pressure to allow delivery of the working volume at a pre-defined minimum rate throughout a withdrawal period (Muhammed et al., 2022). In salt caverns, this pressure is especially important to maintain the structural and mechanical integrity of the cavern and to minimize the inwards closure of the cavern walls by salt creep (e.g. Kumar et al., 2021; Kumar & Hajibeygi, 2021; De Bresser et al., 2022). Therefore, the working gas volume is the actual storage volume for repeated injection and withdrawal. It is important to note that the capacity of a storage facility generally refers to the working volume, and not to the cushion and working volume together.

Caverns are often considered a better option for underground storage than depleted oil or gas fields and aquifers. The advantages of underground storage in a salt cavern are:

- Lower cushion gas volume requirements, therefore higher percentage working volume. Relative to the total volume, the average percentage of working volume for a salt cavern is 70-80%, although it can vary due to

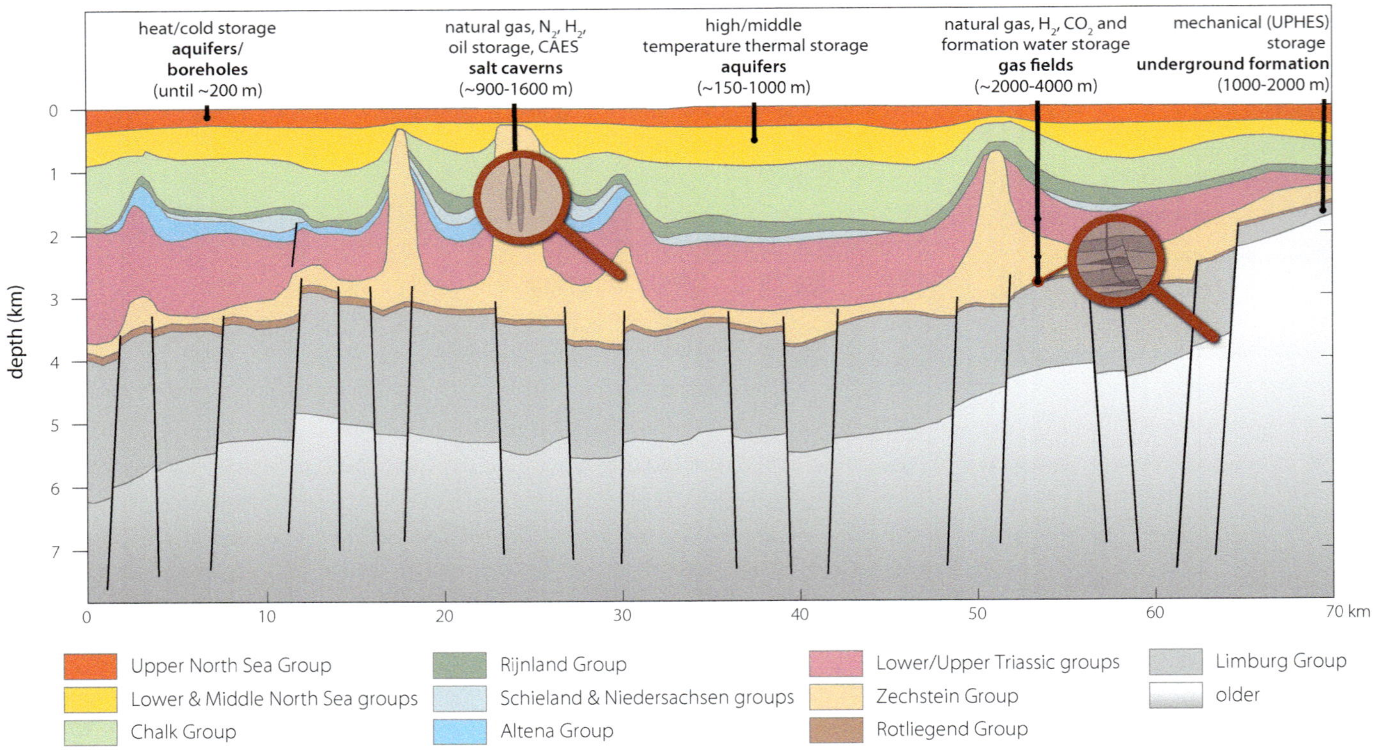

Figure 20.4. *A representative cross-section through the Dutch subsurface with the different storage technologies that are (or can be) applied to particular geological groups.*

site-specific factors such as depth and required storage capacity. In depleted gas fields the percentage is usually ~50-60%, whereas in aquifers it is ~20% (Matos et al., 2019).

- High sealing capacity, mechanical stability and chemically inert nature of the rock salt (in the absence of non-salt rock intercalations).
- A high level of technical maturity for the storage of many types of gasses.
- Injection and withdrawal through the (single) well used to make the salt cavern (by solution mining) is less expensive than the multi-well drilling required in depleted gas fields and aquifers.
- Eventually higher injection and withdrawal rates for short- and medium-term storage.

There are also disadvantages of underground storage in salt caverns:

- Limited availability of large underground salt formations.
- Limited storage volume capacity of hundreds of millions of cubic metres per cavern (relative to billions in a typical gas field/aquifer), although storage in salt caverns, depleted fields or aquifers can be complementary.
- Restrictions on the number of salt caverns that can be developed in a salt structure in view of surface subsidence and cavern stability issues.
- Limitations to the daily volumes that can be injected or withdrawn to preserve the structural integrity of the cavern.
- The shallower depth of some caverns (to minimize cavern volume loss due to salt creep) may increase surface effects (e.g. subsidence).
- Environmental issues related to the water necessary for cavern leaching and the disposal of the produced brine.
- Salt cavern long-term abandonment issues.

Depleted oil and gas fields are the second preferred subsurface storage medium due to their proven ability to contain large volumes of natural gas for millions of years, unlike storage in aquifers which may still require testing to identify potential issues regarding containment (e.g. Tarkowski, 2019; Heinemann et al., 2021; Zivar et al., 2021). Besides the choice of the type of underground reservoir, its location (onshore or offshore) can also be a decisive factor based on the distance to production and supply sites, surface limitations, legal aspects and societal acceptance.

Underground Natural Gas Storage (UGS)

Underground natural gas storage (UGS) is currently one of the most prominent technologies being used to flexibly match energy supply and demand on daily to seasonal timescales around the world. In Europe there is ~100 x 10^9 m^3 (1096 TWh) of working gas capacity in natural gas storage sites, which corresponds to 24% of the annual demand (Cihlar et al., 2021). Most of these storage sites are in porous reservoirs and salt caverns (ESTMAP, 2014; Cihlar et al., 2021). In the Netherlands, the current capacity for underground natural gas storage (working gas volume) is ~14 x 10^9 Nm^3 (~139 TWh), accounting for 34% of the annual domestic demand (CBS, 2022). Natural gas is stored in four natural gas fields and one cluster of salt caverns (Fig. 20.5). The four UGS sites in porous reservoirs are Norg, Grijpskerk, Alkmaar (since 1997) and Bergermeer (since 2014). While the Bergermeer and Alkmaar natural gas fields were significantly depleted prior to their conversion to UGS sites, the Norg and Grijpskerk conversion came after a relatively short period of conventional gas production, and thus their yet-to-be-produced natural gas has been used as cushion gas. The sandstone reservoirs of Norg, Grijpskerk and Bergermeer belong to the Upper Rotliegend Group, and the carbonate reservoir of Alkmaar to the Zechstein Group (Bouroullec & Geel, 2025, this volume). The Upper Rotliegend Group is the most widely used for UGS because of its excellent reservoir properties with clean, well-sorted eolian and/or fluvial sandstones with thicknesses up to a few hundred metres (Juez-Larré et al., 2016). Uplifted horst blocks provide good structural traps and the thick Zechstein salt layers form an excellent top seal. The Zechstein carbonate reservoir at the Alkmaar UGS site is thinner (50 m thick on average) and presents a more complex reservoir as it is part of a marine carbonate platform where the best reservoir properties (permeability) are often found in areas affected by karstification, fracturing and/or faulting. The reservoir is sealed by anhydrites and salts of the same rock unit and is bounded laterally by a combination of fault- and dip-closure.

The only UGS site in a cluster of salt caverns is at Zuidwending, which has been in service since 2011. It consists of a group of six vertical cylindrical salt caverns created by solution mining in a large salt pillar of the Zechstein Z2 Formation (Zechstein Group). The pillar extends from a depth of 2700 up to ~200 m below surface. Four of the caverns have a geometrical volume of 600,000 m^3 (radius ~25 m x height 300 m) and two of them of ~10^6 m^3 (radius ~25 m x height 500 m). They are located at depths between 1550 and 1050 m and are operated at pressures between 50 and 190 bar (Table 20.2). The reported annual loss of cavern volume due to salt creep is very small (~0.4%/yr) and the predicted subsidence at the surface is estimated to be around 1 mm per year (Energystock, 2017).

Table 20.2 shows the storage capacity and operating ranges for each of the five UGS sites. Norg, Grijpskerk and Alkmaar store low-caloric gas from the Groningen field or pseudo-Groningen gas produced from the mixture of high-caloric gas and nitrogen. The low-caloric gas from these three UGS sites is used primarily to provide flexibility to the residential sector for heating purposes (www.pbl.nl/kev). This replaces the former swing production capacity of the Groningen field (Fig. 20.5). Zuidwending also stores Groningen-type gas, but this gas is only traded on the spot market, while the high-caloric gas stored in Bergermeer is mainly used for international trade. There is a salt cavern storage facility in Gronau-Epe (Germany) near Enschede that is connected to the Dutch high-pres-

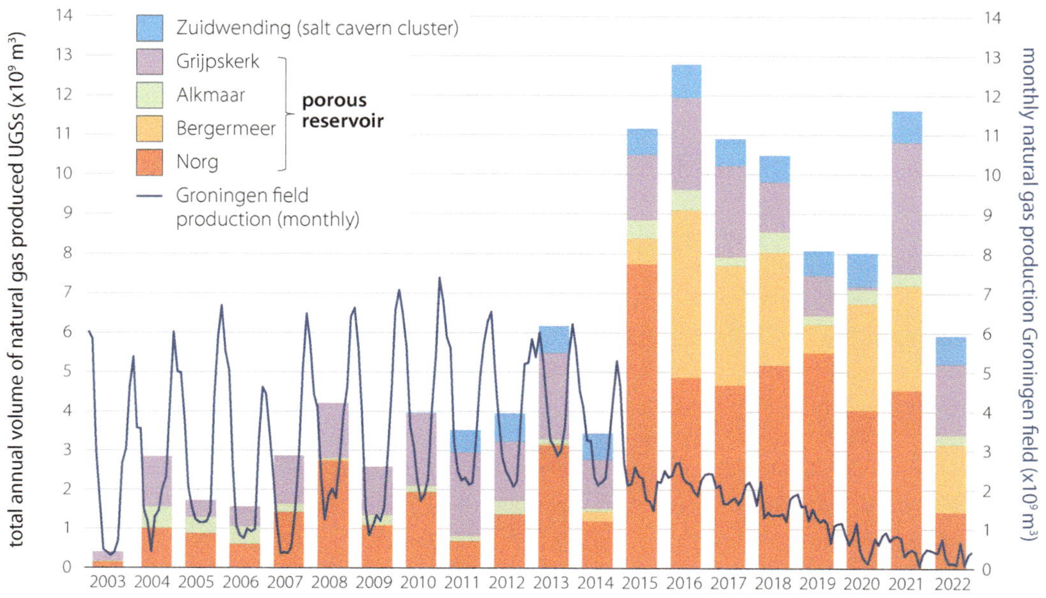

Figure 20.5. Volumes of natural gas withdrawn (discharged) from the various UGSs in the Netherlands. The corresponding monthly production from the Groningen field is plotted as a continuous blue line (MEACP, 2023).

Table 20.2 Rates of withdrawal and injection (single well/facility) and storage capacities of the current underground natural gas storage facilities in the Netherlands. Volumes of working gas and cushion gas for natural gas were extracted from the storage licenses (www.nlog.nl).

Storage natural gas (realized)	Units	Grijpskerk	Norg (Langelo)	Alkmaar	Bergermeer	Zuidwending
Reservoir type	-	Partly depleted gas field	Partly depleted gas field	Depleted gas field	Depleted gas field	Salt caverns (6 caverns)
Geological reservoir	-	Slochteren Fm (Upper Rotliegend Group)	Slochteren Fm And Lower Ten Boer Member-Silverpit Fm (Upper Rotliegend Group)	Zechstein Z3 Member (Zechstein Group)	Slochteren Fm (Upper Rotliegend Group)	Zechstein Z2 (Stassfurt) Fm (Zechstein Group)
Lithology	-	sandstone	sandstone/ conglomerate	dolomite/ limestone	sandstone	salt
Top reservoir depth	m (TVD)	3200	3000	2000	2200	1050
Reservoir thickness (avg.)	m	180	150	50	200	n.a.
Seal	-	Zechstein (salt, claystone)	Zechstein (salt, claystone)	Zechstein (salt, claystone)	Zechstein (salt, claystone)	Zechstein (salt)
Max. withdrawal rate (facility)	-	~60 x 10^6 m^3/d	96 x 10^6 m^3/d	~1.5 x 10^6 m^3/hr	57 x 10^6 m^3/d	1.8 x 10^6 m^3/hr
Max. injection rate (facility)	-	-	51 x 10^6 m^3/d (498 GWh/d*)	0.15 x 10^6 m^3/hr (1 GWh/hr*)	42 x 10^6 m^3/d (455 GWh/d*)	27.12 x 10^6 m^3/hr (11 GWh/h)
Volume at max. working pressure	x10^9 Sm^3	10.8	29.4	3.8	16	0.76
Working Gas Volume	x10^9 Sm^3	2.5 (24.4 TWh*)	6.3 (61.5 TWh*)	0.56 (5.5 TWh*)	4–6 (43–65 TWh**)	0.38 (3.7 TWh*)
Cushion Gas Volume	x10^9 Sm^3	8.3 (89.9 TWh**)	23.0 (249 TWh**)	3.3 (32.2 TWh*)	10–12 (108–130 TWh**)	0.37 (3.6 TWh*)
wv: cv ratio	-	0.3	0.27	0.18	0.3-0.6	1.02

* Low caloric natural gas (35.17 MJ/Nm^3); ** High-caloric natural gas (39 MJ/Nm^3).

sure gas transmission system. This facility provides some additional storage capacity for natural gas to meet the demand in the Netherlands (www.uniper.energy/energy-storage-uniper, 2022).

A study on the maximum theoretical natural gas storage performance of onshore depleted gas fields in the Netherlands, including the five UGS sites, shows that the largest storage capacities and highest withdrawal rates mainly belong to the reservoirs from the Upper Rotliegend and Lower Germanic Trias groups (Fig. 20.6; Juez-Larré et al., 2016; Van Gessel et al., 2018). These results clearly reveal two main things. One is that the four UGS sites in porous reservoirs rank high in their storage performance relative to other depleted gas fields, and the other is that their individual performances are far greater than those of the six salt caverns at Zuidwending (Fig. 20.6). However, in most cases their estimated cushion volumes are very large, as it is the case for Norg, Bergermeer and Grijpskerk (Fig. 20.6, Table 20.2). Initially, this was not a problem for the storage sites of Norg and Grijpskerk, as the original gas-initially-in-place was used as cushion gas. In order to reduce the cushion gas volume, these storage sites could be operated at lower working pressures, however this would reduce their performance. Alternatively, other onshore or offshore depleted gas fields, with better wv:cv ratios, could be developed for underground gas storage in combination with some clusters of salt caverns.

Based on the projected natural gas annual consumption in the Netherlands up to 2030, natural gas imports are expected to increase up to ~20 x 10^9 m^3 (www.pbl.nl/kev; Fig. 20.1). Developments since 2022 regarding security of gas supply are driving Europe to move away from importing Russian natural gas (Patrahau et al., 2022). With climate goals in mind, it is important to note that much of the imported natural gas has a significantly higher Greenhouse Gas (GHG) footprint compared to the natural

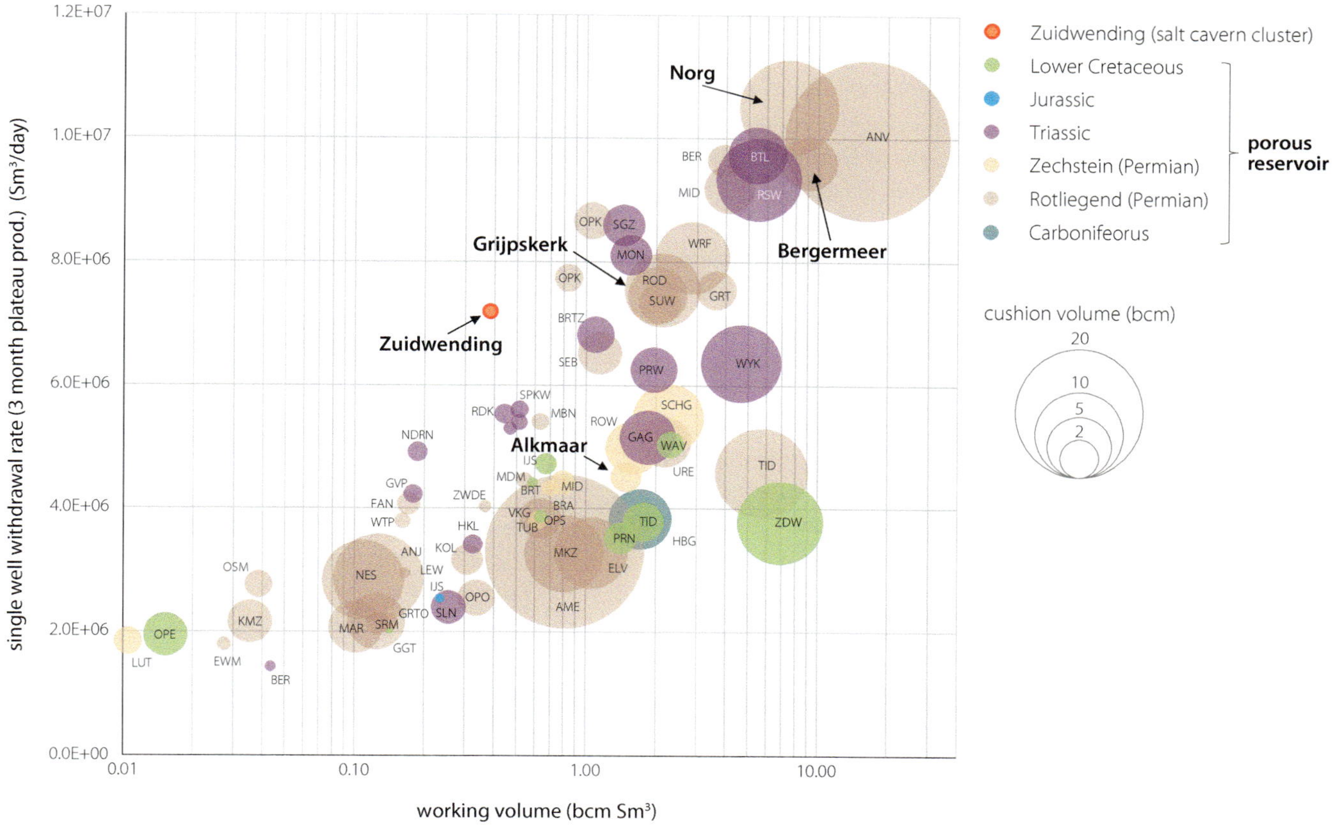

Figure 20.6. Working volume plotted against single well 3-month withdrawal rates calculated by Juez-Larré et al. (2016). The size of the corresponding cushion volume is given by the bubble size. The codes identifying the reservoirs shown in this plots are: AME = Ameland-East; ANJ = Anjum; ANV = Annerveen; BER = Bergen; BRA = Boerakker; BRT = Barendrecht; BRTZ = Barendrecht-Ziedewij; BTL = Botlek; ELV = Eleveld; EWM = Engwierum; FAN = Faan; GAG = Gaag; GGT = Grootegast; GRT = Groet; GRTO = Groet-East; GVP = Geestvaartpolder; HBG = Hardenberg; HKL = Hekelingen; IJS = IJsselmonde; KMZ = Kommerzijl; KOL = Kollumerland; LEW = Leeuwarden 101 RO; LUT = De Lutte; MAR = Marum; MBN = Middelburen; MDM = Middenmeer; MID = Middelie; MON = Monster; MKZ = Munnekezijl; NDRD = Noorderdam; NES = Nes; OPE = Opeinde; OPK = Oude Pekela; OPO = Opende-East; OPS = Opeinde-Zuid; OSM = Oostrum; PRN = Pernis; PRW = Pernis-West; RDK = Reedijk; ROD = Roden; ROW = Rossum-Weerselo; RSW = Roswinkel; SCHG = Schoonebeek-gas; SEB = Sebaldeburen; SGZ = 's-Gravenzande; SLN = Sleen; SPKW = Spijkenisse-West; SRM = Schermer; SUW = Suawoude; TID = Tietjerksteradeel; TUB = Tubbergen; URE = Ureterp; VKG = Vinkega; WAV = Wanneperveen; WRF = Warffum; WTP = Witterdiep; WYK = De Wijk; ZDW = Zuidwal; ZWDE = Zuidwending-East.

gas produced in the Netherlands (Hammond & O'Grady, 2017; Juez-Larré et al., 2018). This is because the energy required for transport of pipeline gas and regasification of LNG leads to higher GHG emissions (Patrahau & Van Geuns, 2021). All of this reinforces the need to implement renewables and to use UGS sites to secure supply for as long as natural gas remains an important source of energy.

Underground Hydrogen Storage (UHS)

Hydrogen is expected to play an important role in our future energy system, as it has the potential to take over (part of) the role of natural gas in securing energy supply (Den Ouden,, 2020). Hydrogen in pure form can be found in geological environments such as ancient continental shields, but it is not found in such large and concentrated quantities as natural gas (Prinzhofer et al., 2018). Global hydrogen is currently primarily produced by steam methane reforming (SMR: $CH_4 + H_2O + \text{heat} \rightarrow CO + 3H_2$), but also by partial oxidation of heavier hydrocarbons and coal gasification (Gupta, 2009). Hydrogen from SMR is termed as 'grey hydrogen', because the CO_2 that is produced from the reaction of CO and steam ($CO + H_2O \rightarrow H_2 + CO_2$) is released to the atmosphere. If the CO_2 produced is captured and (permanently) stored in underground reservoirs or reused, this hydrogen becomes 'blue hydrogen'. Unless in the near future there is a successful upscaling and widespread commercialization of 'turquoise' hydrogen production via methane pyrolysis, which has carbon as by-product instead of CO_2 (Upham et al., 2017), CO_2 storage may well be associated with (blue) hydrogen production.

Hydrogen can also be produced through electrolysis. This is a process that uses electricity to split water into hydrogen and oxygen. Where renewable electricity is generated from wind and solar farms or nuclear power plants – with no or fewer CO_2 emissions – the hydrogen produced is referred to as 'green hydrogen' and 'pink hydrogen', respectively (Gahleitner, 2013). This type of production gives hydrogen the potential to act as a bridge between electricity and the gas grids, and hence to replace part of the role that natural gas plays in the energy system. Today, the efficiency of converting electricity into hydrogen through electrolysis is between 63-81% (Ajanovic et al., 2022) and the re-electrification in fuel cells or combined cycle gas power plants has an efficiency of ~50 and ~60%, respectively (Johansson et al., 2022).

To date, in the Netherlands, hydrogen is mainly used as a feedstock in the chemical industry and for the production of ammonia and methanol and to crack heavier crudes in refineries. It is not yet clear how a market for hydrogen as energy carrier will develop, even if the production, transport and storage are technically feasible and the overall societal benefits outweigh the overall societal costs (Mulder et al., 2019; Staffell et al., 2019). In Europe, hydrogen accounts for less than 2% of the current energy consumption (EnTEC, 2022), and although demand for hydrogen storage is expected to increase from ~23 x 10^9 m^3 (70 TWh) in 2030 to around ~150 x 10^9 m^3 (450 TWh) in 2050 (Cihlar et al., 2021), the actual future demand will depend on factors such as the level of production and demand, the interconnection between national and international hydrogen markets and the availability of other sources of flexibility (Breitschopf et al., 2022).

There are only four operational UHS facilities for pure hydrogen worldwide. In all four cases, the hydrogen is stored in salt caverns, with storage volumes ranging between 10 and 100 million m^3 (Evans & Holloway, 2009; Kruck et al., 2013; Table 20.3). This hydrogen is stored as a feedstock for industry in the event of a supply chain disruptions, which means that storage cyclicity is low (Panfilov, 2016). If hydrogen storage is to be used to balance energy supply and demand, much higher injection and withdrawal rates and an increase in storage cyclicity are expected. Fast-cycling operation brings with it specific challenges that require further research, particularly with regard to geomechanical stability of salt caverns and well integrity (Groenenberg et al., 2020a; Eising, 2021; Ugarte & Salehi, 2021; De Bresser et al., 2022; IEA, 2022a). The total estimated storage capacity for hydrogen in all existing salt caverns in Europe is 50 TWh, which would be insufficient to meet future storage needs (Cihlar et al., 2021). Even with the development of new clusters of salt caverns, the time required for their commissioning, and the limited geographical distribution in a handful of countries in Europe, clearly indicates the need for storage sites in depleted gas fields and aquifers.

Underground storage of hydrogen in porous reservoirs is in many ways similar to underground natural gas storage (Tarkowski, 2019). However, due to hydrogen's smaller molecular size and different chemical properties, further research is still required to demonstrate that hydrogen can be safely stored in porous reservoirs with caprocks that have sufficient sealing properties (Kruck et al., 2013; Heinemann et al., 2020; Zivar et al., 2021,IEA, 2022a; Muhammed et al., 2022). UHS is not expected to cause more significant surface subsidence or induced seismicity than UGS, if operated within the same pressure ranges (Van Eijs et al., 2006; Muntendam-Bos et al., 2008, 2022; Van den Bogert et al., 2016; Van Wees et al., 2017). Hydrogen has a lower volumetric energy density (~10.8 MJ/m^3) relative to natural gas (~39 MJ/m^3) and a lower compressibility (by a factor 0.8-0.9; H_2/natural gas), which implies that the required storage volumes and/or the wv:cv ratio will have to be greater than for natural gas if the same amount of energy is to be provided at the same pressure (Juez-Larré et al., 2020, 2023). In contrast, hydrogen can be withdrawn (discharged) 2 to 3 times faster than natural

Table 20.3. *Overview of current operational UHS in salt caverns outside the Netherlands.*

	Clemens Dome (USA)	Moss Bluff (USA)	Spindletop (USA)	Teeside (UK)
Geology	salt pilar	salt pilar	salt pilar	layered salt
Operator	Conoco Phillips	Praxair	Air Liquide	Sabic Petrochemicals
Start year	1983	2007	2016	1972
Geometrical volume (m^3)	580,000	566,000	906,000	3 x 70,000
Average depth (m)	1000	1200	1340	365
Pressure range (bar)	70–135	55–152	68–202	46
Working volume (10^6 m^3)	27.3 (82 GWh*)	41.5 (124 GWh*)	92.6 (277 GWh*)	9.12 (27 GWh*)

* Estimated energy content of the working volume assuming a low heating value (LHV) of 10.8 MJ/Nm^3 for hydrogen (source: DBI-GUT, 2017).

Table 20.4. Storage capacity parameters of candidate fields for natural gas and hydrogen, including theoretical capacity, effective capacity, working volume (wv) and corresponding energy content (see Fig. 20.7 for location). The Groningen natural gas field was excluded from this study due to its large volume. Theoretical capacity corresponds to the total storage capacity (Gas-Initially-In-Place, GIIP) that meets primary technical preconditions for (safe) storage. The effective capacity fulfills techno-economic criteria for efficient and cost-effective storage. The part of the effective capacity that is available for production/injection is known as the working volume. The working volumes obtained were converted to thermal energy content (TWh_t), using the energy factor 3.0E-9 TWh_t/m^3 (10.79 MJ/m^3) for hydrogen. OFF NH & OFF SH = proximal (nearshore) offshore North Holland and South Holland; OFF distal = offshore distal. From Juez-Larré et al. (2019).

Province/area name (code)	Natural gas			Hydrogen	
	# fields (theoretical capacity) in x10^9 m^3	# fields (wv/effective capacity) in x10^9 m^3	Energy content wv in TWh_t	wv in x10^9 m^3	Energy content wv in TWh_t
Friesland (FR)	39 (216)	21 (34/125)	372	29	88
Groningen (GR)	27 (191)	11 (12/63)	130	10	31
Drenthe (DR)	19 (128)	10 (20/82)	217	17	51
Overijssel (OV)	7 (22)	3 (4/16)	45	3	10
North Holland (NH)	15 (38)	11 (16/35)	177	14	42
South Holland (SH)	17 (57)	17 (22/57)	237	19	56
Rest onshore	6 (5)	0 (0/0)	0	0	0
Total onshore	**130 (657)**	**73 (109/378)**	**1178**	**93**	**277**
Offshore (OFF NH)	6 (31)	3 (8/25)	85	7	20
Offshore (OFF SH)	11 (41)	10 (10/37)	104	8	24
Offshore (OFF distal)	69 (564)	54 (53/414)	572	45	134
Total offshore	**86 (636)**	**67 (71/476)**	**761**	**60**	**179**
TOTAL	**216 (1293)**	**140 (180/854)**	**1939**	**153**	**456**
Current UGS	4 (63)	4 (13/63)	134	11	33

gas, as a result of its (8-10 times) lower density and (1.3-3.0 times) lower viscosity. These high withdrawal velocities can partly compensate for the lower volumetric energy content, but in turn can be limited by technical integrity issues, such as erosion in the production tubing and well-bore damage (Juez-Larré et al., 2023).

Today, many screening studies are looking for potential UHS sites around the world (e.g. Mouli-Castillo et al., 2021; Zivar et al., 2021; Muhammed et al., 2022). In addition, various pilot and demonstration projects in depleted fields are underway to increase the technical maturity of the UHS technology and to make it ready for deployment (IEA, 2022a). Town gas, a mixture of gases containing 50-60% hydrogen, was stored safely in salt caverns, depleted gas fields and aquifers in the past (Bourgeois et al., 1979; Kruck et al., 2013; Reitenbach et al., 2015). Two more recent field projects in depleted fields also show that hydrogen can be stored blended with natural gas, with no adverse effects on installations and the environment (hychico.com.ar; www.underground-sun-storage.at). However, there are still accounts of diffusion, dissolution (into formation water), reactions with rocks and interaction with microbes in reservoirs that can degrade hydrogen purity and/or lead to loss of hydrogen, thus impacting recovery (e.g. Amid et al., 2016; Laban, 2020; Dopffel et al., 2021; Thaysen et al., 2021). To meet quality standards, gas cleaning post-withdrawal may be required, which would significantly impact economics (Taylor, 1986; Yousefi, 2021; Yousefi et al., 2023). This is why research is still needed on many aspects of UHS (Heinemann et al., 2020; IEA, 2022a; Muhammed et al., 2022).

Effective UGS and UHS capacity in the Netherlands

Since 2012, the Dutch Ministry of Economic Affairs and Climate has funded several national (pre-)feasibility studies aimed at estimation of the underground storage potential of depleted gas fields and salt caverns; first for natural gas and in recent years also for hydrogen (Fig. 20.7; Juez-Larré et al., 2016; STRONG, 2018; Van Gessel et al., 2018, 2021, 2022; Groenenberg et al., 2020a). Juez-Larré et al. (2019) estimated the effective storage capacity in still producing gas fields with a minimum depth of 1000 m, a permeability higher than 0.1 mD and with no significant amounts of H_2S (<<1 x 10^4 ppm): the result gave a total cumulative effective capacity of 1178 TWh (onshore) and 761 TWh (offshore) for natural gas and 277 TWh (on-

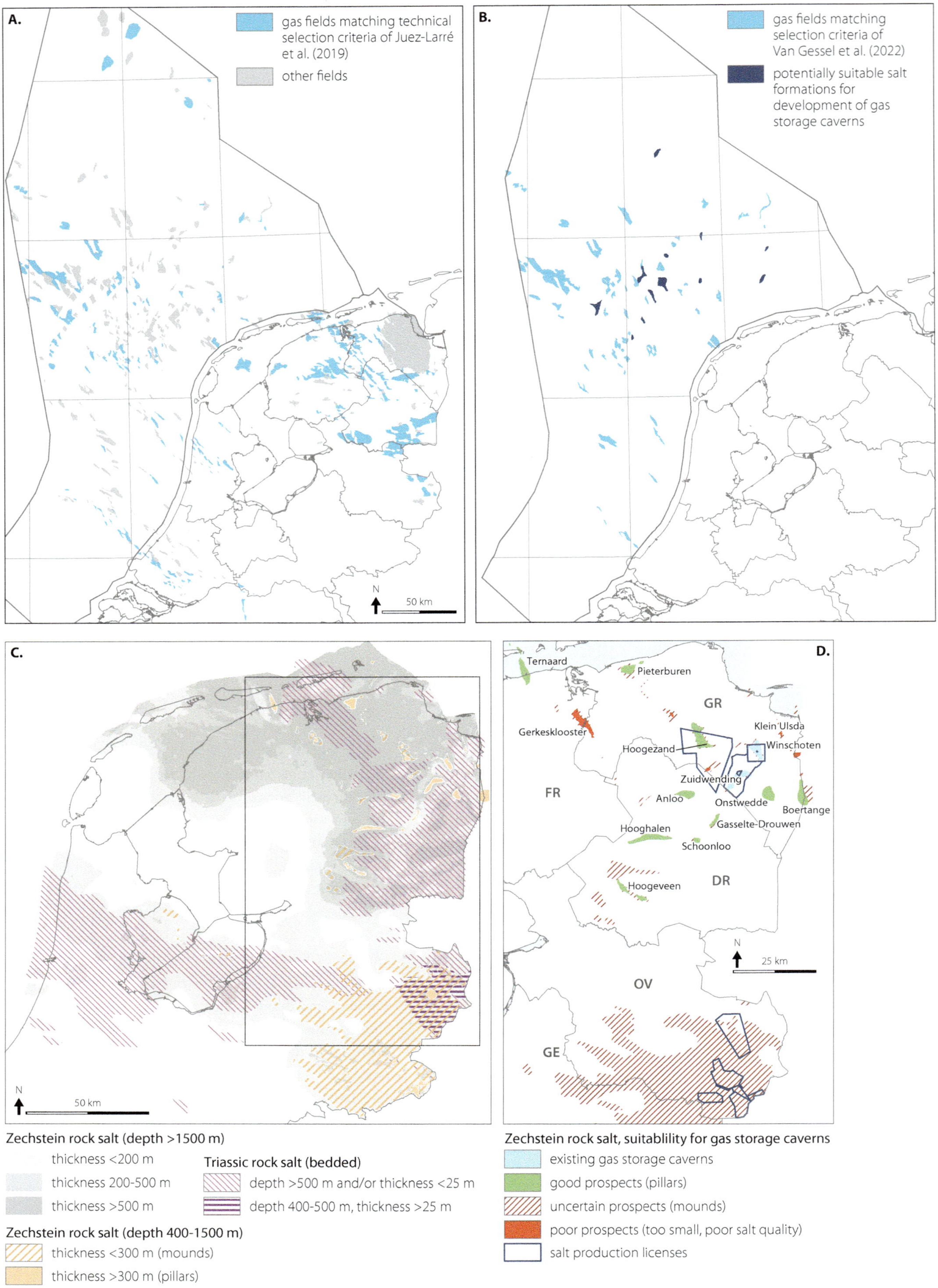
A.
gas fields matching technical selection criteria of Juez-Larré et al. (2019)
other fields
B.
gas fields matching selection criteria of Van Gessel et al. (2022)
potentially suitable salt formations for development of gas storage caverns
N
50 km
C.
D.
Ternaard
Pieterburen
GR
Gerkesklooster
Klein Ulsda
Hoogezand
Winschoten
Zuidwending
FR
Anloo
Onstwedde
Boertange
Hooghalen
Gasselte-Drouwen
Schoonloo
Hoogeveen
DR
25 km
OV
GE
Zechstein rock salt (depth >1500 m)
thickness <200 m
thickness 200-500 m
thickness >500 m
Triassic rock salt (bedded)
depth >500 m and/or thickness <25 m
depth 400-500 m, thickness >25 m
Zechstein rock salt (depth 400-1500 m)
thickness <300 m (mounds)
thickness >300 m (pillars)
Zechstein rock salt, suitablility for gas storage caverns
existing gas storage caverns
good prospects (pillars)
uncertain prospects (mounds)
poor prospects (too small, poor salt quality)
salt production licenses

← ***Figure** 20.7. a) Overview of gas fields matching the technical selection criteria for the storage of hydrogen of Juez-Larré et al. (2019) (onshore and offshore) and (b) Van Gessel et al. (2022) (only offshore and including salt formations). c) Distribution of the Zechstein and Triassic rock salt formations in the north of the Netherlands. d) Detail of (c) showing the suitability of salt caverns for storage of gas. The areas where Zechstein salt pillars have a suitable depth and thickness for the development of salt caverns for gas storage are coloured in green (Juez-Larré et al., 2019; www.nlog.nl/en/rock-salt-0). Provinces in (d): GR = Groningen; FR = Friesland; DR = Drenthe; GE = Gelderland; OV = Overijssel.*

shore) and 179 TWh (offshore) for hydrogen (Table 20.4). Onshore, this capacity can be reduced by up to 60% if surface limitations related to urban, natural and groundwater areas are considered (Van Gessel et al., 2021).

Van Gessel et al. (2018) evaluated the storage capacity of salt caverns in the northeastern part of the Netherlands, where large salt pillars of the Zechstein Group are present (Fig. 20.7d). For the construction of clusters of salt caverns, a depth range between 1000 and 1500 m is considered optimal, with a standard geometric volume of each individual cavern of 600,000 m^3 (radius ~25 m x height ~300 m). This size is similar to that of the salt caverns in the Zuidwending underground gas storage facility (Crotogino et al., 2001). The maximum theoretical number of caverns that could be fitted in each salt pillar was calculated based on the directive described in German salt mining regulations. These regulations specify a minimum distance of 100 m between the cavern wall and the flank and 150 m between the cavern wall and the top of the salt pillar structure (ABVO, 1966) and a distance of 160 to 210 m between neighbouring cavern walls (IfG, 2008). The effective number of caverns was assumed to be a conservative 50% of the theoretical number, resulting in a total of ca. 321 possible caverns onshore, although the actual effective number of potential salt caverns could be much smaller if there are surface limitations and subsidence restrictions. The maximum operational cavern pressure considered was 180 bar, with a working to cushion gas volume ratio of 1:1. For a single cavern, this yields a working volume of about 53 x 10^6 m^3 (0.572 TWh) of natural gas or 45 x 10^6 m^3 (0.135 TWh) of hydrogen. Based on all assumptions above, a total potential working volume of 17.0 x 10^9 m^3 for natural gas (184.3 TWh) and 14.5 x 10^9 m^3 for hydrogen (43.3 TWh) was estimated for the onshore Netherlands in yet to be constructed salt caverns, distributed over the provinces of Groningen, Friesland and Drenthe (Table 20.5, Fig. 20.7d).

As it is the case for the current natural gas storage, studies show that the total cumulative capacity for hydrogen storage onshore and offshore is much larger than the current and expected hydrogen demand in the Netherlands (Van Gessel et al., 2021). In 2030 the demand for hydrogen storage is expected to lie between 42 and 475 GWh (14-16 x 10^6 m^3 hydrogen). For 2050, there are different storage scenarios for storage demands based on different weather patterns: low-case 1.3-4.3 TWh (0.43-1.5 x 10^9 m^3), mid-case 15-26 TWh (5-8.7 x 10^9 m^3) and high-case up to 32.9 TWh (11 x 10^9 m^3). In the event that 25% of the annual consumption would be stored as strategic reserves, as it is currently done for oil in the Netherlands and Europe, the total equivalent storage capacity needed could increase to 51 TWh by 2050. Considering the development time of approximately 3-6 years for the leaching of a single one million cubic metre (geometric volume) salt cavern and assuming 9 sites can be developed simultaneously, a maximum of 60 caverns with a total storage capacity of 15 TWh could be developed between 2030 and 2050 (Van Gessel et al., 2021). Therefore, it is likely that depleted gas fields will be required for hydrogen storage to match hydrogen demand with supply in the Netherlands.

Gasunie has already drawn plans to develop the first large-scale hydrogen storage site in a cluster of salt caverns in the Netherlands. These plans include the development of four new caverns at the current UHS storage site in Zuidwending, with a total storage capacity of 20 kilotonnes (HyStock, 2022), corresponding to ~235 million m^3 H_2 and ~667 GWh. In 2022, a feasibility study was successfully carried out in which the injection and withdrawal of hydrogen in a small salt cavity in an observation well in Zuidwending was tested over short periods. The first cavern is expected to be operational by 2028 (Energystock, 2022).

Hydrogen storage performance for the UHS sites of Grijpskerk, Norg and Alkmaar has also been investigated in detail, as the existing storage facilities could potentially be considered for reuse as UHS (Juez-Larré et al., 2023). The study shows the importance of estimating the storage capacity in terms of volume and energy based on a range of working pressures. This is because hydrogen has a much lower energy density than natural gas (Fig. 20.8). Current research on the possible reuse or retrofit of parts of existing natural gas infrastructures for hydrogen is also an important aspect for the technical development of UHS sites and grids (EBN, 2016; DBI-GUT, 2017; also see www.gasunietransportservices.nl). Another topic for investigation is the societal acceptance of underground activities on land in the Netherlands (e.g. Winters et al., 2020; Mendrinos et al., 2022; Sprenkeling et al., 2022). The possibilities for UHS in the North Sea are currently also being investigated, both for depleted gas fields and salt caverns, as they

Table 20.5. Results of the estimated storage potential of salt caverns (from Juez-Larré et al. (2019), www.nlog.nl). Listed are the number of effective salt caverns, the working volume (wv) and corresponding energy contents for natural gas and hydrogen results are given per salt pillar and grouped per province (see Fig. 20.7 for location). The working volumes obtained were converted to thermal energy content (TWh_t), using the energy factor 3.0E-9 TWh_t/m^3 (10.79 MJ/m^3) for hydrogen.

		Storage natural gas	Storage natural gas	Storage H_2	Storage H_2
Province/ salt pillar	# effective caverns	wv (10^9 m^3)	Energy content (TWh_t)	wv (10^9 m^3)	Energy content (TWh_t)
Groningen	**230**	12.2	132.1	10.4	31.0
Zuidwending	52	2.8	29.9	2.4	7.0
Winschoten	22	1.2	12.6	1.0	3.0
Pieterburen	39	2.1	22.4	1.8	5.3
Onstwedde	66	3.5	37.9	3.00	8.9
Boertange	51	2.7	29.3	2.3	6.9
Friesland	**31**	**1.6**	**17.8**	**1.4**	**4.2**
Ternaard	31	1.6	17.8	1.4	4.2
Drenthe	**60**	**3.2**	**34.4**	**2.7**	**8.1**
Anloo	14	0.7	8.0	0.6	1.9
Hooghalen	37	2.0	21.3	1.7	5.0
Hoogeveen	1	0.1	0.6	0.04	0.1
Schoonloo	8	0.4	4.6	0.4	1.1
Total	**321**	**17.0**	**184.3**	**14.5**	**43.3**

could be an alternative to onshore developments (e.g. NSE, 2022; Van Gessel et al., 2022; poshydon.com; see Fig. 20.7).

Underground Oil Storage (UOS)

The purpose of oil storage sites is to meet strategic and long-term demand. In the case of the European Union all member states are obliged to maintain a strategic oil reserve amounting to 90 days of consumption in critical sectors (Directive 2009/119/EC). As domestic oil production in the Netherlands is only ~5% of the national consumption, most of the oil is imported (CBS, 2022). Oil arrives at the Maasvlakte Oil Terminal (MOT), where it is distributed through pipelines to various surface storage sites and processing facilities in the harbour areas of Rotterdam, Vlissingen and Antwerp. The MOT has 39 oil tanks each with a diameter of 85 m and a height of 22 m, resulting in a total aboveground storage capacity of approximately 4 to 5 x 10^6 m^3.

The Netherlands Petroleum Stockpiling Agency (COVA) is responsible for organizing the storage of strategic petroleum reserves in the event that production and/or trade are interrupted for a longer period of time due to exceptional circumstances such as geopolitical conflicts and trade crises. COVA does not own any storage facilities and therefore its task is to select suitable storage companies and locations at the lowest possible cost for the emergency stock of crude oil for the Dutch state. Therefore, the stock of oil for the Netherlands is stored in several surface facilities in the Netherlands and Belgium. Crude oil is stored in some salt caverns in Etzel, northern Germany (www.storag-etzel.de) and diesel is stored in salt caverns in De Marssteden (Enschede). The Marssteden facility, commissioned in 2016, stores diesel (and home heating oil) in two salt caverns in the Röt Formation and has a total storage capacity of 250,000 m^3 (SodM, 2022). Unlike the deeper salt pillars of the Zechstein Group in the northern part of the Netherlands, the salt caverns at Marssteden were developed horizontally with a height of ~20 m and a diameter of ~130 m in flat salt layers (up to 100 m thick) at depths of 300 to 500 m. In general, salt caverns mined in layered salt formations are much smaller than in salt pillars. This is to ensure cavern stability as the degree of heterogeneity of the salt formation is much higher (Han et al., 2006).

Underground Nitrogen Storage (UNS)

Gas feedstocks like nitrogen can be stored in salt caverns in a similar way as for natural gas and hydrogen. In Heiligerlee (province of Groningen), nitrogen is being stored in a salt cavern of 860,000 m^3 in the Winschoten salt dome. This solution-mined cavern was commissioned in 2010 and was selected as the most suitable cavern among those within the Heiligerlee and nearby Zuidwending cavern clusters (RVO, 2010). The nitrogen stored in this cavern is extracted from air at the Zuidbroek facility (zuidbroek.gasunie.nl). Most of the nitrogen produced from this facility is directly blended with (imported) high-caloric natural gas to produce a Groningen (low-caloric) composition

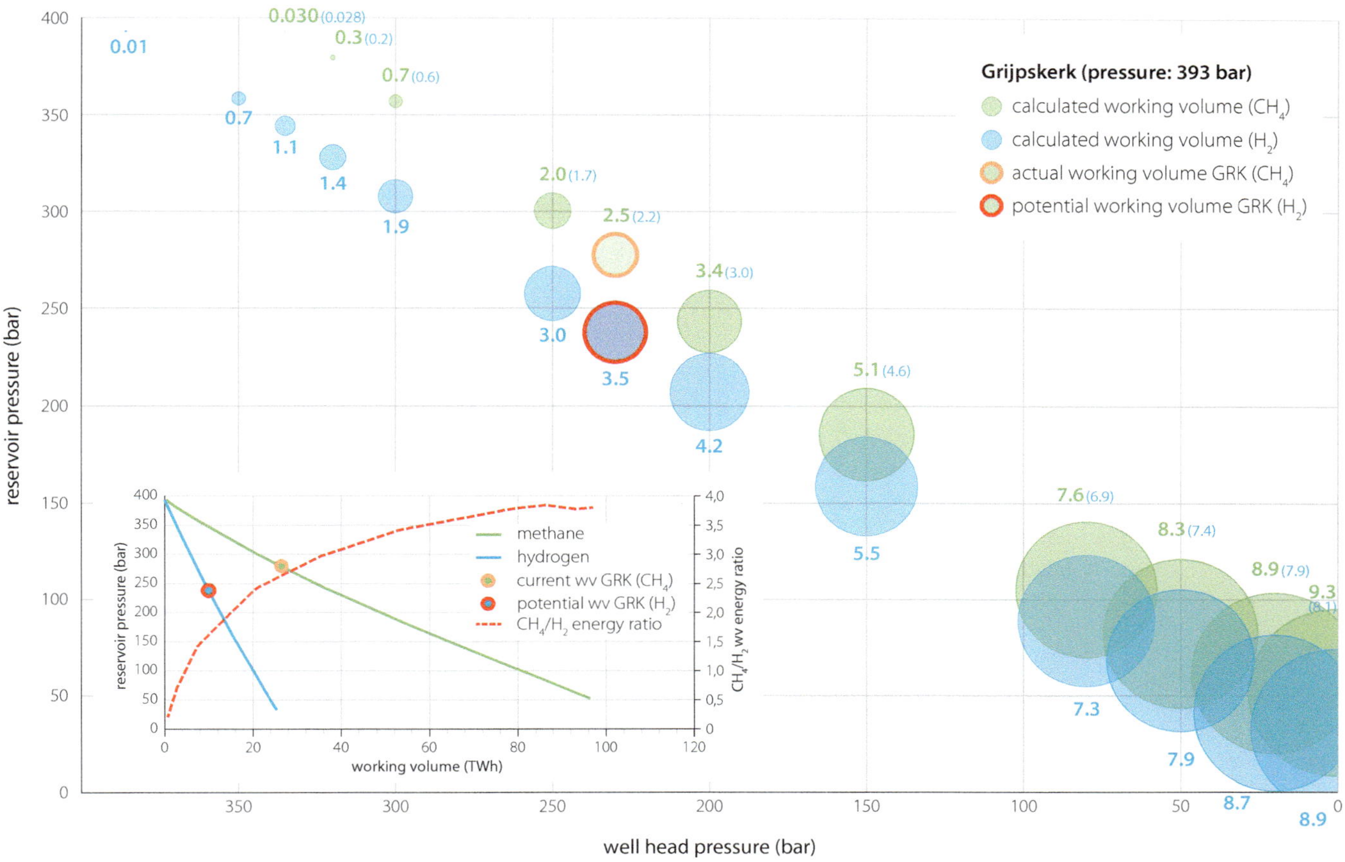

Figure 20.8. Working volumes (wv) of hydrogen (blue bubbles and numbers) and pure methane as proxy for natural gas (green bubbles and numbers) in billion Sm³ calculated by Juez-Larré et al. (2020, 2023) for the Grijpskerk (GRK) reservoir at 393 bar (max. storage pressure), considering a wide range of reservoir depletion pressures (min. storage pressure) and wellhead pressures. A cut-off withdrawal rate of 1 million Sm³/day was enforced. Next to the value for each working volume of methane, the equivalent volume of hydrogen at the same temperature and pressure is indicated in blue numbers between brackets. The graph in the lower left corner shows the corresponding energy content of the calculated working volumes (absolute and ratio) between hydrogen and methane. Modified from Juez-Larré et al. (2023).

gas (also known as pseudo-Groningen gas) suitable for domestic use in the residential sector and for export. The remaining nitrogen is transported via a pipeline (10 km) to the Heiligerlee salt cavern, where approximately 128 x 10^6 m³ can be stored annually, with a minimum available volume of at least 45 x 10^6 m³. This storage facility, along with a recent upgrade of the nitrogen facility at Zuidbroek, was initially realized to partially offset the loss of production capacity from the Groningen field by approximately 7 x 10^9 m³ per year (Gasunie, 2022). There are three other nitrogen facilities in the Netherlands (Ommen, Wieringermeer and Pernis). Although the Ommen and Wieringermeer facilities do not have an underground nitrogen storage, they are directly linked to pseudo-Groningen gas production facilities and are connected to the Grijpskerk and Alkmaar UGS, respectively.

Other gases, such as helium, can also be stored as feedstock. There are very few underground helium storage (UHeS) sites worldwide. Those that are in operation are mostly associated with the natural production of helium as byproduct in natural gas fields, as it is the case in the USA and Russia (e.g. Tade, 1967; Khan et al., 2012; Golovachev & Starokon, 2020). To date, there is no underground helium storage in the Netherlands. The nearest storage facility is in Gronau-Epe in Germany where helium is stored in a salt cavern at a depth of 1300 m. Helium is primarily used to liquefy nearly any other gas, mostly nitrogen and methane, that is, in cryogenic applications.

Compressed Air Energy Storage (CAES)

A CAES system stores electrical energy in mechanical form by compressing air. An electric motor drives a compressor that consumes electricity during off-peak hours when low-cost generating capacity is available (e.g. from nuclear or thermal power stations), or when non-dispatchable electricity is produced from Variable Renewable Energy (VRE) such as wind and solar (Fig. 20.9). The compressed air is stored in underground reservoirs, commonly salt caverns, at high pressure. At discharge, electricity is regenerated by employing compressed air from the reservoir(s) to drive a turbo-expander or a gas turbine, that drives an electricity generator and delivers electricity back to the grid. Because compressed air has a relatively low energy density (2-6 kWh per m^3 of storage space at pressures of 50-200 bar; Fuchs et al., 2015), air volumes in the order of 50-100 x 10^6 m^3 are required to conduct CAES at large-scale (power outputs typically between 100-500 MW and discharge durations of 6-12 hours). This type of storage can play a role in delivering short-to-medium-term and high power flexibility services to the electricity grid (Eckroad & Gyuk, 2003).

There are two main technology concepts: diabatic CAES (D-CAES) and advanced adiabatic CAES (AA-CAES). In a D-CAES system, the heat generated on compression of the air is not stored but is dissipated in the air. Therefore, to enhance discharge power and economics, an external fuel is normally used to heat up the air before driving the turbine. Natural gas is used conventionally, but its combustion generates CO_2 and NOx emissions (indicated with 'exhaust' in Fig. 20.9). Hydrogen is emerging as an alternative, in particular because combustion of hydrogen does not emit CO_2 and it can be produced from renewable electricity. Worldwide, there are two operational D-CAES plants in salt caverns, one in Germany (Huntorf, 321 MW/2.5 GWh, since 1978) and one in the US (McIntosh, 110 MW/2.6 GWh, since 1991) with round-trip efficiencies of 42% and 54%, respectively (Matos et al., 2021). Round-trip efficiencies approaching 60% are deemed feasible with efficient utilization of waste heat of the generation process.

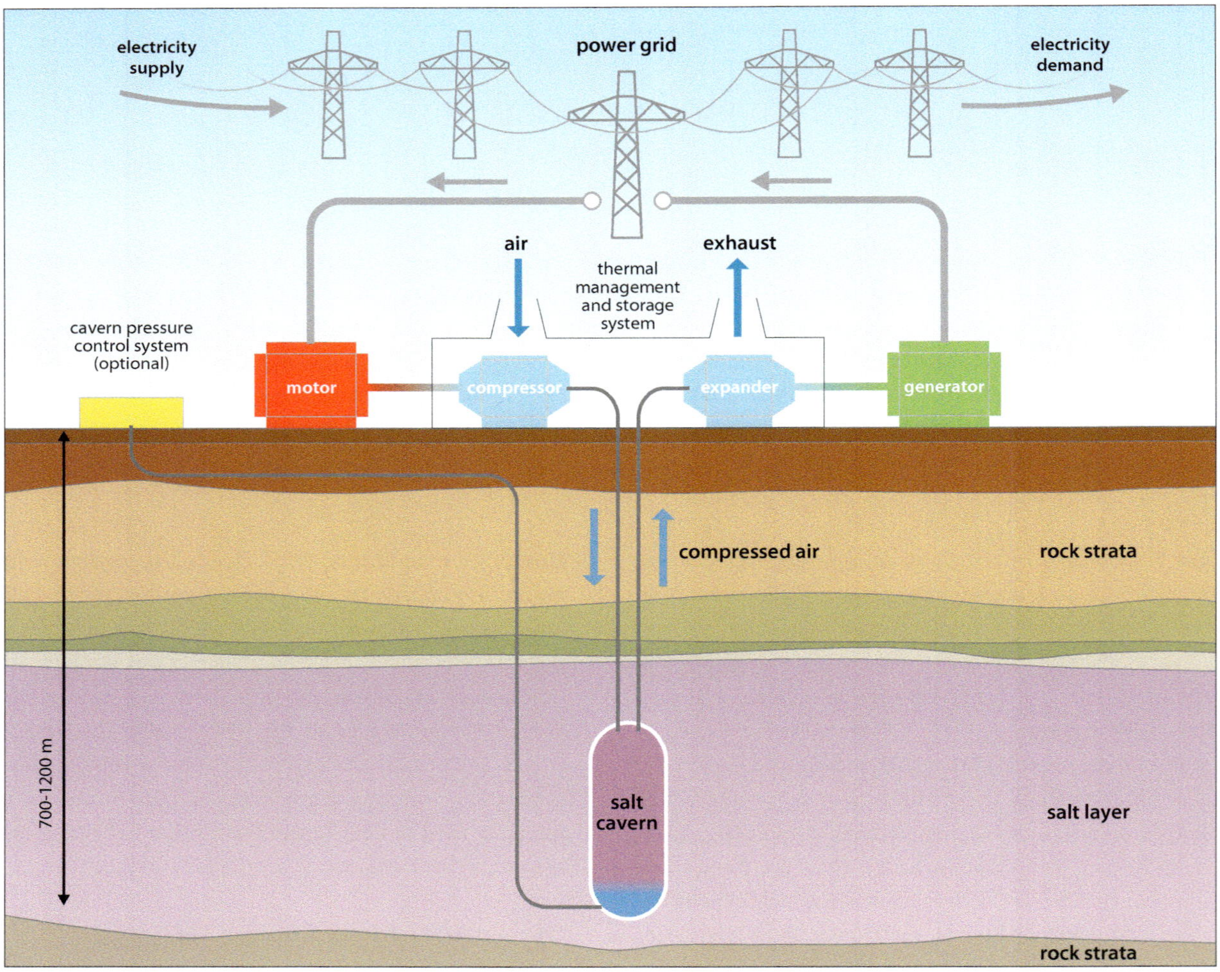

Figure 20.9. Schematic diagram of a CAES system (from Groenenberg et al., 2020b; after storelectric.com). In a diabatic CAES system, a fuel must be burned during (re-)generation of electricity to heat the air in the expander (turbine), which generates flue gases containing e.g. CO_2, NO_x and SO_2 (given natural gas) or water and NO_x (given hydrogen) leaving the system through the exhaust. In an adiabatic CAES system, the thermal storage and management system is capable of storing the heat lost during compression long enough to reuse it to heat the air, thus requiring no fuel.

In an AA-CAES system, the heat produced during compression is stored in a TES (Thermal Energy Storage device) and reused during the discharging process, which eliminates the need to use fuel to heat up the expanding air. With this method, round-trip (power-to-power) efficiencies of up to 70% can be reached. However, efficient thermal storage of heat at the very high temperatures involved (up to 580°C) is challenging and costly and the current level of technological maturity is not high enough for the technique to be commercially applied (King et al., 2021). In 2010, RWE Power and partners General Electric, Zublin and DLR launched the Adele project in Staßfurt – Germany (RWE, 2010) as the world's first large-scale AA-CAES demonstration project. Although the concept and design have been established (200 MW up to 5 hours with a 70% round-trip efficiency), this project was placed on hold in 2016 citing uncertain business conditions. Despite all these issues some major projects are underway on CAES (King et al., 2021).

The Dutch subsurface has a large potential for CAES in salt caverns. To estimate the total CAES capacity onshore, Van Gessel et al. (2018) used the same geometric volume of 0.3-0.6 x 10^6 m^3s for potential development in salt caverns as used for hydrogen storage. In the study, a notional CAES facility was used with a discharge power of 300 MW for a period of 6 hours at assumed operational pressures up to 180 bar, which equals an approximate required energy storage capacity of 1800 MWh (1.8 GWh). With these conditions the total technical potential for developing CAES in salt caverns in the Netherlands was estimated to lie at around 0.58 TWh (Van Gessel et al., 2018). Table 20.6 shows the storage capacity estimated for each province and salt pillar. In recent years various demonstration and commercial projects have been developed, mainly based on the D-CAES concept. This indicates a strong renewed interest in CAES (King et al., 2021), probably sparked by the increasing need for flexibility services to integrate a growing share of variable renewables (wind, solar) into the energy system. In the Netherlands, project developer Corre Energy intends to build a 320-MW D-CAES plant with a storage capacity of 3-4 GWh (Corre Energy, 2022). The project obtained the status of European Project of Common Interest (PCI) in 2017 and receives financial support from the Connected Europe Facility (CEF) fund, which can be considered a recognition of the potential value of this technology in providing flexibility to the increasingly renewables-based European energy system.

Table 20.6. The total number of (effective) caverns and corresponding energy storage for CAES per province and salt pillar (see location in Fig. 20.7). The effective number of caverns is a conservative estimate based on 50% of the theoretical number of caverns that could be mined in a salt structure. From Juez-Larré et al. (2019).

Province/ salt pillars	#Effective salt caverns	Energy storage (TWh)
Groningen	**230**	**0.414**
Zuidwending	52	0.094
Winschoten	22	0.039
Pieterburen	39	0.069
Onstwedde	66	0.119
Boertange	51	0.092
Friesland	**31**	**0.056**
Ternaard	31	0.056
Drenthe	**60**	**0.108**
Anloo	14	0.025
Hooghalen	37	0.067
Hoogeveen	1	0.002
Schoonloo	8	0.014
Total	**321**	**0.577**

Underground Pumped Hydro Energy Storage (UPHES)

This technology is used to store electricity in the form of gravitational potential energy. Large amounts of water are pumped from a lower to a higher elevation basin using low-cost surplus off-peak electric power. In periods of peak electrical demand, the water collected in the upper basin is released into the lower basin through a turbine generating electric power. The round-trip efficiency of a PHES system varies from 60 to 80% (Letcher, 2016). The storage capacity is proportional to the product of the total mass of water and the difference in height between the two basins.

The aboveground variant of the pumped hydro energy storage, located in mountainous areas, is the most widely used electricity storage technology worldwide since the early 20th century (i.e. it has the largest installed capacity). In Europe, about 190 aboveground pump-accumulation systems are currently operational or are under construction (DOE, 2022). The total power of all these systems is approximately 60-70 GW. The average discharge time is 6 to 12 hours. The biggest systems have a capacity of 1 to 2 GW. Aboveground PHES are more difficult to build in areas with lower topographical gradient. In flat regions, a height difference can only be created by placing at least one of the two basins underground. However, the underground excavation and material handling costs, construction risk and the time required to complete the project can hinder the economics of an UPHES project. As a result, most developers around the world are considering utilizing existing subsurface structures such as abandoned mines (e.g. Madlener & Specht, 2020). Preconditions for underground structures are the absence of large fractures and that the UPHES is operated unpressurized to ensure mechanical stability. There are no UPHES in operation today, although many sites and projects have been proposed (e.g. Kitsikoudis et al., 2020). There is only a 500 MW pumped hydro power station in preparation in Paldiski, Estonia (zeroterrain.com).

In the case of the Netherlands, besides the Lievense plan for an aboveground pumped hydro storage in the Markermeer (Rijkswaterstaat, 1982) and the DELTA21 concept next to the Tweede Maasvlakte (www.delta21.nl), there has been a proposal to build a UPHES (OPAC group) in Graetheide in the southeast of the Netherlands (Limburg) since the 1980s. The planned UPHES with a power of 1400 MW could execute about 250 cycles a year, each lasting six hours, storing up to 2.1 TWh annually (Huynen, 2018; Kramer et al., 2020). The design includes a surface water basin (~500 x 500 m) connected to an underground reservoir and machine hall at 1400 m depth, with three mineshafts to collect the water. The underground reservoir could be constructed in the thick succession of the lower Carboniferous (Dinantian) carbonates of the Carboniferous Limestone Group (700 m thick in exploration borehole Geverik-1), as it is a strong and stable rock formation (Müller & Hereth, 1987; Kramers et al., 2011). The construction of shafts, machine hall and tunnels would make use of proven technology. Although underground constructions are common in the Netherlands at shallow depths, the realization of this project at 1400 m depth would be new to the country. The absence of a 'tradition' in hydroelectric energy storage investments, creates an obstacle that has yet to be overcome (Kramer et al., 2020).

Underground Thermal Energy Storage (UTES)

Thermal energy storage (TES) uses the thermal properties of materials to store heat and/or cold. Thermal energy sources are abundant and include industrial waste heat, solar heat, geothermal heat or heat converted from excess wind or solar (photovoltaic) electricity by heat pumps or an electric boiler. Sources of cold are related to industrial processes, low ambient temperatures and surface waters. Where thermal energy storage takes place underground (UTES) in geological formations, it involves significantly larger volumes of sedimentary rock and water than in surface installations. For UTES, water is the most preferred medium because it has one of the highest specific heat capacities (~4.17 MJ/(m^3 °C)), about 3 times larger than that of most rock formations (Cabeza, 2021). The prospect of easily storing large amounts of energy underground is the reason why the number of UTES projects is rapidly expanding in Europe, especially in the Netherlands.

A large variety of UTES concepts are being applied or are in the development phase, among others: Aquifer-TES, Borehole-TES, Tank-TES, Mine-TES and Pit-TES. Other concepts of underground heat storage are still in the conceptual phase, such as depleted oil reservoirs, and will not be further discussed in this chapter (Stricker et al., 2020). The scale at which UTES is applied can vary from the very local (e.g. single house) to city scale (e.g. large heat networks), depending on the temperature ranges and volumes provided by the heat/cold source, the type of demand and the properties of the underground reservoirs. The principles of each of the UTES concepts is briefly described next and is linked to specific underground formations in the Netherlands that have favourable heat storage properties. In particular, many Aquifer-TESs and Borehole-TESs are currently installed in the Netherlands and there is great potential for further development.

Types of Underground Thermal Energy Storage

Aquifer-TES (ATES) is the storage of heat or cold in aquifers typically found at depths of several hundreds of metres. In the Netherlands the Water Act regulates all subsurface activities at depths shallower than 500 m (below ground surface). At greater depths, the Mining Act applies, with strict regulations for ATES, as this type of storage requires specialized drilling equipment and techniques. Storage in an ATES system is carried out through one or several wells that extract groundwater and reinject it into aquifers (Fig. 20.10a). The ATES design needs to meet some requirements such as specific sequences of interbedded sand and clay formations with certain thicknesses and permeabilities. Most of the targeted shallow aquifers are made of unconsolidated sand or gravel formations which are tens of metres thick, with a good permeability (>7 Darcy) and overlain by a layer of clay (aquiclude). Currently, there are more than 2800 ATES systems in operation worldwide, 99% of which are Low-Temperature ATES (<25°C) and 85% of which are located in the Netherlands (Fleuchaus et al., 2018). They can contain between 10,000 and 500,000 m^3 of water (to max. 5 x 10^6 m^3 for multiple systems in a network) with temperatures of up to 25°C (Bloemendal & Hartog, 2018). The power of small- and large-scale systems is 100-300 kW and 5-30 MW, respectively (Fleuchaus et al., 2018). Where water is stored at higher temperatures (up to 90°C), systems typically become cost-effective when the thermal power is more than 5 MW (Kallesøe & Vangkilde-Pedersen, 2019) and where associated storage capacity is more than 10 GWh.

Borehole-TES (BTES) utilizes one or several boreholes with depths from tens of metres up to 250 m. Each borehole contains a 'U' shaped pipe that is inserted into the borehole and then filled with a grouting material to anchor the pipe and to provide good thermal contact with the surrounding formation. The working fluid carrying heat or cold is circulated through the borehole, working as an underground heat exchanger as the heat (or cold) is transferred from the borehole to the surrounding rock formation in a closed loop. Geological formations used for BTES

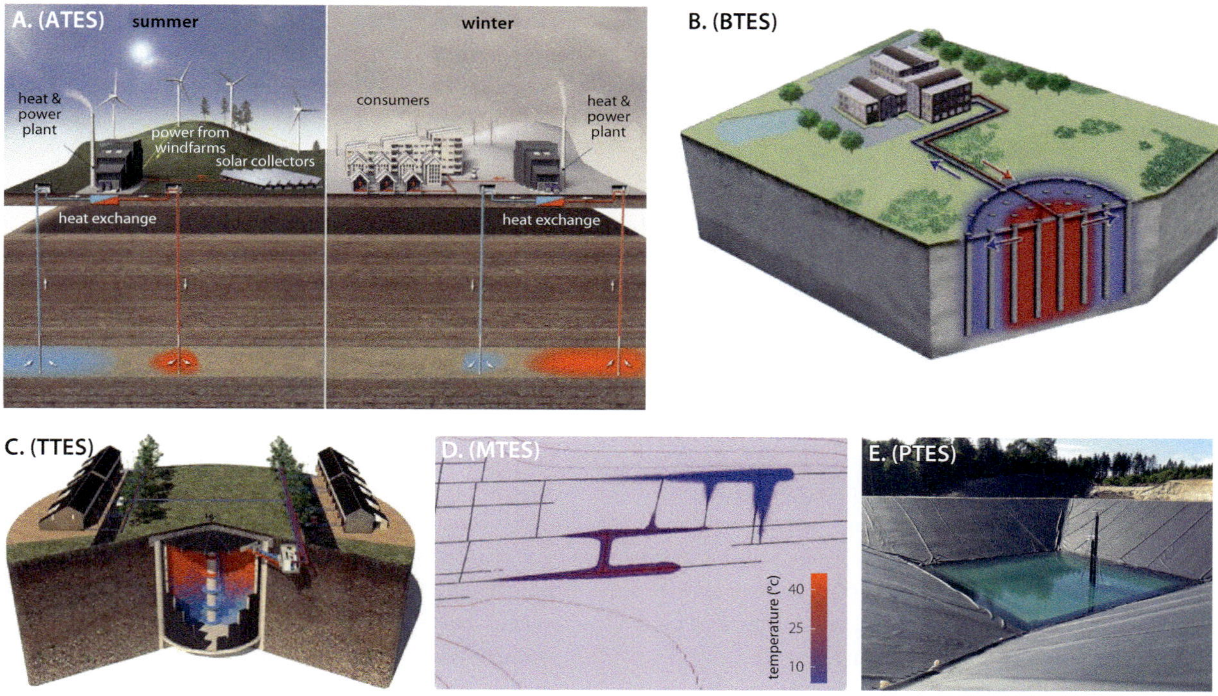

Figure 20.10. *Overview of different UTES storage types: a) Aquifer-TES (© Carsten E. Thuesen, GEUS 2017), b) Borehole-TES (© Underground Energy LLC), c) Tank-TES (© Ecovat), d) Mine-TES (© HEATSTORE 2021a, www.heatstore.eu), e) Pit-TES (© PlanEnergi).*

can range from unconsolidated material to rock (e.g. tight limestone or unfractured crystalline rocks) with or without groundwater, and in order to retain heat efficiently ideally have high heat capacity, low thermal conductivity, limited natural groundwater flow and/or low permeability. Single or small low-temperature BTES systems are typically connected to a single household or utility building. They can provide power of up to 10 kW per individual low-temperature BTES system, while multi-borehole systems can deliver power between 70 and 200 kW. About 75,000 small BTES systems are installed in the Netherlands (wkotool.nl). Large-scale high temperature BTES systems are often designed in radial clusters, with hot water circulating from the inner to the outer boreholes, or vice versa for cold water (Fig. 20.10b). In the Netherlands, an example of a relatively large-scale HT-BTES system is located in Beijum (Groningen), which was built in 1985. It stores solar heat at 60°C with a storage capacity of 600 MWh at a depth of 20 m in the Boxtel and Eem formations (Kallesøe & Vangkilde-Pedersen, 2019). Outside the Netherlands, an example of a large scale BTES system can be found in Brædstrup (Denmark), which stores 400 MWh with a water volume of more than 20,000 m^3 (HEATSTORE, 2021a).

Tank-TES (TTES) systems are (semi-)underground tanks or vessels storing hot water. Small aboveground/underground tanks, with volumes typically up to 12,000 m^3, are commonly used for daily or weekly storage. One of the largest aboveground thermal energy storage tanks is located in Berlin, Germany and has a storage volume of 56,000 m^3 and storage capacity of 2.6 GWh (Vattenfall, 2021). It can run at a maximum thermal output of 200 MW for 13 hours. In the Netherlands, there is a large aboveground storage tank in Diemen, 50 m high and 26 m in diameter, with a storage capacity of 22,000 m^3 that can supply heat to 75,000 homes in Amsterdam, Amstelveen and Almere (Vattenfall, 2013, 2022). Larger tanks may be required for long-term seasonal storage, and although they can be buried underground, they are not designed to take advantage of the thermal properties of subsurface geology. Underground TTES simply needs an easily excavatable soil. Tanks consist of an outer concrete vessel containing a highly insulated inner vessel (Fig. 20.10c). The energy is exchanged by pumping the hot or cold water from the vessel through a centrally located heat exchanger. The heated (or cooled) water is then returned to the vessel through diffusers at different heights, in such a way that the thermal stratification of the water in the vessel is preserved. This allows for high efficiency, as no water is pumped out of the vessel. An example of a large conceptual underground TTES is the one being designed by GroeneWarmte with a diameter of 30-48 m, a depth of 29-54 m and a storage capacity of 1.7 to 8 GWh (Ecovat, 2022).

Mine-TES (MTES) is the storage of excess heat in mine shafts at depths between a couple of hundred metres and up to 2 km, which are naturally or intentionally flooded with groundwater. Hot water (heated at the surface) is injected and maintained at a temperature higher than that of the rock in-situ (Fig. 20.10d). MTES is primarily limited by the availability and integrity of existing shallow-to-deep shallow mine shafts in heat- or cold demand areas. Systems already exist in which water is extracted from deep and shallow flooded mine shafts as a source of heat and cold. Examples of such systems are found in Heerlen, Saxony, Bochum and Essen in eastern Germany (HEATSTORE, 2021b). The Mijnwater project in Heerlen (southeast Netherlands) provides heating and cooling for 350 homes and 9 large utility buildings (mijnwater.com). In Bochum (Germany) a MTES is currently being designed (HEATSTORE, 2021b) and in the city of Kuusikonmäki in Vantaa (Finland), there are plans for the construction of a storage cavern in bedrock at a depth of a 100 m (Cavern-TES). It will store one million cubic metres of water heated up to 140°C by waste heat, and will have a total storage capacity of 90 GWh (Vantaan Energia, 2021).

Pit-TES (PTES) consists of a large shallow pit with walls that slope gently (~27°C) towards the centre and are covered with an impermeable plastic liner (Fig. 20.10e). This technology requires a stable soil to avoid bank instability and a low groundwater table and low natural groundwater flow to reduce heat losses. The water in the pit can reach a maximum temperature of up to 95°C. Like TTES, pit storage is also stratified by temperature and can be used as inter-seasonal storage for heat and cold, but also for daily or weekly use. A PTES usually stores volumes of 6 x 10^4 to 2 x 10^5 m^3. There are no plans yet for the construction of a PTES in the Netherlands. Six PTES systems have already been built in Denmark and more systems are planned (HEATSTORE, 2021b), such as a large one in Aalborg (5 x 10^5 m^3) which has a thermal power potential of between 10 and 40 MW and a storage capacity of 3-12 GWh (Danish Energy Agency, 2018; Fyhn, 2021). Table 20.7 summarizes the requirements and performance indicators for the five types of UTES discussed.

Suitable (hydro)geology for low- to high-temperature ATES and BTES

According to the current Dutch Water Law, the maximum temperature for subsurface injection is limited to 25°C at depths shallower than 500 m. This is labelled low-temperature (LT) and includes most of the ATES and BTES systems installed in the Netherlands so far. Medium (MT: 25-60°C) to high (HT: 60-100°C) temperature ATES and BTES require an exemption from the current legal framework.

ATES and BTES are considered very promising UTES technologies for the Netherlands. From an economic point of view, ATES and BTES have one of the lowest investment costs per unit of energy delivered and require a small surface area. They offer a wide range of market services and storage capacities (Yang et al., 2021), although their current application and future potential are greatly influenced by the suitability of the subsurface and potential interference issues with other activities in the shallow subsurface (e.g. drinking water production).

Future development of MT and HT storage systems can provide larger amounts of hot water to reduce peak demand (currently supplied by natural gas) and/or optimize renewable energy production. In particular, HT-ATES sourced with geothermal, waste or solar heat can be connected to a network to provide a stable supply of heat during the winter. This has an added value when combined

Table 20.7 Summary of the requirements and key performance indicators for different underground thermal energy storage types (high and low temperature). Source: Van der Stoel (2022).

	ATES	BTES	TTES	MTES	PTES
Subsurface requirements	***	**	*	***	*
Pre-investigation	***	***	**	**	**
Surface area[a] [m²]	*	*	** ≤2500	*	*** ≥10^4
Storage volume [m³]	250,000–10^6	>20,000	1000–98,000	25,000–10^6	50,000–500,000
Storage capacity [MWh]	15,000–55,000	>250	100–5000	1000–90,000	5000–40,000
Power [MW]	≥1	0.1–1.5	1–17	≤200	10–40
Investment costs [€/m³]	5–10	20–50	150–280	≥80	25–40
Operation and maintenance[b] [%]	4–5	0.5	0.5	n.a	1
Recovery efficiency [%]	50–80	40–60	70–95	n.a.	70–90
Scalability	***	***	**	*	**

*** = high; ** = medium; * = low
[a] This excludes physical space needed for a technical room, as this is always required. [b] As share of investments cost per year.

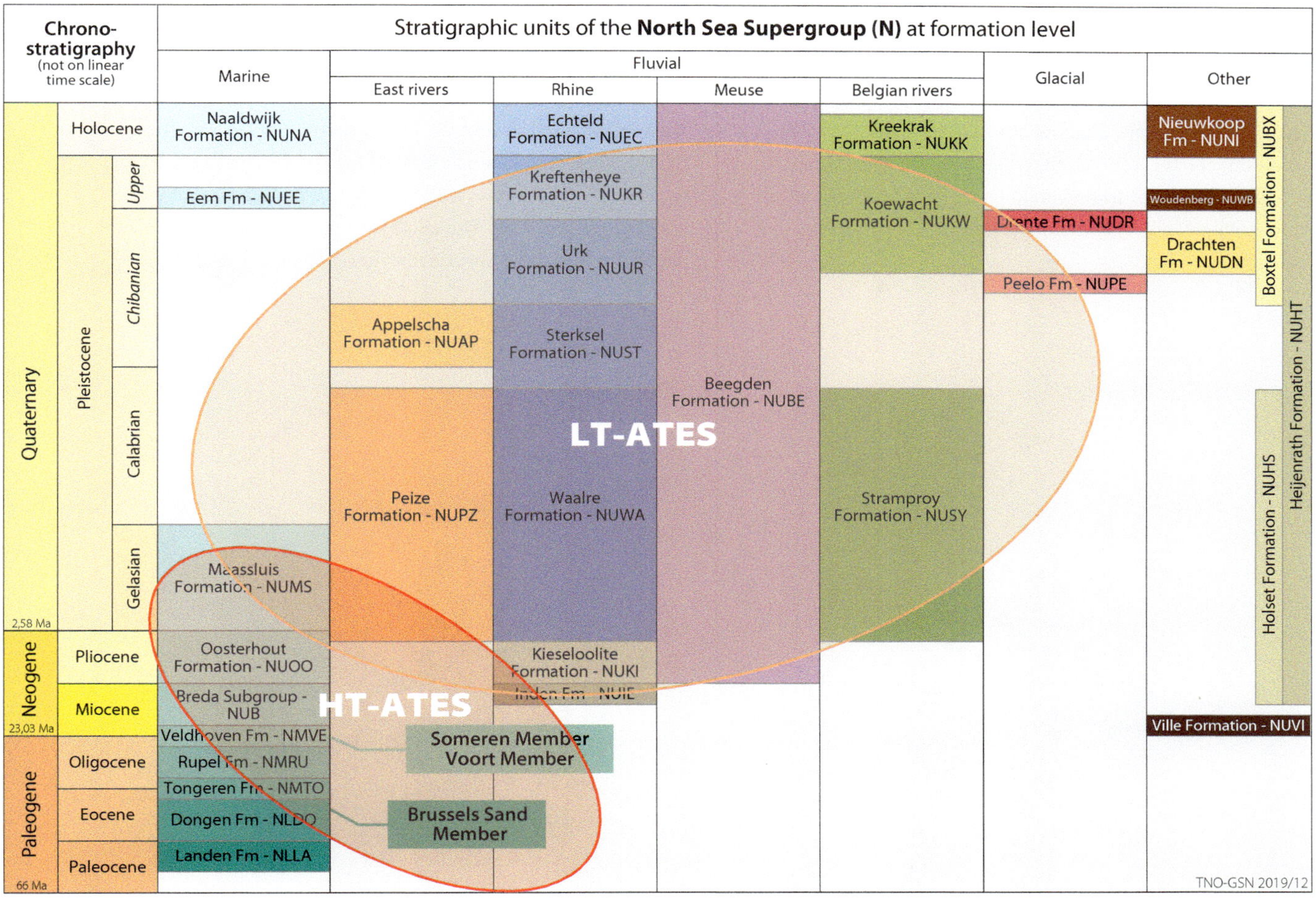

Figure 20.11. Stratigraphic units of the North Sea Supergroup (N) at formation level, with an indication of suitable formations for LT- and HT-ATES (modified from www.dinoloket.nl). Suitable formations for MT-ATES overlap with both LT- and HT-ATES, as this is highly dependent on the applied temperature level and local circumstances.

with geothermal heat, as the geothermal doublet can be used as a baseload, avoiding a suboptimal fluctuating production profile (Van Gessel et al., 2021; Mijnlieff et al., 2025, this volume).

The Dutch subsurface has a great potential for the development of LT-ATES due to the large availability of aquifers with good permeability. These aquifers belong to deltaic depositional systems with marine, coastal and fluvial sediments deposited during the Paleogene, Neogene and Quaternary (Busschers et al., 2025; Munsterman et al., 2025, both in this volume). Successions of coarse-grained unconsolidated sands with high permeability (>30-40 Darcy) and thicknesses greater than 15 m intercalated with clay layers, provide the perfect confined stratified porous medium for the storage of large amounts of low-temperature thermal energy (LT-ATES). The targeted geological formations of the LT-ATES systems are mostly found in fluvial formations (Kreftenheye, Appelscha, Sterksel, Urk, Beegden, Peize- and Waalre, Kieseloolite) and glacial formations (Peelo, Drente; Fig. 20.11). Very few ATES systems have been installed in shallow marine formations such as the Maassluis and Oosterhout formations (Fig. 20.11), as they are known to have greater vertical heterogeneity and to contain an abundance of fine-grained sediments, which tend to migrate towards the producing well causing severe fines production and/or clogging (Cabeza, 2021).

However, the Maassluis, and Oosterhout formations, and the Breda Subgroup seem to have the largest potential for HT-ATES, as they lie deeper (~200-500 m), are hotter and have medium- to fine-grained sediments with relatively low permeability, which can significantly reduce the 'buoyancy flow effect' that causes a reduction in the effective recovery (Marif, 2019).

Technically, HT-ATES systems could be drilled to a depth of at least 700 m with shallow drill rigs, but for greater depths more expensive drill rigs are required. The economic viability of HT-ATES systems also relies on the flow rate that can be achieved (Wesselink et al., 2018). Flow rate depends on reservoir permeability, which in deeper sand reservoirs is expected to be lower due to, among other things, compaction. In addition, some legal aspects will have to be considered, since these systems must comply with the regulations of the Mining Act rather than the Water Law. Shallow aquifers (<50 m) are mostly ruled out for HT-ATES as they lie above the freshwater/to brackish/salt water transition and as such are often part of the (strategic) groundwater reserves.

Subsurface potential maps for existing systems

The potential for LT-ATES and LT-BTES in the Netherlands has already been demonstrated with the large number of systems installed during the last 25 years (Fig. 20.12a and b). HT-ATES uses deeper formations for which the subsurface potential is less well known and the requirements are slightly different. Dinkelman et al. (2020) applied a pre-defined set of subsurface criteria to assess and map

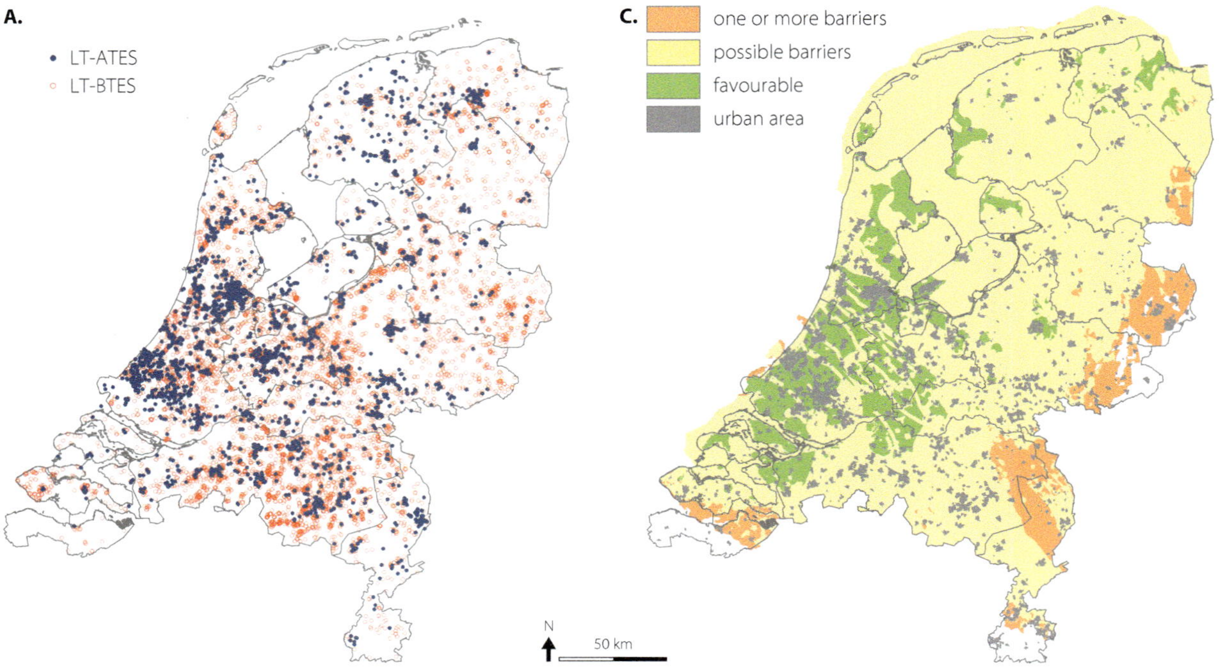

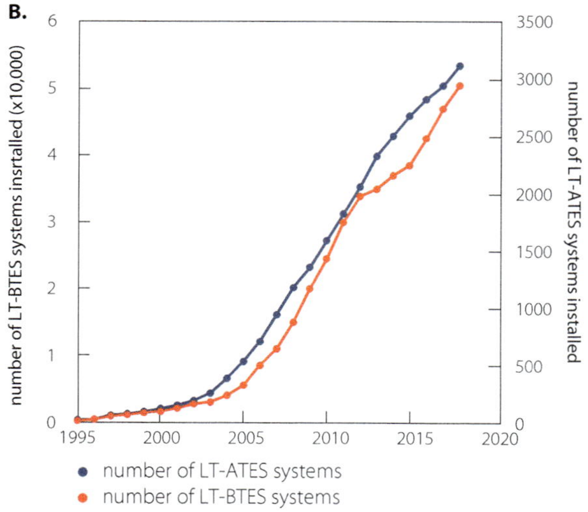

D.

Parameter	Barrier	Possible barrier	Favourable
Lithology	no data available on national scale		
Depth		<50, >500 mbgs*	50-500 mbgs*
Thickness sand layer	<10 m	10-15 m	>15 m
Horizontal hydraulic conductivity – Kh value	<5 m/d		≥5 m/d
Presence of confining cap layer (clay)		thickness <5 m	thickness >5 m
Faults		<1 km	>1 km
Groundwater flow velocity		>25 m/yr	<25 m/yr
Chloride concentration		<1 g/l	>1 g/l

***Figure** 20.12. a) ATES and BTES systems in the Netherlands (wkotool.nl). b) Cumulative amount of LT-ATES and LT-BTES installed (STOWA, 2020). c) Potential map for areas with favourable geological formations for the development of HT-ATES in the Netherlands shown as favourable (green), less favourable due to presence of possible barriers (yellow) and unfavourable (one or more barriers). This is based on the criteria depicted in the table (d) such as depth, thickness, hydraulic conductivity of the aquifer and the presence of a confining layer (aquitard) (Dinkelman et al., 2020). (*) mbgs: metres below ground surface.*

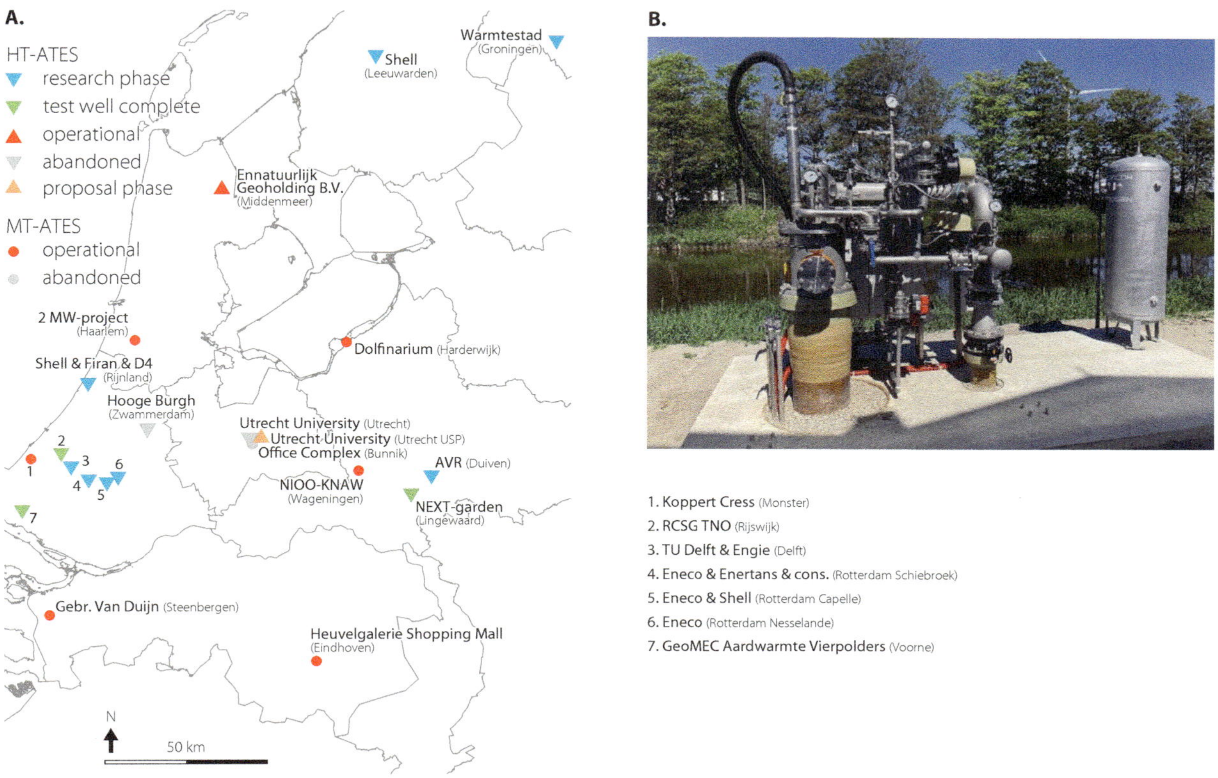

Figure 20.13. *a) Status of MT-ATES and HT-ATES (research) projects in the Netherlands (from publicly available information as of 2023). b) Hot well surface equipment of the HT-ATES installation of Ennatuurlijk Aardwarmte in Middenmeer developed in the HEATSTORE project (source: IF Technology).*

potential obstacles for HT-ATES. The most favourable areas for HT-ATES projects are concentrated in the west of the Netherlands (Fig. 20.12c and d). Less favourable areas are areas where one or more barriers may be present due to less good properties of the subsurface. In addition to the subsurface potential, other factors are important for the successful development of an HT-ATES system. The technical requirements are a good heat recovery efficiency (minimum 60-90%), the availability of a heat source and heat demand (HEATSTORE, 2021b). During the production or recovery period, the produced water temperature will drop over time and a heat pump may be required to raise the temperature of the water produced. Scaling and corrosion issues in the wells can also complicate operations. The economic requirements of a HT-ATES system are determined by the seasonal variations in the volumes and temperatures of heat demand, combined with the presence of heat sources with low marginal costs of production (e.g. waste, geothermal or solar heat; see HEATSTORE, 2021b). The operating temperature of the heating network is especially important, since it often determines the difference in temperature between the hot and cold wells. This, together with the flow rate and the duration determines the energy capacity of the storage project. Storage with high power and/or storage capacity significantly reduces the cost of storage per unit of energy. Experience from previous projects and feasibility studies indicates that the power of a HT-ATES should be at least 5-10 MW (thermal) and with an energy storage capacity equivalent to 2500 annual hours at full load. For the Netherlands, this equates to the annual heat demand of ~1200-2500 households.

Currently, several MT-ATES and HT-ATES projects are in operation (Fig. 20.13a). Two HT-ATES systems were operational already in the Netherlands in the 1980s, but were eventually shut down due to technical failures, lack of economic incentives due to the competition with low natural gas prices and a general mismatch between supply and demand (Bakema & Drijver, 2019). In the Netherlands, the HT-ATES in Middenmeer is an example of a successful project commissioned in 2021 (Fig. 20.13b). At Middenmeer, the operating geothermal system supplies hot water at ~85°C to nearby greenhouses, but in periods of low demand, the excess heat produced is stored in an aquifer located at a depth of 360-380 m in the shallow marine Maassluis Formation (Oerlemans et al., 2022).

The current energy transition is an incentive for the development of new HT-ATES projects. Many research institutes and industry stakeholders around the world are collaborating to investigate, design and implement HT-ATES systems. Examples are the European HEATSTORE project (HEATSTORE, 2021a,b; www.heatstore.eu) and PUSH-IT (www.push-it-thermalstorage.eu). In the Netherlands, initiatives such as the WarmingUP project continue investigating HT-ATES with the aim of removing technical, legal and economic barriers by 2025 (www.warmingup.info).

Long-term (permanent) underground storage

Storage of production water and filtrate residues

In the Netherlands, two main types of water are injected in the underground for permanent storage. One type is the water produced during oil and gas extraction, and another are the filtrate residues left behind after purification of surface water for drinking purposes.

During oil extraction, some of the deep underground water is brought to the surface. To a lesser extent this also occurs during the final phases of natural gas extraction. Although this so-called 'production water' is separated from the oil and gas at the surface, it still contains trace amounts of oil or gas and radioactive elements. It also contains traces of chemicals used during the production process, such as corrosion inhibitors used in wells or glycol or ethanol used in the gas treatment facility. Production water can be managed in two ways, namely by reinjecting it back into the oil or gas field, into a deep nearby aquifer, or by cleaning it and discharging it to the surface. The balance between these options is made in accordance with strict regulations to protect people and the environment (SodM, 2005). For offshore, the cleaning and disposal of production water in the sea takes place in accordance with the OSPAR convention (Karman & Smit, 2019; www.ospar.org). In practice, it seems that when large volumes are involved, reinjection is the most environmentally friendly form of disposal (Brolsma et al. 2017). Once reinjected into the field, the same top layer that sealed off the oil, gas or adjacent aquifers for millions of years ensures the safe and permanent storage of these waters. Two important factors are considered for the reinjection of water in an oil and/or gas field. In the interest of containment and risk of seal fracture the field used for water injection must not be seismically active and no more water can be injected into the field than the original volume of oil and gas.

Reservoirs used for the injection of production water in the Netherlands are mainly located at depths between 0.7 and 3.0 km. They consist of sandstones belonging to the Rijnland Group and Upper Rotliegend Group and carbonates of the Zechstein Group (Hoving et al., 2023). There are currently nine onshore locations where production water is regularly injected into depleted oil/gas fields. These include the Groningen (Borgsweer), Pernis-West, Rotterdam, Schoonebeek, Rossum-Weerselo, Bergermeer, Starnmeer and Nijensleek fields. In the Schoonebeek oil field, the injected steam, which has allowed the production of the high-viscosity oil since 2011, is re-produced as brine at a rate of 6000-12,000 m^3/day. Part of this brine is reinjected into the field and the rest is transported and injected into the Rossum-Weerselo field. Some water was also injected into the Tubbergen and Tubbergen-Mander fields between 2015 and 2016. From all deep reservoir injection activities in the Netherlands since 1972 (Borgsweer), there have been no reports of chemical reactions in the reservoirs or seal layers or waste causing any negative environmental impact in shallower formations (Hoving et al., 2023). The fact that in all cases the aquifers used for water disposal lie much deeper than any drinking water aquifer (>500 m), means that all these activities are in accordance with the General Provisions Act and are managed under it (WABO, 2022).

As for the permanent storage of filtrate residues in the Netherlands, there is currently one underground storage in Andijk. At this location, the surplus water flow after water purification of the IJsselmeer lake is injected for permanent storage in brackish aquifers at a depth of 200-250 m. The injection flow rates are a few tens of m^3/hour. The reservoirs used are the Peize and Waalre formations of the Upper North Sea Group, composed mainly of stacked fining-upwards fluvial sequences. In the past there have been other permanent water storage systems in shallow reservoirs (100-500 m) in Noordbergum, Ridderkerk and Zevenbergen. Despite the shallow depth at which these waters are injected, the Mining Act also takes into account the permanent underground storage of filtrate residues.

Underground storage of CO_2

The Dutch government has a target to reduce greenhouse gas emissions from the Netherlands by at least 49% by 2030 compared to 1990 levels, and 95% by 2050 (MEACP, 2019a) to align with the increasing EU emission reduction targets (EC, 2023). Underground CO_2 storage (carbon sequestration) is nowadays considered one of the technologies that can help to reduce the emissions of CO_2 into the atmosphere. Despite the current transition to renewable energy sources, 80% of the world's energy supply is still derived from fossil fuels (IEA, 2022b). Large amounts of CO_2 can be captured from large point-source emitters. These sites include power plants, various large industrial facilities (e.g. steel and cement manufacturers) and hydrogen production sites. In power plants and industrial facilities CO_2 can be captured 'post-combustion', while in hydrogen production sites CO_2 is captured 'pre-combustion'. CO_2 can be captured using solvent-based (in liquid) or sorbent-based (on solid particles) methods, through cryogenic separation, oxygen-fired combustion or via membrane separation (Ringrose, 2020). CO_2 can be transported to the storage site using pipelines, ships or trucks, ideally in dense or liquid phase, depending on operating pressure and temperature. A CCS project can be combined with carbon capture and utilization (CCU), when CO_2 can be utilized, for instance in the production of high value chemicals, in greenhouse horticulture or in geo-methanation (Zauner et al., 2022).

For long-term geological storage, CO_2 is injected into suitable porous underground reservoirs (e.g. sandstones or carbonates) beneath thick and continuous overburden formations with good sealing properties (e.g. salt or shale). The suitability of the reservoir depends on the storage resources, injectivity and containment characteristics. Reservoir depth for CO_2 storage should be greater than ~800 m in order to make efficient use of pore space when storing CO_2 in the dense (supercritical) phase. Besides depleted gas and oil reservoirs, and saline aquifers, some CO_2 storage projects have used other reservoirs such as coal seams, volcanic rocks (basalt) and underground caverns (CO2RE, 2022). For the Netherlands in particular, depleted natural gas reservoirs, and to some extent also aquifers, are currently considered the most promising storage sites (CATO, 2022).

Several containment mechanisms contribute to the storage of CO_2 in porous reservoirs (Frykman, 2002; Benson et al., 2005; Fig. 20.14). In depleted gas fields, the primary containment is structural trapping. For aquifers, additional important mechanisms are residual trapping (the CO_2 is kept in pores by capillary forces) and dissolution of CO_2 into the brine. On longer timescales, mineral precipitation also plays a role (dissolved CO_2 reacts with the rock matrix and precipitates as a mineral). Lateral and vertical heterogeneity determines whether a free CO_2 plume forms or whether most CO_2 is captured by capillary forces. The lateral extension of the top seal is a critical factor only for trapping the 'free' CO_2 plume. Once dissolved, the CO_2 remains in the reservoir, since brine with dissolved CO_2 is heavier and tends to sink to the bottom (Zhang et al., 2011).

A significant number of research projects on CCS have been carried out in recent decades. In the Netherlands, the large research program known as CATO (CO_2-Afvang, Transport en Opslag, 2004-2014) investigated many of the fundamental aspects of the capture, transport and storage of CO_2 (Lysen et al., 2005; CATO, 2022). The feasibility of CO_2-enhanced coalbed methane production in the former mining area of Limburg was also considered but was ruled out due to high investment costs (Van Bergen et al., 2005; Van Bergen et al. 2025, this volume). Following extensive research and concerning the fact that CCS is technically feasible and deployable (CO2RE, 2022) and combined with the availability of public funding, several large-scale CCS projects are now underway, with injection scheduled

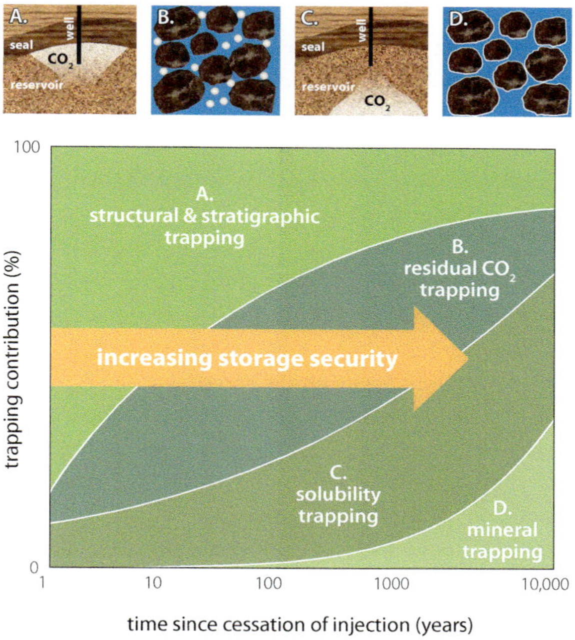

Figure 20.14. *Schematic representation of the contribution of trapping mechanisms for aquifers (modified from www.bigskyco2.org/node/127 and Benson et al., 2005).*

to start in 2025 to 2027. In the following, we describe the development of CCS in the Netherlands with focus on the scientific, technological, economic and societal aspects and the search for potential suitable underground reservoirs onshore and offshore.

In 2004, a small-scale pilot CCS project began in the offshore K12-B natural gas field located 150 km northwest of Amsterdam (Vandeweijer et al., 2021). The natural gas reservoir (Upper Slochteren Member) lies at a depth of around 4000 m and is sealed at the top by the evaporites of the Zechstein Group. Since 1987, the field has been producing natural gas with a relatively high CO_2 content (13%), which was initially separated and vented at the site. The start of the CCS project involved the capture and reinjection of the CO_2 produced at a rate of 0.02 Mt/yr. A few years later, the Shell Pernis refinery in the Rotterdam harbor area proposed a CCS project to inject CO_2 (as by-product of the production of hydrogen) into two nearby depleted natural gas fields, Barendrecht (Cretaceous at 1700 m depth) and Barendrecht-Ziedewij (Triassic at 2700 m depth), both in sandstone reservoirs. The project plan spanned two phases. Phase-I was a 3-year injection into the Barendrecht reservoir and phase-II a 25-year injection in Barendrecht-Ziedewij, at rates of about 0.4 Mt/yr, resulting in a combined (storage) resource estimated at around 10 Mt. Although technically sound, the project encountered strong societal opposition and was cancelled by the Dutch government in November 2010 (Feenstra et al., 2010). Today, only a relatively small amount of CO_2 (0.5 Mt/yr) from Shell's refinery in Pernis and Alco's bioethanol plant in Rotterdam is fed to nearby greenhouses.

After Barendrecht, other onshore natural gas fields were investigated as potential candidates for CCS, but did not develop into projects, mainly due to strong public opposition. CCS research was then redirected to offshore locations, giving rise to the **ROAD** project (Rotterdam Opslag en Afvang Demonstratieproject; Neele et al., 2011; Read et al., 2019). This project aimed to capture CO_2 from coal-fired power plants on the Maasvlakte and store it in the nearshore P18-4 depleted natural gas field, 20 km off the coast (Arts et al., 2012). The project was suspended in 2017, this time due to the collapse of the CO_2 price, which undermined the original business case.

While a combination of economic, political and social factors has led to the cancellation of Dutch CCS projects so far, this trend seems to be changing. The 2015 Paris Agreement and the more recent Dutch Climate Agreement (2019) have led to an increase in political, industrial and societal support for CCS. The focus of CCS projects has shifted from coal-fired power plants (ROAD) to industrial processes that may have no (near-future) alternatives to fossil fuels. A consortium under the name of **Porthos** (Port of Rotterdam CO_2 Transport Hub and Offshore Storage) was formed in 2017 to continue looking for new opportunities for CCS (www.porthosco2.nl). Porthos continues the work done on the ROAD project by updating the transportation and storage design and expanding the set of offshore natural gas fields considered. The goal of this consortium is to enable the capture and storage of 37.5 Mt (2.5 Mt/yr) of CO_2 from the Port of Rotterdam for a period of 15 years starting in 2026, storing it in the offshore block P18 in Lower Germanic Trias Group sandstone reservoirs at ~3500 m depth.

Athos (Amsterdam-IJmuiden CO_2 Transport Hub and Offshore Storage) was another promising offshore CCS project aimed at the large-scale capture and storage of CO_2 generated from major industrial facilities in the Amsterdam area. The decision of one of the partners (TATA steel) to redirect investment towards new clean steel manufacturing technologies involving the use of hydrogen, halted the project in 2021 (ATHOS, 2022). A new initiative named **Aramis** aims to create synergies between different CCS projects (including Porthos) by collecting CO_2 captured by many emitters and transporting it from Rotterdam to several offshore underground storage locations

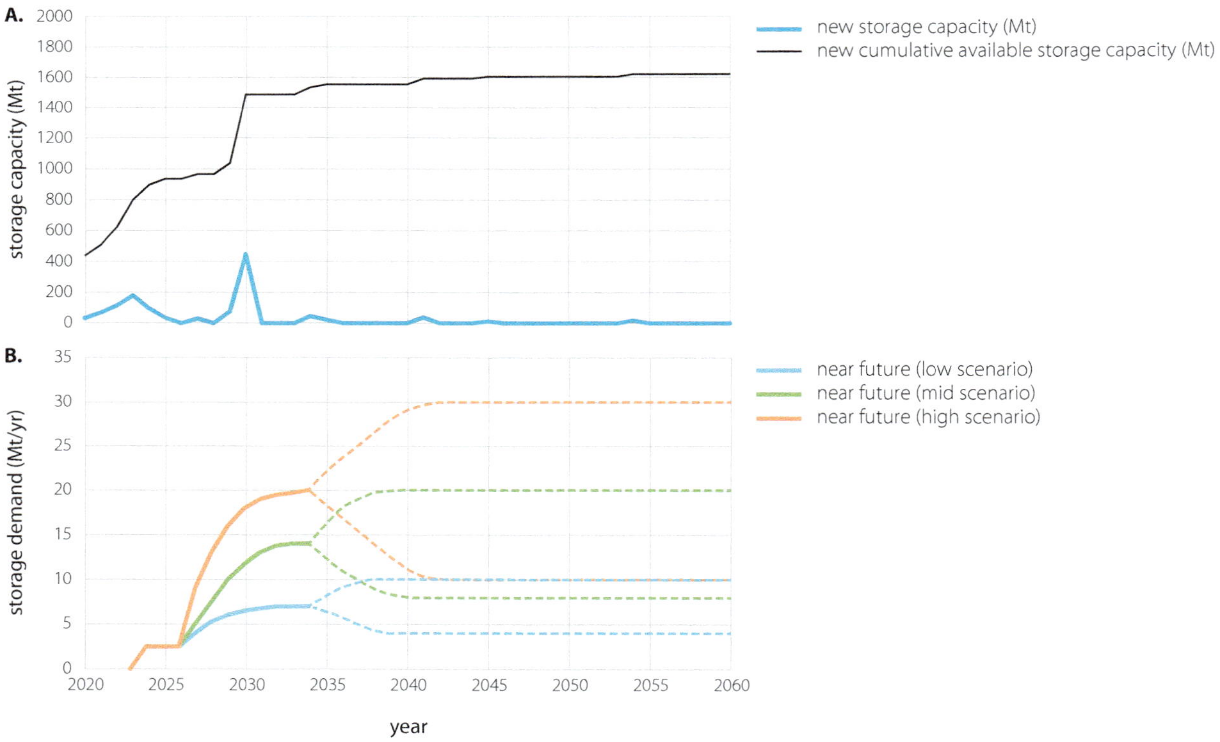

Figure 20.15. *a) Projection of available (cumulative) storage capacity in Megatonnes (Mt). Data from EBN and Gasunie (2018). b) Projection of the possible future demand for storage. Numbers for 2030 are based on the Dutch Climate Agreement (49% emission reduction compared to 1990) with upsides created based on the new EU national climate targets (55-60% reduction as compared to 1990). Long-term targets are uncertain, but are estimated based on aggregate scenarios for 2050.*

(see www.aramis-ccs.com). The project is intended to be open access, potentially connecting several different emitters and storage operators. Possibly, foreign emitters will also be connected through pipelines or ships. This could be the first time multiple emitters and multiple storage locations are connected in one CCS system. The consortium anticipates that most of the depleted gas field candidates for storage are likely to be located in the K and L quadrants of the Dutch central offshore (approximately 150 km from shore), although other candidates may also be included.

In 2018, EBN and Gasunie conducted a portfolio study to estimate the CO_2 storage resources in depleted offshore gas fields in the Netherlands, (EBN & Gasunie, 2018). The results show an estimated storage potential of about 1.6 Gt (Fig. 20.15a), but the total storage resources do not include aquifers, which have yet to be assessed. The ultimately available CO_2 storage capacity will depend on: i) the timing of the CO_2 supply from onshore; ii) the rate of dismantling of offshore infrastructure; and iii) economic feasibility. Despite these uncertainties, the Netherlands' offshore storage resources are significant considering that the Netherlands produces on average 160 Mt CO_2 each year, of which up to 20% (~30 Mt/yr) can be considered for CCS (EBN & Gasunie, 2018). Therefore, the anticipated availability of storage volume in depleted offshore gas fields over the next 20 years is substantially larger than the expected storage demand (Fig. 20.15b). Exactly when storage potential will be released depends on when production from a platform ceases due to economic considerations such as the technically possible production rates, relevant gas prices applicable at the time and operating costs.

The European Commission has implemented various incentives to stimulate CCS (EC, 2022): i) the Emission Trading Scheme (ETS); ii) subsidies for viability studies and demonstration projects of CCS; iii) Emission Performance Standards, since certain facilities are subject to a maximum emission standard per unit of product. Recently, in the Netherlands, the subsidy scheme called 'Stimulation of Sustainable Energy Production and Climate Transition' (SDE++) was introduced to make CCS affordable by compensating for the difference between the cost price and the market value (RVO, 2022). To date (2022), these incentives have resulted in only a limited number of CCS projects and infrastructure.

Many other countries in Europe have shown interest in CCS. The two most prominent countries are the United Kingdom and Norway. Norway already has two fully operational projects (Sleipner and Snøhvit). Apart from small projects for research purposes, not many industrial-scale projects in Europe have progressed to an operational phase, for reasons very similar to those in the Netherlands.

Summary and outlook

The Dutch subsurface has been a major source of raw materials and energy resources in the form of oil, gas and coal, and will continue to play an important role, unlocking new renewable sources and supporting the transition to a renewable energy system with storage. After more than half a century of self-sufficient supply of natural gas, the decline in hydrocarbon production in the Netherlands is increasing the dependence on foreign energy resources and the need for alternative energy supply sources. Meanwhile, the country is embarking on an energy transition largely driven by important global environmental, climate and geopolitical challenges. The ongoing transition is towards a new energy system powered by large-scale variable renewable sources (solar and wind) complemented by energy imports. Along with already existing fluctuations in energy demand, the inevitable variability in energy supply results in an increased need for energy storage. The Dutch subsurface is very suitable, and has a great potential, for storing not only large amounts of energy, but also feedstocks and waste. Therefore, the role of the subsurface in the energy system is shifting from being an energy provider to being more of a service provider for (energy) storage.

This chapter provides an overview of the most relevant state-of-the-art underground storage technologies for the Netherlands. It shows that, compared to the previous edition of this chapter, a lot of research has already been carried out revealing the large potential of the subsurface as an energy storage services provider for fuel-based energy carriers, electricity and heat. In addition, permanent underground storage options will continue to provide very important services to the energy system, beyond the current permanent storage of production water from oil and gas extraction and filtrate residues. The energy storage potential in general and the storage potential for CO_2 in particular are estimated to be large with projects underway or about to be implemented. This will make the subsurface very relevant in the energy transition beyond the era of fossil fuels.

It is evident that the further implementation and upscaling of current and new underground storage technologies cannot be achieved without a good understanding of the subsurface, as well as an adequate insight into energy systems dynamics, commercial viability and social drivers that steer the need for storage development and integration. In the Netherlands, much of the subsurface knowledge comes from long-standing exploration and production activities for coal, oil, gas and salt, or from dedicated surveys (e.g. geothermal) over the past century. The availability of this enormous amount of data and information is of great benefit for the identification and study of a wide range of natural reservoirs as potential candidates for reuse and development as underground storage. This includes (nearly) depleted oil and gas fields, aquifers and engineered underground cavities. It is important to note that in the Netherlands, the geological setting (and the corresponding levels of uncertainty), as well as land use vary greatly from region to region. As a result, related regional strengths, weaknesses, opportunities and threats are also quite variable. With all this in mind, it is essential to investigate whether specific underground storage technologies can be developed and deployed in a safe, responsible and timely manner to meet the scale of storage demand for a particular energy carrier or feedstock and the disposal of any byproduct. In addition, it will be necessary to examine the potential short- and long-term consequences and risks associated with each particular type of underground storage and also how to limit and mitigate them and assess their impact in case of poor management (Dost et al., 2025; Fokker et al., 2025, both in this volume) and potential conflicts between different subsurface users (e.g. hydrogen storage, CAES and CO_2 storage) and surface users (e.g. spatial claims for nature reserves, groundwater and urban areas).

Local and regional authorities and communities should be involved in all phases of the life cycle of (large-scale) underground (energy) projects to ensure societal embeddedness and support and embrace adequate consideration of changing social needs, interests and opinions on how the subsurface should be used. Due to the difficult societal acceptance of the exploration and production of natural resources onshore, we show how alternative sites for underground storage in offshore areas are currently being investigated, as it is the case for CCS and UHS. Further research should focus on the realization of underground storage pilot and demonstration projects, their integration with aboveground storage technologies (as new hybrid energy storage solutions) and the economic, environmental and societal feasibility in the context of the Netherlands and global energy markets.

Acknowledgements

We would like to gratefully acknowledge everyone who has contributed to this chapter with their critical and constructive reviews, which have improved and enriched its content on this broad and rapidly evolving topic of (large-scale) underground (energy) storage in the Netherlands. For all this, we would like to acknowledge: Ingrid Kroon, Harmen Mijnlieff, Frank van Bergen, Kees Geel, Jaap Breunese, Gijs Remmelts, Jos Grunder, Filip Neele, Jasper Griffioen, Jenny Hettelaar, Isis Wetten, Luuk Hunfeld and Thomas Ravestein from the TNO - Geological Survey of the Netherlands; Barthold Schroot and Bastiaan Jaarsma from Energie Beheer Nederland (EBN); Lucia van Geuns from the Hague Center for Strategic Studies (HCSS); Annemarie Muntendam-Bos from the State Supervision of Mines (SodM) and Carla Robledo from the Ministry of Economic Affairs and Climate Policy (MEACP). We also appreciate the review by Susan Riha (Cornell University, USA) and Harry Doust (Vrije Universiteit Amsterdam). The realization of this chapter was supported by TNO and the Advisory Group for the Dutch Ministry of Economic Affairs and Climate Policy.

Digital map data

Spatial data of figures in this chapter for use in geographical information systems can be downloaded here: https://doi.org/10.5117/aup.28164377.

References

ABVO, 1966. Allgemeine Bergverordnung über Untertagebetriebe, Tagebaue und Salinen (ABVO). Sondervorschriften für Salzbergwerke § 224 Sicherheitspfeiler. Niedersachsen (Deutschland) Report No. Nds. MBl Nr. 15/1966: 337 pp.

Afman, M. & Rooijers, F., 2017. Net voor de toekomst. Achtergrondrapport. CE Delft, Report No. 17.3L53.170: 318 pp. https://www.ce.nl

Ajanovic, A., Sayer, M. & Haas, R., 2022. The economics and the environmental benignity of different colors of hydrogen. International Journal of Hydrogen Energy 47 (57): 24136-24154. DOI: 10.1016/j.ijhydene.2022.02.094

Amid, A., Mignard, D. & Wilkinson, M., 2016. Seasonal storage of hydrogen in a depleted natural gas reservoir. International Journal of Hydrogen Energy 41: 5549-5558. DOI: 10.1016/j.ijhydene.2016.02.036

Arts, R.J., Vandeweijer, V.P., Hofstee, C., Pluymaekers, M.P.D., Loeve, D., Kopp, A. & Plug, W.J., 2012. The feasibility of CO2 storage in the depleted P18-4 gas field offshore the Netherlands (the ROAD project). International Journal of Greenhouse Gas Control 11: S10-S20. DOI: 10.1016/j.ijggc.2012.09.010

ATHOS, 2022. Athos-project stopt in huidige vorm na besluit Tata Steel. https://www.gasunie.nl/nieuws/athos-project-stopt-in-huidige-vorm-na-besluit-tata-steel

Bakema, G. & Drijver, B., 2019. HEATSTORE - State of the art HT-ATES in the Netherlands. Evaluation of thermal performance and design considerations for future projects. Project deliverable D1.1: 37 pp.

Benson, S., Cook, P., Anderson, J., Bachu, S., Nimir, H.B., Basu, B., Bradshaw, J., Deguchi, G., Gale, J., Von Goerne, G., Heidug, W., Holloway, S., Kamal, R., Keith, D., Lloyd, P., Rocha, P., Senior, B., Thomson, J., Torp, T., Wildenborg, T., Wilson, M., Zarlenga, F. & Zhou, D., 2005. Underground geological storage. IPCC Special

Report on CarbonD ioxide Capture and Storage: 195-276 pp.
Bergen, F.V., Damen, K.J., Pagnier, H.J., Faaij, A. & Ribberink, H., 2005. Feasibility study of CO2 enhanced coalbed methane production in the Netherlands. In: Wilson, M., Morris, T., Gale, J. & Thambimuthu. K. (eds): Greenhouse Gas Control Technologies 7 (II-2): 2229-2232.
Bloemendal, M. & Hartog, N., 2018. Analysis of the impact of storage conditions on the thermal recovery efficiency of low-temperature ATES systems. Geothermics 71: 306-319. DOI: 10.1016/j.geothermics.2017.10.009
Bos, C.F.M., 2007. Underground storage and sequestration. In: Wong, Th.E., Batjes, D.A.J. & De Jager, J. (eds): Geology of the Netherlands. Royal Netherlands Academy of Arts and Sciences (KNAW, Amsterdam): 335-340.
Bos, C.F.M., 2019. Supply and demand for flexibility in the TPA-grid energy market of the Netherlands until 2050. The potential role of large-scale subsurface energy storage. TNO (Utrecht), Report No. 2019 R10291: 86 pp.
Bourgeois, J.P., Aupaix, N., Bloise, R. & Millet, J.L., 1979. Proposition d'explication de la formation d'hydrogène sulfuré dans les stockages souterrains de gaz naturel par réduction des sulfures minéraux de la roche magasin. Revue d'IFP - Oil & Gas Science and Technology 34: 371-386. DOI: 10.2516/ogst:1979013
Bouroullec, R. & Geel, C.R., 2025. Permian. In: Ten Veen, J.H., Vis, G.-J., De Jager, J. & Wong, Th.E. (eds): Geology of the Netherlands, second edition. Amsterdam University Press (Amsterdam): 127-153. DOI: 10.5117/9789463728362_ch04
Breitschopf, B., Zheng, L., Plaisir, M., Bard, J., Schröer, R., Kawale, D., Koornneef, J.M., Melese, Y., Schaaphok, M., Gorenstein Dedecca, J., Bene, C., Cerny, O. & Gérard, F., 2022. The role of renewable H2 import & storage to scale up the EU deployment of renewable H2. Energy Transition Expertise Centre (ENTEC), Report No. MJ-01-22-112-EN-N: 199 pp. DOI: 10.2833/727785
Breunese, J., Mijnlieff, H.F. & Lutgert, J., 2005. The life cycle of the Netherlands' natural gas exploration: 40 years after Groningen, where are we now? Geological Society, London, Petroleum Geology Conference Series 6 (1): 69-75. DOI: 10.1144/0060069
Brolsma, M.J., De Graeff, J.J., Hack, H.R.G.K., Van Leerdam, A., Van der Pluijm, J.L.P.M., Smal, W. & De Zoeten, G., 2017. Evaluatie verwerking productiewater Schoonebeek. Advies over het Eindrapport alternatievenafweging. Commissie voor de Milieueffectrapportage (MER), MER Adviezen No. 3093-1: 9 pp.
Busschers, F.S., Cohen, K., Wesselingh, F.P., Schokker, J., Bakker, M., Van Balen, R. & Van Heteren, S., 2025. Quaternary. In: Ten Veen, J.H., Vis, G.-J., De Jager, J. & Wong, Th.E. (eds): Geology of the Netherlands, second edition. Amsterdam University Press (Amsterdam): 333-391. DOI: 10.5117/9789463728362_ch10
Cabeza, L.F., 2021. Advances in thermal energy storage systems: Methods and applications. Woodhead Publishing, (Duxford): 775 pp.
CATO, 2022. CO2 capture, transport and storage in the Netherlands. https://www.co2-cato.org.
CBS, 2022. Natural gas, petroleum, coal, supply and consumption (value year 2021). Statistics Netherlands (CBS). https://opendata.cbs.nl/statline/#/CBS/en/
Cihlar, J., Mavins, D. & Van der Leun, K., 2021. Picturing the value of underground gas storage to the European hydrogen system. Gas Infrastructure Europe (GIE, Utrecht): 54 pp.
CO2RE, 2022. CCS facilities database. Global CCS Institute. https://co2re.co/FacilityData
Corre Energy, 2022. Het project CAES Zuidwending. Corre Energy Storage B.V. https://correenergystorage.nl/technology/?lang=en
Correljé, A.F., Van der Linde, J.C. & Westerwoudt, T., 2003. Natural Gas in the Netherlands: From Cooperation to Competition? Clingendael International Energy Programme (CIEP, The Hague): 240 pp.
Crotogino, F., Mohmeyer, K. & Scharf, R., 2001. Huntorf CAES: More than 20 Years of Successful Operation. Nat Gas. 45.
Danish Energy Agency, 2018. Technology data for energy storage. https://ens.dk/en/our-services/projections-and-models/technology-data/technology-data-energy-storage
DBI-GUT, 2017. The effects of hydrogen injection in natural gas networks for the Dutch underground storages (Commissioned by the Ministry of Economic Affairs). Final Report No. FLEXg 16008: 67 pp.
De Bresser, J.H.P., Drury, M.R., Fokker, P.A., Gazzani, M., Hangx, S.J.T., Niemeijer, A.R. & Spiers, C.J., 2022. Mechanical behavior of Salt X. CRC Press/Taylor & Francis Group (Utrecht): 690 pp.
Den Ouden, B., Kerkhoven, J., Warnaars, J., Terwel, R., Coenen, M., Verboon, T., Tiihonen, T. & Koot, A., 2020. Klimaatneutrale energiescenario's 2050. Scenariostudie ten behoeve van de integrale infrastructuuverkenning 2030-2050. Berenschot & Kalavasta (Utrecht), Report No. 61689: 145 pp.
Dinkelman, D., Van Bergen, F. & Veldkamp, J.G., 2020. Geological model, shallow subsurface temperature model and potential maps for HT-ATES applications in the Netherlands. TNO (Utrecht), Report No. TNO 2020 R11686: 120 pp.
DOE, 2022. US Department of Energy - Global Energy Storage Database. gesdb.sandia.gov
Doornenbal, J.C., Kombrink, H., Bouroullec, R., Dalman, R.A.F., De Bruin, G., Geel, C.R., Houben, A.J.P., Jaarsma, B., Juez-Larré, J., Kortekaas, M., Mijnlieff, H.F., Nelskamp, S., Pharaoh, T.C., Ten Veen, J.H., Ter Borgh, M., Van Ojik, K., Verreussel, R.M.C.H., Verweij, J.M. & Vis, G.-J., 2019. New insights on subsurface energy resources in the Southern North Sea Basin area. In: Patruno, S., Archer, S.G., Chiarella, D., Howell, J.A., Jackson, C.A-L. & Kombrink, H. (eds): Cross-border themes in petroleum geology I: The North Sea. Geological Society of London, Special Publications 494: 233-268. DOI: 10.1144/SP494-2018-178
Dopffel, N., Jansen, S. & Gerritse, J., 2021. Microbial side effects of underground hydrogen storage - knowledge gaps, risks and opportunities for successful implementation. International Journal of Hydrogen Energy 46: 8594-8606. DOI: 10.1016/j.ijhydene.2020.12.058

Dost, B., Ruigrok, E., Spetzler, J. & Kruiver, P., 2025. Natural and induced seismicity. In: Ten Veen, J.H., Vis, G.-J., De Jager, J. & Wong, Th.E. (eds): Geology of the Netherlands, second edition. Amsterdam University Press (Amsterdam): 793-823. DOI: 10.5117/9789463728362_ch22

EBN, 2016. Netherlands masterplan for decommissioning and re-use. Energie Beheer Nederland (EBN) in collaboration with NOGEPA and IRO (Utrecht): 49 pp.

EBN & Gasunie, 2018. Transport en opslag van CO2 in Nederland. Verkennende studie door Gasunie en EBN in opdracht van het ministerie van Economische Zaken. Energie Beheer Nederland (EBN)/Gasunie (Utrecht): 88 pp.

EC, 2022. Climate Action. European Commission climate.ec.europa.eu/index_en

EC, 2023. Commission welcomes completion of key 'fit for 55' legislation, putting EU on track to exceed 2030 targets. European Comission (Press Release).

Eckroad, S. & Gyuk, I., 2003. EPRI-DOE handbook of energy storage for transmission & distribution applications. EPRI (Palo Alto, CA) and the U.S. Department of Energy (Washington, DC), Final Report No. 1001834: 512 pp.

Ecovat, 2022. Ecovat Productinformatie. Ecovat Nederland B.V. (Veghel, the Netherlands): 17 pp.

Eising, J., 2021. Risk analysis of worldwide salt cavern storage and its implications for the Dutch cavern storage industry. MSc thesis Vrije Universiteit Amsterdam (the Netherlands): 57 pp.

ENCO, 2020. Possible role of nuclear in the Dutch energy mix in the future. Enconet Consulting Ges.m.b.H. (Vienna, Austria), Final Report No. ENCO-FR-(20)-13: 88 pp.

Energystock, 2017. Aanvraag instemming wijziging opslagplan Zuidwending. Energystock (A Gasunie Company), No. Ol 17.1867: 48 pp.

Energystock, 2022. Ontwikkelen caverne voor waterstofopslag. https://www.energiebufferzuidwending.nl/projecten/ontwikkelen-caverne-voor-waterstofopslag

EP, 2020. Motion for a European parliament resolution on a comprehensive European approach to energy storage. European Parliament (EP), Committee on Industry, Research and Energy, Report No. A9-0130/2020: 23 pp. https://www.europarl.europa.eu/doceo/document/A-9-2020-0130_EN.html#title1

ESTMAP, 2014. First pan-European assessment of underground energy storage potential. Energy Storage Mapping and planning, European Union ENER/C2/2014-640/S12.698827. www.nlog.nl/en/knowledgebase

Evans, D.J. & Holloway, S., 2009. A review of onshore UK salt deposits and their potential for underground gas storage. Geological Society of London, Special Publications 313 (1): 39-80. DOI: 10.1144/SP313.5

Feenstra, C.F.J., Mikunda, T. & Brunsting, S., 2010. What happened in Barendrecht? A case study on the planned onshore carbon dioxide storage in Barendrecht, the Netherlands (; 44 pp.). Energy Research Centre of the Netherlands (ECN)/Global CCS Institute (Petten), Report No. 823136: 44 pp.

Fleuchaus, P., Godschalk, B., Stober, I. & Blum, P., 2018. Worldwide application of aquifer thermal energy storage–A review. Renewable and Sustainable Energy Reviews 94: 861-876. DOI: 10.1016/j.rser.2018.06.057

Fokker, P.A., Candela, T.G.G., Erkens, G., Hanssen, R.F., Kooi, H., Koster, K. & Van Leijen, F.J., 2025. Subsidence. In: Ten Veen, J.H., Vis, G.-J., De Jager, J. & Wong, Th.E. (eds): Geology of the Netherlands, second edition. Amsterdam University Press (Amsterdam): 825-847. DOI: 10.5117/9789463728362_ch23

Frykman, P., 2002. Trapping Diagrams as Case Specific Indicators for Storage Security. Proceedings of the 16th Greenhouse Gas Control Technologies Conference (GHGT-16): 14 pp.

Fuchs, G., Lunz, B., Leuthold, M. & Uwe Sauer, D., 2015. Overview of Nonelectrochemical Storage Technologies. In: Electrochemical Energy Storage for Renewable Sources and Grid Balancing. Elsevier (Durham): 89-102. DOI: 10.1016/B978-0-444-62616-5.00007-3

Fyhn, C.T., 2021. Pit Thermal Energy Storage. Practical experiences from Denmark. Presentation at Energiforsk Slutkonferens Termiska energilager.

Gahleitner, G., 2013. Hydrogen from renewable electricity: An international review of power-to-gas pilot plants for stationary applications. International Journal of Hydrogen Energy 38: 2039-2061. DOI: 10.1016/j.ijhydene.2012.12.010

Golovachev, A.O. & Starokon, I.V., 2020. Underground storage facilities for helium concentrate in porous medium. IOP Conference Series Materials Science and Engineering 860: 012020. DOI: 10.1088/1757-899X/860/1/012020

Groenenberg, R., Juez-Larré, J., Gonçalves, C., Wasch, L., Dijkstra, H., Wassing, B., Orlic, B., Brunner, L., Van der Valk, K., Van der Meulen, T.H. & Kranenburg-Bruinsma, K., 2020a. Techno-economic modelling of large-scale energy storage systems. TNO (Utrecht), Report No. TNO 2020 R12004: 154 pp.

Groenenberg, R., Koornneef, J.M., Sijm, J., Janssen, G., Morales-Espana, G., Van Stralen, J., Hernandez-Serna, R., Smekens, K., Juez-Larré, J., Goncalvez, C., Wasch, L., Dijkstra, H., Wassing, B., Orlic, B., Brunner, L., Van der Valk, K., Unen, M., Van der Meulen, T., Kranenburg-Bruinsma, K., Winters, E., Puts, H., Van Popering-Verkerk, J. & Duijn, M., 2020b. Large-scale energy storage in salt caverns and depleted fields (LSES) - Project findings. TNO (Utrecht), Report No. 2020 TNO R12006; 40 pp.

Guidehouse & Berenschot, 2021. Systeemintegratie wind op zee 2030-2040 [ ]. Rijksdienst voor Ondernemend Nederland (RVO) in samenwerking met Gasunie & Tennet, Eindrapportage: 164 pp.

Gupta, R.B., 2009. Hydrogen fuel: production, transport, and storage. CRC Press/Taylor & Francis Group (Boca Raton, FL): 624 pp.

Hammond, G. & O'Grady, A., 2017. The life cycle greenhouse gas implications of a UK gas supply transformation on a future low carbon electricity sector. Energy 118: 937-949. DOI: 10.1016/j.energy.2016.10.123

Han, G., Bruno, M., Lao, K., Young, J. & Dorfmann, L., 2006. Gas storage and operations in single bedded salt caverns: stability analyses. Paper presented at the SPE Gas Technology Symposium (Calgary, AL): DOI: 10.2118/99520-MS

HEATSTORE, 2021a. Roadmap for flexible energy systems with underground thermal energy storage towards 2050. GEOTHERMICA-ERANET Cofund, Project deliverable No. D6.4: 56 pp.

HEATSTORE, 2021b. The role of UTES in the future EU energy system – a moderated table discussion. HEATSTORE webinar series 21 oct.

Heinemann, N., Alcalde, J., Miocic, J.M., Hangx, S.J.T., Kallmeyer, J., Ostertag-Henning, C., Hassanpouryouzband, A., Thaysen, E.M., Strobel, G.J., Schmidt-Hattenberger, C., Edlmann, K., Wilkinson, M., Bentham, M., Haszeldine, R.S., Carbonell, R. & Rudloff, A., 2020. Enabling large-scale hydrogen storage in porous media – the scientific challenges. Energy & Environmental Science 14(2): 853-864. DOI: 10.1039/doee03536

Heinemann, N., Scafidi, J., Pickup, G., Thaysen, E.M., Hassanpouryouzband, A., Wilkinson, M., Satterley, A.K., Booth, M.G., Edlmann, K. & Haszeldine, R.S., 2021. Hydrogen storage in saline aquifers: The role of cushion gas for injection and production. International Journal of Hydrogen Energy 46: 39284-39296. DOI: 10.1016/j.ijhydene.2021.09.174

Hoving, A., Brautigam, K., Kivits, T., Wilpshaar, M. & Griffioen, J., 2023. Onderzoek naar ondergrondse chemische effecten van waterinjectie in olie- en gasvelden op het Nederlandse vasteland. TNO (Utrecht), Report No. TNO2023_R10940: 58.

Huynen, J.M.H., 2018. Blue battery for green energy. Underground pumped hydro storage. Delft Academic Press: 352 pp.

HyStock, 2022. The HyStock pilot project. Hydrogen conversion https://www.energystock.com/hydrogen-conversion

IEA, 2022a. Hydrogen TCP (Technical Collaboration Platform) Task 42: Underground Hydrogen Storage 2022-2024. International Energy Agency (IEA) https://www.ieahydrogen.org/task/task-42-underground-hydrogen-storage

IEA, 2022b. World Energy Outlook 2022. International Energy Agency (IEA): 522 pp.

IfG, 2008. Rock-mechanical appraisal on the positioning of caverns in a salt dome and safety distances between gas storage caverns. Appendix to the 2012 storage plan of the Zuidwending natural gas storage facility of Energystock. Institut fur Gebirgsmechanik (IfG) GmbH, Confidential report.

Johansson, J., Leirnes, J., Walton, K. & Yilmaz, E., 2022. Hydrogen power and heat with Siemens Energy gas turbines. Reliable and flexible carbon-free energy. Siemens Energy Global GmbH & Co. KG (Erlangen), Report No. PGCM-B10003-00-7600: 24 pp.

Juez-Larré, J., Remmelts, G., Breunese, J.N., Van Gessel, S.F. & Leeuwenburgh, O., 2016. Using underground gas storage to replace the swing capacity of the giant natural gas field of Groningen in the Netherlands. A reservoir performance feasibility study. Journal of Petroleum Science and Engineering 145: 34-53. DOI: 10.1016/j.petrol.2016.03.010

Juez-Larré, J., Van der Gon, H.D., Dellaert, S., Griffioen, J. & Nepveu, M., 2018. Emissie van het broeikasgas methaan gerelateerd aan de olie- en gassector in Nederland met nadruk op exploratie en productie. TNO (Utrecht), Report No. TNO 2018 R11080: 73 pp.

Juez-Larré, J., Van Gessel, S., Dalman, R., Remmelts, G. & Groenenberg, R., 2019. Assessment of underground energy storage potential to support the energy transition in the Netherlands. First Break 37 (7): 57-66. DOI: 10.3997/1365-2397.n0039

Juez-Larré, J., Gonçalves Machado, C. & Groenenberg, R., 2020. Performance assessment of underground gas storage for potential hydrogen storage in the Netherlands. 1st Geoscience & Engineering in Energy Transition Conference, European Association of Geoscientists & Engineers 2020(1): p.1-5. DOI: 10.3997/2214-4609.202021067

Juez-Larré, J., Gonçalves Machado, C., Groenenberg, R.M., Belfroid, S.S.P.C. & Yousefi, S.H., 2023. A detailed comparative performance study of underground storage of natural gas and hydrogen in the Netherlands. International Journal of Hydrogen Energy 48 (74): 28843-28868. DOI: 10.1016/j.ijhydene.2023.03.347

Kallesøe, A.J. & Vangkilde-Pedersen, T., 2019. Underground Thermal Energy Storage (UTES) – state-of-the-art, example cases and lessons learned. GEOTHERMICA –ERA NET Cofund, HEATSTORE Project Report No. D1.1: 130 pp.

Karman, C.C. & Smit, M.G.D., 2019. Whole effluent toxicity data and discharge volumes to assess the likelihood that environmental risks of offshore produced water discharges are adequately controlled. Integrated Environmental Assessment and Management 15 (4): 584-595. DOI: 10.1002/ieam.4139

Khan, S.A., Igoshin, A.I. & Sokhranskiy, V.B., 2012. Gaseous helium storage in salt caverns: influence of physical properties. 10 pp.

King, M., Jain, A., Bhakar, R., Mathur, J. & Wang, J., 2021. Overview of current compressed air energy storage projects and analysis of the potential underground storage capacity in India and the UK. Renewable and Sustainable Energy Reviews 139: 110705. DOI: 10.1016/j.rser.2021.110705

Kitsikoudis, V., Archambeau, P., Dewals, B., Pujades, E., Orban, P., Dassargues, A., Pirotton, M. & Erpicum, S., 2020. Underground pumped-storage hydropower (UPSH) at the Martelange Mine (Belgium): underground reservoir hydraulics. Energies 13: 3512. DOI: 10.3390/en13143512

Kramer, G.J., Arts, T., Urai, J.L., Vrijling, H. & Huynen, J.M.H., 2020. Risk mitigation and investability of a U-PHS project in the Netherlands. Energies 13: 5072. DOI: 10.3390/en13195072

Kramers, L., Van Wees, J.-D., Wassing, B., Krominus, A., Urai, J.L., Kukla, P.A. & Bekendam, R., 2011. Studie naar haalbaarheid van OPAC op de Locatie Graetheide en Nabije Omgeving, TNO (Utrecht), Report No. TNO-060-UT-2011-01251:

Kruck, O., Prelicz, R. & Rudolph, T., 2013. Overview on all Known Underground Storage Technologies for Hydrogen. HyUnder

report No. 3.1: 93 pp.
Kumar, K.R. & Hajibeygi, H., 2021. Multiscale simulation of inelastic creep deformation for geological rocks. Journal of Computational Physics 440: 110439. DOI: 10.1016/j.jcp.2021.110439
Kumar, K.R., Makhmutov, A., Spiers, C.J. & Hajibeygi, H., 2021. Geomechanical simulation of energy storage in salt formations. Scientific Reports 11: 19640. DOI: 10.1038/s41598-021-99161-8
Laban, M.P., 2020. Hydrogen storage in salt caverns. Chemical modelling and analysis of large-scale hydrogen storage in underground salt caverns. MSc thesis Delft University of Technology (the Netherlands): 86 pp.
Letcher, T.M., 2016. Storing Energy, with special reference to renewable energy sources, first edition. Elsevier: 592 pp.
Liang, Y., Ning, Y., Liao, L. & Yuan, B., 2018. Special Focus on Produced Water in Oil and Gas Fields: Origin, Management, and Reinjection Practice. In: Yuan, B. & Wood, D.A. (eds): Formation Damage During Improved Oil Recovery. Gulf Professional Publishing: 515-586. DOI: 10.1016/B978-0-12-813782-6.00014-2
Lysen, E., Faaij, A. & Hendriks, C., 2005. The CATO programme in the Netherlands on CO2 capture, transport and storage. In: Rubin, E.S., Keith, D.W., Gilboy, C.F., Wilson, M., Morris, T., Gale, J. & Thambimuthu, K. (eds): Greenhouse Gas Control Technologies 7. Elsevier Science Ltd (Oxford): 1849-1852. DOI: 10.1016/B978-008044704-9/50225-1
Madlener, R. & Specht, J.M., 2020. An Exploratory Economic Analysis of Underground Pumped-Storage Hydro Power Plants in Abandoned Deep Coal Mines. Energies 13 (21): 5634. DOI: 10.3390/en13215634
Marif, K., 2019. Counteraction of buoyancy flow in high temperature aquifer thermal energy storage systems by applying multiple partially penetrating wells. MSc thesis, Delft University of Technology (the Netherlands): 116 pp.
Matos, C.R., Carneiro, J.F., Silva, P. & Henriques, C.O., 2021. A GIS-MCDA Approach Addressing Economic-Social-Environmental Concerns for Selecting the Most Suitable Compressed Air Energy Storage Reservoirs. Energies 14 (20): 6793. DOI: 10.3390/en14206793
Matos, C.R., Carneiro, J.F. & Silva, P.P., 2019. Overview of large-scale underground energy storage technologies for integration of renewable energies and criteria for reservoir identification. Journal of Energy Storage 21: 241-258. DOI: 10.1016/j.est.2018.11.023
MEACP, 2019a. Klimaatakkoord. Ministry of Economic Affairs and Climate Policy (MEACP, The Hague): 239 pp.
MEACP, 2019b. Wet verbod op kolen bij elektriciteitsproductie. Staatsblad van het Koninkrijk der Nederlanden: 3 pp.
MEACP, 2020. Wetsvoorstel 'Wat na nul' – wetswijzigingen in verband met de definitieve sluiting van het Groningenveld; Minister Eric Wiebes: 3 pp.
MEACP, 2022a. Afkondiging vroegtijdig waarschuwing in Nederland naar aanleiding van de situatie op de Europese gasmarkten Ministry of Economic Affairs and Climate Policy (MEACP, The Hague), Kamerbrief DGKE-E/22263267: 17 pp.
MEACP, 2022b. Gaswinning uit Groningenveld in gasjaar 2022-2023. Kamerbrief PDGGO-DSGG 22457919.
MEACP, 2023. Natural resources and geothermal energy in the Netherlands 2022. Ministry of Economic Affairs and Climate Policy (The Hague): 172 pp. https://www.nlog.nl/en/archive
MEACP, 2024. Wet sluiting Groningenveld gaat per 19 april 2024 in. Rijksoverheid. https://www.rijksoverheid.nl/actueel/nieuws/2024/04/18/wet-sluiting-groningenveld-gaat-per-19-april-in
Mendrinos, D., Karytsas, S., Polyzou, O., Karytsas, C., Nordø, Å.D., Midttømme, K., Otto, D., Gross, M., Sprenkeling, M., Peuchen, R., Geerdink, T. & Puts, H., 2022. Understanding societal requirements of CCS projects: application of the societal embeddedness level Assessment methodology in four national case studies. Clean Technology 4: 893-907. DOI: 10.3390/cleantechnol4040055
Mijnlieff, H., Buijze, L., Rosendaal, E., Schoof, F., Vorage, R. & Van Wees, J.-D., 2025. Geothermal energy. In: Ten Veen, J.H., Vis, G.-J., De Jager, J. & Wong, Th.E. (eds): Geology of the Netherlands, second edition. Amsterdam University Press (Amsterdam): 701-727. DOI: 10.5117/9789463728362_ch19
Mouli-Castillo, J., Heinemann, N. & Edlmann, K., 2021. Mapping geological hydrogen storage capacity and regional heating demands: An applied UK case study. Applied Energy 283: 116348. DOI: 10.1016/j.apenergy.2020.116348
Muhammed, N.S., Haq, B., Al Shehri, D.A., Al-Ahmed, A., Rahman, M.M. & Zaman, E., 2022. A review on underground hydrogen storage: Insight into geological sites, influencing factors and future outlook. Energy Reports 8: 461-499. DOI: 10.1016/j.egyr.2021.12.002
Mulder, M., Perey, P. & Moraga, J.L., 2019. Outlook for a Dutch hydrogen market, economic conditions and scenarios Centre for Energy Economics Research (CEER), University of Groningen, Policy Paper No. 5: 79 pp.
Müller, L. & Hereth, A., 1987. Feasibility Study to the OPAC Project (Study in Three Parts A, B, C). Ingenieurbüro für Tunnel-und Felsbau GmbH (Airning).
Munsterman, D.K., Donders, T.H., Houben, A.J.P., Ten Veen, J.H. & Wesselingh, F.P., 2025. Paleogene-Neogene. In: Ten Veen, J.H., Vis, G.-J., De Jager, J. & Wong, Th.E. (eds): Geology of the Netherlands, second edition. Amsterdam University Press (Amsterdam): 1-2. DOI: 10.5117/9789463728362_ch9
Muntendam-Bos, A.G., Hoedeman, G., Polychronopoulou, K., Draganov, D., Weemstra, C., van der Zee, W., Bakker, R.R. & Roest, H., 2022. An overview of induced seismicity in the Netherlands. Netherlands Journal of Geosciences 101: e1. DOI:10.1017/njg.2021.14
Muntendam-Bos, A.G., Wassing, B.B.T., Geel, C.R., Louh, M. & Van Thienen-Visser, K., 2008. Bergermeer seismicity study. TNO (Utrecht), Report No. TNO-2008-U-R1071/B: 95 pp.

Neeft, E.A.C., Bartol, J., Vuorio, M. & Vis, G.-J., 2025. Geological disposal of radioactive waste. In: Ten Veen, J.H., Vis, G.-J., De Jager, J. & Wong, Th.E. (eds): Geology of the Netherlands, second edition. Amsterdam University Press (Amsterdam): 769-791. DOI: 10.5117/9789463728362_ch21

Neele, F., Wu, H., Hendriks, C. & Brandsma, R., 2011. Planning of CCS development in the Netherlands offshore. Energy Procedia 4: 2756-2763. DOI: 10.1016/j.egypro.2011.02.178

NSE, 2022. Harnessing the North Sea's potential. North Sea Energy (NSE). https://north-sea-energy.eu/en/results/

Oerlemans, P., Drijver, B., Koenen, M., Koornneef, J.M., Dinkelman, D., Bos, W. & Godschalk, B., 2022. First field results on the technical risks and effectiveness of mitigation measures for the full scale HT-ATES demonstration project in Middenmeer. European Geothermal Congress 2022 (Berlin), Conference paper.

Panfilov, M., 2016. 4 - Underground and pipeline hydrogen storage. In: R.B. Gupta, A. Basile & T.N. Veziroğlu (eds): Compendium of Hydrogen Energy. Woodhead Publishing: 91-115. DOI: 10.1016/B978-1-78242-362-1.00004-3

Patrahau, I., Rademaker, M., Van Geuns, L. & Van den Beukel, J., 2022. European tank storage in today's global value chains. What role does it play in our economy? The Hague Centre for Strategic Studies (HCSS, The Hague), Report No.

Patrahau, I. & Van Geuns, L., 2021. Gas supply security in the Netherlands: geopolitical and environmental dilemmas. The Hague Centre for Strategic Studies (HCSS, The Hague): 36 pp.

PBL, 2023. Klimaat- en Energieverkenning 2023. Ramingen van broeikasgasemissies, energiebesparing en hernieuwbare energie op hoofdlijnen. Planbureau voor de Leefomgeving (The Hague), Report No. 5243: 142 pp.

Prinzhofer, A., Tahara Cissé, C.S. & Diallo, A.B., 2018. Discovery of a large accumulation of natural hydrogen in Bourakebougou (Mali). International Journal of Hydrogen Energy 43 (42): 19315-19326. DOI: 10.1016/j.ijhydene.2018.08.193

PRMS, 2018. Petroleum Resources Management System, revised in June 2018. Society of Petroleum Engineers (SPE): 52 pp.

Read, A., Gittins, C., Uilenreef, J., Jonker, T., Neele, F., Belfroid, S., Goetheer, E. & Wildenborg, T., 2019. Lessons from the ROAD project for future deployment of CCS. International Journal of Greenhouse Gas Control 91: 102834. DOI: 10.1016/j.ijggc.2019.102834

Reitenbach, V., Ganzer, L., Albrecht, D. & Hagemann, B., 2015. Influence of added hydrogen on underground gas storage: a review of key issues. Environmental Earth Sciences 73: 6927-6937. DOI: 10.1007/s12665-015-4176-2

Rijkswaterstaat, 1982. Plan Lievense. Waterloopkundige en geohydrologische aspecten bij diverse lokaties van energiebekkens in het Markermeer en IJsselmeer. Ministry of Infrastructure and Water Management, Nota WWNO-N 80-0005: 47 pp.

Ringrose, P., 2020. How to store CO2 Underground: Insights from early-mover CCS projects. Springer Briefs in Earth Sciences: 141 pp.

RVO, 2010. Rijksinpassingplan stikstofbuffer Heiligerlee. Winning, transport en opslag van stikstof in de gemeenten Meterwolde en Oldambt. Rijksdienst voor Ondernemend Nederland (RVO). https://www.rvo.nl/onderwerpen/bureau-energieprojecten/lopende-projecten/stikstofbuffer-heiligerlee

RVO, 2022. Stimulering duurzame energieproductie en klimaattransitie (SDE++). Rijksdienst voor Ondernemend Nederland (RVO). https://www.rvo.nl/subsidies-financiering/sde

RWE, 2010. ADELE - Adiabatic compressed-air energy storage for electricity supply. RWE Power AG: 11 pp.

Scheepers, M., Gamboa Palacios, S., Jegu, E., Pupo Noguieira De Oliveira, L., Rutten, L.W., Van Stralen, J., Smekens, K.E.L. & West, K.J., 2020. Towards a sustainable energy system for the Netherlands in 2050, TNO (Amsterdam), Report No. TNO 2020 P10338: 63 pp.

Sijm, J.P.M., Janssen, G.J.M., Morales-España, G.A., Van Stralen, J., Hernandez-Serna, R. & Smekens, K.E.L., 2020. The role of large-scale energy storage in the energy system of the Netherlands, 2030-2050, TNO (Amsterdam), Report No. TNO 2020 P11106: 136 pp.

SodM, 2005. Convenant Werkgroep injectie. State Supervision of Mines (SodM): 18 pp.

SodM, 2022. De Marssteden opslag. State Supervision of Mines (SodM) https://www.sodm.nl/documenten/vragen-en-antwoorden/meer-informatie-over-ondergrondse-opslag-de-marssteden-enschede

Sprenkeling, M., Geerdink, T., Slob, A. & Geurts, A., 2022. Bridging Social and Technical Sciences: Introduction of the Societal Embeddedness Level. Energies 15 (17): 6252. DOI: 10.3390/en15176252

Staffell, I., Scamman, D., Velazquez Abad, A., Balcombe, P., Dodds, P.E., Ekins, P., Shah, N. & Ward, K.R., 2019. The role of hydrogen and fuel cells in the global energy system. Energy & Environmental Science 12: 463-491. DOI: 10.1039/C8EE01157E

STOWA, 2020. Effecten van bodemenergiesystemen op de grondwaterkwaliteit. Foundation for Applied Water Research (STOWA, Amersfoort): 22 pp

Stricker, K., Grimmer, J.C., Egert, R., Bremer, J., Gholami Korzani, M., Schill, E. & Kohl, T., 2020. The potential of depleted oil reservoirs for high-temperature storage systems. Energies 13: 6510. DOI: 10.3390/en13246510

STRONG, 2018. Structuurvisie Ondergrond (STRONG). Ministry of Infrastructure and Water Management & Ministry of Economic Affairs and Climate Policy (The Hague), Report juni 2018: 110 pp.

Tade, M.D., 1967. Helium storage in Cliffside Field. Journal Petroleum Technology 19 (7): 885-888. DOI: 10.2118/1624-PA

Tarkowski, R., 2019. Underground hydrogen storage: Characteristics and prospects. Renewable and Sustainable Energy Reviews 105: 86-94. DOI: 10.1016/j.rser.2019.01.051

Taylor, J.B., Alderson, J.E.A., Kalyanam, K.M., Lyle, A.B. & Phillips, L.A., 1986. Technical and economic assessment of methods for the storage of large quantities of hydrogen. International Journal of Hydrogen Energy 11 (1): 5-22. DOI: 10.1016/0360-

3199(86)90104-7
Thaysen, E., McMahon, S., Strobel, G.J., Butler, I.B., Ngwenya, B.T., Heinemann, N., Wilkinson, M., Hassanpouryouzband, A., McDermoot, C.I. & Edlmaan, K., 2021. Estimating microbial growth and hydrogen consumption in hydrogen storage in porous media. Renewable and Sustainable Energy Reviews 51: 111481. DOI: 10.1016/j.rser.2021.111481
Ugarte, E.R. & Salehi, S., 2021. A review on well integrity issues for underground hydrogen storage (UHS). Journal of Energy Resources Technology 144: 1-27. DOI: 10.1115/1.4052626
Upham, D.C., Agarwal, V., Khechfe, A., Snodgrass, Z.R., Gordon, M.J., Metiu, H. & McFarland, E.W., 2017. Catalytic molten metals for the direct conversion of methane to hydrogen and separable carbon. Science 358 (6365): 917-921. DOI: 10.1126/science.aao5023
Van Bergen, F., Damen, K., Pagnier, H., Faaij, A. & Ribberink, H., 2005. Feasibility study of CO2 enhanced coalbed methane production in the Netherlands. In: E.S. Rubin, D.W. Keith, C.F. Gilboy, M. Wilson, T. Morris, J. Gale & K. Thambimuthu (eds): Greenhouse Gas Control Technologies 7. Elsevier Science Ltd (Oxford): 2229-2232. DOI: 10.1016/B978-008044704-9/50307-4
Van Bergen, F. & Koster, K., 2025. Peat, coal and coalbed methane. In: J.H. Ten Veen, G.-J. Vis, J. De Jager & Th.E. Wong (eds): Geology of the Netherlands, second edition. Amsterdam University Press (Amsterdam): 577-601. DOI: 10.5117/9789463728362_ch15
Van den Berg, C., Van den Brink, R., Faaij, A. & Peters, R., 2022. Nederlands onafhankelijk van Russisch gas. Opties voor korte en lange termijn. TNO (Utrecht), Report No. TNO 2022 P10971: 28 pp.
Van den Beukel, J. & Geuns, L., 2019. Groningen gas: the loss of a social license to operate. The Hague Center for Strategic Studies, HCSS Geoeconomics: 21 pp.
Van den Beukel, J. & Geuns, L., 2020. The deteriorating outlook for Dutch small natural gas fields. The Hague Centre for Strategic Studies, HCSS Geoeconomics: 17 pp.
Van den Bogert, P.A.J., Van Eijs, R., Solano Viota, J. & Doornhof, D., 2016. Norg UGS fault reactivation study and implications for seismic threat. NAM (Assen), Subsurface Technical Report No. EP201610208045: 58 pp.
Van der Stoel, J.M., 2022. A Techno-Economic Review of Sensible Seasonal Thermal Energy Storage Techniques and a European Assessment of the Potential for Aquifer Storage. MSc thesis, Vrije Universiteit Amsterdam (the Netherlands).
Van Eijs, R.M.H.E., Mulders, F.M.M., Nepveu, M., Kenter, C.J. & Scheffers, B.C., 2006. Correlation between hydrocarbon reservoir properties and induced seismicity in the Netherlands. Engineering Geology 84: 99-111. DOI: 10.1016/j.enggeo.2006.01.002
Van Gessel, S., Breunese, J., Juez-Larré, J., Huijskes, T.D. & Remmelts, G., 2018. Ondergrondse opslag in Nederland - Technische verkenning. TNO (Utrecht), Report No. TNO 2018 R11372: 100 pp.
Van Gessel, S., Huijskes, T.D., Juez-Larré, J. & Dalman, R., 2021. Ondergrondse energieopslag in Nederland 2030-2050: Technische evaluatie van vraag en aanbod. TNO/EBN (Utrecht), Report No. TNO 2021 R11125: 114 pp.
Van Gessel, S., Jaarsma, B., Groenenberg, R., Kleijweg, D., Juez-Larré, J., Klaveren, S., Eikelenboom, W., Remmelts, G., Slooten, R., Boter, E., Huijskes, T. & Koning, M., 2022. Haalbaarheidsstudie ondergrondse waterstofopslag op zee. TNO/EBN (Utrecht), Report No. TNO 2022 R11212: 75 pp.
Van Geuns, L., Juez-Larré, J. & De Jong, S., 2017. Van exporteur naar importeur. De verander(en)de rol van aardgas in Nederland. TNO (Utrecht) & The Hague Centre for Strategic Studies, Whitepaper: 28 pp.
Van Wees, J.-D., Fokker, P.A., Van Thienen-Visser, K., Wassing, B.B.T., Osinga, S., Orlic, B., Ghouri, S.A., Buijze, L. & Pluymaekers, M.P.D., 2017. Geomechanical models for induced seismicity in the Netherlands: inferences from simplified analytical, finite element and rupture model approaches. Netherlands Journal of Geosciences 96 (5): s183-s202. DOI: 10.1017/njg.2017.38
Vandeweijer, V., Hofstee, C., Pelt, W. & Graven, H., 2021. CO2 injection at K12-B, the Final Story. Proceedings 15th Greenhouse Gas Control Technologies Conference, 12. 12.
Vantaan Energia, 2021. World's largest cavern thermal energy storage to be built in Vantaa. https://www.vantaanenergia.fi/en/projects/worlds-largest-cavern-thermal-energy-storage-to-be-built-in-vantaa/
Vattenfall, 2013. Nuon investeert in warmteopslag: Efficiënter produceren met minder CO2. https://group.vattenfall.com/nl/newsroom/archive/nieuws/2013/nuon-investeert-in-warmte-opslag
Vattenfall, 2021. Vattenfall is building Germany's largest district heating heat storage facility in Reuter West. https://group.vattenfall.com/uk/newsroom/News/2021/vattenfall-is-building-germanys-largest-district-heating-heat-storage-facility-in-reuter-west
Vattenfall, 2022. Vattenfall start bouw grote duurzame E-boiler voor stadswarmte. https://group.vattenfall.com/nl/newsroom/persbericht/2022/vattenfall-start-bouw-grote-duurzame-e-boiler-voor-stadswarmte
Verhoef, E.V., Neeft, E.A.C., Bartol, J., Vuorio, M.R., Scholten, C., Buitenhuis, A., Van der Veen, G., Chapman, N.A. & McCombie, C., 2020. Long-term research programme for geological disposal of radioactive waste. COVRA N.V.: 59 pp.
WABO, 2022. General Provisions Act 'WABO' (Wet Algemene Bepalingen Omgevingsrecht). https://rwsenvironment.eu/subjects/general-provisions-0
Wesselink, M., Liu, W., Koornneef, J.M. & Broek, M., 2018. Conceptual market potential framework of high temperature aquifer thermal energy storage. A case study in the Netherlands. Energy 147: 477-489. DOI: 10.1016/j.energy.2018.01.072
Winters, E., Puts, H., Van Popering-Verkerk, J. & Duijn, M., 2020.

Legal and societal embeddedness of large-scale energy storage. TNO (The Hague), Report No. TNO 2020 R11116: 27 pp.
Yang, T., Liu, W., Kramer, G.J. & Sun, Q., 2021. Seasonal thermal energy storage: A techno-economic literature review. Renewable and Sustainable Energy Reviews 139: 110732. DOI: 10.1016/j.rser.2021.110732
Yousefi, H., 2021. Design considerations for developing an underground hydrogen storage facility in porous reservoirs. PhD thesis, University of Twente (the Netherlands): 90 pp.
Yousefi, S.H., Groenenberg, R., Koornneef, J.M., Juez-Larré, J. & Shahi, M., 2023. Techno-economic analysis of developing an underground hydrogen storage facility in depleted gas field: A Dutch case study. International Journal of Hydrogen Energy 48: 28824-28842. DOI: 10.1016/j.ijhydene.2023.04.090
Zauner, A., Fazeni-Fraisl, K., Wolf-Zoellner, P., Veseli, A., Holzleitner, M.-T., Lehner, M., Bauer, S. & Pichler, M., 2022. Multidisciplinary assessment of a novel carbon capture and utilization concept including underground Sun Conversion. Energies 15: 1021. DOI: 10.3390/en15031021
Zhang, Y., Chang, F., Song, Y., Zhao, J., Zhan, Y. & Jian, W., 2011. Density of carbon dioxide + brine solution from Tianjin reservoir under sequestration conditions. Journal of Chemical & Engineering Data 56 (3): 565-573. DOI: 10.1021/je101214e
Zivar, D., Kumar, S. & Foroozesh, J., 2021. Underground hydrogen storage: A comprehensive review. International Journal of Hydrogen Energy 46: 23436-23462. DOI: 10.1016/j.ijhydene.2020.08.138

S220514
S220506
S220510
S200225
S200311
S200187
S200313
S200315
S200224
S200306
S200185
S200237
S200234
S200229
S200341
S200188
S190257
S200236
S190260
S220511
S200245
S200233
S200231
S220563
S220577
S220493

21

Geological disposal of radioactive waste

Erika Neeft, Jeroen Bartol, Marja Vuorio & Geert-Jan Vis*

*Corresponding author: erika.neeft@covra.nl

ABSTRACT

Some types of radioactive waste remain hazardous for thousands of years. Disposal of this long-lived waste in stable geological formation is the best thinkable solution according to the current state of technology and science. This is done by placing packages with radioactive waste in an underground facility, hundreds of metres below the land surface. After closure of a facility, isolation of the waste and containment of the radionuclides are achieved by a disposal system consisting of engineered barriers, the geological formation and surrounding rock formations. This geological disposal of radioactive waste has been studied in the Netherlands for more than four decades. Permian and Triassic rock salt formations are investigated since a facility can be constructed in this rock and its ultralow permeability ensures containment of radionuclides. Upward movement of salt (diapirism) and dissolution due to groundwater flow (subrosion), however, can disrupt the key aspects of rock salt. These natural processes are studied to assess the likelihood of radionuclides from the waste to enter the biosphere. The suitable disposal depth is determined by considering plausible natural processes, such as climate change (including ice ages) that in the future may interfere with underground disposal sites. Paleogene clay formations are an alternative rock type for the disposal of radioactive waste. New techniques have become available to construct underground facilities in these low-permeability argillaceous sediments. Emphasis is now put on obtaining freshly cored material at suitable disposal depths to gain further knowledge on the geotechnical properties of these clays.

<< Emplacing drums with solid processed waste in the Dutch storage facility for Low and Intermediate Level Waste. Forklift trucks are also envisaged for the stacking of waste packages in a future geological disposal facility. Photo: COVRA.

DOI: 10.5117/9789463728362_ch21

Introduction

Every material on earth contains so-called radionuclides, i.e. natural isotopes of elements that release radiation as they change into another element and become more stable. Only when the radionuclide content of a substance exceeds certain thresholds, is it called 'radioactive'. These thresholds, the exemption and clearance levels, are derived for both natural and man-made radionuclides. The materials below these levels can be used, recycled or disposed of without any radiological detriment to health and the environment. The exemption and clearance levels are radionuclide-specific and are regularly updated with latest insights with respect to health damage caused by ionising radiation (EC, 2013).

Radioactivity

Radioactivity has been present on earth since its origin some 4.6 billion years ago. Radioactive potassium, uranium and thorium in rocks and soils are examples of natural radionuclides that have a primordial origin (Cinelli et al., 2019). Some natural radionuclides are continuously generated by neutron capture such as nitrogen-14 (non-radioactive isotope), which in the atmosphere reacts with a neutron to form carbon-14 (radioactive isotope). The neutrons are created from the interaction of cosmic particles, such as high energetic protons from the sun and other stars, with particles in the atmosphere (Kovaltsov et al., 2012). Natural radionuclides can also be generated deep in the earth's crust from neutrons in uranium ores. A very recent source of radionuclides is anthropogenic, for example resulting from neutron irradiation in a nuclear reactor.

Radioactive waste

Radioactive material becomes radioactive waste when no future use is foreseen. Radioactive waste can arise during the generation of nuclear power, production of medicine, research and education, use of diagnostics for human health care and the examination of the integrity of engineered constructions. Radioactive waste can also be collected from facilities in which natural radioactive materials are concentrated, for instance in pipes and casings at oil-field production sites (USGS, 1999).

Radioactive waste needs to be disposed of for a long duration in a safe location disconnected from the biosphere. The design of a final, long-term location for disposal of waste has to take into account the period that waste remains hazardous and needs to consider social accepta-

Radiotoxicity

1

The potential impact of radionuclides on humans is best characterized by their 'radiotoxicity' rather than their radioactivity. Radiotoxicity is a measure of the radiation doses that would result if all the radionuclides in a given amount of waste were to be dissolved in water, and subsequently drunk by humans (Hamstra & Van der Feer, 1981). Although entirely hypothetical, this allows for the comparison of the hazardous nature of different types of radioactive materials. The figure shows a comparison between the radiotoxicity of spent fuel from a plant like the Borsele Nuclear Power Plant and the radiotoxicity of natural uranium ore.

The declining radiotoxicity of spent fuel and vitrified HLW is shown as a function of time after the fuel has been taken out of the reactor or, for vitrified HLW, after it was manufactured, following the reprocessing of the equivalent quantity of spent fuel. The curves are shown normalized to the radiotoxicity of the amount of uranium ore that was originally used to make the fuel. When fuel comes out of a nuclear reactor, it is many thousands of times more radiotoxic than the uranium ore from which it was manufactured, but this diminishes significantly over a period of a some hundreds of years. The 'crossover' time, when spent fuel has a similar level of radiotoxicity to the original ore, is in the order of hundred thousand years. Vitrified HLW has an equivalent 'crossover' time of only about 20,000 years (Gruppelaar et al., 1998). The long time scales are the reason for safety assessments to consider potential impacts on future generations for several millions of years ahead.

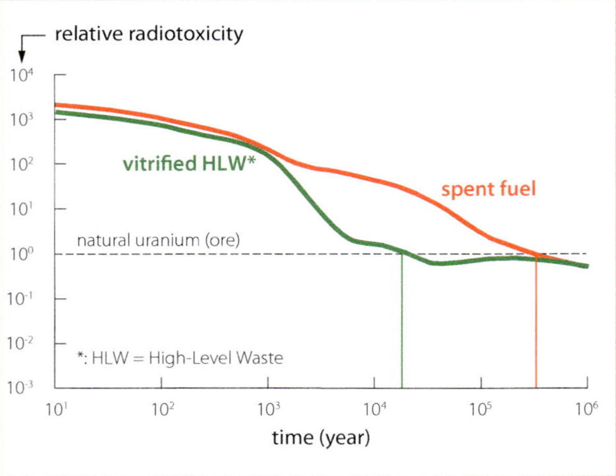

bility. The necessary period for isolation can be determined by the required decay time for the waste to achieve the same radiotoxicity (see Textbox 1) as a natural material (Hamstra & Van der Feer, 1981). It takes about 200,000 years for spent nuclear power fuel to achieve the same radiotoxicity as natural uranium ore, and 20,000 years for the vitrified waste-forms arising from recycling of spent fuel (Gruppelaar et al., 1998). These periods are too long to be managed by society and therefore a solution could be based on isolation by geological barriers. Spent fuel and vitrified waste-forms are called High-Level Waste (HLW). HLW needs to be cooled during storage and requires shielding for safe transportation and storage. Intermediate Level Waste requires shielding only. Low and Intermediate Level Waste (LILW) has the lowest radiation levels. It is possible to stand next to a package LLW without additional precautions.

Disposal of radioactive waste

Radioactive waste can be disposed of in various ways, and the social acceptability of the way in which this is done has changed over time. For example, from 1967 until 1982, it was accepted that the Netherlands deposited packages with solidified LLW at specific locations in the Northeast Atlantic Ocean under strict conditions (NEA, 1985). In 1983, the Dutch government decided to stop offshore disposal of waste packages (Alders et al., 1983; Winsemius, 1983). Since then, radioactive waste is managed in the Netherlands by COVRA, an abbreviation in Dutch for Central Organization For Radioactive Waste. The Dutch radioactive waste policy is based on above-ground storage for at least 100 years, followed by geological disposal of some types of radioactive waste. COVRA has a growing financial provision which guarantees sufficient financial resources to execute the future disposal of waste (see next section). The period of above-ground storage provides time to allow for a part of the waste to decay below the exemption and clearance levels. It also provides time to learn from experiences in other countries, to carry out research and to develop shared disposal facilities that could make disposal of waste more cost-effective (I&E, 2016). Countries with waste inventories larger than the Netherlands decided to build disposal facilities at the surface or near the surface for short-lived LILW. These facilities contain the waste previously destined for ocean dumping. Examples of these facilities are:

- Aube waste disposal facility in France operating since 1992 (international.andra.fr);
- Manche disposal facility in France that operated for 25 years (international.andra.fr);
- Forsmark disposal facility in Sweden operating since 1988 (Vahlund & Andersson, 2015).

Geological disposal

The geological disposal of radioactive waste, especially HLW, is considered the safest and most sustainable long-term solution. Geological disposal involves emplacement of packages with solid waste in a facility constructed in a stable geological environment, which is closed after filling. A geological disposal facility (GDF) is to be constructed at a depth of several hundreds of metres. The depth protects the waste from disturbance by human activities and disruptive natural processes such as erosion and flooding. Although even the most stable deep geological environments will eventually change with the passage of geological time (Ten Veen et al., 2015), the hazard potential of the waste will also decrease by radioactive decay as time passes.

The conceptual basis for geological disposal of radioactive waste is based on a multi-barrier system in which engineered and natural barriers act in concert to isolate the waste and contain the radionuclides (Chapman & Hooper, 2012). Two basic types of facility are considered; a subsurface system of tunnels (galleries) or vertical boreholes penetrating kilometres deep into the subsurface (see Textbox 2). A system of galleries is currently the focus of most studies in the Netherlands.

Around the world there are various examples of more or less successful geological disposal facilities. Between 1967 and 1978 Germany dumped 125,787 packages (124,486 LLW and 1301 ILW) in the former salt mine Asse II, south of Braunschweig (Gerstmann et al., 2002). The former mine proved not to be watertight after so much salt was mined and in 2013 a first-of-its-kind evacuation of all radioactive waste was decided. According to the responsible German Bundesgesellschaft für Endlagerung (BGE) evacuation will start in 2033. This example shows that the use of mines created for the extraction of resources may not provide a suitable base case for the construction of a GDF. In resource mining, often as much material as technically feasible is extracted. This is in contrast with the principle of geological disposal of waste in which the amount of host rock to be removed is minimized. A good example is the operational Waste Isolation Pilot Plant in the USA in which the rock provides the containment of the emplaced waste. This GDF is operational since 1999.

Finland is building a geological disposal facility for spent fuel in Olkiluoto (Fig. 21.1) and Sweden has given permission to construct a geological disposal facility for spent fuel in Forsmark. In both disposal systems, the crystalline Precambrian host rock provides isolation for the facilities built at 400 to 500 metres depth in which the engineered barriers contain the waste.

Disposal of waste in deep boreholes

2

Disposal of waste in deep boreholes drilled from the surface has been the subject of research for several decades. The OPLA programme revised the emplacement method from a borehole filled with air into a borehole filled with brine to reduce the impact of the fall. The borehole would close itself by creep of the salt in which it would be drilled (Anonymous, 1986; OPLA, 1989).

An overview of the current state of science and technology of deep borehole disposal is given by Mallants et al. (2020). Technical advances in drilling for petroleum industry and geothermal energy over the past 20 to 30 years allow for greater depths, required safe distance between multiple boreholes and larger diametres to be realized. Subsurface conditions at a depth of more than a few kilometres favour stagnant or almost stationary pore fluids. This contributes to the isolation potential for disposal and creates a large safety barrier. The great depth of boreholes strongly determines the safety of this type of disposal, even more than the properties of the host rock. The isolation capacity is thus less dependent on a specific rock type, such as rock salt or clay. As a result, many new locations for disposal come into view.

A waste inventory for HLW in the Netherlands estimates that about three boreholes are required (Van de Vate, 2018). However, the dimensions of the packages of the vitrified HLW currently push the limit of feasible boreholes (Chapman, 2019). It is uncertain if these deep boreholes may offer a relatively cheap disposal concept since the estimation of the costs cannot rely on demonstrated techniques yet. Borehole disposal of LILW is not considered due to its large volume and associated envisaged costs. Further research is needed into the effect of the high temperature at depths of 3000-5000 metres which may increase the rate at which engineered barriers alter. Currently the Netherlands does not conduct research into deep boreholes but follows and examines developments through ERDO, the association for multinational radioactive waste solutions.

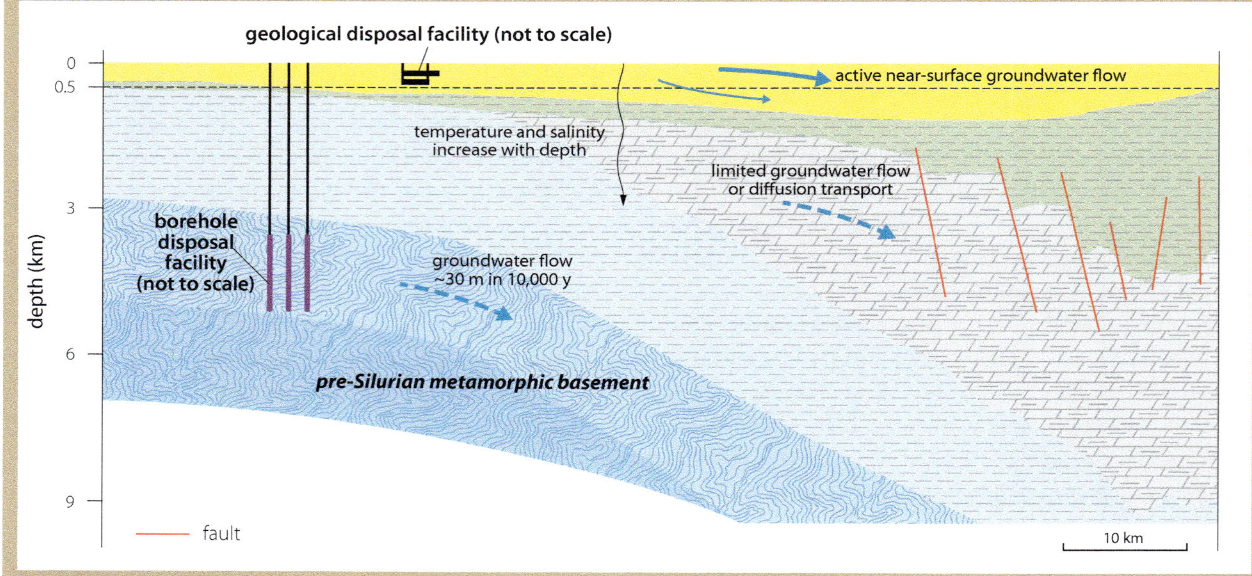

A conceptual model for deep borehole disposal (modified after Chapman & Gibb, 2003).

Research programmes into geological disposal of waste

Extensive research is required before a geological disposal facility can be constructed. The Dutch research for geological disposal of radioactive waste started in the 1970s aiming at waste that would arise from the production of 3500 MWe of nuclear power. This initiative was coordinated by the government that formed the ICK committee, which is a Dutch acronym for Interdepartmental Commission for Nuclear Energy (*Interdepartementale Commissie voor de Kernenergie*). Research in the Netherlands has taken stock of international knowledge. Disposal of radioactive waste in salt deposits was considered the most promising solution in the USA (NRC, 1957) and especially salt domes were considered for the disposal of HLW in the Nether-

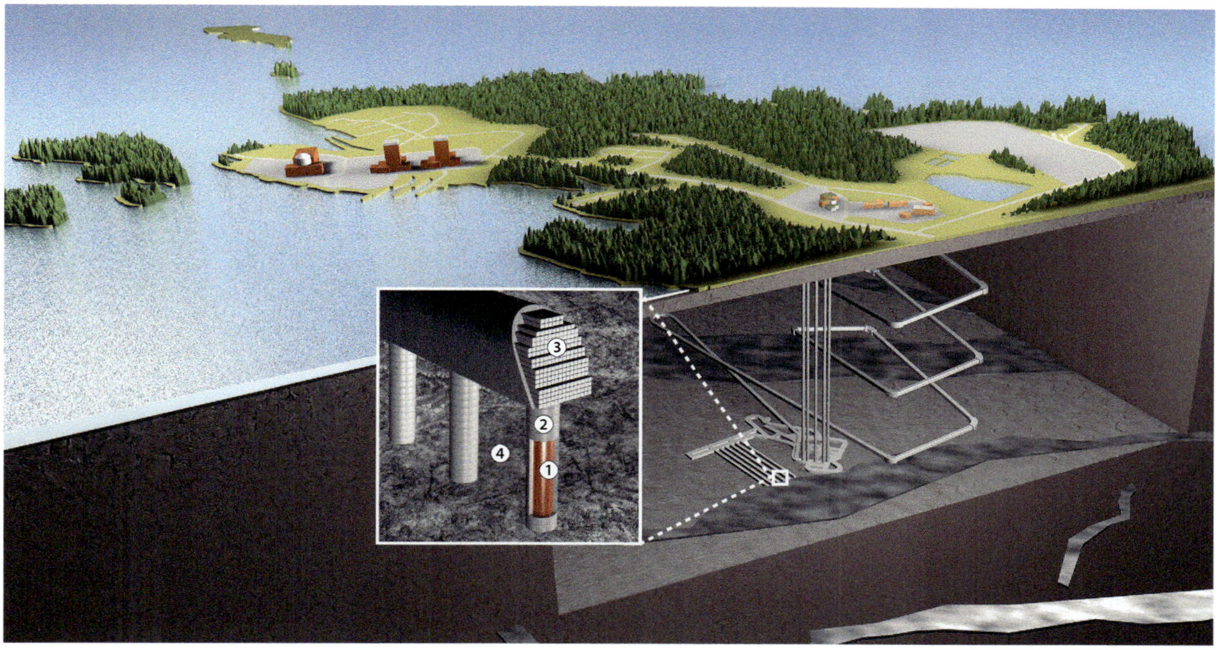

Figure 21.1. Schematic illustration of a repository at the Onkalo site near Olkiluoto Nuclear Power Plant, Finland. Engineered barriers have a highly insoluble ceramic matrix (not shown in the figure). Other components consist of: (1) a reinforced corrosion resistant copper container (diametre 1.05 m), (2) low permeable bentonite in a deposition hole (length 6.6-8.3 m) and (3) an access tunnel (height 6.95 m). In this example the natural barrier is formed by the granitic bedrock (4) at the depth (400-450 m) of the underground facility (POSIVA, 2012; Finnish Energy, 2021). Source: POSIVA.

lands (ICK, 1975; Van Aardenne, 1979). The following sections give an overview of previous research programmes on disposal of radioactive waste in the Netherlands.

Since the publication of the radioactive waste policy for the Netherlands (VROM, 1984), three research programmes focussed on potential host rocks in the subsurface of the Netherlands (Fig. 21.2). Initially these programmes focussed on Zechstein rock salt, but with time attention shifted to poorly indurated clays.

OPLA

The first research programme for geological disposal of radioactive waste was OPLA (Fig. 21.2), a Dutch acronym for onshore disposal (*OPberging te LAnd*). Next to OPLA, the disposal of waste in offshore salt domes, seabed disposal and the possibilities for interim storage of waste were studied in different programmes. OPLA was coordinated by the Geological Survey of the Netherlands (RGD) and financed by the Dutch government (Floor, 2012).
When OPLA started, solidified LLW was dumped in sea (NEA, 1985) and spent research reactor fuel was sent to the USA as part of the larger US nuclear non-proliferation treaty (IAEA, 2017). At that time, exploitation of gas from the Groningen field was in a starting phase and nuclear power was considered an appropriate alternative energy source. The shortage of oil from the Middle East in the 1970s, the so-called oil crisis, stimulated investigations into increasing nuclear power generation in the Netherlands and thereby decreasing the long-term dependency on speculative foreign energy resources. At that time, the nuclear power plants of Borsele (490 MWe) in the southwest and Dodewaard (60 MWe) in the middle of the Netherlands were operational (I&W, 2020). The assumption was that the Netherlands would only need to dispose of limited volumes of vitrified HLW. The envisaged waste inventory was based on the volume of HLW arisen from these two plants and an additional 3000 MWe of nuclear power (OPLA, 1989, 1993). Following protests against nuclear power a discussion about the proposed increase in nuclear power was organized in parallel with OPLA (Van Aardenne et al., 1985).

In 1986 the Soviet nuclear power plant in Chernobyl (Ukraine) exploded and a plume of radionuclides from the reactor reached Europe and the Netherlands. Consequently, the governmental enthusiasm for nuclear power disappeared, despite the fact that the design of this Soviet plant did not comply with European standards.

The OPLA programme focussed on Zechstein salt since the construction method for tunnels in clay was not yet industrialized (OPLA, 1984). In the provinces of Friesland, Groningen, Drenthe and Overijssel (Twente) OPLA investigated 19 salt domes (diapirs) with a cap rock at the top of the dome and a depth of at least 170 metres, 15 salt pillows and 4 horizontally bedded salt formations at a

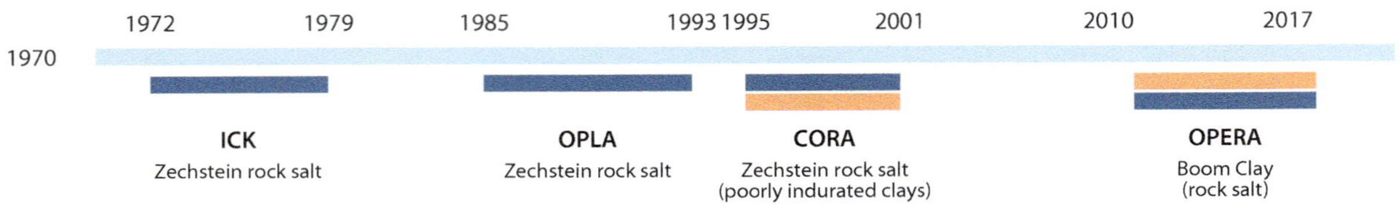

Figure 21.2. *Research activities on host rocks and the three Dutch research programmes (Verhoef et al., 2020). Blue: rock salt, orange: poorly indurated clay. ICK: Interdepartementale Commissie Kernenergie (Interdepartmental Commission for Nuclear Energy); OPLA: OPberging te LAnd (onshore disposal); CORA: Commissie Opberging Radioactief Afval (Commission for Disposal of Radioactive Waste); OPERA: OnderzoeksProgramma Eindberging Radioactief Afval (Research Programme into Geological Disposal of Radioactive Waste).*

maximum depth of 2200 metres. Geohydrological profiles of the sedimentary rock successions at each of these 38 sites were made to model the potential transport of radionuclides from the rock salt into the biosphere (OPLA, 1989). Of each site, the geological history during the Mesozoic and Cenozoic periods was investigated to make quantitative predictions on, for instance, the uplift rate of salt domes (Geluk & Wildenborg, 1988). In the extension of the OPLA programme, 45 sites with rock salt were found to be suitable for disposal of radioactive waste (OPLA, 1993). In the parliamentary evaluation of OPLA, the concept of retrievability was introduced in the Dutch radioactive waste policy to have control over the emplacement of waste packages and closure (Alders, 1993).

CORA

The OPLA research programme (Fig. 21.2) was followed by the CORA programme, a Dutch acronym for Commission for Disposal of Radioactive Waste (*Commissie Opberging Radioactief Afval*). This programme was also coordinated by the Geological Survey of the Netherlands (RGD) and financed by the Dutch government (Floor, 2012). At the start of CORA, solidified LILW was being stored in aboveground facilities as an interim solution. In the meantime, the USA had introduced a fee for accepting spent research reactor fuel (Messick et al., 2006). CORA primarily focused on the technical feasibility of *retrievable* disposal of vitrified HLW. Consequently, the investigations mainly addressed subsurface disposal of radioactive waste in subsurface facilities or in vertical boreholes drilled downward from the facility's gallery floor.

During that time, the Belgian disposal research programme demonstrated the feasibility of the construction of a tunnel in Rupelian clay in Mol (see section Construction of the geological disposal facility in indurated clay). The CORA programme therefore also considered Paleogene and Neogene clay formations (Simmelink et al., 1996). For both the disposal concepts in rock salt and clay an underground facility with short disposal galleries to emplace a single waste package containing vitrified HLW was considered (Barnichon et al., 2000). The facilities were envisaged to be at a depth of 800 metres for rock salt and 500 metres for clay, i.e. below the level of expected future subglacial erosion. The difference in depth reflects the requirement for a sufficiently thick host rock on top of the facility of at least 200 metres for rock salt and 50 metres for clay. The minimum clay thickness was considered to be 100 metres. Rock salt can extend vertically hundreds of metres in a salt dome. In the developed concept, the disposal gallery was to be filled with an easily removable material such as crushed rock (rock salt or claystone) that would also provide shielding against the ionising radiation from the radionuclides in the waste. Retrieval of the waste package was envisaged by removal of the backfill and taking the waste package out (CORA, 2001). The Dutch government became convinced after the evaluation of CORA that the retrievability was technically feasible with these disposal concepts. The government also concluded that COVRA, as an organization that manages all types of radioactive waste, should be involved in research coordination (Geel, 2002).

OPERA

The last research programme was OPERA (Fig. 21.2), a Dutch acronym for Research Programme into Geological Disposal of Radioactive Waste (*OnderzoeksProgramma Eindberging Radioactief Afval*). This third programme was coordinated by COVRA and financed by the government and EPZ, the owner of the nuclear power plant in Borsele. This programme focused on disposal of radioactive waste in a facility hosted in the Boom Member of the Rupel Formation (Vis et al., 2016). This clay had been extensively studied in Belgium but knowledge for disposal of radioactive waste in the Netherlands in such a type of poorly indurated clay was much less developed than for disposal in rock salt.

By the start of OPERA, the local social acceptability for nuclear power had increased. In 2011, cores were allowed to be taken from boreholes in Borsele to investigate the subsurface for a new nuclear power plant. This drilling activity also provided fresh Rupelian clay cores for research into disposal of waste. These cores were therefore also studied in OPERA (e.g. Behrends et al., 2015). However, the project for a new plant was not continued due to financial issues. The amount of waste from nuclear power generation (HLW) to be handled for disposal had increased, since the operational life period of the existing 490 MWe nuclear power plant in Borsele was extended from 30 years to 60 years. This doubling of the operational period did not result in a double volume of vitrified HLW since the radionuclide content per vitrified waste form unit was increased. The envisaged volume for disposal of this type of HLW was 70 m^3 in CORA (CORA, 2001) and 93 m^3 in OPERA (Verhoef et al., 2017). Also, at the start of OPERA, the operational lifetime of the research reactors was increased. The envisaged volume of spent fuel for disposal was increased from 40 m^3 in CORA (CORA, 2001) to 104 m^3 in OPERA (Verhoef et al., 2017). In addition, waste packages with solidified ILW and LLW were envisaged to be emplaced together with HLW in one disposal facility.

Although the disposal concept with a facility at 500 metres and clay as a host rock remained, there were three important differences between the concepts defined in CORA and OPERA. These concern three new features adopted from the Belgian disposal concept for HLW:

1. Only waste packaged in such a way that a worker can handle it without additional shielding will be emplaced
2. More than one waste package will be emplaced in each disposal gallery
3. Engineered concrete supports will also be envisaged for the disposal galleries i.e. not only for the transport tunnels as studied in CORA

The final report of the third Dutch programme OPERA was submitted to the Dutch parliament in 2018 (Verhoef et al., 2017). The European directive to manage radioactive waste (EC, 2011) needed to be implemented in the Dutch policy during the execution of OPERA. This directive required a national programme for the management of radioactive waste and spent fuel (I&E, 2016). From then on, discussions about disposal of radioactive waste have been structured through this national programme (Van Veldhoven-Van der Meer, 2018).

COVRA's long-term research programme

The research programmes in the Netherlands have not been consecutive (Fig. 21.2) since it took time to secure the financial resources for each programme. The financial means for the future construction, operation and closure of a disposal facility have been and will be collected via a fee on the radioactive waste. Since 2018, the research for disposal of radioactive waste is financed directly by the waste fees, so as to secure continuity and prevent dormant phases in research. The research programme integrates all available knowledge in safety cases. Safety cases for disposal facilities in rock salt and poorly indurated clay will be updated each time the government needs to report to the European Commission about the progress made in the national programme, i.e. before 2025, 2035 and further (Verhoef et al., 2020). These safety cases are not site-spe-

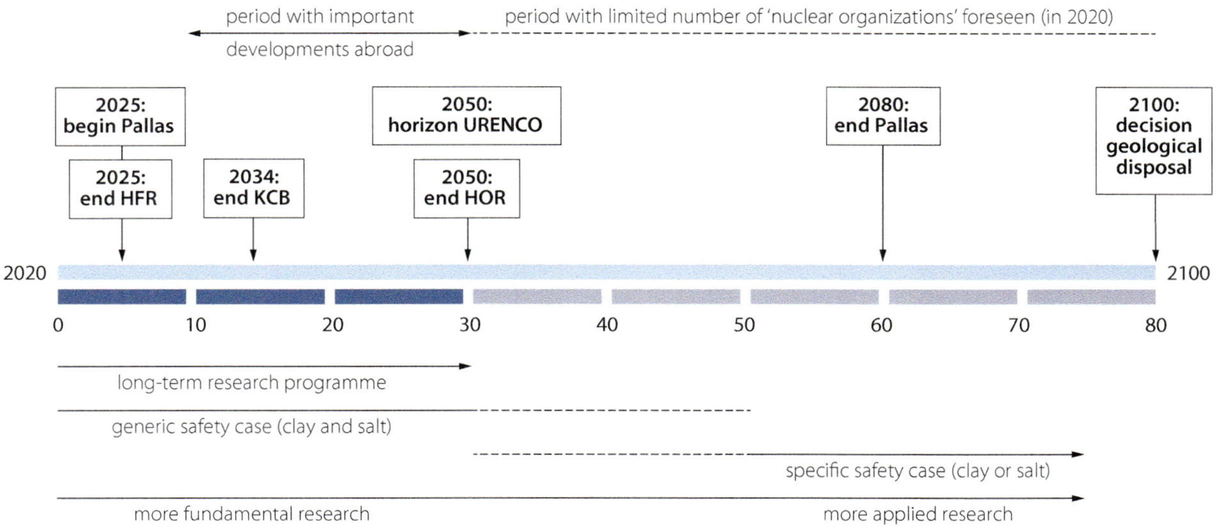

Figure 21.3. Timeline for the future of geological disposal of radioactive waste in the Netherlands (Verhoef et al., 2020). Pallas is a new nuclear plant for medical isotopes; High Flux Reactor (HFR) is the current nuclear plant for medical isotopes, KernCentrale Borsele (KCB) is a nuclear power plant, Hoger Onderwijs Reactor (HOR) is the nuclear plant for education and research, the URanium ENrichment COmpany (URENCO) was planned to be operating in the Netherlands until 2050 in 2020.

cific and more fundamental research will be performed in the next decades (Fig. 21.3). More applied research and site-specific research is currently foreseen after 2050. The current Dutch policy is that a disposal facility should be operational in 2130 (I&E, 2016).

The research for geological disposal of radioactive waste can be subdivided into two main parts. One part studies the feasibility to construct, operate and close an underground facility. The other part assesses the provided safety by a disposal system of engineered and natural barriers after closure.

Geological aspects of subsurface disposal

In the Netherlands two rock types qualify for geological disposal: rock salt and poorly indurated clay both of which have the ability to flow plastically. The low permeability of these rock types (horizontal permeability down to 10^{-19} m^2) ensures containment of radionuclides from the waste. Salt is virtually impermeable to oil, gas and water. Potential host rocks considered are the upper Permian (Lopingian) rock-salt formations of the Zechstein Group and Paleogene and Neogene clay formations. Over the last decades, most attention has been given to the Permian Zechstein salt and the Paleogene Boom Member of the Rupel Formation (the Boom Member was previously known as the Rupel Clay Member and is informally known as Boom Clay). A recent study of the Main Rot Evaporite Member concluded that it appears to be locally not well suited for geological disposal, considering current disposal concepts (Altenburg, 2022).

Disposal of radioactive waste in rock salt

Rock salt has multiple advantageous properties and for that reason is considered for the disposal of radioactive waste (e.g. in USA, United Kingdom, Germany, the Netherlands) and hazardous non-radioactive waste (Chandler facility, Australia). Rock salt has an ultra-low permeability (and is essentially impermeable; e.g. Popp et al., 2001), severely limiting any transport of radionuclides to the surface. Rock salt can also deform plastically and restore its ultra-low permeability when disturbed (e.g. Spiers et al., 1986; Urai et al., 2008). Salt creep can slowly surround other materials such as waste to form a tight, essentially impermeable natural barrier (e.g. Hansen et al., 2016). Additionally, there is substantial hands-on experience with construction in rock salt. Salt mining has existed since the Bronze Age (Reschreiter & Kowarik, 2019) and in the Netherlands salt caverns have been constructed for the storage of gas (Zuidwending salt dome, near Veendam), oil (Marssteden salt dome, Enschede) and nitrogen (Heiligerlee salt dome, near Winschoten). Salt being mined near Enschede using dissolution mining has generated substantial experience with these techniques (Groenenberg et al., 2025, this volume). There is also operational experience with disposal in rock salt: the Waste Isolation Pilot Plant (WIPP) near Carlsbad (New Mexico, USA) has been receiving radioactive waste for disposal since 1999.

Zechstein host rock

Rock salt from the Zechstein Group occurs in large parts of the subsurface of the Netherlands (for more details on Zechstein salt, see Bouroullec & Geel, 2025, this volume; Bouroullec & Ten Veen, 2025, this volume) Two main types of salt formations can be distinguished: bedded and domal salt formations. The characteristics of bedded salt and domal salt are quite different. Due to its uprise, domal salt is relatively pure and homogeneous. The lateral extent of domes is limited (hundreds of metres), and therefore the dome margins are expected to delimit the area useful for a disposal facility. Subrosion by groundwater produces a cap rock at the top of the salt formation and, in some locations, at the margins of the domes. The cap rock may be impermeable or have very low permeability such that it armours the dome against dissolution. Bedded salt is generally less pure than domal salt as it is interbedded with limestone, dolomite, anhydrite, polyhalite and fine-grained siliciclastic beds, which can provide pathways for advective horizontal transport, if reached. Salt beds are typically continuous over large areas.

The Zechstein Group consists of various rock types, including clay, carbonate, anhydrite and rock salt. The spatial distribution of the members containing halite can be consulted online (TNO-GDN, 2023). In the northeastern part of the Netherlands the deposits of the Zechstein Group are present at depths greater than 500 m and reach a thickness of hundreds of metres, mainly due to halokinesis that led to the formation of salt domes and pillars (Fig. 21.4a; Bouroullec & Ten Veen, 2025, this volume). The salt domes can be identified from the thickness map, by assuming a minimum thickness of rock salt in a salt dome of 1300 m.

Salt domes have complex internal geometries and the ductile parts may include brittle inclusions (stringers) consisting of, for example, anhydrite and other impurities such as brine pockets (e.g. Van Gent et al., 2011; Strozyk et al., 2012). The internal structure of salt domes therefore varies and needs to be addressed separately for each individual structure (Geluk, 1998). Several studied salt domes (Harsveldt, 1980, 1986) show the presence of Zechstein Z2 rock salt in the core and younger (Z3) salt along the margins, whereas others show an opposite configuration.

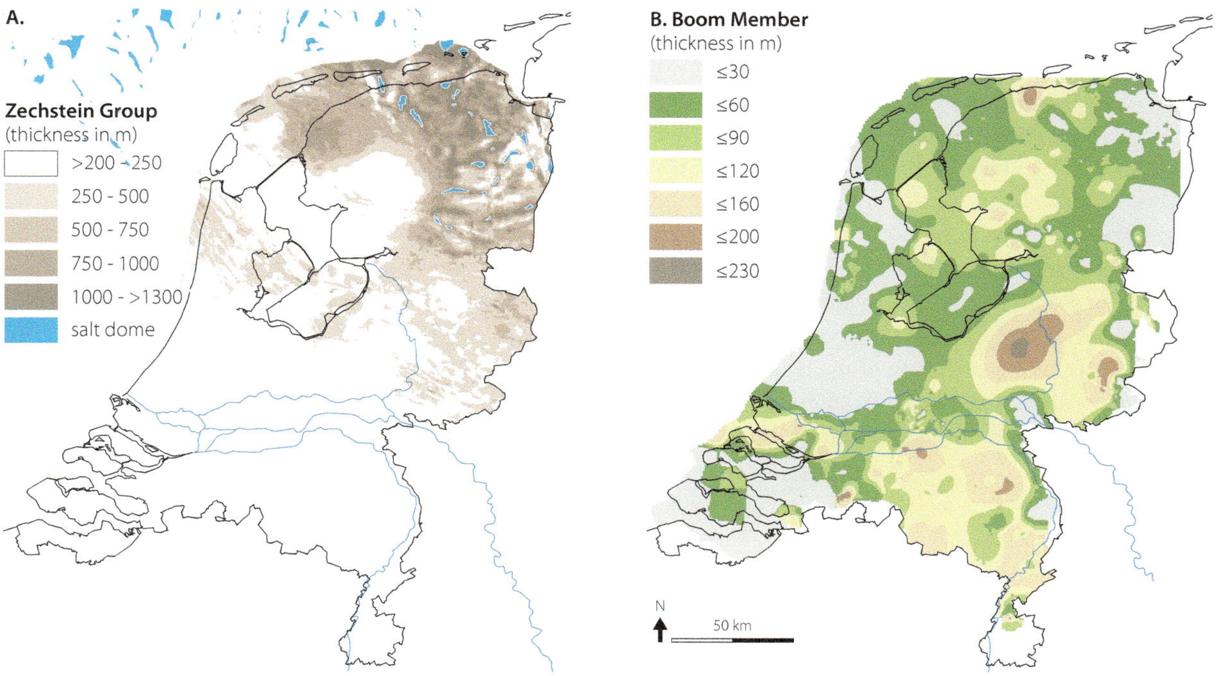

Figure 21.4. *a) Onshore thickness of the Zechstein Group (all salt units and associated lithologies combined), displayed where the top of the group is at a depth greater than 500 m below surface (modified from Hart et al., 2015). b) Onshore thickness of the Boom Member of the Rupel Formation (modified from Vis & Verweij, 2014).*

Disposal of radioactive waste in clay

Clay-rich formations are considered as potential host rocks for disposal of radioactive waste in numerous countries, including France, Switzerland, Belgium, Canada and Germany (Boisson, 2005). In France, the underground research laboratory (URL) in Jurassic (Callovian-Oxfordian) clay in Bure was declared as a public utility in July 2022. This declaration was necessary to apply for a construction licence for a disposal facility for the high- and long-lived intermediate-level waste. For a site selection in Switzerland, the National Cooperative for the Disposal of Radioactive Waste (NAGRA) investigated three regions of about 10 km^2, using seismic surveys and deep boreholes. All these regions have Jurassic (Opalinus) clay with a thickness of about 110 m in the subsurface. In 2022, the Nördlich Lägern region in northern Switzerland was proposed for the construction of an underground disposal facility. The construction of an URL to perform further analyses for this disposal facility is currently foreseen in 2034.

Clay formations have favourable properties such as significant lateral continuity, very low permeability, self-sealing of fractures by cementation for indurated clays and self-healing of fractures by swelling of clay minerals for poorly indurated clays, as well as a capacity to chemically retard radionuclide migration. The analysis of fluids contained in clays provides formation-specific properties and evidence of groundwater system stability and resilience to external perturbation through geologic time (Mazurek et al., 2009). Solute migration is preferentially horizontal due to the lateral continuity in argillaceous sediments. The very low permeability, chemical retardation and self-healing/sealing characteristics contribute to the containment of radionuclides. Examples of hydrothermally formed clays surrounding uranium ores in Canada and Gabon clearly show the effective containment of radionuclides that originated from fission of uranium two billion years ago (e.g. Bentridi et al., 2011).

Poorly indurated clay – Boom Clay

Poorly indurated clay occurring at relevant depth ranges for a geological disposal facility are known from the Paleogene marine clay layers of the Ieper and Asse members, the Oligocene Boom Member, and from the Neogene marine clay layers of the Breda Subgroup and Oosterhout Formation. Up to now, most attention has been given to the (hydro)geological aspects of the Boom Member.

The clay of the Boom Member ('Boom Clay') is present in almost the entire Dutch subsurface at depths of up to about 1500 m (Fig. 21.5; Vis et al., 2016). In the Netherlands, the Boom Clay has a variable thickness and rarely exceeds 125 m (Fig. 21.4b). As such it is much thinner than the Zechstein rock salt. The Boom Clay is generally thinner than 100 m, but occasionally reaches a thickness of more than 160 m in the Roer Valley Graben in the south and the Central Netherlands Basin in the east of the Netherlands (Fig. 21.4b). In the north, some patches with a thickness exceeding 120 m are present. Along the south-western and eastern borders of the country the Boom Clay lies at depths of less than ~50 m from surface. Locally it experienced thinning due to erosion associated with Cenozoic uplift during the Pyrenean and Savian inversion phases. Much like rock salt, the clay is not homogeneous and contains both vertical and lateral grain-size changes. In fact, most Boom Clay samples that were analysed for grain size classify as a silt rather than a clay (Vis et al., 2016). The platy form of the clay particles, however, increases the apparent size of the particles with the technique used and, consequently, the clay fraction in the sample is underestimated (Konert & Vandenberghe, 1997).

Facility locality

The host rock is considered to lie at a suitable depth if the effects of any future tectonic processes and climate change are not expected to affect the integrity of an embedded facility for the duration that the HLW will be hazardous. A geological disposal facility will be constructed at a depth of several hundreds of metres, where other rock formations will act as isolating layers.

The maximum depth of a geological disposal facility depends on how the facility is constructed, operated and closed. Many activities may be performed robotically, but involvement of humans is never excluded. As subsurface temperature increases according to the geothermal gradient, a maximum disposal depth of 1000 metres is considered to provide acceptable underground working conditions.

Predicting the future geological development of a facility in a host rock layer requires knowledge of past processes. Given that geological processes occur intermittently or cyclically, the past can be used as a key to the future. For both rock salt and clay, the research programmes mentioned above have looked at past and future aspects in great detail (e.g. Wildenborg et al., 1990; Remmelts et al., 1993; Ten Veen et al., 2015). We will elaborate on some of these aspects below.

Glaciation

The latitudinal position of the Netherlands makes it susceptible for glaciations as has been extensively documented for the Quaternary (see Busschers et al., 2025, this volume). The geological record of the past half a million years shows that the north of the Netherlands has been impacted more by glaciations than the south. The ice sheets during the Elsterian, Saalian and Weichselian never reached the southern provinces of the country (Westerhoff et al., 2003b; Ten Veen et al., 2015). During the retreat of the Elsterian ice sheet, subglacial erosion scoured sediments up to 600 metres deep in the northern Netherlands (Fig. 21.6a; Keller, 2009; Stackebrandt, 2009; Ten Veen et al., 2015). Glacial basins from the Saalian glaciation are however rarely deeper than 150 metres (Van Dijke & Veldkamp, 1996).

An ice sheet would not only have erosional effects, its physical load also impacts radionuclides released from the engineered barrier system to the host rock. In the case of a clay host rock and an assumed ice-sheet thickness of 1000 metres, the ice load presses the pore water out of the clay (Wildenborg et al., 2000). A recent study estimates that the ice-sheet thickness was around 200 metres in the northern Netherlands during the Saalian glaciation (Ten Veen et al., 2015). Consequently, the ice loading effect is expected to be less than previously assumed (Wildenborg et al., 2000), reducing the rate of enhanced radionuclide transport through clay by a factor of five. Ironically, current global warming may 'postpone' the next glaciation by

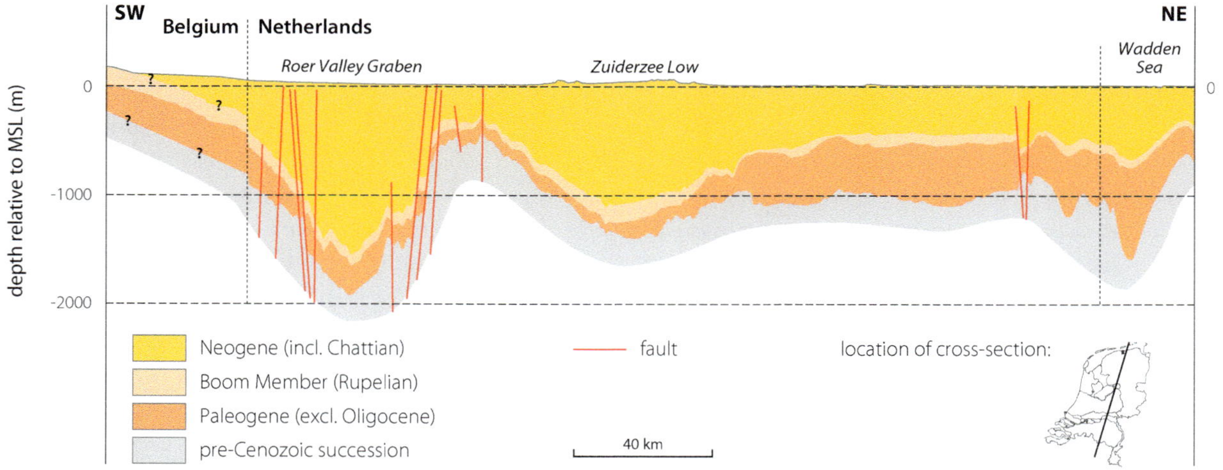

Figure 21.5. Cross-section showing the variable depth of the Boom Member in the Netherlands and the cropping out of this clay in Belgium. The absence of faults in the Zuiderzee Low area probably reflects a lack of data (modified from Vis et al., 2016).

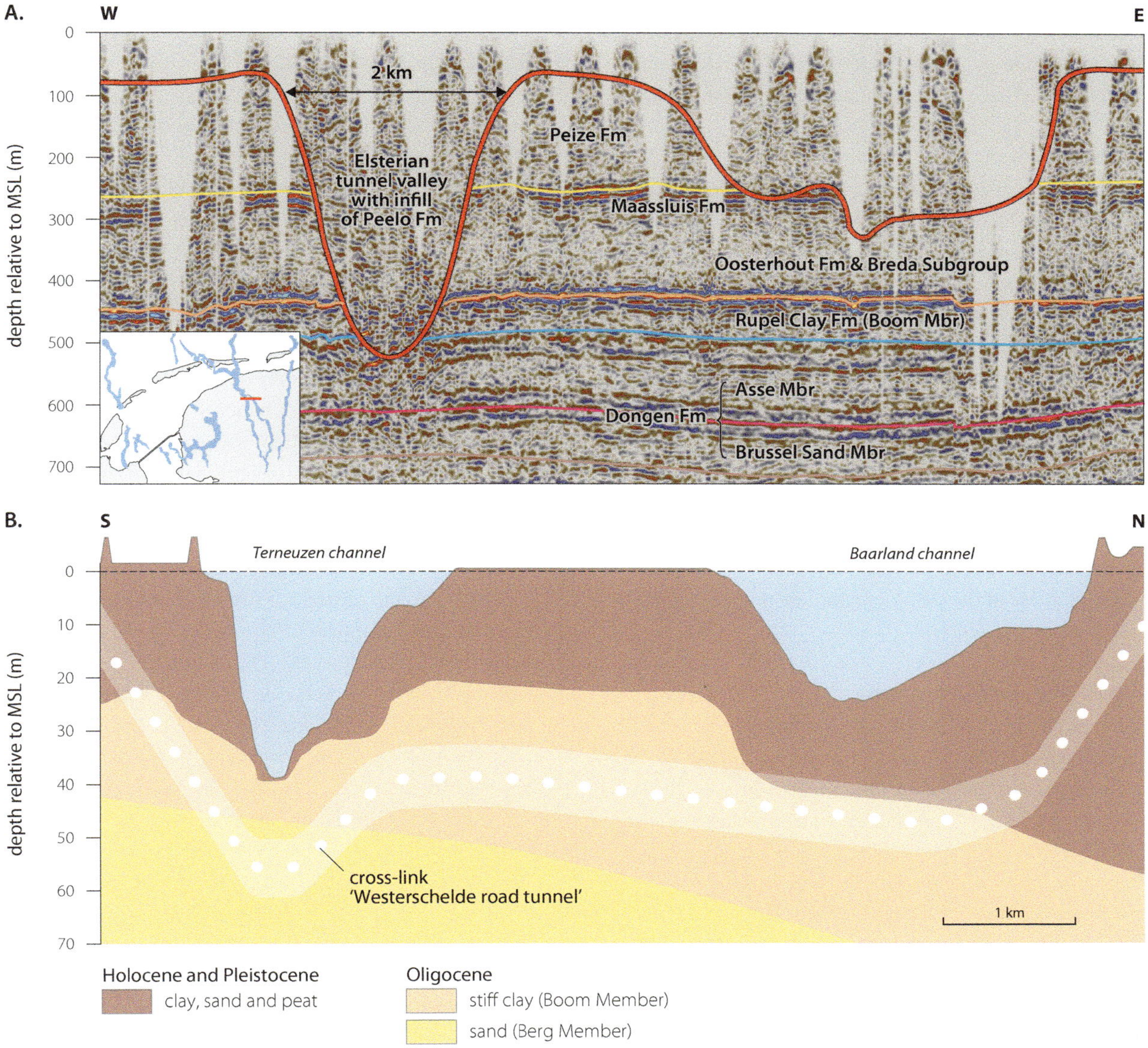

Figure 21.6. a) Erosion down to the Boom Member by tunnel valleys from the Elsterian ice age in the north of the Netherlands (modified from Ten Veen et al., 2015). See inset map for location of tunnel valleys and seismic section. b) Erosion by tidal channels in the southwest of the Netherlands (modified from De Mulder & Ritsema, 2003). Note that the Westerschelde road tunnel is constructed in the clay of the Boom Member.

more than 100,000 years (Ganopolski et al., 2016) such that glacial loading and erosion becomes a negligible hazard for some types of waste such as vitrified HLW. That is the case because this type of waste reaches a radiotoxicity smaller than uranium ore by decay after 20,000 years (Gruppelaar et al., 1998).

The current long-term research programme proposes a minimum disposal depth of 200 metres for the geological isolation of radioactive waste. In the case of a future glaciation which does not result in ice sheet expansion across the Netherlands, permafrost formation is a process that could be expected. Based on a modelling study, the maximum depth penetration of permafrost is expected to range between 140 and 185 metres in the Netherlands (Ten Veen et al., 2015), which is not in the range of the proposed disposal depth.

Channel incision

Incision by tidal channels in which much flow energy is concentrated in a narrow zone is the most powerful type of non-glacial vertical erosion below sea level. Modern examples include the Holocene tidal channels of the 'Pas van Terneuzen' and 'Everingen' (Westerhoff et al., 2003a) in the southwest of the Netherlands (Fig. 21.6b). These channels are up to 100 metres deep and have eroded down into the Boom Clay. The difference in depth between subglacial erosion in the north (up to 600 metres) and tidal channel erosion in the southwest (up to 100 metres), illustrates that suitable disposal depths may differ geographically.

Tectonics

Tectonic processes can affect a geological disposal facility by lifting up its embedding geological layer so much that it gets exposed at the surface, or by hydrological changes affecting fluid flow (Ten Veen et al., 2015). Based on the known pace of tectonic movements in the Netherlands in the past, it is unlikely that exposure of a facility will occur during the next million years (Ten Veen et al., 2015). Tectonic processes and related seismicity can have an effect due to possible changes in groundwater flow patterns through faults and fractures. Fault activity can also be enhanced by loading and unloading of the lithosphere by sediment, water or ice (Ten Veen et al., 2015). Especially differential loading by sediment and ice has been the topic of study in the OPLA-programme related to rock salt (Remmelts et al., 1993).

Hydrology and geochemistry

In natural geological processes, groundwater flow and geochemistry play an important role during burial and diagenesis. As any geological disposal facility is constructed in geological layers that have been affected by these processes for millions of years, any human intervention is expected to have an impact on the natural conditions. After permanent closure of a facility natural processes will take over again and, after a certain period, radionuclides will be released from the engineered barrier. Ideally, the host rock will isolate the disposal facility in such a way that radionuclides will either not escape the host rock at all or only after a delayed period. Due to its physical characteristics, clay has the advantage of being able to geochemically react with radionuclides. For details on recent work on these topics, the reader is referred to recent reports from the Dutch OPERA-programme (Vis & Verweij, 2014; Hart et al., 2015; Griffioen et al., 2017) or studies on the Opalinus Clay in Switzerland (e.g. Bradbury & Baeyens, 2011).

Disposal of radioactive waste in rock salt

Disposal concept for a facility in rock salt

The disposal concept describes how to construct, operate and close a disposal facility. Several disposal concepts have been developed for rock salt (e.g. Hamstra, 1981; Poley, 1999; Grupa & Houkema, 2000). In the most recent concept (Bartol et al., in preparation), it is assumed that a geological disposal facility is constructed in the core of a salt dome belonging to the Zechstein Group (Fig. 21.7). While this concept is for domal salt, this does not exclude a disposal facility in bedded salt.

Construction and operation of the geological disposal facility in rock salt

Within the salt dome, a two-level geological disposal facility is envisaged (Fig. 21.7a), connected to the surface by two vertical shafts. The upper level of the geological disposal facility will be constructed at a depth of at least 750 metres below the surface, i.e. below the observed maximum depth of subglacial erosion. Construction will be done using modern non-explosive mining techniques such as for example a drum miner. The upper level will be used for the disposal of LILW (Fig. 21.7b). For emplacement within the disposal rooms, well-established technologies will be used, similar to the way waste is currently handled and stored at COVRA. Depending on the type of waste, packages will either be placed horizontally in racks (waste in 200 litre drums) or placed on the ground (waste in 1000 litre containers or in Konrad Type II containers) using (heavy) forklift trucks (Fig. 21.7d).

After all waste has been placed within a disposal room, optionally bags containing MgO may be placed next to and on top of the waste packages in order to absorb CO_2 released by decaying organic components in the waste. Open spaces in salt will be constructed sufficiently large to account for salt creep that can be in the order of a few mm per year.

The lower level of the disposal facility at a depth of 850 metres below the surface will contain the HLW (Fig. 21.7c). The construction method will be similar to the upper level. In the most recent disposal concept (Bartol et al., in preparation), HLW will be placed in a single (standardized) overpack at the surface that provides both containment and shielding. Within the geological disposal facility, these overpacks are placed in shallow vertical holes (Fig. 21.7d) half the height of the overpack that should prevent their tipping and consequential damaging. Since an overpack provides shielding, no additional radiation protection is needed within the geological disposal facility. This makes the handling of the overpack within a geological disposal facility easier as it is possible to stand next to it. A disadvantage of using an overpack is that heavy equipment is required for their emplacement. Rather than constructing all the disposal galleries at once, new disposal galleries will be constructed at the time the 'old' disposal galleries have been fully filled. The rock salt removed during the construction of the new galleries will then be used to backfill galleries already filled with waste. There is broad experience and knowledge on the use of granular salt as backfill (e.g. Spiers et al., 1988; Poley, 2000). Seals will be constructed at both ends of a backfilled disposal gallery.

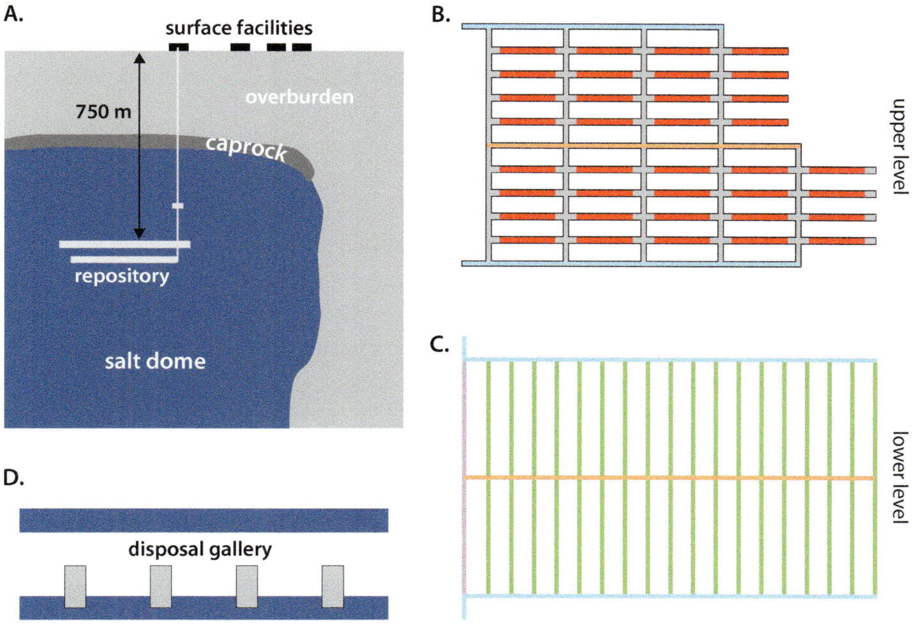

Figure 21.7. a) Schematic cross section through a salt dome, showing a geological disposal facility with two vertical shafts connecting the facility to the surface. b) Plan view of the upper level consisting of disposal rooms (red), a transport tunnel (orange) and ventilation tunnels (light blue). c) Plan view of the lower level consisting of a central transport tunnel (orange), ventilation tunnels (light blue), service tunnel (light purple) and the disposal galleries (green). d) Cross section through a disposal gallery in which waste is emplaced. Grey: standardized overpack, blue: rock salt. None of the figures are at scale.

In the case that HLW needs to be retrieved, the seal and backfill will need to be removed using light machinery or manual mining techniques to avoid damage of the overpacks. Geophysical techniques such as radar could be used to locate overpacks within the backfill (Anonymous, 1986).

Closure of the geological disposal facility in rock salt

At the end of the operational phase, the geological disposal facility will be permanently closed and all the remaining open spaces will be backfilled using granular rock salt. By doing so, no further maintenance of the geological disposal facility will be required as the geological disposal facility provides safety in a completely passive manner. Seals will be placed in the shafts connecting the geological disposal facility with the surface. These seals will likely not only consist of granular rock salt but of a combination of different materials, such as salt and concrete, with each material providing its own water tightness but at different time scales (Bollingerfehr et al., 2013).

Disposal system with rock salt

For the long term evolution of the disposal system in rock salt, both normal and altered scenarios are considered. The normal evolution scenario (Verhoef et al., 2017), in which the degradation of the container and the effects of climate change evolve as expected, is the most likely. The altered evolution refers to one in which either the geological disposal facility is abandoned during the operational phase, during intensified glaciation or unintentional human intrusion takes place (Verhoef et al., 2017). Here we only discuss the normal evolution and unintentional human intrusion.

Normal evolution

In the normal evolution scenario of the disposal system, the engineered barriers (overpacks and seals) will initially provide containment. Over a period of less than 10^3 years, these barriers will degrade and lose their containment function. Concurrently, the backfill with its ultra-low permeability will take over the containment function of the engineered barriers. In the normal evolution, the waste and therefore the radionuclides are effectively trapped within rock salt for a period of $<10^5$ years. It is only at longer geological time scales ($>10^5$ years) that geological processes such as subrosion and diapirism could result in the release of radionuclides into the biosphere. While both processes could result in the release of radionuclides, they will not necessarily affect the safety of the disposal system as these processes are expected to be very slow. Hence, it is expected that it will take more than 10^6 years before any release is expected due to subrosion and diapirism. By that time, the hazardous potential of the waste will have diminished significantly. The process of subrosion may lead to salt dissolution by groundwater flow around and over a salt dome and in this case the natural barrier function will be reduced. Subrosion could eventually result in the total destruction of the natural barrier and consequently the release of radionuclides into the groundwater. When released, these radionuclides will gradually move upward with the groundwater through the subsurface and could eventually reach the surface and enter rivers, lakes and wells. In this scenario, humans could become exposed to the radionuclides in various ways, for example, through drinking water from contaminated wells or by using contaminated ground- or river water for the irrigation of crops.

Diapirism is the upward movement of salt with respect to the mother (bedded) salt layer (Hudec & Jackson, 2007). In case the upward moving salt contains a geological disposal facility, the travel time that radionuclides need to reach the surface will decrease. Over very long time-scales, diapirism could even bring the radioactive waste mingled in salt to the surface (Prij et al., 1993). Salt domes piercing the surface are well known from Iran (e.g. Waltham, 2008), but also occur in the North Sea in block K9 (Laban, 2010). If a repository is brought to the surface by diapirism, exposure to humans may occur by direct radiation and inhalation of contaminated dust.

Both subrosion and diapirism rates vary between salt domes, within a single salt dome and also with time. Diapirism rates are high during periods of tectonic activity (Hudec & Jackson, 2007) while subrosion rates can increase during periods of glaciation (Geluk et al., 1993). Overall, the subrosion rates vary between 0.1 and 1.0 mm/year (Zirngast, 1996; Köthe et al., 2007; Lauwerier, 2022) while diapirism rates vary between 0.01 and 0.1 mm/year (Zirngast, 1996; Köthe et al., 2007; De Gans & Duin, 2010; Lauwerier, 2022). If a relatively high subrosion rate of 1 mm/year is assumed (Köthe et al., 2007; Lauwerier, 2022) and if there are 200 metres of salt between the repository and the overburden, it will take about 200,000 years before radionuclides would be released into the groundwater. If a disposal facility is built at a depth of 850 metres (Bartol et al., in preparation) and a high diapirism rate of 0.5 mm/year is assumed, it will take about 1.7 million years before the facility reaches the surface. In both cases the radioactivity of the waste will have decreased significantly due to natural decay by the time any of the material reaches the biosphere. The waste will have a lower level of radiotoxicity than uranium ore.

In reality a combination of different processes and their interactions should be considered. For example, in a scenario that combines subrosion and diapirism (diapirism-subrosion scenario), subrosion will diminish the salt barrier around the waste, while at the same time the salt dome moves upwards gradually, reducing the travel time of the radionuclides to the surface. As the radioactivity level of waste decays over time, the combination of subrosion and diapirism probably does not pose a problem.

Unintentional human intrusion

Special attention needs to be given to human intrusion and, specifically to inadvertent human intrusion, as salt is currently considered a resource and is used for storage of, for example, gas and nitrogen. It is unlikely that in the near future unintentional human intrusion will occur since the location and content of a geological disposal facility are always registered.

In the far future, inadvertent human intrusion in a geological disposal facility constructed in a salt dome could happen, for example during the construction of a cavern for gas storage. It could also happen for a geological disposal facility built in a bedded salt formation when exploration boreholes are drilled through the same formation. While inadvertent human intrusion cannot be excluded, it is possible to either reduce the consequences or reduce the probability of occurrence (Mobbs, 2012). For the disposal concept discussed here (Bartol et al., in preparation), a two-level geological disposal facility design was selected. Having a two-level geological disposal facility minimizes its footprint and thus decreases the probability of human intrusion. Also, cuttings are currently examined during drilling. If this is also done in the future, workers will be warned as they may encounter material from the upper level of the facility containing lower level waste. Other measures that could be taken are the erection of monuments on the surface or subsurface markers like steel or concrete slabs buried just above the salt dome to indicate that there is a geological disposal facility below the surface (Mobbs, 2012). For the WIPP facility in the USA, such warning signs and surface markers have been suggested (Trauth et al., 1993).

Disposal of radioactive waste in poorly indurated clay

Disposal concept for a facility in poorly indurated clay

The disposal concept developed in OPERA has recently been updated (Fig. 21.8) for the inventory of waste collected until 2130 (Burggraaff et al., 2022). In this concept (scenario 1), all types of waste are placed at the same depth. The eventual design of the facility will be determined by the characteristics of the site. For example, if multiple stacked clay formations are present, the deeper one could be used for waste packages of HLW and the shallower ones for lower level waste. Due to the geometry of clay layers (i.e. large lateral extent and relatively limited

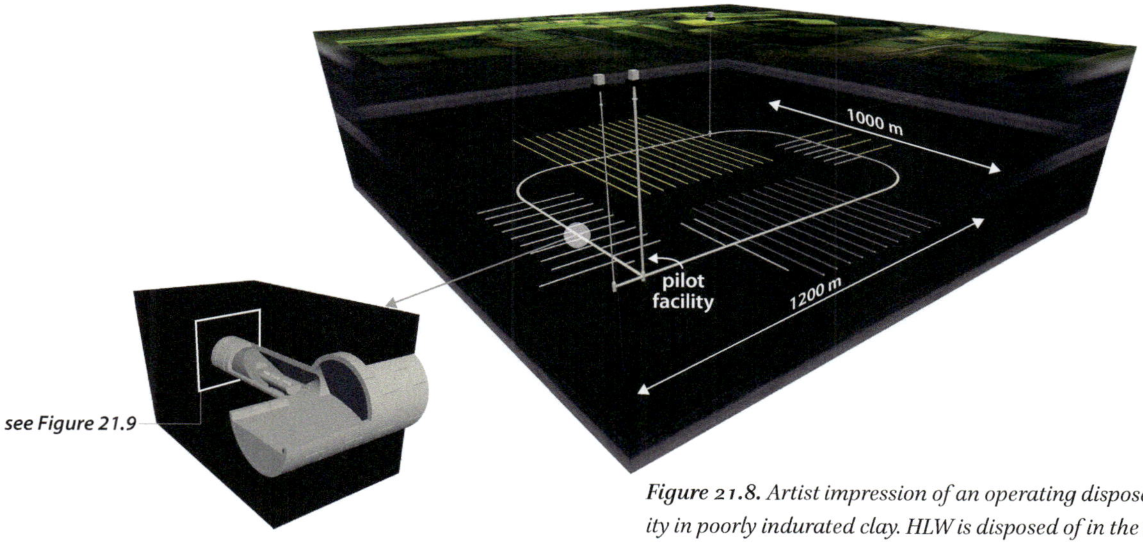

Figure 21.8. Artist impression of an operating disposal facility in poorly indurated clay. HLW is disposed of in the section with short tunnels (pilot facility). LILW is disposed of in the other three sections. Illustration: COVRA.

thickness), galleries will be constructed horizontally, also to minimize vertical disturbance of the clay layer.

Before the operational phase commences, a pilot facility is envisaged which consists of a comparable layout to that proposed for disposal of HLW but containing only a single HLW package. This facility will be equipped with multiple sensors to monitor the impact of the waste package on the behaviour of the engineered barriers and on the in-situ host-rock conditions. The pilot facility will be open to the public.

Construction of the geological disposal facility in poorly indurated clay

Two shafts will connect surface facilities with the underground facility (Fig. 21.8). They will ensure the circulation of fresh air through the facility. The additional rescue shaft has a smaller diameter. The construction will use proven technology such as a tunnel boring machine (TBM) that uses concrete segments to prevent tunnel collapse. This technique was used for the Westerschelde tunnel (Fig. 21.6b) and for constructing the underground facility in the Boom Clay in Mol (Belgium) at 225 m depth (EURIDICE). The number of transport tunnels, their connections with each other and the length of disposal galleries will be continuously updated by assessing the impact of operational events or hazards such as vehicle fire.

The construction of an underground facility requires knowledge about the geotechnical properties of the poorly indurated clay at depth. This knowledge is scarce. For research into the technical feasibility as well as post-closure safety, experiments with clay host rock material are required. Sediment cores of indurated clays are available at the TNO core storage in Zeist (e.g. Koenen & Griffioen, 2014). However, these samples are stored under ordinary air and dry conditions. The clays undergo so-called core shed diagenesis (such as drying and cracking, alteration of minerals and precipitation of salts) that alters their initial state and makes them less suitable for specific studies of the clay. Rather, geotechnical properties are determined from freshly cored and well-preserved clay. To date, experimental work on freshly cored clay is limited to the Boom Clay taken from the town of Blija in 1998 (e.g. Barnichon et al., 2000), the town of Borsele in 2011 (e.g. Behrends et al., 2015) and from the city of Delft in 2022 (Vardon et al., 2022).

Operation of the geological disposal facility in poorly indurated clay

For safety reasons, the civil engineering activities for constructing the facility need to be separated from the radiologically controlled activities. The construction of a facility at a single depth should be completed before emplacement of the waste packages starts.

During the operational phase, the methodology for waste package emplacement strongly depends on the ionising radiation at the surface of the package and also on the mass and dimensions of the package. The largest volume of waste stored at COVRA's premises is packaged in such a way that a worker can handle it without additional shielding. This waste is handled in the storage facilities using electric forklift trucks. A similar handling methodology is foreseen for the emplacement of waste packages in future disposal galleries.

Excavating the tunnels causes damage to the host rock. When using tunnel boring machines and concrete liners, the clay is shielded from activities during the operational phase. Associated ventilation is expected to have no impact on clay properties and drying out and fracturing by shrinkage will be limited.

The construction of a gallery raises the hydraulic conductivity in the vicinity of the tunnel and causes pore water pressure to decrease. The extent of the increase of the conductivity depends on the diametre of the gallery, the overcut used for the excavation and time. After a few months, the conductivity in Boom Clay at 1.5 metres from the interface between concrete and clay for a gallery with a diametre of 4.6 metre was measured be to less than three times the virgin hydraulic conductivity (Bernier et al., 2007). After 8 years at the same position, this conductivity was measured to be less than two times this virgin conductivity (Berckmans et al., 2013). Measurements of the conductivity near the interface between concrete and clay have shown to be similar to conductivities further away from this interface for a smaller gallery (Dizier et al., 2017). Consequently, a hydraulic conductivity equal to the virgin clay host rock may be assumed from the start of the post-closure phase everywhere in the clay formation if an appropriate construction technique is used and the diametre of the disposal gallery is limited.

Closure of the geological disposal facility in poorly indurated clay

A stepwise closure of the disposal facility is foreseen. Galleries in which the emplacement of waste is completed will be backfilled with a material that can be easily removed in order to facilitate the retrieval of a waste package should this be necessary. Foamed concrete has been proposed as a backfill (Verhoef et al., 2014) and studied in detail (Mladenovic et al., 2019). Further closure of the disposal facility should prevent the shafts and ramp from acting as preferential pathways for radionuclide transport in the post-closure phase. The transport galleries would be backfilled with a mixture of excavated clay and grout to mitigate the acidification by oxidation of pyrite within the clay. Most lining material from the underground facility towards the surface would be removed.

Disposal system with poorly indurated clay

After closure of the geological disposal facility, the engineered barriers and the host rock contribute to containment of radionuclides. Many engineered barriers in this disposal system are composed of cementitious materials, manufactured with transport properties that are lower than those in the clay host rock. This limits preferential pathways for radionuclide transport to the host rock itself or to interfaces. After closure, joints between concrete segments are expected to be the preferential pathways for radionuclides into the clay host rock as well as the preferential paths for ingress of clay pore water into the facility (Neeft et al., 2019). The natural barriers are the poorly indurated clay formation and the surrounding rock formations. The low hydraulic conductivity of the clay contributes to the containment of radionuclides. In view of the saline groundwater conditions at disposal depths, the dominant retardation process of radionuclides in the clay needs to be assessed.

Characteristics of natural barriers in the disposal system

In general, there are two transport mechanisms in clay: diffusion and flow. Diffusion is driven by the concentration gradient of the diffusing ions, while flow is driven by a pressure gradient and requires permeability (Hasenpatt et al., 1989). The permeability of plastic clay decreases with increased loading (Wildenborg et al., 2000). In an experiment introducing a load equivalent to 500 metres burial depth, a permeability value of 1.65×10^{-19} m^2 was measured from a core of Boom Clay (Harrington et al., 2017). This permeability value is so low that pore water can be assumed to be stagnant and transport will be diffusion-dominated.

Radionuclides can be prevented from migrating by retardation. This can take place through co-precipitation and chemisorption. Co-precipitation is the simultaneous precipitation of a radionuclide with non-radioactive isotopes of the same element or different elements. An example is the precipitation of carbonates with calcium and uranium. Co-precipitation is a poorly investigated process in clay barriers. Chemisorption is a kind of adsorption that involves a chemical reaction between the surface and the adsorbate and is considered to be the main retardation mechanism in studies performed on Boom Clay from the Belgian research facility in Mol (due to the bicarbonate (HCO_3^-) type of water (De Craen et al., 2004; Bruggeman & Maes, 2017). Because the pore water chemistry is more saline in Paleogene clay host rocks in the Netherlands, the effectiveness of chemisorption as a retardation mechanism may be different from the Belgian example.

The migration of radionuclides in the clay host rock can be investigated by studying past migration of natural radionuclides and non-radioactive analogues or by using artificial radionuclides. Chemical interactions between engineered barriers such as concrete and clay host rock are expected to have a negligible impact on migration.

Evolution of a disposal cell with vitrified HLW

Setting up future evolution scenarios of disposal cells provides insight into the long-term processes that are likely to occur. For an example a cell with vitrified HLW with a highly insoluble waste matrix is used (Fig. 21.9). Initially all the engineered barriers are undamaged (Fig. 21.9a). A carbon steel overpack surrounds the waste package and

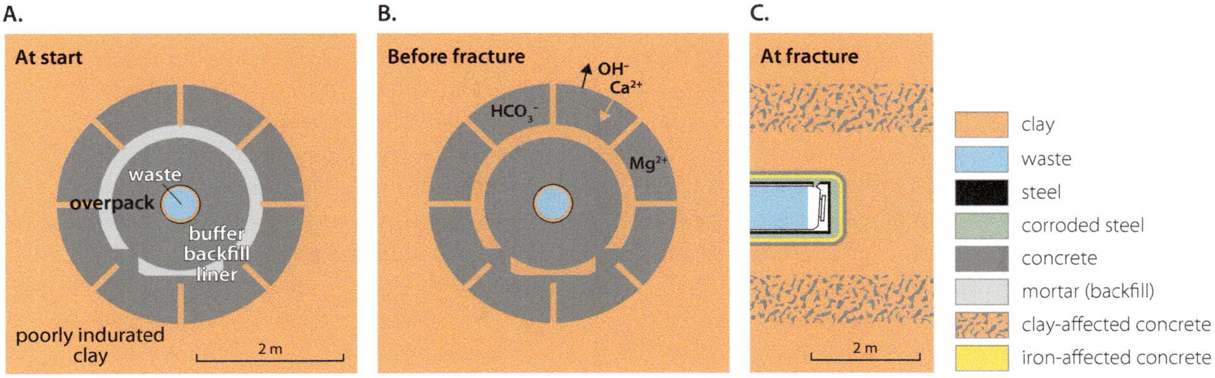

***Figure 21.9.** a) Cross section through a closed horizontal disposal gallery in clay with emplaced high-level waste at the start of the post-closure phase. b) After some time when no fracturing of engineered barriers has occurred. c) Longitudinal section of this disposal gallery after fracturing of the overpack. Note that the scale in (a) and (b) is different than in (c).*

prevents contact between the waste and pore water. The surrounding concrete buffer limits corrosion of the carbon steel overpack by ensuring a high pH, chemically reducing conditions and a limited diffusion of soluble iron. The cementitious backfill (mortar) eliminates the void volume within the disposal cell for sufficient heat dissipation. The concrete segments of the disposal gallery limit the ingress of clay pore water during the initial post-closure phase.

During the first 100 years after closure, the temperature at the interface between waste and steel is calculated to be in the range of 80°C to 94°C, but the temperature of the host rock at the outer interface of the concrete segments remains below 50°C. After 1500 years, the temperature of the host rock will have cooled down to virgin conditions (Neeft et al., 2019). Oxygen from air trapped in the disposal cell will be consumed by corrosion of steel and concrete. Additionally, anaerobic corrosion of the carbon steel overpack takes place (Fig. 21.9b). Archeological analogues of man-made steel or iron embedded in soil or water indicate that anaerobic corrosion rates of steel are at least 100 times lower than aerobic corrosion rates (Crossland, 2005). Over time, the concrete in the vicinity of the (corroding) steel alters by uptake of (dissolved) iron, forming a 'loosely bound material' (Atkins et al., 1991).

The penetration of solutes from the clay pore water also alters the concrete. Magnesium, sulphate, chloride and bicarbonate ions from the pore water will react with the minerals in the cementitious materials. A progressive degradation of concrete is expected (e.g. Atkinson et al., 1985; Dauzeres, 2016). The unaffected parts of the concrete have a high pH and will still have sufficient strength to prevent deformation. After several thousands of years, when the waste no longer heats the host rock, the combination of the lithostatic pressure on the engineered barrier system, the loss in thickness of the carbon steel overpack and the reduction in thickness of the concrete will lead to fracturing of the steel overpack (Fig. 21.9c).

Fracturing allows for contact between pore water and vitrified HLW and will generate a siliceous hydration zone and products precipitating on top of this zone. The alteration mechanisms of vitrified waste are expected to be the same as for basaltic glass (a natural volcanic glass) except that the precipitated product is named palagonite. This is a mixture of clay minerals and zeolites. Palagonite occurs as rims of varying thickness around glass fragments (Lutze et al., 1987). Chemisorption of remaining radionuclides takes place through these precipitated products. The glass alteration is therefore anticipated to have a very small impact on the potential release of radionuclides such as uranium, plutonium and americium (Van Iseghem et al., 1992). An example of a radionuclide that will be released during glass alteration is ^{79}Se. The alteration rates of basaltic glass have been estimated to be 0.1 µm per 1000 years (Lutze et al., 1987). The half-life of ^{79}Se is 327,000 years. Consequently, most of the remaining ^{79}Se is expected to decay within the vitrified waste form and will not be released to the surroundings.

The description of the future evolution of a disposal cell containing vitrified HLW illustrates that processes having an impact on the clay host rock are likely to be limited to minor alteration of the host rock in the vicinity of the disposal facility.

Outlook

Dutch policy aims for radioactive waste to be disposed of in an underground facility by the year 2130 (Fig. 21.3). The timeline for geological disposal of radioactive waste in the Netherlands considers changes in the type and amount of radioactive waste and progress in the knowledge on how to construct and operate disposal facilities. The timeline (Fig. 21.3) is updated every 5-10 years, thereby taking new developments into account. The global political situation and (foreign) experience on the generation of radioactive waste and its disposal influence national public acceptance of modes of disposal. Changing public attitudes will impact the timeline and associated activities. In the Netherlands small but continuous financial resources are available for research into disposal of radioactive waste. This research is used for well-informed adjustments to the timeline. The resources are also used to inform the general public about the current state-of-the art of waste disposal.

The decision-making process for an underground facility requires information on the host-rock formations. In future, such information will be collected from boreholes drilled in the area of selected sites. Until then, COVRA participates in scientific drilling activities in the Netherlands to obtain non-site specific information for a safety case of disposal of radioactive waste in clay. For a disposal facility in salt, the main focus for the near future will be on fundamental research, including the understanding of the long-term evolution of the disposal system with the host rock. In the far future, the focus will shift towards applied research that would, for example, include a more in-depth study of salt domes and bedded salt in the Netherlands.

Acknowledgements

The reviewers J. Prij, L. van de Vate, A.F.B. Wildenborg and J.H. ten Veen made valuable suggestions to improve this chapter. The Master students B.T.M. van Esser and H. Smit and COVRA's deputy director E.V. Verhoef are thanked for their comments on our revised version of this chapter and M.H. Barker, a native speaker in the English language, for his critical review.

Digital map data

Spatial data of figures in this chapter for use in geographical information systems can be downloaded here: https://doi.org/10.5117/aup.28164467.

References

Alders, J.G.M., 1993. Opbergen van afval in de diepe ondergrond. Parliamentary papers, session year 1992-1993. Volkshuisvesting, Ruimtelijke Ordening en Milieubeheer (The Hague, the Netherlands). Report No. 23163: 10 pp.

Alders, J.G.M., Lambers-Haquebard, J.J. & Lankhorst, P.A., 1983. Motie van het lid Alders c.s. Parliamentary papers, session year 1982-1983. Volkshuisvesting, Ruimtelijke Ordening en Milieubeheer (The Hague), Report No. 17600: 1 pp.

Altenburg, R.J., 2022. Storage and disposal potential of the Triassic Röt Formation in the east of the Netherlands: Subsurface mapping and facies interpretation. MSc thesis Utrecht University (the Netherlands): 52 pp

Anonymous, 1986. Locatie-onafhankelijke studie inzake de aanleg, bedrijfsvoering en afsluiting van mogelijke faciliteiten voor de definitieve opberging van radioactief afval in steenzoutformaties in Nederland. OPLA 25 overview report. Van Hattum & Blankevoort, Koninklijke Volker Stevin: 351 pp.

Atkins, M., Beckley, N., Carson, S., Cowie, J., Glasser, F.P., Kindness, A., Macphee, D., Pointer, C., Rahman, A., Jappy, J.G., Evans, P.A., McHugh, G., Natingley, N.J. & Wilding, C. (1991). Medium-active waste form characterization: the performance of cement-based systems. Task 3 Characterization of radioactive waste forms. A series of final reports (1985-89) No 1 (EUR--13542). Commission of the European Communities (CEC): 164 pp.

Atkinson, A., Goult, D.J. & Hearne, J.A., 1985. An assessment of the long-term durability of concrete in radioactive waste repositories. Materials Research Society Symposium Proceedings 50: 239-246. DOI: 10.1557/PROC-50-239

Barnichon, J.D., Neerdael, B., Grupa, J. & Vervoort, A., 2000. Terughaalbare opslag op diepte van 500m in de Boomse kleiformatie (TRUCK II). SCK-CEN Report No. R-3409: 236 pp.

Bartol, J., Vuorio, M., Neeft, E.A.C., Verhoef, E., McCombie, C., Chapman, N., in preparation. COPERA Salt 2024. A Conditional Safety Case & Feasibility study. COVRA N.V.

Behrends, T., Van der Veen, I., Hoving, A. & Griffoen, J., 2015. Geochemical characterization of Rupel (Boom) clay material: pore water composition, reactive minerals and cation exchange capacity. COVRA N.V, Report OPERA-PU-TNO521: 44 pp.

Bentridi, S.-H., Gall, B., Gauthier-Lafaye, F., Seghour, A. & Medjadi, D.-H., 2011. Inception and evolution of Oklo natural nuclear reactors. Comptes Rendus Geoscience 343: 738-748. DOI: 10.1016/j.crte.2011.09.008

Berckmans, A., Boulanger, D., Brassinnes, S., Capouet, M., Depaus, C., Dorado Lopez, E., Gambi, A., Gens, R., Sillen, X., Van Baelen, H., Van Geet, M., Van Marcke, P., Wacquier, W., Wouters, L., Harvey, L. & Wickham, S., 2013. Development and Demonstration (RD&D) Plan for the geological disposal of high-level and/or long-lived radioactive waste including irradiated fuel if considered as waste. State-of-the-art report as of December 2012. Belgian Agency for Radioactive Waste and Enriched Fissile Materials (ONDRAF/NIRAS) (Brussels, Belgium), No. NIROND-TR 2013-12 E: 411 pp.

Bernier, F., Li, X.L., Bastiaens, W., Ortiz, L., Van Geet, M., Wouters, L., Frieg, B., Blümling, P., Desrues, J., Viaggiani, G., Coll, C., Chanchole, S., De Greef, V., Hamza, R., Malinsky, L., Vervoort, A., Vanbrabant, Y., Debecker, B., Verstraelen, J., Govaerts, A.,

Wevers, M., Labiouse, V., Escoffier, S., Mathier, J.-F., Gastaldo, L. & Bühler, C., 2007. Fractures and self-healing within the excavation disturbed zone in clays (SELFRAC). Directorate-General for Research Euratom, No. EYR-22585: 56 pp.

Boisson, J.-Y., 2005. "Catalogue of Characteristics of Argillaceous rocks", Introductory brochure + specific databases in a CD-ROM (Excel versions) per formation (Brochure and CD-Rom Including Database). OECD/NEA/RWMC/IGSC/Clay Club (Paris, France), No. NEA 4436: 72 pp.

Bollingerfehr, W., Buhmann, D., Filbert, W., Krone, J., Lommerzheim, A., Mönig, J., Mrugalla, S., Müller-Hoeppe, N., Weber, J.R. & Wolf, J., 2013. Status of the safety concept and safety demonstration for an HLW repository in salt - Summary report. GRS-BGR-DEBTec (Peine), Report No. TEC-15-2013-AB: 152 pp.

Bouroullec, R. & Geel, C.R., 2025. Permian. In: Ten Veen, J.H., Vis, G.-J., De Jager, J. & Wong, Th.E. (eds): Geology of the Netherlands, second edition. Amsterdam University Press (Amsterdam): 127-153. DOI: 10.5117/9789463728362_ch04

Bouroullec, R. & Ten Veen, J.H., 2025. Salt Tectonics. In: Ten Veen, J.H., Vis, G.-J., De Jager, J. & Wong, Th.E. (eds): Geology of the Netherlands, second edition. Amsterdam University Press (Amsterdam): 457–491 . DOI: 10.5117/9789463728362_ch12

Bradbury, M.H. & Baeyens, B., 2011. Predictive sorption modelling of Ni(II. Applied Clay Science 52 (1-2): 27-33. DOI: 10.1016/j.clay.2011.01.022

Bruggeman, C. & Maes, N., 2017. Radionuclide migration and retention in Boom. SCK-CEN (Mol, Belgium), Report No. ER-0345: 195 pp.

Burggraaff, E., Welbergen, J. & Verhoef, E.V., 2022. Nationale radioactief afval inventarisatie. COVRA N.V.: 27 pp.

Busschers, F.S., Cohen, K., Wesselingh, F.P., Schokker, J., Bakker, M., Van Balen, R. & Van Heteren, S., 2025. Quaternary. In: J.H. Ten Veen, G.-J. Vis, J. De Jager, J. & Th.E. Wong (eds): Geology of the Netherlands, second edition. Amsterdam University Press (Amsterdam): 333-391. DOI: 10.5117/9789463728362_ch10

Chapman, N. & Gibb, F., 2003. A truly final waste management solution: Is very deep borehole disposal a realistic option for High-Level Waste or fissile materials? Radwaste Solutions 26-37.

Chapman, N.A., 2019. Who might be interested in a deep borehole disposal facility for their radioactive waste? Energies 12(8): 1542. DOI: 10.3390/en12081542

Chapman, N.A. & Hooper, A., 2012. The disposal of radioactive waste underground (review paper). Proceedings of the Geologists' Association 123: 46-63. DOI: 10.1016/j.pgeola.2011.10.001

Cinelli, G., De Cort, M. & Tollefsen, T., 2019. European Atlas of Natural Radiation. Publications Office of the European Union (Luxembourg), Report No. OP KJ-02-19-425-EN-N: 190 pp. DOI:10.2760/46388

CORA, 2001. Terugneembare berging, een begaanbaar pad? Onderzoek naar de mogelijkheden van terugneembare berging van radioactief afval in Nederland [Eindrapport]. Commissie Opberging Radioactief Afval, Ministerie van Economische Zaken (The Hague): 110 pp.

Crossland, I., 2005. Long-term corrosion of iron and copper. Proceedings - 10th International Conference on Environmental Remediation and Radioactive Waste Management, ICEM'05: 1402-1408.

Dauzeres, A., 2016. Geochemical and physical evolution of cementitious environments in an aggressive environment. In: Vehmas, T. & Holt, E. (eds): Cebama Deliverable D1.03 - WP1 Experimental studies - State of the art literature review: 136-174.

De Craen, M., Swennen, R. & Keppens, E., 1999. Petrography and geochemistry of septarian carbonate concretions from the Boom Clay Formation (Oligocene, Belgium). Geologie En Mijnbouw 77 (1): 63-76. DOI: 10.1023/A:1003468328212

De Craen, M., Wang, L., Van Geet, M. & Moors, H., 2004. Geochemistry of Boom Clay pore water at the Mol site. SCK-CEN (Mol, Belgium), Scientific Report No. BLG-990: 173 pp.

De Gans, W. & Duin, E., 2010. Steenzout, oppervlaktevormen en landijs. Grondboor & Hamer 64 (4/5): 120-126.

De Mulder, F.J. & Ritsema, I., 2003. Deel 1 - Duurzaam gebruik en beheer van de ondergrond. In: De Mulder, F.J., Geluk, M.C., Ritsema, I., Westerhoff, W.E. & Wong, Th.E. (eds): De ondergrond van Nederland. Nederlands Instituut voor Toegepaste Geowetenschappen-TNO (Utrecht, the Netherlands): 9-118.

Dizier, A., Chen, G., Li, X.L. & Rypens, J., 2017. The PRACLAY Heater test after two years of the stationary phase. EIG EURIDICE (Mol, Belgium), Report No. EUR_PH_17_043: 65 pp.

Dufour, F.C., 2000. Groundwater in the Netherlands – Facts and figures. Netherlands Institute of Applied Geoscience TNO - National Geological Survey (Delft): 96 pp.

EC, 2011. Council Directive 2011/70/Euratom of 19 July 2011 establishing a Community framework for the responsible and safe management of spent fuel and radioactive waste. Official Journal of the European Union L199/48-56 02.08.2011.

EC, 2013. Council directive 2013/39/EURATOM of 5 December 2013 laying down basic safety standards for protection against the dangers arising from exposure to ionising radiation and repealing Directives 89/618/Euratom, 90/641/Euratom, 96/29/Euratom, 97/43/Euratom and 2003/122/Euratom. Official Journal of the European Union L13/1-75 17.10.2014.

Finnish Energy, 2021. Safely in the ancient bedrock. First in the world - final disposal of spent nuclear fuel to begin in Finland. Finnish Energy (Helsinki, Finland): 8 pp.

Floor, P., 2012. Dutch Earth Sciences. Development and impact. Royal Geological and Mining Society of the Netherlands 1912-2012 Centenary Volume. Koninklijk Nederlands Geologische Mijnbouwkundig Genootschap (KNGMG): 273 pp.

Ganopolski, A., Winkelmann, R. & Schellnhuber, H.J., 2016. Critical insolation–CO_2 relation for diagnosing past and future glacial inception. Nature 529: 200-203. DOI: 10.1038/nature16494

Geel, P.L.B.A., 2002. Radioactief afvalbeleid (Parliamentary Papers Session Year 2002-2003. Volkshuisvesting, Ruimtelijke Ordening en Milieubeheer (The Hague, the Netherlands), Report No. 28674: 1-15 pp.

Geluk, M.C., 1998. Internal tectonics of salt structures. Journal of Seismic Exploration 7: 237-251.

Geluk, M.C., Damman, A.H., Klaver, G.T., Obdam, A. & Baker, J.H., 1993. Evaluation of salt bodies and their overburden in the Netherlands for the disposal of radioactive waste - (C) Caprock formation and subrosion. Rijks Geologische Dienst (Haarlem, the Netherlands), OPLA Report No. 30.012C/ERC: 47 pp.

Geluk, M.C. & Wildenborg, A.F.B., 1988. Geologische inventarisatie en ontstaansgeschiedenis van zoutvoorkomens in Noord- en Oost-Nederland. Rijks Geologisch Dienst (Haarlem, the Netherlands), Report No. 10568: 272 pp.

Gerstmann, U., Meyer, H. & Tholen, M., 2002. Bestimmung des nuklidspezifischen Aktivitätsinventars der Schachtanlage Asse – Abschlußbericht. GSF - Forschungszentrum für Umwelt und Gesundheit, Report No. FE Nr. 76277: 68 pp.

Griffioen, J., Koenen, M., Meeussen, H., Cornelissen, P., Peters, L. & Jansen, S., 2017. Geochemical interactions and groundwater transport in the Rupel Clay. A generic model analysis. COVRA N.V., Report No. OPERA-PU-TNO522: 109 pp.

Griffioen J, Verweij H, Stuurman R., 2016. The composition of groundwater in Palaeogene and older formations in the Netherlands. A synthesis. Netherlands Journal of Geosciences 95(3):349-372. DOI:10.1017/njg.2016.19

Groenenberg, R., Den Hartogh, M. & Fokker, P., 2025. Salt Production. In: Ten Veen, J.H., Vis, G.-J., De Jager, J. & Wong, Th.E. (eds): Geology of the Netherlands, second edition. Amsterdam University Press (Amsterdam): 603-625. DOI: 10.5117/9789463728362_ch16

Grupa, J.B. & Houkema, M., 2000. Terughaalbare opberging van radioactief afval in diepe zout en kleiformaties. Modellen voor een veiligheidsstudie. Nuclear Research & consultancy Group (NRG) (Petten, the Netherlands), Report No. 21082/00.33017/P: 196 pp.

Gruppelaar, H., Kloosterman, J.L. & Konings, R.J.M., 1998. Advanced technologies for the reduction of nuclear waste. Energy Research Foundation ECN (Petten, the Netherlands), Report No. R-98-008: 103 pp.

Hamstra, J., 1981. The importance of scenario analysis in the iterative process of achieving' as low as reasonably achievable' solutions for future consequences of underground disposal. Proceedings of the Workshop on Radionuclide Release Scenarios for Geologic Repositories (Paris).

Hamstra, J. & Van der Feer, Y., 1981. Nieuwe ICRP-normen en de ondergrondse opberging van radioactief afval. Energiespectrum 5 (4): 98-104.

Hamstra, J. & Verkerk, B., 1981. The Dutch geologic radioactive waste disposal project. Final report Commission of the European Communities - Nuclear Science, Report No. EUR 7151: 128 pp.

Hansen, F.D., Kuhlman, K.L. & Sobolik, S.R., 2016. Considerations of the Differences between Bedded and Domal Salt Pertaining to Disposal of Heat-Generating Nuclear Waste. Sandia National Lab. (Albuquerque, USA), Report No. SAND2016-6522R: 47 pp. DOI: 10.2172/1333710

Harrington, J.F., Cuss, R.J., Wiseall, A.C., Daniels, K.A., Graham, C.C. & Tamayo-Mas, E., 2017. Scoping study examining the behaviour of Boom Clay at disposal depths investigated in OPERA. COVRA N.V., Report No. OPERA-PU-BGS523 & 616: 215 pp.

Harsveldt, H.M., 1978. Salt resources in The Netherlands as surveyed mainly by AKZO. In: Coogan, A.H. & Hauber, L. (eds) Contr. Fifth International Symposium on Salt, vol. 1, Northern Ohio geological Society: 65-81.

Harsveldt, H.M., 1986. The Netherlands. In: Dunning, F.W & Evans, A.M. (eds): Mineral deposits of Europe. The Institute of Mining and Metallurgy and The Mineralological Society (London) 3, 113-116.

Hart, J., Prij, J., Vis, G.-J., Becker, D.-A., Wolf, J., Noseck, U. & Buhmann, D., 2015. Evaluation of current knowledge for building the Safety Case for salt based repositories. COVRA N.V., Report No. OPERA-PU-NRG221A: 308 pp.

Hasenpatt, R., Degen, W. & Kahr, G., 1989. Flow and diffusion in clays. Applied Clay Science 4 (2): 179-192. DOI: 10.1016/0169-1317(89)90007-0

Hudec, M.R. & Jackson, M.P.A., 2007. Terra infirma: Understanding salt tectonics. Earth-Science Reviews 82: 1-28. DOI: 10.1016/j.earscirev.2007.01.001

IAEA, 2017. Available reprocessing and recycling services for research reactor fuel. Report No. NW-T-1.11: 85 pp.

ICK, 1975. Radioactieve Afvalstoffen in Nederland bij een vermogen aan Kernenergiecentrales van 3500 MWe. Interdepartementale Commissie Kernenergie (ICK), Sub-commissie Radioactieve Afvalstoffen (Leidschendam, the Netherlands): 75 pp.

I&E, 2016. The national programme for the management of radioactive waste and spent fuel. Parliamentary Papers, session year 2015-2016. Ministry of Infrastructure and the Environment (I&E) (The Hague, Netherlands). Report 25422-no. 149: 61 pp.

I&W, 2020. Joint Convention on the safety of spent fuel management and on the safety of radioactive waste management - National Report of the Kingdom of the Netherlands for the Seventh Review Meeting (25 May - 4 June 2021). Ministry of Infrastructure and Water management: 130 pp.

Keller, S., 2009. Glacial channel systems and their significance for the long-term safety of potential geologic repository sites for high-level radioactive waste in north Germany.Bundesanstalt für Geowissenschaften und Rohstoffe (BGR) (Hannover, Germany): 24 pp.

Koenen, M. & Griffioen, J., 2014. Mineralogical and geochemical characterization of the Boom Clay in the Netherlands. COVRA N.V., Report No. OPERA-PU-TNO521-1: 106 pp.

Konert, M. & Vandenberghe, J.F., 1997. Comparison of layer grain size analysis with pipette and sieve analysis: a solution for the underestimation of the clay fraction. Sedimentology 44 (12): 523-535. DOI: 10.1046/j.1365-3091.1997.d01-38.x

Kooijman, G., 1996. Westerscheldetunnel. Cement 10: 7.

Köthe, A., Hoffmann, N., Krull, P., Zirngast, M. & Zwirner, R., 2007. Description of the Gorleben site part 2: Geology of the overburden and adjoining rock of the Gorleben salt dome. Bundesanstalt für Geowissenschaften und Rohstoffe (BGR) (Hannover, Germany): 220 pp.

Kovaltsov, G.A., Mishev, A. & Usoskin, I.G., 2012. A new model of cosmogenic production of radiocarbon 14C in the atmospshere. Earth and Planetary Science Letters 337-338: 114-120. DOI: 10.1016/j.epsl.2012.05.036

Laban, C., 2010. Zout nabij de zeebodem. Grondboor & Hamer 64 (4/5): 111-113.

Lauwerier, W., 2022. The evolution of the Zechstein salt diapirs in the north-eastern Netherlands. Internship report for COVRA N.V.: 62 pp.

Lutze, V., Grambow, B., Ewing, R.C. & Jercinovic, M.J., 1987. The use of natural analogues in the long-term extrapolation of glass corrosion processes. In: Côme, B. Chapman & N.A. (eds): Natural Analogues in Radioactive Waste Disposal: 142-152.

Mallants, D., Travis, K., Chapman, N.A., Brady, P.V. & Griffiths, H., 2020. The state of science and technology of deep borehole disposal of nuclear waste. Energies 13: 7. DOI: 10.3390/en13040833

Mazurek, M., Alt-Epping, P., Bath, A., Gimmi, T. & Waber, H.N., 2009. Natural Tracer Profiles Across Argillaceous Formations: The CLAYTRAC Project. Nuclear Energy Agency (NEA) Organisation for Economic Co-Operation and Development (OECD), NEA Report No. 6253: 365 pp.

Messick, C.E., Taylor, J.L. & Leach, W.E. (2006). Status of the United States foreign research reactor spent nuclear fuel acceptance program (INIS-XA-C--120). International Atomic Energy Agency (IAEA), Report No. INIS-XA-C—120: 344 pp.

Mladenovic, A., Neeft, E.A.C., Deissmann, G., Dähn, R., Geng, G., Koskowski, G. & Markku, L., 2019. ILW: Report describing the selected experiments and the existing/expected experimental results. Deliverable D2.11 of the HORIZON 2020 Project EURAD No. EC Grant agreement no. 847593: 51 pp.

Mobbs, S., 2012. Planning to 'remember to forget'? Mineralogical Magazine 76: 3491-3496. DOI: 10.1180/minmag.2012.076.8.63

NEA, 1985. Review of the continued suitability of the dumping site for radioactive waste in the North-East Atlantic. Nuclear Energy Agency (NEA), OECD (Paris, France): 100 pp.

Neeft, E.A.C., Weetjens, E., Vokal, A., Leivo, M., Cochepin, B., Martin, C., Munier, I., Deissmann, G., Montoya, V., Poskas, P., Grigaliuniene, D., Narkuniene, A., García, E., Samper, J., Montenegro, L. & Mon, A., 2019. Treatment of chemical evolution in National Programmes. Deliverable D2.4 of the HORIZON 2020 Project EURAD No. EC Grant agreement no. 847593: 233 pp.

NRC, 1957. The Disposal of Radioactive Waste on Land. Report on Waste Disposal of the Division of Earth Sciences. National Academy of Sciences - National Research Council (NRC) (Washington DC, USA), Report No. 519: 142 pp.

OPLA, 1984. Proposal for a research programme on geological disposal of radioactive waste in the Netherlands. Ministerie van Economische Zaken (The Hague, the Netherlands): 103 pp.

OPLA, 1989. Onderzoek naar geologische opberging van radioactief afval in Nederland. Eindrapport Fase 1. Ministerie van Economische Zaken (The Hague, the Netherlands): 130 pp.

OPLA, 1993. Onderzoek naar geologische berging van radioactief afval - Eindrapport Aanvullend onderzoek van fase 1. Ministerie van Economische Zaken (The Hague, the Netherlands).

Poley, A.D., 1999. Concept ontwerp terughaalbare opslag van radioactief afval in diepe boorgaten in steenzout (Torad-B). Nuclear Research & consultancy Group (NRG), Report No. 21514/00.31223/C: 54 pp.

Poley, A.D., 2000. BAMBUS project - Nederlandse bijdrage. Nuclear Research & consultancy Group (NRG), Report No. 20038/00.30991/C: 72 pp.

Popp, T., Kern, H. & Schulze, O., 2001. Evolution of dilatancy and permeability in rock salt during hydrostatic compaction and triaxial deformation. Journal of Geophysical Research 106: 4061-4078. DOI: 10.1029/2000JB900381

POSIVA, 2012. Facility Description 2012. Summary Report of the Encapsulation Plant and Disposal Facility Designs. Posiva Oy: 140 pp.

Prij, J., Blok, B.M., Laheij, G.M.H., Van Rheenen, W., Slagter, W., Uffink, G.J.M., Uijt de Haag, P., Wildenborg, A.F.B. & Zandstra, D.A., 1993. PROSA - PRObabilistic Safety Assessment - final report Energieonderzoek Centrum Nederland (ECN) (Petten, the Netherlands), OPLA Report No. OPLA-1A: 428 pp. (excl. Appendices)

Remmelts, G., Muyzert, E., Van Rees, D.J., Geluk, M.C., De Ruyter, C.C. & Wildenborg, A.F.B., 1993. Evaluation of salt bodies and their overburden in the Netherlands for the disposal of radioactiwe waste - (B) Salt movement. Rijks Geologisch Dienst (Haarlem, the Netherlands), OPLA Report No. 30.012B/ERB: 87 pp.

Reschreiter, H. & Kowarik, K., 2019. Der bronzezeitliche Bergbau in Hallstatt. Neue Lebensbilder zum Salzbergwerk. Archaeologia Austriaca 103: 99-136. DOI: 10.1553/archaeologia103s99

Savage, D., 2014. An Assessment of the Impact of the Long Term Evolution of Engineered Structures on the Safety-Relevant Functions of the Bentonite Buffer in a HLW Repository. National Cooperative for the Disposal of Radioactive Waste (NAGRA) (Switzerland), Technical Report No. NTB-13-02: 62 pp.

Simmelink, H.J., Heidema, A.H., Hoogendoorn, A. & Pagnier, H.J.M., 1996. Kartering slecht-doorlatende laagpakketten van Tertiaire formaties (Project 'CAR' - Fase 1). Rijks Geologische Dienst (Heerlen, the Netherlands). Report No. GB 2514: 24 pp.

Spiers, C.J., Peach, C.J., Brzesowsky, R.H., Schutjens, P.M.T.M.,

Liezenberg, J.L. & Zwart, H.J., 1988. Long-term rheological and transport properties of dry and wet salt rocks. Commission of the European Communities (CEC) (Luxembourg), Report No. EUR-11848: 161 pp.

Spiers, C.J., Urai, J.L., Lister, G.S., Boland, J.N. & Zwart, H.J., 1986. The influence of fluid-rock interaction on the rheology of salt rock. Commission of the European Communities (CEC) (Luxembourg), Report No. EUR-10399: 141 pp.

Stackebrandt, W., 2009. Subglacial channels of Northern Germany - a brief review. Zeitschrift der deutschen Gesellschaft für Geowissenschaften 160: 203-210. DOI: 10.1127/1860-1804/2009/0160-0203

Strozyk, F., Van Gent, H., Urai, J.L. & Kukla, P.A., 2012. 3D seismic study of complex intra-salt deformation: An example from the Upper Permian Zechstein 3 stringer, western Dutch offshore. Geological Society of London, Special Publications 363: 489-501. DOI: 10.1144/sp363.23

Ten Veen, J.H., Govaerts, J., Beerten, K., Ventra, D. & Vis, G.-J., 2015. Future evolution of the geological and geohydrological properties of the geosphere. COVRA N.V., Report OPERA-PU-TNO412: 118 pp.

TNO-GDN, 2023. Zechstein Group. Stratigraphic Nomenclature of the Netherlands, TNO – Geological Survey of the Netherlands. http://www.dinoloket.nl/en/stratigraphic-nomenclature/zechstein-group

Trauth, K.M., Hora, S.C. & Guzowski, R.V., 1993. Expert judgment on markers to deter inadvertent human intrusion into the Waste Isolation Pilot Plant (No. SAND92-1382; 349 pp.). Sandia National Laboratories (Albuquerque (USA)): 349 pp.

Urai, J.L., Schléder, Z., Spiers, C.J. & Kukla, P.A., 2008. Flow and transport properties of salt rocks. In: R. Littke, U. Bayer, D. Gajewski & S. Neldskamp (eds): Dynamics of Complex Intracontinental Basins: The Central European Basin System. Springer Berlin (Heidelberg, Germany): 277-290. DOI:10.1007/978-3-540-85085-4_5

USGS, 1999. Naturally occurring radioactive materials (NORM) in produced water and oil-field equipment - an issue for the energy industry (No. Fact Sheet FS-142-99). U.S. Geological Survey.

Vahlund, F. & Andersson, E., 2015. Safety analysis for SFR Long-term safety. Main report for the safety assessment SR-PSU. Revised edition. Swedish Nuclear Fuel and Waste Management Co (Stockholm, Sweden): 500.

Van Aardenne, G.M.V., 1979. Opzetnota. Maatschappelijke discussie over de toepassing van kernenergie voor elektriciteitsopwekking (Parliamentary papers, session year 1978-1979 No. 15100, nr. 15). (The Hague, Netherlands).

Van Aardenne, G.M.V., Winsemius, P., De Koning, J., Brinkman, L.C. & Van den Broek, H., 1985. Regeringsstandpunt met betrekking tot Eindrapport van de Maatschappelijke Discussie Energiebeleid (Parliamentary papers, session year 1984-1985 No. 18830, nr. 1 and 2). (The Hague, the Netherlands).

Van de Vate, L., 2018. Eindberging hoogradioactief afval met diepe boorgaten bij de kerncentrale Borssele. GEOVAT (Castricum, the Netherlands): 30 pp. https://www.laka.org/docu/catalogus/publicatie/1.01.4.30/75_eindberging-hoogradioactief-afval-met-diepe-boorga.

Van Dijke, J.J. & Veldkamp, A., 1996. Climate-controlled glacial erosion in the unconsolidated sediments of northwestern Europe, based on a genetic model for the tunnel valley formation. Earth Surface Processes and Landforms 21: 327-340. DOI: 10.1002/(SICI)1096-9837(199604)21:4%3C327::AID-ESP540%3E3.0.CO;2-P

Van Gent, H., Urai, J.L. & De Keijzer, M. de, 2011. The internal geometry of salt structures - A first look using 3D seismic data from the Zechstein of the Netherlands. Journal of Structural Geology 33: 292-311. DOI: 10.1016/j.jsg.2010.07.005

Van Iseghem, P., Berghman, K., Lemmens, K., Timmermans, W. & Wang, L., 1992. Laboratory and in-situ interaction between simulated waste glasses and clay Task 3 Characterization of radioactive waste forms. A series of final reports (1985-89). No 21 (No. EUR-13607; 127 pp.). Commission of the European Communities (CEC) (Luxembourg): 127 pp.

Van Veldhoven-Van der Meer, S., 2018. Opwerking van radioactief materiaal (Parliamentary Papers Session Year 2017-2018 No. 25422, nr. 217). (The Hague, the Netherlands).

Vardon, P.J., Laumann, S.J., Schmiedel, T., Vargas Meleza, L., Barnhoorn, A., Abels, H.A., Drijkoningen, G.G., Dieudonné, A.A.M., Beernink, S.T.W. & Van den Berg, J.J., 2022. Drilling report. Delftse Hout multipurpose research borehole DAPGEO-02. Delft University of Technology (Delft): 177 pp. 02 DOI: 10.4233/uuid:fde70c00-cf65-4174-9cd3-89585a5e61bd

Verhoef, E.V., De Bruin, A.M.G., Wiegers, R.B., Neeft, E.A.C. & Deissmann, G., 2014. Cementitious materials in OPERA disposal concept in Boom Clay. COVRA N.V., Report No. OPERA-PG-COV020: 20 pp.

Verhoef, E.V., Neeft, E.A.C., Bartol, J., Vuorio, M.R., Scholten, C., Buitenhuis, A., Van der Veen, G., Chapman, N.A. & McCombie, C., 2020. Long-term research programme for geological disposal of radioactive waste. COVRA N.V.: 59 pp.

Verhoef, E.V., Neeft, E.A.C., Chapman, N.A. & McCombie, C., 2017. OPERA Safety case. COVRA N.V.: 147 pp.

Verweij, J.M., Vis, G.-J. & Imberechts, E., 2016. Spatial variation in porosity and permeability of the Rupel Clay Member in the Netherlands. Netherlands Journal of Geosciences 95(3): 253-268. DOI: 10.1017/njg.2016.28

Vis, G.-J. & Verweij, J.M., 2014. Geological and geohydrological characterization of the Boom Clay and its overburden. COVRA N.V., Report No. OPERA-PU-TNO411: 86 pp.

Vis, G.-J., Verweij, J.M. & Koenen, M., 2016. The Rupel Clay Member in the Netherlands: towards a comprehensive understanding of its geometry and depositional environment. Netherlands Journal of Geosciences 95 (3): 221-251. DOI: 10.1017/njg.2016.25

VROM, 1984. Radioactive Waste Policy in the Netherlands. An outline of the government's position. Ministry of Housing, Physical Planning and Environment 21.

Waltham, T., 2008. Salt terrains of Iran. Geology Today 24: 188-194. DOI: 10.1111/j.1365-2451.2008.00686.x

Wang, L., Jacques, D. & De Cannière, P., 2010. Effects of an alkaline plume on the Boom Clay as a potential host formation for geological disposal of radioactive waste. SCK-CEN (Mol, Belgium), Report No. ER-28: 193 pp.

Westerhoff, W.E., Geluk, M.C. & De Mulder, F.J., 2003a. Deel 2 - Geschiedenis van de ondergrond. In: De Mulder, F.J., Geluk, M.C., Ritsema, I., Westerhoff, W.E. & Wong, Th.E. (eds): De ondergrond van Nederland. Nederlands Instituut voor Toegepaste Geowetenschappen-TNO (Utrecht):119-246.

Westerhoff, W.E., Wong, Th.E. & De Mulder, E.F.J., 2003. Opbouw van de ondergrond. In: De Mulder, E.F.J., Geluk, M.C., Ritsema, I.L., Westerhoff, W.E. & Wong, Th.E. (eds): De ondergrond van Nederland. Wolters-Noordhoff (Groningen/Houten): 247-352.

Wildenborg, A.F.B., De Mulder, E.F.J., Bosch, J.H.A., Hillen, R., Schokking, F. & Van Gijssel, K., 1990. Nadere ordening van zoutstructuren ex fase 1 met de nadruk op (peri-)glaciale verschijnselen in de komende 100.000 jaar. Rijks Geologisch Dienst, Report No. 30.010: 77 pp.

Wildenborg, A.F.B., Orlic, B., De Lange, G., Leeuw, C.S., Zijl, W., Van Weert, F., Veling, E.J.M., De Cock, S., Thimus, J.F., Lehnen-de Rooij, C. & Den Haan, E.J., 2000. Transport of RAdionuclides disposed of in Clay of Tertiary ORigin (TRACTOR). Netherlands Institute of Applied Geoscience TNO, TNO-report NITG 00-223-B: 223 pp.

Winsemius, P., 1983. Brief van de Minister van Volkshuisvesting, Ruimtelijke Ordening en Milieubeheer. Parliamentary papers, session year 1982-1983. Volkshuisvesting, Ruimtelijke Ordening en Milieubeheer (The Hague, the Netherlands), Report No. 17600: 2 pp.

Yuan, J., Vardon, P.J., Hicks, M.A., Hart, J. & Fokker, P.A., 2017. Technical feasibility of a Dutch radioactive waste repository in Boom Clay: plugs and seals. COVRA N.V., Report No. OPERA-PU-TUD321b: 36 pp.

Zirngast, M., 1996. The development of the Gorleben salt dome (northwest Germany) based on quantitative analysis of peripheral sinks. Geological Society, London, Special Publications 100 (1): 203-226. DOI: 10.1144/GSL.SP.1996.100.01.13

22 Natural and induced seismicity

Bernard Dost, Elmer Ruigrok, Jesper Spetzler & Pauline Kruiver*

*Corresponding author: bdost36@gmail.com

ABSTRACT

Natural seismicity in the Netherlands is mainly associated with large faults in the south of the country. These faults are part of the Roer Valley Rift System, the northwestern extension of the Rhine rift system. Their style of faulting, as concluded from the analysis of seismicity, is dominantly normal with a small strike-slip component and is in accordance with the current stress pattern. South of the Roer Valley Rift System, smaller earthquakes suggest the reactivation of strike-slip faults and thrust faults along the Brabant Massif. The effects of natural seismicity have been moderate on the whole and include structural damage to buildings in epicentral areas. Hazard analysis based on historical seismicity suggests a maximum expected magnitude of M_L = 6.3. Paleoseismological investigations indicate that events that may reach one magnitude higher, occur at a recurrence interval of at least 2 to 3 kyr along the major fault zones.

In the north of the Netherlands no significant natural activity has been detected, but since 1986 induced events, most of which coincide with producing gas fields have been recorded. Initially, small gas fields started to show seismicity and from 2003 onward seismicity in the large Groningen gas field became dominant. The largest recorded event near Huizinge in 2012 (M_L= 3.6) caused considerable damage to buildings in the region, due to its shallow depth of 3 km. After the Huizinge event, monitoring in Groningen was intensified and a large research programme was started, leading to a significant increase in knowledge on the origin and effects of seismicity in the region. Induced seismicity was also recorded in the vicinity of salt mining, geothermal activities and possibly water injection.

<< The main fault of the Peel Boundary Fault Zone, footwall to the left (Uden, southeast Netherlands). Iron-oxide and silt precipitation accentuate structures. Total fault offset during the last 15,000 years is about 2.5 m. Photo: Ronald van Balen.

DOI: 10.5117/9789463728362_ch22

Introduction

In the Netherlands a clear separation can be made between natural earthquakes in regions where the seismicity is caused by tectonic processes and induced earthquakes in regions where the seismicity is caused by human intervention. Natural onshore seismicity is mainly restricted to the south of the country, where faults are active. Induced earthquakes are thus far mainly restricted to the north and are mostly related to the exploitation of gas fields. Small events have been associated with salt mining near Veendam and have also been recorded near a geothermal well near Venlo. The mining of coal in the past has not resulted in noticeable seismic effects. A compilation of historical seismicity in the Netherlands was given by Houtgast (1991), but no comprehensive and up-to-date description is presently available in the literature. A regularly updated list of earthquakes in the Netherlands and immediate surroundings is maintained by the *Koninklijk Nederlands Meteorologisch Instituut* (KNMI, Royal Netherlands Meteorological Institute) and available on the internet (www.knmi.nl/nederland-nu/seismologie/aardbevingen) The largest recorded natural event, the 1992 Roermond earthquake (M_L= 5.8), occurred in the Roer Valley Rift System and was discussed extensively in Van Eck & Davenport (1994). Seismic events on the Netherlands continental shelf occur less frequently and are restricted to the Dutch Central Graben. Offshore events close to producing gas fields have been detected (KNMI earthquake list) and are suspected to be induced. Following the largest recorded induced event, the 2012 Huizinge earthquake (M_L= 3.6) both observations and research related to induced seismicity were intensified.

The earthquakes in the Netherlands should be viewed in a regional framework. An overview of natural seismicity in western Germany was given by Hinzen et al. (2021), while Camelbeeck et al. (2007) and Vanneste et al. (2013) carried out similar studies for Belgium. The northwestern part of Germany shows mainly events of induced origin (Brandes et al., 2019) and is of interest since these events were detected and located by the KNMI network. Seismological research aims to gain insight into the causes of the earthquakes, their effects at the surface and the ground motions that can be expected in the future.

The objective of this chapter is to give an overview of 1) historic natural and induced seismicity in the Netherlands, 2) the evolving monitoring network, and 3) the insights and main findings of research into seismicity in the Netherlands. Muntendam-Bos et al. (2022) published an overview of induced seismicity in the Netherlands, which gives complimentary information.

Natural seismicity

Natural seismicity in the Netherlands is limited to the southern part of the country, where well-known active fault systems are present.

Observed earthquakes in the Roer Valley Rift System

Most of the natural earthquakes are located in the Roer Valley Rift System, an active rift system located in the Lower Rhine Embayment in the south of the Netherlands, northeastern Belgium and western Germany (Houtgast et al., 2002; Vanneste et al., 2013). The system comprises the Roer Valley Graben and the adjacent Campine and South Limburg blocks to the southwest and the Peel and Venlo blocks to the northeast (Fig. 22.1). The majority of the earthquakes is confined to the Roer Valley Graben, which is bounded by two active fault zones: the Peel Boundary Fault Zone to the northeast and the Feldbiss Fault Zone to the southwest. The main difference between these faults or fault zones is that the former is relatively narrow, up to 1 km, while the latter is relatively wide, up to 5 km (Geluk et al., 1994; Houtgast & Van Balen, 2000). The earthquakes in this area are of limited magnitude, but occasionally powerful enough to cause damage. In the literature, these earthquakes have been associated with known geological structures. Immediately after the 1932 Uden (M_L= 5.0) earthquake, Jongmans & Van Waterschoot van der Gracht (1932) explained this event by movement along the Peel Boundary Fault. Also, Escher (1940) mentioned the association between earthquakes and the main Roer Valley Graben boundary faults.

Peel Boundary Fault

The most recent, significant earthquake associated with the Peel Boundary Fault occurred on April 13, 1992, had a magnitude M_L= 5.8 and was located just south of Roermond (Paulssen et al., 1992; Camelbeeck et al., 1994). It was felt over an area larger than 600,000 km^2 in parts of the Czech Republic, Switzerland, France and England (Haak et al., 1994), and was the strongest event in the Netherlands over the last 100 years. Its maximum observed intensity, a measure of the effects of an earthquake at the surface, was I_o= VII on the EMS-98 scale (damaging). The total damage, mainly resulting from failing chimneys and parapets (Bouwkamp, 1994) and damaged home contents, was estimated at about € 128 million at 1992 value (Berz, 1994). An event of similar size occurred in 1951 near Euskirchen (Germany; Berg, 1953), 80 km southeast of the Roermond epicentre and was associated with the Erft fault. This fault and the Rurrand fault form the southward extension of the Peel Boundary Fault in Germany (Vanneste et al., 2013). In November 1932, the second largest recorded earthquake in the Netherlands, M_L= 5.0, occurred along the Peel Boundary Fault near Uden (I_0= VII, broken chimneys, cracks in walls, floors and ceilings), 64 km northwest of the 1992 Roermond epicentre (Van Dijk, 1934; Gees, 1937). There is little evidence of seismic activity along the fault further to the northwest. Two reports of earthquakes located near Tiel in 1928 and 1932, approximately 90 km north-northwest of Roermond (Houtgast, 1991), are based on doubtful interpretations. The 1928 event was not recorded instrumentally and was reported by only a few people in a large area. The 1932 event, recorded in De Bilt (DBN) in the Netherlands and Uccle (UCC) in Belgium, suggests a source near the Dutch-Belgian border based on S-P time intervals. A small (M_L= 3.0) event was recorded near Tiel in 2004, but significant recent seismic activity along the Peel Boundary Fault appears to terminate to the northwest near Uden. The largest events since 1900 along the Peel Boundary Fault are listed in Table 22.1.

Aftershocks of the Roermond event enabled a detailed study of the fault plane. Camelbeeck et al. (1994) reported more than 200 aftershocks. Hypocentre locations

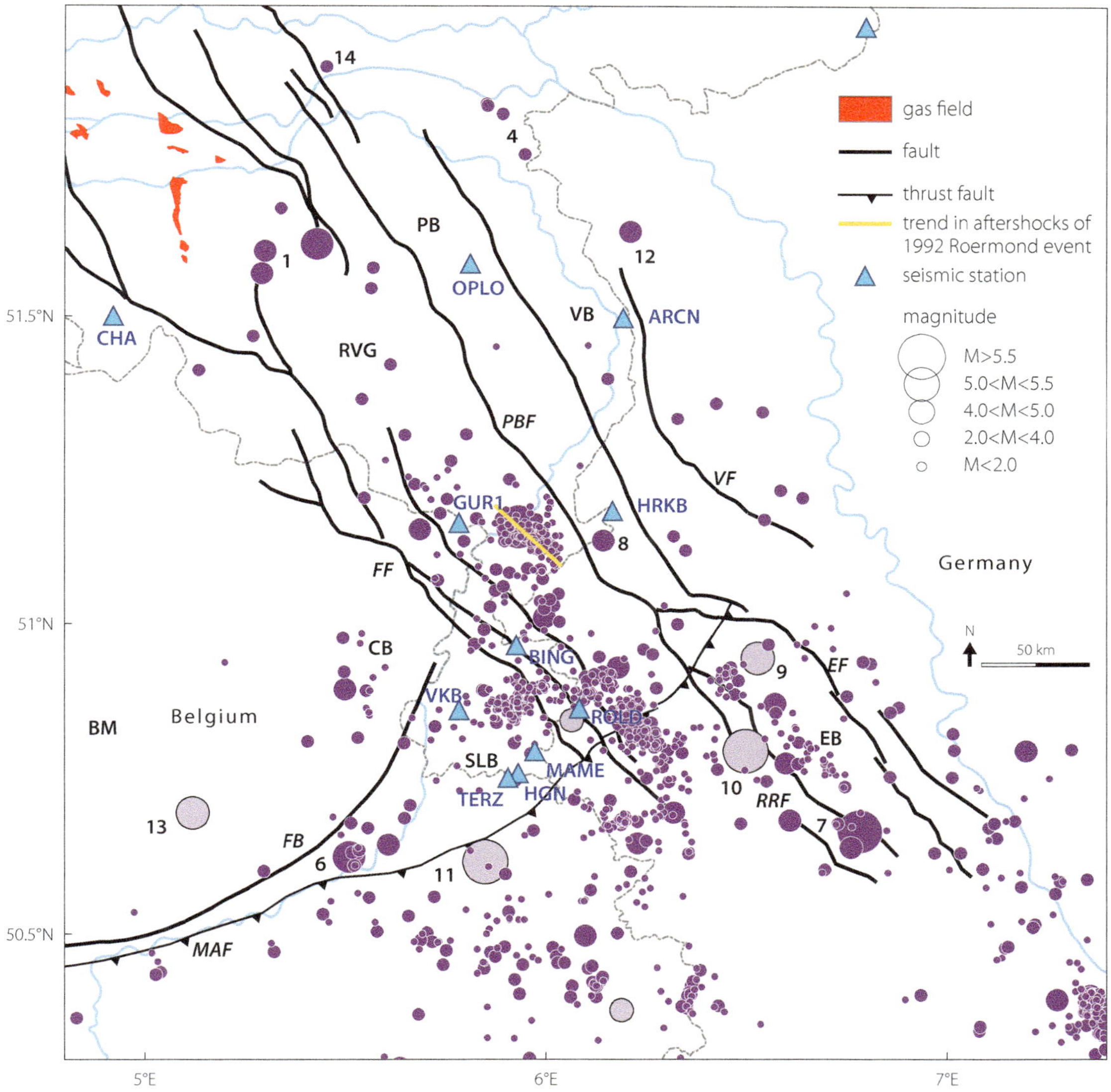

Figure 22.1. Epicentre map of natural seismicity in and around the Roer Valley Graben for the period 1690-2021. Earthquakes are shown as purple circles, scaled according to magnitude; events before 1930 in light purple, those after 1930 in dark purple. Localities mentioned in the text: 1: Uden; 2: Roermond; 3: Alsdorf region; 4: Nijmegen; 5: Voerendaal; 6: Liège; 7: Euskirchen; 8: Heinsberg; 9: Tollhausen; 10: Düren; 11: Verviers; 12: Goch; 13: Hannut, 14: Tiel. Tectonic structures: BM: Brabant Massif; RVG: Roer Valley Graben; PB: Peel Block; VB: Venlo Block; CB: Campine Block; EB: Erft Block; SLB: South Limburg Block; PBF: Peel Boundary Fault; FF: Feldbiss Fault; VF: Viersen Fault; FB: Faille Bordière; MAF: Midi-Aachen thrust Fault; EF: Erft Fault; RRF: Rurrand Fault. Source: KNMI (seismicity data) and TNO (fault data).

Table 22.1. Earthquakes of magnitude $M_L \geq 4$ since 1900 related to the Peel Boundary Fault and its extension into Germany. Source for Table 22.1 and 22.2: KNMI.

Event location	Date	Origin time (UTC)	Latitude (°N)	Longitude (°E)	Depth (km)	M_L
Roermond	1992-04-13	01:20:03	51.162	5.933	15	5.8
Euskirchen (GER)	1951-03-14	09:46:55	50.669	6.785	28	5.8
Uden	1932-11-20	23:36:55	51.630	5.412	16	5.0
Alsdorf (GER)	2002-07-22	05:45:05	50.873	6.212	13	4.9
Roermond	1935-01-04	04:06:13	51.140	6.145	15	4.3
Veghel	1932-11-28	05:41:35	51.612	5.632	15	4.0

Table 22.2. Earthquakes of magnitude $M_L \geq 4$ since 1900 related to the Feldbiss Fault and its extension into Germany.

Event location	Date	Origin time (UTC)	Latitude (°N)	Longitude (°E)	Depth (km)	M_L
Boxtel	1932-11-23	03:08:01	51.577	5.268	10	4.5
Koningsbosch/ Heinsberg (GER)	1971-02-18	23:41:25	51.106	5.997	17	4.4
Stramproy (BEL)	1960-06-25	14:29:15	51.158	5.687	15	4.2
Alsdorf (GER)	1992-04-14	01:06:46	50.935	6.182	15	4.1

could be calculated for 55 of them. Although the scatter in locations is high due to a limited station coverage, an outline of the fault surface becomes clear and corresponds to an average dip of 70° to the southwest. Ahorner (1994) showed similar results in a comparable study, in which less scatter was obtained. For the Uden event one foreshock and eight aftershocks were recorded (Van Dijk, 1934). However, due to the limited instrumentation at that time, the detection threshold was much higher than for the Roermond event and no details on the faulting mechanism could be obtained. Figure 22.1 suggests a difference in trend between the strike of the Peel Boundary Fault and the epicentre locations of the aftershocks south of Roermond (Camelbeeck et al., 1994). This implies that the fault is not planar and that the deeper structure of the fault is relatively complex. Another feature of the Roermond aftershock sequence is that two clusters of aftershocks are evident: one close to the main shock and the other 40 km to the southeast in the vicinity of Alsdorf in Germany, suggesting a triggered effect along the Feldbiss Fault Zone (Prinz et al., 1994). The first aftershock in the distant cluster occurred only three hours after the main shock. No aftershocks are located to the northwest, where the Uden earthquake took place 60 years before. The most recent significant earthquake in the Roer Valley Graben, the July 22, 2002 Alsdorf event, occurred near this remote aftershock region of the Roermond earthquake. Its local magnitude is M_L= 4.9, which makes it the strongest event in the region since ten years. Its depth of 13 km and the location seem to correspond to the Feldbiss Fault Zone, but the fault-plane solution of the event (Hinzen & Reamer, 2007) shows a normal fault dipping to the southwest suggesting a connection with the Rurrand Fault.

Feldbiss Fault Zone

Based on the seismic activity observed in the 20th century, the Feldbiss Fault Zone shows on average events of lower magnitude than the Peel Boundary Fault (Table 22.2). It is worth mentioning that a recent relocation of the 1971 Koningsbosch event shifts its location into Germany, close to Heinsberg (Fig. 22.1). This is in accordance with the findings of Ahorner & Pelzing (1983). For some events that are located in the middle of the Roer Valley Graben and that have no source mechanism solutions available it is not clear whether they belong to the Peel Boundary Fault or to the Feldbiss Fault Zone.

Area south of the Feldbiss Fault Zone

The seismicity south of the Roer Valley Graben is spread over a relatively wide area. The largest event in the 20th century is that of magnitude M_L= 5.0 on November 8, 1983 at Liège in Belgium (Camelbeeck & De Becker, 1985; Fig. 22.1). This shallow event, only 6 km deep, is connected to the eastern prolongation of the Brabant strike-slip zone (Ahorner, 1985), between the Midi-Aachen Thrust Fault and the Faille Bordière. The South Limburg Block is characterized by events that originate in the lower crust at depths of 20 to 24 km (Table 22.3; KNMI earthquake list).

The region around Voerendaal, just south of the Roer Valley Graben, shows an anomalously shallow, swarm-like seismicity at depths between 3 and 8 km. A series of events were recorded during more than one month, starting December 7, 1985, the largest of them having a magnitude of M_L= 3.0. After a period of 15 years, a new swarm of shallow events started in the same area on December 20, 2000. This swarm lasted until January 2002 and a total of 139 events with magnitudes ranging between M_L= 0.1 and 3.9 were recorded by a seismograph network that had been significantly improved with respect to detection capability. The largest event of this swarm occurred about halfway its period of activity. The activity is most probably connected to the E-W oriented Kunrade Fault (Fig. 22.2), which shows a scarp at the surface just south of Voerendaal. The shallow occurrence of the events explains the relatively high maximum intensities, ranging from I_0= III (weak) for M_L= 2.0, to I_0= VII (damaging) for M_L= 3.9. The strongest event caused some damage to structures, including broken or fallen chimneys and roof tiles. This kind of damage is comparable to that caused by the much stronger Uden and Roermond events, which occurred at greater depths, although affecting a much smaller area. The combination of magnitude and depth largely determines the intensity at the surface. Although the Voerendaal swarm is listed as natural seismicity, since the location is close to the Roer Valley Rift System, rising minewater after closure of the coalmines in the region could have triggered the seismicity (Caro Cuenca et al., 2013; Projectgroup GS-ZL, 2016; Hinzen et al., 2021).

Observed earthquakes in other areas

Instrumentally recorded natural seismicity onshore outside the region discussed above is limited to a few locations near Nijmegen, ca. 70 km north of Roermond, where two events of magnitude M_L= 3.2 and 3.0 were recorded in 1972 and 1979 (Fig. 22.1). It is unclear whether they can be associated with the Viersen Fault of the Roer Valley Rift System. A M_L= 4.3 event occurred near Goch (Germany) in 2011, 55 km north of Roermond (Hinzen, 2019). Its location is east of the Viersen fault and the mechanism is a strike-slip movement (Hinzen et al., 2021). A detailed interpretation of the source location and mechanism has not yet been published. In the area south of Bergen op Zoom, close to the Belgian-Dutch border, a small event of magnitude M_L= 2.1, possibly connected to the Antwerpen Fault, took place on June 15, 1988. Two more events on February 15, 2008 with M_L= 2.3 and 1.9 were located in the same area.

The North Sea area is subject to large events, as demonstrated by the 1931 Dogger Bank event, M_L= 6.0, that was felt all over the Netherlands (Gees, 1937). An overview of the seismicity in the area is given by Ringdal (1983). Recent seismicity in the Netherlands continental shelf is lim-

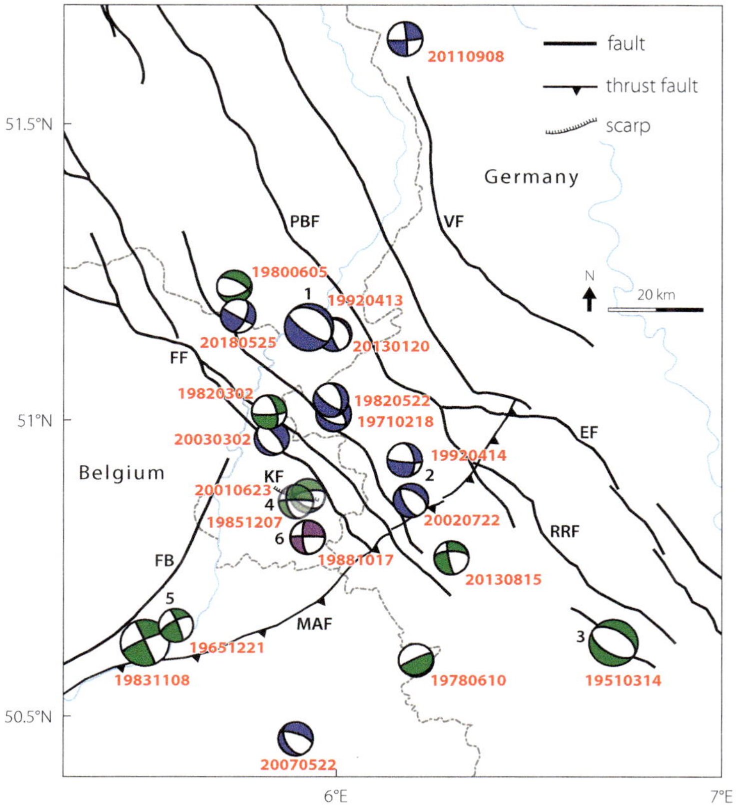

Figure 22.2. Overview of focal mechanisms in and around the Roer Valley Graben, represented as equal-area projections of the lower focal hemisphere. Quadrants of compressional P-wave onsets are filled. Abbreviation of fault names as in Fig. 22.1; KF = Kunrade Fault. Localities of events: 1 = Roermond; 2 = Alsdorf; 3 = Euskirchen; 4 = Voerendaal; 5 = Liège; 6 = Gulpen. Event depth (h) ranges are given in colour: 5 < h < 10 km green, 10 < h < 20 km blue, h > 20 km magenta. Events of $M_L \geq 5.0$ are shown as larger beachballs. Minimum magnitude M_L=2.0. Source: KNMI (seismicity data) and TNO (fault data).

Macroseismic intensities, mapping the effects of felt earthquakes

1

Before the 1930s there were only a few seismographs in operation in the Netherlands and the neighbouring countries. The main source of information we have on events prior to the 1930s, is published reports in newspapers or letters of what people felt during the event. An intensity scale was developed to structure the reported severeness of the felt effects, including damage to structures.

Mercalli (1902) published a first version of a widely used intensity scale, which was modified over time and still in use as the Mercalli-Cancani-Sieberg scale (MCS) or in a slightly modified version: the Modified Mercally Intensity (MMI) scale (Wood & Neumann, 1931). In Europe the European Macroseismic Scale (EMS-98) is currently the standard.

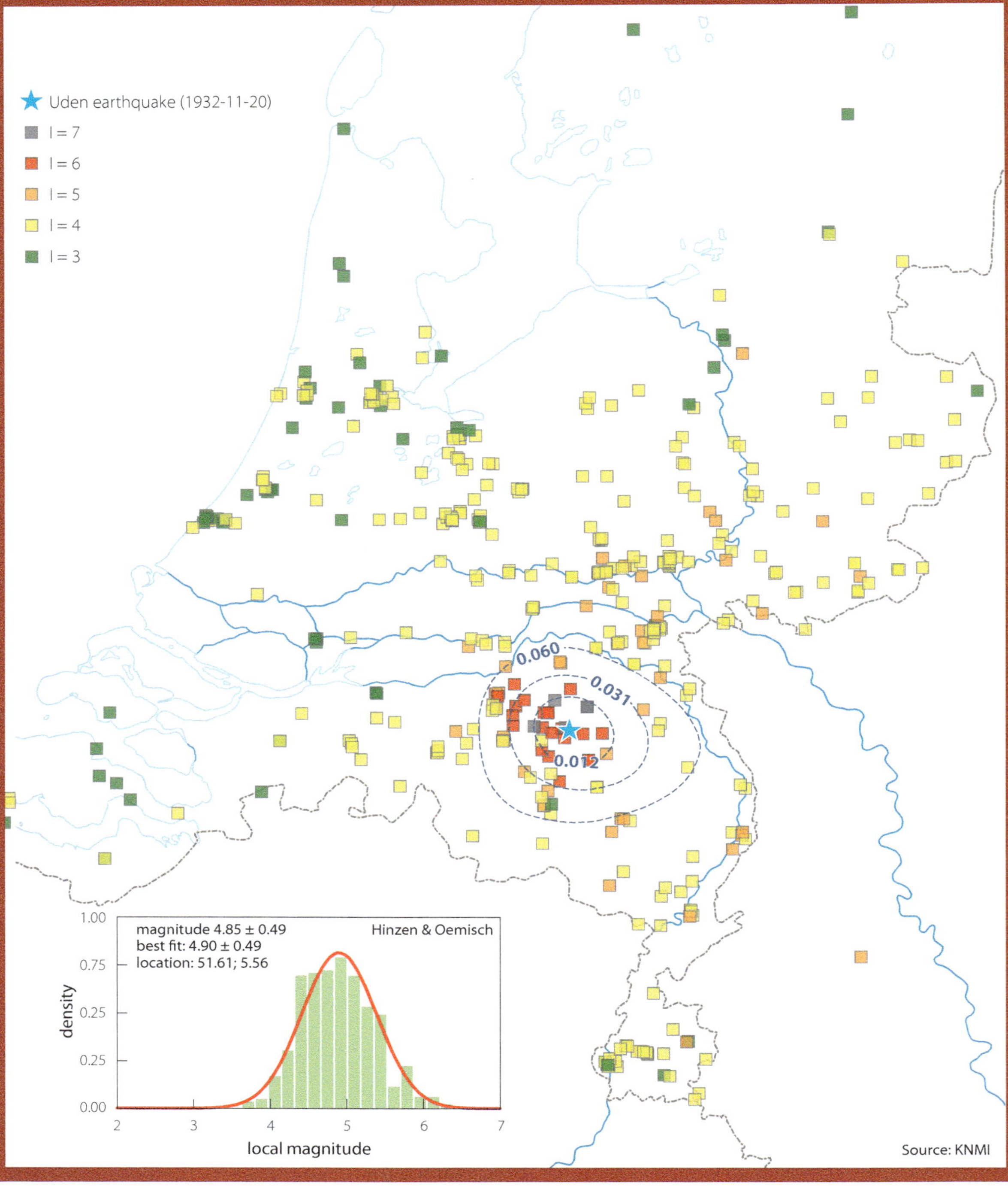

The first macroseismic map for the Netherlands was published in 1934 (KNMI, 1934). The map, showing the effects from the 1932-11-20 Uden event, contains MCS intensity values based on over 400 letters sent to the KNMI and reports in newspapers. The original map reports only intensity values, which was later shown in a contoured map form by Gees (1937). The interpretation of the written reports in terms of intensities and their further analysis is subjective and results are hard to reproduce. The development of standard procedures for the interpretation of intensity data (Bakun & Wentworth, 1997, 1999) and their application to the Lower Rhine Embayment (Hinzen & Oemisch, 2001) changed this situation. Although the intensity assessment is still subjective, the analysis can now be reproduced and formal error estimates of the uncertainty in magnitude and location can be given. The map shows a modern analysis of the 1932-11-20 event, based on a re-interpretation of the original intensity data according to the EMS-98 and calculations on a grid of trial epicentres. Besides the intensity values, the map shows in dashed blue contours the 50% (0.012), 80% (0.031) and 95% (0.060) confidence levels of the location. The best location is shown as a cyan star and magnitudes calculated for the best location are presented in the histogram.

A key parameter in the analysis is the intensity attenuation law used. In order to sample the epistemic uncertainty, multiple attenuation laws will be applied for a final solution. The method is calibrated by comparison with events from the past 50-60 years, that have been well located.

ited to on average one event per year along existing faults. For this region the lowest magnitude event that can be detected and located is estimated at M_L= 2.0-2.5 (see Fig. 22.6).

Instrumentation and location accuracy

The KNMI started seismological observations in De Bilt (station code DBN) near Utrecht, in 1904 on an experimental basis (Houtgast et al., 1987). In 1908 it started continuous recording, and a historical archive of analogue seismograms has been kept and maintained since then. In 1926 a second station was installed in Heerlen (HEE, E-W component only) in order to investigate the local seismicity in the south of the country. Instruments used at that time were heavy and not very portable. In 1951 a third station was added near Witteveen (WIT) in the north (Drenthe), followed since 1970 by additional stations in the south: Ravensbosch (RSB), Epen (ENN), Kerkrade (KRN) and Valkenburg (VKB). All data from these stations were recorded on paper. In the 1970s, experiments were carried out with digital recording and a modernization of the network was planned. In 1993, the Heimansgroeve (HGN) station near Epen started its operation after careful site selection and became the new low-noise reference station for the Netherlands. The station is equipped with state-of-the-art, very broad-band sensors and digital recording (Dost & Haak, 2002). At present four additional broad-band digital stations are operational in South Limburg in addition to HGN: Valkenburg (VKB), Rolduc (ROLD), Mamelis (MAME) and Terziet (TERZ). The northern part of Limburg is monitored by three broad-band stations: Oploo (OPLO), Arcen (ARCN) and Herkenbosch (HRK). The Winterswijk (WTSB) station, an upgrade of the 1974 station WTS in east Gelderland, completes the picture to the north. Apart from the KNMI instrumentation, a set of mobile stations, the NARS array, has been operated by Utrecht University with variable configuration since 1983 (Dost et al., 1984). In the period 1989-1993 the NARS stations have been deployed in the Netherlands, Belgium and Germany and from 2008 to 2013 across the Netherlands (seismologie.sites.uu.nl/research-projects/nars/netherlands). In Belgium, the station density increased after the Liège earthquake of 1983, while at the same time more stations were added in Germany by the University of Köln and the Geological Survey of Nordrhein-Westfalen.

Before the 1930s, the epicentre locations of most earthquakes were identified based on seismic intensity data, due to the low density of seismograph stations at the time. Errors in these locations can easily be in the order of several kilometres, especially if the epicentral area was thinly populated. Assessment of the depth of the events remained problematic until the 1980s, when the number of seismological instruments increased. A good example is the Euskirchen 1951 event, for which Berg (1953) calculated a depth of 28 km based on instrumental data, while Sponheuer (1958) and Van Gils & Zaczek (1978) calculated a shallow depth around 9 km based on intensity data. Since the 1980s the average error in epicentre location is at least 1 km, while the range in depth accuracy is in the order of several kilometres. An effort is underway to calculate formal uncertainties for all events in the catalog. In those cases where a good correlation between waveforms of different events recorded at the same station exists, an increase in resolution can be obtained, improving the error in the relative epicentre location to 0.1 km. Up to now however, such improved resolutions by application of relative location methods, have only been realized for induced earthquakes near two gas fields (Bergermeer and Roswinkel; see section on 'Induced earthquakes').

Although the station density has increased in the south of the Netherlands, routine data analysis is still based on an average crustal seismic velocity model for the Netherlands. Until 2001, the velocity model of Ahorner (1983) was used to locate earthquakes in the southern Netherlands, later this model was slightly adjusted and is still used by KNMI. Reamer & Hinzen (2004) derived a new one-dimensional model, based on a digital version of the seismic phase database from the Bensberg network in western Germany. After reprocessing the Bensberg dataset for 1995-2018, Hinzen et al. (2021) published the latest version of their catalogue. A joint inversion technique using a combined phase database from Bensberg, De Bilt and Uccle may further refine the velocity model(s) and improve the location accuracy of events in the region.

Magnitude

Magnitude is a measure of the strength of an earthquake at its source, while intensity describes the effects of an earthquake at various locations at the surface. Intensity varies with epicentral distance, so the maximum intensity is used to characterize an earthquake. Richter's (1935) definition of the local magnitude (M_L) has been used in different ways by different agencies and has not always been applied to older data. Instead magnitudes, especially of older events, were often estimated from reported maximum intensities and may therefore vary by up to one magnitude unit. In his overview of earthquakes in the Netherlands, Visser (1943) reported no magnitudes, only maximum intensities. Houtgast (1991) estimated magnitudes for events prior to 1900 from the maximum intensity by simple relations originally derived for southern California (Båth, 1979, p. 127). De Crook (1994) derived and used an empirical relation between intensity and local magnitude M_L for the Netherlands, based on 42 observed tectonic earthquakes.

Hinzen & Oemisch (2001) applied a more sophisticated method to estimate location and magnitude directly from the raw seismic intensity data and used recent events (including Roermond 1992) as a training set to calibrate the method. This method provides an objective and reproducible way to interpret data and was used to re-evaluate older events in the northern Rhine area. Results for pre-1900 events will be re-evaluated using this or a similar method in future updates of the catalogue of events in the Netherlands. Richter's local-magnitude M_L scale is based on earthquake recordings from a standard short-period horizontal seismograph (Wood-Anderson). The M_L scale saturates for events larger than magnitude 6.5 (McCalpin, 1996), since the dominant energy of the radiated seismic waves for these magnitudes occurs at lower frequency than the frequency pass band of the seismograph. An alternative is to use the moment magnitude M_w (Hanks & Kanamori, 1979), which is a measure based on the seismic moment, defined as the product of the average displacement over the fault zone, the area of the fault surface and the average shear rigidity of the faulted rocks. M_w can be derived from the displacement spectra of seismograms and does not saturate. For the 1992 Roermond earthquake, where the local magnitude M_L is expected to be close to the moment magnitude M_w, since saturation of the M_L scale is not yet relevant it appears that the local magnitude (5.8) is significantly larger than the moment magnitude (5.4). This inconsistency is not yet resolved and may cause confusion between different reported magnitudes of events.

KNMI uses its own attenuation relation for M_L calculation, which is close to the relation published by Reamer and Hinzen (2004). M_L values reported by Bensberg, Royal Obervatory of Belgium (ROB, Camelbeeck, 1985) and KNMI are in general agreement. Reamer & Hinzen (2004) and Hinzen et al. (2021) made a comparison between M_w and M_L based on broad-band seismograms. Their results support the conclusion that $M_L > M_w$ for events in the Lower Rhine Embayment, with a tendency for the difference M_L-M_w to increase for larger events.

Source mechanisms

Apart from the locations of earthquakes, other source parameters can be derived from seismograms. The fault mechanism, i.e. the direction of movement along the fault plane, combined with an accurate location, makes it possible to relate events to faults and to gain insight into the source processes. An overview of determined source mechanisms in and around the southern Netherlands is shown in Figure 22.2.

The Roer Valley Graben is dominated by normal faulting and subsides with respect to its flanks. The hypocentral depths of earthquakes in the graben range from 5 to 20 km, indicating that the boundary faults extend downwards into the lower crust. In the South Limburg Block, events occur mainly at depths between 20 and 24 km (Table 22.3), and the mechanism appears to change from predominantly normal faulting to strike-slip faulting (Camelbeeck, 1994). However, since these deep events are usually of low magnitude, their source mechanisms are not well constrained. This possible change in mechanism would imply a different cause for these earthquakes, which may be related to large strike-slip faults in Belgium (Camelbeeck & Van Eck, 1994) or to differences in stress regime between the upper and lower crust. Further south, in the Hautes Fagnes region in eastern Belgium and western Germany, small earthquakes with probable thrust mechanisms (e.g. Ahorner, 1983) may indicate the local reactiva-

Table 22.3. Deep earthquakes (≥20 km) since 1970 south of the Feldbiss Fault (M_L ≥2.0). Source: KNMI.

Event location	Date	Origin time (UTC)	Latitude (°N)	Longitude (°E)	Depth (km)	M_L
Gulpen	1988-10-17	19:39:54	50.813	5.922	23	3.5
Verviers (BEL)	1972-02-17	04:03:32	50.671	5.970	22	3.1
Valkenburg	1981-12-20	10:38:53	50.852	5.922	23	2.7
Sippenaeken (BEL)	1995-07-18	12:51:23	50.752	5.927	23	2.4
Maastricht	2011-05-09	16:34:22	50.840	5.725	20	2.3
Valkenburg	1995-03-30	20:29:08	50.846	5.878	20	2.2
Maastricht	1990-10-18	19:30:29	50.813	5.647	21	2.2

tion of SW-NE striking thrust faults like the Midi-Aachen Thrust Fault south of the Brabant Massif (Hance et al., 1999; Mansy et al., 1999).

For the Voerendaal events, different source solutions have been made. For the largest 1985 event, Houtgast (1991) interpreted a normal fault of 270° strike and 80°N, dip whereas Camelbeeck (1994) interpreted a strike-slip movement for the same event. A similar controversy applies to solutions for the largest shocks in the 2000-2002 swarm, although the normal-faulting solution is best constrained and here accepted to be typical for these events. The Voerendaal events thus appear to be connected to the Kunrade Fault, which deviates from the general NW-SE trends in the Roer Valley Graben and which shows an E-W strike. Plenefisch & Bonjer (1997) carried out a stress inversion for the entire Rhine Graben based on 98 focal mechanisms of small to moderate events. Although no definite distinction could be made between strike-slip and normal faulting regimes, a counter-clockwise rotation of the horizontal stress axes of 30° could be observed between the southern Rhine Graben (160°) and the Roer Valley Graben (130°). A separation of the total dataset for the southern Rhine Graben into upper-crust (5-15 km) and lower-crust (15-23 km) parts showed a dominant strike-slip regime in the upper crust, and a normal faulting regime in the lower crust. A similar study for the northern Rhine Graben (Hinzen, 2003) suggested the opposite: a dominant normal faulting regime in the shallow crust (<12 km) and a strike-slip regime in the deep crust.

Seismotectonic setting

Ziegler (1992, 1994) interpreted the evolution of the Roer Valley Rift System as part of the European Cenozoic Rift System (ECRIS). Two factors control the seismotectonic movements in the rift: first the present NW-SE compressive regime, characterized by an orientation of the maximum horizontal stress of 145° and controlled by push from the Mid-Atlantic Ridge and by the collision of Africa and Europe (Grüenthal & Stromeyer, 1992; Goes et al., 2000). A second factor is the existence of major fault zones (Ahorner, 1994) of post-Hercynian or even older origin that have been reactivated during later tectonic phases (De Jager et al., 2025, this volume). Müller et al. (1997) discussed the observation that source mechanisms vary in western Europe over short distances from strike-slip to normal faulting, while the general strike of the faults does not change. This is hard to understand if the regional stress field remains the same and can only be explained by a decoupling of the brittle upper crust from the lithospheric mantle by a ductile lower crust. It also explains the different faulting regimes in the upper and lower crust interpreted by Plenefisch & Bonjer (1997). The occurrence of deep and shallow events south of the Feldbiss Fault Zone may also be explained by this decoupling, although an alternative explanation may be that a deeper set of faults with a different strike exists. Unfortunately, we know little about the deeper geology of South Limburg (>0.5 km), and from the deep (>20 km) events, only the largest event (Gulpen, 1988) has a well-determined source mechanism (strike-slip). If decoupling is assumed, the model needs to explain why normal faulting is dominant in the upper crust and strike-slip in the lower crust, as opposed to the findings in the southern Rhine Graben. Hinzen (2003) explained the difference by interpreting the shallow extensional regime as a local effect superimposed on the regional strike-slip regime. Camelbeeck & Van Eck (1994) argued that the deep Gulpen event and the shallow 1983, M = 5.0, Liège event, both strike-slip events, may belong to the same tectonic region ('Liège-Gulpen zone'), which may indeed be an alternative explanation.

The Roer Valley Rift system has been studied in detail with respect to its recent and long-term subsidence and uplift history (Zijerveld et al., 1992; Geluk et al., 1994; Van den Berg et al., 1994; Houtgast & Van Balen, 2000) and rheological and geomechanical modelling (Dirkzwager et al., 2000). Subsidence rates differ through time and may be related to changes in fault activity (slip rate). Long term subsidence rates based on geological measurements in the Roer Valley Graben show that they were not constant; periods of acceleration and deceleration occurred during the last 8 Myr (Geluk et al., 1994; Michon et al., 2003). This has to be taken into account in the interpretation of paleo-seismicity (e.g. Houtgast, 2003). Geomechanical modelling showed that the boundary faults could be reactivated at reduced friction angles. Details of this modelling are given by Dirkzwager (2002), who modelled the subsidence in and around the graben during such reactivation. Rheological models of a layered crust were developed by Cloetingh & Burov (1996) and used by Mayer et al. (1997) to explain the difference in behaviour between upper crust (strike-slip) and lower crust (normal faulting) for the southern Rhine Graben.

In general terms the earthquakes in the Roer Valley Rift System can be correlated with the major fault zones. However, little is known about the extent of the faults at depth, although seismicity (in particular the Roermond aftershock sequence) indicates a downward continuation at more or less constant dip. Ahorner (1983) combined seismicity data with seismic reflection data and showed that shallow (5 km deep) thrust faults are active to the southeast of station HGN. Information from deep seismic reflection studies in Germany and Belgium (DEKORP-BASIN Research Group, 1991) has confirmed his interpretations. In the Netherlands, only limited information is available from deep seismic surveys (>5 km, Duin et al., 1995). They provide little information on the deep structure of South Limburg. For most of the region south of the Feldbiss Fault Zone, the structure below 200-300 m is largely unknown (NITG-TNO, 1999). This means that the relation between seismicity and geology in the region around Voerendaal cannot yet be established.

Historical and paleo-earthquakes

Evidence for the occurrence of historical earthquakes comes mainly from written sources. For the 19th century these sources are in general clear, but for earlier centuries the help of historians to interpret them is required (Alexandre, 1985, 1994), and the accuracy of reconstructed epicentre locations (10-20 km) and magnitudes (1-2 units) becomes limited. A few 19th century events are worth mentioning (Fig. 22.1). On September 26, 1878, an event with an estimated magnitude M_L= 5.6 occurred near Tollhausen in Germany. This event is possibly connected to the northern boundary fault of the Erft Block (Hinzen & Oemisch, 2001). One year earlier, on June 24, 1877, an event of magnitude M_L= 4.4 occurred near Herzogenrath and was preceded in 1873 by a magnitude M_L= 4.3 event (Hinzen & Oemisch, 2001). Both were most probably connected to the Feldbiss Fault Zone. Other events felt in the southern part of the Netherlands and studied in detail are the ones near Hannut (23-02-1828, M_w= 5.1; Camelbeeck et al., 2021), which can be related to the Brabant Massif and the ones near Hautes Fagnes (3-12-1828, M_w= 4.2; Knuts et al., 2016). In the 18th century a large earthquake of estimated magnitude M_L= 6.4, which occurred near Düren (February 18, 1756), was the largest historical earthquake in the Roer Valley Rift System. And at the end of the 17th century a major event, possibly of magnitude M_L= 6.8, occurred near Verviers in Belgium (September 18, 1692; Hinzen & Oemisch, 2001), in the area of the eastern Ardennes, west of the Roer Valley Rift System. In a detailed study of the Verviers event the moment magnitude was estimated at between M_w= 6 and 6.5 (Alexandre et al., 2008).

As we have at most just a reasonable impression of the larger seismic events over the last 500 years, the question arises: did we miss any of the large events in the Netherlands that are expected to occur on average once every 2000 or 3000 years? Paleoseismic research is directed at solving this question. A first survey for paleo-earthquakes along the Peel Boundary Fault has been carried out (Demanet et al., 2001; Van den Berg et al., 2002). It aimed to collect evidence for sudden displacements where the fault zone reaches the surface. Such displacements can be found in trenches of a few metres deep, oriented perpendicular to the fault plane. Since the preparation of a trench is time-consuming, geophysical methods have been applied to locate the fault at the surface. Near Neer, northwest of Roermond, a trench was prepared intersecting the Peel Boundary Fault. It showed geological evidence for two, or possibly three paleo-earthquakes. The strongest events are dated ca. 15 ka BP with a time interval between them of ca. 2 ka. Their moment magnitudes are estimated at M_w= 6.2 to 6.5, which is one magnitude unit larger than the 1992 Roermond event. More recently a new trench was explored over the Peel Boundary Fault, 32 km NNW of the Neer trench (Van Balen et al., 2019). Two paleo events have been found that are likely similar to the two events found at Neer, suggesting a surface rupture of at least 32 km. Similar results are reported from surveys in Belgium along the Feldbiss Fault Zone (Camelbeeck & Meghraoui, 1998). An overview of research on paleoseismicity in the region is given by Camelbeeck et al. (2007).

Natural seismic hazard and risk

Based on a catalogue of earthquakes in the region, De Crook (1993) carried out a probabilistic hazard assessment for the Netherlands in terms of the probability of intensity recurrence. Important model parameters in this assessment are the definition of the seismotectonic zonation, the relation between maximum observed intensity and number of events, and the intensity-distance attenuation relation. The results of the assessment were presented in a series of maps showing the probability of occurrence per year for the expected intensity. The 1992 Roermond earthquake corroborated the assessment nicely (De Crook, 1994). However, engineering models are better in handling ground-acceleration estimates than intensities. De Crook (1996) presented first rough estimates of expected horizontal peak ground accelerations, based on empirical intensity-peak ground acceleration relations used in Germany. De Vos (2010) provided an update and evaluated the effects of different source zonations, Ground Motion Prediction Equations (GMPE's) and calculation methods.

The statistical relation between the logarithm of the cumulative annual number of events and magnitude, usually referred to as the frequency-magnitude (FM) relation, shows for the southern Netherlands a characteristic linear trend for magnitudes between $M_L = 2.5$ and 4.5. The slope of the regression line, called the b-value, is used to estimate mean return periods for events, while the nonlinear part of the FM curve may be used to estimate the maximum expected magnitude. De Vos (2010) determined a b-value of 0.77 ± 0.03 (based on M_L) for the seismicity in the south of the Netherlands, which is comparable to the b-value of 0.74 determined by Ahorner (1983) for the Rhenish Massif, implying that the seismicity in the south of the Netherlands is comparable to the seismicity in the larger region. De Vos (2010) further estimated the maximum expected magnitude for the south of the Netherlands at $M_L = 6.3 \pm 0.2$.

The use of a dense seismometer network allows us to calculate locations, magnitudes and source mechanisms. The reconstruction of true ground motion in the epicentral area, however, is less accurate since attenuation and site effects on a local scale are not well known. This realization motivated KNMI to install two accelerometers in the epicentral area after the first earthquake near Voerendaal (December 2000), where a new earthquake swarm could be expected. Accelerometers are capable of recording strong motions where seismometers are saturated. Later, two more accelerometers were added to study local variations in ground motion. Recently, Van Ginkel et al. (2019, 2022) developed a nationwide site-response map for the Netherlands, which will be used in future hazard and risk calculations. De Vos (2010) found that source zonation and GMPE's are the most important factors in the hazard calculations. Different versions of both parameters should be taken into account and implemented in a logic tree, similar to the development in the hazard and risk assessment for induced events in Groningen, and this will be implemented in a further hazard update for natural seismicity.

Measured accelerations are used in modelling studies to determine the vulnerability of houses and infrastructure. This relates to earthquake risk, which is equal to the product of hazard and vulnerability. Vulnerability is defined as the probability of damage as a result of ground motion. In times when economic- and insurance losses from earthquakes have increased dramatically (Allman & Smolka, 2001), risk evaluation is gaining importance.

Induced earthquakes

Development of seismicity and monitoring networks

The northern provinces of the Netherlands were long considered to be aseismic (Houtgast, 1991; Haak, 1993). However, since 1986 seismic activity has been recorded, mainly in and around producing gas reservoirs in these provinces. By April 2022, over 1800 induced earthquakes had been recorded. Twenty-two of these were of $M_L \geq 3.0$ (Table 22.4). In the evolution of induced seismicity, three periods can be distinguished. The first one is the period 1986-2004 when seismicity outside the Groningen gas field was dominant (Fig. 22.3a) and when the largest events occurred near Roswinkel (Drenthe, $M_L = 3.4$, 1997)

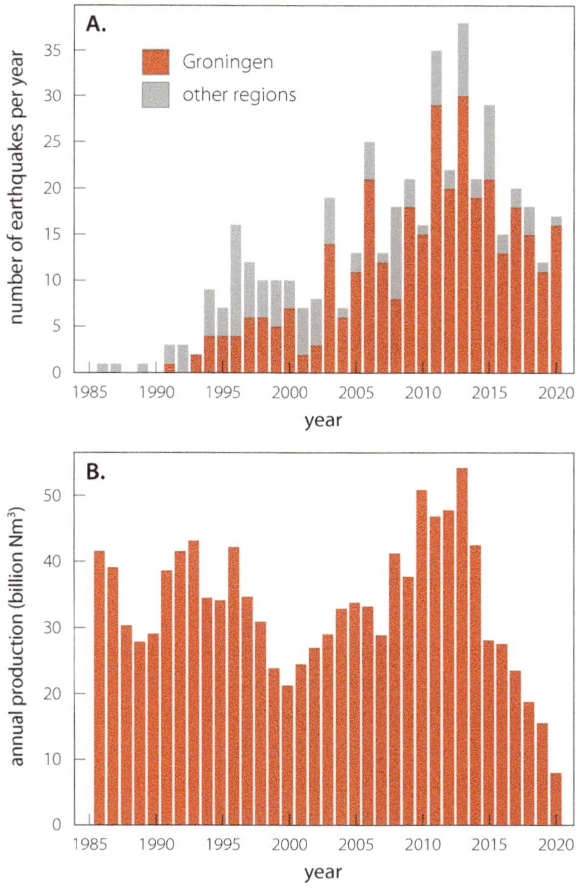

Figure 22.3. a) Histogram of the annual number of induced earthquakes in the Netherlands for $M_L \geq 1.5$; b) Annual gas production from the Groningen gas field. Source: KNMI.

Table 22.4. Induced earthquakes of magnitude $M_L \geq 3.0$ located in the Netherlands (before 01-04-2022). Source: KNMI.

Event location	Date	Origin time (UTC)	Latitude (°N)	Longitude (°E)	Depth (km)	M_L
Huizinge	2012-08-16	20:33:28	53.345	6.672	3	3.6
Alkmaar/Bergen	2001-09-09	06:58:12	52.651	4.713	2	3.5
Westeremden	2006-08-08	05:04:00	53.350	6.697	3	3.5
Roswinkel	1997-02-19	21:53:50	52.832	7.038	2	3.4
Zeerijp	2018-01-08	14:00:52	53.363	6.751	3	3.4
Westerwijtwerd	2019-05-22	03:49:00	53.328	6.652	3	3.4
Roswinkel	1998-07-14	12:12:02	52.833	7.053	2	3.3
Bergen	1994-09-21	01:12:58	52.658	4.705	2	3.2
Roswinkel	2000-10-25	18:10:34	52.832	7.052	2	3.2
Bergen	2001-09-10	04:30:15	52.653	4.712	2	3.2
Loppersum	2008-10-30	05:54:29	53.337	6.720	3	3.2
Garrelsweer	2011-06-27	15:48:09	53.303	6.787	3	3.2
Zandeweer	2013-02-07	23:19:08	53.389	6.667	3	3.2
Garrelsweer	2021-11-16	00:46:48	53.309	6.751	3	3.2
North Sea	2011-10-09	04:33:42	53.281	3.885	3	3.1
Hellum	2015-09-30	18:05:37	53.234	6.834	3	3.1
Bergen	1994-08-06	18:02:19	52.654	4.711	2	3.0
Garrelsweer	2003-10-24	01:52:41	53.295	6.792	3	3.0
Stedum	2003-11-10	00:22:38	53.325	6.690	3	3.0
Zeerijp	2009-05-08	05:23:11	53.354	6.762	3	3.0
Garrelsweer	2013-07-02	23:03:55	53.294	6.785	3	3.0
't Zandt	2014-02-13	02:13:14	53.357	6.782	3	3.0

and near Alkmaar (North Holland, $M_L = 3.5$, 2001). From 2004 until 2014 seismicity in the large Groningen gas field became increasingly dominant, coinciding with an increase in gas production from that field (Fig. 22.3b). At the end of this period the largest recorded event ($M_L = 3.6$) took place, causing damage to buildings in the Groningen region. The period from 2014 until 2021 is characterized by an apparent gradual decrease in the seismic activity rate in the Groningen field following the rapid decrease in gas production from the field. However, variation between the annual number of events is high and a recent statistical analysis showed that there is only an indication for a decrease in annual seismicity of $M_L \geq 1.5$ events. Statistical proof cannot be demonstrated due to the limited duration of the dataset (KNMI, 2021). In addition, seismic events have been detected in the neighbourhood of other mining activities (geothermal operations and salt mining). In this chapter we discuss the characteristics of seismicity in these periods and the development of the monitoring network.

1986-2004

Network development

The first event, at Assen in December 1986, with $M_L = 2.8$, was followed nearly one year later by an earthquake a few kilometres south of Hooghalen ($M_L = 2.5$; Fig. 22.4a). At the time, it was not clear what caused these earthquakes, but a connection with the production of natural gas was already suggested. Similar relations between earthquakes and producing gas fields were noted in France (Grasso & Wittlinger, 1990) and Canada (Wetmiller, 1986). The Ministry of Economic Affairs, responsible for the granting of concessions to the gas-producing companies, asked KNMI to monitor the region. A small-scale seismic network covering an area of 400 km² was installed around Assen in 1988. The network consisted of six vertical-component seismic stations, located at the surface. At the end of 1989, a third event ($M_L = 2.7$) occurred in North Holland near Purmerend, some 100 km to the west (Fig. 22.4a). The problem appeared to be more widespread and the need for more effective monitoring was felt. Although the detection threshold of seismic events in 1986 in the north of the Netherlands was estimated at M_L= 2.5 and smaller events could have occurred, widely felt tremors mentioned by Van der Voort & Vanclay (2015) do have an acoustic origin (Haak, 1985).

Noise conditions in the northern Netherlands do not allow the detection of small seismic events with sensors located at the surface, due to the thick sedimentary cover. Within the Assen array the detection threshold was estimated at magnitude M_L= 1.7. In order to obtain a similar or even improved threshold over a larger area, borehole seismometers were installed in 1995 after extensive testing in a site near Finsterwolde (FSW), east of the Groningen gas field. In boreholes the seismic noise decreases rapidly with depth (Carter et al., 1991; Ruigrok & Dost, 2019) and a reduction by a factor of almost ten in the frequency band between 1 and 10 Hz was achieved in Finsterwolde, showing for the first time dozens of minor earthquakes in the Groningen field (Haak, 1993). In total, ten such borehole stations, seven in the provinces of Groningen and Drenthe (ENM, WDB, ZLV, VLW, ENV, HWF, VBG) and three around Alkmaar in North Holland (OTL, PPB, WMH, Fig. 22.6a), were installed in 1995. The boreholes are 200 m deep, have no casing, and contain minimally four levels of 3-component, low frequency, 4.5 Hz geophones, electronically modified to a 1-Hz corner frequency (Dost et al., 2017). The total network shows a detection and location threshold of magnitude M_L= 1.5, the magnitude of completeness (Mc), although much smaller earthquakes have been detected close to some stations.

At locations where earthquakes were felt, accelerometers were installed in buildings. These instruments are able to record large vibrations, while the boreholes are designed to detect also small events. A total of 19 sites was equipped with an accelerometer at the surface in this period (1986-2004). At 6 borehole sites an additional surface sensor was added in order to evaluate site amplification effects.

Evolution of seismicity

The induced earthquakes in Groningen and Drenthe are all located in or near producing gas fields (Fig. 22.4a). The average error in location in the horizontal plane is ca. 1 km. For the Groningen field the locations cluster around large faults in the producing reservoir at a depth of ca. 3 km. Remarkably, the southeastern part of the field showed almost no seismic activity. Other gas fields with significant seismic activity are Annerveen, south of the Groningen field, several NNW-SSE trending smaller fields to the west, and the Emmen and Roswinkel fields in south-east Drenthe.

The activity around Roswinkel gained special attention. In the period 1992-2006, a total of 40 earthquakes occurred with magnitudes between M_L= 0.8 and 3.4. Events of magnitude 2.0 and higher were felt by people in the village, and minor damage was reported for the events of largest magnitude. Since the seismograms of these events have a high coherency, the locations could be obtained at a high resolution in the horizontal plane, in the order of 0.1 km or less, but the depths remain uncertain. Figure 22.5 shows the locations of the Roswinkel events, the gas field and the faults at the top of the reservoir, which has an anticlinal shape that was formed during the Early Triassic (Roest et al., 1998). The events clearly follow both the WSW-ENE, and the NW-SE striking fault system. Both systems are explained by a complex system dominated by wrench faulting (Roest et al., 1998; Frikken, 1999). The source mechanisms for these events are assumed to be very similar since the waveforms are highly coherent. There are two possibilities: the preferred solution from a seismological point of view is a low-angle thrust, while a steep reverse fault may be an alternative. The latter case would imply reverse reactivation of a normal fault at the top of the reservoir. In reflection seismic profiles vertical faults and shallow low-angle thrust faults are both visible.

To improve depth resolution of seismic events one could add stations in the area or deploy sensors in well bores close to the reservoir. In 2001 the *Nederlandse Aardolie Maatschappij* (NAM) initiated a 'downhole microseismic' experiment near Roswinkel, with its reservoir at a depth of approximately 2 km. A total of 29 minor events with magnitudes between M_L= -6 and -2 could be located. They confirmed the depth of 2 km and in map view they follow the same pattern as the larger events detected in the KNMI shallow-borehole and surface stations (Fig. 22.5). A later downhole experiment in the Annerveen field was less successful due to technical problems. Nevertheless, ten events were recorded with magnitudes less than M_L= 0.9. These experiments show that microseismic events already delineate active fault segments before larger events are observed.

In North Holland, between Alkmaar and Bergen, two earthquakes of magnitude M_L= 3.0 and 3.2 occurred in 1994 (KNMI, 1994a,b). They were located along a NW-SE trending fault in the Bergermeer gas field (Roest & Mulders, 2000; Baisch et al., 2016). In 2001, after seven years of quiescence, the area was confronted with two more events with magnitudes of M_L= 3.2 and 3.5. Analysis showed that the four events were located along a fault that runs from the reservoir to the southeast (Haak et al., 2001). This fault is a normal fault, which is being reactivated in reverse direction by compaction of the gas producing reservoir in the upper block.

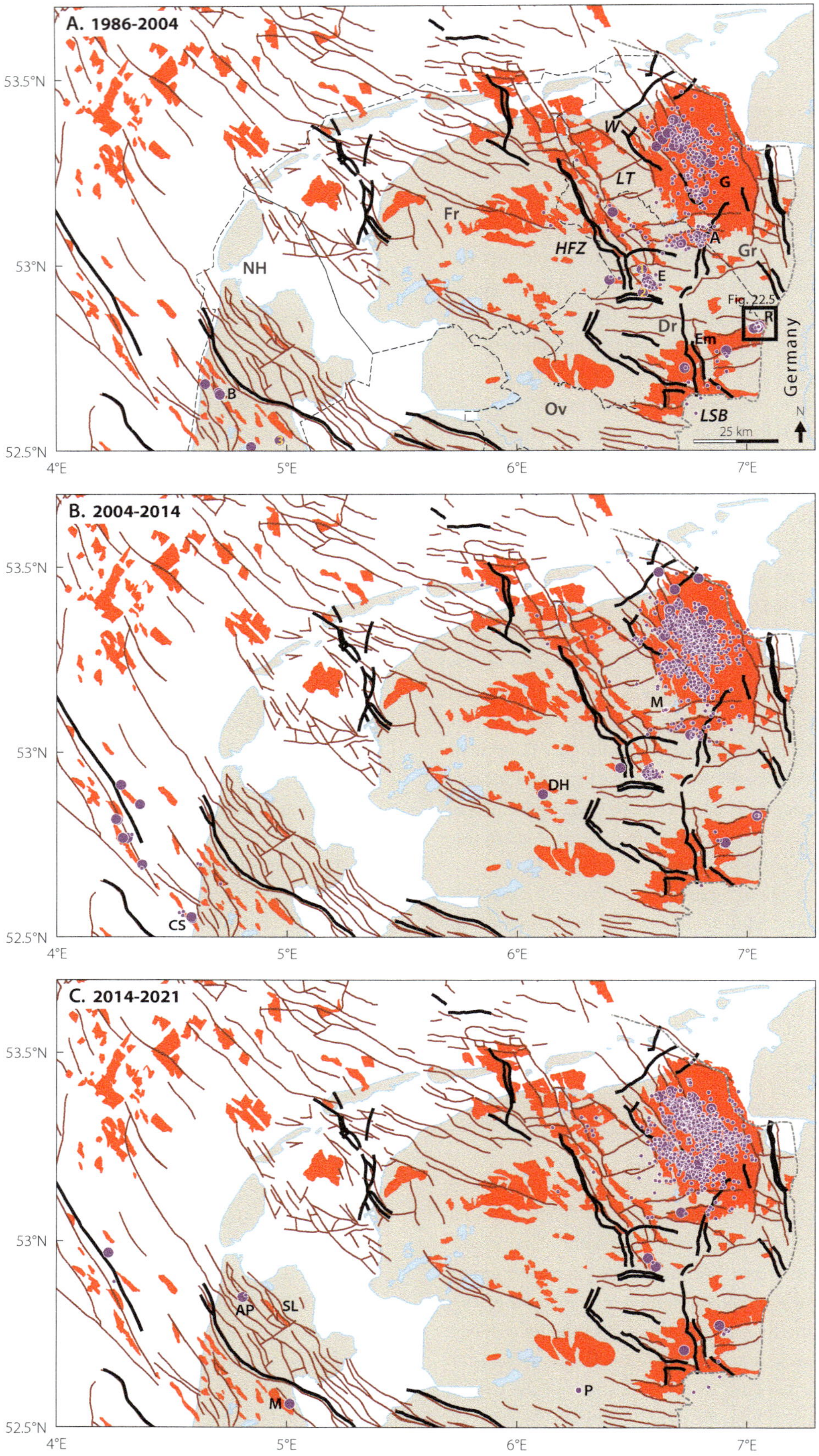
A. 1986-2004
53.5°N
53°N
52.5°N
4°E
5°E
6°E
7°E
W
LT
G
Fr
HFZ
A
Gr
NH
E
Fig. 22.5
R
Dr
Em
Germany
B
Ov
LSB
25 km
N
B. 2004-2014
53.5°N
53°N
52.5°N
4°E
5°E
6°E
7°E
M
DH
CS
C. 2014-2021
53.5°N
53°N
52.5°N
4°E
5°E
6°E
7°E
AP
SL
M
P

← ***Figure** 22.4. Overview of seismicity in the north of the Netherlands. Gas (and oil) fields in red, seismicity in purple (large circles for $M_L \geq 2.0$ and small circles for $M_L < 2.0$) for three periods. a) 1986-2004; provinces are marked with blue borders and blue letters (Gr=Groningen, Fr=Friesland, Dr=Drenthe, NH=North Holland, Ov=Overijssel), gas fields: A=Annerveen, B=Bergermeer, G=Groningen, R=Roswinkel, E=Eleveld, Em=Emmen, structural elements: W=West-Groningen Fault Zone, HFZ= Hantum Fault Zone, LT= Lauwerszee Trough, LSB=Lower Saxony Basin. Locations: 1=Assen, 2=Hooghalen, 3=Purmerend. The area of Fig. 22.5 near Roswinkel is shown as a black box; b) 2004-2014; gas fields: M=Midlaren, CS=Castricum Sea, DH=Den Hoeve (Weststellingwerf/Noordwolde); c) 2014-2021 with locations AP=Anna Paulowna, P=Punthorst, gas fields: M=Middellie, SL=Slootdorp. Base Zechstein faults in brown, base North Sea Supergroup in black. Source: KNMI (seismicity data) and TNO (fault data).*

2004-2014

Network development

After 2004, the accelerometer network was gradually expanded with 11 sites. In 2009/2010, two more borehole stations were added to the west of Groningen (SUH, NIW) and one (SPY) in the central part of the Groningen field (Fig. 22.6a; Dost et al., 2012). In addition, a new monitoring network was installed in the west of Friesland near the city of Harlingen, with the aim of monitoring earthquakes due to the local production of salt at 3 km depth. So far no earthquakes have been detected at this location. In 2013, the existing accelerometers in Groningen were upgraded and transformed into a real-time continuously recording network, later called the B-network (Dost et al., 2017; Ntinalexis et al., 2019). At the Bergermeer site, gas production ceased and the field was transformed into a gas storage facility. A pilot project for monitoring micro-seismicity in the field using a deep borehole with sensors at reservoir level (~2 km) was carried out in early 2010, followed by continuous monitoring later that year. At the end of 2013 NAM installed two deep borehole systems at the Groningen field (Stedum-1 and Zeerijp-1), both equipped with sensors installed at reservoir depth (~3 km; Pickering, 2015; Zhou & Paulssen, 2017; Dando et al., 2019).

Figure 22.6a shows the contours of the magnitude of completeness (Mc) for 2013, calculated using estimated noise levels at 200 m depth in the boreholes and for the accelerometers at the surface. While the majority of the gas fields are located within the Mc= 1.5 contour, exceptions include the gas fields in the southwest of the Netherlands, where no seismicity was recorded.

Evolution of seismicity

At the end of 2003, two M_L= 3.0 events occurred in Groningen. Until that time the maximum recorded magnitude event in the area was M_L= 2.7. Seismic activity rate in the region was considerably lower compared to other gas fields in the Netherlands (Van Eck et al., 2006; Dost et al., 2012). Dost & Kraaijpoel (2013) showed a change in activity rate around 2003. This was corroborated by Nepveu et al. (2016), who clearly showed, based on a statistical analysis, an initial constant activity rate and a change in seismic activity around 2003. Hints on the non-stationary character of seismic activity started to become evident, but could not yet be related to processes in the Groningen gas field. In 2012 the largest recorded induced event, M_L= 3.6, was located near the village of Huizinge in the province of Groningen. This event caused considerable damage in the region and instigated a re-assessment of the gas production levels and the relation between gas production and seismicity.

Bourne et al. (2014) showed that the spatial variation of seismicity in Groningen followed the developing compaction contours and demonstrated an exponential increase of the number of seismic events per unit volume change linked to reservoir compaction. They constructed a first model to forecast annual total moment release over time and produced event density maps for specific production scenarios. The location of energy release, defined by the centre of gravity of the seismic moment release, showed small spatial variations in time (Dost et al., 2017).

Since the average distance between faults in the Groningen gas field is less than the uncertainty in the earthquake location, the increased activity could not be attributed to specific faults in the reservoir. Relative location was expected to improve location accuracy and therefore a cluster analysis was carried out. Unfortunately, only a limited number of highly correlated waveforms could be found (Jagt et al., 2017). Relocation of the clusters indicated a focusing of seismicity along some faults and resulted in some cases to up to 2-3 km distance between the original and final interpreted epicentre.

Kraaijpoel & Dost (2013) investigated the influence of the high-velocity salt structure (the Zechstein Group) with an average, but highly variable thickness of 1 km that overlies the reservoir in the Groningen field. They found that the irregular salt layer causes significant defocusing of seismic energy and introduces mode conversions and multiples. Due to the increased complexity of the waveforms, phase identification is more uncertain and location accuracy is affected.

Seismicity outside the Groningen field was scattered among the NW-SE trending smaller fields in North Holland, around the Eleveld gas field near Assen and in the southeast of Drenthe (Roswinkel/Emmen). A swarm of

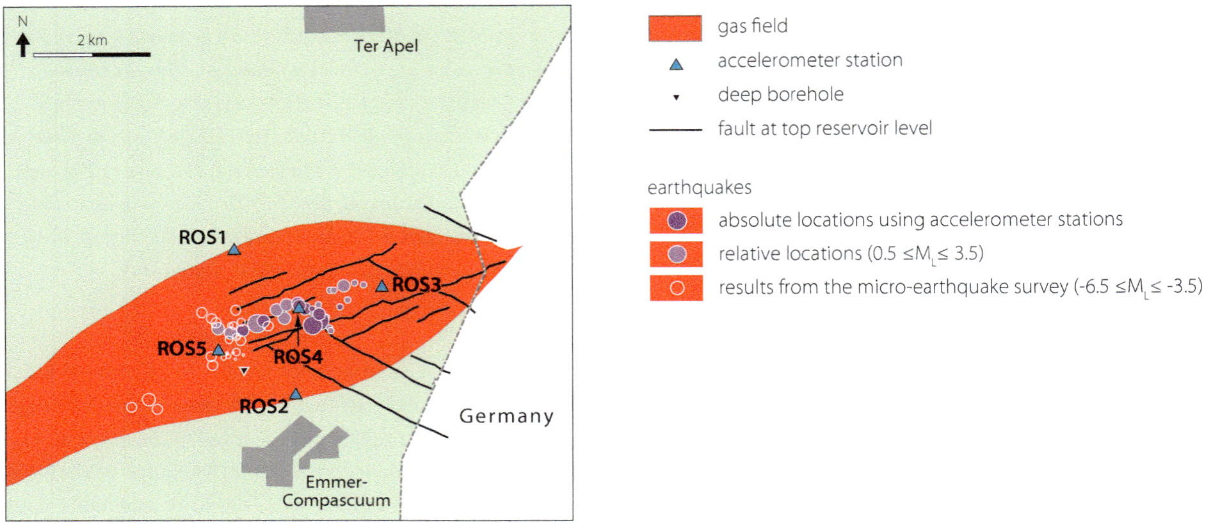

***Figure** 22.5. Map of the Roswinkel area; see Fig. 22.4 for location. Source: KNMI (seismicity data) and TNO (fault data).*

41 small events ($0.2 \leq M_L \leq 1.4$) occurred southwest of the Groningen field in 2009, close to the Midlaren gas field (Dost et al., 2012). At the same time a drilling operation where losses were reported was being carried out 4-5 km from the event location, which may have initiated the seismicity, since once drilling was terminated, the swarm also ended. Increased seismic activity was recorded near offshore gas fields west of Alkmaar/Bergen. In 2013 events were recorded near Castricum (CS in Fig. 22.4b), the two largest of them (M_L= 2.0 and 2.5) were felt in the area. Although the epicentres suggest a connection to the Castricum Sea gas field, uncertainty of the location is large (~ ±3 km in EW direction) and no unique cause could be found (TNO, 2016a). Production from the Castricum Sea gas field stopped in 2004, so the events could represent a delayed response. Losses had been reported a few months prior to the events from an onshore drilling site at 5-8 km distance, which may provide an alternative explanation and there may actually be a natural component as well. Analysis of deep borehole data near the Bergermeer gas storage facility showed microseismicity (M< 1.0) during the first phase of re-pressurization from 30 bar to 90 bar (2010-2015) with the largest two events at M= ~0.7 (Taqa, 2021). These microseismic events are connected to main faults in the reservoir (Baisch et al., 2016). In 2009 an event was recorded near De Hoeve in the province of Friesland (DH in Fig. 22.4b). The location was close to the Weststellingwerf- and Noordwolde gas fields. Although the Weststellingwerf gas field was not in production at the time, it was used for water injection (Dost et al., 2012), so although there is uncertainty in the location, this may have been a first event in the Netherlands related to water injection.

2014-present

Network development

After the Huizinge event in 2012 the need for more intensive monitoring in Groningen was felt. Plans were developed to reduce the distance between borehole stations in the network from an average of 20 km to 4-5 km. In 2015-2016 an additional 69 borehole stations were installed in the Groningen area (G-network, Dost et al., 2017). These borehole stations are all equipped with an additional accelerometer at the surface in free-field conditions. In addition, 10 free-field accelerometers were added west of the Groningen gas field during 2017 and 2018 and 4 broadband sensors (G81-84) were added at 100 m depth, co-located with existing G stations (G81/G22; G82/G39; G83/G56 and G84/G64). These broadband sensors pick up energy outside the frequency range of the geophones in the boreholes (<0.2 Hz), which is relevant for characterizing larger magnitude earthquakes.

Apart from this real-time monitoring network, other networks became operational. A temporary 'household' network of more than 300 accelerometers was installed in 2014-2016 in buildings spread over the Groningen gas field (Ntinalexis et al., 2019) and was operated by the Dutch technology institute TNO. In addition to these monitoring networks, a flexible network consisting of 450 portable geophones was installed as temporary dense arrays at various locations over the Groningen gas field, enabling detailed studies of shear-wave velocity characterization (e.g. Spica et al., 2018; Chmiel et al., 2019; Kruiver et al., 2021).

At the end of 2018, a calibration error was detected in the surface accelerometers of the G-network and subsequently corrected (Dost et al., 2019). The affected data were not yet used in the development of hazard models,

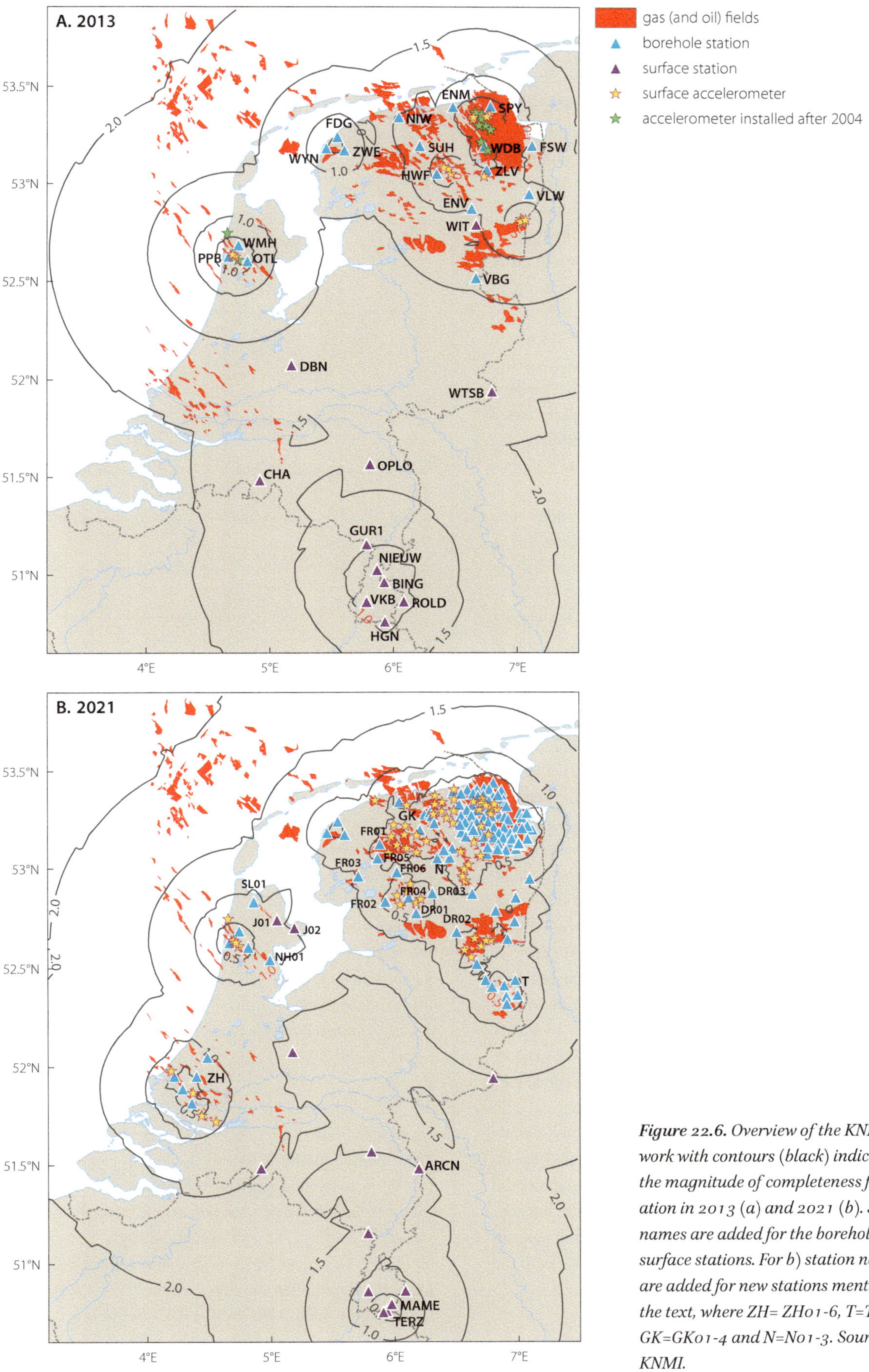

Figure 22.6. Overview of the KNMI network with contours (black) indicating the magnitude of completeness for situation in 2013 (a) and 2021 (b). Station names are added for the borehole and surface stations. For b) station names are added for new stations mentioned in the text, where ZH= ZH01-6, T=T01-8, GK=GK01-4 and N=N01-3. Source: KNMI.

but a recent study of the relation between local and moment magnitudes required correction (Dost et al., 2018, 2019).

The Mc contours for the extended network (Fig. 22.6b) show a significant improvement relative to 2013, lowering the values for the Groningen field from Mc=1.5 to 0.5. All gas fields are now located within the Mc=1.5 contour. An alternative way to estimate Mc is using the cumulative Frequency-Magnitude (FM) curve. This curve generally shows a linear relation for a certain magnitude range but deviation from linearity towards larger and lower M values. Mc can be estimated from the deviations from linearity at the lower magnitude range. Results for Groningen indicated a value of Mc=1.2 from 1996 to 2014, followed by a decrease to Mc=0.5 in 2016 (Paleja & Bierman, 2016; Dost et al., 2017), in line with the estimates based on average noise levels. An example of recordings in the new network is given in Figure 22.7 for the 2019 M= 3.4 Westerwijtwerd event.

Outside Groningen, borehole stations were added next to gas storage facilities (Norg, with station name N01-3, and Grijpskerk: GK01-4) and near smaller producing gas fields, mainly in Drenthe (DR01-3), Friesland (FR01-6) and in North Holland (J01-2, SL01 and NH01, see Fig. 22.6b). In the eastern part of Overijssel province, south of Drenthe, borehole stations were installed to monitor possible seismicity due to wastewater injections (T01-8). In North Holland two small events took place in 2015 outside the monitoring networks at ~5 km from the Slootdorp gas field. Additional stations, one borehole (SL01) and one surface station (J02), were installed to improve location accuracy in the region. More to the south, but also outside the monitoring network, a M_L= 2.5 event occurred close to the Middellie gas field in 2018, which led to the installation of borehole station NH01. In South Holland geothermal operations started and a regional borehole network was installed to lower the magnitude threshold (ZH01-6, Fig. 22.6b) and to establish the natural seismicity baseline. In addition to these surface and shallow borehole installations, whose data are delivered in real time to KNMI, other local surface networks have been installed by the industry. For geothermal operations, a local network was installed in 2014 around Venlo (Californië geothermal site) by Qcon (Baisch & Vörös, 2018). Buijze et al. (2019) made an inventory of seismicity related to geothermal operations worldwide and concluded that the seismic potential for operations in sandstone target formations is low, but moderate in faulted carbonate rocks. These findings can be used for future planning of monitoring networks in the Netherlands.

Evolution of seismicity

In January 2014 gas production in the central part of the Groningen field near Loppersum was strongly reduced. Van Thienen-Visser & Breunese (2015) suggested that seismic activity decreased in the centre of the field, but they concluded that it was too early to support this with statistical evidence. Later, a statistical study by Nepveu et al. (2016) concluded there is a preference for a decreasing activity rate after January 2014 and Paleja & Bierman (2016) show good evidence for such decrease in activity rate in the Loppersum area. For the entire field a decrease in activity rate was found for M ≥ 1.0, but not for M ≥ 1.5.

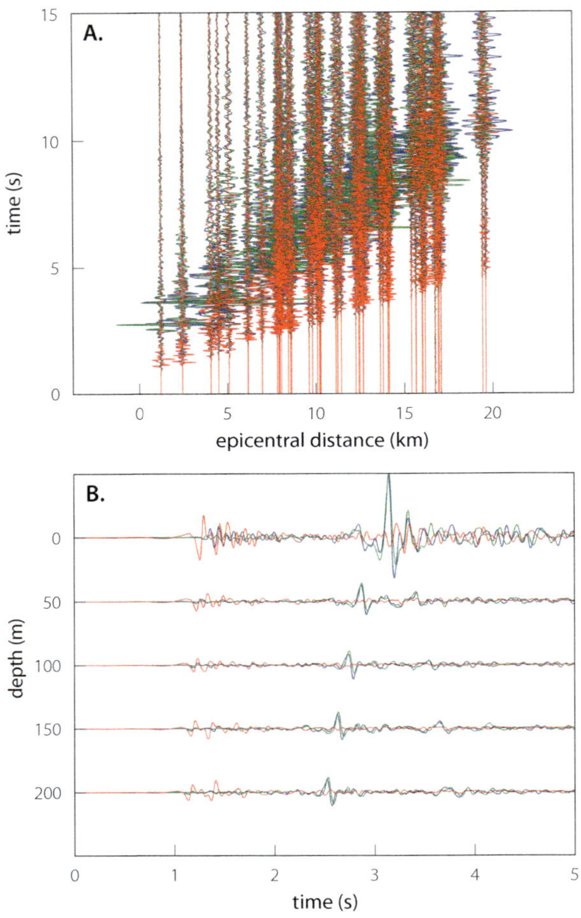

Figure 22.7. Seismic recordings of the Westerwijtwerd M=3.4 event on May 22, 2019. This was one of the largest earthquakes that was induced at the Groningen gas field. It was recorded over most of the KNMI seismic network. a) Three-component recording at 200 m depth over the G-network, within a 20 km epicentral distance. Normalization has been applied in order to show the waveforms at different distances with similar amplitudes. b) Three-component recording at the nearest borehole (G17) from the epicentre without normalization. The red, blue and green wiggles show the vertical, radial and transverse components, respectively. All seismograms have been band-pass filtered between 2 and 20 Hz. Source: KNMI.

Production was further reduced during the following years (Fig. 22.3b) and was eventually stopped on April 17th 2024. Seismicity in the Groningen gas field seems to be slowly declining since 2014, while continuing to show a large annual variation. Statistical analysis of the dataset up to 2021 shows an indication of a decrease in annual seismicity rate, but not yet a statistically significant decrease (KNMI, 2021). Temporal changes in seismicity were studied by Dost et al. (2017) by measuring changes in the slope of the FM curve (b-value) and the activity rate. Within the uncertainty boundaries, the b-value did not change over time, but the activity rate increased up until 2014. The centre of gravity of the seismic moment released per year moved from Loppersum in 2010-2014 towards the south-east in 2015 and 2016 (Dost et al., 2017).

Outside the Groningen gas field two small events were recorded in the north of North Holland near Anna Paulowna (AP in Fig. 22.4c). The epicentre was determined at ~5 km distance from a producing gas field. Due to the limited seismic network in the region, uncertainty in the location was at least 5 km in NW-SE direction. As with the Castricum Sea event, a direct connection with gas production could not be established (TNO, 2016b).

A consequence of the increasing station density is that small very shallow events were detected in the vicinity of salt domes and caverns near Heiligerlee and Zuidwending (Ruigrok et al., 2018, 2019). These events can be distinguished from deeper events and led to the construction of additional local monitoring networks that contribute to the real-time data collection at KNMI. The local network around Venlo captured a M_L= 1.7 induced event, which was located close to a geothermal site but at a much larger depth (4.5-6.5 km), well below the operations. Spetzler et al. (2018) re-assessed the location and showed that the inferred depth depends strongly on the Vp/Vs ratio used. A depth dependent ratio provides a shallow depth of 3.2 km, close to the geothermal operation.

Earthquake location and seismic structure

The smaller interstation spacing in Groningen, combined with the availability of a detailed 3D seismic velocity model (Romijn, 2017) allowed for a higher accuracy of the hypocentre location of events. Spetzler & Dost (2017) investigated the application of new location methods and found a significant improvement resulting in a hypocentre accuracy of better than 0.5 km. One remarkable result of this study was the possibility that small events (M < 1.0) are located in the Zechstein overburden. Further development of the location method, however, showed that these locations did not appear in the improved algorithms and should be viewed as artifacts. Fig. 22.8 shows the relocated seismicity related to the Groningen gas field for the period 2014-2021, which follows closely the known existing fault patterns. Willacy et al. (2018, 2019, 2020) applied full-waveform modelling using a 3D velocity model for Groningen to 200 events recorded in 2015-2018. Their results also show a close correspondence between hypocentre location and known faults at an even higher accuracy (~0.1 km in the horizontal plane and ~0.3 km in depth). Application of a probabilistic location method (Smith et al., 2020) gives similar results. In addition, Paap et al. (2018) demonstrated the feasibility of using full waveform 3D modelling for Groningen and found significant effects of the 3D geological structure, while Dando et al. (2019) showed the complexity of the waveforms recorded in the deep boreholes. Zhou & Paulssen (2017) also used the deep boreholes to derive P- and S-velocities from ambient noise. They also found ~4% azimuthal anisotropy at reservoir depth, as well as hints for temporal variations in P- and S-velocities using deconvolution interferometry (Zhou & Paulssen, 2020).

Source mechanism, relation to existing faults

If events are connected to existing faults, their moment tensor solution should also be in line with the fault geometries known from 3D seismic surveys in the area. Initial results for Groningen, based on first motion analysis, showed mainly normal faulting along steeply (60-70°) dipping faults with a strike of around 290°, close to the average strike and dip of faults in the region (Kraaijpoel & Dost, 2013). Willacy et al. (2018, 2019, 2020) calculated moment tensor solutions and also found dominant normal faulting and a strong correlation with existing faults in the gas reservoir. Kühn et al. (2022) demonstrated the applicability to Groningen of a probabilistic inversion technique, based on full waveform modelling and local 1D models derived from the 3D model. Results for events M > 2.0 recorded in the period 2016-2020 (Dost et al., 2020) show similar results as Willacy et al. (2018, 2019). In addition to the double couple solutions, a significant negative volume change was found for some events (Dost et al., 2020). These results allowed for a detailed interpretation of source location and movement along known faults in the Groningen gas field, which was facilitated by the improved interpretation of faults in the field and their extension into the Carboniferous (Kortekaas & Jaarsma, 2017). From the analysis of 13 M > 2.0 events (2016-2020) it was found that most events occur along faults with a small throw and that reactivated faults in the reservoir often continue downwards into the Carboniferous (Dost et al., 2020). For several events in Groningen, Ameri et al. (2020) found frequency-dependent rupture directivity effects, in good agreement with the existing faults in the reservoir.

Using data from the deep boreholes installed by NAM, Pickering (2015) determined accurate locations of small

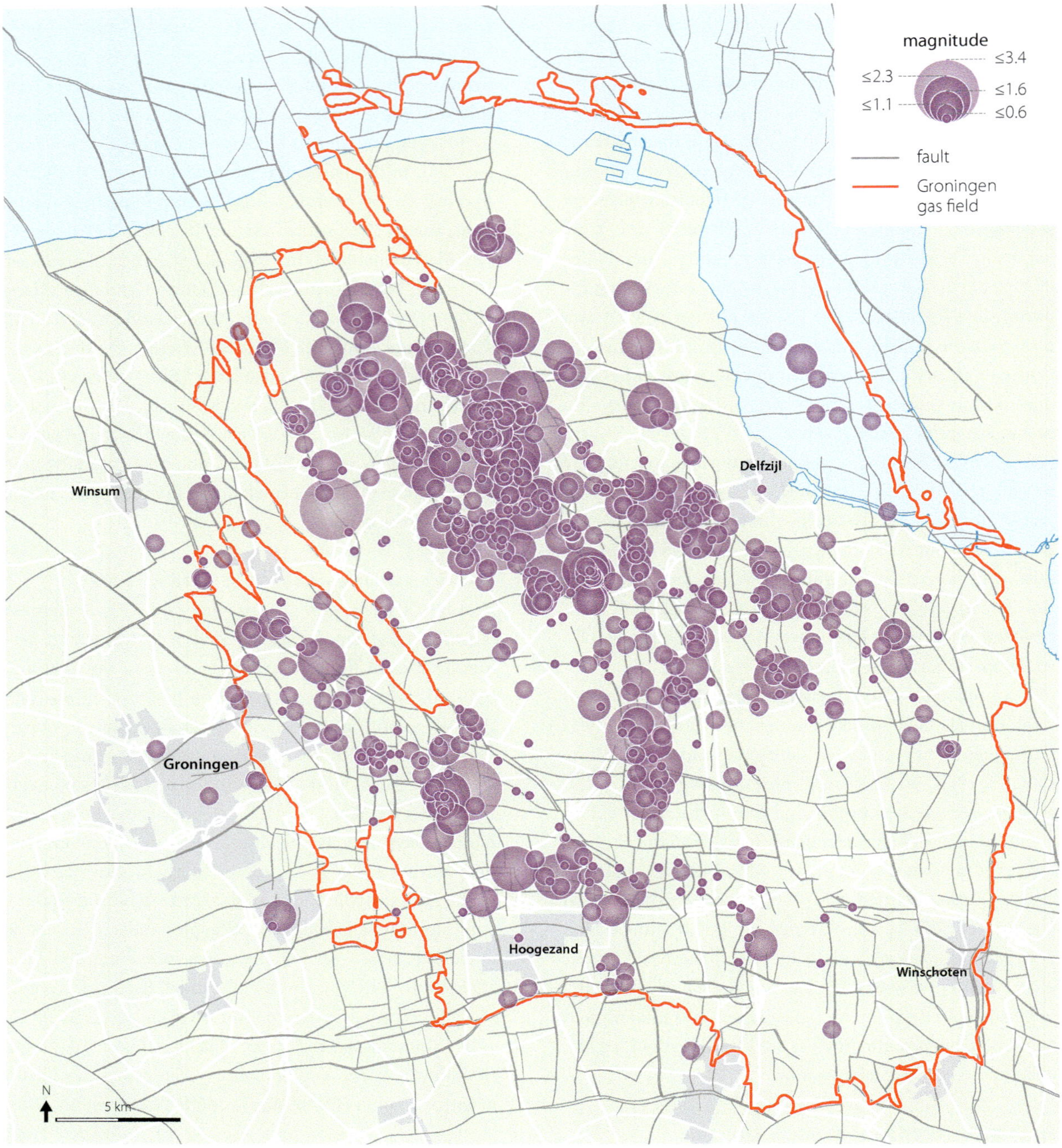

Figure 22.8. *Overview of relocated seismicity (01-2014 until 04-2022) related to the Groningen gas field (area bounded by red contours). Events shown as purple circles, scaled according to magnitude. Known faults shown as thin black lines. Source: KNMI (seismicity data) and TNO (fault data).*

local events and found that all events are confined to a depth within 100 m of the reservoir. This is an important finding, since the depth determination from inversion of shallow borehole data was uncertain and fixed at reservoir level in the inversion. Events could be analysed within a region of 2-3 km from the deep boreholes. Later, Willacy et al. (2019) also used the deep boreholes to locate events and determine their mechanism by full waveform modelling and so corroborated the findings of Pickering (2015).

Aftershocks

Important for hazard studies is the estimate of aftershocks in the dataset. Based on analysis of inter-event times and frequency-magnitude statistics, Luginbuhl et al. (2018) found an almost complete absence of aftershocks in Groningen, while using more complex models Bourne et al. (2018) and Post et al. (2021) demonstrated a 10-20% and 18-36% aftershock contribution, respectively for the same area. Muntendam-Bos (2020) made a distinction between repeated events occurring at short distances from parent events, but separated in time more than expected for foreshock sequences. She concluded that true foreshocks only account for ~5% of the Groningen dataset.

Geomechanical modelling and laboratory measurements

Geomechanical modelling studies provide insight in the causes for induced seismicity in an aseismic region (e.g., Roest & Kuilman, 1994; Roest et al., 1998; Buijze et al., 2017; Van Wees et al., 2017, 2018; Van den Bogert, 2018). The modelling is based on assumptions for rock and fault behaviour, which are often only poorly known at the pressures and temperatures at the depths where seismicity is generated. Wentinck (2018a,b) carried out a kinematic and dynamic rupture modelling for the January 8, 2018 M_L=3.4 Zeerijp event. Spiers et al. (2017) discuss a research programme where this knowledge gap is intended to be filled with special focus on the Groningen gas field. Hunfeld et al. (2017, 2019) report on velocity weakening in basal Zechstein deposits, resulting in a preference for slip at the top of the reservoir and a high potential for faults that cross-cut the caprock of the reservoir. Hunfeld et al. (2021) show that for the larger induced events in Groningen, dynamic weakening plays a role.

Seismotectonic framework

The induced earthquakes in the Netherlands seem to be correlated with faults within or at the boundaries of producing gas fields. A reactivation of existing zones of weakness is, as in the case of natural earthquakes, considered the most likely mechanism. Therefore, the tectonic framework is briefly described below.

In the northeastern Netherlands the tectonic structure is dominated by the Lauwerszee Trough, which is bounded by the NW-SE trending Hantum and West Groningen fault zones, which both contain faults that were active during late Carboniferous and early Permian times (RGD, 1995; Frikken, 1999; De Jager et al., 2025, this volume). Fig. 22.4a shows that the Groningen field is situated east of the Lauwerszee Trough on the Groningen Block, while the smaller fields to the west are near the Hantum Fault Zone. Within the Trough, a series of WSW-ENE trending faults shows dominantly strike-slip movements (Frikken, 1999). Over time, only a limited number of small events has occurred near one of these strike-slip faults. They are too small and data coverage is insufficient to allow source mechanism determination. Only a few events are located west of the Hantum Fault Zone. South of the Eleveld gas field the tectonic structure becomes dominated by the E-W trending faults of the Lower Saxony Basin, active in the Permian (De Jager et al., 2025, this volume).

The region around Alkmaar belongs structurally to the Central Netherlands Basin and offshore to the Broad Fourteens Basin. These basins form, together with the Roer Valley Graben, a set of Mesozoic extensional basins (De Jager et al., 2025, this volume).

Induced seismic hazard and risk

Up to the end of 2003, over 330 induced events have been recorded in the Netherlands. At that time a first Probabilistic Seismic Hazard Assessment (PSHA) for induced seismicity was carried out (Van Eck et al., 2004, 2006). Since 2003, such an assessment has been a requirement for each new mining concession in the Netherlands.

Important building blocks for the PSHA are an estimate of the seismic activity, derived from a FM relation for the region, a Ground Motion Prediction Equation (GMPE), providing estimates of spectral acceleration, Peak Ground Acceleration (PGA) and/or Peak Ground Velocity (PGV) for seismic events in a specified magnitude range, and an estimate of the maximum probable magnitude. Initially, induced seismicity showed a stationary trend and at first a simple GMPE for small-magnitude, shallow source events in the Netherlands was constructed in terms of PGA and PGV (Dost et al., 2004). The slope of the linear part of the FM distribution of all recorded induced events (b-value) was estimated at 0.8 and the maximum probable magnitude (M_{max}) of 3.9 was estimated for the Netherlands from Monte Carlo modelling of the FM curve (Van Eck et al., 2004, 2006).

An update of this study (Dost et al., 2012), based on a dataset of 640 events recorded up to the end of 2010, did not result in significant changes in the estimated hazard parameters except for the b-value, which was found to differ between the Groningen field (b=1.0) and the smaller gas fields (b=0.6). In addition, a first indication of non-stationarity of the seismicity of the Groningen field was found, and the authors concluded that this should be taken into account in future studies. Early 2013 Dost & Kraaijpoel (2013) and Muntendam-Bos & De Waal (2013) noted that in the expanding dataset an estimate of the M_{max} for Groningen could not be obtained anymore, due to the lack of a clear deviation from linearity in the FM curve at larger magnitudes. Comparison with literature studies on induced seismicity in comparable situations provided a new proposed value for M_{max} of 5.0 (Dost & Kraaijpoel, 2013).

After the M_L= 3.6 Huizinge event an international research programme was launched by NAM, the organization operating the Groningen gas field, to develop an integrated hazard and risk assessment for the Groningen field (Van Elk et al., 2017). This programme included the development of detailed reservoir models (e.g. Visser & Solano Viota, 2017), seismological models able to predict induced seismicity for specific production scenarios (Bourne &

Oates, 2017) and GMPE's (Bommer et al., 2017a,c), later called Ground Motion Models (GMMs). For the risk component a separate model was constructed, including an evaluation of the fragility of buildings in the region (Crowley et al., 2017; Van Elk et al., 2017, 2019). As models evolved over time, the seismicity database for Groningen expanded and knowledge increased. An overview of the results of hazard and risk calculations for the Groningen field can be found in Van Elk et al. (2017, 2019) and in annual reports to the Ministry of Economic Affairs.

Seismological model

The seismological model is designed to predict the future seismicity rate. For Groningen, development started with a model linking seismic moment to reservoir volume change (Bourne et al., 2014, 2015) and evolved to a complex statistical model of Coulomb stress failures (Bourne et al., 2018; Bourne & Oates, 2020). The latest versions accurately predict changes in seismicity rate with changes in production from increasing to decreasing production. These source models are implemented in the hazard and risk calculations. Limbeck et al. (2021) report on further development of this model. Alternative models have been developed and applied to the Groningen field (Dempsey & Suckale, 2017; Richter et al., 2020) or to parts of the Groningen field (Candela et al., 2019). The latter two models are based on a rate-and-state model.

The sudden decrease of production in the Loppersum area early 2014 allowed investigations of the effect of poroelastic pressure changes on seismicity. Postma and Jansen (2017) concluded that the effect of changes in production rates on the number or magnitude of earthquakes is small. In contrast, Van Wees et al. (2018) found that rates of pressure change are important in explaining the response of the medium in terms of seismicity. Kühn et al. (2022) give a review of all available models for Groningen and discuss future directions of development. These alternative source models, or parts of source models that need further development, are expected to be implemented in the hazard and risk calculations for Groningen in the near future.

Ground motion model

The Groningen Ground Motion Model (GMM) developed over the years from a simple model with average field-wide site amplification and based on point-source simulations to zone-specific nonlinear site amplification models and extended source simulations (Bommer et al., 2017a-c). In total 7 versions of the GMM for Groningen have been developed (GMM v1-v7). The latest version is described in Bommer et al. (2021). Details of the model development have been published, for instance on modelling of the shallow subsurface (Hofman et al., 2017; Kruiver et al., 2017a,b; Rodriguez-Marek et al., 2017; Noorlandt et al., 2018; Van Ginkel et al., 2019), simulations (Edwards et al., 2019), derivation of a damping model (Ruigrok et al., 2022) and ground motion model development (Bommer et al., 2016a, 2017a; Stafford et al., 2017, 2019). The development of GMMs for Groningen resulted in a significant reduction in the uncertainties and in a general reduction of the hazard and risk estimates.

Since the magnitudes of induced events were calculated as M_L, using a regional attenuation model, and GMM's are usually developed in moment magnitude M_w. Dost et al. (2018, 2019) investigated the relation between magnitudes M_L and M_w. For $M > 2.5$, which is the lower limit used in GMM development for Groningen, both magnitudes are almost identical and M_w=M_L was adopted. For smaller events, Dost et al. (2018, 2019) provide a conversion relation between these two metrics.

The model is based on recorded data from a limited magnitude range ($2.5 < M_L < 3.6$). Since the hazard calculations also include higher magnitudes, simulations have been carried out to estimate their contribution. One of the key parameters in the simulations is an estimate of the stress-drop. In GMM v5, stress-drop measurements were carried out after deconvolving the records to a reference level at 800 m depth and show large values (0.5-100 MPa; Bommer et al., 2017a). Ameri et al. (2020) estimated stress-drop parameters from spectral inversion of Groningen shallow borehole records. They found values in the range 0.1-1 MPa. In the latest version of the GMM (v7), Bommer et al. (2021) inverted stress parameters from surface sensors and found values comparable to Ameri et al. (2020), i.e. in the range 0.5-2.5 MPa.

Maximum magnitude

In 2016, a workshop was organized by NAM, inviting international experts to reassess M_{max} for Groningen. The outcome was a proposed distribution of values, which was subsequently implemented in PSHA calculations (Bommer & Van Elk, 2017). The distribution shows a peak between M_{max}= 4.5 and 5, but also allows values up to M= 7. Zoller & Holschneider (2016) and Beirlant et al. (2019) calculated the M_{max} for Groningen from a pure statistical perspective and found lower values of M_{max}= 4.4 and M_{max}= 3.85-4.5 (90% confidence level), respectively. Zöller & Holschneider (2016) explicitly mention that M_{max} is a constant physical parameter of the medium, while the maximum expected magnitude is time dependent.

Deterministic hazard analysis

Apart from the probabilistic seismic hazard approach for Groningen, a need for seismic hazard and risk assessment for the smaller gas fields was seen. Because of the lack of data, a deterministic approach was chosen with the aim of estimating the probability of earthquakes occurring before any events are recorded (Van Eijs et al., 2006). A correlation between reservoir parameters and observed seismicity was investigated. Although the number of observed induced earthquakes for the small gas fields was limited, three key parameters were identified: pressure drop, fault density of the reservoir and stiffness ratio between the overlying- and reservoir rock. An update of this study showed similar results (Van Thienen-Visser et al., 2012). Based on these studies, a guideline for qualitative seismic risk analysis for small gas fields was developed by State Supervision of Mines (Muntendam-Bos et al., 2015; SodM, 2016; Van Thienen-Visser et al., 2018).

Empirical relations

In addition to the complex PSHA model for Groningen, the need was felt to derive an easy-to-use empirical PGV relation for the field purely based on recorded events without simulation, thus restricting the magnitude to M≤ 3.6. A first version for 2.5≤ M ≤3.6 was published in 2016 (Bommer et al., 2016b), followed by an updated version one year later in which the threshold was lowered from M= 2.5 to M= 1.8 (Bommer et al., 2017b). The discovery of a calibration error in the G-network influenced the empirical PGV and a further update was issued in 2019 (Bommer et al., 2019). Both updates also include the extension of the database with newly recorded events. Bommer et al. (2021) updated the empirical PGV, now including a site amplification term.

After the reduction of production from the Groningen gas field and the foreseen rapid closure of the field, focus shifted to the smaller gas fields. Stakeholders wished to evaluate the possible damage in the neighbourhood of these fields. The Technical Committee on Ground Movement (TCBB, 2018) advised the government to install accelerometers close to them in order to assess possible damage, instead of making a representative measurement of the building quality before the start of gas production. In response, Ruigrok & Dost (2020) slightly modified the empirical PGV model for Groningen (Bommer et al., 2019) and designed a method for the construction of PGV confidence regions around induced events for the smaller gas fields.

Synthesis

In a low-seismicity region like the Netherlands, where moderate natural earthquakes occur at intervals of 50 to 100 years, long-term continuous monitoring is required to study details of fault behaviour. Due to the increased station quality and -density installed over the past 40 years, the earthquake hypocentre locations and mechanisms can now be studied in detail, not only in the south of the country where the seismicity is natural, but also for the induced events in the north. In both regions the seismicity is related to existing faults.

The 2012 M_L= 3.6 Huizinge event and the extension of the monitoring network in Groningen in 2015-2016 increased the scientific interest in induced seismicity in the Netherlands and has resulted in a large number of international publications, covering different aspects of our understanding of the causes and effects of induced seismicity in the region. Many of the results are built into the hazard and risk calculations that are required by the Mining Act and are evaluated by the regulator, who advises the Ministry of Economic Affairs on gas production for the Groningen gas field as well as in the smaller fields. Some key parameters in hazard calculations are difficult to quantify, such as M_{max}. For Groningen an expert judgement was made, resulting in a distribution of values, while for the smaller gas fields individual deterministic qualitative estimates are made. The knowledge gained from the Groningen-related studies can be used in other parts of the Netherlands. This is not limited to the smaller gas fields, but can also be applied to other mining activities, like salt mining, waste-water injection, geothermal operations and Carbon Capture and Storage (CCS) activities. Most of these other activities are located in relatively small areas, except for existing geothermal projects and prospects, which are widely distributed over the Netherlands. This is a challenge for the current monitoring network.

Catalogs of natural (tectonic) and induced seismicity for the Netherlands can be found on www.knmi.nl/kennisen-datacentrum/dataset/aardbevingscatalogus and access to the digital waveform data can be obtained through rdsa.knmi.nl/dataportal.

Acknowledgements

We thank Hanneke Paulssen, Kris Vanneste, Karin Van Thienen-Visser and the editor Jan de Jager for their detailed and constructive comments.

Digital map data

Spatial data of figures in this chapter for use in geographical information systems can be downloaded here: https://doi.org/10.5117/aup.28164503.

References

Ahorner, L., 1983. Historical Seismicity and Present-Day Micro-earthquake Activity of the Rhenish Massif, Central Europe. In: Fuchs, K. et al. (eds): Plateau Uplift. Springer Verlag (Heidelberg): 198-221. DOI: 10.1007/978-3-642-69219-2_27

Ahorner, L., 1985. The general pattern of seismotectonic dislocations in central Europe as the background for the Liege earthquake on November 8, 1983. In: Melchior, P. (ed.): Seismic Activity In Western Europe. NATO ASI series, Reidel (Dordrecht): 41-56. DOI: 10.1007/978-94-009-5273-7_4

Ahorner, L., 1994. Fault-plane solutions and source parameters of the 1992 Roermond, the Netherlands, mainshock and its stronger aftershocks from regional seismic data. Geologie en Mijnbouw 73: 199-214. DOI: 10.1023/A:1009779025108

Ahorner, L. & Pelzing, R., 1983. Seismotektonische Herdparameter von digital registrierten Erdbeben der Jahre 1981 und 1982 in der westlichen Niederrheinischen Bucht. Geologisches Jahrbuch. E26 : 35-63.

Alexandre, P., 1985. Catalogue des séismes survenus au Moyen Age en Belgique et dans les regions voisines. In: Melchior, P. (Ed.): Seismic activity in Western Europe. Reidel (Dordrecht): 189-203. DOI: 10.1007/978-94-009-5273-7_15

Alexandre, P., 1994. Historical seismicity of the lower Rhine and Meuse valleys from 600 to 1525: a new critical review. Geologie en Mijnbouw 73: 431-438. DOI: 10.1007/BF03019480

Alexandre, P., Kusman, D., Petermans, T. & Camelbeeck, T., 2008. The 18 September 1692 earthquake in the Belgian Ardenne and its aftershocks, In: J. Fréchet et al. (eds) Historical seismology: interdisciplinary studies of past and recent earthquakes, modern approaches in solid earth sciences. Springer: 209-230. DOI: 10.1007/978-1-4020-8222-1_10

Allman, A. & Smolka, A., 2001. Increasing Loss Potential in Earthquake Risk - A Reinsurance Perspective. In: Camelbeeck, T. (ed.): Proceedings of the Workshop: Evaluation of the Potential for large earthquakes in regions of present day low seismic activity in Europe, Luxemburg.

Ameri, G., Martin, C. & Oth, A., 2020. Ground-motion attenuation, stress drop, and directivity of induced events in the Groningen gas field by spectral inversion of borehole records. Bulletin of the Seismological Society of America 110: 2077-2094. DOI: 10.1785/0120200149

Baisch, S., Koch, C., Vörös, R. & Rothert, E., 2016. Induced seismicity in the Bergermeer field: Hypocenter relocation and interpretation. Qcon report for Taqa. Available at https://nl.taqa.com/publications

Baisch, S. & Vörös, R., 2018. Earthquakes near the Californie Geothermal site: August 2015-November 2018. Qcon GmbH, archive no. CLGG005: 41 pp.

Bakun, W. H. & Wentworth, C. M., 1997. Estimating earthquake location and magnitude from seismic intensity data. Bulletin of the Seismological Society of America. 87, 1502-1521. DOI: 10.1785/BSSA0870051502

Bakun, W. H. & Wentworth, C. M., 1999. Erratum to Estimating earthquake location and magnitude from seismic intensity data. Bulletin of the Seismological Society of America 89: 557. DOI: 10.1785/BSSA0890050557

Båth, M., 1979. Introduction to Seismology. Birkhauser Verlag (Basel): 428 pp.

Beirlant, J., Kijko, A., Reynkens, T., Einmahl, J.H.J., 2019. Estimating the maximum possible earthquake magnitude using extreme value methodology: the Groningen case. Natural Hazards 98: 1091-1113. DOI: 10.1007/s11069-017-3162-2

Berg, H., 1953. Das Rheinlandbeben bei Euskirchen vom 14. Maerz 1951. Geofisica Pura e Applicata 24: 1-12.

Berz, G., 1994. Assessment of the losses caused by the 1992 Roermond earthquake, the Netherlands (extended abstract). Geologie en Mijnbouw 73: 281.

Bommer, J.J. & van Elk, J., 2017. Comment on "The maximum possible and the maximum expected earthquake magnitude for production-induced earthquakes at the gas field in Groningen, the Netherlands" by Gert Zöller and Matthias Holschneider. Bulletin of the Seismological Society of America 107: 1564-1567. DOI: 10.1785/0120170020

Bommer J.J., Dost, B., Edwards, B., Stafford, P.J., van Elk, J., Doornhof, D. & Ntinalexis, M., 2016a. Developing an application-specific ground-motion model for induced seismicity. Bulletin of the Seismological Society of America 106: 158-173. DOI: 10.1785/0120150135

Bommer, J.J., Stafford, P.J. & Ntinalexis, M., 2016b. Empirical Ground-Motion Prediction Equations for Peak Ground Velocity from Small-Magnitude Earthquakes in the Groningen Field Using Multiple Definitions of the Horizontal Component of Motion. NAM report, November 2016: 17 pp.

Bommer, J.J., Stafford, P.J., Edwards, B., Dost, B., van Dedem, E., Rodriguez-Marek, A., Kruiver, P., van Elk, J., Doornhof, D. & Ntinalexis, M., 2017a. Framework for a ground-motion model for induced seismic hazard and risk analysis in the Groningen gas field, The Netherlands. Earthquake Spectra 33(2): 481-498. DOI: 10.1193/100916EQS202M

Bommer, J.J., Stafford, P.J. & Ntinalexis, M., 2017b. Empirical Ground-Motion Prediction Equations for Peak Ground Velocity from Small-Magnitude Earthquakes in the Groningen Field Using Multiple Definitions of the Horizontal Component of Motion: Updated Model for Application to Smaller Earthquakes. NAM report, November 2017: 16 pp.

Bommer, J.J., Dost, B., Edwards, B., Kruiver, P.P., Ntinalexis, M., Rodriguez-Marek, A., Stafford, P.J. & van Elk, J., 2017c. Developing a model for the prediction of ground motions due to earthquakes in the Groningen gas field. Netherlands Journal of Geosciences 96: s203-s213. DOI: 10.1017/njg.2017.27

Bommer, J.J., Stafford, P.J. & Ntinalexis, M., 2019b. Updated Empirical GMPEs for PGV from Groningen Earthquakes. Rev 1, 4 March 2019: 15 pp.

Bommer, J.J., Edwards, B., Kruiver, P.P., Rodriguez-Marek, A., Stafford, P.J., Dost, B. & Ntinalexis, M., Ruigrok, E. & Spetzler, J.,

2021. V7 ground-motion model for induced seismicity in the Groningen gas field; with assurance letter, NAM report, Van Elk, J. & Uilenreef, J. (eds): 268 pp.

Bourne, S.J. & Oates, S.J., 2017. Development of statistical geomechanical models for forecasting seismicity induced by gas production from the Groningen field. Netherlands Journal of Geosciences 96: s175-s182. DOI: 10.1017/njg.2017.24

Bourne, S. J. & Oates, S. J., 2020. Stress-dependent magnitudes of induced earthquakes in the Groningen gas field. Journal of Geophysical Research: Solid Earth 125: e2020JB020013. DOI: 10.1029/2020JB020013

Bourne, S.J., Oates, S.J., van Elk, J. & Doornhof, D., 2014. A seismological model for earthquakes induced by fluid extraction from a subsurface reservoir, Journal of Geophysical Research 199: 8891-9015. DOI: 10.1002/2014JB011063

Bourne, S. J., Oates, S.J., Bommer, J.J., Dost, B., Van Elk, J. & Doornhof, D., 2015. A Monte Carlo method for probabilistic hazard assessment of induced seismicity due to conventional natural gas production. Bulletin of the Seismological Society of America 105: 1721-1738. DOI: 10.1785/0120140290

Bourne, S.J., Oates, S.J. & van Elk, J., 2018. The exponential rise of induced seismicity with increasing stress levels in the Groningen gas field and its implications for controlling seismic risk. Geophysical Journal International 213: 1693-1700. DOI: 10.1093/gji/ggx575

Bouwkamp, J., 1994. The 1992 Roermond earthquake, the Netherlands: earthquake engineering, Geologie en Mijnbouw 73: 291-298.

Brandes, C., Plenefisch, T., Tanner, D.C., Gestermann, N. & Steffen, H., 2019. Evaluation of deep crustal earthquakes in northern Germany - Possible tectonic causes. Terra Nova 31: 83-93. DOI: 10.1111/ter.12372

Buijze, L., Van den Bogert, P.A.J., Wassing, B.B.T., Orlic, B. & ten Veen, J., 2017. Fault reactivation mechanisms and dynamic rupture modelling of depletion-induced seismic events in a Rotliegend gas reservoir. Netherlands Journal of Geosciences 96: s131-s148. DOI: 10.1017/njg.2-17.27

Buijze, L., van Bijsterveldt, L., Cremer, H., Paap, B., Veldkamp, H., Wassing,, B.B.T., van Wees, J-D., van Yperen, G.C.N., ter Heege, J.H. & Jaarsma, B., 2019. Review of induced seismicity in geothermal systems worldwide and implications for geothermal systems in the Netherlands. Netherlands Journal of Geosciences 98, e13. DOI: 10.1017/njg.2019.6

Camelbeeck, T., 1985. Some notes concerning the seismicity in Belgium - Magnitude scale - detection capability of the Belgian seismological stations. In: Melchior, P. (ed.): Seismic Activity in Western Europe. NATO ASI series, Reidel (Dordrecht): 99-108.

Camelbeeck, T., 1994. Mecanisme au foyer des tremblements de terre et contraintes tectoniques: le cas de la zone intraplaque belge. PhD thesis University of Louvain (Belgium): 344 pp.

Camelbeeck, T. & De Becker, M., 1985. The earthquakes of Liege of November 8, 1983 and December 21, 1965. In: Melchior, P. (ed.): Seismic Activity in Western Europe. NATO ASI series, Reidel (Dordrecht): 233-248. DOI: 10.1007/978-94-009-5273-7_19

Camelbeeck, T. & Meghraoui, M., 1998. Geological and geophysical evidence for large paleo-earthquakes with surface faulting in the Roer Graben (northwest Europe). Geophysical Journal International 132: 347-362. DOI: 10.1046/j.1365-246X.1998.00428.x

Camelbeeck, T. & Van Eck, T., 1994. The Roer Valley Graben earthquake of 13 April 1992 and its seismotectonic setting. Terra Nova 6: 291-300.

Camelbeeck, T., Van Eck, T., Pelzing, R., Ahorner, L., Loohuis, J., Haak, H.W., Hoang-trong, P. & Hollnack, D., 1994. The 1992 Roermond earthquake, the Netherlands, and its aftershocks. Geologie en Mijnbouw 73: 181-197.

Camelbeeck, T., Vanneste, K., Alexandre, P., Verbeek, K., Petermans, T., Rosset, P., Everaerts, M., Warnant, R. & Van Camp, M., 2007. Relevance of active faulting and seismicity studies to assessments of long-term earthquake activity and maximum magnitude in intraplate northwest Europe, between the Lower Rhine Embayment and the North Sea. In: Stein, S. and Mazotti, S. (eds): Continental Intraplate Earthquakes: Science, Hazard and Policy Issues. Geological Society of America Special Paper 425: 193-224. DOI: 10.1130/2007.2425(14)

Camelbeeck, T., Knuts, E., Alexandre, P., Lecocq, T. & Van Noten, K., 2021. The 23 February 1828 Belgian earthquake: a destructive moderate event typical of the seismic activity in Western Europe, Journal of Seismology 25: 369-391. DOI: 10.1007/s10950-020-09977-6

Candela, T., Osinga, S., Ampuero, J.-P., Wassing, B., Pluymaekers, M., Fokker, P.A., van Wees, J.-D., de Waal, H.A. & Muntendam-Bos, A.G., 2019. Depletion-induced seismicity at the Groningen gas field: Coulomb rate-and-state models including differential compaction effect. Journal of Geophysical Research: Solid Earth 124: 7081-7104. DOI: 10.1029/2018JB016670

Caro Cuenca, M., Hooper, A.J. & Hanssen, R.F., 2013. Surface deformation induced by water influx in the abandoned coal mines in Limburg, The Netherlands observed by satellite radar interferometry. Journal of Applied Geophysics 88: 1-11. DOI: 10.1016/j.jappgeo.2012.10.003

Carter, J.A., Barstow, N., Pomeroy, P.W., Chael, E.P. & Leahy, P.A., 1991. High-Frequency Seimic Noise as a Function of Depth. Bulletin of the Seismological Society of America 81: 1101-1114.

Chmiel, M., Mordret, A., Boué, P., Brenguier, F., Lecocq, T., Courbis, R., Hollis, D., Campman, X., Romijn, R., & Van der Veen, W., 2019. Ambient noise multimode Rayleigh and Love wave tomography to determine the shear velocity structure above the Groningen gas field, Geophysical Journal International 218: 1781-1795. DOI: 10.1093/gji/ggz237

Cloetingh, S. & Burov, E., 1996. Thermomechanical structure of European continental lithosphere: Constraints from rheological profiles and EET estimates. Geophysical Journal International 124: 695-723.

Crowley, H., Pinho, R., Polidoro, B., & Van Elk, J., 2017. Developing fragility and consequence models for buildings in the Groningen field, Netherlands Journal of Geosciences 96: s247-s257.

Dando, B.D.E., Oye, V., Näsholm, S.P., Zülsdorff, L., Kühn, D. & Wuestefeld, A., 2019. Complexity in microseismic phase identification: full waveform modelling, traveltime computations and implications for event locations within the Groningen gas field, Geophysical Journal International 217: 620-649.

De Crook, Th., 1993. Probabilistic seismic hazard assessment for the Netherlands. Geologie en Mijnbouw 72: 1-13.

De Crook, Th., 1994. Earthquake hazard for Roermond, the Netherlands. Geologie en Mijnbouw 73: 425-429.

De Crook, Th., 1996. A seismic zoning map confirming to Eurocode 8, and practical earthquake relations for the Netherlands. Geologie en Mijnbouw 75: 11-18.

De Jager, J., 2007. Geological development. In: Wong, Th.E., Batjes, D.A.J. & De Jager, J. (eds): Geology of the Netherlands. Royal Netherlands Academy of Arts and Sciences (KNAW, Amsterdam): 5-26.

De Jager, J., Van Ojik, K. & Smit, J., 2025. Geological development. In: Ten Veen, J.H., Vis, G.-J., De Jager, J. & Wong, Th.E. (eds): Geology of the Netherlands, second edition. Amsterdam University Press (Amsterdam): 21-51. DOI: 10.5117/9789463728362_ch01

De Vos, D., 2010. Probabilisitic Seismic Hazard Assessment for the Southern part of the Netherlands. MSc thesis Utrecht University (the Netherlands): 61 pp.

DEKORP Research Group, 1991. Results of the DEKORP 1 (BELCORP-DEKORP) deep seismic reflection studies in the western part of the Rhenish Massif. Geophysical Journal International 106: 203-228. DOI: 10.1111/j.1365-246X.1991.tb04612.x

Demanet, D., Evers, L.G., Teerlynck, H., Dost, B. & Jongmans, D., 2001. Geophysical investigation across the Peel boundary fault (The Netherlands) for a paleoseismological study. Netherlands Journal of Geosciences / Geologie en Mijnbouw 80 (3/4): 119-127. DOI: 10.1017/S0016774600023787

Dempsey, D. & Suckale, J., 2017. Physics based forecasting of induced seismicity at Groningen gas field, the Netherlands, Geophysical Research Letters 44: 7773-7782. DOI: 10.1002/2017GL073878

Dirkzwager, J.B., 2002. Tectonic modelling of vertical motion and its near surface expression in the Netherlands. Ph. D. thesis Vrije Universiteit Amsterdam (the Netherlands): 156 pp.

Dirkzwager, J.B., Van Wees, J.D., Cloetingh, S.A.P.L., Geluk, M.C., Dost, B. & Beekman, F., 2000. Geo-mechanical and rheological modelling of upper crustal faults and their near surface expression in the Netherlands. Global and Planetary Change 27: 67-88. DOI: 10.1016/S0921-8181(00)00009-000009-0)

Dost, B. & Haak, H.W., 2002. A comprehensive description of the KNMI seismological instrumentation. KNMI (De Bilt), Technical Report 245: 60 pp.

Dost, B. and Kraaijpoel, D., 2013. The August 16, 2012 earthquake near Huizinge (Groningen), KNMI (De Bilt), Scientific Report: 26 pp.

Dost, B., Van Eck, T. & Haak, H., 2004. Scaling of peak ground acceleration and peak ground velocity recorded in the Netherlands. Bollettino di Geofisica Teorica ed Applicata 45: 153- 168.

Dost, B., Van Wettum, A. & Nolet, G., 1984. The NARS Array. Geologie en Mijnbouw 63: 381-386.

Dost, B., Goutbeek, F., van Eck, T. & Kraaijpoel, D., 2012. Monitoring induced seismicity in the
North of the Netherlands: status report 2010, KNMI (De Bilt), Scientific report WR 2012-03: 46 pp.

Dost, B., Ruigrok, E. and Spetzler, J., 2017. Development of seismicity and probabilistic hazard assessment for the Groningen gas field, Netherlands Journal of Geosciences 96: s235-s245. DOI: 10.1017/njg.2017.20

Dost, B., Edwards, B., and Bommer J.J., 2018. The relationship between M and ML: A review and application to induced seismicity in the Groningen gas field, the Netherlands, Seismological Research Letters 89: 1062-1074. DOI: 10.1785/02201700247

Dost, B., Edwards, B. & Bommer, J.J., 2019. Erratum: The relationship between M and ML: A review and application to induced seismicity in the Groningen gas field, the Netherlands. Seismological Research Letters 90(4): 1660-1662. DOI: 10.1785/0220190062

Dost, B., Van Stiphout, A., Kühn, D., Kortekaas, M., Ruigrok, E., and Heimann, S., 2020. Probabilistic 910 Moment Tensor Inversion for Hydrocarbon-Induced Seismicity in the Groningen Gas Field, the Netherlands, Part 2: Application. Bulletin of the Seismological Society of America 110 (5): 2112-2123. DOI: DOI: 10.1785/0120200076.

Duin, E., Rijkers, R. & Remmelts, G., 1995. Deep seismic reflections in the Netherlands, an overview. Geologie en Mijnbouw 74: 191-197.

Edwards B., Zurek, B., van Dedem, E., Stafford, P.J., Oates, S.,van Elk, J., de Martin, B., and Bommer, J.J., 2019. Simulations for the development of a ground motion model for induced seismicity in the Groningen gas field, The Netherlands. Bulletin of Earthquake Engineering 17: 4441- 4456. DOI: 10.1007/s10518-019-00605-5

Escher, B.G., 1940. Algemeene Geologie. N.V. Wereldbibliotheek (Amsterdam): 525 pp.

Frikken, H.W., 1999. Reservoir-geological aspects of productivity and connectivity of gas fields in the Netherlands. PhD thesis Technische Universiteit Delft (the Netherlands) (the Netherlands): 92 pp.

Gees, R.-H., 1937. Die Wellenausbreitung der Erdbeben vom 20. November 1932 (Nordbrabant) und 7. Juni 1931 (Doggerbank). Zeitschrift fur Geophysik 13: 159-179.

Geluk, M.C., Duin, E.J.Th., Dusar, M., Rijkers, R.H.B., Van den Berg, M.W. & Van Rooijen, P., 1994. Stratigraphy and tectonics of the Roer Valley Graben. Geologie en Mijnbouw 73(2-4): 129-141.

Goes, S., Loohuis, J.J.P., Wortel, M.J.R. & Govers, R., 2000. The effect of plate stresses and shallow mantle temperatures on tectonics of northwestern Europe. Global and Planetary Change 27: 23-38. DOI: 10.1016/S0921-8181(00)00003-700003-7)

Grasso, J.-R. & Wittlinger, G., 1990. Ten years of monitoring over a gas field. Bulletin of the Seismological Society of America 80: 450-473. DOI: 10.1785/BSSA0800020450

Gruenthal, G. & Stromeyer, D., 1992. The recent crustal stress field in central Europe: trajectories and finite element modelling. Journal of Geophysical Research 97: 11805-11820. DOI: 10.1029/92JB00543

Haak, H.W., 1985. The seismic event of November 9, 1983. Earthquake or sonic boom, In: P. Melchior (ed.) Seismic Activity in Western Europe, Springer Dordrecht: 215-222.

Haak, H.W. (ed.), 1993. Summary of the Final report on a multidisciplinary study of the relation between Gasproduction and earthquakes in the northern part of the Netherlands. KNMI (De Bilt): 16 pp.

Haak, H.W., Van Bodegraven, J.A., Sleeman, R., Verbeiren, R., Ahorner, L.,Meidow, H., Gruenthal, G.,Hoang-Trong, P.,Musson, R.M.W., Henni, P., Schenkova, Z. & Zimova, R., 1994. The macroseismic map of the 1992 Roermond earthquake, the Netherlands. Geologie en Mijnbouw 73: 265-270.

Haak, H.W., Dost, B. & Goutbeek, F.H., 2001. Seismische analyse van de aardbevingen bij Alkmaar op 9 en 10 september en Bergen aan Zee op 10 october 2001. KNMI (De Bilt), Technical Report 239: 24 pp.

Hance, L., Dejonghe, L., Ghysel, P., Laloux, M. & Mansy, J.L., 1999. Influence of heterogeneous lithostructural layering on orogenic deformation in the Variscan Front Zone (eastern Belgium). Tectonophysics 309: 161-177. DOI: 10.1016/S0040-1951(99)00136-000136-0)

Hanks, T.C. & Kanamori, H., 1979. A moment magnitude scale. Journal of Geophysical Research 84: 2348-2350. DOI: 10.1029/JB084iB05p02348

Hinzen, K.-G., 2003. Stress field in the Northern Rhine area, Central Europe, from earthquake fault plane solutions. Tectonophysics 377: 325-356. DOI: 10.1016/S0040-1951(03)00106-000106-0)

Hinzen, K.-G. , 2019. Engineering seismological parameters of two ML 4.3 intraplate earthquakes in the northern rhine area, Germany, Seismological Research Letters 90: 1258-1267. DOI: 10.1785/0220180265

Hinzen, K.-G. & Oemisch, M., 2001. Location and Magnitude from Seismic Intensity Data of Recent and Historic Earthquakes in the Northern Rhine Area, Central Europe. Bulletin of the Seismological Society of America 91: 40-56. DOI: 10.1785/012000007

Hinzen, K.-G. and Reamer, S. K., 2007. Seismicity, seismotectonics, and seismic hazard in the northern Rhine area. In: Stein, S. and Mazotti, S. (eds): Continental Intraplate Earthquakes: Science, Hazard and Policy Issues. Geological Society of America Special Paper 425: 225-242. DOI:10.1130/2007.2425(15)

Hinzen, K.-G., Reamer, S.K. & Fleischer, C., 2021. Seismicity in the northern Rhine area (1995-2018), Journal of Seismology 25: 351-367. DOI: 10.1007/s10950-021-09909-5

Hofman, L.J., Ruigrok, E., Dost, B., and Paulssen, H., 2017. A shallow seismic velocity model for the Groningen area in the Netherlands. Journal of Geophysical Research: Solid Earth 122: 8035-8050. DOI: 10.1002/2017JB014049

Houtgast, G., 1991. Catalogus Aardbevingen in Nederland. KNMI (De Bilt): 166 pp.

Houtgast, G., Van Bodegraven, J.A. & Ritsema, A.R., 1987. Historical seismograms from the Netherlands. Gerlands Beitrage zur Geophysik 96: 385-394.

Houtgast, R.F., 2003. Quaternary tectonic and fluvial evolution of the Roer Valley Rift System, the Netherlands. PhD thesis Vrije Universiteit Amsterdam (the Netherlands): 176 pp.

Houtgast, R.F. & Van Balen, R.T., 2000. Neotectonics of the Roer Valley Rift System, the Netherlands. Global and Planetary Change 27 (1-4): 131-146. DOI: 10.1016/S0921-8181(01)00063-7

Hunfeld, L.B., Niemeijer, A.R. & Spiers, C.J., 2017. Frictional properties of simulated fault gouges from the Seismogenic Groningen gas field under in-situ P-T-chemical conditions. Journal of Geophysical Research: Solid Earth 122: 8969-8989. DOI: 10.1002/2017JB014876

Hunfeld, L.B., Chen, J., Niemeijer, A.R. & Spiers, C.J., 2019. Temperature and gas/brine content affect seismogenic potential of simulated fault gauges derived from Groningen gas field caprock. Geochemistry, Geophysics, Geosystems 20: 2827-2847. DOI: 10.1029/2019GC008221

Hunfeld, L. B., Chen, J., Niemeijer, A.R., Ma, S. & Spiers, C.J., 2021. Seismic slip-pulse experiments simulate induced earthquake rupture in the Groningen gas field. Geophysical Research Letters 48: e2021GL092417. DOI: 10.1029/2021GL092417

Jagt, L., Ruigrok, E., and Paulssen, H., 2017. Relocation of clustered earthquakes in the Groningen gas field. Netherlands Journal of Geosciences 96: s163-s173.

Jongmans, W.J. & Van Waterschoot van der Gracht, W.J.A.M., 1932. Enkele voorlopige beschouwingen omtrent oorzaak en beteekenis van de in November 1932 in Nederland waargenomen aardbevingen. In: Jaarverslag over 1931 van het Geologisch Bureau voor het Nederlandsche Mijngebied te Heerlen: 51-53.

KNMI, 1994a. Seismische analyse van de aardbeving bij Alkmaar op 6 augustus 1994. KNMI (De Bilt), Technical Report 166: 19 pp.

KNMI, 1994b. Seismische analyse van de aardbeving bij Alkmaar op 21 september 1994. KNMI (De Bilt), Technical Report 167: 26 pp.

KNMI, 2021. Advies met betrekking tot de voorgestelde operationele strategie Groningenveld gasjaar 2021-22. Retreived from https://www.rvo.nl/sites/default/files/2021/06/Advies-KNMI-gaswinning-Groningen-vaststellingsbesluit-2021-2022.pdf

Knuts, E., Camelbeeck, T. & Alexandre, P. , 2016. The 3 December

1828 moderate earthquake at the border between Belgium and Germany. Journal of Seismology 20: 419-437. DOI: 0.1007/s10950-015-9535-7

Kortekaas, M. & Jaarsma, B., 2017. Improved definition of faults in the Groningen field using seismic attributes. Netherlands Journal of Geosciences 96(5): s71-s85.

Kraaijpoel, D. & Dost, B., 2013. Implications of salt-related propagation and mode conversion effects on the analysis of induced seismicity. Journal of Seismology 17: 95-107.

Kruiver, P.P., van Dedem, E., Romijn, R., de Lange, G., Korff, M., Stafleu, J., Gunnink, J.L., Rodriguez-Marek, A., Bommer, J.J., van Elk, J. & Doornhof, D., 2017a. An integrated shear-wave velocity model for the Groningen gas field. The Netherlands. Bulletin of Earthquake Engineering 15: 3555-3580. DOI: 10.1007/s10518-017-0105-y

Kruiver, P.P., Wiersma, A., Kloosterman, F.H., de Lange, G., Korff, M., Stafleu, J., Busschers, F.S., Harting, R., Gunnink, J.L., Green, R.A., van Elk, J. & Doornhof, D., 2017b. Characterisation of the Groningen subsurface for seismic hazard and risk modelling. Netherlands Journal of Geosciences 96: s215-s233. DOI: 10.1017/njg.2017.11

Kruiver, P.P., Pefkos, M., Meijles, E., Aalbersberg, G., Campmans, X., van der Veen, W., Martin, A., Ooms-Asshoff, K., Bommer, J.J., Rodriguez-Marek, A., Pinho, R., Crowley, H., Cavalieri, F., Correia, A.A. & Van Elk, J. , 2021. Incorporating dwelling mounds into induced seismic risk analysis for the Groningen gas field in the Netherlands. Bulletin of Earthquake Engineering: DOI:v10.1007/s10518-021-01225-7

Kühn, D., Hainzl, S., Dahm, T., Richter, G. & Vera Rodriguez, I., 2022, A review of source models to further the understanding of the seismicity of the Groningen field. Netherlands Journal of Geosciences. 2022;101: e11. DOI:10.1017/njg.2022.7

Limbeck, J., Bisdom, K., Lanz, F. , Park, T., Barbaro, E., Bourne, S., Kiraly, F., Bierman, S., Harris, C., Nevenzeel, K., Den Bezemer, T. and Van Elk, J., 2021. Using machine learning for model benchmarking and forecasting of depletion-induced seismicity in the Groningen gas field. Computational Geosciences 25: 529-551. DOI: 10.1007/s10596-020-10023-0

Luginbuhl, M., Rundle, J.B. and Turcotte, D.L., 2018. Natural time and nowcasting induced seismicity at the Groningen gas field in the Netherlands. Geophysical Journal International 215: 753-759. DOI: 10.1093/gji/ggy315

Mansy, J.L., Everaerts, M. & De Vos, W., 1999. Structural analysis of the adjacent Acadian and Variscan fold belts in Belgium and northern France from geophysical and geological evidence. Tectonophysics 309(1-4): 99-116. DOI: 10.1016/S0040-1951(99)00134-1

Mayer, G., Mai, P.M., Plenefisch, T., Echtler, H., Lueschen, E., Wehrle, V., Mueller, B., Bonjer, K.-P., Prodehl, C. & Fuchs, K., 1997. The deep crust of the Southern Rhine Graben: reflectivity and seismicity as images of dynamic processes. Tectonophysics 275: 15-40. DOI: 10.1016/S0040-1951(97)00014-000014-0)

McCalpin, J.P., 1996. Paleoseismology. Academic Press (San Diego): 583 pp.

Mercalli, G., 1902. Sulle modificazioni proposte alla scala sismica De Rossi-Forel. Società Tipografica Modenese.

Michon, L., Van Balen, R., Merle, O. and Pagnier, H., 2003. The Cenozoic evolution of the Roer Valley Rift System integrated at a European scale. Tectonophysics 367 (1-2): 101-126. DOI: 10.1016/S0040-1951(03)00132-X00132-X

Müller, B., Wehrle, V., Zeyen, H. & Fuchs, K., 1997. Short-scale variations of tectonic regimes in the western European stress province north of the Alps and Pyrenees. Tectonophysics 275: 199-219. DOI: 10.1016/S0040-1951(97)00015-100015-1)

Muntendam-Bos, A.G., 2020. Clustering characteristics of gas-extraction induced seismicity in the Groningen gas field, Geophysical Journal International 221: 879-892. DOI: 10.1093/gji/ggz525

Muntendam-Bos, A.G. & de Waal, J.A., 2013. Reassessment of the probability of higher magnitude earthquakes in the Groningen gas field. State Supervison of Mines, report: 35 pp.

Muntendam-Bos, A.G., Roest, J.P.A. & de Waal, J.A., 2015. A guideline for assessing seismic risk induced by gas extraction in the Netherlands. The Leading Edge 43: 672-677., DOI: 10.1190/tle34060672.1

Muntendam-Bos, A.G., Hoedeman, G., Polychronopoulou, K., Draganov, D., Weemstra, C., van der Zee, W., Bakker, R.R. & Roest, H., 2022. An overview of induced seismicity in the Netherlands. Netherlands Journal of Geosciences 101: e1. DOI:10.1017/njg.2021.14

Nepveu, M., Van Thienen-Visser, K. & Sijacic, D., 2016. Statistics of seismic events at the Groningen field. Bulletin of Earthquake Engineering, 14: 3343-3362. DOI: 10.1007/s10518-016-0044-5

NITG-TNO, 1999. Geological atlas of the subsurface of the Netherlands (1 : 250,000) - Explanation to Map sheet XV: Sittard-Maastricht. Netherlands Institute of Applied Geoscience TNO - National Geological Survey (Utrecht): 127 pp.

Noorlandt, R., Kruiver, P.P., de Kleine, M.P.E., Karaoulis, M., de Lange, G., Di Matteo, A., von Ketelhodt, J., Ruigrok, E., Edwards, B., Rodriguez-Marek, A., Bommer, J.J., van Elk, J. & Doornhof, D., 2018. Characterisation of ground motion recording stations in the Groningen gas field. Journal of Seismology 22: 605-623. DOI: 10.1007/s10950-017-9725-6

Ntinalexis, M., Bommer, J.J., Ruigrok, E. , Edwards, B., Pinho, R., Dost, B., Correia, A.A., Uilenreef, J., Stafford, P.J. & Van Elk , J., 2019. Ground-motion networks in the Groningen field: usability and consistency of surface recordings. Journal of Seismology 23: 1233-1253. DOI: 10.1007/s10950-019-09870-x

Paap, B., Kraaijpoel, D., Bakker, M. & Gharti, H.M., 2018. Wave propagation modelling of induced earthquakes at the Groningen gas production site. Geophysical Journal International 214: 1947-1960. DOI: 10.1093/gji/ggy225

Paleja, R. & Bierman, S., 2016. Measuring changes in earthquake occurrence rates in Groningen, Update October 2016. Shell

report SR.16.12820: 37 pp.
Paulssen, H., Dost, B. & Van Eck, T., 1992. The April 13, 1992 earthquake of Roermond (The Netherlands); first interpretation of the NARS seismograms. Geologie en Mijnbouw 71: 91-98.
Pickering, M., 2015. An estimate of the earthquake hypocenter locations in the Groningen gas field. NAM report. Retrieved from https://www.nam.nl/over-nam/onderzoeksrapporten
Plenefisch, T. & Bonjer, K.-P., 1997. The stress field in the Rhine Graben area inferred from earthquake focal mechanisms and estimation of frictional parameters. Tectonophysics 275: 71-97. DOI: 10.1016/S0040-1951(97)00016-400016-4)
Post, R.A.J., Michels, M.A.J., Ampuero, J.-P., Candela, T., Fokker, P.A., van Wees, J.-D., van der Hofstad, R.W. & van den Heuvel, E.R., 2021. Interevent-time distribution and aftershock frequency in non-stationary induced seismicity. Scientific Reports 11: 3540. DOI: 10.1038/s41598-021-82803-2
Postma, T. & Jansen, J-D. , 2017. The small effect of poroelastic pressure transients on triggering of production-induced earthquakes in the Groningen natural gas field. Journal of Geophysical Research: solid earth 123: 401-417. DOI: 10.1002/2017JB014809
Prinz, D., Hollnack, D. & Wohlenberg, J., 1994. The seismic activity near Aachen following the 1992 Roermond earthquake, the Netherlands. Geologie en Mijnbouw 73: 235-240.
Projectgroup GS-ZL, 2016. Na-ijlende gevolgen steenkolenwinning Zuid-Limburg; final report on the results of the working group 5.2.7. – small earthquakes. https://zoek.officielebekendmakingen.nl/blg-796002.pdf
Reamer, S.K. & Hinzen, K.-G., 2004. An Earthquake Catalog for the Northern Rhine Area, Central Europe (1975-2002). Seismological Research Letters 75: 713-725. DOI: 10.1785/75.6.713
RGD, 1995. Geological Atlas of the subsurface of The Netherlands (1 : 250 000). Explanation to map sheet III Rotumeroog-Groningen. Rijks Geologische Dienst (Haarlem): 113 pp.
Richter, C.F., 1935. An Instrumental Earthquake Magnitude Scale. Bulletin of the Seismological Society of America 25: 1-32.
Richter, G., Hainzl, S., Dahm, T. & Zöller, G., 2020. Stress-based, statistical modelling of the induced seismicity at the Groningen gas field, the Netherlands. Environmental Earth Sciences 79: 252. DOI: 10.1007/s12665-020-08941-4
Ringdal, F., 1983. Seismicity of the North Sea area. In: Ritsema, A.R. & Gurpinar, G. (eds): Seismicity and Seismic Risk in the Offshore North Sea Area. Reidel (Dordrecht): 53-75.
Rodriguez-Marek A., Kruiver, P.P., Meijers, P., Bommer, J.J., Dost, B., van Elk, J. & Doornhof, D., 2017. A Regional Site-Response Model for the Groningen Gas Field. Bulletin of the Seismological Society of America 103: 3149-3163. DOI: 10.1785/0120170020
Roest, J.P.A. & Kuilman, W., 1994. Geomechanical Analysis of Small Earthquakes at the Eleveld Gas Reservoir. In: Eurock '94. Balkema (Rotterdam): 573-580.
Roest, J.P.A. & Mulders, F.M.M., 2000. Modelleren van bewegingen en het spanningsveld bij gasreservoirs. In: Technische workshop Geïnduceerde Aardbevingen in Noord-Nederland. KNMI (De Bilt): 12-18.
Roest, J.P.A., Mulders, F.M.M. & Kuilman, W., 1998. Geomechanical modeling of the Roswinkel gas field. Delft University of Technology, Report No. TA/IG/98/15: 27 pp.
Romijn, R. , 2017. Groningen velocity model 2017 – Groningen full elastic velocity model. NAM report: 12 pp.
Ruigrok, E. & Dost, B., 2019. Seismic monitoring and site-characterization with near-surface vertical arrays, Proceedings 25th European Meeting of Environmental and Engineering Geophysics, Held at Near Surface Geoscience Conference and Exhibition 2019. DOI: 10.3997/2214-4609
Ruigrok, E., and Dost, B., 2020. Advice on the computation of peakground-velocity confidence regions for events in gas fields other than the Groningen gas field. KNMI Report No. TR-386: 56 pp. Available at https://cdn.knmi.nl/knmi/pdf/bibliotheek/knmipubTR/TR386.pdf
Ruigrok, E., Spetzler, J., Dost, B. & Evers, L., 2018, Winschoten events, 19-11-2017. KNMI Report No. TR-368. Available at https://cdn.knmi.nl/knmi/pdf/bibliotheek/knmipubTR/TR368.pdf
Ruigrok, E., Spetzler, J., Dost, B. & Evers, L., 2019, Veendam event, 09-01-2019. KNMI Report No. TR-373. Available at https://cdn.knmi.nl/knmi/pdf/bibliotheek/knmipubTR/TR373.pdf
Ruigrok, E., Rodriguez-Marek, A., Edwards, B., Kruiver, P.P., Dost, B. & Bommer, J., 2022, Derivation of a near-surface damping model for the Groningen gas field. Geophysical Journal International 230: 776-795. DOI: 10.1093/gji/ggac069
Smith, J.D., White, R.S., Avouac, J.-P. and Bourne, S., 2020. Probabilistic earthquake locations of induced seismicity in the Groningen region, the Netherlands. Geophysical Journal International 222: 507-516. DOI: 10.1093/gji/ggaa186
SodM, 2016. Methodiek voor risicoanalyse omtrent geïnduceerde bevingen door gaswinning – tijdelijke leidraad voor adressering Mbb 24.1P. State Supervision of Mines, Report version 1.2, 1 February 2016.
Spetzler, J. & Dost, B., 2017. Hypocentre estimation of induced earthquakes in Groningen. Geophysical Journal International 209: 453-465. DOI: 10.1093/gji/ggx020
Spetzler, J., Ruigrok, E., Dost, B. & Evers, L. ,2018. Hypocenter estimation of detected event near Venlo on September 3rd 2038, KNMI Report No. TR-369: 14 pp.
Spica, Z.J., Perton, M., Nakata, N., Liu, X. & Beroza, G.C., 2018. Shallow VS imaging of the Groningen area from joint inversion of multimode surface waves and H/V spectral ratios. Seismological Research Letters 89:, 1720-1729. DOI:10.1785/0220180060
Spiers, C.J., Hangx, S.J.T., Niemeijer, A.R. , 2017. New approaches in experimental research on rock and fault behaviour in the Groningen gas field. Netherlands Journal of Geosciences 96: s55-s69. DOI: 10.1017/njg.2017.32

Sponheuer, W., 1958. Die Tiefen der Erdbebenherde in Deutschland auf Grund macroseismischer Berechnungen. Annali di Geofisica XI: 157-167. DOI: 10.4401/ag-5819

Stafford P.J., Rodriguez-Marek A., Edwards B., Kruiver P.P., Bommer J.J., 2017. Scenario dependence of linear site effect factors for short-period response spectral ordinates. Bulletin of the Seismological Society of America 107(6): 2859-287. DOI: 10.1785/0120170084

Stafford, P.J., Zurek, B.D., Ntinalexis, M., Bommer J.J., 2019. Extensions to the Groningen ground-motion model for seismic risk calculations: component-to-component variability and spatial correlation, Bulletin of Earthquake Engineering 17: 4417-4439. DOI: 10.1007/s10518-018-0425-6

Taqa, 2021. Gasopslag Bergermeer Microseismische monitoring: Referentie rapport (in Dutch). Retrieved from https://nl.taqa.com/wp-content/uploads/sites/23/2022/01/Jaarrapportage-Microseismische-Monitoring-GSB.pdf

TCBB, 2018. Representatieve nulmeting aan gebouwen, letter to the minister of Economic Affairs and Climate change. Retrieved from https://tcbb.nl/file/download/4b-13ddb2-b5de-48e8-b57e-93da60f251bc/1550755802domus-18236399-v5-22082018_nulmeting_ezk.pdf

TNO, 2016a. Evaluatie mogelijke oorzaak aardbevingen nabij Castricum-Zee in oktober-November 2013, TNO (Utrecht) Report No. R11210: 52 pp.

TNO, 2016b. Evaluatie mogelijke oorzaak aardbevingen bij Anna Paulowna in juni 2015, TNO (Utrecht), Report No. R11209: 56 pp.

Van Balen, R.T., Bakker, M.A.J., Kasse, C., Wallinga, J. & Woolderink, H.A.G., 2019. A Late Glacial surface rupturing earthquake at the Peel Boundary fault zone, Roer Valley Rift System, the Netherlands. Quaternary Science Reviews 218: 254-266. DOI: 10.1016/j.quascirev.2019.06.033

Van den Berg, M.W., Groenewoud, W., Lorenz, G.K., Lubbers, P.J., Brus, D.J. & Kroonenberg, S.B., 1994. Patterns and velocities of recent crustal movements in the Dutch part of the Roer Valley rift system. Geologie en Mijnbouw 73: 157-168.

Van den Berg, M., Vanneste, K., Dost, B., Lokhorst, A., Van Eijk, M. & Verbeeck, K., 2002. Paleoseismic investigations along the Peel Boundary Fault: geological setting, site selection and trenching results. Netherlands Journal of Geosciences / Geologie en Mijnbouw 81: 39-60. DOI: 10.1017/S0016774602000050

Van den Bogert, P.A.J., 2018. Depletion-induced fault slip and seismic rupture - 2D Geomechanical models for the Groningen field, The Netherlands. Shell Global Solutions International B.V. (Rijswijk).

Available at http://www.nam.nl/feiten-en-cijfers/onderzoeksrapporten.

Van der Voort, N. & Vanclay, F., 2014. Social impacts of earthquakes caused by gas extraction in the Provinceof Groningen, The Netherlands. Environmental Impact Assessment Review 50: 1-15. DOI: DOI: 10.1016/j.eiar.2014.08.008

Van Dijk, G., 1934. Die Erdbeben in Noordbrabant von 20-28 November 1932. In: Seismische Registrierungen in de Bilt 20: 44-52.

Van Eck, T. & Davenport, C.A. (eds), 1994. Special Issue: Seismotectonics and seismic hazard in the Roer Valley Graben; with emphasis on the Roermond earthquake of April 13, 1992. Geologie en Mijnbouw 73 (2-4): 91-442.

Van Eck, T., Goutbeek, F., Haak, H. & Dost, B., 2004. Seismic hazard due to small shallow induced earthquakes. KNMI Scientific report, WR 2004-01: 87 pp.

Van Eck, T., Goutbeek, F., Haak, H., and Dost, B., 2006. Seismic hazard due to small-magnitude, shallow-source induced earthquakes in The Netherlands. Engineering Geology 87: 105-121. DOI: 10.1016/j.enggeo.2006.06.005

Van Eijs, R.M.H.E., Mulders, F.M.M., Nepveu, M., Kenter, C.J. & Scheffers, B.C., 2006. Correlation between hydrocarbon reservoir properties and induced seismicity in the Netherlands. Engineering Geology 84: 99-111. DOI: 10.1016/j.enggeo.2006.01.002

Van Elk, J., Doornhof, D., Bommer, J.J., Bourne, S.J., Oates, S.J., Pinho, R., & Crowley, H., 2017. Hazard and risk assessment for induced seismicity in Groningen. Netherlands Journal of Geosciences 96: s259-s269. DOI: 10.1017/njg.2017.37

Van Elk, J., Bourne, S.J., Oates, S.J., Bommer, J.J., Pinho, R., Crowley, H., 2019. A Probabilistic Model to Evaluate Options for Mitigating Induced Seismic Risk. Earthquake Spectra 35: 537-564. DOI: 10.1193/050918EQS118M

Van Gils, J.M. & Zaczek, Y., 1978. La séismicité de la Belgique et son application en génie parasismique. Annales Travaux Publics de Belgique 6: 1-38.

Van Ginkel, J., Ruigrok, E. & Herber, R., 2019. Assessing soil amplification in Groningen, the Netherlands. First Break 37: 33-38. DOI: 10.3997/1365-2397.2019026

Van Ginkel, J., Ruigrok, E., Stafleu, J. & Herber, R., 2022, Development of a seismic site-response zonation map for the Netherlands. Natural Hazards and Earth System Sciences 22: 41-63. DOI: 10.5194/nhess-22-41-2022

Van Thienen-Visser, K. & Breunese, J.N., 2015. Induced seismicity of the Groningen gas field: History and recent developments. The Leading Edge: 664-671. DOI: 10.1190/tle34060664.1

Van Thienen-Visser, K., Nepveu, M. & Hettelaar, J., 2012. Deterministische hazard analyse voor geïnduceerde seismiciteit in Nederland. TNO Report No. 2012 R10198: 19 pp.

Van Thienen-Visser, K., Roholl, J.A., van Kempern, B.M.M. & Muntendam-Bos, A.G., 2018. Categorizing seismic risk for the onshore gas fields in the Netherlands. Engineering Geology 237: 198-207. DOI: 10.1016/j.enggeo.2018.05.013

Van Wees, J.-D., Fokker, P.A., Van Thienen-Visser, K., Wassing, B.B.T., Osinga, S., Orlic, B., Ghouri, S.A., Buijze, L., Pluymaekers, M., 2017. Geomechanical models for induced seismicity in the Netherlands: inferences from simplified analytical, finite element and rupture model approaches. Netherlands Journal of Geosciences 96 (5): s183-s202. DOI: 10.1017/njg.2017.38

Van Wees, J.-D., Osinga, S., van Thienen-Visser, K. & Fokker, P., 2018. Reservoir creep and induced seismicity: inferences from geomechanical modeling of gas depletion in the Groningen field. Geophysical Journal International 212: 1487-1497. DOI: 10.1093/gji/ggx452

Vanneste, K., Camelbeeck, T. & Verbeeck, K. ,2013. A model of composite seismic sources for the Lower Rhine Graben, northwest Europe. Bulletin of the Seismological Society of America 103: 984-1007. DOI: 10.1785/0120120037

Visser, C.A. & Solano Viota, J.L., 2017. Introduction to the Groningen static reservoir model. Netherlands Journal of Geosciences 96: s39-s46. DOI: 10.1017/njg.2017.25

Visser, S.W., 1943. Seismologie. Noorduyn's wetenschappelijke reeks, no 11 (Gorinchem): 151 pp.

Wentinck, H.M., 2018a. Dynamic modelling of large tremors in the Groningen field using extended seismic sources. NAM report.

Wentinck, H.M., 2018b. Kinematic modelling of large tremors in the Groningen field using extended seismic sources: First results related to the Zeerijp tremor ML 3.4. NAM report.

Wetmiller, R.J., 1986. Earthquakes near Rocky Mountain House, Alberta, and their relationship to gas production facilities. Canadian Journal of Earth Sciences 23: 172-181. DOI: 10.1139/e86-020

Willacy, C., van Dedem, E., Minisini, S., Li, J., Blokland, J.W., Das, I. & Droujinine, A., 2018. Application of full waveform event location and moment tensor inversion for Groningen induced seismicity. The Leading Edge 37: 92-99. DOI: 10.1190/tle37020092.1

Willacy, C., van Dedem, E., Minisini, S., Li, J., Blokland, J.W., Das, I. & Droujinine, A., 2019. Full-waveform event location and moment tensor inversion for induced seismicity. Geophysics 84: KS39-KS57. DOI: 10.1190/geo2018-0212.1

Willacy, C., Blokland, J.W. & van Dedem, E., 2020. Automating event location monitoring for induced seismicity. The Leading Edge 39: 505-512. DOI: 10.1190/tle39070505.1

Wood, H. O. & Neumann, F., 1931, Modified Mercalli Intensity Scale of 1931. Bulletin of the Seismological Society of America 21: 277-283.

Zhou, W. & Paulssen, H. (2017). P and S velocity structure in the Groningen gas reservoir from noise interferometry. Geophysical Research Letters 44: 11,785-11,791. DOI: 10.1002/2017GL075592

Zhou, W. & Paulssen, H., 2020. Compaction of the Groningen gas reservoir investigated with train

Noise. Geophysical Journal International 223: 1327-1337. DOI: 10.1093/gji/ggaa194

Ziegler, P.A., 1992. European Cenozoic rift system. Tectonophysics 208 (1-3): 91-111. DOI:10.1016/0040-1951(92)90338-7

Ziegler, P.A., 1994. Cenozoic rift system of western and central Europe: an overview. Geologie En Mijnbouw 73(2-4): 99-127.

Zijerveld, L., Stephenson, R., Cloetingh, S., Duin, E. & Van den Berg, M.W., 1992. Subsidence analysis and modeling of the Roer Valley Graben (SE Netherlands). Tectonophysics 208: 159-171. DOI: 10.1016/0040-1951(92)90007-790007-7)

Zöller, G. & M. Holschneider, M., 2016. The maximum possible and the maximum expected earthquake magnitude for production-induced earthquakes at the gas field in Groningen, the Netherlands, Bulletin of the Seismological Society of America 106: 2917-2921. DOI: 10.1785/0120160220

23

Subsidence

Peter A. Fokker, Thibault G.G. Candela, Gilles Erkens, Ramon F. Hanssen, Henk Kooi, Kay Koster & Freek J. van Leijen*

*Corresponding author: Peter.Fokker@tno.nl

ABSTRACT
Subsidence is a complex problem, both in a technical sense and in governance. This is particularly the case in the Netherlands, which is a low-lying and densely populated country where various causes of subsidence interfere with each other. Coping with subsidence in the Netherlands started already 1000 years ago. This long history of subsidence, however, along with its slow manifestation, has resulted in a tendency to adaptation rather than mitigation. There is a growing awareness that this focus on adaptation is actually excluding alternative solutions. Potentially cheaper or more effective options may be unknown and not even considered. At the same time, Dutch society is becoming more aware of the severity of human-induced subsidence as it is one of the most prominent current geological hazards. What is needed therefore, is a sound knowledge base facilitating the exploration of solutions outside the traditional way of thinking. Here we present the different knowledge and governance issues at stake. We start with the description of the natural processes that cause subsidence, and the human-induced causes like groundwater management and exploitation of deep geological resources. Then we elaborate how subsidence can be estimated from measurements. We pay specific attention to the utilization of modern ensemble-based techniques to integrate multiple models and data. The objective is to avoid deterministic predictions and instead produce a range of subsidence forecasts with confidence intervals that are in agreement with observational data and their uncertainties. Finally, we describe how technical knowledge can be integrated in decision making by estimating the costs and benefits of different scenarios, thereby offering an array of options for decision makers. Subsidence will keep playing a role in shaping the future of the Netherlands. Human-induced subsidence will continue with new subsurface activities directed towards the energy transition. Incorporating the grim sea level rise predictions, the issue becomes even more serious. It is therefore of paramount importance to maintain and further develop the current knowledge position and to develop proactive mitigation activities.

<< A row of houses built at the end of the 19th century along the Oostsingel in Schiedam (west of Rotterdam). The deformed facades are caused by subsidence. Photo: Vincent Basler Fotografie.

DOI: 10.5117/9789463728362_ch23

Introduction

The Netherlands is a country in the delta of Rhine and Meuse, with low elevation, dense population, and intense surface use. A movement of the surface downward with respect to the ever rising sea level therefore poses considerable risks for flood safety and has economic repercussions. Some places in the Netherlands are moving upward by both natural and anthropogenic processes, but in many parts of the country the surface has been, and still is going down. Subsidence as a result of human activities commenced locally as early as the Late Iron Age (Vos, 2015). Extensive and systematic human-induced subsidence started roughly 1000 years ago with intensifying land cultivation. The population was forced to react to subsidence with innovative measures: canalization, embankments, wind-driven and mechanical artificial drainage, giving up land, changing land use, and reconstruction of infrastructure in cities (Fig. 23.1). These interventions were focused on short-term adaptation to the consequences of subsidence. The interventions never resulted in decreasing subsidence rates. In fact, some interventions, for instance those that enabled drainage of soft soils, even increased the subsidence.

The impact of subsidence in the Netherlands is slowly being unravelled and quantified (Van den Born et al., 2016). At the same time, over the last decades scientific research has contributed to an increased understanding of the processes leading to subsidence. Technological development in measuring techniques and the ongoing development of numerical predictive models have further enhanced the capability to quantify subsidence. As an example, it has become clear that the greatest contribution to subsidence in the Netherlands is human-induced and thus could theoretically be avoided. Subsidence will therefore continue until an effective policy to reduce the rates has been developed and deployed. These developments have sparked a discussion in the Netherlands on whether the traditional adaptive approach to ongoing subsidence is still the most optimal one. There is a societal demand for conscious decision making and a long term subsidence strategy.

A successful subsidence strategy requires a broad societal support and a justified confidence in its outcome. It is important to understand the processes leading to measured subsidence and to be able to quantify them. A critical check in this regard is, of course, the agreement between modelled surface movements and the movements estimated from measurements. Moreover, movement measurements need to be used to improve the knowledge of the subsidence processes and the model parameters. Such knowledge is required to define economic and effective governance. Finally, to close the loop, feedback is required

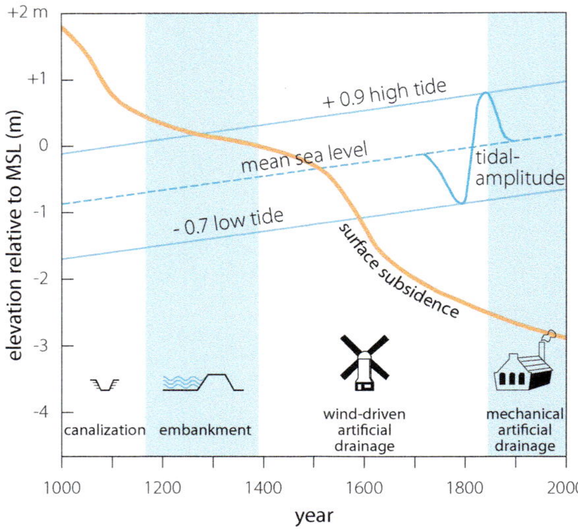

Figure 23.1. Cartoon of 1000 years of subsidence in the coastal zone of the Netherlands showing the response of the Dutch to ongoing subsidence since drainage of the coastal and deltaic areas started (modified after Van de Ven, 1993). Some of these responses, e.g. the construction of windmills and mechanical drainage, have prolonged and further intensified subsidence.

by monitoring the policy results and incorporating lessons learnt in an ongoing process. Such a policy cycle is essential to ensure continual learning and improve the effectiveness of measures to counter subsidence.

This chapter summarizes the current state of knowledge about subsidence in the Netherlands. The processes causing subsidence, both natural and human-induced, are at the basis of this knowledge. Therefore, we start with a summary of the next two sections. Natural causes of surface movement are compaction, isostasy and tectonics. The most important human-induced causes of subsidence in the Netherlands are groundwater regulation and extraction of deep natural resources like gas and salt. Besides subsidence, uplift also occurs locally. The largest uplift is observed in Limburg, where the hydraulic heads are increasing since the abandonment of the coal mines. The fourth section explores the ways surface movement can be estimated from measurements. The classical levelling methods are described as well as modern satellite techniques. The next section presents the methods that combine data and models to increase the quality and accuracy of understanding. Stochastic methods are at the core of many such methods, since they provide efficient ways to not only provide best estimates, but also confidence ranges of parameter values and model predictions. Such knowledge is critical input for decision making, and we subsequently outline the current practice and effective ways forward. We consider a policy cycle that integrates measurements, understanding, models, economic considerations, governance and monitoring in a large feedback loop as the optimal approach. The final section presents a panoramic view on what is currently available and what we can still expect to come.

Natural subsidence processes

Natural subsidence occurs independently of human influence and is an expression of geological deformation processes. With its location on the southern margin of the North Sea Basin with its thick succession of Mesozoic and Cenozoic sedimentary deposits, natural land motion has played a central role in the geological history of the Netherlands. *Tectonic*, *isostatic* and *compaction* contributions are commonly distinguished (Kooi et al., 1998; Fokker et al., 2018). Tectonic subsidence is the expression of faulting and folding within the crust and cooling of the lithosphere. Isostatic subsidence is caused by changes in buoyancy equilibrium of the lithosphere-mantle system and involves the flow of mantle rock. Compaction refers to compression and porosity reduction of sedimentary layers. Long-term (2.5 Ma-present) average subsidence rates for the individual components were inferred by Kooi et al. (1998) using backstripping of Cenozoic stratigraphy, yielding low subsidence rates on the order of 10^{-2}-10^{-1} mm/yr (Fig. 23.2). These rates are presumed to be relative to the centre of mass of the earth.

Figure 23.2 provides a reasonable estimate for the long term, but magnitudes and spatial patterns of historic or present-day tectonic subsidence could be quite different over shorter time periods. To assess the rates over shorter, more recent time periods, other methods and considerations are generally needed. Tectonic deformation, for instance, tends to be irregular in space and time when faulting is involved. Studying detailed stratigraphic records in the Roer Valley Rift System, Houtgast & Van Balen (2000) recognized periods of high and low relative displacement rates on parts of the Peel Boundary Fault and the Feldbiss Fault zones during the Quaternary. The 1992 'Roermond-earthquake' further attests to natural fault activity in the Netherlands in very recent times.

To estimate present-day natural compaction in the western coastal provinces, Kooi (2000) modelled delayed pore-water expulsion and associated compaction (consolidation) of low-permeable Cenozoic clay layers in response to Holocene sediment loading. This process differs from the backstripping technique used for the long-term subsidence which assumes equilibrium compaction based on porosity-depth relationships. The latter approach neglects Darcian flow and thus implies that only present-day sediment loading would cause present-day compaction. Inferred current rates generally are higher than the long-term compaction estimates of Figure 23.2, of up to 0.12 mm/yr to the west of Rotterdam. The younger dune system represents a geologically recent prominent load. Kooi (2008) calculated that a current subsidence rate of 2 mm/yr could theoretically exist in parts of the western coastal provinces.

The long-term isostasy component obtained by backstripping assumes equilibrium of the lithosphere-mantle

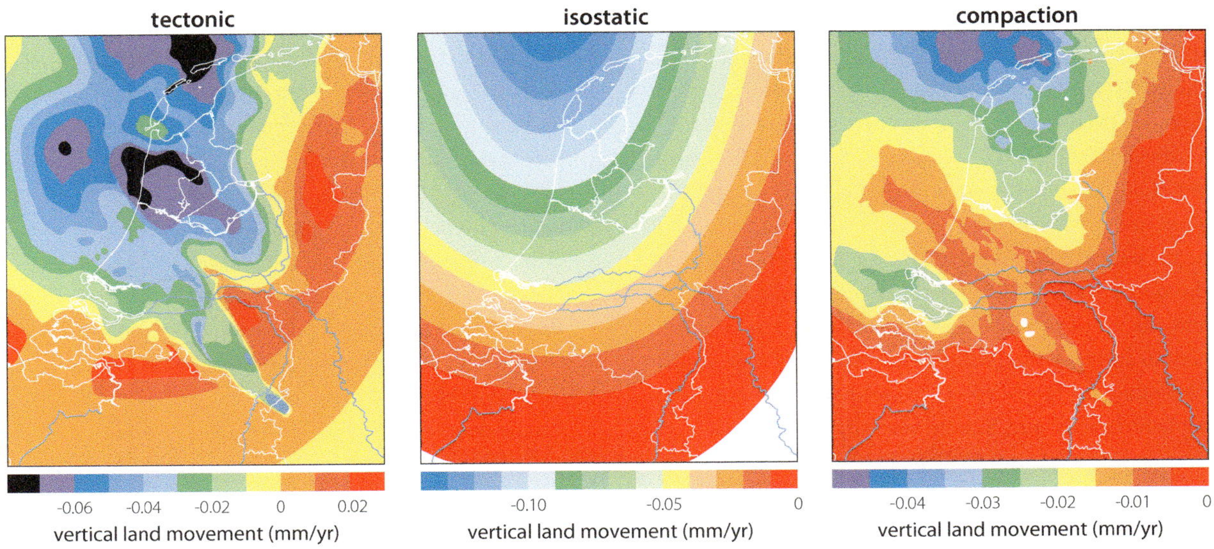

Figure 23.2. *Separation of tectonic, isostatic and compaction contributions to vertical land movement for the Quaternary (ca. 2.6 Ma – present) constructed by 3D back-stripping of Cenozoic stratigraphy of the Netherlands and the Southern North Sea Basin (modified after Kooi et al., 1998). Negative values indicate subsidence.*

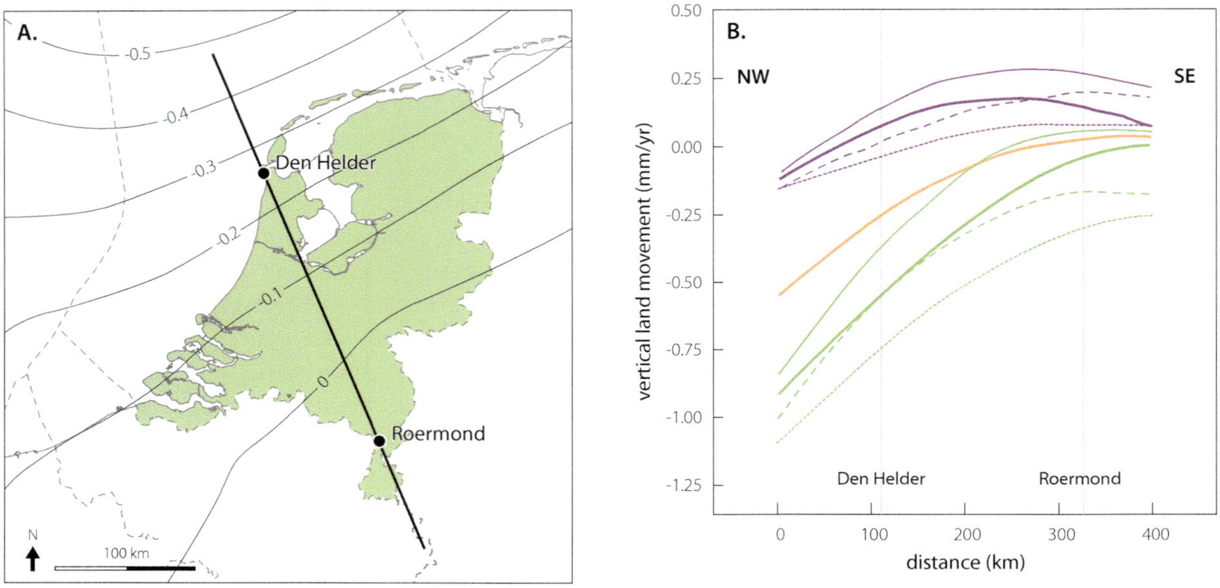

Figure 23.3. Current isostatic vertical land movement (in mm/yr) from GIA modelling of relative sea-level data (modified after Kooi et al., 1998). a) Result from the best-fit earth model. b) Results for the profile in the left panel for the best-fit earth model (orange line) and other earth models that provide a good fit to Fennoscandian relative sea-level data.

system in response to the load of Quaternary deposits. By contrast, current isostatic vertical land movement in the Netherlands is dominated by the dynamic response to changes in ice and water mass on the surface of the earth since approximately the Late Glacial Maximum, ca. 20 ka BP (Kooi et al., 1998; Vermeersen et al., 2018). Various Glacial Isostatic Adjustment (GIA) models that describe the dynamic lithosphere-mantle system response to mass changes on ice-age timescales (resolution ~1 kyr), and that are constrained by relative sea-level data in the central and northern North Sea, indicate that the present isostatic subsidence component includes a tilting of the onshore part of the Netherlands in a northwest to northerly direction (Fig. 23.3; Kooi et al., 1998; studies discussed by Vermeersen et al., 2018). This is consistent with independent Holocene relative sea-level reconstructions from Belgium to the Wadden Sea (Koster et al., 2017; Meijles et al., 2018) which show subsidence rates that increase towards the north. However, the predicted present-day tilting rate and the absolute vertical movement rates are rather uncertain and vary significantly within and between studies, depending on the adopted ice-sheet reconstruction, lithosphere rigidity and mantle viscosity (Kooi et al., 1998; Vermeersen et al., 2018). Riva et al. (2017) showed that post-1900 melting of glaciers and the Greenland and Antarctic ice sheets cause an additional elastic vertical uplift of several tenths of mm/yr. This component of land movement, which varies temporally over timescales less than a decade is not included in the GIA models.

Human-induced subsidence processes

Human-induced subsidence (and uplift, which we regard as 'positive' subsidence) processes are primarily related to the extraction and injection of subsurface substances and the results are largely irreversible. Subsidence can be considered as an unwanted side effect, as these processes themselves are not intended to cause changes in surface elevation.

Human activities that cause subsidence can in general be divided into deep (ca. >500 m), middle-deep (ca. >20 m and <500 m) and shallow activities (ca. <20 m). The deep and middle-deep depths are roughly corresponding to depth intervals of hydrocarbon extraction, geothermal production and salt mining, groundwater pumping and heat storage (Fig. 23.4). Near-surface activities can be many: phreatic groundwater lowering, excess overburden by man-made ground, peat mining, and others.

While subsidence is the accumulation of effects of different origin, both anthropogenic and natural, it is the total effect of surface movement that is of interest to the societal stakeholders. In the following, we will discuss deep, middle-deep, and shallow sources of subsidence.

Deep sources of subsidence

Hydrocarbon extraction

Gas or oil production involves the extraction of hydrocarbons from a reservoir. As a result, the pore pressure drops. The net forces between the sand grains constituting the matrix increase, and the rock will compact. This mechanical response is propagated to the surface and causes subsidence (Geertsma, 1973; Fokker & Orlic, 2006; Fjaer et al., 2008). Models are deployed to quantify this behaviour. Such models start with assessing the pressure distribution of the reservoir fluid, which is controlled by the production profiles and the reservoir flow properties. The outcomes are fed into geomechanical models that describe the mechanical response upon pressure changes. Here, the complete subsurface must be considered: the reservoir compaction, the overburden response as well as the basement properties influence the resulting subsidence. Compaction at a certain location at the reservoir level is distributed over distances at the surface comparable to the depth of the source. Complications can occur due to several factors. When aquifers that are connected to a reservoir are also depleted, they also compact and cause subsidence (Fokker et al., 2018). Furthermore, the mechanical behaviour of the subsurface may be nonlinear or viscoelastic, which results in a time-dependent effect and the persistence of the subsidence even when the activity that caused compaction has ceased (Marketos et al., 2016). This time dependence is difficult to isolate when production is still ongoing. In addition, other processes like water or gas injection can interfere (Hettema et al., 2002). Modelling and forecasting the subsidence, therefore, is not trivial. An important step in quantifying subsidence causes is the matching of modelled subsidence and subsidence estimated from measurements. A section is dedicated to this later in this chapter

Geothermal energy production

Geothermal energy production also involves pumping of

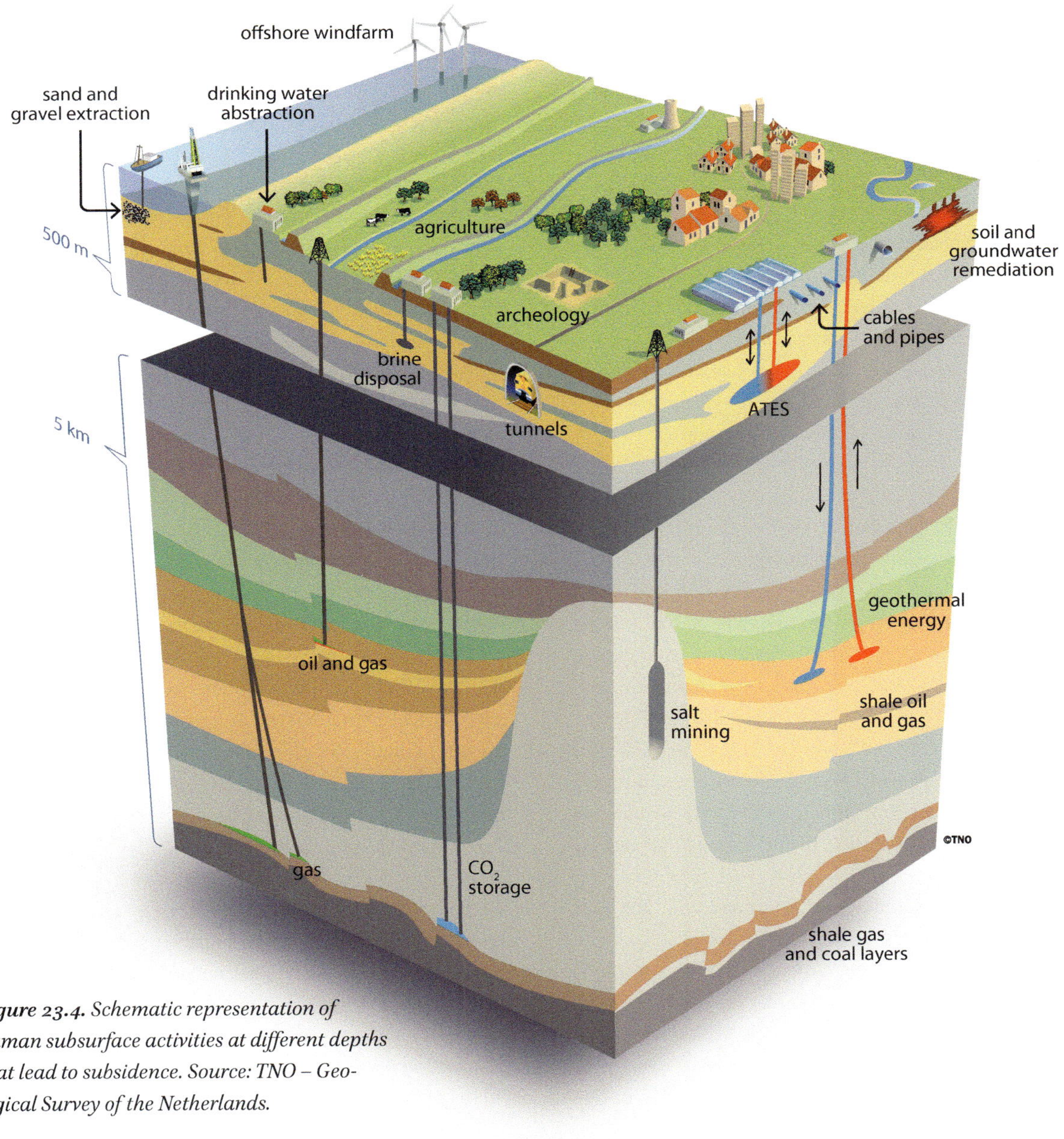

Figure 23.4. Schematic representation of human subsurface activities at different depths that lead to subsidence. Source: TNO – Geological Survey of the Netherlands.

fluids, viz. water or steam into and out of the subsurface. This may also be accompanied by pore pressure changes, compaction and associated surface movement. An additional subsidence effect of geothermal energy production is caused by the heat that is extracted. Even if the pressure is kept constant by injecting as much water as is being produced, progressive cooling around the injector wells may cause substantial thermoelastic stresses resulting in compaction of the subsurface strata and associated subsidence (Grant & Bixley, 2011). Modelling the subsidence effect of geothermal energy production proceeds along similar lines as modelling the effect of hydrocarbon production, with the addition of thermoelastic effects.

Salt mining

The process of salt mining is described by Groenenberg et al. (2025, this volume). Water is injected into the rock salt layers, salt dissolves and a cavern filled with brine develops, from which the saturated brine is produced. The visco-elastic mechanical behaviour of the rock salt causes cavern convergence as the pressure in the cavern is smaller than the stress in the formation around it. The cavern boundaries move inwards, and the surface subsides. Even after cessation of the salt production, this driving force may cause continued subsidence. In addition, the temporal behaviour of the surface subsidence and the shape of the subsidence bowl may be complicated functions of time and space, due to the complex constitutive rock salt behaviour (Breunese et al., 2003). In the recent past, shallow salt mining has also resulted in accelerated subsidence leading to overburden collapse, with sinkholes as a result (see Groenenberg et al., 2025, this volume and references therein).

Coal mining

In the south of the province of Limburg, mining activities caused local depressions above the mined coal seams, 50 to 200 m in diameter and 0.5 to 10 m deep. Most of these depressions were formed during the peak of coal mining in the 20th century. After closure, when the groundwater was allowed to rise, land rising has been observed (Caro Cuenca et al., 2013; Zhao & Konietzky, 2021). However, also some depressions were formed in the decades after the closing of the mines (Vis et al., 2020). A quantitative description based on first principles is not available, but it is clear that their formation is due to sinkholes related to coal mining that are formed by a combination of rock deformation, groundwater- and sediment flow and collapsed mine roofs. In urban areas this form of subsidence is hazardous, because the formation of sinkholes can suddenly accelerate, resulting in foundation instability and therefore collapse of buildings. This is exemplified by a near-collapse of a shopping mall in the city of Heerlen in 2011 (Chang & Hanssen, 2014). Natural sinkholes in Upper Cretaceous chalk are discussed in Van Lochem et al. (2025, this volume).

Middle-deep sources of subsidence

Groundwater abstraction

Sandy aquifers at depths between tens to hundreds of meters below the surface are an important source of freshwater supply for the Netherlands. These aquifers are confined by clayey and sometimes peaty aquitards. Aquifer water pressure drops if more groundwater is extracted than is being replenished by inflow into the aquifer. These pressure drops prograde into bounding aquitards, resulting in aquitard compaction by increasing effective stress and creep processes, and consequently in subsidence. Due to a lack of dedicated subsidence monitoring, knowledge of groundwater-abstraction-induced subsidence in the Netherlands is extremely poor and mostly based on theoretical calculations. By using InSAR data, Kooi (2018) found that subsidence rates in the surroundings of De Lier (province of South Holland) may have locally increased by about 5 mm/yr in the period 2003-2011 due to groundwater abstractions that supplied greenhouses with irrigation water. The same study also showed that model predictions of the subsidence are fraught with uncertainty since the subsidence depends not only on the distribution of compressible clay and peat layers in the Holocene cover layer, but also on the details of the vertical permeability distribution within the Holocene strata. Both of these are variable and uncertain. Another example, using forward modelling of subsidence rates based on groundwater abstraction data and geomechanical schematization of stacked sandy aquifer systems at Nijverdal and Boerhaar (province of Overijssel), reported subsidence rates of 0.2 to 1.6 and 0.4 to 2.3 mm/yr respectively (Valeda, 2021). The upper 100 m of both sites primarily consist of sand, however at Boerhaar a fine-grained aquitard is present at 40 m below the surface. Compaction of this aquitard explains the difference in compaction rates between the two sites.

Subsidence by aquitard compaction is local and could lead to damage to infrastructure and the built environment when groundwater abstraction sites are located within or near domestic areas. This form of subsidence can be mitigated or even partly reversed with managed aquifer recharge, i.e. by pumping water back into aquifers. Water injection coeval with abstraction, for instance, is commonly used during construction works in the Netherlands. Furthermore, subsidence by groundwater abstraction partly reverses when an abstraction site is abandoned, and natural water pressures are restored. Such elastic rebound is currently being monitored in the northern part of Delft, where century-long industrial groundwater abstraction was progressively reduced and even terminated in recent years.

Aquifer Thermal Energy Storage

Aquifer Thermal Energy Storage (ATES) refers to temporary warming of aquifer groundwater at depths less than 500 m by residual heat from e.g. industry, solar thermal collectors, and geothermal systems. The storage is primarily conducted during the summer season, whereas the warm water is being used to heat-up buildings during the winter season. Furthermore, during the winter season, cold water is stored in aquifers to cool buildings during summer. Conventional ATES involves temperature changes in the subsurface typically ranging from 5-10°C to 15-18°C (Rostampour et al., 2019). Efficiency considerations are currently leading to the development of storage at higher temperatures, up to 90°C (Drijver et al., 2019). The associated thermal contraction and expansion of aquifers, which potentially lead to measurable subsidence or uplift at the location of the injection well and surrounding areas are currently subject of investigation.

Shallow sources of subsidence

Phreatic groundwater-level lowering

The coastal plains and inland peaty and clayey areas of the Netherlands are subsiding on a regional scale because of lowered phreatic groundwater levels. Surface level oscillations are superimposed on the net subsiding trend, as seasonal fluctuations in phreatic groundwater level and soil moisture cause stress changes and hence elastic responses in the upper part of peat layers. Phreatic groundwater levels have been kept at artificially low levels for centuries to enable agriculture and urbanization, and to prevent the area from inundation in general. In pristine conditions groundwater levels in coastal plains gradually grade towards mean high sea level or just above, keeping the substrate waterlogged. But when areas are drained, groundwater levels are artificially low and the exposure of peaty and clayey substrates to the atmosphere make them vulnerable to oxidation and shrinkage processes, while lowering of hydrostatic pressures induces various mechanical processes.

Aeration of peat or organic substrate causes the breakdown of its organic material by microbial communities of bacteria, archaea, and fungi (Stephens & Johnson, 1951). This results in degradation of peat layers, greenhouse gas emissions, and subsidence by volumetric loss (Neller, 1944). This process is irreversible. Van den Akker et al (2008) calculated an average emission of 2.26 t CO_2 $ha^{-1}yr^{-1}$ for each mm of subsidence in the Zegveld area. This is a general number for peat in the Netherlands and it only addresses degradation of organic carbon above the groundwater table. The number has since been used in greenhouse gas emissions reporting (e.g. Arets et al., 2019).

Aeration of clayey substrate results in shrinkage by loss of water captured between platy clay particles. In submerged conditions, the negatively charged surfaces of the clay particles adsorb water by attracting dissolved cations (Barciela-Rial et al., 2020). This water is expelled during periods of evaporation, resulting in shrinkage of the clay and thus strong volumetric loss. This is mostly irreversible, but may be partly countered by swelling processes if a clayey substrate is saturated again. An example of this process is the reclaimed province of Flevoland. There, shrinkage of the clayey substrate caused a surface elevation lowering of 1 to 2 m (Fokker et al., 2019).

Hydrostatic pressure lowering in peaty and clayey substrates leads to consolidation by increasing effective stress, consequential pore space reduction, volumetric loss and subsidence. This process is partly reversible when the effective stress decreases again through an increase of the hydrostatic pressure. Furthermore, these stress changes result in creep processes that compact clayey and peaty substrates superimposed on the compaction related to hydrostatic pressure reduction. Creep comprises the visco-plastic rearrangement and flow of soil particles.

Quantification of expected surface movement can be achieved using soil-mechanics models similar to those used for the quantification of subsidence due to deep causes. Using an estimate of the groundwater level development they calculate a volume loss. In comparison with the deep-seated sources, a simplification is that the connection to surface movement is immediate. The soil-mechanics models, however, are rather complicated. They contain an elastic contribution, but also irreversible, non-elastic compaction and creep, shrinkage, and oxidation. Models for these phenomena often depend heavily on laboratory results and field tests of short duration or tests with different goals than long-term predictions. As a result, it is often difficult to interpret data in the context of modelling results (Fokker et al., 2019).

Anthropogenic brought-up soil

Anthropogenic brought-up soil has been used for centuries in the Netherlands to increase the load bearing capacity of peaty and clayey strata, in order to make these soft substrates suitable for supporting infrastructural works and construction activities in general (Van Laarhoven, 2017; Van Asselen et al., 2018). In historic times, brought-up soil consisted of locally derived sods and debris. Initially, the emplacement of anthropogenic brought-up soil on peaty and clayey substrates induces compaction of underlying strata by increasing effective stress, which results in subsidence (Fig. 23.5; Koster et al, 2016). Therefore, such areas needed the periodic addition of anthropogenic soil to compensate for experienced subsidence (Koster, 2016) and, consequently, many historic urban centres in the coastal plains have meters thick strata of anthropogenic brought-up soil that have been compacting underlying

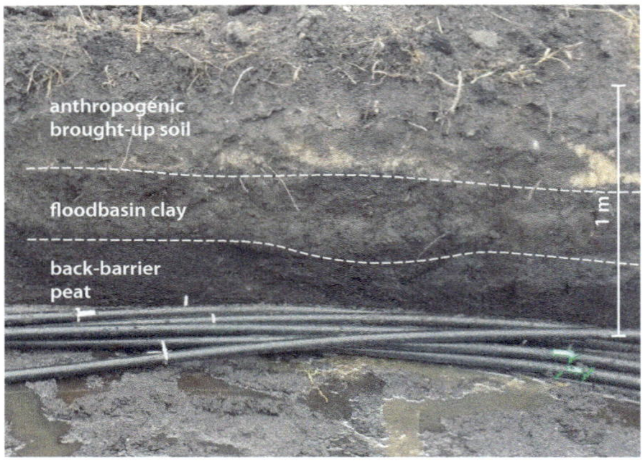

Figure 23.5. Example of anthropogenic brought-up soil in the urbanized area of Amsterdam (from Koster et al., 2018). The overburden of the brought-up soil causes underlying peaty and clayey substrate to compact and leads to subsidence.

peaty and clayey substrates. However, a complex interplay results: effective stress changes by additional overburden have pushed the peaty substrates sometimes meters below local phreatic groundwater levels, thereby hampering oxidation and shrinkage processes, and thus protecting soft substrates from further deterioration (Koster et al., 2018).

Since the industrial revolution, sand is the prime aggregate used to increase load bearing capacities of natural deposits, especially in areas with a peaty substrate (Van der Meulen et al., 2007). Sand has a relatively high specific weight and causes substantial surface elevation lowering when brought-up. Depending on the spatial variability of the substrate this could cause strong differential subsidence on a local scale, potentially resulting in damage to buildings and infrastructure (Nicodemo et al., 2020).

Peat mining

Substantial surface elevation lowering took place between the 12th and late 19th century due to mining of peat bogs for fuel (Van Dam, 2001). Approximately 34% of the volumetric loss of the natural Holocene peaty substrate is due to these activities (Erkens et al., 2016). This lowered the coastal plain substantially and created lakes, which were then partly drained, resulting in shrinkage of their often clayey substrates. Although these activities were terminated more than a century ago, their legacy of a strongly differentially elevated coastal plain and resulting local inundation risk is still a prime cause of the present day strict Dutch phreatic groundwater level management that is the mechanism behind most shallow subsidence processes.

Sinkholes

A relatively new phenomenon in built-up areas in the Netherlands are sinkholes that are formed in man-made ground due to erosion by excess waterflows caused by several processes including burst water pipes. The formation of such sinkholes may result in sudden collapse and instabilities of buildings and infrastructures (e.g. Venmans et al., 2020).

Combined effects of shallow subsidence

For the Netherlands a shallow subsidence prediction map (Fig. 23.6) was published by Erkens et al. (2021). It is based on a combination of results from different forward models targeting the various sources of shallow subsidence for a priori defined scenarios. The map shows that a continuation of the current strategy, the so-called adaptation strategy, will result in up to several decimetres of additional subsidence up to 2100. In the 'Governance' section we show how an alternative strategy can be underpinned with data and information in the form of a scientifically informed policy cycle.

Detecting and monitoring subsidence

Subsidence processes form a subset of all the movements (horizontal and vertical) occurring at the earth's surface and in the objects on it. Such movements exhibit a wide range of variability, both spatially as well as temporally. For example, a small depression in a pavement and a large subsidence bowl due to gas extraction may have the same shape but different dimensions and impact (see Fig. 23.7). Temporally, a surface depression may precede a sinkhole collapse by a few hours, while surface motion due to salt mining will take a long time to develop a significant expression at the surface. The dynamics of peat areas are even more time-variable and can significantly and abruptly vary spatially, e.g. between polders (Morishita & Hanssen, 2015). Some processes even lead to uplift (Hanssen et al., 2008, 2020; Fokker et al., 2016; Conroy et al., 2022).

Surface deformation in real life is usually the combined effect of many processes. The diversity of the earth's surface itself translates almost inevitably into a diversity in subsidence and movement signals.

The consequence of this complexity is that to detect and monitor subsidence using geodetic techniques, (i) the sought 'subsidence' signal needs to be defined specifically for the particular objective at hand, i.e. *'what do we want to know?'*, and subsequently (ii) the most optimal strategy for detecting and monitoring must be selected, i.e. *'how do we measure it?'*. As a consequence, for a particular area of interest one could generate many different subsidence maps, using different techniques, and with different objectives.

The signal of interest

To select an optimal monitoring strategy for the subsidence signal of interest, specific a priori information needs to be established. Obviously, this should include the location of interest as well as the time interval of interest. However, this is not sufficient; the expected extent and smoothness of the signal of interest need to be given both spatially as well as temporally, i.e. how much relative subsidence do we expect between two locations as a function of their distance (see the example in Fig. 23.7). With reference to the smoothness it should be stated whether breakpoints or other spatiotemporal inhomogeneities can be expected. Finally, there needs to be an expectation of the magnitude and type of the signal. This set of expectation parameters is necessary to define the next steps: sampling and parameterization.

Sampling

Subsidence is always estimated as a *double-difference* signal, i.e. it concerns the 3D spatial position of one point relative to another (single spatial difference), as well as changes in time from one epoch to another (single temporal difference). The differences combined are referred

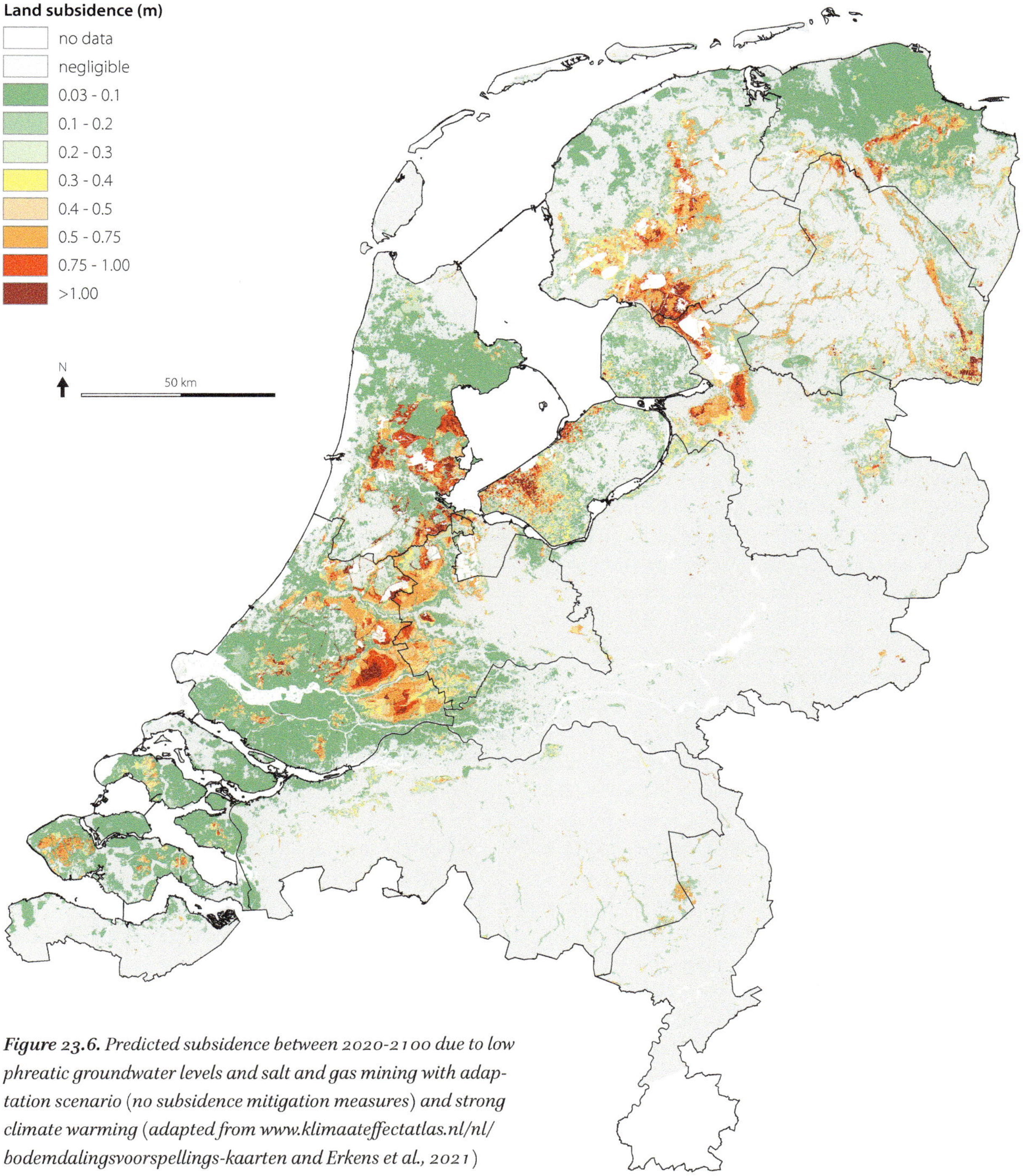

Figure 23.6. *Predicted subsidence between 2020-2100 due to low phreatic groundwater levels and salt and gas mining with adaptation scenario (no subsidence mitigation measures) and strong climate warming (adapted from www.klimaateffectatlas.nl/nl/bodemdalingsvoorspellings-kaarten and Erkens et al., 2021)*

Figure 23.7. Subsidence occurs in many shapes and sizes. a) Subsidence bowl in a pavement. b) Subsidence bowls in the Groningen area, showing the cumulative subsidence between 1992 and 2007.

to as double difference. Once the expected spatial extent of the signal of interest is known, the full area of interest needs to be at least a factor of two wider, in order to incorporate areas that are not affected by the subsidence signal as reference. The expected smoothness drives the spatial sampling. A smooth deformation signal, e.g. due to deep hydrocarbon production, requires a less dense set of measurement points than a deformation signal with very localized characteristics, such as the subsidence preceding a sinkhole collapse. Due to the typical superposition of subsidence contributions, the spatial sampling should also consider the variability of the other signals that will be present in the observations. These can either be disentangled later, or they can be treated as noise in the data. Unless tailor-made benchmarks for a specific driving mechanism are used, the spatial sampling should be sufficiently dense to observe the signal with the highest spatial variability. Expected spatial inhomogeneities need to be considered in the sampling pattern as well.

In addition, the period of interest needs to be defined. This can be related to subsidence processes that are expected in the future, occurred in the past, or are ongoing. Obviously, this choice significantly limits the type of surveying techniques that can be deployed, ranging from advanced electronic systems to information inferred from proxy data. The temporal smoothness of the expected double difference subsidence signal drives the temporal sampling. A subsidence signal that varies smoothly over time, e.g. due to gas production, requires a less frequent sampling than a signal that varies within weeks or days, e.g. the dynamics of organic soils. The temporal sampling should take the potential superposition of signals into account, as the measurements may be susceptible to various components of the superposed signal, including the ones with a higher temporal variability.

Apart from the spatial and temporal characteristics of the subsidence the expected magnitude of the signal is also important to consider, as it influences the choice of geodetic technology that is most suitable.

Parameterization

Parameterization refers to the definition of the parameters of interest and is strongly dependent on the original objective. In some cases the parameter is simply $z(\mathbf{x}, t)$, referring to the elevation of a location with horizontal coordinates $\mathbf{x}=(x, y)$ at an epoch t, expressed in a suitable geodetic datum. This is a so-called 'undifferenced' parameter that gives no information on temporal or spatial changes and for that reason is usually rather meaningless. For example, the actual elevation of a levelling benchmark installed in a wall (see Fig. 23.8), estimated at time t, is of no physical importance for land dynamics even though it is of high importance for several practical procedures in civil engineering. Single-differences are defined either in time as $\Delta_t z(\mathbf{x}, \Delta t)$ or in space as $\Delta_x z(\Delta\mathbf{x}, t)$ with $\Delta t = t_2 - t_1$ and $\Delta\mathbf{x} = \mathbf{x}_p - \mathbf{x}_q$, respectively and with p and q presenting two different locations. These single differences are equally meaningless, because the vertical movement of a single point over a particular time interval, or the elevation difference between two points at a particular epoch does not qualify as subsidence. Only for the *double-difference* parameters $\Delta_{t,x} z(\Delta\mathbf{x}, \Delta t)$ we obtain a physically interpretable variable.

We then parameterize the signal preferably by a minimum number of unknown parameters, for example by a vertical subsidence rate $v_z(\Delta\mathbf{x}_{p,0}, \Delta t_{i,0})$ at every sample point, where $\Delta\mathbf{x}_{p,0} = \mathbf{x}_p - \mathbf{x}_0$ and $\Delta t_{i,0} = t_i - t_0$, and where $\mathbf{x}_0$ and t_0 represent an (arbitrary) reference location and reference epoch, respectively. More elaborate parameterizations, using e.g. seasonal periodicity, acceleration, or higher-degree polynomials, are possible as well, but should only be used when physically plausible. The parameterization can be performed solely on the measurement benchmarks, interpolated to a predefined grid, which would introduce additional interpolation errors, or via a spatio-temporal analytical function. In addition, for satellite techniques there are nuisance parameters that need to be estimated, such as integer ambiguity numbers, atmospheric delays, or scaling parameters (Hanssen 2001).

Absolute and relative deformation

Irrespective of whether the subsidence signal is expressed in terms of a spatially variable velocity or in a more elaborated parameterization, its double-difference nature implies that it is always a *relative* deformation. In fact, the term 'deformation', i.e. a change in 'form', underlines this relativity intrinsically. As such, an *absolute* deformation does not exist, since 'form' is position invariant. Nevertheless, assuming a predefined reference benchmark in the network and a reference epoch in time, and assuming that the position of the reference benchmark is known in a common geodetic datum at time of the reference epoch, it is possible to integrate all double-difference displacement estimates to this spatiotemporal reference system. This effectively yields time series of benchmark positions in the geodetic datum, which are sometimes referred to as quasi-absolute positions.

Benchmarks

After the signal of interest is defined and the sampling and parameterization of the observation system are derived, the next important decision refers to the choice of the benchmarks. This choice is important to ensure reliable, representative, and repeatable observations with sufficient precision and is even more critical than the choice of the surveying technique. There are three options: physical benchmarks, virtual benchmarks, and benchmark-free observations. The first two are Lagrangian, as they follow a physical particle (the benchmark) over time, whereas benchmark-free observations are Eulerian, only reporting the vertical coordinate corresponding to a specific geographic location.

Physical benchmarks are the optimal choice regarding reliability, repeatability and interpretability. Examples are presented in Figure 23.8 and include brass levelling bolts installed in sturdy infrastructure, piles driven down into presumably stable deep strata, masts or buildings carrying antenna, radar reflectors and transponders and integrated geodetic reference stations (IGRS; Hanssen, 2017). Whilst optimal for repeatability and reliability, physical benchmarks are expensive and are often insufficient in terms of spatial sampling. Moreover, they are representative only for the displacement of the monument (object) on which they are placed and not for displacements occurring in strata above the lowest point of the monument, which are often significantly more dynamic. Another drawback is that these benchmarks are subject to loss over time, which is problematic if very long time series are required.

Virtual benchmarks satisfy the condition of repeatability but cannot be pinpointed physically. Such benchmarks are the product of satellite or airborne radar interferometry and are dependent on the earth's scattering conditions. Although such virtual benchmarks have less optimal characteristics compared to physical benchmarks, their advantage lies in their sheer number. There are ~35.000 physical levelling benchmarks in the Netherlands compared to billions of virtual radar benchmarks. This has pros and cons: more benchmarks are favourable for the spatial sampling density and for redundancy, but they are of varying nature and quality, and therefore often harder to interpret unambiguously. Whereas physical benchmarks are installed or selected before measurements take place, virtual benchmarks can only be selected a posteriori, given a set of observations. This enables the estimation of parameters over areas based on archived data.

Benchmark-free subsidence estimates stem from, for instance, repeated airborne laser scanning surveys or from bare-soil levelling campaigns. In these cases the land is

Figure 23.8. *Physical benchmarks. a) the Amsterdam Ordnance Datum NAP ('Normaal Amsterdams Peil') at Noordeinde, Delft. b) A corner reflector for InSAR satellite data. c) Integrated geodetic reference station (IGRS) with GNSS antenna on the mast, airborne laser scanning horizontal platform, levelling benchmark, and two corner reflectors.*

surveyed repeatedly, but typically observations from subsequent campaigns are made at different locations and under different environmental circumstances, which reduces precision and repeatability. Moreover, while the Lagrangian physical or virtual benchmarks guarantee capturing real motion of an object or surface, the Eulerian benchmark-free estimates cannot distinguish between subsidence of the surface and extraction or suppletion of material, e.g. by digging a hole.

Geodetic techniques and observables

Explicit identification of the objectives and of the a priori understanding of the subsidence signal of interest allows for the choice of preferred benchmark types, which leads to a limited set of options for geodetic surveying techniques: (i) spirit or hydrostatic levelling, (ii) continuous or campaign style GNSS (Global Navigation Satellite Systems), (iii) satellite InSAR (Interferometric Synthetic Aperture Radar), (iv) airborne laser scanning, and (v) relative or absolute gravimetry.

Levelling requires a levelling instrument, mounted on a tripod, and operated manually. The principal observable is a spatial single difference in orthometric elevation between two benchmarks (Torge & Müller, 2012). The required double-difference subsidence variable is derived by differencing spatial single-differences between epochs. For high precision and reliability it requires (i) the a priori installation of physical benchmarks (see Fig. 23.8a) representative of the subsidence signal of interest and (ii) manual in-situ land surveying. Cost increases with the area of interest and the temporal sampling.

Using a system of satellites orbiting the earth, GNSS (Global Navigation Satellite Systems) determine the position of a local GNSS antenna. The principal observable of GNSS is the 3D position of the GNSS-antenna reference point, acquired at one position and one epoch (Teunissen & Montenbruck, 2017). The required double-difference subsidence variable is a derived quantity, obtained by differencing either the spatial single-differences between epochs (campaign style), or the temporal single differences between positions (continuous). It requires the a priori installation of physical benchmarks representative of the subsidence signal of interest (see Fig. 23.8c).

Satellite radars emit signals towards the earth and record their subsequent reflections. Using interferometric techniques (InSAR) the change in position of a large number of these reflections can be estimated. For InSAR, the principal observable is the double-difference displacement in the line-of-sight to the radar instrument, derived from the interferometric phase (Hanssen, 2001). The selection of representative virtual benchmarks is performed a posteriori based on the observations. The modulo ambiguity in the phase measurements results in a non-unique solution, which can only be solved by conditioning the parameter estimation using prior knowledge.

Airborne laser scanning is performed from an airplane that sends laser pulses to earth and records the travel time to receive their reflections. The principal observable is the 3D position of a location on the earth's surface (Carter et al., 2007). The required double-difference variable is a derived quantity, obtained by differencing the elevation of two positions for one epoch, and subsequently subtracting two epochs. This results in Eulerian differences.

Gravimetry requires a gravimeter instrument and is operated manually, most often in campaigns. The principal observable is the gravitational acceleration acquired at one position and one epoch (absolute gravimetry), or the spatial single-difference between gravitational accelerations at two positions at one epoch (relative gravimetry; Timmen, 2010). The required double-difference elevation variable is derived by spatio-temporal or temporal differencing, respectively, assuming that there are no mass movements between the epochs, only geometric changes.

Subsidence Estimation

Given that the objectives, benchmarks, and a priori understanding of the signal are known, a set of subsidence parameters can be estimated from various sets of geodetic observations. This involves a mathematical model, consisting of two parts. The functional model describes the relation between the parameters and the observables, while the stochastic model describes the stochastic characteristics of those observables. The choice for these two models is based on the a priori information and assumptions and influences the outcome, as different choices will result in a different product. For example, in a levelling campaign the implicit assumption is that the benchmarks used only exhibit motion that is due to the subsidence signal of interest. For InSAR the implicit assumption is that the phase ambiguities have been successfully resolved.

This dependency on prior knowledge to estimate a subsidence signal introduces a sequential *reciprocity* between the models and the observations, which demands a continuous interaction between physical and observational disciplines.

Extensometers

In-situ observations, obtained for instance with extensometers, may be used to unravel the total subsidence signal. Extensometers are used to measure compaction worldwide (e.g. Poland, 1984). They derive point measurements of vertical movement at various (sub)surface levels at mm-scale accuracy. As such, the contribution of individual layers, and in some cases processes, to the total subsidence can be determined. Traditionally, in the Netherlands, extensometers have been used in geotechnical investigations, for instance for the determination of the subsidence after a building mound has been constructed. With the increasing demand for accurate monitoring of shallow subsidence and for a better understanding of the observed subsidence, extensometers are now also applied in the Dutch (agricultural) peat meadows area. Van Asselen et al. (2020) described the first application of extensometers that are capable of capturing shallow subsidence in peatlands with average rates in the order of mm yr^{-1} and even rates of cm $week^{-1}$, occurring on shorter timescales. Since late 2018, four extensometers at two livestock farms in Rouveen, Overijssel have hourly measured vertical movements of (sub)surface levels. Preliminary results show vertical movements (up and down) in the order of centimetres on the timescale of weeks (Van Asselen et al., 2020). These movements seem to be related to groundwater-level fluctuations, but evapotranspiration is expected to contribute to additional subsidence during the summer period.

Dutch Ground Motion Service (*Bodemdalingskaart*)

An information service containing ground motion has been made publicly available online (Hanssen et al., 2020). Using satellite InSAR data, the displacement time series of 200 million virtual benchmarks have been estimated and the results presented in an interactive portal

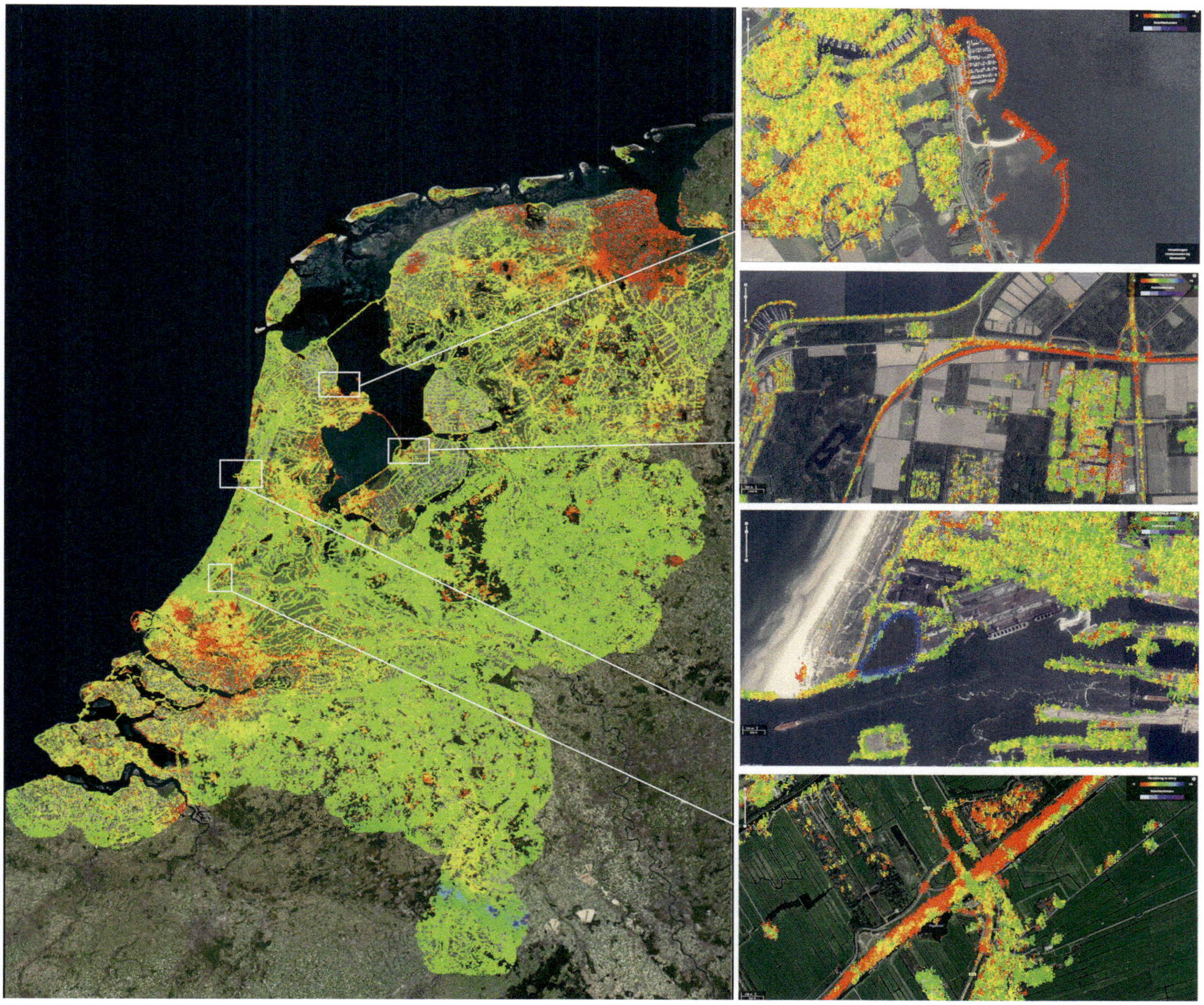

Figure 23.9. *Displacement time series of 200 million virtual benchmarks using satellite InSAR data. The data are presented in an interactive online portal (bodemdalingskaart.nl). Green colours indicate relative stability, blue colours indicate upward soil movement and orange and red colours indicate downward soil movement. Reflections are mainly collected from hard surfaces as illustrated by the inset maps. For a European-scale InSAR viewer: egms.land.copernicus.eu.*

(bodemdalingskaart.nl, see Fig. 23.9). The estimates are based on generic settings and are not optimized for a particular application. Based on the statistical properties of the data, a differentiation has been made between processes in the shallow and the deep subsurface.

Integration: combining detected and modelled subsidence

As outlined in the previous sections, the surface movement constitutes a measurable signature of the processes in the subsurface, which can thus provide information about the subsurface (Maury et al., 1992). This is beneficial, since subsurface parameters are usually associated with large uncertainties. Typical uncertainties include the spatio-temporal distribution of the pressure and temperature changes at depth (e.g. Fokker et al., 2012; Candela et al., 2017), the amount of depletion of aquifers connected to a gas field (Fokker et al., 2016), the value of the compaction coefficient of the reservoir rock and its dependency on porosity and pressure (Van Thienen-Visser & Fokker, 2017), the time dependence of compaction (Fokker et al., 2019), the geomechanical response of the reservoir surroundings up to the surface (Marketos et al., 2016) and other interfering causes such as deep and shallow expressions of compaction (Kroon et al., 2009). These uncertainties immediately propagate into uncertainties in subsidence estimates when forecasting models are deployed.

Inverse modelling or data assimilation can be employed to constrain subsurface knowledge and parameters with

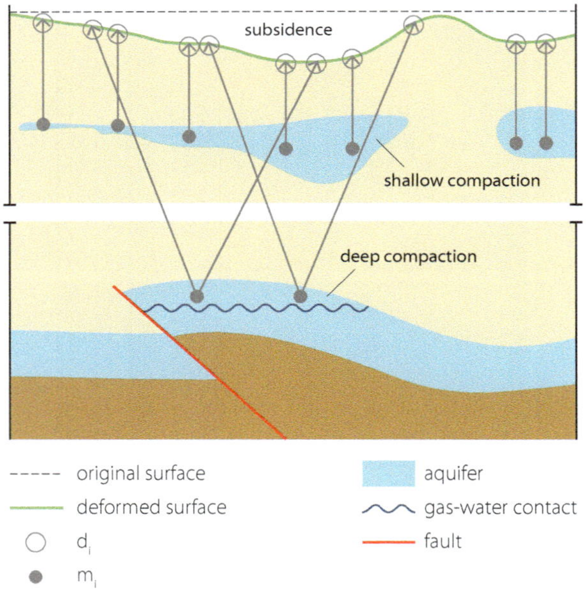

Figure 23.10. Demonstration of compaction sources leading to a cumulative and smoothed subsidence expression.

surface movement data (Tarantola, 2005), although this is not straightforward. It is especially difficult for cases where there is a combination of a spatially distributed source of compaction such as a depleting gas field and a heterogeneous shallow subsurface, where there are strongly nonlinear forward models or interfering causes of subsidence and where there are highly correlated data (see Fig. 23.10). A direct application of inverse models often results in non-unique or unstable solutions. Additional information in the form of prior knowledge is required. We will briefly sketch how inverse modelling of subsidence can be approached in such a way that the result is useful for forecasting and governance.

Linear inversion can be employed if the relationship between driving parameters and estimated surface movement is linear. This is, for instance, the case if the compaction is the unknown parameter and the subsurface response is linearly elastic. The prior knowledge then can be a prior estimate of the compaction estimate (Marchina, 1996; Fokker et al., 2012). The forecast uncertainty for the estimate forms an automatic output of the procedure if the data and model uncertainties are correctly known (Menke, 2018).

A more complicated inversion is usually required. The unknowns may be the subsurface elastic or visco-elastic behaviour or the reservoir permeability that controls the pressure distribution. Classical history matching methods employ a trial-and-error methodology; these may, however, miss important regions of the parameter space (Vasco et al., 2000). When few parameters need to be estimated, an iterative least-squares approach may be appropriate (Vasco et al., 2010). These methods, however, do not result in an automatic estimate of the confidence range of the forecasted subsidence.

Modern estimation algorithms frequently employ an ensemble approach: a large number of model realizations is used. Such models originated in weather forecasting (Mackenzie, 2003) but quickly also found applications in the subsurface domain (Evensen, 2009). One example of an ensemble approach is the particle filter (Van Leeuwen, 2009; Nepveu et al., 2010). Many models are created with limited prior knowledge of the driving parameters, so the ensemble of all realizations should highlight the prior uncertainties of the model parameters in a range of subsidence predictions. The model outcome then is compared with the data and the models that fit the data best are favoured by giving them larger weights than others. A posterior model estimate is finally constructed by means of a weighted average of all prior forecast realizations. The elegance of this approach is that model realizations do not have to be updated: the prior models are used to construct an improved model forecast. However, the problem with the method is that with many measurement points, a tendency towards divergence or ensemble collapse emerges: one single model collects virtually all the weight and the prediction coincides with that model only. This behaviour also renders the calculation of a posterior uncertainty range impossible. An unrealistic number of model realizations would be required to obtain a reasonable number of non-zero posterior weights (Van Leeuwen, 2009).

Current state-of-the-art procedure is to employ data assimilation, whereby both the model forecast and the parameter space are modified in adapting the model ensemble. In data assimilation, the complete ensemble is used to construct an update step. Crucial ingredient in this update step is the correlation between model parameters, model predictions and data. This correlation is used to construct an updated ensemble of model parameters and model predictions that honours the measured data.

Many brands of data assimilation exist. The most common one is the Ensemble Kalman Filter. Every time that new data are collected, the Ensemble Kalman Filter (EnKF) procedure is applied, and the model parameters and predictions are updated. A progressively developing forecast mean and bandwidth are created. The Ensemble Kalman Filter has proven to be very effective in weather forecasting, but also in reservoir engineering approaches (Evensen, 2009; Katterbauer et al., 2014). For subsidence modelling, the Ensemble Smoother has been used more frequently (Baù et al., 2015; Fokker et al., 2016; Zoccarato et al., 2016). This method constructs the new ensemble in a single step (Baù et al., 2015). When applied with nonlinear forward models or non-Gaussian statistics, however, results are inferior to those obtained using EnKF (Tavakoli et al., 2013). Publications by Emerick & Reynolds (2013a,b) and Tavakoli et al. (2013) compare different ensemble techniques and conclude that an Ensemble Smoother with Multiple Data Assimilation (ES-MDA) is a good alternative: computational cost is like that of EnKF but the result of ES-MDA agrees better with the 'true' model in their synthetic cases.

An example application of ES-MDA was given by Fokker et al. (2019) for the case of surface subsidence in the southern Flevopolder. The combined knowledge (data, models, and data assimilation technology) gave a new insight into the processes at work in this reclaimed area. Here, subsidence was mainly the result of clay shrinkage after its first exposure to air. An example of employing a particle filter is NAM's approach to subsidence in the Wadden Sea. All uncertainties present in geology, reservoir dynamics, geomechanical behaviour and subsidence data are stochastically integrated in a large set of models (Van Thienen-Visser et al., 2015; NAM, 2017; TNO, 2017). The forecasts, updated with the modified weights for the various model realizations, were used to assess the subsidence in the Wadden Sea in conjunction with the expected sea level rise and the Wadden Sea morphodynamical development (Van der Spek, 2018; Vermeersen et al., 2018; Wang et al., 2018).

Estimates which are used in inverse models should be as close as possible to the original measurements. An example where this turns out to be important is when Persistent Scatterer InSAR data (PS-InSAR) are employed. These measurements result from the projection of the 3D displacement vector onto the line of sight of the satellite. Using ascending and descending satellite orbits, the displacement vector can be decomposed further, albeit still insufficiently for full decomposition in the east, north and vertical directions (Klemm et al., 2010; Janna et al., 2012; Rucci et al., 2013). The decomposition into vertical and east-west movements, however, introduces the need for a spatial interpolation, because the persistent scatterers identified for ascending and descending satellite orbits are not identical. Thus, such decomposition is allowed only when the deformation signal is sufficiently smooth and sufficiently well-sampled. Interpolation can be avoided by using the line-of-sight data directly in the parameter estimation procedure, as is common practice in the assessment of volcanic activity and earthquakes (Jónsson et al., 2002; Anderssohn et al., 2009; Fokker et al., 2016). Estimates of the purely vertical movement can be made with the updated model ensemble.

Governance

Subsidence is a complex problem. There are many stakeholders with different interests, whilst their level of awareness of the issue is often limited. This may be related to the fact that damage cannot unambiguously be attributed to subsidence, nor to low levels of knowledge and insight into the root cause and possible mediation strategies (Erkens et al., 2015). As a result, as with many problems that root in earth sciences, the process of formulating a subsidence strategy or implementing appropriate measures is often problematic, leading to particularly difficult decision making.

Land subsidence: adaptation or mitigation

In developing coping strategies, there is the option of reducing subsidence (mitigation) or reducing the negative consequences of subsidence (adaptation). Most likely a combination of both will be optimal (Erkens et al., 2015). In coastal areas, where subsidence quickly leads to increased flooding, adaptation often forms the initial hazard management strategy. Adaptation measures may for instance target the increased flood risk. In the Netherlands, the response to human-induced subsidence over the last ca. 1000 years is a prime example of a subsidence adaptation strategy. The stepwise raising of embankments has traditionally been the response to lowering land elevations and the consequently increasing flood risk (Fig. 23.1). Similarly, subsidence that reduced drainage capacity of the surface water and overall wetter conditions have often been addressed by the installation of pumps, which led to further subsidence (e.g. Van de Ven, 1993; Erkens et al., 2016). Even though these actions were mitigation measures to address the increased flood risk, for subsidence they may be considered as adaptation measures. In case of damage to structures by land movement, the traditional approach was to adjust the structure, or completely rebuild. Over the last centuries, the increasing application of pile foundations in the Netherlands also forms an adaptation strategy to cope with the soft soil conditions.

Mitigation measures need to target the root cause of subsidence. For the Netherlands, where there are multiple causes of subsidence (see the previous sections on natural- and human-induced subsidence), this means that, for instance, in case of deep-rooted subsidence, rates could be reduced by diminishing the extraction of hydrocarbons, groundwater or salt. In case of shallow subsidence, measures need to target either the load (weight) on the soft soil or the shallow groundwater level. In both cases, the aim of the mitigation strategy could be to avoid acceleration of subsidence after an intervention or even to actively reduce the current rates. For instance, when applying a load on soft soil with low bearing capacity in the building process, mitigation of subsidence would consist in limiting the processes of consolidation and creep. This can be achieved by using lighter building material, or even by striving to achieve no net increase in loading. Similarly, preventing the temporary lowering of shallow groundwater levels (e.g. during droughts) would limit subsidence, so that on the longer term rates would slowly reduce. Permanently raising groundwater levels would actively reduce subsidence rates, but completely bringing human-induced subsidence to a halt at present is not achievable in the Netherlands. This results from the omnipresence of soft soils combined with the need for groundwater management, in addition to disturbances that cannot be prevented.

The history of coping with subsidence in the Netherlands and lock-in conditions

It is only over the last couple of decades that interest in subsidence mitigation measures has increased. The historically grown dominance of experiences with adaptation measures has led to an optimization of subsidence adaptation strategies. This applies to the technical aspects (for example the building of embankments for flood protection and pumping stations that drain the land), but also to the institutional and financial aspects. Institutions that have implemented these adaptation measures and strategies have strengthened as they have acquired the requisite technical skills and knowledge and the necessary financial resources (Seijger et al., 2018). As a result, over time it has become increasingly easy to implement an adaptation strategy to subsidence problems. This development of a single strategy to manage subsidence problems can be viewed as a path-dependent process (Sydow et al., 2009), where managerial discretion in terms of choices or options depends on the choices made in the past (Fig. 23.11). The end stage of path dependency, when managerial discretion is limited and a single management option fully dominates, is called a lock-in. The Netherlands can thus be regarded as trapped in a dual lock-in condition as the dominating adaptation strategy in terms of the applied technologies and the principal institutions act as constraints to move toward a more long-term sustainable strategy (Seijger et al., 2018) – one that also may include measures to mitigate subsidence (Erkens & Stouthamer, 2020).

The urgency to deal with subsidence with a sustainable long-term strategy is growing. As subsidence continues and cumulative land-level lowering increases, new problems arise and existing problems worsen. Also, subsidence

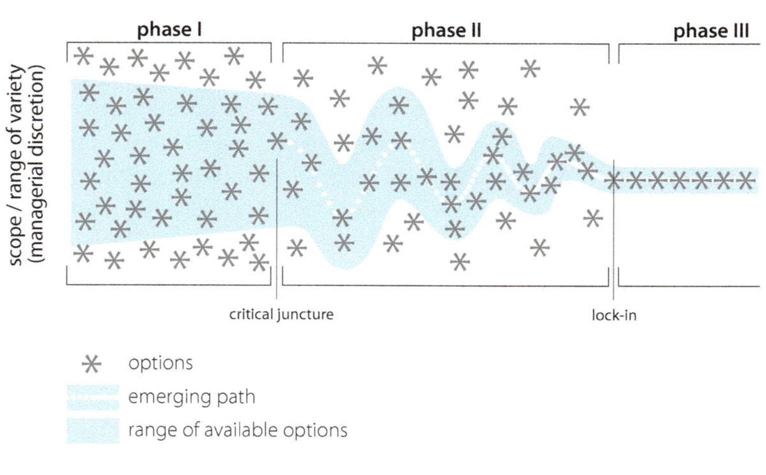

Figure 23.11. Visualization of a lock-in caused by path dependency. This is currently the situation for subsidence in many coastal areas where an adaptation strategy is preferred over a mitigation strategy (from Sydow et al., 2009; reproduced with permission).

and the increasingly grim predictions of future climate warming-induced global sea-level rise are interconnected, as they contribute both to relative sea-level rise. In many coastal areas, including the Netherlands, land elevation is increasingly viewed as an asset (Koster et al., 2020), and therefore losing elevation puts the value at risk (Erkens & Stouthamer, 2020). At the same time, exposure to subsidence is increasing, as population growth, ongoing urbanization and economic growth continues in the Dutch coastal zone, which largely consists of soft soils. As a result of increasing damage to buildings and infrastructure, societal awareness on subsidence and its consequences in the Netherlands has increased. There is a growing awareness that much of the subsidence is in fact human-induced and that it thus can be actively reduced by human action. An additional negative effect of subsidence by peat oxidation is the associated emission of CO_2. In order to comply with legally binding international treaties, the Netherlands is obliged to reduce these greenhouse gas emissions, including those from the drained peatlands and organic soils. This forms an extra incentive to review the current subsidence rates in these peatlands.

Because of the increased urgency, questions arise whether the current subsidence strategies are durable and in fact, these strategies are challenged by stakeholders. The lock-in condition makes it difficult to choose alternative management options, however, as they conflict with existing interests and break with the long-standing traditional approach to managing the subsidence using an adaptation approach. If the societal and financial benefits of alternative management approaches would be fully known a different action perspective could be offered to decision makers. For such alternatives a sound and shared knowledge, based on in-depth understanding of the physical problems, the financial perspectives (what are the costs and benefits of different options?) and governance and legal capabilities (who is responsible and are there sufficient capabilities to implement the management measures?) is required. Only when these conditions are satisfied, will the threshold to act on an alternative management strategy become lower.

Getting out of the lock-in using a sound knowledge base

Realising that in order to emerge from lock-in a sound and shared knowledge base is required to facilitate decision making, it is important to establish which elements must be included in the knowledge base. Based on experience and reviewing best practise cases, Erkens et al. (2015) gathered crucial research elements to facilitate decision making with scientific knowledge. This framework has evolved into the 6M approach to subsidence (Erkens & Stouthamer, 2020), which is a step-wise approach that elaborates along the stages of the policy cycle, with six distinct steps that need to be taken: Measuring, understanding Mechanisms, Modelling, Money, Measures and Monitoring. The policy cycle is a tool that is used to analyse the development of a policy item and has been used for decades in political sciences. In this approach, each of the steps consists of questions that need to be addressed and data that need to be gathered and interpreted. Furthermore, there are governance aspects that often need to be considered in order to answer questions such as regarding the responsibility of parties to act and the ownership and storage of data. In real life short-cuts are often taken, for instance, where measurements from the first step show the urgency of the situation and measures are directly implemented.

In this contribution we have summarized the state-of-the-art concerning the measuring and monitoring of subsidence, the understanding of subsidence mechanisms and the modelling of subsidence in the Netherlands. For policy making and decision making, these steps form the knowledge base (Erkens & Stouthamer, 2020). Here we also argue that policy making and conducting science are coeval activities: it is not recommended to wait with vision and policy forming until the first three steps have been finished and reported. In fact, immediate input from the first three steps will help shaping policy. It is important that lessons learnt are readily available. Preferably, the monitoring data and analysis results should be stored in a central database and freely accessible information platform (Van Daalen et al., 2020; Erkens et al., 2021). Another example of relevant information on the physical system for policy making that needs to be available relatively early in the process is the discrimination of different processes that contribute to subsidence. Often there are both natural and human-induced causes at the same location. For policy development this distinction is important: while it is worthwhile to implement mitigation measures to reduce human-induced subsidence, for natural subsidence only adaptation measures can be taken (Erkens et al., 2015). The proposed measure will often specifically target a single subsidence driver and cause.

While inverse modelling or data assimilation is targeted at the reduction of uncertainty in parameter values and subsidence forecasts, there will always be remaining uncertainty. The great advantage of the ensemble-based methods is that this uncertainty is quantified. It can thus also be propagated into the model forecasts for different governance scenarios. A scientifically underpinned decision making must incorporate the confidence range of the model outcomes in the process.

With subsidence predictions for different management scenarios available, the next step is to perform a cost-benefit analysis for each scenario. Estimating subsidence-related costs (damage) is notoriously complex. Subsidence is a 'hidden threat' because in practice, the actual costs appear on financial sheets as ad hoc investments or planned maintenance schemes but typically are not identified as damage costs related to subsidence and thus remain hidden (Erkens et al., 2015). Dedicated damage estimates can help to raise awareness among policymakers and initiate policy development. For subsidence, being a gradual process, mitigation measures are usually costly in the short term, but cost-effective in the long term (Erkens et al., 2015). Cost-benefit analyses could provide insights into the hidden costs and potential benefits of mitigation measures (usually avoided damage or costs) in a quantitative way. They can also provide a rationale for upfront investment in measures that realize long-term benefits. Often, with subsidence, costs and benefits are different for different stakeholders. This may lead to demands for financial compensation measures or mitigative measures to reduce costs for a certain stakeholder group.

In Gouda, a city in the western peatlands in the Netherlands, Kok and Hommes-Slag (2020) showed that the adaptation strategy that has been followed over the last centuries may be the most beneficial economically, despite the continuation of subsidence. In this scenario, costs made to avoid, reduce and repair the future subsidence-related damage to the structures were calculated at 7-16 M€, but the damage prevented (reducing damp conditions and groundwater nuisance) was approximately 13-20 M€, making this the economically most rational option. This example shows how cost-benefit analyses, based on subsidence model outcomes, may inform decision makers, helping them to unlock the lock-in.

Implementation of measures should follow cost-benefit analysis and informed decision making, with the consequences of remaining uncertainties incorporated. This will often include governance and legal aspects (who is responsible?) and financial aspects (who is paying for and who benefits from the measures?). For various examples around the world, Erkens et al. (2015) showed that the costs of the measures are always high, regardless of the strategy followed. If nothing is done, costs of repairing the subsidence damage (flooding, tilting of structures) are also high. An adaptation strategy is costly, because as subsidence continues adaptation measures need to be taken continuously. A mitigation-focused strategy is costly also because the implementation of measures must be accompanied by investing in development of databases containing measurement data and predictive model outcomes to break out of the lock-in.

The step-wise 6M approach mentioned above is a structural way of facilitating decision making using scientific knowledge prior to, during, and after the decision-making process. Moreover, it is a scheme that offers a framework to exchange best practises and approaches between different subsiding areas around the world.

Summary and outlook

Subsidence is a complex phenomenon and a complex problem to tackle, both in a technical sense and in governance. This is particularly the case in the Netherlands, a low-lying and densely populated country in which various causes of subsidence interfere. The long history of slowly manifesting subsidence in the Netherlands (more than 1000 years) has resulted in a tendency to adapt rather than to mitigate. This 'lock-in' situation is suboptimal, since potentially cheaper and more effective governance options may not be considered. What is needed is a sound knowledge base and a mindset to explore solutions outside the traditional way of thinking. To facilitate this, we have presented several knowledge and governance issues of concern. First, we have described the natural processes that cause subsidence, and the human-induced causes like groundwater management and exploitation of geological resources. After that we have elaborated on how subsidence is detected and monitored. However, well-founded decision making requires quality forecasts for different governance options, and these can only be achieved if data and models are integrated. To this end, modern ensemble-based techniques are available that combine detected and modelled subsidence. An essential element in these methods is the fundamental incorporation of expectation values and expectation bandwidths (or confidence intervals) of the ensuing model parameters and model outcomes. These can then be propagated to forecasts based on management decisions, and with the application of monetary evaluation, the right measures can be proposed and executed. This policy cycle needs to be closed through continuous monitoring and model updating.

The subsidence situation will change in the Netherlands. For the coming decades, it may be expected that subsidence resulting from gas and oil extraction will decrease as greenhouse gas emission mitigation force these activities to diminish or shut down. Conversely, we expect a continuation of surface movements induced by salt production, gas storage (methane or hydrogen) in depleted gas fields or in salt caverns, and geothermal operations including high-temperature thermal storage. In rural areas, greenhouse gas emission reduction targets will no longer allow for substantial CO_2 emissions and the associated shallow subsidence from drained peatlands by 2050. The rate and intensity of peat oxidation must therefore be reduced strongly. This implies the need for a strong reduction in subsidence in these areas as well. For the urban area, a clear aim of subsidence reduction is currently lacking and it is questionable to what extent subsidence will be reduced, if at all. In the longer term, by 2100, following the grim sea level rise predictions, the effects of subsidence may be magnified by absolute sea-level rise, increasing flood risks in the Netherlands. Subsidence will keep playing a role in shaping the future of the Netherlands for decades to come. It is therefore crucial to have a freely accessible central database and information platform for monitoring data and analysis, and to maintain and further develop the current knowledge position.

Acknowledgements

We are indebted to Ger de Lange and Pietro Teatini for their thorough reviews. Their suggestions improved the quality of this chapter. We would also like to thank the editors of this book for providing us the opportunity to include a chapter on subsidence processes in the Netherlands.

Digital map data

Spatial data of figures in this chapter for use in geographical information systems can be downloaded here: https://doi.org/10.5117/aup.28164749.

References

Anderssohn, J., Motagh, M., Walter, T.R., Rosenau, M., Kaufmann, H. & Oncken, O., 2009. Surface deformation time series and source modelling for a volcanic complex system based on satellite wide swath and image mode interferometry: the Lazufre system, central Andes. Remote Sensing of Environment 113: 2062-2075. DOI: 10.1016/j.rse.2009.05.004

Arets, E.J.M.M., Kolk, J.W.H., Hengeveld, G.M., Lesschen, J.P., Kramer, H., Kuikman, P.J. & Schelhaas, M.J., 2019. Greenhouse gas reporting of the LULUCF sector in the Netherlands. Methodological background, update 2019. Wageningen Environmental Research, WOt-technical report No. 146: 111 pp. DOI: 10.18174/472433

Barciela-Rial, M., van Paassen, L. A., Griffioen, J., van Kessel, T. & Winterwerp, J. C. (2020). The effect of solid-phase composition on the drying behavior of Markermeer sediment. Vadose Zone Journal, 19(1): e20028. DOI:10.1002/vzj2.20028

Baù, D., Ferronato, M., Gambolati, G., Teatini, P. & Alzraiee, A., 2015. Ensemble smoothing of land subsidence measurements for reservoir geomechanical characterization. International Journal for Numerical and Analytical Methods in Geomechanics 39: 207-228. DOI: 10.1002/nag.2309

Breunese, J.N., Van Eijs, R.M.H.E., De Meer, S. & Kroon, I.C., 2003. Observation and prediction of the relation between salt creep and land subsidence in solution mining. The Barradeel Case. Paper presented at SMRI Conference (Chester): 1-19.

Candela, T. , Osinga, S., Veer, E. F., Heege, J. H. & P. A. Fokker. 2017. Improving Reservoir Exploitation Using Fast Geomechanical Modelling Coupled With Surface Displacement Data Assimilation. Paper presented at the SPE Reservoir Characterisation and Simulation Conference and Exhibition (Abu Dhabi, UAE). DOI: 10.2118/186028-MS

Caro Cuenca, M., Hooper, A.J. & Hanssen, R.F., 2013. Surface deformation induced by water influx in the abandoned coal mines in Limburg, The Netherlands observed by satellite radar interferometry. Journal of Applied Geophysics 88: 1-11. DOI: 10.1016/j.jappgeo.2012.10.003

Carter, W.E., Shrestha, R.L. & Slatton, K.C., 2007. Geodetic laser scanning. Physics Today 60(12): 41-47. DOI:1063/1.2825070

Chang, L. & Hanssen, R.F., 2014. Detection of cavity migration and sinkhole risk using radar interferometric time series. Remote Sensing of Environment 147: 56-64. DOI: 10.1016/j.rse.2014.03.002

Conroy, P., Van Diepen, S.A.N., Van Asselen, S., Erkens, G., Van Leijen, F.J. & Hanssen, R.F., 2022. Probabilistic Estimation of InSAR Displacement Phase Guided by Contextual Information and Artificial Intelligence. IEEE Transactions on Geoscience and Remote Sensing: 1-1. DOI: 10.1109/TGRS.2022.3203872

Drijver, B., Bakema, G. & Oerlemans, P., 2019. State of the art of HT-ATES in The Netherlands. Proceedings of the European Geothermal Congress, The Hague (The Netherlands): 1-5

Emerick, A. & Reynolds, A.C., 2013a. Ensemble Smoother with multiple data assimilation. Computers & Geosciences 55: 3-15. DOI: 10.1016/j.cageo.2012.03.011

Emerick, A. & Reynolds, A.C., 2013b. Investigation of the sampling performance of ensemble-based methods with a simple reservoir model. Computers & Geosciences 17: 325-350. DOI: 10.1007/s10596-012-9333-z

Erkens, G., Bucx, T., Dam, R., De Lange, G. & Lambert, J., 2015. Sinking coastal cities. Proceedings of the International Association of Hydrological Sciences 372: 189-198. DOI: 10.5194/piahs-372-189-2015

Erkens, G., Kooi, H. & Melman, R., 2021. Actualisatie bodemdalingsvoorspellingskaarten. Rapport Deltares (Utrecht), No. 11206724-002-BGS-0001: 48 pp.

Erkens, G. & Stouthamer, E., 2020. The 6M approach to land subsidence. Proceedings of the International Association of Hydrological Science 382: 733-740. DOI: 10.5194/piahs-382-733-2020

Erkens, G., Van der Meulen, M.J. & Middelkoop, H., 2016. Double trouble: subsidence and CO2 respiration due to 1,000 years of Dutch coastal peatlands cultivation. Hydrogeology Journal 24 (3): 551-568. DOI: 10.1007/s10040-016-1380-4

Evensen, G., 2009. Data assimilation: the Ensemble Kalman filter, second edition. Springer (Berlin-Heidelberg): 307 pp. DOI:10.1007/978-3-642-03711-5

Fjaer, E., Holt, R.M., Horsrud, P., Raaen, A.M. & Risnes, R., 2008. Petroleum related rock mechanics, second edition. Developments in Petroleum Science 53. Elsevier (Amsterdam): 515 pp.
Fokker, P.A. & Orlic, B., 2006. Semi-analytic modeling of subsidence. Mathematical Geology 38: 565-589. DOI: 10.1007/s11004-006-9034-z
Fokker, P.A., Visser, K., Peters, E., Kunakbayeva, G. & Muntendam-Bos, A.G., 2012. Inversion of surface subsidence data to quantify reservoir compartmentalization: A field study. Journal of Petroleum Science and Engineering 96-97: 10-21. DOI: 10.1016/j.petrol.2012.06.032
Fokker, P.A., Wassing, B.B.T., Van Leijen, F.J., Hanssen, R.F. & Nieuwland, D.A., 2016. Application of an ensemble smoother with multiple data assimilation to the Bergermeer gas field, using PS-InSAR. Geomechanics for Energy and the Environment 5: 16-28. DOI: 10.1016/j.gete.2015.11.003
Fokker, P.A., Van Leijen, F.J., Orlic, B., Van der Marel, H. & Hanssen, R.F., 2018. Subsidence in the Dutch Wadden Sea. Netherlands Journal of Geosciences 97 (3): 129-181. DOI: 10.1017/njg.2018.9
Fokker, P.A., Gunnink, J.L., Koster, K. & De Lange, G., 2019. Disentangling and Parameterizing Shallow Sources of Subsidence: Application to a Reclaimed Coastal Area, Flevoland, the Netherlands. Journal of Geophysical Research: Earth Surface 124(5): 1099-1117. DOI: 10.1029/2018JF004975
Geertsma, J., 1973. Land subsidence above compacting oil and gas reservoirs. Journal of Petroleum Technology 25 (6): 734-744. DOI: 10.2118/3730-PA
Grant, M.E. & Bixley, P.F., 2011. Geothermal reservoir engineering, second edition. Elsevier (Amsterdam): 378 pp. DOI: 10.1016/C2010-0-64792-4
Groenenberg, R., Den Hartogh, M. & Fokker, P., 2025. Salt Production. In: Ten Veen, J.H. Vis, G.-J., De Jager, J. & Wong, Th.E. (eds): Geology of the Netherlands, second edition. Amsterdam University Press (Amsterdam): 603-625. DOI: 10.5117/9789463728362_ch16
Hanssen, R.F., 2001. Radar Interferometry: Data Interpretation and Error Analysis. Kluwer Academic Publishers (Dordrecht): 308 pp. DOI: 10.1007/0-306-47633-9
Hanssen, R.F., 2017. A radar retroreflector device and a method of preparing a radar retroreflector device," Jun. 21 2017, international Patent WO2018236215A1. https://Patents.Google.Com/Patent/WO2018236215A1/En.
Hanssen, R.F., Van Leijen, F.J. & Van Zwieten, G., 2008. Bodemdaling Kennemerland waargenomen met satelliet radarinterferometrie (InSAR). In: Barends, F., Dillingh, D., Hanssen, R. & Van Onselen, K. (eds): Bodemdaling langs de Nederlandse kust. Case Hondsbossche en Pettemer Zeewering. IOS press (Amsterdam): 69-99.
Hanssen, R.F., Van Leijen, F.J., Leezenberg, P.B. & Maljaars, H., 2020. De bodemdalingskaart van Nederland.
Hettema, M., Papamichos, E. & Schutjens, P.M.T.M., 2002. Subsidence delay: field observations and analysis. Oil & Gas Science and Technology 57: 443-458. DOI: 10.2516/ogst:2002029
Houtgast, R.F. & Van Balen, R.T., 2000. Neotectonics of the Roer Valley Rift System, the Netherlands. Global and Planetary Change 27 (1-4): 131-146. DOI: 10.1016/S0921-8181(01)00063-7
Janna, C., Castelletto, N., Ferronato, M., Gambolati, G. & Teatini, P., 2012. A geomechanical transversely isotropic model of the Po River basin using PSInSAR derived horizontal displacement. International Journal of Rock Mechanics and Mining Sciences 51: 105-118. DOI: 10.1016/j.ijrmms.2012.01.015
Jónsson, S., Zebker, H., Segall, O. & Amelung, F., 2002. Fault slip distribution of the 1999 Mw 7.1 Hector Mine, California, earthquake, estimated from satellite radar and GPS measurements. Bulletin of the Seismological Society of America 92: 1377-1389. DOI: 10.1785/0120000922
Katterbauer, K., Hoteit, I. & Sun, S., 2014. Data assimilation of InSAR measurements for large scale reservoirs. Paper presented at the SPE Saudi Arabia Section Technical Symposium and Exhibition, Al-Khobar (Saudi Arabia). DOI: 10.2118/172211-MS
Klemm, H., Quseimi, I., Novali, F., Ferretti, A. & Tamburini, A., 2010. Monitoring horizontal and vertical surface deformation over a hydrocarbon reservoir by PSInSAR. First Break 28 (5): 29-37. DOI: 10.3997/1365-2397.2010014
Kok, S. & Hommes-Slag, S., 2020. Cost-benefit analysis of urban subsidence mitigation strategies in Gouda, the Netherlands. Proceedings of the International Association of Hydrological Sciences 382: 761-766. DOI: 10.5194/piahs-382-761-2020
Kooi, H., 2008. Compactiebijdragen aan bodemdaling langs de kust. In: Barends, F., Dillingh, D., Hanssen, R. & Van Onselen, K. (eds): Bodemdaling langs de Nederlandse kust. Case Hondsbossche en Pettemer Zeewering. IOS Press (Amsterdam): 19-28.
Kooi, H., 2000. Land subsidence due to compaction in the coastal area of The Netherlands: The role of lateral fluid flow and constraints from well-log data. Global and Planetary Change 27 (1-4): 207-222. DOI: 10.1016/S0921-8181(01)00067-4
Kooi, H., 2018. Bodemdaling door grondwateronttrekking in het Westland en omgeving. Rapport Deltares (Utrecht), No. 11202399-005: 88 pp.
Kooi, H., Johnston, P., Lambeck, K., Smither, C. & Molendijk, R., 1998. Geological causes of recent (~100 yr) vertical land movement in the Netherlands. Tectonophysics 299: 297-316. DOI: 10.1016/S0040-1951(98)00209-1
Koster, K., 2016. Cone Penetration testing: a sound method for urban archaeological prospection. Archeological Prospection 23 (1): 55-69. DOI: DOI: 10.1002/arp.1531
Koster, K., Erkens, G. & Zwanenburg, C., 2016. A new soil mechanics approach to quantify and predict land subsidence by peat compaction. Geophysical Research Letters 43 (20): 10792-10799. DOI: 10.1002/2016GL070207
Koster, K., Stafleu, J. & Cohen, K.M., 2017. Generic 3D interpolation of Holocene base-level rise and provision of accommo-

dation space, developed for the Netherlands coastal plain and infilled palaeovalleys. Basin Research 29 (6): 775-797.

Koster, K., Stafleu, J. & Stouthamer, E., 2018. Differential subsidence in the urbanised coastal-deltaic plain of the Netherlands. Netherlands Journal of Geosciences 97 (4): 215-227. DOI: 10.1017/njg.2018.11

Koster, K., Stafleu, J., Vos, P.C. & Van der Meulen, M.J., 2020. Can we elevate the subsiding coastal plain of the Netherlands with controlled sedimentation? Proceedings of the International Association of Hydrological Sciences 382: 767-773. DOI: 10.5194/piahs-382-767-2020

Kroon, I.C., Nguyen, B.L., Fokker, P.A., Muntendam-Bos, A.G. & De Lange, G., 2009. Disentangling shallow and deep processes causing surface movement. Mathematical Geosciences 41 (5): 571. DOI: 10.1007/s11004-008-9197-x

Kuikman, P.J., De Groot, W.J.M., Hendriks, R.F.A., Verhagen, J. & De Vries, F., 2003. Stocks of C in soils and emissions of CO2 from agricultural soils in the Netherlands, Alterra-Report 561: 42 pp.

Mackenzie, D., 2003. Ensemble Kalman filters bring weather models up to date. Siam News 36 (8): 10-12.

Marchina, P.J.M., 1996, The Use Of Subsidence Data To Monitor Reservoir Behaviour. Presented at the European Petroleum Conference, Milan (Italy). DOI: 10.2118/36918-MS

Marketos, G., Spiers, C.J. & Govers, R., 2016. Impact of rock salt creep law choice on subsidence calculations for hydrocarbon reservoirs overlain by evaporite caprocks. Journal of Geophysical Research: Solid Earth 121 (6): 4249-4267. DOI: 10.1002/2016JB012892

Maury, V.M.R., Grasso, J.R. & Wittllinger, G., 1992. Monitoring of subsidence and induced seismicity in the Lacq gas field (France): the consequences on gas production and field operation. Engineering Geology 32: 123-135. DOI: 10.1016/0013-7952(92)90041-V

Meijles, E.W., Kiden, P., Streurman, H.J., Van der Plicht, J., Vos, P.C., Gehrels, W.R. & Kopp, R.E., 2018. Holocene relative mean sea-level changes in the Wadden Sea area, northern Netherlands. Journal of Quaternary Sciences 33 (8): 905-923. DOI: 10.1002/jqs.3068

Menke, W., 2018. Geophysical data analysis: Discrete inverse theory, fourth edition. Elsevier, (Amsterdam): 352 pp. DOI: 10.1016/C2011-0-69765-0

Morishita, Y. & Hanssen, R.F., 2015. Deformation parameter estimation in low coherence areas using a multisatellite InSAR approach. IEEE Transactions on Geoscience and Remote Sensing 53(8):249-252. DOI: 10.1109/TGRS.2015.2394394

NAM, 2017. Ensemble Based Subsidence application to the Ameland gas field – long term subsidence study part two (LTS-II) continued study. NAM report EP201710200509: 208 pp.

Neller. J.R., 1944. Oxidation loss of lowmoor peat in fields with different water tables. Soil Science 58: 195-204.

Nepveu, M., Kroon, I.C. & Fokker, P.A., 2010. Hoisting a red flag: an early warning system for exceeding subsidence limits. Mathematical Geosciences 42: 187-198. DOI: 10.1007/s11004-009-9252-2

Nicodemo, G., Peduto, D. & Ferlisi, S., 2020. Building damage assessment and settlement monitoring in subsidence-affected urban areas: case study in the Netherlands. Proceedings of the International Association of Hydrological Sciences 382: 651-656. DOI: 10.5194/piahs-382-651-2020

Poland, J.F., 1984. Guidebook to studies of land subsidence due to ground-water withdrawal. Studies and reports in hydrology, UNESCO. 40 (74). 323 pp.

Riva, R.E.M., Frederikse, T., King, M.A., Marzelon, B. & Van den Broeke, M.R., 2017. Brief communication: The global signature of post-1900 land ice wastage on vertical land motion. The Cryosphere 11: 1327-1332. DOI: 10.5195/tc-11-1327-2017

Rostampour, V., Jaxa-Rozen, M., Bloemendaal, M., Kwakkel, J. & Kevinczky, T., 2019. Aquifer Thermal Energy Storage (ATES) smart grids: Large-scale seasonal energy storage as a distributed energy management solution. Applied Energy 242: 624-639. DOI: 10.1016/j.apenergy.2019.03.110

Rucci, A., Vasco, D.W. & Novali, F., 2013. Monitoring the geologic storage of carbon dioxide using multicomponent SAR interferometry. Geophysical Journal International 193: 197-208. DOI: 10.1093/gji/ggs112

Seijger, C., Ellen, G.J., Janssen, S., Verheijen, E. & Erkens, G., 2018. Sinking deltas: trapped in a dual lock-in of technology and institutions. Prometheus 35(4):1-21. DOI: 10.1080/08109028.2018.1504867

Simon, K.M., Riva, R.E.M., Kleinherenbrink, M. & Frederikse, T., 2018. The glacial isostatic adjustment signal at present day in northern Europe and the Britisch Isles estimated from geodetic observations and geophysical models: Solid Earth 9: 777-795. DOI: 10.5194/se-9-777-2018

Stephens, J.C. & Johnson, L., 1951. Subsidence of organic soils in the Upper Everglades region of Florida. Proceedings of Soil Science Society Florida: 191-237. DOI: 10.5772/intechopen.72925

Sydow, J., Schreyögg, G. & Koch, J., 2009. Organisational Path Dependence: Opening the Black Box. Academy of Management Review 34: 689-709. DOI: 10.5465/AMR.2009.44885978

Tarantola, A., 2005. Inverse problem theory and methods for model parameter estimation. SIAM (Paris): 352 pp. DOI: 10.1137/1.9780898717921

Tavakoli, R., Yoon, H., Delshad, M., ElSheikh, A.H., Wheeler, M.F. & Arnold, B.W., 2013. Comparison of ensemble filtering algorithms and null-space Monte Carlo for parameter estimation and uncertainty quantification using CO2 sequestration data. Water Resources Research 49: 8108-8127. DOI: 10.3997/2214-4609.20141876

Teunissen, P. & Montenbruck, O., 2017. Springer handbook of global navigation satellite systems. Springer Nature (Switzerland): 1268 pp. DOI: 10.1007/978-3-319-42928-1

Timmen, L., 2010. Absolute and relative gravimetry. In: Xu, G. (ed.), Sciences of Geodesy. Springer (Berlin/Heidelberg): 1-48.

TNO, 2017. Ensemble-based Subsidence Interpretation and Prediction ESIP: Technical Reference Manual. TNO (Utrecht), Report No. R11278: 27 pp.

Torge, W. & Müller, J., 2012. Geodesy, fourth edition. Walter de Gruyter (Berlin, Boston): 444 pp. DOI: 10.1515/9783110250008

Valeda, A., 2021. Relation between land subsidence and groundwater abstraction in the Province of Overijssel. MSc thesis Utrecht University (the Netherlands). TNO Report 2021 S11411: 65 pp.

Van Asselen, S., Erkens, G. & De Graaf, F., 2020. Monitoring shallow subsidence in cultivated peatlands. Proceedings of the International Association of Hydrological Sciences 382: 189-194. DOI: 10.5194/piahs-382-189-2020

Van Asselen, S., Erkens, G., Stouthamer, E., Woolderink, H.A.G., Geeraert, R.E.E. & Hefting, M.M., 2018. The relative contribution of peat compaction and oxidation to subsidence in built-up areas in the Rhine-Meuse delta, The Netherlands. Science of the Total Environment 636: 177-191. DOI: 10.1016/j.scitotenv.2018.04.141

Van Daalen, T.M., Fokker, P.A., Bogaard, P.J. & Van der Meulen, M.J., 2020. Why we urgently need a public subsidence information service in the Netherlands. Proceedings of the International Association of Hydrological Sciences, 382: 821-823. DOI: 10.5194/piahs-382-821-2020

Van Dam, P.J., 2001. Sinking Peat Bogs: Environmental Change Holland, 1350-1550. Environmental History 6 (1): 32-45. DOI: 10.2307/3985230

Van de Ven, G.P., 1993. Leefbaar laagland - Geschiedenis van de waterbeheersing en landaanwinning in Nederland. Uitgeverij Matrijs (Utrecht): 455 pp.

Van den Akker, J.J.H., Kuikman, P.J., De Vries, F., Hoving, I., Pleijter, M., Hendriks, R.F.A., Wolleswinkel, R.J., Simões, R.T.L. & Kwakernaak, C., 2008. Emission of CO2 from agricultural peat soils in the Netherlands and ways to limit this emission. In: Farrell, C. & Feehan, J. (eds): 2008. Proceedings of the 13th International Peat Congress After Wise Use – The Future of Peatlands (Tullamore, Ireland): 645-648.

Van den Born, G. J., Kragt, F., Henkens, D., Rijken, B., van Bemmel, B., van der Sluis, S., Polman, N., Bos, E. J., Kuhlman, T., Kwakernaak, C., van den Akker, J., Diogo, V., Koomen, E., de Lange, G., van Bakel, J. & ten Brinke, W. B. M. (Ed.), 2016. Dalende bodems, stijgende kosten: mogelijke maatregelen tegen veenbodemdaling in het landelijk en stedelijk gebied: beleidsstudie. Planbureau voor de Leefomgeving (Den Haag). PBL-publicatie No. 1064: 96 pp.

Van der Meulen, M.J., Van der Spek, A.J.F., De Lange, G., Gruijters, S.H.L.L., Van Gessel, S.F., Nguyen, B.-L., Maljers, D., Schokker, J., Mulder, J.P.M. & Van der Krogt, R.A.A., 2007. Regional Sediment Deficits in the Dutch Lowlands: Implications for Long-Term Land-Use Options. Journal of Soils and Sediments 7 (1): 9-16. DOI: 10.1016/j.quascirev.2004.01.011

Van der Spek, A.J.F., 2018. The development of the tidal basins in the Dutch Wadden Sea until 2100: the impact of accelerated sea-level rise and subsidence on their sediment budget - a synthesis. Netherlands Journal of Geosciences 97 (3): 71-78. DOI: 10.1017/njg.2018.10

Van Laarhoven, S., 2017. Influence of loading history on subsurface architecture and subsidence potential for the historical city of Gouda, The Netherlands. MSc thesis Utrecht University (the Netherlands): 175 pp.

Van Leeuwen, P.J., 2009. Particle filtering in geophysical systems. Monthly Weather Review 137 (12): 4089-4114. DOI: 10.1175/2009MWR2835.1

Van Lochem, H., Vis, G.-J. & Jagt, J.W.M., 2025. Late Cretaceous. In: Ten Veen, J.H., Vis, G.-J., De Jager, J. & Wong, Th.E. (eds): Geology of the Netherlands, second edition. Amsterdam University Press (Amsterdam): 253-291. DOI: 10.5117/9789463728362_ch08

Van Thienen-Visser, K., Breunese, J. N., and A. G. Muntendam-Bos. 2015. Subsidence due to Gas Production in the Wadden Sea: how to Ensure no Harm will be done to Nature. Paper presented at the 49th U.S. Rock Mechanics/Geomechanics Symposium (San Francisco, California)

Van Thienen-Visser, K. & Fokker, P.A., 2017. The future of subsidence modelling: compaction and subsidence due to gas depletion of the Groningen gas field in the Netherlands. Netherlands Journal of Geosciences 96 (5): s105-s116. DOI: 10.1017/njg.2017.10

Vasco, D.W., Karasaki, K. & Doughty, C., 2000. Using surface deformation to image reservoir dynamics. Geophysics 65: 132-147. DOI: 10.1029/2009GL041544

Vasco, D.W., Rucci, A., Ferretti, A., Novali, F., Bissell, R.C., Ringrose, P.S. & Wright, I.W., 2010. Satellite-based measurements of surface deformation reveal fluid flow associated with the geological storage of carbon dioxide. Geophysical Research Letters 37: L03303. DOI: 10.1029/2009GL041544

Venmans, A.M., Op de Kelder, M., De Jong, J., Korff, M. & Houtepen, M., 2020. Reliability of InSAR satellite monitoring of buildings near inner city quay walls. Proceedings of the International Association of Hydrological Sciences 382: 195-199.

Vermeersen, B.L.A., Slangen, A.B.A., Gerkema, T., Baart, F., Cohen, K.M., Dangendorf, S., Duran-Matute, M., Frederikse, T., Grinsted, A., Hijma, M.P., Jevrejava, S., Kiden, P., Kleinherenbrink, M., Meijles, E.W., Palmer, M.D., Rietbroek, R., Riva, R.E.M., Schulz, E., Slobbe, D.C., Simpson, M.J.R., Sterlini, P., Stocchi, P., Van de Wal, R.S.W. & Van der Wegen, M., 2018. Sea-level change in the Dutch Wadden Sea. Netherlands Journal of Geosciences 97 (3): 79-127. DOI: 10.1017/njg.2018.7

Vis, G.-J., Van Linden, E., Van Balen, R.T. & Cohen, K.M., 2020. Depressions caused by localized subsidence in the Netherlands, Belgium and Germany: a link with coal mining? Proceedings of the International Association of Hydrological Sciences 382: 201-205. DOI: 10.5194/piahs-382-469-2020

Vos, P.C., 2015. Origin of the Dutch coastal landscape: long-term landscape evolution of the Netherlands during the Holocene,

described and visualized in national regional and local palaeogeographical map series. PhD thesis Utrecht University (the Netherlands), Barkhuis, Groningen: 369 pp.
Wang, Z.B., Elias, E.P.L. & Van der Spek, A.J.F., 2018. Sediment budget and morphological development of the tidal inlet systems in the Dutch Wadden Sea. Netherlands Journal of Geosciences / Geologie En Mijnbouw 97 (3): 183-214. DOI: 10.1017/njg.2018.8
Zhao, J. & Konietzky, H., 2021. An overview on flooding induced uplift for abandoned coal mines. International Journal of Rock Mechanics and Mining Sciences 148: 104955. DOI: 10.1016/j.ijrmms.2021.104955
Zoccarato, C., Baù, D., Ferronato, M., Gambolati, G., Alzraiee, A. & Teatini, P., 2016. Data assimilation of surface displacements to improve geomechanical parameters of gas storage reservoirs. Journal of Geophysical Research: Solid Earth 121 (3): 1441-1461. DOI: 10.1002/2015JB012090

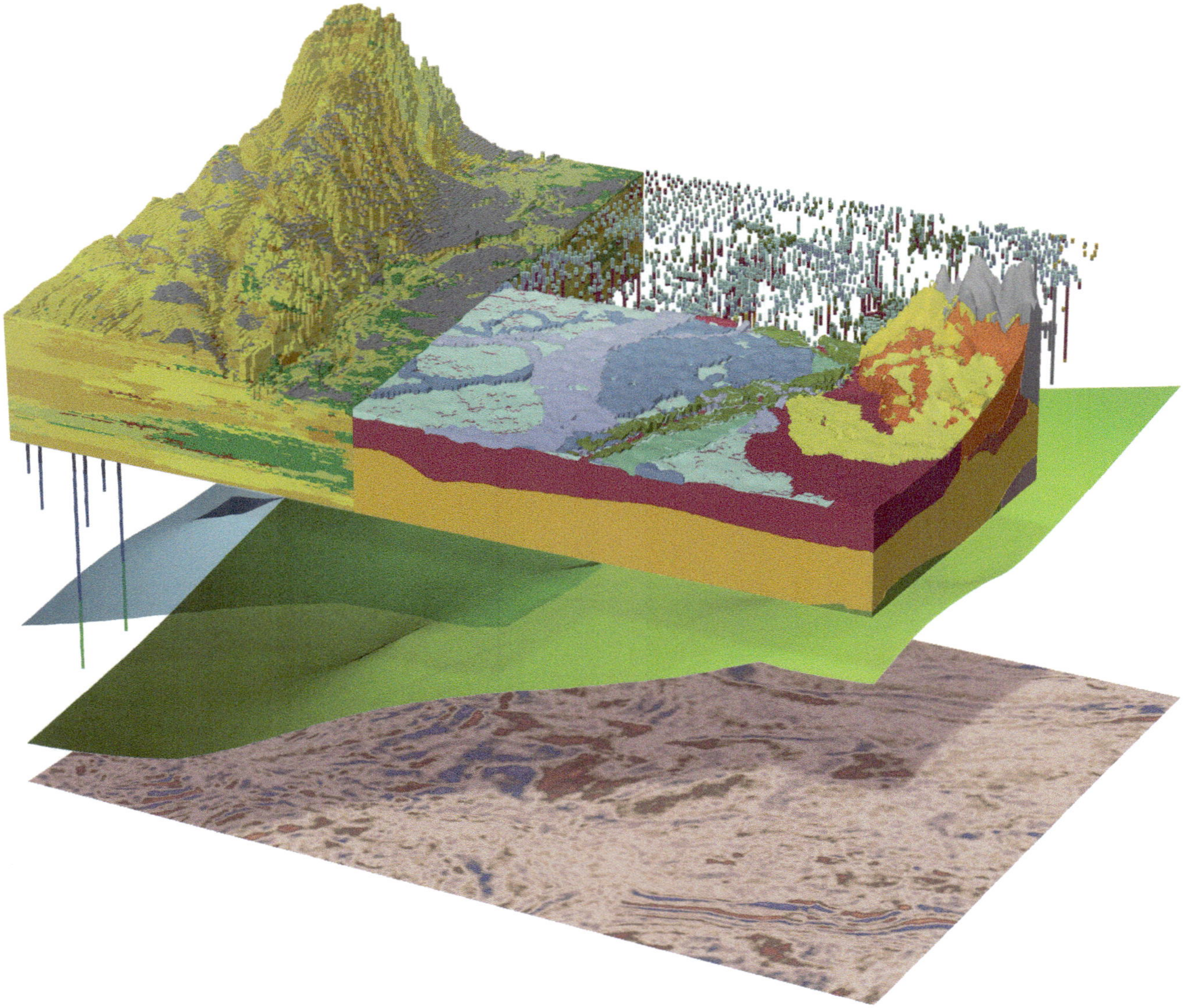

24

Geological subsurface models of the Netherlands

Jan Stafleu, Freek S. Busschers, Michiel J. van der Meulen, Maryke den Dulk, Jan L. Gunnink, Denise Maljers, Jan H. Hummelman, Jeroen Schokker, Ronald W. Vernes, Jelte Stam, Willem Dabekaussen, Johan H. ten Veen, Hans Doornenbal, Romée Kars & Renée de Bruijn*

*Corresponding author: jan.stafleu@tno.nl

ABSTRACT

TNO – Geological Survey of the Netherlands, builds and maintains a suite of 3D national subsurface models. Down to a depth of about 7 km, DGM-deep maps Carboniferous to Neogene seismo-stratigraphic horizons, using exploration data that energy and mining companies must submit to the Survey under the Mining Act. Down to about 500 m, the country is covered by the layer-based models DGM and REGIS II. DGM maps the geometries of Neogene to Quaternary lithostratigraphic units. REGIS II further subdivides DGM into hydraulically parameterized hydrogeological units and is essential to regional groundwater flow models. Down to a depth of 50 m below mean sea level (NAP), the multi-purpose GeoTOP model maps both lithostratigraphy and lithology of the shallow subsurface in a detailed 3D voxel model. A fourth set of models addresses the Quaternary stratigraphic record of the North Sea down to a depth of ca. 30 m below the seabed.

This chapter describes input data and interpretation, modelling approaches and the general application of each of the models. An assessment of model uncertainties is included because the models are constructed using point and/or line observations each of which has a specific spatial density and quality. Next, the effects of the key registry of the subsurface (BRO) on the modelling approach with a focus on quality control is described. Later, the visualization of the models on web portals and in stand-alone software packages is described and is followed by a number of case studies presenting innovative methods. The chapter concludes with perspectives for systematic modelling in the Netherlands.

<< Artist impression showing several Dutch 3D subsurface models (not to scale). The top part of the figure shows a voxel model based on boreholes. At greater depths, lithostratigraphic surfaces and seismostratigraphic horizons are based on boreholes and seismic data.

DOI: 10.5117/9789463728362_ch24

Introduction

TNO – Geological Survey of the Netherlands (TNO-GDN) focuses on the sustainable use and management of the subsurface in the Dutch on- and offshore. One of TNO-GDN's primary tasks is to map the architecture and properties of the subsurface, and to make this data available to the public domain. For the deeper subsurface the management and utilization of information is governed by the Mining Act (Rijksoverheid, 2002), while for the shallow subsurface a new key registry was implemented in 2015 (Van der Meulen et al., 2013; Rijksoverheid, 2015). This key registry is a national database for subsurface data and information, which the Dutch government bodies are obliged to use when making policies or decisions related to the subsurface.

From about the year 2000 onward, following a long history of systematic mapping since the 1960s, TNO-GDN and its predecessor *Nederlands Instituut Toegepaste Geowetenschappen van TNO* (NITG-TNO), replaced its analogue geological mapping projects by digital subsurface modelling procedures (see Van der Meulen et al. (2013) for an overview). Modelling is primarily carried out using borehole and seismic data and, depending on depth, at different resolutions. At present, TNO-GDN runs three subsurface mapping programmes with four national models in total which are all data-driven and stochastic in nature (Stafleu et al., 2019) (Fig. 24.1):

- Down to a depth of ca. 10 km, the Digital Geological Model for the deep subsurface (DGM-deep) programme maps Carboniferous to Neogene seismostratigraphic horizons, using exploration data that energy and mining companies must submit to GDN under the Mining Act.
- Down to a depth of ca. 500 m, the Digital Geological Model for the shallow subsurface (DGM) and the Regional Groundwater Information System II (REGIS II) programmes map the geometries and hydrological characteristics of the Neogene, Pleistocene and Holocene stratigraphic record (Gunnink et al., 2013a). The lithostratigraphic DGM is used for applications requiring geological information. Its primary purpose is to serve as the framework for the REGIS II model, which subdivides DGM lithostratigraphic units into hydraulically parameterized hydrogeological units (Vernes & Van Doorn, 2005; Vernes et al., 2021). REGIS II is the *de facto* hydrogeological standard used in groundwater flow models and other assessments for Dutch water and environmental authorities.
- Down to a depth of 50 m below NAP (Dutch Ordnance Datum, approximately mean sea level - MSL), the GeoTOP programme maps the lithostratigraphy and lithology of the Neogene, Pleistocene and Holocene stratigraphic record in a 3D voxel model (Stafleu et al., 2011a, 2021a). Lithological variation within the lithostratigraphic layers is mapped in great detail. GeoTOP is a multi-purpose model supporting a number of applications involving the shallow subsurface, for example the planning of large infrastructural works, land subsidence predictions, groundwater flow studies and site-response assessments of induced earthquakes. GeoTOP currently covers ca. 70% of the surface area of the Netherlands. In areas where GeoTOP is not yet available, the less detailed voxel model NL3D (Van der Meulen et al., 2013) can be used. Another example of a 3D voxel model with national coverage is TNO-GDN's aggregate resources model (Van der Meulen et al., 2025, this volume).

A fourth set of models addresses the Pleistocene and Holocene stratigraphic records of the North Sea to a depth of ca. 30 m below the seabed. These models are more application oriented than the ones mentioned above. Sand resource models such as DIS (Maljers-Oosterwijk et al., 2014; TNO-GDN, 2022a) and TILES (Hademenos et al., 2019; Van Lancker et al., 2019) are constructed for sand extraction purposes and cover a zone of up-to ca. 12 km seaward of the coastline.

3D modelling at TNO-GDN is primarily data-driven, placing it on the implicit end of the modelling spectrum (Stafleu et al., 2019). In implicit models, geological concepts such as the layering sequence and layer contact types are implemented in specialized modelling software (e.g. Isatis®, Petrel®) and automated modelling workflows (Russell et al., 2019). Furthermore, implicit modelling relies on computer algorithms (e.g. kriging, indicator simulation) to model layer boundaries (tops and bases) and layer properties (e.g. lithology, hydraulic conductivity). It is a complex approach involving many steps that have to be completed each time the model is recreated using new data or new geological insights (Fig. 24.2). However, the emphasis on data and computer algorithms does not imply that explicit geological knowledge is not used. Interpretation of boreholes, wells and seismic data as well as mapping of the spatial extent of geological units still strongly relies on the manual implementation of geological concepts by geologists, as is the case in explicit geological modelling (Russell et al., 2019). This is even the case for the development of automated stratigraphic interpretation routines for GeoTOP, where geologists with programming skills implement geological concepts in Python-code.

The next section describes input data and interpretation, modelling approaches and general application of each of the four models. An assessment of quantification of the model uncertainties is included because the models are constructed using point and/or line observations that each have a specific spatial density and quality, and also because predictions are made using a flow of different stochastic (interpolation) algorithms. The effects of the key registry of the subsurface on the modelling approach with a focus on Quality Control (QC) is described in the third section of this chapter. Subsequently, the presentation of the models on the different web portals and by using stand-alone software packages is described, followed by a number of applications in which innovative methods were used that will be implemented in future updates of the models. The chapter concludes with perspectives for systematic modelling in the Netherlands.

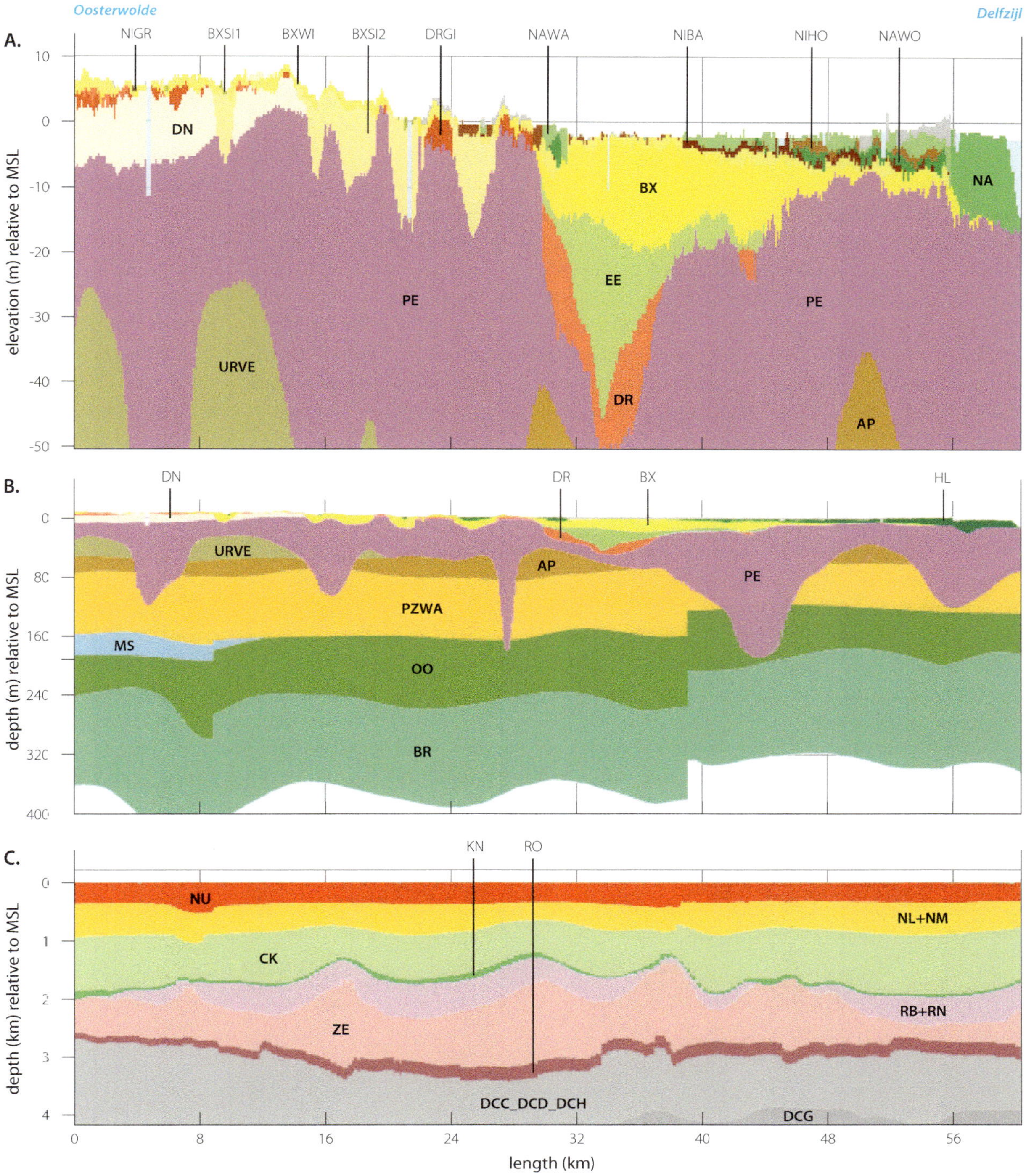

Figure 24.1. Stratigraphic cross-sections through three of the four national subsurface models in the Groningen area (north-eastern Netherlands). a) GeoTOP; b) DGM; c) DGM-deep. The cross-sections were created using the DINO-web portal (www.dinoloket.nl/en). The abbreviations for groups, formations and members are explained on the portal and in Kruiver et al. (2017b, their Table 1). Modified after Kruiver et al. (2017b).

The national models

DGM-model for the deep subsurface

The deep mapping programme was the first systematic modelling effort undertaken by the Geological Survey of the Netherlands. In 1985, the Survey was commissioned to compile a consistent, regional-scale petroleum geological framework. Geological horizons, ranging from Permian to Neogene in age, were mapped, the results of which were first published on paper and later became the constituents of a stacked layer model now referred to as DMG-deep (Figs 24.1c, 24.2; Duin et al., 2006; Kombrink et al., 2012; Van der Meulen et al., 2013). The model is based on 2D and 3D seismic survey data, combined with a variety of well data, and supported by biostratigraphic, petrophysical and geochemical analyses. Attribution of hydrocarbon and later of geothermal reservoirs relies on well data as well as burial history analysis and basin modelling techniques. The latter approach is used to predict maturation levels of source rock, as well as reservoir and seal properties (porosity, permeability, geothermal gradients). The general approach and workflow of the deep mapping programme correspond to those of the hydrocarbon exploration and production industry, but on a regional instead of a reservoir scale.

The modelling of the deep subsurface relies on the combined interpretation of seismic amplitude data and exploration well data (Fig. 24.2). Seismic surveys are ideal for observing the subsurface layer structure including faults at depths of up to several kilometres. A seismic dataset can either be acquired in 2D (cross-section) or 3D (cube) and the penetration depth depends largely on the acquisition parameters of the survey.

TNO-GDN hosts the national geophysical database that contains all seismic data acquired and made public under the Dutch Mining Act. Data delivered to TNO-GDN is so-called post-stack time migrated data, which is the first processing result obtained after stacking and migrating the signal data. In seismic data, depth is recorded in travel time between the generation of the wave at the source and the return of the reflected wave at the receiver. The true depth can only be obtained when the seismic velocity of the rock intervals is known. This velocity depends on many factors (e.g. lithology, fluid content and compaction) and generally increases with depth, but can vary significantly per layer. As velocity measurements are often sparse, the true depth is generally one of the biggest sources of uncertainty in seismic interpretation.

Whereas the seismic data provides information on geometries, well data (especially interpreted well stratigraphy or well tops) help to ground-truth the seismic interpretation by providing lithostratigraphic information in true depth. Depending on the stratigraphic level concerned, 600-3500 wells are available to constrain the seismic interpretation. A good seismic-to-well tie is essential for the interpretation of seismic reflectors that represent geological boundaries. Petrophysical well log data pertaining to sonic velocities and densities, i.e. the two quantities that define acoustic reflection strength, are commonly used to generate synthetic seismograms that can be compared with the real seismic data in order to establish an optimal seismic-to-well tie.

Over several decades, the Geological Survey executes such well-constrained seismic interpretations (TNO-NITG, 2004; Duin et al., 2006; Kombrink et al., 2012) for both nationwide and regional studies, on- and offshore. Over 95% of the publicly available 3D-surveys were used in the interpretation (Fig. 24.3) and all non-confidential wells were consulted. In general, every 10th to 20th inline and crossline in the 3D surveys is interpreted, correspond-

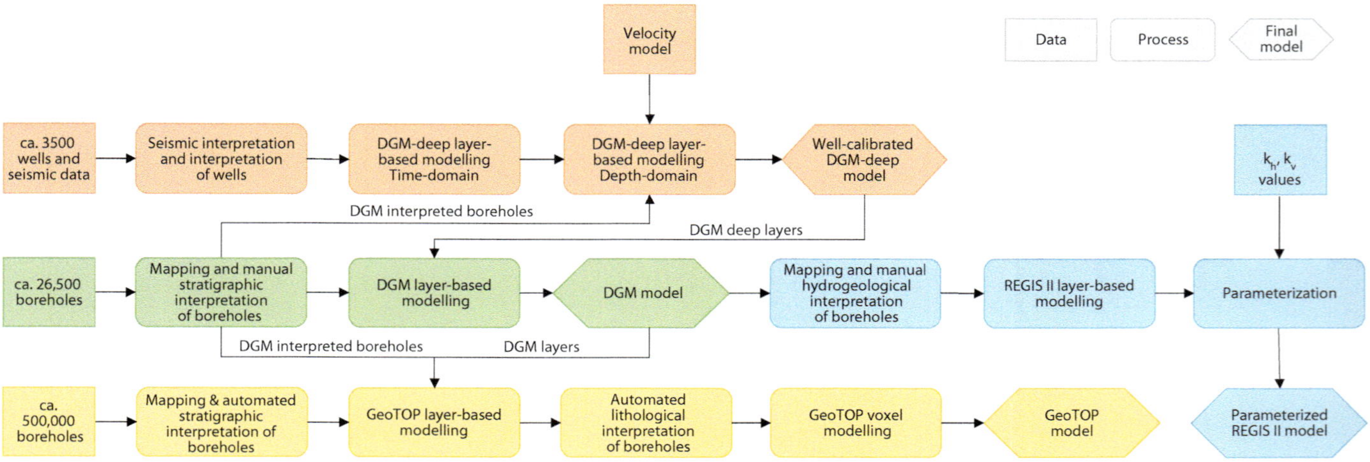

Figure 24.2. Modelling approach for the four national subsurface models with the most important modelling steps involved. Also shown are the interdependencies between the models.

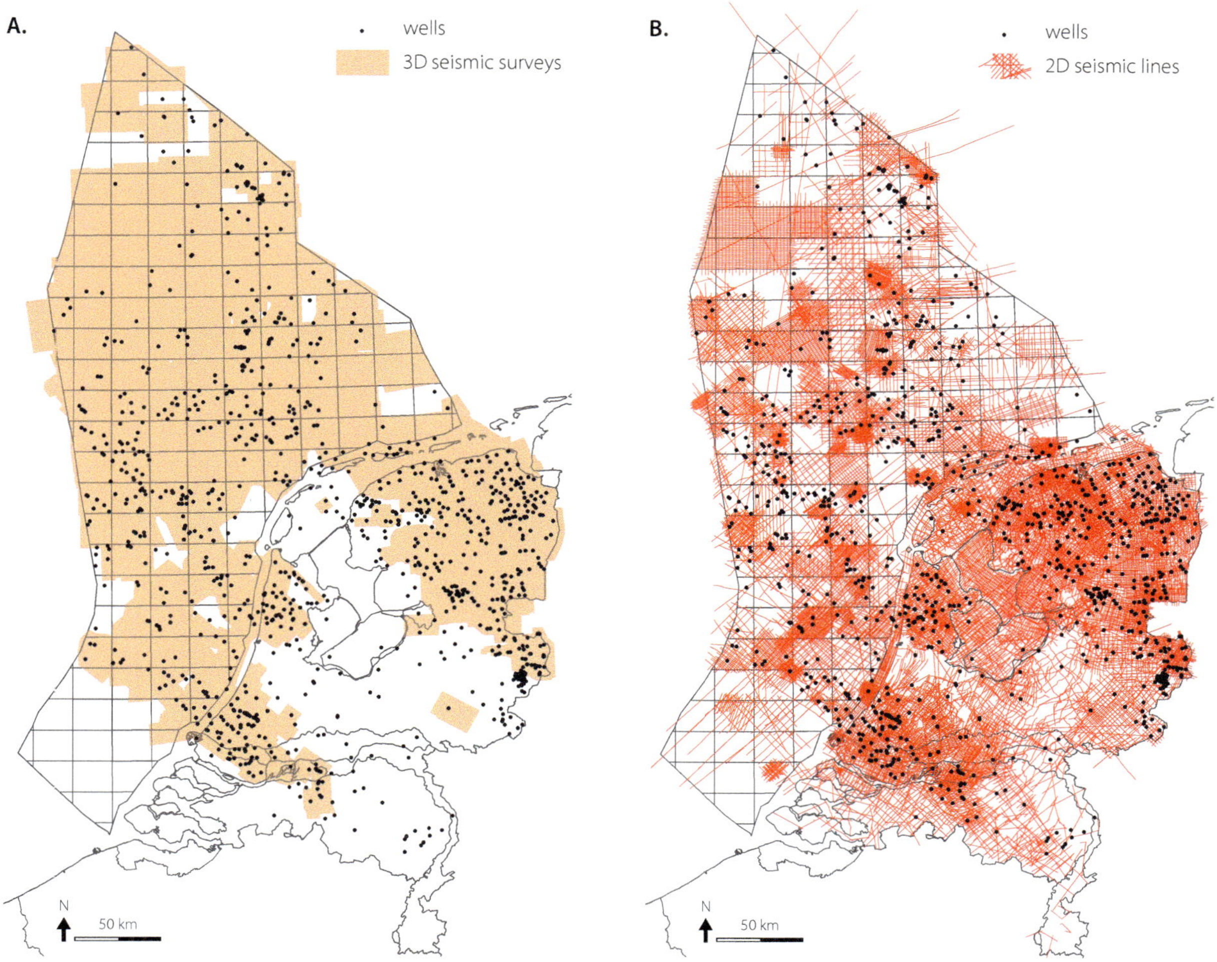

Figure 24.3. *Seismic and well coverage for the DGM-deep model.*

ing to a ca. 250 m resolution. 2D seismic data is used to fill in areas not covered by 3D seismic surveys. In some areas interpretations still originate from the early days of analogue ('paper') 2D seismic interpretations. The interpretations mainly focus on 13 stratigraphic levels that represent the (near) bases of the Paleozoic, Mesozoic, and Cenozoic lithostratigraphic groups (Fig. 24.4) as defined in the Stratigraphic Nomenclature of the Netherlands (TNO-GDN, 2022b). Faults that are evident in seismic data, i.e. faults that have a vertical offset larger than the seismic resolution, are interpreted as well.

Both the seismic interpretations and the well markers are used to produce the DGM-deep model, a stacked layer model of the deep subsurface of the Netherlands represented at 250 x 250 m grid resolution. In every new update of DGM-deep, newly available public data are incorporated. The latest 2019 version, DGM-deep v5, combines on- and offshore data into a single layer model (Fig. 24.4). Faults are modelled separately and are not part of the DGM-deep model. However, the ca. 6000 fault objects that result from the fault-modelling workflow (see Textbox 1) are implicitly used in the layer modelling process by taking into account the fault offset.

The applied modelling workflow initially produces time grids that are generated by interpolating seismic horizon picks using a convergent gridding algorithm in Petrel®. Subsequently, a conversion from the time domain to the depth domain is made using VELMOD 3.1, a velocity model built from well log and check-shot data (Van Dalfsen et al., 2005; Pluymaekers et al., 2017). After time-depth conversion, miss-ties between the generated depth grids and the well marker depths are analysed. Wells with high miss-ties, for instance those that are related to simplified model representations of complex geological features are filtered out. Subsequently, the remaining miss-tie values are interpolated with a kriging algorithm and the resulting well-residual grids are added to the depth converted grids to obtain a well-tied stratigraphic model, i.e. a model that acknowledges all selected well data (see also Kombrink et al., 2012).

In the southwest and easternmost part of the country, where Paleogene and Neogene units are close to the surface, interpreted boreholes used for the DGM-model of the shallow subsurface are used to assist in the construction of the DGM-deep layer model. In these areas, the depth range of the North Sea groups is too shallow to be detected by exploration seismic data. The same applies to the southernmost part of the country where the Upper Cretaceous Chalk Group crops out or is at shallow depths.

A limitation of the modelling approach and the applied grid resolution is that structural complexity is poorly honoured, particularly around salt structures and reverse faults. For example, duplication of stratigraphy, as is the case near overhanging salt structures or reverse faults, is simplified to sub-vertical representations.

In the final modelling step, the model's uncertainty is addressed by calculating standard deviation grids resulting from stochastic simulations of those horizons that are based on seismic interpretation. DGM-deep uses assumptions of uncertainty in data quality as well as geological complexity that are incorporated in different model simulations by a stochastic simulation technique. Quantification of uncertainty in depth of the stratigraphic interfaces is based on the variation that is present between the different model runs (Fig. 24.5).

The DGM-deep model may be regarded as the national 3D geological map suitable for applications requiring geological information on the depth and distribution of consolidated rocks in the deeper subsurface (>1500 m). The lithostratigraphic layers in the model are linked to their

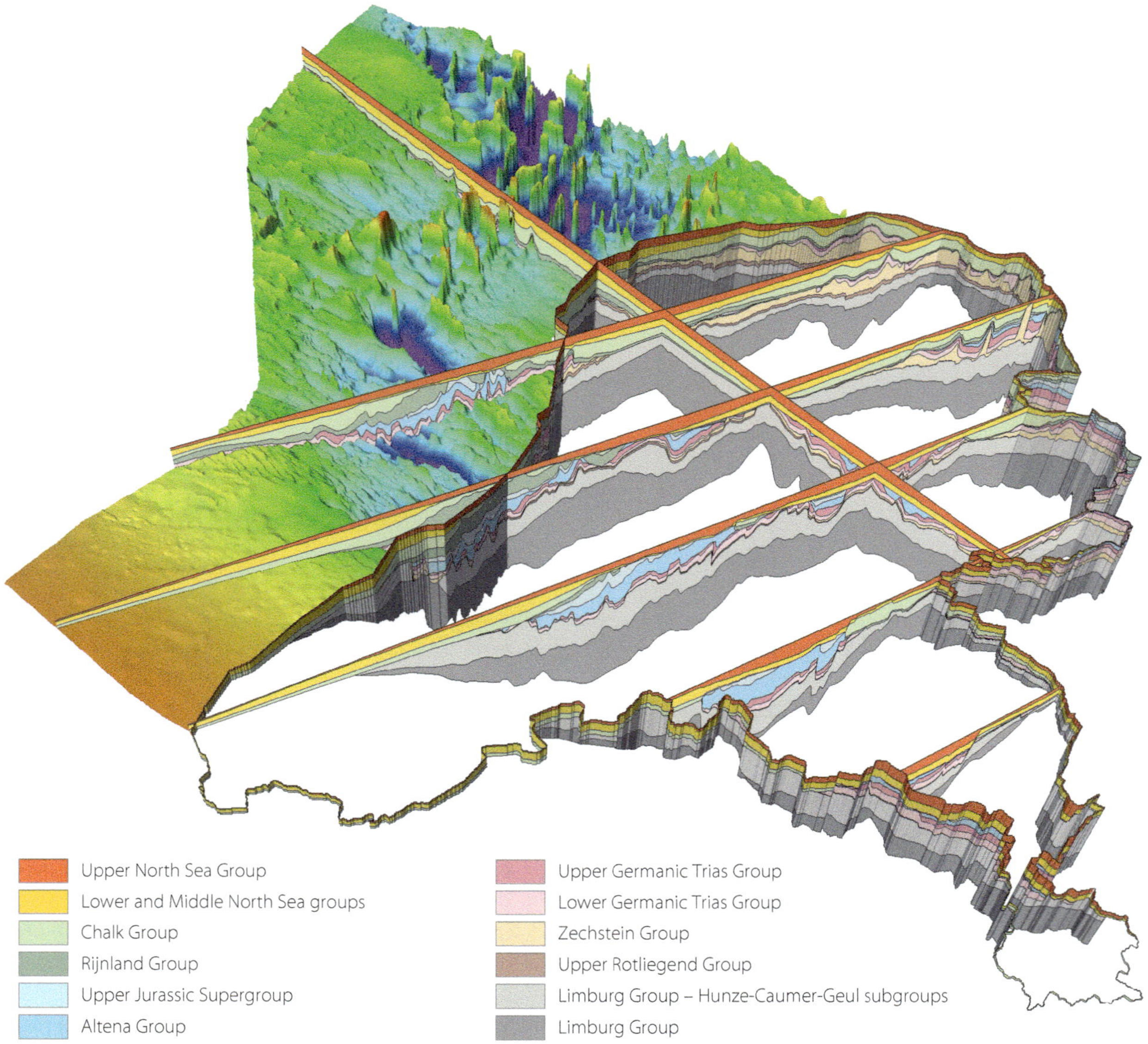

Figure 24.4. Oblique view on the 3D layer model DGM-deep v5 (available at the DINO-web portal, www.dinoloket.nl/en), covering both the on- and offshore domains of the Netherlands. The visualized surface represents the top of the Zechstein Group in the offshore domain. Colour range indicates depth from shallow (red) to deep (purple).

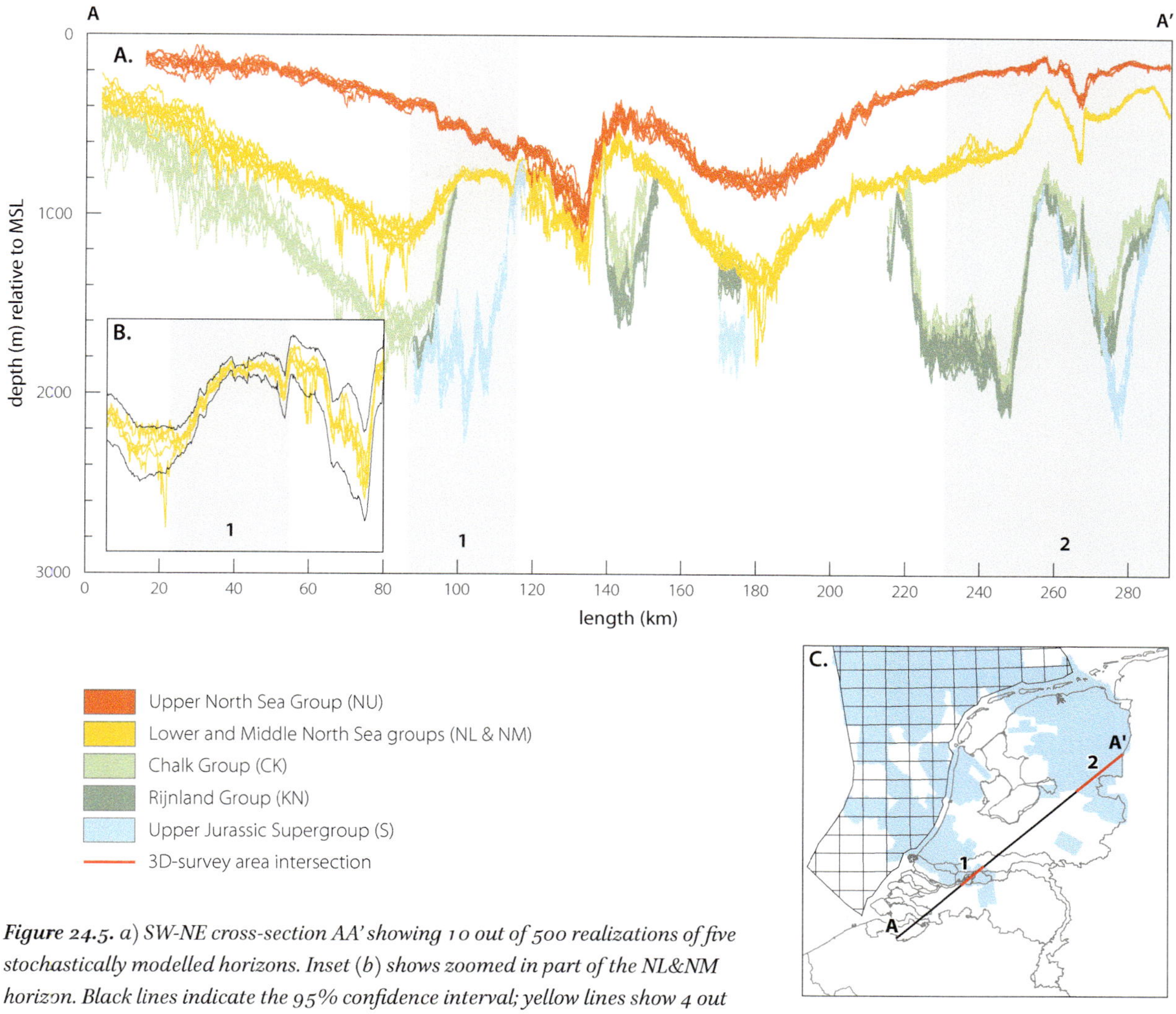

Figure 24.5. *a) SW-NE cross-section AA' showing 10 out of 500 realizations of five stochastically modelled horizons. Inset (b) shows zoomed in part of the NL&NM horizon. Black lines indicate the 95% confidence interval; yellow lines show 4 out of 500 individual realizations. c) Map showing the position of the cross-section AA' with the 3D-survey areas indicated in blue. Red horizontal lines at the top of the cross-section mark 3D-survey areas shown as grey intervals in a).*

definition and description in the Stratigraphic Nomenclature (TNO-GDN, 2022b; see the section about model visualization), allowing users to explore both layer geometry and general rock composition in their area of interest. The model also serves to understand the general geological evolution of the Netherlands, as the presence, thickness and distribution of the units is largely controlled by the main tectonic events (De Jager et al., 2025, this volume). Stratpiler (Kombrink et al., 2012) is a method to automatically extract structural elements from DGM-deep. It groups al possible stratigraphic sequences into specific classes that comply with the definition of highs, platforms and basins that pertain to a certain geological period.

The DGM-deep model does not contain information on individual reservoir- or aquifer layers and cannot be applied directly to study the multivarious potential of the deeper subsurface. The geothermal aquifer potential maps included in ThermoGIS v2.0, however, do include such information and the DGM-deep model was used as a basis for detailing it. The ThermoGIS model will be described in more detail in the section on innovative modelling methods and case studies; a comprehensive description of the construction of the model can be found on the NLOG website (www.nlog.nl/en/welcome-nlog).

DGM-model for the shallow subsurface

Systematic digital mapping of the shallow Dutch subsurface started in 1999 with the development of the Digital Geological Model (DGM; Figs 24.1b, 24.2; Stafleu et al., 2011b, 2019; Gunnink et al., 2013a). DGM is constructed using a set of ca. 26,500 consistently interpreted boreholes. It is a 3D stacked-layer lithostratigraphic model of the entire onshore part of the Netherlands down to a depth of ca. 500 m (with a maximum of 1200 m in the Roer Valley Gra-

Faults

Whereas few faults are visible at the surface, at larger depths (>1 km), the subsurface of the Netherlands is characterized by structural complexity and a dense network of faults dominated by SE-NW orientations (see Ten Veen, 2021; De Jager et al., 2025, this volume). The different tectonic styles in the Netherlands are related to various tectonic phases including the Paleozoic assembly of the Pangea super-continent (Caledonian and Variscan orogenies), the Mesozoic rifting and break-up of Pangea (Cimmerian phases), and the interplay of Alpine collision and Atlantic rifting during the Late Cretaceous and Cenozoic. Consequently, many of the faults have been reactivated during various tectonic phases. At a larger scale, fault zones form tectonic boundaries between different geological domains. At a smaller scale the faults themselves strongly influence or determine the characteristics and kinematic behaviour of rock formations (e.g. fluid flow, tendency to generate earthquakes). In the Netherlands, evident and well-known faults are often linked to the occurrence of natural earthquakes in tectonically active regions. Yet certain subsurface activities (e.g. production and injection) may also trigger earthquakes in regions that are otherwise tectonically stable, induced seismicity in the Groningen gas field being the most renowned example. Furthermore, faults may also be related to other types of hazards such as migration of fluids and gas from deep formations towards shallower groundwater aquifers or the surface environment. Faults are thus a critical component of the TNO-GDN models.

In the deep subsurface, faults can be interpreted based on the offsets of seismic reflectors. Due to the resolution of the seismic data, offsets smaller than 10-20 m are not easily recognizable. At the seismic line, the interpreted fault is represented by a so-called fault stick. In 2D seismic lines the fault sticks are isolated features that in between the seismic lines need to be connected based on the interpreter's judgement. In 3D seismic data, multiple fault sticks can be combined into 3D fault surfaces and therefore fault interpretations can be made with higher confidence.

Based on the interpretation, a fault model is created that selects and groups fault sticks that belong to one fault. In this process, a geologically sensible solution is sought that, as much as possible, honours regional trends. In 3D, the ability to display the seismic amplitude data or seismic attributes horizontally (in time slices) or at geologically sensible surfaces provides adequate means to detect the true orientation and extent of the fault. Where available, fault information at or close to the surface (e.g. Van Balen et al., 2019, 2021) is also used to constrain the location of the modelled fault surface.

3D image of two intersecting seismic lines. Fault interpretations are made on the seismic lines and are represented by so-called 'fault sticks' (in red). The fault sticks are combined into fault planes (light blue planes) to produce a fault model. This involves selecting and (re) grouping fault sticks that belong to one fault and making the interrelationships between faults geologically consistent (green planes are faults). An edge-detection method using the variance (in grey scale) aids in highlighting faults and is here attributed to the base of the North Sea Supergroup surface.

The shallow subsurface (ca. <500 m depth) is not covered adequately by conventional seismic exploration data and very few shallow seismic datasets have been acquired. Therefore, in contrast to the DGM-deep model, faults for the DGM model of the shallow subsurface are inferred from sedimentary thickness and facies changes that emerge from correlation and interpolation of borehole data. To achieve a connection with the deeper subsurface, the uppermost stratigraphic level of the DGM-deep model, i.e. the base of the Upper North Sea Group, is used as a guiding surface for interpreting shallow faults. Contrary to the deep subsurface, in the shallow subsurface faults are represented as vertical planes. This oversimplification is rationalized by the fact that most faults steepen toward the surface. Moreover, this vertical inclination is hardly noticeable at the mapping scales used.

All modelled Dutch faults are collected in the European Fault Database that was developed in the context of the GeoERA-HIKE project (funded by the European Union's Horizon 2020 research and innovation programme) (Ten Veen, 2021). This database builds on a hierarchical classification system (Hintersberger et al., 2018) that ties the mapped faults to tectonic units and to larger- or smaller scale faults. It includes existing (upgraded) fault data and many previously unpublished faults and aligns with the observation that fault systems are segmented and consist of many individual faults. Geometrical fault objects can be retrieved from the GIS-database and, using a vocabulary, are linked to a concept giving it hierarchical and temporal context (Hintersberger et al., 2021). Currently, all modelled faults can be represented as 2D intersection lines with the main stratigraphic horizons (or the land surface) they transect. The database, for example, allows for simple selection of active faults in a certain area and provision of semantic information and fault attributes. The latter are still mainly limited to geometric aspects (length, strike, dip, surface area), fault type (e.g. normal, reversed), timing of fault activity (youngest stratigraphic surface affected) and observation/evaluation method (e.g. seismic interpretation, inferred modelling).

Apart from geological relevance, the fault database can be used to support hazard assessments (natural and induced seismicity, fluid pathways and seals, gas migration, ground motion and subsidence). Furthermore, the database has economic and environmental relevance as it provides the framework for the location of various resources.

ben). It consists of a series of layers, in which each lithostratigraphic unit is represented by rasters for top, base and thickness of the unit (cell size 100 x 100 m). Raster layers are stored in two different formats: Esri's non-proprietary ASCII grid format and ERDAS Imagine's proprietary binary grid format (which has a better performance in GIS applications). The lithostratigraphic units are modelled at formation level. The complex fluvio-deltaic Holocene deposits comprising four formations, six members and eight beds, are represented by one layer only.

The ca. 26,500 borehole descriptions used in DGM are selected from the DINO database (Fig. 24.6). This selection aims at a spatially even distribution of good quality borehole data penetrating the Quaternary and Neogene deposits. Shallow boreholes, which are those less than 10 m deep, are not used in DGM, except in the northern Netherlands where they were used for mapping of the hydrologically important glacial till deposits at the base of the Drente Formation. The selected boreholes are stratigraphically interpreted by manually assigning a lithostratigraphic classification (Westerhoff et al., 2003; TNO-GDN, 2022b). Subsequently, the base of each of the lithostratigraphic units in the boreholes is used for interpolation and modelling. The basic strategy for the lithostratigraphic interpretation is to work from nation-wide to regional-scale cross-sections that, in turn, constitute the geological framework for the final interpretation of individual boreholes. However, these cross-sections are not explicitly used in the interpolation process.

In addition to the interpretation of the boreholes, a map of major faults in the Cenozoic deposits is constructed. The map is a thorough revision of earlier publications, including maps based on seismic data acquired for the exploration of oil and gas. Additional seismic data came from high resolution surveys in the Roer Valley Graben, which is the most prominent tectonic feature in the shallow subsurface of the Netherlands. For every lithostratigraphic unit, the faults that influenced the base of the unit are selected and used as 'barriers' in the interpolation process.

The depths of the base of each lithostratigraphic unit as derived from the borehole data are interpolated to raster surfaces using the block-kriging algorithm (Goovaerts, 1997; Chilès & Delfiner, 2012) as implemented in the geostatistical software-package Isatis®. The top surface follows indirectly from the joined basal surfaces of overlying units when all units are stacked (see 'stacking the units' below). The base surface was chosen because this surface is formed by depositional processes that are linked to the unit itself, whereas the top surface is often a reflection of younger erosive processes. Interpolation is assisted by additional information that includes: (1) maps with the maximum spatial extent of each lithostratigraphic unit; (2) trend surfaces showing geological structures (basins) or trends (dip direction and dip angle); (3) guiding points ('synthetic boreholes') inserted at locations with specific geological features like pinching out or incised channels. An important trend surface at the base of the model (formed by the Breda Subgroup in large parts of the country), is imported from the DGM-deep model. This trend surface is one of the interdependencies of the models as depicted in Fig. 24.2.

The last step in the modelling process consists of stacking the basal surfaces of each unit in a stratigraphically consistent way. In the stacking process, younger basal surfaces may intersect with older ones. In general three types of intersection are possible: (1) the upper unit has eroded the lower units. In this case the lower units are clipped by the upper unit; (2) the upper unit has been deposited against or on top of the relief of the lower unit without eroding the underlying unit. In this case the upper unit is clipped by the lower unit; (3) the intersection is an artefact of the interpolation process occurring between two conformable units. In this case the basal surfaces of the two units are adjusted in such a way that the intersection is removed. The choice of the type of intersection depends on the geological concept. The stacking process is performed within Isatis®, using grid-to-grid operations that are also available in standard GIS software. An impression of the resulting DGM model is shown in Figure 24.7.

The uncertainties in depth and thickness of the stratigraphic units in the DGM model are controlled by many factors (e.g. data quality, interpolation uncertainty). In the current DGM model, model uncertainty is quantified by applying cross-validation (Gunnink et al., 2010). In this approach a data point is left out and the depth of the base of the unit is estimated at that location by interpolation of the remaining data points. This is done sequentially for all datapoints. The difference between the true and the estimated depth at each location is used to quantify a 'regional' standard deviation. The regional uncertainty is lowered near data locations, by a factor derived from the kriging variogram. This method benefits from quantifying the overall magnitude of uncertainty by cross-validation, while retaining the influence of the variogram to determine fine scaled variation in uncertainty at locations in proximity to data points. An updated version of the method described above (Dabekaussen & Hummelman, 2021; Zehner et al., 2021) is currently in use for uncertainty calculations of layer-based modelling at TNO-GDN. The model uncertainty of the stratigraphic units is communicated as standard deviation maps, for example for the base of the Beegden Formation (Fig. 24.8).

DGM may be regarded as the national 3D geological map suitable for applications requiring geological information of the mainly unconsolidated sediments (sand, clay, peat) in the shallow subsurface. The lithostratigraphic layers in the model are linked to their definition and description in the Stratigraphic Nomenclature (TNO-GDN, 2022b) allowing users to explore both layer geometry and general sediment composition in their area of interest. However, the main purpose of the DGM model is to serve as a framework for the REGIS II and GeoTOP models described below.

REGIS II: hydrogeological model

Another important step in digital mapping was the development of the REGIS II hydrogeological model (Regional Groundwater Information System II; Vernes & Van Doorn, 2005; Van der Meulen et al., 2013; Vernes et al., 2021). The model uses the same borehole dataset as used in the DGM model. REGIS II further subdivides the lithostratigraphic

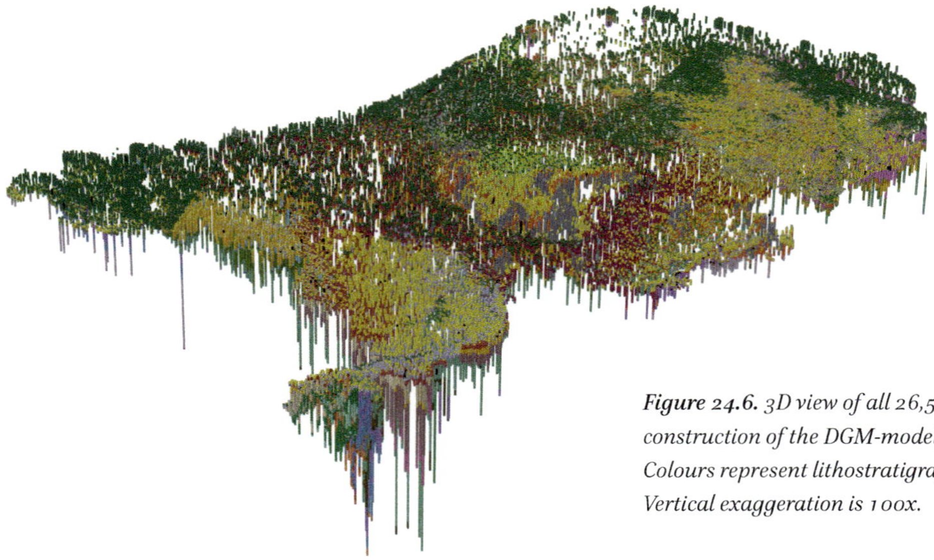

Figure 24.6. 3D view of all 26,500 boreholes used in the construction of the DGM-model of the shallow subsurface. Colours represent lithostratigraphic units at formation level. Vertical exaggeration is 100x.

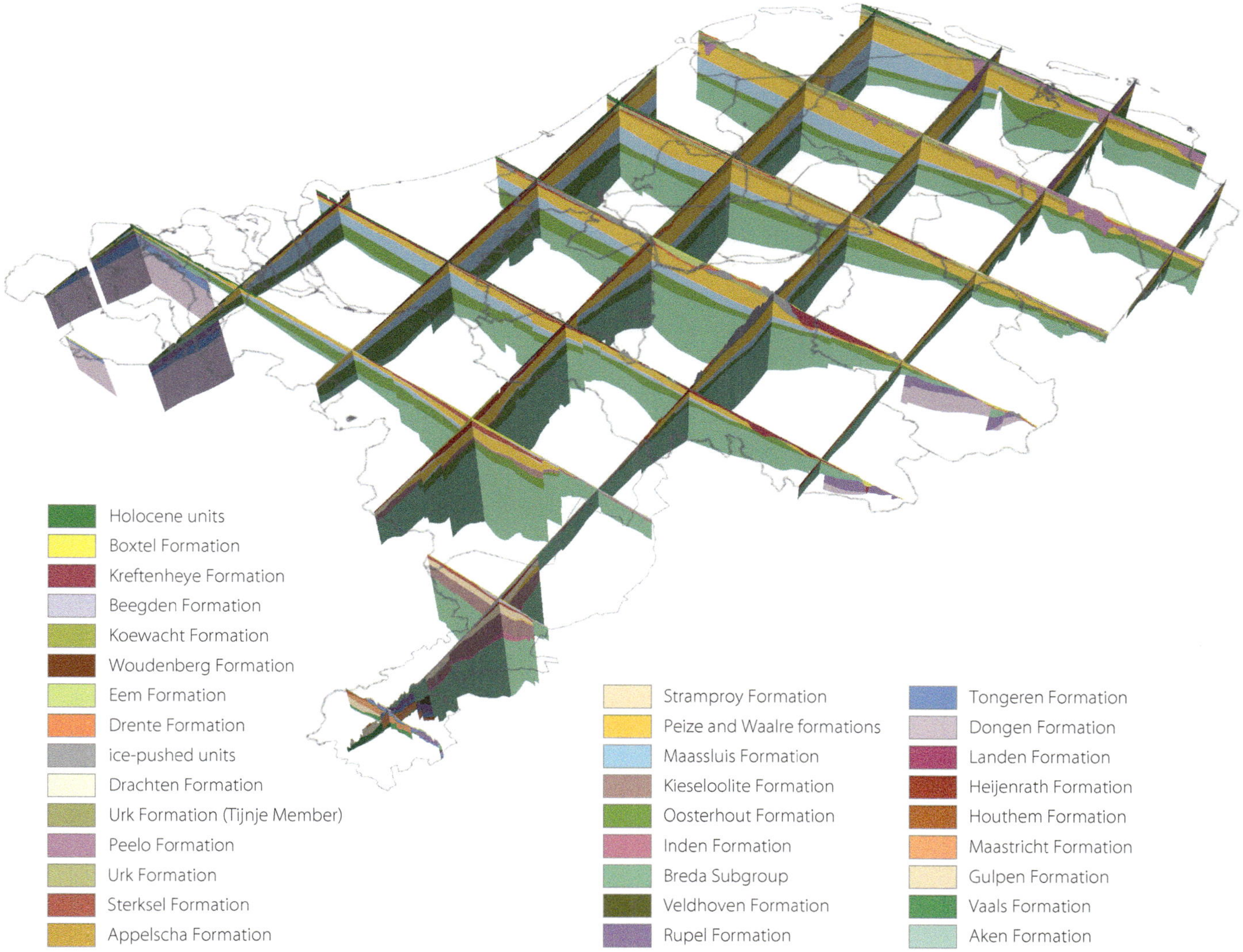

Figure 24.7. *Fence-diagram through the DGM model with an equidistance of 40 km showing the Neogene part of the North Sea Basin infill (Breda Subgroup and younger units) onlapping older Paleogene and Mesozoic formations along the basin shoulder. The Roer Valley Graben locally shows an increase in thickness of the Neogene units. The legend also contains units for other figures in this chapter that are not shown here. Vertical exaggeration is 45x.*

units of DGM into hydrogeological units (aquifers, aquitards and aquicludes). In addition, representative values of hydrological parameters (e.g. hydraulic conductivity) are calculated and assigned to the model hydrogeological units, making it suitable for groundwater flow modelling on a regional scale. Like DGM, REGIS II models the Holocene deposits as a single confining layer. REGIS II is widely used by regional authorities and water supply companies, for example in groundwater flow modelling studies. The reader is referred to Verweij et al. (2025, this volume) for an overview of the hydrogeological setting of the Netherlands.

The REGIS II model workflow uses the lithostratigraphic interpretation of boreholes and the layer-based DGM model as a starting point. The workflow then identifies hydrogeological units in the boreholes and follows with layer-based modelling and parameterization (Figs 24.2, 24.9). The hydrogeological units are interpreted and modelled within the lithostratigraphic units of DGM which remain unaltered during the process, making REGIS II and DGM fully consistent.

In the first step of the hydrogeological interpretation of the boreholes, each lithostratigraphic interval is subdivided into lithological classes: lignite, peat, clay, sandy-clay, clay with an unspecified sand content, fine sand, medium sand, coarse sand, sand with an unspecified grain size, glauconite sand, gravel, shells, chalk and marl, and sandstone and siltstone (Vernes et al., 2021). Stacked borehole intervals with similar lithological classes are aggregated. Subsequently, these aggregated intervals are manually interpreted in terms of hydrogeological units. REGIS II distinguishes six types of hydrogeological units: sand, clay, peat, lignite, chalk and 'complex' (Vernes et al., 2021).

The modelling procedure for the tops and bases of the hydrogeological units is similar to that used in DGM for modelling of the lithostratigraphic units, with two main exceptions. First, the modelling of units within a DGM unit requires the interpolation of two surfaces rather than the base surface only. In most cases top and thickness are interpolated from the borehole data, after which the base is derived by simply subtracting the thickness from the top. In some cases, however, top and base are interpolated and thickness is derived. Second, in many cases it is difficult or even impossible to determine consecutive sandy hydrogeological units within a single geological unit. This is especially the case if these sandy units are not separated by a clayey, peaty or complex unit. Figure 24.10 illustrates this problem in a sandy DGM-unit in which two clay layers 'k1' and 'k2' are present. This problem is solved using a three-step approach where the units that are discriminated by their lithological contrast are modelled first (clayey units 'k1' and 'k2' in Fig. 24.10). The second step involves the construction of extrapolated horizons, which form the continuation of the top of a clayey unit within the extent of the DGM-unit (dashed lines in Fig. 24.10). The percentage thickness distribution of the sandy units at the top of the clayey, peaty and complex units is used to generate artificial data points to be used in the interpolation procedure of these extrapolated horizons. In the third step, the geometry of the sandy units ('z1', 'z2' and 'z3' in Fig. 24.10) is derived by combining the top and base of the clayey units with the extrapolated horizons.

As in DGM, the depths of the top and thickness (or top and base) observed in boreholes are interpolated to raster surfaces using the 'block-kriging' algorithm (Goovaerts, 1997; Chilès & Delfiner, 2012) in Isatis®. The interpolation process is assisted by maps showing the maximum spatial extent of each hydrogeological unit, trend surfaces and guiding points ('synthetic boreholes'). The modelling workflow ultimately subdivides the 34 DGM lithostratigraphic units into 125 hydrogeological units (Fig. 24.11).

The next step in the modelling workflow is the param-

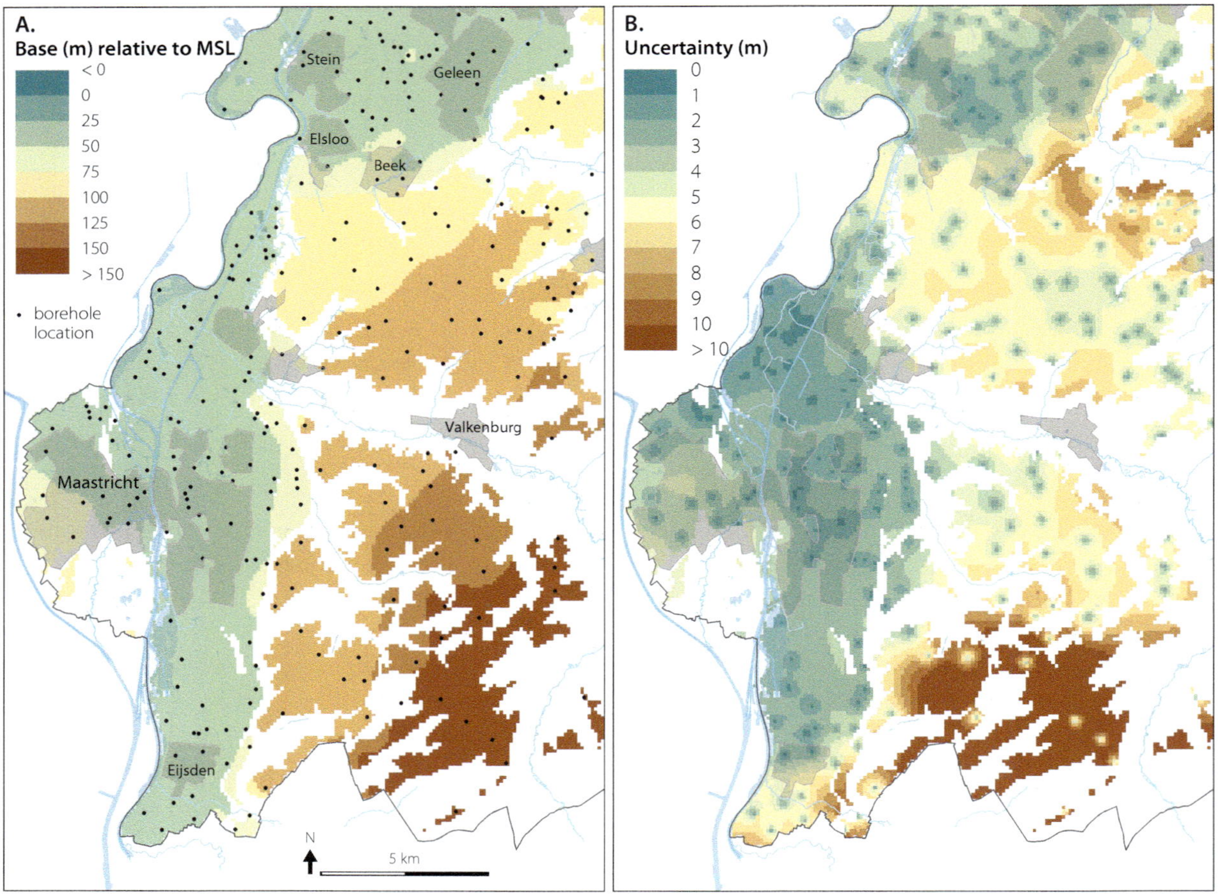

Figure 24.8. Detail of the base Beegden Formation, showing (a) borehole locations and modelled depth in DGM, and (b) the cross-validation based uncertainty. Note how data proximity as well as local geological complexity influence the level of uncertainty (modified after Dabekaussen & Hummelman, 2021).

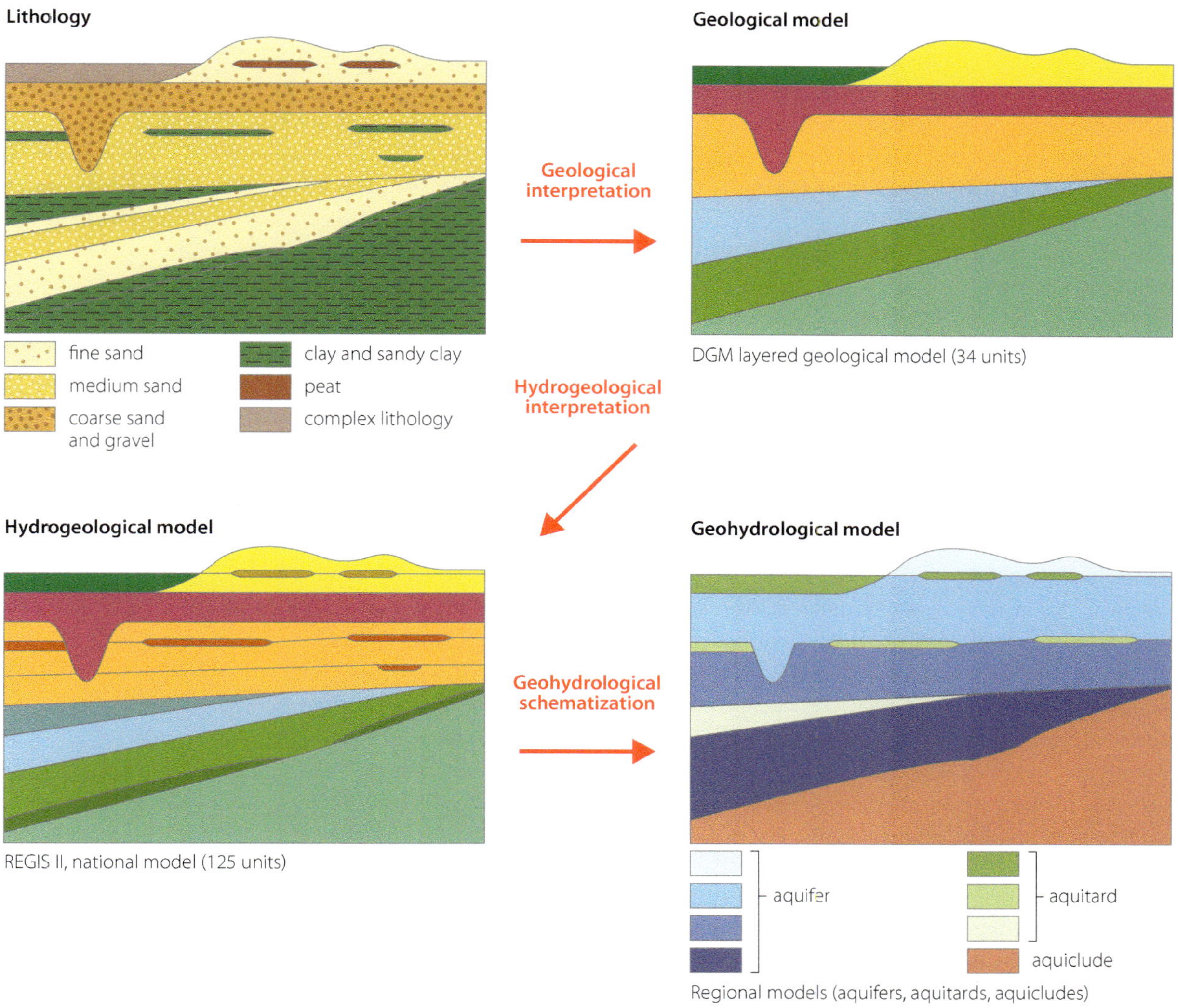

Figure 24.9. The coupled DGM-REGIS II concept: geological interpretation of field observations creates a national lithostratigraphic model (DGM); hydrogeological interpretation of the DGM model produces the REGIS II national hydrogeological model; geohydrological schematization produces regional geohydrological models for national and regional authorities defining aquifers, aquitards and aquicludes (modified after Vernes & Van Doorn, 2005 and Van der Meulen et al., 2013, reprinted with permission).

eterization of the hydrogeological units with hydraulic conductivity values in both the horizontal (k_h) and vertical (k_v) directions (Gunnink et al., 2013b; Vernes et al., 2021). Hydraulic conductivity measurements are only sparsely available, and therefore hydraulic conductivity is correlated with the lithological classes observed in the boreholes and subsequently interpolated to populate the entire volume of each hydrogeological unit.

The procedure involves four steps. In the first step, minimum and maximum hydraulic conductivity values for each combination of lithostratigraphic unit and lithological class are selected from a database comprising a mix of data obtained from measurements on samples taken from boreholes, selected pumping tests, literature and expert judgement. These values are then assigned to each combination of lithostratigraphic unit and lithological class in the boreholes. The second step groups the borehole intervals of the same lithological class within a hydrogeological unit and sums their thicknesses. The third step assumes the conductivities in the grouped lithological class intervals follow a log-normal distribution defined by the minimum (2.5 percentile) and maximum (97.5 percentile) values. Conductivities are then repeatedly (10,000 times) sampled. From the sampled conductivities, the weighted average conductivity at each borehole location is calculated, taking the thickness of each grouped lithological class interval into account. The arithmetic mean for horizontal and the harmonic mean for vertical hydraulic conductivity are used. From the 10,000 weighted average values, a single, upscaled conductivity value for the hydrogeological unit at the location of the borehole is calculated. The last step uses the 'kriging with measurement error' algorithm to interpolate the upscaled values and the corresponding standard deviations to a hydraulic conductivity raster layer covering the entire hydrogeological unit (Fig. 24.12). By repeating this procedure for each hydrogeological unit, the entire model is parameterized.

The procedure outlined above results in upscaled hydraulic conductivity values and their corresponding standard deviations at each borehole location. The standard deviation is used in the interpolation of the borehole data to the hydraulic conductivity raster layer, by applying the 'kriging with measurement error' algorithm. As a result, each resulting hydraulic conductivity layer is accompanied by a standard deviation raster layer representing the model uncertainty (Hummelman et al., 2019; Vernes et al., 2021).

REGIS II's main field of application is groundwater management. The model is available online at www.dinoloket.nl and www.broloket.nl. These portals provide easy, free of charge access to the model, enabling a first analysis and evaluation of the hydrogeological configuration in a specific area through cross-sections, maps and synthetic boreholes. Consulting REGIS II is a first step in obtaining an estimate of subsurface conditions, for example at new drilling locations for monitoring or pumping wells.

Groundwater flow models are an important tool in groundwater management, the scale of which can vary from national to local. Given its regional scale, REGIS II provides the geometry and a first estimate of hydraulic conductivity ranges of the aquifers and aquitards in the Dutch national hydrological model LHM (*Landelijk Hydrologisch Model*; LHM, 2022) as well as for regional groundwater flow models like MIPWA (2022) and IBRAHYM (Vermeulen et al., 2007; Swierstra et al., 2018; nhi.nu/modellen/ibrahym). In addition, REGIS II is often used as a framework for local groundwater flow models where additional local data is used to get a more accurate estimate of the local hydrogeology. Spatial planning of activities in the subsurface requires hydrogeological information as well, for example to assess the potential for thermal energy storage systems, estimate the available fresh groundwater stocks, or regulate the maximum depth of thermal energy storage systems so as to protect groundwater reserves.

GeoTOP: voxel model of the upper tens of metres

GeoTOP is the latest national subsurface model developed at TNO-GDN. GeoTOP schematizes the shallow subsurface in millions of voxels of 100 x 100 x 0.5 m down to a depth of 50 m below mean sea level, which is the main zone of current Dutch subsurface activity (Figs. 24.1a and 24.2; Stafleu et al., 2011a, 2021a; Maljers et al., 2015). GeoTOP models all Holocene formations as well as several Holocene and upper Pleistocene members and beds as separate stratigraphic units by deploying virtually all borehole descriptions available in the national borehole database (ca. 495,000) complemented by ca. 125,000 auger holes of Utrecht University (Cohen, 2017). Each voxel in the model has multiple attributes: the lithostratigraphic unit the voxel belongs to (based on a detailed layer model), the 'most likely' lithological class that is representative for the voxel and the probability of occurrence for each of the lithological classes. Lithological classes are based on the average of 100 equiprobable model realizations. At present, GeoTOP

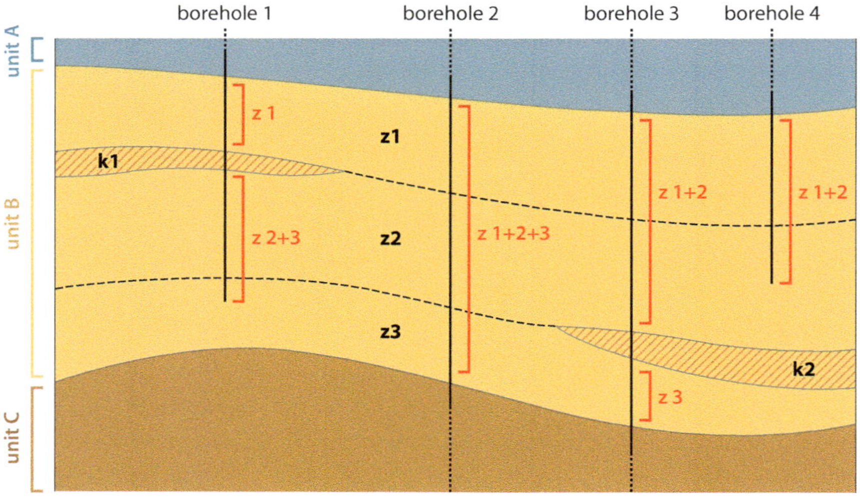

Figure 24.10. *Modelling of two clayey layers and three sandy layers within a single sandy DGM unit. In borehole 1, the topmost sandy layer 'z1' can be distinguished from 'z2' and 'z3' by the intercalating clayey unit 'k1'. In borehole 3, the lowermost sandy layer 'z3' can be distinguished from 'z1' and 'z2' by the intercalating clayey unit 'k2'. In the other boreholes, sandy layers 'z1', 'z2' and 'z3' are modelled by extrapolating the top of each clayey layer. Modified after Hummelman et al. (2019).*

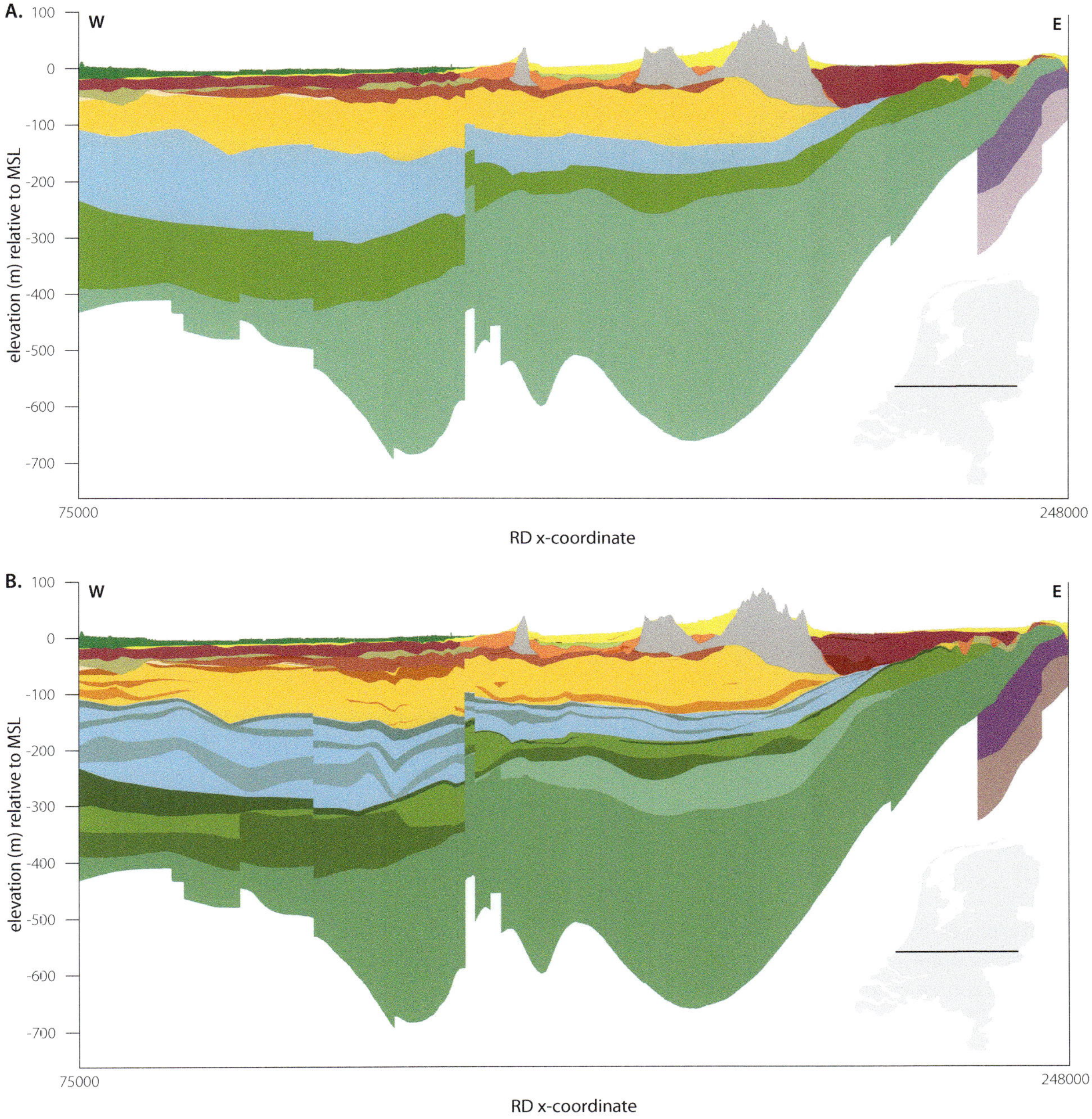

Figure 24.11. Cross-sections through the DGM and REGIS II models: a) DGM model showing lithostratigraphic units (see Fig. 24.7 for legend); b) REGIS II model showing the subdivision in hydrogeological units. Clayey layers are shown in a darker shade of the corresponding lithostratigraphic unit.

covers ca. 70% of the surface area of the Netherlands (including inland waters such as the Wadden Sea) using ca. 415,000 borehole descriptions from the two databases mentioned above.

Given the large number of boreholes, automated stratigraphic interpretation routines (Python scripts) were developed. These routines combine digital maps, stratigraphic rules (e.g. superposition) and lithological criteria (e.g. main lithology, admixtures, grain size and shell content, amongst other criteria) to determine the depth of the top and base of the lithostratigraphic units in borehole descriptions. This procedure delivers a uniform, consistent and reproducible set of interpreted boreholes (Fig. 24.13a).

An example of automated stratigraphic interpretation is the Basal Peat Bed. This thin layer of peat lies on top of the Pleistocene deposits, of which a trend surface is made based on manually labelled boreholes (including the 26,500 boreholes of the DGM model as well as other manually labelled boreholes in the national database). The procedure first selects those boreholes that are within the

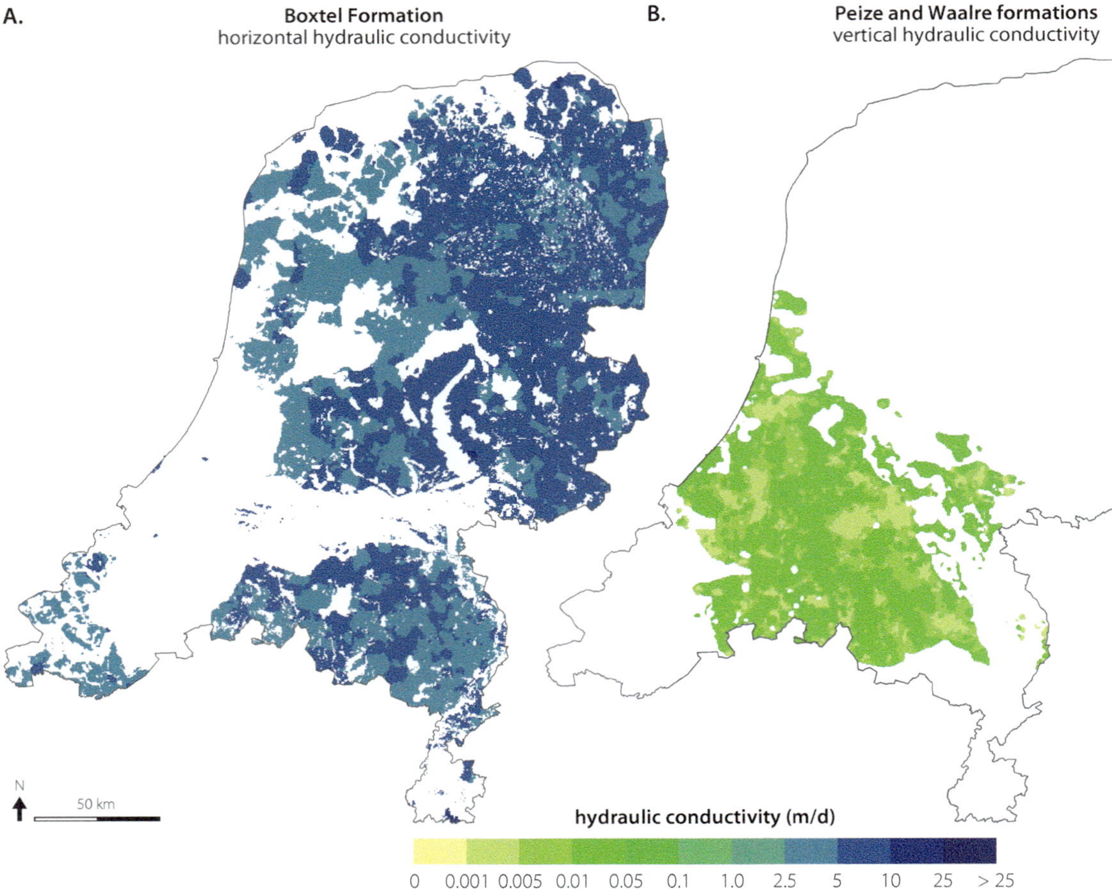

Figure 24.12. *Examples of hydraulic conductivity layers in the REGIS II model: a) sandy unit with horizontal hydraulic conductivity; b) clayey unit with vertical hydraulic conductivity.*

areal extent of the Basal Peat Bed and subsequently examines the description intervals of each borehole from bottom (the deepest interval) to the top (the interval at land surface). The following criteria are then used to identify intervals as part of the Basal Peat Bed:

- The lithology is peat or gyttja;
- The base of the interval is within a buffer of 2 m from the top Pleistocene trend surface;
- The lithology underlying the first interval found is (Pleistocene) sand, or a clay layer with a maximum thickness of 1 m.

The criteria are designed in this way to avoid misinterpretation of other Holocene peat deposits (e.g. the Hollandveen Member).

Another example comprises the small river valley deposits in the southern Netherlands (Boxtel Formation, Singraven Member). These Holocene infills contain peat, clay and clayey sand in an otherwise sandy Pleistocene environment (Fig. 24.14). However, the deeper parts of the infill may consist of fine to coarse grained sand that is difficult to distinguish from the underlying Pleistocene deposits. Therefore, additional criteria are needed. From manually labelled boreholes, the base of the unit is known to lie between 4 to 6 m below land surface. In most cases, the bottom of the unit is simply set at the deepest interval that satisfies this criterion (Fig. 24.14a). In the central part of the Roer Valley Graben however, a more sophisticated procedure can be followed. In this area, the river valleys eroded loam deposits of the Liempde Member (Boxtel Formation; Fig. 24.14b). In this case, the procedure sets the base of the river valley deposits at the deepest interval in a zone of 2 m below a trend surface based on labelled boreholes of the Liempde Member. After the base is determined, the top of the unit representing the Singraven Member is subsequently set at equal to the land surface or, if present, to the base of overlying anthropogenic deposits.

This second example shows that dependencies between the units exist, e.g. the Singraven Member can only be found after the procedures for anthropogenic deposits and the Liempde Member are completed. Fig. 24.14b also shows that defining hard criteria, such as a zone of 2 m, has some limitations: in a manual interpretation, a geologist would probably put the base of the Singraven Member an interval deeper than the automated procedure does.

The automated interpretation procedure works well in the shallow Holocene and upper Pleistocene units, where many good quality boreholes are available (Stafleu et al., 2021a). Furthermore, the shallow units often have sharp lithological contrasts and are therefore easily identifiable. The lithological contrasts between the generally sandy, fluviatile Pleistocene units at deeper levels are more subtle and hence less suitable for automated interpretation routines. The layer modelling of these units therefore relies on the same set of ca. 26,500 manually interpreted boreholes employed by DGM and REGIS II.

In the next modelling step, 2D interpolation techniques, implemented in the geostatistical software package Isatis®, are used to construct surfaces representing the bases of the stratigraphic units (Fig. 24.13b). The interpolation algorithms used (Sequential Gaussian Simulation in older model areas, block kriging in more recently constructed parts of GeoTOP; Goovaerts, 1997; Chilès & Delfiner, 2012) allow for the calculation of a mean depth estimate of each stratigraphic surface and its standard deviation. As in DGM, the interpolation is constrained by maps of the maximal areal extent of the units and is assisted with trend surfaces and guiding points. Subsequently, all surfaces are stacked according to their stratigraphic position, resulting in a consistent, integrated layer-based model with estimates of top and base of each stratigraphic unit (Fig. 24.13b). Top surfaces are derived from the bases of the overlying units.

The surfaces are then used to place each voxel in the model within the correct stratigraphic unit. If the midpoint of a voxel is situated below the top and above the base of a stratigraphic unit, the voxel is assigned to that particular unit. The surfaces of the layer model are also used to assign a stratigraphic unit to borehole intervals that remained uninterpreted in the interpretation step described above. These intervals are from the deeper units that were not part of the automated interpretation routines and do not form part of the 26,500 manually interpreted boreholes.

In the third step, the borehole descriptions are re-analysed and each interval is classified in one of six lithological classes: peat, clay, clayey sand & sandy clay, fine sand, medium sand, coarse sand & gravel. (Fig. 24.13c). Additional classes for gravel and shells are used for stratigraphic units that contain considerable amounts of these lithologies. No classification is applied to the anthropogenic deposits. The classification scheme follows the hydrogeological definition of REGIS II and is suitable for groundwater flow modelling amongst other applications. Compared to the stratigraphic interpretation, the classification is a

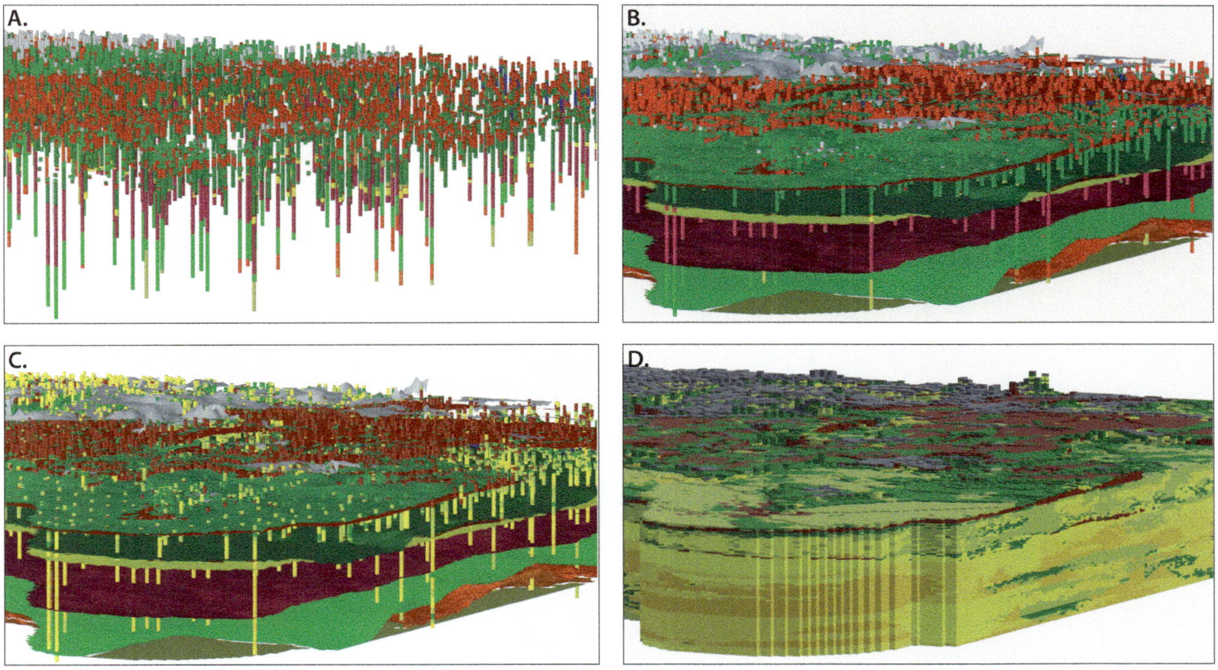

Figure 24.13. The four main modelling steps in the construction of the detailed layer-based model underpinning the GeoTOP voxel model and the voxel modelling itself: a) automated stratigraphic interpretation of borehole descriptions into formations, members and beds; b) 2D interpolation of stratigraphic surfaces; c) subdivision of borehole descriptions into lithological and grain-size classes; d) 3D interpolation of lithological class for each stratigraphic unit separately. In (c) and (d), yellow colours indicate sand in three different grain-size classes, green colour is clay, light green colour is clayey sand and sandy clay, and brown peat. Reproduced from Stafleu et al. (2019).

relatively simple procedure that is implemented in a single Python-script. Figure 24.15 shows examples of lithological class distributions observed in boreholes.

In the fourth modelling step, the voxels in the model are populated with estimates of lithological class. These estimates are calculated from the lithological classes observed in the boreholes using the stochastic simulation technique Sequential Indicator Simulation (Goovaerts, 1997; Chilès & Delfiner, 2012), in Isatis®. Because of the great diversity in composition and heterogeneity of the different stratigraphic units (Fig. 24.15), the simulation is performed for each stratigraphic unit separately.

The simulation results in 100 equally probable realizations of lithological class for each voxel in the model. From these 100 realizations, probabilities of occurrence of the lithological classes are calculated by simply dividing the number of times a particular lithological class is assigned to a voxel by the total number of realizations. In addition, a 'most likely' estimate of lithological class is determined (Fig. 24.13d). Along with the stratigraphic unit, the 'most likely' lithological class is the most used attribute of GeoTOP (Fig. 24.16).

The 100 different but equally probable realizations indicate the degree of uncertainty in the estimates of the lithological class. In the best case, all model runs result in the same value; in the worst case all possible values occur with the same frequency. For an individual voxel, as well as in a 1D vertical voxel stack, the probabilities can be displayed in a bar graph showing the probability distribution and hence the model uncertainty (Fig. 24.22). However, in 2D visualizations, for instance in a vertical cross-section through the voxel model, it is not possible to show all probabilities in a single view: the user will always be presented with one of the probabilities at a time. To solve this visualization problem, the concept of information entropy (Wellmann & Regenauer-Lieb, 2012) is used as a measure of uncertainty in GeoTOP (Stafleu et al., 2021a,b). The information entropy of a voxel is a single value ranging from 0 to 1 that can easily be calculated from each of the probabilities of the lithological classes. An entropy value of 0 represents no uncertainty (all model runs result in the same lithological class), whereas a value of 1 occurs when all lithological classes are equally likely. Values in between 0 and 1 account for both the number of lithological classes with a probability higher than 0 (the more classes, the higher the entropy) and the differences amongst the probabilities (the greater the differences, the lower the entropy).

Figure 24.17 shows the information entropy of the lithological classes in a fence diagram through a small part of the GeoTOP voxel model (Stafleu et al., 2021b). The diagram shows areas that are strongly influenced by single boreholes as vertical zones of low entropy. In addition, homogeneous stratigraphic units dominated by a single

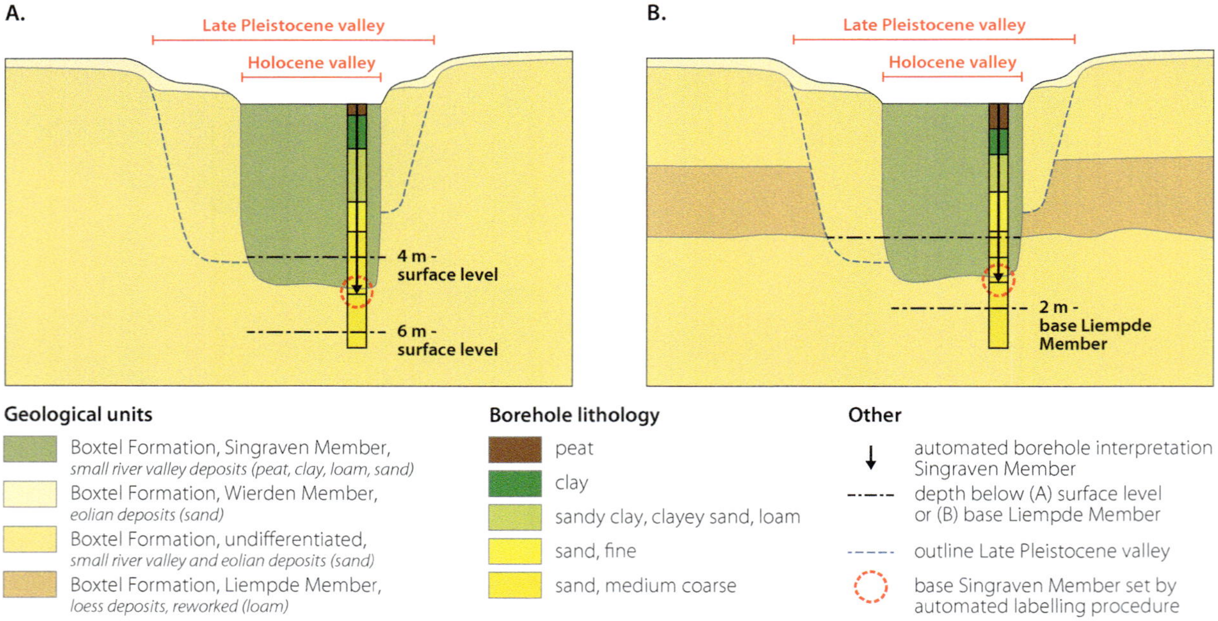

Figure 24.14. Conceptual model illustrating the automated stratigraphic interpretation of a borehole in the small river valley deposits of the Singraven Member (Boxtel Formation): a) in most areas, the base of the unit is simply set at the deepest interval where the base is in between 4 and 6 m below land surface; b) in the Roer Valley Graben, a more sophisticated approach can be used involving a trend surface calculated on the basis of previously interpreted boreholes of the Liempde Member (Boxtel Formation).

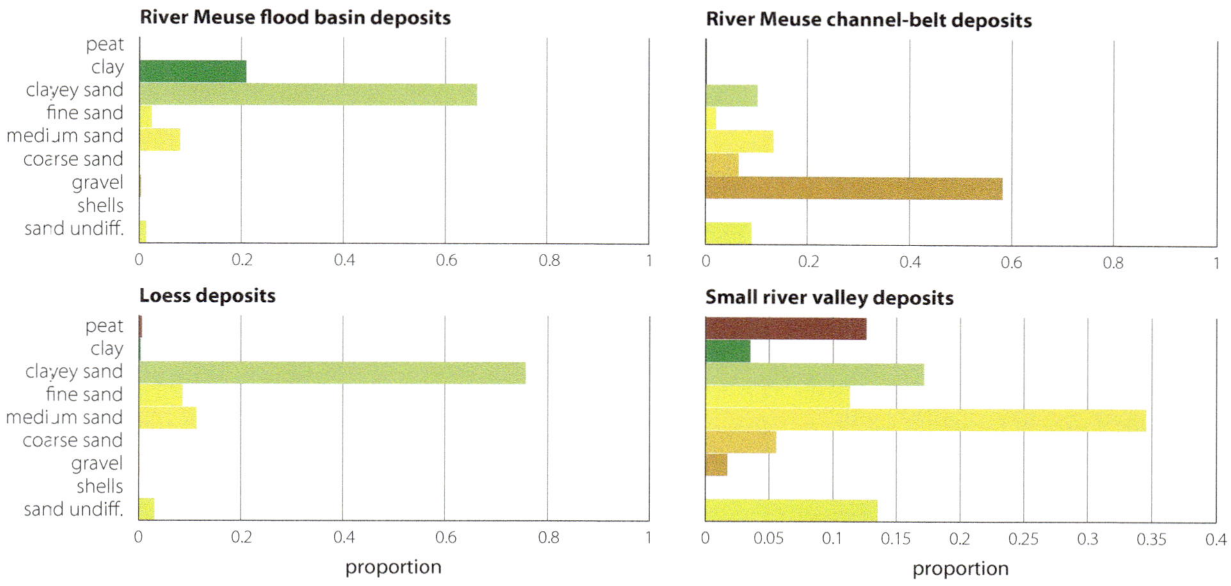

Figure 24.15. Bar graphs of lithological classes in borehole intervals of four stratigraphic units in the GeoTOP model area of the southern Netherlands. Note the diversity in composition and heterogeneity. 'Sand undifferentiated' refers to sandy borehole intervals without grain-size information. In the voxel modelling, these intervals are only used to determine if a voxel is sandy or not.

lithological class can be distinguished (e.g. 'Unit A' in Fig. 24.17, low entropy) from heterogeneous units ('Unit B', high entropy). In general, entropy increases with depth as data density decreases.

GeoTOP is a multi-purpose model supporting applications involving the shallow subsurface, including:

- Evaluations of subsurface geotechnical conditions in the planning stage of large infrastructure projects (Van der Meulen et al., 2013);
- Long-term predictions of land subsidence due to compaction of clay and oxidation of peat (Koster et al., 2018a,b; CAS, 2022; Fig. 24.20);
- Aggregate resource assessments (Van der Meulen et al., 2013; Maljers et al., 2015);
- Defining the lithological characteristics of Holocene channel belt systems (Stafleu & Busschers, 2017);
- Potential environmental impacts of dredging operations (Stafleu et al., 2021a);
- Generalized soil composition map of the upper 8 m, used to assess the suitability of the subsurface for sewage infrastructure (Dambrink et al., 2014; Fig. 24.27c);
- Groundwater flow studies (Van der Meulen et al., 2013);
- Site-response assessment of induced earthquakes (Kruiver et al., 2017a,b; Van Ginkel et al., 2022; see the section on innovative modelling methods and case studies).

In several of these applications, GeoTOP is first parameterized with appropriate property values, like hydraulic conductivity in groundwater flow studies (Gunnink et al., 2013b), shear-wave velocity in the risk-assessment of induced earthquakes (Kruiver et al., 2017a,b) and the organic matter content of peat in advanced land subsidence predictions (Koster et al., 2018a,b). The basic assumption in these three examples is that the spatial variation of the relevant subsurface properties strongly depends on the two main attributes of GeoTOP: the lithostratigraphic unit and the 'most likely' lithological class (Stafleu et al., 2021a). Assigning property values to the voxels in GeoTOP may be simple, by using a lookup-table with a property value for each combination of lithostratigraphic unit and the lithological class, or it may involve a more complex vertical voxel-stack analysis that accounts for depth as a third factor controlling the property value. Some properties (e.g. hydraulic conductivity) require upscaling from the scale of the measurements (samples taken from boreholes) to the scale of the model (individual voxels).

Sand resource models for the North Sea

For the North Sea, two models have been developed to map the offshore sand resources. The DIS model (*Delfstoffen Informatie Systeem*), developed by TNO-GDN and Deltares for the Department of Waterways and Public works (Rijkswaterstaat), schematizes the offshore subsurface in millions of 200 x 200 x 0.5 m blocks down to a depth of 12 m below the seafloor. The model stretches along the entire coast of the Netherlands, starts at a water depth of -20 m NAP (Dutch ordnance datum) and extends 12 Nautical miles offshore. The model and accompanying documentation (in Dutch) is available from the DINOloket web portal (TNO-GDN, 2022a). A second model covers the Belgian and southern Netherlands part of the North Sea (BPNS/sNPNS). This model, developed as part of the Transnational and Integrated Long-term Marine Exploitation Strategies project (TILES) has a resolution of 200 x 200 x 1 m and extends down to a depth of -65 m NAP (Dutch ordnance datum) (Hademenos et al., 2019; Van Lancker et al., 2019).

The modelling approach contains the same four main steps as the GeoTOP model, but there are some differences. In addition to boreholes, seismic data are a major information source in the offshore domain and the modelling approach is using this datatype as described in detail by Hademenos et al. (2019). Step one consists of stratigraphic interpretation of both borehole and seismic data. In contrast to GeoTOP, the interpretation of both datasets is mainly performed manually. In step two, the 2D interpolation of stratigraphic surfaces follows the same principles as applied in GeoTOP. The third step, the subdivision of boreholes into lithological classes, is comparable to the procedure for GeoTOP, but the definition of the classes varies according to the application of the model. In addition to a varying grain-size class definition, classes for the concentration of fines (silt and clay) and shells and gravel (TILES only) are used. These additional classes are of interest when determining the overall sand quality. In the fourth step, 3D interpolation is completed for lithological, mud, shell and gravel classes for each stratigraphic unit separately.

Model-related uncertainty of the offshore aggregate resource models is estimated in the same way as in Geo-

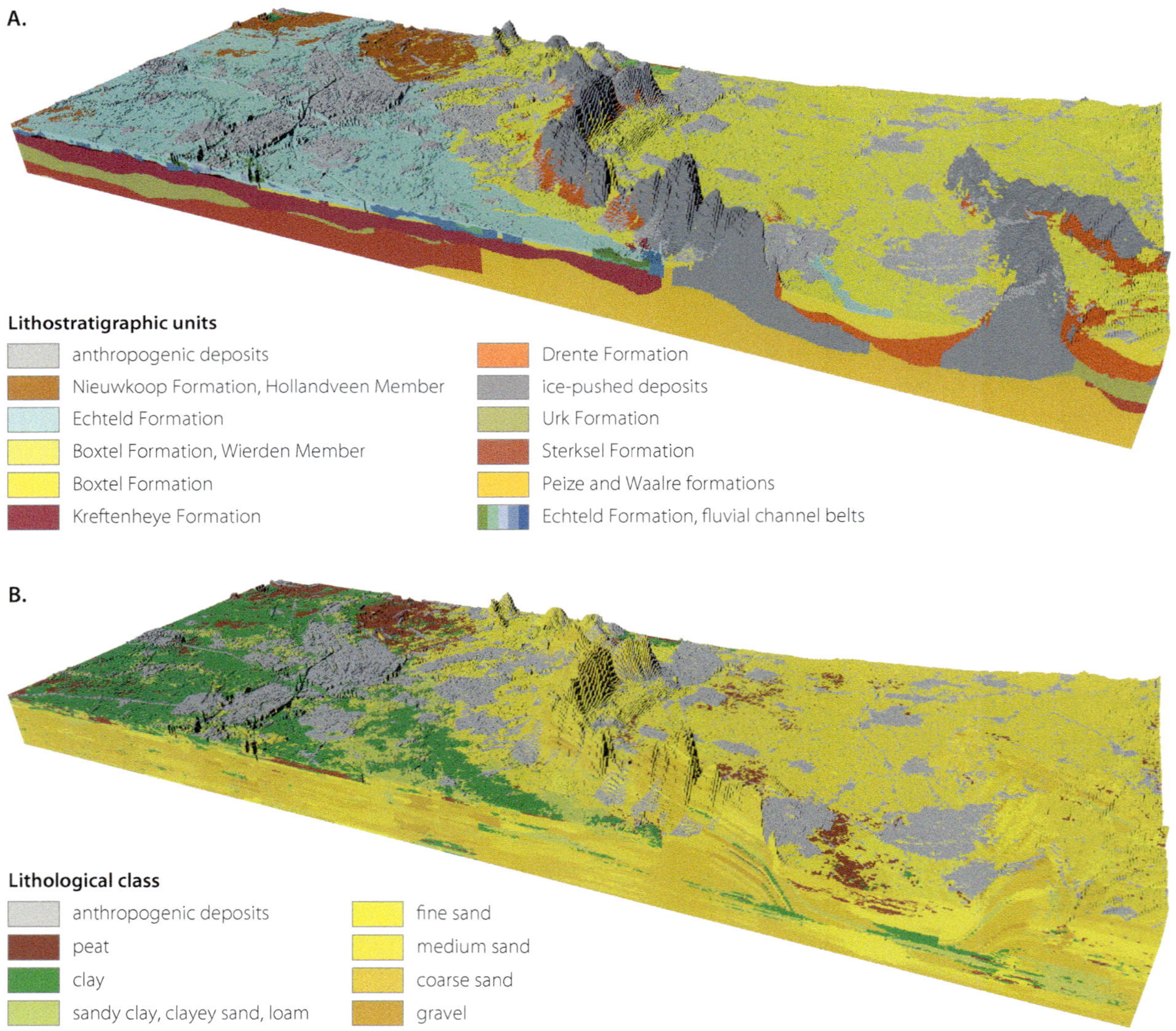

Figure 24.16. *GeoTOP 3D views of the Gelderse Vallei area in the central part of the Netherlands: a) lithostratigraphic units of Holocene and upper Pleistocene formations, members and beds; b) lithological classes. The displayed block measures 62 x 24 km; depth of the base is 50 m below MSL; vertical exaggeration is 75x. Modified after Stafleu & Dubelaar (2016).*

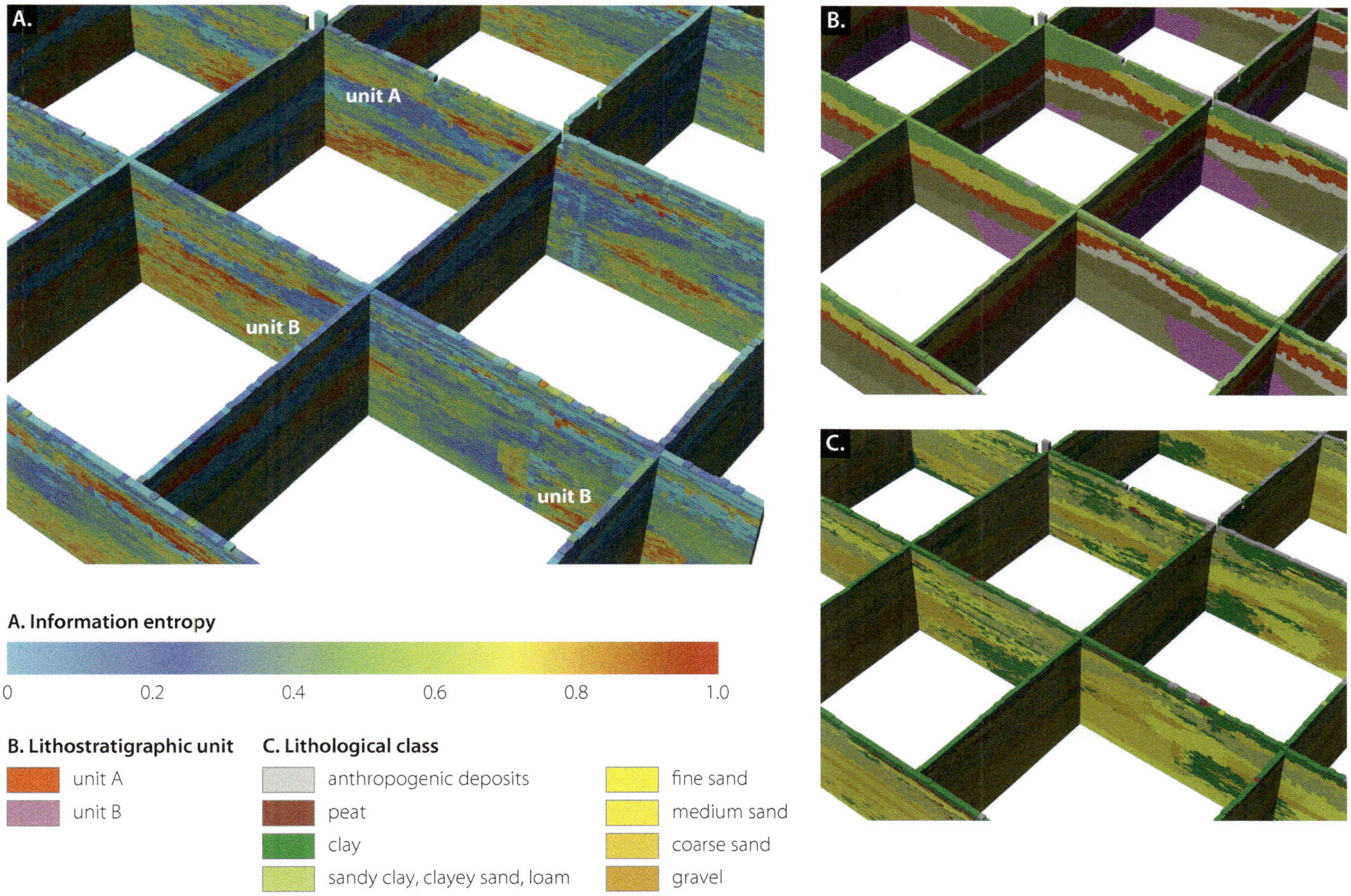

Figure 24.17. a) Information entropy (representing model uncertainty) of lithological class in a fence diagram through part of the GeoTOP voxel model; b) stratigraphic units of the model; c) most likely lithological class. Unit A is dominated by a single lithological class (c) and has a low model uncertainty (a). Unit B is heterogeneous in its lithological composition (c) and has high model uncertainty values (a). Spacing of the fences is 4.5 km. Modified after Stafleu et al. (2021b).

TOP. For the Belgian part of the TILES model, Kint et al. (2020) qualified and quantified data-related uncertainties for positioning, sampling and vintage. Uncertainty on positioning comprises navigational error, on-board and off-deck offsets and underwater drift. Sampling uncertainty arises from the suitability of the instruments to determine the lithological class. Vintage uncertainty is related to the change of the seabed since the moment of sampling. Parameters controlling vintage uncertainty are the age of the boreholes used and the sediment dynamics induced by environmental conditions and human activities (Kint et al., 2020). The three data uncertainties are interpolated and subsequently added as new properties in the voxel model, enabling users to visualize data uncertainty along with the modelled stratigraphy and lithology.

TILES and DIS are developed as decision support systems to enable sustainable management of the offshore sand resources. Both models are intended for regional use and they both show large scale patterns. Figure 24.18 for example shows that the probability to find medium grained sand is much higher further offshore compared to closer to the coast where fine grained sand is more prevalent. From a user's point of view, the greater value of medium grained sand must be weighed against the longer sailing time and higher fuel consumption involved in sand extraction further offshore.

The voxel models form the basis for a range of selection and visualization tools for use in policy development, for example to determine the influence of infrastructure on the short-term sand availability (Fig. 24.19).

Model quality control

In 2015, the Dutch parliament passed a new law, which puts subsurface data and information in the system of so-called key registries (*Basisregistraties*). The key registry of the subsurface (BRO, or *Basisregistratie Ondergrond*), managed by the Geological Survey of the Netherlands, holds subsurface data, including three of the four national

Anthropogenic deposits

2

In most urban areas the natural subsurface has been heavily disturbed by human activity and/or is covered by anthropogenic deposits (or: made ground). These areas do not only include city centres, with their often abundant subsurface archives related to historical urban development (e.g. Jayasena, 2020), but also modern residential and industrial areas, built on extensive sheets of filling sand (see photograph). Other examples include road and railway embankments that consist of natural aggregates or novel anthropogenic materials, coastal reinforcements and dikes, landscaped areas and landfills (see Van der Meulen et al., 2025, this volume) for an overview of the use of different sorts of aggregates over time).

In contrast to past geological maps, the 3D-subsurface model GeoTOP does provide information on the subsurface of urban areas and includes anthropogenic deposits as a separate stratigraphic unit (Schokker et al., 2015). The extent of these deposits has been derived from the Land Use Database of the Netherlands (LGN), a raster dataset in which the country has been subdivided into 39 land use classes. To ensure that highway embankments and dikes appear as continuous structures in GeoTOP, this raster dataset has been supplemented with spatial information from other sources (Stafleu et al., 2020).

However, the modelled thickness of anthropogenic deposits is generally not well constrained and GeoTOP does not give any information on the lithology of these deposits. Also, the composition and properties of the urban subsurface change rapidly over time, which means that data acquired before human interventions took place often do not reflect current subsurface conditions. Nonetheless, the demand for detailed subsurface information in the built environment is high and relates to anything from building stability and ground heat extraction to preserving cultural heritage and mitigating the effects of climate change (Van der Meulen et al., 2013). In preparation for the new Environment and Planning Act, major cities in the Netherlands are currently developing 3D spatial plans to optimize the use of subsurface space, thus potentially releasing valuable space above-ground.

To meet the growing demand for up-to-date, high-resolution subsurface information in urban areas, focus is on improved modelling of anthropogenic deposits. This includes:

- Improved spatial distribution and thickness modelling by incorporating additional data types, like cone penetration test data. To process this information, semi-automated interpretation routines are being employed, including artificial intelligence techniques like neural networks (e.g. Dabekaussen et al., 2020). Furthermore, in future GeoTOP releases of up-to-date information from the key registry BGT (*Basisregistratie Grootschalige Topografie*) will replace LGN data.
- Lithological characterization of anthropogenic deposits. This concerns both reworked natural deposits, as well as recycled non-natural materials (Dijkstra et al., 2019). To better recognize and characterize these deposits, both the lithological description standard (SBB 5.1; Bosch, 2000) and lithostratigraphic nomenclature (TNO-GDN, 2022b) will be updated and extended.

Subsurface disturbance beneath the campus of Vrije Universiteit Amsterdam: sewage and drinking water pipes crossing the local energy system, all embedded in 2 m of filling sand (photo: J. Schokker).

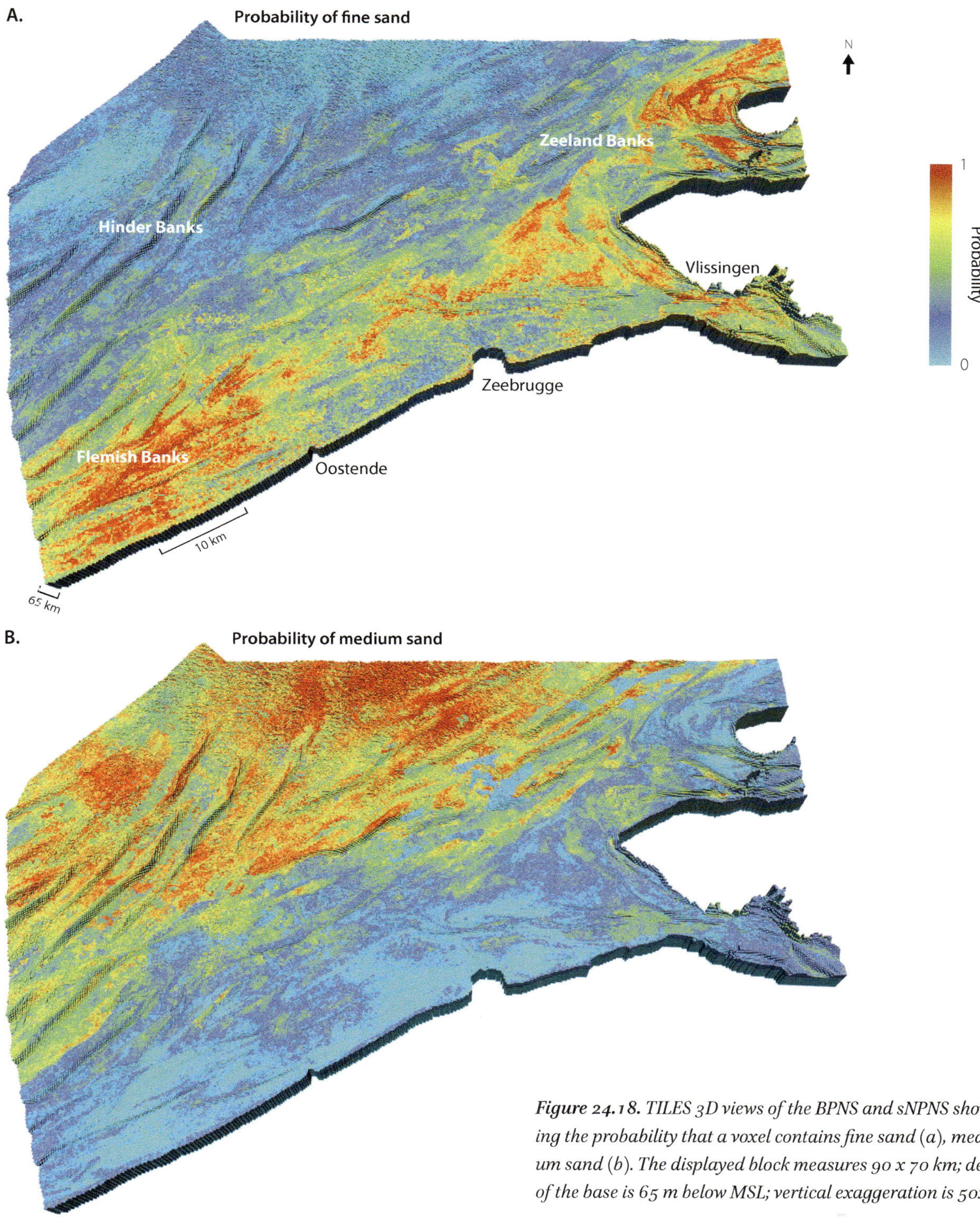

Figure 24.18. TILES 3D views of the BPNS and sNPNS showing the probability that a voxel contains fine sand (a), medium sand (b). The displayed block measures 90 x 70 km; depth of the base is 65 m below MSL; vertical exaggeration is 50x.

subsurface models (DGM, REGIS II and GeoTOP), as well as information on permits and underground infrastructure (Van der Meulen et al., 2013). Public bodies are legally required to consult both the data and the models in the key registry when making decisions related to the subsurface. If, for instance, a regional water authority would like to reinforce one of its dikes, they must first consult the BRO to check that subsurface data underneath the dike is already available. In addition, government agencies have a legal obligation to deliver all subsurface data they acquire and to report any errors they may discover in the data and models in the registry. These legal requirements are expected to lead to an increase in the use of the models. They also impose more stringent demands on the models in terms of their quality as well as in the update frequency, since the user community will expect the data they deliver

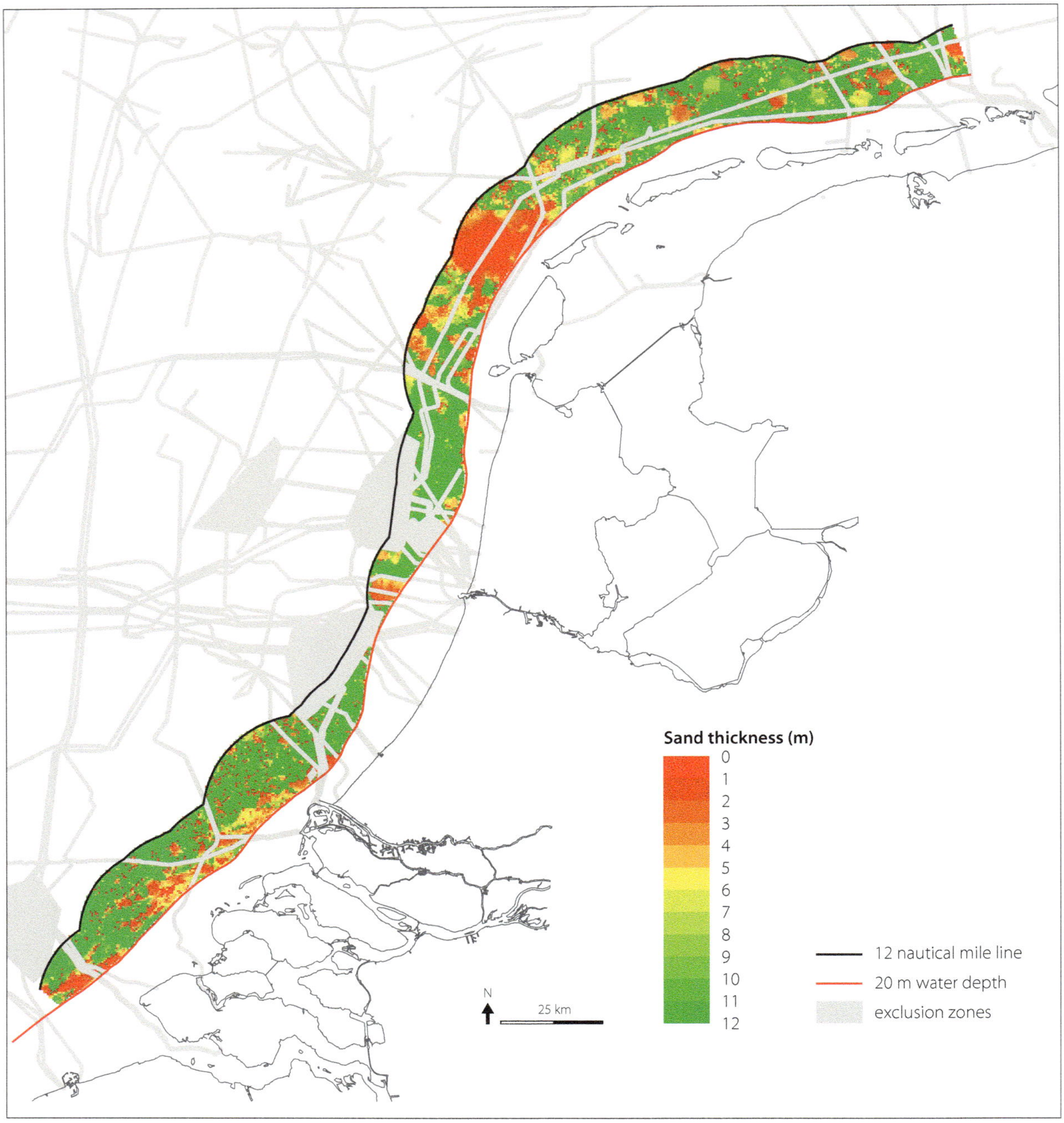

Figure 24.19. Extractable sand in the upper 12 metres below the sea floor outside of exclusion zones (offshore infrastructure, e.g. wind farms, cables and pipelines) according to scenario B2. Modified after Maljers-Oosterwijk et al. (2014) updated using the DIS 2.1 dataset from the DINO web portal (TNO-GDN, 2022a).

to be incorporated in the models as soon as possible (Van der Meulen et al., 2015).

Definitions of quality control usually break down into two elements: first, assessing the extent to which the model meets standards and requirements. This is the *intrinsic quality* of the model. The second element assesses *extrinsic quality*: does the model meet user requirements and expectations, or, in other words: is the model useful and fit for its purpose? Intrinsic and extrinsic quality do not necessarily coincide: a poor-quality model can still be useful, and a detailed, geologically plausible model can still be unfit for a particular application.

Feigenbaum (1991) illustrated the importance of quality control by looking at the costs related to good and poor quality. Good quality is about preventing errors or defects to occur at all, for instance by training staff, providing them with proper work instructions and by certification. In addition, good quality concerns appraisal which involves detecting errors via quality control or audits. Poor quality costs arise from having to do the work again, or to

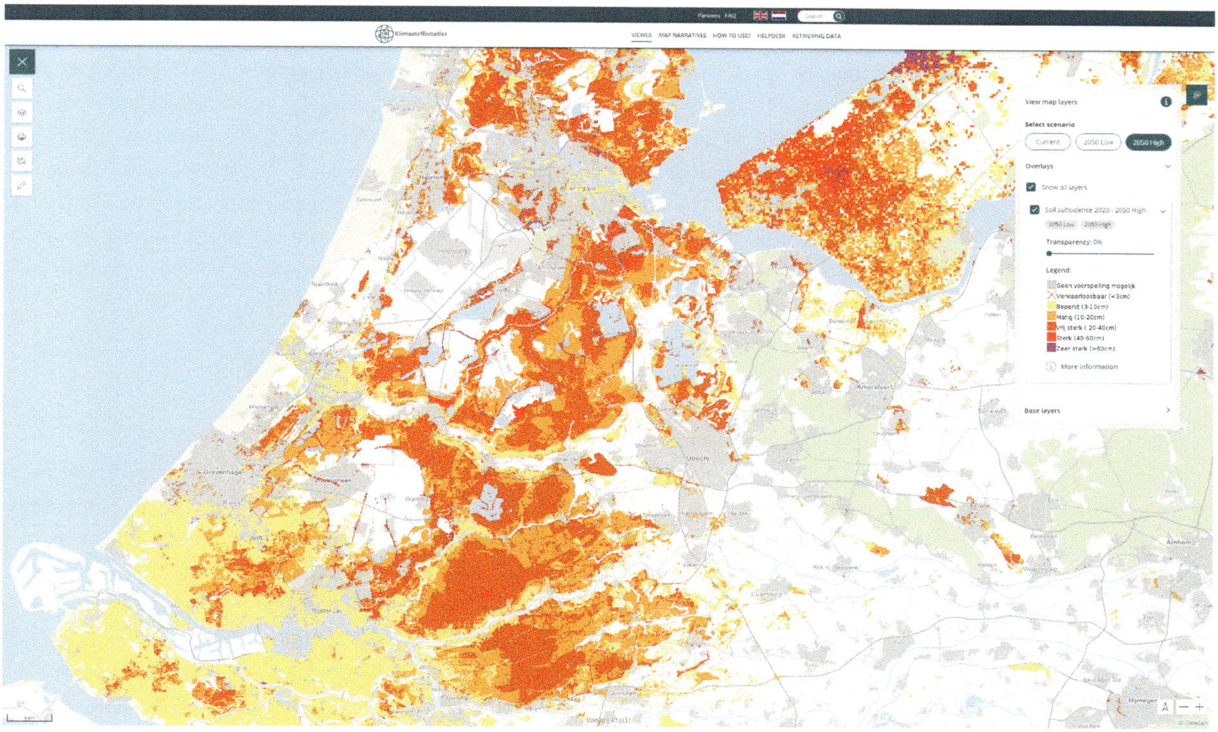

Figure 24.20. Interactive map of predicted land subsidence due to the oxidation of peat in the western and central Netherlands. Screenshot taken from the online Climate Impact Atlas (CAS, 2022).

correct it afterwards. It also comes from dealing with complaints from users and loss of reputation. The sum of the cost of good and poor quality is the total cost of quality. In general, larger investments in good quality lead to even larger savings in poor quality. Therefore, prevention and appraisal lead to a lower total cost of quality and hence lower total model production costs.

Intrinsic quality control

TNO-GDN started to implement a formal round of quality control (QC) of the national models in 2015. Expert geologists, who are not involved in the modelling process itself, inspect the models for defects. They are primarily focusing on intrinsic quality by systematically checking cross-sections, map views and 3D views of the final model. All defects, artefacts and geologically implausible configurations are reported to the modelling team who then address them. Examples of defects include, among others, errors in the mapped extent of geological units, pits and bumps in the layers caused by wrongly interpreted boreholes, faulted layers that are displaced in the wrong direction or have inconsistent displacement, implausible concentrations of a particular lithological class around bad quality boreholes, and artefacts at model area boundaries.

Creating a model is time-consuming and may take several years. This is because the modelling workflow (as outlined in the section on the national models) consists of many components, like drafting a conceptual model or design, mapping the extent of geological units, stratigraphic interpretation of boreholes and the actual modelling of layers and voxels. Defects in the final product may be rooted in one of the components that was created early in the modelling process, potentially one or two years earlier. For example, when an error is detected in the GeoTOP layer model, both the layer model and the voxel model will have to be corrected. Or even worse, if expert geologists disagree with the conceptual model being used, parts of the entire modelling process may have to be redone. Therefore, quality control should not be limited to the final product but should start early in the modelling process by assessing all major components. The modelling team should only start a new stage in the process if the current stage has been approved by the expert geologists. However, this staged approach in quality control makes the modelling process more time-consuming than it already was before QC was implemented. An alternative approach to detect defects early in the modelling process is to work in advance, i.e. to construct a preliminary final model from intermediate model components. For instance, when the modelling team is still working on maps and interpretations, the work-in-progress-result is used to create a first version of the layer model to see if the interpretations lead to implausible configurations. Or, if the team is still working on the layer model, a first draft of the voxel model can be built to see if the subdivision in layers successfully captures the spatial distribution of lithology. Thus, working in advance provides the modelling team with invaluable feedback on the progress of their efforts. This 'fast forward modelling' puts a high demand on the turnaround time of the automated parts of the workflow (for example the interpolation processes implemented in Isatis® and Petrel®).

Another change imposed by the key registry of the subsurface is that users will expect more regular updates of the models. Modelling then becomes a process or a cycle rather than a project with a clear beginning and end. Quality control in turn will focus more on the changes made with respect to the previous version: are these changes reflecting new data and knowledge acquired since the last release, and do they correct for any errors reported by the user community? The focus of intrinsic quality control then shifts from geological plausible configurations to improvements made with respect to the previous version of the model. In other words, quality control will ensure that the new release results in a better model.

Extrinsic quality control

Due to the increasing size of the user community, we foresee another shift from evaluating intrinsic quality towards assessing the usability of the models in applications: will the model deliver the right answer to the users' needs? An example is the map in Fig. 24.20 indicating areas that are prone to land subsidence. The map is largely based on the GeoTOP voxel model and offers an entirely new look from the point of view of an application. Quite often defects in the model become apparent in an application, for instance a very unrealistic amount of subsidence that may have been overlooked by the expert geologist. In this way, applications help to find errors that arise from model configurations that are plausible, but nevertheless wrong. The key registry is another source of extrinsic quality control, as government agencies are required to report any errors they may encounter in the model when using it in their applications.

Model visualization

The models described in this chapter are disseminated free-of-charge via the DINO-web portal (www.dinoloket.nl/en/) in a number of ways, including an online map viewer with the option to create virtual boreholes and vertical cross-sections through the models, and as a series of downloadable GIS products. Full 3D visualization is supported by the freely downloadable SubsurfaceViewer® software, which allows users to download and visualize the models on their desktop computers. A recent development in model visualization is the capability of Esri's GIS-software to visualize large voxel-models such as GeoTOP in 3D. However, many users still prefer to work with the models in a 2D GIS environment. To serve these users, the models are delivered as 2D map products. These map products may be general, e.g. displaying the top, base or thickness of lithostratigraphic units, or they may be tailored to a specific application.

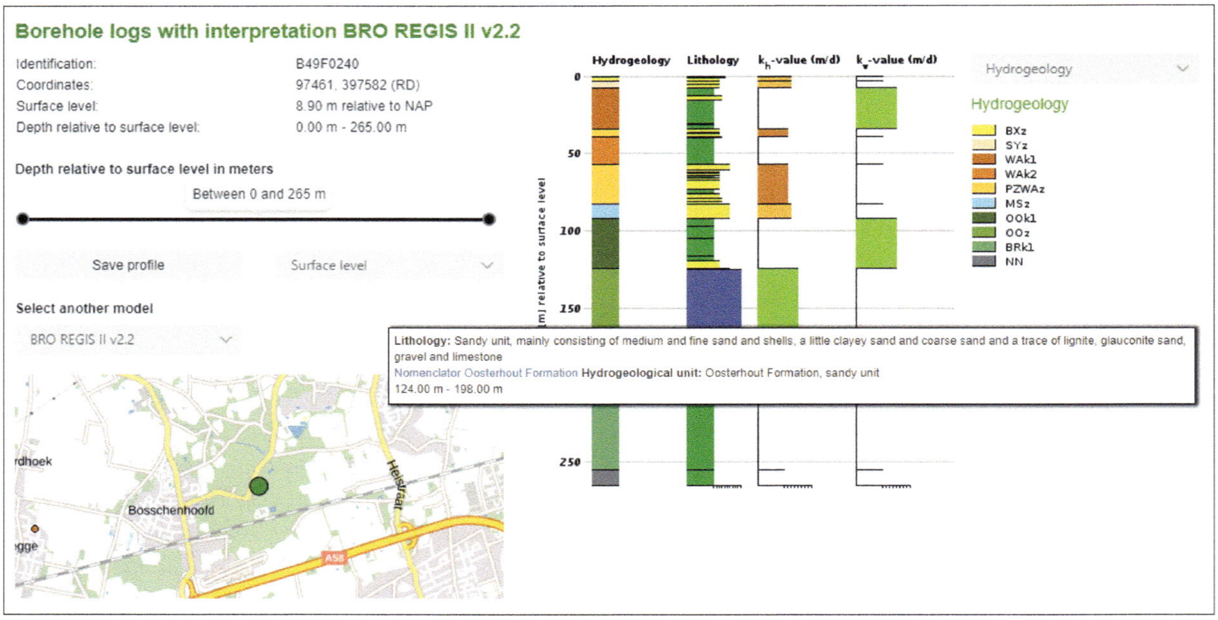

Figure 24.21. Borehole log with hydrogeological interpretation for the REGIS II model as displayed on the DINO web portal. The mouseover text offers a short lithological description of the hydrogeological layer as well as a link to the online Stratigraphic Nomenclature (TNO-GDN, 2022b) of the stratigraphic formation the unit belongs to.

Online viewer

The DINO web portal starts with a topographic map of the Netherlands. After selecting one of the models, a map of the extent of the model as well as the location of the boreholes used in the construction of the model is shown. The user can navigate through the map by zooming and panning or directly go to a location by using search-criteria such as borehole-ID's, (x,y)-coordinates or topographical names. Clicking on a borehole location opens a panel revealing the *borehole log* including the interpretation that is relevant for the selected model (Fig. 24.21). By hovering the mouse over the borehole intervals, the user is presented with short descriptions on the lithology and the interpreted units. In the case of stratigraphic units, a link to the full description of the unit in the online Stratigraphic Nomenclature (TNO-GDN, 2022b) is provided (see Fig. 24.21).

A simple point-and-click tool allows the user to create a *synthetic or virtual borehole* at any location in the model. For the DGM and DGM-deep models, the synthetic borehole shows the vertical succession of geological units at the selected location. In case of REGIS II, hydrogeological units are complemented with vertical plots of conductivity (k_h and k_v) as well as transmissivity of the sandy units and hydraulic resistance of the clayey layers (Fig. 24.21). A synthetic borehole through GeoTOP corresponds to a vertical voxel-stack where voxels are colour-coded with the two main attributes (stratigraphy and lithological class) as well as the probabilities of each lithological class (Fig. 24.22). As with the borehole logs, mouseover texts provide short explanations and, if applicable, a link to the Stratigraphic Nomenclature.

Another tool allows the user to create *vertical cross-sections* at any location similar to those shown in Fig. 24.11. A cross-section through the GeoTOP model offers the opportunity to select and display any of the attributes of the model.

Lastly, *maps* of the (hydro)geological units can be viewed. A map view is accessed via the cross-section by hovering over the appropriate unit. A mouseover text appears, including links to the different maps: depth to base for DGM-deep; top, base and thickness for the three shallow subsurface models and additional maps of hydraulic properties and the associated uncertainties for REGIS II. The map is zoomed in on the location of the cross-section, but the user may zoom or pan to other areas. If applicable, fault trajectories are plotted on top of the maps of the shallow subsurface models.

SubsurfaceViewer®: the 3D geological map

Detailed stratigraphic-lithological models such as GeoTOP may be regarded as 3D geological maps, complementing or replacing their 2D predecessors (Stafleu et al., 2021a).

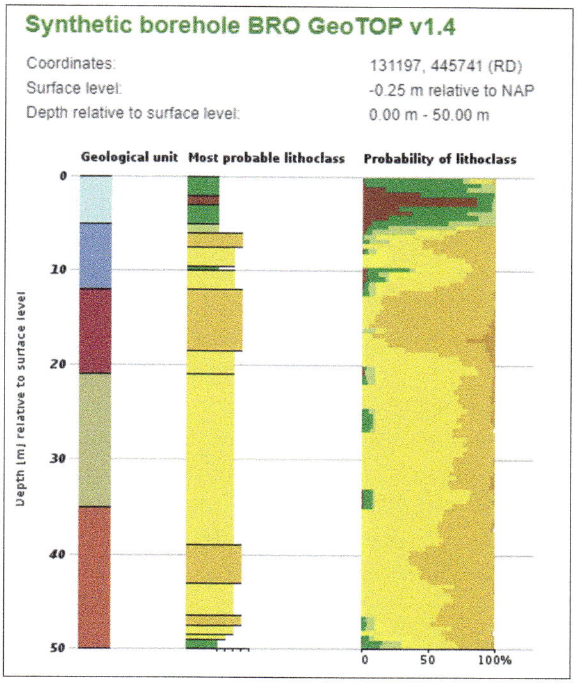

Figure 24.22. Synthetic borehole or vertical voxel-stack taken from the DINOloket web portal showing the vertical succession of voxels at a location selected by the user. The rightmost plot shows areas with a high probability and hence low uncertainty (peat layers close to land surface, coarse sand at ca. 15 m), as well as a zone of high variability in lithological class (from 5 to 12 m). For legend of the lithological classes see Fig. 24.16.

Schokker et al. (2015) used a case study in Amsterdam to demonstrate how 3D visualizations of DGM and GeoTOP create a better understanding of the distribution, thickness and interrelation of the geological units than seen on 2D maps. The SubsurfaceViewer® software, developed by INSIGHT Geologische Software GmbH (Sobisch, 2011), allows for full 3D visualization. The software is offered as a free download on the DINO web portal (www.dinoloket.nl/en) along with predefined cut-outs (usually blocks of 20 x 25 km) of the DGM, REGIS II and GeoTOP models (Stafleu et al., 2013, 2021a). DGM-deep is delivered in two parts: onshore and offshore. The SubsurfaceViewer® includes both a classical map view, a cross-section view and an interactive 3D view (Fig. 24.23). In addition, the SubsurfaceViewer® offers displays of borehole logs as well as synthetic boreholes. Boreholes, cross-sections, geological layers and voxels can all be visualized in 3D, and users can query the voxel model by selecting lithostratigraphic units, lithological classes, and/or probabilities.

NetCDF: 3D arrays

From a computational perspective, both the layer-based raster models and the voxel models may be regarded as

3D-arrays with numbers representing geological units or other attributes. NetCDF (Network Common Data Form; Unidata, 2022) is a data format that supports the creation, access, and sharing of array-oriented scientific data. The online viewer described above draws from NetCDF-files of the models, allowing real-time construction of synthetic boreholes and vertical cross-sections. NetCDF-files of REGIS II and GeoTOP are published on the GDN's OPeNDAP-server (www.dinodata.nl/opendap), allowing users to select an area of interest and download that part of the model as a NetCDF file. The file can then be used as input for software, e.g. in Python-scripts (for users with programming skills) or in off-the-shelf software. An example of the latter is ArcGIS Pro® (Esri, 2020) which has NetCDF-based voxel-functionality, allowing full interactive 3D visualization of GeoTOP (Fig. 24.24). The voxel-functionality includes selections, slicing, the creation of vertical cross-sections and adjusting the vertical exaggeration.

2D map products

GDN's standard model delivery consists of a downloadable dataset (zip-file) with the individual components of the model. The dataset of DGM-deep comprises a set of raster maps in two different ascii-formats: Esri's Ascii-grid and ZMap Plus, the latter format is commonly used in oil and gas Exploration. The layer-based models of DGM and REGIS II, as well as the underlying layer-based model of GeoTOP, are delivered as rasters in Erdas Imagine format, which performs better in desktop GIS software than Ascii-grids. The geometry of DGM, REGIS II and GeoTOP is described in rasters of top, base and thickness and their associated model uncertainties (standard deviations). In case of REGIS II, the download provides additional raster layers of the hydraulic properties (e.g. k_h, k_v) and the associated uncertainty.

Figure 24.25 shows an example of the set of raster maps of a single geological unit in the DGM model. All 2D raster layers can be viewed and analysed in standard GIS software packages. Borehole data are included in tabular form as well as in GIS point features; fault trajectories are included as line features. The GIS data are accompanied by ArcMap documents that serve as a legend presenting the (hydro)geological layers in the correct stratigraphic order and with an appropriate colouring scheme.

The download of GeoTOP is complemented with the voxel-data in a simple Ascii-format comprising the (x,y,z)-coordinates of each voxel-midpoint and the corresponding attribute values of the voxel. To facilitate the use of the 3D voxel-data in a 2D GIS environment, horizontal slices through the two main attributes of the voxel model (stratigraphy and lithological class) are provided as well. These

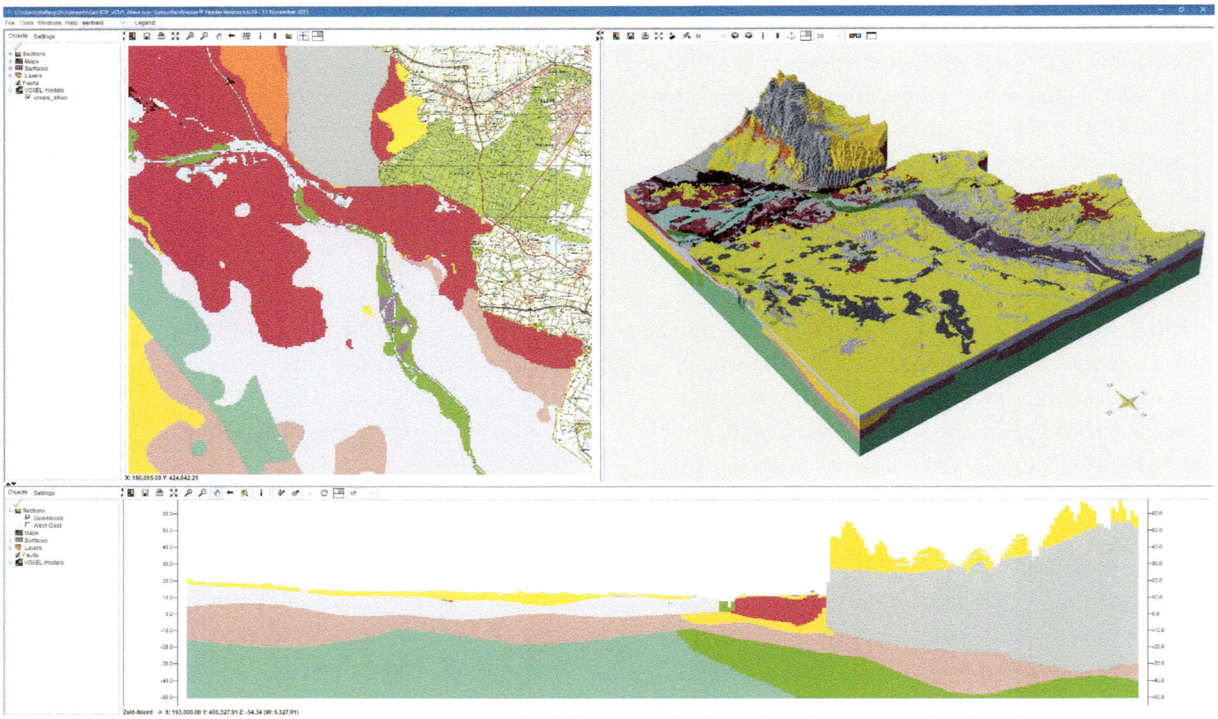

Figure 24.23. Screenshot of the SubsurfaceViewer® with map view (top left), cross-section view (bottom) and interactive 3D view (top right) of a block of 26 x 25 km cut from the GeoTOP model in the area south of Nijmegen. Colours display the lithostratigraphic units the voxels belong to.

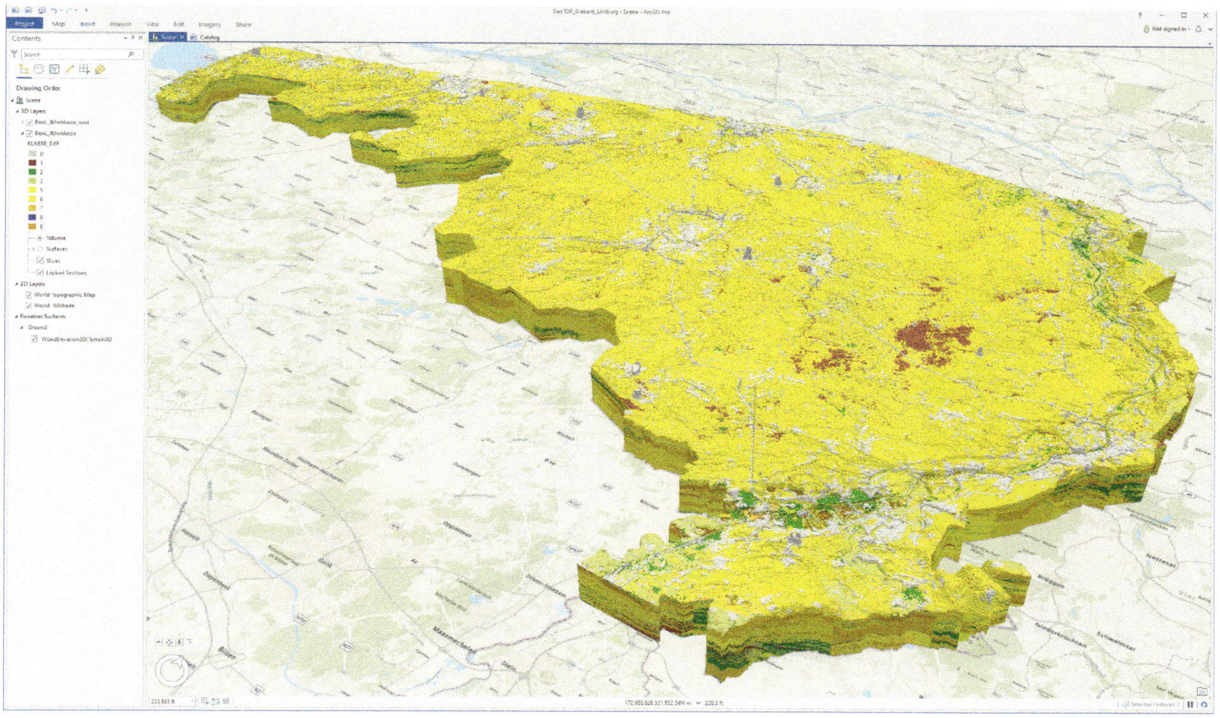

Figure 24.24. *Part of the GeoTOP voxel model as a NetCDF file visualized in ArcGIS Pro®* (*Esri, 2020*).

2D slices show the voxels at a certain depth below NAP (Dutch ordnance datum) or below surface level. Figure 24.26 shows an example of slices through the lithological class model at depths of 0, 2, 4, 6, and 8 m below surface level.

The three subsurface models that are part of the key registry of the subsurface (BRO) are accompanied by formal documentation (in Dutch), including extensive descriptions of the modelling procedure (*Totstandkomingsrapport*) as well as the quality control process (*Kwaliteitstoetsingsdocument*). The data structure of each model is formally described in a data catalogue (*Gegevenscatalogus*), including an UML-based class diagram of the objects (e.g. boreholes, layers, voxels) and definitions of these objects as well as their attributes and mutual relationships.

2D maps from vertical voxel-stack analysis

Other 2D raster maps are created by selectively querying the 3D voxel model by using a vertical voxel-stack analysis (Dambrink et al., 2014, 2015; Stafleu et al., 2021a). Maps based on vertical voxel-stack analysis are a step closer to the user's own area of expertise than the geological models themselves. A vertical voxel-stack analysis processes the voxels at a particular (x,y)-location from the uppermost voxel (at land surface) down to the lowermost one (at the bottom of the model), selecting voxels by their attributes like stratigraphy, lithological classes or depth. The result is a single summary value for the (x,y)-location of the stack.

By repeating this procedure for all voxel-stacks in the model, a 2D raster map is created with cells measuring 100 x 100 m and attributed with the summary value. Three examples are: a surface geological map (Fig. 24.27a), a map showing the cumulative thickness of peat in the Holocene deposits (Fig. 24.27b), and a generalized soil composition map of the upper 8 m (Fig. 24.27c). The examples given are straightforward and relatively easy to calculate. Maps resulting from more complex voxel-stack analyses include a land subsidence prediction map (Fig. 24.20) and a seismic site-response zonation map (Fig. 24.31).

Innovative modelling methods and case studies

Towed Transient Electro-Magnetic technique (tTEM)

Electromagnetic (EM) techniques map electrical resistivity contrasts in the subsurface in three dimensions (Kirsch, 2009). This geophysical technique allows for rapid surveying of large parts of the subsurface and is extensively used in hydrogeological applications. The collected data are processed to obtain an electrical resistivity model that is subsequently interpreted into geology or groundwater quality. Saline groundwater is a good conductor and in coastal areas it is often an important target in EM mapping. Sediments that contain clay minerals are also good conductors and the distinction between clay and sand can

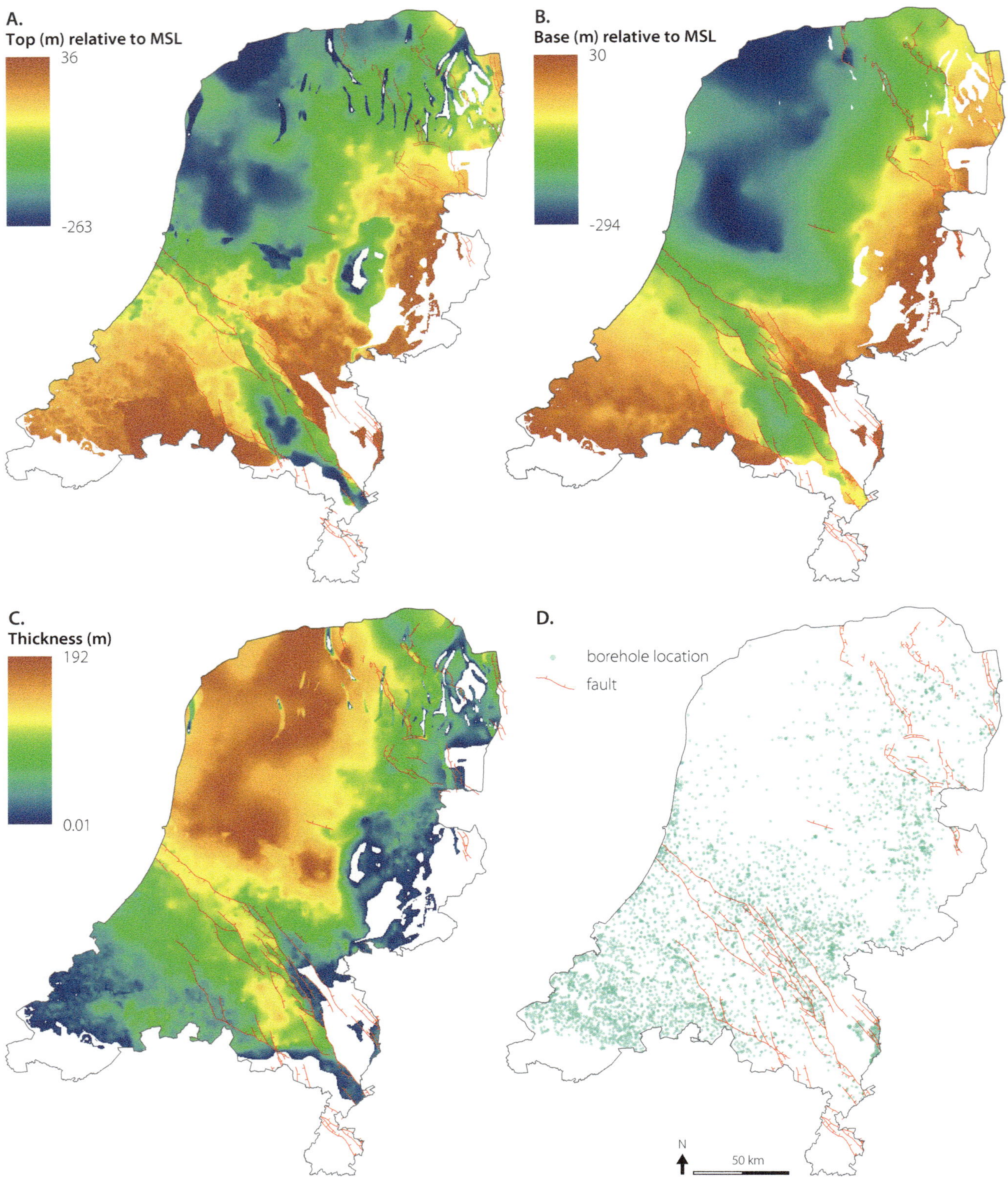

Figure 24.25. *Four raster maps containing the model data of a single geological unit in the DGM model: a) top of the unit; b) base; c) thickness and d) borehole locations where the unit was interpreted. Additional maps of the model uncertainty of top, base and thickness are also available.*

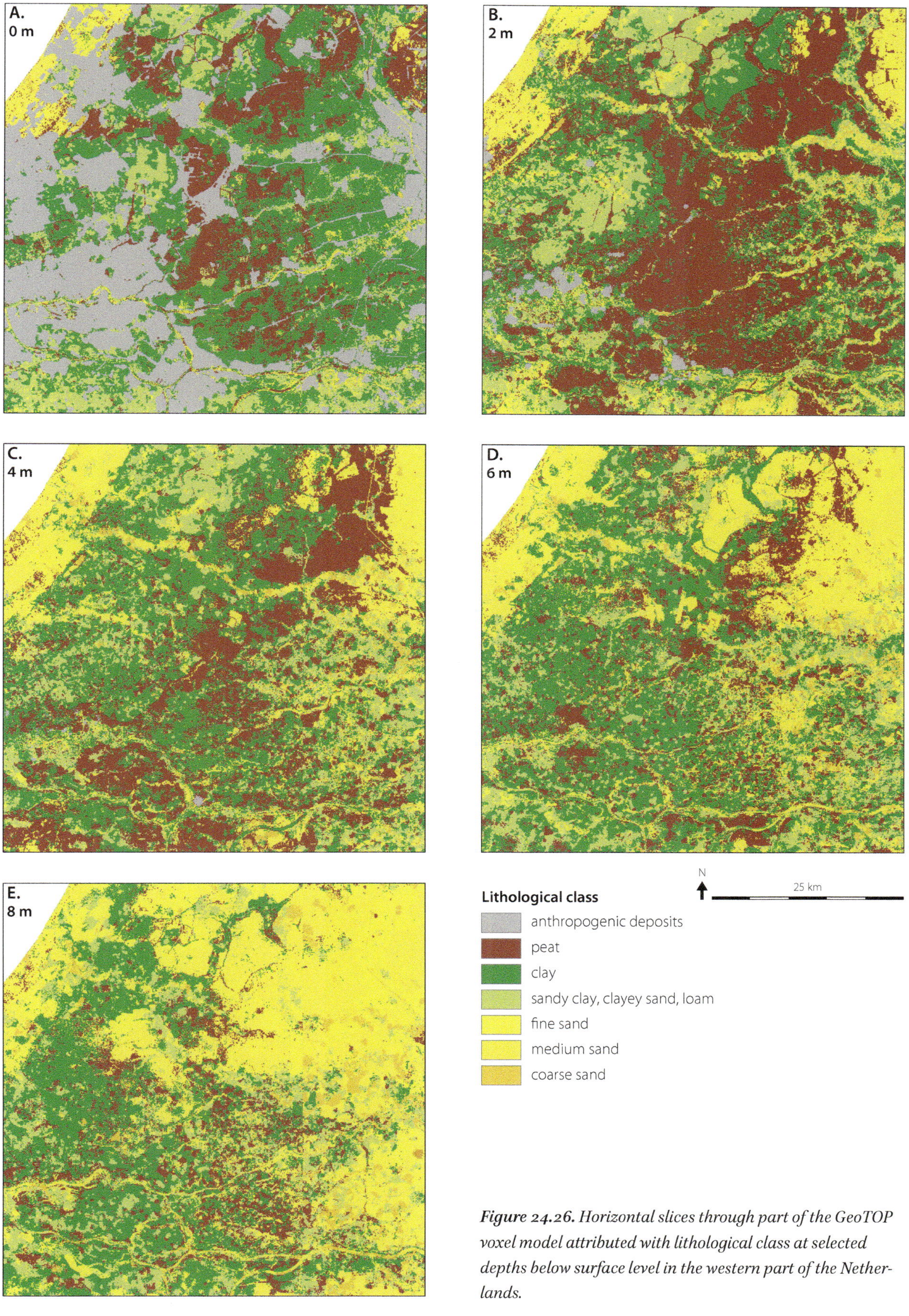

Figure 24.26. Horizontal slices through part of the GeoTOP voxel model attributed with lithological class at selected depths below surface level in the western part of the Netherlands.

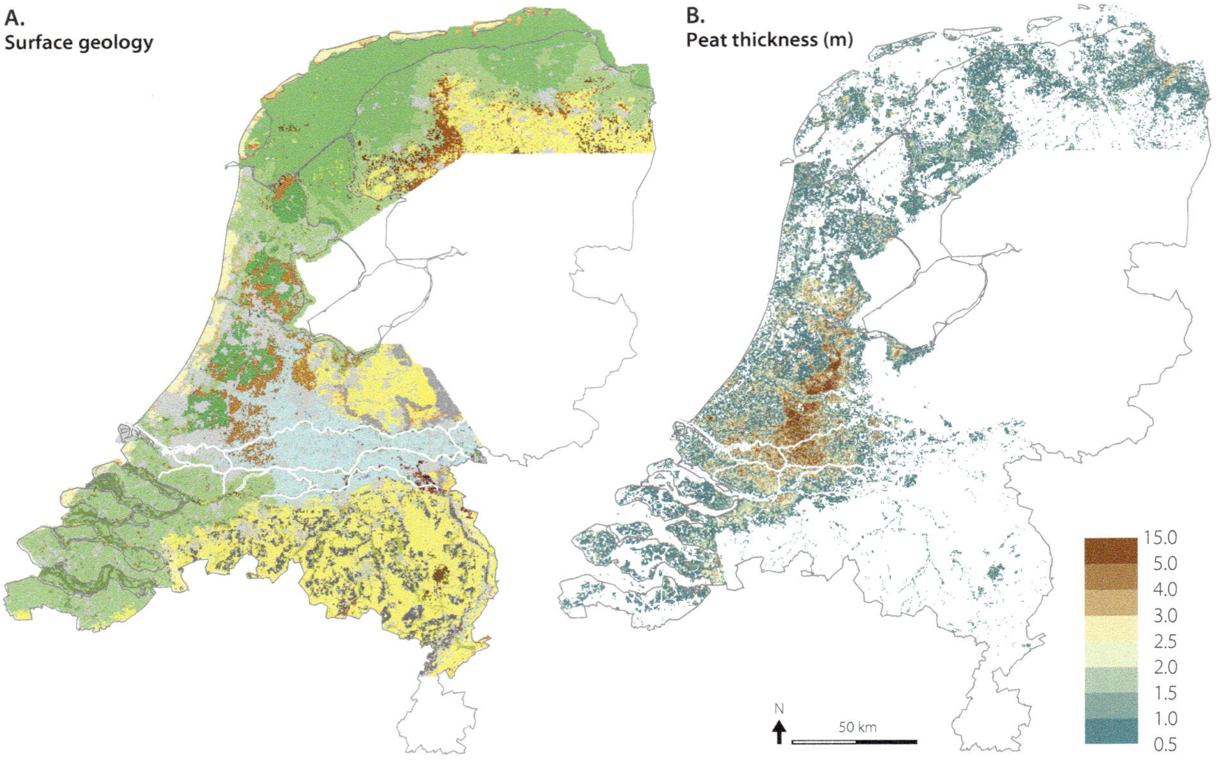

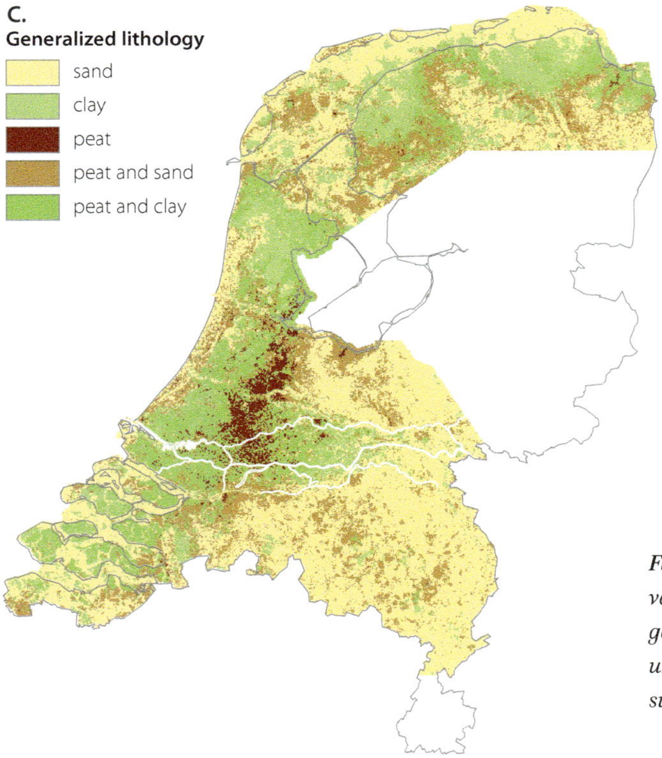

Figure 24.27. *Three examples of 2D maps created by a vertical voxel-stack analysis of the GeoTOP model: a) surface geology; b) cumulative thickness of peat in the Holocene units; and c) generalized lithology of the upper 8 m of the subsurface. Modified after Dambrink et al. (2015).*

often be inferred from the EM signal. The great benefit of the non-invasive EM technique is the ability to measure changes in electrical resistivity (and hence conductivity) with depth, resulting in a 3D model of the subsurface. Expert knowledge of the regional geology and geohydrology, together with existing (borehole) data is of paramount importance for the interpretation. The technique is sensitive to disturbance caused by metal objects (fences, pipelines) and cables that transport electricity, both in the ground and overhead.

The towed Transient Electro-Magnetic technique (tTEM) has been developed to acquire data down to 60-80 m depth by driving the instrument behind an ATV (Sandersen et al., 2021; Fig. 24.28). With a speed of 10-15 km/h, measurements are collected every 5 m. On larger fields, the distance between lines is typically 20 m, resulting in a dense network of data that is inverted into 1D electrical resistivity models, showing the variation of resistivity with depth. Interpolating 1D resistivity models into a 3D model allows for further interpretation in terms of geology and lithology. The tTEM system has been used to map the lateral continuity of clay layers in the province of Drenthe. A field campaign of 2.5 days resulted in almost 40 line-km of data in four different areas. In the results depicted in Figure 24.29 a semi-continuous layer of low resistivity (blue colours) is present, indicating a clay layer. A gas-pipeline is visible in the results causing an interference in the tTEM measurements.

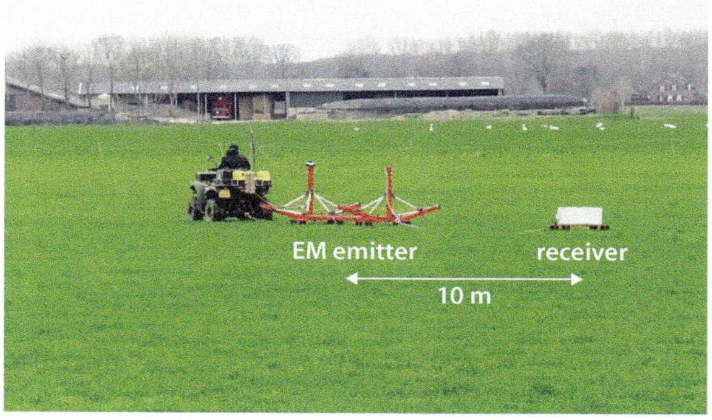

Figure 24.28. A tTEM system, consisting of an EM emitter and receiver towed by an all-terrain vehicle (ATV)..

Site-response assessments

Earthquakes can cause damage to buildings and infrastructure. This holds for natural earthquakes as well as for human-induced earthquakes such as those in the province of Groningen. The composition of the shallow subsurface plays an important role in the extent of the damage caused by earthquakes, because 'weak soils' such as peat, clay and clayey sand/sandy clay can amplify the amplitude of seismic waves. One of the factors related to this 'site-response' is the shear-wave velocity (V_s) of the shallow subsurface, which is often summarized in an average V_s -value for the upper 30 m (V_{s30}).

Kruiver et al. (2017a,b) combined the GeoTOP voxel model of the Groningen area and a set of 60 Seismic Cone Penetration Tests (SCPT's) into a V_s model of the area cov-

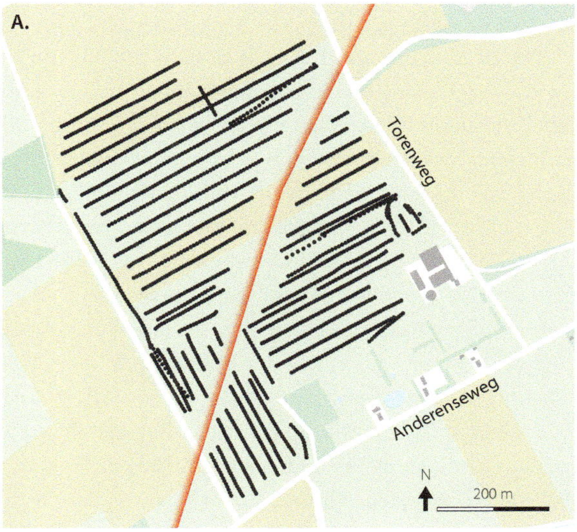

Figure 24.29. A tTEM study on the lateral continuity of clay layers, showing: a) 2D map of tTEM survey lines (black) and a pipeline (red); b) a fence diagram of tTEM results; c) tTEM results showing resistivities under 30 Ω m, indicative for clay. The pipeline interferes with the tTEM measurements. Measurements in the vicinity of the pipeline are therefore discarded.

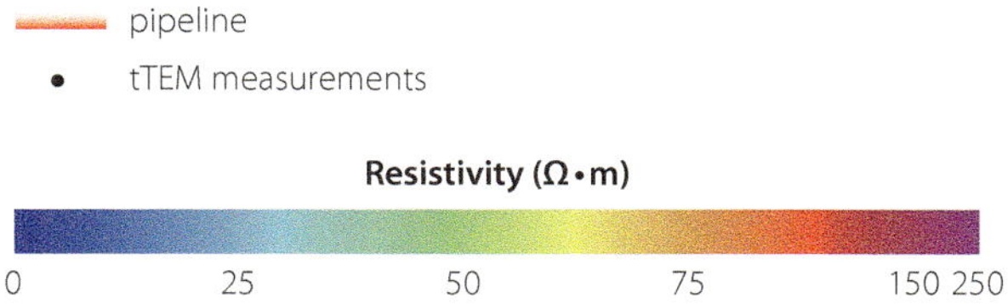

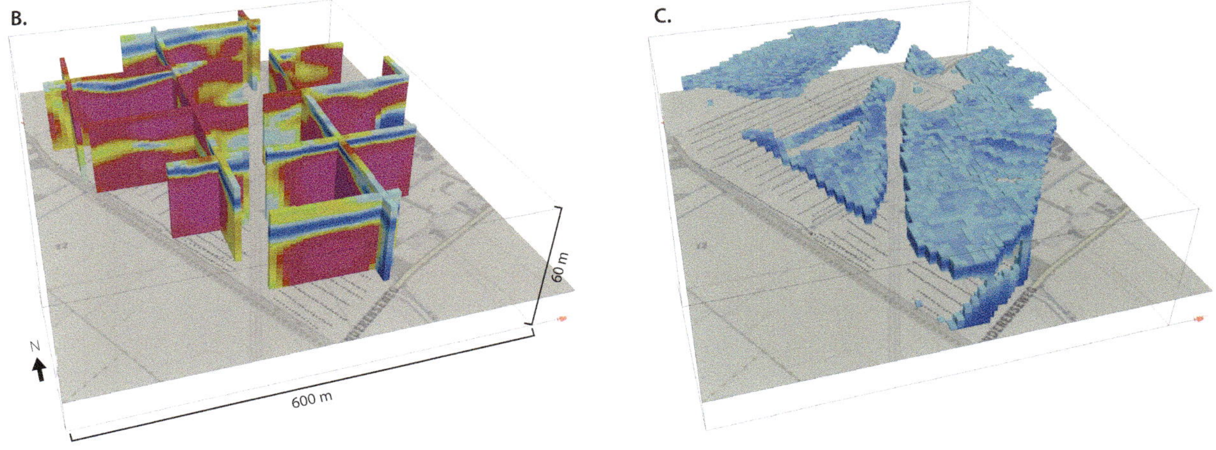

ering the gas field and a buffer zone of 5 km. The SCPT's were used to establish statistical distributions (with mean and standard deviation) of V_s for each combination of lithostratigraphic unit and lithological class. In a number of combinations, it was possible to quantify a depth-dependent distribution (with slope and intercept). These distributions were used to assign a V_s value to each voxel in the model (Fig. 24.30). The resulting 3D V_s model shows how V_s generally increases with depth and contrasts soft Holocene deposits with low V_s with predominantly stiff Pleistocene deposits with high V_s. Vertical voxel stacks attributed with V_s and other soil properties were used as input for site amplification predictions (Kruiver et al., 2017a,b; Stafleu et al., 2021a).

Natural and induced earthquakes may also occur outside the Groningen area. Stafleu et al. (2022) created a preliminary V_{s30} map of the Netherlands based on the V_s measurements from the Groningen area. Van Ginkel et al. (2022) developed a national seismic site-response zonation map of the Netherlands, designed for low-magnitude induced earthquakes. Sediment profiles typical for the Dutch shallow subsurface were categorized into five classes, each representing a level of expected amplification. The mean amplification for each class, as well as its variability, was quantified using data from 66 sites with measured earthquake amplifications and 115 sites with amplifications estimated from ambient vibrations. The zonation map was subsequently derived from GeoTOP by assigning a sediment class to each vertical voxel stack in the model (Fig. 24.31). In areas where GeoTOP is not yet available, TNO-GDN's voxel model NL3D (Van der Meulen et al., 2013) was used.

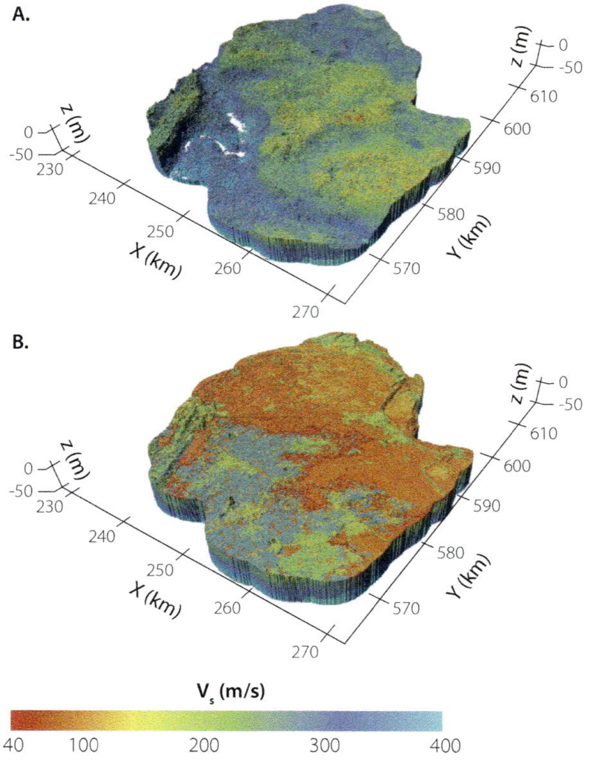

Figure 24.30. GeoTOP model of the Groningen area covering the gas field and a buffer zone of 5 km. Voxels are attributed with shear wave velocity (V_s) derived from V_s measurements in Seismic Cone Penetration Tests (SCPT's): a) the lower part of the model showing the glacial Peelo Formation with relatively high V_s values; b) the full model revealing Holocene deposits with relatively low V_s values. After Kruiver et al. (2017b).

ThermoGIS

ThermoGIS (www.thermogis.nl) is a public, web-based geographic information system supporting companies and the government in developing geothermal energy in the Netherlands (ThermoGIS, 2022). ThermoGIS provides depth, thickness, porosity and permeability maps of many potential aquifers in the Netherlands. Geothermal performance maps are calculated with the use of an integrated, stochastic, techno-economic performance module. The most important outputs of ThermoGIS are geothermal potential maps of the Netherlands that can be consulted in a map viewer (Fig. 24.32).

The DGM-deep model is used as basis for the aquifer model for ThermoGIS v2.2. The depth and thickness of the aquifers within the stratigraphic groups of DGM-deep are constructed using information from a maximum of 3855 wells. For most aquifers, this is considerably more than the number of wells used for the calibration (well-tying) of DGM-deep. ThermoGIS v2.2 presents the geothermal potential for 30 individual aquifers, 5 stacked layers which consist of multiple aquifers within a stratigraphic group and an overview layer that integrates all aquifers (see ThermoGIS (2022) for an overview).

For future ThermoGIS updates it is foreseen that the shallow aquifers of the hydrogeological REGIS II model will be incorporated. This model includes aquifers within the Upper North Sea Group and uses the geological DGM-model for the shallow subsurface as a basis. The ThermoGIS workflow consists of numerous steps of which the three most important ones are: (1) modelling aquifer geometry; (2) estimating aquifer properties and a subsurface temperature model; (3) incorporating technical (doublet) and economic parameters.

Aquifer geometry – Modelling aquifer tops, bases and thicknesses in ThermoGIS has many similarities with modelling the geometry of the hydrogeological units in REGIS II. All aquifers within a DGM-deep unit are modelled with respect to the base of that unit, making ThermoGIS and DGM-deep fully consistent. In a reiterating

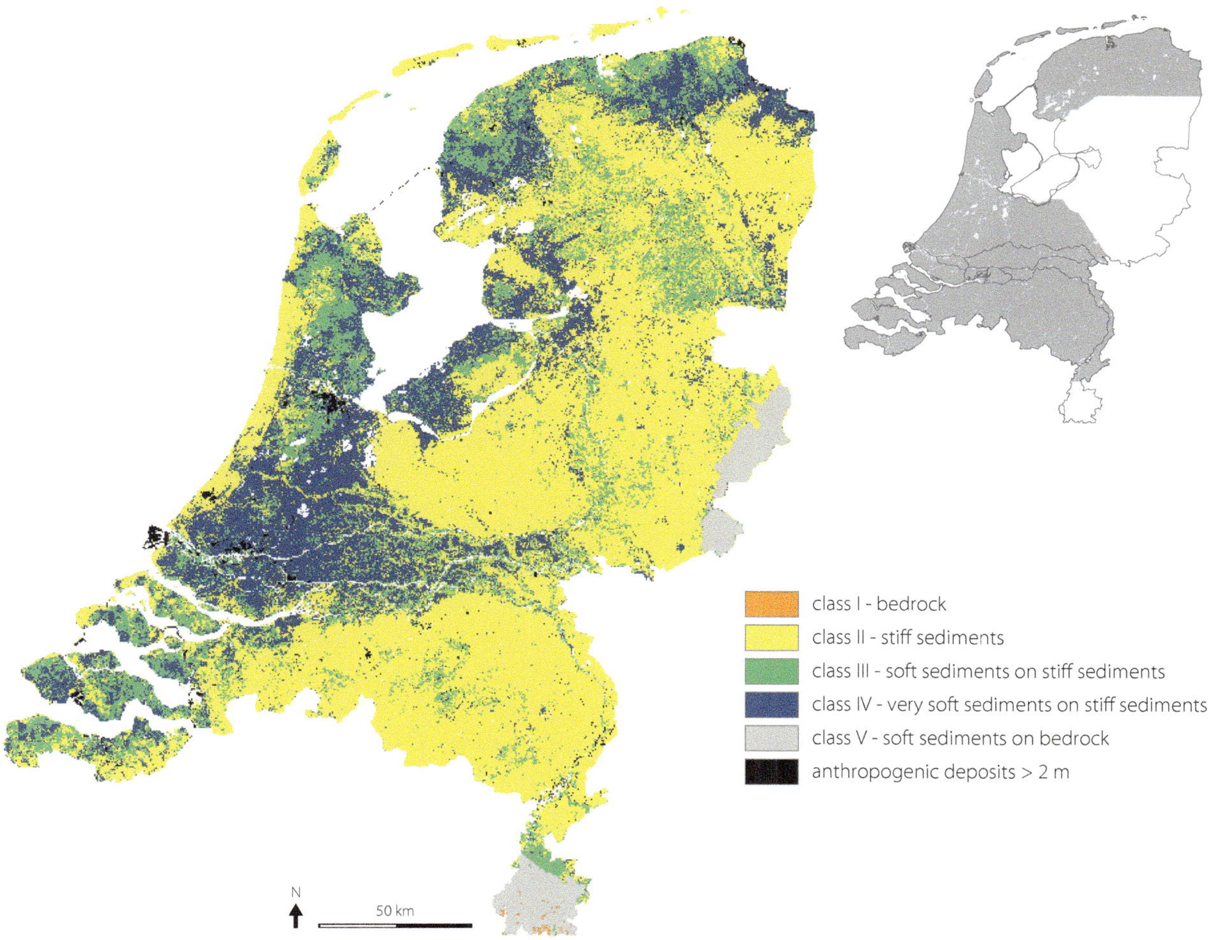

Figure 24.31. *Seismic site-response zonation map for the Netherlands designed for low-magnitude induced earthquakes. Inset map shows the current GeoTOP coverage; areas outside that area are based on the NL3D model. Colours represent the five sediment classes I through V. Areas where anthropogenic deposits exceed 2 m in thickness remain unclassified. After Van Ginkel et al. (2022).*

process, the bases of the aquifers are derived by interpolating the difference between base of the aquifer from well data and the base of the group-level grid and adding it to the latter. Subsequently, the interpolated thickness is added to obtain the top of the aquifer. Multiple aquifers are stacked then between groups' base and top. The interpolation process is assisted by maps with the maximum spatial extent of each aquifer, pinch-out polygons indicating where the thickness converges to zero, and non-pinch-out polygons where the regional thickness trend is retained. In this process, also faults and the overall structural setting of the aquifer are taken into account. For interpolation the convergent gridding algorithm in Petrel® is used. In addition, thicknesses are stochastically modelled using kriging and the resulting variance is subsequently used to calculate P10 and P90 thickness maps. Each aquifer is modelled independently of the position of other aquifers. An integrated and consistent aquifer model is constructed by stacking the aquifers and correcting for intersections.

Aquifer properties and temperature model – For each aquifer, ThermoGIS uses vertically averaged well porosities and permeabilities that are either based on core plug measurements or calculated from petrophysical evaluations and well tests. The porosity-depth relationship is calibrated using burial anomaly maps so as to create country-wide porosity maps. Core-plug derived porosity-permeability relationships (phi-k trends) are used to construct permeability maps and subsequently, the heat in place (HIP) is calculated by using the net thickness, permeability and temperature of the aquifer. The net thickness of the aquifer is determined by multiplying the gross thickness with a net-to-gross ratio. Due to the lack of a reliable and consistent net-to-gross database, a single value was determined for each aquifer based on geological knowledge and available data. The aquifer temperature comes from a 3D temperature model of the Dutch subsurface (Bonté et al., 2012) that was recently updated using new geological and technical results from Békési et al. (2020).

Incorporation of technical (doublet) and economic parameters – In a subsequent step, additional technical (doublet) and economic parameters are incorporated to derive the technical and economic potential and the potential recoverable heat (for more information on these parameters, see Mijnlieff et al., 2025, this volume). The potential maps can be interactively consulted on the ThermoGIS website (ThermoGIS, 2022; Fig. 24.32). The user can either choose for a general overview showing the total potential or a more detailed view where the above-described inputs and outputs of individual aquifers can be considered. In addition, the user can manually alter the technical parameters for site-specific fine-tuning.

High resolution 3D models

Some applications require more detailed information on lithological heterogeneity than is provided by the national subsurface models described in this chapter. An example is the large scale reinforcement by the regional water authority *Hoogheemraadschap De Stichtse Rijnlanden* (HDSR) of the northern Lek River dike. This dike is an important safety measure that keeps a densely populated and economically vital part of the Netherlands safe. The shallow subsurface surrounding the dike consists of a complex alternation of Holocene Rhine sandy channel belt deposits, clayey floodplain deposits and levees or crevasse deposits composed of silt, sand and clays of the Echteld Formation that alternate with peat beds of the Nieuwkoop Formation (Erkens, 2009; Dabekaussen et al., 2019; Fig. 24.33). The Holocene sediments overly a sandy Pleistocene (and locally Early Holocene) substrate of the Kreftenheye Formation with a distinct loam layer at the top (Kreftenheye Formation, Wijchen Member).

Depending on the lithology, different dike-failure mechanisms may play a role. For instance, permeable sand bodies increase the risk of piping, whereas soft soils are prone to macro-instability. Therefore, HDSR needs detailed information on the deposits underneath the dike in order to design location specific and fit-for-purpose reinforcement measures. Incorporating detailed geological information in planning increases confidence in estimating the chance of possible dike failure, potentially leading to an increase of the estimated strength of the dike and thus lower costs by avoiding unnecessary reinforcements.

As more detailed 3D insights into the subsurface below and directly surrounding the dike were needed, the existing GeoTOP dataset was expanded to make a new model with a higher resolution. The required data density was reached by extending the existing GeoTOP dataset (comprising ca. 3600 boreholes) with ca. 1200 additional borehole descriptions from Utrecht University (Cohen, 2017) and by adding ca. 2100 Cone Penetration Test (CPT) results. CPT's, which were not used in the earlier GeoTOP model, are a common geotechnical measurement that determine the strength of the soil. It is a quick and inexpensive measurement that is widely used in soft-sediment

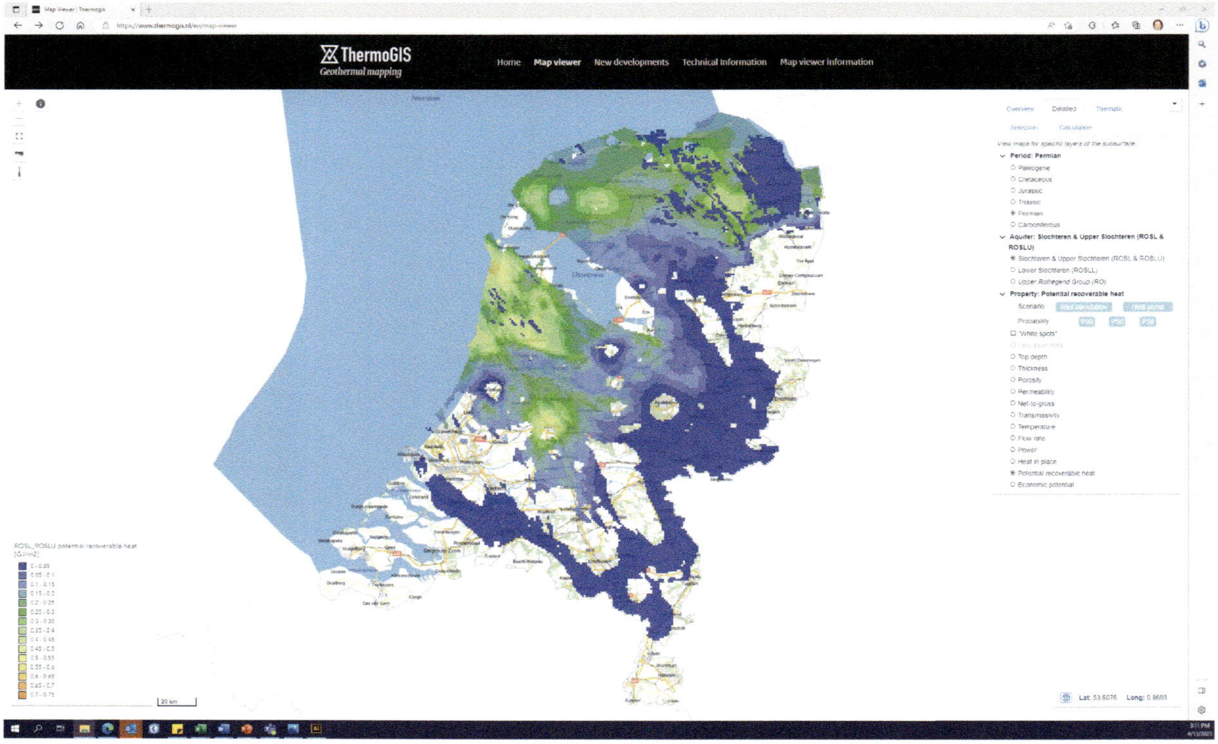

Figure 24.32. *The ThermoGIS Map viewer (ThermoGIS, 2022).*

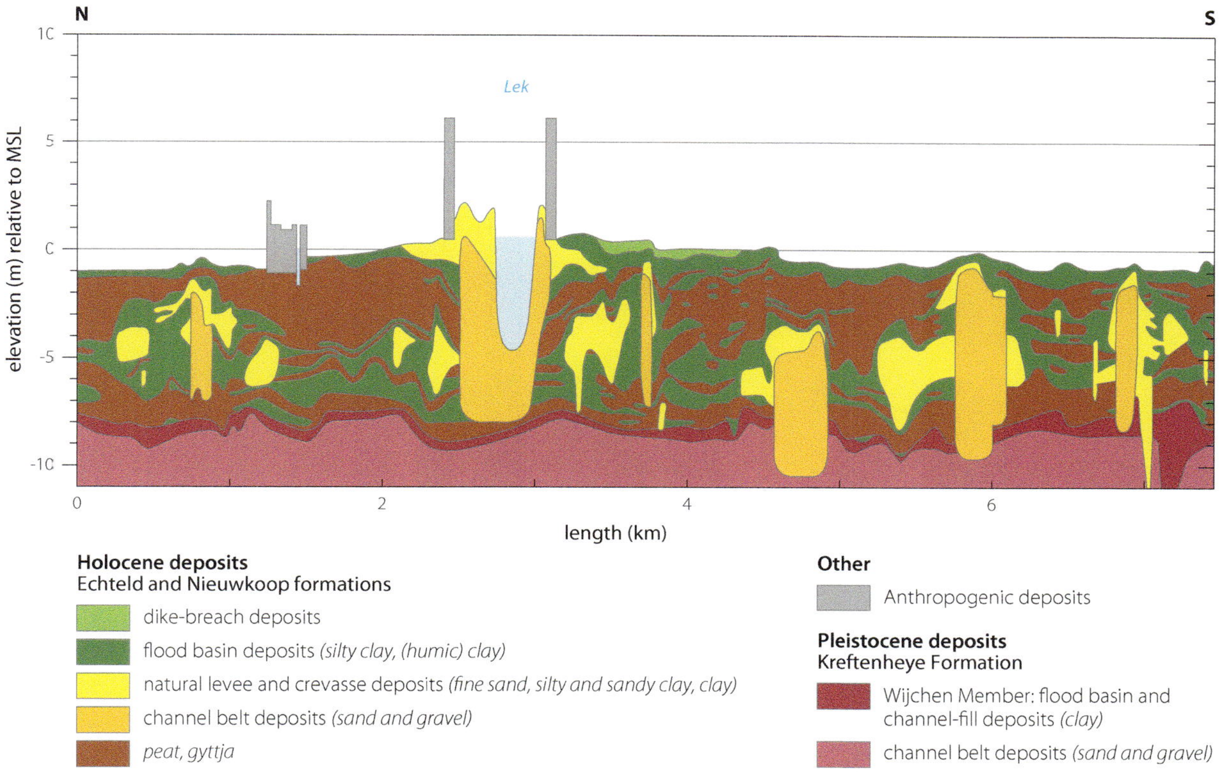

Figure 24.33. Cross-section through the Holocene and uppermost Pleistocene sediments directly surrounding the dike of the Lek River. Modified after Erkens (2009).

environments for infrastructural research (roads, railways, dikes) and in the built environment. This combined dataset allowed us to refine the resolution of GeoTOP from 100 x 100 x 0.5 m to 25 x 25 x 0.25 m.

The existing GeoTOP workflow that was followed (Stafleu et al., 2011a, 2021a) allows borehole information to be processed, but not CPT results, which were therefore translated to lithological information. The characteristics of the CPT measurements (cone resistance, sleeve friction, pore water pressure) can be used as proxies for lithology. To establish the relationship between CPT parameters and lithology, there is the classical approach by Robertson (2010) or machine learning approaches. In the case of the Lek River dike, an Artificial Neural Network (ANN) was employed. More than 200 borehole-CPT pairs within the study area (within a distance of 10 m from each other) were used as a training set for the neural network. This allowed for pairing lithological information from the borehole with the CPT parameters at corresponding depths (Fig. 24.34). The trained neural network was used to translate CPT friction ratio and cone resistance to lithological class and so create 'pseudo' borehole-descriptions at all CPT locations. The translated CPT's, together with the borehole-descriptions were used as input for the layer model and the voxel model in a similar way to the national GeoTOP model.

Comparing the high resolution model with the GeoTOP model (Fig. 24.35) clearly shows the level of detail that can be seen in the new high resolution model. Figure 24.35 demonstrates that spatial distribution and correlation of the different lithological classes is improved with higher data density. Furthermore, the higher resolution allows for a better representation of local sediment bodies, e.g. channel belt and crevasse deposits. The increased insight into the spatial distribution of the various lithologies aids the decision-making process of the regional water authority.

High resolution subsurface modelling, as exemplified above, has the potential to form the basis of many other applications. For instance, the voxels could be populated with geotechnical parameters, opening a way forward for 3D calculations of subsurface characteristics. Further development could also be found in methods to quickly update the model with new data, allowing users to add new measurements to the existing model.

Perspectives for systematic subsurface modelling in the Netherlands

From map sheets to projects

Initiated around the turn of the century, subsurface modelling by TNO-GDN has gradually become as mature an

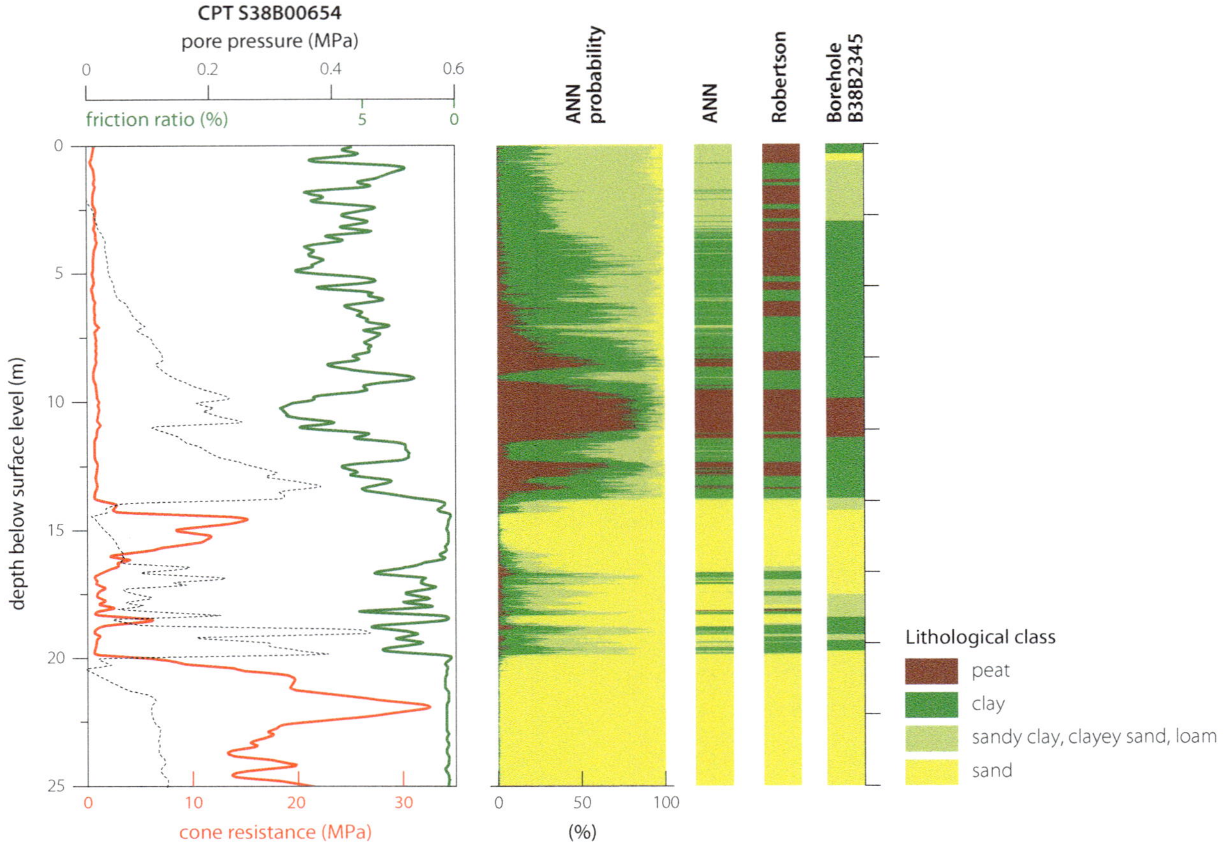

Figure 24.34. Interpretation of Cone Penetration Test (CPT) measurements (cone resistance, sleeve friction, pore pressure and depth below surface level) into lithological classes using an Artificial Neural Network (ANN). Shown is a CPT-borehole pair not used for training of the ANN. The raw ANN output consists of lithological class probabilities, which are used to determine the most likely lithological class. For comparison both a classic interpretation (Robertson, 2010) and lithology as logged in a nearby borehole are shown. Modified after Dabekaussen et al. (2021).

effort as the preceding surface geological mapping programmes it replaces (Van der Meulen et al., 2015). A systematic approach, standards and quality control have been developed and implemented in a suite of modelling programmes that each provide a balance between consistency over time and space on the one hand and the need to innovate on the other, as geological understanding and particularly modelling technology evolve continuously.

TNO-GDN is now entering a next stage. The first rounds of shallow subsurface modelling (GeoTOP, DGM and REGIS II) were similar to traditional mapping projects in having a regional focus. In the surface mapping era, data collection, interpretation and actual map sheet production were connected activities, and planned accordingly. In the first twenty years of subsurface modelling, the Geological Survey primarily relied on data collected in the 20th century and, working area by area was the natural continuation of the decades-old practice of working map sheet by map sheet. In the coming years however, our entire data legacy will have been processed in this way, and updates of the individual models will make use of the new data from the key registry for the subsurface (BRO). Even though it is as yet too early to speculate about rates and volumes, two things are already clear: (1) the key registry is already growing much faster than our survey databases did over the past two decades, and (2) subsurface data will never again be generated map sheet by map sheet. Instead, projects – for example public construction and engineering works, soil quality assessments, and assessments of groundwater and other resources – will define the need for, and be the source of, subsurface data and information. This implies that TNO-GDN will have to update and improve their models in much smaller and more irregularly shaped patches than the original model areas. In a project-centred approach it would make sense to eventually work at variable resolutions instead of the currently employed one-size-fits-all specifications.

The energy transition

After the implementation of the key registry for the sub-

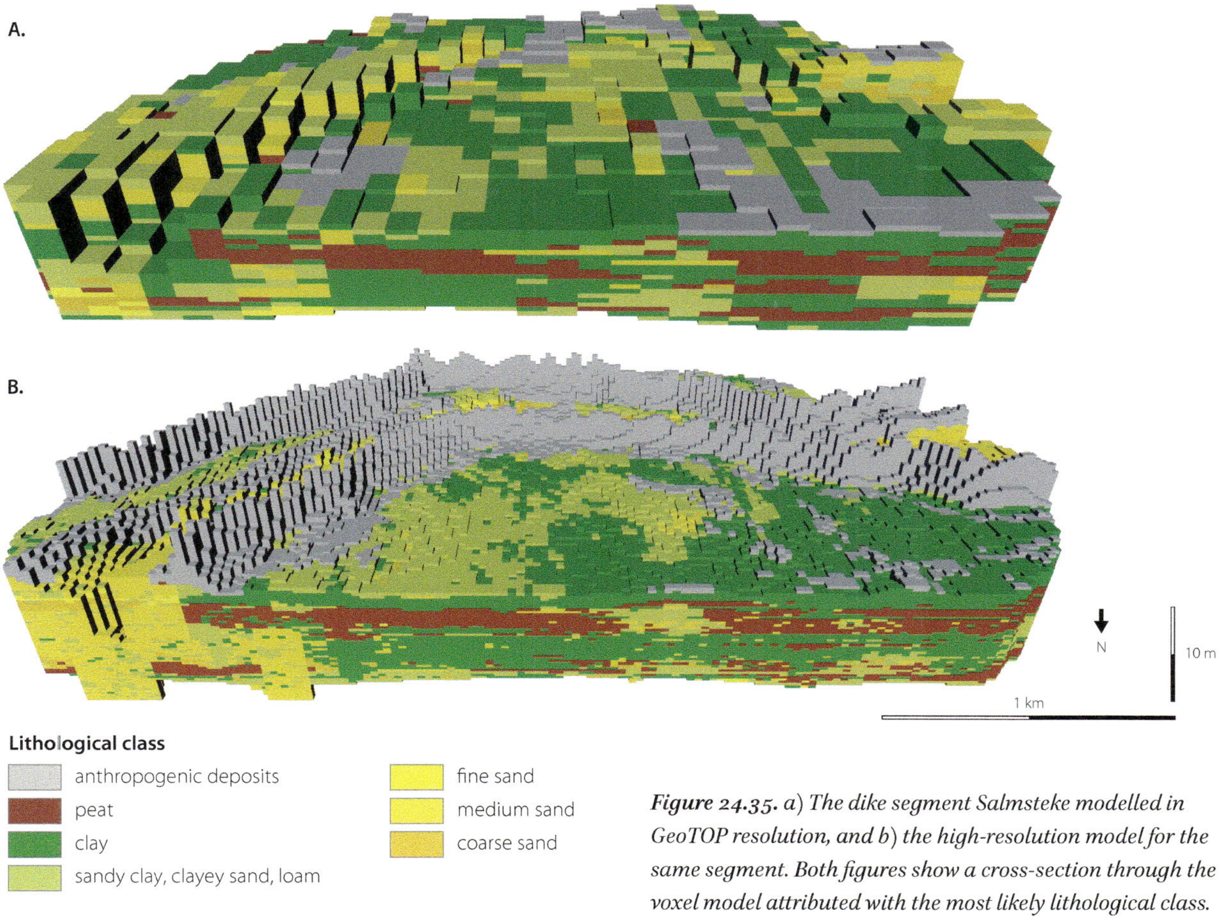

Figure 24.35. a) *The dike segment Salmsteke modelled in GeoTOP resolution, and* b) *the high-resolution model for the same segment. Both figures show a cross-section through the voxel model attributed with the most likely lithological class.*

surface, TNO-GDN's next major effort will be to attune deep subsurface modelling (DGM-deep) to the needs of the energy transition. The observational support, stratigraphic resolution and overall completeness of the survey's current models is more or less inversely proportional to subsurface use intensity. Figure 24.36 defines an 'energy transition blind spot' that exists, roughly, between 250 m down to which groundwater is abstracted, and the 2.5 to 4 km depth range, where most hydrocarbon extraction has been taking place.

This zone is now being explored for geothermal energy, and CO_2 storage potential, and other new forms of subsurface use. At the same time, groundwater is being abstracted from ever greater depths for environmental reasons. All in all, water and energy related uses of the subsurface, which used to be well-separated vertically, are now beginning to interact and compete for underground space, to the extent that for example the integrity of strategic groundwater reserves has become a concern (Anonymous, 2021). To facilitate optimal use of the subsurface, and to inform the associated public decision making, TNO-GDN has to follow suit and map this hitherto underexplored blind spot. This entails covering a larger volume of subsurface in greater detail, for which purpose the budget of TNO-GDN for mining-related information services has approximately been tripled. Geothermal operators generally have smaller exploration budgets than the hydrocarbon multinationals that are currently leaving the country. A successful Dutch energy transition does not only require a larger geological survey effort, it also presents the challenge to create an investment climate that is as favourable for geothermal energy development as it used to be for the hydrocarbon industry. In order to improve the information baseline the Dutch government initiated the SCAN programme, which consists of acquiring seismic and borehole data in areas that were hitherto underexplored because of a low hydrocarbon potential. It represents the first major public investment in exploration since the search for coal at the beginning of the 20th century (Van Waterschoot van der Gracht, 1918).

Contributing to full virtual representations of the physical environment

Guided by the concept of 'digital twinning', i.e. creating virtual representations of physical objects, the geoscience and geospatial communities currently strive to create more fully-fledged representations of the physical environment, combining the atmosphere, hydrosphere, built envi-

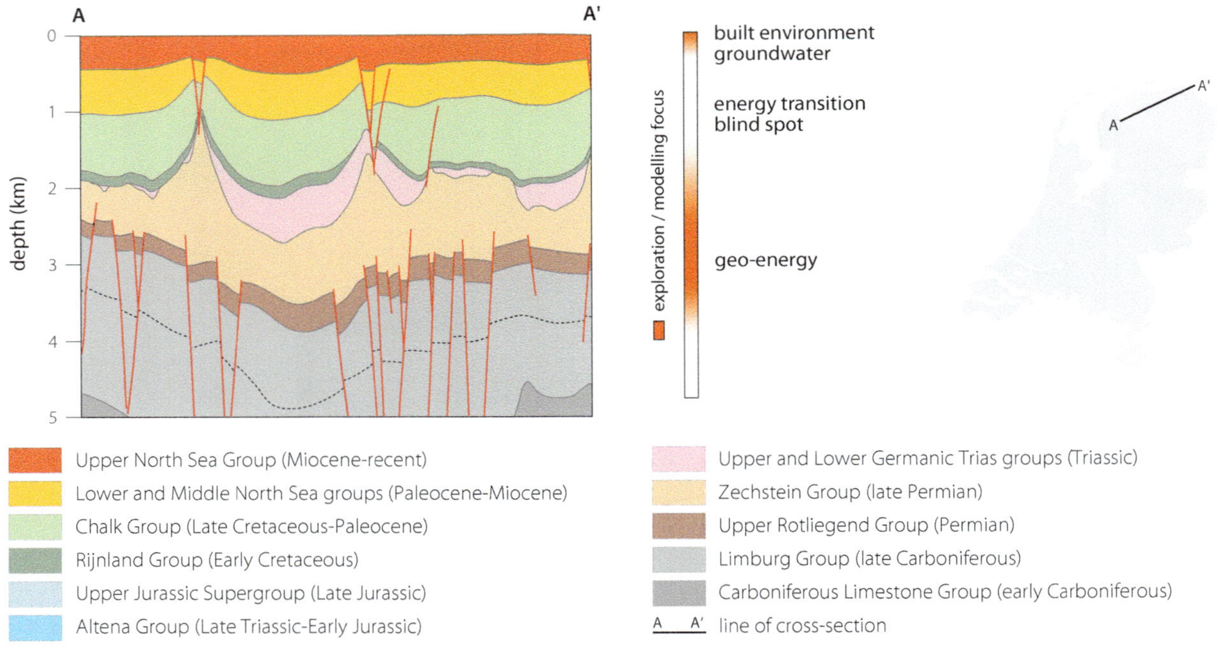

Figure 24.36. An arbitrary cross-section in the northern Netherlands illustrating the relative depth of shallow (hydrogeological, environmental, geo-engineering) and deep (geo-energy) exploration and modelling, and the gap between the two ('energy transition blind spot'). Reprinted with permission from Van der Meulen et al. (2022).

ronment and subsurface compartments (Van der Meulen & Bogaard, 2022). Up till now, science has predominantly captured the world in models, i.e. conceptual and/or mathematical representations of selected aspects and processes. In such a paradigm, it has been valid for TNO-GDN to produce individual models of geological strata or properties. However, a digital twin would represent a section of subsurface as an object with everything relevant included, geologically or otherwise. This could start with the integration of subsurface models that have coinciding or overlapping spatial extents, for example a static hydrogeological model such as REGIS II with groundwater flow and transport models. Obvious next steps, from a hydrological point of view, would be to incorporate bodies and streams of surface water, groundwater wells, pollution plumes, sheet pilings, infiltrations, and so forth. The result would then need to be connected to a virtual above-surface world, constrained by live data, and allowing for realistic human observations and interactions. It goes without saying that the current separate 3D models issued by TNO-GDN will be merged, at least at the layer-model level. Sets of geological units or selected depth ranges from the resulting integrated subsurface models could then be populated with parameters associated with the various applications.

Creating digital twins of the subsurface – or, to be more precise, creating digital twins of the physical environment including the subsurface – involves interactions across various disciplines. They would need to be built together with, and interrogated by, non-geoscientists who cannot be expected to use the geoscience data and models at face value. This, in its turn, calls for an increased effort, by TNO-GDN and the geoscience community in general, to inform the public about the subsurface, geological processes and the extent to which society relies on or affects them. As digital twins aspire to 'mirror' reality, they make much better presentations of the information they hold than visualizations of the abstract models that currently support public decision and policy making. In this way, they can bring scientific data and information much closer to the general public, adding legitimacy to, and trust in, decisions and policies based on them.

Acknowledgements

Ronald van Balen, Holger Kessler and Harry Doust are thanked for reviewing earlier drafts of this chapter.

Digital map data

Spatial data of figures in this chapter for use in geographical information systems can be downloaded here: https://doi.org/10.5117/aup.28165088.

References

Anonymous, 2021. Beleidsnota Drinkwater. Ministry of Instructure and Water Management (The Hague): 68 pp.

Békési, E., Struijk, M., Bonté, D., Veldkamp, H., Limberger, J., Fokker, P.A., Vrijlandt, M. & Van Wees, J.-D., 2020. An updated geothermal model of the Dutch subsurface based on inversion of temperature data. Geothermics 88, 101880, DOI: 10.1016/j.geothermics.2020.101880

Bonté, D., Van Wees, J.-D. & Verweij, J.M., 2012. Subsurface temperature of the onshore Netherlands: new temperature dataset and modelling. Netherlands Journal of Geosciences 91 (4): 491-515. DOI: 10.1017/S0016774600000354

Bosch, J.H.A., 2000. Standaard Boor Beschrijvingsmethode, Versie 5.1, TNO (Utrecht), Rapport No. NITG 00-141-A: 134 pp.

CAS, 2022. Climate Impact Atlas. Climate Adaptation Services (CAS). https://www.klimaateffectatlas.nl/en/

Chilès, J.-P. & Delfiner, P., 2012. Geostatistics – Modeling Spatial Uncertainty. Wiley & Sons (Hoboken, New Jersey): 699 pp. DOI:10.1002/9781118136188

Cohen, K.M., 2017. Laaglandgenese boringendatabase Universiteit Utrecht. DANS. DOI: 10.17026/dans-zcv-knya

Dabekaussen, W., De Bruijn, R., Heerema, C., Kars, R.H., Lourens, A., Schokker, J. & Stafleu, J., 2021. Nieuwe hoge resolutie GeoTOP modellen en update-methodes voor een Sterkere Lekdijk. TNO (Utrecht), Report No. 2021 R12300): 57 pp.

Dabekaussen, W., De Bruijn, R., Kars, R.H., Meijninger, B. & Stafleu, J., 2019. GeoTOP voor Sterke Lekdijk. TNO (Utrecht), Report No. TNO 2019 R11929: 67 pp.

Dabekaussen, W., De Bruijn, R., Kars, R.H., Meijninger, B. & Stafleu, J., 2020. Integrating multiple geotechnical data types with machine learning to construct high-resolution 3D geological models. EGU General Assembly 2020-8500. DOI: 10.5194/egusphere-egu2020-8500

Dabekaussen, W. & Hummelman, H.J., 2021. Uncertainty of the Geological Framework Model of the Netherlands. Workshop Uncertainty Quantification and Visualization in Geological Modelling. Presented at the EAGE Annual conference, Amsterdam, The Netherlands.

Dambrink, R.M., Maljers, D., Schokker, J. & Stafleu, J., 2015. Tailor-made maps generated from 3D geological voxel models (Abstract). 2015 Annual Meeting of the Geological Society of America November 1-4.

Dambrink, R.M., Maljers, D. & Stafleu, J., 2014. Customised geological information derived from multi-purpose 3D voxel models. Presented at 12th Nederlands Aardwetenschappelijk Congres (NAC12) (Veldhoven).

De Jager, J., Van Ojik, K. & Smit, J., 2025. Geological development. In: J.H. Ten Veen, G.-J. Vis, J. De Jager & Th.E. Wong (eds): Geology of the Netherlands, second edition. Amsterdam University Press (Amsterdam): 21-51. DOI: 10.5117/9789463728362_ch01

Dijkstra, J.J., Comans, R.N.J., Schokker, J. & Van der Meulen, M.J., 2019. The geological significance of novel anthropogenic materials: deposits of industrial waste and by-products, Anthropocene 28: 100229. DOI: 10.1016/j.ancene.2019.100229

Duin, E., Doornenbal, J.C., Rijkers, R.H.B., Verbeek, J.W. & Wong, Th.E., 2006. Subsurface structure of the Netherlands; results of recent onshore and offshore mapping. Netherlands Journal of Geosciences 85 (4): 245-276. DOI: 10.1017/S0016774600023064

Erkens, G., 2009. Addendum1. Geological sections across the Lower Rhine valley (Germany) and Rhine-Meuse delta (the Netherlands). In: G. Erkens (ed.): Sediment dynamics in the Rhine catchment - Quantification of fluvial response to climate change and human impact. PhD thesis Utrecht University (the Netherlands), Netherlands Geographical Studies 388: 278 pp.

Feigenbaum, A.V., 1991. Total Quality Control. McGraw-Hill (New York): 3rd ed., Vol. 1, 372 pp.

Goovaerts, P., 1997. Geostatistics for Natural Resources Evaluation. Oxford University Press (New York): 483 pp.

Gunnink, J.L., Maljers, D., Van Gessel, S.F., Menkovic, A. & Hummelman, H.J., 2013a. Digital Geological Model (DGM): a 3D raster model of the subsurface of the Netherlands. Netherlands Journal of Geosciences 92 (1): 33-46. DOI: 10.1017/S0016774600000263

Gunnink, J., Maljers, D. & Hummelman, J., 2010. Quantifying uncertainty of geological 3D layer models, constructed with a-priori geological expertise. Presented at 14th Annual Conference of the International Association for Mathematical Geosciences, IAMG 2010 (Budapest).

Gunnink, J.L., Stafleu, J., Maljers, D. & Hummelman, J., 2013b. Hydraulic parameterization of 3D subsurface models: from measurement-scale to model-scale. In: Russell, H.A.J., Berg, R.C. & Thorleifson, L.H. (eds): Three-Dimensional Geological Mapping, Workshop Extended Abstracts. Annual Meeting of the Geological Society of America (Denver, Colorado): 35-39.

Hademenos, V., Stafleu, J., Missiaen, T., Kint, L. & Van Lancker, V.R.M., 2019. 3D Subsurface characterisation of the Belgian Continenal Shelf: a new voxel modelling approach. Netherlands Journal of Geosciences 98: e1. DOI: 10.1017/njg.2018.18

Hintersberger, E., Iglseder, C., Schuster, R. & Huet, B., 2018. The new database "Tectonic Boundaries" at the Geological Survey of Austria. Jahrbuch Der Geologischen Bundesanstalt 157: 195-207.

Hintersberger, E., Van Ede, R. & Van Gessel, S., 2021. Fault Data Characterization Catalogue. Hazard and Impact Knowledge for Europe Report: 40 pp.

Hummelman, J., Maljers, D., Menkovic, A., Reindersma, R., Stafleu, J. & Vernes, R., 2019. Totstandkomingsrapport Digitaal Geologisch Model (DGM). TNO (Utrecht), Rapport No. R11653: 70 pp. https://www.broloket.nl/sites/default/files/Totstandkomingsrapport-DGM.pdf

Jayasena, R.M., 2020. Graaf- en modderwerk: Een archeologische stadsgeschiedenis van Amsterdam. University of Amsterdam.

pure.uva.nl/ws/files/43045800/11.pdf

Kint, L., Hademenos, V., De Mol, R., Stafleu, J., Van Heteren, S. & Van Lancker, V., 2020. Uncertainty assessment applied to marine subsurface datasets. Quarterly Journal of Engineering Geology and Hydrogeology 54: qjegh2020-028. DOI: 10.1144/qjegh2020-028

Kirsch, R. (ed.), 2009. Groundwater Geophysics – A Tool for Hydrogeology, 2nd ed. Springer Verlag (Berlin – Heidelberg): 548 pp. DOI: 10.1007/978-3-540-88405-7

Kombrink, H., Doornenbal, J.C., Duin, E.J.T., Den Dulk, M., Ten Veen, J.H. & Witmans, N., 2012. New insights into the geological structure of the Netherlands; results of a detailed mapping project. Netherlands Journal of Geosciences 91 (4): 419-446. DOI: 10.1017/S0016774600000329

Koster, K., Stafleu, J. & Stouthamer, E., 2018a. Differential subsidence in the urbanised coastal-deltaic plain of the Netherlands. Netherlands Journal of Geosciences 97 (4): 215-227. DOI: 10.1017/njg.2018.11

Koster, K., Stafleu, J., Cohen, K.M., Stouthamer, E., Busschers, F.S. & Middelkoop, H., 2018b. Three-dimensional distribution of organic matter in coastal-deltaic peat: Implications for subsidence and carbon dioxide emissions by human-induced peat oxidation. Anthropocene 22: 1-9. DOI: 10.1016/j.ancene.2018.03.001

Kruiver, P., Van Dedem, E., Romijn, R., De Lange, G., Korff, M., Stafleu, J., Gunnink, J.L., Rodriguez-Marek, A., Bommer, J.J., Van Elk, J. & Doornhof, D., 2017a. An integrated shear-wave velocity model for the Groningen gas field, The Netherlands. Bulletin of Earthquake Engineering 15: 3555-3580. DOI: 10.1007/s10518-017-0105-y

Kruiver, P.P., Wiersma, A., Kloosterman, F.H., De Lange, G., Korff, M., Stafleu, J., Busschers, F.S., Harting, R., Gunnink, J.L., Green, R.A., Van Elk, J. & Doornhof, D., 2017b. Characterisation of the Groningen subsurface for seismic hazard and risk modelling. Netherlands Journal of Geosciences 96 (5): s215-s233. DOI: 10.1017/njg.2017.11

Maljers - Oosterwijk, D., Vonhögen - Peeters, L., Van Heteren, S., Marges, V., Stafleu, J. & Wiersma, A., 2014. De update van het delfstoffeninformatiesysteem voor het NCP. Deltares (Utrecht), Report No. 1205135-002 (i.o.v Rijkswaterstaat Directie Noordzee): 31 pp.

Maljers, D., Stafleu, J., Van der Meulen, M.J. & Dambrink, R.M., 2015a. Advances in constructing regional geological voxel models, illustrated by their application in aggregate resource assessments. Netherlands Journal of Geosciences 94 (3): 257-270. DOI: 10.1017/njg.2014.46

Maljers, D., Ten Veen, J.H. & Den Dulk, M., 2015b. Description Uncertainty procedure DGM-deep. https://www.dinoloket.nl/en/digital-geological-model-dgm-deep.

Pluymaekers, M.P.D., Doornenbal, J.C. & Middelburg, H., 2017. Velmod-3.1. TNO (Utrecht), Report No. 2017 R11014. https://www.nlog.nl/en/velmod-31

Rijksoverheid, 2002. Wet van 31 oktober 2002, houdende regels met betrekking tot het onderzoek naar en het winnen van delfstoffen en met betrekking tot met de mijnbouw verwante activiteiten. (Mining Act, effective May 2022.). wetten.overheid.nl/BWBR0014168/2022-05-01/0.

Rijksoverheid, 2015. Wet van 30 september 2015, houdende regels omtrent de basisregistratie ondergrond (Key registry of the subsurface, effective May 2022). Retrieved from wetten.overheid.nl/BWBR0037095/2022-05-01/0.

Robertson, P.K., 2010. Soil Behaviour Type from CPT: An Update. 2nd International Symposium on Cone Penetration Testing (Huntington Beach): 575-583.

Russell, H.A.J., Kemp, E.A. & MacCormack, K.E., 2019. Overview of Geological Survey Organizations Contributions on Modelling Approaches. In: K.E. MacCormack, R.C. Berg, H.A.J. Russell, H. Kessler & L.H. Thorleifson (eds): Synopsis of Current Three-Dimensional Geological Mapping and Modelling in Geological Survey Organizations. Alberta Energy Regulator / Alberta Geological Survey (AER/AGS), Special Report 112 (3): 7-18.

Sandersen, P.B.E., Kallesøe, A.J., Møller, I., Høyer, A.-S., Jørgensen, F., Pedersen, J.B. & Christiansen, A.V., 2021. Utilizing the towed Transient ElectroMagnetic method (tTEM) for achieving unprecedented near-surface detail in geological mapping. Engineering Geology 288: 106125. DOI: 10.1016/j.enggeo.2021.106125

Schokker, J., Bakker, M.A.J., Dubelaar, C.W., Dambrink, R.M. & Harting, R., 2015. 3D subsurface modelling reveals the shallow geology of Amsterdam. Netherlands Journal of Geosciences 94 (4): 399-417. DOI: 10.1017/njg.2015.22

Sobisch, H.-G., 2011. One Viewer to Publish 3D Subsurface Models of any Modelling Origin. Abstracts with Programs: GSA Annual Meeting (Minneapolis, Minnesota): 525.

Stafleu, J. & Busschers, F.S., 2017. Analysing Lithological and Grain-Size Trends using a 3D Voxel Model: A Case Study from the Holocene Rhine-Meuse Delta. 79th EAGE Conference & Exhibition, (Paris, France), Workshop Programme: 12-15.

Stafleu, J. & Dubelaar, C.W., 2016. Product Specification Subsurface model GeoTOP. TNO (Utrecht), Report No. 2016 R10133 v1.3: 53 pp.

Stafleu, J., Hummelman, J., De Bruijn, R., Koster, K., Stam, J., Maljers, D. & Schokker, J., 2020. Totstandkomingsrapport GeoTOP - met nieuwe methodiek lagenmodellering, TNO (Utrecht), Report No. 2020 R10758. https://www.broloket.nl/toelichting/geotop

Stafleu, J., Kraaijpoel, D.A. & Osinga, S., 2022. Ideation Challenge 2020: RHINOCEROS: Regional Shear-wave velocIty modelling for seismic site Response – Final scientific report. TNO (Utrecht), Report No. 2022 R10324: 21 pp.

Stafleu, J., Maljers, D., Busschers, F.S., Gunnink, J.L. & Vernes, R.W., 2011a. TNO – Geological Survey of the Netherlands: 3-D Geological Modeling of the Upper 500 to 1,000 Meters of the Dutch Subsurface. In: Berg, R.C., Mathers, S.J., Kessler, H. & Keefer, D.A. (eds): Synopsis of Current Three-dimensional Geo-

logical Mapping and Modeling in Geological Survey Organizations. Illinois State Geological Survey Circular 578: 64-68.
Stafleu, J., Maljers, D., Gunnink, J.L. & Busschers, F.S., 2011b. 3D modelling of the shallow subsurface of Zeeland, the Netherlands. Netherlands Journal of Geosciences 90 (4): 293-310. DOI: 10.1017/S0016774600000597
Stafleu, J., Maljers, D., Busschers, F.S., Schokker, J., Gunnink, J.L. & Dambrink, R.M., 2021a. Models Created as 3-D Cellular Voxel Arrays. In: Turner, A.K., Kessler, H. & Meulen, M.J. (eds): Applied Multidimensional Geological Modeling. DOI: 10.1002/9781119163091.ch11
Stafleu, J., Maljers, D. & Gunnink, J.L., 2021b. Visualisation of uncertainty in voxel models of the shallow subsurface of the Netherlands. Workshop Uncertainty Quantification and Visualization in Geological Modelling, Presented at EAGE Annual Conference (Amsterdam, The Netherlands).
Stafleu, J., Sobisch, H.-G., Maljers, D., Hummelman, J., Dambrink, R.M. & Gunnink, J.L., 2013. Visualization and dissemination of 3D geological property models of the Netherlands. Presented at 2013 Annual Meeting of the Geological Society of America (Denver, Colorado, USA).
Stafleu, J., Van der Meulen, M.J., Gunnink, J.L., Maljers, D., Hummelman, J., Busschers, F.S., Schokker, J., Vernes, R.W., Doornenbal, H., Den Dulk, M. & Ten Veen, J.H., 2019. Systematic 3D subsurface mapping in the Netherlands. In: K.E. MacCormack, R.C. Berg, H.A.J. Russell, H. Kessler & L.H. Thorleifson (eds): Synopsis of Current Three-Dimensional Geological Mapping and Modelling in Geological Survey Organizations. Alberta Energy Regulator / Alberta Geological Survey (AER/AGS), Special Report 112 (3): 179-190.
Swierstra, W., Kerckhoffs, T. & Horn, A., 2018. Aanpassingen en kalibratie stationair model IBRAHYM 2.1, Zuid-Limburg. WAT-BF6136R001F01, Royal Haskoning DHV (Maastricht).
Ten Veen, J.H., 2021. Final fault data collection report and database: Annex 1. Hazard and Impact Knowledge for Europe Report: 329 pp.
ThermoGIS., 2022. ThermoGIS V2.1. https://www.thermogis.nl/en
TNO-GDN, 2022a. Delfstoffen Informatie Systeem, TNO – Geologische Dienst Nederland. https://www.dinoloket.nl/dis
TNO-GDN, 2022b. Stratigraphic Nomenclature of the Netherlands, TNO – Geological Survey of the Netherlands. https://www.dinoloket.nl/en/stratigraphic-nomenclature
TNO-NITG, 2004. Geological Atlas of the Subsurface of the Netherlands - onshore. Netherlands Institute of Applied Geoscience TNO (Utrecht): 104 pp.
Unidata, 2022. Network Common Data Form (NetCDF). https://www.unidata.ucar.edu/software/netcdf/
Van Balen, R.T., Bakker, M.A.J., Kasse, C., Wallinga, J. & Woolderink, H.A.G., 2019. A Late Glacial surface rupturing earthquake at the Peel Boundary fault zone, Roer Valley Rift System, the Netherlands. Quaternary Science Reviews 218: 254-266. DOI: 10.1016/j.quascirev.2019.06.033
Van Balen, R.T., Kasse, C., Wallinga, J. & Woolderink, H.A.G., 2021. Middle to Late Pleistocene faulting history of the Heerlerheide fault, Roer Valley Rift System, influenced by glacio-isostasy and mining-induced displacement. Quaternary Science Review 268: 107111. DOI: 10.1016/j.quascirev.2021.107111
Van Dalfsen, W., Mijnlieff, H.A. & Simmelink, E., 2005. Interval velocities of a Triassic claystone: key to burial history and velocity modeling. Poster 178 presented at 67th EAGE Conference & Exhibition (Madrid, Spain).
Van der Meulen, M.J. & Bogaard, P.J.F., 2022. Medium-term Plan 2022-2025 (MLTP25). TNO – Geological Survey of the Netherlands. (Utrecht, Netherlands): 29 pp.
Van der Meulen, M.J., Dijkstra, J.J., Koopmans, T.P.F., Pietersen, H.S., Stam, J. & Maljers, D., 2025. Surface Mineral Resources. In: J.H. Ten Veen, G.-J. Vis, J. De Jager & Th.E. Wong (eds): Geology of the Netherlands, second edition. Amsterdam University Press (Amsterdam): 679-699. DOI: 10.5117/9789463728362_ch18
Van der Meulen, M.J., Doornenbal, J.C., Gunnink, J.L., Stafleu, J., Schokker, J., Vernes, R.W., Van Geer, F.C., Van Gessel, S.F., Van Heteren, S., Van Leeuwen, R.J.W., Bakker, M.A.J., Boogaard, P.J.F., Busschers, F.S., Griffioen, J., Gruijters, S.H.L.L., Kiden, P., Schroot, B.M., Simmelink, H.J., Van Berkel, W.O., Van der Krogt, R.A.A., Westerhoff, W.E. & Van Daalen, T.M., 2013. 3D geology in a 2D country: perspectives for geological surveying in the Netherlands. Netherlands Journal of Geosciences 92 (4): 217-241. DOI: 10.1017/S0016774600000184
Van der Meulen, M.J., Kessler, H., Ten Veen, J.H., Vernes, R.W. & Maljers, D., 2022. Diving into the energy transition. In: Berg, R.C., Thorleifson, L.H. & MacCormack, K.E. (eds): Three-Dimensional Geological Mapping: Workshop Extended Abstracts. Illinois State Geological Survey (Champaign, IL): 51-54.
Van der Meulen, M.J., Kiden, P., Maljers, D., Stafleu, J., Vernes, R.W., Gunnink, J.L., Westerhoff, W.E. & Van Daalen, T.M., 2015. Systematic Geomodeling: What Does It Actually Imply? (MacCormack, K. E., Thorleifson, L. H., Berg, R. C. & Russel, H. A. J. (eds). Alberta Energy Regulator (Edmonton, AL): 83-88.
Van der Meulen, M.J., Van Gessel, S.F. & Veldkamp, J.G., 2005. Aggregate resources in the Netherlands. Netherlands Journal of Geosciences 84: 379-387. DOI: 10.1017/S0016774600021193
Van Ginkel, J., Ruigrok, E., Stafleu, J. & Herber, R., 2022. Development of a seismic site-response zonation map for the Netherlands. Natural Hazards and Earth System Sciences 22: 41-63. DOI: 10.5194/nhess-22-41-2022
Van Lancker, V., Francken, F., Kapel, M., Kint, L., Terseleer, N., Van den Eynde, D., Hademenos, V., Missiaen, T., De Mol, R., De Tré, G., Appleton, R., Van Heteren, S., Van Maanen, P.-P., Stafleu, J., Stam, J., Degrendele, K. & Roche, M., 2019. Transnational and Integrated Long-term Marine Exploitation Strategies (TILES). Final Report. Belgian Science Policy (Brussels): 75- pp.
Van Waterschoot van der Gracht, W.A.J.M., 1918. Eindverslag over de onderzoekingen en uitkomsten van den Dienst der Rijksopsporing van Delfstoffen in Nederland 1903-1916.

(Amsterdam): 664 pp.
Vermeulen, P., Roelofsen, F. & Veldhuizen, A., 2015. Actualisatie en Kalibratie IBRAHYM Beschrijving activiteiten t.b.v IBRAHYM v2.0 (in Dutch), Deltares (Utrecht), Report No. No. 1206858-000-BGS-0022: 38 pp.
Vermeulen, P., van der Linden, W., Veldhuizen, A., Massop, H., Vermulst, H. & Swierstra, W., 2007. IBRAHYM Grondwater Modelinstrumentarium Limburg. TNO (Utrecht), Report No. 2007-U-R0193/B: 126 pp.
Vernes, R.W., Dabekaussen, W., Gunnink, J.L., Harting, R., De Heer, E., Hummelman, J.H., Menkovic, A., Reindersma, R. & Van der Ven, T.J.M., 2021. Application Theme 2 – Groundwater Evaluations – Case Study 19.4: REGIS II – A 3-D Hydrogeological Model of the Netherlands. In: Turner, A.K., Kessler, H. & Meulen, M.J. (eds): Applied Multidimensional Geological Modeling. DOI: 10.1002/9781119163091.ch19
Vernes, R.W. & Van Doorn, Th.H.M., 2005. Van Gidslaag naar Hydrogeologische Eenheid – Toelichting op de totstandkoming van de dataset REGIS II. TNO (Utrecht), Report NITG 05-038-B: 105 pp.
Verweij, H., Zaadnoordijk, W.J., Oude Essink, G.H.P. & Hartog, N., 2025. Hydrogeology. In: Ten Veen, J.H., Vis, G.-J., De Jager, J. & Wong, Th.E. (eds): Geology of the Netherlands, second edition. Amsterdam University Press (Amsterdam): 627-677. DOI: 10.5117/9789463728362_ch17
Wellmann, J.F. & Regenauer-Lieb, K., 2012. Uncertainties have a Meaning: Information Entropy as a Quality Measure for 3-D Geological Models. Tectonophysics 526-529: 207-216. DOI: 10.1016/j.tecto.2011.05.001
Westerhoff, W.E., Wong, Th.E. & De Mulder, E.F.J., 2003. Opbouw van de ondergrond. In: De Mulder, E.F.J., Geluk, M.C., Ritsema, I.L., Westerhoff, W.E. & Wong, Th.E. (eds): De ondergrond van Nederland. Wolters-Noordhoff (Groningen/Houten): 247-352.
Zehner, B., Ayala, C., Bense, F., Dabekaussen, W., Den Dulk, M., Franek, J., Pueyo, E.L., Malz, A., Müller, C.O. & Stück, H., 2021. Sources of uncertainty in 3D geomodels. 3D Geomodeling for Europe Report (Project number GeoE.171.005). 64 pp.

Contributing authors

Marcel Bakker: TNO – Geological Survey of the Netherlands (marcel.bakker@tno.nl) – Marcel studied physical geography at the University of Amsterdam and holds a PhD at Queen Mary University of London. He works as senior geologist at TNO-GDN with main focus on the Quaternary geology of the Netherlands – both onshore and offshore – and with emphasis on glacial sedimentary systems

Ronald van Balen: TNO – Geological Survey of the Netherlands/Vrije Universiteit Amsterdam (r.t.van.balen@vu.nl) – Ronald is a professor in Quaternary geology and geomorphology working at Vrije Universiteit Amsterdam, chairing the Landscape Dynamics group, and affiliated with TNO-GDN. He is specifically interested in fluvial, glacial and periglacial deposits and landforms, and in neotectonics (including tectonic geomorphology, paleoseismology, and fault hydrology).

Jeroen Bartol: COVRA NV (jeroen.bartol@covra.nl) – Jeroen is responsible for research on geological disposal in rock salt, the safety assessment and cost estimates. Jeroen holds an MSc degree in geophysics and a PhD in tectonophysics from Utrecht University. Since 2018 Jeroen has been working as a researcher at COVRA.

Frank van Bergen: TNO – Geological Survey of the Netherlands (frank.vanbergen@tno.nl) – Frank is a geologist holding a PhD from Utrecht University who worked in research and exploration on coal, petroleum geology and geochemistry, CO_2 storage and geothermal energy both at TNO and in industry. Currently he leads the research facility for well technology at TNO.

Manfred van Bergen: Department of Earth Sciences, Utrecht University (m.j.vanbergen@uu.nl) – Manfred is a retired associate professor in earth sciences at the Faculty of Geosciences of Utrecht University, where he led numerous research projects in the fields of geochemistry, petrology and isotope geology of igneous and hydrothermal systems. He studied active volcanoes and ancient rocks in many parts of the world.

Christian F.M. Bos: TNO – Geological Survey of the Netherlands (cfm.bos@ziggo.nl) – Christian (MSc Petroleum Engineering, TU Delft, 1980) has served 11 years as a reservoir engineer with Shell International, followed by some 30 years as a researcher with TNO. His research includes history matching of reservoir simulation models, exploration and production investment decision-support, CCS, geothermal field development, underground energy storage, and energy system modelling.

Renaud Bouroullec: TNO – Geological Survey of the Netherlands (renaud.bouroullec@tno.nl) – Renaud received his BSc and MSc degrees from UBO (Brest), UC Galway and Université de Rennes and a PhD from Imperial College in 2001. He conducted research at CU Boulder (Colorado), UT Austin (Texas) and CSM (Colorado). He joined TNO-GDN in 2014 as senior scientist and geology scientific lead.

Isaac Brouwers: Department of Earth Sciences, Faculty of Science, Vrije Universiteit Amsterdam (brouwersisaac@gmail.com) – Isaac holds an MSc in earth sciences specializing in geochemistry. Currently he works on research and development for the Netherlands forensic institute, microtrace and materials group by using geochemical methods such as LA-ICP-MS (TOF) for the comparison and classification of polymers for forensic investigations.

Renée de Bruijn: TNO – Geological Survey of the Netherlands (renee.debruijn@tno.nl) – Renée studied physical geography at Utrecht University. After working at other research institutes for a few years, she joined TNO-GDN as a Quaternary geologist and geomodeller. Renée currently focuses on the construction and application of 3D models of the Dutch shallow subsurface.

Loes Buijze: TNO – Geological Survey of the Netherlands (loes.buijze@tno.nl) – Loes is a senior geomechanics expert at TNO-GDN. She specializes in the study of induced seismicity related to gas depletion and sustainable subsurface applications such as geothermal energy and storage, with a focus on understanding the source mechanisms through an integrative approach involving laboratory experiments, modelling, and monitoring.

Freek S. Busschers: TNO – Geological Survey of the Netherlands (freek.busschers@tno.nl) – Freek studied physical geography at Utrecht University and obtained his PhD at Vrije Universiteit Amsterdam. His main focus is the study of the Quaternary North Sea Basin using sedimentary, provenance, modelling and dating techniques. Since 2015 he is project manager and scientific lead of TNO-GDN's Research & Development project.

Thibault G.G. Candela: TNO – Geological Survey of the Netherlands (thibault.candela@tno.nl) – Thibault is geomechanics scientist at TNO-GDN. His research interests include induced subsidence and seismicity.

Kim Cohen: Department of Earth Sciences, Utrecht University (k.m.cohen@uu.nl) – Kim is a Quaternary geologist and geomorphologist employed at Utrecht University. He covers the landscape history of the Lower Rhine in relation to sea-level oscillations, climatic fluctuations, and subsidence regimes. Detailed Rhine-Meuse delta mapping brought many interactions with the archeological field too. Internationally, Cohen is an ICS officer (www.stratigraphy.org).

Willem Dabekaussen: TNO – Geological Survey of the Netherlands (willem.dabekaussen@tno.nl) – Willem obtained his MSc in structural geology at Utrecht University. After working in the geotechnical industry, he joined TNO-GDN, where he has become an expert in 3D modelling, quantifying uncertainties and applying machine learning in subsurface models.

Timothy N. Debacker: Geognostics (tim.debacker@geognostics.com) – Tim is a structural geologist/consultant specialized in potential field data interpretation (gravity and magnetic data), basement mapping, fabric analysis, and soft-sediment deformation, with strong expertise on the lower to middle Paleozoic geology of northwest Europe.

Joris J. Dijkstra: TNO – Geological Survey of the Netherlands (joris.dijkstra@tno.nl) – Joris holds a PhD in environmental geochemistry from Wageningen University. He started his career as environmental chemist at ECN in 1998, and continued it at TNO-GDN in 2018. His research is directed mainly towards the geochemistry of industrial waste and by-products and their interaction with the natural environment.

Dorien Dinkelman: TNO – Geological Survey of the Netherlands (dorien.dinkelman@tno.nl) – Dorien is a research scientist at TNO-GDN in the field of earth sciences and sustainable development, mainly focusing on subsurface simulations, potential mapping and feasibility of high-temperature aquifer thermal energy storage and geothermal energy. She has contributed to national and international heat projects like HEATSTORE and WarmingUP.

Timme H. Donders: Department of Physical Geography, Faculty of Geosciences, Utrecht University (t.h.donders@uu.nl) – Timme is a paleoecologist employed as an associate professor at Utrecht University. Timme is specialized in terrestrial palynology of the Neogene and Quaternary. He focuses on paleoclimate reconstructions and land-sea correlations in temperate to subtropical regions around the globe and integrates multiproxy data with concepts from fluvial sedimentology and ecology.

Hans Doornenbal: TNO – Geological Survey of the Netherlands (hans.doornenbal@ziggo.nl) – Hans has over 40 years of experience in database management, seismic interpretation, structural geology and mapping in various subsurface studies (exploration and production, geothermal, water potential). As project manager at TNO-GDN (retired in 2022) he led a variety of international projects (YEPIC, Geodesa, SPBA, GASH, NAGTEC, EUOGA, GeoERA) and the mapping of the Dutch deep subsurface.

Bernard Dost: Royal Netherlands Meteorological Institute (KNMI) (bdost36@gmail.com) – Bernard studied geophysics at Utrecht University. After completing a PhD in which he developed, installed and operated with a small team a first mobile seismograph network in Europe, he continued his career in observational seismology by setting up the ORFEUS Data Center (ODC). After the 1992 Roermond earthquake he moved to the KNMI including the ODC, at the time when hydrocarbon-induced seismicity started in the north of the Netherlands. Over time his scientific interest shifted from natural to induced seismicity. He is (co)author of 47 scientific papers and retired from KNMI in 2022.

Maryke den Dulk: TNO – Geological Survey of the Netherlands (maryke.dendulk@tno.nl) – Maryke studied geology and obtained a PhD in paleoclimatology at Utrecht University. After working in the petroleum industry, she continued her professional career as a geologist-geomodeller at TNO-GDN with a strong focus on static modelling, seismic interpretation, deep subsurface geology, and mapping in relation to various national- and cross border geomodelling projects (e.g. DGM-deep, H3O).

Gilles Erkens: Deltares/Faculty of Geosciences, Utrecht University (gilles.erkens@deltares.nl) – Gilles works as a Quaternary geologist and subsidence expert at Deltares Research Institute and Utrecht University. He is an expert in land subsidence in soft soil areas in the Netherlands and in deltas and coastal zones around the world. Next to geological aspects, he has a keen interest in the governance of land subsidence.

Marita Felder: Molenaar GeoConsulting (m.felder@casema.nl) – Marita is senior geoscientist specialized in combined sedimentological and petrographic reservoir studies of the Dutch subsurface. She supports PhD students at TU Delft and teaches applied diagenesis at the Technical University of Karlsruhe (KIT) as visiting scientist.

Peter Fokker: Nedmag BV (Veendam), Well Engineering Partners (Hoogeveen), SGS (Voorburg) (p.fokker@nedmag.nl; peter.fokker@we-p.nl) – Peter (1966) works as geomechanics expert and salt mining engineer for mining companies in the Netherlands and abroad for optimizing salt mining or minimizing the environmental consequences. After a PhD at TU Delft on a salt-mining topic (1995), Peter worked for Deltares, Nedmag and Shell (NL and India) until 2017. Now he works as a consultant in mainly salt-related topics and compaction/subsidence prediction for gas fields.

Peter A. Fokker: TNO – Geological Survey of the Netherlands (peter.fokker@tno.nl) – Peter is geomechanics expert working at TNO-GDN, and part-time at Utrecht University and the Technical University of Turin, Italy. He is involved in many geomechanics studies, on topics such as gas-production-induced subsidence, coupled modelling, and induced seismicity.

Cornelis R. Geel: TNO – Geological Survey of the Netherlands (kees.geel@tno.nl) – Kees is a geologist who has worked as a sedimentologist, petrophysicist, seismic interpreter, and reservoir geologist in various staff and consulting positions in the Netherlands and abroad (retired from TNO-GDN in 2024).

Mark C. Geluk†: formerly Shell International Exploration and Production BV – Mark was a regional geologist with special interest in the Permo-Triassic geology of the Permian basins of Europe. After a career at the Geological Survey of the Netherlands (later TNO-GDN) he became principal geoscientist with Shell. Mark wrote an influential PhD thesis on the Triassic of northwest Europe. Unfortunately Mark passed away in 2018.

Serge F. van Gessel: TNO – Geological Survey of the Netherlands (serge.vangessel@tno.nl) – Serge graduated as geologist at Utrecht University and has been working at TNO-GDN since 1998 as a researcher and advisor for the government on the use of the deep subsurface. Among other things, he coordinates international collaboration between companies and universities in the field of underground storage.

Lucia van Geuns: The Hague Centre for Strategic Studies (HCSS) (lucia.vangeuns@gmail.com) – Lucia is Strategic Advisor Energy at HCSS. She is an experienced energy professional with a background in geoscience. Her research focuses on international energy markets and energy transition issues. She was president of the Royal Geological and Mining Society of the Netherlands (KNGMG) from 2013-2019.

Jasper Griffioen: TNO – Geological Survey of the Netherlands/Copernicus Institute of Sustainable Development, Utrecht University (jasper.griffioen@tno.nl) – Jasper is expert researcher at TNO-GDN and professor of water quality management at Utrecht University. He is specialized in environmental geochemistry and hydrology within the framework of sustainable environmental management. Many of his research projects refer to the risks and impact of anthropogenic activities on water resources and soils.

Jan L. Gunnink: TNO – Geological Survey of the Netherlands (jan.gunnink@tno.nl) – Jan is a Quaternary geologist by training, specialized in 3D modelling of the subsurface using geostatistical techniques. His main focus is on integrating various datatypes, including geophysics, for modelling the shallow (~100 m depth) subsurface in order to assist in the current societal challenges: energy transition, groundwater, subsidence, etc.

Ramon F. Hanssen: Faculty of Civil Engineering and Geosciences, TU Delft (r.f.hanssen@tudelft.nl) – Ramon is professor in geodesy and satellite earth observation at TU Delft. He develops methodology and systems for improved parameter estimation and analysis of geodynamical processes using space geodetic technology.

Niels Hartog: KWR, Water Research Institute/Department of Earth Sciences, Faculty of Geosciences, Utrecht University (niels.hartog@kwrwater.nl) – Niels is the principal scientist for the Geohydrology team at KWR Water Research Institute (NL) and KWR's scientific council. He is a guest associate professor in the Environmental Hydrogeology Group at Utrecht University and president of the Netherlands Chapter of the International Association of Hydrogeologists (IAH).

Daan den Hartog Jager: EBN BV (daan.hartogjager-den@ebn.nl) – Daan holds an MSc in geology (Utrecht University, 1985). He was a geoscientist with Shell, both in exploration and production: Shell UK 1986-1991, Brunei Shell 1996-2001, NAM Assen 1991-1996 and 2001-2020. Since 2021 he is a geoscientist with EBN, Utrecht.

Marinus den Hartogh: Nobian Industrial Chemicals BV (marinus.denhartogh@nobian.com) – Marinus (1984) works as technology manager brine fields within the Mining Development & Compliance (MDC) department of Nobian in Hengelo, the Netherlands. Marinus has an MSc degree in geology from Vrije Universiteit Amsterdam. Since 2009 Marinus has worked on various topics at Nobian (and its predecessors), including brine field development, subsidence prevention and production planning and optimization.

Sytze van Heteren: TNO – Geological Survey of the Netherlands (sytze.vanheteren@tno.nl) – Sytze (PhD in geology, Boston University, USA) is a coastal and marine geologist with more than 25 years of experience conducting and supervising projects addressing the North Sea and its shores. Key areas of interest include coastal and marine sedimentology, applied Quaternary geology and transnational seabed mapping.

Alexander J.P. Houben: TNO – Geological Survey of the Netherlands (alexander.houben@tno.nl) – Sander is a geologist and palynologist with extensive experience of the Mesozoic and Cenozoic successions of the greater North Sea area. Over the past few years his interest and experience have also extended to older strata, including the Devonian and lower Carboniferous of northwest Europe.

Alwina Hoving: TNO – Geological Survey of the Netherlands (alwina.hoving@tno.nl) – Alwina is a geochemist at TNO-GDN. She is specialized in the geochemistry of clay minerals. In addition, she works on characterization and parameterization of a wide range of geochemical properties of the Dutch subsurface.

Thijs Huijskes: ONE-Dyas BV (thijs.huijskes@onedyas.com) – Thijs worked as a reservoir engineer at EBN for over 9 years. Currently at One-Dyas, he is working as CCS reservoir engineer, focusing on near-wellbore phenomena, the coupling of reservoir flow and wellbore flow assurance, CO_2 phase-change behaviour and determination of injectivity.

Richard Huis in 't Veld†: Argo Geological Consultants – Richard holds an MSc in geology from Utrecht University (1988), was partner at Argo Geological Consultants since 1987, consultant exploration and production geoscientist for NAM (1989-1995 and 1997-1999), Shell Peru (1996), Wintershall (1999-2020), and ONE-Dyas since 2021. Unfortunately Richard passed away in 2024.

Jan H. Hummelman: TNO – Geological Survey of the Netherlands (jan.hummelman@tno.nl) – Jan joined TNO-GDN as a hydrogeologist in 1983. He is an experienced geomodeller working on regional and cross border (hydro)geological subsurface models. In particular, Jan has extensive experience in the construction of the national geo- and hydrogeological models DGM and REGIS II.

Jan de Jager: De Jager Geological Consultancy (jan.dejager@hotmail.com) – Jan is an experienced exploration geologist with a broad background in regional geology, structural geology, depositional systems and prospect evaluation. Acknowledged expert on the (petroleum) geology of the Netherlands, and author of several key articles on the geology of the Southern North Sea Basin.

John W.M. Jagt: Natuurhistorisch Museum Maastricht (john.jagt@maastricht.nl) – John is curator of paleontology, specializing in Late Cretaceous-early Paleogene macrofossils (echinoderms, molluscs, decapod and thoracican crustaceans), member of various Cretaceous working groups, with over 450 peer-reviewed papers, plus book chapters and (co-)author of around 300 new fossil taxa of genus, species and family rank.

Joaquim Juez-Larré: TNO – Geological Survey of the Netherlands (joaquim.juezlarre@tno.nl) – Joaquim is a senior scientist at TNO-GDN since 2009. He holds an MSc (University Barcelona) and a PhD/PostDoc (Vrije Universiteit Amsterdam) in Geology. During his career, he has worked on topics related to oil/gas/geothermal exploration and production, underground energy storage and energy transition.

Romée Kars: TNO – Geological Survey of the Netherlands (romee.kars@tno.nl) – Romée has a background in Quaternary geology and obtained a PhD in luminescence dating on feldspar sand grains. Currently, at TNO-GDN she is working on various aspects of 3D modelling of the Dutch subsurface.

Ben Kilhams: Shell International Ltd, (b.kilhams@shell.com) – Ben is currently an exploration geoscientist with Shell working on global new ventures. His background includes published work on northwest European Mesozoic intervals with special focus on coals and the impact of igneous systems.

Henk Kooi: Deltares (henk.kooi@deltares.nl) – Henk is a geophysicist and groundwater and subsidence expert working with Deltares – Unit Subsurface and Groundwater Systems. He is involved in national and international applied research and advisory projects in this field and has a leading role in the development of subsidence software at Deltares.

Tjeerd P.F. Koopmans: Mole Underground projects & consultancy (tjeerd.koopmans@mole-underground.nl) – Tjeerd studied geology at Utrecht University. He started working on surface mineral resources in 2001, managing several demand and supply monitoring studies. In 2008 he shifted his focus to the deep subsurface, managing salt cavern development, storage and integrity projects. As of 2020 he works on cavern energy storage.

Janne Koornneef: Department of Earth Sciences, Vrije Universiteit Amsterdam (j.m.koornneef@vu.nl) – Janne is a geochemist studying the earth's mantle and volcanic rocks derived from it. The chemical composition of rocks from all tectonic settings are used to understand the processes that affect the earth's interior composition and evolution over time.

Joris M. Koornneef: TNO – Geological Survey of the Netherlands (joris.koornneef@tno.nl) – Joris has a background in Science & Innovation Studies. He researches the social, technical and economic success/failure factors for energy technologies. He is a senior energy strategy consultant at TNO-GDN with a focus on (offshore) system integration and (underground) energy storage.

Kay Koster: TNO – Geological Survey of the Netherlands (kay.koster@tno.nl) – Kay is a geoscientist at TNO-GDN. He is specialized in Holocene coastal geology and surficial subsidence processes.

Salomon B. Kroonenberg: Anton de Kom University of Suriname/TU Delft (emeritus) (salomonkroonenberg@gmail.com) – Salomon Kroonenberg (1947) holds a PhD from the University of Amsterdam, worked 10 years in Suriname, Swaziland and Colombia and 14 years at Wageningen Agricultural University until his appointment in Delft in 1996. His present research focuses on the geology of the Guiana Shield. In the past he studied Caspian sea level changes related to hydrocarbon exploration in Russia, Azerbaijan and Iran. He wrote 8 books on geological subjects for non-specialists.

Pauline Kruiver: Royal Netherlands Meteorological Institute (KNMI) (pauline.kruiver@knmi.nl) – Pauline joined KNMI as a seismologist in 2020. She studied geophysics (MSc and PhD) at Utrecht University. During her career she worked on various applied geophysics projects including the study of the surface effects of induced seismicity in Groningen. Her current interests are monitoring of seismicity and using monitoring data to obtain a better understanding of the subsurface.

Freek J. van Leijen: Faculty of Civil Engineering and Geosciences, TU Delft (f.j.vanleijen@tudelft.nl) – Freek is a senior researcher at TU Delft in the field of geodesy. He is particularly focusing on methodology development and applications in satellite radar interferometry and geodetic data integration for ground motion analysis.

Henk (J.H.) van Lochem: Argo Geological Consultants (j.h.vanlochem@argo-geoscience.com) – Henk is co-founder of Argo Geological Consultants, working as geoscientist in the North Sea Basin on hydrocarbon and geothermal projects. As consulting geologist he was involved in the exploration and appraisal of the Rembrandt and Vermeer chalk oil fields of Wintershall Noordzee BV and its partners.

Denise Maljers: TNO – Geological Survey of the Netherlands (denise.maljers@tno.nl) – Denise studied physical geography at Utrecht University. After her graduation in 2001, she became a geomodeller at TNO-GDN, contributing to the delivery and development of the survey's national subsurface modelling effort. In 2020, she became head of the department that conducts systematic geological modelling in the Netherlands.

Tom McKie: Shell UK Limited, (tom.mckie@shell.com) – Tom is currently a reservoir geologist with Shell UK working on a wide range of North Sea and global reservoirs. Throughout his career he has maintained a particular interest in dryland continental depositional systems, and particularly those that are encountered in the European Triassic.

Michiel J. van der Meulen: TNO – Geological Survey of the Netherlands (michiel.vandermeulen@tno.nl) – Michiel studied geology at Utrecht University. After obtaining his PhD in 1999, he started working on mineral resources and supplies at Rijkswaterstaat. In 2003, Michiel joined TNO-GDN, where he continued working on minerals, gradually broadening his responsibilities to the survey's full public information and services portfolio.

Harmen Mijnlieff: TNO – Geological Survey of the Netherlands (harmen.mijnlieff@tno.nl) – Harmen is senior geologist at TNO-GDN. He is involved in the advisory work supporting the responsible ministry in drafting and executing their geothermal policy. This includes technical support for the Exploration Guarantee Fund, feed-in tariff scheme, Mining Act amendments, license applications, resource reporting, seismic hazard and risk analysis, spatial planning/interference of surface and subsurface activities and last but not least dissemination of information to the public, government and other geothermal stakeholders.

Nicolaas Molenaar: Molenaar GeoConsulting (nicolaasmolenaar@gmail.com) – Nicolaas is expert petrographer and inorganic geochemist, specialized in siliciclastic and carbonate rocks and their diagenesis. He performs fundamental research on diagenesis, thereby collaborating with the universities of Vilnius, Darmstadt and Karlsruhe. In his own company he carries out petrographic studies for difficult reservoirs and does quality control of industry reports.

Dirk K. Munsterman: TNO – Geological Survey of the Netherlands (dirk.munsterman@tno.nl) – Dirk is a geologist and stratigrapher at TNO – GDN who's research focuses on Mesozoic and Cenozoic stratigraphy of northwest Europe. Dirk studied history at the Nutsseminarium, University of Amsterdam, and sedimentary geology with emphasis on paleoclimate and environment at Utrecht University. He graduated cum laude from both universities. He was project manager, coordinator of the geobiology team and member of the Dutch Stratigraphic Committee.

Erika A.C. Neeft: COVRA NV (erika.neeft@covra.nl) – Erika has been working as a researcher at COVRA since 2010 and coordinated the research programme OPERA. Currently, she is responsible for coordinating the research for a facility in clay and performs safety assessment calculations and cost estimates. She holds an MSc degree in earth sciences from Utrecht University and a PhD in reactor physics from TU Delft.

Susanne Nelskamp: TNO – Geological Survey of the Netherlands (susanne.nelskamp@tno.nl) – Susanne is a geologist with TNO-GDN. Her focus is on regional and structural geology, basin and petroleum system modelling and source rock geochemistry.

Kees van Oijk: EBN BV (kees.ojik-van@ebn.nl) – Kees is an energetic and skilled geoscientist with hands-on experience in a broad spectrum of subsurface resource evaluations and developments (hydrocarbons, CO_2 storage, heat). 'Energising the transition' towards a more sustainable world fuels my ambition to support changing our energy systems and to deliver that support based on expert knowledge of the subsurface.

Gualbert H.P. Oude Essink: Deltares/Department of Physical Geography, Utrecht University (gualbert.oudeessink@deltares.nl) – Gualbert is a hydrogeologist at Deltares & Utrecht University (Associate Professor) researching salt water intrusion and fresh groundwater availability in coastal zones under anthropogenic, paleo and sea level rise conditions in the Netherlands, Vietnam and Egypt. He creates solutions to improve water availability in coastal zones, teaches at IHE Delft and supervises 6 PhDs.

Hans S. Pietersen: Directorate of Public Works and Water Management, Rijkswaterstaat (hans.pietersen@rws.nl) – Hans studied geochemistry at Utrecht University and holds a PhD in civil engineering from TU Delft. In 1991 he started as a building materials consultant and moved to Rijkswaterstaat in 2000, where he first focused on national and EU-level strategic minerals planning before he became advisor international affairs.

Gijs Remmelts: TNO – Geological Survey of the Netherlands (gijs.remmelts@gmail.com) – Gijs was as a geologist with the TNO-GDN (retired in 2024), where he did geoscientific and geotechnical research for policy support to the Dutch government. In addition, he made contributions in petroleum geology, radioactive waste disposal, regional geology and energy transition.

Eveline Rosendaal: EBN BV (eveline.rosendaal@ebn.nl) – Eveline is manager of the business development team Heat Transition at EBN. The team focuses on improving conditions to enable acceleration of geothermal heat development in the Netherlands. This concerns advising on investment conditions, advising on law and regulation amendments and working on local and regional support and acceptance for geothermal heat. She was closely involved in developing the 'Master plan Geothermal Energy' in 2018 and its execution. Eveline has been working at EBN since 2012 in other roles such as team lead Exploration and programme manager Geo-Energy. In the period 1998-2012 she worked in various roles in the oil and gas industry such as petroleum geologist, seismic interpreter and reservoir geophysicist, always with a focus on the Dutch subsurface. Eveline studied structural geology and graduated in 1998 at Utrecht University.

Elmer Ruigrok: Royal Netherlands Meteorological Institute (KNMI)/Utrecht University (elmer.ruigrok@knmi.nl) – Elmer studied applied earth sciences at TU Delft. He completed a PhD and PostDoc at the same university, developing methods to image the deep earth with a combination of seismic interferometry and reflection imaging. Afterwards he focused on induced seismicity during a PostDoc at Utrecht University. Currently he is assistant professor in Utrecht and seismologist at KNMI, working on various topics related to seismicity in the Netherlands. He has supervised 28 student projects and has (co-)authored 50 journal articles.

Jeroen Schokker: TNO – Geological Survey of the Netherlands/Vrije Universiteit Amsterdam (jeroen.schokker@tno.nl) – Jeroen is a senior Quaternary geologist working at TNO-GDN and Vrije Universiteit Amsterdam. He combines sedimentary, biostratigraphic and dating techniques to map and model the shallow subsurface of the Netherlands in 3D. His current focus is on Late Pleistocene climate-related landscape development and anthropogenic deposits.

Frank Schoof: FPP Management BV (frank.schoof@fpp-management.nl) – Frank is former director of *Platform Geothermie* (now part of *Geothermie Nederland*, the sector organization for geothermal energy) and is programme manager for the provinces of North Holland and Flevoland for accelerating the use of geothermal energy in the Amsterdam metropolitan area. He was managing director of the first geothermal project in the built environment in the Netherlands.

Wim Sissingh: Shell International Exploration and Production BV – Wim is a retired review geologist and chief stratigrapher of NAM, Assen.

Jeroen Smit: Institut für Geologie, Mineralogie und Geophysik, Fakultät für Geowissenschaften, Ruhr University Bochum (jeroen.smit@rub.de) – Jeroen is a geoscientist with 20+ years of international experience in structural geology, tectonic modelling and basin geology. Jeroen's work on the North Sea focusses on the long-term, multiphase basin development, linking basin geology to the deep crustal structure, the regional setting and the geodynamic processes at play.

Jesper Spetzler: Royal Netherlands Meteorological Institute (KNMI)/Department of Geoscience and Engineering, TU Delft (jesper.spetzler@knmi.nl) – Jesper holds an MSc in geophysics from the University of Copenhagen and PhD in seismology from Utrecht University. Work experience in the past 20 years includes ultrasonic waveform experiments, time-lapse seismic monitoring, Co_2 sequestration, seismic data processing, geophysical software development, hypocentre location of tectonic and induced earthquakes, seismic hazard analysis, velocity model building and moment tensor inversion. He started at KNMI in 2015 and has until recently worked mostly on the Groningen field.

Job Spijker: National Institute for Public Health and the Environment (RIVM) (job.spijker@rivm.nl) – Job is a senior research scientist at RIVM, with a background in environmental geochemistry. The main focus of his research is on how natural systems are influenced by human activities and about the relations between ecosystems, human health and the environment.

Jan Stafleu: TNO – Geological Survey of the Netherlands (jan.stafleu@tno.nl) – Jan studied geology at Utrecht University and obtained a PhD in seismic modelling from Vrije Universiteit Amsterdam. Jan is an experienced geomodeller focusing on the development and application of the GeoTOP voxel model (since 2006). In addition, he is project manager of TNO-GDN's shallow subsurface mapping and modelling programme.

Jelte Stam: TNO – Geological Survey of the Netherlands (jelte.stam@tno.nl) – Jelte is senior geoscientist at TNO-GDN and holds an MSc from Utrecht University (2010). Before joining TNO in 2014, he worked all over the world in mining and offshore engineering. Currently he focuses on raw materials and subsurface modelling, trying to find innovative solutions for subsurface related challenges.

João P. Trabucho Alexandre: Department of Earth Sciences, Utrecht University (j.trabucho@uu.nl) – João is a Dutch-Portuguese geologist. He is tenured assistant professor in sedimentology at Utrecht University. In 2011, he was awarded his doctorate cum laude at Utrecht University. Before joining the faculty at Utrecht in 2014, João worked as postdoctoral researcher and lecturer at Durham University, UK.

Johan H. ten Veen: TNO – Geological Survey of the Netherlands (johan.tenveen@tno.nl) – Johan is senior research geoscientist at TNO-GDN and holds an MSc from the Vrije Universiteit Amsterdam and a PhD degree from Utrecht University. He has more than 30 years of experience in multidisciplinary basin analysis and skills include structural geology, sedimentology, seismostratigraphy, geomodelling and reservoir studies. As geoscientific lead, many of his current activities are aimed at providing regional understanding of the subsurface potential for energy production or storage of gases, water, radioactive waste and heat.

Ronald W. Vernes: TNO – Geological Survey of the Netherlands (ronald.vernes@tno.nl) – Ronald obtained his MSc at Wageningen University, specialized in hydrogeology and geohydrology, and joined TNO-GDN in 1990. Ronald is a senior hydrogeologist and project manager. He coordinates the hydrogeological mapping of the Netherlands (REGIS II) and initiated the cross-border (hydro)geological mapping with Belgian and German partners (H3O programme).

Roel M.C.H. Verreussel: TNO – Geological Survey of the Netherlands (roel.verreussel@tno.nl) – Roel received his degree in geology with majors in sedimentology and palynology at Utrecht University. After working as a biostratigraphic consultant with the LPP Foundation, Roel joined TNO-GDN where he has been working until the present-day. He is specialized in the regional geology and stratigraphy of the Jurassic.

Hanneke Verweij: Independent hydrogeologist, Delft, the Netherlands/formerly TNO – Geological Survey of the Netherlands, Politecnico di Torino Italy, Utrecht University (hanneke.verweij@outlook.com) – Hanneke is an expert in analyzing pressure and fluid flow systems on different time and spatial scales with focus on regional groundwater and fluid systems in developing sedimentary basins. She is co-chair Europe of IAH_RGFC, editorial board member of Netherlands Journal of Geosciences and DeepNL Programme Committee Member of the Dutch Research Council NWO.

Geert-Jan Vis: TNO – Geological Survey of the Netherlands (geert-jan.vis@tno.nl) – Geert-Jan graduated as Quaternary geologist from Utrecht University (2003) and obtained his PhD at Vrije Universiteit Amsterdam (2009). Geert-Jan is a regional geologist and stratigrapher. He has a broad interest covering the entire rock succession of the Netherlands. Currently his activities focus on the geology of the southeast Netherlands (South Limburg) and bordering regions, including the Paleozoic of that region. The latter driven by the Einstein Telescope project.

Radboud Vorage: Director geothermal project at Koekoekspolder (greenhousegeopower@hotmail.com) – Radboud is engineer by profession, was the initiator of one of the first geothermal enterprises in the Netherlands. The preparation for the first geothermal well in the Slochteren reservoir started in 2009 and the geothermal doublet was realized in 2011. Since the start of the doublet the equivalent of 50 million cubic meters of natural gas was produced by means of geothermal heat. Radboud is board member at the branch organization *Geothermie Nederland*.

Marja Vuorio: COVRA NV (marja.vuorio@covra.nl) – Marja holds a PhD in chemical engineering from Aalto University and has over 15 years of experience in geological disposal. She worked in Posiva, Finland as a project manager and expert on long-term safety issues related to a geological disposal facility. She came to COVRA in 2019 and is responsible for multinational repository research. Marja is secretary of the association for multinational radioactive waste solutions (ERDO).

Jan-Diederik van Wees: TNO – Geological Survey of the Netherlands (jan_diederik.vanwees@tno.nl) – Jan-Diederik is principal scientist at TNO-GDN and professor geothermal exploration at Utrecht University, with over 20 years of relevant research expertise. PhD in tectonics, bridging disciplines, including reservoir characterization and modelling, structural geology and tectonics, numerical methods, decision and risk analysis. He (co)developed many transparent tools and workflows to unlock geothermal energy, including ThermoGIS.

Frank P. Wesselingh: Naturalis Biodiversity Center (frank.wesselingh@naturalis.nl) – Frank is a sedimentary geologist and molluscan paleontologist working on landscape evolution and drivers of biodiversity change. In the past decade his focus has been on Pliocene and Quaternary successions in the North Sea Basin.

Theo E. Wong†: Department of Geosciences, Anton de Kom Universiteit van Suriname – Theo was a Surinamese-Dutch geologist. He was part-time professor at the Anton de Kom University of Suriname. From 1986 to 2007, he worked at the Geological Survey of the Netherlands (later TNO-GDN). Between 1996 and 2008, he was professor at Utrecht University. Unfortunately Theo passed away in 2024.

Willem J. Zaadnoordijk: TNO – Geological Survey of the Netherlands/Department Water Management, Faculty of Civil Engineering and Geosciences, TU Delft (willem_jan.zaadnoordijk@tno.nl) – Willem-Jan is a geohydrologist with experience in groundwater modelling and simulation tool development working on analysis of groundwater head time-series and hydraulic parametrization of subsurface models. He maintains the TNO-GDN website www.grondwatertools.nl and performs research for extension of its groundwater head viewer.

Index